APETALAE 203
CHORIPETALAE 281
SYMPETALAE 626

ACANTHACEAE 798
ACERACEAE 526
AGAVACEAE 191, 174
AIZOACEAE 280
ALISMACEAE 68
AMARANTHACEAE 249, 264
AMARYLLIDACEAE 191
ANACARDIACEAE 521
APIACEAE 606
APOCYNACEAE 651
ARECACEAE 164
ARACEAE 165
ARALIACEAE 605
ARISTOLOCHIACEAE 227
ASCLEPIADACEAE 655
ASTERACEAE 829

BERBERIDACEAE 320
BETULACEAE 214
BIGNONIACEAE 794
BORAGINACEAE 707
BRASICACEAE 325
BROMELIACEAE 166
BURSERACEAE 496
BUXACEAE 521

CACTACEAE 567
CALLITRICHACEAE 521
CAMPANULACEAE 825
CAPPARIDACEAE 355
CAPRIFOLIACEAE 812
CARYOPHYLLACEAE 291
CELASTRACEAE 524
CELTIDACEAE 220
CERATOPHYLLACEAE 303
CHENOPODIACEAE 249
COCHLOSPERMACEAE 557
COMMELINACEAE 166
COMPOSITAE 829
CONVOLVULACEAE 666
CORNACEAE 624
CRASSULACEAE 358
CROSSOSOMATACEAE 371
CRUCIFERAE 325
CUCURBITACEAE 820
CUPRESSACEAE 57
CYPERACEAE 145

ELAEAGNACEAE 586
EPHEDRACEAE 60
EQUISETACEAE 30

ERICACEAE 629
EUPHORBIACEAE 501

FABACEAE 395
FAGACEAE 215
FOUQUIERIACEAE 639

GARRYACEAE 624
GENTIANACEAE 645
GERANIACEAE 484
GRAMINEAE 70
GUTTIFERAE 556

HYDROCHARITACEAE 69
HYDROPHYLLACEAE 696

IRIDACEAE 195

ISOETACEAE 29

JUGLANDACEAE 213
JUNCACEAE 168
JUNCAGINACEAE 67

KOEBERLINIACEAE 558

LABIATAE 731
LAMIACEAE 731
LEGUMINOSAE 395
LEMNACEAE 165
LENNOACEAE 629
LENTIBULARIACEAE 798
LINACEAE 488
LOASACEAE 562
LOGANIACEAE 644
LORANTHACEAE 223
LYTHRACEAE 587

MALPIGHLIACEAE 497
MALVACEAE 536
MARSILEACEAE 33
MARTYNIACEAE 795
MENISPERMACEAE 321
MORACEAE 221

NAJADACEAE 67
NYCTAGINACEAE 270

OLEACEAE 640
ONAGRACEAE 589
OPHIOGLOSSACEAE 31
ORCHIDACEAE 197
OROBANCHACEAE 796
OXALIDACEAE 487

PALMAE 164

PAPAVERACEAE 322
PASSIFLORACEAE 562
PHYTOLACCACEAE 279
PINACEAE 50
PLANTAGINACEAE 802
PLATANACEAE 371
PLUMBAGINACEAE 639
POACEAE 70
POLEMONIACEAE 678
POLYGALACEAE 497
POLYGONACEAE 228
POLYPODIACEAE 34
PONTEDERIACEAE 168
PORTULACACEAE 285
POTAMOGETONACEAE 64
PRIMULACEAE 635

RAFFLESIACEAE 227
RANUNCULACEAE 304, 314
RESEDACEAE 358
RHAMNACEAE 529
ROSACEAE 372
RUBIACEAE 805
RUTACEAE 493

SALICACEAE 207
SALVINIACEAE 32
SANTALACEAE 226
SAPINDACEAE 528
SAPOTACEAE 640
SAURURACEAE 207
SAXIFRAGACEAE 361
SCROPHULARIACEAE 761
SELAGINELLACEAE 28
SIMAROUBACEAE 495
SOLANACEAE 748
SPARGANIACEAE 64
STERCULIACEAE 554

TAMARICACEAE 557
TILIACEAE 536
TYPHACEAE 63

ULMACEAE 220
UMBELLIFERAE 606
URTICACEAE 222

VALERIANACEAE 818
VERBENACEAE 724
VIOLACEAE 559
VITACEAE 534

ZYGOPHYLLACEAE 489

Arizona Flora

Arizona Flora

by THOMAS H. KEARNEY,
ROBERT H. PEEBLES, and collaborators

Second Edition with Supplement by

JOHN THOMAS HOWELL, ELIZABETH McCLINTOCK
and collaborators

University of California Press, Berkeley, Los Angeles, London

UNIVERSITY OF CALIFORNIA PRESS, BERKELEY AND LOS ANGELES, CALIFORNIA
UNIVERSITY OF CALIFORNIA PRESS, LTD., LONDON, ENGLAND

ISBN: 0–520–00637–2

PRINTED IN THE UNITED STATES OF AMERICA

08 07 06 05 04 03 02 01 00
15 14 13 12 11 10 9

THE PAPER USED IN THIS PUBLICATION MEETS THE MINIMUM REQUIREMENTS OF ANSI/NISO Z39.48-1992 (R 1997) (PERMANENCE OF PAPER). ∞

Preface

Arizona Flora by Thomas H. Kearney, Robert H. Peebles, and collaborators was published in August, 1951. In 1959, when the edition was exhausted, it was decided that the work should be reissued with a supplement, since there was scarcely enough new information on the vegetation of the state to justify a complete revision of the text. The two principal authors of *Arizona Flora* have died (Kearney in October, 1956, Peebles in March, 1956), but Dr. Kearney left extensive floristic notes on Arizona plants at the California Academy of Sciences, aside from the three articles he published in *Leaflets of Western Botany* in 1953, 1954, and 1956. These notes and articles have been most helpful in the preparation of much of the material in the supplement.

Many of Dr. Kearney's notes concern the distribution of Arizona plants; but in order to expedite the publication of the present edition, the supplemental notes consist chiefly of plant records and name changes. Moreover, further to expedite the reissue of *Arizona Flora,* these name changes are indicated in the supplement only for the principal entry of the name in the main body of text. No attempt has been made to correct names used in the introductory chapters, in the keys, or in the discussions.

The twenty-three botanists who collaborated with Kearney and Peebles in *Arizona Flora* were asked again to collaborate in the preparation of special groups in the supplement. Many had no additions or corrections to report. We are grateful to the following who have contributed notes that bring their special groups up-to-date:

Benson, Lyman, Pomona College: Genus *Ranunculus.*
Constance, Lincoln, University of California; *see:* Mathias.
Hermann, F. J., United States Department of Agriculture: Genus *Carex* and Family *Juncaceae.*
Hitchcock, C. Leo, University of Washington: Genus *Lathyrus.*
Mathias, Mildred E., and Lincoln Constance, University of California: Family *Umbelliferae.*
Morton, C. V., Smithsonian Institution: Ferns and Fern Allies.
Muller, Cornelius H., University of California: Genus *Quercus.*
Munz, Philip A., Rancho Santa Ana Botanic Garden: Family *Onagraceae.*
Ownbey, Marion, State College of Washington: Genus *Allium.*
Swallen, Jason R., Smithsonian Institution: Family *Gramineae.*
Wheeler, Louis C., University of Southern California: Family *Euphorbiaceae.*

In addition, treatments of particular genera were prepared by:

Barneby, R. C., New York Botanical Garden: Genus *Astragalus.*
Wahl, Herbert A., Pennsylvania State University: Genus *Chenopodium.*

We are grateful also to the following who have sent us special material or have helped us with problems in particular groups: David B. Dunn (*Lupinus*), Alva and Verne Grant (*Gilia*), Charles T. Mason, Jr. (plant records), Herbert L. Mason (*Polemoniaceae*), Reed C. Rollins (*Cruciferae*), and Robert E. Woodson, Jr. (*Asclepias*). Dr. S. F. Blake was working on his treatment of *Compositae* at the time of his death on December 31, 1959, and we have used such material as he had brought together.

The supplement adds one family (*Meliaceae*), 19 genera (7 introduced), and 68 species to the flora of Arizona. Actually there are 83 new plants (63 native, 20 introduced) reported in the supplement, but 15 species are dropped, either removed from the flora entirely or reduced to lower taxonomic categories. The present known flora of Arizona numbers 3,438 species.

JOHN THOMAS HOWELL
ELIZABETH MCCLINTOCK

California Academy of Sciences, San Francisco, California
February 1, 1960

Contents

Introduction: Botanical Exploration 1

PART ONE:

THE PHYSICAL BACKGROUND AND VEGETATION OF ARIZONA

Topography 7

Geology and Soils 10

Climate 11

Vegetation in Relation to Physical Conditions 12

Life Forms 16

PART TWO:

THE FLORA OF ARIZONA

Plan of the Book 21

Geographic Elements of the Flora 23

Taxonomic Composition of the Flora 26

Key to the Phyla 26

Pteridophyta. Ferns and Fern Allies 27

Spermatophyta. Flowering Plants 49

Gymnospermae 50

Angiospermae 62

Monocotyledoneae 62

Dicotyledoneae 203

Series 1. Apetalae 203

Series 2. Choripetalae 281

Series 3. Sympetalae 626

Plates *following page* 22
Literature Consulted 973
Glossary 987
Index 1007
Supplement 1035
Literature Consulted for Supplement 1077
Index to Supplement 1083

Introduction: Botanical Exploration

An Irish botanist, Thomas Coulter, traveled in 1832 from Monterey, California, to the junction of the Gila and Colorado rivers and from there into Sonora. He made a large collection of plants, many of which bear his name, but whether any of these were collected within the limits of the present state of Arizona, which he barely touched, remains wholly uncertain (1, 2).

During the war with Mexico, Lieutenant (afterward General) W. H. Emory traversed southern Arizona, his route following the Gila River from near its source in New Mexico to its confluence with the Colorado. This expedition first brought to the attention of the scientific world some of the remarkable plants of Arizona, notably the saguaro or giant cactus (*Carnegiea gigantea*) and the crucifixion-thorn (*Holacantha Emoryi*).

The naturalists of the first United States–Mexican Boundary Survey, J. M. Bigelow, C. C. Parry, Arthur Schott, George Thurber, and Charles Wright, in the early 1850's, collected extensively in southern Arizona and adjacent Mexico. They discovered many plants new to science, described by the eminent botanists John Torrey, Asa Gray, and George Engelmann in such publications as *Plantae Wrightianae, Plantae Thurberianae,* and *Botany of the United States and Mexican Boundary Survey.* The original boundary left a considerable territory in the Mexican state of Sonora that was acquired by the United States a few years later through the Gadsden Purchase. For this reason many plants collected by the Boundary Survey expeditions and labeled as from Sonora actually were obtained within the present confines of Arizona. This makes it difficult in not a few instances to determine on which side of the present international boundary the types of some of the species were collected, but in these doubtful cases the types came from so near the border as to warrant their citation in this flora of Arizona.

In 1851 Captain L. Sitgreaves, with S. W. Woodhouse as surgeon-naturalist, led an expedition across northern Arizona from the Zuni to the Colorado River which brought back the first considerable collection of plants from that part of the state. Two or three years later J. M. Bigelow, as botanist of Lieutenant A. W. Whipple's survey for a railway to the Pacific, made substantial contributions to knowledge of the flora of northern Arizona. A third expedition, that of Lieutenant J. C. Ives in 1858, ascended the Colorado to what was at that time the limit of navigation and then traversed northern Arizona by way of the Grand Canyon, Hopi Pueblos, and Fort Defiance. J. S. Newberry, who accompanied Ives as geologist and naturalist, obtained the third important collection of plants of western and northern Arizona.

The ornithologists J. C. Cooper and Elliot Coues collected plants in the 1860's, Cooper in the vicinity of Fort Mohave on the Colorado River, Coues chiefly in the country around Prescott. In this decade Charles Smart obtained specimens from along the Verde River and the Mazatzal Mountains, and an indefatigable collector, Edward Palmer, began his explorations in Arizona, which continued in many parts of the state until 1890. The large number of species in the Arizona flora that bear Palmer's name attest the importance of his discoveries. It is regrettable that his earlier collections were so carelessly labeled in Washington that doubt exists in many instances as to whether the plant in question actually was found in Arizona (3).

The forester J. T. Rothrock, as botanist of the Wheeler expeditions (United States geographical surveys west of the 100th meridian), collected extensively in southeastern Arizona, chiefly in 1874. Other collections in that decade were made by Mrs. E. P. Thompson in the "Arizona Strip" near Kanab, Utah, and by P. F. Mohr around Fort Huachuca.

There was much botanical activity in Arizona in the 1880's. Very large collections were made in southern Arizona by Mr. and Mrs. J. G. Lemmon (1880–1882) and C. G. Pringle (1881–1884). The Lemmons collected also in northern Arizona in 1884. Unfortunately the data of locality on their labels are often untrustworthy, and some of their specimens purporting to have been taken in Arizona probably came from California. The Clifton region in eastern Arizona was explored in 1881 by H. H. Rusby, who also made large collections in 1883 in Yavapai County and on the San Francisco Peaks. He obtained the largest representation of the interesting flora of the Peaks that had been assembled up to then. F. H. Knowlton and Edward L. Greene also collected on the San Francisco Peaks, in 1889. E. A. Mearns in 1888 collected many plants in Yavapai and Coconino counties, and two years later, as naturalist of the second United States–Mexican Boundary Survey, along the southern border. Last, but by no means least, Marcus E. Jones began collecting in the state in 1884, continuing his operations there until 1930. He explored the northern counties with great thoroughness and made several expeditions to the southern mountains.

In the 1890's extensive collections were made by A. Davidson, in the Clifton area; by D. T. MacDougal, chiefly in the Flagstaff region; by J. W. Toumey, in many parts of the state; by T. E. Wilcox, in the Huachuca Mountains; by C. A. Purpus, mostly in Yavapai County; and by Walter Hough, in the Hopi country.

During the present century seventy-five or more botanists have collected plants in Arizona. Only some of the largest collections made during this period may be mentioned here. Beginning in 1901 J. J. Thornber, then professor of botany at the University of Arizona, made very large collections of plants in southern Arizona and around Flagstaff. The work was continued actively by his successors at the University, Lyman Benson, R. A. Darrow, F. W. Gould, W. S. Phillips, L. M. Pultz, Kittie F. Parker, and others, who explored all parts of the state. Rose E. Collom collected actively in the Mazatzal Mountains and in Grand Canyon National Park. Large collections of Arizona plants have been made since 1900 by J. C. Blumer, in the Chiricahua and Rincon mountains; Elzada U. Clover, in the Grand Canyon and Havasu Canyon; C. F. Deaver, in the vicinity of Flagstaff and in Havasu Canyon; Alice Eastwood (alone and with J. T. Howell), in the Grand Canyon, in the Navajo country, and in other parts of the state; W. W. Eggleston, mostly in the White Mountains; H. S. Gentry, mostly in southern Arizona; E. A. Goldman, in many parts of the state; L. N. Goodding, in the White Mountains, on the Kaibab Plateau, and in southern Arizona; David Griffiths, chiefly in the White Mountains and the Tucson region; J. T. Howell, in northern Arizona; W. W. Jones, at various localities; E. L. Little, Jr., on the San Francisco Peaks; Aven and Ruth Nelson, in many parts of Arizona; A. A. Nichol, mainly in southwestern Arizona; H. D. Ripley and R. C. Barneby, chiefly in the northern counties; Forrest Shreve, throughout the state; and by the writers and their associates, especially G. J. Harrison, now or formerly stationed at the United States Field Station at Sacaton, Arizona.

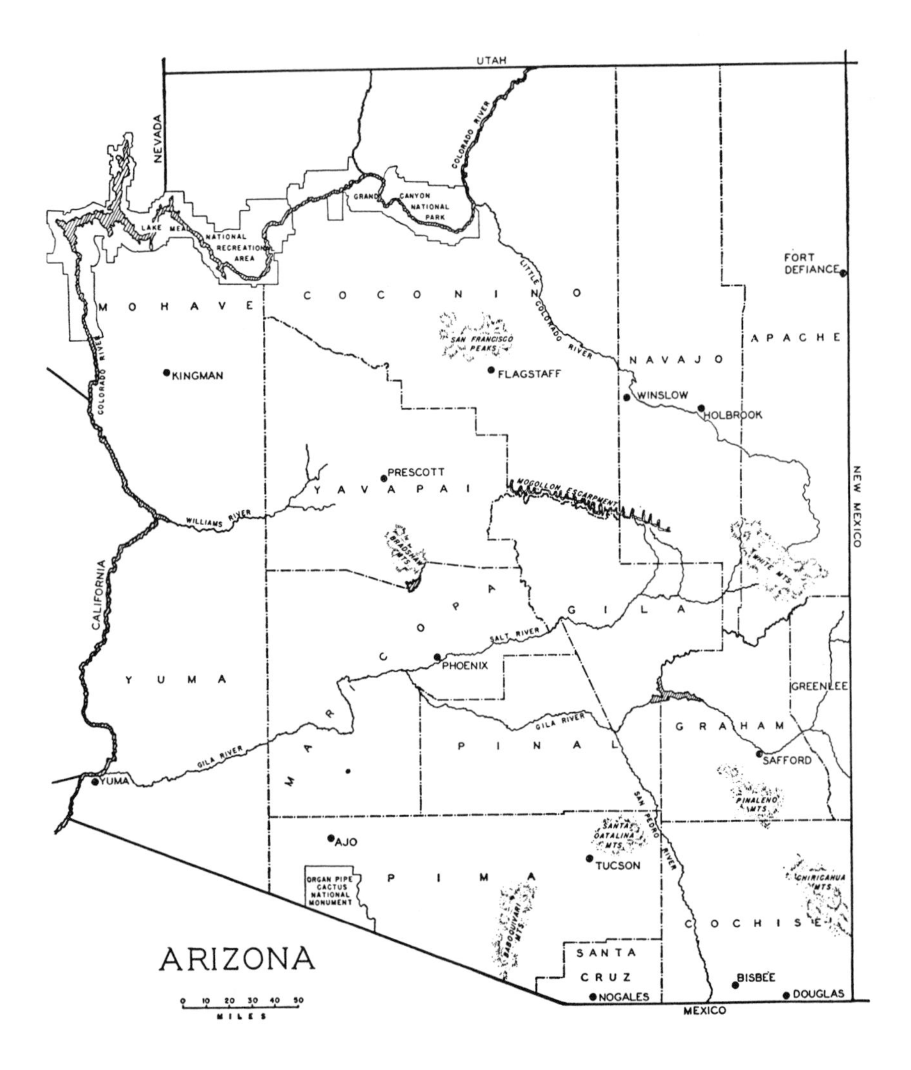
UTAH
NEVADA
COLORADO RIVER
GRAND CANYON NATIONAL PARK
LAKE MEAD NATIONAL RECREATION AREA
FORT DEFIANCE
MOHAVE
COCONINO
SAN FRANCISCO PEAKS
LITTLE COLORADO RIVER
APACHE
NAVAJO
KINGMAN
FLAGSTAFF
WINSLOW
HOLBROOK
PRESCOTT
YAVAPAI
MOGOLLON ESCARPMENT
NEW MEXICO
WILLIAMS RIVER
BRADSHAW MTS.
WHITE MTS.
CALIFORNIA
MARICOPA
GILA
SALT RIVER
PHOENIX
YUMA
GREENLEE
GRAHAM
GILA RIVER
PINAL
SAFFORD
YUMA
PINALENO MTS.
SAN PEDRO RIVER
SANTA CATALINA MTS.
AJO
TUCSON
ORGAN PIPE CACTUS NATIONAL MONUMENT
PIMA
CHIRICAHUA MTS.
BABOQUIVARI MTS.
COCHISE
ARIZONA
SANTA CRUZ
BISBEE
NOGALES
DOUGLAS
MEXICO
0 10 20 30 40 50
MILES

Part I

The Physical Background and Vegetation of Arizona

Topography

This subject as well as "Geology and Soils" and "Climate" are treated briefly and with sole reference to their relationship to the distribution of the plants.

Arizona, with an area of about 113,900 square miles, an extreme length of 400 miles, and an extreme width of approximately 335 miles, is the fifth largest of the United States. It extends in latitude from about 31°20′ to 37° N., and in longitude from 109° to nearly 115° W. The altitudinal range also is great, from a few feet above sea level in the extreme southwestern corner to approximately 12,600 feet at the summit of the San Francisco Peaks in Coconino County.

The northern half of the state is mainly a series of plateaus, some 45,000 square miles in total area, from the Defiance Plateau in the east to the Kaibab Plateau north of the Grand Canyon and the San Francisco or Coconino Plateau south of the Canyon. The elevation of these plateaus is mostly from 5,000 to 9,000 feet. The plateau region is deeply dissected by the Colorado River and its tributaries.

The southern edge of the San Francisco Plateau is marked by one of the most striking topographical features of Arizona, the Mogollon Escarpment or Rim, often designated the Tonto Rim, situated near the north and south center of the state. It extends from near the Verde River in eastern Yavapai County to the White Mountains in southern Apache County. For much of its length the Escarpment presents a nearly vertical wall a thousand feet or so high. West of the Verde River the San Francisco Plateau drops off less abruptly to the south and also to the west, where it descends gradually into the valley of the Colorado River.

Much of the surface of the plateaus is not greatly accidented, except for the canyons mentioned above and such isolated mountains as the Carrizo and Lukachukai ranges in the extreme east, Navajo Mountain, which is mostly in Utah, and the high, volcanic San Francisco Peaks near Flagstaff. Except for these mountains the highest elevation, about 9,000 feet. is reached on the Kaibab Plateau north of the Grand Canyon.

The Grand Canyon is, of course, Arizona's outstanding physical feature, but it is so well known that an attempt at description here would be superfluous. Suffice it to mention that, with an average elevation of about 8,000 feet (maximum 8,800 feet) on the North Rim and about 7,000 feet on the South Rim, it has a maximum depth of more than one mile, the Colorado River at the mouth of Bright Angel Creek being only 2,400 feet above sea level.

The White Mountains in southern Apache County are second in altitude only to the San Francisco Peaks, the summit of Baldy, the culminating peak, having an elevation of about 11,500 feet.

In the southern half of Arizona the surface is broken by numerous isolated mountain ranges, trending for the most part in a northwest-southeast direction and separated by rolling plains, from which they rise abruptly. The mountains are much higher in the southeastern quarter of the state. Beginning at the east the more important ranges are aligned as follows: (1) Gila and Peloncillo; (2) Pinaleno, Dos Cabezas, and Chiricahua; (3) Galiuro, Dragoon, and Mule; (4) Whetstone and Huachuca; (5) Santa Catalina, Rincon, and Santa Rita; (6) Baboquivari. The highest elevation south of the Gila River is the summit of Mount Graham in the Pinaleno Range, with an altitude of 10,500 feet. Elevations of between 9,000 and 10,000 feet are reached in the Chiricahua, Huachuca, Santa Catalina, and Santa Rita mountains.

The Mazatzal Mountains, with a maximum elevation of about 8,000 feet, extend in a somewhat southeasterly direction from near the Mogollon Escarpment to the Salt River near Roosevelt Dam. East of this long, narrow range is the relatively small but floristically important Sierra Ancha, which reaches an altitude of 7,740 feet. It is a meeting ground of northern and southern species, where several of the former reach their southern limit and several of the latter their northernmost extension. The Pinal Mountains, another area of considerable floristic interest, lie between the Mazatzal and Galiuro ranges. Their maximum elevation is 7,850 feet.

The intermontane valleys and plains south of the Gila River, especially along the Mexican boundary and as far west as the Baboquivari Mountains, are relatively high, with elevations mostly between 2,500 and 5,000 feet.

In the southwestern quarter of Arizona, especially as one approaches the Colorado River, the numerous short mountain ranges become progressively lower and are typical desert ranges, very steep and rugged and with little or no timber. They are separated by low desertic plains and valleys. An interesting feature of these desert mountains is the often extensive outwash fans which, from the mouth of each canyon and arroyo, spread out onto the "bajadas," or sloping edges of the plains. This area of low mountains and plains is often distinguished as a geographical province, forming part of the Sonoran Desert, and coördinate with the northern plateau and the more elevated southeastern or Mexican Highland province.

Arizona is not abundantly provided with watercourses. The most important rivers are:

The Colorado, which enters the state from the northeast midway between the eastern and western boundaries, flows through a series of deep canyons culminating in Grand Canyon, and emerges near the northwestern corner of the state, where it now expands into the reservoir called Lake Mead. Thence the Colorado flows almost due south through low desert country, constituting the western boundary of the state, and debouches into the Gulf of California some forty miles below the international boundary. Its principal tributaries in Arizona north of the mouth of the Gila River are the Little Colorado River and the Williams River. The Little Colorado River rises in the White Mountains and flows in a northwesterly direction to join the Colorado near the head of the Grand Canyon, receiving as tributaries from the east the Rio Puerco and Zuni River. The Williams River rises in the mountains of southwestern Yavapai County, flowing almost due west and forming the boundary between Mohave and Yuma counties.

The Salt River is formed by the confluence of the Black and the White rivers, both of which rise in the White Mountains. Its course is generally westerly, and it joins the Gila River a few miles below Phoenix. Its principal tributary is the Verde River, which has its source in Chino Valley in northern Yavapai County and flows almost due south to its confluence with the Salt about twenty-five miles east of Phoenix. The Verde, in turn, has an important tributary in Oak Creek, which rises in southern Coconino County and soon enters a deep canyon. Oak Creek Canyon, with its brilliantly colored cliffs and clear stream at the bottom, is one of the most beautiful and spectacular features of the Arizona landscape.

The Gila River, which enters the state from New Mexico, flows, also in a mainly westerly direction, across the entire width of Arizona, debouching into the Colorado at Yuma. Its most important tributaries are the San Pedro and the Santa Cruz, both of which come into Arizona from the Mexican state of Sonora and flow almost due northward (the Santa Cruz finally northwestward) to their confluences with the Gila.

The courses of the larger rivers are indicated sketchily in figure 1, which shows also the boundaries of the fourteen counties and the location of some of the principal towns.

In addition to the larger rivers, with water flowing permanently along part, at least, of their courses, there are innumerable arroyos and washes in which running water is present only after heavy rains, their sandy beds being dry most of the time.

A very small fraction of the area of Arizona is occupied by lakes, of which the three largest, all artificial, are Lake Mead, on the Colorado River, and the Roosevelt and Coolidge reservoirs in south-central Arizona. There is a chain of small lakes, some of them presumably natural, along the Mogollon Escarpment and a few others on the Kaibab Plateau and in the White Mountain region. A remarkable pond, where aquatic plants are rather numerous, is Montezuma Well in eastern Yavapai County. Although less than five hundred feet in diameter, its craterlike form and considerable depth make it an almost unique feature.

With the exceptions noted, most of the lakes of the state take the form of playas, depressions in which a few feet or inches of water accumulate after heavy rains but which during most of the year present a dry, flat surface almost devoid of vegetation and sometimes covered with a white saline incrustation. These are most numerous in northeastern Arizona, but the largest one is the Willcox Playa in Cochise County, famous for its mirages.

Geology and Soils

Water erosion, although so strikingly exemplified in the Grand Canyon and other canyons of the Colorado River, has been much less effective than wind erosion in shaping the general surface of Arizona. Especially in the northeastern part of the state, on the Navajo Indian Reservation and near by, remarkable specimens of the work of sand-bearing wind, and probably of other agencies, are to be seen in the form of isolated pinnacles, columns, and huge rectangular blocks, such as the celebrated ones in Monument Valley. Shallow archlike recesses are also a conspicuous feature of the landscape. The Chiricahua National Monument, near the southeastern corner of Arizona, offers a striking example of the combined effects of erosion by wind, water, and frost, and of volcanic action, in its grotesque rock forms.

Wind erosion has been active, although with less spectacular results, in all parts of the state. This is evident from the extensive areas of drifting sand, as in the Painted Desert region and in the southwestern quarter near the Mexican boundary. In limited areas these expanses are dotted with low dunes, but Arizona possesses no sand hills comparable in size to those of the Colorado Desert of California.

There has been volcanic activity on a large scale in many parts of the state. The evidences are most conspicuous in the Flagstaff region, where it is estimated that lava flows cover an area of more than two thousand square miles and where the lofty San Francisco Peaks and the much lower Sunset Crater are extinct volcanoes. In southwestern Maricopa, western Pima, and southern Yuma counties considerable areas are covered with lava, these in part being northward extensions of the lava flows of the Pinacate Plateau in northern Sonora.

The stratigraphy of the earth's crust in Arizona is little known except in the Grand Canyon, but there is reason to believe that the succession of strata in the Canyon is not typical of the whole of the state, or perhaps even of the greater part. In the Grand Canyon, Azoic (Archean) schist and gneiss with intrusions of granite are exposed at the bottom. Next above are massive Proterozoic (Algonkian) strata, mostly sandstones and shales, these constituting nearly one half of the entire vertical section. Above the Proterozoic rocks lie Cambrian, Devonian, and Mississippian (Red-Wall Formation) shales, sandstones, and limestones, and upon these are superposed successively Pennsylvanian (Aubrey Group, Coconino and Kaibab formations) sandstones and limestones, then Permian (?), Triassic, and Jurassic sandstones and calcareous shales (Moenkopi Formation etc.). Near the top of the Canyon are almost a thousand feet of Cretaceous shales and sandstones, capped by a thinner deposit of Lower Eocene sandstones, shales, and limestones.

The soils of the northern plateaus of Arizona are derived largely from Kaibab (Aubreyan) limestone toward the west and from Moenkopian sandstones toward the east. The latter soils are characterized by the red and other bright colors that give the Painted Desert its distinction. Much of the soil in the Flagstaff region is of volcanic origin. South of the Mogollon Escarpment the soils are mostly of the Gray Desert type, derived from the granites, gneisses, schists, limestones, and volcanic rocks of the near-by mountains. They are mainly of sandy texture. In some areas, notably around Tucson, the soil is underlain at a slight depth by calcareous hardpan (caliche). Except in wooded areas in the mountains, Arizona soils in the natural state are generally low in humus content and in nitrogen. The brilliant colors of the soils in most of northeastern Arizona are usually lacking in the southern part of the state. Alluvial soils, mostly of more distant origin, of heavier texture, and richer in plant nutrients, are found in the river valleys. Such soils are often more or less saline until reclaimed.

Climate

Arizona, like most arid and semiarid regions, is a land of great climatic contrasts, accentuated here by the wide range of elevation above sea level. There are great differences in temperature and precipitation from place to place and from year to year. Some idea of the magnitude of these variations may be had from the following data[1]:

The mean annual temperature varies from 74.3° at Mohawk, Yuma County (elevation 538 feet) to 42.2° F. at Bright Angel Ranger Station, North Rim of the Grand Canyon (elevation 8,400 feet). The highest absolute maximum temperature, 127° F., was recorded at Fort Mohave and at Parker, in the Colorado River valley at the western edge of the state (elevations 540 and 385 feet respectively). The lowest absolute minimum temperature of record was minus 33° F., at Fort Valley, near Flagstaff (elevation 7,392 feet). There is great daily variation in temperature on most days in most parts of the state, this being generally characteristic of arid and semiarid regions.

The length of the frost-free period is of great importance to vegetation. This varies in Arizona from an annual mean of 348 days at Yuma (elevation 138 feet) to an annual mean of 93 days at Bright Angel Ranger Station (elevation 8,400 feet). The growing season (frost-free period) is generally about three times as long in the low southwestern part of the state as on the high plateaus and mountains.

Arizona has two rainy seasons—winter and late summer. The winter rains are mostly of a steady and penetrating character, as in the Pacific Coast states, whence they usually come. At higher elevations much of the winter precipitation falls as snow. The summer rains, on the other hand, usually take the form of thunderstorms, mostly of brief duration but often violent and resulting in heavy run-off. The summer storms originate in the Gulf of Mexico and Gulf of California. It is most interesting to see how quickly the parched vegetation revives and smaller plants spring up after these storms.

The mean annual precipitation varies from 29.3 inches at Sierra Ancha, Gila County (elevation 5,100 feet), to 3.0 inches in the Yuma Valley (eleva-

[1] Obligingly supplied by Louis R. Juritz, United States Weather Bureau, Phoenix, Arizona.

tion 110 feet). The highest annual precipitation that has been recorded in Arizona was 58.45 inches at Pinal Ranch, Pinal County (elevation 4,520 feet), and the lowest annual of record was 0.44 inches at Wellton, Yuma County (elevation 225 feet). Generally speaking, the driest parts of Arizona are the low western portion and the lower valley of the Little Colorado River in eastern Coconino County, whereas the high plateaus and mountains receive much more precipitation.

Other characteristics of the Arizona climate are low relative humidity of the atmosphere, except during rainy periods, and a very high number of hours during which the sun shines. Only at Phoenix and Yuma have records of annual percentage of possible sunshine been kept during a long enough period to be significant. For these stations the mean percentages are 85 and 90 per cent respectively.

Vegetation in Relation to Physical Conditions

The plant covering of Arizona is extremely varied, as would be expected from the size of the state and the great diversity in topography, altitude, soils, and climate. The life zones range from Arctic-Alpine in two small islands at the summits of Mount Humphreys and Mount Agassiz, highest of the San Francisco Peaks, to Lower Sonoran in the low deserts of the southwestern part, with an infusion of Subtropical in the extreme south.

There are nine principal types of vegetation as named and defined by Forrest Shreve (4). These, in the general order of their altitudinal position, are:

Alpine (11,000 to 12,500 feet)
Northern Mesic Evergreen Forest (6,000 to 11,000 feet)
Western Xeric Evergreen Forest (4,000 to 7,000 feet)
Grassland (5,000 to 7,000 feet)
Desert-Grassland Transition (4,000 to 6,000 feet)
Arizona Chaparral (4,000 to 6,000 feet)
Great Basin Microphyll Desert (3,000 to 6,000 feet)
Arizona Succulent Desert (mostly 3,000 to 3,500 feet)
California Microphyll Desert (below 3,000 feet)

The reader is referred to Shreve for full descriptions of these types and discussions of the physical conditions and more important species of plants characteristic of each. Only brief summaries of Shreve's data are presented here.

Alpine Vegetation

Alpine vegetation, on the San Francisco Peaks, comprises some four dozen species of small herbaceous plants growing among the volcanic rocks and cinders on and near the summits (5).

Northern Mesic Evergreen Forest

Northern Mesic Evergreen Forest represents a southwestern extension of the Rocky Mountain forests and is composed chiefly of several species of pine, Douglas-fir (*Pseudotsuga*) and (mostly above 8,000 feet) two species each of fir (*Abies*) and of spruce (*Picea*). The most extensive areas are on the northern plateaus and in the White Mountains, with smaller colonies in the higher mountains south of the Mogollon Escarpment. What is stated to be the largest continuous forest of ponderosa pine in the United States occurs in Coconino County. The commoner deciduous trees in this type of forest are Gambel oak and aspen. Grasses and other perennial herbs, many with showy flowers, occur in great variety.

Western Xeric Evergreen Forest

Western Xeric Evergreen Forest borders the Northern Mesic Forest at lower altitudes in northern and central Arizona and forms the bulk of the forest on the southeastern mountains. In the latter region the lower elevations are occupied chiefly by several evergreen species of oak, and by pinyon or nut pine (*Pinus cembroides*) and alligator juniper (*Juniperus Deppeana*). Farther north another species of pinyon (*Pinus edulis*) and junipers largely replace the oaks. In central and southeastern Arizona, cypresses (*Cupressus arizonica* and *C. glabra*) are often important components of the Xeric Forest. At higher elevations pines are the dominant trees of this forest type—in the north, ponderosa pine, and in the south, the Arizona, Apache, Chihuahua, and Southwestern White pines. As would be expected from the more arid conditions as compared with those of the Northern Mesic Forest, the smaller associated plants are much more xerophytic, including Cactaceae, *Yucca, Nolina, Dasylirion, Agave,* and various drought-resistant shrubs. Herbaceous plants with showy flowers are numerous but perhaps less so than in the Northern Mesic Forest. Deciduous trees, such as Fremont cottonwood, Arizona walnut, Arizona sycamore, and velvet (Arizona) ash, are confined almost entirely to the banks of streams.

Although covering only a relatively small fraction of the total area of the state, the forests are important in its economy as sources of lumber and firewood and as protection for the watersheds.

Grassland

Grassland was formerly a much more important constituent of the vegetation than it is now. Overgrazing by cattle and sheep and subsequent denudation of the surface of the land have greatly reduced the extent and density of the magnificent stands of grasses that once covered large areas in the northern tier of counties and in central and southeastern Arizona. The remaining Grassland is found mostly along the borders of the forests, extending into them where the trees are not too closely spaced. Species of fescue (*Festuca*), wheat-grass (*Agropyron*), muhly (*Muhlenbergia*), drop-seed (*Sporobolus*), and, most important of all, grama (*Bouteloua*) are among the more valuable grasses of Arizona and are the mainstay of the important grazing industry. Numerous other herbaceous plants and small shrubs usually are intermingled with the grasses.

Desert-Grassland Transition

No sharp distinction can be drawn between this type and the Grassland proper except that it descends to somewhat lower elevations, impinging upon the Great Basin Microphyll and the Arizona Succulent deserts. The constituent grass genera are mostly the same, but there is a larger intermixture of cacti, yuccas, and yuccalike plants.

Arizona Chaparral

Arizona chaparral is very similar in appearance to the better-known chaparral of California and has a few of the same constituent species. It reaches its greatest development in central Arizona (Yavapai and Gila counties). As in California, it is composed mainly of shrubs with thickish evergreen leaves, such as scrub oak (*Quercus turbinella*), manzanitas (*Arctostaphylos* spp.), mountain-mahogany (*Cercocarpus*), and wild-lilac (*Ceanothus Greggii*).

Great Basin Microphyll Desert

This occupies much of northeastern Arizona except the higher mountains and mesas. Roughly, it covers most of the land between the Little Colorado River and the eastern edge of the state. The rainfall is smaller and the plants are more xerophytic than in any other of the vegetation types except the California Microphyll Desert. Small shrubs, mostly with evergreen leaves, including salt-bushes (*Atriplex* spp.) and the leafless joint-firs (*Ephedra* spp.) are characteristic. At somewhat higher elevations sagebrush (*Artemisia tridentata*) sometimes occupies considerable areas. There are many saline playas, where greasewood (*Sarcobatus*) and other salt-tolerating plants are more abundant than elsewhere in Arizona.

Arizona Succulent Desert

Cacti of many species and great diversity of size and form are the outstanding feature of this vegetation type, which reaches its highest development in south-central and southwestern Arizona. These plants range in size from the huge saguaro (*Carnegiea gigantea*) and the many-stemmed organpipe cactus (*Lemaireocereus Thurberi*) to the little pincushion or fishhook cacti (*Mammillaria* spp.), some of which have stems scarcely larger than a baseball. The saguaro, which attains a height of fifty feet or more, is often so abundant as to form "forests," like the well-known one in Saguaro National Monument near Tucson. Other characteristic and extremely abundant constituents are creosote-bush (*Larrea*), with sticky, evergreen, strong-smelling leaves, the gray desert salt-bush (*Atriplex polycarpa*), and bur-sage (*Franseria deltoidea*). Conspicuous also are the palo-verdes (*Cercidium* spp.), with small, deciduous leaves and a wealth of yellow flowers in early spring, ironwood (*Olneya*), and the showy, scarlet-flowered ocotillo (*Fouquieria*). Mesquite (*Prosopis*) is abundant near watercourses, where the Goodding willow (*Salix Gooddingii*) and Fremont cottonwood are also numerous. In the spring months, if the rainfall has been favorable, much of the surface of the Succulent Desert is covered with a multitude of small annual plants with brilliant flowers of every color. These mature their seeds and disappear all too quickly.

California Microphyll Desert

This type of vegetation is found along the western border of Arizona not far from the Colorado River. It represents an eastern extension of the vegetation of the Mojave and Colorado deserts in California, but with elements of the Arizona Succulent Desert, such as saguaro, which barely occur west of the river. It has few distinctive species, but the handsome, magenta-flowered, beavertail cactus (*Opuntia basilaris*) is much more common here than elsewhere in Arizona. The smoke-tree (*Dalea spinosa*) also is characteristic. In Mohave County are found two species of *Yucca* that are especially characteristic of the Mojave Desert in California. These are the weird Joshua-tree (*Y. brevifolia*), which forms small open "forests" in the Sacramento Valley and elsewhere, and the nonarborescent Mohave yucca (*Y. schidigera*). In the alluvial soils of the Colorado and Gila river bottoms flourish forests of cottonwood and willow and extensive thickets of arrow-weed (*Pluchea sericea*), with cat-tail (*Typha*) and bull-rush (*Scirpus*) abundant in the wetter, more open places. The rainfall is lower in this area than in other parts of the state, and the vegetation, except in the stream beds, is correspondingly sparse and stunted.

The three desert types of vegetation cover more than half of the total area of Arizona. Irrigation has made it possible, however, to convert some of it to highly productive cultivation. Most of the agricultural land, which occupies less than two per cent of the total area, borders the Salt, Gila, and Colorado rivers, occupying the alluvial bottoms and also much land that was formerly desert. From a botanist's point of view there is compensation for the removal of part of the original plant cover in the enrichment of the flora with many introduced weeds. Some of these doubtless came in with seeds of crop plants, others, perhaps, with hay shipped to the mining communities in early days.

The relatively few aquatic plants of Arizona are found mostly in small lakes and ponds in the Mesic Evergreen Forest, but some occur at lower elevations—in Montezuma Well and in sloughs and irrigation ditches in the river valleys.

The Grand Canyon, more than a mile in greatest depth, presents a nearly complete cross section of the vegetation of Arizona as well as of the rock strata underlying its surface. In descending into this tremendous chasm from the North Rim, one leaves behind him spruces and firs (and often snow) to pass through most of the vegetation belts described in preceding paragraphs and find cacti, mesquite, and catclaw at the bottom. Here along the river the winters are as warm and the summers as hot as in southern Arizona. The differences in altitude and the complicated topography of Grand Canyon afford a great variety of microclimates; and investigation of the vegetation in different parts of the Canyon offers a fascinating opportunity for students of ecology (6, 7).

Life Forms

Nearly every life form found among North American flowering plants is represented in Arizona.

Chamaephytes, low-growing perennials that include the so-called cushion-plants, are confined largely to the highest elevations, notably the summits of the San Francisco Peaks. Small plants that hug the ground are also numerous in the low, hot Sonoran Desert region, but there they are ephemeral annuals.

There is a great diversity of woody plants throughout the state, ranging in size from the tall pines and Douglas-fir of the high mountains and plateaus to low shrubs (8). The trees are either evergreen (Coniferae and several species of oak) or deciduous (willows, poplars, walnut, sycamore, ash). The deciduous trees occur chiefly along the permanent streams, although the Gambel oak mingles with the pines at higher elevations.

In the Sonoran Desert region, especially, the woody plants are xerophytic, having various means of protection against loss of water by excessive transpiration. Some of these protective devices are: reduction in the size of the leaves; shedding of the leaves in times of drought, as in the palo-verdes and ocotillo (*Cercidium* and *Fouquieria*); thickening of the leaves by development of water-storage tissue; and protective coats of felted hairs, wax, or resin. The creosote-bush (*Larrea*) offers an outstanding example of resin-covered leaves. Some of these desert shrubs have very bizarre forms, notably the crucifixion-thorn (*Holacantha*), which is practically leafless and presents a formidable armature of long, stout, sharp spines. Resembling it in these respects but of smaller stature is the junco or all-thorn (*Koeberlinia*).

Provision for storage of water reaches its highest development in the cacti, which are virtually leafless, the function of leaves being performed by the green outer tissues of the thick stems. The inner tissues of the stems serve as reservoirs in which water is stored during the infrequent rains. Similar but less highly developed arrangements for water storage are found in the thick-leaved century-plants (*Agave*) and the Spanish-bayonets (*Yucca*).

The small annual plants of the desert escape drought entirely by maturing their seeds in a very few weeks after growth starts in early spring.

Even many of the grasses of Arizona are distinctly xerophytic in structure, either suspending growth in dry periods or having relatively thick leaves with the upper surface inrolled so as to protect the water-transpiring stomatal cells.

Halophytic (salt-tolerant) plants are especially abundant at the edges of playas and along the lower courses of the rivers. Notable examples, all belonging to the Chenopodiaceae (Goose-foot Family), are the saltbushes (*Atriplex*), iodine-bush (*Allenrolfea*), greasewood (*Sarcobatus*), and seep-weed (*Suaeda*). These show many of the protections against excessive transpiration of the ordinary xerophytes but in addition are able to endure concentrations of sodium salts in their tissues that would be fatal to most plants.

Many people think of Arizona as all desert, but humus-loving plants, including twenty-two species of terrestrial orchids, are by no means uncommon at higher elevations.

Parasitic flowering plants are well represented by the mistletoes (*Phoradendron* and *Arceuthobium*), dodders (*Cuscuta*), and broomrapes (*Orobanche*). Two rare and little-known parasitic plants, confined in Arizona to the extreme southwestern corner, are *Pilostyles Thurberi,* of the mainly tropical family Rafflesiaceae, and *Ammobroma sonorae,* which belongs to the small American family Lennoaceae.

Only one epiphytic flowering plant is found in the state. This is ball-moss (*Tillandsia recurvata*), of the mainly tropical pineapple family, known only from a limited area in Santa Cruz County, where it finds lodgment on the branches of live oaks but is not parasitic on them. Another life form characteristic of an almost exclusively tropical family, the Palmae, is represented by a small colony of fan-palms (*Washingtonia*) in a canyon of the Kofa Mountains in Yuma County. Insectivorous plants are represented solely by the aquatic bladderwort (*Utricularia vulgaris*), which is rare in Arizona. It has been found in a few small lakes in the White Mountains and in Coconino County.

Part II

The Flora of Arizona

Plan of the Book

The main objective of this work is to afford means for identifying the approximately 3,370 species of flowering plants, ferns, and fern allies known to be growing (or to have grown) without cultivation in the state of Arizona. For this purpose keys are provided to the 132 families, to the genera of each family, to the species of each genus, and to the varieties of many of the species. Formal descriptions are given of the families and genera but not the species. It is realized, of course, that a description of each species would add much to the usefulness of the book but, on the other hand, would make it much too bulky for a single volume of convenient size. As partial compensation for this omission, the characterizations of the species in the keys are usually more ample than is customary. The characters given in the keys, as well as in the description of families and genera, are limited in the main so as to apply only to entities that occur in Arizona.

The nonindented type of key is used throughout, although indented keys are preferred by most botanists. The reason for this is the great saving in the expense of printing and proofreading. In working the keys the user compares the plant in hand with the first paragraph numbered "1" and, if it does not correspond, goes on to second paragraph "1," and so on. Each numbered paragraph ends with the name of a family, genus, or species, or with a number in parentheses. Thus, if the plant to be determined has the characters given in second paragraph "2" and that paragraph ends with the number "3," it is then to be compared with the two paragraphs "3," and so on until a paragraph is reached corresponding with the characters of the plant in hand and ending with the name of a family, genus, or species. The keys are, for the most part, artificial, and the entities often do not appear in the same order in the key and in the text, but each family, genus, or species has the same number in the key and text so that it can be located readily.

The information given for each species comprises: (1) The name, often followed by one or more synonyms in parentheses. Few synonyms are given except those based on Arizona types, or names used in standard works such as Gray's *Synoptical Flora, North American Flora, Das Pflanzenreich,* or in recent monographs and revisions. (2) The distribution of the species in Arizona, stated mostly by counties. If not more than three collections in the state have been recorded, the definite locality, name of the collector, and the collection number are cited. (3) Altitudinal range, in feet, using round numbers because of the uncertainty of precise determinations of altitude. (4)

Habitat. (5) Time of flowering. (6) Citation of the type collection if made in Arizona. (7) General range of the species.

Under many of the species critical taxonomic notes and occasional keys to varieties are given in additional paragraphs. Numerous species are included in this flora which have been reported on doubtful evidence as occurring in Arizona or which are found so close to the borders of the state as to make their eventual discovery there highly probable. In all such instances an asterisk is prefixed to the name of the species.

Monographs and revisions, most of which have been published within the past forty years, are cited after the names of many of the genera by numbers in parentheses. Citations of these publications, correspondingly numbered, are given in the list of "Literature Consulted" (pp. 973–986). In them will be found formal descriptions of the species and other supplemental information. All references to literature are numbered and listed at the end of the book in the order in which they are cited in the text. It has not been thought necessary as a rule to cite such standard works as Gray's *Synoptical Flora of North America, North American Flora,* and *Das Pflanzenreich,* or floras of regions adjacent to Arizona, although they have been drawn upon freely in preparing this work.

Economic value and "human interest" features are mentioned, nearly always under the family, often under the genus, and frequently under individual species. Such matters are dealt with as: food value for man, livestock, and wild animals; utility as timber and for erosion control; medicinal and poisonous properties (including hay-fever relations); and importance as hosts of insect and fungus pests. The utilization of the wild plants by the large Indian population of Arizona is given particular attention.[1]

A list of the publications cited, a glossary of botanical terms, and an index to the names, both scientific and common, of families and genera conclude the volume. The first draft of the glossary was prepared by Myron D. Sutton of Arizona State College, Flagstaff. Species are indexed only if the genus comprises more than ten species occurring in Arizona. In the index the first number after each name indicates the page on which the treatment of the family, genus, or species begins. If it is mentioned in synonymy on other pages, these are indicated by the numbers that follow.

It will be noticed that the taxonomic category next in rank below that of species is designated "subspecies" in some portions of the book, "variety" in other portions. American botanists differ in the use of these words, some employing the term "variety" and regarding it as equivalent to "subspecies," others preferring the latter term and if they also employ the term "variety," treating it as representing a category inferior to that of "subspecies." Several contributors to this work prefer "subspecies," but elsewhere in the book the term "variety" is used to designate the category next in rank below the species unless the treatment is based on a monograph or revision in which the term "subspecies" is used. It is regretted that uniformity in the employment of

[1] The economic information was compiled from various sources, most of which are listed in *Flowering Plants and Ferns of Arizona*, pp. 1037–1040. Several more recent publications also have been consulted.

1. Northern Mesic Evergreen Forest, on the Kaibab Plateau. A burned-over area in which Engelmann spruce is beginning to replace the aspens.

2. Northern Mesic Evergreen Forest, on the San Francisco Peaks, elevation about 8,000 feet. Aspen and ponderosa pine.

3. Northern Mesic Evergreen Forest, on the Defiance Plateau, elevation about 7,500 feet. An open forest of ponderosa pine, with reproduction checked by overgrazing.

4. Western Xeric Evergreen Forest, near Pipe Springs, Mohave County, elevation about 5,000 feet, with Utah junipers and mixed herbaceous vegetation. *Astragalus, Abronia,* and other small plants in foreground.

5. Western Xeric Evergreen Forest, on the Defiance Plateau, elevation 7,000 feet, showing pinyons (*Pinus edulis*) and sagebrush, a common plant association in the Navajo country.

6. Arizona Succulent Desert, in Pinal County, elevation 1,500 feet, showing ironwood (*Olneya*) at the extreme left, saguaros in the background, also little-leaf paloverde and cholla (*Opuntia fulgida*), with bur-sage (*Franseria*) in the foreground.

7. Limber pine (*Pinus flexilis*), on Navajo Mountain, elevation about 8,000 feet.

8. Arizona pine (left) and Chihuahua pine (right), in the Chiricahua Mountains.

9. Arizona cypress, in Chiricahua National Monument, elevation 5,500 feet. An unusually large tree, about 50 feet high.

10. Joint-fir (*Ephedra viridis*), in northern Apache County, elevation about 6,000 feet. A nearly pure stand, making the desert appear at a distance to be covered with grass.

11. Washingtonia palms, in the Kofa Mountains, Yuma County. The self-pruning habit is characteristic of this palm as it grows in Arizona.

12. Bear-grass (*Nolina Bigelovii*), in Palm Canyon, Kofa Mountains, Yuma County.

13. Small-flowered agave (*Agave parviflora*), near Ruby, Santa Cruz County. Distinguished from all other species growing in Arizona by the white leaf markings.

14. Century-plant (*Agave Parryi* var. *Couesii*), near Prescott, Yavapai County. The stalk is 14 feet high.

15. Arizona walnut (*Juglans major*), at base of Patagonia Mountains, Santa Cruz County, elevation about 4,000 feet.

16. Mexican blue oak (*Quercus oblongifolia*), at base of Patagonia Mountains, Santa Cruz County, elevation about 4,000 feet.

17. Emory oak (*Quercus Emoryi*), at base of Patagonia Mountains, Santa Cruz County, elevation about 4,000 feet.

18. Blackbrush (*Coleogyne*), in eastern Coconino County, elevation about 5,500 feet, forming a nearly pure stand.

19. Ironwood (*Olneya testota*), near Gila Bend, Maricopa County. An exceptionally large specimen, about 35 feet high and 3 feet in diameter at the base.

20. Creosote-bush (*Larrea tridentata*), near Picacho, Pinal County. Large specimen, 9 feet high.

21. Night-blooming cereus or reina-da-la-noche (*Peniocereus Greggii*), photographed shortly after sunrise. An exceptionally large plant, the tuberlike root weighing 85 pounds.

22. Saguaro (*Carnegiea gigantea*), Pinal County. A very old plant. This cactus may reach a height of more than 50 feet. The holes in the stem were made originally by woodpeckers and were later appropriated by other birds as nesting sites. A cholla (*Opuntia fulgida*) at left and small-leaf palo-verde in background.

23. Saguaro growing with ocotillo (*Fouquieria*), Pinal County.

24. Flowers of the saguaro or giant cactus, the state flower of Arizona. The white flowers begin to open at night. These were photographed at 9 A.M.

25. Organpipe cactus (*Lemaireocereus Thurberi*), Organ Pipe Cactus National Monument, Pima County, with stems 15 feet high.

26. Pale pink flowers of the organpipe cactus, a night-blooming species, photographed at dawn, Puerto Blanco Mountains, Pima County.

27. Sinita (*Lophocereus Schotti*), pink flowers opening at night, photographed shortly after sunrise, Organ Pipe Cactus National Monument.

28. Hedgehog or strawberry cactus (*Echinocereus pectinatus*), about 8 cm. high, purple flowers, Perilla Mountains, Cochise County.

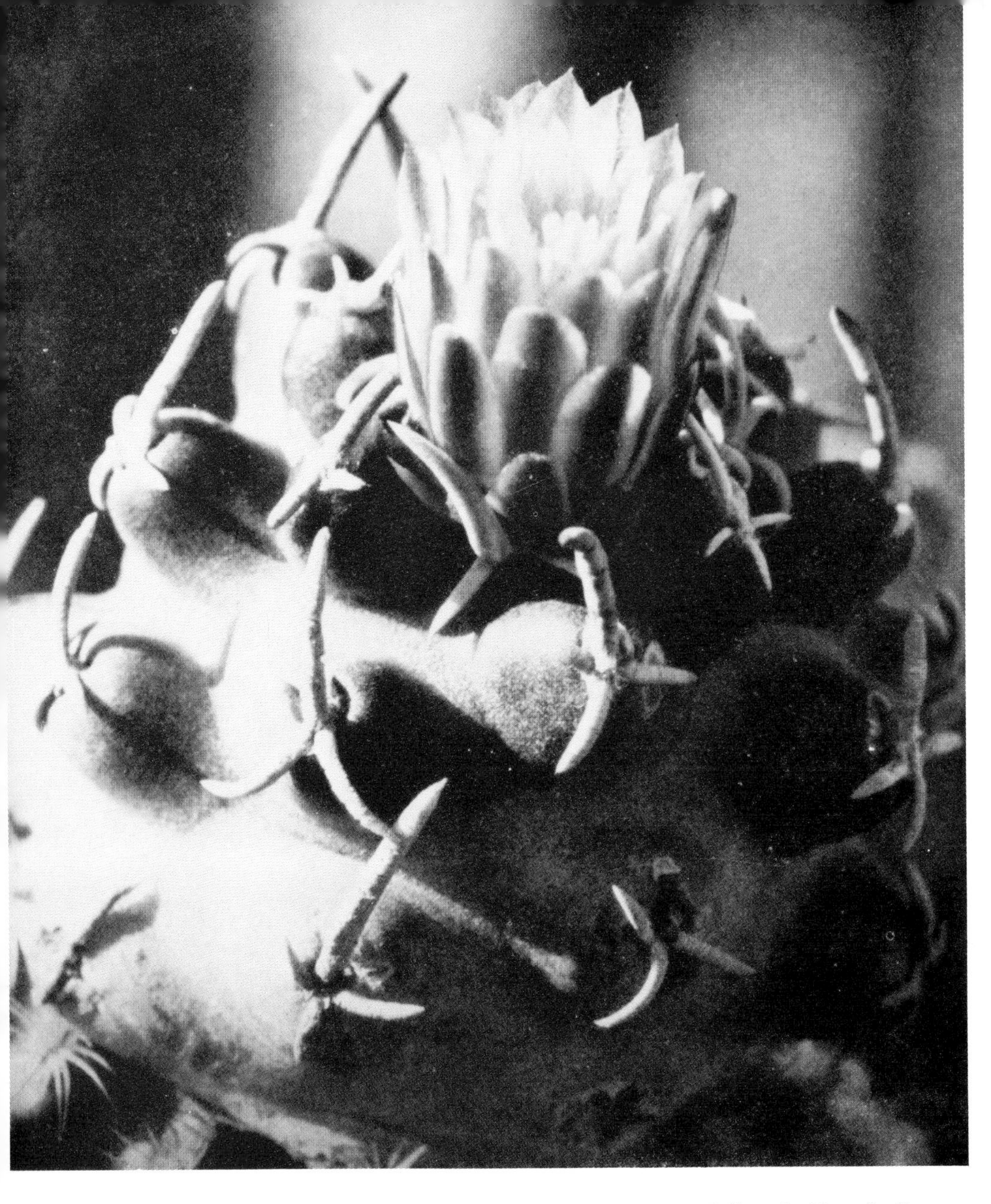

32. Navajo cactus (*Navajoa Peeblesiana*), taken near Holbrook, Navajo County, the only location where this plant is known to grow, and grafted at Sacaton Station. The perianth segments are white with a pink or maroon midstripe.

33. Grafted plant of Navajo cactus (*Navajoa Peeblesiana*) about 3 cm. in diameter. The large tubercles of the stem and the soft covering of the spines are characteristic. This remarkable plant is known only from near Holbrook, Navajo County.

34. Fish-hook or pincushion cactus (*Mammillaria microcarpa*), with pink flowers

35. Fish-hook or pincushion cactus (*Mammillaria Wilcoxii*), with pale pink flowers, Pinaleno Mountains, Graham County.

36. Prickly-pear (*Opuntia aurea*), Pipe Springs, Mohave County. This plant has prostrate stems and pink flowers.

37. Jumping cholla or teddybear cactus (*Opuntia Bigelovii*), with stems about 7 feet high, San Tan Mountains, Pinal County. The heavily armed joints of this cactus break off easily and often cause painful injuries. The flowers are greenish yellow tinged with pink.

38. Cholla (*Opuntia acanthocarpa* var. *ramosa*), about 5 feet high, near Sacaton, Pinal County. The flower buds are cooked and eaten by the Pima Indians.

39. A less spiny cholla (*Opuntia Thornberi*), near Superior, Pinal County.

40. Cholla (*Opuntia spinosior*), about 6 feet high, near Casa Grande, Pinal County. This species commonly has purple flowers, but the petals are sometimes red or yellow.

41. A cholla, presumably a hybrid between *Opuntia spinosior* and *Opuntia fulgida,* common in the usually dry bed of Gila River, Pinal County. This form propagates freely from the fallen joints.

42. Evening primrose (*Oenothera deltoides*), one of the most abundant spring wild flowers of the sandy deserts of western Arizona. The white flowers contrast beautifully with the pink flowers of the sand-verbena, which often grows with it.

43. Palmer's penstemon (*Penstemon Palmeri*), Mingus Mountain, Yavapai County. One of the most beautiful members of this handsome genus. The large pink flowers have a delicate fragance.

44. Ocotillo (*Fouquieria splendens*), sometimes called coach-whip, one of the most conspicuous and abundant of southwestern desert shrubs. After the winter rains, and again during the summer rainy season, the plant puts forth its bright green leaves and tassels of scarlet flowers. During most of the year, however, the bare stems appear lifeless.

45. Desert-willow (*Chilopsis linearis*), at base of Patagonia Mountains, Santa Cruz County. An exceptionally large tree about 35 feet high. The showy catalpa-like flowers are white, usually streaked or spotted with purple.

these terms is impracticable so long as taxonomists continue to differ in their use of them.

It should be stated for the benefit of indexers that no new species, varieties, or combinations are published in this book.

Geographic Elements of the Flora

No part of Arizona constitutes a distinct phytogeographical province, although there is a greater proportion of endemic species than would be expected in an area so open on every side to emigration and immigration. Because of its geographical position, range of altitude, and diversity of climate, this state is a meeting ground for a remarkably large number of floral elements, those of nearly every part of North America being represented. Alone among the states it has both alpine plants, of which a few species occur on the highest mountain summits, and plants of mainly tropical affinity, in the Mexican boundary region.

The 3,370 species of flowering plants and ferns that have been found growing wild in Arizona may be classified very roughly in twelve geographical categories. This does not pretend to be a definitive classification, many of the species not being readily assignable to one or another of these categories, but it may serve as a basis for further study.

Endemic

Very nearly 5 per cent of the species are known only from within Arizona, although further exploration doubtless will discover some of them in adjacent states or in northern Mexico. Of the endemics, forty-six species are confined to northern, twenty-eight to central, and seventy-four to southern Arizona, the remaining sixteen species being more widely distributed in the state. Ten species and two varieties are known only from within the borders of Grand Canyon National Park and its immediate vicinity.

Pacific

Species whose main range lies between British Columbia and Baja California, excluding the desert regions of southeastern California and Baja California. Only about 3 per cent of the Arizona flora has this geographical affinity, but it is remarkable that species known chiefly from coastal California should be found also in south-central Arizona, especially in the Pinal and Mazatzal mountains. Examples are *Dryopteris arguta, Ribes quercetorum, Cercocarpus betuloides, Lupinus succulentus, Rhus ovata, Rhamnus crocea, Fremontia californica,* and *Lonicera interrupta.*

Great Basin

What may be considered a southward prolongation of the Great Basin flora occurs in the drainage basin of the Colorado River from northern Apache County to northern Mohave County at elevations mostly between 4,000 and 7,000 feet. Most of this region is treeless or has a sparse covering of pinyons and junipers. The large genera *Eriogonum* and *Astragalus* are well represented here. Some 6.5 per cent of the Arizona flora is assigned to this category.

Sonoran

Species confined mainly to the deserts of southwestern Arizona, northwestern Sonora, southeastern California, and Baja California at elevations mostly below 3,000 feet. In this category, which comprises roughly 9 per cent of the flora, most of the plants are pronounced xerophytes or ephemeral annuals. The northeastern limit of the Sonoran flora lies in Maricopa and Pinal counties, where notable representatives are *Colubrina californica, Abutilon Palmeri, Horsfordia Newberryi,* and *Beloperone californica.*

Chihuahuan

The center of distribution of this element is the drainage basin of the Rio Grande in western Texas, southern New Mexico, and Chihuahua, but many of the species extend to southeastern Arizona and northeastern Sonora and some into the second tier of Mexican states. Some 1.5 per cent of the mainly Chihuahuan species are found also in the Sonoran deserts, and a few extend as far northwestward as the Colorado River drainage basin. The Chihuahuan province, as here understood, is a region of moderate elevation, mostly from 3,000 to 6,000 feet. This geographical category, next to the Rocky Mountain, is the largest of the Arizona flora, comprising about 18 per cent of the species.

Tropical and Subtropical

Many of the species found in southern Arizona range farther than the southern boundaries of the Chihuahuan and Sonoran provinces, into southern Mexico, Central America, and even South America. A few occur also in the subtropical zone of the southeastern United States and in the West Indies. Species of this category, which comprises about 8.7 per cent of the total number in the flora, are much more numerous than in any other state west of Texas. The relatively few species that are common to Pacific North America and to the temperate regions of Chile and Argentina are not included in this category.

Rocky Mountain

The high plateaus of northern Arizona and the higher mountains of the eastern and central parts of the state are very rich in elements of the Rocky Mountain flora. Many of these species range from the Canadian Rockies to the Sierra Madre in northern Mexico, but no small number are known only from the mountains of New Mexico and Arizona. Assignment of the latter to the Rocky Mountain rather than the Chihuahuan category is sometimes problematic, but such species as are limited to the region north of the Gila or to relatively high altitudes seem more properly to be regarded as Rocky Mountain elements. This is the largest floristic category, comprising about 24 per cent of the total number of species. Nearly one half of these are found also in the higher mountains of the Pacific Coast states (Cascades and Sierra Nevada).

Great Plains

A few species (approximately 1 per cent of the total number in Arizona) have their main center of distribution on the high plains east of the Rocky Mountains, in some instances extending farther eastward into the Prairie region. In Arizona, plants of this category are found mostly on grassy plains in the eastern part. It is noteworthy, however, that three characteristic species of the Great Plains and Prairie districts, buffalo-grass (*Buchloë dactyloides*), larkspur violet (*Viola pedatifida*), and *Houstonia nigricans*, reach their western limit in openings in ponderosa pine forest along the Mogollon Escarpment.

Eastern United States

This category comprises twelve species, most of which are otherwise confined to the southeastern states as far west as central Texas. None of them is known from western Texas and New Mexico. Plants having this remarkably interrupted distribution are: *Ophioglossum Engelmannii, Corallorhiza Wisteriana, Hexalectris spicata, Cerastium texanum, Crotalaria sagittalis, Clitoria mariana, Acalypha ostryaefolia, Viola Rafinesquii, Chimaphila maculata, Trichostema brachiatum, Galium pilosum,* and *Cyclanthera dissecta.*

North American

Species of wider distribution on this continent than those of any of the preceding categories, a few of which are found also in temperate South America. Many of these range across Canada and the northern United States, reaching Arizona only at high elevations. Approximately 13.5 per cent of the Arizona species are assigned to this category.

Transoceanic

Nearly 4.5 per cent of the Arizona species are believed to be indigenous in both the Eastern and the Western hemispheres, or are so well established in both hemispheres that their origin in one or the other cannot be determined with certainty. Most of these are circumpolar (holarctic) and are found in Arizona only in the higher and cooler parts. A much smaller number are common to the tropical and subtropical regions of the two hemispheres. Some of the transoceanic aquatic and marsh plants are of almost world-wide distribution.

Introduced

This category, to which belongs nearly 7 per cent of the Arizona flora, comprises plants brought in by the (mostly unconscious) agency of man, the great majority growing as weeds in cultivated and waste land. Most of these introductions originated in the Mediterranean region or other parts of Eurasia and reached Arizona by way of California. A smaller number came from the New World or the Old World tropics. A few species, natives of central Asia, may have been brought into Arizona with seeds of Turkestan and other Asiatic varieties of alfalfa. Some of these introduced plants have not been seen recently and probably never became established.

Taxonomic Composition of the Flora

The known flora of Arizona comprises 132 families, 907 genera, and 3,370 species of flowering plants, ferns, and fern allies. The ten largest families, with the number of species in each, are as follows:

Compositae—543
Gramineae—369
Leguminosae—287
Scrophulariaceae—112
Cruciferae—105
Cyperaceae—100
Polygonaceae—94
Euphorbiaceae—84
Cactaceae—71
Rosaceae—69

Most of these are among the largest families in many of the local floras of the United States, but the Polygonaceae and Euphorbiaceae are relatively better represented than in the northeastern and northern states. In the number of Cactaceae, however, Arizona is almost unique, no other state except Texas possessing so many species of this highly specialized family. Twelve of the Arizona species of cacti are endemic, so far as is known.

Key to the Phyla

Plants without flowers, producing spores, not seeds . . . . *Pteridophyta*
Plants with flowers, producing seeds . . . . *Spermatophyta*

PTERIDOPHYTA

FERNS AND FERN ALLIES (9)

Contributed by C. V. Morton[1]

Terrestrial, or rarely aquatic, perennial herbs with marked alternation of generations; prothallium (gametophyte, haploid or sexual generation) small, green and thalloid or filamentous, or subterranean and cormlike, or sometimes reduced to a few cells and contained mostly within the spore walls, monoecious or dioecious, bearing antheridia and archegonia; sporophyte (diploid generation) mostly large, diverse in form but always divided into roots, stems, and leaves, these with 1 or several vascular bundles; reproduction by means of spores borne within sporangia, these bivalvate or irregularly dehiscent; spores (haploid) mostly minute, alike, or if unisexual, then the male (microspores) minute and numerous and the female (megaspores) larger and fewer.

KEY TO THE FAMILIES

1. Sporangia borne on the ventral base of the leaf, axillary, or on special peltate sporangiophores; leaves mostly small in comparison with the stem, sessile, simple in structure, mostly with a single vascular bundle (2).
1. Sporangia dorsal or marginal on the leaves; leaves large in comparison with the stem (rhizome), typically stalked, complex in structure, with numerous vascular bundles (veins) (4).
2. Leaves regularly whorled, united into a sheath around the stem, only the apices (teeth) free; stems jointed; sporangia borne on peltate sporangiophores, these grouped into terminal cones .... 3. *Equisetaceae*
2. Leaves not united into sheaths; stems not jointed; sporangia borne on the leaf bases or axillary (3).
3. Stems conspicuously branched, not cormlike; leaves small (not more than 4 mm. long); sporangia axillary .... 1. *Selaginellaceae*
3. Stems unbranched, merely bilobed, cormlike; leaves grasslike, 4 cm. long or more; sporangia large, sunken in the leaf base .... 2. *Isoëtaceae*
4. Stems and leaves fleshy, lacking sclerenchymatous tissue, without scales; leaves not circinate (coiled) in vernation, divided into fertile and sterile parts; leaf bases sheathing; sporangia large, subsessile, with thick walls, lacking an annulus; prothallium subterranean, mycorrhizal .... 4. *Ophioglossaceae*
4. Stems and leaves not fleshy, normally with sclerenchyma; scales present or absent; leaves normally circinate in vernation; leaf bases not sheathing; sporangia short- or long-stalked, with an annulus; prothallium epigaeous or aquatic (5).
5. Plants minute, mosslike, floating in water. Sporangia heterosporous, only one megaspore developing .... 5. *Salviniaceae*
5. Plants larger, not mosslike or floating (6).

[1] The author acknowledges a deep indebtedness to the treatment by Dr. William R. Maxon in *Flowering Plants and Ferns of Arizona*. It does not seem justifiable, however, to attribute the present revision to him because of the extensive alterations that have been made in the descriptions and keys, and especially because of some changed taxonomic concepts.

6. Leaves divided into 4 equal lobes, cloverlike; sori grouped within a closed, hard-walled sporocarp, this borne subbasally on the petiole; sporangia heterosporous 6. *Marsileaceae*

6. Leaves not divided into 4 equal lobes; sori not in sporocarps; sporangia homosporous 7. *Polypodiaceae*

1. SELAGINELLACEAE. Selaginella Family

Small, evergreen, terrestrial, mosslike plants with slender, subdichotomously branched, erect or prostrate stems, these with 1 vascular bundle; leaves minute, numerous, simple, spirally arranged in many ranks, 1-nerved, ciliate; sporophylls more or less similar to the leaves, borne in terminal cones, the microsporophylls and megasporophylls mixed in the same cone; sporangia 1-celled, solitary in the axil of the sporophyll, reniform, subsessile, bivalvate, heterosporous; spores tetrahedral, the megaspores normally 4 to a sporangium, yellowish or chalky white, with a thick, sculptured wall, the microspores minute, numerous, usually red or yellow, spinulose; gametophytes minute, largely contained within the spore walls, lacking chlorophyll.

1. SELAGINELLA

KEY TO THE SPECIES

1. Stems more or less dorsiventral, the leaves of the lower side larger (2).
1. Stems not at all dorsiventral, erect or prostrate, the leaves all alike (4).
2. Terminal seta of the leaves long and slender, flexuous, soon deciduous...1. *S. eremophila*
2. Terminal seta of the leaves shorter and straighter, persistent (3).
3. Plants strongly dorsiventral; leaves attenuate into the seta.............2. *S. arizonica*
3. Plants only slightly dorsiventral; leaves subobtuse, abruptly attenuate into the seta 3. *S. densa*
4. Terminal seta absent or very short; leaves ovate-oblong to narrowly oblong, obtuse 4. *S. mutica*
4. Terminal seta prominent (0.3 to 2 mm. long); leaves subulate-lanceolate (5).
5. Plants bright green, mostly prostrate, freely branching and rooting throughout 5. *S. Underwoodii*
5. Plants bluish green, mostly erect and rooting only in the lower part (6).
6. Leaves whitish-marginate, lance-aciculate, evenly attenuate to a very long, stiff, yellowish-white seta ....6. *S. rupincola*
6. Leaves not marginate, subulate-attenuate to an acutish apex, giving rise rather abruptly to the short, whitish-hyaline seta..............................7. *S. neomexicana*

1. Selaginella eremophila Maxon. Near Tinajas Altas, Yuma County (*Jaeger* in 1934). Southwestern Arizona and southern California.

2. Selaginella arizonica Maxon. Graham, Gila, Maricopa, Pinal, and Pima counties, 2,000 to 4,500 feet, rocky ledges and cliffs, type from the Santa Catalina Mountains (*Shreve* in 1914). Western Texas to Arizona and Sonora.

3. Selaginella densa Rydb. Known in Arizona only from the summit of Mount Baldy, White Mountains, Apache County, 11,470 feet, in crevices of rock, (*Ferriss* in 1913, *Phillips & Reynolds* 2900, *Phillips & Phillips* 3342). British Columbia and Idaho east to Alberta, south to New Mexico and Arizona.

4. Selaginella mutica D. C. Eaton. Tsegi and Betatakin canyons (northern Apache and northern Navajo counties), Grand Canyon (Coconino County), Chiricahua and Mustang mountains (Cochise County), 3,000 to 7,000 feet, damp cliffs. Western Texas to Colorado and Arizona.

5. **Selaginella Underwoodii** Hieron. Kaibab Plateau to the Mogollon Escarpment (Coconino and Gila counties), Pinaleno Mountains (Graham County), Chiricahua Mountains (Cochise County), Santa Rita Mountains (Santa Cruz County), Santa Catalina and Baboquivari mountains (Pima County), 5,000 to 8,000 feet, cliffs and rock ledges. Western Texas to Wyoming and Arizona.

Var. *dolichotricha* Weatherby, differing from the typical variety in having longer terminal setae (0.8 to 1.8 mm. long) and usually better-developed cilia on the leaves, occurs in the Pinaleno, Chiricahua, Huachuca, Rincon, Santa Catalina, and Santa Rita mountains.

A somewhat similar species, *S. Watsoni* Underw., has been found on Navajo Mountain, Utah, near the Arizona border, and may eventually be found in Arizona. It may be distinguished by its more compact habit, congested leaves, and especially by the short, yellowish, smooth terminal seta of the leaves.

6. **Selaginella rupincola** Underw. Chiricahua, Mule, and Dragoon mountains (Cochise County), Patagonia and Santa Rita mountains (Santa Cruz County), Rincon, Santa Catalina, and Coyote mountains (Pima County), 3,000 to 6,000 feet, dry cliffs and ledges. Western Texas to Arizona and northern Mexico.

7. **Selaginella neomexicana** Maxon. Known in Arizona only from a single collection from Paradise, Cochise County. Western Texas to Arizona.

2. ISOËTACEAE (10). Quillwort Family

Plants aquatic or amphibious, the stem (corm) fleshy, flattened, bilobed; leaves simple and grasslike, spirally arranged on top of the corm, entire, glabrous and scaleless, hollow, divided into 4 longitudinal cavities, these with cross walls at irregular intervals; sporangia large, oblong, sunken in the ventral leaf base, divided by transverse partitions; velum (a fold of leaf tissue) partly covering the upper end of the sporangium; megaspores large (0.3 mm. in diameter or more), white and chalky, tetrahedral, marked with wrinkles and small tubercles; microspores small, numerous, bilateral, nearly smooth; gametophytes reduced, dioecious, borne wholly within the spore walls.

The only genus, *Isoëtes,* numbers about eighty species distributed throughout the world but most numerous in the North Temperate Zone. The species are difficult to distinguish, even with the aid of a compound microscope.

1. ISOËTES. Quillwort

1. **Isoëtes Bolanderi** Engelm. Lukachukai Mountains, Apache County (*Goodman & Payson* 2936), Dude Lake, Mogollon Escarpment, Coconino County, 8,000 feet (*Coville* 1053, *Phillips & Reynolds* 2892). Mountain ponds and lakes, British Columbia to Wyoming, Colorado, Arizona, and California.

Var. *pygmaea* (Engelm.) Clute was collected in the Huachuca Mountains by Lemmon; known otherwise only from Mono Pass, California, and Walker Lake, Nevada. It differs from the typical phase of the species in its stouter and somewhat shorter leaves.

3. EQUISETACEAE. Horsetail Family

Deciduous or evergreen plants with deeply subterranean, branched rhizomes; stems green, solitary or fasciculate, jointed, the internodes hollow, prominently ridged, the ridges bearing siliceous tubercles or cross bands, internally with longitudinal vallecular cavities alternating with the ridges and usually also with a series of smaller carinal cavities; stomata prominent, sunken, borne in regular rows or in irregular bands in the depressions between the stem ridges; leaves minute, scalelike, undivided, whorled, fused laterally and forming a sheath around the stem, only the tips free, each leaf continuous with a ridge of the internode below and alternating with the ridges above; branches, if present, whorled, breaking through the bases of the stem sheaths; sporangia borne in terminal cones, these subsessile or long-pedunculate, composed of whorled sporangiophores, these stalked, peltate, the flat, polygonal tips fitting closely together; sporangia elongate, borne on the under side of the sporangiophores, dehiscent by a longitudinal slit; spores spherical, with a thick perispore consisting of 4 spirally wound bands; gametophytes thalloid, epigaeous, dorsiventral, consisting of a round, cushionlike base and several flattened, erect lobes.

1. EQUISETUM. Horsetail

KEY TO THE SPECIES

1. Sterile and fertile stems distinct, the fertile unbranched, soon withering, the sterile with numerous, whorled, 3- or 4-ridged branchlets........................1. *E. arvense*
1. Sterile and fertile stems alike, persistent, mostly unbranched or with a few irregular branches, these with many more than 4 ridges (2).
2. Stem sheaths much longer than broad, widened upwardly, all except the lowest without a dark basal band, the teeth promptly deciduous; stems mostly smoothish, with only slightly developed bands of silex on the ridges....................2. *E. laevigatum*
2. Stem sheaths not much longer than broad, not widened upwardly, with a prominent, dark basal band, the teeth mostly subpersistent or irregularly deciduous; stems rough, with strongly developed cross bands of silex on the ridges..................3. *E. hiemale*

1. Equisetum arvense L. Apache, Navajo, Coconino, Graham, and Gila counties, 5,000 to 8,000 feet, moist soil along streams. Newfoundland and Labrador to Alaska, southward nearly throughout the United States; Greenland; Eurasia.

Forma *ramulosum* (Rupr.) Klinge, with the branchlets again branched, has been found in Gila County (*Little* 4302).

2. Equisetum laevigatum A. Braun (*E. kansanum* Schaffn.). Navajo and Coconino counties south to Pima, Santa Cruz, and Cochise counties, 5,000 to 8,000 feet, marshes, alluvial thickets, and sandy banks, sometimes a weed in cultivated places. Ontario to New York and New Jersey, west to Illinois and Minnesota; North Dakota west to British Columbia and south to Texas and California; Mexico and Guatemala.

A form with more prominent cross bands of silex on the ridges is forma *scabrellum* (Engelm.) Broun.

3. Equisetum hiemale L. Scouring Rush. Apache, Navajo, and Coconino counties to Cochise, Santa Cruz, and Pima counties, 5,000 to 8,000 feet, moist, alluvial soil and springy places. Nearly throughout the United States; Eurasia.

The Arizona plant is var. *elatum* (Engelm.) Morton (*E. prealtum* Raf.).

4. OPHIOGLOSSACEAE (11). Adder's Tongue Family

Rhizomes short, erect, unbranched, glabrous and scaleless, bearing enlarged, hairless, mycorrhizal roots; fronds often solitary, scaleless, typically glabrous, soft and fleshy (lacking sclerenchymatous tissue), the stipe (common stalk) enlarged at base and with a stipular sheath enclosing a bud (embryonic leaf), the blade erect or bent in bud, not circinate, divided into an entire or compound, green, sessile or stalked sterile segment, and a stalked, simple or compound, apparently ventral fertile blade, this with leaf tissue reduced or lacking; veins free or anastomosing; sporangia large, homosporous, lateral, free or sunken in the leaf tissue, thick-walled, lacking an annulus, bivalvate; spores tetrahedral, very numerous, thick-walled; gametophytes small, subterranean, not green, cylindric or flattened, branched or unbranched.

A small family of three genera and about fifty species, widespread throughout the world.

KEY TO THE GENERA

1. Veins free, forked; sterile blades mostly lobate or compound; sporangia free, short-stalked .... 1. *Botrychium*
1. Veins anastomosing; sterile blades entire; sporangia fused with the leaf tissue of fertile blades .... 2. *Ophioglossum*

1. BOTRYCHIUM. Grape-fern

Fronds mostly solitary; stipe short or elongate; sterile blades pinnately or subpalmately compound, sessile or stalked, glabrous or pubescent with simple hairs; fertile blades stalked, pinnate to decompound, the segments minute; veins free, forked; sporangia marginal, not fused with leaf tissue, short-stalked, green when immature.

The species are all excessively rare in Arizona.

KEY TO THE SPECIES

1. Sterile blades glabrous, even in the bud enclosed within the petiolar leaf sheath, pinnate or pinnate-pinnatifid, deciduous (2).
1. Sterile blades slightly pilose, at least in bud, bipinnate to quadripinnatifid (3).
2. Sterile pinnae rounded to obtuse; sterile blades oblong .... 1. *B. Lunaria*
2. Sterile pinnae acuminate; sterile blades deltoid .... 2. *B. lanceolatum*
3. Sterile blades long-stalked, inframedial, evergreen, the stipe much shorter than the blade 3. *B. multifidum*
3. Sterile blades sessile, about medial, deciduous, the stipe equaling or exceeding the blade 4. *B. virginianum*

1. Botrychium Lunaria (L.) Swartz. San Francisco Peaks, 11,000 feet, open spruce-fir forests (*Kearney & Peebles* 12123, *Little* 4741, *Collom* 890). Labrador and Newfoundland to Alaska, southward in varietal forms to Maine, New York, Michigan, Colorado, Arizona, and southern California; Argentina; Greenland; Eurasia; Australia; New Zealand.

2. Botrychium lanceolatum (S. G. Gmel.) Rupr. San Francisco Peaks, 11,000 to 11,500 feet, subalpine meadows (*Little* 4679, 4740, *Clausen & Trapido*). Labrador to Virginia, west to Ontario and south to Ohio and West Virginia; Colorado, Arizona, and Utah, west to Washington and north to Alaska; Eurasia.

3. Botrychium multifidum (S. G. Gmel.) Rupr. White Mountains, Apache County, 8,000 feet (*Goodding & Schroeder* 338, 340–41). Newfoundland south to North Carolina, west to British Columbia and south to Colorado, Arizona, and California; Eurasia.

The Arizona plant is var. *intermedium* (D. C. Eaton) Farwell (*B. silaifolium* Presl, *B. Coulteri* Underw.).

4. Botrychium virginianum (L.) Swartz. Santa Rita Mountains (*Pringle* in 1884). Labrador and Newfoundland south to Florida, west to British Columbia and south to Arizona; Mexico; Eurasia.

Pringle's station has been destroyed, apparently, and this species has not since been found in Arizona.

2. OPHIOGLOSSUM. Adder's Tongue

Leaves 1 or 2, glabrous, the stipe green; blades divided into a simple, entire, elliptic sterile segment, and a ventral, stalked, erect, simple, sharp-tipped, spikelike fertile segment; veins anastomosing; sporangia large, subspherical, fused with the leaf tissue, green when immature, borne in two rows.

KEY TO THE SPECIES

1. Principal veins forming large primary areoles enclosing numerous smaller secondary areoles; sterile segment apiculate . . . . 1. *O. Engelmannii*
1. Principal veins forming smaller areoles not enclosing secondary areoles; sterile segment not apiculate . . . . 2. *O. vulgatum*

1. Ophioglossum Engelmannii Prantl. Huachuca Mountains and near Hereford (Cochise County), Mustang Mountains and Sonoita Valley (Santa Cruz County), 4,000 feet or higher, in damp places, usually in calcareous soil. Virginia to Florida, west to southern Illinois, Kansas, and Arizona; Mexico.

2. Ophioglossum vulgatum L. A single specimen, said to have been collected in the Huachuca Mountains, is cited by Clausen. Nova Scotia south to Florida, west to Ontario and south to Texas and Arizona (?); Mexico; Alaska; Eurasia.

5. SALVINIACEAE. Water-fern Family

Floating aquatics with reduced vascular tissue; rhizomes filiform, branched; leaves small, papillose; plants monoecious, the sporocarps terminal on leaf lobes, the indusium forming the sporocarp wall, the sporangia arising in acropetal succession, the annulus reduced or absent; microspores numerous, aggregated into masses (massulae); megaspores few; gametophytes dioecious, minute, with or without green tissue.

1. AZOLLA (12)

Rhizomes branched dichotomously above every third leaf; roots slender, unbranched, unattached; leaves bilobed, the upper aerial lobes papillose, several cells thick, the lower submersed, larger, 1 cell thick, submembranous; sporocarps borne on the first leaf of a branch, paired at the apex of the reduced bilobed submersed lobe, usually unlike, the microsporocarps large, globose, the wall membranous, 2-layered, the megasporocarp small, subfusiform, the wall thickened at apex; microspores 64, agglutinated into 4 to 6 massulae, these armed with numerous appendages (glochidia) with terminal, retrorse hooks; megasporangia each with a solitary, basal megaspore, this with a bell-shaped basal portion and 3 spongy apical lobes.

The species of this genus cannot be identified definitely without the use of a compound microscope.

1. Azolla filiculoides Lam. Near Tucson (Pima County), Crittenden (Santa Cruz County), Wikieup and mouth of Williams River (southern Mohave County), Alamo Crossing (northern Yuma County), 2,000 to 4,000 feet. Alaska to Guatemala; South America.

The Arizona plants were referred previously to *A. caroliniana* Willd. It is likely that *A. mexicana* Presl also occurs in Arizona, for it has been found in Nevada and Sinaloa. It differs in having megaspores with pitted basal portions rather than reticulate, as in *A. filiculoides*, and in having the glochidia of the massulae with numerous cross partitions.

6. MARSILEACEAE. Pepperwort Family

Rhizomes creeping at or just below the surface, branching, bearing roots at nodes; leaves alternate, in 2 rows, circinate in vernation, long-stalked, the blade divided to base into 4 leaflets, these folded upwardly in bud and at night; veins dichotomously forked, with frequent cross veins; sporocarps apparently axillary at base of leaves (in ours), ovoid, stalked, solitary, bituberculate near insertion of the stalk, thick-walled; sori several within a sporocarp, in 2 rows, each surrounded by a delicate indusium, the receptacle elongate, bearing megasporangia on top and microsporangia on the sides; spores tetrahedral, the microspores numerous, the megaspores 1 to a sporangium; gametophytes dioecious, minute; spermatozoid with a spiral appendage, multiciliate.

The plants of this family are aquatic or semiaquatic and have leaves resembling clover leaves.

1. MARSILEA. Pepperwort

1. Marsilea mucronata A. Br. Navajo, Coconino, Yavapai, Pinal, Cochise, and Pima counties, 1,500 to 7,000 feet. Saskatchewan and Alberta south to Texas and Arizona; Mexico.

Referred by Dr. Maxon to *M. vestita* Hook. & Grev., which has a dense tuft of very slender, light-colored hairs at the bases of the petioles, long, spreading hairs on the young petioles, and dense, more or less spreading hairs on the sporocarps. It ranges from British Columbia to southern California and has outlying stations in Montana and Colorado. *M. mucronata* is evidently very closely allied but is less hairy throughout, the hairs being coarser, those of the petioles few and subappressed, and those of the sporocarps thick, flattish, and closely appressed. (See Weatherby, 13, p. 325.)

7. POLYPODIACEAE. Fern Family

Stems (rhizomes) creeping or erect, sparingly branched, traversed by one or several vascular bundles, bearing hairs or scales, the scales thin, one cell thick, acute or acuminate; leaves large compared with the stems, circinate in vernation, divided into a stalk and a blade, the blade pinnatifid to decompound, glabrous or with various types of indument (wax, hairs, scales); fertile and sterile leaves alike or dissimilar; reproduction by means of spores, these borne in sporangia, which are grouped in sori; sori dorsal or marginal, borne on a vein or vein tip, rotund or elongate, often covered or surrounded by a thin, scalelike structure (indusium); sporangia mostly small, not developing all at once, mostly long-stalked, elastically dehiscent by means of a vertical, interrupted row of thick-walled cells (annulus); sexual generation (gametophyte) minute, the prothallium thalloid, green, dorsiventral, terrestrial, monoecious, the sex organs borne on the under surface.

A large family, widely distributed but most abundant in tropical mountains. The ferns are mostly not important economically, but many are highly prized as ornamentals.

KEY TO THE GENERA

1. Rhizomes (and blades) with hairs only. Sori linear, marginal; stipe bundles several; blades large, coarse, compound, subternate........................... 1. *Pteridium*
1. Rhizome scaly (2).
2. Sori marginal or nearly so, or borne in lines along the veins and then lacking an indusial covering; stipe bundle 1; spores tetrahedral (3).
2. Sori dorsal on the veins, if elongate then covered by a membranaceous indusium; stipe bundles 2 or more at base, often united upwardly; spores bilateral (8)
3. Sporangia borne on the veins on the under side of a reflexed marginal lobe. . 2. *Adiantum*
3. Sporangia not borne *on* a marginal lobe, but often borne on the leaf surface *under* a marginal lobe (4).
4. Sporangia borne in lines on the veins. Blades pentagonal in outline (5).
4. Sporangia submarginal (6).
5. Blades waxy beneath, otherwise glabrous and scaleless; rhizome short, thick 3. *Pityrogramma*
5. Blades pubescent, not waxy; rhizome slender, widely creeping............ 4. *Bommeria*
6. Margin strongly modified and reflexed (except in *P. limitanea* and *P. Jonesii*), elongate and continuous. Pinnae scaleless, glabrous (except in *P. intermedia* and *P. atropurpurea*), sometimes white-waxy beneath; rhizomes multicipital (creeping in *P. intermedia*) .... 5. *Pellaea*
6. Margin reflexed or not, mostly interrupted (7).
7. Vein ends thickened; margin mostly reflexed, sometimes modified; rhizomes multicipital or creeping; pinnae glabrous or mostly hairy or scaly, not waxy; fronds 2- to 4-pinnate .... 6. *Cheilanthes*
7. Vein ends thickened or not; margin not or scarcely reflexed, not modified; rhizomes multicipital; pinnae scaly, hairy, or mostly white- or yellow-waxy; fronds 1- to 4-pinnate .... 7. *Notholaena*
8. Indusium absent. Fronds articulate near base of stipe; blades pinnatifid, the segments entire or merely denticulate. Rhizomes wide-creeping and fronds scattered 8. *Polypodium*
8. Indusium present, or if absent (in *Gymnocarpium*) the blades bipinnate-pinnatfid or tripinnate (9).

9. Veins anastomosing and forming costular areoles, the large, tumid sori borne in a chain-like row close to the midrib.......................................18. *Woodwardia*
9. Veins all free or, in *Phanerophlebia*, casually anastomosing but not forming regular areoles (10).
10. Indusium inferior, i.e. borne under the sporangia and surrounding them, deeply lacerate into filamentous or broad segments..................................9. *Woodsia*
10. Indusium at one side of the sorus, or centrally attached, or absent (11).
11. Indusium attached at a point or absent (12).
11. Indusium with a linear attachment. Stipe bundles 2 at base of stipe (17).
12. Indusium ovate or oblong, hoodlike or convex........................10. *Cystopteris*
12. Indusium orbicular or reniform, or absent (13).
13. Indusium orbicular and centrally attached. Fronds coriaceous, the segments with spinulose teeth (14).
13. Indusium reniform and attached at one side or absent (15).
14. Sori in a single row on either side of the midrib; veins all free........11. *Polystichum*
14. Sori in 2 rows or more; veins often casually anastomosing..........12. *Phanerophlebia*
15. Acicular, unicellular hairs present on the midribs above; segments ciliate; stipe bundles 2, these united below the base of the blade; rhizome scales ciliate. Rhizomes creeping; fronds membranaceous; veins reaching the margin.............13. *Thelypteris*
15. Acicular hairs absent on the midribs above; segments not ciliate; stipe bundles free to the base of the blade; rhizome scales not ciliate, sometimes toothed (16).
16. Lowest pair of pinnae articulate to the rachis; stipe bundles 2; veins reaching the margin; rhizomes slender, wide-creeping; blades membranaceous, deltoid; indusium absent ....14. *Gymnocarpium*
16. Lowest pair of pinnae not articulate; stipe bundles 3 to 7; veins ending short of the margin in elongate hydathodes; rhizomes thick, short-creeping to erect; blades herbaceous to coriaceous, ovate to lanceolate; indusium present..........15. *Dryopteris*
17. Rhizome scales with short, broad cells and dark partition walls; stipe bundles without incurved margins, joining upwardly into an X-shaped bundle; indusia not crossing the veins; blades rather small, mostly thick-herbaceous to subcoriaceous 16. *Asplenium*
17. Rhizome scales with elongate, narrow cells, the walls pale; stipe bundles with incurved margins, joining upwardly into a U-shaped bundle; indusia (especially the apical ones) often crossing the veins; blades large, thin-membranaceous......17. *Athyrium*

1. PTERIDIUM (14). Bracken

Rhizomes wide-creeping, deeply subterranean, provided with septate hairs but not scaly; stipes coarse, with several vascular bundles free to base of blade; blades coarse, deciduous, bipinnate-pinnatifid to tripinnate-pinnatifid; pinnae opposite or nearly so, the pinnules alternate, the segments oblong to linear, with revolute margins; veins free, twice- or thrice-forked; sporangia borne on a continuous submarginal vascular strand connecting the vein ends, protected by an outer false indusium (the recurved, modified margin) and a minute, hyaline, extrorse inner indusium; spores tetrahedral, without perispore.

1. Pteridium aquilinum (L.) Kuhn. Lukachukai Mountains (Apache County) and Kaibab Plateau (Coconino County) to the mountains of Cochise and Pima counties, 5,000 to 8,500 feet, very common in open pine forests; nearly cosmopolitan.

The Arizona plants are referable to var. *pubescens* Underw., which is widely distributed in western and northern North America. This fern is reported to be poisonous to cattle and horses when eaten in large quantities, but the poisonous properties may be eliminated by cooking. The rootstocks and young fronds have been utilized for human food.

2. ADIANTUM. Maidenhair Fern

Woodland ferns with creeping, scaly rhizomes, the scales subclathrate, concolorous, persistent, entire, attenuate; fronds distichous, drooping; stipes and rachises black or red-black, polished, scaleless except at base; bundle 1; blades membranaceous, compound, glabrous, scaleless; veins free, several- to many-times forked, ending in the teeth of the sterile segments; indusium none; sporangia borne on the veins of the lower side of reflexed, modified marginal lobes; spores tetrahedral.

KEY TO THE SPECIES

1. Blades broader than long, the stipes equally forked at apex, the pinnae borne along one side of the 2 curved rachises; segments subrectangular; rhizome scales large, 1 to 2 mm. wide .... 1. *A. pedatum*
1. Blades longer than broad, the stipe not forked, continuous with the rachis; segments flabellate or rhomboid; rhizome scales minute, less than 0.5 mm. broad 2. *A. Capillus-Veneris*

1. Adiantum pedatum L. Oak Creek Canyon and Schnebly Hill (Coconino County), near General Springs (Gila County), Huachuca Mountains (Cochise County). Widely distributed in temperate North America.

The Arizona plants have been referred erroneously to var. *aleuticum* Rupr.

2. Adiantum Capillus-Veneris L. Throughout most of the state, 1,500 to 7,000 feet, mostly on moist cliffs and in springy places. Virginia south to Florida, west to Kentucky, Tennessee, Missouri, Texas, and Arizona; tropical America; Eurasia.

3. PITYROGRAMMA. Gold Fern

Evergreen ferns of dry banks and ledges, with short-creeping, stout, scaly rhizomes, the scales linear-subulate, bicolorous, entire; stipes longer than the blades, red-brown, polished, glabrous and scaleless; blades membranaceous, tripinnate-pinnatifid, deltoid, the lowest pinnae opposite, basiscopically developed, almost equal in size to the remainder of the blade, glabrous and scaleless, golden-waxy beneath; veins free, branched, reaching the unreflexed margin; sporangia following the veins throughout; indusium none.

1. Pityrogramma triangularis (Kaulf.) Maxon. Graham, Gila, Maricopa, Pinal, and Pima counties, 2,000 to 5,000 feet, rock ledges, not uncommon. British Columbia (Vancouver Island) to Nevada and northwestern Mexico.

In Arizona the species is represented by var. *Maxonii* Weatherby (type from the Rincon Mountains, *Blumer* 3271), which occurs also in southern California, Sonora, and Baja California.

4. BOMMERIA

Small evergreen rock ferns with slender, wide-creeping, scaly rhizomes, the scales linear-lanceolate, concolorous, entire, persistent; stipes longer than the blades, light brown, hairy and sparingly scaly; blades tripinnatifid, the main rachis winged throughout, pedate, the basal segments basiscopically developed, the segments pilose above, brown-tomentose and scaly beneath, not waxy; veins free (in ours), forked; sporangia following the veins throughout; indusium none.

1. **Bommeria hispida** (Mett.) Underw. Mountains of Yavapai, Gila, Graham, Pinal, Cochise, Santa Cruz, and Pima counties, 4,000 to 6,000 feet, on shaded cliffs. Western Texas to Arizona and Mexico.

5. PELLAEA. CLIFF-BRAKE

Evergreen rock ferns with stout, short, multicipital rhizomes (long-creeping in *P. intermedia*), the scales linear-subulate, filiform-tipped, concolorous or bicolorous, glabrous, nearly entire; stipes mostly shorter than the blades, wiry, pale brown to dark purplish, with 1 vascular bundle; blades subcoriaceous, subdimorphic, the sterile segments broader than the fertile ones, 1- to 4-pinnate, scaleless, not white- or yellow-waxy (except in *P. limitanea*), glabrous (except in *P. atropurpurea* and *P. intermedia*); veins free, simple or forked; sori exindusiate, submarginal, the sporangia scattered on the tips of the vein branches, confluent with age, protected by the reflexed, modified, continuous, indusiumlike margin (except in *P. Jonesii* and *P. limitanea*); spores tetrahedral, without perispore, often rugose.

KEY TO THE SPECIES

1. Pinnae white-waxy beneath; blades 4-pinnate, the segments minute.......1. *P. limitanea*
1. Pinnae not waxy beneath; blades 1- to 3- pinnate (2).
2. Rhizome scales concolorous (3).
2. Rhizome scales with a dark central stripe (4).
3. Blades 2- to 3-pinnate, the segments small, not over 5 mm. long, roundish...2. *P. Jonesii*
3. Blades 1- to 2-pinnate, the segments mostly over 1 cm. long, oblong....3. *P. atropurpurea*
4. Rachises hairy, deeply channeled on the upper side; rhizome wide-creeping 4. *P. intermedia*
4. Rachises glabrous, terete throughout; rhizomes multicipital (5).
5. Blades triangular-ovate; pinnae mostly with 6 to 10 pairs of ovate to ovate-oblong pinnules ....5. *P. longimucronata*
5. Blades linear; pinnae with 1 to 4 pairs of mostly narrowly oblong segments 6. *P. Wrightiana*

1. **Pellaea limitanea** (Maxon) Morton (*Notholaena limitanea* Maxon). Marble Canyon and Wupatki National Monument (Coconino County), to Cochise, Santa Cruz, Pima, and Yuma counties, 2,000 to 7,000 feet, hillsides and cliffs. Western Texas to southern Utah and Arizona.

Var. *mexicana* (Maxon) Morton, founded on material from Chihuahua, Mexico, is represented by a collection from the Chiricahua Mountains, 6,000 feet (*Blumer* 2390). It differs in its narrower and less-divided blades, with larger segments.

2. **Pellaea Jonesii** (Maxon) Morton (*Notholaena Jonesii* Maxon). Marble Canyon and Havasu Canyon (Coconino County), Meriwitica Canyon (Mohave County), near Superior (Pinal County), Waterman Mountains (Pima County), 2,000 to 4,000 feet, crevices of limestone cliffs. Southern Utah and Arizona to southern California.

3. **Pellaea atropurpurea** (L.) Link. Heber (Navajo County), Havasu Canyon (Coconino County), Mogollon Escarpment (Coconino and Gila counties), and mountains of Graham, Gila, Cochise, Santa Cruz, and Pima counties, 3,500 to 7,000 feet, on cliffs. Vermont and Ontario to South Dakota, south to Florida and Texas, New Mexico, and Arizona; Mexico and western Guatemala.

Var. *simplex* (Butters) Morton (*P. Suksdorfiana* Butters) occurs in Coconino County (Kaibab Plateau, Grand Canyon, San Francisco Peaks, and Walnut Canyon), 6,500 to 9,000 feet. It ranges from British Columbia south to Utah, Mexico, and Arizona. It differs in its smaller size and glabrous rachises.

4. Pellaea intermedia Mett. Mountains of Gila, Cochise, Santa Cruz, and Pima counties, 3,500 to 7,000 feet, dry, rocky slopes and in crevices of limestone ledges. Southwestern Texas to Arizona and Mexico.

5. Pellaea longimucronata Hook. (*P. truncata* Goodding). Grand Canyon (Coconino County) and Grand Wash Cliffs (northern Mohave County), to the mountains of Cochise, Santa Cruz, Pima, and Yuma counties, 2,000 to 6,000 (8,000?) feet, very common among rocks and on cliffs, type of *P. truncata* from the Mule Mountains (*Goodding* 977). Colorado and New Mexico to Nevada, Arizona, and northern Mexico.

6. Pellaea Wrightiana Hook. Near Flagstaff (Coconino County), and mountains of Graham, Gila, Pinal, Cochise, Santa Cruz, and Pima counties, (3,000?) 4,000 to 8,000 feet, among rocks and on cliffs. Southwestern Oklahoma and central Texas, west to Arizona.

This species includes the Arizona specimens formerly referred to *P. ternifolia* (Cav.) Link. These are somewhat intermediate between the two species but seem closer to *P. Wrightiana*.

6. CHEILANTHES (15). Lip Fern

Evergreen, mostly xerophytic rock ferns, with long-creeping or short, multicipital rhizomes, the scales abundant, mostly entire and glabrous, persistent, often with a dark, sclerotic central stripe; stipes shorter than the blades, dark and shining, with 1 vascular bundle; blades uniform, 2- to 4-pinnate, thick-membranaceous or mostly coriaceous, often tomentose or paleaceous, not white- or yellow-waxy; pinnae many pairs, petiolulate, the lower subopposite, the ultimate segments often minute; veins all free, not reaching the margin, clavate at the tip, mostly forked; sori borne on the vein tips, usually protected by the reflexed leaf margin, this sometimes modified (indusiumlike), mostly interrupted; spores tetrahedral, without perispore.

KEY TO THE SPECIES

1. Segments minute, rotund, beadlike (2).
1. Segments not rotund or beadlike (10).
2. Pinnae scaleless, villous or tomentose beneath (3).
2. Pinnae scaly, at least on the midribs beneath (4).
3. Segments hairy above; rhizomes multicipital.............................1. *C. Feei*
3. Segments glabrous above; rhizomes long-creeping.....................2. *C. lendigera*
4. Segments glabrous above. Rhizomes widely creeping (5).
4. Segments hairy above (7).
5. Scales of blade not ciliate, inconspicuously denticulate.................3. *C. Fendleri*
5. Scales long-ciliate, at least at base (6).
6. Rhizome scales narrow, rigid, with strongly sclerotic walls; scales of blade with deeply cordate base with overlapping basal lobes..........................4. *C. Covillei*
6. Rhizome scales broader, thinner, pale brown or only partly sclerotic-walled; scales of blade rounded at base or merely subcordate........................5. *C. Wootoni*
7. Larger scales of blade ciliate at base; rhizome widely creeping.......6. *C. Lindheimeri*
7. Larger scales not ciliate; rhizome multicipital (8).

8. Scales of lower side of pinnae large, whitish, ovate, concealing the segments; segments coarsely villous above . . . . 7. *C. villosa*
8. Scales narrower, not concealing the segments; segments finely villous or tomentulose above (9).
9. Scales of rachis narrowly linear-subulate, pale; ultimate segments mostly less than 1 mm. in diameter . . . . 8. *C. tomentosa*
9. Scales broader, brown; ultimate segments mostly more than 1 mm. in diameter 9. *C. Eatoni*
10. Pinnae scaly on the midribs, glabrous or nearly so. Rhizome creeping . . . . 10. *C. Pringlei*
10. Pinnae scaleless (11).
11. Segments hairy above, white-villous beneath with long hairs. Rhizomes multicipital 11. *C. Parryi*
11. Segments glabrous above, nearly glabrous beneath (12).
12. Rachis terete throughout, hairy on the upper side. Rhizomes creeping . 12. *C. alabamensis*
12. Rachis deeply channeled on the upper side, glabrous (13).
13. Segments not spathulate, not narrowed at base; blades oblong; rhizomes creeping 13. *C. Wrightii*
13. Segments spathulate, narrowed at base; blades subdeltoid; rhizomes multicipital 14. *C. pyramidalis*

1. Cheilanthes Feei Moore. Almost throughout the state, 2,000 to 7,000 feet, dry, rocky slopes and cliffs. Illinois and Wisconsin, west to British Columbia and south to Arkansas, Oklahoma, Texas, New Mexico, Arizona, and southern California.

2. Cheilanthes lendigera (Cav.) Swartz (*Pomatophytum pocillatum* M. E. Jones). Chiricahua, Mule, and Huachuca mountains (Cochise County), Oro Blanco Mountains (Santa Cruz County), Rincon Mountains (Pima County), 5,000 to 6,000 feet, moist, shaded ravines and canyons, locally abundant, the type of *Pomatophytum pocillatum* from the Huachuca Mountains (*Jones* 24690). Western Texas; Mexico to Venezuela and Ecuador.

3. Cheilanthes Fendleri Hook. Walnut Canyon and Slate Mountain (Coconino County), Hualpai Mountain (Mohave County), to the mountains of Greenlee, Cochise, Santa Cruz, and Pima counties, 4,000 to 9,500 feet, dry, rocky slopes and cliffs. Western Texas to Colorado and Arizona.

4. Cheilanthes Covillei Maxon. Black Mountains (Mohave County) to the mountains of Cochise and Pima counties, 2,000 to 5,000 feet, cliffs and rock ledges. Southwestern Utah and Arizona to southern California and Baja California.

5. Cheilanthes Wootoni Maxon. Grand Canyon (Coconino County) to the mountains of Greenlee, Cochise, Santa Cruz, and Pima counties, 3,000 to 9,000 feet, rock ledges and among boulders, common. Oklahoma, southeastern Colorado, and western Texas to Arizona and northern Mexico (Sonora).

6. Cheilanthes Lindheimeri Hook. Prescott (Yavapai County) and mountains of Graham, Gila, Maricopa, Pinal, Cochise, Santa Cruz, and Pima counties, 2,000 to 8,000 feet, dry slopes among rocks, common. Western Texas to Arizona and northern Mexico.

7. Cheilanthes villosa Davenp. Mountains of Maricopa, Cochise, Santa Cruz, Pima, and Yuma counties, 2,000 to 4,000 feet, granitic and limestone slopes and ledges. Western Texas to southern Arizona and northern Mexico.

8. Cheilanthes tomentosa Link. Mountains of Graham, Cochise, Santa Cruz, and Pima counties, 4,000 to 7,500 feet, shaded, rocky places. Virginia south to Georgia, west to Missouri, Oklahoma, Texas, and Arizona; Mexico.

9. Cheilanthes Eatoni Baker. Navajo and Coconino counties to Graham, Cochise, Santa Cruz, and Pima counties, 3,000 to 8,000 feet, dry, rocky slopes and cliffs, often in chaparral. Oklahoma and central Texas to Colorado, Arizona, and Mexico.

Plants with the upper surface of the segments green and sparingly villous are referable to f. *castanea* (Maxon) Correll (*Cheilanthes castanea* Maxon).

10. Cheilanthes Pringlei Davenp. Mazatzal Mountains (Gila County), Chiricahua Mountains (Cochise County), and mountains of Pima County, 3,000 to 5,000 feet, at base of cliffs, type from "southeastern Arizona" (*Pringle* in 1883). Southern Arizona and northern Mexico.

11. Cheilanthes Parryi (D. C. Eaton) Domin (*Notholaena Parryi* D. C. Eaton). Coconino and Mohave counties to Pima and Yuma counties, 1,000 to 6,500 feet, dry, hot situations, crevices of canyon walls and among rocks, very common. Southwestern Utah and Arizona to the desert region of southern California.

Sometimes *C. Parryi* superficially resembles *C. Feei,* from which it may be immediately distinguished by the densely hairy upper surfaces of the segments; these in *C. Feei* are green and very sparingly hairy or mostly glabrate. *C. Feei* is essentially tripinnate, with round segments mostly 1 to 1.5 mm. long and wide; *C. Parryi* is mostly only bipinnate or bipinnate-pinnatifid, and the segments are much longer than wide, often 2.5 to 3.5 mm. long.

12. Cheilanthes alabamensis (Buckl.) Kunze. Mustang and Huachuca mountains (Cochise County), Canille (Canelo) Hills and near Sonoita (Santa Cruz County), 4,500 to 5,000 feet, shaded north slopes, on limestone. Virginia to Georgia, west to Missouri, Oklahoma, Texas, and Arizona; Mexico; Jamaica.

13. Cheilanthes Wrightii Hook. Mountains of Greenlee, Graham, Gila, Maricopa, Pinal, Cochise, Santa Cruz, and Pima counties, 1,000 to 6,000 feet, rocky slopes and ledges. Western Texas to Arizona and northern Mexico.

14. Cheilanthes pyramidalis Fée. Chiricahua, Mule, and Huachuca mountains (Cochise County), about 7,000 feet. Southeastern Arizona, Mexico, and Central America.

A highly variable species, represented in Arizona by var. *arizonica* (Maxon) Broun, type from the Huachuca Mountains (*Goodding* 1327).

7. NOTHOLAENA. Cloak Fern

Evergreen, mostly xerophytic rock ferns, with short, congested, multicipital rhizomes, the scales abundant, mostly entire and glabrous; stipes often longer than the blades, dark and wiry, with 1 vascular bundle; blades uniform, 1- to 4-pinnate, thick-membranaceous or coriaceous, tomentose, scaly, or white- or yellow-waxy, linear to pentagonal; pinnae few to many pairs, the ultimate segments not beadlike; veins all free, not reaching the margin, enlarged at tip or not; sori borne at the vein tips, the leaf margin not much recurved, not modified; spores tetrahedral, without perispore.

This genus contains an artificial assemblage of probably unrelated species and is difficult to define, for it merges with *Pellaea* on the one hand and

Cheilanthes on the other. It has been united outright with *Cheilanthes* by some authors, but it seems best to retain the traditional grouping until a comprehensive survey of the entire group can be made. Two species included in *Notholaena* in *Flowering Plants and Ferns of Arizona* are here removed to *Pellaea*, and one to *Cheilanthes*.

KEY TO THE SPECIES

1. Blades simply pinnate (2).
1. Blades 2- to 4-pinnate (3).
2. Pinnae not scaly, densely tomentose beneath . . . . 1. *N. aurea*
2. Pinnae densely scaly beneath . . . . 2. *N. sinuata*
3. Pinnae hairy above, not waxy beneath, densely scaly . . . . 3. *N. Aschenborniana*
3. Pinnae not hairy above, white- or yellow-waxy beneath (4).
4. Pinnae scaly on the midribs beneath . . . . 4. *N. Grayi*
4. Pinnae scaleless (5).
5. Blades linear-oblong . . . . 5. *N. Lemmoni*
5. Blades ovate to pentagonal (6).
6. Blades merely bipinnatifid, the rachis scarcely if at all free, the primary segments adnate 6. *N. Standleyi*
6. Blades 2- to 4-pinnate (7).
7. Rhizome scales concolorous; stipes reddish brown . . . . 7. *N. californica*
7. Rhizome scales with a dark central stripe; stipes blackish . . . . 8. *N. neglecta*

1. Notholaena aurea (Poir.) Desv. Willow Spring (southern Apache County) and mountains of Gila, Pinal, Cochise, Santa Cruz, and Pima counties, 4,000 to 7,000 feet, dry ledges and rocky slopes. Texas to Arizona; Mexico to Argentina; Jamaica; Hispaniola.

2. Notholaena sinuata (Lag.) Kaulf. Canyons of the Colorado (Coconino and Mohave counties), to the mountains of Graham, Cochise, Santa Cruz, Pima, and Yuma counties, 1,000 to 7,000 feet, dry, rocky slopes and crevices, often on limestone, very common. Western Oklahoma and Texas to southern California; Mexico to Chile; Jamaica; Hispaniola.

KEY TO THE VARIETIES, after Weatherby (13, pp. 314, 315)

1. Pinnae 1 cm. long or more, ovate, commonly subacute and with 4 to 6 pairs of oblong lobes; scales of upper surface of lamina with a narrow central portion or reduced to stellate processes, usually soon deciduous. Rhizome scales pectinate-ciliate or serrulate . . . . *N. sinuata* (typical)
1. Pinnae mostly less than 1 cm. long, very obtuse, entire or with 1 to 3 pairs of broadly ovate lobes; scales of upper surface with a relatively broad central portion, usually persistent until full maturity of the frond. (2).
2. Pinnae oblong, entire or with about 3 pairs of shallow lobes; rhizome scales pectinate-ciliate or serrulate . . . . var. *integerrima*
2. Pinnae subquadrate, nearly or quite as wide as long, with 1 or 2 (3) pairs of lobes; rhizome scales entire or nearly so . . . . var. *cochisensis*

Var. *cochisensis* (Goodding) Weatherby (type from the Huachuca Mountains, *Goodding* 373) occurs throughout most of the range of the species in Arizona where it is the commonest variety, especially in Cochise and Pima counties. There is experimental evidence that this variety is toxic to sheep and goats whereas the typical species is harmless. Var. *integerrima* Hook. has been collected at a few localities in Yavapai, Gila, Pinal, Cochise, and Pima counties.

3. Notholaena Aschenborniana Klotzsch. Mule and Huachuca mountains (Cochise County), Santa Rita Mountains (Santa Cruz or Pima County), about 5,000 feet, dry, rocky slopes. Western Texas to Arizona and Mexico.

4. Notholaena Grayi Davenp. (*N. hypoleuca* Goodding). Mountains of Greenlee, Graham, Cochise, Santa Cruz, and Pima counties, 4,000 to 6,000 feet, dry, rocky slopes, type from southeastern Arizona (*Courtis*), type of *N. hypoleuca* from the Mule Mountains (*Goodding* 1004). Western Texas to Arizona and Mexico.

5. Notholaena Lemmoni D. C. Eaton. Santa Rita, Tumacacori, Rincon, and Santa Catalina mountains (Santa Cruz and Pima counties), about 4,000 feet, type from the Santa Catalina Mountains (*Lemmon* in 1880). Southern Arizona and northern Mexico.

6. Notholaena Standleyi Maxon. Greenlee, Graham, Gila, Maricopa, Pinal, Cochise, Santa Cruz, and Pima counties, 1,000 to 6,500 feet, among rocks, common. Western Oklahoma and Texas to Colorado, Arizona, and Mexico.

7. Notholaena californica D. C. Eaton. Near Congress Junction (Yavapai County), Fish Creek (Maricopa County), Kofa, Tule, and Tinajas Altas mountains (Yuma County), 1,000 to 3,000 feet, dry, rocky slopes and canyons. Arizona, southern California, and Baja California.

8. Notholaena neglecta Maxon. Cochise County, in the Mule Mountains (*Goodding* 1384) and Huachuca Mountains (*Lemmon* in 1882), south faces of limestone cliffs. Southeastern Arizona and northern Mexico.

8. POLYPODIUM. Polypody

Evergreen rock ferns with wide-creeping, scaly rhizomes; fronds scattered, subcoriaceous; stipes bisulcate on the upper side, articulate near base, with 3 vascular bundles near base, these uniting upwardly; blades pinnatisect, with alternate segments, not narrowed toward base, glabrous, scaly or not, the segments entire or remotely toothed; veins not reaching the margin, thickened at tip, once- or twice-forked, sometimes anastomosing; sori round, subterminal on the basal anterior vein fork, exindusiate; spores bilateral, verrucose, without perispore.

KEY TO THE SPECIES

1. Stipes and lower surfaces of blades scaleless; segments remotely toothed; rhizome scales concolorous; veins obvious, all free .... 1. *P. vulgare*
1. Stipes and lower surfaces of blades densely scaly; segments entire; rhizome scales with a dark median band; veins obscure, partly anastomosing .... 2. *P. thyssanolepis*

1. Polypodium vulgare L. Mogollon Escarpment and Oak Creek (southern Coconino County) and mountains of Graham, Gila, Cochise, and Pima counties, 7,000 to 9,000 feet, sides of canyons and ledges in forests. Widely distributed in North America and Eurasia.

The Arizona plant is var. *columbianum* Gilbert (*P. hesperium* Maxon, *P. prolongilobum* Clute, *P. vulgare* var. *perpusillum* Clute). The types of *P. prolongilobum* and *P. vulgare* var. *perpusillum* came from the Santa Catalina Mountains (*Ferriss* in 1910).

2. Polypodium thyssanolepis A. Braun. Chiricahua and Huachuca mountains (Cochise County), Baboquivari Mountains (Pima County), 5,000 to 6,000 feet, among rocks in canyons. Western Texas to Arizona, Mexico, and South America; Jamaica; Hispaniola.

9. WOODSIA

Rock ferns with deciduous, clustered fronds and short, erect, scaly rhizomes, the scales linear-lanceolate, clathrate, subdenticulate, with a dark, sclerotic central stripe; stipes relatively stout, shorter than the blades, persistent, with 2 vascular bundles at base, these uniting upwardly; blades uniform, lanceolate, pinnate-pinnatifid, membranaceous, slightly reduced at base, scaleless (in ours), often pilose or glandular; veins free, several times forked, not reaching the margin, slightly thickened at apex; sori dorsal on the veins, borne at the apex of the anterior fork, supramedial, rotund in outline but the receptacle elongate; indusium inferior, i.e. borne below and surrounding the sporangia, consisting of a small disk with few to many lobes, these broad, or narrow and hairlike; spores large, bilateral, winged.

KEY TO THE SPECIES

1. Fronds bearing long, articulate, many-celled hairs on the midribs and veins beneath 1. *W. scopulina*
1. Fronds lacking articulate hairs, but mostly provided with unicellular or glandular hairs (2).
2. Indusial lobes broad, covering the sporangia when young, lacerate at apex; fronds glanduliferous ....2. *W. Plummerae*
2. Indusial lobes broad only at base, deeply lacerate into filamentous lobes; fronds normally eglandular (3).
3. Indusial lobes elongate, much exceeding the sporangia; leaf margins whitish, with numerous hairlike processes..3. *W. mexicana*
3. Indusial lobes shorter, often concealed at maturity; leaf margins little altered 4. *W. oregana*

1. Woodsia scopulina D. C. Eaton. Recorded for Arizona on the basis of a collection purporting to be from the Huachuca Mountains (*Lemmon* in 1882) and one from "Maricopa" (*Pringle* in 1882). Quebec, south in varietal forms to North Carolina and Tennessee, west to Colorado, Arizona, and California, and north to Alaska.

2. Woodsia Plummerae Lemmon (*W. pusilla* Fourn. var. *glandulosa* (Eaton & Faxon) Taylor). Apache and Coconino counties south to the mountains of Cochise, Santa Cruz, Pima, and Yuma counties, 2,000 to 9,000 feet, shaded ledges and cliffs. Western Texas to Arizona and northern Mexico.

3. Woodsia mexicana Fée (*W. pusilla* Fourn. var. *mexicana* (Fée) Taylor). Lukachukai and White mountains (Apache County), Grand Canyon (Coconino County), and Greenlee County, south to Cochise, Santa Cruz, and Pima counties, 3,500 to 9,500 feet, crevices of cliffs and rocky slopes. Western Texas to Arizona and Mexico.

A form, that described as *W. pusilla* Fourn., with fronds slightly glandular beneath, occurs in the Huachuca Mountains (see Weatherby, 13, p. 308).

4. Woodsia oregana D. C. Eaton. Navajo Mountain, Kaibab Plateau, Grand Canyon, San Francisco Peaks, etc. (Coconino County), 5,500 to 11,500 feet, from the pinyon belt to timber line, rock crevices. Quebec west to British

Columbia, south to New York, Michigan, Oklahoma, New Mexico, Arizona, and southern California.

10. CYSTOPTERIS. Bladder Fern

Delicate, deciduous, woodland ferns with creeping, unbranched, scaly rhizomes, the scales fibrous, concolorous, glabrous, subentire or denticulate; stipes weak, subremote, stramineous or pale brown, shorter than the blades, scaly at base only, with 2 vascular bundles free to base of blade; blades ovate-lanceolate to linear-lanceolate, thin-membranaceous, glabrous or with a few septate hairs on the rachis, scaleless, pinnate-pinnatifid to subtripinnate; veins free, simple or forked, reaching the margin, slightly thickened at apex; sori dorsal on the veins, rotund; indusium thin, extrorse, attached at point, hoodlike or deeply convex, often caducous; spores bilateral, spinulose.

KEY TO THE SPECIES

1. Fronds long-attenuate at apex; basal pinnae the largest; rachis often bulblet-bearing at apex; veins mostly running to the sinuses .... 1. *C. bulbifera*
1. Fronds merely acute or acuminate; basal pinnae shorter than the second pair; rachis not bulblet-bearing; veins mostly running to the teeth .... 2. *C. fragilis*

1. Cystopteris bulbifera (L.) Bernh. Oak Creek (Coconino County), very common on the West Fork, 5,500 to 7,000 feet. Newfoundland south to Georgia and Alabama, west to Manitoba and south to Arkansas, New Mexico, and Arizona.

2. Cystopteris fragilis (L.) Bernh. White Mountains (Apache County), Betatakin (Navajo County), and Grand Canyon (Coconino County), to the mountains of Cochise, Santa Cruz, and Pima counties, 5,000 to 12,000 feet, rich, moist, shaded soil among rocks and around springs.

This is one of the most widely distributed of all ferns, being nearly cosmopolitan. In North America the typical phase ranges from Labrador to Alaska, south to North Carolina, the Great Lakes region, Texas, New Mexico, Arizona, and southern California. Most of the Arizona specimens are referable to var. *tenuifolia* (Clute) Broun, a delicate and beautiful large form. Other varieties occur in the eastern United States.

11. POLYSTICHUM. Holly Fern

Evergreen, terrestrial ferns with erect or ascending, thick, densely scaly rhizomes and fasciculate, coriaceous, erect fronds; stipes much shorter than the blades, persistently scaly, the scales spreading, brown, fibrous, denticulate, mixed with small, appressed, hairlike ones, the vascular bundles 4 or 5, free; blades pinnate or subbipinnate, glabrous, scaly, especially on the rachis and costae; pinnae or pinnules mostly subfalcate, spinulose-toothed; veins free, not reaching the margin, mostly once-forked; sori dorsal on the anterior vein branches, in one row on either side of the midrib (in ours), often confluent with age; indusium central, rotund, centrally attached, persistent; spores bilateral, with perispore.

KEY TO THE SPECIES

1. Pinnae unlobed, with a superior basal auricle .... 1. *P. Lonchitis*
1. Pinnae with 2 or 3 nearly free segments at base .... 2. *P. scopulinum*

1. Polystichum Lonchitis (L.) Roth. West Fork of Oak Creek, Coconino County (*Deaver* 1961), Pinaleno Mountains, Graham County (*Thornber & Shreve* 7767, *Phillips et al.* 2482), in deep shade, about 8,000 feet. Newfoundland to Alaska, southern Ontario, Michigan, and Montana, and in the mountains to Colorado, Arizona, and California; Greenland; Eurasia.

2. Polystichum scopulinum (D. C. Eaton) Maxon. Elden Mountain (Coconino County), 8,500 feet (*Collins* 14, *Lefebure* 859). Quebec; Idaho and Washington; Arizona and California.

12. PHANEROPHLEBIA

Evergreen, terrestrial ferns with short, thick, erect, densely scaly rhizomes and fasciculate, coriaceous fronds; stipes shorter than the blades, bearing large, pale brown, hair-pointed, persistent, spreading scales, the vascular bundles 3, free; fronds imparipinnate, glabrous, scaly, especially on the rachis and costae; pinnae numerous, falcate, unlobed, the lower usually auriculate at the superior base, spinulose-toothed; veins several times forked, free or casually anastomosing but not forming regular areoles, not reaching the margin; sori dorsal on the veins, not confined to the anterior branches, in 2 rows or more on either side of the midrib; indusium central, persistent, centrally attached; spores bilateral, with perispore.

1. Phanerophlebia auriculata Underw. Mountains of Graham, Pinal, Cochise, Pima, and Yuma counties, 2,000 to 7,000 feet, damp, shaded canyon walls, local. Western Texas to Arizona and northern Mexico.

13. THELYPTERIS

Woodland ferns with thick, long-creeping, scaly rhizomes, the scales few, deciduous, ciliate, entire, fibrous; fronds light green, thin-membranaceous; stipes stramineous, nearly scaleless; blades pinnate-pinnatifid, not or only slightly reduced at base, nearly scaleless, the pinnae slightly reduced at base, the midribs pilose above and beneath, the minor axes not decurrent, the segments entire, ciliate, pubescent on the veins; veins simple, reaching the margins, the lowest pair connivent to the sinus; sori dorsal on the veins, in 1 row on either side of the midribs; indusium reniform, persistent, rather small, densely pilose; spores bilateral, with perispore.

1. Thelypteris augescens (Link) Munz & Johnston. Santa Maria River (southwestern Yavapai County), Aravaipa Canyon (Graham and Pinal counties), Santa Catalina Mountains (Pima County).

The Arizona plant is var. *puberula* (Fée) Munz & Johnston (*Dryopteris Feei* C. Chr.), which ranges from central Arizona and southern California south into Mexico.

14. GYMNOCARPIUM. Oak Fern

Delicate, deciduous, woodland ferns with long-creeping, thin, branched, scaly rhizomes, the scales few, concolorous, fibrous, glabrous, not ciliate, entire; fronds scattered, thin-membranaceous; stipes stramineous, slender, longer than the blades, glabrous, sparsely scaly at base, with 2 vascular bundles free to the apex; blades deltoid or pentagonal, bipinnate-pinnatifid to tripinnate-pinnatifid, glabrous and scaleless, sometimes bearing a few glands, the lowest pair of pinnae articulate at base, long-petiolulate, inequilateral; veins free, forked, reaching the margin; sori dorsal on the anterior vein branch, rotund, exindusiate; spores bilateral.

1. Gymnocarpium Dryopteris (L.) Newm. (*Dryopteris Linnaeana* C. Chr., *Dryopteris disjuncta* (Rupr.) Morton). Bonita Creek, White Mountains (Apache County), about 9,000 feet, on steep, shaded slopes (*Goodding* 1222). Newfoundland to Alaska, south to Virginia, Kansas, New Mexico, Arizona, and Oregon; Greenland; Eurasia.

15. DRYOPTERIS. Shield Fern

Evergreen, terrestrial, woodland ferns with thick, erect, densely scaly rhizomes, the scales fibrous, glabrous, entire or toothed, not ciliate; fronds coriaceous, glabrous or with capitate glands, mostly scaly; stipes stout, stramineous, shorter than the blades, with 3 to 7 free vascular bundles; blades bipinnate to tripinnate-pinnatifid, the ultimate segments mostly toothed, not ciliate, the minor axes decurrent on the major ones and forming the sides of dorsal grooves; veins free, simple or mostly forked, not reaching the margin, thickened at apex; sori dorsal on the veins, borne on the anterior vein branch, in 1 row on either side of the midrib; indusium large, reniform, persistent, glabrous (or glandular on the margin); spores bilateral, with perispore.

KEY TO THE SPECIES

1. Fronds subtripinnate-pinnatifid, the blades broadest near base, obviously capitate-glandular, not scaly on the pinnae....................1. *D. patula*
1. Fronds bipinnate or bipinnate-pinnatifid, the blades not broadest at base, not or inconspicuously capitate-glandular, scaly on the midribs of the pinnae and often also on the pinnules (2).
2. Indusium thin, deeply concave with incurved margins and mostly covering the sporangia; pinnules lobed, the lobes ending in 2 to 5 sharp, spreading teeth; scales of the midribs entire or merely denticulate....................2. *D. arguta*
2. Indusium thick, plane or the margins recurved, not covering the sporangia; pinnules scarcely lobed, the teeth incurved; scales of the midribs long-fimbriate..3. *D. Filix-mas*

1. Dryopteris patula (Swartz) Underw. Huachuca Mountains (Cochise County), moist swales in canyons (*Lemmon* in 1882, *Goodding* 1328). Widely distributed in tropical America.

This variable species is represented in the United States (Arizona only) by var. *Rossii* C. Chr., which extends into Mexico.

2. Dryopteris arguta (Kaulf.) Watt. Queen Creek Canyon and Superstition Mountains (Pinal County), Sierra Ancha (Gila County), 4,000 to 5,000 feet, along streams. Extreme southwestern Washington to southern California and Arizona.

3. Dryopteris Filix-mas (L.) Schott. White Mountains (Apache County), and mountains of Coconino, Graham, Gila, Cochise, Santa Cruz, and Pima counties, 6,500 to 10,000 feet, in rich soil among rocks and along streams. Newfoundland to British Columbia, south to Vermont, Michigan, and in the mountains to western Oklahoma, western Texas, New Mexico, Arizona, southern California, and Mexico; Greenland; Eurasia.

From this species, known from the time of the old herbalists as male-fern, is derived the drug aspidium, used extensively as a vermifuge, especially for tapeworm. It is a violent poison, and grave consequences have resulted from overdoses.

16. ASPLENIUM. Spleenwort

Evergreen rock ferns with short-creeping to erect, scaly rhizomes, the scales lanceolate to linear-lanceolate, thin, often iridescent, the cells with dark walls; fronds membranaceous to spongiose-coriaceous; stipes mostly wiry, green to black, glabrous, more or less scaly, with 2 oval vascular bundles at base, these united below the blade into an X-shaped bundle; blades pinnatifid to bipinnate-pinnatifid, mostly with numerous pairs of pinnae, the segments entire to toothed or incised, nearly scaleless; veins free, simple or forked, not reaching the margin; sori dorsal, elongate, the indusia attached at one side, facing distally, thin and hyaline, glabrous, subentire; spores bilateral, with perispore.

KEY TO THE SPECIES

1. Rachis green throughout or dark only in the basal half on the lower side, without 2 ridges on the upper side (2).
1. Rachis dark throughout, with 2 sharp ridges on the upper side. Blades linear, simply pinnate; stipes much shorter than the blades (5).
2. Fronds grasslike, the stipe much longer than the blade, green except at base, the blade consisting of 2 or 3 alternate, linear segments....................1. *A. septentrionale*
2. Fronds fernlike (3).
3. Stipes longer than the blades; blades broadest at base, bipinnate-pinnatifid 2. *A. Adiantum-nigrum*
3. Stipes much shorter than the blades; blades narrowed toward base (4).
4. Blades spongy-coriaceous, merely pinnatifid; rachis scaly beneath......3. *A. Dalhousiae*
4. Blades membranaceous, pinnate-pinnatifid, the ultimate segments incised; rachis not scaly ....4. *A. exiguum*
5. Sori confined to the proximal (basal) side of the pinnae or more numerous there than on the distal side; fertile veins mostly simple......................5. *A. monanthes*
5. Sori more numerous on the distal side than on the proximal side; fertile veins mostly forked (6).
6. Pinnae oval to broadly oblong, over half as wide as long, not or scarcely auriculate, mostly not over 6 mm. long..6. *A. Trichomanes*
6. Pinnae mostly oblong, often less than half as wide as long, subauriculate at the superior base, mostly 7 to 10 mm. long (7).
7. Sori mostly medial between the midrib and margin; some of the fronds narrowed and proliferous at apex...7. *A. Palmeri*
7. Sori mostly supramedial; fronds not proliferous.......................8. *A. resiliens*

1. Asplenium septentrionale (L.) Hoffm. San Francisco Peaks, Elden Mountain, and Mogollon Escarpment (Coconino County), Santa Catalina Mountains (Pima County), 8,000 to 9,000 feet. Black Hills of South Dakota to western Oklahoma, New Mexico, Arizona, southern California, and Baja California; Eurasia.

2. Asplenium Adiantum-nigrum L. Elden Mountain, Coconino County (*Whitehead* 2051, *Whiting & Bradey, Lefebure* 868), about 7,500 feet. Colorado, Utah, and northern Arizona; Eurasia and Africa.

3. Asplenium Dalhousiae Hook. (*A. Ferrissii* Clute, *A. rupium* Goodding, *Ceterach Dalhousiae* C. Chr.). Mule and Huachuca mountains (Cochise County), Baboquivari Mountains (Pima County), 4,000 to 6,000 feet, shaded, moist soil and rocky canyons, type of *A. Ferrissii* from the Huachuca Mountains (*Ferriss* in 1907), type of *A. rupium* from the Mule Mountains (*Goodding* 969). Known otherwise only from northwestern Mexico, Abyssinia, and the mountains of southern Asia.

4. **Asplenium exiguum** Bedd. Conservatory Canyon, Huachuca Mountains, Cochise County (*Lemmon* in 1882), Bear Valley or Sycamore Canyon, Santa Cruz County, about 3,500 feet (*Goodding* in 1937, *Darrow* 2047). Southeastern Arizona and northern Mexico; Asia.

The Arizona specimens collected by Lemmon were distributed as "*Asplenium fontanum*, var.," and Mexican material was the basis of *A. Glenniei* Baker. The American specimens are indistinguishable from the Himalayan *A. exiguum*, a case of sporadic distribution similar to that of *A. Dalhousiae*.

5. **Asplenium monanthes** L. Huachuca Mountains, Cochise County (*Lemmon* in 1882, *Pringle* in 1884, and others), about 8,000 feet, on shaded cliffs. Known in the United States only from South Carolina and Arizona; Mexico to Chile; West Indies; Africa; Hawaiian Islands.

6. **Asplenium Trichomanes** L. Mountains of southern Apache, Coconino, Graham, Gila, Cochise, and Pima counties, 6,000 to 9,000 feet, sheltered crevices of cliffs. Nova Scotia and Quebec south to Georgia and Alabama, west to Alaska and south to Texas, New Mexico, Arizona, and Oregon; Eurasia.

7. **Asplenium Palmeri** Maxon (*A. parvulum* var. *grandidentatum* Goodding). Mule Mountains (Cochise County), mountains of Santa Cruz County (several collections), Baboquivari Canyon (Pima County), moist, sheltered, rocky situations, type of *A. parvulum grandidentatum* from the Mule Mountains (*Goodding* 976). New Mexico, Arizona, Mexico, and Guatemala.

8. **Asplenium resiliens** Kunze. Near Flagstaff (Coconino County), and mountains of Greenlee, Graham, Gila, Cochise, and Yuma counties, 2,000 to 7,000 feet, among boulders and in crevices of cliffs. Southern Pennsylvania south to Florida, west to Illinois, Oklahoma, New Mexico, and Arizona; Jamaica; Mexico.

17. ATHYRIUM. Lady Fern

Large, delicate, terrestrial wood ferns with thick, erect, scaly rhizomes, the scales brown, fibrous, entire; fronds deciduous, thin-membranaceous; stipes stramineous, thick, shorter than the blades, deeply bisulcate above, with 2 vascular bundles at base, these with inflexed margins, uniting upwardly into a single, U-shaped bundle; blades bipinnate-pinnatifiid, narrowed at base and apex, nearly glabrous and scaleless, sparingly glandular, not ciliate; veins free, mostly twice-forked, reaching the margin, running into the teeth; sori dorsal on the veins, the indusia elongate and hooked at tip or subreniform, often crossing the veins, sparingly long-ciliate; spores bilateral, with perispore.

1. **Athyrium Filix-femina** (L.) Roth. White Mountains (Apache County), Pinaleno Mountains (Graham County), Chiricahua and Huachuca mountains (Cochise County), Santa Catalina Mountains (Pima County), 7,000 to 9,500 feet, rich, shaded ground about springs and along streams. Widely distributed in North and South America and Eurasia.

The Arizona specimens belong to var. *californicum* Butters, which ranges from Idaho and western Wyoming to New Mexico, Arizona, and California.

18. WOODWARDIA. Chain Fern

Large, coarse, evergreen, woodland ferns with thick, ascending, scaly rhizomes, the scales brown, fibrous, lanceolate, broadly attached, acuminate, glabrous, entire; fronds coriaceous, fasciculate; stipes stramineous, deeply sulcate, shorter than the blades, scaly at base, elsewhere bearing a few pale, minute, hairlike scales, with 2 large, flat vascular bundles and several smaller accessory bundles; fronds pinnate-pinnatifid, pinnatifid at apex, glabrous, sparingly scaly, the pinnae acuminate, with many pairs of falcate, acuminate segments, the margins sharply spinulose-serrulate; veins anastomosing to form costal and costular areoles, otherwise free or partly uniting to form a series of secondary areoles, the branches reaching the cartilaginous margin; sori elongate, borne on the outer vein of the costular areoles; indusium tumid, persistent, facing inward or at length reflexed; spores bilateral, with perispore.

1. Woodwardia fimbriata J. E. Smith. Mountains of southern Apache, Graham, Gila, Cochise, Santa Cruz, and Pima counties, 5,000 to 7,000 feet, rich soil in canyons, local. British Columbia to Nevada, southern Arizona, and California.

SPERMATOPHYTA

FLOWERING PLANTS

Seed-producing plants, the first phase of the life cycle very brief and concealed; pollen grains (microspores) borne in anther sacs (microsporangia); ovules (macrosporangia) enclosed in an ovary (except in the Gymnospermae), each ovule containing an embryo sac (macrospore); embryo resulting from the union of an egg cell in the embryo sac and a pollen grain, the ovule thereupon developing into a seed containing the embryo, the latter usually consisting of 1 or more leaves (cotyledons), a hypocotyl and radicle, and a terminal bud (plumule).

KEY TO THE CLASSES AND SUBCLASSES

1. Stigma none; ovules and seeds not enclosed, borne on the face of a scale or bract
Class *Gymnospermae*
1. Stigma or stigmas present; ovules and seeds in a closed cavity (ovary)
Class *Angiospermae* (2)
2. Cotyledon usually 1; flower parts commonly in threes; stem not differentiated into bark, wood, and pith (endogenous); veins of the leaves mostly longitudinally parallel (sometimes with netted veinlets between the parallel principal veins)
Subclass *Monocotyledoneae*
2. Cotyledons usually 2; flower parts commonly not in threes; stem differentiated into bark, wood, and pith (exogenous); veins of the leaves seldom parallel (commonly branching at a greater or less angle from the midvein, this alone extending the whole length of the leaf) .... Subclass *Dicotyledoneae* (3)
3. Perianth none, or single or appearing so, with segments all much alike in texture and color .... Series 1. *Apetalae*
3. Perianth present, evidently double or multiple, the outer segments (calyx) and the inner segments (corolla) usually conspicuously different in texture and color (4).
4. Petals distinct or united only at base .... Series 2. *Choripetalae*
4. Petals united well above the base .... Series 3. *Sympetalae*

Gymnospermae

KEY TO THE FAMILIES

1. Stems jointed; leaves reduced to more or less deciduous scales, these distant, opposite or in whorls of 3; fruits in small thin-scaled cones....................10. *Ephedraceae*
1. Stems not jointed; leaves needlelike or linear or if scalelike, then persistent and closely imbricate; fruits in thick-scaled cones, these sometimes berrylike (2).
2. Leaves elongate, narrowly linear or needlelike, arranged spirally or in fascicles; cones usually more or less elongate, dry and woody at maturity................8. *Pinaceae*
2. Leaves (in the adult state) short, scalelike, opposite or in whorls of 3; cones globose, either dry and woody or fleshy at maturity.........................9. *Cupressaceae*

8. PINACEAE. Pine Family

Trees, with resinous wood and foliage; leaves elongate, needle-shaped or narrowly linear, arranged spirally or in fascicles; flowers unisexual, the plants monoecious; perianth none; staminate inflorescences with numerous, spirally arranged stamens; scales of the pistillate inflorescences also arranged spirally, numerous, each scale subtended by a distinct bract and with 2 naked ovules at base; fruiting cones dry, more or less woody.

KEY TO THE GENERA

1. Leaves sheathed at base, at least when young, usually in fascicles of 2 or more, mostly needlelike; cone scales very thick and woody, umbonate on the back; bracts minute, much shorter than the scales; fruit maturing in the second (rarely third) season 1. *Pinus*
1. Leaves not sheathed or fascicled, linear or subulate; cone scales not very thick and woody, not umbonate; bracts relatively large; fruit maturing in the first season (2).
2. Cones erect, the scales falling from the axis at maturity and much longer than the bracts; leaves sessile, flat or somewhat 4-sided.................................4. *Abies*
2. Cones pendulous, the scales persistent on the axis; leaves stalked (3).
3. Branchlets roughened by the persistent, hard, peglike leaf bases; leaves mostly 4-sided, deciduous in drying; bracts shorter than the cone scales, erose-dentate or nearly entire 2. *Picea*
3. Branchlets not roughened by persistent leaf stalks; leaves compressed but strongly ribbed and channeled, persistent in drying; bracts longer than the cone scales, conspicuous, 2-lobed and aristate at apex...................................3. *Pseudotsuga*

1. PINUS (16). Pine

Trees; leaves in fascicles subtended by a sheath, rarely solitary, needle-shaped or narrowly linear; scales of the pistillate flowers in the axils of minute persistent bracts; cones in fruit with thick, woody scales, these umbonate on the back; fruit maturing in the second or third season; seeds winged or wingless.

This genus comprises some of the most valuable timber trees of the world. The ponderosa pine in Arizona, as in most of its range, is by far the most important species economically. It is the only species forming extensive nearly pure stands in readily accessible localities. Lumbering is rated as a five-million-dollar industry in Arizona, and ponderosa pine constitutes about 95 per cent of the total of sawed lumber. Sold locally as "native pine," the wood is heavy, hard, and brittle but not coarse-grained, yellow to reddish brown in color. The sapwood, known to the trade as "Western white pine," is easily worked and is much used for finishing. The wood of this and several other pines that occur in Arizona is used locally for rough construction, poles, fence posts, railway ties, and fuel.

Seeds of all pines are important food of squirrels, other rodents, and birds. The well-flavored seeds of the pinyons or nut pines (*Pinus cembroides, P. edulis, P. monophylla*) are used by the Indians for food, and in recent years those of *P. edulis,* by far the most abundant and widely distributed of the three species, have become an article of commerce. This is an important source of revenue to the Indians of northern Arizona and New Mexico. The seeds are picked up on the ground, taken from the nests of rodents, or extracted by roasting the nearly ripe cones. The resin of *P. edulis* is used by the Indians to waterproof bottles for holding water and to cement the turquoise stones in their jewelry.

Pines suffer considerably from the ravages of bark beetles. Species of the white pine group, with normally 5 needles in the fascicle (*P. aristata, P. flexilis, P. reflexa*), are likely to be attacked by blister rust if that destructive fungus should reach Arizona.

KEY TO THE SPECIES

1. Leaves commonly less than 5 cm. long, usually strongly incurved, the margins entire or very nearly so. Trees small (2).
1. Leaves usually more than 5 cm. long, straight or only slightly incurved (5).
2. Fascicles mostly 5-leaved; bark of young trees white, smooth; sheaths persistent 1 or 2 years, soon revolute; cones 6 to 10 cm. long, commonly at least 1½ times as long as wide, cylindric or cylindric-ovoid, deep brown-purple at maturity; scales with a slender, deflexed, deciduous prickle about 5 mm. long. Leaves densely crowded toward the ends of the branches, stout, deep green....................................4. *P. aristata*
2. Fascicles seldom more than 3-leaved; bark of young trees not white; sheaths soon deciduous, short, revolute; cones usually less than 6 cm. long, little if any longer than wide, broadly ovoid or nearly globose, light brown at maturity; scales usually muticous, the prickle, if any, very stout, strongly deflexed, and not more than 1 mm. long. Seeds edible: Pinyons, nut pines (3).
3. Leaves commonly in threes, seldom more and often less than 1 mm. wide, deep bluish green (at least on young trees) and strongly glaucous on the ventral face, usually very crowded toward the ends of the branches..........................1. *P. cembroides*
3. Leaves commonly in twos or single, commonly more than 1 mm. wide, yellowish green or sometimes moderately glaucous, usually not very crowded (4).
4. Leaves commonly in twos, semiterete, deeply channeled....................2. *P. edulis*
4. Leaves mostly single, terete..3. *P. monophylla*
5. Fascicles 5-leaved; leaf margins entire or very obscurely and remotely serrulate; cones at maturity cylindric or cylindric-ovoid, at least 1½ times as long as wide, 8.5 to 25 cm. long, pendent; scales muticous; bark of young trees silvery gray. Branches often drooping; sheaths deciduous; leaves slender: White pines (6).
5. Fascicles normally 3-leaved (except in *P. ponderosa* var. *arizonica*); leaf margins minutely but distinctly and closely serrulate; cones at maturity ovoid to nearly globose, seldom more and usually less than 1½ times as long as wide; scales bearing a short, stout, deflexed prickle; bark of young trees not silvery gray, that of older trees deeply and narrowly furrowed. Leaves dark yellowish green (7).
6. Cone scales broadly truncate at apex, the tip not or scarcely reflexed; trunk short, branched nearly from the ground; leaves yellowish green................5. *P. flexilis*
6. Cone scales narrowed toward the rounded apex, the tip strongly reflexed; trunk of mature trees well developed and clear of branches to a considerable height, often tapering rapidly; leaves bluish green..6. *P. reflexa*

7. Sheaths soon deciduous; leaves 5 to 12 cm. long, about 1 mm. wide; cones 4 to 7 cm. long, on a stalk 10 to 15 mm. long, this often falling with the cone; prickles of the cone scales gradually deciduous . . . . 7. *P. chihuahuana*
7. Sheaths persistent; leaves 10 cm. long, or longer; cones 5 to 15 cm. long, subsessile, the basal scales usually persistent on the branch after the rest of the cone falls; prickles of the cone scales persistent: Yellow pines (8).
8. Leaves 10 to 20 (25) cm. long, mostly 1 to 1.5 mm. wide; sheaths mostly 10 to 20 mm. long . . . . 8. *P. ponderosa*
8. Leaves usually more than 25 (up to 35) mm. long, about 2 mm. wide; sheaths 20 to 35 mm. long . . . . 9. *P. latifolia*

1. Pinus cembroides Zucc. Big Lue Range (Greenlee County) and from the Chiricahua Mountains to the Baboquivari Mountains (Cochise, Santa Cruz, and Pima counties), 5,000 to 7,500 feet. Western Texas to Arizona and northern Mexico.

Mexican pinyon. The trees attain a height of 15 m. (50 feet) and a trunk diameter of 35 cm. (14 inches), but are usually smaller. The trunk is commonly very short, the crown compact and conic in young trees, wide and rounded with mostly horizontal main branches in older trees. The bark of old trunks is thin, scaly, reddish brown.

2. Pinus edulis Engelm. Widely distributed and abundant in northern and central Arizona, from the Carrizo Mountains (Apache County) to the Kaibab Plateau (Coconino County), southward to the Chiricahua Mountains (Cochise County), Pinal Mountains (Gila County), and Prescott (Yavapai County), 4,000 to 7,000 feet, sometimes occurring in continuous stands of considerable extent. Western Oklahoma and Texas to Wyoming, eastern Utah, Arizona, and Baja California.

Colorado pinyon, nut pine. The trees are commonly straggling, with usually short and often crooked trunks, attaining a height of 10.5 m. (35 feet) and a trunk diameter of 75 cm. (30 inches), but usually smaller. The crowns are broadly conic in young trees, rounded or flat-topped in older trees. The old bark is yellowish brown or reddish brown, irregularly furrowed, and broken superficially into small scales. Pure stands have been likened to an old apple orchard.

M. E. Jones (*Zoe* 2: 251), reduced *P. edulis* to a variety of *P. monophylla,* stating that both single and paired leaves are found frequently on the same individual tree.

3. Pinus monophylla Torr. & Frém. Sporadic in Coconino, Mohave, Yavapai, Greenlee, Graham, and Gila counties, 4,000 to 6,500 feet. Utah and Arizona to California and Baja California.

Single-leaf pinyon. As it occurs in Arizona this pine scarcely differs from the ordinary pinyon (*P. edulis*) except in its solitary leaves, and may be only a variant of that species. *P. monophylla,* in California and Nevada, has thicker and more rigid leaves and larger cones than the Arizona tree.

4. Pinus aristata Engelm. San Francisco Peaks (Coconino County), 9,500 to 12,000 feet. Colorado and northern New Mexico to northern Arizona, Nevada, and California.

Bristle-cone pine, fox-tail pine. This tree reaches a height of 12 m. (40 feet) and a trunk diameter of 75 cm. (30 inches), but is usually smaller. The crown is pyramidal in young trees and in dense stands, but older trees growing in exposed situations are characterized by long, more or less erect upper limbs and long, drooping lower branches. The deep-green leaves are very crowded and appressed at the ends of the branchlets. The young bark is smooth and nearly white, the older bark dull reddish brown and not deeply furrowed.

5. **Pinus flexilis** James. San Francisco Peaks and Navajo Mountain (Coconino County), probably also in the White Mountains (Apache County) and Pinaleno Mountains (Graham County), 8,000 feet or higher. Alberta to western Texas, Arizona, and California.

Limber pine. In Arizona the trees reach a height of at least 15 m. (50 feet) and a trunk diameter of 0.9 m. (3 feet). The trunk is relatively short, and the crown widely branched with drooping limbs. The bark is smooth and grayish white in young trees, but on old trunks it is nearly black and split by deep furrows into wide plates. This species affords a small quantity of sawed lumber in Arizona.

6. **Pinus reflexa** Engelm. (*P. strobiformis* of authors, non Engelm.). White Mountains (Apache County) to the Pinaleno Mountains (Graham County), Chiricahua and Huachuca mountains (Cochise County), and Santa Rita and Santa Catalina mountains (Santa Cruz and Pima counties), 6,500 to 10,000 feet, type from the Santa Rita Mountains. Southern New Mexico, southern Arizona, and northern Mexico.

Southwestern white pine. This tree attains a height of 18 to 30 m. (60 to 100 feet) and a trunk diameter of 0.5 to 0.9 m. (20 to 36 inches). The bark of the trunk is dark gray or dull reddish brown, somewhat deeply and irregularly furrowed and narrowly ridged. The absence of stomata on the backs of the leaves is said always to distinguish this pine from the limber pine.

This pine is closely related to the Mexican *P. Ayacahuite* Ehrenb. (*P. strobiformis* Engelm.) differing apparently only in the shorter seed wings.

7. **Pinus chihuahuana** Engelm. (*P. leiophylla* Schiede & Deppe var. *chihuahuana* (Engelm.) Shaw). White River (southern Apache or Navajo County), Pinaleno Mountains (Graham County), Pinal Mountains (Gila County), Chiricahua Mountains (Cochise County), and west to the Santa Rita and Santa Catalina mountains (Santa Cruz and Pima counties), 5,000 to 7,500 feet, mostly on dry slopes and benches, fairly common in most of its range. Southwestern New Mexico, Arizona, and Mexico.

Chihuahua pine. A relatively small tree, reaching a maximum of 18 m. (60 feet) in height and 0.6 m. (2 feet) in trunk diameter, with widespreading limbs, dark brown, deeply furrowed older bark, and very persistent cones.

8. **Pinus ponderosa** Lawson. Widely distributed in Arizona, from the Carrizo Mountains (Apache County) to the Kaibab Plateau (Coconino County), southward to the Pinaleno Mountains (Graham County), Pinal Mountains (Gila County), and the Prescott region (Yavapai County), sometimes, especially in Coconino County, occurring in nearly pure stands of great extent, 5,500 to 8,000 feet, rarely as low as 3,500 feet or as high as 9,500 feet. United States and Canada from the Rocky Mountains to the states of the Pacific coast.

Ponderosa pine, Western yellow pine. In Arizona this species attains a height of 38 m. (125 feet) and a trunk diameter of 0.9 m. (3 feet) or more. The massive straight trunk, free from branches to a great height in mature trees, and the long, narrowly pyramidal or nearly cylindric crown with upturned branches are characteristic. The bark is gray-brown to black in young trees, warm russet-brown and split into broad plates covered with small concave scales in older trees.

The ponderosa pine is represented in northern and central Arizona by the rather vaguely characterized var. *scopulorum* Engelm. (*P. brachyptera* Engelm.), which differs little from typical *P. ponderosa* of the Pacific coast states. In the mountains of Graham, Cochise, Santa Cruz, and Pima counties the prevailing tree is var. *arizonica* (Engelm.) Shaw (*P. arizonica* Engelm.), type from the Santa Rita Mountains (*Rothrock* 652). In this variety the somewhat slenderer leaves are mostly in fascicles of fives (commonly in threes in var. *scopulorum*), and the cones are prevailingly shorter (5 to 9 cm. in var. *arizonica,* 7 to 15 cm. in var. *scopulorum*). I. M. Johnston (13, p. 331), who recognized *P. arizonica* as a species, distinguished it from *P. ponderosa* in "its somber brownish (rather than russet), more or less asymmetric, frequently stalked cones, weak, non-pungent umbo on the cone-scales, 3–5 needles, usually glaucescent branches and more southern distribution."

9. Pinus latifolia Sarg. (*P. apacheca* Lemmon, *P. Mayriana* Sudworth). Chiricahua, Huachuca, and Dragoon mountains (Cochise County), Santa Rita Mountains (Santa Cruz County), infrequent, 5,000 to 8,000 feet, type of *P. apacheca* from the Chiricahua Mountains (*Lemmon* in 1881), type of *P. latifolia* from the Santa Rita Mountains (*Mayr* in 1887). New Mexico and Arizona (probably also northern Mexico).

Apache pine, Arizona long-leaf pine. The great length of the leaves gives young trees a rather striking resemblance to the long-leaf pine of the Southeastern states (*P. australis*), but mature trees have much the habit of ponderosa pine. The Apache pine is reported to attain a height of 23 m. (75 feet) and a trunk diameter of 75 cm. (30 inches). The bark of the trunks is described as darker-colored than in *P. ponderosa.* The lumber is reported to be of fine quality, but the tree is not sufficiently abundant to have commercial importance.

2. PICEA (17). SPRUCE

Trees; leaves evergreen, narrow, 4-sided, short-stalked, blue-green or whitish, often silvery in young trees, deciduous in drying; branchlets rough with the persistent, peglike bases of the leaves; cones pendulous, cylindric, with large, relatively thin, persistent scales longer than the erose-dentate or nearly entire bracts; seeds small, winged.

On the Kaibab Plateau, in the White Mountains, and on the summit of Mount Graham the Engelmann spruce occurs in extensive stands that afford protection to the headwaters of streams. Present conditions are unfavorable to commercial exploitation in this state, but elsewhere the wood, which is rather weak and knotty, is utilized to some extent, chiefly for making boxes. Both species that occur in Arizona are valuable ornamentals but thrive only in a cool, moist climate. They are in demand for Christmas trees. Some species of spruce are very important as a source of paper pulp.

KEY TO THE SPECIES

1. Young branches and petiolelike leaf bases commonly pubescent or puberulent; leaves not rigid, acute or acutish at apex; cones commonly about 5 cm. long, the scales more or less rounded and distinctly thinner at apex........................1. *P. Engelmanni*
1. Young branches and leaf bases commonly glabrous; leaves rigid, spinescent-acuminate at apex; cones commonly about 8 cm. long, the scales truncate and not distinctly thinner at apex..2. *P. pungens*

1. Picea Engelmanni Parry. Kaibab Plateau, San Francisco Peaks, White Mountains, Pinaleno Mountains, and Chiricahua Mountains (Coconino, Apache, Graham, and Cochise counties), reported also from the Santa Catalina Mountains (Pima County), 8,000 to 12,000 feet, often in dense stands. British Columbia through the Rocky Mountains to New Mexico and Arizona, and through the Pacific coast states to northern California.

Engelmann spruce. Attains a height of 30 m. (100 feet) and a trunk diameter of 90 cm. (3 feet), but is perhaps rarely so large in Arizona. Trunk straight and tapering, the crown narrow-pyramidal, pointed and short in dense stands, but in widely spaced trees much longer, with drooping branches often extending nearly to the ground. Bark of older trunks russet to dark purplish brown or gray-brown, thin, hard, splitting into plates. This species affords some sawed lumber, especially on Mount Graham, where growth conditions are good.

2. Picea pungens Engelm. Kaibab Plateau (Coconino County), Lukachukai and White mountains (Apache County), 7,000 to 11,000 feet, sometimes in dense stands. Wyoming to New Mexico and Arizona.

Blue spruce. This species is less widely distributed in Arizona than Engelmann spruce, which it resembles in habit and appearance, although usually smaller, with foliage of young trees more pronouncedly glaucous and bark rougher and duller-colored. According to Marco (18), the resin canals extend into the upper half of the leaf in *P. pungens,* whereas in *P. Engelmanni* they terminate in the basal half.

3. PSEUDOTSUGA. DOUGLAS-FIR

Tree; leaves evergreen, narrowly linear, obtuse, short-petioled, compressed but strongly ribbed and channeled, not deciduous in drying, their bases not persisting on the branchlets; cones pendulous, ovoid-cylindric, with thin persistent scales shorter than the conspicuous, 2-lobed, aristate bracts.

1. Pseudotsuga taxifolia (Poir.) Britton. Common in the northern and central parts of the state, southward to the Chiricahua, Huachuca, Santa Catalina, and Santa Rita mountains, 6,500 to 10,000 feet, descending to 5,000 feet in canyons. British Columbia to western Texas, Arizona, California, and northern Mexico.

This is one of the most important timber trees of the Pacific Northwestern states, and the wood is in great demand for heavy construction. In Arizona it is of little importance commercially, as it seldom occurs in pure stands, commonly being mixed with ponderosa pine and at higher elevations with spruce. A limited quantity is sawed, and it is also used for rough construction, telephone poles, and railway ties. The wood of young trees is reddish and coarse-grained, hence the trade name "red fir," whereas in older trees it is yellowish and fine-grained, is marketed under the name "Oregon pine," and is used for fine finish work.

In Oregon and Washington this tree attains a height of 90 m. (300 feet) and a trunk diameter of 4.5 m. (14.5 feet), but in Arizona the species reaches no such dimensions, although in favorable situations it is probably the largest tree in the state. The trunk is usually straight, the crown broadly pyramidal, with drooping lower branches, and the bark furrowed, firm but not hard, in age cinnamon-brown and very thick. The tree of the Rocky Mountain region (including Arizona) is var. *glauca* (Mayr) Sudworth (*P. glauca* Mayr), regarded by some authorities as specifically distinct from the Pacific coast form (19, p. 208).

4. ABIES (17). Fir

Trees; the young bark with numerous horizontally elongate resin pockets; leaves evergreen, linear, flat, mostly blunt or notched at apex, or on fruiting branchlets acutish to acuminate; branchlets marked with conspicuous circular scars left by the fallen leaves; cones erect, ovoid-cylindric to nearly globose, with thin deciduous scales much longer than the bracts; seeds winged.

No species of fir occurs in Arizona in sufficient abundance and accessibility to be commercially important. In other Western states the light, soft, straight-grained wood of the white fir (*A. concolor*) is used to a limited extent as saw timber.

KEY TO THE SPECIES

1. Twigs glabrous when 1 year old; resin ducts near the lower leaf epidermis; cones grayish green; bracts of the cone scales with a short triangular tip............1. *A. concolor*
1. Twigs pubescent when 1 year old; resin ducts central; cones dark brown-purple; bracts with a long, subulate tip.......................................2. *A. lasiocarpa*

1. Abies concolor (Gordon & Glendinning) Hoopes. Rim of the Grand Canyon (Coconino County) to the Chiricahua Mountains (Cochise County), Santa Catalina Mountains (Pima County), and Hualpai Mountain (Mohave County), 5,500 to 9,000 feet. Wyoming to Oregon, south to northern Mexico.

White fir. In Arizona, where this is much the commoner species of fir, it attains a height of about 30 m. (100 feet) and the massive trunk a diameter of 1 m. (40 inches). In young trees the crown is symmetrically conic and sharp-pointed but becomes more or less irregular and rounded at the summit. Where the trees are widely spaced, the branches extend nearly to the ground, but in crowded stands half or more of the trunk is often bare. The bark, smooth and brownish gray at first, becomes very thick, hard, deeply furrowed, and ash-colored. A limited quantity of the soft, coarse-grained, light-colored wood is sawed in Arizona.

2. Abies lasiocarpa (Hook.) Nutt. Kaibab Plateau and Navajo Mountain (Coconino County), Escudilla Mountain and White Mountains (Apache County), Pinaleno Mountains (Graham County), also reported from the Santa Catalina Mountains (Pima County), 8,000 feet and upward. Alberta to Alaska, south to northern New Mexico, Arizona, and Oregon.

Alpine fir. Tree up to 27.5 m. (90 feet) high and 0.6 m. (2 feet) in trunk diameter, but much dwarfed at high elevations. The crown is elongate, narrowly conic, pointed, with branches often extending nearly to the base of the tree. The bark, smooth and ashy gray or whitish at first, becomes shallowly furrowed and gray or grayish brown.

The corkbark fir, *A. lasiocarpa* var. *arizonica* (Merriam) Lemmon (*A. arizonica* Merriam) occurs on the San Francisco Peaks (Coconino County), where the type was collected by Merriam in 1896, at elevations of 8,500 feet or higher. It is characterized by the creamy white, thick, and corky bark of the trunk, whereas, in typical *A. lasiocarpa,* the bark is usually gray or grayish brown and thinner. There seem to be no other constant diagnostic characters, however, and trees that are intermediate even in the bark characters are met with occasionally.

9. CUPRESSACEAE. Cypress Family

Trees or shrubs, resinous; leaves, in the adult state, mostly short, scalelike, opposite or in whorls of 3, and closely imbricate; flowers unisexual, the plants monoecious or dioecious; perianth none; pistillate inflorescences of few scales, these opposite or whorled, each with 1 to several ovules and without an evident bract; fruiting cones globose or nearly so, dry and with the scales separating at maturity, or else fleshy and with the scales fused into a berrylike fruit.

KEY TO THE GENERA

1. Cones dry at maturity, the scales woody and finally separating, the cones long-persistent on the branches; seeds numerous under each scale, winged. . . . . 1. *Cupressus*
1. Cones berrylike, often juicy, the scales fleshy and remaining fused at maturity, the cones not long-persistent; seeds 1 or 2 under each scale, not winged. . . . . 2. *Juniperus*

1. CUPRESSUS (20). Cypress

Trees; leaves all alike, small, scalelike, closely imbricate and appressed to the branchlets, usually with a pit on the back containing a resin gland; cones nearly globose, with woody scales that separate at maturity, persistent on the branches several years, long after the seeds have fallen; seeds numerous under each scale, winged.

The moderately soft, light, straight-grained wood is suitable for manufacture of sashes, doors, and blinds. The timber is rarely if ever milled, however, because the stands are of limited extent and usually rather inaccessible. Locally, cypress is cut for small rough construction and for making posts, although there is some question as to the durability of the wood. The Arizona cypress is frequently cultivated as an ornamental and is very attractive, especially in the young state while the foliage is covered with a heavy bloom. Individuals differ greatly in symmetry of growth, and there is much opportunity for selection of desirable specimens for planting. It is reported that the French government has found Arizona cypress useful for reclaiming barren land in France and northern Africa.

KEY TO THE SPECIES

1. Outer bark persistent except in saplings, rough. . . . . 1. *C. arizonica*
1. Outer bark deciduous except on the trunks of very old trees, leaving exposed the smooth, dark purplish-red inner bark. . . . . 2. *C. glabra*

1. Cupressus arizonica Greene (*C. arizonica* var. *bonita* Lemmon). Mountains of Greenlee, Graham (?), Cochise, and Pima counties, 3,000 to 7,000 feet, type from near Clifton (*Greene* in 1880). Western Texas to southern Arizona and northern Mexico.

Rough-bark Arizona cypress. The trees attain a height of 27 m. (90 feet) and a stem diameter of 1.6 m. (5½ feet), but such large specimens are rarely seen where the stands are readily accessible. The trunk branches from near the ground or is well developed. The crown is narrowly pyramidal or broad and flat. The outer bark is thin, dark brownish gray, longitudinally fissured and fibrous, or occasionally checkered somewhat as in *Juniperus Deppeana*.

2. Cupressus glabra Sudworth. Oak Creek and Verde River canyons (Coconino and Yavapai counties), Mogollon Escarpment and Mazatzal Mountains (northwestern Gila and northeastern Maricopa counties), 3,500 to 5,500 feet, type from Verde River Canyon. Known only from central Arizona.

Smooth-bark Arizona cypress. Questionably distinct as a species from *C. arizonica*, the only satisfactory character for distinguishing it being the habit of shedding the outer bark of the trunk until the tree has reached a very advanced age. The exposed inner bark is much like that of manzanita (*Arctostaphylos* spp.). Both *C. arizonica* and *C. glabra* are highly variable and show intergradation in shape of the crown (narrowly conic to broad and rounded), color of the foliage, presence or absence and size, when present, of the dorsal pit and gland of the leaves, and size of the seeds. The ranges of the two entities apparently do not overlap.

2. JUNIPERUS (21, 22). Juniper, Cedar

Evergreen trees or shrubs, monoecious or dioecious; leaves small, in alternate pairs or whorls, subulate and spreading in juvenile form, imbricate and appressed in the mature form (except in *J. communis*); cones berrylike, the scales becoming fleshy and not separating at maturity, or only slightly so at apex of the cone.

Arizona possesses extensive stands of juniper, especially in the central and northern portions, chiefly on well-drained, rather sterile soils. The wood is much used in manufacturing lead pencils, but in Arizona its chief utilization is for fuel and in making fence posts, for which the very durable heartwood adapts it admirably. *J. osteosperma*, with relatively straight branches, is especially suitable for fence posts. The foliage of junipers is browsed when other forage is scarce, but is injurious to livestock if eaten too freely. The berries are eaten greedily by birds and other wild creatures, and formerly were used as food by the Indians of Arizona. Those of *J. communis* are employed elsewhere to give the characteristic flavor to gin and are the source of oil of juniper, which has been used extensively in patent medicines. This species forms a valuable ground cover where it is sufficiently abundant. Juniper is used in various ways for medicinal and ritualistic purposes by the Hopi Indians. The pollen is believed to be responsible for cases of hay fever.

KEY TO THE SPECIES

1. Shrub, usually less than 1 m. high; leaves 5 to 15 mm. long, lanceolate, not thickened or imbricate, the upper surface exposed to view and conspicuously whitened. Branches prostrate or ascending .... 1. *J. communis*
1. Shrubs or trees, 2 to 16 m. high; leaves of the ultimate twigs 1 to 3 mm. long, ovate, thickened, closely imbricate, the upper surface not exposed to view (2).
2. Bark divided into rectangular plates. Foliage usually copiously dotted with resinous exudate; mature fruit mealy or resinous but not succulent; seeds mostly 3 or 4 2. *J. Deppeana*
2. Bark fibrous, longitudinally fissured (3).
3. Leaves entire, paired; seeds 1 to 3, usually 2; unseasoned heartwood reddish or purplish; branchlets flattened, often more or less drooping. Fruits 4 to 6 mm. in diameter, succulent .... 3. *J. scopulorum*
3. Leaves minutely denticulate, usually in pairs but often in whorls of 3; seeds 1 or 2, commonly solitary; heartwood yellowish brown; branchlets not flattened (4).
4. Plants dioecious; mature fruits succulent, 4 to 7 mm. in diameter; trunk almost invariably wanting, the plant spreading, with curved limbs arising at or below ground level; foliage inclined to bunch at ends of the branches; cotyledons 2...... 4. *J. monosperma*
4. Plants monoecious; mature fruits mealy or fibrous, 7 to 18 mm. in diameter; trunk either continuous or branched, but the limbs usually arising above ground level; foliage not inclined to bunch; cotyledons 4 to 6............................ 5. *J. osteosperma*

1. Juniperus communis L. Kaibab Plateau and San Francisco Peaks (Coconino County), Carrizo, White, and Chuska mountains (Apache County), above 8,000 feet. Cooler parts of North America; Eurasia.

Common juniper, also known as dwarf juniper and ground-cedar. The species is represented in Arizona only by a dwarf form, var. *saxatilis* Pall. (var. *montana* Ait.).

2. Juniperus Deppeana Steud. Common in the southeastern and central parts of the state, extending at least as far north as Flagstaff (Coconino County) and west to the Baboquivari Mountains (Pima County), 4,500 to 8,000 feet. Western Texas to Arizona and Mexico.

Alligator juniper. This, the largest of the Arizona species, is usually a tree, exceptionally attaining a height of 20 m. (65 feet). It rarely occurs in pure stands. The Arizona representatives of this species are referrable to var. *pachyphlaea* (Torr.) Martinez (*J. pachyphlaea* Torr.), type from near Flagstaff (*Woodhouse*), but that variety may not be worth maintaining (23, p. 132).

3. Juniperus scopulorum Sarg. (*J. virginiana* L. var. *scopulorum* Lemmon). Kaibab and Coconino plateaus to the Mogollon Escarpment (Coconino and Gila counties), Lukachukai and White mountains (Apache County), and Blue (Greenlee County), 5,000 to 9,000 feet. Alberta and British Columbia to New Mexico, Arizona, and Nevada.

Rocky Mountain juniper. The form with drooping branchlets, called "weeping juniper," is not uncommon in Arizona. The tree is graceful and highly ornamental despite the usually sparse foliage, reaching a height of 6 m. (20 feet) and a trunk diameter of 45 cm. (18 inches). There seems to be no constant character by which this species can be distinguished from *J. virginiana* L. (24), but it may be practicable to separate them on a statistical basis (25).

4. **Juniperus monosperma** (Engelm.) Sarg. Common and well distributed over the state, except in the extreme western and southwestern portions, occasionally forming forests, as on the slope of the Coconino Plateau east of Flagstaff, but more often growing with Utah juniper or other trees, 3,000 to 5,000 feet or somewhat higher. Kansas to Nevada and southward into Mexico.

One-seed juniper. Fruits with exposed seeds, the principal distinguishing character of *J. gymnocarpa* (Lemmon) Cory, are a not uncommon abnormality. Trees occurring in the Ajo Mountains, Pima County (*Goodding* 471–45, 14–46), were described by the collector as having a well-developed trunk, shreddy bark, and fruits red at maturity. They may belong to *J. monosperma* var. *gracilis* Martinez, known definitely only from Mexico.

5. **Juniperus osteosperma** (Torr.) Little (*J. utahensis* (Engelm.) Lemmon, *J. cosnino* Lemmon). Abundant over most of the state north of the Gila River, 3,000 to 7,500 feet, type of *J. cosnino* from near Flagstaff. Southern Idaho to New Mexico, Arizona, and California.

Utah juniper. The most valuable and abundant species of juniper in Arizona, often forming forests below the pinyons and ponderosa pines, such as the one south of Ash Fork, Yavapai County. It grows ordinarily as a small, conical tree 3 to 4.5 m. (10 to 15 feet) high, with a definite trunk. The limbs are not much curved. *J. megalocarpa* Sudworth, with fruit described as 14 to 18 mm. in diameter, seems to be only a large-fruited form of *J. osteosperma.*

10. EPHEDRACEAE. Joint-fir Family

1. EPHEDRA (26, 27). Joint-fir

Xerophytic, dioecious shrubs; branches opposite or whorled, slender, jointed, striate, equisetoid; leaves reduced to scales, paired or ternate, more or less connate; inflorescence (strobile) conelike; staminate flowers with 2 to 8 or more stamens, the filaments united; bracts of the ovulate strobile in several pairs or ternate whorls, in the Arizona species firm or scarious; seeds 1 to 3, hard, angled or subterete.

All Arizona species, with the possible exception of *E. trifurca,* are valuable browse in winter when better forage is lacking, and these plants are reported to be a food plant of the bighorn. *E. Cutleri* has considerable value as a soil binder. A palatable tonic beverage (Mormon tea, Brigham tea) is made from the dried stems and flowers of these plants, which contain certain alkaloids, such as pseudoephedrin, and tannins. The Indians and early white settlers esteemed *Ephedra* for treatment of syphilis and other diseases. The drug ephedrine, commonly administered as an astringent and as a mild substitute for adrenalin, is obtained from *E. sinica,* a Chinese herb. The plants flower in spring.

KEY TO THE SPECIES

1. Leaves 3 at each node; bracts of the fruiting cones clawed, 7 to 10 mm. wide, flexible and scarious; seeds slender, about twice as long as thick (2).
1. Leaves 2 at each node; bracts not clawed, 3 to 5 mm. wide, firm, with narrow scarious margin; seeds plump (3).
2. Leaves 5 to 12 mm. long, persistent but soon shreddy; bracts of the fruiting cones entire-margined, reddish brown, cordate .... 1. *E. trifurca*

2. Leaves 3 to 5 mm. long, tardily deciduous but not becoming shreddy; bracts erose-margined, yellowish, rounded or truncate at base........................2. *E. Torreyana*
3. Base of the leaves brown, persistent. Seeds prevailingly paired (4).
3. Base of the leaves gray, deciduous (5).
4. Peduncles of the ovulate spikes wanting or very short; stems not viscid....3. *E. viridis*
4. Peduncles of the ovulate spikes up to 2 cm. long; stems often viscid........4. *E. Cutleri*
5. Seeds mostly paired, brown, smooth..............................5. *E. nevadensis*
5. Seeds mostly solitary, grayish or light brown, vertically wrinkled......6. *E. fasciculata*

1. **Ephedra trifurca** Torr. Greenlee County to Gila, Yavapai, and Mohave counties, south to the Mexican boundary, up to 4,500 feet, desert or grassland. Southwestern Texas to southern California and adjacent Mexico.

The largest Arizona species, occasionally attaining a height of 4.5 m. (15 feet).

2. **Ephedra Torreyana** Wats. Apache County to Mohave and northern Yavapai counties, mostly 4,000 to 6,000 feet. Southwestern Colorado to Nevada, south to western Texas, Arizona, and Chihuahua.

3. **Ephedra viridis** Coville. Navajo, Coconino, Mohave, and Yavapai counties, 3,000 to 7,000 feet. Southwestern Colorado, Utah, Nevada, Arizona, and California.

4. **Ephedra Cutleri** Peebles (*E. Coryi* Reed var. *viscida* Cutler, *E. viridis* var. *viscida* L. Benson). Apache, Navajo, Coconino, Yavapai, and Gila counties, mostly 5,000 to 6,000 feet, type from west of Rock Point, Apache County (*Cutler* 2209). Southwestern Colorado, southeastern Utah, northern New Mexico, and Arizona.

Similar in many respects to *E. viridis*, but distinguished by the well-developed peduncles and often viscid stems. This species is very common on the Navajo Indian Reservation, often growing so thickly on sandy plains as to resemble a luxuriant growth of grass. *E. arenicola* Cutler is the name for a presumable hybrid between *E. Torreyana* and *E. Cutleri* found near Dinnehotso, Apache County (*Cutler* 2217 and 2221).

5. **Ephedra nevadensis** Wats. Western Coconino County, at head of Hualpai Canyon, Mohave County, at Kingman and Fort Mohave, reported also from the Grand Canyon and from Lake Mead. Oregon, Utah, Arizona, and California.

In typical *E. nevadensis* the young stems are usually smooth and glaucous, the ovulate spikes distinctly stalked, and the seeds are usually paired. In var. *aspera* (Engelm.) L. Benson (*E. aspera* Engelm.), which Cutler (26, p. 398) treated as a distinct species, the stems are commonly yellowish and minutely papillate, the ovulate spikes sessile or subsessile, and the seeds usually solitary. The variety has been collected in Havasu and Hualpai canyons (Coconino County), between Kingman and Oatman and at Goldroad (Mohave County), and in the Harcuvar and Castle Dome mountains (Yuma County), about 3,000 feet.

6. **Ephedra fasciculata** A. Nels. Grand Canyon (Coconino County) and Mohave, Graham, Gila, Maricopa, Pinal, and Yuma counties, up to about 4,000 feet, type from Phoenix (*A. Nelson* 10268). Arizona and southeastern California.

Benson (27) reduced *E. fasciculata* to a synonym of *E. nevadensis* var.

aspera (Engelm.) L. Benson. Although *E. fasciculata* was described originally from sterile branches, Cutler (26, p. 402) stated that the seeds are 8 to 13 mm. long. *E. Clokeyi* Cutler (*E. fasciculata* var. *Clokeyi* (Cutler) Clokey), has seeds only 5 to 8 mm. long but is otherwise very similar to *E. fasciculata* and, according to Cutler (26, p. 404), has about the same distribution in Arizona.

E. californica Wats. is given for Arizona by Cutler (26, p. 396), but the specimen cited by him seems not to belong to this species.

Angiospermae

MONOCOTYLEDONEAE

KEY TO THE FAMILIES

1. Plants strictly aquatic, immersed in or floating on the surface of water; perianth none, or very inconspicuous, except in Hydrocharitaceae (2).
1. Plants terrestrial (in one family epiphytic) or semiaquatic or if strictly aquatic, then with a relatively large and showy perianth (8).
2. Flowers perfect. Stamens 4, with sepallike appendages forming a false perianth; leaves mostly alternate, with or without broad, floating blades; flowers in spikes: genus *Potamogeton* .... 13. *Potamogetonacae*
2. Flowers unisexual (3).
3. Stems very short or none, the whole plant floating. Inflorescence subtended by a spathe (4).
3. Stems elongate, mostly immersed (5).
4. Plants with a short stem and a rosette of rather large leaves; inflorescence few-flowered, axillary, the spathe conspicuous, white; stamens 2, united.............. 21. *Araceae*
4. Plants reduced to small, thalluslike bodies without distinction of stem and leaf; inflorescence 2- or 3-flowered, borne on the edge of the thallus, the spathe minute; stamen 1 .... 22. *Lemnaceae*
5. Leaves alternate, linear; flowers many, in dense, globose heads. Staminate heads above, the pistillate heads becoming burlike in fruit; stamens 3 or more; stems usually partly emersed .... 12. *Sparganiaceae*
5. Leaves opposite or whorled, narrowly linear or filiform; flowers few, not in heads (6).
6. Flowers of both sexes enclosed, in the bud, in a common spathe, the staminate flowers without a perianth. Stamens 1 or 2; pistillate flowers with a very inconspicuous cuplike perianth: genus *Zannichellia*.......................... 13. *Potamogetonaceae*
6. Flowers solitary, the staminate flowers with a perianth (7).
7. Staminate flowers with a double sheathlike perianth, the pistillate flowers naked; stamen 1; leaves serrate or serrulate.............................. 14. *Najadaceae*
7. Staminate and pistillate flowers both with a perianth, that of the pistillate flowers with a long pedicellike tube; stamens 3 to 9; leaves entire or minutely serrulate
17. *Hydrocharitaceae*
8. Perianth none or rudimentary (9).
8. Perianth evident (12).
9. Plants semiaquatic, the stems partly immersed; flowers subtended by inconspicuous bristles or scales; flowers unisexual, the plants monoecious. Inflorescences dense, spicate or capitate, the staminate flowers above (10).
9. Plants mostly terrestrial; flowers subtended by relatively conspicuous chaffy scales (glumes); flowers normally perfect except in *Carex* and a few genera of Gramineae. Leaves with a sheathlike basal portion enclosing the stem (11).
10. Flowers in thick spikes; floral bractlets bristlelike................... 11. *Typhaceae*
10. Flowers in globose heads, the pistillate heads burlike in fruit; floral bractlets scalelike
12. *Sparganiaceae*
11. Stems round or flattened, usually hollow except at the nodes; leaves 2-ranked, the sheath commonly split on one side; flowers each subtended by an outer scale (lemma) with (usually) a 2-keeled inner scale (palea) facing the lemma...... 18. *Gramineae*

11. Stems round or trigonous, usually solid; leaves often 3-ranked, the sheath not split; flowers each subtended by a single scale, the palea wanting.........19. *Cyperaceae*
12. Divisions of the perianth not showy or petallike, greenish or brownish. Plants mostly of wet ground (13).
12. Divisions of the perianth (at least the inner ones) mostly showy and petallike (14).
13. Flowers in elongate, slender, spikelike racemes; perianth segments herbaceous, not sharp-pointed; carpels separating at maturity..................15. *Juncaginaceae*
13. Flowers not in elongate racemes; perianth segments chaffy, glumelike, usually sharp-pointed; carpels not separating, the fruit a capsule..................26. *Juncaceae*
14. Plants trees, with a tall trunk; leaves fan-shaped, many-ribbed, deeply lobed 20. *Palmae*
14. Plants not trees or if treelike, then the leaves not fan-shaped or lobed (15).
15. Plants epiphytic, growing on the branches of trees................23. *Bromeliaceae*
15. Plants terrestrial (rooted in soil) or aquatic (16).
16. Pistils several or many, in a head or ring; ovary 1-celled; flowers perfect or unisexual. Plants aquatic or semiaquatic.................................16. *Alismaceae*
16. Pistil 1; ovary usually 3-celled, or the ovules borne on 3 placentas; flowers mostly perfect (17).
17. Stamens more than 3, commonly 6, some of them (in genus *Commelina*) often imperfect (18).
17. Stamens 3 or fewer (20).
18. Perianth adnate below to the ovary and appearing as if borne upon it 28. *Amaryllidaceae*
18. Perianth free from the ovary or nearly so (19).
19. Segments of the perianth sharply differentiated, the outer ones green, sepallike, the inner ones petallike, fugacious, not bearing glands; flowering stems not from an onionlike bulb .... 24. *Commelinaceae*
19. Segments of the perianth not sharply differentiated or if so (in genus *Calochortus*), then the inner ones bearing a conspicuous fringed gland and the flowering stems from an onionlike bulb...27. *Liliaceae*
20. Perianth very irregular, adnate below to the ovary; stamens 1 or 2; flowers not subtended by spathes...30. *Orchidaceae*
20. Perianth regular or only slightly irregular; stamens 3; flowers subtended by spathelike bracts (21).
21. Ovary superior; plants aquatic or semiaquatic; leaves not equitant..25. *Pontederiaceae*
21. Ovary inferior; plants terrestrial; leaves equitant (in 2 ranks, enfolding one another) 29. *Iridaceae*

11. TYPHACEAE. Cat-tail Family

1. TYPHA (28). Cat-tail

Semiaquatic perennial herbs; flowering stems from creeping rootstocks, tall, terete, not jointed; leaves long, flat, equitant; plants monoecious, the flowers very numerous in a dense, cylindric, continuous or interrupted spike, the staminate flowers above; perianth none; fertile flowers composed of a stipitate ovary, an elongate style, and a stigma, the stipe bearing numerous long hairs; abortive flowers composed of a slender stipe expanded and club-shaped at apex, hairy below.

Plants of marshes and sloughs. Various parts of the plants have been used as food by Indians and other people.

KEY TO THE SPECIES

1. Spike dark brown, usually not interrupted (but sometimes with a considerable interval), usually about 6 times as long as thick, frequently a little thicker toward the apex; bracts none; stigmas lance-ovate, fleshy, usually long-persistent, forming a densely matted mass over the surface of the spike...........................1. *T. latifolia*

1. Spike light brown, interrupted (the interval 0.7 to 4.5 cm. long), usually about 10 times as long as thick, not thicker toward apex; bracts present, with a long stalk and a very small, usually ovate and apiculate blade; stigmas linear, not fleshy, soon deciduous, forming a loosely curly mass over the surface of the spike.........2. *T. domingensis*

1. **Typha latifolia** L. Apache, Navajo, Coconino, Gila, and Maricopa counties (doubtless elsewhere), 3,500 to 7,500 feet. Widely distributed in North America and Eurasia.

2. **Typha domingensis** Pers. Navajo, Coconino, Gila, Maricopa, and Cochise counties, 1,000 to 5,000 (?) feet. Southern United States to southern South America.

Arizona specimens have been referred to *T. angustifolia* L.

12. SPARGANIACEAE. Bur-reed Family

1. SPARGANIUM. Bur-reed

Aquatic or semiaquatic herbs, the stems usually partly emersed; leaves 2-ranked, alternate, sessile, linear; plants monoecious, the flowers many, small, in dense globose heads, these sessile or stalked, the staminate heads above, the pistillate heads burlike in fruit; perianth none or rudimentary, the flowers subtended by scales; stamens 3 or more; fruit nutlike, 1- or 2-seeded.

KEY TO THE SPECIES

1. Achenes sessile or very nearly so, obovoid or cuneate-obpyramidal, truncate or depressed at apex, with a stout conic beak more than half as long as the achene; inflorescence usually branched; fruiting heads at maturity commonly 2 to 3 cm. in diameter 1. *S. eurycarpum*

1. Achenes stipitate, fusiform, pointed at apex; inflorescence not branched; fruiting heads usually less than 2 cm. in diameter............................2. *S. simplex*

1. **Sparganium eurycarpum** Engelm. McNary (Apache County), 7,500 feet, in a marsh (*Peebles & Smith* 12480), Lakeside, Navajo County, 6,500 feet, in shallow water (*Pultz* 1738). Newfoundland to British Columbia, south to Virginia, Arizona, and California.

2. **Sparganium angustifolium** Michx. Kaibab Plateau (Coconino County), about 9,000 feet, "abundant in older ponds" (*Grand Canyon Herb.* 146, *Goodding* 251–48). Widely distributed in the Northern Hemisphere.

The Arizona specimens may belong to *S. multipedunculatum* (Morong) Rydb. (*S. simplex* of American authors), the leaves being rather broad for *S. angustifolium.*

13. POTAMOGETONACEAE. Pondweed Family

Plants aquatic; flowers perfect or unisexual, in spikes or in small axillary clusters; true perianth none, except in the pistillate flowers of *Zannichellia,* a false perianth present in *Potamogeton;* pistils 1 or 4; ovary 1-celled, the ovule solitary.

KEY TO THE GENERA

1. Flowers perfect, in spikes; stamens 4, the anthers sessile, with the connective expanded dorsally into a herbaceous sepallike appendage; pistils 4; fruits sessile; leaves alternate or nearly opposite, floating or immersed......................1. *Potamogeton*

1. Flowers unisexual, in small, umbellike axillary clusters; stamens 1 or 2, the anthers not appendaged; pistil 1; fruit short-stalked; leaves opposite or whorled, all immersed .... 2. *Zannichellia*

1. POTAMOGETON (29, 30). Pondweed

Leaves with membranaceous stipules, mostly alternate, all much alike or the lower ones narrower and immersed and the upper ones with broad, floating blades; flowers perfect, in axillary spikes, these often emersed; perianth technically none but closely simulated by the concave sepallike appendages of the 4 stamens; anthers sessile; pistils 4, the style short or the stigma sessile.

Plants of ponds and sluggish streams, flowering mostly in summer. In irrigation ditches the plants are sometimes so abundant as materially to retard the flow. The species are at best difficult to identify, and the difficulty is especially great with Arizona specimens, which rarely have good fruit. The following treatment is necessarily provisional.

KEY TO THE SPECIES

1. Leaves all much alike and immersed, the blades linear or capillary; spikes mostly immersed (2).
1. Leaves dimorphic, some floating and with broad, thickish, long-petioled blades, others immersed and narrower and thinner; spikes emersed, elongate (5).
2. Fruiting spikes commonly 15 mm. long or longer, conspicuously interrupted; flowers seldom fewer than 10 in the spike........ 1. *P. pectinatus*
2. Fruiting spikes less than 15 mm. long; flowers commonly fewer than 10 in the spike (3).
3. Stipules adnate to the base of the leaf; fruits 1 to 1.5 mm. long, strongly compressed, 3-keeled, suborbicular to subreniform........ 2. *P. diversifolius*
3. Stipules free from the leaf bases; fruits 1.8 to 2.8 mm. long, relatively plump, 1-keeled or the lateral keels obscure, obovoid to suborbicular (4).
4. Leaves usually without basal glands; fruiting spikes subcapitate or densely short-cylindric, 2 to 5 mm. long; fruits suborbicular to obovoid, conspicuously keeled, the keel dentate or sharply undulate........ 3. *P. foliosus*
4. Leaves usually biglandular at base; fruiting spikes cylindric, interrupted, 6 to 12 mm. long; fruits obovoid, often sigmoid, obscurely keeled, the keel broad.... 4. *P. pusillus*
5. Immersed leaves narrowly linear, not more than 2 mm. wide, the nerves 3 to 5, inconspicuous (6).
5. Immersed leaves (except in *P. gramineus*) usually more than 2 mm. wide, linear to ovate-elliptic, the nerves 3 to 19. Floating leaves cuneate to rounded at base (7).
6. Floating leaves 40 to 120 mm. long, 25 to 65 mm. wide, usually cordate at base, the nerves 13 to 37; fruiting spikes 30 to 50 mm. long; fruits usually 3.5 to 5 mm. long 5. *P. natans*
6. Floating leaves 10 to 40 mm. long, 3 to 20 mm. wide, not cordate, the nerves 7 to 15; fruiting spikes not more than 20 mm. long; fruits 1 to 1.5 mm. long... 2. *P. diversifolius*
7. Immersed leaves with petioles 2 to 13 cm. long, acutish but not sharp-pointed or mucronate at apex; fruits 3.5 to 4.3 mm. long, usually reddish, the keels mostly muricate 6. *P. nodosus*
7. Immersed leaves sessile or with petioles up to 4 cm. long, acutish or sharp-pointed, often somewhat mucronate at apex; fruits 1.7 to 3.5 mm. long, usually greenish, the keels scarcely muricate (8).
8. Stem usually much-branched, 0.5 to 1.0 mm. in diameter; immersed leaves seldom more than 10 mm. wide, mostly 7 to 12 times as long as wide, sessile, the nerves mostly 3 to 9; floating leaves 15 to 50 (70) mm. long, 10 to 20 (30) mm. wide, the petioles mostly longer than the blades; fruiting spikes 10 to 25 mm. long, 6 to 8 mm. thick; fruits 1.7 to 2.8 mm. long........ 7. *P. gramineus*

8. Stem simple or once-branched, 1 to 5 mm. in diameter; immersed leaves 15 to 40 mm. wide, mostly 3 to 5 times as long as wide, sessile or petiolate, the nerves mostly 9 to 17; floating leaves 40 to 130 (190) mm. long, 20 to 65 mm. wide, the petioles mostly shorter than the blades; fruiting spikes 25 to 70 mm. long, 8 to 10 mm. thick; fruits 2.5 to 3.5 mm. long.......................................8. *P. illinoensis*

1. Potamogeton pectinatus L. (*P. interior* Rydb.). Yavapai, Gila, Pinal, Maricopa, and Pima counties, 1,000 to 5,000 (?) feet. Almost throughout North America; cosmopolitan.

2. Potamogeton diversifolius Raf. (*P. hybridus* Michx.). South Rim of Grand Canyon, 7,500 feet (*Thornber* 8505, *Collom* in 1941). Widely distributed in North America.

3. Potamogeton foliosus Raf. Coconino, Mohave, Maricopa, Pima, and Santa Cruz counties, 1,000 to 8,500 feet. Throughout most of North America.

Both var. *genuinus* Fernald, with 3- to 5-nerved leaves 1.4 to 2.7 mm. wide, the stipules 7 to 18 mm. long, and var. *macellus* Fernald, with 1- to 3-nerved leaves 0.3 to 1.5 mm. wide, the stipules 3 to 11 mm. long, occur in Arizona. The former has been collected on the San Francisco Peaks, at Mesa (Maricopa County), and in Sonoita Valley (Santa Cruz County). Var. *macellus* has been found in Havasu Canyon (Coconino County), at Canelo (Santa Cruz County), and near Tucson.

4. Potamogeton pusillus L. (*P. panormintanus* Biv.-Bern.). Crater Lake and Summit Ranch, San Francisco Peaks, Coconino County (*Lemmon* 3284, *MacDougal* 274, *Thornber* 2856). Widely distributed in North America and in other parts of the world.

5. Potamogeton natans L. Marsh Lake, White Mountains, Apache County, 9,000 feet (*Goldman* 2453), Greenland Lake, north of the Grand Canyon, 8,500 feet (*Collom* in 1942), Walker Lake, San Francisco Peaks, Coconino County, 8,500 feet (*Knowlton* 288). Newfoundland to New Jersey, west to Alaska and California.

6. Potamogeton nodosus Poir. (*P. americanus* Cham. & Schlecht.). Navajo, Coconino, Yavapai, Gila, and Maricopa counties, 1,500 to 8,000 feet. New Brunswick to British Columbia, south to Virginia, Louisiana, Arizona, and California.

7. Potamogeton gramineus L. Coconino and (?) Yavapai counties. Greenland to Alaska, south to New Jersey, Indiana, Nebraska, New Mexico, Arizona, and California; Eurasia.

Var. *maximus* Morong occurs at Crater Lake, San Francisco Peaks (*Lemmon*), apparently also at Montezuma Well, Yavapai County (*Goodding* 91–47). It has the immersed leaves larger and with usually more numerous nerves than in var. *typicus* Ogden (var. *graminifolius* Fries). Specimens intermediate between the latter and var. *maximus* have been collected at Mormon Lake, Coconino County (*MacDougal* 80).

***8. Potamogeton illinoensis** Morong (*P. Zizii* of American authors, in part). This species, known from New Mexico and California, has not been reported from Arizona, although Ogden (30, p. 187) identified specimens from Montezuma Well, Yavapai County (*Taylor* 78, *Jackson* 52) as *P. gramineus* x *illinoensis*. It is doubtful that they are distinguishable from *P. gramineus* var. *maximus*.

2. ZANNICHELLIA. Horned-pondweed

Stems filiform; leaves very narrow, opposite or whorled, entire or nearly so; plants monoecious, the flowers axillary, those of both sexes enclosed, in bud, in a common spathe, the staminate flower naked, with 1 or 2 stamens, the pistillate flowers few, with an inconspicuous cuplike perianth, a flask-shaped ovary, and an expanded stigma.

1. **Zannichellia palustris** L. Coconino and Mohave counties to Cochise and Pima counties, in ponds, ditches, and slow streams. Cosmopolitan.

14. NAJADACEAE. Naiad Family

1. NAJAS. Naiad

Stems slender, wholly immersed; leaves narrow, serrulate to spinulose-dentate, mostly opposite or whorled; flowers unisexual, axillary, solitary; staminate flowers with a double, sheathlike perianth, the stamen solitary; pistillate flowers naked, the ovary 1, the stigmas 2 to 4, subulate.

1. **Najas marina** L. Near Fort Mohave (Mohave County), Crittenden (Santa Cruz County), near Tucson (Pima County), 2,500 to 4,000 feet. Almost throughout North America; Eurasia.

Both typical *N. marina*, with teeth of the leaves entire, and var. *recurvata* Dudley, with the teeth bidentate at base, probably occur in Arizona.

N. flexilis (Willd.) Rostk. & Schmidt and *N. guadalupensis* (Spreng.) Magnus are likely to be found in Arizona. They differ from *N. marina* in having very narrow, obscurely serrulate leaves. The seeds are shiny and nearly smooth in the former, dull and distinctly reticulate in the latter.

15. JUNCAGINACEAE. Arrow-grass Family

1. TRIGLOCHIN (31). Arrow-grass

Plants herbaceous, rushlike; leaves all basal, long, narrow, semiterete, with membranaceous sheaths; flowers in narrow spikes or spikelike racemes, these terminating long scapes; perianth small, greenish, of 6 concave segments; stamens 3 to 6, the anthers sessile or nearly so; pistil compound, of 3 to 6 carpels, these separating at maturity.

Plants of marshes and shallow ponds. Especially when stunted by drought, these plants may develop hydrocyanic acid and become toxic, particularly to cattle.

KEY TO THE SPECIES

1. Plants mostly less than 30 cm. high, with slender, elongate rootstocks; ligules 2-parted ..... 1. *T. concinna*
1. Plants mostly more than 30 cm. high, with stout, short rootstocks; ligules entire ..... 2. *T. maritima*

1. **Triglochin concinna** Burtt Davy. Holbrook, Navajo County (*Zuck* in 1896), Tuba, Coconino County (*Peebles* 11843). Wyoming, Colorado, and northeastern Arizona to interior Oregon and California; along the coast from British Columbia to Baja California.

The Arizona (and Great Basin) plant is var. *debilis* (M. E. Jones) J. T. Howell, which differs from the typical coastal phase of the species in being more robust, with taller scapes and usually a fibrous-coated rootstock.

2. Triglochin maritima L. Fredonia, northwestern Coconino County (*Hester* 1030). Widely distributed in the Northern Hemisphere.

16. ALISMACEAE. Water-plantain Family

Plants herbaceous, aquatic or semiaquatic; stems scapelike; leaves with sheathing bases and usually with broad blades; flowers perfect or unisexual; perianth regular, with green calyxlike outer segments and white, petallike inner segments; stamens 6 or more; pistils several or many in a ring or a dense head; ovary 1-celled; ovule usually solitary; fruit achenelike, compressed.

KEY TO THE GENERA

1. Carpels in 1 or few series; leaves not sagittate, short-cuneate to subcordate at base; flowers perfect, small, in ample compound panicles....................1. *Alisma*
1. Carpels in several series, forming a dense head in fruit; leaves (some or all of them) commonly sagittate; plants dioecious, or monoecious with the lower flowers pistillate and the upper ones staminate, in verticillate, simple or somewhat compound panicles 2. *Sagittaria*

Representatives of the genera *Echinodorus* and *Lophotocarpus* are also likely to be found in Arizona. The former differs from *Alisma* in having the carpels in several series forming a dense head in fruit. The latter differs from *Sagittaria* in having the lower flowers perfect.

1. ALISMA (32). Water-plantain

Roots fibrous; leaves mostly emersed, with the blades broadly ovate and usually cordate or rounded at base, but the leaves occasionally floating and narrower; inflorescence a large, open panicle; flowers small, perfect; stamens commonly 6; receptacle flat.

Plants growing in shallow water, flowering in summer.

KEY TO THE SPECIES

1. Sepals broadly scarious-margined, in anthesis 3 to 4 mm. long; petals 3.5 to 6 mm. long; stamens much surpassing the ovaries; anthers 0.6 to 0.8 mm. long; style about equaling the ovary in anthesis; achenes 2.2 to 3 mm. long.................1. *A. triviale*
1. Sepals narrowly margined, in anthesis 2 to 2.5 mm. long; petals 1 to 3 mm. long; stamens little surpassing the ovaries; anthers 0.3 to 0.5 mm. long; style much shorter than the ovary; achenes 1.5 to 2 mm. long............................2. *A. subcordatum*

1. Alisma triviale Pursh (*A. brevipes* Greene). North Rim of Grand Canyon, Lake Mary, and Foxboro (Coconino County), Lakeside (southern Navajo County), Sulphur Springs Valley (Cochise County), 4,000 to 8,000 feet. Quebec to British Columbia, south to Maryland, Iowa, Nebraska, and northern Mexico.

2. Alisma subcordatum Raf. Mormon Lake and near Williams (Coconino County), near Prescott (Yavapai County), 5,000 to 7,000 feet. New England and southern Ontario to Florida, Texas, Arizona, and Mexico.

The Arizona specimens of *Alisma* are not easily identified by the characters given in the key, and it is somewhat doubtful that there are 2 species in this state.

2. SAGITTARIA (33). Arrow-head

Flowering stems from rootstocks, these often tuber-bearing; leaves (except in *S. graminea*) mostly emersed, with triangular-ovate, deeply sagittate blades; inflorescence a narrow, verticillate panicle, simple or branched below; flowers relatively large, unisexual; stamens commonly numerous; receptacle elevated.

Plants of shallow, stagnant ponds and mud flats, flowering in summer. One species (*S. latifolia*) produces edible tubers, eaten by the Indians and Chinese in the Pacific coast states. The plant is sometimes called tule-potato in California.

KEY TO THE SPECIES

1. Leaves all or mostly entire, linear, lanceolate, or elliptic..............1. *S. graminea*
1. Leaves all or mostly sagittate, broader (2).
2. Achenes with a conspicuous, horizontal, straight or slightly curved beak. Terminal leaf lobe commonly equaling or longer than the basal lobes; achenes not winged or crested on the faces, the marginal wing broad, especially at top.............2. *S. latifolia*
2. Achenes minutely or inconspicuously beaked (3).
3. Basal leaf lobes not longer (usually shorter) than the terminal lobe; achenes with an erect or suberect beak. Terminal leaf lobe commonly ovate or broadly lanceolate, acute or slightly acuminate; achenes not winged or crested on the faces, the marginal wing broad and rather thick..3. *S. cuneata*
3. Basal leaf lobes much longer than the terminal lobe; achenes with a minute horizontal or ascending beak (4).
4. Leaf blades not more than 25 cm. long, the terminal lobe linear or lanceolate, usually long-acuminate, commonly less than half as long as the basal lobes; scape usually simple; bracts 6 to 8 mm. long; achenes without facial wings........4. *S. longiloba*
4. Leaf blades 20 to 40 cm. long, the terminal lobe ovate to broadly lanceolate, acute or short-acuminate, more than half as long as the basal lobes; scape sometimes branched; bracts 10 to 30 mm. long; achenes with facial wings................5. *S. Greggii*

1. **Sagittaria graminea** Michx. Hooker Cienega, southern Graham County (*Thornber* in 1905). Newfoundland to Saskatchewan, south to Florida, Texas, and southeastern Arizona.

2. **Sagittaria latifolia** Willd. Near Tucson, Pima County, 2,500 feet (*Thornber* 250, 4271). Throughout most of North America.

3. **Sagittaria cuneata** Sheld. Showlow Creek, southern Navajo County, 6,500 feet (*Pultz* 1713), Mormon Lake, Coconino County, 7,000 feet (*Deaver* in 1933), Sacaton, Pinal County, 1,300 feet (*Harrison & Peebles* 1999). Ontario to British Columbia, south to Michigan, New Mexico, Arizona, and California.

4. **Sagittaria longiloba** Engelm. Bonita, Graham County, 4,500 to 5,000 feet (*Shreve* 5217, *Pultz* 1813). Kansas to Colorado, Arizona, and Mexico.

5. **Sagittaria Greggii** J. G. Smith. 40 miles southwest of Tucson, Pima County (*Harrison* 8937), about 3,000 feet. Arizona, California, and Mexico. This entity apparently intergrades with *S. longiloba*, but the material available is too scanty to permit a decision as to the validity of the species.

17. HYDROCHARITACEAE. Frogs-bit Family

1. ANACHARIS. Water-weed

Immersed aquatic herbs with slender, leafy stems and narrow, 1-nerved, entire or serrulate, opposite or whorled leaves; plants dioecious, the solitary flowers each subtended by a 2-cleft axillary spathe; staminate flowers long-pedicelled, with 3 greenish reflexed outer segments and 3 whitish, more or less spreading inner segments; pistillate flowers with a small, 6-parted limb and a very long, slender tube adnate at base to the ovary.

The name *Elodea* for this genus has been proposed for conservation.

KEY TO THE SPECIES

1. Leaves in whorls of 4 or 5, lanceolate, acuminate, 15 to 25 mm. long, about 3 mm. wide, very crowded, concealing the internodes, especially toward the end of the branches; petals of the staminate flowers 6 to 8 mm. long........................1. *A. densa*
1. Leaves opposite or in whorls of 3, linear, acute or obtusish, up to 20 mm. long and 5 mm. wide but usually much smaller, not crowded, most of the internodes apparent; petals of the staminate flowers less than 5 mm. long........................2. *A. canadensis*

1. Anacharis densa (Planch.) Marie-Victorin (*Elodea densa* (Planch.) Casp.). Canelo, Santa Cruz County, in an artificial pond (*Benson* 11445a). Florida, southern Arizona, and California, adventive from South America.

2. Anacharis canadensis (Michx.) Planch. (*Elodea canadensis* Michx.). Big Lake, Apache County (*Peebles* 15155), near Williams, Coconino County (*M. Wetherill* in 1938), 12 miles east of Mayer, Yavapai County (*Pulich* in 1948). Widely distributed in North America.

The specimens cited are without flowers. They are assigned tentatively to var. *Planchonii* (Casp.) Marie-Victorin (*Elodea Planchonii* Casp.), the variety of *A. canadensis* that is most common in the Western states.

18. GRAMINEAE (34). Grass Family

Contributed by Jason R. Swallen

Herbs (woody in *Arundo*); stems (culms) hollow or solid, closed at the nodes; leaves 2-ranked, parallel-veined, composed of a sheath enclosing the culm, and a blade, with a hairy or membranaceous appendage (ligule) between them on the inside; flowers perfect or sometimes unisexual, arranged in spikelets, these consisting of a short axis (rachilla) and 2 to many 2-ranked bracts, the lower 2 bracts (glumes) empty, the succeeding ones (lemmas) bearing in their axils a single flower, and between the flower and the rachilla a 2-nerved bract (palea), the lemma, palea, and included flower constituting the floret; stamens 1 to 6; anthers 2-celled; pistil 1, with 2 (rarely 1 or 3) styles, and usually plumose stigmas; spikelets mostly aggregate in spikes or panicles at the ends of the main culms and branches.

This very large family contains the most valuable of all plants used by man—the cultivated cereals, such as wheat, barley, maize, and rice. It also includes the most important forage plants, many of which, notably blue grass, timothy, and fescue, are extensively grown for hay and pasturage. Nearly all grasses of temperate regions are eaten by grazing animals, but some are more nutritious and palatable than others. Some of the native grasses of Arizona, especially species of the genera *Bouteloua* (grama) and *Hilaria*, are outstanding range plants upon which the maintenance of the cattle and sheep industries of the state is largely dependent. Seeds of many of the native species were used as food by the Indians. The sod-forming habit of many grasses makes them pre-eminent as soil binders. A few of the Arizona species, including the native red sprangle-top (*Leptochloa filiformis*) and the introduced species Johnson grass (*Sorghum halepense*) and Bermuda grass (*Cynodon dactylon*), are troublesome weeds in cultivated lands. The two latter, however, are of considerable forage value, and Bermuda grass, in the long, hot summers of southern Arizona, is the only satisfactory grass for lawns. Several species are commonly cultivated as ornamentals, notably such large plants as the giant-reed (*Arundo*), pampas grass (*Cortaderia*), and silver grass (*Miscanthus*). A tribe of large, woody grasses, the bamboos, comprises species of great utility, especially in eastern Asia, as substitutes for timber in construction. Another grass of very great economic importance is sugar-cane (*Saccharum officinarum*).

Many people are allergic to the pollen of grasses, one of the commonest causes of hay fever. So far as is known, there is little difference within this family in the allergenic toxicity of the pollens, the harmfulness of a particular species depending upon the abundance of the plants and the freedom with which the pollen is shed. Bermuda grass (*Cynodon dactylon*) is by far the most important hay-fever grass in Arizona, Johnson grass (*Sorghum halepense*) probably ranking next.

KEY TO THE GENERA

1. Spikelets in groups of 3 to 5, falling entire, sometimes enclosed in spiny burs: Tribe Zoysieae; see also genus *Cenchrus* (2).
1. Spikelets not arranged in groups (4).
2. Spikelets in groups of 2 to 5, sessile on a short, zigzag rachis, enclosed in small, spiny burs composed of the indurate second glumes of the 2 lower spikelets......52. *Tragus*
2. Spikelets in groups of 3, the central one perfect, the lateral spikelets staminate, not enclosed in burs (3).
3. Rigid perennials; groups of spikelets erect on the stiff axis...............53. *Hilaria*
3. Delicate annual; groups of spikelets nodding on one side of the very slender axis 54. *Aegopogon*
4. Spikelets dorsally compressed, with one terminal fertile floret and a sterile or staminate floret below, the latter usually without a palea, the articulation below the spikelets, either in the pedicel, the rachis, or at base of a cluster of spikelets: subfamily Panicoideae (5).
4. Spikelets more or less laterally compressed; sterile florets, if any, above the fertile ones (except in tribe Phalarideae); articulation above the glumes (except in genera *Leersia, Sphenopholis, Trisetum, Lycurus, Polypogon, Alopecurus*, and *Spartina*): subfamily Festucoideae (25).
5. Fertile lemma and palea firmer than the glumes, usually indurate: tribe Paniceae (6).
5. Fertile lemma and palea thin, hyaline, the glumes indurate. Sterile lemma like the fertile one (16).
6. Spikelets subtended by one or more bristles, these distinct or united at base, forming an involucre or spiny bur (7).
6. Spikelets not subtended by bristles (9).
7. Bristles persistent, the spikelets deciduous...........................79. *Setaria*
7. Bristles deciduous with the spikelets at maturity (8).
8. Bristles not united, slender, plumose toward base..................80. *Pennisetum*
8. Bristles united below, forming a burlike involucre, retrorsely barbed....81. *Cenchrus*
9. Spikelets in open panicles (except in *Panicum obtusum* and *P. texanum*), not conspicuously one-sided on the branches, the pedicels short or elongate (10).
9. Spikelets short-pedicelled on one side of the panicle branches (12).
10. Spikelets covered with silky, pink, spreading hairs much longer than themselves; panicle branches and pedicels flexuous........................78. *Rhynchelytrum*
10. Spikelets not so covered; panicle branches and pedicels straight (11).

11. Fruit dark brown; lemma with more or less prominent white margins, these not inrolled .... 73. *Leptoloma*
11. Fruit pale; margins of the indurate lemma inrolled........ 76. *Panicum*
12. Second glume mucronate, the sterile lemma mucronate or awned........ 77. *Echinochloa*
12. Second glume and sterile lemma awnless (13).
13. Spikelets covered with long, silky hairs........ 71. *Trichachne*
13. Spikelets appressed-pubescent (except sometimes in *Eriochloa*) or glabrous (14).
14. Racemes digitate. First glume present; weedy, decumbent, spreading annuals 72. *Digitaria*
14. Racemes panicled (15).
15. First glume and the rachilla joint forming a swollen ringlike callus below the spikelet; back of the fruit turned away from the rachis........ 74. *Eriochloa*
15. First glume wanting (occasionally present in *Paspalum distichum*); back of the fruit turned toward the rachis........ 75. *Paspalum*
16. Spikelets unisexual, the pistillate ones below, the staminate spikelets above in the same inflorescence .... 91. *Tripsacum*
16. Spikelets in pairs, one sessile, the other pedicellate, both usually fertile, or the sessile spikelet perfect and the pedicellate one sterile or staminate, sometimes much reduced: tribe Andropogoneae (17).
17. Spikelets all alike, surrounded by a conspicuous tuft of soft hairs (18).
17. Spikelets unlike, the sessile one perfect, the pedicellate spikelet usually sterile, but the sessile spikelet staminate and the pedicellate one perfect in genus *Trachypogon* (19).
18. Rachis continuous; fertile lemma awnless........ 82. *Imperata*
18. Rachis breaking into joints at maturity; fertile lemma awned........ 83. *Erianthus*
19. Racemes reduced to 1 or few joints, these peduncled in a subsimple or compound panicle (20).
19. Racemes of several to many joints, solitary, digitate, or aggregate in panicles (21).
20. Pedicellate spikelets staminate........ 85. *Sorghum*
20. Pedicellate spikelet wanting, only the pedicel present........ 86. *Sorghastrum*
21. Spikelets awnless (22).
21. Spikelets, or at least some of them, awned (23).
22. Rachis joint and pedicel distinct; perfect spikelet lanceolate; tufted perennial. Rachis joints and pedicels woolly........ 89. *Elyonurus*
22. Rachis joint and pedicel adnate; perfect spikelet globose; freely branching annual. Racemes short, partly enclosed in the sheaths........ 90. *Hackelochloa*
23. Awns delicate, glabrous, not more than 2.5 cm. long........ 84. *Andropogon*
23. Awns stout, conspicuous, 3 to 12 cm. long or if rather slender, then plumose on the lower part. Racemes solitary (24).
24. Awns hirsute; primary spikelet fertile, the pedicellate one sterile. Lower pairs of spikelets all staminate or neuter........ 87. *Heteropogon*
24. Awns plumose; primary spikelet sterile, the pedicellate one fertile.... 88. *Trachypogon*
25. Spikelets with 2 sterile or staminate florets below the fertile one: tribe Phalarideae (26).
25. Spikelets with no sterile florets below the 1 or more fertile florets (27).
26. Lower florets staminate, as large as the fertile floret........ 68. *Hierochloë*
26. Lower florets sterile, reduced to small scales........ 69. *Phalaris*
27. Glumes wanting. Pedicels articulate below the 1-flowered spikelets; lemma and palea indurate, equal, strongly keeled: tribe Oryzeae........ 70. *Leersia*
27. Glumes present (28).
28. Spikelets sessile or subsessile in spikes or spikelike racemes (29).
28. Spikelets in open or spikelike panicles (47).
29. Spikelets on opposite sides of the continuous or disarticulating rachis: tribe Hordeae (30).
29. Spikelets subsessile on one side of the continuous rachis: tribe Chlorideae (35).
30. Spikelets solitary at each node of the rachis (31).
30. Spikelets (except sometimes in *Elymus*) more than 1 at each node of the rachis (33).
31. First glume wanting except on the terminal spikelet; spikelets placed edgewise to the rachis .... 27. *Lolium*

31. First glume present; spikelets placed flatwise to the rachis (32).
32. Plants perennial; glumes commonly with 2 or more nerves..............22. *Agropyron*
32. Plants annual; glumes 1-nerved, subulate..................................23. *Secale*
33. Spikelets 3 at each node of the rachis, 1-flowered, the lateral ones usually reduced to awns..26. *Hordeum*
33. Spikelets usually 2 at each node of the rachis (34).
34. Rachis usually continuous; glumes acute to aristate, entire..............24. *Elymus*
34. Rachis readily disarticulating; glumes narrow, extending into long divergent awns 25. *Sitanion*
35. Spikes solitary. Plant small, the spikes elongate, curved, about 1 mm. wide 58. *Microchloa*
35. Spikes more than 1, digitate or racemose on a common axis (36).
36. Inflorescence very small, enclosed in broad, crowded sheaths at the ends of the branches 66. *Munroa*
36. Inflorescence distinctly exserted (37).
37. Plants monoecious or dioecious, the spikelets all or mostly unisexual (38).
37. Plants with perfect flowers (39).
38. Plants with unisexual, rarely perfect florets, the staminate and pistillate ones in the same spike..65. *Cathestecum*
38. Plants monoecious or dioecious, the staminate and pistillate florets in separate spikes. Staminate spikes pectinate, the pistillate spikes with the rachis thickened and indurate, this together with the second glume forming a false involucre around the floret..67. *Buchloë*
39. Spikelets 1-flowered with no rudimentary florets above the perfect one (40).
39. Spikelets 2- to several-flowered or, if only 1-flowered, then with 1 or more rudimentary florets above the fertile one (42).
40. Spikes digitate..59. *Cynodon*
40. Spikes racemose (41).
41. Rachilla articulate above the glumes; spikes very slender, elongate, widely spreading 60. *Schedonnardus*
41. Rachilla articulate below the glumes, the spikelets falling entire; spikes short and thick. Glumes unequal, the first shorter, the second longer than the floret..61. *Spartina*
42. Spikes solitary or racemose (43).
42. Spikes digitate or aggregate on a very short axis (44).
43. Lemmas entire or minutely bifid, awnless or 1-awned; spikelets with 2 to several perfect florets..55. *Leptochloa*
43. Lemmas variously lobed or dentate, 3-awned; spikelets with 1 perfect floret and 1 or 2 rudimentary florets above it..64. *Bouteloua*
44. Lemmas awnless (45).
44. Lemmas awned (46).
45. Rachis not prolonged..56. *Eleusine*
45. Rachis prolonged beyond the spikelet into a naked point..........57. *Dactyloctenium*
46. Lemmas 1-awned; spikes digitate..62. *Chloris*
46. Lemmas 3-awned; spikes aggregate on a very short axis but scarcely digitate 63. *Trichloris*
47. Spikelets 1-flowered: tribe Agrostideae (48).
47. Spikelets 2- to several-flowered (64).
48. Fruit indurate, terete, awned; callus well developed, oblique, bearded (49).
48. Fruit thin or firm but not indurate; callus not well developed (52).
49. Awn trifid, the lateral awns sometimes minute........................51. *Aristida*
49. Awn simple, with a distinct line of demarcation between lemma and awn (50).
50. Awn readily deciduous, straight or sometimes bent but not tightly twisted 48. *Oryzopsis*
50. Awn persistent, twice-geniculate (except in *Stipa speciosa*), at least the lower segment tightly twisted (51).
51. Edges of the lemma not meeting, exposing the palea, this projecting from the summit as a minute point; callus short, acute........................49. *Piptochaetium*

51. Edges of the lemma overlapping, enclosing the palea; callus sharp-pointed, usually acuminate .... 50. *Stipa*
52. Callus bearded (53).
52. Callus not bearded (54).
53. Lemma membranaceous, awned from the back..................... 37. *Calamagrostis*
53. Lemma firm, chartaceous, awnless.............................. 38. *Calamovilfa*
54. Panicles narrow, dense, spikelike (55).
54. Panicles open or dense but scarcely spikelike, except in *Polypogon monspeliensis* with long-awned glumes (60).
55. Spikelets in pairs, one perfect, the other staminate or neuter, the pair falling together 42. *Lycurus*
55. Spikelets all alike, not paired (56).
56. Glumes equal, similar, united at base, enclosing the floret; articulation below the glumes, the spikelets falling entire (57).
56. Glumes unequal, dissimilar, not united at base; articulation above the persistent glumes (58).
57. Glumes awnless, pubescent or villous but not hispid-ciliate on the keels...40. *Alopecurus*
57. Glumes abruptly awned, hispid-ciliate on the keels...................... 43. *Phleum*
58. Glumes saccate at base. Panicle dense, spikelike; lemma with a long, delicate awn; palea about as long as the lemma................................ 44. *Gastridium*
58. Glumes not saccate (59).
59. Lemma 3-nerved, mostly awned from the tip or mucronate; caryopsis cylindric or somewhat compressed dorsally, usually not falling readily from the floret 45. *Muhlenbergia*
59. Lemma 1-nerved, awnless; caryopsis obovoid, usually strongly compressed laterally, falling readily from the floret................................ 46. *Sporobolus*
60. Glumes longer than the lemma. Lemma and palea much thinner than the glumes (61).
60. Glumes shorter than the lemma, or the awn tips exceeding the lemma in *Muhlenbergia racemosa* (62).
61. Glumes awnless; palea much shorter than the lemma, sometimes reduced to a small, nerveless scale.. 39. *Agrostis*
61. Glumes long-awned; palea nearly as long as the lemma................ 41. *Polypogon*
62. Lemma densely pubescent on the nerves........................ 47. *Blepharoneuron*
62. Lemma glabrous or, if pubescent, then not densely so on the nerves (63).
63. Lemma 3-nerved, mostly awned from the tip or mucronate; caryopsis cylindric or somewhat compressed dorsally, usually not falling readily from the floret 45. *Muhlenbergia*
63. Lemma 1-nerved, awnless; caryopsis obovoid, usually strongly compressed laterally, falling readily from the floret................................ 46. *Sporobolus*
64. Glumes longer than the lowest floret, usually longer than the spikelet (or shorter in genus *Sphenopholis* with a broadly obovate second glume); lemmas awnless or awned from the back: tribe Aveneae (65).
64. Glumes shorter (except sometimes in *Tridens*) than the lowest floret; lemmas awnless or awned from the tip or from a minutely bifid apex: tribe Festuceae (73).
65. Lemmas awnless or sometimes with a very short awn in genus *Koeleria* (66).
65. Lemmas awned (68).
66. Glumes alike, gradually acuminate; spikelets 5- or 6-flowered.......... 28. *Schismus*
66. Glumes unlike, the first narrow, the second wider, broadened above the middle; spikelets 2- or 3-flowered (67).
67. Second glume broadened above the middle, then abruptly narrowed to an acute tip, the first glume narrower but not conspicuously so; lemmas pale and shining; spikelets articulate above the glumes.................................. 29. *Koeleria*
67. Second glume broadly obovate, obtuse; lemmas firm, not shining; spikelets articulate below the glumes.. 30. *Sphenopholis*
68. Florets 2, one perfect, the other staminate (69).
68. Florets all alike (70).
69. Lower floret staminate, bearing a twisted, geniculate, exserted awn...34. *Arrhenatherum*

69. Lower floret perfect, awnless..35. *Holcus*
70. Spikelets several-flowered. Awns conspicuous, flat, bent...............36. *Danthonia*
70. Spikelets 2-flowered, sometimes with a rudimentary third floret (71).
71. Spikelets large, the glumes more than 1 cm. long.........................33. *Avena*
71. Spikelets small, the glumes much less than 1 cm. long (72).
72. Lemmas keeled, awned from above the middle, the tip minutely bifid.....31. *Trisetum*
72. Lemmas rounded on the back, awned from below the middle, the tip erose 32. *Deschampsia*
73. Plants dioecious; pistillate lemmas with 3 long, twisted, divergent awns, the staminate lemmas mucronate...21. *Scleropogon*
73. Plants with perfect flowers or if dioecious, then the staminate and pistillate spikelets similar in appearance (74).
74. Lemmas divided at summit into 5 or more awns or awnlike lobes (75).
74. Lemmas awnless, 1-awned, or if 3-awned, then the lateral awns minute (77).
75. Rachilla disarticulating between the florets; summit of the lemmas with awns intermixed with awned teeth...18. *Cottea*
75. Rachilla not disarticulating between the florets, all florets falling attached; summit of the lemmas with awns only (76).
76. Glumes 1-nerved; lemmas dissected into an indefinite number of fine, unequal, scaberulous awns...19. *Pappophorum*
76. Glumes 7-nerved; lemmas crowned with 9 flat, equal or subequal, plumose awns 20. *Enneapogon*
77. Tall, stout reeds with plumelike panicles (78).
77. Low or rather tall grasses; panicles not plumelike (79).
78. Lemmas hairy; rachilla naked...14. *Arundo*
78. Lemmas naked; rachilla hairy.......................................15. *Phragmites*
79. Plants dioecious, the staminate and pistillate spikelets similar in appearance. Culms erect from creeping rhizomes; plants of saline soil.................10. *Distichlis*
79. Plants with perfect flowers (80).
80. Spikelets of two kinds, arranged in fascicles, the terminal one 1-flowered, perfect, the others several-flowered, sterile...................................13. *Lamarckia*
80. Spikelets all alike, not arranged in fascicles (81).
81. Lemmas 3-nerved, the nerves usually prominent (82).
81. Lemmas 5- to many-nerved (85).
82. Lemmas glabrous, awnless (83).
82. Lemmas more or less pubescent on the nerves or callus, mucronate or awned (84).
83. Spikelets 3- to several-flowered; lemmas acute or acuminate............7. *Eragrostis*
83. Spikelets 2-flowered; lemmas truncate, strongly nerved, brown..........8. *Catabrosa*
84. Callus densely bearded; lemmas firm, mucronate, the nerves glabrous....9. *Redfieldia*
84. Callus glabrous; lemmas membranaceous, pubescent on the nerves........17. *Tridens*
85. Spikelets with 1 or more empty lemmas below the fertile ones; lemmas rather faintly many-nerved. A tall, broad-leaved grass with a drooping panicle of large, flat spikelets .... 11. *Uniola*
85. Spikelets without empty basal lemmas; lemmas 5- to 9-nerved (86).
86. Lemmas mucronate or awned, keeled at least toward apex (87).
86. Lemmas awnless, usually rounded on the back (89).
87. Spikelets densely clustered toward the ends of the branches. Glumes hispid-ciliate on the keel...12. *Dactylis*
87. Spikelets not clustered, rather evenly distributed in narrow or open panicles (88).
88. Lemmas awned from between the teeth of the minutely bifid apex..........1. *Bromus*
88. Lemmas awnless, or awned from the tip...............................2. *Festuca*
89. Glumes papery, 5-nerved; palea much shorter than the lemma; upper florets reduced to a club-shaped rudiment...16. *Melica*
89. Glumes not papery, 1- to 3-nerved; palea as long as the lemma; upper florets similar to the lower ones (90).
90. Nerves of the lemma converging toward the apex. Leaves with boat-shaped tips 6. *Poa*
90. Nerves of the lemma parallel, not converging toward the apex (91).

91. Nerves faint; plants usually of saline soil. Low annuals or perennials. . 3. *Puccinellia*
91. Nerves prominent; plants of fresh-water marshes (92).
92. Lemmas 7-nerved: styles as long as the stigmas. . . . . 4. *Glyceria*
92. Lemmas 5-nerved; stigmas sessile or nearly so. . . . . 5. *Torreyochloa*

1. BROMUS. Brome

Annuals or perennials with closed sheaths, flat blades, and open or sometimes rather dense panicles; lemmas keeled, or rounded below on the back, the margins not clasping the palea, awned from between the teeth of the minutely bifid apex; palea thin, ciliate on the keels, usually shorter than the lemma, adhering to the fruit.

Bromes in the young stage are relished by all classes of livestock, but at maturity the "beards" cause mouth injuries. Among the best Arizona forage species are the mountain bromes (*B. marginatus* and *B. carinatus*) and fringed brome (*B. ciliatus*). Some of the annual species, collectively known as "chess," such as cheat grass (*B. tectorum*) and foxtail brome (*B. rubens*), are introduced weeds. Smooth brome (*B. inermis*) is cultivated for hay and pasture in the northern plains region, and rescue grass (*B. catharticus*) is cultivated for winter forage in the southern United States.

KEY TO THE SPECIES

1. Spikelets flattened, the glumes and lemmas compressed-keeled (2).
1. Spikelets not conspicuously flattened, the lemmas rounded on the back (5).
2. Lemmas awnless or very short-awned. . . . . 1. *B. catharticus*
2. Lemmas distinctly awned, the awn 3 mm. or longer (3).
3. Lemma hirsute on the margins, the lobes rather prominent; plant annual 2. *B. arizonicus*
3. Lemma glabrous or evenly scabrous or pubescent over the back, the lobes minute; plants perennial (4).
4. Panicle with spreading, relatively slender branches; awn about ½ to ⅔ the length of the lemma, usually more than 7 mm. long. . . . . 3. *B. carinatus*
4. Panicle with stiff, usually erect branches; awn about ⅓ to ½ the length of the lemma, usually less than 7 mm. long. . . . . 4. *B. marginatus*
5. Plants perennial (6).
5. Plants annual (11).
6. Rhizomes present. Lemmas nearly awnless. . . . . 5. *B. inermis*
6. Rhizomes wanting (7).
7. Lemmas (at least the upper portion) glabrous or merely scabrous (8).
7. Lemmas evenly pubescent or pilose on the back, the pubescence sometimes sparse (9).
8. Lemmas villous-ciliate on the margins, pubescent on the lower part of the back, the upper portion glabrous. . . . . 6. *B. ciliatus*
8. Lemmas scabrous to nearly smooth. . . . . 7. *B. texensis*
9. Culms usually less than 60 cm. high; panicles usually less than 10 cm. long, with lax, few-flowered branches; first glume 3-nerved. . . . . 8. *B. anomalus*
9. Culms usually 80 to 120 cm. high; panicles more than 10 cm. long, with relatively stiff branches; first glume 1-nerved (10).
10. Panicle branches short, stiffly ascending or spreading, few-flowered; spikelets appressed, short-pedicelled. . . . . 9. *B. Orcuttianus*
10. Panicle branches long, arcuate-spreading; spikelets long-pedicelled. Lower sheaths glabrous or nearly so. . . . . 10. *B. frondosus*
11. Awns geniculate. Teeth of the lemma aristate. . . . . 21. *B. Trinii*
11. Awns not geniculate (12).
12. Lemmas narrow, acuminate, hyaline toward the tip, the teeth more or less aristiform (13).

12. Lemmas broad, acute to truncate (16).
13. Panicle contracted, more or less dense (14).
13. Panicle open, the branches widely spreading (15).
14. Culms pubescent below the panicle....................................17. *B. rubens*
14. Culms glabrous or merely scabrous below the panicle.............18. *B. madritensis*
15. Panicle branches slender; pedicels flexuous; awns 12 to 14 mm. long..19. *B. tectorum*
15. Panicle branches relatively stout; pedicels straight; awns 3.5 to 5 cm. long 20. *B. rigidus*
16. Panicle dense, the branches appressed (17).
16. Panicle open, the branches spreading or drooping (18).
17. Spikelets glabrous..11. *B. racemosus*
17. Spikelets pubescent..12. *B. mollis*
18. Panicle branches stout, stiffly spreading or drooping; lemmas firm...13. *B. commutatus*
18. Panicle branches slender, flexuous; lemmas rather thin (19).
19. Palea much shorter than the lemma..............................14. *B. japonicus*
19. Palea about as long as the lemma (20).
20. Sheaths pubescent; awns straight or nearly so........................15. *B. arvensis*
20. Sheaths glabrous or nearly so; awns undulate......................16. *B. secalinus*

1. Bromus catharticus Vahl (*B. unioloides* H.B.K.). Apache County to Mohave County, south to Cochise, Santa Cruz, and Pima counties, a weed in waste places, May to September. Southern United States, introduced from South America; West Indies and western South America.

2. Bromus arizonicus (Shear) Stebbins (*B. carinatus arizonicus* Shear). Almost throughout the state, at moderate elevations, type from Tucson (*Pringle* in 1884). Arizona, California, and Baja California.

3. Bromus carinatus Hook. & Arn. Southern Apache County to Coconino County, south to Cochise and Pima counties, 5,500 to 9,000 feet, moist meadows, August to October. Montana to British Columbia, south to New Mexico, Arizona, California, and northern Mexico.

4. Bromus marginatus Nees. Coconino, Mohave, Yavapai, Gila, Pinal, and Pima counties, up to 9,500 feet, roadsides, moist meadows, and rocky hills, sometimes common, March to June. South Dakota to British Columbia, south to northern Mexico.

5. Bromus inermis Leyss. Grand Canyon and Long Valley (Coconino County), Pinaleno Mountains (Graham County), Pinal Creek (Gila County), 7,000 to 10,000 feet. Escaped from cultivation, Minnesota and Kansas to eastern Washington, Oregon, and Arizona, occasional eastward; introduced from the Old World.

6. Bromus ciliatus L. (*B. Richardsoni* Link). Apache and Coconino counties to Cochise and Pima counties, up to 11,000 feet, moist woods and rocky slopes, July to October. Labrador to Alaska, south to Tennessee, Iowa, western Texas, Arizona, and southern California.

Specimens with exceptionally pubescent lemmas were collected in the Huachuca Mountains (*Gould & Haskell* 3323).

7. Bromus texensis (Shear) Hitchc. Huachuca Mountains, Cochise County 7,000 feet (*Gould et al.* 2472, 3324, *Darrow et al.* 1470). Southern Texas, southeastern Arizona, and northern Mexico.

8. Bromus anomalus Rupr. (*B. Porteri* (Coult.) Nash). Apache, Coconino, Yavapai, Greenlee, Graham, Cochise, Pima, and Santa Cruz counties, 2,000 to 10,000 feet, open woods, July to September. Saskatchewan and Idaho to western Texas, Arizona, California, and Mexico.

A form with densely lanate sheaths is var. *lanatipes* (Shear) Hitchc. Arizona specimens that were referred previously to *B. latiglumis* (Shear) Hitchc. and *B. purgans* L. are large specimens of *B. anomalus*.

9. **Bromus Orcuttianus** Vasey. Huachuca Mountains, Cochise County (*Lemmon* in 1883), open woods. Washington to California and Arizona.

10. **Bromus frondosus** (Shear) Woot. & Standl. Coconino, Greenlee, Maricopa, Cochise, and Pima counties, 5,500 to 9,500 feet, rocky hillsides and pine woods. Utah, New Mexico, and Arizona.

11. **Bromus racemosus** L. A weed along road at Grand Canyon (Coconino County). Waste places, Washington to Idaho and Colorado, south to Arizona and California, rare eastward; introduced from Europe.

12. **Bromus mollis** L. Grand Canyon (Coconino County), Roosevelt (Gila County), near Superior (Pinal County), Tucson (Pima County). A weed in fields and waste places; introduced from Europe.

13. **Bromus commutatus** Schrad. Flagstaff and Grand Canyon (Coconino County). Fields and waste places throughout the United States; introduced from Europe.

14. **Bromus japonicus** Thunb. Apache, Navajo, Coconino, Gila, Cochise, and Pima counties. Fields and waste places throughout the United States; introduced from the Old World.

15. **Bromus arvensis** L. Lukachukai Mountains (Apache County), Flagstaff and South Rim of Grand Canyon (Coconino County). A weedy European species introduced in a few scattered localities in the United States.

16. **Bromus secalinus** L. Grand Canyon, Coconino County (*Thornber* 8317). Almost throughout the United States; naturalized from Europe.

17. **Bromus rubens** L. Coconino and Mohave counties to Pima County, roadsides and waste places. Washington to Arizona and California; introduced from Europe.

18. **Bromus madritensis** L. Havasu Canyon, Coconino County (*J. T. Howell* 26443). Open ground and waste places, northern Arizona, Oregon, and California; introduced from Europe.

19. **Bromus tectorum** L. Navajo, Coconino, and Yavapai counties. Waste places throughout the United States except in the southeast; introduced from Europe.

A form with glabrous spikelets is var. *glabratus* Spenner.

20. **Bromus rigidus** Roth. Coconino and Mohave counties to Graham, Cochise, and Pima counties, a weed in waste places. British Columbia and Idaho to Arizona and California, rare eastward; introduced from Europe.

21. **Bromus Trinii** Desv. Mohave, Yavapai, Maricopa, and Pinal counties, dry plains and wooded slopes, March to May. Oregon and Colorado to Arizona and Baja California; introduced from Chile.

Chilean chess. Var. *excelsus* Shear, with larger spikelets, 7- (instead of 5-) nerved lemmas, and the awns spreading but not twisted or bent, has been collected near Lake Mead, Mohave County (*Clover* 6093).

After this book went to press, a paper was published by H. Keith Wagnon (Leafl. West. Bot. 6: 64–69) in which was described a new species of *Bromus*, *B. mucroglumis* Wagnon. Numerous collections in Coconino, Graham, Cochise, Santa Cruz, and Pima counties, Arizona, were cited, in addition to collections that were stated to be intermediate between *B. mucroglumis* and *B. Richardsoni* Link. These citations include specimens that had been referred previously to *B. anomalus*, *B. ciliatus*, etc. Pending the publication of Wagnon's revision of this group of *Bromus*, the writer is unable to evaluate his conclusions and has therefore made no alterations in the above treatment of the genus.

2. FESTUCA. Fescue

Plants annual or perennial; spikelets in narrow or open panicles, few- to several- (rarely 1-) flowered; rachilla disarticulating above the glumes and between the florets; glumes narrow, acute, unequal; lemmas rounded on the back, 5-nerved, usually awned from the tip or rarely from a minutely bifid apex.

Although inclined to become rather tough with age, the fescues are browsed extensively by sheep in the high mountain parks, where Arizona fescue (*F. arizonica*) is particularly abundant. The small annual six-weeks fescue (*F. octoflora*) furnishes winter and spring forage. Sheep fescue (*F. ovina*) and red fescue (*F. rubra*) are cultivated to a limited extent in the eastern United States in lawns and pastures. Meadow fescue (*F. elatior*) is cultivated for hay and pasture in the central United States.

KEY TO THE SPECIES

1. Plants annual (2).
1. Plants perennial (8).
2. Spikelets more than 5-flowered, the florets crowded; lemmas 4 to 5 mm. long, the margins inrolled; awns 2 to 5 mm. long .... 1. *F. octoflora*
2. Spikelets usually fewer than 5-flowered or if more than 5-flowered, then the lemmas 7 to 8 mm. long and the awns 10 to 13 mm. long (3).
3. Panicle narrow, the branches appressed (4).
3. Panicle more open, the branches and often the spikelets spreading or reflexed (5).
4. Lemmas ciliate toward apex; first glume less than half as long as the second 2. *F. megalura*
4. Lemmas not ciliate; first glume barely ⅓ as long as the second .... 3. *F. myuros*
5. Spikelets glabrous (6).
5. Spikelets pubescent (7).
6. Pedicels appressed; spikelets mostly 3- to 5-flowered .... 4. *F. pacifica*
6. Pedicels spreading, especially those on the upper part of the main axis; spikelets 1- or 2-flowered .... 5. *F. reflexa*
7. Pedicels appressed; lower branches of the panicle spreading or reflexed; lemmas pubescent or scabrous-puberulent; glumes glabrous or pubescent .... 6. *F. Grayi*
7. Pedicels and panicle branches finally spreading or reflexed; glumes and lemmas pubescent .... 7. *F. Eastwoodae*
8. Blades flat, 3 to 6 mm. wide, rather thin; lemmas acuminate, sometimes with an awn as much as 2 mm. long (9).
8. Blades involute, less than 3 mm. wide (10).
9. Spikelets mostly not more than 5-flowered .... 8. *F. sororia*
9. Spikelets mostly 8- to 10-flowered .... 9. *F. elatior*
10. Ligule 2 to 4 mm. long, or longer .... 10. *F. Thurberi*
10. Ligule short (11).
11. Culms loosely tufted, decumbent at base; lower sheaths red, fibrillose; blades soft, glabrous or scabrous .... 11. *F. rubra*

11. Culms densely tufted; blades hard and firm, usually scabrous (12).
12. Panicle narrow, the branches appressed; lemmas 4 to 5 mm. long; blades short ..... 12. *F. ovina*
12. Panicle open, the branches ascending or spreading; lemmas about 7 mm. long; blades elongate (13).
13. Awns 2 to 4 mm. long..........13. *F. idahoensis*
13. Awns very short or obsolete..........14. *F. arizonica*

1. Festuca octoflora Walt. Throughout the state, up to 6,500 feet but usually lower, sterile, rocky, open ground, spring-flowering. Southern Canada, throughout the United States, and Baja California.

A form with densely pubescent or hirtellous lemmas is var. *hirtella* Piper.

2. Festuca megalura Nutt. Fort Huachuca (Cochise County), Tucson (Pima County), open ground at low or medium altitudes, spring-flowering. British Columbia and Idaho south to Arizona and Baja California; Pacific slope of South America.

3. Festuca myuros L. Tucson Mountains, Pima County (*Thornber* in 1907). Widely naturalized in the United States, from Europe.

4. Festuca pacifica Piper. Mohave, Yavapai, Gila, Pinal, Maricopa, and Pima counties, open ground and open woods at low altitudes, spring and early summer. British Columbia and Montana, south to New Mexico, Arizona, and Baja California.

5. Festuca reflexa Buckl. Jerome Junction, Yavapai County (*Tidestrom* 933), dry or rocky slopes, April and May. Washington and Utah, south to Arizona and Baja California.

6. Festuca Grayi (Abrams) Piper. Yavapai, Gila, Maricopa, Pinal, and Pima counties, about 4,000 feet, moist or dry situations, usually spring and early summer. Washington to southern California and Arizona.

7. Festuca Eastwoodae Piper. Pinal and Pima counties, open, wooded slopes, usually April and May. Oregon, California, and Arizona, infrequent.

8. Festuca sororia Piper. Coconino, Graham, Cochise, and Pima counties, 7,500 to 11,000 feet, open woods, August and September, type from the Rincon Mountains (*Nealley* 177). Colorado, Utah, New Mexico, and Arizona.

This is the only species of *Festuca* in Arizona with broad, flat blades.

9. Festuca elatior L. McNary, Apache County (*Gould* 5302), Lakeside, Navajo County (*Harrison* 5511), meadows and waste places. Introduced in the cooler regions of North America; native of Eurasia.

10. Festuca Thurberi Vasey. Pinaleno Mountains, Graham County, 9,000 feet (*Shreve* 4334). Dry slopes and rocky hills, Wyoming to New Mexico and Arizona.

11. Festuca rubra L. Baldy Peak (Apache County), North Rim of Grand Canyon and San Francisco Peaks (Coconino County), 8,500 to 11,000 feet, also in Pima County, moist meadows and marshes, July and August. Arctic America, south in the mountains to Georgia, Colorado, Arizona, and California; also in the Old World.

12. Festuca ovina L. Apache and Coconino counties, 7,000 to 12,500 feet, open woods and rocky slopes, July to September. Alaska to Nebraska, New Mexico, Arizona, and California, introduced eastward; circumpolar.

An alpine form with short culms and smooth blades, also found in Arizona, is var. *brachyphylla* (Schult.) Piper.

13. Festuca idahoensis Elmer. White Mountains (Apache County), San Francisco Peaks (Coconino County), open woods and rocky slopes, July and August. British Columbia to Alberta, south to northern New Mexico, Arizona, and central California.

Some of the Arizona specimens are almost intermediate between this and the next species.

14. Festuca arizonica Vasey. Apache, Coconino, Yavapai, Greenlee, Graham, and Pima counties, 2,500 to 9,500 feet, dry plains and open woods, June to August, type from Flagstaff (*Tracy* 118). Colorado, Nevada, New Mexico, and Arizona.

This is the commonest species of *Festuca* in Arizona.

3. PUCCINELLIA (35). Alkali Grass

Dwarf annuals or perennials; spikelets several-flowered; glumes unequal; lemmas firm, obtuse, glabrous or pubescent; palea as long as the lemma or a little shorter.

Plants of moist, usually saline soil.

KEY TO THE SPECIES

1. Lemmas pubescent on the nerves; plant annual .... 1. *P. Parishii*
1. Lemmas glabrous or if pubescent, then the hairs not confined to the nerves; plants apparently perennial (2).
2. Panicle ellipsoid, rather compact, less than 10 cm. long, the branches floriferous nearly to the base; spikelets 3 to 4 mm. long .... 2. *P. fasciculata*
2. Panicle pyramidal, open, mostly 10 to 20 cm. long, the branches naked below; spikelets 4 to 7 mm. long .... 3. *P. airoides*

1. Puccinellia Parishii Hitchc. Shato, Navajo County (*J. T. Howell* 24573), Tuba, Coconino County (*Peebles* 11842), 5,000 to 6,000 feet, marshy ground. Southern California and northern Arizona, rare.

2. Puccinellia fasciculata (Torr.) Bicknell. Winslow, Navajo County (*Reeder & Reeder* 34–45). Nova Scotia to Delaware; Utah and northern Arizona; Europe.

3. Puccinellia airoides (Nutt.) Wats. & Coult. (*P. Nuttalliana* (Schult.) Hitchc.). Near Tsaili Peak, Apache County (*J. Howell, Jr.* 39), Cameron, Coconino County (*J. T. Howell* 24402), 4,000 to 7,500 feet. Minnesota to British Columbia, south to Kansas, New Mexico, Arizona, and California.

The plants from Cameron are very small, with an atypically narrow inflorescence.

4. GLYCERIA (35). Manna Grass

Aquatic perennials with closed sheaths; second glume 1-nerved; lemmas broad, scarious at apex, 7-nerved, glabrous; palea as long as or a little longer than the lemma; caryopsis dark brown or blackish; styles as long as the stigmas, persistent, the stigmas early deciduous.

These are palatable forage grasses, and the seeds have been used as human food in Europe.

KEY TO THE SPECIES

1. Spikelets linear, not compressed, 10 to 15 mm. long; panicles narrow, elongate .... 1. *G. borealis*
1. Spikelets ovate or oblong-ovate, more or less compressed; panicles open, the branches (at least the upper ones) drooping (2).
2. First glume 1.5 mm. long .... 2. *G. grandis*
2. First glume not more than 1 mm. long (3).
3. Culms usually less than 1 meter high; first glume 0.5 mm. long .... 3. *G. striata*
3. Culms 1 to 2 meters high; first glume 1 mm. long .... 4. *G. elata*

1. Glyceria borealis (Nash) Batchelder. Apache, Coconino, Cochise, and Pima counties, 7,500 to 9,000 feet, wet places, often in water, August. Newfoundland to Alaska, southward to Massachusetts, New Mexico, Arizona, and California.

2. Glyceria grandis Wats. White Mountains (Apache County), 7,000 to 7,500 feet (*Nichol* in 1933, *Gould & Robinson* 5079), perhaps also Hooker Cienega, southern Graham County (*Thornber* in 1905). Canada and Alaska to Tennessee, New Mexico, eastern Arizona, and Oregon.

3. Glyceria striata (Lam.) Hitchc. White Mountains (Apache and Navajo counties), Betatakin (Navajo County), Oak Creek (Coconino County), Sierra Ancha (Gila County), up to 9,000 feet, wet places, June to August. Newfoundland to British Columbia, southward to Florida, Arizona, and northern California.

4. Glyceria elata (Nash) Hitchc. Lukachukai and White mountains (Apache County), Oak Creek Canyon (Coconino County), Pinaleno Mountains (Graham County), Chiricahua Mountains (Cochise County), Rincon and Santa Catalina mountains (Pima County), 5,000 to 9,000 feet, wet meadows and moist woods, August. Montana to British Columbia, south in the mountains to New Mexico, Arizona, and southern California.

This species closely resembles *G. striata* but has stouter, taller culms and slightly larger spikelets. The length of the first glume seems undependable as a differentiating character.

5. TORREYOCHLOA

Aquatic perennial with open sheaths; second glume 3-nerved; lemmas 5-nerved; palea as long as the lemma; caryopsis light brown; stigmas sessile or nearly so, persistent.

1. Torreyochloa pauciflora (Presl) Church (*Glyceria pauciflora* Presl). White Mountains, Apache County, 7,000 to 9,000 feet (*Nichol* in 1933, *Gould & Robinson* 4971, 5057). South Dakota to Alaska, south to New Mexico, eastern Arizona, and California.

6. POA. Blue Grass

Annual or perennial, cespitose or rhizomatous grasses with open or contracted panicles; blades with boat-shaped tips; spikelets ovate or oblong, few- to several-flowered; lemmas glabrous or pubescent on the nerves, sometimes pubescent on the internerves, the intermediate nerves usually obscure.

The blue grasses are palatable and nutritious forage grasses. In Arizona, mutton grass (*P. Fendleriana*) and Bigelow blue grass (*P. Bigelovii*) furnish good forage during winter and spring. Kentucky blue grass (*P. pratensis*) is a standard pasture grass and is cultivated in lawns. Canada blue grass (*P. compressa*) is cultivated in pastures in the northeastern United States. Annual blue grass (*P. annua*) is a common weed in lawns.

KEY TO THE SPECIES

1. Plants annual (2).
1. Plants perennial (3).
2. Panicle 7 to 15 cm. long, contracted, rather dense; lemmas webbed at base, pubescent below on the internerves....1. *P. Bigelovii*
2. Panicle 3 to 10 cm. long, open, the branches rather few-flowered, naked toward base; lemmas more or less pubescent on the nerves, glabrous on the internerves, not webbed 2. *P. annua*
3. Rhizomes present: See also *P. Fendleriana* and *P. longiligula* with dense narrow panicles (4).
3. Rhizomes wanting (7).
4. Culms strongly compressed; panicles narrow, rather dense....6. *P. compressa*
4. Culms terete or only slightly compressed; panicles open (5).
5. Lemmas more or less webbed at base, the intermediate nerves glabrous; panicle pyramidal, the lower branches mostly in fives, arcuate-spreading....3. *P. pratensis*
5. Lemmas not webbed (sometimes villous) at base, the intermediate nerves usually pubescent, at least below; panicle narrow, the lower branches not in fives, appressed or ascending (6).
6. Culms not more than 50 cm. long; panicle narrow, the branches appressed or strictly ascending .... 4. *P. arida*
6. Culms up to 100 cm. long; panicle more open, the branches ascending-spreading 5. *P. glaucifolia*
7. Lemmas webbed at base, the web sometimes obscure in *P. interior* (8).
7. Lemmas not webbed but sometimes villous at base (10).
8. Panicle branches slender, reflexed at maturity, the lower branches in twos; culms rather soft at base. Blades relatively short, flat, mostly 2 to 4 mm. wide..7. *P. reflexa*
8. Panicle branches stiffly ascending, spreading, or drooping but not reflexed; culms firm toward base (9).
9. Panicle 10 to 30 cm. long, broad, the branches spreading or drooping; culms decumbent at the reddish base; ligule 3 to 5 mm. long....8. *P. palustris*
9. Panicle less than 10 cm. long, the branches short, ascending; culms erect, densely tufted; ligule less than 1 mm. long....9. *P. interior*
10. Spikelets strongly compressed, the lemmas at maturity distinctly keeled (11).
10. Spikelets not strongly compressed, the lemmas at maturity convex or rounded on the back, scarcely keeled (13).
11. Blades lax or soft, usually flat; spikelets usually 3-flowered....12. *P. rupicola*
11. Blades firm, involute, erect; spikelets 5- or 6-flowered, about 8 mm. long. Plants incompletely dioecious (12).
12. Ligule inconspicuous, less than 1 mm. long....10. *P. Fendleriana*
12. Ligule conspicuous, up to 7 mm. long....11. *P. longiligula*
13. Sheaths more or less scabrous; lemmas glabrous or minutely scabrous...13. *P. nevadensis*
13. Sheaths glabrous; lemmas more or less crisp-pubescent on the lower part of the back 14. *P. Canbyi*

1. Poa Bigelovii Vasey & Scribn. Apache, Coconino, and Mohave counties to Cochise and Pima counties, 1,000 to 3,000 feet, open ground, type from near Tucson (*Pringle*). Oklahoma and western Texas to Nevada, Arizona, southern California, and northern Mexico.

2. Poa annua L. Coconino, Yavapai, Graham, Gila, Maricopa, Pinal, Cochise, Pima, and Yuma counties, 500 to 8,000 feet, open ground, lawns, and waste places. Newfoundland to Alaska, south to Florida, Arizona, and California; also in tropical America at higher altitudes; introduced from Europe.

3. Poa pratensis L. Apache County to Mohave County, south to Cochise, Santa Cruz, and Pima counties, medium to high altitudes, moist meadows and open woods. Canada and throughout the United States, except in dry or hot situations; introduced from Europe.

This, the well-known Kentucky blue grass, is an extremely variable species, differing markedly in appearance depending on the location where it is found. Although the plant is sometimes depauperate and sometimes very tall and coarse, the florets are relatively uniform in having a pronounced cobweb and densely pubescent keel and lateral nerves.

4. Poa arida Vasey. Grand Canyon, Coconino County (*Grant* 5604), White Mountains, Apache County (*Phillips* 3085), Pinaleno Mountains, Graham County (*Moeller* 352), up to 9,500 feet. Manitoba to Alberta, south to Iowa, Texas, and northern Arizona.

Plains blue grass. Grant's specimen is exceptional in having the intermediate nerves of the lemma glabrous.

5. Poa glaucifolia Scribn. & Williams. San Francisco Peaks (Coconino County), moist places at high altitudes, July and August. Wisconsin to New Mexico and Arizona.

6. Poa compressa L. White Mountains (Apache County), near Lakeside (Navajo County), near Flagstaff (Coconino County), Pinaleno Mountains (Graham County), Sierra Ancha (Gila County), 6,500 to 8,500 feet, open ground, meadows, and waste places. Newfoundland to Alaska, south to Georgia, Oklahoma, New Mexico, Arizona, and California; introduced from Europe.

7. Poa reflexa Vasey & Scribn. San Francisco Peaks (Coconino County), 8,000 to 12,000 feet, moist, open meadows and stream banks, late June to August. Montana to eastern British Columbia, south to New Mexico and Arizona.

8. Poa palustris L. Pinaleno Mountains, Graham County (*Peebles et al.* 4413). Meadows and moist, open ground, at low and medium altitudes. Newfoundland and Quebec to Alaska, south to Virginia, Missouri, New Mexico, Arizona, and California; Eurasia.

9. Poa interior Rydb. San Francisco Peaks (Coconino County), 9,000 to 12,000 feet, moist meadows, grassy slopes, and open woods, July and August. Quebec to British Columbia, south to Vermont, Minnesota, New Mexico, and Arizona.

10. Poa Fendleriana (Steud.) Vasey (*P. Fendleriana arizonica* Williams). Apache County to Mohave County, south to Cochise, Santa Cruz, and Pima counties, 3,500 to 10,500 feet, rocky slopes and open woods, April to August, the type of var. *arizonica* from Yavapai Creek (*Rusby* in 1883). South Dakota to Idaho, south to western Texas, Arizona, and southern California.

Usually *P. Fendleriana* is a tufted grass but sometimes produces rhizomes. Typically the lemmas are pubescent on the keel and nerves, but frequently forms occur with the lemmas nearly or entirely glabrous.

11. Poa longiligula Scribn. & Williams (*P. Fendleriana* var. *longiligula* Gould). Navajo, Coconino, Mohave, Yavapai, Gila, Pinal, and Pima counties, 3,500 to 7,500 feet, rocky slopes and open woods, April to June. North Dakota to Oregon, south to New Mexico, Arizona, and California.

This species very closely resembles *P. Fendleriana,* differing in having a long, conspicuous ligule.

12. Poa rupicola Nash. San Francisco Peaks (Coconino County), 11,500 to 12,500 feet, alpine meadows and rocky slopes above timber line, August and September. Montana to Oregon, south to Arizona and California.

Similar in appearance to *P. interior* but usually smaller and the florets without a cobweb at base.

13. Poa nevadensis Vasey. "Arizona," without definite locality (*Pringle* 44, 133). Wet places, Montana to Yukon Territory and Washington, south to Colorado, Arizona, and California.

Pringle's specimens are not the typical form, differing in having shorter and relatively broader blades. The ligule is long, as in the common form.

14. Poa Canbyi (Scribn.) Piper. Mokiak Springs, northern Mohave County, 3,000 feet (*Gould* 1643). Quebec to Yukon, south to western Nebraska, Colorado, northern Arizona, and eastern Oregon.

Poa secunda Presl, an indigenous species closely related to *P. Canbyi* but a smaller plant with more numerous and finer innovations, was collected at Willow Point, Kaibab Plateau (*Goodding* 17–49). *P. bulbosa* L., introduced from Europe, has been found in lawns at Fort Huachuca (*Goodding* 26–50). It has the culms more or less bulbous at base and the spikelets mostly proliferous and converted into bulblets.

7. ERAGROSTIS. Love Grass

Annual or perennial grasses with flat blades and open panicles; spikelets several-flowered, the rachilla continuous, the lemma persistent, the palea deciduous; glumes acute, 1-nerved, shorter than the first floret; lemma 3-nerved, rounded on the back; palea about as long as the lemma.

This genus has comparatively little forage value. Probably the best Arizona species in this regard is plains love grass (*E. intermedia*). Alkali love grass (*E. obtusiflora*) furnishes a large part of the forage locally in extremely saline soils in Cochise County. The seeds of teff (*E. Tef*) are utilized for food in Africa. Stink grass (*E. cilianensis*), with an odor of cockroaches, is reported sometimes to poison horses, both the fresh plant and in hay.

KEY TO THE SPECIES

1. Plants perennial (2).
1. Plants annual (6).
2. Plants with stout, creeping rhizomes bearing hard, closely imbricate scales. Culms firm, wiry, erect or ascending; lemmas erose......................11. *E. obtusiflora*
2. Plants cespitose, without rhizomes (3).
3. Lateral nerves of the lemma obscure. Lemmas rounded on the back, 1.8 to 2 mm. long (4).
3. Lateral nerves of the lemma prominent (5).
4. Panicle 15 to 35 cm. long; glumes acute; lemmas 1.8 to 2.0 mm. long...14. *E. intermedia*
4. Panicle 7.5 to 15 cm. long; glumes obtuse or acutish; lemmas about 1.5 mm. long 15. *E. Lehmanniana*
5. Panicle purple, the branches stiffly spreading; culms 30 to 60 cm. high; lemmas about 1.5 mm. long..12. *E. spectabilis*

5. Panicle lead-colored, the branches slender, drooping; culms usually more than 1 m. high; lemmas about 2.5 mm. long .... 13. *E. curvula*
6. Plants with minute glands on the branches or on the keels of the lemmas (7).
6. Plants not glandular or with a few scattered glands on the sheaths (8).
7. Spikelets 2.5 to 3 mm. wide; keels of the lemmas with a few prominent glands; panicles open, dark gray-green or tawny .... 1. *E. cilianensis*
7. Spikelets 1 to 2 mm. wide; keels of the lemmas without glands; panicles narrow, the branches ascending or appressed, yellowish green .... 2. *E. lutescens*
8. Spikelets about 1 mm. wide (9).
8. Spikelets 1.5 mm. wide, or wider (10).
9. Plants delicate; spikelets 3 to 5 mm. long; lemmas 1 to 1.5 mm. long.... 3. *E. pilosa*
9. Plants rather stout; spikelets 5 to 7 mm. long; lemmas about 2 mm. long 4. *E. Orcuttiana*
10. Panicle narrow, the short branches stiffly ascending or spreading, few-flowered, bearing spikelets nearly to the base .... 5. *E. Barrelieri*
10. Panicle open, diffuse, the branches slender, usually somewhat drooping (11).
11. Spikelets appressed along the main branches of the panicle or the appressed primary branchlets (12).
11. Spikelets on usually slender spreading pedicels (13).
12. Panicle branches simple; spikelets rather distant.................. 6. *E. pectinacea*
12. Panicle branches compound; spikelets somewhat crowded............. 7. *E. diffusa*
13. Spikelets linear; pedicels stiff, widely divergent, longer than the spikelets.... 8. *E. arida*
13. Spikelets ovate to ovate-oblong; pedicels spreading but not stiffly divergent (14).
14. Plants up to 1 meter high; panicles 20 to 40 cm. long, with drooping, many-flowered branches. Blades up to 1 cm. wide .... 9. *E. neomexicana*
14. Plants usually less than 30 cm. high; panicles small, few-flowered, the branches relatively short, spreading but not drooping .... 10. *E. mexicana*

1. Eragrostis cilianensis (All.) Link. Coconino County to Mohave County, south to Graham, Santa Cruz, Pima, and Yuma counties, up to 6,000 feet, a common weed in cultivated ground and waste places, May to October. Throughout the United States except at higher altitudes, southward to Argentina; introduced from Europe.

2. Eragrostis lutescens Scribn. Arizona without definite locality (*Lemmon* 1321, in 1882). Dry ground and sandy shores, rare. Idaho and Washington to Arizona and California.

3. Eragrostis pilosa (L.) Beauv. Above Superior, Pinal County, 2,500 feet (*Peebles et al.* 2333), open ground and waste places. Massachusetts to Colorado and Arizona, south to Argentina; introduced from Europe.

4. Eragrostis Orcuttiana Vasey. Nogales (Santa Cruz County), Tucson and Santa Rita mountains (Pima County), fields and waste places, summer-flowering. Colorado and Arizona to Oregon and California.

5. Eragrostis Barrelieri Daveau. Near Duncan (Greenlee County), near Tombstone (Cochise County), common at Tucson (Pima County), 2,500 to 4,500 feet, waste places. Oklahoma and Texas to California; introduced from Europe.

***6. Eragrostis pectinacea** (Michx.) Nees. Arizona without definite locality (*Palmer* in 1869). Open ground and waste places, Maine to North Dakota, south to Florida, Texas (and Arizona?), rare westward.

Palmer's specimen is small and immature, and may be only an extreme form of *E. diffusa.*

7. Eragrostis diffusa Buckl. Almost throughout the state, up to 7,000 feet, open ground and waste places, July to September. Texas to Nevada, southern California, and northern Mexico, introduced eastward in a few localities.

8. Eragrostis arida Hitchc. Pinal, Cochise, Santa Cruz, and Pima counties, 1,000 to 5,500 feet, dry soil, rocky ground, and waste places, August to October. Missouri (where probably introduced) and Texas to Arizona.

9. Eragrostis neomexicana Vasey. Coconino, Yavapai, Cochise, Santa Cruz, and Pima counties, wet ground, fields, and waste places, mostly 4,000 to 5,500 feet, July to September. Oklahoma to Arizona and Mexico.

10. Eragrostis mexicana (Hornem.) Link. Coconino and Mohave counties, south to Cochise, Santa Cruz, and Pima counties, 4,000 to 8,500 feet, fields and waste places. Texas to Arizona, and southward through Mexico.

11. Eragrostis obtusiflora (Fourn.) Scribn. Common locally in saline soil around Willcox and near the Chiricahua Mountains (Cochise County), about 4,000 feet, April to September. Southern New Mexico, Arizona, and northern Mexico.

12. Eragrostis spectabilis (Pursh) Steud. Apache, Coconino, and Pima counties, sandy or rocky soil, at medium altitudes, September and October. Maine to Minnesota, south to Florida and Arizona.

13. Eragrostis curvula (Schrad.) Nees. Crook National Forest (*Purchase* 283), Pinal Mountains, Gila County, 7,500 feet (*Peebles* 14098). Introduced from Africa, grown for ornament and forage in several localities in the southern part of the United States.

14. Eragrostis intermedia Hitchc. (*E. lugens* of authors, in part, non Nees). Greenlee, Gila, and Yavapai counties to Cochise, Santa Cruz, and Pima counties, 3,500 to 6,000 feet, rocky hills, canyons, and plains, June to September. Missouri to Louisiana and Arizona; Mexico to Costa Rica.

15. Eragrostis Lehmanniana Nees. Near Fort Huachuca, Cochise County (*Goodding* 1121–49), Tucson, Pima County, well established at roadsides (*Gould* 2533). Introduced from South Africa by the Soil Conservation Service.

Another African species, *E. echinochloidea* Stapf, has been introduced into Arizona by the Soil Conservation Service and seems to be spreading from the original plantings.

8. CATABROSA. Brook Grass

Aquatic perennials; spikelets brown, 2-flowered, the florets distant; glumes nerveless, irregularly toothed; lemmas broad, the apex scarious; palea as long as the lemma.

***1. Catabrosa aquatica** (L.) Beauv. The only collection that has been attributed to Arizona (*E. Palmer* 486, in 1877) probably was made in Utah (3). The known range is Newfoundland to Alberta, south of Colorado, Utah, and Oregon; Eurasia.

9. REDFIELDIA. Blow-out Grass

Perennial, with extensive rhizomes; panicle large, the branches capillary; spikelets 3- or 4-flowered, the florets closely imbricate; glumes acuminate, 1-nerved; palea as long as the lemma.

1. Redfieldia flexuosa (Thurb.) Vasey. Hopi Indian Reservation, Navajo County (*Hough* 119), sand hills, August and September. South Dakota to Oklahoma, west to Arizona.

This grass is a valuable sand binder but is very rare in Arizona.

10. DISTICHLIS (36). Salt Grass

Dioecious perennials with creeping, scaly rhizomes, rigid culms, and dense, few-flowered panicles; glumes broad, acute, keeled, 3- to 7-nerved; lemmas closely imbricate, coriaceous; palea usually a little shorter than the lemma.

Plants of saline soils, of low forage value.

1. Distichlis stricta (Torr.) Rydb. (*D. dentata* Rydb.). Apache County to Coconino County, south to Cochise, Pima, and Yuma counties, up to 7,000 feet, May to October. Saskatchewan to Washington, south to Texas, Arizona, California, and Mexico.

11. UNIOLA

Plants perennial, with short, strong rhizomes and rather tall culms; blades broad and flat; panicle drooping, the spikelets large and flat, many-flowered; glumes and lemmas compressed-keeled, the first lemma empty; palea shorter than the lemma.

1. Uniola latifolia Michx. Near Coolidge Dam, Pinal County (*G. Sheets* 762, in 1935). New Jersey to Kansas, south to Florida and Texas; southern Arizona.

It seems very unlikely that this grass is indigenous to Arizona, but no evidence could be obtained that it had been planted at the station mentioned.

12. DACTYLIS. Orchard Grass

Densely tufted perennial with flat blades and open panicles; spikelets subsessile in dense clusters at the ends of the branches; spikelets compressed, few-flowered; glumes unequal, acute, hispid, ciliate on the keel; lemmas keeled, mucronate, ciliate on the keel.

Orchard grass is cultivated in some parts of the United States for hay and pasturage.

1. Dactylis glomerata L. Navajo, Coconino, Graham, and Gila counties, fields, meadows, and waste places. Newfoundland to Alaska, south to Florida, Arizona, and California; introduced from Eurasia.

13. LAMARCKIA. Golden-top

Low annual with dense 1-sided panicles; fertile spikelet with 1 perfect floret and a rudimentary floret raised on a long rachilla joint; lemma bearing a delicate awn just below the apex; sterile spikelets composed of numerous imbricate empty lemmas.

1. Lamarckia aurea (L.) Moench. Roosevelt (Gila County), Canyon Lake (Maricopa County), Tucson (Pima County), open ground and waste places. Texas, Arizona, southern California, and northern Mexico; introduced from the Mediterranean region.

14. ARUNDO. Giant-reed

Tall perennial with broad, linear blades and plumelike panicles; glumes as long as the spikelet, membranaceous, tapering to a slender point; lemma thin, densely pilose, the nerves excurrent, the midnerve extending into a straight awn.

1. **Arundo Donax** L. Occasional along streams and irrigation ditches in southern Arizona. Texas to southern California and northern Mexico; tropical America; introduced from the Old World.

The stout culms are used for lattices, mats, and screens, and in the construction of adobe huts. In Europe the culms are utilized for reeds of clarinets and organ pipes. The plants are useful for windbreaks and for controlling erosion along streams.

15. PHRAGMITES. Reed

Coarse perennial with broad, flat, linear blades and usually large panicles; glumes unequal, the first about half as long as the second; lemmas long-acuminate, glabrous, the summits of all of them about equal; palea much shorter than the lemma.

1. **Phragmites communis** Trin. Apache, Navajo, Coconino, Gila, Maricopa, and Yuma counties, marshes and wet ground along irrigation canals and river banks, July to October. Distributed throughout the world.

In Arizona and Mexico, where it is known as carrizo, shafts of arrows, Indian prayer sticks, weaving-rods, pipestems, mats, screens, cordage, nets, and thatching have been made from the culms of the reed. Both the rootstocks and the seeds have been used as food by Indians of various tribes.

16. MELICA (37). Melic Grass

Perennials with closed sheaths and usually rather narrow panicles of large spikelets; glumes thin, acute or obtuse, nearly as long as the lower floret; lemmas firm with scarious margins, awnless; palea much shorter than the lemma.

Several species are known as onion grass. Two exotic species, *M. altissima* L. and *M. ciliata* L., are sometimes cultivated as ornamentals.

KEY TO THE SPECIES

1. Spikelets not falling entire; pedicels straight. Glumes nearly as long as the spikelet; culms somewhat woody; panicles narrow, rather dense............3. *M. frutescens*
1. Spikelets falling entire; pedicels slender or capillary, recurved (2).
2. Spikelets 3-flowered, 8 to 10 mm. long; lower panicle branches spreading..1. *M. nitens*
2. Spikelets 4- or 5-flowered, 10 to 15 mm. long; panicle branches narrowly ascending or appressed .... 2. *M. Porteri*

*1. **Melica nitens** (Scribn.) Piper. A specimen labeled as from Arizona without definite locality (*Palmer*) is the only basis for including this species in the flora of the state. The known range is from Pennsylvania to Iowa, south to Virginia, Arkansas, and Texas.

2. Melica Porteri Scribn. Coconino, Yavapai, Graham, Gila, Cochise, Santa Cruz, and Pima counties, 2,500 to 8,000 feet, moist, open woods and in canyons, July to October. Colorado, Texas, New Mexico, Arizona, and Mexico.

Var. *laxa* Boyle, with more spreading panicle branches and purple-tinged spikelets, occurs in the mountains of Cochise and Pima counties.

3. Melica frutescens Scribn. Fish Creek (Maricopa County), Superstition Mountains (Pinal County), 1,500 to 5,000 feet, dry hills and canyons, April and May. Arizona, southern California, and Baja California.

17. TRIDENS

Cespitose or stoloniferous perennials with open or contracted panicles; glumes nearly equal; lemmas rounded on the back, the apex toothed or lobed, the midnerve usually excurrent in a short awn, the lateral nerves often excurrent as minute points, all the nerves pubescent.

In *Flowering Plants and Ferns of Arizona* the name *Triodia* was used for these grasses. Although often very abundant, especially fluff grass (*T. pulchellus*), the species of this genus are not important range plants.

KEY TO THE SPECIES

1. Plants widely stoloniferous. Panicles small, capitate, exceeded by the fascicles of leaves 1. *T. pulchellus*
1. Plants cespitose, without stolons (2).
2. Panicle open or loose, not dense or spikelike; spikelets seldom more than 5 (rarely 8) mm. long; lemmas about 2 mm. long, obscurely pubescent.........2. *T. eragrostoides*
2. Panicle narrow or dense; spikelets 5 to 15 mm. long; lemmas 4 to 6 mm. long, conspicuously pubescent (3).
3. Lemmas deeply lobed...3. *T. grandiflorus*
3. Lemmas minutely toothed or subentire (4).
4. Panicle ovoid, 1 to 2 cm. long; lemmas acute, the awn 1 to 2 mm. long......4. *T. pilosus*
4. Panicle elongate, 10 to 25 cm. long; lemmas obtuse, entire or minutely notched, awnless, sometimes mucronate (5).
5. Glumes acute, usually longer than the lowest floret, the second glume 3-nerved 5. *T. elongatus*
5. Glumes obtuse, usually shorter than the lowest floret, the second glume 1-nerved 6. *T. muticus*

1. Tridens pulchellus (H.B.K.) Hitchc. (*Triodia pulchella* H.B.K.). Throughout the state, up to 5,500 feet, mesas and rocky hills, March to October. Utah and Nevada to western Texas, Arizona, and southern California.

2. Tridens eragrostoides (Vasey & Scribn.) Nash (*Triodia eragrostoides* Vasey & Scribn.). Pima County, at Colossal Cave, Rincon Mountains, near Sells, and in the Ajo Mountains, 3,000 to 3,500 feet. Southern Florida and Cuba; Texas; southern Arizona and northern Mexico.

3. Tridens grandiflorus (Vasey) Woot. & Standl. (*Triodia grandiflora* Vasey). Greenlee, Cochise, and Pima counties, 3,000 to 5,500 feet, dry, rocky slopes, August to October. Western Texas to Arizona and northern Mexico.

4. Tridens pilosus (Buckl.) Hitchc. (*Triodia pilosa* (Buckl.) Merr.). Coconino, Mohave, Yavapai, Gila, and Santa Cruz counties, up to 5,500 feet, plains and rocky hills. Kansas to Nevada, south to central Mexico.

5. Tridens elongatus (Buckl.) Nash (*Triodia elongata* (Buckl.) Scribn.). Apache, Coconino, Yavapai, Greenlee, and Cochise counties, 3,500 to 7,000 feet, plains and rocky slopes, August to October. Missouri and Kansas to Arkansas and Arizona.

6. Tridens muticus (Torr.) Nash (*Triodia mutica* (Torr.) Scribn.). Navajo, Coconino, and Mohave counties to Cochise, Santa Cruz, Pima, and Yuma counties, up to 5,500 feet, dry plains, gravelly slopes, and rocky hills, April to October. Southern Colorado and Texas to southeastern California.

18. COTTEA. Cotta Grass

Small, tufted perennial with flat blades and narrow or oblong, rather dense panicles; glumes nearly equal, about as long as the lowest floret, several-nerved; lemmas rounded on the back, villous below, with 9 to 11 prominent nerves extending into teeth or awns; palea a little longer than the body of the lemma.

1. Cottea pappophoroides Kunth. Cochise and Pima counties, 2,000 to 4,000 feet, dry hills and plains, August to October. Western Texas to southern Arizona and central Mexico; Ecuador to Argentina.

19. PAPPOPHORUM. Pappus Grass

Tufted perennials with culms up to 100 cm. long, the nodes glabrous; panicles narrow, dense, 10 to 20 cm. long; spikelets 4- to 6-flowered, the upper florets sterile; rachilla disarticulating above the glumes and sometimes tardily between the florets; glumes nearly equal, 1-nerved; lemmas rounded on the back, many-nerved, dissected above into numerous, more or less unequal, scabrous awns; palea equaling the body of the lemma.

KEY TO THE SPECIES

1. Panicle spikelike, tawny or whitish . . . . 1. *P. mucronulatum*
1. Panicle narrow but loose, pinkish . . . . 2. *P. bicolor*

1. Pappophorum mucronulatum Nees. Cochise and Pima counties, 2,500 to 4,000 feet, open ground and low places on plains, July to September. Texas, Arizona, and northern Mexico; South America.

2. Pappophorum bicolor Fourn. Santa Cruz River at La Noria, Santa Cruz County (*Mearns* 1175), open valley land. Texas, southern Arizona, and Mexico.

20. ENNEAPOGON (38)

Differs from *Pappophorum* in the shorter culms (up to 40 cm. long), pubescent nodes, lead-colored panicles usually less than 5 cm. long, 3-flowered spikelets, 7-nerved glumes, and lemmas bearing 9 equal or nearly equal plumose awns.

1. Enneapogon Desvauxii Beauv. (*Pappophorum Wrightii* Wats.). Navajo County to Mohave County, south to Cochise, Santa Cruz, and Pima counties, 3,000 to 6,000 feet, rocky hills and plains, July to October. Texas to Arizona and Oaxaca; Peru to Argentina.

Reported to have considerable forage value. Cleistogamous spikelets are produced sometimes at the base of the culms.

21. SCLEROPOGON. Burro Grass

Monoecious or dioecious, stoloniferous, perennial, with short, flexuous blades and narrow panicles; staminate spikelets several-flowered, the lemma similar to the glumes, mucronate, the palea obtuse, shorter than the lemma; pistillate spikelets several-flowered, the florets falling together, the lowest one with a sharp, bearded callus, the upper ones reduced to awns; nerves of the lemmas extending into slender, spreading awns.

1. Scleropogon brevifolius Phil. Apache County to Mohave County, south to Cochise and Pima counties, up to about 5,000 feet, mesas, open slopes, and valleys, May to October. Texas, Colorado, Nevada, and Arizona to central Mexico; Argentina.

Although grazed by livestock before reaching maturity, this grass often spreads on heavily grazed ranges at the expense of more palatable species. The pointed awns penetrate clothing and wool.

22. AGROPYRON. Wheat Grass

Perennials, often with creeping rhizomes; culms erect or decumbent; spikes usually erect; glumes equal, firm, acute or awned, usually shorter than the first lemma; lemmas firm, rounded on the back, acute or awned, the awn straight or divergent.

Most of the species furnish forage, and some of them are among the most valuable range grasses of the western United States. In the valleys these grasses may grow in sufficient abundance to produce wild hay. Slender wheat grass (*A. trachycaulum*) has been cultivated in the Northwestern states. Western wheat grass (*A. Smithii*), sometimes erroneously called bluestem, furnishes a good deal of forage in Arizona, and in open depressions is a source of hay. The species with strong, creeping rhizomes are valuable soil binders. Quack grass (*A. repens*) is a troublesome weed. The rhizomes of this species, often adulterated with Bermuda grass, are used in treating urinary disorders.

KEY TO THE SPECIES

1. Plants with creeping rhizomes. Blades firm, strongly nerved (2).
1. Plants normally cespitose and without rhizomes, but apparently sometimes short-rhizomatous in *A. Saundersii* and *A. spicatum* (6).
2. Glumes acuminate or gradually tapering into a short awn, more or less asymmetric, the lateral nerves usually obscure....................................1. *A. Smithii*
2. Glumes acute or abruptly awn-pointed, symmetric, the lateral nerves evident (3).
3. Joints of the rachilla villous (4).
3. Joints of the rachilla glabrous or scaberulous (5).
4. Lemmas glabrous or puberulent, and ciliate......................2. *A. pseudorepens*
4. Lemmas villous..3. *A. dasystachyum*
5. Blades firm, stiff, often involute, not pilose........................4. *A. riparium*
5. Blades lax, flat, usually sparsely pilose above..........................5. *A. repens*
6. Spikelets much compressed, closely imbricate, divergent.............6. *A. desertorum*
6. Spikelets not much compressed or closely imbricate (7).
7. Lemmas awnless or very short-awned. Rachilla internodes villous; glumes broad, nearly as long as the spikelet.......................................7. *A. trachycaulum*
7. Lemmas distinctly awned (8).
8. Awns erect or nearly so (9).

8. Awns divergent, at least when dry (10).
9. Rachis continuous; awns of the glumes less than 1 cm. long.........8. *A. subsecundum*
9. Rachis finally disarticulating; awns of the glumes 2 to 4 cm. long....9. *A. Saundersii*
10. Glumes long-awned, the awns often as long as those of the lemmas; rachis finally disarticulating (11).
10. Glumes acute or with a very short awn; rachis continuous (12).
11. Culms prostrate or decumbent-spreading, often flexuous. Spikes nodding or flexuous 10. *A. Scribneri*
11. Culms erect...11. *A. saxicola*
12. Spike straight, 8 to 15 cm. long; spikelets appressed, the awns abruptly divergent; blades 1 to 2 mm. wide......................................12. *A. spicatum*
12. Spike flexuous, 15 to 30 cm. long; spikelets somewhat spreading, the awns divergent but not abruptly so; blades 4 to 6 mm. wide....................13. *A. arizonicum*

1. Agropyron Smithii Rydb. Apache, Navajo, Coconino, Yavapai, and Pima counties, 3,000 to 7,500 feet, dry hills, moist, open ground, and pine forests, May to September. New York to Alberta and Washington, south to Kentucky, Texas, Arizona, and California.

A form with pubescent lemmas is var. *molle* (Scribn. & Smith) M. E. Jones, and one with densely pubescent sheaths is var. *Palmeri* (Scribn. & Smith) Heller.

2. Agropyron pseudorepens Scribn. & Smith. North Rim of Grand Canyon and San Francisco Peaks (Coconino County), White Mountains (Apache County), about 8,000 feet. Nebraska to British Columbia, south to Texas and Arizona.

3. Agropyron dasystachyum (Hook.) Scribn. Arizona without definite locality (*U. S. Forest Service*, several collections). Plains and sandy shores, Michigan to British Columbia, south to Illinois, Nebraska, Colorado, Arizona, and Nevada.

The Arizona specimens may represent only a form of *A. riparium* with pubescent spikelets.

4. Agropyron riparium Scribn. & Smith. Kaibab Plateau, Coconino County (*Storm* 225, *Darrow* 2920), about 8,500 feet, dry or moist meadows and hills. North Dakota to Alberta and Washington, south to Colorado, northern Arizona, and Nevada.

5. Agropyron repens (L.) Beauv. Kaibab Plateau, South Rim of Grand Canyon, and near Flagstaff (Coconino County), 7,000 to 8,500 feet. Extensively naturalized in North America, from Eurasia.

6. Agropyron desertorum (Fisch.) Schult. Tonto Lake (Apache County), Flagstaff (Coconino County), Sierra Ancha (Gila County). Cultivated as a forage plant under the name of *A. cristatum* and becoming naturalized in some parts of the United States, introduced from Eurasia.

7. Agropyron trachycaulum (Link) Malte (*A. pauciflorum* (Schwein.) Hitchc.). Apache County to Mohave County, south to Cochise and Pima counties, 5,000 to 12,000 feet, moist meadows and open woods, May to October. Labrador to Alaska, south to West Virginia, Kansas, Arizona, and California.

A reduced form, with culms about 30 cm. long, spikes 4 to 5 cm. long, and short-awned glumes, occurs on the San Francisco Peaks at about 12,000 feet (*Wolf* 3126).

8. **Agropyron subsecundum** (Link) Hitchc. White Mountains (Apache County), North Rim of Grand Canyon and San Francisco Peaks (Coconino County), Sierra Ancha (Gila County), 7,000 to 9,000 feet, moist meadows and open woods. Newfoundland to Alaska, south to West Virginia, Missouri, Arizona, and California.

9. **Agropyron Saundersii** (Vasey) Hitchc. Mormon Lake, Coconino County 7,000 feet (*Darrow* 3236). Colorado, Utah, northern Arizona, and California.

Darrow reported that the plants at Mormon Lake showed a slight tendency to produce rhizomes.

10. **Agropyron Scribneri** Vasey. White Mountains (Apache County), San Francisco Peaks (Coconino County), alpine slopes up to 12,000 feet, July to September. Montana and Idaho to New Mexico and Arizona.

11. **Agropyron saxicola** (Scribn. & Smith) Piper. Low, open ground near Prescott, Yavapai County (*Hitchcock* 13195). South Dakota to Washington, south to Arizona and California.

12. **Agropyron spicatum** (Pursh) Scribn. & Smith. Hualpai Mountain (Mohave County), Chiricahua and Huachuca mountains (Cochise County), Bear Valley or Sycamore Canyon (Santa Cruz County), 4,500 to 7,500 feet, rocky slopes, plains, and dry, open woods, June to October. Michigan to Alaska, south to New Mexico, Arizona, and California.

13. **Agropyron arizonicum** Scribn. & Smith. Graham, Cochise, Santa Cruz, and Pima counties, 5,000 to 8,000 feet, rocky slopes and moist stream banks, late July to October, type from the Rincon Mountains (*Nealley* 67). Western Texas to Nevada, Arizona, California, and Chihuahua.

23. SECALE. Rye

Annual, with flat blades and broad, dense, long-awned spikes; spikelets solitary at the nodes, usually 2-flowered; glumes subequal, subulate, rigid, sharply acuminate; lemmas much broader than the glumes, sharply keeled, 5-nerved, hispid-ciliate on the keel and margins, tapering into a long awn.

1. **Secale cereale** L. Coconino and Yavapai counties, occasional at roadsides and in waste ground, but not established. Native of Eurasia.

Rye, the well-known cultivated cereal.

24. ELYMUS. Wild-rye

Cespitose or rhizomatous perennials, with usually broad, flat blades and slender or sometimes dense spikes; spikelets 2- to 6-flowered, more or less dorsiventral to the axis; glumes equal, firm or indurate, somewhat asymmetric; lemmas rounded on the back, awnless or awned from the tip.

These grasses are utilized as forage mainly before maturity, at which time they become too coarse. Mammoth wild-rye (*E. giganteus*) is cultivated occasionally for ornament.

KEY TO THE SPECIES

1. Plants with slender, creeping rhizomes. Spikelets often solitary, rather irregularly placed on the axis; lemmas short-awned .... 1. *E. triticoides*
1. Plants without creeping rhizomes, or these short and stout in *E. cinereus* (2).
2. Glumes narrow or subulate, obscurely nerved, not broadened above the base (3).
2. Glumes rather prominently nerved, broadened above the base. Spikelets awned (4).

3. Spikes large, thick, often compound; spikelets 2 to 4 at each node; culms usually tall and stout .... 2. *E. cinereus*
3. Spikes narrow, slender, loosely flowered; spikelets mostly solitary; culms relatively slender .... 3. *E. salinus*
4. Glumes relatively thin, not indurate at base .... 4. *E. glaucus*
4. Glumes indurate at base, rather prominently bowed out (5).
5. Awns usually flexuous, divergent, implicate; base of the glumes not terete .... 5. *E. canadensis*
5. Awns straight, erect; base of the glumes terete, usually straw-colored... 6. *E. virginicus*

1. **Elymus triticoides** Buckl. Navajo, Coconino, eastern Mohave, Yavapai, Greenlee, and Pima counties, 2,500 to 7,000 feet, rocky hills, canyons, and open woods, May to October. Montana and Washington to Texas, Arizona, California, and Baja California.

2. **Elymus cinereus** Scribn. & Merr. Oak Creek, Yavapai or Coconino County (*Rusby* 908, 909½), near Prescott (*Harrison* 7203), dry plains and slopes, June to September. Saskatchewan to British Columbia, south to New Mexico, central Arizona, and eastern California.

Arizona specimens have been referred to *E. condensatus* Presl.

3. **Elymus salinus** M. E. Jones. North end of Carrizo Mountains, Apache County (*Standley* 7466), Kaibab National Forest, Coconino County, 5,000 feet (*Darrow* 2980), Bright Angel Trail, Coconino County (*Silveus* 1923), dry hills. Wyoming, Idaho, Utah, and Arizona.

4. **Elymus glaucus** Buckl. Coconino, Yavapai, Graham, Gila, Maricopa, and Pima counties, 1,500 to 8,000 feet, open woods, thickets, and along streams, April to September. Ontario and Michigan to Alaska, south through Montana to Arizona and California.

5. **Elymus canadensis** L. Apache, Navajo, and Coconino counties, south to Cochise and Pima counties, 2,500 to 7,000 feet, moist ground along streams, thickets, and open ground, July to October. Quebec to Alaska, south to North Carolina, Mississippi, Texas, Arizona, and California.

A variant with larger, stouter, scarcely nodding spikes, var. *robustus* (Scribn. & Smith) Mackenz. & Bush, has been collected on the Hopi Indian Reservation, at Springerville (Apache County) and near Tucson (Pima County). A form with exceptionally short, straight awns was found in Oak Creek Canyon (*Deaver* 2389).

6. **Elymus virginicus** L. Oak Creek, Yavapai County, about 3,000 feet (*Rusby* 909). Moist ground, low woods and along streams, Newfoundland to Alberta, south to Florida and Arizona.

25. SITANION. Squirrel-tail

Tufted perennial with firm, narrow blades and rather dense, bushy spikes; spikelets 2- to few-flowered; glumes firm, very narrow, extending into long, scabrous, divergent awns, sometimes with a short bristle from the margins; lemmas firm, rounded on the back, minutely bifid, the midnerve extending into a long, divergent awn.

KEY TO THE SPECIES

1. Glumes cleft into 3 or more long, awnlike divisions .... 1. *S. jubatum*
1. Glumes entire or 2-cleft .... 2. *S. Hystrix*

1. Sitanion jubatum J. G. Smith. Grand Canyon, Coconino County (*Clover & Jotter* 4147), Black Mountains, Mohave County (*Gould & Darrow* 4294). Idaho and eastern Washington to Arizona, Nevada, and Baja California.

2. Sitanion Hystrix (Nutt.) J. G. Smith. Apache County to Mohave County, south to Cochise, Santa Cruz, and Pima counties, 2,000 to 11,500 feet, open, sandy ground, rocky hills, and open pine woods, March to September. South Dakota to British Columbia, south to Missouri, Texas, Arizona, and Mexico.

Eaten when young by livestock. The mature awns penetrate the flesh of grazing animals, causing inflammation.

26. HORDEUM (39). Barley

Annuals or perennials with flat blades and dense, bristly spikes; spikelets 1-flowered, 3 at each node of the articulate rachis, the middle one sessile, the lateral ones pedicelled, usually imperfect, sometimes reduced to bristles; glumes narrow or subulate; lemmas with back turned toward the rachis, rounded, obscurely nerved, tapering into an awn.

The wild barleys are grazed before maturity. Some of them are aggressive weeds. The "beards" and the sharp joints of the spikes injure stock by piercing their mouths and nostrils and get into wool, causing sores when they penetrate the skin. Common cultivated barley is *H. vulgare.*

KEY TO THE SPECIES

1. Plants perennial (2).
1. Plants annual (3).
2. Awns 2 to 5 cm. long; spike nodding .... 1. *H. jubatum*
2. Awns mostly less than 1 cm. long; spike erect .... 2. *H. brachyantherum*
3. Glumes ciliate, except the outer ones of the lateral spikelets (4).
3. Glumes not ciliate (5).
4. Internodes of rachis mostly 3 mm. long; prolongation of rachilla of lateral spikelets slender, 2.8 to 3.7 mm. long .... 3. *H. leporinum*
4. Internodes of rachis mostly 2 mm. long; prolongation of rachilla of lateral spikelets rather stout, 1 to 2.2 mm. long .... 4. *H. Stebbinsii*
5. Glumes of the central spikelet and first glume of the lateral spikelets dilated above the base .... 5. *H. pusillum*
5. Glumes not dilated, setaceous or reduced to awns (6).
6. Spikes 1 to 4 (usually 2 to 3) cm. long; awns of glumes and lemmas stiffly ascending, rather stout, subequal; pedicels of florets of lateral spikelets short and thick .... 6. *H. Hystrix*
6. Spikes 5 to 10 cm. long; awns of glumes and lemmas slender, somewhat flexuous and spreading, the awn of the fertile floret usually exceeding those of the glumes; pedicels of florets of lateral spikelets relatively long and slender .... 7. *H. arizonicum*

1. Hordeum jubatum L. Apache, Navajo, Coconino, and Maricopa counties, up to 7,500 feet, a common weed in moist, open ground, along ditches and in waste places, June to September. Newfoundland to Alaska, south to Maryland, Missouri, Texas, Arizona, and California.

Fox-tail barley, sometimes called squirrel-tail, a name belonging properly to *Sitanion.* A form with awns 1.5 to 3 cm. long is var. *caespitosum* (Scribn.) Hitchc.

2. Hordeum brachyantherum Nevski (*H. nodosum* of authors, non L.). Apache, Coconino, Greenlee, Maricopa, Cochise, and Pima counties, up to

10,000 feet, meadows and open ground, June to October. Montana to Alaska, south to New Mexico, Arizona, and California; introduced in a few localities in the Eastern states.

3. Hordeum leporinum Link (*H. murinum* of authors, in part, non L.). Coconino and Cochise counties, a weed in cultivated land and waste places, mostly April to June. Massachusetts to Georgia; British Columbia to western Texas, Arizona, and California; introduced from Europe.

4. Hordeum Stebbinsii Covas (*H. murinum* of authors, in part, non L.). Coconino, Mohave, Yavapai, Maricopa, Cochise, and Pima counties, a weed in cultivated land and waste places, mostly April to June. Idaho and Washington; New Mexico to California and Baja California; Old World.

5. Hordeum pusillum Nutt. Apache, Navajo, and Yavapai counties south to Cochise and Pima counties, open ground and waste places, March to June. Delaware to Washington, south to Florida, Arizona, and California; South America.

6. Hordeum Hystrix Roth (*H. Gussonianum* Parl.). Coconino, Yavapai, and Pinal counties, a weed along ditches, May and June. British Columbia to Utah, Arizona, and California, also Maine and New Jersey; introduced from Europe.

7. Hordeum arizonicum Covas. Pinal, Maricopa, and Pima counties, roadsides and waste places, April and May, type from Camp Lowell (*Thornber* 536). Southern Arizona and southeastern California.

The name *Hordeum adscendens* has been misapplied to this species.

27. LOLIUM. Rye Grass

Annuals or perennials with flat blades and usually long, slender spikes; spikelets several-flowered, placed edgewise to the rachis, the first glume wanting; lemmas rounded on the back, obtuse, acute, or awned.

Perennial rye grass (*L. perenne*) and Italian rye grass (*L. multiflorum*) are cultivated as forage grasses and in lawns. Darnel (*L. temulentum*), which is presumably the "tares" of the Bible, contains in its grains a narcotic poison, but it is uncertain whether the toxin is elaborated by the plant itself or is due to fungus infection. All the Arizona species are introductions from the Old World.

KEY TO THE SPECIES

1. Plants annual (2).
1. Plants perennial. Glume much shorter than the spikelet (3).
2. Glume equaling or longer than the spikelet; lemmas awned........... 1. *L. temulentum*
2. Glume shorter than the spikelet; lemmas awnless..................... 2. *L. rigidum*
3. Lemmas awnless or nearly so... 3. *L. perenne*
3. Lemmas, some or all of them, awned.................................... 4. *L. multiflorum*

1. Lolium temulentum L. Maricopa, Pinal, Pima, and Yuma counties, a weed in cultivated fields and along ditches. Massachusetts and Wisconsin, south to Georgia and Texas; Washington to Arizona and California.

2. Lolium rigidum Gaudin. Hualpai Mountain, Mohave County, 5,000 feet, apparently established (*Buzan* in 1939). Here and there in the United States; introduced from Europe.

3. Lolium perenne L. Coconino, Pinal, Cochise, Santa Cruz, and Pima counties, meadows, lawns, and waste places. Newfoundland to Alaska, south to Virginia, Arizona, and California.

4. Lolium multiflorum Lam. Coconino, Maricopa, Cochise, Santa Cruz, and Pima counties, open ground and waste places. Range about same as that of *L. perenne.*

28. SCHISMUS

Low annual with short, narrow blades and small, rather dense panicles; spikelets several-flowered; glumes acute, subequal, nearly as long as the spikelets; lemmas broad, rounded on the back, bidentate, pilose; palea broad, hyaline, the nerves near the margins.

KEY TO THE SPECIES

1. Lemma rounded and shallowly notched at apex, usually hairy only near the margins 1. *S. barbatus*
1. Lemma deeply notched, the lobes acute, sparsely to copiously long-hairy on the back 2. *S. arabicus*

1. Schismus barbatus (L.) Thell. Gila, Yavapai, and Mohave counties, south to Santa Cruz, Pima, and Yuma counties, 1,000 to 4,000 feet, growing in great abundance on the open desert, where it is an important spring range grass. Arizona and California; introduced from the Old World.

2. Schismus arabicus Nees. Near Mesa (Maricopa County), near Tucson (Pima County), 1,000 to 2,500 feet, sandy soil, well established. Occasional in southern Arizona, southern Nevada, and California; introduced from western Asia.

29. KOELERIA. June Grass

Tufted perennial with narrow blades and shining spikelike panicles; spikelets 2- to 4-flowered, the rachilla prolonged beyond the florets as a slender bristle; glumes dissimilar, the first narrow, 1-nerved, the second broadened above the middle, 3- to 5-nerved; lemmas pale, shining, acute, awnless.

1. Koeleria cristata (L.) Pers. Apache County to Mohave and Yavapai counties, south to Cochise and Pima counties, up to 8,500 feet, dry plains, rocky ground, and pine woods, May to October. Ontario to British Columbia, south to Delaware, Missouri, Louisiana, Arizona, California, and Mexico.

The plant affords summer forage in the mountains.

30. SPHENOPHOLIS. Wedge-scale, Wedge Grass

Slender perennials with flat blades and narrow, often dense, panicles; spikelets 2- or 3-flowered; pedicel disarticulating below the glumes; rachilla produced beyond the upper floret; first glume narrow, acute, the second broadly obovate; lemmas firm, awnless, the first lemma usually a little longer than the second glume.

KEY TO THE SPECIES

1. Panicle dense, often spikelike, erect; second glume very broad, obtuse....1. *S. obtusata*
1. Panicle rather loose, nodding, never spikelike; second glume subacute..2. *S. intermedia*

1. Sphenopholis obtusata (Michx.) Scribn. Apache, Navajo, Coconino, Yavapai, Pinal, Cochise, and Pima counties, moist ground and open woods at medium altitudes, May to July. Maine to British Columbia, south to Florida, Arizona, California, and Mexico.

2. Sphenopholis intermedia (Rydb.) Rydb. Fort Apache, Navajo County (*Palmer* 577), Huachuca Mountains, Cochise County (*Palmer* 467, *Darrow* 2589), about 6,000 feet, moist ground, mostly in shade. Newfoundland to British Columbia, south to Florida and Arizona.

31. TRISETUM

Tufted perennials with flat blades and open or spikelike, usually shining panicles; spikelets 2- or 3-flowered; rachilla prolonged beyond the upper floret, usually villous; glumes acute, the second a little longer than the first; lemmas 2-toothed, bearing from just below the cleft a bent exserted awn.

KEY TO THE SPECIES

1. Plants annual; spikelets disarticulating below the glumes; rachilla pubescent 1. *T. interruptum*
1. Plants perennial; spikelets disarticulating above the glumes; rachilla villous (2).
2. Panicle dense, spikelike, more or less interrupted below................2. *T. spicatum*
2. Panicle open, rather densely flowered but not spikelike................3. *T. montanum*

1. Trisetum interruptum Buckl. Mescal (Cochise County), Santa Rita Mountains (Santa Cruz or Pima County), and at several localities in Pima County, 3,000 to 4,000 feet, in open ground. Texas, Colorado, New Mexico, and Arizona.

2. Trisetum spicatum (L.) Richt. Summit of Baldy Peak (Apache County), San Francisco Peaks (Coconino County), alpine slopes and meadows, 10,500 to 12,000 feet. Arctic America, south to Connecticut, Pennsylvania, Minnesota, and in the mountains to New Mexico, Arizona, and California; high mountains of Mexico and South America; alpine regions of the Old World.

3. Trisetum montanum Vasey. Pinaleno Mountains, Graham County, 9,500 feet (*Kearney & Peebles* 9970, *Darrow* in 1942). Mountain slopes and meadows, Colorado, Utah, New Mexico, and Arizona.

32. DESCHAMPSIA. Hair Grass

Annuals or perennials with narrow or open, shining panicles; spikelets 2-flowered, the hairy rachilla prolonged beyond the upper floret, sometimes bearing a rudimentary floret; glumes equal, acute; lemmas thin, erose-truncate with a straight or bent and twisted awn from at or below the middle.

KEY TO THE SPECIES

1. Plants annual; panicle open, the stiffly ascending capillary branches usually in twos 1. *D. danthonioides*
1. Plants perennial; panicle narrow or open, the slender branches appressed or drooping (2).
2. Glumes usually longer than the florets; panicle usually narrow, as much as 30 cm. long, the branches appressed; blades filiform, lax........................2. *D. elongata*
2. Glumes shorter than the florets; panicle open, nodding, 10 to 25 cm. long, the branches drooping; blades firm, flat or folded............................3. *D. caespitosa*

1. **Deschampsia danthonioides** (Trin.) Munro. "Sulphur Val., S. E. Ariz.," probably Sulphur Springs Valley, Cochise County (*Lemmon* 386, 387, in 1881). Dry or moist, open ground, Alaska to Montana, south to Arizona and Baja California; Texas; Chile.

If Lemmon's collections are labeled correctly as to locality, the plant was probably introduced from California, perhaps with alfalfa seed.

2. **Deschampsia elongata** (Hook.) Munro. Virgin Mountains (northern Mohave County), Pinaleno Mountains (Graham County), Santa Catalina Mountains (Pima County), 4,000 to 9,000 feet, moist or dry, open ground. Alaska to Wyoming, Arizona, California, and Mexico; southern Argentina and Chile.

3. **Deschampsia caespitosa** (L.) Beauv. Apache, Coconino, and Cochise counties, up to 9,500 feet, damp or wet mountain meadows, June to September. Greenland to Alaska, south to North Carolina, Illinois, North Dakota, New Mexico, Arizona, and California; southern Argentina and Chile; also in the Old World.

This species affords excellent forage in mountain meadows.

33. AVENA. Oat

Annuals with relatively broad blades and open panicles of large spikelets; spikelets 2- or 3-flowered; rachilla villous; glumes equal, several-nerved, exceeding the florets; lemmas indurate, bidentate, hairy, bearing a dorsal, bent or twisted awn.

The wild oats are often troublesome weeds but occasionally are useful for hay and pasturage. The long-awned fruits are injurious to the mouths of animals. The commonly cultivated oat is *Avena sativa,* which is found occasionally at roadsides in Arizona but has not become established. It is readily distinguished from the 2 species of wild oats by having the awn either straight or wanting.

KEY TO THE SPECIES

1. Teeth of the lemma acute; pedicels slender but not capillary............1. *A. fatua*
1. Teeth of the lemma setaceous; pedicels curved, capillary..............2. *A. barbata*

1. **Avena fatua** L. Apache, Coconino, and Mohave counties to Cochise and Pima counties, in waste places. Maine to Pennsylvania, Missouri, South Dakota, New Mexico, Arizona, and California; introduced from Europe.

2. **Avena barbata** Brot. Tucson, Pima County (*Toumey* 747). A weed in waste places, Washington to California and Arizona; introduced from Europe.

34. ARRHENATHERUM

A rather tall perennial with flat blades and long, narrow panicles; spikelets 2-flowered, the lower flower staminate, the upper one perfect; rachilla disarticulating above the glumes, produced beyond the florets; glumes rather broad and papery, the first 1-nerved, the second a little longer, 3-nerved; lemmas 5-nerved, with a hairy callus, the lower lemma bearing near the base a twisted, geniculate, exserted awn, the upper lemma bearing near the apex a short, straight, slender awn.

1. **Arrhenatherum elatius** (L.) Presl. Near Flagstaff, Coconino County (*Whiting* 1211), Prescott, Yavapai County (*E. W. Phillips* in 1945), Pinaleno Mountains, Graham County (*Thornber & Shreve* 7837).

Tall-oat grass, extensively naturalized in North America, from Europe, sometimes grown for hay.

35. HOLCUS. Velvet Grass

Tufted perennial with flat, velvety-pubescent blades and compact panicles; spikelets 2-flowered, the pedicel disarticulating below the glumes; glumes nearly equal, longer than the florets; first floret perfect, the lemma awnless; second floret staminate, the lemma bearing a short, recurved awn from the back near the apex.

1. **Holcus lanatus** L. Flagstaff, Sedona, and Oak Creek Canyon (Coconino County), 4,500 to 7,000 feet, meadows and moist places. Widely distributed in North America; introduced from Europe.

Occasionally cultivated in meadows in the eastern United States. It is reported that the plant, either fresh or wilted, sometimes develops hydrocyanic acid.

36. DANTHONIA. Oat Grass

Tufted perennials with small, open panicles of rather large spikelets; spikelets several-flowered; glumes equal, broad, papery, exceeding the florets; lemma rounded on the back, bifid, the lobes acute, with a stout, flat, twisted, geniculate awn from between the lobes.

KEY TO THE SPECIES

1. Panicle narrow, the pedicels appressed; glumes about 15 mm. long; lemmas 7 to 8 mm. long ....1. *D. intermedia*
1. Panicle open, the pedicels spreading or somewhat reflexed; glumes mostly 15 to 20 mm. long; lemmas 8 to 10 mm. long................................2. *D. californica*

1. **Danthonia intermedia** Vasey. White Mountains (Apache County), Mogollon Escarpment (Coconino County), Pinaleno Mountains (Graham County), 7,500 to 8,500 feet. Wet meadows and pine forest in northern or alpine regions. Quebec to Alaska, south to Michigan, New Mexico, Arizona, and California.

2. **Danthonia californica** Boland. Pinaleno Mountains, Graham County, 9,000 feet (*Moeller* 348). Montana to British Columbia, south to Colorado, Arizona, and California.

37. CALAMAGROSTIS. Reed Grass

Erect, rhizomatous perennials with firm, flat or loosely involute blades and spikelike or narrow and open panicles; glumes equal, acute or acuminate; rachilla prolonged beyond the floret, hairy; lemma shorter than the glumes, the callus bearded, the midnerve exserted as an awn.

Several species of this genus are important forage grasses in other regions. Blue-joint reed grass (*C. canadensis*) is a source of much of the wild hay in Wisconsin and Minnesota. In some parts of the United States northern reed grass (*C. inexpansa*) is grazed by horses and cattle.

KEY TO THE SPECIES

1. Panicle nodding, rather loose and open. Callus hairs copious, about as long as the lemma .... 1. *C. canadensis*
1. Panicle erect, dense or spikelike (2).
2. Blades flat; glumes 4 to 6 mm. long; panicle moderately dense.......... 2. *C. scopulorum*
2. Blades involute; glumes 3 to 4 mm. long; panicle spikelike, more or less interrupted below. Culms scabrous below the inflorescence.......... 3. *C. inexpansa*

1. Calamagrostis canadensis (Michx.) Beauv. Kaibab Plateau and San Francisco Peaks (Coconino County), White Mountains (Apache County), Pinaleno Mountains (Graham County), Rincon Mountains (Pima County), 7,500 to 11,000 feet, wet places and open woods. Greenland to Alaska, south to North Carolina, Kansas, New Mexico, Arizona, and California.

2. Calamagrostis scopulorum Jones. North side of Grand Canyon, Coconino County, about 7,000 feet (*Eastwood & Howell* 7075, *Collom* 1584, 1585). Moist soil, Montana, Wyoming, Colorado, Utah, New Mexico, and Arizona.

3. Calamagrostis inexpansa Gray. North Rim of Grand Canyon and Mogollon Escarpment (Coconino County), White Mountains (Apache County), Willow Spring (Apache or Coconino County), wet meadows above 5,000 feet, July to September. Greenland to Alaska, south to Massachusetts, Indiana, Nebraska, New Mexico, Arizona, and California.

38. CALAMOVILFA. SAND-REED

Coarse, tall perennial with stout, creeping rhizomes, long, attenuate blades, and large, open panicles; glumes firm, 1-nerved, the first glume half as long, the second one as long as the floret; lemma chartaceous, acute, awnless, the callus densely bearded with hairs nearly half as long as the lemma.

1. Calamovilfa gigantea (Nutt.) Scribn. & Merr. Apache, Navajo, and Coconino counties, about 5,000 feet, sand hills, July to October. Kansas and Utah to Texas and Arizona.

This grass is a valuable sand binder. The reedlike stems are used by the Hopi Indians for various purposes.

39. AGROSTIS. BENT GRASS, BENT

Slender perennials with flat or involute blades and dense or very open panicles of small spikelets; glumes equal, acute, longer than the floret; lemma much thinner than the glumes, awnless; palea minute or sometimes nearly as long as the lemma.

Many of the species are important forage grasses. Red-top bent grass (*A. alba*) and creeping bent grass (*A. palustris*) are utilized for pastures and lawns. Spike bent grass (*A. exarata*) is valuable for forage in the western United States.

KEY TO THE SPECIES

1. Palea present, at least half as long as the lemma (2).
1. Palea wanting or only a small, nerveless scale, rarely longer (4).
2. Glumes scabrous over the back; palea nearly as long as the lemma; panicle dense, lobed, the branches short, densely flowered.......... 1. *A. semiverticillata*
2. Glumes scabrous only on the keel; palea about half as long as the lemma; panicle open or sometimes contracted but not lobed, the branches loosely flowered (3).

3. Culms erect; rhizomes present; panicle open, the branches ascending.......2. *A. alba*
3. Culms usually decumbent at base; rhizomes none; panicle contracted, the branches appressed ....3. *A. palustris*
4. Panicle contracted, the short, densely flowered branches appressed, floriferous from the base or nearly so..4. *A. exarata*
4. Panicle open, the branches floriferous from or above the middle (5).
5. Panicle very diffuse, the branches very scabrous, elongate, drooping........5. *A. scabra*
5. Panicle not diffuse, the branches glabrous or nearly so, short, usually ascending 6. *A. idahoensis*

1. **Agrostis semiverticillata** (Forsk.) C. Chr. (*A. verticillata* Vill., *Polypogon semiverticillatus* Hylander). Apache County to Mohave County, south to Cochise, Santa Cruz, and Pima counties, 6,500 feet or lower, wet ground, May to October. Texas to Washington, Arizona, and California, south to Guatemala; Peru to Argentina and Chile; introduced from the Old World.

2. **Agrostis alba** L. Apache, Navajo, Coconino, Yavapai, Greenlee, and Cochise counties, moist ground and waste places. Throughout the cooler regions of North America; apparently introduced from the Old World.

3. **Agrostis palustris** Huds. Yavapai, Gila, and Pinal counties, in damp places along streams and ditches, June to August, introduced in Arizona. Newfoundland to Maryland and British Columbia, south to Arizona and central California; Eurasia.

4. **Agrostis exarata** Trin. Apache County to Coconino County, south to Cochise and Pima counties, up to 8,000 feet, moist ground, June to September. Nebraska to Alberta and Alaska, south to Mexico.

5. **Agrostis scabra** Willd. Apache, Navajo, Coconino, Yavapai, Graham, Cochise, and Pima counties, moist ground up to 9,500 feet, May to September. Newfoundland to Alaska, south to Maryland, Illinois, Nebraska, New Mexico, Arizona, and California, rarely in the Southeastern states.

Tickle grass. This plant has been referred by authors to *A. hiemalis* (Walt.) B.S.P., which is a distinct species confined to the Southeastern states.

6. **Agrostis idahoensis** Nash (*A. filiculmis* Jones). White Mountains, Apache County (*Phillips* 3286), Kaibab Plateau, Coconino County (*Jones* in 1894, the type of *M. filiculmis*), up to 9,500 feet, wet meadows. Montana to Washington, south to Arizona and California.

40. ALOPECURUS. Fox-tail

Semiaquatic annuals or perennials with flat blades and soft, spikelike panicles; glumes equal, united at base, ciliate on the keel; lemma about as long as the glumes, with margins united at base, bearing from below the middle a straight or geniculate, included or exserted awn; palea wanting.

All the species are palatable and nutritious, but these grasses are not common enough in Arizona to be important.

KEY TO THE SPECIES

1. Awns straight, included or only slightly longer than the glumes. Plant perennial 1. *A. aequalis*
1. Awns geniculate, twisted below, much longer than the glumes (2).
2. Plants perennial; anthers 1.5 mm. long...........................2. *A. geniculatus*
2. Plants annual; anthers about 0.5 mm. long.......................3. *A. carolinianus*

1. Alopecurus aequalis Sobol. Apache, Coconino, and Yavapai counties, 5,000 to 9,000 feet, bogs and wet ground, June to August. Greenland to Alaska, south to Pennsylvania, Kansas, New Mexico, Arizona, and California; Eurasia.

2. Alopecurus geniculatus L. Apache, Coconino, Yavapai, and Pima counties, 2,000 to 9,000 feet, marshes and wet ground, July to September. Newfoundland to British Columbia, south to Virginia, Arizona, and California; Eurasia.

3. Alopecurus carolinianus Walt. Foxboro (Coconino County), Payson (Gila County), Rincon Mountains (Pima County), about 5,000 feet, wet ground, May and June. New Jersey to British Columbia, south to Florida, Arizona, and California.

41. POLYPOGON

Decumbent annuals or perennials with flat, scabrous blades, and dense, narrow or spikelike panicles; glumes equal, entire or minutely lobed, awned from the tip or from between the lobes, the awns slender, usually longer than the glumes; lemma thin, hyaline, about half as long as the glumes.

KEY TO THE SPECIES

1\. Plants annual; glumes minutely lobed, the awns very slender, (4) 6 to 8 mm. long; panicles very dense, spikelike ....1. *P. monspeliensis*

1\. Plants perennial; glumes not lobed, the awn not more than 5 mm. long; panicles moderately dense (2).

2\. Glumes abruptly narrowed above, the awn 2.5 to 5 mm. long...........2. *P. interruptus*

2\. Glumes gradually tapering into a short awn, this 1 to 2 mm. long........3. *P. elongatus*

1. Polypogon monspeliensis (L.) Desf. Throughout the state, up to 8,000 feet, waste places, along ditches, April to October. New Brunswick to Alaska, south to Virginia, mostly near the coast, and common in the Western states from Washington to Nebraska, south to Texas, Arizona, and California; introduced from Europe.

Sometimes known as rabbit-foot grass.

2. Polypogon interruptus H.B.K. Apache, Navajo, Coconino, Santa Cruz, and Pima counties, up to 7,500 feet, along ditches and streams, May to September. British Columbia to California, east to Louisiana; Mexico to Argentina.

3. Polypogon elongatus H.B.K. Madera Canyon, Santa Rita Mountains, Santa Cruz or Pima County (*Silveus* 3488). Along ditches, streams, and in wet places, Arizona; Mexico to Argentina.

42. LYCURUS. Wolf-tail

Rather low, slender, tufted perennial with short, narrow blades and narrow, bristly spikelike panicles; spikelets in pairs, the lower one sterile; glumes equal, the first usually 2-awned, the second 1-awned; lemma longer than the glumes, tapering into a slender awn, pubescent on the margins; palea acute, nearly as long as the lemma, pubescent.

1. Lycurus phleoides H.B.K. Apache County to Coconino County, south to Cochise, Santa Cruz, and Pima counties, 4,000 to 6,500 feet, dry, rocky hills and plains, July to October. Colorado and Utah to Texas, Arizona, and Mexico; Ecuador to Argentina.

Sometimes called Texas-timothy, a common and important forage grass in Arizona.

43. PHLEUM. Timothy

Densely tufted perennials with flat blades and dense, cylindric panicles; glumes equal, abruptly awned, keeled, the keels ciliate; lemma much shorter than the glumes, hyaline, truncate; palea narrow, a little shorter than the lemma.

Common timothy (*P. pratense*) is a very valuable hay plant in the northeastern United States. It has been demonstrated that under favorable conditions this grass can be established on depleted Western ranges.

KEY TO THE SPECIES

1. Culms mostly more than 50 cm. high, erect from a swollen bulblike base; panicle narrow, several times longer than wide....................................1. *P. pratense*
1. Culms 20 to 50 cm. high, from a decumbent, somewhat creeping base; panicle usually not more than twice as long as wide, bristly..........................2. *P. alpinum*

1. Phleum pratense L. Apache County to Coconino County, south to Graham and Pima counties, usually in moist soil. Escaped from cultivation throughout the United States; introduced from the Old World.

2. Phleum alpinum L. White Mountains (Apache County), North Rim of Grand Canyon and San Francisco Peaks (Coconino County), up to 12,000 feet, wet meadows, July to September. Greenland to Alaska, south to New Hampshire, Michigan, and in the Western states to New Mexico, Arizona, and California; Eurasia.

44. GASTRIDIUM. Nit Grass

Small annual grasses; foliage scanty; panicles dense, spikelike, shining; spikelets narrow, 1-flowered, the rachilla prolonged behind the palea as a minute bristle; glumes unequal, somewhat swollen at base, long-pointed; lemma much shorter than the glumes, broad, truncate, pubescent, with a delicate, somewhat geniculate awn; palea about equaling the lemma.

1. Gastridium ventricosum (Gouan) Schinz & Thell. Foothills of the Rincon Mountains, Pima County (*Thornber* in 1905). Here and there in the United States, fields and waste ground; introduced from Europe.

45. MUHLENBERGIA (40). Muhly

Annuals or (usually) perennials, tufted or rhizomatous, with simple or branching culms and narrow or open panicles; glumes usually much shorter than the lemma, sometimes as long as the lemma in robust species with narrow panicles; lemma firm, 3-nerved, with a very short, usually minutely pilose callus, awned or awnless.

This, the largest genus of grasses in Arizona, comprises several species of high forage value. Among the most important of these on Arizona ranges are mountain muhly (*M. montana*), New Mexican muhly (*M. pauciflora*), ring muhly or ring grass (*M. Torreyi*), and spike muhly (*M. Wrightii*). Sand-hill muhly (*M. pungens*) furnishes forage in sandy areas in the valley of the Little Colorado River and is used by the Hopi in making brushes. Bush muhly (*M. Porteri*) seeks the protection of shrubs, and cattle eat it mainly in winter when other grasses become scarce. Aparejo grass (*M. utilis*) has been used to stuff improvised pack saddles.

KEY TO THE SPECIES

1. Plants annual (2).
1. Plants perennial (14).
2. Lemmas awnless (3).
2. Lemmas awned or if merely mucronate (in *M. filiformis*), then the panicle narrow, with appressed branches (6).
3. Pedicels capillary, very flexuous; anthers 1 to 1.4 mm. long. Glumes long-pilose, usually about half as long as the floret....................................1. *M. sinuosa*
3. Pedicels thicker or if filiform, then straight and rather stiff; anthers 0.2 to 0.5 mm. long (4).
4. Pedicels relatively stout, at least the lateral ones shorter than the spikelets. Anthers 0.2 mm. long..2. *M. Wolfii*
4. Pedicels filiform or capillary, much longer than the spikelets (5).
5. Panicle very diffuse, 4 to 10 cm. wide, the branches spreading or reflexed; glumes glabrous; spikelets about 1 mm. long.............................3. *M. fragilis*
5. Panicle narrower or, if as much as 4 cm. wide, then the branches ascending; glumes pilose, at least apically; spikelets 1.2 to 1.5 mm. long............4. *M. minutissima*
6. Panicle open, the branches ascending or spreading (7).
6. Panicle narrow, the branches appressed (9).
7. Pedicels thick, short; glumes glabrous; awn of the lemma 10 to 30 mm. long 8. *M. microsperma*
7. Pedicels slender; glumes pilose; awn 1 to 2 mm. long (8).
8. Spikelets 1.5 to 1.8 mm. long; pedicels spreading, mostly much longer than the spikelets .... 6. *M. texana*
8. Spikelets 2 to 2.5 mm. long; pedicels appressed, mostly much shorter than the spikelets 7. *M. eludens*
9. Lemma merely mucronate. Culms very slender, usually erect...........10. *M. filiformis*
9. Lemma awned (10).
10. Glumes entire, erose, or sometimes minutely bifid. Awn of the lemma 10 to 30 mm. long (11).
10. First glume minutely to deeply bifid, or the second glume 3-toothed (12).
11. Glumes abruptly acute or acuminate, commonly awned..............5. *M. pectinata*
11. Glumes obtuse, 1 to 2 mm. long. Cleistogamous spikelets often present in the lower, reduced sheaths..9. *M. appressa*
12. First glume entire, the second 3-toothed........................11. *M. pulcherrima*
12. First glume minutely to deeply bifid, the second acuminate or setaceous (13).
13. Glumes as long as the floret; lemma 3 to 3.5 mm. long, the awn 2 to 10 mm. long 12. *M. depauperata*
13. Glumes shorter than the floret; lemma 4 to 5 mm. long, the awn 10 to 20 mm. long 13. *M. brevis*
14. Plants with prominent scaly creeping rhizomes: see also *M. polycaulis* (15).
14. Plants tufted; rhizomes wanting: see also *M. glauca* (28).
15. Panicle open; spikelets on slender, rather long, usually spreading pedicels (16).
15. Panicle narrow, often condensed; spikelets short-pedicelled (18).
16. Panicle branches 3 to 5, divided into fascicles of capillary, finally spreading, very scabrous branchlets; spikelets 4 to 5 mm. long; lemma awned, the awn 1 to 1.5 mm. long; blades short, sharp-pointed, involute.......................14. *M. pungens*
16. Panicle branches numerous, the branchlets not arranged in fascicles; spikelets 1 to 2 mm. long; lemma acute or mucronate; blades flat, relatively soft (17).
17. Ligule 1 to 2 mm. long, auriculate.............................15. *M. arenacea*
17. Ligule minute, without auricles...................................16. *M. asperifolia*

18. Blades flat, at least some of them more than 3 mm. wide, lax, spreading (19).
18. Blades involute or if flat, then less than 2 mm. wide (22).
19. Callus hairs copious, as long as the lemma. Panicles silky, often tinged with purple 24. *M. andina*
19. Callus hairs rather sparse, not more than half as long as the lemma (20).
20. Glumes awned, the awns exceeding the awnless floret; panicles compact, bristly 25. *M. racemosa*
20. Glumes acuminate or awn-pointed but not exceeding the lemma; panicles not bristly (21).
21. Lemma awnless; panicles compactly flowered. Blades mostly less than 10 cm. long 26. *M. mexicana*
21. Lemma awned, the awn 5 to 10 mm. long; panicles rather loosely flowered 27. *M. sylvatica*
22. Culms 1 to 3 meters high, woody at base, as much as 6 mm. thick, freely branching at the middle and upper nodes....................................17. *M. dumosa*
22. Culms not more (usually much less) than 60 cm. high (23).
23. Blades 5 to 15 cm. long, flat; culms 30 to 60 cm. high, slender, branching from the lower nodes; awn of the lemma 0.5 to 6 mm. long..................23. *M. glauca*
23. Blades mostly much less than 5 cm. long (24).
24. Culms widely creeping. Blades conspicuously recurved-spreading (25).
24. Culms erect or decumbent at base, but not widely creeping (26).
25. Spikelets about 3 mm. long..18. *M. repens*
25. Spikelets about 2 mm. long..19. *M. utilis*
26. Culms nodulose-roughened. Glumes about half as long as the floret; ligule 2 to 3 mm. long .... 20. *M. Richardsonis*
26. Culms glabrous or pubescent, but not nodulose (27).
27. Lemma mucronate or short-awned; plants forming dense cushions, the leaves crowded toward base...21. *M. Thurberi*
27. Lemma with an awn 1 to 5 mm. long; plants forming loose bunches, the leaves not crowded toward base...22. *M. curtifolia*
28. Panicle open, or at least loose, the branches naked at base (29).
28. Panicle narrow, usually dense or spikelike, the branches floriferous from the base (35).
29. Culms wiry, freely branching, geniculate, widely spreading............38. *M. Porteri*
29. Culms erect or sometimes decumbent at base, but not widely spreading (30).
30. Blades rarely more (usually much less) than 8 cm. long. Culms rather freely branching at the lower nodes (31).
30. Blades elongate, much more than 8 cm. long (33).
31. Blades flat or folded, with conspicuous, firm, white midnerve and margins 39. *M. arizonica*
31. Blades involute. Culms loosely tufted, sometimes forming large cushions, erect from a decumbent base (32).
32. Panicle mostly less than 15 cm. long; blades 1 to 4 cm. long, curled or falcate 40. *M. Torreyi*
32. Panicle mostly more than 20 cm. long; blades commonly 5 to 8 cm. long 41. *M. arenicola*
33. Basal sheaths strongly compressed-keeled; glumes nearly as long as to somewhat longer than the lemma...44. *M. Emersleyi*
33. Basal sheaths rounded on the back; glumes much shorter than the lemma (34).
34. Panicle pale or tawny, the slender branches flexuous, ascending or spreading; glumes about half as long as the lemma; awn of the lemma 18 to 25 mm. long 42. *M. xerophila*
34. Panicle dark purple, the capillary branches finally spreading; glumes not more than ¼ as long as the lemma; awn of the lemma 10 to 15 mm. long..........43. *M. rigida*
35. Awn less than 5 (or occasionally as much as 10) mm. long, or the lemma sometimes awnless (36).
35. Awn usually at least 10 mm. long (41).
36. Panicle spikelike (37).
36. Panicle narrow but not spikelike, moderately dense to rather open (39).

37. Blades usually flat, 1 to 3 mm. wide; panicles 5 to 15 cm. long; glumes awn-pointed, distinctly shorter than the lemma....28. *M. Wrightii*
37. Blades involute; panicles 10 to 30 cm. long; glumes acute or obtuse (38).
38. Glumes ⅔ as long as to nearly equaling the awnless lemma....29. *M. rigens*
38. Glumes about ½ as long as the usually awned lemma....45. *M. dubia*
39. Glumes nearly as long as the awnless lemma. Spikelets 2 to 3 mm. long 30. *M. longiligula*
39. Glumes ½ to ⅔ as long as the awned lemma (40).
40. Lemma 2.5 to 3 mm. long; second glume usually sharply 3-toothed at apex 31. *M. filiculmis*
40. Lemma 3.5 to 4 mm. long; second glume more or less erose at apex....32. *M. dubioides*
41. Second glume 3-toothed. Lemma often yellowish; leaves mostly basal, the sheaths becoming flat and loose....33. *M. montana*
41. Second glume entire (42).
42. Old sheaths flat and more or less coiled at base of the plant. Spikelets about 5 mm. long; glumes nearly as long as the lemma....34. *M. virescens*
42. Old sheaths never flat or coiled (43).
43. Lemma 4 mm. long, scaberulous; glumes about half as long as the lemma, acuminate or awn-tipped....37. *M. pauciflora*
43. Lemma 2.5 to 3 mm. long, pilose on the lower part; glumes at least ⅔ as long as the lemma (44).
44. Culms loosely tufted, hard and wiry at base; floret loosely villous toward base 35. *M. polycaulis*
44. Culms usually densely tufted, slender, not wiry at base; floret densely pilose at base 36. *M. monticola*

1. Muhlenbergia sinuosa Swallen. Apache, Coconino, Cochise, and Pima counties, 3,500 to 7,000 feet. New Mexico and Arizona.

Specimens of *M. sinuosa, M. fragilis,* and *M. minutissima* were included previously under *Sporobolus microspermus* (Lag.) Hitchc.

2. Muhlenbergia Wolfii (Vasey) Rydb. Southern Apache (or northern Greenlee), Coconino, and Pima counties. Colorado, New Mexico, Arizona, and Chihuahua.

Specimens of this species were included previously under *Sporobolus ramulosus* (H.B.K.) Kunth.

3. Muhlenbergia fragilis Swallen. Apache, Coconino, Yavapai, Gila (?), Santa Cruz, and Pima counties. Western Texas to Arizona and central Mexico.

4. Muhlenbergia minutissima (Steud.) Swallen. Apache, Navajo, Coconino, Yavapai, Gila, and Pima counties, 4,000 to 8,000 feet. Montana to Washington, south to northern Mexico.

5. Muhlenbergia pectinata C. O. Goodding. Mule Mountains (Cochise County) and Bear Valley or Sycamore Canyon (Santa Cruz County), wet places below and on the face of cliffs, September. Southeastern Arizona to Jalisco.

A delicate, spreading annual with narrow panicles and ciliate lemmas, related to *M. ciliata* (H.B.K.) Kunth.

6. Muhlenbergia texana Buckl. Mule and Huachuca mountains (Cochise County), Bear Valley or Sycamore Canyon (Santa Cruz County), 4,000 to 8,000 feet. Western Texas to southern Arizona and northern Mexico.

7. Muhlenbergia eludens C. G. Reeder. Cochise County, in Rucker Canyon, 6,500 feet, and in the Huachuca Mountains, Canelo Hills, Santa Cruz County. New Mexico, Arizona, and northern Mexico.

8. **Muhlenbergia microsperma** (DC.) Kunth. Grand Canyon (Coconino County) and Mohave County, south to Santa Cruz, Pima, and Yuma counties, 5,000 feet and lower, rocky slopes and canyons, February to May and at other times after rains. Arizona and southern California to Guatemala; Colombia and Venezuela to Peru.

9. **Muhlenbergia appressa** C. O. Goodding. Graham, Gila, Maricopa, Pinal, and Pima counties, 3,000 to 4,000 feet, canyons and slopes, March and April, type from Devil's Canyon, Pinal or Gila County (*Harrison & Kearney* 1493). Known only from southern Arizona.

Closely related to *M. microsperma,* differing in having narrow panicles with closely appressed branches and larger spikelets 4.5 to 6 mm. long.

10. **Muhlenbergia filiformis** (Thurb.) Rydb. Apache and Coconino counties, up to 8,000 feet, meadows and wet places in the mountains. South Dakota and British Columbia, south to Kansas, New Mexico, Arizona, and California. Pull-up muhly.

11. **Muhlenbergia pulcherrima** Scribn. Apache County at Big Cienaga (*Schroeder* in 1940) and Phelps Botanical Area, White Mountains (*Phillips* 3368), 9,000 to 9,500 feet. Arizona and northern Mexico.

12. **Muhlenbergia depauperata** Scribn. Coconino, Yavapai (?), and Cochise counties, dry, gravelly soil and open grassland at medium altitudes, August to October, type from Arizona (*Pringle*). Texas, Colorado, New Mexico, Arizona, and Mexico.

13. **Muhlenbergia brevis** C. O. Goodding. Bowie, Cochise County (*Jones* in 1884), 3,500 to 4,000 feet, open ground, August to October. Colorado and Texas to Arizona, south to the Federal District, Mexico.

This species is related to *M. depauperata,* but differs in having lemmas 4 to 5 mm. long, awns 10 to 20 mm. long, and glumes much shorter than the spikelet.

14. **Muhlenbergia pungens** Thurb. Apache, Navajo, and Coconino counties, 5,000 to 7,000 feet, common on sandy mesas, July to October. South Dakota and Wyoming to New Mexico and Arizona.

15. **Muhlenbergia arenacea** (Buckl.) Hitchc. Cochise County (several localities), Nogales (Santa Cruz County), Baboquivari Mountains (Pima County), dry mesas. Texas to southern Arizona and Sonora.

16. **Muhlenbergia asperifolia** (Nees & Mey.) Parodi. Apache County to Coconino County, south to Pima County, dry hills or moist ground at medium altitudes, May to October. Wisconsin to British Columbia, south to Mexico; southern South America.

17. **Muhlenbergia dumosa** Scribn. Gila, Maricopa, Pinal, Santa Cruz, Pima, and Yuma counties, rocky canyon slopes and valleys at low altitudes, March to May, type from the Santa Catalina Mountains (*Pringle* in 1884). Southern Arizona and northwestern Mexico.

18. **Muhlenbergia repens** (Presl) Hitchc. (*M. abata* Johnst.). Apache, Coconino, Yavapai, Cochise, Santa Cruz, and Pima counties, dry, open, rocky or sandy ground, April to September. Texas to Arizona and Mexico.

19. **Muhlenbergia utilis** (Torr.) Hitchc. Monkey Springs Reservoir near Sonoita, Santa Cruz County (*Darrow & Haskell* 2276). Texas, Nevada, Arizona, southern California, and Mexico.

20. Muhlenbergia Richardsonis (Trin.) Rydb. Vicinity of Flagstaff and Kaibab Plateau (Coconino County), 7,000 to 8,500 feet, low, open ground. New Brunswick to Alberta, south to South Dakota, New Mexico, Arizona, California, and Baja California.

21. Muhlenbergia Thurberi Rydb. (*M. curtifolia Griffithsii* Scribn.). Apache County, about 6,000 feet, dry slopes and sandy ground, July to September, the type of *M. curtifolia Griffithsii* from Canyon de Chelly (*Griffiths* 5837). Texas, New Mexico, and Arizona.

22. Muhlenbergia curtifolia Scribn. Apache and Coconino counties, 7,500 to 8,000 feet, moist or rocky, open slopes, July to September, rare. Utah, Nevada, and Arizona.

23. Muhlenbergia glauca (Nees) Mez. (*M. huachucana* Vasey). Mountains of Cochise County, Bear Valley or Sycamore Canyon (Santa Cruz County), 4,000 to 7,000 feet, dry, rocky slopes, type of *M. huachucana* from the Huachuca Mountains (*Lemmon* 2915). Western Texas to California and Mexico.

24. Muhlenbergia andina (Nutt.) Hitchc. Apache, Navajo, Coconino, and Graham counties, 6,000 to 9,000 feet, moist, open ground, August and September. Montana and Washington to New Mexico, Arizona, and California.

25. Muhlenbergia racemosa (Michx.) B.S.P. Apache and Coconino counties, 4,000 to 8,000 feet, moist ground in canyons and meadows, June to September. Newfoundland to British Columbia, south to Maryland, Kentucky, Oklahoma, and Arizona.

26. Muhlenbergia mexicana (L.) Trin. Chiricahua Mountains, Cochise County, in springy soil along a creek, 5,500 feet (*Blumer* 1784). Maine and Quebec to Washington, south to North Carolina, Arkansas, New Mexico, and Arizona.

Forma *setiglumis* (Wats.) Fern., with awns of the lemmas 4 to 10 mm. long, was collected at Big Bug, Yavapai County (*Toumey* in 1891).

27. Muhlenbergia sylvatica Torr. Grapevine Creek, Grand Canyon (*Toumey* 169). Moist woods, Maine to South Dakota, south to Alabama, Texas, and Arizona.

28. Muhlenbergia Wrightii Vasey. Apache, Coconino, Yavapai, and Pima counties, plains and open slopes up to 8,500 feet, July to October. Colorado and Utah to northern Mexico.

29. Muhlenbergia rigens (Benth.) Hitchc. Apache and Coconino counties to Cochise, Santa Cruz, and Pima counties, 2,500 to 7,000 feet, open slopes, canyons, and forests, July to October. Texas to southern California and northern Mexico.

Deer grass. Ivan M. Johnston (13, p. 392) has segregated specimens from southern Arizona and adjacent territory as *M. mundula* Johnst., which he distinguishes from what he considers typical *M. rigens* in having the spike well exserted and with lower branches only 5 to 15 mm. long. Also the glumes are distinctly shorter than the lemma, whereas in typical *M. rigens* they are nearly as long. Collections at Mormon Lake (*Rice* in 1921), Flagstaff (*Schallert* in 1943), and in the Sierra Ancha (*Gould & Hudson* 3770) also correspond with *M. mundula*.

30. Muhlenbergia longiligula Hitchc. (*Epicampes distichophylla* var. *mutica* Scribn.). Apache, Navajo, and Coconino counties to Cochise, Santa Cruz, and Pima counties, up to 9,000 feet, canyons and rocky slopes, July to October, type from the Santa Rita Mountains (*Pringle* in 1884), type of *E. distichophylla* var. *mutica* also from Arizona (*Toumey* 740).

31. Muhlenbergia filiculmis Vasey. V. T. Ranch, Kaibab Plateau, Coconino County (*Goodding* in 1935). Wyoming, Colorado, New Mexico, and northern Arizona.

32. Muhlenbergia dubioides C. O. Goodding. Santa Rita and Santa Catalina mountains (Pima County), type from Box Canyon (*Silveus* 3490). Known only from these localities.

33. Muhlenbergia montana (Nutt.) Hitchc. (*M. gracilis* var. *major* Vasey). Apache, Coconino, Yavapai, Graham, Gila, Cochise, and Santa Cruz counties, 5,000 to 8,500 feet, dry mesas and rocky hills, August and September, type of *M. gracilis* var. *major* from Mount Graham (*Rothrock* 744). Montana and central California to western Texas, Arizona, and southern Mexico.

34. Muhlenbergia virescens (H.B.K.) Kunth. Coconino, Gila, Cochise, and Pima counties, 5,000 to 9,500 feet, rocky hills and mesas, April to June. New Mexico and Arizona to central Mexico.

35. Muhlenbergia polycaulis Scribn. Chiricahua, Dragoon, and Huachuca mountains (Cochise County), near Sonoita and Bear Valley or Sycamore Canyon (Santa Cruz County), Santa Rita, Santa Catalina, and Baboquivari mountains (Pima County), 4,000 to 5,000 feet, rocky slopes. Texas, southern Arizona, and Mexico.

36. Muhlenbergia monticola Buckl. Havasu Canyon, Coconino County (*Clover* 7151) and Pinal, Cochise, Santa Cruz, and Pima counties, 5,000 to 8,000 feet, rocky hills and canyons, August to October. Texas to Arizona and central Mexico.

37. Muhlenbergia pauciflora Buckl. Navajo, Coconino, Yavapai, Graham, Cochise, Santa Cruz, and Pima counties, 4,000 to 7,500 feet, dry, rocky slopes and canyons, August to October. Western Texas to Colorado, Arizona, and northern Mexico.

38. Muhlenbergia Porteri Scribn. Navajo County to Mohave County, south to Cochise, Pima, and Yuma counties, 2,000 to 5,500 feet, mesas and rocky slopes, August to October. Colorado and Nevada to western Texas, Arizona, California, and northern Mexico.

39. Muhlenbergia arizonica Scribn. Cochise, Pima, and Santa Cruz counties, 3,500 to 5,500 feet, dry, rocky hills and open woods, August to October, type from Arizona near the Mexican boundary (*Pringle*). Arizona and northwestern Mexico.

40. Muhlenbergia Torreyi (Kunth) Hitchc. Apache, Navajo, Coconino, Mohave, and Yavapai counties, 5,000 to 7,000 feet, July to September. Western Kansas and Colorado to Texas and Arizona.

41. Muhlenbergia arenicola Buckl. Mohave, Yavapai, Cochise, Pima, and Santa Cruz counties, 4,000 to 7,000 feet, dry mesas and open ground, often forming "fairy rings," August to October. Western Kansas and Colorado to Arizona and northern Mexico.

42. Muhlenbergia xerophila C. O. Goodding. Pima and Santa Cruz counties, canyons, September to November, type from Bear Valley or Sycamore Canyon (*L. N. Goodding* M262). Known only from southern Arizona.

43. Muhlenbergia rigida (H.B.K.) Kunth. Greenlee, Cochise, Santa Cruz, and Pima counties, 5,000 to 6,000 feet, rocky slopes, September and October. Texas, New Mexico, Arizona, and Mexico.

44. Muhlenbergia Emersleyi Vasey (*M. Vaseyana* Scribn.). Southern Apache County to Yavapai County, south to Cochise, Santa Cruz, and Pima counties, rocky slopes and canyons at medium altitudes, July to October, type from southern Arizona (*Emersley*), type of *M. Vaseyana* also from Arizona (*Rothrock* 282). Texas to Arizona and Mexico.

Bull grass. An awnless form with more open, fewer-flowered panicles (*Epicampes subpatens* Hitchc.) has been collected on Hualpai Mountain, Mohave County (*Nichol* in 1937), at Prescott (*Thornber* 8644), and in the Baboquivari Mountains, Pima County (*Goodding* 451–45).

***45. Muhlenbergia dubia** Fourn. A fragmentary specimen, doubtfully referred to this species, was collected in Tonto National Forest (*Read* 98). Texas, New Mexico, central Arizona (?), and Mexico.

A *Muhlenbergia*, possibly *M. Schreberi* Gmelin, was discovered recently as a garden weed at Kingman, Mohave County (*J. T. Howell* 26333). It has decumbent stems rooting at the nodes and rather broad flat blades. It differs from *M. racemosa,* to which it comes closest in the preceding key, in having very small awnless glumes and a short-awned lemma.

46. SPOROBOLUS. Drop-seed

Annuals or perennials with open or spikelike panicles of small spikelets; glumes equal or (usually) unequal, the second often as long as the lemma; lemma membranaceous, 1-nerved, awnless; palea as long as the lemma; fruit free from the lemma and palea.

The perennial species are useful forage plants. Two large, coarse bunch grasses, alkali-sacaton (*S. airoides*) and sacaton (*S. Wrightii*), are important elements of some Arizona ranges, but they are grazed mainly when more succulent grasses are not available. Alkali-sacaton thrives on the plains and can withstand rather strongly saline soil conditions. Sacaton reaches its highest development in the bottom lands in the southeastern part of the state. Both species are utilized for hay. Black drop-seed (*S. interruptus*), abundant on the plateau south of Flagstaff, is one of the most valuable forage grasses in its range. Sand drop-seed (*S. cryptandrus*) furnishes some forage. The seeds of several species have been used for food by the Indians of Arizona.

KEY TO THE SPECIES

1. Plants annual (2).
1. Plants perennial (5).
2. Panicles usually included in the sheaths, narrow, few-flowered, spikelike (3).
2. Panicles exserted, open, many-flowered (4).
3. Spikelets 3 to 5 mm. long; lemma pubescent........................1. *S. vaginiflorus*
3. Spikelets 2 to 3 mm. long; lemma glabrous........................2. *S. neglectus*
4. Blades 4 to 7 cm. long, mostly 2 to 5 mm. wide; spikelets 1.5 to 1.7 mm. long, appressed 3. *S. pulvinatus*
4. Blades 1 to 2 cm. long, not more than 2 mm. wide; spikelets 1.8 to 2 mm. long, spreading 4. *S. patens*

5. Spikelets 4 to 6 mm. long (6).
5. Spikelets 1.5 to 3 mm. long (7).
6. Panicle contracted, more or less included in the sheaths; second glume shorter than the glabrous lemma....................5. *S. asper*
6. Panicle open, exserted; second glume about as long as the lemma......6. *S. interruptus*
7. Summit of the sheaths glabrous on the back and sides, usually with long, rather stiff hairs in the throat (8).
7. Summit of the sheaths conspicuously villous on the back, or sides, or both (10).
8. Pedicels elongate, capillary....................7. *S. texanus*
8. Pedicels short (9).
9. Panicle loose, 1 to 2 times as long as broad, the branches naked below, the branchlets spreading; blades mostly involute....................8. *S. airoides*
9. Panicle relatively dense, 30 to 60 cm. long, the numerous, relatively short, crowded, densely flowered branches floriferous nearly to base, the branchlets appressed; blades usually flat....................9. *S. Wrightii*
10. Panicle open, the branches spreading, naked at base (11).
10. Panicle contracted, spikelike (13).
11. Panicle branches loosely flowered, the branchlets and pedicels implicate, spreading 10. *S. flexuosus*
11. Panicle branches densely flowered, the branchlets and short pedicels appressed (12).
12. Plants closely tufted at base; culms up to 100 cm. long.............11. *S. cryptandrus*
12. Plants with clustered knotty rhizomes at base; culms seldom more than 50 cm. long, erect, slender. Blades short, involute, spreading....................12. *S. Nealleyi*
13. Spikelets 2 to 2.5 mm. long; culms slender, usually less than 1 meter high 13. *S. contractus*
13. Spikelets 2.5 to 3 mm. long; culms robust, 1 to 2 meters high..........14. *S. giganteus*

1. Sporobolus vaginiflorus (Torr.) Wood. Fort Huachuca and Willcox (Cochise County), sandy soil or open waste ground. Maine and Ontario to Minnèsota and Nebraska, south to Georgia, Texas, and Arizona.

2. Sporobolus neglectus Nash. Coconino County at Garland Prairie, Kaibab National Forest, 7,000 feet (*Darrow* 3069), and at Black Springs, Coconino National Forest, in deep, black, gravelly loam (*Talbot* C-13). Quebec and Maine to North Dakota, south to Maryland, Tennessee, and Texas; Washington to Arizona.

3. Sporobolus pulvinatus Swallen. Apache, Graham, Maricopa, Pinal, Cochise, and Pima counties, 1,000 to 5,500 feet, open, sandy plains and roadsides, type from Adamana, Apache County (*Griffiths* 5107). Texas to Arizona and northern Mexico.

Arizona specimens of this species have been referred to *S. pyramidatus* (Lam.) Hitchc. (*S. argutus* (Nees) Kunth) but differ in being annual, with short, flat blades, and in having smaller spikelets, with the second glume and lemma abruptly acute or subobtuse.

4. Sporobolus patens Swallen. Willcox, Cochise County (*Silveus* 354, the type). Known only by the type collection.

A slender, erect annual with short, flat blades 1 to 2 cm. long and rather delicate, loosely flowered panicles with long-pedicelled spikelets 1.8 to 2 mm. long.

5. Sporobolus asper (Michx.) Kunth. South of Beklohito (Beclabito?), Apache County, 6,000 feet, in red, sandy loam (*J. Howell, Jr.* 56). Vermont to Iowa, North Dakota, Utah, and Washington, south to Louisiana and Arizona.

6. **Sporobolus interruptus** Vasey. Navajo, Coconino, Yavapai, and Gila counties, 5,000 to 7,000 feet, dry, rocky hillsides, July to September, type from near Flagstaff (*Rusby* 885). Known only from Arizona.

7. **Sporobolus texanus** Vasey. Bright Angel Trail, Grand Canyon (*Thornber* 8232). Kansas and Colorado to Texas, New Mexico, and northern Arizona.

8. **Sporobolus airoides** Torr. Apache County to Coconino County, south to Cochise, Santa Cruz, and Pima counties, 1,000 to 5,500 feet, often in saline soil, June to October. South Dakota to eastern Washington, south to Texas, Arizona, and southern California.

9. **Sporobolus Wrightii** Munro. Navajo, Coconino, Cochise, Santa Cruz, and Pima counties, 2,000 to 5,500 (7,500?) feet, dry, sandy, open ground, rocky slopes, and river banks, July to October, type from Pantano, Pima County (*Pringle*). Western Texas to southern California and central Mexico.

Specimens from the northern counties are less typical, resembling *S. airoides*.

10. **Sporobolus flexuosus** (Thurb.) Rydb. Apache, Navajo, Coconino, Yavapai, Graham, Cochise, and Pima counties, 2,500 to 5,500 feet, dry or moist, open, sandy soil, June to October. Western Texas to southern Utah, Arizona, southeastern California, and northern Mexico.

11. **Sporobolus cryptandrus** (Torr.) Gray. Apache County to Mohave County, south to Cochise, Santa Cruz, Pima, and Yuma counties, 150 to 7,000 feet, dry, open, sandy ground and rocky slopes, April to September. Maine and Ontario to Alberta and Washington, south to North Carolina, Indiana, Louisiana, Arizona, and Mexico.

12. **Sporobolus Nealleyi** Vasey. Northeastern corner of Apache County, open pine forest on a volcanic plateau (*Gould & Phillips* 4812, 4847), near Winslow, Navajo County, in gypseous soil (*Darrow* 2684), 5,000 to 8,500 feet. Western Texas to northeastern Arizona.

13. **Sporobolus contractus** Hitchc. Apache County to Mohave County, south to Cochise, Santa Cruz, and Pima counties, 1,500 to 6,000 feet, dry mesas, bluffs, and sand hills, August to October, type from Camp Lowell, Pima County (*Pringle*). Colorado to Nevada, south to western Texas, Arizona, southeastern California, and northern Mexico.

14. **Sporobolus giganteus** Nash. Apache, Navajo, Coconino, Pinal, Cochise, and Pima counties, 2,500 to 6,000 feet, dry, open ground, June to October. Western Texas to Arizona.

47. BLEPHARONEURON

Densely tufted perennial with slender, flat or involute, more or less flexuous blades, and narrow, open panicles; glumes subobtuse, nearly equal or the second a little longer and broader; lemma broad, abruptly pointed, densely pubescent on the nerves; palea slightly longer than the lemma, densely villous between the nerves.

1. **Blepharoneuron tricholepis** (Torr.) Nash. Apache County to Mohave County, south to Cochise and Pima counties, 2,500 to 10,000 feet, dry or moist, open woods, July to October. Colorado to Utah, south to Texas, Arizona, and Mexico.

This grass, sometimes known as pine drop-seed, is valuable for forage.

48. ORYZOPSIS (41). Rice Grass

Slender perennials with flat or involute blades and narrow or open panicles; glumes equal, gradually or abruptly acuminate; lemma firm, terete, glabrous or villous, with a short, rather blunt callus, and a short, straight or weakly geniculate, deciduous awn.

Indian rice grass (*O. hymenoides*), sometimes known as Indian-millet, furnishes much forage on arid sandy plains in northern Arizona. In former times it was cut for hay, and the seeds were utilized for food by the Indians. The other species are palatable but less abundant.

KEY TO THE SPECIES

1. Spikelets 3 to 4 mm. long; lemma glabrous or rarely pubescent, 2 to 2.5 mm. long. Panicle branches slender, finally spreading, spikelet-bearing toward the ends...1. *O. micrantha*
1. Spikelets 6 to 10 mm. long; lemma densely villous, 3 to 5 mm. long (2).
2. Panicle diffuse, the branches in pairs, the branchlets and pedicels divaricately spreading, flexuous; lemma 3 mm. long..................................2. *O. hymenoides*
2. Panicle rather narrow, the branches stiffly ascending, the branchlets and pedicels usually appressed; lemma 5 mm. long....................................3. *O. Bloomeri*

1. Oryzopsis micrantha (Trin. & Rupr.) Thurb. Navajo, Coconino, Mohave, Yavapai, Maricopa, and Pima counties, 2,000 to 7,000 feet, rocky slopes and dry, open woods, June to August. Saskatchewan to Montana, south to New Mexico and Arizona.

2. Oryzopsis hymenoides (Roem. & Schult.) Ricker. Apache, Navajo, Coconino, Mohave, Yavapai, and Pima counties, dry, open woods and sandy plains at medium altitudes, June to August. Manitoba to British Columbia, south to Texas, Arizona, California, and northern Mexico.

3. Oryzopsis Bloomeri (Boland.) Ricker. Betatakin, Navajo County (*J. T. Howell* 24532), Grand Canyon, Coconino County (*Silveus* 1927, *Hawbecker* in 1935), 3,500 to 7,000 feet. Montana to eastern Washington, south to New Mexico, Arizona, and California.

According to B. Lennart Johnson (42, p. 608), *O. Bloomeri* consists of hybrids between *O. hymenoides* and several species of *Stipa,* the plant collected by Silveus being *O. hymenoides* × *S. robusta.*

Oryzopsis miliacea (L.) Benth. & Hook., an Old World species, has been collected in the Sierra Ancha, Gila County (*Dibbern* in 1948), where it is reported to have been planted by the U. S. Forest Service and may become established. It has also been found recently in Havasu Canyon, Coconino County, growing as a weed in fields (*J. T. Howell* 26516). This species can be distinguished from *O. micrantha* by its broad leaves (5 mm. or more wide) and the short awn of the lemma (about 4 mm. long).

49. PIPTOCHAETIUM. Pinyon Rice Grass

Densely tufted perennial with narrow or involute blades and open few-flowered panicles; glumes subequal, acute; fruit obovate, dark brown, asymmetric, glabrous or hispid; awn deciduous or persistent, more or less geniculate, often twisted below; palea firm except near the margins, its apex projecting above the lemma as a minute point.

1. Piptochaetium fimbriatum (H.B.K.) Hitchc. Greenlee, Graham, Cochise, Santa Cruz, and Pima counties, 4,000 to 7,000 feet, rocky hills, limestone cliffs, and open woods. Western Texas, Arizona, and Mexico.

This grass is reported to make excellent forage. Specimens from the Mexican border region were referred by I. M. Johnston (13, p. 396) to var. *confine* Johnst., distinguished from the typical plant of central Mexico in having green (not purple), evidently nerved glumes.

50. STIPA. Needle Grass

Tufted perennials with usually involute blades and narrow or sometimes open panicles; spikelets disarticulating above the glumes, the articulation oblique, leaving a sharp, bearded callus on the floret; glumes equal, thin, narrow, longer than the floret; lemma indurate, terete, terminating in a prominent, geniculate awn, this twisted below.

These plants, known variously also as feather grass, spear grass, and porcupine grass, make good forage while young, but some of the species when mature are injurious, especially to sheep, on account of the sharp-pointed fruits, which penetrate the flesh and injure the mouths and eyes of grazing animals. They are a nuisance in wool and damage the hides. Sleepy grass (*S. robusta*) in certain districts in New Mexico has been reported as having a narcotic effect on horses and slightly on sheep but not on cattle. The toxic principle is unknown. An Old World species, esparto (*S. tenacissima*), is used in the manufacture of fine paper and cordage.

KEY TO THE SPECIES

1. Terminal segment of the awn plumose, flexuous, up to 12 cm. long....1. *S. neomexicana*
1. Terminal segment of the awn not plumose (2).
2. Lower segment of the once-geniculate awn conspicuously plumose, the hairs 5 to 8 mm. long .... 2. *S. speciosa*
2. Lower segment of the awn glabrous, scabrous, or pubescent, but not plumose (3).
3. Lemma densely villous with white hairs 3 to 4 mm. long. Awn once-geniculate 3. *S. coronata*
3. Lemma pubescent or if sometimes villous toward the apex, then the hairs less than 3 mm. long (4).
4. Panicle open, the slender branches ascending or spreading (5).
4. Panicle narrow, usually rather dense, the branches appressed, floriferous from the base (7).
5. Ligule about 2 mm. long; mature lemma dark brown, 7 to 8 mm. long, papillose-pubescent. Callus acute, 1 mm. long; awn 2.5 to 3 cm. long, the terminal segment straight 4. *S. Pringlei*
5. Ligule usually 3 mm. long or longer; mature lemma pale (6).
6. Mature lemma 8 to 12 mm. long, glabrous or sparsely pubescent above the callus. Callus acuminate, 3 mm. long; awn 10 to 15 cm. long, the terminal segment long and flexuous........5. *S. comata*
6. Mature lemma 5 to 7 mm. long, pubescent........6. *S. eminens*
7. Upper portion of the lemma conspicuously villous, the hairs 2 to 2.5 mm. long (8).
7. Upper portion of the lemma not conspicuously villous, if appressed-pubescent then the hairs usually not more than 1 mm. long (9).
8. Apex of the lemma with lobes 0.8 to 1.5 mm. long, the lemma evenly villous all over 7. *S. lobata*
8. Apex of the lemma not lobed or obscurely so, the lemma conspicuously villous above, less so below........8. *S. Scribneri*

9. Awn 4 to 6 cm. long, obscurely geniculate, the terminal segment flexuous....9. *S. arida*
9. Awn 2 to 3 cm. long, twice-geniculate, the terminal segment straight (10).
10. Sheaths villous at the mouth; panicle 10 to 30 cm. long, the lower nodes villous (11).
10. Sheaths glabrous at the mouth; panicle 5 to 15 cm. long, rather narrow, the lower nodes glabrous (12).
11. Glumes thin, papery; plants rather slender, mostly less than 1 m. high; panicle rather slender, open..10. *S. viridula*
11. Glumes firm; plants robust, usually more than 1 m. high; panicle larger, more compact 11. *S. robusta*
12. Hairs at apex of the lemma about as long as the others; awn mostly more than 2 cm. long ....12. *S. columbiana*
12. Hairs at apex of the lemma longer than the others; awn mostly 1.5 to 2 cm. long. Blades slender, involute, crowded toward the base of the plant............13. *S. Lettermani*

1. **Stipa neomexicana** (Thurb.) Scribn. Apache County to Mohave County, south to Cochise and Pima counties, 3,500 to 7,000 feet, dry, sandy or rocky hills and plains, May to August. Western Texas and Colorado to Utah and Arizona.

2. **Stipa speciosa** Trin. & Rupr. Apache County to Mohave County, south to Pima and Yuma counties, 3,000 to 7,000 feet, dry, rocky hills and canyons, April to June. Colorado and Nevada to Arizona and Baja California; southern South America.

3. **Stipa coronata** Thurb. Only a variety of this species, var. *depauperata* (Jones) Hitchc., is found in Arizona, in the Grand Canyon, Coconino County (*Hitchcock* 13062, *Clover* 4159, *Darrow* in 1942), growing in large clumps on slopes, June and July. This variety, which ranges from Utah and Nevada to Arizona and southern California, differs from the species in the once- rather than twice-geniculate awn. *Stipa coronata*, excluding the variety, is confined to the Coast Ranges of California from Monterey to Baja California.

4. **Stipa Pringlei** Scribn. Apache, Coconino, Graham, Gila, Cochise, and Pima counties, 5,000 to 9,000 feet, rocky, mostly wooded, slopes and openings in pine forests, June to September. Texas to Arizona and Chihuahua.

5. **Stipa comata** Trin. & Rupr. Apache County to eastern Mohave County, and in Pima County, 3,500 to 8,500 feet, dry hills, open woods, and sandy soil, often with juniper, May to August. Indiana to Yukon Territory, south to Texas, Arizona, and California.

Needle-and-thread. Var. *intermedia* Scribn. & Tweedy, in which the third segment of the awn is relatively short and straight, has been found in Apache, Coconino, and Yavapai counties. Specimens collected in Sitgreaves National Forest, Coconino County (*Darrow* 3289) and at Perkinsville, Yavapai County (*Goodwin* 365) are aberrant in their relatively short ligules and long glumes.

6. **Stipa eminens** Cav. Graham, Cochise, and Pima counties, 3,500 to 5,500 feet, rocky hills, June to October. Texas to Arizona and central Mexico.

7. **Stipa lobata** Swallen. Mustang Mountains, Cochise County, 5,000 feet (*Darrow* 3622). Western Texas to Arizona.

8. **Stipa Scribneri** Vasey. Grand Canyon (Coconino County), dry, rocky banks along Bright Angel Trail (*Hitchcock* 10448) and Kaibab Trail to Roaring Springs (*Eastwood & Howell* 991), 5,000 to 7,000 feet, May to September. Colorado, Utah, New Mexico, and Arizona.

A collection in the Grand Canyon (*Silveus* 1925) is more or less intermediate between *S. Scribneri* and *S. lobata.*

9. Stipa arida Jones. Navajo and Coconino counties, 3,000 to 7,000 feet, dry, rocky hills and canyons, May and June. Southwestern Colorado, Utah, northern Arizona, and southeastern California.

10. Stipa viridula Trin. Southeast of Greasewood, Apache County, 7,000 feet (*Shreve* 9017). New York to Alberta, south to Kansas, New Mexico, and northeastern Arizona.

11. Stipa robusta Scribn. Apache, Coconino, and Yavapai counties, 6,000 to 8,000 feet, dry plains, hills, and open woods, June to September. Colorado to western Texas, Arizona, and northern Mexico.

12. Stipa columbiana Macoun. Whisky Creek (southern Navajo County), Kaibab Plateau (Coconino County), Santa Catalina Mountains (Pima County), 5,000 to 8,000 feet, and reported by L. N. Goodding from Mingus Mountain (Yavapai County), dry plains and open woods at medium and higher altitudes. Wyoming to Yukon Territory, south to Texas, northern Arizona, and California.

A variant differing from the species in being larger, with broader blades and longer, denser panicles, var. *Nelsonii* (Scribn.) Hitchc., was collected in Grand Canyon National Park (*Merkle* in 1937, *Grant* 5718, *Collom* in 1935). Mrs. Collom's specimen is exceptionally robust, with a panicle 30 cm. long.

13. Stipa Lettermani Vasey. Franks Lake, Kaibab Plateau, Coconino County (*Forest Service* 62594), Huachuca Mountains, Cochise County, 7,000 feet (*Gould & Haskell* 3386). Open ground and open woods at upper altitudes, Wyoming to Montana and Oregon, south to New Mexico, northern Arizona, and California.

51. ARISTIDA. Three-awn

Tufted annuals or perennials with firm, usually involute blades and narrow or open panicles; glumes equal or unequal, acute, acuminate, or awn-tipped; lemma indurate with a sharp, bearded callus, 3-awned, the lateral awns sometimes much reduced, the base sometimes undivided, twisted, forming a column.

Several species of this genus are very abundant in Arizona, but their forage value is relatively small. The purple three-awn (*A. purpurea*) and other species are grazed in the immature stage. The sharp-pointed fruits, like those of *Stipa,* are troublesome to livestock and become entangled in wool.

KEY TO THE SPECIES

1. Lemma articulate with the column of the awn. Awns nearly equal; plants perennial (2).
1. Lemma not articulate (3).
2. Culms pubescent. . . . . 1. *A. californica*
2. Culms glabrous. . . . . 2. *A. glabrata*
3. Lateral awns minute or wanting. Panicle open, the branches stiffly spreading to drooping, naked at base (4).
3. Lateral awns nearly as long as the central awn (5).
4. Column of the awn twisted at base; panicle branches often drooping. . . . 3. *A. Orcuttiana*
4. Column of the awn not twisted; panicle branches rather stiffly spreading. . 4. *A. ternipes*
5. Plants annual (6).

5. Plants perennial (7).
6. Awns mostly 4 to 7 cm. long, terete, spreading........................5. *A. oligantha*
6. Awns 10 to 15 mm. long, flattened at base........................6. *A. adscensionis*
7. Panicle open, the branches naked at base (8).
7. Panicle narrow, often rather dense, the branches ascending or appressed, at least some of them floriferous nearly to the base (11).
8. Panicle branches stiffly ascending. Panicle narrow, stiffly erect, 10 to 20 cm. long 10. *A. pansa*
8. Panicle branches stiffly and abruptly spreading at base (9).
9. Branchlets divaricate and implicate................................7. *A. barbata*
9. Branchlets appressed (10).
10. Summit of the lemma narrowed into a twisted neck 2 to 5 mm. long....8. *A. divaricata*
10. Summit of the lemma not twisted................................9. *A. hamulosa*
11. Glumes nearly equal or the first sometimes a little longer (12).
11. Glumes unequal, the first half as long as the second or (in *A. glauca* and occasionally in *A. Fendleriana*) as much as ⅔ as long (13).
12. Column of the awn distinctly twisted, 3 to 5 mm. long; panicle narrow, the branches appressed, floriferous nearly to the base........................16. *A. arizonica*
12. Column of the awn straight or obscurely twisted; panicle somewhat open, the branches rather distant, stiffly ascending, naked at base....................17. *A. Parishii*
13. Lemma narrowed into a slender beak 5 to 6 mm. long. Awns 1.5 to 2.5 cm. long, widely spreading ....11. *A. glauca*
13. Lemma not narrowed as above (14).
14. Panicle branches very slender, more or less flexuous; lemma conspicuously scabrous in lines; awns terete at base....................................12. *A. purpurea*
14. Panicle branches stiffly ascending or appressed, or sometimes rather lax in *A. longiseta,* but then the lemma nearly glabrous and the awns flattened toward base (15).
15. Culms rather stout, 30 to 60 cm. high; panicle 15 to 20 cm. long, densely flowered; awns 2 cm. long..15. *A. Wrightii*
15. Culms rather slender, 20 to 30 cm. high; panicle mostly 10 to 15 cm. long, relatively few-flowered; awns 2 to 8 cm. long (16).
16. Leaves crowded toward the base in a dense cluster; lemma scabrous toward the summit, the awns 2 to 5 cm. long; panicle rather stiff..................13. *A. Fendleriana*
16. Leaves not crowded toward the base; lemma glabrous or nearly so, the awns 6 to 8 cm. long; panicle lax ....14. *A. longiseta*

1. Aristida californica Thurb. Fort Mohave, Mohave County (*Cooper*), and, in Yuma County, common along roadsides on desert plains and mesas, apparently flowering as favorable conditions permit. Southwestern Arizona, southern California, and northern Mexico.

2. Aristida glabrata (Vasey) Hitchc. Mohave, Maricopa, Pinal, Santa Cruz, Pima, and Yuma counties, dry ground and mesas, common in the foothills of the Santa Rita Mountains, apparently flowering whenever conditions are favorable. Southern Arizona to Baja California.

3. Aristida Orcuttiana Vasey. Navajo, Yavapai, Graham, Gila, Pinal, Cochise, Santa Cruz, and Pima counties, 4,000 to 7,000 feet, dry, rocky hills and canyons, August and September. Texas to southern California and northwestern Mexico.

4. Aristida ternipes Cav. Mohave, Yavapai, Graham, Gila, Pinal, Cochise, Santa Cruz, Pima, and Yuma counties, up to 6,000 feet, rocky hills and mesas, August to October. New Mexico and Arizona to northern South America and West Indies.

The name spider grass is sometimes applied to this species. Var. *minor*

(Vasey) Hitchc. is a smaller plant with less diffuse panicles, the shorter branches usually stiffly spreading. It is more common than the species.

5. Aristida oligantha Michx. McNary (Apache County), Cedar Wash (Navajo County), Coconino National Forest (Coconino County), Sierra Ancha (Gila County), 5,500 to 7,500 feet, dry, open ground. Massachusetts to South Dakota, south to Florida and Texas; Oregon to California and Arizona.

6. Aristida adscensionis L. Southern Apache County to Mohave County, south to the southern boundary, up to 6,000 feet but usually lower, dry mesas, deserts, and rocky slopes, flowering whenever conditions are favorable. Western Missouri and Texas to California and southward; warmer parts of the Old World. Six-weeks three-awn.

7. Aristida barbata Fourn. Navajo, Coconino, Yavapai, Cochise, and Pima counties, dry slopes and mesas at medium altitudes, May to October. Western Texas to Arizona and central Mexico.

8. Aristida divaricata Humb. & Bonpl. Apache, Navajo, and Coconino counties to Greenlee, Cochise, Santa Cruz, and Pima counties, mostly 5,000 to 7,000 feet, rocky hills, July to September. Kansas to southern California, south to Texas and Guatemala. Poverty three-awn.

9. Aristida hamulosa Henr. Grand Canyon (Coconino County), and Graham, Gila, Pinal, Cochise, Santa Cruz, and Pima counties, 2,500 to 5,500 feet, dry slopes and mesas, June to September, type from Tucson (*Toumey*). Western Texas to southern California, south to Guatemala.

10. Aristida pansa Woot. & Standl. Coconino, Cochise, and Santa Cruz counties, up to 5,000 feet, dry plains and open ground, July to October. Texas to Arizona.

11. Aristida glauca (Nees) Walp. Navajo, Coconino, Mohave, Yavapai, Cochise, Santa Cruz, Pima, and Yuma counties, 1,000 to 5,000 feet, dry, rocky slopes and plains, March to September. Western Texas to Utah, Nevada, Arizona, and southern California, south to central Mexico.

12. Aristida purpurea Nutt. Mohave and Yavapai counties to Graham, Cochise, Santa Cruz, Pima, and Yuma counties, up to 5,000 feet, dry, rocky hills and plains. Arkansas and Kansas to Utah, Arizona, southern California, and northern Mexico.

Var. *laxiflora* Merr., with capillary branches bearing 1 or 2 spikelets, is found in Cochise County (*Griffiths* 1880).

13. Aristida Fendleriana Steud. Apache County to Mohave County, south to Cochise, Maricopa, and Pinal counties, 3,500 to 7,000 feet, mesas, dry hills, and open, rocky ground, May to August. South Dakota to Montana, south to Texas, Utah, Arizona, and southern California.

14. Aristida longiseta Steud. Apache County to Mohave County, south to Cochise and Pima counties, 3,000 to 7,000 feet, dry hills and plains, April to October. Texas to Colorado and Arizona. Red three-awn.

Two varieties are recognized: var. *rariflora* Hitchc., which differs in the few-flowered panicles with long, capillary, flexuous branches; and var. *robusta* Merr., which differs in being a larger plant with a rather dense panicle, the branches relatively short and stiffly ascending. The former has been collected in Yavapai, Cochise, and Pima counties, and the latter in Coconino, Mohave, and Pima counties.

15. Aristida Wrightii Nash. Mohave, Pinal, Maricopa, Cochise, Santa Cruz, Pima, and Yuma counties, 1,000 to 5,000 feet, dry plains and rocky slopes, flowering apparently whenever conditions are favorable. Texas, Colorado, and Utah, to southern California and central Mexico.

16. Aristida arizonica Vasey. Navajo, Coconino, Graham, and Cochise counties, 4,500 to 8,000 feet, dry plains, rocky slopes, and open woods, May to September, type from Arizona (*Rusby* 875). Colorado and western Texas to Arizona, and southward through Mexico.

17. Aristida Parishii Hitchc. Mohave, Yavapai, Gila, Maricopa, Pinal, Pima, and Yuma counties, 1,500 to 4,000 feet, dry, rocky hills, February to May. Arizona and southern California.

52. TRAGUS. Bur Grass

Low annuals with flat blades, the spikes closely arranged on a slender axis; spikes subsessile, falling entire, composed of 2 to 5 spikelets on a short zigzag rachis; first glumes small or wanting, the second glumes of the 2 lower spikelets bearing stout, hooked spines along each side; lemma and palea thin, the lemma flat, the palea convex.

KEY TO THE SPECIES

1. Spikes or burs subsessile, 2 to 3.5 mm. long, scarcely exceeding the spines 1. *T. Berteronianus*
1. Spikes or burs pedicelled, 4 to 5 mm. long, projecting beyond the spines. .2. *T. racemosus*

1. Tragus Berteronianus Schult. Navajo, Mohave, Graham, Pinal, Cochise, Santa Cruz, and Pima counties, up to 5,500 feet, dry, open ground, September and October. Texas to Arizona, south to Argentina; warmer parts of the Old World.

***2. Tragus racemosus** (L.) All. Campus of the University of Arizona (Pima County), probably only cultivated. Waste ground, occasional from Maine to North Carolina; Texas; introduced from the Old World.

53. HILARIA

Stiff perennials with solid culms, narrow blades, and usually dense, narrow, terminal spikes; spikelets arranged in groups of three, the central spikelet perfect, 1-flowered, the lateral ones staminate, 2-flowered, the groups falling entire; glumes usually firm, asymmetric, those of the lateral spikelets bearing an awn on one side from about the middle, those of the fertile spikelet variously divided into awns or lobes; lemma and palea hyaline, nearly equal, the lemma 3-nerved, shallowly lobed, awnless or short-awned.

On the Arizona range this genus is second in importance only to the gramas (*Bouteloua* spp.). Although not so palatable as the latter, the plants are better able to withstand close grazing and trampling. Curly-mesquite (*H. Belangeri*) abounds on dry, open foothills. Galleta (*H. Jamesii*) is perhaps the most characteristic and important forage grass on the Navajo Indian Reservation, and is inferior only to blue grama (*Bouteloua gracilis*) on the higher grasslands in the northern part of the state. Its place is taken in the southeastern grassland by tobosa (*H. mutica*). This species and big galleta (*H. rigida*) attain fullest development in depressions or on heavy, alluvial soil. *H. Jamesii* is used by the Hopi Indians as a fill for basketry and in making ceremonial articles.

KEY TO THE SPECIES

1. Culms tufted, sending out slender, wiry stolons, the internodes 5 to 25 cm. long .... 1. *H. Belangeri*
1. Culms not stoloniferous, erect from a stout decumbent or rhizomatous base (2).
2. Culms felty-pubescent .... 2. *H. rigida*
2. Culms glabrous or puberulent, the nodes pubescent or villous (3).
3. Glumes of the lateral spikelets narrowed toward the apex .... 3. *H. Jamesii*
3. Glumes of the lateral spikelets conspicuously widened toward the subhyaline apex .... 4. *H. mutica*

1. Hilaria Belangeri (Steud.) Nash. Greenlee County to Yavapai and Coconino (?) counties, south to Cochise, Santa Cruz, and Pima counties, 1,500 to 6,000 feet, mesas and foothills, March to October. Texas to Arizona, southeastern California, and northern Mexico.

Var. *longifolia* (Vasey) Hitchc., with erect culms, elongate blades, and without rhizomes, is found on rocky hills around Tucson and in the Galiuro Mountains (Pinal County).

2. Hilaria rigida (Thurb.) Benth. Mohave, Yavapai, Pinal, Maricopa, western Pima, and Yuma counties, up to 4,000 feet, deserts, plains, and rocky hills, February to September, the type (*Cooper* 2230) probably from Fort Mohave. Southern Utah and Nevada to Arizona, southern California, Baja California, and Sonora.

3. Hilaria Jamesii (Torr.) Benth. Apache, Navajo, Coconino, Mohave, and Cochise counties, 4,500 to 7,000 feet, dry hills, rocky canyons, and sandy plains, May to September. Wyoming to Nevada, south to Texas, Arizona, and California.

Collections in Monument Valley, Navajo County (*J. T. Howell* 24695) and in northeastern Apache County (*Gould & Phillips* 4807) have exceptionally stout culms with unusually villous nodes.

4. Hilaria mutica (Buckl.) Benth. Greenlee County to Mohave County, south to Cochise, Santa Cruz, Pima, and Yuma counties, 2,000 to 6,000 feet, dry mesas and hills, May to October. Texas to Arizona and northern Mexico.

54. AEGOPOGON (43)

Low, delicate, decumbent, spreading annual with flat, narrow blades and loose racemes of spreading spikelets; spikelets in groups of 3, the central one perfect, the lateral spikelets staminate or neuter, falling entire; glumes membranaceous, toothed at apex, the midnerve usually extending into a delicate awn; lemma and palea thin, longer than the glumes, the nerves usually extending into awns.

1. Aegopogon tenellus (DC.) Trin. Cochise, Santa Cruz, and Pima counties, open ground in the mountains, August and September, 5,000 to 7,000 feet. Southern Arizona to northern South America.

Var. *abortivus* (Fourn.) Beetle, with awns inconspicuous or wanting, occurs in the Chiricahua and Mule mountains (Cochise County).

55. LEPTOCHLOA. Sprangle-top

Annuals or perennials with flat blades and numerous spikes scattered along the common axis; spikelets few- to several-flowered, the upper floret reduced to a small awnless rudiment; glumes 1-nerved, the second usually longer and broader than the first; lemmas acute to obtuse, pubescent or pilose on the nerves and sometimes on the internerves.

KEY TO THE SPECIES

1. Plants perennial; lemmas notched at apex, awnless, the lobes broad, obtuse...1. *L. dubia*
1. Plants annual; lemmas mucronate or awned from between the teeth of a minutely bifid apex, if awnless, then not with broad obtuse lobes (2).
2. Sheaths sparsely papillose-pilose; spikelets 1 to 2 mm. long, 3- or 4-flowered
2. *L. filiformis*
2. Sheaths glabrous or scabrous; spikelets 3 to 12 mm. long, 5- to 12-flowered (3).
3. Lemmas 1.5 to 2.5 mm. long, mucronate only, scarcely narrowed toward the apex, the lateral nerves excurrent..5. *L. uninervia*
3. Lemmas 2 to 5 mm. long, awned, the awns sometimes minute (4).
4. Lemmas 2 mm. long, acute, viscid on the back; panicle usually less than 10 cm. long, tinged with purple..3. *L. viscida*
4. Lemmas 4 to 5 mm. long, acuminate, not viscid; panicle more than 10 cm. long, not tinged with purple..4. *L. fascicularis*

1. Leptochloa dubia (H.B.K.) Nees. Yavapai, Graham, Gila, Maricopa, Pinal, Cochise, Santa Cruz, and Pima counties, mostly 3,000 to 5,000 feet, rocky hills and open ground, July to October. Oklahoma to Arizona, Mexico, and southern Florida; Argentina.

This species affords good grazing and is sometimes cut for hay.

2. Leptochloa filiformis (Lam.) Beauv. Apache, Navajo, and Coconino counties, south to Cochise, Santa Cruz, Pima, and Yuma counties, 1,000 to 5,000 feet, cultivated land and along streams, May to September. Virginia to eastern Kansas, south to Florida, Texas, Arizona, southern California, and Mexico; Argentina.

3. Leptochloa viscida (Scribn.) Beal. Maricopa, Pinal, Cochise, Santa Cruz, Pima, and Yuma counties, 1,000 to 3,500 feet, open ground and waste places, June to October, type from Tucson (*Pringle* in 1881). Western Texas (El Paso) to California and northern Mexico.

4. Leptochloa fascicularis (Lam.) Gray. Navajo, Graham, Gila, Pinal, Cochise, and Pima counties, 1,500 to 5,000 feet, along ditches and in moist, waste places, often in brackish marshes, July to October. Throughout the United States, south to Argentina.

5. Leptochloa uninervia (Presl) Hitchc. & Chase. Mohave, Pinal, Maricopa, Pima, and Yuma counties, along ditches, roadsides, and moist, waste places. Mississippi to Colorado, Arizona, southern California, and northern Mexico; Peru to Argentina.

56. ELEUSINE. Goose Grass, Yard Grass

Annual; culms compressed, with 2 to several (rarely only 1) digitate spikes, frequently with 1 or 2 spikes shortly below the apex; spikelets few- to several-flowered, compressed, sessile, closely imbricate, in 2 rows on one side of the rather broad, flattened rachis; glumes unequal, acute, 1-nerved, shorter than the first lemma; lemmas acute, strongly 3-nerved, keeled; seeds finely rugose, loosely enclosed in a thin pericarp.

1. **Eleusine indica** (L.) Gaertn. Tucson, Pima County (*Thornber* in 1934), Yuma, Yuma County (*E. Palmer*). Roadsides, lawns, and waste ground, extensively naturalized in the United States; from Eurasia.

57. DACTYLOCTENIUM. Crowfoot Grass

Decumbent, spreading annual with short, broad blades and 2 to several thick, digitate, ascending or spreading spikes, the rachis extending beyond the spikelets; spikelets compressed, 3- to 5-flowered, horizontally spreading; glumes subequal, the second bearing a short, stout, spreading awn; lemmas thin, acute, 3-nerved, awnless.

1. **Dactyloctenium aegyptium** (L.) Richt. A weed at and near Tucson (Pima County). North Carolina to Florida, California, and tropical America; introduced from the Old World.

58. MICROCHLOA

Small, densely tufted perennial; culms slender, flattened; leaves mostly basal, the blades about 1 mm. wide, more or less convolute; spikes solitary, very slender, curved, 3 to 8 cm. long; spikelets 1-flowered, borne in 2 rows on one side of the narrow, flattened rachis; glumes persistent, membranaceous, keeled, acute or acutish, nearly equal, awnless; lemma shorter, hyaline, muticous, the keel and often the margin ciliate; palea nearly as long as the lemma.

1. **Microchloa Kunthii** Desv. Huachuca Mountains, Cochise County, about 5,500 feet (*Pultz et al.* 1400, *Darrow* 2525). Southeastern Arizona and Mexico to Argentina.

59. CYNODON. Bermuda Grass

Stoloniferous perennial with narrow, often short blades and few to several, slender, digitate spikes; spikelets 1-flowered, the rachilla prolonged beyond the spikelet in a naked stipe; glumes subequal, the first lunate, the second lanceolate; lemma acute, awnless, pubescent on the nerves; palea as long as the lemma.

A very abundant grass in the irrigated valleys of southern Arizona, where it is hard to eradicate except by frequent cultivation or shading. Lawns and old pastures usually become Bermuda-grass sod. Indeed, a large proportion of the lawns are planted with this grass, although its abundant production of pollen makes it the commonest cause of hay fever in Arizona. The bulk of the world supply of Bermuda-grass seed is produced near Yuma.

1. **Cynodon dactylon** (L.) Pers. Throughout the state, mostly at low altitudes, lawns and waste places. New Hampshire to Michigan, south to Florida, Arizona, and southern California; introduced in America.

60. SCHEDONNARDUS. Tumble Grass

Slender, freely branching perennial, with few to several stiffly spreading spikes distant on a slender, trigonous axis; spikelets sessile, appressed, in 1 row on each of 2 sides of a trigonous rachis; glumes abruptly narrowed into stiff awn points, the second longer than the first; lemma 3-nerved, acuminate, a little longer than the glumes, awnless.

1. **Schedonnardus paniculatus** (Nutt.) Trel. Apache, Navajo, Coconino, Yavapai, and Cochise counties, up to 7,000 feet, on plains. Illinois to Saskatchewan and Montana, south to Texas and Arizona; Argentina.

61. SPARTINA. Cord Grass

Rather coarse perennial with strong, scaly rhizomes and several ascending or spreading spikes, these racemose on a common axis, the rachis produced beyond the spikelets; spikelets 1-flowered, disarticulating below the glumes; first glume shorter, the second longer than the floret; lemma firm but thinner than the glumes, keeled, subobtuse; palea as long as or longer than the lemma, with thin, very wide margins.

1. **Spartina gracilis** Trin. Apache and Navajo counties, at medium altitudes, plains and in saline soil, August. Saskatchewan to British Columbia, south to Kansas, Arizona, and California.

62. CHLORIS

Annuals or perennials, sometimes stoloniferous, with several digitate spikes; spikelets with 1 (exceptionally 2) perfect florets, the rachilla prolonged beyond the floret, bearing a club-shaped rudiment composed of 1 or more reduced sterile lemmas; fertile lemma 3-nerved, awned from the back just below the apex.

C. virgata, sometimes known as feather finger grass, is a common annual weed, especially in alfalfa fields. It furnishes appreciable quantities of forage on the cattle ranges in the southeastern counties, where in favorable seasons it is cut for hay. Rhodes grass (*C. Gayana*) is grown to a very limited extent in the irrigated districts of southern Arizona as a pasture and hay crop.

KEY TO THE SPECIES

1. Plants annual; margins of the lemma short-ciliate on the lower part, long-ciliate on the upper third, the hairs up to 4 mm. long .... 1. *C. virgata*

1. Plants perennial; margins of the lemma scaberulous to rather evenly ciliate or, if longer-ciliate above, then the hairs much less than 4 mm. long (2).

2. Spikes widespreading. Plant not stoloniferous; culms slender, tufted; rudiment composed of 1 reduced floret (3).

2. Spikes ascending (4).

3. Spikes mostly 3 to 6 cm. long; spikelets crowded, 2 to 2.5 mm. long; rudiment broad, triangular .... 2. *C. latisquamea*

3. Spikes mostly 8 to 10 cm. long; spikelets rather distant, 2.5 to 3 mm. long; rudiment oblong .... 3. *C. verticillata*

4. Spikes digitate; spikelets crowded; lemma 3 mm. long, the margin hispid above, the awn 1 to 5 mm. long; cleistogamous spikelets none; stolons stout, leafy...4. *C. Gayana*

4. Spikes racemose on a short axis; spikelets not crowded; lemma 5 to 6 mm. long, the margin scaberulous, the awn up to 12 mm. long; cleistogamous spikelets borne on underground branches; stolons none .... 5. *C. chloridea*

1. Chloris virgata Swartz. Navajo, Coconino, and Mohave counties to Greenlee, Cochise, Santa Cruz, Pima, and Yuma counties, a common weed in cultivated land and waste places. Nebraska to Texas, west to Nevada, Arizona, and southern California, introduced in a few Eastern localities; tropical America; also in the Old World.

2. Chloris latisquamea Nash. Santa Cruz River at La Noria, Santa Cruz County (*Mearns* 1205). Texas, Arizona, and northeastern Mexico.

3. Chloris verticillata Nutt. Gila County (*Gould* 3710). Missouri to Colorado, south to Louisiana and Arizona.

4. Chloris Gayana Kunth. Pinal and Pima counties, escaped from cultivation near Sacaton and Tucson. North Carolina and Florida, west to southern California; tropical America; introduced from Africa.

5. Chloris chloridea (Presl) Hitchc. Western Pima County, between Sells and Santa Rosa (*Thornber & Freeman* 8062) and at Vamori (*Goodding* 194–45). Southern Texas, southern Arizona, and Mexico.

A specimen of *Chloris cucullata* Bisch. in the herbarium of the California Academy of Sciences is labeled "Southern Pacific railroad, Arizona" (*Eastwood* in 1909). This species is not known otherwise from west of the Pecos Valley, New Mexico. It is readily distinguished from any of the foregoing species by the inflated, broadly triangular-truncate rudimentary florets and by the very short awns (about 1 mm. long).

63. TRICHLORIS

Tufted, leafy perennial with several narrowly ascending spikes crowded on a short axis; spikelets 2-flowered, the upper one reduced; glumes acuminate, 1-nerved, persistent; lemmas rounded on the back, obscurely 3-nerved, 3-awned; palea broad, slightly exceeding the lemma.

1. Trichloris crinita (Lag.) Parodi (*T. mendocina* (Phil.) Kurtz). Graham, Pinal, Maricopa, Cochise, and Pima counties, up to 4,000 feet, mesas and rocky hills. Texas to Arizona and northern Mexico; southern South America.

A rather large, showy grass, rarely cultivated as an ornamental.

64. BOUTELOUA. Grama

Cespitose or sometimes stoloniferous annuals or perennials with slender culms and 1 to many short, 1-sided spikes, these racemose on a short or often elongate axis; spikelets with 1 fertile floret, and 1 or 2 rudimentary florets above it; fertile lemma 3-nerved, variously lobed or dentate at apex, the nerves usually excurrent in short awns; rudiment reduced to 3 awns, glumaceous and lobed, or dentate with 3 usually conspicuous awns.

This is Arizona's most important genus of forage grasses. Rothrock grama (*B. Rothrockii*) and side-oats grama (*B. curtipendula*) are sometimes cut for hay in the southeastern part of the state. Blue grama (*B. gracilis*) is highly valued and often predominant in "short-grass" areas north of the Mogollon Escarpment. When too heavily grazed, these areas are encroached upon by noxious weeds and grasses of lower palatability. Three annual gramas furnish a large quantity of forage, but this is of poorer quality and shorter duration than that produced by the perennial species. Of these, six-weeks grama (*B. barbata*) and needle grama (*B. aristidoides*) are abundant in the foothills and plains of the southern counties, and mat grama (*B. simplex*) is important on the northern plateaus.

Species of *Bouteloua* harbor the telial and uredial stages of a fungus (*Puccinia Stakmanii* Presley), the aecial stage of which causes a destructive rust disease of cotton in southern Arizona. These grasses should be eradicated in the near vicinity of cotton fields.

KEY TO THE SPECIES

1. Spikes deciduous, falling entire; spikelets not pectinate or obscurely so (2).
1. Spikes persistent; spikelets pectinate (7).
2. Plants annual; spikes very narrow, abruptly spreading above, the rachis sharp-pointed at base. Spikelets appressed....................10. *B. aristidoides*
2. Plants perennial; spikes relatively broad, the rachis not sharp-pointed at base (3).
3. Spikes 20 or more. Spikelets appressed to deflexed...............15. *B. curtipendula*
3. Spikes fewer than 15 (4).
4. Second glume hairy (5).
4. Second glume glabrous (6).
5. Spikes 3 to 8, rhomboid, the margins of the rachis densely ciliate; spikelets obscurely pectinate....................11. *B. chondrosioides*
5. Spikes 10 to 13, triangular, the margins of the rachis not conspicuously ciliate; spikelets not pectinate....................12. *B. eludens*
6. Base of the plant hard and rhizomatous; sheaths usually broad and conspicuous 13. *B. radicosa*
6. Base of the plant comparatively soft, not rhizomatous; sheaths inconspicuous 14. *B. filiformis*
7. Plants annual (8).
7. Plants perennial (10).
8. Spike solitary, ascending or spreading, curved....................1. *B. simplex*
8. Spikes 4 to 10, finally spreading, straight (9).
9. Second glume papillose-hispid on the keel; rachis papillose-hispid-ciliate..2. *B. Parryi*
9. Second glume scabrous on the keel; rachis not ciliate....................3. *B. barbata*
10. Rachis produced beyond the spikelets, pointed; second glume tuberculate-hispid (11).
10. Rachis terminating in a spikelet, this often rudimentary; second glume glabrous, scabrous, or pubescent, sparsely papillose-pilose in *B. gracilis* (12).
11. Culms glabrous ....4. *B. hirsuta*
11. Culms retrorsely hirsute below the nodes....................5. *B. glandulosa*
12. Spikes normally 2, rarely 1 or 3; second glume sparsely papillose-pilose..6. *B. gracilis*
12. Spikes 3 to 8; second glume not at all papillose-pilose (13).
13. Culms felty-pubescent, wiry, straggling, sparingly stoloniferous.........7. *B. eriopoda*
13. Culms glabrous, erect, cespitose, not stoloniferous (14).
14. Culms branching; awns of the fertile lemma 2 to 3 mm. long. Plant often appearing annual .... 8. *B. Rothrockii*
14. Culms simple; awns of the fertile lemma about 5 mm. long................9. *B. trifida*

1. Bouteloua simplex Lag. Apache, Coconino, and Yavapai counties, up to 7,000 feet, dry plains and open woods, August and September. Texas to Colorado, Utah, Arizona, and Mexico; Ecuador to Argentina and Chile. Mat grama.

2. Bouteloua Parryi (Fourn.) Griffiths. Separation Canyon (Mohave County) and Pinal, Cochise, and Pima counties, mostly 3,000 to 5,000 feet, mesas and rocky hills, August to October. New Mexico, Arizona, and northern Mexico.

3. Bouteloua barbata Lag. (*B. micrantha* Scribn. & Merr.). Apache County to Mohave County, south to Cochise, Pima, and Yuma counties, 1,000 to 5,500 feet, dry mesas and rocky hills, July to October, type of *B. micrantha* from Fort Lowell (*Griffiths* 1556). Texas to southern Utah, Arizona, southeastern California, and Mexico; Argentina. Six-weeks grama.

4. Bouteloua hirsuta Lag. Apache County to Mohave County, south to Cochise, Santa Cruz, and Pima counties, 1,000 to 6,500 feet, dry, rocky hills and mesas, August to October. Wisconsin and South Dakota to Texas, Arizona, southern California, and Mexico. Hairy grama.

5. Bouteloua glandulosa (Cerv.) Swallen. Near Ruby, Santa Cruz County (*Goodding et al.* 3582), rocky hills and plains. Arizona and Mexico.

6. Bouteloua gracilis (H.B.K.) Lag. Apache County to Mohave County, south to Cochise, Santa Cruz, and Pima counties, 1,000 to 5,000 (rarely 8,000) feet, dry plains, July to October. Wisconsin to Manitoba and Alberta, south to Missouri, Texas, Arizona, southern California, and Mexico. Blue grama.

7. Bouteloua eriopoda Torr. Apache County to Mohave County, south to Cochise and Pima counties, 3,000 to 5,500 feet, dry hills, mesas, and open ground, July to October. Colorado and Texas to southern Utah, Arizona, and Mexico. Black grama.

8. Bouteloua Rothrockii Vasey. (*B. polystachya* (Benth.) Torr. var. *major* Vasey). Mohave and Yavapai counties to Graham, Cochise, Santa Cruz, Pima, and Yuma counties, up to 5,000 feet, mesas and rocky foothills, August to October, type from Sonoita Valley (*Rothrock* 347). Arizona, southern California, and northern Mexico.

Rothrock grama, very similar in appearance to *B. barbata* and sometimes difficult to distinguish before the perennial habit has become apparent. The awns are longer than is usual in *B. barbata*, making the spikes wider, and the stems are usually less branched (Agnes Chase, personal communication).

9. Bouteloua trifida Thurb. Grand Canyon (Coconino County), and Mohave County to Santa Cruz, Pima, and Yuma counties, mostly 2,000 to 4,000 feet, mesas and dry, rocky hillsides, March to June, and sometimes October. Texas to southwestern Utah, Nevada, Arizona, and northwestern Mexico.

10. Bouteloua aristidoides (H.B.K.) Griseb. Greenlee County to western Coconino and Mohave counties, south to the southern border, up to 5,500 feet, deserts, dry mesas, and rocky hillsides, June to October. Texas to southern California and northern Mexico; Argentina.

Needle grama. Var. *arizonica* M. E. Jones, type from Tucson (*Thornber* 177) with much longer, curved spikes, and a nonciliate rachis, occurs in Graham, Pinal, Santa Cruz, and Pima counties.

11. Bouteloua chondrosioides (H.B.K.) Benth. Cochise, Santa Cruz, and Pima counties, 2,000 to 5,000 feet, rocky hills, August to October. Western Texas, southern Arizona, and Mexico; Costa Rica and Honduras. Spruce-top grama.

12. Bouteloua eludens Griffiths. Nogales and Atascosa Mountain (Santa Cruz County), Santa Rita and Santa Catalina mountains (Pima County), rocky hills and desert grassland, August and September, type from the Santa Rita Mountains (*Griffiths* 7269). Arizona and Sonora.

13. Bouteloua radicosa (Fourn.) Griffiths. Apache, Greenlee, Pinal, Cochise, Santa Cruz, and Pima counties, 3,500 to 7,000 feet, rocky hills and canyons, August to October. New Mexico to southern California and Mexico. Purple grama.

14. Bouteloua filiformis (Fourn.) Griffiths. Mohave, Greenlee, Graham, Pinal, Cochise, Santa Cruz, and Pima counties, below 5,000 feet, mesas and rocky foothills, July to October. Texas to Arizona, south to Guatemala; Panama, Colombia, and Venezuela. Slender grama.

15. Bouteloua curtipendula (Michx.) Torr. Throughout the state, up to 7,000 feet, dry hills and mesas, April to October. Maine and Ontario to Montana, south to Maryland, Alabama, Texas, Arizona, and southern California; Mexico to El Salvador; Colombia to Argentina. Side-oats grama.

65. CATHESTECUM

Stoloniferous perennial with relatively short, flat blades and several V-shaped or rhomboid, spreading spikes arranged on opposite sides of the slender, flattened axis; spikes falling entire, consisting of 3 spikelets, the lateral ones 2-flowered, staminate or sterile, the central spikelet 3-flowered, the lowest floret pistillate or rarely perfect, the upper florets staminate or sterile; lemmas dissimilar, the lower ones cleft about ¼ their length, awned from between the lobes, the awns equaling or slightly exceeding them, the upper lemmas deeply cleft, their awns villous in the lower part, extending as much as 3 mm. beyond the lobes.

* **1. Cathestecum erectum** Vasey & Hack. "Southern Arizona" without definite locality (*E. Palmer* in 1869). Dry hills and plains, western Texas, southern Arizona (?), and northern Mexico.

66. MUNROA. False Buffalo Grass

Freely branching, widely decumbent, spreading annual with short, firm, pungent blades and short spikes hidden in the crowded sheaths at the ends of the branches; spikes composed of 2 or 3 spikelets, the lower one or two 3- or 4-flowered, the upper spikelet 2- or 3-flowered; glumes of the lower 1 or 2 spikelets equal, 1-nerved, those of the upper spikelet much shorter, the first about half as long as the second; lemmas 3-nerved, excurrent in short awns, the central awn longer and stouter than the lateral awns.

1. Munroa squarrosa (Nutt.) Torr. Apache County to Mohave County, south to Cochise, Gila, and Yavapai counties, 4,000 to 6,000 feet, open plains and hills, June to October. Alberta to Texas and Arizona.

The plants often have a white weblike covering, the remains of egg cases of a woolly aphid.

67. BUCHLOË. Buffalo Grass

Stoloniferous perennial with short, slender culms and flat, narrow blades; plants monoecious or dioecious, sometimes with perfect flowers; staminate spikes 1 to 4, pectinate, the spikelets 2- or 3-flowered, the glumes acute, the second glume about twice as long as the first, the lemmas acute or subobtuse, 3-nerved, awnless; pistillate spikes 1 or 2, falling entire, the short, thickened, indurate rachis and second glumes forming a false involucre around the spikelets, the spikelets 1-flowered, the first glume thin, acuminate, sometimes obsolete, the lemma indurate, 3-toothed, with margins overlapping and enclosing the palea, the palea firm, about as long as the lemma.

1. Buchloë dactyloides (Nutt.) Engelm. North of Ash Fork, in southwestern Coconino County, 5,500 feet, sandy soil (*Darrow* 3092), rocky, limestone soil on ridge running from Promontory to Tonto Creek, Gila County (*Forest Service* 58246). Minnesota to Montana, south to Iowa, western Louisiana, Texas, and Arizona.

A dominant forage grass in the Great Plains region, but very rare in Arizona.

68. HIEROCHLOË. Sweet Grass

Slender, erect, sweet-smelling perennial, with slender, creeping rhizomes, flat blades, and small, open panicles of bronze-colored spikelets; spikelets with 1 terminal fertile floret and 2 staminate florets, these falling attached to the fertile one; staminate lemmas as long as the glumes, boat-shaped, hispidulous; fertile lemma a little shorter than the others, indurate, awnless.

1. Hierochloë odorata (L.) Beauv. Apache, Navajo, Coconino, and Pima counties, wet places, 7,000 feet or higher, June and July. Labrador to Alaska, south to New Jersey, Oregon, and in the mountains to New Mexico and Arizona; Eurasia.

Sweet grass, also known as holy grass, vanilla grass, and seneca grass, is used by the Indians in some parts of the United States for making fragrant baskets.

69. PHALARIS. Canary Grass

Annuals or perennials with flat blades and erect, spikelike, sometimes interrupted panicles; spikelets laterally compressed, with 1 fertile floret and 1 or 2 much-reduced sterile florets below the fertile one; fertile lemma coriaceous, shorter than the glumes.

The canary seed of commerce is obtained from *P. canariensis*. A form of *P. arundinacea*, reed canary grass (var. *picta*) is cultivated as a garden ornamental under the name ribbon grass.

KEY TO THE SPECIES

1. Spikelets in groups of 7, a fertile one surrounded by 6 sterile ones, the group falling entire .... 1. *P. paradoxa*
1. Spikelets all alike, the groups not falling entire (2).
2. Plants perennial with creeping rhizomes. Panicle interrupted below, the branches spreading in anthesis .... 2. *P. arundinacea*
2. Plants annual (3).
3. Glumes broadly winged. Panicle ovate or ovate-oblong (4).
3. Glumes wingless or nearly so (5).
4. Sterile lemma 1; fertile lemma 3 mm. long .... 3. *P. minor*
4. Sterile lemmas 2; fertile lemma 4 to 6 mm. long .... 4. *P. canariensis*
5. Panicle mostly 2 to 6 cm. long, tapering to each end; glumes 5 to 6 mm. long 5. *P. caroliniana*
5. Panicle mostly 6 to 15 cm. long, subcylindric; glumes 3.5 to 4 mm. long .... 6. *P. angusta*

1. Phalaris paradoxa L. Tempe and Mesa, Maricopa County, roadsides and fields (*Davis* in 1935, *Streets* in 1947). Occasional in the United States; introduced from the Mediterranean region.

Var. *praemorsa* (Lam.) Coss. & Dur. has been collected at Yuma (*Forbes* in 1906). It is distinguished by the outer glumes of all the sterile spikelets in the panicle being clavate. In the species only those in the lower part of the panicles are deformed.

2. Phalaris arundinacea L. Crater Lake, Coconino County, 8,000 feet (*MacDougal* 834). Moist places, New Brunswick to southeastern Alaska, south to North Carolina, Kentucky, Oklahoma, New Mexico, Arizona, and northeastern California; Eurasia.

3. Phalaris minor Retz. Tempe and near Glendale, Maricopa County (*McLellan & Stitt* 765, *Gould* 3497), near Tucson, Pima County (*Thornber* in 1913). Introduced in a few scattered localities in the United States, rather common in California; native of the Mediterranean region.

4. Phalaris canariensis L. Near Government Hill, Gila County (*Forest Service* 39185). Waste places, Nova Scotia to Alaska, south to Virginia, Texas, Arizona, and California; introduced from the Mediterranean region.

5. Phalaris caroliniana Walt. Graham, Gila, and Yavapai counties to Cochise, Pima, and Yuma counties, 1,000 to 6,000 feet, moist ground, April to August. Virginia to Colorado, south to Florida and Texas, west to Arizona, California, and Oregon.

6. Phalaris angusta Nees. Sacaton, Pinal County (*Harrison* 5266). Open ground at low altitudes. Mississippi, Louisiana, Texas, Arizona, and California; southern South America.

70. LEERSIA. Cut Grass

Slender, erect or decumbent perennial with creeping rhizomes and flat, scabrous blades; spikelets 1-flowered, laterally compressed, disarticulating from the pedicel; glumes wanting; lemma chartaceous, boat-shaped, hispid; palea as long as the lemma but much narrower, the margins firmly held by the margins of the lemma.

1. Leersia oryzoides (L.) Swartz. Greenlee and Pima counties, a weed along irrigation ditches. Quebec and Maine to eastern Washington, south to northern Florida, Arizona, and southern California; Europe.

71. TRICHACHNE. Cotton-top, Cotton Grass

Perennial, with flat blades and slender, erect or ascending racemes forming a silky inflorescence; first glume minute; second glume and sterile lemma equal, covering the fruit, conspicuously villous with long hairs; fertile lemma acuminate, brown, with broad, hyaline margins.

KEY TO THE SPECIES

1. Spikelets (excluding the hairs) about 4 mm. long, the hairs tawny; lemma lanceolate, acuminate; culms up to 1.5 m. long; blades usually more than 5 mm. wide; panicle up to 30 cm. long . . . . 1. *T. insularis*
1. Spikelets about 3 mm. long, the hairs white or purplish; lemma obovate, abruptly pointed; culms up to 1 m. long; blades not more than 5 mm. wide; panicle seldom more than 15 cm. long . . . . 2. *T. californica*

1. Trichachne insularis (L.) Nees. Huachuca Mountains (Cochise County), Patagonia Mountains (Santa Cruz County), Santa Catalina and Baboquivari mountains (Pima County), 3,000 to 4,000 feet, in mesquite-catclaw association on rocky slopes, September. Florida to southern Texas; southern Arizona; West Indies and Mexico to Argentina.

The blades in the Arizona specimens are narrower than in most specimens from tropical America.

2. Trichachne californica (Benth.) Chase. Southern Navajo County, Grand Canyon (Coconino County), and Mohave County, to Cochise, Santa Cruz, Pima, and Yuma counties, 1,000 to 6,000 feet, mesas and rocky hills in open ground. Texas to Colorado, Arizona, and Mexico.

Cotton-top furnishes a considerable quantity of palatable forage for a short period following spring or summer rain.

72. DIGITARIA. Crab Grass

Decumbent, spreading annual with flat blades and rather slender, ascending or spreading, digitate racemes; spikelets usually in pairs on one side of a flat, winged rachis; first glume evident; second glume shorter than the sterile lemma, exposing the fruit, more or less pubescent; fertile lemma cartilaginous, pale, with hyaline margins.

1. Digitaria sanguinalis (L.) Scop. Havasu Canyon (Coconino County) and Greenlee, Gila, Maricopa, Pinal, Cochise, Santa Cruz, and Pima counties, up to 6,000 feet. A common weed in temperate and tropical regions of the world.

Where it is sufficiently abundant this grass supplies forage of good quality.

73. LEPTOLOMA. Fall Witch Grass

Slender perennial, felty-pubescent at base, with branching, brittle culms, flat blades, and diffuse panicles; first glume minute or obsolete; second glume and sterile lemma nearly equal, appressed-hairy on the internerves and margins; fertile lemma elliptic, acute, brown, with thin, hyaline margins.

1. Leptoloma cognatum (Schult.) Chase. Cochise, Santa Cruz, and Pima counties, up to 5,500 feet, rocky slopes and gravelly plains, August and September. New Hampshire to Minnesota, south to Florida, Texas, southern Arizona, and northern Mexico.

The mature panicles break away and become tumbleweeds. The species is a fairly palatable forage plant.

74. ERIOCHLOA. Cup Grass

Erect or decumbent annuals with flat blades and several to numerous erect or ascending racemes, approximate on a common axis; spikelets usually solitary in two rows on one side of a narrow rachis; first glume and rachilla joint forming a cuplike callus below the spikelet; second glume and sterile lemma equal, longer than the fruit, the lemma sometimes enclosing a palea or a staminate flower; fertile lemma indurate, minutely rugose, mucronate or awned.

KEY TO THE SPECIES

1. Pedicels with numerous erect hairs, the hairs at least half as long as the spikelets. Leaves and spikelets villous....................................1. *E. Lemmoni*

1. Pedicels scabrous, or short-pubescent, or (in *E. gracilis*) with a few long hairs (2).

2. Second glume and sterile lemma awned; spikelets, including the awns, 7 to 10 mm. long ..... 2. *E. aristata*
2. Second glume and sterile lemma awnless or mucronate; spikelets not more than 6 mm. long (3).
3. Fruit 3 to 4 mm. long, apiculate to cuspidate ..... 3. *E. gracilis*
3. Fruit 2 to 2.5 mm. long, with an awn 0.5 to 1 mm. long (4).
4. Spikelets 3 to 3.5 mm. long, the second glume merely acute; rachis slender, scabrous only ..... 4. *E. procera*
4. Spikelets, including the awn of the second glume, 4.5 to 5 mm. long; rachis relatively stout, pubescent ..... 5. *E. contracta*

1. Eriochloa Lemmoni Vasey & Scribn. Cochise, Santa Cruz, and Pima counties, 4,000 to 5,000 feet, rocky, grassy slopes, August and September, type from the Huachuca Mountains (*Lemmon* 2910). Southern Arizona and northern Mexico.

2. Eriochloa aristata Vasey. Vail, Tucson, and Gunsight (Pima County). Arizona, California, and northern Mexico.

3. Eriochloa gracilis (Fourn.) Hitchc. (*E. Lemmoni* var. *gracilis* Gould). Yavapai, Graham, Gila, Maricopa, Pinal, Cochise, Santa Cruz, and Pima counties, 2,500 to 5,000 feet, moist, open ground, June to October. Texas, New Mexico, and Arizona to central Mexico.

A smaller plant, with more crowded, less acuminate spikelets, and fertile lemma as long as the second glume and sterile lemma, is var. *minor* (Vasey) Hitchc. This species is of some value as a forage plant.

***4. Eriochloa procera** (Retz.) C. E. Hubbard. Campus of the University of Arizona, Tucson, Pima County (*Griffiths* 1516), probably only cultivated. Introduced in Cuba from Asia.

5. Eriochloa contracta Hitchc. Bear Valley or Sycamore Canyon, Santa Cruz County, about 3,500 feet (*Kearney & Peebles* 14484), Menenger Dam, western Pima County (*Gooding & Lusher* 134–45), Yuma (Yuma County), 140 feet (*W. M. Wootton* in 1950). Open ground, ditches, and wet places, Kansas to Louisiana and Arizona.

75. PASPALUM (44)

Perennials, with flat blades and 2 to several spikelike racemes paired or racemose on a common axis; spikelets solitary or paired in 2 rows on one side of the rachis; first glume usually wanting; second glume and the sterile lemma equal, covering the fruit; fertile lemma indurate, smooth, usually obtuse.

These grasses are good forage plants. Dallis grass (*P. dilatatum*) has been cultivated as a pasture grass in the southern United States and elsewhere. Knot grass (*P. distichum*) serves as a soil binder along streams but sometimes causes trouble by clogging irrigation ditches.

KEY TO THE SPECIES

1. Racemes 2, paired; first glume often developed; plants often with extensively creeping stolons ..... 1. *P. distichum*
1. Racemes 1 to 5, never paired; first glume obsolete; plants cespitose, without stolons (2).
2. Inflorescences both terminal and axillary; sterile lemma 2-nerved. Spikelets 2.1 to 2.2 mm. long, glabrous to densely pubescent with short hairs ..... 2. *P. stramineum*
2. Inflorescences terminal only; sterile lemma with 3 or more nerves (3).

3. Culms rather stout; pedicels flattened; spikelets 2.8 to 3.8 mm. long; glume and sterile lemma acuminate, pubescent with long, silky hairs....................3. *P. dilatatum*
3. Culms slender; pedicels 3-angled; spikelets 2 mm. long; glume and sterile lemma acutish, the glume minutely pubescent, the lemma glabrous.....................4. *P. Virletii*

1. Paspalum distichum L. Navajo County to Mohave County, south to Cochise, Santa Cruz, Pima, and Yuma counties, at low altitudes, moist ground along streams and ditches, June to September. New Jersey to Florida, west to California and northwest to Idaho and Washington; Mexico and the West Indies to Argentina and Chile; seacoasts of the Eastern Hemisphere.

2. Paspalum stramineum Nash. Cochise and Santa Cruz counties, about 4,000 feet, sandy, open ground, June to September. Indiana to Minnesota, south to Texas, Arizona, and northwestern Mexico.

3. Paspalum dilatatum Poir. Springerville, Apache County (*McGinnies* in 1932), Tucson and vicinity, Pima County (*Hitchcock* 3474, *Darrow et al.* 3483). Introduced from South America.

4. Paspalum Virletii Fourn. Bear Valley or Sycamore Canyon, Santa Cruz County, 3,500 feet (*Darrow & Haskell* 2226). Southeastern Arizona and northern Mexico.

76. PANICUM (45)

Annuals or perennials, with spikelets in open or condensed panicles; first glume minute to more than half as long as the spikelet; second glume and the sterile lemma equal, usually covering the fruit, the sterile lemma sometimes enclosing a staminate flower; fertile lemma indurate, typically obtuse.

Bulb panicum (*P. bulbosum*), vine-mesquite (*P. obtusum*), and switch grass (*P. virgatum*) yield forage and are sometimes cut for hay. Vine-mesquite is an excellent plant for controlling too rapid erosion in gullies but is sometimes looked upon as a weed. The seeds of several species were used for food by the Indians of Arizona.

KEY TO THE SPECIES

1. Plants annual (2).
1. Plants perennial (11).
2. Fruit transversely rugose. Spikelets arranged in spikelike racemes (3).
2. Fruit smooth and shining (5).
3. Spikelets dark brown, strongly reticulate-veined, glabrous..........1. *P. fasciculatum*
3. Spikelets green, reticulate-veined only at the apex, usually pubescent (4).
4. Spikelets 3.5 to 3.8 mm. long; sheaths and blades glabrous to papillose-hispid 2. *P. arizonicum*
4. Spikelets 5 to 6 mm. long; sheaths and blades soft-pubescent............3. *P. texanum*
5. First glume short, truncate ....4. *P. dichotomiflorum*
5. First glume ⅓ to ¾ as long as the spikelet, acute or acuminate (6).
6. Panicle more or less drooping (7).
6. Panicle erect (8).
7. Spikelets less than 3.5 mm. long......................................5. *P. sonorum*
7. Spikelets 4.5 to 5 mm. long ....6. *P. miliaceum*
8. Panicle usually more than half the height of the culm, the branches stiffly spreading 7. *P. capillare*
8. Panicle not more than ⅓ the height of the culm, the branches ascending (9).
9. First glume subacute or blunt, about ⅓ the length of the spikelet....8. *P. stramineum*
9. First glume acuminate, usually more than half the length of the spikelet (10).
10. Spikelets 4 mm. long...9. *P. pampinosum*

10. Spikelets not more than 3.3 mm. long.......................... 10. *P. hirticaule*
11. Basal leaves distinctly different from those of the culm, forming a winter rosette. Culms at first simple, later becoming much branched (12).
11. Basal leaves similar to those of the culm, not forming a winter rosette (14).
12. Spikelets 3.2 to 3.3 mm. long, glabrous or sparsely pubescent.... 11. *P. Scribnerianum*
12. Spikelets 1.6 to 1.8 mm. long, pubescent. Ligule of conspicuous hairs 4 to 5 mm. long (13).
13. Blades glabrous on the upper surface, with rather conspicuous, firm, white margins 12. *P. tennesseense*
13. Blades pilose on the upper surface, sometimes near the base only...... 13. *P. huachucae*
14. Plants cespitose; rhizomes and stolons wanting (15).
14. Plants rhizomatous or with widely creeping stolons (18).
15. Fruit transversely rugose (16).
15. Fruit smooth (17).
16. Culms distinctly bulbous at base.............................. 14. *P. bulbosum*
16. Culms not bulbous .... 15. *P. plenum*
17. Spikelets 3 to 3.7 mm. long; leaves crowded toward base, the blades becoming curled or twisted with age.. 16. *P. Hallii*
17. Spikelets 4 to 4.2 mm. long; leaves not crowded toward base, the blades not curled 17. *P. lepidulum*
18. Plants without rhizomes. First glume obtuse; fruit as long as the spikelet; culms erect from a knotty crown, with long, widely spreading stolons, the nodes of the stolons densely bearded... 18. *P. obtusum*
18. Plants rhizomatous. Tall, long-leaved grasses (19).
19. Panicle densely many-flowered; spikelets 2 to 2.5 mm. long; first glume obtuse, not more than half as long as the spikelet; fruit nearly equaling the spikelet 19. *P. antidotale*
19. Panicle open, relatively few-flowered; spikelets 3.5 mm. or longer; first glume acute; fruit much shorter than the spikelet (20).
20. Spikelets densely villous, 6 to 7 mm. long; nodes densely bearded.... 20. *P. Urvilleanum*
20. Spikelets glabrous, 3.5 to 5 mm. long; nodes glabrous................. 21. *P. virgatum*

1. Panicum fasciculatum Swartz. Pinal and Pima counties, in open, sandy ground and waste places, mostly around Tucson. Arkansas and Louisiana to Arizona, south through Mexico and Central America to Argentina.

This species typically has broad blades and open panicles with spikelets 2.1 to 2.5 mm. long. Only var. *reticulatum* (Torr.) Beal, with narrow blades, narrow condensed panicles, and spikelets 2.6 to 3 mm. long, is found in Arizona.

2. Panicum arizonicum Scribn. & Merr. Greenlee, Gila, and Mohave counties to Cochise, Santa Cruz, Pima, and Yuma counties, 1,000 to 5,500 feet, sandy ground and open, rocky slopes, July to October, type from Camp Lowell, Pima County (*Pringle* 465). Texas to southern California and Mexico.

3. Panicum texanum Buckl. Introduced at Phoenix, Maricopa County (*Thornber* in 1912), and at Wilmot, Pima County (*Thornber* 39, in 1903). Moist, open ground, Texas and northern Mexico, introduced in a few localities eastward.

4. Panicum dichotomiflorum Michx. Tempe, Maricopa County (*McLellan & Stitt* 576). Moist ground, a weed in waste places, Maine to Nebraska, south to Florida and Texas, occasionally introduced westward.

5. Panicum sonorum Beal. Yuma, Yuma County (*Forbes & Skinner* in 1903). Southwestern Arizona and northern Mexico.

6. Panicum miliaceum L. Douglas, Cochise County (*W.W. Jones* in 1930), Tucson, Pima County (*Thornber* in 1903). Here and there in the United States, occasional at roadsides and in waste ground but not naturalized; introduced from the Old World.

Broom-corn millet or proso, cultivated as a forage plant in the United States and for human food in Europe.

7. Panicum capillare L. Apache, Navajo, and Coconino counties, south to Cochise, Pima, and Yuma counties, 2,000 to 7,000 feet, moist, open ground, July to October. Prince Edward Island to British Columbia, south to New Jersey, Missouri, Texas, Arizona, and California.

Old-witch grass, a tumbleweed. Only var. *occidentale* Rydb. (*P. barbipulvinatum* Nash) occurs in Arizona, differing from the species in the shorter, less pubescent leaves crowded toward the base and more exserted panicles of somewhat larger spikelets.

8. Panicum stramineum Hitchc. & Chase. Tucson and Vamori (Pima County), moist, sandy plains, August and September. Southern Arizona and northwestern Mexico.

9. Panicum pampinosum Hitchc. & Chase. Big Lue Range (Greenlee County), near Hilltop (Gila County), Chiricahua Mountains (Cochise County), and in Pima County at Wilmot (*Thornber* 193, the type collection) and in the Tucson Mountains. Texas to Arizona and Mexico.

10. Panicum hirticaule Presl. Throughout the state, 1,000 to 7,500 feet, dry, open ground, waste places, and along streams, June to September. Arkansas and western Texas to southern California; Mexico and the West Indies to Argentina.

11. Panicum Scribnerianum Nash. Apache and Coconino counties, south to Cochise, Santa Cruz, and Pima counties, 4,000 to 6,000 feet, open, sandy ground and among rocks, June to September. Maine to British Columbia, south to Maryland, Tennessee, Texas, Arizona, and northern California.

12. Panicum tennesseense Ashe. Santa Catalina Mountains, Pima County (*Harrison & Kearney* 7256). Open ground and borders of woods, Maine to Minnesota, south to Georgia, Texas, and northeastern Mexico; also at a few localities in Colorado, Utah, New Mexico, and Arizona.

13. Panicum huachucae Ashe. Grand Canyon (Coconino County), Sierra Ancha (Gila County), Swisshelm and Huachuca mountains (Cochise County), Santa Catalina Mountains (Pima County), 4,000 to 6,500 feet, open ground, type from the Huachuca Mountains (*Lemmon* in 1882). Nova Scotia to Montana, south to Florida and Texas; also westward in a few scattered localities.

Var. *fasciculatum* (Torr.) Hubbard, which is more slender and less pubescent, with thin, lax, spreading blades sparsely pilose on the upper surface, occurs near Tucson (*Toumey* 781). It has about the same general range as the species.

14. Panicum bulbosum H.B.K. Apache, Navajo, and Coconino counties to Cochise, Santa Cruz, and Pima counties, 4,500 to 8,000 feet, moist canyons and open woods, July to October. Western Texas to Arizona and Mexico.

Var. *minus* Vasey is a smaller, slender form with narrow blades, and spikelets 2.8 to 3.2 mm. long. The variety is as common as the species and has the same range.

15. Panicum plenum Hitchc. & Chase. Hualpai Mountain (Mohave County), and Cochise, Santa Cruz, and Pima counties, at medium altitudes, along streams and on rocky hills, August and September. Texas to Arizona and Mexico.

16. Panicum Hallii Vasey. Coconino, Yavapai, Greenlee, Gila, Pinal, Cochise, Santa Cruz, and Pima counties, 3,500 to 5,500 feet, dry prairies and rocky hills, June to October. Texas to Arizona and Mexico.

17. Panicum lepidulum Hitchc. & Chase. Nogales, Santa Cruz County (*Peebles et al.* 4621), Quinlan Mountains, Pima County (*Gould et al.* 2834). New Mexico, southern Arizona, and Mexico.

These are the only typical specimens of this species from the United States. A collection in the Santa Catalina Mountains (*Griffiths* 7063) was doubtfully referred to this species by Hitchcock and Chase in their revision of *Panicum.*

18. Panicum obtusum H.B.K. Throughout the state, 1,000 to 6,000 feet, low, open ground, May to October. Missouri to Colorado, south to Texas, Arizona, and central Mexico.

19. Panicum antidotale Retz. Near Douglas, Cochise County (*W. W. Jones* in 1944, *Gould & Haskell* 4546). Introduced from the Old World tropics by the United States Soil Conservation Service and well established at roadsides.

***20. Panicum Urvilleanum** Kunth. Fort Mohave, Mohave County (*Lemmon* 4665). Sandy deserts, western Arizona (?) and southern California; Argentina, Chile.

The data of locality on Lemmon's labels are notoriously undependable, and this collection may have been made in the Mojave Desert, California.

21. Panicum virgatum L. Navajo, Coconino, Yavapai, Gila (?), and Pima counties, 2,000 to 7,000 feet, moist canyons and open, sometimes rocky ground, June to September. Quebec and Maine to Montana, south to southern Mexico.

77. ECHINOCHLOA. Cock-spur

Slender or stout annuals, with flat blades and few to several spikelike racemes along a common axis; spikelets hispid, densely arranged on one side of the rachis; first glume acute, about half as long as the spikelet; second glume and sterile lemma equal, pointed, the lemma often with a long, conspicuous awn.

These grasses are readily grazed by livestock. Forms of barnyard grass (*E. Crusgalli*) and jungle-rice (*E. colonum*) are cultivated in Asia and Africa for the seeds, which are used for human food.

KEY TO THE SPECIES

1. Culms slender, 20 to 40 cm. high; blades 3 to 5 mm. wide; spikelets 2 mm. long, arranged in about 4 rows, awnless, weakly hispid-scabrous on the nerves..........1. *E. colonum*
1. Culms stout, up to 150 cm. high; blades 5 to 15 mm. wide; spikelets 3 mm. long, irregularly fascicled, awned or awnless, strongly papillose-hispid on the nerves, the internerves hispid-scabrous ....2. *E. Crusgalli*

1. Echinochloa colonum (L.) Link. Yavapai, Graham, Pinal, Maricopa, Cochise, Santa Cruz, Pima, and Yuma counties, up to 5,500 feet, moist, weedy places, May to October. Virginia to Missouri, southward and southwestward to Florida, Arizona, and southeastern California; tropical regions of both hemispheres; introduced in America.

2. Echinochloa Crusgalli (L.) Beauv. Apache County to Coconino County, south to the southern border, 150 to 7,000 feet, moist ground along ditches and in waste places, July to September. New Brunswick to Washington, south to Florida, Arizona, California, and northern Mexico; temperate and subtropical regions of both hemispheres.

Var. *zelayensis* (H.B.K.) Hitchc. differs from the species in having an awnless or mucronate, sterile lemma. Var. *mitis* (Pursh) Peterm. differs from the preceding in having mostly simple, more spreading racemes and less strongly hispid spikelets. Both varieties occur throughout most of the range of the species in Arizona.

78. RHYNCHELYTRUM. Natal Grass

Slender annual, with open or somewhat contracted but not dense panicles of silky spikelets on short, capillary pedicels; first glume minute, the second glume short-stipitate; sterile lemma equaling the second glume; fertile lemma shorter than the spikelet, smooth, obtuse, the margins thin, not inrolled; palea well developed in both florets.

1. Rhynchelytrum roseum (Nees) Stapf & Hubb. (*Tricholaena rosea* Nees). Pima County near Tucson (*Nichol* in 1939) and in the Santa Catalina Mountains (*Parker & Haskell* 5903). Adventive in Florida, Texas, and southern Arizona; from South Africa.

A handsome grass with long, purplish-pink, silky hairs on the spikelets, sometimes grown as an ornamental and cultivated for forage in the Southeastern states.

79. SETARIA. Bristle Grass

Annuals or perennials, with flat blades and spikelike or somewhat open panicles; spikelets subtended by one or more scabrous bristles, the spikelets deciduous, the bristles persistent; first glume broad, less than half as long as the spikelet; second glume and the sterile lemma equal or the glume a little shorter; fruit smooth or transversely rugose.

Plains bristle grass (*S. macrostachya*) is Arizona's best forage species of *Setaria*. Foxtail-millet (*S. italica*), which is grown in the United States for hay, has been cultivated for food by primitive peoples in Europe since prehistoric times.

KEY TO THE SPECIES

1. Bristles below each spikelet more than 5. Panicle dense, cylindric, spikelike; fruit evidently rugose (2).
1. Bristles below each spikelet 1 to 3 (3).
2. Plants annual; leaves often spirally twisted..........................1. *S. lutescens*
2. Plants perennial, the culms from short knotty rootstocks; leaves seldom twisted 2. *S. geniculata*
3. Bristles more or less retrorsely scabrous..........................3. *S. verticillata*
3. Bristles antrorsely scabrous (4).

4. Plants annual (5).
4. Plants perennial (7).
5. Fertile lemma coarsely transverse-rugose. Panicle loosely flowered......4. *S. Liebmanni*
5. Fertile lemma finely cross-lined or nearly smooth (6).
6. Panicle rather loose, tapering toward apex..........................5. *S. Grisebachii*
6. Panicle dense, cylindric, scarcely tapering.............................6. *S. viridis*
7. Spikelets 3 mm. long; blades villous...............................7. *S. villosissima*
7. Spikelets 2 to 2.5 mm. long; blades scabrous or pubescent (8).
8. Panicle spikelike, interrupted, with branches usually not more than 1 cm. long, appressed; blades mostly less than 1 cm. wide............................8. *S. macrostachya*
8. Panicle rather loose, the lower branches spreading, up to 3 cm. long; blades flat, up to 1.5 cm. wide..9. *S. Scheelei*

1. **Setaria lutescens** (Weigel) Hubbard (*S. glauca* (L.) Beauv.). Apache, Coconino, Gila, Maricopa, Cochise, Pima, and Yuma counties, 500 to 7,500 feet, cultivated ground, waste places, and along streams. New Brunswick to South Dakota, south to Florida and Texas; occasional from British Columbia to California, Arizona, and New Mexico; introduced from Europe.

2. **Setaria geniculata** (Lam.) Beauv. Bear Valley or Sycamore Canyon, Santa Cruz County, 3,500 feet, locally abundant in sand (*Goodding* in 1939, *Darrow & Haskell* 2225). Massachusetts to Florida, west to California, south to South America.

3. **Setaria verticillata** (L.) Beauv. Havasu Canyon (Coconino County), Kingman (Mohave County), Huachuca Mountains (Cochise County), Tucson (Pima County), up to 7,000 feet, cultivated and waste land. Massachusetts to North Dakota, south to Alabama and Missouri, occasional westward to California; introduced from Europe.

4. **Setaria Liebmanni** Fourn. Nogales (Santa Cruz County) and at several localities in Pima County, open, sandy or rocky soil. Arizona to Nicaragua.

5. **Setaria Grisebachii** Fourn. Navajo and Coconino counties, south to Cochise, Santa Cruz, and Pima counties, 2,000 to 6,000 feet, sandy or stony ground, June to October. Texas to Arizona and Mexico.

6. **Setaria viridis** (L.) Beauv. Apache, Coconino, Pinal, Cochise, Pima, and Yuma counties, 2,000 to 8,000 feet, fields and open woods. Temperate regions of both hemispheres; introduced from Europe.

7. **Setaria villosissima** (Scribn. & Merr.) Schum. Arizona, without definite locality (*Emersley* 19 and 21 C, in 1890). Texas and Arizona.

8. **Setaria macrostachya** H.B.K. Navajo and Coconino counties, south to Cochise, Santa Cruz, Pima, and Yuma counties, 2,000 to 7,000 feet, dry, rocky soil, May to October. Texas to Colorado, Arizona, and Mexico.

9. **Setaria Scheelei** (Steud.) Hitchc. Near base of the Baboquivari Mountains, Pima County (*Harrison & Kearney* 8018). Texas and Arizona.

Foxtail-millet (*S. italica* (L.) Beauv.) is occasionally seen at roadsides in Arizona but has not become established. It can be distinguished from *S. viridis* by the usually large and heavy, often lobed or interrupted panicles and by the fruit falling freely from the glumes and sterile lemma.

80. PENNISETUM

Perennial, tall, with tufted stems and narrow, elongate blades; inflorescence a long cylindric, bristly, spikelike panicle; spikelets solitary or in groups of 3, subtended by unequal, purplish bristles 2 to 4 cm. long, these plumose below and united at the very base; fertile lemma chartaceous, smooth, thin-margined, enclosing the palea.

1. Pennisetum setaceum (Forsk.) Chiov. (*P. Ruppelii* Steud.). Santa Catalina Mountains, Pima County, 4,500 feet, in disturbed soil at roadside (*Pultz* 1901). Cultivated as an ornamental under the name fountain-grass; native of Ethiopia.

81. CENCHRUS. Sand-bur

Decumbent or geniculate, spreading annuals, with flat blades and rather dense, spikelike racemes of burs; burs composed of numerous coalescing bristles enclosing 2 to 4 spikelets, falling entire; first glume usually ½ to ⅔ as long as the spikelets; second glume and sterile lemma equal, subacute or acuminate; fruit acuminate, about as long as the second glume and the sterile lemma.

The plants make good forage when young but become troublesome after maturity. The burs are especially obnoxious when mixed with hay, their barbed spines being painful to human beings and animals, difficult to extract, and sometimes causing inflammation and infection.

KEY TO THE SPECIES

1. Burs with a ring of slender bristles at base; spikelets usually 4 in each bur 1. *C. echinatus*
1. Burs with no ring of slender bristles at base; spikelets usually 2 in each bur 2. *C. pauciflorus*

1. Cenchrus echinatus L. Maricopa, Cochise, Pima, and Yuma counties, a troublesome weed in irrigated land and cultivated fields. Open ground and waste places, South Carolina to Florida and Arizona; Mexico and the West Indies to Argentina.

2. Cenchrus pauciflorus Benth. Navajo and Coconino counties, south to Cochise, Santa Cruz, and Pima counties, up to 5,500 feet, open, sandy ground, July to September. Maine and Ontario to Washington, south through Mexico and the West Indies to Argentina.

82. IMPERATA. Satin-tail

Slender, erect perennials from hard, scaly rhizomes, with linear blades narrowed toward base to the thickened midrib, and narrow, terminal, silky panicles; spikelets all alike, paired, awnless, unequally pedicellate on a continuous rachis, surrounded by long, silky hairs; glumes about equal, membranaceous; sterile lemma, fertile lemma, and palea thin and hyaline.

1. Imperata brevifolia Vasey (*I. Hookeri* Rupr. ex Hack.). Coconino, Mohave, Yavapai, Santa Cruz, and Pima counties, up to 4,500 feet, rocky canyons, May to August. Western Texas to Utah, Nevada, Arizona, California, and northern Mexico.

83. ERIANTHUS. Plume Grass

Coarse perennials with long, narrow blades and large, silvery panicles; spikelets all alike, in pairs along a slender axis, one sessile, the other pedicellate; rachis disarticulating below the spikelets; glumes coriaceous, with long, silky hairs at base; sterile and fertile lemma hyaline, the latter with a short, slender awn.

1. **Erianthus ravennae** (L.) Beauv. Sparingly naturalized along the Arizona Canal, Maricopa County (*Harrison* 5835). Introduced from Europe.

Occasionally cultivated for ornament under the name ravenna grass.

84. ANDROPOGON. BLUE-STEM, BROOM-SEDGE

Perennials, with flat or folded blades and few to numerous racemes, these solitary, paired, digitate, or several to numerous and approximate on a short or somewhat elongate axis; sessile spikelet perfect, the pedicellate one staminate and similar to the sessile one, or sterile and much reduced; glumes of the fertile spikelet coriaceous, the first rounded, flat, or concave on the back, several-nerved; sterile lemma shorter than the glumes, empty, hyaline; fertile lemma hyaline, narrow, entire or bifid, usually bearing from the tip a bent and twisted awn.

The blue-stems, sometimes known as beard grass, furnish a good deal of the summer forage in the hills and mountains of southern Arizona, but the plants tend to become woody as they approach maturity. Large, coarse-stemmed species with digitate inflorescences (*A. Hallii, A. Gerardi*) are known as turkey-foot.

KEY TO THE SPECIES

1. Racemes few to numerous, approximate on a relatively long axis; culms sparingly branched from the base only (2).
1. Racemes solitary, paired, or digitate; culms branching toward the summit (3).
2. Spikelets 5 to 6 mm. long; panicle short-exserted or partly enclosed in the sheath; racemes relatively few on a short axis; nodes densely bearded........7. *A. barbinodis*
2. Spikelets about 4 mm. long; panicle usually long-exserted; racemes numerous on a relatively long axis; nodes glabrous or appressed-hispid...............8. *A. saccharoides*
3. Racemes solitary on each peduncle (4).
3. Racemes paired or digitate on each peduncle (6).
4. Racemes flexuous; internodes of the rachis relatively slender. Sterile pedicel and spikelet usually spreading..3. *A. scoparius*
4. Racemes straight; internodes of the rachis relatively thick (5).
5. First glume of the sessile spikelet pilose; rachis joints and sterile pedicels pilose 1. *A. hirtiflorus*
5. First glume of the sessile spikelet glabrous; sterile pedicel glabrous or ciliate only near apex .... 2. *A. cirratus*
6. Pedicellate spikelet much reduced, the pedicel not always developed; racemes 2, slender, partly enclosed in the narrow spathe, aggregate in a dense silky inflorescence 6. *A. glomeratus*
6. Pedicellate spikelet staminate, similar to the sessile one; racemes 5 to 10 cm. long, 2 to 6 on each peduncle, long-exserted (7).
7. Rhizomes short or wanting; racemes inconspicuously hairy, pale or tinged with purple; awn of the sessile spikelet 1 to 2 cm. long..........................4. *A. Gerardi*
7. Rhizomes well developed; racemes conspicuously villous with golden hairs; awn of the sessile spikelet rarely more than 0.5 cm. long........................5. *A. Hallii*

1. **Andropogon hirtiflorus** (Nees) Kunth. Coconino, Graham, Cochise, Santa Cruz, and Pima counties, 4,000 to 7,500 feet, canyons and rocky slopes, June to October. Western Texas to Arizona and Mexico.

The species is represented in Arizona by var. *feensis* (Fourn.) Hack.

2. **Andropogon cirratus** Hack. Coconino, Yavapai, Greenlee, Gila, Pinal, Cochise, Santa Cruz, and Pima counties, 2,000 to 7,500 feet, canyons and rocky slopes, August to October. Western Texas to Arizona, southern California, and northern Mexico.

3. **Andropogon scoparius** Michx. Navajo, Coconino, and Pima counties, at medium altitudes, dry plains and rocky hills, July to September. Quebec and Maine to Alberta and Idaho, south to Florida and Arizona.

Little blue-stem. Var. *neomexicanus* (Nash) Hitchc. is a variant with nearly straight racemes, the rachis and pedicels conspicuously villous. It is found in Apache, Navajo, Coconino, Yavapai, Cochise, Santa Cruz, and Pima counties.

4. **Andropogon Gerardi** Vitm. (*A. furcatus* Muhl.). White Mountain region (Apache County), Lakeside (Navajo County), Walnut Canyon (Coconino County), White River Crossing (Gila County), 4,000 to 7,500 feet, dry, rocky hills, prairies, and open woods. Maine and Quebec to Saskatchewan and Montana, south to Mexico.

5. **Andropogon Hallii** Hack. Apache, Navajo, Coconino, and Cochise counties, at medium altitudes, sandy plains, dry hills, and open pine woods, July to September. North Dakota and eastern Montana to Texas, Wyoming, Utah, and Arizona.

6. **Andropogon glomeratus** (Walt.) B.S.P. Coconino, Mohave, Maricopa, Pinal, and Santa Cruz counties, up to 4,500 feet, moist ground, July to September. Massachusetts to Florida, west to Kentucky, Arizona, and southern California; Mexico and West Indies to Panama.

7. **Andropogon barbinodis** Lag. (*A. saccharoides* var. *barbinodis* Hack.). Navajo and Coconino counties, south to Greenlee, Graham, Cochise, Santa Cruz, and Pima counties, 1,000 to 6,000 feet, open, sandy or gravelly ground and rocky slopes, May to October. Oklahoma and Texas to California and Mexico.

8. **Andropogon saccharoides** Swartz. Navajo and Coconino counties, south to Santa Cruz, Pima, and Yuma (?) counties, 2,000 to 5,500 feet, prairies and rocky slopes, June to October. Missouri to Colorado, Alabama, Arizona, and southern California; Mexico and the West Indies to Argentina. Silver blue-stem.

85. SORGHUM

Annual or perennial grasses with flat blades and open panicles of short, few-flowered racemes; sessile spikelet ovate, with a twisted, geniculate awn, the glumes indurate; pedicellate spikelet lanceolate, awnless, the glumes membranaceous.

Johnson grass (*S. halepense*), because of the large size of the plant, rank growth, and difficulty of eradication, is a very costly pest in the irrigated valleys of southern Arizona, where it thrives to the detriment of all summer field crops. The seeds were formerly eaten by the Pima Indians. The plant produces much pollen, to which many persons are allergic. Many varieties of *S. vulgare* Pers. are cultivated for grain, ensilage, or sirup. Sudan grass (*S. sudanense*) yields large quantities of forage on irrigated lands. Under certain conditions the sorghums (including Johnson grass) cause prussic-acid poisoning in livestock.

KEY TO THE SPECIES

1. Plants perennial, the culms from extensively creeping, scaly rhizomes.....1. *S. halepense*
1. Plants annual..2. *S. sudanense*

1. Sorghum halepense (L.) Pers. Throughout the state, up to 6,000 feet, a weed in waste places, fields, and along irrigation ditches, flowering April to November. Massachusetts to Wyoming, south to Florida, Texas, Arizona, and southern California; Mexico and the West Indies to Argentina and Chile; naturalized from the Old World.

2. Sorghum sudanense (Piper) Stapf (*S. vulgare* var. *sudanense* (Piper) Hitchc.). Sudan grass has been found growing spontaneously at the United States Field Station, Sacaton (Pinal County), and is reported by Clover to have been introduced by the United States Indian Service in Havasu Canyon (Coconino County), but does not appear to have become naturalized in Arizona. It is readily distinguished from the cultivated grain sorghums by its slender culms 1 to 2 meters high, narrow blades, and more open panicle.

86. SORGHASTRUM. Indian Grass

A rather tall, tufted, rhizomatous perennial, with flat, narrow blades and narrow, rather dense terminal panicles of 1- to 3-jointed racemes; sessile spikelet perfect, the glumes coriaceous, brownish, the first glume sparsely hirsute, the fertile lemma hyaline, extending into a tightly twisted, once-geniculate awn; pedicellate spikelet wanting, only the hairy pedicel present.

1. Sorghastrum nutans (L.) Nash. Apache, Navajo, Coconino, and Cochise counties, at medium altitudes, dry slopes and canyons, August to October. Quebec and Maine to Manitoba and North Dakota, south to Florida, Arizona, and Mexico.

This grass affords fairly good forage.

87. HETEROPOGON. Tangle-head

Annuals or perennials, with flat blades and solitary terminal racemes; lower few pairs of spikelets alike, staminate, awnless, the remaining sessile spikelets fertile, long-awned, the pedicellate spikelets staminate like the lower ones; rachis continuous below, bearing the fertile spikelets above, disarticulating at base of each joint, the joint forming a sharp, barbed callus below the fertile spikelet; glumes of the fertile spikelet dark brown, coriaceous, the first glume enclosing the second one; glumes of the staminate spikelet membranaceous, broad, obscuring the fertile spikelets; lemmas hyaline, the fertile one with a long, stout, twisted, geniculate awn.

The mature fruits of tangle-head are injurious to sheep, but in the young stage the plants make good forage. Sweet tangle-head (*H. melanocarpus*) owes its common name to the fragrance of the fresh foliage.

KEY TO THE SPECIES

1. Plants annual; culms usually more than 1 meter high; first glume of the staminate spikelet with a row of conspicuous glands on the back..............1. *H. melanocarpus*
1. Plants perennial; culms slender, usually less than 1 meter high; first glume of the staminate spikelet glandless, usually sparsely papillose-hispid toward apex..2. *H. contortus*

1. Heteropogon melanocarpus (Ell.) Benth. Cochise, Santa Cruz, and Pima counties, 3,000 to 5,500 feet, desert grassland, stream banks, and oak woods. Georgia, Florida, Alabama, and Arizona; tropical and semitropical regions of both hemispheres.

2. Heteropogon contortus (L.) Beauv. Mohave, Yavapai, Pinal, Cochise, Santa Cruz, Pima, and Yuma counties, also reported from the Grand Canyon, 1,000 to 5,500 feet, sandy plains and rocky slopes and canyons, September and October, occasionally January to April. Texas to Arizona; warmer regions of both hemispheres.

88. TRACHYPOGON. Crinkle-awn

Densely tufted perennial with flat or subinvolute blades and solitary terminal racemes; spikelets in pairs on a continuous rachis, the sessile one staminate, awnless, persistent, the pedicellate spikelet perfect, with a long, geniculate, villous awn, its pedicel obliquely disarticulating, forming a sharp-pointed callus below the spikelet; first glume firm, rounded on the back, obtuse.

1. Trachypogon secundus (Presl) Scribn. Mohave, Graham, Gila, Pinal, Cochise, Santa Cruz, and Pima counties, 1,000 to 6,000 feet, rocky hills, September and October. Southern Texas, New Mexico, and Arizona, south to Panama.

T. Montufari (H.B.K.) Nees, to which the Arizona specimens were referred previously, is a distinct species that does not occur in the United States.

89. ELYONURUS

Rather slender, erect perennial with involute blades and solitary, erect, woolly racemes; spikelets in pairs at each joint of the tardily disarticulating rachis, the sessile spikelet perfect, the pedicellate one staminate, similar to the sessile one; first glume firm, flattened, acute, the margins enclosing the second glume; lemmas thin, hyaline, awnless.

1. Elyonurus barbiculmis Hack. (*E. barbiculmis parviflorus* Scribn.). Mohave, Cochise, Santa Cruz, and Pima counties, mostly 4,000 to 6,000 feet, rocky slopes and canyons, July to October, type from Arizona (*Lemmon* 2926), type of var. *parviflorus* also from Arizona (*Griffiths* 1849). Western Texas to Arizona and northern Mexico.

This grass, where sufficiently abundant, affords good grazing.

90. HACKELOCHLOA

Erect or spreading, freely branching annual with short, solitary, exserted or partly included racemes; spikelets in pairs, awnless, very dissimilar, the sessile one perfect, the pedicellate one staminate, conspicuous; first glume of the sessile spikelet broad, alveolate, the margins clasping the fused rachis joint and pedicel of the staminate spikelet.

1. Hackelochloa granularis (L.) Kuntze. Cochise, Santa Cruz, and Pima counties, sandy plains and waste places, August to October. Georgia and Florida to Louisiana, New Mexico, and Arizona; tropics of both hemispheres, introduced in America.

91. TRIPSACUM (46). Gama Grass

Robust perennials with broad, flat blades and monoecious terminal and axillary inflorescences, the pistillate part below, the staminate part above on the same rachis; staminate spikelets 2-flowered, in pairs, one sessile, the other sessile or pedicellate; pistillate spikelets solitary on opposite sides at each joint of the thick, hard, articulate lower part of the rachis, sunken in hollows, consisting of 1 perfect floret and a sterile lemma.

This genus is of interest mainly on account of its relationship to Indian-corn or maize.

1. **Tripsacum lanceolatum** Rupr. (*T. Lemmoni* Vasey). Mule Mountains and Huachuca Mountains (Cochise County), 15 miles from Patagonia and in Bear Valley or Sycamore Canyon (Santa Cruz County), 3,500 to 6,000 feet, rocky hills and canyon bottoms, September and October, type of *T. Lemmoni* from the Huachuca Mountains (*Lemmon* 2932). Southern Arizona to Guatemala.

19. CYPERACEAE. Sedge Family

Plants herbaceous, grasslike or rushlike; flowering stems round or trigonous (seldom flat), usually solid, from fibrous roots or from rootstocks, these sometimes bulblike or bearing tubers; leaves with narrow, sometimes terete blades, the sheaths closed, the blades sometimes wanting; flowers subtended by 2-ranked or spirally imbricate scales or glumes (some of these often empty), arranged in spikelets (these in *Carex* often called spikes); perianth rudimentary or wanting; stamens 1 to 3; ovary 1-celled; ovule 1; fruit an achene.

A family of large size but very little economic importance, the commonly tough and wiry herbage giving the plants small forage value as compared with that of the grasses. The fact that most of the gregarious species occur in marshes makes their value as soil binders negligible.

KEY TO THE GENERA

1. Achene enclosed in a perigynium (a saclike, often beaked envelope with an apical opening through which the styles protrude); flowers none of them perfect, the staminate and pistillate ones in separate spikes or at opposite ends of the same spike . . 8. *Carex*
1. Achene not enclosed in a perigynium; flowers all perfect or if some of them imperfect, then the staminate and pistillate ones not sharply segregated as above (2).
2. Spikelets with scales in 2 rows, the spikelets often strongly flattened. Empty basal scales not more than 2; perianth bristles none . . . . 2. *Cyperus*
2. Spikelets with scales mostly spirally imbricate around the axis in several rows, the spikelets terete or not strongly flattened (3).
3. Base of the style noticeably enlarged (4).
3. Base of the style not noticeably enlarged (6).
4. Thickened base deciduous with the rest of the style. Spikelets usually several, in capitate or umbellate clusters; perianth bristles none; achenes biconvex or lenticular 5. *Fimbristylis*
4. Thickened base persistent after the rest of the style has fallen (5).
5. Stems leafless or their leaves reduced to sheaths; spikelet solitary; perianth bristles usually present; style base conic or bulbous, often large; achenes biconvex to trigonous 4. *Eleocharis*
5. Stems usually bearing 1 or more leaves with filiform blades; spikelets solitary or several in cymose clusters; perianth bristles none; style base apiculate, small; achenes trigonous . . . . 6. *Bulbostylis*

6. Stamen 1 (rarely 2); perianth represented by a hyaline scale, this often bifid, sometimes wanting. Achene somewhat obovoid; plant dwarf, with filiform stems and leaves; spikelets not more than 3 mm. long .... 1. *Hemicarpha*
6. Stamens 2 or 3; perianth, if any, represented by 1 or more bristles (7).
7. Flowers all perfect; spikelets with not more than 2 empty basal scales; perianth bristles usually present; achene trigonous, lenticular, or plano-convex; leaf margins not spinulose .... 3. *Scirpus*
7. Flowers often partly staminate, only the one or two uppermost ones perfect; spikelets with 3 or more empty basal scales; perianth bristles none; achene ovoid or globose; leaf margins spinulose-serrate .... 7. *Cladium*

1. HEMICARPHA (47)

Plants annual, dwarf, with tufted, slender stems; spikelets terminal, solitary or in clusters of 2 to 4, not more than 3 mm. long, terete, subtended by 1 to 3 leaflike bracts; flowers all perfect, each subtended by a mucronate glume, the glumes spirally imbricate; perianth usually present, consisting of a hyaline scale; stamen 1 (rarely 2).

1. Hemicarpha micrantha (Vahl) Pax. Cochise, Santa Cruz, and Pima counties, 2,500 to 4,000 feet, in wet sand along streams. Maine to Washington, southward to South America.

The species is represented in Arizona mainly by var. *minor* (Schrad.) Friedland, characterized by having the mucro shorter than the body of the glume and the perianth scale shorter than the achene and usually more or less bifid. Specimens from Chiricahua National Monument, Cochise County (*Clark* 8599) and from Camp Lowell, Pima County (*Thornber* 4284), with the glume exceptionally long-mucronate, seem referable to var. *aristulata* Coville. Another collection at Camp Lowell, however (*Rothrock* 715), was identified by Friedland as var. *Drummondii* (Nees) Friedland (*H. Drummondii* Nees), described as having the perianth scale equaling or exceeding the achene and not deeply excised. This variety has been collected also in the Huachuca Mountains (*Goodding* 614–49).

2. CYPERUS (48, 49, 50, 51). Flat-sedge

Plants annual or perennial; stems trigonous; leaves mostly basal, with 1 or more apical leaves forming an involucre subtending the inflorescence; spikelets in simple or compound umbels or heads, often flat, with the scales in 2 ranks; stamens 1 to 3; achene lenticular or trigonous, not subtended by bristles.

Most of the species inhabit marshes and banks of streams, but a few occur in pine woods. They flower in summer. Nut-grass (*C. rotundus*) and chufa (*C. esculentus*) are troublesome weeds in irrigated land in southern Arizona and are difficult to eradicate because of the nutlike underground tubers by which they propagate. These tubers are greatly relished by pigs and are used as human food in some parts of the Old World.

KEY TO THE SPECIES

1. Style branches and (or) stigmas 2; achenes lenticular (2).
1. Style branches and (or) stigmas 3; achenes trigonous (4)
2. Spikelets turgid; achenes compressed dorsally, with a side facing the rachilla; flowering stems from a horizontal, creeping rootstock. Inflorescences small, dense, sub-

capitate, appearing lateral, the longest bract of the involucre erect and resembling a continuation of the stem; scales (glumes) closely imbricate, whitish, often blotched with reddish brown . . . . 1. *C. laevigatus*

2. Spikelets flat; achenes compressed laterally, with an edge facing the rachilla; rootstock none or if present, then usually shorter and less horizontal (3).

3. Inflorescence open, the rays up to 20 cm. long and often branched; scales loosely imbricate, with broad, white, hyaline margins, erose at apex; achenes obovoid, black, nearly equaling to longer than the scale . . . . 2. *C. albomarginatus*

3. Inflorescence subcapitate, the rays less than 1 cm. long; scales closely imbricate, not broadly hyaline-margined, light brown to dark brown, entire at apex; achenes ellipsoid, yellowish brown or purplish brown, not more than ⅔ as long as the scale 3. *C. niger*

4. Rachilla of the spikelet disarticulating into several or many 1-fruited joints. Inflorescence usually large and compound, with some of the rays elongate, but sometimes contracted; spikelets several- to many-flowered, terete, very slender, not more than 1 mm. wide; scales persistent on the rachilla joints; stamens 3 . . . . 4. *C. odoratus*

4. Rachilla continuous with the rachis or articulated only at base (5).

5. Stamen 1, rarely 2 (6).

5. Stamens 3, rarely only 2 (10).

6. Scales less than 1 mm. long, rounded or truncate at apex, the lateral nerves indistinct. Plants annual; scales hyaline-margined, 3-nerved, caducous . . . . . 5. *C. difformis*

6. Scales 1 mm. or longer, acuminate to aristate at apex, the lateral nerves prominent (7).

7. Plants often strong-scented when dry; scales 7- to 9-nerved, terminating in a spreading or recurved awn ⅓ or more as long as the body of the scale. Plants annual, reddish brown at base . . . . 6. *C. aristatus*

7. Plants not strong-scented; scales 3-nerved, terminating in a cusp or mucro, or merely acuminate (8).

8. Involucral bracts scarcely, if at all, surpassing the inflorescence; spikelets reddish brown, not crowded, up to 18 mm. long. Plants annual; spikelets digitately radiate 7. *C. amabilis*

8. Involucral bracts much surpassing the inflorescence; spikelets yellowish or pale brown, usually crowded in the head, not more than 10 mm. long (9).

9. Plants perennial, with a tuberiferous rootstock, this sometimes very short; stems much thickened and fibrous-coated at base; leaf sheaths nearly black; involucral bracts more or less reflexed; inflorescence reduced to a single head; achenes broadly obovoid, obtuse and apiculate at apex, becoming nearly black . . . . 8. *C. seslerioides*

9. Plants annual or biennial; rootstock none; stems scarcely thickened or fibrous-coated basally; leaf sheaths often reddish brown or purplish brown at base; involucral bracts erect or ascending; inflorescence with 2 to 5 rays, rarely reduced to a single head; achenes ellipsoid or narrowly obovoid, acutish at apex, brown or purplish brown . . . . 9. *C. acuminatus*

10. Rachilla wings none or if present, then not more than 0.2 mm. wide, flat, and not closely clasping the achene. Plants with short, thick stolons (11).

10. Rachilla wings conspicuously developed (15).

11. Spikelets widespreading or the lower ones reflexed, more or less 2-ranked, rarely crowded; secondary prophyllum originating at apex of the conspicuously swollen basal callosity. Scales 2 to 3 mm. long, about 2 mm. wide, usually reddish brown; achenes 1.7 to 2.2 mm. long . . . . 10. *C. spectabilis*

11. Spikelets ascending, several-ranked, usually crowded; secondary prophyllum originating at base of the callosity (12).

12. Leaves, bracts, and stems (in Arizona specimens) scaberulous or hirtellous; scales 1.5 to 2.5 mm. long, 1 to 1.5 mm. wide, scarcely mucronate, 7- to 9-nerved; achenes 1.2 to 1.7 mm. long, 0.5 to 0.8 mm. wide . . . . 11. *C. manimae*

12. Leaves and bracts glabrous, or scaberulous only on the margins and midnerve; stems smooth or sparsely scaberulous (sometimes copiously so in *C. Schweinitzii*); scales 2 to 3.5 mm. long, 2 to 3 mm. wide, with a distinct mucro or cusp; achenes 1.5 to 2.8 mm. long, 0.8 to 1.2 mm. wide (13).

13. Rays of the inflorescence wanting or nearly so, the 1 to 4 spikes crowded into a dense, sessile, usually 3- or 4-lobed head; spikelets 3- to 8-flowered; mucro of the scales spreading, often curved. Scales 9- to 11-nerved, the mucro up to 0.5 mm. long; achenes broadly obovoid..12. *C. Fendlerianus*
13. Rays well developed; spikelets commonly 8- to 16-flowered; mucro erect or nearly so, straight (14).
14. Stems usually smooth; scales 9- to 11-nerved, the mucro not more than 0.3 mm. long; achenes broadly obovoid, 1.5 to 1.9 mm. long, subtruncate at apex, cuneate at base 13. *C. Rusbyi*
14. Stems usually scaberulous; scales 11- to 15-nerved, the mucro of the upper scales up to 1.0 mm. long; achenes ellipsoid or narrowly obovoid, 2.0 to 2.8 mm. long, acutish at each end..14. *C. Schweinitzii*
15. Achenes not more than ⅗ as long as the scales; spikelets usually more than 5-flowered. Rachilla wings hyaline, scalelike (16).
15. Achenes half as long as to nearly equaling the scales; spikelets 1- to 5-flowered or sometimes (in *C. hermaphroditus*) 7-flowered. Rootstocks or stolons, if present, thick, more or less woody, not more than 3 cm. long (19).
16. Plants annual or biennial, without rootstocks or stolons; rachilla wings caducous (17).
16. Plants perennial, with elongate stolons, these often tuberous-thickened at the distal end; rachilla wings persistent (18).
17. Spikes elongate, the rachis longer than the spikelets, the latter not more than 10 mm. long; scales not more than 1.5 mm. long, 3- to 5-nerved, the edges gold- or copper-colored at maturity; inflorescence usually ample, with elongate rays, but sometimes contracted .... 15. *C. erythrorhizos*
17. Spikes short, the rachis shorter than the spikelets, the latter up to 20 mm. long; scales 2 to 3.5 mm. long, 7- to 11-nerved, the edges red or mahogany-colored; inflorescence relatively small, open or contracted..............................16. *C. Parishii*
18. Stolons numerous, very slender, herbaceous; scales golden brown, prominently 7- to 9-nerved .... 17. *C. esculentus*
18. Stolons usually few, relatively thick, becoming somewhat woody; scales reddish brown to nearly black, faintly to rather prominently 5- to 7-nerved........18. *C. rotundus*
19. Spikelets 3 to 4 mm. wide; wings of the rachilla thickish, not scalelike, not clasping the achene; scales cuspidate, the cusp up to 1 mm. long, often excurved. Scales conspicuously spreading, straw-colored, the margins whitish...........19. *C. Wrightii*
19. Spikelets not more than 2 mm. wide; wings of the rachilla very thin (hyaline), scalelike, clasping the achene; scales muticous or short-mucronate, or (in *C. flavus*) sometimes cuspidate (20).
20. Leaves and bracts more or less scabrous; spikelets turgid, the scales more or less inflated; inflorescence open, the longest rays 6 to 20 cm. long; achenes oblong or ellipsoid. Stems scabrous to nearly smooth; spikelets mostly spreading or deflexed. 20. *C. Mutisii*
20. Leaves and bracts glabrous or obscurely scabrous on the margins and midnerve; spikelets compressed, the scales scarcely inflated, or if the spikelets turgid and the scales inflated (in *C. flavus*), then the inflorescence contracted, with the longest ray not more than 5 cm. long and the achenes obovoid (21).
21. Scales with nerves crowded toward the keel, the nerveless marginal portion broad; leaves not more than 3 mm. wide; spikelets conspicuously subulate at apex, the sterile terminal scale strongly involute. Rays usually well developed; spikelets 1- to 5-flowered; fertile scales and rachilla wings tightly clasping the achenes....21. *C. uniflorus*
21. Scales with nerves rather evenly spaced, the nerveless marginal portion narrow; leaves up to 6 or 7 mm. wide; spikelets obtuse to acuminate (exceptionally subulate) at apex (22).
22. Scales about 1 mm. wide; spikes 10 to 15 mm. wide; spikelets 3- to 7-flowered. 22. *C. hermaphroditus*
22. Scales 1.5 to 2 mm. wide; spikes 4 to 10 (exceptionally 15) mm. wide; spikelets 1- to 5-flowered (23).
23. Rays wanting or not more than 5 cm. long; spikelets 1- to 3-flowered; scales usually strongly inflated; achenes obovoid..................................23. *C. flavus*

23. Rays well developed, the longest 6 to 12 cm. long; spikelets 3- to 5-flowered; scales not strongly inflated; achenes ellipsoid . . . . 24. *C. Pringlei*

1. Cyperus laevigatus L. Mouth of Williams River (Mohave County), San Bernardino Ranch (Cochise County), Quitobaquito (Pima County), Alamo and near Planet (northern Yuma County), 400 to 4,000 feet, moist, sandy flats. Texas to California, southward to Argentina; Eastern Hemisphere.

2. Cyperus albomarginatus Mart. & Schrad. Huachuca Mountains and Canille (Canelo) Hills (Cochise County), western Pima County, near Ruby (Santa Cruz County) 2,500 to 4,500 feet, rocky slopes and along washes. Virginia to South Carolina (Florida?), westward to southern Arizona and southward to Paraguay.

3. Cyperus niger Ruiz & Pavon. Apache, Navajo, Yavapai, Greenlee, Gila, Cochise, Santa Cruz, and Pima counties, 3,500 to 7,000 feet, in wet soil. Texas to California, southward to South America.

The species is represented in Arizona (and in the United States) by var. *capitatus* (Britton) O'Neill (var. *castaneus* (Wats.) Kükenth.), which differs from the typical and other forms of the species in its relatively light-colored (chestnut-brown) glumes and strongly apiculate achenes.

4. Cyperus odoratus L. (*C. ferax* L. C. Rich.). Mohave, Yavapai, and Gila counties to Cochise, Pima, and Yuma counties, 4,500 feet or lower, common along streams and ditches. Massachusetts to Florida, westward to California and southward to South America; Eastern Hemisphere.

The seeds of *C. odoratus*, as reported by E. Palmer, were eaten by the Cocopah Indians.

5. Cyperus difformis L. (*C. lateriflorus* Torr.). Near Topock, Mohave County (*Monson* in 1948), also "east of Santa Cruz, Sonora," about 10 miles south of the Arizona boundary (*Wright* 1950, the type of *C. lateriflorus*). New Mexico, Arizona, California, and Mexico; tropics of the Eastern Hemisphere.

6. Cyperus aristatus Rottb. (*C. inflexus* Muhl.). Navajo and Coconino counties, south to Cochise, Santa Cruz, and Pima counties, 2,500 to 7,500 feet. Throughout most of North America; South America; tropics of the Eastern Hemisphere.

The plant when dry has a strong odor of the bark of slippery elm (*Ulmus fulva*).

7. Cyperus amabilis Vahl. Sonoita Valley and foothills of the Patagonia Mountains, Santa Cruz County, 4,000 to 6,000 feet (*Rothrock* 599, *Harrison & Peebles* 4733). Southern Arizona to South America.

The Arizona plant is var. *macrostachyus* Kükenth., characterized by having the scales cuspidate, whereas in typical *C. amabilis* they are merely short-mucronate.

8. Cyperus seslerioides H.B.K. Chiricahua and Huachuca mountains (Cochise County), Sonoita Valley (Santa Cruz County), about 6,500 feet. Texas, southern Arizona, and Mexico.

9. Cyperus acuminatus Torr. & Hook. (*C. cyrtolepis* Torr. & Hook.). Coconino (?), Graham, Gila, Cochise, and Pima counties, 5,000 feet or lower. Illinois to Georgia and Louisiana, westward to Washington, Oregon, and California.

10. Cyperus spectabilis Link. Sierra Ancha, Gila County (*Gould & Hudson* 3745), Santa Rita Mountains, Pima or Santa Cruz County (*Peebles et al.* 4543), 4,000 to 5,000 feet. Western Texas, southern Arizona, and Mexico; Peru.

11. Cyperus manimae H.B.K. Huachuca Mountains, Cochise County, moist, rocky slopes in oak forest (*Gentry* 3358), Sonoita Valley, Santa Cruz County (*Rothrock* 600, 601), 4,500 to 6,000 feet. Southern Arizona to Costa Rica.

The Arizona plants are var. *asperrimus* (Liebm.) Kükenth., which differs from typical *C. manimae* only in having the stems scabrous or densely hirtellous at apex (stems glabrous in the species).

12. Cyperus Fendlerianus Boeckl. Apache County to Coconino County, southward to Cochise, Santa Cruz, and Pima counties, 4,500 to 9,500 feet. Colorado and western Texas to Nevada, Arizona, and northern Mexico.

Common in pine forests. It is said that the underground parts were eaten by the Apache Indians.

13. Cyperus Rusbyi Britton (*C. Fendlerianus* var. *debilis* (Britton) Kükenth.). Graham, Gila, Pinal, Cochise, Santa Cruz, and Pima counties, 4,000 to 6,500 feet. Western Texas to Arizona and Mexico.

Grows on dry slopes, usually with grasses, commonly at lower elevations than *C. Fendlerianus.*

***14. Cyperus Schweinitzii** Torr. Not known definitely to occur in Arizona, but the species has been collected at Kanab, Utah, just north of the Arizona boundary, and is to be looked for in the Kaibab region. Southeastern Canada to Idaho, southward to New Jersey, Texas, New Mexico, and Utah.

15. Cyperus erythrorhizos Muhl. Separation Canyon, Mohave County (*Clover* 4236), Picacho Lake, Pinal County, 1,500 feet (*Taylor* in 1934), Yuma, Yuma County (*Fochtman* 28). New Hampshire to Florida, west to Washington, Oregon, and California.

16. Cyperus Parishii Britton. Near Prescott (Yavapai County), near Phoenix (Maricopa County), Sulphur Springs Valley (Cochise County), 1,000 to 5,500 feet. New Mexico to southern California.

17. Cyperus esculentus L. Apache, Navajo, and Coconino counties to Cochise, Santa Cruz, and Pima counties (probably also Yuma County), up to 6,000 feet. Widely distributed in North America; cosmopolitan.

Chufa, yellow nut-grass. Common in wet soil, often a weed in cultivated fields and pastures. The tubers of this and the next species are relished by hogs, which often are employed to suppress these plants.

In addition to the more common, typical *C. esculentus,* with spikelets 0.5 to 1.5 cm. long, there occur in Arizona var. *angustispicatus* Britton and var. *macrostachyus* Boeckl., both with spikelets 2 to 3 cm. long. In the former they are less than 2 mm. wide, tapering to slender points, and in the latter 2 to 3 mm. wide, uniformly linear and rounded at apex.

18. Cyperus rotundus L. Salt and Gila River valleys and at Tucson (Pinal, Maricopa, Pima, and Yuma counties). Widely distributed in the warmer parts of North America; probably of Old World origin.

Nut-grass. A common and troublesome weed in cultivated land and pastures.

19. Cyperus Wrightii Britton. Mountains of Cochise, Pima, and Santa Cruz counties, 3,500 to 5,000 feet, rocky slopes and rich soil in wooded canyons. Southern New Mexico, southern Arizona, and northern Mexico.

A peculiar form, with exceptionally loose spikes and exceptionally long spikelets, was collected between Nogales and Ruby, Santa Cruz County (*Kearney & Peebles* 14906).

20. Cyperus Mutisii (H.B.K.) Griseb. Pinaleno Mountains (Graham County), Huachuca Mountains (Cochise County), Santa Rita, Santa Catalina, and Baboquivari mountains (Pima County), near Nogales (Santa Cruz County), 4,000 to 7,000 feet. Southern Arizona to Central America; Jamaica.

Var. *asper* (Liebm.) Kükenth. (*C. asper* O'Neill) also has been collected in the Baboquivari Mountains (*Harrison* 9053). It differs from typical *C. Mutisii* in the more scabrous stems, leaves, and bracts, more slender stems, often narrower leaves, usually shorter rays, and commonly unbranched spikes, whereas in typical *C. Mutisii* the spikes usually have 1 to several lateral branches.

21. Cyperus uniflorus Torr. & Hook. Beaver Creek, Yavapai County (*Purpus* 8294). Arkansas, Oklahoma, Texas, New Mexico, Arizona, and northern Mexico.

22. Cyperus hermaphroditus (Jacq.) Standl. A specimen from the Baboquivari Mountains, Pima County (*Jones*), was referred to this species by Horvat (50, p. 48). Northern Mexico to Central America; South America.

23. Cyperus flavus (Vahl) Nees (*C. subambiguus* Kükenth.). Near Globe and Miami (Gila County), mountains of Cochise, Pima, and Santa Cruz counties, 4,000 to 6,000 feet. Southern Arizona to South America; West Indies.

24. Cyperus Pringlei Britton. Rincon, Santa Catalina, and Baboquivari mountains (Pima County), about 4,000 feet, type from the Santa Catalina Mountains (*Pringle* in 1884). Southern Arizona and northern Mexico.

The roots of this and the preceding species are strongly aromatic, with an odor of camphor.

3. SCIRPUS (52). BULRUSH

Plants (in Arizona) perennial, with terete or trigonous, mostly leafy, flowering stems from creeping rootstocks; spikelets solitary or in heads or compound panicles; scales spirally imbricate, all fertile or the lowest scales empty; flowers perfect; stamens 3; achene plano-convex to trigonous, usually subtended by 1 or more bristles representing the perianth.

Plants of marshes, stagnant ponds, stream banks, and ditches. The tough stems of an Old World species (*S. lacustris* L.) are used for making mats and cane chair seats. The bases of the stalks of one species (*S. acutus* ?) are eaten raw by the Hopi Indians.

KEY TO THE SPECIES

1. Stems leafy, 3-edged; inflorescence subtended by 2 or more ascending or spreading, long, leafy bracts (2).
1. Stems not leafy; inflorescence subtended by a single erect bract, this appearing like a continuation of the stem. Achene lenticular or plano-convex, rarely trigonous; style normally bifid (4).

2. Stolons often bearing thick, woody tubers; inflorescence compact or dense; spikelets few to many, 10 to 25 (35) mm. long. Achene lenticular or plano-convex; style bifid 1. *S. paludosus*
2. Stolons not tuberiferous; inflorescence usually loose, umbelliform; spikelets many, seldom more than 5 mm. long (3).
3. Inflorescence usually only once compound, the clusters of spikelets commonly dense; achene obtusely trigonous; style trifid................................2. *S. pallidus*
3. Inflorescence usually 2 or more times compound, the clusters loose; achene lenticular or plano-convex; style bifid.....................................3. *S. microcarpus*
4. Stems 3-edged, seldom more than 1 m. long; spikelets few (rarely more than 15), sessile (5).
4. Stems mostly terete, up to 5 m. long, stout; spikelets several to many, mostly in pedunculate glomerules (6).
5. Sides of the relatively stout stems somewhat concave; involucral bract thick, rigid; spikelets commonly obtuse; scales mucronate or short-cuspidate..........4. *S. Olneyi*
5. Sides of the slender, wiry stems flat; involucral bract slender, more or less flexuous; spikelets commonly acute; scales long-cuspidate or aristate.........5. *S. americanus*
6. Branchlets of the inflorescence glabrous; stems often 3-edged toward apex; perianth bristles 2 to 4, flattened, hirtellous to short-plumose.................6. *S. californicus*
6. Branchlets scabrous-puberulent; stems round; bristles 4 to 6, terete or nearly so, scabrous or hirtellous (7).
7. Scales equaling or little longer than the mature achene; spikelets ovoid, dark reddish brown; roots fibrous...7. *S. validus*
7. Scales longer than the achene; spikelets subcylindric, usually pale brown or reddish; roots swollen, spongy...8. *S. acutus*

1. **Scirpus paludosus** A. Nels. Apache, Navajo, and Mohave counties to Pinal and Maricopa counties, 1,000 to 5,000 feet. Widely distributed in the United States and Canada; Argentina; Hawaiian Islands.

2. **Scirpus pallidus** (Britton) Fern. White Mountains, Apache County (*Griffiths* 5422). Minnesota to Texas, westward to the Pacific Coast.

3. **Scirpus microcarpus** Presl. Northeastern Apache, southern Coconino, Yavapai, Gila, Cochise, and Pima counties, 4,000 to 7,500 feet. Alaska to New Mexico, Arizona, and California.

4. **Scirpus Olneyi** Gray. Apache and Coconino counties southward to Cochise and Pima counties, 1,000 to 7,000 feet. Widely distributed in North America.

S. Olneyi is perhaps not specifically distinct from the South American *S. chilensis* Nees & Meyen, to which it was referred by Beetle (52v.30, p. 397).

5. **Scirpus americanus** Pers. Apache County to Mohave County, south to Cochise and Maricopa counties, 1,000 to 6,000 feet. Widely distributed in North America; South America.

This wiry-stemmed plant occurs often in saline marshes. Beetle (52v.30, pp. 398–400) cites no specimens of typical *S. americanus* from Arizona. He cites a collection from Coconino County (*Hanson* 249) of var. *polyphyllus* (Boeckl.) Beetle, distinguished by having relatively short and less erect stems, several blade-bearing sheaths, a trifid style, and somewhat larger anthers. This variety has been found also at Shato (Shonto), Navajo County (*J. T. Howell* 24423). Collections from Navajo, Mohave, and Maricopa counties were referred by Beetle to var. *longispicatus* Britton, which differs from var. *polyphyllus* in having stricter stems, blades borne on only 1 or 2 basal sheaths, and a bifid style.

6. Scirpus californicus (C. Meyer) Steud. Topock and Fort Mohave (Mohave County), near Phoenix (Maricopa County), Arivaca (Pima County), near Yuma (Yuma County), 100 to 3,500 feet. South Carolina and Florida to California, south to Argentina and Chile.

7. Scirpus validus Vahl. Coconino, Yavapai, Gila, Pinal, and Cochise counties, up to 5,000 feet. Almost throughout the United States and Canada; Mexico; South America.

8. Scirpus acutus Muhl. Apache, Navajo, and Coconino counties, south to Cochise and Pima counties, 2,500 to 9,000 feet. Widely distributed in Canada and the United States.

Very similar to *S. validus*. Both species are known in California as tule. According to E. Palmer the Indians of Arizona used the roots of this or a related species as food.

4. ELEOCHARIS (53). SPIKE-RUSH

Contributed by Henry K. Svenson

Plants leafless, glabrous throughout; inflorescence a single spikelet, bractless, with spiral to distichous scales, the lowermost frequently sterile; stamens 3 or 2; style with 3 or 2 branches, fimbriate; achenes trigonous to biconvex, usually with a prominent style base (tubercle), this articulated or constricted at base; surface of the achene smooth to reticulate.

This genus differs from *Fimbristylis* in the glabrous, always single-headed inflorescence; and from single-headed species of *Scirpus* in the fimbriate style, bractless inflorescence, and firm (not spongy) achene surface. The plants grow mostly in wet soil, in marshes, springy places, and along streams.

KEY TO THE SPECIES

1. Achenes biconvex (lenticular); styles normally bifid but frequently trifid in *E. Engelmanni* and *E. montana* (2).
1. Achenes trigonous to nearly terete; styles trifid (5).
2. Plants annual. Style base (tubercle) much less than 1/3 as high as the achene (3).
2. Plants perennial, with creeping rootstocks, often stoloniferous. Style base much narrower than the achene, deltoid to elongate: species difficult to distinguish (4).
3. Spikelets ovate to lanceolate; mature achene brown; style base very low, about as wide as the truncate achene and not differentiated in color........1. *E. Engelmanni*
3. Spikelets globose; mature achene black; style base whitened, usually pointed, less than 1/3 as wide as the achene ....2. *E. caribaea*
4. Culms septate, i. e. with bars extending across the culm, these sometimes very obscure; upper sheaths horizontal at summit, with a conspicuous rigid mucro; achenes pitted-reticulate. Spikelets 1 to 2 cm. long, commonly pale brown, with numerous rigid, acute scales..3. *E. montana*
4. Culms not septate; upper sheaths usually oblique, rarely mucronate; achenes smooth. Empty basal scales of the spikelet 1 to 3; culms soft, compressed; spikelets narrow-cylindric or lanceolate, 1 to 3 cm. long; scales pale brown to dark brown 4. *E. macrostachya*
5. Style base confluent with the apex of the achene. Achene trigonous, green to olivaceous, obovoid to elongate, reticulate or slightly verrucose under high magnification, not with longitudinal ribs and cross bars (6).
5. Style base not confluent with the apex of the achene, usually constricted at base (8).
6. Culms capillary, 2 to 5 (rarely 7) cm. long, few-flowered; achenes 1 to 1.5 mm. long, without bristles. Minute tubers frequently present among the roots: var. *anachaeta* 5. *E. parvula*

6. Culms 1 mm. or more wide, at least 20 cm. long; achenes 2 to 3 mm. long (7).
7. Achenes elongate, strongly reticulate; spikelets not proliferous; tubers frequently present: var. *Suksdorfiana* .... 6. *E. pauciflora*
7. Achenes obovoid, usually with a smooth (almost greasy) surface; spikelets frequently proliferous; tubers none .... 7. *E. rostellata*
8. Achenes not cross-ribbed, trigonous, dark brown when mature; plants perennial, with creeping rootstocks (9).
8. Achenes cross-ribbed between the longitudinal ribs, terete to obscurely trigonous; plants with capillary culms, annual, or perennial with filiform rootstocks. Lowest scale fertile (10).
9. Spikelets linear to narrowly lanceolate, acuminate; scales acute or acutish, appressed, dark brown (frequently mottled with white); achene smooth or nearly so and appearing somewhat greasy under magnification .... 8. *E. Parishii*
9. Spikelets ovoid to oblong, usually obtuse; scales very obtuse, dull brown; achene pitted-reticulate under magnification .... 9. *E. montevidensis*
10. Anthers 0.8 to 1.2 mm. long. Scales with a greenish center and reddish-brown sides; bristles none; plant perennial .... 10. *E. acicularis*
10. Anthers less than 0.5 mm. long (11).
11. Plants perennial, with light-green, spongy culms and filiform, creeping rootstocks; bristles exceeding the achenes, rarely lacking .... 13. *E. radicans*
11. Plants annual, forming tufts (i.e., cespitose); bristles none (12).
12. Achenes elliptic to fusiform; cross bars about 30 in a longitudinal series.... 11. *E. bella*
12. Achenes ovoid; cross bars about 15 in a longitudinal series .... 12. *E. cancellata*

1. Eleocharis Engelmanni Steud. White Mountains (Apache County) and at several localities in Coconino County. Maine to Arizona and California.

Represented in Arizona by var. *monticola* (Fern.) Svenson, characterized chiefly by pale spikelets.

2. Eleocharis caribaea (Rottb.) Blake. Rye Creek near the Mazatzal Mountains, Gila County (*Harrison & Kearney* 8369), Quitobaquito, western Pima County (*Darrow et al.* 2403). Throughout the world, in tropical and subtropical regions.

3. Eleocharis montana (H.B.K.) Roem. & Schult. Pima County, Santa Catalina Mountains (*Pringle* in 1881, *Thornber* 309, etc.), Rincon Mountains (*Peebles & Kearney* 10463). Florida and Arizona to Argentina.

The species is represented in Arizona by var. *nodulosa* (Roth) Svenson (*E. nodulosa* Schult.), which ranges from Florida to southern Arizona and southward into tropical America.

4. Eleocharis macrostachya Britton. Apache County to Coconino County, south to Graham, Pima, and Yuma counties, 150 to 6,500 feet. Missouri to British Columbia, south to Mexico.

The most variable and abundant species in northern and central Arizona.

*__5. Eleocharis parvula__ (Roem. & Schult.) Link. Not definitely known from Arizona, but var. *anachaeta* (Torr.) Svenson, with characters as given in the key, is likely to be found there. This variety ranges from South Dakota to Idaho, southward to Venezuela.

6. Eleocharis pauciflora (Lightf.) Link. Alpine, Apache County, 9,000 feet (*Benson* 9538), Flagstaff, Coconino County (*Hitchcock* in 1915), wet, often calcareous places. North America; northern Eurasia; Andes of South America.

The Arizona specimens belongs to the well-marked var. *Suksdorfiana* (Beauverd) Svenson, which closely simulates *E. rostellata*.

7. Eleocharis rostellata Torr. Grand Canyon, Coconino County (*Wooton* 2012), Fort Huachuca, Cochise County (*Lemmon* 2907), wet, saline places. North and South America; West Indies.

8. Eleocharis Parishii Britton. Widely distributed in Arizona from desert areas to mountain meadows, Apache County to Mohave County, south to Cochise and Pima counties. Oregon to Arizona, California, and northern Mexico.

9. Eleocharis montevidensis Kunth. Huachuca and Chiricahua mountains (Cochise County), near Tucson (Pima County), probably also at Cameron (Coconino County), wet, sandy places. South Carolina to California and Mexico; South America.

10. Eleocharis acicularis (L.) Roem. & Schult. White Mountains (Apache County), and widely distributed in Coconino County, wet soil, 7,000 to 9,000 feet, perhaps also in Maricopa and Pima counties at much lower elevations. North America and Eurasia.

11. Eleocharis bella (Piper) Svenson. Alpine (Apache County), Flagstaff and Mormon Lake (Coconino County), Rucker Valley (Cochise County), 5,500 to 9,000 feet. Montana to Washington, southward to New Mexico and Arizona.

***12. Eleocharis cancellata** Wats. Not definitely known from Arizona, but C. Wright's No. 1937, labeled as from New Mexico, may have been collected in what is now Arizona. New Mexico to central Mexico.

13. Eleocharis radicans (Poir.) Kunth. Grand Canyon, Coconino County (*Tracy* 363), near Camp Verde, Yavapai County (*Goodding* 56–47), San Pedro River, Cochise County (*Toumey* in 1894). Michigan, southern United States, and Arizona; West Indies and South America.

5. FIMBRISTYLIS

Plants annual or perennial; spikelets terete, usually in heads or umbellike inflorescences; scales spirally imbricate; achenes (in the Arizona species) lenticular or biconvex, the apical tubercle deciduous.

KEY TO THE SPECIES

1. Plants perennial; stems wiry, 30 to 50 cm. long. Spikelets in open, simple, or compound umbellike inflorescences; style pubescent..........................1. *F. thermalis*
1. Plants annual or if perennial, then the stems not wiry, less than 20 cm. long (2).
2. Spikelets in small, dense heads subtended by long leaflike bracts; style glabrous or minutely puberulent. Stems densely tufted, not more than 15 cm. long....2. *F. Vahlii*
2. Spikelets few, in loose umbels with unequal rays, or solitary; style pubescent (3).
3. Umbels mostly simple; spikelets commonly more than 5 mm. long; achenes fully 1 mm. long, the ribs noticeably tuberculate..........................3. *F. Baldwiniana*
3. Umbels compound; spikelets less than or barely 5 mm. long; achenes less than to barely 1 mm. long, not or obscurely tuberculate..........................4. *F. alamosana*

1. Fimbristylis thermalis Wats. Grand Canyon, Coconino County, foot of Bright Angel Trail. Nevada, northern Arizona, and California.

2. Fimbristylis Vahlii (Lam.) Link. "Between Tucson and Silver Lake," 2,500 feet (*Blake* in 1903), Yuma, in sand along the Colorado River, "very abundant, forming small dense cushions" (*Thornber* in 1912). North Carolina to Florida, westward to California.

3. **Fimbristylis Baldwiniana** (Schult.) Torr. Bowie, Cochise County (*Jones* 4317), Bear Valley (Sycamore Canyon), Santa Cruz County (*Kearney & Peebles* 14476), September. Pennsylvania to Florida, west to Kansas and southeastern Arizona.

4. **Fimbristylis alamosana** Fern. Bear Valley (Sycamore Canyon), Santa Cruz County (*Goodding* 6633, *Kearney & Peebles* 14475), in moist sand, August and September. Southern Arizona and Sonora.

6. BULBOSTYLIS

Plants annual or perennial; stems and leaves filiform; spikelets terete, in small umbels or heads, or solitary; scales spirally imbricate; achenes trigonous, more or less distinctly rugose transversely, the apical tubercle persistent.

KEY TO THE SPECIES

1. Plants perennial; stems becoming hardened and bulblike at base; achenes 1 to 1.2 mm. long, slate-colored at maturity. Stems up to 40 cm. long; spikelets in a small cluster, or solitary, at the apex of the stems....................1. *B. juncoides*
1. Plants annual; stems not hardened and bulblike; achenes less than or barely 1 mm. long, straw-colored or pale brown at maturity (2).
2. Spikelets usually 2 or more at the apex of the stems, usually none basal; stems up to 25 cm. long....................2. *B. capillaris*
2. Spikelets usually solitary, both apical and at base of the stems among the leaves; stems seldom more than 10 cm. long....................3. *B. Funckii*

1. **Bulbostylis juncoides** (Vahl) Kükenth. Yavapai, Cochise, Santa Cruz, and Pima counties, 3,000 to 6,000 feet, in moist sand. Western Texas to Arizona and Mexico; South America.

The Arizona plant is var. *ampliceps* Kükenth., with more expanded inflorescence than in typical *B. juncoides*. Young plants simulate *B. capillaris*, but the species may be distinguished by the slate color of the mature achenes.

2. **Bulbostylis capillaris** (L.) C. B. Clarke. Yavapai, Greenlee, Gila, Cochise, Santa Cruz, and Pima counties (2,500?) 4,500 to 6,000 feet. Widely distributed in North America.

Several of the Arizona specimens have basal or subbasal spikelets, as in *B. Funckii*, but these are seen occasionally in plants of *B. capillaris* from eastern North America.

3. **Bulbostylis Funckii** (Steud.) C. B. Clarke. Hualpai Mountain (Mohave County), Pinal Mountains (Gila County), Chiricahua and Huachuca mountains (Cochise County), Santa Catalina and Baboquivari mountains (Pima County), 5,000 to 7,500 feet. New Mexico and Arizona, southward to South America; West Indies.

7. CLADIUM. Saw-grass

Plants perennial, tall; stems leafy; spikelets terete, fusiform, in small clusters forming an ample cymose panicle; scales imbricate, only the uppermost ones usually subtending perfect flowers, the lowest scales empty; anthers elongate, aristate-acuminate; achenes obovoid, without an apical tubercle.

1. **Cladium californicum** (Wats.) O'Neill. Bottom of the Grand Canyon (Coconino County), near Columbine (Emory) Falls (Mohave County). Northern Arizona, southern Nevada, and southern California to Central America.

8. CAREX.[1] SEDGE

Contributed by F. J. Hermann

Plants grasslike, perennial, monoecious or sometimes dioecious; culms mostly trigonous; leaves 3-ranked, the upper ones (bracts) elongate, or short, or wanting; spikes 1 to many, sessile or peduncled, either wholly pistillate, wholly staminate, androgynous (staminate flowers above), or gynaecandrous (staminate flowers below); flowers solitary in the axils of the scales; perianth none; stamens 3 (rarely 2); pistil single, the style 1, the stigmas 2 or 3; achene trigonous, lenticular, or plano-convex, completely surrounded by the perigynium.

The members of this genus that occur in Arizona are generally found in the mountains, seldom growing below 5,000 feet except in shady canyons. The southwestern part of the state, comprising southern Mohave, southwestern Yavapai, western Maricopa, and western Pima counties, together with the whole of Yuma County, does not contain a single *Carex* as far as is known. Of the forty-nine species known to occur in the state, several have been collected in only one locality. More intensive collecting of this genus will without doubt extend the known range of these species and will result in the addition of other species to the recorded flora.

KEY TO THE SPECIES

1. Spike one. Stigmas 3; achenes trigonous (2).
1. Spikes more than 1 (4).
2. Plants dioecious; spike entirely staminate or entirely pistillate.......32. *C. curatorum*
2. Plants monoecious; spike androgynous (3).
3. Leaves flattened, canaliculate, 1.5 to 2.5 mm. wide toward base; lowest scale usually conspicuously awned; culms stout, sharply angled, scabrous above..27. *C. oreocharis*
3. Leaves acicular, 0.25 to 0.5 mm. wide toward base; lowest scale rarely awned; culms slender, bluntly angled or subterete, smooth........................28. *C. filifolia*
4. Stigmas prevailingly 3; achenes mostly trigonous (5).
4. Stigmas prevailingly 2; achenes mostly lenticular (18).
5. Perigynia pubescent (6).
5. Perigynia glabrous or (in *C. Meadii*) granular or (in *C. specuicola*) strongly papillose and often sparsely scabrous above (9).
6. Pistillate spikes many- (25 to 75) flowered; achenes with concave sides; tall (40 to 100 cm.) plants of wet habitats....................................35. *C. lanuginosa*
6. Pistillate spikes few- (3 to 10) flowered; achenes with sides convex above; low (5 to 40 cm.) plants of dry habitats (7).
7. Plants with no basal spikes; leaves about half the length of the fertile culms; staminate spike 12 to 30 mm. long....................................29. *C. leucodonta*
7. Plants with basal spikes; leaves usually longer or but little shorter than the fertile culms; staminate spike 3 to 12 mm. long (8).
8. Bracts leaflike; culms not strongly filamentose at base; beak of perigynium 0.75 to 1.5 mm. long..30. *C. Rossii*
8. Bracts squamiform or none; culms strongly filamentose at base; beak of perigynium 0.5 to 0.75 mm. long...31. *C. geophila*
9. Style articulated to the achene, at length withering; spikes (except in *C. Meadii* and *C. specuicola*) blackish (10).

[1] Based upon the treatment by J. W. Stacey in *Flowering Plants and Ferns of Arizona*, pp. 168–175. The artificial key to the Arizona species has been adapted from that of Mackenzie (*North American Flora* 18: 9–478), to whom credit should be given for the most modern treatment of the genus in North America.

9. Style continuous with the achene, not withering; spikes not blackish (13).
10. Lateral spikes linear, the lower ones nodding on long peduncles............36. *C. bella*
10. Lateral spikes linear-oblong to ovoid, the peduncles all erect (11).
11. Scales black, with a conspicuous white-hyaline margin, equaling or shorter than the perigynia..37. *C. albo-nigra*
11. Scales brown or copper-colored (12).
12. Perigynia dark purple, ovate-suborbicular to obovate, 2.3 to 5 mm. wide. Scales exceeding the perigynia..38. *C. chalciolepis*
12. Perigynia yellowish green or brownish, broadly elliptic or obovate, not more than 2.5 mm. wide (12a).
12a. Terminal spike staminate, long-peduncled; perigynia strongly many-nerved 34a. *C. Meadii*
12a. Terminal spike usually gynaecandrous, short-peduncled or sessile; perigynia nerveless or finely few-nerved..39. *C. specuicola*
13. Leaves not septate-nodulose; perigynia nerveless or obscurely several-nerved (14).
13. Leaves septate-nodulose; perigynia strongly ribbed (15).
14. Perigynia squarrose-spreading at maturity, becoming inflated; style abruptly bent 44. *C. spissa*
14. Perigynia appressed-ascending, little or not at all inflated; style straight..45. *C. ultra*
15. Pistillate scales acuminate or short-awned, usually dark-colored and conspicuous in the spikes; ribs of the perigynia coarse (16).
15. Pistillate scales abruptly long- and rough-awned, usually pale and often with only the awns visible in the spikes; ribs of the perigynia fine and close (17).
16. Culms sharply trigonous below the spikes, rough; perigynia appressed or ascending; teeth of perigynia long or the perigynia gradually long-beaked; ligule much longer than wide; lower sheaths fragile, becoming strongly filamentose; rootstocks without long, horizontal stolons..48. *C. vesicaria*
16. Culms bluntly trigonous below the spikes, smooth; perigynia ascending to squarrose at maturity; teeth of perigynia short or the perigynia abruptly short-beaked; ligule slightly if at all longer than wide; lower sheaths not fragile, not becoming filamentose; rootstocks with long, horizontal stolons......................49. *C. rostrata*
17. Perigynia 5 to 7 mm. long, the beak 2 mm. long; pistillate scales with small bodies 46. *C. hystricina*
17. Perigynia 4 to 5 mm. long, the beak 1.5 mm. long; pistillate scales with large bodies 47. *C. Thurberi*
18. Lateral spikes (at least the lower ones) peduncled (19).
18. Lateral spikes sessile, short (25).
19. Terminal spike gynaecandrous..39. *C. specuicola*
19. Terminal spike staminate (20).
20. Perigynia few, pulverulent or golden yellow at maturity (21).
20. Perigynia many, not pulverulent or golden yellow at maturity (22).
21. Perigynia whitish-pulverulent at maturity; pistillate scales appressed....33. *C. Hassei*
21. Perigynia golden yellow or brownish at maturity; pistillate scales widely spreading at maturity..34. *C. aurea*
22. Beak of perigynium prominent, 0.5 to 1 mm. long (23).
22. Beak of perigynium rudimentary, not more than 0.3 mm. long. Perigynia nerveless or few-nerved; pistillate scales usually blunt (never markedly acuminate) the pale midrib not reaching the apex (24).
23. Perigynia 1.5 to 2 (rarely 2.5) mm. long, lightly several-nerved or almost nerveless, the beak entire; scales obtuse or acutish, not awned................40. *C. Kelloggii*
23. Perigynia 3 to 3.5 mm. long, strongly many-ribbed, the beak bidentate; pistillate scales acuminate, the pale midrib usually extended beyond the apex as a short awn 41. *C. nebraskensis*
24. Perigynia nerveless, 2.25 to 2.75 mm. long........................42. *C. aquatilis*
24. Perigynia conspicuously few-nerved, 3 to 3.5 mm. long..................43. *C. senta*
25. Perigynia not wing-margined or thin-edged, elliptic, whitish-puncticulate. Spikes gynaecandrous .... 13. *C. canescens*

25. Perigynia wing-margined or thin-edged, usually broadest below the middle, not whitish-puncticulate (26).
26. Culms arising singly or few together from long-creeping rootstocks. Terminal spike androgynous, or the plants tending to be dioecious (27).
26. Culms from loosely to densely cespitose, the rootstocks sometimes short-prolonged, but not long-creeping (31).
27. Perigynia wing-margined, flattened, plano-convex........................5. *C. siccata*
27. Perigynia not wing-margined, plano-convex or unequally biconvex (28).
28. Culms obtusely trigonous; leaves narrowly involute or canaliculate. Rootstocks slender (29).
28. Culms sharply trigonous; leaves flat or channeled (30).
29. Perigynia with a long (1.75 mm.) beak; plants dioecious or nearly so; heads large (10 to 25 mm. wide)..1. *C. Douglasii*
29. Perigynia with a short (0.5 to 0.75 mm.) beak; plants monoecious, the spikes androgynous; heads relatively small (5 to 10 mm. wide)..................2. *C. Eleocharis*
30. Rootstocks slender; lower sheaths light brownish; perigynia chestnut-colored, unequally biconvex, 1.75 to 2.25 mm. long............................3. *C. simulata*
30. Rootstocks stout; lower sheaths dark brown or black; perigynia blackish in age, plano-convex, 3 to 4 mm. long.......................................4. *C. praegracilis*
31. Spikes androgynous or wholly pistillate, never gynaecandrous (32).
31. Spikes gynaecandrous (38).
32. Perigynia not more, usually less than 3.5 mm. long (33).
32. Perigynia not less, usually more than 3.5 mm. long (36).
33. Beak of the perigynium usually about ⅓ as long as the body; perigynia more or less spongy at base; spikes relatively few (34).
33. Beak of the perigynium as long as or longer than the body; perigynia scarcely spongy at base; spikes many (35).
34. Perigynia 2.5 to 3.5 mm. long, broadest well above the base, abruptly contracted into the beak, with raised margins.................................7. *C. occidentalis*
34. Perigynia about 3.5 mm. long, broadest at base, tapering into the beak, sharp-edged but without raised margins..................................11. *C. chihuahuensis*
35. Scales obtusish to short-awned; perigynia 2.5 to 3.5 mm. long, broadest near the middle, tapering into the beak; leaves 1 to 2 mm. wide...................9. *C. agrostoides*
35. Scales long-awned; perigynia 2 to 2.5 mm. long, broadest below the middle, abruptly contracted into the beak; leaves 2 to 4 mm. wide................10. *C. vulpinoidea*
36. Perigynia 3.5 to 4 mm. long, serrulate above the middle, tapering or abruptly contracted into the beak. Inflorescence compound, more than 2.5 cm. long, the spikes numerous...8. *C. alma*
36. Perigynia 4 to 5 mm. long, not serrulate below the beak, tapering into the beak (37).
37. Stems 25 to 35 cm. long, much surpassing the leaves; leaves not more than 3 mm. wide; beak about ⅓ as long as the body of the perigynium, obscurely bidentate; plants of dry habitats..6. *C. Rusbyi*
37. Stems 30 to 120 cm. long, shorter than to surpassing the leaves; leaves 4 to 8 mm. wide; beak about as long as the body of the perigynium, evidently bidentate; plants of wet habitats..12. *C. stipata*
38. Perigynia at most thin-edged (39).
38. Perigynia wing-margined (41).
39. Perigynia spreading at maturity; culms wiry; ligule wider than long....14. *C. interior*
39. Perigynia appressed; culms not wiry; ligule much longer than wide (40).
40. Perigynia shallowly bidentate, 3.5 to 4 mm. long, 1.5 mm. wide......15. *C. leptopoda*
40. Perigynia deeply bidentate, 4 to 4.5 mm. long, 1 to 1.3 mm. wide.......16. *C. Bolanderi*
41. Bracts conspicuously surpassing the head......................26. *C. athrostachya*
41. Bracts not conspicuously surpassing the head (42).
42. Perigynia with the beak flattened (43).
42. Perigynia with the beak slender and nearly terete, at least near the tip (44).
43. Perigynia lanceolate to narrowly ovate-lanceolate, 2 mm. wide or less, about 3 times as long as wide, shallowly bidentate..........................24. *C. scoparia*

43. Perigynia broadly ovate to suborbicular, 2.5 mm. wide or more, less than twice as long as wide, strongly bidentate........25. *C. brevior*
44. Perigynia 3 to 3.5 mm. long, plano-convex, ovate........21. *C. subfusca*
44. Perigynia 3.5 to 8 mm. long, flattened (45).
45. Perigynia 3.5 to 5 mm. long, nerved on both faces; culms generally slender and tall, 3 to 10 dm. high; scales dark reddish or dull brown to brownish black. Plants found usually at 8,000 feet altitude or less (46).
45. Perigynia 6 to 8 mm. long, or if only 4.5 to 6 mm. long (in *C. Haydeniana*), then usually nerveless ventrally; culms generally stouter and lower, 1 to 5 dm. high; scales brownish black or blackish (47).
46. Perigynia strongly margined, ovate, appressed; scales dark chestnut to brownish black 17. *C. festivella*
46. Perigynia very narrowly margined, lanceolate-ovate, ascending-spreading; scales dull brown........18. *C. microptera*
47. Scales with very narrow or almost no hyaline margins (48).
47. Scales with broad, white, hyaline margins (49).
48. Perigynia narrowly lanceolate, usually finely many-nerved ventrally, the beaks appressed; culms stiff, erect........19. *C. ebenea*
48. Perigynia generally ovate and nerveless ventrally, the beaks ascending; culms slender, ascending or decumbent........20. *C. Haydeniana*
49. Perigynia ventrally nerved, 2.25 mm. wide, brownish green to yellowish brown or reddish brown; usually only the lowermost bract conspicuous, the others rudimentary 22. *C. petasata*
49. Perigynia ventrally nerveless, 2.5 to 3 mm. wide, light green or straw-colored; usually several of the setaceous bracts well developed and conspicuous........23. *C. Wootoni*

1. Carex Douglasii Boott. Kaibab Plateau and San Francisco Peaks (Coconino County), near Chloride (Mohave County), 5,000 feet and higher, apparently rare. Manitoba to British Columbia, south to New Mexico, Arizona, and California.

2. Carex Eleocharis Bailey. Lukachukai Mountains (Apache County), Kayenta (Navajo County), Grand Canyon (Coconino County), 6,000 to 8,000 feet, rare. Manitoba to Yukon, south to Iowa, New Mexico, Arizona, and Oregon.

3. Carex simulata Mackenz. White Mountains, Apache County, 7,000 to 9,500 feet (*Palmer* 553, *Phillips* 3121a?, 3125?). Montana to Washington, south to New Mexico, Arizona, and California.

Phillips's specimens are too immature for certain identification.

4. Carex praegracilis W. Boott. Apache County to Mohave County, south to Cochise and Pima counties, 3,000 to 9,500 feet, rather common. Manitoba to Yukon, south to central Mexico and California; Michigan; South America.

This is a very variable species and one of the most widespread in the western United States. It is sometimes confused with *C. siccata* Dewey, but that species has slender, brown rootstocks, whereas *C. praegracilis* has stout, blackish ones.

5. Carex siccata Dewey. Apache, Coconino, Yavapai, Greenlee, and Cochise counties, 8,000 to 11,500 feet, rather common. Maine to Mackenzie and Washington, south to New Jersey, Illinois, New Mexico, and Arizona.

6. Carex Rusbyi Mackenz. Graham County to Coconino County, south to Cochise, Santa Cruz, and Pima counties, 7,000 to 9,500 feet, fairly common, especially in the southeastern part of the state, type from Yavapai County (*Rusby* 859). New Mexico and Arizona.

7. Carex occidentalis Bailey. Apache County to Coconino County, south to Cochise and Pima counties, 6,500 to 9,500 feet, common, especially in Coconino County. Wyoming and Utah to New Mexico and Arizona.

8. Carex alma Bailey. Gila, Maricopa, and Cochise counties, 2,000 to 6,000 feet, not rare. Arizona, Nevada, and California.

9. Carex agrostoides Mackenz. Gila, Maricopa, Cochise, and Pima counties, 4,000 to 5,000 feet. New Mexico, Arizona, and northern Mexico.

This species has about the same range as *C. alma* and is related to it, but may be distinguished by its lighter-colored spikes and much lighter-colored perigynia.

10. Carex vulpinoidea Michx. White Mountains, Apache County (*Griffiths* 5426); also collected somewhere in Arizona by Rothrock in 1874. Newfoundland to British Columbia, south to Florida, Texas, Arizona, and Oregon.

11. Carex chihuahuensis Mackenz. Pinaleno Mountains, Graham County, about 6,000 feet (*Maguire & Maguire* 11483, 11494). Southeastern Arizona and Chihuahua.

12. Carex stipata Muhl. Apache, Navajo, Coconino, Graham, and Gila counties, up to 9,000 feet. Newfoundland to Alaska, south to North Carolina, New Mexico, Arizona, and California; Japan.

13. Carex canescens L. White Mountains, Apache County, up to 9,500 feet (*Goodding* 1154, *Phillips* 3137, 3283). Throughout the cooler parts of North America; South America; Eurasia; Australia.

14. Carex interior Bailey. Apache County, in the Lukachukai Mountains (*Peebles* 14380) and at Willow Spring (*Palmer* 548), 7,000 to 8,000 feet. Labrador to British Columbia, south to Pennsylvania, Kansas, northern Mexico, and northern California.

15. Carex leptopoda Mackenz. Greer (Apache County), San Francisco Peaks (Coconino County), Santa Catalina Mountains (Pima County), 8,000 to 10,000 feet, not common. Idaho to British Columbia, south to Arizona and California.

16. Carex Bolanderi Olney. Coconino, Pinal, Cochise, Santa Cruz, and Pima counties. Montana to British Columbia, south to New Mexico, Arizona, and California.

17. Carex festivella Mackenz. Apache, Coconino, Greenlee, Graham, and Cochise counties, 8,500 to 9,500 feet, rather common. Manitoba to British Columbia, south to northern Mexico and California.

18. Carex microptera Mackenz. North Rim of the Grand Canyon, Coconino County, 8,000 feet (*Collom* 1590), perhaps also in the Pinaleno Mountains, Graham County (*Thornber & Shreve* 7757). Wyoming to British Columbia, south to Arizona, Nevada, and California.

19. Carex ebenea Rydb. Common on the San Francisco Peaks (Coconino County), also summit of Baldy Peak (Apache County), 10,500 to 12,000 feet. Montana to Utah and south to New Mexico and Arizona.

20. Carex Haydeniana Olney. North Rim of Grand Canyon and San Francisco Peaks, Coconino County (*Collom* in 1943, *Little* 4771), White Mountains, Apache County (*Phillips* 3248 etc.), 8,000 to 11,500 feet. Alberta to Oregon, south to Arizona and California.

21. Carex subfusca W. Boott. Coconino, Yavapai, Gila, Graham, Pinal, and Pima counties, 3,500 to 9,500 feet, common and variable. Arizona to Oregon and California.

22. Carex petasata Dewey. Kaibab Plateau, North Rim of Grand Canyon, and San Francisco Peaks (Coconino County), White Mountains (Apache County), 8,000 to 9,500 feet. Saskatchewan to British Columbia, south to Colorado, Arizona, Nevada, and Oregon.

23. Carex Wootoni Mackenz. Apache, Coconino, Graham, and Pima counties, 7,000 to 11,500 feet. New Mexico and Arizona.

May be at once recognized and differentiated from related species (except the next) by the fact that the scales are broadly hyaline-margined.

24. Carex scoparia Schkuhr. McNary, Apache County, 7,500 feet (*Peebles & Smith* 12477). Newfoundland and British Columbia, south to South Carolina, New Mexico, Arizona, and Oregon.

25. Carex brevior (Dewey) Mackenz. Near Lakeside, southern Navajo County, in water and wet soil, 6,500 feet (*Pultz* 1678). Maine to British Columbia, south to Virginia, Tennessee, Texas, eastern Arizona, and Oregon.

26. Carex athrostachya Olney. Coconino County, 6,500 to 8,500 feet, not rare. Saskatchewan to Alaska, south to Colorado, Arizona, and California.

27. Carex oreocharis Holm. Summit Ranch, San Francisco Peaks (*Thornber* 2955). Colorado and Arizona.

28. Carex filifolia Nutt. Near Ganado, Apache County, 6,500 feet, hillsides among junipers (*Gould & Phillips* 4877), San Francisco Peaks (Coconino County), rare in Arizona. Manitoba to Yukon, south to Texas, Arizona, and California.

29. Carex leucodonta Holm. Cochise, Santa Cruz, and Pima counties, 6,000 to 7,000 feet, in pine forests, type from the Huachuca Mountains (*Lemmon* 2904). Southern Arizona and northern Mexico.

30. Carex Rossii Boott. Coconino County, at the Grand Canyon (*Eastwood & Howell* 888, erroneously reported by J. W. Stacey as *C. pityophila* Mackenz., *Leafl. W. Bot.* 1: 209. 1936) and on the San Francisco Peaks (*M. E. Jones* in 1890). Northern Michigan and South Dakota to Yukon, south to Colorado, northern Arizona, and California.

31. Carex geophila Mackenz. Chuska and White mountains (Apache County), Pinaleno Mountains (Graham County), apparently also near Prescott (Yavapai County), 5,000 (?) to 8,000 feet. New Mexico and Arizona.

32. Carex curatorum Stacey. Known only from the type locality, Kaibab Trail to Roaring Springs, Grand Canyon (types *Eastwood & Howell* 1100, 1101).

33. Carex Hassei Bailey. Navajo and Coconino counties, about 8,000 (?) feet, rare. Mackenzie and Yukon, south to Utah, northern Arizona, and Baja California.

34. Carex aurea Nutt. Apache and Coconino counties, more common than *C. Hassei,* but rare in Arizona. Newfoundland to British Columbia, south to Pennsylvania, Indiana, New Mexico, northern Arizona, and California.

34a. Carex Meadii Dewey. Hermit Creek, Grand Canyon, Coconino County (*Eastwood* 6004). New Jersey to Saskatchewan, southward to Georgia, Texas, and Arizona.

35. Carex lanuginosa Michx. Apache County to Mohave and Yavapai counties, up to 8,500 feet, rather common. New Brunswick to British Columbia, south to Tennessee, Texas, Arizona, and California.

36. Carex bella Bailey. Apache, Navajo, Coconino, and Graham counties, 9,000 to 12,000 feet. Colorado, Utah, New Mexico, Arizona, and northern Mexico.

This is one of the most frequently collected species of *Carex* in Arizona, not because it is the most common species but because of its striking beauty, whence the specific name *bella.*

37. Carex albo-nigra Mackenz. San Francisco Peaks (Coconino County) up to 12,000 feet, common. Alberta to Washington, south to northern Arizona and California.

38. Carex chalciolepis Holm. San Francisco Peaks (Coconino County) about 11,000 feet, common. Wyoming, Colorado, Utah, and northern Arizona.

This species and *C. albo-nigra* are often found growing together, and specimens of both are sometimes mounted on the same sheet. They are easily differentiated, as the heads of *C. albo-nigra* are nearly black whereas the heads of *C. chalciolepis* are brownish or copper-colored.

39. Carex specuicola J. T. Howell. Near Inscription House, Coconino County (*J. T. Howell* 24609, the type collection). Known only from this collection.

This species is unusual but by no means unique in having both 2-branched styles, with lenticular achenes, and 3-branched styles, with trigonous achenes. Although the former condition is the more common one, the species seems to belong to section *Atratae.*

40. Carex Kelloggii W. Boott. Wood's Canyon, Mogollon Escarpment, Coconino County, about 7,500 feet (*J. F. Arnold* 33, in 1949). Alberta to Alaska, south to Colorado, Arizona, and California.

41. Carex nebraskensis Dewey. Apache, Coconino, and Mohave counties, up to 11,500 feet, rather common. South Dakota to British Columbia, south to Kansas, New Mexico, Arizona, and California.

42. Carex aquatilis Wahl. White Mountains, Apache County (*Phillips* 3247), perhaps also in the Chiricahua Mountains, Cochise County (*Kusche* in 1927). Greenland to northern Alaska, south to Quebec, New Mexico, Arizona, and California; Eurasia.

43. Carex senta Boott. Apache, Coconino, Maricopa, and Cochise counties, 2,000 feet or higher. Arizona and California.

*__44. Carex spissa__ Bailey. "Arizona (Pringle)" (cited by L. H. Bailey, *Proc. Amer. Acad.* 22: 71). Arizona (?), southern California, and northern Baja California.

The Pringle specimen probably was mislabeled. There are three different handwritings on the sheet, and the word "Arizona" has evidently been added later. Specimens from the Huachuca Mountains (*Lemmon* in 1882), in the Dudley Herbarium, Stanford University, identified by K. K. Mackenzie as "*Carex spissa* Bailey (*C. ultra* Bailey)," are probably part of the type collections of *C. ultra.*

45. Carex ultra Bailey. Apache, Pinal, Cochise, and Santa Cruz counties, types from southern Arizona (*Lemmon* 2901, 2902). New Mexico and southern Arizona.

46. Carex hystricina Muhl. Apache, Navajo, Coconino, and Maricopa counties, up to about 6,000 feet, rather common. New Brunswick to Washington, south to Virginia, Texas, Arizona, and California.

47. Carex Thurberi Dewey. Coconino, Gila, Cochise, Santa Cruz, and Pima counties, 4,000 to 6,000 feet, common, especially in the southern counties. Arizona to Guatemala; West Indies.

Related to *C. hystricina,* but that species does not seem to extend farther south in Arizona than Maricopa County.

48. Carex vesicaria L. Wood's Canyon, Mogollon Escarpment, Coconino County, 7,500 feet (*J. F. Arnold* 31, in 1949). Newfoundland to British Columbia, south to Delaware, Missouri, New Mexico, Arizona, and California; Eurasia.

49. Carex rostrata Stokes. Apache and Coconino counties, also reported from Pima County, 7,500 to 9,500 feet. Greenland to Alaska, south to Delaware, Indiana, New Mexico, Arizona, and California.

20. PALMAE. Palm Family

A very large, chiefly tropical family, including many species that are grown as ornamentals and many economically valuable plants. The best-known of the latter are the coconut palm (*Cocos nucifera*) and the date palm (*Phoenix dactylifera*), the latter extensively cultivated for its fruit in southern Arizona.

1. WASHINGTONIA. California Palm

Tree; leaves clustered at apex of the columnar trunk around the terminal bud, the petioles long, stout, with hooked marginal spines, the blades very large, fan-shaped, many-ribbed, splitting longitudinally into numerous narrow segments, these fibrous-margined; inflorescences axillary, subtended by spathes, many-flowered, drooping; flowers perfect or unisexual; perianth segments 6, in 2 series; style and stigma 1; ovary 3-celled; fruit a small drupe with thin, dry flesh.

1. Washingtonia filifera Wendl. Kofa Mountains, Yuma County, about 2,500 feet. Southwestern Arizona, southeastern California, and northern Baja California.

Although California palms are extensively planted in the southern part of the state, the approximately sixty-five individuals growing in small lateral canyons or pockets in the walls of a larger canyon in the Kofa Mountains are the only plants of this species known in the wild in Arizona. The trees here reach a maximum height of about 30 feet (9 m.), and the trunk soon becomes naked. The inflorescences, including the peduncle, reach a length of 12 feet (3.6 m.). The self-pruning habit of these palms as they grow in Arizona may warrant recognition as a variety, but apparently there are no other differences from the Californian phase of the species.

21. ARACEAE. ARUM FAMILY

Most of the members of this large family are tropical. Many, including the familiar calla (*Zantedeschia aethiopica*), are cultivated as ornamentals. Some are important sources of human food in tropical countries, notably taro (*Colocasia esculenta*).

1. PISTIA. WATER-LETTUCE

Plants aquatic; stem very short; leaves in a rosette floating on the water surface, the petioles short, the blades broadly obovate; inflorescence axillary, few-flowered, enclosed in a small white spathe; plants monoecious, the staminate flowers above; perianth none; pistil solitary; ovary 1-celled.

1. **Pistia Stratiotes** L. Yuma, in an irrigation canal (*Sister Mary Noel* 44), probably adventive. Southeastern United States and widely distributed in the tropics.

22. LEMNACEAE. DUCKWEED FAMILY

Plants minute, floating on the surface of ponds and slow streams, thallus-like, without differentiation of stem and leaf; plants monoecious; flowers (rarely produced) borne on the edge of the thallus in a minute spathe; perianth none; stamen and pistil each solitary; fruit a utricle.

KEY TO THE GENERA

1. Rootlets several; thallus several-nerved............................1. *Spirodela*
1. Rootlet solitary; thallus few-nerved, often very obscurely so............2. *Lemna*

Species of the genus *Wolffiella,* characterized by the absence of rootlets and spathes and by the elongate, very thin thalli, are to be looked for in Arizona.

1. SPIRODELA

Thallus broadly obovate, 7- to 15-nerved; anthers longitudinally dehiscent; ovules 2; fruit rarely seen.

1. **Spirodela polyrhiza** (L.) Schleiden. Showlow Creek, southern Navajo County, 6,500 feet (*Pultz* 1785), Huachuca Mountains (*Lemmon* in 1882). Almost cosmopolitan.

2. LEMNA. DUCKWEED

Thallus orbicular or obovate to oblong or lanceolate, 1- to 5-nerved, or almost nerveless; anthers transversely dehiscent; ovules 1 to 6; fruit ovoid.

KEY TO THE SPECIES

1. Thalli with a stalklike base, 6 to 10 mm. long, oblong or lanceolate, remaining connected in chains, mostly immersed..1. *L. trisulca*
1. Thalli sessile or nearly so, not more than 5 mm. long, soon separating, floating (2).
2. Lower surface of the thallus strongly gibbous and much paler than the upper surface. Thallus suborbicular or obovate..................................2. *L. gibba*
2. Lower surface of the thallus flat or nearly so (3).
3. Thallus symmetric or nearly so, elliptic-obovate to suborbicular, obscurely 3-nerved 3. *L. minor*
3. Thallus asymmetric (4).
4. Thallus obliquely obovate...4. *L. perpusilla*
4. Thallus oblong to obovate-oblong..................................5. *L. valdiviana*

1. **Lemna trisulca** L. Willow Spring, southern Apache County, 7,000 feet (*Palmer* 531). Nova Scotia to New Jersey, Texas, and west to the Pacific coast; in all continents except South America.

2. **Lemna gibba** L. Mohave, Yavapai, Maricopa, Pinal, Cochise, Santa Cruz, and Pima counties. Nebraska to Texas, Arizona, and California; almost cosmopolitan.

3. **Lemna minor** L. Apache, Navajo, Yavapai, Maricopa, Pinal, Cochise, and Pima counties. In the greater part of North America; almost cosmopolitan.

4. **Lemna perpusilla** Torr. Between Ruby and Nogales, Santa Cruz County, 4,000 feet (*Kearney & Peebles* 14486, *Gould & Robbins* 3607), western Pima County (*Goodding* 412–45). Widely distributed in North America.

Arizona plants have been referred to *L. minima* Phil. The thalli are described as 3-nerved in *L. perpusilla,* 1-nerved in *L. valdiviana,* but this distinction apparently does not always hold.

5. **Lemna valdiviana** Phil. Oak Creek Canyon, Coconino or Yavapai County (*Goodding* 13), East Eagle Creek, Greenlee County, 5,500 feet (*Gould* 5230), foothills near Tucson, Pima County (*Phillips* 2587). Widely distributed in North and South America.

23. BROMELIACEAE. Pineapple Family

The cultivated pineapple (*Ananas sativa*) is the best-known member of this family, which also includes Spanish-moss (*Tillandsia usneoides*) and other epiphytic plants, and many that are cultivated as ornamentals.

1. TILLANDSIA

Plants epiphytic (growing on the branches of trees but not parasitic); leaves 2-ranked, crowded, awl-shaped from an enlarged base, scurfy-canescent; inflorescence spicate; flowers perfect, regular, the perianth segments 6, the inner ones petallike, violet-colored; stamens 6; stigmas 3, spreading; fruit a prismatic capsule.

1. **Tillandsia recurvata** L. Santa Cruz County, in Bear Valley or Sycamore Canyon (*Phillips* in 1910), and in Flux Canyon, Patagonia Mountains (*Bartram, Peebles & Bartram* 10611), 4,000 to 5,000 feet. Florida to Texas; southeastern Arizona; widely distributed in tropical America.

Ball-moss. This is the only epiphytic flowering plant in Arizona. In Flux Canyon it grows on the branches of live oaks (*Quercus Emoryi* and *Q. Toumeyi*). The Arizona specimens have exceptionally short peduncles.

24. COMMELINACEAE. Spiderwort Family

Plants herbaceous; leaves alternate, with sheathing bases; flowers perfect, regular or irregular, subtended by bracts, these spathelike or leaflike; outer perianth segments sepallike, the inner ones petallike, showy, fugacious; stamens 6, all perfect or some of them sterile; ovary superior; capsule 3-celled.

KEY TO THE GENERA

1. Inflorescences subtended by a single bract, this spathelike; flowers irregular; fertile stamens usually 3; filaments naked . . . . 1. *Commelina*

1. Inflorescences subtended by several bracts, these not spathelike; flowers regular; stamens 6, all perfect; filaments bearded below........................2. *Tradescantia*

1. COMMELINA (54). DAYFLOWER

Stems weak, erect to procumbent; bract strongly compressed, the two halves folded together; flowers irregular, the petallike inner perianth segments blue or white, unequal, the lateral ones larger; sterile stamens with 4-lobed, empty anthers.

Some of the species are said to have the property of stopping blood flow.

KEY TO THE SPECIES

1. Floral bracts 3 to 6 cm. long, distinct, long-acuminate (the tip usually equaling or longer than the body of the bract), glabrous, puberulent, or sparsely pubescent with short, mostly appressed hairs; roots tuberous-thickened; stems simple to much-branched, usually erect but sometimes decumbent; all the petals blue..........1. *C. dianthifolia*
1. Floral bracts not more than 3 cm. long, connate below, acute or short-acuminate (the tip much shorter than the body of the bract), pubescent with short, subappressed hairs and (especially toward base) with long, more spreading, flaccid hairs; roots thick but scarcely tuberlike; stems much-branched, decumbent or spreading; one petal white 2. *C. erecta*

1. Commelina dianthifolia Delile. Apache and Coconino counties, southward to Cochise, Santa Cruz, and Pima counties, 3,500 to 9,500 feet, commonly in pine woods, August and September. New Mexico, Arizona, and nearly throughout Mexico.

2. Commelina erecta L. Cochise, Santa Cruz, and Pima counties, 3,500 to 5,000 feet, May to September. New York to Colorado, south to Florida, Texas, and Arizona.

The Arizona plant is var. *crispa* (Wooton) Palmer & Steyermark (*C. crispa* Wooton).

2. TRADESCANTIA (55). SPIDERWORT

Plants perennial, with thickened roots; outer one or more of the bracts leaflike, the inner ones scarious; inner perianth segments purple, all alike.

These plants were used by the Indians as pot herbs, and the tuberlike roots of *T. pinetorum* are said to have been eaten also.

KEY TO THE SPECIES

1. Roots partly tuberous-thickened, fascicled at base of the stem or borne on a creeping rootstock; stems slender, hispidulous, commonly unbranched; sheaths pubescent or puberulent, especially on the margins; corolla not more than 2 cm. in diameter 1. *T. pinetorum*
1. Roots thick but not tuberlike; rootstock none; stems stout, glabrous, often branched; sheaths glabrous; corolla often more than 2 cm. in diameter........2. *T. occidentalis*

1. Tradescantia pinetorum Greene. Apache, Navajo, and Coconino counties to Cochise, Santa Cruz, and Pima counties, often in pine woods, 4,500 to 9,500 feet, August and September. New Mexico and Arizona, probably also in northern Mexico.

Flowers handsome, the inner perianth segments violet or purple.

2. Tradescantia occidentalis (Britton) Smyth. Apache, Navajo, and Coconino counties, south to Graham and Pima counties, 2,500 to 7,000 feet, April to September. Wisconsin to Montana, Texas, and Arizona.

Var. *scopulorum* (Rose) Anderson & Woodson (*T. scopulorum* Rose) is about as frequent and as widely distributed in Arizona as typical *T. occidentalis*, which has the sepals and pedicels more or less glandular-pubescent whereas they are glabrous in the variety. The type of *T. scopulorum* was collected in the Santa Catalina Mountains, Pima County (*Pringle* in 1884).

25. PONTEDERIACEAE. Pickerel-weed Family

1. HETERANTHERA. Mud-plantain

Plants herbaceous, aquatic or semiaquatic; leaves alternate, narrow and grasslike or with broad blades; inflorescence subtended by a spathe, 1- or 2-flowered; flowers perfect, somewhat irregular, the perianth corollalike, of 6 segments; stamens 3, inserted on the perianth; ovary superior, fusiform, 1-celled or incompletely 3-celled; fruit a many-seeded capsule.

KEY TO THE SPECIES

1. Leaves linear, grasslike; perianth yellow................................1. *H. dubia*
1. Leaves with long, stout petioles and elliptic to lance-ovate blades rounded at base; perianth commonly blue, sometimes white............................2. *H. limosa*

1. Heteranthera dubia (Jacq.) MacMillan. Verde River near Camp Verde, Yavapai County (*Goodding* 80–47), irrigation canals in the Salt River Valley, Maricopa County (*Loomis* 5537). Ontario to Washington, North Carolina, and Arizona.

Water-stargrass. This plant sometimes grows so profusely in canals and ditches as to obstruct the flow of water, causing considerable expense for removal. It is strictly aquatic.

2. Heteranthera limosa (Swartz) Willd. Near Bonita, Graham County (*Pultz* 1812), San Bernardino Ranch, Cochise County (*Mearns* 609), Bear Valley or Sycamore Canyon, Santa Cruz County (*Haskell* 1014), about 4,000 feet, mud flats, summer and early autumn. Virginia to South Dakota, southward to Florida, Louisiana, and southeastern Arizona; tropical America.

26. JUNCACEAE. Rush Family

Contributed by F. J. Hermann

Plants herbaceous, mostly perennial, grasslike, with narrow sheathing leaves; perianth regular, the 6 divisions glumelike; stamens 3 to 6; ovary 1-celled or 3-celled; stigmas 3; fruit a capsule.

KEY TO THE GENERA

1. Plants glabrous; leaf sheaths open; ovary usually more or less 3-celled; ovules many, attached to the axis or walls of the ovary (the placentas axial or parietal)..1. *Juncus*
1. Plants with long, soft hairs (these sometimes very few); leaf sheaths closed; ovary 1-celled; ovules 3, attached to the base of the ovary (the placentas basal)...2. *Luzula*

1. JUNCUS. Rush

Principally perennial, grasslike herbs of wet habitats, with glabrous, pithy or hollow, usually simple, stems; leaves glabrous, the sheaths open, the blades terete or flattened, sometimes wanting; inflorescence cymose, paniculate, or glomerate, often unilateral; flowers small, greenish or brownish, glumaceous; capsule 3-celled with a central placenta, or 1-celled with parietal placentas; seeds numerous, reticulate or ribbed, sometimes appendaged.

Plants bearing mature fruit and the persistent perianth and stamens are essential for identification of most of the species in this genus. Occasionally *Juncus saximontanus* is sufficiently abundant in moist meadows to become a principal ingredient of the "grasses" cut for hay, and other species may be locally so plentiful as to have an appreciable forage value. Otherwise the rushes of Arizona are of no economic importance.

KEY TO THE SPECIES

1. Inflorescence appearing lateral, the involucral bract terete, stiffly erect, resembling a continuation of the stem. Leaves all basal or nearly so, never septate (2).
1. Inflorescence obviously terminal, or else the leaves septate and the involucral bracts flat or channeled along the upper side; involucral bracts leaflike, not strictly erect or resembling a continuation of the stem (7).
2. Flowers 1 to 3 (rarely 4 or 5); seeds conspicuously tailed; low alpine plant, 5 to 20 cm. high .... 1. *J. Drummondii*
2. Flowers many; seeds not tailed (or slightly so in *J. Cooperi* and *J. acutus*); plants usually taller, of the Sonoran and Transition zones (3).
3. Stems relatively slender, not very rigid; inflorescence not glomerate; flowers each with 2 bracteoles at base in addition to the bractlet at base of the pedicel (4).
3. Stems very coarse, stout and rigid; flowers in headlike clusters arranged in open panicles, from the axil of a single bractlet but without bracteoles (6).
4. Stems compressed; leaf blades usually present. Perianth usually greenish or straw-colored .... 4. *J. mexicanus*
4. Stems terete; leaf blades none or reduced to filiform rudiments (5).
5. Perianth segments 2 to 3 mm. long, straw-colored to pale brown; stamens 3; anthers not longer than the filaments; capsule obovoid, obtuse or retuse.......... 2. *J. effusus*
5. Perianth segments 3.5 to 5 mm. long, purplish brown; stamens 6; anthers much longer than the filaments; capsule narrowly ovoid, acute, mucronate.......... 3. *J. balticus*
6. Perianth segments acutish to acuminate, narrowly scarious-margined, greenish or straw-colored, 4 to 6 mm. long, nearly equaling the narrowly ovoid, acute or acuminate capsule.......... 5. *J. Cooperi*
6. Perianth segments (at least the inner ones) obtuse or truncate, broadly scarious-margined, brown, 2 to 4 mm. long, much shorter than the subglobose, obtuse, mucronate capsule.......... 6. *J. acutus*
7. Leaves transversely flattened (inserted with the flat surface facing the stem), or in age involute, not septate (8).
7. Leaves terete or ensiform (flattened and inserted with one edge facing the stem), not transversely flattened (16).
8. Flowers in heads, not bracteolate, i. e., with only the bractlet at base of the pedicel (9).
8. Flowers not in heads, inserted singly on the branches of the inflorescence, each with a pair of bracteoles at base in addition to the bractlet at base of the pedicel (11).
9. Stamens 3; perianth 2 to 3 mm. long.......... 12. *J. marginatus*
9. Stamens 6; perianth 5 to 6 mm. long (10).
10. Outer perianth segments equaling or slightly exceeding the inner ones; auricles of the leaf sheaths 0.5 to 1.5 mm. long; leaf blades flat; anthers cream-colored or pale yellow .... 13. *J. longistylis*
10. Outer perianth segments shorter than the inner ones; auricles 1.5 to 3 mm. long; leaf blades channeled; anthers brownish.......... 14. *J. macrophyllus*
11. Inflorescence more than half the height of the plant; low annuals. Flowers scattered along the loosely forking branches (12).
11. Inflorescence much less than half the height of the plant; perennials (13).
12. Capsule oblong, 3 to 4.5 mm. long; perianth 4 to 6 mm. long.......... 7. *J. bufonius*

12. Capsule subglobose to broadly ovoid; perianth 3 to 4 mm. long......8. *J. sphaerocarpus*
13. Capsule completely 3-celled, retuse....................................9. *J. confusus*
13. Capsule 1-celled, with septa extending halfway to the center, acutish to obtuse, not retuse (14).
14. Auricles at summit of the sheaths very thin, white and scarious, conspicuously produced beyond the point of insertion, 1 to 3.5 mm. long. Bracteoles blunt......10. *J. tenuis*
14. Auricles firm, not conspicuously produced beyond the point of insertion (15).
15. Bracteoles blunt; auricles cartilaginous, yellow, very rigid and glossy, especially the short produced portion: var. *Dudleyi*................................10. *J. tenuis*
15. Bracteoles acuminate to aristate; auricles with the very slightly produced portion membranaceous, not rigid, easily broken..............................11. *J. interior*
16. Leaves terete, the septa complete (17).
16. Leaves ensiform, the septa incomplete (21).
17. Capsule subulate; stamens 6 (18).
17. Capsule oblong or ovoid to obovoid; stamens 3 or 6 (19).
18. Plants low, 10 to 40 cm. high; leaves erect or ascending; flowers 3 to 4 mm. long; inner perianth segments equaling or exceeding the outer ones.............15. *J. nodosus*
18. Plants taller, 40 to 100 cm. high; leaves divaricate; flowers 4 to 5 mm. long; inner perianth segments shorter than the outer ones.........................16. *J. Torreyi*
19. Perianth greenish or straw-colored; rhizome very short or obsolete; stamens 3; capsule ovate-lanceolate in outline.....................................17. *J. acuminatus*
19. Perianth chestnut-colored or dark brown; rhizomes well developed; stamens 6 (rarely only 3 in *J. Mertensianus*); capsule oblong, ovate, or obovate in outline (20).
20. Anthers shorter than the filaments; heads usually solitary; styles included; capsule obovoid, obtuse. Plants alpine..............................18. *J. Mertensianus*
20. Anthers much longer than the filaments; heads usually 2 or more; styles exserted; capsule oblong to ovoid, acutish.......................................19. *J. badius*
21. Stamens 3; bract ensiform, more than half the length of the inflorescence 20. *J. ensifolius*
21. Stamens 6; bract narrower, less than half the length of the inflorescence (22).
22. Perianth segments equal in length, very narrow and thin, often shorter than the oblong, acute capsule, spreading, overlapping only near the base, thus exposing about ¾ of the capsule; blades of the larger leaves 7 to 12 mm. wide; stems stout 21. *J. xiphioides*
22. Perianth segments unequal, the inner ones shorter, the segments broader and firmer in texture, usually exceeding the oblong-obovoid capsule, appressed, overlapping most of their length, thus exposing little of the capsule; blades of the larger leaves seldom more than 5 mm. wide; stems relatively slender. Valves of the capsule more rigid than in *J. xiphioides* (23).
23. Seeds tailed; styles long-exserted.......................................22. *J. Tracyi*
23. Seeds not tailed; styles usually little if at all exserted............23. *J. saximontanus*

1. Juncus Drummondii E. Mey. Represented from Arizona by a single collection from the San Francisco Peaks, Coconino County, 11,500 feet (*Little* 4758). Alaska to California, northern Arizona, and New Mexico.

2. Juncus effusus L. The typical form of this European species is not known from America. Two varieties occur in Arizona: (1) var. *brunneus* Engelm., characterized by a dark-brown perianth, this firm to almost rigid in texture, appressed to and from slightly shorter than to slightly exceeding the capsule, its segments with narrow, scarious, not or scarcely involute margins. This variety has been collected near Baker Butte, southern Coconino County, 7,500 feet (*Darrow* 3264), in the Rincon Mountains, Pima County, 7,500 feet (*Nealley* 158), and perhaps in the Mazatzal Mountains, Gila County (*Collom* 1620, immature). It ranges from British Columbia to California and southern Arizona. (2) var. *exiguus* Fern. & Wieg., with a pale-brown perianth thin in texture, spreading, about 1½ times as long as the capsule, and the segments with broadly scarious, more or less involute margins, has been found on Pinal Peak, Gila County, 7,500 feet (*Smith* 14066) and in the Santa Catalina Mountains, Pima County (*Thornber* 7518). This variety is limited to California and southern Arizona.

3. Juncus balticus Willd. A Eurasian species represented in Arizona by var. *montanus* Engelm. Common throughout the state in moist habitats, 3,500 to 9,500 feet, July to September, and ranging from Kansas to Alaska, Arizona, and California. Wire rush.

4. Juncus mexicanus Willd. Apache, Navajo, Coconino, Yavapai, Cochise, and Santa Cruz counties, 3,000 to 7,000 feet, frequent, especially in slightly saline soils, July and later. Texas, Arizona, California, and Mexico.

***5. Juncus Cooperi** Engelm. Saline flats of the Colorado and Mojave deserts, California, and of southern Nevada; reported also from southern Utah and to be looked for in western Arizona.

6. Juncus acutus L. A European rush of saline habitats, represented in Arizona by var. *sphaerocarpus* Engelm., which has been collected in the Grand Canyon (Coconino County), at Topock (Mohave County), northwest of Hillside (Yavapai County), and at the mouth of Williams River (northern Yuma County), ranging to southern California and Baja California.

7. Juncus bufonius L. Throughout Arizona, mostly 2,000 to 3,000 feet, stream banks and dried-up pools, April to July. Nearly throughout North America; cosmopolitan. Toad rush.

Var. *halophilus* Buch. & Fern., of brackish soils, has been collected at Hance's Ranch, Coconino (?) County (*Wooton* in 1892).

8. Juncus sphaerocarpus Nees. Coconino, Yavapai, and Pima counties, 4,000 to 7,000 feet, borders of pools and streams, July and August. Idaho and Oregon to Arizona and southern California; Old World.

9. Juncus confusus Coville. Kaibab Plateau and north wall of Grand Canyon, about 8,000 feet (*Kearney & Peebles* 13732, *Eastwood & Howell* 7027, *Goodding* 259–48). Montana to Washington, south to Arizona and California.

10. Juncus tenuis Willd. (*J. macer* S. F. Gray). Pinaleno Mountains (Graham County), Chiricahua Mountains (Cochise County), Rincon Mountains (Pima County), 7,000 to 8,000 feet, June to September. Almost throughout North America; Chile, Argentina, and Brazil.

Var. *Dudleyi* (Wieg.) Hermann (*J. Dudleyi* Wieg.) occurs in Coconino, Yavapai, Maricopa, Pinal, and Cochise counties, 1,000 to 6,500 feet, and ranges from Newfoundland to Washington, south to Mexico and California. It differs from typical *J. tenuis* in the thicker, firmer, yellowish auricles of the leaf sheaths but intergrades freely with typical *J. tenuis.*

11. Juncus interior Wieg. Coconino County to Graham, Cochise, and Pima counties, 1,000 to 7,000 feet, June to September. Illinois and Missouri to Washington, south to Texas and Arizona.

KEY TO THE VARIETIES

1. Perianth 3 to 4 mm. long, equaling the capsule.................. *J. interior* (typical)
1. Perianth 4 to 5 mm. long, exceeding the capsule (2).

2. Bracteoles lanceolate, acuminate; perianth segments erect, rigid, lanceolate, acuminate, with narrow, relatively opaque, hyaline margins.....................var. *arizonicus*
2. Bracteoles broadly ovate, acute to abruptly aristate; perianth segments spreading, not rigid, broadly ovate, acuminate, with broad, transparent, scarious margins and brown lateral bands bordering the green center.........................var. *neomexicanus*

Typical *J. interior* occurs in Coconino and Yavapai counties and in the Huachuca and Santa Catalina mountains (Cochise and Pima counties), about 7,000 feet. Var. *arizonicus* (Wieg.) Hermann (*J. arizonicus* Wieg.) is known from Coconino, Yavapai, Graham, Maricopa, Gila, Pinal, Cochise, and Pima counties, 1,000 to 7,000 feet. Var. *neomexicanus* (Wieg.) Hermann (*J. neomexicanus* Wieg.) is present in Coconino, Gila, Cochise, and Pima counties, 4,000 to 7,000 feet. The type of the unimportant *J. arizonicus* var. *curtiflorus* Wieg. was collected at Flagstaff (*MacDougal* 305), and that of *J. neomexicanus* Wieg. at "Huachuca" (*Palmer* 461b).

12. Juncus marginatus Rostk. Pinaleno Mountains (Graham County), Rincon and Santa Catalina mountains (Pima County), 2,500 to 4,500 feet, along streams, June to August. Maine to Ontario and Nebraska, southward to Florida, Louisiana, and Mexico.

Typical *J. marginatus,* with dull, lusterless capsules and blunt inner perianth segments, is known in Arizona only from a collection at Camp Lowell, Pima County (*Parish* in 1884). A commoner plant in this state is the southwestern var. *setosus* Coville (*J. setosus* Small), with glossy capsules and aristate inner perianth segments.

13. Juncus longistylis Torr. Apache, Navajo, Coconino, Yavapai, Greenlee, Graham, and Gila counties, montane meadows, 4,500 to 9,500 feet, June to September. Minnesota to British Columbia, south to New Mexico, Arizona, and California.

The commoner phase in Arizona is var. *scabratus* Hermann, type from near Prescott (*Peebles et al.* 2712), which has the vegetative parts (particularly the pedicels and the ends of the leaves) strongly scabrous and the auricles tending to be prolonged, free, and acute.

14. Juncus macrophyllus Coville. Yavapai, Maricopa, and Pinal counties, rare, on damp slopes below 5,500 feet, July and August. Southern California to Arizona and Baja California.

***15. Juncus nodosus** L. To be sought in the northern counties, having been collected at Farmington, northwestern New Mexico, and in the Charleston Mountains, southern Nevada. Nova Scotia to British Columbia, south to Virginia, New Mexico, and Nevada.

16. Juncus Torreyi Coville. Navajo County to Mohave County, south to Pima and Yuma counties, very common in wet soil below 5,000 feet, July and August. Massachusetts to Washington, south to Alabama, Texas, Arizona, and California.

17. Juncus acuminatus Michx. Lakeside (southern Navajo County), Oro Blanco (Pajarito) Mountains (Santa Cruz County), and locally common in the Santa Catalina and Rincon mountains (Pima County), 3,000 to 6,500 feet, March to October. Maine to British Columbia, south to Georgia, Arizona, and Oregon.

18. Juncus Mertensianus Bong. Kaibab Plateau and North Rim of Grand Canyon, Coconino County, 8,500 to 9,000 feet (*Jones* 6056, *Eastwood & Howell* 7031, *Goodding* 216–48). Alaska to New Mexico, northern Arizona, and California.

19. Juncus badius Suksd. (*J. truncatus* Rydb.). Extreme northwestern New Mexico, in the Chuska Mountains, a range extending into Apache County, Arizona (*Carter* in 1935), Kaibab Plateau to the Mogollon Escarpment (Coconino County), 6,500 to 9,000 feet, collected once also in Yavapai County, June to September. Wyoming to Washington, south to New Mexico and northern Arizona.

20. Juncus ensifolius Wikstr. Square Lake, ? County, about 5,000 feet (*P. A. South* in 1910). Saskatchewan to Alaska, south to Arizona and California.

21. Juncus xiphioides E. Mey. Montezuma Well and Bradshaw Mountains (Yavapai County), Ashdale (Maricopa County), common in the Santa Rita Mountains (Santa Cruz and Pima counties), 3,500 feet and higher, June to August. Arizona, California, and northern Baja California.

22. Juncus Tracyi Rydb. Lukachukai Mountains (Apache County), north side of Grand Canyon and near Fredonia (Coconino County), 5,000 to 8,000 feet. Montana and Idaho to northern Arizona and Nevada.

23. Juncus saximontanus A. Nels. (*J. xiphioides* var. *montanus* Engelm., *J. parous* Rydb.). Grand Canyon (Coconino County), and Yavapai, Greenlee, Graham, Cochise, and Pima counties, 4,000 to 9,500 feet, July to October. Colorado to British Columbia, south to New Mexico, Arizona, and California.

The range in Arizona, as stated above, is that of typical *J. saximontanus*. Forma *brunnescens* (Rydb.) Hermann (*J. brunnescens* Rydb.) occurs throughout the state, except in the low, western portion, and is by far the commonest rush in Arizona. It has numerous (usually more than 10) heads, which are few- (5- to 12-) flowered and average 5 to 6 mm. in diameter; whereas in typical *J. saximontanus* the few (seldom more than 10) heads are many- (15- to 25-) flowered and average 7 to 10 mm. in diameter.

2. LUZULA. Wood-rush

Plants perennial, leafy-stemmed; leaves flat, channeled, or involute; flowers in heads, spikes, or loose cymes, these forming dense or open compound inflorescences; flower bracteolate; stamens 6.

Plants eaten by livestock, and sometimes a fairly important element in mountain pastures.

KEY TO THE SPECIES

1. Flowers on slender pedicels in a loose, somewhat drooping, many-flowered, cymose panicle; herbage glabrous except for a few long hairs near the throat of the sheath; perianth segments about 2 mm. long, shorter than or barely equaling the capsule ... 1. *L. parviflora*

1. Flowers crowded, subsessile, in few headlike or spikelike glomerules; herbage sparsely villous with long, loose hairs; perianth segments longer than the capsule (2).

2. Inflorescence erect; glomerules capitate, borne on elongate branches; leaves flat; perianth segments about 3 mm. long .... 2. *L. multiflora*

2. Inflorescence nodding; glomerules short-spicate, sessile or nearly so, forming an interrupted spike or a small, compact panicle of spikes; leaves channeled, often involute; perianth segments about 2 mm. long .... 3. *L. spicata*

1. Luzula parviflora (Ehrh.) Desv. San Francisco Peaks (Coconino County), Baldy Peak, White Mountains (Apache County), 9,500 to 11,500 feet, June to September. Labrador to Alaska, southward to New York, New Mexico, Arizona, and California; Eurasia.

The name *Luzula parviflora* probably covers a species complex, but until the group has received a complete modern revision it seems best to include the forms under this name.

2. Luzula multiflora (Retz.) Lejeune. Rincon and Santa Catalina mountains (Pima County), about 7,500 feet. Throughout temperate North America; Eurasia.

3. Luzula spicata (L.) DC. & Lam. San Francisco Peaks (Coconino County), 10,000 to 12,000 feet. Greenland to Alaska, New Mexico, northern Arizona, and California; Eurasia.

27. LILIACEAE. Lily Family

Plants perennial, herbaceous or in a few genera woody, scapose or with leafy stems; flowers mostly perfect, regular or nearly so, often showy; stamens commonly 6; ovary superior or partly inferior; stigmas usually 3; fruit a capsule or a berry.

Some of the handsomest and showiest plants of Arizona, notably the lemon lily (*Lilium Parryi*), the mariposas (*Calochortus* spp.), and species of *Yucca*, belong to this family. A substitute for soap is made from the leaves of the yuccas, and the strong fibers obtained from them were much used by the Indians, who also ate the fruits of the fleshy-fruited species. The bulbs of wild onions (*Allium* and related genera) were eaten by the natives of Arizona. The Liliaceae are mostly innocuous, but a few species—death-camas (*Zigadenus*) and false-hellebore (*Veratrum*)—are poisonous. The most important cultivated food plants of this family are onion and asparagus, the latter occasionally growing wild as an escape from gardens.

KEY TO THE GENERA

1. Plants with a large, woody caudex, this mainly subterranean, or largely above ground and trunklike; leaves numerous, in large rosettes at apex of the caudex or of its branches, narrow, elongate, mostly rigid and spine-tipped (2).

1. Plants herbaceous, without a large, woody caudex; leaves not in large rosettes, never rigid or spine-tipped (4).

2. Flowers seldom less than 2 cm. long, all perfect; capsules large, not lobed or winged; seeds numerous in each cell of the capsule, flattened. Leaf margins usually filiferous 16. *Yucca*

2. Flowers much less than 1 cm. long, all or many of them unisexual; capsules small, 3-lobed or winged; seed solitary in each cell of the capsule, turgid. Inflorescences with small scarious bracts (3).

3. Leaves not very rigid, the margins not spiny (sometimes serrulate); plants incompletely dioecious, some of the flowers perfect; capsule 3-celled, 3-lobed; seeds round 17. *Nolina*

3. Leaves very rigid, the margins armed with sharp curved spines; plants completely dioecious; capsule 1-celled, winged; seeds trigonous. Staminate flowers in dense catkinlike spikes..18. *Dasylirion*

4. Styles 3, distinct. Flowers in racemes or panicles; perianth greenish or whitish (5).

4. Style 1 (often wanting in *Calochortus*) or the styles united, at least toward base (6).

5. Leaves linear or narrowly lanceolate; perianth segments each with a gland near the base 1. *Zigadenus*
5. Leaves broadly elliptic or ovate; perianth segments not gland-bearing....2. *Veratrum*
6. Perianth segments very unlike, the 3 outer ones much narrower, sepallike, the 3 inner ones broad, petallike, bearing a large, variously fringed or bordered gland near the base; fruit a septicidal capsule (dehiscing through the partitions)....15. *Calochortus*
6. Perianth segments all alike or nearly so, none bearing glands; fruit a loculicidal capsule (dehiscing between the partitions), or berrylike (7).
7. Fruits berrylike, indehiscent; flowering stems from (usually horizontal) rootstocks (8).
7. Fruits capsular, dehiscent; flowering stems from bulbs, corms, short vertical rootstocks, or tuberlike roots (12).
8. Stems much-branched, with very slender branches; leaves small, scalelike.19. *Asparagus*
8. Stems simple or sparingly branched; leaves large, with broad blades (9).
9. Flowers not nodding, in terminal racemes or panicles; perianth rotate, the segments not more than 7 mm. long..20. *Smilacina*
9. Flowers nodding, axillary or subaxillary or, if terminal, in very few-flowered umbellike clusters; perianth campanulate or tubular, the segments 8 mm. long or longer (10).
10. Perianth tubular, the segments united below; stems simple.........21. *Polygonatum*
10. Perianth campanulate, the segments distinct or nearly so; stems usually dichotomously branched (11).
11. Flowers terminal; peduncle or pedicel not jointed or bent; perianth segments not recurved; fruit 3-lobed, depressed-globose, papillate...................22. *Disporum*
11. Flowers not terminal; peduncles filiform, jointed and abruptly twisted or bent; perianth segments becoming recurved; fruit entire, ellipsoid, smooth..........23. *Streptopus*
12. Stems scaly but scarcely bulbous at base; roots more or less tuberous-thickened. Leaves long and narrow, all basal or nearly so; flowers in racemes; perianth segments distinct or very nearly so; capsules transversely rugose (13).
12. Stems bulbous or cormlike at base; roots not tuberous-thickened, or exceptionally so in *Hesperocallis* (14).
13. Raceme loose, elongate; perianth orange; anthers straight or nearly so..3. *Anthericum*
13. Raceme rather dense; perianth whitish with green veins; anthers becoming strongly incurved .... 4. *Eremocrinum*
14. Plants caulescent, the stems more or less leafy (15).
14. Plants scapose, the leaves (except in *Triteleiopsis*) all basal. Flowers in umbels, these subtended by scarious bracts, or the flower solitary (17).
15. Bulbs tunicate, the coats thin and dry; leaf margins conspicuously undulate; flowers in scarious-bracted racemes.......................................5. *Hesperocallis*
15. Bulbs with fleshy scales; leaf margins not undulate; flowers solitary or few, terminal or subterminal (16).
16. Perianth segments 5 to 9 cm. long, bright yellow or orange-red, usually with darker spots .... 13. *Lilium*
16. Perianth segments not more than 2 cm. long, dull yellow, veined and mottled with brownish purple...14. *Fritillaria*
17. Perianth segments distinct to the base or nearly so, pink or whitish (18).
17. Perianth segments united below into a funnelform tube (19).
18. Ovules 2 in each cell of the ovary; seeds 1 or 2 in each cell; plants with odor of onions 6. *Allium*
18. Ovules several and seeds usually more than 2 in each cell; plants without odor of onions 7. *Nothoscordum*
19. Filaments partly united into a tube, this with toothlike lobes between the anthers 9. *Androstephium*
19. Filaments not united into a tube (20).
20. Umbels 2- or 3-flowered, or the flower solitary; perianth salverform; stigma large, somewhat funnelform...10. *Milla*
20. Umbels several- to many-flowered; perianth funnelform or funnelform-campanulate; stigma small (21).
21. Leaves borne on the lower part of the stem; perianth with internal, transverse, scalelike

appendages alternating with the perianth lobes. Anthers basifixed; capsule stipitate; seeds elongate . . . . . 8. *Triteleiopsis*
21. Leaves arising from the bulb; perianth without transverse scales (22).
22. Perianth (in the Arizona species) blue-violet; anthers basifixed; capsule sessile or nearly so; seeds elongate . . . . . 9. *Dichelostemma*
22. Perianth (in the Arizona species) bright yellow fading purplish; anthers versatile; capsule stipitate; seeds not elongate . . . . . 10. *Triteleia*

1. ZIGADENUS. Death-camas

Plants glabrous; flowering stems from bulbs, subscapose; leaves narrow, grasslike; flowers in racemes or narrow panicles; perianth segments distinct or united below, each with a gland near the base; styles 3, distinct; ovary superior or partly inferior; fruit a 3-lobed capsule.

These plants are poisonous (*Z. elegans* perhaps only slightly so) and sometimes cause heavy loss of sheep in spring and early summer when other forage is scarce. Cattle and horses seldom eat death-camas. The toxic principle (zygadenin?) is found in all parts of the plant, even in the seeds.

KEY TO THE SPECIES

1. Perianth segments not more than 5 mm. long, pale yellow, abruptly contracted into the claw and often subcordate at base, the glands obovate with the upper margin not sharply defined; stamens much-exserted; ovary superior . . . . . 1. *Z. paniculatus*
1. Perianth segments not less (commonly more) than 5 mm. long, not abruptly contracted, more or less cuneate at base, the glands obcordate with a sharply defined upper margin; stamens not or only moderately exserted; ovary partly inferior (2).
2. Branches of the inflorescence erect or ascending; pedicels commonly ascending, seldom decurved; perianth yellowish white . . . . . 2. *Z. elegans*
2. Branches of the inflorescence spreading; pedicels divergent, often decurved; perianth greenish white, sometimes tinged with purple . . . . . 3. *Z. virescens*

1. Zigadenus paniculatus (Nutt.) Wats. Apache County to Coconino County, 5,500 to 7,500 feet, May. Montana to Washington, New Mexico, northern Arizona, and California.

Sometimes known as sand-corn.

2. Zigadenus elegans Pursh. Coconino County and southern Apache County, 5,000 to 10,000 feet, rich soil in pine woods, July and August. Saskatchewan to Alaska, south to New Mexico and Arizona.

3. Zigadenus virescens (H.B.K.) Macbr. (*Z. porrifolius* Greene). White Mountains (Apache County), Huachuca Mountains (Cochise County), 6,500 to 11,000 feet, rich soil in coniferous forests, July to September. New Mexico and Arizona to Central America.

Z. porrifolius is, perhaps, varietally distinct.

2. VERATRUM. False-hellebore, Corn-lily

A coarse, pubescent herb with tall, leafy stems from thick rootstocks; leaves clasping, broad, strongly veined; flowers numerous in an ample panicle, many of them unisexual; perianth glandless, greenish yellow; anthers cordate; styles 3, distinct; fruit a 3-lobed capsule.

1. Veratrum californicum Durand. White Mountains (Apache County), near Lakeside (Navajo County), Mogollon Mesa (Coconino County), Pinaleno Mountains (Graham County), Chiricahua Mountains (Cochise County),

7,500 to 9,500 feet, bogs and wet meadows, July and August. Montana to Washington, south to New Mexico, Arizona, and California.

This plant is known in New Mexico as skunk-cabbage. The root and young shoots, which contain an alkaloid, veratrin, are poisonous to stock, although seldom eaten. The flowers are poisonous to insects, sometimes causing heavy losses in honeybees.

3. ANTHERICUM. Crag-lily, Amber-lily

Roots thick, cylindric, fascicled; stems scapose; leaves narrow, grasslike; flowers in a slender raceme; perianth orange-yellow, the segments narrow, distinct or nearly so; fruit an oblong capsule.

1. **Anthericum Torreyi** Baker. Apache, Navajo, and Coconino counties, south to Cochise, Santa Cruz, and Pima counties, 6,000 to 9,000 feet, commonly in pine woods, August. New Mexico, Arizona, and southward.

The Arizona plants differ from typical *A. Torreyi* in having nearly smooth filaments, according to Johnston (56, p. 59).

4. EREMOCRINUM

Roots tuberous-thickened; stems scapose, not more than 30 cm. long; leaves long and narrow; flowers few, in a short, rather dense raceme; perianth whitish with green veins, the segments distinct or nearly so; anthers becoming strongly incurved; fruit a capsule.

1. **Eremocrinum albomarginatum** Jones. West of Carrizo, Apache County (*Peebles & Smith* 13578), north of Kayenta and in Monument Valley, Navajo County (*Peebles & Fulton* 11950, *Sampson* 192), 5,000 to 5,500 feet, sandy soil, June. Southern Utah and northern Arizona.

The star-lily or sand-lily, *Leucocrinum montanum* Nutt., is to be looked for in northern Arizona. It is readily distinguishable from *Eremocrinum* by having the peduncles originating underground and the perianth segments united below into a long tube.

5. HESPEROCALLIS. Desert-lily

Roots often slightly thickened and coated with sand; flowering stems from a large, tunicate bulb, with few, large, undulate-margined leaves; flowers large, in a scarious-bracted raceme; perianth segments whitish with a green central stripe, 5 to 6 cm. long, united at base; fruit a capsule.

1. **Hesperocallis undulata** Gray. Western part of Maricopa and Pima counties and in Mohave and Yuma counties, mostly below 2,000 feet, in sandy soil of deserts, February to May, types from Jessup Rapids (*Newberry*) and Fort Mohave (*Cooper*). Southwestern Arizona and southeastern California, probably in northwestern Sonora.

The bulbs are eaten by the Indians of southwestern Arizona. This is one of the showiest of the Arizona desert wild flowers.

6. ALLIUM (57). Onion

Contributed by Marion Ownbey

Flowering stems scapose, from a tunicate bulb, this sometimes borne on a short rootstock; leaves sheathing, usually narrowly linear; flowers regular, perfect, in a terminal simple umbel; perianth persistent, more or less accrescent, rose-purple to nearly white, the segments distinct or nearly so; ovary usually completely 3-celled, often crested; ovules 2 in each cell (in the Arizona species).

Such important culinary herbs as onion, garlic, leek, and chives belong to this genus. The bulbs of the Arizona native species were formerly much utilized by the Indians for food and seasoning, being consumed raw or after heating in ashes. Sometimes they were stored for use in winter.

KEY TO THE SPECIES

1. Outer bulb scales persisting as reticula of coarse anastomosing fibers (2).
1. Outer bulb scales without fibers or with parallel fibers, never fibrous-reticulate (5).
2. Leaves usually 2 per scape; alveoli on seeds not pustuliferous. Bracts of spathe 3- to 5-nerved; ovary crested, usually conspicuously, with 6 flattened processes: low, desert species .... 4. *A. macropetalum*
2. Leaves 3 or more per scape; alveoli on seeds pustuliferous: tall, mountain species (3).
3. Bracts of the spathe 3- to 5-nerved; ovary conspicuously crested with 6 flattened, lacerate processes; bulbs stipitate (short-rhizomatous) at base.......3. *A. Plummerae*
3. Bracts of the spathe mostly 1-nerved; ovary inconspicuously crested with 6 knoblike processes (or sometimes crestless?); bulbs not stipitate at base (4).
4. Umbel wholly floriferous; flowers fertile.............................1. *A. Geyeri*
4. Umbel with most of the flowers replaced by bulbils; flowers usually sterile 2. *A. rubrum*
5. Perianth segments of the inner whorl obscurely serrulate-denticulate, long-acuminate with recurved tips; outer segments similar but broader, longer, and usually entire; ovary and capsule not prominently crested; outer bulb scales cellular-reticulate, the meshes about equally long and wide, relatively large, with thick walls 5. *A. acuminatum*
5. Perianth segments of both outer and inner whorls entire; other characteristics never combined as above (6).
6. Ovary and capsule never crested. Outer bulb scales with rectangular cells in regular vertical rows, the cell walls sometimes obscure even under strong magnification (7).
6. Ovary and capsule conspicuously crested (9).
7. Bulbs elongate, terminating stout, *Iris*-like rhizomes; outer bulb scales with persistent, parallel fibers, striate with elongate cells in regular vertical rows....6. *A. Gooddingii*
7. Bulbs ovoid or subspherical, without *Iris*-like rhizomes; outer bulb scales without fibers, obscurely or not at all striate (8).
8. Bulbs about 1 cm. thick, solitary, proliferating from the base by means of slender *Agropyron*-like rhizomes.................................7. *A. glandulosum*
8. Bulbs about 2 cm. thick, often clustered, sometimes stipitate (short-rhizomatous) at base, but without long, slender rhizomes.........................8. *A. Kunthii*
9. Umbel nodding; stamens exserted; bulb elongate, stipitate (short-rhizomatous) at base. Outer bulb scales striate with elongate cells in regular, vertical rows...9. *A. cernuum*
9. Umbel erect; stamens included; bulb ovoid or subspherical, not stipitate at base (10).
10. Leaf solitary, terete, surpassing the scape in length. Alveoli on seeds minutely roughened (11).
10. Leaves 2 or more per scape, flattened and channeled. Outer bulb scales cellular-reticulate (12).
11. Perianth pale pink or nearly white, 8 to 12 mm. long, commonly much less than twice as long as the stamens; pedicels slender, longer than the flowers; outer bulb scales, or some of them, usually with distinct or indistinct, contorted, cellular reticulations 10. *A. nevadense*
11. Perianth usually bright pink, 12 to 20 mm. long, about twice as long as the stamens; pedicels stout, mostly shorter than the flowers; outer bulb scales without reticulations 11. *A. Parishii*
12. Scape low; outer bulb scales dark brown, the meshes vertically elongate or contorted,

the walls not sinuous; perianth segments pink, acute; pedicels erect or ascending in fruit; alveoli on seeds smooth .... 12. *A. Bigelovii*

12. Scape taller; outer bulb scales grayish or brownish, the meshes usually transversely elongate, the walls very sinuous; perianth segments purplish, acuminate; outer pedicels reflexed in fruit; alveoli on seeds pustuliferous .... 13. *A. Palmeri*

1. **Allium Geyeri** Wats. (*A. funiculosum* A. Nels.). White Mountains (southern Apache and Navajo counties), Coconino and Gila counties, Huachuca Mountains (Cochise County), Santa Rita Mountains (Santa Cruz County), Rincon Mountains (Pima County), 5,000 to 10,000 feet, commonly in pine woods, June to August. Western Texas to southeastern Arizona, northward through New Mexico and Colorado to southern Wyoming, disjunct to eastern Washington and adjacent Idaho, with outlying stations in western Montana, southern Alberta, and the Black Hills of South Dakota.

From one isolated mountain range to another the geographical races of this species may differ slightly in their morphology. These differences, however, are not of the order of those which usually distinguish taxonomic entities.

2. **Allium rubrum** Osterh. (*A. sabulicola* Osterh.). White Mountains (Apache and Greenlee counties), up to 9,500 feet, July and August. Eastern Arizona and New Mexico, northward through the Rocky Mountain region to southern Alberta and eastern Oregon, with an outlying station on Vancouver Island, British Columbia.

Under *A. rubrum* are grouped a number of asexual, bulbiliferous races assumed to have been derived from *A. Geyeri*. There is often a certain amount of morphological parallelism between the race of *A. Geyeri* and that of *A. rubrum* occurring in the same region, but this is not always the case. Even when they occur in the same region they grow in separate colonies, and there is no evidence that the transformation from *A. Geyeri* into *A. rubrum* takes place very often.

3. **Allium Plummerae** Wats. Chiricahua and Huachuca mountains (Cochise County), Baboquivari Mountains (Pima County), 5,500 to 9,000 feet, June to September, type from Tanner's Canyon, Huachuca Mountains (*Lemmon* 2893). Southeastern Arizona and adjacent northern Mexico.

A. Plummerae is similar in habit to *A. cernuum* Roth but differs in its fibrous-reticulate bulb scales, erect umbel, and included stamens. It seems more closely related to *A. Geyeri* but differs from this also in several important characteristics.

4. **Allium macropetalum** Rydb. (*A. deserticola* (Jones) Woot. & Standl.). Apache County to Coconino and Yavapai counties, south to Cochise and Pima counties, 1,000 to 7,000 feet, March to June. Western Colorado and eastern Utah to Arizona, New Mexico, and western Texas.

One of the most widely distributed of the wild onions of Arizona and sometimes very abundant, covering the ground in places with its orchid-pink flowers.

5. **Allium acuminatum** Hook. Coconino and Mohave counties to Graham, Cochise, and Pinal counties, 2,000 to 7,000 feet, April to June. Montana (?) to British Columbia, south to Arizona and California.

Flowers deep lavender-pink or rose-pink. The 2 or 3 slender leaves usually wither by the time the flowers are open and readily break off.

6. Allium Gooddingii Ownbey. White Mountains, Apache County, at Bonita Creek (*Goodding* 1233, the type collection) and Phelps Botanical Area, 9,500 feet (*Phillips* 3092). Known only from these collections.

Related to *A. brevistylum* Wats. of the Rocky Mountains but confused previously with *A. Plummerae,* from which it is very distinct.

7. Allium glandulosum Link & Otto (*A. rhizomatum* Woot. & Standl.). Mule Mountains, Cochise County (*Goodding* 993, 1003), August. Western Texas to southeastern Arizona and far south into Mexico.

The Arizona plant, which has been described as *A. rhizomatum* (type locality Gila Hot Springs, New Mexico), seems only to represent the Mexican *A. glandulosum* (type locality near Mexico City) at its northern limits. Typical *A. glandulosum* has deep-red flowers. The flowers in Arizona plants are pale, with the color confined mostly to the midribs. This appears to be the principal difference, and on the basis of the material examined does not seem to warrant specific segregation. The glands to which the specific epithet alludes are septal glands evidently, which are not apparent on herbarium specimens representing either extreme.

8. Allium Kunthii Don. Cochise and Pima counties, 4,000 to 5,000 feet, August and September. Western Texas to southeastern Arizona, south to Central America.

Flowers cream-colored, fading pink. This species closely resembles *A. glandulosum,* from which it differs most conspicuously in the absence of slender rhizomes and in the larger, often clustered bulbs. The reported flowering of this species in April and May seems to have been based on specimens of *Nothoscordum texanum.*

9. Allium cernuum Roth. Apache, Navajo, and Coconino counties to Cochise, Santa Cruz, and Pima counties, mostly in pine forests, 5,000 to 8,500 feet, July to October. In mountainous or cool regions, almost throughout temperate North America.

Nodding onion. The commonest woodland species of the state, distinguished by its slender scapes, recurved at apex, and deep- to pale-pink or nearly white flowers. Two well-marked geographical variants occur in Arizona. Var. *neomexicanum* (Rydb.) Macbr. (*A. neomexicanum* Rydb.), with relatively taller scapes, broader, thinner, nearly plane leaves, and white or nearly white inner bulb scales, is found from the Santa Rita Mountains (Pima and Santa Cruz counties) through the Huachuca and Chiricahua mountains (Cochise County) to western Texas and northern Mexico. Var. *obtusum* Cockerell, with shorter scapes, narrower, thicker, concave-convex leaves, and bright-pink or red inner bulb scales, occurs in the White Mountains (Apache, Navajo, and Greenlee counties), Tunitcha Mountains (Apache County), and in the vicinity of Flagstaff (Coconino County). This is the Rocky Mountain phase of the species.

10. Allium nevadense Wats. Northwestern Coconino and northern Mohave counties at Grand Canyon, Peach Springs, between Hackberry and Peach

Springs, and Pagumpa, in gravelly or stony (calcareous) soil, 4,000 to 7,000 feet, April to June. Southern Idaho and southeastern Oregon through Utah and Nevada to northwestern Arizona and eastern California.

The usual form in Arizona is var. *cristatum* (Wats.) Ownbey (*A. cristatum* Wats.), in which the contorted reticulations on the outer bulb scales are indistinct or lacking. Only a collection at Pagumpa (*Jones* 5082) approaches typical *A. nevadense* in this character.

11. Allium Parishii Wats. Chemehuevi, Mohave County (*Jones* in 1903), Kofa Mountains, Yuma County (*Nichol* in 1937), rocky slopes, 3,000 to 4,000 feet, April. Western Arizona and southeastern California.

The specimens from the Kofa Mountains have trifid stigmas but otherwise agree closely with other collections of this species including the type.

12. Allium Bigelovii Wats. 10 miles south of Tonto Basin (Gila County), Walnut Grove and Ash Fork (Yavapai County), 1½ miles west of Burro Creek (Mohave County), 1,500 to 5,000 feet, April and May. Southwestern New Mexico and Arizona.

In *Flowering Plants and Ferns of Arizona* (p. 192) this name applies, at least in part, to specimens of *A. Palmeri,* to which true *A. Bigelovii* shows no particularly close affinity.

13. Allium Palmeri Wats. Apache, Navajo, Coconino, Mohave, Yavapai, Gila, Santa Cruz, and Pima counties, 4,000 to 7,500 feet, often in pine forests, May and June. Southern Utah, eastern Nevada, northwestern New Mexico, and Arizona.

The bulbs of this species often proliferate from the base by means of long, slender rhizomes, each terminated by a daughter bulb. They differ in this way from bulbs of *A. bisceptrum* Wats., which bear a cluster of bulblets at the base. Often, however, the rhizomes have been detached and lost from herbarium specimens, so that it cannot be determined if the species is constant in this respect.

7. NOTHOSCORDUM

Very similar to *Allium,* differing chiefly in having more than 2 ovules in each cell of the ovary and in the absence of onionlike odor.

1. Nothoscordum texanum Jones. Gila, Pinal, Cochise, Santa Cruz, and Pima counties, 4,000 to 6,000 feet, growing in the open on hillsides and plains in shallow, hard, gravelly soil, April and May. Western Texas and southern Arizona, probably also in southwestern New Mexico and northern Mexico.

The type (*Jones* in 1930) was collected in Cochise County, Arizona, near Rodeo, New Mexico. The flowers are yellowish white, tinged with purple externally, and are somewhat fragrant.

8. TRITELEIOPSIS (60, p. 98)

Stems leafy at base; lèaves several, crowded, bearing bulblets in the axils; umbel many-flowered; involucral bracts scarious, acuminate; perianth funnelform, purplish blue, with transverse scalelike appendages between the stamens; anthers basifixed, the slender filaments adnate below to the perianth; stigma small, entire or nearly so; ovary stipitate; seeds elongate.

1. Triteleiopsis Palmeri (Wats.) Hoover (*Brodiaea Palmeri* Wats.). Western base of the Gila Mountains, Yuma County, 250 feet (*Ripley & Barneby* in 1949). Known otherwise only from Baja California.

According to Barneby (personal communication) the stout scape is about 2 feet high and the leaves are broadly lanceolate. At the Arizona station it grows on dunes with *Hesperocallis* and *Oenothera*.

9. DICHELOSTEMMA (59)

Stems naked, the few leaves arising directly from the bulb; umbel few- to many-flowered, subcapitate or if more open, then the rays unequal; involucral bracts large, ovate or lance-ovate; perianth blue-violet, without internal scales, the lobes less than twice as long as the campanulate tube; stamens dimorphic, the inner (fertile) ones with filaments completely adnate to the perianth and winged on each side of the anther, the anthers basifixed, appressed to the style; stigma small, 3-lobed; ovary sessile or nearly so; seeds elongate.

In *Flowering Plants and Ferns of Arizona* (pp. 192–193), this and the following genus were referred to *Brodiaea* (58).

1. Dichelostemma pulchellum (Salisb.) Heller (*Brodiaea capitata* Benth.). Coconino and Mohave counties to Cochise, Santa Cruz, and Pima (doubtless also Yuma) counties, 5,000 feet or (usually) lower, common and abundant, February to May (rarely late summer). Southwestern New Mexico to Oregon, California, Baja California, and Sonora.

Bluedicks, covena, grass-nuts. The violet-colored flowers of this species are conspicuous on the mesas and open slopes in early spring. The bulbs were eaten by the Pima and Papago Indians. The common form in Arizona is the poorly characterized var. *pauciflorum* (Torr.) Hoover, with narrower, paler-colored bracts, these usually shorter than the pedicels.

10. TRITELEIA (60)

Stems naked; leaves 1 or 2; umbel with rays all more or less elongate and nearly equal; involucral bracts small, lanceolate or oblong; perianth bright yellow fading purplish, with lobes much more than twice as long as the turbinate tube; stamens all alike, the filaments not winged, the anthers versatile, sagittate (in the Arizona species), not appressed to the style; stigma obscurely lobed; ovary short-stipitate; seeds not elongate.

1. Triteleia Lemmonae (Wats.) Greene (*Brodiaea Lemmonae* Wats.). Springerville (Apache County), Flagstaff and Mogollon Escarpment (Coconino County), northern Gila County, 5,000 to 7,000 feet, in partial shade of pines, May to August, type from Oak Creek, Coconino County (*Lemmon* in 1884). Known only from Arizona.

11. ANDROSTEPHIUM

Flowering stems from a bulb, scapose; leaves all basal, few, narrow, grass-like; perianth funnelform, the segments united below; filaments partly united into a tube, this with short lobes or teeth between the anthers; capsule 3-celled, obtusely 3-angled.

1. Androstephium breviflorum Wats. Northern Arizona, Apache County to Mohave County, in sandy soil, 2,000 to 7,000 feet, March and April. Western Colorado to northern Arizona and southeastern California.

12. MILLA. Mexican-star

Flowering stems from a tunicate bulb; leaves all basal, narrow, grasslike; flowers solitary or in umbellike clusters of 2 or 3; perianth large, salverform, the lobes about 2 cm. long, white with a green midvein; capsule sessile, somewhat obovoid.

1. Milla biflora Cav. Cochise, Santa Cruz, and Pima counties, near the Mexican boundary, 4,000 to 7,000 feet, mostly in open woods of oak or pine, August and September. Southern New Mexico and Arizona, southward to Oaxaca, Mexico.

13. LILIUM. Lily

Flowering stems tall, leafy, from thick-scaled bulbs; leaves mostly in whorls, linear or lanceolate; flowers large, 5 to 9 cm. long, yellow or orange, often spotted with brown or purple; perianth funnelform or campanulate, the segments distinct; anthers linear, versatile; stigma capitate, 3-lobed; capsule 3-celled, loculicidal.

KEY TO THE SPECIES

1. Perianth clear lemon-yellow, usually finely spotted inside with darker color, the segments lanceolate, very acuminate, tapering gradually into the claws, 8 to 10 cm. long 1. *L. Parryi*

1. Perianth red or reddish orange, coarsely purple-spotted inside, the segments lanceolate to ovate and somewhat rhombic, obtuse or acutish, abruptly contracted into the claws, 5 to 7 cm. long .... 2. *L. umbellatum*

1. Lilium Parryi Wats. Huachuca Mountains (Cochise County), Santa Rita Mountains (Santa Cruz County), rich soil along streams, rare, May to July. Southern Arizona and southern California.

Lemon lily. One of the handsomest and showiest plants of Arizona, with fragrant flowers. It has been collected by Palmer, Pringle, and Goodding, and recently by Darrow and by Gould, at an elevation of 6,500 feet in the Huachuca Mountains.

***2. Lilium umbellatum** Pursh. A specimen collected by E. Palmer in 1869, labeled "Arizona" without definite locality, seems to be the only basis for including this lily in the flora of the state. The known range of the species is Ohio to Alberta, south to Arkansas and New Mexico.

14. FRITILLARIA (61). Fritillary

Flowering stem from a thick-scaled bulb, leafy, bearing 1 or few nodding flowers; leaves alternate, linear; perianth campanulate, of 6 distinct segments not more than 2 cm. long, greenish yellow mottled with maroon; style 3-cleft, the linear lobes introrsely stigmatic; fruit a 3-valved capsule.

Several species of this genus are cultivated as ornamentals. A California species, *F. biflora,* is known as mission-bells.

1. Fritillaria atropurpurea Nutt. Navajo, Coconino, and Gila counties, rich soil in woods, 6,000 to 8,500 feet, April to June. North Dakota to Washington, south to New Mexico, Arizona, and California.

15. CALOCHORTUS (62). MARIPOSA

Flowering stems from tunicate bulbs, subscapose, simple or sparingly branched; leaves few, narrow, alternate; flowers solitary, or few in a terminal inflorescence, large and showy; outer perianth segments green, sepallike, glandless, the inner segments larger, petallike, each with a large, hairy gland near the base; fruit a 3-angled or 3-winged septicidal capsule.

The mariposas are among the most beautiful of Arizona wild flowers, their petals exhibiting a range of colors from pale yellow to nearly scarlet and from whitish to deep lavender. The rather large bulbs were eaten by the Hopi and Navajo Indians. It is said that the Mormon pioneers of Utah also used the bulbs of the sego-lily (*C. Nuttallii*) as food in times of scarcity.

KEY TO THE SPECIES

1. Stem usually decumbent and flexuous, often branched; inflorescence, if more than 1-flowered, with a distinct main axis (monochasial); petal gland not depressed (rarely slightly so), not surrounded by a membranaceous border, transversely elongate to nearly circular .... 1. *C. flexuosus*

1. Stem erect or nearly so, usually simple; inflorescence subumbellate; petal gland more or less depressed, surrounded by a membranaceous border (2).

2. Petal gland not transversely elongate, the bordering membrane broad, usually continuous (3).

2. Petal gland usually elongate transversely, the bordering membrane narrow, discontinuous. Petals whitish to deep lavender-purple, the hairs thickened apically and often branched (4).

3. Hairs of the petals near the gland slender; petals whitish, purple, or yellow; stem often bearing bulblets toward the base .... 2. *C. Nuttallii*

3. Hairs of the petals thickened apically; petals yellow to nearly scarlet; stem rarely bulbiferous .... 3. *C. Kennedyi*

4. Anthers obtuse or (exceptionally) acutish; petal gland only slightly elongate transversely, or nearly circular; stem often bulbiferous toward the base...... 4. *C. ambiguus*

4. Anthers usually acute or acuminate; petal gland distinctly elongate transversely; stem rarely bulbiferous........ 5. *C. Gunnisoni*

1. Calochortus flexuosus Wats. Navajo, Coconino, Mohave, Yavapai, and Gila counties, 2,000 to 7,000 (8,500?) feet, open ground or in chaparral, April to June. Southwestern Colorado to southeastern California and central Arizona.

Petals pale purple to nearly white. The bent or twisted stems are characteristic.

2. Calochortus Nuttallii Torr. & Gray. Apache County to northern Mohave, eastern Yavapai, and northern Gila counties, 4,500 to 8,000 feet, mesas, slopes, and open pine forests, May to July. Western North Dakota to eastern Oregon, south to Nebraska, northern New Mexico and Arizona, and eastern California.

Sego-lily, the state flower of Utah. The typical plant, with whitish to lavender-blue petals, is known in Arizona from the Kaibab Plateau and Grand Canyon (Coconino County), and from the mountains of northern Mohave and northern Gila counties. Var. *aureus* (Wats.) Ownbey (*C. aureus*

Wats.), with lemon-yellow petals, occurs in Apache, Navajo, and eastern Coconino counties, and is known elsewhere only from northwestern New Mexico and southern Utah.

3. Calochortus Kennedyi Porter. Yavapai and Mohave counties, south to Cochise, Pima, Santa Cruz, and Yuma counties, 5,000 feet or (usually) lower, March to May. Arizona, Nevada, southern California, and northern Sonora.

Desert mariposa. In favorable seasons and localities this plant gives a gorgeous display of color, rivaling that of the California-poppy (*Eschscholtzia*). The usual color of the petals is a deep, rich orange, but in var. *Munzii* Jepson they are clear yellow. This form is occasional in most parts of the range of the species.

4. Calochortus ambiguus (Jones) Ownbey. Navajo County to eastern Mohave County, south to Cochise, Santa Cruz, and Pima counties, 3,000 to 8,000 feet, dry slopes, chaparral, and open pine forests, April to August, type from near Flagstaff (*Jones* in 1884). New Mexico and Arizona.

The most widely distributed and generally the most abundant of the Arizona mariposas, resembling *C. Nuttallii* in general appearance.

5. Calochortus Gunnisoni Wats. Lukachukai and White mountains (Apache County), Keet Seel (northern Navajo County), San Francisco Peaks and near Flagstaff (Coconino County), near Montezuma Castle (Yavapai County), 6,500 to 8,000 feet, June to August. Western South Dakota and Montana to New Mexico and Arizona.

16. YUCCA (63). SOAP-WEED, SPANISH-BAYONET, DATIL

Large plants with a thick, branching, mainly subterranean caudex, or a distinct trunk above ground; leaves numerous, clustered at the ends of the branches, narrow, elongate, commonly spine-tipped; flowers large, perfect, numerous or many in terminal racemes or panicles; perianth segments rather thick, whitish; ovary 3-celled; fruit dry or fleshy, dehiscent or indehiscent; seeds many, flat.

The yuccas are an important resource of the Indians of the Southwest. The buds, flowers, and emerging flower stalks are eaten raw or boiled, or the flower stalks roasted like mescal. The large, pulpy fruits of the baccate species (*Y. baccata*, etc.) are eaten raw or roasted, dried for winter use, or ground into meal, and the seeds also are used for food. A fermented beverage was made from these fruits. Fiber from the leaves furnishes material for rope, mats, sandals, baskets, and cloth. The roots, known as amole, have saponifying properties and are used as a sort of soap and as a laxative. The seeds are the natural food of the larvae of small moths, which pollinate the flower by gathering the pollen into a mass that is pushed into the tube of the stigma.[1]

KEY TO THE SPECIES

1. Fruit indehiscent (2)
1. Fruit dehiscent (7).
2. Plants treelike, the stems commonly branched high above the base, up to 9 or even 12 m. high (only 3 to 6 m. in var. *Jaegeriana*); leaves clustered at the ends of the branches, less than 50 cm. long; inflorescence, including the scape, not more than 50 cm. long;

[1] In the following treatment Mrs. McKelvey's monograph (63) has been followed in the main.

outer perianth segments erect at anthesis; fruit semicapsular, soon becoming dry, spreading. Leaves rigid, fibrous, the margins corneous, sharply toothed
1. *Y. brevifolia*

2. Plants not treelike, branched only toward the base or sometimes (in *Y. Schottii*) arborescent and branched as high as 2 m. above the base, up to 5.5 m. high (in *Y. Schottii*) but usually much lower; leaves all basal or, in caulescent plants, borne up to the base of the scape, usually more than 50 cm. long; inflorescence, including the scape, seldom less than 60 cm. long (except in *Y. Schottii*); perianth segments all more or less spreading at anthesis; fruit fleshy, becoming pendent (3).

3. Pistil less than 4 cm. long at anthesis; perianth with the united portion not, or barely, surpassing the base of the ovary; filaments attached at the very base of the perianth (4).

3. Pistil 4.5 cm. or longer at anthesis; perianth with the united portion evidently surpassing the base of the ovary; filaments attached somewhat above the base of the perianth (5).

4. Leaf margins thin, without free fibers; ovary not more than 7 mm. in diameter at anthesis; scape extremely short .... 2. *Y. Schottii*

4. Leaf margins thick, with numerous detaching fibers; ovary 7 to 12 mm. in diameter at anthesis; scape 15 cm. or longer .... 3. *Y. schidigera*

5. Plants acaulescent or with short, procumbent stems, these solitary or forming small, usually open clumps; filaments (at anthesis) with the tip forming a more or less conspicuous angle to the lower portion. Inflorescence not or little surpassing the foliage, somewhat flattened at apex; leaves often twisted, the marginal fibers broad and coarse
4. *Y. baccata*

5. Plants caulescent (sometimes obscurely so in *Y. arizonica*), with erect or ascending stems 1 to 2 m. long, these forming large, crowded clumps; filaments not conspicuously angled (6).

6. Inflorescence many-branched, surpassing the foliage for most of its length, broadest near the apex; free marginal fibers numerous on young leaves, almost wanting on old leaves; bracts tough, somewhat persistent.......................... 5. *Y. arizonica*

6. Inflorescence few-branched, surpassing the foliage for not more than half of its length, broadest near the middle; free marginal fibers wanting on young leaves, numerous on old leaves; bracts becoming thin and brittle...................... 6. *Y. Thornberi*

7. Leaf margins very sharp, finely denticulate; stigmas capitate, bright green, covered with white, translucent, fleshy hairs; filaments usually as long as or longer than the pistil, finely papillose, thickest near the middle; anthers with kidney-shaped cells; capsules loculicidal .... 7. *Y. Newberryi*

7. Leaf margins thin, paperlike, more or less filiferous; stigmas more or less lobed, white, papillose; filaments commonly shorter than the pistil, more or less pubescent, thickest toward apex (more or less clavate); anthers sagittate or hastate; capsules mainly septicidal but also more or less loculicidal (8).

8. Style bright green, darker than the ovary at anthesis. Plants acaulescent or short-stemmed; leaves rather soft and flexible, 10 to 25 mm. in greatest width
8. *Y. Harrimaniae*

8. Style white or pale green, paler than the ovary at anthesis (9).

9. Scape elongate; inflorescence paniculate, borne high above the foliage (10).

9. Scape short; inflorescence simply racemose (rarely with a very few short or abortive branches), starting below or just above the foliage (12).

10. Plants at maturity arborescent, with 1 to several elongate stems, not forming clumps; capsules 5 to 8 cm. long .... 9. *Y. elata*

10. Plants at maturity acaulescent or short-stemmed, forming clumps; capsules (except in *Y. kanabensis*) less than 6 cm. long (11).

11. Inflorescence branched most of its length; leaves up to 70 cm. long, usually concavo-convex for most of their length; pistil 2.5 to 3.2 cm. long; capsules 5 to 5.7 cm. long, the walls strong and persistent................................ 10. *Y. utahensis*

11. Inflorescence commonly branched only toward base; leaves up to 45 cm. long, usually plano-convex for most of their length; pistil 2 to 2.5 cm. long; capsules about 4.5 cm. long, the walls thin and fragile................................ 11. *Y. verdiensis*

12. Capsules 3.5 to 5 (rarely nearly 6) cm. long, the walls thin and fragile. Inflorescence including the scape) less than 1.5 m. long........................12. *Y. angustissima*
12. Capsules 5 cm. or longer, the walls tough and persistent (13).
13. Inflorescence (including the scape) about 1 m. long...................13. *Y. navajoa*
13. Inflorescence (including the scape) 2 to 3 m. long..................14. *Y. kanabensis*

1. Yucca brevifolia Engelm. Northern Mohave County to southwestern Yavapai and northern Yuma counties, up to about 3,500 feet, deserts, March to May. Southwestern Utah, southern Nevada, northwestern Arizona, and southeastern California.

Joshua-tree. There are fine "forests" of these treelike yuccas in Mohave County, south of Pierce Ferry and in the Sacramento Valley. In var. *Jaegeriana* McKelvey (var. *Wolfei* Jones), which occurs in northwestern Mohave County (*Jones* 5008, *McKelvey* 4160), the plant is only 3 to 6 m. high, branching near the ground, and the leaves are not more than 22 cm. long. For a description of this variety see Munz (64).

2. Yucca Schottii Engelm. Greenlee (?), southeastern Pinal, Cochise, Santa Cruz, and Pima counties, 4,000 to 7,000 feet, hillsides and canyons, April to August, type from near Tubac, Santa Cruz County (*Schott* in 1855). Southwestern New Mexico, southeastern Arizona, and northern Mexico.

Hoary yucca. A large, handsome species, often cultivated.

3. Yucca schidigera Roezl (*Y. mohavensis* Sarg.). Western Mohave County, up to 3,500 feet, deserts, March and April. Southern Nevada, northwestern Arizona, southeastern California, and Baja California. Mojave yucca.

4. Yucca baccata Torr. Apache County to Mohave, Yavapai, and Gila counties, 3,000 to 8,000 feet, often with pinyon and juniper, April to July. Southwestern Colorado, southern Utah and Nevada, and southwestern Texas to southeastern California.

Blue yucca, banana yucca, datil. Very abundant in central and northern Arizona. In typical *Y. baccata* the plant is acaulescent or has a few procumbent stems. Var. *vespertina* McKelvey, a densely cespitose plant with many short, procumbent to nearly erect stems and with the flowers more attenuate and stipelike at base, is common in Mohave and Yavapai counties, type from near Peach Springs (*McKelvey* 2167).

5. Yucca arizonica McKelvey (? *Y. confinis* McKelvey). Gila (?), Pinal, Cochise, Santa Cruz, and Pima counties, mostly below 4,000 feet, May, type from near Tubac (*Schott* in 1855). Southern Arizona and doubtless also northern Mexico.

Mrs. McKelvey (personal communication, November 9, 1947) suggested that *Y. confinis*, type from near Douglas (*McKelvey* 2099), may be only a juvenile form of *Y. arizonica*.

6. Yucca Thornberi McKelvey. Southwestern Gila, southeastern Pinal, Cochise, and Pima counties, 3,000 to 5,000 feet, type from foothills of the Rincon Mountains (*McKelvey* 1627). Known only from southern Arizona.

Y. Thornberi, *Y. confinis*, and *Y. arizonica* were included under the combination *Y. baccata* Torr. var. *brevifolia* (Schott) Benson & Darrow (*Amer. Jour. Bot.* 30: 234. 1943).

7. Yucca Newberryi McKelvey. Southern bank of the Colorado River from Kanab Creek to Lake Mead (Mohave County), 1,000 to 2,000 feet, April to June, type from New Water Point (*McKelvey* 4087). Known only from northwestern Arizona.

Very closely related to the Californian *Y. Whipplei* Torr. ("Our Lord's Candle"), to which the Arizona plant was referred previously. The flowers are reported to be fragrant.

8. Yucca Harrimaniae Trel. Carrizo Mountains, Apache County (*Standley* 7314). Colorado, Utah, northeastern New Mexico, and northeastern Arizona.

9. Yucca elata Engelm. Graham, Gila, Yavapai, eastern Maricopa, Cochise, Santa Cruz, and Pima counties, 1,500 to 6,000 feet, grassland and desert, May to July, lectotype from Camp Grant, Graham County (*Rothrock* 382). Southwestern Texas to central and southern Arizona.

Soaptree yucca, palmilla. According to J. M. Webber (personal communication) the stems reach a height of 9 m. (30 feet). *Y. elata* has proved useful as emergency ration for livestock during periods of drought. The chopped stems, mixed with concentrated food such as cottonseed meal, are palatable and nourishing. The plants sprout from the roots but cannot be cropped too often because of the slow rate of growth. The roots, known as amole, are detergent and have been used in the manufacture of shampoo soap. A substitute for jute has been manufactured from the leaf fibers. Indians eat the young flower stalks and lower part of the stem.

10. Yucca utahensis McKelvey. Grand Canyon, Coconino County (*Eastwood*), near Congress Junction, Yavapai (?) County (*Peebles et al.* 4127). Southwestern Utah, northern and central Arizona, and (?) southern Nevada.

11. Yucca verdiensis McKelvey. Southern Navajo, Coconino, Yavapai, and Gila counties, 3,000 to 6,000 feet, type from between Clarkdale and Cottonwood, Yavapai County (*McKelvey* 2752). Known only from central Arizona.

12. Yucca angustissima Engelm. Apache, Coconino, Mohave, and Yavapai counties, 3,000 to 7,000 (?) feet, May and June, type from "deserts of the Colorado River region" (*Bigelow* in 1854). Northwestern New Mexico, northern and central Arizona, and (?) southwestern Utah.

13. Yucca navajoa J.M. Webber (*Y. Baileyi* Woot. & Standl., in part, *Y. Standleyi* McKelvey). Apache, Navajo, and eastern Coconino counties, 4,500 to 7,500 feet, sandy plains, hillsides, and pine forests, May and June. Colorado, southeastern Utah, northwestern New Mexico, and northern Arizona.

According to Webber (65) this species develops from 1 to 44 rosettes of leaves, the average number being 10.

14. Yucca kanabensis McKelvey. Eastern Coconino County to northeastern Mohave County, 4,500 to 6,000 feet. Southwestern Utah and northern Arizona.

17. NOLINA (66). Bear-grass

Plants with a large, woody caudex, this subterranean or forming a distinct trunk above ground; leaves numerous, clustered, long and narrow; flowers perfect or unisexual, very numerous, in ample terminal panicles; capsule 3-lobed; seed 1 in each cell, turgid.

It is reported that sheep and goats are sometimes poisoned by these plants.

KEY TO THE SPECIES

1. Plants acaulescent; leaves not more than 12 mm. wide; fruit 5 to 8 mm. wide, not or but moderately inflated, the seed becoming exposed (2)
1. Plants caulescent; leaves 15 to 35 mm. wide; fruit up to 15 mm. wide, greatly inflated, the seed not exposed. Trunk 1 to 2 m. high, clothed with dead leaves (3).
2. Fruit not inflated, 5 to 7 mm. wide, the seed soon exposed and protruding....1. *N. texana*
2. Fruit moderately inflated, 7 to 8 mm. wide, the seed exposed only after dehiscence of the fruit, or not protruding..2. *N. microcarpa*
3. Leaf margins not serrulate, shreddy; leaves nearly flat, usually not more than 25 mm. wide; perianth segments about 3 mm. long; fruit 10 to 12 (seldom 15) mm. wide 3. *N. Bigelovii*
3. Leaf margins sharply serrulate; leaves concave ventrally, keeled dorsally, up to 35 mm. wide; perianth segments about 4 mm. long; fruit 12 to 15 mm. wide......4. *N. Parryi*

1. Nolina texana Wats. (*N. affinis* Trel., *N. caudata* Trel.). Cochise and Santa Cruz counties. Texas to southeastern Arizona and northern Mexico.

The Arizona plants belong to var. *compacta* (Trel.) Johnst. (*N. erumpens compacta* Trel.), which is more robust and has a longer, more densely and rigidly branched inflorescence than in typical *N. texana* (I. M. Johnston, *Jour. Arnold Arbor.* 24: 90; and 25: 64).

2. Nolina microcarpa Wats. Nearly throughout the state, common in the central and southeastern counties, usually in exposed situations on mountainsides, about 3,000 to 6,500 feet, May and June, type from "Rocky Canyon," southern Arizona (*Rothrock* 278). Western Texas to Arizona and northern Mexico.

Sacahuista. The caudex and emerging flower stalk apparently were prepared for food by the Indians in the same way that the corresponding parts of *Agave* and *Yucca* were used. The leaves are browsed in times of drought.

3. Nolina Bigelovii (Torr.) Wats. Mohave, western Yavapai, and Yuma counties, not uncommon, up to at least 3,500 feet, hills and canyons, type from William River (*Bigelow* in 1853–1854). Western Arizona, southeastern California, Baja California, and Sonora.

4. Nolina Parryi Wats. (*N. Bigelovii* var. *Parryi* L. Benson). Grand Canyon, Coconino County (*Eastwood* 6050), and in Mohave County. A collection at Fort Whipple, Yavapai County (*Coues & Palmer* in 1865), is cited by Trelease (66, p. 423). Western Arizona, southern California, and Baja California.

There seems to be intergradation between *N. Bigelovii* and *N. Parryi* in Arizona. Specimens from Mohave County, from the Grand Canyon National Monument (*Laws* in 1944), and from between Kingman and Oatman (*Eastwood* 18402) have the flat leaves and smaller flowers of the former but the serrulate leaf margins of the latter. It is doubtful, in fact, that true *N. Parryi* occurs in Arizona (see Munz & Roos, *El Aliso* 2: 221–227).

18. DASYLIRION. Sotol

Plants dioecious, with a thick, woody, mostly subterranean caudex; leaves in large clusters, very rigid, the margins armed with sharp, curved spines; flowers in large terminal panicles, these on the staminate plants composed of dense catkinlike spikes; capsule 1-celled, 3-winged.

1. Dasylirion Wheeleri Wats. Greenlee, Graham, Gila, Pinal, Cochise, Santa Cruz, and Pima counties, 4,000 to 6,000 feet, May to August, types from southern Arizona (*Rothrock* 329, 655). Western Texas to Arizona and northern Mexico.

Cattle feed upon the short, round heads of plants that have been burned or split open. Bighorns are reported to browse these plants. The heads contain much sugar and have been used in the manufacture of alcohol. From the roasted hearts the natives prepared an article of food similar to that obtained from mescal (*Agave*) and also a potent beverage generally known as sotol.

19. ASPARAGUS

Stems simple, fleshy and scaly in the juvenile (edible) state, becoming tall, slender, much-branched; leaves minute, scalelike, with filiform branchlets in their axils; flowers commonly solitary, nodding, small; perianth campanulate, greenish white; fruit a few-seeded berry, red at maturity.

1. Asparagus officinalis L. The well-known garden vegetable, occasionally escaping from cultivation, as at Sacaton, Pinal County, but scarcely naturalized. Native of Europe.

20. SMILACINA (67)

Flowering stems from horizontal rootstocks, unbranched, leafy; leaves alternate, mostly sessile or subsessile, with ample lanceolate to ovate blades; flowers small, whitish, in terminal racemes or panicles; perianth rotate; ovary 3-celled; fruit a few-seeded berry.

KEY TO THE SPECIES

1. Inflorescence a many-flowered panicle; perianth segments not more than 2.5 mm. long, shorter than the stamens; berries red with purplish dots; leaves oval, ovate, or lance-ovate . . . . 1. *S. racemosa*
1. Inflorescence a few-flowered raceme; perianth segments 3 to 7 mm. long, longer than the stamens; berries green with vertical, dark-blue (exceptionally red) stripes, becoming nearly black; leaves lanceolate or oblong-lanceolate . . . . 2. *S. stellata*

1. Smilacina racemosa (L.) Desf. Apache, Coconino, Greenlee, Graham, Gila, Cochise, and Pima counties, 6,000 to 10,000 feet, rich soil in coniferous forests, May to July. Throughout most of temperate North America.

False-Solomon-seal. The commoner plant in Arizona is var. *amplexicaulis* (Nutt.) Wats., with leaves sessile or subsessile and often slightly clasping, acute or short-acuminate. This intergrades with typical *S. racemosa,* in which the leaves are short-petioled. The latter occurs in the Chiricahua and Huachuca mountains (Cochise County) and in the Santa Catalina Mountains (Pima County). Var. *cylindrata* Fern. is a reduced form with a relatively small and narrow panicle.

2. Smilacina stellata (L.) Desf. Kaibab Plateau to Oak Creek (Coconino County), White Mountains (Apache County), Pinaleno Mountains (Graham County), Chiricahua and Huachuca mountains (Cochise County), Santa Catalina Mountains (Pima County), 7,500 to 10,000 feet, rich woods, May and June. Throughout most of temperate North America; Europe. Starflower.

21. POLYGONATUM (68). SOLOMON-SEAL

Flowering stems from horizontal, knotted rootstocks; herbage glabrous; leaves alternate, somewhat glaucous beneath, the blades elliptic, the petioles short and broad; flowers solitary or in clusters of 2 or 3, borne on strongly decurved pedicels and peduncles; perianth greenish white, 12 to 23 mm. long, the tube cylindric, the limb slightly expanded, the lobes erect; stamens inserted on the perianth tube.

1. Polygonatum cobrense (Woot. & Standl.) Gates (*Salomonia cobrensis* Woot. & Standl.). Southern Navajo County at Fort Apache (*Hoyt* in 1893) and near Whiteriver (*Phillips & Humphrey* 3049), Galiuro Mountains, Graham or Pinal County (*Wood* in 1939). Western New Mexico and eastern Arizona.

22. DISPORUM

Herbage pubescent, puberulent, or glabrescent; flowering stems from rootstocks, leafy, dichotomously branched; leaves broad, sessile or clasping; flowers terminal, solitary or 2 or 3 in an umbellike cluster, nodding, the perianth yellowish white; fruit somewhat lobed, its surface papillate.

1. Disporum trachycarpum (Wats.) Benth. & Hook. Kaibab Plateau, San Francisco Peaks, Elden Mountain, and Bill Williams Mountain (Coconino County), White Mountains (Apache County), Huachuca Mountains (Cochise County), Santa Catalina Mountains (Pima County), 7,500 to 9,500 feet, rich woods, May and June. Manitoba and British Columbia to New Mexico and Arizona.

The Arizona specimens are mostly referable to var. *subglabrum* Kelso, the type of which was collected near Flagstaff (*MacDougal* 64).

23. STREPTOPUS. TWISTED-STALK

Herbage glabrous; flowering stems from rootstocks, leafy, branched; leaves broad, cordate-clasping; flowers lateral, extra-axillary, solitary or in pairs, nodding, the perianth greenish white; fruit entire, smooth.

1. Streptopus amplexifolius (L.) DC. Baldy Peak, White Mountains, Apache County (*Goodding* 625, *Peebles & Smith* 12510), 10,000 to 11,000 feet, springy places in forests, July and August. Greenland to Alaska, south to North Carolina, New Mexico, Arizona, and California; Eurasia.

28. AMARYLLIDACEAE. AMARYLLIS FAMILY

Plants herbaceous or somewhat woody at base; flowering stems scapose, from a bulb, a short rootstock, or a large, woody caudex; flowers perfect, regular or nearly so; perianth segments united below into a tube adnate to the ovary (the ovary inferior); stamens 6, inserted on the perianth; ovary 3-celled; style 3-lobed; fruit a capsule.

This family includes many handsome plants that are cultivated as ornamentals, notably species of *Amaryllis* and of *Narcissus* (daffodils, jonquils, and others).

KEY TO THE GENERA

1. Plants with a large, woody caudex; leaves very thick, rigid, spine-tipped and often spiny-margined; flowers numerous, in elongate racemes or panicles; perianth fleshy 3. *Agave*
1. Plants without a woody caudex, the flowering stems from a bulb or a cormlike rootstock; leaves grasslike, not rigid or spine-tipped; flowers solitary or few in an umbellike inflorescence; perianth not fleshy (2).
2. Herbage glabrous; scape from a large tunicate bulb, 1-flowered, not much shorter than to surpassing the leaves; perianth funnelform........................1. *Zephyranthes*
2. Herbage hairy; scape from a small cormlike rootstock, often bearing more than 1 flower, much shorter than the leaves; perianth spreading........................2. *Hypoxis*

1. ZEPHYRANTHES. Zephyr-lily

Plants herbaceous, glabrous; flowering stems from a tunicate bulb; leaves narrow, somewhat fleshy; flower solitary, large, subtended by a spathelike bract; perianth funnelform, white or tinged with pink; anthers versatile; capsule subglobose, 3-lobed, dehiscent by valves.

1. Zephyranthes longifolia Hemsl. San Bernardino Ranch, Tombstone, and Fort Huachuca (Cochise County), foothills of the Rincon, Santa Catalina, and Santa Rita mountains (Pima County), 4,000 to 6,000 feet, in gravelly soil on hillsides, June and July. Western Texas to Arizona and Mexico.

2. HYPOXIS (69). Goldeye-grass

Plant herbaceous, pubescent; flowering stems from a cormlike rootstock; leaves narrow, grasslike, not fleshy; flowers few, subumbellate; perianth segments spreading, green outside, yellow within; anthers erect, sagittate; capsule somewhat elongate, irregularly dehiscent.

1. Hypoxis mexicana Schult. Known in Arizona only from a collection in the Huachuca Mountains (*Lemmon* 2891), September. Arizona to southern Mexico.

3. AGAVE (70, 71)

Flowering stems from a more or less woody caudex or a short, erect rootstock; leaves succulent, numerous, imbricate, forming a basal rosette; scapes tall and stout, ending in an elongate, bracted raceme or panicle; flowers numerous; perianth tubular or funnelform; stamens exserted, the anthers versatile; capsule thick-walled, many-seeded.

The names century-plant and mescal are applied to the large, paniculate species, and some of the small species are known as lechuguilla and amole. In Mexico and Central America species of *Agave* are cultivated for the fibers known in commerce as henequen and sisal. Other species yield intoxicating beverages. Pulque is the fermented juice, and mescal or tequila is a potent drink obtained by distillation of mash made from the caudex. The name "mescal" is applied also to the food obtained by roasting the caudex and emerging flower stalk. This food was of importance to the Indians of Arizona, who had access to such large species as *A. Palmeri*, *A. chrysantha*, and *A. Parryi*. Numerous pits in which mescal and sotol were roasted are still in evidence on the mountainsides, and even at the present time a small quantity of mescal is made by the Papagos. The juice of certain Mexican species is said to be emmenagogic, laxative, and diuretic, also toxic to livestock. The fresh juice of *A. Parryi* irritates the skin of some persons.

KEY TO THE SPECIES

1. Inflorescence spicate or appearing so, narrowly cylindric (2).
1. Inflorescence amply paniculate. Scapes 3 to 8 m. high; leaves usually more or less constricted near the base, with prickles on the margins (6).
2. Leaf margins beset with prickles; leaves 2 to 4 (6.5) cm. wide; scapes (including inflorescence) usually more than 2 m. long; capsules 25 to 30 mm. long (3).
2. Leaf margins filiferous, occasionally serrulate toward base; leaves commonly narrower, more or less falcate; scapes 1 to 2 m. long; capsules 10 to 25 mm. long (4).
3. Plants cespitose, the rosettes several; leaves more or less incurved; scapes 2 to 4 m. long. Flowers yellow, 22 to 30 mm. long including the ovary; filaments inserted near the middle of the short, broadly funnelform perianth tube...............1. *A. utahensis*
3. Plants not cespitose, the rosette solitary; leaves straight, ascending-spreading; scapes 4.5 to 7.5 m. long. Flower characters unknown....................2. *A. kaibabensis*
4. Filaments inserted at base of the perianth tube, the latter cylindric and much longer than the rounded lobes; leaves with coarse marginal fibers and conspicuous glaucous markings. Leaves flat or even slightly convex on the upper surface except near the apex, oblong-linear, 5 to 10 cm. long, 7 to 10 mm. wide, dark green; flowers somewhat red-tinged, glaucous, 12 mm. long including the ovary; capsules globose, 9 to 12 mm. long, about 10 mm. in diameter.................................3. *A. parviflora*
4. Filaments inserted at or near the summit of the perianth tube, the latter shorter than the lobes; leaves with fine marginal fibers and with glaucous markings only when immature (5).
5. Flowers 18 to 25 mm. long (including the ovary), greenish or pale yellow; perianth tube broadly funnelform, 5 mm. long; leaves subulate to linear; capsules 11 to 15 mm. long. Leaves 10 to 25 mm. wide, 15 to 35 cm. long, concave on the upper surface 4. *A. Toumeyana*
5. Flowers 30 to 50 mm. long (including the ovary), yellow; perianth tube narrowly funnelform, 8 to 11 mm. long; leaves linear; capsules 10 to 25 mm. long........5. *A. Schottii*
6. Filaments inserted at base of the perianth lobes; leaves rarely more than 45 cm. long (7).
6. Filaments inserted near the middle of the perianth tube; leaves usually 40 to 80 (up to 150) cm. long. Leaves lanceolate to linear, $\frac{1}{12}$ to $\frac{1}{4}$ as wide as long (8).
7. Flower buds greenish; scapes rarely stout; panicle lanceolate in outline or reduced and confined to the upper part of the scape, the branches ascending; clusters comparatively few-flowered, hemispheric or subglobose; leaves often with conspicuous transverse markings, the marginal prickles commonly stout, up to 10 mm. long; flowers pale yellow. Leaves not closely imbricate, lanceolate or oblong-lanceolate, attenuate-acuminate, 30 to 50 cm. long, the end spine slender...................6. *A. deserti*
7. Flower buds strongly charged with red; scapes stout; panicle elliptic in outline, occupying $\frac{1}{3}$ to $\frac{1}{2}$ or more of the total length of the scape, the branches horizontal; clusters many-flowered, horizontally flattened; leaves without conspicuous markings, the marginal prickles commonly 3 to 5 mm. long; flowers yellow............7. *A. Parryi*
8. End spine stout, 12 to 15 mm. long; leaves linear, not greatly thickened at base, the naked acumination, including the spine, 30 to 35 mm. long; inflorescence proliferous, the branches strongly ascending, the bulbils long-persistent. Leaves up to 65 cm. long, dark green, strongly concave toward apex; flowers greenish yellow 8. *A. Murpheyi*
8. End spine slender, 20 to 50 mm. long; leaves lanceolate to linear, greatly thickened at base, the naked acumination, including the spine, 50 to 150 mm. long; inflorescence rarely proliferous (9).
9. Flowers greenish or yellowish, more or less tinged with purple; leaves green; prickles usually not more than 5 mm. long, at the middle of the leaf margin 7 to 20 mm. apart; panicle branches horizontal, twice as long as the flower clusters; flowers not especially numerous or congested..9. *A. Palmeri*

9. Flowers yellow, not purple-tinged; leaves glaucous; prickles stouter, 7 to 10 mm. long, at the middle of the leaf margin 20 to 35 mm. apart; panicle branches ascending, about as long as the flower clusters; flowers congested, occasionally as many as 300 in one cluster .... 10. *A. chrysantha*

1. Agave utahensis Engelm. (? *A. Newberryi* Engelm., *A utahensis* var. *discreta* Jones). Coconino and Mohave counties, 3,000 to 7,500 feet, May to July, type of *A. Newberryi* from Peacock Spring, Mohave County (*Newberry* in 1858), type of var. *discreta* from near Oatman (*Jones* in 1930). Southern Utah and northern Arizona to southeastern California.

2. Agave kaibabensis McKelvey. Kaibab Plateau on the north side of Grand Canyon, Coconino County (*McKelvey* 4381, the type collection), also within the Canyon. Known only from northern Arizona.

The difference in habit as compared with *A. utahensis* is well shown by Mrs. McKelvey's photographs (*Jour. Arnold Arbor.* 30 : 227–230, pl. 1 & 2).

3. Agave parviflora Torr. Mountainous region west of Nogales (Santa Cruz County) along the Mexican boundary, type from Pajarito Mountains (*Schott*), Arivaca to Ruby, 4,500 feet (*Peebles & Fulton* 11444), June and July. Arizona, Chihuahua, and Sonora.

4. Agave Toumeyana Trel. Pinal Mountains (Gila or Pinal County), Fish Creek Hill (eastern Maricopa County), Superstition Mountains (Pinal County), 2,000 to 5,000 feet, May to July. Known only from south-central Arizona.

5. Agave Schottii Engelm. (*A Mulfordiana* Trel.) Gila, Pinal, Cochise, Pima, and Santa Cruz counties, exposed mountainsides, 4,000 to 7,000 feet, May to October, type from the Pajarito Mountains (*Schott* in 1855), type of *A. Mulfordiana* from the Rincon Mountains (*Toumey* in 1894). Southwestern New Mexico, southern Arizona, and northern Sonora.

In typical *A. Schottii* the leaves are yellowish green, 6 to 9 (rarely 12) mm. wide, 15 to 35 cm. long, concave on the upper surface, and the capsules are 10 to 15 or even 25 mm. long. Var. *Treleasei* (Toumey) Kearney & Peebles is known from one station on the southern slopes of the Santa Catalina Mountains, 6,500 feet, where Toumey found the type growing with typical *A. Schottii*, and from the Ajo Mountains, 3,000 feet. The stamens are inserted as in the typical plant, and the variety differs only in the character of the leaves, which are dark green, 15 to 25 mm. wide, 20 to 40 cm. long, and nearly flat on the upper surface.

6. Agave deserti Engelm. North side of the Grand Canyon, Coconino County (*Schellbach* in 1944), western Yavapai and southern Mohave counties, southward to western Pima and northern Yuma counties, 500 to 3,500 feet, June. Arizona, southern California, and Baja California.

In Arizona the species does not colonize appreciably, although the plants produce offsets freely.

7. Agave Parryi Engelm. Southern Coconino County and Yavapai County to Cochise, Santa Cruz, and Pima counties, 4,500 to 8,000 feet, June to August, type probably from Graham County (*Rothrock* 274). New Mexico, Arizona, and northern Mexico.

In the typical plant the leaves are compactly imbricate, broadly oblong,

⅓ to ⅖ as wide as long, 10 to 15 cm. wide, 30 to 40 cm. long, with a stout end spine and the flowers 50 to 60 mm. long. Var. *Couesii* (Engelm.) Kearney & Peebles (*A. Couesii* Engelm.) is found in Yavapai and Gila counties, sometimes with typical *A. Parryi*, the type from Date Creek (*Coues & Palmer* 253). This is a somewhat smaller plant, with flowers 25 to 50 mm. long and leaves less closely imbricate, oblong to lanceolate, ⅕ to ⅓ as wide as long. In the robust var. *huachucensis* (Baker) Little (*A. huachucensis* Baker), type from the Huachuca Mountains (*Pringle* in 1884) and known only from that vicinity, the leaves are up to 35 cm. wide and up to 65 cm. long, and the flowers up to 75 mm. long.

8. **Agave Murpheyi** Gibson. Paradise Valley (Maricopa County), Roosevelt and Tonto Basin (Gila County), Queen Creek, near Superior, the type locality (Pinal County), March and April. Known only from Arizona.

9. **Agave Palmeri** Engelm. Graham, Gila, Cochise, Santa Cruz, and Pima counties, in the mountains, 3,500 to 7,500 feet, June to August. New Mexico, southern Arizona, and Sonora.

The original description was based on several Arizona collections (*Schott* in 1855, *Palmer* in 1869, *Rothrock* in 1874).

10. **Agave chrysantha** Peebles (*A. Palmeri* var. *chrysantha* Little). Gila, eastern Maricopa, Pinal, and Pima counties, 3,000 to 7,000 feet, common along the Apache Trail, June to August, type from the Pinal Mountains (*Peebles & Harrison* 5543). Known only from Arizona.

29. IRIDACEAE. Iris Family

Perennial herbs; flowering stems from rootstocks or bulbs; leaves long and narrow, 2-ranked, equitant (folded together lengthwise and enfolding one another); flowers perfect, regular or nearly so, subtended by spathelike bracts; stamens 3, inserted on the perianth; style 3-cleft; ovary inferior; fruit a 3-celled, dehiscent, many-seeded capsule.

KEY TO THE GENERA

1. Sepals recurved, longer than the erect petals; style branches petallike, opposite to and overarching the stamens; filaments distinct; flowers about 8 cm. in diameter . . 1. *Iris*
1. Sepals and petals alike, all spreading; style branches not petallike; filaments more or less united; flowers not more than 6 cm. in diameter (2).
2. Flowering stems from a tunicate bulb; style 3-branched, the branches 2-cleft or 2-parted . . . . 2. *Nemastylis*
2. Flowering stems from a short rootstock, this often nearly obsolete; style entire or if 3-branched, then the branches entire . . . . 3. *Sisyrinchium*

1. IRIS (72). Flag, Fleur-de-lis

Flowering stems from a thick, mostly horizontal, more or less branched rootstock; leaves long and rather narrow; flowers large and showy, violet; perianth segments united below into a tube, the 3 inner ones narrower; stigmas below the tips of the style branches.

Some of the exotic species have been brought to a high state of perfection as cultivated ornamentals. Orris-root powder, used extensively in toilet preparations, is obtained from certain European species. The leaves and rootstocks of the Arizona species are reputed to be poisonous to livestock.

1. Iris missouriensis Nutt. Apache, Navajo, and Coconino counties, south to the Chiricahua and Huachuca mountains (Cochise County), 6,000 to 9,500 feet, wet meadows, May to September. North Dakota to British Columbia, south to New Mexico, Arizona, and California.

Rocky Mountain iris. Var. *arizonica* (Dykes) R. C. Foster (*I. arizonica* Dykes) has been collected "near Brome on the road from Prescott" (*McKelvey* 1251), in the Chiricahua Mountains (*Blumer* 1556, the type collection), and in the Huachuca Mountains (*Pringle* in 1886). As compared with typical *I. missouriensis* it has longer and broader leaves (up to 75 cm. long and 12 mm. wide), these equaling the stems and not or only slightly glaucous, shorter pedicels (not more than 10 cm. long), entire (not emarginate) petals, and shorter filaments (about 1 cm. long). A collection in the White Mountains, Apache County (*Ferris* 1247), was referred by Foster (72, pp. 66–67) to var. *pelogonus* (Goodding) R. C. Foster. This variety is characterized by thick, rigid leaves, pedicels not more than 10 cm. long, and a relatively short ovary (not more than 15 mm. long). A collection near Flagstaff, Coconino County (*Fulton & Osborn* 7136), also seems to be of this variety.

2. NEMASTYLIS (73)

Flowering stem from a tunicate bulb, often branched; leaves narrow, grasslike; flowers solitary or very few in a cluster, rather large; perianth violet, the segments all alike; filaments united below into a tube; capsule oblong-ellipsoid, dehiscent near the apex.

1. Nemastylis tenuis (Herb.) Baker. Huachuca Mountains (Cochise County), about 6,000 feet, open, grassy pine woods. Western Texas and southeastern Arizona to Guatemala.

The Arizona plant belongs to var. *Pringlei* (Wats.) R. C. Foster (*N. Pringlei* Wats.).

3. SISYRINCHIUM

Flowering stems from a short rootstock or apparently from a cluster of fibrous roots; leaves long, narrow, grasslike; flowers in few-flowered terminal umbels; perianth blue, violet, or yellow, the segments all alike or nearly so; capsule globose to oblong, 3-valved.

The blue-flowered species are called blue-eyed-grass.

KEY TO THE SPECIES

1. Stems branched above, leafy, stout, 30 cm. or longer; leaves usually more than 5 mm. wide; perianth orange, the segments 15 mm. or longer; capsules oblong, more than 10 mm. long. Filaments united about ⅓ of their length; anthers erect, about 6 mm. long; style 3-branched..1. *S. arizonicum*

1. Stems commonly unbranched, scapelike; leaves all basal or nearly so, less than 5 mm. wide; perianth segments not more than 15 mm. long; capsules broadly ellipsoid or subglobose, less than 10 mm. long (2).

2. Perianth blue or violet; filaments united into a tube; anthers erect, less than 2 mm. long; style not cleft..2. *S. demissum*

2. Perianth yellow or orange, with brown veins; filaments united only near the base; anthers versatile; style cleft (3).

3. Pedicels erect or ascending; perianth 8 to 12 mm. long; anthers 2 to 3 mm. long 3. *S. longipes*

3. Pedicels spreading or recurved; perianth 3 to 5 mm. long; anthers 0.5 to 1.5 mm. long 4. *S. cernuum*

1. Sisyrinchium arizonicum Rothr. White Mountains (Apache County), Mogollon Mesa (Coconino County), Chiricahua and Huachuca mountains (Cochise County), Santa Catalina Mountains (Pima County), 6,000 to 9,500 feet, in coniferous woods, July and August, type from Willow Spring, southern Apache County (*Rothrock* 238). New Mexico and Arizona, doubtless also in northern Mexico.

2. Sisyrinchium demissum Greene. Apache and northern Greenlee counties to Coconino and Yavapai counties, 5,000 to 9,500 feet, in wet meadows and springy places, June to September, type from Bill Williams Mountain, Coconino County (*Greene* in 1889). Western Kansas (?) to Arizona.

The typical plant has leaves 1 to 2 mm. wide and a perianth not more than 10 mm. long; whereas in var. *amethystinum* (Bicknell) Kearney & Peebles (*S. amethystinum* Bicknell) the leaves are commonly 3 to 4 mm. wide and the perianth up to 15 mm. long. In Arizona the variety appears to be the more common and more widely distributed plant, ranging from the San Francisco Peaks and the White Mountains to the Huachuca Mountains (Cochise County) and the Rincon Mountains (Pima County). The type of *S. amethystinum* was collected at the last-mentioned locality (*Nealley* 153).

S. macrocarpon Bicknell, known only by the type collection at Willow Spring, Apache County (*Palmer* 490a), is perhaps only an exceptionally large-fruited variant of *S. demissum* with capsules 6 to 7 mm. long as compared with the ordinary length of 4 to 5 mm.

3. Sisyrinchium longipes (Bicknell) Kearney & Peebles (*Hydastylus longipes* Bicknell). San Francisco Peaks (Coconino County), White Mountains (Apache County), Pinaleno Mountains (Graham County), Chiricahua Mountains (Cochise County), Rincon and Santa Catalina mountains (Pima County), 7,500 to 9,500 feet, in springy places and open spruce and pine woods, July to September, type from the San Francisco Peaks (*Knowlton* 34). Arizona and northern Mexico.

Varies greatly in size of the plant and in width and relative length of the leaves.

4. Sisyrinchium cernuum (Bicknell) Kearney (*Hydastylus cernuus* Bicknell). Happy Valley Road, Cochise County, 4,000 feet, along a stream, March (*Pultz & Phillips* 1573). Southeastern Arizona and Mexico.

30. ORCHIDACEAE. Orchis Family

Perennial herbs, some of them parasitic or saprophytic and without green coloring matter; flowering stems from bulbs, corms, rootstocks, or more or less thickened roots; perianth very irregular, the 3 outer segments sepallike and similar, the lowest of the 3 inner segments (lip) usually very unlike and larger than the other 2, sometimes saccate or spurred; stamen or stamens united with the style in a column; pollen grains (except in *Cypripedium*) coherent in 2 or more masses (pollinia), these attached at base to a viscid gland and united above by elastic threads; ovary inferior; capsule 3-valved; seeds minute, very numerous.

This large family, mainly tropical, includes some of the handsomest of all flowers. The species most extensively grown as ornamentals are mainly tropical and epiphytic. The Arizona species are all terrestrial (rooted in soil) and have relatively inconspicuous flowers except the lady's-slipper (*Cypripedium*) and *Calypso*. All orchids are interesting, however, because of the complicated structure of their flowers, which are specialized for pollination by various insects. It is remarkable that notwithstanding the enormous number of seeds produced, these plants are usually rare and seldom abundant. Aside from the value of some of the species in the florist trade, the family is of little importance economically except that extract of vanilla is manufactured from the pods of certain climbing species of tropical America.

KEY TO THE GENERA

1. Plants without green coloring matter, yellowish or purplish brown; inflorescence not spirally twisted; roots none; rootstock thick, corallike. Leaves reduced to sheathing scales; flowers several or numerous, in a raceme (2).
1. Plants with green, well-developed leaves or if not so, then the inflorescence spirally twisted; roots present; rootstock, if any, not corallike (3).
2. Lip entire, erose, or with a pair of short lobes near the base, with a small callus or fold on each side of the midvein.......................................7. *Corallorhiza*
2. Lip 3-lobed toward the apex, with several longitudinal winglike crests...10. *Hexalectris*
3. Fertile anthers 2; lip an inflated sac 2 to 5 cm. long..................1. *Cypripedium*
3. Fertile anther 1; lip not more than 2 cm. long (4).
4. Spur of the lip distinct, saccate to elongate-clavate.....................2. *Habenaria*
4. Spur of the lip none or obscure (5).
5. Lip not saccate or strongly concave (6).
5. Lip saccate or strongly concave, at least toward the base (7)
6. Flowering stem not from a bulb; leaves 2; lip longer than the sepals, short-stalked, broadest at the retuse or obcordate apex..................................5. *Listera*
6. Flowering stem from a bulb; leaf solitary; lip shorter than the sepals, sessile, broadest near the base, entire or nearly so at apex..............................8. *Malaxis*
7. Leaf solitary, this and the scape arising from a corm; flower solitary, the lip conspicuously bearded, much larger than the other perianth segments............9. *Calypso*
7. Leaves several; corm none; flowers several or numerous; lip beardless, not (or not much) larger than the other segments (8).
8. Plants appearing scapose, the foliage leaves mostly in a basal rosette and the stem leaves mostly reduced to sheathing bracts; flowers glandular-pubescent; lip strongly concave except the tip..6. *Goodyera*
8. Plants caulescent, the stems leafy except in *Spiranthes parasitica;* flowers glabrous; lip concave only toward base (9).
9. Flowers greenish or purplish, in a loose raceme; sepals and petals distinct, spreading; column not beaked...3. *Epipactis*
9. Flowers ochroleucous, in a spirally twisted, spikelike raceme; sepals and petals united or connivent, forming a galea over the column; column slender-beaked..4. *Spiranthes*

1. CYPRIPEDIUM. Lady's-slipper

Flowering stems from a cluster of somewhat thickened roots, leafy; herbage glandular-pubescent; leaves large, oval, sessile, strongly nerved; flower usually solitary; lip large, yellow, saccate, the other perianth segments long and narrow, greenish or brownish purple; stamens 3, the upper one sterile and somewhat petaloid, covering the summit of the style; stigma broad, slightly 3-lobed.

1. **Cypripedium Calceolus** L. White Mountains, Apache County (*Zuck* in 1907, *Goodding* 1122, *Schroeder* in 1938), in moist soil, probably in shade,

June and July. Newfoundland to British Columbia, south to Georgia, Arizona, and Washington; Europe.

It is the large-flowered form of this species, var. *pubescens* (Willd.) Correll (*C. pubescens* Willd.), that occurs in Arizona. Some persons who come into contact with the glandular hairs of this plant suffer severely from a form of dermatitis resembling ivy poisoning.

2. HABENARIA (74)

Roots clustered, tuberous-thickened; stems leafy; flowers small, greenish or yellowish, several or many in elongate, bracted, spikelike racemes; lip spreading or drooping, with a tubular or saccate spur at base; sacs of the solitary anther divergent.

Plants of bogs and rich, moist woods, sometimes known as rein-orchis and bog-orchid.

KEY TO THE SPECIES

1. Spur slender, 2 or more times as long as the lip. Lip commonly linear, strongly arcuate at base, with a small basal or subbasal tubercle or cushionlike callus......1. *H. limosa*

1. Spur not more than 1½ times as long as the lip (2).

2. Spur strongly inflated (saccate). Stems leafy; inflorescence elongate, usually loosely flowered; spur usually much shorter than the linear to elliptic-lanceolate lip, rarely more than ⅔ as long..2. *H. saccata*

2. Spur cylindric or clavate (moderately enlarged below), never saccate (3).

3. Raceme usually relatively short and densely flowered; lip not callose-thickened, commonly lanceolate and longer than the more or less clavate spur........3. *H. hyperborea*

3. Raceme usually elongate and loosely flowered; lip callose-thickened toward base, linear or lance-linear, usually shorter than the slender, cylindric or subclavate spur 4. *H. sparsiflora*

1. **Habenaria limosa** (Lindl.) Hemsl. (*H. Thurberi* Gray, *Limnorchis arizonica* Rydb.). Mountains of Cochise and Pima counties, 7,000 to 8,000 feet, rich, moist woods and cool, springy places, June to September, type of *Habenaria Thurberi* from south of Babocomari, Cochise County (*Thurber* 925), type of *Limnorchis arizonica* from the Rincon Mountains, Pima County (*Nealley* 78). New Mexico, southern Arizona, and Mexico.

2. **Habenaria saccata** Greene. Lukachukai Mountains (Apache County), White Mountains (Apache or Greenlee County), Pinaleno Mountains (Graham County), 8,500 to 9,500 feet, July to September. Alberta to Alaska, south to New Mexico and Arizona.

3. **Habenaria hyperborea** (L.) R. Br. Lukachukai and White mountains (Apache and Greenlee counties), 7,500 to 9,500 feet, in rich, moist woods, July and August. Newfoundland to Alaska, south to New Jersey, Arizona, and Oregon.

4. **Habenaria sparsiflora** Wats. Apache, Navajo, Coconino, and Graham counties, 5,000 to 9,000 feet, June to October. Colorado and New Mexico to Washington, northern Arizona, and southern California.

A small-flowered variant, var. *laxiflora* (Rydb.) Correll, occurs on the Navajo Indian Reservation (Navajo County?) and in the Grand Canyon (Coconino County), 4,500 to 7,000 feet.

3. EPIPACTIS. Helleborine

Flowering stem tall, leafy, from a creeping rootstock; inflorescence a few-flowered, bracted raceme; flowers greenish or purplish, the lip saccate only near the base, the larger terminal portion deltoid-ovate, inconspicuously crested.

1. Epipactis gigantea Dougl. Navajo, Coconino, and northeastern Mohave counties to Cochise and Pima counties, 3,000 to 8,000 feet, in moist soil, April to July. Montana to British Columbia, south to western Texas, Arizona, and California.

4. SPIRANTHES. Lady's-tresses

Plants from a cluster of tuberous-thickened roots, mostly leafy-stemmed; leaves with well-developed blades or reduced to sheathing scales; inflorescence a twisted spike with yellowish-white flowers in 1 to 3 ranks; lip concave toward the base, the distal portion expanded, erose.

KEY TO THE SPECIES

1. Stem leaves reduced to thin sheathing scales; scapes slender, glandular-pubescent above. Spike slender, few-flowered, the flowers in 1 rank; lip dentate........1. *S. parasitica*
1. Stem leaves well developed, lanceolate or oblanceolate, green; scapes stout (2).
2. Inflorescence glabrous or nearly so; flowers crowded, commonly in 3 ranks, less than 1 cm. long; leaves seldom more than 1 cm. wide....................2. *S. Romanzoffiana*
2. Inflorescence villous; flowers not crowded, commonly in 1 rank, 1 cm. long or longer; leaves 2 cm. or wider...3. *S. michuacana*

1. Spiranthes parasitica A. Rich. & Gal. Santa Catalina Mountains, Pima County (*Thornber & Lloyd* 4196, *Phillips* 2403, 2420), 8,000 to 8,500 feet, June and July. Southern Arizona and Mexico.

Phillips's specimens were collected on north slopes in deep shade under Douglas-firs.

2. Spiranthes Romanzoffiana Cham. Kaibab Plateau and North Rim of Grand Canyon (Coconino County), White Mountains (Apache County), Pinaleno Mountains (Graham County), 8,500 to 9,500 feet, in bogs and wet meadows, August and September. Newfoundland to Alaska, south to New York, New Mexico, Arizona, and California.

3. Spiranthes michuacana (La Llave & Lex.) Hemsl. Cochise and Pima counties, Rucker Valley or Huachuca Mountains (*Lemmon* 477), Cochise (Dragoon?) Mountains (*Barneby* 2653), Santa Catalina Mountains, 7,000 feet (*Shreve* 5416), rocky canyons and slopes, September and October. Western Texas, southeastern Arizona, and nearly throughout Mexico.

The floral bracts are remarkably large, attaining a length of 3 cm.

5. LISTERA. Tway-blade

Small plants with fibrous roots and a pair of nearly opposite, broadly ovate leaves; flowers inconspicuous, greenish yellow, in a short raceme; lip wedge-shaped, broadest at apex, slightly toothed or angled on each side at base.

1. Listera convallarioides (Swartz) Nutt. Santa Catalina Mountains (Pima County), about 8,000 feet, July to September. Newfoundland to Alaska, south to New England, Arizona, and California.

6. GOODYERA. Rattlesnake-plantain

Flowering stem scapose, from a cluster of thickened roots; foliage leaves in a basal rosette, somewhat fleshy, sometimes mottled or striped with white; inflorescence a somewhat 1-sided, bracted spike; flowers small, whitish, glandular-pubescent; upper sepal united with the petals, forming a galea; column ending in a forked or bidentate beak.

KEY TO THE SPECIES

1. Scapes up to 30 cm. long but usually shorter; leaves (including petiole) 2 to 4 cm. long, seldom mottled or striped; spikes usually loosely flowered; lip deeply saccate, the margin flaring or revolute..........1. *G. repens*
1. Scapes up to 45 cm. long; leaves 4 to 10 cm. long, often white-striped; spikes commonly densely flowered; lip shallowly saccate, the margin involute........2. *G. oblongifolia*

1. Goodyera repens (L.) R. Br. Phelps Meadow, White Mountains, Apache County, 9,500 to 10,000 feet (*Phillips* 3326). Widely distributed in the Northern Hemisphere.

The plants of the Rocky Mountain region are often referred to var. *ophioides* Fern., but the collection cited was identified by Donovan S. Correll as typical *G. repens.*

2. Goodyera oblongifolia Raf. (*G. decipiens* (Hook.) F. T. Hubbard). Lukachukai and White mountains (Apache and Greenlee counties), Kaibab Plateau, Elden Mountain, and Bill Williams Mountain (Coconino County), Pinaleno Mountains (Graham County), 8,000 to 9,500 feet, rich woods, July to September. Nova Scotia to Alaska, south to New Hampshire, New Mexico, Arizona, and California.

7. CORALLORHIZA. Coral-root

Plants without chlorophyll, the leaves reduced to sheathing scales; rootstock thick, branching, corallike; flowers in racemes; lateral sepals united with the base of the column, often forming a short spur or projection, this adnate to the summit of the ovary.

KEY TO THE SPECIES

1. Longer perianth segments commonly 10 to 15 mm. long; lip somewhat darker-colored than the other segments, conspicuously striate (as are the other segments), oval or oblong-ovate, entire; spur none..........1. *C. striata*
1. Longer perianth segments mostly less than 10 mm. long; lip contrasting sharply in color with the other (brown) segments, whitish and more or less spotted with crimson, broadly ovate, obovate, or suborbicular, erose-denticulate; spur or a saclike protuberance present (2).
2. Spur usually evident and partly free; lip with a pair of elongate folds, 3-lobed, the lateral lobes much smaller than the midlobe, the other perianth segments obtuse or acutish .... 2. *C. maculata*
2. Spur a mere swelling adnate to the ovary; lip with a pair of slender basal callosities, not lobed, the other segments acuminate..........3. *C. Wisteriana*

1. Corallorhiza striata Lindl. Pinaleno Mountains (Graham County), Pinal Mountains and Sierra Ancha (Gila County), Chiricahua Mountains (Cochise County), Santa Catalina Mountains (Pima County), 7,000 to 9,000 feet, in deep shade of pine and spruce forests, July. Quebec to British Columbia, Michigan, Arizona, and California.

2. Corallorhiza maculata Raf. Lukachukai and White mountains (Apache County), North Rim of Grand Canyon and San Francisco Peaks (Coconino County), south to the Pinaleno Mountains (Graham County) and the Santa Catalina Mountains (Pima County), 6,000 to 10,000 feet, coniferous forests, June and July. Nova Scotia to Alaska, south to Florida, New Mexico, Arizona, and California.

3. Corallorhiza Wisteriana Conrad. North Rim of Grand Canyon and San Francisco Peaks (Coconino County), Sierra Ancha and Pinal Mountains (Gila County), Santa Catalina Mountains (Pima County), 6,000 to 8,000 feet, May. Pennsylvania to Florida and Texas; Arizona.

8. MALAXIS. Adders-mouth

Small plants with flowering stems from a corm, bearing a solitary rounded-ovate clasping leaf; flowers inconspicuous, in a terminal raceme or spike; sepals and petals distinct, spreading; lip cordate or lobed at base.

KEY TO THE SPECIES

1. Inflorescence a short, often corymbiform, loose raceme; leaves deeply cordate at base. Pedicels equaling or longer than the flowers........................1. *M. corymbosa*
1. Inflorescence elongate, not corymbiform; leaves not, or very slightly, cordate at base (2).
2. Flowers 3 to 4 mm. long, sessile, greenish yellow, in a dense, elongate spike 3 to 4 mm. wide; perianth segments oblong to ovate............................2. *M. Soulei*
2. Flowers commonly more than 4 mm. long, on pedicels half to equally as long as themselves; inflorescence loosely racemose; perianth segments linear or narrowly lanceolate (3).
3. Flowers greenish, 10 to 12 mm. long..................................3. *M. tenuis*
3. Flowers brownish purple, 4 to 6 mm. long.......................4. *M. Ehrenbergii*

1. Malaxis corymbosa (Wats.) Kuntze. Chiricahua and Huachuca mountains (Cochise County), Santa Catalina Mountains (Pima County), 6,500 to 7,500 feet, August, type from Tanners Canyon, Huachuca Mountains (*Lemmon* 2882). Southern Arizona to Guatemala.

2. Malaxis Soulei L. O. Williams (*Microstylis montana* Rothr.). White Mountains (Apache County), Mogollon Escarpment (Coconino County), Pinaleno Mountains (Graham County), fairly common in the mountains of Cochise, Santa Cruz, and Pima counties, 6,000 to 9,500 feet, pine woods, July to September, type from Mount Graham (*Rothrock* 264). Western Texas to Arizona, south to Panama.

3. Malaxis tenuis (Wats.) Ames. Santa Catalina Mountains, Pima County (*Thornber* in 1903, *Peebles et al.* 2518), 7,000 feet. New Mexico, Arizona, and Mexico.

4. Malaxis Ehrenbergii (Reichb. f.) Kuntze. (*Microstylis purpurea* Wats.). Huachuca Mountains (Cochise County), Santa Catalina Mountains (Pima County), about 7,500 feet, moist, mossy places, August, type of *Microstylis purpurea* from the Huachuca Mountains (*Lemmon* 2881). New Mexico and Arizona to Guatemala.

9. CALYPSO

Stem from a corm, low, 1-leaved and 1-flowered; leaf broadly ovate or suborbicular; flower showy; lip boat-shaped, pendent, more than 1 cm. long, purple, striped and mottled with darker color, bearded with long, usually bright-yellow hairs.

1. Calypso bulbosa (L.) Oakes. North Rim of Grand Canyon and San Francisco Peaks (Coconino County), White Mountains (Apache and Greenlee counties), 8,500 to 10,000 feet, spruce and fir forests, June to August. Labrador to Alaska, south to New England, Michigan, Arizona, and California; Europe.

A remarkably beautiful little plant.

10. HEXALECTRIS

Plants without chlorophyll, the leaves reduced to sheathing scales; flowering stem from a thick corallike rootstock; flowers in a raceme, the perianth 1.5 cm. or longer; sepals and petals purplish brown, striate; lip cuneate-obovate to suborbicular, 3-lobed, with lamellate crests, not saccate.

KEY TO THE SPECIES

1. Raceme few- (not more than 8-) flowered; crests of the lip scalloped and more or less interrupted .... 1. *H. Warnockii*
1. Raceme several- (up to 12-) flowered; crests entire .... 2. *H. spicata*

1. Hexalectris Warnockii Ames & Correll. Chiricahua National Monument, Cochise County (*Fish* 20, in 1939). Western Texas, northern Mexico (?), and southeastern Arizona.

2. Hexalectris spicata (Walt.) Barnhart (*Corallorhiza arizonica* Wats.). Santa Rita Mountains, Santa Cruz or Pima County (*Pringle* 353, the type of *C. arizonica, Thornber* in 1903), rich soil in woods among rocks, June and July. Virginia to Missouri, Florida, and Texas; Arizona.

Crested coral-root. The plant resembles a large-flowered *Corallorhiza*. The flowers, as observed in the southeastern United States, have a delicious odor of violets.

H. grandiflora (A. Rich. & Gal.) L. O. Williams (*H. mexicana* Greenm.) is also to be sought in southern Arizona. It resembles *H. spicata* in the entire crests of the lip but has a more deeply lobed lip with only 2 to 4 short, lateral crests or folds, whereas *H. spicata* has 5 or 6 crests, longer and more central.

DICOTYLEDONEAE

Series 1. Apetalae (Families 31 to 48)

KEY TO THE FAMILIES[1]

1. Flowers unisexual, those of one or both sexes in catkins or catkinlike inflorescences. Plants mostly trees or shrubs (2).
1. Flowers perfect or, if unisexual, then not in catkins or catkinlike inflorescences (8).

[1] Other families, belonging properly to series Choripetalae and Sympetalae, are included in this key because some of their members are apetalous or have an apparently uniseriate perianth.

2. Pistillate flowers solitary or in few-flowered clusters, the staminate flowers in catkins (3).
2. Pistillate (and usually also the staminate) flowers in catkins or catkinlike inflorescences (5).
3. Fruit small, achenelike, enveloped by a winged calyx; leaves narrow, entire, fleshy: genus *Sarcobatus*..44. *Chenopodiaceae*
3. Fruit large, nutlike; leaves large, not fleshy (4).
4. Leaves pinnately compound; nut with a thick, hard shell; cotyledons 2-lobed 33. *Juglandaceae*
4. Leaves simple but sometimes deeply lobed; nut (acorn) with a relatively thin shell, partly embraced by a cuplike involucel; cotyledons entire.............35. *Fagaceae*
5. Plants monoecious, the pistillate flowers subtended by conspicuous bracts; mature catkins dry..34. *Betulaceae*
5. Plants dioecious or if monoecious, then the flowers not conspicuously bracted and the pistillate catkins fleshy at maturity (6).
6. Perianth none. Fruit a capsule; seeds with a conspicuous tuft of silky hairs 32. *Salicaceae*
6. Perianth present, at least in the staminate flowers (7).
7. Carpels numerous and fused into a compound, fleshy fruit, or dry and subtended by conspicuous, papery bracts; leaves thin, usually lobed...............37. *Moraceae*
7. Carpels remaining distinct, not subtended by conspicuous, papery bracts; leaves thick, not lobed: genus *Garrya*.......................................101. *Cornaceae*
8. Ovary inferior or appearing so, wholly or partly adnate to the perianth or (in Nyctaginaceae and Elaeagnaceae) very closely enveloped by it in fruit (9).
8. Ovary superior, free or very nearly free from the perianth when the latter is present (20).
9. Plants aquatic, the stems wholly or partly immersed in water. Leaves entire and in whorls, or the immersed ones finely dissected; flowers axillary or in interrupted terminal spikes, minute...98. *Haloragaceae*
9. Plants not aquatic or if so (in a few genera of Umbelliferae), then the leaves orbicular and peltate or the flowers in umbels or dense heads (10).
10. Perianth really double but appearing single and corollalike. Calyx with tube wholly adnate to the ovary and the limb obsolete or reduced to a mere border or teeth: families belonging mainly to series Choripetalae and Sympetalae (11).
10. Perianth in 1 series, calyxlike or corollalike (14).
11. Fruit a pair of contiguous 1-seeded carpels (12).
11. Fruit achenelike, not paired (13).
12. Flowers in umbels, heads, or spikes; leaves alternate or basal, mostly compound 100. *Umbelliferae*
12. Flowers in cymes or solitary in the axils; leaves appearing whorled, simple: genus *Galium* .... 127. *Rubiaceae*
13. Corolla spurred; anthers distinct; flowers not in involucrate heads but sometimes in rather dense clusters: genus *Plectritis*......................129. *Valerianaceae*
13. Corolla not spurred; anthers often connate; flowers in heads subtended by an involucre 132. *Compositae*
14. Plants parasitic on the stems of trees and shrubs (15).
14. Plants terrestrial, autophytic or (exceptionally) root-parasitic (16).
15. Stems well developed, often much-branched; leaves with well-developed blades or reduced to small scales; fruit a 1-seeded berry...................39. *Loranthaceae*
15. Stems almost none outside the bark of the host, the visible plant reduced to a flower subtended by a few imbricate bracts; fruit several-seeded........42. *Rafflesiaceae*
16. Perianth very irregular, the segments united below into a tube. Flowers 2 to 3 cm. long 41. *Aristolochiaceae*
16. Perianth regular or nearly so (17).
17. Fruit of 2 partly distinct, beaked carpels: genus *Heuchera*........60. *Saxifragaceae*
17. Fruit of a single carpel (18).
18. Ovules 2 or more; ovary truly inferior, crowned in fruit by the persistent perianth limb 40. *Santalaceae*

18. Ovule solitary; ovary technically superior but in fruit very closely enveloped by the perianth tube and appearing inferior (19).
19. Plants herbaceous or suffrutescent; pubescence not stellate; perianth with a campanulate, salverform, or funnelform limb, this usually corollalike in texture; fruit dry 46. *Nyctaginaceae*
19. Plants shrubby or arborescent; pubescence stellate-scurfy; perianth of the fertile flowers cylindric or urn-shaped, not corollalike, its base fleshy in fruit 95. *Elaeagnaceae*
20. Pistils commonly more than 1. Stamens usually numerous; fruits indehiscent (achenes) or dehiscent on one suture (follicles) .... 52. *Ranunculaceae*
20. Pistil 1, simple or compound (21).
21. Style and (or) stigma 1, or sometimes 2 in genus *Trianthema* (22).
21. Styles and (or) stigmas 2 or more, exceptionally 1 in family Euphorbiaceae (34).
22. Plants aquatic, the stems immersed; leaves in whorls, finely dissected 51. *Ceratophyllaceae*
22. Plants not aquatic or if so, then the leaves not finely dissected (23).
23. Fruit a circumscissile, crested, several-seeded capsule: genus *Trianthema* 48. *Aizoaceae*
23. Fruit otherwise (24).
24. Plants herbaceous, seldom suffrutescent (25).
24. Plants shrubs or trees (30).
25. Fruit berrylike. Flowers in terminal and axillary racemes; perianth segments 4: genus *Rivina* .... 47. *Phytolaccaceae*
25. Fruit not berrylike (26).
26. Ovary 2- or more-celled; fruit a capsule (27).
26. Ovary 1-celled; fruit a utricle or an achene (28).
27. Petals and stamens free from the calyx; capsule compressed, obcordate, 2-celled: genus *Lepidium* .... 56. *Cruciferae*
27. Petals and stamens borne on the calyx throat; capsule turgid, 2- to 4-celled 96. *Lythraceae*
28. Leaves pinnately compound; stamens usually 2. Achene enclosed in a 4-winged hypanthium: genus *Sanguisorba* .... 63. *Rosaceae*
28. Leaves simple; stamens usually more than 2 (29).
29. Stipules present or if absent, then the flowers subtended by an involucre of soft, green, leaflike bracts .... 38. *Urticaceae*
29. Stipules none; floral bracts hyaline, scarious, or rigid .... 45. *Amaranthaceae*
30. Stamens 10 or more, or if fewer than 10, then the leaves bipinnate (31).
30. Stamens 5 or fewer; leaves never bipinnate (32).
31. Leaves simple; flowers solitary or in few-flowered fascicles: genera *Cercocarpus*, *Coleogyne* .... 63. *Rosaceae*
31. Leaves compound; flowers in heads, spikes, or spikelike racemes: subfamily Mimosoideae and genus *Parryella* .... 64. *Leguminosae*
32. Ovary 4- or 5-celled; fruit a capsule; stamens 5, the filaments united below. Perianth segments large, petallike, bright yellow: genus *Fremontia* .... 85. *Sterculiaceae*
32. Ovary 2- to 4-celled; fruit not a capsule; stamens distinct (33).
33. Calyx well developed, the throat lined with a disk bearing the stamens; leaves simple, alternate or fascicled; fruit a drupe .... 81. *Rhamnaceae*
33. Calyx inconspicuous or wanting; disk none; leaves simple or pinnate, opposite or fascicled; fruit a samara or a drupe .... 108. *Oleaceae*
34. Stems woody, the plants shrubs or trees (35).
34. Stems herbaceous or at most slightly suffrutescent (42).
35. Fruit a drupe (36).
35. Fruit otherwise (37).
36. Style none; stigmas elongate, spreading or recurved, plumose or tomentose; ovary 1-celled; drupe with 1 stone; leaves very unequal at base and very scabrous above: genus *Celtis* .... 36. *Ulmaceae*

36. Style present but often very short; stigmas not recurved or plumose; ovary 2- or 3-celled; drupe with 1 to 4 stones; leaves not conspicuously unequal at base, nor very scabrous: genera *Condalia, Rhamnus*..........................81. *Rhamnaceae*
37. Fruit a pair of samaras, united at base, each with a large, 1-sided wing; leaves palmately lobed or divided, or pinnate with few leaflets................79. *Aceraceae*
37. Fruit otherwise; leaves simple but deeply palmately lobed in certain Euphorbiaceae (38).
38. Fruit an acornlike capsule with usually only 1 large seed. Leaves entire, coriaceous; flowers unisexual, in dense axillary clusters......................76. *Buxaceae*
38. Fruit otherwise (39).
39. Ovary 2- to 4-celled (exceptionally 1-celled); fruit a capsule, commonly dehiscent longitudinally. Flowers unisexual, in *Dodonaea* by abortion (40).
39. Ovary 1-celled; fruit not dehiscent longitudinally (41).
40. Capsule not winged; leaves not viscid, entire to palmately parted..74. *Euphorbiaceae*
40. Capsule conspicuously 2- to 4-winged; leaves somewhat viscid, narrow, entire: genus *Dodonaea* .... 80. *Sapindaceae*
41. Flowers perfect, subtended by a cylindric, turbinate, or cup-shaped gamophyllous involucre; stamens 9; fruit a 3-angled or 3-winged achene: genus *Eriogonum* 43. *Polygonaceae*
41. Flowers perfect or unisexual, not with an involucre as in *Eriogonum*, but the pistillate flowers often closely subtended by a pair of accrescent bractlets; stamens 5 or fewer; fruit a utricle..44. *Chenopodiaceae*
42. Inflorescence of many small, naked flowers in a dense cylindric spike subtended by a conspicuous involucre of white petallike bracts, simulating a single large, *Anemone*-like flower...31. *Saururaceae*
42. Inflorescence otherwise (43).
43. Plants aquatic, the stems wholly or partly immersed, not jointed; stipules none; fruit at maturity separating into 4 nutlets..........................75. *Callitrichaceae*
43. Plants not aquatic or if so, then the stems jointed and the stipules united into a sheath; fruit not separating into nutlets (44).
44. Ovary completely or incompletely 2- or more-celled, or occasionally 1-celled in families Aizoaceae and Euphorbiaceae (45).
44. Ovary 1-celled (47).
45. Fruit a depressed-globose, several-seeded berry. A coarse herb with large leaves: genus *Phytolacca* .... 47. *Phytolaccaceae*
45. Fruit a capsule (46).
46. Flowers perfect, with a perianth..................................48. *Aizoaceae*
46. Flowers unisexual, with or without a perianth..................74. *Euphorbiaceae*
47. Fruit a many-seeded, longitudinally dehiscent capsule: genus *Sagina* 50. *Caryophyllaceae*
47. Fruit a 1-seeded achene or utricle (48).
48. Flowers subtended by a gamophyllous, calyxlike involucre, or each flower subtended by a single bract, or else the stems jointed and the stipules united into a sheath; fruit an achene, usually trigonous but sometimes lenticular......43. *Polygonaceae*
48. Flowers not subtended as in the foregoing, but (in genus *Atriplex*) the pistillate ones by a pair of appressed, accrescent bractlets; stems not jointed; stipules none, or not united into a sheath; fruit not trigonous (49).
49. Stipules present; flowers perfect, in axillary clusters or terminal cymes. Plants small, with spreading, prostrate, or densely cespitose stems...........50. *Caryophyllaceae*
49. Stipules none; flowers perfect or unisexual (50).
50. Plants often more or less succulent; pubescence often mealy; floral bracts not hyaline or scarious, seldom rigid...............................44. *Chenopodiaceae*
50. Plants not succulent; pubescence never mealy but sometimes closely appressed and the hairs stellate; floral bracts hyaline, scarious, or rigid........45. *Amaranthaceae*

(Keys to the families of series Choripetalae and Sympetalae will be found on pages 281 to 285 and 626 to 629, respectively).

31. SAURURACEAE. Lizard-tail Family

1. ANEMOPSIS. Yerba-mansa

Plants herbaceous, perennial; leaves mostly basal, large, somewhat fleshy, cordate at base; inflorescence a dense, cylindric spike, subtended by large, white, petallike bracts; flowers perfect, very numerous, small, without a perianth.

1. Anemopsis californica (Nutt.) Hook. & Arn. Grand Canyon (Coconino County), Yavapai County, and Cochise to Yuma counties, 2,000 to 5,500 feet, wet, saline soil, May to August. Western Texas to Utah, Arizona, and California, southward into Mexico.

The Arizona plants belong to var. *subglabra* Kelso. The inflorescence with its white bracts suggests the single flower of an *Anemone*. An infusion of the root was used by Californians of Spanish descent and by the Pima Indians as a remedy for various ailments.

32. SALICACEAE. Willow Family

Trees or large shrubs, dioecious; leaves deciduous, alternate, simple; flowers naked (perianth none), those of both sexes in catkins, appearing before or with the leaves; ovary 1-celled; stigmas 2 to 4; fruit a 1-celled, 2- to 4-valved capsule; seeds minute, subtended by silky hairs.

KEY TO THE GENERA

1. Winter buds covered by several scales; scales of the catkins laciniate or fimbriate; flowers borne on broad or cup-shaped disks; stigmas elongate, the lobes slender or dilated ... 1. *Populus*
1. Winter buds covered by 1 scale; scales of the catkins entire or merely dentate; flowers without disks; stigmas short ... 2. *Salix*

1. POPULUS (75). Cottonwood, Poplar, Aspen

Trees, with more or less resinous buds; leaves mostly long-petioled, the blades mostly deltoid or ovate, sometimes lanceolate; stipules minute, caducous; catkins long and drooping; stamens numerous.

The familiar cottonwood (*P. Fremontii*) is conspicuous along streams throughout the state except in the higher mountains, and is planted everywhere as a shade tree. The trees grow rapidly and propagate readily from branch cuttings. The great quantities of pollen produced by the staminate trees and the copious quantities of downy seeds shed by the pistillate ones are distinct disadvantages to the use of the cottonwood in dooryards. Another drawback is the frequent mutilation that results from windstorms. The wood is light and tough and not durable, but has been used to some extent for fence posts. The Pima Indians use the twigs of this species for basket material and formerly ate the uncooked catkins. The inner bark was esteemed by the Indians as an antiscorbutic. Cattle browse on branches within their reach. Beavers cut the stems of poplars for dams, and the bark is their principal food. Poplar pollen is responsible for some cases of hay fever in Arizona.

The aspen (*P. tremuloides*) occurs on the higher mountains, mingled with conifers or in pure stands of considerable extent, conspicuous because of the nearly white bark of the trunks and the brilliant yellow of the foliage in autumn. Stands of aspen usually quickly repopulate forest areas denuded by fire, in this way retarding soil erosion and conserving soil moisture. Aspen stands are usually transient, eventually being replaced by conifers. Aspen has limited value as a timber tree and as browse for livestock. The wood is used chiefly for paper pulp and in making containers for butter and other food, as it imparts no flavor. An infusion of the inner bark was formerly used in treatment of intermittent fever and is regarded by the Indians as an antiscorbutic.

KEY TO THE SPECIES

1. Leaves considerably longer than wide, finely crenulate or serrulate (except sometimes on vigorous shoots), gradually or not very abruptly acuminate; petioles terete, shorter (commonly much shorter) than the blades (2).
1. Leaves nearly as wide as, to considerably wider than, long; petioles flattened laterally (3).
2. Petioles usually less than 1/3 as long as the blades, these lanceolate or ovate-lanceolate, more than twice (commonly 3 to 7 times) as long as wide, gradually acuminate at apex and often at base, the marginal teeth extending very nearly to the apex
1. *P. angustifolia*
2. Petioles commonly more than 1/3 (often half or more) as long as the blades, these ovate or lance-ovate, commonly less than twice as long as wide, somewhat abruptly acuminate at apex, rounded or broadly short-cuneate at base, the marginal teeth not extending very nearly to the apex..2. *P. acuminata*
3. Stigma lobes broad, more or less flattened, crenate; leaves broadly deltoid or subrhombic-deltoid, crenate or crenate-dentate (usually coarsely so), abruptly and commonly broadly acuminate at apex, truncate or very short-cuneate at base, not conspicuously paler beneath...3. *P. Fremontii*
3. Stigma lobes slender; leaves broadly ovate to rhombic-suborbicular, crenulate or serrulate with numerous teeth, abruptly very short-acuminate or apiculate at apex, short-cuneate, truncate, or rounded at base, usually conspicuously paler beneath
4. *P. tremuloides*

1. Populus angustifolia James. Apache, Coconino, and Yavapai counties, south to Cochise and Pima counties, 5,000 to 7,000 feet, along streams. South Dakota to Alberta, south to Nebraska, New Mexico, Arizona, and Chihuahua.

Narrow-leaf cottonwood. A small tree, reaching a height of about 15 m. (50 feet) and a trunk diameter of 45 cm. (18 inches). The crown is usually narrowly pyramidal, with slender, narrowly ascending branches.

2. Populus acuminata Rydb. Apache, Navajo, Yavapai, and Greenlee counties, 5,000 to 7,000 feet, along streams. Nebraska to Assiniboia, south to New Mexico and Arizona.

Lance-leaf cottonwood. This species reaches about the same size as *P. angustifolia*, from which it is distinguished chiefly by the broader, more spreading crown, stouter branches, and broader leaves. Some of the Arizona specimens seem to be nearly intermediate. A variant with pubescent branchlets and bud scales (var. *Rehderi* Sarg.) is of frequent occurrence throughout the range of the species in Arizona.

3. Populus Fremontii Wats. Throughout the state, along streams, 6,000 feet or lower. Western Texas to Nevada, Arizona, California, and northern Mexico.

Fremont cottonwood. The trees are frequently 15 m. (50 feet), sometimes even 30 m. (100 feet), high, and the trunk diameter may reach 1.2 m. (4 feet). The main branches are large, the crown wide and flat-topped, and the older bark gray-brown, thick, deeply furrowed.

Many Arizona specimens have been referred to *P. arizonica* Sarg., but if I. M. Johnston's interpretation of that species is correct, its occurrence in this state is doubtful (56, pp. 434–435). A common form in Arizona, with persistently pubescent branchlets, petioles, and leaf veins, is *P. Fremontii* var. *pubescens* Sarg. Var. *Macdougalii* (Rose) Jepson (*P. Macdougalii* Rose), which is even more pubescent and has bluish-green foliage, is common in the lower valley of the Colorado River. Var. *Toumeyi* Sarg., which occurs in Pima County and at the Grand Canyon (*Eastwood* 6002), and probably elsewhere in Arizona, is described as having very shallowly subcordate leaves and an exceptionally large disk; and var. *Thornberi* Sarg., of southern Arizona and New Mexico, is described as having leaves with more numerous serratures, ellipsoidal capsules, a smaller disk, and shorter pedicels (see Sargent, Charles Sprague. *Man. Trees North Amer.* ed. 2, 131 (1922). Several specimens from eastern Arizona approach *P. Wislizeni* (Wats.) Sarg. in their long pedicels, but they have the relatively broad capsules of *P. Fremontii.*

4. Populus tremuloides Michx. Apache to Mohave counties, south to the mountains of Cochise and Pima counties, 6,500 to 9,000 feet. Labrador to Alaska, south to New Jersey, Missouri, Arizona, and northern Mexico.

Quaking aspen. The species is represented in Arizona by the golden aspen, var. *aurea* (Tidestrom) Daniels (*P. aurea* Tidestrom), characterized by the more intense autumnal coloration of the foliage (golden or orange). The trees reach a height of 24 m. (80 feet) and a diameter of 75 cm. (30 inches) but usually are much smaller. The slightest breeze causes movement of the leaves, hence the common name quaking aspen.

2. SALIX (75, 76, 77, 77a). WILLOW

Trees or shrubs; bud scale with an inner membrane, this adherent, or loose and then giving the appearance of 2 scales; leaves mostly short-petioled with mostly lanceolate blades; stipules large or small, persistent or caducous; catkins not pendulous; stamens 2 to 10.

The genus is a difficult one, and the Arizona species require further study. Most of the specimens in herbaria are incomplete, as leaves, staminate flowers, and fruit are needed for accurate determination of many of the species. The following treatment should be regarded as only provisional.[1]

Willows grow mostly along streams and are valuable for their shade, especially on the stock ranges in the southern part of the state, where *S. Gooddingii*, Arizona's largest willow, is widely distributed and common. The foliage is eaten by livestock, and the bark by beavers. Small roots develop in abundance, and thus the willows are important agents in checking soil erosion along watercourses. In high mountains *S. Scouleriana*, in particular, invades burned forest areas, holding the soil and acting as a nurse for the

[1] The writers are indebted to Carleton R. Ball for helpful suggestions, but he is not responsible for the treatment here presented.

reproduction of conifers in much the same way as the aspen. The Pima and other Indians of Arizona used willow twigs for basket material. Timbers bound with willow withes have been found in cliff dwellings. The drug salicin, obtained from the bark of various species, has tonic, antiperiodic, and febrifugal properties.

KEY TO THE SPECIES

1. Petioles very short (not more than 3 mm. long), or none; leaves seldom more and usually much less than 7 mm. wide (rarely 12 mm. in *S. exigua*), the margins entire or remotely denticulate. Stamens 2; leaves linear-lanceolate or narrowly oblanceolate, sericeous, at least beneath (2).
1. Petioles more than 3 mm. long or if shorter, then the larger leaves more than 7 mm. wide, or closely serrate or serrulate (3).
2. Leaves oblanceolate, sessile or very nearly so, at maturity not more than 3 cm. long; capsules silky-villous, then glabrate.............................1. *S. taxifolia*
2. Leaves linear-lanceolate, short-petioled, at maturity 5 cm. long or longer; capsules glabrous .... 2. *S. exigua*
3. Margins of the leaves entire or obscurely serrulate (4).
3. Margins of the leaves serrate or serrulate (10).
4. Leaves rounded or subcordate at base, green but paler beneath....14. *S. pseudocordata*
4. Leaves acutish to attenuate at base, white or whitish beneath (5).
5. Capsules silky-pubescent (6).
5. Capsules glabrous. Leaves lanceolate or oblanceolate, acute to acuminate at apex, glabrous and often shiny above (8).
6. Leaves broadly obovate, rounded to acutish at apex, glabrous above...3. *S. Scouleriana*
6. Leaves not broadly obovate (7).
7. Leaves not more than 4 times as long as wide, elliptic, oval, oblong-lanceolate, or narrowly obovate, rounded to acute at apex, the upper surface persistently pubescent 4. *S. Bebbiana*
7. Leaves 5 or more times as long as wide, narrowly lanceolate or oblanceolate, more or less acuminate at apex, the upper surface glabrous, or pubescent only along the midvein ....5. *S. Geyeriana*
8. Shrub up to 4 m. high; branchlets usually dark purple (plum-colored) and very glaucous ....6. *S. irrorata*
8. Shrubs or trees; branchlets yellowish to dark brown, not or but slightly glaucous (9).
9. Leaves oblanceolate to linear and acute or short-acuminate; stamens 2, the filaments glabrous ....7. *S. lasiolepis*
9. Leaves prevailingly lanceolate or oblong-lanceolate and long-acuminate; stamens more than 2, the filaments hairy toward the base........................8. *S. laevigata*
10. Lower leaf surface green, slightly paler than the upper surface but seldom distinctly glaucous (11).
10. Lower leaf surface decidedly paler than the upper, usually glaucous (13).
11. Petioles and leaf bases glandular....................................9. *S. caudata*
11. Petioles and leaf bases not glandular (12).
12. Shrub; twigs brown; leaves seldom more than 4 times as long as wide, rounded or subcordate at base, obtuse to short-acuminate at apex..........14. *S. pseudocordata*
12. Tree; twigs yellowish; leaves much more than 4 times as long as wide, attenuate at base, long-acuminate at apex..................................10. *S. Gooddingii*
13. Leaves all, or some of them, conspicuously and sharply long-acuminate at apex. Stamens 3 or more; filaments hairy toward the base; usually trees (14).
13. Leaves rounded to short-acuminate at apex (17).
14. Petioles slender, those of the larger leaves usually 10 mm. long or longer; leaves commonly not more than 3 times as long as wide, not shiny above....11. *S. amygdaloides*
14. Petioles stout, usually less than 10 mm. long; leaves commonly at least 4 times as long as wide, often shiny above (15).
15. Margins of the leaf blades and the petioles near the apex bearing conspicuous yellowish glands; branchlets and the upper surface of the leaves very shiny....12. *S. lasiandra*

15. Margins of the leaf blades and the petioles not or not conspicuously glandular; branchlets and the upper surface of the leaves not or only moderately shiny (16).
16. Leaves commonly broadly lanceolate (less than 6 times as long as wide), usually only moderately acuminate, glaucous but ordinarily not silvery white beneath 8. *S. laevigata*
16. Leaves commonly narrowly lanceolate (6 or more times as long as wide), very long- and sharp-acuminate, silvery white beneath..................13. *S. Bonplandiana*
17. Bases of the leaves rounded or subcordate (seldom cuneate). Stamens 2, the filaments glabrous (18).
17. Bases of the leaves cuneate or attenuate (sometimes rounded in *S. laevigata*), the leaves often shiny above and very glaucous beneath (19).
18. Bark of the twigs yellow or brown; leaves oblong-lanceolate to linear-lanceolate, yellowish green and usually not glossy above, moderately glaucous beneath....15. *S. lutea*
18. Bark of the twigs reddish brown to plum-colored; leaves elliptic (often broadly so), dark green and slightly glossy above, strongly glaucous beneath. Stipules large, usually persistent...16. *S. pseudomonticola*
19. Leaves prevailingly oblanceolate to broadly obovate (20).
19. Leaves prevailingly lanceolate or oblong-lanceolate (21).
20. Capsules hairy; leaves commonly obovate.........................3. *S. Scouleriana*
20. Capsules glabrous; leaves commonly oblanceolate...................7. *S. lasiolepis*
21. Branchlets commonly dark purple (plum-colored) and very glaucous; filaments glabrous 6. *S. irrorata*
21. Branchlets yellow to dark brown, not or but slightly glaucous; filaments hairy toward base ....8. *S. laevigata*

1. **Salix taxifolia** H.B.K. Cochise, Santa Cruz, and Pima counties, 3,500 to 6,000 feet, along streams and washes in the foothills. Western Texas to Arizona, south to Guatemala.

Yew-leaf willow. A relatively slow-growing large shrub or tree, attaining a height of 12 m. (40 feet) and a trunk diameter of 45 cm. (18 inches), but in Arizona usually much smaller. It affords excellent browse for livestock. The very small-leafed var. *microphylla* (Schlecht. & Cham.) Schneid. has been collected in the Canelo (Canille) Hills, Santa Cruz County (*Goodding* 1057–49).

2. **Salix exigua** Nutt. Almost throughout the state, up to 9,500 feet, along streams. Saskatchewan to British Columbia, south to northern Mexico.

Coyote willow. Usually a shrub and not more than 4.5 m. (15 feet) high, often forming thickets. A very narrow-leaved variant, var. *stenophylla* (Rydb.) Schneid., is found here and there in Arizona. Var. *nevadensis* (Wats.) Schneid., characterized by the presence of 2 glands on the glabrous ovary, is known from Coconino, Graham, Gila, and Cochise counties.

3. **Salix Scouleriana** Barratt. Lukachukai and White mountains (Apache County), Kaibab Plateau, Grand Canyon, San Francisco Peaks, and Mount Elden (Coconino County), Pinaleno Mountains (Graham County), Chiricahua Mountains (Cochise County), Santa Catalina Mountains (Pima County), 7,000 to 10,000 feet. South Dakota to Alaska, south to New Mexico, Arizona, and California.

Scouler willow, sometimes known as fire willow. In Arizona more often a shrub but occasionally attaining a height of 9 m. (30 feet). It frequently invades burned-over areas in forests.

4. **Salix Bebbiana** Sarg. North Rim of Grand Canyon, San Francisco Peaks, and Mormon Lake (Coconino County), White Mountains (Apache

County), Lakeside (Navajo County), perhaps also summit of Mount Graham (Graham County), 8,000 to 11,000 feet, in coniferous forests, chiefly along streams. Canada and Alaska, south to Pennsylvania, New Mexico, and Arizona.

Bebb willow. Usually a shrub in Arizona, up to 4.5 m. (15 feet) high.

5. **Salix Geyeriana** Anderss. Apache, Navajo, Coconino, and Yavapai counties, 5,000 to 7,000 feet. Colorado and Montana to British Columbia, south to northern Arizona and California.

6. **Salix irrorata** Anderss. White Mountains (Apache County), Chiricahua Mountains (Cochise County), Rincon and Santa Catalina mountains (Pima County), 6,000 to 7,500 feet (and probably higher), along streams. Western Texas to Colorado and Arizona.

A shrub, seldom more than about 4 m. (12 feet) high.

7. **Salix lasiolepis** Benth. Apache, Navajo, and Coconino counties, south to the mountains of Cochise and Pima counties, 4,000 to 7,500 feet, along streams. Idaho and Washington to Arizona, California, and northern Mexico.

Arroyo willow. Usually a shrub but sometimes treelike and up to 9 m. (30 feet high). The prevailing form in Arizona is the relatively narrow-leaved var. *Bracelinae* Ball (77a).

8. **Salix laevigata** Bebb. Navajo, Coconino, Mohave, Yavapai, Gila, and Cochise counties, 2,000 to 7,000 feet, along streams. Southwestern Utah to Arizona and California.

Red willow. Usually arborescent, reaching a height of 12 m. (40 feet) and a trunk diameter of 0.6 m. (2 feet). A variant with pubescent young twigs, petioles, and bases of the midveins, and often very large leaves, var. *araquipa* (Jepson) Ball, has been collected in the Grand Canyon, Havasu Canyon, and Oak Creek Canyon (Coconino County) and near Fort Verde (Yavapai County).

9. **Salix caudata** (Nutt.) Heller. Reported by Goodding from Springerville, Apache County. South Dakota to British Columbia, south to New Mexico, Arizona, and California.

The variant reported as occurring in eastern Arizona is var. *Bryantiana* Ball & Bracelin, which has a range nearly coextensive with that of the species.

10. **Salix Gooddingii** Ball (*S. nigra* Marsh var. *vallicola* Dudley). Throughout most of the state, along streams, up to 7,000 feet but usually much lower. Western Texas to California and northern Mexico.

Goodding willow. Very similar to the black willow of the eastern United States (*S. nigra*). Along the lower courses of the Colorado and Gila rivers it forms veritable forests, growing with cottonwood (*Populus Fremontii*), attaining a height of 13.5 m. (45 feet) and a trunk diameter of 75 cm. (30 inches) or more. The typical form, with pubescent ovaries and young capsules, is restricted mainly to the Colorado River Valley, the trees in other parts of Arizona belonging mostly to var. *variabilis* Ball, with glabrous ovaries and capsules (77a).

11. **Salix amygdaloides** Anderss. Tunitcha Mountains (Apache County), Rincon Mountains (Pima County), up to 7,000 feet, along streams, apparently rare in Arizona. Quebec to British Columbia, south to New York, Texas, Arizona, and Oregon.

Peach-leaf willow. Usually treelike but seldom more than 9 m. (30 feet) high and 30 cm. (12 inches) in trunk diameter.

Benson and Darrow (8, pp. 106–108) stated that *Salix Wrightii* Anderss. (*S. amygdaloides* Anderss. var. *Wrightii* (Anderss.) Schneid.) is present in Arizona, but its occurrence farther west than New Mexico is questionable. Arizona specimens that had been identified as *S. Wrightii* were referred by C. R. Ball to *S. Gooddingii.*

12. Salix lasiandra Benth. White Mountains (Apache, Navajo, and Greenlee counties), Hopi Indian Reservation (Navajo County), Havasu Canyon (western Coconino County), Tonto Basin (Gila County), 5,000 to 6,500 feet, along streams. Colorado to Yukon, south to New Mexico, Arizona, and California.

Pacific willow. A shrub or small tree, rarely reaching a height of 12 m. (40 feet).

13. Salix Bonplandiana H.B.K. Yavapai, Greenlee, Gila, Santa Cruz, and Pima counties, 3,000 to 5,000 feet. New Mexico and Arizona to Guatemala.

A handsome tree, usually not more than 7.5 m. (25 feet) but sometimes 15 m. (50 feet) high. Arizona plants are commonly referred to var. *Toumeyi* (Britton) Schneid. (*S. Toumeyi* Britton), described as having narrower, less deeply dentate leaves and shorter catkins, but according to Johnston (56, p. 433) the variety is scarcely distinguishable from typical *S. Bonplandiana.*

14. Salix pseudocordata Anderss. White Mountains (Apache County), Betatakin (Navajo County). Montana, Idaho, and Oregon to Arizona and California.

15. Salix lutea Nutt. Apache County to Coconino and Yavapai counties, 4,000 to 6,000 feet (and probably higher). Nebraska to Alberta, south to New Mexico, Arizona, and California.

Yellow willow, usually a shrub. Var. *ligulifolia* Ball, now regarded by Dr. Ball as a distinct species (*S. ligulifolia* Ball), type from Fort Apache (*Coville* 1977), with narrower leaves than in typical *S. lutea* and brown rather than yellow bark of the often very pubescent twigs, is much more common in Arizona than typical *S. lutea.*

16. Salix pseudomonticola Ball. White Mountains, Apache County (*Coville* 2009). Saskatchewan and Alberta to Colorado and Arizona.

A shrub, not more than 3 m. (10 feet) high. The specimen above cited has been referred (77, p. 450) to var. *padophylla* (Rydb.) Ball.

33. JUGLANDACEAE. Walnut Family

This family is outstanding in that many of its members produce edible nuts, the best-known of these being the Persian ("English") and black walnuts, produced by species of *Juglans,* and the hickory nuts and pecans, produced by species of *Carya.*

1. JUGLANS. Walnut

A tree, or occasionally shrubby, with strong-scented, pinnately compound leaves; leaflets commonly 9 to 13, large, lanceolate or ovate-lanceolate, acuminate, serrate; staminate flowers in long, drooping catkins; pistillate flowers solitary, or very few in a cluster; fruit a large, usually nearly globose, hard-shelled nut enclosed in a finally dry husk; cotyledons 2-lobed.

1. Juglans major (Torr.) Heller (*J. rupestris* Engelm. var. *major* Torr.). Almost throughout the state, 3,500 to 7,000 feet, along streams, commonly with cottonwood (*Populus Fremontii*) and buttonwood or sycamore (*Platanus Wrightii*). New Mexico, Arizona, and northern Mexico.

Arizona walnut. A tree up to 15 m. (50 feet) high and 1.2 m. (4 feet) in trunk diameter but usually smaller, with mostly widely spreading branches. The native walnut or nogal is a fine shade tree. The small, thick-shelled nuts are eaten by the Indians in New Mexico and probably in Arizona. Other species of walnut have been employed medicinally and as insecticides, and it is probable that the native species has similar properties. The wood is reported to be durable but is little used.

It is questionable whether *J. major* should be maintained as a species. Specimens from the Pinal Mountains, from Lowell (Cochise County), and from near Tucson approach *J. microcarpa* Berlandier (*J. rupestris* Engelm.) in their narrower, more finely serrate leaflets. *J. elaeopyren* Dode, based on a collection in the Santa Rita Mountains (*Pringle* in 1881), was characterized as having a more elongate nut than in *J. major* and *J. microcarpa*. According to D. T. MacDougal (personal communication), "*J. elaeopyren* is to be found near the lower limits of yellow pine about 7,000 feet, in the Santa Catalinas. Its individuality is quite clear."

34. BETULACEAE. Birch Family

Monoecious trees or large shrubs; leaves alternate, simple; flowers appearing with or before the leaves, those of both sexes in catkins (aments), staminate catkins pendulous; pistillate flowers subtended by conspicuous bracts and (or) bractlets; style 2-branched; ovary 2-celled; fruit a 1-seeded nutlet.

KEY TO THE GENERA

1. Perianth absent in the staminate flowers, present in the pistillate flowers; nutlets wingless, each enclosed in a large, bladderlike, papery bractlet.............1. *Ostrya*
1. Perianth present in the staminate flowers, absent in the pistillate flowers; nutlets winged or thin-edged, not enclosed in a bladderlike bractlet (2).
2. Pistillate catkins solitary, their bracts remaining thin, deciduous with or soon after the nutlets, deeply 3-lobed, the midlobe longer.........................2. *Betula*
2. Pistillate catkins usually several in a racemelike cluster, their bracts becoming thick and woody, long-persistent on the branch after the nutlets have fallen, shallowly 3- to 5-lobed...3. *Alnus*

1. OSTRYA. Hop-hornbeam

Bark scaly, grayish; leaves ovate, sharply double-serrate; staminate flowers without a perianth, the stamens often numerous; nutlets wingless, each enclosed in a greatly enlarged, thin-walled bractlet.

1. Ostrya Knowltoni Coville. Grand Canyon and San Francisco Peaks (Coconino County), 5,000 to 7,000 feet, reported also from northern Yavapai County, type from Grand Canyon (*Toumey* 272). Southeastern Utah and northern Arizona.

Knowlton hop-hornbeam. Plant sometimes treelike, attaining a height of 7.5 m. (about 25 feet) but usually smaller and shrubby. Bark ashy-gray. The fruiting "cones" resemble those of hops.

2. BETULA. Birch

Bark smooth, dark copper-colored (in the Arizona species), with conspicuous lenticels; herbage resinous; leaves broadly ovate, rhombic, or suborbicular, sharply serrate; staminate flowers with a small perianth; nutlets wing-margined.

1. Betula fontinalis Sarg. Tunitcha and Lukachukai mountains (Apache County), northern Navajo County, North Rim of the Grand Canyon (Coconino County), 7,000 to 8,000 feet, mostly along streams, often forming thickets. South Dakota to British Columbia, south to Nebraska, New Mexico, northern Arizona, and California.

Water birch. A small, finely branched tree, reaching a height of 9 m. (30 feet) and a stem diameter of 35 cm. (14 inches) but perhaps never so large in Arizona. The plant affords good browse for sheep and goats.

3. ALNUS. Alder

Bark becoming scaly and dark gray or brown; leaves elliptic to ovate, doubly serrate-dentate; staminate flowers with a small perianth; nutlets thin-margined, scarcely winged.

Alders are browsed to some extent by livestock and tend to check erosion along watercourses in the mountains, where they often form thickets.

KEY TO THE SPECIES

1. Leaves ovate or oblong-ovate, rounded, truncate, or subcordate at base, deeply and doubly serrate-dentate, often somewhat lobed; stamens 4.............1. *A. tenuifolia*
1. Leaves elliptic or ovate-oblong (exceptionally ovate), acutish or short-cuneate at base, shallowly and doubly serrate-dentate, seldom lobed; stamens 1 to 3 (usually 2) 2. *A. oblongifolia*

1. Alnus tenuifolia Nutt. Tunitcha and White mountains (Apache County), Pinaleno Mountains (Graham County), Rincon Mountains (Pima County), 7,000 to 9,500 feet, mostly along streams. Yukon to New Mexico, Arizona, and California.

Thin-leaf alder. A large shrub or small tree up to 7.5 m. (25 feet) high, with reddish-brown bark on the older trunks.

2. Alnus oblongifolia Torr. Apache and Coconino counties south to Graham and Pima counties, 5,000 to 7,500 feet, mostly along streams in the mountains. New Mexico, Arizona, and northern Mexico.

New Mexican alder, Arizona alder. A tree, up to 18 m. (60 feet) high and 1 m. (3 feet) in trunk diameter, with grayish-brown bark on the older trunks. More widely distributed and abundant in Arizona than *A. tenuifolia.*

35. FAGACEAE. Beech Family

Beeches (species of *Fagus*), chestnuts (species of *Castanea*), and oaks are the best-known members of this family of exclusively woody plants.

1. QUERCUS. Oak

Contributed by Cornelius H. Muller

Trees or shrubs, monoecious; leaves alternate, simple, petioled, the blades entire, toothed, or lobed, persistent or deciduous; stipules associated with the buds, ligulate, often caducous; staminate flowers in flaccid, pendulous aments, the perianth about 5-lobed, the stamens 5 to 10, free; pistillate flowers solitary or clustered, subsessile or peduncled, enclosed in an involucre of numerous flat scales, the perianth 6-lobed, the ovary 3-carpellate, 1-celled, the ovules 6 (5 abortive), the styles 3, short; fruit a nut (acorn), 1-seeded, partly enveloped by an involucre (cup) of flat or basally thickened scales, maturing in 1 or 2 seasons.

The shrubby or scrub oaks, such as *Q. turbinella,* are the principal elements of the chaparral on exposed mountainsides in southern and central Arizona. In this region they are considered the chief reserve winter feed for cattle and in addition are of inestimable value in retarding soil erosion. Some of the tree oaks of southern Arizona (such as *Q. arizonica, Q. Emoryi*) are important browse plants. At higher elevations, especially in northern Arizona, the deciduous white oaks (principally *Q. Gambelii*), are abundant and provide ground cover as well as browse for livestock and deer. Although livestock may suffer poisoning when feeding exclusively on oak browse, the foliage is fairly nutritious if supplemented with other feed. The fattening effect of acorns, especially for swine, is well known. They are an important food of birds, squirrels, and other wild animals. The Indians gathered acorns for food, generally roasting them and often making a meal, which they mixed with meat or fat. Although the Arizona oaks afford valuable shade on cattle ranges, they rarely attain sufficient size for logging. The wood is used locally for fuel, fence posts, and mine props. Oak bark is one of the principal sources of tanning material.

KEY TO THE SPECIES

1. Shell of the fruit tomentose within; abortive ovules lateral or apical; cup scales thin and flat or if thickened, then covered with a dense, golden tomentum; stigmas short and broad, or elongate and spatulate (2).

1. Shell of the fruit not tomentose within; abortive ovules basal; cup scales much thickened basally, narrowed at apex; stigmas short and broad. Leaves, if toothed or spinescent, not aristate; bark commonly gray and scaly or flaking: subgenus *Lepidobalanus,* White oaks (4).

2. Cup scales basally thickened, covered with a dense, golden tomentum, the cup large, heavy, loosely fitting the acorn; stigmas short and broad; abortive ovules lateral; leaves glaucous, dull, usually yellow and resinous-pubescent or glaucous beneath, ovate, lobed (rarely entire), the lobes mucronate to long-aristate, the margin markedly crisped, concave beneath: subgenus *Protobalanus,* Intermediate oaks 10. *Q. Palmeri*

2. Cup scales thin, flat, thinly scurfy-pubescent with gray or buff hairs, or glabrate, the cup medium-sized, closely fitting the acorn; stigmas elongate, spatulate; abortive ovules apical; leaves not glaucous or resinous, lanceolate, usually not long-aristate, the margins flat or revolute, not crisped or concave: subgenus *Erythrobalanus,* Black oaks (3).

3. Leaves flat, broadly lanceolate, usually distinctly cordate at base and subentire or with a few short teeth, green on both sides, with a small tuft of stellate tomentum on both sides of the base of the midrib beneath, otherwise glabrate..........11. *Q. Emoryi*

3. Leaves revolute, usually narrowly lanceolate or attenuate, cuneate or rarely rounded at base, entire or with several short teeth or small lobes above the middle, glaucous-green and glabrate or nearly so above, densely white-tomentose beneath 12. *Q. hypoleucoides*
4. Leaves deciduous in autumn, deeply incised or crenate, the lobes round, not mucronate (5).
4. Leaves persistent until spring, entire, mucronate-serrate or acutely shallow-lobed (6).
5. Leaves shallowly crenate-lobed, about 3 to 4.5 cm. long................8. *Q. undulata*
5. Leaves deeply incised with usually narrow lobes, 8 to 15 (rarely only 3 to 5) cm. long 9. *Q. Gambelii*
6. Veins slightly impressed above, markedly prominent beneath, the leaves usually mucronate, sometimes sparingly toothed (7).
6. Veins not impressed, usually somewhat prominent above, not markedly prominent beneath, the leaf margins entire or toothed (8).
7. Leaves concave beneath, obovate to suborbicular, the base strongly cordate; fruit long-stalked .... 1. *Q. reticulata*
7. Leaves flat, narrowly obovate, oblanceolate, or oblong, the base usually cordate, occasionally rounded to subcuneate; fruit subsessile..................3. *Q. arizonica*
8. Leaves oblong, rounded at both ends or cordate at base, glabrous, very glaucous, the margins entire, or occasionally crenate on young shoots..........2. *Q. oblongifolia*
8. Leaves not oblong or rounded at both ends or if so, then not glabrous or glaucous, the margins entire or toothed (9).
9. Leaves medium-sized, 2 to 4 cm. wide, 3 to 6 cm. long, more or less undulate-crisped, roughly short-pubescent or scabrous to the touch (10).
9. Leaves usually small, 0.5 to 2 cm. wide, 1 to 5 cm. long, entire or if toothed, then flat, glabrate or soft-hairy (11).
10. Leaves oblong, pungently lobed, strongly crisped with the lobes appearing twisted, often harsh and scabrous like sandpaper...............................7. *Q. pungens*
10. Leaves ovate, mucronately shallow-lobed, somewhat crisped, moderately rough-pubescent .... 8. *Q. undulata*
11. Leaves narrowly elliptic to lanceolate. Bark roughly scaly or flaky on branches 5 years old; leaves usually very small, acute, entire or rarely with a few short teeth, sparsely pubescent beneath with soft, white hairs, shiny above...............5. *Q. Toumeyi*
11. Leaves broadly elliptic, broadly lanceolate, or ovate (12).
12. Leaves entire or with a few mucronate teeth, relatively large, elliptic to broadly lanceolate or ovate, gray-green or blue-green, the upper surface shining and sparsely stellate-pubescent, the lower surface perceptibly roughened with stellate hairs. Leaves usually obtuse but sometimes acute and mucronate at apex......4. *Q. grisea*
12. Leaves almost aristately many-toothed, small, broadly elliptic to ovate, glaucous, the upper surface dull, nearly glabrous, the lower surface with white or yellow, resinous pubescence .... 6. *Q. turbinella*

1. Quercus reticulata Humb. & Bonpl. (*Q. diversicolor* Trel.). Southern Coconino County to Cochise, Santa Cruz, and Pima counties, 4,000 to 6,500 (exceptionally 8,500?) feet, mostly in canyons, nowhere very abundant. Western Texas to central and southern Arizona and Mexico.

Net-leaf oak. Not very closely related to any other species in the United States. It occurs as a low or tall tree or a shrub, from 2 to 12 m. (6.5 to 40 feet) high. The typically broadly obovate leaves, with several subaristate teeth about the round apex and strongly reticulate venation beneath, make this species readily distinguishable, even in sterile condition, from other Southwestern oaks.

2. Quercus oblongifolia Torr. Mohave County and mountains of Cochise, Santa Cruz, and Pima counties, up to about 6,000 feet, common in the foothills of desert mountain ranges. Arizona and northern Mexico.

Mexican blue oak. Commonly a small tree 5 to 8 m. (16 to 26 feet) high, with a broadly spreading crown, but it may mature as a shrub at higher altitudes.

3. Quercus arizonica Sarg. Southern Coconino County and Yavapai County to Greenlee, Graham, Cochise, Santa Cruz, and Pima counties, up to about 6,000 (exceptionally 7,500) feet, common. Western Texas, New Mexico, Arizona, and northern Mexico.

Arizona white oak. A well-marked species in Arizona although it grades into *Q. grisea* Liebm. in New Mexico. It becomes a tree up to 18 m. (60 feet) high, with a trunk 1 m. (3 feet) in diameter, or it may mature as a large shrub.

4. Quercus grisea Liebm. White Mountains (Apache County) and Grand Canyon (Coconino County) to Cochise and Santa Cruz counties. Texas to Arizona and northern Mexico.

Gray oak. A small or large tree, reaching a height of 20 m. (65 feet) in moist, protected situations. It is one of the most difficult of Arizona oaks to distinguish, intergrading freely in this area with both *Q. arizonica* and *Q. turbinella,* although the much more abundant typical forms of the 3 species are clearly distinct. The species reaches the northwestern limit of its range in Arizona and is not an important element of the flora there, whereas in New Mexico and western Texas it is dominant in the "encinal" over large areas.

5. Quercus Toumeyi Sarg. Greenlee, Cochise, and Santa Cruz counties, 3,500 to 7,000 feet, type from the Mule Mountains, Cochise County (*Toumey* in 1899). Southwestern New Mexico, southern Arizona, and northern Mexico.

Toumey oak. A small tree usually less than 10 m. (33 feet) high, or a shrub 1 or 2 m. high. Its characteristically very small, entire, yellowish-green leaves, shiny above, mucronately acute, and densely disposed on the branches, make it easily distinguishable from all other species in its range.

6. Quercus turbinella Greene (*Q. subturbinella* Trel.). Almost throughout the state, up to 8,000 feet but commonly lower. Colorado and western Texas to California and northern Mexico.

Shrub live oak. This species centers in Arizona, where it is the most abundant element of the chaparral, growing as a shrub or at most a small tree up to 4 m. (13 feet) high. The species was described from California specimens. Trelease's treatment of the Arizona material as distinct apparently was based on supposed differences in geographical distribution, but the species has a nearly continuous range from California to western Texas and in the absence of morphological difference *Q. subturbinella* must be reduced to synonymy. By reason of frequent misidentification *Q. turbinella* in Arizona and New Mexico has been much confused with *Q. pungens* Liebm., to which it is not related and bears not even a superficial resemblance. The species is rather close to *Q. grisea* Liebm. and occasionally is confused with toothed-leaved forms of that species, but *Q. turbinella* is readily recognized by its glaucous leaves.

7. Quercus pungens Liebm. Cochise County, 4,000 to 5,000 feet, sometimes on limestone. Western Texas to southeastern Arizona.

A low shrub, 0.5 to 2 m. (2 to 6.5 feet) high. The distinctive leaves, pungently lobed and often markedly crisped and scabrous, clearly distinguish this species from all others except occasional forms of *Q. Vaseyana* Buckl. of Texas.

8. **Quercus undulata** Torr. (*Q. Fendleri* Liebm., *Q. subobtusifolia* Camus). Navajo, Coconino, Mohave, Gila, and Cochise counties, commonly 6,000 to 8,000 feet. Colorado, Utah, Texas, New Mexico, and Arizona.

Wavyleaf oak. A low shrub, usually not more than 0.3 to 2 m. (1 to 6.5 feet) high, but attaining a height of 4 m. (13 feet) in moist, protected spots. *Q. subobtusifolia* represents the deciduous-leaved phase of *Q. undulata.* The narrow, shallowly lobed leaves and obligate shrub habit of *Q. undulata* and its tendency to frequent open drier and lower sites readily distinguish it from the arboreal (sometimes shrubby) *Q. Gambelii,* the latter having broad, deeply lobed leaves.

9. **Quercus Gambelii** Nutt. (*Q. submollis* Rydb.). Throughout the state except the extreme western and southwestern parts, commonly 5,000 to 8,000 feet, often forming thickets, type of *Q. submollis* from southern Arizona (*Wilcox* 191a, in part). Colorado to Nevada, south to northern Mexico.

Gambel oak, Rocky Mountain white oak. In Arizona *Q. Gambelii* ranges from a shrub 2 m. (6.5 feet) or less in height to a tree 15 m. (50 feet) high in protected sites. The deeply lobed, deciduous leaves of this species make it readily recognizable among Arizona oaks. In a species so variable in size, lobing, and pubescence of the leaves, the fine distinctions of Rydberg are untenable.

10. **Quercus Palmeri** Engelm. (*Q. Wilcoxii* Rydb.). Oak Creek Canyon (Coconino County) and Hualpai Mountain (Mohave County) to the mountains of Greenlee, Cochise, and Pima counties, 3,500 to 7,000 feet, often forming thickets, type of *Q. Wilcoxii* from Fort Huachuca (*Wilcox* in 1892). Arizona, California, and northwestern Mexico.

Palmer oak. A shrub or small tree, commonly 2 to 4 m. (6 to 13 feet) high. It cannot be confused with other species of oak in Arizona, but its specific distinction from *Q. chrysolepis* Liebm. of California is open to question. The superficial resemblance of its pungently toothed, glaucous leaves, resinous beneath, to descriptions of *Q. turbinella* might be confusing until one becomes familiar with both species. The scales of the fruiting cups are dependable criteria.

11. **Quercus Emoryi** Torr. Southern Coconino County to Cochise, Santa Cruz, and Pima counties, 3,000 to 8,000 feet, dry foothills and moist canyons. Western Texas to Arizona and northern Mexico.

Emory oak, known locally as blackjack oak and bellota, is one of the most abundant oaks in the "encinal" areas of the Mexican border region. It forms a shrub, a small tree, or a large tree up to 15 m. (50 feet) high and 0.8 m. (2.5 feet) in trunk diameter.

12. **Quercus hypoleucoides** Camus (*Q. hypoleuca* Engelm. non Miq.). Mountains of Greenlee, Graham, Cochise, Santa Cruz, and Pima counties, 5,000 to 7,000 (exceptionally 8,500?) feet, on slopes and in canyons. Western Texas to southern Arizona and northern Mexico.

Silverleaf oak. This is a widely distributed and fairly common shrub or small tree.

36. ULMACEAE. Elm Family

Elms (species of *Ulmus*) are the typical representatives of this family.

1. CELTIS. Hackberry

Trees or shrubs; leaves simple, alternate, unequal at base, very rough above; flowers perfect or unisexual, axillary, solitary or in small clusters; perianth 5- or 6-parted; style none, the 2 stigmas elongate, spreading or recurved, plumose; ovary 1-celled; fruit a globose drupe with thin flesh and a hard-shelled stone, mostly yellow to dull red at maturity.

Hackberries are browsed by cattle when other forage is scarce, and the sweet but rather dry and insipid fruits are relished by birds and small desert mammals. The Papago and probably other Indians gather the fruits for food. The wood is sometimes used for fence posts.

KEY TO THE SPECIES

1. Plant a spiny, intricately branched shrub; leaves not more than 2 cm. wide, elliptic to oblong-ovate, rounded to acutish at apex, entire or sparingly crenate-dentate, not reticulate-veined; herbage puberulent.................................1. *C. pallida*
1. Plants large shrubs or small trees, not spiny or intricately branched; leaves more than 2 cm. wide, ovate or lance-ovate, acute to sharply acuminate at apex, obliquely cuneate to obliquely cordate at base, serrate-dentate or entire, more or less prominently reticulate-veined beneath; twigs, petioles, and lower surface of the leaves (especially the veins) short-pilose (usually sparsely so) to nearly glabrous. Bark warty 2. *C. reticulata*

1. Celtis pallida Torr. (*C. Tala* Gillies var. *pallida* (Torr.) Planch.). Greenlee, Gila, Pinal, Maricopa, Cochise, and Pima counties, 1,500 to 3,500 feet, foothills and mesas, often forming dense thickets. Western Texas to Arizona and northern Mexico.

Granjeno, desert hackberry. This is reportedly of value as a honey plant and for erosion control.

2. Celtis reticulata Torr. (*C. Douglasii* Planch., *C. laevigata* Willd. var. *reticulata* L. Benson). Almost throughout the state, 2,500 to 6,000 feet, usually along streams. Oklahoma and Colorado to Arizona and northern Mexico.

Netleaf hackberry, palo-blanco, sugar-berry. The species is highly variable in the shape, thickness, and pubescence of the leaves and prominence of the veins, also in length of the pedicels, but there is so much intergradation that recognition, even of varieties, is difficult. *C. brevipes* Wats. is a form with pedicels somewhat shorter than to about equaling the petioles. At the other extreme are specimens from the Grand Canyon, referred by Sargent to *C. Douglasii* Planch., with serrate, long-acuminate, more or less cordate leaves and pedicels up to 2 cm. long and 2 to 4 times as long as the petioles. *C. Douglasii* is reported to have brownish-purple fruits, whereas in typical *C. reticulata* they are usually orange-colored at maturity.

An extraordinarily thick-leaved form with strongly cordate leaf bases was collected at Fish Creek, eastern Maricopa County (*Eastwood* 15956). Another extreme form, with rounded-cordate, merely acute leaves only about 2 cm. long, short pedicels, and rugose fruits, has been found in the Ajo Mountains, western Pima County (*Goodding* 476–45). Mr. Goodding reported that it is common there, and constant in its characters. It is highly probable that more than one species is included here under *C. reticulata,* but pending thorough revision of the North American species of *Celtis* no other treatment seems practicable.

37. MORACEAE. Mulberry Family

Small trees, shrubs, or nearly herbaceous twining plants, mostly dioecious; leaves simple but often deeply lobed, alternate or opposite; flowers inconspicuous, those of both sexes, or the pistillate ones only, in catkins; perianth present, calyxlike.

KEY TO THE GENERA

1. Plant a small tree; pistillate and staminate flowers both in catkins; perianths of the pistillate flowers becoming thick, succulent, and fused, the infloresence as a whole becoming a juicy, oblong, aggregate fruit..............................1. *Morus*
1. Plant climbing, perennial, herbaceous or suffrutescent; staminate flowers in loose panicles, the pistillate flowers in short, catkinlike spikes; bracts of the pistillate inflorescence foliaceous, imbricate, becoming much enlarged and membranaceous in fruit ....2. *Humulus*

1. MORUS. Mulberry

A tree or large shrub; leaves alternate, crenate-serrate or palmately lobed (especially on young shoots); perianth 4-parted; stigmas 2, linear, spreading.

Mulberry leaves are the favorite food of the silkworm.

1. Morus microphylla Buckl. Havasu Canyon (Coconino County), where perhaps introduced, and Greenlee County to Yavapai County, south to Cochise, Santa Cruz, and Pima counties, 3,500 to 5,000 feet, usually along streams. Western Texas to Arizona and northern Mexico.

Texas mulberry. The stems seldom exceed 4.5 m. (15 feet) in height. The tart, palatable fruits are gathered by the Papago Indians and are greedily eaten by birds and other animals. There is great variation in the shape of the leaves, and largely on this basis several segregate species founded on Arizona types were described by E. L. Greene. These are: *M. confinis* (type *Pringle* in 1881, Santa Rita Mountains), *M. crataegifolia* (type *Hough* in 1905, Blue River, Greenlee County), *M. radulina* (type *Fernow* in 1896, Beaver Creek, Yavapai County), and *M. grisea* (type *Palmer* in 1869, Hell Canyon, Yavapai County). The last is an exceptionally pubescent form.

2. HUMULUS. Hop

Plant twining, perennial, herbaceous or nearly so; leaves opposite, palmately 3- to 7-lobed; perianth 5-parted in the staminate flowers, entire in the pistillate flowers; stigmas 2, filiform.

1. Humulus americanus Nutt. Apache, Navajo, and Coconino counties, south to Graham and Pima counties, 5,500 to 9,500 feet, mountain woods. Rocky Mountain region, south to New Mexico and Arizona.

The pistillate inflorescences of the closely related European species (*H. Lupulus* L.) are used in the brewing of malt beverages to impart the bitter flavor and in medicine as a tonic and soporific. The Southwestern plant apparently was utilized by the Indians, the Apache name being said to mean "to make bread with it." Rare cases of dermatitis from contact with the plant have been reported.

38. URTICACEAE. Nettle Family

Plants herbaceous; leaves simple, opposite or alternate, with or without stipules; inflorescences axillary; flowers perfect or unisexual; perianth small, calyxlike; petals none; stamens commonly 4, opposite the perianth segments; ovary 1-celled; fruit an achene.

Several important fiber plants, notably hemp (*Cannabis sativa*) and ramie (*Boehmeria nivea*), belong to this family.

KEY TO THE GENERA

1. Segments of the pistillate flowers distinct or nearly so, in pairs, the outer ones much smaller, the inner ones enveloping the achene; plants armed with stinging, bristle-like hairs; leaves opposite .... 1. *Urtica*
1. Segments of the pistillate flowers united much of their length in a tubular or bell-shaped, 4-lobed perianth; plants without stinging hairs; leaves alternate .... 2. *Parietaria*

1. URTICA. Nettle

Stems tall; leaves opposite, with stipules; herbage more or less canescent or tomentose, usually also hispid with stinging hairs; inflorescences elongate, spikelike; stigma sessile or nearly so; achene compressed.

All the species flower in Arizona in summer (July to September). Contact with nettles is very irritating to the skin because of the toxin released by the stinging hairs.

KEY TO THE SPECIES

1. Plants annual; inflorescences unbranched; petioles of the lower leaves more than half as long as to longer than the blades, these broadly ovate (at least the lower ones), often nearly as wide as long, cordate to subcuneate at base, the marginal teeth broadly triangular-ovate, obtuse or acutish, directed only slightly upward .... 1. *U. gracilenta*
1. Plants perennial; inflorescences branched; petioles mostly less than half as long as the blades, these commonly lanceolate and much longer than wide, rounded to cuneate (exceptionally cordate) at base, the marginal teeth triangular, acute, directed strongly upward (2).
2. Stipules linear or narrowly lanceolate, acuminate; herbage green, sparsely to copiously (but not densely) pubescent as well as hispid .... 2. *U. gracilis*
2. Stipules oblong or broadly lanceolate, obtuse or acutish; stem and the lower surface of the leaves usually densely grayish pubescent as well as hispid .... 3. *U. Serra*

1. Urtica gracilenta Greene. Sierra Ancha (Gila County), Chiricahua and Huachuca mountains (Cochise County), Bear Valley (Santa Cruz County), Baboquivari Mountains (Pima County), 4,000 to 8,000 feet. New Mexico and Arizona.

The leaves are bright green on both surfaces, and the shorter hairs are sometimes spreading and straight, sometimes subappressed and curly.

2. Urtica gracilis Ait. Lukachukai and White mountains (Apache County), Kaibab Plateau (Coconino County), Pinaleno Mountains (Graham County), Chiricahua and Huachuca mountains (Cochise County), up to 9,000 feet, in springy places and along streams. Almost throughout temperate North America.

F. J. Hermann (78) concluded that *U. gracilis* is not distinct as a species from *U. dioica* L.

3. Urtica Serra Blume (*U. Breweri* Wats.). Kaibab Plateau and Grand Canyon (Coconino County), about 7,500 feet. Idaho and Oregon to New Mexico, Arizona, California, and Mexico.

In *Flowering Plants and Ferns of Arizona* (p. 230) this species was mistaken for *U. holosericea* Nutt. A specimen from the Carrizo Mountains, Apache County (*Standley* 7509), seems to be intermediate between *U. gracilis* and *U. Serra,* the herbage being sparsely pubescent but the stipules broad and obtusish.

2. PARIETARIA. Pellitory

Plants annual; leaves alternate, without stipules, lanceolate to ovate; inflorescences small, glomerate, subtended by leafy bracts; style short, the stigma tufted; achene enclosed by the persistent calyx.

In Arizona these insignificant plants are in flower almost throughout the year.

KEY TO THE SPECIES

1. Involucre 2 to 3 times as long as the flowers; stems simple or sparingly branched, erect or ascending; leaves commonly lanceolate or ovate-lanceolate...... 1. *P. pensylvanica*
1. Involucre commonly less than twice as long as the flowers; stems diffusely branched from the base; leaves commonly broadly ovate or suborbicular, sometimes oblong or lanceolate 2. *P. floridana*

1. Parietaria pensylvanica Muhl. Grand Canyon (Coconino County), and Mohave, Yavapai, Gila, Maricopa, Pinal, and Pima counties, 1,500 to 4,000 (?) feet. Throughout most of temperate North America.

2. Parietaria floridana Nutt. Throughout much of the state, 5,000 feet or lower. Southeastern United States to southern California.

Parietaria obtusa Rydb. apparently is a variant with obtuse bracts and sepals.

39. LORANTHACEAE. Mistletoe Family

Plants dioecious, parasitic on trees and shrubs, with or without chlorophyll; stems jointed, brittle when dry; leaves opposite; perianth calyxlike, small, with 2 to 4 lobes or teeth; stamens inserted on the perianth, the anthers sessile; ovary inferior, 1-celled; fruit a 1-seeded berry.

Nearly all members of this mainly tropical family are parasitic shrubs. The plants of both Arizona genera sap the vitality of the host tree and when abundant, cause considerable damage in forests. Birds are very fond of the berries of *Phoradendron,* and through their agency the viscid seeds are transferred from tree to tree. The Papago Indians dry the berries of *P. californicum* in the sun and store them as food supply. The Hopi are reported to use *P. juniperinum* medicinally. The berries of some species are reputedly poisonous and contain a principle similar in action to epinephrin. The European mistletoe (*Viscum album*) and some of the American species of *Phoradendron* are popular for Christmas decoration because of their white or whitish, pearllike berries.

KEY TO THE GENERA

1. Fruits sessile, globose or nearly so, whitish or reddish; flowers in axillary spikes; pistillate perianth normally 3-toothed; anthers 2-celled.......................1. *Phoradendron*
1. Fruits on short, often curved pedicels, longer than wide, greenish, bluish, or purplish; flowers not in spikes; pistillate perianth normally 2-toothed; anthers 1-celled 2. *Arceuthobium*

1. PHORADENDRON (79, 80). MISTLETOE

Plants with or without chlorophyll, parasitic on various trees and shrubs; leaves with well-developed blades or reduced to scales; style short, the stigma usually captitate.

KEY TO THE SPECIES

1. Leaves reduced to triangular scales not more than 2 mm. long (2).
1. Leaves with well-developed blades seldom less than 1 cm. long (3).
2. Stems not crowded, the branches relatively slender, terete, flexuous; leaf scales only slightly connate; pistillate spikes with 3 or more nodes, often tomentulose; berries usually red ....1. *P. californicum*
2. Stems crowded, the branches stout, obscurely quadrangular, rather rigid; leaf scales strongly connate; pistillate spikes with only 1 node; berries whitish. .2. *P. juniperinum*
3. Pistillate spikes with 2 flowers at each node, commonly only 2-flowered; leaves linear-spatulate to oblong, not more (usually much less) than 10 mm. wide, sessile or nearly so; spikes in flower not more than 5 mm. long.......................3. *P. Bolleanum*
3. Pistillate spikes with more than 2 (usually 6 or more) flowers at each node; leaves broadly elliptic, ovate, obovate, or nearly orbicular, commonly more than 10 mm. wide, distinctly petiolate; spikes more than 5 mm. long (4).
4. Fruits glabrous (the perianth sometimes puberulent), 4 to 5 mm. in greatest diameter when mature; herbage glabrous to variously pubescent; leaves commonly more than 3 cm. long...4. *P. flavescens*
4. Fruits puberulent at apex (the pubescence not confined to the perianth), about 3 mm. in greatest diameter when mature; herbage closely puberulent; leaves rarely more than 3 cm. long ....5. *P. Coryae*

1. Phoradendron californicum Nutt. Grand Canyon (Coconino County) and common in the semidesert southern and western parts of the state, 4,000 feet or lower. Southern Utah and Arizona to southern California and northern Mexico.

Parasitic chiefly on leguminous shrubs and trees (*Acacia, Prosopis, Cercidium*), but occasionally on Rhamnaceae, rarely on *Larrea*. Var. *distans* Trelease, with spikes more elongate and whorls of fruit more widely spaced than in the typical phase of the species, is common in Arizona.

2. Phoradendron juniperinum Engelm. Apache, Navajo, and Coconino counties, south to Cochise and Pima counties, 4,000 to 7,000 feet. Southwestern Colorado and southern Utah to western Texas, Arizona, and northern Mexico.

Parasitic on several species of juniper, rarely on pinyon (*Pinus monophylla*). The plant has yellowish or light-brown stems and resembles a large *Arceuthobium*.

3. Phoradendron Bolleanum (Seem.) Eichler. Coconino County to Cochise, Santa Cruz, and Pima counties, 3,500 to 7,000 feet. Western Texas to Oregon, California, and northern Mexico.

KEY TO THE VARIETIES

1. Herbage stellate-tomentulose; leaves linear-spatulate................var. *capitellatum*
1. Herbage glabrous or sparsely hispidulous, the hairs not stellate; leaves oblong or somewhat oblanceolate (2).
2. Plant compact; leaves sessile, 10 to 15 mm. long, seldom more than 4 mm. wide var. *densum*
2. Plant open; leaves subpetiolate, 15 to 30 mm. long, 5 to 10 mm. wide....var. *pauciflorum*

Typical *P. Bolleanum* is not known in Arizona. Var. *capitellatum* (Torr.) Kearney & Peebles (*P. capitellatum* Torr.) is the commonest variety in Arizona, occupying the entire range of the species in the state. It parasitizes *Juniperus*. Var. *densum* (Torr.) Fosberg (*P. densum* Torr.) is known near Flagstaff, Coconino County (*Hedgcock* 4915, in part, *Deaver* 1596), and from the Mazatzal Mountains, Gila County (*Peebles* 11557). In Arizona it is parasitic on *Cupressus* and sometimes on *Juniperus*. Var. *pauciflorum* (Torr.) Fosberg (*P. pauciflorum* Torr.) occurs in southern Coconino, Gila, and Pima counties. Its hosts are *Cupressus* in Coconino and Gila counties, *Abies* in the Santa Catalina Mountains.

4. Phoradendron flavescens (Pursh) Nutt. Yavapai County to Graham, Cochise, Santa Cruz, and Pima counties, doubtless also in Yuma County. New Jersey to Florida, west to California, south to central Mexico.

KEY TO THE VARIETIES

1. Herbage nearly glabrous, or glabrescent..........................var. *macrophyllum*
1. Herbage copiously to densely pubescent, at least when young (2).
2. Herbage stellate-pubescent or short-hirtellous, the hairs usually branched or tufted var. *pubescens*
2. Herbage velvety-pubescent, the hairs simple, not tufted..................var. *villosum*

This is the most conspicuous and largest-leaved of the Arizona mistletoes, forming great masses, especially on cottonwood (*Populus Fremontii*) but also parasitizing sycamore, ash, hackberry, walnut, willow, and mesquite. The commonest variety in Arizona is var. *pubescens* Engelm. (*P. Engelmanni* Trel.), but var. *macrophyllum* Engelm. (*P. macrophyllum* Cockerell, *P. macrophyllum* var. *circulare* Trel. and var. *Jonesii* Trel., *P. coloradense* Trel., *P. longispicum* Trel.) is not infrequent. Var. *villosum* (Nutt.) Engelm. (*P. villosum* Nutt., *P. flavescens* var. *tomentosum* Engelm.) may also occur, but if so it is much less common than in California.

5. Phoradendron Coryae Trel. Southern Coconino County to Graham, Cochise, Santa Cruz, and Pima counties, 3,500 to 8,500 feet, types from the Chiricahua Mountains (*Blumer* 1514, 1516). Western Texas to Arizona and Sonora.

A very common species in Arizona, parasitic on several evergreen species of oak, rarely on barberry (*Berberis haematocarpa*).

2. ARCEUTHOBIUM (81). SMALL-MISTLETOE

Plants without chlorophyll, parasitic only on conifers; leaves reduced to connate scales; flowers solitary or several in the axils of the scales.

A. vaginatum sometimes causes considerable damage in stands of young ponderosa pine.

KEY TO THE SPECIES

1. Stems slender, about 1 mm. in diameter at base, seldom more than 3 cm. long, commonly scattered along the stem of the host plant. Plants greenish, flowering in spring and early summer .... 1. *A. Douglasii*
1. Stems relatively stout, 2 to 5 mm. in diameter at base, usually more than 3 cm. long, commonly clustered (2).
2. Plants mostly yellowish, flowering in spring and early summer; stems seldom less than 3 mm. in diameter at base .... 2. *A. vaginatum*
2. Plants olive green, brownish, or sometimes yellowish, flowering in late summer; stems often less than 3 mm. in diameter at base........................ 3. *A. campylopodum*

1. Arceuthobium Douglasii Engelm. Apache, Navajo, and Coconino counties, south to Greenlee, Graham, and Pima counties, up to 8,500 feet (perhaps higher). Western Canada to New Mexico, Arizona, California, and Mexico. Parasitic on Douglas-fir (*Pseudotsuga*).

2. Arceuthobium vaginatum (H.B.K.) Eichler. Apache, Navajo, and Coconino counties, south to Cochise and Pima counties, 5,500 to 8,000 feet. Colorado and Utah to Mexico.

Parasitic on pines, the presumably typical phase on *Pinus latifolia, P. chihuahuana,* and *P. ponderosa* var. *arizonica* in the southern counties; whereas forma *cryptopodum* (Engelm.) Gill (*A. cryptopodum* Engelm.) infests *P. ponderosa* var. *scopulorum* in central and northern Arizona.

3. Arceuthobium campylopodum Engelm. Throughout most of the state in various forms, 4,500 feet or higher. Western Canada and Alaska to Mexico.

The typical phase of the species has not been reported from Arizona. Forma *abietinum* (Engelm.) Gill (*A. Douglasii* var. *abietinum* Engelm.) parasitic on firs (*Abies* spp.) has been collected in Coconino County, at the Grand Canyon (*Gill* in 1934) and on the San Francisco Peaks (*Knowlton* 40), also in the Pinaleno Mountains, Graham County (*Stouffer & Gill* in 1934). Forma *divaricatum* (Engelm.) Gill (*A. divaricatum* Engelm.), with the old stems often conspicuously divaricate, is common in the northern and central portions (Apache to Coconino and Yavapai counties), parasitic on pinyons (*Pinus edulis*). Forma *microcarpum* (Engelm.) Gill (*A. Douglasii* var. ? *microcarpum* Engelm.) is frequent, ranging from the Kaibab Plateau (Coconino County) and the White Mountains (Apache County) to the Pinaleno Mountains (Graham County), parasitic on spruce (*Picea*). The type collection of this form is from the White Mountains (*Gilbert* 112). Forma *Blumeri* (A. Nels.) Gill (*A. Blumeri* A. Nels.) occurs in the mountains of Graham, Cochise, and Pima counties, parasitic on *Pinus reflexa*. The type of *A. Blumeri* was collected by Blumer in the Chiricahua Mountains. Forma *cyanocarpum* (A. Nels.) Gill has been collected in Arizona only on the San Francisco Peaks, 10,000 feet, on *Pinus aristata* (*Leiberg* 5884), and at Blue Summit, Greenlee County, on *Pinus flexilis* (*Gill* in 1934).

40. SANTALACEAE. Sandalwood Family

Many members of this family are root parasites. The Old World sandalwoods (species of *Santalum*), esteemed for the exquisite fragrance of their wood, are the best-known representatives.

1. COMANDRA. Bastard-toadflax

Plant a root parasite, small, herbaceous or nearly so; leaves alternate, narrow, entire; flowers perfect, small, whitish or pinkish, in terminal corymbose clusters; perianth urn-shaped, 4- or 5-cleft; stamens borne on a fleshy disk in the perianth tube; ovary inferior; ovules 2 to 4; fruit nutlike, indehiscent, 1-seeded, crowned by the persistent perianth teeth.

1. Comandra pallida A. DC. Apache County to Mohave County, south to Cochise, Santa Cruz, and Pima counties, 4,000 to 9,500 feet, usually in pine woods, April to August. Minnesota to British Columbia, south to Texas, Arizona, and California.

Plants of many different families have been reported as hosts.

41. ARISTOLOCHIACEAE. Birthwort Family

1. ARISTOLOCHIA

Plant (the Arizona species) herbaceous, perennial; stems from a thick root, trailing; leaves alternate, triangular-hastate, often purplish; flowers perfect, axillary, solitary; perianth irregular, tubular below, dull greenish and brown-purple; ovary inferior; fruit a dehiscent 4- to 6- (usually 5-) valved capsule; seeds numerous, horizontal, flat.

This is by far the largest genus of its family. It includes numerous large, woody climbers, especially in tropical America, some of which have very large flowers, these attaining, in the pelican-flower (*A. grandiflora*) a diameter of 20 inches and, with the taillike appendage, a length of 3 feet.

1. Aristolochia Watsoni Woot. & Standl. Greenlee, Gila, and southern Mohave counties to Cochise, Santa Cruz, Pima, and Yuma counties, 2,000 to 4,500 feet, usually among shrubs, April to October. Western Texas to southern Arizona.

Known locally as Indian-root and reported to have been used medicinally, especially as a remedy for snake bite, by Indians and by the white settlers. The flowers, as in many tropical species, have a fetid odor and are so constructed as to insure insect pollination.

42. RAFFLESIACEAE. Rafflesia Family

1. PILOSTYLES

Plant dioecious (?), parasitic, with only the small, brown flowers and a few subtending, imbricate bracts visible outside the bark of the host; flowers unisexual, the perianth of 4 or 5 distinct segments similar to the subtending bracts, with a thick central column expanded at apex into a fleshy disk, the latter bearing under its margin in staminate flowers the anthers, in pistillate flowers the ring-shaped stigma; ovary inferior; fruit a many-seeded capsule.

1. Pilostyles Thurberi Gray. Yuma County, April, type from near the mouth of the Gila River (*Thurber* in 1850). Southwestern Arizona and southeastern California.

Parasitic on the stems of shrubby species of the leguminous genus *Dalea,* especially *D. Emoryi.* It is noteworthy that the family to which this tiny plant belongs includes a plant having the largest of all known flowers, *Rafflesia Arnoldii,* a native of Sumatra.

43. POLYGONACEAE. Buckwheat Family

Plants herbaceous or woody; stems jointed in some of the genera; leaves simple, entire, with or without stipules; flowers small, regular, commonly perfect; perianth not differentiated into a calyx and a corolla but usually corollalike; stamens 2 to 9; pistil 1, the ovary superior, 1-celled; style cleft or parted, the divisions 2 or 3; fruit a trigonous or lenticular achene.

This family includes buckwheat (*Fagopyrum*) and rhubarb (*Rheum*), also several common weeds of the genera *Rumex* (dock) and *Polygonum* (knotweed).

KEY TO THE GENERA

1. Flowers subtended by a campanulate, turbinate, or cylindric involucre, this composed of more or less united bracts (2).
1. Flowers not subtended by an involucre of the foregoing character (4).
2. Teeth or lobes of the involucre not bristly or spiny at tip................5. *Eriogonum*
2. Teeth or lobes of the involucre ending in bristles or spines (3).
3. Involucre subtending usually only 1 flower, its teeth tipped with hooked or straight spines or bristles; bracts not connate-perfoliate..........................3. *Chorizanthe*
3. Involucre subtending 2 or more flowers, its teeth tipped with straight spines or bristles; upper bracts connate-perfoliate, forming a cup-shaped disk............4. *Oxytheca*
4. Stems without swollen joints; leaves opposite or basal, without sheathlike stipules (ocreae); flowers each subtended by a bract, the bracts not sheathlike (5).
4. Stems with swollen joints; leaves alternate or basal, with sheathlike stipules; bracts of the inflorescence sheathlike, more or less scarious (6).
5. Stems leafy; leaves opposite; flowers solitary in the axils; bracts enlarged in fruit, 2-lobed, bigibbous on the back....................................1. *Pterostegia*
5. Stems subscapose, the leaves all or nearly all basal; flowers in subglobose, sessile heads; bracts not becoming enlarged, entire, not gibbous on the back..........2. *Nemacaulis*
6. Perianth segments 4 or 6, the outer ones spreading or reflexed, remaining small, the inner ones usually erect and enlarged in fruit (7).
6. Perianth segments 5 (exceptionally 4), all similar and erect in fruit (8).
7. Leaves lanceolate to oblong-ovate; perianth segments 6, one or more of the inner ones often with a large dorsal tubercle; style 3.............................6. *Rumex*
7. Leaves reniform; perianth segments 4; styles 2...........................7. *Oxyria*
8. Achene enclosed by the fruiting perianth or, if exserted, then the leaves narrow and attenuate at base ....8. *Polygonum*
8. Achene long-exserted; leaves broadly deltoid, hastate or cordate.........9. *Fagopyrum*

1. PTEROSTEGIA

Plant herbaceous, annual; stems slender, weak, diffusely branched, often prostrate; leaves opposite, petioled, cuneate-obovate or fan-shaped, usually deeply lobed; flowers solitary, subtended by a single, folded, 2-lobed, dentate bract, this in fruit enlarged, reticulate-veined, bigibbous on the back.

1. Pterostegia drymarioides Fisch. & Meyer. Grand Canyon (Coconino County) and in Mohave, Yavapai, Pinal, Maricopa, Pima, and Yuma counties, 3,500 feet or lower, February to April. Oregon to Utah, Arizona, and Baja California.

2. NEMACAULIS

Plant small, annual, subscapose; stems prostrate or ascending, with few spreading branches; larger leaves all basal or nearly so, spatulate, white-woolly, the stem leaves greatly reduced and bractlike; floral bracts woolly above, involving the flowers in a mass of wool, oblong, spatulate, or obovate, forming a whorl but not united; perianth 6-cleft, yellowish or pinkish, 1 mm. long; achenes ovoid.

1. Nemacaulis denudata Nutt. Yuma County, at Mohawk (*Alice Bird*) and near Bouse (*L. Benson* 10910), 1,500 feet or lower, in sandy soil. Southwestern Arizona, southeastern California, Baja California, and northwestern Sonora.

The plant resembles some of the annual species of *Eriogonum* but is readily distinguished by the combination of distinct involucral bracts and conspicuously woolly flower heads.

3. CHORIZANTHE (82)

Plants annual, herbaceous, dichotomously branched; foliage leaves in a basal rosette, soon disappearing, the stem leaves mostly bractlike, opposite or in threes; flowers (usually solitary) subtended by a tubular or funnelform involucre; stamens commonly 9.

Plants of the more desertic parts of the state.

KEY TO THE SPECIES

1. Bracts divergently 3-lobed; involucre small, with 3 rather broad divaricate spurs at base. Plant often reddish; stem glandular-puberulent, especially toward the base, slender, diffusely branched, not becoming rigid; foliage leaves all basal, oblanceolate or spatulate; teeth of the bracts and involucre straight .... 1. *C. Thurberi*
1. Bracts entire; involucre not spurred (2).
2. Involucres 6-toothed, the tube 6-ribbed, the teeth strongly hooked at apex; stems very brittle, soon disarticulating; foliage leaves all basal, narrowly oblanceolate, strigose or villous, the stem leaves reduced to subulate bracts.................. 2. *C. brevicornu*
2. Involucres with fewer than 6 teeth, the tube either 3-ribbed (angled) or not ribbed; stems not very brittle; stem leaves, at least the lower ones, very like the basal leaves, the blades woolly beneath (3).
3. Tube of the involucre 3-angled, much shorter than the teeth, with prominent transverse ridges, the teeth straight. Leaves mostly broadly elliptic or suborbicular, shorter than the petioles; teeth of the involucre very unequal, leaflike, the spiny tips (and those of the bracts) straight .... 3. *C. rigida*
3. Tube of the involucre not angled, cylindric, nearly equaling to longer than the teeth, cross-corrugate, the teeth hooked (4).
4. Involucral teeth 5, unequal, one much larger than the others and often leaflike; basal leaves oblanceolate; tube of the involucre at maturity 4 to 6 mm. long, much longer than the teeth .... 4. *C. Watsoni*
4. Involucral teeth 3, equal or nearly so; basal leaves ovate or suborbicular; tube of the involucre 2 to 3 mm. long, about as long as the teeth.................. 5. *C. corrugata*

1. Chorizanthe Thurberi (Gray) Wats. (*Centrostegia Thurberi* Gray). Mohave, Yavapai, Graham, Gila, and Maricopa counties, 4,000 feet or lower, in sandy soil, March to May. Arizona, Nevada, and California.

2. Chorizanthe brevicornu Torr. Graham, Gila, Yavapai, and Mohave counties to Pima and Yuma counties, 2,500 feet or lower, March to May, type from the Gila River (*Parry*). Arizona, Nevada, and California.

3. Chorizanthe rigida (Torr.) Torr. & Gray (*Acanthogonum rigidum* Torr.). Mohave County to Pima and Yuma counties, 2,500 feet or lower, March to May, type from Williams River (*Bigelow*). Arizona, California, and Baja California.

The blackened plants are persistent on the desert long after they have died.

4. Chorizanthe Watsoni Torr. & Gray. Apparently only once collected in Arizona, at Chloride, Mohave County (*Jones* in 1903). Washington to western Arizona and California.

5. Chorizanthe corrugata (Torr.) Torr. & Gray. Western Maricopa and Pima counties to Mohave and Yuma counties, 1,500 feet or lower, type from near Fort Yuma (*Thomas*). Arizona, California, and Baja California.

C. leptoceras (Gray) Wats. (*Centrostegia leptoceras* Gray) was collected at Needles, California (*Meiere* in 1917), and is to be looked for in Mohave County, Arizona. It is characterized by very deeply cleft, 2- or 3-flowered involucres, the segments subulate and sharply aristate, the short tube bearing 6 spreading, hooked spines.

4. OXYTHECA

Plant annual, small, slender-stemmed, dichotomously branched; leaves oblanceolate, the basal ones in a rosette, the stem leaves opposite; upper bracts connate-perfoliate, these and the involucre teeth tipped with long, stiff bristles.

1. Oxytheca perfoliata Torr. & Gray. Mohave County, at Fort Mohave (*Lemmon* in 1884), and north of Chloride (*Kearney & Peebles* 11201), 500 to 2,000 feet, April. Arizona, Nevada, and California.

5. ERIOGONUM (83). WILD-BUCKWHEAT

Plants annual or perennial, herbaceous or shrubby; leaves alternate or whorled, simple, entire, the larger ones often in a basal rosette and the stem leaves often bractlike; flowers several or many in a campanulate, turbinate, or cylindric, toothed or lobed involucre, the involucres arranged in cymose, racemose, or glomerate inflorescences, or sometimes solitary in the forks of the branches; perianth corollalike; stamens 9; fruit an achene, usually trigonous.

This large genus is extraordinarily well represented in Arizona in number both of species and of individuals. Often the roads are bordered for miles by plants of one species, notably skeleton-weed (*E. deflexum*) in the low semidesert areas and *E. Wrightii* in the foothills and mountains. The shrubby species known collectively as buckwheat-brush furnish browse for livestock, and flat-top buckwheat-brush or California-buckwheat (*E. fasciculatum*) is a valued honey plant.

KEY TO THE SPECIES

1. Plants annual or biennial, without a woody caudex (2).
1. Plants perennial, the caudex, at least, thickened and woody (29).
2. Involucres all sessile or subsessile, axillary, solitary (3).
2. Involucres (except occasionally in *E. Hookeri*) all or some of them peduncled (9).

3. Herbage pilose or sericeous with appressed or subappressed hairs, not tomentose or lanate; involucres turbinate-campanulate, cleft to the middle or farther, not strongly ribbed, more or less concealed by the subtending leaves and bracts (4).
3. Herbage (at least the under side of the basal leaves) white-tomentose or lanate; involucres cylindric or narrowly turbinate, short-toothed, strongly ribbed. Outer perianth lobes wedge-shaped or fan-shaped, truncate or very obtuse, constricted near base (6).
4. Stems very leafy to the apex, the upper leaves not greatly reduced; perianth lobes strongly differentiated, the outer ones spreading, fan-shaped and somewhat cucullate (hooded), the inner ones erect, much narrower, oblong. Plant small, many-branched from the base; perianth pink or yellowish, white-hirtellous at base and on margins of the outer segments.................................1. *E. Darrovii*
4. Stems not very leafy above, the upper leaves much reduced; perianth lobes scarcely differentiated, all oblong or ovate (5).
5. Plants leafy-stemmed, the stem leaves gradually smaller than the basal ones and subtended by small, triangular bracts; stems widespreading or decumbent; involucres cleft to about the middle; perianth yellow, hispidulous, often glandular 2. *E. divaricatum*
5. Plants scapose, the stem leaves all reduced to small bracts; stems erect or ascending; involucres cleft nearly to the base; perianth pink, glabrous or nearly so 3. *E. puberulum*
6. Stems not more, usually much less, than 30 cm. long, diffusely many-branched; involucres not more than 1.5 mm. long, loosely and rather sparsely lanate-tomentose. Herbage lanate-tomentose; stem leaves greatly reduced, bractlike (7).
6. Stems seldom less than 30 cm. long, the branches few, ascending; involucres 2 to 3 mm. long, glabrous except on the margins. Perianth pink or whitish (8).
7. Plants intricately branched, the branches more or less incurved at maturity; perianth usually pale yellow...4. *E. nidularium*
7. Plants more open, the branches spreading, or ascending at a rather wide angle; perianth whitish or pink..5. *E. densum*
8. Plants scapose, the scapes glabrous; stem leaves reduced to bracts, the basal leaves orbicular, subcordate; involucres relatively few, 2.5 to 3 mm. long, cylindric-turbinate, their margins glabrous or very nearly so.......................6. *E. vimineum*
8. Plants leafy-stemmed, the stems white-tomentose; lower stem leaves large, oblong-obovate or oblanceolate, cuneate or attenuate at base; involucres very numerous, about 2 mm. long, short-turbinate, their margins white-lanate......7. *E. polycladon*
9. Plants (except rarely in *E. maculatum*) leafy-stemmed, the lower stem leaves like the basal ones but smaller (10).
9. Plants scapose or subscapose, the stem leaves all (or nearly all) reduced to small, usually scalelike bracts, the large, foliaceous leaves all or mostly confined to the basal rosette (13).
10. Stem leaves subtended by small, triangular bracts; peduncles seldom more than 2 cm. long; involucre and perianth glandular-puberulent. Plants more or less white-floccose; stems sharply angled; involucre hemispheric, many-flowered; perianth white or pink, the outer segments more or less inflated...................8. *E. maculatum*
10. Stem leaves not subtended by bracts; peduncles often at least 3 cm. long; involucre and perianth not glandular (11).
11. Involucres parted or divided, glabrous; perianth yellow, hispidulous, with narrow, attenuate lobes; herbage glabrous. Leaves oblanceolate to linear; peduncles filiform 9. *E. salsuginosum*
11. Involucres deeply cleft but not parted or divided, pubescent; perianth white, pink, or yellowish, glabrous, with broad outer lobes, the inner lobes longer and much narrower; herbage more or less pubescent (12).
12. Leaves villous, those of the lower stem oblong-lanceolate to broadly ovate, flat; outer perianth lobes nearly orbicular, cordate, not saccate..............10. *E. Abertianum*
12. Leaves white-lanate beneath, all linear-lanceolate or linear-oblanceolate, more or less revolute; outer perianth lobes ovate, bisaccate at base..........11. *E. pharnaceoides*

13. Perianth more or less whitish-hispidulous, yellow (sometimes fading pinkish). Peduncles filiform; scapes usually much-branched (14).
13. Perianth not whitish-hispidulous. Basal leaves white-lanate, at least beneath (19).
14. Outer perianth lobes saccate-dilated below; basal leaves lanate-tomentose beneath. Involucres 0.5 to 1.5 mm. long, glabrous or merely ciliolate........12. *E. Thomasii*
14. Outer perianth lobes not saccate-dilated; basal leaves mostly villous or loosely floccose, lanate beneath only in *E. Clutei* (15).
15. Glandular pubescence present (sometimes very sparse) on the scapes and often the peduncles, the latter often geniculate; involucres deeply cleft, often to below the middle, glabrous or glandular-puberulent. Scapes not inflated; lower bracts often well developed, up to 10 mm. long..........................13. *E. glandulosum*
15. Glandular pubescence none; peduncles not geniculate; involucres not cleft to below the middle, glabrous (16).
16. Internodes of the scape (the first and sometimes one or more higher) usually strongly inflated toward apex. Basal leaves villous, often sparsely so; scapes usually 40 cm. or more long, sparingly branched, the branches ascending at a narrow angle; involucres usually 1 to 2 mm. long and with 4 or more flowers, cleft to about the middle; plants usually perennial....................................14. *E. inflatum*
16. Internodes of the scape not inflated, or but slightly so (17).
17. Floccose pubescence none, the basal leaves villous, often sparsely so; scapes usually diffusely many-branched; involucres 0.5 to 1 mm. long..............15. *E. trichopes*
17. Floccose pubescence present, at least on the basal leaves; scapes few-branched, the branches ascending; involucres 1 to 1.5 mm. long (18).
18. Scapes not inflated, not glaucous, these and the leaves loosely floccose; branches of the scape ascending at a narrow angle; leaves green above, usually obovate and narrowed at base, 2 to 6 cm. long; involucres 1- to 3-flowered; perianth less than 2 mm. long 16. *E. Ordii*
18. Scapes (the first internode) slightly inflated, glaucous, glabrous; branches of the scape ascending at a wide angle; leaves grayish-floccose above, densely white-lanate beneath, reniform, 1 to 1.5 cm. long; involucres 4- or 5-flowered; perianth 2.5 to 3 mm. long ....17. *E. Clutei*
19. Perianth puberulent. Peduncles very slender, usually elongate; involucres broadly campanulate, often about 10-flowered; perianth yellow, often with a red central stripe and fading pinkish, the outer lobes not saccate-dilated (20).
19. Perianth glabrous or very nearly so (21).
20. Involucres and bracts glandular-puberulent; outer perianth lobes oblong-obovate or oblanceolate; basal leaves obovate to nearly orbicular, not cordate..18. *E. pusillum*
20. Involucres glabrous or merely ciliolate; outer perianth lobes oblong-lanceolate; basal leaves round-cordate or reniform..............................19. *E. reniforme*
21. Peduncles often slender but not filiform, seldom more and usually less than 10 mm. long, sometimes none. Scapes glaucous (22).
21. Peduncles filiform, often more than 10 mm. long (26).
22. Involucres hemispheric, mostly sessile or subsessile, glabrous; perianth yellowish, fading pinkish. Outer perianth lobes nearly orbicular, cordate; inflorescence branches somewhat flexuous or zigzag...................................20. *E. Hookeri*
22. Involucres turbinate-campanulate, distinctly peduncled, the peduncles usually more than 1 mm. long; perianth white or pink (23).
23. Outer perianth lobes obovate, truncate, often emarginate, narrowed at base; peduncles usually at least 5 mm. long. Involucres glabrous (24).
23. Outer perianth lobes ovate to suborbicular, rounded at apex, more or less cordate at base; peduncles usually less than 5 mm. long (25).
24. Peduncles usually more or less deflexed or decurved; involucres often many- (more than 10-) flowered; outer perianth lobes more or less fiddle-shaped; scapes usually erect or ascending...21. *E. cernuum*
24. Peduncles spreading; involucres usually few- (about 5-) flowered; outer perianth lobes fan-shaped; scapes spreading.............................22. *E. rotundifolium*
25. Involucres, peduncles, and bracts normally glandular-puberulent; scapes with wide-spreading branches ....23. *E. Parryi*

25. Involucres, peduncles, and bracts glabrous; scapes with ascending or spreading branches 24. *E. deflexum*

26. Glandular pubescence present on involucres, peduncles, and bracts; involucres hemispheric, 2 to 2.5 mm. long, usually many- (more than 10-) flowered; outer perianth lobes fan-shaped and somewhat hooded (cucullate), greatly dilated from the narrow claw. Scapes floccose below; peduncles often 20 mm. long; perianth white or pinkish .... 25. *E. Thurberi*

26. Glandular pubescence none or obscure; involucres turbinate, not more than 1.5 mm. long, few- (seldom more than 8-) flowered; outer perianth lobes not fan-shaped or hooded (27).

27. Scapes 10 to 20 cm. long, diffusely many-branched, glabrous or nearly so, as are also the bracts. Leaves glabrescent above; perianth yellowish, fading pinkish, the outer lobes oblong, not constricted .... 26. *E. Wetherillii*

27. Scapes usually 30 cm. or longer, with fewer, ascending branches, more or less villous or tomentose at the nodes, as are also the lower bracts (28).

28. Plants biennial (?); scapes leafy and with very short internodes toward base; leaves sparsely lanate to densely whitish-tomentose on both faces, their margins crispate; outer perianth lobes narrow, fiddle-shaped .... 27. *E. capillare*

28. Plants annual; scapes not leafy above the basal rosette; leaves green and glabrescent above, white-lanate beneath; outer perianth lobes oblong, not constricted 28. *E. subreniforme*

29. Stems noticeably woody above ground. Involucres (except sometimes in *E. Simpsoni*) mostly sessile or nearly so (30).

29. Stems not, or scarcely, woody above ground (42).

30. Perianth silky-villous, white or pink; cymes (or their ultimate divisions) dense, subcapitate; leaves pilose or villous above, lanate beneath, narrow, strongly revolute. Plants shrubby; leaves often fascicled; involucres silky-villous, many-flowered 49. *E. fasciculatum*

30. Perianth glabrous; cymes more or less open; leaves lanate or glabrate above (31).

31. Branches of the inflorescence rigid, divaricate, more or less spinescent, scabrous or puberulent; involucres glabrous or glabrescent. Low, intricately branched shrubs; stems usually conspicuously white-lanate in the lower axils; inflorescences seldom conspicuously cymose; involucres sessile, turbinate, few-flowered, 1 to 2 mm. long, the teeth very obtuse or truncate; leaves usually caducous, lanceolate or oblanceolate, the petioles short (32).

31. Branches of the inflorescence not rigid, divaricate, and spinescent or if so, then more or less villous or lanate and the involucres also lanate (33).

32. Inflorescence branches sharply quadrangular, deeply sulcate, obscurely puberulent, not papillate .... 50. *E. sulcatum*

32. Inflorescence branches terete or rounded-quadrangular, not deeply sulcate, pronouncedly scabrous-papillate .... 51. *E. Howellii*

33. Inflorescence not conspicuously cymose. Involucres lanate, racemosely disposed and often crowded on the ultimate divisions of the inflorescence; stems lanate; leaves linear-lanceolate, elliptic, or oblanceolate, not more than 6 mm. wide; perianth white or pink (34).

33. Inflorescence conspicuously cymose (36).

34. Involucres divergent, the teeth obtuse or acutish, narrowly scarious-margined; ultimate divisions of the inflorescence short, usually conspicuously secund. Stems woody only toward base, much-branched, dichotomously or trichotomously so at the lower nodes 52. *E. leptocladon*

34. Involucres appressed to the branchlets, the teeth acute, not noticeably scarious-margined; ultimate divisions of the inflorescence not conspicuously secund (35).

35. Plants suffrutescent, few-branched; involucres 2 to 3 mm. long, scattered or crowded; perianth 2.5 to 3.5 mm. long, the outer lobes broadly obovate .... 44. *E. Wrightii*

35. Plants shrubby, intricately many-branched; involucres 1 to 1.5 mm. long, scattered along the branches; perianth 1.5 mm. long, the outer lobes oblanceolate or narrowly obovate 45. *E. Pringlei*

36. Stems woody only near base, lanate, at least below the inflorescence (37).
36. Stems woody almost throughout. Leaves densely lanate beneath (38).
37. Plants not at all scapose, with well-developed leaves nearly to the inflorescence, these caducous; ultimate branches of the inflorescence flexuous, widespreading or even decurved; involucres glabrous or glabrescent except on the margin, their teeth obtuse, scarious-margined .... 42. *E. Plumatella*
37. Plants subscapose, with well-developed leaves only near base, these persistent; ultimate branches of the inflorescence stiff, stout, ascending; involucres lanate, their teeth acute, not noticeably scarious-margined, often dark-colored.......... 48. *E. Jonesii*
38. Leaves oblong-lanceolate to broadly ovate, seldom less than 6 mm. wide, cuneate to subcordate at base, irregularly undulate or crispate, not or only slightly revolute; petioles usually well developed, often half as long as the blade. Involucres more or less lanate (39).
38. Leaves linear or lanceolate, seldom more than 4 mm. wide, acute at base, strongly revolute; petioles usually very short, rarely ⅓ as long as the blade. Perianth white or pink (40).
39. Inflorescence branches ascending, not rigid, or spinescent, or glutinous; perianth white or pink.. 53. *E. corymbosum*
39. Inflorescence branches divaricate, rigid, sometimes almost spinescent, sometimes slightly glutinous; perianth yellow, white, or pink........................ 54. *E. aureum*
40. Inflorescence glabrous or obscurely puberulent, relatively dense, broomlike, the numerous branches erect or ascending at a narrow angle; leaves 2 to 5 cm. long, very strongly revolute, the glabrous upper surface often completely hiding the white-lanate lower surface; involucres glabrous or obscurely ciliolate, cylindric-turbinate; perianth 3 to 4 mm. long, the outer lobes oblanceolate.................. 55. *E. leptophyllum*
40. Inflorescence more or less lanate, relatively open, not broomlike, the few branches ascending at a wide angle; leaves seldom more than 2 cm. long, less strongly revolute, the upper surface rarely completely glabrous, the white-lanate lower surface nearly always evident; involucres loosely lanate or glabrescent, turbinate or turbinate-campanulate; perianth 2 to 2.5 mm. long, the outer lobes obovate to nearly orbicular (41).
41. Stems erect or ascending, few-branched from the base, the inflorescence branches 2.5 to 4 cm. long; leaves 10 to 20 (30) mm. long; perianth white or pinkish, rarely bright pink. Leaves sparsely floccose or green and glabrescent above; involucres, especially those in the forks, often pedunculate.......................... 56. *E. Simpsoni*
41. Stems decumbent or spreading, many-branched from the base, the inflorescence branches not more than 2 cm. long; leaves 5 to 10 mm. long; perianth bright pink to pale pink 57. *E. Mearnsii*
42. Achenes conspicuously winged, with membranaceous wings. Hairs of the herbage long, fine, not closely appressed; stems tall, stout, usually leafy, sparingly branched; leaves oblanceolate, tapering into long petioles; involucres not strongy ribbed (43).
42. Achenes not conspicuously winged (45).
43. Perianth sericeous, yellow; inflorescence relatively short, compact, short-branched; herbage copiously to densely grayish-pubescent; achenes 4 to 5 mm. long, winged above the middle. Plants sometimes subscapose, with reduced stem leaves; involucres rather densely sericeous................................... 29. *E. hieracifolium*
43. Perianth glabrous; inflorescence elongate, open, long-branched; herbage green, loosely pubescent to glabrescent; achenes 6 to 8 mm. long, winged nearly to the base (44).
44. Achenes 6 to 7 mm. long, usually considerably more than half as wide; involucres sericeous; perianth yellow..................................... 30. *E. alatum*
44. Achenes 7 to 8 mm. long, about half as wide; involucres glabrous or sparsely sericeous; perianth purplish or yellowish.................................... 31. *E. triste*
45. Peduncles filiform or nearly so, 5 to 20 mm. long; larger leaves usually more or less cordate and orbicular or nearly so. Plants scapose or subscapose; stems tall, sparingly branched; involucres glabrous; perianth yellow or fading pinkish (46).
45. Peduncles not filiform; larger leaves (except sometimes in *E. ovalifolium*, *E. racemosum*, and *E. Jonesii*) not cordate, usually considerably longer than wide (47).

46. Perianth whitish-hispidulous, the lobes not fiddle-shaped; scapes usually with 1 or more internodes conspicuously inflated; well-developed leaves all basal, rather sparsely villous or hirsute..14. *E. inflatum*
46. Perianth nearly or quite glabrous, the lobes fiddle-shaped; scapes not inflated; well-developed leaves at several of the crowded lower nodes of the stem, sparsely lanate to white-tomentose on both surfaces............................27. *E. capillare*
47. Bracts subtending the divisions of the inflorescence large, verticillate, resembling the basal leaves but somewhat smaller; perianth attenuate at base, stipitate. Involucres turbinate, many-flowered, membranaceous, villous or sericeous, not strongly ribbed (48).
47. Bracts much smaller than the basal leaves, usually inconspicuous but occasionally (in *E. racemosum*) large and foliaceous; perianth not attenuate-stipitate (52).
48. Perianth cream-colored or pinkish, densely silky-villous...............32. *E. Jamesii*
48. Perianth bright yellow (49).
49. Stems usually only once-forked and bearing a single whorl of foliaceous bracts; inflorescence compact, often subcapitate; involucres few (50).
49. Stems more than once-forked and bearing more than 1 whorl of foliaceous bracts; inflorescence more open, never subcapitate; involucres often numerous (51).
50. Involucres short-toothed or subtruncate; perianth densely silky-villous...33. *E. flavum*
50. Involucres deeply cleft, the lobes usually reflexed; perianth glabrous..34. *E. umbellatum*
51. Perianth silky-villous; involucres dentate, the teeth broad, erect or somewhat revolute 35. *E. Bakeri*
51. Perianth glabrous; involucres deeply cleft, usually to the middle or farther, the lobes reflexed ....36. *E. cognatum*
52. Plants pulvinate or cespitose; stems not more than 30 cm. long (53).
52. Plants not pulvinate or cespitose; stems often more than 30 cm. long. Involucres mostly sessile; perianth glabrous (57).
53. Involucres glabrate or very sparsely villous, solitary on very short bractless scapes; herbage sparsely pubescent except the densely tomentose lower leaf surfaces. Stems 5 to 15 cm. long; leaves oblanceolate, strongly revolute, 2 to 6 mm. long, not more than 1 mm. wide; perianth glabrous, whitish, the segments with a reddish median stripe ....38. *E. Ripleyi*
53. Involucres villous or tomentose, clustered on the scapes or (in *E. lachnogynum*) solitary on the cyme branches; herbage densely pubescent (54).
54. Scapes seldom less than 10 cm. long (55).
54. Scapes not more than 8 cm. long. Perianth villous, white, pink, or yellowish (56).
55. Involucres sessile or subsessile in a capitate inflorescence; scapes usually not branched; leaves oval to nearly orbicular; perianth glabrous, pale yellow, cream-colored, or pinkish, the outer lobes broadly obovate, truncate or emarginate, subcordate at base 37. *E. ovalifolium*
55. Involucres solitary on the more or less elongate cyme branches; scapes sparingly branched above; leaves lanceolate, narrowly elliptic, or oblanceolate; perianth densely white-villous, yellow, the outer lobes ovate or obovate, obtuse. Herbage silky-villous 39. *E. lachnogynum*
56. Leaves elliptic or lanceolate, acute, villous-sericeous..............40. *E. villiflorum*
56. Leaves mostly obovate or oblanceolate, obtuse, lanate-tomentose......41. *E. Shockleyi*
57. Involucres glabrous or glabrescent except on the margin, the teeth obtuse or truncate (58).
57. Involucres lanate on most of their surface. Stems lanate-tomentose; perianth white or pink (59).
58. Plants not scapose, the stems leafy up to the inflorescence, the leaves mostly caducous; stems lanate, at least below the inflorescence; involucres about 2 mm. long, more or less ciliate; perianth whitish; inflorescence compound-cymose, several-branched, the branches flexuous, the branchlets divergent or decurved............42. *E. Plumatella*
58. Plants scapose, the large leaves all basal or nearly so, persistent; scapes glabrous except at the nodes; involucres 3 to 3.5 mm. long, glabrous or obscurely ciliolate; perianth yellow; inflorescence dichotomous or trichotomous, the branches ascending 43. *E. Thompsonae*

59. Leaves short-petioled, lanceolate, elliptic, or oblanceolate, acute at base, not more than 6 mm. wide; plants not scapose, the stems leafy well above the base. Inflorescence simple or with few branches ascending at a narrow angle, racemiform or the involucres glomerate toward the ends of the branches........44. *E. Wrightii*

59. Leaves long-petioled, ovate or oblong-ovate, rounded, subcordate, or occasionally cuneate at base, 10 mm. or wider; plants scapose or subscapose, the well-developed leaves all, or nearly all, basal (60).

60. Inflorescence compound-cymose, many-branched, the branches ascending-spreading, stout; involucres not more than 2 mm. long, much surpassing the nearly distinct bracts ....48. *E. Jonesii*

60. Inflorescence racemiform, elongate, simple or with very few, nearly erect branches; involucres 3 to 5 mm. long, not or but little surpassing the strongly connate bracts (61).

61. Scapes tomentose, not inflated; involucres 4 to 5 mm. long, several- to many-flowered; perianth cream-colored or pinkish..............................46. *E. racemosum*

61. Scapes glabrous, somewhat glaucous, inflated; involucres about 3 mm. long, few-flowered; perianth greenish or bright red..........................47. *E. zionis*

1. Eriogonum Darrovii Kearney. Near Ryan, Kaibab National Forest, northwestern Coconino County, 6,500 feet, locally abundant with sagebrush and *Cowania* (*Darrow* 2998, the type collection), west of Slide Tank, Kaibab Plateau (*Goodding* 197–49), August and September. Known only from these collections.

2. Eriogonum divaricatum Hook. Apache, Navajo, and Coconino counties, 4,500 to 6,000 feet, May to September. Colorado, Utah, and Arizona.

***3. Eriogonum puberulum** Wats. Not known definitely to occur in this state, but has been collected in Utah and Nevada very near the Arizona boundary.

4. Eriogonum nidularium Coville (*E. vimineum* subsp. *nidularium* Stokes). Mohave County, 3,500 feet or lower, April and May. Idaho to Arizona and southern California.

5. Eriogonum densum Greene (*E. vimineum* var. *densum* Stokes). Coconino and Mohave counties to Pima and Cochise counties, 1,000 to 6,500 feet, common, May to October. Utah and Nevada to New Mexico and Arizona.

Very common at roadsides in the foothills and mountains. Closely related to *E. nidularium* Coville.

6. Eriogonum vimineum Dougl. Coconino and Mohave counties to Gila and Maricopa counties, 3,000 to 7,000 feet, May to October. Idaho and Washington to Arizona and California.

Broom eriogonum. Represented in Arizona by subsp. *juncinellum* (Gandoger) Stokes. The type of *E. juncinellum* Gandoger was collected at the Grand Canyon (*MacDougal* 182).

7. Eriogonum polycladon Benth. (*E. vimineum* subsp. *polycladon* Stokes). Coconino and Mohave counties to Cochise, Santa Cruz, and Pima counties, 2,500 to 5,500 feet, June to November. Western Texas to Arizona and northern Mexico.

Sorrel eriogonum. So common at roadsides and in washes, especially in southern Arizona, as to color the landscape in places with its tall, gray stems and pink flowers.

8. Eriogonum maculatum Heller (*E. angulosum* Benth. subsp. *maculatum* Stokes). Yavapai and Mohave counties to Graham, Pinal, Maricopa, and Pima counties, 4,500 feet or (usually) lower, April to June. Idaho and eastern Washington to Arizona and (?) Baja California.

The remarkable inflation of the perianth segments distinguishes this from all other Arizona species of *Eriogonum*. Arizona specimens have been referred to *E. angulosum* Benth., which is confined to California, according to Howell (84). A collection of the related *E. gracillimum* Wats. labeled as from "Peach Springs, Arizona" (*Tracy & Evans* in 1887), is more likely to have come from California, the species being unknown otherwise outside that state.

9. Eriogonum salsuginosum (Nutt.) Hook. Apache and Navajo counties, 5,000 to 6,000 feet, apparently rare, June to September. Wyoming to Utah, New Mexico, and northeastern Arizona.

10. Eriogonum Abertianum Torr. Coconino and Yavapai counties to Greenlee, Graham, Cochise, Santa Cruz, and Pima counties, 1,500 to 7,000 feet, foothills and mountains, very common, March to September. Western Texas to Arizona and Chihuahua.

Several varieties have been distinguished by Fosberg (85). Of these the 3 following are credited to Arizona: var. *neomexicanum* Gandoger (*E. pinetorum* Greene), characterized by a paniculate inflorescence with inconspicuous, not leafy bracts, found in southern Greenlee, southern Yavapai, Graham, Cochise, and Pima counties; var. *villosum* Fosberg with the inflorescence leafy, not paniculate, found in Gila, Cochise, Santa Cruz, and Pima counties; and var. *Gillespiei* Fosberg (*E. cyclosepalum* Greene var. *Gillespiei* Johnst.), similar to var. *villosum* in the type of inflorescence but with the upper leaves more reduced and bractlike, and the basal leaves shorter-petioled and attenuate (instead of truncate or cordate) at base, the type from Apache Gap, Pinal County, 2,500 feet (*Gillespie* 8797).

11. Eriogonum pharnaceoides Torr. (*E. arizonicum* Gandoger). Apache, Navajo, and Coconino counties, south to Cochise and Pima counties, 4,500 to 7,000 feet, frequent in pine woods, July to October, type of *E. arizonicum* Gandoger from near Bill Williams Mountain, Coconino County (*MacDougal* 311). Utah, New Mexico, and Arizona.

12. Eriogonum Thomasii Torr. Graham County to Mohave County, south to Pima and Yuma counties, 3,000 feet or lower, February to May, type from near Fort Yuma (*Thomas*). Southern Utah to southeastern California, Arizona, and Sonora.

A small annual, common in sandy soil in the drier parts of the state. What appears to be an extraordinarily robust form of this species, with stems about 40 cm. high and 5 mm. in diameter, was collected at Burro Point on Lake Mead, Mohave County (*Clover* 8286).

13. Eriogonum glandulosum Nutt. (*E. flexum* Jones, *E. trichopes* subsp. *glandulosum* Stokes). Moenkopi, Coconino County, 4,500 feet, the type locality of *E. flexum* (*Jones* in 1890), this being apparently the only collection of *E. glandulosum* in Arizona. Colorado, Utah, and Arizona.

14. Eriogonum inflatum Torr. & Frém. Apache County to Mohave County, south to Graham, Pima, and Yuma counties, 3,500 feet or lower (occasionally considerably higher), March to October. Utah and Arizona to southern California and Baja California.

Desert-trumpet, bladder-stem, Indianpipe-weed. A familiar plant on the rocky foothills and lower slopes of desert mountains, remarkable for the inflated stem. A form with noninflated scapes and few-flowered involucres was collected in Havasu Canyon, western Coconino County (*Clover* 5127).

15. Eriogonum trichopes Torr. Mohave County to Graham, Cochise, Pima, and Yuma counties, 4,000 feet or lower, flowering almost throughout the year. Colorado and Utah to Arizona, California, and northwestern Mexico.

Often extremely abundant on the deserts and low hills, sometimes covering large areas in nearly pure stand, as in northern Maricopa County and northwest of Kingman, Mohave County. The many hairlike branches of the inflorescence and very small, few-flowered involucres are distinctive.

***16. Eriogonum Ordii** Wats. The type (*Lemmon* 4189, in 1884) is labeled as having been collected at Fort Mohave, Arizona, but in view of the known distribution of the species in California, the absence of later collections from Arizona, and the uncertainty of Lemmon's data of locality, it is improbable that *E. Ordii* really occurs in this state.

17. Eriogonum Clutei Rydb. Known apparently only from the type collection at Cameron, Coconino County, "common in the driest places" (*Clute* 71a).

18. Eriogonum pusillum Torr. & Gray. (*E. reniforme* subsp. *pusillum* Stokes). Hackberry and Kingman (Mohave County), Wickenburg (Maricopa County), 2,000 to 3,000 feet, April and May. Southern Utah, Arizona, and southeastern California.

19. Eriogonum reniforme Torr. & Frém. Mohave, Pima, and Yuma counties, 2,500 feet or lower, March to June. Nevada to Arizona and Baja California.

20. Eriogonum Hookeri Wats. (*E. deflexum* subsp. *Hookeri* Stokes). Near Fredonia, north of Cameron, and Wupatki National Monument (Coconino County), Topock (Mohave County), 500 to 4,500 feet, August and September, apparently rare in Arizona. Utah, Nevada, and northern Arizona.

21. Eriogonum cernuum Nutt. Apache County to western Coconino and northern Yavapai counties, 3,000 to 8,000 feet, locally common, July to October. Saskatchewan and Alberta, south to Nebraska, northern Arizona, and California.

Nodding eriogonum. A collection on Mount Trumbull, Mohave County (*K. F. Parker et al.* 6283), belongs to subsp. *tenue* Stokes, having the peduncles spreading or curved upward and the stems leafy a short distance above the base.

22. Eriogonum rotundifolium Benth. (*E. cernuum* subsp. *rotundifolium* Stokes). Greenlee, Graham (?), and Cochise counties, 3,500 to 4,000 feet, on plains, rare in Arizona. Colorado to Texas, Arizona, and Chihuahua.

23. Eriogonum Parryi Gray (*E. deflexum* subsp. *Parryi* Stokes). Northern Mohave County, near Littlefield, about 1,500 feet (*Goodman & Hitchcock* 1661), and at Burro Point, Lake Mead (*Clover* 8286). Southern Utah and northwestern Arizona.

The Clover collection is exceptionally robust.

24. Eriogonum deflexum Torr. (*E. turbinatum* Small). Coconino and Mohave counties, south to western Cochise, Pima, and Yuma counties, 4,000 feet and lower, flowering most of the year, type of *E. turbinatum* from Tucson (*Pringle* in 1884). Utah to Arizona and Baja California.

Skeleton-weed. Very abundant in the semidesert parts of the state, conspicuous at roadsides, often continuously for many miles. An extremely drought-resistant plant, flourishing when nearly all other herbaceous plants of the desert have disappeared.

The species is extremely variable in size of the plant, angle of the branches and peduncles, and size and shape of the perianth segments. Tall, strictly branched plants have been referred to subsp. *exaltatum* (Jones) Stokes (*E. exaltatum* Jones), and if the plants are of this character and the outer perianth segments are oblong-ovate (longer than wide) and not deeply cordate, to subsp. *insigne* (Wats.) Stokes (*E. insigne* Wats.). The latter is not uncommon in northern Mohave County.

25. Eriogonum Thurberi Torr. (*E. panduratum* Wats., *E. cernuum* subsp. *Thurberi* Stokes). Mohave County to Graham, Cochise, Santa Cruz, Pima, and Yuma counties, 4,000 feet or usually lower, in sandy soil, March to June. Arizona, California, and Baja California.

One of the commonest of the small, annual, spring-flowering species in semidesert areas. *E. cernuum* subsp. *viscosum* Stokes appears to be merely a densely glandular form of *E. Thurberi*. The type of *E. panduratum* was collected by Lemmon, probably in Arizona.

26. Eriogonum Wetherillii Eastw. Apache County to eastern Coconino County, 3,500 to 6,000 feet, May to July. Utah, New Mexico, and northeastern Arizona.

27. Eriogonum capillare Small (*E. arizonicum* Stokes). San Carlos, Roosevelt, and near Winkelman (Gila County), Patagonia Mountains (Santa Cruz County), 2,000 to 3,000 feet, rare, June to October, type of *E. capillare* from San Carlos (*Ebert* in 1893), type of *E. arizonicum* Stokes from Arizona, without definite locality (*Orcutt* 186). Known only from south-central Arizona.

28. Eriogonum subreniforme Wats. Apache, Navajo, and Coconino counties, 4,500 to 6,000 feet, June to September. Southern Utah, New Mexico, and northern Arizona.

29. Eriogonum hieracifolium Benth. White Mountains (Apache and Navajo counties), 5,000 to 7,000 feet, June to August. Western Texas to eastern Arizona.

At Fort Apache it grows among junipers and live-oaks (*Quercus Emoryi*), at higher altitudes probably among pines.

30. Eriogonum alatum Torr. Apache County to Cochise County, westward to Coconino, Yavapai, and Gila counties, 5,500 to 9,500 feet, July to September. Nebraska to Texas, Utah, and Arizona.

Winged eriogonum. A common plant in open pine forests, with rather tall wandlike stems. The large root was reported by Mrs. Collom to be used medicinally by the Indians.

31. Eriogonum triste Wats. (*E. alatum* subsp. *triste* Stokes). Apache

County to Coconino County, 7,000 to 8,000 feet, July to September. Colorado and Utah to New Mexico and Arizona.

Probably not specifically distinct from *E. alatum.*

32. Eriogonum Jamesii Benth. Navajo and Gila counties to Cochise and Pima counties, 5,000 to 9,000 feet, among rocks in pine and oak woods, July to October. Kansas to Colorado, south to Texas and Arizona.

Sometimes known as antelope-sage. Easily distinguished from other species of its group (species 33 to 36) by the cream-colored (not bright-yellow) perianth. Mrs. Collom reported that the roots are used medicinally by Indians.

33. Eriogonum flavum Nutt. Kaibab Plateau and North Rim of the Grand Canyon, Coconino County, 8,000 to 8,500 feet (*Goodman & Hitchcock* 1643, *Kearney & Peebles* 13696, *Goodding* 245–48), among pines and in open meadows, August (probably other months). Manitoba and Alberta to Colorado and Arizona.

34. Eriogonum umbellatum Torr. Navajo Creek, Kaibab Plateau, and South Rim of the Grand Canyon (Coconino County), Virgin Mountains and Pagumpa Springs (Mohave County), 5,000 to 9,000 feet, April to September. Wyoming to Washington, northern Arizona, and California.

Sulphur eriogonum or sulphur-flower, names equally applicable to all of these closely related, showy, yellow-flowered species.

35. Eriogonum Bakeri Greene (*E. Jamesii* subsp. *Bakeri* Stokes). Hopi Indian Reservation (Navajo County), throughout Coconino County, Hualpai Mountain (Mohave County), 5,000 to 8,000 feet, in pine forests, June to September. Colorado, Utah, New Mexico, and northern Arizona.

E. arcuatum Greene (*E. Jamesii* subsp. *flavescens* (Wats.) Stokes var. *arcuatum* (Greene) Stokes), a smaller plant with less-branched inflorescence, is more or less intermediate between *E. flavum* and *E. Bakeri.* It occurs here and there within the range of the latter in Arizona.

36. Eriogonum cognatum Greene (*E. Ferrissii* A. Nels., *E. umbellatum* subsp. *cognatum* Stokes). Navajo, Coconino, Mohave, Gila, and northern Yavapai counties, 5,000 to 8,500 feet, common in pine forests in Coconino County, July to September, type of *E. cognatum* from near the San Francisco Peaks (*Greene* in 1889), type of *E. Ferrissii* from Betatakin, Navajo County (*Clute* 10c). Definitely known only from Arizona.

Perhaps too near *E. stellatum* Benth. Both species appear like luxuriant developments of *E. umbellatum* Torr.

37. Eriogonum ovalifolium Nutt. Carrizo Mountains (Apache County) to the Kaibab Plateau (Coconino County) and Pagumpa Springs (Mohave County), 5,000 to 7,000 feet, April to June, apparently rather rare in Arizona. Alberta to New Mexico, northern Arizona, and California.

Subsp. *vineum* (Small) Stokes (*E. vineum* Small), a large form with the perianth white or pink, occurs on the Kaibab Plateau (Coconino County) and in northeastern Mohave County, in the sagebrush-grass association.

38. Eriogonum Ripleyi J. T. Howell. Near Frazier Well, western Coconino County, about 6,000 feet, among pinyons (*Ripley & Barneby* 5226, the type collection, 8445). Known only from these collections.

39. Eriogonum lachnogynum Torr. Petrified Forest, Apache County (*Ward* in 1901, *A. S. Hitchcock* 8), June, rare in Arizona. Kansas to Texas, Colorado, and northeastern Arizona.

*__40. Eriogonum villiflorum__ Gray. Apparently not yet collected in Arizona, but the type came from southern Utah not far from the Arizona line.

41. Eriogonum Shockleyi Wats. Apache, Navajo, and Coconino counties, 4,500 to 6,000 feet, sometimes (always?) in gypseous soil, June to August. Utah, Nevada, and northern Arizona.

Plants densely cespitose, forming hummocks about a foot in diameter, the perianth yellowish white. The Arizona plants apparently belong to subsp. *candidum* (Jones) Stokes (*E. pulvinatum* Small), characterized by a deeply cleft involucre.

42. Eriogonum Plumatella Dur. & Hilg. (*E. Palmeri* Wats.). Mohave County, road to Searchlight Ferry, also near Kingman and southeastward to the Aquarius Mountains, about 3,500 feet, in dry thickets. Southern Utah and western Arizona to southern California.

Arizona possesses both typical *E. Plumatella,* with lanate inflorescence branches and involucres copiously lanate-ciliate, and var. *Jaegeri* (Munz & Johnst.) Stokes, with inflorescence branches glabrous except at the nodes and involucres sparsely ciliate.

43. Eriogonum Thompsonae Wats. West of Pipe Springs, Mohave County, 4,500 feet, sandy soil in the pinyon-juniper-sagebrush association, August (*K. F. Parker et al.* 6242). Otherwise known only by the type collection at Kanab, Utah (*Thompson* in 1873).

44. Eriogonum Wrightii Torr. (*E. trachygonum* Torr. subsp. *Wrightii* Stokes). Almost throughout the state, 3,000 to 7,000 feet, often very abundant, June to October. Colorado to Texas, Arizona, California, and northern Mexico.

The plant affords fair browse for cattle and is, according to Nichol, the most important deer-browse plant in the state. Mrs. Collom reported that the flowers yield a fine, almost colorless honey. A variant with involucres glomerate toward the ends of the branches (these more scattered in typical *Wrightii*) is frequent in southern Arizona and has been collected on Hualpai Mountain (Mohave County). It is subsp. *glomerulum* Stokes, type from western Gila County (*Eastwood* 16643).

45. Eriogonum Pringlei Coult. & Fish. "Hills near Maricopa," Pinal County (*Pringle* in 1882, the type collection), Sierra Estrella (Maricopa County), Ajo Mountains (Pima County), Tinajas Altas (Yuma County), on dry, rocky slopes, flowering as late as October. Reported to occur also in southern California.

Closely related to *E. Wrightii,* but a larger, shrubbier plant with smaller flowers and involucres.

46. Eriogonum racemosum Nutt. Apache County to Hualpai Mountain (Mohave County), 5,000 to 9,000 feet, common in pine forests, June to October. Colorado and Utah to western Texas and Arizona.

Red-root eriogonum.

47. Eriogonum zionis J. T. Howell. Both rims of the Grand Canyon, Coconino County, 7,000 to 8,000 feet (*Bryant* in 1941, *Collom* in 1941). Southern Utah and northern Arizona.

The Arizona variant, with vivid carmine perianths up to 5 mm. long, is var. *coccineum* J. T. Howell, type from Point Sublime (*Bryant* in 1941). In typical *zionis* the perianths are greenish yellow and not more than 4 mm. long.

48. Eriogonum Jonesii Wats. (*E. lanosum* Eastw.). Coconino County and near Peach Springs (Mohave County), 4,000 to 7,000 feet, August and September, type of *E. Jonesii* from Cosnino (*Jones* in 1884), type of *E. lanosum* from near Canyon Diablo (*Eastwood* 15747). Southern Utah and northern Arizona.

49. Eriogonum fasciculatum Benth. Coconino, Mohave, and western Gila counties to Pima and Yuma counties, 1,000 to 4,500 feet, dry, rocky slopes, March to June. Arizona to Nevada and California.

Flat-top buckwheat-brush, California-buckwheat, represented in Arizona by var. *polifolium* (Benth.) Torr. & Gray (*E. polifolium* Benth.), a xerophilous shrub up to about 1 m. (3 feet) high, with involucres in dense headlike clusters and whitish or pinkish, slightly fragrant flowers.

50. Eriogonum sulcatum Wats. (*E. Heermanni* Dur. & Hilg. subsp. *sulcatum* Stokes). Prospect Valley, Coconino County (*Goldman* 2291), Pagumpa Springs, Mohave County (*Jones* 5089p), 4,000 to 5,000 feet, September. Utah, Nevada, and Arizona.

51. Eriogonum Howellii Stokes. Lees Ferry, Canyon of the Little Colorado River, Kaibab Plateau, and Grand Canyon (Coconino County), Kingman (Mohave County), 2,500 to 8,500 feet, dry, rocky places, August and September. Nevada, northern Arizona, and eastern California.

The type of var. *subracemosum* Stokes, to which all the Arizona specimens probably belong, was collected 17 miles west of Cameron, Coconino County (*Kearney & Peebles* 12818). Specimens collected by M. E. Jones in 1903 and labeled "8 miles south of Vail" (Pima County) are very like *E. Howellii.* The locality as stated is almost certainly erroneous.

52. Eriogonum leptocladon Torr. & Gray. (*E. pallidum* Small, *E. effusum* Nutt. subsp. *leptocladon* Stokes). Apache County to Mohave County, 3,500 to 5,500 feet, sandy soil, July to September, type of *E. pallidum* from in or near the Hopi Indian Reservation, Navajo County (*Hough* 30). Utah, New Mexico, and northern Arizona.

53. Eriogonum corymbosum Benth. (*E. effusum* subsp. *corymbosum* Stokes). Northern Navajo County to western Coconino County, 5,000 to 7,500 feet. Colorado to Nevada, New Mexico, and northern Arizona.

The Arizona specimens are not quite typical, approaching *E. aureum* and perhaps should be referred to that species.

54. Eriogonum aureum Jones (*E. microthecum* Nutt. subsp. *aureum* Stokes). Apache County to Mohave County, 4,500 to 8,000 feet, often very abundant, August to October. Colorado to Nevada, New Mexico, and northern Arizona.

The color of the perianth varies from bright yellow to nearly white, sometimes in the same colony. A more pubescent, slightly glutinous form is var. *glutinosum* Jones, type from Holbrook, Navajo County (*Rusby* in 1883).

55. Eriogonum leptophyllum (Torr.) Woot. & Standl. Apache, Navajo, and eastern Coconino counties, 5,000 to 6,000 feet, August to October. Northwestern New Mexico and northeastern Arizona.

56. Eriogonum Simpsoni Benth. (*E. Macdougalii* Gandoger, *E. effusum* subsp. *Simpsoni* Stokes). Apache County to Mohave and Yavapai counties, 3,500 to 7,500 feet, June to September. Colorado to Nevada, New Mexico, and Arizona.

One of the commonest species in the pine forests of northern Arizona. Perhaps only a variety of *E. microthecum* Nutt., differing chiefly in its strongly revolute leaves. *E. Macdougalii* (type *MacDougal* 176, Grand Canyon) is a form with exceptionally long peduncles.

57. Eriogonum Mearnsii Parry. Navajo, Coconino, and Yavapai counties, 3,000 to 6,500 feet, August to October. Known only from Arizona.

The typical phase, with leaves green and glabrous above, is known only from the type collection (Fort Verde, Yavapai County, *Mearns* 179). Much more common is var. *pulchrum* (Eastw.) Kearney & Peebles (*E. pulchrum* Eastw.) with leaves floccose-pubescent and whitish above. The type of *E. pulchrum* was collected at Crater Mound, Coconino County (*Eastwood* 15746).

6. RUMEX (86). Dock, Sorrel

Plants herbaceous, mostly perennial; leaves alternate, simple, the stipules united into a cylindric, more or less caducous sheath (ocrea); flowers perfect or unisexual; perianth calyxlike, 6-parted, the 3 inner segments (valves) becoming enlarged and modified in fruit; stamens 6; stigmas 3; fruit a trigonous achene.

Most of the species are coarse plants, and several are weeds introduced from the Old World. *R. triangulivalvis* and *R. violascens* are reported to be eaten freely by livestock. The majority flower in summer. Several species are used in Europe as potherbs, and American Indians made meal from the seeds. Identification of specimens without mature fruit is often difficult.

KEY TO THE SPECIES

1. Leaves hastately lobed; plants usually dioecious....................1. *R. Acetosella*
1. Leaves not hastate; plants usually monoecious or the flowers perfect (2)
2. Stems with axillary shoots, erect to procumbent (3).
2. Stems usually without axillary shoots and erect (7).
3. Leaves ovate-lanceolate, broadest below the middle; valves usually more than 4 mm. long in fruit, their margins entire or nearly so (4).
3. Leaves usually narrower, lanceolate or linear-lanceolate; valves 3 to 4 mm. long, their margins entire, or minutely crenulate or denticulate toward base (5).
4. Valves (1 or all) often bearing a prominent grainlike callosity.........2. *R. altissimus*
4. Valves without a callosity or 1 valve bearing a very narrow, elongate callosity (essentially a thickened midnerve)....................................3. *R. ellipticus*
5. Valves without callosities, about 3 mm. long......................4. *R. californicus*
5. Valves all, or 1 of them, bearing a callosity, this much narrower than the valve (6).
6. Achenes about 2.5 mm. long; valves 4 to 5 mm. long.................5. *R. mexicanus*
6. Achenes about 2 mm. long; valves about 3 mm. long.............6. *R. triangulivalvis*

7. Callosities wanting on the valves, but the midnerve sometimes thickened (8).
7. Callosities present on at least one of the valves (10).
8. Valves in fruit 10 mm. long or longer, deeply cordate, usually reddish; stipular sheaths firm, persistent; leaves thickish .... 7. *R. hymenosepalus*
8. Valves in fruit 4 to 5 mm. long; sheaths delicate, deciduous; leaves thinnish. Stems up to nearly 2 m. long (9).
9. Principal lateral leaf veins straight, spreading at nearly a right angle to the midvein; leaves rounded, cuneate, or cordate at base, up to 50 cm. long and nearly half as wide 8. *R. orthoneurus*
9. Principal lateral veins arcuate-ascending; leaves more or less cordate at base, usually less than half as wide as long .... 9. *R. occidentalis*
10. Valves entire or if denticulate, then the leaf margins usually distinctly crisped (11).
10. Valves sharply dentate or denticulate toward base; leaf margins flat or nearly so (12).
11. Leaves small, flat, truncate or the lowest cordate at base; valves 2.5 to 3 mm. long, scarcely broader than the thick callosity; inflorescence conspicuously leafy, much interrupted .... 10. *R. conglomeratus*
11. Leaves large, crisped or undulate on the margin, commonly narrowed (seldom truncate) at base; valves 4 to 6 mm. long, much broader than the callosity; inflorescence not leafy and interrupted, or so only toward the base .... 11. *R. crispus*
12. Plants perennial; basal leaves not more than 2½ times as long as wide, cordate at base. Valves usually sharply and conspicuously dentate .... 12. *R. obtusifolius*
12. Plants mostly annual; basal leaves 3 to 6 times as long as wide (13).
13. Valves 2.5 to 3 mm. long, short-dentate or denticulate; leaves oblanceolate or obovate, commonly about 3 times as long as wide .... 13. *R. violascens*
13. Valves not more than 2 mm. long, mostly long-dentate; leaves linear-lanceolate, usually more than 3 times as long as wide .... 14. *R. fueginus*

1. Rumex Acetosella L. Apache, Navajo, Coconino, Graham, Gila, Cochise, and Pima counties, 5,500 to 8,000 feet, occasional. Naturalized almost throughout temperate North America; native of Eurasia.

Sheep sorrel. An abundant and troublesome weed of fields and pastures in some parts of the United States. The herbage is very sour to the taste, containing much oxalic acid.

2. Rumex altissimus Wood. Coconino, Graham, Gila, Cochise, and Pima counties, 3,000 to 7,000 feet. Eastern and central United States to Arizona.

3. Rumex ellipticus Greene. Pinal Mountains, Gila County (*Peebles et al.* 4432), Tucson, Pima County (*Toumey* 343a), probably more widely distributed in the state. Texas to Arizona.

Scarcely more than a variety of *R. altissimus*.

4. Rumex californicus Rech. f. North Rim of the Grand Canyon, 9,000 feet (*Collom* in 1940), Prescott, Yavapai County (*Peebles et al.* 8861), Tucson, Pima County (*Toumey* 343c). Arizona and California.

The Arizona specimens are not typical, resembling *R. triangulivalvis* except in the absence of callosities on the valves.

5. Rumex mexicanus Meisn. Coconino County, North Rim of the Grand Canyon, 8,000 to 9,000 feet (*Collom* in 1939 and No. 1483), and near Flagstaff (*Schallert*). New Mexico, northern Arizona, and Mexico.

One of Mrs. Collom's specimens (No. 1483) resembles *R. californicus* in the absence of callosities on the valves, but these are about 4 mm. long and wide and lack the attenuate tips characteristic of *R. californicus*.

6. Rumex triangulivalvis (Danser) Rech. f. Coconino County to Graham and Pima counties, 1,500 to 8,500 feet. Quebec to British Columbia, south to New Mexico, Arizona, and California.

Arizona specimens previously identified as *R. mexicanus* were referred to this species by Rechinger. It is doubtful that this entity is more than a small-fruited variety of *R. mexicanus*.

7. Rumex hymenosepalus Torr. (*R. arizonicus* Britton). Navajo, Coconino, and Mohave counties, south to Graham, Santa Cruz, and Pima counties, 6,000 feet or lower, common and conspicuous in sandy stream beds and fields, March and April, type of *R. arizonicus* from Fort Verde, Yavapai County (*Mearns* 300). Wyoming to Utah, western Texas, Arizona, northern Mexico, and California.

Canaigre, wild-rhubarb. The high tannin content has aroused interest in canaigre, but attempts to cultivate the plant have not been financially successful hitherto. The United States Department of Agriculture is conducting experiments to this end. The petioles make a good substitute for rhubarb in pies. Indians and Mexicans use the leaves of this and other species of *Rumex* for greens and eat the petioles roasted or stewed with sugar. The Papago Indians roast rather than boil canaigre leaves, as with other greens that are sufficiently succulent, probably because of the frequent scarcity of water. The Hopi and Papago Indians used the roots for treating colds and sore throat, and a dye was formerly obtained from them by the Navajos.

8. Rumex orthoneurus Rech. f. Chiricahua Mountains, Cochise County, 8,000 to 8,500 feet (*Blumer* 1449, the type collection, *Kearney & Peebles* 15114). Known definitely only from these collections.

9. Rumex occidentalis Wats. Phelps Botanical Area, White Mountains, Apache County, 9,500 feet, in a wet meadow (*Phillips* 3230, *Phillips & Kearney* 3403). Labrador to British Columbia, south to Texas, New Mexico, Arizona, and California.

The Arizona specimens are immature; but although the leaves are exceptionally long and narrow, they seem closer to *R. occidentalis* than to any other species.

10. Rumex conglomeratus Murr. Occasional in Maricopa, Pinal, and Pima counties, along ditches and streams. Here and there in the United States; naturalized from Europe.

11. Rumex crispus L. Apache and Coconino counties to the southern border of the state, up to 8,000 feet, along streams and ditches. Naturalized in most of temperate North America; native of Eurasia.

Curly-leaf dock. The plant is reputed to have medicinal value and is sometimes used as a potherb.

12. Rumex obtusifolius L. Chiricahua and Huachuca mountains (Cochise County), Santa Catalina Mountains (Pima County), up to 8,000 feet. Extensively naturalized in North America; native of Europe.

13. Rumex violascens Rech. f. Mohave and Gila counties to Pima and Yuma counties. Texas to California and Mexico.

14. Rumex fueginus Phil. Springerville to Fort Apache, Apache or Navajo County (*Eggleston* 15755). Canada and most of the United States; South America.

7. OXYRIA. Mountain-sorrel

Plants resembling small plants of *Rumex,* differing in having reniform leaves, only 4 perianth segments, 2 stigmas, and lenticular achenes.

1. Oxyria digyna (L.) Hill. San Francisco Peaks (Coconino County), 10,000 to 12,000 feet, July to September. Greenland to Alaska, south in the mountains to New Hampshire and northern Arizona; Eurasia.

8. POLYGONUM (87). Knotweed, Smartweed

Plants herbaceous, annual or perennial, a few species semiaquatic; stems jointed, often swollen at the nodes; leaves alternate, simple, entire, the stipules united in a sheath (ocrea), this usually cylindric, often lacerate or fringed; flowers small, mostly perfect, in spikes or racemes, or scattered in the leaf axils; pedicels jointed; perianth cleft or parted, with 4 or 5 lobes; stamens 5 to 9; stigmas 2 or 3; achene lenticular or trigonous.

Many of the species are weeds but are not troublesome in Arizona. The knotweeds (section *Avicularia*) tend to become abundant on overgrazed land. Some of the smartweeds or lady's-thumbs (section *Persicaria*) have acrid juice that is irritating to the skin, eyes, and nostrils.

KEY TO THE SPECIES

1. Stems twining; outer perianth segments winged or strongly keeled. Leaves (at least the lower ones) ovate, cordate-sagittate; flowers greenish white, in loose axillary and terminal racemes or racemiform panicles: section *Bildерdykia*......1. *P. Convolvulus*
1. Stems not twining; outer perianth segments not winged (2).
2. Leaves with a hingelike joint at point of attachment to the sheath; flowers in axillary clusters; bracts of the inflorescence with well-developed blades. Leaves oblong, elliptic, lanceolate, or linear; flowers greenish, whitish, or pinkish: section *Avicularia* (3).
2. Leaves without a distinct joint at the point of attachment; flowers in terminal (sometimes also axillary) spikelike racemes, these sometimes forming panicles; bracts of the inflorescence reduced to sheaths (8).
3. Flowers crowded toward the ends of the branches, the inflorescences appearing like terminal spikes. Stems mostly erect (4).
3. Flowers scattered along the stems in small axillary clusters (5).
4. Stems slender, usually less than 15 cm. long; floral leaves little reduced; achenes minutely roughened..2. *P. Kelloggii*
4. Stems stout, 30 to 100 cm. long; floral leaves greatly reduced except near the base of the spike; achenes smooth, shiny............................3. *P. argyrocoleon*
5. Stems decumbent or prostrate. Upper leaves not greatly reduced; achenes dark brown, the surface dull and minutely roughened........................4. *P. aviculare*
5. Stems usually erect or ascending (6).
6. Pedicels mostly reflexed or deflexed; perianth 3 to 4 mm. long. Achenes black, the surface (sometimes only the angles) shiny and smooth.............5. *P. Douglasii*
6. Pedicels erect or ascending; perianth not more than 3 mm. long (7).
7. Stems usually slender; upper leaves reduced to subulate bracts; achene black, acutely angled, shiny and smooth...................................6. *P. sawatchense*
7. Stems stout; upper leaves often reduced but not subulate; achenes minutely roughened, not very shiny..7. *P. ramosissimum*
8. Rootstock usually much thickened and bulblike; basal leaves well developed. Inflorescence solitary; plants alpine or subalpine; flowers white or tinged with pink: section *Bistorta* (9).
8. Rootstock, if any, not bulblike; basal leaves none or soon disappearing: section *Persicaria* (10).

9. Inflorescence narrowly cylindric, 5 to 8 mm. wide, usually viviparous (bearing bulblets) below .... 8. *P. viviparum*
9. Inflorescence broadly cylindric or somewhat ovoid, 10 to 20 mm. wide, not viviparous 9. *P. bistortoides*
10. Inflorescences often solitary, all terminal or nearly so; plants aquatic or semiaquatic. Flowers bright pink (11).
10. Inflorescences usually several, axillary as well as terminal; plants not aquatic, but often growing in wet soil (12).
11. Leaves obtuse or acute, commonly widest near the middle; inflorescence seldom more than 3 cm. long, more than 10 mm. wide........ 10. *P. amphibium*
11. Leaves acuminate, commonly widest near the base; inflorescence 3 to 10 cm. long, seldom more and often less than 10 mm. wide........ 11. *P. coccineum*
12. Sheaths without marginal bristles (occasionally short-ciliate when young). Flowers pink or pinkish (13).
12. Sheaths with marginal bristles, these sometimes almost wanting in *P. fusiforme* (15).
13. Inflorescences nodding or drooping, usually numerous, slender and elongate 12. *P. lapathifolium*
13. Inflorescences erect or nearly so (14).
14. Leaves glabrous, or strigose beneath; perianth usually bright pink 13. *P. pensylvanicum*
14. Leaves floccose-tomentose beneath; perianth pale pink........ 14. *P. incanum*
15. Inflorescences loose, more or less interrupted, narrowly cylindric; perianth greenish, glandular-punctate .... 15. *P. punctatum*
15. Inflorescences dense, usually not interrupted; perianth pink, not or very obscurely glandular (16).
16. Internodes normally fusiform-inflated; inflorescences commonly more than 3 cm. long, narrowly cylindric, not more than 6 mm. wide........ 16. *P. fusiforme*
16. Internodes not normally inflated; inflorescences commonly less than 3 cm. long, broadly cylindric or somewhat ovoid, usually more than 6 mm. wide.. 17. *P. Persicaria*

1. Polygonum Convolvulus L. Apache, Navajo, and Coconino counties to Cochise and Pima counties, 2,500 to 8,000 feet, roadsides, May to August. Widely distributed in the United States; naturalized from Eurasia.

Corn-bind. The only climbing plant of this family in Arizona.

2. Polygonum Kelloggii Greene. Kaibab Plateau and North Rim of Grand Canyon, Coconino County, 8,000 to 9,000 feet (*Mead* 967, *Collom* in 1950), July and August. Wyoming to British Columbia, south to Colorado, northern Arizona, and California.

The Mead collection was referred previously to *P. Watsoni* Small, but the flowers have only 3 anther-bearing stamens (See L. C. Wheeler, *Rhodora* 40: 309-317).

3. Polygonum argyrocoleon Steud. Mohave, Maricopa, Pinal, and Yuma counties, 100 to 3,500 feet, roadsides, April to October. Texas, Arizona, and California; naturalized from central Asia.

The plant resembles *P. ramosissimum* Michx., but the inflorescences are more spicate. This species has become abundant in Maricopa and Pinal counties.

4. Polygonum aviculare L. Navajo and Coconino counties to Cochise and Pima counties, 1,000 to 8,000 feet, roadsides, April to October. Widely distributed in North America; naturalized from Eurasia.

5. Polygonum Douglasii Greene. Apache, Coconino, and Graham counties, 7,000 to 9,500 feet, June to September. Saskatchewan to British Columbia, south to New Mexico, Arizona, and California.

6. Polygonum sawatchense Small. Apache County to Mohave County, south to Cochise and Pima counties, 5,500 to 9,500 feet, common in dry pine woods, June to September. South Dakota to Washington, New Mexico, Arizona, and California.

7. Polygonum ramosissimum Michx. Flagstaff, Coconino County (*Jones* 3991), White Mountains, Apache County (*Griffiths* 5378), Beaver Creek, Yavapai County (*Toumey* 346½), August. Widely distributed in North America.

***8. Polygonum viviparum** L. Inclusion of this species in the Arizona flora rests only upon the doubtful basis of a collection by E. Palmer in 1869, without locality except "Arizona." Greenland to Alaska, south to New Mexico (and Arizona?); Eurasia.

9. Polygonum bistortoides Pursh. Apache, Coconino, Greenlee, and Graham counties, 8,500 to 11,000 feet, June to September. Montana to British Columbia, New Mexico, Arizona, and California.

A conspicuous plant in wet mountain meadows, with rather tall, erect stems and dense spikes of white flowers. Most of the Arizona specimens belong to the glabrous-leaved var. *oblongifolium* (Meisn.) St. John.

10. Polygonum amphibium L. White Mountains (Apache County), Lakeside (Navajo County), Tuba and Foxboro (Coconino County), 5,000 to 9,000 feet, in ponds, August and September. Canada and Alaska to New Jersey, New Mexico, eastern Arizona, and California; Europe.

The American plant was referred by Fernald (*Rhodora* 48: 49) to var. *stipulaceum* Coleman (*P. natans* (Michx.) Eaton).

11. Polygonum coccineum Muhl. Apache, Navajo, and Coconino counties, south to Santa Cruz and Pima counties, 2,500 to 7,000 feet, ponds, ditches, and marshes, July to September. Maine to Alaska, south to Virginia, Mexico, Arizona, and California.

E. L. Greene published *Persicaria fistulosa, P. ophiophila,* and *P. Rothrockii,* based upon Arizona types, but these seem to be merely forms of *Polygonum coccineum* or of *P. amphibium.*

12. Polygonum lapathifolium L. Navajo and Coconino counties to Cochise and Pima counties, 1,000 to 6,000 feet, along streams, April to October. Throughout most of North America; probably introduced from Europe.

Persicaria granulata Greene is probably a synonym. It was based upon a collection on the Verde River (*Smart* in 1867).

13. Polygonum pensylvanicum L. Near McNary (Apache County), near Tuba and Mormon Lake (Coconino County), Santa Maria Mountains (Yavapai County), near Tucson (Pima County), 2,500 to 7,000 feet, wet places, summer. Nova Scotia to Minnesota, south to Florida, Arizona, and Mexico.

14. Polygonum incanum F. W. Schmidt. White Mountains (southern Apache and northern Greenlee counties), 7,500 to 8,500 feet, wet meadows, August and September. Introduced sparingly into the United States from Europe.

Identification of the Arizona plants as *P. incanum* is somewhat doubtful.

15. Polygonum punctatum Ell. Navajo County and western Coconino County, south to Cochise, Santa Cruz, and Pima counties, 2,500 to 5,000 (?)

feet, along streams, July to October. Throughout most of the United States and south to Guatemala.

16. Polygonum fusiforme Greene. Topock (Mohave County), near Tucson (Pima County), near Yuma (Yuma County), 150 to 3,000 feet, July to October. Southern Arizona and southeastern California.

The specimens from near Tucson (Sabino Canyon) have somewhat less pointed achenes than in the type of the species, which was collected near Needles, California.

17. Polygonum Persicaria L. Billie Creek (southern Navajo County), Coconino County (several localities), Prescott (Yavapai County), Huachuca Mountains (Cochise County), 5,000 to 7,000 feet, marshy places along streams, July to September. Extensively naturalized in North America; from Europe.

9. FAGOPYRUM. Buckwheat

An annual glabrous herb; leaves alternate, petioled, with broad hastate blades; flowers in terminal or axillary racemes, these usually forming panicles; perianth greenish white, 5-parted; stamens 8; style 3-parted; achene trigonous.

1. Fagopyrum sagittatum Gilib. (*F. esculentum* Moench). Near Patagonia, Santa Cruz County, at roadside (*Peebles et al.* 5613). An escape from cultivation in many parts of the United States, but scarcely naturalized.

44. CHENOPODIACEAE. Goose-foot Family

Plants herbaceous or shrubby; leaves simple, without stipules; flowers perfect or unisexual, not showy; perianth, when present, with 1 to 5 segments; stamens as many as or fewer than the perianth segments, opposite to them; styles 2 or 3; ovary 1-celled; fruit a 1-seeded utricle or achene.

This large family includes the cultivated sugar beet, garden beet, and spinach. Many of the species are weeds, and some of them are valuable for browse and grazing. Species of *Atriplex* and other genera are characteristic plants of strongly saline and alkaline soils, often taking up so much sodium chloride from the soil solution as to give the herbage a distinctly salty taste. The pollen of many Chenopodiaceae is believed to cause hay fever in Arizona. Russian-thistle (*Salsola*) is the worst offender, but the salt-bushes (*Atriplex*) are also very important because of their abundance over large areas.

KEY TO THE GENERA

1. Embryo spirally coiled. Leaves narrow, entire, either thick and fleshy or spine-tipped (2).
1. Embryo not spirally coiled, circular to horseshoe-shaped, or conduplicate (4).
2. Flowers without bractlets, the plants monoecious, the staminate flowers in catkinlike spikes, naked; perianth of the pistillate flowers confluent with the ovary 13. *Sarcobatus*
2. Flowers with a pair of bractlets and a perianth, mostly perfect, never in catkinlike spikes; perianth free from the ovary (3).
3. Bractlets minute, scalelike, much shorter than the perianth; fruiting perianth scarcely enlarged, connivent, not winglike, remaining fleshy; plants mostly perennial, often suffrutescent; leaves soft and fleshy, subterete........................14. *Suaeda*
3. Bractlets narrow, elongate, equaling or longer than the perianth; fruiting perianth greatly enlarged, spreading, winglike, dry, scarious; plants annual; leaves strongly

spine-tipped. Plants becoming hard and prickly, the intricately branched stem breaking off at the surface of the ground and becoming a tumble-weed........15. *Salsola*

4. Leaves reduced to small scales; stems appearing jointed; flowers in dense, continuous, cylindric, fleshy spikes, perfect.................................12. *Allenrolfea*
4. Leaves with well-developed blades; stems not appearing jointed; flowers not in dense, continuous, fleshy spikes (5).
5. Perianth segments strongly imbricate, nearly distinct; individual flowers relatively conspicuous; leaves opposite.....................................1. *Nitrophila*
5. Perianth segments, if present, not or only slightly imbricate; individual flowers inconspicuous; leaves all or most of them alternate (6).
6. Fruit at maturity naked. Flowers perfect, in long, slender spikes, without bractlets, borne in the axils of conspicuous, scarious-margined bracts but not enclosed by them 11. *Corispermum*
6. Fruit at maturity enclosed by the perianth or by the enlarged bractlets (7).
7. Pubescence of the herbage pilose, villous, or lanate, the hairs simple and slender. Flowers mostly perfect (8).
7. Pubescence of the herbage wholly or partly of stellate, glandular, or inflated hairs (the last collapsing and scurflike when dry), seldom none. Perianth, if any, not spiny or horizontally winged (10).
8. Perianth in fruit not winged, bearing a dorsal tubercle or spine on each segment. Plants annual; stem tall, much-branched; leaves narrow, entire...........9. *Bassia*
8. Perianth in fruit with conspicuous, scarious, horizontal wings (9).
9. Leaves coarsely sinuate-dentate; perianth wing annular, continuous or nearly so 3. *Cycloloma*
9. Leaves entire; perianth wing of 5 usually distinct, wedge-shaped segments 10. *Kochia*
10. Flowers mostly perfect, without bractlets, with a perianth; plants herbaceous (11).
10. Flowers unisexual (or some of them perfect in *Eurotia*), the pistillate flowers with bractlets and usually without a perianth; plants mostly woody, at least at base. Bractlets becoming enlarged and enclosing the fruit (12).
11. Perianth segments and stamens 3 to 5; upper leaves of the inflorescence usually much reduced ....2. *Chenopodium*
11. Perianth segment and stamen one; upper leaves of the inflorescence little reduced 4. *Monolepis*
12. Pubescence entirely or chiefly of simple inflated hairs, these collapsing and scurflike when dry (13).
12. Pubescence of branched, scarcely inflated hairs. Bractlets united to the middle or higher; seeds vertical (14).
13. Bractlets distinct, at least near the apex; seeds all vertical, or both vertical and horizontal on the same plant......................................5. *Atriplex*
13. Bractlets united up to the depressed apex; seeds all horizontal..............6. *Zuckia*
14. Fruiting bractlets very thin and flat, glabrous or scurfy-pubescent, winged, more or less retuse at apex; herbage puberulent or glabrate....................7. *Grayia*
14. Fruiting bractlets forming a 2-beaked tube, densely long-villous, not winged; herbage conspicuously and densely stellate-pubescent........................8. *Eurotia*

1. NITROPHILA

Glabrous perennial herbs; stems numerous, tufted, not more than 30 cm. long; leaves opposite, fleshy, narrow, semiterete; flowers perfect, axillary, conspicuous for the family, pink or white; perianth segments and stamens usually 5; style longer than the ovary, persistent.

1. Nitrophila occidentalis (Moq.) Wats. Gila Crossing, Pinal County (*Plumb* 75, *Peebles* 13232), Quitobaquito, Pima County (*Nichol* in 1939), about 1,000 feet, moist, saline soil, April and May. Oregon to Arizona, California, and northwestern Mexico.

2. CHENOPODIUM (88, 89). GOOSE-FOOT

Plants herbaceous, annual or perennial, often mealy, sometimes glandular-pubescent; leaves alternate, entire to pinnatifid; flowers green, perfect, in glomerules, these axillary or forming spikes or panicles; perianth herbaceous or fleshy, with 2 to 5 lobes or segments; stamens 2 to 5.

Plants mostly weedlike. Several of the species occurring in Arizona are said to be eaten freely by sheep and cattle. The Indians use the leaves for greens and the seeds of certain species for making mush and cakes, sometimes mixing them with corn meal. *C. ambrosioides* yields oil of chenopodium, distilled from the leaves and stems, which is a powerful anthelmintic.

KEY TO THE SPECIES

1. Herbage, or at least the calyx, glandular-pubescent and (or) resinous-glandular, often more or less lanuginose, not farinose. Plants strong-scented; leaves (some or all of them) coarsely toothed or pinnatifid; seeds mostly horizontal (2).
1. Herbage glabrous or farinose, the hairs at first blisterlike, or (in *C. glaucum*) sometimes elongate (4).
2. Flowers in glomerules, these disposed in panicles of spikes; odor fetid 1. *C. ambrosioides*
2. Flowers solitary or loosely clustered, disposed in dichotomous cymes; odor not fetid (3).
3. Herbage glabrous or sparsely puberulent; branches becoming somewhat spinescent; perianth resinous-granular, the lobes tuberculate- or corniculate-appendaged 2. *C. incisum*
3. Herbage copiously glandular-pilose; branches not spinescent; perianth not or scarcely granular, the lobes not appendaged............................3. *C. Botrys*
4. Seeds all vertical, or partly vertical and partly horizontal (5).
4. Seeds nearly always horizontal (7).
5. Perianth at maturity fleshy and bright red; inflorescence almost bractless, or leafy-bracted below. Leaves hastate and dentate, or the upper ones entire, glabrescent; flowers in dense glomerules, these disposed in elongate spikes......4. *C. capitatum*
5. Perianth not fleshy; inflorescences mostly bracteate (6).
6. Leaves densely white-farinose beneath, sharply sinuate-dentate. Inflorescence often sparsely pilose ....5. *C. glaucum*
6. Leaves glabrous or glabrescent on both surfaces, irregularly sinuate-dentate to pinnatifid, rarely entire...6. *C. rubrum*
7. Seeds rugose or alveolate (8).
7. Seeds smooth or nearly so (11).
8. Surface of the seeds with irregular markings (undulate ridges). Leaves thickish, deltoid to rhombic-ovate, coarsely and often sharply dentate or laciniate; seeds sharp-edged, the pericarp closely adherent...............................7. *C. murale*
8. Surface of the seeds regularly and finely alveolate. Pericarp adherent (9).
9. Herbage densely farinose, whitish; stems much-branched from the base, stout, the branches decumbent to ascending-spreading; leaves scarcely longer than wide, thickish; plants very ill-scented.................................8. *C. Watsoni*
9. Herbage moderately farinose to glabrescent, green; stems usually not branched from the base, the branches mostly erect or ascending at a narrow angle; leaves usually considerably longer than wide, thin; plants not or only moderately ill-scented (10).
10. Stems slender; glomerules of flowers small and rather widely spaced, the inflorescence open; perianth not closely investing the fruit, spreading at full maturity; leaves entire, or subhastate with 1 pair of short teeth.............9. *C. arizonicum*
10. Stems stout; glomerules relatively large and crowded, the inflorescence usually dense; perianth closely investing the fruit, more or less connivent even at full maturity; leaves often with several teeth.................................10. *C. Berlandieri*

11. Leaves linear, narrowly lanceolate, or narrowly oblong (sometimes broader in *C. pratericola*), entire to slightly 3-lobed or hastate (12).
11. Leaves rhombic, deltoid, or broadly lanceolate (14).
12. Leaves linear, entire, 1-nerved. Herbage densely farinose; perianth lobes closely investing the utricle; pericarp adherent............................11. *C. leptophyllum*
12. Leaves narrowly lanceolate or oblong, entire or slightly 3-lobed or hastate, the lower ones 3-nerved (13).
13. Herbage copiously and coarsely farinose; perianth lobes at maturity erect or spreading and exposing the utricle; pericarp adherent; plants with a rank odor.....12. *C. hians*
13. Herbage sparsely (rarely copiously) farinose to glabrescent; perianth lobes closely investing the utricle; pericarp free; plants without a rank odor....13. *C. pratericola*
14. Leaves about as wide as long, deltoid to rhombic-orbicular. Herbage sparsely to densely farinose; pericarp free; seeds 0.8 to 1.2 mm. in diameter..14. *C. Fremontii*
14. Leaves usually noticeably longer than wide, usually oval to rhombic-ovate, seldom lanceolate (15).
15. Plants farinose throughout; pericarp free; seeds 1 mm. in diameter....15. *C. albescens*
15. Plants farinose chiefly on the leaves; pericarp adherent; seeds 1.3 to 1.5 mm. in diameter 16. *C. album*

1. Chenopodium ambrosioides L. Pinal, Maricopa, Cochise, and Yuma counties, occasional at roadsides. Throughout most of the United States; introduced from tropical America.

Spanish-tea, Mexican-tea. The species is represented in Arizona by subsp. *euambrosioides* Aellen, of which two varieties, var. *typicum* (Speg.) Aellen, with bracteate glomerules, and var. *anthelminticum* (L.) Aellen, with bractless glomerules, probably occur in this state.

2. Chenopodium incisum Poir. (*C. graveolens* Willd. non Lag. & Rodr.). Apache, Navajo, and Coconino counties, south to Cochise, Santa Cruz, and Pima counties, 5,000 to 9,000 feet, usually in pine woods, August and September. Western Texas and Colorado to Arizona, southward to Central America; South America.

The Arizona plant is var. *neomexicanum* Aellen, type from the Santa Catalina Mountains (*Harrison* 3026). The plant has a strong but not unpleasant odor, and turns bright red in autumn.

3. Chenopodium Botrys L. Yavapai and Maricopa counties, 1,000 to 5,500 feet, at roadsides. Extensively naturalized in North America; from the Old World.

Jerusalem-oak, feather-geranium.

4. Chenopodium capitatum (L.) Asch. (*Blitum capitatum* L.). Kaibab Plateau, Grand Canyon, San Francisco Peaks, and Flagstaff (Coconino County), White Mountains (Apache and Greenlee counties), 7,000 to 10,000 feet, in rich, moist soil, June to September. Quebec to Alaska, southward to New Jersey, New Mexico, Arizona, and Oregon; Europe.

Strawberry-blite.

5. Chenopodium glaucum L. Apache and Navajo counties, 5,000 to 6,000 feet, in saline soil, July to September. North Dakota to Alaska, southward to Missouri, New Mexico, Arizona, and Oregon.

The range as stated is that of subsp. *salinum* (Standl.) Aellen (*C. salinum* Standl.), which apparently is the only form occurring in Arizona. The typical plant (subsp. *euglaucum* Aellen), with often shorter petioles and less sharply dentate blades, is widely distributed in North America, probably adventive from Europe.

*6. **Chenopodium rubrum** L. Not known to occur in Arizona, but has been collected in northwestern New Mexico. Widely distributed in Canada and the United States; probably introduced from Eurasia.

7. **Chenopodium murale** L. Coconino County to Cochise, Pima, and Yuma counties, 150 to 8,000 feet, flowering throughout the year. Widely distributed in North America; naturalized from Europe.

Nettle-leaf goose-foot, a common winter weed in waste places in southern Arizona.

8. **Chenopodium Watsoni** A. Nels. Navajo County to eastern Mohave County, south to Greenlee and Santa Cruz counties, 1,000 to 7,000 feet, March to September. South Dakota and Montana to New Mexico and Arizona.

9. **Chenopodium arizonicum** Standl. Pinal, Cochise, and Pima counties, 4,000 to 8,000 feet, type from the Santa Rita Mountains (*Griffiths* 5982). Utah, Arizona, and Mexico.

10. **Chenopodium Berlandieri** Moq. Apache County to Mohave County, near Elgin (Santa Cruz County), Baboquivari Mountains (Pima County), probably elsewhere in the state, up to 7,000 (9,500?) feet. Virginia to Florida, west to Washington, Oregon, and California, south to Mexico.

This polymorphic species is represented in Arizona by subsp. *Zschackei* (Murr) Zobel var. *typicum* (Ludwig) Aellen, with the lateral lobes and greatest width of the leaf mostly at or above the middle, and subsp. *pseudo-petiolare* Aellen, with leaves broadest below the middle and distinctly 3-lobed.

11. **Chenopodium leptophyllum** Nutt. (*C. inamoenum* Standl.). Apache, Navajo, and Coconino counties (probably elsewhere), 5,000 to 7,000 feet, August and September. Wyoming to Oregon, south to Texas, New Mexico, and Arizona.

12. **Chenopodium hians** Standl. Near Jacob Lake, Grand Canyon, Flagstaff, and Wupatki National Monument (Coconino County), 5,000 to 7,000 feet. Wyoming to Nevada, south to New Mexico and Arizona.

A plant collected on the north side of the Grand Canyon (*Eastwood & Howell* 7102) has the open calyx of *C. hians* but otherwise resembles *C. Fremontii*. It may have been a hybrid.

13. **Chenopodium pratericola** Rydb. (*C. leptophyllum* of authors, non Nutt.). Apache, Navajo, and Coconino counties, south to Cochise and Pima counties, 1,500 to 8,000 feet, May to September. New Jersey and Virginia to Washington, Oregon, California, and Mexico; Argentina.

There occur in Arizona subsp. *eupratericola* Aellen, a tall, erect plant with linear, lanceolate, or narrowly oblong leaves, the lower ones often somewhat 3-lobed; and subsp. *desiccatum* (A. Nels.) Aellen (*C. desiccatum* A. Nels.), a low, usually diffusely branched plant with oblong or oval, entire leaves.

14. **Chenopodium Fremontii** Wats. Apache County to Mohave County, south to Cochise, Santa Cruz, and Pima counties, 2,500 to 9,000 feet, June to September. North Dakota to British Columbia, south to New Mexico, Arizona, and Mexico.

A common plant in chaparral and pine forests, furnishing feed for cattle in autumn. Var. *incanum* Wats. (*C. incanum* Heller), which differs from the typical phase of the species in being a lower, more diffusely branched and spreading, more densely farinose plant, with leaves thicker and more pointed at base, is also common in Arizona.

15. Chenopodium albescens Small (*C. Berlandieri* subsp. *Zschackei* var. *glaucoviride* Aellen). A collection at Willow Spring, Apache County (*E. Palmer* 586) was cited by Aellen (88, p. 58). Iowa, Texas, and Arizona.

The Arizona specimen apparently is not typical of the species.

16. Chenopodium album L. Sacaton (Pinal County), perhaps also in Walnut Canyon (Coconino County), at Prescott (Yavapai County), and in the White Mountains (Apache County). Naturalized almost throughout North America; from Europe.

Lambs-quarters, a weed of cultivated and waste land. This species is often confused with *C. Berlandieri* subsp. *Zschackei,* which is much more common in Arizona.

3. CYCLOLOMA

Annual herbs with diffusely branched stems; leaves coarsely sinuate-dentate; inflorescence a panicle of interrupted spikes; flowers sessile; perianth 5-lobed, in fruit developing a thin, horizontal, irregularly denticulate wing; stamens 5; utricle depressed; seed horizontal.

1. Cycloloma atriplicifolium (Spreng.) Coult. Navajo and Coconino counties, also near Tempe (Maricopa County), and near Bisbee (Cochise County), 1,000 to 6,000 feet. Indiana to Manitoba, south to Texas and Arizona.

Plant becoming a tumbleweed. The Indians made mush and cakes from the ground-up seeds.

4. MONOLEPIS. Patata, Patota

Annual, slightly succulent, nearly glabrous herbs; stems low, diffuse or prostrate; flowers in small axillary clusters; perianth reduced to a single, persistent segment; stamen 1; utricle compressed laterally.

1. Monolepis Nuttalliana (Schult.) Greene. Apache County to Mohave County, south to Cochise, Pima, and Yuma counties, commonly 3,000 feet or lower, abundant in southern Arizona, January to April. Manitoba and Alberta, south to Texas, Sonora, and California.

The plant affords good spring pasturage for cattle and is used as greens by the Indians, who also make pinole from the seeds. The leaves are usually subhastately lobed, but specimens collected at Maricopa, Pinal County (*Eastwood* 6310), have narrow, entire leaves. They somewhat resemble *M. spathulata* Gray, but the latter has more spatulate and longer-petioled leaves.

5. ATRIPLEX (90). Salt-bush, Orache

Plants annual or perennial, herbaceous or shrubby, mostly mealy; leaves commonly alternate, entire to sinuate-dentate; plants monoecious or dioecious, the flowers in glomerules, these axillary or (the staminate ones especially) forming terminal spikes or panicles; perianth of the staminate flowers 3- to

5-parted; pistillate flowers without a perianth, subtended by 2 connivent bractlets, these in fruit enlarged, usually more or less connate, often dentate or tuberculate.

Most of the species are very salt-tolerant and flower in summer and early autumn. The shrubby species are good browse, especially in winter. Some of the herbaceous species also are grazed by livestock. The salty taste of the plants probably increases their palatability. The fruits are very nutritious. The Indians used to depend on the salt-bushes as a source of meal, which was made from the parched seeds and, like the pinole made from mesquite pods, sometimes was drunk mixed with water. The leaves and young shoots were used for greens, the Hopi Indians usually boiling them with meat. This tribe is reported to have used the ashes of *A. canescens* as a substitute for baking powder. The pollen of many species is a cause of hay fever.

KEY TO THE SPECIES

1. Fruiting bractlets fleshy-thickened and bright red at maturity. Plants perennial; stems prostrate or nearly so, sometimes woody below, much-branched; leaves whitish-scurfy beneath, glabrate above, oblong or obovate-oblong, shallowly dentate or entire; bractlets deltoid-cuneate, coarsely few-toothed............3. *A. semibaccata*
1. Fruiting bractlets not becoming fleshy or bright red (2).
2. Stems herbaceous or occasionally somewhat woody; plants mostly annual, commonly monoecious, the staminate and pistillate flowers in the same or in separate clusters (3).
2. Stems distinctly woody, at least near the base; plants perennial, mostly dioecious (11).
3. Leaves triangular-hastate, commonly becoming glabrate and green on both surfaces. Lowest leaves often opposite; fruiting bractlets acute or acutish, rounded-deltoid, often hastate, united only at base................................1. *A. patula*
3. Leaves not distinctly hastate, sometimes subhastate, usually remaining scurfy and whitish, grayish, or yellowish, at least on the lower surface (4).
4. Fruiting bractlets (often those in the same axil) commonly distinctly dimorphic, some of them wedge-shaped, about 3 mm. long, broadly truncate and emarginate or denticulate at apex and with smooth faces, others rounded-triangular or suborbicular, larger, with dentate margins and prominently crested faces; leaves more or less cordate at base, broadly triangular to suborbicular, the margins entire 4. *A. saccaria*
4. Fruiting bractlets not noticeably dimorphic; leaves not cordate at base (5).
5. Bractlets orbicular, their margins herbaceous and evenly dentate all around (6).
5. Bractlets not orbicular (sometimes semiorbicular in *A. argentea*), their margins not herbaceous and evenly dentate all around (7).
6. Stems erect or ascending, often slightly woody below; leaves oblanceolate; margins of the bractlets dissected nearly to the base, the teeth slender........7. *A. elegans*
6. Stems decumbent to nearly prostrate; leaves obovate or broadly spatulate; margins of the bractlets dissected about halfway to the base, the teeth relatively broad 8. *A. fasciculata*
7. Leaves strongly 3-nerved from the base with long, ascending, lateral nerves, entire, whitish-farinose beneath; fruiting bractlets conspicuously differentiated into a 2- to several-crested basal portion and a smooth, truncate and apiculate, or acute, apical portion. Plants sometimes with arachnoid as well as scurfy pubescence 6. *A. Powellii*
7. Leaves not strongly 3-nerved from the base or if so, then the lateral nerves short or spreading, or the margins of the blades dentate; fruiting bractlets not conspicuously differentiated basally and apically (8).
8. Staminate flowers in conspicuous naked or nearly naked terminal panicles; leaves seldom less than 3 times as long as wide; fruiting bractlets 2 to 3 mm. long (9).

8. Staminate flowers not in conspicuous, naked panicles; leaves less than 3 times as long as wide; bractlets more than 3 mm. long at maturity (10).

9. Leaves obtuse to acute at apex, commonly widest above the middle, densely white-farinose beneath when mature; fruiting bractlets strongly ribbed but usually not tuberculate on the faces, broadly deltoid or somewhat wedge-shaped, coarsely dentate .... 9. *A. Wrightii*

9. Leaves sharply acute or acuminate at apex, widest at or below the middle, sparsely farinose or glabrate beneath when mature; fruiting bractlets usually tuberculate on the faces .... 10. *A. Serenana*

10. Leaves oblong-lanceolate to ovate, conspicuously and sometimes acutely dentate; fruiting bractlets rhombic or deltoid, acutish to short-acuminate at apex, cuneate at base, dentate, strongly 3-nerved, usually short-crested on the faces. Leaves sometimes green and glabrescent on both surfaces when mature .... 2. *A. rosea*

10. Leaves rounded-triangular or triangular-subhastate, entire or inconspicuously repand-dentate; fruiting bractlets obovate-cuneate or semiorbicular, more or less truncate at apex, laciniate, not strongly 3-nerved, usually conspicuously long-crested on the faces .... 5. *A. argentea*

11. Fruits conspicuously 4-winged, the wings much broader than the body of the fruit and extending to its base. Plants shrubby; glomerules of flowers mostly aggregated in leafy or nearly naked panicles of spikes terminating the branches; leaves with entire margins, linear, oblong-lanceolate, or spatulate (12).

11. Fruits not 4-winged (14).

12. Leaves commonly less than 3 mm. wide; fruiting bractlets 4 to 8 mm. long; herbage whitish. Bractlets with free tips commonly much surpassing the wings, the latter coarsely dentate to laciniate .... 20. *A. linearis*

12. Leaves commonly more than 3 mm. wide; fruiting bractlets 6 to 20 mm. long; herbage gray or yellowish green (13).

13. Leaves usually broadest above the middle and less than 1 cm. wide; free tips of the fruiting bractlets commonly not surpassing the wings; shrubs 0.4 to 1.5 m. high .... 21. *A. canescens*

13. Leaves usually broadest at the middle and 1 cm. wide or wider; free tips of the fruiting bractlets commonly surpassing the wings; undershrubs less than 0.4 m. high .... 22. *A. Garrettii*

14. Bractlets of the fruits with entire to denticulate margins, rounded-deltoid, orbicular, or reniform, the faces smooth (15).

14. Bractlets of the fruits normally with dentate to laciniate margins, the faces often tuberculate or crested (18).

15. Leaves deeply and rather sharply dentate or laciniate; plants not spinescent. Foliage silvery white; fruiting bractlets distinct, thin, strongly compressed, entire .... 16. *A. hymenelytra*

15. Leaves sometimes subhastate, the margins otherwise entire; plants commonly spinescent, the ends of the branches becoming indurate and sharp-pointed (16).

16. Glomerules of flowers small and dense, axillary or in short, leafy spikes, the pistillate flowers sometimes solitary; fruiting bractlets 5 to 15 mm. long, united only at base, entire or denticulate; leaves oblanceolate, oval, ovate, obovate, or suborbicular, never subhastate-deltoid; plants usually rounded and compact, less than 1 m. high .... 17. *A. confertifolia*

16. Glomerules of flowers in slender, dense or interrupted, elongate, often drooping spikes, the spikes commonly aggregated in large, open terminal panicles, these leafy or nearly naked toward apex; fruiting bractlets not more than 6 mm. long, sometimes united up to the middle, usually crenulate; leaves oblong to deltoid-ovate; plants commonly tall, 1 to 3 m. high (17).

17. Branches terete; leaves usually ovate or deltoid, often subhastate, rounded to cuneate at base; fruiting bractlets 2.5 to 4 mm. long .... 18. *A. lentiformis*

17. Branches angled; leaves oblong or oval, not at all hastate, acute at base; fruiting bractlets 4 to 5 mm. long .... 19. *A. Griffithsii*

18. Leaves with sinuate or sinuate-dentate (seldom entire) margins; fruiting bractlets pedicellate, 6 to 14 mm. long. Bractlets with laciniate margins and the faces bearing long hornlike processes; plants suffrutescent; foliage usually silvery white 11. *A. acanthocarpa*
18. Leaves with entire margins; fruiting bractlets not more than 6 mm. long (19).
19. Plants intricately and rigidly branched shrubs; stems 0.5 to 2 m. high; leaves mostly less than 1 cm. long. Fruiting bractlets normally deeply dentate on nearly the whole margin, including the truncate apex .... 15. *A. polycarpa*
19. Plants suffrutescent; stems not more than 0.5 m. high; leaves seldom less than 1 cm. long (20).
20. Apical portion of the fruiting bractlets broadly deltoid, with margins entire or denticulate .... 14. *A. corrugata*
20. Apical portion of the fruiting bractlets truncate and dentate, or entire and beaklike, the faces bearing a few short, blunt appendages, or nearly smooth (21).
21. Bractlets united only at the truncate or subcordate base; herbage silvery 12. *A. obovata*
21. Bractlets united above the broadly cuneate base; herbage yellowish.... 13. *A. Jonesii*

1. **Atriplex patula** L. Winslow, Navajo County (*Griffiths* 5034), Moenkopi, Coconino County (*Clute* 63), 4,500 to 5,000 feet, reported to be common in Moenkopi Wash. Throughout most of the Northern Hemisphere.

Represented in Arizona by var. *hastata* (L.) Gray, which is widely distributed in Canada and the United States.

2. **Atriplex rosea** L. Apache County to Coconino County, south to Pima County, 1,000 to 7,000 feet. Extensively naturalized in the western United States; from Europe.

Red-scale, red orache. A common weed in waste places in parts of northern Arizona.

3. **Atriplex semibaccata** R. Br. Graham, Maricopa, Cochise, Pima, and Yuma counties, 1,000 to 3,000 feet. New Mexico, Arizona, and California; naturalized from Australia.

Australian salt-bush. Introduced into the United States many years ago, now a common weed in southern Arizona. The low plants help bind the soil along irrigation ditches, crowd out undesirable weeds, and furnish forage for domestic animals, particularly sheep.

4. **Atriplex saccaria** Wats. Apache County to Coconino County, 4,500 to 6,000 feet. Wyoming and Utah to New Mexico and Arizona.

Plants apparently sometimes dioecious.

5. **Atriplex argentea** Nutt. Apache, Navajo, and Coconino counties, about 5,000 feet. North Dakota to New Mexico, northern Arizona, and California.

This plant is a tumbleweed. The species is represented in Arizona by subsp. *typica* Hall & Clements var. *Caput-Medusae* (Eastw.) Fosberg (*A. Caput-Medusae* Eastw.), in which the faces of the fruiting bractlets are covered with long, often twisted, hornlike processes.

6. **Atriplex Powellii** Wats. Apache County to eastern Coconino County, 5,000 to 6,000 feet, type grown from "Arizona seeds" obtained by the Powell Expedition. South Dakota and Montana to New Mexico and Arizona.

Griffiths reported that this species covers extensive areas of denuded land in the Little Colorado River region. The plants apparently are sometimes dioecious.

7. Atriplex elegans (Moq.) D. Dietr. Greenlee, Maricopa, Pinal, Cochise, Santa Cruz, Pima, and Yuma counties, 3,500 feet or lower. Western Texas to southern California and northern Mexico.

A plant of weedlike habit, very common at roadsides and in waste land in southern Arizona and freely grazed by cattle. The Pima Indians boiled it with other food, sometimes with the flower buds of *Opuntia*. *A. elegans* is a host of the beet leaf hopper in Arizona.

Var. *Thornberi* Jones (*A. Thornberi* Standl.) has the faces of the bractlets prominently crested. This variety is known only from Tucson, Pima County (*Thornber, Jones* in 1903).

8. Atriplex fasciculata Wats. (*A. elegans* var. *fasciculata* Jones). Maricopa, Pinal, Pima, and Yuma counties, 2,500 feet or lower. Southern Arizona and southeastern California.

9. Atriplex Wrightii Wats. Coconino County to Cochise and Pima counties, 1,000 to 7,000 feet. Southern New Mexico, Arizona, and Sonora.

Common, especially in southern Arizona, in similar habitats as are occupied by *A. elegans*. It is usually a larger plant than the latter. This is one of the species held by the Indians in particular esteem as a potherb.

***10. Atriplex Serenana** A. Nels. Not known to occur in Arizona, but has been collected at Needles, California, on the Colorado River.

11. Atriplex acanthocarpa (Torr.) Wats. Gila River Valley near Safford (Graham County), about 3,500 feet, in saline soil. Western Texas to southern Arizona and northern Mexico.

The Arizona specimens referred doubtfully to this species have the fruiting bractlets less united, thinner, and less spongy in texture than in typical *A. acanthocarpa*. Apparently they represent a transition to *A. obovata*, with which they were growing.

12. Atriplex obovata Moq. Near Safford (Graham County), 2,500 to 3,500 feet. Western Texas to southern Arizona and northern Mexico.

According to Johnston (56, p. 148) the Arizona plants are greener and more slender than the typical Mexican plant.

13. Atriplex Jonesii Standl. (*A. sabulosa* Jones non Rouy). Apache County to eastern Coconino County, 3,500 to 5,000 feet, in dry, saline soil, type from Winslow, Navajo County (*Jones* 4109). Northwestern New Mexico and northeastern Arizona.

This species has been united with *A. obovata* (90, p. 322), but it has a different geographical distribution and appears to be specifically distinct. The plants are readily browsed by cattle, sheep, and goats.

***14. Atriplex corrugata** Wats. Not known definitely to occur in Arizona, but has been collected in northwestern New Mexico near the Arizona state line.

15. Atriplex polycarpa (Torr.) Wats. Mohave, Graham, Maricopa, Pinal, Pima, and Yuma counties, 3,500 feet or lower, the type presumably from Williams River (*Bigelow*). Arizona to Nevada, southern California, and northwestern Mexico.

Desert salt-bush, cattle-spinach, all-scale, commonly known in Arizona by the confusing names "sage" and "sagebrush." In the deserts of southwestern

Arizona this plant covers vast areas of moderately saline or nonsaline soil, in pure stands or associated with cresote-bush (*Larrea*), the bushes often symmetrically rounded and evenly spaced. It is by far the most important native forage plant of that region, which, however, is too arid to support many cattle.

16. Atriplex hymenelytra (Torr.) Wats. Western Maricopa, Mohave, and Yuma counties, below 1,000 feet, in dry, sandy or stony soil, the type presumably from Williams River (*Bigelow*). Southwestern Utah to southeastern California and northwestern Mexico.

Desert-holly. The silvery, evergreen leaves are gathered for Christmas decorations and winter bouquets, and are sometimes dyed or gilded.

17. Atriplex confertifolia (Torr. & Frém.) Wats. Apache County to Mohave County, also in eastern Graham County, 2,500 to 5,500 feet, dry plains and mesas, usually forming small hummocks. North Dakota to Oregon, south to Chihuahua, northern Arizona, and California.

Shad-scale, sheep-fat, spiny salt-bush. In the northern part of the state shad-scale often occurs over large areas in pure stands, crowding out nearly all other plants. It has the ability to resist overgrazing. *A. collina* Woot. & Standl., type from the Carrizo Mountains, Apache County (*Standley* 7481), differs from typical *A. confertifolia* in having dentate fruiting bractlets, but there seems to be complete intergradation.

18. Atriplex lentiformis (Torr.) Wats. Coconino and Mohave counties, south to Pima and Yuma counties, 4,000 feet or (usually) lower, in moist or dry, saline soil, type from along the Colorado River (*Woodhouse?*). Southern Utah and Nevada to Sonora and California.

Quail-brush, lens-scale, sometimes known in Arizona as white-thistle. This is the largest and showiest salt-bush in Arizona, reaching a height of 10 feet where the water table is high. It is especially abundant in the low western and southwestern parts of the state. Cattle browse the plants, and the forage is palatable.

19. Atriplex Griffithsii Standl. (*A. lentiformis* subsp. *Griffithsii* Hall & Clements). Known only from near Willcox, Cochise County, Arizona (*Griffiths* 1895, the type collection, *Darrow* in 1937, *K. F. Parker* 6563).

20. Atriplex linearis Wats. (*A canescens* subsp. *linearis* Hall & Clements). Pinal, Maricopa, Pima, and Yuma counties, 2,500 feet or lower, in dry, saline soil. Arizona, southeastern California, Sonora, and Baja California.

21. Atriplex canescens (Pursh) Nutt. Throughout the state, 6,500 feet or lower, commonly in sandy, sometimes in saline, soil. South Dakota to Oregon, south to northern Mexico.

Four-wing salt-bush, cenizo, chamiso, chamiza, often erroneously called shad-scale and sagebrush. This plant is adapted to very diverse soil and climatic conditions, and is found in association with creosote-bush, sagebrush, pinyon, and sometimes ponderosa pine. It is highly prized as a browse plant, and the fruits are so relished by livestock that reproduction is often greatly hindered. The plant is deep-rooted and should be useful for erosion control.

22. Atriplex Garrettii Rydb. (*A. canescens* subsp. *Garrettii* Hall & Clements). Lees Ferry and vicinity (Coconino County), about 3,500 feet. Western Colorado, eastern Utah, and northern Arizona.

Atriplex Nuttallii Wats. Both subsp. *typica* Hall & Clements and subsp. *cuneata* (A. Nels.) Hall & Clements have been reported from Arizona, but there seems to be no satisfactory evidence of their occurrence in this state. The species bears considerable resemblance to *A. Jonesii* Standl.

6. ZUCKIA

A low, erect shrub, pubescent with inflated, whitish hairs; leaves alternate, petioled, entire; plants dioecious, the pistillate flowers without perianth, bibracteolate, sessile, solitary or in small clusters, forming short, interrupted, naked panicles of spikes; fruiting bractlets accrescent, united except for a small apical orifice, 6-keeled, 2 of the keels broader and winglike; stigmas 2, filiform, exserted; seed horizontal.

1. **Zuckia arizonica** Standl. Adamana (Apache County), Chalcedony Park (Petrified Forest?), Navajo County (*Zuck* 39, the type collection), near Tuba (Coconino County), about 5,000 feet. Northeastern Arizona (and eastern Utah?).

7. GRAYIA. Hop-sage

Low, branched shrubs with stiff, divergent branches; leaves oblanceolate, slightly fleshy; flowers unisexual, in glomerules, these forming terminal or axillary spikes or panicles; fruits closely subtended by a pair of conspicuous, thin, flat-winged, connivent bractlets, these united to the middle or higher; seeds vertical.

The spiny hop-sage (*G. spinosa*) is an excellent browse plant, relished by all livestock, but not sufficiently abundant in Arizona to be of much importance.

KEY TO THE SPECIES

1. Fruiting bractlets at maturity glabrous, not carinate, more than 6 mm. wide; plant often spiny .... 1. *G. spinosa*
1. Fruiting bractlets at maturity scurfy-pubescent, carinate, not more than 6 mm. wide; plant not spiny .... 2. *G. Brandegei*

1. **Grayia spinosa** (Hook.) Moq. Keam Canyon (Navajo County) and at several localities in Mohave County, 2,500 to 6,000 feet, flowering in spring. Wyoming to Washington, south to Arizona and California.

2. **Grayia Brandegei** Gray. Petrified Forest (Apache County), about 5,000 feet, in clay soil (*Hall* 11167). Colorado, Utah, and northeastern Arizona.

Hall's specimen has the bractlets broader and more retuse than in typical *G. Brandegei*.

8. EUROTIA. Winter-fat

Plants shrubby or suffrutescent, stellate-tomentose; leaves alternate, entire, linear; flowers unisexual or perfect, in axillary clusters and terminal spike-like inflorescences; perianth 4-parted; stamens 4; fruiting bractlets united into a villous, 2-beaked tube.

1. **Eurotia lanata** (Pursh) Moq. Apache County to Mohave County, south to Cochise and Pima counties, 2,000 to 7,000 feet, dry plains and mesas, usually among grasses, May to October. Saskatchewan to Washington, south to Texas, Arizona, California, and Mexico.

One of the most valuable native forage plants, especially as winter feed for sheep. As implied by the vernacular name, livestock fatten well on

Eurotia. The plant is also known (erroneously) as white sage. Mrs. Collom reported that the Indians used the plant medicinally, applying the powdered root to burns and treating fever with a decoction of the leaves. The plant avoids very saline soil. It is of value for controlling soil erosion.

Var. *subspinosa* (Rydb.) Kearney & Peebles (*E. subspinosa* Rydb.) occurs throughout the range of the species in Arizona and is apparently the only form present in the southern counties. As compared with typical *E. lanata* it has normally more woody stems, more spreading branches, and hairs with few or no elongate rays, but there is intergradation in all characters.

9. BASSIA

Annual, herbaceous; stems tall, much-branched; herbage loosely villous; leaves narrow, entire; flowers perfect, without bractlets, in open leafy-bracted panicles of short spikes; perianth lobes each bearing a dorsal tubercle or spine.

1. **Bassia hyssopifolia** (Pall.) Kuntze (*Echinopsilon hyssopifolius* Moq.). Navajo, Coconino, Mohave, Maricopa, and Yuma counties, flowering in summer. A weed, here and there in the western United States; introduced from Asia.

Smother-weed. This species is now abundant in Maricopa County and other parts of Arizona. According to L. L. Stitt (personal communication) it is a late-summer host of species of *Lygus*, insects that cause great damage to various crop plants.

10. KOCHIA

Plants annual or perennial, herbaceous or suffrutescent; leaves mostly alternate, narrow, entire; flowers perfect or pistillate, axillary, sessile, solitary or in small clusters; perianth subglobose, with incurved lobes, in fruit with 5 wedge-shaped, horizontal, distinct or confluent wings.

KEY TO THE SPECIES

1. Plants perennial, suffrutescent; stems numerous, mostly unbranched, not more than 0.5 m. high; leaves sessile, terete or nearly so, somewhat fleshy..........1. *K. americana*
1. Plants annual; stems much-branched, up to 1.5 m. high; leaves petiolate, flat, lanceolate, thin ....2. *K. scoparia*

1. **Kochia americana** Wats. Apache and Navajo counties, 5,000 to 6,000 feet, on dry, open plains, often in saline soil, June to August. Wyoming to California, northwestern New Mexico, and northeastern Arizona.

This species, sometimes known as green-molly and red-sage, was reported by Griffiths as sufficiently abundant in the valley of the Little Colorado River to furnish winter feed for sheep. Both the glabrate typical form and var. *vestita* Wats. (*K. vestita* Rydb.), which has densely and permanently sericeous-villous herbage, occur in Arizona.

2. **Kochia scoparia** (L.) Schrad. Tucson (*Thornber* in 1913), apparently not established. Here and there in the United States, an escape from gardens; native of Eurasia.

Summer-cypress. A very hairy form of this species, presumably var. *subvillosa* Moq., has been collected in waste land at Cameron, Coconino County (*J. T. Howell* 24403, *Deaver* 2510).

11. CORISPERMUM. Bug-seed

Plants annual, herbaceous, with branching stems; herbage glabrous or sparsely pubescent; leaves alternate, entire, narrow, 1-nerved; flowers perfect, in narrow, loose, terminal spikes, each flower subtended by a leaflike, scarious-margined bractlet broader than the foliage leaves; perianth usually of only 1 segment, this deciduous; stamens 1 to 3; styles 2, persistent; utricle narrowly winged; seed vertical.

These plants when mature behave as tumbleweeds.

KEY TO THE SPECIES

1. Spikes slender, loosely flowered; lower bractlets narrower than the fruit; fruit 2 to 3 mm. long..1. *C. nitidum*
1. Spikes stout, dense; lower bractlets usually as wide as or wider than the fruit; fruit 3.5 to 4.5 mm. long..2. *C. hyssopifolium*

1. Corispermum nitidum Kit. Apache County to eastern Mohave and Yavapai counties, 5,000 to 7,000 feet, usually in sandy soil, July to September. A widely distributed weed in the United States; naturalized from Europe.

2. Corispermum hyssopifolium L. Holbrook, Navajo County (*Zuck* 23, in 1903), Flagstaff, Coconino County (*Schallert* in 1943), Prescott, Yavapai County (*Thornber* 8612, in 1916), 5,000 to 7,000 feet. Ontario to Washington, south to Missouri, Arizona, and northern Mexico; introduced from Eurasia.

12. ALLENROLFEA. Iodine-bush

Plants fleshy, much-branched, woody toward the base; stems constricted at intervals, appearing jointed; leaves reduced to triangular scales; flowers perfect, in threes in the axils of spirally arranged bracts, forming dense, cylindric spikes.

1. Allenrolfea occidentalis (Wats.) Kuntze. Almost throughout the state, 5,000 feet or lower, July to November. Oregon to western Texas, Sonora, and Baja California.

A reliable indicator of strongly saline soil, to which the plant is mainly confined. Unpalatable to livestock and eaten only when other feed is lacking. The abundant pollen is said to cause hay fever in some persons. Other common names of this plant are pickle-weed and chico.

13. SARCOBATUS. Grease-wood

Plants monoecious, shrubby, up to 2.4 m. (8 feet) high, with spreading, rigid branches; leaves fleshy, entire, narrow; staminate flowers in catkinlike spikes, without a perianth, each subtended by a peltate, stipitate bractlet; pistillate flowers solitary or in small clusters, each flower enclosed in a perianth, this with a turbinate tube and a spreading, winglike limb much enlarged in fruit.

1. Sarcobatus vermiculatus (Hook.) Torr. Apache, Navajo, and Coconino counties, south to Pinal and Maricopa counties, 1,000 to 6,000 feet, in saline, usually moist soil, June to September, most abundant in northeastern Arizona, where it often covers large areas in pure stands or mixed with *Suaeda*. North Dakota to Alberta, south to western Texas, New Mexico, Arizona, and California.

A valuable browse plant, the young shoots and leaves being eaten by cattle and sheep in winter and spring; but bloating, and perhaps poisoning by the oxalates in the sap, may result from eating this forage too freely. The plant often indicates the presence in the soil of alkali carbonates, "black alkali." Various articles, such as planting sticks, are made from the wood by the Hopi Indians and it is used as fuel in their "kivas."

14. SUAEDA. Seep-weed, Quelite-salado

Plants mostly perennial, herbaceous or suffrutescent; leaves alternate, fleshy, mostly terete or subterete; flowers perfect or some of them unisexual, axillary, solitary or in small clusters; perianth 5-lobed or 5-parted, fleshy, enclosing the utricle; stamens 5.

These plants are indicators of moderate to excessive soil salinity. They are browsed to some extent when other feed is scarce. The young plants are used for greens by the Pimas and other Indians, and are sometimes eaten with cactus fruits. Pinole was made from the seeds. The dried leaves were applied to sores by the Hopi.

The taxonomy of the genus is perplexing.

KEY TO THE SPECIES

1. Perianth lobes (at least some of them) corniculate-appendaged or winged, unequal; plants annual, glabrous or very nearly so; leaves of the inflorescence broadly lanceolate or ovate-lanceolate. Stems erect and little-branched, or low and spreading 1. *S. depressa*
1. Perianth lobes not appendaged or winged, often carinate or cucullate, equal; plants perennial, suffrutescent, green or glaucous; leaves all linear or lance-linear, subterete or somewhat flattened (2).
2. Young stems and leaves copiously soft-pubescent; branches stout, commonly short and ascending at a narrow angle................................2. *S. suffrutescens*
2. Young stems and leaves commonly glabrous, sometimes pubescent; branches slender, often flexuous, commonly elongate and spreading or ascending at a wide angle 3. *S. Torreyana*

1. Suaeda depressa (Pursh) Wats. Near Holbrook, Navajo County (*Thornber* in 1929), near Tuba, Coconino County (*Darrow* 2762). Minnesota to Saskatchewan, south to western Texas, New Mexico, Arizona, and California.

2. Suaeda suffrutescens Wats. Navajo County and eastern Coconino County, 3,000 to 5,000 feet, March to July. Western Texas to Arizona and Chihuahua.

3. Suaeda Torreyana Wats. Apache County to Mohave County, south to Cochise, Pima, and Yuma counties, 5,000 feet or lower, very common, July to September. Alberta to Oregon, south to northern Mexico.

This species is closely allied to the Old World *S. fruticosa* (L.) Forsk., to which many of the Arizona specimens have been referred. The plants sometimes reach a height of 2.4 m. but are usually smaller. Var. *ramosissima* (Standl.) Munz (*Dondia ramosissima* Standl.) has been collected at Lees Ferry, eastern Coconino County (*E. W. Nelson* 62, the type collection), and near Yuma, where it grows on sand hills. It differs from typical *S. Torreyana* in the copiously short-pilose herbage.

15. SALSOLA

Annual much-branched herbs, becoming hard and prickly; leaves awl-shaped, spine-tipped; flowers perfect, each subtended by 2 bractlets, sessile, axillary; perianth 5-parted, in fruit with horizontal, scarious, dorsal wings; stamens commonly 5; styles 2; seed horizontal.

1. Salsola Kali L. Throughout the state. The Arizona form is var. *tenuifolia* Tausch (*S. pestifer* A. Nels.), which is extensively naturalized in the western United States; from Eurasia. It differs from the typical, maritime phase of the species in being a larger, more bushy plant with longer, narrower leaves and (usually) smaller fruiting calyces; but specimens with fruiting calyces 10 mm. in diameter have been collected in Arizona.

Russian-thistle. Abundant along roads in some of the irrigated districts of southern Arizona and on overgrazed ranges in the northern part of the state. In early spring the young plants are readily eaten by livestock, and the dead plants are eaten in winter after softening by rains. In case of need, good ensilage can be made from the mature plants, which otherwise are unpalatable. Hay sometimes is made of the young plants. The plant is a typical tumbleweed, breaking off at the surface of the ground when mature and piling up along fences. One of the Hopi Indian names signifies "white man's plant."

45. AMARANTHACEAE. Amaranth Family

Stems herbaceous or slightly woody below; leaves simple, entire, without stipules; flowers small, unisexual or some of them perfect, commonly in dense, bracteate heads or spikes; perianth and bracts hyaline, scarious, or rigid; stamens usually as many as the perianth segments, the filaments commonly more or less united; stigmas 1 to 3; ovary 1-celled; fruit a utricle, circumscissile or bursting irregularly.

KEY TO THE GENERA

1. Leaves alternate; anthers 4-celled, appearing 2-celled after dehiscence. Plants annual, without lanate pubescence (2).
1. Leaves opposite; anthers 2-celled, often appearing 1-celled after dehiscence (3).
2. Pistillate flowers with a perianth, not concealed by the bracts, these narrow, not cordate 1. *Amaranthus*
2. Pistillate flowers without a perianth, more or less concealed by broad, cordate, spine-tipped, scarious-margined bracts .... 2. *Acanthochiton*
3. Stamens adnate to the perianth, perigynous. Sepals united below, the free portion glabrous, the tube soft-pubescent .... 3. *Brayulinea*
3. Stamens free from the perianth, hypogynous (4).
4. Perianth segments united into a tube, this in fruit hardened and longitudinally crested, winged, or bearing longitudinal rows of spines .... 5. *Froelichia*
4. Perianth segments distinct or united only near the base, not forming an appendaged tube in fruit (5).
5. Flowers in few-flowered, rather loose axillary glomerules, these subtended by leaves with the bases becoming more or less hardened and united, forming a turbinate involucre; herbage densely stellate-pubescent .... 4. *Tidestromia*
5. Flowers in dense heads or spikes, these naked or if subtended by leaves, their bases not becoming hardened and united; pubescence not stellate. Bracts of the inflorescence thin, scarious, white, yellowish, or pink (6).

6. Stigma capitate. Perianth segments distinct, hirsute with jointed, barbed hairs 6. *Alternanthera*
6. Stigma with 2 or 3 subulate or filiform lobes (7).
7. Flowers perfect, in globose or ovoid heads, these terminal or both terminal and axillary, solitary or in dense subcapitate clusters . . . . 7. *Gomphrena*
7. Flowers unisexual, in a loose panicle of numerous slender spikes . . . . 8. *Iresine*

1. AMARANTHUS. Amaranth

Plants herbaceous, annual; leaves alternate, petioled, pinnate-veined; flowers mostly unisexual, small, commonly subtended by a bract and 2 bractlets; perianth segments 2 to 5, distinct; stamens 2 to 5; utricle 2- or 3-beaked by the persistent styles.

These plants are commonly known as pig-weed. Their seeds, produced in great abundance, are an important food supply for birds, such as the common dove, whitewing, and quail. Indians of several tribes gathered the young leaves for greens and the seeds for meal. In southern Arizona careless-weed (*A. Palmeri*), abundant in river bottoms, is sometimes cut for hay and is relished by stock in both the green and dry state. Most of the species are weeds of cultivated land and roadsides, flowering in summer. The pollen of the amaranths, notably of the very common *A. Palmeri*, is an important cause of hay fever in Arizona.

KEY TO THE SPECIES

1. Perianth segments of the pistillate flowers broadly spatulate, with a flabelliform or obovate blade considerably wider than the claw. Apex of the segments obtuse, truncate, or emarginate, often apiculate, exceptionally acutish (2).
1. Perianth segments of the pistillate flowers mostly linear, lanceolate, oblong, or elliptic, but occasionally narrowly spatulate and with a blade distinctly wider than the claw. Plants chiefly monoecious (5).
2. Utricle not circumscissile, narrow, nearly equaling the calyx. Plants monoecious, the staminate flowers often very few; leaves lanceolate or narrowly oblong, not more than 3 cm. long . . . . 1. *A. obcordatus*
2. Utricle circumscissile, subglobose, considerably shorter than the calyx (3).
3. Plants dioecious; inflorescence naked or nearly so; bracts and sepals becoming rigid and spinose. Petioles slender, equaling or longer than the broad, rhombic-ovate blades; flowers mostly in elongate, compound spikes, these often panicled; perianth segments often denticulate . . . . 2. *A. Palmeri*
3. Plants monoecious, the staminate flowers sometimes very few; inflorescence leafy, at least below; bracts and perianth segments not becoming rigid and spinose (4).
4. Axillary flower clusters much shorter than the petioles; leaves linear-lanceolate to ovate-lanceolate; bracts shorter than the flowers; perianth segments of the pistillate flowers usually with fimbriate margins . . . . 3. *A. fimbriatus*
4. Axillary flower clusters mostly equaling or longer than the petioles; leaves linear to elliptic; bracts equaling or longer than the flowers; perianth segments of the pistillate flowers with entire or denticulate margins . . . . 4. *A. Torreyi*
5. Utricle not circumscissile. Leaves ovate, up to 8 cm. long; flowers in a rather small panicle of slender spikes; perianth segments 3 . . . . 5. *A. viridis*
5. Utricle circumscissile (6).
6. Flowers all in small, glomerate or racemiform, leafy-bracted, axillary clusters (7).
6. Flowers mostly in terminal and axillary, compound spikes, these often clustered in a terminal, leafy or nearly naked panicle. Leaves mostly rhombic-ovate; species of similar appearance, difficult to distinguish (8).
7. Stems commonly prostrate and purplish; bracts ovate or oblong-ovate, cuspidate, little surpassing the perianth segments . . . . 6. *A. graecizans*

7. Stems commonly erect or ascending and whitish; bracts lanceolate or subulate, aristate, much surpassing the perianth segments.............................7. *A. albus*
8. Perianth segments of the pistillate flowers (or some of them) acute to aristate-acuminate at apex, erect, shorter than to slightly surpassing the utricle (9).
8. Perianth segments of the pistillate flowers mostly obtuse or truncate at apex (11).
9. Seeds usually whitish and dull, but sometimes dark red or black, shining; perianth segments of the pistillate flowers mostly oblong or ovate, acute to aristate 8. *A. leucocarpus*
9. Seeds black or dark brown, shining; perianth segments of the pistillate flowers lanceolate, tapering into a terminal awn or cusp (10).
10. Bracts not more than twice as long as the perianth segments; seeds orbicular 9. *A. hybridus*
10. Bracts 2 to 3 times as long as the perianth segments; seeds broadly oval or obovate in outline ....10. *A. Powellii*
11. Plants glabrous or very nearly so; stems slender, usually less than 50 cm. long; spikes slender, more or less interrupted, leafy. Perianth segments of the pistillate flowers about equaling the utricle..................................11. *A. Wrightii*
11. Plants villous, at least in the inflorescence; stems stout, usually more than 50 cm. long; spikes ample, usually dense (12).
12. Utricle considerably surpassing the perianth segments, the latter 1.5 mm. long, remaining erect ....12. *A. cruentus*
12. Utricle equaling to considerably shorter than the perianth segments, the latter 3 mm. long, spreading at maturity...................................13. *A. retroflexus*

1. Amaranthus obcordatus (Gray) Standl. Cochise and Pima counties, 2,000 to 3,500 feet, apparently not common. Western Texas to Arizona and northern Mexico.

2. Amaranthus Palmeri Wats. Coconino, Yavapai, and Greenlee counties, south to Cochise, Santa Cruz, Pima, and Yuma counties, 5,500 feet or lower. Kansas to Texas, Arizona, California, and central Mexico.

Careless-weed, red-root, quelite, bledo. A tall, coarse, weedy plant, abundant in river bottoms and irrigated land.

3. Amaranthus fimbriatus (Torr.) Benth. Western Coconino County and Mohave County, south to Greenlee, Gila, Cochise, Pima, and (probably) Yuma counties, below 4,000 feet, commonly in sandy washes, type from the Gila River (*Schott*). Southern Utah and Nevada to Arizona and northwestern Mexico.

The typical plant, with fimbriate perianth segments, is much the more common in Arizona, but var. *denticulatus* (Torr.) Uline & Bray (*A. venulosus* Wats.), with these entire or denticulate, has been collected at Tucson, Pima County (*Toumey, Thornber*), and at Casa Grande, Pinal County (*Jones*).

4. Amaranthus Torreyi (Gray) Benth. (*A. Pringlei* Wats.). Cochise, Santa Cruz, and Pima counties, 3,500 to 5,500 feet. Western Texas to Arizona and northern Mexico.

5. Amaranthus viridis L. (*A. gracilis* Desf.). Sacaton (Pinal County), a casual introduction from the tropics, probably not established in Arizona.

6. Amaranthus graecizans L. (*A. blitoides* Wats.). Navajo, Coconino, and Mohave counties, south to Cochise, Santa Cruz, and Pima counties, 1,000 to 8,000 feet, a common roadside weed. Widely distributed in North America.

7. Amaranthus albus L. (*A. graecizans* of authors, non L.). Apache County to Coconino County, south to Cochise and Pima counties, 1,500 to 8,000 feet. Widely distributed in North America.

A characteristic tumbleweed, common at roadsides and in fields. Typical *A. albus* has the herbage glabrous, or puberulent but not viscid. Var. *pubescens* (Uline & Bray) Fern. (*A. graecizans* var. *pubescens* Uline & Bray), which in Arizona apparently is confined to the northern counties, has viscid-pubescent herbage. The type of the variety was collected at Flagstaff (*Jones* 3978).

8. Amaranthus leucocarpus Wats. (*A. caudatus* of authors). "Pueblo," Apache or Navajo County (*Stevenson* 7), the type of *A. leucocarpus* grown from "Arizona seeds" collected by the Powell Expedition, without definite locality. Occasional in the United States; introduced from tropical America.

Love-lies-bleeding, often grown in old-fashioned gardens.

9. Amaranthus hybridus L. Jacob Lake and Havasu Canyon (Coconino County), near Prescott (Yavapai County), Salt River Valley (Maricopa County), where a common weed in cultivated land, near Sacaton (Pinal County), 1,000 to 8,000 feet. Widely distributed in North America.

10. Amaranthus Powellii Wats. Apache, Navajo, and Coconino counties, south to Cochise and Yavapai counties, 3,000 (?) to 8,000 feet, the type "cultivated from seeds brought from Arizona by Col. Powell." Wyoming to Oregon, Chihuahua, and Arizona.

Very similar to *A. hybridus,* but "commonly it may be distinguished by being a more slender and lower plant with much simpler less floriferous inflorescences, having stiffer somewhat longer bracts, and a green rather than tawny color" (Johnston, 56, p. 157).

11. Amaranthus Wrightii Wats. San Francisco Peaks and Wupatki National Monument, Coconino County (*Knowlton* 205, *D. J. Jones* 65), reported also from Fort Verde, Yavapai County (*Mearns* 277). Southern Colorado, New Mexico, and Arizona.

12. Amaranthus cruentus L. Hopi Indian Reservation, Navajo County (*Hough* 44, *Mrs. Colton* in 1933), possibly also in the Pinaleno Mountains, Graham County, 8,000 feet (*Shreve* 5367). Occasional in the United States; introduced from tropical America.

The Hopi are reported to use the seeds for coloring corn bread pink for certain ceremonies.

13. Amaranthus retroflexus L. Apache County to Coconino and Yavapai counties, 5,000 to 7,000 feet. Southern Canada to northern Mexico.

A collection at Fort Apache, southern Navajo County (*Palmer* 587), was identified by Johnston (56, p. 157) as var. *salicifolius* Johnst., characterized by more slender habit than in typical *A. retroflexus,* and lanceolate leaves often 3 to 4 times as long as wide.

2. ACANTHOCHITON

Plants annual, glabrous; stems striped green and white, erect, branched; plants commonly dioecious, the staminate flowers in glomerules forming spikes, with a perianth of 5 segments and bractless, the pistillate flowers without a perianth, subtended by cordate bracts, these becoming spiny.

1. Acanthochiton Wrightii Torr. Holbrook, Navajo County (*Zuck* in 1897), Hopi Indian Reservation and Little Colorado River region (*Hough* 60). Western Texas to Arizona and Chihuahua.

While young the plants are relished by livestock. They are also eaten by the Indians of northern Arizona, both as greens and when dried and stored. This plant is a host of the beet leaf hopper.

3. BRAYULINEA

Plants perennial; stems numerous, much branched, prostrate from a thick root, forming mats; leaves opposite, ovate, of very unequal size in the pair, lanate-pubescent beneath, as is the inflorescence; flowers perfect, in dense axillary glomerules; perianth 5-lobed; utricle indehiscent.

1. Brayulinea densa (Humb. & Bonpl.) Small. Greenlee, Graham, Gila, Cochise, Santa Cruz, and Pima counties, 2,500 to 6,000 feet, May to October. Western Texas to Arizona, south to South America.

4. TIDESTROMIA

Plants annual or perennial, pubescent with branched hairs; stems erect to prostrate, branched, herbaceous, or woody toward the base; leaves mostly opposite; flowers perfect, in small axillary clusters, the perianth yellow; stamens 5, the filaments united, with intervening staminodia sometimes present; utricle compressed.

KEY TO THE SPECIES

1. Plants annual; stems procumbent or prostrate, radiating from the root; larger leaves broadly obovate to spatulate, the veins not prominent beneath; perianth commonly 2 to 3 mm. long; staminodia minute or wanting..................1. *T. lanuginosa*
1. Plants perennial; stems erect, ascending, or decumbent, often woody below; larger leaves suborbicular, ovate, elliptic, or oblong, the veins prominent beneath; perianth less than 2 mm. long; staminodia triangular, nearly half as long as the filaments of the fertile stamens..................2. *T. oblongifolia*

1. Tidestromia lanuginosa (Nutt.) Standl. Throughout the state, 5,500 feet or lower, June to October. Western Kansas to Utah, south to northern Mexico.

The whitish mats of this plant are conspicuous soon after summer rains on the deserts in southern Arizona, and are well adapted for checking the blowing of sandy soils. This is one of the host plants of the beet leaf hopper.

2. Tidestromia oblongifolia (Wats.) Standl. Mohave and Yuma counties, 5,000 feet or lower, in sandy soil, June to October. Arizona, Nevada, and southeastern California.

Plants often shrubby, up to about 0.6 m. high.

5. FROELICHIA. Snake-cotton

Plants herbaceous, annual or perennial, lanate-tomentose; stems erect, sparingly branched; flowers perfect, subtended by glabrous, dark-colored bracts, in somewhat elongate glomerules, these forming terminal panicles or interrupted spikes; stamens 5, the filaments united; utricle indehiscent.

The plants are relished by livestock.

KEY TO THE SPECIES

1. Plants annual; stems slender, simple or branched at base, leafy only near the base, commonly not more than 30 cm. long; leaves thin, lanceolate, 1 cm. wide or narrower 1. *F. gracilis*

1. Plants perennial, with a thick, woody root; stems stout, often sparingly branched above the base, commonly leafy well above the base, often much more than 30 cm. long; leaves thick, commonly oblanceolate and 1 to 2 cm. wide....................2. *F. arizonica*

1. Froelichia gracilis (Hook.) Moq. Navajo and Yavapai counties southeastward to Cochise County, 4,500 to 5,500 feet, summer. Iowa to Colorado, south to Chihuahua and Arizona.

2. Froelichia arizonica Thornber. Gila, Cochise, Santa Cruz, and Pima counties, 3,500 to 5,500 feet, dry, grassy plains and rocky slopes, late summer, type from the Santa Rita Mountains (*Griffiths & Thornber* 73). Western Texas, southern Arizona, and northern Mexico.

6. ALTERNANTHERA

Plants herbaceous, perennial, with a thick, woody, vertical root; stems prostrate or procumbent, forming mats; leaves opposite, oval or obovate, those of the pair very unequal; flowers perfect, in short axillary spikes, with conspicuous white bracts; perianth segments 5, pubescent with stiff, jointed hairs, these minutely barbed at apex.

1. Alternanthera repens (L.) Kuntze (*Achyranthes repens* L.). Cochise, Santa Cruz, and Pima counties, 2,500 to 5,500 feet, summer. South Carolina to Arizona, south to tropical America.

A roadside weed, resembling *Brayulinea densa,* but with larger leaves and flower spikes.

7. GOMPHRENA. GLOBE-AMARANTH

Plants herbaceous; stems leafy or scapose; flowers perfect, in globose or ovoid heads, conspicuously subtended by white, pink, or yellowish scarious bracts and bractlets; perianth 5-lobed or 5-parted.

The plants, which grow on dry plains and slopes, usually with grasses, are eaten freely by cattle and probably other livestock.

KEY TO THE SPECIES

1. Bractlets crested along the keel with a laciniate-dentate crest. Stems leafy; leaves elliptic, ovate, or obovate; spikes usually solitary, commonly subtended by 2 or more leaves; bractlets thin but firm, yellowish white or pink......................1. *G. nitida*

1. Bractlets not crested (2).

2. Plants annual or perennial, not cespitose; stems leafy, 15 to 60 cm. long; leaves commonly narrow, elliptic, lanceolate, or oblanceolate; spikes subtended by 2 or more leaves, usually clustered; bractlets thin but firm (scarious), strongly carinate, entire, cream-colored, pale orange, or pink...............................2. *G. sonorae*

2. Plants perennial with a deep, woody root, cespitose; stems usually scapelike, not more than 15 cm. long; leaves oblanceolate, obovate, or nearly orbicular; spikes not subtended by leaves, usually solitary; bractlets very thin and soft (hyaline), not strongly carinate, often denticulate at apex, white.........................3. *G. caespitosa*

1. Gomphrena nitida Rothr. Cochise, Santa Cruz, and Pima counties, 4,000 to 6,000 feet, August and September, type from the Chiricahua Mountains (*Rothrock* 520). Western Texas to southeastern Arizona and northern Mexico.

2. Gomphrena sonorae Torr. Gila County to Cochise, Santa Cruz, and Pima counties, 3,000 to 5,500 feet, August and September. Arizona and northern Mexico.

3. Gomphrena caespitosa Torr. Coconino County to Greenlee, Cochise, Santa Cruz, and Pima counties, 3,500 to 5,000 feet, April to August. New Mexico, Arizona, and northern Mexico.

Sometimes known as ball-clover. *G. viridis* Woot. & Standl. appears to be merely a greener, less pubescent form.

8. IRESINE. Blood-leaf

Plants dioecious, perennial, herbaceous; stem tall, erect, leafy; leaves opposite, petioled, with thin, broad blades; flowers small, white, in loose terminal panicles of spikes; calyx 5-parted; utricle indehiscent.

1. Iresine heterophylla Standl. Pinal, Cochise, Santa Cruz, and Pima counties, 3,500 to 4,500 feet, usually among trees and bushes, late summer. Western Texas and southern Arizona to central Mexico.

46. NYCTAGINACEAE (91). Four-o'clock Family

Plants annual or perennial, herbaceous or suffrutescent; leaves (in the Arizona genera) opposite, often very unequal in the pair; flowers perfect, subtended by bracts, these distinct or united in a calyxlike involucre; perianth with the free upper portion corollalike and campanulate, funnelform, or salverform, the lower part of the tube persistent, becoming hardened, and closely investing the fruit, forming an anthocarp; ovary 1-celled, appearing inferior but technically superior; ovule solitary; anthocarp usually grooved, ribbed, or winged.

Species of *Bougainvillea* and of *Mirabilis* (four-o'clock) are well-known cultivated ornamentals.

KEY TO THE GENERA

1. Cotyledon 1, by abortion; stigma narrowly linear, this and the stamens included in the perianth tube (2).
1. Cotyledons 2; stigma globose or hemispheric (3).
2. Limb of the perianth 5-lobed; wings of the fruit thickish, opaque, interrupted above and below the body of the fruit....................................10. *Abronia*
2. Limb of the perianth 4- or 5-lobed; wings of the fruit thin, nearly transparent, conspicuously reticulate-veined, continuous around the body of the fruit 11. *Tripterocalyx*
3. Wings of the fruit 3 to 5, conspicuous, scarious (4).
3. Wings of the fruit none or coriaceous, sometimes (in genus *Boerhaavia*) the angles narrowly winglike, subscarious (5).
4. Free portion of the perianth tubular-funnelform; stamens 5 or 6, attached to the lower part of the perianth tube....................................1. *Selinocarpus*
4. Free portion of the perianth broadly campanulate; stamens 2 or 3, free from the perianth ....2. *Ammocodon*
5. Fruit strongly compressed, oval or obovate in outline, the (usually dentate) margins commonly strongly inflexed over the dorsal (outer) face................6. *Allionia*
5. Fruit not compressed, terete or angled, without inflexed, dentate margins (6).
6. Floral bracts more or less united into a calyxlike involucre. Stamens and pistil more or less exserted (7).
6. Floral bracts distinct, remaining small (8).
7. Fruits not strongly 5-angled (sometimes noticeably 5-ribbed), usually not constricted at base; involucre in fruit scarcely enlarged, remaining leaflike in texture 4. *Mirabilis*
7. Fruits strongly 5-angled longitudinally, constricted at base; involucre in fruit greatly enlarged, thin, conspicuously veined..............................5. *Oxybaphus*

8. Flowers solitary, axillary; bracts persistent........................3. *Acleisanthes*
8. Flowers in umbels, cymes, or racemes; bracts usually soon deciduous (9).
9. Fruits with not more than 5 angles or ribs; free portion of the perianth campanulate to nearly rotate, with scarcely any tube............................7. *Boerhaavia*
9. Fruits 10-ribbed; free portion of the perianth funnelform, with a distinct tube (10).
10. Fruits narrowly clavate, wingless, conspicuously beset with wartlike, stipitate glands 8. *Commicarpus*
10. Fruits biturbinate, with a rigid, horizontal, median, annular wing, glandless. Ribs of the fruit frequently winged or verrucose........................9. *Anulocaulis*

1. SELINOCARPUS

Plants perennial; stems low, diffusely branched; leaves thickish, ovate; flowers few, axillary and solitary, or in short-stalked terminal leafy clusters; free portion of the perianth 3 cm. long, or longer; stamens 5 or 6; fruit conspicuously winged.

*1. **Selinocarpus diffusus** Gray. Not known definitely to occur in Arizona, but has been collected near St. George, Utah. Western Texas to southern Nevada.

2. AMMOCODON

Plants perennial, with the habit of *Selinocarpus;* leaves broadly ovate; flowers several or numerous, in simple or compound cymes, these usually long-stalked; free portion of the perianth not more than 5 mm. long; stamens 2 or 3; fruit conspicuously winged.

1. **Ammocodon chenopodioides** (Gray) Standl. (*Selinocarpus chenopodioides* Gray). Duncan, Greenlee County (*Davidson* in 1900), Chiricahua Mountains, Cochise County (*Lemmon* in 1881), Tucson, Pima County (*Thornber* 259), 2,500 to 4,000 feet, late summer. Western Texas to southern Arizona and Chihuahua.

3. ACLEISANTHES

Plants perennial; stems low and spreading from a woody root; leaves deltoid-ovate to narrowly lanceolate, usually acuminate; flowers axillary, solitary, the subtending bracts small, distinct, persistent; perianth salverform, the free portion 8 cm. long or longer, the tube very slender; stamens 2 to 5, often exserted; fruit angled or ribbed.

1. **Acleisanthes longiflora** Gray. Salt River Mountains, Maricopa County (*Bailey* in 1913), Table Top Mountain, Pinal County (*Harrison & Kearney* 7299), Plomosa Mountains, Yuma County (*Peebles & Fulton* 8515), 2,000 to 3,000 feet, among rocks, April to August. Texas to California and northern Mexico.

Plant night-blooming, the flowers fragrant. It is known in California as yerba-de-la-rabia, and sometimes as angel-trumpet.

4. MIRABILIS. Four-o'clock

Plants herbaceous, mostly perennial; leaves mostly petioled, deltoid-ovate to narrowly lanceolate, usually acuminate; flowers solitary or several in the calyxlike involucre; perianth salverform or funnelform; stamens 3 to 5; fruit smooth or tuberculate, sometimes 5-ribbed.

KEY TO THE SPECIES

1. Involucre subtending more than 1 flower. Perianth purplish red (2).
1. Involucre subtending a single flower (3).
2. Perianth not more than 1 cm. long; involucre subrotate, 3-flowered; perianth campanulate-funnelform; stamens 3, the filaments distinct; herbage copiously viscid-villous to glabrate .... 1. *M. oxybaphoides*
2. Perianth 4 to 6 cm. long; involucre campanulate, usually more than 3-flowered; perianth tubular-funnelform; stamens 5, the filaments connate at base; herbage puberulent, commonly not viscid... 2. *M. multiflora*
3. Perianth about 1 cm. long, less than 3 times as long as the involucre, salverform-campanulate, white or pinkish; fruit smooth, longitudinally striate but not angled or ribbed .... 3. *M. Bigelovii*
3. Perianth at least 3 cm. long, 3 or more times as long as the involucre, funnelform or elongate-salverform; fruit rugose-tuberculate, obtusely angled or ribbed (4).
4. Stamens not conspicuously exserted; perianth bright red or purplish red, not more than 6 cm. long... 4. *M. Jalapa*
4. Stamens conspicuously exserted; perianth white or tinged with purple, 7 to 17 cm. long 5. *M. longiflora*

1. Mirabilis oxybaphoides Gray (*Quamoclidion oxybaphoides* Gray). Apache, Navajo, and Coconino counties, 6,000 to 8,000 feet, August and September, apparently rare in Arizona. Colorado and Utah to western Texas and northern Arizona.

2. Mirabilis multiflora (Torr.) Gray (*Quamoclidion multiflorum* Torr.). Almost throughout the state, 2,500 to 6,500 feet, on hillsides and mesas, often among rocks and shrubs, April to September. Colorado and Utah to northern Mexico.

A handsome plant with large magenta-purple flowers and dark-green foliage. According to Mrs. Collom the powdered root is used as a remedy for stomach-ache. It is reported that the Hopi Indians eat the root to induce visions.

3. Mirabilis Bigelovii Gray (*Hesperonia Bigelovii* Standl., *H. glutinosa gracilis* Standl.). Coconino and Mohave counties to Pima and Yuma counties, 3,000 feet or lower, rocky slopes, March to October, type from the Grand Canyon (*Gray* in 1885), type of *H. glutinosa gracilis* from Sabino Canyon, Pima County (*Toumey* 471c). Southern Utah to Arizona and southeastern California.

A straggling, weak-stemmed plant with pale-pink flowers and viscid, pilose or villous herbage. A less pubescent variety, with scabrous-puberulent stems (var. *retrorsa* (Heller) Munz, *Hesperonia retrorsa* Standl.) occurs in Mohave, Gila, and Pinal counties, probably elsewhere.

4. Mirabilis Jalapa L. Cave Creek, Chiricahua Mountains (*Harrison & Kearney* 6132), probably an escape from cultivation. Native of tropical America. This is the well-known four-o'clock of old-fashioned gardens.

5. Mirabilis longiflora L. Apache and Yavapai counties to Cochise, Santa Cruz, and Pima counties, 2,500 to 7,000 feet, rich soil among trees and shrubs, August and September. Western Texas to Arizona and far southward in Mexico.

Plant remarkable for the very long and slender perianth tube of the white or pinkish flowers. There occur in Arizona both the typical plant, with short-villous, very viscid stems and sessile or subsessile upper leaves, and, more commonly, var. *Wrightiana* (Gray) Kearney & Peebles, which has puberulent, scarcely viscid stems and all the leaves usually distinctly petioled.

5. OXYBAPHUS

Plants perennial, herbaceous; stems tall and erect, or low and decumbent; leaves usually thickish; involucre calyxlike, gamophyllous, becoming enlarged and papery in fruit; flowers 1 to 5 in each involucre; perianth campanulate or short-funnelform, slightly oblique; fruits more or less obovoid, 5-angled.

KEY TO THE SPECIES

1. Leaves not more than 4 times as long as wide, short-cuneate, truncate, or cordate at base; petioles mostly elongate, well differentiated from the blade. Stems pilose or villous above, or throughout, with soft, weak hairs, often also glandular; fruit short-pilose (2).
1. Leaves 5 or more times as long as wide, narrowly linear to broadly lanceolate, acutish to long-attenuate at base; petioles short or almost none. Stems normally tall, erect, and not branched below the inflorescence (3).
2. Stems low, decumbent or nearly prostrate, much-branched, viscid, and pilose or villous to the base; leaves deltoid-ovate to nearly orbicular, not or scarcely longer than wide; perianth pale pink or purplish, pubescent............................1. *O. pumilus*
2. Stems normally tall and erect, commonly not branched below the inflorescence, glabrous below or puberulent in lines (seldom on the whole surface); leaves elongate-deltoid (sometimes deltoid-ovate or ovate-lanceolate), commonly 1.5 to 4 times as long as wide; perianth purplish red, rarely pale pink, often nearly glabrous....2. *O. comatus*
3. Perianth bright red, funnelform, 3 to 4 times as long as the copiously strigose or short-pilose involucre. Stems glaucous, glabrous or inconspicuously strigose, only the peduncles bearing spreading hairs..............................3. *O. coccineus*
3. Perianth pink or purple, usually pale-colored, campanulate, about twice as long as the involucre (4).
4. Stems, involucre, perianth, and fruit glabrous or sparsely strigose. Leaves narrow, elongate, attentuate at both ends, thick...............................4. *O. glaber*
4. Stems usually glandular, and pilose or villous, above; involucre pilose or villous, often glandular; perianth more or less pubescent; fruit copiously strigose or pilose 5. *O. linearis*

1. Oxybaphus pumilus (Standl.) Standl. (*Allionia pumila* Standl., *A. pachyphylla* Standl.). Coconino, Mohave, Gila, and Yavapai counties, 3,000 to 7,500 feet, June to September, type of *A. pumila* from Kingman, Mohave County (*Lemmon* in 1884), type of *A. pachyphylla* from the Grand Canyon (*Toumey* 485). New Mexico, Arizona, Nevada, and southeastern California.

A collection in the Santa Catalina Mountains, Pima County, south of the main area of *O. pumilus* in Arizona (*Toumey* 484), is referred doubtfully to this species. It has an exceptionally deeply cleft involucre, with narrowly triangular teeth.

2. Oxybaphus comatus (Small) Weatherby (*Allionia comata* Small, *A. pratensis* Standl., *A. melanotricha* Standl.). Coconino County to Greenlee, Cochise, Santa Cruz, and Pima counties, 3,500 to 9,000 feet, May to October, types of *A. pratensis* and *A. melanotricha* from the Chiricahua Mountains (*Blumer* 1384 and 1385, respectively). Western Texas and Arizona, south into Mexico.

3. Oxybaphus coccineus Torr. (*Allionia coccinea* Standl., *A. gracillima scabridata* (Heimerl) Standl.). Southern Apache and Coconino counties to Cochise, Santa Cruz, and Pima counties, 4,000 to 6,500 feet, open pine woods and grassy slopes, May to August, type of *A. gracillima scabridata* from the Santa Rita Mountains (*Pringle*). New Mexico, Arizona, and Sonora.

The showiest of the Arizona species, with brilliant, carmine-red flowers.

4. Oxybaphus glaber Wats. (*Allionia glabra* Kuntze). Hopi Indian Reservation, Navajo County (*Zuck* in 1897, *Hough* 53), August. Kansas to Utah, south to Chihuahua and northeastern Arizona.

The Indians are reported to have treated wounds with an infusion of the leaves.

5. Oxybaphus linearis (Pursh) Robins. (*Allionia linearis* Pursh). Apache County to Mohave County, south to Cochise and Pima counties, 4,500 to 9,500 feet, often in pine woods, July to September. South Dakota and Montana to northern Mexico.

Occurs in two forms, about equally common throughout the area of distribution in Arizona. These are: (1) the typical form with linear to narrowly lanceolate, sessile or subsessile leaves; (2) var. *decipiens* (Standl.) Kearney & Peebles (*Allionia decipiens* Standl.), with broader, sometimes ovate-lanceolate, short-petioled leaves.

6. ALLIONIA

Plants herbaceous, annual or perennial, usually glandular-pubescent; stems prostrate; leaves petioled, very unequal in the pair, oblong to broadly ovate; involucres solitary on axillary peduncles, 3-flowered; perianth campanulate-rotate; fruit flattened, the dorsal (seemingly the inner) face bearing 2 rows of stipitate glands.

KEY TO THE SPECIES

1. Fruit with a broad entire or subentire wing on the inner side, the margin as in *A. incarnata;* perianth 6 mm. long . . . . 1. *A. cristata*
1. Fruit not winged on the inner side; perianth more than 6 mm. long (2).
2. Plants perennial; margin of the fruit strongly incurved, usually with about 6 broadly triangular teeth, seldom entire . . . . 2. *A. incarnata*
2. Plants annual; margin of the fruit spreading or moderately incurved, with more numerous, relatively slender, gland-tipped teeth . . . . 3. *A. Choisyi*

1. Allionia cristata (Standl.) Standl. (*Wedelia cristata* Standl.). Known only from the type collection at Holbrook, Navajo County (*Zuck* in 1896).

2. Allionia incarnata L. Throughout most of the state, 6,000 feet or (usually) lower, April to October. Colorado and Utah to southern Mexico; South America.

Trailing-four-o'clock. A conspicuous plant on open plains, mesas, and slopes, with long, trailing stems and showy, rose-purple (occasionally white) flowers.

3. Allionia Choisyi Standl. (*A. glabra* Standl., non Kuntze). Apache, Navajo, Coconino, Yavapai, and Cochise counties, 4,000 to 6,000 feet, July to October. Western Texas to Arizona, south to Oaxaca.

7. BOERHAAVIA. Spiderling

Plants herbaceous, annual or perennial; stems usually branched, often with a viscid band around each internode; leaves petioled, usually very unequal in the pair, linear-lanceolate to nearly orbicular; flowers small, mostly in terminal racemes or cymes, the perianth limb campanulate to nearly rotate; fruits obpyramidal or clavate, the ribs sometimes winged, the furrows between the ribs rugose.

The plants usually grow where exposed to full sunlight, but sometimes in open chaparral, flowering in late summer and autumn.

KEY TO THE SPECIES

1. Fruit pubescent, the hairs more or less spreading; plants perennial, with a woody caudex; perianth carmine or dark red (2).
1. Fruit glabrous or with strigose hairs in the furrows; plants annual; perianth not carmine or dark red (3).
2. Flowers solitary on long, slender pedicels; plants glabrous or obscurely puberulent, not glandular ....1. *B. gracillima*
2. Flowers in glomerules, sessile or on short pedicels; plants usually densely glandular-puberulent in the inflorescence, often more or less hirsute below......2. *B. coccinea*
3. Flowers in elongate racemes, these forming a cymose or paniculate inflorescence (4).
3. Flowers not in racemes, the inflorescence cymose or cymose-paniculate (7).
4. Bracts persistent, more than half as long as the fruit; fruit 4- (rarely 5-) angled. Herbage viscid-villous..3. *B. Wrightii*
4. Bracts deciduous, much less than half as long as the fruit; fruit 5-angled (5).
5. Flowers crowded; bracts broadly ovate or obovate, usually much surpassing the ovary at anthesis; stems conspicuously viscid-villous below the inflorescence..4. *B. spicata*
5. Flowers not crowded; bracts lanceolate or ovate-lanceolate, not or scarcely surpassing the ovary; stems not conspicuously villous. Species of very similar appearance (6).
6. Fruit with narrow ridges and open furrows, these conspicuously transverse-rugose 5. *B. Torreyana*
6. Fruit with broad ridges and narrow or nearly closed furrows, these scarcely or not conspicuously transverse-rugose ....6. *B. Coulteri*
7. Inflorescence glandular-villous; bracts large, equaling or surpassing the fruit, persistent. Flowers short-pedicelled or nearly sessile, in dense glomerules, these borne on long, slender peduncles; fruit with narrow ridges and very broad furrows 7. *B. purpurascens*
7. Inflorescence glabrous or puberulent, the internodes often ringed with a viscid band; bracts very small, much shorter than the fruit (8).
8. Fruit not winged; cymules not dense, umbelliform or racemiform, the flowers borne on pedicels often more than 2 mm. long (9).
8. Fruit conspicuously winged; cymules dense, the flowers borne on pedicels not more than 2 mm. long (10).
9. Stems erect, often more than 50 cm. long; ultimate divisions of the inflorescence subumbellate to subracemose, the pedicels of unequal length; fruits 3 to 4 mm. long 8. *B. erecta*
9. Stems erect or decumbent, usually less than 50 cm. long; ultimate divisions of the inflorescence umbellate, the pedicels subequal; fruits commonly 2 to 2.5 mm. long 9. *B. intermedia*
10. Wings of the fruit 4 (rarely 3); fruit abruptly contracted into a short, winged stipe, the body coarsely transverse-rugose; stems decumbent or procumbent 10. *B. pterocarpa*
10. Wings 5; fruit not stipitate, the body smooth or nearly so; stems erect or ascending 11. *B. megaptera*

1. **Boerhaavia gracillima** Heimerl. Pinaleno Mountains (Graham County), Huachuca Mountains (Cochise County), Santa Catalina, Baboquivari, and Comobabi mountains (Pima County), 2,500 to 4,500 feet, April to September. Western Texas to Arizona and Mexico.

The specimens from Pima County have longer, narrower, and thinner leaves than those from the Huachuca Mountains. A remarkable plant from the Santa Catalina Mountains (*Thornber* in 1910) has the pubescence of the fruits nonglandular, as in *B. gracillima,* but the flowers are often in pairs. In the stiff, apparently erect stems it differs from all species of this group.

2. **Boerhaavia coccinea** Mill. (*B. caribaea* Jacq.). Mohave and Yavapai counties to Cochise, Santa Cruz, and Pima counties, up to 7,000 feet but usually lower, common at roadsides and in fields, April to November. Widely distributed in tropical and subtropical America.

With its long, trailing stems and viscid herbage this is sometimes an annoying weed in gardens.

3. **Boerhaavia Wrightii** Gray. Coconino and Mohave counties to Cochise, Pima, and Yuma counties, 4,000 feet or lower, July to September. Western Texas to Nevada, Arizona, and northwestern Mexico.

A collection near the base of the Gila Mountains, Yuma County (*Harrison & Kearney* 6257), approaches *B. triquetra* Wats. of Baja California in its turbinate, truncate, narrowly wing-angled fruits, and in having only 2 stamens, but is nearer *B. Wrightii* in other characters. This and other specimens from Yuma County may represent an undescribed species.

4. **Boerhaavia spicata** Choisy. Pinal and Pima counties, 1,500 to 2,500 feet. Arizona and northern Mexico.

5. **Boerhaavia Torreyana** (Wats.) Standl. Navajo, Coconino, Yavapai, Gila, and Maricopa counties, 1,000 to 5,000 feet. Western Texas and Coahuila to southern California.

6. **Boerhaavia Coulteri** (Hook.f.) Wats. Coconino and Mohave counties to Graham, Cochise, Pima, and Yuma counties, 500 to 5,000 feet. Arizona, southern California, Sonora, and Baja California.

One of the commonest and most widely distributed species of *Boerhaavia* in Arizona.

7. **Boerhaavia purpurascens** Gray. Gila, Pinal, Cochise, Santa Cruz, and Pima counties, 3,500 to 5,500 feet, often in chaparral and on limestone. Western Texas to Arizona and Sonora.

8. **Boerhaavia erecta** L. (*B. Thornberi* Jones). Yavapai, Cochise, Santa Cruz, Pima, and Yuma counties, 1,000 to 5,000 feet, type of *B. Thornberi* from Tucson (*Thornber* 10). Widely distributed in tropical and subtropical America.

9. **Boerhaavia intermedia** Jones (*B. universitatis* Standl., *B. erecta* var. *intermedia* Kearney & Peebles). Mohave, Gila, Cochise, and Pima counties (doubtless elsewhere), 1,000 to 4,500 feet, type of *B. universitatis* from Tucson (*Thornber* in 1903). Western Texas to southeastern California and northern Mexico.

This species intergrades with *B. erecta,* but most of the specimens are readily distinguishable.

10. Boerhaavia pterocarpa Wats. Cochise and Pima counties, 2,500 to 4,500 feet, at roadsides, type from Apache Pass, Cochise County (*Lemmon* in 1881). Arizona and Sonora.

11. Boerhaavia megaptera Standl. Pima County from Tucson westward, 2,500 to 4,500 feet, type from the Tucson Mountains (*Thornber* 162). Known only from southern Arizona.

8. COMMICARPUS

Plants suffrutescent; stems leafy, long and weak, usually supported on other plants; leaves petioled, subequal in the pair, broadly ovate, usually more or less cordate; flowers in umbels; perianth less than 5 mm. long, yellowish white; fruits narrowly clavate, bearing stipitate glands, these rarely wanting.

1. Commicarpus scandens (L.) Standl. Greenlee, Maricopa, Pinal, Cochise, and Pima counties, 2,000 to 4,500 feet, March to October. Western Texas, Arizona, and southward; widely distributed in tropical America.

9. ANULOCAULIS

Plants herbaceous, perennial, with a tuberous-thickened root, subscapose; stems usually with a broad viscid band on each internode; leaves mostly basal, leathery, broadly ovate to suborbicular, more or less cordate; perianth funnelform, white, 2 to 3 cm. long; filaments purple; fruits biturbinate, glandless, with a horizontal, thickish, median wing, the longitudinal ribs also often winged or verrucose.

1. Anulocaulis leisolenus (Torr.) Standl. Havasu Canyon (Coconino County), Detrital Wash (Mohave County), near Fort Verde (Yavapai County), about 3,500 feet, in calcareous soil. Western Texas to southern Nevada and central Arizona.

The basal leaves attain a length and width of 15 cm.

10. ABRONIA. Sand-verbena

Plants herbaceous, annual or perennial, often viscid-pubescent; stems mostly decumbent or prostrate, leafy or scapose; leaves petioled, lanceolate, elliptic, or ovate; flowers numerous, in heads subtended by conspicuous, distinct, thin bracts; perianth funnelform-salverform, the tube slender and elongate, the limb small; fruits deeply lobed or winged.

The sand-verbenas are attractive plants, with bright-pink to white, more or less fragrant flowers. They are conspicuous in spring in sandy, open places, often in dry beds of streams.

KEY TO THE SPECIES

1. Perianth purplish red; plants annual; bracts linear-lanceolate to ovate-lanceolate, attenuate-acuminate, herbaceous or not conspicuously scarious, greenish. Leaves prevailingly ovate (2).

1. Perianth with a white or pink limb and a greenish tube; plants perennial, the root often thickened and woody; bracts elliptic to suborbicular or obovate, obtuse to short-acuminate, conspicuously scarious, whitish. Fruits commonly with winglike lobes (3).

2. Stems copiously to densely viscid-villous; fruits mostly broadly winged, the central cavity not extending to the edge of the wings, the wings often surpassing the beak of the fruit and rounded or acutish at apex.........................1. *A. villosa*

2. Stems viscid-puberulent (commonly also sparsely villous); fruits mostly with winglike lobes, the central cavity extending to the edge of the lobes, the lobes considerably surpassed by the beak of the fruit and truncate or subtruncate at apex
2. *A. angustifolia*

3. Plants acaulescent or nearly so, cespitose; petioles much longer than the blades. Herbage sparsely villous and often hirtellous, commonly glabrescent in age; leaves elliptic-lanceolate to elliptic-ovate; peduncles slender, scapelike, 7 to 15 cm. long. . . . 3. *A. nana*

3. Plants leafy-stemmed; petioles mostly shorter to little longer than the blades (4).

4. Perianth 13 to 18 mm. long. Herbage viscid-puberulent and sparsely short-pilose to glabrescent, not villous . . . . 4. *A. pumila*

4. Perianth 18 to 25 mm. long (5).

5. Stems viscid-villous, at least above, usually copiously so; bracts oval, commonly narrowly so, acute or acuminate at apex; fruits thick-walled, olive or brownish, mostly biturbinate (the lobes narrowed at apex), up to 10 mm. long, considerably longer than wide . . . . 5. *A. fragrans*

5. Stems viscid-puberulent or glabrate, sometimes also sparsely villous; bracts broadly oval, suborbicular, or obovate, rounded and often apiculate at apex; fruits thin-walled, straw-colored, mostly simply turbinate (the lobes truncate at apex), seldom more than 6 mm. long, little if any longer than wide . . . . 6. *A. elliptica*

1. Abronia villosa Wats. Mohave, Maricopa, Pima, and Yuma counties, 1,500 feet or lower, February to May, type from "Arizona" (*Wheeler Expedition*). Arizona, southern California, and Sonora.

2. Abronia angustifolia Greene (*A. lobatifolia* Standl.). Graham, Maricopa, Pinal, and Cochise counties, 1,000 to 4,000 feet, March to July (occasionally September), type of *A. lobatifolia* from "Arizona" (*Palmer* in 1869). Western Texas and Chihuahua to Arizona.

The species is represented in Arizona by var. *arizonica* (Standl.) Kearney & Peebles (*A. arizonica* Standl.), which is usually more pubescent and broader-leaved than typical *A. angustifolia.* Specimens without fruit are difficult to distinguish from *A. villosa.* The type of *A. arizonica* was obtained in Arizona (*Vasey* in 1882, without definite locality).

3. Abronia nana Wats. (*A. nana* var. *lanciformis* Jones). Coconino and Mohave counties, especially in the Grand Canyon region, 3,000 to 5,500 feet, April to August, type of var. *lanciformis* from Hackberry (*Jones* 4689). Southern Utah and Nevada and northern Arizona.

Specimens having a bright-pink perianth limb have been collected at the Grand Canyon (*Collom* in 1940).

4. Abronia pumila Rydb. Northern Navajo and eastern and northern Coconino counties, 2,500 to 7,000 feet, August and September. Southern Utah and Nevada, and northern Arizona.

A specimen from the Grand Canyon (*Collom* in 1940) is atypical, having exceptionally large leaves and oblanceolate floral bracts, these less than 3 mm. wide.

5. Abronia fragrans Nutt. North of Ganado, Apache County (*Peebles* 13497), east of Tuba, Coconino County (*Harrison & King* 8715), 5,000 to 6,000 feet. South Dakota to Idaho, south to Chihuahua and northern Arizona.

6. Abronia elliptica A. Nels. (*A. fragrans* var. *elliptica* Jones, *A. ramosa* Standl.). Apache County to Mohave County, 2,000 to 7,000 feet, common in northeastern Arizona, April to September, type of *A. ramosa* from Holbrook (*Ward* in 1901). Wyoming to northern New Mexico and Arizona.

Perhaps only varietally distinct from *A. fragrans* Nutt., with which it seems to intergrade in Arizona.

11. TRIPTEROCALYX

Plants similar to *Abronia* but often with a 4-lobed perianth limb and with the body of the fruit harder, spindle-shaped, completely surrounded by the 2 to 4 broad, thin, nearly transparent, conspicuously reticulate-veined wings.

The fruits, with their large fish-scale-like wings, are conspicuous.

KEY TO THE SPECIES

1. Perianth less than 2 cm. long, sparsely pubescent or glabrate, the limb greenish or pink 1. *T. micranthus*
1. Perianth more than 2 cm. long, copiously pubescent, the limb pink outside, white within 2. *T. Wootonii*

1. Tripterocalyx micranthus (Torr.) Hook. Mohave County, at Pierce Ferry, 1,500 feet (*Jones* 5077av), and Pipe Springs, 4,500 feet (*Peebles* 13068), April to June. North Dakota and Montana to Kansas, Nevada, and Arizona.

The Arizona plant is *T. pedunculatus* (Jones) Standl. (*Abronia micrantha* var. *pedunculata* Jones), which seems at most to be only varietally distinct from *T. micranthus*.

2. Tripterocalyx Wootonii Standl. Apache County to Coconino County, 4,500 to 6,000 feet, May to September. Northern New Mexico and Arizona.

A specimen collected by C. B. Carter (No. 202, in 1926) is labeled as from near Texas Canyon (Cochise County), 3,500 feet, but the occurrence of this species so far to the south seems doubtful.

47. PHYTOLACCACEAE. Pokeberry Family

Plants perennial, herbaceous or nearly so; stems leafy; leaves alternate, petioled, entire; flowers mostly perfect, small, in racemes; perianth of 4 or 5 distinct or nearly distinct segments, whitish; ovary superior; style 1, or styles several; fruit berrylike, juicy, of 1 or several 1-seeded carpels.

KEY TO THE GENERA

1. Perianth segments and stamens commonly 4; fruit 1-seeded.................. 1. *Rivina*
1. Perianth segments normally 5, stamens 9 to 12; fruit several-seeded...... 2. *Phytolacca*

1. RIVINA. Rouge-plant, Pigeon-berry

Plant herbaceous or barely suffrutescent; stems erect, up to 1 m. high; leaves ovate, acute or acuminate; racemes terminal and axillary; perianth white or pinkish; style and stigma 1; fruit red or yellow at maturity.

1. Rivina humilis L. Greenlee County to Maricopa County, south to Cochise, Santa Cruz, and Pima counties, 1,500 to 4,500 feet, ravines and canyons, mostly in shade, June to October. Florida to Arizona, south into tropical America.

In Mexico a red dye is obtained from the fruits, and the leaves are reported to be used medicinally. The plant is worth cultivating as an ornamental.

2. PHYTOLACCA. Pokeberry

Stems tall, usually branched, very juicy; leaves large, ovate-lanceolate, acute or acuminate at both ends; racemes terminal but appearing opposite the leaves as growth of the stem continues; styles and stigmas several, the styles recurved; fruit depressed-globose, dark purple (nearly black) at maturity.

1. Phytolacca americana L. Chiricahua Mountains (Cochise County), near Patagonia (Santa Cruz County), 4,000 to 6,000 feet, August, perhaps introduced from farther east. Maine and Ontario to Florida, Texas, and Arizona.

Also known as poke-weed and scoke. The roots and berries contain a bitter substance, probably saponin, and are poisonous, but the succulent young shoots are esteemed as a potherb and are harmless when thoroughly cooked. The root has been used medicinally.

48. AIZOACEAE. Carpet-weed Family

Plants annual, often succulent, with low or prostrate, usually branched stems; leaves opposite or nearly so, or in whorls, entire; flowers perfect, axillary, solitary or in small clusters, inconspicuous; perianth segments 5, distinct or united below; petals none; ovary superior, 1- to 5-celled; fruit a capsule.

The largest genus of this family is *Mesembryanthemum,* of which several species, including the ice-plant (*M. crystallinum*) are in cultivation as ornamentals.

KEY TO THE GENERA

1. Capsule dehiscent by valves; perianth of 5 distinct segments, these not appendaged .... 1. *Mollugo*
1. Capsule circumscissile; perianth 5-lobed, the lobes usually with a thick dorsal ridge ending in a free, hornlike tip, the margins thin and petallike (2).
2. Styles 3 to 5; ovary 3- to 5-celled; seeds many........................2. *Sesuvium*
2. Styles 1 or 2; ovary 1- or 2-celled; seeds few........................3. *Trianthema*

1. MOLLUGO

Plants not fleshy; stems prostrate or erect; leaves narrow, mostly in whorls of 3 to 6; sepals white within; stamens 3 to 5; capsules longitudinally dehiscent, 3- to 5-celled.

KEY TO THE SPECIES

1. Stems commonly prostrate, radiating from the root, not filiform, green; stem leaves mostly oblanceolate and more than 1 mm. wide; seeds reniform, red-brown, very shiny, with several parallel ribs along the back and sides...........1. *M. verticillata*
1. Stems erect or ascending, filiform, straw-colored; stem leaves linear or linear-lanceolate, commonly about 1 mm. wide; seeds irregularly obovoid, dark brown, scarcely shiny, finely reticulate..2. *M. Cerviana*

1. Mollugo verticillata L. Yavapai, Greenlee, Cochise, Santa Cruz, and Pima counties, 2,500 to 5,000 feet, commonly in sandy soil, August to October. Widely distributed in North America; Eastern Hemisphere.

Carpet-weed, sometimes called Indian-chickweed.

2. Mollugo Cerviana (L.) Seringe. Coconino County to Cochise and Pima counties, 1,500 to 7,000 feet, sandy soil, August to October. Texas to California, south to tropical America; Eastern Hemisphere.

2. SESUVIUM. SEA-PURSLANE

Plants fleshy; leaves opposite, narrow, usually oblanceolate; stipules none; petioles more or less connate at base; perianth 5-lobed, the lobes purplish within, the tube turbinate; stamens numerous, inserted on the perianth tube.

1. Sesuvium verrucosum Raf. (*S. sessile* Robins.). Maricopa, Pinal, and Yuma counties, 1,500 feet or lower, saline soil, March to November. Arkansas to California, south to tropical America.

3. TRIANTHEMA

Plants somewhat fleshy; leaves with stipules, opposite, those of the pair very unequal in size, obovate, cuneate at base; perianth purplish within, the lobes concave, with a hornlike dorsal appendage; stamens 6 to 10; capsule crested.

1. Trianthema Portulacastrum L. Havasu Canyon (Coconino County), and Greenlee, Maricopa, Pinal, Cochise, and Pima counties, 1,000 to 4,000 feet, June to October. Widely distributed in tropical and subtropical America; Eastern Hemisphere.

A common weed in irrigated land in southern Arizona, but easily controlled by cultivation. Locally known as pigweed, a name applied to several unrelated plants. This plant is a host of the beet leaf hopper in Arizona.

Series 2. Choripetalae (Families 49 to 101)

KEY TO THE FAMILIES[1]

1. Corolla distinctly irregular (2).
1. Corolla regular or nearly so (11).
2. Leaves compound, rarely reduced to a single leaflet (3).
2. Leaves simple, sometimes palmately lobed or parted (4).
3. Sepals 2; corolla conspicuously spurred, not pealike; leaves decompound, with numerous narrow segments: genus *Corydalis*............................55. *Papaveraceae*
3. Sepals or calyx lobes more than 2; corolla not or inconspicuously spurred, usually pealike (reduced to 1 petal in *Amorpha*); leaves variously compound......64. *Leguminosae*
4. Carpels normally more than 1, distinct, in fruit becoming several-seeded follicles. Leaves palmately cleft to parted; flowers showy, mostly blue or bluish, very irregular, the sepals larger and usually more highly colored than the petals: genera *Delphinium*, *Aconitum* ....52. *Ranunculaceae*
4. Carpel solitary or if more than one, then the carpels united to form a single fruit (5).
5. Stamens many more than 12. Plants herbaceous; leaves palmately cleft or parted; flowers large and showy; fruit a large, thick-walled capsule....89. *Cochlospermaceae*
5. Stamens not more than 12 (6).
6. Plants large shrubs; flowers appearing before the leaves, somewhat pealike. Leaves entire, round-cordate; fruit a flat 2-valved pod: genus *Cercis*......64. *Leguminosae*
6. Plants herbaceous, or low shrubs; flowers appearing with the leaves, not pealike (7).
7. Petals and stamens borne on the throat of the calyx, this gamophyllous, asymmetric, and enlarged in fruit: genus *Cuphea*..........................96. *Lythraceae*
7. Petals and stamens free from the calyx, this of distinct or nearly distinct sepals (8).
8. Stipules present; stamens 5, the anthers connivent or slightly cohering, the filaments distinct or almost none..91. *Violaceae*
8. Stipules none or reduced to glands; stamens more or fewer than 5, the anthers distinct, the filaments more or less united (9).
9. Petals 5, two of them fleshy and glandlike; fruit spiny, indehiscent: genus *Krameria* 64. *Leguminosae*

[1] Two families that belong properly to series Sympetalae are included in this key because some of their members have the petals distinct.

9. Petals usually fewer than 5, none glandlike; fruit not spiny (10).
10. Stamens 3; petals 2; capsule turgid, 4-lobed........................58. *Resedaceae*
10. Stamens 6 or 8; petals commonly 3; capsule more or less compressed, entire or merely notched at apex..73. *Polygalaceae*
11. Ovary inferior, at least the lower part distinctly adnate to the calyx tube or hypanthium (12).
11. Ovary superior, free from the calyx or very nearly so (25).
12. Stems very thick and succulent, flat or cylindric, with spines, or barbed bristles, or both, these borne on cushionlike areoles; leaves none, or greatly reduced and terete; perianth segments and stamens indefinitely numerous................94. *Cactaceae*
12. Stems not succulent or, if so, then not areolate; leaves mostly well developed; petals not more, usually fewer than 10 (13).
13. Flowers very small, in umbels or dense round heads, or, if in slender interrupted spikes, then the leaves orbicular and peltate (14).
13. Flowers not in umbels or dense round heads, if the inflorescence subcapitate, then the plant armed with barbed, stinging hairs (15).
14. Styles 5; fruit a several-seeded berry..............................99. *Araliaceae*
14. Styles 2; fruit of 2 closely contiguous, 1-seeded carpels............100. *Umbelliferae*
15. Herbage very rough-pubescent, the hairs commonly barbed, sometimes stinging 93. *Loasaceae*
15. Herbage not rough-pubescent, or the hairs never barbed or stinging (16).
16. Stems herbaceous or merely suffrutescent (17).
16. Stems woody (21).
17. Plants aquatic; flowers minute; leaves (at least the immersed ones) finely dissected 98. *Haloragaceae*
17. Plants not aquatic or, if so, then the flowers showy; leaves not finely dissected (18).
18. Stems and leaves more or less succulent; fruit a circumscissile capsule: genus *Portulaca* 49. *Portulacaceae*
18. Stems and leaves not succulent; fruit not circumscissile (19).
19. Styles and (or) stigmas 2 to 4; ovary of 2 to 4 partly distinct or completely united carpels; fertile stamens 5 or 10..............................60. *Saxifragaceae*
19. Style 1; ovary entire; stamens of some other number (20).
20. Stamens 2, 4, or 8; fruit a 2- to 5-celled capsule, or sometimes indehiscent and nutlike 97. *Onagraceae*
20. Stamens 3; fruit 1-celled, large and gourdlike: genus *Apodanthera*..130. *Cucurbitaceae*
21. Fruit dry, follicular or capsular (22).
21. Fruit a more or less fleshy drupe, pome, or berry (23).
22. Stamens 8 or more..60. *Saxifragaceae*
22. Stamens 5 or fewer..81. *Rhamnaceae*
23. Flowers small, in many-flowered compound cymes; leaves opposite, simple, entire; fruit a drupe, the stone containing 1 or 2 seeds: genus *Cornus*...........101. *Cornaceae*
23. Flowers in relatively few-flowered racemes or corymbs or if in many-flowered compound cymes, then the leaves alternate and pinnate; fruit usually several-seeded (24).
24. Stamens not more than 5; ovary 1-celled; fruit a berry; leaves simple, palmately lobed: genus *Ribes*..60. *Saxifragaceae*
24. Stamens usually more than 5; ovary normally 2- or more-celled; fruit a pome (applelike), or sometimes appearing berrylike; leaves simple or pinnately compound 63. *Rosaceae*
25. Anthers opening by terminal valves or pores (26).
25. Anthers opening longitudinally (28).
26. Plants shrubs or undershrubs; leaves compound; stamens 6; ovary 1-celled; fruit a berry..53. *Berberidaceae*
26. Plants herbaceous or nearly so; leaves simple; stamens 8 or more; ovary several-celled; fruit a capsule (27).
27. Flowers somewhat irregular, the petals of unequal width and the stamens dimorphic; leaves palmately cleft or parted; stems from a large, tuberlike root; stamens numerous ....89. *Cochlospermaceae*

27. Flowers nearly or quite regular; leaves not lobed (sometimes reduced to scales and the plant without chlorophyll); root not tuberlike; stamens 8 to 12 103. *Ericaceae*
28. Flowers in dense globose heads, very numerous and minute; plant a tree; leaves large, palmately lobed...61. *Platanaceae*
28. Flowers not in dense heads, or the plant not a tree with palmately lobed leaves (29).
29. Flowers with a more or less thickened disk under or surrounding the ovary, this often bearing the stamens, the disk sometimes obsolete in Aceraceae. Plants mostly woody (30).
29. Flowers (except sometimes in Saxifragaceae and Euphorbiaceae) without a disk, or this rudimentary, or represented by distinct glands or scales (43).
30. Stems climbing or trailing, with tendrils, herbaceous or suffrutescent; flowers relatively large; calyx throat bearing a conspicuous, fringed corona.........92. *Passifloraceae*
30. Stems not climbing or trailing or, if so, then the flowers small and the calyx throat not with a fringed corona (31).
31. Leaves reduced to small scales. Plants shrubby or arborescent; flowers small (32).
31. Leaves with well-developed blades (33).
32. Plants dioecious, with rigid, intricate, spine-tipped branches; leaves not crowded, deciduous; flowers solitary or glomerate; fruit a circle of 5 or more, nearly distinct, dry, 1-seeded drupes, long-persistent.........................70. *Simaroubaceae*
32. Plants with perfect flowers, the branches not rigid or spine-tipped; leaves crowded, overlapping, persistent; flowers in elongate spikelike inflorescences; fruit a 3- to 5-valved capsule...88. *Tamaricaceae*
33. Herbage punctate with translucent oil glands. Plants mostly shrubby or arborescent; fruit a 2- or 3-lobed capsule or a flat, nearly orbicular samara..........69. *Rutaceae*
33. Herbage not glandular-punctate or if obscurely so (in Burseraceae), then the fruit drupaceous (34).
34. Fruit a pair of laterally winged samaras. Leaves simple and palmately lobed, or palmately divided, or pinnate with few leaflets; trees or large shrubs.....79. *Aceraceae*
34. Fruit otherwise (35).
35. Plants vinelike, with tendrils, the stems climbing or trailing. Fruit a berry..82. *Vitaceae*
35. Plants not vinelike, tendrils none (36).
36. Pistil 1, with 3 styles, these distinct or nearly so. Leaves compound or simple; flowers small, paniculate; fruit a small drupe..........................77. *Anacardiaceae*
36. Pistils 1 to many, each with 1 style, or the styles more or less united, or the stigma sessile (37).
37. Leaves compound (38).
37. Leaves simple (40).
38. Fruits dry achenes or follicles, or the achenes or drupelets aggregated in a fleshy, compound fruit, or enclosed in an enlarged hypanthium; plants herbaceous or shrubby 63. *Rosaceae*
38. Fruits otherwise; plants large shrubs or small trees (39).
39. Fruit dry, drupelike, trigonous; bark and foliage strong-scented......71. *Burseraceae*
39. Fruit fleshy, globose, with translucent pulp; bark and foliage not strong-scented 80. *Sapindaceae*
40. Stamens more than 10, usually numerous (41).
40. Stamens 10 or fewer. Pistil 1, or exceptionally 2 in genus *Glossopetalon* (42).
41. Fruits follicular (dehiscent on one suture only); flowers solitary; leaves entire, without stipules; pistils 2 to 5; seeds with a fimbriate aril; plants shrubby 62. *Crossosomataceae*
41. Fruits various, if follicular then the inflorescence several- to many-flowered; leaves entire to deeply lobed, often with stipules; pistils 1 to numerous; seeds not arillate; plants nearly herbaceous to arborescent............................63. *Rosaceae*
42. Stamens (or in *Glossopetalon* the longer ones) alternate with the petals; fruit capsular or follicular...78. *Celastraceae*
42. Stamens opposite the petals; fruit a somewhat fleshy drupe, or a capsule 81. *Rhamnaceae*
43. Filaments united, at least at base (44).

43. Filaments distinct (52).
44. Carpels adnate to a central column from which they become more or less detached at maturity; stamens 10, the filaments united only at base............65. *Geraniaceae*
44. Carpels not adnate to and separating from a central column or if so (in Malvaceae), then the stamens usually indefinitely numerous and the filaments united well above the base (45).
45. Leaves compound, often sensitive (46).
45. Leaves simple (47).
46. Stamens more conspicuous than the small petals; leaves bipinnate, with several or numerous, small leaflets; fruit a flat, 2-valved pod or a loment, with several 1-seeded segments: subfamily *Mimosoideae*............................64. *Leguminosae*
46. Stamens less conspicuous than the petals; leaves digitate, with 3 or more wedge-shaped, obcordate leaflets; fruit a turgid 5-celled capsule.................66. *Oxalidaceae*
47. Flowers mostly unisexual. Pubescence (if any) usually of stellate hairs or the hairs dolabriform (2-armed and affixed at the middle); stamens several or many: genera *Croton* and *Ditaxis*....................................74. *Euphorbiaceae*
47. Flowers all or some of them perfect (48).
48. Pubescence (if any) of simple hairs. Filaments united only at base (49).
48. Pubescence, at least partly, of forked or stellate hairs (50).
49. Fertile stamens 5; ovary 4- or 5- (or apparently 8- or 10-) celled; leaves not punctate 67. *Linaceae*
49. Fertile stamens numerous; ovary 3-celled, or 1-celled with 3 placentas; leaves punctate 86. *Guttiferae*
50. Leaves opposite, entire; stems trailing or twining, woody below; fruit nutlike or a pair of samaras..72. *Malpighiaceae*
50. Leaves alternate; stems not trailing or twining; fruit otherwise (51).
51. Stamens much more numerous than the calyx lobes; anthers 1-celled; fruit of several or numerous finally separating carpels, or a several-celled capsule.....84. *Malvaceae*
51. Stamens (the fertile ones) of the same number as the calyx lobes; anthers 2- or 3-celled; fruit a 1- to 5-celled capsule..........................85. *Sterculiaceae*
52. Pistils 2 or more, distinct or united at base (53).
52. Pistil 1, simple or compound (57).
53. Sepals united below into a hypanthium, this surrounding at least the base of the ovary. Plants herbaceous; leaves compound: genera *Potentilla, Agrimonia*....63. *Rosaceae*
53. Sepals distinct or (in Crassulaceae) united at base (54).
54. Stems twining; flowers unisexual, small. Sepals and petals 6, each in 2 series; fruit a drupe ....54. *Menispermaceae*
54. Stems not twining; flowers (except in genus *Thalictrum*) mostly perfect (55).
55. Plants evidently succulent or if only slightly so, then the stems appearing winged when dry. Pistils 3 to 5; fruits follicular.......................59. *Crassulaceae*
55. Plants not succulent; stems not appearing winged (56).
56. Fruit of achenes or of several-seeded follicles...................52. *Ranunculaceae*
56. Fruit of several torulose carpels, these at maturity breaking transversely into indehiscent, 1-seeded segments. Leaves simple, entire; flowers solitary, on long peduncles; sepals 3; petals 6, in 2 series: genus *Platystemon*................55. *Papaveraceae*
57. Pistil evidently compound, the ovary more or less lobed and the carpels tending to separate at maturity, at least toward apex (58).
57. Pistil simple or if compound, then the ovary entire and the carpels not separating (60).
58. Styles and stigmas 5, the styles more or less united. Leaves palmately lobed, or pinnate with pinnatifid leaflets.......................................65. *Geraniaceae*
58. Styles and (or) stigmas fewer than 5 (59).
59. Stigmas 2 to 4; leaves simple..................................60. *Saxifragaceae*
59. Stigma 1, entire or obscurely lobed; leaves compound...........68. *Zygophyllaceae*
60. Leaves compound (61).
60. Leaves simple, entire to pinnatifid or if 1 to 3 times pinnate (genus *Descurainia*) then the stamens 6, tetradynamous (65).
61. Fruit berrylike; flowers small, in racemes or panicles (62).

61. Fruit a dry, 2-valved pod or a several-jointed loment (63).
62. Stems erect; leaves large, decompound; stamens numerous: genus *Actaea* 52. *Ranunculaceae*
62. Stems trailing or climbing, with tendrils; leaves digitate, with 3 to 7 large leaflets; stamens 5: genus *Parthenocissus*.................................82. *Vitaceae*
63. Sepals 2 or 3, distinct, deciduous; leaves ternately decompound. Stamens numerous: genus *Eschscholtzia*..55. *Papaveraceae*
63. Sepals or calyx lobes 4 or 5; leaves otherwise (64).
64. Leaves digitately 2- to 5-foliolate; plants herbaceous..............57. *Capparidaceae*
64. Leaves pinnate or bipinnate, the leaflets often numerous; plants woody or herbaceous: subfamilies *Mimosoideae* and *Caesalpinioideae*...................64. *Leguminosae*
65. Plants shrubby or arborescent. Sepals or calyx lobes 4 to 6 (66).
65. Plants herbaceous, or sometimes suffrutescent (68).
66. Leaves with well-developed blades, these narrow, emarginate at apex, otherwise entire, lepidote beneath; plant with rigid but not spiny branches. Fruit berrylike: genus *Atamisquea* ....57. *Capparidaceae*
66. Leaves reduced to small scales; plants very spiny (67).
67. Petals and stamens 5; fruit a 5-valved capsule: genus *Canotia*......78. *Celastraceae*
67. Petals 4; stamens 8; fruit a globose 2-celled berry..............90. *Koeberliniaceae*
68. Sepals united most of their length, the calyx cylindric to campanulate (69).
68. Sepals distinct or nearly so (70).
69. Stamens 10, these and the petals free from the calyx; styles 2 to 5..50. *Caryophyllaceae*
69. Stamens 4 to 6, these and the petals borne on the throat of the calyx; style 1 96. *Lythraceae*
70. Petals more numerous than the sepals, the latter commonly 2 or 3 (71).
70. Petals not more numerous than the sepals, the latter (except in Elatinaceae) 4 or 5 (72).
71. Plants mostly succulent; leaf margins entire; styles or style branches slender, introrsely stigmatic..49. *Portulacaceae*
71. Plants not succulent; leaf margins not entire; styles very short or obsolete, the stigmas large, terminal, more or less united...........................55. *Papaveraceae*
72. Stamens 6, tetradynamous (2 shorter, 4 longer, barely so in genus *Stanleya*). Fruit a 2-valved, more or less dehiscent pod or else 1-celled, flat, indehiscent, and containing a single seed...56. *Cruciferae*
72. Stamens of some other number, not tetradynamous (73).
73. Fruit a long, slender, 2-valved pod. Petals yellow; stamens commonly 10; style 1 83. *Tiliaceae*
73. Fruit otherwise (74).
74. Stamens more than 10; petals yellow or salmon-colored; leaves punctate 86. *Guttiferae*
74. Stamens 10 or fewer; petals whitish or pink; leaves not punctate (75).
75. Plants scapose, the leaves nearly all basal; staminodia (aborted stamens) present, alternating with the 5 fertile stamens. Flowers solitary, on long 1-leaved scapes: genus *Parnassia*...60. *Saxifragaceae*
75. Plants leafy-stemmed; staminodia none (76).
76. Ovary commonly 1-celled; sepals and petals 4 or 5; plants not aquatic 50. *Caryophyllaceae*
76. Ovary 2- to 4-celled; sepals and petals 2 to 4; plants semiaquatic........87. *Elatinaceae*

49. PORTULACACEAE. Portulaca Family

Plants mostly small and herbaceous (seldom suffrutescent), annual or perennial, often succulent; leaves simple, entire; flowers perfect, regular or very nearly so; sepals commonly 2; petals mostly 4 or 5 or (in *Lewisia*) the sepals and petals more numerous; stamens few or numerous; style cleft or divided, the branches introrsely stigmatic; ovary superior or partly inferior, 1-celled; fruit a circumscissile or longitudinally dehiscent capsule.

KEY TO THE GENERA

1. Capsule circumscissile (2).
1. Capsule valvate, splitting downward from the top (3).
2. Calyx of distinct sepals, free from the ovary; capsule circumscissile near the base and splitting upward longitudinally; plants subacaulescent, the leaves mostly basal 3. *Lewisia*
2. Calyx 2-lobed, the tube adherent to the ovary; capsule circumscissile near the middle, the calyx lobes coming away with the top of the capsule; plants caulescent, with leafy stems . . . . 7. *Portulaca*
3. Stigmas and valves of the capsule 2; inflorescence more or less scorpioid. Plants annual 4. *Calyptridium*
3. Stigmas and valves of the capsule 3; inflorescence not scorpioid (4).
4. Sepals deciduous. Plants perennial . . . . 1. *Talinum*
4. Sepals persistent (5).
5. Petals 5 to 7; stem leaves alternate; stamens 5 to 12, usually more numerous than the petals. Plants annual . . . . 2. *Calandrinia*
5. Petals and stamens 5; stem leaves opposite (6).
6. Plants perennial, with a large globose corm; stem leaves 1 pair, not connate; ovules 6 5. *Claytonia*
6. Plants annual, or perennial with runners ending in bulblets; stem leaves several pairs or, if one pair, then connate-perfoliate; ovules 3 . . . . 6. *Montia*

1. TALINUM (92)

Plants perennial, glabrous; stems leafy or scapose; leaves alternate, the blades broad and flat to narrow and nearly terete; inflorescence paniculate or cymose; stamens 5 to numerous; style 3-cleft; ovary superior.

KEY TO THE SPECIES

1. Lower leaves 1 to 5 cm. wide, flat; inflorescence a many-flowered, elongate, open panicle, its leaves mostly reduced to small bracts. Petals commonly pink, not more than 6 mm. long; flowering stems 25 cm. long or longer, from a thick, tuberous root 1. *T. paniculatum*
1. All the leaves less than 5 mm. wide; inflorescence a few-flowered cyme, or the flowers solitary in the leaf axils (2).
2. Flowers in terminal cymes. Petals pink (3).
2. Flowers solitary, or in very few-flowered axillary cymes (5).
3. Stamens 10 or more; stems and scapes spreading, less than 10 cm. long 2. *T. validulum*
3. Stamens 4 to 8; stems and scapes erect or ascending, commonly at least 10 cm. long (4).
4. Sepals obtuse to short-acuminate, not or obscurely cuspidate . . . . 3. *T. parviflorum*
4. Sepals bearing a stout, erect, apical-dorsal, hornlike cusp . . . . 4. *T. Gooddingii*
5. Stems not more than 10 cm. long; leaves crowded, very thick, less than 15 mm. long, commonly spatulate; pedicels not reflexed in fruit; petals lavender or rose-pink. Plant with aspect of *Sedum* . . . . 5. *T. brevifolium*
5. Stems commonly more than 10 cm. long; leaves not crowded or very thick, 10 to 60 mm. long, linear or narrowly lanceolate; pedicels reflexed in fruit; petals yellow or orange (6).
6. Petals yellow; capsules globose or nearly so; stems woody below, with exfoliating bark 6. *T. angustissimum*
6. Petals orange or copper-colored; capsules ovoid; stems herbaceous or barely suffrutescent 7. *T. aurantiacum*

1. Talinum paniculatum (Jacq.) Gaertn. Greenlee County to Cochise, Santa Cruz, and Pima counties, 3,500 to 5,500 feet, rich soil among rocks in partial shade, July to September. Florida to Arizona, south to tropical America.

Easily distinguished from the other Arizona species of this genus by its relatively tall, leafy stems, broad, flat leaves, and large, open panicle of small flowers.

2. Talinum validulum Greene. Coconino County, in the Tusayan (now the Kaibab) National Forest (*Hill* in 1912, the type collection), and at Rattlesnake Tanks, 6,000 feet (*Leiberg* 5966). Known only from Arizona.

3. Talinum parviflorum Nutt. Showlow (Navajo County), near Inscription House (Coconino County), near Prescott (Yavapai County), Sierra Ancha (Gila County), Huachuca Mountains (Cochise County), 4,500 to 7,000 feet, in pine forest and with sagebrush, July and August. Minnesota and North Dakota to Arkansas, New Mexico, and Arizona.

4. Talinum Gooddingii P. Wilson. Anderson Mesa (Coconino County), Boyles, San Francisco River, Greenlee County (*Goodding* 1282, the type collection), Nogales (Santa Cruz County), 4,000 to 7,000 feet, open, rocky slopes, August. Known only from Arizona.

The specimens from Nogales have shorter and stouter cusps of the sepals than in the type collection. Traces of such appendages are occasionally seen in specimens of *T. parviflorum* from farther east, and these entities apparently are not distinguishable by the shape of the capsule as indicated in the key in *North American Flora.*

5. Talinum brevifolium Torr. Navajo and Coconino counties, 5,000 to 7,000 feet, May to September, type from along the Little Colorado River (*Woodhouse*). Southern Utah, New Mexico, and northeastern Arizona.

6. Talinum angustissimum (Gray) Woot. & Standl. Hereford (Cochise County), between Benson and Tucson and near Sells (Pima County), 2,500 to 4,000 feet, August and September. Western Texas to southern Arizona.

7. Talinum aurantiacum Engelm. Greenlee, Graham, Gila, Pinal, Cochise, Santa Cruz, and Pima counties, 4,000 to 5,000 feet, plains and rocky slopes, often among grasses, July to September. Western Texas to southern Arizona and northern Mexico.

Arizona's largest-flowered and showiest species. Indians in Arizona cooked and ate the roots, which often become very large and more or less woody.

2. CALANDRINIA. Rock-purslane

Small, somewhat fleshy annual herbs; leaves alternate; inflorescence umbelliform or racemiform; sepals 2, persistent; petals 3 to 7; style 3-cleft; capsule longitudinally dehiscent.

KEY TO THE SPECIES

1. Leaves oblong-spatulate, very obtuse; flowers in rather compact, umbelliform, lateral and terminal clusters; sepals glabrous; petals white, shorter than the sepals; capsules obtuse .... 1. *C. ambigua*
1. Leaves narrowly to rather broadly lanceolate or oblanceolate, often subrhombic, acutish or obtuse; flowers in loose, leafy terminal racemes; sepals usually more or less pubescent with thick, white hairs; petals rose-red (exceptionally white), mostly longer than the sepals; capsules acuminate.......... 2. *C. ciliata*

1. Calandrinia ambigua (Wats.) Howell. Near Yuma, Yuma County (*L. Swingle* in 1916, *Harrison & Peebles* 5056, *McMurry & Phillips* 2670), sandy soil, February and March. Deserts of southwestern Arizona and southeastern California.

2. Calandrinia ciliata (Ruiz & Pavon) DC. (*C. arizonica* Rydb., ?*Calyptridium depressum* A. Nels.). Graham, Gila, Maricopa, Pinal, Santa Cruz, and Pima counties, 1,500 to 4,000 feet, February to April, type of *C. arizonica* from near Tucson (*Pringle* in 1881). British Columbia to Baja California and Arizona; South America.

Red-maids, one of the common, early-spring wild flowers of the southern deserts. The Arizona plants belong to var. *Menziesii* (Hook.) Macbr. *Calyptridium depressum*, type from the Pena Blanca Mountains, Santa Cruz County (*R. & A. Nelson* 1203), appears to be a somewhat aberrant form of *Calandrinia ciliata.*

3. LEWISIA

Plants small, perennial, somewhat fleshy, scapose, with thick roots; leaves mostly basal, narrow; flowers rather showy; sepals distinct; petals pink or white.

Bitter-root (*L. rediviva*) is the state flower of Montana. The roots of this species were eaten by the Indians.

KEY TO THE SPECIES

1. Scapes jointed much below the middle, bearing 2 or more, subulate, scarious bracts just below the joint, the upper portion (pedicel) becoming detached on drying; leaves subterete; sepals or sepallike bracts 6 to 8; petals 12 or more, 15 to 30 mm. long; stamens 20 or more....................1. *L. rediviva*

1. Scapes not evidently jointed or if so, then the pedicel more or less persistent; leaves not terete, somewhat fleshy; bracts and sepals 2 each; petals 9 or fewer; stamens 15 or fewer (2).

2. Bracts closely subtending the sepals and similar to them; scapes nodeless; leaves oblanceolate or spatulate; sepals not glandular-denticulate; petals 12 to 24 mm. long; stamens 10 to 15....................2. *L. brachycalyx*

2. Bracts remote from and much narrower than the sepals, borne at a distinct node near the middle of the scape; leaves linear or linear-oblanceolate; sepals usually glandular-denticulate at apex; petals 8 to 10 mm. long; stamens 5 to 8..........3. *L. pygmaea*

1. Lewisia rediviva Pursh. Eastern boundary of Grand Canyon National Park (Coconino County), near the Navahopi entrance, about 7,000 feet (*Brueck* in 1943, *Collom* in 1947), May. Montana and British Columbia to northern Arizona and northern California.

2. Lewisia brachycalyx Engelm. (*Oreobroma brachycalyx* Howell). Apache, Navajo, Coconino, Gila, and Yavapai counties, 5,000 to 8,000 feet, among oaks, junipers, and pines, March to May. Southern Utah, Arizona, and southern California.

3. Lewisia pygmaea (Gray) Robins. (*Oreobroma pygmaeum* Howell). Tsegi and Big Lake (Apache County), Kaibab Plateau and North Rim of the Grand Canyon (Coconino County), 8,000 to 9,000 feet, June to August. Montana to Washington, south to New Mexico, northern Arizona, and California.

Specimens with entire or nearly entire sepals, collected in the White Mountain region, Apache County (*Phillips & Kearney* 3435), approach *L. nevadensis* (Gray) Robins.

4. CALYPTRIDIUM

Small, glabrous, somewhat succulent, annual herbs; stem leaves alternate, petiolate, spatulate; flowers small, ephemeral, in scorpioid racemes; petals 2 to 4, pink or white; stamens 1 to 3.

KEY TO THE SPECIES

1. Inflorescences relatively loose, the flowers not closely imbricate; sepals 1.5 to 2.5 mm. long; capsules more or less falcate, acutish or obtusish, usually 3 or more times as long as the sepals..1. *C. monandrum*
1. Inflorescences dense, the flowers closely imbricate; sepals 2.5 to 4 mm. long; capsules straight or nearly so, very obtuse, not more than twice as long as the sepals 2. *C. Parryi*

1. Calyptridium monandrum Nutt. Mohave County to Greenlee, Pinal, Santa Cruz, Pima, and (doubtless) Yuma counties, 450 feet or lower, common on desert plains and slopes, March to May. Arizona, California, and Baja California. Sand-cress.

2. Calyptridium Parryi Gray. Canoa to Arivaca, and Rosemont, Pima County (*Griffiths* 3556, 4125), Santa Rita Mountains (*Darrow* in 1937), Nogales, Santa Cruz County (*Peebles & Fulton* 11454), 3,500 to 5,000 feet, March and April. Southern Arizona and southern California.

The Arizona plant is var. *arizonicum* J. T. Howell (type *Griffiths* 4125), which has the pedicels not jointed to the rachis and the seeds smooth and shiny whereas in typical *C. Parryi* the pedicels appear to be jointed to the rachis and the seeds are more or less tuberculate.

5. CLAYTONIA. Spring-beauty

Plants small, glabrous, perennial, with a large corm; stem leaves 1 pair, opposite, narrowly lanceolate; flowers few, in loose racemes, rather conspicuous; sepals 2, persistent; petals and stamens 5, the petals pale pink; style 3-cleft; ovules 6; capsule 3-valved.

1. Claytonia rosea Rydb. Grand Canyon, Rogers Lake, and Oak Creek Canyon (Coconino County), Sierra Ancha and Mazatzal Mountains (Gila County), Santa Catalina Mountains (Pinal County), 5,500 to 7,000 feet, moist, rich soil, preferring shade on northern exposures, February to May. Wyoming and Utah to New Mexico and Arizona.

The Arizona specimens have longer leaves than those from the Rocky Mountains (up to 10 cm. long) and perhaps constitute a distinct variety. *C. rosea* is perhaps not specifically distinct from *C. lanceolata* Pursh.

6. MONTIA

Small, glabrous plants of diverse habit, annual or perennial; stem leaves opposite, in 1 or more pairs; flowers small and inconspicuous, in loose, simple or compound racemes; sepals 2; petals and stamens 5; style 3-cleft; ovules 3; capsule 3-valved.

KEY TO THE SPECIES

1. Plants perennial, with runners ending in a bulblet; stems commonly decumbent and rooting at the lower nodes; leaves none of them long-petioled, the stem leaves several pairs, obovate to oblong-spatulate, narrowed at base; racemes axillary (often appearing terminal); petals 6 to 8 mm. long..........................1. *M. Chamissoi*

1. Plants annual, with fibrous roots; stems erect or ascending; basal leaves very long-petioled, the stem leaves 1 pair, broader than long, connate-perfoliate, forming a nearly orbicular or somewhat angled disk subtending the inflorescence; racemes terminal; petals 3 to 5 mm. long . . . . 2. *M. perfoliata*

1. Montia Chamissoi (Ledeb.) Durand & Jackson. Apache County to Coconino, Yavapai, and Gila counties, 6,000 to 9,500 feet, wet ground in forests, June to August. Alaska to Iowa, New Mexico, Arizona, and California.

2. Montia perfoliata (Donn) Howell. Coconino and Mohave counties to Santa Cruz and Pima counties, 2,500 to 7,500 feet, along brooks and around springs, February to May. South Dakota to British Columbia, south to Arizona and California.

Miners-lettuce, Indian-lettuce, easily recognized by the saucer-shaped disk formed by the stem leaves, has been used as a salad plant and potherb by both white people and Indians. A variant with narrower basal leaves, var. *parviflora* (Dougl.) Jepson (*Claytonia parviflora* Dougl.), often grows with the typical plant and is equally common in Arizona. A depauperate form occasionally found in the state is var. *depressa* (Gray) Jepson (*Limnia humifusa* Rydb.). Several Arizona specimens have been identified as *Limnia utahensis* Rydb.

7. PORTULACA

Small, more or less succulent plants, annual or perennial, with diffuse or ascending, leafy stems; sepals 2, united below; petals and stamens inserted on the calyx, the petals mostly 5, the stamens often numerous; ovary partly inferior; capsule opening by an apical lid.

All the species prefer dry soil and full sunlight, growing on plains and mesas. Some of the Arizona species were used by the Indians as potherbs, and mush and bread were made from the seeds. The popular garden annual (*P. grandiflora* Hook.) is a native of southern South America.

KEY TO THE SPECIES

1. Capsule rim expanded into a circular, membranaceous wing. Plants annual; herbage glabrous; stems commonly erect or ascending; leaves flat, lanceolate or spatulate; petals yellow with a reddish base . . . . 1. *P. coronata*
1. Capsule rim not winged (2).
2. Leaf axils and inflorescence glabrous or inconspicuously short-pilose; leaves obovate-cuneate or spatulate, thick but flat. Plants annual; stems commonly decumbent or prostrate; petals yellow, less than 5 mm. long (3).
2. Leaf axils and inflorescence conspicuously villous with long, white, kinky hairs; leaves linear or narrowly oblanceolate, subterete (4).
3. Seeds (at low magnification) conspicuously and sharply granulate (almost echinate), very nearly to fully 1 mm. in greatest diameter . . . . 2. *P. retusa*
3. Seeds minutely and not sharply granulate, usually distinctly less than 1 mm. in diameter 3. *P. oleracea*
4. Plants perennial, the taproot tuberous-thickened; stems commonly erect or ascending, sometimes slightly woody at base; petals 7 to 12 mm. long, copper-colored 4. *P. suffrutescens*
4. Plants annual, the taproot slender to rather stout but not tuberous-thickened; stems commonly decumbent or ascending-spreading; petals seldom more than 6 mm. long (5).
5. Petals yellow to bronze, 2 to 2.5 mm. long; capsule diameter 1.5 to 2 mm. . . . 5. *P. parvula*
5. Petals red-purple, 3 mm. or longer; capsule diameter 2.5 to 3.5 mm. . . . . . . 6. *P. mundula*

1. Portulaca coronata Small (*P. lanceolata* Engelm. non Haw.). Greenlee, Graham, Gila, Pinal, Cochise, Santa Cruz, and Pima counties, 2,500 to 6,000 feet, August and September. Western Texas to Arizona and Mexico.

2. Portulaca retusa Engelm. Navajo, Coconino, Graham, Gila, Pinal, Cochise, and Santa Cruz counties, 1,500 to 5,500 feet, sometimes in saline soil, August to October. Arkansas and Texas to Utah and Arizona.

Difficult to distinguish from *P. oleracea* except by the seed characters.

3. Portulaca oleracea L. Apache, Navajo, Coconino, Yavapai, Greenlee, and Santa Cruz counties, 4,000 to 8,500 feet, late summer. Widely distributed in both the Eastern and the Western hemispheres.

Common purslane, pusley. An abundant garden weed in the eastern United States, but rare in Arizona.

4. Portulaca suffrutescens Engelm. Greenlee County to Yavapai County, south to Cochise, Santa Cruz, and Pima counties, 3,000 to 5,500 feet, July to September. Arkansas and Texas to Arizona and northern Mexico.

Arizona's showiest species, the flowers sometimes 3 cm. wide.

5. Portulaca parvula Gray. A collection at "El Sauz" (San Simon Valley), Cochise County (*Hayes* 72) is cited by Johnston (*Jour. Arnold Arbor.* 29: 194), and one at Camp Grant, Graham County, 5,000 feet (*Shreve* 4419), with very small capsules but no flowers, probably belongs here. Oklahoma, western Texas, southeastern Arizona, and northern Mexico.

6. Portulaca mundula Johnst. Navajo and Coconino counties to Cochise, Santa Cruz, and Pima counties, 1,500 to 7,000 feet, August and September. Missouri and Kansas to Texas, New Mexico, Arizona, and northern Mexico.

Arizona specimens have been referred to *P. parvula,* which apparently is quite rare in this state.

50. CARYOPHYLLACEAE. Pink Family

Plants herbaceous or slightly suffrutescent, annual or perennial; leaves opposite, entire, with or without stipules; flowers perfect, regular, mostly with both a calyx and corolla but sometimes apetalous; stamens 10 or fewer; styles 2 to 5, sometimes united nearly to the apex; ovary superior, 1-celled; fruit a many-seeded capsule, dehiscent by valves, at least at apex, or a 1-seeded achene or utricle.

This family includes those favorites of gardeners and florists, the pinks and carnations, but in many of the genera the plants are insignificant with small, inconspicuous flowers.

KEY TO THE GENERA

1. Fruit a 1-seeded achene or utricle, enclosed in the calyx tube; petals minute or none. Stipules scarious; sepals more or less united (2).

1. Fruit a several-seeded capsule, valvate, at least at apex; petals usually present (4).

2. Plants perennial, pulvinate-cespitose; stipules large and conspicuous, not or not much shorter than the leaf blades; flowers in terminal, cymose clusters....9. *Paronychia*

2. Plants annual; stipules relatively small, much shorter than the blades; flowers in axillary clusters (3).

3. Sepals with conspicuous, thin, white blades and green claws, these united below into an elongate, turbinate tube....................................8. *Achyronychia*

3. Sepals greenish, without differentiation of claw and blade, united only toward the base .... 10. *Herniaria*
4. Sepals united for most of their length, forming a more or less inflated gamosepalous calyx, this 5-toothed at apex; petals with long claws, usually with appendages forming a corona in the throat of the corolla (5).
4. Sepals distinct to the base or nearly so; petals not clawed or appendaged (7).
5. Calyx terete or 5-angled, inconspicuously veined; petals with or without a corona. Herbage glabrous or obscurely puberulent; styles 2........ 13. *Saponaria*
5. Calyx with 10 or more conspicuous, longitudinal nerves or ribs; petals with coronalike appendages, these sometimes reduced to small scales (6).
6. Styles normally 3; ovary and capsule stipitate, or at least narrowed at base; blade of the petal broader than the claw........ 11. *Silene*
6. Styles 4 or 5; ovary and capsule sessile or very nearly so, broad at base; blade of the petal not or very little wider than the apical part of the claw........ 12. *Lychnis*
7. Style 1, 3- cleft or 3-toothed. Capsule 3-valved (8).
7. Styles 3 to 5, usually distinct to the base (9).
8. Petals usually present, although sometimes very small; sepals not recurved or rigid 6. *Drymaria*
8. Petals minute or none; sepals recurved, becoming rigid and almost spinose 7. *Loeflingia*
9. Stipules present, scarious, conspicuous; petals pink, entire, sometimes rudimentary 5. *Spergularia*
9. Stipules none; petals white, occasionally purple-tinged, rarely wanting (10).
10. Styles alternate with the sepals and of the same number, usually 5........ 3. *Sagina*
10. Styles opposite the sepals or, if fewer in number, then opposite the outer sepals (11).
11. Capsule elongate, cylindric, often curved, dehiscent at apex, with twice as many teeth as there are styles; styles usually 5........ 2. *Cerastium*
11. Capsule short, ovoid or oblong, straight, splitting into as many valves as there are styles, the valves often 2-cleft; styles usually 3 (12).
12. Petals bifid, often deeply so, rarely none........ 1. *Stellaria*
12. Petals entire or slightly emarginate........ 4. *Arenaria*

1. STELLARIA. Starwort

Plants small, herbaceous, annual or perennial, usually with tufted stems; flowers solitary in the axils or in few-flowered terminal cymes, inconspicuous; sepals and petals commonly 5; styles usually 3, sometimes partly united; fruit a dehiscent capsule.

KEY TO THE SPECIES

1. Plants glandular-pubescent, at least in the inflorescence, perennial; petals 6 to 8 mm. long, about twice as long as the sepals, retuse or bifid (cleft not more than halfway to the base). Flowering stems from creeping, often tuberous-thickened rootstocks, sharply 4-angled; leaves lanceolate, attenuate-acuminate, 3 cm. long or longer, closely sessile........ 1. *S. Jamesiana*
1. Plants not glandular, or the pedicels obscurely so in *S. media;* petals not more than 5 mm. long, not or scarcely surpassing the sepals, 2-parted (2).
2. Leaves distinctly petioled, at least the basal ones. Plants annual (3).
2. Leaves all sessile or subsessile. Pedicels elongate, filiform (4).
3. Stems procumbent, pubescent in lines; leaves mostly long-petioled, the blades ovate; petioles ciliate; flowers axillary, solitary on long, soon deflexed, puberulent pedicels; sepals pubescent, not shiny, broadly lanceolate, not setose-tipped, not conspicuously scarious-margined .... 2. *S. media*
3. Stems erect or ascending, glabrous or sparsely pubescent below, not in lines; leaves (except the lowest) sessile, lanceolate; flowers in terminal, scarious-bracted cymes; sepals glabrous, shiny, narrowly lanceolate, setose-acuminate, conspicuously scarious-margined .... 3. *S. nitens*

4. Petals none or rudimentary; plants annual; leaves thin, lanceolate to elliptic-oblong; flowers numerous, in terminal, subumbellate inflorescences. Pedicels spreading or deflexed .... 4. *S. umbellata*
4. Petals equaling or surpassing the sepals; plants perennial, from creeping rootstocks; leaves linear or linear-lanceolate, acuminate, 1 to 3 cm. long; flowers solitary or in very few-flowered cymes (5).
5. Leaves ascending, somewhat shiny, broadest near the base; pedicels erect or ascending, straight or nearly so.. 5. *S. longipes*
5. Leaves spreading, not shiny, broadest near the middle; pedicels becoming widespreading or deflexed and usually curved.. 6. *S. longifolia*

1. Stellaria Jamesiana Torr. Navajo and Coconino counties, 7,000 to 8,500 feet, April to July. Wyoming to Idaho, Texas, northern Arizona, and California.

Represented in Arizona by a relatively narrow-leaved form (*Alsine Curtisii* Rydb.).

2. Stellaria media (L.) Cyrillo. North Rim of Grand Canyon (Coconino County), Mazatzal Mountains (Gila County), Tempe (Maricopa County), Warren (Cochise County), Tucson (Pima County). In most parts of the United States; naturalized from Europe.

Chickweed, a common weed of lawns and gardens.

3. Stellaria nitens Nutt. Grand Canyon (Coconino County) and Gila, Pinal, Maricopa, and Pima counties, 1,500 to 4,000 feet, February and March. Montana to British Columbia, south to Arizona and California.

4. Stellaria umbellata Turcz. San Francisco Peaks (Coconino County), 11,000 to 12,000 feet, July and August. Montana to Oregon, south to New Mexico, northern Arizona, and California; Siberia.

5. Stellaria longipes Goldie. White Mountains (Apache County), Chiricahua Mountains (Cochise County), 8,500 to 9,500 feet, coniferous forests and wet meadows, May to July. Greenland to Alaska, south to New Mexico, Arizona, and California; northern Asia.

6. Stellaria longifolia Muhl. White Mountains, Apache County (*Schroeder* in 1938, *Phillips* 3241), Buck Springs Ranger Station, Coconino County, 7,500 feet (*Collom* 783), July. Canada to Maryland and Arizona.

2. CERASTIUM

Plants herbaceous, annual to perennial, mostly pubescent; flowers in terminal cymes; sepals and petals commonly 5, the petals 2-lobed or 2-cleft, rarely wanting; capsules elongate, dehiscent only near the apex.

The names powder-horn, suggested by the shape of the capsule, and mouse-ear chickweed are sometimes applied to these plants.

KEY TO THE SPECIES

1. Capsule teeth at maturity strongly revolute from the tip; leaves oblanceolate or obovate. Plants annual; pedicels commonly considerably longer than the straight capsules
1. *C. texanum*
1. Capsule teeth not revolute from the tip, but often with inflexed margins; leaves prevailingly linear to ovate-oblong, or the lowest sometimes oblanceolate (2).
2. Petals not surpassing the sepals. Stems low, more or less matted; herbage copiously pubescent, more or less viscid.. 2. *C. vulgatum*
2. Petals commonly considerably surpassing the sepals (3).

3. Plants annual or at most biennial (rarely perennial?). Capsules often curved, usually nodding (4).
3. Plants perennial (5).
4. Pedicels in fruit little longer (often shorter) than the capsule, straight or slightly and gradually curved; capsule less than twice as long as the calyx....3. *C. brachypodum*
4. Pedicels in fruit longer (often much longer) than the capsule, usually strongly curved or hooked toward the apex; capsule 2 or more times as long as the calyx 4. *C. nutans*
5. Stem leaves mostly linear or linear-lanceolate, often rather stiff; petals 2 to 3 times as long as the calyx; stems erect or ascending........................5. *C. arvense*
5. Stem leaves mostly oblong-lanceolate to ovate, not stiff; petals commonly less than twice as long as the calyx (6).
6. Leafy basal shoots elongate, producing flowering branches from alternate axils; leaves acute ....6. *C. sordidum*
6. Leafy basal shoots short; leaves obtuse to acutish..................7. *C. beeringianum*

1. Cerastium texanum Britton. Lukachukai Mountains (Apache County), southern Coconino and Greenlee counties to Cochise, Santa Cruz, and Pima counties, 1,500 to 8,500 feet, moist, partly shaded places, March to May, also late summer. Texas and Arizona.

2. Cerastium vulgatum L. Southern Apache and Navajo counties, near Tucson (Pima County), 2,500 to 8,000 feet, May to September. Widely distributed in the United States; naturalized from Europe.

3. Cerastium brachypodum (Engelm.) Robins. Apache, Navajo, Coconino, Greenlee, and Cochise counties, 7,000 to 9,500 feet, May to August. South Dakota to Alberta, south to Virginia, Missouri, Arizona, Oregon, and northern Mexico.

Specimens collected in the Huachuca Mountains (*Goodding* 1296, *Darrow* 2600) are apparently perennial.

4. Cerastium nutans Raf. Apache and Gila counties to Cochise, Santa Cruz, and Pima counties, 7,000 to 9,500 feet, March to October. Nova Scotia to British Columbia, south to North Carolina, Arizona, and nearly throughout Mexico.

Both typical *C. nutans*, viscid-villous with spreading hairs, and var. *obtectum* Kearney & Peebles (*C. sericeum* Wats., non Pourr.) occur in Arizona. The latter has the stems and leaves, at least near the base of the plant, sericeous with long nonglandular hairs. Many Arizona specimens are intermediate in character of the pubescence. The type of *C. sericeum* Wats. was obtained in the Huachuca Mountains (*Lemmon* in 1882).

5. Cerastium arvense L. Apache County to Coconino County, 7,000 to 9,500 feet, wet meadows, June and July. Labrador to Alaska, south to Georgia, New Mexico, northern Arizona, and California; Europe.

Flowers showy for the genus, with pure white petals. The representative of this polymorphic species in Arizona is probably *C. scopulorum* Greene.

6. Cerastium sordidum Robins. Chiricahua Mountains, Cochise County (*Blumer* 1428, *Goodman & Hitchcock* 1189), Santa Rita Mountains, Santa Cruz County (*Darrow* in 1937), 7,500 to 9,000 feet, pine forest, May and June. Southeastern Arizona and Chihuahua.

Easily recognizable by the long basal shoots bearing oblanceolate, petioled leaves, these up to 5 cm. long and 1.5 cm. wide. The petals are sometimes twice as long as the sepals.

7. Cerastium beeringianum Cham. & Schlecht. San Francisco Peaks (Coconino County), 11,000 to 12,000 feet, July and August. Quebec to Alaska, south to New Mexico, northern Arizona, and California.

Cerastium adsurgens Greene was based upon a collection on the San Francisco Peaks (*Greene* in 1889). From Greene's description it would appear to be nearer *C. vulgatum* than any other of the species included in this flora.

3. SAGINA. Pearl-wort

Plants herbaceous, annual or perennial, small and inconspicuous; stems slender, diffuse, often matted; leaves narrowly linear; sepals and petals 4 or 5, the petals entire, sometimes minute or obsolete; capsules dehiscent to the base, 4- or 5-valved.

KEY TO THE SPECIES

1. Plants perennial (?), with leafy rosettes clustered at the base; pedicels and calyx glabrous; sepals spreading in fruit, about half as long as the mature capsule
1. *S. saginoides*

1. Plants annual, without basal rosettes; pedicels and calyx (in the Arizona plants) sparsely glandular-pilosulous; sepals erect in fruit, about ⅔ as long as the mature capsule
2. *S. occidentalis*

1. Sagina saginoides (L.) Karst. San Francisco Peaks, Coconino County, 11,500 feet (*Knowlton* 134). Greenland to Alaska, south to New Mexico, Arizona, and California; Eurasia.

The Arizona specimens belong to var. *hesperia* Fern., with sepals not more than 2 mm. long.

2. Sagina occidentalis Wats. Between Superior and Miami (Gila (?) County), Rincon and Santa Catalina mountains (Pima County), March to May. Idaho to British Columbia, south to Arizona and California.

The Arizona specimens are referred doubtfully to this species, having glandular hairs on the pedicels and calyx, and the pedicels shorter and the calyx narrower and less rounded at base than in typical *S. occidentalis*. They seem to approach *S. decumbens* (Ell.) Torr. & Gray of the eastern United States.

4. ARENARIA (93, 94, 95, 96). Sandwort

Contributed by Bassett Maguire

Plants mostly perennial; stems slender, often tufted; flowers mostly in cymes, these open or congested; sepals 5, commonly ribbed or keeled; petals 5, entire or nearly so; stamens normally 10; styles 3; capsules longitudinally dehiscent, the valves 3, entire or 2-cleft.

Where sufficiently abundant these plants are said to furnish excellent forage but do not withstand heavy grazing.

KEY TO THE SPECIES

1. Valves of the capsule not cleft (2).
1. Valves of the capsule cleft. Plants perennial (5).
2. Plants annual, diffusely branched from near the base; leaves often more than 1 cm. long, linear-subulate to nearly filiform, not rigid or pungent; plants of low elevations
1. *A. Douglasii*

2. Plants perennial, more or less cespitose; leaves less than 1 cm. long, linear-subulate, often slightly pungent; plants of medium to high elevations (3).
3. Sepals obtuse, not surpassing and usually shorter than the petals........4. *A. obtusiloba*

3. Sepals acute or acuminate, surpassing the petals (4).
4. Plants wholly glandular-puberulent; seeds pale reddish brown, 0.4 to 0.7 mm. in diameter 2. *A. rubella*
4. Plants wholly glabrous; seeds dark reddish brown, 0.7 to 1.0 mm. in diameter 3. *A. filiorum*
5. Leaves lanceolate or broader, not or scarcely rigid or pungent. Glandular pubescence none (6).
5. Leaves linear-subulate or acicular, more or less rigid or pungent. Plants grasslike, more or less cespitose, with a subligneous caudex (7).
6. Pedicels divergent after anthesis, often 4 times as long as the fruit; plants green, sparsely puberulent or glabrate; sepals at anthesis mostly 3 to 3.5 mm. long. Stems lax, diffusely branched, 20 to 40 cm. long.................................5. *A. confusa*
6. Pedicels erect or ascending, seldom more than 3 times as long as the fruit; plants grayish and densely puberulent, to green and glabrate; sepals 3 to 6 mm. long....6. *A. saxosa*
7. Sepals obtuse; capsules mostly 7 to 9 mm. long.........................7. *A. aberrans*
7. Sepals acute or acuminate; capsules mostly less than 7 mm. long (8).
8. Petals (except in var. *Parishiorum*) conspicuously surpassing the sepals; plants more or less suffrutescent...10. *A. macradenia*
8. Petals shorter than to moderately surpassing the sepals; plants with only the caudex ligneous (9).
9. Plants bluish green; caudex branches relatively few and thick, the old leaves densely and closely marcescent; leaves not rigid or very sharply pungent; inflorescence glandular-puberulent ....8. *A. Fendleri*
9. Plants light green or yellowish; caudex branches relatively numerous and thin, wiry, the old leaves spreading, soon disintegrating; leaves very rigid and sharply pungent; inflorescence glandular-puberulent or glabrous. Leaves more or less spreading, commonly not more than 2 cm. long.................................9. *A. Eastwoodiae*

1. Arenaria Douglasii Fenzl. Southern Yavapai, Gila, and Maricopa counties, chiefly in the vicinity of the Mazatzal Mountains, 2,500 to 3,500 feet, sandy soil, March to June. South-central Arizona, and from southern Oregon to Baja California.

2. Arenaria rubella (Wahlenb.) J. E. Smith (*A. propinqua* Richards.). San Francisco Peaks (Coconino County), 11,000 to 12,500 feet, July and August. Circumpolar, arctic Eurasia and North America south to Quebec, southern Colorado, northern Arizona, southern Nevada, and California.

Arizona specimens have been referred to the European *A. verna* L.

3. Arenaria filiorum Maguire. Kaibab Plateau and North Rim of Grand Canyon (Coconino County), 7,000 to 8,500 (12,000?) feet, gravelly or stony soil, June to August. South-central Utah, southern Nevada, and northern Arizona.

4. Arenaria obtusiloba (Rydb.) Fern. San Francisco Peaks, 11,500 to 12,000 feet (*Knowlton* 128, *Little* 4654), August. Alberta, British Columbia, and Washington to northern New Mexico, Arizona, and California.

5. Arenaria confusa Rydb. Navajo, Coconino, and Yavapai counties to Greenlee, Cochise, Santa Cruz, and Pima counties, 5,500 to 9,500 feet, pine forests, July to September. Colorado, Utah, Texas, New Mexico, Arizona, and California.

The commonest species at moderate elevations in central and southern Arizona. It is most closely related to and possibly not specifically distinct from the Eastern *A. lanuginosa* (Michx.) Rohrb., and apparently intergrades in Arizona with *A. saxosa*.

6. **Arenaria saxosa** Gray (*A. polycaulos* Rydb.). White Mountains (Apache and Greenlee counties), North Rim of Grand Canyon and San Francisco Peaks (Coconino County), Pinaleno Mountains (Graham County), Chiricahua and Huachuca mountains (Cochise County), Santa Catalina Mountains (Pima County), 7,000 to 12,000 feet, commonly in coniferous forest, May to September. Southern Colorado and Utah, New Mexico, Arizona, and northern Mexico.

KEY TO THE VARIETIES

1. Plants compact, the stems more or less matted and seldom more than 10 cm. long; leaves small, oblong-lanceolate; sepals 4 to 6 mm. long............... *A. saxosa* (typical)
1. Plants more open, the stems erect or ascending, seldom less than 15 cm. long; leaves longer and relatively narrower; sepals 3 to 4 mm. long (2).
2. Herbage cinereous-puberulent; stems mostly stiffly erect; leaves erect or strictly ascending ....var. *cinerascens*
2. Herbage green, sparsely pubescent or glabrate; stems relatively lax; leaves spreading var. *Mearnsii*

The distribution in Arizona, as given above, is that of the more typical plant. Var. *cinerascens* Robins. ranges from the Lukachukai Mountains (Apache County) and the Kaibab Plateau (Coconino County) to the mountains of Cochise and Pima counties, 7,000 to 9,000 feet, type from the Huachuca Mountains (*Lemmon*). Var. *Mearnsii* (Woot. & Standl.) Kearney & Peebles (*A. Mearnsii* Woot. & Standl.) is found in the mountains of Graham, Cochise, Santa Cruz, and Pima counties, 4,000 to 9,000 feet. There is intergradation among all forms of this species.

7. **Arenaria aberrans** Jones. Coconino, Mohave, Yavapai, and Gila counties, 5,500 to 9,000 feet, oak and pine forests, May to July, type from Mount Dellenbaugh (*Cottam* 4159). Known only from northern and north-central Arizona.

Arizona specimens were referred previously to *A. capillaris* Poir., a closely related species, and to *A. aculeata* Wats.

8. **Arenaria Fendleri** Gray. Apache County to Mohave County, and Huachuca Mountains (Cochise County), 4,000 to 12,000 feet, common in pine forests, April to September. Wyoming and Utah to western Texas and Arizona.

KEY TO THE VARIETIES

1. Sepals acuminate, 6 to 7.5 mm. long; inflorescence moderately glandular (2).
1. Sepals acute, 4 to 6 (7) mm. long; inflorescence densely glandular (3).
2. Leaves 3 to 6 cm. long...var. *Fendleri*
2. Leaves 1 to 2 (3) cm. long. Plants of reduced stature, alpine............var. *Porteri*
3. Leaves 2 to 4 (6) cm. long; sepals 5 to 6 (7) mm. long. Plants of strict habit var. *brevifolia*
3. Leaves 1 to 2 cm. long; sepals 4 to 5 (6) mm. long. Plants of reduced stature var. *Tweedyi*

Arenaria Fendleri is a polymorphic species, having given rise to a number of freely intergrading varietal populations. The major and typical population (var. *Fendleri* Maguire) is represented throughout the range of the species in Arizona. Var. *Porteri* Rydb. and var. *Tweedyi* (Rydb.) Maguire are known in Arizona only from the San Francisco Peaks, the latter at an elevation of about 12,000 feet. Var. *brevifolia* (Maguire) Maguire is represented from the Tunitcha and White mountains (Apache County), Kaibab Plateau, Grand Canyon, and San Francisco Peaks (Coconino County), and Chiricahua Mountains (Cochise County).

9. Arenaria Eastwoodiae Rydb. Apache County to Mohave and Yavapai counties, 3,500 to 8,000 feet, sandy soil, plains and mesas, tending to make small hummocks, June to September. Colorado, Utah, New Mexico, and northern Arizona.

The Hopi Indians are reported to use the plant as an emetic. The typical glabrous plant (var. *Eastwoodiae* Maguire) is less common in Arizona than the glandular-puberulent var. *adenophora* Kearney & Peebles, type from Tuba, Coconino County (*Kearney & Peebles* 11856). Specimens from Prescott (*Eastwood* 8800, 17649a) are intermediate between var. *adenophora* and *A. aberrans* Jones.

10. Arenaria macradenia Wats. Coconino, Mohave, and Yavapai counties, dry, rocky slopes and ledges, 2,000 to 5,500 feet, April to July. Southwestern Utah and northwestern Arizona to southern California and Baja California.

KEY TO THE SUBSPECIES AND VARIETIES

1. Sepals 4 to 5.5 mm. long; stems not or but slightly woody at base......subsp. *Ferrisiae*
1. Sepals 5.5 to 6.5 mm. long; stems woody at base: subsp. *macradenia* (2).
2. Petals conspicuously surpassing the sepals; stem leaves 5 or more pairs; innovations (sterile shoots) seldom produced................................var. *macradenia*
2. Petals equaling or barely surpassing the sepals; stem leaves fewer than 5 pairs; innovations frequently produced......................................var. *Parishiorum*

The typical and more extensively distributed plant (subsp. *macradenia* Maguire var. *macradenia* Maguire) is a suffrutescent desert plant frequently exceeding 6 dm. in height. Usually at higher altitudes, between 5,000 and 7,000 feet, the species is represented by a less robust and less woody variant, subsp. *Ferrisiae* Abrams. Var. *Parishiorum* Robins. has been collected in Grand Wash, Mohave County (*Maguire & Holmgren* 20665). Specimens with copiously glandular-puberulent stems, intermediate between subsp. *Ferrisiae* and var. *Parishiorum*, were obtained near Burro Creek, southeastern Mohave County (*Crooks & Darrow* in 1937, *Gould & Darrow* 4247).

Arenaria congesta Nutt. has generally been credited to Arizona on the basis of a collection by Newberry and one by Palmer in 1869, these belonging to the indefinite var. *subcongesta* Wats., but no definite locality was assigned to them. No recent or further confirmatory material has been collected in the state. *A. congesta* is an exceedingly variable and widespread species with congested, subcongested, or umbellate inflorescences. Its natural range, with the exception of the Charleston Mountains, southern Nevada, occupies the Great Basin, Rocky Mountains, and Sierra-Cascade ranges from California, Nevada, Utah, and Colorado northward to Washington, Idaho, and Montana.

5. SPERGULARIA (97). SAND-SPURRY

Plants annual, somewhat fleshy, usually glandular-pubescent, with spreading or prostrate, branched stems; leaves narrow, nearly terete, often fascicled; sepals 5; petals usually 5, sometimes wanting; styles 3; capsule longitudinally dehiscent to the base.

1. **Spergularia marina** (L.) Griseb. Eastern Coconino, southern Yavapai, Gila, Maricopa, Pinal, Pima, and Yuma counties, 1,000 to 3,500 feet, moist, strongly saline soil, March to June. Throughout most of the Northern Hemisphere.

The Arizona specimens are of the smooth-seeded form (*S. leiosperma* (Kindb.) F. Schmidt).

6. DRYMARIA (98)

Plants annual, with slender, usually tufted stems; stipules present but sometimes caducous; sepals 5; petals 2- to 5-cleft or -parted; stamens usually 5; capsule longitudinally dehiscent.

KEY TO THE SPECIES

1. Stem leaves nearly as wide as to wider than long, elliptic to suborbicular, at least 4 mm. wide (2).
1. Stem leaves much longer than wide, linear, narrowly lanceolate, or oblanceolate, not more than 2 mm. wide. Flowers in open, mostly terminal cymes (3).
2. Plants bright green, glandular-puberulent; leaves thin, subreniform, wider than long, abruptly short-acuminate or apiculate; flowers in long-peduncled, dense or rather loose clusters; sepals lanceolate, setose-acuminate....................1. *D. Fendleri*
2. Plants glaucous, glabrous; leaves thickish, ovate or broadly elliptic, as long as or longer than wide, obtuse; flowers in sessile or subsessile axillary clusters; sepals ovate, obtuse or apiculate..2. *D. pachyphylla*
3. Cauline leaves fascicled, appearing verticillate; mature capsules 3 to 5 mm. long; petals fimbriate-appendaged ....3. *D. sperguloides*
3. Cauline leaves opposite; mature capsules not more than 2.5 mm. long; petals merely bifid (4).
4. Petals surpassing the sepals; herbage puberulent or glabrescent. Stems 6 to 20 cm. long, erect, branching above; herbaceous portion of the sepals obtuse........4. *D. effusa*
4. Petals not surpassing and usually shorter than the sepals; herbage glabrous or very nearly so (5).
5. Herbaceous portion of the sepals acute; stems erect, up to 20 cm. long, branching above; leaves linear, not more than 1 mm. wide; seeds minutely muriculate....5. *D. tenella*
5. Herbaceous portion of the sepals obtuse; stems decumbent or spreading, not more than 5 cm. long, branching from the base; leaves oblong or oblanceolate, 1.5 to 3 mm. wide; seeds minutely tessellate.................................6. *D. depressa*

1. **Drymaria Fendleri** Wats. Coconino, Yavapai, Greenlee, Cochise, and Santa Cruz counties, 5,000 to 6,000 feet, thickets, August to November. New Mexico, Arizona, and Mexico.

2. **Drymaria pachyphylla** Woot. & Standl. Chiricahua Mountains, Cochise County (*Jones* in 1931). Western Texas to southeastern Arizona and northern Mexico.

The plant is very poisonous to cattle and sheep, as has been demonstrated by feeding experiments, but fortunately animals avoid it when other feed is obtainable.

3. **Drymaria sperguloides** Gray (*Mollugophytum sperguloides* Jones). Apache County to Mohave County, south to Cochise, Santa Cruz, and Pima counties, 4,000 to 8,000 feet, open places, August and September. Western Texas to Arizona and northern Mexico.

4. **Drymaria effusa** Gray. Cochise, Santa Cruz, and Pima counties, 6,000 to 7,500 feet, dry pine woods and among rocks, August and September. Southeastern Arizona and northern Mexico.

5. **Drymaria tenella** Gray. Coconino County to Cochise, Santa Cruz, and Pima counties, 6,000 to 8,000 feet, dry pine woods, August and September. New Mexico, Arizona, and northern Mexico.

6. **Drymaria depressa** Greene. White Mountains (Apache and Greenlee counties), Pinaleno Mountains (Graham County), 7,500 to 9,500 feet, openings in pine forests, August and September. New Mexico and eastern Arizona.

7. LOEFLINGIA

Plants annual, small; stems branched from the base, stiff; leaves subulate, rigid; flowers axillary, inconspicuous; petals 3 to 5, sometimes wanting; stamens 3 to 5; stigmas 3; capsule longitudinally 3-valved.

1. **Loeflingia squarrosa** Nutt. Maricopa, Pinal, and Pima counties, 1,000 to 3,000 feet, March and April. Southern Arizona, California, and Baja California.

8. ACHYRONYCHIA

Plants annual, glabrous or nearly so; stems branched, prostrate; flowers in axillary clusters; perianth segments in 1 series, the blades white, petallike, the claws united; stamens 10 to 15, few of them fertile; style 2-cleft; fruit 1-seeded, indehiscent.

1. **Achyronychia Cooperi** Gray. Mohave, Maricopa, and Yuma counties, 2,000 feet or lower, February to April. Western Arizona, southern California, and Baja California.

An attractive little plant, forming mats on the desert sand, with bright-green foliage and snow-white flowers.

9. PARONYCHIA (99). NAILWORT

Plants perennial, with an almost woody caudex, pulvinate-cespitose; herbage scabrous-puberulent; flowering stems short; leaves crowded, linear, thickish, cuspidate, the stipules conspicuous, white, sharply acuminate; calyx lobes hooded toward the apex, cuspidate; petals none; stamens usually 4 or 5; fruit a utricle.

1. **Paronychia Jamesii** Torr. & Gray. South Rim of Grand Canyon, 7,500 feet, on limestone (*Ripley & Barneby* 8476), Crater Mound, Coconino County, about 5,500 feet (*Eastwood & Howell* 6918), Chiricahua Mountains, Cochise County (*Lemmon*), May to September. Nebraska and Wyoming to Oklahoma, Texas, New Mexico, Arizona, and northern Mexico.

10. HERNIARIA. BURST-WORT

Plants annual, small and inconspicuous, hispidulous; stems branching from the base, spreading or prostrate; flowers minute, greenish, in dense axillary clusters; perianth segments in 1 series; fruit 1-seeded, indehiscent.

1. **Herniaria cinerea** DC. Sentinel, Maricopa County (*Orcutt* 117), near Casa Grande, Pinal County (*Harrison* 7518), March. Southern Arizona and California; introduced from southern Europe.

11. SILENE (100). Catchfly, Campion

Plants herbaceous, annual or perennial; herbage often viscid; flowers in racemes, cymes, cymose panicles, or solitary; calyx gamophyllous, campanulate or cylindric, more or less inflated, longitudinally nerved; petals 5, often deeply notched, with scalelike or fringed appendages; stamens 10; ovary stipitate, the stipe, together with the filaments and petal bases, forming a carpophore; capsule dehiscent by apical teeth.

KEY TO THE SPECIES

1. Stems very short, densely cespitose in cushionlike mats; leaves crowded, narrowly linear. Herbage glabrous or puberulent, not glandular; flowers about 1 cm. long; petals normally purplish pink..1. *S. acaulis*
1. Stems elongate, leafy, not or very loosely cespitose; leaves not crowded (2).
2. Plants annual; petals entire to shallowly cleft (3).
2. Plants perennial; petals commonly deeply cleft (4).
3. Herbage puberulent, hirtellous, or glabrate; upper internodes usually with a sharply defined viscid area; leaves mostly acute, linear to oblong-lanceolate, the lowest oblanceolate; inflorescence cymose-paniculate; capsules oblong-ovoid......2. *S. antirrhina*
3. Herbage villous; internodes not viscid; leaves very obtuse, apiculate, all broadly spatulate; inflorescence racemose, 1-sided; capsules broadly ovoid........3. *S. gallica*
4. Petals cardinal red, greatly surpassing the calyx, laciniate with several lobes, rarely merely bifid. Leaves oblong-lanceolate, oblanceolate, or obovate........4. *S. laciniata*
4. Petals white, greenish, pink, or purplish, moderately or only slightly surpassing the calyx, emarginate or bifid (5).
5. Corolla, including the carpophore, less than 12 mm. long; inflorescence very leafy, the floral bracts not much reduced; claws of the petals barely equaling the calyx; carpophore glabrous, less than 2 mm. long. Petals white; calyx 5 to 8 mm. long; appendages (segments of the corona) 0.1 to 0.5 mm. long.......................5. *S. Menziesii*
5. Corolla, including the carpophore, usually more than 12 mm. long or, if smaller, then the floral bracts usually considerably reduced, or the petals well exserted, or the carpophore pubescent and more than 2 mm. long (6).
6. Appendages of the petals 0.1 to 0.5 mm. long, nearly as wide as long; calyx papery-membranaceous, 7 to 11 mm. long, very sparsely glandular-puberulent; petal blades 2 to 4 mm. long, shallowly bilobed, otherwise entire; carpophore finely puberulent
6. *S. rectiramea*
6. Appendages mostly more than 0.5 mm. long, usually distinctly longer than wide; calyx not papery, mostly more than 10 mm. long, copiously pubescent; petal blades 3 to 8 mm. long, deeply bilobed, often with lateral teeth on the 2 main lobes; carpophore puberulent to woolly (7).
7. Stems freely branched; inflorescence loose and elongate, with numerous lateral branches; calyx tubular, 8 to 11 mm. long; petal blades about 3 mm. long, the appendages scarcely 1 mm. long; carpophore about 1.5 mm. long........................7. *S. Thurberi*
7. Stems usually simple; inflorescence usually simple and narrow, but sometimes compound; calyx campanulate-tubular, 10 to 18 mm. long; petal blades 4.5 to 7.5 mm. long, the appendages 1 to 2.5 mm. long; carpophore 2 to 5 mm. long
8. *S. Scouleri*

1. Silene acaulis L. San Francisco Peaks (Coconino County), 11,500 to 12,000 feet, July to September. Circumpolar, and alpine in both the Eastern and the Western hemispheres.

Moss campion. An attractive little plant, forming mosslike cushions studded with bright-pink flowers, among rocks above timber line. The species is represented in Arizona by subsp. *subcaulescens* (F. N. Williams) Hitchcock & Maguire, with larger flowers and narrower petals than in the typical northern plant.

2. Silene antirrhina L. Coconino and Mohave counties, south to Cochise, Santa Cruz, Pima, and Yuma counties, 6,000 feet or lower, March to May. Throughout temperate North America. Sleepy catchfly.

3. Silene gallica L. (*S. anglica* of authors, non L.). Canyon Lake, Maricopa County (*Peebles* 3904), Santa Rita Mountains and near Tucson (*Thornber*), April and May. Naturalized in the eastern United States and in Arizona and California; from Europe.

4. Silene laciniata Cav. Apache, Navajo, and Coconino counties, south to Cochise, Santa Cruz, and Pima counties, 5,500 to 9,000 feet, mostly in pine forests, July to October. Western Texas to California and Mexico.

Mexican campion. Arizona's showiest species of *Silene*, with large flowers of a brilliant cardinal red. The Arizona specimens belong to the relatively broad-leaved subsp. *Greggii* (Gray) Hitchcock & Maguire.

5. Silene Menziesii Hook. Lukachukai Mountains, Apache County (*Goodman & Payson* 2829), Betatakin, Navajo County, on a "sandy flat" (*Wetherill* in 1935), Greenland Lake, Kaibab Plateau (*Collom* in 1939), 7,000 to 9,000 feet, June and July. Canada to Missouri, northern New Mexico and Arizona, and California.

6. Silene rectiramea Robins. South Rim of the Grand Canyon (Coconino County), 6,500 to 7,000 feet, May and June (*MacDougal* 181, the type collection, *H. E. Bailey* in 1935, *Grant* 5728, *Hawbecker* 288). Known only from these collections.

7. Silene Thurberi Wats. Cochise County, Swisshelm Mountains (*Lemmon*), Chiricahua Mountains (*Eggleston* 11003, *Kearney & Harrison* 6196), 5,000 to 6,000 feet. Southeastern Arizona and northern Mexico.

8. Silene Scouleri Hook. Apache, Coconino, Yavapai, Graham, Cochise, and Pima counties, 5,000 to 9,500 feet, July to September. Montana to British Columbia, south to northern Mexico.

The species is represented in Arizona by subsp. *Pringlei* (Wats.) Hitchcock & Maguire (*S. Pringlei* Wats.) which is confined largely to New Mexico, Arizona, and northern Mexico. Four intergrading varieties occur in Arizona according to Hitchcock and Maguire (100, pp. 27, 28). The Arizona plants referred previously to *S. Hallii* Wats. (*S. Scouleri* subsp. *Hallii* Hitchcock & Maguire) presumably all belong to subsp. *Pringlei*.

KEY TO THE VARIETIES OF SUBSP. *PRINGLEI*

1. Plants completely eglandular.......................................var. *eglandulosa*

1. Plants glandular above, the pedicels and calyx usually more or less viscid (2).

2. Stem leaves usually lanceolate, mostly spreading; lower pedicels usually 1-flowered and mostly only 1 to 2 cm. long.......................................var. *concolor*

2. Stem leaves linear or linear-lanceolate, strict; lower pedicels or inflorescence branches mostly elongate, 2 to 10 cm. long (3).

3. Stems 20 to 40 cm. long; stem leaves 3 to 5 pairs, linear, 3 to 7 cm. long, 2 to 4 mm. wide; petal blades oblong, cleft to the middle, usually without lateral teeth var. *leptophylla*

3. Stems mostly taller; stem leaves 4 to 10 pairs, linear-lanceolate, up to 20 cm. long, 2 to 20 mm. wide; petal blades more nearly obovate, usually with lateral teeth var. *typica*

Var. *leptophylla* is known definitely only from the Kaibab Plateau and the Grand Canyon, type from V. T. Meadow (*C. L. Hitchcock et al.* 4540), whereas the other varieties are more southern in their range.

The night-flowering catchfly, *Silene noctiflora* L., a European species, was collected at Flagstaff, Coconino County (*Whiting* 814), but there is no evidence that it has become established in Arizona. It is annual, viscid-pubescent, and has rather large, deeply cleft, white petals.

12. LYCHNIS. Campion

Plants perennial, densely puberulent; stems erect, usually branched only in the inflorescence; leaves narrow, oblanceolate to linear; flowers few, in a cymose panicle; calyx gamophyllous, longitudinally nerved; petals scarcely surpassing the calyx, with very narrow white or purplish-pink blades, these entire or nearly so; styles commonly 5.

1. Lychnis Drummondii (Hook.) Wats. Kaibab Plateau, Grand Canyon, and San Francisco Peaks (Coconino County), 7,000 to 11,500 feet, June and July. Manitoba to British Columbia, south to New Mexico and northern Arizona.

13. SAPONARIA. Soapwort

Plants herbaceous, annual or perennial, glabrous or nearly so; leaves sessile or nearly so, lanceolate to ovate; flowers in terminal, compound cymes or corymbs, showy; calyx not conspicuously ribbed; petals pink or red, with or without appendages; stamens 10; styles 2.

KEY TO THE SPECIES

1. Plants annual; flowers in a broad, open, flat-topped, corymbiform panicle; calyx ovoid, strongly 5-angled; petals rose-red, not appendaged....................1. *S. Vaccaria*
1. Plants perennial; flowers in compact cymes, these forming a more or less elongate panicle; calyx cylindric (rarely ovoid), not angled; petals normally pale pink, appendaged with long, subulate teeth at the junction of the blade and claw
2. *S. officinalis*

1. Saponaria Vaccaria L. (*Vaccaria vulgaris* Host). Tucson (Pima County), April and May. A field weed here and there in the United States; naturalized from Europe. Scarcely established in Arizona.

Cow soapwort. The seeds contain saponin and may be somewhat poisonous.

2. Saponaria officinalis L. Flagstaff (Coconino County), Hualpai Mountain (Mohave County), Prescott (Yavapai County), well established in waste land, midsummer. Widely distributed in the United States; naturalized from Europe.

Bouncing-bet. The leaves and roots contain saponin and form a lather in water. The plant has been used medicinally as a tonic, etc.

51. CERATOPHYLLACEAE. Hornwort Family

1. CERATOPHYLLUM. Hornwort

Plants aquatic, immersed; stems slender, much-branched; leaves whorled, dissected into very narrow, dentate segments; flowers unisexual, minute, axillary, sessile, subtended by a many-cleft involucre simulating a perianth; stamens numerous; pistil with a 1-celled ovary and an elongate, filiform style.

1. Ceratophyllum demersum L. Southern Navajo and Coconino counties, southeastern Mohave County, Maricopa, Pinal, and Cochise counties, 2,000 to 6,500 feet. Almost throughout North America; Europe.

52. RANUNCULACEAE. Crowfoot Family

Plants mostly perennial, herbaceous or suffrutescent; leaves alternate or opposite, simple or compound; flowers mostly perfect, regular or irregular; perianth either in one series and then usually corollalike, or of both a calyx and a corolla; stamens often numerous; pistils usually several or numerous; ovary superior, 1-celled, the ovules 1 to many; fruit either several-seeded and dehiscent on one suture only (follicular), or 1-seeded and indehiscent (achenes), or several-seeded and berrylike.

This large and diverse family contains some of the showiest of Arizona wild flowers, notably the columbines, larkspurs, and monkshoods (species of *Aquilegia, Delphinium,* and *Aconitum*). Many beautiful garden flowers belong to this family, notably, in addition to the 3 genera already mentioned, species of *Ranunculus, Clematis, Anemone,* and the peonies (*Paeonia* spp.).

KEY TO THE GENERA

1. Carpels with 2 or more ovules; fruit follicular or (in *Actaea*) berrylike (2).
1. Carpels with a solitary ovule, in fruit achenes (7).
2. Flowers irregular, large and showy, commonly blue or bluish (3).
2. Flowers regular (4).
3. Upper sepal extended into a conspicuous, cylindric spur..............5. *Delphinium*
3. Upper sepal expanded into a helmet-shaped hood......................6. *Aconitum*
4. Petals with a relatively small, erect limb, produced below into a long, tapering, hollow spur, this clavate-thickened at the end.............................4. *Aquilegia*
4. Petals very small and not spurred, or none (5).
5. Flowers solitary or in very few-flowered cymes; sepals 4 to 10, showy, flat, somewhat persistent; plants subacaulescent; leaves simple, broad, cordate..........1. *Caltha*
5. Flowers numerous or many, in simple or few-branched racemes; sepals 3 to 5, not showy, somewhat cucullate, deciduous at anthesis; plants caulescent; leaves large, decompound (6).
6. Fruit berrylike, commonly red when ripe; raceme short....................2. *Actaea*
6. Fruit of 1 to 3 dry, thin-walled follicles; raceme elongate.............3. *Cimicifuga*
7. Petals present (except sometimes in *Myosurus*). Leaves commonly alternate, or all basal (8).
7. Petals none or rudimentary (9).
8. Sepals spurred at base, the spur usually elongate, scarious; petals very small or none; receptacle becoming greatly elongate, the fruits in a slender, cylindric spike; plants scapose ....9. *Myosurus*
8. Sepals not spurred; petals usually present, often showy; receptacle not becoming greatly elongate, the fruits in a conic, ovoid, or hemispheric head; stems commonly leafy ....11. *Ranunculus*
9. Sepals large and showy, petallike, somewhat persistent (10).
9. Sepals small, less conspicuous than the stamens, caducous (11).
10. Sepals 5 or more; head of achenes ovoid to cylindric, woolly, the achenes without long tails ....7. *Anemone*
10. Sepals commonly 4; head of achenes globose, the achenes with long plumose tails 8. *Clematis*
11. Leaves simple, palmately lobed or parted; outer filaments flat, somewhat petaloid; anthers oval or ovate, about 1 mm. long..........................10. *Trautvetteria*
11. Leaves decompound; filaments all filiform; anthers narrowly linear, much more than 1 mm. long...12. *Thalictrum*

1. CALTHA. Marsh-marigold

Plants perennial, herbaceous, glabrous; leaves simple, mostly basal, with broad, cordate blades; flowers regular, rather showy, the perianth segments all alike, petallike, purplish outside, white within; stamens numerous; pistils several, becoming several-seeded follicles in fruit.

1. Caltha leptosepala DC. Apache, Coconino, and Graham counties, 7,500 to 11,000 feet, wet meadows, June to September. Montana to Alaska, south to New Mexico, Arizona, and Washington.

Known as elks-lip in New Mexico.

2. ACTAEA. Bane-berry

Plants herbaceous, perennial; leaves few, very large, the lower leaves long-stalked, ternately compound, with large serrate-dentate leaflets; flowers in short, simple, terminal racemes, with 3 to 5 caducous, petallike sepals, and 4 to 10 small petals, these less conspicuous than the numerous stamens; pistil 1; fruit a several-seeded berry.

The root and berries of the European *A. spicata* (probably also those of *A. arguta*) are poisonous, and deaths of children eating the berries have been reported.

1. Actaea arguta Nutt. Lukachukai and White mountains (Apache County), San Francisco Peaks (Coconino County), southward to the mountains of Cochise and Pima counties, 7,000 to 10,000 feet, May to July. South Dakota to Alaska, south to New Mexico, Arizona, and California.

The Arizona plant seems to be *A. viridiflora* Greene, of which the type was collected on the San Francisco Peaks (*Greene* in 1889). It may be at least varietally distinct.

3. CIMICIFUGA. Bugbane

Racemes long, sometimes branched; sepals commonly 5, petallike, cucullate; petals none (in the Arizona species); stamens numerous; pistils usually more than 1, becoming, in fruit, dry, several-seeded follicles; plants otherwise resembling *Actaea.*

The name snakeroot is sometimes given these plants. The underground parts, at least in *C. racemosa* of eastern North America, contain a drug—cimicifugin—which has been used in treatment of uterine disorders and of rheumatism. The Eastern species were also reputed beneficial in cases of snake bite. The botanical name of the genus, of which "bugbane" is a translation, suggests that the plant was used as an insecticide.

1. Cimicifuga arizonica Wats. Bill Williams Mountain and Oak Creek Canyon (Coconino County), Sierra Ancha (Gila County), 5,500 to 7,000 feet, rich soil in wooded ravines, July and August, type from Bill Williams Mountain (*Lemmon* 3275). Known only from central Arizona.

The long, white, spikelike racemes, borne on tall stems above the large leaves, make this a conspicuous plant when in flower.

4. AQUILEGIA (101). Columbine

Plants perennial, herbaceous; leaves ternately decompound; flowers regular, solitary or few on long peduncles, large and showy; sepals 5, petallike, soon deciduous; petals with a small blade and a long spur; pistils 5, becoming in fruit many-seeded follicles with long, slender beaks.

KEY TO THE SPECIES

1. Leaflets viscid-pubescent beneath. Flowers usually nodding, cream-colored, often tinged with pink; sepals considerably longer than the petal blades; spurs slender, 15 to 30 mm. long .... 1. *A. micrantha*
1. Leaflets not viscid (2).
2. Flowers nodding; spurs, and usually the sepals, red; spurs less than 3 cm. long (3).
2. Flowers erect; spurs and sepals not red; spurs 3 cm. or longer (5).
3. Sepals little longer than the petal blades, usually greenish or yellowish, at least toward apex; stamens not more than 6 mm. longer than the petal blades ...... 2. *A. elegantula*
3. Sepals considerably longer than (up to twice as long as) the petal blades, usually red throughout; stamens about 10 mm. longer than the petal blades (4).
4. Petal blades 4 to 5 mm. long; sepals 8 to 10 (rarely 20) mm. long, usually about half as long as the spurs; basal leaves mostly biternate .... 3. *A. desertorum*
4. Petal blades 7 to 8 mm. long; sepals 12 to 20 mm. long, ⅔ as long as to nearly equaling the spurs; basal leaves usually triternate .... 4. *A. triternata*
5. Spurs and sepals blue to white, the sepals obtuse to acute; stamens shorter to somewhat longer than the petal blades .... 5. *A. caerulea*
5. Spurs and sepals yellow, the sepals usually acuminate; stamens longer (usually much longer) than the petal blades (6).
6. Petioles 5 to 20 cm. long; flowers clear yellow; spurs 4 to 7 cm. long .. 6. *A. chrysantha*
6. Petioles up to 30 cm. long; flowers pale yellow; spurs 9 to 15 cm. long ... 7. *A. longissima*

1. Aquilegia micrantha Eastw. Northern Navajo and Coconino counties, 6,500 to 7,000 feet, rock ledges, June to September. Southwestern Colorado, southeastern Utah, and northeastern Arizona.

2. Aquilegia elegantula Greene. Lukachukai Mountains, northern Apache County, 9,000 feet (*Goodman & Payson* 2867), apparently also near Flagstaff, Coconino County (*Thornber* in 1930). Southwestern Colorado and southeastern Utah to northern Mexico.

3. Aquilegia desertorum (Jones) Cockerell. Northern Navajo County and Coconino County, 7,000 to 8,000 feet, July, type from Flagstaff (*Jones* in 1884). Known only from northern Arizona.

4. Aquilegia triternata Payson. White Mountains (southern Apache County), and mountains of Cochise, Santa Cruz, and Pima counties, 4,000 to 10,000 feet, chiefly in coniferous forests, May to October, type from the Chiricahua Mountains (*Goodding* 2325). Western Colorado, western New Mexico, and eastern Arizona.

5. Aquilegia caerulea James. Tunitcha and Lukachukai mountains (Apache County), Kaibab Plateau, North Rim of Grand Canyon, and San Francisco Peaks (Coconino County), 8,000 to 11,000 feet, June and July. Southwestern Montana to northern New Mexico and northern Arizona.

The very beautiful typical phase of the species, with blue sepals and spurs and white petal blades, is the state flower of Colorado. The Arizona specimens probably all are referrable to subsp. *pinetorum* (Tidestrom) Payson (*A. pinetorum* Tidestrom), which has paler-colored flowers and longer, more slender spurs than in typical *caerulea*. The type of *A. pinetorum* came from the Kaibab Plateau, Coconino County (*Tidestrom* 2328).

6. Aquilegia chrysantha Gray. Apache County to Mohave County, south to Cochise, Santa Cruz, and Pima counties, 3,000 to 11,000 feet, rich, moist soil, chiefly in the pine belt, but descending lower along streams, April to September. Southern Colorado to New Mexico, Arizona, and northern Mexico.

With its large, long-spurred, canary-yellow flowers, this is one of the handsomest plants of the state. It is by far the most abundant and widely distributed of the Arizona columbines and is exceptional in its wide altitudinal range.

7. Aquilegia longissima Gray. Huachuca Mountains, Cochise County (*Breninger* in 1897), Baboquivari Mountains, Pima County (*Gilman* 93, *Gentry* 3418). Western Texas, southern Arizona, and northeastern Mexico.

5. DELPHINIUM (102). Larkspur

Contributed by Joseph Ewan

Perennial herbs; leaves palmately divided; inflorescence racemose; sepals 5, irregular, the upper one produced into a spur; petals in 2 unequal and unlike pairs, the upper ones firm with an oblique blade, the lower ones wholly membranaceous, more or less ligulate and notched; follicles many-seeded; seeds with wings or angles, or enveloped in a papery pellicle.

The seeds, usually unnoticed when ripe, are the most reliable evidence for positive identification and define the subgeneric groups. Larkspurs in general react readily to variations in soil, shade, and moisture, and numerous natural hybrids are known. The low arenicolous species are spring- or early-summer-flowering and occur in large or small colonies. The tall hydrophilous species are summer- and autumn-flowering and occur as close stands in mountain meadows or partly shaded ravines. The plants contain delphinine and other toxic alkaloids. The extent and exact nature of larkspur poisoning is as yet little known, but the tall meadow species are often deadly to cattle, apparently less so to sheep and horses.

KEY TO THE SPECIES

1. Stems mostly less than 70 cm. high, usually not uniformly leafy throughout, the leaves clustered at or near the base; plants of plains, deserts, foothills, and dry pine forests, usually in sandy or gravelly soil (2).

1. Stems (except in *D. tenuisectum*) mostly 1 m. or more high, usually leafy throughout; plants of ravines, meadows, and valley floors, in heavy, often wet soil. Flowers dark blue or purple (5).

2. Sepals whitish or pale lavender-blue, the spur slender, 15 to 20 mm. long; lower petals white, conspicuously white-pilose, deeply notched at apex into cuneate-acuminate lobes, the sinus 4 to 6 mm. deep; stems puberulent with fine, curling, white hairs
1. *D. virescens*

2. Sepals azure blue to intense royal blue, the spur 10 to 12 (or 15) mm. long; lower petals bluish purple to dark purple, shallowly to deeply notched into ovate lobes, the sinus not more than 4 mm. deep; stems glabrous, often glaucous, rarely glandular-hairy, but not closely and evenly puberulent (3).

3. Sinuses of the lower petals 3 to 4 mm. deep; leaves chiefly strictly basal, the primary segments obovate or even subspatulate; flowers mostly clear dark blue or royal blue
2. *D. scaposum*

3. Sinuses of the lower petals less than 3 mm. deep; leaves not strictly basal, the primary segments linear to broadly cuneate-oblanceolate; flowers blue-purple to pale blue or azure blue (4).
4. Stems narrowing at ground level, easily disarticulating from the tuberlike roots; flowers dark blue, the upper petals bicolored, prominently venulose; plants of pine forests 3. *D. Nelsoni*
4. Stems not narrowing at ground level, persistent upon the woody-fibrous roots; flowers variously light blue, the upper petals pallid or if somewhat bicolored, then not conspicuously venulose; plants of canyons and deserts.................... 4. *D. amabile*
5. Racemes dense; leaves nearly circular in outline. Leaves velvety gray-puberulent beneath, 4 to 10 cm. wide, the primary segments cuneate-obovate, with short, barely acute teeth 5. *D. geraniifolium*
5. Racemes loose, or interrupted, or elongating early; leaves more or less pentagonal in outline (6).
6. Sepals lance-acuminate; rachis of the raceme lustrous-hairy, with glandular spreading hairs; follicles subglabrous, dark-venulose.......................... 6. *D. Barbeyi*
6. Sepals ovate; rachis nonglandular; follicles puberulent (7).
7. Leaves more or less dimorphic, glabrous. Basal leaves broadly and shallowly lobed, the upper ones palmatisect into pinnatifid divisions, the ultimate segments linear; racemes not conspicuously bracteate; follicles 12 to 15 mm. long............ 8. *D. scopulorum*
7. Leaves all much alike, finely puberulent (8).
8. Stem 1 to 2 m. high; leaves rather coarsely dissected; racemes open, not conspicuously bracteate; sinus of the lower petals 1 mm. deep; follicles oblong, 10 to 14 mm. long 7. *D. andesicola*
8. Stems usually less than 1 m. high; leaves finely dissected; racemes interrupted-spicate, usually conspicuously leafy-bracteate below; sinus of the lower petals 3 to 4 mm. deep; follicles ovate, 9 to 11 mm. long (oblong and 13 to 26 mm. long in the typical Mexican plant).. 9. *D. tenuisectum*

1. Delphinium virescens Nutt. Pinaleno, Chiricahua, Huachuca, Tumacacori, and Rincon mountains and adjacent valleys (Graham, Cochise, Santa Cruz, and Pima counties), 6,000 feet or lower, open flats, valley floors, and gentle foothill slopes, May. Widely distributed in the Great Plains region.

This larkspur is notable for its deeply notched, white-pilose petals and long, often geniculate spurs. The species is represented in Arizona and southern New Mexico by subsp. *Wootoni* (Rydb.) Ewan (*D. Wootoni* Rydb.), with bluish-white flowers fading buff and the lower petals more exserted than in typical *D. virescens.*

A hybrid, *D. virescens* subsp. *Wootoni* × *D. scaposum* (*D. confertiflorum* Wooton), occurs in grassy openings in spruce-poplar forests at Big Lake, about 9,000 feet, Apache County (*Gould & Robinson* 5123). The plants somewhat suggest a browsed *D. scaposum* that has subsequently grown a compact, spicate raceme. The stems of the hybrid are leafy, the leaves gray with a close-appressed puberulence on both surfaces. Flowers borne on bracteate pedicels, often nodding, the sepals dull blue or bluish purple, the spur vertical, the lower petals merely retuse, the sinus closed. Known otherwise only from McKinley and Socorro counties, New Mexico.

2. Delphinium scaposum Greene. Coconino, Mohave, Yavapai, Graham, Gila, Maricopa, Pinal, and Pima counties, 5,000 (rarely 8,000) feet or lower, locally frequent on open deserts and gravelly mesas, March to June. Southwestern Colorado to southern Nevada, New Mexico, and Arizona.

The typically scapose habit and deep-royal-blue flowers (except in northern

Arizona plants) are distinctive. Plants of the Colorado River and plateau drainages have lighter-blue flowers, more diffuse racemes, and follicles 18 to 20 mm. long, as compared with 10 to 13 mm. long in the type. This larkspur is used by the Hopi Indians, who call it "tcoro'si." It is reported that they grind the flowers with corn to make a blue meal, "blue pollen," for the flute altar and also use the plant as an emetic in one of their rituals. At least in parts of its range this species is known to be poisonous to cattle.

3. Delphinium Nelsoni Greene. Defiance Plateau (Apache County), Betatakin (Navajo County), Kaibab Plateau to the Mogollon Escarpment (Coconino County), 6,000 to 8,500 feet, pine forests, June. South Dakota to Idaho, south to Colorado, northern Arizona, and Nevada.

This species is represented in Arizona by forma *pinetorum* (Tidestrom) Ewan (*D. pinetorum* Tidestrom), of which the type was collected on the Kaibab Plateau (*Tidestrom* 2375). As compared with typical *D. Nelsoni,* it is usually a smaller plant, with more slender stems and reduced leaves, ashy-puberulent.

4. Delphinium amabile Tidestrom. Coconino and Mohave counties to Graham and Pima counties, 5,000 feet or lower, rocky knolls and desert mesas, February to May. Southwestern Utah and Arizona to southern California.

This is the most xerophytic of North American larkspurs. The reduction in vegetative parts enables the plants to withstand withering heat or prolonged droughts. The species has developed many local races. The range in Arizona, as given above, is that of the typical plant. Subsp. *apachense* (Eastw.) Ewan (*D. apachense* Eastw.), type from Roosevelt, Gila County (*Eastwood* 17144), occurs in most of the range of the species in Arizona and is commoner than the typical phase. It has sinuses of the lower petals 2 to 2.5 mm. deep, the lobes distally spreading, leaves rather conspicuous at flowering time, flowers clear sky blue, in loose racemes; whereas in typical *D. amabile* the sinuses of the lower petals are 1.5 to 2 mm. deep, the lobes not noticeably spreading, leaves withering, not noticeable at flowering time, flowers mostly deep blue, in moderately dense racemes. The size and form of the leaves vary greatly in the subspecies in response to varying ecologic factors.

5. Delphinium geraniifolium Rydb. Apache, Navajo, and Coconino counties, 7,000 to 9,500 feet, mountain parks, summer and autumn, type from "Charles" (Clark) Valley (*Rusby* in 1883). Known only from northern Arizona.

The plant is of a strict, often simple, habit with flowers densely massed in a turretlike spike. Occasional cream-colored variants may be noted among the usual purplish-blue-flowered plants. The leaves recall *Geranium* or *Paeonia* in their distinctive contour and lobation.

6. Delphinium Barbeyi Huth. Baldy Peak, White Mountains, Apache County (*Goodding* 605), apparently scarce, in subalpine parks. Wyoming, Colorado, Utah, New Mexico, and eastern Arizona.

The lower flowers in the short raceme are subtended by broad, leaflike, few-parted bracts, the stems usually hollow and succulent.

7. Delphinium andesicola Ewan. Chiricahua, Huachuca, and Santa Rita mountains (Cochise, Santa Cruz, and Pima counties), 5,000 to 9,500 feet, in swales on open, pine-forested slopes, summer and autumn, type from Barfoot Park, Chiricahua Mountains (*Blumer* 136). Known only from southeastern Arizona. The plants are browsed by deer.

Subsp. *amplum* Ewan (*D. sierrae-blancae* Wooton subsp. *amplum* Ewan) occurs in the White, Pinaleno, and Chiricahua mountains (southern Apache, Greenlee, Graham, and Cochise counties), the type from 6 miles south of Hannigan Meadow, White Mountains, Greenlee County (*Kearney & Peebles* 12274). It has shorter and broader ultimate divisions of the leaves than in typical *D. andesicola,* and the stems are less pubescent below.

8. Delphinium scopulorum Gray. East Eagle Creek (Greenlee County), Huachuca Mountains (Cochise County), Santa Rita Mountains (Pima or Santa Cruz County), 5,500 to 6,500 feet, stream flats in the oak-sycamore belt and lower edge of the pine belt, late summer. Southwestern New Mexico and southeastern Arizona.

9. Delphinium tenuisectum Greene. Apache County, at Big Lake, 9,000 feet (*Phillips* 2827), and 15 miles east of Fort Apache, 6,500 feet (*Gould & Robinson* 4947), openings in deciduous woods and meadow borders. New Mexico, eastern Arizona, and northern Mexico.

The Arizona plant belongs to subsp. *amplibracteatum* (Wooton) Ewan (*D. amplibracteatum* Wooton), which has shorter stems, less finely dissected leaves, and much smaller follicles than typical *D. tenuisectum.*

6. ACONITUM. Monks-hood

Plants herbaceous, perennial; stems tall, leafy; leaves palmately lobed; flowers in simple or branched racemes, perfect, very irregular; sepals 5, the upper one much the largest, helmet-shaped; petals small, varying in number; stamens numerous; pistils 3 to 5, these becoming several-seeded, short-beaked follicles.

Showy plants with a general resemblance to larkspur, but different in the structure of the large, normally dark-blue or violet flowers. All parts of the plants contain aconitine and other alkaloids. The European monks-hood, *A. Napellus,* is the source of the powerful heart stimulant aconite. *A. columbianum* is reputed to be poisonous to livestock. It is stated that the plants are most toxic in the preflowering stage.

KEY TO THE SPECIES

1. Inflorescence often paniculate (somewhat branched below), the peduncles and branches mostly spreading, or ascending at a wide angle; hood sepal rarely less than 2 cm. long. Herbage villous and somewhat viscid, to glabrous................1. *A. columbianum*
1. Inflorescence racemiform, the peduncles erect or ascending at a narrow angle; hood sepal not more than 1.5 cm. long. Leaves deeply, narrowly, and acutely dissected, the upper ones copiously short-pilose, at least on the upper surface.........2. *A. infectum*

1. Aconitum columbianum Nutt. (*A. arizonicum* Greene). Apache, Coconino, Cochise, Santa Cruz, and Pima counties, 5,000 to 9,500 feet, rich, moist soil along mountain brooks, June to September, type of *A. arizonicum* from the Santa Rita Mountains (*Pringle* in 1881). British Columbia to Montana, south to New Mexico, Arizona, and California.

2. Aconitum infectum Greene. Known only from the San Francisco Peaks (Coconino County), 9,500 to 11,000 feet, July and August, type collected by MacDougal (No. 396).

Similar forms occur elsewhere in the range of *A. columbianum,* of which *A. infectum* seems scarcely more than a good variety.

7. ANEMONE

Plants herbaceous, perennial, with erect, scapelike stems; leaves basal and in an involucrelike pair or whorl subtending the inflorescence, pedately parted or divided; flowers solitary on long peduncles, regular, showy; perianth segments 5 to 10, all alike, petaloid; stamens and pistils numerous; achenes compressed, in a dense, ovoid or cylindric head.

These attractive plants are sometimes known as wind-flower. Some of the Old World species, notably *A. japonica,* are popular garden flowers.

KEY TO THE SPECIES

1. Sepals 8 to 10, linear or narrowly elliptic; stems from a short, tuberlike, sometimes forked root, up to 40 cm. long; herbage sparsely soft-pubescent or nearly glabrous. Leaves ternate or biternate, the ultimate divisions usually cleft or coarsely toothed; fruiting head cylindric or ovoid-cylindric; sepals pink or purplish....1. *A. tuberosa*

1. Sepals 5 to 8, elliptic to ovate; stems from an elongate, more or less woody caudex; herbage copiously pubescent (2).

2. Stems commonly not more than 30 cm. long; pubescence soft, spreading, the hairs long and very fine; leaves pedately several-parted, the divisions all narrow, lanceolate or oblanceolate, entire or few-cleft, the uppermost leaves subsessile or very short-petioled; sepals purplish outside, yellowish or purplish within; fruiting head globose or short-ovoid, not more than 15 mm. long; styles usually deciduous............2. *A. globosa*

2. Stems commonly more than 30 cm. long; pubescence subappressed; leaves pedately 3-parted, the main divisions broadly cuneate-obovate, several-cleft and several-toothed, the uppermost leaves with petioles nearly to quite as long as the blades; sepals whitish; fruiting head cylindric, 20 mm. long or longer; styles (at least the bases) persistent 3. *A. cylindrica*

1. Anemone tuberosa Rydb. Coconino and Mohave counties to Greenlee, Cochise, Santa Cruz, and Pima counties, 2,500 to 5,000 feet, among rocks on mesas and foothills, February to April. Texas to Utah, Arizona, and California.

A pretty spring wild flower. Perhaps not specifically distinct from *A. sphenophylla* Poepp. of southern South America.

2. Anemone globosa Nutt. San Francisco Peaks (Coconino County), 10,500 to 12,000 feet, July. Western Canada to New Mexico, northern Arizona, and California.

3. Anemone cylindrica Gray. Apache and Coconino counties, 6,500 to 7,500 feet, rich soil along streams, June and July. New Brunswick to British Columbia, south to New Jersey, New Mexico, and Arizona.

8. CLEMATIS (103)

Plants perennial; stems woody below, weak, usually climbing by means of tendrillike petioles; leaves opposite, the pairs scattered along the stem, pinnate or bipinnate; flowers perfect or unisexual, regular; perianth in one series, the segments thin and petallike or thick and leathery, the petals none or rudimentary; stamens and pistils numerous; achenes in globose heads, with long, persistent, plumose styles.

Several species are grown extensively as ornamentals, and some of the exotic species are among the handsomest of cultivated climbing plants. The species of the section *Viorna* are known collectively as leather-flower.

KEY TO THE SPECIES

1. Flowers in panicles of cymes, numerous, mostly unisexual. Sepals cream-colored or ochroleucous, spreading (2).
1. Flowers solitary, very few, perfect (3).
2. Leaflets divergently cleft or parted, not more than 5 cm. long; tails of the mature carpels commonly more than 5 (up to 10) cm. long; herbage grayish-pubescent, usually copiously so..1. *C. Drummondii*
2. Leaflets with ascending lobes and teeth, 3 to 7 cm. long; tails of the carpels not more (usually less) than 5 cm. long; herbage loosely pubescent to nearly glabrous, green 2. *C. ligusticifolia*
3. Sepals thin, spreading, violet or purple, glabrous or inconspicuously pubescent; stamens spreading. Leaves ternate or biternate; leaflets few-toothed or cleft: section *Atragene* 3. *C. pseudoalpina*
3. Sepals thick, erect or connivent, dull brown or purplish brown, conspicuously white-lanate, at least on the margins; stamens erect: section *Viorna* (4).
4. Leaves bipinnate or occasionally ternate, the ultimate divisions linear-lanceolate to nearly filiform, seldom more than 6 mm. wide; stems erect, commonly unbranched and 1-flowered; herbage villous to glabrate. Tails of the achenes plumose 4. *C. hirsutissima*
4. Leaves pinnate or bipinnate, the ultimate divisions oblong-lanceolate to nearly orbicular, usually more than 6 mm. wide; stems more or less twining, commonly branched and bearing more than 1 flower, the flowers solitary at the ends of the branchlets; herbage glabrous to sparsely pubescent (5).
5. Tails of the achenes about 3 cm. long, glabrous or sparsely pubescent toward the base 5. *C. Bigelovii*
5. Tails of the achenes 4 to 5 cm. long, plumose throughout..................6. *C. Palmeri*

1. Clematis Drummondii Torr. & Gray. Southern Yavapai, Greenlee, Gila, Maricopa, Pinal, Cochise, and Pima counties, seldom above 4,000 feet, among shrubs, usually in comparatively open ground, March to September. Texas to Arizona and northern Mexico.

2. Clematis ligusticifolia Nutt. Apache, Navajo, and Coconino counties, south to Cochise, Santa Cruz, and Pima counties, 3,000 to 8,000 feet, commonly along streams, May to September. Western Canada and North Dakota to New Mexico, Arizona, and California.

Apparently intergrades or hybridizes in Arizona with *C. Drummondii*, and is extremely variable in the shape, size, and dentation of the leaflets. *C. neomexicana* Woot. & Standl., a variant with broad, rounded teeth and lobes, occurs in the Pinal Mountains (*Toumey* 44) and Chiricahua Mountains (*Blumer* 1510). An extremely small-leaved form, with leaflets not more than 2.5 cm. long and equally wide, was collected in the White Mountains (*Zuck* in 1896). An extraordinary form collected at Rock Point, Apache County (*Peebles* 13518), has elongate, nearly entire leaflets 5 to 10 mm. wide.

Occasionally grown as an ornamental, this plant was formerly used by the Indians as a remedy for sore throat and colds, and it is stated that the crushed roots were placed in the nostrils of tired horses to revive them.

3. Clematis pseudoalpina (Kuntze) A. Nels. Apache, Coconino, and northern Greenlee counties, 7,000 to 8,500 feet, rich soil in forests, apparently rare in Arizona. South Dakota and Montana to New Mexico and Arizona.

4. Clematis hirsutissima Pursh. Coconino County, 7,000 to 8,500 feet, April to July. Montana to Washington, south to New Mexico, Arizona, and Oregon.

The typical phase of the species is confined in Arizona to the North Rim of the Grand Canyon and the Kaibab Plateau. On the South Rim and in the vicinity of Flagstaff occurs var. *arizonica* (Heller) Erickson (*C. arizonica* Heller), which has more spreading petioles, narrower, almost filiform leaflets, and mostly smaller flowers, the sepals usually not more than 2.5 cm. long. The type of *C. arizonica* came from Walnut Canyon (*MacDougal* 343). Although Erickson has been followed in treating *C. arizonica* as a variety of *C. hirsutissima,* it may be specifically distinct.

5. Clematis Bigelovii Torr. Fort Apache, Navajo County (*Hoyt* in 1893). New Mexico and eastern Arizona.

6. Clematis Palmeri Rose. Known in Arizona only from the type collection at Fort Apache, Navajo County, 5,000 feet (*Palmer* 600, in 1890), and from a collection at Black River, White Mountains (*Goodding* 4417). Western New Mexico and eastern Arizona.

More material is required to permit a conclusion as to whether *C. Palmeri* is specifically distinct from *C. Bigelovii.*

9. MYOSURUS. Mouse-tail

Plants annual, dwarf; leaves all basal, narrowly linear or oblanceolate, entire; flowers solitary on slender scapes, inconspicuous; sepals commonly 5, deciduous; petals rudimentary or none; stamens 5 to numerous; pistils many; achenes in a dense, slender, cylindric spike.

Plants of wet soil, along streams and around springs.

KEY TO THE SPECIES

1. Achenes when mature roundish, with a dorsal cup or border nearly surrounding the base of the beak, the cup often larger than the body of the achene (2).
1. Achenes when mature more or less quadrangular, without cup or border, keeled dorsally from base to apex, the beak subulate, not strongly flattened laterally (3).
2. Border of the achene deep, cup-shaped, suborbicular, erose or nearly entire, corky-thickened, the beak stout, triangular, strongly flattened laterally; plants up to 15 cm. high .... 1. *M. cupulatus*
2. Border of the achene shallow, relatively thin, the beak elongate, subulate; plants not more than 3 cm. high .... 2. *M. Egglestonii*
3. Back of the achene scarcely wider on each side than the very prominent keel, the latter prolonged into a beak at least half as long as the body of the achene .. 3. *M. aristatus*
3. Back of the achene distinctly wider on each side than the relatively low keel, the latter prolonged into a beak much less than half as long as the body of the achene, or the beak sometimes obsolete .... 4. *M. minimus*

1. Myosurus cupulatus Wats. Greenlee County to Mohave County, south to Cochise, Santa Cruz, and Pima counties, 2,500 to 5,000 feet. February to May, type from between the San Francisco and Gila rivers, Greenlee County (*Greene* in 1880). New Mexico to California and Sonora.

2. Myosurus Egglestonii Woot. & Standl. Kaibab Plateau, Grand Canyon, and Leroux Spring near Flagstaff (Coconino County), Young to Payson (northern Gila County), 2,000 to 8,000 feet. New Mexico and Arizona.

3. Myosurus aristatus Benth. Kaibab Plateau and Mormon Lake (Coconino County), Santa Rita Mountains and probably also Santa Catalina Mountains (Pima County), March to June. Nebraska to British Columbia, south to New Mexico, Arizona, and California; South America.

4. Myosurus minimus L. Navajo, Coconino, Yavapai, Pinal, Cochise, and Pima counties, 1,000 to 7,000 feet, March and April. Canada to Florida, Arizona, and California; Europe and Africa.

10. TRAUTVETTERIA

Plants perennial, herbaceous; stems tall, branched; leaves alternate, large, deeply palmately lobed, the lobes incised and serrate; flowers in corymbose panicles, perfect, white; perianth in 1 series (petals none), of 3 to 5 segments, these concave, caducous; stamens and pistils numerous; achenes in heads, sharply angular, not ribbed, or the ribs not extending to the apex of the achene.

1. Trautvetteria grandis Nutt. Head of Black River, White Mountains, Apache County, near springs (*Goodding* 1206), July. Idaho to British Columbia, south to New Mexico, eastern Arizona, and California.

11. RANUNCULUS (104). BUTTERCUP, CROWFOOT

Contributed by Lyman Benson

Plants herbaceous; basal leaves entire, 3-lobed, 3-parted, 3-divided, or pinnately compound, the cauline leaves alternate or rarely opposite; flowers from terminal buds; sepals 5 or rarely 3, seldom persistent in fruit; petals 5, or rarely 6 to 26, yellow, seldom white or red, each bearing a scale-covered or rarely naked nectariferous pit at base of the blade; pistils 5 to many, the single ovule attached at base of the ovary.

KEY TO THE SPECIES

1. Achenes roughly transversely ridged; petals not glossy, white, the claws sometimes yellow; plants aquatic or rarely palustrine, with finely dissected, submerged leaves: subgenus *Batrachium* (2).

1. Achenes not transversely ridged; petals usually glossy, yellow or (in one species) red; plants terrestrial or palustrine, rarely aquatic, if so, then with simple entire leaves (3).

2. Pedicels not recurved at fruiting time; leaves usually petioled, the first divisions arising usually, but not always, well above the nondilated stipular leaf bases (the ends of these not free), usually collapsing when withdrawn from the water, not circinate; achenes about 10 to 20 or up to 40; dissected leaves repeatedly trichotomous .... 1. *R. aquatilis*

2. Pedicels recurved at the bases at fruiting time; leaves usually sessile, the first divisions arising usually within the dilated, stipular leaf bases (the ends of these often free), usually not collapsing when withdrawn from the water, circinate; achenes mostly 30 to 45 or 80; dissected leaves usually once or twice trichotomous 2. *R. circinatus*

3. Sepals persistent in fruit; fruits utricular; petals red. Leaves compound, dissected into lingulate leaflets: subgenus *Crymodes*.......... 3. *R. juniperinus*

3. Sepals not persistent when the fruit is mature; fruits not utricular; petals yellow (4).
4. Wall of the fruit striate, the nerves 3 or more on each face, often branched, the ovary wall very thin and usually fragile. Plants perennial, palustrine, stoloniferous, with simple, crenate leaves and elongate fruiting heads each bearing 50 to 300 achenes: subgenus *Cyrtorhyncha* ....4. *R. Cymbalaria*
4. Wall of the fruit not striate or nerved, thick and firm: subgenus *Euranunculus* (5).
5. Styles lacking, the achenes practically beakless. Plants annual, of brackish streams and marshes; juice very acrid; stem markedly fistulous; fruiting head elongate, bearing 40 to 300 achenes: section *Hecatonia*......................5. *R. sceleratus*
5. Styles present, the achenes beaked (6).
6. Leaves all entire or dentate, serrulate, or undulate. Dorsoventral dimension of the achene not more than twice or thrice the lateral dimension: section *Flammula* (7).
6. Leaves all deeply crenate, lobed, parted, or divided (9).
7. Nectary scale ciliate along the distal margin, the adjacent surface of the petal sparsely pubescent; receptacle short-pubescent; achenes about 3 to 3.2 mm. long, pubescent, each with a flat, winged stalk. Stems scapose; herbage glabrous; roots stout; petals 7 to 12 mm. long..6. *R. oreogenes*
7. Nectary scale and petal glabrous; receptacle glabrous; achenes 1.2 to 2.5 mm. long, glabrous, sessile (8).
8. Basal leaves lanceolate, oblanceolate, or linear, never ovate, the cauline leaves similar to the basal..7. *R. Flammula*
8. Basal leaves (of Arizona plants) ovate or cordate-ovate, the cauline leaves similar, or lanceolate or oblanceolate..............................8. *R. hydrocharoides*
9. Nectary scale not forming a pocket, free laterally for at least ⅔ of its length; achene discoid, the dorsoventral dimension 3 to 15 times the lateral dimension; sepals rarely with lavender or purple pigment: section *Chrysanthe* (10).
9. Nectary scale forming a pocket, attached to the petal laterally for all or most of its length; achene not discoid, plump, the dorsoventral dimension only 1 to 2.5 times the lateral dimension; sepals nearly always tinged dorsally with lavender or purple: section *Epirotes* (13).
10. Petals 8 to 18; achene beaks 3 to 5 mm. long, straight or at least not regularly curved. Plants perennial, with stout stems up to 0.9 m. long; leaves simple (and parted) or compound, 3.5 to 15 cm. long, 4 to 10 cm. wide...................9. *R. macranthus*
10. Petals 5; achene beaks 1 to 2 mm. long, curving above (11).
11. Receptacle glabrous, in fruit not more than 2.5 times its length in anthesis; head of achenes hemispheroidal or globose. Plants usually annual, the herbage glabrous or sparsely hirsute; stem leaves usually larger than the basal ones......10. *R. uncinatus*
11. Receptacle hairy, in fruit at least 3 times its length in anthesis; head of achenes ovoid or cylindroid (12).
12. Sepals equaling or a little shorter than the petals, the latter 3 to 7 mm. long; plants perennial; stems rooting at the nodes, at least in Arizona specimens. Head of achenes ovoid, 7 to 9 mm. long...11. *R. Macounii*
12. Sepals about twice as long as the petals, the latter 2 to 3 mm. long; plants annual; stems never rooting, erect......................................12. *R. pensylvanicus*
13. Achenes swollen or broadened at base and therefore oblong. Fruiting receptacle and head of achenes cylindroid or ovoid; herbage and receptacle glabrous; basal leaves cleft and again once-lobed, or the middle division entire........13. *R. Eschscholtzii*
13. Achenes obovoid, flattened-obovoid, or discoid, not swollen or flat-stiped at base (14).
14. Nectary scale and the petal glabrous (15).
14. Nectary scale ciliate (except possibly rarely in *R. arizonicus*), the adjacent surface of the petal sometimes hairy also (16).
15. Petals narrow, mostly 4 to 6 or rarely 8 mm. long, 2 to 4 mm. wide; sepals glabrous or thinly appressed-pubescent; some of the basal leaves on each plant crenate, occasional leaves 3-lobed or -divided...............................14. *R. inamoenus*
15. Petals none or, when present, broadly cuneate-obovate, about 9 to 10 mm. long, at least 5 mm. wide; sepals pilose-tomentose; basal leaves proximally cordate
15. *R. pedatifidus*

16. Petals obovate, cuneate-obovate, or obovate-deltoid, nearly or fully as wide as long; fruiting receptacle ovoid to ovoid-cylindroid; dead leaf bases only moderately separating into fibers; plants of mountain meadows............16. *R. cardiophyllus*
16. Petals oblanceolate-obovate, usually 2 to 3 times as long as wide; fruiting receptacle slender; dead leaf bases represented by numerous brown fibers, the basal leaves not persisting at flowering time; plants of dry situations..............17. *R. arizonicus*

1. Ranunculus aquatilis L. Apache County to Coconino and eastern Yavapai counties, 4,500 to 9,000 feet, ponds and streams in the pine and sagebrush belts. Almost throughout northern and middle North America; Eastern Hemisphere.

An aquatic perennial, flowering in late spring and summer. The variety occurring in Arizona is var. *capillaceus* DC. (*R. trichophyllus* Chaix).

2. Ranunculus circinatus Sibth. Coconino County and northern Gila County, 5,000 to 7,000 feet, occasional in ponds, lakes, and streams in the pine and perhaps in the oak belt. Almost throughout North America except the extreme West.

An aquatic perennial, flowering in late spring and summer. The variety occurring in North America is var. *subrigidus* (W. Drew) L. Benson.

3. Ranunculus juniperinus Jones. Virgin Mountains, Mohave County (*Goodding* 2135), 5,000 to 6,000 feet, rocky slopes in the juniper-pinyon belt, May. Utah to eastern Nevada and northwestern Arizona.

4. Ranunculus Cymbalaria Pursh. Across the state in northern (Williams River and northward) and east-central Arizona, 500 or mostly 5,000 to 8,000 feet, saline soil along streams and about springs in the pine, sagebrush, juniper-pinyon, and oak belts. Alberta to Vancouver Island and south to Kansas, Arizona, California, and Baja California.

A stoloniferous perennial with erect flowering stems, flowering in late spring and summer. It is called desert crowfoot. The Arizona plant is var. *saximontanus* Fern.

5. Ranunculus sceleratus L. Sacaton (Pinal County), about 1,200 feet, (seeds probably having been carried down the Gila River from the New Mexico highlands), in wet places, May to August. Canada and Alaska, south to Florida, New Mexico, and Arizona; Eurasia.

The species is represented in Arizona by the Western var. *multifidus* Nutt. (*R. eremogenes* Greene). The acrid juice of *R. sceleratus* is said to have been used by beggars to induce sores. It is also reported that livestock eating this plant may develop severe intestinal inflammation. The active principle (anemonal) is reputed to be a cardiac poison. The species is too rare in Arizona to make these undesirable properties important there.

6. Ranunculus oreogenes Greene. Coconino County, from the Kaibab Plateau to Rogers Lake and Oak Creek Canyon, 7,000 to 8,000 feet, March to May. Southeastern Utah to southwestern Colorado and northern Arizona.

A synonym is *R. Collomae* L. Benson, type from South Rim of Grand Canyon (*Collom* in 1941). *R. oreogenes* was named earlier from a small collection of fruiting plants from Colorado. A collection of *R. oreogenes* from the North Rim of the Grand Canyon (*McHenry* 2076) was referred previously to *R. glaberrimus* Hook. var. *ellipticus* Greene.

7. Ranunculus Flammula L. Lukachukai Mountains, Apache County, 9,000 feet (*Goodman & Payson* 2934), Kaibab Plateau (several collections), May to August. Canada and Alaska, south to New Jersey, New Mexico, northern Arizona, and California; Eurasia.

A creeping perennial. The Western plant occurring in Arizona is var. *ovalis* (Bigel.) L. Benson.

8. Ranunculus hydrocharoides Gray. White Mountains (Apache County) and San Francisco Peaks (Coconino County) to the Chiricahua and Huachuca mountains (Cochise County), 7,000 to 9,500 feet, marshes, streams, and springs in the pine belt, June to September. Southwestern New Mexico; Arizona; above Owens Lake, California; Mexico.

An aquatic or palustrine perennial, represented in Arizona by both the typical plant, with entire basal leaves and ovate or cordate-ovate stem leaves, and petals markedly longer than the sepals, and by forms tending toward the Mexican var. *stolonifer* (Hemsl.) L. Benson (*R. stolonifer* Hemsl.), with dentate basal leaves, lanceolate or oblanceolate stem leaves, and petals about equal to the sepals. Plants approaching the variety have been collected in the White Mountains (Apache and Greenlee counties) and on the San Francisco Peaks (Coconino County).

9. Ranunculus macranthus Scheele. Eastern Arizona from the White Mountains (Apache County and southern Navajo County) to the Huachuca Mountains and Hereford (Cochise County), 4,000 to 7,500 feet, in pine forests and wet meadows, June to August. Western and southern Texas, eastern Arizona, and almost throughout the highlands of Mexico.

The showiest buttercup in Arizona, worthy of cultivation as an ornamental.

10. Ranunculus uncinatus D. Don. North Rim of Grand Canyon, Coconino County, about 8,000 feet (*Collom* in 1940). Western Montana to Alaska, south to Colorado, northern Arizona, and California; at an unknown locality in Mexico (Mociño & Sessé).

A synonym is *R. Bongardi* Greene var. *tenellus* (Nutt.) Greene. The plant is reported by Mrs. Collom as very abundant at Kanabownits Spring, sometimes forming nearly pure stands.

11. Ranunculus Macounii Britton. San Francisco Peaks and Mogollon Escarpment (Coconino County) to the White Mountains (Apache County), 6,000 to 8,000 feet, usually in pine forests, July and August. Labrador to Alaska, south to Minnesota, New Mexico, Arizona, and Goose Lake, California.

Subpalustrine perennial, creeping in mud and, at least in the Arizona plants, rooting adventitiously.

12. Ranunculus pensylvanicus L. f. Vicinity of Flagstaff (Coconino County), East Fork of White River (Apache or Navajo County), near Lakeside (southern Navajo County), Camp Verde (Yavapai County), 3,000 to 8,000 feet, woods or bottom lands, summer. Newfoundland to Alaska, south to New Jersey, Missouri, New Mexico, Arizona, and Washington; China.

13. Ranunculus Eschscholtzii Schlecht. San Francisco Peaks near timber line, 11,000 feet (*Kearney & Peebles* 12155), July and August. Alaska to Colorado, northern Arizona, and the Sierra Nevada, California.

Var. *eximius* (Greene) L. Benson (*R. eximius* Greene) has been collected on the San Francisco Peaks (*Whiting & Sanders*). It is distinguished by cleft, rather than parted, basal leaves, with the lobes sharply acute; scarious leaf bases 1 to 2 or 2.5 (instead of 1 to 1.5) cm. long; and petals 8 to 12 or 17 (instead of 7 to 10) mm. long and 6 to 11 or 19 (instead of 5 to 10) mm. wide.

14. Ranunculus inamoenus Greene. Lukachukai Mountains (Apache County), Kaibab Plateau, San Francisco Peaks and vicinity, and Potato Lake (Coconino County), Sierra Ancha (Gila County), 6,500 to 9,500 feet, coniferous forests, April to August. Rocky Mountains, from Alberta to New Mexico and Arizona.

KEY TO THE VARIETIES

1. Achenes about 1.5 mm. long and 1.3 mm. thick dorsoventrally, the beak 0.8 to 0.9 mm. long, recurved; petals 2.5 to 5 (8) mm. long, 2 to 4 mm. wide; stipular leaf bases 1.5 to 2.5 cm. long....................var. *typicus*
1. Achenes about 2 mm. long and 1.5 to 1.7 mm. thick dorsoventrally, the beak usually 1.5 to 2 mm. long, not curved, hooked at tip; petals 6 to 9 mm. long, 2.5 to 7 mm. wide; stipular leaf bases 2.5 to 3 cm. long....................var. *subaffinis*

The type of a plant described as *R. affinis* R.Br. var. *micropetalus* Greene, a form with small flowers and fruits, was collected on the San Francisco Peaks (*Greene* in 1889). This variety is not recognized by the writer. There occurs also as an endemic on the San Francisco Peaks, near timber line at 10,000 to 12,000 feet, var. *subaffinis* (Gray) L. Benson, the type of which was collected on Mount Agassiz (*Lemmon* 4152).

15. Ranunculus pedatifidus J. E. Smith. San Francisco Peaks, Coconino County, 12,000 feet (*Little* 4632). Greenland to Yukon, south to Colorado and northern Arizona; northern Asia.

The Arizona plant and similar collections from southern Utah may represent an undescribed variety. It is apetalous and has stems not more than 15 cm. long; the basal leaves are 1 to 2 cm. long, deeply lobed or cleft into simple divisions, the upper lobes erect, the lower ones divergent or somewhat reflexed.

16. Ranunculus cardiophyllus Hook. Kaibab Plateau, Grand Canyon, and San Francisco Peaks (Coconino County) and occasional southeastward to the White Mountains (Apache County), 7,000 to 9,500 feet, pine forests, flowering mostly in June and July. Chiefly in the Rocky Mountains, from Alberta to northern New Mexico, Arizona, and northeastern Washington.

KEY TO THE VARIETIES

1. Herbage pilose, often densely so; stems not scapose, 2 to 4 mm. in diameter, branching from near the base, 1- to 5- (8-) flowered; basal leaves round-cordate, 1 to 6 cm. long, 1 to 5 cm. wide; petals cuneate-obovate, 8 to 15 mm. long, 6 to 13 mm. wide; achenes 50 to 125 in a cylindroid head, this (8) 10 to 15 mm. long and (5) 7 to 9 mm. in diameter, the beak curved....................var. *typicus*
1. Herbage glabrous or moderately pilose; stems scapose, 1 to 1.5 or 2.5 mm. in diameter, usually unbranched below, 1- to 3- or 4-flowered; basal leaves cordate or long-ovate, rarely subsagittate, 1.5 to 4.5 cm. long, 1 to 4 cm. wide; petals obovate-obdeltoid, 7 to 15 mm. long, 5.5 to 13 mm. wide; achenes 20 to 50 in an ovoid head, this 5 to 6 mm. long and 4 to 6 mm. in diameter, the beak straight....................var. *subsagittatus*

In addition to typical *R. cardiophyllus*, var. *subsagittatus* (Gray) L. Benson is common in northern Arizona, ranging from the North Rim of the Grand

Canyon (where it is uncommon and poorly defined) and the San Francisco Peaks to the White Mountains (Apache and Greenlee counties). The type of this variety was collected by Lemmon in De La Vergne Park (now Fort Valley), near Flagstaff. Forms intermediate between this variety and typical *R. cardiophyllus* are as numerous as the extreme forms in the area about Flagstaff where the ranges overlap.

17. Ranunculus arizonicus Lemmon. Greenlee County to the Chiricahua, Huachuca, and Patagonia mountains (Cochise and Santa Cruz counties), 5,000 to 7,000 feet, dry situations, summer, type from Rucker Valley, Cochise County (*Lemmon* 585). Southwestern New Mexico and southeastern Arizona; northwestern Mexico.

A species remarkable in this genus for its habitat in dry situations.

12. THALICTRUM (105). MEADOW-RUE

Plants herbaceous, perennial; leaves alternate, large, twice or thrice ternate with numerous cleft or shallowly lobed leaflets, the basal leaves long-stalked; flowers mostly unisexual, small, greenish or yellowish, in terminal panicles; perianth of 4 or 5 caducous segments, these all alike; petals none; stamens numerous; pistils few, becoming asymmetric achenes in fruit, these with prominent longitudinal ribs extending from base to apex.

KEY TO THE SPECIES

1. Stem leaves sessile or very nearly so, the leaflets longer than wide, rather thick and rigid, distinctly pubescent beneath; filaments white, slightly dilated toward apex; achenes moderately asymmetric.................................1. *T. dasycarpum*
1. Stem leaves petioled (except sometimes the uppermost), the leaflets usually wider than long, thin, not rigid, minutely puberulent or glabrate beneath; filaments yellow, filiform; achenes very asymmetric, gibbous on the ventral side..........2. *T. Fendleri*

1. Thalictrum dasycarpum Fisch. & Lall. Showlow, Navajo County, 6,000 feet, July (*Hough*). Ontario to British Columbia, south to Ohio, Louisiana, Texas, and Arizona.

Var. *hypoglaucum* (Rydb.) Boivin is distinguished by Boivin (105, pp. 482, 483) as having thinner leaves, often longer stigmas, and a more elongate receptacle than in the typical plant. Specimens from the White Mountains, Apache County (*Lead* 1511) and from Oak Creek, Coconino (?) County (*Rusby* in 1883) are referred by Boivin to this variety.

2. Thalictrum Fendleri Engelm. Apache County to Mohave County, south to Cochise, Santa Cruz, and Pima counties, 5,000 to 9,500 feet, mostly in pine forests, common, April to August. Wyoming to Oregon, south to northern Mexico.

Extremely variable in size of leaflets, these from 4 to 30 mm. wide. Var. *Wrightii* (Gray) Trel. occurs in the Santa Rita Mountains, Pima County, and probably elsewhere in southern Arizona. It is distinguished from typical *T. Fendleri* by Boivin (105, p. 457) in having the herbage more frequently glabrous and the achenes with only 1 lateral nerve or with the central nerve much thicker than the others, whereas in the typical plant the herbage is pubescent and all 3 lateral nerves of the achene are conspicuous.

53. BERBERIDACEAE. Barberry Family

1. BERBERIS. Barberry, Holly-grape

Shrubs or undershrubs; wood and inner bark yellow; leaves compound, pinnate or palmately trifoliolate, the leaflets thick, evergreen, spiny-toothed, more or less conspicuously reticulate-veined; inflorescence racemose or subcorymbose; flowers perfect, regular, yellow; sepals, petals, and stamens 6, the sepals and petals each in 2 series; anthers opening by 2 apical valves; ovary superior, 1-celled; fruit a few-seeded berry.

These handsome plants are secondary hosts of the black stem rust of cereals, which restricts their otherwise considerable value for planting as ornamentals. Some (perhaps all) of the species contain berberine, which has limited use as a drug, and the Indians are said to use the root as a tonic. The plants are reputed to be sometimes poisonous to livestock. The juicy berries of *B. repens* and *B. haematocarpa* are excellent for making jelly and are eaten by birds and various mammals. A brilliant yellow dye is obtainable from the roots. All the Arizona species belong to the holly-grapes, subgenus *Mahonia* or *Odostemon*.

KEY TO THE SPECIES

1. Leaves palmately trifoliolate. Leaflets with 1 or 2 pairs of large teeth (2).
1. Leaves pinnate, the larger ones usually with 5 or more leaflets (3).
2. Foliage usually very glaucous; leaflets lanceolate, long-acuminate, the terminal one 1.5 to 2 (exceptionally 2.5) cm. wide (including the teeth), longer than the petiole; filaments not toothed; berries red when ripe....................1. *B. trifoliolata*
2. Foliage not glaucous; leaflets rhombic-oblong or triangular, short-acuminate, the terminal one 2.5 to 3.5 cm. wide, usually equaling or somewhat shorter than the petiole; filaments conspicuously bidentate, with spurlike teeth; berries blue-black when ripe 2. *B. Harrisoniana*
3. Inflorescence elongate, racemose, many-flowered, usually dense at anthesis; berries ovoid or ellipsoid, blue-black and very glaucous when ripe. Leaves not or only moderately glaucous, the leaflets broadly oblong-ovate to nearly orbicular (4).
3. Inflorescence usually short and subcorymbose, relatively few-flowered, loose; berries globose or nearly so, not or only slightly glaucous (5).
4. Stems above ground very short or almost none, seldom more than 10 cm. long; leaflets with numerous, commonly 10 or more, small and slender teeth, seldom strongly reticulate ....3. *B. repens*
4. Stems above ground usually 20 cm. long or longer; leaflets with usually fewer than 10 coarse teeth, commonly strongly reticulate......................4. *B. Wilcoxii*
5. Berries at maturity light yellow, rarely red, becoming dry and more or less inflated; leaflets usually distinctly reticulate, moderately glaucous, ovate or broadly oblong, the terminal one short-acuminate, seldom more than 2.5 cm. long or more than twice as long as wide..5. *B. Fremontii*
5. Berries at maturity blood red, remaining juicy; leaflets not or obscurely reticulate, very glaucous, mostly lanceolate or oblong-lanceolate, the terminal one long-acuminate, commonly 3 or more (up to 11) cm. long and 2 to 5 times as long as wide 6. *B. haematocarpa*

1. Berberis trifoliolata Moric. Pima County, in the Santa Catalina Mountains (*Vasey* in 1881) and Rincon Mountains, 3,000 feet (*Benson* 11325, *Frost* 487). Texas to southern Arizona and northern Mexico.

Known in Texas as algerita. The Arizona plants presumably all belong to *Mahonia trifoliolata* var. *glauca* Johnst., described as differing from typical *B. trifoliolata* in the opaque, glaucous, minutely papillate leaves.

2. Berberis Harrisoniana Kearney & Peebles. Palm Canyon, Kofa Mountains, Yuma County (*Peebles & Loomis* 6768, the type collection), Pitahaya Canyon, Ajo Mountains (*Nichol* in 1939), 2,500 to 3,500 feet, rocky slopes, February and March. Known only from southwestern Arizona.

In the Nichol specimen, which has leaves only, the terminal leaflets are longer (up to 5 cm. long) and more attenuate at base than in the type.

3. Berberis repens Lindl. Apache, Navajo, Coconino, Graham, Gila, and Yavapai counties, 5,000 to 8,500 feet, chiefly in coniferous forests, April to June. Wyoming to British Columbia, New Mexico, Arizona, and California.

Creeping mahonia. The spreading rootstocks of this low shrub make it an excellent ground cover, protective against erosion.

4. Berberis Wilcoxii Kearney. Near Jerome Junction, Yavapai County, and in the mountains of Graham, Gila, Cochise, Santa Cruz, and Pima counties, 5,500 to 8,000 feet, April, type from the Huachuca Mountains (*Wilcox* in 1894). New Mexico, Arizona, and Sonora.

A straggling shrub, frequently 2 m. high, the flowers fragrant. The plant seems to be related to *B. dictyota* Jepson, of California.

5. Berberis Fremontii Torr. Apache County to eastern Mohave and northern Yavapai counties, 4,000 to 7,000 feet, often with pinyon and juniper, April to July. Colorado, Utah, New Mexico, and northern Arizona.

In the valley of the Little Colorado River this is a roundish shrub 6 to 8 feet high, but it is reported as occasionally reaching tree size in the Grand Canyon National Park. Various articles are made from the wood by the Hopi Indians.

6. Berberis haematocarpa Wooton (*B. Nevinii* Gray var. *haematocarpa* L. Benson). Southern Apache County, Grand Canyon (Coconino County), and Hualpai Mountain (Mohave County), south to Cochise, Santa Cruz, Pima, and Yuma counties, 4,500 feet or lower, common, usually with scrub oak and other chaparral plants, February to May. Western Texas, New Mexico, Arizona, and Mexico.

Red mahonia. Mrs. Collom reported that the flowers are fragrant and that a delicious red jelly can be made from the fruits. The ranges of this species and its near relative *B. Fremontii* scarcely overlap in Arizona.

54. MENISPERMACEAE. Moon-seed Family

1. COCCULUS. Snail-seed

Plants dioecious, woody, climbing; leaves alternate, simple, with entire, thickish, semievergreen, lanceolate to ovate blades; flowers regular, small, in axillary panicles; sepals, petals, and stamens 6, the sepals and petals each in 2 series, the 3 inner sepals larger than the other perianth segments; anthers completely or incompletely 4-celled; pistils 3 to 6 in fertile flowers; fruit a dark-purple drupe, with a flattened stone.

1. Cocculus diversifolius DC. Pima County, westward to the Baboquivari and Santa Rosa mountains, 3,500 to 5,000 feet, in thickets, May to August. Southern Texas, southern Arizona, and far south in Mexico.

55. PAPAVERACEAE. Poppy Family

Plants herbaceous, of diverse character; leaves simple or decompound; flowers perfect, regular or irregular, solitary or in small clusters; sepals 2 or 3, caducous; petals 4 to 6, distinct or the inner ones cohering at apex; stamens 6 to numerous; pistil 1, or the pistils several; fruits various.

The outstanding plants of this family are the true poppies (genus *Papaver*), comprising several species of great value as ornamentals, one of them (*P. somniferum*) being the source of the drug opium. This species was observed in 1932 growing wild at a roadside near Picacho, Pinal County, but did not become established. *Eschscholtzia californica,* the state flower of California, is a deservedly popular plant in flower gardens and is usually known as California-poppy.

KEY TO THE GENERA

1. Flowers very irregular, 1 or both of the outer petals spurred at base, the smaller inner petals united at apex and enclosing the anthers and stigma; stamens 6. Leaves decompound: subfamily *Fumarioideae*......................................5. *Corydalis*
1. Flowers regular, the petals all alike, distinct; stamens numerous: subfamily *Papaveroideae* (2).
2. Herbage, sepals, and capsules prickly; leaves large, sinuate-dentate or sinuate-pinnatifid; sepals with hornlike appendages; stems leafy....................3. *Argemone*
2. Herbage, sepals, and capsules not prickly (or the leaves somewhat so in *Arctomecon*); leaves not sinuate or pinnatifid; sepals not appendaged; plants scapose or subscapose (3).
3. Fruit of several separating carpels, these becoming torulose or moniliform 1. *Platystemon*
3. Fruit a single capsule (4).
4. Leaves compound, ternately dissected; sepals united into a conic, acuminate cap, this pushed off as the flower expands..............................2. *Eschscholtzia*
4. Leaves simple; sepals not united into a cap.........................4. *Arctomecon*

1. PLATYSTEMON. Cream-cups

Plants annual; herbage pilose or subhirsute; leaves linear or narrowly lanceolate, entire; flowers solitary, long-stalked; sepals 3; petals 6, in 2 series, cream-colored; pistils 6 or more, connivent or coherent in a circle, separating in fruit; stigmas subulate-filiform.

The peculiar fruits have been compared to tiny ears of corn with the husks removed.

1. Platystemon californicus Benth. Yavapai, Mohave, Graham, Gila, Maricopa, Pinal, and Pima counties, 1,500 to 4,500 feet, moist ground along streams, March to May. Southern Utah, Arizona, and California.

There is much variation, especially in the pubescence of the herbage and fruits, the latter varying from glabrous to bristly-hirsute. Several segregate species were described by E. L. Greene (*Pittonia* 5: 176, 190. 1903), of which the following are based on Arizona types: *P. arizonicus* (type *Pringle* in 1882, Santa Catalina Mountains), *P. confinis* (type *Toumey* 47b, Bradshaw Mountains), and *P. mohavensis* (type *Jones* in 1884, Hackberry). There seem to be no satisfactory characters for distinguishing these entities, even as varieties.

2. ESCHSCHOLTZIA. Gold-poppy, California-poppy

Plants annual; leaves ternately dissected, smooth, glaucous; flowers solitary on long peduncles, or in small, loose clusters at the ends of the branches; petals orange or yellow; receptacle concave around the base of the pistil, often with a spreading outer rim and an erect, scarious, inner rim; fruit a slender 1-celled, 2-valved, several-seeded, longitudinally ribbed capsule.

The juice of *E. californica* Cham. is reported to be mildly narcotic and to have been used by the Indians of California in alleviating toothache.

KEY TO THE SPECIES

1. Outer rim of the hypanthium distinct, usually at least 0.5 mm. wide, nearly as wide as to wider than the scarious inner rim, after anthesis more or less cartilaginous and flaring or revolute; cotyledons bifid. Stems leafy and branching above the base, or the plants nearly acaulescent; petals 15 to 30 mm. long; mature seeds dark-colored, rugose-reticulate .... 1. *E. mexicana*
1. Outer rim of the hypanthium indistinct or nearly obsolete, less than 0.5 mm. wide, usually narrower than the inner rim, after anthesis not flaring; cotyledons entire (2).
2. Plants acaulescent or nearly so, very glaucous; leaf segments numerous and crowded, very narrow, often almost filiform; petals 10 to 25 mm. long; stamens usually more than 15; mature seeds at least partly covered with a thick, gray, deeply pitted, outer coat .... 2. *E. glyptosperma*
2. Plants branching and leafy well above the base, slightly to rather pronouncedly glaucous; leaf segments relatively few and not crowded; petals less than 10 mm. long; stamens usually fewer than 15; mature seeds dark-colored, with grayish reticulations, not deeply pitted........ 3. *E. minutiflora*

1. Eschscholtzia mexicana Greene. Throughout the state except the northeastern portion, 4,500 feet or lower, on plains and mesas, February to May. Western Texas to southern Utah, southeastern California, and northern Sonora.

In favorable springs the landscape is colored over extensive areas by the showy orange-colored (rarely white or pink) flowers. It is reported that in southern Arizona the plants are grazed by cattle in winter and early spring when other feed is scarce.

E. mexicana is closely related to *E. californica* Cham., differing apparently only in being a smaller, more scapose, probably always annual plant and in having a narrower, sometimes nearly obsolete, outer rim of the hypanthium. E. L. Greene (*Pittonia* 5: 260–262) also published as species, based on Arizona types, *E. aliena, E. arizonica, E. Jonesii,* and *E. paupercula,* but these do not seem sufficiently distinct from *E. mexicana.*

2. Eschscholtzia glyptosperma Greene. Mohave and Yuma counties, 2,000 feet or lower, in sandy soil of deserts, February to May. Southwestern Utah and western Arizona to California.

3. Eschscholtzia minutiflora Wats. (*E. micrantha* Greene). Mohave County to Pinal, Maricopa, Santa Cruz, Pima, and Yuma counties, 4,500 feet or lower, commonly in sandy soil, February to May. Southern Utah to southeastern California and western Arizona.

3. ARGEMONE. Prickle-poppy, Chicalote

Perennial, herbaceous, glaucous, rather coarse, prickly plants, with yellow sap; stems erect, leafy; flowers large; sepals usually 3, bearing a rigid, hornlike, spine-tipped appendage; petals 4 to 6, white or pale yellow, contrasting with the numerous orange-colored stamens; stigma large, with radiating lobes; capsule dehiscent apically by valves.

The white-flowered species are handsome and, except for the prickles, resemble the Matilija-poppy (*Romneya*) of California. The plants grow in dry soil in fields and at roadsides and are decidedly drought-resistant. They are unpalatable to livestock. An abundance of these plants on cattle range is an indication of excessive overgrazing. The acrid yellow juice of *A. mexicana* has been used to treat cutaneous diseases.

KEY TO THE SPECIES

1. Petals pale yellow to orange. Herbage sparsely spiny, without fine bristles; horns of the sepals without lateral spines .... 1. *A. mexicana*
1. Petals white, occasionally tinged with pink (2).
2. Horns of the sepals long and slender, dilated only at base, without lateral spines or with a very few slender ones near the base; valves of the capsule not becoming indurate-thickened, rather sparsely spiny; herbage with few or no short fine bristles 2. *A. intermedia*
2. Horns of the sepals short and stout, dilated well above the base, usually with several lateral spines and bristles, these sometimes extending nearly to the apex of the horn; valves of the capsule becoming indurate-thickened, usually copiously to densely spiny; herbage usually with numerous short, fine bristles .... 3. *A. platyceras*

1. Argemone mexicana L. Occasional near Tucson (Pima County), April and May; probably introduced from tropical America.

2. Argemone intermedia Sweet. Yavapai, Maricopa, Pinal, Cochise, Santa Cruz, and Pima counties, 1,500 to 5,000 feet, flowering most of the year. South Dakota and Wyoming to Arizona and northern Mexico.

3. Argemone platyceras Link & Otto. Apache County to Mohave County, south to Cochise, Santa Cruz, and Pima counties, 1,500 to 8,000 feet, flowering almost throughout the year. Nebraska and Wyoming to Arizona and Mexico.

Specimens with numerous short, fine bristles approach var. *hispida* (Gray) Prain.

4. ARCTOMECON. Desert-poppy

Plants herbaceous, biennial or perennial, with a stout taproot; leaves mostly basal, hirsute or hispid with long hairs, wedge-shaped, commonly dentate or 3-lobed at apex; flowers on long peduncles, large; sepals 2 or 3, caducous; petals 4 or 6, white or pale yellow; capsule oblong, ovoid, or obovoid, 3- to 6-valved, dehiscent apically.

KEY TO THE SPECIES

1. Herbage sparsely hispid-hirsute; petals 4, white; stems up to 25 cm. long, usually leafy; leaves deeply 3-toothed at apex .... 1. *A. humilis*
1. Herbage copiously shaggy-villous; petals usually 6, yellow; stems up to 60 cm. long, nearly naked; leaves shallowly toothed at apex .... 2. *A. californica*

1. Arctomecon humilis Coville. North of Wolf Hole, Mohave County, 2,500 feet, May (*Peebles & Parker* 14749). Southwestern Utah and northwestern Arizona.

A handsome plant, the petals pure white.

2. Arctomecon californica Torr. & Frém. Near Pierce Ferry, northern Mohave County (*Hester* in 1941). Southern Nevada and northwestern Arizona.

5. CORYDALIS (106)

Plants herbaceous, biennial or short-lived perennial; herbage glabrous, glaucous; leaves dissected into numerous small segments; flowers very irregular, in spikelike racemes; sepals 2; corolla yellow, conspicuously spurred; stamens 6; capsules elongate, cylindric, usually curved, 2-valved, more or less torulose; seeds numerous, black, shining.

The plants contain several alkaloids and are said to be poisonous to sheep, less so to cattle, if eaten freely.

1. Corydalis aurea Willd. Throughout the state except the extreme western part, 1,500 to 9,500 feet, February to June (sometimes also late summer). Quebec to Alaska, south to Pennsylvania, Arizona, northern Mexico, and California.

Two subspecies, both occurring practically throughout the range of the species in Arizona, are recognized by Ownbey. These are: subsp. *aurea* G. B. Ownbey (*C. Wetherillii* Eastw., *C. Jonesii* Fedde), with racemes usually surpassed by the leaves, fruits spreading or pendent, and seeds not ring-margined; and subsp. *occidentalis* (Engelm.) G. B. Ownbey (*C. montana* Engelm., *C. Jonesii* var. *stenophylla* Fedde, *C. pseudomicrantha* var. *Griffithsii* Fedde), with racemes usually surpassing the leaves, fruits erect, often incurved, and seeds usually ring-margined. Intermediate forms are of frequent occurrence. The type of *C. Wetherillii* came from the Grand Canyon (*A. Wetherill* in 1897), that of *C. Jonesii* from near Pagumpa, Mohave County (*Jones* 5085), that of *C. Jonesii* var. *stenophylla* from Flagstaff (*MacDougal* 105), and that of *C. pseudomicrantha* var. *Griffithsii* from the Santa Rita Mountains (*Griffiths* 3846).

56. CRUCIFERAE.[1] Mustard Family

Plants herbaceous or (in *Lepidium*) sometimes suffrutescent, annual or perennial; leaves alternate, commonly simple and entire to pinnatifid (pinnate to tripinnate in *Descurainia*); flowers perfect, regular; sepals and petals 4, the petals rarely wanting; stamens usually 6 and usually tetradynamous (1 pair shorter); pistil 1, the stigma entire or 2-lobed, the ovary superior, sessile or stipitate; fruit a capsule (silique or silicle), commonly 2-celled with a thin longitudinal partition and more or less dehiscent, or fruit 1-celled and indehiscent in a few genera.

This large family, chiefly of temperate regions, includes such well-known garden vegetables and culinary herbs as cabbage, cauliflower, turnip, radish, cress, and mustard. Many of the species found in Arizona are weeds of Old World origin. Few of them have any considerable value as forage, but species of *Lepidium, Lesquerella,* and *Descurainia,* although avoided while the plants are green, are relished, especially by horses, when the capsules are ripe. It has been reported that hay containing mature plants of *Sisymbrium, Brassica, Camelina,* and *Conringia* may cause disorders in livestock because of the oil of mustard in the seeds. Plants of certain genera (*Stanleya* and *Descurainia*) are used as potherbs by the Indians.

[1] Dr. Reed C. Rollins has obligingly given many useful suggestions on the treatment of this family.

There is much difference of opinion concerning the limits of genera in the Cruciferae, and any treatment of them is necessarily tentative. Several of the segregate genera that were recognized in *Flowering Plants and Ferns of Arizona* are here restored to *Thelypodium* and *Sisymbrium* respectively, although this disposition makes these genera somewhat heterogeneous aggregates.

KEY TO THE GENERA

1. Fruits 1-celled, 1-seeded, indehiscent, thin and flat, orbicular or nearly so. Pedicels strongly decurved in fruit (2).
1. Fruits 2-celled, normally containing 2 or more seeds and normally dehiscent (capsular) (3).
2. Pubescence, if any, of simple hairs; fruits with conspicuous, commonly perforate wing margins, not bearing hooked hairs........................24. *Thysanocarpus*
2. Pubescence partly of branched hairs; fruits wingless, bearing more or less hooked hairs 25. *Athysanus*
3. Capsules strongly compressed contrary to the very narrow partition, not more than twice as long as wide (4)
3. Capsules not compressed contrary to the partition, but sometimes appearing so in dried specimens (10).
4. Petals brown-purple, 15 to 20 mm. long. Plants perennial, the herbage densely stellate-pubescent; capsules obovate or wedge-shaped, often constricted between base and apex ....16. *Lyrocarpa*
4. Petals white to yellow, sometimes purple-tinged, less than 15 mm. long (5).
5. Capsules 2-lobed (didymous). Herbage stellate-pubescent (6).
5. Capsules not 2-lobed, at most obcordate (7).
6. Plants not lepidote; capsules flat; seed solitary in each cell..............17. *Dithyrea*
6. Plants silvery-lepidote; capsules inflated; seeds commonly more than 1 in each cell 18. *Physaria*
7. Ovule and seed solitary in each cell; pubescence, if any, of simple hairs..6. *Lepidium*
7. Ovules (and usually the seeds) 2 or more in each cell; pubescence, if any, at least partly of stellate hairs (8).
8. Leaves all entire or merely dentate. Herbage glabrous; stem leaves auriculate-clasping 7. *Thlaspi*
8. Leaves (the basal ones) usually pinnatifid (9).
9. Plants glabrous or stellate-puberulent; stems very slender; leaves mostly petioled, entire or few-toothed; capsules somewhat turgid, elliptic, entire at apex 20. *Hutchinsia*
9. Plants more or less hirsute below; stems relatively stout; stem leaves sessile, auriculate, the basal ones pinnatifid; capsules flat, obdeltoid, notched at apex......21. *Capsella*
10. Capsules not more than twice as long as wide (11).
10. Capsules more than twice as long as wide (14).
11. Capsules flat, strongly compressed parallel to the broad partition. Herbage commonly with stellate or forked hairs..23. *Draba*
11. Capsules more or less turgid (12).
12. Valves of the obovoid capsules with a distinct central nerve extending from base to apex. Leaves entire or denticulate...22. *Camelina*
12. Valves of the ovoid, ellipsoid, or globose capsules without a distinct central nerve, or the nerve not extending to the apex (13).
13. Seeds plump; herbage glabrous or sparsely pubescent with simple hairs; leaves pinnate or pinnatifid; aquatic or marsh plants................................14. *Rorippa*
13. Seeds flat; herbage densely stellate-pubescent or lepidote; leaves mostly entire (the basal ones lyrate-pinnatifid in one species); dry-land plants........19. *Lesquerella*
14. Capsules long-stipitate, the stipe (gynophore) at least 10 mm. long. Sepals in anthesis divaricate or reflexed; petals narrow, yellow or ochroleucous, 8 mm. or longer; pods elongate, slender, terete or subterete..................................1. *Stanleya*
14. Capsules sessile or if stipitate, then the stipe much less than 10 mm. long (15).

15. Calyx closed or nearly so in anthesis, usually somewhat flask-shaped, the sepals spreading, if at all, only at the tip (16).
15. Calyx open in anthesis, not flask-shaped (21).
16. Herbage sparsely pubescent with gland-tipped hairs; capsules not longitudinally dehiscent, the lower portion torulose, with transverse partitions between the seeds, the upper portion seedless, beaklike. Plants annual; petals pink or lilac 30. *Chorispora*
16. Herbage glabrous or nearly so, often glaucous; capsules longitudinally dehiscent, not torulose or transversely partitioned (17).
17. Petal blades flat or nearly so, much wider than the claws; anthers soon becoming strongly curved or coiled; stigma entire or slightly 2-lobed, with lobes parallel to the placentas and septum..8. *Sisymbrium*
17. Petal blades strongly crisped, channeled, or cucullate, little if any wider than the claws; anthers straight or moderately curved; stigma entire or, if 2-lobed, then with lobes at a right angle to the placentas and septum, hence over the valves (18).
18. Valves of the mature capsules remaining attached to the frame (replum) toward the apex. Stem leaves not clasping; flowers inconspicuous; capsules conspicuously beaked, becoming reflexed, somewhat compressed, less than 2 mm. wide; seeds winged 4. *Streptanthella*
18. Valves of the mature capsules separating completely from the frame (19).
19. Capsules at maturity strongly compressed, 3 mm. or wider (if narrower, then the inflorescence bracteate, diffusely paniculate); seeds flat, conspicuously winged. Stem leaves clasping, cordate or sagittate at base........................5. *Streptanthus*
19. Capsules terete or nearly so, about 2 mm. wide; seeds not, or only narrowly, winged (20).
20. Stems slender, flexuous; stem leaves auriculate; petals ochroleucous, sometimes purple-veined; style evident; stigma entire or shallowly lobed; capsules reflexed 2. *Thelypodium*
20. Stems stout, often inflated; stem leaves not auriculate; petals purple or purplish; style none; stigma deeply lobed; capsules erect or somewhat ascending 3. *Caulanthus*
21. Mature capsules strongly compressed parallel to the partition, flat or twisted or if subterete (in *Arabis glabra*), then erect, crowded, and not more than 1 mm. wide. Pubescence, if any, usually at least partly of stellate or forked hairs (22).
21. Mature capsules terete or tetragonal, not strongly compressed (except in *Diplotaxis*), but sometimes appearing so in immature dried specimens (23).
22. Capsules not more than 12 mm. long, often twisted; petals white or yellow 23. *Draba*
22. Capsules much more than 12 mm. long, not twisted; petals white, ochroleucous, or purplish pink ....27. *Arabis*
23. Capsules becoming strongly reflexed. Herbage more or less pubescent..2. *Thelypodium*
23. Capsules not strongly reflexed but sometimes recurved-spreading (24).
24. Beak of the capsule stout, indehiscent, extending much beyond the valves. Leaves (at least the lower ones) commonly lyrate-pinnatifid (25).
24. Beak of the capsule none, or slender, or not more than 3 mm. long (occasionally longer in *Erysimum, Conringia,* and *Cardamine*) (26).
25. Capsules moderately compressed, the beak not tapering, 2 to 3 mm. long; seeds in 2 rows...10. *Diplotaxis*
25. Capsules terete or only slightly compressed, with a distinctly tapering (conic or subulate) beak usually more than 3 mm. long; seeds mostly in 1 row........11. *Brassica*
26. Petals deeply lobed...13. *Dryopetalon*
26. Petals entire or nearly so (27).
27. Pubescence partly of forked or stellate hairs (28).
27. Pubescence of simple hairs, or none (31).
28. Leaves pinnate or very deeply pinnatifid; pubescence often partly glandular 26. *Descurainia*
28. Leaves entire to moderately pinnatifid; glandular hairs none (29).

29. Pubescence closely appressed, harsh; petals 6 mm. long or longer, yellow, orange, or maroon; stigma large, deeply 2-lobed. Capsules tetragonal, rigid....28. *Erysimum*
29. Pubescence not closely appressed; petals 6 mm. long or shorter, white, whitish, or pink; stigma small, entire or nearly so (30).
30. Herbage finely and softly pubescent; stems lax; petals white; capsules not more than 2 cm. long, not rigid, terete, with a relatively long, very slender beak 9. *Halimolobos*
30. Herbage coarsely and harshly pubescent; stems somewhat rigid; petals purplish pink; capsules 4 to 6 cm. long, rigid, tetragonal, very short-beaked........29. *Malcolmia*
31. Stem leaves pinnately cleft to pinnate (32).
31. Stem leaves entire or merely dentate. Capsules elongate (35).
32. Capsules rounded-tetragonal, elongate. Petals yellow, 3 to 5 mm. long....12. *Barbarea*
32. Capsules terete or obscurely tetragonal (33).
33. Capsules 3 cm. or longer or, if less than 3 cm. long, then closely appressed to the stem; dry-land plants, or weeds of fields and roadsides..................8. *Sisymbrium*
33. Capsules less than 3 cm. long, not closely appressed; plants chiefly of moist soil or aquatic (34).
34. Petals yellow or white; capsules spreading or ascending................14. *Rorippa*
34. Petals whitish, 2 to 3 mm. long; capsules erect, very slender, less than 1 mm. wide 15. *Cardamine*
35. Petals yellow. Herbage glabrous (36).
35. Petals white, pink, or purple. Plants mostly perennial (37).
36. Plants perennial; stem leaves narrow, not clasping; petals somewhat channeled or crisped; capsules terete or nearly so; anthers deeply sagittate......8. *Sisymbrium*
36. Plants annual; stem leaves broad, clasping; petals flat; capsules sharply tetragonal; anthers not or scarcely sagittate.................................31. *Conringia*
37. Stem leaves conspicuously petiolate, deeply cordate; herbage glabrous or soft-pilose; anthers not sagittate..15. *Cardamine*
37. Stem leaves sessile or nearly so, not cordate but often auriculate or sagittate; herbage glabrous or very nearly so; anthers sagittate (38).
38. Stigma conic, often pointed, entire or nearly so; outer sepals strongly gibbous at base; petals 10 to 15 mm. long, purplish pink; stem leaves not clasping 8. *Sisymbrium*
38. Stigma not conic or pointed; outer sepals not or not strongly gibbous; petals less than 10 mm. long or, if longer, violet-purple; stem leaves often clasping (39).
39. Capsules torulose, the septum with a central strip of elongate cells appearing, at low magnification, like a broad yellowish midvein; stigma lobes, if any, at a right angle to the septum, hence over the valves............................2. *Thelypodium*
39. Capsules not torulose, the septum undifferentiated; stigma lobes, if any, parallel to the septum ....8. *Sisymbrium*

1. STANLEYA (107). Desert-plume

Plants perennial, herbaceous or slightly suffrutescent, rather coarse; stems tall, stout; leaves entire or pinnatifid; flowers large for the family, in elongate, terminal racemes; petals yellow or cream-colored, with long claws; stamens not or scarcely tetradynamous, the anthers long and narrow, often curved; capsules long-stipitate, slender, nearly terete.

The plants grow usually on seleniferous soils, hence are probably poisonous to livestock. *S. pinnata* is an outstanding seleniferous plant. The Indians used the plants as a potherb and made mush with the seeds.

KEY TO THE SPECIES

1. Petals with the claw glabrous on the inner face, the blade about 1 mm. wide; leaves entire or nearly so. Herbage glabrous; leaves 3 to 10 cm. wide; petals yellow to nearly white...1. *S. elata*

1. Petals with the claw densely pubescent on the inner face, the blade 1.5 mm. wide or wider; lower stem leaves usually pinnatifid (2).
2. Plants suffrutescent; petal blades oblong, 1.5 to 3 mm. wide, bright yellow . . 2. *S. pinnata*
2. Plants herbaceous; petal blades obovate, 4 to 10 mm. wide, pale yellow or whitish 3. *S. albescens*

1. Stanleya elata Jones. North of Tuba, Coconino County (*Jaeger* in 1927, cited by Rollins). Northern Arizona to southeastern California.

What seems to be an exceptionally narrow-leaved form of this species was collected at Wikieup, southeastern Mohave County (*Gentry* 5962).

2. Stanleya pinnata (Pursh) Britton. Apache County to Mohave, Yavapai, Gila, and Yuma counties, 2,500 to 6,000 feet, dry plains and mesas, May to September. North Dakota to Idaho, south to Texas, Arizona, and California.

The Arizona specimens belong to var. *typica* Rollins.

3. Stanleya albescens Jones. Moenkopi (*Jones* in 1890, the type collection) and north of Cameron (Coconino County), Hopi Villages and Coal Mine Canyon (Navajo County), 4,000 to 5,000 feet. Colorado, Utah, New Mexico, and northern Arizona.

2. THELYPODIUM (108, pp. 260–282)

Plants annual or perennial; herbage glabrous or pubescent, the hairs simple or branched; leaves entire to pinnatifid; inflorescences simple or compound; petals white or purplish-pink; capsules long and slender, sessile or short-stipitate, terete or nearly so, erect to reflexed.

KEY TO THE SPECIES

1. Sepals in anthesis spreading or reflexed. Herbage glabrous; leaves, at least the lower ones, usually coarsely toothed, cleft, or pinnatifid; racemes dense in flower; petals white or pinkish, 5 to 7 mm. long; anthers sagittate, soon becoming curved or coiled; stigma entire or very nearly so; capsules usually short-stipitate, almost filiform, spreading or upcurved . . . . 1. *T. Wrightii*
1. Sepals in anthesis erect or ascending (2).
2. Calyx in anthesis closed or nearly so, more or less flask-shaped, the sepals erect, or spreading only at the tip; stem leaves auriculate-clasping. Stems more or less flexuous; herbage glabrous; racemes open, few-flowered; petals ochroleucous, sometimes purple-veined; capsules reflexed . . . . 2. *T. Cooperi*
2. Calyx open in anthesis, not flask-shaped; stem leaves not clasping (3).
3. Pubescence partly of forked or stellate hairs. Leaves entire or merely dentate (4).
3. Pubescence of simple hairs or none (5).
4. Sepals greenish; petals whitish, 2 to 3 mm. long; capsules erect . . . . 3. *T. micranthum*
4. Sepals and petals purple, the petals 4 to 5 mm. long; capsules strongly reflexed 4. *T. longifolium*
5. Herbage sparsely hirsute; stem leaves pinnatifid; racemes few-flowered, soon becoming very open; capsules becoming reflexed, not torulose . . . . 5. *T. lasiophyllum*
5. Herbage glabrous; stem leaves usually entire; racemes many-flowered, dense; capsules spreading or curved upward, more or less torulose. Pedicels noticeably widened and flattened at base . . . . 6. *T. integrifolium*

1. Thelypodium Wrightii Gray (*Stanleyella Wrightii* Rydb.). Navajo, Coconino, and Mohave counties to Greenlee, Cochise, Gila, Pinal, and Pima counties, 2,500 to 7,000 feet, rich soil among pines and chaparral, March to November. Colorado, Utah, New Mexico, and Arizona.

A conspicuous plant because of the tall stems and numerous racemes of white flowers.

2. Thelypodium Cooperi Wats. (*Caulanthus Cooperi* Payson). Kingman and Chloride to the Colorado River (Mohave County), Harcuvar Mountains (Yuma County), 3,500 feet or lower, March and April. Western Arizona, southern Nevada, and southern California.

This plant superficially resembles *Streptanthella longirostris,* but has auriculate-clasping stem leaves, and capsules with the valves separating completely from the partition at maturity.

3. Thelypodium micranthum (Gray) Wats. (*Pennellia micrantha* Nieuwland). Yavapai, Greenlee, Graham, Gila, Pinal, Cochise, Santa Cruz, and Pima counties, 3,000 to 7,000 feet, often among bushes, July to September. Colorado and Utah to Mexico.

4. Thelypodium longifolium (Benth.) Wats. (*Lamprophragma longifolium* O. E. Schulz). Apache County to Coconino County, south to Cochise and Pima counties, 5,500 to 9,500 feet, dry pine forests, July to September. New Mexico, Arizona, and Mexico.

Var. *catalinense* Jones, "from the Catalina and Santa Rita Mountains," does not seem to be well marked.

5. Thelypodium lasiophyllum (Hook. & Arn.) Greene (*Microsisymbrium lasiophyllum* O. E. Schulz). Grand Canyon (Coconino County), and Mohave, Graham, Maricopa, Pinal, Cochise, Pima, and Yuma counties, 3,500 feet or lower, common in the western and southern deserts, usually among bushes, February to April. Washington to Arizona, California, and Baja California.

Payson referred all the Arizona specimens examined by him to *Caulanthus lasiophyllus* var. *utahensis* (Rydb.) Payson.

6. Thelypodium integrifolium (Nutt.) Endl. (*T. lilacinum* Greene, *T. rhomboideum* Greene). Painted Desert (Navajo or Coconino County), Grand Canyon and Havasu Canyon (Coconino County), near Peach Springs (Mohave County), near Pine (Gila County), 2,500 to 5,500 feet, July and August. Western Nebraska to Washington, south to New Mexico, Arizona, and California.

3. CAULANTHUS (108, pp. 283–309)

Plants herbaceous, annual or perennial; herbage glabrous or nearly so; stems stout, more or less inflated; stem leaves sessile or short-petioled; calyx somewhat flask-shaped, closed at anthesis or nearly so; petals (and often the sepals) purple or purplish, with channeled, crisped, or cucullate blades little wider than the claws; stigma deeply lobed; capsules erect or strongly ascending, elongate, terete.

1. Caulanthus crassicaulis (Torr.) Wats. Navajo Mountain, Kaibab Plateau, and South Rim of the Grand Canyon (Coconino County), about 7,000 feet, May and June. Idaho to northern Arizona and California.

Squaw-cabbage, so called because the plant was cooked and eaten by Indians of the Great Basin region.

Caulanthus sulfureus Payson was based on a collection near Tucson, Pima County (*Griffiths* 4058). The type seems to be an immature specimen of *Brassica campestris.*

4. STREPTANTHELLA

Plants annual or biennial, glabrous, usually glaucous; stems slender, simple or sparingly branched; stem leaves linear-lanceolate or oblanceolate, narrowed at base; flowers small, the petals with crisped or channeled blades little wider than the claws; capsules conspicuously beaked, becoming reflexed, somewhat flattened, the valves at maturity not separating toward the apex; seeds winged.

1. **Streptanthella longirostris** (Wats.) Rydb. Apache County to Mohave County, south to western Pima and Yuma counties, up to 7,000 feet but usually much lower, commonly in sandy soil, January to June. Wyoming to Oregon, New Mexico, Arizona, and California.

A collection near Dome, Yuma County (*Monnet* 11031) has been referred to var. *derelicta* J. T. Howell, a nonglaucous variety with pinnately parted leaves.

5. STREPTANTHUS. Twist-flower

Plants annual or biennial, glabrous, glaucous; leaves entire to lyrate-pinnatifid; stem leaves clasping, with a cordate or sagittate base; inflorescence racemose or paniculate; calyx flask-shaped, closed or nearly so at anthesis, ochroleucous, yellow, brown-purple, or violet; petals with long claws and narrow, crisped or channeled blades, these brown-purple, at least the veins; capsules strongly compressed; seeds flat, winged.

KEY TO THE SPECIES

1. Stems slender, diffusely branched above; inflorescence an open panicle, bracteate, the bracts ovate or suborbicular; flowers not more than 8 mm. long; capsules about 1 mm. wide, 5 to 7.5 cm. long, becoming reflexed or pendulous................1. *S. Lemmoni*
1. Stems stout, simple or few-branched; inflorescence racemose, elongate, ebracteate; flowers 10 mm. long or longer; capsules 3 mm. wide or wider, erect or ascending (2).
2. Leaves thick and firm, the upper ones broadly ovate, obtuse or acutish, entire, the basal ones rather sharply dentate; calyx brown-purple; petal blades about 1 mm. wide; herbage very glaucous..2. *S. cordatus*
2. Leaves rather thin, the upper ones lanceolate, oblong, or ovate-lanceolate, acute or acuminate, the basal ones commonly lyrate-pinnatifid; calyx violet, whitish, or yellow; petal blades 1.5 to 2 mm. wide; herbage moderately glaucous (3).
3. Sepals violet-purple; basal leaves lyrate-pinnatifid....................3. *S. carinatus*
3. Sepals whitish to yellow; basal leaves nearly entire to lyrate-pinnatifid..4. *S. arizonicus*

1. **Streptanthus Lemmoni** Wats. Known only from the type collection, Santa Catalina Mountains, Pima County, 5,000 feet (*Lemmon* 27).

2. **Streptanthus cordatus** Nutt. Navajo, Coconino, and Mohave counties, 2,000 to 8,000 feet, March to May. Wyoming to Oregon, New Mexico, northern Arizona, and California.

3. **Streptanthus carinatus** Wright. Santa Rita Mountains, Pima County, 4,500 feet, limestone slope (*Darrow & Haskell* 3425). Western Texas and southern Arizona.

4. **Streptanthus arizonicus** Wats. Greenlee County to Yavapai County, south to Cochise and Pima counties, 1,500 to 4,500 feet, January to April, type from mountains of southern Arizona (*Pringle* in 1881). Arizona and Chihuahua.

The calyx is normally ochroleucous. A variant with the calyx bright yellow when fresh but soon fading whitish, differing also in having a longer style and pods more attenuate at apex, var. *luteus* Kearney & Peebles, occurs in western Pima County, in the Ajo Mountains (*Kearney* 10813, the type) and the Growler Mountains.

It is questionable that *S. arizonicus* is more than varietally distinct from *S. carinatus.*

6. LEPIDIUM (109). PEPPER-GRASS

Plants annual or perennial, herbaceous or suffrutescent; stems usually much-branched; leaves pinnatifid to entire; flowers mostly very small, in dense racemes, these becoming elongate and open in fruit; fruits 2-celled, more or less dehiscent except in one species, orbicular or elliptic, often emarginate, strongly flattened at a right angle to the partition; seed solitary in each cell.

Cress (*L. sativum*), a European species, is grown for greens, and cress-seed oil is obtained from the seeds. The Arizona species are mostly weeds of roadsides and fields. The seeds of *L. Fremontii* and other species were used by the Arizona Indians as food and for flavoring.

KEY TO THE SPECIES

1. Leaves, at least the upper ones, perfoliate or with clasping bases (2).
1. Leaves none of them perfoliate or clasping. Fruits dehiscent (4).
2. Stem leaves (the upper ones) perfoliate, broadly ovate to nearly orbicular; capsules flat. Plants annual; upper leaves entire, the lower ones finely dissected; petals only slightly surpassing the sepals; style in fruit less than 1 mm. long; fruits tardily dehiscent .... 1. *L. perfoliatum*
2. Stem leaves sagittate-clasping, oblong, oblanceolate, or obovate; capsules turgid, at least toward base (3).
3. Plants perennial; basal leaves dentate; petals about twice as long as the sepals; style in fruit 1 to 2 mm. long; fruits indehiscent........ 2. *L. Draba*
3. Plants annual; basal leaves entire to pinnatifid; petals only slightly longer than the sepals; style in fruit less than 1 mm. long; fruits tardily dehiscent.... 3. *L. campestre*
4. Petals 2 to 3 mm. long, with broad blades and long, narrow claws, much surpassing the sepals; style in fruit 0.3 mm. or longer (5).
4. Petals less than 2 mm. long, linear or spatulate, often rudimentary or wanting; style in fruit less than 0.3 mm. long, usually not surpassing the notch of the capsule. Plants chiefly annual or biennial, the stems never woody (7).
5. Stems of mature plants woody well above the base; capsules at maturity 4 to 7 mm. wide, broadly obovoid or suborbicular. Herbage glabrous, slightly glaucous; leaves linear, very narrowly lanceolate, or oblanceolate, entire or pinnatifid with very few, narrow, entire lobes........ 4. *L. Fremontii*
5. Stems not woody or only slightly so at base; capsules less than 4 mm. wide (6).
6. Leaves pinnatifid or pinnately toothed; stems soft-pilose or villous, seldom merely puberulent .... 5. *L. Thurberi*
6. Leaves, at least the uppermost ones, entire; stems puberulent or glabrous, rarely short-pilose .... 6. *L. montanum*
7. Pubescence of stiff, spreading hairs; pedicels conspicuously flattened, commonly about twice as wide as thick. Plants diffusely branched from the base; stems decumbent or spreading, seldom erect; basal leaves deeply incised or pinnatifid, the stem leaves coarsely toothed, incised, or pinnatifid; capsules commonly short-hirsute or hispid, at least on the margin........ 7. *L. lasiocarpum*
7. Pubescence soft, or more or less appressed, or minute, rarely none; pedicels seldom conspicuously flattened or as much as twice as wide as thick (8).

8. Stems prostrate or strongly decumbent; basal leaves often bipinnatifid; stem leaves mostly pinnately parted or divided; sepals usually persistent after the petals have fallen, often until the fruit is nearly mature. Capsules mostly glabrous 8. *L. oblongum*

8. Stems commonly erect or ascending; lower leaves coarsely toothed or incised, the basal ones sometimes pinnatifid; upper stem leaves narrow, often entire; sepals usually deciduous with the petals (9).

9. Petals usually well developed, often surpassing the sepals; pedicels spreading soon after anthesis; capsules usually glabrous..........................9. *L. medium*

9. Petals shorter than the sepals, commonly not more than half as long, sometimes obsolete; pedicels remaining erect or ascending long after anthesis; capsules glabrous or sparsely pubescent. Stems puberulent, pilose, or short-villous......10. *L. densiflorum*

1. Lepidium perfoliatum L. Common in gardens at Fredonia (Coconino County), also at the Grand Canyon and Flagstaff, and at Tucson (Pima County). A weed in many parts of the United States; introduced from Europe.

2. Lepidium Draba L. (*Cardaria Draba* Desv.). Snowflake (Navajo County), Peeples Valley and Jerome Junction (Yavapai County), and reported as occurring elsewhere in Arizona, local at roadsides and in fields and pastures. Introduced from Europe. Hoary cress.

3. Lepidium campestre (L.) R. Br. Mesa, Maricopa County (*T. J. Smith* in 1945). A weed of waste ground, widely distributed in the United States; introduced from Europe.

4. Lepidium Fremontii Wats. Havasu Canyon (Coconino County) and Mohave and Yuma counties, 3,000 feet or lower, dry, sandy soil of plains and mesas, March to May. Southwestern Utah and western Arizona to southeastern California.

The only really shrubby species in Arizona, attaining a height of 1 m. The fragrant flowers are showy for the genus.

5. Lepidium Thurberi Wooton. Greenlee County to Yavapai County, south to Cochise, Santa Cruz, and Pima counties, 5,000 feet or lower, February to November. New Mexico and Arizona.

A common roadside weed in central and southern Arizona, conspicuous because of its relatively large, pure white flowers.

6. Lepidium montanum Nutt. Apache County to Mohave and Yavapai counties, 3,000 to 7,000 feet, April to September. Colorado to Oregon, south to Texas, Arizona, and California.

KEY TO THE VARIETIES

1. Leaves all entire. Stems and sepals rather densely hairy; leaves thick, acute var. *integrifolium*

1. Leaves (at least some of them) lobed, parted, or divided (2).

2. Stems usually glabrous below the inflorescence, or very nearly so (3)

2. Stems more or less pubescent below the inflorescence (4).

3. Plants entirely glabrous...var. *glabrum*

3. Plants (the basal leaves, or the inflorescence, or both) somewhat pubescent..var. *Jonesii*

4. Plants suffrutescent; upper stem leaves entire.........................var. *alyssoides*

4. Plants herbaceous to the base; upper stem leaves often lobed, parted, or divided var. *canescens*

A collection at Fort Verde, Yavapai County (*Mearns* 309), was referred doubtfully by Hitchcock to var. *integrifolium* (Nutt.) C. L. Hitchc. Var. *glabrum* C. L. Hitchc. occurs at Dinnehotso (Apache County) and in the Grand Canyon–Flagstaff region, type from Grand Canyon (*Eastwood* 5826). Var. *Jonesii* (Rydb.) C. L. Hitchc. ranges from Apache to Mohave and Yavapai counties. Var. *alyssoides* (Gray) Jones is known from Holbrook (Navajo County) and from Crater Mound, House Rock Valley, and edge of Kaibab Plateau (Coconino County). Var. *canescens* (Thell.) C. L. Hitchc., the commonest variety in this state, is found throughout the range of the species in Arizona.

7. Lepidium lasiocarpum Nutt. Almost throughout the state, 4,000 feet or lower, commonly in sandy soil, January to April. Southwestern Colorado to Arizona and California, southward into Mexico.

KEY TO THE VARIETIES

1. Hairs of the capsules with swollen (pustular) bases; stem leaves lobed to pinnatifid var. *Wrightii*
1. Hairs of the capsules not swollen at base, or the stem leaves entire to dentate (2).
2. Stems glabrous to rather soft-pubescent; pedicels usually glabrous on the lower side var. *georginum*
2. Stems hirsute or hispid; pedicels usually pubescent on both sides..........var. *typicum*

Var. *Wrightii* (Gray) C. L. Hitchc. has been collected in Mohave, Maricopa, Pinal, Santa Cruz, and Pima counties. Var. *georginum* (Rydb.) C. L. Hitchc. is occasional in Santa Cruz and Pima counties. Var. *typicum* C. L. Hitchc. occurs abundantly almost everywhere in Arizona.

8. Lepidium oblongum Small. Sacaton and Maricopa (Pinal County), Redington and vicinity of Tucson (Pima County), 1,000 to 3,000 feet, waste ground. Arkansas to Texas, Arizona, and California; probably introduced from South America.

9. Lepidium medium Greene (*L. virginicum* var. *medium* C. L. Hitchc.). Apache County to Coconino County, south to Cochise and Pima counties, 7,500 feet or lower, February to August. Montana to British Columbia, south to Texas, Arizona, and California.

Resembles the common peppergrass of the eastern United States (*L. virginicum* L.) but the Arizona specimens apparently always have incumbent (not accumbent) cotyledons. The species occurs in two forms that are about equally common in Arizona: the typical plant with upper parts of the stem and pedicels glabrous, and var. *pubescens* (Greene) Robinson (*L. hirsutum* Rydb.) with stems and pedicels puberulent or cinereous-pilose throughout. Specimens that apparently belong to var. *pubescens* but with the capsules also pubescent, have been collected in the Grand Canyon (*Clover* 6307a) and in the Ajo Mountains, Pima County (*Benson* 10680).

10. Lepidium densiflorum Schrad. Apache, Navajo, Coconino, Mohave, Yavapai, and Gila counties, 5,000 to 8,000 feet. Widely distributed in the United States and Canada.

The common form in Arizona is var. *Bourgeauanum* (Thell.) C. L. Hitchc., which differs from typical *L. densiflorum* in having distinctly flattened pedicels and capsules about 3 mm. long.

7. THLASPI (110, 111)

Plants herbaceous, annual or perennial, glabrous; stem leaves auriculate-clasping; flowers small, the petals white or tinged with purple; capsules 2-celled, dehiscent, flattened at a right angle to the partition; seeds 2 or more in each cell.

KEY TO THE SPECIES

1. Plants annual; stems commonly 30 cm. long or longer, usually branched; stem leaves usually dentate; capsules orbicular or nearly so, rounded or slightly cuneate at base, at least 10 mm. wide at maturity, broadly winged all around, deeply notched at apex; style minute, much shorter than the notch............................1. *T. arvense*
1. Plants perennial; stems not more (usually much less) than 30 cm. long, not branched; stem leaves entire or denticulate, the basal leaves entire to dentate; capsules wedge-shaped, not more than 6 mm. wide, not winged, or obscurely so near the truncate or very shallowly notched apex; style elongate, much longer than the notch. Sepals usually purplish..2. *T. Fendleri*

1. Thlaspi arvense L. Keam Canyon, Navajo County, 6,500 feet (*Peebles & Smith* 13413), Ryan, Coconino County (*Goodding* 155–49). A troublesome weed in some parts of the United States; naturalized from Europe. Penny-cress.

2. Thlaspi Fendleri Gray (*T. purpurascens* Rydb.). Apache County to Hualpai Mountain (Mohave County), south to Cochise, Santa Cruz, and Pima counties, 4,000 to 12,000 feet, mostly in coniferous forests, February to August, type of *T. purpurascens* collected "in Arizona" without definite locality (*Palmer* 571). Colorado, Utah, New Mexico, and Arizona.

Sometimes called wild-candytuft. A collection at Greer, Apache County (*Goodding* 1187), was referred by Payson (110, p. 152) to *T. glaucum* A. Nels. var. *typicum* Payson. According to Maguire (111, p. 468) *T. glaucum* is not specifically distinct from *T. Fendleri*. Dwarf specimens with short, crowded, corymbiform inflorescences, occurring at elevations of 11,000 to 12,000 feet on the San Francisco Peaks, Coconino County (*Smith* 12023, *Little* 4618), are referrable to var. *coloradense* (Rydb.) Maguire (111, p. 469). Three species of *Thlaspi*, based on Arizona types, were described by Aven Nelson (*Amer. Jour. Bot.* 32: 287–288). These are: *T. australe*, type from the Baboquivari Mountains (Pima County); *T. prolixum*, type from the San Carlos Indian Reservation (Gila County); and *T. stipitatum*, type from the Pena Blanca Mountains (Santa Cruz County). None of these seems to be specifically distinct from *T. Fendleri*, but *T. stipitatum*, with capsules conspicuously narrowed at base and appearing substipitate, may be worthy of recognition as a variety. It has been collected also in the Pinaleno Mountains (Graham County) and in the Santa Catalina and Santa Rita mountains (Pima and Santa Cruz counties) at elevations of 4,000 to 5,000 feet.

8. SISYMBRIUM (112)

Plants of diverse appearance, annual or perennial; stems simple or branched; leaves entire, dentate, or pinnatifid, the stem leaves sometimes with auriculate-clasping bases; flowers in racemes; petals yellow or purple; style short or obsolete; stigma entire, or 2-lobed with lobes over the partition and placentas; capsules terete, slender.

It is doubtful that all the following species belong properly to this genus.

KEY TO THE SPECIES

1. Stem leaves auriculate-clasping. Petals purple, flat or nearly so (2).
1. Stem leaves not auriculate-clasping (3).
2. Petals 10 to 15 mm. long, rich violet-purple; anthers becoming curled or coiled; capsules distinctly stipitate, the stipe 2 mm. long or longer. Stem tall and stout; basal leaves dentate .... 1. *S. ambiguum*
2. Petals less than 10 mm. long, pale purple; anthers becoming curved and somewhat twisted; capsules nearly sessile, the stipe not more than 1 mm. long...... 2. *S. elegans*
3. Petals purplish pink, 10 to 15 mm. long; outer sepals strongly gibbous at base; stigma entire, conic, often pointed. Leaves entire or denticulate.......... 3. *S. linearifolium*
3. Petals yellow or whitish, less than 10 mm. long; outer sepals not strongly gibbous; stigma usually distinctly 2-lobed, flat or depressed (4).
4. Plants perennial, with creeping rootstocks; leaves all or nearly all entire and linear or narrowly lanceolate, the basal ones sometimes pinnatifid; petals more or less crisped or somewhat cucullate; capsules erect or ascending, very slender, 3 to 7 cm. long. Anthers deeply sagittate........ 4. *S. linifolium*
4. Plants annual or biennial; leaves all or nearly all pinnate or pinnatifid; petals flat or nearly so; capsules eventually spreading (5).
5. Plants glabrous or very sparsely hirsute near the base; divisions of the upper stem leaves never filiform; capsules at maturity 3 to 5 cm. long, not rigid; anthers scarcely sagittate .... 5. *S. Irio*
5. Plants hirsute up to the inflorescence, usually very sparsely so above; divisions of the upper stem leaves very narrow, often nearly filiform; capsules at maturity more than 5 (up to 10) cm. long, becoming rather rigid; anthers often pronouncedly sagittate 6. *S. altissimum*

1. Sisymbrium ambiguum (Wats.) Payson. Coconino, Mohave, and Yavapai counties, 2,000 to 6,000 feet, March to June, type from Long Valley, Coconino County (*Newberry*). Also in southern Utah.

A coarse plant, but very showy when in flower.

***2. Sisymbrium elegans** (Jones) Payson. Not known definitely to occur in Arizona, but has been collected at Kanab, Utah, very near the Arizona state line. Colorado and Utah.

3. Sisymbrium linearifolium (Gray) Payson (*Hesperidanthus linearifolius* Rydb.). Apache, Navajo, and Coconino counties, south to Cochise, Santa Cruz, and Pima counties, 2,500 to 9,500 feet, common in chaparral and pine forests, May to September. Colorado to Arizona and northern Mexico.

4. Sisymbrium linifolium Nutt. Carrizo Mountains, Apache County (?) possibly collected on the New Mexico side of the state line (*Matthews* in 1892), eastern and western slopes of the Kaibab Plateau (*Goodding & Gunning* 2959, *Goodding* 61–49). Montana to British Columbia, Utah, and northern Arizona.

5. Sisymbrium Irio L. Western Coconino and Mohave counties to Cochise, Santa Cruz, Pima, and Yuma counties, 4,500 feet and lower, winter and early spring. Introduced from Europe.

An extremely abundant weed in irrigated sections of southern Arizona, disappearing on the advent of hot weather, of possible value for green manure.

6. Sisymbrium altissimum L. Navajo, Coconino, Mohave, and Yavapai counties, a common roadside weed, 5,000 to 7,000 feet, occasional farther south and lower, April to September. Introduced from Europe.

Tumble-mustard. A bad weed in grain fields in some parts of the United States.

Sisymbrium officinale (L.) Scop., an introduced weed common throughout most of the United States, is to be looked for in Arizona. It is readily distinguished from all the foregoing species of *Sisymbrium* by its closely appressed, awl-shaped capsules not more than 15 mm. long. The petals are pale yellow.

9. HALIMOLOBOS (113)

Plants perennial, soft-canescent with mostly forked or stellate hairs; stems diffusely branched above; leaves deeply sinuate or pinnatifid; petals bright white; capsules terete, very slender, less than 1 mm. in diameter.

1. Halimolobos diffusus (Gray) O. E. Schulz. Gila, Cochise, and Pima counties, 4,000 to 7,000 feet, mostly in crevices of rocks, July to September. Western Texas to southern California and northern Mexico.

10. DIPLOTAXIS

Plants annual or perennial, glabrous or sparsely hispid; leaves coarsely toothed or pinnatifid; flowers rather few, in open racemes; petals yellow; capsules erect or ascending, long and narrow, somewhat flattened; seeds in 2 rows.

KEY TO THE SPECIES

1. Plants annual or biennial; stems scapose or subscapose; leaves coarsely toothed, or pinnatifid with broadly oblong or triangular lobes; petals about 6 mm. long; pedicels in fruit up to 22 mm. long..1. *D. muralis*
1. Plants perennial; stems leafy nearly to the inflorescence; leaves usually deeply pinnatifid with elongate, linear or narrowly oblong lobes; petals usually more than 6 mm. long; pedicels in fruit up to 40 mm. long..........................2. *D. tenuifolia*

1. Diplotaxis muralis (L.) DC. Tuba, Coconino County, "common about buildings" (*Clute* 111), Tucson, Pima County, "streets, spreading rapidly" (*Thornber* 7545, in 1913). In waste ground here and there in the United States; introduced from Europe.

2. Diplotaxis tenuifolia (L.) DC. Benson, Cochise County, "common in railway yard and spreading" (*Thornber* 5365, 8219, 9202). Introduced from Europe.

11. BRASSICA

Plants annual or biennial, glabrous and glaucous, or sparsely and stiffly pubescent; leaves petioled, or sessile and clasping, at least the lower ones lyrate-pinnatifid; flowers rather large, in elongate racemes, the petals yellow or whitish; capsules conspicuously beaked, commonly torulose, dehiscent below, the beak indehiscent; seeds in 1 or 2 rows in each cell.

This genus includes the cultivated cabbage, cauliflower, turnip, and mustard as well as the weedy species enumerated here. Oil of mustard, used medicinally as a skin stimulant (rubefacient), is obtained from the seeds of *B. nigra* and *B. juncea*. Most of the species make palatable greens.

KEY TO THE SPECIES

1. Seeds in 2 rows in each cell; valves of the capsules strongly keeled; petals whitish. Beak nearly as long as to equaling the rest of the capsule, broad and flat......1. *B. Eruca*
1. Seeds in 1 row; valves not strongly keeled; petals yellow (2).
2. Beak half or more as long as the rest of the capsule, flat or conspicuously angled, usually containing 1 seed in an indehiscent cell. Leaves not clasping at base (3).
2. Beak usually less than half as long as the rest of the capsule, conic, seedless (4).

3. Leaves petioled, pinnatifid; pedicels more than 5 (often 10) mm. long; capsules ascending on spreading pedicels, the dehiscent part of the capsule bristly, shorter than or equaling the beak .... 2. *B. hirta*
3. Leaves nearly sessile, the upper ones merely toothed; pedicels 3 to 5 mm. long; capsules and pedicels strongly ascending, the dehiscent part of the capsule usually smooth, longer than the beak .... 3. *B. Kaber*
4. Upper leaves clasping, glaucous and glabrous, the lower ones usually with scattered hairs. Pedicels spreading; capsules 4 to 9 cm. long, stout, ascending, with a stout beak forming ⅛ to ⅓ of the total length of the capsule .... 4. *B. campestris*
4. Upper leaves not clasping, not or only slightly glaucous, often sparsely hirsute (5).
5. Pedicels in fruit less than 5 mm. long, erect; capsules appressed to the stem, 1 to 2 cm. long, not more than 2 mm. in diameter, somewhat quadrangular .... 5. *B. nigra*
5. Pedicels in fruit more than 5 mm. long, spreading; capsules not appressed to the stem, 3 to 5.5 cm. long, 2 to 3.5 mm. in diameter, nearly terete, but the valves with a stout midnerve .... 6. *B. juncea*

1. Brassica Eruca L. (*Eruca sativa* Mill.). Near Tucson, Pima County (*Thornber* 7330, 7339). Fields and waste land, here and there in the United States; introduced from Europe.

Rocket-salad or roquette, sometimes cultivated as a salad plant and potherb.

2. Brassica hirta Moench (*Sinapis alba* L.). Base of the Mazatzal Mountains, Gila County, 4,000 feet (*Collom* 907). Adventive in the United States; from Eurasia. White mustard.

3. Brassica Kaber (DC.) L. C. Wheeler (*Sinapis arvensis* L.). Roosevelt, Gila County (*Peebles et al.* 5205), near Glendale, Maricopa County (*Gould* 3505), near Tucson, Pima County (*Thornber* in 1913), roadsides and waste ground. Widely distributed in the United States; introduced from Europe.

Charlock. The identification of the Arizona specimens as *B. Kaber* is somewhat questionable.

4. Brassica campestris L. Kingman (Mohave County), Mesa (Maricopa County), Sacaton and south of Florence (Pinal County), occasional at roadsides, probably an escape from cultivation. Native of Eurasia.

5. Brassica nigra (L.) Koch (*Sinapis nigra* L.). Coconino, Pinal, Maricopa, Cochise, and Pima counties, occasional at roadsides. Widely distributed in the United States; naturalized from Europe. Black mustard.

6. Brassica juncea (L.) Cosson (*Sinapis juncea* L.). Flagstaff (Coconino County), Prescott (Yavapai County), near Tucson (Pima County). Widely distributed in the United States; introduced from Asia. Indian mustard.

The garden radish (*Raphanus sativus* L.) and wild radish (*R. Raphanistrum* L.) are likely to be found in waste places in Arizona. They are easily distinguished from *Brassica* by their indehiscent fruits with spongy cross partitions between the seeds, the garden radish also by its white or pink petals.

12. BARBAREA. Winter-cress

Plants commonly biennial, glabrous; stems erect, branched above, angled; leaves lyrate-pinnatifid; flowers in elongate racemes, the petals yellow; capsules elongate, 4-sided; seeds in 1 row in each cell.

1. Barbarea orthoceras Ledeb. White Mountains (Apache County), Buck Springs (Coconino County), Camp Grant (Graham County), Chiricahua

Mountains (Cochise County), up to 9,500 feet, spring and early summer. Labrador to Alaska, south to New Hampshire, Colorado, Arizona, California, and Mexico; Eurasia.

Var. *dolichocarpa* Fern., with less crowded, spreading or ascending (rather than erect) capsules, has been collected near Flagstaff.

13. DRYOPETALON (114)

Plants annual; stems branching above, hispid below; leaves lyrate-pinnatifid, the segments broad, coarsely toothed; sepals white-margined, strongly gibbous at base; petals bright white, pinnately cleft; stigma sessile, slightly 2-lobed; capsules 2.5 to 4 cm. long, terete, spreading or somewhat recurved.

1. Dryopetalon runcinatum Gray. Greenlee, Gila, Pinal, Cochise, Santa Cruz, and Pima counties, 2,000 to 4,000 feet, commonly in moist rock crevices in canyons, February to May. New Mexico, southern Arizona, and northern Mexico.

14. RORIPPA

Plants annual or perennial; stems usually branched; leaves simple or pinnate; flowers in racemes, these rather short and dense at first, becoming loose and elongate; petals yellow or white; capsules globose, to elongate and narrowly cylindric; seeds commonly in 2 rows, very small, turgid.

The plants all prefer wet ground, and one species, the true water-cress (*R. Nasturtium-aquaticum*), is semiaquatic.

KEY TO THE SPECIES

1. Petals clear white, surpassing the sepals; leaves pinnate; plants perennial, semiaquatic. Leaflets entire or nearly so, the terminal one much the largest, broadly ovate or suborbicular; capsules cylindric or subclavate, 8 to 20 mm. long 1. *R. Nasturtium-aquaticum*

1. Petals yellow or whitish; leaves pinnatifid or sometimes pinnate; plants terrestrial, in moist soil (2).

2. Sepals considerably shorter than the petals; plants commonly perennial, with rootstocks. Leaf segments mostly acute or acutish, the terminal one lanceolate or oblong; capsules seldom less than 3 times as long as wide, acute or acutish at apex; stigma much thicker than the style (3).

2. Sepals not or only slightly shorter than the petals (except sometimes in *R. curvisiliqua*); plants commonly annual (4).

3. Leaves often pinnate, the segments mostly toothed, incised, or laciniate; inflorescence and capsules glabrous; style not more than 1 mm. long; capsules slender; seeds often in 1 row..2. *R. sylvestris*

3. Leaves pinnatifid, the segments entire or nearly so; inflorescence and capsules scurfy-papillate; style more than 1 mm. long; capsules rather thick, usually curved; seeds in 2 rows..3. *R. sinuata*

4. Capsules very slender, linear, more or less curved, often 1 cm. or longer, acute or acutish at apex. Stigma very small, not thicker than the very short style, or sessile; leaf segments commonly acute....................................4. *R. curvisiliqua*

4. Capsules thick, straight or nearly so, usually much less than 1 cm. long, very obtuse or truncate at apex (5).

5. Pedicels 2 to 4 mm. long, mostly longer than the globose capsules. Plants low, diffuse, glabrous or very sparsely pilose; leaf segments obtuse............5. *R. sphaerocarpa*

5. Pedicels mostly 4 mm. long or longer; capsules longer than wide (6).

6. Stems villous or hirsutulous, commonly sparsely so; capsules not more than twice as long as wide, ovoid or ellipsoid, mostly shorter than the pedicels. Stigma scarcely thicker than the style .... 6. *R. hispida*
6. Stems glabrous or very nearly so; capsules usually more than twice as long as wide, cylindric or ovoid-cylindric, mostly longer than the pedicels (7).
7. Stigma thicker than the style, distinctly 2-lobed; leaf segments numerous, often acutish .... 7. *R. islandica*
7. Stigma not or scarcely thicker than the style, entire or very nearly so; leaf segments few, obtuse .... 8. *R. obtusa*

1. Rorippa Nasturtium-aquaticum (L.) Schinz & Thell. (*Nasturtium officinale* R. Br.). Apache County to Mohave County, south to Cochise, Santa Cruz, and Pima counties, 1,500 to 7,000 feet, springs, brooks, and ponds, April to August. Naturalized throughout temperate North America; from Europe.

Water-cress, sometimes cultivated for salads and garnishings.

2. Rorippa sylvestris (L.) Besser. Near Flagstaff, Oak Creek, and Rogers Lake (Coconino County), near Fort Huachuca (Cochise County). Extensively naturalized in North America; from Europe. Yellow cress.

3. Rorippa sinuata (Nutt.) A. S. Hitchc. Apache County to Coconino County, 5,000 to 6,000 feet, May and June. Illinois to Saskatchewan, south to Texas, northern Arizona, and California.

Represented in Arizona only by a form with scurfy-papillate inflorescence and capsules (*Nasturtium trachycarpum* Gray).

4. Rorippa curvisiliqua (Hook.) Bessey. Billie Creek (southern Navajo County), Flagstaff, San Francisco Peaks, and Bill Williams Mountain (Coconino County), Beaver Creek (Yavapai County), bed of the Gila River (Pinal County), up to 7,000 feet, June to August. Montana and Wyoming to British Columbia, south to Arizona and California.

The Arizona specimens are referred doubtfully to this species, having longer, more slender, and less curved capsules than most specimens of *R. curvisiliqua* from the Pacific coast states. The specimen from Billie Creek (*Pultz* 1694) has bright-yellow flowers.

5. Rorippa sphaerocarpa (Gray) Britton. San Francisco Peaks, Flagstaff, Mormon Lake, and Oak Creek (Coconino County), Tonto Basin (Gila County), 5,000 to 8,000 feet, July to September. Illinois to Wyoming, south to Texas, Arizona, and California.

6. Rorippa hispida (Desv.) Britton (*R. islandica* var. *hispida* (Desv.) Butters & Abbe). Guthrie, Greenlee County (*Davidson* 315), Arizona, without locality (*Palmer* in 1869), Gila River bed near Sacaton, Pinal County, 1,300 feet (*Harrison & Peebles* 1742, 2006). New Brunswick to Alaska, south to New Mexico and Arizona.

7. Rorippa islandica (Oeder) Borbás (*R. palustris* (L.) Besser). North side of the Grand Canyon and Foxboro (Coconino County), White Mountains (Apache County), Winkelman (Pinal County), along the Colorado River (Yuma (?) County), up to 9,500 feet. Widely distributed in North America and Eurasia.

Palmer noted on a label: "Used for greens by the Cocopa Indians."

8. Rorippa obtusa (Nutt.) Britton. White Mountains (Apache and Greenlee counties) in meadows, Grand Canyon and Baker Butte (Coconino County), bed of the Gila River near Sacaton (Pinal County), where doubtless a stray from the mountains. Michigan to Washington, south to Texas, Arizona, and California.

15. CARDAMINE. Bitter-cress

Plants perennial; herbage glabrous or soft-pilose; leaves simple, conspicuously petiolate, broadly ovate or suborbicular, deeply cordate, shallowly sinuate-dentate; petals white, 8 to 10 mm. long; capsules ascending on long pedicels, conspicuously beaked; seeds in 1 row in each cell.

1. **Cardamine cordifolia** Gray. San Francisco Peaks (Coconino County), White Mountains (Apache County), 9,000 to 11,000 feet, July and August. Wyoming and Idaho, south to New Mexico and Arizona.

C. parviflora L., a normally annual species, has been reported from Havasu Canyon, Coconino County. It is a plant of very different appearance from *C. cordifolia*, being much smaller, with slender stems, the leaves pinnate with narrow leaflets, and the petals only 2 to 3 mm. long.

16. LYROCARPA (115)

Plants perennial, stellate-canescent; stems weak and straggling, usually supported on shrubs; leaves lyrate-pinnatifid; petals subulate, brown-purple, 1.5 to 2 cm. long; capsules irregularly obovoid-triangular, obcordate, often constricted below the apex, flattened contrary to the partition.

1. **Lyrocarpa Coulteri** Hook. & Harv. Western parts of Maricopa, Pinal, and Pima counties and southern Yuma County, 3,000 feet or lower, among mesquite and other bushes in partial shade, March and August. Arizona, southeastern California, Sonora, and Baja California.

The Arizona plants belong to var. *typica* Rollins. The flowers are fragrant.

17. DITHYREA. Spectacle-pod

Plants annual, stellate-canescent; stems leafy, erect or decumbent; leaves sinuate-dentate to nearly entire; petals white or yellowish; capsules didymous (the 2 cells side by side), strongly flattened contrary to the partition, wider than long; seed solitary in each cell.

KEY TO THE SPECIES

1. Stems decumbent or spreading; herbage yellowish green; leaves ovate or oblong-ovate, shallowly sinuate-dentate; petals 8 to 12 mm. long, yellowish; pedicels in fruit 2 to 3 mm. long; fruit rather deeply notched both above and below, canescent 1. *D. californica*

1. Stems erect; herbage gray or whitish; stem leaves linear-lanceolate to ovate-lanceolate, deeply sinuate-dentate to nearly entire; petals 5 to 8 mm. long, white; pedicels in fruit seldom less than 10 mm. long; fruit notched below, truncate or very shallowly notched above, canescent or glabrous.............................2. *D. Wislizeni*

1. **Dithyrea californica** Harv. Mohave, western Pima, and Yuma counties, 2,500 feet or lower, sandy soil, plains and mesas, February to April, sometimes autumn-flowering. Southern Nevada, western Arizona, California, and Baja California.

2. **Dithyrea Wislizeni** Engelm. Apache County to Mohave County, south to Cochise, Pinal, Maricopa, and Yuma counties, 1,000 to 6,000 feet, sandy soil, often along streams, February to October. Colorado and Utah to Arizona and northern Mexico.

Occasional in Arizona is var. *Griffithsii* (Woot. & Standl.) Payson, with glabrous capsules but not consistently with narrow, entire leaves. It is said that *D. Wislizeni* is used by the Hopi Indians in treating wounds.

18. PHYSARIA (116). Twin-pod

Plants small, cespitose, more or less silvery-lepidote; leaves mostly basal, or the basal leaves much larger than the stem leaves, orbicular to spatulate; petals narrow, yellow; capsules didymous, bladderlike, deeply notched at apex; seeds usually 2 or more in each cell.

KEY TO THE SPECIES

1. Style 2 to 3 mm. long; replum of the capsule lanceolate................1. *P. Newberryi*
1. Style 6 to 8 mm. long; replum oblong.............................2. *P. Chambersii*

1. Physaria Newberryi Gray. Apache, Navajo, and Coconino counties, 5,000 to 7,000 feet, May, type from Tegua, Hopi Villages, Navajo County (*Newberry* in 1858). Utah, Nevada, New Mexico, and northern Arizona.

2. Physaria Chambersii Rollins. North side of Grand Canyon (*Eastwood & Howell* 1028), near Pipe Springs, Mohave County, 5,000 feet (*Peebles & Parker* 14709), April and May. Utah, Nevada, and northwestern Arizona.

19. LESQUERELLA (117). Bladder-pod

Plants annual or perennial; herbage stellate-canescent; leaves largely basal, entire to sinuate-dentate; flowers in loose racemes, the petals yellow, or white tinged with purple; pedicels elongate in fruit; styles persistent; capsules inflated, globose or ovoid, the valves nerveless; seeds 2-rowed in each cell.

The roots of one species, perhaps *L. intermedia,* are reported to be used by the Hopi Indians as an antidote for rattlesnake venom.

KEY TO THE SPECIES

1. Plants annual. Herbage green and sparsely pubescent to densely silvery-canescent; leaves lanceolate, oblanceolate, or spatulate, entire (exceptionally sinuate-dentate), the basal leaves sometimes lyrate, the stem leaves 2 to 12 mm. wide; petals 6 to 8 mm. long, bright yellow, sometimes fading reddish; pedicels in fruit spreading, becoming sigmoid; capsules globose or nearly so, about 4 mm. in diameter, usually somewhat longer than the style, glabrous or pubescent, short-stipitate to sessile in the calyx..1. *L. Gordoni*
1. Plants perennial (2).
2. Petals at anthesis white, sometimes purple-veined, commonly fading purple; pedicels in fruit ascending to simply recurved, not sigmoid; capsules glabrous. Leaves green and sparsely pubescent to canescent, the basal ones much larger than the stem leaves, usually lyrate; petals 6 to 10 mm. long; pedicels slender, up to 2 cm. long in fruit; capsules sessile in the calyx or nearly so, about 6 mm. in diameter......2. *L. purpurea*
2. Petals at anthesis yellow, sometimes fading reddish or whitish; pedicels in fruit not simply recurved, or if so, then the capsules pubescent (3).
3. Capsules glabrous, shorter to somewhat longer than the style. Fruiting inflorescence often elongate; leaves oblanceolate or spatulate, commonly entire; pedicels in fruit up to 20 mm. long, erect or strongly ascending, straight or nearly so; petals 8 to 10 mm. long...3. *L. Fendleri*
3. Capsules pubescent (4).
4. Pedicels in fruit simply recurved (not sigmoid), up to 15 mm. long; capsules usually pendent; blades of the petals not more than twice as wide as the claws. Leaves linear or narrowly oblanceolate, the lower ones up to 10 cm. long..........4. *L. ludoviciana*

4. Pedicels in fruit not recurved, either straight, simply curved upward, or sigmoid; capsules not pendent; blades of the petals at least twice as wide as the claws. Species difficult to distinguish (5).
5. Basal leaves usually forming a distinct rosette, some of them commonly broadly oblanceolate to nearly orbicular and 4 mm. or more wide (6).
5. Basal leaves not forming a distinct rosette, linear or narrowly oblanceolate, mostly less than 4 mm. wide (8).
6. Blades of the basal leaves abruptly contracted into the petiole; fruiting racemes short and dense. Capsules ovoid or ellipsoid............................5. *L. Wardii*
6. Blades of the basal leaves tapering into the petiole; fruiting racemes usually elongate and loose (7).
7. Stems erect or ascending; basal rosette not dense; stem leaves not crowded 6. *L. rectipes*
7. Stems decumbent or prostrate; basal rosette dense; stem leaves often crowded 7. *L. cinerea*
8. Styles in fruit 2.5 to 6 mm. long; basal leaves mostly 3 cm. long or longer, usually involute 8. *L. intermedia*
8. Styles in fruit seldom more than 2 mm. long; basal leaves not more than 2.5 cm. long, usually flat ....9. *L. arizonica*

1. Lesquerella Gordoni (Gray) Wats. Greenlee County to Mohave and Yavapai counties, south to Cochise, Pima, and Yuma counties, 5,000 feet or lower, dry plains and mesas, February to May. Oklahoma to Utah, Arizona, and California, south to northern Mexico.

Extensive desert areas are colored in spring with the bright-yellow flowers of this plant. It is reported to afford good forage, probably after the capsules mature. Throughout most of the range of the species in Arizona, often growing with the typical plant, is var. *sessilis* Wats. (*L. Palmeri* Wats.), characterized by pubescent, sometimes sessile capsules, type of *L. Palmeri* from Arizona (*Palmer* in 1872).

2. Lesquerella purpurea (Gray) Wats. Coconino County to Cochise and Pima counties, 1,500 to 5,000 feet, usually in partial shade of bushes, January to May. Texas to Arizona and northern Mexico.

Easily distinguished from all other Arizona species by its white or purplish flowers.

3. Lesquerella Fendleri (Gray) Wats. Apache and Navajo counties to Cochise, Santa Cruz, and Pima counties, 4,000 to 7,000 feet, plains and mesas, April to June. Kansas to Utah, south to northern Mexico.

4. Lesquerella ludoviciana (Nutt.) Wats. "Arizona" without definite locality (*Palmer* in 1869), Joseph City, Navajo County, 5,000 feet (*Wooton* in 1892). Minnesota and the Dakotas to Kansas, Utah, and Arizona.

5. Lesquerella Wardii Wats. Pleasant Valley, Kaibab Plateau, Coconino County (*Goodding* 165–49). Southern Utah and Nevada and northern Arizona.

6. Lesquerella rectipes Woot. & Standl. Navajo Experiment Station and White Mountains (Apache County), Kaibab Plateau and near Flagstaff (Coconino County), 6,000 to 8,000 feet, June. Colorado, Utah, northwestern New Mexico, and northern Arizona.

7. Lesquerella cinerea Wats. Willow Springs Pass, Kaibab Plateau, and near Flagstaff (Coconino County), 6,000 to 7,000 feet, near Burro Creek (southeastern Mohave County), about 2,500 feet, April to August, type from Arizona without definite locality (*Palmer* in 1869). Known only from Arizona.

8. Lesquerella intermedia (Wats.) Heller. Apache County to Coconino, eastern Mohave, and Yavapai counties, 5,500 to 7,000 feet, April to August. Colorado, Utah, New Mexico, and Arizona.

The commonest species in northeastern Arizona.

9. Lesquerella arizonica Wats. Coconino, Mohave, and northern Yavapai counties, 3,500 to 7,000 feet, rocky slopes and mesas, April and May, type from Juniper Mesa, Yavapai County (*Palmer* 16). Southern Utah to central Arizona.

This species apparently intergrades with *L. intermedia*. Specimens collected near Prescott (*Eastwood* 8867) have glabrous capsules, although otherwise resembling *L. arizonica*. A specimen collected on Hualpai Mountain (*Barneby* 2979) has the habit and appearance of *L. arizonica*, but the capsules are 7 to 8 mm. long and the styles about 3.5 mm. long. A reduced form, with leafless scapes, var. *nudicaulis* Payson, is known only from the type collection on the "Buckskin Mountains," northern Coconino County (*Jones* in 1890).

20. HUTCHINSIA

Plants annual, small, glabrous or stellate-puberulent; stems very slender, branched, spreading; leaves mostly petioled, narrow, entire or few-toothed; petals small, whitish; capsules 4 mm. long or less, not strongly compressed, ellipsoid, entire at apex.

1. Hutchinsia procumbens (L.) Desv. Pierce Spring, Mohave County, 1,500 feet (*Jones* 5077b), April. Labrador to British Columbia, south to Colorado, northern Arizona, and Baja California; Asia.

21. CAPSELLA. Shepherds-purse

Plants annual; root leaves in a rosette, lyrate-pinnatifid, the stem leaves dentate or entire, auricled at base; flowers in elongate racemes, very small, the petals white; capsules flat, wedge-shaped, dehiscent; seeds numerous in each cell.

1. Capsella Bursa-pastoris (L.) Medic. Coconino, Yavapai, Graham, Gila, Maricopa, Pinal, Cochise, Santa Cruz, and Pima counties, a weed of waste land and lawns. Extensively naturalized in North America; from Europe.

22. CAMELINA. False-flax

Plants annual; stems erect, rather strict; leaves entire or denticulate, the stem leaves auriculate-clasping; flowers in elongate racemes, small, the petals yellow; capsules obovoid, strongly margined, turgid but somewhat compressed; seeds numerous in each cell.

Weeds of fields and waste land; introduced from Europe. An oil, somewhat resembling linseed oil, is obtained from the seeds.

KEY TO THE SPECIES

1. Body of the capsule 6 to 8 mm. long, more than 3 times as long as the persistent style 1. *C. sativa*

1. Body of the capsule 4 to 6 mm. long, less than 3 times as long as the style . 2. *C. microcarpa*

1. Camelina sativa (L.) Crantz. Yuma (*H. Brown* in 1905). Almost throughout the United States.

2. Camelina microcarpa Andrz. Grand Canyon and Flagstaff (Coconino County), near Mesa (Maricopa County), Chiricahua Mountains (Cochise County), 1,000 to 8,000 feet. Widely distributed in the United States. Doubtfully distinct as a species from *C. sativa.*

23. DRABA (118)

Contributed by C. Leo Hitchcock

Plants of diverse habit, annual, biennial, or perennial, with scapose or leafy stems, usually pubescent with simple or forked hairs; leaves entire or dentate; racemes short and corymbose to elongate; petals white or yellow; capsules (silicles) 2-celled, dehiscent, strongly compressed parallel to the partition and flat, or elongate and often twisted; seeds numerous, in 2 rows in each cell.

KEY TO THE SPECIES

1. Plants evidently perennial, the stems usually tall and leafy. Petals yellow, often fading to white, usually as much as 4 mm. long; style evident, persistent; capsules elongate, often twisted (2).

1. Plants winter annuals, except *D. crassifolia,* a small, scapose biennial or short-lived perennial (7).

2. Plants scapose, the leaves in basal rosettes, grayish-hirsute with stalked cruciform hairs; capsules densely pubescent . . . . 1. *D. asprella*

2. Plants not truly scapose, some of the leaves cauline; capsules often glabrous (3).

3. Leaves ciliate with simple or forked hairs, mostly basal; stems glabrous above; capsules glabrous. Leaves narrowly oblanceolate, 1.5 to 8 cm. long, 2 to 7 mm. wide; style 1 to 1.5 mm. long . . . . 2. *D. Standleyi*

3. Leaves usually rather uniformly pubescent, many of them cauline; stems mostly pubescent throughout; capsules glabrous or pubescent (4).

4. Stem leaves dentate to denticulate, pubescent with cross-shaped, closely appressed hairs, not ciliate on the petioles; styles 1 to 3.5 mm. long . . . . 3. *D. spectabilis*

4. Stem leaves usually entire but if dentate, then the petioles ciliate and the hairs forked to cross-shaped but not appressed; styles often less than 1 mm. long (5).

5. Basal leaves oblanceolate, 3 to 8 cm. long, their bases matted on the caudices; pubescence rather sparse, hence all the leaves bright green. Styles 1 to 3 mm. long 5. *D. petrophila*

5. Basal leaves otherwise; whole plant usually more or less pallid, the pubescence dense (6).

6. Styles 0.5 to 1.5 mm. long; pedicels in fruit ascending to erect . . . . 4. *D. aurea*

6. Styles 1.5 to 3.5 mm. long; pedicels in fruit spreading or ascending-spreading 6. *D. Helleriana*

7. Petals yellow at anthesis, often fading to whitish; plants of elevations at or above 6,000 feet (8).

7. Petals white or whitish at anthesis; plants winter annuals of elevations usually less than 6,000 feet (9).

8. Stems leafy nearly or quite to the inflorescence; whole plant, including the capsules, copiously pubescent . . . . 7. *D. rectifructa*

8. Stems scapose (or with one leaf); plants sparsely pubescent, the capsules nearly always glabrous . . . . 8. *D. crassifolia*

9. Stems leafy up to the inflorescence; basal leaves petioled; capsules 2 to 5 mm. long; leaves all entire or some of them sparingly dentate . . . . 9. *D. brachycarpa*

9. Stems not leafy up to the inflorescence; basal leaves sessile or nearly so; capsules 4 to 14 mm. long; some or all the leaves usually coarsely few-toothed (10).

10. Inflorescence and pedicels glabrous . . . . 10. *D. reptans*

10. Inflorescence and pedicels pubescent . . . . 11. *D. cuneifolia*

1. Draba asprella Greene. Coconino and Yavapai counties, perhaps also in Gila County, 5,000 to 8,000 feet, pine forests, local, commencing to flower in February, type from Lynx Creek, Yavapai County (*Rusby* in 1883). Known only from Arizona.

KEY TO THE VARIETIES

1. Hairs of the capsules long and stiff, simple or, if forked, then with the branches unequal ..... var. *typica*
1. Hairs of the capsules short and fine, once-forked to stellate (2).
2. Petioles and lower portion of the scapes with many long, coarse hairs; pubescence of the leaves coarse and stiff ..... var. *stelligera*
2. Petioles and lower portion of the scapes with chiefly branched hairs; pubescence of the leaves soft and very dense ..... var. *kaibabensis*

Var. *typica* C. L. Hitchc. is known definitely only from Yavapai County. Var. *stelligera* O. E. Schulz occurs in Coconino County from the Kaibab Plateau to Oak Creek Canyon, type from Flagstaff (*Purpus* 16, in 1902). Var. *kaibabensis* C. L. Hitchc. is found on the Kaibab Plateau and North Rim of Grand Canyon, where the type was collected by Eastwood & Howell (No. 1061).

2. Draba Standleyi Macbr. & Payson. Apache Pass and Chiricahua Mountains (Cochise County), 5,000 to 8,500 feet. Western Texas to southeastern Arizona.

3. Draba spectabilis Greene. Lukachukai Mountains, northern Apache County, 7,000 feet (*Peebles* 14388). Colorado, Utah, and northeastern Arizona.

The collection cited belongs to var. *typica* C. L. Hitchc.

4. Draba aurea Vahl. White Mountains (Apache County), San Francisco Peaks (Coconino County), mountains of Cochise County, 5,000 to 12,000 feet, July and August. Widely distributed in the cooler parts of the Northern Hemisphere.

The Arizona specimens mostly belong to var. *leiocarpa* (Payson & St. John) C. L. Hitchc., with relatively small flowers, short, glabrous capsules, and short, fruiting styles, but approaches to the typical phase in some of these characters also occur.

5. Draba petrophila Greene. Mountains of Graham, Cochise, Santa Cruz, and Pima counties, 4,000 to 9,000 feet, chiefly in rock crevices, type from the Santa Rita Mountains (*Pringle* in 1884). Southern Arizona and northern Sonora.

Typical *D. petrophila* and var. *viridis* (Heller) C. L. Hitchc. (*D. viridis* Heller) have about the same range in Arizona. The variety (type from Fort Huachuca, *Wilcox* in 1893) has longer styles and stiff, simple or forked hairs on the surface of the capsules, whereas in typical *D. petrophila* the capsules are glabrous or merely ciliate.

6. Draba Helleriana Greene. White Mountains (Apache and Greenlee counties), San Francisco Peaks (Coconino County), and mountains of Graham, Cochise, and Pima counties, usually among rocks in coniferous forests, 6,000 to 11,500 feet, July to September. Colorado (?), New Mexico, Arizona, and northern Mexico.

KEY TO THE VARIETIES

1. Hairs of the stem leaves almost all simple or once-forked . . . . var. *bifurcata*
1. Hairs of all the leaves largely 4-rayed (2).
2. Plants evidently perennial and of several years duration, with thick crowns of marcescent leaf bases; leaves oblong to oblanceolate, the pubescence soft. Leaves mostly entire, silvery green . . . . var. *Blumeri*
2. Plants biennial or short-lived perennial, the bases not thickened by marcescent leaf bases; leaves ovate to obovate, the hairs rather stiff . . . . var. *patens*

Var. *bifurcata* C. L. Hitchc. is known only from southeastern Arizona westward to the Santa Catalina Mountains, type from the Chiricahua Mountains (*Blumer* 1608). Var. *Blumeri* C. L. Hitchc. occurs on the San Francisco Peaks and the White, Pinaleno, Chiricahua, Huachuca, and Santa Rita mountains, type from the Chiricahua Mountains (*Blumer* 1465). Var. *patens* (Heller) O. E. Schulz (*D. patens* Heller) has been collected on the San Francisco Peaks and in the White Mountains.

7. Draba rectifructa C. L. Hitchc. Jacob Lake to the Grand Canyon (Coconino County), White Mountains (Apache County), 8,500 to 9,000 feet, open coniferous forest, August. Colorado, Utah, and northern New Mexico and Arizona.

8. Draba crassifolia Graham (*D. albertina* Greene). San Francisco Peaks, 10,000 to 12,000 feet, July. Montana to British Columbia, south to Colorado, northern Arizona, and California.

9. Draba brachycarpa Nutt. Devil's Canyon, Pinal or Gila County (*Porter* 802), between Payson and Pine, Gila County (*Eastwood* 17195a, with pubescent fruits), February to May. Virginia to Kansas, south to Florida, Texas, and central Arizona.

10. Draba reptans (Lam.) Fern. Lukachukai Mountains and Navajo Experiment Station (northern Apache County), Payson to Pine (Gila County), Pinaleno Mountains (Graham County), Santa Catalina Mountains (Pima County), 3,000 to 6,500 feet. Massachusetts to Georgia, west to Washington, California, and northern Sonora.

Easily distinguished from *D. cuneifolia* by the more nearly entire leaves and the glabrous inflorescence. Most of the Arizona plants are var. *stellifera* (O. E. Schulz) C. L. Hitchc., characterized by prevailingly stellate hairs on the upper leaf surface, the capsules being either glabrous or hispidulous; but var. *typica* C. L. Hitchc. forma *micrantha* (Nutt.) C. L. Hitchc., characterized by having the hairs of the herbage mostly unbranched and the capsules hispidulous, has been collected in the Lukachukai Mountains (*Eastwood & Howell* 6782).

11. Draba cuneifolia Nutt. Throughout the state, 1,000 to 7,000 feet, very common in sandy soil, February to May. Illinois to Washington, south to Florida, Texas, California, and northern Mexico.

KEY TO THE VARIETIES

1. Stems leafy nearly to the inflorescence; racemes elongate, usually forming at least half the total height of the plant; capsules hispidulous (the hairs simple), oval or somewhat obovate, 3 to 4 mm. wide; pedicels nearly as long as to longer than the fruit var. *platycarpa*

1. Stems leafy only below the middle; racemes forming less than half the height of the plant or if half or more, then the capsules stellate-pubescent or glabrous; capsules narrowly elliptic to linear, commonly less than 3 mm. wide; pedicels usually shorter than the fruit, often only half as long (2).
2. Hairs of the capsules and herbage mostly branched; racemes forming about half the length of the stem; style evident in fruit, slender.................... var. *integrifolia*
2. Hairs of the capsules simple or the capsules glabrous, the hairs of the lower part of the stem also, in part, often simple; racemes forming usually less than half the length of the stem; style not evident .... var. *typica*

Var. *platycarpa* (Torr. & Gray) Wats. (*D. platycarpa* Torr. & Gray) occurs near Granite Reef Dam, Maricopa County, and in Cochise and Pima counties, 3,500 to 5,000 feet. Var. *integrifolia* Wats. (*D. sonorae* Greene) is occasional in Mohave, Yavapai, Graham, Maricopa, Pinal, Pima, and Yuma counties. Var. *typica* C. L. Hitchc. is found throughout the range of the species in Arizona and is much the commonest form.

24. THYSANOCARPUS. Lace-pod, Fringe-pod

Plants annual, the herbage usually nearly glabrous; stems slender, erect, simple or sparingly branched, leafy; leaves sessile, usually auriculate-clasping, nearly entire to deeply dentate, the basal ones sometimes pinnatifid; flowers small, in elongate racemes; fruits flat, indehiscent, 1-seeded, orbicular or nearly so, winged, the wing usually with radiating nerves and regularly incised or perforate.

The dainty and peculiar fruits make these otherwise insignificant plants singularly attractive. The taxonomy of the genus is confused, and the following disposition of the Arizona material is tentative.

KEY TO THE SPECIES

1. Fruits 3 to 4 mm. wide, glabrous .... 1. *T. laciniatus*
1. Fruits 4.5 to 7 mm. wide, glabrous to copiously pubescent, the wing regularly and conspicuously notched or perforate .... 2. *T. amplectens*

1. Thysanocarpus laciniatus Nutt. Grand Canyon (Coconino County), Mohave County (several localities), Prescott (Yavapai County), 3,500 to 5,000 feet. Arizona and California.

The Arizona plants seem to belong to var. *crenatus* (Nutt.) Brewer, having fruits with well-defined rays, notched or perforate between the rays, and leaves mostly entire to coarsely dentate. Specimens from between Kingman and Chloride, Mohave County (*Kearney & Peebles* 11176, *Darrow* 3744), approach typical *T. laciniatus* in the rayless, subentire wing, but the fruits are pubescent and exceptionally large.

2. Thysanocarpus amplectens Greene. Apache and Coconino counties, south to Cochise, Santa Cruz, Pima, and Yuma counties, rarely above 4,000 feet, preferring moist, sandy soil, January to May. Southwestern New Mexico and Arizona.

Greene's description of *T. amplectens* needs considerable amplification to include all the Arizona specimens here included under that species. *T. filipes* Greene, based on a collection at Clifton, Greenlee County (*Davidson* in 1899), may be a form of this rather variable species.

25. ATHYSANUS

Plants annual, small, pubescent with partly forked hairs; stems filiform-branched from at or near the base; racemes elongate, 1-sided; flowers on reflexed or recurved pedicels, the petals minute or wanting; fruits indehiscent, 1-seeded, wingless.

1. Athysanus pusillus (Hook.) Greene. Grand Canyon (Coconino County), and Mohave, southern Yavapai, Graham, Gila, Maricopa, and Pinal counties, 5,000 feet or lower, March to May. Idaho to British Columbia, Arizona, and California.

The type of var. *glabrior* Wats., a nearly glabrous form, came from Fort Mohave (*Lemmon* in 1884).

26. DESCURAINIA (119). TANSY-MUSTARD

Plants annual, stellate-pubescent and sometimes glandular; stems often tall, leafy, simple or sparingly branched; leaves deeply pinnatifid or once to thrice pinnate, the segments mostly small; racemes terminal, becoming elongate; flowers small, the petals yellow or whitish; capsules dehiscent, 2-celled, slender, elongate, terete or nearly so; seeds many, small, in 1 or 2 rows in each cell.

Some of the species are reported to be used by the Indians as greens and for making pinole (gruel or mush) from the parched and ground seeds. Among the Mexicans the seeds are used in poultices for wounds. The plants grow mainly in open ground, flowering chiefly in early spring.

KEY TO THE SPECIES

1. Capsules clavate or subclavate, spreading. Leaves bipinnate or the upper ones simply pinnate . . . . 1. *D. pinnata*
1. Capsules not clavate, or somewhat so in *D. obtusa* var. *brevisiliqua* (2).
2. Seeds in 2 rows in each locule of some or all the capsules. Herbage canescent 2. *D. obtusa*
2. Seeds in 1 row in each locule (3).
3. Leaves twice or thrice pinnate, with narrow segments; capsules containing 20 or more seeds, 10 to 30 mm. long . . . . 3. *D. Sophia*
3. Leaves simply pinnate, the leaflets often deeply incised; capsules often containing fewer than 20 seeds (4).
4. Capsules 3 to 7 mm. long, attenuate at apex and tipped with the prominent style 4. *D. californica*
4. Capsules 8 to 15 mm. long, not attenuate at apex, the style short or obsolete 5. *D. Richardsonii*

1. Descurainia pinnata (Walt.) Britton. Throughout the state, the most abundant and widely distributed species, up to 7,000 feet. Southeastern United States to Mackenzie, south to Arizona, California, and northern Mexico.

KEY TO THE SUBSPECIES

1. Fruiting pedicels spreading at an angle of about 45 (30 to 70) degrees. Herbage densely canescent . . . . subsp. *Paysonii*
1. Fruiting pedicels spreading at an angle of about 75 (60 to 90) degrees (2).
2. Leaf segments narrowly oblong to linear. Plants usually short and branched below; petals mostly less than 2 mm. long, yellow or whitish . . . . subsp. *halictorum*
2. Leaf segments ovate to oblanceolate (3).

3. Herbage canescent; petals whitish to pale yellow; capsules mostly 8 to 10 mm. long; racemes pubescent. Plants often wide-branching from the base...... subsp. *ochroleuca*
3. Herbage green, moderately pubescent; petals yellow; capsules mostly about 6 mm. long; racemes glabrous .... subsp. *glabra*

A collection at Betatakin, Navajo County (*Peebles & Fulton* 11911), is cited by Detling under his subsp. *Paysonii*. The specimens in question have much the appearance of *D. obtusa* and do not resemble other forms of *D. pinnata*. Subsp. *halictorum* (Cockerell) Detling and subsp. *glabra* (Woot. & Standl.) Detling occur throughout most of the range of the species in Arizona. Subsp. *ochroleuca* (Wooton) Detling has been collected in Apache, Coconino, Yavapai, Gila, Maricopa, Pinal, and Pima counties.

2. Descurainia obtusa (Greene) O. E. Schulz. Apache County to Mohave County, south to Greenlee and Pima counties. New Mexico to southern California and northern Mexico.

KEY TO THE SUBSPECIES

1. Plants glandular, at least in the inflorescence...................... subsp. *adenophora*
1. Plants not glandular (2).
2. Capsules 5 to 9 mm. long .... subsp. *brevisiliqua*
2. Capsules 10 to 20 mm. long .... subsp. *typica*

Subsp. *adenophora* (Woot. & Standl.) Detling is known from Apache, Navajo, Coconino, and Yavapai counties. Subsp. *brevisiliqua* Detling occurs at Flagstaff and near Sunset Crater (Coconino County), near Prescott (Yavapai County) and has been reported from Fort Mohave (Mohave County). Subsp. *typica* Detling ranges from Apache to Mohave counties, south to Greenlee and Pima counties.

3. Descurainia Sophia (L.) Webb. Coconino, Mohave, Graham, Gila, Pinal, Cochise, Santa Cruz, and Pima counties, an abundant roadside weed in some localities. Extensively naturalized in the United States; from Eurasia.

4. Descurainia californica (Gray) O. E. Schulz. Lukachukai Mountains and Tsegi Canyon (northern Apache County), Kaibab Plateau and Grand Canyon (Coconino County), 6,000 to 8,500 feet. Wyoming to Oregon, south to New Mexico, northern Arizona, and California.

5. Descurainia Richardsonii (Sweet) O. E. Schulz. Apache, Coconino, and Greenlee counties, 6,500 to 9,500 feet. Great Lakes region to Yukon, south to northern Mexico and Baja California.

Represented in Arizona by the glandular subsp. *viscosa* (Rydb.) Detling, which has been found in the White Mountains (Apache and Greenlee counties) and at Walnut Canyon (Coconino County), also by the nonglandular subsp. *incisa* (Engelm.) Detling, known from the Carrizo Mountains and near Round Rock and Rock Point (Apache County), Betatakin (Navajo County), and Walnut Canyon (Coconino County).

27. ARABIS (120). ROCK-CRESS

Contributed by Reed C. Rollins

Biennial or perennial herbs with acrid juice; caudex simple or multicipitally branched; young plants rosulate; leaves entire or dentate, the basal ones petiolate, the cauline leaves sessile except in *A. tricornuta;* inflorescence racemose; siliques narrowly linear, flattened or at least slightly compressed parallel to the septum; valves 1-nerved at least below; seeds numerous, winged or rarely wingless; cotyledons accumbent.

Six of the ten Arizona species have a northerly distribution in the Sierra Nevada, Great Basin, or Rocky Mountains and are rare in Arizona. Two species, *A. gracilipes* and *A. tricornuta,* are endemic to Arizona. *A. perennans* is widely distributed in the state and is apparently the only species of the genus commonly encountered.

KEY TO THE SPECIES

1. Mature siliques strictly erect, numerous, mostly crowded, and appressed to the rachis; petals white to ochroleucous, rarely pinkish (2).
1. Mature siliques spreading or deflexed, few, not crowded; petals white to purple (4).
2. Siliques subterete; stigma markedly expanded, conspicuously bilobed; seeds biseriate, narrowly winged or rarely wingless; petals ochroleucous................1. *A. glabra*
2. Siliques decidedly flattened parallel to the septum; stigma unexpanded, entire or inconspicuously bilobed; seeds uniseriate or biseriate, winged all around; petals usually white, rarely pinkish (3).
3. Stems sparsely to densely hirsute below; siliques about 1 mm. wide; seeds uniseriate; petals less than 6 mm. long ....2. *A. hirsuta*
3. Stems glabrous throughout or appressed-pubescent at base; siliques about 2 mm. wide; seeds biseriate; petals up to 1 cm. long ....3. *A. Drummondi*
4. Lower cauline leaves petiolate; petals white; sepals glabrous; style about 1 mm. long 4. *A. tricornuta*
4. Lower cauline leaves sessile; petals pink to purple; sepals at least sparsely pubescent; stigma sessile or nearly so (5).
5. Siliques finely pubescent, strongly reflexed but the pedicels not geniculate, straight, 2.5 to 3 mm. wide; seeds broadly winged, biseriate; petals 1.5 to 2 cm. long. Leaves all narrow, entire or very nearly so ....5. *A. pulchra*
5. Siliques glabrous, widely spreading to pendulous, arcuate, usually less than 2.5 mm. wide; seeds narrowly winged or nearly wingless, uniseriate or biseriate; petals less than 1 cm. long (6).
6. Cauline leaves numerous (30 to 80 on each stem), closely imbricate; stems robust, densely leafy below, single or rarely 2 from a simple base; pedicels 2 to 4 cm. long 6. *A. gracilipes*
6. Cauline leaves fewer (usually fewer than 15 on each stem), scarcely imbricate; stems mostly slender, not densely leafy below, 1 to several from a simple or branched base; pedicels less than 2 cm. long (7).
7. Pubescence of lower stems and basal leaves dendritic, fine to coarse but not setaceous; leaf margins not ciliate; seeds winged, uniseriate (8).
7. Pubescence of lower stems and basal leaves coarse, simple or forked, setaceous; leaf margins ciliate; stems hirsute below with simple trichomes; seeds narrowly winged to wingless, biseriate or imperfectly so (9).
8. Basal leaves narrowly oblanceolate, acute, entire, densely pubescent with minute, dendritic trichomes; pubescence of lower part of stem closely appressed; flowering pedicels sparsely pubescent, the fruiting pedicels arched strongly downward....7. *A. lignifera*
8. Basal leaves broadly oblanceolate to broader, obtuse, at least the lower ones dentate to repand, pubescent with coarse, dendritic trichomes; pubescence of lower part of stem spreading, coarse; flowering pedicels glabrous, the fruiting pedicels widely spreading 8. *A. perennans*
9. Siliques 2 to 4 cm. long; plants mostly less than 3 dm. high; cauline leaves small, mostly remote; basal leaves hirsute with simple trichomes, or glabrous........9. *A. pendulina*
9. Siliques 4 to 6 cm. long; plants 2.5 to 6 dm. high; cauline leaves fairly ample, the lower ones somewhat imbricate; basal leaves hirsute on the blade surfaces with forked trichomes, ciliate with simple or forked trichomes.....................10. *A. Fendleri*

1. Arabis glabra (L.) Bernh. North Rim of the Grand Canyon (Coconino County), to the mountains of Yavapai and Gila counties, 4,500 to 8,000 feet, open scrub, pinelands, grassy slopes, or rarely near streams, May to July. Southern Quebec to North Carolina, central Arizona, and California; Europe and Asia.

Many European and some American botanists treat this species in the segregate genus *Turritis* on account of the subterete siliques and yellowish flowers, but so many of the characters coincide with those ordinarily attributed to *Arabis* that such a classification is misleading. The species is certainly native in many areas of the United States, but it tends to become weedy and may have been recently transported to some localities.

2. Arabis hirsuta (L.) Scop. Plateau and mountain areas of northern Arizona (Apache and Coconino counties), 6,000 to 9,000 feet, open slopes, under small trees or shrubs, and in grassy meadows, May to July. Quebec to Yukon Territory, south to Georgia, northern Arizona, and California; Eurasia.

Arabis hirsuta is a rather variable species, with several varieties present in North America. The typical variety is not found in North America, and only var. *pycnocarpa* (Hopkins) Rollins is known from Arizona.

3. Arabis Drummondi Gray. Apparently rare in the Lukachukai Mountains (northern Apache County) and in the plateau region of northern Coconino County, 7,000 to 9,000 feet, coarsely granitic or sandy soils, floor of pine forests, or open slopes in spruce-fir forests, May to September. Southern Labrador to British Columbia, south to New Jersey, northern Arizona, and California.

4. Arabis tricornuta Rollins. Huachuca and Chiricahua Mountains (Cochise County), Rincon and Santa Rita mountains (Pima and Santa Cruz counties), 7,000 to 9,000 feet, July to September, type from the Rincon Mountains (*Blumer* 3478). Known only from southern Arizona.

This species is of particular interest because of the superficial resemblance at flowering time to *Thelypodium micranthum*. The flowers are small and in long, loose, very narrow racemes.

5. Arabis pulchra Jones. The species is represented in Arizona by var. *pallens* Jones (*A. formosa* Greene), which has been collected at Kayenta and Betatakin (Navajo County) and on the Kaibab Plateau and in the Grand Canyon (Coconino County), about 6,000 feet, sandy hillsides and open knolls among sagebrush, pinyon, and juniper, May to July. This variety occurs in the Colorado River drainage area in western Colorado, eastern Utah, and northeastern Arizona.

Other varieties of *Arabis pulchra* have not been collected in Arizona, but *A. pulchra* var. *munciensis* Jones is known from a locality between St. George and the Beaver Dam Mountains in adjacent Utah. This variety is distinguished from var. *pallens* by having spreading pedicels, pendulous siliques, and smaller, purple instead of white or pink flowers. Among Arizona species of *Arabis, A. pulchra* is unique in having relatively large, showy flowers and finely pubescent siliques.

6. Arabis gracilipes Greene (*A. arcuata* (Nutt.) Gray var. *longipes* Wats.). Coconino, Mohave, Gila, and Yavapai counties, hot, sandy canyons and lower mountain slopes (rarely up to 8,000 feet), seldom collected, April to June, type of *A. gracilipes* from near Flagstaff (*N. C. Wilson* in 1893), type of *A. arcuata* var. *longipes* from near Fort Mohave (*Lemmon* in 1884). Known only from Arizona.

Arabis gracilipes is related to both *A. perennans* and *A. Fendleri* but is readily separated from them by the robust habit, numerous large cauline leaves, very long, filiform pedicels, and usually solitary stems.

7. Arabis lignifera A. Nels. South of Witch Well, Apache County (*Ripley & Barneby* 8424), 12 miles east of Keam Canyon, Navajo County (*Peebles & Smith* 13438), the only collections known from Arizona, about 6,000 feet, sandy or stony soils, often among sagebrush, pinyon, or juniper, April and May. Wyoming to Idaho, south to Colorado, northeastern Arizona, and Nevada.

This species has something of the habit of *A. perennans* but differs markedly in having short, rigid, recurved pedicels, fewer stems, and a very fine, dense indument on the basal leaves and lower portions of the plant. The species is abundant in the basin ranges to the north.

8. Arabis perennans Wats. (*A. eremophila* Greene, *A. recondita* Greene). Apache County to Mohave County, south to Graham and Pima counties, 2,000 to 8,000 feet, mostly on lower mountain slopes and in hot canyons, February to October, type from the Santa Catalina Mountains, Pima County (*Pringle* in 1881), type of *A. eremophila* from Peach Springs, Mohave County (*Baker* in 1899), type of *A. recondita* from Diamond Creek, Mohave County (*Wilson* in 1893). Western Colorado to Arizona, southern California, northern Mexico, and Baja California.

This, the earliest-flowering and most abundant species of the genus in Arizona, often develops a ligneous footlike extension of the caudex between the functional basal leaves and the ground surface. This foot appears to be formed as a result of the shedding of basal leaves during successive seasons and the occurrence of new leaves at a higher level on the caudex each succeeding year.

9. Arabis pendulina Greene. Coconino County (*Wiegand* 658), and on the Kaibab Plateau (*Goodding* 197–49), dry washes and rocky and gravelly hillsides in the upper pinyon-sagebrush areas and at slightly higher elevations in the mountains. Utah, Nevada, and northern Arizona.

This species is most closely related to *Arabis Fendleri* but is much less robust, with several to numerous slender stems. It has smaller, remote, usually nonclasping cauline leaves and shorter siliques than *A. Fendleri*.

10. Arabis Fendleri (Wats.) Greene. White Mountains (Apache County) and northern Navajo County to northern Mohave County, 5,000 to 8,000 feet, open pine woods, April to June. Colorado and New Mexico to Nevada and northern Arizona.

Arabis Fendleri is much like *A. perennans* in habit of growth, but the latter flowers much earlier and ordinarily occurs at lower altitudes.

28. ERYSIMUM

Plants annual, biennial, or perennial, rather coarse, with appressed, harsh pubescence; leaves entire or repand-dentate; petals 6 mm. long or longer, yellow, orange, or maroon; capsules elongate, 4-sided, rigid, erect to recurved-spreading, the valves strongly keeled; seeds in 1 row in each cell.

KEY TO THE SPECIES

1. Pods divaricate, often somewhat curved, about 1 mm. wide; leaves repand-dentate. Petals not more than 8 mm. long, pale yellow............................1. *E. repandum*
1. Pods ascending or erect, straight or nearly so, commonly more than 1 mm. wide; leaves entire or sparingly dentate (2).
2. Petals more than 10 mm. long, yellow, orange, or maroon...............2. *E. capitatum*
2. Petals less than 10 mm. long, pale yellow.........................3. *E. inconspicuum*

1. Erysimum repandum L. White Mountains (Apache ? County), Grand Canyon (Coconino County) to northern Yavapai County, also Gila River bed (Pinal County), March to June. Ohio to Oregon and Arizona; introduced from Europe.

2. Erysimuum capitatum (Dougl.) Greene (*E. elatum* Nutt., *E. Wheeleri* Rothr.). Throughout the state except the extreme western portion, 2,500 to 9,500 feet, March to September, types of *E. Wheeleri* from Camp Grant and Mount Graham (*Rothrock*). Saskatchewan to Washington, south to New Mexico, Arizona, and California.

Western-wallflower, a showy plant, with flowers resembling those of the cultivated wallflower. The petals are usually bright yellow, but in the variant described as *E. Wheeleri* they are orange or maroon, and the plant is usually found in coniferous forests at higher altitudes, rarely below 7,000 feet. This entity is perhaps entitled to varietal rank.

3. Erysimum inconspicuum (Wats.) MacMillan. Kaibab Plateau, Grand Canyon, Havasu Canyon, Flagstaff, Walnut Canyon, and Winona (Coconino County). Ontario and Alaska to Kansas, northern Arizona, and Oregon.

29. MALCOLMIA

Annual; herbage roughly pubescent with mostly forked hairs; stems several, sparingly branched above; leaves petioled, oblong or lanceolate, coarsely few-toothed; petals purplish pink, the blades much shorter than the claws; capsules ascending-spreading, tetragonal, slender, 4 to 6 cm. long, with a short, conic beak.

1. Malcolmia africana (L.) R. Br. Near Wolf Hole and Fredonia (northern Mohave County), 3,500 to 4,500 feet, April and May. Utah, southern Nevada, and northwestern Arizona; introduced from the Mediterranean region.

Matthiola bicornis (Sibth. & Smith) DC., of the eastern Mediterranean region, has been collected at Tucson (*Thornber* 7424). The collector reported it as "becoming common on the mesas," in 1914, but apparently it has since disappeared. It somewhat resembles *Malcolmia africana,* but the flowers are much larger (15 to 20 mm. long), and the long, slender, divaricate capsules are 2-pronged at apex. It is much less common in cultivation than the common stock (*Matthiola incana*).

30. CHORISPORA

Plants annual; stems sparingly branched; herbage sparsely pubescent with short, gland-tipped hairs; leaves dentate or denticulate, the lower ones sometimes pinnatifid; flowers rather large, in elongate, loose racemes; calyx narrowly cylindric, only the tips of the sepals spreading; petals with long, erect claws and narrow, spreading, pink or lilac blades; anthers sagittate; stigma minute; capsules slender, elongate, the lower portion torulose, with transverse partitions between the seeds, the upper portion (more or less half the length of the pod) seedless, beaklike.

1. Chorispora tenella (Pall.) DC. Copper Basin Road near Prescott, Yavapai County (*Dearing* 5819), Oak Creek Canyon, Coconino County (*Story & Deaver* in 1950). Here and there in the western United States; introduced from southern Asia.

31. CONRINGIA. Hares-ear-mustard

Plants glabrous, annual; stems tall, leafy; stem leaves sessile, broad, entire or denticulate, with clasping bases; petals pale yellow; capsules long and narrow, 4-sided, spreading; seeds in 1 row in each cell.

1. Conringia orientalis (L.) Dum. Keam Canyon, Navajo County, 6,500 feet (*Peebles & Smith* 13406), Mescal, Cochise County (*Thornber* 4928). Here and there in the United States; introduced from Europe.

57. CAPPARIDACEAE. Caper Family

Plants herbaceous or woody, sometimes ill-scented; leaves alternate, simple or palmately compound; flowers perfect, regular or nearly so, commonly in bracted racemes; sepals and petals usually 4; stamens 6 or more, usually much exserted; ovary commonly 1-celled and stipitate; fruit a 2-valved capsule, or indehiscent.

The pickled flower buds of the European *Capparis spinosa* are the familiar condiment capers. It is reported that the Hopi and other Indians use young plants of *Cleome* and *Wislizenia* as potherbs.

KEY TO THE GENERA

1. Plants shrubby; leaves simple, entire or merely emarginate; fruit berrylike .... 5. *Atamisquea*
1. Plants herbaceous; some or all of the leaves compound; fruit not berrylike (2).
2. Fruits didymous, indehiscent, the carpels divaricate, 1-seeded, becoming coriaceous. Leaves trifoliolate, rarely simple; flowers numerous, small; petals yellow. 3. *Wislizenia*
2. Fruit not didymous, a 2-valved, thin-walled, capsule, several- to many-seeded (3).
3. Stamens 12 or more; herbage glandular-villous, very clammy. Leaves trifoliolate; petals whitish .... 4. *Polanisia*
3. Stamens 6; herbage not glandular (4).
4. Capsules slender, cylindric, at least 1 cm. long; leaflets 3 or more; petals pink or yellow .... 1. *Cleome*
4. Capsules broad, more or less triangular or quadrangular, not more than 5 mm. long; leaflets 1 to 3; petals yellow .... 2. *Cleomella*

1. CLEOME. Spider-flower

Herbage not glandular; stems mostly tall and branched; leaflets 3 to 5, entire or serrulate; petals clawed; stamens commonly 6, inserted on a receptacle above the petals; capsule elongate, many-seeded, long-stipitate, more or less torulose.

KEY TO THE SPECIES

1. Sepals distinct or very nearly so, soon deciduous; leaves all very short-petioled or subsessile; leaflets 3, narrowly linear, not more (usually less) than 3 cm. long; petals 4 to 5 mm. long, pink or white; capsules 1 to 1.5 cm. long............... 1. *C. sonorae*
1. Sepals united below, persistent; leaves all distinctly petioled; leaflets 3 or more, lanceolate, oblanceolate, or oblong, usually more than 3 cm. long; petals usually not less than 6 mm. long; capsules seldom less than 2 cm. long (2)
2. Petals purplish pink or white; leaflets 3.............................. 2. *C. serrulata*
2. Petals yellow; leaflets commonly more than 3 (3).
3. Capsules less than 4 cm. long, the stipes not more than 1.5 cm. long; petals light yellow, 4 to 7 mm. long; longer filaments 10 to 15 mm. long..................... 3. *C. lutea*
3. Capsules 4 to 6 cm. long, the stipes up to 2.5 cm. long; petals golden yellow, 10 to 13 mm. long; longer filaments 20 to 30 mm. long......................... 4. *C. Jonesii*

1. Cleome sonorae Gray. "Low subsaline grounds west of the Chiricahui Mts., Sonora" (now Arizona), the type collection (*Wright*), Willcox, "alkaline sink" 4,000 feet (*Thornber* 2231, *Pultz* 1038), August and September. Colorado, New Mexico, and southeastern Arizona.

2. Cleome serrulata Pursh. Apache County to Mohave, Yavapai, Greenlee, and Gila counties, 4,500 to 7,000 feet, chiefly at roadsides, June to September. Saskatchewan to Kansas, Arizona, and Oregon.

Rocky Mountain bee-plant. Grows as a weed and appears as if introduced in Arizona. Sometimes called skunk-weed because of the unpleasant odor of the crushed herbage. The seeds are eaten by doves.

3. Cleome lutea Hook. (*Peritoma breviflorum* Woot. & Standl.). Apache County to Coconino and Yavapai counties, also occasional in Gila, Pinal, Santa Cruz, and Pima counties, 2,000 to 6,000 feet, mostly along streams, May and June (and probably later). Nebraska to Washington, south to New Mexico, Arizona, and eastern California.

Yellow bee-plant. *Peritoma breviflorum* is a form with relatively small flowers and pods and short stipes.

4. Cleome Jonesii (Macbr.) Tidestrom (*C. lutea* var. *Jonesii* Macbr.). Coconino, Yavapai, and Mohave counties to Santa Cruz and Pima counties, 1,000 to 4,000 feet, sandy soil along streams, May to September, type from the Verde River Valley, Yavapai County (*W. W. Jones* 168). Known only from Arizona.

Usually a larger plant than *C. lutea,* the stems reaching a height of about 2 m. Although the ranges of the two entities in Arizona overlap to a greater extent than was formerly supposed, there seems to be very little intergradation.

2. CLEOMELLA (121)

Plants annual; stems erect and branching above, or diffuse; leaves trifoliolate, or the upper ones reduced to 1 leaflet; pedicels subtended by small bracts or by leaves of normal size; petals sessile or nearly so, yellow; stamens 6, inserted on a short receptacle; capsules seldom longer than broad, the valves concave, hemispheric or conic; seeds usually few.

These plants mostly prefer saline soil.

KEY TO THE SPECIES

1. Leaflets obovate, less than 3 times as long as wide, pubescent, very obtuse, somewhat succulent; flowers not in definite racemes, crowded at the ends of the stems and

branches in the axils of not greatly reduced leaves; stems diffusely branched from the base; stipules conspicuous, white, filiform-laciniate; stipes not more than 6 mm. long, becoming reflexed..1. *C. obtusifolia*

1. Leaflets linear, oblong, or oblanceolate, 3 or more times as long as wide, glabrous; flowers in definite terminal racemes and subtended by greatly reduced leaves or bracts; stems normally erect; stipules inconspicuous, often reduced to a mere bristle; stipes spreading or somewhat deflexed (2).

2. Stipes in fruit 7 to 20 mm. long; leaflets commonly oblanceolate; inflorescence becoming elongate, often 5 cm. long or longer, the bracts small, much shorter than the fruiting pedicels; stems up to 80 cm. long; capsules 3 to 6 mm. long, the valves irregularly conic; seeds broadly ovoid..2. *C. longipes*

2. Stipes in fruit not more than 10 mm. long; leaflets linear or linear-lanceolate; inflorescence not conspicuously elongate, commonly not more than 3 cm. long, the bracts large, often equaling the pedicels; stems 20 to 40 cm. long; capsules 2 to 3 mm. long, rhombic, the valves broadly deltoid; seeds obovoid, often mottled
3. *C. plocasperma*

*1. **Cleomella obtusifolia** Torr. & Frém. Specimens labeled: "Mohave Desert, Arizona" (*Lemmon* in 1884) possibly were collected at Fort Mohave on the Colorado River, Mohave County. Southeastern California.

Plant known as Mojave-stinkweed. Doves are said to relish the seeds.

2. **Cleomella longipes** Torr. "West of the Chiricahui Mountains" (*Wright* 857, the type collection), saline playa near Willcox, Cochise County, about 4,000 feet (*Thornber* in 1905, *Pultz* 1045), July to September. Western Texas to southeastern Arizona and Sonora.

*3. **Cleomella plocasperma** Wats. There seems to be no definite evidence of the occurrence of this species in Arizona. Its known range is Utah to California and Oregon.

A collection from "near Fort Mohave in western Arizona" (*Cooper*) was referred by Watson (in King, *Geol. Expl. 40th Par.* 5: 33) to *C. parviflora* Gray. This species, known definitely only from Nevada and southeastern California, is distinguished from *C. plocasperma* by its very short stipes, these much shorter than the capsule.

3. WISLIZENIA. Jackass-clover

Plants annual; stems erect, much-branched; leaves trifoliolate, the leaflets elliptic, oblanceolate, or obovate; flowers very numerous, small, the petals yellow; stamens 6; fruits long-stipitate, 1.5 to 2.5 mm. long, 2-seeded, the rounded apex more or less rugose to tuberculate-dentate.

1. **Wislizenia refracta** Engelm. (*W. scabrida* Eastw.). Navajo, Coconino, and Mohave counties, south to Graham, Cochise, and Pima counties, reported from Greenlee County, 1,000 to 6,500 feet, usually in sandy soil, often very abundant, March to October, types of *W. scabrida* from near Tucson (*Lemmon* in 1880, *Pringle* in 1881). Western Texas to southern California.

Conspicuous in late summer at roadsides and in stream beds. The prevailing form in northeastern Arizona is var. *melilotoides* (Greene) Johnst., type from Holbrook (*Rusby* 581), characterized by obovate leaflets, these thicker and broader than in typical *W. refracta*, and by a more compact and leafy appearance.

4. POLANISIA. Clammy-weed

Plants annual; stems erect, branched; herbage glandular-pubescent and strong-scented; leaves trifoliolate, the leaflets elliptic or lanceolate; petals whitish or pale yellow; stamens numerous, the filaments long-exserted, purple; capsules sessile or nearly so, elongate, somewhat flattened, tipped by the slender style; seeds numerous.

1. Polanisia trachysperma Torr. & Gray. Navajo and Coconino counties to Graham, Cochise, Santa Cruz, and Pima counties, 1,000 to 6,500 feet, usually in sandy stream beds, May to October. Saskatchewan to British Columbia, south to Texas, New Mexico, and Arizona.

5. ATAMISQUEA

Shrubby, with rigid, brittle branches, ill-scented; leaves coriaceous, dark green and glabrate above, silvery-lepidote beneath, linear or narrowly oblong, the margins entire, the apex emarginate; flowers solitary or in small fascicles; fertile stamens 6; fruit oval or subglobose, about 8 mm. long, berrylike.

1. Atamisquea emarginata Miers. Quitobaquito, southwestern Pima County (*Harbison* 26181), 1,000 to 1,500 feet. Southern Arizona, Sonora, Sinaloa, and Baja California; Argentina.

The plant is reported to reach a height of 6 m. (20 feet), but is doubtless smaller in Arizona.

58. RESEDACEAE. Mignonette Family

1. OLIGOMERIS

Plants annual, slightly succulent, glabrous; leaves numerous, alternate or fascicled, linear, entire; flowers in slender terminal spikes, small, very irregular, the 2 white petals and the 3 stamens borne on one side; sepals 4; capsule turgid, 4-lobed, apically dehiscent.

1. Oligomeris linifolia (Vahl) Macbr. Maricopa, Pinal, Pima, Mohave, and Yuma counties, 2,500 feet or lower, rather common in dry, sandy, sometimes saline soil, March to June. Texas to California and northern Mexico; Asia and Africa.

59. CRASSULACEAE. Orpine Family

Plants herbaceous, usually succulent, annual or perennial, glabrous or puberulent; leaves simple, entire; flowers perfect, regular or nearly so; sepals and petals mostly 4 or 5, the petals distinct, or united below; stamens as many or twice as many as the petals, sometimes borne on the corolla; ovaries 3 to 5, superior, becoming distinct, or nearly distinct, 1- to several-seeded follicles.

Many members of this family are cultivated as ornamentals and are very popular with fanciers of succulent plants. Species of the genus *Sedum* are favorites in rock gardens.

KEY TO THE GENERA

1. Plants annual, small; leaves opposite; flowers minute, in axillary and terminal glomerules ... 3. *Tillaea*
1. Plants perennial; leaves alternate; flowers larger, in cymes, racemes, or open panicles (2).

2. Petals distinct or very nearly so, white, pink, or pale yellow; stems not scapose, leafy up to the inflorescence; cauline leaves often crowded, not conspicuously smaller than the basal ones; basal leaves not in a well-marked rosette..................1. *Sedum*
2. Petals united below, the lower part of the corolla tubular or funnelform, yellow, orange, or red, or marked with those colors; stems scapose or subscapose; cauline leaves scattered, much smaller than the basal leaves or reduced to scales; basal leaves in a conspicuous rosette..2. *Echeveria*

1. SEDUM (122). Stonecrop

Plants succulent, perennial; stems erect or decumbent, leafy; leaves simple, entire or dentate, terete or flat, often crowded and imbricate; inflorescence cymose, terminal or axillary; stamens usually 10; pistils 4 or 5, distinct or nearly so; follicles several- to many-seeded.

KEY TO THE SPECIES

1. Flowers in axillary racemes or cymes, these forming a spikelike or headlike inflorescence. Plants glabrous; stems erect, up to 30 cm. long, usually numerous; leaves entire or dentate, flat; petals pink or white; carpels erect, usually with spreading tips
1. *S. rhodanthum*
1. Flowers in terminal cymes (2).
2. Petals yellow, narrow, sharply acuminate. Plants with creeping rootstocks and with short, leafy, sterile shoots at the base, the flowering stems erect or nearly so; leaves mostly linear or lanceolate, the basal ones commonly papillate; cymes mostly narrow and compact, with ascending branches; mature carpels erect or spreading only at the tip, narrowed gradually into the style..........................2. *S. stenopetalum*
2. Petals white, sometimes tinged with purple (3).
3. Leaves subterete when fresh, not noticeably papillate; mature carpels strongly divaricate. Leaves mostly linear or lanceolate; cyme branches commonly spreading; petals acute to acuminate...3. *S. stelliforme*
3. Leaves flat or nearly so when fresh, minutely to prominently papillate; mature carpels erect, or the tips (persistent styles) somewhat spreading (4).
4. Basal leaves 1 to 4 mm. wide, minutely papillate; leaves of the flowering stems ascending or spreading, oblong-oblanceolate, spatulate, or oblong-lanceolate, obtuse or rounded (rarely acute) at apex, light green; floral bracts oblong-oblanceolate; sepals usually obtuse, 4 to 5 mm. long............................4. *S. Cockerellii*
4. Basal leaves 0.5 to 2 mm. wide, prominently papillate; leaves of the flowering stems spreading or somewhat recurved, linear, narrowly oblong, or narrowly elliptic, acute or acutish at apex, usually dark green; floral bracts linear or narrowly oblong; sepals acute or acutish, 4 to 9 mm. long..................................5. *S. Griffithsii*

1. Sedum rhodanthum Gray (*Clementsia rhodantha* Rose). San Francisco Peaks (Coconino County), Baldy Peak (Apache County), 9,000 to 12,000 feet, July to September. Montana to Utah, New Mexico, and Arizona.

2. Sedum stenopetalum Pursh. Coconino County, North Rim of the Grand Canyon (*Grand Canyon Herb.* 625, 626), Deboschibeko, 6,500 feet (*Darsie* in 1933), June to August. Alberta to Nebraska, New Mexico, northern Arizona, and California.

The Arizona specimens may belong to *S. lanceolatum* Torr. rather than to *S. stenopetalum* (see R. T. Clausen, *Cact. and Succ. Jour.* 20:143–146).

3. Sedum stelliforme Wats. (*S. Topsentii* Raymond-Hamet). Apache, Graham, and Cochise counties, 7,000 to 9,500 feet, July to September, type of *S. stelliforme* from the Huachuca Mountains (*Lemmon* 2702), type of *S. Topsentii* from the Chiricahua Mountains (*Blumer* 2150). New Mexico, Arizona, and Chihuahua.

4. Sedum Cockerellii Britton (*S. Wootoni* Britton). Mountains of Apache, Coconino, Gila, and Graham counties, 5,000 to 11,500 feet, usually on rocks in partial shade, often among mosses, June to October. Southern Colorado, New Mexico, and Arizona.

5. Sedum Griffithsii Rose (*S. Wootoni* var. *Griffithsii* Kearney & Peebles). Mountains of southern Apache, Graham, Cochise, Santa Cruz, and Pima counties, 3,500 to 9,000 feet, in habitats similar to those of *S. Cockerellii,* type from the Santa Rita Mountains (*Griffiths* 6061). Southwestern New Mexico, eastern and southern Arizona, and northern Mexico.

Clausen and Uhl (122, p. 45) state: "Herbarium specimens are frequently difficult to distinguish from *S. Cockerellii.*" The characterizations of these 2 species in the key are drawn from the descriptions by Clausen and Uhl based on living plants. These authors indicate (122, p. 33) a difference in number of chromosomes, *n* being 16 in *S. Cockerellii* and 14 in *S. Griffithsii.*

2. ECHEVERIA

Plants perennial, succulent; stems scapose; leaves of the stem alternate, bractlike, the basal ones much larger, in a rosette; flowers in cymes or open panicles; calyx lobes and petals 5, the petals united below into a cylindric or funnelform tube.

KEY TO THE SPECIES

1. Corolla broadly campanulate or rotate-campanulate, the lobes spreading, at least as long as the tube, pale or greenish yellow and transversely banded or irregularly blotched with red; stamens in age often spreading or deflexed outside the corolla: subgenus *Graptopetalum* (2).

1. Corolla cylindric or tubular-funnelform, the lobes erect or slightly spreading at apex, bright yellow or red, not banded or blotched; stamens not exserted: subgenus *Dudleya* (3).

2. Basal leaves not more than 3.5 cm. long; scapes (including the inflorescence) less than 25 cm. long; inflorescence a broad, very open cyme; sepals oblong, spatulate, or obovate, very obtuse; valves of the carpels gradually attenuate into the styles 1. *E. Rusbyi*

2. Basal leaves up to 6 cm. long; scapes and inflorescence up to 30 cm. long; inflorescence an elongate panicle, broad and much-branched, or narrow and subracemose; sepals narrowly deltoid to oblong-lanceolate, acute or acuminate; valves of the carpels abruptly tipped with the styles.................................2. *E. Bartramii*

3. Basal leaves abruptly acuminate, ovate-lanceolate to rhombic-ovate, usually widest above the base, the larger ones 2 to 4 cm. wide; calyx 4 to 5 mm. long, green when fresh; corolla 10 to 12 mm. long, deep red to apricot yellow fading reddish, the tube about equaling the lobes and longer than the calyx................3. *E. pulverulenta*

3. Basal leaves gradually long-acuminate, lanceolate, widest at or near the base, not more (usually less) than 2 cm. wide; calyx 5 to 8 mm. long, bright red when fresh; corolla 12 to 17 mm. long, clear yellow, the tube shorter than the lobes and shorter than the calyx..4. *E. Collomae*

1. Echeveria Rusbyi (Greene) Nels. & Macbr. (*Graptopetalum Rusbyi* Rose, *G. Orpettii* E. Walther). Greenlee, Graham, Gila, Maricopa, Pinal, and Pima counties, 2,500 to 5,000 feet, open places among rocks, May, type from the San Francisco River region, Greenlee County (*Rusby* in 1881), type of *Graptopetalum Orpettii* from near Superior, Pinal County (*Howard*). Known only from Arizona.

2. Echeveria Bartramii (Rose) Kearney & Peebles (*Graptopetalum Bartramii* Rose). Patagonia and Tumacacori mountains (Santa Cruz County), Baboquivari Mountains (Pima County), 4,000 to 5,000 feet, growing with scrub oaks, September to February, type from the Patagonia Mountains (*Bartram* in 1924). Known only from southern Arizona and northern Mexico.

The maintenance of *Graptopetalum* as a genus has been advocated recently by Moran (123).

3. Echeveria pulverulenta Nutt. (*Dudleya pulverulenta* Britt. & Rose). Maricopa, Mohave, and Yuma counties, 500 to 2,500 feet, on cliffs, April, October, and December. Southern Nevada, western Arizona, southern California, northern Sonora, and Baja California.

The Arizona plant is *E. pulverulenta* subsp. *arizonica* (Rose) Clokey (*Dudleya arizonica* Rose, *Echeveria arizonica* Kearney & Peebles), a desert variant with fewer and smaller leaves and less pulverulent herbage than in typical *E. pulverulenta* of coastal California. The type of *Dudleya arizonica* came from Yucca, Mohave County (*Bly* in 1921).

4. Echeveria Collomae (Rose) Kearney & Peebles (*Dudleya Collomae* Rose). Yavapai, Gila, Maricopa, and Pinal counties, 2,000 to 6,000 feet, among rocks, March to May, type from near Payson, Gila County (*Collom* in 1924). Known only from central Arizona.

This, the showiest and commonest *Echeveria* of Arizona, is handsome in flower and well worthy of cultivation. It is closely related to several California species and was presumably included under *Dudleya Parishii* Rose in *North American Flora* (22 : 41).

3. TILLAEA. Pigmy-weed

Plants annual, not more than 10 cm. high, glabrous, soon becoming reddish brown; stems slender, branched, appearing winged when dry; leaves opposite, connate-perfoliate; flowers minute, in small axillary clusters; sepals, petals, and stamens usually 4; seeds 1 or 2 in each follicle.

1. Tillaea erecta Hook. & Arn. Southwestern Yavapai, Maricopa, Pinal, and Pima counties, 1,500 to 4,000 feet, March and April. Oregon to Baja California and southern Arizona; Chile.

60. SAXIFRAGACEAE. Saxifrage Family

Plants perennial, herbaceous or shrubby; leaves mostly simple, alternate, opposite, or mostly basal; flowers perfect or some of them unisexual, regular or nearly so, in racemes, corymbs, cymes, or panicles, or solitary, usually with a well-developed hypanthium in the throat of which the petals and stamens are inserted; sepals or calyx lobes and the petals usually 5, the petals rarely wanting; stamens 4 to many; pistil 1, simple or compound, or of 2 nearly distinct carpels; ovary inferior, or nearly free from the calyx; fruit follicular, capsular, or baccate.

The best-known members of this family are the currants and goose-berries. It includes many species, both herbaceous and shrubby, that are highly esteemed as cultivated ornamentals.

KEY TO THE GENERA

1. Plants herbaceous (2).
1. Plants shrubby (5).
2. Fertile stamens 10. Staminodia none (3).
2. Fertile stamens 5. Plants scapose, the leaves mostly basal (4).
3. Styles 2; capsule 2-celled and 2-beaked, the carpels sometimes nearly distinct; petals entire or nearly so .... 1. *Saxifraga*
3. Styles 3; capsule 1-celled, 3-valved, not beaked; petals usually laciniate or deeply dentate 3. *Lithophragma*
4. Staminodia none; flowers small, in racemes or cymose panicles; leaves round-reniform, crenate or dentate; carpels 2, united below, distinct and divergent above 2. *Heuchera*
4. Staminodia in clusters alternating with the fertile stamens; flowers relatively large, solitary on long peduncles; leaves oval or ovate, acute or short-cuneate at base, entire; carpels 3 or 4, completely united .... 4. *Parnassia*
5. Leaves alternate, those of the branchlets usually appearing fascicled, more or less deeply palmately lobed; ovary wholly adnate to the calyx; fruit a juicy berry; stems often spiny or bristly .... 9. *Ribes*
5. Leaves opposite, entire or merely toothed; upper part of the ovary free; fruit a dry capsule; stems unarmed (6).
6. Sepals and petals 4 (rarely 5). Leaves entire or shallowly few-toothed (7).
6. Sepals and petals 5. Stamens 10, the filaments not lobed (8).
7. Petals not clawed; stamens many more than 8 (commonly 20 or more), the filaments terete, not lobed; capsule urceolate or obovoid, abruptly beaked...... 5. *Philadelphus*
7. Petals long-clawed; stamens 8, the filaments broad and flat, 2-lobed, the anthers borne between the lobes; capsule ovoid or conic, attenuate-beaked .... 7. *Fendlera*
8. Leaves broad, thin, deeply and regularly crenate, long-petioled .... 6. *Jamesia*
8. Leaves narrow, thickish, entire, sessile or nearly so .... 8. *Fendlerella*

1. SAXIFRAGA. Saxifrage

Plants herbaceous; stems leafy or scapelike, erect, decumbent, or prostrate; flowers perfect, regular, solitary, or in simple or compound cymes; calyx lobes and petals 5; stamens 10; carpels 2, united below or nearly distinct; ovary nearly free, or partly inferior; follicles beaked, divergent, many-seeded.

Plants mostly of relatively high altitudes, in moist soil of coniferous forests, usually among rocks. Three arctic-alpine species are found near the summit of the highest mountains in Arizona.

KEY TO THE SPECIES

1. Foliage leaves all basal. Flowers numerous, borne on scapes, these usually naked below the inflorescence, the latter small-bracteate (2).
1. Foliage leaves not all basal, the stems leafy below the inflorescence, but the cauline leaves sometimes much reduced (4).
2. Inflorescence a very open panicle; leaf blades thin, commonly much shorter than the petioles, orbicular or nearly so, more or less cordate at base, deeply crenate-dentate with numerous teeth .... 1. *S. arguta*
2. Inflorescence a contracted panicle of cymules; leaf blades thickish, commonly longer than the petioles, oblong, ovate, or rhombic, truncate or cuneate at base, shallowly crenate or crenate-dentate (3).
3. Leaves lanate beneath with long, reddish hairs; petioles copiously ciliate with similar hairs; cymules lax, few-flowered, the inflorescence becoming relatively open 2. *S. eriophora*
3. Leaves glabrous or sparsely pubescent beneath; petioles sparsely ciliate; cymules dense, subcapitate, the inflorescence remaining contracted. Scapes up to 30 cm. long; inflorescence up to 10 cm. long, the cymules solitary to several........ 3. *S. rhomboidea*

4. Leaves entire, conspicuously ciliate with thick, bristlelike hairs. Plants with slender, elongate, naked stolons ending in a tuft of small leaves; leaves of the flowering stems scarcely reduced; flowering stems commonly solitary, erect, very leafy, bearing one or very few flowers; petals yellow, broadly obovate, 6 to 9 mm. long. . 4. *S. flagellaris*
4. Leaves deeply dentate or lobed, not bristly-ciliate, the hairs, if any, soft and slender, or glandular (5).
5. Plants densely cespitose, not bulbiferous at base, copiously glandular-pubescent; flowering stems several or numerous; basal leaves flabelliform-cuneate, deeply, narrowly, and acutely 3-lobed at apex; petioles very short or none. . . . . .5. *S. caespitosa*
5. Plants not cespitose, bulbiferous at base, glabrous or sparsely pubescent; flowering stems few or solitary, 1- to 3-flowered; basal leaves suborbicular or reniform, palmately several-lobed or coarsely crenate; petioles elongate. . . . . .6. *S. debilis*

1. Saxifraga arguta D. Don (*S. punctata* L. var. *arguta* Engl. & Irmsch.). Baldy Peak, Apache County (*Goodding* 626, *Peebles & Smith* 12514, *Phillips* 3345), 11,000 feet, July and August. Montana to British Columbia, New Mexico, eastern Arizona, and California.

2. Saxifraga eriophora Wats. Pinaleno Mountains (Graham County), Huachuca Mountains (Cochise County), Santa Rita, Rincon, and Santa Catalina mountains (Santa Cruz and Pima counties), 5,000 to 8,500 feet, March to May, type from the Santa Catalina Mountains (*Lemmon* in 1881). Southern New Mexico and southern Arizona.

Less pubescent specimens collected in the Mazatzal Mountains, Gila County (*Collom* 46), seem to be intermediate between this and the next species.

3. Saxifraga rhomboidea Greene. Apache County to Coconino and Gila counties, 5,500 to 11,000 feet, usually in shade but sometimes in open grassland, April to July. Montana to New Mexico and Arizona.

A small variant, with scapes not more than 10 cm. long and inflorescence reduced to 1 to 3 cymules, is var. *franciscana* (Small) Kearney & Peebles (*Micranthes franciscana* Small). It occurs on the San Francisco Peaks, 11,500 to 12,000 feet, type *Mearns* 26, in 1887.

4. Saxifraga flagellaris Willd. San Francisco Peaks (Coconino County), 10,000 to 12,000 feet, July to September. Circumpolar, south in the Rocky Mountains to New Mexico and northern Arizona.

5. Saxifraga caespitosa L. San Francisco Peaks, 11,500 to 12,000 feet, July and August. Circumpolar, southward in the mountains of North America to Colorado, northern Arizona, and Oregon.

The Arizona plants belong to subsp. *exaratoides* (Simmons) Engl. & Irmsch. var. *Lemmonii* Engl. & Irmsch., of which the type was collected on Mount Agassiz (*Lemmon* in 1884).

6. Saxifraga debilis Engelm. San Francisco Peaks, 10,000 to 12,000 feet, July and August. Montana to Colorado, northern Arizona, and California.

2. HEUCHERA (124). Alum-root

Perennial herbs with scapose stems from a large, somewhat woody, scaly caudex; leaves nearly all basal, long-petioled, orbicular or broadly ovate, cordate, dentate; flowers in narrow racemes or cymose panicles, slightly irregular, with a well-developed hypanthium; petals small, rarely wanting; ovary partly inferior, 1-celled; fruit of 2 divergent, beaked follicles, these several-seeded.

The rootstocks have astringent properties, as indicated by the common name, and have been used by hunters and others in cases of diarrhea caused by drinking alkali water.

KEY TO THE SPECIES

1. Stamens surpassing the sepals; pistils exserted (sometimes tardily). Styles slender, elongate, gradually expanded below into the conic beaks of the carpels; flowers pink or pinkish; petals much surpassing the sepals, with narrow blades and long, slender claws (2).
1. Stamens not equaling the sepals; pistils included (3).
2. Filaments noticeably widened and flattened at base; stamens inserted at or slightly below the level of insertion of the petals; petioles finely glandular-puberulent, sometimes also sparsely hirtellous. Inferior part of the ovary at anthesis not longer than wide ....1. *H. rubescens*
2. Filaments not noticeably widened and flattened at base; stamens usually inserted well below the level of insertion of the petals; petioles commonly villous or hirsute with long hairs, sometimes merely glandular-puberulent..................2. *H. versicolor*
3. Flowers deep pink to carmine; inflorescence relatively short and expanded, the lower branches seldom less than 2 and up to 6 cm. long; hypanthium funnelform, becoming urceolate. Petals much shorter than the sepals.....................3. *H. sanguinea*
3. Flowers greenish or yellowish, the petals often whitish; inflorescence elongate and contracted, the lower branches usually less than 2 cm. long; hypanthium broadly turbinate, or campanulate (4).
4. Sepals broadly triangular, more or less spreading; petals commonly surpassing the sepals, the blade usually at least twice as wide as the short claw; pistils commonly shorter than the tube of the hypanthium......................4. *H. parvifolia*
4. Sepals oblong or ovate, erect or incurved at tip; petals, when present, shorter than the sepals, the blade usually not much wider than the claw; pistils equaling or surpassing the tube of the hypanthium (5).
5. Flowers hexamerous, glandular-puberulent, sometimes also slightly hirtellous; petals none, or much reduced...5. *H. Eastwoodiae*
5. Flowers normally pentamerous; petals present (6).
6. Flowers merely glandular-puberulent; petioles hirsute............6. *H. novomexicana*
6. Flowers copiously white-hirsute, also puberulent; petioles finely puberulent 7. *H. glomerulata*

1. Heuchera rubescens Torr. (*H. Clutei* A. Nels.). Hidden Spring, Navajo Mountain, Coconino County (*Clute* 80, the type collection of *H. Clutei*). Utah to Oregon, northern Arizona, and California.

2. Heuchera versicolor Greene (*H. Sitgreavesii* Rydb.). Apache County to Hualpai Mountain (Mohave County), south to Cochise, Santa Cruz, and Pima counties, 6,500 to 12,000 feet, moist, shaded rocks in coniferous forest, May to October, type of *H. Sitgreavesii* from between Williams and Flagstaff, Coconino County (*Woodhouse* in 1851). Western Texas and southern Utah to Arizona, California, and northern Mexico.

An attractive plant, suitable for rock gardens in cool climates. A depauperate form (forma *pumila* Rosendahl et al.) is found on the San Francisco Peaks at 10,000 to 12,000 feet. Specimens from near the summit of the San Francisco Peaks (*Taylor* in 1926, *Wolf* 3118) were referred doubtfully to *H. alpestris* Rosendahl et al. (124, p. 106). The material is scarcely adequate to permit a conclusion as to whether they can be distinguished satisfactorily from *H. versicolor* forma *pumila*. Having practically the same range in

Arizona as typical *H. versicolor* but less common is var. *leptomeria* (Greene) Kearney & Peebles (*H. leptomeria* Greene), which has the inferior portion of the ovary narrower.

3. Heuchera sanguinea Engelm. Southern Apache County to Cochise, Santa Cruz, and Pima counties, 4,000 to 8,500 feet, moist, shaded rocks, March to October. Arizona and northern Mexico.

Coral-bells. The showiest of the Arizona species, with bright-pink to carmine flowers, often cultivated as an ornamental. Distinguished by its more pubescent calyx, with longer hairs, is var. *pulchra* (Rydb.) Rosendahl (*H. pulchra* Rydb.), occasional throughout the range of the species, type from the Santa Catalina Mountains (*Griffiths* in 1901). This variety is not well marked.

4. Heuchera parvifolia Nutt. The typical plant, widely distributed in the Rocky Mountains, apparently is not found in Arizona, but two varieties occur. These are: (1) var. *arizonica* Rosendahl et al., Coconino County, on both sides of the Grand Canyon and south to Oak Creek Canyon, and in the Pinaleno Mountains (Graham County), 7,000 to 9,000 feet, type from the Grand Canyon (*Eastwood* 5775); and (2) var. *flavescens* (Rydb.) Rosendahl et al. (*H. flavescens* Rydb.), found in the White Mountains (Apache County), in Navajo County, and on the Kaibab Plateau (Coconino County), 7,000 to 11,500 feet. In var. *arizonica* the petioles are pilose, villous, or hirsute, and the disk of the hypanthium is distinctly developed; whereas, in var. *flavescens*, the petioles are commonly glabrous or finely puberulent, although sometimes villous-hirsute, and the disk is nearly obsolete.

5. Heuchera Eastwoodiae Rosendahl, Butters & Lakela. Yavapai and Gila counties, 5,000 to 6,000 feet, and perhaps higher, May to August, type from Senator Mine near Prescott (*Eastwood* 17659). Known only from central Arizona.

6. Heuchera novomexicana Wheelock. White Mountains (Apache or Greenlee County) 8,000 feet (*Nichol* in 1939), Coconino County, near Flagstaff (*Leiberg* 5538), and at Deboschibeko (*Darsie* in 1933), 6,000 to 7,000 feet, June. New Mexico and Arizona.

7. Heuchera glomerulata Rosendahl, Butters, & Lakela. Pinaleno Mountains, Graham County, 4,500 feet (*Eggleston* 19913), Chiricahua Mountains, Cochise County (*Stone* 385, the type collection, *Darrow & Phillips* 2496), about 6,500 feet, shaded, rocky slopes, May. Known only from southeastern Arizona.

3. LITHOPHRAGMA. Woodland-star

Plants herbaceous, perennial, small, with slender rootstocks bearing bulblets; stems slender; leaves mostly basal, round, variously cleft or divided; flowers few, rather showy, in a narrow raceme; petals white or pink, clawed, deeply cleft; stamens 10, not exserted; ovary partly inferior.

1. Lithophragma tenellum Nutt. Coconino and Gila counties, 5,000 to 8,000 feet, pine forests, April to June. Alberta to New Mexico, north-central Arizona, and California.

A very attractive little plant. The Arizona specimens have been referred to *L. australe* Rydb. and *L. brevilobum* Rydb., but it is doubtful that these entities are sufficiently distinct from *L. tenellum*. The type of *L. australe* came from Cedar Creek (*Newberry* in 1858).

4. PARNASSIA. Grass-of-Parnassus

Plants herbaceous, perennial, with a short rootstock; leaves in a basal rosette, except a single one borne on the scape (usually below the middle), entire, several-veined from near the base; petals oval, conspicuously veined, 6 to 10 mm. long; staminodia 5 to 7 in the cluster, united below; fruit a 1-celled, many-seeded capsule.

The popular name is misleading, as these plants are not at all grasslike.

1. Parnassia parviflora DC. Near Greer, Apache County, 9,000 feet (*Eggleston* 17098), Oak Creek, Coconino County (*Deaver* in 1928, *Thornber* in 1930), in wet meadows, August. Canada to New Mexico and eastern Arizona.

5. PHILADELPHUS (125). Mock-orange

Shrubs with branching stems and exfoliating bark; leaves opposite, thickish, entire or dentate, white-sericeous beneath; flowers solitary or very few in the cluster, rather large and showy; petals white, broad and rounded; stamens numerous; ovary partly inferior.

The Old World *P. coronarius* and other species are favorite ornamental shrubs, often known by the inappropriate name "syringa." *P. Lewisii* Pursh is the state flower of Idaho.

1. Philadelphus microphyllus Gray. Apache County to Coconino County, south to Cochise, Santa Cruz, and Pima counties, 5,000 to 9,000 feet, rocky slopes and canyons in the chaparral and pine belts, June and July. Southwestern Wyoming to western Texas, southeastern California, and northern Mexico.

The Mexican bighorn or "mountain sheep" is reported to browse these plants.

KEY TO THE SUBSPECIES

1. Stamens usually about 32. Leaves seldom more than 15 mm. long; hairs of the leaves and calyx mostly appressed, not matted, or the upper leaf surface hirsute and the hairs erect or nearly so; calyx copiously pubescent; filaments usually free or united only at base...subsp. *stramineus*
1. Stamens usually more than 32, often 40 or more (2).
2. Filaments mostly united, sometimes nearly to the tips; calyx moderately to copiously pubescent, often silvery. Leaves commonly more than 15 mm. long (3).
2. Filaments mostly distinct or nearly so; calyx glabrate to moderately pubescent, not silvery. Hairs of the leaves and calyx appressed (4).
3. Upper leaf surface strigose-pubescent; hairs of the calyx mostly appressed, not matted subsp. *argenteus*
3. Upper leaf surface hirsute, the hairs erect or nearly so; hairs of the calyx either matted and of two kinds, or not matted and all alike....................subsp. *argyrocalyx*
4. Leaves seldom more than 16 mm. long; petals 9 to 11 mm. long.......subsp. *occidentalis*
4. Leaves seldom less than 17 mm. long; petals 11 to 17 mm. long..........subsp. *typicus*

Specimens with relatively few stamens, apparently referrable to subsp. *stramineus* (Rydb.) C. L. Hitchc., have been collected on the north side of

Grand Canyon (*Eastwood* 5890, *Hawbecker* in 1935) and near Patagonia, Santa Cruz County (*Harrison* 7180). The Harrison collection was referred by Hitchcock (125, p. 43) to subsp. *argenteus,* but the stamens are 32, with the filaments free or very nearly so. It extends the previously known range of this subspecies (southern Nevada to northern Lower California) considerably eastward. Subsp. *argenteus* (Rydb.) C. L. Hitchc. occurs in the mountains of Graham, Cochise, and Pima counties and also in northern Mexico. This and the next are much the commonest of the subspecies in Arizona. Subsp. *argyrocalyx* (Wooton) C. L. Hitchc. ranges from southern Coconino County to the mountains of Graham, Cochise, and Pima counties, occurring also in southern New Mexico and northern Mexico. Hitchcock pointed out (125, p. 46) that the Arizona specimens are not typical, approaching subsp. *argenteus.* Subsp. *occidentalis* (A. Nels.) C. L. Hitchc. is represented by two collections in northern Apache County (*Standley* 7323, *Goodman & Payson* 2843), and these are not typical, approaching subsp. *typicus* (125, pp. 52, 53). Several collections in the Grand Canyon, Coconino County, were considered by Hitchcock (125, p. 53) to be intermediate between subsp. *occidentalis* and subsp. *stramineus.* The center of distribution of subsp. *typicus* C. L. Hitchc. is in New Mexico and southern Colorado, but Hitchcock (125, p. 49) stated that it extends to Apache County, Arizona.

6. JAMESIA. Cliff-bush

Shrubby, often 2 m. or more high, with shreddy bark; leaves large, thin, ovate, crenate, bright green above, whitish-pubescent beneath; flowers numerous, rather small, in dense cymes; petals white or pink; filaments broad and flat, not lobed; styles 3 to 5.

1. Jamesia americana Torr. & Gray. Mountains of Graham, Cochise, and Pima counties, 7,500 to 9,500 feet, coniferous forests, along streams, and on the walls of canyons, June and July. Wyoming to New Mexico and southeastern Arizona.

Sometimes cultivated as an ornamental.

7. FENDLERA

Straggling shrubs, usually 1 to 2 (sometimes 3) m. high, widely branched; leaves opposite, linear-lanceolate to narrowly ovate, entire, thickish, the margins often revolute; flowers large and showy, mostly solitary, the petals white or tinged with purple; fruit a 4-celled capsule.

1. Fendlera rupicola Gray. Apache County to Hualpai Mountain (Mohave County), south to Cochise, Santa Cruz, and Pima counties, 4,000 to 7,000 feet, dry, rocky and gravelly slopes, March to June. Southern Colorado to western Texas and Arizona.

Browsed by goats, deer, bighorns, and to a less extent by cattle when other feed is scarce. A beautiful shrub when in flower, worthy of more extensive cultivation as an ornamental. In northeastern Arizona the prevailing variety is var. *falcata* (Thornber) Rehder (*F. falcata* Thornber), with narrow, usually strongly revolute leaves, these green and sparsely pubescent to glabrate on both surfaces. In the southeastern mountains (Graham, Cochise, and Pima counties) var. *Wrightii* Gray (*F. tomentella* Thornber) is about as common as the typical plant. Compared with the latter it has the leaves more densely pubescent and white beneath, also usually narrower and more revolute. The type of *F. tomentella* came from Blue River, Greenlee County (*Hough* 470).

8. FENDLERELLA

Small, much-branched shrubs, 0.5 to 1 m. high; herbage and calyx strigose; leaves small, lanceolate, elliptic, or oblanceolate, entire; flowers inconspicuous, in cymose clusters, the petals white; hypanthium turbinate; capsule longer than the calyx.

1. Fendlerella utahensis (Wats.) Heller (*Whipplea utahensis* Wats.). Coconino County and (probably) northern Mohave County, common in and about the Grand Canyon, 5,000 to 8,000 feet, dry, open pine woods, June to September. Southern Utah, northern Arizona, and southern California.

The distribution of typical *F. utahensis* is given above. In southern Arizona, southern New Mexico, and northern Mexico occurs var. *cymosa* (Greene) Kearney & Peebles (*F. cymosa* Greene), which has the leaves commonly narrower and more pointed than in the typical plant. The variety occurs in the mountains of Graham, Cochise, Santa Cruz, and Pima counties, 4,500 to 7,000 feet, often on limestone, May to September. The type of *F. cymosa* was collected in the Santa Rita Mountains, Pima County (*Pringle* in 1884). The plants are browsed by deer.

9. RIBES. Currant, Goose-berry

Shrubs, usually straggling, often spiny; leaves alternate, or appearing fascicled because of the much-shortened internodes, broad, rounded, commonly lobed; flowers in fascicles or somewhat elongate racemes, sometimes solitary, usually inconspicuous; petals much reduced, these and the stamens 4 or 5, inserted on the calyx throat; ovary inferior; berry globose or nearly so, juicy, several-seeded, crowned by the persistent calyx lobes.

These shrubs harbor one stage of the white pine blister rust (*Cronartium ribicola*) and are being exterminated wherever commercially important stands of white pines occur. The fruits, fresh or dried, of some of the wild species are eaten by the Indians, and are sometimes used for making jelly. They are much liked by birds. *R. cereum* was used by the Hopi to alleviate stomach-ache. Both domestic animals and deer browse the plants.

KEY TO THE SPECIES

1. Fruit disarticulating from the pedicel; spines at the stem nodes none or, if present, then the hypanthium very shallow: subgenus *Ribesia* (2).

1. Fruit not disarticulating; nodal spines present; hypanthium elongate: subgenus *Grossularia* (7).

2. Spines present, usually numerous and conspicuous, often clustered. Leaves seldom more than 3 cm. wide, deeply 5-lobed or 5-parted, copiously pubescent on both surfaces to nearly glabrous; inflorescence few-flowered; flowers dull pink or red; berries bright red, glandular-bristly............................1. *R. montigenum*

2. Spines none (3).

3. Anthers without an apical gland but sometimes apiculate (4).

3. Anthers with a cup-shaped apical gland. Leaves shallowly lobed with rounded lobes, crenate-dentate (5).
4. Hypanthium glabrous, yellow (as are the calyx lobes), tubular-funnelform, equaling or longer than the calyx lobes; berries at maturity smooth, red, black, or yellow, 6 to 8 mm. in diameter; leaves deeply 3-lobed, the lobes commonly sparingly crenate-dentate or shallowly cleft, cuneate to subcordate at base, up to 5 cm. wide. Flowers very showy for the genus..2. *R. aureum*
4. Hypanthium glandular-pubescent, greenish white or pinkish (as are the calyx lobes), turbinate, shorter than the calyx lobes; berries at maturity glandular-pubescent, black (sometimes with a bloom), 8 to 12 mm. in diameter; leaves shallowly to rather deeply 3- to 5-lobed with acutish lobes, dentate with numerous teeth, cordate at base, up to 9 cm. wide but commonly smaller..........................3. *R. Wolfii*
5. Flowers 14 to 17 mm. long from the base of the hypanthium; herbage not or but sparsely resinous-granular, hispidulous and puberulent, the hairs often gland-tipped; leaves commonly at least 5 cm. wide, rather deeply lobed, cordate; hypanthium cylindric-campanulate, less than twice as long as wide; berry at maturity black, glandular-bristly ....4. *R. viscosissimum*
5. Flowers seldom more than 10 mm. long; herbage usually noticeably resinous-granular, often puberulent or short-pilose, fragrant; leaves not more (commonly much less) than 4 cm. wide, shallowly lobed, cuneate to subcordate at base; hypanthium tubular, more than twice as long as wide; berry at maturity bright red, smooth or glandular-pubescent, not bristly (6).
6. Bracts cuneate-obovate, deeply toothed or lobed at the (commonly) truncate or rounded apex ....5. *R. cereum*
6. Bracts rhombic, often narrowly so, entire or denticulate at the acutish apex 6. *R. inebrians*
7. Styles hairy below; hypanthium, calyx lobes, and ovary usually glabrous. Stems sometimes with a few short bristles; spines few or none, seldom as much as 10 mm. long; leaves up to 6 cm. wide; flowers (including the ovary) 7 to 9 mm. long; calyx lobes 1 to 2 times as long as the hypanthium; berry at maturity wine-colored, smooth, about 8 mm. in diameter..7. *R. inerme*
7. Styles not hairy, glabrous or puberulent; hypanthium and calyx lobes pubescent (8).
8. Ovary densely bristly; leaves up to 4 cm. wide; berry densely spiny, at maturity 10 to 15 mm. in diameter (excluding the spines), dark purple. Stems without bristles, glabrous or puberulent when young; nodal spines stout, up to 12 mm. long, often somewhat curved; flowers (including the ovary) up to 18 mm. long; calyx lobes narrow, longer than the hypanthium, often twice as long...........8. *R. pinetorum*
8. Ovary not bristly, the hairs, if any, soft, glandular or nonglandular; leaves not more than 2 cm. wide; berry not spiny, less than 10 mm. in diameter (9).
9. Hypanthium as wide as or wider than long; berry at maturity yellow (sometimes purple), copiously soft-pubescent; herbage almost tomentose; stems without bristles; spines straw-colored, mostly straight and slender, up to 2 cm. long. Flower when dry (including the ovary) not more than 7 mm. long..........9. *R. velutinum*
9. Hypanthium longer than wide; berry black or dark red, commonly glabrous; herbage glabrous or moderately pubescent; stems sometimes bristly but commonly not so; spines yellowish brown to brownish gray (10).
10. Calyx lobes and hypanthium white; hypanthium 4 to 6 mm. long....10. *R. leptanthum*
10. Calyx lobes and hypanthium yellow; hypanthium 2 to 3 mm. long....11. *R. quercetorum*

1. Ribes montigenum McClatchie. White Mountains (Apache County) and mountains of Coconino County, 6,500 to 11,500 feet, June to August. Montana to British Columbia, south to New Mexico, Arizona, and California. Gooseberry currant.

2. Ribes aureum Pursh. Navajo and Yavapai counties to Greenlee, Cochise, Santa Cruz, and Pima counties, 4,000 to 6,000 feet, March to June. South Dakota to Assiniboia and Washington, south to New Mexico, Arizona, and California.

Golden currant, the showiest species growing in Arizona, with bright-yellow, fragrant flowers. The plant is often cultivated as an ornamental, and at least some of the collections in Arizona probably were escapes from gardens.

3. Ribes Wolfii Rothrock. Kaibab Plateau (Coconino County), Tunitcha and White mountains (Apache County), Pinaleno Mountains (Graham County), 8,500 to 11,500 feet, moist woods and springy places, May to August. Colorado, Utah, New Mexico, and Arizona.

4. Ribes viscosissimum Pursh. Grand Canyon, both rims, Coconino County, 7,000 to 8,000 feet (*H. & V. Bailey* 1025, 3387), May to July. Montana to British Columbia, Colorado, northern Arizona, and California. Sticky currant.

5. Ribes cereum Dougl. Apache, Navajo, Coconino, and Yavapai counties, 5,500 to 9,000 feet, common in pine forest, sometimes on cliffs, May to July. Montana to British Columbia, south to Arizona and California. Wax currant.

6. Ribes inebrians Lindl. Apache County to Mohave County, south to Greenlee and Yavapai counties, 5,000 to 9,000 feet, May to August. South Dakota and Nebraska to Idaho, New Mexico, Arizona, and California.

Squaw currant. Closely resembles *R. cereum* and intergrades completely with it in Arizona, being probably not more than varietally distinct. It is reported that the berries are eaten by the Hopi Indians, although causing illness.

7. Ribes inerme Rydb. (*R. divaricatum* Dougl. var. *inerme* (Rydb.) McMinn, *Grossularia inermis* Cov. & Britt.). Kaibab Plateau, North Rim of Grand Canyon, and Lake Mary (Coconino County), Greer (Apache County), 7,000 to 8,500 feet. Montana to British Columbia, south to New Mexico, Arizona, and California.

White-stem gooseberry. H. J. Fulton reported that at Lake Mary the plants were more or less trailing bushes not more than 1 m. high. They were atypical at this locality in having the ovary and hypanthium sparsely glandular-hirsute.

8. Ribes pinetorum Greene (*Grossularia pinetorum* Cov. & Britt.). Apache, Navajo, and Coconino counties, south to Cochise and Pima counties, 7,000 to 10,000 feet, coniferous forests, April to September. New Mexico and Arizona.

Orange goose-berry. The most abundant species in the mountains of southern Arizona and the handsomest wild goose-berry of the state, with reddish flowers and large, densely prickly berries, these purple at maturity.

9. Ribes velutinum Greene (*Grossularia velutina* Cov. & Britt.). Coconino County, on both sides of the Grand Canyon, 6,500 to 8,000 feet, April and May. Utah and northern Arizona to Oregon and California.

10. Ribes leptanthum Gray (*Grossularia leptantha* Cov. & Britt.). Apache, Navajo, and Coconino counties, south to Cochise (and Pima?) counties, 6,000 to 9,500 feet, along streams, May and June. Colorado, Utah, New Mexico, and Arizona.

11. Ribes quercetorum Greene (*Grossularia quercetorum* Cov. & Britt.). Superstition Mountains (Pinal County), Sierra Estrella (Maricopa County), Ajo Mountains (Pima County), 3,500 to 4,500 feet, steep, rocky slopes and canyon walls, growing in the Sierra Estrella with scrub oaks, *Vauquelinia*, and *Penstemon microphyllus*, November to April. South-central Arizona, California, and Baja California.

A shrub, barely 1 m. high, tending to form low thickets. The Arizona specimens approach *R. leptanthum* in length of the hypanthium but have the yellow flowers of *R. quercetorum*.

61. PLATANACEAE. Plane-tree Family

1. PLATANUS. Sycamore, Button-wood, Plane-tree

Trees, with the outer bark flaking off and exposing the smooth, whitish, inner bark; buds enclosed in the dilated bases of the petioles; leaves large, alternate, palmately lobed; plants monoecious, the flowers very many in dense globose heads; sepals and petals minute; pistils 3 or 4, distinct; fruit a 4-sided achene, with a basal tuft of long hairs.

1. Platanus Wrightii Wats. (*P. racemosa* Nutt. var. *Wrightii* L. Benson). Southern parts of Coconino and Mohave counties to Greenlee, Cochise, Santa Cruz, and Pima counties, 2,000 to 6,000 feet, along streams, April and May, type from near the San Pedro River (*Wright* 1880). The Mogollon Escarpment coincides approximately with the northern limit of this species in Arizona. New Mexico, Arizona, and northern Mexico.

Arizona sycamore. A large, spreading tree, attaining a height of 24 m. (80 feet), with beautifully arched, white-barked branches. The roots effectively bind the soil, preventing excessive erosion. This tree is perhaps not specifically distinct from the California sycamore (*P. racemosa* Nutt.). In *P. Wrightii* the fruiting heads are usually stalked and the leaf lobes relatively long, whereas *P. racemosa* usually has sessile heads and short leaf lobes, but intergradations are reported to occur (126, p. 34).

62. CROSSOSOMATACEAE. Crossosoma Family

1. CROSSOSOMA

Rough-barked shrubs; leaves without stipules, alternate, narrow, entire, thickish, smooth and somewhat glaucous; flowers solitary, rather large; sepals 5; petals 5, white; stamens numerous, borne on a disk in the turbinate hypanthium; pistils 2 to 5 (seldom reduced to 1), becoming distinct, several-seeded follicles, these with thickish, reticulate-veined walls; seeds with a fleshy, fimbriate aril.

KEY TO THE SPECIES

1. Follicles seldom less than 9 mm. long, oblong, ovoid, or somewhat obovoid, usually conspicuously beaked....1. *C. Bigelovii*
1. Follicles about 6 mm. long, broadly ovoid, abruptly short-beaked......2. *C. parviflorum*

1. Crossosoma Bigelovii Wats. Southern parts of Yavapai, Mohave, and Gila counties to Pinal, Maricopa, Pima, and Yuma counties, 1,500 to 4,000 feet, dry, rocky slopes and cliffs, February to May (rarely September), type from near the mouth of Williams River (*Bigelow*). Southern and western Arizona, southeastern California, and northwestern Mexico.

A straggling shrub, up to 2.4 m. (8 feet) high, with astringent bark and white flowers, these sometimes tinged with purple. Worthy of cultivation on account of the delicious fragrance of the flowers. In the eastern part of its range var. *glaucum* (Small) Kearney & Peebles (*C. glaucum* Small) largely replaces typical *C. Bigelovii.* The variety is distinguished by broader and more glaucous follicles, but intergradation between the two phases is complete. The type of *C. glaucum* is *Palmer* 560, Arizona without definite locality.

2. Crossosoma parviflorum Robins. & Fern. Known certainly only from the type specimen collected in the Grand Canyon (*Gray* in 1885), with fruit only, but collections in Mohave County at Quartermaster Canyon, Lake Mead, 5,000 feet (*Nichol* in 1939), and in the Black Mountains, between Kingman and Oatman (*Eastwood* 18420), seem also to belong here. The material is insufficient to permit a conclusion as to whether *C. parviflorum* is specifically distinct from *C. Bigelovii.*

63. ROSACEAE. Rose Family

Plants herbaceous or woody; leaves alternate (except in *Coleogyne*), simple or compound, usually with stipules; flowers mostly perfect, regular or nearly so; sepals partly united; petals commonly 5, occasionally none; stamens commonly numerous, rarely fewer than 5, nearly always borne on the throat of the calyx (hypanthium) or on a disk surrounding the ovary or ovaries; pistils 1 to many; ovary 1- to 5-celled, free from or adnate to the calyx; ovules 1 to several; fruit various.

This large and very diverse family includes many of the most important cultivated fruits, such as the apple, pear, peach, plum, cherry, apricot, almond, strawberry, raspberry, and blackberry. The flowers and foliage are usually attractive, often beautiful, and many plants of this family, first and foremost the roses, are highly prized as cultivated ornamentals. Many of the species in Arizona are important browse plants, both for domestic animals and deer, and the fruits supply much food for birds and other wild animals. Most plants of the Rosaceae are harmless, but a few are reputed to be somewhat poisonous.

KEY TO THE GENERA

1. Carpel solitary; fruit a dry or fleshy, usually 1-seeded drupe (plumlike). Calyx more or less persistent at base of the fruit; plants small trees or large shrubs; leaves simple; flowers white or greenish, in racemes or corymbs, or solitary in the leaf axils 21. *Prunus*

1. Carpels more than one or, if solitary, then the fruit an achene (2).

2. Ovary inferior, enclosed in and adnate to the calyx tube (hypanthium), the latter becoming more or less fleshy. Fruit a pome (applelike); calyx lobes more or less persistent at apex of the fruit; plants shrubby or treelike; petals white (3).

2. Ovary superior; calyx tube not fleshy and enclosing the pistils or, if so, then not adnate to them (5).

3. Leaves pinnate. Flowers small, in short, broad, compound, many-flowered cymes 6. *Sorbus*

3. Leaves simple (4).

4. Plants unarmed; ovary with both complete and false partitions, the cells twice as many as the number of styles; flowers relatively large, in racemes or corymbose fascicles .... 7. *Amelanchier*

4. Plants armed with strong spines; ovary without false partitions, the cells of the same number as the styles; flowers relatively small, in corymblike cymes 8. *Crataegus*
5. Calyx tube (hypanthium) enclosing the numerous carpels, becoming fleshy. Fruit simulating a pome, crowned by the persistent calyx lobes................20. *Rosa*
5. Calyx tube not enclosing the carpels, or not becoming fleshy (6).
6. Fruits becoming juicy and more or less edible (7).
6. Fruits not becoming juicy (8).
7. Plants shrubby, often with prickly stems; receptacle not becoming greatly enlarged and fleshy, bearing a conic, ovoid, or nearly globose mass of more or less juicy drupelets ....9. *Rubus*
7. Plants herbaceous, unarmed, acaulescent; receptacle becoming greatly enlarged and very fleshy, the numerous dry achenes embedded in its surface..........10. *Fragaria*
8. Carpels becoming dehiscent capsules or follicles, containing usually more than one seed (9).
8. Carpels becoming indehiscent 1-seeded achenes, or sometimes tardily dehiscent in *Holodiscus* (12).
9. Seeds with a thin, elongate, terminal wing; leaves evergreen, simple. Plants large shrubs or small trees..4. *Vauquelinia*
9. Seeds not winged; leaves not evergreen or, if so, then compound (10).
10. Leaves more or less evergreen, bipinnate; stipules persistent; herbage scurfy with stellate hairs and somewhat viscid...........................3. *Chamaebatiaria*
10. Leaves deciduous, simple; stipules deciduous or none; herbage not scurfy or viscid (11).
11. Plants shrubs; leaves large, rounded-cordate, usually shallowly 3-lobed; flowers in corymbs; fruits much-inflated................................1. *Physocarpus*
11. Plants undershrubs; leaves small, spatulate, entire; flowers in dense spikelike inflorescences; fruits not inflated..............................2. *Petrophytum*
12. Plants herbaceous above ground, the caudex often somewhat woody. Leaves pinnately or digitately compound (13).
12. Plants shrubby (16).
13. Hypanthium 4-winged or strongly 4-angled; bracts and margins of the calyx lobes scarious; petals none; flowers in dense cylindric spikes............19. *Sanguisorba*
13. Hypanthium not winged; bracts and calyx lobes herbaceous; petals present, sometimes minute; flowers not in dense spikes (14).
14. Hypanthium bearing numerous hooked bristles. Stems tall; leaves pinnate, with several much smaller leaflets interspersed among the large ones; flowers in slender, elongate, spikelike racemes; petals yellow; pistils 2.................18. *Agrimonia*
14. Hypanthium not bristly (15).
15. Style wholly deciduous from the mature achene.......................11. *Potentilla*
15. Style (at least the lower portion) persistent on the achene, often plumose....12. *Geum*
16. Flowers in panicles, very numerous, very small. Leaves simple, deeply cleft or coarsely toothed ....5. *Holodiscus*
16. Flowers solitary or in few-flowered cymes or fascicles (17).
17. Style wholly deciduous from the mature achene; leaves pinnately compound. Bark exfoliating; flowers showy, the petals bright yellow.................11. *Potentilla*
17. Style persistent; leaves simple but sometimes deeply cleft or pinnatifid (18).
18. Petals normally none (exceptionally present in *Coleogyne*); leaves entire or merely dentate. Pistil solitary (19).
18. Petals present; leaves usually deeply cleft or pinnatifid (20).
19. Pistil not enclosed in a sheath; style terminal, not twisted or bent, in fruit greatly elongate and plumose nearly to the apex.......................15. *Cercocarpus*
19. Pistil at anthesis enclosed in a thin sheathlike prolongation of the disk; style lateral, much twisted and bent, villous only near the base.................16. *Coleogyne*
20. Style not greatly elongate or plumose in fruit, stout, beaklike; petals spatulate, cream-colored or yellow; pistil solitary or rarely the pistils 2 or 3.......17. *Purshia*
20. Style greatly elongate and plumose in fruit; petals broadly obovate or nearly orbicular, white; pistils several or numerous (21).

21. Bark exfoliating; bractlets present, alternating with the sepals; achenes with purplish tails .... 13. *Fallugia*
21. Bark not or but tardily exfoliating; bractlets none; achenes with white tails 14. *Cowania*

1. PHYSOCARPUS. Ninebark

Plants small shrubs, with bark exfoliating in strips; leaves simple, petioled, palmately lobed; flowers in terminal corymbs; petals 5, these and the numerous stamens inserted on the throat of the calyx; pistils 1 to 5, short-stipitate, becoming inflated few-seeded capsules dehiscent on both sutures.

1. Physocarpus monogynus (Torr.) Coult. White Mountains (Apache County), Pinaleno Mountains (Graham County), Chiricahua Mountains (Cochise County), 8,000 to 9,500 feet, pine and spruce forests, June and July. South Dakota to Texas and eastern Arizona.

2. PETROPHYTUM. Rock-mat

Small undershrubs, or the numerous stems woody only at base, forming mats; leaves small, spatulate, entire; flowers many, in very dense, simple or sparingly branched, spikelike inflorescences terminating the few-bracted scapes; stamens numerous; pistils few; fruits capsular (finally dehiscent on both sutures).

1. Petrophytum caespitosum (Nutt.) Rydb. (*Spiraea caespitosa* Nutt.). Navajo, Coconino, northeastern Mohave, Yavapai, and northern Gila counties, also Huachuca and Mustang mountains (Cochise County), 5,000 to 8,000 feet, common on both rims of the Grand Canyon, dry rock ledges, often on limestone, June to October. South Dakota and Montana to New Mexico, Arizona, and California.

Var. *elatius* (Wats.) Tidestrom, with longer, frequently branched inflorescences and longer floral bracts, intergrades freely with typical *P. caespitosum* and has the same distribution in Arizona. An unusual form, with spreading rather than appressed hairs on the leaves, was found in Havasu Canyon (*J. T. Howell* 26596).

3. CHAMAEBATIARIA. Fern-bush

Shrubs, commonly 1.5 to 2 m. high, very leafy, aromatic; leaves more or less evergreen, much dissected, the ultimate segments very small; flowers numerous, small, panicled; petals white; stamens many; pistils 5; fruit follicular.

1. Chamaebatiaria Millefolium (Torr.) Maxim. Navajo, Coconino, eastern Mohave, and northern Yavapai counties, 4,500 to 8,000 feet, often with pinyons and junipers, July to November. Idaho to Arizona and California.

Browsed by sheep, goats, and deer, apparently not by cattle. Also known as tansy-bush and desert-sweet.

4. VAUQUELINIA

Large shrubs or small trees; leaves evergreen, simple, lanceolate, serrate; flowers numerous, in flat-topped, cymose panicles; petals white; stamens 15 or more; pistils 5, connate at base, becoming somewhat woody follicles.

1. Vauquelinia californica (Torr.) Sarg. Mountains of Gila, Maricopa, Pinal, Cochise, and Pima counties, 2,500 to 5,000 feet, often among live-oaks, sometimes on limestone, May to July, type from along the Gila River (*Emory*). Arizona and northern Mexico.

Arizona-rosewood. The wood is close-grained, hard, and heavy, but the trunks probably are too small to warrant exploitation.

5. HOLODISCUS (127). Rock-spiraea

Shrubs up to 3 m. high, much branched; leaves deciduous, simple, cuneate at base, often decurrent on the petiole, coarsely crenate, white-sericeous to sparsely villous beneath; flowers small, in ample terminal panicles, or the inflorescence reduced to a small raceme; calyx and petals cream-colored; stamens many; pistils 5.

The names cream-bush, foam-bush, mountain-spray, and ocean-spray are also applied to these plants.

1. Holodiscus dumosus (Nutt.) Heller. Apache County to Mohave County, south to Cochise, Santa Cruz, and Pima counties, 5,500 to 10,000 feet, commonly in pine or spruce forests, often on cliffs, June to September. Wyoming to Utah, south to western Texas, New Mexico, Arizona, and northern Mexico.

A beautiful shrub, with aromatic foliage and feathery panicles of small flowers. The species is represented in Arizona by 2 varieties. Var. *typicus* Ley, a shrub usually about 1 m. high, with small inflorescences and leaves only 1 to 2 cm. long, the lower surface sparsely villous to glabrate and usually obviously resinous-granular, occurs in northern Arizona, especially in and near the Grand Canyon. Var. *australis* (Heller) Ley (*H. australis* Heller), a shrub usually 2 to 3 m. high, with usually ample inflorescences and leaves up to 5 cm. long, the lower surface white with a dense sericeous-tomentose pubescence, is found chiefly in the southern counties, but occasionally as far north as the San Francisco Peaks. These varieties were referred previously (under *H. discolor*) to var. *glabrescens* (Greenm.) Heller and var. *dumosus* (Nutt.) Dippel respectively. Specimens from near Shato, Navajo County (*Eastwood & Howell* 6589), and from the Kaibab Plateau, Coconino County (*Collom* in 1940), have the ample inflorescence of var. *australis*, but the leaves are about as wide as long and are green and merely loosely villous beneath.

Ley (127, pp. 283, 284) referred a specimen from the Kaibab Plateau (*Jones* 6056b) to *H. microphyllus* Rydb. var. *typicus* Ley, characterized by villous lower leaf surfaces, and a specimen from the Huachuca Mountains (*Hilend* in 1931) to *H. microphyllus* var. *sericeus* Ley, characterized by densely white-sericeous lower leaf surfaces. It seems probable that these specimens are more properly referrable to *H. dumosus* vars. *typicus* and *australis* respectively and that *H. microphyllus* should not be included in the Arizona flora.

6. SORBUS (128). Mountain-ash

Shrubs, commonly about 2.5 m. high; leaves odd-pinnate, the leaflets numerous, lanceolate, sharply serrate; flowers small, in terminal compound cymes; petals white; stamens many; pistil one, compound, the styles usually 3, the ovary inferior, usually 3-celled; fruit a small, berrylike pome.

The plants are favorite cultivated ornamentals because of their attractive foliage and highly colored, usually bright red, mature fruits, which are much relished by birds and other wild animals. Some species are small trees. The bark of *Sorbus americana,* of eastern North America, is used medicinally for its tonic, astringent, and antiseptic properties.

1. Sorbus dumosa Greene. Apache, Navajo, and Coconino counties, south to Cochise and Pima counties, 8,000 to 10,000 feet, moist, rich soil in coniferous forests, June and July, type from the San Francisco Peaks (*Greene* in 1889). New Mexico and Arizona.

A specimen from Lukachukai Pass, Apache County (*Eastwood & Howell* 6790), without flowers or fruit, seems referrable to *S. scopulina* Greene on the character of the winter buds, these being glossy, glutinous, and only sparsely pubescent.

7. AMELANCHIER (129, 130, 131). SERVICE-BERRY

Large shrubs or small trees; leaves simple, petioled, serrate, dentate, or nearly entire; flowers rather large, in few-flowered racemes or fascicles, seldom solitary; calyx lobes narrow, reflexed; petals oblong or narrowly obovate, white; stamens numerous; pistil 1, compound; styles 2 to 5, distinct or more or less connate; ovary inferior; fruit a small several-celled pome; seed solitary in each cell.

The small, applelike fruits were used by Indians and by early European explorers in North America, especially when dried or cooked in various ways, and they make excellent jelly. They are greedily devoured by birds and other animals. The plants are said to afford good browse for sheep, goats, and deer, and in spring for cattle. These plants are hosts of the cedar-apple fungus.

The taxonomy of this genus is in great confusion, and the following treatment of the Arizona plants is necessarily tentative. Authorities differ widely in their interpretations of the species in the Rocky Mountain region, and pending extensive further study in the field it seems best to maintain the treatment substantially as in *Flowering Plants and Ferns of Arizona.* G. N. Jones (131, pp. 86–93) gave only one species for Arizona. This is *A. utahensis* Koehne, under which *A. Bakeri* Greene, *A. Covillei* Standley, *A. Goldmanii* Woot. & Standl., *A. Jonesiana* C. K. Schneid., *A. mormonica* C. K. Schneid., *A. nitens* Tidestrom, *A. oreophila* A. Nels., and *A. rubescens* Greene were cited as synonyms. *A. polycarpa* Greene was made a synonym of *A. pumila* Nutt., but Jones did not include Arizona in the range of that species.

KEY TO THE SPECIES

1. Styles normally 2 or 3; fruit at maturity orange or yellow, rather dry and mealy. Twigs rigid, the bark soon becoming gray; leaves commonly finely pubescent or subtomentose, at least beneath; hypanthium, calyx lobes, and the top of the ovary commonly pubescent; styles distinct to the base or nearly so (2).
1. Styles normally 4 or 5; fruit at maturity dark plum purple, glaucous, more or less juicy. Leaves broadly oval, ovate, or suborbicular, broadly rounded or truncate at apex, glabrate or loosely villous beneath (3).
2. Leaves elliptic or oblong-ovate, finely and usually bluntly toothed, often somewhat narrowed to the obtuse or acutish apex............................1. *A. utahensis*

2. Leaves mostly suborbicular, coarsely and often sharply toothed, broadly rounded or truncate and often retuse at apex .... 2. *A. Bakeri*
3. Top of the ovary and the calyx glabrous from the beginning, as is the whole plant. Bark of the twigs dark reddish brown, becoming gray; leaves coarsely serrate 3. *A. polycarpa*
3. Top of the ovary and usually also the calyx persistently pubescent (4).
4. Leaves broadly crenate-dentate, often nearly to the base, cordate or subcordate at base, usually soon glabrate on both surfaces. Bark of the twigs dark reddish brown, finally becoming gray; leaves 3 to 5 cm. long and about equally wide. . 4. *A. Goldmanii*
4. Leaves serrate-dentate, rounded, truncate, or short-cuneate (rarely subcordate) at base, persistently pubescent, at least on the lower surface (5).
5. Hypanthium at anthesis commonly villous, often copiously so; calyx lobes pubescent on both surfaces; leaves seldom more than 2.5 cm. long, becoming thickish, more or less persistently pubescent on both surfaces, grayish green above, the margin finely toothed, usually not to far below the middle; bark of the twigs soon becoming gray .... 5. *A. oreophila*
5. Hypanthium at anthesis glabrous or nearly so; calyx lobes commonly glabrate externally; leaves mostly not less than 3 cm. long, remaining thin, glabrate and bright green above, the margin coarsely toothed, often nearly to the base; bark of the twigs more persistently reddish brown .... 6. *A. mormonica*

1. Amelanchier utahensis Koehne. Apache County to Mohave County, 2,000 to 7,500 feet, dry, rocky slopes, common in and around the Grand Canyon, April and May. Colorado to Nevada, New Mexico, and northern Arizona.

Specimens with the herbage, ovary, and usually the calyx glabrous and the leaves often shiny above are referrable to subsp. *Covillei* (Standl.) Clokey (*A. Covillei* Standl., *A. nitens* Tidestrom), which is apparently common in the Grand Canyon.

2. Amelanchier Bakeri Greene. Apache (?), Navajo, Coconino, Yavapai, and Gila counties, 4,000 to 7,000 feet, April and May. Colorado, New Mexico, and Arizona.

Commonly about 2.5 m. (8 feet) high.

3. Amelanchier polycarpa Greene. North Rim of the Grand Canyon (*Eastwood & Howell* 1047a, 7069). Wyoming to New Mexico and northern Arizona.

The specimens cited have larger leaves than as described by Greene for *A. polycarpa,* and the blades are deeply serrate, almost laciniate. Other specimens from the Grand Canyon (*Eastwood* 5811, *Darrow* in 1938), with smaller, less deeply incised leaves, seem more typical.

4. Amelanchier Goldmanii Woot. & Standl. North Rim of the Grand Canyon, San Francisco Peaks, and Flagstaff (Coconino County), Lukachukai Mountains and Willow Spring (Apache County), 7,000 to 9,500 feet, May and June. New Mexico and Arizona.

A. Goldmanii may not be distinct from *A. Jonesiana* C. K. Schneid., the latter name having priority. It is doubtful that this and the 2 following entities are more than varieties of the species, widely distributed in western North America, that is known commonly as *A. alnifolia* Nutt. but whose proper name may be *A. florida* Lindl. (McVaugh, 129, p. 92).

5. Amelanchier oreophila A. Nels. (*A. utahensis* subsp. *oreophila* Clokey, *A. florida* var. *oreophila* R. J. Davis). Apache, Navajo, Coconino, Gila, and Yavapai counties, 4,000 to 8,000 feet, April and May. Montana to New Mexico, Arizona, and Nevada.

In the absence of mature fruit this entity is likely to be confused with *A. Bakeri,* but the latter has usually more coarsely toothed leaves.

6. Amelanchier mormonica C. K. Schneid. Apache, Navajo, Coconino, and Gila counties, 6,500 to 9,500 feet, April to June, also Bear Valley (Sycamore Canyon), Santa Cruz County, about 4,000 feet (*Goodding* 6488), type from Mormon Lake, Coconino County (*MacDougal* 102). Wyoming to New Mexico and Arizona.

This shrub attains a height of at least 3.6 m. (12 feet). *A. mormonica* seems to be intermediate between *A. Goldmanii* and *A. oreophila.*

Peraphyllum ramosissimum Nutt., the so-called squaw-apple, is to be sought in northern Arizona. It is closely related to *Amelanchier* but has the flowers solitary or 2 or 3 (rarely 5) in a cluster, the petals pink, and the leaves narrowly oblanceolate or elliptic, entire or serrulate.

8. CRATAEGUS. Hawthorn

Shrubs or small trees, armed with strong, sharp thorns; leaves simple, petioled, serrate to shallowly lobed, strongly veined, sparsely pubescent beneath or glabrate; flowers in several-flowered corymbs; petals rather small, round, usually white; stamens few to numerous; ovary inferior, 1- to 5-celled; fruit a nearly globose, thin-fleshed pome, almost filled by the large, bony seeds.

KEY TO THE SPECIES

1. Spines few, not more than 2.5 cm. long; leaves elliptic, about twice as long as wide, not or scarcely lobed, tapering at base............................1. *C. rivularis*
1. Spines numerous, 3 to 5 cm. long; leaves ovate, less than twice as long as wide, often distinctly lobed, rather abruptly contracted at base..............2. *C. erythropoda*

1. Crataegus rivularis Nutt. Clear Creek Canyon, southeast of Winslow, Navajo County, about 5,000 feet (*Vaughn & Benham* 6463). Wyoming and Idaho, south to New Mexico and northern Arizona.

2. Crataegus erythropoda Ashe. Bonita Creek, White Mountains (Apache County), Oak Creek Canyon (Coconino County), near Calva (Graham County), 4,500 to 6,000 feet, along streams in canyons. Wyoming to New Mexico and Arizona.

The specimens from Oak Creek Canyon approach *C. Wootoniana* Eggleston in their rather deeply lobed leaves.

9. RUBUS (132)

Plants shrubby, of diverse habit and appearance, often prickly; leaves simple or pinnate; flowers in few-flowered racemes or corymbs, sometimes solitary, the petals white or tinged with pink; stamens numerous; pistils several or numerous, on a convex receptacle, becoming 1-seeded drupelets aggregated in a more or less fleshy fruit.

This large genus includes the cultivated blackberries and raspberries. In Arizona, *R. strigosus* var. *arizonicus,* a raspberry, and *R. arizonensis,* a blackberry, produce edible fruits, good for eating raw and often used locally for making jam and jelly. The fruits are much relished by birds and other wild animals. The most ornamental of the species, with large white flowers, are the thimble-berries, *R. neomexicanus* and *R. parviflorus,* which are reported to be extensively browsed by deer.

KEY TO THE SPECIES

1. Stems unarmed, not prickly or bristly; bark exfoliating; leaves simple, 3- to 5-lobed; petals 15 to 30 mm. long; styles short, club-shaped. Fruit red, not very juicy, hemispheric or flatter (2).
1. Stems prickly or at least bristly; bark not exfoliating; leaves compound; petals less than 15 mm. long; styles elongate (3).
2. Leaves 5 to 20 cm. wide, the lobes acute or acuminate; flowers usually in loose clusters of 3 or more, seldom solitary; styles glabrous; drupelets capped by a hard, pubescent cushion ....1. *R. parviflorus*
2. Leaves 3 to 10 cm. wide, the lobes acute or obtuse; flowers solitary, or in twos; styles hairy; drupelets not capped by a pubescent cushion..............2. *R. neomexicanus*
3. Sepals shorter than the petals; fruit globose or nearly so; drupelets glabrous at maturity or very nearly so. Stems, petioles, and leaf veins, beneath, armed with short, flattened, reflexed or recurved prickles; leaves of the shoots usually pedately 5-foliolate, those of the flowering stems mostly 3-foliolate (4).
3. Sepals equaling or longer than the petals, caudate-acuminate; fruit hemispheric or somewhat higher; drupelets pubescent. Leaflets sharply double-serrate, white-tomentose beneath (5).
4. Stems prostrate, only the flowering branches erect; lower leaf surface green, sparsely pubescent or glabrescent, the veins only slightly prominent; inflorescence short-cymose, 1- to 5-flowered; fruit dark red at maturity, the drupelets large 3. *R. arizonensis*
4. Stems erect, ascending, or recurved; lower leaf surface permanently white-tomentose, the veins prominent; inflorescence cymose-paniculate, elongate, several- to many-flowered; fruit black at maturity...............................4. *R. procerus*
5. Spines slender or reduced to bristles, straight or nearly so; inflorescence and hypanthium spinose-bristly, more or less glandular; leaves all pinnate, on the shoots 5- to 7- (rarely 9-) foliolate, on the flowering branches usually 5- (sometimes 3-) foliolate; terminal leaflet ovate or ovate-lanceolate; mature fruit bright red 5. *R. strigosus*
5. Spines stout, laterally flattened, those of the branchlets strongly curved; inflorescence and hypanthium not bristly, the pubescence soft; leaves of the shoots often pedately 5-foliolate, those of the flowering branches 3-foliolate; terminal leaflet broadly ovate; mature fruit dark red to nearly black...........................6. *R. leucodermis*

1. Rubus parviflorus Nutt. Lukachukai, Tunitcha, and White mountains (Apache County), Pinaleno Mountains (Graham County), Chiricahua Mountains (Cochise County), 8,000 to 9,500 feet, July to September. Michigan to Alaska, south to New Mexico, eastern Arizona, and California.

The Arizona plants presumably belong to var. *parvifolius* (Gray) Fern. The specimens from the Pinaleno and Chiricahua mountains are merely suffrutescent, but the plant is a shrub up to 2 m. high in favorable situations.

2. Rubus neomexicanus Gray. Coconino County to Graham, Cochise, Santa Cruz, and Pima counties, 5,000 to 9,000 feet, May to September. New Mexico, Arizona, and northern Mexico.

A collection in the Huachuca Mountains (*Jones* 24918) was cited by Bailey (132, p. 917) under *R. exrubicundus* (Woot. & Standl.) Bailey, which is distinguished from *R. neomexicanus* in having the leaves pubescent beneath only on the veins, and the hairs (as also on the branches and petioles) more appressed than in *R. neomexicanus*. It is doubtful that this entity is more than varietally distinct.

3. Rubus arizonensis Focke (*R. oligospermus* Thornber). Yavapai, Greenlee, Graham, Gila, Maricopa, Pinal, Cochise, Santa Cruz, and Pima counties,

3,500 to 5,000 feet, often along streams in partial shade, March to May, type of *R. arizonensis* from the Santa Rita Mountains (*Pringle* in 1881), type of *R. oligospermus* from the Santa Catalina Mountains (*Pringle* in 1881). Western Texas, central and southern Arizona, and northern Mexico.

Arizona dew-berry. The trailing habit of the plant makes it a good ground cover, protecting the soil against erosion.

4. Rubus procerus P. J. Muell. Grand Canyon and Oak Creek Canyon (Coconino County), Sierra Ancha (Gila County), about 6,000 feet.

This is an introduced species, probably the one cultivated under the name Himalaya-berry, and although apparently growing wild at the stations mentioned, it can scarcely be regarded as established in the Arizona flora. The identity of the species is uncertain.

5. Rubus strigosus Michx. Apache County to Mohave County, south to Cochise and Pima counties, 7,000 to 11,500 feet, rich soil in pine and spruce forests, June and July. Widely distributed in the cooler parts of North America.

This species is a progenitor of some of the cultivated raspberries. It is represented in Arizona by var. *arizonicus* (Greene) Kearney & Peebles (*Batidaea arizonica* Greene), type from the San Francisco Peaks (*Greene* in 1889), which seems to differ from typical *R. strigosus* only in having usually more numerous leaflets, up to 9 on the shoots, mostly 5 on the flowering branches.

6. Rubus leucodermis Dougl. Along Oak Creek, Coconino County, about 5,500 feet (*Goldman* 2171, *Fulton* 4356, *Shreve* 4794). Montana to British Columbia, south to New Mexico, central Arizona, and California.

The Arizona specimens have a few glandular hairs on the petioles, pedicels, and calyx (character of *R. bernardinus* (Greene) Rydb.), but such hairs are found also on specimens of *R. leucodermis* from northern California.

10. FRAGARIA. Strawberry

Plants herbaceous, acaulescent; scapes from short rootstocks, these emitting long runners; leaves long-petioled, digitately trifoliolate, the leaflets obovate or wedge-shaped, coarsely toothed; flowers rather large, the petals broad and rounded, white; stamens many; pistils numerous, becoming achenes embedded in pits on the surface of the enlarged fleshy receptacle.

The fruits of the wild strawberries, although edible, are too small to be of much use to man, although doubtless they are relished by birds and other animals.

KEY TO THE SPECIES

1. Hairs of the scapes (and commonly of the petioles) appressed or ascending; leaves commonly somewhat glaucous, thickish; leaflets obovate-oblong to nearly spatulate, long-cuneate at base, few-toothed, mainly toward the apex; fruit deeply pitted, the seeds partly buried in the flesh..1. *F. ovalis*
1. Hairs of the scapes and petioles soon spreading or somewhat reflexed; leaves not glaucous, thin; leaflets rhombic-ovate or obovate, short-cuneate at base, toothed to well below the middle; fruit shallowly pitted, the seeds superficial........2. *F. bracteata*

1. Fragaria ovalis (Lehm.) Rydb. Apache, Navajo, and Coconino counties, south to Cochise and Pima counties, 7,000 to 11,000 feet, common in coniferous forests, May to October. Wyoming to New Mexico and Arizona.

Plants more or less intermediate between this species and *F. bracteata* have been collected on the San Francisco Peaks, in Oak Creek Canyon, and in the White, Chiricahua, and Santa Catalina mountains.

2. Fragaria bracteata Heller. White Mountains (Apache County), Pinaleno Mountains (Graham County), Chiricahua and Huachuca mountains (Cochise County), 7,000 to 9,500 feet, May to September. Montana to British Columbia, south to New Mexico, Arizona, and California.

11. POTENTILLA (133, 134, 135). CINQUEFOIL

Plants biennial or perennial, mostly herbaceous (one species shrubby); stems leafy or scapose, often from a somewhat woody, branched caudex or rootstock; inflorescence cymose (the flowers seldom solitary); sepallike bractlets usually present, alternating with the sepals; petals commonly yellow but sometimes whitish, red, or purple; stamens 5 to many, inserted at base of the receptacle or on the margin of the flat to turbinate hypanthium; pistils few to numerous, the style basal to terminal; fruits of achenes.

Some of the species are grazed by sheep.

KEY TO THE SPECIES[1]

1. Receptacle stipitate, conic-columnar; calyx usually without bractlets. Plants small, subscapose, the flowering stems from a woody caudex, this often branched; herbage more or less glandular-pubescent; leaflets 2 to 5 pairs, broad, deeply toothed or cleft; hypanthium campanulate; petals yellow; stamens 5; pistils borne at the summit of the hairy receptacle, the styles subterminal: section *Purpusia* 1. *P. Osterhoutii*

1. Receptacle sessile, flat or shallowly saucer-shaped; calyx with bractlets (2).

2. Styles lateral; petals yellow or yellowish (3).

2. Styles terminal or nearly so (7).

3. Achenes silky-villous; plants shrubby, the bark shreddy. Leaves pinnately 3- to 7-foliolate, sericeous, the leaflets linear or narrowly oblanceolate, entire; flowers large and showy, the petals bright yellow, 10 mm. long or longer, much surpassing the bractlets and sepals; stamens and pistils numerous: section *Dasiphora*........2. *P. fruticosa*

3. Achenes glabrous; plants herbaceous (4).

4. Plants caulescent, viscid-villous. Stems 30 cm. long or longer, leafy; basal leaves pinnate, with 5 to 9 broadly ovate or obovate, coarsely toothed leaflets; stamens and pistils numerous; style attached near the base of the achene: section *Drymocallis* (5).

4. Plants acaulescent or nearly so, not viscid (6).

5. Herbage conspicuously villous; leaflets thickish, densely pubescent to glabrate above, the terminal one oval to rhombic-obovate; inflorescence dense, strict; sepals at anthesis 5 to 9 mm. long..3. *P. arguta*

5. Herbage not conspicuously villous; leaflets thin, sparsely pubescent to glabrate above, the terminal one obovate; inflorescence dense or rather open; sepals at anthesis 4 to 6 mm. long..4. *P. glandulosa*

6. Plants without runners; leaves green on both surfaces, digitately trifoliolate; leaflets cuneate-obovate, 3- to 5-dentate at the truncate apex; petals ochroleucous or pale yellow, shorter than the sepals; stamens usually 5: section *Sibbaldia*....5. *P. Sibbaldi*

6. Plants with runners; leaves silvery-sericeous, at least beneath, pinnately multifoliolate with much smaller leaflets interspersed with the large ones; leaflets oblong, elliptic, or obovate, coarsely toothed; petals bright yellow, much longer than the sepals; stamens numerous: section *Argentina*..........................6. *P. Anserina*

---

[1] Dr. David D. Keck has made helpful suggestions in regard to some of the Arizona species.

7. Stamens 5, inserted at a considerable distance from the base of the receptacle and from the pistils; pistils 1 to 5. Leaves pinnate; hypanthium with a flat or saucer-shaped peripheral expansion, on which the stamens are inserted; leaves elongate, narrow, with very many small leaflets; stems erect, 30 cm. long or longer: section *Ivesia* (8).
7. Stamens seldom fewer than 10, inserted close to the pistils, these 5 or more: section *Eupotentilla* (9).
8. Leaflets cleft at apex; petals dark purple........................7. *P. multifoliolata*
8. Leaflets parted or divided; petals yellow..........................8. *P. sabulosa*
9. Petals rose red or dark red; plants caulescent. Stems erect or nearly so, seldom less than 30 cm. long; basal leaves digitate or nearly so, with 5 to 7 leaflets 9. *P. Thurberi*
9. Petals yellow (10).
10. Inflorescence usually many-flowered, conspicuously leafy, its leaves not bractlike although usually smaller than the other stem leaves; plants annual or biennial; flowers inconspicuous, the petals seldom surpassing the sepals. Leaves digitate, or the lower ones pinnate with a short rachis; leaflets coarsely toothed or shallowly cleft (11).
10. Inflorescence few-flowered or if many-flowered, then not conspicuously leafy, its leaves mostly much reduced; plants perennial; flowers commonly showy; petals usually surpassing the sepals (13).
11. Petals not much shorter than the sepals; achenes commonly finely rugose; hairs of the stem mostly rather stiff, long, and spreading or widely ascending. Leaves all trifoliolate or the lower ones pinnately 5-foliolate.................10. *P. norvegica*
11. Petals much shorter than the sepals; achenes usually smooth; hairs of the stem soft, mostly short and subappressed (12).
12. Branches erect, or ascending at a narrow angle; cymes elongate, racemiform; pubescence partly glandular. Leaves all digitate, trifoliolate..............11. *P. biennis*
12. Branches spreading, or ascending at a wide angle; cymes not racemiform; pubescence not glandular..12. *P. rivalis*
13. Basal leaves pinnate, the rachis distinct. Plants caulescent; inflorescence several- to many-flowered (14).
13. Basal leaves digitate or nearly so, the rachis, if any, very short. Leaflets 7 (rarely 9) or fewer (18).
14. Leaves at base of the plant bipinnate, the primary divisions parted or divided into linear, lanceolate, or oblanceolate lobes. Pubescence strigose or villous; flowering stems decumbent, seldom more than 20 cm. long; cymes few-flowered; pedicels elongate, slender, often somewhat arcuate; bractlets nearly equaling the ovate-lanceolate, acuminate sepals; petals considerably longer than the sepals........13. *P. plattensis*
14. Leaves at base of the plant simply pinnate, the primary divisons at most pinnatifid (15).
15. Leaflets shallowly toothed toward the apex, usually entire below the middle, strigose to sericeous beneath...14. *P. crinita*
15. Leaflets deeply toothed or cleft more than half of their length, usually nearly to the base. Stems seldom less than 25 cm. long; inflorescence open, often many-flowered; petals distinctly surpassing the sepals (16).
16. Basal leaves obscurely pinnate, the rachis very short (3 to 8 mm. long). Leaflets 5 to 9, crowded, green above, usually conspicuously white-tomentose as well as sericeous beneath ....15. *P. pulcherrima*
16. Basal leaves noticeably pinnate, the rachis 10 to 50 mm. long (17).
17. Herbage grayish green and loosely villous or substrigose; leaflets cleft nearly to the midvein ....16. *P. pensylvanica*
17. Herbage silvery-sericeous, the hairs appressed or subappressed; leaflets seldom cleft more than halfway to the midvein.............................17. *P. Hippiana*
18. Plants caulescent, the flowering stems erect or ascending, seldom less and usually more than 20 cm. long; inflorescence several- (usually more than 5-) to many-flowered (19).
18. Plants subacaulescent, strongly cespitose, the flowering stems spreading or prostrate, less than 20 cm. long; inflorescence few- (usually not more than 5-) flowered (20).

19. Leaves white-tomentose beneath, the upper surface dark green; anthers mostly 1 mm. long .... 15. *P. pulcherrima*
19. Leaves silky-strigose to glabrate beneath, the two surfaces not contrasting strongly in color; anthers mostly 0.5 mm. long .... 18. *P. diversifolia*
20. Leaves more or less white-tomentose as well as sericeous beneath, green and sericeous above; styles rather short and thick .... 19. *P. concinna*
20. Leaves not white-tomentose beneath; styles elongate, slender (21).
21. Plants conspicuously and densely silky-villous, more or less granuliferous 20. *P. viscidula*
21. Plants sparsely villous or subhirsute, often also glandular-puberulent, especially in the inflorescence (22).
22. Leaflets of the basal leaves 5 (occasionally 7); inflorescence usually with numerous tack-shaped, glandular hairs .... 21. *P. subviscosa*
22. Leaflets of the basal leaves 3; inflorescence viscid but without tack-shaped hairs 22. *P. albiflora*

1. Potentilla Osterhoutii (A. Nels.) J. T. Howell (*Purpusia Osterhoutii* A. Nels., *P. arizonica* Eastw.). Grand Canyon (Coconino County), both rims, 6,500 to 7,500 feet, Schnebly Hill (southern Coconino County), crevices of rocks, June to September, the types of *Purpusia Osterhoutii* and *P. arizonica* (*Osterhout* 7103, *Eastwood* 5662) from the Grand Canyon. Northern Arizona, southern Nevada, and southeastern California.

2. Potentilla fruticosa L. (*Dasiphora fruticosa* Rydb.). San Francisco Peaks and vicinity of Flagstaff (Coconino County), White Mountains (Apache County), 7,000 to 9,500 feet, along streams and in wet meadows, June to August. Widely distributed in the cooler parts of the Northern Hemisphere.

Bush cinquefoil, the only shrubby species in Arizona, browsed by sheep, goats, and deer, sometimes until the plants are stunted. Rated as an excellent erosion-control species in localities to which it is adapted. Very handsome when flowering.

3. Potentilla arguta Pursh. White Mountains, Apache County, on the White River (*Hough* in 1918) and on Bonito Creek, in meadows (*Goodding* 1219), Clear Creek Canyon, Navajo County (*Goodding* 6473), June and July. New Brunswick to Mackenzie, south to the District of Columbia, New Mexico, and eastern Arizona.

The Arizona plants belong to subsp. *convallaria* (Rydb.) Keck (*P. convallaria* Rydb., *Drymocallis convallaria* Rydb.). The petals are white.

4. Potentilla glandulosa Lindl. North Rim of Grand Canyon, near Williams, Oak Creek Canyon, and Mormon Lake (Coconino County), Sierra Ancha and Mazatzal Mountains (Gila County), 5,000 to 8,000 feet, wet places, May and June. South Dakota to British Columbia, south to New Mexico, central Arizona, and California.

The Arizona plants belong to subsp. *arizonica* (Rydb.) Keck (*Drymocallis arizonica* Rydb., *Potentilla Macdougalii* Tidestrom), the type of which was collected at Mormon Lake (*MacDougal* 64). The petals are pale yellow or cream-colored.

5. Potentilla Sibbaldi Hall. f. (*Sibbaldia procumbens* L.). San Francisco Peaks (Coconino County), 11,000 to 12,000 feet, July and August. Circumpolar and alpine in the Northern Hemisphere.

6. Potentilla Anserina L. (*Argentina Anserina* Rydb.). White Mountains (Apache and Greenlee counties), Flagstaff and vicinity (Coconino County), wet mountain meadows, May to August. Widely distributed in the cooler parts of the Northern Hemisphere.

Silverweed. The sweetish roots are reported to be edible either raw or cooked. The typical plant, with leaves green above, occurs in the White Mountains, and var. *concolor* Ser. (*Argentina argentea* Rydb.), with leaves silvery-sericeous on both surfaces, occurs in the Flagstaff region and was reported from the Kaibab Plateau.

7. Potentilla multifoliolata (Torr.) Kearney & Peebles (*Comarella multifoliolata* Rydb., *Ivesia multifoliolata* Keck). San Francisco Peaks, Mormon Lake, Oak Creek, Mogollon Escarpment (Coconino County), 6,000 to 7,500 feet, in washes, June to August, type collected in "western New Mexico" (now Arizona) by Woodhouse. Known only from Arizona.

*__8. Potentilla sabulosa__ Jones (*Ivesia sabulosa* Keck). Not known definitely from Arizona, but occurs in southern Utah and Nevada.

9. Potentilla Thurberi Gray. Apache, Navajo, and Coconino counties, south to Cochise and Pima counties, 6,000 to 9,000 feet, rich soil in coniferous forests, July to October. New Mexico, Arizona, and northern Mexico.

The typical plant, with leaflets rather sparsely sericeous beneath, and var. *atrorubens* (Rydb.) Kearney & Peebles (*P. atrorubens* Rydb.), with leaflets densely silvery-sericeous beneath, are about equally common. The type of *P. atrorubens* was collected on Mount Graham (*Rothrock* 399). Var. *sanguinea* (Rydb.) Kearney & Peebles (*P. sanguinea* Rydb.), with basal leaves subpinnate rather than strictly digitate as in the other forms, has been collected only in Coconino County, at Walnut Canyon (*MacDougal* 331, the type collection) and in Oak Creek Canyon (*Fulton* 9661).

10. Potentilla norvegica L. (*P. monspeliensis* L.). Widely distributed in Coconino County, White Mountains (Apache and Greenlee counties), 7,000 to 9,500 feet, wet meadows, June to August. Almost throughout North America; probably introduced from Europe.

11. Potentilla biennis Greene. Kaibab Plateau (*Goodding* 296–49), South Rim of Grand Canyon (*Toumey* 98, *Whiting* 5114), June and July. Saskatchewan to British Columbia, south to Colorado, northern Arizona, and Baja California.

12. Potentilla rivalis Nutt. San Francisco Peaks (Coconino County), Del Rio (Yavapai County), Gila River bed, Sacaton (Pinal County), the last doubtless a stray from the mountains, April to July. Illinois to Washington, south to Arizona, northern Mexico, and California.

Var. *millegrana* (Engelm.) Wats. (*P. millegrana* Engelm.), a smaller and more diffuse plant with basal leaves all digitately trifoliolate (these usually pinnately 5- to 7-foliolate in typical *P. rivalis*), has been collected near Flagstaff (*Thornber* in 1920).

13. Potentilla plattensis Nutt. (*Ivesia pinnatifida* Wats., *Potentilla arizonica* Greene). Flagstaff, Coconino County (*Lemmon* 3200, the type collection of *Ivesia pinnatifida* and of *P. arizonica*), White Mountains, Apache

County (*Goodding* 1192, *Kearney & Peebles* 12421), 7,000 to 8,500 feet, moist, grassy meadows, July and August. Saskatchewan to Alberta, south to New Mexico, Arizona, and Nevada.

The Arizona specimens apparently differ from most of the specimens from farther east and north only in the more villous herbage.

14. Potentilla crinita Gray. Apache County to Coconino County, also in the Huachuca Mountains (Cochise County), 6,000 to 8,000 feet, commonly in pine forests, April to September. Colorado, Utah, New Mexico, and Arizona.

Var. *Lemmoni* (Wats.) Kearney & Peebles (*Ivesia Lemmoni* Wats., *Potentilla Lemmoni* Greene), occurs on the Kaibab Plateau, at the Grand Canyon (both rims), and near Flagstaff (Coconino County), type from Oak Creek Canyon (*Lemmon* in 1884). It differs from typical *P. crinita* in having the leaflets toothed only at or very near the apex, and petals usually not surpassing the sepals. A form with leaflets entire or merely emarginate at apex was collected at Flagstaff (*Goodwin* in 1947).

15. Potentilla pulcherrima Lehm. Apache County to Coconino County, 8,000 to 11,500 feet, open coniferous forests, July and August. Alberta and British Columbia to New Mexico, northern Arizona, and eastern Nevada.

This species apparently intergrades with *P. Hippiana*.

16. Potentilla pensylvanica L. Lukachukai Mountains, Apache County, 9,500 feet (*Goodman & Payson* 2852), San Francisco Peaks, Coconino County (*Thornber* in 1929). New England and Canada to Kansas, New Mexico, and northern Arizona.

17. Potentilla Hippiana Lehm. Apache, Navajo, and Coconino counties, Pinaleno Mountains (Graham County), Chiricahua Mountains (Cochise County), 7,000 to 11,500 feet, grassy meadows and open coniferous forests, June to September. Saskatchewan and Alberta, south to New Mexico, Arizona, and Nevada.

The typical plant has the leaves almost equally sericeous on both surfaces. In *P. Hippiana* var. *diffusa* Lehm. (*P. propinqua* Rydb.), which has much the same distribution in Arizona as the typical plant and is at least equally abundant, the upper surface of the leaves is green and much less sericeous than the lower surface. What seems to be an extreme form of this variety, from near McNary, Apache County (*Deaver* 1769), has stems about 0.5 meter high and the leaflets green on both surfaces, sparsely strigose above, thinly sericeous beneath.

18. Potentilla diversifolia Lehm. Kaibab Plateau, Grand Canyon, and San Francisco Peaks (Coconino County), White Mountains (Apache County), Santa Catalina Mountains (Pima County), 8,000 to 12,000 feet, July to September. Western Canada to New Mexico, Arizona, and California.

The Arizona specimens belong to the sparsely pubescent or glabrate form (*P. glaucophylla* Lehm.). A remarkably luxuriant plant, with stems about 40 cm. long and leaves 5 cm. long, was collected on the North Rim of the Grand Canyon (*Collom* in 1941). David D. Keck (personal communication) mentioned that this collection is aberrant also in having the leaflets with a scanty tomentum, in addition to the usual strigose pubescence, and the calyces copiously glandular.

19. Potentilla concinna Richards. White Mountains (Apache County), Betatakin (Navajo County), Kaibab Plateau and San Francisco Peaks (Coconino County), 7,000 to 12,000 feet, May and June. Canada to Colorado, Utah, New Mexico, and Arizona.

Some of the Arizona specimens, with leaflets sparsely sericeous above, correspond to *P. concinnaeformis* Rydb., type from Mount Agassiz (*Lemmon* 3294). Others, with a densely sericeous upper surface of the leaflets, seem to belong to *P. modesta* Rydb. It is doubtful that these entities are distinct as species from *P. concinna.*

20. Potentilla viscidula Rydb. Huachuca Mountains (Cochise County), Santa Rita Mountains (Santa Cruz County), 8,000 feet and probably higher, rock crevices, May and June, type from the Santa Rita Mountains (*Pringle* in 1881). Southwestern Arizona, Chihuahua, and southeastern California.

Perhaps too nearly related to *P. Wheeleri* Wats., to which the Arizona plants were referred in *Flowering Plants and Ferns of Arizona.*

21. Potentilla subviscosa Greene. Navajo and Coconino counties to Graham, Cochise, and Pima counties, 6,500 to 12,000 feet, coniferous forests and mountain meadows, April to June. New Mexico and Arizona.

More abundant in Arizona than the typical plant is var. *ramulosa* (Rydb.) Kearney & Peebles (*P. ramulosa* Rydb.), which is distinguished by the less deeply and more coarsely incised leaflets, and often by a less strictly acaulescent habit of growth. The type of *P. ramulosa* was collected in Arizona (*Lemmon* 399).

22. Potentilla albiflora L. Williams. Known only from the Pinaleno Mountains, Graham County (*Goodding* 1045, the type collection), 7,500 to 9,500 feet, abundant on rocky slopes and in open coniferous forest, May to August.

The petals were described by Williams as white, but they are pale yellow when fresh.

12. GEUM. Avens

Plants herbaceous, perennial; flowering stems from a thick rootstock or caudex, leafy or subscapose; leaves pinnate or deeply pinnatifid; flowers solitary or in few-flowered cymes, relatively large; petals yellow, sometimes tinged with purple or pink; stamens and pistils numerous; achenes tipped by the long, often plumose, sometimes jointed, persistent styles.

KEY TO THE SPECIES

1. Stems leafy, the lower stem leaves not much smaller than the basal ones; leaves with few divisions, the terminal one much the largest; styles conspicuously geniculate above the middle, the upper section promptly deciduous, the persistent lower section becoming sharply hooked (2).

1. Stems subscapose, all their leaves greatly reduced, the basal leaves with many divisions, the terminal one not much larger than the other upper ones; styles not conspicuously geniculate, straight or somewhat curved toward the apex, the upper section persistent or tardily deciduous, glabrous (3).

2. Upper section of the style hirsute, the lower section glabrous or sparsely pubescent toward base, not glandular..1. *G. strictum*

2. Upper section of the style short-pubescent near base, or glabrate, the lower section glandular-puberulent ....2. *G. macrophyllum*

3. Styles not greatly elongate in fruit, glabrous; hypanthium turbinate, acute at base; plants not conspicuously pubescent............................3. *G. turbinatum*

3. Styles greatly elongate in fruit, plumose below the short terminal section; hypanthium not turbinate, broad, rounded, and more or less depressed at base; plants conspicuously hirsute. Leaflets cuneate, few-toothed or cleft................4. *G. triflorum*

1. Geum strictum Ait. Apache, Navajo, Coconino, and Yavapai counties, 5,500 to 9,500 feet, rich soil in pine forests, July and August. Widely distributed in the cooler parts of the Northern Hemisphere.

The Arizona specimens belong to var. *decurrens* (Rydb.) Kearney & Peebles (*G. decurrens* Rydb.), the common form of the Rocky Mountain region, with the upper leaf segments more or less decurrent on the rachis, type from Baker Butte (*Mearns* 59, in 1887).

2. Geum macrophyllum Willd. Lukachukai and White mountains (Apache County), Kaibab Plateau (Coconino County), Pinaleno Mountains (Graham County), 7,000 to 9,500 feet, rich soil along streams and openings in pine forests, July to September. Throughout most of the cooler part of North America.

The Rocky Mountain (and Arizona) plant is var. *perincisum* (Rydb.) Raup (*G. perincisum* Rydb.).

3. Geum turbinatum Rydb. San Francisco Peaks (Coconino County), 10,000 to 12,000 feet, June to September. Montana to New Mexico, northern Arizona, Nevada, and Oregon.

It is doubtful that *G. turbinatum* is specifically distinct from the arctic *G. Rossii* (R. Br.) Ser.

4. Geum triflorum Pursh. Apache, Navajo, Coconino, and northern Gila counties, 6,000 to 9,500 feet, pine forests and open hillsides, May to August. Alberta and British Columbia, south to New Mexico, Arizona, and California.

Oldman-whiskers, grandfathers-beard. An attractive plant with pink flowers and silvery-plumose tails to the fruits, said to make good forage for sheep. The Arizona plant is var. *ciliatum* (Pursh) Fassett (*Geum ciliatum* Pursh, *Sieversia ciliata* G. Don). E. L. Greene described *Erythrocoma grisea* (type from the San Francisco Peaks, *Leiberg* 25, in 1901), *E. arizonica* (types from Mormon Lake and Bill Williams Mountain, *MacDougal* 65, *Rusby*), and *E. tridentata* (type from Willow Spring, *Palmer* 506, in 1890), but these appear to be merely forms of *G. triflorum* var. *ciliatum*.

13. FALLUGIA. Apache-plume

Plants much-branched, somewhat straggling shrubs, 1 to 2 m. high; branches slender, with white bark; leaves more or less evergreen, fascicled, obovate-cuneate, pinnately cleft or divided; flowers large, commonly solitary; hypanthium hemispheric, bearing 5 narrow bractlets alternating with the sepals; petals broad, white; stamens and pistils numerous; achenes with long, purplish, plumose, persistent styles.

1. Fallugia paradoxa (D. Don) Endl. Navajo, Coconino, and Mohave counties to Greenlee, Graham, Cochise, and Pima counties, 3,500 to 8,000 feet, common, often in chaparral, April to October. Southern Colorado and western Texas to southeastern California and northern Mexico.

This handsome shrub affords fairly good browse for cattle and sheep and is of value as a soil binder. The Hopi Indians use an infusion of the leaves as a stimulant of hair growth.

14. COWANIA (136). Cliff-rose

Plants shrubby or arborescent, often resinous and strong-smelling, the stems usually erect and rather stiff; leaves evergreen, thick, either obovate-cuneate and pinnately cleft or parted, or entire and narrowly spatulate, with revolute margins, loosely lanate or glabrate above, densely white-lanate beneath; flowers solitary at the ends of the branchlets, rather large; hypanthium funnelform to campanulate; petals broadly obovate, white or pale yellow; pistils few, the achenes with long, plumose, whitish, persistent styles.

KEY TO THE SPECIES

1. Shrubs up to 7.5 m. (25 feet) high, with stiffly ascending branches and reddish-brown to dark-gray bark; herbage usually viscid; leaves cuneate-obovate, pinnately veined, deeply 3-cleft, usually conspicuously glandular-punctate; pedicels and hypanthium commonly beset with stipitate glands; hypanthium funnelform to campanulate 1. *C. mexicana*
1. Shrubs up to 0.75 m. (3 feet) high, with ascending-spreading branches and pale-gray bark; herbage whitish-tomentose, not viscid; leaves oblanceolate, mostly 1-veined and entire but occasionally with 1 or 2 very short, subapical teeth, minutely and obscurely glandular-punctate; pedicels and hypanthium without stipitate glands; hypanthium narrowly funnelform. . . . . 2. *C. subintegra*

1. Cowania mexicana D. Don. Apache County to Mohave County, south to Cochise, Santa Cruz, and Pima counties, 3,000 to 8,000 feet, very common on dry slopes and mesas, especially in the juniper-pinyon association, often on limestone, April to September. Southern Colorado to southeastern California, south to central Mexico.

Sometimes called quinine-bush. This is one of the most important winter browse plants for cattle, sheep, and deer despite the bitter taste of the foliage. It is reported that strips of the inner bark were braided together by the aborigines of Utah and Nevada and used for clothing, sandals, rope, and mats. The plant is used by the Hopi Indians as an emetic and as a wash for wounds, and the wood formerly for making arrows. Under favorable circumstances the shrub attains a height of 7.5 m. (25 feet) but is ordinarily much smaller. The flowers are fragrant.

The Arizona plants belong mostly to var. *Stansburiana* (Torr.) Jepson (*C. Stansburiana* Torr., *C. Davidsonii* Rydb.), which differs from typical *C. mexicana* in having the primary leaf lobes toothed or cleft (not entire); and in having a funnelform (not campanulate) hypanthium, and stipitate glands on the pedicels and hypanthium. The type of *C. Davidsonii* Rydb. came from Blue River, Greenlee County (*Davidson* 754).

2. Cowania subintegra Kearney. Near Burro Creek, southeastern Mohave County, 2,500 feet (*Darrow & Crooks* 3, *Darrow & Benson* 10890, the type collection). Known only from the type locality.

According to Lyman Benson the trunk is 2.5 to 3.5 cm. thick at base but continues only a few centimeters above ground.

15. CERCOCARPUS. Mountain-mahogany

Shrubs or small trees; leaves simple, fascicled, thickish, entire or dentate, linear to obovate, often prominently veined beneath; flowers solitary or in small fascicles, inconspicuous, with small, yellowish sepals and no petals; stamens numerous; pistil 1, the ovary superior; hypanthium sheathlike in fruit, enclosing the slender, villous achene, the long, persistent, plumose style exserted.

The plants are sometimes known locally as deer-browse, and certain species are important elements of the chaparral in central and southern Arizona, useful in protecting the soil against erosion and affording excellent browse for cattle, sheep, and goats, as well as for deer and bighorns. Cases have been reported of hydrocyanic-acid poisoning of animals eating the leaves of *C. montanus*. The Hopi Indians are reported to use the bark of one species to dye leather red-brown. The wood is hard, and that of some species was used by the Indians for making digging sticks and is occasionally used for making tool handles. The sharp-pointed basal end of the achene and the corkscrewlike tail enable it to penetrate the ground, as in *Stipa* and *Erodium*.

KEY TO THE SPECIES

1. Leaves evergreen, coriaceous, resinous, linear or elliptic, acute at both ends, revolute, entire, the lateral veins not very prominent beneath (2).
1. Leaves deciduous, often thickish but scarcely coriaceous, not noticeably resinous, obovate-cuneate or oblanceolate, rounded or truncate at apex, flat or slightly revolute, usually dentate, at least at apex, the lateral veins very prominent beneath (3).
2. Leaf margins slightly to rather strongly revolute but with much of the lower leaf surface exposed; leaves elliptic, 10 to 30 mm. long, 4 to 10 mm. wide; style in fruit 4 to 7 cm. long........1. *C. ledifolius*
2. Leaf margins very strongly revolute, with little of the lower surface exposed except the midvein; leaves narrowly linear, 5 to 15 mm. long, 2 to 3 mm. wide; style in fruit 2.5 to 5 cm. long........2. *C. intricatus*
3. Leaves entire, or toothed only at or very near the apex, commonly oblanceolate or spatulate, thickish........3. *C. breviflorus*
3. Leaves toothed well below the apex, mostly obovate (4).
4. Leaves thickish, finely dentate with triangular teeth, commonly 2 to 3 times as long as wide, pale green or grayish beneath........4. *C. betuloides*
4. Leaves thin, coarsely dentate with ovate teeth, commonly less than twice as long as wide, white or whitish beneath........5. *C. montanus*

1. Cercocarpus ledifolius Nutt. Kaibab Plateau, both rims of Grand Canyon, and Navajo Mountain (Coconino County), 6,000 to 9,500 feet, April to June. Montana to Washington, south to Colorado, northern Arizona, and California.

Curl-leaf mountain-mahogany. H. F. Loomis reported that the plants observed by him were treelike, widely branching, about 4.5 m. high and 0.6 m. in trunk diameter. A height of 6 m. (20 feet) in Grand Canyon National Park has been reported. The Arizona plants belong to var. *intercedens* C. K. Schneid., which is more or less intermediate between *C. ledifolius* var. *typicus* C. K. Schneid. and *C. intricatus*.

2. Cercocarpus intricatus Wats. (*C. ledifolius* var. *intricatus* Jones). Navajo, Coconino, and northeastern Mohave counties, especially in and near the Grand Canyon, 3,000 to 7,000 feet, April to June. Utah, Nevada, northern Arizona, and southern California.

Little-leaf mountain-mahogany. A much-branched shrub, up to about 2.5 m. (8 feet) high. The typical plant (*C. intricatus* var. *typicus* C. K. Schneid.)

with herbage and hypanthium more or less appressed-pubescent, then glabrescent, is rather rare in Arizona. Much more common, especially in the Grand Canyon region, is var. *villosus* C. K. Schneid. (*C. arizonicus* Jones), with young branches villous-tomentose and the upper leaf surface and hypanthium pilose with curly hairs, sometimes permanently so. The type of *C. arizonicus* was collected at Willow Spring, Coconino County (*Jones* in 1890).

3. Cercocarpus breviflorus Gray. Southern Apache, Navajo, Coconino, and Yavapai counties to Cochise and Pima counties, 5,000 to 8,000 feet, common in chaparral on dry slopes and mesas, March to November. Western Texas to Arizona and Mexico.

A shrub, sometimes treelike and up to about 4.5 m. (15 feet) high. Nearly typical *C. breviflorus* has been collected in the Chiricahua and Huachuca mountains (Cochise County) and in the Baboquivari Mountains (Pima County). Much more common in Arizona is var. *eximius* C. K. Schneid. (*C. eximius* Rydb.) which has more spreading pubescence, larger and more distinctly dentate leaves, and usually a longer hypanthium and style.

4. Cercocarpus betuloides Nutt. (*C. Douglasii* Rydb.). Apache County to Mohave County, south to Cochise and Pinal counties, 3,000 to 6,500 feet, mostly in chaparral, March to July. Arizona, Oregon, and California.

Birch-leaf mountain-mahogany. A large shrub or small tree, sometimes attaining a height of about 6 m. (20 feet). Probably the most important of the Arizona species as a browse plant. Intergradations with *C. breviflorus* occur.

5. Cercocarpus montanus Raf. Apache, Coconino, Yavapai, and Gila counties, 4,500 to 7,000 feet, with pinyons and junipers. South Dakota and Montana to Kansas, New Mexico, and Arizona.

Alder-leaf mountain-mahogany. The Arizona plants belong to var. *flabellifolius* (Rydb.) Kearney & Peebles (*C. flabellifolius* Rydb.), which apparently differs from typical *C. montanus* only in the more appressed pubescence of the leaves and hypanthium.

Floyd L. Martin, in his recently published revision of *Cercocarpus* (*Brittonia* 7: 91–111), has reduced all the Arizona plants to varieties of *C. montanus* and of *C. ledifolius*.

16. COLEOGYNE. Black-brush

Small to rather large shrubs with rigid spinescent branches and strigose-pubescent herbage; leaves opposite, crowded, narrowly spatulate, entire; flowers solitary; sepals 4, yellow; petals usually none; stamens numerous; disk at base of the hypanthium with a sheathlike prolongation, this hairy at the throat and enclosing the solitary achene.

1. Coleogyne ramosissima Torr. Apache County to Mohave and Yavapai counties, 3,000 to 6,500 feet, in well-drained, usually gravelly soils on open plains and mesas, sometimes in pure stands to the exclusion of other shrubs, March to May. Southwestern Colorado to northern Arizona and southeastern California.

Sometimes erroneously called burro-brush. The plants are browsed by sheep and goats, to a lesser extent by cattle, and withstand heavy browsing successfully. The plant is normally apetalous, but a specimen collected in Mohave County (*Kearney & Peebles* 11219) had 2 pale-yellow obovate petals opposite to and considerably longer than the outer sepals.

17. PURSHIA. Antelope-brush

Shrubs, intricately branched, erect or sprawling; leaves fascicled, small, tomentose, wedge-shaped, 3-toothed at apex; flowers solitary at the ends of the branchlets; sepals and petals 5, the petals yellow; stamens many; disk none; pistil usually solitary (sometimes 2, rarely 3); style stout, beaklike, persistent; achene enveloped only at base by the turbinate or funnelform hypanthium.

1. Purshia tridentata (Pursh) DC. Apache County to Coconino County, 4,000 to 9,000 feet, open slopes and mesas, and coniferous forests, April to June. Montana to British Columbia, south to New Mexico, northern Arizona, and California.

Also known as bitter-brush. A very important browse plant for sheep and cattle in regions where it is more abundant than in Arizona. The often prostrate stems, rooting where they touch the ground, doubtless give this plant value for control of soil erosion.

18. AGRIMONIA. Agrimony

Plants herbaceous, perennial; stems erect, leafy, usually branched above; leaves alternate, with conspicuous stipules, odd-pinnate, the leaflets alternately large and small, serrate-dentate; flowers numerous, small, in long slender spikelike racemes; hypanthium obconic to hemispheric, with hooked bristles; petals yellow; stamens 5 to 15; pistils 2; achenes 1 or 2.

KEY TO THE SPECIES

1. Bristles of the hypanthium (at least the outer ones) strongly reflexed at maturity; leaflets oblong or ovate-oblong, scarcely acuminate at apex, thin, sparsely pubescent to glabrate beneath .... 1. *A. gryposepala*
1. Bristles all erect, or the outer ones ascending-spreading at maturity; leaflets lanceolate or oblong-lanceolate (seldom broader), acuminate at apex, usually thickish and copiously pubescent beneath .... 2. *A. striata*

1. Agrimonia gryposepala Wallr. Coconino County, at Walnut Canyon (*Leiberg* 5791), and Oak Creek (*Fulton* 7332, *Deaver* 459), about 6,000 feet, August and September, apparently rare in Arizona. Almost throughout North America.

2. Agrimonia striata Michx. Apache County to Coconino County, south to Cochise and Pima counties, 6,500 to 8,500 feet, frequent in rich soil in pine forests, July to October. Nova Scotia to British Columbia, south to West Virginia, New Mexico, and Arizona.

19. SANGUISORBA. Burnet

Plants herbaceous; leaves pinnate, the leaflets coarsely toothed to pinnatifid; flowers perfect or unisexual, small, in dense, cylindric, bracted spikes; sepals 4; petals none; stigma brushlike; ovary superior, 1-celled; hypanthium in fruit 4-angled or -winged, enclosing the achene or achenes.

KEY TO THE SPECIES

1. Leaflets pinnately parted, the segments narrow; stamens 2 or 4; pistil 1; hypanthium 4-winged in fruit, the faces smooth or finely reticulate ....1. *S. annua*
1. Leaflets coarsely toothed; stamens numerous in the staminate flowers; pistils 2; hypanthium strongly 4-angled in fruit, the faces coarsely reticulate or alveolate..2. *S. minor*

1. Sanguisorba annua Nutt. San Carlos Indian Reservation, Gila County (*A. & R. Nelson* 1840), near Superior, Pinal County (*Harrison* 1877), April and May. Kansas, Arkansas, Texas, and Arizona.

2. Sanguisorba minor Scop. West of Glendale, Maricopa County, in an alfalfa field (*Gould* 3509). Introduced here and there in the United States; from Eurasia.

20. ROSA. Rose

Shrubs, with prickly (and often also bristly) stems; leaves alternate, odd-pinnate, with conspicuous stipules adnate to the petioles; flowers large and showy, fragrant, solitary or in few-flowered clusters; petals broad, normally pink; stamens numerous, inserted on an annular disk in the hypanthium, the latter globose or ellipsoid, often constricted below the throat, enclosing, but not adhering to, the numerous pistils, berrylike and usually bright red at maturity.

The plants are browsed. They probably have some value in controlling soil erosion where they form thickets, but this is seldom the case in Arizona. The fruits are much eaten by birds and other animals. Like nearly all species of this genus, those of Arizona are beautiful in flower and fruit.

It is doubtful that all the species here enumerated are valid. Erlanson (137) relegated *R. Fendleri*, *R. neomexicana*, *R. arizonica*, and *R. granulifera* to synonymy under *R. Woodsii* Lindl. and regarded *R. manca* as "a dwarf hexaploid of the southern Rocky Mountain region, closely allied to *R. nutkana*."

KEY TO THE SPECIES

1. Leaflets wedge-shaped, coarsely toothed at and near the apex, commonly less than 1 cm. long, nearly as wide as long; young stems stellate-pubescent, or copiously glandular-hispidulous, or both, the older stems with numerous long, nearly straight prickles, commonly also bristly; flowers solitary, bractless....................1. *R. stellata*
1. Leaflets not wedge-shaped, but often cuneate at base, toothed well below the apex, commonly more than 1 cm. long, distinctly longer than wide; stems not stellate-pubescent; inflorescence commonly of 2 or more flowers and bracteate (2).
2. Prickles of the stem straight or nearly so, normally very slender. Hypanthium globose or nearly so...2. *R. Fendleri*
2. Prickles mostly recurved, normally rather stout (3).
3. Hypanthium ellipsoid, noticeably longer than wide, strongly constricted below the calyx lobes ....3. *R. neomexicana*
3. Hypanthium globose or nearly so, not strongly constricted (4).
4. Leaflets glabrous on both surfaces.....................................4. *R. manca*
4. Leaflets puberulent and more or less granuliferous beneath.............5. *R. arizonica*

1. Rosa stellata Wooton. Both sides of the Grand Canyon (Coconino County) at Dutton Point, Powell Plateau (*Ferriss* in 1908), and Mesa Eremita (*Hawbecker* in 1935), about 6,500 feet, dry, rocky places, June to September. Western Texas, southern New Mexico, and northern Arizona.

In Ferriss's specimen the young twigs are pubescent with soft, stellate hairs and also closely beset with short-stalked glands. Hawbecker's specimen lacks the stellate hairs, having only glandular pubescence, and is therefore technically *R. mirifica* Greene (*R. stellata mirifica* Cockerell), but these specimens are otherwise very similar.

2. Rosa Fendleri Crépin. Apache, Coconino, and Yavapai counties, also in the Chiricahua and Huachuca mountains (Cochise County), 5,500 to 9,000 feet, June to August. Minnesota to British Columbia, south to northern Mexico.

3. Rosa neomexicana Cockerell. White River (Apache ? County), Kaibab Plateau, Grand Canyon, and Pinchot Ranger Station (Coconino County), 6,500 to 7,500 feet, June and July. Colorado, Utah, New Mexico, and Arizona.

The Arizona specimens are scarcely typical, seemingly approaching *R. arizonica.*

***4. Rosa manca** Greene. Given for Arizona in *North American Flora* (22: 518). Specimens collected in Coconino County, at Rainbow Lodge, 6,500 feet (*Peebles & Smith* 13941), and on the San Francisco Peaks, 7,000 to 9,000 feet (*Heiser* 1573, part), have the leaflets glabrous beneath but do not correspond otherwise with Greene's description. Known certainly only from Colorado and Utah.

5. Rosa arizonica Rydb. Apache, Navajo, and Coconino counties, south to Cochise and Pima counties, 4,000 to 9,000 feet, along streams and in pine forests, in partial shade, May to July, type from near Flagstaff (*MacDougal* 110). New Mexico and Arizona.

The most abundant and widely distributed wild rose in Arizona. A form with usually double-serrate leaflets, these copiously granuliferous beneath, and sepals more constantly bearing stipitate glands, is found occasionally in northern and central Arizona but seems to be absent in the southernmost counties. It is var. *granulifera* (Rydb.) Kearney & Peebles (*R. granulifera* Rydb.), the type of which was collected west of Holbrook, Navajo County (*Zuck* in 1896).

21. PRUNUS. Plum, Cherry

Small trees or large shrubs; leaves alternate or fascicled, simple; flowers usually perfect, in racemes or corymbs, or solitary in the leaf axils; calyx free from the ovary, 5-lobed, with a disk at the bottom; petals 5, these and the numerous stamens inserted on the calyx; pistil 1, the ovary 1-celled; fruit a dry or fleshy drupe, with 1 bony seed.

The plants are browsed, but cases have been reported of hydrocyanic-acid poisoning of cattle and sheep. The foliage is usually supposed to be more dangerous when wilted, but this is disputed. The fruits are much eaten by birds and other animals, and preserves are sometimes made from those of the choke-cherries (subgenus *Padus*).

Seedlings of the cultivated peach (*P. persica*) are met with occasionally, growing without cultivation, but the species has not become established as a member of the Arizona flora.

KEY TO THE SPECIES

1. Drupe pubescent, the exocarp almost dry, splitting on one side; leaves entire or sparingly and irregularly dentate, sessile or nearly so, fascicled, linear-spatulate, puberulent. Plants polygamo-dioecious; flowers mostly solitary in the leaf axils, sessile or nearly so; petals 2 to 3 mm. long; plants shrubby, divaricately branched: subgenus *Emplectocladus* .... 1. *P. fasciculata*
1. Drupe glabrous, the exocarp juicy; leaves regularly crenate or serrate, petioled (2).
2. Flowers few, in short, corymbiform inflorescences; pedicels commonly villous; leaves crenate or crenulate, with conspicuously gland-tipped teeth. Petals 4 to 6 mm. long: subgenus *Cerasus*.. 2. *P. emarginata*
2. Flowers numerous, in elongate racemes; pedicels glabrous or puberulent; leaves serrate or serrulate, the teeth not conspicuously glandular, commonly appressed or incurved: subgenus *Padus* (3).
3. Calyx deciduous long before maturity of the fruit; leaves mostly rounded or subcordate at base; petals about 5 mm. long.................................. 3. *P. virginiana*
3. Calyx persistent until maturity of the fruit; leaves mostly acute or acutish at base; petals about 3 mm. long.. 4. *P. virens*

1. Prunus fasciculata (Torr.) Gray (*Emplectocladus fasciculatus* Torr.). Grand Canyon and Havasu Canyon (Coconino County), Mohave County (many localities), near Wickenburg (Maricopa County), Kofa Mountains (Yuma County), 5,000 feet or lower, dry plains and slopes, often forming thickets, March. Southern Utah and Nevada, Arizona, and southern California.

Desert-almond, browsed by sheep and goats.

2. Prunus emarginata (Dougl.) D. Dietr. Coconino County and Hualpai Mountain (Mohave County) to Cochise and Pima counties, 5,000 to 9,000 feet, mostly in pine forests, April to June. Montana to British Columbia, south to New Mexico, Arizona, and California.

Bitter cherry. In Arizona there occur both the typical plant, with leaves oval or obovate, mostly obtuse or rounded at apex, sometimes sparsely pubescent beneath, and var. *crenulata* (Greene) Kearney & Peebles (*Cerasus crenulata* Greene), with leaves elliptic or oblanceolate, mostly acute or acutish at apex, commonly glabrous beneath. The variety has been collected in the White Mountains (Apache County), on the San Francisco Peaks and Bill Williams Mountain (Coconino County), and in the Pinaleno Mountains (Graham County).

3. Prunus virginiana L. Apache, Navajo, and Coconino counties, south to Greenlee and Gila counties, 4,500 to 8,000 feet, coniferous forests, April to June. Canada to Georgia, New Mexico, Arizona, and California.

Common choke-cherry. The tree reaches a height of 7.5 m. (25 feet) and a trunk diameter of 20 cm. Represented in Arizona by var. *demissa* (Nutt.) Torr. (*Prunus demissa* D. Dietr.), the Western choke-cherry, with twigs and lower leaf surface pubescent when young, the mature fruit dark red, and by var. *melanocarpa* (A. Nels.) Sarg. (*Prunus melanocarpa* Rydb.), the black Western choke-cherry, with twigs and lower leaf surface glabrous or nearly so, the mature fruit nearly black.

4. Prunus virens (Woot. & Standl.) Shreve (*Padus virens* Woot. & Standl.). Apache County to Mohave County, south to Cochise, Santa Cruz, and Pima counties, 4,500 to 6,000 feet, common, usually along streams, April and May. Western Texas to Arizona and northern Mexico.

Southwestern choke-cherry. Usually a large shrub, up to 6 m. (20 feet) high, sometimes arborescent, with semievergreen foliage, handsome in flower. The typical, glabrous plant is much more common in Arizona than var. *rufula* (Woot. & Standl.) Sarg. (*Padus rufula* Woot. & Standl.), the Gila choke-cherry, which has the young twigs, petioles, and midribs pubescent with more or less tawny pubescence.

64. LEGUMINOSAE. Pea Family

Herbs, shrubs, or small trees; leaves alternate, mostly compound (occasionally reduced to 1 leaflet, simple in 2 genera); flowers mostly perfect, commonly irregular (except in *Mimosoideae*); petals distinct or partly united (especially the 2 lowest or keel petals in papilionaceous flowers), commonly 5, rarely only 1 or none; stamens few or many, frequently 10, distinct, or more often with the filaments variously united; pistil 1, the ovary superior, usually 1-celled; fruit commonly a dehiscent, 2-valved pod (legume) or a loment with several indehiscent segments, or entire and indehiscent in a few genera.

This huge family includes many plants used as human food, such as beans and peas, many forage plants of first importance, such as clovers and alfalfa, and valuable timber trees, especially in the tropics. Some members (logwood, indigo) are (or were) important commercially as dye plants. Species of several tropical genera are used locally as fish poisons and contain rotenone, a substance now finding wide application as an insecticide. In Arizona the native Leguminosae are of great value as range forage and as soil-binding plants, useful for controlling erosion. Several species of the genera *Astragalus* and *Oxytropis*, the so-called loco-weeds, and certain species of *Astragalus* that prefer soils rich in selenium are the cause of fatal poisoning in domestic animals.

KEY TO THE SUBFAMILIES

1. Corolla valvate in bud, regular or very nearly so. Leaves bipinnate, the leaflets usually small; flowers small, in many-flowered heads, spikes, or spikelike racemes; calyx 4- or 5-lobed; petals 4 or 5, small and inconspicuous; stamens conspicuously exserted, the filaments distinct or united below . . . . 1. *Mimosoideae*
1. Corolla imbricate in bud, more or less irregular, the petals unlike in size or shape, or both (2).
2. Uppermost petal internal in bud, enveloped by the lateral ones, the corolla therefore not papilionaceous, moderately to strongly irregular . . . . 2. *Caesalpinioideae*
2. Uppermost (odd or banner) petal external in bud, the corolla commonly very irregular (papilionaceous), with the 2 lowest (keel) petals often more or less united. Petals commonly 5 (1 in *Amorpha*, none in *Parryella*); pods mostly dehiscent (indehiscent in a few genera), commonly 1-celled but sometimes 2-celled . . . . 3. *Papilionoideae*

1. MIMOSOIDEAE. Mimosa Subfamily

KEY TO THE GENERA

1. Filaments united below, more than 1 cm. long. Stamens very numerous; plants unarmed; flowers in heads (2).
1. Filaments distinct to the base or very nearly so, less than 1 cm. long (3).
2. Plants undershrubs, or herbaceous above ground; pods less than 10 mm. wide, elastically dehiscent from the apex, the valves not separating from the margins; seeds not more than 6 mm. long . . . . 1. *Calliandra*

2. Plants large shrubs; pods 15 to 30 mm. wide, not elastically dehiscent from the apex, the valves finally separating from the margins; seeds 8 to 10 mm. long..2. *Lysiloma*
3. Stamens many more than 10. Plants suffrutescent to large shrubs or small trees; flowers in heads or spikes....................3. *Acacia*
3. Stamens not more than 10 (4).
4. Anthers gland-tipped. Pods indehiscent, flat or spirally coiled; plants large shrubs or small trees; flowers in spikes....................6. *Prosopis*
4. Anthers not gland-tipped (5).
5. Valves of the pod finally separating from the persistent margins, often in sections; plants shrubby; stems (and often the pods) prickly; flowers in round heads or spike-like racemes....................4. *Mimosa*
5. Valves of the pod not separating from the margins; plants suffrutescent; stems and pods unarmed; flowers in round heads....................5. *Desmanthus*

1. CALLIANDRA. False-mesquite

Plants perennial, herbs or low shrubs, not prickly; flowers in heads, pink or white; stamens many, long and conspicuous; pods with thick riblike margins, the valves recurved after dehiscence.

These plants, especially *C. eriophylla,* are also known as mesquitilla and fairy-duster. All the Arizona species of *Calliandra* are valuable as soil binders and are palatable to livestock, but *C. eriophylla,* because of its relatively large size, resistance to browsing, and abundance, far exceeds the other species in economic value and is highly palatable to deer.

KEY TO THE SPECIES

1. Stems herbaceous above ground, rarely more than 20 cm. long (2).
1. Stems woody above ground, commonly more than 20 cm. long (3).
2. Plants glabrous or sparsely pilose; pinnae 1 to 5 pairs; leaflets not more (usually fewer) than 12 pairs, 7 to 14 mm. long, 3 to 6 mm. wide, obliquely oval or oblong, conspicuously veined....................1. *C. reticulata*
2. Plants pilose, usually copiously so; pinnae 4 to 9 pairs; leaflets 10 to 18 pairs, rarely more than 5 mm. long, 1 to 2 mm. wide, linear or narrowly oblong, seldom conspicuously veined....................2. *C. humilis*
3. Branches slender, not rigid, erect or ascending at a narrow angle, moderately woody; leaflets few (not more than 7 pairs), not imbricate, spreading, thin, bright green, glabrous or pubescent, 5 to 12 mm. long, 2 to 5 mm. wide; flowers glabrous; stamens not more than 15 mm. long; pods glabrous or inconspicuously puberulent..3. *C. Schottii*
3. Branches stout, rigid, often divaricate, very woody; leaflets numerous (usually more than 7 pairs), imbricate, ascending, rather thick, grayish-pubescent, not more than 7 mm. long, 1 to 2 mm. wide; flowers pubescent; stamens about 20 mm. long; pods densely and conspicuously pubescent....................4. *C. eriophylla*

1. Calliandra reticulata Gray (*C. humilis* var. *reticulata* L. Benson). Southern Navajo and Coconino counties to Greenlee, Cochise, Santa Cruz, and Pima counties, 4,500 to 8,000 feet, commonly in dry pine forests, May to September. New Mexico, Arizona, and Mexico.

2. Calliandra humilis Benth. Navajo and Coconino counties to Greenlee, Cochise, Santa Cruz, and Pima counties, 4,000 to 9,000 feet, dry soil among oaks or pines, June to August. Western Texas to Arizona and northern Mexico.

More common than *C. reticulata,* especially in central and northern Arizona. Intergrades with that species, but most of the specimens are readily distinguishable.

3. Calliandra Schottii Torr. Chiricahua Mountains (Cochise County), Santa Catalina and Baboquivari mountains (Pima County), 3,000 to 5,000 feet, rocky slopes, ascending to the pinyon belt, August. Southern Arizona and northern Mexico.

Stems much-branched, up to 1.2 m. high. It is reported that the root system is extraordinarily well developed, and that the stems are often frozen back but the plants soon recover.

4. Calliandra eriophylla Benth. Greenlee County to southern Mohave County, south to Cochise, Santa Cruz, Pima, and Yuma counties, 5,000 feet or lower, very common on dry, gravelly slopes and mesas, February to May. Western Texas to southeastern California and Mexico.

A straggling shrub, up to 1.2 m. high.

2. LYSILOMA

Large, spreading, unarmed shrubs, with dense foliage; leaves large, with numerous pinnae and many leaflets; flowers in heads, white; stamens very numerous; pods large, flat.

1. Lysiloma Thornberi Britt. & Rose. Pima County, on rock ledges along Chimenea (Chimney) Creek in the foothills of the Rincon Mountains, about 3,500 feet, May and June, type collected by Thornber in 1926. Known only from the type locality.

A very rare and handsome shrub with hard, brittle, dark-brown wood and feathery canopied foliage, well worth cultivating as an ornamental. As seen by the writers at Chimenea Creek, it has the appearance of being very long-lived and occasionally frozen back.

L. Thornberi is doubtfully distinct from *L. microphylla* Benth., to which species it was referred by Benson and Darrow (8, p. 156). There is considerable variation in pubescence in Arizona specimens, the leaflets on some plants being nearly glabrous, on others pubescent on both surfaces.

3. ACACIA (138, 139)

Plants shrubs, small trees, or sometimes nearly herbaceous, spiny or unarmed; flowers whitish to bright yellow; stamens numerous, the anthers minute, barely 0.1 mm. long.

Several of the Arizona species, especially the nearly herbaceous *A. angustissima*, are browsed by cattle and horses, and the pods are eaten with relish. Many exotic, arborescent species are in cultivation as ornamentals in the warmer parts of the United States. Gum arabic and tanbark are among the useful products obtained from various species of this large genus.

KEY TO THE SPECIES

1. Inflorescence racemose or spicate. Corolla and stamens ochroleucous or cream-colored; pods flat or compressed, thin-walled, oblong, 15 to 20 mm. wide; small trees or large shrubs: section *Senegalia* (2).

1. Inflorescence capitate (3).

2. Branches unarmed or bearing slender, straight spines; pinnae 5 to 12 pairs; leaflets very numerous, linear, cuspidate, usually glabrous or merely ciliolate; pods very flat, not twisted or torulose..1. *A. millefolia*

2. Branches usually armed with stout, curved spines; pinnae 1 to 3 pairs; leaflets 2 to 7 pairs, oblong or obovate, obtuse or retuse, pilose, rarely glabrous; pods not very flat, often somewhat twisted and torulose............................2. *A. Greggii*
3. Flowers distinctly pedicelled; calyx undulate or very shallowly dentate; corolla and stamens white or whitish; pods rather promptly dehiscent; stems woody below or almost entirely herbaceous above the crown, unarmed, commonly pilose or hirsute. Leaflets glabrous or sparsely ciliate, seldom pilose; pods flat, thin-walled, not or scarcely torulose: section *Acaciella*......................................3. *A. angustissima*
3. Flowers sessile or nearly so; calyx deeply dentate; corolla and stamens bright yellow; pods indehiscent or tardily dehiscent; shrubs or small trees, usually armed with straight or nearly straight, commonly slender spines (4).
4. Pods turgid, woody, somewhat curved, not conspicuously torulose, less than 10 times as long as wide; leaflets 10 to 25 pairs: section *Vachellia*............4. *A. Farnesiana*
4. Pods compressed, not woody, strongly torulose, more than 10 times as long as wide; leaflets 6 to 12 pairs: section *Acaciopsis* (5).
5. Plants not, or scarcely, viscid, usually pilose; pinnae seldom fewer and often more than 3 pairs; bracts and calyx teeth ciliate; pods 6 to 12 cm. long......5. *A. constricta*
5. Plants very viscid, glabrous or nearly so; pinnae 1 to 3 pairs; bracts and calyx teeth gland-tipped, not ciliate; pods usually not more than 6 cm. long......6. *A. vernicosa*

1. Acacia millefolia Wats. Cochise and Pima counties, 3,500 to 5,000 feet, July and August, apparently rare. Southern Arizona and northern Mexico.

2. Acacia Greggii Gray. Grand Canyon and Havasu Canyon (Coconino County), and Mohave County to Greenlee, Cochise, Pima, and Yuma counties, 4,500 feet or lower, often forming thickets along streams and washes, April to October. Texas to southern Nevada, Arizona, southeastern California, and northern Mexico.

Catclaw acacia, devils-claw. A common, often abundant, large shrub or small tree, reaching a height of 6 m. (20 feet) and a stem diameter of 20 cm., the sapwood yellow, the heartwood reddish brown. The new foliage is relished by cattle in early spring, otherwise catclaw is valuable chiefly as a reserve food in times of drought or on depleted ranges. The Arizona Indians made meal of the pods, using it as mush and cakes. The flowers are one of the most important sources of honey for bees kept on the desert. The wood is very strong and is used locally for making doubletrees and singletrees as well as for firewood. This is probably the most heartily disliked plant in the state, the sharp, strong prickles tearing the clothes and lacerating the flesh.

3. Acacia angustissima (Mill.) Kuntze. Greenlee County to eastern Yavapai County, south to Cochise, Santa Cruz, and Pima counties, 3,000 to 6,500 feet, dry, rocky slopes, usually in chaparral, May to September. Missouri to Texas and Arizona, southward to Guatemala; southern Florida.

White-ball acacia, a very handsome plant, with feathery foliage and panicled round heads of cream-colored flowers, easily cultivated. The habit of growth is well adapted to protecting soil against erosion. The stems reach a height of about 1.5 m.

A widely distributed species, extremely variable in stature, pubescence, number of pinnae and of leaflets, and venation of the leaflets. The commonest phase in Arizona is var. *hirta* (Nutt.) Robins. (*A. hirta* Nutt., *A. suffrutescens* Rose, *A. angustissima* subsp. *suffrutescens* Wiggins). Certain Arizona collections were referred by Wiggins (138, p. 234) and Benson (139) to *A. cuspidata* Schlecht. (*A. texensis* Torr. & Gray, *A. angustissima* var. *cuspidata*

L. Benson), characterized by fewer pinnae and leaflets; but in Arizona material there seems to be complete intergradation in these characters with *A. angustissima* var. *hirta*. A more distinct entity is *A. Lemmoni* Rose (*A. angustissima* subsp. *Lemmonii* Wiggins), known only from the Huachuca Mountains (type *Lemmon* in 1882). This differs from most specimens of var. *hirta* in its larger, acute, strongly pinnately veined leaflets, but the plant described as *Acaciella Shrevei* Britt. & Rose, also from the Huachuca Mountains (type *Shreve* 5064), is nearly intermediate. Another extreme variation occurs in the Baboquivari Mountains (*Jones* 25024, part, *Goodding* 4321). This has the pods up to 15 mm. wide, and broad and rounded at apex, whereas in most specimens of var. *hirta* (including another from the Baboquivari Mountains numbered by Jones 25024) the pods are less than 10 mm. wide.

4. Acacia Farnesiana (L.) Willd. Near Tucson and in canyons on the west slope of the Baboquivari Mountains (Pima County), reported as occurring also near Ruby (Santa Cruz County), 2,500 to 4,000 feet, April to November. Florida to southern California, southward to Argentina.

Sweet acacia, huisache. Extensively cultivated as an ornamental because of the exquisite fragrance of the flowers, which are used in France in the manufacture of perfumery. In Arizona a small tree up to 6 m. (20 feet) high, rare.

5. Acacia constricta Benth. Greenlee, Gila, and Yavapai counties to Cochise, Pima, and Yuma counties, 2,500 to 5,000 feet, shallow caliche soil on dry slopes and mesas, May to August (occasionally November). Texas to Arizona and Mexico.

Mescat acacia, white-thorn, abundant over large areas in southeastern Arizona. A pretty shrub, especially when covered with the small, orange-yellow balls of very fragrant flowers. The foliage is not palatable to livestock, although the pods are eaten. The plants are mostly well armed with long, slender, straight, white spines, but var. *paucispina* Woot. & Standl., with few or no spines, is occasionally found.

6. Acacia vernicosa Standl. (*A. constricta* var. *vernicosa* L. Benson). Cochise and eastern Pima counties, 3,500 to 5,000 feet, usually on limestone, June to September. Western Texas to southeastern Arizona and northern Mexico.

A species generally similar to *A. constricta* but readily distinguished by the characters given in the key.

4. MIMOSA

Plants shrubby, usually armed; leaves with numerous small leaflets; flowers small, sessile, whitish or pink, in many-flowered heads or spikes; stamens not more than 10; anthers 0.3 to 0.4 mm. long; fruits with or without prickles, in some species breaking up into 1-seeded sections.

These plants are mostly very attractive in flower, and the flowers are commonly fragrant.

KEY TO THE SPECIES

1. Inflorescence elongate, spicate or racemose; flowers purplish pink; valves of the fruits jointed, the segments separating at maturity (2).
1. Inflorescence capitate; flowers pale pink or whitish; valves of the fruits not jointed, usually prickly-margined (3).

2. Young stems, leaves, and flowers glabrous or inconspicuously pubescent; spines few or none; pinnae 1 to 4 pairs; leaflets 2 to 5 pairs, elliptic to obovate, rounded and often mucronulate at apex, 5 to 10 mm. long, at least half as wide; flowers short-pedicelled; fruits very flat, thin-walled, glabrous, unarmed, not or scarcely torulose
1. *M. laxiflora*
2. Young stems, leaves, and flowers sericeous or villous; spines numerous; pinnae 5 or more pairs; leaflets 7 or more pairs, linear-lanceolate or narrowly oblong (the terminal one sometimes obovate), acute or acutish at apex, 3 to 6 mm. long, less than half as wide; flowers sessile; fruits not very flat, thickish-walled, tomentose, strongly torulose ....2. *M. dysocarpa*
3. Spines of the stem stout, conspicuously flattened and broadened to far above the base, strongly recurved toward the apex; leaves, including the petiole, 1.5 to 4 cm. long, the pinnae crowded, the internodes of the rachis seldom more than 5 mm. long; leaflets 2 to 3.5 mm. long, not evidently pinnate-veined; fruits 3 to 4 mm. wide, acute or acuminate at apex, more or less curved. Calyx and corolla pubescent
3. *M. biuncifera*
3. Spines of the stem relatively slender, conspicuously broadened only near the base, often straight or nearly so; leaves, including the petiole, 5 to 10 cm. long, the pinnae not crowded, the internodes of the rachis commonly at least 6 mm. long; leaflets 3 to 6 mm. long, distinctly pinnate-veined; fruits 5 mm. wide or wider, rounded and often apiculate at apex, straight or nearly so........................4. *M. Grahami*

1. Mimosa laxiflora Benth. Known in Arizona only from the Papago Indian Reservation and Ajo Mountains, western Pima County, about 2,500 feet, rocky slopes and along washes, locally abundant with *Acacia Greggii, Cercidium, Simmondsia,* and *Carnegiea,* August. Southern Arizona and northern Mexico.

2. Mimosa dysocarpa Benth. Cochise, Santa Cruz, and Pima counties, 3,500 to 6,500 feet, commonly along arroyos and washes, May to September. Western Texas to southern Arizona and northern Mexico.

Arizona's showiest and handsomest *Mimosa,* with long spikes of purplish-pink flowers. Occupying the same region and commoner than the typical plant is var. *Wrightii* (Gray) Kearney & Peebles (*M. Wrightii* Gray), type from the Sonoita (*Wright* 1041). This has the pods normally somewhat narrower and unarmed (sparsely prickly in typical *M. dysocarpa*) and leaflets glabrous or glabrate above, but the correlation between these characters is low.

3. Mimosa biuncifera Benth. Southern Apache County to Hualpai Mountain (Mohave County), south to Cochise, Santa Cruz, and Pima counties, 3,000 to 6,000 feet, dry soil on mesas and rocky slopes, commonly in chaparral, May to August. Western Texas to Arizona and northern Mexico.

Wait-a-bit, wait-a-minute, often also called catclaw, but this name belongs properly to *Acacia Greggii.* A common, straggling shrub in south-central Arizona, reaching, exceptionally, a height of 2.4 m. (8 feet). It often forms dense thickets of considerable extent, making a very efficient soil binder. When other food is scarce, it furnishes forage for livestock. It is reputed to be a good honey plant.

Var. *glabrescens* Gray, much less pubescent than the typical plant, occurs throughout the range of the species in Arizona, type from the Sonoita (*Wright* 1039).

4. Mimosa Grahami Gray. Cochise, Santa Cruz, and Pima counties, 4,000 to 6,000 feet, dry slopes, April to August, type from "between the San Pedro and the Sonoita" (*Wright* 1042). Southwestern New Mexico, southeastern Arizona, and northern Mexico.

About equally common in Arizona are the typical plant and var. *Lemmoni* (Gray) Kearney & Peebles (*Mimosa Lemmoni* Gray), which seems to differ only in the copious pubescence of the herbage, flowers, and pods (these glabrous or sparsely pubescent in typical *M. Grahami*). Numerous specimens are intermediate. The type of *M. Lemmoni* was collected in the Huachuca Mountains (*Lemmon* 2692).

5. DESMANTHUS

Plants herbaceous or shrubby, unarmed; leaflets very numerous, small; flowers sessile, in axillary heads, whitish; stamens 10; pods slender, elongate.

KEY TO THE SPECIES

1. Stems decumbent or spreading; stipules 1 to 2 mm. long, sometimes wanting; leaves 3 to 5 cm. long, the leaflets ciliate, at least when young; peduncle not more than 3 cm. long; pods few or solitary..1. *D. Cooleyi*
1. Stems erect or ascending; stipules commonly more than 2 mm. long; leaves 4 to 10 cm. long, the leaflets glabrous; peduncle 2 to 5 cm. long; pods usually several or numerous (2).
2. Pinnae 1 to 7 pairs; petiolar gland oblong or elliptic, usually borne just below the lowest pair of pinnae and not subtended by stipels........................2. *D. virgatus*
2. Pinnae 7 to 13 pairs; petiolar gland suborbicular, borne near the base of the petiole, subtended by a pair of subulate stipels........................3. *D. bicornutus*

1. Desmanthus Cooleyi (Eaton) Trel. Apache, Navajo, and Coconino counties, south to Cochise, Santa Cruz, and Pima counties, 3,500 to 7,500 feet, dry slopes, mesas, and plains, May to September. Nebraska to Arizona and northern Mexico.

2. Desmanthus virgatus (L.) Willd. Guadalupe Canyon, Cochise County, about 4,000 feet (*Goodding* 8330, A9734), Baboquivari Mountains, Pima County (*Gilman* 136). Widely distributed in tropical and subtropical America.

Gilman No. 136 is the type collection of *D. Covillei* (Britt. & Rose) Wiggins var. *arizonicus* B. L. Turner (139a, p. 129).

3. Desmanthus bicornutus Wats. Near Ruby, Santa Cruz County, about 4,000 feet (*Goodding* in 1938), August and September. Southeastern Arizona, Sonora, and Chihuahua.

6. PROSOPIS (140, 141). MESQUITE

Shrubs or small trees, usually armed with straight spines; leaves with 2 to 4 pinnae, and numerous narrow leaflets; flowers in cylindric spikes, small, greenish yellow, somewhat fragrant; fruits indehiscent, compressed but somewhat turgid, or spirally coiled.

KEY TO THE SPECIES

1. Fruits not coiled, compressed, more or less constricted between the seeds, much more than 4 cm. long; leaflets commonly many more than 9 pairs; spines, if any, yellowish, often stout..1. *P. juliflora*
1. Fruits tightly spirally coiled, 2 to 4 cm. long; leaflets 5 to 9 pairs; spines white, slender. Herbage sparsely to copiously grayish-pubescent..................2. *P. pubescens*

1. Prosopis juliflora (Swartz) DC. Coconino County (Little Colorado River, bottom of the Grand Canyon) to Mohave County, southward to Cochise, Santa Cruz, Pima, and Yuma counties, 5,000 (exceptionally 6,000) feet or lower, very common, chiefly along streams and where the water table is relatively high, March to August. Southern Kansas to southeastern California and Mexico; West Indies.

Two intergrading varieties occur in Arizona: (1) var. *velutina* (Woot.) Sarg. (*P. velutina* Woot.), the more common one, with pubescent foliage and leaflets less than 15 mm. long; (2) var. *Torreyana* L. Benson, with glabrous or glabrate foliage and leaflets commonly more than 15 mm. long. Benson (141) mentioned certain specimens from Cochise County as possibly referrable to typical *P. juliflora*. Burkart (140) apparently excluded from *P. juliflora* the plants of the southwestern United States, referring them, at least as to var. *velutina*, to *P. articulata* Wats.

Common mesquite, known also as honey mesquite. A large shrub or small tree along watercourses, reaching a height of 9 m. (30 feet) or more and a trunk diameter of nearly 1 m. Scattered as a smaller shrub on grasslands and lower mountain slopes with much of the trunk underground. It is reported that the roots sometimes penetrate to a depth of 60 feet. The foliage, and particularly the pods, are eaten by livestock. The sapwood is yellow, the heavy, reddish-brown heartwood hard and slow-burning. With the exception of *Olneya* mesquite is the best firewood obtainable in the semidesert region. Trees cut to the ground sprout again. The wood is used for fence posts, and the heartwood is said to take a fine polish. Mesquite increases rapidly on overgrazed grassland in southeastern Arizona and is considered a serious range pest under such circumstances.

This plant has been a mainstay of existence to the aborigines of the Southwest. When cultivated crops failed, the Indians subsisted mainly upon mesquite beans. Pinole, a meal made from the long, sweet pods, prepared in the form of cakes and in other ways, was a staple food with the Pimas and still is eaten by them to some extent. Fermented pinole was a favorite intoxicating drink. The gum that exudes from the bark was used to make candy, to mend pottery, and as a black dye. The inner bark furnished the Indians material for basketry and coarse fabrics, as well as medicine to treat a variety of disorders. Under normal conditions large quantities of excellent honey are obtained from the flowers of mesquite, which is rated by beemen as the most valuable honey plant of the state. The pollen is reported to be responsible for cases of hay fever.

2. Prosopis pubescens Benth. (*P. odorata* Torr. & Frém., *Strombocarpa pubescens* Gray). Mohave County to Cochise, Pima, and Yuma counties, 4,000 feet or (usually) lower, flood plains, often in saline soil, May (and doubtless later). Western Texas to southern Nevada, southern California, and northern Mexico.

Screwbean, also known as screwpod mesquite, and tornillo. A shrub or small tree up to 6 m. (20 feet) high. Not generally so abundant as the true mesquite, but the sweet pods were used by the Indians in much the same way. Livestock browse on the leaves and the pods. The bark of the roots was used by the Pima Indians to treat wounds. Fence posts and handles for tools are obtained locally from the screwbean.

2. CAESALPINIOIDEAE. Senna Subfamily

KEY TO THE GENERA

1. Corolla pink or purple; leaves simple, entire (2).
1. Corolla yellow; leaves compound (3).
2. Fruits indehiscent, turgid, globose or ovoid, spiny; plants small straggling shrubs or perennial herbs; leaves small, linear or narrowly lanceolate; flowers borne on the branchlets .... 7. *Krameria*
2. Fruits dehiscent, flat, oblong, not spiny; plants large shrubs or small trees; leaves large, round-cordate; flowers borne on the old wood .... 8. *Cercis*
3. Leaves once-pinnate. Plants herbaceous or shrubby .... 9. *Cassia*
3. Leaves twice-pinnate, sometimes appearing simply pinnate because of the shortening of the primary rachis (4).
4. Filaments 3 or 4 times as long as the petals, bright red, very conspicuous. Plants shrubby; pods 5 cm. long or longer, broad, flat .... 13. *Caesalpinia*
4. Filaments not or but slightly surpassing the petals (5).
5. Plants unarmed, herbaceous or somewhat woody; pods not torulose ... 12. *Hoffmanseggia*
5. Plants spiny, large shrubs or small trees; pods often torulose. Young bark green (6).
6. Rachis of the pinnae flattened, 10 cm. long or longer; leaflets alternate; inflorescence an elongate raceme up to 20 cm. long; calyx lobes strongly imbricate in bud 10. *Parkinsonia*
6. Rachis of the pinnae terete, not more than 4 cm. long; leaflets opposite; inflorescence a short raceme or a corymb; calyx lobes valvate or induplicate-valvate in bud 11. *Cercidium*

7. KRAMERIA. Ratany

Straggling low shrubs or perennial herbs, the herbage grayish-pubescent and sometimes glandular; leaves alternate, simple, without stipules; flowers very irregular, purplish, in racemes or solitary and axillary; petals 5, smaller than the sepals, the upper 3 petals long-clawed, the others reduced to nearly orbicular fleshy glands; stamens 3 or 4, more or less united; fruits globose or nearly so, indehiscent, thick-walled, spiny, 1-seeded.

These plants, sometimes called chacate, are supposed to be root parasites. With the exception, perhaps, of *K. lanceolata,* they are relished by livestock. *K. Grayi,* sometimes called crimson-beak, is very drought-resistant. The powerfully astringent rhatany root of commerce is obtained from a South American species. The Papago Indians treat sore eyes with an infusion of the twigs of *K. parvifolia,* and a dye used to color wool and other materials is obtained from the roots. All the species grow on dry plains and mesas. The genus is peculiar in this family in the structure of the flowers and fruits.

KEY TO THE SPECIES

1. Stems herbaceous above ground, from a thick, woody caudex, prostrate or nearly so; spines of the fruit not barbed. Leaves and flowers more or less secund; upper petals with broad blades, more or less united below; stipitate glands none .... 1. *K. lanceolata*
1. Stems woody, the plants straggling shrubs; spines of the fruit rarely without barbs (2).
2. Herbage densely sericeous-tomentose, the pubescence not closely appressed, soft; branchlets spinescent; upper petals narrow, distinct to the base; spines of the fruit with barbs in one series at the apex of the spine; stipitate glands none .... 2. *K. Grayi*
2. Herbage densely to sparsely strigose, the pubescence rather harsh; branchlets rigid but scarcely spinescent; upper petals broad, more or less united toward the base; spines of the fruit with scattered barbs below the apex, rarely barbless; stipitate glands present or absent .... 3. *K. parvifolia*

1. Krameria lanceolata Torr. Cochise, Santa Cruz, and Pima counties, 2,500 to 5,000 feet, May to August. Kansas and Arkansas to southeastern Arizona and Mexico.

2. Krameria Grayi Rose & Painter. Mohave, Maricopa, Pinal, Pima, and Yuma counties, 4,000 feet or (usually) lower, April to September. Western Texas to southern Nevada, Arizona, southern California, and northern Mexico. White ratany.

3. Krameria parvifolia Benth. Coconino County (bottom of the Grand Canyon) and Mohave County to Graham, Cochise, Santa Cruz, Pima, and Yuma counties, 5,000 feet or lower, April to October. Texas to southern Nevada, Arizona, southern California, and northern Mexico.

Range ratany. The species occurs in Arizona in 2 nearly equally common phases: (1) var. *glandulosa* (Rose & Painter) Macbr. (*K. glandulosa* Rose & Painter), with stipitate glands on the pedicels and often elsewhere; (2) var. *imparata* Macbr. (*K. imparata* Britton), without glands.

8. CERCIS (142). Redbud

Large shrubs or small trees, the herbage glabrous; leaves large, simple, round-cordate, entire; flowers appearing before the leaves, in scattered fascicles on the old wood, irregular, purplish red; stamens 10, distinct; pods flat, dehiscent, thin-walled, several-seeded.

Several species very like that of Arizona are grown as ornamentals. The astringent bark of *C. canadensis* has been used as a remedy for diarrhea and dysentery.

1. Cercis occidentalis Torr. Near Kayenta (Navajo County), Grand Canyon and Havasu Canyon (Coconino County), Pagumpa Springs (Mohave County), Superstition Mountains (Pinal County), 4,000 to 4,500 feet, March and April. Southern Utah and Nevada, Arizona, and California.

In Arizona the stems reach a height of 3.5 m. (12 feet) and have smooth gray bark.

9. CASSIA. Senna

Plants annual or perennial, herbaceous or shrubby; leaves pinnate, the leaflets few or numerous; flowers moderately irregular, in racemes or panicles, or solitary in the leaf axils, the petals yellow; stamens 5 or 10, often unequal, some of them sometimes sterile; fruits dehiscent or indehiscent, flat or turgid, the walls thin, or thick and almost woody.

A very large genus, chiefly tropical and subtropical, of which a few species are cultivated as ornamentals in the warmer parts of the United States. The cathartic drug senna is obtained from certain Old World species.

KEY TO THE SPECIES

1. Fruits indehiscent or tardily and not elastically dehiscent; stems usually 1 m. long or longer (2).
1. Fruits more or less promptly and elastically dehiscent; stems usually much less than 1 m. long. Plants herbaceous or barely suffrutescent (4).
2. Stipules none; leaflets caducous; fruits turgid, not more than 4 cm. long. Plants shrubby; leaf rachis spinescent; leaflets few and distant, less than 1 cm. long, thickish, oval or oblong-ovate; flowers in loose terminal panicles.................... 1. *C. armata*

2. Stipules present; leaflets persistent; fruits compressed, 6 cm. or longer (3).
3. Plants shrubby, with rigid branches; petiole not gland-bearing; leaflets 2 or 3 pairs, less than 1 cm. long, oval or obovate; inflorescences few-flowered, axillary and terminal; fruits up to 12 cm. long, black and shiny at maturity..........2. *C. Wislizeni*
3. Plants herbaceous or at most suffrutescent; petiole bearing a large gland near the base; leaflets 4 to 8 pairs, 2 cm. or longer, lanceolate or oblong-lanceolate; inflorescences usually many-flowered, terminal; fruits up to 25 cm. long, brown and not shiny at maturity..3. *C. leptocarpa*
4. Rachis of the leaves normally eglandular; petals not conspicuously veined (5).
4. Rachis usually bearing 1 or more subulate glands, these situated between the opposite leaflets; petals conspicuously veined. Leaflets oblong, elliptic, or obovate, 6 mm. wide, or wider; plants perennial; pubescence of the herbage and pods soft; seeds gray or olive brown, dull, rugose (7).
5. Leaflets 2 pairs, 6 mm. wide, or wider, obovate or oval, thin, ciliate; petioles normally eglandular; petals nearly equal, about 5 mm. long; pods sparsely long-hirsute; seeds obovate, smooth, black and shiny. Plants annual; stems glandular-hirsute..4. *C. Absus*
5. Leaflets 6 or more pairs, not more than 3 mm. wide, narrowly oblong or oblanceolate, mucronate or cuspidate; petioles bearing a cup-shaped, usually stipitate gland; petals conspicuously unequal; pods appressed-pubescent; seeds not black and shiny (6).
6. Plants perennial; stems decumbent or prostrate, glabrous or nearly so; leaflets 6 to 9 pairs, glabrous; pedicels seldom less than 10 mm. long; longest petal not less than 12 mm. long; seeds obovoid, smooth.................................5. *C. Wrightii*
6. Plants annual; stems erect or ascending, pubescent with short, subappressed and longer, spreading hairs; leaflets usually more than 9 pairs, long-ciliate; pedicels seldom more than 5 mm. long; longest petal 5 to 7 mm. long; seeds irregularly triangular or quadrangular, rugose...6. *C. leptadenia*
7. Leaflets 1 pair; flowers all axillary, solitary or in twos on peduncles surpassing the petioles. Pubescence all appressed or partly spreading; pods obliquely cuspidate 7. *C. bauhinioides*
7. Leaflets 2 or more pairs; flowers in terminal, leafy panicles, as well as in axillary clusters (8).
8. Leaflets 2 or 3 pairs; pubescence usually closely appressed; anthers orange; pods 2 to 3 cm. long, with the cusp central, erect or nearly so, 2 to 4 mm. long......8. *C. Covesii*
8. Leaflets 4 to 8 pairs; pubescence not closely appressed; anthers red; pods 3.5 to 6 cm. long, with the cusp usually somewhat lateral and ascending, 1 to 2 mm. long 9. *C. Lindheimeriana*

1. Cassia armata Wats. Yucca, Mohave County (*Meire* in 1917), about 2,000 feet, and at Lake Mead, on the Nevada side (*Clover* 8222), February to October. Deserts of western Arizona, southern Nevada, and southeastern California.

A shrub about 1 m. high, leafless most of the year.

2. Cassia Wislizeni Gray. Cochise County, 4,000 to 5,000 feet, dry slopes and mesas, usually on limestone, June to September. Southern Texas to southeastern Arizona and northern Mexico.

A much-branched shrub up to 1.5 m. (5 feet) high, with dark-colored bark and prominent lenticels.

3. Cassia leptocarpa Benth. Cochise, Pinal, Santa Cruz, and Pima counties, 2,500 to 5,500 feet, along streams and washes, July to September. New Mexico and southern Arizona to South America.

A handsome but rather coarse plant, with large terminal panicles of bright-yellow flowers, the foliage ill-smelling. The Arizona plants mostly belong to var. *glaberrima* Jones (*C. Gooddingii* A. Nels.), being less pubescent and with narrower leaflets than in typical *C. leptocarpa,* but a collection in the Baboquivari Mountains (*Kearney & Peebles* 10384) is nearly typical. The types of var. *glaberrima* were from southern Arizona, and that of *C. Gooddingii* from the Huachuca Mountains (*Goodding* 2431).

4. Cassia Absus L. Bear Valley (Sycamore Canyon), Santa Cruz County, 4,000 feet (*Darrow & Haskell* 2014), western slope of the Baboquivari Mountains, Pima County, 5,000 feet (?), granitic slopes and ridges (*Gilman* B212, B234), September. Widely distributed in tropical America; supposed to have been introduced from the Old World.

The presence of this plant in such remote localities in Arizona is remarkable.

5. Cassia Wrightii Gray (*Chamaecrista Wrightii* Woot. & Standl.). Santa Cruz County (probably elsewhere in southern Arizona), 3,500 to 5,000 feet, open chaparral, August, type from the Sonoita (*Wright* 1034). Southeastern Arizona and northeastern Sonora.

Partridge-pea, sensitive-pea, names applied generally to the subgenus *Chamaecrista.*

6. Cassia leptadenia Greenm. (*Chamaecrista leptadenia* Cockerell). Greenlee, Graham, Gila, Pinal, Cochise, Santa Cruz, and Pima counties, 3,500 to 5,000 feet, dry plains and mesas, August to October. Western Texas to Arizona and Mexico.

Very similar to *C. nictitans* L. of the eastern United States, but the plant has ciliate leaflets, narower pods, and stems usually with long, spreading hairs as well as shorter, more appressed ones.

7. Cassia bauhinioides Gray. Coconino, Mohave, Yavapai, and Greenlee counties to Cochise, Santa Cruz, and Pima counties, 2,000 to 5,500 feet, dry, rocky slopes and mesas, April to August. Texas to Arizona and Mexico.

The common phase in the state is var. *arizonica* Robins., type from the Mule Mountains, Cochise County (*Toumey* in 1894), with hairs of the stem appressed or subappressed, but the typical phase, with spreading hairs, likewise occurs.

8. Cassia Covesii Gray. Distribution in Arizona and habitat same as for *C. bauhinioides,* but usually at somewhat lower altitudes, mostly 1,000 to 3,000 feet, April to October, types from Camp Grant and near Prescott (*Coues, Palmer*). New Mexico to Nevada, Arizona, California, and northwestern Mexico.

9. Cassia Lindheimeriana Scheele. Cochise County, 4,500 to 5,500 feet, dry mesas and foothills, June to September. Texas to southeastern Arizona and northern Mexico.

10. PARKINSONIA (143). JERUSALEM-THORN

Large shrubs or small trees, attaining a height of 12 m. (40 feet) and a trunk diameter of 30 cm. (1 foot); bark smooth, yellowish green, becoming brown; leaves bipinnate but simulating a pair of elongate, simply pinnate leaves (the common rachis almost none), the rachis of the pinnae broad, flat, up to 60 cm. long, the leaflets many, caducous; flowers moderately irregular, showy, bright yellow, in elongate racemes; pods turgid, nearly terete, torulose.

1. Parkinsonia aculeata L. (*P. Thornberi* Jones). Castle Dome Mountains (Yuma County), foothills of the Coyote and Baboquivari mountains (Pima County), occasional in sandy soil along washes, also south of Tucson, where perhaps an escape from cultivation, 2,000 to 3,000 feet, May (probably other months), type of *P. Thornberi* from south of Tucson. Florida, southern Texas, and southern Arizona to South America.

A very attractive, rapid-growing plant, much used for ornamental planting in warmer parts of the United States. Sometimes known as horse-bean.

11. CERCIDIUM (143). PALO-VERDE

Large shrubs or small trees, attaining a height of about 8 m. (25 feet); young bark smooth, green; rachis of the pinnae short, terete; flowers showy, yellow, in corymbose fascicles; pods more or less torulose.

The Arizona palo-verdes of the genus *Cercidium* are very common and characteristic plants of the lower and drier parts of the state and are a glorious sight when in full flower. They are leafless in the dry season but always conspicuous because of the green bark. The wood is soft and brittle and burns very quickly, giving off an unpleasant odor and leaving few coals. The pods are fairly palatable but are not much eaten by livestock except during prolonged droughts. It is reported that the plants are browsed by bighorns. The Indians ate the seeds sometimes in the form of meal. The flowers are said to yield good honey.

KEY TO THE SPECIES

1. Young bark yellowish green; leaves appearing simply pinnate, the common rachis nearly obsolete; leaflets of each pinna commonly 4 to 8 pairs, very small, not more than 3 mm. long; flowers not more than 10 mm. long, pale yellow, the odd petal often whitish; pod turgid, ending in a flat triangular or sword-shaped beak 1. *C. microphyllum*

1. Young bark bluish green; leaves evidently bipinnate, the common rachis short but evident; leaflets of each pinna 1 to 3 pairs, 4 to 8 mm. long; flowers 12 to 17 mm. long, bright yellow, all the petals alike in color; pod flat, with a short, triangular beak, or almost beakless..2. *C. floridum*

1. Cercidium microphyllum (Torr.) Rose & Johnston. Northern Mohave County to Cochise, Pima, and Yuma counties, 4,000 feet or lower, dry, rocky hillsides and mesas, April and May, types from the Colorado River and Williams River. Southern and western Arizona, southeastern California, Sonora, and Baja California.

Little-leaf palo-verde. Ends of the leafy branchlets spinescent.

2. Cercidium floridum Benth. (*C. Torreyanum* (Wats.) Sarg.). Greenlee, Gila, Pinal, and Pima counties to Mohave and Yuma counties, 3,500 (rarely 6,000) feet or lower, common along washes and on flood plains, less frequently on dry lower slopes, March to May (occasionally August to October). Range outside Arizona same as that of *C. microphyllum*.

Blue palo-verde. Leafy branchlets not or not strongly spine-tipped, the rudimentary branchlets transformed into spines. A handsomer plant when in flower than *C. microphyllum* because of the larger, more highly colored flowers. It appears to be less xerophytic and usually grows larger, having a better water supply. It is somewhat earlier-flowering than *C. microphyllum*.

12. HOFFMANSEGGIA (144)

Plants perennial, herbaceous or shrubby; leaves bipinnate, the leaflets small; flowers in terminal or lateral racemes, moderately irregular; petals yellow; stamens 10, sometimes red; pods flat.

KEY TO THE SPECIES

1. Plants shrubby; stems erect, virgate, densely puberulent, up to 1.2 m. long; leaves soon deciduous, the pinnae 3, the terminal one much the longest. Racemes elongate; pods lunate, 1.5 to 2 cm. long, 5 to 7 mm. wide........................1. *H. microphylla*
1. Plants herbaceous or suffrutescent; stems decumbent or spreading, seldom more than 30 cm. long; leaves persistent, the pinnae 5 or more, the terminal one not conspicuously longer (2).
2. Leaflets dotted beneath with conspicuous, black glands, the stems, flowers, and pods similarly punctate; pods asymmetrically lunate, conspicuously wider above the middle. Root large, woody, fusiform....................................2. *H. Jamesii*
2. Leaflets, etc. not punctate with black glands; pods not or not conspicuously wider above the middle (3).
3. Flowering stems from a thick, woody taproot; inflorescence not glandular; petals short-clawed, glandless; pods strongly falcate, often forming nearly a semicircle. Herbage finely appressed-pubescent..........................3. *H. drepanocarpa*
3. Flowering stems from creeping rootstocks; inflorescence conspicuously glandular; petals long-clawed, the claws bearing subulate glands; pods moderately falcate to nearly straight ....4. *H. densiflora*

1. Hoffmanseggia microphylla Torr. Southern Yuma County, in and around the Mohawk, Gila, and Tinajas Altas mountains, 1,000 feet or lower, dry, sandy or rocky, mesas and slopes, March to October. Southwestern Arizona, southeastern California, Sonora, and Baja California.

Very different in appearance from the other Arizona species and much more xerophytic.

2. Hoffmanseggia Jamesii Torr. & Gray. Apache, Navajo, and Coconino counties, also in the Galiuro Mountains (Graham or Pinal County), at several localities in Cochise County, and near Sonoita (Santa Cruz County), 4,500 to 6,000 feet, dry plains, mesas, and dunes, sometimes with pinyons and junipers, May to August. Kansas and Colorado to Texas and Arizona.

3. Hoffmanseggia drepanocarpa Gray. Coconino, Mohave, Yavapai, Gila, and Cochise counties, 3,000 to 5,000 feet, May to September. Colorado and western Texas to Arizona and Chihuahua.

4. Hoffmanseggia densiflora Benth. Navajo County to Mohave County, south to Graham, Cochise, Pima, and Yuma counties, 5,000 feet or lower, April to September. Kansas to Arizona, southern California, and central Mexico.

Hog-potato, camote-de-raton. Common at roadsides in the irrigated districts, often forming large colonies and becoming a troublesome weed in cultivated fields and pastures, difficult to eradicate, especially on heavy soils. The tuberous enlargements of the roots make valuable hog feed, and after roasting were used for food by the Indians. The plant is considered a good soil binder.

The species is closely related to the South American *H. falcaria* Cav., to which it was reduced by Fisher (144). He described several varieties under

H. falcaria, of which the following were stated to occur in Arizona: var. *stricta* (Benth.) Fisher, var. *demissa* (Gray) Fisher, var. *Pringlei* Fisher (type from near Tucson, *Pringle* in 1881, part), and var. *capitata* Fisher (type from near Tucson, *Pringle* in 1881, part).

13. CAESALPINIA

Shrubs, up to 3 m. (10 feet) high; leaves bipinnate, large, with many small leaflets; flowers in terminal racemes, large and showy, the petals yellow; stamens and pistil red, the stamens with very long filaments; pods large, flat.

1. Caesalpinia Gilliesii Wall. (*Poinciana Gilliesii* Hook.). Occasionally escaped from cultivation, and perhaps naturalized here and there in Mohave, Gila, Maricopa, Cochise, and Pima counties, April to September; introduced from South America.

Bird-of-paradise-flower. Much grown as an ornamental, especially by the Mexican population. Plant ill-smelling, very showy because of the extraordinarily long, red filaments.

3. PAPILIONOIDEAE. Bean Subfamily

KEY TO THE GENERA

1. Filaments all distinct to the base or very nearly so (2).

1. Filaments all, or 9 of them, united, at least near the base or if distinct (*Parryella*), then attached at base to the calyx, and the corolla wanting (3).

2. Leaves pinnate with numerous leaflets; stipules very small; corolla white, blue, or lilac; pods torulose; plants small herbs or good-sized shrubs............14. *Sophora*

2. Leaves digitately trifoliolate; stipules large, foliaceous; corolla yellow; pods not torulose; plants tall herbs....................................15. *Thermopsis*

3. Anthers strongly differentiated, some very small and versatile (dorsifixed), others much larger and basifixed. Plants herbaceous (4).

3. Anthers not sharply differentiated, all of approximately the same size or if sharply differentiated (*Glycyrrhiza, Stylosanthes, Zornia*), then the pods prickly, or else the inflorescence conspicuously bracted, the corolla yellow, and the pods jointed or ending in a hook (5).

4. Leaves unifoliolate or pinnately trifoliolate; corolla yellow or orange; pods much inflated, bladderlike; plants annual..............................16. *Crotalaria*

4. Leaves palmately compound with usually more than 3 leaflets; corolla violet, purple, or whitish; pods not bladderlike; plants annual or perennial...........17. *Lupinus*

5. Rachis of the leaf extended into a simple or forked tendril, this (in *Lathyrus*) sometimes reduced to a soft bristle or rudimentary. Plants herbaceous; stems weak; leaves pinnate, with several leaflets; corolla purple, pink, or whitish (6).

5. Rachis of the leaf not ending in a tendril but greatly prolonged in place of a terminal leaflet in certain species of *Astragalus* (7).

6. Style with a tuft of hairs on the back or all around, just below the stigma; stamen tube oblique at apex...49. *Vicia*

6. Style hairy along the inner side only, with no apical tuft; stamen tube not oblique 50. *Lathyrus*

7. Fruit (loment) of several indehiscent 1-seeded segments, these often separating at maturity, or if reduced to 1 seed-bearing segment, then ending in a wing (*Nissolia*), or a hook (*Stylosanthes*), or the plant a spiny shrub (8).

7. Fruit not segmented, but sometimes constricted between the seeds (14).

8. Corolla pink or purple, exceptionally white (9).

8. Corolla yellow or orange (11).

9. Plants intricately branched, spiny, glabrous shrubs; leaves small, unifoliolate. Fruit of 1 to several segments, these not separating at maturity...............43. *Alhagi*

9. Plants herbaceous above ground or suffrutescent, not spiny; leaves pinnately compound, sometimes reduced to one leaflet in *Desmodium* (10).
10. Leaflets more than 3, without stipels............................42. *Hedysarum*
10. Leaflets 1 or 3, with stipels..48. *Desmodium*
11. Anthers all alike; inflorescences not conspicuously bracteate, few-flowered, axillary (12).
11. Anthers conspicuously different, some large and basifixed, others small and versatile; inflorescences conspicuously bracteate. Plants herbaceous; stems decumbent to erect (13).
12. Stems twining or prostrate, often suffrutescent; leaflets usually 5, rather large, ovate or ovate-lanceolate; segments of the fruit 1 to 3, the terminal segment winged 44. *Nissolia*
12. Stems erect; leaflets many, small, linear; segments of the fruit commonly more than 3, all alike..45. *Aeschynomene*
13. Leaves trifoliolate; bracts of the inflorescence not valvelike, often trifoliolate, sharply cuspidate; fruit 1- or 2-jointed, tipped by the hooked persistent style 46. *Stylosanthes*
13. Leaves bifoliolate; bracts valvelike, a closely appressed pair partly hiding each flower, entire, conspicuously veined, not sharply cuspidate; fruit several-jointed, without a hooked tip..47. *Zornia*
14. Leaves (and usually the calyx) glandular-punctate, sometimes obscurely so (15).
14. Leaves not glandular-punctate (22).
15. Stems trailing or twining; flowers few, in axillary racemes or fascicles; corolla bright yellow; pods completely dehiscent, strongly compressed. Leaves pinnately trifoliolate 55. *Rhynchosia*
15. Stems erect or ascending; flowers numerous, in mostly terminal racemes or spikes; corolla not bright yellow, sometimes ochroleucous or pale yellow; pods not or very tardily dehiscent (16).
16. Pods prickly. Plants tall, leafy-stemmed; leaves pinnate with numerous leaflets; flowers in dense, spikelike racemes; corolla whitish................41. *Glycyrrhiza*
16. Pods not prickly (17).
17. Ovule solitary; leaves digitately trifoliolate to 5-foliolate. Plants perennial, herbaceous ....23. *Psoralea*
17. Ovules 2 or more; leaves mostly pinnate with several or many leaflets, seldom digitately trifoliolate (18).
18. Petals none or only one. Plants shrubby; leaves pinnate with numerous leaflets, these stipellate; flowers in elongate, spikelike racemes (19).
18. Petals 5 (20).
19. Leaflets very numerous; stipels glandular, persistent; corolla none; stamens distinct, attached at base to the calyx....................................24. *Parryella*
19. Leaflets several; stipels subulate, caducous; corolla of 1 petal, this dark violet; stamens united at base..25. *Amorpha*
20. Stamens 5. Plants herbaceous; flowers in dense, cylindric spikes; corolla white, pink, or purplish...28. *Petalostemum*
20. Stamens 9 or 10 (21).
21. Plants shrubby; corolla white; leaflets numerous, with subulate stipels; flowers in elongate, spikelike racemes; style bearing a large sessile gland near the apex; pods thin, flat, long-exserted from the calyx..........................26. *Eysenhardtia*
21. Plants small herbs to large shrubs, when shrubby the corolla not white; leaflets few or numerous, without stipels or these mostly glandlike; flowers in loose or dense heads, spikes, or racemes; style not gland-bearing; pods turgid, included in the calyx, or long-exserted in a few shrubby species...............................27. *Dalea*
22. Margins of the leaflets denticulate to serrate, rarely entire in *Trifolium*. Leaves trifoliolate; plants herbaceous (23).
22. Margins of the leaflets entire (25).
23. Pods scythe-shaped to spirally coiled......................................18. *Medicago*
23. Pods straight or nearly so (24).

24. Flowers in elongate, rather loose racemes; corolla yellow or white; plants strongly aromatic when dry....................19. *Melilotus*
24. Flowers in heads or short dense spikes; corolla white, pink, or purple; plants not noticeably aromatic when dry....................20. *Trifolium*
25. Plants shrubby or arborescent (26).
25. Plants herbaceous or suffrutescent (31).
26. Leaves trifoliolate; corolla about 5 cm. long, scarlet; banner petal narrow, strongly keeled, much longer than the other petals; pods very large, coriaceous; seeds bright red....................53. *Erythrina*
26. Leaves multifoliolate; corolla much less than 5 cm. long, not scarlet; banner petal not strongly keeled, not much longer than the other petals; pods not very large, not coriaceous; seeds not bright red (27).
27. Calyx in bud closely subtended by 2 caducous bractlets; corolla yellowish; outer wall of the pod separating from the inner wall. Plants low shrubs....................33. *Diphysa*
27. Calyx without bractlets; corolla purple, pink, or white; outer wall of the pod not separating from the inner wall (28)
28. Herbage with dolabriform hairs (hairs attached at the middle); pods subglobose, 1-seeded, tardily dehiscent; plants low shrubs. Leaflets numerous....22. *Indigofera*
28. Herbage without dolabriform hairs; pods elongate, several-seeded, rather promptly dehiscent; plants large shrubs or small trees (29).
29. Inflorescence not glandular; leaflets not cuspidate. Plants small trees; corolla purple and white, about 12 mm. long; pods torulose....................32. *Olneya*
29. Inflorescence glandular; leaflets cuspidate (30).
30. Corolla 20 to 25 mm. long, purplish pink; pods not torulose, hispid and usually glandular....................31. *Robinia*
30. Corolla not more than 15 mm. long, white or tinged with pink, rarely yellow; pods torulose, puberulent....................34. *Coursetia*
31. Filaments all or some of them broad and flat, or dilated just below the anther. Flowers solitary in the leaf axils, or in few-flowered umbellike clusters; leaflets few; corolla usually yellow....................21. *Lotus*
31. Filaments all filiform or if flat, then narrow (32).
32. Leaves pinnately trifoliolate. Corolla purple or purplish (33).
32. Leaves all or some of them with more than 3 leaflets, or reduced to a single leaflet, or sometimes (in *Astragalus*) the leaflets none and the rachis greatly prolonged (36).
33. Style bearded (34).
33. Style not bearded (35).
34. Corolla 5 to 6 cm. long, lavender-purple, the keel slightly incurved at apex; stems erect or ascending....................51. *Clitoria*
34. Corolla not more than 2 cm. long, the keel strongly incurved, curled, or spirally coiled; stems usually trailing or twining....................56. *Phaseolus*
35. Stems not or barely twining; flowers solitary or in pairs in the axils; corolla deep red-purple....................52. *Cologania*
35. Stems more or less twining; flowers in bracted racemes; corolla pale......54. *Galactia*
36. Calyx closely subtended by a pair of small, caducous bractlets. Plants tall; leaflets numerous, linear, narrowly oblong, or narrowly elliptic; flowers in axillary racemes (37).
36. Calyx without bractlets (38).
37. Plants annual; pods very long and slender, completely dehiscent; corolla pale yellow, usually with purple markings....................37. *Sesbania*
37. Plants perennial, with rootstocks, resembling some species of *Astragalus;* pods bladder-like, indehiscent or tardily and irregularly dehiscent; corolla dull red..38. *Swainsona*
38. Pods with cross partitions between the seeds, narrow, elongate, flat, completely dehiscent; seeds more or less quadrate (39).
38. Pods without cross partitions; seeds commonly rounded, often reniform, quadrate in a few species of *Astragalus* (40).
39. Leaflets several, broad and thin; flowers several, in loose, long-stalked racemes; seeds not constricted; plants perennial....................35. *Cracca*

39. Leaflet 1, narrowly linear, elongate; flowers 1 or 2 in the leaf axils, short-pedicelled; seeds constricted at the middle; plants annual................36. *Sphinctospermum*
40. Sutures of the pods not introverted, the pods completely 1-celled, flat, 10 or more times as long as wide (41).
40. Sutures 1 or both of them introverted, the pods often more or less completely 2-celled, usually much less than 10 times as long as wide (42).
41. Stipules not spinescent; lateral veins of the leaflets conspicuous........29. *Tephrosia*
41. Stipules spinescent; lateral veins inconspicuous or obsolete................30. *Peteria*
42. Keel not beaked or if so, then the plant caulescent and the stems decumbent or procumbent ....39. *Astragalus*
42. Keel with a prominent, erect or ascending beak; plants acaulescent, the scapes erect or ascending...40. *Oxytropis*

14. SOPHORA

Shrubs or perennial herbs; leaves pinnate, with several or many leaflets; flowers in racemes, the petals white, blue, or lilac; stamens 10, distinct to the base or very nearly so; pods flat or turgid, moderately torulose to moniliform, tardily dehiscent.

The herbage and seeds of some (perhaps all) of the species are poisonous to livestock when eaten in quantity. The herbaceous species, which resemble in habit some species of *Astragalus,* colonize freely by means of horizontal roots and thus are efficient soil binders, especially in the sandy areas of northeastern Arizona. Several shrubby and arboreous species from different parts of the world are grown as ornamentals, and the native *S. arizonica* and *S. formosa,* with their large wisteria-colored flowers, are well worth cultivating.

KEY TO THE SPECIES

1. Plants shrubby, up to 3.5 m. high; leaves coriaceous, evergreen; seeds red; pods flat, about 10 mm. wide, moderately torulose. Herbage sericeous to glabrescent; corolla lilac or violet (2).
1. Plants herbaceous, seldom more than 30 cm. high; leaves not coriaceous or evergreen; seeds not red; pods turgid, about 5 mm. wide, usually moniliform (3).
2. Leaflets not more (usually less) than 10 mm. wide; upper calyx teeth 1 to 1.5 mm. long; corolla about 22 mm. long, lilac, the banner petal obovate-oblong (distinctly wider above the middle), about ⅔ as wide as long, the keel petals with claws less than half as long as the blades................................1. *S. arizonica*
2. Leaflets up to 12 mm. wide; upper calyx teeth 2.5 to 3 mm. long; corolla about 16 mm. long, violet, the banner petal broadly oval (not wider above the middle), ⅘ as wide as long, the keel petals with claws ½ to ⅗ as long as the blades........2. *S. formosa*
3. Leaflets narrowly linear, 15 to 25 mm. long, pubescent on both surfaces; corolla blue; herbage velvety-tomentose..................................3. *S. stenophylla*
3. Leaflets oblong or oblong-obovate, commonly not more than 10 mm. long, glabrous or glabrescent above; corolla white or ochroleucous; herbage sericeous....4. *S. sericea*

1. Sophora arizonica Wats. Mohave County, "Cactus Pass and on White Cliff Creek" (*Bigelow,* the type collection), southeast of Yucca, foothills of Hualpai Mountain, Big Sandy River, 60 miles southeast of Kingman, 2,000 to 4,000 feet, dry, rocky hillsides and banks of arroyos, with *Quercus turbinella* and *Canotia,* March. Known only from western Arizona.

2. Sophora formosa Kearney & Peebles. Graham County, foothills of the Pinaleno Mountains (*Maguire* 10993, the type collection), about 3,500 feet, habitat similar to that of *S. arizonica,* associated with *Quercus turbinella, Yucca, Dasylirion, Prosopis, Larrea,* and *Fouquieria,* April. Known only from southeastern Arizona.

Both this species and the closely related *S. arizonica* are very local, but are fairly abundant at the stations where they occur. They appear to be relict species, barely holding their own under present conditions.

3. Sophora stenophylla Gray. Hopi Indian Reservation, Betatakin, and Snowflake (Navajo County) 5,500 to 7,000 feet, reported also from Canyon Diablo (Coconino County), June, type from Oraibi (*Newberry* in 1858). Southern Utah, New Mexico, and northeastern Arizona.

4. Sophora sericea Nutt. Apache County to Yavapai and eastern Mohave counties, also in Cochise and Pima counties, 4,000 to 7,000 feet, often growing in dense colonies, sometimes in cultivated land, April to June. South Dakota and Wyoming to Texas, southern Utah, and Arizona.

15. THERMOPSIS (145). Golden-pea

Plants herbaceous, perennial; stems erect, branching, leafy; leaves trifoliolate, the leaflets large, lanceolate to ovate or somewhat rhombic, the stipules large, foliaceous; flowers large, in rather dense, terminal racemes, the petals bright yellow; stamens 10, incurved, distinct; pods sessile or nearly so, pubescent, at least when young.

1. Thermopsis pinetorum Greene. Apache County to Coconino County, south to Cochise, Gila, and Yavapai counties, 6,000 to 9,500 feet, common in pine forests, April to July. Colorado, Utah, New Mexico, and Arizona.

A showy plant when in flower, reported to be unpalatable to cattle. The plants spread by rootstocks, forming patches. Doubtfully distinct as a species from *T. montana* Nutt.

16. CROTALARIA (146). Rattle-box

Plants herbaceous or suffrutescent; leaves unifoliolate or trifoliolate, with stipules; flowers in lateral or terminal racemes, these few- to many-flowered; calyx somewhat bilabiate; petals yellow, the keel curved or bent; stamens dimorphic; pods much inflated, many-seeded, oblong to nearly globose.

The plants are palatable to livestock, but the seeds of some of the species are known to be poisonous.

KEY TO THE SPECIES

1. Leaves unifoliolate, the leaflet linear, lanceolate, elliptic, oval, or the lowest obovate; stems simple, or branched only near the base; herbage (including the upper leaf surface) loosely villous or subhirsute with long, more or less spreading hairs; peduncles lateral, 1- to 3-flowered; calyx 6 to 9 mm. long; corolla pale yellow, scarcely surpassing the calyx; pods sometimes black at maturity................1. *C. sagittalis*
1. Leaves trifoliolate, the leaflets commonly oblanceolate or obovate; stems branched well above the base, rarely simple; herbage strigose-pubescent, the upper leaf surface glabrous; peduncles often appearing terminal as well as lateral, 1- to many-flowered; calyx 3 to 4 mm. long; corolla orange-yellow, much surpassing the calyx; pods reddish brown at maturity....................................2. *C. pumila*

1. Crotalaria sagittalis L. Cochise, Santa Cruz, and Pima counties, 4,500 to 6,000 feet, sandy soil, usually along brooks, August to October. New England to South Dakota and Texas, southern Arizona, and south to Panama.

Two varieties occur in Arizona. These are: (1) var. *Blumeriana* Senn, type from the Chiricahua Mountains (*Blumer* 1772), differing from typical *C. sagittalis* in its shorter stems, shorter and relatively broader leaflets, inconspicuous or obsolete stipules, and smaller pods; and (2) var. *fruticosa* (Mill.) Fawc. & Rend. (*C. Pringlei* Gray), which is suffruticose and has uniformly linear leaves. The second variety is known in Arizona only from the type collection of *C. Pringlei* in the Santa Catalina Mountains (*Pringle* 276).

2. Crotalaria pumila Ortega (*C. lupulina* H.B.K.). Greenlee, Graham, Cochise, Santa Cruz, and Pima counties, 4,000 to 6,000 feet, preferring sandy soil, June to October. Florida; Texas to southern Arizona and Mexico.

Varies greatly in size of the plant and of the flowers, and in the number of flowers in the racemes, these 1 to many. The petals are often tinged or streaked with reddish color.

17. LUPINUS (147). Lupine

Plants herbaceous (the Arizona species), annual or perennial; stems leafy or subscapose; leaves digitately compound, with 4 to 15 leaflets; flowers in terminal racemes or spikes, often showy; calyx strongly bilabiate; petals blue, purple, or white, the keel usually curved and enclosed by the connivent wing petals; stamens dimorphic; pods more or less compressed, sometimes constricted between the seeds.

Several species are cultivated as ornamentals, and most of those occurring in Arizona are handsome plants. The seeds of *L. albus* are used as food in Europe. It is known that some of the American species, including *L. sparsiflorus,* common in Arizona, contain alkaloids that are poisonous to livestock, especially to sheep. The seeds are especially toxic, the pods less so, and the herbage is relatively harmless, often containing so little of the dangerous alkaloids that sheep graze the plants without ill effect. Blue-bonnet (*L. subcarnosus* or *L. texensis*) is the state flower of Texas.

The taxonomy of this genus is difficult, and authorities differ greatly in their interpretations, particularly of the perennial species.

KEY TO THE SPECIES

1. Plants annual or biennial (2).
1. Plants perennial but sometimes (especially in *L. huachucanus*) appearing annual when young. Ovules and seeds normally more than 2 (12).
2. Cotyledons petioled after germination; pods mostly 3- to 5-seeded, oblong (3).
2. Cotyledons sessile after germination; pods mostly 2- (exceptionally 3-) seeded (7).
3. Flowers 8 mm. or longer; keel usually ciliate (4).
3. Flowers 4 to 7 mm. long; keel naked. Herbage villous or hirsute (6).
4. Leaflets mostly truncate and often slightly retuse at apex; flowers verticillate, 10 to 15 mm. long; keel ciliate toward base, both above and below. Stems strigose or glabrate, stout, more or less fistulose; leaflets wedge-shaped, widest at apex, thickish, glabrous above, sparsely strigose beneath; corolla violet-blue...... 1. *L. succulentus*
4. Leaflets acute to rounded at apex; flowers not distinctly verticillate, 8 to 10 mm. long; keel ciliate toward base on the lower side only, or sometimes naked (5).
5. Plants not succulent, drying green; stems slender, or stout but scarcely fistulose; leaflets narrowly lanceolate or oblanceolate, seldom more than 4 mm. wide, acute or acutish at apex, thin, usually sparsely strigose above; corolla when fresh violet-blue, occasionally white.. 2. *L. sparsiflorus*

5. Plants somewhat succulent, drying brownish; stems stout, fistulose; leaflets broadly oblanceolate, 5 to 12 mm. wide, rounded and mucronate or mucronulate at apex, thickish, glabrous or glabrate above; corolla when fresh pale purplish pink, sometimes drying violet....3. *L. arizonicus*
6. Leaflets 4 to 9 mm. wide, rounded at apex; inflorescences not or but little surpassing the foliage; flowers not verticillate; petals pale purple, the keel tip darker purple 4. *L. concinnus*
6. Leaflets 2 to 3 mm. wide, acutish to obtuse; inflorescences mostly surpassing the foliage; flowers more or less verticillate; petals violet-blue and white....5. *L. bicolor*
7. Racemes in flower dense, often subcapitate, usually less than 2 cm. long; pods not noticeably constricted between the seeds, ovate or lance-ovate in outline, villous all over (8).
7. Racemes in flower loose, seldom less than 2 cm. long or if the racemes rather dense, then the pods distinctly constricted between the seeds (9).
8. Plants acaulescent or nearly so, the stems scarcely 1 cm. long, the leaves mostly basal; lower calyx lip more than twice as long as the (often obsolete) upper lip; pods broadly ovate....6. *L. brevicaulis*
8. Plants caulescent, the stems 3 cm. long or longer, branched, leafy; lower calyx lip less than twice as long as the upper lip, the lips often nearly equal; pods ovate or lance-ovate....7. *L. Kingii*
9. Pods oblong to oblong-lanceolate in outline, usually distinctly constricted between the seeds, hirsute all over. Plants more or less caulescent; stems and the lower leaf surface hirsute (10).
9. Pods ovate in outline, not noticeably constricted between the seeds, smooth or scaly on the sides, villous or hirsute on the upper (ventral) suture (11).
10. Leaflets obtuse at apex; racemes slightly if at all surpassing the foliage; pedicels hirsute; calyx hirsute all over, the upper lip 3 to 5 mm. long....8. *L. pusillus*
10. Leaflets acutish at apex; racemes usually considerably surpassing the foliage; pedicels glabrous or nearly so; calyx glabrous or nearly so toward base, the upper lip not more than 3 mm. long....9. *L. rubens*
11. Herbage loosely pilose; plants acaulescent or nearly so; racemes 5 to 12 cm. long, greatly surpassing the foliage; pedicels and calyx tube glabrous or nearly so 10. *L. odoratus*
11. Herbage conspicuously silky-villous; plants more or less caulescent but short-stemmed; racemes usually less than 5 cm. long, not or but slightly surpassing the foliage; pedicels and calyx tube villous....11. *L. Shockleyi*
12. Petals whitish, ochroleucous, or violet; beak of the keel long, attenuate, almost at a right angle to the lower part and usually exserted from the wings at anthesis, the keel sparsely ciliate toward base. Herbage sparsely strigose, appearing glabrous; stems tall, stout, erect; racemes 10 cm. long or longer, many-flowered, the flowers large....12. *L. Parishii*
12. Petals normally violet to pale purple; beak of the keel shorter and less attenuate, not or barely exserted at anthesis, the keel usually ciliate toward apex (13).
13. Plants acaulescent and cespitose or, if caulescent, then the stems not more than about 20 cm. long. Herbage copiously villous-hirsute with long spreading hairs; leaflets 5 to 7, acute at both ends; corolla 7 to 9 mm. long, the banner and wings violet-blue, the tip of the keel purple....13. *L. huachucanus*
13. Plants caulescent, the stems leafy, usually more than 20 cm. long. Species difficult to distinguish and apparently largely confluent (14).
14. Banner petal sericeous in the center of the back (15).
14. Banner petal glabrous or with a few scattered hairs (17).
15. Flowers 12 to 15 mm. long. Stems hirsute; leaves green, the longer leaflets 40 to 80 mm. long....14. *L. barbiger*
15. Flowers less than 10 mm. long (16).
16. Stems, petioles, and pedicels soft-pilose, the hairs short and spreading; leaflets silvery-sericeous on both surfaces, densely so beneath, obtuse and mucronate, not more than 25 mm. long; flowers 6 to 8 mm. long; calyx scarcely gibbous 15. *L. Osterhoutianus*

16. Stems, petioles, and pedicels strigose or sericeous, the hairs appressed; leaflets green and glabrous or glabrescent above to silvery-sericeous on both surfaces, obtuse to very acute, up to 50 mm. long; flowers 8 to 9 mm. long; calyx usually strongly gibbous or even short-spurred on the upper side at base..............16. *L. argenteus*
17. Flowers 5 to 7 mm. long; racemes usually conspicuously verticillate. Herbage sericeous, the hairs of the stem and petioles loosely appressed or partly spreading 17. *L. Hillii*
17. Flowers 8 mm. or longer; racemes not conspicuously verticillate, or sometimes so in *L. argenteus* and *L. Lemmonii* (18).
18. Hairs of the stems and petioles appressed or subappressed (19).
18. Hairs of the stems and petioles, at least the longer ones, spreading or ascending-spreading (20).
19. Herbage green and sparsely to densely pubescent, the leaflets glabrescent to densely silvery-sericeous above: northern................................16. *L. argenteus*
19. Herbage, including the upper surface of the leaflets, densely sericeous: southern 18. *L. Lemmonii*
20. Leaflets sericeous above; long-petioled basal leaves usually present at anthesis. Stems copiously to densely pubescent; flowers 8 to 10 mm. long, the corolla normally violet 19. *L. Palmeri*
20. Leaflets sparsely strigose or glabrescent above; basal leaves usually disappearing before anthesis (21).
21. Stems copiously pilosulous, with a few longer, very soft hairs, all of the hairs more or less spreading; calyx strongly gibbous or short-spurred at base on the upper side. Flowers 8 to 11 mm. long; larger leaflets 40 to 50 mm. long and up to 12 mm. wide ....20. *L. Cutleri*
21. Stems sparsely to somewhat copiously pubescent, the longer hairs rather stiff; calyx not strongly gibbous (22).
22. Flowers 12 to 14 (seldom only 10) mm. long, the corolla pale lavender to reddish purple; stems hirsute with long, spreading hairs and pilosulous with very short hairs, these also more or less spreading: southern..................21. *L. Blumeri*
22. Flowers 9 to 11 (rarely 14) mm. long, the corolla normally violet; stems scarcely hirsute, the longer hairs shorter and ascending, the short hairs appressed: northern 22. *L. Sitgreavesii*

1. Lupinus succulentus Dougl. Roosevelt and north of San Carlos (Gila County), Canyon Lake and Camp Creek (Maricopa County), 1,500 to 4,500 feet, open slopes and mesas, March and April. Arizona, California, and Baja California.

The plants have extraordinarily large root nodules. The Arizona plants are relatively small-flowered.

2. Lupinus sparsiflorus Benth. Greenlee County to Mohave County, south to the Mexican boundary, 4,500 feet or lower, mesas and foothills, preferring sandy soil, January to May. Nevada and Arizona to California, Sonora, and Baja California.

In favorable seasons this handsome lupine colors extensive areas with the rich violet of its flowers. Arizona plants, with flowers not more (usually less) than 1 cm. long, were referred by Smith (147, p. 121) to var. *arizonicus* (Wats.) C. P. Smith.

3. Lupinus arizonicus Wats. (*L. concinnus* var. *arizonicus* Wats., *L. sparsiflorus* var. *barbatulus* Thornber). Mohave and Yuma counties, eastward to the western part of Maricopa, Pinal, and Pima counties, 3,000 feet or lower, usually in sandy washes, January to May. Western Arizona, southeastern California, Sonora, and Baja California.

This species occurs usually at lower altitudes and in deeper sand than *L. sparsiflorus,* the two species seldom commingling and when they do, rarely intergrading, in Arizona at any rate. Watson's description of *L. concinnus* var. *arizonicus* seems clearly applicable to the plant published subsequently as *L. sparsiflorus* var. *barbatulus* Thornber, type from the Colorado River Valley (*Palmer* 88, in 1876), but Watson's description of *L. arizonicus* as a species (with citation of *L. concinnus* var. *arizonicus* Wats. as a synonym) seems to include both *L. sparsiflorus* and the plant here under consideration.

4. Lupinus concinnus Agardh. Southern Apache County to Mohave County, south to the Mexican boundary, 5,000 feet or (usually) lower, March to May. New Mexico to southern Nevada, Arizona, California, and northern Mexico.

Very abundant in the sandy, desert areas in spring. The commoner plant in Arizona is the relatively small-flowered var. *Orcuttii* (Wats.) C. P. Smith (*L. Orcuttii* Wats., *L. micensis* Jones), but Smith (147, p. 122) identified numerous Arizona specimens as typical *L. concinnus.*

5. Lupinus bicolor Lindl. Mazatzal Mountains and north of San Carlos (Gila County), near Oracle (Pinal County), 2,500 to 4,500 feet, March to May. Vancouver Island to southern California; perhaps introduced in Arizona.

The Arizona plants belong apparently to var. *Pipersmithii* (Heller) C. P. Smith, having a nonciliate keel, although Smith (147, p. 124) identified a collection from the Mazatzal Mountains (*Eastwood* 17343) as var. *microphyllus* (Wats.) C. P. Smith.

6. Lupinus brevicaulis Wats. Apache County to Mohave County, south to Cochise, Santa Cruz, and Pima counties, 3,000 to 7,000 feet, dry slopes and mesas, April to July. Colorado to Oregon, New Mexico, Arizona, and California.

7. Lupinus Kingii Wats. (*L. capitatus* Greene). Apache, Navajo, Coconino, and Yavapai counties, perhaps also in Cochise and Pima counties, 6,000 to 8,000 feet, usually in open pine forest, June to September, type of *L. capitatus* from near Flagstaff (*Lemmon* in 1884). Colorado, Utah, New Mexico, and Arizona.

8. Lupinus pusillus Pursh. Apache County to northeastern Mohave County, 4,500 to 8,000 feet, sandy plains, May and June. Saskatchewan to Washington, south to Kansas, New Mexico, and northern Arizona.

The Arizona specimens belong mostly to the relatively small-flowered var. *intermontanus* (Heller) C. P. Smith, which intergrades completely with the typical plant.

9. Lupinus rubens Rydb. Navajo County to Mohave County, 2,000 to 5,000 feet, April to June. Southern Utah and northern Arizona to southeastern California.

10. Lupinus odoratus Heller. Near Lake Mead, Peach Springs, Hackberry, Kingman, and Chloride (Mohave County), 3,000 to 4,500 feet, April and May. Western Arizona, Nevada, and southeastern California.

The Arizona specimens are not typical, belonging to var. *pilosellus* C. P. Smith. The handsome flowers are fragrant.

11. Lupinus Shockleyi Wats. Kingman to Peach Springs, Mohave County (*Lemmon* in 1884, *Eastwood* 18443), 3,500 to 5,000 feet. Western Arizona, Nevada, and southeastern California.

The Eastwood collection is less typical, approaching *L. odoratus* in the more sparsely pubescent herbage, more elongate peduncles, and nearly glabrous pedicels and calyx.

12. Lupinus Parishii Eastw. Williams (Coconino County), Prescott (Yavapai County), 5,000 to 7,000 feet, at Prescott along a stream in partial shade, May to July. Central Arizona and California.

Most of the Arizona specimens have a whitish or ochroleucous corolla, whereas *L. Parishii* was described from violet-flowered California specimens. The Arizona plant was identified by C. P. Smith (147, p. 139) as *L. latifolius* var. *Parishii* C. P. Smith, but Alice Eastwood (personal communication) regards *L. Parishii* as more nearly related to *L. polyphyllus* Lindl. than to *L. latifolius* Agardh, notwithstanding the absence of cilia on the keel in *L. polyphyllus.*

13. Lupinus huachucanus Jones (*L. platanophilus* Jones). Chiricahua and Huachuca mountains (Cochise County), Santa Rita Mountains (Pima or Santa Cruz County), 5,000 to 6,000 feet, pine woods, March to May, type of *L. huachucanus* from the Santa Rita Mountains (*Jones* in 1903), type of *L. platanophilus* from the Huachuca Mountains (*Jones* 26135). Southern Arizona and Chihuahua.

The plant somewhat resembles *L. concinnus,* but is readily distinguished by its ciliate keel and racemes surpassing the foliage. A specimen from the Chiricahua Mountains (*Price* in 1894), with exceptionally narrow leaflets, was referred by Smith (147, p. 124) to *L. chihuahuensis* Wats., but that species, although evidently related to *L. huachucanus,* differs from it in its much more sparsely pubescent herbage, interrupted-verticillate inflorescence, larger flowers, and appressed or subappressed pubescence of the pedicels and calyx.

14. Lupinus barbiger Wats. Kaibab Plateau to the South Rim of the Grand Canyon (Coconino County), 7,000 to 8,500 feet, June to August. Colorado, Utah, and northern Arizona.

L. barbiger was given as a synonym of *L. sericeus* Pursh by Smith (147, p. 138).

15. Lupinus Osterhoutianus C. P. Smith. Near the Grand Canyon, Coconino County (*Osterhout* 6971, in part, the type collection, not examined), North Rim of the Grand Canyon (*Behle* in 1942). Known only from northern Arizona.

Behle's specimen, identified from Smith's description, has bright-pink flowers, but is probably merely a color variation.

16. Lupinus argenteus Pursh. Mountains of Apache County and from Navajo Mountain and the Kaibab Plateau to the Flagstaff region (Coconino County), 7,000 to 10,000 feet, mostly in open coniferous forests, June to October. North Dakota and Montana, south to New Mexico and northern Arizona.

A widely distributed and polymorphic aggregate, including probably several entities that have been segregated as different species. Five varieties, all based on types from the Flagstaff region, were described by Smith (147, pp. 133–135). Intensive field study is needed to determine whether most of these are more than individual variations. Many of the specimens now considered as belonging to *L. argenteus* were referred formerly to *L. aduncus* Greene, a species that probably does not occur in Arizona (Alice Eastwood, personal communication). A form with densely silvery-sericeous herbage occurs on the Kaibab Plateau (*Eastwood & Howell* 1112, 6390, 6432, 6440). Specimens from Navajo Mountain (*Darsie* in 1933, *Peebles & Smith* 13954) with green, sparsely pubescent herbage and exceptionally long and narrow leaflets (35 to 50 mm. long, 4 to 5 mm. wide) were referred previously to *L. Greenei* A. Nels. Similar specimens have been collected in the Tunitcha Mountains, northern Apache County (*Goodman & Payson* 2923).

17. Lupinus Hillii Greene. Kaibab Plateau (Coconino County) to southern Apache and northern Gila counties, 6,000 to 9,000 feet, often very abundant in pine forests in the Flagstaff region, May to September, type from the Coconino National Forest (*Hill* in 1911). Known only from northern and central Arizona.

This species resembles *L. Palmeri* except in the smaller and usually more numerous and crowded flowers. Greene (*Leaflets* 2:236) stated: "The flowers are the smallest known among those of perennial lupines." The type of *L. ingratus* Greene var. *arizonicus* C. P. Smith, from the Grand Canyon (*Eggleston* 15664), seems scarcely distinct from *L. Hillii*. Specimens collected between Young and Payson, northern Gila County (*Peebles & Smith* 13283) resemble *L. Hillii* in the small flowers and definitely verticillate inflorescence but *L. Palmeri* in the leaves and in the pubescence. They may represent an interspecies hybrid.

18. Lupinus Lemmonii C. P. Smith. Chiricahua National Monument, Apache Pass, and Sulphur Springs (Cochise County), Santa Rita Mountains (Pima County), 4,000 to 4,500 feet, type from Sulphur Springs (*Lemmon* in 1881). Known only from southern Arizona.

Rothrock's specimens were referred by Smith to *L. Greenei* A. Nels., but they correspond closely to Smith's description of *L. Lemmonii* (147, p. 125). The stems are apparently slightly woody toward the base. Except in the closely appressed pubescence this species greatly resembles *L. Palmeri*.

19. Lupinus Palmeri Wats. Kaibab Plateau (Coconino County), Hualpai Mountain (Mohave County), and southward and southeastward through Yavapai and Gila counties to southern Navajo, Cochise, and Pima counties, 4,000 to 8,000 feet, mostly in pine forests, April to October, type from near Prescott (*Palmer* 754). New Mexico and Arizona.

The commonest and most widely distributed of the perennial lupines of Arizona. Two varieties, based on types from Coconino County, were described by Smith (147, p. 126). A pink-flowered form is found near Flagstaff.

L. Marcusianus C. P. Smith (147, p. 137) evidently is closely related to *L. Palmeri*, differing chiefly in the longer, narrower, more acute leaflets. The stem hairs are mostly appressed, but this condition is occasional in otherwise typical *L. Palmeri*.

20. Lupinus Cutleri Eastw. Apache County, in sandy soil northwest of Fort Defiance (*Cutler* 2141, the type collection) and in the Lukachukai Mountains (*Peebles* 14384, *Shreve* 9007), about 7,000 feet, June. Known only from northeastern Arizona.

This species resembles *L. aduncus* Greene in the spurred calyx but differs markedly in other characters.

21. Lupinus Blumeri Greene. Pinaleno, Chiricahua, Huachuca, and Santa Rita mountains (Graham, Cochise, and Santa Cruz counties), 6,000 to 9,000 feet, April to September, type from the Chiricahua Mountains (*Blumer* 1357). Apparently endemic in southeastern Arizona.

With the exception of *L. Parishii* this is the largest-flowered of the perennial Arizona lupines and is very showy in flower according to Blumer. The leaflets are up to 15 mm. wide.

22. Lupinus Sitgreavesii Wats. Lukachukai and White mountains (Apache County), San Francisco Peaks and vicinity (Coconino County), 6,500 to 9,500 feet, open coniferous forests, July to October, type from the San Francisco Peaks (*Sitgreaves* in 1851). Reported to occur also in southern Utah and western New Mexico.

This species seems ill-defined, differing from *L. argenteus,* as here interpreted, chiefly in the more spreading pubescence. What appears to be a glabrate form of *L. Sitgreavesii,* although referred by Smith (147, p. 136) to *L. alpestris* A. Nels., was collected between Springerville and Cooley Ranch, Apache or Navajo County (*Ferris* 1273).

A probably undescribed species of *Lupinus* related to *L. aridus* Dougl. and *L. lepidus* Dougl. has been collected recently between Frazier Well and Hualpai Hilltop, Coconino County (*J. T. Howell* 26610). The plant is an acaulescent or low-caulescent perennial with a deep taproot. In the preceding key it would come next to *L. huachucanus* Jones but is readily distinguished therefrom by the appressed or subappressed pubescence of the herbage and the larger corolla (10 to 12 mm. long).

18. MEDICAGO. Medick

Plants herbaceous, annual or perennial; leaves trifoliolate, the leaflets dentate; flowers small, in axillary racemes or heads; corolla yellow or violet; stamens diadelphous (1 separate from the other 9), the anthers all alike; pods indehiscent, strongly curved or spirally coiled.

All the species are natives of the Old World. Alfalfa (*M. sativa*) is the outstanding cultivated forage plant of the western United States, and the other weedlike species afford nutritious and palatable feed, although not sufficiently abundant in Arizona to be important. *M. hispida* is much used in California as a winter cover crop and green-manure crop.

KEY TO THE SPECIES

1. Plants perennial; stems erect; corolla violet. Flowers numerous, in rather dense racemes; pods spirally coiled, several-seeded, unarmed, usually glabrous..........1. *M. sativa*
1. Plants usually annual; stems procumbent; corolla yellow. Pods strongly veined (2).
2. Pods kidney-shaped, coiled in 1 plane, 1-seeded, pubescent, without marginal prickles or tubercles, black at maturity; flowers rather numerous, in ovoid spikes, these becoming oblong in fruit..2. *M. lupulina*

2. Pods spirally coiled, several-seeded, with marginal prickles or tubercles, straw-colored or brown at maturity; flowers few, in subcapitate clusters, or solitary (3).
3. Herbage and pods glabrous or nearly so; stipules broad, laciniate; marginal prickles of the pods long or short, or reduced to tubercles..................... 3. *M. hispida*
3. Herbage and pods copiously pubescent; stipules narrow, entire or dentate; marginal prickles usually long... 4. *M. minima*

1. Medicago sativa L. Here and there at roadsides, an occasional escape from cultivation but apparently nowhere naturalized in the state. Alfalfa, lucerne.

2. Medicago lupulina L. Navajo, Coconino, Yavapai, Graham, Gila, and Pima counties, 2,500 to 8,000 feet, occasional at roadsides and in lawns. Extensively naturalized in North America.

Black medick, none-such. A troublesome weed in lawns in some parts of the United States. Although usually annual this plant shows a tendency to perennial growth in Arizona.

3. Medicago hispida Gaertn. Yavapai, Gila, Maricopa, Pinal, Cochise, and Pima counties, occasional in waste land, March to May.

Bur-clover. Extensively naturalized in California and elsewhere in the United States. A specimen collected at Eloy, Pinal County (*Peebles et al.* 6498), with the marginal prickles reduced to low tubercles, belongs to var. *confinis* (Koch) Burnat (*M. apiculata* of authors, non Willd.).

4. Medicago minima (L.) Grufberg. Oak Creek Canyon, Coconino County (*J. T. Howell* 24385). A rare adventive in Texas, Arizona, and California.

19. MELILOTUS. Sweet-clover

Plants annual or biennial, with leafy, branched stems; herbage fragrant when dried; leaves trifoliolate, the leaflets dentate; flowers small, in elongate narrow racemes; petals white or yellow; stamens diadelphous, the anthers all alike; pods small, ovoid or globose; seeds 1 or few.

Coumarin, the substance that gives the strong odor to these plants, makes them distasteful to livestock at first, but animals acquire a liking for them, especially for the cured hay. White sweet-clover (*M. albus*) is now cultivated as a forage plant in some parts of the United States. Sour-clover (*M. indicus*) is extensively grown in the Southwest as a winter cover crop and occasionally escapes from cultivation. These plants, especially *M. albus*, do well on moderately saline soil. They are excellent honey plants. The Arizona species are all introductions from Eurasia.

KEY TO THE SPECIES

1. Flowers not more than 2.5 mm. long; stems commonly less than 1 m. long. Corolla yellow; pods alveolate-rugose.. 1. *M. indicus*
1. Flowers not less than 4 mm. long; stems often more than 1 m. long (2).
2. Corolla yellow, the banner not or but slightly surpassing the wings....... 2. *M. officinalis*
2. Corolla white, the banner considerably surpassing the wings................ 3. *M. albus*

1. Melilotus indicus (L.) All. Coconino, Mohave, Yavapai, Maricopa, Pinal, Cochise, Pima, and Yuma counties, occasional at roadsides, along ditches, and in fields, April to September. Sour-clover.

2. Melilotus officinalis (L.) Lam. Shato (northern Navajo County), Kaibab Plateau to Oak Creek Canyon (Coconino County), near Prescott (Yavapai County), Pinal Mountains (Gila County), July to October. Yellow sweet-clover.

3. Melilotus albus Desr. Navajo, Coconino, Yavapai, Graham, Pinal, Cochise, and Pima counties, summer. White sweet-clover.

20. TRIFOLIUM. Clover

Plants herbaceous, annual, biennial, or perennial, leafy-stemmed or (a few species) subscapose; leaves mostly trifoliolate, the leaflets denticulate to deeply serrate; flowers in dense globose heads or short spikes, these frequently subtended by an involucre, the flowers often reflexed after anthesis; corolla mostly rose red or purple to nearly white, usually persistent; pods small, usually terete, indehiscent or tardily dehiscent; seeds 1 or few.

The clovers, although palatable and nutritious to livestock and beneficial to the soil, are not sufficiently abundant in Arizona to be of much economic importance. Most of the native species grow in wet soil along streams in the pine belt. Several of them are sod-forming, but in the situations where they grow they can have little importance in protecting against erosion. The introduced red clover (*T. pratense*) and alsike clover (*T. hybridum*) are important forage plants elsewhere in the United States and in Europe. White clover (*T. repens*), likewise an introduced species, is a useful constituent of pastures and lawns in cooler climates, and the flowers yield honey of the highest quality. The plants and seeds of the native species are reported to have been used as food by the Arizona Indians.

KEY TO THE SPECIES

1. Heads subtended by 2 or 3 broad membranaceous bracts (modified stipules), these often bearing reduced trifoliolate blades. Plants subacaulescent, the flowering stems numerous from a much-branched, woody caudex; herbage and calyces villous; scapes up to 5 cm. long........1. *T. andinum*
1. Heads not so subtended, the involucral bracts, if present, not bearing blades (2).
2. Involucre manifest, the bracts in a whorl and usually more or less united (3).
2. Involucre greatly reduced, or obsolete (9).
3. Pubescence copious, at least on the peduncles and involucres. Leaflets obcordate to oblanceolate; involucre saucer-shaped, the bracts united at least 1/3 of their length, entire, ovate, scarious-margined, aristate; heads 8 to 12 mm. wide, very dense; flowers not reflexed; corolla not, or but slightly, surpassing the calyx....2. *T. microcephalum*
3. Pubescence sparse or none (4).
4. Heads normally not more than 10 mm. wide and not more than 10-flowered. Stems prostrate to ascending, slender; leaflets oblanceolate or obovate; involucre 1/2 to 2/3 as long as the calyces, with usually obovate, deeply toothed lobes; lower flowers becoming strongly reflexed; calyx teeth usually dark purple; corolla 6 to 9 mm. long, much surpassing the calyx........3. *T. variegatum*
4. Heads normally more than 10 mm. wide and more than 10-flowered (5).
5. Stipules lanceolate, long-attenuate, entire or with a few short teeth; leaflets oblanceolate or obovate, obtuse, truncate, or slightly retuse at apex; peduncles (and sometimes the petioles and upper portion of the stems) often sparsely pilose; involucre cleft nearly to the base, the divisions entire, subulate-setaceous to narrowly lanceolate........4. *T. pinetorum*
5. Stipules broad, oblong-lanceolate to ovate, deeply and sharply toothed to laciniate; leaflets prevailingly elliptic, oblanceolate, or linear-lanceolate, commonly acutish to

acuminate at apex; peduncles normally glabrous; involucre with an entire, cup-shaped basal portion forming usually at least ⅓ of the length of the involucre, the divisions relatively broad and deeply cleft, with setaceous teeth (6).

6. Heads (in pressed specimens) 1 to 2 cm. wide; corolla less than 12 mm. long (7).
6. Heads 2 to 4 cm. wide; corolla 12 mm. or longer (8).
7. Leaflets narrowly linear, lanceolate, or oblanceolate, often conspicuously cuspidate, spinulose-serrate, often deeply so .... 5. *T. lacerum*
7. Leaflets oblong-lanceolate or elliptic, not conspicuously cuspidate, serrulate 6. *T. arizonicum*
8. Leaflets prevailingly elliptic, oblanceolate, or (in the lower leaves) obovate, usually less than 3 cm. long; stems not fistulose .... 7. *T. Fendleri*
8. Leaflets prevailingly linear-lanceolate, mostly 3 to 4 cm. long; stems fistulose 8. *T. fistulosum*
9. Involucre usually manifest although very small (10).
9. Involucre none, but the minute bractlets subtending the individual flowers of *T. Rusbyi* (and perhaps other species) sometimes simulating a greatly reduced involucre (11).
10. Plants caulescent. Stems prostrate; leaflets cuneate-obovate, glabrous or glabrate; petioles, peduncles, and calyx villous; involucre minute, the bracts seldom more than 1 mm. long; flowers strongly reflexed .... 9. *T. amabile*
10. Plants acaulescent or nearly so. Leaflets often more than 3, elliptic to obovate, doubly dentate, mucronate; involucre minute, almost obsolete, all the bracts less than 2 mm. long, truncate or erose, entirely scarious; calyx copiously villous .. 10. *T. subcaulescens*
11. Plants annual or biennial (12).
11. Plants perennial, but sometimes short-lived (14).
12. Leaflets sericeous or villous, up to 25 mm. long; heads 10 to 15 mm. long; calyx villous with long hairs; flowers not noticeably reflexed after anthesis. Corolla dark purple and whitish .... 11. *T. albopurpureum*
12. Leaflets glabrous or nearly so, seldom more than 12 mm. long; heads not more than 10 mm. long; calyx glabrous or nearly so; flowers strongly reflexed after anthesis (13).
13. Calyx teeth much longer than the tube; corolla pink to reddish purple 12. *T. gracilentum*
13. Calyx teeth not or little longer than the tube; corolla yellow .... 13. *T. dubium*
14. Leaves (the uppermost pair) subtending the sessile or subsessile head; flowers not becoming reflexed. Stipules very large and conspicuously veined; heads at anthesis seldom less than 2 cm. in diameter; corolla rose-colored .... 14. *T. pratense*
14. Leaves not subtending the head, the latter long-peduncled; flowers (the lower ones) becoming reflexed after anthesis (15).
15. Calyx teeth glabrous or nearly so, less than twice as long as the tube; peduncle glabrous or nearly so; corolla white or tinged with pink (16).
15. Calyx teeth villous-plumose, 2 to 3 times as long as the tube; peduncle sparsely to densely villous, especially toward apex; corolla deep pink (17).
16. Stems repent and often rooting at the nodes; leaflets obovate, notched at apex 15. *T. repens*
16. Stems erect or ascending; leaflets oval or ovate, rounded at apex .... 16. *T. hybridum*
17. Heads 10 to 15 mm. long from the apex of the peduncle; leaflets elliptic to narrowly obovate, the larger ones seldom more than 2 cm. long, sparsely sericeous or villous beneath .... 17. *T. Rusbyi*
17. Heads seldom less than 20 mm. long; leaflets narrowly lanceolate, the larger ones 3 cm. or longer, densely sericeous beneath with long hairs, at least when young. Leaflets 5 to 10 times as long as wide, sharply acuminate at apex, the marginal teeth incurved 18. *T. neurophyllum*

1. Trifolium andinum Nutt. North Rim of Grand Canyon (*Mead* in 1929, *Eastwood & Howell* 941, *Collom* in 1945), about 8,000 feet, June. Wyoming and Utah to northern Arizona.

2. Trifolium microcephalum Pursh. Pinal and Mazatzal mountains (Gila County), near Superior (Pinal County). Montana to British Columbia, south to Arizona and Baja California.

3. Trifolium variegatum Nutt. Near Kirkland, Yavapai County (*Harrison & Peebles* 4194), Santa Rita Mountains, Pima County (*Thornber* in 1905). Montana to British Columbia, south to Arizona and Baja California.

4. Trifolium pinetorum Greene (*T. longicaule* Woot. & Standl., *T. Wormskjoldii* var. *longicaule* L. Benson). Kaibab Plateau and North Rim of Grand Canyon (Coconino County), mountains of Apache, Graham, Cochise, Santa Cruz, and Pima counties, 6,500 to 9,000 feet, moist soil in coniferous forests, June to October. New Mexico and Arizona.

The glabrate plant, with few-flowered heads, found on the Kaibab Plateau, probably should be distinguished, at least as a variety.

5. Trifolium lacerum Greene. Coconino County to Cochise and Pima counties, 2,000 to 8,000 feet, wet places, March to August. New Mexico and Arizona.

6. Trifolium arizonicum Greene. Near Flagstaff, Coconino County (type *Rusby* in 1883), about 7,000 feet, August. Northern Arizona and southeastern California.

7. Trifolium Fendleri Greene. Apache County to Coconino County, south to Greenlee and Gila counties, 7,000 to 9,500 feet, moist soil in coniferous forests and meadows, June to September. Colorado, Utah, New Mexico, and Arizona.

8. Trifolium fistulosum Vaughan. Cochise County, in the Huachuca Mountains (*Lemmon* in 1882) and Sulphur Springs Valley (*Thornber* 2249), July and August. Southeastern Arizona and Mexico.

This and the 4 preceding species are related, perhaps too closely, to the Mexican *T. Ortegae* Greene (*T. involucratum* Ortega) and to *T. Wormskjoldii* Lehm. (*T. Wildenovii* Spreng.) of the Pacific coast states.

9. Trifolium amabile H.B.K. Huachuca Mountains (Cochise County), about 6,000 feet, wet, sandy soil near springs and brooks, August to October. Southeastern Arizona to Central America.

10. Trifolium subcaulescens Gray. Apache County to Coconino County, 7,000 to 10,000 feet, pine and spruce-fir forests, May and June, type from near Fort Defiance, Apache County (*Newberry* in 1858). Colorado, New Mexico, and northern Arizona.

Flowers pink or whitish. Leaflets often with a whitish, rounded-triangular mark, as in red clover (*T. pratense*). Perhaps not specifically distinct from *T. gymnocarpon* Nutt. (148).

11. Trifolium albopurpureum Torr. & Gray. Gila and Maricopa counties, 3,000 to 4,000 feet, March to May. Arizona and California. Rancheria clover.

12. Trifolium gracilentum Torr. & Gray. Gila, Maricopa, Pinal, and Pima counties, 2,500 to 4,000 feet, March to May. Washington to Arizona and California.

13. Trifolium dubium Sibth. Tucson (Pima County), "a common weed in lawns" (*Thornber* 5419, etc.). Naturalized from Europe. Shamrock.

14. Trifolium pratense L. Occasionally growing wild in Arizona, as at Lakeside (Navajo County) and McNary (Apache County); introduced from Europe. Red clover, extensively cultivated in the eastern United States.

15. Trifolium repens L. Roadsides and lawns, near Lakeside (Navajo County), Grand Canyon (Coconino County), Kingman (Mohave County), Prescott (Yavapai County), Chiricahua Mountains (Cochise County). Introduced from Europe. White clover, often used in lawn mixtures.

16. Trifolium hybridum L. North Rim of the Grand Canyon, Coconino County (*Collom* in 1940). Here and there in the United States; naturalized from Europe. Alsike clover.

17. Trifolium Rusbyi Greene (*T. longipes* Nutt. var. *pygmaeum* Gray). White Mountains (Apache County), Kaibab Plateau, San Francisco Peaks, Bill Williams Mountain (Coconino County), 7,000 to 8,500 feet, coniferous forests, June to September, types of *T. Rusbyi* from on or near the San Francisco Peaks (*Lemmon, Rusby*), type of *T. longipes* var. *pygmaeum* from Bill Williams Mountain (*Newberry* in 1858). Colorado to Arizona, probably also in Oregon and northern California.

18. Trifolium neurophyllum Greene. White Mountains, Apache and Greenlee counties (*Goodding* 1076, *Kearney & Peebles* 12423), 8,000 to 8,500 feet, August. New Mexico and eastern Arizona.

Easily distinguished from most of the Arizona clovers by the very large flower heads.

21. LOTUS (149). DEER-VETCH

Plants annual or perennial, herbaceous or suffrutescent; stems leafy; leaves pinnately compound, but sometimes appearing digitate by shortening of the rachis, the leaflets 3 or more (rarely only 1), entire; flowers axillary, solitary or in few-flowered umbellike clusters, sessile or pedunculate; corolla yellow to reddish orange (in one species whitish, fading pink); filaments all or some of them flattened, at least near the apex; pods narrow, subterete or somewhat flattened, several-seeded.

These plants are known also as deer-clover and red-and-yellow-pea. Most of the species are grazed or browsed by domestic animals and deer, and the forage is rated as of fair to good quality. *L. Wrightii* and *L. alamosanus* are considered excellent for control of erosion. A decoction of the leaves and flowers of *L. rigidus* is reported to have been used as a tonic by the early settlers, under the name of Hills tea.

Many of the Arizona specimens have been identified by Ottley (149) as interspecies hybrids, but in default of cytogenetic evidence these conclusions need substantiation.

KEY TO THE SPECIES

1. Stipules herbaceous or membranaceous. Plants perennial; flowers umbellate, on peduncles much surpassing the leaves; corolla yellow; pods glabrous (2).
1. Stipules glandlike or obsolete (4).
2. Stipules herbaceous, resembling and nearly as large as the leaflets; calyx turbinate, firm. Leaves trifoliolate; umbels compact, 3- to 12-flowered............1. *L. corniculatus*
2. Stipules membranaceous, much smaller than the leaflets; calyx cylindric or campanulate, thin (3).
3. Stems stout, erect or nearly so; leaflets 7 to 11, narrowly lanceolate or elliptic to obovate, 1 to 3 cm. long; corolla 10 to 14 mm. long, purple-veined; pods about 2 mm. wide 2. *L. oblongifolius*
3. Stems slender, procumbent; leaflets 3 to 5, broadly obovate, less than 1 cm. long; corolla 5 to 7 (11?) mm. long; pods 1 to 1.5 mm. wide....................3. *L. alamosanus*

4. Corolla whitish, often fading pink; leaves mostly trifoliolate; calyx teeth subulate, much longer than the tube. Plants annual, more or less pubescent; stems erect or ascending, branched . . . . 4. *L. Purshianus*
4. Corolla yellow or orange; leaves (some or all of them) usually with more than 3 leaflets; calyx teeth not, or not much, longer than the tube (5).
5. Plants annual, flowering in spring; corolla not more (usually less) than 7 mm. long. Stems decumbent to prostrate (6).
5. Plants perennial; corolla not less than 8 mm. long (9).
6. Pods indehiscent, with a beak (persistent style) nearly as long as to longer than the slender body and strongly incurved toward the apex like a fishhook. Umbels nearly sessile . . . . 5. *L. hamatus*
6. Pods dehiscent, with a straight or moderately curved beak much shorter than the body (7).
7. Herbage soft-villous, usually copiously so; flowers solitary, sessile or subsessile; pods oblong, 2.5 to 4 mm. wide, villous with long hairs. Calyx teeth much longer than the tube . . . . 6. *L. humistratus*
7. Herbage strigose with short hairs, often glabrescent; flowers in small clusters or solitary, the inflorescence distinctly stalked; pods linear, usually less than 2.5 mm. wide, strigose or glabrescent (8).
8. Rachis of the leaves usually at least 1 mm. wide; pubescence not very closely appressed; leaflets thickish, slightly succulent, cuneate-oblanceolate to rather narrowly obovate, rounded or truncate and often emarginate at apex; calyx teeth shorter than the tube 7. *L. tomentellus*
8. Rachis of the leaves usually less than 1 mm. wide; pubescence closely strigose, sparse, the plant often glabrate; leaflets very thin, broadly obovate, rounded to acutish and often mucronulate at apex; calyx teeth equaling or longer than the tube 8. *L. salsuginosus*
9. Stems rigid, erect or ascending, somewhat woody below, the internodes commonly more than twice as long as the leaves. Leaves 3- to 5-foliolate, the rachis distinct in leaves with more than 3 leaflets; leaflets mostly obtuse or truncate, sometimes emarginate; peduncles much surpassing the leaves; corolla 15 to 25 mm. long; herbage and pods glabrescent . . . . 9. *L. rigidus*
9. Stems not rigid, herbaceous above the caudex, the internodes usually much less than twice as long as the leaves (10).
10. Peduncles usually shorter than the leaves or obsolete; leaves appearing digitate, the rachis usually obsolete. Leaflets (at least those of the upper leaves) acute or acutish; corolla 10 to 20 mm. long; pods commonly strigose at maturity. . . . . . 10. *L. Wrightii*
10. Peduncles much longer than the leaves; leaves pinnate, the rachis (except in *L. utahensis*) usually distinct (11).
11. Stems, leaves, and pods copiously villous. Stems decumbent or nearly prostrate; leaflets broadly cuneate-obovate to narrowly oblanceolate, rounded and apiculate to acute at apex, up to 15 but usually less than 10 mm. long, 2 to 6 mm. wide; calyx teeth shorter to somewhat longer than the tube, villous; corolla 12 to 18 mm. long; pods 20 to 30 mm. long, 2.5 to 3.5 mm. wide. . . . . 11. *L. Greenei*
11. Stems, leaves, and pods strigose or sericeous (12).
12. Pubescence silvery; pods 4 to 5 mm. wide, 15 to 25 mm. long. Stems decumbent or nearly prostrate; leaflets obovate or oblanceolate, obtuse, sometimes emarginate, not more and usually less than 15 mm. long, 3 to 6 mm. wide; calyx teeth commonly shorter than the tube; corolla 12 to 18 mm. long. . . . . 12. *L. Mearnsii*
12. Pubescence not silvery; pods not more and usually less than 4 mm. wide, 20 to 40 mm. long (13).
13. Leaflets 4 to 13, prevailingly linear-lanceolate or narrowly oblanceolate and acute, only those of the lowest leaves more than 2 mm. wide; calyx teeth subulate-setaceous, equaling or somewhat longer than the tube, villous; corolla up to 18 mm. long. Stems commonly erect . . . . 13. *L. oroboides*
13. Leaflets 3 to 6 (rarely more), oblanceolate, obtuse or acutish, the larger ones usually 3 mm. wide or wider; calyx teeth subulate or narrowly lanceolate, shorter than to equaling the tube, pubescent with more or less appressed hairs; corolla seldom more than 15 mm. long (14).

14. Stems erect or ascending; leaves mostly sessile; peduncles mostly 2- or 3-flowered
14. *L. utahensis*

14. Stems decumbent to ascending; leaves mostly short-petioled; peduncles 1- or 2-flowered
15. *L. neomexicanus*

1. Lotus corniculatus L. Skull Valley, Yavapai County (*Thornber* 8798, in 1916). Adventive here and there in the United States; from Eurasia. Birdsfoot-trefoil.

2. Lotus oblongifolius (Benth.) Greene. Huachuca Mountains, Cochise County (*Lemmon* 2670). Southern Arizona, southern California, and northern Mexico.

The presence of this species of the section *Hosackia* in southeastern Arizona is remarkable, and an error in the locality as stated on the labels of Lemmon's specimens would be suspected were it not that a very similar specimen collected near Colonia García, Chihuahua (*Townsend & Barber* 159), was identified by Greene as *L. Torreyi* Greene (*L. oblongifolius* var. *Torreyi* Ottley), to which Lemmon's specimens also are referrable.

3. Lotus alamosanus (Rose) Gentry. Bear Valley or Sycamore Canyon (Santa Cruz County), about 4,000 feet, reported to be abundant, flowering April and May. Southern Arizona, Sonora, and Durango.

4. Lotus Purshianus (Benth.) Clements & Clements (*L. americanus* (Nutt.) Bisch., non Vell.). Apache, Navajo, Yavapai, Gila, and Cochise counties, 5,000 to 7,500 feet, open pine forests, July to October. Minnesota to Washington, New Mexico, Arizona, and California. Spanish-clover.

5. Lotus hamatus Greene. Robles Ranch near Tucson, "on flats" (*Goodding* 1672), about 2,500 feet, April, probably introduced from southern California.

6. Lotus humistratus Greene. Throughout most of the state except the northeastern portion, 5,000 feet or (usually) lower, very common on sandy deserts, March to June. New Mexico to California and northern Mexico.

7. Lotus tomentellus Greene. Mohave, Graham, Maricopa, Pinal, Pima, and Yuma counties, 3,000 feet or lower, very common on sandy deserts, March to May. Southern Nevada, Arizona, southeastern California, and northwestern Mexico.

8. Lotus salsuginosus Greene. Mohave, Yavapai, Maricopa, Pinal, Pima, and Yuma counties, 3,000 feet or lower, common on dry hills and mesas, February to May. Arizona, California, and Baja California.

The Arizona specimens all belong to var. *brevivexillus* Ottley (*L. humilis* Greene), with plants and flowers smaller than in typical *L. salsuginosus.*

9. Lotus rigidus (Benth.) Greene. Southern Apache County to Mohave County, south to Graham, Pima, and Yuma counties, 5,500 feet or lower, dry, rocky slopes, February to May. Southern Utah and Nevada, Arizona, southeastern California, and Baja California.

This is Arizona's most xerophytic *Lotus*. Presumable hybrids with *L. oroboides* from the Santa Catalina Mountains, with *L. Wrightii* from many parts of the state, with *L. Greenei* from the Santa Catalina and Baboquivari mountains, and with *L. Mearnsii* from Mohave, Gila, Maricopa, and Pinal counties were mentioned by Ottley (149, pp. 108, 116, 120).

10. Lotus Wrightii (Gray) Greene. Apache County to Mohave County, south to Cochise and Pima counties, 4,500 to 9,000 feet, very common in dry, open pine forests, May to September. Southwestern Colorado and southern Utah to New Mexico, Arizona, and southeastern California.

Hybrids with *L. Mearnsii* from Coconino, Mohave, Yavapai, and Gila counties were mentioned by Ottley (149, p. 122).

11. Lotus Greenei (Woot. & Standl.) Ottley (*Hosackia mollis* Greene non Nutt., *H. Greenei* Wiggins). Graham, Pinal, Cochise, Santa Cruz, and Pima counties, 3,000 to 5,000 feet, mesas and rocky hillsides, March to May (sometimes August), type of *H. mollis* from the Huachuca Mountains (*Lemmon* 2669). Southern New Mexico, southern Arizona, and northern Mexico.

Hybrids of *L. Greenei* with *L. oroboides* from Cochise, Santa Cruz, and Pima counties were mentioned by Ottley (149, p. 119).

12. Lotus Mearnsii Britton. Navajo, Coconino, Mohave, and Yavapai counties, 3,000 to 7,000 feet, grassland and dry mesas and slopes, March to August, type from Fort Verde, Yavapai County (*Mearns* 342). Known only from Arizona.

13. Lotus oroboides (H.B.K.) Ottley (*L. puberulus* (Benth.) Greene). Southern Coconino County to Cochise, Santa Cruz, and Pima counties, 5,000 to 8,000 feet, mostly in pine woods, April to September. Western Texas to Arizona and Mexico.

14. Lotus utahensis Ottley. Coconino and Mohave counties, mostly north of the Colorado River, 3,500 to 9,000 feet, April to August. Southern Utah and northern Arizona.

This species is nearly related to *L. Wrightii,* differing chiefly in the elongate peduncles. The plants now referred here were treated previously as *L. longebracteatus* Rydb., but Ottley (149, p. 108) concluded that the type of the latter and also of *Hosackia rigida* var. *nummularia* Jones represent hybrids between *L. utahensis* and *L. rigidus.* Similar plants of more or less intermediate character are not uncommon where the ranges of these species overlap in northern Arizona.

15. Lotus neomexicanus Greene. Greenlee and Graham counties, 3,500 to 5,000 feet, March to July. Western New Mexico and eastern Arizona.

According to Ottley (149, pp. 117, 118) *L. neomexicanus* is probably only an atypical form of *L. Greenei,* but Greene's description (*Pittonia* 2:144) does not correspond to *L. Greenei* while applying rather closely to specimens from Greenlee and Graham counties in the herbarium of the University of Arizona, here referred tentatively to *L. neomexicanus.*

22. INDIGOFERA. Indigo

Low shrubs, up to 1.2 m. (4 feet) high; herbage canescent with appressed hairs, these attached at the middle; leaves pinnate, the leaflets numerous; flowers in axillary racemes, small; corolla white or pink; pods small, nearly globose, 1-seeded, tardily dehiscent.

Before the invention of synthetic dyes the blue dye indigo was obtained from *Indigofera tinctoria,* an Old World species, and from the Mexican *I. suffruticosa.* Indigo was an important crop at one time in the southeastern United States. Arizona's single native species is browsed but is not abundant enough to be important. It is considered useful for erosion control.

1. Indigofera sphaerocarpa Gray (*Amorpha ovalis* Jones). Cochise, Santa Cruz, and Pima counties, 4,000 to 6,000 feet, July and August, type of *Amorpha ovalis* from the Huachuca Mountains (*Jones* 25027). Southeastern Arizona and northern Mexico.

23. PSORALEA. Scurf-pea

Plants perennial, herbaceous; flowering stems from a thick tuberlike taproot or from branched rootstocks, scapose or leafy; herbage glandular-punctate; leaves digitately 3- to 5-foliolate; inflorescences axillary or appearing terminal in the scapose species; calyx regular or irregular; pods small, ovoid, indehiscent, 1-seeded.

The tuberous-rooted species, known as bread-root, were used for food by the Indians and early white settlers. *P. lanceolata* is an admirable soil binder in sandy areas in northern Arizona, but is sometimes a troublesome weed in fields because it propagates by creeping rootstocks. Its presence is often an indication of overgrazing. *P. tenuiflora* has been reported to poison horses and cattle.

KEY TO THE SPECIES

1. Flowering stems from a very thick, tuberous, rounded or fusiform taproot; plants conspicuously pubescent, often subacaulescent, the main stem not more than 10 cm. long; leaves prevailingly 5-foliolate; leaflets broadly obovate to nearly orbicular; inflorescence conspicuously bracteate, very dense; pods regularly or irregularly cirsumscissile near the middle, the beak equaling or longer than the body of the pod: section *Pediomelum* (2).

1. Flowering stems from branched rootstocks, the roots not tuberous-thickened; plants not conspicuously pubescent, strongly caulescent, the main stem seldom less than 20 cm. long; leaves mostly trifoliolate; leaflets narrowly obovate, oblanceolate, or linear; inflorescence inconspicuously bracteate, loose or moderately dense; pods indehiscent, the beak much shorter than the body of the pod, the surface glandular-warty: section *Psoralidium* (5).

2. Lowest calyx lobe much more than twice as wide as the others, spatulate or obovate; seeds transversely wrinkled. Petioles appressed-pubescent; leaflets cuneate-obovate or rhombic; inflorescence 2 to 4 cm. long; corolla not more than 10 mm. long 1. *P. castorea*

2. Lowest calyx lobe about twice as wide as the others; seeds smooth (3).

3. Hairs of the petiole and peduncle all appressed or ascending; corolla about 20 mm. long; inflorescence about 2 cm. long. Leaflets cuneate-obovate.......2. *P. megalantha*

3. Hairs of the petiole and peduncle (the longer hairs) spreading or retrorse; corolla 10 to 17 mm. long; inflorescence 2 to 7 cm. long (4).

4. Plants acaulescent; leaflets grayish green above, copiously pubescent, the punctate glands more or less concealed by the hairs; banner noticeably to greatly surpassing the longest calyx lobe...3. *P. mephitica*

4. Plants short-caulescent; leaflets bright green above, sparsely pubescent or glabrescent, the glands conspicuous; banner barely to moderately surpassing the longest calyx lobe 4. *P. epipsila*

5. Corolla whitish, the tip of the keel often purple; pods subglobose, rounded and abruptly beaked at apex; leaves all trifoliolate. Leaflets oblanceolate to nearly filiform 5. *P. lanceolata*

5. Corolla violet; pods ovoid, somewhat tapering into the beak; lower leaves often 4- or 5-foliolate (6).

6. Leaves (except those near the base of the flowering stems), reduced to small, subulate scales; flowers few, distant, subsessile; pods white-sericeous; plants shrubby 6. *P. juncea*

6. Leaves well developed, mostly trifoliolate; flowers numerous, usually distinctly pedicelled, the pedicels 1 to 4 mm. long; pods glabrous; plants herbaceous........7. *P. tenuiflora*

1. **Psoralea castorea** Wats. Northern Mohave County, at Beaver Dam and Pagumpa Springs, 2,000 to 4,000 feet, sandy soil, April and May, type from between Beaver Dam, Arizona, and St. Thomas, Nevada (*Palmer*). Southern Utah and northwestern Arizona to southeastern California.

*2. **Psoralea megalantha** Woot. & Standl. Not known to occur in Arizona, but has been collected near the borders in New Mexico and Utah.

3. **Psoralea mephitica** Wats. Apache County to Mohave County, south to Greenlee, western Gila, and eastern Yavapai counties, 3,000 to 5,500 feet, April to June. Southern Utah and Arizona.

The type of var. *retrorsa* (Rydb.) Kearney & Peebles (*Pediomelum retrorsum* Rydb.) was collected at Peach Springs, Mohave County (*Lemmon* in 1884). The variety differs from the typical plant in its larger leaflets (up to 4 cm. long), longer peduncles and inflorescences, and larger corollas.

4. **Psoralea epipsila** Barneby. Kaibab Plateau (Buckskin Mountains), Coconino County (*Jones* in 1890), June. Southern Utah and northern Arizona.

5. **Psoralea lanceolata** Pursh (*P. micrantha* Gray). Apache County to Coconino County, 5,500 to 7,500 feet, sandy soil, open mesas and pine forests, May to September. Saskatchewan, Alberta, and Washington, south to Missouri, Texas, and northern Arizona.

Lemon scurf-pea, lemon-weed. *P. micrantha* tends to have narrower leaflets than typical *P. lanceolata,* but there seem to be no other differences.

*6. **Psoralea juncea** Eastw. This species occurs in southeastern Utah about 15 miles from the Arizona state line and is to be looked for in northeastern Arizona.

The plant was described by Alfred R. Purchase (personal communication) as a "symmetric rounded bush."

7. **Psoralea tenuiflora** Pursh. Apache County to Mohave County, south to Cochise, Santa Cruz, and Pima counties, 4,000 to 7,500 feet, dry slopes and plains, often in open pine forest, May to September. North Dakota and Montana to Arizona and northern Mexico.

The prevailing form in Arizona is var. *Bigelovii* (Rydb.) Macbr. (*Psoralidium Bigelovii* Rydb.), which seems to differ from typical *P. tenuiflora* only in its broader leaflets.

24. PARRYELLA

Much-branched shrubs with glandular-punctate herbage; leaves pinnate, the leaflets numerous; flowers small, yellowish, in terminal, elongate, often branched, spikelike racemes; petals none; stamens 9 or 10, distinct, attached at base to the calyx; pods small, conspicuously gland-dotted, 1-seeded.

These plants, which reach a height of about 1 m. and are slightly aromatic, occur sporadically in sandy soil in northern Arizona, often forming low dunes. They are well adapted to reduce wind erosion and are sometimes planted for this purpose. They are employed for making baskets and brooms by the Hopi Indians, who also are reported to use the seeds in treating toothache and the plant as an insecticide. Tests by the Bureau of Plant Industry showed the insecticidal value of *Parryella* to be small.

KEY TO THE SPECIES

1. Herbage inconspicuously strigose; leaflets linear-filiform to narrowly elliptic, not more than 2 mm. wide, 5 to 15 mm. long; stipules minute to about 1 mm. long; racemes loose, commonly more than 2 (up to 13) cm. long; calyx 3 to 4 mm. long, eglandular or sparsely glandular-punctate, nearly or quite glabrous except the often conspicuously white-ciliate margin; pods oblong-obovoid, 2.5 to 3 mm. wide..........1. *P. filifolia*
1. Herbage densely appressed-pubescent; leaflets broadly oval to nearly orbicular, 3 to 5 mm. wide, not more than 6 mm. long; stipules 2 to 5 mm. long; racemes dense, 1 to 2 cm. long; calyx 4 to 6 mm. long, conspicuously glandular-punctate, copiously pubescent on the whole surface externally, densely so on the lobes; pods ovoid or somewhat obovoid, 4 to 5 mm. wide....................................2. *P. rotundata*

1. Parryella filifolia Torr. & Gray. Apache County to Coconino County (a doubtful record from Fort Verde, Yavapai County), 4,000 to 6,000 feet, often on rolling, treeless, sandy plains with *Aplopappus heterophyllus,* June to September. New Mexico and northern Arizona.

2. Parryella rotundata Wooton. Winslow, Navajo County (*Wooton* in 1892, the type collection), Willow Spring, Coconino County (*Jones* in 1890, *Ripley & Barneby* 4876, 8495), 4,500 to 5,000 feet. Known only from northern Arizona.

Jones's collection was distributed by him as *Dalea nummularia* Jones. The plant was described by R. C. Barneby (*Leaflets West. Bot.* 4: 9) as growing "in crevices of hard red sandstone pavement, forming low gnarled bushes of rounded outline and only two or three decimeters high." It was stated to be much earlier-flowering than *P. filifolia,* which grew on near-by dunes.

25. AMORPHA (150). FALSE-INDIGO, INDIGO-BUSH

Shrubs, with gland-dotted herbage; leaves odd-pinnate, with numerous, rather large leaflets; flowers small, in dense terminal spikelike racemes; petal 1 (the banner or standard only), dark violet; stamens 9 or 10, united below; pods small, asymmetrically clavate or obovoid, gland-dotted, with 1 or 2 seeds, indehiscent or tardily dehiscent.

Sometimes called spice-bush in Arizona. The plants are unpalatable to livestock but are effective in controlling erosion. They are sometimes cultivated as ornamentals.

KEY TO THE SPECIES

1. Twigs, petioles, and leaf rachis bearing glands, these more or less spinelike; calyx lobes triangular-lanceolate, at least half as long as the tube, all acute or acutish, gland-tipped; pods pubescent; herbage soft-pilose......................1. *A. californica*
1. Twigs etc. without spinelike glands; calyx lobes deltoid, much less than half as long as the tube, the upper ones obtuse; pods glabrous; herbage strigose or glabrescent, exceptionally pilose..2. *A. fruticosa*

1. Amorpha californica Nutt. Yavapai, Graham, Cochise, and Pima counties, 5,000 to 6,500 feet, mostly along streams, apparently rare in Arizona, June. Arizona, California, and Baja California.

Plant sometimes known as stinking-willow and mock-locust.

2. Amorpha fruticosa L. Apache County to Coconino County, south to Cochise, Santa Cruz, and Pima counties, 2,500 to 6,500 feet, rich soil in canyons and along streams, fairly common, May and June. Ohio to Manitoba, south to Florida, Arizona, California, and northern Mexico.

The plants reach a height of 3 m. (10 feet). The Arizona variety is var. *occidentalis* (Abrams) Kearney & Peebles (*A. occidentalis* Abrams), with more elongate spikes and these usually fewer on the branches than in most of the Eastern specimens. E. J. Palmer (150, pp. 184, 185), who recognized *A. occidentalis* as a species, described under it two varieties, a glabrescent one with leaflets truncate to emarginate at apex (var. *emarginata* E. J. Palmer, *A. emarginata* Eastw.), type from Fish Creek, eastern Maricopa County (*Eastwood* 8745), and a copiously pubescent one (var. *arizonica* (Rydb.) E. J. Palmer, *A. arizonica* Rydb.), type from the Huachuca Mountains, Cochise County (*Goodding* 136).

26. EYSENHARDTIA. Kidney-wood

Plants shrubby or arborescent; leaves pinnate, the leaflets numerous, glandular-punctate; flowers many, small, in spikelike terminal racemes; corolla white, nearly regular; stamens 10, diadelphous; pods indehiscent, 1-seeded, flat, long-exserted.

The wood of some of the species is reputed to have diuretic properties and is fluorescent when soaked in water. The plants are browsed by cattle, horses, goats, and deer notwithstanding the rather disagreeable odor of the herbage.

1. Eysenhardtia polystachya (Ortega) Sarg. Graham, Pinal, Cochise, Santa Cruz, and Pima counties, 3,500 to 6,000 feet, usually among rocks in canyons, May to August. Southwestern New Mexico, southern Arizona, and Mexico.

27. DALEA (151). Indigo-bush, Pea-bush

Glandular-punctate herbs or shrubs; leaves odd-pinnate, rarely unifoliolate or digitate; bracts deciduous; calyx 5-toothed, persistent; petals clawed; stamens 9 or 10, rarely 7 or 8; pod small, indehiscent.

Most of the perennial species are ornamental, especially when flowering, and a number of them contribute to the forage value of the stock ranges. The Indians of southwestern Arizona formerly dyed basket material with an extract from the glandular twigs of *Dalea Emoryi*. The Hopi ate the roots of *D. terminalis* as a sweet. *D. scoparia* has been suggested for use in control of soil erosion.

KEY TO THE SPECIES

1. Petals inserted on the calyx at base of the stamen tube. Plants shrubby or arborescent (2).
1. Petals (the paired ones) manifestly inserted on the stamen tube, the scars visible on the tube after the petals fall (10).
2. Leaves unifoliolate (exceptionally trifoliolate), in one species caducous. Flowers dark blue (3).
2. Leaves several-foliolate. Plants commonly more or less spinescent (5).
3. Calyx externally glabrous or nearly so, shining. Arborescent shrubs with spinescent branches and persistent linear leaves; flowers racemose, not crowded; glands in each interval between the calyx ribs 4 or fewer, small, sometimes wanting. .3. *D. Schottii*
3. Calyx pubescent (4).
4. Leaves caducous; calyx strigose-canescent, with 1 large gland in each interval between the calyx ribs; plants arborescent, the branches spiny; flowers in racemes 4. *D. spinosa*

4. Leaves persistent; calyx villous, with 2 to 4 glands in each interval; plants shrubby, 0.5 to 2 m. high, unarmed; flowers in dense subglobose or ovoid-cylindric spikes 12 to 15 mm. wide. Branches and peduncles finely retrorse-strigose; leaves linear-spatulate or oblanceolate, exceptionally trifoliolate............................7. *D. scoparia*
5. Branches when young retrorsely hairy and conspicuously punctate with orange-colored glands, spinescent; leaflets 2 to 4 mm. long. Calyx lobes shorter than the tube; corolla pink or purple (6).
5. Branches not retrorsely hairy; leaflets 4 to 15 mm. long (8).
6. Calyx glabrous externally, the ribs rather prominent. Calyx lobes ciliate, the upper ones obtuse or rounded, the lowest lobe acute; leaflets 5 to 9, elliptic or slightly obovate ....8. *D. Thompsonae*
6. Calyx copiously pubescent externally, the ribs not prominent (7).
7. Racemes dense; calyx lobes acute, lance-subulate; leaflets 7 to 13, orbicular to oblanceolate ....5. *D. polyadenia*
7. Racemes lax, few-flowered; calyx lobes obtuse or rounded, the upper ones nearly as wide as long; leaflets 11 to 19, oblong to oblanceolate................9. *D. Whitingi*
8. Branches velutinous-canescent, unarmed; flowers purple, in capitate spikes. Shrubs about 1 m. high; leaflets 1 to 13, more or less serrate, the terminal one 10 to 25 mm. long, usually 2 or 3 times longer than the lateral leaflets..............6. *D. Emoryi*
8. Branches strigose, often canescent but not velutinous; flowers indigo, in loose racemes (9).
9. Calyx lobes dissimilar (the upper ones broader), deltoid to triangular-lanceolate, usually shorter than the tube...1. *D. Fremontii*
9. Calyx lobes all alike, lance-subulate, equaling the tube..................2. *D. amoena*
10. Flowers evidently pedicelled, but often very shortly so; bracts caducous (11).
10. Flowers sessile; bracts often more persistent (17).
11. Calyx lobes broad, not setaceous. Stems not conspicuously glandular (12).
11. Calyx lobes setaceous from a deltoid base (14).
12. Stems prostrate or nearly so, entirely herbaceous. Calyx densely pilose, the lobes foliaceous, lanceolate, nearly twice as long as the tube; leaflets 15 to 29; flowers purple and white...10. *D. calycosa*
12. Stems strongly ascending or erect, often somewhat suffrutescent (13).
13. Herbage and calyx glabrous; calyx lobes ovate, obtuse, shorter than the tube; leaflets of the stem leaves 13 to 23, those of the branches 3 to 9; paired petals purple, the banner yellowish...11. *D. diffusa*
13. Herbage and calyx strigose-canescent; calyx lobes triangular, acute, shorter than or equaling the tube; leaflets 15 to 33; petals purple and white..........12. *D. Parryi*
14. Plants annual; stem erect, branching above the middle, sparsely glandular, glabrous; stamen tube long-exserted.......................................17. *D. Lagopus*
14. Plants perennial, but flowering the first year and thus often appearing annual; stems prostrate or weakly ascending, conspicuously glandular, pubescent; stamen tube not exserted (15).
15. Racemes 4 to 8 cm. long, 15 to 20 mm. wide; leaflets 9 to 11, mostly oblong, 8 to 15 mm. long; flowers purplish...................................13. *D. lachnostachys*
15. Racemes 1 to 5 cm. long, 10 to 15 mm. wide; leaflets 7 to 15, obovate, 3 to 8 mm. long; flowers white and purple (16).
16. Calyx 3 to 5 mm. long, usually surpassed by the corolla; wing petals commonly retuse and with a gland in the apical notch.............................14. *D. mollis*
16. Calyx 5 to 8 mm. long, usually surpassing the corolla; wing petals usually entire 15. *D. neomexicana*
17. Plants annual, except sometimes *D. Lemmoni.* Herbage glabrous or glabrescent; stems slender, usually erect; calyx lobes subulate or setaceous from a deltoid base, villous or plumose; color of the flowers not known for all these species (18).
17. Plants perennial (24).
18. Leaflets many (11 to 51), oblong or oblanceolate, truncate, often retuse (19).
18. Leaflets few, 3 to 11 (21).

19. Stamen tube twice as long as the calyx. Calyx tube slit dorsally to below the middle; spikes 3 to 10 cm. long, 12 to 14 mm. wide; stem erect, branching above the middle 17. *D. Lagopus*
19. Stamen tube exserted, but less than twice as long as the calyx (20).
20. Calyx tube glabrous; bracts glabrous; spikes 1 to 3 cm. long; plants usually branching at base..16. *D. urceolata*
20. Calyx tube villous; bracts usually pilose on the margins and toward the base; spikes 2 to 5 cm. long; stem erect, branching above the middle..............18. *D. leporina*
21. Leaflets 3 to 5, narrowly linear or filiform, usually much longer than the rachis, this obsolete in trifoliolate leaves. Calyx copiously villous............19. *D. filiformis*
21. Leaflets 3 to 11, narrowly oblanceolate or elliptic, shorter than the rachis, or only the terminal leaflet longer (22).
22. Calyx lobes much longer than the tube, conspicuously plumose; bracts lanceolate, long-acuminate, long-ciliate.......................................21. *D. Lemmoni*
22. Calyx lobes equaling to somewhat longer than the tube, moderately plumose; bracts ovate or obovate, shortly acuminate, ciliate or glabrous on the margin (23).
23. Spikes subglobose; corolla yellowish, tinged with pink............20. *D. brachystachys*
23. Spikes oblong; corolla whitish or purplish........................22. *D. polygonoides*
24. Spikes usually elongate and relatively slender (3 to 8 cm. long and seldom more than 7 mm. wide). Plants entirely herbaceous, copiously pubescent except the glabrous calyx tube; stems prostrate; leaflets 7 to 15, obovate; bracts broadly obovate, abruptly acuminate, conspicuously glandular-punctate; flowers purple 23. *D. terminalis*
24. Spikes stouter, if less than 7 mm. wide, then not elongate (25).
25. Flowers yellow, usually drying pink. Leaves and stems strigose or sericeous, the stems entirely herbaceous from a woody caudex; calyx villous, the lobes setaceous from a deltoid base, plumose, longer than the tube (26).
25. Flowers purple or white (29).
26. Leaves digitately trifoliolate. Stems usually about 10 cm. long, decumbent 24. *D. Jamesii*
26. Leaves pinnate, the leaflets 5 to 7 (27).
27. Upper leaves reduced; peduncles usually 5 to 10 cm. long. Spikes 2 to 5 cm. long, about 1.5 cm. wide..25. *D. aurea*
27. Upper leaves scarcely reduced; peduncles not more than 2 cm. long, often obsolete (28).
28. Spikes 15 to 20 mm. wide; bracts lanceolate, attenuate; leaflets usually acute 26. *D. Wrightii*
28. Spikes 10 to 15 mm. wide; bracts ovate, short-acuminate; leaflets usually obtuse 27. *D. nana*
29. Bracts conspicuously scarious-margined and glandular-punctate. Leaflets 3 to 7, glabrous; calyx lobes setaceous, plumose, longer than the tube; corolla purple; stems herbaceous from a woody caudex...........................28. *D. pogonathera*
29. Bracts not conspicuously scarious-margined or punctate (30).
30. Bracts densely villous dorsally toward the base, glabrate above, conspicuously ciliate. Calyx lobes pilose, shorter than the tube; spikes ovoid, compact, 6 to 8 mm. wide; leaflets 17 to 41, linear-lanceolate, 4 to 10 mm. long; stems herbaceous from a woody caudex; corolla white, fading pink.............................29. *D. Lumholtzii*
30. Bracts either glabrous or pubescent, but not as in *D. Lumholtzii* (31).
31. Paired petals inserted below the middle of the stamen tube, long-clawed; spikes capitate or subcapitate, rarely ovoid or short-cylindric; plants shrubby or suffrutescent. Calyx lobes long, plumose; flowers purple (32).
31. Paired petals inserted above the middle of the stamen tube, short-clawed; spikes commonly cylindric, always dense; plants herbaceous above the caudex. Leaflets many (34).
32. Branches and leaves glabrous; leaflets thickish, in dried specimens strongly involute or conduplicate, 7 to 15 in number, 1.5 to 2.5 mm. long; calyx tube 3 to 4 mm. long, the lobes 4 to 10 mm. long, setaceous; flowers normally not crowded...30. *D. formosa*

32. Branches and leaves pubescent or if glabrous, then the leaflets thin, plane, and mostly larger, and the calyx shorter; flowers crowded (33).
33. Leaflets 5 to 7, abundantly sericeous-strigose, in dried specimens mostly involute or conduplicate ....31. *D. Greggii*
33. Leaflets 9 to 21, not sericeous, plane or somewhat involute............32. *D. Wislizeni*
34. Flowers purple; calyx lobes lance-subulate, commonly longer than the tube. Paired petals oval or elliptic, the wings very short-clawed; stems and leaves glabrous 33. *D. Pringlei*
34. Flowers white, sometimes fading purple; calyx lobes deltoid-subulate, shorter than or barely equaling the tube (35).
35. Herbage glabrous; claws of the wing petals very short. Paired petals obliquely oblong; banner often drying purple....................................34. *D. Grayi*
35. Herbage pubescent; claws of the wing petals longer, ¼ to ⅓ as long as the blade (36).
36. Stems from a branched caudex, pilose to densely villous, relatively stout, up to 60 cm. long; leaflets oblong-elliptic or obovate; peduncles usually longer than the spikes; bracts long-attenuate, surpassing the calyx......................35. *D. albiflora*
36. Stems from creeping rootstocks, sericeous, slender, seldom more than 30 cm. long; leaflets linear to oblanceolate; peduncles usually shorter than the spikes; bracts shorter than or barely equaling the calyx..........................36. *D. Ordiae*

1. **Dalea Fremontii** Torr. Northern Mohave County, up to 3,000 feet, April to June. Southern Utah to southeastern California, and Arizona.

The species is represented in Arizona mainly by var. *minutifolia* (Parish) L. Benson (var. *Johnsoni* (Wats.) Munz), which has 5 to 11 linear or linear-oblanceolate leaflets. A collection at Wolf Hole, northern Mohave County (*Peebles & Parker* 14751), approaches the typical phase of the species, with 1 to 7 oblong or elliptic leaflets.

2. **Dalea amoena** Wats. Coconino and Mohave counties, 500 to 5,000 feet, occasional, April and May, type from northern Arizona. Southern Utah and Nevada, and northern Arizona.

In *D. amoena*, which is known from only a few collections, the hairs of the calyx are short and closely appressed. Var. *pubescens* (Parish) Peebles (*D. Fremontii* var. *pubescens* (Parish) L. Benson) occurs in the vicinity of Lees Ferry, Coconino County, the type locality (type, *Jones* 3076). It is distinguished by spreading, usually longer hairs on the calyx.

3. **Dalea Schottii** Torr. Yuma County, infrequent, type from along the Colorado River (*Schott*). Southwestern Arizona, southern California, and Baja California.

4. **Dalea spinosa** Gray. Western Mohave, Yuma, western Maricopa, and western Pima counties, 1,500 feet or lower, in sandy washes, April to June, type from along the Gila River (*Emory*). Arizona, southern California, Baja California, and Sonora.

Smoke-tree, smoke-thorn. The names refer to the normally gray appearance of the tree, which changes when it is covered with a profusion of violet or indigo-blue flowers.

*5. **Dalea polyadenia** Torr. Not known from Arizona but collected not far from the northern boundary in southern Utah at St. George (*Siler* in 1875). Southwestern Utah to the Mojave Desert, California.

6. **Dalea Emoryi** Gray. Yuma County, especially abundant on sandy mesas

near Yuma, where the type was collected by Emory, below 500 feet, flowering in spring (occasionally in autumn). Southern California, Baja California, Arizona, and Sonora.

This, and occasionally *D. Schottii* and perhaps other species, are hosts of the rare parasitic plant *Pilostyles Thurberi.*

7. Dalea scoparia Gray. Willcox Flat, Cochise County (*Shreve* 4256), near Leupp, eastern Coconino County (*Whiting* 3321). Western Texas to eastern Arizona and northern Mexico.

8. Dalea Thompsonae (Vail) L. O. Williams. Known only from the type, which was collected in northern Arizona, probably near Kanab, Utah (*Mrs. Thompson* in 1872).

Rydberg should have included this species in his segregate genus *Psorothamnus* (*North Amer. Flora* 24: 45), the petals being inserted on the hypanthium.

9. Dalea Whitingi Kearney & Peebles. Known only from the type locality, Wupatki National Monument, Coconino County (*Whiting & Jones* in 1938, the type collection, *D. J. Jones* in 1939), and from southeastern Utah.

10. Dalea calycosa Gray. Graham, Pinal, Cochise, Santa Cruz, and Pima counties, 4,000 to 5,000 feet, April to September, type from along the San Pedro River (*Wright* 994). New Mexico, southern Arizona, and northern Mexico.

11. Dalea diffusa Moric. Patagonia Mountains, Santa Cruz County, 4,500 feet (*Kearney & Peebles* 10054). Southeastern Arizona, Mexico, and Guatemala.

The Mexican name is *escoba colorada.* Apparently only a single plant of this species has been found in Arizona.

12. Dalea Parryi Torr. & Gray (*D. angulata* Jones). Gila County to Mohave County, south to Cochise, Pima, and Yuma counties, 4,000 feet or lower, March to June (October), type of *D. Parryi* from Fort Mohave (*Cooper*), type of *D. angulata* from the Baboquivari Mountains (*Jones* 24881). Arizona, southern California, Sonora, and Baja California.

A xerophytic perennial 1 to 2.5 feet high, with more or less woody stems, common on low deserts.

13. Dalea lachnostachys Gray. Cochise, Santa Cruz, and Pima counties, 3,500 to 5,000 feet, September. Western Texas, southern New Mexico, Arizona, and Chihuahua.

14. Dalea mollis Benth. Mohave, western Pima, and Yuma counties, up to 3,000 feet but usually lower, sandy or rocky soil, January to April. Arizona, southern California, Sonora, and Baja California.

15. Dalea neomexicana (Gray) Cory. Mohave, Maricopa, Pinal, Cochise, Pima, and Yuma counties, up to 4,000 feet, sandy deserts and dry grassland, December to May. Western Texas to southern Nevada, southeastern California, and northern Mexico.

There occur in Arizona both the typical phase, with wing petals oblong and rounded or retuse at apex, and subsp. *mollissima* (Rydb.) Wiggins (*Parosela mollissima* Rydb.), with wing petals ovate and somewhat narrowed at apex. Although most of the plants can be distinguished from *D. mollis* by the characters given in the key, some specimens possessing the large calyx of *D. neomexicana* have the wing petals notched and even with a gland in the notch.

16. Dalea urceolata Greene. Hilltop, Apache Indian Reservation (*Harrison* 4895), Flagstaff (*Carter* in 1927), September and October. New Mexico and Arizona.

17. Dalea Lagopus (Cav.) Willd. Bear Valley (Sycamore Canyon), Santa Cruz County about 3,500 feet (*Goodding* in 1939), September. Southern Arizona, southern Mexico, and central America.

18. Dalea leporina (Ait.) Kearney & Peebles. Near the San Francisco Peaks (Coconino County), near Prescott (Yavapai County), Big Lue Range (Greenlee County), near the Mogollon Escarpment (Gila County), Chiricahua, Huachuca, and Santa Rita mountains (Cochise and Santa Cruz counties), 5,500 to 8,000 feet, June to October. New Mexico and Arizona to Guatemala.

This species and *D. alopecuroides* Willd. are perhaps too closely related. The latter is distinguished by the entirely pubescent, mostly pale bracts, and the wholly straw-colored stamen tube, whereas in *D. leporina* the darker-colored bracts are pubescent only below and on the margins, and the stamen tube is tinged with purple. A specimen from the White Mountains, Apache County (*Gould & Robinson* 5091), resembles *D. leporina* except in its less hairy calyx, with shorter and broader teeth.

19. Dalea filiformis Gray. White Mountains (Apache or Navajo County), Hualpai Mountain (Mohave County), and in Yavapai, Greenlee, Cochise, Santa Cruz, and Pima counties, 3,500 to 8,000 feet, mountains and grassland, August and September. New Mexico, Arizona, and northern Mexico.

20. Dalea brachystachys Gray. Chiricahua Mountains, Cochise County (*Lemmon* in 1881), Santa Rita Mountains, Santa Cruz or Pima County, 5,000 feet (*Thornber* 213), September, type from "between the San Pedro and the Sonoita" (*Wright* 990). New Mexico, southeastern Arizona, and Mexico.

21. Dalea Lemmoni Parry. Cochise, Santa Cruz, and Pima counties, about 5,000 feet, September, type from near Fort Bowie (*Lemmon* in 1881). Southern Arizona and Mexico.

22. Dalea polygonoides Gray. Apache and Coconino counties, south to Cochise and Pima counties, 5,500 to 9,000 feet, pine forest and grassland, August to October. New Mexico, Arizona, and Chihuahua.

Var. *anomala* (Jones) Morton (*D. Hutchinsoniae* Jones var. *anomala* Jones, *D. polygonoides* var. *laevituba* Kearney & Peebles), with a glabrous calyx tube, is more common in Arizona, especially in the southern counties, than the typical plant, with a villous calyx tube.

23. Dalea terminalis Jones. Navajo, Coconino, Mohave, and Graham counties, 2,000 to 5,000 feet, sandy soil, May to September. Western Texas to southern Utah, Arizona, and Chihuahua.

The essentially glabrous calyx tube distinguishes this species from the more eastern *Dalea lanata* Spreng.

24. Dalea Jamesii (Torr.) Torr. & Gray. Pima, Santa Cruz, and Cochise counties, 5,000 feet, grassland, infrequent. Kansas and Colorado to southern Arizona and northern Mexico.

25. Dalea aurea Nutt. Navajo County, at Fort Apache (*Palmer* 611) and Cibecue Creek (*Thornber* in 1905), June. South Dakota and Wyoming to Texas, Coahuila, and Arizona.

26. Dalea Wrightii Gray. Navajo, Cochise, Santa Cruz, and Pima counties, 3,500 to 5,000 feet, grassland, May to October. Western Texas to Arizona and Mexico.

27. Dalea nana Torr. Southern Navajo, Yavapai, Cochise, Santa Cruz, and Pima counties, 3,500 to 5,000 feet, grassland, May to September. Kansas to Texas, Arizona, and Mexico.

In the typical plant the leaflets are sericeous on both surfaces and commonly obtuse. More common in Arizona is var. *carnescens* (Rydb.) Kearney & Peebles (*Parosela carnescens* Rydb.), with leaflets green and glabrate above and obtuse to acutish, and stems usually stouter.

28. Dalea pogonathera Gray. Greenlee, Graham, Pinal, Cochise, Santa Cruz, and Pima counties, 2,500 to 6,000 feet, grassland and hills, March to September. Texas to southern Arizona and Mexico.

29. Dalea Lumholtzii Robins. & Fern. (*Parosela arizonica* Vail). Santa Catalina and Baboquivari mountains (Pima County), Pajarito Mountains (Santa Cruz County), 3,500 to 7,000 feet, September and October, type of *P. arizonica* from near Tucson (*Eastwood*). Southern Arizona and Sonora.

The foliage has fragrance similar to that of lemon-verbena (*Aloysia triphylla*).

30. Dalea formosa Torr. Navajo, southern Coconino, and Yavapai counties to Graham, Cochise, Santa Cruz, and Pima counties, 2,000 to 6,500 feet, hills and mountains, March to June (September). Colorado, southern Utah, western Texas, New Mexico, Arizona, and northern Mexico.

31. Dalea Greggii Gray. Cochise, Pima, and Santa Cruz counties, rocky hills, 2,500 to 5,000 feet, February to May. Southern Arizona and Mexico.

A handsome plant, well worth cultivating, with straight, more or less woody stems and globose heads of rose-purple flowers. There is doubt that the Arizona plant so called is true *D. Greggii* (H. S. Gentry, personal communication).

32. Dalea Wislizeni Gray. Greenlee, Graham, Pinal, Cochise, Santa Cruz, and Pima counties, 3,000 to 6,000 feet, rocky hills, April to October. New Mexico, southeastern Arizona, and Mexico.

Typical *D. Wislizeni* has the flower spikes at the ends of elongate branches, and the herbage more or less villous. Var. *sanctae-crucis* (Rydb.) Kearney & Peebles (*Parosela sanctae-crucis* Rydb.), growing in Santa Cruz County and in the Baboquivari Mountains (Pima County), differs seemingly only in the glabrous or sparsely pubescent herbage. Var. *sessilis* Gray with spikes mainly at the ends of short lateral branches, occurs in Greenlee, Pinal, Cochise, Pima, and Santa Cruz counties.

33. Dalea Pringlei Gray. Cochise, Santa Cruz, and Pima counties, 2,500 to 5,000 feet, hills and mountains, April to October, type from the Santa Catalina Mountains (*Pringle* in 1881). Southern Arizona, Sonora, and Chihuahua.

34. Dalea Grayi (Vail) L. O. Williams (*D. laevigata* Gray, non Moç. & Sessé). Graham, Gila, Cochise, Santa Cruz, and Pima counties, 3,500 to 5,500 feet, hills and mountains, May to September, types of *D. laevigata* Gray from the Chiricahua Mountains and vicinity (*Wright* 989). New Mexico, Arizona, Sonora, and Durango.

35. Dalea albiflora Gray (*Thornbera albiflora* Rydb., *T. villosa* Rydb.). Oak Creek (southern Coconino County) and Hualpai Mountain (Mohave County) to Greenlee (?), Graham, Cochise, Santa Cruz, and Pima counties, 3,500 to 7,500 feet, April to October, types of *D. albiflora* from the San Pedro and Babocomari (*Wright* 987), type of *Thornbera villosa* from the Santa Rita Mountains (*Griffiths & Thornber* 130). New Mexico, Arizona, and northern Mexico.

Thornbera villosa (*T. albiflora* subsp. *villosa* (Rydb.) Wiggins) is a robust plant with very villous stems.

36. Dalea Ordiae Gray. Southern Apache County to Cochise County, west to the Sierra Ancha (Gila County) and Baboquivari Mountains (Pima County), 5,000 to 7,500 feet, among pines and junipers and on open, grassy slopes, August to October, type from Cochise County (*Lemmon* in 1881). Southwestern New Mexico, southern Arizona, and adjacent Mexico.

After the foregoing treatment was prepared, a paper by Howard Scott Gentry, entitled "Studies in the Genus Dalea," was published (*Madroño* 10: 225–250). The Arizona specimens that were referred previously to *D. Greggii* Gray were assigned to a new species, *D. pulchra* Gentry, type from the Santa Catalina Mountains (*Gould & Robbins* 3534). A new species, *D. tentaculoides* Gentry, was described from specimens collected in Santa Cruz County and in the Baboquivari Mountains, type from Sycamore Canyon (Bear Valley), Santa Cruz County (*Darrow* in 1941). It is characterized by "elongate tentacle-like glands on the bracts and calyx lobes." The only form of *D. Wislizeni* Gray that Gentry recognized as occurring in Arizona is subsp. *sessilis* (Gray) Gentry (var. *sessilis* Gray), of which *Parosela sanctae-crucis* Rydb. was cited as a synonym.

28. PETALOSTEMUM. Prairie-clover

Plants herbaceous, annual or perennial, often deep-rooted; herbage gland-dotted; leaves odd-pinnate, the leaflets several or numerous; flowers in dense spikes, only slightly irregular; petals white, lilac, or rose pink, the claws of the wings and keel adnate to the stamen tube; stamens 5; pods small, included in the calyx, containing 1 or 2 seeds.

The plants are attractive but of small economic importance. The name of the genus is often spelled *Petalostemon.*

KEY TO THE SPECIES

1. Calyx tube glabrous or sparsely puberulent, the lobes ciliolate. Plants perennial; herbage glabrous; leaflets 3 to 9, commonly 5, linear-oblanceolate; calyx lobes about ⅓ as long as the prominently 10-ribbed tube; petals white; blade of the banner cordate or reniform, wider than long .... 1. *P. candidum*

1. Calyx tube pilose or villous (2).

2. Plants annual. Stems slender, glabrous; leaflets 3 to 5, linear, 1 to 3 cm. long, glabrous; spikes long-peduncled, slender, about 6 mm. wide in fruit; bracts conspicuous, dark-colored, elliptic, ovate, or rhombic, subulate-tipped; corolla lilac, about 4 mm. long; blade of the banner shorter than the claw .... 2. *P. exile*

2. Plants perennial (3).

3. Corolla white; blade of the banner equaling or longer than the claw, quadrilateral. Leaflets 3 to 7; stem glabrous or sparsely pilose, rather conspicuously glandular-punctate; peduncles 10 to 20 cm. long; spikes 10 to 12 mm. wide .... 3. *P. flavescens*

3. Corolla rose-colored or purplish; blade of the banner shorter than the claw (4).
4. Stems and leaves sparsely villous or glabrate; stems very leafy almost to the inflorescence, with additional leaves often fascicled in the axils; leaflets commonly 5, narrowly linear, about 1 mm. wide, mucronate, strongly involute; peduncles commonly not more than 2 cm. long; calyx tube densely silky-villous..........4. *P. purpureum*
4. Stems and leaves glabrous; stems not very leafy, naked for a considerable distance below the inflorescence; leaflets 5 to 9, elliptic to obovate, 2 to 4 mm. wide, not strongly involute; peduncles up to 15 cm. long; calyx tube loosely pilose or villous 5. *P. Searlsiae*

1. Petalostemum candidum (Willd.) Michx. Apache County to Coconino County, south to Cochise, Santa Cruz, and Pima counties, 3,000 to 7,000 feet, mesas and openings in pine forests, rather common, May to September. Indiana to Saskatchewan and Montana, south to Mississippi, Texas, Arizona, and northern New Mexico.

White prairie-clover. The species is represented in Arizona by the Western var. *oligophyllum* (Torr.) Hermann (*P. oligophyllus* Torr. ex Smyth, *P. occidentale* (Gray) Fern., *P. sonorae* Rydb.), which has the leaflets usually smaller and narrower than in *P. candidum* as it occurs farther east. The plant is reported to be used by the Hopi Indians as an emetic.

2. Petalostemum exile Gray. Greenlee, Cochise, Santa Cruz, and Pima counties, 5,000 to 7,000 feet, mostly among pines, September. Southern New Mexico and Arizona, and northern Mexico.

3. Petalostemum flavescens Wats. Navajo and Coconino counties, 5,000 to 8,000 feet, apparently rare, June. Southern Utah and northernmost Arizona.

The corolla was described as yellow by Watson but is white in fresh specimens. The crushed foliage is lemon-scented.

4. Petalostemum purpureum (Vent.) Rydb. Near Prescott, Yavapai County (*Peebles et al.* 4253), at roadside, July. Indiana to Saskatchewan, south to Texas and central Arizona, where perhaps introduced from farther east.

5. Petalostemum Searlsiae Gray (*P. Rothrockii* Rydb.). Coconino, eastern Mohave, and northern Yavapai counties, 3,000 to 7,000 feet, April to June. Southern Utah and Nevada, northern Arizona, and southern California.

The type of *P. Rothrockii* was collected in Arizona (*Rothrock* in 1874, without definite locality). It differs from most specimens of *P. Searlsiae* in the broader, more abruptly acuminate bracts, but intergradations occur.

Petalostemon pilosulus Rydb. was described from a specimen collected by Nealley (no. 237), probably in the Rincon Mountains, Pima County. The type appears to be a form of *Dalea albiflora.*

29. TEPHROSIA (151a)

Plants perennial, herbaceous or suffrutescent; stems either sympodial and the racemes appearing lateral, opposite the leaves, or monopodial and the racemes terminal and axillary; herbage not punctate; leaves odd-pinnate, the leaflets few to numerous; corolla whitish or purple; stamens more or less united; pods linear, flat, several-seeded.

Although some of the species of this genus are a source of rotenone, an extensively used insecticide, the Arizona species apparently contain little or none of this substance. They are, however, suspected of being poisonous.

KEY TO THE SPECIES

1. Racemes few-flowered, very open; corolla about 8 mm. long; style glabrous. Herbage appressed-pubescent; leaflets linear to oblanceolate....................1. *T. tenella*
1. Racemes many-flowered; corolla more than 10 mm. long; style barbate on the inner (upper) side (2).
2. Stems finely strigose-canescent; leaflets glabrous above; racemes elongate, narrow, loosely flowered, many of them axillary; pods glabrous or nearly so, 5 to 6.5 mm. wide ....2. *T. leiocarpa*
2. Stems with spreading as well as short appressed hairs; leaflets pubescent on both surfaces; racemes at first short, broad, and dense, mostly terminal; pods hirsutulous, 3.5 to 4.5 mm. wide..3. *T. Thurberi*

1. Tephrosia tenella Gray. Pinal, Cochise, Santa Cruz, and Pima counties, 3,000 to 6,000 (?) feet, slopes and mesas, April to September, type from along the San Pedro River (*Wright* 966). Texas to southern Arizona and northern Mexico.

2. Tephrosia leiocarpa Gray. Cochise (?), Santa Cruz, and Pima counties, 4,500 to 5,500 feet, August and September, type from along the Sonoita (*Wright* 965). Southern Arizona and Mexico.

A roundish shrub about 1 m. high.

3. Tephrosia Thurberi (Rydb.) C. E. Wood. Cochise, Santa Cruz, and Pima counties, 4,500 to 7,000 feet, dry slopes among oaks and pines, July to September. Southern Arizona and northern Mexico.

In *Flowering Plants and Ferns of Arizona* species 1 and 3 were referred respectively to *T. purpurea* (L.) Pers. and *T. leucantha* H.B.K.

30. PETERIA

Plants herbaceous, perennial; stems low, slender, rather stiff; leaves pinnate, the leaflets many, small, the stipules spiny; flowers in long-stalked loose racemes, the banner recurved, the keel incurved; pods narrow, several-seeded.

The tuberous rootstocks of *P. scoparia,* known in Texas as camote-de-monte, are reported to be edible.

KEY TO THE SPECIES

1. Leaflets not more than 2 mm. wide, acute or acutish at apex, soon deciduous; stem freely branched above the base; calyx strigose, only slightly glandular-puberulent, the teeth deltoid, 2 to 4 mm. long; corolla 12 to 16 mm. long.........1. *P. scoparia*
1. Leaflets 4 mm. or wider, obtuse, rounded, or slightly retuse and often mucronate at apex; stem simple or very sparingly branched above the base; calyx more or less hirsute, copiously glandular-pubescent, the teeth subulate, (3) 5 mm. long, or longer; corolla (12) 16 to 25 mm. long................................2. *P. Thompsonae*

1. Peteria scoparia Gray. Petrified Forest, Apache County (*Toole & Gooding* in 1936), July and August. Western Texas to northeastern Arizona.

2. Peteria Thompsonae Wats. Northern Coconino County and northeastern Mohave County near the Utah state line (*Kelly* in 1932, *Cottam et al.* 4199), 4,000 feet, sometimes with *Coleogyne.* Southern Utah and northern Arizona.

Mrs. Kelly's specimen is exceptionally small-flowered.

31. ROBINIA. Locust

Large shrubs or small trees, thorny; leaves odd-pinnate, the leaflets numerous, rather large; flowers many, in rather dense racemes, large and showy, fragrant, the corolla purplish pink; pedicels glandular; stamens diadelphous, 9 of them with the filaments united below into a tube; pods flat, 2-valved, several-seeded, hispid and usually glandular.

1. Robinia neomexicana Gray. Navajo, Coconino, and Mohave counties, south to Greenlee, Cochise, Santa Cruz, and Pima counties, 4,000 to 8,500 feet, common and often abundant, mostly in canyons and in coniferous forests, often with *Quercus Gambelii,* May to July. Southern Colorado to southern Nevada, western Texas, New Mexico, Arizona, and northern Mexico.

New Mexican locust. The plant reaches a height of 7.5 m. (25 feet) and, being very handsome in flower, is sometimes cultivated as an ornamental. The flowers are relished by cattle, and the foliage is browsed by both cattle and deer. The bark, roots, and seeds are reported, however, to be poisonous. The Hopi Indians are said to use the plant as an emetic and in treating rheumatism. The habit of forming thickets and of sprouting freely from stumps and roots makes this plant valuable for control of erosion. Durable fence posts are made from the trunks.

Two forms of this species are common in Arizona, var. *luxurians* Dieck (*R. luxurians* Rydb.) and var. *subvelutina* (Rydb.) Kearney & Peebles (*R. subvelutina* Rydb.), type from the Natanes Plateau (*Goodding* 1092). The former is more common in the southern counties, the latter in northern and central Arizona. Both are supposed to be distinguished from typical *R. neomexicana* by the presence of gland-tipped hairs on the pods. In var. *luxurians* the pubescence of the herbage is finer and more appressed than in var. *subvelutina,* and the petioles are usually glandular-hispid in the latter variety, not so in var. *luxurians.* But there is so much intergradation among all forms of this species that the attempt to maintain even varieties is scarcely worth while.

32. OLNEYA. Tesota

Trees, attaining a height of 9 m. (30 feet) and a trunk diameter of 45 cm. (1.5 feet), the branches armed with spines, the bark thin and scaly; leaves pinnate, the leaflets 8 to 24 (commonly 11 to 15), grayish-pubescent; flowers in short racemes; corolla about 12 mm. long, purple and white; pods glandular-pubescent, usually several-seeded and torulose.

1. Olneya Tesota Gray. Maricopa, Pinal, Pima, and Yuma counties, 2,500 feet or lower, common along washes in the foothills, May and June. (A collection by B. E. Fernow labeled "Pinery Creek" (Cochise County) must have come from an altitude of 4,000 feet or more if correctly labeled as to locality.) The type was collected along the Gila River in Arizona. Southern Arizona, southeastern California, Sonora, and Baja California.

Known commonly in Arizona as ironwood, or palo-de-hierro. A desert tree with very handsome flowers, limited to warm locations, and for this reason used as an indicator in selecting sites for citrus orchards. The foliage is evergreen except in very cold winters. The wood is brittle, hard, remarkably heavy, and burns very slowly, making good coals. The ironwood has been used so extensively for firewood that it is unusual to find a large tree that has not been cut back to the stump. The wood was used by the Indians for arrowheads and for tool handles. Experiments have been made to utilize it commercially, but it is too hard for ordinary woodworking tools. The hard, sharp spines of the branches do not prevent desert-bred horses from eating the foliage with evident relish, and the trees are browsed by bighorns. The seeds are an important food of desert animals and formerly were eaten parched by the Pima Indians.

33. DIPHYSA

Plants low shrubs; branches of the inflorescence armed with short, weak prickles; leaves pinnate, the leaflets numerous, thin, oval to suborbicular, 1 to 1.5 cm. long; flowers rather large, in axillary racemes; corolla yellow or yellowish; pods 1-celled, the outer wall separating from the inner wall, forming 2 elongate bladderlike cavities paralleling the seed cavity.

*1. **Diphysa Thurberi** (Gray) Rydb. Montezuma Camp, southeast of the Huachuca Mountains (*Lemmon* 2659). Southeastern Arizona (?) and northern Mexico.

34. COURSETIA

Shrubs up to 6 m. (20 feet) high, unarmed; leaves pinnate, the leaflets numerous, thin, oval; flowers appearing with the leaves; inflorescences racemose, axillary, glandular; corolla white or tinged with pink, with a yellow center (rarely entirely yellow); pods linear, torulose, 2-valved.

1. **Coursetia microphylla** Gray. Maricopa, Pima, and Yuma counties, 4,000 feet or lower, canyons and dry, rocky slopes, locally abundant, March and April, type from the Santa Catalina Mountains (*Pringle* in 1881). Southern Arizona and northern Sonora.

This species is closely related to *C. glandulosa* Gray and is perhaps only a good variety of the latter. The mature leaflets attain a length of more than 2 cm. but are usually much smaller. The plant is often browsed. The stems sometimes are heavily encrusted with orange-colored lac, resulting from infestation by an insect of the genus *Tachardia*. This lac was used by the Papago Indians to seal jars containing saguaro sirup and is reported to be used by the Mexicans in treating colds and fever.

35. CRACCA

Plants herbaceous, perennial, small; herbage sericeous or villous; leaves pinnate, the leaflets several, broad, thin; flowers few, in loose, long-stalked racemes; corolla with ochroleucous or pale-yellow wings and keel, the banner usually more or less purple; pods linear, flat, with cross partitions between the seeds, completely dehiscent; seeds rounded-quadrate.

It is reported that the plants are heavily grazed but quickly recover.

1. **Cracca Edwardsii** Gray (*Benthamantha Wrightii* Rydb.). Cochise, Santa Cruz, and Pima counties, 4,000 to 6,000 feet, common on rocky slopes, often with live-oaks, July to September, type of *B. Wrightii* from "between the San Pedro and the Sonoita" (*Wright* 963, in part). Southern Arizona and northern Mexico.

Var. *glabella* Gray (*Benthamantha glabella* Rydb.) occurs in the Huachuca and Patagonia mountains (Cochise and Santa Cruz counties), type probably from Santa Cruz County (*Wright* 963, in part). Normally it differs from typical *C. Edwardsii* in having flowering stems from a cormlike caudex (instead of an elongate, woody rootstock), some of the roots tuberous-thickened, the pubescence of the herbage looser and more spreading, the leaflets more broadly obovate and more obtuse at apex, and the banner petal yellowish, often purple-veined (instead of red or purple). Rydberg may have been justified in giving this entity specific rank, although apparently it intergrades with typical *C. Edwardsii.*

36. SPHINCTOSPERMUM

Small annual herbs; stem slender, erect, sparingly branched; herbage sparsely strigose; leaves reduced to 1 long, narrowly linear leaflet; flowers axillary, solitary or in pairs, short-pedicelled, small and inconspicuous; pods as in *Cracca;* seeds quadrate, sharply constricted at the middle.

1. **Sphinctospermum constrictum** (Wats.) Rose. Santa Cruz and Pima counties, 2,500 to 4,000 feet, not common, open, sandy places, July to September. Southern Arizona, Sonora, and Baja California.

The hourglass-shaped seeds are very distinctive.

37. SESBANIA

Plants annual, glabrous; stems tall, erect, sparingly branched, leafy; leaves bright green, elongate, pinnate, the leaflets numerous, narrow, oblong or elliptic; flowers in axillary few-flowered racemes; corolla pale yellow, usually streaked and spotted with brown-purple; pods long, very slender, dehiscent, with cross partitions between the numerous oblong seeds.

1. **Sesbania macrocarpa** Muhl. (*Sesban exaltatus* (Raf.) Rydb., *S. sonorae* Rydb.). Bottom lands along the Colorado River, in southern Mohave and Yuma counties, and occasionally escaped from cultivation elsewhere in irrigated districts of southern Arizona, August to October. Missouri to Florida, Texas, southwestern Arizona, southeastern California, and northwestern Mexico.

Sometimes known as Colorado-River-hemp. Often planted as a soil-improvement crop on farms in irrigated sections of the Southwest and as a cover crop in citrus orchards. It is a fiber plant, producing lustrous, smooth, and very strong threads, which are used by the Yuma Indians for nets and fish lines. The stems reach a height of 3 m. (10 feet).

38. SWAINSONA

Plants herbaceous, perennial, glabrous or nearly so; flowering stems from creeping rootstocks, tall, erect, leafy, strictly branched above; leaves pinnate, the leaflets numerous, thin, narrowly elliptic or oblanceolate, about 1 cm. long; flowers in loose axillary racemes, large and showy; corolla dull red; style bearded around the apex and along the inner side; pods large, bladderlike, nearly globose, long-stipitate, with thin, papery walls, indehiscent or tardily and irregularly dehiscent.

1. **Swainsona Salsula** (Pall.) Taubert. Navajo County, at Winslow (*Peebles* 9595, *Shreve* 9024), and Holbrook to Winslow (*McKelvey* 4565), May and June. Here and there in the western United States; introduced from Asia.

Related species are reputed to be poisonous. The plant might easily be mistaken for a large, bladdery-fruited *Astragalus*.

39. ASTRAGALUS (152, 153). MILK-VETCH

Plants of very diverse habit, mostly herbaceous perennials, but a few species annual and at least 1 species suffrutescent; herbage usually pubescent, the hairs either basifixed or 2-armed (dolabriform); stems leafy or scapose, erect to prostrate; leaves pinnate, rarely reduced to 1 leaflet; inflorescences commonly racemose, sometimes umbellike or capitate; keel petals usually arched or bent; pods diverse, dehiscent or indehiscent, with papery to leathery or woody walls, 1-celled or more or less completely 2-celled by introversion of 1 or both sutures.

This is the largest genus of flowering plants in Arizona. Many of the species look much alike and can be distinguished satisfactorily only by the characters of the fruit. A few species with prostrate stems probably have a limited value for control of erosion. The name loco-weed is applied to some of these plants, and species with bladderlike pods are called rattle-weed. The former name implies that the species is one of those containing the poisonous constituent causing the well-known and often fatal loco disease of livestock, especially of horses. Species known to cause this disease that are found in Arizona are: *A. allochrous, A. lentiginosus* (several varieties), *A. nothoxys, A. Thurberi,* and *A. Wootoni,* probably also *A. arizonicus.* Fortunately these plants are seldom eaten when better forage is available, but it is said that animals may acquire the habit of eating loco-weed.

Other species, known collectively as poison-vetch, prefer soils rich in selenium, taking up sufficient quantities of this toxic element to make them poisonous to animals. Most of the Arizona species have not been proved to be injurious, and some apparently are grazed with impunity, but all are under suspicion until positive evidence of their harmlessness is forthcoming. Beath et al. (154) listed the following species occurring in Arizona, all characterized by a rank, disagreeable odor, as dangerously seleniferous: *A. Beathii, A. confertiflorus, A. Haydenianus, A. moencoppensis, A. Pattersoni, A. praelongus,* and *A. Preussii.* Species that have been examined for selenium with negative results are: *A. calycosus, A. humistratus, A. Layneae, A. lentiginosus* (vars. *diphysus* and *palans*), *A. nothoxys, A. Nuttallianus,* and *A. Wootoni.*

The key to the species is a purely artificial one. The genus is classified mainly by the characters of the fruit, but as specimens without fruit are often collected, the key has been constructed primarily on the vegetative and flower characters. These are to be understood as follows: Stem length is measured from the surface of the ground to the base of the uppermost petiole, excluding the peduncles. Length and width of the leaflets are those of the largest leaflet on the individual plant. Flower length is measured externally from the base of the calyx to the tip of the banner, allowing for curvature of the flower. The length of the calyx teeth relative to the tube is determined by the longest tooth. Shape of the pods refers to the outline, as if a flat object, like a leaf, were being described. Dolabriform (malpighiaceous) hairs, as contrasted with basifixed hairs (155), are 2-armed, having the point of attachment somewhere between the base and the middle, with one arm often much shorter than the other. They are pointed at both ends and are usually stiff and closely appressed. Dolabriform hairs in place, when touched with the point of a needle, can be pivoted around the point of attachment. Basifixed hairs are often mingled with the dolabriform hairs.[1]

KEY TO THE SPECIES

1. Plants acaulescent or nearly so, the leaves all basal or subbasal and the internodes mostly concealed by the crowded leaf bases and stipules, or else the stems very short, forming mats or cushions close to the ground (2).
1. Plants caulescent, with leaves scattered along the stems, the upper internodes not concealed, the plants not forming close mats or cushions (23).
2. Flowers not more, usually less than 10 mm. long (3).
2. Flowers (except occasionally in *A. confertiflorus*) not less, usually more than 10 mm. long (9).
3. Keel long-acuminate, the beaklike apical portion forming nearly a right angle with the claws. Flowering stems from a much-branched caudex; stipules conspicuously connate around the stem; pods usually mottled (4).
3. Keel not long-acuminate (5).
4. Herbage copiously strigose to sericeous with rather stiff dolabriform hairs; flowers 6 to 8 mm. long, solitary or in loose 2- or 3-flowered racemes; corolla drying deep violet; pods 1-celled, moderately inflated, asymmetrically and narrowly obovate, barely 10 mm. long, strigose................................25. *A. sesquiflorus*
4. Herbage grayish-pilose with more or less spreading, soft, curly, basifixed hairs; flowers 9 to 10 mm. long, in subcapitate 2- to 5-flowered racemes; corolla drying whitish or pale purple; pods 2-celled, greatly inflated, nearly symmetric, oval, 12 to 15 mm. long, villous....................................58. *A. striatiflorus*
5. Plants pulvinate (cushionlike), with many, very short, creeping, leafy stems; leaflets 1 to 4 pairs, less than 4 mm. long; inflorescences loose, 1- to 3-flowered; flowers less than 6 mm. long; pods not more than 5 mm. long. Corolla whitish or pale purple; pods 1-celled, ovate or oblong-ovate (6).
5. Plants not pulvinate, acaulescent or nearly so; leaflets more numerous and larger; inflorescences compact or dense, normally with more than 3 flowers; flowers 6 mm. or longer; pods mostly more than 5 mm. long (7).
6. Herbage closely strigose, the hairs mostly dolabriform; leaflets 1 to 2 mm. long; flowers about 4 mm. long; pods 3 to 4 mm. long..............24. *A. cremnophylax*
6. Herbage hirsute, the hairs basifixed; leaflets up to 3.5 mm. long; flowers 5 to 5.5 mm. long; pods 4 to 5 mm. long............................26. *A. micromerius*
7. Leaflets glabrous above; herbage green, minutely and rather sparsely strigose; flowers about 6 mm. long; pods 2-celled. Leaflets 9 to 13; pods 8 to 10 mm. long, furrowed along the lower suture................................67. *A. hypoxylus*
7. Leaflets persistently pubescent above; herbage silvery-strigose or sericeous; flowers mostly 8 to 9 mm. long; pods 1-celled (8).
8. Hairs of the herbage dolabriform; leaflets prevailing oblanceolate or narrowly obovate; pods 12 to 18 mm. long, strongly ribbed dorsally, furrowed ventrally 22. *A. accumbens*
8. Hairs mostly basifixed; leaflets prevailingly elliptic or lanceolate; pods 5 to 7 mm. long, strongly ribbed on both sutures........................23. *A. gilensis*
9. Leaflets narrowly linear, usually at least 5 times as long as wide. Hairs of the herbage some or all dolabriform; racemes many-flowered; calyx campanulate, the teeth

[1] Rupert C. Barneby has obligingly rendered much assistance in the preparation of the following treatment.

½ to ⅔ as long as the tube; corolla whitish or tinged with purple; pods 1-celled, 8 to 12 mm. long .... 8. *A. confertiflorus*

9. Leaflets elliptic, oval, or oblanceolate to broadly obovate, not more than 4 times as long as wide (10).

10. Hairs, at least some of them, dolabriform (11).

10. Hairs all basifixed (13).

11. Wing petals deeply notched; pods 2-celled. Herbage silvery-sericeous; leaflets 11 or fewer; inflorescences usually considerably surpassing the leaves; pods oblong, 10 to 15 mm. long, sericeous .... 60. *A. calycosus*

11. Wing petals entire; pods 1-celled, the dorsal suture rarely slightly inflexed (12).

12. Corolla bright purple, the banner 13 to 28 mm. long; hairs mostly attached near the middle; pods 2 to 5 cm. long, oblong, often acute or acuminate at base, usually strongly curved .... 14. *A. amphioxys*

12. Corolla whitish or purplish, the banner 12 to 17 mm. long; hairs mostly attached near one end; pods usually less than 2 cm. long, commonly ovate to nearly orbicular, obtuse at base, moderately curved .... 16. *A. castaneaeformis*

13. Herbage and pods densely shaggy-villous; leaflets 21 or more; pods 2-celled. Leaflets mostly obtuse, less than 3 times as long as wide; pods oblong to broadly ovate, moderately curved to nearly straight (14).

13. Herbage and pods strigose, sericeous, or short-pilose or if shaggy-villous (in *A. desperatus* and *A. Newberryi*), then the leaflets 15 or fewer and the pods 1-celled (16).

14. Inflorescences not, or barely, surpassing the leaves, not more than 10-flowered; hairs of the herbage subappressed and relatively short; pods broadly ovate, little longer than wide. Leaflets broadly ovate, less than 10 mm. long and about equally wide; pods 2-celled to the apex, 15 to 20 mm. long .... 70. *A. Matthewsii*

14. Inflorescences usually considerably surpassing the leaves and more than 10-flowered; hairs longer and more spreading; pods oblong to ovate, distinctly longer than wide (15).

15. Herbage and pods yellowish- or brownish-pubescent, at least in dried specimens; leaflets often more than 15 mm. long; pods 2-celled throughout, oblong or oblong-ovate, obtuse and apiculate to acute at apex, 8 to 15 mm. long, 4 to 6 mm. wide .... 68. *A. Bigelovii*

15. Herbage and pods whitish-pubescent; leaflets seldom more than 12 mm. long; pods 1-celled at the tip, ovate, short-acuminate at apex, 15 to 20 mm. long, more than 6 mm. wide .... 69. *A. Thompsonae*

16. Calyx campanulate, the teeth about half as long as the tube; pods with thin but firm walls. Hairs of the herbage stiff; inflorescences usually considerably surpassing the leaves; pods 1-celled (17).

16. Calyx more or less cylindric, the teeth less than half as long as the tube; pods (in most of the species) with thick, leathery walls (19).

17. Inflorescence dense, not more than 3 cm. long; leaflets 4 to 8 mm. wide, oval or obovate; pods asymmetrically suborbicular, straight or nearly so, 5 to 6 mm. long. Calyx and pods villous-hirsute .... 11. *A. troglodytus*

17. Inflorescence usually loose, up to 7 cm. long; leaflets 2 to 3 mm. wide, elliptic, lanceolate, or oblanceolate; pods oblong or elliptic, strongly curved, 8 mm. or longer (18).

18. Calyx and pods shaggy-villous, the hairs with pustulate bases; pods 8 to 15 mm. long, acute at apex, sometimes mottled .... 27. *A. desperatus*

18. Calyx and pods strigose, not pustulate; pods 15 to 20 mm. long, acuminate at apex, not mottled .... 28. *A. musimonum*

19. Calyx 6 to 7 (9) mm. long; pods 2-celled, linear-lanceolate. Leaflets 10 to 20 mm. long, broadly oval to obovate or suborbicular; petals whitish, the keel (sometimes other petals) purple-tipped; pods 30 to 50 mm. long, strongly curved, sometimes into a nearly complete circle .... 59. *A. Layneae*

19. Calyx usually more than 7 mm. long; pods 1-celled, oblong-lanceolate to ovate (20).

20. Herbage densely silky-villous, the hairs mostly appressed and more or less matted; leaflets 3 to 9 (13); pods shaggy-villous or tomentose, usually densely so and with hairs more than 1 mm. long .... 17. *A. Newberryi*

20. Herbage strigose, sericeous, or (exceptionally) short-pilose; leaflets seldom fewer than 11; pods strigose (rarely glabrate) or pilose with incurved or curly hairs, these not more than 1 mm. long (21).
21. Leaflets rather distant, more or less rhomboidal, usually acute; pods conspicuously mottled .... 12. *A. zionis*
21. Leaflets approximate, seldom rhomboidal, acute or obtuse; pods rarely mottled (22).
22. Racemes usually surpassed by the leaves; leaflets mostly acute, equally pubescent on both surfaces .... 13. *A. argophyllus*
22. Racemes usually surpassing the leaves; leaflets mostly obtuse, sometimes emarginate, usually less pubescent (often glabrescent) on the upper surface.... 15. *A. tephrodes*
23. Mature leaflets 4 mm. or more wide (24).
23. Mature leaflets not more, commonly less, than 4 mm. wide (60).
24. Flowers less than 10 mm. long (25).
24. Flowers (except occasionally in *A. lentiginosus*) 10 mm. or longer (38).
25. Calyx teeth (except occasionally in *A. Haydenianus*) less than ⅔ as long as the tube. Leaflets (except occasionally in *A. wingatanus*) glabrous or glabrescent above; pods more or less pendulous (26).
25. Calyx teeth (except occasionally in *A. allochrous, A. insularis*, and *A. Nuttallianus*) ⅔ as long as to longer than the tube (28).
26. Leaflets 9 to 13 (exceptionally 15 or 17); flowers erect or spreading; pods sessile in the calyx, compressed laterally. Herbage rather sparsely pubescent; pods 1-celled, flat, oblong or elliptic, glabrous .... 3. *A. wingatanus*
26. Leaflets 15 or more; flowers deflexed; pods stipitate, more or less compressed dorsoventrally (at a right angle to the sutures). Corolla whitish, the keel sometimes purple-tipped (27).
27. Herbage sparsely to copiously strigose; racemes compact or dense; flowers 8 to 12 mm. long; calyx tube more or less distended on the upper (ventral) side; pods 1-celled, 5 to 11 mm. long, deeply grooved ventrally............. 7. *A. Haydenianus*
27. Herbage glabrous or nearly so; racemes loose; flowers 6 to 8 mm. long; calyx tube not noticeably distended; pods incompletely 2-celled, 15 to 20 mm. long, not deeply grooved ventrally.. 52. *A. Rusbyi*
28. Leaflets glabrous or glabrescent above, the herbage otherwise sparsely to rather copiously strigose (29).
28. Leaflets (except occasionally in *A. Nuttallianus* and *A. Sileranus*) persistently pubescent above (31).
29. Flowering stems from slender, creeping rootstocks; pods nearly symmetric and straight, not inflated, imperfectly 2-celled, elliptic or oblong....... 51. *A. cobrensis*
29. Flowering stems not from creeping rootstocks; pods very asymmetric, strongly curved on the lower side, greatly inflated, 1-celled. Herbage sparsely to rather copiously strigose (30).
30. Leaflets 11 to 19; flowers 6 to 10 mm. long; pods spreading or pendulous, 25 to 40 mm. long, 10 to 15 mm. wide.............................. 33. *A. allochrous*
30. Leaflets 17 to 27; flowers 6 to 7 mm. long; pods erect, 12 to 15 mm. long, less than 10 mm. wide.. 41. *A. Palmeri*
31. Herbage and pods pilose or villous, the hairs more or less spreading. Pods 1-celled (32).
31. Herbage and pods strigose or sericeous, the hairs mostly appressed (33)
32. Plants perennial, with a more or less compact, somewhat woody caudex; stem sparingly branched, decumbent to prostrate; leaflets mostly distinctly notched at apex; inflorescence usually compact or dense; pods greatly inflated, ovate to suborbicular, not falcate, usually mottled.................................. 37. *A. Sileranus*
32. Plants annual, with a taproot; stems freely branched, erect, ascending, or decumbent only toward base; leaflets rounded, truncate, or obscurely notched at apex; inflorescence loose; pods moderately inflated, oblong-ovate, falcate, not mottled 42. *A. sabulonum*
33. Inflorescence umbellike or subcapitate; pods flat, usually 2-celled. Plants annual; leaflets mostly narrowly elliptic; inflorescences 2- to 5-flowered; pods linear, more or less curved, 10 to 20 mm. long........................... 63. *A. Nuttallianus*

33. Inflorescence distinctly racemose; pods more or less inflated, 1-celled (34).
34. Racemes usually more than 10-flowered, the flowers mostly more than 6 mm. long (35).
34. Racemes usually fewer-flowered, the flowers 4 to 7 mm. long. Plants annual; herbage copiously strigose or sericeous; inflorescences not or but slightly (seldom considerably) surpassing the leaves; pods 10 to 20 mm. long (36).
35. Plants evidently perennial; stems less than 40 cm. long; herbage grayish-sericeous; stipules connate opposite the petiole; inflorescences shorter than to but slightly surpassing the subtending leaves; pods only slightly inflated, symmetric, 8 to 10 mm. long, little surpassing the calyx........................10. *A. sophoroides*
35. Plants often appearing annual, at most short-lived perennial; stems up to 60 cm. long; herbage sparsely to rather copiously strigose; stipules distinct; inflorescences usually surpassing the subtending leaves; pods greatly inflated, very asymmetric, 25 to 40 mm. long, greatly surpassing the calyx....................33. *A. allochrous*
36. Leaflets 7 to 9, notched at apex, elliptic or oblong; pods 3-edged, strigose 38. *A. triquetrus*
36. Leaflets 11 to 21, not or obscurely notched; pods 2-edged (37).
37. Herbage and pods white-sericeous, the hairs not closely appressed; corolla whitish or purplish; pods about 6 mm. wide..............................39. *A. aridus*
37. Herbage and pods copiously strigose, the hairs closely appressed; corolla purple; pods 8 to 11 mm. wide...40. *A. insularis*
38. Calyx teeth (except occasionally in *A. allochrous*) ⅔ as long as to longer than the tube (39).
38. Calyx teeth (except occasionally in *A. Haydenianus*) less than ⅔ as long as the tube (42).
39. Plants evidently perennial; inflorescences shorter than to but slightly surpassing the leaves (40).
39. Plants often appearing annual, at most short-lived perennial; inflorescences often surpassing the leaves. Herbage nearly glabrous to whitish-sericeous, the hairs basifixed; pods more or less inflated, seldom less than 15 mm. long, greatly surpassing the calyx (41).
40. Herbage grayish-sericeous, with dolabriform hairs; flowers not more than 11 mm. long; pods sessile in the calyx, only slightly inflated, 8 to 10 mm. long, little surpassing the calyx, symmetric, the walls firm, not mottled............10. *A. sophoroides*
40. Herbage green, glabrous or nearly so, the few hairs basifixed; flowers 13 to 18 mm. long; pods stipitate, greatly inflated, 25 to 40 mm. long, greatly surpassing the calyx, asymmetric (more strongly curved dorsally than ventrally), the walls papery, usually conspicuously mottled.................................31. *A. oophorus*
41. Leaflets glabrous to strigose above; flowers not more than 12 mm. long; petals sharply bent, the blades forming nearly a right angle to the claws; pods 1-celled, greatly inflated, 25 to 40 mm. long, nearly straight on the upper (ventral) suture, strongly curved on the lower suture, the walls papery, not mottled..........33. *A. allochrous*
41. Leaflets glabrous to whitish-sericeous above; flowers up to 20 mm. long; petals more or less curved but not sharply bent at a right angle; pods more or less completely 2-celled, slightly to strongly inflated, 10 to 30 mm. long, more or less curved on both sutures, the walls papery to somewhat leathery, often mottled 56. *A. lentiginosus*
42. Leaflets (except occasionally in *A. tephrodes*) persistently pubescent above (43).
42. Leaflets (except occasionally in *A. ensiformis*) glabrous or glabrescent above (48).
43. Dolabriform hairs present. Herbage grayish- to silvery-sericeous; inflorescences not or but slightly surpassing the leaves (44).
43. Dolabriform hairs none. Pods greatly surpassing the calyx (45).
44. Flowers not more than 11 mm. long; corolla whitish or tinged with purple; pods little surpassing the calyx, 8 to 10 mm. long, straight.................10. *A. sophoroides*
44. Flowers 15 mm. or longer; corolla purple; pods greatly surpassing the calyx, 15 to 40 mm. long, falcate. Plants normally acaulescent or nearly so......14. *A. amphioxys*
45. Stipules (at least the lower ones) evidently connate around the stem; calyx saccate-ventricose at base; pods short-stipitate. Calyx teeth very short; pods 1-celled, considerably inflated, broadly elliptic, ovate, or obovate, the walls somewhat leathery 29. *A. famelicus*

45. Stipules distinct; calyx at most somewhat oblique at base; pods sessile in the calyx (46).

46. Plants distinctly caulescent, the leafy stems seldom less than 30 cm. long; herbage glabrous or strigose to white-sericeous, the hairs mostly appressed; pods with papery to somewhat leathery walls, often mottled. Corolla purple or whitish 56. *A. lentiginosus*

46. Plants short-caulescent (normally acaulescent or nearly so), the leaves mostly basal or subbasal and the stems usually much less than 30 cm. long; herbage copiously and rather loosely pubescent, many of the hairs often more or less spreading; pods with thick, leathery to almost woody walls, not mottled (47).

47. Petals purple, except toward base; leaflets elliptic to obovate; calyx 8 to 15 mm. long; pods 1-celled, oblong to ovate, slightly to moderately curved, rounded at base, 17 to 40 mm. long .... 15. *A. tephrodes*

47. Petals whitish, the keel (sometimes other petals), purple-tipped; leaflets broadly oval or obovate to suborbicular; calyx 6 to 9 mm. long; pods 2-celled, linear-lanceolate, strongly curved (sometimes into a nearly complete circle), attenuate at base, 30 to 50 mm. long .... 59. *A. Layneae*

48. Flowers 18 to 23 mm. long; pods 2-celled, globose or nearly so, glabrous, about 2 cm. long, the walls thick and fleshy, becoming spongy. Stems decumbent, at least toward base; leaflets elliptic .... 72. *A. crassicarpus*

48. Flowers less than 18 mm. long, or the pods otherwise (49).

49. Plants short-caulescent or nearly acaulescent, the stems (excluding the peduncles) not more and usually much less than 25 cm. long. Calyx teeth about ⅕ as long as the tube (50).

49. Plants strongly caulescent, the stems seldom less and usually much more than 25 cm. long (51).

50. Leaflets 7 to 11, mostly obovate or oblanceolate, these and the stems glabrous or nearly so; flowers 12 to 18 mm. long; pods 1-celled, long-stipitate (the stipe 2 to 3 times as long as the calyx), greatly inflated, oval or ovate, straight or nearly so, the walls papery but rather firm .... 44. *A. ampullarius*

50. Leaflets 13 to 19, mostly elliptic or oval, these (at least beneath) and the stems rather copiously strigose; flowers 12 to 13 mm. long; pods 2-celled, sessile in the calyx, flat, oblong, more or less curved, the walls leathery.................. 61. *A. ensiformis*

51. Stipules (at least the lower ones) evidently connate around the stem; calyx saccate-ventricose at base. Pods stipitate (52).

51. Stipules distinct or nearly so; calyx, at most, somewhat oblique at base (53).

52. Racemes usually more than 25-flowered; corolla 8 to 12 mm. long, whitish; pods oblong or narrowly elliptic, 5 to 11 mm. long, compressed dorsoventrally (at a right angle to the sutures) .... 7. *A. Haydenianus*

52. Racemes fewer than 25-flowered; corolla 12 to 15 mm. long, purplish pink; pods broadly elliptic, ovate, or somewhat obovate, 20 to 25 mm. long, considerably inflated 29. *A. famelicus*

53. Plants malodorous; stems becoming rather hard and woody; pods with leathery to almost woody walls. Herbage glabrous or strigose; pods nearly terete, oblong-lanceolate to oblong-ovate, straight or nearly so (54).

53. Plants not malodorous; stems herbaceous; pods with papery to somewhat leathery walls (58).

54. Leaflets linear-lanceolate to narrowly elliptic; pods erect. Corolla whitish; pods sessile in the calyx or nearly so, about 3 times as long as wide .... 48. *A. Pattersoni*

54. Leaflets broadly elliptic, oval, or obovate; pods ascending or spreading (55).

55. Corolla whitish, with keel often purple-tipped; stems up to 100 (150?) cm. long; racemes in flower usually compact. Pods 2 to 3 times as long as wide, rounded at base, sessile in the calyx or nearly so.......................... 49. *A. praelongus*

55. Corolla usually purple; stems up to 70 cm. long; racemes in flower often loose. Lower leaves often with obcordate leaflets (56)

56. Leaflets sparsely to copiously strigose beneath. Pods sessile in the calyx or very short-stipitate, 2 to 3 times as long as wide .... 47. *A. Crotalariae*

56. Leaflets glabrous or glabrescent (57).

57. Pods 2 to 5 times as long as wide, usually ascending and more or less stipitate, not septate .... 45. *A. Preussii*
57. Pods 4 to 6 times as long as wide, spreading or reflexed, sessile or subsessile in the calyx, with a septum about 2 mm. wide .... 46. *A. Beathii*
58. Pods 1-celled but with a narrow septum, erect or nearly so, straight or very slightly curved, long-stipitate, the stipe considerably to greatly surpassing the calyx. Herbage sparsely strigose to glabrate; leaflets 19 or more; flowers 14 to 19 mm. long; pods oblong, 15 to 20 mm. long .... 55. *A. eremiticus*
58. Pods more or less completely 2-celled, spreading or ascending-spreading, more or less curved, sessile in the calyx or nearly so (59).
59. Corolla purple or whitish; pods more or less inflated, oblong-lanceolate to nearly orbicular, 1 to 6 times as long as wide.. 56. *A. lentiginosus*
59. Corolla violet, at least when dry; pods strongly compressed, linear or linear-lanceolate, 5 to 8 times as long as wide .... 57. *A. Bryantii*
60. Flowers (except occasionally in *A. arizonicus, A. confertiflorus, A. lancearius,* and *A. striatiflorus*) 10 mm. or longer (61).
60. Flowers (except occasionally in *A. moencoppensis, A. pinonis,* and *A. Sileranus*) less than 10 mm. long (86).
61. Calyx teeth (except occasionally in *A. allochrous*) ⅔ as long as to longer than the tube (62).
61. Calyx teeth (except occasionally in *A. arizonicus* and *A. convallarius*) less than ⅔ as long as the tube (67).
62. Terminal leaflet decurrent on the rachis.. 35. *A. niveus*
62. Terminal leaflet jointed to the rachis (63).
63. Stems strongly decumbent to prostrate; keel sharply bent toward apex, with a conspicuous beaklike prolongation at a right angle to the claws. Stipules conspicuously connate around the stem (64).
63. Stems erect, ascending, or decumbent only toward base; keel moderately incurved, without a conspicuous beaklike prolongation. Pods 1-celled (65).
64. Herbage strigose or sericeous with mostly appressed dolabriform hairs (rarely loosely villous with curly, mostly basifixed hairs); flowering stems usually much more than 3 cm. long; leaflets linear-lanceolate or narrowly elliptic (seldom oblanceolate), the apex acutish to acuminate; racemes commonly more than 5-flowered; pods 1-celled, compressed, more or less curved, commonly strigose or sericeous. . 20. *A. humistratus*
64. Herbage grayish-pilose with soft, curly, basifixed hairs; flowering stems not more than 3 cm. long; leaflets oblanceolate or obovate, the apex obtuse, truncate, or slightly notched; racemes 3- to 5-flowered; pods 2-celled, greatly inflated, not curved, villous 58. *A. striatiflorus*
65. Plants annual or, at most, short-lived perennial, the flowering stems from a more or less thickened taproot; herbage sparsely to moderately (rarely copiously) strigose, the hairs basifixed; pods very asymmetric (straight on the upper, curved on the lower suture), greatly inflated, the walls papery. Stems commonly at least 30 cm. long; leaflets 11 or more, broadly linear to elliptic; flowers 6 to 12 mm. long 33. *A. allochrous*
65. Plants evidently perennial, the flowering stems from a branched more or less woody caudex; herbage copiously strigose to silvery-sericeous, the hairs at least partly dolabriform; pods nearly symmetric and straight, little inflated, the walls firm (66).
66. Leaflets narrowly linear, mostly acute or acutish at apex, up to 25 mm. long; inflorescences usually much surpassing the leaves.. 8. *A. confertiflorus*
66. Leaflets broadly linear to narrowly elliptic, mostly obtuse at apex, up to 18 mm. long; inflorescences shorter than to barely surpassing the leaves........ 10. *A. sophoroides*
67. Leaflets (except occasionally in *A. nothoxys*) glabrous or glabrescent above (68).
67. Leaflets (except occasionally in *A. arizonicus, A. lonchocarpus,* and *A. tephrodes*) persistently pubescent above (75).
68. Leaflets narrowly linear. Stems erect, ascending, or decumbent only toward base, or (exceptionally in *A. confertiflorus*) strongly decumbent (69).
68. Leaflets mostly broader (71).

69. Petals sharply bent, the blades spreading at a right angle to the claws; terminal leaflet often greatly elongate and rachislike, or the whole leaf reduced to a rachis. Flowers up to 11 mm. long; pods more or less pendulous, straight or nearly so, more or less flattened laterally, 15 to 40 mm. long, 3 to 5 mm. wide...... 4. *A. convallarius*
69. Petals moderately curved, the blades ascending; terminal leaflet not greatly elongate or rachislike (70).
70. Herbage sericeous; stipules connate around the stem; leaflets not more than 5 times as long as wide; flowers 9 to 15 mm. long; pods not winged, straight, 8 to 12 mm. long, the walls firm but not woody............................ 8. *A. confertiflorus*
70. Herbage glabrous or sparsely strigose; stipules distinct; leaflets 5 to 10 times as long as wide; flowers 15 to 20 mm. long; pods conspicuously winged, crescent-shaped or even more strongly curved, 20 to 40 mm. long, the walls becoming woody 18. *A. tetrapterus*
71. Flowers 18 to 23 mm. long; pods 2-celled, globose or nearly so, glabrous, about 2 cm. long, the walls thick and fleshy, becoming spongy. Stems decumbent, at least toward base; leaflets elliptic .... 72. *A. crassicarpus*
71. Flowers smaller, or the pods otherwise (72).
72. Keel abruptly contracted into a short, slender, beaklike acumination; pods 2-celled. Plants spring-flowering; pods compressed, linear or lance-linear.... 64. *A. nothoxys*
72. Keel not beaked; pods 1-celled, sometimes with a narrow septum (73).
73. Stipules distinct or nearly so; pods erect. Herbage glabrous or nearly so; flowers 15 to 20 mm. long, the corolla whitish; pods 20 to 25 mm. long, thick-walled, oblong-lanceolate .... 48. *A. Pattersoni*
73. Stipules (at least the lower ones) connate around the stem; pods more or less pendulous (74).
74. Herbage whitish-sericeous; calyx, at most, somewhat oblique at base; corolla dull yellow, tinged with purple; pods lanceolate, about 12 mm. long, thin-walled 21. *A. albulus*
74. Herbage grayish-strigose; calyx saccate-ventricose at base; corolla purplish pink; pods broadly elliptic, ovate, or somewhat obovate, 20 to 25 mm. long, thickish-walled 29. *A. famelicus*
75. Stems short, the leaves mostly basal (plants normally acaulescent or nearly so); calyx 7 to 15 mm. long, cylindric. Herbage commonly strigose to sericeous; flowers seldom less than 15 mm. long, the corolla normally purple; pods 1-celled, 17 to 40 mm. long, the walls leathery (76).
75. Stems elongate, the leaves mostly suprabasal; calyx not more than 8 mm. long, mostly campanulate or cylindric-campanulate (77).
76. Hairs dolabriform; pods lanceolate or oblong-lanceolate, 4 to 5 times as long as wide, usually attenuate-acuminate at both ends, more or less falcate..... 14. *A. amphioxys*
76. Hairs basifixed; pods oblong to ovate, seldom more than 3 times as long as wide, acute to short-acuminate at apex, rounded at base, straight or nearly so... 15. *A. tephrodes*
77. Keel with a beaklike prolongation at a right angle to the claws, this acute or acutish; stems strongly decumbent to procumbent; hairs of the herbage mostly dolabriform; pods 2-celled. Herbage copiously strigose to silvery-sericeous; leaflets linear or lance-linear; pods compressed, linear or lance-linear.............. 62. *A. arizonicus*
77. Keel not with a beaklike prolongation or if somewhat beaked (in *A. convallarius*), then the stems erect, ascending, or decumbent only toward base, and the hairs basifixed; pods 1-celled (78).
78. Calyx ventricose-saccate at base. Pods more or less pendulous (79).
78. Calyx not ventricose-saccate, at most somewhat oblique at base. Leaflets narrowly oblong or oblanceolate to nearly filiform (80).
79. Plants malodorous; stems coarse, often swollen; leaflets linear to nearly filiform, the terminal one not articulated to the rachis, often greatly elongate and rachislike, or the whole leaf reduced to a rachis; flowers 15 to 20 mm. long; pods long-stipitate, compressed .... 19. *A. lonchocarpus*
79. Plants not malodorous; stems wiry, not swollen; leaflets elliptic, oblanceolate, or obovate, the terminal one articulated, never rachislike; flowers 9 to 14 mm. long; pods short-stipitate, inflated .... 29. *A. famelicus*

80. Hairs of the herbage mostly dolabriform; pods 8 to 12 mm. long. Herbage silvery-sericeous; terminal leaflet little longer than the others, articulated to the rachis; flowers 9 to 15 mm. long; pods erect or ascending, not stipitate...8. *A. confertiflorus*
80. Hairs basifixed; pods 15 to 40 mm. long (81).
81. Flowers 15 to 20 mm. long; pods conspicuously winged, strongly curved, turgid, spreading or ascending, the walls becoming woody; leaflets often sharply cuspidate. Herbage glabrous or sparsely strigose; terminal leaflet seldom greatly elongate; pods sessile in the calyx ....18. *A. tetrapterus*
81. Flowers not more, usually less than 15 mm. long; pods not winged, straight or nearly so, more or less compressed or bladdery-inflated, mostly pendulous (except perhaps in *A. niveus*), the walls remaining thin; leaflets not cuspidate (82)
82. Banner suborbicular, strongly arcuate, forming at least a right angle with the calyx. Terminal leaflet (except in var. *foliolatus*) not articulated to the rachis and often greatly elongate and rachislike; pods sessile in the calyx..........4. *A. convallarius*
82. Banner obovate, moderately arcuate (83).
83. Stems abundantly leafy, with short, sterile branches in the lower axils; leaflets densely and permanently silky-canescent, the terminal one not elongate or rachislike; flowers purple, the wings white-tipped; pods greatly inflated................35. *A. niveus*
83. Stems sparsely leafy, without sterile branches; leaflets sparsely strigose, the terminal one often greatly elongate and rachislike; flowers ochroleucous; pods not greatly inflated (84).
84. Terminal leaflet articulated to the rachis, not elongate; pods long-stipitate 5. *A. canovirens*
84. Terminal leaflet not articulated, often greatly elongate and rachislike; pods sessile in the calyx or nearly so (85).
85. Pods flat, strongly compressed laterally..........................2. *A. lancearius*
85. Pods subterete, compressed dorsoventrally, if at all..................6. *A. kaibensis*
86. Calyx teeth (except occasionally in *A. convallarius* and *A. Egglestonii*) less than ⅔ as long as the tube (87).
86. Calyx teeth (except occasionally in *A. allochrous, A. ceramicus, A. cobrensis,* and *A. Nuttallianus*) ⅔ as long as to longer than the tube (98).
87. Leaflets persistently pubescent above. Pods more or less pendulous (88).
87. Leaflets (except occasionally in *A. Hartwegii, A. nothoxys,* and *A. wingatanus*) glabrous or glabrescent above (91).
88. Flowers 8 mm. or longer. Herbage strigose-canescent; pods 1-celled, straight or nearly so (89).
88. Flowers 5 to 7 mm. long. Leaflets seldom fewer (often more) than 13, the terminal one not greatly elongate or rachislike (90).
89. Leaflets usually fewer than 13, the terminal one (except in var. *foliolatus*) often elongate and rachislike, or the whole leaf reduced to a rachis; pedicels commonly about half as long as the calyx; calyx teeth 1 mm. long or shorter; pods linear or narrowly oblanceolate....................................4. *A. convallarius*
89. Leaflets usually 13 or more, the terminal one not greatly elongate or rachislike; pedicels much shorter than the calyx; calyx teeth 1 to 1.5 mm. long; pods oblong-lanceolate ....30. *A. pinonis*
90. Herbage copiously strigose; corolla purple; pods 1-celled, asymmetrically ovate, greatly inflated...40. *A. insularis*
90. Herbage glabrous or sparsely strigose; corolla whitish or purple-tinged; pods 2-celled, linear, not inflated, strongly curved............................54. *A. recurvus*
91. Leaflets narrowly linear to nearly filiform, the terminal one often greatly elongate and rachislike, or the whole leaf reduced to a rachis. Herbage rather copiously strigose; racemes loose, few-flowered; pods 1-celled, more or less flattened laterally, 3 to 5 mm. wide, more or less pendulous.......................4. *A. convallarius*
91. Leaflets broader, the terminal one never greatly elongate or rachislike (92).
92. Keel abruptly contracted into a short, slender, beaklike acumination. Stems strongly decumbent to prostrate; herbage green, sparsely to rather copiously strigose; leaflets elliptic, oblanceolate, or obovate, 3 to 4 mm. wide, rounded, truncate, or retuse at

apex; flowers 8 to 12 mm. long; pods 2-celled, linear or lance-linear, compressed, deeply grooved dorsally..64. *A. nothoxys*

92. Keel not abruptly beaked, but sometimes prolonged at a right angle to the claws and gradually attenuate (93).

93. Flowers ascending or spreading; leaflets linear to narrowly oblanceolate; pods 1-celled (94).

93. Flowers deflexed, the pedicels decurved at or soon after anthesis; leaflets elliptic, oblong, oblanceolate, or obovate; pods more or less completely 2-celled. Herbage glabrous or sparsely (seldom copiously) strigose; flowers 5 to 8 mm. long; corolla whitish or purple-tinged; pods pendulous (95).

94. Herbage sparsely to copiously strigose; racemes in flower loose, surpassing the leaves; calyx teeth about half as long as the tube; keel moderately incurved toward apex, obtuse; pods flat, oblong or elliptic, 3 to 4 times as long as wide, glabrous, exceptionally mottled..3. *A. wingatanus*

94. Herbage strigose-canescent; racemes in flower usually compact, not surpassing the leaves; calyx teeth ¼ to ½ as long as the tube; keel strongly bent toward apex, acute or acutish; pods greatly inflated, oval or ovate, not more than twice as long as wide, strigose, conspicuously mottled with brownish purple.....36. *A. subcinereus*

95. Racemes in flower compact or dense, usually more than 20-flowered; pods not more than 15 mm. long, sessile in the calyx.......................65. *A. Hartwegii*

95. Racemes in flower loose, seldom more than 20-flowered; pods 15 to 20 mm. long (96).

96. Corolla purple-tinged; pods strongly curved, sessile or subsessile in the calyx. Leaves ascending-spreading; leaflets 1 to 2 mm. wide; pods finely strigose..54. *A. recurvus*

96. Corolla whitish; pods straight or nearly so, distinctly stipitate (97).

97. Leaves ascending-spreading; pods sparsely strigose, the stipe often considerably surpassing the calyx..52. *A. Rusbyi*

97. Leaves nearly erect; pods glabrous, the stipe usually shorter than the calyx 53. *A. Egglestonii*

98. Leaflets (except occasionally in *A. cobrensis, A. Hartwegii,* and *A. Wootoni*) glabrous or glabrescent above (99).

98. Leaflets (except occasionally in *A. Nuttallianus* and *A. Sileranus*) persistently pubescent above (109).

99. Leaflets not more than 3 times as long as wide, obovate or oblanceolate, more or less conspicuously notched at apex (100).

99. Leaflets usually more than 3 times as long as wide, narrowly linear to elliptic or somewhat oblanceolate, entire or nearly so at apex (101).

100. Plants perennial, the flowering stems from a branched caudex or rootstock; inflorescences distinctly racemose; flowers 5 to 7 mm. long; pods elliptic or oblong, 9 to 13 mm. long, the walls veiny but scarcely rugose.................51. *A. cobrensis*

100. Plants annual; inflorescences densely short-spicate or subcapitate; flowers 4 to 5 mm. long; pods ovate, not more than 4 mm. long, the walls transversely rugose with hard ridges..71. *A. didymocarpus*

101. Stems strongly decumbent to prostrate; stipules conspicuously connate around the stem; hairs of the herbage (except in var. *crispulus*) mostly dolabriform. Plants perennial, the flowering stems from a well-developed, more or less woody caudex; herbage strigose or sericeous; leaflets usually acute at apex; racemes short and compact; keel sharply bent toward apex and prolonged at a right angle to the claws 20. *A. humistratus*

101. Stems erect, ascending, or decumbent only toward base; stipules not or not conspicuously connate; hairs basifixed (102).

102. Plants annual or, at most, short-lived perennial; pods not mottled or (in *A. Wootoni*) exceptionally so. Pods greatly inflated, thin-walled, 1-celled but with the ventral suture sometimes introverted (103).

102. Plants evidently perennial or if flowering the first year and appearing annual (*A. subcinereus*), then the pods conspicuously mottled (105).

103. Stems up to 60 cm. long; racemes usually surpassing the subtending leaves; peduncles up to 10 cm. long. Pods 25 to 40 mm. long, very asymmetric, nearly straight on the upper (ventral) side, strongly curved on the lower side...........33. *A. allochrous*

103. Stems up to 30 cm. long; racemes usually not surpassing the leaves; peduncles not more than 6 cm. long (104).

104. Racemes 5- to 10-flowered; pods 15 to 25 mm. long, broadly ovate, very asymmetric, much more strongly curved on the lower (dorsal) side than on the upper (ventral) side ....34. *A. Wootoni*

104. Racemes 10- to 25-flowered; pods less than 15 mm. long, nearly orbicular, symmetric or nearly so..43. *A. Thurberi*

105. Flowering stems from a deep rootstock, copiously white-strigose, or -sericeous; pods conspicuously mottled with brown-purple spots, greatly inflated, papery-walled, 20 mm. or longer, 1-celled....................................36. *A. subcinereus*

105. Flowering stems from a well-developed caudex, glabrate to rather copiously strigose; pods not mottled, moderately inflated, firm-walled, 15 mm. or shorter (106).

106. Leaflets usually more than 15, linear, elliptic, or oblanceolate; racemes in flower compact or dense; flowers deflexed, the pedicels decurved at or soon after anthesis; pods pendulous, 2-celled (107).

106. Leaflets usually fewer than 15, narrowly linear; racemes in flower loose; flowers and pods ascending or spreading (108).

107. Flowers 6 to 7 mm. long; calyx teeth much shorter than to about ⅔ as long as the tube; banner petal considerably longer than the wings and keel; pods more than 10 mm. long, the hairs closely appressed........................65. *A. Hartwegii*

107. Flowers about 5 mm. long; calyx teeth nearly as long as the tube; banner not or but slightly longer than the other petals; pods less than 10 mm. long, the hairs not closely appressed...66. *A. vaccarum*

108. Racemes usually more than 10-flowered; flowers 6 to 10 mm. long; corolla purple; pods 1-celled, 6 to 7 mm. long, ovate-lanceolate................9. *A. moencoppensis*

108. Racemes not more than 8-flowered, very loose; flowers about 5 mm. long; corolla whitish or purple-tinged; pods incompletely 2-celled, 10 to 15 mm. long, oblanceolate-elliptic 50. *A. Brandegei*

109. Leaflets sharply spine-tipped. Stems simple to much-branched; leaflets usually 5, subulate or narrowly lanceolate; flowers axillary, solitary or in very few-flowered racemes or fascicles, these usually shorter than the subtending leaves; corolla whitish or purplish; pods 1-celled, lanceolate, 6 to 10 mm. long, 2- to 4-seeded 1. *A. Kentrophyta*

109. Leaflets not spine-tipped (110).

110. Leaflets usually distinctly to deeply notched at apex (111).

110. Leaflets (except occasionally in *A. sabulonum*) entire or nearly so at apex (114).

111. Plants perennial, the flowering stems from a more or less branched, woody caudex or rootstock. Herbage and pods pilose, the hairs more or less spreading; leaflets oblanceolate to obovate (112).

111. Plants annual. Stems not more than 30 cm. long; flowers 4 to 6 mm. long (113).

112. Flowering stems up to 60 cm. long, strongly decumbent or prostrate; racemes compact or dense; flowers 7 to 9 mm. long; pods 1-celled, greatly inflated, oval, ovate, or nearly orbicular, usually mottled.............................37. *A. Sileranus*

112. Flowering stems not more than 20 cm. long, erect or ascending; racemes loose; flowers 5 to 7 mm. long; pods incompletely 2-celled, little-inflated, elliptic or oblong, not mottled ....51. *A. cobrensis*

113. Stems whitish-strigose or sericeus; leaflets 6 to 12 mm. long, oblong, elliptic, or oblanceolate; inflorescences loose, distinctly racemose, few-flowered; pods 1-celled, several-seeded, inflated, 3-angled, somewhat curved, about 15 mm. long, the walls thin, not rugose..38. *A. triquetrus*

113. Stems inconspicuously strigose; leaflets not more than 5 mm. long, oblanceolate or obovate; inflorescences dense, short-spicate or subcapitate, usually more than 10-flowered; pods 2-celled, usually 2-seeded, not inflated, not angled or curved, not more than 4 mm. long, the walls thickish, with hard transverse ridges 71. *A. didymocarpus*

114. Plants annual or, at most, short-lived perennial, without a well-developed caudex (115).

114. Plants evidently perennial. Pods 1-celled (118).
115. Flowers solitary or in subumbellate or subcapitate clusters of 2 to 5; pods usually 2-celled, linear, not inflated. Corolla whitish or purplish, the keel tip usually dark purple; pods 10 to 20 mm. long, curved......................63. *A. Nuttallianus*
115. Flowers several or numerous, in racemes; pods 1-celled, oblong-ovate or broader, inflated (116).
116. Herbage green, sparsely to rather copiously strigose; stems up to 60 cm. long; pods 25 to 40 mm. long. Leaflets broadly linear to elliptic; pods strongly inflated, very asymmetric, strigose.......................................33. *A. allochrous*
116. Herbage grayish, copiously pubescent; stems up to 30 cm. long; pods 10 to 20 mm. long (117).
117. Hairs closely appressed; leaflets linear to elliptic; pods strigose, 8 to 11 mm. wide, very asymmetric (straight on the upper, strongly curved on the lower side) 40. *A. insularis*
117. Hairs somewhat spreading; leaflets oblong, elliptic, oblanceolate, or obovate; pods villous, less than 8 mm. wide, moderately asymmetric, somewhat curved on both sides ....42. *A. sabulonum*
118. Flowering stems borne at intervals on long slender creeping rootstocks; stipules distinct or weakly connate; pods stipitate, moderately to greatly inflated, 12 to 35 mm. long. Stems usually less than 30 cm. long; herbage sparsely strigose to densely strigose-canescent; racemes not more than 3 cm. long; flowers 6 to 8 mm. long; pods 1-celled, thin-walled, conspicuously mottled with brown-purple 32. *A. ceramicus*
118. Flowering stems from a branched, more or less woody caudex; stipules strongly connate around the stem; pods sessile in the calyx, little-inflated, 7 to 12 mm. long. Herbage copiously strigose to sericeous (exceptionally villous), the hairs mostly dolabriform (119).
119. Keel obtuse, not prolonged at a right angle to the claws; stems erect, ascending, or decumbent only toward base; pods symmetric or nearly so, straight, erect or ascending. Herbage silvery-sericeous; racemes usually 10- to 20-flowered, shorter than to barely surpassing the leaves; corolla whitish or purplish; pods 8 to 10 mm. long ....10. *A. sophoroides*
119. Keel obtuse to acuminate, conspicuously prolonged at a right angle to the claws; stems strongly decumbent to prostrate; pods often very asymmetric, more or less curved, spreading (120).
120. Flowering stems up to 60 cm. long; racemes compact, usually several-flowered; leaflets linear-lanceolate or narrowly elliptic (seldom oblanceolate), acutish to acuminate (exceptionally obtuse) at apex; corolla dull yellow and (or) purple; pods ovate or obovate, seldom mottled..........................20. *A. humistratus*
120. Flowering stems not more than 6 cm. long; racemes loose, 1- to 3-flowered; leaflets narrowly elliptic, obtuse to very acute or cuspidate at apex; corolla violet, at least when dry; pods narrowly obovate, usually mottled.............25. *A. sesquiflorus*

1. Astragalus Kentrophyta Gray. Navajo and Coconino counties, especially in the Grand Canyon, 3,500 to 8,500 feet, July to September. Saskatchewan to Alberta and eastern Oregon, south to western Nebraska, New Mexico, northern Arizona, and Nevada.

Very different in appearance from any other of the Arizona species of *Astragalus*, with hard, almost woody stems, sharply spine-tipped leaflets and stipules, and small axillary flowers. This species is considered excellent for control of erosion.

The Arizona plants belong to var. *elatus* Wats. (*A. Kentrophyta* var. *coloradoensis* Jones, *A. impensus* (Sheld.) Woot. & Standl., *Kentrophyta impensa* Rydb., *K. coloradensis* Rydb.). The type of *A. Kentrophyta* var. *coloradoensis* came from Lees Ferry, Coconino County (*Jones* in 1890).

2. Astragalus lancearius Gray (*A. episcopus* Wats., *Homalobus lancearius* Rydb.). Coconino County (?) and northern Mohave County, 2,000 to 5,500 feet, type of *A. lancearius* from Beaver Dam, Mohave County (*Palmer* in 1877). Southern Utah and northern Arizona.

3. Astragalus wingatanus Wats. (*Homalobus wingatanus* Heller). Apache, Navajo, Coconino, and northern Gila counties, 6,000 to 6,500 feet, May to August. Southern Colorado and Utah, northwestern New Mexico, and northeastern Arizona.

4. Astragalus convallarius Greene (*A. junciformis* A. Nels., *Homalobus junceus* Nutt.). Navajo and Coconino counties, 5,000 to 6,000 feet, April and May. Montana and Idaho to northern Arizona.

The relatively small flowers (8 to 11 mm. long), sharply bent so that the blades of the petals form a right angle with the claws, distinguish *A. convallarius* from all related "junciform" *Astragali* of Arizona. The species is represented in this state by two well-marked varieties. These are: (1) var. *xiphoides* Barneby, characterized by leaflets decurrent on (not articulated to) the rachis, the terminal leaflet often greatly elongate and rachislike, or the whole leaf reduced to a rachis, and by strongly compressed pods 4 to 5 mm. wide; and (2) var. *foliolatus* Barneby, with leaflets articulated to the rachis, the terminal one not greatly elongate or rachislike, and with the pods somewhat narrower (3 to 4.5 mm. wide) and less flattened when mature. Var. *xiphoides* has been collected near Holbrook, Navajo County (*Hough* 37, *Ripley & Barneby* 5246, the type collection), the latter collection on slopes of clay hills. Var. *foliolatus* is known only from near Frazier Well, western Coconino County (*Ripley & Barneby* 5229, the type collection), where it is reported to be not infrequent among junipers on sandy soil overlying limestone, and from near Peach Springs, Mohave County (*Ripley & Barneby* 7017).

5. Astragalus canovirens (Rydb.) Barneby (*Homalobus canovirens* Rydb). Near Sweetwater, Apache County, about 6,000 feet (*Peebles & Smith* 13557), May and June. Southwestern Colorado, southeastern Utah, northwestern New Mexico, and northeastern Arizona.

6. Astragalus kaibensis Jones (*Lonchophaca kaibensis* Rydb.). Known only from House Rock Valley, northern Coconino County, about 5,000 feet (*Jones* in 1890, the type collection, *Ripley & Barneby* 4856).

This species, notwithstanding the tendency to dorsoventral compression of the pods, is considered by R. C. Barneby (personal communication) to be closely related to *A. lancearius* Gray.

7. Astragalus Haydenianus Gray (*Diholcos scobinatulus* (Sheld.) Rydb.). House Rock, Coconino County (*Jones* 25427). Wyoming to New Mexico, Nevada, and northern Arizona.

This species is easily distinguished by its dense, many-flowered racemes and pendulous, transversely ribbed, longitudinally furrowed pods. It is common on, if not confined to, seleniferous soils and hence is doubtless toxic to animals.

8. Astragalus confertiflorus Gray (*Cnemidophacos confertiflorus* Rydb.). Near Rock Point (Apache County), near Holbrook (Navajo County), south of Lees Ferry (Coconino County), northern Mohave County, 5,000 to 6,000 feet, May. Colorado, Utah, Nevada, New Mexico, and Arizona.

Prefers soils of relatively high selenium content and is probably poisonous.

9. Astragalus moencoppensis Jones (*Cnemidophacos moencoppensis* Rydb.). Navajo County and eastern Coconino County, 3,500 to 5,500 feet, May and June, type from Willow Spring, Coconino County (*Jones* in 1890). Southern Utah and northeastern Arizona.

10. Astragalus sophoroides Jones (*Cnemidophacos sophoroides* Rydb.). Eastern Coconino County, 4,500 to 5,000 feet, May and June, type from Willow Spring (*Jones* in 1890). Known only from northern Arizona.

11. Astragalus troglodytus Wats. (*Cnemidophacos troglodytus* Rydb.). Kaibab Plateau, San Francisco Peaks, Mogollon Escarpment near Oak Creek, and Flagstaff to Ash Fork (Coconino County), 6,500 to 7,500 (?) feet, among pines, April to June, type from the San Francisco Peaks (*Lemmon* in 1884). Known only from northern Arizona.

The flowers apparently are red when fresh, a color unique among the Arizona species. The plant is characterized also by subacaulescent habit and bracts strongly reflexed after anthesis.

12. Astragalus zionis Jones (*Xylophacos zionis* Rydb.). Betatakin (Navajo County), Navajo Mountain, near Navajo Bridge, and Grand Canyon (Coconino County), up to 7,500 feet, sandstone ledges, May and June. Southern Colorado and Utah and northern Arizona.

The Arizona plants, as compared with typical Utah specimens, tend to have more copious pubescence, shorter and less acute leaflets, shorter peduncles, and less arcuate pods (153 II, p. 149).

***13. Astragalus argophyllus** Nutt. (*Xylophacos argophyllus* Rydb.). Nagle Ranch, northern Coconino County, about 7,000 feet (*Jones* 6054h). Montana and Idaho to Colorado, northern Arizona, and Nevada.

The Jones collection was referred doubtfully by Barneby to this species (var. *pephragmenoides* Barneby). The variety differs from typical *A. argophyllus* in the paler-colored, usually smaller corolla, in the more spreading hairs, and in the relatively slender rootstock.

14. Astragalus amphioxys Gray (*Xylophacos amphioxys* Rydb.). Northern Apache County to northern Mohave County, 2,000 to 7,500 feet, April to June. Western Texas to southern Nevada and northern Arizona.

The species is quite variable, especially in the shape of the pods. Specimens with mottled pods are occasional. Caulescent forms have been collected near Kayenta and Winslow (Navajo County) and in the Grand Canyon and Havasu Canyon (Coconino County). In one of these collections (*Cook & Johnson* in 1934, Grand Canyon) the stems reach a length of 12.5 cm., with 4 well-developed internodes, and the plant is aberrant also in the acute leaflets of the upper leaves, and the black-hairy calyx. Var. *melanocalyx* (Rydb.) Tidestr. (*Xylophacos melanocalyx* Rydb.), distinguished chiefly by the black-hairy calyx, is found mostly in northern Mohave County on sandy plains, 2,000 to 4,000 feet. Var. *vespertinus* (Sheldon) Jones (*A. vespertinus* Sheldon) occurs in Monument Valley, near the Utah state line (*J. T. Howell* 24729), and has been collected also at Kanab, Utah. It differs from typical

A. amphioxys in the larger flowers (banner 22 to 28 mm. long) and in the more exserted claws of the keel. A variant characterized by very slender stems, petioles, and peduncles, broad stipules, thin, sparsely strigose leaflets, and large flowers is known only from the Grand Canyon (*Eastwood* 5711, 5712).

15. Astragalus tephrodes Gray (*Xylophacos tephrodes* Rydb.). Apache County to Mohave County, south to Graham, Gila, and eastern Maricopa counties, 3,500 to 8,000 feet, mostly in pine forest, April to July. Western Texas to southern Nevada, Arizona, and northern Mexico.

Var. *typicus* Barneby is rare in Arizona, being known only from a few collections in Greenlee and Yavapai counties. It has relatively small flowers (calyx tube 3.5 to 8 mm. long, banner 11 to 17 mm. long), relatively thin-walled, short pods (seldom more than 15 mm. long), and commonly conduplicate leaflets. Much more common and widely distributed in the state is var. *brachylobus* (Gray) Barneby (*A. pephragmenus* Jones, *A. intermedius* Jones, *A. phoenicis* Jones, *A. remulcus* Jones, *A. remulcus* var. *chloridae* Jones, *A. chloridae* Tidestrom, *Xylophacos pephragmenus* etc. Rydb.). This variety has the calyx tube 7 to 10 mm. long, the banner 17 to 22 mm. long, the pod commonly more than 15 mm. long, with rigid, woody valves when dry, and the leaflets flat. The type of *A. pephragmenus* came from the Pinal Mountains (*Jones* in 1890), that of *A. remulcus* from Bangharts (Del Rio), Yavapai County (*Rusby* 576), that of *A. intermedius* (and *A. phoenicis*) from "Arizona" (*Palmer*), and that of *A. remulcus* var. *chloridae* from Chloride, Mohave County (*Jones* in 1903). The last-mentioned is typically an extreme variant, with inflorescences greatly surpassing the leaves, the peduncles 15 to 25 cm. long, and the relatively sparse hairs of the herbage and pods straighter and more appressed than in other forms of *A. tephrodes*. Specimens from the Mazatzal Mountains are of more or less intermediate character. Var. *brachylobus* as a whole is extremely diverse in the quantity and quality of the pubescence, the hairs of the leaves and pods varying from straight and closely appressed to incurved or curly and more or less spreading.

For a further account of the synonymy and of the variation in *A. tephrodes*, especially in var. *brachylobus*, see Barneby (153 VII, pp. 463–471).

16. Astragalus castaneaeformis Wats. (*Xylophacos castaneaeformis* Rydb.). Coconino and northern Gila counties, 6,000 to 8,000 feet, among pines, often in rocky places, May, type from Williams (*Lemmon* in 1884). Known only from Arizona and Utah.

Distinguished from most of the related acaulescent species by its short, broad pods. There is, however, considerable variation in the shape and size of the pods. An extreme form, with pods only half as wide as long and rather strongly curved, should perhaps be distinguished as a variety. The species is represented in Arizona by var. *typicus* Barneby and in Utah by var. *consobrinus* Barneby.

17. Astragalus Newberryi Gray (*Xylophacos Newberryi* Rydb.). Apache County to Mohave, Gila, and Yavapai counties, 2,000 to 7,000 feet, dry, stony mesas, March and April, type from northern Arizona (*Newberry* in 1858). Idaho and Oregon to New Mexico, Arizona, and southeastern California.

A. Newberryi is peculiar among the acaulescent species of Arizona in having the pods densely villous with long, soft hairs. Var. *Blyae* (Rose) Barneby (*Xylophacos Blyae* Rose, *Astragalus Blyae* Tidestrom), which is known only in Mohave County, Arizona, types from near Kingman (*Mrs. Bly* in 1925 and 1927), differs from typical *A. Newberryi* in the smaller flowers, not more than 18 mm. long (22 to 31 mm. in the species), and in the usually more numerous leaflets. Specimens of more or less intermediate character have been found at Hackberry (*Kearney & Peebles* 11299), in Toroweap Valley (*Laws* in 1944), and southwest of Fredonia (*Hester* in 1945).

18. Astragalus tetrapterus Gray (*Pterophacos tetrapterus* Rydb.). Near Bonelli Ferry, on the Colorado River near the mouth of Virgin River, Mohave County (*Jones* in 1894). Southern Utah and Nevada and northwestern Arizona.

The remarkable winged pods distinguish this from all other species of *Astragalus* in Arizona. The plant is reported to be poisonous.

19. Astragalus lonchocarpus Torr. (*Lonchophaca macrocarpa* (Gray) Rydb.). Lukachukai Mountains, Apache County, 6,500 feet (*Goodman & Payson* 2815). Colorado to Nevada, south to northwestern New Mexico and northeastern Arizona.

20. Astragalus humistratus Gray (*Batidophaca humistrata* Rydb.). Apache County to Mohave County, south to Cochise and Pima (?) counties, 5,000 to 9,000 feet, mostly in pine and pinyon forests, May to September. Southern Colorado to southern Nevada, New Mexico, Arizona, and northern Mexico.

The prostrate stems, closely hugging the ground, make this a good plant for control of erosion.

KEY TO THE VARIETIES

1. Pods nearly symmetric, linear-lanceolate, oblong-lanceolate, or somewhat oblanceolate, strongly curved, deeply sulcate on the dorsal (lower) suture (2).
1. Pods very asymmetric, commonly oblong-ovate or oblong-obovate, moderately curved, not or scarcely sulcate dorsally. Herbage strigose or sericeous, the hairs largely dolabriform; walls of the pods coriaceous (3).
2. Herbage strigose or sericeous with straight hairs, many of these dolabrifrom; leaflets 8 to 18 mm. long, glabrous to persistently pubescent above; pods 10 to 20 mm. long, falcate, the walls coriaceous............................*A. humistratus* (typical)
2. Herbage loosely villous with curly hairs, these basifixed or very nearly so; leaflets 4 to 7 mm. long, persistently pubescent above; pods less than 10 mm. long, crescent-shaped, the walls membranaceous.................................var. *crispulus*
3. Leaflets glabrous or glabrescent above. Herbage sparsely to moderately pubescent; corolla 6 to 8 mm. long; pods ovate or oblong-ovate, 4 to 5 mm. wide var. *Hosackiae*
3. Leaflets persistently pubescent above (4).
4. Herbage silvery-sericeous; leaflets 5 to 16 mm. long; peduncles 2 to 8 cm. long; racemes usually with 8 or more flowers; corolla 7 to 12 mm. long; pods commonly more or less obovate or oblanceolate, 3 to 5 mm. wide.........................var. *sonorae*
4. Herbage sparsely to rather copiously strigose; leaflets 3 to 6 mm. long; peduncles 1.5 to 3 (6) cm. long; racemes usually with 6 or fewer flowers; corolla 6 to 7 mm. long; pods oblong-lanceolate to ovate-lanceolate, about 3 mm. wide..........var. *tenerrimus*

Typical *A. humistratus* is rather rare in Arizona, but has been collected in Apache, Navajo, Coconino, Greenlee, Gila, and Cochise counties. Var. *crispulus* Barneby, remarkably different from the other varieties in the character of the pubescence, is known only from the type collection near Nutrioso, southern Apache County (*Ripley & Barneby* 5074). Var. *Hosackiae* (Greene) Jones (*Astragalus Hosackiae* Greene, *Batidophaca Hosackiae* Rydb.) occurs in Navajo, Coconino, and Gila counties, type from Flagstaff (*Rusby* in 1883). Var. *sonorae* (Gray) Jones (*Astragalus sonorae* Gray, *Batidophaca sonorae* Rydb., *B. humivagans* Rydb.) ranges from Apache County to Mohave County, southward to Cochise and Pima (?) counties, characteristically in the pinyon zone (Barneby). The type of *A. sonorae* came from "between the San Pedro and the Sonoita" (*Wright* 1005), and the type of *B. humivagans* from Mokiak Pass, northern Mohave County (*Palmer* 108). Var. *tenerrimus* Jones (*Batidophaca tenerrima* Rydb., *Astragalus sonorae* var. *tenerrimus* Kearney & Peebles) is a reduced form peculiar to the Kaibab Plateau at 8,500 to 9,000 feet elevation, where the type was collected (*Jones* in 1894).

21. Astragalus albulus Woot. & Standl. (*Batidophaca albula* Rydb.). Apache County, St. Johns to Springerville (*Marsh* 14245, 14248), Richville Valley, sedimentary badlands (*Gentry* 3869), 5,500 to 7,000 feet, August and September. Western New Mexico and eastern Arizona.

According to C. L. Porter (*Jour. Colo.-Wyom. Acad. Sci.* 3:38) this species is more closely related to Rydberg's genera *Diholcos* and *Cnemidophacos* than to *Batidophaca*, where Rydberg placed it in *North American Flora*.

***22. Astragalus accumbens** Sheldon (*Batidophaca accumbens* Rydb.). No Arizona specimens have been identified definitely as of this species, but it has been collected at Fort Wingate, New Mexico, about 30 miles east of the Arizona state line.

23. Astragalus gilensis Greene (*Batidophaca gilensis* Rydb.). Apache and Navajo counties, 6,000 to 8,500 feet, in sandy or gravelly soil, often under pines, May to September. Western New Mexico and eastern Arizona.

The corolla is pink and white when fresh, drying blue.

24. Astragalus cremnophylax Barneby. South Rim of Grand Canyon near El Tovar, Coconino County, Arizona, 7,000 feet, in fissures of limestone pavement (*Jones* in 1903, *Ripley & Barneby* 8473, the type collection), April to June. Known only from this locality.

This remarkable little plant was referred previously to *A. humillimus* Gray.

25. Astragalus sesquiflorus Wats. (*Batidophaca sesquiflora* Rydb.). Northern Navajo and Coconino counties, 7,000 to 10,000 feet, sandstone ledges, May and June. Southern Utah and northern Arizona.

The Arizona plants differ from typical Utah specimens in their more copious and looser pubescence, smaller, usually more obtuse leaflets, and smaller calyx with relatively shorter teeth (R. C. Barneby, personal communication).

***26. Astragalus micromerius** Barneby. Tunitcha Mountains, New Mexico (*Goodman & Payson* 3218). Known definitely only from northwestern New Mexico, but since the Tunitcha Mountains are partly in Arizona the species almost certainly will be found in this state.

27. Astragalus desperatus Jones (*Batidophaca desperata* Rydb.). Holbrook (Navajo County), Moenkopi, 50 miles south of Lees Ferry, and Kaibab National Forest (Coconino County), 4,000 to 5,000 feet, April and May. Colorado, Utah, and northern Arizona.

Var. *conspectus* Barneby, which differs from typical *desperatus* in its denser inflorescence and larger flowers, with relatively narrower calyx and petals, occurs along the Little Colorado River near Holbrook (*Ripley & Barneby* 8451, the type collection).

***28. Astragalus musimonum** Barneby. Mokiak Pass, Mohave County, 5,500 feet (*Ripley & Barneby* 4321). Southern Nevada and (?) northern Arizona.

The Arizona specimens were immature and identified with some uncertainty by R. C. Barneby.

29. Astragalus famelicus Sheldon (*Pisophaca famelica* Rydb.). Apache County to Coconino and Gila counties, 5,500 to 8,500 feet, mostly in open pine forests, very common in the Flagstaff region, also in the Santa Catalina Mountains (*Lemmon* in 1881), June to September. Southern Colorado, New Mexico, and Arizona.

30. Astragalus pinonis Jones (*Pisophaca pinonis* Rydb.). South Canyon, Kaibab Plateau, Coconino County, pinyon-juniper flats (*Goodding* 92–49). Utah, Nevada, and northern Arizona, 5,000 to 6,500 feet, April to July, usually (R. C. Barneby, personal communication), with sage-brush, on limestone.

31. Astragalus oophorus Wats. (*Phaca oophora* Rydb.). Coconino, Mohave, and northern Yavapai counties, 2,500 to 7,000 feet, May. Colorado to Oregon, southeastern California, and northern Arizona.

Very showy in fruit, with large, bladderlike, thin-walled pods heavily mottled with reddish brown. The Arizona plant is var. *caulescens* Jones (*A. artipes* Gray), characterized by a cream-colored corolla and pods that are strongly convex on the dorsal side only, whereas in typical *A. oophorus* the banner is purple, the wings white, and the pods are more nearly symmetric. The type of *A. artipes* was collected at Mokiak Pass, northern Mohave County (*Palmer* in 1877).

32. Astragalus ceramicus Sheldon (*A. pictus* Gray, non Steud.). Apache County to Coconino County, 5,000 to 7,000 feet, in sand, May and June. Western South Dakota to Idaho, south to Nebraska, New Mexico, and northeastern Arizona.

This plant is easily recognized by the habit of spreading by slender, creeping rootstocks, the narrow leaflets, and the conspicuously brown-mottled pods. The sweet rootstocks are eaten in spring by Hopi Indian children.

Typical *A. ceramicus* has distinctly dolabriform hairs on the herbage, the leaves more or less spreading, with 4 to 14 lateral leaflets, the terminal leaflet usually with a distinct blade not more than twice as long as the lateral leaflets, the racemes with 5 to 15 flowers, and the pods not more than half as wide as long. In var. *filifolius* (Gray) Hermann (*A. ceramicus* var. *imperfectus* Sheldon, *A. mitophyllus* Kearney), which is less common in Arizona, the hairs are mostly basifixed, the leaves erect, the lateral leaflets fewer (often none), the terminal leaflet very long and rachislike, the racemes fewer-flowered, and the pods more inflated, often ⅔ as wide as long; but many specimens are intermediate in these characters. An extreme variant with pods only ¼ to ⅓ as wide as long (*A. pictus* Gray var. *angustus* Jones) also occurs in Arizona.

33. Astragalus allochrous Gray (*Phaca allochroa* Rydb.). Coconino and Mohave counties to Graham, Cochise, Santa Cruz, and Pima (doubtless also Yuma) counties, 1,500 to 7,000 feet, very common on plains and mesas, occasional in oak-walnut woodland, March to May, type from Wickenburg, Maricopa County (*Palmer* 588). Southern New Mexico and Arizona.

One of the most conspicuous species in the state because of the large size of the plant and of the bladdery pods. It causes loco disease in horses.

34. Astragalus Wootoni Sheldon (*Phaca Wootoni* Rydb.). Equally common in Arizona as *A. allochrous* and having about the same geographic and altitudinal range, habitat, and time of flowering. Western Texas to southeastern California and northern Mexico.

This plant is known definitely to produce loco disease in horses, cattle, and sheep. It is smaller in all its parts than *A. allochrous* but is otherwise very similar. Most of the Arizona plants belong to var. *typicus* Barneby, but var. *endopterus* Barneby occurs in the valley of the Little Colorado River at Cameron (*Ripley & Barneby* 8491, the type collection) and below Tanners Crossing (*Ward* in 1901). The variety is distinguished as follows (153X, p. 498): Leaflets 4 or 5 pairs (mostly 6 to 9 pairs in var. *typicus*); peduncles up to 2 cm. long, very slender, borne on nearly the whole length of the stem (up to 5 cm. long, stout, borne in the median and upper axils, in var. *typicus*); pods membranaceous, mottled, the ventral suture introverted as a wing 2.5 to 3.5 mm. wide (chartaceous, seldom mottled, the ventral suture filiform or the wing less than 1 mm. wide in var. *typicus*).

***35. Astragalus niveus** (Rydb.) Barneby (*A. Peirsonii* Munz & McBurney, *Phaca nivea* Rydb.). Not yet reported from Arizona, but it occurs in eastern Imperial County, California, growing on sand hills with *Ammobroma*, and extends to northeastern Baja California and northwestern Sonora.

36. Astragalus subcinereus Gray (*Phaca subcinerea* Rydb.). Apache County to Coconino County, 5,000 to 7,000 feet, sandy soil, May to August, type from Mokiak Pass (*Palmer* in 1877). Southern Utah and northern Arizona.

37. Astragalus Sileranus Jones (*Phaca Silerana* Rydb.). Kaibab Plateau and North Rim of Grand Canyon (Coconino County), 7,000 to 9,000 feet, openings in coniferous forests, June to August. Southern Utah and northern Arizona.

When fresh the corolla is yellowish with a purple-tipped keel.

38. Astragalus triquetrus Gray (*Phaca triquetra* Rydb.). Beaver Dam, Mohave County, 2,000 feet (*Peebles & Parker* 14678), sandy soil, April and May. Extreme southern Nevada and adjacent Arizona.

39. Astragalus aridus Gray (*Phaca arida* (Gray) Rydb.). Southwestern Arizona (*Gilman* 1145), near Fortuna Station, Yuma County (*Ripley & Barneby* 10091), sandy deserts, March and April. Southwestern Arizona, southeastern California, and Baja California.

40. Astragalus insularis Kellogg (*Phaca insularis* Rydb.). Near Bouse, Yuma County, 1,500 feet, on sand dunes (*Benson & Darrow* 10916), March and April. Southwestern Arizona, southeastern California, and Baja California.

The Arizona plants belong to var. *Harwoodii* Munz & McBurney, which has larger pods than in the type of the species.

41. Astragalus Palmeri Gray (*Phaca Palmeri* Rydb.). Graham County, at Camp Grant, 5,000 feet (*Palmer* in 1867, the type collection), and San Carlos River (*Mohr* in 1873). Known only from southeastern Arizona.

42. Astragalus sabulonum Gray (*Phaca sabulonum* Rydb.). Navajo, Coconino, and Mohave counties, chiefly along the Little Colorado and Colorado rivers, from Monument Valley and Leupp to near the mouth of Virgin River, also in Detrital Valley (northern Mohave County), 1,000 to 4,500 feet, sometimes with *Artemisia tridentata,* April to June. Southern Utah and Nevada to northwestern New Mexico, Arizona, southeastern California, and Sonora.

43. Astragalus Thurberi Gray (*Phaca Thurberi* Kearney). Graham, Cochise, Santa Cruz, and Pima counties, 3,000 to 5,000 feet, common on dry, rocky slopes and mesas, March to May. Western New Mexico, southern Arizona, and Sonora.

This species is definitely known to cause loco disease. It is a small plant with nearly globose, thin-walled pods.

***44. Astragalus ampullarius** Wats. (*Phaca ampullaria* Rydb.). Known only from southern Utah, but likely to be found in extreme northern Arizona, the type having been collected at Kanab, a few miles north of the state line.

45. Astragalus Preussii Gray (*Phaca Preussii* Rydb.). Apache County to northern Mohave County, 2,000 to 6,000 feet, April to June. Western Colorado to northern Arizona and southeastern California.

Most of the Arizona specimens belong to var. *latus* Jones (var. *arctus* Sheldon, *Jonesiella arcta* Rydb.). This entity, perhaps specifically distinct, has the leaflets elliptic to narrowly obovate, the corolla whitish, and the pods usually pronouncedly stipitate. In the typical phase of the species, which has been collected near Navajo Bridge (*Peebles* 14647), and in var. *laxiflorus* Gray (*Phaca laxiflora* Rydb.), type from Beaver Dam (*Palmer* 104, in 1877), the leaflets are usually broadly obovate and the corolla usually purple. The pods are very short-stipitate in var. *laxiflorus*. This species is reported to prefer seleniferous soils and is probably poisonous.

46. Astragalus Beathii C. L. Porter. Near Cameron, eastern Coconino County (*Goodding* 34–39, the type collection), about 4,000 feet, March to May. Known only from this vicinity.

47. Astragalus Crotalariae (Benth.) Gray (*Phaca Crotalariae* Benth.). Near Yuma, Yuma County, 200 feet, in sandy soil (*Benson* 4198, *Crowder* in 1947), February to April. Southwestern Arizona, California, and Baja California.

48. Astragalus Pattersoni Gray (*Jonesiella Pattersoni* Rydb.). Near Fredonia, Coconino County, about 5,000 feet (*Ripley & Barneby* 4361). Colorado and northern Arizona.

49. Astragalus praelongus Sheldon (*A. Rothrockii* Sheldon ?, *Jonesiella praelonga* Rydb., *J. Mearnsii* Rydb.). Apache County to Mohave and Yavapai counties, 3,000 to 6,500 feet, common, May to July, type of *J. Mearnsii* from Fort Verde, Yavapai County (*Mearns* 163). Southern Utah and Nevada, New Mexico, and Arizona.

A coarse, ill-smelling plant with cream-colored flowers, resembling *A. Preussii* but with walls of the large pods thicker, almost woody. This plant, preferring (if not confined to) seleniferous soils, has been proved by experiments with sheep to be very toxic. Specimens apparently of this species but with exceptionally broad leaflets and pods only about 2 cm. long have been collected in Havasu Canyon (Coconino County).

50. Astragalus Brandegei Porter (*Atelophragma Brandegei* Rydb.). Banghart's Ranch (Del Rio), Yavapai County (*Rusby* 572), about 4,500 feet, May and June. Colorado, Utah, New Mexico, and central Arizona.

51. Astragalus cobrensis Gray (*Atelophragma cobrense* Rydb.). Metcalf (Greenlee County), Miami to Potato Patch (Pinal County), Chiricahua Mountains (Cochise County), April and May. Southwestern New Mexico and southeastern Arizona.

The typical phase of the species, with strigose pubescence and glabrous upper surface of the leaflets, occurs in Greenlee County. The collections from Pinal and Cochise counties, with more copious and more spreading pubescence and leaflets sparsely pubescent above, belong to var. *Maguirei* Kearney, type from the Chiricahua Mountains (*Maguire et al.* 11079).

52. Astragalus Rusbyi Greene (*Atelophragma Rusbyi* Rydb.). San Francisco Peaks, Flagstaff, and north of Williams (Coconino County), Mount Trumbull (Mohave County), 6,500 to 8,000 feet, May to September, type from Mount Humphreys (*Rusby* 573). Known certainly only from northern Arizona.

53. Astragalus Egglestonii (Rydb.) Kearney & Peebles (*Tium Egglestonii* Rydb.). White Mountains, Apache County, 6,500 to 9,000 feet, August. New Mexico and eastern Arizona.

Very similar to *A. Rusbyi* and scarcely more than a variety of that species.

54. Astragalus recurvus Greene (*Tium recurvum* Rydb.). Coconino, Gila, and Yavapai counties, 5,000 to 7,500 feet, May and June, type from "mountains of northern Arizona" (*Rusby* in 1883). Utah and Arizona.

Similar in most characters to the 2 preceding species, but the pods sessile or nearly so.

55. Astragalus eremiticus Sheldon (*Tium eremiticum* Rydb.). Mokiak Pass, Peach Springs, Chloride, and Pagumpa Springs (Mohave County), 3,000 to 5,500 feet, April and May. Idaho and Oregon to northwestern Arizona and Nevada.

56. Astragalus lentiginosus Dougl. (*Cystium lentiginosum* Rydb.). Almost throughout the state in one or another of its numerous varieties, 200 to 7,000 feet, plains and mesas, usually in sandy soil, February to June, sometimes later. Idaho and Washington, south to New Mexico, Arizona, California, and adjacent Mexico.

KEY TO THE VARIETIES

(after Barneby, 153 IV, pp. 75–80)

1. Racemes short and dense, with the axis not exceeding 4 cm. in fruit (2).
1. Racemes loose and open or, if rather dense in fruit, then the axis of the raceme more than 4 cm. long (6).
2. Pods ovoid, strongly inflated, more than 6 mm. in greatest diameter, or, if less, then not exceeding 13 mm. in length (3).

2. Pods lanceolate, barely inflated, not exceeding 6 mm. in greatest diameter and 3 times longer than wide (4).
3. Pods coriaceous or firmly chartaceous; peduncles usually strict........a. var. *diphysus*
3. Pods membranaceous; peduncles widely spreading..................b. var. *oropedii*
4. Petals purple; pods spreading. Stems erect..........................c. var. *palans*
4. Petals whitish or pinkish; pods strongly ascending (5).
5. Stems prostrate, flexuous; petals white or pinkish.................d. var. *Wilsonii*
5. Stems erect or ascending; petals ochroleucous......................e. var. *maricopae*
6. Pods not exceeding 6 mm. in greatest diameter (7).
6. Pods exceeding 6 mm. in greatest diameter; ovoid, inflated (11).
7. Herbage silvery-sericeous or tomentulose........................f. var. *borreganus*
7. Herbage green, strigose or glabrous (8).
8. Pods membranaceous or thin-chartaceous, narrowly ovoid............g. var. *australis*
8. Pods decidedly coriaceous, lanceolate or lance-falcate (9).
9. Corolla ochroleucous..e. var. *maricopae*
9. Corolla evidently purple (10).
10. Pods strongly ascending, in very open racemes; petals drying violet..h. var. *mokiacensis*
10. Pods spreading, in rather compact racemes; petals drying purple or whitish c. var. *palans*
11. Pods coriaceous, glabrous..................................h. var. *mokiacensis*
11. Pods membranaceous or chartaceous or if firm, then evidently pubescent (12).
12. Pods chartaceous, rather firm, pubescent with incurved, curly hairs...i. var. *stramineus*
12. Pods membranaceous, glabrous or strigulose (13).
13. Leaves 6 to 8 cm. long; leaflets 6 to 12 mm. long, glabrous..............j. var. *vitreus*
13. Leaves more than 8 cm. long; leaflets 10 to 20 mm. long (14).
14. Keel 10 to 11 mm. long; petals ochroleucous, white, or faintly purple-tinged; pods greatly inflated..k. var. *yuccanus*
14. Keel 12 to 15 mm. long; petals purple; pods often little inflated......g. var. *australis*

a. Var. *diphysus* (Gray) Jones (*Astragalus diphysus* Gray, *A. Macdougalii* Sheldon) is the common variety in Apache, Navajo, Coconino, and northern Yavapai counties, perhaps also in Graham and Cochise counties, mostly at elevations of 4,500 to 7,000 feet, and is found also in New Mexico and southwestern Colorado. The type of *A. Macdougalii* was collected in Walnut Canyon, near Flagstaff (*MacDougal* 438).

b. Var. *oropedii* Barneby is known only from the North Rim of the Grand Canyon and the Kaibab Plateau (Coconino County), type from North Kaibab Trail, Grand Canyon (*Eastwood & Howell* 7064). A specimen from the same vicinity (*Collom* 1451) has the short, dense inflorescence of var. *oropedii* but approaches var. *mokiacensis* in the size of the corolla (18 mm. long).

c. Var. *palans* Jones (*Astragalus palans* Jones) is chiefly a plant of the canyons of the Colorado River and its tributaries in northern Arizona, western Colorado, and southern Utah. It intergrades with var. *diphysus* and is highly variable.

d. Var. *Wilsonii* (Greene) Barneby (*Astragalus Wilsonii* Greene) is found principally in the Flagstaff region (Coconino County) and in eastern Yavapai County, and apparently is confined to Arizona, type from "northern Arizona" (*N. C. Wilson* in 1893).

e. Var. *maricopae* Barneby is known only from Maricopa County, where it has been collected near Tempe (*Harrison* 1790, the type collection), at Cave Creek Dam (*Peebles et al.* 3684), and at Fish Creek (*Peebles et al.* 5238).

f. Var. *borreganus* Jones (*Astragalus Arthu-Schottii* Gray, *A. agninus* Jepson) occurs near the Colorado River from Lake Mead to the Mexican boundary (western Mohave and Yuma counties), and in adjacent California, Sonora, and Baja California. It occurs at lower elevations in Arizona than any other variety of this species.

g. Var. *australis* Barneby occurs at moderate elevations in Hualpai Valley, Mohave County (*Benson* 10180), and in southern Greenlee, Graham, Cochise, and Pima counties, extending to southern New Mexico and northern Mexico, type from near Robles, Pima County (*A. & R. Nelson* 1537).

h. Var. *mokiacensis* (Gray) Jones (*Astragalus mokiacensis* Gray) is found in northern Mohave County, apparently also in Houserock Valley (Coconino County), extending into southern Nevada, type from Mokiak Pass (*Palmer* in 1877).

i. Var. *stramineus* (Rydb.) Barneby (*Cystium stramineum* Rydb.) occurs in extreme northern Mohave County, also in southern Utah and Nevada.

j. Var. *vitreus* Barneby is found in the northern part of Coconino and Mohave counties, also in southern Utah.

k. Var. *yuccanus* Jones (*Cystium yuccanum* Rydb., *Astragalus yuccanus* Tidestrom) is known only from western Arizona, ranging from central Mohave County to Yuma County and the western parts of Yavapai, Maricopa, and Pima counties, type from Yucca, Mohave County (*Jones* 3886). It intergrades with vars. *diphysus* and *australis*.

57. Astragalus Bryantii Barneby. Grand Canyon, Coconino County, at the mouth of Hermit Creek (*Eastwood* 5991), at the head of Phantom Creek (*Bryant* in 1939, the type collection), and on Clear Creek Trail (*Collom* in 1945), December to April. Known only from the Grand Canyon.

***58. Astragalus striatiflorus** Jones. Known definitely only from southern Utah, where it has been collected east of Kanab, very near the Arizona boundary, growing in sand and on rock ledges at an elevation of 5,500 feet, May and June.

59. Astragalus Layneae Greene (*Hamosa Layneae* Rydb.). Northern Mohave County, from near Lake Mead to Kingman, sandy soil, April. Northwestern Arizona, southern Nevada, and southeastern California.

Unmistakable when in fruit, the long, somewhat prominently veined, thick-walled pods being curved so as often to form a nearly complete circle.

60. Astragalus calycosus Torr. (*Hamosa calycosa* Rydb.). Apache County to eastern Mohave and northern Yavapai counties, 3,000 to 8,000 feet, common on rocky slopes and mesas, often with juniper and pinyon, April and May. Wyoming and Idaho to New Mexico, northern Arizona, and eastern California.

The silvery-sericeous foliage and the purple-and-white flowers make this an attractive little plant. Most of the Arizona specimens are referrable to var. *scaposus* (Gray) Jones (*A. scaposus* Gray, *Hamosa scaposa* Rydb.), type from Mokiak Pass (*Palmer* in 1877), with short, deltoid calyx teeth; but specimens from the Kaibab Plateau (*Jones* in 1890) and the Grand Canyon (*Collom* in 1940) have the relatively long and slender calyx teeth of typical *A. calycosus*. A very depauperate form of *A. calycosus*, with leaflets only 2 to 3 mm. long, is found on both rims of the Grand Canyon (*Collom* in 1946 and 1947). Extraordinarily large plants of var. *scaposus,* with scapes up to 25 cm. long, racemes about 20-flowered, and as many as 11 leaflets, were collected at Cottonwood (Yavapai County), in mesquite grassland (*Darrow* in 1941).

61. Astragalus ensiformis Jones (*Hamosa ensiformis* Rydb.). Northern Mohave County, head of Grand Wash near Pagumpa Springs (*Jones* 5095a, the type collection) and Mokiak Pass (*Ripley & Barneby* 4311), 4,000 to 5,500 feet, April and May. Southern Utah and northern Arizona.

62. Astragalus arizonicus Gray (*Hamosa arizonica* Rydb.). Coconino and Mohave counties to Graham, Cochise, Santa Cruz, and Pima counties, 4,500 feet and lower, common on plains and mesas, March to May, type from near Camp Grant, Graham County (*Palmer*). New Mexico, Arizona, and Sonora.

The radiating, nearly prostrate stems, narrow, silvery-sericeous leaflets, and dingy-purple flowers are characteristic. The species has been reported to cause loco disease, but apparently this has not been proved experimentally.

63. Astragalus Nuttallianus DC. (*Hamosa Nuttalliana* Rydb.). Almost throughout the state, 100 to 4,000 (rarely 7,000) feet, very common on dry plains, mesas, and slopes, February to May (occasionally late summer). Arkansas and Texas to California and northern Mexico.

Arizona's most common annual species, occurring in 2 almost equally abundant forms, one with smooth pods (*Hamosa Emoryana* Rydb.) and the other with pubescent pods (var. *trichocarpus* Torr. & Gray, *Hamosa austrina* Small). A variant with 1-celled pods but otherwise apparently indistinguishable from *A. Nuttallianus* is var. *imperfectus* (Rydb.) Barneby (*Hamosa imperfecta* Rydb.). It is the prevailing form in northern Mohave County.

64. Astragalus nothoxys Gray (*Hamosa nothoxys* Rydb., *H. Gooddingii* Rydb.). Graham, Gila, Maricopa, Pinal, Cochise, Santa Cruz, and Pima counties, 1,500 to 6,000 feet, common on slopes and mesas, often with live-oaks, March to May, type of *A. nothoxys* from southern Arizona (*Thurber*), type of *H. Gooddingii* from the Huachuca Mountains (*Goodding* 1299). Southwestern New Mexico, southern Arizona, and northern Mexico.

Sheep loco. Plant handsome in flower, sometimes covering the ground in openings in oak woods with its purplish-pink flowers that change to violet in drying. Experiments with sheep and cattle have proved this plant to be as toxic as *A. lentiginosus* var. *diphysus,* although it is reported to be eaten readily by deer.

The slender, erect or ascending, beaklike acumination of the keel distinguishes this species from all other *Astragali* of Arizona. Because of this character it was transferred to *Oxytropis* by M. E. Jones, but later Jones restored it to the genus *Astragalus* (152, p. 271). In growth habit it is markedly different from the Arizona species of *Oxytropis.*

65. Astragalus Hartwegii Benth. (*Hamosa Hartwegii* Rydb.). Babocomari Creek, Cochise County (*Lemmon* 2638). Southeastern Arizona and northern Mexico.

66. Astragalus vaccarum Gray (*Hamosa vaccarum* Rydb.). Flagstaff, Coconino County (*Thornber* in 1928). Southwestern New Mexico, Arizona, and Mexico.

The specimen cited resembles the type of *A. vaccarum* from southwestern New Mexico except that the dried corollas are violet-blue, the petals are less arcuate, and the banner, wings, and keel are more nearly of equal length. If really collected at Flagstaff it represents a considerable extension of range northwestward.

67. Astragalus hypoxylus Wats. (*Hamosa hypoxyla* Rydb.). Known certainly only from the type collection in the Huachuca Mountains, Cochise County (*Lemmon* 2656).

Plant acaulescent, loosely cespitose, the leaflets small, obovate, obtuse, the inflorescence subcapitate.

68. Astragalus Bigelovii Gray. Cochise, Santa Cruz, and Pima counties, 4,000 to 5,500 feet, dry slopes and mesas, April and May. Texas to southern Arizona and northern Mexico.

Reputed to cause loco disease, but the dried plants have been fed to cattle in large quantity without ill effect. The range as given in the first paragraph is that of var. *typicus* Barneby. Var. *mogollonicus* (Greene) Barneby (*A. mogollonicus* Greene), which occurs on the Coconino Plateau, south of the Grand Canyon, 6,000 to 7,500 feet, extending southeastward to the Mogollon Mountains, New Mexico, differs in having a pubescence of shorter, more appressed, and more matted hairs and somewhat smaller leaflets, flowers, and pods.

69. Astragalus Thompsonae Wats. Apache County to Coconino County, 3,500 to 7,000 feet, usually in sandy soil, sometimes in pine forest, April to June. Colorado, Utah, Nevada, and northeastern Arizona.

From all forms of *A. Bigelovii* this species may be distinguished by its incompletely 2-celled pods, the septum extending not quite to the apex.

***70. Astragalus Matthewsii** Wats. (*A. Bigelovii* var. *Matthewsii* Jones). Not known definitely to occur in Arizona, but the type was collected at Fort Wingate, New Mexico, about 30 miles east of the Arizona state line.

Resembles the 2 preceding very hairy, acaulescent species, but the pubescence is more appressed and the hairs are shorter. The pods are more broadly ovoid than in *A. Thompsonae* and are completely 2-celled.

71. Astragalus didymocarpus Hook & Arn. Pinal, Maricopa, and Pima counties, 1,000 to 2,500 feet, February to April. Southern Arizona, southeastern California, and Baja California.

An inconspicuous plant, differentiated from all Arizona's other species by the hard and sharp transverse ridges of the small pods. The species is represented in Arizona by var. *dispermus* (Gray) Jeps. (*A. dispermus* Gray), type from Wickenburg, Maricopa County (*Palmer* in 1876).

72. Astragalus crassicarpus Nutt. (*Geoprumnon crassicarpum* Rydb.). Near Springerville and Nutrioso, southern Apache County, on open mesas, about 7,500 feet (*Ferris* 1232, *Ripley & Barneby* 5049, 8430), May and June. Manitoba and Saskatchewan, south to Tennessee, Missouri, New Mexico, and eastern Arizona.

Ground-plum, Indian-pea, buffalo-bean. The round, fleshy fruits, which distinguish this from all other species of *Astragalus* in Arizona, were eaten, raw or cooked, by the Indians of the prairie and plains states and by white settlers.

What seems to be an undescribed species of *Astragalus,* erroneously referred in a former publication to *A. ensiformis,* was collected near Kayenta, Navajo County (*Peebles & Fulton* 11928). It is a small plant characterized as follows: leaflets oblanceolate or obovate, 5 to 6 mm. long; flowers few, in a congested raceme about 1.5 cm. long, the peduncles 3 to 4 cm. long; calyx 6 to 7 mm. long; corolla about 13 mm. long, light purple, the keel tipped with dark purple, obtuse and slightly incurved at apex, nearly as long as the wings and somewhat shorter than the broadly obovate, slightly notched banner, the wing petals and keel petals with large, obtuse, reflexed auricles; pods 2-celled, linear-lanceolate, moderately curved, 15 to 25 mm. long. According to R. C. Barneby (personal communication) this plant apparently is related to *A. naturitensis* Payson and *A. desperatus* Jones.

40. OXYTROPIS

Plants perennial, herbaceous, acaulescent; leaves all basal, pinnate, the leaflets numerous; racemes spikelike, elongate or subcapitate; flowers white or purple, the keel with a prominent erect or ascending beak; pods indehiscent, 1-celled or imperfectly 2-celled.

These plants resemble some of the acaulescent species of *Astragalus.* The genus is a very weak one and is merged with *Astragalus* by some authorities.

KEY TO THE SPECIES

1. Herbage silvery-sericeous, the hairs all basifixed; leaflets 7 to 11, not more than 1 cm. long; inflorescences subcapitate, dense, not more than 2.5 cm. long, seldom more than 8-flowered; flowers about 12 mm. long; pods broadly ovate, much inflated ..... 1. *O. oreophila*

1. Herbage green to white-sericeous, with at least some of the hairs dolabriform; leaflets usually more than 11 and more than 1 cm. long; inflorescences elongate, loose, many-flowered; flowers 15 to 25 mm. long; pods oblong or oblong-lanceolate, not greatly inflated ..... 2. *O. Lambertii*

1. Oxytropis oreophila Gray. Kaibab Plateau (Coconino County), 7,500 to 8,500 feet, open meadows, July. Idaho to northern Arizona, southern Nevada, and southeastern California.

The Arizona specimens represent a rather luxuriant phase of *O. oreophila,* with larger leaves and longer and more numerously flowered scapes than the pulvinate form that grows at higher elevations in Nevada and California (R. C. Barneby, personal communication).

2. Oxytropis Lambertii Pursh (*O. bilocularis* A. Nels., *Aragallus Knowltoni* Greene). Apache County to Coconino County, and in the Chiricahua and Huachuca mountains (Cochise County), 5,000 to 8,000 feet, open, sandy situations and pine forests, June to September, type of *O. bilocularis* from the Huachuca Mountains (*Goodding* 2411). Western Canada, south to Texas, New Mexico, and Arizona.

The Arizona plants belong to var. *Bigelovii* Gray, which has broader leaflets and thinner-walled pods than typical *O. Lambertii. Aragallus Knowltoni,* a form with exceptionally broad leaflets, was described from a collection on the San Francisco Peaks (*Knowlton* 44).

O. Lambertii, a showy plant when in flower, is one of the most dangerous of the loco-weeds because it is eaten readily by horses, cattle, and sheep, especially when grass is scarce, often with fatal effect. It seems to be a pronouncedly habit-forming plant.

41. GLYCYRRHIZA. Licorice

Plants perennial, herbaceous; roots stout; stems tall, erect, very leafy; leaves pinnate, the leaflets numerous, narrow; racemes many-flowered, dense, spikelike; corolla whitish; alternate anthers smaller; pods few-seeded, indehiscent, covered with hooked prickles.

G. lepidota, which is a good soil binder but sometimes a bad weed on fertile soils, contains in its sweet roots practically as much crude glycyrrhizin as the imported licorice root of *G. glabra* L., which is used in the manufacture of tobacco, confections, and fire-extinguisher compounds as well as by druggists.

1. **Glycyrrhiza lepidota** (Nutt.) Pursh. Apache, Navajo, Coconino, and Yavapai counties, 2,000 to 7,000 feet, common along the Colorado River near Lees Ferry, May to July. Ontario and New York to Washington, New Mexico, Arizona, and California.

42. HEDYSARUM (156). Sweet-vetch

Plants perennial, herbaceous; stems erect, leafy; leaves pinnate, the leaflets numerous, finely punctate; flowers in axillary stalked racemes, rather large and showy, the corolla rose-purple; fruit (loment) flat, several-jointed, deeply indented above and below between the seeds, the segments rounded, reticulate-veined.

The roots of North American species are eaten by Indians and Eskimos and are said to taste like carrots when cooked. These plants make excellent forage but are too rare in Arizona to be important.

1. **Hedysarum boreale** Nutt. Apache County to Coconino County, 6,000 to 7,000 feet, June and July. Saskatchewan and Alberta, south to Oklahoma, Nevada, and northern Arizona.

The Arizona plant is var. *typicum* Rollins.

43. ALHAGI. Camel-thorn

Intricately branched, spiny, glabrous shrubs; leaves small, reduced to a single leaflet; flowers numerous, in panicles of racemes, the corolla purplish pink; fruit (loment) few-jointed, of 1 to several segments, these not separating at maturity.

1. **Alhagi camelorum** Fisch. Dinnehotso, Snowflake, and near Holbrook (Navajo County), along the Little Colorado River in scattered colonies sometimes of considerable extent (Coconino County), near Gillespie Dam (western Maricopa County), along streams and canals, in fields, and on rocky hillsides, May to July. Introduced into the southwestern United States from Asia.

Of great value as a browse plant in the desert regions of Asia, but a dangerous introduction since it is extremely difficult to eradicate from cultivated fields, having deep and extensive rootstocks. In Persia and Afghanistan an exudate, similar to the drug manna that is obtained from *Fraxinus Ornus,* is collected from the camel-thorn. The identification of the Arizona specimens as *A. camelorum* is perhaps questionable.

44. NISSOLIA

Plants with twining or trailing stems, often suffrutescent; leaves pinnate, the leaflets commonly 5; flowers few, axillary, in short racemes or fascicles, the corolla yellow; fruit (loment) few-jointed, sometimes reduced to a single segment, the terminal segment winged, samaralike.

KEY TO THE SPECIES

1. Stems prostrate or ascending; leaflets not more than 15 mm. long; calyx tube pubescent, longer than the triangular teeth; fruit turgid, commonly 2- or 3-seeded and deeply constricted between the seeds, straight, the terminal winglike segment smaller (often much smaller) than the others.................................1. *N. Wislizeni*
1. Stems twining; leaflets up to 30 mm. long; calyx tube glabrous, equaling or shorter than the subulate teeth; fruit flat, commonly 1- or 2-seeded and scarcely constricted between the seeds, somewhat falcate, the terminal wing much larger than the body of the fruit.................................2. *N. Schottii*

1. Nissolia Wislizeni Gray. Cochise County, about 5,000 feet, July and August. Southeastern Arizona and Mexico.

2. Nissolia Schottii (Torr.) Gray. Pima County, chiefly in the Rincon, Santa Catalina, Tucson, and Baboquivari mountains, 2,500 to 4,000 feet, July to September. Southern Arizona and northern Mexico.

45. AESCHYNOMENE. Sensitive Joint-vetch

Plants herbaceous and annual (in Arizona); stems tall, leafy, erect; leaves pinnate with many linear leaflets, these sensitive; flowers in short axillary racemes, the corolla yellow; fruit (loment) narrow, flat, of 3 or more segments, deeply indented between the seeds below but not above.

1. Aeschynomene americana L. Huachuca Mountains (Cochise County), Canelo Hills and Bear Valley or Sycamore Canyon (Santa Cruz County), 4,000 to 5,000 feet, locally abundant, September. Widely distributed in tropical America.

The Arizona specimens are rather exceptional in their sparse pubescence and few-jointed loments (segments 2 to 5). The roots are thickly beset with bacterial nodules.

46. STYLOSANTHES. Pencil-flower

Perennial herbs; leaves digitately trifoliolate, with sheathing stipules, the leaflets conspicuously veined; flowers small, in interrupted terminal spikes with leaflike bracts, or some axillary, the corolla yellow; fruit (loment) of 2 segments, the terminal one reduced and infertile, dehiscent at apex, tipped by the persistent hooked style.

***1. Stylosanthes biflora** (L.) B.S.P. This species, of which the known range is from New York to Kansas, Florida, and Texas, is included doubtfully in the Arizona flora on the basis of a specimen in the Gray Herbarium collected in "Arizona or New Mexico" (*W. F. Parish* 314), identified by M. L. Fernald as var. *hispidissima* (Michx.) Pollard & Ball.

It is improbable that this plant really occurs in Arizona. On the label of a duplicate of the collection cited, in the Dudley Herbarium, Stanford University, the collector's name has been changed to "S. B. Parish."

47. ZORNIA

Plants herbaceous, perennial; leaves bifoliolate (in the Arizona species), the stipules sagittate; flowers in axillary and terminal spikes or some of them solitary in the axils; bracts very different from the foliage leaves, paired, connivent, nearly enclosing the flower; corolla orange yellow, the keel incurved; fruit (loment) flat, several-jointed.

1. Zornia diphylla (L.) Pers. Cochise, Santa Cruz, and Pima counties, 4,000 to 6,000 feet, dry, rocky slopes and mesas, August to October. Southern Arizona to Central America.

A form collected at Nogales, also occasional in Mexico and Central America, has longer, much narrower, and longer-acuminate leaflets, less prominently veined and more acuminate bracts, and less copious pubescence than in the normal phase of the species. It may be referable to var. *leptophylla* Benth., described from Brazil.

48. DESMODIUM (157). Tick-clover

Plants annual or perennial, herbaceous or suffrutescent; stems erect to prostrate; leaves trifoliolate or (in a few species) reduced to 1 leaflet; flowers in terminal or axillary racemes, these simple or compound, the corolla purplish pink or sometimes white; fruit (loment) flat, of several 1-seeded segments, these indehiscent or nearly so.

The name beggar-ticks is also sometimes used for these plants. The joints of the pods stick tightly to clothing and to the hair of animals. There seems to be little evidence that the plants are relished by livestock although, in view of the great abundance of some of the species in Arizona and the presumable absence of any poisonous or disagreeable component, they would seem likely to be important range plants. Perhaps the fact that most of the species grow where grasses are abundant and are at the height of their growth in late summer when the grasses are at their best accounts for this seeming neglect. Several of the Arizona tick-clovers are recommended for erosion control.

KEY TO THE SPECIES

1. Leaves all unifoliolate; plants perennial; segments of the fruit normally not contorted (2).
1. Leaves trifoliolate or if some or all of them unifoliolate, then the plants annual and the segments of the fruit normally somewhat contorted (3).
2. Flowering stems more or less woody below; leaflet very long and narrow, not more than 5 mm. wide..1. *D. angustifolium*
2. Flowering stems not woody; leaflet much shorter and broader, 12 to 28 mm. wide 2. *D. psilophyllum*
3. Bracts very conspicuous before anthesis, densely imbricate at the ends of the branches, mostly ovate and 8 to 10 mm. long, attenuate-acuminate. Stems procumbent to nearly erect, usually densely pubescent, the hairs uncinate; leaflets ovate-lanceolate to broadly rhombic-ovate; fruit usually pubescent, the segments normally contorted, about 3 mm. long..3. *D. intortum*
3. Bracts not very conspicuous or densely imbricate or if so, then not more than 7 mm. long, or narrowly lanceolate or subulate (4).
4. Fruit nearly or quite as deeply notched above as below, hence appearing moniliform (5).

4. Fruit less deeply notched above than below, the segments normally not contorted. Plants perennial, often with a woody caudex (10).
5. Plants perennial. Stems diffuse or procumbent; leaflets oblong-lanceolate or ovate-lanceolate to ovate; fruit glabrous or puberulent, the segments 3 mm. long or shorter, normally not contorted.......................................4. *D. retinens*
5. Plants annual or (*D. psilocarpum*) sometimes perennial (6)
6. Segments of the fruit orbicular to elliptic, not contorted, the margins flat or nearly so. Leaves all much alike and trifoliolate (7).
6. Segments of the fruit rhombic or the terminal one triangular, all or some of them slightly to strongly contorted, the margins involute or revolute. Fruit pubescent, at least on the sutures, the terminal 1 or 2 segments sometimes glabrous (8).
7. Leaflets ovate, the larger ones 15 to 50 mm. wide; segments of the fruit at maturity 6 to 10 mm. long, sparsely pubescent, at least on the sutures.......5. *D. psilocarpum*
7. Leaflets linear or linear-lanceolate, not more than 5 mm. wide; segments of the fruit at maturity less than 5 mm. long, glabrous or nearly so.................6. *D. Rosei*
8. Terminal segment of the fruit triangular, not contorted, distinctly larger than the others, 7 to 8.5 mm. long; leaves all trifoliolate, the leaflets of much the same shape throughout the plant and longer than wide.....................7. *D. scopulorum*
8. Terminal segment much like the others in shape and contortion, if larger, then not triangular, none of the segments more than 4 mm. long; leaves differentiated, the basal ones with leaflets mostly wider than long, often unifoliolate (9).
9. Fruit usually sessile or nearly so, the segments 3 to 5 (rarely only 2), moderately contorted; leaflets of the upper leaves lanceolate to ovate; bracts usually persistent. Some or all of the upper leaves rarely unifoliolate............8. *D. neomexicanum*
9. Fruit distinctly stipitate, the segments 2 to 4, strongly contorted (the edges alternately involute and revolute); leaflets of the upper leaves linear or linear-lanceolate; bracts often early deciduous...9. *D. procumbens*
10. Stems diffuse or procumbent; leaves dark green above, pale or somewhat glaucous beneath. Bracts conspicuous before anthesis, 6 mm. long or shorter, imbricate, ovate or lance-ovate, attenuate-acuminate; pubescence uncinate (11).
10. Stems erect or ascending; leaves lighter green but not glaucous beneath (12).
11. Segments of the fruit 5 to 8 mm. long, very pubescent; leaflets ovate, often conspicuously reticulate beneath, the terminal one less than twice as long as wide; bracts ovate-lanceolate, not closely imbricate......................10. *D. Grahami*
11. Segments of the fruit not more than 4 mm. long, sparsely pubescent or glabrous; leaflets oblong-lanceolate, not conspicuously reticulate, the terminal one seldom less than 3 times as long as wide; bracts ovate, closely imbricate....11. *D. batocaulon*
12. Bracts conspicuous before anthesis, imbricate, lanceolate or ovate-lanceolate, 4 to 7 mm. long; leaves nearly sessile; fruit subsessile or very short-stipitate. Terminal leaflet lanceolate, 5 or more times as long as wide...............12. *D. arizonicum*
12. Bracts not conspicuous or noticeably imbricate, subulate or narrowly lanceolate; lower leaves with petioles 1.5 cm. long or longer; fruit rather long-stipitate (13).
13. Leaflets obtuse, the terminal one 2 to 3 times as long as wide, oblong, ovate, or obovate; hairs of the fruit not uncinate............................13. *D. cinerascens*
13. Leaflets acute or acutish, the terminal one up to 5 times as long as wide, oblong-lanceolate; hairs of the fruit uncinate.........................14. *D. Metcalfei*

1. Desmodium angustifolium (H.B.K.) DC. Southwestern Cochise County, Patagonia Mountains and Bear Valley or Sycamore Canyon (Santa Cruz County), Santa Catalina and Baboquivari mountains (Pima County), 3,500 to 5,000 feet, dry, rocky slopes with live-oaks and grasses, August and September. Southern Arizona to northern South America.

The Arizona plants belong to var. *gramineum* (Gray) Schubert (*D. gramineum* Gray), the type of which was collected "on the Sonoita," probably in southwestern Cochise County (*Wright* 1009), and which is limited to southern Arizona and northern Mexico. *D. angustifolium* is easily distinguished from all Arizona's other species of *Desmodium* by the somewhat woody stems and long, narrow, grasslike, unifoliolate leaves.

2. Desmodium psilophyllum Schlecht. (*D. Wrightii* Gray). Bear Valley or Sycamore Canyon (Santa Cruz County), Santa Catalina and Baboquivari mountains (Pima County), about 3,500 feet, August to October. Texas, southern Arizona, and Mexico.

3. Desmodium intortum (Mill.) Urban (*D. sonorae* Gray). The type of *D. sonorae* was collected "on the Sonoita," probably in what is now Cochise County (*Wright* 1014). The species apparently has not been collected since in Arizona. It is widely distributed in tropical America.

4. Desmodium retinens Schlecht. (*D. Wislizeni* Engelm., *D. Wislizeni* var. *venulosum* Jones). Chiricahua Mountains, Cochise County (*Blumer* 1188), Santa Catalina Mountains, Pima County (*Pringle* 14), Santa Rita Mountains, Santa Cruz or Pima County (*Jones* in 1903, the type collection of *D. Wislizeni* var. *venulosum*). Southern Arizona and Mexico.

5. Desmodium psilocarpum Gray. Near Nogales (Santa Cruz County), Santa Catalina and Baboquivari mountains (Pima County), about 4,000 feet, type from near Santa Cruz, Sonora (*Wright* 1016), August and September. Southern Arizona and northern Mexico.

6. Desmodium Rosei Schubert. Navajo and Coconino counties to Greenlee, Cochise, Santa Cruz, and Pima counties, 3,500 to 6,500 feet, common on slopes and mesas, often in grassland, August and September. New Mexico, Arizona, and northern Mexico.

7. Desmodium scopulorum Wats. (*D. Wigginsii* Schubert). Baboquivari Mountains, Pima County (*Jones* 24903, 24904 in part), September. Southern Arizona and northern Sonora.

8. Desmodium neomexicanum Gray (*D. Bigelovii* Gray, *Meibomia parva* Schindler). Yavapai, Gila, and Greenlee counties, to Cochise, Santa Cruz, and Pima counties, 3,500 to 6,000 feet, grassy slopes and mesas, very common, August and September, type of *D. Bigelovii* from along the San Pedro River, Cochise County (*Wright* 1012), type of *Meibomia parva* from Paradise, Cochise County (*Blumer* 1675). Western Texas to Arizona and Mexico; South America.

A form collected in Bear Valley or Sycamore Canyon, Santa Cruz County (*Kearney & Peebles* 14466), has some of the upper leaves as well as the basal ones unifoliolate. It resembles the type of *Desmodium annuum* Gray, collected "on the Sonoita," probably in southwestern Cochise County (*Wright* 1009a), except that the latter has all the leaves unifoliolate.

9. Desmodium procumbens (Mill.) A. S. Hitchc. Greenlee, Gila, Cochise, Santa Cruz, and Pima counties, 3,500 to 6,500 feet, not common, August and September. Southern Arizona to South America.

The Arizona plants belong to var. *exiguum* (Gray) Schubert (*D. exiguum* Gray), the type of which was collected "in mountain ravines on the Sonoita" (*Wright* 1010), probably in southwestern Cochise County.

10. Desmodium Grahami Gray. Navajo, Coconino, and Greenlee counties to Cochise and Pima counties, 4,500 to 8,000 feet, often in pine woods, August and September. Texas to Arizona and Mexico.

11. Desmodium batocaulon Gray. Greenlee, Graham, Cochise, Santa Cruz, and Pima counties, 3,500 to 6,500 feet, common, often in pine woods, June to September, type from the San Pedro (*Wright* 1013). Southern New Mexico and Arizona and northern Mexico.

12. Desmodium arizonicum Wats. Gila, Cochise, Santa Cruz, and Pima counties, 4,000 to 8,000 feet, frequent, especially in dry pine woods, July to September, type from southern Arizona (*Lemmon* 540). Southern New Mexico and Arizona and northern Mexico.

13. Desmodium cinerascens Gray (*Meibomia Canbyi* Schindler). Cochise, Santa Cruz, and Pima counties, 4,000 to 6,000 feet, frequent on dry, sunny slopes, August and September, type from between Babocomari, Arizona, and Santa Cruz, Sonora (*Wright* 1017), type of *Meibomia Canbyi* from the Santa Catalina Mountains (*Pringle* in 1881). Southern Arizona and northern Mexico.

The numerous stems, hard and almost woody toward base, often form large clumps and reach a height of 1.5 m. (5 feet).

14. Desmodium Metcalfei (Rose & Painter) Kearney & Peebles. A collection at Cave Creek, Chiricahua Mountains, Cochise County (*Eggleston* 11027), was identified by A. K. Schindler as of this species, otherwise known only from southwestern New Mexico. A specimen from Sedona, southern Coconino County (*R. S. Beal Jr.* in 1947), with stems at least 75 cm. high, apparently belongs to this species but has flowers only.

49. VICIA. Vetch

Plants herbaceous, annual or perennial; stems leafy, weak, climbing or trailing; leaves pinnate, ending in a tendril; flowers axillary, solitary or in racemes; stamens all, or 9 of them, united below; style with a subapical tuft of hairs; pods narrow, flat, 2-valved, dehiscent.

Most of the vetches are excellent forage plants, and several Old World species are much cultivated in the United States for green manure, hay, and as cover crops in orchards. The native species do not withstand close grazing.

KEY TO THE SPECIES

1. Flowers mostly in twos in the upper axils; peduncle almost none; corolla 10 to 20 mm. long, deep rose-purple; calyx teeth somewhat shorter to somewhat longer than the tube, setaceous-acuminate. Leaflets mostly oblong-oblanceolate, truncate or emarginate, mucronate or cuspidate........1. *V. sativa*

1. Flowers in peduncled racemes or, if solitary, then the peduncle well developed and the corolla not more than 10 mm. long; calyx teeth much shorter than the tube, except sometimes in *V. leucophaea* (2).

2. Corolla 15 to 25 mm. long, purple. Plants glabrous or sparsely (rarely copiously) pubescent; leaflets mucronate or cuspidate; racemes 2- to several-flowered..2. *V. americana*

2. Corolla not more than 10 mm. long (3).

3. Peduncle bearing several or many (usually 10 or more) flowers. Corolla 5 to 7 mm. long, cream-white, often with the banner purple-veined and the keel purple-tipped
3. *V. pulchella*

3. Peduncle bearing 1 or 2 flowers (4).

4. Herbage and calyx sparsely to copiously villous; pods sericeous; corolla 8 to 10 mm. long
4. *V. leucophaea*

4. Herbage and calyx sparsely pubescent with mostly appressed hairs, or glabrate; pods glabrous; corolla commonly less than 8 mm. long........5. *V. exigua*

1. **Vicia sativa** L. An occasional escape from cultivation, as at Sacaton (Pinal County), but apparently not established anywhere in Arizona. Common in the eastern United States; naturalized from Europe.

2. **Vicia americana** Muhl. (*V. perangusta* Greene, *V. hypolasia* Greene). Apache County to Coconino County, south to Cochise and Pima counties, 5,000 to 10,000 feet, common, especially in pine forests, May to September, type of *V. perangusta* from the Tusayan (now the Kaibab) National Forest (*Read* in 1912), type of *V. hypolasia* from the Chiricahua Mountains (*Blumer* 1348). Canada to Virginia, New Mexico, Arizona, and California.

Flowers large and handsome for the genus. A polymorphic species, several forms of which have been segregated as species. About equally common in Arizona are: (1) the typical phase, with relatively thin and broad leaflets, these 4 to 10 mm. wide, rounded to acutish at apex; and (2) var. *linearis* (Nutt.) Wats., with usually rather thick and prominently veined leaflets 1 to 4 mm. wide. Var. *truncata* (Nutt.) Brewer, with leaflets truncate and often emarginate and denticulate at apex, is less common but not rare. There is intergradation among all these entities.

3. **Vicia pulchella** H.B.K. (*V. melilotoides* Woot. & Standl.). Apache County to Coconino County, south to Cochise, Santa Cruz, and Pima counties, 6,000 to 8,500 feet, frequent in pine forests, July to September. Western Texas to Arizona and throughout Mexico.

The stems often are so thickly matted as to be subject to mildew.

4. **Vicia leucophaea** Greene. Southern Apache County to Cochise, Santa Cruz, and Pima counties, 5,500 to 8,000 feet, pine forests, July to September. Western New Mexico and southeastern Arizona.

5. **Vicia exigua** Nutt. Greenlee County to Mohave County, south to the Mexican boundary, 4,000 feet or lower, common among bushes on slopes and in canyons, March to May. Western Texas to Oregon and California.

Vicia villosa Roth, a European species often cultivated as a green-manure crop, has been found as a weed in a yard at Douglas, Cochise County (*Mrs. F. Conrey* in 1950). It has the flowers in dense racemes, the corollas violet.

50. LATHYRUS. Peavine

Contributed by C. Leo Hitchcock

Plants perennial, herbaceous; stems erect or weakly climbing, not winged in the Arizona species; leaves pinnate, commonly ending in a tendril, the leaflets usually fewer, larger, thicker, and more prominently veined than in *Vicia;* stipules large and conspicuous; flowers large and showy, commonly in axillary racemes, the corolla purple to nearly white; style hairy along the inner side; pods much as in *Vicia.*

The best-known member of this genus is the sweet-pea (*L. odoratus*), a native of Sicily. The native species of Arizona bear a general resemblance to this favorite of gardens. Apparently the plants are less palatable to livestock than are the vetches.

KEY TO THE SPECIES

1. Tendrils bristlelike, simple, not prehensile; flowers white, 11 to 14 mm. long. Leaflets mostly 4 or 6 .... 5. *L. arizonicus*
1. Tendrils prehensile, simple or branched, or corollas purplish or over 14 mm. long (2).
2. Corollas white, usually over 15 mm. long; leaflets seldom less (usually more) than 5 mm. wide........ 3. *L. laetivirens*
2. Corollas bluish or purplish, or less than 15 mm. long, or the leaflets narrower (3).
3. Corollas 12 to 16 mm. long. Leaflets seldom more (often much less) than 5 mm. wide 4. *L. graminifolius*
3. Corollas well over 16 mm. long (4).
4. Banner deeply cordate, 20 to 25 mm. long, the blade nearly as wide as the combined length of blade and claw; calyx teeth scarcely half as long as the tube 2. *L. brachycalyx*
4. Banner shallowly cordate, 20 to 30 mm. long, the blade not much more than half as wide as the total length of the petal; calyx teeth but little shorter than the tube 1. *L. eucosmus*

1. Lathyrus eucosmus Butters & St. John (*L. ornatus* Nutt. apud White, *L. decaphyllus* Pursh apud Britton). Apache, Navajo, Coconino, Yavapai, Greenlee, and (?) Santa Cruz counties, 5,000 to 7,000 feet, often in open, dry woods of pinyon and of ponderosa pine, May to August. Colorado, Utah, New Mexico, and Arizona.

Stems 1.5 to 4 dm. tall; leaflets 6 to 8 (10), thick and coriaceous; tendrils well developed on the upper leaves; flowers rose to purple, 2 to 3 cm. long; calyx 10 to 15 mm. long, the teeth but slightly shorter than the tube; banner narrowly obcordate, 25 to 30 mm. long, about ⅔ as wide.

A handsome plant, with larger and showier flowers than in any other of the Arizona species. Its habit of spreading by horizontal rootstocks makes it a useful soil binder.

2. Lathyrus brachycalyx Rydb. Apache, Navajo, and Coconino counties (several collections at the Grand Canyon), in the sagebrush-juniper area, and reported from the Dos Cabezas Mountains (Cochise County). Utah and northern Arizona.

Stems 2 to 4.5 dm. tall; leaflets (4) 6 to 10 (12), linear or linear-lanceolate; tendrils well developed; flowers 2 to 2.5 cm. long, pinkish orchid; calyx 7 to 9 mm. long, the teeth not much more than half as long as the tube; banner 20 to 25 mm. long, as wide or wider than long, deeply obcordate. The Arizona plants may belong to another (unpublished) species.

3. Lathyrus laetivirens Greene. Coconino and Yavapai counties, about 6,000 to 7,000 feet. Colorado, Utah, and Arizona.

Stems 1.5 to 6 dm. tall; leaflets 4 to 10 (12); tendrils well developed; flowers white to pale pink, 15 to 22 mm. long. Perhaps not distinct as a species from *L. leucanthus* Rydb. In *Flowering Plants and Ferns of Arizona* the Arizona specimens were erroneously referred to *L. parvifolius* Wats.

4. Lathyrus graminifolius (Wats.) White. Coconino County and White Mountains (Apache County) to Cochise, Santa Cruz, and Pima counties, 4,000 to 9,000 feet, common, chiefly in pine forests, April to September. Western Texas to Arizona and northern Mexico.

Stems 2 to 6 dm. tall; leaflets (4) 6 to 10, linear, always at least 10 times as long as wide; tendrils well developed; flowers 12 to 16 mm. long, bluish to purple (never white?).

5. Lathyrus arizonicus Britton. Apache and Coconino counties, south to Cochise and Pima counties, 8,000 to 11,000 feet, chiefly in coniferous forests, May to October, type from the Mogollon Escarpment (*Mearns* 57). Southern Colorado and southeastern Utah to north-central Mexico.

Stems 1 to 4 dm. tall; leaflets usually 2 or 4, sometimes 6, rarely 8, elliptic to linear-elliptic, but less than 10 times as long as wide; tendrils bristlelike, unbranched, not at all prehensile; flowers 11 to 14 mm. long, white (pink-penciled).

51. CLITORIA. BUTTERFLY-PEA

Plants herbaceous, perennial, glabrous or nearly so; stems usually erect or ascending; leaves pinnately trifoliolate, the leaflets large, ovate-lanceolate or oblong-lanceolate, the stipules and stipels subulate, persistent; flowers axillary, mostly solitary, sometimes 2 or 3 in a cluster, 5 to 6 cm. long; calyx tubular, 5-toothed; corolla with an erect banner much larger than the other petals; pods narrowly oblong, several-seeded.

1. Clitoria mariana L. Sierra Ancha (Gila County), and mountains of Cochise, Santa Cruz, and Pima counties, from the Chiricahua to the Baboquivari mountains, 4,000 to 6,000 feet, rich soil among live-oaks and junipers, July and August. New Jersey to Florida and Texas; southern Arizona.

This handsome plant is outstanding among the Leguminosae of Arizona in the very large size of its lilac-colored corolla. It is also remarkable as an example of interrupted distribution, being apparently absent in the area between central Texas and southeastern Arizona. It is infrequent in Arizona except in the Chiricahua Mountains and in Santa Cruz County, where it is reported to be locally abundant.

52. COLOGANIA

Plants perennial, herbaceous; stems trailing to nearly erect, scarcely twining; leaves pinnately trifoliolate or occasionally 5-foliolate; flowers axillary, short-pedicelled, solitary or in clusters of 2 or 3; calyx tubular, 4-toothed; corolla reddish purple, 2 to 3 cm. long; pods flat, narrowly oblong, several-seeded.

KEY TO THE SPECIES

1. Pubescence of the herbage and pods conspicuously shaggy-villous, that of the pods tawny; petioles very short; leaflets 3, broadly obovate, the terminal one 18 to 35 mm. wide; corolla 2.5 to 3 cm. long .... 1. *C. Lemmoni*
1. Pubescence of the herbage and pods appressed or subappressed, grayish; petioles elongate; leaflets 3 to 5, linear, linear-lanceolate, narrowly oblong, or elliptic, the terminal one not more (commonly much less) than 10 mm. wide, up to 15 cm. long; corolla 2 to 2.5 cm. long .... 2. *C. longifolia*

1. Cologania Lemmoni Gray. Chiricahua and Huachuca mountains (Cochise County), Rincon and Santa Catalina mountains (Pima County), 6,000 to 7,500 feet, openings in pine forests, July and August, type from the Huachuca Mountains (*Lemmon* 2681). Southern Arizona and northern Mexico.

2. Cologania longifolia Gray. Apache County to Coconino County, south to Cochise, Santa Cruz, and Pima counties, 4,000 to 9,000 feet, rich soil in coniferous forests, July to September, co-type from along Sonoita Creek (*Wright* 961). Western Texas to Arizona and northern Mexico.

A very variable complex, possibly separable into more than one species. The genus needs revision.

53. ERYTHRINA (158, 159). CORAL-TREE

Plants shrubby or arborescent, with prickly stems and petioles; leaves pinnately trifoliolate, with large, fan-shaped leaflets; flowers in axillary or terminal racemes, large and showy; corolla bright red, the banner petal much longer than the others, about 5 cm. long, strongly keeled; stamens partly exserted; pods large, thick-walled, torulose; seeds several, large, normally bright red.

1. Erythrina flabelliformis Kearney. Mountains of Cochise, Santa Cruz, and Pima counties, 3,000 to 5,500 feet, fairly common on dry, rocky slopes, spring and sometimes late summer. Southwestern New Mexico, southern Arizona, and northern Mexico.

This plant, known as Western coral-bean, Indian-bean, and chilicote, reaches a height of 4.5 m. (15 feet) and a trunk diameter of 25 cm. (10 inches), but is usually much smaller. The wood is brittle, the bark light brown with longitudinal white lines. The flamelike flowers appear mainly in spring before the leaves. The plant has been cultivated locally, but scarcely can be recommended as an ornamental on account of its sensitiveness to frost and the long period when it is leafless. The plant is said to be browsed, but the attractive scarlet seeds contain poisonous alkaloids. In Mexico necklaces are made of them.

54. GALACTIA

Perennial, herbaceous or sometimes slightly suffrutescent; stems long, commonly twining; leaves pinnately trifoliolate, the stipules small, caducous, the leaflets with stipels, elliptic, oblong, or oblong-lanceolate; flowers in axillary, peduncled, bracted racemes; calyx 4-cleft; corolla pale purple and greenish yellow, the wings and keel united below; pods linear, several-seeded.

The plants are of some value as forage but do not withstand heavy grazing. They make an excellent ground cover.

1. Galactia Wrightii Gray. Greenlee, Graham, Gila, Pinal, Cochise, Santa Cruz, and Pima counties, 3,000 to 6,000 feet, common on dry slopes, often in oak chaparral, July to September. Western Texas to southern Arizona and northern Mexico.

Var. *mollissima* Kearney & Peebles, with herbage densely soft-villous (hairs more spreading than in typical *G. Wrightii*), has been collected in the Pinaleno Mountains (Graham County), at Fish Creek (Maricopa County), in the Perilla and Chiricahua mountains (Cochise County), and in the Patagonia Mountains (Santa Cruz County), type from near Patagonia (*Peebles et al.* 4657).

55. RHYNCHOSIA. Rosary-bean

Perennial herbs with trailing or weakly twining stems; leaves pinnately trifoliolate, without stipels; flowers small, in axillary few-flowered racemes or fascicles; corolla yellow, the keel more or less falcate; pods flat, more or less asymmetric, completely dehiscent; seeds 1 or 2.

The plants afford excellent ground cover and may be useful for erosion control.

KEY TO THE SPECIES

1. Stems usually retrorsely puberulent; leaflets seldom less than twice as long as wide, linear or narrowly elliptic to ovate-lanceolate (exceptionally ovate), rounded to acutish at apex, the veins often very prominent beneath; pods puberulent or short-pilose, asymmetrically oblanceolate and slightly lunate........................... 1. *R. texana*
1. Stems short-hirsute and slightly viscid; leaflets normally less than twice as long as wide, ovate or ovate-lanceolate, acute or short-acuminate at apex, the veins not very prominent beneath; pods short-pilose and sparsely hirsute, scarcely asymmetric, elliptic .... 2. *R. rariflora*

1. Rhynchosia texana Torr. & Gray. Greenlee County to Yavapai County, south to Cochise, Santa Cruz, and Pima counties, 3,500 to 5,500 feet, fairly common on dry plains and mesas, May to September. Western Texas to Arizona and Mexico.

In specimens from southern Arizona the hairs of the stem are appressed or subappressed and retrorse, whereas specimens from central Arizona have spreading or ascending hairs. An unusual form, with twining stems and exceptionally large, thin leaflets, was collected in the foothills of the Rincon Mountains and at Vamori, Pima County (*Kearney & Peebles* 8752, *Goodding* 188–45).

2. Rhynchosia rariflora Standl. Bear Valley or Sycamore Canyon (Santa Cruz County), about 4,000 feet, grassy slopes, September and October. Southern Arizona and Chihuahua.

The corolla in the Arizona specimens is somewhat smaller than in the type from southwestern Chihuahua.

56. PHASEOLUS (160). Bean

Plants herbaceous, annual or perennial, often with very large roots; stems usually long, and trailing or weakly twining; leaves pinnately trifoliolate, with stipels; flowers axillary, mostly in racemes; corolla purplish pink to deep purple, or brick red, the keel strongly incurved, curled, or spirally coiled; pods flat, linear to broadly ovate or obovate in outline, completely dehiscent; seeds rounded, rather large.

Several exotic species are widely cultivated for human food, notably the common or string bean (*P. vulgaris*) and the lima bean (*P. lunatus*). The scarlet-runner (*P. multiflorus*) is often grown as an ornamental climbing plant. All the native beans improve the soil, make a good ground cover, and provide excellent forage for livestock. The form of *P. acutifolius* known as tepary bean is grown to such an extent by the Papago Indians that they have been nicknamed "bean people," and it is also planted by white farmers in the Southwest both for the beans and for increasing the fertility of the soil. *P. Metcalfei* produced abundant forage in an experimental planting in New Mexico. The beans of this and probably other species were eaten by the Apache Indians.

KEY TO THE SPECIES

1. Flowering stems from a deep-seated, globose tuber, erect or slightly twining, seldom more than 30 cm. long; calyx not subtended by bractlets (bracts present at base of the pedicel only). Leaflets lanceolate, entire, less than 1 cm. wide; peduncles 1- or 2-flowered; corolla purple, 15 to 20 mm. long; pods linear, 2.5 to 3.5 cm. long 1. *P. parvulus*
1. Flowering stems not from a globose tuber, trailing or twining, usually more than 30 cm. long; calyx subtended by a pair of bractlets, these often deciduous long before anthesis (2).
2. Keel strongly incurved but not curled or coiled. Stems rooting at the nodes; herbage sparsely pilose; leaflets linear-lanceolate, entire, thickish, strongly reticulate; peduncles long, nearly filiform; inflorescences short, few-flowered; pods linear, elongate 2. *P. leiospermus*
2. Keel more or less curled or coiled (3).
3. Plants annual, the root not thickened; pods 3 to 7 cm. long; leaflets linear and nearly entire to triangular-lanceolate or -ovate and very shallowly subhastately lobed on one or both sides. Pods 4 to 10 mm. wide, tipped with a slender persistent style 3. *P. acutifolius*
3. Plants perennial, with a more or less thickened taproot or if annual, then the pods less than 3 cm. long and the leaflets more deeply lobed (4).
4. Pods linear, less than 5 mm. wide (5).
4. Pods broader than linear, commonly more than 5 mm. wide (6).
5. Pubescence soft, often very dense, the hairs mostly appressed; leaflets nearly entire to deeply lobed; corolla deep purple, 15 to 20 mm. long; pods usually spreading, 6 to 10 cm. long, 3 to 4.5 mm. wide, gradually long-acuminate....4. *P. atropurpureus*
5. Pubescence villous or subhirsute, the hairs spreading or retrorse; corolla brick red, drying purplish, not more than 15 mm. long; pods reflexed, not more than 3 cm. long and 3 mm. wide..5. *P. heterophyllus*
6. Leaflets broadly rounded-ovate, entire or very nearly so, the veins reticulate and somewhat prominent beneath; corolla deep rose purple; seeds smooth or nearly so (7).
6. Leaflets linear to triangular-ovate or, if broader, then deeply lobed, the veins not noticeably reticulate and prominent beneath; corolla purplish pink. Pods not more (usually less) than 8 mm. wide, commonly broadest above the middle (8).
7. Pods 7 to 10 mm. wide, short-stipitate; bractlets 1 mm. long or shorter....6. *P. ritensis*
7. Pods 12 to 18 mm. wide, sessile in the calyx or nearly so; bractlets 1.5 mm. long or longer 7. *P. Metcalfei*
8. Pods persistently pubescent; style in fruit stout, about 1 mm. long; seeds smooth or nearly so; peduncles often more than 10 (up to 25) cm. long. Taproot thick and woody, up to 3 cm. in diameter; leaflets (at least the terminal one) deeply lobed, broadly deltoid, up to 5 cm. long..8. *P. Grayanus*
8. Pods glabrescent; style in fruit slender, 1.5 to 2.5 mm. long; seeds more or less rugose; peduncles mostly less than 10 cm. long (9).
9. Leaflets entire or merely subhastate, lanceolate to triangular-ovate, much longer than wide; plants evidently perennial..............................9. *P. angustissimus*
9. Leaflets usually distinctly to deeply lobed, broadly deltoid, seldom much longer than wide; plants annual or, at most, short-lived perennial................10. *P. Wrightii*

1. Phaseolus parvulus Greene (*Alepidocalyx parvulus* Piper). White, Chiricahua, Huachuca, Santa Rita, Rincon, and Santa Catalina mountains (Apache, Cochise, Santa Cruz, and Pima counties), 6,500 to 8,000 feet, rich soil in coniferous forests, August and September. Southwestern New Mexico and southern Arizona.

2. Phaseolus leiospermus Torr. & Gray (*Strophostyles pauciflora* Wats., *S. leiosperma* Piper). Between Nogales and Ruby, Santa Cruz County (*Kearney & Peebles* 14901). Indiana to South Dakota, southward to Mississippi, Texas, Arizona (?), and Mexico.

The specimen cited is referred doubtfully to *P. leiospermus*. The flowers were brick red, drying purplish. The fruit was not seen.

3. Phaseolus acutifolius Gray. Greenlee, Graham, Pinal, Cochise, Santa Cruz, and Pima counties, 3,000 to 6,000 feet, August to October. Western Texas to southern Arizona and Mexico.

A broad-leaved variant, var. *latifolius* Freeman, cultivated under the name tepary bean and occasionally found growing wild in Arizona, was thought by Vavilov to have originated in southern Mexico or Central America, but W. W. Mackie (*Hilgardia* 15: 5) regarded it as indigenous in Arizona. It is very drought-resistant. The common wild form in Arizona, with much narrower leaflets, is var. *tenuifolius* Gray (*P. tenuifolius* Woot. & Standl.).

4. Phaseolus atropurpureus DC. Baboquivari and Coyote mountains, Pima County (*Gilman* B166, *Peebles* 8802, *Hester* in 1937), about 4,000 to 5,000 feet, August and September. Southern Arizona to Central America.

The Arizona specimens are exceptionally thin-leaved and only moderately sericeous. The long stems clamber over bushes.

5. Phaseolus heterophyllus Willd. (*P. macropoides* Gray). Southern Navajo, Greenlee, Cochise, Santa Cruz, and Pima counties, 4,000 to 7,500 feet, fairly common on dry plains and mesas, often with grasses, August to October. Western Texas to Arizona, south to Central America.

Distinguished from most of the Arizona species by the brick-red color of the fresh flowers. The more or less typical plant (*P. macropoides*) has the leaflets oblong to rhombic-ovate and commonly lobed, and the pods usually short-pilose. In var. *rotundifolius* (Gray) Piper (*P. rotundifolius* Gray), which has about the same distribution in Arizona as the typical plant, the leaflets are nearly orbicular and mostly entire and the pods are commonly villous. The type of *P. rotundifolius* was collected west of the Chiricahua Mountains, Cochise County (*Wright* 954).

6. Phaseolus ritensis Jones. White Mountains (Apache County) and mountains of Graham, Cochise, Santa Cruz, and Pima counties, 4,500 to 7,000 feet, common among live-oaks and at the lower limit of the pine belt, July to September, type from the Santa Rita Mountains (*Jones* in 1903). Southern Arizona and Mexico.

7. Phaseolus Metcalfei Woot. & Standl. Southern Navajo County and Yavapai County to Cochise and Pima counties, about 6,000 feet, July to September. Southwestern New Mexico and Arizona.

Closely related to *P. ritensis* and perhaps only varietally distinct.

8. Phaseolus Grayanus Woot. & Standl. (*P. Wrightii* var. *Grayanus* Kearney & Peebles). White Mountains (Apache County), Pinaleno and Galiuro mountains (Graham County), Chiricahua and Huachuca mountains (Cochise County), Santa Rita Mountains (Santa Cruz County), Santa Catalina Mountains (Pima County), 5,000 to 8,500 feet, mostly in pine forest, July to September. Western Texas to southern Arizona and northern Mexico.

This entity is related, perhaps too closely, to *P. scabrellus* Benth. The characters distinguishing *P. Grayanus* from *P. Wrightii* were pointed out by Oliver A. Norvell (personal communication). Specimens from southern Coconino and northern Gila counties, with very shallowly lobed leaflets, may constitute a variety of *P. Grayanus*.

9. Phaseolus angustissimus Gray. Apache County to northern Mohave County, south to Cochise and Pima counties, 3,500 to 7,000 feet, common on mesas, mostly among trees and shrubs, May to October. New Mexico and Arizona.

The typical plant has linear-lanceolate leaflets, these entire or merely angulate at base. More common in Arizona is var. *latus* Jones (*P. dilatatus* Woot. & Standl.), with oblong-lanceolate leaflets, these usually subhastately lobed at base. The type of var. *latus*, which intergrades completely with the typical plant, was collected along the Little Colorado River near Winslow, Navajo County (*Jones* in 1890).

10. Phaseolus Wrightii Gray. Eastern Maricopa, Pinal, and Pima counties, 1,000 to 4,000 feet, rocky slopes and canyons, flowering throughout the year. Western Texas to southern Arizona and northern Mexico.

It is questionable that *P. Wrightii* is distinct as a species from *P. filiformis* Benth. Specimens from Fish Creek, Maricopa County (*Eastwood* 8755), and from Bates Well and Quitobaquito, western Pima County (*Harbison* 26158, *Peebles* 14554a), annual and very slender-stemmed, closely resemble specimens of *P. filiformis* from Baja California and Sonora.

65. GERANIACEAE. Geranium Family

Plants herbaceous; leaves opposite or basal, with stipules; flowers perfect, regular, in cymose or umbellike clusters; sepals and petals 5, the sepals persistent, the petals whitish to rose purple, deciduous; stamens 10, all or only 5 of them fertile, the filaments distinct or united toward base; pistil compound, the 5 carpels at first united to the central column, separating at maturity, long-beaked by the persistent styles.

The so-called geraniums of gardens belong to the mainly South African genus *Pelargonium*.

KEY TO THE GENERA

1. Carpels with thick, not spindle-shaped bodies, the tails (persistent styles) not bearded within, becoming recoiled, thus bringing the carpels toward the apex of the column, but not spirally twisted; stamens 10, all anther-bearing................1. *Geranium*
1. Carpels with slender, spindle-shaped bodies, sharp-pointed at base, the tails bearded within, becoming spirally twisted, not bringing the carpels upward; anther-bearing stamens 5, the alternate filaments scalelike or obsolete..................2. *Erodium*

1. GERANIUM (161). Cranesbill

Leaves palmately lobed or parted; stamens all fertile, 5 of the filaments usually longer than the others; styles persistent, becoming recoiled in fruit.

The plants are reported to afford fairly good forage for sheep. The rootstock of *G. maculatum* L. of the eastern United States is used medicinally as an astringent, and the Arizona species may have the same property. These plants grow in rich soil, mostly in coniferous forests.

KEY TO THE SPECIES

1. Plants annual or biennial, without a thick caudex, the taproot slender; petals not, or scarcely, surpassing the sepals, pale pink. Plants pilose, usually glandular-pubescent in the inflorescence; leaves 5-parted, the divisions cleft into linear or narrowly oblong lobes....................1. *G. carolinianum*
1. Plants perennial, with a thick caudex, the taproot stout, woody; petals surpassing the sepals, usually considerably (2).
2. Petals pilose within not more than ¼ of their length. Plants more or less cespitose, the stems often decumbent, the caudex usually branched; petals lavender; inflorescence somewhat loosely cymose; pedicels glandular (3).
2. Petals pilose within ⅓ to ½ of their length (4).
3. Petioles of the basal leaves and the lower internode of the stem glandular-pubescent 2. *G. Parryi*
3. Petioles of the basal leaves and the lower internode of the stem finely retrorsely pubescent to nearly glabrous, not glandular....................3. *G. Fremontii*
4. Petals 7 to 9 mm. long, white or pinkish (5).
4. Petals 10 to 20 mm. long or if slightly less than 10 mm., then not white (6).
5. Stems, petioles, and pedicels copiously glandular-villosulous; mature stylar column 20 to 25 mm. long, glandular-pubescent; stylodia 4 to 5 mm. long........4. *G. lentum*
5. Stems and petioles sparsely pilose, not glandular or the pedicels sometimes glandular-villous; mature stylar column 14 to 18 mm. long, puberulent or pilose; stylodia 2 to 3 mm. long....................5. *G. Wislizeni*
6. Pedicels glandular (rarely glandless in *G. eremophilum*). Stylar column more or less glandular (7).
6. Pedicels not glandular, retrorsely short-pubescent. Plants completely eglandular (8).
7. Stylodia 3 to 4 mm. long; petals white or purple-tinged; leaves sharply incised, the lobes and teeth acute or acuminate. Pedicels usually copiously glandular-villosulous, the trichomes with usually purplish glands; filaments 6 to 9 mm. long 6. *G. Richardsonii*
7. Stylodia 4.5 to 9 mm. long; petals pink or lavender; leaves usually not sharply incised, the lobes and teeth obtuse to acute. Stems erect or decumbent, cespitose, freely branched, the first internode 1 to 8 cm. long....................7. *G. eremophilum*
8. Petals pale purple to deep rose purple; stylodia 5 to 8 mm. long; stems erect or decumbent, profusely branching....................8. *G. caespitosum*
8. Petals white or purple-tinged; stylodia 3 to 4 mm. long; stems erect, not profusely branching....................6. *G. Richardsonii*

1. Geranium carolinianum L. Gila, Maricopa, and Pima counties (probably elsewhere), 3,000 to 5,500 feet, April and May. Canada to northern Mexico.

2. Geranium Parryi (Engelm.) Heller. Coconino and northern Mohave counties, 7,000 to 8,500 feet, June and July. Wyoming, Colorado, Utah, and northern Arizona.

3. Geranium Fremontii Torr. (*G. furcatum* Hanks, *G. atropurpureum* Heller var. *furcatum* Kearney & Peebles). Apache County (White Mountains) and Coconino County, perhaps also in Pima County (Santa Catalina Mountains), up to 9,000 feet, coniferous forests, June to September, type of *G. furcatum* from the Grand Canyon (*Allen* in 1897). Wyoming, Colorado, New Mexico, and Arizona.

4. Geranium lentum Woot. & Standl. Johnsons Basin, Cochise (?) County (*Wooton* in 1892), apparently also in the Pinaleno Mountains, Graham County, 8,500 feet (*Kearney & Peebles* 9864, immature). Western Texas to southeastern Arizona and Mexico.

5. Geranium Wislizeni Wats. Chiricahua and Huachuca mountains (Cochise County) and near Ruby (Santa Cruz County), 3,500 to 6,000 feet, August and September. Texas, southern Arizona, and northern Mexico.

The stylar column is pilose rather than puberulent in Arizona specimens.

6. Geranium Richardsonii Fisch. & Trautv. Apache County to Coconino County, south to Cochise and Pima counties, 6,500 to 11,500 feet, common in coniferous forests, April to October. South Dakota to British Columbia, southward to New Mexico, Arizona, and southern California.

7. Geranium eremophilum Woot. & Standl. Apache County to Hualpai Mountain (Mohave County), south to Cochise and Pima counties, 4,500 to 9,000 feet, June to October. New Mexico and Arizona.

This entity is perhaps not specifically distinct from *G. caespitosum*. The two apparently intergrade both in flower color and the presence or absence of glandular hairs.

8. Geranium caespitosum James (*G. atropurpureum* Heller). Apache County to Coconino County, south to the mountains of Cochise and Pima counties, 5,000 to 9,000 feet, very common in pine forests, May to September. Colorado and Utah, south to western Texas, New Mexico, Arizona, and Mexico.

2. ERODIUM. Heron-bill

Plants annual; leaves palmately lobed or pinnate; stamens 5 or if 10, then the alternate ones rudimentary; bodies of the carpels spindle-shaped, sharp-pointed at base, the persistent styles pubescent on the inner face, becoming spirally twisted at maturity.

Both of the Arizona species are excellent spring forage plants but alfilaria or filaree (*E. cicutarium*) is especially important because of its great abundance. It is believed that alfilaria was introduced into the Southwest at an early date by the Spaniards. The plants usually survive only a few weeks in the more arid desert regions, but livestock continue to feed on the dried stems. The corkscrewlike "tails" of the fruits are tightly twisted when dry but uncoil when moist, and as a result the sharp-pointed fruits penetrate the soil as if driven by an auger, as happens also with the fruits of *Stipa* and other grasses and mountain-mahogany (*Cercocarpus*).

KEY TO THE SPECIES

1. Leaves nearly to quite as wide as long, palmately lobed to almost divided, often cordate at base; herbage canescent with short appressed hairs; sepal tips not appendaged; petals commonly more than 6 mm. long .... 1. *E. texanum*
1. Leaves much longer than wide, pinnate, the leaflets pinnatifid and their segments often cleft; herbage sparsely glandular-villous; sepal tips with 1 or 2 white bristlelike appendages; petals not more, commonly less than 6 mm. long .... 2. *E. cicutarium*

1. Erodium texanum Gray. Grand Canyon, Coconino County, and Yavapai, Gila, Pinal, and Pima counties, also reported from Greenlee County, 1,000 to 4,500 feet, plains and mesas, February to April. Texas to southern Utah and southeastern California.

2. Erodium cicutarium (L.) L'Hér. Throughout the state, up to 7,000 feet, common and often very abundant on plains and mesas, February to July. Extensively naturalized in the United States; from Europe.

66. OXALIDACEAE. Wood-sorrel Family

1. OXALIS. Wood-sorrel

Plants herbaceous, mostly perennial with creeping rootstocks or bulbs, caulescent or acaulescent, the sap acid; leaves digitately compound, the leaflets 3 or more, wedge-shaped, obcordate; flowers perfect, regular; sepals 5; petals 5, yellow or purplish pink; stamens 10, the filaments united at base, 5 of them longer than the others; pistil of 5 united carpels, the ovary superior; fruit a dehiscent 5-celled capsule.

Several species are grown as ornamentals. The pleasantly acid leaves are refreshing when chewed and in salads, but should be eaten sparingly because of their high content of oxalic acid. These plants are sometimes known as sour-grass and sheep-sour.

KEY TO THE SPECIES

1. Petals yellow; plants caulescent; bulb none; sepals not bearing callosities; capsules pubescent. Leaflets 3: section *Xanthoxalis* (2).
1. Petals purplish pink, often drying violet; plants acaulescent, from a scaly bulb; sepals bearing apical callosities; capsules glabrous: section *Ionoxalis* (5).
2. Stems decumbent to erect (3).
2. Stems prostrate, creeping (4).
3. Hairs of the stems, petioles, and pedicels spreading or retrorse; stems conspicuously pilose or villous; capsules little shorter to longer than the pedicels........1. *O. pilosa*
3. Hairs mostly appressed or subappressed-ascending; stems not conspicuously pubescent; capsules commonly shorter than the pedicels. Leaflets ciliate, glabrous or nearly so above, usually sparsely strigose beneath............................2. *O. stricta*
4. Rootstocks and taproot rather thick, more or less woody; leaflets sparsely to copiously strigose, often on both surfaces...................................3. *O. albicans*
4. Rootstocks and taproot slender, not woody; leaflets nearly glabrous or sparsely strigose 4. *O. corniculata*
5. Leaflets 4 or more, longer than wide, broadly to narrowly wedge-shaped, entire or notched; longer filaments not appendaged, or with the appendage completely adnate to the filament. Outer bulb scales several-nerved.......................5. *O. Grayi*
5. Leaflets 3, wider than long, obreniform or V-shaped, the notch very broad; longer filaments appendaged, the appendage usually with a free tip (6).
6. Outer bulb scales 3- to several-nerved; leaflets 1 to 3 cm. wide, with broadly ovate lobes usually as wide as or wider than long; scapes mostly 10 to 20 cm. long; callosities of the sepals 0.3 to 0.5 mm. long, seldom longer...................6. *O. Metcalfei*
6. Outer bulb scales normally with more than 3 nerves; leaflets 2.5 to 5 cm. wide, with oblong-ovate lobes usually longer than wide; scapes 15 to 35 mm. long; callosities of the sepals 0.5 to 0.8 mm. long...7. *O. amplifolia*

1. Oxalis pilosa Nutt. Havasu Canyon, Coconino County (*J. T. Howell* 26539), Sierra Ancha, Gila County (*Eastwood* 17584), Ashdale, Maricopa County (*Peebles* 11627), May. Southern Arizona, California, and Sonora.

2. Oxalis stricta L. Coconino, Yavapai, Graham, Maricopa, Cochise, Santa Cruz, and Pima counties, 2,500 to 6,000 feet, usually along streams, April to September. Throughout much of North America.

3. Oxalis albicans H.B.K. (*O. Wrightii* Gray). Apache County to Coconino County, south to Cochise, Santa Cruz, Pima, and Yuma counties, 2,500 to 7,500 feet, common, preferring moist soil and partial shade, March to November. Western Texas to Arizona and Mexico.

4. Oxalis corniculata L. (*O. repens* Thunb.). Tempe, Maricopa County, common in lawns (*McLellan & Stitt* 1360), Tucson, Pima County (*Mrs. Thornber* 5479). Extensively naturalized in the United States; from the Old World.

5. Oxalis Grayi (Rose) Knuth. Apache, Navajo, Coconino, Yavapai, Gila, and Cochise counties, 5,000 to 9,500 feet, in pine or deciduous woods, July and August. New Mexico, Arizona, and northern Mexico.

Plants collected in the Patagonia Mountains, Santa Cruz County (*Peebles et al.* 5599), are referred here doubtfully. They have exceptionally large leaves, the sepals bear a light-brown apical gland in addition to the ordinary callosities, and the filaments are appendaged on the back.

6. Oxalis Metcalfei (Small) Knuth. Greenlee County to Coconino County, south to Cochise, Santa Cruz, and Pima counties, 5,500 to 9,000 feet, common in rich soil in coniferous and deciduous forests, July to September. New Mexico and Arizona.

Very close to *O. violacea* L., but the leaflets are more deeply obcordate and the bulb scales less distinctly and prominently 3-nerved than in most specimens of that species.

7. Oxalis amplifolia (Trel.) Knuth. Cochise, Santa Cruz, and Pima counties, 3,500 to 8,000 feet, rich soil in shade, August, type from the Santa Rita Mountains (*Pringle* 300, in 1881). Western Texas to southern Arizona.

Specimens collected in Santa Cruz County (*Peebles et al.* 4656, 5612) differ from Trelease's description in having the sepals acutish instead of "very obtuse" and in having 2 (not 4) callosities, these sometimes deeply lobed at base.

67. LINACEAE. Flax Family

1. LINUM. Flax

Plants herbaceous, annual or perennial; stems slender, commonly erect and branched; leaves simple, sessile, mostly alternate; inflorescence cymose-paniculate or racemose; flowers regular, perfect; sepals and petals 5, the sepals persistent, the petals deciduous, blue or yellow; stamens 5, sometimes with 5 additional rudiments, the filaments united at base; styles 2 to 5, distinct or more or less united; fruit a 4- to 10-valved capsule.

The outstanding species, of Old World origin, is flax (*L. usitatissimum*), from which linen (from stem fibers) and linseed oil (from seeds) are manufactured. One of Arizona's yellow-flowered species, *L. neomexicanum,* is reported to be poisonous to horses and sheep. Cyanogen is thought to be the toxic constituent.

KEY TO THE SPECIES

1. Petals sky blue, 1 to 2 cm. long; sepals glandless; stigmas introrse (2).
1. Petals yellow or orange; sepals (at least the inner ones) with marginal glands; stigmas terminal, capitate (3).
2. Plants annual; stigmas considerably longer than wide..............1. *L. usitatissimum*
2. Plants perennial; stigmas slightly longer than wide.....................2. *L. Lewisii*
3. Styles distinct to the base; sepals not or scarcely aristate, the outer ones entire or with few glandular teeth. Inflorescence elongate, narrow, interrupted-racemiform; plants

annual or biennial; outer sepals lanceolate or oblong-lanceolate; stems erect, the branches ascending at a narrow angle; capsule subglobose-ovoid. . 3. *L. neomexicanum*

3. Styles united nearly to the apex; sepals spinulose-aristate, the outer ones with numerous glandular teeth. Plants mostly annual; stems often branching from near the base, the branches ascending-spreading (4).

4. Pedicels and stems densely puberulent; angles of the stem not winged 4. *L. puberulum*

4. Pedicels and stems glabrous or obscurely puberulent; angles of the stem narrowly winged . . . . 5. *L. aristatum*

1. Linum usitatissimum L. Near Florence, Pinal County, at roadside (*Arnberger* 240, in 1949). Here and there in the United States; adventive from Europe. Cultivated flax.

2. Linum Lewisii Pursh. Apache County to Mohave County, south to Cochise and Pima counties, 3,500 to 9,500 feet, common on open mesas and in coniferous forests, March to September. Saskatchewan and Alaska to northern Mexico.

This handsome plant closely resembles cultivated flax, but is perennial. Occasional specimens have nearly white flowers. It is stated that the Indians in some of the Western states used the long fibers of the stems for making cordage.

3. Linum neomexicanum Greene. Apache County to Hualpai Mountain, Mohave County, south to Cochise and Pima counties, 4,500 to 9,000 feet, common in pine forests, June to September. New Mexico, Arizona, and northern Mexico.

4. Linum puberulum (Engelm.) Heller. Apache County to eastern Mohave County, south to Cochise, Santa Cruz, and Pima counties, 3,500 to 6,500 feet, April to October. Colorado, Utah, New Mexico, and Arizona.

Scarcely more than a variety of *L. rigidum* Pursh, differing chiefly in its denser puberulence and in having the apex of the pedicel not or only slightly cupulate.

5. Linum aristatum Engelm. Apache County to Hualpai Mountain, Mohave County, 4,500 to 8,000 feet, plains and mesas, mostly in sandy soil, May to September. Colorado and western Texas to northern Arizona and Mexico.

Var. *australe* (Heller) Kearney & Peebles (*L. australe* Heller) has the same range in Arizona as the typical plant and also extends farther south, to the Rincon Mountains (Pima County). As compared with typical *L. aristatum* it has shorter outer sepals, more distinctly dentate inner sepals, more diffusely branched stems, and usually serrulate leaves, the leaves being commonly entire in the typical plant. *L. aristatum* is reported to be used by the Hopi Indians in cases of childbearing.

68. ZYGOPHYLLACEAE. Caltrop Family

Plants herbaceous or shrubby; leaves alternate or opposite, compound or pinnately dissected; flowers perfect, regular or nearly so, 4- to 6-merous; stamens more numerous than the petals, in 2 whorls; style 1 or the styles completely united; ovary 2- to 12-celled, sometimes splitting in fruit into as many nutlets.

KEY TO THE GENERA

1. Leaves once or twice pinnately or subternately dissected. Branched perennial herbs; petals white or pale yellow . . . . 1. *Peganum*
1. Leaves compound, the leaflets entire (2).
2. Plants shrubby or suffrutescent; leaves digitately bifoliolate or trifoliolate (3).
2. Plants herbaceous; leaves pinnately 6-foliolate or more. Petals yellow or orange (4).
3. Leaves trifoliolate; stipules spiny; petals purple; filaments unappendaged; fruit smooth 2. *Fagonia*
3. Leaves bifoliolate; stipules not spiny; petals yellow; filaments with scalelike appendages; fruit villous . . . . 3. *Larrea*
4. Fruit flat, radiate, breaking up into 5 nutlets, each with 2 strong dorsal spines and containing 2 or more seeds, these separated by transverse septa . . . . 4. *Tribulus*
4. Fruit hemispheric or higher, not radiate, breaking up into more than 5 nutlets, these not spiny, 1-seeded . . . . 5. *Kallstroemia*

1. PEGANUM

Branching perennial herbs, the herbage glabrous or nearly so; leaves alternate, deeply dissected, with long, narrow segments; sepals and petals 4 or 5, the sepals elongate, often deeply cleft; filaments more or less dilated at base; fruit a 2- to 4-celled, many-seeded capsule.

1. Peganum Harmala L. Tucson (*K. F. Parker* 5820), probably only a casual escape from a garden. Adventive from the Old World.

This characteristic plant of the North African and Asiatic deserts has been collected also near Deming, New Mexico (*Hershey* in 1940).

P. mexicanum Gray was recorded by Gray (*Pl. Wright.* 2: 106. 1853) as having been collected on "sides of the Chiricahua Mountains, Sonora," now Cochise County, Arizona, but the writer is informed that the collection in question (*Wright* 1428) was actually made in Hudspeth County, Texas. A. Davidson (*Bull. S. Calif. Acad. Sci.* 4: 35. 1905) listed *P. mexicanum* as occurring in the Clifton area (Greenlee County), but his specimen, so labeled, is *Kallstroemia parviflora* Norton. *Peganum mexicanum* is known definitely only from southwestern Texas and northeastern Mexico. It is distinguished from *P. Harmala* by the much narrower leaf segments, 4 pale-yellow petals, filaments only slightly dilated at base, and peduncles decurved in fruit, whereas in *P. Harmala* the petals are 5 and white, the filaments conspicuously dilated at base, and the peduncles erect.

2. FAGONIA

Plants suffrutescent, puberulent to nearly glabrous; leaves digitately trifoliolate, the stipules spinescent; flowers small, solitary in the leaf axils; petals purplish pink; fruit of 5 united carpels.

1. Fagonia californica Benth. (*F. laevis* Standl., *F. longipes* Standl.). Southern Coconino County, Graham County (?), western parts of Pinal, Maricopa, and Pima counties, and throughout Yuma County, usually not above 2,500 feet, frequent on dry, rocky slopes and mesas, January to April (sometimes October). Southern Utah to southern California, Sonora, and Baja California.

F. laevis (*F. californica* subsp. *laevis* (Standl.) Wiggins), type from Yuma, *Jones* in 1906), is a very nearly glabrous form. *F. longipes* (type from Williams River, *Palmer* 58, in 1876) has exceptionally long pedicels.

3. LARREA. Creosote-bush

Plants much-branched shrubs, up to 3.5 m. (11.5 feet) high; leaves evergreen, thick, glutinous, strong-scented, the leaflets 2, oblong to obovate, united at base; flowers axillary, solitary; petals yellow; filaments each adnate to a conspicuous, 2-cleft, often laciniate scale; capsule 5-celled, densely white-villous.

1. Larrea tridentata (DC.) Coville (*Covillea tridentata* Vail). Havasu Canyon and Lava Falls (Coconino County), and from Mohave County to Greenlee, Cochise, Santa Cruz, Pima, and Yuma counties, 5,000 feet or lower, dry plains and mesas, flowering from time to time throughout the year, but most profusely in spring. Western Texas to southern Utah, Arizona, California, and northern Mexico.

Closely related to *L. divaricata* Cav. of southern South America and perhaps not specifically distinct, in which case the name *divaricata* has priority. Often erroneously called greasewood in Arizona and California. An outstanding xerophyte and a very important element of the perennial desert flora in southern and western Arizona. The plants cover thousands of square miles, often in nearly pure stand, usually with remarkably little variation in size. The Pima Indians formerly used the leaves in decoction as an emetic and to poultice sores. Small quantities of lac are found on the branches as a resinous incrustation. This was used for fixing arrow points and mending pottery. Creosote-bush has a strong characteristic odor, especially noticeable when the foliage is wet. The plant ordinarily is not touched by livestock, although it is reported that sheep, especially pregnant ewes, have been killed by partaking of it. This plant is reported to cause dermatitis in exceptional persons who are allergic to it.

4. TRIBULUS. Caltrop

Plants annual, with long, prostrate stems radiating from the root; leaves pinnate, the leaflets 8 to 12; flowers small, axillary, solitary, peduncled; petals yellow; fruits flat, of 5 nutlets, each dorsally armed with 2 strong spines and containing 2 or more seeds separated by transverse partitions.

1. Tribulus terrestris L. Here and there throughout the state, 7,000 feet or lower, often very abundant at roadsides and in fields, March to October. Extensively naturalized in the United States; from southern Europe.

Plant now commonly known as puncture-vine, also as bull-head and burnut. A troublesome annual weed because of the numerous spiny fruits each plant produces. Although seeds are produced in great quantity and the fruits are readily disseminated by means of furred animals and automobile tires, the weed is easily controlled by cultivation. The spines of the fruit are incapable of puncturing serviceable automobile tires but do penetrate bicycle tires. Hay containing the fruits may cause troublesome sores in the mouths of livestock.

5. KALLSTROEMIA

Plants very similar to *Tribulus*, but the fruit more convex, breaking up at maturity into 8 to 12 nutlets, these 1-seeded and merely tuberculate on the back.

KEY TO THE SPECIES

1. Petals 15 to 30 mm. long, orange; sepals 8 to 15 mm. long; beaks of the mature fruit slender, slightly conic at base; leaflets up to 25 mm. long. Stems hirsute-hispid, usually also copiously short-pilose; beaks of the fruit 8 to 11 mm. long, much longer than the nutlets, usually at least twice as long........................1. *K. grandiflora*
1. Petals not more and usually less than 12 mm. long, orange-yellow, often fading whitish; sepals less than 8 mm. long; beaks of the fruit stout, strongly conic at base; leaflets seldom more than 15 mm. long (2).
2. Beaks of the mature fruit 4 to 6 mm. long, commonly longer than the nutlets; petals 6 to 12 mm. long. Sepals persistent until after maturity of the fruit; nutlets usually prominently tuberculate on the back............................2. *K. parviflora*
2. Beaks of the fruit not more than 3 mm. long, shorter than the nutlets; petals not more than 6 mm. long (3).
3. Plants conspicuously hirsute, many of the hairs long and spreading; sepals persistent; beaks of the fruit copiously strigose-pubescent. Nutlets with low, transversely elongate (ridgelike) tubercles....................................3. *K. hirsutissima*
3. Plants not conspicuously hirsute, most of the hairs appressed; sepals usually deciduous before maturity of the fruit; beaks of the fruit glabrous or nearly so 4. *K. californica*

1. Kallstroemia grandiflora Torr. Greenlee County to Yavapai County, south to Cochise, Santa Cruz, Pima, and Yuma counties, 5,000 feet or lower, common on open plains and mesas, February to September, type from along the Gila River (*Emory*). Texas to Arizona and Mexico.

This plant, often locally known as Arizona-poppy, Mexican-poppy, and summer-poppy, is one of the most attractive annuals in the southern part of the state. The large flowers, rich orange in color, superficially resemble those of the California-poppy (*Eschscholtzia*). There is remarkable variation in size and shape of the anthers, apparently uncorrelated with other characters. The anthers are usually nearly orbicular and 1 to 2 mm. in diameter, but in specimens collected at San Bernardino Ranch, Cochise County (*Mearns* 1982), they are elliptic and up to 4 mm. long.

2. Kallstroemia parviflora Norton (*K. laetevirens* Thornber). Navajo County to Mohave County, south to Greenlee, Cochise, Santa Cruz, and Pima counties, 1,000 to 5,000 feet, plains and mesas, August to October. Mississippi to Arizona and northern Mexico.

3. Kallstroemia hirsutissima Vail. San Bernardino Ranch and Benson (Cochise County), near Elgin (Santa Cruz County), also (less typical) at Tucson (Pima County), 2,500 to 4,500 feet, apparently rare, August to October. Kansas and Colorado to Texas, southeastern Arizona, and northern Mexico.

4. Kallstroemia californica (Wats.) Vail. Apache, Coconino, and Yavapai counties to Santa Cruz, Pima, and Yuma counties, 7,000 feet or (usually) lower, plains and mesas, commonly in sandy soil, May to October. Southern Colorado to Arizona, southeastern California, and Mexico.

The typical plant, commonly with 4 to 7 pairs of leaflets and relatively elevated and sharp dorsal tubercles on the carpels, seems to be confined to the southern part of the state, from Graham County to Yuma County. In the north and central portions the prevailing variety is var. *brachystylis* (Vail) Kearney & Peebles (*K. brachystylis* Vail), with only 3 to 5 (rarely 6) pairs of leaflets and lower, blunter dorsal tubercles. There is, however, much intergradation.

69. RUTACEAE. Rue Family

Plants large shrubs or small trees to nearly herbaceous, more or less strong-scented; herbage glandular-punctate; leaves simple or digitately compound; flowers perfect or unisexual, regular; sepals and petals commonly 4 or 5, the stamens as many or twice as many, borne on a fleshy, hypogynous disk; ovary superior, 2- to 5-celled; fruit various.

The most important members of this family from an economic standpoint are the citrus fruits—orange, grapefruit, and lemon.

KEY TO THE GENERA

1. Leaves simple, linear or narrowly spatulate; petals erect; fruit a deeply 2-lobed capsule; ovules 5 or more in each carpel. Plants small shrubs or nearly herbaceous ... 2. *Thamnosma*
1. Leaves digitately compound; petals spreading; fruit not deeply lobed; ovules 1 or 2 in each carpel (2).
2. Plants small shrubs; herbage and capsules glandular-pustulate; leaves with usually more than 3 linear leaflets; petals white; fruit turgid, of 2 or 3 carpels....1. *Choisya*
2. Plants large shrubs or small trees; herbage glandular-punctate, not pustulate; leaves trifoliolate, the leaflets lanceolate or broader; petals greenish or yellowish; fruit flat, broadly winged, 1- or 2-celled..3. *Ptelea*

1. CHOISYA (162). Star-leaf, Zorillo

Contributed by Cornelius H. Muller

Aromatic shrubs, often prominently glandular; leaves opposite, 3- to 7-foliolate or more, the leaflets digitate, linear, prominently glandular on the margins and petioles; flowers solitary in the axils but crowded toward the tips of the stems; sepals 5, glandular and ciliate; petals 5, at least 1 cm. long, white, glabrous; stamens 10, in 2 series of alternating, unequal filaments; ovary of 5 united pubescent carpels with glabrous, glandular tips; styles connate above, the lobed stigma capitate; fruit of 2 or 3 pubescent and glandular carpels, the glandular tips migrating to form dorsal protuberances, the persistent style bases forming apical points; seeds oval or reniform, reticulately pitted.

The Arizona species belong to the subgenus *Astrophyllum,* regarded by some botanists as a distinct genus.

KEY TO THE SPECIES

1. Herbage appressed-hirtellous or partly glabrate; leaflets narrowly linear, 1 to 3 (rarely 4.5) mm. wide, strongly revolute, the petioles always less than ⅓ the length of the leaflets; glands prominent...1. *C. arizonica*
1. Herbage spreading-pilose; leaflets broadly linear, 3 to 5 mm. wide or rarely narrower, slightly revolute, flat and willowlike, the petioles always half as long as the leaflets or longer; glands scarcely prominent..................................2. *C. mollis*

1. Choisya arizonica Standl. (*C. dumosa* (Torr.) Gray var. *arizonica* L. Benson). Graham, Cochise, Santa Cruz, and Pima counties, 3,000 to 5,500 feet, canyons and rocky slopes, usually on limestone, April to July, type from the Santa Rita Mountains (*Pringle* in 1884). Known only from southeastern Arizona.

An attractive shrub, 3 to 6 feet high. Specimens from the Galiuro Mountains (Pinal County), Mustang Mountains (Santa Cruz County), Rincon

Mountains (Pima County), and near Klondyke (Graham County) approach *C. mollis* in their long petioles and rather broad leaflets.

2. Choisya mollis Standl. (*C. dumosa* var. *mollis* L. Benson). Mountains of Santa Cruz County near Nogales and Ruby, 3,500 to 5,000 feet, dry slopes and sides of canyons, type collected by Schott in the Pajarito Mountains (Santa Cruz County or adjacent Sonora). Known definitely only from southern Arizona.

A shrub 3 to 5 feet high, with leaflets resembling the leaves of *Salix taxifolia.*

2. THAMNOSMA

Plants shrubby to herbaceous or nearly so; leaves simple, alternate, narrow, entire, persistent or soon deciduous; flowers in small cymose or racemose clusters; corolla cylindric to campanulate; capsule deeply 2-lobed.

KEY TO THE SPECIES

1. Plants shrubby, more or less spinescent; herbage yellowish; leaves mostly linear-spatulate, few, soon deciduous; corolla cylindric-funnelform or slightly urceolate, 8 to 14 mm. long; petals dark blue; stipe of the capsule longer than the calyx; seeds 4 to 6 mm. long, smooth or somewhat rugose.............................1. *T. montana*
1. Plants suffrutescent or nearly herbaceous, not spinescent; herbage glaucous-green; leaves linear to nearly filiform, numerous, persistent; corolla campanulate, 3 to 5 mm. long; petals yellowish or brownish purple; stipe of the capsule shorter than the calyx or none; seeds not more than 2 mm. long, tuberculate..........................2. *T. texana*

1. Thamnosma montana Torr. & Frém. Grand Canyon (Coconino County), Mohave, Yavapai, Gila, Pinal, Maricopa, and Yuma counties, 4,500 feet or lower, desert mesas and slopes, frequent and locally abundant, February to April. Southern Utah and Nevada, southeastern California, Arizona, Sonora, and Baja California.

Turpentine-broom, so called because of the appearance and strong odor of the plant. Reported to have been used by the Indians as a tonic and in treatment of gonorrhea.

2. Thamnosma texana (Gray) Torr. (*Rutosma purpurea* Woot. & Standl.). Greenlee County to Coconino County, south to Cochise, Santa Cruz, and Pima counties, 2,000 to 5,000 feet, dry, rocky slopes and mesas, March to June. Colorado and western Texas to southern Arizona and northern Mexico.

3. PTELEA. Hop-tree

Shrubs or small trees; leaves commonly trifoliolate, the leaflets lanceolate or ovate, somewhat rhombic; flowers small, perfect or unisexual, in compound cymes; calyx lobes, petals, and stamens 4 or 5; fruit flat, nearly orbicular, winged, mostly 2-celled, indehiscent.

The plants have a strong, somewhat disagreeable odor and are not eaten by livestock. The fruits are reported to have been used in brewing as a substitute for hops and in making bread. Most of the numerous forms that have been described as species in this genus appear to be only individual variations. There is great diversity in the shape and size of the fruits as well as of the leaflets, but there appears to be little correlation between the characters of the two organs.

KEY TO THE SPECIES

1. Bark of the twigs straw-colored to light olive-colored; leaves yellowish green, paler but not glaucous beneath, often somewhat shiny above, rather thick and firm, glabrous or sparsely pubescent beneath; leaflets prevailingly rhombic-lanceolate but sometimes rhombic-ovate, the terminal one commonly 3 or more times as long as wide 1. *P. pallida*
1. Bark of the twigs brown or dark purple (commonly mahogany-colored or plum-colored); leaves bright green or bluish green, often glaucous beneath, not shiny above, thin, glabrate or permanently soft-pubescent beneath; leaflets prevailingly rhombic-ovate but sometimes rhombic-lanceolate, commonly less than 3 times as long as wide 2. *P. angustifolia*

1. Ptelea pallida Greene. Coconino, Mohave, and Yavapai counties, 2,000 to 7,000 feet, especially abundant in and near the Grand Canyon, April to June, type from Peach Springs, Mohave County (*Greene* in 1889).

This seems to be the plant interpreted by Wooton and Standley (*Flora of New Mexico*, p. 389) as *P. angustifolia* Benth. The latter was described from Mexican specimens, whereas *P. pallida* is mainly a species of the high plateaus of central and northern Arizona. Additional names published by Greene (163, pp. 71–73), based on types from the Grand Canyon and vicinity, are *P. argentea, P. elegans, P. lutescens, P. nitida, P. saligna,* and *P. triptera.* What may be a form of *P. pallida* with exceptionally broad leaflets is *P. straminea* Greene (163, p. 70), type from the Virgin Mountains, Mohave County (*Purpus* 6165).

2. Ptelea angustifolia Benth. (*P. jucunda* Greene). Apache County to Hualpai Mountain (Mohave County), south to Cochise, Santa Cruz, and Pima counties, 3,500 to 8,500 feet, commonly in the pine belt, mostly in canyons, May and June. New Mexico, Arizona, and Mexico.

Some persons suffer dermatitis as a result of contact with this plant.

P. angustifolia may be not specifically distinct from the Eastern *P. trifoliata* L. (and var. *mollis* Torr. & Gray) differing chiefly in its usually narrower leaflets and the thicker wings and relatively large body of the fruits, which are more often emarginate at base and apex. Additional, probably synonymous, names published by Greene (163, pp. 60–65) on the basis of Arizona types are *P. attrita, P. crenata, P. similis,* and *P. tortuosa.* Typical *P. angustifolia* has the foliage and twigs more or less permanently pubescent. A glabrate form, usually with narrower leaflets, is var. *cognata* (Greene) Kearney & Peebles (*P. cognata* Greene, 163, p. 62), the type of which was collected in the Huachuca Mountains (*Wilcox* 477). *P. betulifolia* Greene (163, p. 64), based on a specimen collected at Fort Bowie (*Fisher* in 1894), seems to be a synonym. This variety is more common than typical *P. angustifolia* in the southern part of the state and begins to flower as early as March.

70. SIMAROUBACEAE. Simarouba Family

1. HOLACANTHA. Crucifixion-thorn

Large, dioecious shrubs, of grotesque appearance, intricately branched, with short, stout, sharply spinose branches; leaves reduced to small deciduous scales; flowers small, greenish yellow; petals commonly 7 or 8; stamens 12 or more in staminate flowers; fruit a ring of 5 or more nearly distinct, divergent, 1-seeded, drupelike carpels.

1. Holacantha Emoryi Gray. Pinal, Maricopa, Pima, and Yuma counties, 2,000 feet or lower, frequent but not abundant on desert plains, June and July, type from between Tucson and the Gila River (*Emory*). Southern Arizona, southeastern California, and northern Mexico.

The fruits persist for years, so it is usually possible to identify each preceding season's fruit clusters by the degree of weathering. The plant attains a height of 3.5 m. (12 feet).

71. BURSERACEAE. Torch-wood Family

1. BURSERA (164). Elephant-tree

Shrubs or small trees, unarmed, strongly aromatic; young bark smooth and brown, the older bark exfoliating; leaves alternate, pinnate, deciduous; flowers small, solitary or in very few-flowered clusters; calyx lobes and petals 3 to 5; stamens 6 to 10, these and the petals inserted at base of a ring-shaped disk; ovary 3-celled; fruit drupelike, trigonous, with 1 large, bony seed.

A resin called copal is obtained in Mexico from many of the species, including *B. fagaroides*. It is used for cement and varnish and for treating bites of scorpions. It is burned as incense in the churches and was formerly so employed in the Aztec and Mayan temples.

KEY TO THE SPECIES

1. Leaflets lanceolate, acute or acutish at apex, 15 to 40 mm. long, 4 to 15 mm. wide; young bark gray-brown, the old bark exfoliating in large, thin sheets........1. *B. fagaroides*
1. Leaflets narrowly oblong, oval, or spatulate (the terminal one sometimes nearly orbicular), obtuse at apex, 2 to 8 mm. long, 1 to 2 mm. wide; young bark red-brown, the old bark exfoliating in flakes..............................2. *B. microphylla*

1. Bursera fagaroides (H.B.K.) Engler (*B. odorata* T. S. Brandeg.). Known in Arizona only on dry limestone cliffs near Fresnal, western foothills of the Baboquivari Mountains, Pima County (*Gilman* 3529, etc.) about 4,000 feet, July. Southern Arizona and Mexico (including Baja California).

The few plants at Fresnal, which reach a height of 4.5 m. (15 feet), have the appearance of being survivors from a time when conditions were more favorable to the species at this locality. The young bark is bright green, and the crushed herbage has an odor of tangerine skin. The old bark can be removed in translucent sheets resembling parchment. The branches yield quantities of gum and, when cut from the tree, dry very slowly.

2. Bursera microphylla Gray. Mountains of southwestern Arizona from the Salt River Mountains (Maricopa County) and western Pima County to the Gila and Tinajas Altas mountains (Yuma County), 2,500 feet or lower, locally abundant on arid, rocky slopes, July, type from the Tule Mountains (*Schott*). Southwestern Arizona, southeastern California, and northwestern Mexico.

The trees reach a height in Arizona of 6 m. (20 feet) and a trunk diameter of 0.3 m. (1 foot). The crooked branches taper rapidly, and their shape suggests the trunk of an elephant. The plant cannot withstand much cold. The bark contains tannin and was gathered in Sonora for export. In that region the gum was used for treating venereal diseases.

72. MALPIGHIACEAE Malpighia Family

Plants suffrutescent; stems trailing or twining; leaves simple, opposite, entire; hairs of the herbage appearing simple but affixed at the middle (dolabriform); flowers mostly perfect, axillary or terminal, solitary or in small clusters, some of them often cleistogamous and apetalous; sepals 5, all or 4 of them bearing a pair of conspicuous fleshy glands externally at base; petals 5, somewhat unequal, abruptly contracted into claws; stamens 5 or 6, some of them often sterile, the filaments stout, united at base; style and stigma 1; ovary superior, 3-lobed.

KEY TO THE GENERA

1. Carpels each with a conspicuous, scarious, dorsal wing, the wings divergent; leaves lance-linear; petals yellow, entire or denticulate........................1. *Janusia*
1. Carpels not winged, nutlike, irregularly trigonous, strongly keeled on the back, the edges margined or tuberculate; leaves oblong-lanceolate, oval, or ovate; petals orange, fimbriate ....2. *Aspicarpa*

1. JANUSIA

Stems slender, twining, often tangled; leaves short-petioled, narrow; fruit a pair of samaras (exceptionally 3).

1. Janusia gracilis Gray. Grand Canyon (Coconino County), and Greenlee and Mohave counties, south to Cochise, Santa Cruz, Pima, and Yuma counties, 1,000 to 5,000 feet, dry, rocky slopes, April to October. Western Texas to Arizona and northern Mexico (including Baja California).

2. ASPICARPA

Stems commonly trailing; leaves sessile or short-petioled, oblong-lanceolate to ovate; flowers dimorphic, the petaliferous ones mostly in terminal clusters, the cleistogamous flowers axillary, long-peduncled to nearly sessile; fruit a pair of strongly keeled, usually tuberculate nutlets.

1. Aspicarpa hirtella Rich. Cochise, Santa Cruz, and Pima counties, 4,000 to 5,500 feet, commonly in chaparral, August and September. Southern Arizona and Mexico.

The Arizona specimens all appear to belong to one species, although identified variously as *A. longipes* Gray, *A. humilis* (Benth.) Small, and *A. hirtella*. As these species were defined by Niedenzu (*Pflanzenreich* IV. 141: 559–560) the description of *A. hirtella* seems to apply best to the Arizona plant.

73. POLYGALACEAE. Milk-wort Family

Plants annual or perennial, herbaceous or suffruticose; leaves alternate, opposite, or verticillate, sessile or short-petioled, simple, entire; flowers perfect, very irregular; sepals 5, the 2 inner ones more or less petaloid; petals normally 3, often united at base, the 2 upper ones partly adnate to the stamen tube, the lower petal (keel) boat-shaped, often crested or beaked; stamens 6 to 8, the filaments united below, the anthers opening by an apical or subapical pore; fruit a flat or flattish 1- or 2-celled capsule; seeds usually carunculate (appendaged around the hilum).

KEY TO THE GENERA

1. Ovary 2-celled; capsule flat, dehiscent; seeds usually caruneulate..........1. *Polygala*
1. Ovary 1-celled; capsule slightly turgid, indehiscent; seeds not caruneulate..2. *Monnina*

1. POLYGALA. Milk-wort

Herbs or undershrubs; flowers mostly in narrow terminal racemes, sometimes few or solitary in the leaf axils; petals united below, forming a dorsally cleft tube; capsule 2-celled, thin and flat, dehiscent; seeds usually with a caruncle, this usually umbrella-shaped or veillike, with a thickened center (umbo) and a scarious margin.

These dainty plants are interesting because of the unusual structure of their flowers and seeds. Most of the species are reported to be distasteful to grazing animals, but the somewhat woody *P. macradenia* is browsed occasionally.

KEY TO THE SPECIES

1. Keel petal not crested or beaked, at most apiculate. Flowers not more than 6 mm. long: *P. acanthoclada*, an intricately branched shrub with scattered, yellowish flowers, these solitary or in fascicles of not more than 4, might be sought here (2).
1. Keel petal crested or beaked (9).
2. Leaves and capsules conspicuously dotted with large, sessile glands; leaves all alike; flowers axillary, mostly solitary; plants suffrutescent. Herbage copiously short-pilose; leaves very numerous, less than 10 mm. long, linear to oblong-lanceolate; wing petals purple, pubescent, the keel greenish yellow................1. *P. macradenia*
2. Leaves and capsules not gland-dotted; leaves more or less dimorphic, the lower ones broader and shorter than the upper ones; flowers few or somewhat numerous in loose, terminal racemes; plants herbaceous, of very similar appearance. Capsules ciliolate (3).
3. Caruncle of the seed with the depth of the entire or denticulate, scarious, marginal portion less than the height of the umbo (4).
3. Caruncle of the seed with the depth of the lobed or lobulate, scarious, marginal portion equaling or exceeding the height of the umbo. Wing petals greenish purple or violet (6).
4. Capsules persistently more or less pubescent on the sides; middle and upper leaves linear-lanceolate or linear-elliptic, 2 to 3 mm. wide, the lower leaves oblanceolate 2. *P. piliophora*
4. Capsules glabrous on the sides when mature; middle and upper leaves linear or lance-linear (5).
5. Middle and upper leaves scalelike, 5 to 10 mm. long, 1 to 1.5 mm. wide..3. *P. Barbeyana*
5. Middle and upper leaves not scalelike, 15 to 30 mm. long, 1.5 to 3 mm. wide. Wing petals greenish, tinged with purple............................4. *P. racemosa*
6. Capsules glabrous on the sides at maturity; caruncle not veillike, covering only the tip of the seed, the scarious portion not much deeper than the height of the umbo (7).
6. Capsules pubescent on the sides at maturity (seldom glabrescent); caruncle veillike, covering ¼ or more of the seed, the scarious portion much deeper than the height of the umbo (8).
7. Upper leaves less than 25 mm. long, the lowest leaves less than 10 mm. long; stems less than 20 cm. long; capsules oblong-oval........................5. *P. reducta*
7. Upper leaves mostly 20 to 35 mm. long, the lowest leaves seldom less than 10 mm. long; stems commonly at least 20 (up to 45) cm. long; capsules broadly oval, ovate, or slightly obovate..6. *P. longa*
8. Hairs of the stem all incurved or subappressed; middle and upper leaves up to 12 mm. wide, but usually much narrower.................................7. *P. obscura*

8. Hairs of the stem all widespreading; middle and upper leaves not more than 6 mm. wide .... 8. *P. orthotricha*
9. Crest of the keel none, the keel ending in a conic or cylindric, entire beak, this almost obsolete in *P. acanthoclada;* racemes terminal or lateral, few-flowered except sometimes in *P. Tweedyi* (10).
9. Crest of the keel conspicuous, fimbriate; racemes terminal, several- to many-flowered. Leaves (except the lowest) linear or linear-lanceolate (13).
10. Branches not or only slightly spinescent. Stems less than 20 cm. long (11).
10. Branches strongly spinescent (12).
11. Herbage pilose; stems few, relatively stout; leaves 3 to 12 mm. wide; wing petals 8 to 10 mm. long, rose-purple, the keel with a very stout and blunt, yellow beak 9. *P. Rusbyi*
11. Herbage puberulent; stems numerous, slender; leaves commonly not more than 3 mm. wide; wing petals 4 to 5 mm. long, the keel with a relatively slender beak 10. *P. Tweedyi*
12. Plants suffrutescent; stems not more than 15 cm. long; herbage puberulent or glabrescent; leaves 3 to 6 mm. wide; wing petals 8 to 12 mm. long, purplish-pink, the keel conspicuously beaked; capsules more than 5 mm. long .... 11. *P. subspinosa*
12. Plants shrubby, intricately branched; stems up to 90 cm. long; herbage puberulent or short-pilose; leaves seldom more than 3 mm. wide; wing petals not more than 5 mm. long, yellowish, sometimes tipped with purple, the keel inconspicuously or obscurely beaked; capsules 4 to 5 mm. long .... 12. *P. acanthoclada*
13. Plants annual; stems not rushlike, freely branched above the base, very slender, rarely more than 30 cm. long; flowers in loose racemes, not more than 2.5 mm. long, rose purple to nearly white. Capsules not winged; lower leaves in whorls of 4 or 5 13. *P. glochidiata*
13. Plants perennial; stems rushlike, not or sparingly branched above the base, angled, sulcate, often more than 30 cm. long; flowers in spikelike racemes (these rather dense at first), more than 2.5 mm. long, whitish, the petals often with greenish or purplish veins (14).
14. Capsules not winged, at most narrowly margined on both cells; leaves (at least the lowest) often in whorls, mostly linear. Plants often with short, very leafy basal shoots .... 14. *P. alba*
14. Capsules with a scarious wing on the upper cell; leaves all alternate, scattered, linear-acicular (15).
15. Stems usually few, glabrous or nearly so, sparsely leafy; racemes elongate, relatively loose; wing of the capsule broad .... 15. *P. hemipterocarpa*
15. Stems usually many, puberulent, the leaves crowded; racemes relatively short and dense; wing of the capsule narrow .... 16. *P. scoparioides*

1. Polygala macradenia Gray. Gila, Maricopa, Pinal, Cochise, Santa Cruz, Pima, and Yuma counties, 1,500 to 4,500 feet, dry, rocky slopes, April to July. Western Texas to southern Arizona and northern Mexico.

2. Polygala piliophora Blake. Known only from the type collection in the Huachuca Mountains, Cochise County (*Wilcox* in 1894), August.

3. Polygala Barbeyana Chodat. Specimens from Beaver Creek, Yavapai County (*Rusby* 527, *Purpus* 8290), were identified by S. F. Blake as *P. Barbeyana,* a species known otherwise only from northern Mexico. They resemble *P. racemosa* Blake, although somewhat small-fruited for that species.

4. Polygala racemosa Blake. Near Bisbee and near Benson, Cochise County (*Mearns* 1014, part, *Carlson* in 1915, *Harrison & Kearney* 5816), about 5,000 feet, May and June. Southeastern Arizona and northern Mexico.

5. Polygala reducta Blake. Pinal, Cochise, and Santa Cruz counties, 4,000 to 5,500 feet, May. Southeastern Arizona and northern Mexico.

6. **Polygala longa** Blake. Santa Cruz and Pima counties, 3,000 to 5,000 feet, spring and autumn. Western Texas to southern Arizona and northern Mexico.

This species apparently intergrades with *P. reducta.*

7. **Polygala obscura** Benth. Southern Navajo, Greenlee, Graham, Gila, Pinal, Cochise, Santa Cruz, and Pima counties, 3,500 to 6,000 feet, often among live-oaks, July to September. Western Texas to Arizona and Mexico.

Arizona's most widely distributed and commonest species.

8. **Polygala orthotricha** Blake. Chiricahua Mountains (Cochise County), Santa Rita Mountains (Santa Cruz or Pima County), August, type from the Santa Rita Mountains (*Pringle* in 1884). Known only from southern Arizona.

The character of the pubescence alone seems to differentiate this entity from *P. obscura.*

9. **Polygala Rusbyi** Greene. Peach Springs to Kingman (Mohave County), also Yavapai County, April to July, type from near Prescott (*Rusby* in 1883). Known only from central Arizona.

Flowers larger and showier than in any other Arizona species.

10. **Polygala Tweedyi** Britton (*P. Lindheimeri* Gray var. *parvifolia* Wheelock, *P. arizonae* Chodat). Chiricahua, Dos Cabezas, and Huachuca mountains (Cochise County), Santa Rita Mountains (Santa Cruz or Pima County), about 5,000 feet, limestone ledges, June to September, type of *P. Lindheimeri* var. *parvifolia* (and of *P. arizonae*) from the Santa Rita Mountains (*Pringle* in 1884). Oklahoma and western Texas to southern Arizona and northern Mexico.

11. **Polygala subspinosa** Wats. Near Pipe Springs, Mohave County, 5,000 feet (*Peebles & Parker* 14714). Western Colorado to northwestern Arizona and southern California.

The keel is yellow, the other petals purple.

12. **Polygala acanthoclada** Gray. Navajo, eastern and northern Coconino, and southwestern Yavapai counties, 2,500 to 5,000 feet, June. Southwestern Colorado to northern Arizona and southeastern California.

Arizona's most shrubby species, with stems up to 1 m. (3 feet) high, intricately branched, tending to form hummocks. Both the typical plant, with spreading hairs and puberulent sepals, and var. *intricata* Eastw., with incurved or reflexed, matted hairs and glabrous or nearly glabrous sepals, occur in Arizona.

13. **Polygala glochidiata** H.B.K. Mule Mountains, Cochise County (*Goodding* in 1939), southern slopes of the Santa Rita Mountains, Santa Cruz County (*Goodding* in 1937), November. Southern Arizona, Mexico, and widely distributed in tropical America.

14. **Polygala alba** Nutt. Apache County to Coconino County, south to Cochise, Santa Cruz, and Pima counties, 5,000 to 7,500 feet, May to September. South Dakota to Washington, south to southern Mexico.

Var. *suspecta* Wats., with most of the leaves whorled (only the lowest whorled in the typical plant), has been collected in the Huachuca Mountains (*Jones* in 1903).

15. Polygala hemipterocarpa Gray. Cochise, Santa Cruz, and Pima counties, 4,000 to 7,000 feet, grassy or stony slopes, May and June, type from the Sonoita (*Wright* 937). Texas to southern Arizona and Mexico.

Specimens from Empire Ranch, eastern Pima County (*Thornber* in 1905), and from Santa Cruz County (*Benson* 10425) seem intermediate between this and *P. scoparioides*, the stems being glabrous but the capsule very narrowly winged.

16. Polygala scoparioides Chodat. Yavapai, Gila, Cochise, Santa Cruz, and Pima counties, 3,500 to 5,000 feet, rocky mesas and slopes, March to October, type from the Santa Rita Mountains (*Pringle* in 1884). Western Texas to Arizona and northern Mexico.

2. MONNINA

Plants annual; stems slender, leafy, erect, few-branched; herbage minutely puberulent; leaves short-petioled, lanceolate, acuminate; flowers very small, in narrow terminal racemes; corolla pale blue.

1. Monnina Wrightii Gray. Southern Apache (or Navajo), Yavapai, Greenlee, Cochise, Santa Cruz, and Pima counties, 4,000 to 7,500 feet, sometimes on limestone with *Ceanothus*, September and October. New Mexico, Arizona, and northern Mexico.

74. EUPHORBIACEAE. Spurge Family

Contributed by Louis C. Wheeler

Herbs, shrubs, or trees, often with milky juice; leaves alternate, opposite, or whorled, simple or rarely compound; stipules present or absent; flowers unisexual; calyx and corolla present or absent; stamens 1 to indefinitely numerous; ovary superior, mostly 3-locular, sometimes 1- to 4-locular; ovules pendulous, 1 or 2 per locule; styles as many as the locules, distinct or partly connate, often divided; fruit capsular, the carpels usually dehiscent by 2 elastic valves, sometimes tardily dehiscent or even indehiscent; seeds with or without a caruncle, the testa crustaceous, the endosperm copious, oily, the embryo straight or curved.

A very large family, chiefly tropical and subtropical. The sap is often poisonous, and some of these plants, notably castor-bean, are used medicinally as purgatives. The most valuable of the rubber plants of the world, *Hevea brasiliensis*, belongs to this family, and members of certain other genera also are sources of rubber. Cassava, an important food of tropical countries, and tapioca are obtained from the roots of species of *Manihot*. Wood oil or tung oil, valuable in varnishes because of its quick-drying property, is obtained from the seeds of species of *Aleurites*, a mainly Asiatic genus. Castor oil, chemically altered into a drying oil, is of increasing importance in paints.

KEY TO THE GENERA

1. Leaves palmately lobed (2).
1. Leaves not palmately lobed (5).
2. Stamens indefinitely numerous; leaves peltate. Petals absent............8. *Ricinus*

2. Stamens 8 to 10; leaves not peltate (3).
3. Petals present. Stinging hairs absent; filaments united. . . . . 9. *Jatropha*
3. Petals absent (4).
4. Herbage and capsule bearing harsh, stinging hairs; filaments united; calyx petaloid 10. *Cnidoscolus*
4. Herbage and capsule glabrous; filaments distinct, attached around a fleshy central disk; calyx not petaloid. . . . . 11. *Manihot*
5. Plants clothed with stellate, sometimes scalelike hairs (6).
5. Plants with the hairs simple or malpighiaceous (affixed at the middle), or the plants glabrous (7).
6. Leaves entire or nearly so; anthers turned in and down in the bud; staminate calyx lobes 5 or if 4, then the carpels 2 or 3, not carinate; seeds carunculate. . . . . . 3. *Croton*
6. Leaves crenate; anthers erect in bud; staminate calyx lobes 3 or 4; carpels 3, carinate, obviously so when young, only at the apex at maturity; seeds not carunculate 5. *Bernardia*
7. Flowers seemingly perfect, the cluster consisting of a central pistillate flower with mostly 5 radial glomerules of several to few or even 1 staminate flower each, all surrounded by a persistent gamophyllous involucre bearing 1 to 5 often petaloid-appendaged glands on the rim. . . . . 14. *Euphorbia*
7. Flowers plainly unisexual (8).
8. Petals present; flowers monoecious or if dioecious, then the plants clothed with malpighiaceous hairs; filaments united into a column; stamens 8 to 12 (9).
8. Petals none; flowers monoecious or dioecious; malpighiaceous hairs never present; filaments distinct or if shortly united, then the stamens only 2 (10).
9. Seeds not carunculate; flowers monoecious or dioecious; some malpighiaceous hairs usually present (always present if the flowers are dioecious); petals entirely distinct; anthers about as wide as long. . . . . 4. *Ditaxis*
9. Seeds carunculate; flowers monoecious; hairs of the ordinary simple type or absent; petals more or less connivent; anthers obviously longer than wide. . . . . 9. *Jatropha*
10. Plants shrubby with rigid, divaricate branches and small, fascicled, entire leaves; flowers dioecious. Ovules 2 in each locule. . . . . 2. *Tetracoccus*
10. Plants herbaceous, often perennial or if shrubby, then the leaves not fascicled or entire; flowers monoecious, or sometimes dioecious in *Reverchonia* (11).
11. Ovules and seeds 2 in each locule; stigmas subsessile; plants glabrous, annual herbs; stamens 2; staminate sepals 4. . . . . 1. *Reverchonia*
11. Ovules and seeds 1 in each locule; styles of appreciable length; plants not glabrous annuals or if so, then the stamens more than 2 or the calyx only 2-lobed (12).
12. Pistillate flowers subtended by accrescent, foliaceous bracts; styles dissected into filiform segments; anther locules linear, flexuous, attached only at the tip. . 6. *Acalypha*
12. Pistillate flowers not subtended by foliaceous bracts; styles entire; anther locules subglobose or reniform, attached along their sides (13).
13. Calyx 2- to 6-lobed; herbage with harsh, often stinging, hairs. Plants slender and often twining. . . . . 7. *Tragia*
13. Calyx 2-lobed; herbage glabrous (14).
14. Plants herbaceous, sometimes woody at base; styles 3; ovary and capsule 3-locular; staminate flowers solitary in the axil of each bract. . . . . 12. *Stillingia*
14. Plants shrubby; styles 2, rarely 3; ovary and capsule 2- or rarely 3-locular; staminate flowers several in the axil of each bract. . . . . 13. *Sapium*

1. REVERCHONIA

Glabrous annual herbs; leaves simple, alternate, entire, linear-spatulate to spatulate, petioled; stipules small, thin, triangular-subulate; flowers monoecious or dioecious, axillary; staminate sepals 4; stamens 2, the filaments distinct; pistillate sepals 6; glandular disk present; ovary 3-locular; ovules 2 in each locule; seeds not carunculate.

1. Reverchonia arenaria Gray. Hopi Indian Reservation, Navajo County (*Hough* 39, *Whiting* in 1935), Leupp, eastern Coconino County, 5,000 feet (*Oakley* 373), July and August. Utah and northeastern Arizona to Texas and northern Mexico.

This is supposed to be the only euphorbiaceous genus outside the Australian region having the cotyledons no wider than the radicle. Another unique feature is that the seeds are attached below the middle rather than near the apex. The seeds are reported to be used by the Hopi Indians in treatment of hemorrhage and also for oiling their stone griddles.

2. TETRACOCCUS

Rigid, divaricately branched shrubs; leaves small, spatulate to obovate-spatulate, petiolate, fascicled on very short lateral branchlets; flowers dioecious; staminate flowers 1 to 5 on the branchlets, pedicelled, the sepals 5 or 6, the stamens 5 to 8, the filaments distinct, attached around the lobed central disk, a rudimentary pistil present; pistillate flowers solitary on the short lateral branchlets, with short, thick pedicels, the ovary 3- or occasionally 4-locular, the styles 3 or occasionally 4, entire, spatulate, the ovules 2 in each cell; seeds carunculate, mostly solitary by abortion of 1 ovule and then dorso-ventrally compressed, or laterally compressed when both seeds develop.

1. Tetracoccus Hallii T. S. Brandeg. (*Halliophytum Hallii* Johnst.). Rawhide Mountains (southern Mohave County), Williams River near Alamo, Sheep Tanks, and Kofa Mountains (Yuma County), 500 to 2,500 feet, along sandy washes, flowering in spring. Southwestern Arizona and southeastern California.

3. CROTON

Herbs or shrubs; leaves alternate, petiolate, simple, the stipules mostly obsolete; flowers monoecious or dioecious; inflorescence racemose, the staminate flowers above and the pistillate below in monoecious species; calyx 4- or 5-lobed; petals present or absent in the staminate flowers, the stamens several to many; petals sometimes present but usually absent or rudimentary in the pistillate flowers, the ovary 3- or sometimes 2-locular, the ovules solitary, the styles 3 or sometimes 2, 1- to 4-times bifid; capsules 3- or sometimes 1-seeded; seeds small to medium, carunculate.

These plants are odorous and probably more or less poisonous. The powerful purgative oil of croton is obtained from an Asiatic species, *C. Tiglium* L.

KEY TO THE SPECIES

1. Flowers mostly monoecious; staminate flowers petalliferous; style branches usually only 6 or if 12, then the leaf margins bearing stalked glands (2).
1. Flowers usually dioecious; staminate flowers apetalous; style branches 12 or more and leaves devoid of stalked glands (7).
2. Shrubs (3).
2. Herbs, annual or suffruticose-perennial (5).
3. Leaf margins and the obvious laciniate stipules bearing stipitate glands; style branches about 12 .... 1. *C. ciliato-glandulosus*
3. Leaf margins and the minute stipules devoid of glands (4).

4. Mature leaves with only scattered stellae beneath; sepals subglabrous except for a few simple hairs at the tips .... 2. *C. sonorae*
4. Mature leaves stellate-tomentose beneath; sepals stellate-tomentose outside 3. *C. fruticulosus*
5. Suffruticose perennial; anthers about 1.5 mm. long .... 4. *C. corymbulosus*
5. Annual; anthers about 0.5 mm. long (6).
6. Ovary 3-celled; capsule 3-seeded; styles 3; staminate calyx 5-lobed 5. *C. Lindheimerianus*
6. Ovary 2-celled; capsule 1-seeded; styles 2; staminate calyx 4-lobed 6. *C. monanthogynus*
7. Perennial; upper surfaces of leaves densely covered with overlapping stellae, hence grayish .... 7. *C. californicus*
7. Annual; upper surfaces of leaves with mostly scattered, rarely overlapping stellae, hence green .... 8. *C. texensis*

1. Croton ciliato-glandulosus Ortega. Bear Valley or Sycamore Canyon, Santa Cruz County, 3,500 feet, abundant on rocky west slopes with *Mimosa* and *Anisacanthus* (*Darrow & Haskell* 2215). Southern Arizona and Baja California to Central America.

2. Croton sonorae Torr. Western part of Pinal and Pima counties, dry, rocky slopes, 2,000 to 3,000 feet, August. Southern Arizona and Mexico.

3. Croton fruticulosus Engelm. Guadalupe Canyon, southeastern Cochise County, 4,500 feet, abundant in canyon bottoms under oaks (*Darrow et al.* 3569). Southern and western Texas to southeastern Arizona and northern Mexico.

4. Croton corymbulosus Engelm. (*C. eremophilus* Woot. & Standl.). Greenlee, Cochise, Santa Cruz, and Pima counties, 2,500 to 6,000 feet, common on dry, rocky slopes, May to October, type from Camp Bowie, Cochise County (*Rothrock* 506). Texas to southern Arizona and northern Mexico.

The leaves are reported to be used in domestic medicine in Texas.

5. Croton Lindheimerianus Scheele. The solitary specimen seen from Arizona, perhaps introduced, was collected at Beaver Creek, Yavapai County (*C. A. Purpus* 8258). Kansas to Texas and northern Mexico.

6. Croton monanthogynus Michx. In Arizona known only from the Santa Cruz River at La Noria, Santa Cruz County (*Mearns* 1167); perhaps introduced from farther east. Florida to Illinois and Texas.

7. Croton californicus Muell.Arg. Rillito Creek (Pima County), Verde River (Maricopa County ?), Parker and Yuma (Yuma County), 200 to 2,500 feet, February to October. Southwestern Arizona, California, and Baja California.

The Arizona plants are best referred to var. *mohavensis* Ferguson. The varieties of this species are ill-defined and vague.

8. Croton texensis (Klotzsch) Muell.Arg. Apache County to Coconino County, south to Cochise, Santa Cruz, Pima, and Yuma counties, 500 to 7,000 feet, very common, roadsides, fields, and dry stream beds, May to November. Illinois to Wyoming, south to Arkansas, Arizona, and Mexico.

This plant is sometimes called dove-weed, the seeds being a favorite food of that bird. The plant gives off a disagreeable odor, so strong as to be noticeable to passing motorists. Hay containing this plant is reported to have poisoned cattle. The Hopi Indians use it as an emetic and an eyewash.

Broad-leaved specimens have been referred to *C. luteovirens* Woot. & Standl. A practically glabrous form was collected at Seneca, Gila County (*Darrow* in 1940).

4. DITAXIS

Herbs or shrubs with mostly malpighiaceous hairs; leaves simple, alternate, exstipulate; flowers mostly monoecious, borne in bracteate axillary racemes, or sometimes solitary; calyx 5-lobed; petals 5 or sometimes wanting in the pistillate flowers; glands 5, alternating with the petals; stamens 8 to 12 (usually 10), the filaments united below into a column (androphore), the anthers in 2 whorls, the lower with 5 or 6, the upper with 3, 5, or 6 anthers; ovary 3-locular, the ovules solitary; styles 3, bifid, distinct or partly connate; seeds not carunculate, small.

KEY TO THE SPECIES

1. Gland-tipped teeth present on the leaves, bracts, and pistillate sepals. . 1. *D. adenophora*
1. Gland-tipped teeth lacking (2).
2. Claws of the staminate petals entirely free from the column of united filaments (androphore); stigmas not flattened, sometimes subclavate. Leaves petiolate (3).
2. Claws of the staminate petals united to the androphore; stigmas flattened and dilated (4).
3. Pistillate petals ovate-lanceolate, with a few short hairs on the back scarcely extending beyond the tip of the petal, or the margin shortly ciliate; styles united only at base; glands in the staminate flowers mostly thickish, obtuse; seeds shallowly faveolate 6. *D. neomexicana*
3. Pistillate petals obovate-cuneate, with abundant coarse hairs on the back extending markedly beyond the tip of the petal; styles united ⅓ to ½ of their length; glands in the staminate flowers thin, acute; seeds nearly smooth. . . . . 7. *D. serrata*
4. Leaves sessile; anther whorls so close together as to be scarcely distinguishable; seeds ellipsoid to subspheroid, 3.4 to 4.6 mm. long (5).
4. Leaves petiolate; anther whorls obviously distinct; seeds definitely truncate at base (6).
5. Pistillate petals wanting; sepals strigose outside; glands linear. . . . . . 2. *D. mercurialina*
5. Pistillate petals present; sepals glabrous outside; glands not longer than wide 3. *D. cyanophylla*
6. Little-branched shrubs 1 to 2 m. high; lateral branches more than 3 mm. thick except at tip, pithy; seeds trigonous-pyramidal, truncate, 4 mm. or more long; leaves with few hairs or none. . . . . 4. *D. Brandegei*
6. Freely branching low shrubs, less than 1 m. high; main stem rarely more than 3 mm. thick; seeds ovoid-truncate, not more than 3 mm. long; leaves thickly covered with hairs, often silvery. . . . . 5. *D. lanceolata*

1. Ditaxis adenophora (Gray) Pax & Hoffmann. Agua Caliente (western Maricopa County), mouth of Williams River (Yuma-Mohave counties), March. Western Arizona, southeastern California, and Sonora.

2. Ditaxis mercurialina (Nutt.) Coult. Fort Apache (Navajo County). Kansas and Texas to eastern Arizona.

3. Ditaxis cyanophylla Woot. & Standl. Slate Mountain, Flagstaff, and Walnut Canyon (Coconino County), Coyote Spring near Springerville (Apache County), 6,000 to 6,500 feet, pine forests. New Mexico to southern Nevada and northern Arizona.

The flowers impart a purple color to water on boiling.

4. Ditaxis Brandegei (Millsp.) Rose & Standl. Tule Tank and Gila Mountains (Yuma County), dry slopes, April to July. Southwestern Arizona to Sonora (?) and Baja California.

The plant occurring in Arizona is var. *intonsa* Johnston, where it attains a height of 1.8 m. The flowers impart a strong purple color to water on boiling.

5. Ditaxis lanceolata (Benth.) Pax & Hoffmann (*Argythamnia sericophylla* Gray). Southern Yavapai County to Pima and Yuma counties, 350 to 3,000 feet, dry, rocky slopes, February to September. Southwestern Arizona to southeastern California, Sonora, and Baja California.

Plants from excessively dry sites show a tendency to be either wholly staminate or wholly pistillate.

6. Ditaxis neomexicana (Muell.Arg.) Heller. Grand Canyon (Coconino County) and Mohave County to Greenlee, Cochise, Pima, and Yuma counties, 1,000 to 4,000 feet, February to September. Texas to southern Arizona, Sonora, and Baja California.

Leaves lanceolate, acute, and serrulate or entire.

7. Ditaxis serrata (Torr.) Heller. Southern Yuma County, 200 to 300 feet, sandy soil, March to October. Southwestern Arizona, southeastern California, Sonora, and Baja California.

Leaves mostly broadly cuneate-spatulate, with truncate, coarsely toothed apices.

5. BERNARDIA

Deciduous shrubs; leaves slightly greener above, petiolate, alternate; stipules small, fleshy, acuminate; flowers dioecious; staminate flowers 2 to several in the axil of each bract of the short, lateral, racemiform panicles, the calyx lobes 3 or 4, the stamens 3 to 6, the filaments distinct; pistillate flowers solitary, terminal, the sepals 5, the ovary 3-locular, the ovules solitary, the styles 3, short, verrucose, bifid; seeds medium-sized, not carunculate.

1. Bernardia incana Morton. Grand Canyon (Coconino County) and Mohave County, to Cochise, Pima, and Yuma counties, 1,500 to 5,000 feet, dry, rocky slopes, April to October, type from the Tucson Mountains (*Pringle* in 1884). Arizona, southeastern California, and Baja California.

Specimens with exceptionally large and relatively narrow leaves (up to 27 mm. long and ½ as wide) were collected in Havasu Canyon, Coconino County (*Clover* 7030).

6. ACALYPHA

Herbs or shrubs; leaves alternate, simple, petioled; stipules small; inflorescence of terminal or axillary spikes or spiciform racemes, entirely staminate, or staminate above and pistillate below, or wholly pistillate; staminate flowers several to numerous in the axil of each bract, pedicelled, the sepals 4, the stamens 6 to 8, the filaments distinct, the anther locules linear; pistillate flowers sessile, 1 or 2 in the axil of each bract, the sepals 3, the carpels 3 or rarely 2, the ovules solitary, the styles 3, distinct, dissected into filiform segments; capsule usually 3-celled; seeds small, ovoid, with small caruncles.

KEY TO THE SPECIES

1. Plants herbaceous, annual (2).
1. Plants shrubby or suffrutescent (4).
2. Pistillate bracts pectinately toothed; seeds 2 to 2.3 mm. long.........1. *A. ostryaefolia*
2. Pistillate bracts shallowly toothed; seeds 1.2 to 1.4 mm. long (3).
3. Pistillate bracts obscurely veined, evenly toothed all around..............2. *A. indica*
3. Pistillate bracts conspicuously veined, the middle tooth markedly prolonged 3. *A. neomexicana*
4. Leaf bases cordate to truncate; pistillate bracts with 9 to 17 small obtuse teeth; inflorescence heavily covered with stipitate glands.....................4. *A. Pringlei*
4. Leaf bases broadly cuneate; pistillate bracts with not more than 6 large acute teeth; inflorescence with a few glands on the pistillate bracts only........5. *A. Lindheimeri*

1. Acalypha ostryaefolia Riddell. Cochise, Santa Cruz, and Pima counties, 3,000 to about 5,500 feet, rich woods, not common, August and September. New Jersey to Florida, west to southern Arizona and Mexico.

2. Acalypha indica L. Paradise and Cave Creek, Chiricahua Mountains (Cochise County), about 5,000 feet. Introduced from the Old World tropics.

3. Acalypha neomexicana Muell.Arg. Greenlee County to Yavapai County, south to Cochise, Santa Cruz, and Pima counties, 2,500 to 7,500 feet, August to October. New Mexico, Arizona, and northern Mexico.

4. Acalypha Pringlei Wats. Quijotoa and Ajo mountains and Organ Pipe Cactus National Monument (western Pima County), 1,500 to 3,500 feet, locally abundant on rocky slopes, June to December. Southern Arizona and Sonora.

A small shrub up to about 1 m. high.

5. Acalypha Lindheimeri Muell.Arg. Mountains of Cochise, Santa Cruz, and Pima counties, 4,500 to 5,500 feet, June to October. Texas to southern Arizona.

Var. *major* Pax & Hoffmann represents a large-leaved, subglabrous variation that is too ill-defined for recognition, at least with the specimens available. The neat foliage and the spikes of crimson flowers make *A. Lindheimeri* an attractive plant.

7. TRAGIA. Nose-burn

Slender, often twining, perennial herbs clothed with stinging hairs, these sometimes sparse; leaves alternate, stipulate, petiolate, simple or compound, serrate; flowers monoecious, borne in bracteate racemes disposed terminally or laterally but not in the axils; staminate flowers above, 2 to many, the sepals and stamens 3 to 5; pistillate flowers below, 1 or 2, the sepals 6, the ovary normally 3-locular, the ovules solitary, the styles 3, more or less united below, entire, rugulose to strongly papillose; seeds small, spheroidal, not carunculate.

KEY TO THE SPECIES

1. Leaves, except the uppermost, compound, with 3 leaflets, laciniately toothed 1. *T. laciniata*
1. Leaves simple, toothed but more shortly so (2).
2. Styles 1.9 to 3.8 mm. long, nearly smooth; racemes with 2 to 4 staminate flowers; stamens mostly 4 or 5, sometimes only 3............................2. *T. stylaris*
2. Styles shorter, papillose; racemes with 6 to many staminate flowers; stamens 3 3. *T. nepetaefolia*

1. Tragia laciniata (Torr.) Muell.Arg. Bear Valley or Sycamore Canyon, Sonoita Creek, and Nogales (Santa Cruz County), 3,500 to 4,000 feet, August and September, type collected "on the Sonoita" (*Wright* 1795). Southern Arizona and Sonora.

2. Tragia stylaris Muell.Arg. Apache County to Mohave County and northern Gila County, also in the mountains of Cochise County, 5,000 to 7,000 feet, April to June. Colorado and Texas to California and northern Mexico.

3. Tragia nepetaefolia Cav. Apache County to Coconino County, south to Cochise, Santa Cruz, Pima, and Yuma counties, 2,500 to 7,000 feet, March to November. Texas to Arizona and Mexico.

8. RICINUS. Castor-bean

Glabrous herbs, shrubs, or trees; leaves long-petioled, mostly 7- to 10-lobed; stipules membranaceous, sheathing, caducous; inflorescence racemose or paniculate, the pistillate flowers above, the staminate below; staminate calyx mostly 5-lobed, the filaments united into many dendritic androphores; pistillate calyx 5-lobed, caducous, the ovary 3-locular, mostly echinate, the ovules solitary, the styles 3, bifid, papillose, red; capsule mostly echinate; seeds large, carunculate.

1. Ricinus communis L. Pretty well established in eastern Yavapai and eastern Pinal counties, also near Yuma, flowering throughout most of the year. Introduced from the Old World tropics.

A poisonous plant, well known as the source of castor oil, mainly used in making Turkey red oil and related substances for the dyeing industry and, after chemical alteration, as a drying oil in paints. The plant is sometimes cultivated as an ornamental and to provide shade for fowls.

9. JATROPHA

Perennial herbs, or shrubs; leaves simple, alternate; stipules small or none; flowers monoecious or dioecious, borne in terminal or lateral cymes, or the pistillate flowers sometimes solitary; calyx 5-lobed; petals 5, more or less connivent into a tube; stamens 8 to 10, the filaments united below into a column, the anthers in two whorls, the lower whorl of 5 and the upper one of 3 to 5 anthers; ovary 1- to 3-locular, the ovules solitary; styles 1 to 3, entire or shortly bifid; seeds large, carunculate.

KEY TO THE SPECIES

1. Leaves palmately lobed; perennial herbs............................1. *J. macrorhiza*
1. Leaves not palmately lobed except on seedlings or suckers; shrubs (2).
2. Leaves obovate-obcordate to spatulate, 5 to 11 mm. long, the petioles about 1 mm. long 2. *J. cuneata*
2. Leaves about as broad as long, widest near the base, mostly 2 cm. or more long at maturity, the petioles 1 cm. or more long (3).
3. Leaves glabrous, cordate-deltoid, the apex acuminate, the margin crenate 3. *J. cardiophylla*
3. Leaves canescent at least beneath, orbicular-reniform, the margin entire 4. *J. cinerea*

1. Jatropha macrorhiza Benth. McNary (Apache County), Cochise, Santa Cruz, and Pima counties, 3,500 to 7,500 feet, mesas and plains, May to October. Southern New Mexico, Arizona, and Mexico.

The species is represented in Arizona by var. *septemfida* Engelm. (*J. arizonica* Johnst.), type of *J. arizonica* from the Santa Rita Mountains (*Pringle* in 1882). The large, thick root is said to be strongly purgative.

2. Jatropha cuneata Wiggins & Rollins. Southwestern Pima and southern Yuma counties, 1,000 to 2,000 feet, dry mesas and slopes, July and August. Southwestern Arizona, Sonora, and Baja California.

A much-branched shrub up to 2 m. high. The related *J. spathulata* (Ortega) Muell.Arg., (sangre-de-drago) is reported to have been employed in Mexico medicinally and for the manufacture of various articles from the tough, flexible stems.

3. Jatropha cardiophylla (Torr.) Muell.Arg. Southwestern Maricopa County and from the Rincon Mountains to the Organ Pipe Cactus National Monument (Pima County), 2,000 to 3,000 feet, dry plains, mesas, and foothills, July and August. Southern Arizona and Sonora.

Sangre-de-Cristo, sangre-de-drago. A handsome shrub with shining, green, heart-shaped leaves. The roots contain both tannin and a red dye and were used by the natives of Arizona and Mexico for tanning hides. The clear sap coagulates quickly on contact with air and can be used for stanching the flow of blood from slight wounds. As much as 3 per cent of rubber has been extracted from the dry stems.

4. Jatropha cinerea (Ortega) Muell.Arg. (*J. canescens* (Benth.) Muell. Arg.). Organ Pipe Cactus National Monument, western Pima County, 1,000 to 1,500 feet, June to September. Southern Arizona to Sinaloa and Baja California.

10. CNIDOSCOLUS (165)

Perennial herbs; leaves long-petioled, the lobes attenuate and slenderly toothed; stipules thin, lacerate; cymes terminal on the stems and branches; staminate calyx white, petaloid, 5-lobed, the stamens 10, the staminodia 3, filiform, the filaments united into a column with a ring of hairs at base, the anthers in 2 whorls of 5, the glands united with the androphore just beneath the ring of hairs; pistillate calyx white, petaloid, 5-merous, the sepals distinct, caducous, the ovary 3-locular, the ovules solitary, the styles 3, connate below, twice bifid above; seeds large, carunculate.

1. Cnidoscolus angustidens Torr. Cochise, Santa Cruz, and Pima counties, 2,500 to 5,000 feet, rocky slopes, May to September. Southern Arizona, Sonora, and Baja California.

Mala-mujer. A handsome plant with transparent, stinging hairs from conspicuous, white, pustulate bases.

11. MANIHOT (166)

Leaves long-petioled, 5- to 7-lobed; stipules small, subulate; inflorescence racemose; staminate flowers several to many, at the distal end of the raceme, the calyx 5-lobed, slightly inflated in bud, the stamens 10, alternately long and short; pistillate flowers few to several at base of the raceme, the calyx caducous, the ovary 3-locular, the ovules solitary; seeds large, carunculate.

Both species are rather rare in Arizona and are interesting chiefly because of their kinship with *M. Glaziovii,* the tree that produces the ceara rubber of commerce, and with *M. esculenta,* from which cassava, tapioca, and other foods are derived. *M. carthaginensis* has been cultivated in Bahia, Brazil, for the starch obtained from the fleshy roots. The oil of the seeds is both emetic and cathartic, hence similar to castor oil.

KEY TO THE SPECIES

1. Primary lobes of the leaves broadly lobed toward the apex..............1. *M. Davisiae*
1. Primary lobes of the leaves lobed only at or below the middle, narrow and tapering to the apex..2. *M. angustiloba*

1. Manihot Davisiae Croizat. Santa Catalina and Baboquivari mountains (Pima County), 3,500 to 4,000 feet, August, type from the Santa Catalina Mountains (*Lemmon* in 1883). Known only from southern Arizona.

2. Manihot angustiloba (Torr.) Muell.Arg. Mountains of Cochise, Santa Cruz, and Pima counties, 3,000 to 5,000 feet, June to September. Southern Arizona to southern Mexico.

12. STILLINGIA

Leaves alternate to subopposite, simple, exstipulate; staminate flowers borne in terminal or axillary spikes, the calyx 2-lobed, the stamens 2; pistillate flowers 1 to 6 at base of the spikes, or axillary, the styles 3, long, entire; seeds ovoid or ovoid-ellipsoid, small, the caruncle minute or none.

KEY TO THE SPECIES

1. Plants annual, or perhaps sometimes perennial; leaves 3-nerved, mostly elliptic-cuneate, the apex attenuate, the margins with sharp teeth from base to apex....1. *S. spinulosa*
1. Plants perennial; leaves 1-nerved, linear or nearly so, the margins entire or with a few teeth near the base or the apex but not both (2).
2. Low, rounded, leafy bushes; inflorescence little overtopping the foliage; some of the leaves with a few slender teeth near the base; filaments exserted 1 to 1.7 mm. beyond the calyx at maturity..................................2. *S. paucidentata*
2. Strictly erect with few, sparsely leafy stems; inflorescence generally overtopping the foliage; leaves entire or rarely minutely serrulate near the apex; filaments exserted 0.4 to 0.6 mm. beyond the calyx at maturity.....................3. *S. linearifolia*

1. Stillingia spinulosa Torr. Southwestern Yuma County, 150 to 1,500 feet, sandy deserts, February to October. Southern Nevada to southwestern Arizona and southeastern California.

2. Stillingia paucidentata Wats. Known in Arizona only from the type collection, "Colorado Valley, near mouth of Williams River" (*Palmer* 517, in 1876). Western Arizona and southeastern California.

3. Stillingia linearifolia Wats. Near Yucca and Topock (Mohave County), western Pima (?) and southern Yuma counties, 500 to 2,000 feet, April and October. Western Arizona, southern California, and Baja California.

A collection in western Pima or southern Yuma County (*Goodding & Hardies* 9588) has much-branched stems 0.6 m. long, from a stout, woody root.

13. SAPIUM

Shrubs or small trees with milky sap; leaves alternate, coriaceous, serrulate; stipules triangular, oblique; staminate flowers borne in terminal spikes, the calyx 2-lobed, the stamens 2, the filaments shortly united; pistillate flowers 1 or 2 at base of the spikes, or axillary; seeds not carunculate, subspheroidal, large.

1. Sapium biloculare (Wats.) Pax. Near Gila Bend (southwestern Maricopa County), and from western Pima County to the Tinajas Altas Mountains (southern Yuma County), 1,000 to 2,000 feet, locally abundant along sandy washes and on rocky slopes, March to November. Southwestern Arizona, Sonora, and Baja California.

The plant attains a height of 4.5 m. (15 feet) in Arizona. The flowers are very fragrant. According to Pringle it was called yerba-de-fleche and was used by the Apaches to poison their arrows. The sap is reputedly very poisonous and causes sore eyes when introduced in the smoke of burning wood. The natives of Mexico are said to have used the juice to stupefy fish. The seeds often are infested with larvae of a small moth, which cause them to move about, roll over, or even jump a little, like the famous Mexican jumping beans obtained from *Sebastiana Pavoniana.*

14. EUPHORBIA (167, 168, 169). SPURGE

Annual or perennial herbs; leaves simple, alternate, opposite, or whorled; flowers monoecious, borne in cyathia simulating simple flowers; pistillate flower solitary in the center of the cyathium, pedicellate, naked, the ovary 3-locular, the styles 3, usually bifid; staminate flowers in 5 glomerules, 1 to several per glomerule, naked, consisting of a pedicel jointed to the filament, the glomerules opposite the lobes of the involucre; both kinds of flowers surrounded by a hypanthium or calyxlike involucre bearing on its margin 1 to 5 nectariferous glands of various shapes alternating with the lobes of the involucre, with petaloid appendages often extending from beneath the glands; fruit a 3-locular, 3-seeded, elastically dehiscent, usually nodding capsule.

The milky, acrid juice of some of the species causes dermatitis in susceptible persons and in horses. The fresh plants are rarely eaten by livestock, but when present in hay they are reported to be toxic to cattle. *E. hirta* is an official drug plant, used in treating asthma and bronchitis. Certain species, such as *E. albomarginata,* are known as rattlesnake-weed and, by the Mexicans, as golondrina. They are popularly supposed to be efficacious in treating snake bite, and the root of *E. albomarginata* is said to have been used as an emetic by the Pima Indians. Prostrate species are useful soil binders. The showy *E. marginata,* snow-on-the-mountain, is often cultivated as an ornamental, and its juice has been used in Texas in branding cattle. The cultivated poinsettia (*E. pulcherrima*), with bright-red floral bracts, is a favorite Christmas plant.

KEY TO THE SPECIES

1. Glands of the involucre without petaloid appendages, deeply cupped, or lunate if the leaves below the inflorescence are opposite; leaves essentially symmetric (2).
1. Glands of the involucre with petaloid appendages or if unappendaged, then the leaves all strictly opposite, but not decussate, and with inequilateral bases (17).
2. Involucral glands either deeply cupped, or concealed by the inflexed linear segments of the margin. Stem never branching into a symmetrical 3- to several-rayed inflorescence: subgenus *Poinsettia* (3).
2. Involucral glands flat or convex, never concealed. Leaves alternate below or if opposite, then decussate, in a single whorl beneath the pleiochasium (cyme resembling an umbel), opposite or whorled in the symmetrically forking inflorescence: subgenus *Esula* (6).
3. Stems few to several from a thickened perennial root; cyathia in compact terminal cymes subtended by colored floral leaves, the flowering stems otherwise bearing only reduced bracteal leaves, the leafy shoots appearing later; capsules 6 to 7 mm. long .... 1. *E. radians*
3. Stems solitary from slender annual roots or (in *E. eriantha*) the stems becoming thickened and woody and the plant apparently perennial; terminal cymes subtended by leaves usually undifferentiated in color from the ordinary foliage; shoots not dimorphic; capsules 5 mm. long or shorter (4).
4. Capsules scarcely lobed, plainly longer than wide; seeds markedly wider than thick; caruncle obvious and stipitate; glands concealed by 5 to 7 inflexed strigose segments .... 2. *E. eriantha*
4. Capsules strongly 3-lobed, plainly wider than long; seeds not markedly flattened dorsoventrally; caruncle minute and sessile, or wanting; glands naked (5).
5. Leaves mostly opposite throughout, mostly slightly to coarsely many-toothed; stems mostly strigose; seeds mostly 2.6 to 2.8 (rarely up to 3.1) mm. long.... 3. *E. dentata*
5. Leaves alternate between the first pair of secondary leaves and those at the stem tip, essentially entire or if serrulate, then also with 1 or 2 pairs of upwardly projecting lobes; stems not strigose; seeds mostly 3 to 3.4 (rarely only 2.7) mm. long 4. *E. heterophylla*
6. Involucral glands entire, elliptic; capsule verrucose to papillate; seeds reticulate, brown to nearly black; leaves serrulate (7).
6. Involucral glands either dentate or with horns and lunate; capsule smooth or slightly rugulose, glabrous or hairy, never verrucose or papillate; seeds not obviously reticulate but often mottled, rugulose, or pitted, usually grayish; leaves nearly always entire (8).
7. Plants mostly 10 to 35 cm. high; capsules verrucose; terminal pleiochasium in mature plants mostly ¼ to ⅓ of the total length of the plant............... 5. *E. spathulata*
7. Plants mostly 28 to 60 cm. high; capsules papillate, the papillae up to 0.5 mm. long; terminal pleiochasium $1/10$ to ⅕ of the total length of the mature plant.... 6. *E. alta*
8. Stem leaves decussate; stem commonly unbranched below the inflorescence; capsule 7 to 15 mm. long, spongy when fresh............................ 7. *E. Lathyris*
8. Stem leaves alternate; stems usually several in perennial species; capsule 5 mm. long or shorter, never spongy (9).
9. Stems robust, strictly erect, mostly 2 to 5 mm. thick at the base, mostly 30 to 55 cm. high, with densely leafy sterile branches on the upper half below the whorled leaves; stem leaves narrowly oblong to oblong-lanceolate, glabrous, glaucous, subsessile; capsule 4 to 5 mm. long; seeds truncate at base, often shallowly pitted 8. *E. Chamaesula*
9. Stems mostly either shorter or more slender, without sterile leafy branches; stem leaves various, often pubescent or papillate; capsule often shorter; seeds rounded at base (10).
10. Stem leaves either elliptic and tapering equally to the acute apex and the petioled base, or (at least the upper ones) widest well below the middle; leaf epidermis not papillate (11).

10. Stem leaves either orbicular to elliptic-oblong and obtuse, or obviously widest above the middle; leaf epidermis often papillate (14).
11. Stem leaves, at least the upper ones, widest near the base, subcordate, subsessile 14. *E. robusta*
11. Stem leaves elliptic and tapering equally to the acute apex and the petioled base. Stems mostly slender (about 1.5 mm. thick), mostly numerous and often sinuous (12).
12. Glands irregularly toothed all along the margin, without horns much exceeding the teeth ....11. *E. incisa*
12. Glands with sharp horns about ½ as long as the gland or longer, otherwise essentially entire (13).
13. Stems more than 20 cm. long; rays commonly repeatedly branched, often forming a dense inflorescence in age....................................9. *E. brachycera*
13. Stems only about 20 cm. long; rays commonly only once- or twice-forked 10. *E. odontadenia*
14. Lobes of the capsule crested. Seeds with a white outer coat, deeply pitted, the caruncle high, conical..15. *E. Peplus*
14. Lobes not crested (15).
15. Rays of the pleiochasium mostly 3; first pair of floral leaves often deltoid, usually longer than wide; leaf epidermis not papillate..................9. *E. brachycera*
15. Rays of the pleiochasium mostly 5; floral leaves all broadly rounded; leaf epidermis usually papillate (16).
16. Stem leaves mostly oblong to suborbicular; floral leaves wider than long 12. *E. Palmeri*
16. Stem leaves spatulate to oblanceolate; floral leaves about as wide as long 13. *E. lurida*
17. Leaves alternate, opposite, or even whorled, symmetric at base; stipules glandlike or none: subgenus *Agaloma* (18).
17. Leaves all strictly opposite, usually strongly inequilateral at base; stipules mostly well developed, always evident in species with symmetric leaves: subgenus *Chamaesyce* (21).
18. Stems branched above into a 3- (rarely 4-) rayed, essentially symmetric pleiochasium; floral leaves with broad, white margins..........................16. *E. marginata*
18. Stems often forking above, never branching into a cyme with more than 2 branches; floral leaves green throughout (19).
19. Plants perennial with a thickened root; seeds smooth, ovoid.........17. *E. Plummerae*
19. Plants annual with slender roots; seeds definitely angled (20).
20. Leaves all entire; appendages all symmetrically bifid to the base; seeds often dull black ....18. *E. bilobata*
20. Leaves, at least some of them, serrulate to serrate; appendages entire to irregularly toothed; seeds brown to sordid white.........................19. *E. exstipulata*
21. Ovary and capsule bearing hairs (22).
21. Ovary and capsule glabrous (33).
22. Plants perennial but flowering the first year; staminate flowers 16 to 60 (rarely as few as 15 in *E. melanadenia*); appendages never markedly unequal; involucres never urceolate (23).
22. Plants annual (except *E. arizonica* with an urceolate involucre); staminate flowers up to 12, or if as many as 15, then 1 pair of the appendages more than twice as long as the other (26).
23. Cyathia borne in dense glomerules, a few sometimes also solitary in the upper forks; leaves often serrate.......................................25. *E. capitellata*
23. Cyathia solitary in the forks and at the tips of the branches; leaves always entire (24).
24. Seeds scarcely angled, narrowly ovoid, encircled by 4 or 5 rounded ridges 34. *E. pediculifera*
24. Seeds quadrangular, at most slightly wrinkled (25).
25. Herbage with short, straight, spreading hairs; appendages as narrow as the (usually) pink or red glands: var. *hirtella*...............................32. *E. polycarpa*

25. Herbage with appressed, curly hairs; appendages usually much wider than the dark-purple glands..33. *E. melanadenia*
26. Involucres urceolate; appendages essentially equal, more than twice as wide as the glands. Herbage pilose, somewhat viscid (27).
26. Involucres obconic to campanulate; appendages less than twice as wide as the glands, or one pair more than twice as long as the other (28).
27. Appendages entire to crenulate; hairs mostly clavate; plants perennial..37. *E. arizonica*
27. Appendages deeply parted into a few attenuate segments; hairs tapering; plants annual ....38. *E. setiloba*
28. Cyathia borne in dense, axillary and terminal, leafless glomerules........26. *E. hirta*
28. Cyathia solitary or on short, leafy, lateral branchlets (29).
29. Appendages (the 2 proximal ones) greatly prolonged, often concealing the capsule 43. *E. indivisa*
29. Appendages (proximal and distal) without marked disparity in size (30).
30. Seeds punctately pitted and mottled, depressed-truncate at base, sharply acute at apex; styles entire, or sometimes emarginate; capsule more heavily pubescent at base than elsewhere..44. *E. stictospora*
30. Seeds not punctately pitted or mottled, obtuse at base, not sharply acute at apex; styles bifid; pubescence of the capsule either uniform, or more abundant on the angles (31).
31. Glands without appendages; seeds smooth; leaves entire; plants usually glabrous or if not, then the pubescence short, straight, and spreading..........42. *E. micromera*
31. Glands appendiculate; seeds variously ridged; leaves often serrulate; plants always somewhat covered with long and weak, crisped or appressed hairs (32).
32. Seeds with rounded transverse ridges; capsules evenly strigose...........45. *E. supina*
32. Seeds with sharp transverse ridges; capsules with crisped, spreading hairs confined mainly to the external angles...............................46. *E. Chamaesyce*
33. Leaves linear, symmetric; herbage glabrous; plants annual, mostly erect (34).
33. Leaves rarely linear, inequilateral; herbage sometimes pubescent; plants annual or perennial, often prostrate (37).
34. Capsules sharply 3-angled, 1.3 to 1.4 mm. long; staminate flowers 5 to 12 per cyathium; involucres 0.5 to 1 mm. in diameter; delicate plants with capillary ultimate branchlets (35).
34. Capsules roundly 3-lobed to bluntly 3-angled, 2 to 2.5 mm. long; staminate flowers numerous; involucres 1.4 to 2 mm. in diameter; plants coarser (36).
35. Seeds smooth; involucres 0.5 to 0.7 mm. in diameter; appendages usually longer than the glands; ultimate branchlets about 0.1 mm. in diameter; longest leaves usually shorter than 1 cm...35. *E. gracillima*
35. Seeds transversely wrinkled or ridged; involucre 0.9 to 1 mm. in diameter; appendages not longer than the glands; ultimate branchlets 0.15 to 0.25 mm. in diameter; longest leaves rarely as short as 1 cm...........................36. *E. revoluta*
36. Leaves entire, less than 3 cm. long; appendages narrow, never reddish; seeds smooth, ovoid-triangular ....22. *E. Parryi*
36. Leaves sharply but remotely serrulate, up to 6 cm. long; appendages often reddish in age, 1 to 2.8 mm. long; seeds with 2 (or 3) low transverse ridges, the dorsal and lateral angles slightly winged................................23. *E. florida*
37. Capsules more than 3 mm. long (38).
37. Capsules less than 3 mm. long (39).
38. Capsules about 4 mm. long; seeds nearly flat on the inner face, rounded on the outer; leaves entire, up to 1 cm. long.............................20. *E. platysperma*
38. Capsules 3.1 to 3.3 mm. long; seeds quadrangular; leaves serrulate at least at the apex, 1 to 4 cm. long.................................24. *E. trachysperma*
39. Stipules united into a glabrous, membranaceous scale; nodes often rooting; leaves entire; herbage glabrous; seeds smooth (40).
39. Stipules not united into a glabrous, membranaceous scale; nodes not rooting; leaves sometimes serrulate; herbage sometimes pubescent; seeds often wrinkled or ridged (41).

40. Plants perennial, common; staminate flowers 12 or more..........29. *E. albomarginata*
40. Plants annual, rare; staminate flowers 5 to 10........................30. *E. serpens*
41. Capsules at least 2 mm. long (42).
41. Capsules less than 2 mm. long (45).
42. Leaves (the longer ones) considerably more than 15 mm. long......27. *E. hyssopifolia*
42. Leaves not more than 15 mm. long, mostly shorter (43).
43. Plants pilose; leaves serrulate......................................47. *E. serrula*
43. Plants entirely glabrous; leaves entire (44).
44. Seeds ovoid or obscurely angled; glands circular or nearly so, the appendages absent: var. *arenicola*...21. *E. ocellata*
44. Seeds quadrangular; glands transversely oblong, the appendages present 31. *E. Fendleri*
45. Cyathia in dense terminal glomerules, a few also sometimes solitary in the upper axils. Plants perennial, but flowering the first year...............25. *E. capitellata*
45. Cyathia solitary in the axils and at the tips of the branches (46).
46. Seeds with definite, regular transverse ridges, these usually passing through the angles (47).
46. Seeds smooth to strongly wrinkled but never with definite transverse ridges (49).
47. Plants erect; leaves mostly 15 mm. or longer; seeds with shallow, concave depressions separated by low, narrow ridges..............................27. *E. hyssopifolia*
47. Plants mostly prostrate; leaves rarely up to 15 mm. long; seeds with rounded transverse ridges as wide as or wider than the valleys between (48).
48. Seeds radially ovate; capsule widest below the middle; herbage always glabrous 40. *E. glyptosperma*
48. Seeds radially oblong-ovate to oblong; capsule widest at the middle; herbage (at least the stems) often pubescent.................................41. *E. Abramsiana*
49. Leaves always entire; stems wingless (50).
49. Leaves mostly serrulate, if entire then the stems more or less winged (51).
50. Plants perennial but flowering the first year; glands transversely elongate, the appendages present but often narrow.................................32. *E. polycarpa*
50. Plants annual; glands circular or nearly so, the appendages absent...42. *E. micromera*
51. Herbage more or less pubescent; leaves often wider below the middle and tapering to the apex; seeds reticulately wrinkled to nearly smooth, the testa dark gray under the whitish outer coat................................28. *E. vermiculata*
51. Herbage wholly glabrous; leaves widest at or above the middle; seeds more or less wrinkled but never reticulately so, the testa tan to brown under the whitish outer coat ....39. *E. serpyllifolia*

1. **Euphorbia radians** Benth. (*Poinsettia radians* Klotzsch & Garcke). Cochise, Santa Cruz, and Pima counties, about 5,000 feet, desert grassland. Western Texas to southern Arizona and Mexico.

2. **Euphorbia eriantha** Benth. (*Poinsettia eriantha* Rose & Standl.). Graham, Maricopa, Pinal, Pima, and Yuma counties, 300 to 3,500 feet, dry, hot slopes and canyons, February to October. Texas to southeastern California and northern Mexico.

3. **Euphorbia dentata** Michx. (*Poinsettia dentata* Klotzsch & Garcke). Southern Apache County and Yavapai County, south to Cochise, Santa Cruz, and Pima counties, 3,000 to 8,000 feet, August to October. Eastern United States to Utah, Arizona, and Mexico.

Var. *cuphosperma* Engelm., which occurs in most of the range of the species in Arizona, differs from the species in having sharply quadrangular seeds, strigose capsules, and shallowly toothed to entire, often lanceolate to linear leaves, whereas the typical form has ovoid seeds, glabrous capsules, and generally coarsely toothed, mostly ovate-lanceolate to obovate-cuneate leaves.

Var. *gracillima* Millsp. is unworthy of recognition. It is intermediate between var. *cuphosperma* and the species. The type came from Bowie (*Jones* in 1884).

4. **Euphorbia heterophylla** L. (*Poinsettia heterophylla* Klotzsch & Garcke). Cochise, Santa Cruz, and Pima counties, 2,500 to 5,000 feet, August to October. Southeastern United States to southern Arizona, south to tropical America.

Painted spurge. The plants referred here differ in the angled seeds from the plants to which the name is usually applied. The Arizona plants may represent another, perhaps undescribed, species, but decision must await a careful revision of a group of plastic tropical species. The floral leaves are often partly colored pink or red. A race with narrowly linear, entire leaves (var. *graminifolia* (Michx.) Engelm.) occurs in the Baboquivari Mountains, Pima County (*Kearney & Peebles* 10413), near Ruby, Santa Cruz County (*Darrow* in 1939), and in western Texas.

5. **Euphorbia spathulata** Lam. (*E. dictyosperma* Fisch. & Meyer, *Tithymalus dictyospermus* Heller). White Mountains (Apache County) and Yavapai, Gila, Pinal, Cochise, and Pima counties, 1,500 to 7,500 feet, March to August, perhaps also on the North Rim of the Grand Canyon (*Eastwood & Howell* 962). Widely distributed in the United States and northern Mexico.

6. **Euphorbia alta** Norton (*Tithymalus altus* Woot. & Standl.). White Mountains (Apache County), Chiricahua and Huachuca mountains (Cochise County), Santa Rita, Rincon, and Santa Catalina mountains (Santa Cruz and Pima counties), up to 8,000 feet, June to October. New Mexico, Arizona, and Mexico.

7. **Euphorbia Lathyris** L. (*Tithymalus Lathyris* Hill). Near Sonoita, Santa Cruz County, "where introduced to repel rodents in an orchard" (*Goodding*), Tucson, Pima County (*Nicholson* in 1948). Here and there in the United States; adventive from Europe.

8. **Euphorbia Chamaesula** Boiss. (*Tithymalus Chamaesula* Woot. & Standl.). Apache, Navajo, and Coconino counties to Cochise and Pima counties, 5,500 to 8,500 feet, June to August. New Mexico, Arizona, and Mexico.

9. **Euphorbia brachycera** Engelm. (*Tithymalus brachycerus* Small). Cave Creek (Maricopa County), Chiricahua, Mule, and Huachuca mountains (Cochise County), Rincon Mountains (Pima County), 4,500 to 7,000 feet, May to September. Texas to southern Arizona and Chihuahua.

10. **Euphorbia odontadenia** Boiss. Segi Canyon, Navajo County (*Whitehead* in 1916). Otherwise known from the region of El Paso, Texas.

An uncertain entity closely allied to *E. incisa.*

11. **Euphorbia incisa** Engelm. (*E. schizoloba* Engelm., *Tithymalus schizolobus* Norton). Navajo County to Mohave County, south to Cochise and Pima counties, 3,000 to 9,000 feet, February to August, type from the Cerbat Mountains, Mohave County (*Newberry* in 1858). Nevada, Arizona, and California.

Var. *mollis* (Norton) L. C. Wheeler (*E. yaquiana* Tidestrom) occurs in Yavapai, Graham, Gila, Cochise, and Pima counties, type from the Santa Catalina Mountains (*Pringle* in 1881). It is distinguished from the species in being pubescent throughout and usually more robust, intergrading with *E. Palmeri* var. *subpubens* and with *E. robusta.*

12. Euphorbia Palmeri Engelm. (*Tithymalus Palmeri* Abrams). Williams, Flagstaff, and San Francisco Peaks (Coconino County), Sierra Ancha and Mazatzal Mountains (Gila County), 5,000 to 8,000 feet, May and June. Utah and Arizona to California.

Var. *subpubens* (Engelm.) L. C. Wheeler has a wider range in Arizona than the typical form, occurring in the mountains of Coconino, Yavapai, Gila, and Pima counties, type from Prescott (*Palmer* 112, in 1876). It differs from the species mainly in being pubescent. The specimens from Gila and Pima counties are not typical.

13. Euphorbia lurida Engelm. (*E. lurida* var. *Pringlei* Norton, *Tithymalus luridus* Woot. & Standl.). Apache, Coconino, Yavapai, Greenlee, Graham, Cochise, and Pima counties, 3,500 to 7,500 feet, April to August, type of *E. lurida* from the San Francisco Peaks (*Newberry*), type of var. *Pringlei* from the Santa Rita Mountains (*Pringle* in 1881). Southern Utah, New Mexico, and Arizona.

14. Euphorbia robusta (Engelm.) Small (*Tithymalus robustus* Small). Camp Grant (Graham County), Mazatzal Mountains (Gila County), Organ Pipe Cactus National Monument (Pima County), apparently also north side of Grand Canyon (Coconino County), 2,500 to 6,000 feet. South Dakota, Montana, and Utah to New Mexico and Arizona.

A copiously soft-pubescent form was collected at Camp Grant (*Goodding* 1042).

15. Euphorbia Peplus L. (*Tithymalus Peplus* Gaertn.). Campus of the University of Arizona, Tucson, Pima County (*Thornber* 8081, in 1915, *Streets* in 1945). A weed of waste ground, here and there in the United States; introduced from Europe.

16. Euphorbia marginata Pursh (*Dichrophyllum marginatum* Klotzsch & Garcke). Gila Valley (Graham County?), Fort Huachuca (Cochise County). Montana south to southern Mexico, widely introduced east of the Mississippi River.

Snow-on-the-mountain. Three different collectors found this in Arizona before 1900, but no recent collections seem to have been made.

17. Euphorbia Plummerae Wats. Known in Arizona only from the type collection from Tanner Canyon, Huachuca Mountains, Cochise County (*Lemmon* 2874). Southeastern Arizona, Chihuahua, and Sonora.

18. Euphorbia bilobata Engelm. (*Zygophyllidium bilobatum* Standl.). Yavapai, Greenlee, Gila, Pinal, Cochise, and Santa Cruz counties, 3,500 to 6,000 feet, August to October. Western Texas to Arizona.

19. Euphorbia exstipulata Engelm. (*Zygophyllidium exstipulatum* Woot. & Standl.). Coconino, Mohave, Yavapai, Cochise, and Pima counties, 4,000 to 6,000 feet, September. New Mexico, Arizona, and northern Mexico.

***20. Euphorbia platysperma** Engelm. (*E. eremica* Jepson). An extremely rare species known only from the type, collected by Edward Palmer in 1869 "near the mouth of the Colorado River, Arizona," and from two collections in the Colorado Desert, California. The plant may not occur in Arizona, but should be sought in the Yuma region.

21. Euphorbia ocellata Dur. & Hilg. (*Chamaesyce ocellata* Millsp.). Virgin River, Mohave County, about 1,500 feet (*Purpus* 6187). Southern Idaho to southern California.

The Arizona plant is var. *arenicola* (Parish) Jepson, which has the median leaves ovate-lanceolate and acuminate, instead of ovate-deltoid-falcate and blunt or mucronulate.

22. Euphorbia Parryi Engelm. (*E. flagelliformis* Engelm., *Chamaesyce Parryi* Rydb.). Apache County to northern Mohave County, south to Graham and Cochise counties, 1,500 to 5,500 feet, sandy soil, May to September. Colorado and western Texas to California and Chihuahua.

23. Euphorbia florida Engelm. (*Chamaesyce florida* Millsp.). Yavapai Graham, Gila, Pinal, Cochise, Santa Cruz, and Pima counties, 2,000 to 5,000 feet, August to October, type from west of the Chiricahua Mountains, Cochise County (*Wright* in 1851). Arizona and northwestern Mexico.

24. Euphorbia trachysperma Engelm. (*Chamaesyce trachysperma* Millsp.). Maricopa (Pinal County), valley of the San Pedro River (Cochise County), Tucson (Pima County), near La Paz (Yuma County), very rare, 300 to 4,000 feet, type from along the San Pedro River (*Wright* 1832). Southern Arizona and Sonora.

A specimen collected at St. David, Cochise County (*Gould & Haskell* 4492), has the involucral glands unappendaged.

25. Euphorbia capitellata Engelm. (*E. pycnanthema* Engelm., *E. Rusbyi* Greene, *Chamaesyce capitellata* Millsp.). Throughout the state except the lower western desert and northeastern portion, 1,500 to 5,000 feet, March to October, type of *E. capitellata* from San Bernardino, Cochise County (*Wright* in 1851), type of *E. Rusbyi* from northern Arizona (*Rusby* in 1883). Western Texas to Arizona and northern Mexico.

26. Euphorbia hirta L. (*Chamaesyce hirta* Millsp.). Cochise, Santa Cruz, and Pima counties, 3,000 to 5,000 feet, August to October. Florida, Texas, and southern Arizona, south to Argentina.

27. Euphorbia hyssopifolia L. (*E. Jonesii* Millsp., *Chamaesyce hyssopifolia* Small). Yavapai, Graham, Gila, Pinal, and Maricopa counties, where infrequent, Santa Cruz, Cochise, and eastern Pima counties, where common, 1,000 to 6,000 feet, August to November, type of *E. Jonesii* from Bowie, Cochise County (*Jones* 4247). Southern Florida; western Texas to Arizona, south to South America.

28. Euphorbia vermiculata Raf. (*Chamaesyce vermiculata* House). Mountains of Apache, Navajo, Gila, and Cochise counties, about 6,000 feet, May to October. Nova Scotia to Pennsylvania and west to Michigan; probably introduced in British Columbia; New Mexico and Arizona.

Chamaesyce Rothrockii Millsp., type from Crittenden, Santa Cruz County (*Rothrock* 672), is intermediate between *E. vermiculata* and *E. maculata* L., as are some of the collections included above. No true *E. maculata* has been seen from Arizona.

29. Euphorbia albomarginata Torr. & Gray (*Chamaesyce albomarginata*

Small). Throughout the state except the extreme northern, northeastern, and southwestern portions, 1,000 to 6,000 feet, often common on clay and loam flats, February to October. Oklahoma to California and northern Mexico.

30. Euphorbia serpens H.B.K. (*Chamaesyce serpens* Small). Santa Cruz River at La Noria, Santa Cruz County, about 5,000 feet (*Mearns* 1192). Ontario and Montana, south to South America.

31. Euphorbia Fendleri Torr. & Gray (*Chamaesyce Fendleri* Small). Apache County to Mohave and Yavapai counties, also at Dragoon (Cochise County), 4,000 to 7,000 feet, May to September. Nebraska to Texas, west to California.

Var. *chaetocalyx* Boiss. (*Chamaesyce chaetocalyx* Woot. & Standl.), characterized by narrower, acute leaves and obtuse and crenate petaloid appendages, whereas in typical *E. Fendleri* the leaves are about as wide as long and obtuse and the appendages are narrowly deltoid and entire, occurs in most of the range of the species in Arizona, intergrading with the typical form in Navajo and Coconino counties.

32. Euphorbia polycarpa Benth. (*Chamaesyce polycarpa* Millsp.). Mohave, southern Yavapai, Maricopa, Pinal, Pima, and Yuma counties, 500 to 3,000 feet, sandy or gravelly plains and mesas, flowering throughout the year. Nevada to Sonora, California, and Baja California.

A pubescent variant, var. *hirtella* Boiss. (*Chamaesyce tonsita* Millsp.), occurs in Maricopa, Mohave, and Yuma counties, intergrading with the species, which is glabrous.

33. Euphorbia melanadenia Torr. (*Chamaesyce melanadenia* Millsp.). In the entire southwestern part of the state, eastward to Gila and Pima counties, and in the White Mountain region (Apache County), 500 to 5,000 feet, common on dry, sunny hillsides, often among shrubs, flowering throughout the year. Arizona, southern California, Sonora, and Baja California.

The plants are generally erect or ascending, but on open flats, particularly in disturbed soil, they may be nearly prostrate.

34. Euphorbia pediculifera Engelm. (*E. vermiformis* M. E. Jones, *Chamaesyce pediculifera* Rose & Standl.). Southern Yavapai, Graham, Maricopa, Pinal, Cochise, Santa Cruz, Pima, and Yuma counties, 500 to 4,000 feet, flowering throughout the year, type of *E. vermiformis* from Ajo, Pima County (*Jones* 24856). Arizona, southeastern California, and northwestern Mexico.

Specimens from the deserts at low altitude often simulate *Euphorbia melanadenia* in vestiture and leaf size, but the ovoid-oblong seeds, encircled by 4 or 5 ridges, of *E. pediculifera* are readily distinguishable from the quadrangular and smooth, or nearly smooth, seeds of *E. melanadenia.*

35. Euphorbia gracillima Wats. (*Chamaesyce gracillima* Millsp.). Tucson Mountains and Papago Indian Reservation (Pima County), 2,000 to 2,500 feet, August to October. Southern Arizona and northwestern Mexico.

36. Euphorbia revoluta Engelm. (*Chamaesyce revoluta* Small). Coconino and Mohave counties to Graham, Cochise, Santa Cruz, and Pima counties, 3,000 to 6,000 feet, August to October. Colorado to Arizona and Chihuahua.

37. Euphorbia arizonica Engelm. (*E. versicolor* Greene, *Chamaesyce versicolor* Norton, *C. arizonica* Arthur). Grand Canyon (Coconino County), Yucca

(Mohave County), southward and southeastward to Greenlee, Graham, Gila, Maricopa, Santa Cruz, Pima, and Yuma counties, 1,000 to 4,000 feet, February to October, type of *E. arizonica* from Sonora (*Schott* in 1856), type of *E. versicolor* from "San Francisco Mountains," Greenlee (?) County (*Greene* in 1880). Texas to southeastern California and northern Mexico.

38. Euphorbia setiloba Engelm. (*Chamaesyce setiloba* Millsp.). Greenlee, Graham, and Cochise counties to Mohave and Yuma counties, 200 to 5,000 feet, March to October, type from near Fort Yuma. Western Texas to California and northwestern Mexico.

39. Euphorbia serpyllifolia Pers. (*Chamaesyce serpyllifolia* Small). Apache County to Mohave County, south to Cochise, Santa Cruz, and Pima counties, 3,000 to 7,000 feet, May to October. Widely distributed in western North America from Canada to Mexico.

40. Euphorbia glyptosperma Engelm. (*Chamaesyce glyptosperma* Small). Apache County to Coconino County, 5,000 to 7,500 feet, June to September. New Brunswick to British Columbia, south to Texas and northeastern Arizona.

41. Euphorbia Abramsiana L. C. Wheeler. Yavapai, Maricopa, Pinal, Cochise, Pima, and Yuma counties, 150 to 3,000 feet, August to November. Arizona, southeastern California, and northwestern Mexico.

42. Euphorbia micromera Boiss. (*E. pseudoserpyllifolia* Millsp., *Chamaesyce micromera* Woot. & Standl.). Navajo and Coconino counties to Cochise, Pima, and Yuma counties, 500 to 5,000 feet, August to October, type from near the San Pedro River, Cochise County (*Wright* in 1851), type of *E. pseudoserpyllifolia* from the Gila River (*P. Mohr* in 1873). Western Texas to Utah, southeastern California, and northern Mexico.

43. Euphorbia indivisa (Engelm.) Tidestrom (*Chamaesyce indivisa* Millsp.). Greenlee County to Yavapai County, south to Cochise, Santa Cruz, and Pima counties, 2,500 to 5,000 feet, August to October. Western Texas to Arizona and northern Mexico.

44. Euphorbia stictospora Engelm. (*Chamaesyce stictospora* Small). Mountains of Greenlee, Cochise, Santa Cruz, and Pima counties, 3,500 to 5,500 feet, August to October. South Dakota and Wyoming to Arizona, Durango, and Zacatecas.

45. Euphorbia supina Raf. (*E. maculata* of authors, non L.). Near Metcalf (Greenlee County), Devil's Canyon (Gila County), Litchfield (Maricopa County), Douglas (Cochise County), Tucson (Pima County), 1,000 to 5,500 feet, July to September. Eastern United States to North Dakota and Texas, introduced here and there farther west.

46. Euphorbia Chamaesyce L. (*E. prostrata* Ait., *Chamaesyce prostrata* Small). Pinal, Cochise, Santa Cruz, and Pima counties, 1,500 to 4,500 feet, June to October. South Carolina and Florida to southern Arizona, south into tropical America.

47. Euphorbia serrula Engelm. (*Chamaesyce serrula* Woot. & Standl.). Peach Springs (Mohave County), near Duncan (Greenlee County), Tucson (Pima County), and at several localities in Cochise County, 2,500 to 5,000 feet, September. Western Texas to Arizona and northern Mexico.

75. CALLITRICHACEAE. Water-starwort Family

1. CALLITRICHE. Water-starwort

Aquatic plants with slender stems and small, opposite, entire leaves, these often crowded at the ends of the stems; flowers minute, axillary, perfect or unisexual, without a perianth, subtended by 2 saccate bracts; stamen and pistil 1; stigmas 2, filiform, caducous; fruit at maturity separating into 4 nutlets, these 1-seeded.

1. Callitriche palustris L. Chuska Mountains (Apache County), Kaibab Plateau and Grand Canyon (Coconino County), Young to Payson (Gila County), Tucson (Pima County), in streams and ponds. Widely distributed in the Northern Hemisphere.

Further study may show that more than one species of *Callitriche* exists in Arizona. A collection in Sabino Canyon, Pima County (*Eastwood* in 1930), was identified tentatively by J. T. Howell as *C. Bolanderi* Hegelm., which may not be sufficiently distinct from *C. heterophylla* Pursh. In the Eastwood collection the fruits are nearly orbicular, whereas in *C. palustris* they are longer than wide.

76. BUXACEAE. Box Family

This small family is typified by box (*Buxus sempervirens*), a well-known ornamental shrub.

1. SIMMONDSIA. Jojoba, Deer-nut

Evergreen, dichotomously branched shrubs; leaves opposite, simple, entire, thick, leathery; flowers unisexual (the plants perhaps always dioecious), apetalous, in dense axillary clusters, small, yellowish; sepals 4 to 6; stamens 10 to 12; styles 3; ovary superior, 3-celled; fruit an acornlike capsule, with usually only 1 large seed.

1. Simmondsia chinensis (Link) Schneid. (*S. californica* Nutt.). Greenlee County to southern Yavapai County, south to Cochise, Pima, and Yuma counties, 1,000 to 5,000 feet, common, often abundant, dry slopes and along washes, December to July. Southern Arizona, southern California, Sonora, and Baja California.

Also known as goat-nut, wild-hazel, coffee-bush, and quinine-plant. It is a rather handsome shrub, seldom more than 2 m. (6.5 feet) high in Arizona, and is the best browse plant within its range, both for domestic and for many wild animals. The "nuts" are rich in an edible oil (chemically, a liquid wax). This reputedly has medicinal virtues and is used in small quantities in the manufacture of hair oil. It has been discovered recently that this substance can be substituted for beeswax in the manufacture of electrical insulation, phonograph records, and varnishes. In early days the Indians and white settlers made a substitute for coffee from the fruits. The "nuts" may be eaten raw or parched but are too bitter with tannin in their natural state to please the ordinary palate.

77. ANACARDIACEAE. Cashew Family

Rhus is much the largest genus of this family, which includes also the tropical tree (*Anacardium occidentale*) that yields the cashew nuts of commerce.

1. RHUS (170). Sumac

Shrubs, the sap usually acrid and resinous, sometimes poisonous; leaves alternate, either simple and entire or compound; flowers regular, perfect or unisexual, mostly 5-merous, small, greenish, yellowish, or whitish, in axillary or terminal panicles, with a ring-shaped or cup-shaped disk around the ovary; stamens 5; styles 3; ovary superior, 1-celled; fruit a small, 1-seeded drupe.

All the Arizona species are ornamental. The fruits, except in poison-ivy (*R. radicans*), are thin-fleshed, sweet, pleasantly acid, and can be used to make a refreshing beverage. They are an important food of birds and other wild animals. From the latex of an Asiatic species, *R. verniciflua*, the fine lacquer of China and Japan is manufactured.

KEY TO THE SPECIES

1. Fruits yellowish white when mature, shiny, glabrous or nearly so, never glandular-pubescent. Plants finely pubescent or glabrate; stems erect, ascending, or climbing by aerial rootlets; leaves deciduous, trifoliolate, middle leaflet long-stalked; leaflets up to 10 cm. long, oblong-lanceolate to ovate, usually coarsely few-toothed; inflorescences loose, paniculate, appearing after the leaves; petals greenish white. 1. *R. radicans*

1. Fruits red when mature, glandular-pubescent (2).

2. Leaves simple, evergreen, leathery, entire or very nearly so (3).

2. Leaves compound, sometimes reduced to a single leaflet in *R. trilobata* (4).

3. Leaves broadly ovate, acute or short-acuminate, often conduplicate, bright green above; petioles usually more than 1 cm. long.................................2. *R. ovata*

3. Leaves broadly oblong or oval, obtuse or acutish, flat, dark green above with conspicuous whitish veins; petioles usually less than 1 cm. long............3. *R. Kearneyi*

4. Leaflets not more than 3, coarsely crenate, the middle one sessile or nearly so, often cleft; leaves deciduous, not leathery. Inflorescences dense, spikelike, appearing before the leaves (except in one variety); petals yellow.......................4. *R. trilobata*

4. Leaflets more than 3, or else the leaves evergreen and leathery (5).

5. Flowers very numerous, in naked terminal inflorescences; leaflets seldom fewer than 11, whitish beneath, conspicuously serrate. Stems (below the inflorescence) and leaves glabrous or glabrescent; leaves deciduous, the leaflets oblong-lanceolate, 4 to 10 cm. long ....5. *R. glabra*

5. Flowers not very numerous, often in axillary as well as terminal inflorescences; leaflets 9 or fewer, paler but not whitish beneath, entire (6).

6. Leaves evergreen, the rachis not winged; leaflets 3 to 5, leathery, somewhat shiny above, glabrate, petiolulate, 2 to 6 cm. long, 1 to 3 cm. wide, acute or short-acuminate at apex; flowers appearing after the leaves.......................6. *R. choriophylla*

6. Leaves deciduous, the rachis winged; leaflets 5 to 9, not leathery or shiny, pilose, sessile, less than 2 cm. long, 2 to 6 mm. wide, rounded to acutish at apex; flowers appearing before the leaves..7. *R. microphylla*

1. Rhus radicans L. (*Toxicodendron radicans* Kuntze). Apache County to Coconino County, south to Cochise, Santa Cruz, and Pima counties, 3,000 to 8,000 feet, common in rich soil in ravines and canyons, April to September. Throughout most of North America.

Poison-ivy, poison-oak. A variable, often climbing plant, containing a nearly nonvolatile oil, urushiol, that causes painful swelling and eruption of the skin with many persons. The milky juice is poisonous when taken internally.

The Arizona plants are referrable to var. *Rydbergii* (Small) Rehder, which differs from typical *R. radicans* of eastern North America "in being always an upright shrub, in the sinuate-dentate thickish leaflets, glabrous or pubescent on the veins beneath, and in its larger flowers and fruits" (Alfred Rehder, *Jour. Arnold Arbor.* 20: 416). E. L. Greene (*Leaflets* 1: 123–124) published *Toxicodendron arizonicum, T. laetevirens,* and *T. pumilum,* based on Arizona types, but these are probably only ecotypes or individual variants. A collection from near Bisbee (*Goodding* 46), with seeds constricted at the side, may represent *T. radicans* var. *divaricatum* (Greene) Barkley (*T. divaricatum* Greene), known otherwise only from Sonora and Baja California.

2. Rhus ovata Wats. Southern Coconino County and Hualpai Mountain (Mohave County) to western Graham, Gila, and eastern Maricopa counties, 3,000 to 5,000 feet, slopes and mesas, common in chaparral, March and April (sometimes autumn). Central Arizona, southern California, and Baja California.

Sugar sumac, sugar-bush, sometimes called mountain-laurel in Arizona. A handsome shrub, cultivated as an ornamental in California, with bright-green, somewhat shiny, leathery, entire leaves, deep-red flower buds, and cream-colored flowers. The shrubs reach a height of at least 4.5 m. (15 feet). The old bark is very shaggy.

3. Rhus Kearneyi Barkley. Tinajas Altas, southern Yuma County (*Goldman* 2311, *Kearney & Harrison* 6573, the type collection) 1,000 to 1,500 feet, dry cliffs. Known only from this locality.

Reported by L. N. Goodding to reach a height of 5.5 m. (18 feet).

4. Rhus trilobata Nutt. Throughout the state, 2,500 to 7,500 feet, very common on slopes and in canyons, often in chaparral, March to June (August). Saskatchewan to Washington, south to Mexico.

Skunk-bush, squaw-bush. A polymorphic species with aromatic foliage, yellow flowers, and bright-red fruits. The plants are browsed. The Indians used the pliable stems in basketry and ate the berries, also using them as a mordant in dyes.

KEY TO THE VARIETIES

1. Terminal leaflet considerably longer than wide; fruit about 6 mm. in diameter or smaller ..... *R. trilobata* (typical)
1. Terminal leaflet nearly as wide as long; fruit usually more than 7 mm. in diameter (2).
2. Pedicels usually more than 5 mm. long; plants flowering in midsummer, after maturity of the leaves ..... var. *racemulosa*
2. Pedicels usually less than 4 mm. long; plants flowering in spring, with or before the unfolding of the leaves (3).
3. Young branches densely soft-pilose; leaves tomentose, at least beneath ... var. *pilosissima*
3. Young branches and leaves puberulent or sparsely pubescent, to glabrescent (4).
4. Leaves unifoliolate or the lateral leaflets very small, the single or terminal leaflet often shallowly lobed ..... var. *simplicifolia*
4. Leaves trifoliolate, the lateral leaflets not greatly reduced (5).
5. Leaflets commonly glabrescent, the terminal one not deeply lobed ..... var. *anisophylla*
5. Leaflets somewhat pubescent, the terminal one deeply lobed ..... var. *quinata*

The typical phase of the species occurs in Apache, Coconino, and Gila counties but is not common in Arizona. The distribution of the varieties is as follows: Var. *racemulosa* (Greene) Barkley, remarkable for its summer-flowering habit, occurs in the mountains of Greenlee, Cochise, and Pima counties, 5,000 to 6,000 feet, type from Fort Huachuca (*Wilcox* 378). Var. *pilosissima* Engler (*Schmaltzia Emoryi* Greene), a very hairy plant, is found in the central and southern counties as far north as Sedona (Coconino County). Var. *simplicifolia* (Greene) Barkley ranges from Apache County to Mohave County and is especially frequent in the Grand Canyon, where the type of *Schmaltzia simplicifolia* Greene was collected (*Greene* in 1889). Var. *anisophylla* (Greene) Jepson occurs throughout most of the range of the species in Arizona. Var. *quinata* Jepson has been found at the Grand Canyon, on the Mogollon Escarpment, and near Prescott. The last 2 varieties are not easily distinguishable. Greene (*Leaflets* 1: 136–138) published also, based on Arizona types, *Schmaltzia botryoides, S. cissodes, S. elegantula, S. hirtella, S. puncticulata,* and *S. trinervata,* all of which were referred by Barkley to *Rhus trilobata* vars. *anisophylla* and *simplicifolia.*

5. **Rhus glabra** L. (*R. cismontana* Greene). Apache County to Coconino County, south to Cochise and Pima counties, 5,000 to 7,000 feet, common in rich soil, often forming thickets, ascending to the pine belt, June to August. Canada to Florida, New Mexico, Arizona, and Chihuahua.

Scarlet sumac. In the fall the leaves turn bright red, giving a brilliant touch of color to the forest. The acidulous fruits have astringent and refrigerant properties. Greene published also *R. albida, R. elegantula,* and *R. calophylla,* based upon Arizona types, but these appear to be mere forms of *R. glabra.*

6. **Rhus choriophylla** Woot. & Standl. Cochise, Santa Cruz, and Pima counties, 4,000 to 6,000 feet, rocky slopes (often on limestone), sometimes growing with *Cupressus arizonica* and *Juniperus Deppeana,* July to September. Southern New Mexico, southeastern Arizona, and Sonora.

Very close to *R. virens* Lindh., differing chiefly in the usually larger, glabrate leaflets and in the more frequent occurrence of axillary inflorescences. This shrub attains a height of about 2 m. (7 feet).

7. **Rhus microphylla** Engelm. Greenlee, Cochise, and eastern Pima counties, 3,500 to 6,000 feet, dry mesas and slopes, March to May. Western Texas to southeastern Arizona and northern Mexico.

A much-branched shrub, up to 1.8 m. (6 feet) high.

A specimen without flowers or fruit, labeled as collected at Metcalf, Greenlee County (*Davidson* 626), appears to be *R. lanceolata* (Gray) Britton, although the leaflets are scarcely falcate. It differs from all the foregoing species in its numerous long, narrow, entire, thickish leaflets, these up to 7 cm. long and barely 1 cm. wide. The known range of *R. lanceolata* is Oklahoma to southeastern New Mexico, south into eastern Mexico.

78. CELASTRACEAE. Bitter-sweet Family

Plants of diverse habit, more or less woody; leaves simple, alternate or opposite, sometimes reduced to scales; flowers small, regular, usually perfect; calyx lobes and petals 4 to 6; stamens as many or twice as many as the petals, inserted on or below the margin of a basal disk, this rudimentary or wanting in *Canotia;* style 1 or the stigma sessile; fruit capsular or follicular.

KEY TO THE GENERA

1. Leaves reduced to small deciduous scales; seeds with a thin terminal wing. Plants large shrubs or small trees .... 4. *Canotia*
1. Leaves with well-developed blades; seeds without a terminal wing (2).
2. Plants creeping undershrubs; leaves mostly opposite; flowers reddish brown 1. *Pachystima*
2. Plants intricately branched shrubs; leaves alternate; petals whitish (3).
3. Herbage scabrous, yellowish; leaves persistent, very thick, with cartilaginous margins, elliptic to nearly orbicular; flowers in panicles, these mostly terminal; stamens commonly 5; fruit indehiscent, symmetric, short-cylindric, rounded and abruptly apiculate at apex .... 2. *Mortonia*
3. Herbage not scabrous, glabrous or puberulent, pale green, somewhat glaucous; leaves deciduous, not very thick, elliptic or oblanceolate; flowers axillary, scattered along the branchlets, often solitary; stamens commonly 10; fruit dehiscent on one suture only (follicular), asymmetric, ovoid, pointed at apex .... 3. *Glossopetalon*

1. PACHYSTIMA. Box-leaf

Undershrubs with creeping stems, glabrous or nearly so; leaves mostly opposite, thickish, evergreen, somewhat shiny above, the margins serrate or serrulate; flowers minute, 4-merous, axillary, solitary or in very few-flowered clusters; fruit a 2-celled, finally dehiscent capsule.

The name of the genus was published as *Paxistima.*

1. Pachystima Myrsinites (Pursh) Raf. Apache County to Coconino County, south to Cochise and Gila counties, 6,000 to 9,000 feet, coniferous forests, May to July. Canada to New Mexico, Arizona, and California.

Myrtle box-leaf, mountain-lover, Oregon-boxwood. A low and inconspicuous ground cover in the mountains. Relished by deer but not much eaten by domestic livestock.

2. MORTONIA

Plants yellowish, scurfy-scabrous shrubs up to 1.8 m. (6 feet) high, with many stiff, nearly erect, very leafy branches; leaves alternate, small, leathery, commonly elliptic; flowers small, 5-merous, whitish, in narrow panicles; style short, the stigma 5-lobed; ovary 5-celled.

1. Mortonia scabrella Gray. Southern Gila, Cochise, and Pima counties, 3,000 to 5,500 feet, dry plains, mesas, and slopes, often on limestone, March to September, co-type from along the San Pedro (*Wright*). Western Texas to southern Arizona and northern Mexico.

Var. *utahensis* Coville (*M. utahensis* A. Nels.) has been collected in Havasu Canyon (Coconino County) and in northern Mohave County. It differs from typical *M. scabrella* in little but the larger leaves (up to 20 mm. long), but has a different distribution—southwestern Utah to southeastern California.

3. GLOSSOPETALON (171, 172). Grease-bush

Spinescent, often intricately branched, glabrous or puberulent shrubs, the young bark green; leaves alternate, small, simple, entire, short-petioled; flowers axillary, solitary or in small clusters, the petals 4 to 6, whitish, inserted under the edge of a fleshy disk; stamens 6 to 10; stigma sessile; ovary superior, 1-celled; fruit an asymmetric follicle.

The plants are browsed by sheep and deer.

KEY TO THE SPECIES

1. Stipules present, frequently adnate to the persistent, glandular-thickened, often dark-colored base of the petiole; stems 1 to 2 m. high . . . . 1. *G. nevadense*
1. Stipules absent; stems less than 1 m. high . . . . 2. *G. spinescens*

1. Glossopetalon nevadense Gray (*Forsellesia nevadensis* Greene). Coconino County and eastern Mohave County, especially in and near the Grand Canyon, 3,500 to 6,500 feet, dry, rocky slopes, often on limestone, March to June. Idaho, south to northern Arizona and eastern California.

The Arizona plants are usually glabrous or nearly so, but the pubescent (typical) phase also occurs.

2. Glossopetalon spinescens Gray (*Forsellesia spinescens* Greene). Chiricahua, Dos Cabezas, and Mustang mountains (Cochise County and probably Santa Cruz County), 4,500 to 5,500 feet, limestone cliffs. Western Texas to southeastern Arizona and northern Mexico.

4. CANOTIA

Large shrubs or small trees, with numerous rigid, mostly spine-tipped, suberect, rushlike branches and yellowish-green bark; leaves reduced to small deciduous scales; flowers in small axillary clusters, inconspicuous, without a disk, or this rudimentary; ovary superior; fruit a somewhat woody 5-valved capsule, each valve splitting above into 2 long, slender points.

An anomalous plant in this family.

1. Canotia holacantha Torr. Havasu Canyon (western Coconino County) and Mohave County to Greenlee, Cochise, Pima, and Yuma counties, 2,000 to 4,500 feet, dry slopes and mesas, abundant in places, May to August, type from Williams River (*Bigelow*). Southern Utah, Arizona, southeastern California, and northwestern Mexico.

Sometimes known locally as palo-verde, a name properly belonging to *Parkinsonia* and *Cercidium*. This striking and peculiar plant is a characteristic feature of the landscape in large areas of southern and western Arizona. The trees attain a height of at least 4.5 m. (15 feet). The living branches burn readily as if resinous.

79. ACERACEAE. Maple Family

1. ACER. Maple

Trees or large shrubs; leaves opposite, simple or palmately or pinnately compound; flowers small, perfect or unisexual, in cymose or racemose, mostly axillary inflorescences; petals present or absent; stamens 4 to 12, borne on a ring-shaped disk, or the disk sometimes obsolete; styles 2; ovary 2-celled; fruit a pair of laterally winged samaras, these united at base.

All the maples are ornamental, but the genus is of minor importance in Arizona. In autumn the leaves change color and give great beauty to the mountain scenery. It is reported that maple sugar is sometimes made from the sap of *A. grandidentatum*. The plants are browsed by livestock and deer.

KEY TO THE SPECIES

1. Leaves pinnately compound, 3- to 5-foliolate, the terminal leaflet long-stalked 1. *A. Negundo*
1. Leaves simple or palmately trifoliolate, the terminal leaflet sessile or short-stalked (2).
2. Leaves thin, glabrous, with numerous acute teeth; inflorescence long-stalked 2. *A. glabrum*
2. Leaves thickish, usually persistently pubescent beneath, with few obtuse teeth; inflorescence nearly sessile . . . . 3. *A. grandidentatum*

1. **Acer Negundo** L. Apache County to Mohave County, south to Cochise and Pima counties, 3,500 to 8,000 feet, common along streams, April and May. Throughout most of the United States.

Box-elder. A rapid-growing tree, reaching a height of 15 m. (50 feet) and a trunk diameter of 0.8 m. (2.5 feet). It is often planted in dooryards and along streets but is short-lived and subject to damage by windstorms. The trunk is short, the crown broad, rounded, and dense, the bark pale grayish brown, and the wood soft and light-colored. The species is represented in Arizona by var. *interius* (Britton) Sarg., which differs from typical Eastern *A. Negundo* in its somewhat thicker, commonly more pubescent leaves. There is intergradation, however, and the form described as var. *arizonicum* Sarg. (type from the Santa Catalina Mountains, *Rehder* 463) approaches the Eastern tree in its thin, very sparsely pubescent leaves.

2. **Acer glabrum** Torr. Apache and Coconino counties, south to Cochise and Pima counties, 7,000 to 9,000 feet, rich soil in coniferous forests, May and June. South Dakota to Alaska, south to New Mexico, Arizona, and California.

Rocky Mountain maple, a shrub or small tree up to 10 m. (33 feet) high and 30 cm. (1 foot) in trunk diameter, with smooth, gray bark. Three varieties, distinguished by Keller (173), are reported to occur in Arizona. These are: var. *typicum* (Wesmael) Keller, on the San Francisco Peaks and, probably elsewhere; var. *diffusum* (Greene) Smiley (*A. diffusum* Greene), North Rim of the Grand Canyon (*Eastwood & Howell* 1066, *H. & V. Bailey* 1217), also Elden Mountain, Coconino County (*Deaver* in 1945); and var. *neomexicanum* (Greene) Kearney & Peebles (*A. neomexicanum* Greene) in the mountains of Apache, Coconino, Graham, Cochise, and Pima counties. In var. *neomexicanum* the leaves are mostly 3-parted, whereas in the other varieties they are usually merely lobed. Var. *diffusum* is a small-leaved form (the leaves not more than 5 cm. wide), with usually whitish instead of reddish or grayish twigs.

3. **Acer grandidentatum** Nutt. Grand Canyon and Oak Creek Canyon (Coconino County), Hualpai Mountain (Mohave County), southeast and south to the mountains of Greenlee, Cochise, and Pima counties, 4,500 to 7,000 feet, mostly with coniferous trees, April. Montana and Idaho to western Texas, New Mexico, and Arizona.

Big-tooth maple. Tree up to 15 m. (50 feet) high and 30 cm. (1 foot) in diameter, with smooth, gray or brownish bark. The wood makes excellent fuel. Var. *brachypterum* (Woot. & Standl.) E. J. Palmer (*A. brachypterum* Woot. & Standl.), with smaller fruits and shorter, few-toothed or nearly entire leaf lobes, is found chiefly in the southeastern part of the state but also near Prescott, Yavapai County.

80. SAPINDACEAE. Soapberry Family

Small trees or shrubs; leaves pinnate or simple; inflorescences terminal or lateral; flowers small, perfect or unisexual, with or without petals; disk present or wanting; stamens 5 to 10; style 1 or the styles united; ovary 2- to 4-celled; fruit various.

KEY TO THE GENERA

1. Leaves pinnate with numerous leaflets, not viscid; flowers in large terminal panicles; petals present; fruit berrylike, 1- to 3-seeded........................1. *Sapindus*
1. Leaves simple, viscid when young; flowers in small lateral clusters; petals none; fruit a dry, winged capsule..2. *Dodonaea*

1. SAPINDUS. Soapberry

Small trees or large shrubs; leaves pinnate, with numerous lanceolate or oblong-lanceolate leaflets; flowers small, whitish, in broad many-flowered panicles, with a basal disk; sepals and petals 4 or 5, the petals hairy and crested at base of the blade; stamens 8 to 10, the filaments hairy; fruits with amber-colored, translucent pulp.

1. Sapindus Saponaria L. Southern Coconino County and southeastern Mohave County to Greenlee (?), Graham, Cochise, Santa Cruz, and Pima counties, 2,500 to 5,000 feet, along streams, May to August. Kansas and Louisiana to Arizona and northern Mexico.

Soapberry. The trees very often attain a height in Arizona of 6 m. (20 ft.), perhaps more. The fruits, which often hang on the trees long after ripening, have been used in the southwestern United States and in Mexico for washing clothes. They are poisonous, containing a high percentage of saponin, and cause dermatitis in some persons. According to Mrs. Collom, soapberry is not considered safe browse for cattle, and they seldom touch it.

The Arizona plant is var. *Drummondii* (Hook. & Arn.) L. Benson (*S. Drummondii* Hook. & Arn.). The distribution of this variety is given in the first paragraph. The range of the species extends to South America.

2. DODONAEA (174). Hopbush

Shrubs, with viscid foliage; leaves simple, short-petioled, narrow, entire, linear to oblanceolate; flowers yellowish, in small lateral racemes or corymbs, apetalous; fruits conspicuous, dry, with 2 to 4 broad wings.

1. Dodonaea viscosa Jacq. Southern Yavapai County to Cochise, Santa Cruz, and Pima counties, 2,000 to 5,000 feet, fairly common on dry, rocky slopes and in canyons, often on limestone, frequently with *Simmondsia* and *Fouquieria*, February to October. Widely distributed in the warmer parts of the world.

Sometimes called switch-sorrel. The glutinous leaves and the bark have been used to treat various diseases. The attractive winged fruits have been used as a substitute for hops. In Arizona a shrub up to 3.5 m. (12 feet) high. The plant appears to be unpalatable to livestock and tends to increase on overstocked ranges. It is doubtless harmful, since it contains saponin, and is used in some countries as a fish poison.

According to Sherff (174, p. 214) the Arizona plant is *D. viscosa* var. *linearis* (Harv. & Sond.) Sherff forma *arizonica* (A. Nels.) Sherff (*D. arizonica* A. Nels.). The type of *D. arizonica* came from eastern Maricopa County (*A. Nelson* 11276).

81. RHAMNACEAE. Buck-thorn Family

Shrubs or small trees; leaves simple; flowers small, perfect or unisexual, regular or nearly so, 4- or 5-merous, with or without petals, with a disk in the calyx throat on which the stamens often are borne; ovary 2- to 4-celled, superior or partly inferior.

Most of these plants are browsed by domestic animals and deer.

KEY TO THE GENERA

1. Fruits drupelike, with a single stone, this 1- or 2-celled. Leaves alternate or fascicled; branches rigid, often spiny; ovary superior (2).
1. Fruits capsular or drupelike, with 2 to 4 distinct or nearly distinct stones (3).
2. Leaves petioled, evidently pinnate-veined, the margins flat or nearly so ...... 1. *Condalia*
2. Leaves sessile, not evidently pinnate-veined, the margins strongly revolute 2. *Microrhamnus*
3. Ovary superior (free from the calyx); fruit drupelike, more or less fleshy, the cells indehiscent (4).
3. Ovary partly inferior (adnate below to the calyx); fruit capsulelike, dry at maturity, the cells dehiscent (5).
4. Flowers sessile, in small panicles terminating the branchlets; petals present; style 3-lobed 3. *Sageretia*
4. Flowers pedicellate, axillary, solitary or in small clusters; petals sometimes absent; style 2-lobed .... 4. *Rhamnus*
5. Calyx lobes petaloid; petals white, bluish, or lavender-pink, long-clawed, hooded, often spreading away from the stamens .... 5. *Ceanothus*
5. Calyx lobes not petaloid; petals greenish or yellowish, short-clawed, standing close to the stamens .... 6. *Colubrina*

1. CONDALIA

Shrubs or small trees with rigid, usually spiny branches; leaves alternate or fascicled, pinnately veined, often conspicuously so; flowers axillary, in small fascicles or solitary; petals present or absent; ovary superior; fruit a drupe with 1 stone.

KEY TO THE SPECIES

1. Leaves elliptic to ovate; petals present; fruit globose or nearly so, not beaked, at maturity dark blue with a bloom, 6 to 8 mm. in diameter. Spines numerous, stout, divaricate or slightly decurved; leaves commonly at least 5 mm. wide, entire or dentate, the veins (except the midvein) slender and rather inconspicuous .... 1. *C. lycioides*
1. Leaves spatulate or obovate; petals none; fruit ovoid, ellipsoid, or somewhat obovoid, often beaked with the persistent style, at maturity black without bloom, less than 5 mm. in diameter (2).
2. Fruits nearly sessile; branchlets stout, very rigid, spiny throughout, relatively short; leaves oblanceolate or obovate, 4 to 6 mm. wide, with relatively slender and widely spaced, inconspicuous lateral veins .... 2. *C. mexicana*
2. Fruits distinctly pedicelled; branchlets slender, moderately rigid, spiny toward the apex, relatively long; leaves spatulate, with broad, conspicuous lateral veins, these occupying much of the surface (3).
3. Shrubs usually not more than 2 m. high, very dense, the branches ascending, often crooked; leaves very numerous, crowded, 3 to 10 (commonly about 5) mm. long including the petiole, 1 to 2 (rarely 3) mm. wide, with veins occupying nearly the whole of

the lower surface; bark of the branchlets commonly dark gray; pedicels commonly shorter than the fruit . . . . 3. *C. spathulata*

3. Shrubs up to 5 m. high, comparatively open, the branches divaricate, nearly straight; leaves relatively few, not crowded, 7 to 12 mm. long, 2 to 5 mm. wide, with veins more widely spaced; bark of the branchlets light gray or brown; pedicels nearly as long as to longer than the fruit . . . . 4. *C. globosa*

1. Condalia lycioides (Gray) Weberb. Coconino and Mohave counties to Greenlee, Cochise, Santa Cruz, Pima, and Yuma counties, 1,000 to 5,000 feet, common on dry plains, mesas, and slopes, often forming thickets, May to September. Western Texas to southeastern California and northern Mexico.

Gray-thorn. The typical plant, with glabrous or glabrate leaves, rare in Arizona, has been collected near Douglas and San Bernardino, Cochise County (*Peebles* 11694, 11703). The species is represented in this state mainly by var. *canescens* (Gray) Trel., type from the Gila River Valley (*Rothrock* 331), which has the leaves permanently canescent. The plants reach a height of 2.5 m. (8 feet), but are usually smaller. They are not or seldom browsed. Birds relish the insipid, dark-blue fruits and find refuge in the impenetrable growth. The Pima Indians treated sore eyes with a decoction of the roots, which also have been used as a substitute for soap.

2. Condalia mexicana Schlecht. Cochise, Santa Cruz, and Pima counties, 3,000 to 4,500 feet, dry slopes and canyons, usually with other large shrubs, July to October. Southern Arizona and Mexico.

Mexican blue-wood. The plants reach a height of 3 m. (10 feet).

3. Condalia spathulata Gray. Southwestern Gila County and eastern Pinal County to Cochise and Pima counties, 2,500 to 5,000 feet, common on dry mesas and "bajadas," July to September. Western Texas to southern Arizona and northern Mexico.

Squaw-bush. Shrub up to 3 m. (10 feet high) but usually 1.5 to 2 m. The stones of the black fruits have a soapy taste and are only slightly bitter.

4. Condalia globosa Johnst. Western Pima and southern Yuma counties, 1,000 to 2,500 feet, common on dry, sandy plains and along washes, March. Southwestern Arizona, southeastern California (?), northwestern Sonora, and Baja California.

Represented in Arizona by var. *pubescens* Johnst. A large shrub or small tree, reaching a height of 4.5 to 6 m. (15 to 20 feet) and a trunk diameter of 0.6 m. (2 feet). The fruits have extremely bitter stones. The ranges of *C. globosa* and the nearly related *C. spathulata* scarcely overlap in Arizona but specimens from the Ajo Mountains, 3,500 feet (*Nichol* in 1939, etc.), are more or less intermediate.

2. MICRORHAMNUS

Intricately branched, glabrous or nearly glabrous shrubs, up to 1.5 m. (5 feet) high, the branchlets ending in sharp, slender spines; young bark whitish, splitting, old bark dark gray; leaves alternate or fascicled, sessile, small, linear-oblanceolate, the margins strongly revolute, leaving only the broad midrib exposed beneath; flowers solitary, 5-merous, small, yellowish, pedicelled; petals cucullate; ovary immersed in the fleshy disk; fruit a slender, asymmetric, 1-celled, 1-seeded drupe, 6 to 9 mm. long, crowned by the more or less persistent style.

1. **Microrhamnus ericoides** Gray. Santa Rita, Santa Catalina, and Baboquivari mountains, Pima County (*Griffiths* 4200, *Rose* 12127, *Jones* in 1931), March and April. Western Texas to southern Arizona and northern Mexico.

3. SAGERETIA

Straggling shrubs with slender, somewhat spiny, divaricate branches; leaves nearly opposite, bright green, shining; flowers in glomerules, these forming open, leafy panicles; petals present, whitish; ovary superior; fruit a somewhat fleshy drupe with 3 stones.

1. **Sageretia Wrightii** Wats. Greenlee County to eastern Maricopa County, south to Cochise, Santa Cruz, and Pima counties, 3,000 to 5,000 feet, in canyons among rocks, March and September. Western Texas, southern Arizona, and Mexico.

4. RHAMNUS (175). Buck-thorn

Shrubs or small trees, not spiny; leaves mostly alternate; flowers perfect or unisexual, greenish, axillary, in small fascicles or solitary, with or without petals; calyx free from the ovary; fruit a drupe with 2 to 4 stones.

All the Arizona species are ornamental. The laxative drug cascara is obtained from the bark of *R. Purshiana,* a species of the Pacific coast states. *R. crocea* is an alternate host of the crown-rust disease of oats. The plants, especially the more or less evergreen forms, are of some value as browse in winter, and the fruits are eaten by wild pigeons and other birds.

KEY TO THE SPECIES

1. Bud scales present; flowers solitary or in small axillary fascicles without a common peduncle, mostly 4-merous and unisexual; style exserted (2).
1. Bud scales none; flowers in pedunculate cymes, mostly 5-merous and perfect; style included. Leaf margins serrulate to entire; fruits black or nearly so at maturity (3).
2. Leaves evergreen, with spinose-dentate margins, oval to suborbicular, commonly truncate or emarginate at apex; petals usually wanting; fruits bright red at maturity
1. *R. crocea*
2. Leaves deciduous, serrulate or crenulate, lanceolate, elliptic, or narrowly ovate, obtuse to acute at apex; petals present, small; fruits black at maturity ....2. *R. Smithii*
3. Leaves evergreen, thickish, whitish-tomentose beneath, oblong-lanceolate to broadly elliptic, commonly less than 3 cm. wide; fruits usually 2-seeded.......3. *R. californica*
3. Leaves deciduous, thin, green on both surfaces, sparsely to copiously pubescent beneath but not tomentose, broadly elliptic, ovate, or obovate, commonly more than 3 cm. wide; fruits usually 3-seeded ....4. *R. betulaefolia*

1. **Rhamnus crocea** Nutt. Southern Coconino County, Toroweap Valley and Hualpai Mountain (Mohave County), to Greenlee, Cochise, and Pima counties, 3,000 to 7,000 feet, common in chaparral and in open coniferous forest, March to May (rarely October). Arizona, California, and Baja California.

Red-berry buck-thorn. A handsome shrub, with bright-green hollylike leaves and bright-red fruits, attaining a height of 4 m. (13 feet) in California, but perhaps never so large in Arizona. It is browsed by deer and bighorns. Apache Indians ate the fruits with meat.

The species is represented in Arizona mainly by var. *ilicifolia* (Kellogg)

Greene, the holly-leaf buck-thorn, which has leaves as described in the key, whereas in typical *R. crocea* they are smaller, narrower, and merely serrulate to almost entire. A specimen collected in the Ajo Mountains, Pima County (*Nichol* in 1939), approaches typical *R. crocea.*

2. Rhamnus Smithii Greene. White Mountains, Apache County (*Schroeder* in 1938). Southwestern Colorado, western Texas, New Mexico, and eastern Arizona.

The Arizona plant is subsp. *fasciculata* (Greene) C. B. Wolf, with leaves obtuse at apex and finely pubescent above.

3. Rhamnus californica Esch. Southern Coconino County and Hualpai Mountain (Mohave County), to Graham, Cochise, and Pima counties, 3,500 to 6,500 feet, common in chaparral and open coniferous forests, May and June. New Mexico and Arizona to southern Oregon, California, and Baja California.

California buck-thorn, coffee-berry, pigeon-berry. The plants sometimes attain a height of 5.5 m. (18 feet), but are usually smaller. The species is represented in Arizona by subsp. *ursina* (Greene) Wolf (*R. ursina* Greene), which has the leaves whitish-tomentose beneath, whereas in typical *R. californica* they are green and glabrous or nearly so beneath. *R. Blumeri* Greene (type from the Chiricahua Mountains, *Blumer* 1290, in part) was thought by Wolf to be a hybrid between this variety and *R. betulaefolia.*

4. Rhamnus betulaefolia Greene. Southern Apache and southern Coconino counties to Cochise and Pima counties, 3,500 to 7,500 feet, mostly along streams in the live-oak and pine belts, May and June. New Mexico, eastern Arizona, and northern Mexico.

Birch-leaf buck-thorn. Usually a smaller shrub than *R. californica,* reaching a height of 2.5 m. (8 feet). The typical plant, with elliptic or oblong leaves, is limited in Arizona to the south-central and southern counties. In the northern part of the state is found var. *obovata* Kearney & Peebles, the type of which was collected on Navajo Mountain, Coconino County (*Peebles & Smith* 13930). This variety is apparently common in and near the Grand Canyon and Havasu Canyon, and extends into southern Utah and Nevada, thus being well separated geographically from the main area of *R. betulaefolia.* The variety is characterized by more or less obovate leaves with thicker, more prominent veins.

5. CEANOTHUS (176)

Much-branched shrubs with commonly rigid or spinescent branches; leaves alternate or opposite, entire or dentate, conspicuously veined; flowers perfect, with hooded, long-clawed petals, in cymose panicles; calyx adnate to the lower part of the ovary; fruit a 3-celled capsule.

These plants afford good browse for cattle, sheep, and especially for deer. They are esteemed as honey plants. It is reported that an infusion of the bark of *C. integerrimus* and perhaps other species occurring in Arizona has been employed as a tonic and that the flowers form a lather in water. Many species of this genus are cultivated as ornamentals and are often known as wild-lilac or mountain-lilac. The seeds are discharged with considerable force when the capsules open.

KEY TO THE SPECIES

1. Leaves opposite, pinnately veined with several broad veins; stipules thick, commonly persistent (at least the lower portion). Leaves not more than 2.5 cm. long, thick, pilose or tomentulose, at least beneath, when young; inflorescences small, not or but little surpassing the leaves . . . . 1. *C. Greggii*
1. Leaves alternate, palmately 3-nerved; stipules thin, usually soon deciduous except for the thickened basal scar (2).
2. Inflorescences terminating elongate branches, amply thyrsoid-paniculate, 5 to 15 cm. long, naked or nearly so, greatly surpassing the leaves; twigs not rigid, glabrous or obscurely puberulent; leaves (the larger ones) 3 to 6 cm. long, glabrous or sparsely pilose on the veins beneath, bright green above, somewhat paler beneath, broadly elliptic or ovate, the margins entire or very nearly so . . . . 2. *C. integerrimus*
2. Inflorescences axillary or terminating short branches, not more than 4 cm. long, commonly leafy; twigs rigid, copiously pubescent or puberulent; leaves commonly less than 3 cm. long (3).
3. Plants unarmed; leaves mostly oval to suborbicular, broadly rounded at apex, commonly rounded or subcordate at base, usually less than twice as long as wide, with often denticulate or serrulate margins, puberulent, then glabrescent, not whitish beneath 3. *C. Martini*
3. Plants usually spiny; leaves narrowly to broadly elliptic (rarely suborbicular), narrowed and obtuse or acute at apex, often narrowed at base, usually at least twice as long as wide, with entire or nearly entire margins, permanently pubescent and often whitish beneath . . . . 4. *C. Fendleri*

1. Ceanothus Greggii Gray. Coconino and Mohave counties to Greenlee (?), Graham, Cochise, and Pima counties, 3,000 to 7,000 feet, very common in chaparral, March to May (occasionally September). Western Texas to Utah, Nevada, southern California, and Mexico.

The plants are seldom more than 1.5 m. (5 feet) high, with petals commonly whitish but frequently bluish or pinkish. Arizona specimens vary greatly in pubescence and in the shape of the leaves. The latter are commonly relatively short and broad (var. *vestitus* (Greene) McMinn), reaching an extreme in var. *orbicularis* Kelso, the type of which came from Hackberry, Mohave County (*Goldman* 2946). Occasional specimens from throughout the range of the species in Arizona, however, approach typical *C. Greggii*, having narrower, elliptic or oblanceolate leaves, these commonly entire or sparingly denticulate. Specimens with some of the leaves spinose-dentate, collected in Mohave, Yavapai, and western Maricopa counties, are referable to var. *perplexans* (Trel.) Jepson.

2. Ceanothus integerrimus Hook. & Arn. Southern Coconino County to Cochise and Pima counties, 3,500 to 7,000 feet, chaparral and open coniferous forest, preferring shade, May to October. New Mexico and Arizona to Washington, Oregon, and California.

Deer-brush. Rated in California as one of the most valuable browse plants for livestock. Although widely distributed in Arizona, it is nowhere abundant. It is a handsome plant, with large, bright-green leaves and feathery panicles of whitish flowers. The shrubs reach a height of about 2.5 m. (8 feet), but are usually smaller.

The Arizona plants belong to var. *californicus* (Kellogg) Benson (*C. myrianthus* Greene), with usually broader leaves than in typical *C. integerrimus*, these 3-veined from the base. The type of *C. myrianthus* came from near Fort Huachuca (*Palmer* in 1890).

3. Ceanothus Martini Jones. Kaibab Plateau and north side of Grand Canyon (Coconino County), 7,500 to 8,500 feet, May and June. Utah, Nevada, and northern Arizona.

4. Ceanothus Fendleri Gray. Throughout the state except the extremely desert portions, 5,000 to 10,000 feet, very common in pine forests, tending to form low thickets, April to October. Colorado and Utah to western Texas, New Mexico, Arizona, and northern Mexico; reported also from South Dakota and Wyoming.

Buck-brush, deer-brier. Its great abundance gives this plant outstanding importance as browse for livestock and deer. It is more spinescent than the other Arizona species, seldom more than 1 m. (3 feet) tall and at high altitudes often less than 0.5 m. The typical phase of the species has the leaves sparsely puberulent to glabrous above, grayish-pubescent or tomentose beneath. There occur also in Arizona var. *venosus* Trel. with both surfaces of the leaves pubescent, often sericeous, and var. *viridis* Gray, with leaves glabrous on both surfaces or with hairs only along the veins. These entities intergrade freely.

6. COLUBRINA

Shrubs with divaricate, subspinescent branches; leaves alternate, petioled, entire or denticulate, oval or obovate, pinnately few-veined; flowers axillary, solitary or in small fascicles, inconspicuous, greenish or yellowish; petals present, hooded, short-clawed; calyx adnate to the lower part of the ovary; fruit a 3-celled drupelike capsule.

1. Colubrina californica Johnst. (*C. texensis* (Torr. & Gray) Gray var. *californica* L. Benson). Eastern Maricopa, Pinal, western Pima, and Yuma counties, 2,000 to 3,000 feet, dry slopes and along washes, June to August. Southern Arizona to southeastern California and Baja California.

In favorable situations the plants are 3 m. (10 feet) high, but are usually smaller. The Arizona specimens have smaller fruits than those from the type locality in Baja California.

82. VITACEAE. Grape Family

Stems woody, climbing or trailing, with prominent nodes and usually with tendrils; leaves alternate, simple or compound, with the long petioles dilated at base; flowers small, greenish, perfect or unisexual, in cymose panicles, 4- or 5-merous, often with a disk; petals deciduous at anthesis; stamens opposite the petals; pistil 1, the ovary superior, 2-celled; fruit berrylike; seeds bony.

The cultivated grapes (species of *Vitis*) are the only economically very important members of this family.

KEY TO THE GENERA

1. Leaves simple, often shallowly lobed; petals partly united and connivent at anthesis ..... 1. *Vitis*
1. Leaves compound or very deeply lobed; petals distinct and spreading at anthesis (2).
2. Leaves digitately trifoliolate to 7-foliolate; flowers 5-merous; disk obsolete ..... 2. *Parthenocissus*
2. Leaves deeply 3-lobed to trifoliolate; flowers 4- or 5-merous; disk evident ..... 3. *Cissus*

1. VITIS (177). GRAPE

Stems climbing or trailing; leaves large, rounded, cordate, often shallowly lobed, persistently pubescent or glabrate; flowers rarely perfect; petals more or less coherent at apex until they fall; berries globose or nearly so, juicy, few-seeded.

1. Vitis arizonica Engelm. Navajo, Coconino, and Mohave counties to Greenlee, Cochise, Santa Cruz, and Pima counties, 2,000 to 7,500 feet, common along streams and in canyons, often climbing on trees, April to July. Western Texas to southern Utah, Arizona, and northern Mexico.

Canyon grape. The berries are of good quality for jelly and grape juice and are much eaten by birds. They are also eaten, both fresh and dried, by the Indians. The vines are useful in checking soil erosion along creeks. The leaves when chewed allay thirst.

In addition to typical *V. arizonica*, with permanently more or less pubescent leaves, the nearly glabrous var. *glabra* Munson occurs throughout the range of the species in Arizona. Specimens with leaves more deeply and irregularly toothed than is usual, collected in Chevelon Canyon, Navajo or Coconino County (*Zuck* in 1896), and near Flagstaff (*Munson & Hopkins* in 1889), may be *V. Treleasei* Munson, which was reported by Trelease (in A. Gray *Syn. Fl.* 1^1: 423) from the Bradshaw Mountains (Yavapai County), but this entity is perhaps not specifically distinct from *V. arizonica*. An exceptionally pubescent form was collected near Lake Mead, Mohave County (*Clover* 4294).

A remarkable plant, with deeply 3-lobed, very coarsely few-toothed, sharply long-acuminate leaves, collected in Chiricahua National Monument, Cochise County (*O. M. Clark* 8142), may belong to another species. The leaf shape is like that of *V. palmata* Vahl as described and figured by Bailey (177, pp. 224–226), but the young growth is very pubescent. *V. californica* Benth. is "apparently in Arizona" (Bailey 177, p. 211), but no collections were cited.

2. PARTHENOCISSUS. VIRGINIA-CREEPER

Stems climbing or trailing; tendrils often expanded at apex into disks that adhere to bark or other surfaces; leaves large, digitately compound, the leaflets 3 to 7, deciduous; flowers 5-merous, the petals spreading at anthesis; berries thin-fleshed; seeds 1 to 4.

The Virginia-creeper (*P. quinquefolia*) of the eastern United States is often cultivated as an ornamental climber. The handsome foliage is beautifully colored in autumn.

1. Parthenocissus inserta (Kerner) K. Fritsch (*P. vitacea* (Knerr) A. S. Hitchc.). Apache County to Coconino County, south to Cochise and Pima counties, 3,000 to 7,000 feet, May to September. Ohio to Wyoming, New Mexico, and Arizona.

P. vitacea is distinguished from *P. quinquefolia* (L.) Planch. by its fewer-branched tendrils without adherent disks, or the disks only weakly developed, and by its mostly lateral, regularly dichotomous inflorescences with subequal branches; whereas in *P. quinquefolia* the inflorescences are chiefly terminal and more paniculate with unequal branches.

3. CISSUS

Stems trailing or clambering, from large tubers; tendrils not expanded at apex; leaves digitately trifoliolate or deeply 3-lobed, somewhat fleshy, persistent; flowers 4-merous, the petals spreading at anthesis; ovary 2-celled; berries 1- to 4-seeded.

1. **Cissus trifoliata** L. Gila, Santa Cruz, and Pima counties, 3,000 to 5,000 feet, not common, usually among rocks and clambering over bushes, July and August. Southern Arizona, Mexico, and widely distributed in tropical America.

The plant has a rank, disagreeable odor suggesting that of jimson-weed (*Datura*). The tubers are reputedly poisonous, and contact with the plant causes dermatitis in susceptible persons.

83. TILIACEAE. Linden Family

The best-known members of this family are the linden trees, species of *Tilia*.

1. CORCHORUS

Plants herbaceous (in Arizona); leaves alternate, simple, serrate; flowers perfect, regular, solitary or few on short peduncles opposite the leaves; petals 4 or 5, yellow; stamens commonly 10, distinct; ovary superior, 2- to 5-celled; capsule long and slender, longitudinally dehiscent.

Two Old World species furnish the important fiber jute.

1. **Corchorus hirtus** L. Apparently the only record, presumably for Arizona, is a collection from "sandy plains near the Mexican boundary" (*Pringle* in 1884), August. Southern Arizona, Mexico, and widely distributed in tropical America.

Pringle's specimens are small and evidently juvenile. They probably belong to var. *glabellus* Gray.

84. MALVACEAE. Mallow Family

Plants herbaceous or shrubby, with more or less mucilaginous sap, usually pubescent with stellate or forked hairs; leaves simple, alternate, petioled, stipulate; flowers regular, perfect; calyx often subtended by a calyxlike involucel (calyculus, epicalyx); petals 5, hypogynous, convolute in the bud, asymmetric, more or less united at base to the stamen column; stamens numerous, monadelphous; anthers 1-celled, reniform; pollen grains large, spiny; carpels 3 or more, 1-celled; style usually several-branched; fruit a loculicidal capsule, or in most of the genera the mature carpels separating from one another and from the receptacle; seeds often pubescent.

This family is notable as the one to which the cotton plants (*Gossypium* spp.) belong. Cotton is an important crop in Arizona, and a peculiar variety (*Gossypium hopi* Lewton), distantly related to upland cotton (*G. hirsutum* L.), was cultivated by the Hopi Indians from prehistoric until very recent times. There is, however, no record of its growing wild anywhere in the state. The garden vegetable okra (*Hibiscus esculentus*) also belongs to the Malvaceae. Many plants of this family have showy flowers, and numerous species of *Hibiscus, Malva,* and *Abutilon,* also the hollyhock (*Althaea rosea*), are cultivated as ornamentals. The marshmallow (*Althaea officinalis*), a European plant with thick, mucilaginous roots, is used medicinally as well as in making the well-known confection.

KEY TO THE GENERA

1. Fruit a loculicidal capsule; carpels 3 to 5, not separating at maturity from one another or from the axis. Flowers mostly solitary and axillary; ovules and seeds several in each carpel (2).
1. Fruit a schizocarp, of several or many carpels, these at maturity usually separating from one another and from the axis, but sometimes connate and sometimes remaining for a time attached to the axis by a threadlike prolongation of the midrib of the carpel (3).
2. Style branches 5, elongate, finally spreading; stigmas capitate; calyx 5-lobed; seeds reniform; petals yellow or purple; plants suffruticose, or herbaceous above the caudex 12. *Hibiscus*
2. Style not branched, clavate and ribbed toward the apex; calyx entire or shallowly dentate; seeds angulate-obovoid; petals white; plants shrubby..........13. *Gossypium*
3. Style branches filiform, longitudinally and introrsely stigmatic. Carpels reniform or subreniform, indehiscent, muticous, 1-ovulate; petals mauve or white; plants herbaceous (4).
3. Style branches terminating in a capitate or truncate stigma (6).
4. Flowers in elongate racemes; involucel none; stamens in 2 whorls, one apical, the other subapical, the filaments more or less united in groups (phalanges). Plants perennial; leaves, at least the upper ones, palmately parted or divided, the lobes narrow 7. *Sidalcea*
4. Flowers axillary, solitary or in small clusters, or in a loose terminal cymose inflorescence; involucel present, the bractlets narrow; stamens not in 2 whorls, the filaments not in phalanges (5).
5. Plants annual, the roots not thickened; leaves crenate, often shallowly cleft with broad, rounded lobes; peduncles short; petals not more than 15 mm. long; filaments all apical or subapical on the stamen column..............................8. *Malva*
5. Plants perennial, the taproot thick; leaves pedately 5-cleft, the lobes also deeply cleft; peduncles elongate; petals 20 to 30 mm. long; filaments extending well below the apex of the stamen tube...9. *Callirhoë*
6. Carpels sharply differentiated into a more or less reticulate indehiscent basal portion and a smooth dehiscent apical portion, the 2 portions separated by a pronounced ventral notch, or the upper portion expanded and winglike (7).
6. Carpels not sharply differentiated into a reticulate basal and a smooth apical portion or if so (in species of *Sida*), then the portions not separated by a distinct notch and the apical portion not expanded and winglike (8).
7. Apical portion of the carpel much wider than the basal portion, thin-walled, winglike, muticous, the notch indistinct or none; involucel none; petals yellow or pink; plants shrubby ....3. *Horsfordia*
7. Apical portion of the carpel not or not much wider than the basal portion, not winglike, the 2 portions separated by a (usually deep) notch; involucel usually present; petals bright red, orange, or mauve; plants herbaceous or suffruticose.......4. *Sphaeralcea*
8. Ovules, and usually the seeds, 2 or more in each carpel or if the ovule solitary, then the carpels membranaceous and much inflated; carpels dehiscent to the base or nearly so, not reticulate (9).
8. Ovule solitary; carpels not membranaceous or much inflated, indehiscent, or dehiscent only toward apex (11).
9. Involucel of 3 narrow bractlets; column antheriferous to far below the apex; petals pink or lavender. Carpels hirsute with long simple hairs, also stellate-canescent..5. *Iliamna*
9. Involcel none; column antheriferous only at and near the apex; petals mostly yellow or orange (10).

10. Fruit cylindric or ovoid; carpels not conspicuously inflated, usually mucronate to aristate at apex, the walls firm-chartaceous to coriaceous in fruit .... 1. *Abutilon*
10. Fruit globose; carpels conspicuously inflated, rounded and muticous at apex, the walls thin and papery in fruit .... 2. *Gayoides*
11. Involucel present; petals orange-yellow, purple, or purplish; ovule erect or ascending. Carpels indehiscent .... 6. *Malvastrum*
11. Involucel none or if present, then the petals yellowish white; ovule pendulous or resupinate-horizontal (12).
12. Lateral walls of the carpels mostly persistent, firm, and intact until maturity; carpels muticous to aristate at apex, not umbonate or spurred on the back, indehiscent or apically dehiscent; plants mostly perennial .... 10. *Sida*
12. Lateral walls of the carpels fragile, breaking up before maturity or if more persistent, then very thin and becoming more or less lacerate; carpels muticous at apex, often umbonate, angled, or spurred on the back; plants annual .... 11. *Anoda*

1. ABUTILON. Indian-mallow

Plants herbaceous, suffrutescent, or shrubby, canescent or tomentose with short, stellate hairs, or hirsute with longer, simple hairs; leaves crenate or dentate, not or obscurely lobed; flowers solitary in the axils, or in leafy panicles; involucel none; corolla usually orange or yellow; fruit truncate-cylindric or ovoid, the carpels with smooth sides, dehiscent nearly to the base when mature; ovules 2 or more in each carpel.

KEY TO THE SPECIES

1. Plants annual, tall; carpels usually more than 10, with long, divergent awns 1. *A. Theophrasti*
1. Plants perennial; carpels seldom more than 10 (2).
2. Carpels seldom more than 5 (3).
2. Carpels mostly 7 or more (6).
3. Herbage sparsely pubescent with long, rather stiff, mostly simple hairs; carpels aristate, the awns 2 to 3 mm. long, hispid .... 2. *A. Thurberi*
3. Herbage finely or minutely pubescent, the hairs mostly stellate; carpels muticous to short-cuspidate (4).
4. Stems woody well above the caudex; herbage finely and densely puberulent; leaves often considerably longer than wide, gradually acuminate; petals white or pink, with a dark-crimson spot at base .... 3. *A. Pringlei*
4. Stems not woody above the caudex or only near the base; pubescence coarser or more sparse; leaves little if any longer than wide, abruptly acuminate; petals not spotted (5).
5. Leaves thickish, usually densely soft-canescent or tomentulose on both surfaces, the larger ones usually 4 cm. long or longer; flowers usually numerous, in panicles; petals orange-yellow, 6 to 10 mm. long; carpels muticous or nearly so .... 4. *A. incanum*
5. Leaves thin, less pubescent, dark green above, the larger ones about 3 cm. long or shorter; flowers few, scattered, mostly solitary and axillary; petals pink or red, 4 to 6 mm. long; carpels mucronate or short-cuspidate .... 5. *A. parvulum*
6. Stems more or less woody above the caudex, leafy nearly to the apex; flowers solitary in the axils, or in small axillary and terminal clusters; leaves not more (usually less) than 10 cm. long, mostly shallowly cordate, the basal sinus open. Carpels cuspidate (7).
6. Stems herbaceous above the caudex, nearly naked above; flowers in long, small-bracted, terminal panicles; leaves usually at least 10 cm. long, deeply cordate, the basal sinus often closed and up to 3 cm. deep in the lower leaves. Stems stout, erect, usually 90 cm. or longer; petals light orange or yellow, much surpassing the calyx; calyx and carpels stellate-canescent (8).
7. Leaves shallowly cordate, the sinus usually much less than 1 cm. deep; petals light orange-yellow, little surpassing the calyx; herbage, calyx, and carpels stellate-canescent or tomentulose .... 6. *A. californicum*

7. Leaves more deeply cordate, the sinus usually 1 to 2 cm. deep; petals rich orange, 2 to 2½ times as long as the calyx; calyx and carpels (and usually the stems) villous; leaves velvety-tomentose .... 7. *A. Palmeri*
8. Stems and petioles sparsely to copiously hirsute with long, spreading or reflexed hairs; carpels mucronate or cuspidate .... 8. *A. sonorae*
8. Stems and petioles glabrous or puberulent; carpels muticous or nearly so . . 9. *A. reventum*

1. Abutilon Theophrasti Medic. Near Glendale, Maricopa County, in a cotton field (*Kearney* 5986), probably nowhere established in Arizona. Common in the eastern United States; naturalized from Asia.

Leaves large, cordate, velvety-pubescent; carpels much surpassing the calyx, villous.

2. Abutilon Thurberi Gray. Near mouths of canyons on the western slope of the Baboquivari Mountains (Pima County), in partial shade, about 3,500 feet (*Gilman* B35, *Peebles* 8987, 9060), August to October. Southern Arizona and northern Sonora.

Calyx and carpels long-hirsute. Stems from creeping rootstocks.

3. Abutilon Pringlei Hochr. Yavapai, Greenlee, Graham, Pinal, Maricopa, and Pima counties, 2,000 to 4,000 feet, common on dry slopes, March to October, type from the Tucson Mountains, Pima County (*Pringle* in 1884). Southern Arizona and northern Sonora.

This species apparently intergrades with *A. incanum.*

4. Abutilon incanum (Link) Sweet. Mohave, Yavapai, Maricopa, Pinal, Cochise, Pima, and Yuma counties, 1,000 to 4,000 feet, common on dry slopes, April to October. Central Texas to southern Arizona and Mexico.

Fibers extracted from the stems are reported to be used in Mexico for making rope.

5. Abutilon parvulum Gray. Navajo County to Mohave County, south to Cochise and Pima counties, 2,500 to 5,000 feet, rather common on dry plains and slopes, sometimes with scrub oak, April to October. Colorado and Texas to Arizona, California, and northern Mexico.

6. Abutilon californicum Benth. (*A. Lemmoni* Wats.). Yavapai, Graham, Pinal, Maricopa, Cochise, Santa Cruz, and Pima counties, 2,000 to 4,000 feet, frequent on dry, rocky slopes, March to September. Southern Arizona and Mexico.

A shrub, up to 2 m. high, the leaves varying greatly in size according to the moisture supply. The flowers open in the evening, according to C. G. Pringle, who collected the type of *A. Lemmoni* in the Santa Catalina Mountains.

7. Abutilon Palmeri Gray (*A. Parishii* Wats.). Along Salt River (eastern Maricopa County), Santa Catalina Mountains to the Growler Mountains (Pima County), 1,000 to 3,000 feet, dry, rocky slopes, March to May, types of *A. Parishii* from the Santa Catalina Mountains and vicinity (*Parish, Pringle,* in 1884). Southern Arizona, Sonora, and Baja California.

8. Abutilon sonorae Gray. Greenlee, Graham, Gila, Pinal, Cochise, Santa Cruz, and Pima counties, 3,000 to 4,500 feet, not infrequent in rich soil near streams, August to October, type (*Wright* 899) collected "on the Sonoita," probably in southwestern Cochise County. Southern Arizona and Mexico.

9. Abutilon reventum Wats. Santa Catalina and Baboquivari mountains (Pima County), 3,000 to 4,000 feet, rich soil near streams, April to September. Southern Arizona and Mexico.

Plant very similar in appearance to *A. sonorae*. Petals bright yellow, not orange as in most of the species.

2. GAYOIDES

Plants herbaceous or merely suffrutescent, the stems often weak and vine-like; leaves ovate, cordate, often prominently reticulate-veined; flowers small, on slender axillary peduncles; involucel none; petals yellow or orange; fruit globose; carpels rounded and muticous, becoming greatly inflated, the walls thin and papery, normally hirsute with long hairs.

1. Gayoides crispum (L.) Small (*Abutilon crispum* Sweet). Pinal, Maricopa, Pima, and Yuma counties, 3,500 feet or lower, common on dry slopes, flowering almost throughout the year. Florida; southern Arizona and California; and widely distributed in tropical America.

3. HORSFORDIA

Plants shrubby, densely stellate-canescent or tomentose, the pubescence yellowish; leaves denticulate or crenulate, thick, truncate or subcordate at base; flowers axillary, solitary or in few-flowered clusters, often assembled in leafy panicles; involucel none; carpels 8 to 12, the winged apical portion green, bluish, or purplish, often slightly erose, promptly dehiscent.

KEY TO THE SPECIES

1. Leaves ovate-lanceolate; petals yellow, about 8 mm. long..............1. *H. Newberryi*
1. Leaves ovate, usually broadly so; petals pink, 10 to 15 mm. long............2. *H. alata*

1. Horsfordia Newberryi (Wats.) Gray. Pinal, Maricopa, Pima, and Yuma counties, 1,000 to 3,500 feet, dry, rocky hillsides, March to October, type from the Purple Hills, Yuma County (*Newberry* in 1858). Southern Arizona, southeastern California, Sonora, and Baja California.

Stems commonly 1 to 2 m. high.

2. Horsfordia alata (Wats.) Gray. Tule Well and Tinajas Altas and Gila mountains (southern Yuma County), 500 to 1,500 feet, sandy washes, March to October. Southwestern Arizona, southeastern California, Sonora, and Baja California.

The stems reach a height of 3 m. and a diameter at base of 5 cm. The petals turn blue in drying.

4. SPHAERALCEA (178). GLOBE-MALLOW

Plants herbaceous or suffrutescent, mostly perennial, the herbage stellate-pubescent; leaves shallowly dentate to pedately dissected; inflorescences racemose or paniculate; calyx nearly always tribracteolate; corolla usually red (grenadine); fruit hemispheric to truncate-conic; carpels often remaining attached to the axis after maturity by a threadlike extension of the dorsal nerve; ovules and seeds 1 to 3 in each carpel.

Several of the species flower in spring and again after summer rains. They are known locally in Arizona as hollyhock and sore-eye-poppy, the latter name doubtless a translation of the Mexican *mal de ojos.* The Pima Indian name of these plants signifies "a cure for sore eyes." The Hopi Indians are reported to employ certain species as a remedy for disorder of the bowels and to use the mucilaginous stems, under the name of "kopona," as a substitute for chewing gum. The plants are browsed to some extent by sheep, goats, and bighorns. At least one species in southern Arizona is a host of the fungus, *Phymatotrichum omnivorum,* that causes the destructive root rot of cotton and other cultivated plants. Many of the species are difficult to identify, especially in the absence of mature fruit.

KEY TO THE SPECIES

1. Annual or biennial; petals orange. Leaves shallowly lobed; carpels very thin-walled, the reticulate basal part forming ⅔ or more of the carpel and conspicuously wider than the unreticulate apical part (2).
1. Perennial; petals usually red (grenadine), sometimes pink or white (3).
2. Plants densely yellowish-canescent, appearing scurfy; leaves thick and firm, crenulate; carpels rather deeply notched .... 1. *S. Orcuttii*
2. Plants grayish-pubescent, usually rather sparsely and loosely so; leaves thin and soft, coarsely crenate; carpels shallowly notched .... 2. *S. Coulteri*
3. Carpels with thick, coriaceous walls, the reticulate part forming ⅔ or more of the carpel, conspicuously wider than the unreticulate part. Leaves, at least the lower ones, 3-parted or divided (4).
3. Carpels with thinner, scarious to chartaceous walls, the reticulate part forming less than ⅔ of the carpel, not conspicuously wider than the unreticulate part (5).
4. Herbage, and calyx silvery-lepidote; upper leaves entire, linear to narrowly oblanceolate; carpels 7 to 9.. 15. *S. leptophylla*
4. Herbage and calyx canescent or more coarsely pubescent; all of the leaves deeply cleft, parted, or divided; carpels 10 to 14 .... 16. *S. coccinea*
5. Walls of the carpels very thin, scarious, the reticulate part with very small, transparent areolas (6).
5. Walls of the carpels thicker, chartaceous, the reticulate part with coarser, often opaque areolas (7).
6. Inflorescence narrow, many-flowered, thyrsoid-glomerate; leaves mostly considerably longer than wide; herbage usually green .... 3. *S. Emoryi*
6. Inflorescence an open, relatively few-flowered panicle; leaves scarcely longer than wide; herbage usually whitish-canescent .... 5. *S. laxa*
7. Reticulate part of the carpel usually rugose or muricate on the back, the reticulations prominent and coarse (8).
7. Reticulate part of the carpel smooth or nearly so on the back, the reticulations less prominent and finer (10).
8. Fruit hemispheric or nearly so; carpels usually muticous or mucronulate, about ⅔ as wide as high; plants often suffrutescent; leaves suborbicular, shallowly lobed; inflorescence usually an open panicle.............................. 4. *S. ambigua*
8. Fruit truncate-conic; carpels usually cuspidate, less than ⅔ as wide as high; plants not suffrutescent; leaves considerably longer than wide; inflorescence racemiform or narrowly thyrsoid (9).
9. Inflorescence a long, many-flowered thyrse with usually more than 3 flowers to the node; caudex well developed; stems erect or nearly so, seldom less than 60 cm. long; pedicels usually persistent .... 3. *S. Emoryi*
9. Inflorescence racemiform or subthyrsoid with 1 to 3 flowers to the node; caudex usually poorly developed; stems often originating as root shoots, commonly decumbent at base, not more than 50 cm. long; pedicels detaching promptly at maturity of the fruit 11. *S. subhastata*

10. Inflorescence usually relatively few-flowered and open, not thyrsoid-glomerate; carpels about half as wide as high (11).
10. Inflorescence usually many-flowered and relatively dense or if few-flowered and racemiform, then the carpels much more than half as wide as high (12).
11. Calyx not conspicuously more pubescent than the herbage, this commonly densely whitish-canescent or tomentose; inflorescence an open, long-branched panicle; carpels usually acutish, cuspidate, and rather prominently reticulate with semitransparent areolas; leaves shallowly and broadly 3-lobed to almost 3-parted..........5. *S. laxa*
11. Calyx conspicuously more pubescent than the herbage, this usually green and sparsely pubescent; inflorescence typically narrow and short-branched; carpels mostly very obtuse, muticous or mucronate, and faintly to rather prominently reticulate with opaque areolas; leaves commonly pedately parted or divided, the lobes usually relatively narrow ....6. *S. Rusbyi*
12. Fruit hemispheric or nearly so; carpels ⅗ to fully as wide as high (13).
12. Fruit truncate-conic; carpels ½ to ⅗ as wide as high (15).
13. Leaves not pedate, shallowly lobed near the middle with broad, rounded lobes, broadly deltoid or suborbicular, more or less cordate, thickish, the veins prominent beneath 12. *S. parvifolia*
13. Leaves pedately cleft, parted, or divided (14).
14. Inflorescence a narrow, interrupted, many-flowered thyrse; midlobe of the leaves 10 to 20 mm. wide...13. *S. grossulariaefolia*
14. Inflorescence racemiform or subthyrsoid, few-flowered; midlobe of the leaves not more than 5 mm. wide..14. *S. digitata*
15. Leaves about equally long and wide, shallowly lobed, usually cordate; pubescence grayish or whitish..12. *S. parvifolia*
15. Leaves much longer than wide or if not so, then either pedate, or cuneate at base, or the pubescence yellowish (16).
16. Leaves pedately cleft or parted ....7. *S. Wrightii*
16. Leaves not pedate (17).
17. Leaves merely angulate or toothed near the base, linear-lanceolate or oblong-lanceolate, not more than ⅓ as wide as long. Pubescence grayish ....10. *S. angustifolia*
17. Leaves more or less distinctly lobed, ovate or oblong-ovate, more than ⅓ as wide as long (18).
18. Pubescence yellowish, scurfy; leaves shallowly lobed, the lateral lobes usually broad and rounded; petals 10 to 17 mm. long; column 6 to 8 mm. long; hairs of the stem very short, many-rayed ....8. *S. incana*
18. Pubescence grayish or whitish; leaves often deeply cleft, the lateral lobes triangular and often acutish; petals 8 to 13 mm. long; column 4 to 6 mm. long or if the petals and the column longer, then the hairs of the stem usually relatively long and few-rayed 9. *S. Fendleri*

1. Sphaeralcea Orcuttii Rose. Southern Yuma County, 500 feet or lower, abundant, roadsides and fields, usually in sandy soil, March to May. Southwestern Arizona, southeastern California, Sonora, and Baja California.

Plant annual or biennial, with tall, wandlike stems and very numerous small flowers.

2. Sphaeralcea Coulteri (Wats.) Gray. Pinal, Maricopa, Pima, and Yuma counties, 2,500 feet or lower, roadsides, fields, and mesas, abundant, usually in sandy soil, January to May (occasionally later). Southwestern Arizona, southeastern California, and northwestern Mexico.

A specimen labeled as from Flagstaff (*Toumey* in 1892) probably was collected in southern Arizona.

3. Sphaeralcea Emoryi Torr. Mohave, Graham, Pinal, Maricopa, Cochise, Pima, and Yuma counties, mostly 2,500 feet or lower, roadsides and fields, flowering mainly in spring, type from near the mouth of the Gila River (*Emory* in 1846). Western and southern Arizona, southern Nevada, southeastern California, and northern Baja California.

KEY TO THE VARIETIES

1. Leaves not lobed, at most angulate or subhastate at base (2).
1. Leaves distinctly 3-lobed to 3-parted or even 3-divided (3).
2. Leaves ½ to ¾ as wide as long; carpels usually rather coarsely reticulate *S. Emoryi* (typical)
2. Leaves ⅓ to ½ as wide as long; carpels very finely reticulate............var. *nevadensis*
3. Carpels relatively thick-walled and coarsely reticulate, with opaque areolas var. *variabilis*
3. Carpels thin-walled and finely reticulate, with transparent areolas............var. *arida*

A highly variable species, intergrading or perhaps hybridizing frequently with *S. ambigua*. The corolla is normally grenadine red but varies to white, pink, or lavender. Typical *S. Emoryi* is rather rare and is found chiefly in the Colorado River region. Var. *nevadensis* Kearney has been collected at Silver Lake, Navajo County (*Griffiths* 2698). Much the commonest variety is var. *variabilis* (Cockerell) Kearney (*S. variabilis* Cockerell), which is especially abundant along roadsides in the irrigated districts, type from Phoenix (*Cockerell* in 1899). Var. *arida* (Rose) Kearney (*S. arida* Rose) has been collected on the California side of the Colorado River, near Yuma, and probably occurs in southwestern Arizona.

4. Sphaeralcea ambigua Gray. Mohave, Gila, Pinal, Maricopa, Pima, and Yuma counties, 3,500 feet or lower, dry, rocky slopes and edges of sandy washes, flowering almost throughout the year. Southwestern Utah to southern California, Sonora, and northern Baja California.

Desert-mallow, apricot-mallow. This is the most xerophytic of the Arizona species of *Sphaeralcea*. The stems become woody below and are often very numerous, 100 or more from a single root. It is one of the largest-flowered species, the petals attaining a length of 3 cm. The inflorescence is typically an open, long-branched panicle, but plants with a narrower, more thyrsoid inflorescence are not infrequent. A variant, of local occurrence throughout most of the range in Arizona, is var. *rosacea* (Munz & Johnston) Kearney, distinguished by a mauve- rather than apricot- or grenadine-colored corolla. Specimens suggesting hybridization with *S. Rusbyi* and with *S. parvifolia* have been collected, the former in western Gila County and the latter in Mohave County.

5. Sphaeralcea laxa Woot. & Standl. Navajo and Coconino counties (local), southward to Graham, Cochise and Pima counties, 2,000 to 6,000 feet, usually on limestone soils, in the open, March to November. Western Texas to Arizona and northeastern Sonora.

On caliche soils near Tucson this is the most abundant species of the genus. It varies from shade forms with thin, bright-green, shallowly lobed leaves to forms with thick, whitish-tomentose, often deeply dissected leaves. The open, relatively few-flowered inflorescence and dark-purple anthers are distinctive.

6. Sphaeralcea Rusbyi Gray. Navajo, Coconino, Yavapai, Graham, Gila, Pinal, and Maricopa counties, 2,500 to 6,000 feet, well-drained slopes, often in openings in pine forests, April to September, type from Prescott, Yavapai County (*Rusby* 537). Southern Utah to central Arizona and southeastern California.

In its typical phase this species is distinguished from its nearest relative, *S. laxa,* by its sparse pubescence of long-rayed hairs and usually narrow inflorescence. A luxuriant, highly variable phase, with usually less deeply pedate leaves, larger flowers, and often more copious and shorter pubescence, is var. *gilensis* Kearney, type from Devil's Canyon, Pinal County (*Kearney & Smith* 9011). This often occurs at somewhat lower elevations and is especially abundant in the Pinal Mountain region and along the Gila River east of Florence.

7. **Sphaeralcea Wrightii** Gray. Camp Grant, Graham County (*E. Palmer* 19, etc.), Douglas and Sulphur Springs Valley, Cochise County (*Carlson* in 1915, *Price* in 1894), apparently rare. Western Texas to southeastern Arizona and Chihuahua.

None of the Arizona specimens is typical, having less distinctly pedate leaves than the type of the species.

8. **Sphaeralcea incana** Torr. Apache County to western Coconino County, 4,000 to 6,000 feet, sandy or gravelly mesas and slopes, summer and autumn. Western Texas to northern Arizona and Chihuahua.

The tall, wandlike stems and very numerous flowers make this a conspicuous plant, distinguished from its nearest relative, *S. Fendleri,* by the yellowish-green color of the foliage. It is characteristically a plant of the open, whereas *S. Fendleri* commonly inhabits forests or thickets. The more frequent form in Arizona is var. *cuneata* Kearney, with narrower leaves, these cuneate, not subcordate or truncate, at base.

9. **Sphaeralcea Fendleri** Gray. Apache County to Coconino County, south to Cochise, Santa Cruz, and Pima counties, 3,000 to 8,000 feet, commonly in pine forests, sometimes among oaks at lower elevations, summer and autumn. Southern Colorado to western Texas, Arizona, and northern Mexico.

This is the characteristic species of the ponderosa pine belt. It is highly variable in leaf shape and in the reticulation of the carpels, which often is scarcely perceptible (*S. leiocarpa* Woot. & Standl.).

KEY TO THE VARIETIES

1. Hairs of the stem 0.6 to 0.8 mm. long, 4- to 8-rayed; petals 15 to 20 mm. long, usually pink and drying lavender or violet but occasionally grenadine............var. *venusta*
1. Hairs not more than 0.4 mm. long, 10- to 18-rayed; petals less than 15 mm. long, usually grenadine red (2).
2. Herbage whitish-velutinous, especially the lower leaf surfaces............var. *albescens*
2. Herbage green or grayish, not velutinous (3).
3. Leaves cleft nearly to the base, the lateral lobes ⅔ as long as the midlobe, spreading, obtuse ....var. *tripartita*
3. Leaves more shallowly cleft, the lateral lobes usually less than half as long as the midlobe, usually ascending, acutish to obtuse (4).
4. Leaves thin and rather sparsely pubescent, 5 to 9 cm. long, usually less than half as wide, shallowly lobed ....var. *elongata*
4. Leaves usually thickish and more or less canescent, 3 to 6 cm. long, usually more than half as wide, often deeply cleft ....*S. Fendleri* (typical)

Var. *venusta* Kearney, the showiest variety, is of rather frequent occurrence in the live-oak belt and at the lower edge of the pine belt, usually along streams, in the mountains of Graham, Cochise, Santa Cruz, and Pima counties, occurring also in northern Chihuahua, type from the Santa Catalina Mountains (*Pringle* in 1881). In the Huachuca Mountains occurs a form with the large flowers of this variety but with relatively short hairs and grenadine-colored petals. Var. *albescens* Kearney is known only from Santa Cruz County, Arizona, where it grows in thickets along Sonoita Creek (type *Harrison & Kearney* 9092). Var. *tripartita* (Woot. & Standl.) Kearney has been collected in Havasu Canyon, Coconino County (*Clover* 7003), and in the Pinaleno Mountains, Graham County (*Kearney & Peebles* 9776), occurring also in New Mexico. Var. *elongata* Kearney, known in Arizona only by 2 collections in Havasu Canyon (*Clover* 5137, 5141), is fairly frequent in the mountains of New Mexico.

10. Sphaeralcea angustifolia (Cav.) G. Don. Navajo and Coconino counties, south to Cochise, Santa Cruz, and Pima counties, 3,000 to 7,000 feet, roadsides and edges of cultivated fields, May to October. Western Kansas and Colorado to western Texas, Arizona, and Mexico.

The species is represented in Arizona by var. *cuspidata* Gray (*S. cuspidata* Britton), distinguished from the typical plant by its usually denser pubescence, narrower leaves with more pronounced basal teeth, grenadine or grenadine-pink petals, and less connate, narrower, mucronate or cuspidate carpels. In the vicinity of Tucson a form with broader and more distinctly lobed leaves approaches var. *lobata* (Wooton) Kearney (*S. lobata* Wooton).

11. Sphaeralcea subhastata Coult. Apache County to Mohave County, south to Cochise and Pima counties, common on treeless plains and mesas, usually in heavy, relatively impermeable soils, March to October. Western Texas to Arizona and northern Mexico.

A highly variable species, characterized by low growth (stems less than 50 cm. long) and racemose or, in one variety, subthyrsoid, relatively few-flowered inflorescences.

KEY TO THE VARIETIES

1. Calyx lobes usually deltoid and acute, about equaling the fruit; carpels 3/5 to 3/4 as wide as high. Leaves deeply cleft to almost divided, the lateral divisions at least 1/3 as long as the midlobe .... var. *pumila*
1. Calyx lobes usually lanceolate and acuminate, conspicuously surpassing the fruit; carpels 1/2 to 3/5 as wide as high (2).
2. Flowers usually 2 or 3 at each node of the inflorescence; leaves deeply cleft to almost divided, the lateral lobes or divisions usually 1/4 or more as long as the midlobe var. *thyrsoidea*
2. Flowers usually solitary at each node, or more than 1 only at the lowest nodes; leaves subhastately angled or toothed or, if more deeply cleft, then the lateral lobes usually less than 1/4 as long as the midlobe (3).
3. Carpels at maturity usually strongly connate and not separable without tearing; leaves usually about half as wide as long .... var. *connata*
3. Carpels at maturity free or nearly so; leaves usually less than half as wide as long *S. subhastata* (typical)

Var. *pumila* (Woot. & Standl.) Kearney (*S. pumila* Woot. & Standl.) is found in the northern and southeastern counties and in New Mexico. Var.

thyrsoidea Kearney is known only from Pima County, Arizona, at elevations of 2,000 to 3,000 feet, commonly growing in depressions where water collects at times, type from west of Tucson (*Harrison* 6825). Var. *connata* Kearney occurs in Navajo, Coconino, and Mohave counties at elevations of 3,000 to 7,000 feet, also in northwestern New Mexico. Typical *S. subhastata,* common in western Texas and New Mexico, is known in Arizona only from Cochise and Pima counties.

12. Sphaeralcea parvifolia A. Nels. (*S. arizonica* Heller). Apache County to Mohave County, south to Yavapai and Gila counties, mostly 4,000 to 7,000 feet, dry slopes and mesas, usually in the open, April to October, type of *S. arizonica* from Flagstaff (*MacDougal* 120). Western Colorado to New Mexico, Arizona, and eastern California.

An extremely abundant plant in north-central and northern Arizona, the roadsides often being colored red with its flowers. This species superficially resembles *S. incana,* but is distinguished by the whitish or grayish (not yellowish) herbage, less virgate inflorescence, smaller, more rounded leaves, flatter fruit, and less cuspidate carpels. From forms of *S. ambigua,* with contracted inflorescence, *S. parvifolia* is distinguishable by its more numerous and smaller flowers, fruit equaling or surpassing the calyx, and less galeate and less prominently reticulate carpels.

13. Sphaeralcea grossulariaefolia (Hook. & Arn.) Rydb. Apache County to Mohave, Yavapai, Gila, and Pinal counties, 3,000 to 6,000 feet, mesas and slopes, often among junipers and pinyons, occasionally descending along streams to lower elevations, April to October. Idaho and Washington to New Mexico, Arizona, and California.

This species is represented in Arizona chiefly by var. *pedata* (Torr.) Kearney (*S. pedata* Torr.), which differs from the typical plant in its higher, narrower, more pointed carpels, these often mucronate or cuspidate and 2-seeded. It resembles forms of *S. Rusbyi,* but may usually be distinguished by its finer and denser pubescence and shorter carpels. At the southern limit of its range in Arizona specimens suggesting hybridization with *S. Rusbyi* have been collected.

14. Sphaeralcea digitata (Greene) Rydb. Apache County to eastern Coconino County, 4,000 to 7,000 feet, well-drained slopes, often among junipers or pines. Western Texas to southern Utah, northern Arizona, and northern Chihuahua.

15. Sphaeralcea leptophylla (Gray) Rydb. Apache County to eastern Coconino County, 4,000 to 6,000 feet, dry, rocky hills and mesas, in shallow soils, May to September. Southwestern Colorado, southern Utah, New Mexico, northeastern Arizona, and Chihuahua.

The silvery-lepidote pubescence and the very narrow, simple or 3-divided, otherwise entire leaves, distinguish this from all other species of the genus.

16. Sphaeralcea coccinea (Pursh) Rydb. Apache County to Coconino County, 5,000 to 8,000 feet, dry plains and mesas, often associated with grasses, sometimes among pinyons and junipers, summer and autumn. Saskatchewan and Alberta, south to Texas, New Mexico, and northeastern Arizona.

KEY TO THE VARIETIES

1. Primary lateral divisions of the leaves less than ⅔ as long as the midlobe; dehiscent part of the carpel usually ascending, deltoid, and acutish, forming about ⅓ of the whole var. *elata*
1. Primary lateral divisions at least ⅔ as long as the midlobe; dehiscent part of the carpel usually horizontal, irregularly quadrilateral, forming less (often much less) than ⅓ of the whole (2).
2. Herbage usually densely whitish-canescent; leaves with very narrow divisions var. *dissecta*
2. Herbage usually green and less densely pubescent; leaves with relatively broad divisions *S. coccinea* (typical)

The typical plant and var. *elata* (Baker f.) Kearney and var. *dissecta* (Nutt.) Garrett are found pretty much throughout the range of the species in Arizona. Specimens with distinctly cuspidate carpels, referred doubtfully to var. *elata,* may be hybrids with *S. subhastata.*

5. ILIAMNA (179). WILD-HOLLYHOCK

Plants herbaceous above the caudex; herbage sparsely to copiously hirsute, the hairs mostly stellate; stems tall, leafy; leaves rather shallowly 3- to 7-cleft, with broad, triangular lobes; flowers in long, interrupted, thyrsoid panicles; involucel of 3 narrow, persistent bractlets; column stellate-hirsute below; carpels not reticulate, dehiscent to the base or nearly so, remaining attached to the axis by threads; ovules and seeds usually 3.

1. Iliamna grandiflora (Rydb.) Wiggins (*Sphaeralcea grandiflora* Rydb.). South Canyon, Kaibab Plateau, Coconino County, in damp places (*P. Mead* in 1929), Sierra Ancha, Gila County about 6,500 feet, in oak woodland (*Gould & Martin* 4451), summer. Southwestern Colorado, northern New Mexico, and Arizona.

A showy plant, with lavender petals 2 to 3 cm. long and large, maplelike leaves. The Sierra Ancha plant has exceptionally hirsute stems.

6. MALVASTRUM

Plants annual or perennial, herbaceous or shrubby, sparsely to densely stellate-pubescent, or hispid; leaves ovate to orbicular, crenate or palmately cleft; flowers axillary, or in terminal, bracted inflorescences; calyx subtended by an involucel of narrow bractlets; petals yellow or purple; carpels few, compressed or somewhat turgid, rugose on the sides, sometimes tuberculate on the back, indehiscent.

KEY TO THE SPECIES

1. Plants shrubby; leaves ovate, crenate-dentate, sparsely to densely stellate-canescent; petals orange-yellow; carpels rather thick, with coriaceous walls, radiately rugose on the sides, with 2 conical processes or cusps on the back .... 1. *M. bicuspidatum*
1. Plants herbaceous, annual; leaves orbicular or nearly so; petals purple or nearly white; carpels thin-walled, not appendaged on the back (2).
2. Leaves merely coarsely crenate; stems and petioles hispid with long, mostly simple hairs; petals 15 to 20 mm. long, conspicuously spotted at base, lilac or mauve, drying violet-purple; carpels many more than 15, thin, flat, black at maturity, reticulate near the edges .... 2. *M. rotundifolium*
2. Leaves palmately cleft, with rounded lobes; pubescence mostly stellate; petals not more than 6 mm. long, not spotted, lavender or whitish; carpels not more than 15, somewhat turgid, not becoming black, transversely rugose .... 3. *M. exile*

1. Malvastrum bicuspidatum (Wats.) Rose. Eastern Maricopa, Pinal, and Pima counties, 3,000 to 5,000 feet, occasional on rocky slopes, March to October. Southern Arizona and Mexico.

Closely related to the South American *M. scabrum* (Cav.) Gray and *M. scoparium* (L'Hér.) Gray. The Arizona plant was referred by Gray to the latter species.

2. Malvastrum rotundifolium Gray (*Eremalche rotundifolia* Greene). Mohave, western Maricopa, Pinal, and Yuma counties, 100 to 1,500 feet, dry, sandy soil, often in washes, March to May, type from Fort Mohave (*Cooper*). Western Arizona, southern Nevada, and southern California.

A showy plant when in flower.

3. Malvastrum exile Gray (*Eremalche exilis* Greene). Mohave, Yavapai, Greenlee (?), Graham, Pinal, Maricopa, Cochise, Pima, and Yuma counties, up to 4,000 feet, common at roadsides and in fields, February to May, type from Pyramid Canyon, Mohave County (*Newberry* in 1858). Southwestern Utah to southern Arizona and California.

Reported to be used as food by the Pima Indians in times of scarcity. The plant affords considerable grazing in southern Arizona in early spring.

7. SIDALCEA (180). CHECKER-MALLOW, PRAIRIE-MALLOW

Plants perennial, herbaceous above the caudex, sparsely pubescent; leaves palmately lobed or cleft, the lower ones orbicular or nearly so, sometimes merely crenate; flowers showy, short-pedicelled, in terminal, bracted racemes; involucel none; petals 1.5 to 2.5 cm. long; style branches filiform, introrsely papillate; carpels few, indehiscent, smooth or reticulate on the sides.

1. Sidalcea neomexicana Gray. Apache County to Coconino and Yavapai counties, also in the Chiricahua and Huachuca mountains (Cochise County), 5,000 to 9,500 feet, frequent in wet meadows and along streams, June to September. Wyoming and Idaho to northern Mexico and California.

A handsome plant, with stems up to about 1 m. long and large, commonly mauve-colored flowers. Reported to be sometimes used as greens by the Indians.

8. MALVA. MALLOW, CHEESE-WEED

Plants annual or biennial, sparsely pubescent or glabrate; leaves orbicular or reniform; flowers small, axillary, solitary or in small cymules, short-pedicelled, involucellate; petals white or pink; style branches filiform, introrsely papillate; fruit depressed, disklike, the carpels numerous, compressed, reniform, indehiscent.

The plants are reported to be boiled and eaten by the Indians in times of scarcity. Both species are natives of the Old World, extensively naturalized in the United States.

KEY TO THE SPECIES

1. Petals not more than 6 mm. long, the claws glabrous; carpels glabrous or pubescent, sharply and radiately rugose on the sides, reticulate on the back......1. *M. parviflora*
1. Petals 7 to 15 mm. long, the claws bearded; carpels puberulent but otherwise smooth or nearly so ....2. *M. neglecta*

1. Malva parviflora L. Mohave, Coconino, Graham, Maricopa, Pinal, Cochise, Pima, and Yuma counties, a common field weed, March to September.

2. Malva neglecta Wallr. (*M. rotundifolia* of authors, non L.). Navajo, Coconino, Yavapai, Graham, and Gila counties, July to September.

9. CALLIRRHOË. Poppy-mallow

Herbage hirsute or hispid, with both simple and branched hairs; leaves palmately parted, the lobes toothed or pinnatifid; flowers large and showy, solitary on long axillary peduncles; involucel of 3 bractlets; petals wedge-shaped, erose-denticulate at apex; carpels numerous, deeply reticulate, with a short, inflexed beak.

1. Callirhoë involucrata (Torr. & Gray) Gray. Petrified Forest, Apache (?) County (*Meiere* in 1917). Probably introduced from farther east, the previously known range being from Minnesota to Texas and northeastern New Mexico.

10. SIDA

Plants mostly perennial, herbaceous or suffrutescent, more or less pubescent with forked, stellate, or scalelike hairs; flowers axillary, solitary or in small cymules, these sometimes assembled in terminal leafy panicles; involucel usually none; carpels indehiscent or dehiscent only part way from the apex, more or less rugose and often reticulate on the sides.

KEY TO THE SPECIES

1. Calyx greatly enlarged in fruit, papery, veiny, the segments appearing cordate at base; taproot long, tuberlike, often fusiform; herbage hispid (usually sparsely so) with long, mostly few-rayed hairs; petals not or scarcely surpassing the calyx 1. *S. physocalyx*

1. Calyx not or but slightly enlarged in fruit, not papery or veiny, the segments not appearing cordate; taproot sometimes stout but not tuberlike or fusiform; herbage puberulent, canescent, or lepidote (if long hairs also present, these simple, not stiff); petals considerably surpassing the calyx (2).

2. Flowering stems from elongate rootstocks, decumbent or prostrate; herbage conspicuously whitish stellate-canescent or lepidote; leaves very oblique at base; petals white or ochroleucous when fresh, often fading pink. Peduncles axillary, mostly 1-flowered, commonly decurved or sigmoid after anthesis; petals 10 to 20 mm. long; carpels reticulate on the sides, muticous or nearly so (3).

2. Flowering stems not from rootstocks; herbage not conspicuously whitish-canescent or lepidote; leaves not or scarcely oblique at base; petals orange-yellow when fresh. Carpels more or less dehiscent at apex, rugose or reticulate on the sides (4).

3. Herbage densely whitish-canescent with short, stellate hairs; leaves broadly deltoid or suborbicular, wider than long, rounded at apex, rather regularly dentate; involucel of 1 to 3 subulate bractlets, usually persistent until or after anthesis; petals ochroleucous when fresh; carpels indehiscent .... 2. *S. hederacea*

3. Herbage sparsely to densely silvery-lepidote with scalelike hairs; leaves triangular-lanceolate to triangular-ovate, longer than wide, acutish to acuminate at apex; involucel none; petals white when fresh; carpels dehiscent at apex......... 3. *S. lepidota*

4. Stems decumbent or prostrate, these and the petioles sparsely hirsute with long, spreading, very slender, simple hairs (these rarely absent), also finely stellate-canescent or puberulent, and with short, glandular hairs; flowers axillary, solitary, on long, slender peduncles; calyx usually both long-hirsute and puberulent.......... 4. *S. procumbens*

4. Stems erect or ascending or if somewhat decumbent, then the flowers clustered at the ends of the stems and branches; stems, petioles, and pedicels finely stellate-canescent or puberulent, without long simple hairs; calyx not hirsute (5).

5. Plants perennial, without a definite axis, the stems diffuse, several or numerous from a woody root; leaves linear or narrowly oblong (the lowest sometimes ovate), serrate; petals fading pink; carpels muticous or short-mucronate, rugulose only at the edges, scarcely differentiated apically and basally .... 5. *S. neomexicana*
5. Plants normally with a definite axis, this and the branches erect or nearly so, virgate; leaves narrowly lanceolate to ovate, crenate or crenate-dentate; petals not fading pink; carpels with a smooth dehiscent apical portion sharply differentiated from the reticulate indehiscent basal portion (6).
6. Taproot slender, not woody; plants annual, herbaceous; leaves narrowly lanceolate, not more than 8 mm. wide, acuminate at apex, finely crenate-dentate; flowers mostly in small clusters in the upper axils and at the ends of the main stem and branches; peduncles usually less than 1 cm. long; carpels commonly bicuspidate or biaristate, the basal portion rugose-tuberculate.................................... 6. *S. spinosa*
6. Taproot thick, woody; plants perennial, often suffrutescent; leaves ovate or oblong-ovate, mostly 10 to 20 mm. wide, obtuse or acutish at apex, coarsely crenate; flowers solitary, axillary (by reduction of the upper leaves the inflorescence sometimes appearing as an elongate, terminal raceme); peduncles 1 to 3 cm. long; carpels bimucronate, the basal portion coarsely reticulate .... 7. *S. tragiaefolia*

1. Sida physocalyx Gray (*S. hastata* St. Hil., non Willd.). Cochise, Santa Cruz, and Pima counties, 2,500 to 5,000 feet, rather frequent in rich soil in canyons, also at Beaver Creek, Yavapai County (*Purpus* 57), March to October. Texas to Arizona and northern Mexico; South America.

Readily distinguished from all other species of *Sida* in Arizona by the inflated, strongly 5-angled calyx and the large, tuberlike root.

2. Sida hederacea (Dougl.) Torr. Apache County to Coconino, Graham, Pinal, and Yuma counties, 100 to 5,000 feet, moist, often saline soil, usually near streams, May to October. Oklahoma and Texas to Washington, California, Arizona, and Mexico.

The fruits are extensively parasitized by insects, and the seeds seldom mature. Known in New Mexico as meloncilla. Often a troublesome weed on heavy soils.

3. Sida lepidota Gray. Mohave, Yavapai, Pinal, Maricopa, Cochise, and Pima counties, 1,000 to 6,000 feet, roadsides, March to October. Western Texas and southern Colorado to Arizona and Mexico.

The typical plant, with triangular-ovate, dentate, not or barely hastate leaves, has been collected near Duncan (Greenlee County), at Safford (Graham County), and at several localities in Cochise County. Var. *sagittaefolia* Gray is more common in Arizona. It has lanceolate or oblong-lanceolate, asymmetrically hastate or subsagittate leaves with 1 to 3 pairs of teeth near the base, the margins otherwise entire or nearly so.

4. Sida procumbens Sw. (*S. diffusa* H.B.K.). Greenlee, Graham, Gila, Pinal, Cochise, Santa Cruz, and Pima counties, 2,500 to 6,000 feet, common on plains and mesas in dry, sandy soils, April to October. Southern United States and West Indies to northern South America.

5. Sida neomexicana Gray. Cochise, Santa Cruz, and Pima counties, 4,000 to 6,000 feet, plains in the open and among trees in canyons, June to October. Western Texas to southern Arizona and Mexico.

6. Sida spinosa L. Cochise, Santa Cruz, and Pima counties, 3,500 to 4,000 feet, dry, sandy plains, late summer. Southern United States to Argentina.

The Arizona plants belong to the narrow-leaved var. *angustifolia* (Lam.) Griseb. (*S. angustifolia* Lam.).

7. Sida tragiaefolia Gray. Graham, Pima, and Santa Cruz counties, 2,500 to 3,500 feet, dry, rocky slopes, late summer and autumn. Western Texas, southern Arizona, and northeastern Mexico.

Plant suffrutescent, the stems up to 1.8 m. (6 feet) tall.

11. ANODA (181)

Plants sparsely hirsute to densely puberulent or tomentose; leaves, especially the upper ones, often hastate at base; involucel none; fruit depressed, hemispheric or disklike; carpels usually umbonate or spurred on the back, with fragile lateral walls, these usually breaking up before maturity, the inner layer forming a saclike envelope of the seed or becoming closely adherent to the seed coat, the dorsal wall more persistent.

KEY TO THE SPECIES

1. Fruit a flattened disk of radiating carpels, these conspicuously long-hirsute or hispid, with an elongate dorsal spur. Calyx spreading in fruit (2).

1. Fruit hemispheric or somewhat lower, but not flat and disklike; carpels puberulent to short-hirsute, rounded, umbonate, angled, or short-spurred on the back (3).

2. Petals purple; herbage sparsely hirsute with long, mostly simple, spreading or retrorse hairs, often also puberulent, sometimes glabrate; carpels 9 to 20, hispid, not reticulate on the back; leaves truncate or short-cuneate at base .... 1. *A. cristata*

2. Petals orange-yellow, purplish at base; herbage densely puberulent or short-pilose, a few long, simple hairs often also present, the pubescence slightly viscid; carpels 10 to 12, hirsute, strongly reticulate on the back with black veins; lower leaves cordate or subcordate at base .... 2. *A. Wrightii*

3. Inner layer of the carpel wall separating from the outer at maturity as a reticulate, loose, saclike envelope of the seed. Lower leaves broadly cordate-angulate or shallowly lobed (4).

3. Inner layer not separating or, if so, then becoming closely adherent to the seed and arilliform. Carpels more or less umbonate or gibbous on the back; stems tall, virgate; inflorescence elongate, often nearly leafless above; pubescence often somewhat viscid (5).

4. Upper leaves mostly elongate and hastate; petals orange-yellow; carpels prominently angled or short-spurred on the back .... 3. *A. crenatiflora*

4. Upper leaves narrowly 3- to 5-lobed; petals purple; carpels rounded on the back 4. *A. reticulata*

5. Petals 8 to 12 mm. long, more than twice as long as the calyx; lateral walls of the carpels persistent but fragile and becoming torn; herbage velvety-tomentose, the stems also villous, at least below, with spreading, simple hairs; leaves all broadly ovate, cordate. Petals orange-yellow, often fading pink or purplish; leaves up to 10 cm. wide, crenate-dentate, long-acuminate at apex .... 5. *A. abutiloides*

5. Petals not more than 6 mm. long, about twice as long as the calyx; lateral walls of the carpel breaking up before maturity; herbage puberulent or glabrescent; upper leaves narrow, more or less hastate (6).

6. Carpels 8 or 9, strongly 3-nerved on the back; petals purple .... 6. *A. Thurberi*

6. Carpels 5 to 7, 1-nerved on the back; petals orange-yellow, often fading pink or purplish 7. *A. pentaschista*

1. Anoda cristata (L.) Schlecht. (*A. arizonica* Gray). Southern Apache County to Yavapai County, south to Cochise, Santa Cruz, and Pima counties, 3,500 to 6,500 feet, common in moist meadows and along streams, August to

October, type of *A. arizonica* from "S. Arizona" (*Lemmon* 599). Western Texas to southern Arizona, southward to South America.

Highly variable in size of plant and shape of the leaves, these being usually narrowly to broadly triangular in outline and coarsely crenate, often hastate, the basal ones sometimes digitately several-lobed. The variant with lobed leaves is var. *digitata* (Gray) Hochr. (*A. arizonica* var. *digitata* Gray), type from "S. Arizona" (*Lemmon* 517), which intergrades freely with forms having less dissected leaves.

2. **Anoda Wrightii** Gray. Cave Creek, Chiricahua Mountains, Cochise County, about 5,000 feet, rich soil in pine forests, September (*Harrison & Kearney* 6179). Southwestern New Mexico, southeastern Arizona, and Mexico.

3. **Anoda crenatiflora** Ortega. Tumacacori Mission, Santa Cruz County, about 3,000 feet, in thickets (*Harrison* 8146), August and September. Southern Arizona and Mexico.

4. **Anoda reticulata** Wats. Santa Catalina Mountains, Pima County (*Pringle* in 1881, the type collection), Bear Valley (Sycamore Canyon), Santa Cruz County (*Kearney & Peebles* 14453). Known only from southern Arizona.

5. **Anoda abutiloides** Gray. Pima and Santa Cruz counties, 3,500 to 5,000 feet, dry slopes and rich soil in canyons, May to October, type from the Santa Catalina Mountains (*Pringle* in 1882). Apparently known only from southern Arizona, but doubtless also in Sonora.

This plant bears a marked resemblance to *Abutilon sonorae* in its large, deeply cordate, long-pointed, velvety-pubescent leaves. In structure of the carpels it forms a link with the genus *Sida*, the lateral walls, although fragile, being more persistent than in any other species of *Anoda*.

6. **Anoda Thurberi** Gray. Cochise County (several localities), 4,000 to 5,500 feet, sandy plains and on limestone, September and October, types from southern Arizona (*Thurber, Wright, Lemmon*). Southern New Mexico, southeastern Arizona, and northern Mexico.

7. **Anoda pentaschista** Gray (*Sidanoda pentaschista* Woot. & Standl.). Greenlee, Maricopa, Pinal, Cochise, and Pima counties, 1,000 to 4,000 feet, roadsides and fields, June to September. Western Texas to southern Arizona and Mexico.

12. HIBISCUS. Rose-mallow

Plants perennial, often shrubby; leaves merely crenate or dentate, or pedately cleft; flowers axillary, solitary, the petals 2 cm. long or longer; involucel usually present; fruit a loculicidal capsule, the carpels 5; seeds several in each locule, long-hairy.

KEY TO THE SPECIES

1. Petals lavender; bractlets less than half as long as the calyx, often caducous; stems densely grayish-canescent or -tomentose with short, stellate hairs. Plants suffrutescent or shrubby .... 1. *H. denudatus*

1. Petals yellow with a large red basal spot; bractlets nearly equaling to longer than the calyx, persistent; stems strigose or hispid with long, simple or forked hairs (2).

2. Stems distinctly woody above the caudex, evenly strigose with forked hairs; pedicels usually disarticulating at maturity of the fruit; seeds completely covered with long hairs .... 2. *H. Coulteri*

2. Stems scarcely woody above the caudex, finely pubescent in 1 or 2 lines, also hispid with long, simple or forked hairs; pedicels not disarticulating; seeds naked or nearly so in the center .... 3. *H. biseptus*

1. Hibiscus denudatus Benth. Pinal, Maricopa, Cochise, Pima, and Yuma counties, January to October. Western Texas to southern California and northern Mexico.

Two rather well-marked phases occur in Arizona: the typical phase, a shrub reaching a height of 1 m., with the involucel commonly little developed, growing in sandy washes, 2,000 feet or lower, in Pinal, Maricopa, and Yuma counties; and var. *involucellatus* Gray (*H. involucellatus* Woot. & Standl.), an undershrub, commonly not more than 0.5 m. high, with a better-developed and more persistent involucel, the prevailing variety on shallow, caliche soils in Greenlee (?), Graham, Cochise, and Pima counties, 2,000 to 4,000 feet.

2. Hibiscus Coulteri Harv. (*H. Coulteri* var. *brevipedunculata* Jones). Southern Yavapai, western Gila, Pinal, Maricopa, Pima, and Yuma counties, 1,500 to 4,500 feet, rather common on rocky slopes and sides of canyons, flowering throughout the year, type of var. *brevipedunculata* (*sic*) from Congress Junction (*Jones* in 1903). Western Texas to southern Arizona and Mexico.

A straggling shrub, the stems up to 1.2 m. long.

3. Hibiscus biseptus Wats. Pinal, Maricopa, and Pima counties, 3,000 to 4,500 feet, not infrequent on rocky slopes in canyons, April to October. Southern Arizona and northern Mexico.

Arizona specimens differ from the typical phase in having the fine pubescence in only one line and the longer hairs mostly forked.

13. GOSSYPIUM. Cotton

Plants shrubby; branchlets and petioles quadrangular; leaves pedately 3- to 5-parted, with lanceolate or lance-elliptic, attenuate-acuminate lobes; extrafloral nectaries present on the leaves and at apex of the peduncles; flowers mostly solitary and axillary, borne on short branches; involucel present, persistent; petals white or whitish, 2 to 3 cm. long; fruit a 3- or 4-celled loculicidal capsule; seeds turbinate-angulate, rather sparsely villous.

1. Gossypium Thurberi Todaro (*Thurberia thespesioides* Gray). Graham, Gila, Pinal, Maricopa, Cochise, Santa Cruz, and Pima counties, reported also from the Bradshaw Mountains (Yavapai County), 2,500 to 5,000 (rarely 7,000) feet, rather common on rocky slopes and sides of canyons, late summer and autumn. Southern Arizona and northern Mexico.

A handsome shrub, known in Sonora as algodoncillo (little cotton), reaching a height of 4.2 m. (14 feet). Petals normally spotless, but plants with faint crimson basal spots are not rare. The plant is interesting because a subspecies of the cotton boll weevil breeds in the capsules. The form of this insect of which *G. Thurberi* is the normal host also occasionally attacks nearby cultivated cotton, consequently the United States Department of Agriculture endeavored at one time to eradicate the plant where it grew near areas of cotton cultivation.

85. STERCULIACEAE. Cacao Family

Plants herbaceous to treelike, the pubescence wholly or partly of forked or stellate hairs; leaves alternate, simple; flowers perfect, regular or nearly so; calyx usually 5-lobed; petals free or united with the stamen tube, absent in one genus; fertile stamens 5, the filaments more or less united below, with staminodia sometimes also present, the anthers 2- or 3-celled; fruit a capsule, or an achenelike follicle.

The most important plant of this family is *Theobroma Cacao,* a native of tropical America, from the seeds of which cocoa and chocolate are obtained.

KEY TO THE GENERA

1. Sepals large, bright yellow; petals none; plants large shrubs or small trees. . 1. *Fremontia*
1. Sepals relatively small and inconspicuous; petals present; plants small shrubs or herbs (2).
2. Petals reddish or purplish, with broad, hooded blades abruptly contracted into slender claws; stamens 10, the 5 fertile ones alternating with staminodia, the anthers 3-celled; stigma capitate, larger than the style . . . . 4. *Ayenia*
2. Petals orange, with flat, spatulate or obovate blades tapering gradually into relatively broad claws; stamens 5, all fertile, the anthers 2-celled; stigma minute, not or scarcely larger than the style (3).
3. Flowers solitary in the leaf axils, on elongate peduncles; calyx tube not strongly ribbed; petals 6 to 8 mm. long; anthers hispidulous; styles 5, not contorted; fruit much larger than the calyx, 5-celled, bladderlike; seeds several in each cell. . . . . 2. *Hermannia*
3. Flowers numerous, sessile, in dense glomerules aggregated into small panicles; calyx tube strongly 10-ribbed, turbinate; petals about 4 mm. long; anthers glabrous; style 1, somewhat contorted; fruit enclosed in the calyx, 1-celled, not inflated; seed solitary 3. *Waltheria*

1. FREMONTIA (182)

Large evergreen shrubs or small trees; leaves thickish, usually palmately lobed, truncate or subcordate at base, the lower surface whitish or yellowish, scurfy-tomentose with minute, stellate hairs; flowers solitary, extra-axillary, showy; sepals large, bright yellow, each with a hairy gland at base; petals none; capsule 4- or 5-celled.

1. Fremontia californica Torr. (*Fremontodendron californicum* (Torr.) Coville). Peeples Valley (Yavapai County), near Payson, in the Mazatzal Mountains, and at the junction of Rock and Pinto creeks (Gila County), Cliff Dweller Canyon (Pinal County), reported to occur also in the Bradshaw Mountains (Yavapai County), rare and local, 3,500 to 6,000 feet, on dry, usually north slopes in canyons, May. Central Arizona, California, and Baja California.

A handsome plant when in flower, frequently planted in California as an ornamental, sometimes known as flannel-bush and California slippery elm. The bark is said to have the same properties as that of the true slippery elm (*Ulmus fulva*) and to be used for the same purpose, that is, to relieve irritation of the throat. Cattle browse this plant.

The Arizona specimens were referred by Harvey to *Fremontia californica* var. *typica* M. Harvey.

2. HERMANNIA

Plants herbaceous or slightly woody at base, loosely pubescent or glabrate; leaves deltoid or oblong-ovate, serrulate; flowers small, axillary, pendulous; calyx 5-cleft, the lobes longer than the tube; petals orange; anthers about 2.5 mm. long; styles sparsely hispidulous; capsule with stout teeth along the edges of the valves.

1. Hermannia pauciflora Wats. Santa Catalina and Tucson mountains (Pima County), about 3,000 feet, dry slopes, February to April, type from the Santa Catalina Mountains (*Pringle* in 1881). Southern Arizona and Sonora.

3. WALTHERIA

Plants herbaceous or suffrutescent, tomentose or canescent; leaves elliptic to broadly ovate, shallowly dentate; flowers in dense axillary clusters; calyx 5-toothed, the teeth shorter than the tube; petals pale yellow, fading reddish; anthers about 1 mm. long; style bearded; fruit 1-seeded, achenelike, pubescent.

1. Waltheria americana L. (*W. detonsa* Gray). Cochise County (?), Bear Valley or Sycamore Canyon (Santa Cruz County), Santa Catalina, Baboquivari, and Quinlan mountains (Pima County), 3,000 to 4,500 feet, canyons and slopes, flowering most of the year, type of *W. detonsa* from southern Arizona (*Wright* 904). Southern Arizona, Mexico, and widely distributed in tropical America.

4. AYENIA

Herbs or undershrubs, resembling certain Euphorbiaceae; flowers inconspicuous; petals with the claw much longer than the blade; staminodia appearing as teeth or lobes at the margin of the expanded apex of the stamen column; capsule 5-celled, conspicuously warty.

KEY TO THE SPECIES

1. Plants small, intricately branched shrubs; stamen column short, wholly funnelform; staminodia large, fleshy, crenate; blades of the petals deltoid-reniform, not adnate to the summit of the column, not appendaged dorsally; ovary and fruit very short-stipitate or subsessile in the calyx. Leaves ovate, crenate-dentate....1. *A. microphylla*
1. Plants suffrutescent, or herbaceous above the woody caudex; stamen column slender, elongate, abruptly expanded into the funnelform apical portion; staminodia small, toothlike; blades of the petals not reniform, adnate to the summit of the column, bearing a hornlike dorsal appendage; ovary and fruit distinctly stipitate..........2. *A. pusilla*

1. Ayenia microphylla Gray. Near Vail and in the Rincon and Tucson mountains (Pima County), Table Top Mountain (Pinal County), 2,000 to 3,000 feet, dry, rocky slopes and mesas, spring and late summer. Western Texas, southern Arizona, and northern Mexico.

A small shrub, less than 0.5 m. high.

2. Ayenia pusilla L. Grand Canyon (Coconino County), and from Greenlee, Gila, and Yavapai counties south to Cochise, Santa Cruz, and Pima counties, 2,000 to 4,000 feet, dry, rocky slopes, March to October. Southern Florida, southern Texas, southern Arizona, southeastern California, and widely distributed in tropical and subtropical America.

There is much variation in the shape of the leaves, which vary from narrowly lanceolate or oblong-lanceolate to ovate, with crenate-dentate to serrate margins.

86. GUTTIFERAE

1. HYPERICUM. St. Johns-wort

The Arizona species perennial, glabrous herbs; leaves opposite, simple, sessile, entire, glandular-punctate; flowers perfect, regular, the sepals and petals 4 or 5, the petals yellow or salmon-colored; stamens numerous, usually in 3 to 5 clusters with the filaments united below; ovary 3-celled or with 3 placentas; capsule dehiscent, many-seeded.

The common St. Johns-wort (*H. perforatum* L.), a European plant extensively naturalized in the United States, contains a photosensitizing principle that causes blistering and loss of hair in white-skinned horses, cattle, and sheep if exposed to strong sunlight after eating the plant. It is not known whether the native species have this effect.

KEY TO THE SPECIES

1. Stems erect from creeping rootstocks, 20 to 70 cm. long; leaves 10 to 35 mm. long, black-dotted along the margin; flowers several or numerous, in leafy terminal panicles; petals yellow, 7 to 14 mm. long . . . . 1. *H. formosum*
1. Stems procumbent, often forming mats, rooting at the lower nodes, 3 to 25 cm. long; leaves 4 to 12 mm. long, not black-dotted; flowers few, in terminal cymes, sometimes solitary, occasionally axillary; petals salmon-colored, 2 to 4 mm. long 2. *H. anagalloides*

1. Hypericum formosum H.B.K. Apache County to Coconino County, south to Cochise and Pima counties, 5,000 to 9,500 feet, moist soil in coniferous forests, July to September. Wyoming to Arizona, southern California, and Mexico.

2. Hypericum anagalloides Cham. & Schlecht. North Rim of Grand Canyon and V. T. Park, Coconino County, 8,000 feet, low ground (*Grand Canyon Herb.* 1013, *Goodding* 228–48). Montana to British Columbia, northern Arizona, and southern California.

87. ELATINACEAE. Water-wort Family

1. ELATINE (183, 184). Water-wort

Plants small, herbaceous, annual, semiaquatic, glabrous; stems slender, rooting at the nodes, seldom more than 5 cm. long; leaves opposite, simple, entire, with stipules, linear to spatulate; flowers axillary, minute, 2- to 4-merous; fruit an irregularly dehiscent capsule, with several to many sculptured seeds.

These insignificant little plants grow in or around shallow ponds.

KEY TO THE SPECIES

1. Seeds relatively long and slender, with more than 15 transverse ridges (and rows of pits) 1. *E. americana*
1. Seeds relatively short and thick, with 9 to 15 transverse ridges . . . . 2. *E. brachysperma*

1. Elatine americana (Pursh) Arn. San Francisco Peaks, Coconino County (*MacDougal* 273), northeastern Yavapai County (*Peebles* 14410), near Bonita, Graham County (*Shreve* 5219), 5,000 to 8,000 feet. Widely distributed in North America.

The collections cited may represent *E. triandra* Schkuhr rather than *E. americana.*

2. Elatine brachysperma Gray (*E. triandra* var. *brachysperma* Fassett). Near the San Francisco Peaks, Coconino County (*Lemmon* 3313), White Mountains, Apache County, 9,500 feet (*Phillips* 3308), Young to Payson, Gila County, 5,500 feet (*Peebles & Smith* 13295). Ohio to Oregon, south to Texas, Arizona, and southern California.

88. TAMARICACEAE. Tamarix Family

1. TAMARIX (184a)

Large shrubs or small trees with slender branches, these covered when young by the small, imbricate, scalelike leaves; flowers in slender spikes terminating the branchlets, usually perfect, regular, small, with 4 or 5 sepals and petals; stamens borne on a fleshy, lobed, hypogynous disk; styles 3 to 5; fruit a 3- to 5-valved capsule; seeds many, usually with a tuft of hairs at one end.

1. Tamarix pentandra Pall. Abundant along streams in most of the state at altitudes not above 5,000 feet, often forming extensive thickets, March to August. Naturalized from Eurasia.

Tamarix. A handsome plant, with deep-pink to very nearly white flowers, from which much honey is obtained in Arizona. In some places the plant is looked upon with favor as a control for too rapid soil erosion. It is seldom browsed by livestock but is used by cattle as a hiding place in the river bottoms. It can be grown on saline soils. The Arizona plants were referred previously to *T. gallica* L.

Athel (*T. aphylla* (L.) Karst.) is an evergreen tree from Africa frequently planted in Arizona for wind breaks and shade but almost never occurring spontaneously. The wood is brittle and weak, fragrant when burning. The trees grow rapidly but are objectionable on account of the expense of removal, which often is necessitated by danger of the heavy limbs falling and the fact that other plants cannot be grown to best advantage in the dense shade and against the competition of the shallow feeding roots.

89. COCHLOSPERMACEAE. Cochlospermum Family

1. AMOREUXIA (185)

Plants herbaceous, with short flowering stems from a large tuberlike root; leaves long-petioled, palmately lobed or parted with more or less wedge-shaped lobes; flowers few in a terminal raceme, large and somewhat irregular, the petals of unequal width and the numerous stamens dimorphic, one set with long, incurved filaments and purple anthers, the other set with shorter, straight filaments and yellow anthers; petals orange, all but the lowest bearing 1 or 2 large red spots; anthers opening by terminal pores; style and stigma 1; ovary 3-celled; capsule large, with a thick outer wall and a thin inner wall, these separating at maturity.

The flowers are handsome, and the capsules are singular on account of the hyaline endocarp, through which the seeds may be seen as through a window after the exocarp falls away or is removed. The fruits are said to be used as food in Sonora and Chihuahua, and the roots were roasted and eaten by the Indians of southern Arizona. They are reported to taste like carrots or parsnips.

KEY TO THE SPECIES

1. Stems and petioles puberulent; leaves glabrous; capsule broadly ovoid, less than 4 cm. long, short-acuminate, puberulent; seeds reniform, the outer coat close, hirsutulous; cotyledons oblong, at least twice as long as wide ....1. *A. palmatifida*

1. Stems and petioles short-pilose; leaves pilose on the veins beneath; capsule ellipsoid, 4 to 7 cm. long, long-acuminate, copiously short-pilose; seeds globose, the outer coat loose, pilose; cotyledons nearly orbicular ....2. *A. Gonzalezii*

1. Amoreuxia palmatifida Moç. & Sessé. Cochise, Santa Cruz, and Pima counties, near the Mexican boundary, 3,500 to 5,500 feet, rocky slopes and mesas, July and August. Southern Arizona and Mexico.

2. Amoreuxia Gonzalezii Sprague & Riley. Santa Rita Mountains, Pima County (*Wooton* in 1914). Southern Arizona and northwestern Mexico.

90. KOEBERLINIACEAE. Junco Family

1. KOEBERLINIA. Junco, All-thorn

Intricately branched, very thorny shrubs with green bark; leaves reduced to small scales; inflorescences lateral, few-flowered, umbellike or short-racemose; flowers perfect, regular; petals 4, somewhat hooded; stamens 8, the filaments thickened at the middle; ovary 2-celled, stipitate; fruit a globose berry with 2 to 4 seeds.

1. Koeberlinia spinosa Zucc. Greenlee, Graham, Gila, and Pinal counties, and from Cochise County west to near Tucson (Pima County), 2,500 to 5,000 feet, hillsides and mesas, May to July. Western Texas to southeastern Arizona and northern Mexico.

Sometimes called crown-of-thorns, crucifixion-thorn, and corona-de-Cristo. These very strongly armed plants repel livestock and no doubt assist in controlling soil erosion. Wherever abundant they are considered a range pest. They tend to form thickets.

The distribution in Arizona, as given in the first paragraph, is that of typical *K. spinosa*. Var. *tenuispina* Kearney & Peebles occurs farther west, in western Maricopa and Yuma counties, also in southeastern California and northwestern Sonora. Normally it differs from the typical plant in having longer and more slender branches, bluish-green rather than yellowish-green bark, longer and relatively narrower sepals, and longer petals and filaments. It flowers in March and April, hence earlier than typical *K. spinosa,* prefers sandier soil, and grows at lower elevations, 2,000 feet or lower. The plant is usually more open and taller, reaching a height of 4.5 m. (15 feet), whereas the typical plant seldom exceeds 1.8 m. (6 feet) in height. Further study may show this entity to be a distinct species, but there is considerable variation and some overlapping with typical *K. spinosa* in most of the characters.

91. VIOLACEAE. Violet Family

Plants herbaceous, annual or perennial; leaves simple, with stipules, alternate or all basal; flowers 5-merous, irregular, the lower petal concave, saccate or spurred; stamens 5, the anthers connivent or slightly coherent, the filaments very short or none; fruit a 3-valved capsule.

Two favorite garden plants, the pansy (*Viola tricolor*) and the sweet violet (*V. odorata*), belong to this family. Both are natives of the Old World. Many North American species of *Viola* have beautiful flowers.

KEY TO THE GENERA

1. Sepals not auriculate at base; lower petal concave or slightly saccate at base, constricted at the middle; plants caulescent, the stems leafy; flowers axillary, solitary or in small clusters .... 1. *Hybanthus*
1. Sepals dilated or auriculate at base; lower petal produced at base into a spur or deep sac, not noticeably constricted at the middle; plants caulescent or acaulescent; flowers solitary, on long, 2-bracted peduncles .... 2. *Viola*

1. HYBANTHUS (186)

Plants annual or perennial; stems leafy; leaves linear to rhombic-lanceolate; flowers axillary, solitary or few in a cluster, inconspicuous, greenish or purplish.

KEY TO THE SPECIES

1. Plants perennial; leaves sessile or subsessile, linear, lanceolate, or oblanceolate (the lowest ones sometimes obovate), not more than 6 mm. wide; corolla 3 to 4 mm. long, the lower petal little longer than the others, with a cucullate blade .... 1. *H. verticillatus*
1. Plants annual; leaves distinctly petioled, rhombic-lanceolate, up to 20 mm. wide; corolla 6 to 9 mm. long, the lower petal at least twice as long as the others, with a flat blade 2. *H. attenuatus*

1. Hybanthus verticillatus (Ortega) Baill. Greenlee County to eastern Mohave County, south to Cochise, Santa Cruz, and Pima counties, 3,500 to 5,500 feet, dry plains and mesas, May to September. Kansas and Colorado to Arizona and Mexico.

2. Hybanthus attenuatus (Humb. & Bonpl.) G. K. Schulze. Mule and Huachuca mountains (Cochise County), Santa Rita and Baboquivari mountains (Pima County), about 4,000 feet, rich soil in canyons, August and September. Southern Arizona to Central America.

2. VIOLA (187). Violet

Plants annual or perennial, caulescent or acaulescent; leaves lanceolate to obovate or round-reniform, entire to palmately parted; flowers on long peduncles, all or some of them large and showy, the corolla violet, yellow, or whitish, the lower petal spurred or deeply saccate at base; inconspicuous cleistogamous flowers also often present; stamens closely surrounding the ovary.

KEY TO THE SPECIES

1. Plants annual; stipules nearly as large as the blades, lyrate-pinnatifid. Corolla pale blue or whitish, the lateral petals bearded .... 1. *V. Rafinesquii*
1. Plants perennial; stipules much smaller than the blades, not lyrate-pinnatifid (2).

2. Plants strictly acaulescent, the leaves and scapes from rootstocks, these vertical to horizontal. Herbage glabrous, puberulent, or sparsely short-pilose; corolla normally violet, the spur rather short and thick (3).
2. Plants caulescent or soon becoming so (5).
3. Leaves palmately twice-parted or -divided, the divisions narrow.........2. *V. pedatifida*
3. Leaves undivided, the margins crenulate or denticulate (4).
4. Peduncles usually surpassing the leaves, often considerably; leaves round-reniform 3. *V. nephrophylla*
4. Peduncles scarcely surpassing the leaves; leaves ovate, short-cuneate to subcordate at base 4. *V. umbraticola*
5. Petals yellow above, often dark brown or dark purple on the back. Plants subacaulescent when young, becoming short-caulescent (6).
5. Petals violet or white, sometimes yellow at base (8).
6. Leaves all entire; spur puberulent externally. Herbage cinereous-puberulent, especially on the leaf veins ....5. *V. charlestonensis*
6. Leaves (or some of them) dentate or denticulate; spur glabrous externally (7).
7. Herbage green, sparsely pubescent to nearly glabrous; leaves somewhat elongate, entire to sparingly denticulate; upper petals commonly yellow on the back; capsule glabrous 6. *V. Nuttallii*
7. Herbage usually grayish and copiously pubescent; leaves relatively short and broad, mostly coarsely dentate; upper petals commonly brown or purple on the back; capsule puberulent ....7. *V. purpurea*
8. Petals white, purple-veined and usually purple-tinged on the back; stems mostly erect and tall (commonly 20 to 30 cm. long); spur saccate, much less than ⅓ as long as the petal blade. Leaves triangular-ovate to suborbicular, deeply cordate, crenate, glabrous or puberulent ....8. *V. canadensis*
8. Petals violet; stems mostly procumbent and short (seldom more than 10 cm. long); spur cylindric, half as long to as long as the petal blade, often hooked. Leaves denticulate or crenulate; petals bearded; stipules sharply serrate or laciniate..........9. *V. adunca*

1. Viola Rafinesquii Greene. Collom Camp, Mazatzal Mountains, western Gila County, 4,000 feet (*Collom* 1636), perhaps introduced. New York to Georgia, west to Michigan and Texas; eastern Colorado; south-central Arizona.

Field-pansy, the nearest relative of the cultivated pansy that is presumably indigenous in North America.

2. Viola pedatifida G. Don. Maverick, Apache Indian Reservation, about 7,500 feet (*Collins* in 1949), Mogollon Escarpment, Coconino County (?) (*Mearns* 44), pine forest, May and June. Ohio to Saskatchewan, south to Oklahoma, New Mexico, and central Arizona.

Larkspur violet. The presence of this plant in pine forests in central Arizona is remarkable. The species belongs mainly to prairies and plains east of the Rocky Mountains.

3. Viola nephrophylla Greene. Apache and Coconino counties, south to Cochise and Pima counties, 5,000 to 9,500 feet, rich soil in coniferous forests and wet meadows, April to June. Newfoundland to British Columbia, south to New Mexico, Arizona, and California.

Represented in Arizona chiefly by var. *arizonica* (Greene) Kearney & Peebles (*V. arizonica* Greene), characterized by sparsely pubescent or at least ciliate leaves, these glabrous in the typical plant. The type of *V. arizonica* was collected near Fort Verde, Yavapai County (*Mearns* in 1888).

4. Viola umbraticola H.B.K. Santa Catalina Mountains, Pima County (*Thornber* 5593, etc., *Peebles & Harrison* 2223), about 7,500 feet, ponderosa pine forest, June and July. Southern Arizona and Mexico.

The Arizona plants have been identified by M. S. Baker as var. *glaberrima* Becker, with herbage glabrous or obscurely puberulent (densely short-pubescent in typical *V. umbraticola*).

5. Viola charlestonensis Baker & Clausen. "Jacob's Pool," Arizona (*Jaeger*). Known otherwise only from Zion National Park, Utah, and the Charleston Mountains, southern Nevada.

Doubt is expressed by the authors of the species (*Madroño* 8 : 60) that the specimen cited really was collected in Arizona, but if the locality was Jacob Lake on the Kaibab Plateau (elevation about 8,000 feet) the range extension is not extraordinary.

6. Viola Nuttallii Pursh. White Mountains (Apache County), Kaibab Plateau (Coconino County), Pagumpa Springs (Mohave County), Ash Fork and Fort Verde (Yavapai County), 5,000 feet or higher, often with sagebrush, April and May. South Dakota to British Columbia, south to Kansas, Colorado, Arizona, and California.

7. Viola purpurea Kellogg. Gila County, on North Peak, Mazatzal Mountains, 6,000 feet (*Collom* 48), and in the Sierra Ancha (*Crooks et al.* in 1939). Montana to British Columbia, south to Colorado, central Arizona, and California.

Mrs. Collom's collection was annotated by Baker and Clausen as "a very luxuriant form."

8. Viola canadensis L. (*V. muriculata* Greene). Apache County to Coconino County, south to Cochise and Pima counties, 6,000 to 11,500 feet, rich, moist soil in coniferous forests, April to September, type of *V. muriculata* from the San Francisco Peaks (*Greene* in 1889). New Brunswick to British Columbia, south to South Carolina, Alabama, New Mexico, and Arizona.

V. muriculata is a more pubescent form, with leaves puberulent on both surfaces, often slightly scabrous above, and the stems also usually puberulent. It intergrades completely with typical *V. canadensis*. Specimens collected at base of the San Francisco Peaks and on the Mogollon Escarpment (*Mearns* 19, 94) were described as *V. canadensis* var. *scariosa* Porter, characterized as having exceptionally large, scarious stipules. A specimen from the Huachuca Mountains (*Goodding* 460) was identified by Milo S. Baker as *V. rugulosa* Greene (*V. canadensis* var. *Rydbergii* (Greene) House), which is stated to differ from *V. canadensis* chiefly in the production of underground runners (Baird, 187, p. 52).

9. Viola adunca J. E. Smith. Near Flagstaff, Coconino County (*MacDougal* 132), White Mountains, Apache County (*Phillips* 3277), 7,000 to 9,500 feet, coniferous forests. New Brunswick to Alaska, south to New Mexico, Arizona, and California.

92. PASSIFLORACEAE. Passionflower Family

Contributed by Ellsworth P. Killip

1. PASSIFLORA (188). Passionflower

Plants herbaceous or suffrutescent, the stems trailing or climbing, with tendrils opposite the leaves, these alternate, deeply lobed; flowers perfect, regular, normally 5-merous, on 1-flowered axillary peduncles, these often in pairs; receptacle expanded and disklike at base, prolonged into a column on which the ovary is borne; calyx throat bearing a conspicuous, single or multiple, fringed corona; petals present or absent; stamens 5, the filaments united below into a tube enveloping, and adnate to, the stalk of the 1-celled ovary; styles usually 3, the stigmas capitate; fruit a many-seeded berry.

The Arizona passionflowers are relatively inconspicuous, but some of the species of this chiefly tropical genus are highly prized cultivated ornamentals with very handsome large flowers. Two species, *P. quadrangularis* (the granadilla) and *P. edulis*, are cultivated for their fruits, the latter especially in Australia. The fruits are eaten fresh or are used for flavoring ices and for making fruit sirup.

KEY TO THE SPECIES

1. Leaves deeply 2-lobed, the lobes ascending, the margin entire; stipules narrowly linear or setaceous; seeds transversely sulcate; plants glabrous..............1. *P. mexicana*
1. Leaves 3- to 5-lobed, the lateral lobes divaricate, the margin dentate, denticulate, or sinuate; stipules semiovate or pinnatisect; seeds reticulate; plants pubescent nearly throughout (2).
2. Petioles biglandular near the apex; bracts setaceous; stipules semiovate; corona 1-ranked; ovary glabrous; herbage hispidulous.......................2. *P. bryonioides*
2. Petioles glandless; bracts deeply pinnatisect, with filiform divisions; stipules pinnatisect; corona in several ranks; ovary pilose; herbage grayish-villous...3. *P. foetida*

1. Passiflora mexicana Juss. Pinal, Cochise, Santa Cruz, and Pima counties, 2,500 to 5,000 feet, usually along streams, sometimes on dry mesas, July and August. Southern Arizona and Mexico.

2. Passiflora bryonioides H.B.K. Near Ruby, Santa Cruz County (*Harrison* 5644, *Goodding* in 1936, *Darrow* in 1938), about 4,000 feet, August and September. Southern Arizona and Mexico. Petals white, the corona purple.

3. Passiflora foetida L. Canyons on west side of the Baboquivari Mountains (Pima County), 4,000 to 5,500 feet, thickets, August and September. Almost throughout tropical and subtropical America.

Represented in Arizona by var. *arizonica* Killip, known only from Arizona and Sonora, type from the Baboquivari Mountains (*Harrison* 4774). The plant has a rank, disagreeable odor. The flowers are apparently vespertine and have a lilac-colored corona.

93. LOASACEAE. Loasa Family

Plants annual or perennial, herbaceous or suffrutescent; herbage rough-pubescent, the hairs often barbed, sometimes stinging; stems mostly brittle and with pale exfoliating bark; leaves alternate, simple but often deeply pinnatifid; flowers perfect, regular or nearly so, 5- (seldom 4-) merous, in terminal inflorescences or some of them solitary in the forks of the branches; calyx tube enclosing and adnate to the ovary (ovary inferior); petals yellow or whitish; stamens few to many; style 1, entire or 3- to 5-cleft; ovary 1-celled; fruit a capsule, indehiscent or tardily and irregularly dehiscent, 1- to many-seeded.

This family is remarkable for the diversity and peculiar structure of the hairs.

KEY TO THE GENERA

1. Ovary of 1 carpel, 1-celled, containing a single, pendulous ovule; stamens 5 or fewer (2).
1. Ovary of more than one carpel, each carpel containing several to many ovules borne on parietal placentas; stamens 10 or more (3).
2. Filaments very short, linear; anthers large, 2-celled, the connective produced into a hyaline spoon-shaped body longer than the anther cells..................1. *Cevallia*
2. Filaments elongate, filiform; anthers small, 4-celled, the connective not conspicuously produced ....2. *Petalonyx*
3. Carpels usually 3; placentas narrow or flat, not projecting far into the cavity of the ovary; ovules in 1 or 2 rows on the placenta; leaves not cordate at base....3. *Mentzelia*
3. Carpels commonly 5; placentas thick, more or less circular in cross section, projecting far into the cavity of the ovary and connected with the ovary wall by a thin plate; ovules in several rows on the placenta; leaves subcordate at base, round-ovate, crenate-dentate, rather thick...4. *Eucnide*

1. CEVALLIA

Plants herbaceous, perennial, canescent, also hispid with stinging hairs; leaves sinuate-pinnatifid; flowers small, in dense, narrow-bracted heads; calyx tube short, the calyx lobes and the petals similar and seemingly in 1 series, long, narrow, erect, plumose with white hairs.

1. Cevallia sinuata Lag. Greenlee, Graham, Pinal, Cochise, and Pima counties, 2,500 to 5,000 feet, dry mesas and slopes, June to November. Western Texas to southeastern Arizona and northern Mexico.

A noteworthy plant because of the stinging hairs and the peculiar structure of the flowers.

2. PETALONYX. SANDPAPER-PLANT

Plants woody, at least at base, the herbage scabrous; leaves entire or dentate; flowers small, in short, broad-bracted spikes, the petals white or whitish, the slender claws more or less coherent.

KEY TO THE SPECIES

1. Leaves broad at base, sessile; plants woody only at base. Herbage gray, very scabrous, densely hispidulous with retrorse hairs; leaves lanceolate to ovate, often few-toothed; floral bracts triangular-ovate, denticulate toward base; petals 4 to 6 mm. long
1. *P. Thurberi*
1. Leaves narrowed at base or short-petioled; plants woody well above the base (2).
2. Leaves sessile or nearly so, lance-oblong, 3 to 6 mm. wide, entire, green, muricate-scabrous, not shiny; inflorescences subcapitate in flower, elongate in fruit; floral bracts ovate-cordate, entire, obtuse or acutish, densely soft-pubescent; petals 4 to 5 mm. long
2. *P. linearis*
2. Leaves short-petioled, ovate to lanceolate, commonly more than 6 mm. wide, denticulate or dentate, somewhat shiny; inflorescences densely paniculate; floral bracts ovate or lanceolate, not cordate, often denticulate, more or less acuminate, rough-pubescent; petals 7 to 10 mm. long (3).

3. Petals 7 to 8 mm. long; leaves ovate, few-dentate, usually coarsely so, rounded, truncate, or short-cuneate at base; stems herbaceous above . . . . 3. *P. nitidus*

3. Petals about 10 mm. long; leaves lanceolate to ovate, finely crenate, cuneate at base; stems almost wholly woody . . . . 4. *P. Parryi*

1. Petalonyx Thurberi Gray. Mohave, Maricopa, Pinal, Pima, and Yuma counties, up to 5,000 feet but usually much lower, common in sandy soil, often in dry stream beds, April to October, type from the Gila River valley (*Thurber*). Arizona, Nevada, southeastern California, and northwestern Mexico.

2. Petalonyx linearis Greene. Yuma County near Laguna Dam on the Colorado River (*Jones* in 1906, *Harrison & Peebles* 5060) and at Tinajas Altas (*Goodding* in 1938), March and April. Southwestern Arizona, southeastern California, and Baja California.

3. Petalonyx nitidus Wats. Yucca, Mohave County, about 2,000 feet (*Jones* 4483), May. Western Arizona, southern Utah and Nevada, and southeastern California.

4. Petalonyx Parryi Gray. Western Coconino and northern Mohave counties, up to 3,000 feet, dry washes, April and May. Southern Utah and Nevada, and northern Arizona.

A rounded shrub about 3 feet high.

3. MENTZELIA (189). Stick-leaf

Plants annual or perennial, herbaceous or suffrutescent; herbage scabrous, without stinging hairs; flowers in terminal cymose inflorescences, or some of them solitary in the forks of the branches, small or large, cream-colored, yellow, or orange; stamens usually numerous, the outer ones often petallike.

As the common name implies, fragments of the leaves and stems stick readily to clothing and to the hair of animals. Many of the species are handsome in flower.

KEY TO THE SPECIES

1. Outer filaments cleft at the dilated apex, the anther borne on a slender prolongation of the filament between the 2 subulate or triangular-lanceolate teeth or lobes; petals ochroleucous; flowers closely subtended by deeply laciniate bracts. Stems stout, white or whitish, seldom more than 30 cm. long, branching at or near the base; petals 2 to 4 cm. long: section *Bicuspidaria* (2).

1. Outer filaments not cleft at apex; petals yellow or orange, often fading whitish; flowers not bracteate or, if loosely so, then the bracts not deeply laciniate, often entire (3).

2. Bracts at anthesis concealing the calyx tube, mostly broadly ovate, the lower portion of different texture from the upper, scarious or thinly membranaceous, whitish, the teeth usually more than 8 and shorter than the width of the undivided central portion of the bract . . . . 1. *M. involucrata*

2. Bracts at anthesis not concealing the calyx tube, lanceolate, green and of uniform texture throughout, the teeth 6 to 8, at least as long as the width of the undivided central portion of the bract . . . . 2. *M. tricuspis*

3. Stems not conspicuously whitish, these and the capsules hispid with stout, glochidiate hairs 0.5 to 1.0 mm. long; leaves thin, deep green, irregularly dentate and often subhastately lobed near the base; petals orange-yellow, 6 to 10 mm. long. Capsules elongate, clavate or narrowly obconic: section *Eumentzelia* (4).

3. Stems becoming conspicuously whitish, scabrous-puberulent, short-pilose, or glabrate; leaves thickish, light green or yellowish green, not subhastate; petals yellow (5).

4. Petioles of the lower leaves more than half as long as the blade; blades broadly ovate, less than twice as long as wide; filaments of the outer stamens petaloid-dilated toward the apex . . . . 3. *M. aspera*
4. Petioles less than half as long as the blade; blades lanceolate to ovate, often much more than twice as long as wide; filaments not dilated . . . . 4. *M. asperula*
5. Capsules clavate or narrowly obconic, more than 5 times as long as wide; seeds pendulous, thick, faceted, not winged; hairs not glochidiate, the longer ones very slender; plants annual; outer filaments not or scarcely dilated. Leaves mostly pinnatifid: section *Trachyphytum* or *Acrolasia* (6).
5. Capsules turbinate or broadly obconic, seldom more than 3 times as long as wide; seeds horizontal, thinly lenticular, winged; glochidiate hairs usually present, at least on the calyx; plants mostly biennial or perennial; outer filaments more or less dilated: section *Bartonia* or *Nuttallia* (8).
6. Seeds grooved on the angles . . . . 5. *M. affinis*
6. Seeds not grooved, or but slightly so on one of the angles. Leaves deeply pinnatifid, or the upper ones few-toothed or entire (7).
7. Petals 8 mm. long or longer . . . . 6. *M. nitens*
7. Petals less than 8 mm. long . . . . 7. *M. albicaulis*
8. Petals pale yellow when fresh, 15 to 20 mm. long. Stems up to 1.5 m. long, stout, strict, not branched below; leaves sinuate-dentate, the lower ones up to 20 cm. long; capsules 20 to 30 mm. long, commonly about 3 times as long as wide, acutish at base 8. *M. Rusbyi*
8. Petals bright yellow when fresh (9).
9. Stems usually tall (40 cm. or longer) and erect, seldom branched from the base, herbaceous; leaves mostly linear or oblong, sinuate-dentate to pinnatifid, rarely nearly entire; petals 10 to 20 mm. long; capsules 10 to 20 mm. long, acutish to rounded at base . . . . 9. *M. pumila*
9. Stems low (seldom more than 30 cm. long), commonly decumbent, diffusely branched from the base, often suffrutescent; leaves oblong, ovate, or somewhat obovate, shallowly toothed to nearly entire; petals 8 to 10 mm. long; capsules 9 to 12 mm. long, rounded at base . . . . 10. *M. puberula*

1. Mentzelia involucrata Wats. Western part of Maricopa, Pinal, and Pima counties, and in Yuma County, 3,000 feet or lower, dry, sandy soil, February to April. Southwestern Arizona, southeastern California, and northwestern Sonora.

This and the following species are similar in appearance, short-stemmed, with large pale-yellow or cream-colored flowers.

2. Mentzelia tricuspis Gray. Western Mohave County, 2,000 feet or lower, dry, sandy soil, March and April. Southern Utah to western Arizona and southeastern California.

3. Mentzelia aspera L. Santa Cruz and Pima counties, about 4,000 feet, thickets, August. Widely distributed in tropical and subtropical America.

4. Mentzelia asperula Woot. & Standl. Cochise, Santa Cruz, and Pima counties, 4,000 to 6,000 feet, slopes and mesas, usually among shrubs, sometimes on limestone, August to October. Western Texas to southern Arizona and northern Mexico.

5. Mentzelia affinis Greene. Canyon Lake (Maricopa County), base of Galiuro Mountains (Pinal County), near Tucson (Pima County), 1,500 to 2,500 feet, March to June. Southern Arizona and California.

Arizona specimens have smaller petals than most of those from California and are scarcely distinguishable from *M. albicaulis* except by the seed character.

6. Mentzelia nitens Greene. White Mountains (Apache County) and Mohave, southern Yavapai, Maricopa, Pinal, Pima, and Yuma counties, 3,000 feet or lower, February to May. Southern Utah, Arizona, and southeastern California.

Differs from *M. albicaulis* chiefly in its larger flowers, and is connected with it by *M. nitens* var. *Jonesii* (Urban & Gilg) J. Darl., a relatively small-flowered variant with petals not more than 12 mm. long and the upper leaves mostly entire or few-toothed. This variety occurs in Arizona throughout the range of *M. nitens* and intergrades completely with the typical plant. Resembling var. *Jonesii* but with slenderer stems and narrower leaves is var. *leptocaulis* J. Darl., known only from the type collection on Williams River (*Palmer* 157).

7. Mentzelia albicaulis Dougl. Throughout the state, 7,000 feet or lower, very common in sandy soil on plains and along washes, February to August. Wyoming to Washington, south to New Mexico, Arizona, and California.

Coextensive in range in Arizona with the typical plant and almost equally abundant is var. *Veatchiana* (Kellogg) Urban & Gilg, which has petals 4 to 6 mm. long, whereas in typical *M. albicaulis* they are not more than 4 mm. long. This variety seems scarcely distinguishable from *M. albicaulis* var. *gracilis* J. Darl. Parched meal made from the seeds of *M. albicaulis* was eaten by the Arizona Indians.

8. Mentzelia Rusbyi Wooton. Coconino County, 6,000 to 8,500 feet, mostly in pine forests, July to September, type from Bellemont, Coconino County (*Rusby* 614). Wyoming to New Mexico and northern Arizona.

Plant coarser, more robust, and larger-leaved than in any other of the Arizona species.

9. Mentzelia pumila (Nutt.) Torr. & Gray. Throughout the state, 100 to 8,000 feet, abundant in dry stream beds and along roads, February to October. Wyoming and Utah to southeastern California and northern Mexico.

KEY TO THE VARIETIES

1. Leaves sparingly and shallowly lobed to nearly entire; stems seldom more than 40 cm. long .... var. *integra*
1. Leaves sinuate-dentate to pinnatifid; stems usually more than 40 cm. long (2).
2. Petals 15 to 20 mm. long, obtuse (often appearing acute in dried specimens); outer filaments narrow .... var. *multiflora*
2. Petals 10 to 15 mm. long, acute or short-acuminate; outer filaments broad *M. pumila* (typical)

Typical *M. pumila*, including the scarcely distinguishable var. *procera* (Woot. & Standl.) J. Darl., occurs throughout the state, as does also var. *multiflora* (Nutt.) Urban & Gilg (*M. multiflora* Gray). Specimens that may be referable to *M. multiflora* var. *integra* Jones (*M. integra* Tidestrom) have been collected in the Grand Canyon and in northern Mohave County. The relationship of the latter variety to *M. puberula* J. Darl. needs clarification. All the varieties intergrade so completely that it is scarcely worth while to maintain them.

10. Mentzelia puberula J. Darl. Detrital Wash (Mohave County), Gila Mountains (Yuma County), apparently also in Havasu Canyon (western

Coconino County), 500 to 1,000 (2,000?) feet, dry rocky slopes, March and sometimes later. Western Arizona and southeastern California.

An undershrub, not more than 0.5 m. high.

A large, apparently somewhat woody plant of the section *Eumentzelia*, with thin, broad, shallowly lobed leaves and turbinate capsules about 1 cm. long, was collected in the Mule Mountains, Cochise County (*Goodding* 448). Better material is required for its identification.

4. EUCNIDE. Rock-nettle

Plants herbaceous, hispid with long, slender, sharp-pointed stinging hairs and shorter, segmented and barbed ones; leaves broadly ovate, subcordate, crenate-dentate; flowers few, in terminal cymes, the petals pale yellow or cream-colored, about 5 cm. long; filaments of the stamens all alike, none of them petaloid.

1. Eucnide urens Parry. Along the Colorado River from Grand Canyon (Coconino County) to Ehrenberg (northern Yuma County), 300 to 2,500 feet, dry, rocky slopes and canyons, April to September, type from the Colorado River valley (*Bigelow*). Southern Utah and Nevada to western Arizona and southeastern California.

Plant remarkable for the large size of the flowers and the stinging hairs.

94. CACTACEAE (190, 191, 191a, 192, 192a, 193). Cactus Family

The Arizona species mostly succulent, perennial, spiny and xerophytic plants with mucilaginous or rarely milky juice, characterized by complex cushionlike organs (areoles) from which spines, branches, or flowers arise; stems of 1 or more joints, these flattened, cylindric, or globose, often tuberculate or ribbed; leaves wanting or rudimentary; flowers perfect, mostly regular and solitary; perianth segments several to many, more or less united at base, inserted on a hypanthium; stamens numerous, inserted within the hypanthium tube; ovary inferior, 1-celled, the ovules parietal, numerous; style solitary; stigma lobes several; fruit several- to many-seeded, fleshy or dry, indehiscent or dehiscent, spiny, scaly, or smooth.

The cactus family is characteristic of the desert regions of Arizona, where the plants are abundant and conspicuous. The outstanding species in scenic appeal is the gigantic saguaro, the largest succulent in the United States.

KEY TO THE GENERA

1. Areoles furnished with glochids (barbed bristles); spines barbed or scabrous 15. *Opuntia*
1. Areoles not furnished with glochids; spines not barbed or scabrous (2).
2. Flowers borne in the axil of the tubercle or at base of the groove, at some distance from the spiniferous areole. Tubercles distinct, disposed in spiral rows....14. *Mammillaria*
2. Flowers borne at apex of the tubercle, contiguous with or actually on the spiniferous areole (3).
3. Flowers nocturnal, borne on the spiniferous areoles; stems greatly elongate, many times longer than thick; plants usually more or less branched above the base (4).
3. Flowers diurnal, not borne on the spiniferous areoles; stems not greatly elongate; plants cespitose or the stems unbranched (8).
4. Stems not more than 3 cm. in diameter, densely puberulent; spines less than 10 mm. long; flowers salverform, white; plants rarely 2 m. high; root tuberous. Fruit red (5).

4. Stems very stout, glabrous; spines more than 10 mm. long; flowers funnelform; plants arborescent, 4 to 17 m. high; roots not tuberous (6).
5. Branches 1.5 to 3 cm. thick, strongly angled with 3 to 6 prominent ribs; root large, carrot-shaped; flowers 12 to 20 cm. long; fruit 12 to 15 cm. long.......1. *Peniocereus*
5. Branches 5 to 8 mm. thick, not angled, deeply striate; ribs 6 to 9, low, broad, flattened; roots clustered; flowers 7 to 15 cm. long; fruit 4 to 5 cm. long...........2. *Wilcoxia*
6. Plants with a massive, continuous trunk more than 30 cm. thick, this usually producing several smaller, high branches; flowers white. Ribs 12 to 24 in number, 5 to 10 cm. apart; flowers 10 to 12 cm. long, borne in crownlike clusters at the ends of the branches; fruit ovoid, 6 to 9 cm. long, naked or sparsely spiny..........3. *Carnegiea*
6. Plants with several to many stems, these of about equal size, less than 20 cm. thick, mainly produced from the base of the plant; flowers pink (7).
7. Spines similar all along the stem; ribs 12 to 17 in number, 2 to 4 cm. apart; flowers 6 to 7.5 cm. long; fruit globose, 5 to 7.5 cm. in diameter, densely spiny 4. *Lemaireocereus*
7. Spines twisted and conspicuously longer on the upper (floriferous) part of the stem; ribs 5 to 7 in number, 6 to 10 cm. apart; flowers 3 to 4 cm. long, often 2 or more at an areole; fruit globose, 2 to 3 cm. in diameter, unarmed............5. *Lophocereus*
8. Hypanthium spiny; flowers lateral..................................6. *Echinocereus*
8. Hypanthium devoid of spines, commonly scaly; flowers terminal (9).
9. Fruit fleshy, persisting without drying for more than one season, finally dehiscing by means of a basal orifice; spines annulate. Stem conspicuously ribbed (10).
9. Fruit thin-walled, dry, dehiscent by vertical fissure or rarely by a basal orifice, persisting only a few weeks; spines not annulate. Plants small, rarely 50 cm. high (11).
10. Hypanthium and fruit densely woolly with long hairs from the axils of the scales, the scales linear-lanceolate to subulate, acuminate, entire; spines stout, curved but not hooked; stems commonly 20 to 30 cm. long.......................7. *Echinocactus*
10. Hypanthium and fruit glabrous, the scales orbicular-ovate, obtuse, finely ciliate; spines densely puberulent; stem normally solitary, 0.5 to 3 m. high..........8. *Ferocactus*
11. Stem not ribbed, the distinct tubercles arranged in spiral rows. Spines not hooked (12).
11. Stems ribbed (14).
12. Central spines strongly flattened and grooved, flexible, twisted and curved, papery 12. *Toumeya*
12. Central spines not flattened or flexible or papery (13).
13. Plants up to 15 cm. high; spines horny; outer perianth segments conspicuously fringed 11. *Utahia*
13. Plants very small, 3 cm. high; spines spongy except for a small hard core; perianth segments not fringed...13. *Navajoa*
14. None of the spines hooked or flattened...........................9. *Echinomastus*
14. One or more of the lower central spines strongly hooked, the upper central erect, not hooked, often flattened....................................10. *Sclerocactus*

1. PENIOCEREUS. Night-blooming Cereus, Reina de la Noche

Plants small; stems slender, several arising from a large, carrot-shaped root, strongly angled with few prominent ribs, densely puberulent, weakly armed with short, dark spines; flowers large, salverform, white, nocturnal; fruit ovoid, red, weakly spiny.

1. Peniocereus Greggii (Engelm.) Britt.& Rose. (*Cereus Greggii* Engelm.). Southern Mohave County to Graham, Cochise, Pima, and Yuma counties, commonly at low altitudes with *Larrea,* June and July. Western Texas, southern New Mexico, Arizona, and northern Mexico.

The beautiful, white, fragrant flowers last only one night. The root ordinarily weighs from 5 to 15 pounds, although occasional specimens are known to weigh as much as 87 pounds. It is reported that the Indians formerly utilized the root for food.

2. WILCOXIA

Plants small, branching; trunk arising from a cluster of dahlialike roots; branches very slender with broad, flat ribs and narrow grooves, densely puberulent; areoles minute, weakly armed with bristlelike spines; flowers salverform, nocturnal.

1. Wilcoxia Diguetii (Weber) Peebles (*Cereus Diguetii* Weber, *Neoevansia Diguetii* Marshall). Near the Sonora boundary in southern Pima County (*George Lindsay, George Bean*) and in Sonora, Mexico in the vicinity of Sonoita (*J. Whitman Evans*), flowering in late July. Southern Arizona and Sonora.

3. CARNEGIEA. Saguaro, Giant Cactus

Plants arborescent, tall and massive, the continuous, columnar trunk usually bearing several erect branches well above the base; stems stout, prominently many-ribbed, heavily armed with stout, straight spines; flowers spineless, funnelform, white, nocturnal; fruit oblong-obovoid, sparsely and weakly spiny.

1. Carnegiea gigantea (Engelm.) Britt. & Rose (*Cereus giganteus* Engelm.). Yavapai and Mohave to Graham, Santa Cruz, Pima, and Yuma counties, up to 3,500 or exceptionally 4,500 feet (the highest recorded station 5,100 feet), warm situations in well-drained soil, common, flowering May and June, type from along the Gila River in southern Arizona (*Emory*). Arizona, Sonora, and very locally in southeastern California.

Designated the Arizona state flower, this massive succulent is the largest cactus in the state, occasionally attaining a height of more than 50 feet and developing as many as 50 arms. Large individuals are believed to be from 150 to 200 years old. The beautiful white flowers are not fragrant but have an odor like that of ripe melon.

The saguaro has contributed substantially toward the subsistence of the Pima and Papago Indians, furnishing materials for food and shelter. The great capacity for water storage, combined with slow rate of growth, enables the plant to fruit annually, more or less irrespective of drought. The fruit, or pitahaya of the early Spaniards, matures in June and July. The watermelon-red pulp is eaten fresh or stored in the form of sirup and preserves. At the annual harvest an intoxicating beverage is prepared by allowing the juice to ferment. The Papago Indians make a sort of butter from the seeds. The whitewing dove, a favorite game bird of Arizona sportsmen, feeds largely on saguaro seeds during the fruiting season.

4. LEMAIREOCEREUS. Organpipe Cactus

Plants large, arborescent, many-branched from the base, the trunk very short or wanting; stems many-ribbed, spiny; flowers pink-tinged, nocturnal, funnelform, the globose ovary spiny; fruit globose, densely spiny.

1. Lemaireocereus Thurberi (Engelm.) Britt. & Rose (*Cereus Thurberi* Engelm.). Southern Pinal, southern Maricopa, and western Pima counties, up to about 3,000 feet, flowering May to July. Arizona, Sonora, and Baja California.

The flowers open shortly after sunset and close the following day. The inner perianth segments are pale pink in the center, white along the margin and toward the base. The large, extremely spiny fruits are highly esteemed by the Papago Indians. The old name, pitahaya dulce, denotes the sweetness of the pulp. A large area in the desert south of Ajo, Pima County, has been set aside as a national monument for the preservation of the organpipe cactus and the sinita.

5. LOPHOCEREUS. Sinita

Plants large but not arborescent, many-branched; trunk wanting; stems simple, prominently few-ribbed, the spines longer on the upper (floriferous) part; flowers small, pink, 2 or more at an areole; fruit small, globose, unarmed.

1. **Lophocereus Schottii** (Engelm.) Britt. & Rose (*Cereus Schottii* Engelm.). Western Pima County, 1,000 to 2,000 feet, flowering April to August. Arizona, Sonora, and Baja California.

The odorless, pink flowers open soon after sunset and wither the following morning.

Rathbunia alamosensis (Coult.) Britt. & Rose (*Cereus alamosensis* Coult.), which occurs in Sonora, has been reported dubiously from western Pima County. This species has weak, usually reclining stems and scarlet flowers.

6. ECHINOCEREUS. Hedgehog or Strawberry Cactus

Plants small, rarely 0.5 m. high, cespitose or simple; stems 1-jointed, cylindric or ovoid, erect or strongly ascending, rarely decumbent; spines not hooked; flowers funnelform or subcampanulate, arising from a rupture of the epidermis immediately above mature spiniferous areoles, diurnal, lasting several days; hypanthium scales small, acute, entire, rather persistent; fruit with large and readily detachable spine clusters, thin-fleshed, juicy, edible; seeds black, tuberculate, the tubercles more or less confluent.

KEY TO THE SPECIES

1. Flowers scarlet or crimson; plants cespitose .... 1. *E. triglochidiatus*

1. Flowers purple; plants cespitose or simple (2).

2. Areoles narrowly elliptic; radial spines closely appressed and pectinate, not more than 12 mm. long, red or pink .... 2. *E. pectinatus*

2. Areoles oblong to circular; radial spines not pectinate. Central spines never wanting (3).

3. Central spines 2 to 6 or more, all well developed, more or less curved or twisted, the lower ones deflexed, commonly flattened and angled toward the base. Stems stout, up to 50 cm. long; ribs 11 to 18 .... 3. *E. Engelmannii*

3. Central spine solitary, terete, often accompanied by shorter superposed accessory centrals (4).

4. Central spine curved toward the base, deflexed; spines straw-colored, monochromatic, translucent. Accessory centrals 1 to 4 or wanting; stems stout, up to 50 cm. long; ribs 13 to 16 .... 5. *E. Ledingii*

4. Central spine not curved near the base; spines commonly opaque (5).

5. Ribs 12 to 22, usually 14 to 17; principal central spine porrect or deflexed, reddish, white, or rarely straw-colored, translucent. 1 or 2 accessory centrals on at least some of the areoles .... 4. *E. Boyce-Thompsoni*

5. Ribs 8 to 13, usually not more than 10; central spines dark brown or ashy gray 6. *E. Fendleri*

1. Echinocereus triglochidiatus Engelm. Throughout the state in various forms, but the typical phase, which more closely resembles var. *gonacanthus* than any other Arizona variety but is smaller and often lacks any central spines, has not been definitely located in Arizona.

KEY TO THE VARIETIES

1. Stems obviously green, not obscured by the spines; central spine usually solitary, conspicuously 6- or 7-angled . . . . var. *gonacanthus*
1. Stems green or grayish green, more or less obscured by the spiny armament (2).
2. Plants with few stems, these stout, 20 to 50 cm. long. Central spines 3 or 4, stout, straight, often angled, ashy gray or darkened . . . . var. *polyacanthus*
2. Plants strongly proliferous; stems seldom more than 20 cm. long (3).
3. Stems seldom more than 50; spines straight or nearly so, never flexuous, not or scarcely angled, often clear yellow; central spines 1 to 4, acicular . . . . var. *melanacanthus*
3. Stems often very numerous (500 or more) in large, hemispheric mounds; spines curved or even flexuous, angled, ashy gray; central spine usually solitary. . . . . var. *mojavensis*

Var. *gonacanthus* (Engelm. & Bigel.) Engelm. & Bigel. occurs near St. Johns, Apache County, and in New Mexico. Var. *polyacanthus* (Engelm.) L. Benson (*E. polyacanthus* Engelm., *E. arizonicus* Orcutt) is found in Graham, Pinal, Cochise, and Pima counties, mountains, 3,000 to 6,000 feet, flowering April and May, also in New Mexico and Mexico. Var. *melanacanthus* (Engelm.) L. Benson (*E. coccineus* Engelm.) is confined to the eastern half of the state, mountains, 4,000 to 9,000 feet, common, flowering May to July, and occurs also in Colorado, Utah, and New Mexico. Var. *mojavensis* (Engelm. & Bigel.) L. Benson (*E. mojavensis* Rümpler) is found in Navajo, Coconino, and Mohave counties, 4,000 to 6,000 feet, rocky situations and sandy plains, and extends to southern Utah, Nevada, and southeastern California.

2. Echinocereus pectinatus (Scheidw.) Engelm. The typical phase of the species, chiefly Mexican, has been collected in Arizona in the Perilla Mountains, Cochise County (*Harlan* in 1939). Var. *rigidissimus* (Engelm.) Engelm. (*E. rigidissimus* Rose), the rainbow cactus, is much more common in Arizona, occurring in Cochise, Pima, and Santa Cruz counties, 4,000 to 6,000 feet, rocky situations, flowering June to August. Also in northern Sonora.

Typical *E. pectinatus* is characterized by the presence of 2 to 7 (usually 4 or 5) central spines. Var. *rigidissimus* is distinguished by the absence of central spines and the presence of horizontal bands of color on the stems. The flowers are indistinguishable from those of typical *E. pectinatus*.

3. Echinocereus Engelmannii (Parry) Rümpler. Coconino and Mohave to Pima and Yuma counties, up to 5,000 feet, common, flowering February to May. Southern Utah to Sonora and Baja California.

A common and variable species. Ordinarily spines of more than one color occur at the same areole, and they may be white, brown, black, or yellow, opaque or translucent. Var. *Nicholii* L. Benson, characterized by yellow spines and lavender rather than purple flowers, occurs in western Pima County. Var. *decumbens* (Clover & Jotter) L. Benson (*E. decumbens* Clover & Jotter), which forms dense masses of elongate, slender, decumbent stems, occurs in Marble Canyon, Coconino County (type locality), and Palm Canyon, Yuma County.

4. Echinocereus Boyce-Thompsoni Orcutt (*E. Fendleri* var. *Boyce-Thompsoni* L. Benson). Coconino, Yavapai, Gila, Graham, and Pinal counties, 2,000 to 5,000 feet, flowering April and May. This abundant and well-defined species is endemic to Arizona.

Var. *Bonkerae* (Thornber & Bonker) Peebles (*E. Fendleri* var. *Bonkerae* L. Benson), distinguished by spines not more than 10 mm. long and stems not obscured by the armament, occurs in the Pinal and Santa Catalina mountains.

5. Echinocereus Ledingii Peebles. Pinaleno Mountains (Graham County), the type locality, 4,500 to 6,000 feet, flowering May. Known only from Arizona.

This very well-marked species was reported by Nichol as occurring also in Cochise and Pima counties, but specimens have been collected only in the Pinaleno Mountains.

6. Echinocereus Fendleri (Engelm.) Rümpler. Widely distributed in Arizona in one form or another. Utah, New Mexico, and Arizona.

KEY TO THE VARIETIES

1. Central spine curved and, except in age, strongly ascending, flexible, 2.5 to 4.5 cm. long; accessory central spines wanting; radials straight or curved; stem rarely more than 15 cm. long, flaccid .... *E. Fendleri* (typical)
1. Central spine straight, porrect; accessory centrals usually present on some of the areoles; stem rarely less than 15 cm. long, firm (2).
2. Spines 1 to 2.5 cm. long, stout, rigid; stems 1 to 5 in number, 8 to 25 cm. long var. *rectispinus*
2. Spines 2.5 to 6 cm. long, flexible; stems as many as 15 in number, 25 to 45 cm. long var. *robustus*

Typical *E. Fendleri* ranges from Apache to Coconino and Yavapai counties, 4,500 to 7,500 feet, not common, flowering May and June, also in Utah and New Mexico. Var. *rectispinus* (Peebles) L. Benson (*E. rectispinus* Peebles) is found in Graham, Pinal, Cochise, Pima, and Santa Cruz counties, 3,000 to 5,000 feet, grassland and hills, rather common, flowering April, type from Santa Cruz County, also in southern New Mexico. Var. *robustus* (Peebles) L. Benson (*E. robustus* Peebles) occurs in Graham, Cochise, and Pima counties, 2,000 to 3,000 feet, flowering in April, type from Pima County, also in adjacent Sonora.

7. ECHINOCACTUS

Plants low, simple or cespitose; stems ribbed, the low tubercles armed with stout annulate spines, these not hooked; flowers funnelform; hypanthium tube densely woolly with hairs produced in the axils of the linear to subulate scales; fruit densely woolly, fleshy, persistent, finally dehiscent by a basal orifice.

KEY TO THE SPECIES

1. Stem usually solitary, glaucous; ribs about 8, broad, obtuse, not obscured by the glabrous spines; flowers pink .... 1. *E. horizonthalonius*
1. Stems several to many; ribs 13 to 21, narrow and acute, partially obscured by the spines; flowers yellow (2).
2. Seeds tessellate; spines glabrate; hypantnium scales up to 3 cm. long 2. *E. xeranthemoides*
2. Seeds papillose; spines densely puberulent; hypanthium scales very short 3. *E. polycephalus*

1. Echinocactus horizonthalonius Lemaire. Silver Bell, Pima County, 2,500 feet, flowering May and June. Western Texas, southern New Mexico, Arizona, and Mexico. There is some reason to suspect that the Arizona plants were imported by mine laborers.

2. Echinocactus xeranthemoides Engelm. ex Coult. Coconino and Mohave counties. Southern Utah and Arizona.

3. Echinocactus polycephalus Engelm. & Bigel. Mohave and Yuma counties, desert regions at low altitudes, flowering February and March. Arizona, southern California, and northwestern Sonora.

Niggerhead cactus. In this species and the foregoing the plants are normally densely cespitose, forming mounds.

8. FEROCACTUS. Barrel Cactus, Bisnaga

Plants large, the simple stem cylindric, ribbed, armed with stout, annulate, often hooked spines; flowers funnelform; scales of the hypanthium orbicular to ovate, imbricate, ciliate, not woolly in the axils; fruit glabrous, fleshy, persistent, dehiscent by a basal orifice.

KEY TO THE SPECIES

1. Central spines usually brown or gray and hooked; stem thick, barrellike, the circumference approximating the length; inner perianth segments with a broad, purplish median stripe (2).
1. Central spines pink, red, or rarely straw-colored; stem of mature plants thinner, not barrellike; inner perianth segments with a narrow, reddish median stripe, or entirely yellow. Bristlelike spines normally borne on the margins of the areoles (3).
2. Bristlelike spines wanting; spines variable, often not hooked, not rarely as thick as wide ... 1. *F. Covillei*
2. Bristlelike spines borne on the margin of the areole; spines hooked, flattened ... 2. *F. Wislizeni*
3. Central spines porrect, commonly twisted and hooked................3. *F. acanthodes*
3. Central spines commonly deflexed or ascending, usually not much twisted, rarely hooked ... 4. *F. Lecontei*

1. Ferocactus Covillei Britt. & Rose (*Echinocactus Covillei* Berger). Pinal, Maricopa, and Pima counties, 1,500 to 4,500 feet, flowering June to August. Arizona and northern Sonora.

2. Ferocactus Wislizeni (Engelm.) Britt. & Rose (*Echinocactus Wislizeni* Engelm.). Greenlee and Cochise to Maricopa and Pima counties, up to at least 4,500 feet, flowering July to September. Western Texas, southern New Mexico, Arizona, and northern Mexico.

This and the preceding species grow very large. A specimen of *F. Wislizeni* that had attained a height of 11 feet was called to our attention by F. A. Thackery.

3. Ferocactus acanthodes (Lemaire) Britt. & Rose (*Echinocactus acanthodes* Lemaire). Yuma County, up to 3,000 feet, April to June. Southern Nevada to Arizona and Baja California.

4. Ferocactus Lecontei (Engelm.) Britt. & Rose (*Echinocactus Lecontei* Engelm.). Mohave, Maricopa, Pinal, Pima, and Yuma counties, up to 3,500 feet, flowering April and May. Southern Utah to Baja California, Arizona, and Sonora.

F. Lecontei should perhaps be treated as a variety of *F. acanthodes,* but there are several points of difference in the spine characters.

9. ECHINOMASTUS

Plants globose or cylindric, tuberculate-ribbed; spines straight or slightly curved, not hooked; flowers from nascent areoles; hypanthium scales broad, erose; fruit dry, bearing a few chartaceous scales, dehiscent by vertical fissure.

KEY TO THE SPECIES

1. Central spines 4 to 9, all alike, straight or slightly curved; flowers yellow or magenta ... 1. *E. Johnsoni*
1. Central spines 1 to 4 (if more than 1, then dissimilar), straight; flowers pink (2).
2. Central spines about 4, one very short and porrect, the upper ones appressed and otherwise like the radials ... 2. *E. intertextus*
2. Central spines 1 or 2, rarely several, the upper one long and erect, the lower porrect and commonly much shorter ... 3. *E. erectocentrus*

1. Echinomastus Johnsoni (Parry) Baxter (*Echinocactus Johnsoni* Parry, *Ferocactus Johnsoni* Britt. & Rose). Mohave and Yuma counties, up to 3,000 feet, flowering in April. Southwestern Utah to California and Arizona.

Beehive cactus. The flowers are magenta in the typical form, greenish yellow in *Echinocactus Johnsoni* var. *lutescens* Parish, the more common form in Arizona. The color phases scarcely deserve recognition.

2. Echinomastus intertextus (Engelm.) Britt. & Rose (*Echinocactus intertextus* Engelm.). Sonoita, Santa Cruz County (*Peebles & Loomis* SF 186), Fort Huachuca, Cochise County (*Goodding*). Southwestern Texas to Arizona and northern Mexico.

3. Echinomastus erectocentrus (Coult.) Britt. & Rose (*Echinocactus erectocentrus* Coult.). Pinal, Cochise, and Pima counties, 1,000 to 4,500 feet, March and April, type from Cochise County. Known only from Arizona.

10. SCLEROCACTUS

Plants small, globose or cylindric, undulate-ribbed, usually very spiny; spines straight or curved, at least 1 of the centrals hooked; flowers from nascent areoles, funnelform to subcampanulate; hypanthium scales few, small, erose, usually woolly in the axils; fruit thin-walled, dry, few-scaled or naked, dehiscent by a basal pore; seeds black, tuberculate.

KEY TO THE SPECIES

1. Upper central spines not very different from the other centrals. Flowers greenish yellow or purple, 2.5 to 4 cm. long, rarely larger; style puberulent, rarely if ever glabrous; stem ovoid, usually cylindric, densely armed ... 1. *S. parviflorus*
1. Upper central spines conspicuously dissimilar to the other centrals (2).
2. Upper central spine erect-appressed, strongly flattened, rapidly diminishing from a base 2.5 mm. wide; flowers purple or greenish yellow, 3 to 3.5 cm. long; stem globose or ovoid, sparsely armed. Style puberulent ... 2. *S. Whipplei*
2. Upper central spine not strictly erect, curved, flattened but slender, more or less prominently ribbed on the upper surface; flowers purple; stem ovoid to cylindric (3).
3. Style glabrous; flowers 5 to 7.5 cm. long ... 3. *S. polyancistrus*
3. Style puberulent; flowers 4 to 5 cm. long ... 4. *S. intermedius*

1. Sclerocactus parviflorus Clover & Jotter (*S. havasupaiensis* Clover). Navajo and Coconino counties, not uncommon, type from "Forbidding Canyon," Coconino County. Known only from Arizona.

2. Sclerocactus Whipplei (Engelm. & Bigel.) Britt. & Rose (*Echinocactus Whipplei* Engelm. & Bigel.). Apache County, Ganado to Chinle (*King* in 1938) and Lithodendron Creek (*Bigelow* in 1853, the type). Reported also from southwestern Colorado and southern Utah.

Var. *pygmaeus* Peebles is a very small form, sparsely armed and lacking the characteristic flattened and appressed upper central spine of typical *S. Whipplei*. Known only from the type collection in Apache County (*Peebles & Smith* SF 1054).

***3. Sclerocactus polyancistrus** (Engelm. & Bigel.) Britt. & Rose (*Echinocactus polyancistrus* Engelm. & Bigel.). Not definitely known to occur in Arizona. Utah, Nevada, and California.

4. Sclerocactus intermedius Peebles. Sweetwater, Apache County (*Peebles & Smith* SF 1059) and 9 miles southwest of Pipe Springs, Mohave County (*Peebles & Parker* 14712, the type). Known only from these collections.

11. UTAHIA

Plants small, globose; tubercles distinct, prominent, spirally arranged, crowded toward base of stem; spines rather stout, straight or somewhat curved but not hooked; flowers from nascent tubercles, small, yellowish, the outer perianth segments imbricate, oblong to lanceolate, fimbriate, the inner segments oblanceolate, entire; ovary short; fruit oblong, thin-walled, dry, dehiscent by vertical fissure; seeds obovoid, black, tuberculate, with a large hilum.

1. Utahia Sileri (Engelm.) Britt. & Rose (*Echinocactus Sileri* Engelm.). Locally common in the vicinity of Pipe Springs, Mohave County, the type locality. Known certainly only from northwestern Arizona.

The spines at the base of old plants are whitened, shreddy, and greatly weathered.

12. TOUMEYA

Plants small, cylindric; tubercles distinct, spirally arranged, in age shedding the spines; spines long, thin, flexible, curved; flowers borne on nascent tubercles, white; perianth segments lanceolate, acuminate, entire; scales of the hypanthium entire or erose, naked or sparsely hairy in the axils; ovary small, hemispheric; fruit small, globose, dry; seeds black, tessellate.

1. Toumeya papyracantha (Engelm.) Britt. & Rose (*Echinocactus papyracanthus* Engelm.). Near Showlow, Navajo County, 6,500 feet (*Bumstead* in 1935), where apparently not extremely rare, flowering in May. Northern New Mexico and Arizona.

13. NAVAJOA (194)

Plants very small, the stems globose, prominently tuberculate, not ribbed; spines few, slightly curved, glabrous, internally spongy except for a small ligneous core; flowers borne on young tubercles, campanulate; inner perianth segments white with a pink stripe, subacute; outer segments broadly oblong, very obtuse, maroon with white margin.

1. Navajoa Peeblesiana Croizat (*Toumeya Peeblesiana* Marshall). Discovered by Mr. Whittaker of the Arizona Highway Department in the vicinity of Holbrook, Navajo County, the type locality. Not known from any other station and apparently very rare.

14. MAMMILLARIA. Fishhook or Pincushion Cactus

Small or low plants, the stems short, solitary to numerous, ovoid, cylindric, turbinate, or hemispheric, furnished with crossing spiral rows of distinct nipplelike tubercles; spines straight or hooked; flowers borne in the axils of the tubercles or on the tubercles at the base of a groove, funnelform, diurnal, lasting several days; hypanthium scales usually ciliate, not extending to the ovary; fruit a smooth berry; seeds usually pitted.

KEY TO THE SPECIES

1. Tubercles grooved from apex to middle or base; flowers borne at apex of the stem, or laterally in *M. recurvata* (2).
1. Tubercles not grooved, or shortly and indistinctly grooved at apex; flowers borne well below the apex of the stem, or apically in *M. Wilcoxii* (6).
2. Spines yellow, the central one more or less curved or even hooked; seeds red or reddish brown, delicately reticulate or tessellate, shining. Flowers yellow; berry green (3).
2. Spines dark brown, whitish, or brown-tipped, straight; seeds red, pitted, 1 to 2.5 mm. long (4).
3. Tubercles short, crowded; central spines 1 to 2 cm. long, curved-reflexed, the radials shorter, appressed; stems several to many, obscured by spines, 10 to 20 cm. thick, broadly cylindric, the apex depressed; berry about 8 mm. long; seeds 1.5 mm. long. Flowers borne well below the apex of the stem .... 1. *M. recurvata*
3. Tubercles 15 to 30 mm. long, widely spaced except toward the base of the stem, often with large glands in the grooves; central spines 3 to 4 cm. long, stout, curved or hooked; stems solitary or several, not obscured by spines; berry 4 to 6 cm. long; seeds 3.5 mm. long .... 2. *M. robustispina*
4. Flowers 2.5 to 3.5 cm. long, pink or yellowish; stems solitary or few, ovoid or cylindric. Central spines white with dark tips; inner perianth segments narrowly lanceolate 3. *M. chlorantha*
4. Flowers 3.5 to 5.5 cm. long, pink; stems solitary, few, or numerous in congested mounds, globose to ovoid (5).
5. Inner perianth segments oblanceolate, pale pink; central spines white with dark-brown tips .... 4. *M. aggregata*
5. Inner perianth segments lanceolate or linear-oblanceolate, darker pink; central spines mostly dark brown .... 5. *M. arizonica*
6. Juice milky; stem turbinate or hemispheric; spines not hooked; seeds small, red, translucent, vertically rugose, the deep depressions or pits elongate and about as wide as the ridges (7).
6. Juice clear; stem cylindric, ovoid, or hemispheric; some of the spines hooked (rarely so in *M. Oliviae*); seeds black or dark brown, commonly opaque, when pitted the depressions circular or somewhat angled but not elongate (8).
7. Outer perianth segments erose or fimbriate; flowers greenish yellow; radial spines 9 to 15 12. *M. Macdougalii*
7. Outer perianth segments entire, somewhat crispate; flowers pink; radial spines about 20 13. *M. Heyderi*
8. Central spines rarely hooked, usually straight and short. Stem ovoid to cylindric; berry clavate, scarlet, 20 to 25 mm. long; seeds black, pitted .... 10. *M. Oliviae*
8. Central spines hooked (9).
9. Seeds coarsely rugose and finely reticulate, dark brown, 2 mm. long, the hilum greatly enlarged, corky, light brown. Stem ovoid to cylindric; berry obovoid-clavate, scarlet, 10 to 20 mm. long; central spines 1 to 4 at an areole, 1 or all of them hooked 6. *M. tetrancistra*

9. Seeds smaller, pitted, the hilum not enlarged (10).
10. Spines yellow. Stem hemispheric; berry obovoid, scarlet or red, 5 to 7 mm. long 7. *M. Mainae*
10. Spines not yellow, in various combinations of white and dark reddish brown (11).
11. Plants colonizing freely; stem rarely more than 3 cm. in diameter and 12 cm. long, flaccid. Berry obovoid to clavate, scarlet, 7 to 13 mm. long ....8. *M. fasciculata*
11. Plants solitary; stem rarely less than 5 cm. in diameter (12).
12. Berry scarlet, clavate, 12 to 25 mm. long; perianth segments spreading, pink with white margin; stem ovoid to cylindric, firm; spines glabrous ....9. *M. microcarpa*
12. Berry green or purplish, ovoid to subglobose, 10 to 15 mm. in diameter, borne at or near the apex of the stem; perianth segments erect with spreading tips, pink, or sometimes straw-colored or greenish white; stem hemispheric to cylindric, flaccid; spines glabrous or pubescent ....11. *M. Wilcoxii*

1. Mammillaria recurvata Engelm. (*Coryphantha recurvata* Britt. & Rose). Pajarito Mountains, Santa Cruz County, about 4,500 feet, flowering July. Arizona and Sonora.

A single plant may have as many as 50 stout stems.

2. Mammillaria robustispina Schott ex Engelm. (*Coryphantha robustispina* Britt. & Rose). Cochise, Pima, and Santa Cruz counties, 2,500 to 6,000 feet, flowering July. Arizona and northern Sonora.

The identification of a specimen from Cochise County (*Loomis & Peebles* SF 200) as *M. Engelmannii* Cory (*Coryphantha Muehlenpfordtii* (Poselger) Britt. & Rose) may be in error. Since herbarium material was not preserved, the reported occurrence of this species in Arizona cannot be confirmed.

3. Mammillaria chlorantha Engelm. (*M. deserti* Engelm., *Coryphantha chlorantha* Britt. & Rose). Near Wolf Hole, Mohave County (*Peebles & Parker* 14759). Southwestern Utah, southeastern California, and Arizona.

4. Mammillaria aggregata Engelm. (*M. vivipara* var. *aggregata* L. Benson, *Coryphantha aggregata* Britt. & Rose). Southeastern quarter of the state, 3,000 to 7,000 feet, common in rocky situations, flowering May. New Mexico and Arizona.

5. Mammillaria arizonica Engelm. (*M. vivipara* var. *arizonica* L. Benson, *Coryphantha arizonica* Britt. & Rose). Apache County to Mohave, Gila, and Yavapai counties, mostly north of the Mogollon Escarpment, common, 5,000 to 8,000 feet, flowering May and June, type from northern Arizona. Southern Utah and Arizona.

Plants grown at low altitude in southern Arizona retain the characteristically narrow, deep-pink perianth segments and dark spines.

6. Mammillaria tetrancistra Engelm. (*Phellosperma tetrancistra* Britt. & Rose). Mohave, southern Yavapai, Maricopa, Pinal, Pima, and Yuma counties, low desert regions. Southwestern Utah, southern Nevada, Arizona, and southern California.

Reported by Clover and Jotter (192a) from Marble Canyon and Grand Canyon, Coconino County.

7. Mammillaria Mainae K. Brandeg. (*Neomammillaria Mainae* Britt. & Rose). Santa Cruz and Pima counties, 2,000 to 4,000 feet, flowering July. Arizona and Sonora.

This species is not common in Arizona. It is found on rocky hills, but more often on plains with ironwood and mesquite.

8. Mammillaria fasciculata Engelm. (*Neomammillaria fasciculata* Britt. & Rose). Graham (?), Pinal, and Pima counties, 1,200 to 3,000 feet, flowering May and June, type from along the Gila River, Graham County (?). Arizona and Sonora.

This little fishhook cactus has great tolerance for saline soil and seems to show preference for deep soils. Although often abundant, the plants usually grow under shrubbery and are not easily found.

9. Mammillaria microcarpa Engelm. (*Neomammillaria microcarpa* Britt. & Rose, *N. Milleri* Britt. & Rose). Western and southern Arizona, up to 4,500 feet, common in both heavy and well-drained soil, flowering June and July, type from Pinal County. Southern Utah to western Texas, Arizona, California, and northern Mexico.

A widespread and variable species. *Neomammillaria Milleri,* type from near Phoenix, is one of the stout forms.

10. Mammillaria Oliviae Orcutt (*Neomammillaria Oliviae* Britt. & Rose). Cochise and Pima counties, 3,500 to 4,500 feet, type from Pima County. Arizona and Sonora.

A neat plant, hoary with a dense covering of short, white, prevailingly straight (not hooked) spines.

11. Mammillaria Wilcoxii Toumey (*Neomammillaria Wilcoxii* Britt. & Rose, *N. viridiflora* Britt. & Rose). Hualpai Mountain (Mohave County) and in Pinal, Graham, Cochise, and Santa Cruz counties, 2,500 to 5,000 feet or higher, flowering May and June, type from Arizona. Southern New Mexico and Arizona.

The original description of *M. Wilcoxii* is incomplete, and Toumey's plant may never be identified with certainty. The flowers are usually pink, and the berries are always greenish and subglobose, not scarlet and clavate. *Neomamillaria viridiflora* is a form with very pale-pink or greenish-white flowers known only from the vicinity of Superior, Pinal County.

12. Mammillaria Macdougalii Rose (*M. Heyderi* var. *Macdougalii* L. Benson, *Neomammillaria Macdougalii* Britt. & Rose). Cochise, Santa Cruz, and Pima counties, 3,000 to 5,500 feet, rocky situations, flowering March and April, type from Pima County. Known only from Arizona, but the species probably occurs in Sonora.

13. Mammillaria Heyderi Muehlenpfordt (*Neomammillaria Heyderi* Britt. & Rose). Bisbee, Cochise County (*Peebles* SF 922), also reported from the Peloncillo Mountains by Benson (193, p. 264), flowering May. Texas, New Mexico, Arizona, and northern Mexico.

15. OPUNTIA

Shrubs or herbaceous perennials with short-jointed stems; joints flattened or terete, often tuberculate but never ribbed; leaves small, fleshy, subulate, caducous; areoles furnished with glochids (barbed bristles); spines minutely barbed, at least at the tip, never hooked; floriferous and spiniferous areoles combined in one organ; flowers diurnal; hypanthium scales caducous, resembling the leaves; hypanthium tube short; perianth segments usually broad, entire, spreading; fruit indehiscent, fleshy or dry, glochidiate, often spiny; seeds bony, compressed or angled, pallid.

The flat-jointed kinds are known as prickly-pear, and those with edible fruits as tuna. Species with cylindric joints are called cane cactus, the spinier ones cholla. Flowers remain fresh only one day, at least in the Arizona species. Some opuntias increase rapidly and become pestiferous on range land subjected to prolonged overgrazing. On the other hand, some of the species provide emergency feed for cattle. Cattle can feed on opuntias for a limited time without undue injury provided the spines are burned or the joints macerated.

The genus presents many taxonomic difficulties, and satisfactory classification of the Arizona species is far from an accomplished fact. Diagnostic characters are often greatly altered by environment. Natural hybrids are not rare.

KEY TO THE SPECIES

1. Joints flattened. Spines not sheathed (2).
1. Joints not flattened (17).
2. Fruit dry. Plants small or low, rarely more than 0.5 m. high; areoles 5 to 15 mm. apart; joints inclined to be transversely marked or wrinkled (3).
2. Fruit fleshy. Flowers yellow (10).
3. Fruit not spiny (4).
3. Fruit spiny. Joints inclined to be tuberculate (5).
4. Branches erect, 1-jointed; areoles deeply depressed; joints unarmed, densely puberulent 1. *O. basilaris*
4. Branches prostrate or low, several-jointed; areoles depressed or level with the surface of the joint; joints unarmed or somewhat spiny, glabrous .... 2. *O. aurea*
5. Joints readily detached, not more than 5 cm. long, often ovoid or subglobose, spiny or unarmed. Plants very small and low, with 1 to several branches, these 1- to several-jointed; flowers yellow .... 3. *O. fragilis*
5. Joints firmly attached, 5 to 20 cm. long (6).
6. Areoles less than 10 mm. apart; spines appressed or strongly deflexed, 1 to 3 cm. long; flowers pale yellow. Joints usually suborbicular, spiny on most of the surface 4. *O. polyacantha*
6. Areoles mostly more than 10 mm. apart; spines not appressed, 2 to 20 cm. long; plants of the same species with either yellow or pink flowers (7).
7. Spines bristlelike, obscuring the surface of the joint, those at base of old joints flexuous and up to 20 cm. long .... 6. *O. ursina*
7. Spines not bristlelike, often numerous but not obscuring the surface of the joint (8).
8. Spines 4 to 12 cm. long, 4 or more at an areole. Joints obovate, 8 to 15 cm. long, not thickened, usually heavily armed over the entire surface .... 5. *O. hystricina*
8. Spines 2 to 6 cm. long (9).
9. Spines 4 or more at an areole, slender, acicular; joints 8 to 20 cm. long, elliptic or oblong, armed over the entire surface, usually thickened, sometimes subterete... 7. *O. erinacea*
9. Spines 1 to 4 at an areole, stout; joints 5 to 12 cm. long, elliptic to obovate, spiny only on the upper half, often somewhat thickened .... 8. *O. rhodantha*
10. Roots tuberous; plants small or low, at most not more than 0.5 m. high; joints rarely more than 15 cm. long, transversely marked or wrinkled. Spines 1 to 3 at an areole, mostly confined to the upper areoles; joints usually lead green......... 9. *O. plumbea*
10. Roots fibrous; plants commonly large, bushy; joints mostly more than 20 cm. long, not transversely marked or wrinkled (11).
11. Areoles 15 to 30 mm. apart; joints prevailingly orbicular, usually 6 to 10 mm. thick (12).
11. Areoles 30 to 50 mm. apart; joints usually broadest above the middle, seldom less than 10 mm. thick (14).

12. Principal spines deflexed or appressed, yellow, translucent, subsetaceous, 2 to 4 cm. long; joints yellowish green, usually crowded and orbicular, the areoles soon abundantly filled with large glochids .... 10. *O. chlorotica*
12. Principal spines spreading, never appressed, opaque, reddish brown or darker, coarse, 4 to 7 cm. long; joints purple-tinged (13).
13. Central spines wanting, or few and confined almost entirely to the upper margin; plants of erect bushy habit; joints crowded, orbicular or ovate .... 11. *O. santa-rita*
13. Central spines never wanting, produced from the sides as well as the margin of the joint; plants of rather open habit; joints not crowded, orbicular or obovate 12. *O. macrocentra*
14. Joints yellowish green, the spines short and few or wanting. Plants bushy (15).
14. Joints commonly gray-green, with more numerous and longer spines, obovate (16).
15. Joints oblanceolate, asymmetric, spineless or nearly so; fruits with a long stipelike base 13. *O. laevis*
15. Joints orbicular-obovate, symmetric, spiny on the upper margin and sometimes more or less so on the sides, the spines 2 to 3 cm. long; fruits not contracted into a stipelike base 14. *O. gilvescens*
16. Branches prostrate or decumbent; joints 15 to 25 cm. long; principal spines not appreciably flattened or angled, straight, spreading, often spirally twisted, 5 to 6 cm. long, reddish brown, variegated, or whitened; center of flower darkened 15. *O. phaeacantha*
16. Branches ascending or erect, or the lowermost prostrate, forming a bushy plant; joints commonly more than 25 cm. long; principal spines stout, more or less angled or flattened, slightly curved, 2 to 4 cm. long, the lower or lowest deflexed and whitened, those in the lateral areoles arranged in a characteristic figure resembling a bird's foot; center of flowers only slightly darkened .... 16. *O. Engelmannii*
17. Spines sheathed only at the tip, strongly flattened, scabrous; plants low, creeping, the stems few-jointed; joints more or less clavate. Base of the fruit contracted (18).
17. Spines completely invested with a loose, papery sheath, minutely barbed but not scabrous; plants not creeping (except in some forms of *O. Whipplei*), bushy or arborescent, the stems many-jointed; joints cylindric (20).
18. Joints mostly 3 to 7 cm. long .... 17. *O. Parishii*
18. Joints mostly more than 7 cm. long (19).
19. Tubercles extremely prominent, 25 to 45 mm. long, 8 to 11 mm. wide, 10 to 15 mm. high, always distinct .... 18. *O. Stanlyi*
19. Tubercles smaller, 5 to 30 mm. long, about 5 mm. wide, 5 to 10 mm. high, distinct or somewhat confluent .... 19. *O. Kunzei*
20. Ultimate joints rarely more than 1.5 cm. thick; spines usually fewer than 6 at an areole (21).
20. Ultimate joints more than 1.5 cm. thick; spines 6 to 30 at an areole (26).
21. Joints tessellate with somewhat rhombic tubercles; fruit dry, bristly. Joints gray, often purple-tinged .... 20. *O. ramosissima*
21. Joints not tessellate; fruit fleshy (22).
22. Mature fruit conspicuously tuberculate (except in var. *enodis*), 2.5 to 4 cm. long. Plants erect with a distinct trunk, or low and creeping; joints prominently tuberculate; principal spines 1 to 4; flowers greenish yellow .... 31. *O. Whipplei*
22. Mature fruit not tuberculate (23).
23. Flowers purplish. Spines 1 to 6, commonly 4, at an areole; tubercles long, narrow, rather prominent but not crowded; plants openly branched shrubs, about 1 m. high, without a definite trunk; fruit 2 to 2.5 cm. long, scarlet-tinged........... 22. *O. tetracantha*
23. Flowers yellow (24).
24. Flowers greenish yellow; principal spines 1 to 4 at an areole; joints strongly tuberculate, the tubercles crowded and only about twice as long as wide...... 31. *O. Whipplei*
24. Flowers bronze yellow; spines usually solitary at the areoles of the ultimate joints; tubercles low, not prominent, not crowded (25).
25. Fruits 25 to 40 mm. long, greenish yellow, obovoid or clavate, often proliferous; ultimate joints 7 to 10 mm. in diameter.................................. 21. *O. arbuscula*

25. Fruits 10 to 18 mm. long, scarlet, obovoid; joints 3 to 5 mm. in diameter. Plants less than 1 m. high, sparingly branched except toward base .... 23. *O. leptocaulis*
26. Fruit not persisting for more than one season (27).
26. Fruit persisting for more than one season, fleshy. Plants 2 to 4 m. high (30).
27. Fruit fleshy, bristly with numerous setaceous, readily detached spines; joints 3 to 5 cm. thick, clavate, impenetrably armed, readily detached. Tubercles nearly as broad as long, somewhat 4-sided; spines and sheaths straw-colored; flowers greenish yellow, the outer segments pink-tinged .... 24. *O. Bigelovii*
27. Fruit dry, the spines stout, firmly attached; joints less than 3 cm. thick (28).
28. Ultimate joints mostly less than 10 cm. long; tubercles about twice as long as wide, not strongly compressed except in new growth; plants typically with a well-developed trunk and very dense crown, 1 to 1.5 m. high. Flowers greenish yellow 25. *O. echinocarpa*
28. Ultimate joints mostly 15 to 30 cm. long; tubercles twice or thrice as long as wide, strongly compressed; trunk rarely well developed (29).
29. Joints strongly armed; spines numerous, the principal ones more than 2.5 cm. long, conspicuously sheathed; plants 1.5 to 3 (rarely up to 5) m. high..26. *O. acanthocarpa*
29. Joints rather weakly armed; spines less than 2.5 cm. long, the sheaths not conspicuous; plants 1 to 2 m. high. Flowers yellow, red, or rarely purple......... 27. *O. Thornberi*
30. Mature fruit evidently tuberculate, solitary. Joints 1.5 to 3 cm. thick; tubercles 6 to 15 mm. long; spines 10 to 15 mm. long; flowers purple, occasionally red or yellow 30. *O. spinosior*
30. Mature fruit slightly or not at all tuberculate (31).
31. Joints readily detached, impenetrably armed, 3 to 5 cm. thick, pale green; spines and sheaths straw-colored; fruits proliferous, suspended in chainlike clusters; flowers pink 28. *O. fulgida*
31. Joints not readily detached, very spiny but not impenetrably armed, about 2.5 cm. thick, elongate, usually purplish; spines dark-colored; fruits solitary or sparingly proliferous; flowers commonly purple, occasionally red or yellow............ 29. *O. versicolor*

1. Opuntia basilaris Engelm. & Bigel. Mohave, Yavapai, Maricopa, and Yuma counties, up to 3,000 feet, flowering March and April, type from Mohave County. Southern Utah to southern California, Arizona, and Sonora.

Beavertail cactus. The typical form is a handsome plant, with large, velvety, ash-gray joints and magenta flowers with somewhat crispate perianth segments.

2. Opuntia aurea Baxter. Northern Mohave County, type from Pipe Springs, about 5,000 feet, flowering May. Eastern California, southern Nevada, southwestern Utah, and Arizona.

The flowers are either yellow or pink. Plants at the type locality exhibit an amazing variety of forms, possibly from hybridization with some such species as *Opuntia rhodantha*. The fruit is not known to the writer but is assumed to be dry and spineless.

3. Opuntia fragilis (Nutt.) Haw. Apache County to Coconino County, 6,500 to 7,500 feet, with pines, flowering June. Wisconsin to British Columbia, south to Texas and Arizona.

4. Opuntia polyacantha Haw. Apache County to Coconino County, 6,000 to 7,000 feet, flowering June. Alberta to North Dakota, Washington, Texas, and Arizona.

A widely ranging and variable species. The typical plant appears to be rare in Arizona, but var. *trichophora* (Engelm. & Bigel.) Coult., characterized by long, flexuous, bristlelike spines at the base of old joints, is not uncommon in the northeastern part of the state.

5. Opuntia hystricina Engelm. & Bigel. (*O. erinacea* var. *hystricina* L. Benson). Apache County to Coconino County, 4,500 to 6,500 feet, common, flowering May and June, type from Coconino County. New Mexico and Arizona.

A plant found in the vicinity of Navajo Bridge, Coconino County, is characterized by extremely coarse white spines and rather large joints, but it does not seem to be properly referable to *O. hystricina.* On sandy soil between Chinle and Ganado, Apache County, *O. hystricina* has become pestiferous in its great abundance. In that locality the yellow and pink color phases are equally represented.

6. Opuntia ursina Weber (*O. erinacea* var. *ursina* Parish). Near Littlefield, Mohave County, 2,500 feet (*Peebles* SF 1007). Southern Nevada, Arizona, and southeastern California.

7. Opuntia erinacea Engelm. & Bigel. Northern Apache County and Mohave County, 3,000 to 5,000 feet, flowering May and June, reported also from Grand Canyon and Havasu Canyon by Clover and Jotter (192a). Southern Utah and Nevada, Arizona, and eastern California.

8. Opuntia rhodantha Schumann (*O. erinacea* var. *rhodantha* L. Benson). Apache County to Mohave County, 4,500 to 7,000 feet, common, flowering May and June. Western Nebraska, Colorado, Utah, and Arizona.

As observed in Arizona, the flowers are prevailingly pink or apricot pink.

9. Opuntia plumbea Rose (*O. delicata* Rose, *O. Loomisii* Peebles). Apache County to Mohave, Yavapai, Cochise, and Santa Cruz counties, 4,000 to 7,500 feet, common in juniper and pine forests, flowering May and June, type from Gila County, the types of *O. delicata* and *O. Loomisii* from Santa Cruz and Yavapai counties respectively. New Mexico and Arizona.

The flowers are yellow when fresh, but in the afternoon often change to shades of red or orange.

The inclusion of *O. Rafinesquei* Engelm. in *Flowering Plants and Ferns of Arizona* was due to misidentification of several collections. The species is not known to occur in Arizona.

10. Opuntia chlorotica Engelm. & Bigel. Throughout most of the state except in the northeastern quarter, common, 2,000 to 6,000 feet, flowering in spring, type from Mohave County. Southern Utah to Baja California, New Mexico, Arizona, and Sonora.

Pancake-pear, silver-dollar cactus.

11. Opuntia santa-rita (Griffiths & Hare) Rose (*O. chlorotica* var. *santa-rita* Griffiths & Hare). Cochise, Pima, and Santa Cruz counties, 2,000 to 4,000 feet, flowering in spring, type from "Celero" (Tucson?) Mountains. Arizona and probably Sonora.

Santa-rita cactus. The combination of bright-yellow flowers and purple joints is very attractive. The plant is often grown for ornament. This species is closely related to *O. Gosseliniana* Weber of Sonora and Baja California, which has joints well armed on the sides, long, brown principal spines and the secondary spines, when present, strongly deflexed or appressed. Specimens collected by A. A. Nichol in western Pima County perhaps should be referred to *O. Gosseliniana.*

12. Opuntia macrocentra Engelm. Greenlee County to Gila, Cochise, and Pima counties, 2,000 to 5,000 feet, flowering April and May. Western Texas, southern New Mexico, Arizona, and northern Mexico.

13. Opuntia laevis Coult. Gila, Pinal, Cochise, and Pima counties, 2,500 to 4,500 feet, infrequent in protected situations, flowering April and May, type from Pima County, reported also from the Grand Canyon by Clover and Jotter (192a). Known only from Arizona.

Spineless cactus. The plants attain a height of 3 or 4 m.

14. Opuntia gilvescens Griffiths (*O. flavescens* Peebles). Pima and Santa Cruz counties, 2,500 to 4,500 feet, flowering April and May, type from Pima County, type of *O. flavescens* from near Sells, Pima County.

This species is related to *O. laevis,* but differs in many respects and occurs on open grassland or mesas.

15. Opuntia phaeacantha Engelm. (*O. procumbens* Engelm. & Bigel., *O. angustata* Engelm. & Bigel., *O. tortispina* Engelm. & Bigel.). Throughout most of the state, 1,000 to 7,500 feet, common and variable, flowering April to June. Colorado, Utah, and Arizona to Texas and northern Mexico.

A rare, pink-flowered form, or perhaps a distinct species, occurs in the vicinity of House Rock Canyon, Coconino County (*Peebles* SF 997 and SF 1066).

A herbarium specimen collected in a garden at Yavapai Point, Grand Canyon (*Peebles* in 1935) consists of a fertile, tuberculate fruit and a photograph of two suborbicular joints. This rarity is an undescribed species, or perhaps the true *O. mojavensis* Engelm. & Bigel., which Engelmann figured with a tuberculate although sterile fruit.

The writer has seen no local material of *O. tenuispina* Engelm., although the species was reported to occur in Arizona.

16. Opuntia Engelmannii Salm-Dyck (*O. discata* Griffiths,? *O. canada* Griffiths). Common or abundant throughout most of the state, 1,000 to 6,500 feet or somewhat higher, flowering April to June. Texas to Arizona and Mexico.

In Arizona the plants are somewhat variable but usually recognizable, 0.5 to 1.5 m. high, the joints obovate, 20 to 35 cm. long. A very robust form, perhaps a tetraploid, is common in the southeastern counties. This form attains a spread of 2 to 3 m., and the joints are orbicular or slightly rhombic, 30 to 45 cm. long. The types of *O. discata* and *O. canada* were found in Pima County. *O. Engelmannii* has become a pest on some cattle ranges.

17. Opuntia Parishii Orcutt (*Corynopuntia Parishii* Knuth). 35 miles north of Hackberry, Mohave County (*Evans* in 1935). Nevada, Arizona, and southern California.

Reported from Apache and Navajo counties is *O. clavata* Engelm., which is distinguished from *O. Parishii* by white spines. *O. pulchella* Engelm., characterized by purple flowers and flattened but slender spines, has been reported from Arizona and is to be looked for in northern Mohave County.

18. Opuntia Stanlyi Engelm. (*Corynopuntia Stanlyi* Knuth). Greenlee, Graham, Gila, and Pinal counties, common, 2,000 to 5,000 feet, flowering June. Southwestern New Mexico and Arizona.

19. Opuntia Kunzei Rose (*O. Stanlyi* var. *Kunzei* L. Benson). Pima and Yuma counties, 500 to 2,000 feet, flowering May and June, type from Pima County. Arizona, southeastern California, and Sonora.

Typically all the tubercles are distinct, the joints clavate, 2 to 3.5 cm. thick, 8 to 10 or 15 cm. long, and the tubercles 10 to 30 mm. long, 5 to 10 mm. high. Var. *Wrightiana* (Baxter) Peebles (*Grusonia Wrightiana* Baxter, *Opuntia Wrightiana* Peebles, *O. Stanlyi* var. *Wrightiana* L. Benson) is a more robust form in which some of the tubercles are confluent, the joints subcylindric, 10 to 20 cm. long, 3 to 4 cm. thick, and the tubercles about 5 mm. high.

20. Opuntia ramosissima Engelm. (*Cylindropuntia ramosissima* Knuth). Mohave, Maricopa, western Pima, and Yuma counties, 500 to 2,000 feet, flowering May to September. Southwestern Utah and Nevada to Sonora and California.

21. Opuntia arbuscula Engelm. (*Cylindropuntia arbuscula* Knuth). Maricopa, Pinal, Pima, and Yuma counties, 1,000 to 3,000 feet, flowering May and June, type from Pinal County. Arizona and Sonora.

Pencil cholla. Typically a small, arborescent shrub with compactly branched crown and well-developed trunk, but sometimes the plant is less than 1 m. high and openly branched. The Papago Indians utilized the young joints of the pencil cholla and similar species as a boiled vegetable, but probably only in times of want.

The obscure tubercles and large fruits of *Opuntia vivipara* Rose, which is known only from a few plants in the vicinity of Tucson, Pima County, indicate relationship with *O. arbuscula*. On the other hand, *O. vivipara* resembles *O. tetracantha* in having 1 to 4 spines at an areole, these only 2 to 4 cm. long, and purplish flowers.

22. Opuntia tetracantha Toumey (*Cylindropuntia tetracantha* Knuth). Cochise, Pinal, and Pima counties, infrequent, flowering May, type from Pima County, reported also from Diamond Creek, Mohave County, by Clover and Jotter (192a). Known only from Arizona.

23. Opuntia leptocaulis DC. (*Cylindropuntia leptocaulis* Knuth). Southern Arizona, 1,000 to 5,000 feet, flowering May and June. Texas, New Mexico, Arizona, and Mexico.

Sometimes called Christmas cactus. Common and variable, especially in respect to development of the spines, but the small, scarlet fruits, very slender stems, and small size of the plant mark the species well.

24. Opuntia Bigelovii Engelm. (*Cylindropuntia Bigelovii* Knuth, *O. abyssi* Hester). Western and southern Arizona, up to 3,000 feet, common on talus slopes, flowering February to May, type from Mohave County. Southwestern Utah and southern Nevada to Sonora and Baja California.

Jumping cholla, teddybear cactus. The plants occur in abundance on warm slopes of desert mountains. The combination of barbed spines and densely armed, easily detached joints has earned profound respect for this formidable cholla.

25. Opuntia echinocarpa Engelm. & Bigel. (*Cylindropuntia echinocarpa* Knuth). Mohave and Yuma counties, up to 3,000 feet, flowering April, type from Mohave County. Southwestern Utah and southern Nevada to Arizona and Baja California.

26. Opuntia acanthocarpa Engelm. & Bigel. (*Cylindropuntia acanthocarpa* Knuth). Mohave, Yavapai, Maricopa, Pinal, Pima, and Yuma counties, common, 500 to 3,500 feet, flowering in spring, type from Mohave County, reported also from Havasu Canyon by Clover and Jotter (192a). Southwestern Utah and southern Nevada to Sonora and southern California.

The typical form is a robust plant of open habit, commonly 2 or 3 m. high, with the younger branches ascending at acute angles and the joints at least 2.5 cm. thick. In Pinal and Pima counties the typical form is replaced by var. *ramosa* Peebles, a bushy plant 1 or 2 m. high, with joints less than 2.5 cm. thick and red, yellow, or variegated flowers. The Pima Indians use the flower buds of var. *ramosa* for food. The product, which is prepared by a steaming process, keeps well and is eaten as needed, usually in combination with pinole or with saltbush greens.

27. Opuntia Thornberi Thornber & Bonker (*O. acanthocarpa* var. *Thornberi* L. Benson). Eastern Mohave County fide Benson (193) and southern Yavapai County to Graham and Pima counties, 1,500 to 3,500 feet, foothills and detrital slopes, flowering April and May, type from southern Arizona. Known only from Arizona.

28. Opuntia fulgida Engelm. (*Cylindropuntia fulgida* Knuth). Southern Arizona, up to 4,000 feet, common, flowering June to August. Arizona to Sinaloa.

Cholla. Var. *mammillata* (Schott) Coult., distinguished by less densely armed joints, occurs with the typical plant but often replaces the latter at the higher elevations. Cattle relish the fruits of both forms and will eat the viciously armed joints if the spines are scorched. In fact a few cattle acquire a taste for this cholla and actually eat the spiny joints as they occur in nature. The reticulate stems are used in the manufacture of small articles of furniture, such as picture frames and lamp stands.

29. Opuntia versicolor Engelm. (*Cylindropuntia versicolor* Knuth). Pinal and Pima counties, 1,000 to 4,000 feet, common, flowering May, type from Pima County. Arizona and northern Mexico.

Deerhorn or staghorn cholla.

30. Opuntia spinosior (Engelm. & Bigel.) Toumey (*O. Whipplei* var. *spinosior* Engelm. & Bigel., *Cylindropuntia spinosior* Knuth). Yavapai County to Greenlee, Cochise, and Pima counties, 1,000 to 5,000 feet, occasionally higher, common, flowering May and June, type from southern Arizona. Western New Mexico, Arizona, and northern Mexico.

An apparent hybrid between *O. spinosior* and *O. fulgida* is rather abundant in the bed of the Gila River between Florence and Casa Blanca, Pinal County. The hybrid plants propagate freely by means of fallen joints but produce very little seed.

31. Opuntia Whipplei Engelm. & Bigel. (*Cylindropuntia Whipplei* Knuth, *O. hualpaensis* Hester). Apache County to Mohave, Yavapai, and Gila counties, 3,500 to 6,500 feet, common, flowering June and July. Southwestern Colorado, southern Utah, western New Mexico, and Arizona.

Var. *enodis* Peebles, characterized by nontuberculate, shallowly umbilicate fruits, is known only from the type collection on Hualpai Mountain, Mohave County, about 4,000 feet (*Kearney & Peebles* SF 883). The fruits of *O. Whipplei* are utilized by the Hopi Indians for food and for seasoning food.

95. ELAEAGNACEAE. Oleaster Family

Shrubs or small trees, scurfy-pubescent with stellate hairs; leaves alternate or opposite, simple, entire; flowers small, apetalous, regular, perfect or unisexual, axillary solitary or in small clusters; perianth 4-toothed; stamens 4 to 8; ovary 1-celled, 1-ovulate, technically superior but the perianth tube becoming fleshy and closely investing the achene, the whole structure simulating a drupe.

KEY TO THE GENERA

1. Leaves alternate; flowers perfect or the plants polygamous; stamens 4.....1. *Elaeagnus*
1. Leaves opposite; flowers unisexual, the plants dioecious; stamens normally 8 ..... 2. *Shepherdia*

1. ELAEAGNUS. Oleaster

Shrubs or small trees, up to about 7.5 m. (25 feet) high, often thorny; leaves lanceolate or oblong-lanceolate, bright green above, closely silvery-lepidote beneath; flowers pale yellow, very fragrant; fruit oval, yellow, silvery-lepidote.

1. Elaeagnus angustifolia L. Oak Creek Canyon, southern Coconino County, 5,500 feet (*Kearney & Peebles* 12208), an escape from cultivation, apparently nowhere naturalized in Arizona.

Russian-olive, native of the Old World, often cultivated as an ornamental in the United States.

2. SHEPHERDIA. Buffalo-berry

Dioecious shrubs; leaves persistent or deciduous; stamens alternating with the teeth of a fleshy disk; fruits globose or ellipsoid.

KEY TO THE SPECIES

1. Leaves persistent, thick, often subcordate at base, not more than 3 cm. long; fruits almost globose, scurfy. Leaves rounded-oval, ovate, or suborbicular, silvery-lepidote above, densely yellowish or whitish lepidote-tomentose beneath......1. *S. rotundifolia*
1. Leaves deciduous, thin, not subcordate at base, up to 5 cm. long; fruit more or less elongate, not scurfy (2).
2. Plants somewhat thorny, 1 to 6 m. high; leaves oblong, cuneate at base, copiously to densely silvery-lepidote on both surfaces; fruits ovoid-ellipsoid, usually scarlet, edible ..... 2. *S. argentea*
2. Plants not thorny, 1 to 2 m. high; leaves elliptic-oblong to ovate, usually rounded at base, sparsely lepidote to nearly glabrous above, silvery- and rusty-lepidote beneath; fruits ellipsoid, yellowish red, unpalatable..............................3. *S. canadensis*

1. Shepherdia rotundifolia Parry. Navajo, Coconino, and eastern Mohave counties, 5,000 to 8,000 feet, May and June. Southern Utah and northern Arizona.

Round-leaf buffalo-berry. An evergreen shrub, commonly about 1 m. high, with silvery leaves and a compact habit of growth, common and even abundant in certain areas and pretty well distributed throughout northeastern Arizona, apparently preferring steep slopes (sides of mesas and badly gullied areas). The ripe fruit contains a sweet, watery, pale-yellow juice.

2. Shepherdia argentea (Pursh) Nutt. Houserock Valley, Coconino County, 5,000 feet, along a stream (*Darrow* 3014). Canada to Kansas, New Mexico, northern Arizona, and California.

Silver buffalo-berry. The berries are eaten both raw and cooked, and are said to make a delicious jelly.

3. Shepherdia canadensis (L.) Nutt. White Mountains, Apache County at Twelve Mile Creek, head of the Little Colorado River, and at Marsh Lake, about 9,000 feet, Kaibab Plateau (Coconino County). Newfoundland to Alaska, south to New York, New Mexico, Arizona, and Oregon. Russet buffalo-berry.

96. LYTHRACEAE. Loosestrife Family

Plants (in Arizona) herbaceous, annual or perennial; leaves simple, entire, opposite or alternate, without stipules; flowers perfect, regular or irregular, axillary or in terminal racemes or spikes; calyx with tube enclosing but free from the ovary, often with toothlike appendages in the sinuses; petals and stamens borne on the throat of the calyx; stamens 4 to 12; style 1, the stigma entire or lobed; ovary 2- to 4-celled; fruit a capsule; seeds numerous.

A well-known member of this family is the ornamental shrub crape-myrtle (*Lagerstroemia indica*), cultivated extensively in southern Arizona. The plant from which is obtained the cosmetic dye henna (*Lawsonia inermis*) also belongs to the Lythraceae.

KEY TO THE GENERA

1. Calyx broad, campanulate or turbinate in flower, hemispheric to globose in fruit, not prominently ribbed; petals usually inconspicuous, sometimes wanting. Leaves opposite; flowers axillary, solitary or in small, short-stalked glomerules (2).
1. Calyx narrow, cylindric, tubular, or subclavate, with several longitudinal ribs; petals usually conspicuous, normally rose purple (3).
2. Capsules regularly and longitudinally dehiscent; leaves attenuate at base, short-petioled or subsessile . . . . 1. *Rotala*
2. Capsules bursting irregularly; leaves some or all of them auriculate-clasping at base 2. *Ammannia*
3. Plants perennial; herbage glabrous or obscurely puberulent; leaves mostly alternate, sessile or subsessile; flowers nearly regular; calyx almost symmetric, narrowly cylindric or subclavate in fruit . . . . 3. *Lythrum*
3. Plants annual; herbage glandular-hispid; leaves mostly opposite, petioled; flowers irregular, the upper petals larger than the others; calyx oblique, turgid in fruit, hispid 4. *Cuphea*

1. ROTALA

Small annual herbs, glabrous or nearly so; leaves opposite, narrow; flowers regular or nearly so, axillary, mostly solitary, small, bibracteolate; calyx with appendages shorter to longer than the proper teeth; petals 4, or wanting; ovary ellipsoid; valves of the capsule minutely transverse-striate.

KEY TO THE SPECIES

1. Bractlets usually shorter than the calyx; appendages half as long as to about equaling the calyx teeth; ovary subovoid-globose . . . . 1. *R. ramosior*

1. Bractlets often twice as long as the calyx; appendages often 3 times as long as the calyx teeth; ovary ellipsoid..2. *R. dentifera*

1. Rotala ramosior (L.) Koehne. Apache Pass, Cochise County (*Lemmon* 493). Widely distributed in North America; South America; Philippine Islands.

***2. Rotala dentifera** (Gray) Koehne. No specimens from Arizona have been seen, but the type was collected near Santa Cruz, Sonora, a few miles south of the international boundary, at "margins of pools and mountain streams" (*Wright* 1063). Apparently known only from Sonora.

2. AMMANNIA

Annual herbs, differing from *Rotala* chiefly by the characters given in the key to genera and in the plants being usually larger, with the flowers in few-flowered, axillary cymes.

KEY TO THE SPECIES

1. Capsules equaling or somewhat surpassing the calyx lobes; flowers distinctly pedicelled, the pedicels often elongate ....1. *A. auriculata*
1. Capsules not or barely equaling the calyx lobes; flowers sessile or subsessile, the pedicels rarely up to 4 mm. long ....2. *A. coccinea*

***1. Ammannia auriculata** Willd. (*A. Wrightii* Gray). Not known definitely to occur in Arizona, but was collected near Santa Cruz, Sonora, and along the San Pedro River, possibly in Arizona (*Wright* 1062, type collection of *A. Wrightii*). Missouri to Texas, Arizona (?), and southward; almost throughout the warmer parts of the world.

According to Koehne (*Pflanzenreich* IV. 216:46) all American specimens of *A. auriculata* belong to var. *arenaria* (H.B.K.) Koehne.

2. Ammannia coccinea Rottb. Maricopa, Cochise, Pima, and Yuma counties, 4,000 feet or lower, infrequent in wet soil or shallow water, late summer and autumn. New Jersey to Washington, south to Florida, Arizona, and California; Mexico; West Indies; South America.

3. LYTHRUM. Loosestrife

Plants herbaceous or nearly so, perennial, stoloniferous; stems erect, very leafy, with exfoliating bark; leaves alternate or nearly opposite, sessile, the stem leaves linear-lanceolate, the leaves of the basal shoots oblong or somewhat oblanceolate; flowers mostly solitary in the axils, short-pedicelled, forming slender leafy spikes; petals normally rose purple.

1. Lythrum californicum Torr. & Gray. Apache, Yavapai, and Mohave counties to Cochise, Santa Cruz, and Pima counties, 500 to 5,500 feet, frequent in wet soil along streams and in bogs, June to August. Texas to southern Nevada, Arizona, California, and Mexico.

4. CUPHEA

Plants annual; stems erect, leafy, usually branched; leaves slender-petioled, broadly lanceolate to ovate, thin; flowers axillary, also in terminal, leafy racemes or narrow panicles; petals 6; stamens 10 to 12.

Several exotic species are cultivated as ornamentals, the best-known one being *C. platycentra*, the cigar-flower, with a scarlet, black-and-white-tipped calyx.

1. Cuphea Wrightii Gray. Graham, Cochise, Santa Cruz, and Pima counties, 4,000 to 6,000 feet, rich soil in canyons, August to October. Southern Arizona to Panama.

A variant with small, very narrow petals, these grayish with very short hairs, is var. *nematopetala* Bacigalupi. This was based on a collection near Bowie, Cochise County (*Jones* in 1884), and was collected by Jones also in the Huachuca Mountains.

97. ONAGRACEAE. Evening-primrose Family

Contributed by Philip A. Munz

Herbs or rarely shrubs, with simple, alternate or opposite leaves; stipules none; flowers perfect, axillary or in terminal racemes, the parts mostly in twos or fours; hypanthium adnate to the ovary and usually prolonged beyond it; sepals 4 (sometimes 2 or 5); petals 4 (sometimes 2 or 5), inserted at summit of the hypanthium; stamens as many or twice as many as the petals, borne at the summit of the hypanthium; ovary inferior, 4- (sometimes 2- or 5-) celled; style 1; stigma 4-lobed, capitate or discoid; fruit a capsule, rarely nutlike.

This family is notable for the beauty of the flowers in many of the genera. Numerous species of *Fuchsia, Zauschneria, Clarkia, Godetia,* and *Oenothera* are highly prized as cultivated ornamentals. The evening-primroses with large white flowers (*Oenothera deltoides* etc.) are among the handsomest of the desert plants, and in favorable springs make a glorious display, often growing in great profusion with sand-verbena (*Abronia*).

KEY TO THE GENERA

1. Sepals persistent, divided down to the ovary (2).
1. Sepals deciduous after flowering (3).
2. Sepals 5; petals 5, about 1 cm. long .... 1. *Jussiaea*
2. Sepals 4; petals 4 and minute, or lacking .... 2. *Ludwigia*
3. Flowers 2-merous. Fruit indehiscent, obovoid, usually with hooked hairs .... 10. *Circaea*
3. Flowers 4-merous (4).
4. Seeds with tufts of hairs (coma) at one end (5).
4. Seeds without coma (6).
5. Hypanthium 2 to 3 cm. long, funnelform, with a transverse row of scales within, about midway of its length; flowers scarlet .... 3. *Zauschneria*
5. Hypanthium less than 1 cm. long, without scales within, sometimes lacking; flowers not scarlet .... 4. *Epilobium*
6. Fruit nutlike, indehiscent .... 9. *Gaura*
6. Fruit a capsule, dehiscent (7).
7. Ovary 2-celled; hypanthium not prolonged beyond the ovary. Flowers minute; stems with capillary branches .... 8. *Gayophytum*
7. Ovary 4-celled; hypanthium prolonged beyond the ovary (8).
8. Anthers usually versatile, attached near the middle; petals yellow or white, rarely red except on aging .... 7. *Oenothera*
8. Anthers innate, attached near the base, erect; petals white or pink (9).
9. Petals distinctly clawed, the claw at least ¼ as long as the blade .... 5. *Clarkia*
9. Petals not or scarcely clawed, the claw not more than 1/10 as long as the blade .... 6. *Godetia*

1. JUSSIAEA (195)

Plants perennial, herbaceous, aquatic; leaves alternate; flowers 5-merous, solitary in the leaf axils; ovary 5-celled, many-ovulate; capsule cylindric-clavate.

1. Jussiaea repens L. South of Flagstaff (Coconino County), Pecks Lake (eastern Yavapai County), near Tempe (Maricopa County), 1,000 to 7,000 feet, shallow ponds, May to October. Widely distributed in the warmer parts of the world.

Yellow water-weed. The species is represented in Arizona by var. *peploides* (H.B.K.) Griseb. (*J. californica* (Wats.) Jepson), which ranges from Oregon to South America.

2. LUDWIGIA (196). Seed-box

Plants perennial, herbaceous, inhabiting marshes and wet places; leaves opposite (in the Arizona species); flowers 4-merous; petals often lacking; stamens 4; ovary flattened at the broad apex; capsule short, 4-valved.

1. Ludwigia palustris (L.) Ell. Pima County at Rillito Creek (*Lemmon* 176) and near Arivaca (*Thornber* 2913), also reported by Shreve from the Santa Catalina Mountains. Nova Scotia to California and southern Arizona, south to Guatemala.

Marsh-purslane. Characterized by longitudinal green bands on the fruit and by the 4 persistent sepals. The Arizona plant is var. *americana* (DC.) Fern. & Griscom.

3. ZAUSCHNERIA (197). Hummingbird-trumpet

Plants perennial, herbaceous, somewhat woody at base and with shredding bark; leaves sessile or nearly so, more or less fascicled, elliptic to elliptic-lanceolate, somewhat coriaceous, usually denticulate, pallid, more or less villous; flowers mostly 3 to 4 cm. long, scarlet, horizontal, fuchsialike.

1. Zauschneria latifolia (Hook.) Greene. Southern Coconino, Greenlee, and Gila counties, south to the Mexican border, west to the Ajo Mountains (Pima County), damp places on rocky slopes and in canyons, 2,500 to 7,000 feet, June to December.

The Arizona plant is var. *arizonica* (Davidson) Hilend (*Z. arizonica* Davidson), type from Metcalf, Greenlee County (*Davidson* 365). This variety occurs in southwestern New Mexico, southern Arizona, and northern Sonora.

4. EPILOBIUM. Willow-weed

Plants annual or perennial, herbaceous; leaves sessile or short-petioled; flowers axillary or in terminal racemes or panicles; hypanthium either short or not prolonged beyond the ovary; sepals 4; petals 4, usually notched, white, pink, or purplish; stamens 8, the alternate ones shorter; capsule elongate, subcylindric to clavate, 4-celled, loculicidal.

KEY TO THE SPECIES

1. Hypanthium not prolonged beyond the ovary; flowers large; petals 8 to 18 mm. long, entire, spreading .... 1. *E. angustifolium*
1. Hypanthium prolonged beyond the ovary; flowers smaller; petals 2 to 6 mm. long, notched, ascending (2).
2. Plants annual, of dry situations; stems with an exfoliating epidermis (3).
2. Plants usually perennial, mostly of moist situations; epidermis of the stems not exfoliating (4).
3. Stems glabrous except in the upper parts, usually more than 30 cm. long; leaves usually alternate, with fascicles in the axils; hypanthium 1 to 3 mm. long.... 2. *E. paniculatum*
3. Stems puberulent throughout, less than 30 cm. long; leaves mostly opposite, without fascicles; hypanthium scarcely 1 mm. long .... 3. *E. minutum*
4. Rootstocks bearing turions, i.e., globose or ovoid winter buds with fleshy overlapping scales (5).
4. Rootstocks not bearing turions (6).
5. Leaves lanceolate, appressed-erect, sharply denticulate................ 4. *E. Halleanum*
5. Leaves lance-ovate to ovate, spreading, subentire to denticulate...... 5. *E. saximontanum*
6. Stems mostly 10 to 20 cm. tall, simple above, with a few pairs of opposite leaves (7).
6. Stems mostly 30 cm. or taller, usually freely branching, especially above (8).
7. Leaves sessile, suberect, oblong or linear .... 6. *E. oregonense*
7. Leaves more or less distinctly petioled and spreading, ovate to elliptic-ovate 7. *E. Hornemanni*
8. Inflorescence glandular-pubescent .... 8. *E. adenocaulon*
8. Inflorescence whitish-pubescent, not glandular .... 9. *E. californicum*

1. Epilobium angustifolium L. Kaibab Plateau and San Francisco Peaks (Coconino County) to the mountains of southern Apache, Greenlee, and Graham counties, 7,000 to 11,500 feet, damp places, July to September. Widely distributed in the Northern Hemisphere.

Fire-weed, blooming-sally. The large, rose to almost lilac flowers, strongly perennial habit, and elongate capsules opening to expose masses of dingy coma are characteristic. The young shoots are said to make a palatable potherb.

2. Epilobium paniculatum Nutt. A polymorphous species of wide distribution in the western United States, occurring in Arizona in the following forms: (1) forma *adenocladon* Hausskn., with capsules and pedicels glandular-puberulent and petals rose to lilac, 5 to 7 mm. long, in Apache, Coconino, Yavapai, and Gila counties, dry, open and disturbed places, 5,000 to 8,500 feet, August to October; (2) forma *subulatum* Hausskn., with capsules and pedicels glabrous and petals as in the preceding, in Coconino and Yavapai counties; (3) forma *Tracyi* (Rydb.) St. John, with petals white, 2 to 3 mm. long, and capsule subglabrous, on the Kaibab Plateau, 7,500 to 8,000 feet (*Jones* 6054i, *Kearney & Peebles* 13662).

3. Epilobium minutum Lindl. Gila County, between Young and Payson, 5,500 feet (*Peebles & Smith* 13273), between Miami and Superior (*A. & R. Nelson* 1905), May. Montana to British Columbia, south to Arizona and central California.

4. Epilobium Halleanum Hausskn. (*E. Drummondii* Hausskn.). Kaibab Plateau, Grand Canyon, and Flagstaff (Coconino County), White Mountains (Apache and Greenlee counties), 8,000 to 10,000 feet, damp places, July and August. Colorado to British Columbia and Arizona.

5. Epilobium saximontanum Hausskn. (*E. ovatifolium* Rydb.). Lukachukai and White mountains (Apache County), San Francisco Peaks (Coconino County), Pinaleno Mountains (Graham County), 8,000 to 11,500 feet, moist places, July to September. Colorado to Alberta and Arizona.

6. Epilobium oregonense Hausskn. Phelps Botanical Area, White Mountains, Apache County, 9,500 feet, open woods and wet meadows (*Phillips* 3209, 3244). British Columbia to Idaho, eastern Arizona, Nevada, and southern California.

7. Epilobium Hornemanni Reichenb. San Francisco Peaks, Coconino County, 9,500 to above 10,000 feet (*Leiberg* 5720, *Collom* 955). Arctic America to Arizona and California; Europe.

8. Epilobium adenocaulon Hausskn. (*E. novomexicanum* Hausskn.). Widely distributed in Arizona, 4,000 to 9,500 feet, moist places, June to September. Eastern United States to British Columbia, Arizona, and Mexico.

Occasional specimens with narrow, petioled leaves, very slender stems, and an exceedingly close, appressed pubescence in the upper parts may be referred to the uncertain var. *perplexans* Trelease.

9. Epilobium californicum Hausskn. (*E. Fendleri* Hausskn.). Widely distributed in Arizona, 5,000 to 9,500 feet, moist places, June to October. Washington to New Mexico, Arizona, and Baja California.

Intergrades freely with E. *adenocaulon* and is perhaps only a variety of it.

5. CLARKIA (198)

Plants annual, herbaceous; leaves few, subopposite, lance-ovate to elliptic; inflorescence spicate; buds nodding; flowers rose purple; capsules 1 to 3 cm. long, 2 to 4 mm. thick, on short pedicels; seeds densely cellular-pubescent.

1. Clarkia rhomboidea Dougl. Gila and Pima counties, 3,000 to 6,000 feet, rather dry and often disturbed places, May and June. South Dakota to Washington, Arizona, and Baja California.

The rose-purple petals may or may not be dotted.

6. GODETIA (199)

Plants annual, herbaceous; leaves linear to spatulate, the lower ones usually deciduous, the secondary ones borne in fascicles; hypanthium with an inner ring of hairs; seeds with a fimbriate upper margin.

KEY TO THE SPECIES

1. Buds erect; hypanthium with the inner ring of hairs about 1/3 way from the base; petals lavender to purple; sepals usually distinct in anthesis; capsule 2 to 3 mm. thick
   1. *G. quadrivulnera*
1. Buds nodding; hypanthium with the inner ring of hairs near the summit; petals whitish; sepals united in anthesis; capsule 1 to 1.5 mm. thick .... 2. *G. epilobioides*

1. Godetia quadrivulnera (Dougl.) Spach. Gila County, Pinal Mountains (*Harrison* 1897), and Mazatzal Mountains (*Collom*), Pima County, Tucson (*Thornber* 2530), April and May. Washington to Arizona and Baja California.

2. Godetia epilobioides (Nutt.) Wats. Pinal, Maricopa, and Pima counties, 1,500 to 3,000 feet, damp and disturbed places, March to May. Southern Arizona to central California and Baja California.

7. OENOTHERA (200, 200a). EVENING-PRIMROSE, SUN-DROPS

Plants herbaceous; leaves alternate or basal; flowers yellow or white, rarely red except in age, when commonly so; hypanthium deciduous after flowering; sepals and petals 4; stamens 8, equal or unequal; capsule straight, curved, or coiled, membranaceous or woody, 4-celled, 4-valved, dehiscent; seeds many, naked.

The roots of *O. biennis,* and probably of the related *O. Hookeri,* are sometimes eaten. When cooked, they are reported to taste like salsify (oyster plant).

KEY TO THE SPECIES

1. Stigma with 4 linear lobes; flowers mostly vespertine (2).
1. Stigma capitate, discoid, or 4-toothed; flowers mostly diurnal (21).
2. Capsule sharply 4-angled or -winged, at least in the upper part, rather short for its thickness. Plants perennial (3).
2. Capsule terete or round-angled, elongate, not winged (8).
3. Flowers white to red; capsule club-shaped, the lower part narrow and sterile, the upper part thicker, ribbed or winged. Seeds in more than 2 rows in each cell: subgenus *Hartmannia* (4).
3. Flowers yellow; capsule body proper ovoid or prismatic (6).
4. Buds nodding; lower sterile part of the capsule cylindric and sessile; petals 2.5 to 4 cm. long, white to pink .... 15. *O. speciosa*
4. Buds erect; lower sterile part of the capsule tapering toward the base; petals 0.5 to 1.5 cm. long (5).
5. Petals rose, 5 to 10 mm. long; hypanthium 4 to 8 mm. long .... 16. *O. rosea*
5. Petals white to pink, 10 to 15 mm. long; hypanthium 10 to 20 mm. long. 17. *O. Kunthiana*
6. Seeds in 1 row in each cell, with corky tubercles; capsule broadly winged throughout its length. Hypanthium 5 to 15 cm. long; petals 3 to 5 cm. long; herbage pubescent, often grayish-tomentose: subgenus *Megapterium* .... 20. *O. brachycarpa*
6. Seeds in 2 rows in each cell, granular, with a winglike margin around the obtuse summit; capsule winged most widely in the upper half: subgenus *Lavauxia* (7).
7. Tube of the hypanthium 4 to 8 cm. long; sepals and petals 1 to 2 cm. long; wings of the capsule 2 to 5 mm. wide; herbage more or less pubescent................. 18. *O. flava*
7. Tube of the hypanthium 12 to 18 cm. long; sepals and petals 3 to 4 cm. long; wings of the capsule 1 to 3 mm. wide; herbage bright green, nearly glabrous 19. *O. taraxacoides*
8. Seeds sharply angled, in 2 rows in each cell of the capsule. Capsule somewhat cylindric or fusiform, gradually tapering upward; flowers yellow, opening in the evening: subgenus *Euoenothera* or *Onagra* (9).
8. Seeds not sharply angled (11).
9. Petals 10 to 14 mm. long, not very red in age. Free tube of the hypanthium 2 to 4 cm. long 3. *O. procera*
9. Petals 25 to 45 mm. long, reddish or purplish in age (10).
10. Free tube of the hypanthium 3 to 5 cm. long; leaves more or less sinuate-dentate 1. *O. Hookeri*
10. Free tube of the hypanthium 8 to 12 cm. long; leaves sinuate-serrulate to almost entire 2. *O. longissima*
11. Plants tufted or almost acaulescent; seeds with a deep furrow along the raphe. Capsule cylindric or lance-ovoid, thick-walled, usually ridged, often tuberculate: subgenus *Pachylophis* (12).
11. Plants definitely caulescent; seeds not with a deep raphal groove (14).

12. Plants perennial; sepals 25 to 35 mm. long; anthers 10 to 14 mm. long. Petals white, aging pink .... 12. *O. caespitosa*
12. Plants annual; sepals less than 30 mm. long; anthers not more than 10 mm. long (13).
13. Sepals 15 to 28 mm. long; petals yellow, aging red; anthers 5 to 10 mm. long 13. *O. primiveris*
13. Sepals 5 to 15 mm. long; petals white, aging pink; anthers 2 to 7 mm. long 14. *O. cavernae*
14. Capsule slightly enlarged in the upper part (or sometimes slightly narrowed in *O. albicaulis*); seeds in 2 rows in each cell, with shallow pits in regular rows: subgenus *Raimannia* (15).
14. Capsule slightly narrowed toward the apex, woody or membranaceous, cylindric; seeds in 1 row in each cell, not pitted. Flowers white: subgenus *Anogra* (17).
15. Flowers yellow; buds erect. Leaves entire to deeply laciniate; petals 5 to 15 mm. long 4. *O. laciniata*
15. Flowers white; buds nodding (16).
16. Plants perennial from slender underground rootstocks; hypanthium with conspicuous white hairs in the throat; petals 7 to 11 mm. long; capsule 8 to 20 mm. long 5. *O. coronopifolia*
16. Plants annual or winter annual; hypanthium not long-hairy in the throat; petals 15 to 40 mm. long; capsule 20 to 40 mm. long .... 6. *O. albicaulis*
17. Plants coarse spring or winter annuals; basal leaves rhombic, 2 to 8 cm. long; capsules woody, 2 to 7 cm. long, with exfoliating epidermis. Buds often shaggy. .7. *O. deltoides*
17. Plants perennial or biennial; basal leaves smaller and narrower; capsules not woody (18).
18. Upper portion of the plant conspicuously pilose with long, spreading hairs; hypanthium 3 to 5 cm. long; capsules quite erect, 2 to 3 cm. long. Sepals 2 to 3 cm. long 8. *O. neomexicana*
18. Upper portion of the plant not conspicuously pilose with long, spreading hairs, or if so, then also pallid with fine, appressed hairs; hypanthium 1.5 to 3.5 cm. long; capsules spreading, deflexed, or coiled (19).
19. Plants essentially glabrous; capsules usually contorted; anthers 5 to 10 mm. long 9. *O. pallida*
19. Plants (except in *O. runcinata* var. *brevifolia*) canescent to hoary; capsules spreading or curved, sometimes contorted; anthers 3 to 6 mm. long (20).
20. Underground system a running rootstock .... 10. *O. runcinata*
20. Underground system a taproot .... 11. *O. trichocalyx*
21. Hypanthium about 1 mm. long, lined with a lobed, orange disk; seeds with purple dots. Stigma capitate; capsules linear, often strongly refracted: subgenus *Eulobus* 21. *O. leptocarpa*
21. Hypanthium usually longer, not lined with a disk; seeds not purple-dotted (22).
22. Stigma discoid, somewhat shallowly 4-lobed; flowers yellow (23).
22. Stigma capitate; flowers yellow or white (26).
23. Tube of the hypanthium 5 to 15 mm. long; stamens of 2 lengths; sepals distinctly keeled, especially toward the tip: subgenus *Calylophis* .... 25. *O. serrulata*
23. Tube of the hypanthium 25 to 50 mm. long; stamens subequal; sepals not keeled: subgenus *Salpingia* (24).
24. Plants densely strigose or pilose, with a peculiar olive-gray-green color, low and densely tufted, 5 to 20 cm. high; leaves typically 5 to 15 mm. long. Petals 13 to 22 mm. long; sepals 7 to 12 mm. long, with free tips 1 to 2 mm. long .... 22. *O. lavandulaefolia*
24. Plants greenish or if grayish, then with longer leaves, not so densely tufted, often quite woody, 10 to 40 cm. high (25).
25. Stems subglabrous to minutely glandular-pubescent; leaf margins not crisped or wavy 23. *O. Hartwegii*
25. Stems conspicuously pubescent or even pilose; leaf margins tending to be crisped or wavy 24. *O. Greggii*
26. Capsule cylindric or tapering toward the tip, sessile or nearly so: subgenus *Sphaerostigma* (27).

26. Capsule cylindric or clavate, distinctly pedicelled: subgenus *Chylismia* (32).
27. Flowers yellow, often drying greenish, borne in the axils of the foliage leaves (28).
27. Flowers white, often drying pinkish, borne in terminal spikes (29).
28. Cauline leaves linear to lance-linear, 1 to 3 mm. wide; petals 2 to 3 mm. long; capsules not 4-angled ....30. *O. contorta*
28. Cauline leaves broadly lanceolate, 5 to 10 mm. wide; petals 3 to 7 mm. long; mature capsules 4-angled ....31. *O. micrantha*
29. Capsules terete, cylindric, linear, not thickened in the lower portion, usually scarcely if at all coiled, not noticeably attenuate at tip (30).
29. Capsules not strictly cylindric, somewhat enlarged at base and attenuate at tip, curved or bent (31).
30. Petals 5 to 7 mm. long, suborbicular; style exceeding the corolla; hypanthium 4 to 6 mm. long; capsules refracted or spreading, occasionally coiled............26. *O. refracta*
30. Petals 3 mm. long, spatulate; style shorter than the corolla; hypanthium 2.5 to 3 mm. long; capsules divaricately spreading......................27. *O. chamaenerioides*
31. Leaves chiefly basal, subglabrous, lance-ovate to oblanceolate; stems glabrous or nearly so, the epidermis exfoliating promptly; capsule 15 to 25 mm. long..28. *O. decorticans*
31. Leaves well distributed, glandular-pubescent to glandular-villous, ovate to oblong-ovate; stems glandular-pubescent to glandular-villous, the epidermis exfoliating tardily, if at all; capsule 10 to 15 mm. long.................................29. *O. Boothii*
32. Leaves orbicular-cordate, well distributed along the stem, not at all pinnatifid, commonly glandular-pubescent. Plants rather coarse; flowers yellow, becoming bright red in age; anthers glabrous; capsules on rather short pedicels, coarse-cylindric 32. *O. cardiophylla*
32. Leaves ovate, oblong, or lanceolate, mostly near the base of the plant, commonly pinnatifid (33).
33. Capsules linear, elongate, usually more than 2 cm. long (34).
33. Capsules somewhat clavate, usually less than 2 cm. long (36).
34. Stems slender, commonly freely branched above; pedicels capillary, 10 to 25 mm. long; capsules 1 to 3.5 cm. long; anthers glabrous in some forms. Petals 1.5 to 9 mm. long 35. *O. multijuga*
34. Stems coarse, commonly branched only near the base; pedicels short, usually 3 to 15 mm. long; capsules 2 to 9 cm. long, widely spreading; anthers hairy (35)
35. Petals bright yellow, 12 to 15 mm. long; stem usually with conspicuous spreading hairs, especially near the base; sepals pilose; capsules 5 to 9 cm. long, 2 to 3 mm. thick; hypanthium with a swelling within on each rib at the upper edge of the pubescent portion ....33. *O. brevipes*
35. Petals pale yellow, 8 to 12 mm. long; stem usually ashy-strigose, often without conspicuous spreading hairs; sepals strigillose to subglabrous; capsules 2 to 5 cm. long, 1 to 2 mm. thick; hypanthium without inner swellings..............34. *O. pallidula*
36. Branches in well-developed plants arising freely throughout the plant, capillary; capsules 3 to 9 mm. long; anthers oblong to linear-oblong, glabrous; style not exceeding the petals. Petals 3 to 7 mm. long.................................38. *O. Parryi*
36. Branches in well-developed plants few to several and arising at base of the plant only, not capillary; capsules 10 to 25 mm. long; anthers linear, with scattered white hairs; style exceeding the petals (37).
37. Stems slender; flowers few, not crowded; leaves ovate, subentire; petals usually less than 4 mm. long..36. *O. scapoidea*
37. Stems somewhat coarse; flowers crowded in close terminal clusters; leaves usually pinnatifid; petals 4 to 7 mm. long............................37. *O. clavaeformis*

1. Oenothera Hookeri Torr. & Gray. This species is represented in Arizona by two subspecies, both growing in damp places, 3,500 to 9,500 feet, flowering July to October. These are: subsp. *hirsutissima* (Gray) Munz, with hirsute as well as strigose pubescence, muricate stems, and papillose sepals; and subsp. *Hewettii* Cockerell (*O. irrigua* Woot. & Standl.), with mostly appressed and ashy pubescence, not or scarcely muricate stems, and not or scarcely papillose sepals. Subsp. *hirsutissima* occurs in Apache and Coconino counties, south to Cochise and Pima counties, and ranges from Colorado and Utah to northern Mexico. Subsp. *Hewettii* is known from Navajo, Coconino, Graham, and Gila counties, and ranges from Kansas to Nevada, south to northern Mexico. These subspecies intergrade with each other and with other forms of *O. Hookeri.*

2. Oenothera longissima Rydb. Two subspecies occur in Arizona, in damp and springy places, 4,000 to 8,000 feet, flowering July to September. These are: subsp. *typica* Munz, with pubescence ashy-strigose, stems not muricate, and inflorescence not glandular-pubescent; and subsp. *Clutei* (A. Nels.) Munz (*O. Clutei* A. Nels.), with pubescence more or less hirsute, especially above, stems somewhat muricate, and inflorescence glandular-pubescent. Subsp. *typica* is found in northern Navajo and Coconino counties, also in Utah and Nevada. Subsp. *Clutei* occurs in northern Coconino and Mohave counties, also in southern Utah and Nevada and southeastern California, the type from Navajo Mountain (*Clute* 4, in 1919).

3. Oenothera procera Woot. & Standl. Oak Creek, Coconino County (*Fulton* 7329), Ryan Ranch, southern Apache County (*Harrison* 4843). Southern Colorado, New Mexico, and Arizona.

Doubtfully distinct from *O. strigosa* (Rydb.) Mackenz. & Bush in its smaller flowers and less shaggy pubescence.

4. Oenothera laciniata Hill. Almost throughout the state except the extreme western part, 1,500 to 9,500 feet, disturbed and fairly damp places, commonly in pine forest, May to October.

Represented in Arizona by var. *pubescens* (Willd.) Munz, which ranges from Texas to Arizona, south to Ecuador. The flowers open in the evening.

5. Oenothera coronopifolia Torr. & Gray (*Anogra coronopifolia* Britton). Coconino County (several collections), Peach Springs (Mohave County), Benson (Cochise County), 5,500 to 8,000 feet, usually in dry soil on plains, June to August. South Dakota to Kansas, Utah, and Arizona.

6. Oenothera albicaulis Pursh (*Anogra albicaulis* Britton). Widely distributed in Arizona, 2,500 to 7,500 feet, rather dry, grassy and disturbed places, March to July. South Dakota and Montana to Chihuahua and Arizona.

7. Oenothera deltoides Torr. & Frém. (*O. trichocalyx* of authors, non Nutt.).

KEY TO THE VARIETIES

1. Hairs of the sepals, hypanthium, and upper stems spreading, not closely appressed (2).
1. Hairs of the sepals, hypanthium, etc. closely appressed, or both appressed and spreading (3).
2. Petals 2 to 4 cm. long; uppermost leaves usually not deeply divided, rarely pinnatifid; buds with straight hairs 1 to 1.5 mm. long; capsules 2.5 to 7 cm. long, 2 to 3 mm. thick at base .... var. *typica*
2. Petals usually less than 2 cm. long; uppermost leaves deeply sinuate-dentate or pinnatifid, the undissected portion only 3 to 4 mm. wide; buds with curly hairs 2 or more mm. long; capsules 1.5 to 3 cm. long, 3 to 5 mm. thick at base........ var. *Piperi*
3. Sepals with both appressed and scattered spreading hairs, and with a purple spot at the base of each spreading hair; cauline leaves usually deeply and regularly pectinate-pinnatifid; capsules very slender, 1.5 to 2.5 mm. thick at base........ var. *arizonica*

3. Sepals with appressed hairs only, without purple spots on the buds; leaves subentire to shallowly and coarsely dentate; capsules usually 2.5 to 5 mm. thick at base (4).
4. Sepals with free tips 1 to 2 mm. long; buds usually 4-angled toward the tip var. *decumbens*
4. Sepals without free tips; buds not at all 4-angled toward the tip.......... var. *cineracea*

Var. *typica* Munz occurs in Mohave and Yuma counties, usually below 1,000 feet, sandy, open places, February and March, ranging into the deserts of California and Baja California. Var. *Piperi* Munz has been collected at Fredonia, Coconino County (*Jones,* in 1929), and ranges from eastern Oregon to eastern California, Nevada, and Arizona. Var. *arizonica* Munz, type from near Tucson (*Thornber* 509), occurs in Pinal, Maricopa, and Pima counties, 500 to 2,500 feet, February to May, and is known only from Arizona. A collection on Hualpai Mountain, Mohave County, 5,000 feet (*Kearney & Peebles* 12668), is also to be referred here. A variant of var. *arizonica,* forma *floccosa* Munz, with the long hairs on the sepals so closely set as to make the buds appear like tufts of wool, is known from Phoenix and Tempe (Maricopa County). Var. *decumbens* (Wats.) Munz occurs in northern Mohave County, 2,000 feet or lower, March and April, ranging into Utah and Nevada. Var. *cineracea* (Jepson) Munz is found in Yuma County, below 1,000 feet, March and April, ranging into southeastern California and northwestern Sonora.

8. Oenothera neomexicana (Small) Munz (*Anogra neomexicana* Small). Oak Creek Canyon (Coconino County), 5,000 feet, Greenlee and Graham counties, 8,000 to 9,500 feet, dry, open places, July and August. New Mexico and Arizona.

9. Oenothera pallida Lindl. (*Anogra pallida* Britton). Apache County to Mohave and Yavapai counties, 650 to 7,500 feet, dry, open places, May to September. Eastern Washington to New Mexico and Arizona.

In Arizona many plants intergrade with the next species, but to the north it is much more distinct in its more glabrous and less deeply divided leaves.

10. Oenothera runcinata (Engelm.) Munz (*Anogra runcinata* Woot. & Standl.).

KEY TO THE VARIETIES

1. Herbage glabrous. Leaves pinnatifid................................ var. *brevifolia*
1. Herbage pubescent (2).
2. Leaves deeply dentate; herbage hirsute as well as strigillose........... var. *leucotricha*
2. Leaves deeply sinuate-dentate to runcinate-pinnatifid, hairy-strigillose, without longer hairs .... var. *typica*

Var. *typica* Munz occurs in Apache, Navajo, Coconino, Yavapai, Greenlee, Cochise, and Pinal counties, 4,000 to 7,000 feet, dry plains and hills, May to September, ranging to Texas and Chihuahua. Var. *brevifolia* (Engelm.) Munz has been collected in the Lukachukai Mountains, Apache County (*Goodman & Payson* 2900a), and at Clifton, Greenlee County (*Davidson* 240), ranging into New Mexico, Texas, and Chihuahua. Var. *leucotricha* (Woot. & Standl.) Munz apparently is represented by a single collection in the state, on Hualpai Mountain, Mohave County, 4,500 feet (*Kearney & Peebles* 12650), being known otherwise from central New Mexico.

11. Oenothera trichocalyx Nutt. Known in Arizona only from Beaver Dam Creek, Mohave County (*Maguire* 4886). Wyoming, Colorado, Utah, and northwestern Arizona.

This species, characterized by conspicuously hairy buds and deeply sinuate leaves, is very much like *O. runcinata* var. *leucotricha* in its capsules and pubescence but has a taproot instead of creeping rootstocks.

12. Oenothera caespitosa Nutt. (*Pachylophis caespitosa* Raimann).

KEY TO THE VARIETIES

1. Capsule pedicelled, linear-cylindric, straight, symmetric; plants usually caulescent, villous-hirsute throughout. Leaves sinuate-pinnatifid; capsule 3 to 4 cm. long, tuberculate var. *marginata*
1. Capsule sessile, ovoid, often curved and asymmetric; plants acaulescent (2).
2. Capsule tuberculate. Herbage pubescent with soft, somewhat curly, spreading hairs var. *Jonesii*
2. Capsule not tuberculate (3).
3. Plants nearly glabrous except for the canescent or villous leaf margins and veins; capsule glabrous or somewhat villous, about 2 cm. long....................var. *montana*
3. Plants finely cinereous throughout with appressed hairs; capsule appressed-canescent var. *australis*

The typical glabrous form of the species is not known in Arizona, but 4 varieties are found there: Var. *marginata* (Nutt.) Munz is well distributed throughout the state, up to 7,500 feet, dry, stony slopes, April to August, ranging from Colorado and Arizona to Washington and California. Var. *Jonesii* Munz, otherwise known only from Utah, has been found north of Cameron, Coconino County (*Ripley & Barneby* 8498). Var. *montana* (Nutt.) Durand, occurs in Apache, Navajo, and Coconino counties, 4,000 to 6,500 feet, May and June, ranging from Montana and Oregon to New Mexico and Arizona. Var. *australis* (Woot. & Standl.) Munz is known from several localities in Cochise County and is found also in New Mexico. A glabrescent form, approaching var. *longiflora* (Heller) Munz, was found on the north side of the Grand Canyon (*Eastwood & Howell* 1073a).

13. Oenothera primiveris Gray. Greenlee and Mohave counties, south to Cochise, Santa Cruz, Pima, and Yuma counties, below 4,500 feet, dry, open deserts, March to May. Texas to Nevada and California.

Plant cespitose, with yellow flowers and leaves peculiarly pinnatifid into lanceolate or ovate lobes with rounded teeth or lobes. A caulescent form, var. *caulescens* Munz, occurs in southern Yuma County, where it seems to be the prevailing form of the species.

14. Oenothera cavernae Munz. Coconino County, north and west of Navajo Bridge, at the Grand Canyon, and in Havasu Canyon, about 4,000 feet, April to June. Southern Nevada and northern Arizona.

The Arizona specimens differ from the typical Nevada plant in having much larger flowers and purple spots on the calyx.

15. Oenothera speciosa Nutt. (*Hartmannia speciosa* Small). Reported from Fort Verde, Yavapai County, by Britton (*Trans. N. Y. Acad. Sci.* 8: 67) as having been collected by *Mearns* (No. 330), and at one time introduced on the campus of the University of Arizona at Tucson (*Thornber* in 1904). Missouri and Kansas to Louisiana, Texas, and northern Mexico.

The large white to rose-colored flowers are very handsome.

16. Oenothera rosea Ait. (*Hartmannia rosea* G. Don). Pinal, Cochise, Santa Cruz, and Pima counties, 1,000 to 5,500 feet, occasional in river bottoms and canyons, April to August. Texas and southern Arizona to Bolivia.

17. Oenothera Kunthiana (Spach) Munz (*Hartmannia Kunthiana* Spach). Fort Huachuca, Cochise County (*Wilcox* in 1892). Western Texas to southeastern Arizona and central Mexico.

18. Oenothera flava (A. Nels.) Garrett (*Lavauxia flava* A. Nels.). Navajo, Coconino, Mohave, Yavapai, and Gila counties, 2,500 to 9,000 feet, fairly damp flats and meadows, April to September. Canada to New Mexico, Arizona, California, and Mexico.

19. Oenothera taraxacoides (Woot. & Standl.) Munz (*Lavauxia taraxacoides* Woot. & Standl.). Apache, Navajo, Coconino, Graham, and Gila counties particularly in the White Mountains, 5,000 to 9,500 feet, sandy or wet soil in pine forests, May to August. Western Texas to Arizona and Chihuahua.

20. Oenothera brachycarpa Gray (*Lavauxia brachycarpa* Britton). Navajo, Coconino, Cochise, and Pima counties, 4,000 to 6,000 feet, rare on dry slopes and benches, May to July. Idaho to Texas, Arizona, and Chihuahua.

Represented in Arizona by var. *Wrightii* (Gray) Léveillé (*Oenothera Wrightii* Gray, *Lavauxia Wrightii* Small).

21. Oenothera leptocarpa Greene (*Eulobus californicus* Nutt.). Greenlee, Yavapai, and Mohave counties south to Pima and Yuma counties, below 4,500 feet, dry slopes, plains, and washes, February to June. Western Arizona and southern California to Sonora and Baja California.

The cruciferlike aspect of this plant is striking.

22. Oenothera lavandulaefolia Torr. & Gray (*Galpinsia lavandulaefolia* Small). 10 miles southeast of Tuba, Coconino County, 5,500 feet (*Peebles & Smith* 13363), near Oracle, Pinal County (*Peebles & Loomis* 7117), these being apparently the only collections in Arizona of var. *typica* Munz, which ranges from Wyoming to Texas and Arizona.

Hypanthium and calyx strigose-canescent in var. *typica*. Var. *glandulosa* Munz, with hypanthium and calyx glandular, occurs in Navajo, Coconino, and Mohave counties, 5,000 to 7,500 feet, dry slopes and flats, May to July, ranging to Nevada, Colorado, and Texas.

23. Oenothera Hartwegii Benth. (*Galpinsia Hartwegii* Britton).

KEY TO THE VARIETIES

1. Fascicles of small leaves present in the main axils; free tips of the calyx lobes 3 to 12 mm. long; petals rounded at apex, not noticeably rhomboidal; larger leaves widespreading ....var. *Toumeyi*
1. Fascicles of small leaves usually not evident; free tips of the calyx lobes 1 to 4 mm. long; petals rhomboidal; leaves ascending (2).
2. Plants nearly glabrous throughout; leaves 3 to 6 mm. wide; hypanthium (in pressed specimens) 10 to 18 mm. wide at apex; petals 20 to 25 mm. long........var. *Fendleri*
2. Plants minutely glandular-pubescent throughout, to nearly glabrous; leaves 1 to 8 mm. wide; hypanthium 5 to 8 mm. wide at apex; petals 10 to 20 mm. long......var. *typica*

Var. *typica* Munz occurs in Apache, Navajo, Coconino, Cochise, and Santa Cruz counties, 4,500 to 7,500 feet, occasional on dry mesas, May to September, ranging from Texas to Arizona and Zacatecas. Var. *Fendleri* Gray occurs rarely in Navajo, Coconino, and Gila counties, ranging from Oklahoma and Texas to Arizona. Var. *Toumeyi* (Small) Munz (*Galpinsia Toumeyi* Small), type from the Chiricahua Mountains (*Toumey* 197), which intergrades freely with the other varieties, is found in the Chiricahua, Huachuca, and Santa Rita mountains (Cochise, Santa Cruz, and Pima counties), 5,000 to 9,000 feet, rocky places, June to September, occurring also in New Mexico, Sonora, and Chihuahua.

24. Oenothera Greggii Gray (*Galpinsia Greggii* Small). This species is represented in Arizona by var. *lampasana* (Buckl.) Munz, which occurs from the White Mountains (Apache County) to the mountains of Pinal, Cochise, Santa Cruz, and Pima counties, 4,000 to 7,000 feet, occasional on limestone hills, April to June, ranging from Oklahoma and Texas to southeastern Arizona.

The characteristic phase has spreading hairs, the leaves crinkled-wavy, 5 to 9 mm. wide, the hypanthium 25 to 40 mm. long, and the petals 15 to 30 mm. long. Some Arizona collections with narrower leaves and hairs closely appressed (none spreading) approach var. *Pringlei* Munz of northern Mexico.

25. Oenothera serrulata Nutt. (*Meriolix serrulata* Raf.). Carrizo and Willow Spring (Apache County), Forestdale and Fort Apache (Navajo County), 5,000 to 7,000 feet. Assiniboia to Texas and eastern Arizona, ranging widely over the plains east of the Rocky Mountains.

26. Oenothera refracta Wats. (*Sphaerostigma refractum* Small, *Oenothera deserti* Jones). Mohave, western Pima, and Yuma counties, below 4,000 feet, frequent on open deserts, March to May, type from gravelly hills on the Colorado (*Bigelow*). Southern Utah and western Arizona to California.

27. Oenothera chamaenerioides Gray (*Sphaerostigma chamaenerioides* Small). Grand Canyon (Coconino County), and Yavapai and Mohave counties to Greenlee, Santa Cruz, Pima, and Yuma counties, below 5,500 feet, open deserts, February to May. Texas to Utah, Arizona, and California.

28. Oenothera decorticans (Hook. & Arn.) Greene (*Sphaerostigma decorticans* Small). Mohave, Maricopa, Pima, and Yuma counties, up to 2,500 feet, open deserts, March to May. Utah and Nevada to western Arizona and southeastern California.

The Arizona plant is var. *condensata* Munz.

29. Oenothera Boothii Dougl. (*Sphaerostigma Boothii* Walp.). Coconino County, San Francisco Peaks and vicinity (*Knowlton* 200, *Leiberg* 5808) and Wupatki National Monument, 5,000 feet, in cinder soil (*D. Jones* in 1939). Eastern Washington to northern Arizona and California.

In Jones's specimen the bark is more exfoliated than is usual in this species.

30. Oenothera contorta Dougl. (*Sphaerostigma contortum* Walp.). Fort Huachuca, Cochise County (*Wilcox* in 1892–1893).

Represented in Arizona by var. *epilobioides* (Greene) Munz, characterized by almost linear leaves, small yellow flowers, and narrow cylindric capsules. This variety occurs in southern Arizona, California, and Baja California.

31. Oenothera micrantha Hornem. (*Sphaerostigma micranthum* Walp.). Grand Canyon (Coconino County), and in Yavapai, Mohave, Gila, Pinal, Maricopa, and Pima counties, below 4,500 feet, deserts, March to May.

The species is represented in Arizona by var. *exfoliata* (A. Nels.) Munz (*Sphaerostigma pallidum* Abrams), which occurs in Arizona and southeastern California.

32. Oenothera cardiophylla Torr. (*Chylismia cardiophylla* Small). Var. *typica* Munz, with hypanthium 5 to 10 mm. long, petals 3 to 8 mm. long, and style 12 to 16 mm. long, occurs in Pinal and Yuma counties, at low altitudes, desert washes and rocky slopes, February to April, ranging also into the deserts of California and Baja California, type from near Fort Yuma. Var. *splendens* Munz & Johnston (var. *longituba* Jepson, *Chylismia arenaria* A. Nels.), with hypanthium 20 to 35 mm. long, petals 13 to 25 mm. long, and style 30 to 60 mm. long, is more common than var. *typica* in Yuma County, especially in the Gila and Tinajas Altas mountains, and is found also in adjacent California. The type of *C. arenaria* came from near Yuma (*A. Nelson* 11140).

33. Oenothera brevipes Gray (*Chylismia brevipes* Small). Mohave, western Pima, and Yuma counties, below 4,500 feet, dry washes and desert plains, February to May. Nevada, western Arizona, and southeastern California.

34. Oenothera pallidula (Munz) Munz. Northern Mohave County to western Maricopa and southern Yuma counties, at low altitudes, open deserts and washes, March to May. Utah and Nevada to western Arizona and southeastern California.

Plants from Arizona have hairier stems than those from farther north and west.

35. Oenothera multijuga Wats. (*Chylismia multijuga* Small).

KEY TO THE VARIETIES

1. Calyx lobes 6 to 7 mm. long; petals 7 to 9 mm. long; anthers hairy..........var. *typica*
1. Calyx lobes not more than 4 mm. long; petals not more than 5 mm. long; anthers essentially glabrous (2).
2. Leaves all basal; calyx lobes 3 to 4 mm. long; petals 3 to 5 mm. long....var. *parviflora*
2. Leaves usually present on the lower third of the stem; calyx lobes 1 to 1.5 mm. long; petals 1.5 to 2 mm. long..var. *orientalis*

Var. *typica* Munz is found in western Coconino County and in Mohave County, up to 5,500 but mostly below 2,500 feet, washes and canyons, April to June, ranging into Nevada and southern Utah. Var. *parviflora* (Wats.) Munz occurs in the Grand Canyon (Coconino County) and near Mokiak Springs and Lake Mead (northern Mohave County), ranging to Death Valley, California, and St. George, Utah. An unusual, almost completely glabrous form occurs in the Grand Canyon (*Eastwood & Howell* 7107, *Collom* in 1940), flowering as late as September; and a specimen with capsules only 8 mm. long was collected in Havasu Canyon (*Deaver* 1520). Var. *orientalis* Munz occurs rarely in Coconino and Mohave counties, about 4,000 feet, and ranges from western Colorado and eastern Utah to northern Arizona.

36. Oenothera scapoidea Nutt. (*Chylismia scapoidea* Small). Apache, Navajo, and eastern Coconino counties, 5,000 to 5,500 feet, May and June.

The Arizona plants belong to var. *seorsa* (A. Nels.) Munz, which ranges from Wyoming to Oregon and northern Arizona.

37. Oenothera clavaeformis Torr. & Frém. (*Chylismia clavaeformis* Heller).

KEY TO THE VARIETIES

1. Petals yellow; stems villous with spreading hairs. Leaves often much divided .... var. *Peirsonii*
1. Petals white; stems glabrate to finely pubescent (2).
2. Hypanthium and calyx lobes strigillose .... var. *aurantiaca*
2. Hypanthium and calyx lobes glandular-puberulent .... var. *Peeblesii*

Var. *aurantiaca* (Wats.) Munz is found in Mohave, Greenlee, Graham, Gila, Maricopa, Pinal, Pima, and Yuma counties, up to 4,500 but usually below 3,000 feet, washes and sandy places, February to April and occasionally in autumn, type from Fort Mohave (*Cooper* in 1861), ranging from southwestern Utah to Arizona and California. Var. *Peeblesii* Munz occurs in Yavapai, Pinal, Maricopa, Cochise, Pima, and Yuma counties, below 3,500 feet, washes and dry, sandy places, March to May, known only from south-central Arizona, type from Casa Grande (*Peebles & Harrison* 3537). Var. *Peirsonii* Munz has been collected between Growler Pass and Tule Well, Pima or Yuma County (*Kearney & Peebles* 10859), and near Laguna Dam, Yuma County (*McMurry & Phillips* 2653), occurring also in the Colorado Desert of California and in northern Baja California.

38. Oenothera Parryi Wats. (*Chylismia Parryi* Small). Northwestern Coconino County and northern Mohave County, 1,000 to 4,500 feet, April to June. Southwestern Utah and northwestern Arizona.

Petals bright yellow.

8. GAYOPHYTUM (201)

Annuals, with slender stems and filiform branches; leaves alternate, entire; flowers very small; petals white or pink; stamens 8, the alternate ones much reduced and usually sterile.

KEY TO THE SPECIES

1. Capsules not torulose, subsessile; plants branched only at base or sparingly above; upper leaves well developed .... 3. *G. racemosum*
1. Capsules torulose, pedicelled; plants freely branched above the base, repeatedly dichotomous; upper leaves bractlike (2).
2. Petals 0.5 mm. long; capsules 2 to 5 mm. long, commonly shorter than the usually deflexed pedicels; plants usually quite glabrous .... 1. *G. ramosissimum*
2. Petals 1 to 1.5 mm. long; capsules 5 to 12 mm. long, exceeding the (usually) erect or ascending pedicels; plants usually puberulent above .... 2. *G. Nuttallii*

1. Gayophytum ramossimum Torr. & Gray. Chuska Mountains (Apache County), and in Coconino County, 6,500 to 9,000 feet, open places in pine forests, June to August. Colorado to Washington and northern Arizona.

2. Gayophytum Nuttallii Torr. & Gray. Chuska and Lukachukai mountains (Apache County), Coconino County, Prescott (Yavapai County), 6,000 to 9,500 feet, open places in pine woods, July to September. South Dakota to Washington, New Mexico, Arizona, and southern California.

Var. *intermedium* (Rydb.) Munz, which approaches *G. ramosissimum* in having deflexed pedicels, occurs in the Lukachukai Mountains (*Eastwood & Howell* 6783).

3. Gayophytum racemosum Torr. & Gray. Coconino County, 6,500 to 8,500 feet, open places in pine forests, June to September. Colorado to Washington, northern Arizona, and southern California.

9. GAURA (202)

Annual or perennial herbs, with red to white, somewhat irregular, 4-merous flowers; stamens declined, all fertile, 8 in number; stigma deeply lobed, with a cuplike indusium at base; fruit woody, small, indehiscent, 1- to 4-seeded.

KEY TO THE SPECIES

1. Flowers small, the petals 1.5 to 2 mm. long, the sepals 1.5 to 3 mm. long; anthers oval, 1 mm. long; tall, weedy biennials or winter annuals, the stem mostly simple below and branched above ....1. *G. parviflora*
1. Flowers larger, the petals 4 to 6 mm. long, the sepals 4 to 8 mm. long; anthers linear-oblong, 2 to 5 mm. long; low perennials, the stem much-branched from the base (2).
2. Body of the fruit ovoid-pyramidal, widest near the narrow base and winged almost throughout, transversely wrinkled on each face; plants 40 to 90 cm. high; basal leaves sinuate-dentate, 4 to 10 cm. long..2. *G. gracilis*
2. Body of the fruit widest at or above the middle, winged only beyond this point, the basal portion terete and thick, not transversely wrinkled; plants mostly 10 to 40 cm. high; basal leaves subentire, 1.5 to 3.5 cm. long..............................3. *G. coccinea*

1. Gaura parviflora Dougl. Navajo County to Mohave County, south to Cochise, Pima, and Yuma counties, 1,000 to 6,000 feet, waste and disturbed places, June to October. Mississippi Valley to Washington and northern Mexico; Argentina.

Var. *typica* Munz has the ovary and capsule glabrous and the hypanthium minutely puberulent. Forma *glabra* Munz, with the hypanthium also glabrous, has been collected in Arizona at Kirkland, Yavapai County (*Peebles et al.* 4244), and in Ramsey Canyon, Huachuca Mountains (*Jones* 24941). Var. *lachnocarpa* Weatherby, with the hypanthium, ovaries, and capsules short-pubescent, occurs almost throughout the range of the species in Arizona, mostly below 3,000 feet, March to June.

2. Gaura gracilis Woot. & Standl. Nearly throughout the state except the most western portion, especially abundant about Flagstaff and in the mountains of Cochise County, 2,500 to 9,000 feet, canyons and along roadsides, May to October. Western Texas to Arizona and northern Mexico.

Both var. *typica* Munz, with hypanthium and calyx strigillose to glabrous, and forma *glandulosa* (Woot. & Standl.) Munz, with gland-tipped hairs on the hypanthium and calyx, are found in Arizona, the latter perhaps at somewhat higher altitudes.

3. Gaura coccinea Nutt. Throughout the state except the southwestern desert portion, 2,000 to 8,000 feet, dry flats and plains, April to September. Southern Canada to Mexico.

KEY TO THE VARIETIES

1. Capsules about 10 mm. long, 1.5 to 2 mm. wide; anthers 4 to 5 mm. long....var. *arizonica*
1. Capsules 5 to 7 mm. long, 2.5 to 3 mm. wide; anthers 2 to 4 mm. long (2).
2. Stem and leaves glabrous..var. *glabra*
2. Stem and leaves more or less pubescent (3).

3. Leaves linear, subentire..var. *parvifolia*
3. Leaves (at least the lower ones) lanceolate and sinuate-dentate (4).
4. Main leaves mostly plane, oblong, crowded; flowers crowded; inflorescence not peduncled; plants 10 to 30 cm. high..var. *typica*
4. Main leaves mostly waved and crisped, acutish, not crowded; flowers not crowded; inflorescence with a short peduncle; plants 30 to 40 cm. high...........var. *epilobioides*

Var. *typica* Munz has been found in most parts of Arizona and ranges from southern Canada to Texas, Arizona, and eastern California. Var. *epilobioides* (H.B.K.) Munz has much the same distribution in the state and ranges from Texas to Arizona and Mexico. Var. *parvifolia* (Torr.) Torr. & Gray is known from Apache, Coconino, Yavapai, Cochise, and Santa Cruz counties, and ranges from Kansas and Colorado to northern Mexico. Var. *glabra* (Lehm.) Torr. & Gray has the same general range within and outside the state as has var. *typica*. Var. *arizonica* Munz occurs in Mohave, Yavapai, Gila, Cochise, Santa Cruz, and Pima counties, 2,000 to 6,000 feet, type from Globe (*Eastwood* 8657), and ranges into New Mexico and Coahuila.

10. CIRCAEA. Enchanters-nightshade

Plants low, delicate, herbaceous, perennial, with short, slender rootstocks and small tubers; leaves thin, opposite, petioled; flowers small, racemose; sepals and petals 2; fruit 1-celled, 1-seeded, indehiscent, pear-shaped, bristly with hooked hairs.

1. Circaea pacifica Asch. & Magn. Greer (Apache County), Oak Creek (Coconino County), Pinaleno Mountains (Graham County), 6,000 to 9,000 feet, rich soil in shady ravines, June to September. Rocky Mountains to British Columbia, Arizona, and southern California.

Very doubtfully distinct from *C. alpina* L.

98. HALORAGACEAE. Water-milfoil Family

Plants herbaceous, perennial, aquatic, the stems wholly or partly immersed; leaves commonly in whorls; flowers minute, perfect or unisexual; petals when present usually 4; stamens 1 to 8; ovary inferior, 1- to 4-celled; fruit indehiscent.

KEY TO THE GENERA

1. Leaves (at least the immersed ones) pinnatifid to capillary-dissected; flowers mostly unisexual; stamens more than 1; ovary 2- to 4-celled.................1. *Myriophyllum*
1. Leaves all entire; flowers mostly perfect; stamen 1; ovary 1-celled.........2. *Hippuris*

1. MYRIOPHYLLUM. Water-milfoil

Immersed leaves (often all the leaves) pinnately dissected into capillary divisions; flowers axillary or in interrupted terminal spikes, often emersed; petals present or absent; stamens 4 or more; fruit 4-celled, deeply 4-lobed.

KEY TO THE SPECIES

1. Leaves all alike, in whorls of 4 to 6, pinnatifid, the segments linear-filiform, not more than 5 mm. long; flowers axillary................................1. *M. brasiliense*
1. Leaves dimorphic, the immersed ones in whorls of 3 or 4, pinnate, the segments capillary, commonly at least 10 mm. long, the emersed (floral) leaves bractlike, entire or merely dentate, not or but slightly surpassing the flowers; flowers in interrupted terminal spikes..2. *M. exalbescens*

1. **Myriophyllum brasiliense** Cambess. In a pond near Sacaton, Pinal County (*Peebles* 10607), apparently well established, perhaps introduced by migrating wild fowl. Native of South America.

Parrot-feather, commonly grown in aquaria and ponds.

2. **Myriophyllum exalbescens** Fern. (*M. spicatum* L. var. *exalbescens* (Fern.) Jepson). Apache, Navajo, and Coconino counties, 5,000 to 9,000 feet, ponds and lakes, June to August. Canada to Florida, northern Arizona, and California.

2. HIPPURIS. MARES-TAIL

Stems usually partly emersed, erect, not branched; leaves all simple, entire, in whorls of 6 or more; flowers axillary, sessile; calyx not lobed, almost completely adnate to the ovary; petals none; style filiform, in a groove formed by the lobes of the single anther.

1. **Hippuris vulgaris** L. White Mountains (Apache County) at Marsh Lake (*Goldman* 2450) and Big Lake (*Darrow* in 1940), 9,000 feet. Widely distributed in the cooler parts of the Northern Hemisphere.

99. ARALIACEAE. GINSENG FAMILY

English ivy (*Hedera Helix* L.) is a well-known member of this family.

1. ARALIA

Plants perennial, herbaceous or shrubby; leaves large, compound, with large leaflets; flower small, regular, perfect or unisexual, in umbels, these forming terminal panicles; calyx adnate to the ovary; petals and stamens 5 each, borne on the calyx; styles 5, separate or partly united; fruit berry-like, several-seeded.

The underground parts are more or less spicy-aromatic. The roots of North American and Asiatic species of a related genus, *Panax*, commonly known as ginseng, are highly esteemed for medicinal purposes in China. The aromatic root of *A. racemosa* is one of the ingredients of root beer. An allied species, *A. cordata* Thunb., is a favorite vegetable in Japan, where it is known as udo.

KEY TO THE SPECIES

1. Shrubs, up to 2.5 m. high; leaves mostly simply pinnate, with 5 to 9 leaflets, these crenulate or crenate with obtuse or acutish, often glandular-mucronulate teeth, light green or yellowish green, commonly thickish, persistently and usually copiously puberulent beneath, lance-ovate or oblong-ovate; peduncles and pedicels sparsely puberulent or glabrate .... 1. *A. humilis*

1. Large herbs, 1 to 2 m. high; leaves (all except the uppermost) ternate or biternate, the ultimate divisions pinnate with 3 to 5 leaflets, these doubly crenate-serrate with setose-cuspidate teeth, deep green, thin, puberulent only on the veins beneath when mature, lance-ovate to broadly ovate; peduncles and pedicels copiously puberulent 2. *A. racemosa*

1. **Aralia humilis** Cav. Cochise, Santa Cruz, and Pima counties, 3,500 to 5,000 feet, canyons, July to October. Southern Arizona and Mexico.

The stems reach a height of 3 m. (10 feet) and a diameter at base of 5 cm., and the older bark is rough.

2. Aralia racemosa L. (*A. bicrenata* Woot. & Standl., *A. arizonica* Eastw.). Apache, Navajo, and Coconino counties to Cochise and Pima counties, 5,000 to 9,500 feet, rich soil in coniferous forests, preferring shade, July and August, type of *A. arizonica* from the Chiricahua Mountains (*Kusche* in 1929). Canada to Georgia, Arizona, and northern Mexico.

American spikenard. The plants reach a height of 2 m. (7 feet). The phase occurring in Arizona (*A. bicrenata* Woot. & Standl.) should perhaps be regarded as a variety, having more deeply serrate margins of the leaflets and a more ample inflorescence than in most specimens from the eastern United States. The fruits are eaten by various birds.

100. UMBELLIFERAE (203). Parsley Family

Contributed by Mildred E. Mathias and Lincoln Constance

Plants herbaceous, annual or perennial, with commonly hollow stems; leaves alternate or basal, usually compound, with usually sheathing petioles; flowers small, regular, in simple or compound umbels, or the umbels sometimes proliferous or capitate; rays sometimes subtended by bracts forming an involucre; umbellets usually subtended by bractlets forming an involucel; calyx tube wholly adnate to the ovary, the calyx teeth obsolete or small; petals 5, usually with an inflexed tip; stamens 5, inserted on an epigynous disk; ovary inferior, 2-celled, with 1 anatropous ovule in each cell; styles 2, sometimes swollen at base to form a stylopodium; fruit consisting of 2 mericarps united by their faces (commissure), each mericarp with 5 ribs, 1 down the back (dorsal rib), 2 on the edges near the commissure (lateral ribs), and 2 between the dorsal and lateral ribs (intermediate ribs), with oil tubes usually present in the intervals (spaces between the ribs), and on the commissural surface; mericarps 1-seeded, splitting apart at maturity, usually suspended from the summit of a slender prolongation of the axis (carpophore); embryo small, the endosperm cartilaginous.

Among the useful Umbelliferae are garden vegetables, such as carrot, parsnip, celery, and parsley, and condiment plants, such as caraway, fennel, dill, anise, and coriander. On the other hand, some of these plants are very poisonous, notably the water-hemlocks (*Cicuta* spp.) and poison-hemlock (*Conium maculatum*).

The genera are arranged in the sequence followed by Mathias and Constance (203).

KEY TO THE GENERA

1. Inflorescence capitate, not distinctly umbellate, terminal on the branches or scapes, or borne on short peduncles in the forks of the branches (2).
1. Inflorescence a distinct umbel (rarely spicate), more or less spreading, not capitate or if nearly so (in *Torilis*), then the inflorescences scattered along the stem, opposite the leaves, sessile or short-peduncled (3).
2. Fruit winged, not squamose .... 24. *Cymopterus*
2. Fruit not winged, ribless, variously squamose .... 34. *Eryngium*
3. Leaves simple; umbels simple, proliferous, or compound (4).
3. Leaves variously compound; umbels compound, or rarely simple by reduction (7).
4. Ovary and fruit covered with stellate hairs; foliage more or less stellate-pubescent; leaves stipulate .... 2. *Bowlesia*

4. Ovary and fruit glabrous; foliage glabrous; leaves not stipulate (5).
5. Leaves reduced to hollow, cylindric, jointed petioles....................22. *Lilaeopsis*
5. Leaves with a definite blade (6).
6. Leaves long-petioled, orbicular or roundish-reniform, peltate or nonpeltate; plants aquatic or subaquatic..1. *Hydrocotyle*
6. Leaves sessile or nearly so, oblong, ovate, or obovate, the stem leaves perfoliate-clasping; plants not aquatic..14. *Bupleurum*
7. Ovary and fruit armed with bristles, callose teeth, or papillae (8).
7. Ovary and fruit not armed, sometimes pubescent (15).
8. Ovary and fruit linear or linear-oblong, several times longer than wide; oil tubes absent or very small. Fruit hispid with straight bristles (9).
8. Ovary and fruit ovate to oblong, not more than twice as long as wide; oil tubes present in mature fruit (10).
9. Plants perennial; leaves ternate-pinnate, the leaflets large; beak of the fruit shorter than the body..6. *Osmorhiza*
9. Plants annual; leaves pinnately decompound, the leaflets small; beak of the fruit much longer than the body..7. *Scandix*
10. Stems and leaves variously hispid (11).
10. Stems and leaves glabrous or somewhat roughened, never hispid. Involucre mostly absent; bractlets of the involucel linear or filiform and entire, or absent; fruit armed with short bristles, callose teeth, or papillae (13).
11. Inflorescences subcapitate, with short rays, scattered along the stem, sessile or on very short peduncles opposite the leaves; bracts of the involucre (when present) and bractlets of the involucel small, linear; fruit papillate as well as bristly........8. *Torilis*
11. Inflorescences with more or less elongate rays, terminating the stem and branches, the peduncles well developed; bracts and bractlets foliaceous, the bractlets usually pinnately divided; fruit not papillate (12).
12. Fruit flattened laterally, the bristles hooked, not barbed; stylopodium conic; calyx teeth prominent ....9. *Caucalis*
12. Fruit flattened dorsally, the bristles barbed at tip; stylopodium and calyx teeth absent 10. *Daucus*
13. Involucel absent; umbels sessile. Fruit papillate......................3. *Apiastrum*
13. Involucel present; umbels peduncled (14).
14. Fruit covered with short bristles; seed face more or less sulcate........4. *Spermolepis*
14. Fruit covered with callose teeth; seed face plane to somewhat concave..5. *Ammoselinum*
15. Ribs of the fruit not prominently winged (or somewhat so in *Ligusticum*); ovary and fruit terete in cross section or somewhat flattened laterally (16).
15. Ribs of the fruit (some or all of them) prominently winged, the fruit more or less flattened dorsally (24).
16. Petals conspicuously unequal; fruit subglobose, not constricted at the commissure 18. *Coriandrum*
16. Petals equal; fruit orbicular to oblong, more or less constricted at the commissure (17).
17. Flowers yellow; plants with an aniselike odor.......................11. *Foeniculum*
17. Flowers not yellow; plants without an aniselike odor (18).
18. Ribs or the whole surface of the fruit corky; plants of marshes, or aquatic (19).
18. Ribs of the fruit not corky; plants of dry ground or moist meadows (21).
19. Involucre inconspicuous or absent; some or all of the leaves ternate-pinnate..21. *Cicuta*
19. Involucre conspicuous; leaves simply pinnate (20).
20. Ribs filiform, the pericarp forming a continuous corky covering; stylopodium conic 19. *Berula*
20. Ribs of the fruit corky, equal; stylopodium depressed.....................20. *Sium*
21. Ribs narrowly winged; leaflets variously incised....................17. *Ligusticum*
21. Ribs prominent but not winged; leaflets entire to lobed (22).
22. Involucre present, the bracts scarious-margined; stems usually spotted or streaked with purple; basal leaves large, ternately, then pinnately decompound; fruit with crenulate ribs ....15. *Conium*

22. Involucre usually absent; stems not purple-spotted; leaves once or twice pinnate or, if ternately decompound, then not large; fruit with entire ribs (23).
23. Plants from slender, elongate roots; umbels axillary or terminal; calyx lobes inconspicuous .... 13. *Apium*
23. Plants from fascicled tuberous or fusiform roots; umbels terminal; calyx lobes conspicuous........ 16. *Perideridia*
24. Fruit with both lateral and dorsal wings developed, or the dorsal ribs prominent (25).
24. Fruit with lateral wings only, these more or less prominent, the dorsal ribs filiform or obsolete (34).
25. Plants acaulescent or short-caulescent, low, subscapose (26).
25. Plants caulescent, mostly tall, the stems leafy, often stout (30).
26. Calyx teeth prominent, linear-lanceolate, acuminate (27).
26. Calyx teeth evident to obsolete, when present then ovate to deltoid (28).
27. Plants low, mostly less than 15 cm. high, acaulescent, puberulent; fruit moderately compressed .... 23. *Oreoxis*
27. Plants more than 15 cm. high, usually caulescent, essentially glabrous; fruit strongly compressed .... 25. *Pteryxia*
28. Fruit not strongly compressed, the ribs narrowly corky-winged........ 12. *Aletes*
28. Fruit strongly compressed, the ribs broadly winged (29).
29. Plants acaulescent to subcaulescent; peduncles glabrous or (rarely) scaberulous 24. *Cymopterus*
29. Plants more or less caulescent; peduncles noticeably hirtellous-pubescent at base of the umbel .... 26. *Pseudocymopterus*
30. Stylopodium absent (31).
30. Stylopodium present, conic (32).
31. Calyx teeth linear-lanceolate, attenuate, often unequal........ 25. *Pteryxia*
31. Calyx teeth deltoid or ovate........ 26. *Pseudocymopterus*
32. Plants annual, with a strong, aniselike odor; involucels usually wanting; flowers yellow 29. *Anethum*
32. Plants perennial, not anise-scented; involucels usually present; flowers white, pink, or purplish (33).
33. Plants glabrous except for the puberulent inflorescence; stems from a cluster of fleshy roots; rays subequal; ovaries glabrous........ 27. *Conioselinum*
33. Plants somewhat pubescent; stems from a taproot; rays unequal; ovaries sparsely hispidulous........ 28. *Angelica*
34. Leaf segments filiform to linear; plants acaulescent or if caulescent, then the stems slender (35).
34. Leaf segments broad, entire, toothed, or lobed; plants strongly caulescent, the stems stout (36).
35. Peduncles conspicuously hirtellous-pubescent at base of the umbel 26. *Pseudocymopterus*
35. Peduncles glabrous, or pubescent throughout........ 31. *Lomatium*
36. Plants without stem leaves, or the divisions linear, 3 mm. or less wide; stylopodium absent .... 31. *Lomatium*
36. Plants with stem leaves, the divisions mostly ovate, more than 5 mm. wide; stylopodium present (37).
37. Outer petals of the inflorescence radiant and often 2-cleft; herbage tomentose; oil tubes not extending to the base of the fruit........ 33. *Heracleum*
37. Outer petals not differentiated; herbage glabrous to somewhat pubescent; oil tubes extending to the base of the fruit (38).
38. Ovaries sparsely hispidulous; dorsal ribs of the fruit narrowly winged. Flowers white, pink, or purplish........ 28. *Angelica*
38. Ovaries glabrous; dorsal ribs not winged (39).
39. Flowers white or purple; stems from fascicled tubers........ 30. *Oxypolis*
39. Flowers yellow; stems from taproots........ 32. *Pastinaca*

1. HYDROCOTYLE. Water-pennywort

Plants glabrous, perennial, with slender, floating or creeping stems or rootstocks; leaves round, peltate or nonpeltate; flowers in a simple axillary umbel or an interrupted spike; calyx teeth minute; corolla white, greenish, or yellow; stylopodium depressed to conspicuously conic; fruit transversely ovate to orbicular, 1 to 3 mm. long, strongly flattened laterally, the ribs slender or obsolete, the oil-bearing cells conspicuous to obsolete, the seed face plane to concave.

Plants of wet ground, or aquatic.

KEY TO THE SPECIES

1. Leaves peltate, orbicular, shallowly 8- to 14-lobed; inflorescence an interrupted simple or branched spike, the flowers borne in scattered verticils..............1. *H. verticillata*
1. Leaves not peltate, roundish-reniform, 5- or 6-lobed to about the middle; inflorescence a simple umbel ....2. *H. ranunculoides*

1. **Hydrocotyle verticillata** Thunb. Mohave, Yavapai, Gila, Maricopa, Cochise, Santa Cruz, and Pima counties, May to August. Massachusetts to Florida and the West Indies, west to southern Utah, Arizona, and California.

In the typical plant the inflorescence is often bifurcate, and the fruit sessile or subsessile. Var. *triradiata* (A. Rich.) Fern., distinguished from the species by a rarely bifurcate inflorescence and pedicellate fruit with pedicels up to 10 mm. long, has been collected in Mohave, Yavapai, and Pima counties.

2. **Hydrocotyle ranunculoides** L. f. San Pedro River valley, Cochise County (*Toumey* in 1894), Crittenden, Santa Cruz County (*Thornber* 2889), Tucson, Pima County (*Toumey* in 1894). Pennsylvania and Delaware to Florida, west to Washington and Arizona, south to tropical America.

2. BOWLESIA

Plants annual, stellate-pubescent, prostrate or suberect, dichotomously branching; leaves suborbicular, palmately 5- to 7-lobed; peduncles axillary, shorter than the leaves; umbels simple, few-flowered; involucral bracts small; calyx teeth prominent; corolla greenish white; stylopodium depressed-conic; fruit broadly ovate, 1 to 1.5 mm. long, stellate-pubescent, turgid, narrowed at the commissure, the ribs and oil tubes obsolete, the seed face plane or convex.

1. **Bowlesia incana** Ruiz & Pavon (*B. septentrionalis* Coult. & Rose). Havasu Canyon (Coconino County) and Mohave County to Graham, Pima, and (probably) Yuma counties, 1,000 to 3,500 feet, common, usually among bushes, February to May, type of *B. septentrionalis* from near Tucson (*Zuck*). Southern Texas to Arizona and central California; Sonora; Baja California; South America.

3. APIASTRUM

Plants annual, slender, glabrous; stems sometimes simple, usually dichotomously or trichotomously branching; leaves 2- or 3-ternate, the segments subfiliform to linear; umbels compound, sessile in the axils or opposite the upper leaves; involucre and involucel none; calyx teeth obsolete; corolla white; stylopodium minute and depressed; fruit ovate or cordate, 1 to 1.5 mm. long, papillate-roughened, the ribs inconspicuous, the oil tubes solitary in the intervals, 2 on the commissure, the seed face concave to sulcate.

1. Apiastrum angustifolium Nutt. Santa Rita Range Reserve and Coronado National Forest, Pima County (*Griffiths* 429, 474, 516). Southern Arizona, California, and Baja California.

4. SPERMOLEPIS

Plants annual, caulescent, alternately branching, glabrous; leaves ternately decompound, the segments filiform; peduncles axillary and terminal, exceeding the leaves; umbels compound; involucre absent; bractlets of the involucel few, filiform, shorter than the pedicels, glabrous or callose-toothed; calyx teeth obsolete; corolla white; stylopodium low-conic; fruit ovoid, 1.5 to 2 mm. long, laterally compressed, covered with short, echinate bristles, the oil tubes 1 to 3 in the intervals, 2 on the commissure, the seed face sulcate.

1. Spermolepis echinata (Nutt.) Heller. Graham, Gila, Maricopa, Pinal, Cochise, and Pima counties, 1,000 to 5,000 feet, February to May. Missouri to Louisiana, west to southern Arizona, southern California, and Coahuila.

5. AMMOSELINUM

Plants annual, caulescent, branching, more or less roughened; leaves ternate-pinnately dissected, the segments linear; peduncles axillary and terminal, up to 4 cm. long; umbels compound; involucre mostly absent; bractlets of the involucel few, linear, acute, somewhat callose-toothed, about equaling the pedicels; calyx teeth obsolete; corolla white; stylopodium low-conic; fruit oblong-ovate, 3 to 5 mm. long, somewhat constricted toward the apex, subcordate at base, laterally compressed, covered with callose teeth, the ribs corky, acute to rounded, the oil tubes 3 in the intervals, 2 on the commissure, the seed face concave.

1. Ammoselinum giganteum Coult. & Rose. Mesas near Phoenix (*Pringle* 28, the type collection), Maricopa, Pinal County (*Dewey* in 1894), Eloy, Pinal County (*Peebles et al.* 6496), March. Southern Arizona, California, and Coahuila.

6. OSMORHIZA (204). SWEET-ROOT, SWEET-CICELY

Plants perennial, caulescent, glabrous to hirsute; leaves biternate or ternate-pinnate, the leaflets broad, variously toothed or lobed, distinct; peduncles exceeding the leaves; umbels compound; involucre usually absent; calyx teeth obsolete; corolla white to greenish yellow; stylopodium conic; fruit linear to linear-oblong, 10 to 20 mm. long, cylindric or clavate, more or less attenuate at base, slightly compressed laterally or not at all, the ribs inconspicuous, the oil tubes obsolete in mature fruit, the seed face sulcate.

The sweetish roots have a strong, aniselike flavor.

KEY TO THE SPECIES

1. Involucels present but more or less deciduous, the bractlets usually well developed at anthesis. Fruits narrowed toward apex, 12 to 20 mm. long..........1. *O. brachypoda*
1. Involucels absent or rudimentary (2).
2. Rays and pedicels spreading-ascending; fruits linear-oblong, cylindric, narrowed toward the apex, 12 to 27 mm. long..........2. *O. chilensis*
2. Rays and pedicels divaricate; fruits clavate, widest near the apex, seldom more than 10 (up to 15) mm. long..........3. *O. obtusa*

1. **Osmorhiza brachypoda** Torr. Mazatzal Mountains, Gila County (*Harrison* 7815, 7830, *Collom* 866), April and May. Central Arizona and California.

2. **Osmorhiza chilensis** Hook. & Arn. (*O. nuda* Torr.). Oak Creek Canyon (Coconino County), Mogollon Escarpment (Gila County), Pinaleno Mountains (Graham County), Santa Catalina Mountains (Pima County), 6,000 (?) to 7,500 feet, May and June. Newfoundland, Quebec, and New Hampshire, to Alaska, Arizona, and California; southern South America.

3. **Osmorhiza obtusa** (Coult. & Rose) Fern. Apache and Coconino counties to Cochise and Pima counties, 7,000 to 10,000 feet, May to October. Labrador and Newfoundland to the Rocky Mountains, south to northern Arizona; southern South America.

7. SCANDIX. Venus' Comb

Plants annual, caulescent, the stems branched, the herbage hispid; leaves pinnately decompound, the ultimate divisions small; umbels terminal and lateral, simple or compound, few-rayed; involucre wanting or of a single foliaceous bract; bractlets of the involucel lobed or dissected; calyx teeth minute or obsolete; petals white; fruit linear or narrowly oblong, with a slender beak much longer than the body, the ribs filiform, prominent; oil tubes solitary in the intervals, very small or obscure.

1. **Scandix Pecten-Veneris** L. Mescal, western Cochise County, along railway (*Thornber* in 1906), probably not established. Widely distributed in the United States; introduced from Eurasia.

8. TORILIS. Hedge-parsley

Plants annual, the herbage hispid or hispidulous; leaves pinnate, the leaflets bipinnately dissected; peduncles much shorter than and borne opposite to the leaves; umbels subcapitate, the rays very short; involucre absent, or of 1 bract; involucel of small, linear bractlets; calyx teeth triangular, persistent; corolla white; stylopodium short; fruit ovoid, 3 to 5 mm. long, somewhat flattened laterally, sharply papillate, the primary ribs inconspicuous, the secondary ribs prominent and bearing conspicuous, hooked prickles, these often wanting on the inner fruits of the umbel, the oil tubes solitary and inconspicuous in the intervals, 2 and conspicuous on the commissure, the seed face sulcate.

1. **Torilis nodosa** (L.) Gaertn. Tempe (Maricopa County), common in lawns, summer (*Judd* in 1942). Here and there in the United States; naturalized from the Mediterranean region.

9. CAUCALIS

Plants annual, caulescent, branching, hispid; leaves pinnately decompound, the segments short, linear to filiform; peduncles axillary and terminal; umbels compound or occasionally simple by reduction; involucre of foliaceous bracts shorter than the rays; bractlets of the involucel foliaceous, entire or pinnately divided, mostly shorter than the fruits; calyx teeth prominent; corolla white; stylopodium conic; fruit oblong, 3 to 7 mm. long, laterally compressed, the carpels with 5 filiform, bristly, prominent ribs and 4 prominent, winged, secondary ribs bearing hooked bristles, the oil tubes solitary in the intervals, 2 on the commissure, the seed face deeply sulcate.

1. Caucalis microcarpa Hook. & Arn. Grand Canyon (Coconino County) and Mohave County to Gila, Santa Cruz, and Pima counties, 4,500 feet or lower, April and May. Idaho to British Columbia, south to Arizona and Baja California.

10. DAUCUS. Carrot

Plants annual, caulescent, variously hispid; leaves 3- or 4-pinnatisect, the leaflets small, linear; peduncles terminal; umbels compound; involucre foliaceous, equaling or exceeding the rays, pinnately decompound into short, linear or lanceolate segments; bractlets of the involucel linear, about equaling the pedicels; calyx teeth obsolete; corolla white; stylopodium none; fruit oblong, 3 to 5 mm. long, somewhat flattened dorsally, the primary ribs slender, bristly, the secondary ribs with a single row of prominent, barbed bristles, the oil tubes solitary in the intervals, 2 on the commissure, the seed face plane to concave.

1. Daucus pusillus Michx. Greenlee County to Yavapai and Mohave counties, south to Cochise, Santa Cruz, Pima, and (doubtless) Yuma counties, 4,000 feet or lower, March to May. South Carolina to Florida, west to Missouri, Arizona, Washington, California, and Mexico.

This humble relative of the cultivated carrot (*D. Carota* L.) is known in California as rattlesnake-weed. It is reported that the Navajo Indians ate the roots both raw and cooked.

D. Carota, a common weed in many parts of the United States, is to be looked for in Arizona. It is a taller plant than *D. pusillus*, with longer-rayed umbels, the central flower of the umbellets usually pink or purple.

11. FOENICULUM. Fennel

Plants perennial, caulescent, branching, glabrous, with a strong odor of anise; stem tall; leaves large, pinnately decompound, the segments filiform, glaucous; peduncles axillary and terminal, exceeding the leaves; umbels compound; involucre and involucel none; calyx teeth obsolete; corolla yellow; stylopodium conic; fruit oblong, 3.5 to 4 mm. long, laterally compressed, the ribs slender, acute, the oil tubes solitary in the intervals, 2 on the commissure, the seed face slightly concave.

1. Foeniculum vulgare Mill. Mule Mountains, Cochise County (*Harrison & Kearney* 8284). Waste ground in various parts of North America, a common weed in coastal California; naturalized from Europe.

The blanched leaf stalks are popular in southern Europe as a substitute for celery.

12. ALETES

Plants perennial, low, cespitose, acaulescent; herbage glabrous; leaves pinnate or bipinnate, oblong in outline, the leaflets linear-oblong to ovate, entire or toothed; umbels compound, open; involucre usually absent; involucel of several linear bractlets; calyx teeth deltoid-ovate; petals yellow; stylopodium none; fruit oblong, 4 to 8 mm. long, the wings corky, the oil tubes solitary in the intervals, 2 on the commissure.

1. Aletes Macdougali Coult. & Rose (*Oreoxis Macdougali* Rydb.). Coconino County, at the Grand Canyon (*MacDougal* 192, the type collection, *Eastwood* 5713) and Oak Creek (*Deaver* 1383), about 7,000 feet, April to June. Southwestern Colorado, southeastern Utah, northwestern New Mexico, and northern Arizona.

Aletes acaulis (Torr.) Coult. & Rose, ranges in the Rocky Mountains from Colorado to New Mexico, western Texas, and Coahuila. In the herbarium of the University of California there is a specimen of this species bearing the label "Gila River" and collected by Mohr. This is probably an error, since Dr. Mohr collected the same species in Colorado. *A. acaulis* has the inflorescence 8- to 15-rayed and the bractlets connate at base, whereas in *A. Macdougali* the rays are 4 to 8 and the bractlets are distinct.

13. APIUM. Celery

Plants annual or perennial, glabrous, branched above; leaves pinnate to ternate-pinnately decompound; umbels compound (or simple by reduction), axillary and terminal, sessile to short-pedunculate; involucre and involucel none; calyx teeth absent or inconspicuous; corolla white; stylopodium short-conic; fruit ovate to suborbicular, 1 to 3 mm. long, laterally compressed, somewhat constricted at the commissure, the ribs obtuse, conspicuous, the oil tubes solitary in the intervals. 2 on the commissure, the seed face more or less plane.

KEY TO THE SPECIES

1. Plants annual; leaves pinnately or ternate-pinnately decompound, the segments linear to filiform .... 1. *A. leptophyllum*
1. Plants perennial; leaves pinnate, the segments ovate to suborbicular or cuneate .... 2. *A. graveolens*

1. Apium leptophyllum (Pers.) F. Muell. Tempe and Phoenix (Maricopa County), St. David (Cochise County), Tucson (Pima County), a troublesome weed in lawns, March to September. North Carolina to Florida and California, occasionally introduced elsewhere in the United States; West Indies; Mexico to South America.

2. Apium graveolens L. Coconino County to Cochise and Pinal counties, naturalized in Havasu Canyon (Coconino County), moist soil, May to October.

Common celery, more or less naturalized throughout North America; from Eurasia.

14. BUPLEURUM

Plants annual, caulescent, the herbage glabrous and glaucous; leaves entire, the basal ones oblong- to obovate-lanceolate, often subpetiolate, the upper cauline leaves ovate, perfoliate; umbels axillary and terminal, usually compound; involucre wanting; involucel of several large, broadly ovate to obovate bractlets, these united at base and much longer than the flowers; calyx teeth obsolete; petals yellow; fruits short-pedicelled, oblong-oval, smooth, the ribs filiform, the oil tubes obscure or wanting.

1. **Bupleurum rotundifolium** L. Tucson, Pima County (*Toumey* 196), probably not established. A weed of waste ground, occasional in the eastern and central United States; introduced from the Mediterranean region.

15. CONIUM. Poison-hemlock

Plants biennial, tall; herbage glabrous; stems branched, usually streaked or spotted with purple; leaves decompound, the leaflets pinnatifid or pinnately incised; umbels compound; involucre and involucels small, the bracts and bractlets scarious-margined; calyx teeth obsolete; corolla white; stylopodium prominent; fruit broadly ovate, somewhat flattened laterally, the ribs prominent, crenulate, the oil tubes very small and numerous, the seed face deeply sulcate.

1. **Conium maculatum** L. Prescott (Yavapai County), Duncan (Greenlee County), Safford and Aravaipa Canyon (Graham County), ditch banks and waste ground, summer. Extensively naturalized in the United States; from Eurasia.

The herbage is malodorous, and all parts of the plant are poisonous.

16. PERIDERIDIA

Plants perennial from a solitary tuber or from fascicled tuberous or fusiform roots, usually caulescent, glabrous; leaves mostly pinnate or ternate-pinnate, the leaflets linear to lanceolate; peduncles exceeding the leaves; umbels compound; involucre usually absent; involucel usually present, the bractlets linear to lanceolate; calyx teeth prominent; corolla white or pink; stylopodium conic; fruit orbicular to oblong, 2 to 3.5 mm. long, laterally compressed, the ribs inconspicuous, the oil tubes 1 to 4 in the intervals, 2 to 6 on the commissure, the seed face plane to broadly concave.

KEY TO THE SPECIES

1. Leaves pinnate or bipinnate; bractlets of the involucel (when present) linear-acuminate; fruit orbicular to suborbicular, 2 to 3 mm. long, the oil tubes solitary in the intervals
1. *P. Gairdneri*

1. Leaves ternate, rarely biternate, the uppermost entire; bractlets of the involucel narrowly lanceolate; fruit ovoid to oblong, 2.5 to 3.5 mm. long, the oil tubes 2 to 4 in the intervals
2. *P. Parishii*

1. **Perideridia Gairdneri** (Hook. & Arn.) Mathias (*Carum Gairdneri* Gray). White Mountains (Apache County), Flagstaff, Mogollon Escarpment, and Bill Williams Mountain (Coconino County), Springer Mountain (southern

Navajo County), 7,000 feet or higher, moist soil. South Dakota to New Mexico, west to British Columbia and southern California.

Yampa, wild-caraway. The tuberous roots have a pleasant, nutty flavor and were used by the Indians as food. The seeds were used as seasoning.

2. Perideridia Parishii (Coult. & Rose) Nels. & Macbr. (*Eulophus Parishii* Coult. & Rose, *E. Parishii* var. *Rusbyi* Coult. & Rose). North Rim of the Grand Canyon to Oak Creek, Coconino County, 6,500 to 8,000 feet, moist soil in pine woods, July to September, type of *E. Parishii* var. *Rusbyi* from Bill Williams Mountain (*Rusby* 629). Northern Arizona, California, and Nevada.

17. LIGUSTICUM

Plants perennial, caulescent, glabrous or puberulent; leaves once- or twice-ternate, then once- or twice-pinnate, the leaflets mostly distinct, ovate, more or less incised; peduncles axillary or often verticillate, exceeding the leaves; umbels compound; involucre absent or occasionally of a solitary, deciduous, linear bract; involucel wanting, or of several linear bractlets; calyx teeth small or wanting; corolla white; stylopodium depressed-conic; fruit oblong, 5 to 8 mm. long, glabrous, terete, the lateral and dorsal ribs narrowly winged, the oil tubes 4 to 6 in the intervals, 8 to 10 on the commissure, the seed face concave.

1. Ligusticum Porteri Coult. & Rose. Mountains of Apache, Coconino, Yavapai (?), Graham, and Cochise counties, 6,500 to 11,500 feet, June to August. Wyoming to Arizona and Chihuahua.

Chuchupate, osha. A palatable forage plant.

18. CORIANDRUM. Coriander

Plants annual, caulescent, glabrous; leaves biternate to ternate-pinnate, the leaflets mostly cuneate, doubly lobed at apex, the upper leaflets linear; peduncles axillary or terminal; umbels compound; involucre usually absent; bractlets of the involucel few, inconspicuous, lanceolate; calyx teeth acute, unequal; corolla white or roseate, the petals conspicuously unequal; stylopodium conic; fruit subglobose, about 3 mm. long, not constricted at the commissure, the ribs filiform to obscure, the oil tubes wanting, the seed face concave.

1. Coriandrum sativum L. Huachuca Mountains, Cochise County (*Thornber* 7238), near Tucson, Pima County (*Thornber* in 1903, *G. E. Wilcox* in 1905), probably only a chance escape from cultivation. Sparingly naturalized from Europe in various parts of North America.

19. BERULA. Water-parsnip

Plants perennial from fascicled, fibrous roots, caulescent, glabrous, aquatic; leaves simply pinnate, the segments linear to ovate, entire to variously lobed; peduncles axillary and terminal, exceeding the leaves; umbels compound; involucre conspicuous, of 6 to 8 unequal linear or lanceolate bracts; involucel of 4 to 8 conspicuous, lanceolate or linear bractlets; calyx teeth inconspicuous; corolla white; stylopodium depressed-conic; fruit oval to orbicular, 1.5

to 2 mm. long, laterally compressed, emarginate at base, the ribs slender, inconspicuous, scarcely raised above the surface, the oil tubes numerous and somewhat contiguous in the innermost layer of the mericarp immediately surrounding the seed, the seed face plane.

1. **Berula erecta** (Huds.) Coville (*B. pusilla* (Nutt.) Fern.). Apache County to Coconino County, south to Cochise and Santa Cruz counties, 4,000 to 7,000 feet, wet places, June and probably later. Widely distributed in North America; Eurasia.

20. SIUM. Water-parsnip

Plants perennial from fascicled, fibrous roots, caulescent, aquatic, glabrous; leaves simply pinnate, the leaflets lanceolate to linear, finely to coarsely serrate or incised, mostly acute; peduncles exceeding the leaves; umbels compound; involucre of 6 to 10 unequal, lanceolate or linear bracts, these reflexed in fruit; involucel of 4 to 8 linear-lanceolate bractlets, these shorter than the flowers; calyx teeth minute; corolla white; stylopodium depressed; fruit oval to orbicular, 2 to 3 mm. long, laterally compressed, the ribs prominent, corky, the oil tubes solitary in the intervals, 2 to several on the commissure, the seed face plane.

The roots and even the herbage were eaten by Indians in some parts of North America.

1. **Sium suave** Walt. (*S. cicutaefolium* Schrank). Near Tuba, Coconino County, in a marsh, 5,000 feet (*Kearney & Peebles* 12857), September. Widely distributed in North America.

21. CICUTA (205). Water-hemlock

Plants perennial from a vertical or horizontal, short or elongate, tuberous base bearing fibrous or fleshy-fibrous roots, caulescent, glabrous; leaves 1- to 3-pinnate, the leaflets linear-lanceolate to ovate-lanceolate, distinct or some of them confluent, remotely to coarsely serrate or incised; peduncles exceeding the leaves; umbels compound; involucre of few bracts or none; bractlets of the involucel several, ovate-lanceolate to linear, acute to acuminate; calyx teeth evident; corolla white or greenish white; stylopodium low-conic; fruit orbicular to oval, 2 to 4 mm. long, slightly compressed laterally, definitely constricted at the commissure, the ribs low, broad and corky, wider than the reddish-brown or homochromous intervals, the oil tubes solitary in the intervals, 2 on the commissure, the seed face plane.

Plants violently toxic to warm-blooded animals, especially the roots and young growth. No antidote is known, but in human beings the use of emetics has proved effective. Symptoms are vomiting, colicky pains, staggering, unconsciousness, and convulsions. The poisonous principle, cicutoxin, affects the nerve centers. These plants should be eradicated from ranges because they have caused the loss of much livestock.

1. **Cicuta Douglasii** (DC.) Coult. & Rose (*C. grandifolia* Greene). Apache, Navajo, Greenlee, and Coconino counties, 6,000 to 9,000 feet, wet ground, August, type of *C. grandifolia* from Mormon Lake (*Pearson* 140). Montana and Alaska to Arizona, California, and Chihuahua.

22. LILAEOPSIS

Plants perennial, glabrous; stems creeping and rooting in the mud; leaves reduced to hollow, cylindric petioles, transversely septate, elongate when growing in water; peduncles shorter than the leaves; umbels simple; involucre of few small bracts; calyx teeth minute; corolla white; stylopodium depressed; fruit subglobose or slightly compressed laterally, 2 to 2.5 mm. long, the lateral ribs very thick and corky, the dorsal ribs filiform, the oil tubes solitary in the intervals, 2 on the commissure, the seed face somewhat convex.

1. Lilaeopsis recurva A. W. Hill. Huachuca Mountains, Cochise County (*Lemmon* 2895, etc., *Goodding* 824–49), Santa Cruz River valley near Tucson, Pima County (*Pringle* in 1881, the type collection). Known only from southern Arizona.

23. OREOXIS

Plants perennial from slender, elongate roots, low, acaulescent, puberulent; leaves oblong, pinnate or bipinnate, the segments short, linear, mostly distinct; peduncles exceeding the leaves; umbels compound; involucre usually absent; involucel of several linear bractlets about equaling the flowers; calyx teeth conspicuous; corolla yellow; stylopodium absent; fruit oblong, 3 to 6 mm. long, laterally compressed, the seed dorsally compressed, the lateral and dorsal wings of the fruit obovate in cross section, the oil tubes usually solitary in the intervals, 2 to 4 on the commissure, the seed face plane or slightly concave.

1. Oreoxis alpina (Gray) Coult. & Rose. Keet Seel, Navajo Indian Reservation (*J. Howell, Jr.* 69), without definite locality (*Palmer* in 1869). Wyoming to northern New Mexico and northeastern Arizona, in the mountains.

24. CYMOPTERUS (206)

Plants perennial from long taproots, low, acaulescent or subacaulescent, glabrous or pubescent; leaves pinnate to ternate-pinnately decompound, the segments narrow, short, entire to pinnately lobed; peduncles shorter than to exceeding the leaves; umbels compound, or globose with the rays fused into a disk; involucre absent or present; bractlets of the involucel usually conspicuous, usually dimidiate, foliaceous to scarious; calyx teeth inconspicuous; corolla white, yellow, or purple; stylopodium absent; fruit ovate to oblong, 4 to 18 mm. long, more or less compressed dorsally, the lateral wings present, the dorsal wings usually present, the oil tubes small, 1 to 24 in the intervals, 2 to 22 on the commissure, the seed face slightly to deeply concave.

Some of the species are known as corkwing and wafer-parsnip.

KEY TO THE SPECIES

1. Rays of the umbel fused to form a discoid inflorescence; involucel bractlets scarious and paleaceous, obscure .... 1. *C. megacephalus*
1. Rays of the umbel distinct, 0.2 to 9 cm. long; involucel bractlets not paleaceous (2).
2. Involucre conspicuous, scarious, the bracts sometimes united; bractlets of the involucel conspicuous, scarious (3).
2. Involucre wanting or inconspicuous, never scarious; bractlets of the involucel rarely scarious and inconspicuous, or else foliaceous and conspicuous (5).

3. Bractlets of the involucel purple or greenish white, conspicuously many-nerved; pedicels less than 1 mm. long, or obsolete . . . . 4. *C. multinervatus*
3. Bractlets of the involucel white or whitish, few-nerved; pedicels 3 to 12 mm. long (4).
4. Umbels more or less spreading, the mature rays 10 to 50 mm. long; fruit ovate-oblong to oblong, the wings mostly narrower than the body of the fruit . . . . 2. *C. bulbosus*
4. Umbels densely globose, the mature rays 4 to 10 mm. long; fruit ovate, the wings much broader than the body of the fruit . . . . 3. *C. purpurascens*
5. Bractlets of the involucel inconspicuous, not foliaceous; oil tubes 1 to 8 in the intervals 7. *C. purpureus*
5. Bractlets of the involucel conspicuous, foliaceous; oil tubes 3 to 17 in the intervals (6).
6. Plants subcaulescent, with a well-developed stem (pseudoscape); leaf segments usually longer than wide . . . . 5. *C. Fendleri*
6. Plants acaulescent, the pseudoscape absent and the leaves borne at the apex of the taproot; leaf segments wider than long . . . . 6. *C. Newberryi*

1. Cymopterus megacephalus Jones. Eastern Coconino County (several localities), about 5,000 feet, gravelly or shaly soil, April to June, type from mouth of Moenkopi Wash (*Jones* in 1890). Known only from northern Arizona.

2. Cymopterus bulbosus A. Nels. Apache, Navajo, and Graham counties, 2,500 to 6,000 feet, April to June. Southwestern Wyoming to western Texas and eastern Arizona.

3. Cymopterus purpurascens (Gray) Jones. Apache County to eastern Mohave County, 3,500 to 7,000 feet, April and May, type from Oraibi (*Newberry* in 1858). Southern Idaho to central Arizona and southeastern California.

4. Cymopterus multinervatus (Coult. & Rose) Tidestrom (*Phellopterus multinervatus* Coult. & Rose). Apache, Coconino, and Mohave counties to Cochise, Santa Cruz, and Pima counties, 3,000 to 7,000 feet, March to May, type from Peach Springs, Mohave County (*Lemmon* in 1884). Southern Utah to Texas, northern Mexico, Arizona, and southeastern California.

5. Cymopterus Fendleri Gray. Apache County to Coconino County, 5,000 to 6,000 feet, May and June. Utah to Chihuahua and northern Arizona. Chimaya.

6. Cymopterus Newberryi (Wats.) Jones. Apache, Navajo, and Coconino counties, 4,500 to 7,000 feet, usually in sand, April to June, type from the "Flax" (Little Colorado) River (*Newberry* in 1858). Southern Utah and northern Arizona.

The sweet roots are eaten by Hopi children in spring.

7. Cymopterus purpureus Wats. Apache County to northwestern Mohave County, 4,000 to 7,000 feet, May and June. Southwestern Colorado, Utah, northwestern New Mexico, and northern Arizona.

Cymopterus Jonesii Coult. & Rose, of southwestern Utah, may be found in Arizona.

25. PTERYXIA

Plants perennial from a slender taproot, caulescent, glabrous or nearly so; leaves bipinnate or tripinnate, the segments narrow, acute; peduncles exceeding the leaves; umbels compound; involucre usually absent; bractlets of the involucel linear; calyx teeth conspicuous, persistent; corolla yellow; stylopodium absent; fruit ovate to oblong, compressed dorsally, the lateral and dorsal wings linear in cross section, the oil tubes 1 to 8 in the intervals, 3 to 15 on the commissure, the seed face somewhat concave.

KEY TO THE SPECIES

1. Stems unbranched or with few branches below; cauline leaves few or none; herbage glabrous throughout; young fruit glabrous; mature fruit 4.5 to 7 mm. long
1. *P. petraea*
1. Stems branched above; cauline leaves several to many; stem hirtellous at base of the umbel; young fruit puberulent; mature fruit 3 to 4 mm. long.........2. *P. Davidsoni*

1. Pteryxia petraea (Jones) Coult. & Rose. Both rims of the Grand Canyon (*Cook & Johnson* 2063, *Eastwood* 5833, *Eastwood & Howell* 1006). Southern Idaho and southeastern Oregon to northern Arizona and California.

2. Pteryxia Davidsoni (Coult. & Rose) Mathias & Constance (*Pseudocymopterus Davidsoni* Mathias). Greenlee County, among rocks in a moist creek bed near Clifton (*Davidson* in 1900, the type collection), Garfield (*Davidson*), 45 miles north of Clifton, 8,000 feet (*Kearney & Peebles* 12240, 12241), August. Western New Mexico and southeastern Arizona.

26. PSEUDOCYMOPTERUS

Plants perennial with a long, slender taproot, subcaulescent or caulescent, glabrous or pubescent; leaves pinnate to tripinnate, the segments filiform, linear, or lanceolate, short or elongate; peduncles exceeding the leaves, hirtellous at base of the umbel; umbels compound; involucre mostly absent; bractlets of the involucel usually conspicuous; calyx teeth conspicuous; corolla purple or yellow; stylopodium absent; fruit oblong to ovate-oblong, 3 to 7 mm. long, compressed dorsally or subterete, the lateral wings present, the dorsal wings similar to the laterals or absent through abortion, the wings mostly thin, sublinear in cross section, the oil tubes 1 to 5 in the intervals, 2 to 6 on the commissure, the seed face plane.

1. Pseudocymopterus montanus (Gray) Coult. & Rose (*Peucedanum Lemmoni* Coult. & Rose). Apache County to Coconino County, south to Cochise and Pima counties, 5,500 to 12,000 feet, common in pine woods and grassland, May to October, type of *Peucedanum Lemmoni* from the Huachuca Mountains (*Lemmon* 392). Southern Wyoming to western Utah, south to northern Mexico.

This species is exceedingly variable in vegetative characters, such as leaf division, shape and size of the ultimate segments, and height of growth. These characters are governed to a great extent by the habitat of the individual plant. The flower color sometimes varies on the same plant from yellow through orange-purple to purple. In some individuals the dorsal wings of the fruit are absent through abortion.

27. CONIOSELINUM

Plants perennial from a cluster of fleshy roots, caulescent, glabrous or the inflorescence puberulent; leaves bipinnate or ternate-pinnate, the leaflets ovate to ovate-lanceolate, acute, laciniately toothed or pinnatifid; peduncles exceeding the leaves; umbels compound; involucre mostly absent; involucel wanting, or of 1 to several, more or less elongate, narrow bractlets; calyx teeth obsolete; corolla white; stylopodium low-conic; fruit oval, 3 to 6 mm. long, compressed dorsally, the lateral wings broad, corky, the dorsal ribs slender, winged, the oil tubes 1 or 2 in the intervals, 4 on the commissure, the seed face plane or slightly concave.

KEY TO THE SPECIES

1. Leaves 10 to 20 cm. long; leaflets laciniately pinnatifid; peduncles stout; bractlets of the involucel 2 to 8 mm. long; rays 10 to 20; fruit 4 to 6 mm. long, 2 to 3 mm. wide
1. *C. scopulorum*

1. Leaves 6 to 14 cm. long; leaflets coarsely pinnatifid and toothed; peduncles slender; bractlets of the involucel 2 to 5 mm. long; rays 5 to 11; fruit 3 mm. long, 2 mm. wide
2. *C. mexicanum*

1. Conioselinum scopulorum (Gray) Coult. & Rose. East Fork of White River (Navajo County?) and mountains of Graham, Cochise, and Pima counties, 4,500 to 9,000 feet, August and September. Wyoming to New Mexico and Arizona.

2. Conioselinum mexicanum Coult. & Rose. Huachuca Mountains, Cochise County (*Jones* in 1903), Santa Rita Mountains, Pima County (*Kearney & Peebles* 10497), September and October. Southern Arizona and Chihuahua.

28. ANGELICA

Plants perennial, caulescent, somewhat pubescent; leaves pinnate to incompletely bipinnate, the leaflets lanceolate to ovate-lanceolate, acute, distinct, coarsely to remotely serrate; peduncles exceeding the leaves; umbels compound; involucre and involucel usually absent; calyx teeth obsolete; corolla white or pinkish; stylopodium conic; fruit oblong-oval, 3 to 6 mm. long, sparsely hispidulous when young, glabrous at maturity, compressed dorsally, the lateral wings broad, the dorsal ribs slender, winged, the oil tubes 1 or 2 in the intervals, several on the commissure, the seed face slightly concave.

1. Angelica pinnata Wats. Lukachukai Mountains, Apache County (*Goodman & Payson* 3194, *Eggleston* 17129), Kaibab Plateau, Coconino County, in partial shade along a stream (*Goodding* 384–48). Wyoming and Utah to New Mexico and northern Arizona.

29. ANETHUM. Dill

Plants annual from slender subfusiform roots, caulescent, branching, glabrous, with a strong odor of anise; leaves pinnately decompound, the segments filiform, distinct; peduncles axillary and terminal, exceeding the leaves; umbels compound; involucre and involucel mostly absent; calyx teeth obsolete; corolla yellow; stylopodium conic, the margin crenulate; fruit ovate, about 4 mm. long, compressed dorsally, the lateral wings much narrower than the body of the fruit, the dorsal wings obsolete, the oil tubes solitary in the intervals, 2 to 4 on the commissure, the seed face plane.

1. Anethum graveolens L. Sacaton, Pinal County (*King* 1832), Tucson (*Thornber* in 1903, *Streets* in 1938), an occasional escape from gardens; introduced from Europe.

Cultivated as a condiment plant.

30. OXYPOLIS

Plants perennial, glabrous, caulescent, from fascicled tubers; leaves with dilated sheaths, oblong in general outline, simply pinnate or ternate; leaflets few, broadly ovate to narrowly lanceolate, crenate-dentate or serrate, rarely incised; peduncles 8 to 20 cm. long; inflorescence of loose compound umbels; involucre and involucel wanting; calyx teeth conspicuous; corolla white or purple; stylopodium conic; fruit oblong to oval, 3 to 5 mm. long, glabrous, strongly compressed dorsally, the dorsal ribs filiform, the lateral ones broadly thin-winged, the oil tubes large, solitary in the intervals, 2 to 6 on the commissure, the seed face plane.

1. Oxypolis Fendleri (Gray) Heller. Phelps Botanical Area, White Mountains, Apache County, 9,500 feet, stream side in partial shade (*Phillips & Humphrey* 3106, *Phillips & Kearney* 3426). Wyoming to New Mexico, southeastern Utah, and eastern Arizona.

31. LOMATIUM (207). Biscuit-root, Indian-root

Plants perennial, with moniliform tubers or long taproots, acaulescent to tall-caulescent, glabrous or pubescent; leaves ternately or pinnately decompound, the segments filiform to ovate; peduncles equaling or exceeding the leaves; umbels compound; involucre of few bracts or absent; bractlets of the involucel filiform to obovate, foliaceous to subscarious, distinct to connate, rarely none; calyx teeth small or obsolete; corolla greenish white, yellow, or purple; stylopodium absent; fruit ovate to linear, 4.5 to 16 mm. long, compressed dorsally, the lateral wings present, the dorsal ribs absent or filiform, the wings thin or corky-thickened, the oil tubes small or obscure, the seed face plane or slightly concave.

Many of the species have edible roots that the Indians ate raw, cooked, or ground into flour. The plants are grazed by livestock.

KEY TO THE SPECIES

1. Stems tall, leafy; calyx teeth usually obsolete; wings of the fruit corky-thickened; oil tubes obscure .... 1. *L. dissectum*

1. Stems short or the plants acaulescent; calyx teeth small; wings of the fruit thin; oil tubes 1 to several in the intervals, 2 to several on the commissure (2).

2. Plants with elongate, moniliform, tuberous roots; fruit narrowly oblong, 10 to 15 mm. long, 2 to 5 mm. wide, the wings less than half the width of the body; rays and pedicels strict, suberect .... 2. *L. leptocarpum*

2. Plants with more or less thickened, elongate taproots, sometimes with very deep-seated tubers; fruit orbicular to broadly oblong, with broader wings; rays and pedicels more spreading (3).

3. Fruiting pedicels 10 to 17 mm. long; plants glabrous; leaves narrowly oblong, 10 to 20 cm. long; fruit oblong........ 6. *L. Parryi*

3. Fruiting pedicels mostly less than 10 mm. long; plants more or less pubescent; leaves broadly oblong to ovate, 5 to 15 cm. long; fruit suborbicular to oblong-obovate (4).

4. Involucel bractlets with a conspicuous, scarious margin, never tomentose or villous; flowers white . . . . 3. *L. nevadense*
4. Involucel bractlets not conspicuously scarious-margined, more or less tomentose or villous; flowers yellow or purple (5).
5. Plants more or less villous throughout; petioles shorter than the blades; flowers yellow or purplish-tinged . . . . 4. *L. Macdougali*
5. Plants hoary-pubescent, never villous; petioles longer than the blades; flowers purple 5. *L. mohavense*

1. Lomatium dissectum (Nutt.) Mathias & Constance (*Leptotaenia dissecta* Nutt.). Southern Navajo, southern Coconino, and northern Gila counties, 4,000 to 6,000 feet, in chaparral, April to September. Colorado to British Columbia, south to Arizona and California.

The species is represented in Arizona by var. *multifidum* (Nutt.) Mathias & Constance (*Leptotaenia multifida* Nutt.). The plant is sometimes known as carrot-leaf and wild-carrot. It is reported that it is palatable to livestock and that the Indians roasted and ate the large roots.

2. Lomatium leptocarpum (Torr. & Gray) Coult. & Rose. Grand Canyon (*Jones*), abundant under pines in camp grounds on the North Rim of the Grand Canyon (*Mathias* 736, *Eastwood & Howell* 902). Northwestern Colorado to northern Idaho, northern Arizona, and northeastern California.

3. Lomatium nevadense (Wats.) Coult. & Rose. Coconino and Mohave counties to Greenlee, Graham, Gila, Maricopa, and Pima counties, 3,000 to 7,000 feet, mesas and rocky slopes, March to May. Western Utah to Oregon and eastern California, south to western New Mexico, Arizona, and Sonora.

KEY TO THE VARIETIES

1. Ovary and fruit puberulent . . . . *L. nevadense* (typical)
1. Ovary and fruit glabrous (2).
2. Petioles not prominently scarious-margined; wings narrower than the body of the fruit var. *Parishii*
2. Petioles prominently scarious-margined; wings broader than the body of the fruit. Dorsal ribs evident . . . . var. *pseudorientale*

Typical *L. nevadense* occurs near Flagstaff and in Yavapai, Gila, eastern Maricopa, and Pima counties. Var. *Parishii* (Coult. & Rose) Jepson (*Cogswellia decipiens* Jones), the commonest and most widely distributed variety in Arizona, occupies the entire range of the species in Arizona. The type of *C. decipiens* came from Hualpai Mountain (*Jones* in 1903). Var. *pseudorientale* (Jones) Munz is nearly as widely distributed as var. *Parishii* but is apparently less common. The type was collected in Skull Valley, Yavapai County (*Jones* in 1903).

4. Lomatium Macdougali Coult. & Rose. Coconino, eastern Mohave, and Yavapai counties, 4,500 to 8,000 feet, March to June, type from Mormon Lake (*MacDougal* 84). Western Wyoming to central Oregon, south to central Arizona.

***5. Lomatium mohavense** Coult. & Rose. Peach Springs, Mohave County, 5,000 feet (*Lemmon* in 1884). Deserts of southern California and adjacent Nevada.

In view of the unreliability of the data on the labels of Lemmon's specimens, further evidence of the presence of this species in Arizona is required.

6. Lomatium Parryi (Wats.) Macbr. South end of Navajo Mountain, Coconino County, 7,500 to 8,500 feet (*Peebles & Smith* 13963), May and June. Mountains of southeastern Utah to northern Arizona and eastern California.

Other species of this genus that may be found in Arizona are: *Lomatium Nuttallii* (Gray) Macbr., which has been collected in northwestern New Mexico; *L. latilobum* (Rydb.) Mathias, known from southeastern Utah; and *L. scabrum* (Coult. & Rose) Mathias, known from southwestern Utah.

32. PASTINACA. Parsnip

Plants perennial from thick taproots, caulescent, branching, glabrous or nearly so; stem stout, fluted; leaves pinnate, the leaflets oblong to ovate, coarsely serrate and pinnately lobed or divided; peduncles axillary and terminal; umbels compound; involucre none, or of a few narrow, deciduous bracts; involucel absent; calyx teeth obsolete; corolla yellow; stylopodium depressed-conic; fruit obovate to orbicular, 5 to 6 mm. long, strongly flattened dorsally, the dorsal ribs filiform, the lateral wings narrow, the oil tubes solitary in the intervals, 2 to 4 on the commissure, the seed face plane.

1. Pastinaca sativa L. Snowflake (Navajo County), Williams (Coconino County), Huachuca Mountains (Cochise County). The parsnip has escaped from gardens and established itself widely in North America but has scarcely become naturalized in Arizona.

33. HERACLEUM. Cow-parsnip

Plants perennial, with fascicled, fibrous roots, caulescent, tomentose; stem tall, stout; leaves ternately compound, the leaflets 3, large, ovate to orbicular, sharply serrate and lobed, petiolulate; peduncles axillary and terminal; umbels compound; involucre of narrow, entire, deciduous bracts; bractlets of the involucel small, linear; calyx teeth obsolete; corolla white, the petals obcordate, the marginal ones of the umbel much larger; stylopodium thick, conic; fruit broadly obovate or obcordate, 5 to 11 mm. long, strongly flattened dorsally, more or less pubescent, the dorsal and intermediate ribs filiform, the lateral wings broad, the oil tubes solitary in the intervals, visible on the dorsal surface and extending from the apex to about the middle of the mericarp, 2 on the commissure, the seed face plane.

1. Heracleum lanatum Michx. (*H. maximum* Bartr., nom. illegit.). Pinaleno Mountains (Graham County), Santa Catalina Mountains (Pima County), McClintock Canyon, Black Mesa (Apache or Navajo County), 7,500 to 9,000 feet, July and August. Widely distributed in the United States and Canada.

A coarse plant, reported to be relished by livestock. The young leaves and stems were eaten by the Indians. The root appears to be somewhat stimulant and carminative and has been used in epilepsy. It is reported that the Apache Indians used it medicinally. Contact with the wet foliage is said to cause dermatitis in susceptible persons.

34. ERYNGIUM (208). Eryngo, Button-snakeroot

Plants perennial, caulescent, glabrous; stems simple or branched; leaves lanceolate to oblanceolate and reticulate-veined, or linear and parallel-veined, entire to pinnatifid; peduncles exceeding the leaves; inflorescence a dense bracteate head; bracts linear to ovate-lanceolate, entire to spinulose-serrate; floral bractlets usually entire; calyx lobes lanceolate to ovate, obtuse or acute; corolla white or blue; stylopodium absent; fruit ovoid, flattened laterally, covered with hyaline scales or tubercules, the ribs obsolete, the oil tubes several, inconspicuous, the seed face plane.

KEY TO THE SPECIES

1. Leaves elongate, linear, parallel-veined, usually entire...........1. *E. sparganophyllum*
1. Leaves shorter, lanceolate to oblanceolate, reticulate-veined, crenate to spinose-serrate or pinnatifid (2).
2. Basal leaves obscurely crenate or serrate, not spinose; heads cylindric-ovoid, amethystine 2. *E. phyteumae*
2. Basal leaves spinose-serrate or pinnatifid; heads ovoid, not amethystine (3).
3. Plants from a cylindric taproot; lower cauline leaves pinnatifid to bipinnatisect; inflorescence paniculately branched, the heads comate; bracts linear-lanceolate to lanceolate, entire or with 1 or 2 pairs of lateral spines near the middle, commonly yellowish above ....3. *E. heterophyllum*
3. Plants from a fascicle of fibrous or fleshy roots; lower cauline leaves spinose-serrate; inflorescence successively trifurcate, the heads not comate; bracts broadly lanceolate to oblanceolate, spinose-serrate with 2 or 3 pairs of teeth, silvery-white above 4. *E. Lemmoni*

1. Eryngium sparganophyllum Hemsl. Agua Caliente ranch, near Tucson, 3,000 feet, marshy ground (*Shreve* in 1908, *Thornber* in 1910). New Mexico to southern Arizona and Mexico.

2. Eryngium phyteumae Delar. (*E. discolor* Wats.). In water, Huachuca Mountains, Cochise County (*Lemmon* in 1882, the type of *E. discolor*). Southeastern Arizona and Mexico.

3. Eryngium heterophyllum Engelm. (*E. Wrightii* Gray). Cochise, Santa Cruz, and eastern Pima counties, 4,000 to 6,500 feet, plains and along water courses, August to October. Western Texas to southeastern Arizona and Mexico.

Known locally as Mexican-thistle. The flowers are pale blue.

4. Eryngium Lemmoni Coult. & Rose. Cochise County, chiefly in the Chiricahua and Huachuca mountains, 6,000 to 9,000 feet, August to October, type from the Chiricahua Mountains (*Lemmon* 17). Southeastern Arizona and northern Mexico.

101. CORNACEAE. Dogwood Family

Small trees or shrubs; leaves opposite, simple, without stipules, entire or nearly so; flowers perfect or unisexual, regular, small; petals (when present), and stamens 4; pistil 1, the styles 1 or 2, the ovary inferior, 1- or 2-celled; fruit a drupe or berrylike.

KEY TO THE GENERA

1. Flowers in catkins, unisexual, apetalous; styles 2. Plants dioecious...........1. *Garrya*
1. Flowers in flat-topped, compound cymes, perfect, with petals; style 1.........2. *Cornus*

1. GARRYA. Silk-tassel

Large evergreen shrubs, the branchlets quadrangular; leaves short-petioled, thick; flowers of both sexes in loose or dense catkinlike spikes, the staminate ones in clusters of 3, the pistillate flowers solitary in the axils of the bracts; calyx with tube adnate to the ovary, the limb reduced to 4 teeth or lobes in the staminate flowers, obsolete or nearly so in the pistillate flowers; fruit berrylike, dry or somewhat juicy.

Sometimes called quinine-bush and coffeeberry-bush. The leaves are bitter, and an alkaloid—garryin—obtained from 1 or 2 species of the genus, is used medicinally. The plants are sometimes browsed by cattle, deer, and bighorns. Both of the Arizona species occur usually in the chaparral association on dry slopes and in canyons.

This genus is often referred to a separate family, Garryaceae.

KEY TO THE SPECIES

1. Inflorescences spiciform, dense (especially the pistillate ones); bracts not leaflike, triangular, strongly connate; fruits ellipsoid-ovoid, acutish at apex, densely whitish-sericeous; leaves entire or nearly so, when mature copiously grayish-sericeous to glabrescent above, densely sericeous-tomentose beneath, the larger ones commonly 5 cm. long or longer . . . . 1. *G. flavescens*
1. Inflorescences of both sexes loose; bracts leaflike, not triangular, connate only at base; fruits globose, rounded at apex, when ripe glabrous or very nearly so and dark blue with a slight bloom; leaves more or less callose-denticulate, when mature glabrous or pulverulent on both faces (sometimes very sparsely appressed-pubescent beneath with straight hairs), the larger ones commonly less than 5 cm. long . . . . 2. *G. Wrightii*

1. Garrya flavescens Wats. (*G. mollis* Greene). Coconino, Mohave, Gila, Yavapai, Cochise, and Yuma counties, 2,500 to 7,000 feet, January to April, type of *G. mollis* from Oak Creek Canyon, Coconino County (*Pearson* 339). Western Texas, Arizona, southern Utah and Nevada, California, and northern Mexico.

This shrub attains a height of at least 1.8 m. (6 feet). The leaves are sometimes 25 cm. long. The bark is grayish green.

2. Garrya Wrightii Torr. Grand Canyon (Coconino County) and Greenlee, Gila, Cochise, Santa Cruz, and Pima counties, 3,000 to 8,000 feet, dry slopes, March to August. Western Texas to Arizona and northern Mexico.

The plants reach a height of 3 m. (10 feet). Rubber, in small quantity, has been extracted from this shrub.

2. CORNUS. Dogwood

Shrubs, with purplish-red young bark; leaves ovate, entire, finely appressed-pubescent, whitish beneath; flowers small; calyx limb of 4 minute teeth, the tube wholly adnate to the ovary; petals 4, white; fruit a drupe, normally white when mature.

1. Cornus stolonifera Michx. (*C. sericea* L. subsp. *stolonifera* (Michx.) Fosberg). Apache County to Coconino County, south to Cochise and Pima counties, (2,500?) 5,000 to 9,000 feet, along streams, often with willows and alder, May to July. Canada and Alaska, south to the District of Columbia, New Mexico, Arizona, and California.

Red-osier dogwood, so called because the bark resembles that of some willows. The fruits are very attractive to birds.

Series 3. Sympetalae (Families 102 to 132)

KEY TO THE FAMILIES[1]

1. Corolla distinctly irregular (2).
1. Corolla regular or nearly so (17).
2. Ovary inferior (3).
2. Ovary superior (free from the calyx) (6).
3. Stamens distinct, borne on the corolla (4).
3. Stamens united by the filaments, the anthers, or both (5).
4. Plants woody; calyx limb usually evident, as teeth or lobes; fruit berrylike, several-seeded: genus *Lonicera* .... 128. *Caprifoliaceae*
4. Plants herbaceous; calyx limb obsolete or represented by plumose bristles; fruit achene-like, 1-seeded .... 129. *Valerianaceae*
5. Flowers not in heads; calyx limb well developed, the lobes 3 to 5; stamens free from the corolla or nearly so; fruit a turgid, many-seeded capsule: subfamily Lobelioideae 131. *Campanulaceae*
5. Flowers in heads subtended by a specialized involucre, rarely solitary; calyx limb obsolete or represented by a pappus of bristles, awns, or scales; stamens borne on the corolla; fruit an achene, 1-seeded.............................. 132. *Compositae*
6. Stamens more or less united by the filaments; corolla not strongly sympetalous (7).
6. Stamens distinct, the filaments attached to the corolla; corolla strongly sympetalous, more or less bilabiate (9).
7. Leaves compound, rarely reduced to 1 leaflet; stamens commonly 10. Petals usually 5; fruit a 2-valved or indehiscent pod, or a loment composed of 2 or more indehiscent, 1-seeded segments .... 64. *Leguminosae*
7. Leaves simple; stamens fewer than ten (8).
8. Petals 5, 2 of them fleshy and glandlike; stamens 3 or 4; fruit an indehiscent, spiny pod: genus *Krameria* .... 64. *Leguminosae*
8. Petals normally 3, not glandlike; stamens 6 or 8; fruit a flattened, 1- or 2-celled capsule .... 73. *Polygalaceae*
9. Fruit at maturity separating into 2 or 4 dry nutlets or if a fleshy drupe (in genus *Lantana*), then the flowers in headlike, involucrate clusters (10).
9. Fruit capsular, not separating into nutlets, somewhat fleshy and long-beaked in genus *Proboscidea* (11).
10. Ovary entire, or longitudinally grooved, or slightly 4-lobed; style apical; fruit of 2 or 4 nutlets, or drupelike.. 117. *Verbenaceae*
10. Ovary 4-lobed or 4-parted, the style arising between the lobes; fruit of (normally) 4 nutlets .... 118. *Labiatae*
11. Plants root parasites, without chlorophyll, the leaves reduced to fleshy scales 123. *Orobanchaceae*
11. Plants not or only weakly parasitic, with chlorophyll; leaves with well-developed blades (12).
12. Plants aquatic, all but the scapes immersed; leaves finely dissected, these and the stems bearing bladders closed by lids............................ 124. *Lentibulariaceae*
12. Plants terrestrial; leaves not finely dissected or bearing bladders (13).
13. Anther-bearing stamens 5; capsule usually 3-celled............... 114. *Polemoniaceae*
13. Anther-bearing stamens 4 or 2; capsule 2-valved (14).
14. Ovary 1-celled; fruit large, somewhat fleshy, ending in a long, hooked beak 122. *Martyniaceae*
14. Ovary 2-celled; fruit not with a long, hooked beak (15).
15. Capsule long and slender; seeds comose or winged................ 121. *Bignoniaceae*
15. Capsule relatively short; seeds not comose, seldom winged (16).
16. Stems usually terete; seeds commonly indefinitely numerous, the cotyledons narrow, the seed stalks not persisting on the placenta................... 120. *Scrophulariaceae*

[1] Several families belonging properly to series Choripetalae are included in this key also, for the reason that some of their members have the petals partly united.

16. Stems more or less 4-angled; seeds few (usually only 2 in each cell), the cotyledons broad and heart-shaped, the seed stalks (except in genus *Elytraria*) persisting on the placenta as hardened, hooklike projections....................125. *Acanthaceae*
17. Ovary inferior or partly so (18).
17. Ovary superior (free from the calyx) or very nearly so (26).
18. Flowers in heads subtended by a specialized involucre, rarely solitary; calyx limb represented by a pappus of bristles, awns, or scales, sometimes obsolete. Stamens more or less united; fruit an achene.............................132. *Compositae*
18. Flowers not in heads with a specialized involucre; calyx limb usually well developed, not pappuslike, except in Valerianaceae (19).
19. Perianth segments and stamens indefinitely numerous; stems very succulent, with spines, or barbed bristles, or both, these borne on cushionlike areoles; leaves wanting or rudimentary ....94. *Cactaceae*
19. Corolla lobes few (usually 5); stamens 10 or fewer; stems not succulent or if so, then not areolate and spiny; leaves well developed, except in Lennoaceae (20).
20. Plants with tendrils, the stems trailing or climbing; anthers more or less united; flowers unisexual, the plants monoecious or dioecious..................130. *Cucurbitaceae*
20. Plants without tendrils; anthers distinct; flowers mostly perfect (21).
21. Stamens free from the corolla or very nearly so (22).
21. Stamens borne on the corolla (23).
22. Plants shrubby; corolla cylindric or urceolate; anthers opening by apical pores; fruit a berry: genus *Vaccinium*...103. *Ericaceae*
22. Plants herbaceous; corolla campanulate or rotate; anthers opening longitudinally; fruit a capsule: subfamily Campanuloideae........................131. *Campanulaceae*
23. Calyx limb reduced to bristles, these elongate and plumose in fruit..129. *Valerianaceae*
23. Calyx limb otherwise (24).
24. Fruit an apically 5-valved, many-seeded capsule: genus *Samolus*......104. *Primulaceae*
24. Fruit not a capsule or if so, then not 5-valved (25).
25. Fruit dry, a 2-celled capsule, or achenelike, or separating into 2 to 4 usually closed carpels ....127. *Rubiaceae*
25. Fruit berrylike or drupelike or if dry and achenelike, then the stems creeping and the flowers in pairs on long peduncles...........................128. *Caprifoliaceae*
26. Stamens more or less united by the filaments or (in Solanaceae) by the anthers (27).
26. Stamens distinct (31).
27. Leaves bipinnate. Flowers in heads or spikes; stamens much more conspicuous than the small corolla: subfamily Mimosoideae..........................64. *Leguminosae*
27. Leaves simple, sometimes pedately cleft or parted (28).
28. Flowers unisexual; seeds caruncullate: genus *Jatropha*.............74. *Euphorbiaceae*
28. Flowers perfect; seeds not caruncullate (29).
29. Stamens numerous; corolla only slightly sympetalous. Fruit of several separating carpels, or a several-celled capsule...............................84. *Malvaceae*
29. Stamens 5; corolla weakly to strongly sympetalous (30).
30. Ovaries 2, connected only by the common stigma; pollen grains cohering in masses (pollinia); filaments united into a column; fruit a pair of follicles, or 1 by abortion 112. *Asclepiadaceae*
30. Ovary 1, usually 2-celled; pollen grains not cohering in pollinia; filaments distinct, the anthers connivent; fruit a berry: genus *Solanum*..................119. *Solanaceae*
31. Filaments free from the corolla, or very nearly so (32)
31. Filaments attached to the corolla (34).
32. Leaves bipinnate; flowers in dense heads or spikes; stamens much more conspicuous than the small corolla: subfamily Mimosoideae...................64. *Leguminosae*
32. Leaves simple; flowers not in dense heads or spikes; stamens less conspicuous than the corolla (33).
33. Anthers awned, or opening by pores, or both; plants shrubby, or else saprophytic, without chlorophyll, and the leaves reduced to scales; corolla urceolate or subglobose 103. *Ericaceae*

33. Anthers not awned, opening longitudinally; plants herbaceous or suffrutescent, always with chlorophyll; leaves well developed; corolla salverform: genus *Plumbago* 105. *Plumbaginaceae*
34. Plants without chlorophyll, parasitic; leaves reduced to scales (35).
34. Plants with chlorophyll and well-developed leaves (36).
35. Root parasites, the thick, succulent stems mostly subterranean; flowers small, very numerous, crowded on an expanded saucer-shaped receptacle; corolla lobes and stamens 6 or more . . . . . 102. *Lennoaceae*
35. Stem parasites, the slender, yellow stems twining; flowers in cymose clusters; corolla lobes and stamens not more than 5: genus *Cuscuta* . . . . . 113. *Convolvulaceae*
36. Fruit at maturity separating into 2 to 4 dry nutlets or (by abortion) only 1 (37).
36. Fruit not separating into nutlets, but (in 1 genus of Convolvulaceae) of 2 partly distinct utricles (39).
37. Stamens and corolla lobes 5; stems not quadrangular . . . . . 116. *Boraginaceae*
37. Stamens 2 or 4, if 4, then in 2 pairs; stems often quadrangular (38).
38. Ovary entire, or longitudinally grooved, or slightly 4-lobed; style apical or very nearly so; nutlets 2 or 4 . . . . . 117. *Verbenaceae*
38. Ovary 4-lobed or 4-parted; style arising between the lobes; nutlets normally 4 118. *Labiatae*
39. Fruit a 1- or 2-seeded drupe; plants shrubs or small trees, usually thorny or spiny (40).
39. Fruit not a drupe, if berrylike, then several-seeded; plants various (41).
40. Fertile stamens 5, with 5 petaloid staminodia; corolla with a pair of lobelike appendages in each sinus of the 5 true lobes; flowers in axillary clusters . . . . . 107. *Sapotaceae*
40. Fertile stamens 4; staminodia none; corolla not appendaged; flowers in terminal, headlike clusters: genus *Lantana* . . . . . 117. *Verbenaceae*
41. Stamens 10 or more. Leaves simple, entire (42).
41. Stamens fewer than 10, seldom more than 5 (43).
42. Plants herbaceous, succulent, with a rosette of relatively large basal leaves, the stem leaves bractlike, persistent; carpels normally 5, distinct or nearly so: genus *Echeveria* 59. *Crassulaceae*
42. Plants thorny shrubs, with long, whiplike branches; stem leaves not bractlike, deciduous; carpels 3, united . . . . . 106. *Fouquieriaceae*
43. Fruit a samara, or a didymous capsule. Leaves simple or pinnate; stamens 2 or 4 108. *Oleaceae*
43. Fruit otherwise (44).
44. Ovaries 2, united only by the common style or stigma; fruit a pair of elongate follicles or (by abortion) 1; seeds often comose . . . . . 111. *Apocynaceae*
44. Ovary 1, sometimes deeply parted; fruit a single capsule or berry, or a pair of partly distinct utricles; seeds not comose (45).
45. Ovary 1-celled, or sometimes imperfectly 2-celled by introflection of the placentas (46).
45. Ovary 2-celled or more or if 2-celled only at base (in genus *Limosella*), then the plants small semiaquatic herbs with small solitary flowers, a 5-lobed corolla, and 4 stamens (48).
46. Stamens opposite the corolla lobes. Style 1; stigma entire . . . . . 104. *Primulaceae*
46. Stamens alternate with the corolla lobes (47).
47. Style 1 and entire, or none and the stigma sessile; herbage usually glabrous; leaves simple and entire; inflorescence not scorpioid. Stigma 1 and more or less distinctly 2-lobed, or the stigmas 2 . . . . . 110. *Gentianaceae*
47. Styles 2, or single and 2-cleft; herbage usually pubescent; leaves simple or compound, if simple, then the blades entire to pinnatifid; inflorescence commonly scorpioid 115. *Hydrophyllaceae*
48. Fruit a circumscissile capsule. Flowers small, in (usually dense) terminal spikes; plants scapose, the foliage leaves all basal . . . . . 126. *Plantaginaceae*
48. Fruit not circumscissile (49).
49. Plants shrubby; flowers small, in dense, axillary clusters, these forming interrupted, leafy spikes . . . . . 109. *Loganiaceae*

49. Plants not shrubby, or the inflorescence otherwise (50).
50. Stigmas 3. Style 1, often 3-cleft; fruit a capsule..................114. *Polemoniaceae*
50. Stigmas 1 or 2 or (in family Convolvulaceae) sometimes 4 (51).
51. Ovules 2 in each ovary cell. Stems often twining; fruit a capsule or (in *Dichondra*) of 2 partly distinct utricles....................................113. *Convolvulaceae*
51. Ovules more than 2 in each cell (52).
52. Styles 2, distinct or partly united. Inflorescence often scorpioid..115. *Hydrophyllaceae*
52. Style 1, entire (53).
53. Leaves commonly alternate, sometimes fascicled; plants leafy-stemmed; stamens 4 or 5, equal in number to the corolla lobes and all perfect; inflorescence never spicate; or if (in genus *Petunia*) the leaves often opposite and 1 of the stamens rudimentary, then the flowers small, solitary, extra-axillary and the corolla narrowly funnelform 119. *Solanaceae*
53. Leaves, at least the lower ones, opposite, or the plant subscapose, with leaves mostly in a basal rosette; fertile stamens 2 or 4, fewer than the corolla lobes; or if (in genus *Verbascum*) the stamens and corolla lobes both 5, then the flowers many, in elongate, terminal spikes or spikelike racemes........................120. *Scrophulariaceae*

102. LENNOACEAE. Lennoa Family

1. AMMOBROMA (209). Sand-root, Sand-food

Plants without chlorophyll, root parasites; stems thick, succulent, subterranean; leaves reduced to scales; flowers regular, perfect, small, very numerous on an expanded receptacle; sepals filiform, plumose-hairy; corolla funnelform, its lobes normally 6 to 9; stamens borne on the corolla.

1. Ammobroma sonorae Torr. Southern Yuma County along the Mexican boundary (*Harrison & Kearney* 8435), below 500 feet, drifting sand, April. Southwestern Arizona, southeastern California, and northwestern Sonora.

One of the most remarkable plants of the Arizona flora. Only the saucer-shaped receptacle is normally seen above the ground. This is commonly 3.5 to 12.5 cm. in diameter, about the color of the surrounding sand, and thickly studded with tiny, violet-colored flowers opening in successive circles. The long, succulent, underground stems are attached to the roots of various desert shrubs. They were formerly much used as food by a Western group of Papago Indians, the "Sand Papagos," who ate them raw, roasted, or ground into meal.

103. ERICACEAE (210). Heather Family

Plants of various habit, herbaceous to treelike, with or without chlorophyll, often evergreen; flowers perfect, regular or nearly so; corolla sympetalous or the petals nearly distinct; stamens free from the corolla or nearly so; stamens 8 to 12, the anthers mostly opening by terminal pores, often awned; style 1; ovary superior or inferior, 4- to 10-celled; fruit dry or juicy.

This large family, chiefly of the cooler parts of the world, includes some of the most beautiful of cultivated trees and shrubs, notably the rhododendrons and azaleas, and the heathers (species of *Erica*). The blueberries and cranberries, belonging to the genus *Vaccinium*, are the most important economic plants of the heather family.

KEY TO THE GENERA

1. Plants saprophytic, without green coloring matter, yellowish brown or red; leaves reduced to scales. Anthers not beaked; style erect, stout (2).
1. Plants autophytic and green, the leaves with well-developed blades, except sometimes in *Pyrola picta* (3).
2. Petals distinct; anthers not awned, opening at apex; ovary borne on a deeply toothed disk; seeds not winged but with taillike extensions of the coat..........1. *Monotropa*
2. Petals united below; anthers with 2 long dorsal awns, opening lengthwise; disk none; seeds with a hyaline wing much larger than the body..................2. *Pterospora*
3. Petals distinct; plants small, herbaceous or barely suffrutescent; anthers not awned, opening by pores and often tubular-beaked at the apparent apex (4).
3. Petals united; plants shrubby or arborescent; anthers with or without awns, opening by apical pores or slits (7).
4. Disk present at base of the ovary; collar at base of the stigma obscure or wanting (5).
4. Disk wanting; collar at base of the stigma evident (6).
5. Inflorescence an elongate, 1-sided raceme; filaments not dilated; style elongate, the stigma only moderately expanded...................................3. *Ramischia*
5. Inflorescence short, corymbose or cymose, rarely 1-flowered; filaments greatly dilated toward base; style very short, the stigma very broad and thick.........4. *Chimaphila*
6. Flowers several; petals converging at anthesis; style often declined; stigma lobes very short; valves of the capsule sparsely arachnoid-pubescent on the edges when opening 5. *Pyrola*
6. Flower solitary; petals spreading at anthesis; style straight; stigma lobes elongate, triangular, narrow; valves of the capsule not pubescent on the edges.........6. *Moneses*
7. Leaves thin, deciduous; ovary wholly inferior; fruit a more or less juicy berry, crowned by the persistent calyx teeth.......................................10. *Vaccinium*
7. Leaves thick, evergreen; ovary superior; fruit not juicy (8).
8. Low, heatherlike shrubs; leaves linear, strongly revolute; corolla campanulate; anthers not awned; fruit a capsule.......................................7. *Phyllodoce*
8. Tall shrubs or arborescent; leaves broad and flat; corolla cylindric-urceolate; anthers with 2 dorsal awns; fruit indehiscent (9).
9. Fruit berrylike, only slightly fleshy, the surface granular-tessellate, glabrous..8. *Arbutus*
9. Fruit drupelike, dry, the surface not granular, glabrous or glandular-pilose 9. *Arctostaphylos*

1. MONOTROPA. Pinesap

Plants saprophytic, yellowish brown or red; stems beset with scalelike leaves; flowers soon nodding; calyx of 3 to 5 distinct, bractlike sepals; petals distinct; anthers more or less reniform; style erect, stout; fruit a dehiscent 4- or 5-celled capsule.

1. Monotropa latisquama (Rydb.) Hultén (*Hypopitys latisquama* Rydb.). Coconino County to Cochise and Pima counties, 7,000 to 9,000 feet, rich soil in shade of pines, firs, and aspens, July and August. Widely distributed in the Northern Hemisphere.

The plant is pink or red and somewhat fragrant in drying, with stems up to 30 cm. long and petals at least 10 mm. long.

2. PTEROSPORA. Pine-drops

Plants with a superficial resemblance to *Monotropa,* but with a more elongate inflorescence, smaller, more numerous, pendulous flowers, a 5-parted calyx, and a sympetalous corolla.

1. Pterospora andromedea Nutt. Apache and Coconino counties to Cochise and Pima counties, 6,000 to 9,500 feet, coniferous forests, June to August. Canada to Pennsylvania, Arizona, California, and northern Mexico.

3. RAMISCHIA. Sidebells-pyrola

Plants small, herbaceous, perennial; well-developed leaves mostly basal or nearly so, thin, ovate or oblong-ovate, serrulate; corolla campanulate, the greenish-white petals and the stamens somewhat connivent.

1. Ramischia secunda (L.) Garcke (*Pyrola secunda* L.). Mountains of Apache, Coconino, Graham, and Pima counties, 7,000 to 9,500 feet, coniferous forests, July and August. Widely distributed in the Northern Hemisphere.

4. CHIMAPHILA. Pipsissewa

Plants herbaceous or nearly so; stems somewhat leafy; leaves evergreen, thick and leathery, tending to form whorls; inflorescences corymbose or umbelliform, the flowers nodding; petals pink or whitish, waxlike, spreading.

KEY TO THE SPECIES

1. Leaves lanceolate or ovate, whitish-mottled along the veins; dilated portion of the filaments conspicuously villous..1. *C. maculata*
1. Leaves oblanceolate or spatulate, not mottled; dilated portion of the filaments glabrous or merely ciliolate..2. *C. umbellata*

1. Chimaphila maculata (L.) Pursh. Chiricahua Mountains, Cochise County, 6,500 to 7,000 feet (*Darrow & Phillips* 2500, *Phillips* 2856), Santa Catalina Mountains (Pima County), about 8,000 feet, among rocks in pine forest (*Harrison & Kearney* 8100), August. Massachusetts and Ontario, south to Georgia, Arizona, and Mexico.

Represented in Arizona by var. *dasystemma* (Torr.) Kearney & Peebles (*C. dasystemma* Torr.), which differs from most of the Eastern specimens of *C. maculata* in its shorter and broader leaves, these 2 to 4 cm. long, ⅓ to ⅔ as wide as long.

2. Chimaphila umbellata (L.) Nutt. Coconino County, White Mountains (Apache County), Pinaleno Mountains (Graham County), Chiricahua Mountains (Cochise County), 6,500 feet or higher, coniferous forest, July. Widely distributed in the Northern Hemisphere.

Represented in Arizona by var. *acuta* (Rydb.) Blake (*C. acuta* Rydb.), which may be distinguished from other forms of *C. umbellata* by its narrowly oblanceolate, acute or acutish, few-toothed leaves, not prominently veined beneath. The type of *C. acuta* was collected on the Mogollon Escarpment, head of Tonto Basin (*Mearns* 136).

This plant is used as an ingredient of root beer.

5. PYROLA. Wintergreen, Shinleaf

Small, herbaceous perennials, more or less scapose; leaves mostly basal or nearly so, evergreen, usually with well-developed blades; flowers several, nodding, in terminal racemes; corolla nearly globose; anthers commonly reversed at anthesis, the basal pores appearing apical.

These plants are not aromatic. The aromatic wintergreen or checker-berry of eastern North America, from which oil of wintergreen is derived, belongs to a different genus, *Gaultheria*.

KEY TO THE SPECIES

1. Leaves more or less whitish-mottled along the veins, ovate or oval, thick, 3 to 8 cm. long, usually longer than the petiole, denticulate to nearly entire; anther cells narrowed or beaklike at the apparent apex..1. *P. picta*
1. Leaves uniformly green above (2).
2. Blades commonly shorter than the petiole, elliptic to orbicular, thickish; anther cells narrowed or beaklike at the apparent apex............................2. *P. virens*
2. Blades equaling or longer than the petiole, elliptic or obovate-oval, thin; anther cells not or very slightly narrowed at the apparent apex.......................3. *P. elliptica*

1. **Pyrola picta** J. E. Smith. North Rim of Grand Canyon and near Flagstaff (Coconino County), Pinaleno Mountains (Graham County), 8,000 to 9,500 feet, July and August. Montana to British Columbia, south to New Mexico, Arizona, and California.

A form with the leaves and the chlorophyll content of the plant greatly reduced, *P. picta* forma *aphylla* (J. E. Smith) Camp (*P. aphylla* J. E. Smith), has been collected in "northern Arizona," probably near Kanab, Utah (*Mrs. Thompson* in 1872), and at Baker Butte, southern Coconino County (*Mearns*) (Ref. 211).

2. **Pyrola virens** Schweigg. (*P. chlorantha* Swartz). Lukachukai and White mountains (Apache County) and Kaibab Plateau (Coconino County), to the Chiricahua and Huachuca mountains (Cochise County) and Santa Catalina Mountains (Pima County), 6,500 to 10,000 feet, coniferous forests, July and August. Canada to the District of Columbia, New Mexico, Arizona, and California; Europe.

A variant with mostly elliptic or oblong-ovate leaves, var. *saximontana* Fernald, probably also occurs in Arizona.

3. **Pyrola elliptica** Nutt. Pinaleno Mountains, Graham County (*Shreve* 5257), Santa Catalina Mountains, Pima County (*Peebles et al.* 2524), 7,000 to 9,000 feet, reported also from the Grand Canyon, July. Canada to the District of Columbia, New Mexico, and Arizona.

6. MONESES

Resembles *Pyrola* except in the characters given in the key to the genera.

1. **Moneses uniflora** (L.) Gray. San Francisco Peaks (Coconino County), Baldy Peak (Apache County), 10,000 to 11,500 feet, July and August. Widely distributed in the cooler parts of the Northern Hemisphere.

One of the most attractive of the Arizona high-mountain plants.

7. PHYLLODOCE. Mountain-heath

Dwarf shrubs; leaves evergreen, crowded, linear, 5 to 12 mm. long, articulated with the stem, the margins revolute; flowers in umbellike clusters at the ends of the branches; corolla campanulate, 4 to 6 mm. long; stamens 7 to 12, the anthers not appendaged; fruit a capsule, dehiscing from the apex.

1. **Phyllodoce empetriformis** (J. E. Smith) D. Don. South Kaibab Trail, Grand Canyon, Coconino County (*McKee* in 1928, No. 1108 of Grand Canyon Herbarium). Yukon and Alaska to Colorado, northern Arizona, and California.

The occurrence of this montane plant so far south and at so low an elevation (7,000 feet or lower) is remarkable.

8. ARBUTUS. Madroño, Madrone

Usually trees, up to 15 m. (50 feet) high, with smooth, thin, exfoliating bark; leaves alternate, thick, evergreen, somewhat shiny above, entire to serrate; flowers in terminal racemes or panicles; corolla urn-shaped, white or pink, the lobes short; stamens 10; fruit berrylike, the surface tessellate-warty.

1. Arbutus arizonica (Gray) Sarg. Mountains of Graham, Cochise, and Pima counties, 4,000 to 8,000 feet, often with live-oaks, April to September. Southwestern New Mexico, southeastern Arizona, and northern Mexico. Arizona madroño.

9. ARCTOSTAPHYLOS (212, 213). Manzanita

Plants shrubby or suffruticose; leaves mostly alternate, thick, evergreen; flowers in terminal racemes or panicles, much like those of *Arbutus;* fruit globose, several-seeded, the surface (in the Arizona species) smooth or nearly so.

A. pungens and *A. Pringlei* are characteristic plants of the chaparral association in Arizona. The wood is very hard, and the smooth, mahogany-colored bark is distinctive. The plants are seldom browsed except by goats. Birds, bears, and other animals eat the fruits, and a delicious jelly can be made from the unripe fruits of *A. pungens. A. Uva-ursi* is employed for treating disorders of the urinary tract, and it is reported that a decoction of the leaves of *A. pungens* has been used locally in Arizona as a remedy for stomach trouble.

KEY TO THE SPECIES

1. Plants with creeping, much-branched stems, forming large mats close to the ground; leaves spatulate, obtuse or retuse . . . . 1. *A. Uva-ursi*
1. Plants large shrubs; stems normally erect or ascending, 1 m. long or longer; leaves elliptic, lanceolate, or ovate, exceptionally oblanceolate or obovate (2).
2. Herbage, pedicels, and calyx glandular-pilose; ovary (and often the mature fruit) pubescent; bracts commonly more than 3 mm. long, thin, often pink. Leaves commonly rounded or subcordate at base . . . . 2. *A. Pringlei*
2. Herbage puberulent, subtomentose, or glandular, at least when young; calyx and ovary glabrous, or the calyx ciliolate; bracts commonly only 2 to 3 mm. long, thickish, firm (3).
3. Leaves pale green or bluish green, mostly elliptic or lanceolate, usually acute or acutish and pungent at apex and narrowed or subcuneate at base, seldom more than 1.5 cm. wide; branches of the inflorescence subtomentose, not glandular . . . . 3. *A. pungens*
3. Leaves bright green, broadly ovate or suborbicular, usually obtuse or rounded at apex and rounded or truncate at base, commonly 2 to 3 cm. wide; branches of the inflorescence glandular-pubescent . . . . 4. *A. patula*

***1. Arctostaphylos Uva-ursi** (L.) Spreng. Included in the Arizona flora on the doubtful basis of a collection labeled only "Arizona" (*Palmer* in 1869). Widely distributed in the cooler parts of the Northern Hemisphere. Known as bear-berry, sand-berry, and kinnikinnick.

2. Arctostaphylos Pringlei Parry. Southern Coconino County and Hualpai Mountain (Mohave County) to the Pinaleno Mountains (Graham County), Chiricahua Mountains (Cochise County), and Rincon and Santa Catalina mountains (Pima County), 4,000 to 6,500 feet, common on dry slopes, often with cypress, juniper, and pinyon, April to June. Arizona, southern California, and Baja California.

A large shrub, often 2 m. (6 feet) high.

3. Arctostaphylos pungens H.B.K. Apache County to northern Mohave County, south to Cochise, Santa Cruz, and Pima counties, 3,500 to 8,000 feet, abundant on dry slopes, often with *A. Pringlei,* February to May. New Mexico, southern Utah, Arizona, southern California, and Mexico.

Point-leaf manzanita. A smaller shrub than *A. Pringlei,* with a more pronounced tendency to form dense thickets, these often nearly impenetrable because of the rigid, crooked stems. Exceptionally broad-leaved specimens, e.g. a collection at the Grand Canyon (*Tidestrom* 2350), resemble the next species except in the absence of glandular pubescence.

4. Arctostaphylos patula Greene. Navajo Indian Reservation (*Vorhies* in 1916), Navajo Mountain and North Rim of the Grand Canyon (Coconino County), 7,000 to 8,500 feet, coniferous forests, May and June. Colorado to northern Arizona and California.

Green-leaf manzanita. In Arizona a low shrub, commonly not more than 1 m. (3 feet) high, thicket-forming, the stems rooting where they touch the soil. Reported to be very resistant to fire. The foliage is brighter green than in the other species.

10. VACCINIUM (214). WHORTLEBERRY, BLUEBERRY

The Arizona species low shrubs, less than 0.5 m. high; twigs yellowish or tinged with red, acutely angled; leaves ovate, oval, oblanceolate, or obovate, serrulate, rounded to acutish at apex, 1 to 3 cm. long; corolla about 4 mm. long; fruit black-purple, not glaucous, up to 8 mm. in diameter, juicy.

1. Vaccinium oreophilum Rydb. White Mountains (Apache County), Pinaleno Mountains (Graham County), Chiricahua and Huachuca mountains (Cochise County), Santa Catalina Mountains (Pima County), 8,000 to 11,000 feet, open woods and hillsides, June and July. Canada to New Mexico and Arizona.

Rocky Mountain whortleberry. On Baldy Peak, above 9,000 feet, this plant is an important element of the ground cover, in spruce forests.

Camp (214) referred tentatively most of the Arizona specimens to *V. Myrtillus* L., a species widely distributed in North America and Eurasia, and gave as a synonym *V. oreophilum* Rydb., in part. He identified one collection from Baldy Peak, Apache County (*Goodding* 1134), as *V. geminiflorum* H.B.K., a chiefly Mexican species. This is a smaller plant with much smaller leaves than most of the Arizona specimens here referred to *V. oreophilum.* The Goodding collection is closely matched by two others from Baldy Peak (*Bailey* 1442, *Coville* 1997).

104. PRIMULACEAE. Primrose Family

Plants herbaceous, annual or perennial, scapose or with leafy stems; leaves simple, entire or shallowly dentate; inflorescence various; flowers perfect, regular, mostly 5-merous; corolla sympetalous but sometimes cleft nearly to the base; stamens distinct, inserted on the corolla opposite its lobes; style and stigma 1; ovary superior or partly inferior, 1-celled; fruit a capsule.

The family includes some highly ornamental plants, notably several species of *Primula* that are much cultivated under the names "primrose" and "cowslip," and the well-known cyclamen.

KEY TO THE GENERA

1. Ovary partly inferior, its base enveloped by and adnate to the base of the calyx tube 3. *Samolus*
1. Ovary superior, not adnate to the calyx tube (2).
2. Plants caulescent, the stems leafy. Corolla rotate (3).
2. Plants acaulescent or nearly so. Flowers in umbels or solitary at the apex of the scape; corolla white or purplish pink, often with a different-colored eye (5).
3. Flowers numerous, in leafy panicles; capsule dehiscent longitudinally. Corolla yellow 4. *Lysimachia*
3. Flowers few, solitary, axillary; capsule circumscissile, globose. Plants small, annual (4).
4. Corolla normally red, longer than the calyx; leaves opposite; filaments bearded 5. *Anagallis*
4. Corolla whitish, shorter than the calyx; leaves alternate; filaments not bearded 6. *Centunculus*
5. Corolla with entire, reflexed lobes and a dark eye, showy; stamens inserted in the throat of the corolla, exserted, the filaments united at least at base, the anthers connivent around the pistil........7. *Dodecatheon*
5. Corolla with erect or spreading lobes; stamens inserted in the tube of the corolla, included, the filaments distinct, the anthers not connivent (6).
6. Plants perennial; flowers relatively large and showy........1. *Primula*
6. Plants annual; flowers small and inconspicuous........2. *Androsace*

1. PRIMULA (215). Primrose

Plants scapose, often tufted; calyx tube elongate, angled; corolla surpassing the calyx at anthesis, often with obcordate lobes, the limb pink or reddish purple, the throat open, greenish or yellowish; capsule opening apically by valves or teeth.

KEY TO THE SPECIES

1. Scapes stout, up to 55 cm. long; leaves entire (exceptionally dentate); pedicels and calyx copiously to densely glandular-puberulent, not at all mealy; corolla limb 2 cm. or more wide, the lobes shallowly notched or nearly entire, the tube not or but slightly surpassing the calyx........1. *P. Parryi*
1. Scapes slender, not more than 25 cm. long; leaves denticulate or dentate; pedicels and calyx not glandular-puberulent, more or less white-mealy; corolla limb usually less than 2 cm. wide, the lobes distinctly notched, the tube surpassing the calyx (2).
2. Leaves more or less puberulent but not white-mealy beneath, the teeth conspicuously glandular; involucral bracts lanceolate or lance-ovate, thin; corolla limb more than 1 cm. wide........2. *P. Rusbyi*
2. Leaves more or less white-mealy beneath, at least when young, the teeth not conspicuously glandular; bracts subulate or narrowly lanceolate, distinctly thickened at base; corolla limb not more than 1 cm. wide (3).

3. Petioles shorter than the blades; involucral bracts 4 to 10 mm. long; calyx 6 to 9 mm. long, longer than the capsule . . . . 3. *P. specuicola*
3. Petioles about equaling the blades; bracts 2 to 4 mm. long; calyx 3 to 4 mm. long, shorter than the capsule . . . . 4. *P. Hunnewellii*

1. Primula Parryi Gray. San Francisco Peaks (Coconino County), Baldy Peak (Apache County), 10,000 to 12,000 feet, among rocks, often along brooks, June to August. Montana and Idaho to New Mexico and Arizona.

The plant is showy but rather coarse, and the flowers have an odor of carrion.

2. Primula Rusbyi Greene. Mountains of Graham, Cochise, Santa Cruz, and Pima counties, 7,500 to 10,500 feet, usually on damp, mossy ledges, May to September. New Mexico and southeastern Arizona.

3. Primula specuicola Rydb. North side of the Grand Canyon, Coconino County, May and June (*Eastwood & Howell* 1024). Southeastern Utah and northern Arizona.

4. Primula Hunnewellii Fern. North Rim of the Grand Canyon, Coconino County, limestone cliffs, August (*Hunnewell* 10883, the type). Known only by the type collection.

Doubtfully distinct as a species from *P. specuicola*. The Eastwood & Howell specimens cited under the latter species seem to be intermediate, having petioles about as long as the blades and involucral bracts only 3 to 6 mm. long, but the calyx is 5 to 6 mm. long and surpasses the capsule.

2. ANDROSACE (216, 217). Rock-jasmine

Small, often cespitose, scapose plants, the foliage leaves basal; flowers inconspicuous, slender-pedicelled, in umbels subtended by an involucre of small, nearly distinct bracts; corolla funnelform, constricted at the throat, 5-ridged within, white or whitish, not or little surpassing the calyx at anthesis.

KEY TO THE SPECIES

1. Involucral bracts lance-ovate or somewhat obovate; plants of low altitudes, flowering in early spring . . . . 1. *A. occidentalis*
1. Involucral bracts lanceolate or subulate; plants of relatively high altitudes, flowering in summer . . . . 2. *A. septentrionalis*

1. Androsace occidentalis Pursh. Coconino, Mohave, Yavapai, and Greenlee (?) counties, south to Cochise, Santa Cruz, and Pima counties, 1,000 to 5,000 feet, chiefly along streams, February to April. Manitoba to British Columbia, south to Indiana, Texas, New Mexico, and Arizona.

Var. *arizonica* (Gray) St. John (*A. arizonica* Gray), type from the Santa Catalina Mountains (*Pringle* 330), with larger, green and foliaceous calyx lobes, these more spreading than in the typical plant or even slightly recurved, is found throughout the southern part of the range of the species in Arizona. It often grows with the typical plant and intergrades with it.

2. Androsace septentrionalis L. Mountains of northern Apache County and Kaibab Plateau (Coconino County), south to the mountains of Cochise and Pima counties, 7,000 to 12,000 feet, usually in springy places in coniferous forests, April to September. Widely distributed in the cooler parts of the Northern Hemisphere.

KEY TO THE VARIETIES

1. Plants with usually only the central scape well developed, the others shorter or wanting; umbels more or less compact; pedicels rather strictly ascending to nearly erect. Herbage and calyces puberulent with numerous reddish, glandular hairs........var. *glandulosa*
1. Plants usually with several or many scapes of nearly equal length; umbels more or less diffuse; pedicels spreading or ascending-spreading (2).
2. Herbage and calyces nearly glabrous, the scapes puberulent only toward base; calyx lobes slender-subulate to acerose..................................var. *subulifera*
2. Herbage and calyces more or less puberulent, the scapes densely so; calyx lobes narrowly triangular ....var. *puberulenta*

Var. *glandulosa* (Woot. & Standl.) St. John occurs in the Chuska and White mountains (Apache County) and in the Flagstaff region (Coconino County). Var. *subulifera* Gray (*A. diffusa* Small) has been collected in the Lukachukai and White mountains, Apache County (*Goodman & Payson* 2903, *Phillips* 3371). Var. *puberulenta* (Rydb.) Knuth is the most abundant and widely distributed variety in Arizona, ranging from the Kaibab Plateau (Coconino County) and the White Mountains (Apache County) to the mountains of Cochise and Pima counties. A dwarf form of this variety that occurs near the summit of the San Francisco Peaks has been mistaken for var. *subumbellata* A. Nels.

3. SAMOLUS. WATER-PIMPERNEL

Plants mostly perennial, glabrous or nearly so; stems leafy, at least below; leaves broad, entire; flowers small, pedicelled, in loose racemes or panicles, the petals white; capsule globose, opening apically by valves.

KEY TO THE SPECIES

1. Leaves crowded near the base of the stem, broadly spatulate; pedicels bractless, ascending, rather stiff; corolla 3 mm. or longer.........................1. *S. ebracteatus*
1. Leaves scattered along the stem, oval, elliptic, or somewhat obovate; pedicels bearing a small bract, spreading, lax; corolla less than 3 mm. long (2).
2. Stems erect or ascending, bearing several or numerous leaves; basal leaves commonly oval or elliptic; inflorescences several- to many-flowered; bract borne near the middle of the pedicel; calyx lobes commonly shorter than the tube..........2. *S. floribundus*
2. Stems mostly procumbent, stolonlike, bearing few leaves; basal leaves obovate or spatulate; inflorescences few-flowered; bract borne usually near the base of the pedicel; calyx lobes equaling or longer than the tube.........................3. *S. vagans*

***1. Samolus ebracteatus** H.B.K. Not known definitely from Arizona, but has been collected along the Muddy River, Nevada, near the northwestern border. Florida to Texas; southern Nevada; Mexico; West Indies.

2. Samolus floribundus H.B.K. Navajo, Coconino, Yavapai, Greenlee, Pinal, and Pima counties, 1,000 to 5,000 feet, wet soil along streams, April to August. Throughout most of North America; South America.

3. Samolus vagans Greene. Cochise, Santa Cruz, and Pima counties, 3,500 to 6,000 feet, in wet sand, May to October, type from the Chiricahua Mountains (*Blumer* 1546). Apparently known only from southeastern Arizona.

4. LYSIMACHIA. LOOSE-STRIFE

Plants perennial; stems leafy; leaves opposite or appearing whorled, the petioles ciliate, the blades lanceolate or ovate-lanceolate; corolla rotate, yellow; stamens borne on a ring at base of the corolla, the 5 fertile ones alternating with 5 staminodia.

1. **Lysimachia ciliata** L. (*Steironema ciliatum* Raf.). Apache, Navajo, Coconino, and Gila counties, 6,000 to 7,500 feet, moist, rich soil, July to September. Canada to Georgia, New Mexico, Arizona, and Washington.

Represented in Arizona by var. *validula* (Greene) Kearney & Peebles (*Steironema validulum* Greene), which differs from most Eastern specimens of *L. ciliata* in its relatively narrow leaves, these at most obscurely ciliolate. The type of *S. validulum* was collected near Flagstaff, Coconino County (*Lemmon* in 1884).

5. ANAGALLIS. Pimpernel

Stems low, spreading or procumbent; leaves ovate, sessile; peduncles surpassing the leaves; corolla rotate, with scarcely any tube, salmon-colored to scarlet, the lobes denticulate and bearing stalked glands on the margin.

1. **Anagallis arvensis** L. Mesa (Maricopa County), near Fort Huachuca (Cochise County), Sacaton (Pinal County), Tucson (Pima County), probably elsewhere, a weed in lawns, April and later. Widely distributed in North America; naturalized from Europe.

The flowers close quickly at the approach of storms, hence the English name "poorman's-weatherglass."

6. CENTUNCULUS. Chaff-weed

Stems short, tufted, ascending, leafy; flowers small, nearly sessile; corolla with a rotate limb and a short tube, usually persistent at the apex of the capsule.

1. **Centunculus minimus** L. Graham, Cochise, Santa Cruz, and Pima counties, about 4,000 feet, wet soil along streams, apparently rare in Arizona, but easily overlooked. Illinois and Minnesota to British Columbia, south to Florida, Texas, Arizona, and California; Europe.

7. DODECATHEON. Shooting-star

Perennial, acaulescent herbs, with a short rootstock, the herbage glabrous or nearly so; leaves in a basal rosette, petioled; scapes elongate, bearing 1 to many showy flowers in a terminal umbel; calyx and corolla deeply cleft, the lobes narrow, becoming reflexed; capsule ovoid, dehiscent by valves.

Handsome plants, sometimes cultivated as ornamentals.

KEY TO THE SPECIES

1. Leaves thin, bright green, ovate or ovate-oblong, not or scarcely more than twice (rarely 3 times) as long as wide, sinuate to coarsely dentate, rounded, truncate, or subcuneate at base, abruptly contracted into petioles often longer than the blades; corolla lobes white .... 1. *D. Ellisiae*

1. Leaves thickish, dull green, oblanceolate, more (often much more) than twice as long as wide, entire or subsinuate, tapering into petioles shorter than the blades; corolla lobes pink, drying purple (2).

2. Flowers 5-merous; filaments yellow, short but evident; stigma little wider than the style 2. *D. radicatum*

2. Flowers 4-merous; filaments purple, very short or nearly obsolete; stigma about twice the diameter of the style .... 3. *D. alpinum*

1. Dodecatheon Ellisiae Standl. White Mountains (Apache County), Pinaleno Mountains (Graham County), Santa Catalina Mountains (Pima County), 8,000 to 9,500 feet, rich, moist soil in coniferous forests, June to August. New Mexico and southern Arizona.

2. Dodecatheon radicatum Greene. Pinaleno Mountains, Graham County, 9,000 feet (*Phillips & Haskell* 2952). South Dakota and Wyoming to New Mexico, eastern Arizona, and northern Mexico.

3. Dodecatheon alpinum (Gray) Greene. White Mountains (Apache County), Lakeside (Navajo County), 6,500 to 9,500 feet, moist meadows, June to September. California, Nevada, and eastern Arizona.

The writers are indebted to Henry J. Thompson for pointing out the distinguishing characters of this species.

105. PLUMBAGINACEAE. Plumbago Family

1. PLUMBAGO

Plants suffrutescent; leaves alternate, simple, entire; flowers nearly sessile, in panicles of spikelike racemes, perfect, regular, commonly 5-merous; calyx tubular, beset with stipitate glands; corolla sympetalous, salverform, with a long, slender tube; stamens 5, distinct, free or nearly free from the corolla, borne opposite its lobes; style 1; stigmas 5, slender; fruit a slender 1-seeded capsule.

P. capensis, a South African species with sky-blue flowers, is often cultivated as an ornamental in the warmer parts of the United States. The roots and leaves of *P. scandens* are reported to cause dermatitis in susceptible persons.

1. Plumbago scandens L. Pinal and Pima counties, 2,500 to 4,000 feet, in canyons, May to September. Southern Florida, southern Arizona, and widely distributed in tropical America.

Flowers whitish or tinged with blue.

Limonium californicum (Boiss.) Heller occurs in Vegas Wash, on the Nevada side of the Colorado River near Lake Mead, and is to be looked for on the Arizona side. It is easily recognizable by its scapose habit, large, fleshy leaves in a basal rosette, much-branched scapes with the small violet flowers clustered at the ends of the branchlets and subtended by chaffy bractlets, and nearly distinct petals. Certain species of *Limonium* (or sea-lavender) are cultivated for winter bouquets.

106. FOUQUIERIACEAE. Ocotillo Family

1. FOUQUIERIA. Ocotillo

Large, thorny shrubs with numerous long, whiplike, unbranched stems; petioles of the short-lived primary leaves becoming thorns, bearing in their axils the fascicles of secondary leaves; flowers perfect, regular, in dense, terminal panicles, showy, bright red; corolla tubular, 5-lobed; stamens 10 or more, exserted, the thickened bases of the filaments lightly adnate to the base of the corolla tube; fruit an incompletely 3-celled capsule; seeds flat, winged, the wing disintegrating into hairlike filaments.

1. Fouquieria splendens Engelm. Southern Apache, Coconino, and northern Mohave counties to Cochise, Santa Cruz, Pima, and Yuma counties, 5,000 feet or lower (reported to occur at 6,500 feet in the Chiricahua Mountains), very common on dry mesas and slopes, April to June, sometimes later. Western Texas to southeastern California and northern Mexico.

Sometimes called slimwood and coach-whip. It is one of the oddest and most conspicuous of Arizona plants and is very attractive in flower. The ocotillo drops its leaves as soon as the soil dries, but as quickly refoliates after a good rain, except in winter when temperatures are low. Cuttings root readily, and it is not uncommon to see living fences or hedges of this plant. The straight, thorny stems are set in the ground thickly to build coyote-proof runs and corrals for fowl, and also are utilized by the Indians for constructing crude huts and outhouses. The Coahuila Indians of southern California are said to eat the flowers and capsules. It is reported that belt dressing of good quality can be manufactured from the wax that coats the stems. According to Mrs. Collom, the Apache Indians relieve fatigue by bathing in a decoction of the roots and also apply the powdered roots to painful swellings.

107. SAPOTACEAE. Sapote Family

The highly esteemed tropical American fruit sapodilla is produced by *Achras Zapota* L., a tree that also yields chicle, from which chewing gum is manufactured.

1. BUMELIA (218)

Large shrubs with spiny branches; leaves mostly fascicled on the branchlets, oblanceolate- or obovate-cuneate, rusty-lanate beneath; flowers small, perfect, regular, in axillary fascicles; corolla white, with lobelike appendages in the sinuses; stamens 10, of these 5 sterile and petallike; fruit a drupe, usually 1-seeded.

1. Bumelia lanuginosa (Michx.) Pers. Cochise, Santa Cruz, and Pima counties, 3,500 to 6,000 feet, forming thickets along streams, June and July. Georgia and Florida to Illinois, Kansas, southern Arizona, and northern Mexico.

The Arizona plant is var. *rigida* Gray (*B. rigida* Small), which has smaller, more strongly cuneate leaves than in the typical plant. In Arizona it is a shrub, ordinarily 2.4 to 3 m. (8 to 10 feet) high. The flowers are very fragrant. Gum exuded from the bark is reported to be used for chewing by children in Texas.

108. OLEACEAE. Olive Family

Trees, shrubs, or herbs of diverse habit; leaves simple or pinnate, alternate or opposite; flowers regular, perfect or unisexual, with or without a corolla; stamens 2 to 4, borne on the corolla tube when a corolla is present; style 1 or the stigma sessile; ovary superior, 2-celled; fruit various.

The best-known members of this family are olive, ash, and lilac.

KEY TO THE GENERA

1. Leaves pinnately compound or if unifoliolate, then broad and ovate, oval, or suborbicular; fruit a samara with a conspicuous, mainly terminal, flat wing. Corolla present in only 1 species .... 1. *Fraxinus*
1. Leaves simple; fruit not winged (2).
2. Flowers often unisexual, appearing before the leaves; corolla none or rudimentary; fruit a 1-seeded drupe .... 2. *Forestiera*
2. Flowers perfect, appearing after the leaves; corolla large; fruit a membranaceous, didymous capsule with 2 to 4 seeds in each cell .... 3. *Menodora*

1. FRAXINUS (219). Ash

Trees or large shrubs; leaves opposite, commonly pinnate, petioled; flowers in racemes or panicles, mostly unisexual, apetalous or with a 4-parted corolla; stamens commonly 2, with large anthers; fruit dry, with a large, flat, terminal wing, indehiscent; seeds 1 or 2.

Some of the North American ashes are important timber trees, but the species occurring in Arizona do not grow large enough to make the wood valuable. The herbage is of limited value as browse. The Arizona ash (*F. velutina* var. *glabra*) is an exceptionally fine shade tree and is planted extensively in dooryards and along streets in southern Arizona. The pollen, notably of *F. velutina,* is believed to be an important cause of hay fever.

KEY TO THE SPECIES

1. Twigs evidently quadrangular; wing of the fruit extending nearly or quite to the base of the thin, strongly compressed body. Twigs and foliage glabrous or glabrate (rarely decidedly pubescent in *F. Lowellii*); corolla none; fruit elliptic or obovate, obtuse or truncate and often retuse at apex (2).
1. Twigs terete or rounded-quadrangular; wing of the fruit subterminal, not extending nearly to the base of the body, except sometimes in *F. cuspidata* (3).
2. Leaves 1- or 3-foliolate, the single or terminal leaflet broadly ovate or orbicular, truncate or short-cuneate at base, commonly obtuse or retuse at apex, the margin entire to crenate; plants commonly shrubs .... 1. *F. anomala*
2. Leaves 3-, 5-, or 7-foliolate; leaflets oblong-lanceolate, elliptic, or ovate, the terminal one usually obovate, cuneate at base, the margin commonly crenate-serrate; plants small trees .... 2. *F. Lowellii*
3. Corolla present, 4-parted, greenish white, 10 to 15 mm. long; body of the fruit thin, strongly compressed; leaflets 1 to 5 (rarely 7). Leaflets not more than 4 cm. long, lanceolate to broadly ovate, acutish to acuminate or sometimes obtuse or emarginate, the margin entire or nearly so; fruit spatulate-oblong or oblong-lanceolate, the wing obtuse or truncate and often emarginate at apex .... 3. *F. cuspidata*
3. Corolla none; body of the fruit thick, more or less terete; leaflets commonly 5 or more (4).
4. Old leaves persistent until the flowers appear; rachis narrowly winged; leaflets not more, usually much less, than 4 cm. long, oblanceolate, narrowly obovate, or elliptic, obtuse or acutish at apex, inconspicuously crenate or dentate; shrubs or small trees; wing of the fruit much longer than the thick body, narrowly elliptic or oblanceolate. Herbage glabrous or nearly so; leaflets coriaceous, sessile .... 4. *F. Greggii*
4. Old leaves deciduous before the flowers appear; rachis not winged; leaflets commonly more than 4 cm. long, lanceolate to ovate, mostly acuminate and serrate or serrulate; normally small trees; wing equaling or shorter (seldom longer) than the body of the fruit .... 5. *F. velutina*

1. Fraxinus anomala Torr. (*F. anomala* var. *triphylla* Jones). Apache County to the Cerbat and Hualpai mountains (Mohave County) and southeastern Yavapai (and northwestern Gila?) counties, common in the Grand Canyon, 2,000 to 6,000 feet, April, type of var. *triphylla* from Pagumpa, Mohave County (*Jones* 5082W). Western Colorado and New Mexico to northern Arizona and southeastern California.

Single-leaf ash. A shrub, sometimes 6 m. (20 feet) high.

2. Fraxinus Lowellii Sarg. Coconino County and eastern Mohave County to western Gila and eastern Maricopa counties (perhaps also in the Santa Rita Mountains), 3,000 to 6,500 feet, mostly along streams, March to May, type from Oak Creek Canyon (*Rehder* 53). Known only from Arizona.

Lowell ash, commonly shrubby, reaching a height of at least 7.5 m. (25 feet).

3. Fraxinus cuspidata Torr. Western Navajo, Coconino, and northern Mohave counties, 5,500 to 7,000 feet, April to June. Western Texas to northern Arizona and northern Mexico.

Flowering ash, the only Arizona *Fraxinus* having a corolla, commonly a large shrub, the flowers fragrant. The species is represented in this state by var. *macropetala* (Eastw.) Rehder (*F. macropetala* Eastw.), type from the Grand Canyon (*Wooton* 1102). The leaflets are usually fewer and broader than in typical *F. cuspidata.*

4. Fraxinus Greggii Gray. Near Ruby, Santa Cruz County, 4,000 to 5,000 feet, steep, rocky slopes (*Goodding* in 1936). Western Texas, southern Arizona, and northern Mexico.

Gregg ash, a handsome shrub or small tree with smooth, iron-gray bark and neat foliage. Further study may show the Arizona plant to be at least a good variety. It has tomentulose twigs, and some of the leaves are 7-foliolate. As compared with specimens of *F. Greggii* from Texas and Mexico, the leaflets are broader and more distinctly crenate, the body of the fruit is thinner, longer, more slender, and less obtuse, and the wing extends farther toward the base of the fruit.

5. Fraxinus velutina Torr. (*F. Standleyi* Rehder). Apache County to Coconino County, south to Cochise, Santa Cruz, and Pima counties, 2,000 to 7,000 feet, along streams, March to May. Western Texas to southern California and northern Mexico.

KEY TO THE VARIETIES

1. Herbage glabrous. Leaflets usually distinctly stalked, often coarsely serrate. .var. *glabra*
1. Herbage pubescent, usually copiously so (2).
2. Leaflets subsessile or very short-stalked. . . . . *F. velutina* (typical)
2. Leaflets distinctly stalked. . . . . var. *Toumeyi*

Velvet ash, usually a tree, reaching a height of at least 9 m. (30 feet), extremely variable. Typical *F. velutina* is less common in Arizona than the varieties. Var. *glabra* Rehder, the Arizona ash, type from near Tucson (*Wooton* in 1911), occasionally approaches var. *coriacea* (Wats.) Rehder in the thickness of the leaflets. This and var. *Toumeyi* (Britton) Rehder (*F. attenuata* Jones, in part, *F. Toumeyi* Britton), type from near Tucson (*Toumey* in 1895), occur throughout the range of the species in Arizona.

2. FORESTIERA. Adelia

Large, much-branched shrubs; leaves simple, opposite or fascicled at the ends of the branchlets; flowers very small, apetalous or nearly so, in lateral clusters; calyx minute or none; fruit an ellipsoid or narrowly ovoid, thin-fleshed drupe with a bony seed.

These plants are sometimes known as wild-olive. The tough wood of *F. neomexicana* is said to have been used by the Hopi for making digging sticks.

KEY TO THE SPECIES

1. Leaves with entire or slightly sinuate, more or less revolute margins, lanceolate, oblanceolate, or narrowly oblong, not more (commonly less) than 6 mm. wide, usually 3 or more times as long as wide, commonly pubescent; filaments 3 to 5 mm. long, the anthers dark purple; drupes very asymmetric, ellipsoid or subclavate, about twice as long as thick .... 1. *F. phillyreoides*

1. Leaves with crenulate or serrulate margins (rarely nearly entire), ovate, obovate, oblanceolate, or rhombic-elliptic, commonly more than 6 mm. wide, usually not more than twice (exceptionally 3 times) as long as wide; filaments usually not more than 3 mm. long, the anthers yellow; drupes symmetric or nearly so, ellipsoid or narrowly ovoid, commonly less than twice as long as thick .... 2. *F. neomexicana*

1. Forestiera phillyreoides (Benth.) Torr. (*F. Shrevei* Standl.). Southern Maricopa, Cochise, Santa Cruz, Pima, and Yuma counties, 2,000 to 4,500 feet, dry, rocky slopes, often forming thickets, December to March, type of *F. Shrevei* from the Ajo Mountains (*Shreve* 6201). Southern Arizona and Mexico.

Reaches a height of at least 2.5 m. (8 feet). What seems to be a form of this species, collected by L. N. Goodding on the summit of the Ajo Mountains, Pima County, has drupes 7.5 to 9.5 mm. long, whereas they usually do not exceed 8 mm. in this species.

2. Forestiera neomexicana Gray. Carrizo Mountains (Apache County) to Hualpai Mountain (Mohave County), south to Graham, Gila, and Yavapai counties, 2,000 to 7,000 feet, March to May. Colorado and Utah to western Texas, California, and northern Mexico.

Commonly 1.8 to 2.5 m. (6 to 8 feet) high, varying greatly in leaf shape. Var. *arizonica* Gray (*F. arizonica* Rydb.), which has the twigs and sometimes the leaves copiously and often persistently soft-pilose (these glabrous or obscurely puberulent in the typical plant), occurs throughout most of the range of the species in Arizona. The type of var. *arizonica* was collected near Prescott by E. Palmer. This variety seems to connect *F. neomexicana* and *F. pubescens* Nutt.

3. MENODORA (220)

Perennial herbs or small shrubs; leaves simple, entire, mostly alternate; flowers showy, the corolla large, sympetalous, rotate-campanulate, yellow or white; capsules didymous, thin-walled, circumscissile or indehiscent; seeds commonly 2 in each cell.

The plants are reported to be highly palatable to livestock, in some localities constituting a significant proportion of the total forage.

KEY TO THE SPECIES

1. Stems woody throughout, spiny; larger leaves not more than 12 mm. long; corolla white (often brownish outside), the lobes about 2 mm. wide, shorter than the tube; capsules not circumscissile .... 1. *M. spinescens*
1. Stems herbaceous or woody toward base, not spiny; larger leaves 15 mm. long or longer; corolla yellow, the lobes 2.5 to 5 mm. wide, longer than the tube; capsules circumscissile (2).
2. Calyx lobes normally 5 or 6, additional ones, if present, smaller; herbage and calyx glabrous or nearly so; leaves of the inflorescence mostly reduced to small subulate bracts .... 2. *M. scoparia*
2. Calyx lobes 7 or more; herbage and calyx usually scabrous-puberulent; leaves of the inflorescence small but commonly foliaceous........................... 3. *M. scabra*

1. Menodora spinescens Gray. Detrital Valley, Mohave County, 2,500 feet (*L. Benson* 10153). Southern Nevada, northwestern Arizona, and southeastern California.

2. Menodora scoparia Engelm. Mohave, Yavapai, Gila, Pinal, Cochise, Pima, and Yuma counties, 3,000 to 5,000 feet, dry slopes and mesas, April to September. Arizona, southeastern California, and northwestern Mexico.

What appears to be an exceptional form of this species, collected in the Ajo Mountains, Pima County (*Goodding* 4392), is a low, dense shrub, woodier than is usual in *M. scoparia.*

3. Menodora scabra Gray. Apache County to Mohave County, south to Cochise, Pima, and Yuma counties, 1,500 to 7,500 feet, dry mesas and slopes, March to September. Western Texas to southern Utah, Arizona, southeastern California, and northern Mexico.

In the southern part of the state a variant with woodier, more branched, longer stems, var. *ramosissima* Steyerm., is commoner than the typical plant. The stems are sometimes nearly 1 m. long in this variety. Similar to var. *ramosissima* except in its longer corolla tube is var. *longituba* Steyerm., known only from the type collection in the Mazatzal Mountains (*Smart* 213). Steyermark (220, p. 139) cited specimens from the vicinity of Tucson under var. *laevis* (Woot. & Standl.) Steyerm., a glabrescent plant with broader leaves and corolla lobes than in typical *M. scabra.*

109. LOGANIACEAE. Logania Family

A member of this family, *Strychnos Nux-vomica,* is the source of the powerful drug and poison, strychnine.

1. BUDDLEJA

Shrubs; leaves opposite or whorled; flowers small, in dense axillary clusters, these often forming leafy interrupted spikes; flowers perfect, regular, normally 4-merous; stamens borne on the corolla tube, alternate with the lobes, the anthers sessile or nearly so; style and stigma 1; ovary 2-celled; fruit a capsule, apically dehiscent.

Several exotic species are cultivated as ornamental plants, the best-known one being *B. Davidii,* known as butterfly-bush and summer-lilac. *B. sessiliflora* is used medicinally in Mexico.

KEY TO THE SPECIES

1. Herbage and inflorescence usually only moderately tomentose to glabrate; leaves distinctly petioled, rather thin, with veins not or scarcely impressed above and not very prominent beneath, lanceolate to rhombic-ovate, acuminate at apex, entire or serrate, commonly at least 5 cm. long and 15 mm. wide; glomerules of flowers usually about 1 cm. in diameter.......1. *B. sessiliflora*
1. Herbage and inflorescence densely and conspicuously lanate-tomentose; leaves sessile or nearly so, thick, with veins deeply impressed above and very prominent beneath, linear or narrowly oblong-elliptic, obtuse or acutish at apex, crenate, seldom more than 3 cm. long and 6 mm. wide; glomerules of flowers usually much less than 1 cm. in diameter 2. *B. utahensis*

1. Buddleja sessiliflora H.B.K. (*B. Pringlei* Gray). Valley of the Santa Cruz River and foothills of the Santa Catalina and Quinlan mountains, Pima County, up to 3,000 feet, flowering in spring, type of *B. Pringlei* from near Tucson (*Pringle* in 1883). Southern Arizona and Mexico.

2. Buddleja utahensis Coville. Virgin Narrows, northern Mohave County, 2,500 feet, limestone cliffs (*Cottam* 8499). Southwestern Utah to southeastern California and northwestern Arizona.

110. GENTIANACEAE. Gentian Family

Plants herbaceous, annual or perennial, mostly glabrous; leaves simple, entire, commonly opposite, sessile; flowers solitary or in simple or compound cymes, perfect, regular or the calyx somewhat irregular, the corolla sympetalous; stamens borne on the corolla, alternate with its lobes, distinct or nearly so; style 1 or none; stigmas 1 or 2; ovary superior, 1-celled; fruit a capsule.

KEY TO THE GENERA

1. Corolla with 1 or 2 conspicuous, fringed, glandular pits on the inner side and toward the base of each lobe.......3. *Swertia*
1. Corolla lobes without glands or pits, or these not fringed (2).
2. Lobes of the corolla each bearing a conspicuous divaricate spur; corolla yellow.4. *Halenia*
2. Lobes of the corolla without spurs; corolla not yellow, but sometimes ochroleucous in *Gentiana* (3).
3. Corolla pink, without folds or fringes, salverform; anthers becoming spirally twisted after anthesis.......1. *Centaurium*
3. Corolla not pink; anthers not becoming twisted (4).
4. Stamens inserted in the corolla tube; corolla with folds or plaits between the lobes, or fringed in the throat, or the margins of the lobes fringed or erose-dentate (if otherwise, then the corolla ochroleucous and not more than 1 cm. long); style short and stout, or none; inflorescence narrow and leafy, or the flowers solitary......2. *Gentiana*
4. Stamens inserted in the corolla throat; corolla 2 to 3 cm. long, without folds or fringes, the lobes entire; style elongate; inflorescence an open, small-bracted panicle, or the flowers (in small plants) solitary and terminal.......5. *Eustoma*

1. CENTAURIUM

Plants annual or biennial; stems branched; calyx and corolla limb 4- or 5-parted, the calyx lobes keeled, the corolla salverform; stamens inserted in the throat of the corolla; stigma 2-lobed.

The name "canchalagua" is sometimes used for these plants.

KEY TO THE SPECIES

1. Corolla lobes 5, at least 7 mm. long, nearly as long as the tube; stems up to 60 cm. long, slender or stout, with usually more than 6 pairs of leaves. Leaves narrowly lanceolate or oblanceolate to broadly elliptic.................................1. *C. calycosum*
1. Corolla lobes 4 or 5, not more than 5 mm. long; stems seldom more than 30 cm. long, slender, rarely with more than 6 pairs of leaves. Basal leaves forming a rosette, conspicuously wider than the stem leaves; plants rather lax, the branches, peduncles, and often the leaves ascending-spreading...............................2. *C. nudicaule*

1. Centaurium calycosum (Buckl.) Fern. (*C. arizonicum* (Gray) Heller). Almost throughout the state, 150 to 6,000 feet, moist soil, March to November. Western Texas to southern Utah, Nevada, Arizona, and northern Mexico.

There is much variation, especially in the width of the leaves and of the corolla lobes, but there seems to be no satisfactory basis for the distinction of varieties.

2. Centaurium nudicaule (Engelm.) Robins. Santa Catalina and Baboquivari mountains (Pima County), 3,000 to 4,000 feet, along streams, April and May, type from the Santa Catalina Mountains (*Pringle* in 1881). Southern Arizona and Baja California.

Specimens collected at Willow Spring, southern Apache County (*Palmer* in 1890), and in northeastern Pinal County (*Smith* 12994) seem to be intermediate between this species and *C. exaltatum* (Griseb.) W. F. Wight. The latter species has been reported from Arizona but probably does not occur there.

2. GENTIANA. Gentian

Plants annual or perennial; stems mostly erect, simple or sparingly branched; flowers terminal or axillary, solitary or in cymose clusters, these often forming narrow leafy panicles; corolla cylindric, funnelform, or salverform, usually with folds, these often extended into teeth or fringed appendages between the lobes; stamens attached to the corolla tube; stigmas 2, or the stigma 2-lobed.

The medicinal use of gentian root is of great antiquity, but the drug is very mild and is probably only a stimulant of gastric secretions. Many of the species are very beautiful in flower.

KEY TO THE SPECIES

1. Flowers terminal and solitary; corolla 9 to 15 mm. long. Plants annual, alpine; stems not more than 12 cm. long; corolla lobes entire or nearly so (2).
1. Flowers clustered or if terminal and solitary, then the corolla more than 20 mm. long (3).
2. Leaves appressed to the stem, these and the calyx lobes with conspicuous, white, scarious margins; flower nearly sessile, or the peduncle not more than 1 cm. long; corolla salverform with a long, slender tube and spreading lobes, whitish or greenish purple, with broad, emarginate folds in the sinuses.........................1. *G. Fremontii*
2. Leaves not appressed or scarious-margined; flower borne on a peduncle 2 to 8 cm. long; corolla nearly cylindric, the tube about as wide as the throat, the lobes erect, pale blue (sometimes ochroleucous?) without folds in the sinuses, fringed in the throat 2. *G. tenella*
3. Corolla folded in the sinuses, the terminal portion of the folds free and bifid or irregularly dentate or laciniate. Plants perennial; leaves thickish; corolla usually violet (4).
3. Corolla without folds in the sinuses (7).

4. Flowers solitary or few (not more than 5), terminal and subterminal; corolla campanulate-funnelform, 30 to 40 mm. long, rich violet with a green band externally between each pair of folds extending nearly to the apex, the folds 2-cleft at apex and nearly equaling the lobes .... 3. *G. Parryi*
4. Flowers usually several or numerous in a more or less elongate inflorescence, some of them usually axillary well below the apex of the stem; corolla narrowly funnelform or subcylindric (5).
5. Calyx irregular, spathaceous, cleft on 1 or both sides, toothless or with 1 or 2 subulate teeth much less than ⅓ as long as the tube .... 4. *G. Forwoodii*
5. Calyx nearly regular, not spathaceous, with 5 well-developed lobes at least ⅓ as long as the tube. Corolla nearly closed in anthesis (6).
6. Inflorescence relatively short and dense, seldom constituting more than ⅓ of the length of the stem, not conspicuously leafy-bracted, the floral leaves erect or narrowly ascending, not surpassing the flowers; stem leaves (excluding the basal and floral ones) lanceolate, usually not more than 5 times as long as wide; calyx lobes oblong-lanceolate, equaling or somewhat longer than the tube; corolla 25 to 35 mm. long, the lobes acutish to acuminate .... 5. *G. affinis*
6. Inflorescence elongate and rather loose below, usually constituting more than ⅓ (sometimes ½) of the length of the stem, conspicuously leafy-bracted, the floral leaves spreading and much surpassing the flowers; stem leaves lance-linear, 7 or more times as long as wide; calyx lobes subulate, from much shorter to longer than the tube; corolla 20 to 25 mm. long, the lobes obtuse and apiculate .... 6. *G. Bigelovii*
7. Lobes of the corolla with erose or fimbriate margins; flowers usually 4-merous, solitary or not more than 3 on each stem or branch; corolla 25 mm. or longer, not fimbriate internally (8).
7. Lobes of the corolla with entire or nearly entire margins; flowers usually 5-merous, numerous, in terminal and lateral clusters; corolla not more than 25 mm. long, fimbriate internally (except in *G. microcalyx*). Plants annual or biennial; stems quadrangular, the angles often narrowly winged (10).
8. Plants perennial; flowers sessile or nearly so; corolla lobes long-fimbriate from below the apex nearly to the base; filaments bearded below. Stems not more than 15 cm. long; flowers closely invested by the bractlike uppermost pair of leaves, occasionally also 1 or 2 closely approximate axillary flowers borne on short, bractless peduncles .... 7. *G. barbellata*
8. Plants annual; flowers on naked peduncles at least 2 cm. long (usually much longer); corolla lobes erose and fimbriate at and near the apex; filaments naked (9).
9. Stems from base to the uppermost leaves not more than 25 cm. long, seldom with more than 3 pairs of leaves; peduncle up to 15 cm. long but usually much shorter; calyx lobes shorter to somewhat longer than the tube; corolla 25 to 45 (rarely 50) mm. long .... 8. *G. thermalis*
9. Stems to the uppermost leaves 20 to 50 cm. long, often with 4 or more pairs of leaves; peduncle 10 to 20 cm. long; calyx lobes much longer than the tube; corolla 45 to 60 mm. long .... 9. *G. grandis*
10. Calyx without a distinct tube, parted or cleft very nearly to the base, the lobes strikingly dissimilar, 2 of them foliaceous, about ⅓ longer and at least twice as wide as the others. Corolla pale blue or ochroleucous .... 10. *G. heterosepala*
10. Calyx with a distinct but short tube, the lobes unequal in length and width but not strikingly dissimilar (11).
11. Corolla 20 to 25 mm. long, ochroleucous. Calyx lobes lanceolate, sharply acuminate; plants strict, the branches nearly erect .... 11. *G. Wrightii*
11. Corolla less than 20 mm. long (12).
12. Calyx spathaceous (split down one side), the tube thin-scarious, much like the corolla in texture, whitish or purplish, much longer than the thick, subulate, green teeth, these less than 1 mm. long. Corolla conspicuously fimbriate internally .... 12. *G. Wislizeni*
12. Calyx not spathaceous, the tube not like the corolla in texture, much shorter than the teeth or lobes (13).

13. Plants open, the branches ascending at a relatively wide angle; calyx 2 to 3 mm. long; corolla ochroleucous or occasionally violet, naked internally.......13. *G. microcalyx*
13. Plants strict, the branches erect or nearly so; calyx 5 to 10 mm. long; corolla ochroleucous to violet, fimbriate internally..........................14. *G. strictiflora*

1. Gentiana Fremontii Torr. White Mountains, southern Apache County (*Darrow* in 1940, *Phillips* 3070), San Francisco Peaks (*Toumey* 490, in 1892), 9,000 feet or higher, summer. Alberta to New Mexico, northern Arizona, and California.

2. Gentiana tenella Rottb. (*G. monantha* A. Nels.). San Francisco Peaks, 11,500 to 12,000 feet (*Knowlton* 131, *Little* 4728), August. Colorado and Idaho to Alaska, northern Arizona and California; circumpolar.

3. Gentiana Parryi Engelm. White Mountains (Apache and Greenlee counties), Kaibab Plateau and North Rim of Grand Canyon (Coconino County), 8,500 to 11,500 feet, alpine and subalpine meadows, August and September. Wyoming and Utah to New Mexico and eastern Arizona.

4. Gentiana Forwoodii Gray. De Motte Park, Kaibab Plateau, 9,000 feet (*Jones* 6056a), September. Alberta to Colorado and northern Arizona.

5. Gentiana affinis Griseb. Apache, Navajo, Coconino, Gila, and northern Greenlee counties, 7,000 to 9,500 feet, mountain meadows, August to October. Saskatchewan to British Columbia, south to Colorado, northern Arizona, and California.

Pleated gentian. Flowers dark blue or violet.

6. Gentiana Bigelovii Gray. Mountains of Graham, Cochise, and Pima counties, about 7,500 feet, August and September. Colorado, New Mexico, and southeastern Arizona.

The Arizona plant, which is very similar to *G. interrupta* Greene, was referred by Gray to *G. Bigelovii* but seems at least varietally distinct from typical *G. Bigelovii* of northern New Mexico.

7. Gentiana barbellata Engelm. San Francisco Peaks, 10,500 to 12,000 feet (*Knowlton* 126, *Brady* 725/2490, *Little* 4780), August and September. Wyoming and Utah to New Mexico and northern Arizona.

8. Gentiana thermalis Kuntze (*G. elegans* A. Nels.). Pinaleno Mountains (Graham County), 8,000 to 9,000 feet, September. Mackenzie to New Mexico and Arizona.

Western fringed gentian.

9. Gentiana grandis (Gray) Holm. Huachuca and Santa Rita mountains (Cochise and Santa Cruz counties), 5,000 to 7,500 feet, rich soil in shade along brooks, September and October, type from near Babocomari, Cochise County (*Thurber*). Known only from southern Arizona.

Arizona's most beautiful gentian, the very large, fringed corolla with a violet-purple limb and white throat veined with purple. It is doubtfully distinct from the Mexican *G. superba* Greene.

10. Gentiana heterosepala Engelm. Kaibab Plateau, Grand Canyon, and San Francisco Peaks (Coconino County), White Mountains (Apache and Greenlee counties), 7,000 to 11,000 feet, June to September. Colorado, Utah, New Mexico, and Arizona.

11. Gentiana Wrightii Gray. Known in Arizona only from the type collection, in Santa Cruz County near Santa Cruz, Sonora (*Wright* 1659), flowering in autumn. Southern Arizona and northern Mexico.

12. Gentiana Wislizeni Engelm. White Mountains, Apache County (*Rothrock* 799), Chiricahua Mountains, Cochise County (*Blumer* 1414, *Eggleston* 10819), 7,000 to 8,000 feet, openings in pine forest, September and October. Southeastern Arizona and northern Mexico.

13. Gentiana microcalyx Lemmon. Chiricahua and Huachuca mountains (Cochise County), Rincon, Santa Rita, and Santa Catalina mountains (Santa Cruz and Pima counties), 5,000 to 7,500 feet, frequent in rich soil in canyons, August to October, type from the Chiricahua Mountains (*Lemmon*). Known only from southern Arizona.

14. Gentiana strictiflora (Rydb.) A. Nels. Kaibab Plateau and vicinity of Flagstaff (Coconino County), White Mountains (Apache and Greenlee counties), Pinaleno Mountains (Graham County), 7,000 to 11,500 feet, mountain meadows and forests, August and September. Saskatchewan to Alaska, south to New Mexico, Arizona, and California.

The taxonomy of the entities closely related to *G. Amarella* L. is confused. The Arizona specimens are probably referable to *G. scopulorum* (Greene) Tidestrom (*Amarella scopulorum* Greene) if the latter be considered as specifically distinct from *G. strictiflora*.

3. SWERTIA (221, 222). Elkweed, Green-gentian

Plants perennial; stem leaves mostly opposite or whorled; flowers in cymose panicles; corolla rotate, deeply 4- or 5-lobed, the lobes bearing, on the inner side toward base, 1 or 2 large, pitted glands (foveae), these with fringed margins; filaments often united at base; stigma 2-lobed, nearly sessile or the style elongate; seeds flat, usually margined or winged.

KEY TO THE SPECIES

1. Glandular pits 2 on each corolla lobe; leaves gradually reduced in size, but even the uppermost foliaceous, thickish but not leathery, not white-margined (2).
1. Glandular pits 1 on each corolla lobe; leaves of the inflorescence small and bractlike, the lower leaves leathery, conspicuously white-margined. Corolla greenish with purple dots or streaks, the pits with conspicuous, fringed crests (3).
2. Stems relatively slender, few-leaved; leaves opposite or sometimes alternate; corolla usually 5-lobed, blue, sometimes streaked or spotted, the pits oval to suborbicular, sparsely fringed, not more than 1 mm. long; stigma sessile or nearly so...1. *S. perennis*
2. Stems stout, many-leaved; leaves mostly whorled; corolla 4-lobed, greenish flecked with purple, the pits elliptic, copiously fringed, 3.5 mm. or longer; style evident.2. *S. radiata*
3. Stems seldom more than 0.5 m. long, usually branched from at or near the base, the branches often numerous; lower stem leaves whorled, the basal ones with conspicuously crisped margins; corolla lobes acute to sharply acuminate at apex, the pit much narrower than the lobe, widened and notched at apex................3. *S. albomarginata*
3. Stems up to 1 m. long, seldom branched near the base, the branches usually few; stem leaves opposite, the basal ones with flat or somewhat undulate margins; corolla lobes rounded and apiculate or abruptly acuminate at apex, the pit nearly as broad as the lobe, entire at apex, 2-lobed below..............................4. *S. utahensis*

1. Swertia perennis L. (*S. scopulina* Greene). White Mountains (Apache County), 9,000 to 10,000 feet, mountain meadows, August and September. Alaska to New Mexico, eastern Arizona, and California; Eurasia.

In Arizona specimens the corolla is deep slate blue, sometimes white-veined beneath.

2. Swertia radiata (Kellogg) Kuntze (*Frasera speciosa* Griseb., *F. speciosa* var. *scabra* Jones). Apache County to Coconino County, south to Cochise and Pima counties, 5,000 to 10,000 feet, common in rich soil in open pine forests, May to August, type of *F. speciosa* var. *scabra* from Pine, Gila County (*Jones* in 1890). South Dakota to Washington, south to New Mexico, Arizona, California, and northeastern Mexico.

Deers-ears. A large-leaved conspicuous plant with stout stems up to 1.8 m. (6 feet) high. Most of the Arizona specimens have the leaves pubescent beneath, but var. *macrophylla* (Greene) St. John (*Frasera macrophylla* Greene), with glabrous leaves, also occurs.

3. Swertia albomarginata (Wats.) Kuntze (*Frasera albomarginata* Wats.). Apache County to Mohave County, also, growing with *Larrea* and *Flourensia*, near Johnson (Cochise County), 4,500 to 7,000 feet, May to September. Colorado, Utah, Arizona, and southern California.

4. Swertia utahensis (Jones) St. John (*Frasera paniculata* Torr.). Apache County to northeastern Mohave County, north of the Colorado River, 4,000 to 7,500 feet, June to September. Southern Utah and Nevada, and northern New Mexico and Arizona.

4. HALENIA (223). Spur-gentian

Small, annual herbs; stems erect, leafy, usually simple; flowers in panicles of cymes; corolla light yellow, without glands, folds, or fringe, the lobes erect, usually extended at base into nectariferous spurs; stigmas 2, sessile, persistent; capsule somewhat flattened.

1. Halenia recurva (J. E. Smith) Allen (*H. Rothrockii* Gray). White Mountains (southern Apache County) and Mogollon Escarpment (southern Coconino County) to the Chiricahua Mountains (Cochise County), 7,500 to 10,000 feet, rich, moist soil in coniferous forests and mountain meadows, August and September, type of *H. Rothrockii* from Mount Graham (*Rothrock* 733). Arizona and northern Mexico.

5. EUSTOMA

Plants perennial, glaucous; stems leafy, erect or nearly so, up to 0.7 m. (2.5 feet) high; leaves oblong, oblong-lanceolate, or slightly oblanceolate; calyx lobes keeled, long-acuminate; corolla deeply campanulate, violet, the lobes 2 to 3 cm. long, about twice as long as the tube; stigma very large, with broad, flattened lobes; capsule ellipsoid, obtuse, 2-valved.

1. Eustoma exaltatum (L.) Griseb. Mohave, Yavapai, Gila, Maricopa, and Pima counties, 500 to about 2,500 feet, along ditches and in beds of streams, June to September. Florida, southern Arizona, southern California, and Mexico.

Catchfly-gentian, a very showy plant, rare in Arizona.

111. APOCYNACEAE. Dogbane Family

Plants perennial, herbaceous or slightly woody at base, commonly with acrid, milky juice; leaves simple, entire, opposite or alternate; flowers perfect, regular, sympetalous, 5-merous; stamens distinct, borne on the corolla; pistils 2, the ovaries superior, distinct, the styles or the stigmas united; fruit a pair of elongate follicles or by abortion 1; seeds often with coma.

This mainly tropical family comprises many handsome plants, including the well-known oleander (*Nerium Oleander*). Not a few of the Apocynaceae are poisonous, and some of them are important rubber plants. This substance occurs in small quantity in native species of *Amsonia* and *Apocynum*.

KEY TO THE GENERA

1. Anthers not produced at base, free from the stigma; leaves all, or mostly, alternate; corolla normally either blue or bright yellow; seeds naked, or comose at both ends (2).
1. Anthers produced basally into a sterile appendage below the polliniferous portion, connivent around and more or less adherent to the stigma; leaves opposite; corolla whitish or pink; seeds with apical coma (3).
2. Corolla not constricted at the throat, bright yellow, the lobes longer than the tube; seeds with coma at both ends..1. *Haplophyton*
2. Corolla constricted at or near the throat, blue (exceptionally whitish), the lobes shorter than the tube; seeds naked..2. *Amsonia*
3. Flowers few, large and showy; corolla salverform, with a long tube and throat; stamens inserted in the throat of the corolla; ovary superior................3. *Macrosiphonia*
3. Flowers numerous, small; corolla campanulate, cylindric, or urceolate; stamens inserted near the base of the corolla; ovary partly inferior....................4. *Apocynum*

1. HAPLOPHYTON. Cockroach-plant, Hierba-de-la-cucaracha

Stems branched, woody below; leaves mostly alternate, bright green, lanceolate or lance-ovate; flowers showy, terminal, solitary or in clusters of 2 or 3, the corolla bright yellow, salverform, with broad lobes longer than the tube; follicles slender, terete; seeds long, slender, black, with a pappuslike, deciduous tuft of long, white hairs at each end.

Superficially these plants somewhat resemble *Menodora*. An extract of the dried leaves mixed with molasses is an effective poison for such insects as cockroaches, flies, mosquitoes, fleas, and lice.

1. Haplophyton Crooksii L. Benson. Pinal, Cochise, Santa Cruz, and Pima counties, 2,000 to 4,500 feet, rocky slopes and canyons, July to October, type from the Santa Catalina Mountains (*Crooks & Darrow* in 1939). Western Texas to southern Arizona and northern Mexico.

Arizona specimens have been referred until recently to *H. cimicidum* A.DC., which has larger leaves and seeds and differs also in the character of the seed surface. *H. cimicidum* ranges from central Mexico to Guatemala (224).

2. AMSONIA (225)

Plants herbaceous; stems leafy, commonly erect, often numerous; leaves alternate or appearing whorled, sessile or short-petioled; flowers in terminal cymose panicles; corolla salverform, pale blue or whitish; seeds numerous, in 1 series, cylindric.

The Arizona species grow in the open or among shrubs, often along streams and washes, preferring sandy soil.

KEY TO THE SPECIES

1. Follicles torulose (constricted between the seeds on the edges as well as on the faces), more or less compressed. Corolla lobes 4 to 7 mm. long (2).
1. Follicles not torulose (3).
2. Leaves lanceolate to rhombic-ovate, up to 2 cm. wide; corolla tube and throat 7 to 10 mm. long; herbage and follicles glabrous to lanate-tomentose............1. *A. tomentosa*
2. Leaves linear or linear-lanceolate, less than 1 cm. wide; corolla tube and throat 10 to 20 mm. long; herbage and follicles glabrous....................2. *A. Eastwoodiana*
3. Lobes of the corolla very nearly as long as the tube, 5 to 7 mm. long. Herbage glabrous; leaves broadly ovate to ovate-oblong, 3 to 5 cm. long, 1.5 to 3 cm. wide, slightly glaucous; calyx 2 mm. long..3. *A. Jonesii*
3. Lobes of the corolla not more than half as long as the tube (4).
4. Corolla tube very slender below, swollen toward the apex, slightly constricted just below the limb but without a distinct neck, 16 to 18 mm. long; corolla lobes about 10 mm. long, half as long as the tube. Herbage glabrous; leaves slightly dimorphic, the upper ones almost filiform, shorter and more crowded than the middle ones..4. *A. grandiflora*
4. Corolla tube relatively stout, swollen near the middle, usually strongly constricted below the limb and with a distinct neck; corolla lobes less than 10 mm. long, rarely more than ⅓ as long as the tube (5).
5. Follicles compressed, regularly but shallowly impressed on the faces between the seeds, about 6 mm. wide at maturity, glabrous; seeds fully 10 mm. long. Herbage soft-villous with spreading hairs, rarely glabrate; leaves short-petioled, the upper ones lanceolate, 5 mm. wide or wider; corolla lobes 3 to 4 mm. long................5. *A. Kearneyana*
5. Follicles terete or nearly so, not impressed between the seeds or irregularly and obscurely so, 3 to 4 mm. wide; seeds (so far as known) seldom more than 8 mm. long (6).
6. Leaves linear-lanceolate to ovate-lanceolate, the largest ones sometimes 2 cm. wide; calyx lobes long-ciliate; herbage glabrous to villous. Corolla lobes 4 to 8 mm. long 6. *A. hirtella*
6. Leaves mostly linear, the lower ones narrowly lanceolate or oblanceolate, all much less than 1 cm. wide; calyx and herbage glabrous or very nearly so (7).
7. Corolla lobes 3 to 4 mm. long, ⅕ to ⅓ as long as the tube, the latter 10 to 15 mm. long 7. *A. Palmeri*
7. Corolla lobes 7 to 12 mm. long, fully ⅓ as long as the tube, the latter 16 to 18 mm. long 8. *A. Peeblesii*

1. Amsonia tomentosa Torr. & Frém. Navajo, Coconino, and Mohave counties, 100 to 5,000 feet, April to June. Southern Utah and Nevada, northern and western Arizona, and southern California.

Typical *A. tomentosa*, with broad leaves, lanate herbage, and more or less pubescent follicles, is known from several stations in southern Mohave County. A glabrous variant (*A. brevifolia* Gray) has been collected at Mokiak Pass, northern Mohave County (*Palmer* 302, the type collection), at Cameron (Coconino County), and on the south side of the Grand Canyon. In addition to these glabrous plants, some of which approach *A. Eastwoodiana* in their relatively narrow leaves, a narrow-leaved but pubescent variant, var. *stenophylla* Kearney & Peebles, occurs in Monument Valley (Navajo County) and along the Colorado and Little Colorado rivers (eastern Coconino County), about 5,000 feet. This variety is nearly intermediate between *A. tomentosa* and *A. arenaria* Standl., the latter being a species of western Texas, southern New Mexico, and Chihuahua.

2. Amsonia Eastwoodiana Rydb. Near Holbrook (Navajo County), Lees Ferry, House Rock Valley, and Grand Canyon (Coconino County), 3,000 to 5,000 feet, April to June. Southern Utah and northeastern Arizona.

3. Amsonia Jonesii Woodson. Navajo Spring, western edge of the Kaibab Indian Reservation, Pagumpa Springs, and Pipe Springs (Coconino and Mohave counties), 4,000 to 5,000 feet, April and May. Colorado, Utah, and northern Arizona.

4. Amsonia grandiflora Alexander. Near Patagonia and Ruby (Santa Cruz County), 4,000 to 4,500 feet, April and May, type from near Patagonia (*Peebles & Harrison* 6986). Southern Arizona and northern Mexico.

5. Amsonia Kearneyana Woodson. Plains and mouths of canyons along the western front of the Baboquivari Mountains (Pima County), about 3,500 feet, March and April, type collected by F. A. Thackery (No. 55). Known only from this well-isolated locality, which is farther southwest than any other recorded station of *Amsonia* in Arizona.

In *North American Flora* (29 : 129) Woodson reduced this species to synonymy under *A. Palmeri,* but in the characters of the mature follicles it seems quite distinct.

6. Amsonia hirtella Standl. (*A. pogonosepala* Woodson, *A. arizonica* A. Nels., *A. biformis* A. Nels.). Greenlee County to southern Mohave County, south to Cochise, Pinal, and Maricopa counties, 1,500 to 5,000 feet, March and April, type of *A. pogonosepala* from near Clifton, Greenlee County (*Rusby* 256), type of *A. arizonica* from south of Ash Fork, Yavapai County (*Nelson* 10247), types of *A. biformis* from near Duncan, Greenlee County (*Nelson* 11278, 11279). Western Texas to Arizona and Chihuahua.

Typical *A. hirtella,* with copious pubescence and linear-lanceolate leaves, has been collected along the Salt River in eastern Maricopa County (*Peebles* 11655). Much more common in Arizona is a glabrous or nearly glabrous form with some of the leaves usually broadly lanceolate (*A. pogonosepala* Woodson, *A. hirtella* var. *pogonosepala* Wiggins).

7. Amsonia Palmeri Gray. Lees Ferry (Coconino County) and in Mohave and Yavapai counties, 2,500 to 4,500 feet, March to June (occasionally September). Known only from Arizona, where the type was collected by E. Palmer, without definite locality.

8. Amsonia Peeblesii Woodson. Near St. Johns, Apache County (*Ripley & Barneby* 8449), eastern Coconino County, 4,500 to 6,000 feet, April to June, type from near Leupp (*Peebles* 9568). Known only from northern Arizona. Resembles *A. Eastwoodiana* except in the characters of the fruit.

3. MACROSIPHONIA

Plants low, suffrutescent, puberulent or glabrate; leaves opposite, ovate or elliptic-ovate; flowers mostly terminal, solitary or in twos or threes, sessile or nearly so; corolla funnelform, the limb 1.5 to 2.5 cm. wide; anthers connivent and cohering with the large stigma.

1. Macrosiphonia brachysiphon (Torr.) Gray (*M. dulcis* A. Nels.). Cochise, Santa Cruz, and eastern Pima counties, 4,000 to 5,500 feet, dry mesas and slopes, July to September, type of *M. dulcis* from the Huachuca Mountains (*Goodding* 2413). Southern New Mexico and Arizona and northern Mexico.

The showy white flowers, opening in the evening, are said to be used medicinally in Mexico. The plant is browsed by livestock (L. N. Goodding, personal communication) but, like most plants of this family, it is suspected of being poisonous.

4. APOCYNUM (226). Dogbane, Indian-hemp

Plants herbaceous; stems leafy, commonly erect; leaves large, opposite (exceptionally in whorls of 3); flowers small for the family, in usually dichotomous, compound cymes; corolla cylindric, urceolate, or campanulate, pink or whitish, the tube with internal basal appendages; anthers connivent and adherent to the stigma; follicles long, slender, terete.

The American aborigines used the bark of *Apocynum* for cordage. A cardiac stimulant similar although inferior to digitalis has been obtained from the root of *A. cannabinum.* The medicinal properties of these plants appear to have been known to the Indians. Some if not all of the species develop cyanogen and are toxic to cattle, horses, and sheep in both the green and the dry states.

KEY TO THE SPECIES

1. Corolla whitish or greenish, cylindric or narrowly urceolate, seldom more than 3 mm. long or more than twice as long as the calyx, the lobes erect or nearly so; leaves ascending or somewhat spreading, slightly paler beneath (2).
1. Corolla normally pink or striped with pink, campanulate or broadly urceolate, usually at least 4 mm. long and 2 to 3 times as long as the calyx, the lobes more or less spreading; leaves widespreading or drooping, whitish or conspicuously paler beneath. Leaves commonly ovate or oblong-ovate, deep green above (4).
2. Calyx lobes less than 2 mm. long. Corolla cylindric; plant glabrous throughout; leaves mostly short-petioled, the blades narrowed at base .... 1. *A. Suksdorfii*
2. Calyx lobes 2 mm. or longer (3).
3. Leaves of the main stem distinctly petioled, narrowed at base, oblong-lanceolate to ovate; herbage pubescent or glabrous .... 2. *A. cannabinum*
3. Leaves of the main stem sessile or subsessile, rounded or subcordate at base, lanceolate or oblong-lanceolate; herbage glabrous throughout .... 3. *A. sibiricum*
4. Corolla 5 mm. long or longer, broadly campanulate, the lobes conspicuously spreading, often recurved at apex .... 4. *A. androsaemifolium*
4. Corolla 4 to 5 mm. long, narrowly campanulate or urceolate, the lobes moderately or slightly spreading, not recurved .... 5. *A. medium*

1. Apocynum Suksdorfii Greene. Apache County to Coconino and Yavapai counties, Santa Catalina Mountains (Pima County), 3,000 to 7,000 feet, May to August. Colorado to Washington, south to New Mexico and Arizona.

Var. *angustifolium* (Wooton) Woodson, with lanceolate leaves not more than 2 cm. wide, is more common in Arizona than the typical plant, which has oblong-lanceolate leaves, some of them commonly at least 2.5 cm. wide.

A. Jonesii Woodson, known only by the type collection at Flagstaff, Arizona (*M. E. Jones* in 1884), is closely related to *A. Suksdorfii,* differing chiefly in having somewhat broader calyx lobes and a somewhat more urceolate corolla.

2. Apocynum cannabinum L. Apache, Coconino (?), Yavapai, and Cochise counties, up to 7,500 feet. Southern Canada and almost throughout the United States.

Typical *A. cannabinum,* with pubescent herbage, has been collected near McNary, Apache County (*Goodman & Hitchcock* 1318). Var. *glaberrimum* A.DC. apparently is more common in Arizona.

3. Apocynum sibiricum Jacq. Willow Spring, southern Apache County (*Palmer* 511), White Mountains, Navajo County (*Hough* 109), Clemenceau, Yavapai County (*W. W. Jones* 79), perhaps also in the Santa Catalina Mountains, Pima County (*Thornber* 5732), June and July. Throughout most of North America.

The species is represented in Arizona by var. *salignum* (Greene) Fern., which has a narrower corolla tube than in typical *A. sibiricum.*

4. Apocynum androsaemifolium L. Apache County to Coconino County, south to Cochise and Pima counties, 7,000 to 9,500 feet, openings in pine forests, June to August. Canada, south to Georgia and Arizona.

The most attractive of the species occurring in Arizona and also the most widely distributed and abundant. Most of the Arizona specimens are presumably typical, with leaves pubescent beneath, but var. *glabrum* Macoun (*A. ambigens* Greene), with leaves glabrous beneath, has been collected in the White Mountains, Apache County (*Goodding* 1210), and at Flagstaff, Coconino County (*Thornber* in 1930).

5. Apocynum medium Greene (*A. abditum* Greene). Apache, Navajo, Coconino, Yavapai, Gila, and Cochise counties, 5,000 to 7,500 feet, May to August, type of *A. abditum* from the Coconino National Forest (*Pearson* 235). Throughout much of North America.

The typical plant, with leaves pubescent beneath, is much more common in Arizona than var. *floribundum* (Greene) Woodson, with leaves entirely glabrous. The variety has been collected at the Grand Canyon, near Prescott, and near McNary Junction (southern Navajo County).

112. ASCLEPIADACEAE. Milkweed Family

Plants perennial, herbaceous or somewhat woody, mostly with milky sap; leaves simple, the margins entire or somewhat undulate or crisped; flowers perfect, regular, 5-merous, of highly specialized structure, usually in umbels or cymes; stamens and style coherent in a column, this adnate to the base of the corolla, a corona of distinct or united, often hoodlike appendages usually present between the corolla and the column, and adnate to one or the other, or both; anthers commonly winged, often scarious-tipped; pollen in pear-shaped masses (pollinia), usually 1 in each anther cell, these attached in pairs to the summit of the column; ovaries 2, united only by the common stigma; fruit a pair of follicles (or one by abortion); seeds usually with a pappuslike crown of fine bristles or hairs.

The flowers of this family rival those of the Orchidaceae in complexity of structure and in the manner in which they are adjusted for cross-pollination by insects. The plants are probably somewhat poisonous to livestock but are scarcely eaten except when other forage is unavailable. Rubber has been found in the latex of Arizona species of *Asclepias, Funastrum,* and *Lachnostoma.*

The genera represented in Arizona are reduced to 5 by Woodson (227). He gave cogent reasons for his treatment, but it seems more practical in the present publication to retain all the Arizona genera recognized in *Flowering Plants and Ferns of Arizona* except that *Gomphocarpus, Acerates,* and *Asclepiodora* are here combined with *Asclepias.* A supplementary key to the genera, as defined by Woodson, has been inserted below for convenient comparison, and the new combinations published by him are included in the synonymy.

KEY TO THE GENERA

1. Stems not twining, commonly erect; flowers in terminal and (or) lateral clusters. Leaves never sagittate or deeply cordate at base; anthers scarious-tipped, usually conspicuously so (2).
1. Stems twining; flowers lateral, solitary or clustered (3).
2. Corolla not bearded (at most puberulent) within; segments of the corona strongly concave, at least at base, usually crested or appendaged within.............2. *Asclepias*
2. Corolla bearded within with stiff, white hairs; segments of the corona flat or slightly concave, not crested within...8. *Pherotrichis*
3. Corona none, the campanulate corolla and the column unappendaged. Stems filiform; leaves narrowly linear, almost filiform; flowers yellowish, less than 3 mm. long 1. *Astephanus*
3. Corona present, sometimes reduced to subulate teeth (4).
4. Anthers not or very inconspicuously scarious-tipped (5).
4. Anthers conspicuously scarious-tipped. Corolla rotate or broadly campanulate, deeply lobed, the tube very short (6).
5. Corolla greenish or brownish, rotate-campanulate or funnelform-campanulate; segments of the corona with a median, winglike or subulate, internal crest; stems puberulent or pilose with soft, more or less retrorse hairs.........................7. *Gonolobus*
5. Corolla whitish, funnelform-campanulate; segments of the corona not crested or appendaged within; stems hirsute with stiff, spreading hairs............9. *Lachnostoma*
6. Corolla rotate, the lobes short, ovate, flat; corona appearing double, consisting of 5 large, turgid bodies united at base to an annulus in the corolla throat. Inflorescences pedunculate, umbellike ....5. *Funastrum*
6. Corolla very open-campanulate, the lobes elongate, lanceolate or strap-shaped, recurved at apex, more or less pubescent within; corona not appearing double, the segments lanceolate or subulate, at least above (7).
7. Leaves long-petioled, triangular-ovate, cordate, thin; inflorescences short-racemose or corymbiform, pedunculate, several-flowered; corolla 8 mm. long or longer, whitish, with strap-shaped lobes; segments of the corona subulate from a broad, ovate or subquadrate base, nearly equaling to slightly surpassing the corolla lobes 6. *Mellichampia*
7. Leaves short-petioled or subsessile, linear or narrowly lanceolate, not cordate, thickish; inflorescences umbellike, nearly sessile, few-flowered or the flowers solitary; corolla about 3 mm. long, with lanceolate lobes; segments of the corona subulate or lanceolate, much shorter than the corolla lobes. Corolla white-pilose within (8).
8. Corolla lobes valvate in the bud; segments of the corona subulate; stigma truncate, rounded, or merely apiculate....................................3. *Metastelma*
8. Corolla lobes slightly imbricate in the bud; segments of the corona lanceolate or triangular-subulate; stigma long-beaked...........................4. *Basistelma*

KEY TO THE ARIZONA GENERA AS DEFINED BY WOODSON (227)

1. Pollinia strictly pendulous, their faces uniformly flattened or rounded, uniformly fertile to the attachment of the translators: Tribe Asclepiadeae (2).
1. Pollinia usually horizontal or essentially so, occasionally ascending or descending, but 1 or both faces more or less excavated, and with a sterile hyaline margin or indentation near the attachment of the translators: Tribe Gonolobeae (4).
2. Erect or decumbent perennial herbs; pollinia very strongly flattened. Corona of 5 cucullate, calceolate, or clavate hoods, usually with an internal horn or crest..1. *Asclepias*[1]

[1] Including *Gomphocarpus* (of American authors), *Acerates, Asclepiodora.*

2. Lianas or twining undershrubs; pollinia broadly rounded or only slightly compressed on the faces (3).
3. Corona of 5 distinct or united, laminate to filiform scales, occasionally compounded or with internal processes, rarely wholly lacking........................2. *Cynanchum*[2]
3. Corona of 5 closed, inflated vesicles joined at the bases by a fleshy ring adnate to the corolla throat ....3. *Sarcostemma*[3]
4. Anthers relatively simple, not conspicuously vesicular, nor with dorsal appendages 4. *Matelea*[4]
4. Anthers with spreading, more or less laminate, fleshy dorsal appendages....5. *Gonolobus*[5]

1. ASTEPHANUS

Glabrous herbs with slender, twining stems; leaves narrowly linear; flowers small, yellowish, in axillary clusters of 3 to 5; corolla without a corona, campanulate, dull yellow, the lobes inflexed at apex; follicles long-acuminate; seeds rough-granulate.

1. **Astephanus utahensis** Engelm. (*Cynanchum utahense* Woodson). Mohave County, at Hardyville (*Palmer* 440) and near Yucca (*Jones* in 1884), 500 to 2,000 feet, April to June. Southern Utah and northwestern Arizona to southeastern California.

2. ASCLEPIAS. Milkweed

Stems herbaceous or woody below, commonly erect, never twining, the sap usually milky; leaves commonly opposite or whorled, sometimes early-deciduous; inflorescences terminal or lateral, usually many-flowered; corolla lobes commonly reflexed in anthesis; hoods usually with a hornlike internal crest; follicles ovoid, lanceolate, or fusiform.

A. subverticillata (*A. galioides* of authors) and perhaps other species contain a glucoside that is poisonous to livestock, especially to sheep, but the plants are seldom eaten. Appreciable quantities of rubber are found in the sap of some of the Arizona species. The Hopi Indians are reported to cook the young shoots and leaves of *A. speciosa* with meat. *A. subverticillata* is considered by the Hopi to increase the flow of milk in women. Milkweeds contain a substance—asclepain—that can be used as a substitute for papain for tenderizing meat. The possibility of using the coma or seed hairs of certain species in place of kapok as a filling for life rafts has been under investigation by the United States Department of Agriculture.

KEY TO THE SPECIES

1. Hoods (segments of the corona) not appendaged within or if sometimes so (in *A. auriculata*), then the stems tall and wandlike and the leaves mostly alternate, narrowly linear, and elongate (2).
1. Hoods crested within with a winglike or hornlike appendage (4).
2. Leaves all opposite, oval or ovate, whitish-tomentose beneath; flowers dark red; hoods concave only near the base, strap-shaped above, much surpassing the column, with long, narrow auricles at base; anther wings widest at base..........1. *A. hypoleuca*

[2] Including *Astephanus* (of American authors), *Metastelma, Basistelma, Mellichampia.*
[3] Including *Funastrum.*
[4] Including *Gonolobus* (of American authors in large part), *Pherotrichis, Rothrockia.*
[5] Including *Lachnostoma.*

2. Leaves mostly alternate or subopposite, oblong or narrower, not whitish-tomentose beneath; flowers greenish or purplish; hoods concave most of their length, not surpassing the column, with short, broad auricles at base; anther wings widest above the base (3).
3. Mass of the anthers and stigma nearly globose, usually considerably surpassing the hoods; column evident below the hoods but very short; hoods broadly ovate or the upper part rectangular, truncate, with large, winglike, basal auricles; leaves mostly alternate, narrowly linear, elongate; herbage glabrous or sparsely pubescent 2. *A. Engelmanniana*
3. Mass of the anthers and stigma longer than wide, usually very slightly surpassing the hoods; column concealed by the insertions of the hoods; hoods oblong-lanceolate, obtuse, with small, concealed basal auricles; leaves mostly nearly opposite, lanceolate, elliptic, or oblong; herbage tomentulose........................3. *A. viridiflora*
4. Corolla lobes ascending or spreading in anthesis; anther wings narrowed at base, broadest near the middle; hoods arched and hollow at apex. Stems usually decumbent; leaves thickish, lanceolate to linear; flowers large, in terminal umbels 4. *A. capricornu*
4. Corolla lobes mostly reflexed in anthesis; anther wings widened down to the base, usually triangular; hoods involute or complicate, not arched at apex (5).
5. Stems naked or nearly so, rushlike, the filiform leaves soon deciduous. Follicles fusiform; stems glaucous, more or less puberulent above (6).
5. Stems leafy, not rushlike, the leaves persistent (7).
6. Stems 10 mm. or more in diameter at base, decidedly woody below; corolla lobes 5 to 6 mm. long, white tinged with purple; hoods shorter than the anthers, spreading, not dilated at apex ....5. *A. albicans*
6. Stems less than 10 mm. in diameter at base, not noticeably woody above the caudex; corolla lobes 7 to 10 mm. long, pale green; hoods at least twice as long as the anthers, erect, dilated at apex..6. *A. subulata*
7. Corolla lobes orange or scarlet; stems villous or hirsute. Leaves very numerous, linear- to oblong-lanceolate ....7. *A. tuberosa*
7. Corolla lobes whitish, greenish, or purplish; stems not villous or hirsute, except in *A. Lemmoni* (8).
8. Leaves seldom more than 15 mm. wide, ovate-lanceolate or narrower, usually more than 3 times as long as wide (9).
8. Leaves more than 15 mm. wide, lance-oblong or broader, usually less than 3 times as long as wide (19).
9. Fruiting pedicels normally erect or ascending and straight, but sometimes sigmoid-curved. Leaves linear or linear-lanceolate, not more than 5 mm. wide; follicles fusiform; column about 1 mm. long (10).
9. Fruiting pedicels deflexed or decurved (12).
10. Hoods surpassing the column; leaves distinctly short-petioled, rather flaccid, mostly in pairs; corolla pinkish white.................................8. *A. angustifolia*
10. Hoods shorter than to equaling the column; leaves sessile or short-petioled, not flaccid, mostly in whorls of 3 or 4; corolla greenish white, sometimes tinged with purple (11).
11. Leaves narrowly linear to nearly filiform, not more than 3 mm. wide, the margins more or less revolute.......................................9. *A. subverticillata*
11. Leaves linear-lanceolate, 4 mm. or wider, flat.....................10. *A. fascicularis*
12. Plants suffrutescent, puberulent. Leaves narrowly linear or almost filiform, with more or less revolute margins (13).
12. Plants herbaceous above the caudex (14).
13. Hoods with long, subulate, recurved, hispidulous tips; leaves opposite, not crowded or rigid; umbels seldom more than 5-flowered; follicles fusiform........11. *A. macrotis*
13. Hoods not subulate-attenuate; leaves alternate or falsely verticillate, crowded, rather rigid; umbels usually more than 10-flowered; follicles ovate-lanceolate to broadly ovate in outline..12. *A. Linaria*
14. Hoods conspicuously dentate at the truncate apex; plants obscurely puberulent or glabrate; leaves narrowly linear, not more than 4 mm. wide; corolla lobes pale green; follicles fusiform, about 10 cm. long.......................13. *A. quinquedentata*

14. Hoods not conspicuously dentate at apex; plants more or less floccose-tomentose, at least on the young parts; leaves linear to ovate; corolla lobes usually purple or purplish; follicles ovate to lance-ovate in outline, considerably less than 10 cm. long (15).
15. Leaves at most obscurely ciliolate; umbels mostly lateral, few-flowered; hoods shorter than to equaling the anthers (16).
15. Leaves conspicuously white-ciliate with curved hairs, at least when young; umbels terminal; hoods nearly equaling to much surpassing the anthers. Umbels usually sessile or nearly so and closely subtended by 2 or more long leaves (17).
16. Leaves lanceolate, the lower ones 6 to 15 mm. wide near the base; umbels distinctly stalked; hoods much shorter than the anthers, strongly angulate-toothed in front; follicles conspicuously longitudinally striate with darker-colored stripes 14. *A. brachystephana*
16. Leaves narrowly linear; umbels sessile or nearly so; hoods equaling the anthers, not toothed in front; follicles not striate. Stems slender, 10 to 20 cm. long...15. *A. Cutleri*
17. Stems 2.5 to 5 cm. long; umbels 3- or 4-flowered; corolla lobes broadly ovate, 3 to 4 mm. long; hoods strongly cucullate, with an orbicular body and triangular wings nearly as large as the body, the whole structure much broader than high, the horn represented by a broadly ovate plate somewhat shorter than the body of the hood...16. *A. uncialis*
17. Stems 5 to 15 cm. long; umbels mostly more than 10-flowered; corolla lobes oblong-ovate, 4.5 to 6 mm. long; hoods quadrate-ovate, subauriculate near the base, the horn broadly triangular with a somewhat exserted subulate tip. Stems from a vertical, tuberlike root (18).
18. Herbage, pedicels, and calyx puberulent or short-pilose with subappressed hairs; leaves prevailingly lanceolate, seldom more than 10 mm. wide; hoods 3 to 4 mm. long, rounded at apex..17. *A. involucrata*
18. Herbage, pedicels, and calyx tomentose with crispate hairs, sometimes sparsely so; leaves (at least the lower ones) ovate-lanceolate to nearly orbicular, often considerably more than 10 mm. wide; hoods 2 to 3 mm. long, truncate at apex 18. *A. macrosperma*
19. Anther wings broadest considerably above the base; umbels all lateral; hoods stipitate, at least 3 times as long as the anthers, about 8 mm. long, strongly compressed laterally, the apical portion dilated................................19. *A. nyctaginifolia*
19. Anther wings broadest at base; umbels usually both terminal and lateral; hoods not stipitate (20).
20. Leaves acutish to acuminate at apex (21).
20. Leaves obtuse, truncate, or retuse, often mucronate, sometimes abruptly short-acuminate, at apex (23).
21. Follicles bearing soft, subulate processes, densely tomentose; hoods attenuate-acuminate. Corolla lobes purple, 8 to 10 mm. long; hoods 9 mm. or longer, 3 or more times as long as wide ....20. *A. speciosa*
21. Follicles without subulate processes; hoods not attenuate-acuminate (22).
22. Leaves acute or acutish, the margin not cartilaginous or erose-denticulate; corolla lobes purplish; hoods 5 to 6 mm. long, 2 to 3 times as long as the anthers......21. *A. Hallii*
22. Leaves cuspidate-acuminate, the margin cartilaginous and erose-denticulate; corolla lobes pale green; hoods 3 to 4 mm. long, not more than 1.5 times as long as the anthers 22. *A. erosa*
23. Leaves 2 to 4 pairs; stems seldom more than 20 cm. long. Leaves broadly ovate to suborbicular, or even wider than long (24).
23. Leaves usually more than 4 pairs; stems seldom less than 30 cm. long (25).
24. Stems 10 cm. long or shorter, the plant often subacaulescent; leaves mostly crowded near the base of the stem, lanate-tomentose at least when young; peduncles longer (often much longer) than the leaves; corolla lobes purplish, 4 to 5 mm. long; hoods purplish, much surpassing the anthers........................23. *A. nummularia*
24. Stems commonly 15 to 20 cm. long; leaves well distributed along the stem, glabrous or ciliolate; umbels sessile or short-peduncled; corolla lobes pale yellow, about 10 mm. long; hoods pink, shorter than to barely surpassing the anthers....24. *A. cryptoceras*

25. Herbage more or less villous with delicate, flaccid, pluricellular hairs; hoods about 3 times as long as the anthers, 7 mm. long, with a thin, oblique, attenuate tip. Leaves very large, up to 25 cm. long and 15 cm. wide; corolla lobes 8 to 10 mm. long 25. *A. Lemmoni*
25. Herbage puberulent or glabrate; hoods not more than twice as long as the anthers, 3 to 4 mm. long, not attenuate at apex, their sides petaloid, whitish or yellow (26).
26. Plants more or less glaucous; leaves thin, oblong or elliptic, 2 to 3 times as long as wide, sessile and often cordate-clasping; peduncles nearly as long as to longer than the leaves; corolla lobes 8 to 10 mm. long .... 26. *A. elata*
26. Plants not glaucous; leaves thick, ovate, obovate, or nearly orbicular, usually not more than 1.5 times as long as wide, short-petioled; peduncles much shorter than the leaves; corolla lobes 7 to 8 mm. long .... 27. *A. latifolia*

1. Asclepias hypoleuca (Gray) Woodson (*Gomphocarpus hypoleucus* Gray). White Mountains (Apache County) and mountains of Cochise, Santa Cruz, and Pima counties, 6,000 to 9,500 feet, open pine forests, July and August, type from the Santa Rita Mountains (*Pringle* in 1881). Southern New Mexico and Arizona and northern Mexico.

2. Asclepias Engelmanniana Woodson (*Acerates auriculata* Engelm., *Asclepias auriculata* Holz. non H.B.K.). Navajo and Coconino counties to Cochise and Pima counties, 4,000 to 7,000 feet, usually in open pine forests, June to September. Nebraska to Texas and Arizona.

Most of the Arizona specimens belong to var. *Rusbyi* (Vail) Kearney (*Acerates Rusbyi* Vail), which differs chiefly in the presence of a minute to well-developed hornlike crest on the inner face of the hoods, whereas typical *A. Engelmanniana* usually shows no trace of this appendage. The type of *Acerates Rusbyi* was collected along Oak Creek, Yavapai County (*Rusby* in 1883).

3. Asclepias viridiflora Raf. (*Acerates viridiflora* Eaton). Near Flagstaff (Coconino County), Prescott (Yavapai County), 5,500 to 7,000 feet, June and July. Massachusetts to Saskatchewan, south to Florida and Arizona.

The Arizona plants have the relatively long and narrow leaves of *Asclepias viridiflora* var. *lanceolata* (Ives) Torr.

Acerates bifida Rusby is known only by a single specimen in the Gray Herbarium. It differs from *A. viridiflora*, in which the hoods are entire except for the basal auricles, in having the hoods parted into 2 lanceolate divisions. Gray presumed that the plant was collected in Yavapai County, Arizona, but a letter from Rusby to Gray, filed with the type specimen, leaves it uncertain even that it was collected in the state of Arizona.

4. Asclepias capricornu Woodson (*Asclepiodora decumbens* (Nutt.) Gray). Almost throughout the state, 3,000 to 9,000 feet, common on dry plains and slopes, sometimes in openings in pine forests, April to August. Kansas and Arkansas to Nevada and Arizona, south to northern Mexico.

Antelope-horns. Flowers greenish yellow and maroon, slightly fragrant. Abandonment of grazing grounds for sheep because of the prevalence of this plant has been reported. The species is represented in Arizona by subsp. *occidentalis* Woodson, which has narrower and basally more acute leaves than in typical *A. capricornu.*

5. Asclepias albicans Wats. Western Maricopa County and Yuma County, 2,000 feet or lower, dry, granitic or volcanic slopes, March and April. Southwestern Arizona, southeastern California, and northwestern Mexico.

Stems exceptionally as many as 50 from 1 root, reaching a height of 3 m. (10 feet) and a diameter of 2 cm.

6. Asclepias subulata Decne. Mohave, Gila, Maricopa, Pinal, and Yuma counties, 3,000 feet or lower, locally abundant on dry slopes, mesas, and plains, April to October. Southern Arizona, southeastern California, and northwestern Mexico.

The sap contains an appreciable quantity of rubber, latex sometimes constituting as much as 6 per cent of the dry weight. The plant is reported to be used as a laxative by Pima Indians. *A. subulata* and *A. albicans* are the most xerophytic of the Arizona species of *Asclepias.*

7. Asclepias tuberosa L. Apache County to Coconino County, south to Cochise, Santa Cruz, and Pima counties, 4,000 to 8,000 feet, mostly in open places in pine forests, May to September. Throughout much of the United States.

Butterfly-weed, pleurisy-root. The plant is very showy in flower and is notable in the genus in not having milky sap. The flowers are normally orange-red, but vary to pale orange or even yellow. The roots are said to be used locally in treatment of affections of the lungs.

The Arizona plants belong to subsp. *interior* Woodson, characterized by mostly ovate to oblong-lanceolate leaves, these usually acute to acuminate at apex.

8. Asclepias angustifolia Schweig. Huachuca Mountains (Cochise County), Bear Valley or Sycamore Canyon (Santa Cruz County), Santa Catalina Mountains (Pima County), 4,000 to 5,000 feet, June and July. Southern Arizona and Mexico.

This plant was included in *Flowering Plants and Ferns of Arizona* under the name *Asclepias linifolia* H.B.K.

9. Asclepias subverticillata (Gray) Vail (*A. galioides* of authors, non H.B.K.). Apache County to Coconino and Yavapai counties, south to Cochise, Santa Cruz, and Pima counties, 2,500 to 8,000 feet, plains and mesas, common at roadsides, May to September. Kansas and Colorado to Texas, Arizona, and northern Mexico.

Poison milkweed, horsetail milkweed. Very poisonous to livestock, especially to sheep, but fortunately the plant is unpalatable and eaten only in the absence of better-liked forage. This dangerous weed should be eradicated from pastures and ranges wherever practicable.

10. Asclepias fasicularis Decne. (*A. mexicana* of authors, non Cav.). Flagstaff, Coconino County (*Jones* 4063, *Goodwin* 200), Rincon and Santa Catalina mountains, Pima County (*Thornber*). Idaho and Washington to northern Arizona and California.

11. Asclepias macrotis Torr. Cochise and Pima counties, 4,000 to 6,000 feet, limestone ridges, June to August. Western Texas to southeastern Arizona.

12. Asclepias Linaria Cav. Graham, Gila, Maricopa, Pinal, Cochise, Santa Cruz, and Pima counties, 1,500 to 6,000 feet, common on dry, rocky slopes and mesas, February to October. Southern Arizona and Mexico.

13. Asclepias quinquedentata Gray. Huachuca Mountains (Cochise County), Santa Rita and Rincon mountains (Pima County), June to August. Western Texas to southern Arizona.

14. Asclepias brachystephana Engelm. Coconino, Yavapai, Graham, Cochise, Santa Cruz, and Pima counties, 3,500 to 7,000 feet, dry plains and mesas, May to August. Kansas and Wyoming to Arizona and northern Mexico.

15. Asclepias Cutleri Woodson. Near Rock Point, Apache County (*Cutler* 2177, the type collection), 27 miles west of Carrizo, Apache County (*Peebles & Smith* 13581), about 5,000 feet, in sand, rare. Southeastern Utah and northeastern Arizona.

16. Asclepias uncialis Greene. White Mountains near Springerville, Apache County (*Ellis* 9). Wyoming to New Mexico and eastern Arizona.

17. Asclepias involucrata Engelm. Apache County to Mohave County, south to Cochise, Santa Cruz, and Pima counties, 3,500 to 7,000 feet, common on dry plains and mesas, sometimes with pine, April to June. Southern Utah, New Mexico, and Arizona.

18. Asclepias macrosperma Eastw. Apache County to eastern Coconino County, 5,000 to 6,000 feet, May and June. Southern Utah, northwestern New Mexico, and northeastern Arizona.

Intergrades with the preceding species, but well marked in its typical phase.

19. Asclepias nyctaginifolia Gray. Mohave County to Cochise, Santa Cruz, and Pima counties, 1,500 to 5,000 feet, common on plains and mesas, often in sandy washes, May to September. Arizona and southern California.

20. Asclepias speciosa Torr. Apache, Navajo, Coconino, Greenlee, and Gila counties, 6,000 to 9,000 feet, mostly in open coniferous forests, June to August. Saskatchewan to British Columbia, south to New Mexico, Arizona, and California.

A large, showy, coarse plant with dull-pink flowers.

21. Asclepias Hallii Gray (*A. lonchophylla* Greene). San Francisco Peaks, Coconino County (*Purpus* 30, the type collection of *A. lonchophylla*). Colorado and northern Arizona.

22. Asclepias erosa Torr. (*A. demissa* Greene). Mohave and Yuma counties, 3,500 feet or lower, usually at roadsides and in washes, May to October, types from the Gila River valley (*Schott, Thurber*). Southern Utah and western Arizona to southeastern California and northwestern Mexico.

Desert milkweed. The stems grow in clumps, reaching a height of 1.8 m. (6 feet). This is one of the most promising sources of rubber among plants native in the United States. *A. demissa,* based on an Arizona type (*Loew* in 1875), appears to be merely an extremely depauperate form.

23. Ascelpias nummularia Torr. Cochise, Santa Cruz, and Pima counties, 3,000 to 5,000 feet, dry mesas and slopes, March to June. Western Texas to southern Arizona.

24. Asclepias cryptoceras Wats. Near Pipe Springs, Mohave County, 5,000 feet (*Peebles & Parker* 14705). Utah to Oregon, northwestern Arizona, and California.

25. Asclepias Lemmoni Gray. Mountains of Cochise and Pima counties, 4,000 to 6,000 feet, mostly in open pine woods, July and August, type from the Chiricahua or the Huachuca Mountains (*Lemmon* in 1881). Known only from southern Arizona.

One of the largest-leaved of the Arizona species, with pale-pink hoods.

26. Asclepias elata Benth. Graham, Gila, Cochise, Santa Cruz, and Pima counties, 4,000 to 6,000 feet, openings in pine forests, August and September. New Mexico, Arizona, and Mexico.

Flowers greenish yellow, faintly fragrant.

27. Asclepias latifolia (Torr.) Raf. Apache County to Coconino and Yavapai counties, 3,000 to 7,000 feet, plains and mesas, often abundant along roadsides, June to August. Nebraska to Utah, Texas, and Arizona.

3. METASTELMA

Stems twining (often around one another), slightly woody at base; leaves narrow, rather thick; flowers small, solitary or in few-flowered lateral umbels; corolla 5-parted, white-pubescent inside, the segments of the corona not hooded, narrow, inserted at base of the column, surpassing the stigma; follicles slender, long-acuminate.

1. Metastelma arizonicum Gray. Pinal, Maricopa, and Pima counties, 1,500 to 4,500 feet, dry, rocky slopes, flowering almost throughout the year, type from near Tucson (*Pringle*). Known only from southern Arizona.

4. BASISTELMA

Plants very similar to *Metastelma arizonicum* except in the characters given in the key to genera.

***1. Basistelma angustifolium** (Torr.) Bartlett. Not known definitely to occur in Arizona, but the type (*Wright* 1677) was collected at Santa Cruz, Sonora, only a few miles south of the Arizona border.

5. FUNASTRUM. Climbing-milkweed

Stems twining; leaves opposite, linear to cordate-ovate or sagittate; flowers numerous, in lateral umbels; corolla campanulate-rotate, deeply lobed, the lobes twisted, the corona appearing double; follicles fusiform, attenuate-acuminate, smooth or warty.

The whitish, yellowish, or purplish flowers are fragrant. It is reported that the Papago Indians ate the fruits raw or cooked.

These plants have been referred by Woodson (227) and R. W. Holm (227a) to the genus *Sarcostemma*.

KEY TO THE SPECIES

1. Peduncles shorter than the leaves; leaves with usually crisped margins. Herbage copiously cinereous-puberulent; leaves thickish, narrowly to broadly lanceolate, hastate or sagittate at base; follicles 9 to 16 cm. long, 1 to 2 cm. in greatest width, long-acuminate at apex, acute or short-attenuate at base, smooth .... 1. *F. crispum*

1. Peduncles equaling or longer than the leaves; leaves not crisped (2).

2. Leaves ovate-lanceolate to broadly ovate, cuspidate-acuminate at apex (often abruptly so), cordate or sagittate at base (usually deeply so), thin; flowers white or whitish; follicles 1 to 1.5 cm. in greatest width, finely ridged. Herbage sparsely short-pubescent or glabrate .... 2. *F. cynanchoides*

2. Leaves narrowly linear to broadly lanceolate, acuminate at apex, acute and entire to pronouncedly hastate at base; flowers yellowish or purplish; follicles barely 1 cm. in greatest width, not noticeably ridged (3).

3. Herbage and follicles canescent-pilose with short, spreading hairs; corolla 3 to 4 mm. long, greenish yellow; segments of the corona longer than wide; follicles acutish at base; leaves narrowly linear, less than 1 cm. wide, the base not at all hastate or auriculate .... 3. *F. hirtellum*
3. Herbage glabrous or sparsely puberulent with subappressed hairs (rarely sparsely pilose); corolla 5 to 6 mm. long, purplish; segments of the corona as wide as or wider than long; follicles short-attenuate at base; leaves narrowly linear to lanceolate, up to 2 cm. wide but usually much narrower, the base acute and entire, angled, or auriculate 4. *F. heterophyllum*

1. **Funastrum crispum** (Benth.) Schlechter (*Sarcostemma crispum* Benth.). Greenlee, Gila, Pinal, Cochise, Santa Cruz, and Pima counties, 4,000 to 6,000 feet, in canyons among shrubs, April to July. Western Texas to southern Arizona and Mexico.

2. **Funastrum cynanchoides** (Decne.) Schlechter (*Sarcostemma cynanchoides* Decne.). Havasu Canyon (western Coconino County) and Yavapai and Greenlee counties to Cochise and Pima counties, 1,500 to 4,500 feet, along streams and washes, climbing over bushes, May to September. Western Texas to southern Arizona and northern Mexico.

Var. *subtruncatum* (Robins. & Fern.) Macbr., with leaves truncate or subcordate at base and attenuate at apex (these deeply cordate at base and apiculate or short-acuminate at apex in the typical plant), is occasional in Arizona.

3. **Funastrum hirtellum** (Gray) Schlechter (*Sarcostemma hirtellum* (Gray) R. Holm). Western Mohave County, chiefly in the valley of the Colorado River, 2,000 feet or lower, April, type from near Fort Mohave (*Cooper*). Southern Nevada, western Arizona, and southeastern California.

4. **Funastrum heterophyllum** (Engelm.) Standl. (*Sarcostemma cynanchoides* ssp. *Hartwegii* (Vail) R. Holm). Grand Canyon (Coconino County) to Yucca (Mohave County), south to Graham, Cochise, Pima, Santa Cruz, and Yuma counties, 5,500 feet or lower, common along washes, climbing over small trees and shrubs, March to October, type from near Fort Yuma. Western Texas to southeastern California and Mexico.

6. MELLICHAMPIA

Stems twining, retrorsely short-pilose in a definite line; leaves long-petioled, triangular-ovate, cordate; inflorescences short-racemose or corymbiform; corolla whitish, broadly campanulate, the lobes strap-shaped, 8 to 13 mm. long, strongly recurved at apex, the segments of the corona subulate from a broad base, shorter than to slightly longer than the corolla lobes; anthers round, conspicuously scarious-tipped; stigma 2-lipped; pods oblong-lanceolate in outline.

1. **Mellichampia sinaloensis** (T. S. Brandeg.) Kearney & Peebles (*Cynanchum sinaloense* Woodson). Along the Sonoita, Nogales to Patagonia, Santa Cruz County (*Peebles et al.* 4654), Baboquivari Mountains, Pima County (*Proctor, Goodding,* in 1936), August. Southern Arizona and western Mexico.

The cream-white flowers are fragrant.

7. GONOLOBUS. Angle-pod

Stems trailing or weakly climbing; leaves opposite, subhastate to sagittate-cordate at base; flowers lateral (not axillary), solitary or in umbellike clusters; corolla rotate-campanulate or funnelform-campanulate, greenish or dull purple; stigma truncate or depressed.

KEY TO THE SPECIES

1. Flowers solitary or in pairs, nearly sessile, with scarcely any common peduncle; herbage canescent-puberulent, the hairs subappressed; leaves less than 2 cm. long, not more than 1 cm. wide, deltoid-oblong, obtuse or acutish at apex, shallowly hastate or subhastate at base; corolla 3 to 4 mm. long, rotate-campanulate with spreading lobes and a very short tube; internal crests of the corona segments attached to the column; follicles fusiform, not more than 1 cm. wide, sparsely warty..........1. *G. parvifolius*

1. Flowers in peduncled umbels, seldom solitary; herbage pilose, the hairs somewhat spreading; leaves up to 5 cm. long and 3 cm. wide, triangular-ovate, attenuate-acuminate at apex, sagittate-cordate at base; corolla more than 5 (commonly about 10) mm. long, funnelform-campanulate with nearly erect lobes and a well-developed tube; internal crests of the corona segments free from the column; follicles ovoid, smooth
2. *G. productus*

1. Gonolobus parvifolius Torr. (*Vincetoxicum parvifolium* Heller, *Matelea parvifolia* Woodson). Gila County to Cochise, Pinal, Pima, and eastern Yuma counties, 2,000 to 5,000 feet, dry slopes and mesas, March to October. Western Texas and southern Arizona.

2. Gonolobus productus Torr. (*Vincetoxicum productum* Vail, *Matelea producta* Woodson). Yavapai, Gila, Cochise, and Santa Cruz counties, apparently rare in Arizona, May to August. Western Texas to Arizona.

8. PHEROTRICHIS

Stems erect or ascending from a thick, woody, tuberlike root; herbage and calyx hispid-hirsute; leaves broadly oblong or ovate, rounded or subcordate at base; flowers in lateral, sessile or subsessile umbels; corolla rotate-campanulate, the lobes lineate-veined, the segments of the corona truncate; stigma capped by a large, globose appendage.

1. Pherotrichis Balbisii (Decne.) Gray (*P. Schaffneri* Gray, *Matelea Balbisii* Woodson). Huachuca Mountains, Cochise County (*Lemmon* 2816), September. Southern Arizona and northern Mexico.

9. LACHNOSTOMA

Stems twining; leaves opposite, petioled, triangular-ovate; inflorescences lateral, few-flowered; corolla whitish, conspicuously reticulate with green veins, cleft about halfway, the lobes oblong-ovate, the tube villous within; stigma depressed, not surpassing the stamens, with a broad 5-angled disk; follicles large, lance-ovate in outline, strongly angled and finely ridged longitudinally.

1. Lachnostoma arizonicum Gray (*Gonolobus arizonicus* Woodson). Bear Valley or Sycamore Canyon (Santa Cruz County), Rincon, Santa Catalina, and Baboquivari mountains (Pima County), 3,500 to 4,500 feet, along streams in canyons, May to September, type from the Santa Catalina Mountains (*Lemmon* 3036, in 1883). Known only from southern Arizona.

Rothrockia cordifolia Gray (*Matelea cordifolia* Woodson) has been attributed to Arizona, apparently by confusion with *Lachnostoma arizonicum*, from which it may be distinguished by the larger, much more deeply cleft corolla, this glabrous within and not conspicuously veined, by the elevated stigma much surpassing the stamens, and by the smooth follicles. It occurs in Sonora not far from the southern boundary of Arizona.

113. CONVOLVULACEAE. Convolvulus Family

Plants herbaceous or suffrutescent, mostly with twining or trailing stems, in one genus without chlorophyll and parasitic; leaves alternate, simple but sometimes deeply lobed or parted, in one genus reduced to minute scales; flowers perfect, regular, mostly 5-merous, often showy, axillary, solitary or in cymes; sepals imbricate, distinct or partly united; pistil of 2 united or partly distinct carpels, the styles 1 or 2, often cleft, the ovary superior, 2-celled; fruit a capsule or a pair of utricles.

The morning-glories and other favorite ornamentals belong to this family, of which the most important member economically is the sweet-potato (*Ipomoea Batatas*).

KEY TO THE GENERA

1. Plants without green coloring matter, parasitic on the stems of various hosts; leaves reduced to small scales. Stems twining; flowers small; corolla white or whitish, usually with fimbriate or dentate appendages within............................1. *Cuscuta*
1. Plants with green coloring matter, autophytic; leaves with well-developed blades (2).
2. Ovary deeply 2-lobed; fruit utricular. Styles 2; stems creeping, rooting at the nodes; leaves reniform, wider than long; flowers solitary, small and inconspicuous; pedicels after anthesis often strongly revolute or sigmoid-curved................2. *Dichondra*
2. Ovary not lobed; fruit capsular (3).
3. Corolla imbricate in the bud, white. Styles 2, entire; stigmas capitate.........4. *Cressa*
3. Corolla plicate-convolute in the bud (4).
4. Styles 2, distinct to the base or nearly so, each 2-cleft. Stigmas linear-filiform or slender-clavate; stems not twining; corolla rotate-campanulate or broadly funnelform 3. *Evolvulus*
4. Style 1 or, if 2-cleft, then the divisions entire (5).
5. Stigma 1, globose or nearly so, entire or lobed............................7. *Ipomoea*
5. Stigmas 2, more or less elongate (6).
6. Style entire; stigmas ovate or oblong; stems not or scarcely twining....5. *Jacquemontia*
6. Style 2-cleft at apex, or entire; stigmas linear-filiform to ovate; stems mostly twining 6. *Convolvulus*

1. CUSCUTA (228, 229). Dodder

Contributed by T. G. Yuncker

Plants leafless and rootless, herbaceous, parasitic; stems yellowish, filiform, twining; flowers small (mostly 2 to 6 mm. long), sessile or short-pedicellate, in few- to many-flowered cymose clusters, commonly 5-merous but regularly 3- or 4-merous in a few species; perianth parts mostly united; stamens inserted in the throat of the corolla, alternating with the lobes; appendages commonly present at base of the corolla opposite the stamens, these scalelike, more or less toothed, fringed, or fimbriate; ovary 2-celled, the styles 2, the stigmas (in the Arizona species) capitate; fruit a capsule, this remaining closed, or opening with a regular or irregular line of circumscission near the base; embryo acotyledonous, filiform or more or less enlarged at one end.

Upon emergence from the seed the slender, elongate seedling coils about an available host to which it becomes firmly attached by means of its sucker-like organs (haustoria). A few species have been shown to possess small amounts of chlorophyll and are thereby partly autophytic. Although certain dodders show a preference in the choice of host, most of them grow readily upon various plants. Those which parasitize economically important crops sometimes cause considerable damage. This is especially true in fields of clover and alfalfa, where dodder seeds are commonly introduced with those of the host.

KEY TO THE SPECIES

1. Capsules not circumscissile, i.e. not separating in a regular line of cleavage, when forcibly separated either coming away entirely from the receptacle or breaking very irregularly (2).
1. Capsules circumscissile, i.e. easily separating near the base in a more or less regular line of cleavage (10).
2. Flowers mostly 3- or 4-parted (3).
2. Flowers mostly 5-parted (4).
3. Perianth membranaceous, the lobes obtuse; corolla lobes not inflexed at tip; scales oblong, reaching the filaments and free from the corolla tube above. Corolla when withered remaining at top of the capsule........................1. *C. Cephalanthi*
3. Perianth fleshy-papillate, the lobes acute; corolla lobes erect with inflexed tips; scales reduced to lateral wings along the stamen attachment..................5. *C. Coryli*
4. Infrastamineal scales lacking. Perianth parts acute to acuminate; calyx lobes triangular-ovate to sublanceolate; corolla lobes lanceolate, reflexed ....3. *C. californica*
4. Infrastamineal scales present (5).
5. Perianth fleshy-papillate; corolla lobes commonly erect, with inflexed tips. Scales prominent and mostly free from the corolla tube, at least above.............4. *C. indecora*
5. Perianth not fleshy-papillate; corolla lobes various, but not as in *C. indecora* (6).
6. Corolla lobes triangular or lanceolate, acute to acuminate (7).
6. Corolla lobes ovate or suborbicular, obtuse (9).
7. Scales prominent, commonly exserted; corolla lobes triangular to sublanceolate, acute, reflexed, with inflexed tips; capsules mostly depressed-globose, 2- to 4-seeded
2. *C. campestris*
7. Scales included; corolla lobes lanceolate, acute to acuminate; capsules globose-conic, mostly 1-seeded (8).
8. Flowers 2 to 3 mm. long; corolla lobes ovate-lanceolate; scales attached to the corolla tube most of their length; anthers oval, the filaments well developed......6. *C. salina*
8. Flowers 3 to 4 mm. long; corolla lobes lanceolate; scales commonly free; anthers oval-oblong, subsessile ....7. *C. nevadensis*
9. Flowers about 2 mm. long, subsessile, in few-flowered glomerules; calyx lobes orbicular, broadly overlapping; capsules conic, mostly 1-seeded. Margins of the perianth lobes denticulate ....8. *C. denticulata*
9. Flowers mostly much larger, pedicellate, in many-flowered, cymose clusters; calyx lobes ovate, obtuse; capsules globose-ovoid, mostly 3- or 4-seeded..........9. *C. Gronovii*
10. Styles subulate, commonly much thicker at base. Calyx lobes commonly more or less carinate (11).
10. Styles more slender, mostly about equally thick throughout, but sometimes slightly thicker near the base (12).

11. Flowers mostly 3 to 5 mm. long; styles pronouncedly subulate, becoming subconic in fruit . . . . 10. *C. mitraeformis*
11. Flowers mostly about 3 mm. long; styles subulate but not becoming conic in fruit. Perianth lobes more or less irregularly denticulate, thickened medianally to form a carina . . . . 11. *C. erosa*
12. Scales dentate only toward apex (13).
12. Scales fimbriate (14).
13. Flowers whitish when dry, more or less granulate- or scabrous-papillate; calyx lobes broad, ovate-deltoid, short-acute; scales bridged below the middle. . 13. *C. odontolepis*
13. Flowers reddish when dry, smooth; calyx lobes triangular, acute, commonly somewhat thickened in the center to form a low carina; scales bridged near the middle 14. *C. dentatasquamata*
14. Calyx ½ to ¾ as long as the cylindric corolla tube; scales reaching to about the middle of the tube. Calyx lobes triangular to sublanceolate, acute, often carinate 16. *C. tuberculata*
14. Calyx equaling or surpassing the campanulate corolla tube; scales reaching the filaments. Capsules mostly quickly and definitely circumscissile (15).
15. Lobes of the calyx triangular-ovate, obtuse, commonly carinate; pedicels short or almost none . . . . 12. *C. applanata*
15. Lobes of the calyx triangular-ovate-lanceolate, acute to acuminate; pedicels definite, often as long as or longer than the flowers, these in loose, umbellate cymes 15. *C. umbellata*

1. Cuscuta Cephalanthi Engelm. Apparently rare in Arizona, where collected in the southern part of the state by Wright without indication of definite locality, probably in Cochise County. Massachusetts to Oregon and south to Mexico, infrequent westward.

Host genera numerous, including *Salix, Spiraea, Vicia, Dracocephalum, Teucrium, Cephalanthus, Solidago, Aster, Sonchus.* This species is often confused with *C. Gronovii,* to which it bears some resemblance. It is distinguished by the mostly 3- or 4-parted flowers and the persistence of the corolla at top of the capsule instead of about it or at base, as with *C. Gronovii.* The capsule is subglobose rather than conic as in that species.

2. Cuscuta campestris Yuncker. Coconino, Yavapai, and Pima (probably also Graham and Maricopa) counties. Widely distributed throughout the range of the genus.

Hosts numerous, mostly herbaceous, including grasses. If this species has any host preference it is for clover, alfalfa, and other legumes, and probably its wide distribution is due to association of its seeds with those of such economically important hosts. Pedicels mostly shorter than the flowers, which are yellow when dry and mostly about 2 mm. long; calyx lobes broadly ovate to oval-ovate, almost enclosing the corolla tube; corolla campanulate, enlarging about the base of the rapidly developing capsule, which becomes up to 4 mm. wide; scales abundantly fringed; styles slender, scarcely subulate; capsules mostly wider than long.

3. Cuscuta californica Hook. & Arn. Topock, Mohave County (*Eastwood* 8907). Common in the Pacific coast states, especially California, but apparently rare in Arizona.

Host plants include: *Eriogonum, Abronia, Dalea, Foeniculum, Asclepias, Franseria.* An attractive species, easily recognized by the lanceolate, reflexed corolla lobes and the absence of infrastamineal scales.

4. Cuscuta indecora Choisy. Coconino, Yavapai, Graham, Gila, Maricopa, Pinal, and Pima counties. Southern and western United States; Mexico, West Indies, and South America.

This species occurs on a great variety of both woody and herbaceous hosts, including *Crossosoma, Acacia, Prosopis, Sapindus, Condalia, Datura, Solidago, Aster, Hymenoclea, Baccharis,* and *Pluchea.* Though named *indecora* the plant is ordinarily attractive with its abundant, white, fleshy, and more or less papillate flowers. Styles about as long as the somewhat pointed ovary, becoming divaricate on the globose capsule, which is enveloped by the withered corolla.

5. Cuscuta Coryli Engelm. Grand Canyon, Coconino County (*Eggert* in 1886). Eastern United States to Montana and northern Arizona, more common eastward.

This dodder occurs on a great variety of woody and herbaceous hosts, including *Salix, Rhus, Ceanothus, Daucus, Stachys, Symphoricarpos, Solidago, Aster, Helianthus,* and *Chrysanthemum.* Flowers about 2 mm. long on pedicels longer or shorter than the flowers, or the flowers originating endogenously; calyx lobes triangular-ovate, about reaching the sinuses of the corolla or somewhat longer; corolla lobes triangular-ovate or more or less lanceolate; scales mostly reduced to toothed wings along the line of attachment of the filaments but sometimes free and bifid or toothed; stamens nearly as long as the corolla lobes; styles shorter than or equaling the globose-ovoid ovary, becoming divergent in fruit; capsule depressed-globose, enveloped by the withered corolla.

6. Cuscuta salina Engelm. Pinal and Pima counties, probably elsewhere in Arizona, on saline soils. British Columbia to Arizona and southern California.

Host plants include *Atriplex, Suaeda, Allenrolfea, Salsola, Nitrophila,* and *Cressa.* Flowers 2 to 3 mm. long, narrowly campanulate, mostly pedicellate, with pedicels of different lengths, forming close or loose cymose inflorescences; calyx enclosing the corolla tube; corolla lobes mostly spreading to reflexed and equaling the tube; anthers oval, on short filaments; scales oblong, shallowly fringed, free from the corolla tube at the upper end only.

***7. Cuscuta nevadensis** Johnst. This species occurs in southern Nevada and is to be looked for in northwestern Arizona.

8. Cuscuta denticulata Engelm. Near Topock and Lake Mead (Mohave County), Hope (Yuma County). Southern Utah to southern California and western Arizona.

This dodder occurs in Arizona mostly on *Larrea,* but also on *Coleogyne, Euphorbia, Sphaeralcea, Tamarix, Nicotiana,* and various Compositae. Stems very slender; flowers small, mostly in 2- or 3-flowered clusters; calyx yellow and more or less glistening in dry specimens, with large and conspicuous cells, almost enclosing the corolla tube; corolla becoming urceolate, the lobes oval-oblong, commonly obtuse; scales about reaching the anthers, oblong, denticulate; embryo with an enlarged globose, knoblike end, a character found otherwise only in the closely allied *C. nevadensis* and *C. Veatchii.*

9. Cuscuta Gronovii Willd. Grand Canyon, Coconino County (*Eggert* in 1886). Eastern and central United States, where this is the commonest species of dodder, to northern Arizona.

It occurs on a very wide range of both woody and herbaceous hosts, with no apparent preference. Flowers 2 to 4 mm. long, in loose or somewhat dense, paniculate cymes; calyx mostly shorter than the corolla tube; corolla lobes spreading to reflexed, shorter than the campanulate tube; scales commonly oblong, about reaching the stamens; styles commonly about equaling the ovary; capsule enveloped by the withered corolla.

10. Cuscuta mitraeformis Engelm. Cave Creek, Chiricahua Mountains, Cochise County (*Kearney & Harrison* 6176), about 6,000 feet, on *Lupinus.* Southeastern Arizona and Mexico.

Stems coarse; flowers on short pedicels, forming compact, globular clusters; calyx lobes about as long as the corolla tube, ovate, obtuse, more or less unequal, irregular, the larger lobes often strongly and unevenly carinate; corolla lobes ovate, obtuse, about as long as or exceeding the campanulate tube; scales oblong, mostly somewhat truncate and bifid, or less commonly ovate, as long as the tube, and deeply fringed; styles shorter than the conic ovary, becoming widely divergent; capsule 5 to 8 mm. long, enveloped by the withered corolla.

11. Cuscuta erosa Yuncker. Mountains of Pima County. Southern Arizona and northern Sonora.

Host plants include *Amaranthus, Gomphrena, Kallstroemia, Abutilon, Ipomoea, Siphonoglossa, Anisacanthus,* and *Franseria.* Pedicels mostly shorter than the flowers; calyx lobes orbicular, membranaceous, denticulate on the margin, fleshier in the median part, nearly distinct; corolla lobes erect to reflexed, about as long as or slightly shorter than the campanulate tube, ovate-oblong, obtuse; scales broad, fringed, about equaling the corolla tube, bridged at about the middle; styles longer than the globose ovary, becoming divergent in fruit; capsule globose, thin toward the base, bearing the withered corolla about the middle or at top.

12. Cuscuta applanata Engelm. Coconino, Cochise, and Santa Cruz counties, type from "Arizona Territory south of the Gila River" (*Wright* 541). New Mexico, Arizona, and Mexico.

Hosts various, including *Boerhaavia* and *Ambrosia.* Flowers in dense clusters; corolla lobes oblong to ovate-lanceolate, spreading; scales exserted, fringed; capsules globose-depressed, thin, readily circumscissile.

13. Cuscuta odontolepis Engelm. Santa Rita Mountains (Santa Cruz or Pima County), on various hosts including *Amaranthus,* type from "near a deserted rancho on a rocky hillside in Arizona" (*Wright* 1624). Southern Arizona and Sonora.

Flowers 4 to 5 mm. long, short-pedicellate, forming rather large, dense clusters; corolla cylindric-campanulate, the lobes ovate-lanceolate, acute; scales oblong or subspatulate, dentate near the apex only; styles slender, mostly longer than the ovary; capsule globose.

14. Cuscuta dentatasquamata Yuncker. Florida Canyon, Santa Rita Mountains, Pima County (*Kearney & Peebles* 10580), on *Bouvardia.* A rare species, known elsewhere only from Los Pinitos, Sonora, the type locality.

Flowers pedicellate, somewhat fleshy; calyx deep, the lobes equaling or exceeding the corolla tube; corolla campanulate, the lobes triangular, acute, shorter than the tube, spreading; scales oblong, dentate; styles slender, slightly subulate, about equaling or longer than the depressed-globose ovary; capsule depressed-globose, thin, somewhat irregularly circumscissile.

15. Cuscuta umbellata H.B.K. Pinal, Cochise, Pima, and Yuma counties. Southern United States to Arizona, West Indies, Mexico, and northern South America.

On a variety of herbaceous hosts, including *Polygonum, Atriplex, Suaeda, Alternanthera, Amaranthus, Boerhaavia, Trianthema, Sesuvium, Kallstroemia, Tribulus,* and *Euphorbia.* Flowers in compound cymes, the ultimate divisions of these umbellate, of 3 to 7 flowers; calyx turbinate, yellow and shining when dry; corolla lobes lanceolate, acute to acuminate, reflexed; styles longer than the globose ovary; capsule depressed-globose, surrounded by the withered corolla. Specimens bearing unusually large flowers (4 to 6 mm. long) have been distinguished as var. *reflexa* (Coult.) Yuncker.

16. Cuscuta tuberculata T. S. Brandeg. Pinal, Pima, and Yuma counties. Southwestern New Mexico, southern Arizona, and northwestern Mexico.

It occurs most commonly on *Boerhaavia,* but is found occasionally on other hosts, such as *Euphorbia.* Flowers cylindric, on slender pedicels; calyx thickened and keeled toward the base; corolla papillose in the basal, calyx-enveloped part; stamens somewhat shorter than or equaling the corolla lobes; styles slender, longer than the ovary, exserted; capsule globose, enveloped and surmounted by the withered corolla.

2. DICHONDRA

Plants perennial, more or less sericeous; stems creeping, rooting at the nodes; leaves petioled, round-reniform; flowers small, solitary on bractless peduncles, the corolla broadly campanulate, whitish; pistils distinct or nearly so; utricles 1- or 2-seeded.

These plants are very efficient soil binders but are rare in Arizona.

KEY TO THE SPECIES

1. Stems and petioles relatively stout, the petioles straight or nearly so; leaves densely silvery-sericeous on both surfaces, with a very shallow sinus or nearly truncate at base; corolla very villous externally. Peduncles stout, about 5 mm. long, strongly decurved after flowering .... 1. *D. argentea*

1. Stems and petioles slender, the petioles mostly curved; leaves bright green and sparsely sericeous above, with a deep, broad or narrow sinus; corolla glabrous or sparsely villous externally (2).

2. Leaves silvery-sericeous beneath, seldom more than 2 cm. wide; peduncles filiform, commonly more than 1 cm. long .... 2. *D. repens*

2. Leaves green and sparsely sericeous beneath, commonly more than 2 (up to 5) cm. wide; peduncles relatively stout, mostly less than 1 cm. long .... 3. *D. brachypoda*

1. Dichondra argentea Willd. Foothills near Bisbee, Cochise County (*Harrison* 8256), about 5,500 feet, late summer. Western Texas to southeastern Arizona and Mexico.

2. Dichondra repens Forst. Bear Valley or Sycamore Canyon, Santa Cruz

County, about 3,500 feet (*Goodding* 6620, *Darrow & Haskell* 2217). Widely distributed in tropical America.

The collections cited belong to var. *sericea* (Swartz) Choisy.

3. Dichondra brachypoda Woot. & Standl. Near San Bernardino, Chiricahua Mountains, and Rucker Canyon, Cochise County (*Goodding* 6629, *Clark* 8533, *Gould & Haskell* 4610), 4,000 to 6,000 feet. Western Texas to southeastern Arizona and northern Mexico.

3. EVOLVULUS (230)

Small, pubescent, perennial herbs; stems numerous, erect or diffuse, never twining; flowers solitary or few in a cluster, subsessile to long-pedunculate; corolla rotate-campanulate or broadly funnelform, cream-colored, purple, or sky blue; styles 2, each 2-cleft.

The plants are sun-loving, growing on dry plains and mesas, often among grasses.

KEY TO THE SPECIES

1. Stems rarely more than 15 cm. long, spreading or decumbent; upper leaves only slightly reduced; flowers mostly solitary; peduncles or pedicels much shorter than the subtending leaves, often decurved in fruit (2).

1. Stems commonly 30 cm. or longer, erect or ascending; upper leaves greatly reduced; flowers one or few on very slender, bracted peduncles longer than the subtending leaves. Inflorescence a very open, leafy, terminal panicle; corolla rotate, azure blue or occasionally white (3).

2. Sepals linear or narrowly lanceolate; corolla rotate-campanulate, lavender, drying violet-purple; leaves appressed or narrowly ascending, closely imbricate along the stem, at least above, densely villous-sericeous on both surfaces .... 1. *E. pilosus*

2. Sepals lanceolate or ovate-lanceolate; corolla rotate, cream-colored or azure blue; leaves spreading, not closely imbricate along the stem, commonly glabrous on the upper surface .... 2. *E. sericeus*

3. Corolla not more than 7 mm. in diameter .... 3. *E. alsinoides*

3. Corolla 8 to 20 mm. in diameter .... 4. *E. arizonicus*

1. Evolvulus pilosus Nutt. Apache County to Coconino County, south to Cochise, Santa Cruz, and Pima counties, 3,000 to 5,500 feet, March to July. North Dakota and Montana to Texas and Arizona.

2. Evolvulus sericeus Swartz. Navajo and Yavapai counties to Greenlee, Cochise, Santa Cruz, and Pinal counties, 3,500 to 5,500 feet, May to September. Texas to southeastern California, south to Argentina; West Indies.

The common phase in Arizona is var. *discolor* (Benth.) Gray (*E. Wilcoxianus* House), type of *E. Wilcoxianus* from Fort Huachuca (*Wilcox* 96). This has the upper leaf surface green and glabrate, and the corolla usually cream-colored. The phase with leaves sericeous on both surfaces is occasional in Cochise County.

3. Evolvulus alsinoides L. Gila, eastern Maricopa, Cochise, Santa Cruz, and Pima counties, 2,500 to 5,000 feet, April to September. Widely distributed in tropical and subtropical regions of both the Eastern and the Western hemispheres.

The Arizona plant is var. *acapulcensis* (Willd.) Van Ooststroom, which intergrades in Arizona with *E. arizonicus*, differing chiefly in its smaller corolla.

4. Evolvulus arizonicus Gray. Graham County to Yavapai County, south to Cochise, Santa Cruz, and Pima counties, 3,500 to 5,000 feet, April to October, type from Sonora near the border of Arizona. Southwestern New Mexico, Arizona, and northern Mexico; Argentina.

One of Arizona's most beautiful wild flowers, with deep sky-blue corollas. The typical plant has the hairs of the herbage all or nearly all short and appressed. Almost equally abundant in the 3 southern counties is var. *laetus* (Gray) Van Ooststroom (*E. laetus* Gray), with many of the hairs long and spreading, but intergrading with typical *E. arizonicus*. The type of *E. laetus* was collected in the Santa Rita Mountains (*Pringle* in 1881).

4. CRESSA

Plants small, perennial, herbaceous, silky-villous; stems erect or spreading, very leafy; leaves sessile, narrowly elliptic, entire; flowers small, solitary in the upper axils, rather crowded; corolla whitish, the lobes becoming reflexed.

1. Cressa truxillensis H.B.K. Eastern Coconino County and Mohave, Maricopa, Pinal, and Yuma counties, 100 to 4,000 feet, strongly saline soil, May to August. Texas to southern Utah and southern California, south to tropical America.

5. JACQUEMONTIA

Plants annual or perennial, often suffruticose, pubescent with forked or stellate hairs; leaves petioled, entire, rounded or subcordate at base; flowers long-peduncled, solitary or in small, loose inflorescences, the corolla funnelform, blue or lavender; sepals all alike or the outer ones much broader than the inner ones.

KEY TO THE SPECIES

1. Plants annual (in Arizona); stems herbaceous throughout; herbage loosely soft-pilose; sepals nearly alike in size and shape, lanceolate or lance-ovate, acuminate; corolla narrowly funnelform, 8 to 10 mm. long, deep blue.................... 1. *J. Palmeri*
1. Plants perennial; stems woody at least toward base; herbage finely canescent or subtomentose, the hairs mostly appressed; outer sepals much wider than the inner ones, broadly ovate or suborbicular, acutish or apiculate; corolla broadly funnelform, 20 to to 25 mm. long, pale lavender.................................... 2. *J. Pringlei*

1. Jacquemontia Palmeri Wats. Canyons on the western slope of the Baboquivari Mountains (Pima County), about 4,000 feet, September and October Southern Arizona, Sonora, and Baja California.

2. Jacquemontia Pringlei Gray. Tanque Verde, Santa Catalina, Tucson, and Ajo mountains (Pima County), south end of Gila Mountains (Yuma County), also a doubtful record from Cochise County, 3,000 to 4,500 feet, canyons and oak woodland, July to September, type from the Santa Catalina Mountains (*Pringle*). Southern Arizona to Central America.

6. CONVOLVULUS. BIND-WEED

Plants mostly perennial, herbaceous; stems (in the Arizona species) twining or trailing; leaves more or less lobed; corolla broadly funnelform, white or pink; stigmas more or less elongate; capsule normally 2-celled or imperfectly 4-celled.

C. tricolor L., a low annual with erect or spreading stems, narrow entire leaves, and small but showy, parti-colored flowers, is sometimes cultivated as an ornamental. Specimens of what appears to be this species were collected at Tucson by Toumey in 1892, but there is no evidence that the plant has become established in Arizona.

KEY TO THE SPECIES

1. Bracts larger than the sepals, inserted immediately below the calyx and enclosing it; stigmas oblong or oval. Herbage glabrous or soft-pilose; leaves deltoid-hastate; bracts ovate, obtuse or apiculate; corolla 3 to 6 cm. long, white or pink..........1. *C. sepium*
1. Bracts much smaller than the sepals, remote from or not closely subtending the calyx; stigmas linear, filiform, or slender-clavate (2).
2. Leaves shallowly sagittate or hastate, with short, entire or sparingly dentate, triangular-ovate basal lobes. Herbage glabrous or sparsely soft-pilose; leaves oblong or oblong-ovate, commonly very obtuse; bracts seldom more than 5 mm. long, lanceolate, oblanceolate, or narrowly oblong; corolla white or striped with pink, 1.5 to 2.5 cm. long 2. *C. arvensis*
2. Leaves hastately lobed, some of them usually with elongate, linear or lanceolate lobes (3).
3. Herbage and calyx sericeous, rarely glabrate; peduncles seldom more than twice as long as the subtending leaf; bracts 2 to 3 mm. long, subulate; corolla not more than 2 cm. long, usually pink, drying purplish; anthers 2 mm. long................3. *C. incanus*
3. Herbage and calyx glabrous, often slightly glaucous; peduncles usually 3 or more times as long as the subtending leaf; bracts up to 25 mm. long, linear-lanceolate; corolla 3 to 4 cm. long, pale pink or white; anthers 3 to 4 mm. long..............4. *C. linearilobus*

1. Convolvulus sepium L. Lakeside, Navajo County (*Harrison* 5506), near Williams, Coconino County (*Ripley & Barneby* 7042), 6,000 to 7,000 feet, June and July. Throughout most of temperate North America; Eurasia.

Hedge bind-weed. The plant is sometimes a troublesome weed. The Arizona plants, with relatively short, prostrate or barely twining stems and relatively small corollas (3 to 4 cm. long), probably are referable to *C. interior* House, which seems to be scarcely more than a variety of *C. sepium*.

2. Convolvulus arvensis L. Apache County to Coconino County, south to Cochise, Santa Cruz, and Pima counties, roadsides and fields, May to July. Extensively naturalized in North America; from Europe.

Field bind-weed. A troublesome weed, difficult and expensive to eradicate, considered in California the worst weed in the state. An antihemorrhagic substance has been discovered in this plant.

3. Convolvulus incanus Vahl. Apache County to eastern Mohave County, south to Cochise and Pima counties, 3,000 to 6,000 feet, common on dry slopes and mesas, May to October. Nebraska and Colorado to Texas and Arizona.

Stems trailing or clambering over bushes.

4. Convolvulus linearilobus Eastw. Mohave, Yavapai, and Gila counties, chiefly in the Mazatzal and Hualpai mountains, 3,500 to 5,000 feet, slopes and banks, often among oaks, May to October, type from the Mazatzal Mountains (*Eastwood* 17264). Known only from central Arizona.

This species is related to *C. longipes* Wats. of Nevada and southern California. The numerous long stems often form tangled masses. The narrowly lobed leaves and the large, pale pink or white flowers are distinctive.

7. IPOMOEA (231). MORNING-GLORY

Plants herbaceous, annual or perennial; stems erect, trailing, or twining; leaves entire to pedately parted; flowers solitary or in few-flowered clusters; outer sepals commonly larger than the inner ones; corolla mostly funnelform, sometimes salverform, the limb entire or very shallowly lobed; capsule globose, 2- to 4-valved; seeds commonly 4.

Several species of this large genus are favorite ornamentals, cultivated under the names morning-glory and cypress-vine.

KEY TO THE SPECIES

1. Corolla bright red. Stems twining; herbage glabrous or nearly so; corolla salverform or the elongate tube narrowly funnelform, the limb not more and usually less than 15 mm. wide .... 1. *I. coccinea*
1. Corolla pink, purple, blue, or white (2).
2. Leaves entire, obtuse or short-cuneate at base, 6 or more times as long as wide. Plant perennial, glabrous; stems stout, semiprostrate, not twining; leaves linear to oblong-lanceolate, often 10 cm. or longer; sepals broad, very obtuse, scarious-margined; corolla white with a pink throat, broadly funnelform, 6 to 10 cm. long. . 2. *I. longifolia*
2. Leaves variously toothed or lobed or, if entire, then cordate at base, much less than 6 times as long as wide (3).
3. Plants perennial with a tuberous-thickened root. Herbage glabrous or nearly so; stems not twining or very weakly so (4).
3. Plants annual or, if perennial, then the root not tuberous-thickened except in *I. heterophylla* (8).
4. Sepals setaceous-caudate; corolla 5 to 8 cm. long, the tube elongate, narrow, rather abruptly expanded into the throat; leaves usually sparsely strigose. Tuber elongate; leaves deeply sagittate to pedately lobed, the lobes divergent, lanceolate, linear, or oblong; calyx not warty or very obscurely so at base.......... 3. *I. Thurberi*
4. Sepals not setaceous-caudate; corolla not more than 5 cm. long, the tube gradually expanded into the throat; leaves glabrous (5).
5. Leaves laciniate-dentate at the broad apex, otherwise entire, obovate-cuneate. Tuber globose or nearly so; petioles less than 5 mm. long; calyx conspicuously warty 4. *I. egregia*
5. Leaves pedately parted or divided, the segments elongate, narrowly linear or filiform, not more than 2 mm. wide (6).
6. Sepals not or not conspicuously warty; petioles 10 to 20 mm. long; corolla 4 to 5 cm. long. Tuber elongate.......... 5. *I. Lemmoni*
6. Sepals conspicuously warty; petioles seldom more than 5 mm. long; corolla not more than 3 cm. long (7).
7. Tuber elongate; sepals 5 to 6 mm. long; peduncle and pedicel together usually not much longer than the calyx.......... 6. *I. muricata*
7. Tuber usually globose or nearly so; sepals 7 to 9 mm. long; peduncle and pedicel considerably longer than the calyx, often twice as long.......... 7. *I. Plummerae*
8. Leaves pedately 5- to 9-parted or divided, the lobes linear or almost filiform, the midlobe rarely more than 4 mm. wide. Sepals conspicuously scarious-margined (9).
8. Leaves entire or angulate-lobed or, if pedately parted, then the midlobe not less than 6 mm. wide (except sometimes in the upper leaves of *I. barbatisepala*). Stems twining (11).
9. Corolla 5 to 8 cm. long; herbage entirely glabrous; stems twining. Calyx and pedicel often verrucose; corolla trumpet-shaped, with a long, narrow tube..... 8. *I. tenuiloba*
9. Corolla not more than 4 cm. long; herbage sparsely and inconspicuously hirsute, or glabrate; stems erect to procumbent, sometimes feebly twining at apex (10).
10. Calyx and pedicel glabrous or puberulent; sepals 3 to 7 mm. long; corolla with tube and throat not more than 10 mm. long, the limb less than 10 mm. wide. Sepals often crested or warty on the midrib.......... 9. *I. costellata*

10. Calyx and pedicel hirsute; sepals 8 to 12 mm. long; corolla with tube and throat 20 to 25 mm. long, the limb 25 to 40 mm. wide..........................10. *I. leptotoma*
11. Outer sepals conspicuously dilated toward base, attenuate-acuminate toward apex, not scarious-margined, the inner ones much narrower, lanceolate, scarious-margined toward base; herbage and calyx sericeous with long, very fine, more or less appressed hairs, these not pustulate at base; capsule 5-celled. Leaves (rarely entire?) pedately 5- to 7-parted, the mid-lobe strongly constricted at base; corolla 6 to 9 cm. long; root tuberous-thickened ....11. *I. heterophylla*
11. Outer sepals not conspicuously dilated toward base, much like the inner ones; herbage and calyx not sericeous; capsule 2- or 3-celled (12).
12. Calyx hispid, the hairs stout, conspicuously subulate, mostly 2 to 3 mm. long. Herbage glabrous; petioles usually sparsely beset with elevated, pricklelike warts; leaves pedately 3- to 5-parted, the basal lobes usually deeply cleft; sepals narrowly lanceolate; corolla about 2 cm. long...............................12. *I. barbatisepala*
12. Calyx villous, hirsute, or glabrous, the hairs, if any, very slender, not conspicuously subulate (13).
13. Leaves all entire, cordate (14).
13. Leaves (some or all of them) angulate-lobed or more deeply cleft (15).
14. Herbage and calyx glabrous; peduncles mostly 1-flowered; calyx and pedicel rugose-verrucose; sepals ovate, 4 to 5 mm. long at anthesis, scarious-margined to the apex; corolla 2 to 3 cm. long...13. *I. cardiophylla*
14. Herbage and calyx more or less hirsute; peduncles mostly 2- to several-flowered; calyx and pedicel not rugose-verrucose but the bases of the hairs often enlarged; sepals lanceolate or somewhat spatulate, 8 to 25 mm. long at anthesis, not scarious-margined to the apex; corolla 4 to 6 cm. long.............................14. *I. purpurea*
15. Corolla not more than 2 cm. long, narrowly funnelform; calyx at anthesis less than 10 mm. long, the sepals ovate or lance-ovate, scarious-margined, usually conspicuously so, sparsely villous, long-ciliate, or glabrous; herbage glabrous or sparsely pubescent; leaves deeply 3-lobed, or some of them entire and deeply cordate........15. *I. triloba*
15. Corolla 2.5 to 4 cm. long, broadly funnelform; calyx at anthesis seldom less than 10 mm. long, the sepals lanceolate, not or not conspicuously scarious-margined, copiously villous-hirsute; herbage more or less villous-hirsute; leaves angulate-lobed to pedately 5-parted ....16. *I. hirsutula*

1. Ipomoea coccinea L. (*Quamoclit coccinea* Moench). Navajo and Coconino counties to Greenlee, Cochise, Santa Cruz, and Pima counties, 2,500 to 6,000 feet, hillsides and canyons, May to October. Western Texas to Arizona and south into tropical America.

Star-glory. Easily distinguished from all the other Arizona species by the scarlet, narrowly trumpet-shaped corolla. Typical *I. coccinea,* with cordate or subsagittate, otherwise nearly entire leaves, is less common in Arizona than var. *hederifolia* (L.) Gray, with some or all of the leaves deeply 3-lobed to pedately 5-parted.

2. Ipomoea longifolia Benth. Cochise and Santa Cruz counties, 4,000 to 6,000 feet, plains and mesas, usually with grasses, July and August. Oklahoma to southern Arizona and Mexico.

The trailing stems up to 3 m. long, elongate, narrow, entire leaves, and large, white, pink-throated corolla make this species easily distinguishable.

3. Ipomoea Thurberi Gray. Cochise, Santa Cruz, and Pima counties, 4,000 to 5,000 feet, plains and mesas, August and September, type from southern Arizona near Santa Cruz, Sonora (*Thurber* 966). Southern Arizona and Sonora.

Stems trailing, the purple flowers opening in the evening.

4. Ipomoea egregia House (*I. cuneifolia* Gray, non Meisn.). Known only from the type collection in the Huachuca Mountains, Cochise County (*Lemmon* 2837), September.

5. Ipomoea Lemmoni Gray. Known only from the type collection in the Huachuca Mountains (*Lemmon* 2840), August.

6. Ipomoea muricata Cav. (*I. patens* (Gray) House). Cochise and Santa Cruz counties, 5,000 to 6,000 feet, July to September. New Mexico and southern Arizona to northern South America.

7. Ipomoea Plummerae Gray. Apache County to Coconino County, south to the Pinaleno Mountains (Graham County), Chiricahua Mountains (Cochise County), and Santa Catalina Mountains (Pima County), 5,000 to 9,000 feet, mostly in coniferous forests, August and September, type from Arizona (*Lemmon* 2839). Arizona and northern Mexico.

Corolla pink, the tuber reported to be edible.

8. Ipomoea tenuiloba Torr. Chiricahua Mountains, Cochise County, 6,000 feet (*Blumer* 2138, *Clark* 8680), August and September. Western Texas to southeastern Arizona and northern Mexico.

9. Ipomoea costellata Torr. (? *I. futilis* A. Nels.). Navajo, Yavapai, and Greenlee counties to Cochise, Santa Cruz, and Pima counties, 3,500 to 6,000 feet, common on dry, grassy plains and mesas, sometimes in pine forest, July to October, type of *I. futilis* (not examined) from near Prescott (*Hanson* 1016). Western Texas to Arizona and Mexico.

10. Ipomoea leptotoma Torr. Cochise, Santa Cruz, and Pima counties, 3,000 to 4,500 feet, common on dry, grassy plains and mesas, June to October, type from Sonora near the Arizona border. New Mexico, southern Arizona, and Mexico.

An attractive plant with rather large pink (seldom white) corollas, apparently hybridizing occasionally with *I. costellata*. Specimens with noticeably hirsute stems belong to var. *Wootoni* Kelso (*I. leptotoma* f. *Wootonii* Wiggins) of which the type was collected in the Santa Rita Mountains (*Wooton* in 1914).

11. Ipomoea heterophylla Ortega. Near Bisbee, Tombstone, Hereford, and Gleeson (Cochise County), western slope of the Baboquivari Mountains (Pima County), 3,500 to 4,500 feet, mesas and plains, August and September. Western Texas, southern Arizona, and northern Mexico.

The corolla is purple. Specimens from Cochise County resemble Mexican specimens, but those from the Baboquivari Mountains probably represent a distinct variety, having the leaf lobes narrower and more attenuate at both ends, the calyx with shorter and less dense pubescence, and the sepals with broader bases and more abrupt tips. A collection from near Fort Huachuca (*Lemmon* 2835), with entire or merely angulate leaves, is the type of *I. Lindheimeri* Gray var. *subintegra* House (231, p. 196). It is probably an unusual form of *I. heterophylla*, the outer sepals being much broader than the inner ones.

12. Ipomoea barbatisepala Gray. Greenlee, Cochise, Santa Cruz, and Pima counties, 3,000 to 5,000 feet, canyons, climbing on shrubs, August and September. Western Texas, southern Arizona, and Mexico.

The corolla is normally purplish pink, sometimes white.

*13. **Ipomoea cardiophylla** Gray. House (231, p. 258) gives the range as "western Texas to Arizona and Mexico," but cites no Arizona collections, and none has been seen by us.

14. **Ipomoea purpurea** (L.) Roth. Occasional in cultivated fields in central and southern Arizona, July to September. Widely distributed in the United States; naturalized from tropical America.

One of the cultivated morning-glories, in places a troublesome field weed.

15. **Ipomoea triloba** L. Valley of the Santa Cruz River near Tucson, Pima County (*Pringle* in 1884, *Thornber* in 1912). Southern Florida, southern Arizona, Mexico, and southward.

Pringle's specimen differs from the common phase of the species in having glabrous leaves, sepals, and capsules and nearly glabrous stems.

16. **Ipomoea hirsutula** Jacq. f. (*I. desertorum* House). Southern Navajo, Yavapai, and Greenlee counties to Cochise, Santa Cruz, and Pima counties, 1,000 to 5,500 feet, fields and roadsides, exceptionally in pine forests, July to November. Western Texas to central and southern Arizona, south to Central America.

Sometimes a weed in cultivated land, especially in cotton fields in Graham County. A common variant, with elongate sepal tips, is *I. desertorum* House. A similar variation is found occasionally also in *I. purpurea*. The type of *I. desertorum* was collected at Tucson (*Thornber* 29).

114. POLEMONIACEAE (232). Phlox Family

Plants annual or perennial, herbaceous or suffrutescent; leaves simple or compound; flowers perfect, mostly regular, 5-merous; stamens separately attached to the sympetalous corolla, inserted equally or unequally; ovary superior, mostly 3-celled; style usually 3-cleft; fruit a longitudinally dehiscent, usually 3-celled capsule.

An almost wholly American family comprising many plants with beautiful flowers. Species of the genera *Phlox, Gilia,* and *Polemonium* and the climbing *Cobaea scandens* are garden favorites.

KEY TO THE GENERA

1. Leaves compound, pinnate, the leaflets lanceolate or narrowly elliptic to nearly orbicular; plants perennial, herbaceous; calyx entirely herbaceous and green, accrescent but not becoming ruptured in fruit. Leaves alternate, the leaflets numerous, the apical ones often confluent; flowers in cymose clusters; corolla regular, blue, violet, or yellow, exceptionally white .... 12. *Polemonium*

1. Leaves simple, entire to deeply pinnately or palmately dissected or, if appearing compound, then the divisions narrower, or the plants annual, or the calyx tube more or less scarious between the ribs and becoming ruptured in fruit (2).

2. Corolla distinctly irregular (bilabiate), pale violet; plants suffrutescent perennials, with exfoliating bark; leaves lanceolate to ovate, serrate. Calyx scarious, except the ribs; stamens long-exserted, the filaments strongly declined and incurved. 11. *Loeselia*

2. Corolla regular or if distinctly irregular, then the plants annual, or the leaves or their divisions narrowly linear (3).

3. Calyx tube not becoming ruptured at maturity, not carinate between the ribs. Plants annual; leaves alternate; flowers in leafy-bracted glomerules; corolla funnelform or funnelform-salverform (4).

3. Calyx tube splitting between the lobes at maturity, usually carinate between the ribs (5).
4. Leaves and bracts entire, not pungent; calyx lobes equal or nearly so, herbaceous, not pungent; plants relatively tall, unbranched or with ascending branches...3. *Collomia*
4. Leaves and bracts mostly pinnatifid, rigid and pungent; calyx lobes unequal, rigid and pungent; plants low, often diffusely branched from the base..........4. *Navarretia*
5. Leaves entire and mostly opposite; flowers diurnal; calyx cylindric-campanulate, the tube usually not, or but slightly, longer than the lobes; corolla salverform or (in one species of *Phlox*) funnelform-salverform, pink or white (6).
5. Leaves not entire and opposite or, if so, then the flowers mostly vespertine, the calyx cylindric with tube usually considerably longer than the lobes, and the corolla funnelform-salverform, or else the corolla open-campanulate and yellow (7).
6. Plants (the Arizona species) perennial; flowers relatively large and showy; calyx little-changed after anthesis; corolla tube not flaring at base; seeds unchanged when wet 1. *Phlox*
6. Plants annual; flowers small; calyx more or less accrescent; corolla tube flaring at base; seeds mucilaginous when wet.......................................2. *Microsteris*
7. Teeth or lobes of the leaves tipped with long, slender, white bristles, the basal teeth or lobes reduced to similar bristles. Plants small, annual, with several spreading stems or branches; leaves pinnately toothed or pinnatifid; calyx teeth subequal, bristle-tipped; corolla regular to distinctly bilabiate, pale blue, lavender, or whitish...10. *Langloisia*
7. Teeth or lobes of the leaves, if any, not tipped with long bristles or if so, then none of them reduced to bristles (8).
8. Inflorescences conspicuously cobwebby with long, very fine, white hairs, closely subtended by leafy bracts, dense, subcapitate; calyx lobes usually unequal, bristle-tipped; anthers usually cordate or sagittate. Plants annual; stems low, erect or diffusely branched; leaves entire or pinnately parted near the base with 2 to 4 narrow lateral lobes; corolla regular to distinctly bilabiate, bright blue to whitish, the tube usually much longer than the calyx...5. *Eriastrum*
8. Inflorescences not conspicuously cobwebby or if sometimes so (in *Gilia achilleaefolia*), then not bracteate, the calyx lobes equal or nearly so and the anthers not cordate or sagittate (9).
9. Plants suffrutescent; corolla regular, white or tinged with pink, funnelform-salverform. Stems very leafy; leaves palmately (exceptionally subpinnately) parted or divided, the divisions very narrow (10).
9. Plants herbaceous or if suffrutescent, then the corolla normally blue or violet and either (in *Gilia rigidula*) regular and rotate-campanulate with a very short tube or (in *G. multiflora*) more or less bilabiate, with a slender, elongate tube (11).
10. Calyx angles strongly carinate, the scarious junction membranes conspicuous; stems woody well above the caudex; leaves mostly alternate and fascicled, rigid, appressed or narrowly ascending, the divisions subulate, very pungent; corolla tube considerably surpassing the calyx...6. *Leptodactylon*
10. Calyx angles not strongly carinate, the junction membranes inconspicuous or obsolete; stems not or scarcely woody above the caudex; leaves opposite (appearing whorled), flaccid or somewhat stiff, spreading, the divisions narrowly linear to almost filiform, scarcely pungent; corolla tube not or but slightly surpassing the calyx 7. *Linanthastrum*
11. Flowers scattered, axillary or borne in the forks of the stem, solitary or in small clusters, sessile or pedicelled; leaves opposite (except often in *L. demissus*); plants annual, small, spring-flowering ....8. *Linanthus*
11. Flowers (except in a few species) aggregated in definite inflorescences, these open-paniculate, thyrsoid-glomerate, or more or less capitate; leaves alternate or basal, or (in *G. filiformis*) the lowest leaves opposite; plants annual, biennial, or perennial, flowering in spring or summer...9. *Gilia*

1. PHLOX

Contributed by Edgar T. Wherry

Plants perennial (the Arizona species), herbaceous or somewhat woody; leaves opposite, entire; inflorescence cymose, often reduced to a solitary flower; corolla salverform or rarely funnelform; stamens 5, irregular; ovules 1 to 3 in each carpel.

Flowering occurs twice a year, in spring and occasionally again after summer rains. The species are in many cases not well defined. Some have been renamed more than once, and names are often applied to specimens not closely related to the type material on which those names were originally founded.

The phloxes are popular garden plants, especially where a mass of blooms is desired. Several of the native species of Arizona have been brought under cultivation. The flowers of *Phlox* are relished by sheep.

KEY TO THE SPECIES

1. Shoots tending to be short and branched, and the inflorescence to be simple: depressed phloxes (2)
1. Shoots tending to be elongate and little-branched, and the inflorescence to be compound: elate phloxes (7).
2. Pubescence more or less glandular (3).
2. Pubescence wholly eglandular (4).
3. Glandular pubescence extending throughout the herbage and even onto the corolla tube 1. *P. gladiformis*
3. Glandular pubescence limited to the inflorescence herbage; corolla tube glabrous 2. *P. caespitosa*
4. Leaves thickish, gray-green, more or less acerose (5).
4. Leaves thinnish, soft, scarcely acerose (6).
5. Hairs fine; calyx intercostally carinate........................3. *P. austromontana*
5. Hairs coarse; calyx intercostally flat...............................4. *P. griseola*
6. Herbage bright (rarely gray-) green, glabrate to moderately pubescent; leaves linear-subulate, the larger ones often more than 1 mm. wide; calyx intercostally somewhat carinate ....5. *P. diffusa*
6. Herbage gray-green, canescent to arachnoid-tomentose, rarely glabrate; leaves subulate, rarely more than 1 mm. wide; calyx intercostally flat.................6. *P. Hoodii*
7. Corolla funnelform. Leaves small and sparse.......................7. *P. tenuifolia*
7. Corolla salverform (8).
8. Underground parts chiefly long slender rootstocks, terminating in clusters of evergreen leaves, from which arise the flowering shoots of the next season. Leaves mostly narrowly elliptic and obtusish; calyx intercostally flat.................8. *P. Cluteana*
8. Underground parts chiefly taproots; leaves deciduous, or a few cauline ones evergreen (9).
9. Styles short, not equaling the sepals (10).
9. Styles elongate, exceeding the sepals (11).
10. Woody tissue well developed; corolla lobes notched.................9. *P. Woodhousei*
10. Woody tissue little developed; corolla lobes entire.......................10. *P. nana*
11. Corolla tube 20 to 25 mm. long.................................11. *P. Stansburyi*
11. Corolla tube 12 to 18 mm. long (12).
12. Calyx intercostally flat to slightly carinate. Leaves thick, short-acuminate to acutish 12. *P. amabilis*
12. Calyx intercostally carinate, moderately to strongly so (13).
13. Leaves tending to be short-acuminate and thick.........................13. *P. Grayi*
13. Leaves tending to be long-acuminate and thin......................14. *P. longifolia*

***1. Phlox gladiformis** (Jones) E. Nels. Although not yet actually collected in Arizona, this *Phlox* occurs so near the state borders in both Nevada and Utah that it seems likely to be found there. It is a cespitose plant characterized by having the herbage and even the corolla tube densely covered with gland-tipped hairs that emit a musky odor.

***2. Phlox caespitosa** Nutt. This *Phlox* grows in Nevada, Utah, and Colorado not far from the Arizona line, and its occurrence in Arizona is to be expected. The plants are pulvinate, with linear-subulate, coarsely ciliate leaves.

3. Phlox austromontana Coville. Northern and central Arizona, also in Graham County, and in the Harquahala Mountains (Maricopa or Yuma County), up to 8,000 feet, rocky slopes. Idaho and Oregon to northwestern New Mexico, Arizona, California, and Mexico.

This species forms grayish-green, prickly cushions and mats, more or less pubescent with fine eglandular hairs, the flowers pink or white.

KEY TO THE SUBSPECIES

1. Plants spreading, with long-decumbent stems; corolla tube 12 to 18 mm. long; styles 4.5 to 6 mm. long. Longest leaves 15 to 30 mm. long.................... subsp. *prostrata*
1. Plants compact; corolla tube 8 to 14 mm. long; styles 2.5 to 6 mm. long (2).
2. Stems short-decumbent or erect; longest leaves 12 to 20 mm. long........... subsp. *vera*
2. Stems short; longest leaves 8 to 12 mm. long........................... subsp. *densa*

Subsp. *prostrata* (E. Nels.) Wherry (*P. acerba* A. Nels.) occurs in Coconino, Mohave, and southern Graham counties, type of *P. acerba* from Oak Creek Canyon, Coconino County (*A. & R. Nelson* 2119). Subsp. *vera* Wherry occurs in Coconino, Yavapai, and Mohave counties. Subsp. *densa* (Brand) Wherry is found in Apache, Navajo, Coconino, Yavapai, and Mohave counties.

4. Phlox griseola Wherry. Northwestern corner of Mohave County, 4,000 feet, dry, sandy barrens, early spring. Eastern Nevada, western Utah, and northwestern Arizona.

This species was treated previously as an aberrant phase of *P. Covillei* E. Nels.

5. Phlox diffusa Benth. Northern Apache and Coconino counties, 6,000 to 9,000 feet, plateaus and canyon rims, on rock ledges and gravelly slopes, late spring. Idaho and Washington to northern Arizona and southern California. Reaches the North Rim of Grand Canyon but apparently does not cross the Canyon.

The Arizona plants belong to subsp. *subcarinata* Wherry, which forms mats densely covered with soft, bright-green or exceptionally grayish, linear-subulate leaves, producing an abundance of white to pink flowers.

6. Phlox Hoodii Richards. Northwestern corner of Mohave County, 4,000 feet, dry, sandy barrens, early spring. Western North America.

This widespread species reaches its southern limit in Arizona. The material from north of Beaver Dam seems to represent a copiously pubescent form of subsp. *canescens* (Torr. & Gray) Wherry.

7. Phlox tenuifolia E. Nels. (*P. gilioides* A. Nels.). Graham, Gila, Maricopa, Pinal, Cochise, and Pima counties, 1,500 to 5,000 feet, rocky slopes, flowering in spring and rarely again in autumn, type from the Santa Catalina Mountains (*Pringle* in 1881), type of *P. gilioides* also from the Santa Catalina Mountains (*Hanson* 1087). Known only from Arizona.

When growing in the open, the plants form tufts of slender, woody-based stems up to 75 cm. long, bearing sparse, small leaves. In partial shade the stems are elongate, supporting themselves on other shrubs and attaining a length of 1.2 m. The white flowers are unique in the genus in having the corolla tube funnelform. In some colonies they are sweet-scented, in others unpleasantly musky.

8. Phlox Cluteana A. Nels. Lukachukai Mountains (Apache County), Keet Seel (Navajo County), Navajo Mountain (Coconino County), 6,000 to 10,000 feet, pine forests, late spring and early summer. Southern Utah and northeastern Arizona.

The slender rootstocks creep through the humus of the forest floor, producing numerous clusters of evergreen leaves. The flowers are large and brilliantly phlox purple.

9. Phlox Woodhousei (Gray) E. Nels. (*P. Woodhousei oculata* A. Nels.). White Mountains (southern Apache and Navajo counties) and Grand Canyon (Coconino County), south to Greenlee or Graham County, 3,500 to 8,000 feet, open woods on rocky slopes, flowering in spring and autumn, type from west of the present Fairview, Coconino County (*Woodhouse* in 1851), type of var. *oculata* from Cooley Lake, Apache County (*A. Nelson* 10357). Almost endemic in Arizona, extending but a short distance into New Mexico.

This dwarf shrub is easily recognized by its thick, oblong, acutish or obtusish leaves and bright-pink flowers with deeply notched corolla lobes and short styles.

10. Phlox nana Nutt. Chiricahua Mountains and vicinity (Cochise County), 5,000 to 6,000 feet, in chaparral, May and June. Western Texas, New Mexico, southeastern Arizona, and Chihuahua.

The species is represented in Arizona by subsp. *glabella* (Gray) Brand (*P. triovulata* Thurb.), distinguished by eglandular pubescence. A collection in Rucker Valley (*Lemmon* 415) was named by Brand *P. Nelsonii,* but his diagnostic character—lobes of the calyx 1½ times as long as the tube—is too inconstant and trivial for nomenclatorial recognition. *P. nana* is a herbaceous perennial, varying in habit from one season to another, with purple to white corollas larger than in any other *Phlox* of Arizona.

11. Phlox Stansburyi (Torr.) Heller. Northern Navajo County to northern Mohave County, and in Cochise County, 4,000 to 6,500 feet, dry soil, often with sagebrush, spring. Utah, Nevada, New Mexico, northern Arizona, and eastern California.

This species name has been applied to all sorts of dissimilar phloxes, but the type material is characterized by having the corolla tube 20 to 25 mm. long, and there seems to be no reason for expanding the definition of the species to cover the shorter-tubed *P. Cluteana, P. longifolia,* and *P. amabilis.*

12. Phlox amabilis Brand. Endemic in southern Coconino, Yavapai, and Graham counties, type collected by E. Palmer on a trip from Prescott to Cottonwood Valley. The number 391 was assigned to several different entities he obtained.

APETALAE 203
CHORIPETALAE 281
SYMPETALAE 626

ACANTHACEAE 798
ACERACEAE 526
AGAVACEAE 191, 174
AIZOACEAE 280
ALISMACEAE 68
AMARANTHACEAE 249, 264
AMARYLLIDACEAE 191
ANACARDIACEAE 521
APIACEAE 606
APOCYNACEAE 651
ARECACEAE 164
ARACEAE 165
ARALIACEAE 605
ARISTOLOCHIACEAE 227
ASCLEPIADACEAE 655
ASTERACEAE 829

BERBERIDACEAE 320
BETULACEAE 214
BIGNONIACEAE 794
BORAGINACEAE 707
BRASICACEAE 325
BROMELIACEAE 166
BURSERACEAE 496
BUXACEAE 521

CACTACEAE 567
CALLITRICHACEAE 521
CAMPANULACEAE 825
CAPPARIDACEAE 355
CAPRIFOLIACEAE 812
CARYOPHYLLACEAE 291
CELASTRACEAE 524
CELTIDACEAE 220
CERATOPHYLLACEAE 303
CHENOPODIACEAE 249
COCHLOSPERMACEAE 557
COMMELINACEAE 166
COMPOSITAE 829
CONVOLVULACEAE 666
CORNACEAE 624
CRASSULACEAE 358
CROSSOSOMATACEAE 371
CRUCIFERAE 325
CUCURBITACEAE 820
CUPRESSACEAE 57
CYPERACEAE 145

ELAEAGNACEAE 586
EPHEDRACEAE 60
EQUISETACEAE 30

ERICACEAE 629
EUPHORBIACEAE 501

FABACEAE 395
FAGACEAE 215
FOUQUIERIACEAE 639

GARRYACEAE 624
GENTIANACEAE 645
GERANIACEAE 484
GRAMINEAE 70
GUTTIFERAE 556

HYDROCHARITACEAE 69
HYDROPHYLLACEAE 696

IRIDACEAE 195

ISOETACEAE 29

JUGLANDACEAE 213
JUNCACEAE 168
JUNCAGINACEAE 67

KOEBERLINIACEAE 558

LABIATAE 731
LAMIACEAE 731
LEGUMINOSAE 395
LEMNACEAE 165
LENNOACEAE 629
LENTIBULARIACEAE 798
LINACEAE 488
LOASACEAE 562
LOGANIACEAE 644
LORANTHACEAE 223
LYTHRACEAE 587

MALPIGHLIACEAE 497
MALVACEAE 536
MARSILEACEAE 33
MARTYNIACEAE 795
MENISPERMACEAE 321
MORACEAE 221

NAJADACEAE 67
NYCTAGINACEAE 270

OLEACEAE 640
ONAGRACEAE 589
OPHIOGLOSSACEAE 31
ORCHIDACEAE 197
OROBANCHACEAE 796
OXALIDACEAE 487

PALMAE 164

PAPAVERACEAE 322
PASSIFLORACEAE 562
PHYTOLACCACEAE 279
PINACEAE 50
PLANTAGINACEAE 802
PLATANACEAE 371
PLUMBAGINACEAE 639
POACEAE 70
POLEMONIACEAE 678
POLYGALACEAE 497
POLYGONACEAE 228
POLYPODIACEAE 34
PONTEDERIACEAE 168
PORTULACACEAE 285
POTAMOGETONACEAE 64
PRIMULACEAE 635

RAFFLESIACEAE 227
RANUNCULACEAE 304, 314
RESEDACEAE 358
RHAMNACEAE 529
ROSACEAE 372
RUBIACEAE 805
RUTACEAE 493

SALICACEAE 207
SALVINIACEAE 32
SANTALACEAE 226
SAPINDACEAE 528
SAPOTACEAE 640
SAURURACEAE 207
SAXIFRAGACEAE 361
SCROPHULARIACEAE 761
SELAGINELLACEAE 28
SIMAROUBACEAE 495
SOLANACEAE 748
SPARGANIACEAE 64
STERCULIACEAE 554

TAMARICACEAE 557
TILIACEAE 536
TYPHACEAE 63

ULMACEAE 220
UMBELLIFERAE 606
URTICACEAE 222

VALERIANACEAE 818
VERBENACEAE 724
VIOLACEAE 559
VITACEAE 534

ZYGOPHYLLACEAE 489

Phlox amabilis is a low plant with thick, oblong leaves, and when, as is often the case, its corolla lobes are deeply notched, it bears a striking resemblance to *P. Woodhousei.* In the former, however, the stamens and styles are nearly as long as the corolla tube, whereas in the latter they are much shorter than the tube.

13. Phlox Grayi Woot. & Standl. (*P. visenda* A. Nels). White Mountains (Apache County) and northern Navajo and Coconino counties, south to Santa Cruz County, 3,000 to 9,000 feet, dry, gravelly slopes, spring and occasionally autumn, type of *P. visenda* from Grandview, Coconino County (*A. Nelson* 10219). Southern Utah and Nevada to western New Mexico, Arizona, and eastern California.

This species was combined previously with *P. amabilis,* but further study has shown the desirability of keeping them distinct. Under the present interpretation *P. amabilis* is restricted to entities with more or less oblong leaves, flat calyx membranes, and notched corolla lobes; whereas the name *P. Grayi* is applied to the more widespread entities that have more or less lanceolate to ovate leaves, carinate calyx membranes, and subentire corolla lobes. *P. Stansburyi* subvar. *microcalyx* Brand, based on another entity included under Palmer's No. 391, seems to be a minor variant of this.

14. Phlox longifolia Nutt. Apache, Coconino, Mohave, Yavapai, and Pima counties, 4,000 to 7,500 feet, dry, sandy or rocky slopes, spring. Wyoming to British Columbia, south to New Mexico, Arizona, and California.

KEY TO THE SUBSPECIES

1. Herbage eglandular. Leaves short.................................subsp. *cortezana*
1. Herbage glandular in the inflorescence (2).
2. Leaves (the larger ones) 50 to 70 mm. long, rather narrow and thin......subsp. *longipes*
2. Leaves (none of them) more than 45 mm. long.........................subsp. *compacta*

Subsp. *cortezana* (A. Nels.) Wherry, referred previously to subsp. *humilis* (Dougl.) Wherry but much larger-leaved, has been collected at Black Rock Spring, northern Mohave County (*Jones* 5098j). Subsp. *longipes* (Jones) Wherry is found in the Grand Canyon region (Coconino County) and in Mohave County. Subsp. *compacta* (Brand) Wherry, which intergrades with *P. Grayi,* occurs throughout the range of the species in Arizona.

2. MICROSTERIS

Small, annual herbs; leaves mostly opposite, entire; flowers small, axillary or in small terminal clusters; corolla salverform, pink or white, the limb 2 to 3 mm. wide, the slender tube not or but little surpassing the calyx; stamens inserted at different levels; ovules usually solitary in each cell of the ovary; seeds mucilaginous in water.

1. Microsteris gracilis (Hook.) Greene (*M. Macdougalii* Heller, *Gilia gracilis* Hook., *Phlox gracilis* Greene). Apache County to Mohave County, south to Santa Cruz and Pima counties, 3,000 to 7,000 feet, moist soil around springs and along streams, February to May, type of *M. Macdougalii* from Flagstaff (*MacDougal* 42). Montana to Alaska, south to New Mexico, Arizona, and Baja California; South America.

Authorities differ as to the status of *Microsteris*. Mason (233) believed that it belongs to *Phlox*, whereas Wherry (234) preferred to maintain it as a genus.

3. COLLOMIA (235)

Plants annual; leaves sessile, entire, linear to lanceolate, or the floral ones sometimes ovate; flowers in dense, terminal, leafy-bracted heads; corolla regular, funnelform-salverform, the tube long and slender; stamens unequally inserted; seeds mucilaginous in water.

KEY TO THE SPECIES

1. Corolla purplish pink or whitish, 10 to 18 mm. long, the tube little surpassing the calyx, the throat narrow and the limb small; calyx lobes subulate or narrowly lanceolate, acute or acuminate .... 1. *C. linearis*
1. Corolla cream, salmon pink, or apricot, 15 to 30 mm. long, the tube much surpassing the calyx, the throat and limb ample; calyx lobes ovate-lanceolate, obtusish or acutish .... 2. *C. grandiflora*

1. Collomia linearis Nutt. White Mountains (Apache County), Kaibab Plateau to Oak Creek (Coconino County), 6,500 to 9,500 feet, July and August. New Brunswick to British Columbia, south to New Mexico, Arizona, and California.

2. Collomia grandiflora Dougl. North Rim of Grand Canyon (Coconino County), mountains of Gila County, 3,500 to 8,000 feet, coniferous forests and open slopes, May to July. Montana to British Columbia, south to western Colorado, Arizona, and California.

4. NAVARRETIA

Plants small, annual, the stems often diffusely branched from the base; leaves alternate, these and the bracts mostly pinnatifid, with very slender, rigid, and pungent segments; flowers small, in leafy-bracted glomerules; calyx lobes unequal, rigid and pungent; corolla narrowly funnelform, white or yellow, the tube shorter than the calyx; stamens exserted, the anthers oval.

Valuable information on this genus has been supplied by Beecher Crampton.

KEY TO THE SPECIES

1. Stems usually erect, often diffusely branched; herbage glandular-puberulent; corolla usually yellow, 5 to 8 mm. long; stigmas 3-cleft; capsule with relatively thick walls, dehiscing by valves from the base upward....... 1. *N. Breweri*
1. Stems depressed, often densely tufted; herbage sparsely villous, not glandular; corolla white, 3 to 6 mm. long; stigmas entire to 2-cleft; capsule thin-walled, disintegrating on dehiscence (2).
2. Sinus membranes of the calyx truncate at summit; corolla lobes 1-veined throughout, the vein unbranched; stamens inserted at middle of the throat; stigma entire or minutely 2-lobed .... 2. *N. minima*
2. Sinus membranes V-shaped at summit; corolla lobes 1-veined at base, but the vein branched above into 2 or 3 veinlets; stamens inserted at base of the throat; stigma deeply 2-cleft .... 3. *N. propinqua*

1. Navarretia Breweri (Gray) Greene (*Gilia Breweri* Gray). Navajo Mountain, Coconino County (*Peebles & Smith* 13957), about 8,000 feet, June. Wyoming to Idaho and Oregon, south to Colorado, northern Arizona, Nevada, and California.

2. Navarretia minima Nutt. (*Gilia minima* Gray). Near Williams, Coconino County, about 7,000 feet (*Kearney & Peebles* 14006, *Ripley & Barneby* 7034), mesas and open pine forests, July. Eastern Washington to northern Arizona and California.

The Arizona collections cited are not typical, showing an approach to the characters of *N. propinqua* (Beecher Crampton, personal communication).

3. Navarretia propinqua Suksd. (*N. intertexta* Hook. var. *propinqua* (Suksd.) Brand). Navajo and Coconino counties, about 7,000 feet, in meadows, July. North and South Dakota to Washington, south to Wyoming, northern Arizona, Nevada, and California.

Apparently commoner in Arizona than *N. minima*, 4 collections having been reported.

5. ERIASTRUM (236)

Small annuals, the stems mostly erect, sometimes diffusely branched; leaves alternate, entire or pinnately parted near the base with few narrow lobes; flowers in dense, leafy-bracted heads, these conspicuously cobwebby with long, fine, white hairs; calyx lobes more or less unequal, bristle-tipped; corolla nearly regular to distinctly bilabiate, bright blue to whitish, the tube usually much longer than the calyx; anthers usually sagittate or cordate.

KEY TO THE SPECIES

1. Anthers 0.5 to 1.0 mm. long; corolla regular or nearly so, 6 to 8 (12) mm. long, the lobes usually much shorter than the tube; stamens commonly included.......1. *E. diffusum*
1. Anthers 1.2 to 2.2 mm. long; corolla more or less irregular, 8 to 20 mm. long, the lobes often nearly equaling the tube; stamens often exserted...............2. *E. eremicum*

1. Eriastrum diffusum (Gray) Mason (*Gilia filifolia* Nutt. var. *diffusa* Gray, *Huegelia diffusa* Jepson). Northern Coconino and Mohave counties to Graham, Cochise, Santa Cruz, and Pima counties, 1,000 to 5,500 feet, plains and mesas, March to June. Western Texas to southern Utah and Nevada, California, and northwestern Mexico.

Subsp. *Jonesii* Mason, with larger corollas and anthers and longer filaments than in typical *E. diffusum*, occurs in Gila and Pima counties.

2. Eriastrum eremicum (Jepson) Mason (*Huegelia eremica* Jepson, *Gilia eremica* Craig). Coconino and Mohave counties to Graham, Santa Cruz, Pima, and Yuma counties, up to 5,000 feet, common on dry plains and mesas, March to June. Southern Utah and Nevada, Arizona, and southeastern California.

Subsp. *Yageri* (Jones) Mason (*Gilia eremica* var. *arizonica* Craig), with more nearly regular corollas and less divided leaves than in typical *E. eremicum*, is more common in Arizona than the latter. The form described as *Gilia eremica* var. *zionis* Craig is intermediate between the species and the subspecies. These plants are sometimes so abundant as to color large areas with their sky-blue flowers.

6. LEPTODACTYLON

Plants suffrutescent, the stems woody well above the caudex, very leafy; leaves mostly alternate, fascicled in the axils, rigid, appressed or narrowly ascending, palmately or subpinnately parted or divided, the segments subulate, very pungent; flowers solitary or in small terminal clusters; calyx angles strongly carinate, the scarious junction membranes conspicuous; corolla regular, funnelform-salverform, white or tinged with pink, the tube well exserted from the calyx.

1. Leptodactylon pungens (Torr.) Nutt. ex Rydb. (*Gilia pungens* Benth.). Navajo and Coconino counties, 6,500 to 7,500 feet, May and June. Montana to British Columbia, south to New Mexico, northern Arizona, and Baja California.

The Arizona plants probably all belong to the short-leaved subsp. *brevifolium* (Rydb.) Wherry.

7. LINANTHASTRUM (237)

Plants perennial, with a woody caudex, the stems only slightly woody above ground, very leafy; leaves opposite (appearing whorled), flaccid or somewhat rigid, spreading, palmately parted or divided, the segments narrowly linear to nearly filiform; inflorescences cymose, open or dense; calyx angles moderately carinate, the junction membranes inconspicuous or obsolete; corolla regular, funnelform-salverform, white or tinged with pink, the tube not or but slightly surpassing the calyx.

1. Linanthastrum Nuttallii (Gray) Ewan (*Gilia Nuttallii* Gray, *Siphonella Nuttallii* Heller). Apache, Navajo, Coconino, Yavapai, Greenlee, Gila, and Pima counties, 5,500 to 8,000 feet, mostly in open pine forests, July to November. Idaho and Washington to Chihuahua, Arizona, and Baja California.

Arizona possesses both typical *L. Nuttallii* and subsp. *floribundum* (Gray) Ewan (*Gilia floribunda* Gray), the latter having the leaves with fewer divisions and the pedicels longer than in the typical plant.

8. LINANTHUS

Small desert annuals, flowering in spring; leaves mostly opposite (except in *L. demissus*), entire or palmately parted with narrow divisions; flowers solitary or in few-flowered clusters, but not aggregated in definite heads or panicles; corolla campanulate, funnelform, or somewhat salverform; stamens inserted equally; ovules few to many in each locule.

KEY TO THE SPECIES

1. Calyx campanulate, cleft nearly to the base, the lobes very unequal, broadly scarious-margined; corolla broadly campanulate, the tube very short; leaves mostly alternate. Stems diffusely branched; flowers diurnal, short-pedicelled or subsessile; corolla whitish, or tinged or marked with purple, 5 to 8 mm. long..............1. *L. demissus*

1. Calyx turbinate or cylindric, not cleft nearly to the base, the lobes seldom noticeably unequal; corolla funnelform or somewhat salverform, the tube well developed; leaves opposite (2).

2. Leaves palmately parted, the divisions linear or oblanceolate, seldom more than 1 cm. long; stems usually diffusely branched; flowers diurnal; calyx turbinate; corolla broadly funnelform, 8 to 15 mm. long, the limb usually bright yellow......2. *L. aureus*

2. Leaves entire, or parted with few, almost filiform divisions, these usually more than 1 cm. long; stems usually erect and sparingly branched but sometimes diffusely branched; flowers mostly vespertine; calyx cylindric; corolla funnelform-salverform, the limb whitish or tinged outside with brownish purple (3).

3. Corolla 15 to 30 mm. long, the lobes equaling or longer than the tube; stamens inserted near base of the corolla tube; seeds with a loose, white testa..........3. *L. dichotomus*
3. Corolla usually less than 15 mm. long, the lobes much shorter than the tube; stamens inserted high in the corolla tube; seeds reddish brown, the testa close, not white 4. *L. Bigelovii*

1. Linanthus demissus (Gray) Greene (*Gilia demissa* Gray, *G. Dactylophyllum* Torr.). Havasu Canyon (western Coconino County) and Mohave County, south to Pima and Yuma counties, up to 2,000 feet, desert sands, locally abundant, March to May, type from Diamond Creek, Mohave County (*Newberry* in 1858). Southern Utah, Arizona, and southeastern California.

There is a question whether the older name, *G. Dactylophyllum*, was properly published (see H. L. Mason, *Madroño* 9:253).

2. Linanthus aureus (Nutt.) Greene (*Gilia aurea* Nutt., *G. Ashtonae* A. Nels). Almost throughout the state, 2,000 to 6,000 feet, common on dry plains and mesas, March to June. Western Texas to southern Nevada and southeastern California.

The plants are sometimes so abundant as to color extensive areas with their bright-yellow flowers. A rather rare form, *G. aurea* var. *decora* Gray, has the corolla limb whitish. *G. Ashtonae* (type from Tortilla Flat, now Canyon Lake, Maricopa County, *A. & R. Nelson* 1768) is a form with elongate pedicels (up to 22 mm. long) occasional in western Gila and eastern Maricopa counties.

3. Linanthus dichotomus Benth. (*Gilia dichotoma* Benth.). Eastern Mohave, Maricopa, and Pinal counties (probably elsewhere), 1,000 to 3,500 feet, dry mesas and slopes, not common, March to May. Southern Arizona, California, and Baja California.

Known in California as evening-snow. The very fragrant flowers open in the evening, as in the next species.

4. Linanthus Bigelovii (Gray) Greene (*Gilia Bigelovii* Gray). Grand Canyon (Coconino County) and Mohave, Greenlee, Graham, Gila, Pima, and Yuma counties, 3,500 feet or lower, common on dry mesas and slopes, February to May. Western Texas to southern Utah, California, and Baja California.

Corolla limb cream-colored streaked with crimson, the tube mahogany-colored within. Specimens with stipitate glands on the herbage and calyx are not uncommon in Arizona. This is a character of *L. Jonesii* (Gray) Greene (*Gilia Jonesii* Gray, *G. Bigelovii* var. *Jonesii* Brand). According to Jepson (*Fl. Calif.* 3^2: 205), *L. Jonesii* also differs from *L. Bigelovii* in having a yellow or yellowish corolla and deeply notched seeds. It is uncertain whether the Arizona plants possess these latter characters.

9. GILIA (238, 239, 240)

Plants of diverse habit, annual to perennial, mostly herbaceous, but 2 species normally or occasionally suffrutescent; leaves alternate or basal, or (in *G. filiformis*) the lower leaves opposite, entire to bipinnatifid; calyx and corolla 5-lobed; corolla mostly regular but in a few species slightly bilabiate, tubular-funnelform to salverform or open-campanulate, usually with a well-developed, often elongate, tube; stamens 5, inserted in the corolla throat; ovules 1 to many in each locule; capsule 3-celled and 3-valved.

The genus *Gilia* remains a very heterogeneous assemblage, even after elimination of the segregate genera *Navarretia, Eriastrum, Leptodactylon, Linanthastrum, Linanthus,* and *Langloisia,* all of which were included in *Gilia* in *Flowering Plants and Ferns of Arizona* (pp. 717–726). But until the whole genus has been thoroughly revised, it seems unwise to attempt further segregation. In the key that follows, the names of subgenera recognized by Mason and Grant (240) have been inserted. Their treatment is limited, however, to species that are found in California, and the position of several Arizona species (*G. rigidula, G. subnuda, G. multiflora, G. polyantha,* and *G. longiflora*) remains uncertain. The first of these species was included in his section *Giliastrum* by Brand (*Pflanzenreich* IV. 250 : 147), and the others are here included provisionally in subgenus *Ipomopsis* since they had been placed in that group by Brand (*Ibid.* pp. 112–119).

KEY TO THE SPECIES

1. Corolla open-campanulate, the tube very short, much shorter than the lobes; flowers scattered, not in definite inflorescences (2).
1. Corolla tubular-funnelform to salverform, the tube well developed, usually at least as long as the lobes; flowers (except in *G. gilioides*) aggregated in definite inflorescences (3).
2. Plants perennial, suffrutescent; leaves more or less distinctly pinnatifid, the lobes few, acicular, rigid, and pungent; flowers large, the corolla normally bright blue: section *Giliastrum* .... 1. *G. rigidula*
2. Plants annual; leaves entire and narrowly linear, or rarely with a few lobes, not rigid and pungent; flowers small, the corolla yellow or yellowish. Branches and pedicels very slender, divaricate: subgenus *Tintinabulum*.................. 2. *G. filiformis*
3. Seeds numerous in each locule; leaves coarsely and sharply few-toothed or shallowly lobed, oblong-ovate to nearly orbicular. Plants annual; flowers in an open cymose panicle; corolla pink at anthesis, broadly funnelform, the limb 4 to 5 mm. wide, the tube shorter than the calyx: subgenus *Gilmania*.................... 3. *G. latifolia*
3. Seeds 1 to several in each locule; leaves otherwise (4).
4. Ovule 1 (exceptionally 2) in each locule; corolla tubular-funnelform, normally bright blue or violet. Plants annual; leaves (the lower ones) coarsely and irregularly pinnatifid, the upper leaves entire or 3-parted; flowers scattered or in loose, few-flowered glomerules; corolla tubular-funnelform, the limb 2 to 3 mm. wide, the tube and throat 5 to 10 mm. long, much longer than the calyx; stamens unequally inserted on the corolla throat: subgenus *Greenianthus*.......................... 4. *G. gilioides*
4. Ovules 2 or more in each locule; corolla not tubular-funnelform, or not bright blue or violet (5).
5. Inflorescences capitate or subcapitate, leafy-bracteate. Plants annual, small; stems decumbent or if erect, then usually diffusely branched; corolla not more than 10 mm. long, whitish or bluish; stamens inserted in or just below the corolla sinuses: subgenus *Elaphocera* (6).
5. Inflorescences more open, or not leafy-bracteate (8).
6. Corolla regular or nearly so, 2 to 5 mm. long; stamens included. Leaves dentate to pinnatifid .... 5. *G. polycladon*
6. Corolla somewhat bilabiate, (4) 6 to 10 mm. long; stamens exserted and more or less declined (7).
7. Leaves mostly pinnatifid, with 1 or 2 pairs of divergent lobes............ 6. *G. pumila*
7. Leaves all entire, or the lowest occasionally with 1 or 2 teeth.......... 7. *G. Gunnisoni*
8. Inflorescenes densely capitate to loosely subcapitate. Plants annual; stems erect or ascending, leafy; lower leaves mostly pinnately to tripinnately dissected, the ultimate divisions very narrow; corolla funnelform, not more than 10 mm. long: subgenus *Capitata* (9).

8. Inflorescences open-paniculate, or elongate-thyrsoid with the flowers more or less glomerate at the ends of the branches (10).

9. Flower clusters densely capitate, many-flowered; peduncles long; calyx usually lanate 8. *G. achilleaefolia*

9. Flower clusters loosely subcapitate, few-flowered; peduncles long or short; calyx (in Arizona plants) glabrous to glandular-villous....................9. *G. multicaulis*

10. Inflorescence an open panicle; corolla funnelform, pale pink, lavender, whitish, or yellowish at anthesis, the tube and throat not more (usually much less) than 15 mm. long. Plants annual; stems scapelike; larger leaves mostly in a basal rosette, rather regularly pinnately toothed to bipinnatifid, the upper leaves mostly reduced and bractlike: subgenus *Eugilia* (11).

10. Inflorescence thyrsoid-glomerate, usually narrow and elongate or if more diffusely paniculate, then the corolla salverform and normally pale blue, or funnelform and bright coral pink: subgenus *Ipomopsis* (15).

11. Panicle branches spreading or ascending-spreading; basal leaves rather shallowly pinnately incised, the lobes not or but little longer than the width of the rachis; herbage glandular-puberulent. Panicle broad and very open; corolla narrowly funnelform, 4 to 6 mm. long, the limb 2 to 3 mm. wide.........................10. *G. leptomeria*

11. Panicle branches mostly erect or ascending; basal leaves pinnatifid or bipinnatifid, the lobes usually much longer than the width of the rachis; herbage glabrous to arachnoid-tomentose below, usually glandular-puberulent only in the inflorescence. Species difficult to distinguish (12).

12. Corolla tube longer than the throat, often 2 to 3 times as long; lower leaves coarsely to somewhat finely pinnatifid, the lobes seldom more than twice as long as wide, often coarsely toothed; calyx rupturing very tardily. Upper leaves usually coarsely few-toothed; tube and throat of the corolla 7 to 10 mm. long, the limb 2 to 7 mm. wide 11. *G. scopulorum*

12. Corolla tube shorter than to somewhat longer than the throat (rarely twice as long); lower leaves usually finely pinnatifid or bipinnatifid, the lobes or divisions usually more than twice as long as wide, often very narrow; calyx rupturing rather promptly (13).

13. Stem leaves palmately 3- to 5-lobed, the basal leaves pinnately 5- to 9-lobed, the lobes very narrow; corolla yellow or cream-colored, the lobes sometimes purple-tipped 12. *G. ochroleuca*

13. Stem leaves entire or few-toothed, the basal leaves pinnately or bipinnately lobed or divided, the segments usually numerous; corolla pink, lavender, or whitish when fresh (14).

14. Corolla broadly funnelform, the tube and throat 3 to 6 times as long as the calyx, the limb 6 to 11 mm. wide, purplish pink, usually drying violet..........13. *G. tenuiflora*

14. Corolla narrowly funnelform, the tube and throat barely surpassing to about twice (rarely 3 times) as long as the calyx, the limb 2 to 6 (8) mm. wide, whitish or pale pink, seldom drying violet....................................14. *G. sinuata*

15. Flowers in open, long-branched panicles. Stamens included (16).

15. Flowers in narrow, usually elongate thyrses, the glomerules borne on short branches. Plants mostly biennial or perennial; stems leafy; leaves pinnately parted into narrow segments or sometimes (in *G. multiflora*) entire (18).

16. Corolla funnelform, coral pink, the tube and throat 12 to 20 mm. long; plants biennial or perennial; herbage copiously glandular-puberulent; stems scapelike, the larger leaves in a basal rosette, the stem leaves much reduced; leaves entire to shallowly and coarsely pinnatifid; flowers short-pedicelled, mostly in small glomerules at ends of the panicle branches. Stamens with the filaments much shorter than the anthers 15. *G. subnuda*

16. Corolla salverform, pale blue or white; plants annual or biennial; herbage glabrous or sparsely pilose, the inflorescence often minutely glandular-puberulent; stems leafy; lower leaves pinnate or pinnatifid with few, nearly filiform segments, the uppermost leaves often entire; flowers long-pedicelled (17).

17. Tube and throat of the corolla 20 to 50 mm. long, the lobes usually obtuse or abruptly short-acuminate .... 16. *G. longiflora*
17. Tube and throat of the corolla 10 to 15 mm. long, the lobes acute........ 17. *G. laxiflora*
18. Corolla somewhat irregular (bilabiate); stamens more or less declined. Herbage copiously pubescent with both short and glandular, and longer, nonglandular, white hairs; corolla normally pale blue or lavender with darker spots or streaks in the throat, the tube and throat not more than 15 mm. long; stamens conspicuously exserted (19).
18. Corolla regular or very nearly so; stamens erect (20).
19. Tube and throat of the corolla 7 to 15 mm. long, 1½ to 2½ times as long as the calyx and 1½ to 3 times as long as the corolla lobes; plants often suffrutescent 18. *G. multiflora*
19. Tube and throat of the corolla 4.5 to 6 mm. long, shorter than to only slightly longer than the calyx (rarely 1½ times as long) and shorter than to 1½ times as long as the corolla lobes; plants herbaceous............................ 19. *G. polyantha*
20. Corolla normally pink to bright red, the lobes acutish (rarely obtuse) to long-acuminate 20. *G. aggregata*
20. Corolla blue, violet, or deep purple, the lobes rounded, more or less emarginate and apiculate (21).
21. Stamens not or only slightly exserted, the filaments short; corolla 12 to 20 mm. long; style usually glabrous.. 21. *G. Macombii*
21. Stamens well exserted, the filaments elongate; corolla (25?) 30 to 40 mm. long; style sparsely short-pilose .... 22. *G. Thurberi*

1. Gilia rigidula Benth. Apache, Navajo, Cochise, and Santa Cruz counties, 4,500 to 6,500 feet, dry plains and mesas, April to September. Kansas and Colorado to Arizona and Mexico.

The species is represented in Arizona by var. *acerosa* Gray (*G. acerosa* Britton). The bright-blue flowers are very attractive.

2. Gilia filiformis Parry. Mohave County, at Fort Mohave (*Lemmon* in 1884) and Yucca (*Jones* 3909), 500 to 2,000 feet, April and May. Utah, Nevada, western Arizona, and southeastern California.

3. Gilia latifolia Wats. Mohave and Yuma counties, 2,000 feet or lower, sandy soil, March and April. Southwestern Utah, western Arizona, and southeastern California.

4. Gilia gilioides (Benth.) Greene. Gila and Yavapai counties to Pima County, 3,000 to 5,000 feet, moist soil along streams, March to May. Nevada and Oregon to Arizona and California.

5. Gilia polycladon Torr. Navajo, Coconino, Mohave, Greenlee, and Pinal counties, 2,000 to 5,000 feet, plains and mesas, April to June. Colorado and western Texas to southeastern California and northern Mexico.

Two different-looking forms, possibly specifically distinct, are represented by Arizona specimens identified on technical characters as *G. polycladon*.

6. Gilia pumila Nutt. Apache and Navajo counties, 5,000 to 6,500 feet, plains, May to August. Kansas to Wyoming, Texas, and northeastern Arizona.

7. Gilia Gunnisoni Torr. & Gray. Apache County to Coconino County, 4,500 to 7,000 feet, plains and mesas, sandy or heavy soil, May to September. Colorado, Utah, New Mexico, and northern Arizona.

Very similar to *G. pumila* and apparently intergrading with it. The herbage of *G. Gunnisoni* is usually described as glabrous or nearly so, but some of the Arizona specimens are decidedly pubescent, in this respect resembling *G. depressa* Jones.

8. Gilia achilleaefolia Benth. Oracle, Pinal County (*Thornber* 4526), near Tucson, Pima County (*Thornber,* several collections), March to May. California and Baja California; probably introduced in Arizona.

The collections cited belong to subsp. *staminea* (Greene) Mason & Grant, differing from typical *G. achilleaefolia* in having the corolla tube shorter than the throat, and the stamens exserted.

9. Gilia multicaulis Benth. Mazatzal Mountains, Gila County, about 4,000 feet (*Peebles* 11592, *Collom* 899), May. Central Arizona, California, and Baja California.

The plant of the Mazatzal Mountains, which is entirely glabrous and has very small flowers in few-flowered clusters, seems to belong to var. *alba* Milliken. It may have been introduced from California. Specimens purporting to have been collected at Clifton, Greenlee County, by Mrs. J. C. Reynolds have more compact, glandular-villous inflorescences, the herbage also being sparsely villous. They strikingly resemble specimens of subsp. *Nevinii* (Gray) Mason & Grant from Guadalupe Island and islands off the coast of California.

10. Gilia leptomeria Gray. Northern Apache County to northern Mohave County, 4,000 to 6,000 feet, common on rocky slopes, April to June. Wyoming to eastern Washington, south to New Mexico, northern Arizona, and California.

Corolla lobes lavender. Subsp. *rubella* (Brand) Mason & Grant (*G. hutchinsifolia* Rydb.), with more deeply dissected leaves, has been reported as occurring in northern Arizona (240 p. 214).

11. Gilia scopulorum Jones. Coconino and Mohave counties to Pinal and Yuma counties, 2,500 feet or lower, dry, rocky slopes, March and April. Southern Utah and Nevada, Arizona, and southeastern California.

12. Gilia ochroleuca Jones. Beaver Dam Creek, Mohave County (*Maguire et al.* 4923). Northwestern Arizona, eastern California, and western Nevada.

The species is represented in Arizona by subsp. *transmontana* Mason & Grant, of which the collection cited is the type. The subspecies has a narrow inflorescence with virgate branches and somewhat wider leaf divisions than in typical *G. ochroleuca.*

13. Gilia tenuiflora Benth. (*G. arenaria* Benth., ? *G. flavocincta* A. Nels.). Apache, Coconino, and Mohave counties, south to Santa Cruz, Pima, and Yuma counties, 200 to 4,000 feet, very common on plains and in washes, preferring sandy soil, February to May, type of *G. flavocincta* (not examined) from Canyon Lake, Maricopa County (*Nelson* 11228). Arizona, California, and Baja California.

14. Gilia sinuata Dougl. (*G. inconspicua* Dougl., non Sweet). Almost throughout the state, up to 7,000 feet but usually much lower, very common, usually in open, sandy places, February to June. Wyoming to eastern Washington, south to western Texas, Arizona, southern California, and northern Mexico.

This and the preceding species are extremely variable, especially in pubescence and in the degree of dissection of the basal leaves. Intergradations occur, but the great majority of the Arizona specimens can be assigned readily to one or the other species.

15. Gilia subnuda Torr. Apache County to eastern Coconino County, 5,000 to 8,000 feet, sandy or rocky hills, June and July. Utah, Nevada, New Mexico, and northern Arizona.

Very attractive when in flower, the corollas of a beautiful coral pink or watermelon pink. The Arizona plants belong to subsp. *superba* (Eastw.) Brand (*G. superba* Eastw.), characterized by herbage glandular-viscid throughout and leaves sparingly toothed or shallowly lobed.

16. Gilia longiflora (Torr.) G. Don. Almost throughout the state, 1,000 to 8,000 feet, common on dry plains and mesas, often on limestone soil, March to October. Colorado and western Texas to Arizona and Chihuahua.

The plant is conspicuous and attractive, with very long, salverform, pale-blue to nearly white (rarely variegated) corollas. A remarkable form, with caudate-acuminate corolla lobes, was collected in Segi (Laguna) Canyon, Navajo County (*Cutler* 3022).

17. Gilia laxiflora (Coult.) Osterh. (*G. Macombii* var. *laxiflora* Coult.). San Simon, Cochise County (*Bolton* in 1914). Colorado, southeastern Utah, western Texas, and southeastern Arizona.

18. Gilia multiflora Nutt. Apache County to Mohave County, south to Cochise and Pima counties, 4,000 to 9,000 feet, common on dry slopes, usually among pines, July to October. New Mexico to southern Nevada and Arizona.

The leaves vary from all entire to all pinnatifid. An exceptionally woody variant, with all (not merely a few) of the hairs glandular, was collected in the Santa Catalina Mountains (*Kearney & Peebles* 10330). Medicinal use of the plant by the Indians has been reported.

This species and the following one (*G. polyantha*) approach the genus *Loeselia* in their slightly bilabiate corollas and exserted and declined stamens (238).

19. Gilia polyantha Rydb. Widely distributed in Coconino County, 6,500 to 9,000 feet, August and September. Southern Colorado, New Mexico, and northern Arizona.

The Arizona plant is var. *Whitingi* Kearney & Peebles, type from the Grand Canyon (*Whiting* 1072–5200), which has a pale-violet corolla with usually rather narrow, oblanceolate or obovate lobes; whereas, in typical *G. polyantha*, the corolla is whitish, sometimes purple-dotted, and the lobes are broadly elliptic or oval.

20. Gilia aggregata (Pursh) Spreng. Apache County to Mohave County, south to Graham, Gila, and Pima counties, 5,000 to 8,500 feet, mostly in open coniferous forests, May to September. Montana to British Columbia, south to New Mexico, Arizona, and California.

Skyrocket. One of the showiest wild flowers of Arizona. The corolla is normally brilliant red but sometimes pink or even pale orange. The plant is sometimes cultivated in gardens. It is reported to be browsed by livestock and deer, and the flowers attract hummingbirds.

KEY TO THE VARIETIES

1. Tube and throat of the corolla 10 to 15 mm. long, little longer than the oblong-lanceolate to oblong-oval, acutish (rarely obtuse) to short-acuminate lobes; stamens included, the filaments shorter than the anthers. Corolla normally bright red...... var. *arizonica*

1. Tube and throat of the corolla more than 15 mm. long, much longer than the lanceolate, caudate-acuminate lobes; stamens usually exserted, the filaments commonly longer than the anthers (2).
2. Corolla purplish pink, the tube and throat 35 to 40 mm. long..........var. *macrosiphon*
2. Corolla normally bright red, the tube and throat usually less than 35 mm. long *G. aggregata* (typical)

Typical *G. aggregata* ranges in Arizona from the northern border to the Sierra Ancha (Gila County) and the Bradshaw Mountains (Yavapai County). The species is highly variable, the most distinct of the variants in Arizona being: (1) Var. *arizonica* (Greene) Fosberg (*Callisteris arizonicus* Greene, *Gilia arizonica* Rydb.), which occurs in Coconino County and northeastern Mohave County, usually among pines or junipers, often on volcanic soil, type from near Flagstaff (*MacDougal* 148). In var. *arizonica* the stems are usually shorter than in other forms of *G. aggregata*. A yellow-flowered form of this variety was collected on the North Rim of the Grand Canyon (*Christensen* in 1949). (2) Var. *macrosiphon* Kearney & Peebles (*G. tenuituba* Rydb. forma *macrosiphon* Wherry) is found in the Pinaleno Mountains (Graham County) and the Santa Catalina Mountains (Pima County), type from the Santa Catalina Mountains (*Peebles et al.* 2522). This handsome variety has a purplish-pink corolla, the lobes with darker-purple, lineate spots. It differs from *G. tenuituba* Rydb. in the corolla color and in its more southern distribution. Specimens having an exceptionally long, slender corolla tube, but usually with the bright-red flower color of *G. aggregata*, from the north side of the Grand Canyon, are referable to *G. tenuituba*, which does not seem to be more than a variety of *G. aggregata* although it was maintained as a species by Wherry (239).

21. Gilia Macombii Torr. Cochise and Santa Cruz counties, 4,000 to 8,000 feet, stony slopes and openings in pine forests, August to October. Southern Arizona and northern Mexico.

22. Gilia Thurberi Torr. Mountains of Cochise, Santa Cruz, and Pima counties as far westward as the Baboquivari Mountains, 4,000 to 6,500 feet, open slopes and canyons (April), August to October. Southern New Mexico, southern Arizona, and northern Mexico.

Arizona's showiest species (excepting *G. aggregata*) and worthy of cultivation. The corolla varies from dark bluish purple to claret color, and shows considerable variation in the diameter of the tube and the width and apiculation of the lobes. There is also much variation in the length of the anthers.

10. LANGLOISIA

Small, annual, desert plants with several stems or branches, these usually decumbent or prostrate; leaves pinnately toothed or pinnatifid, the teeth or lobes tipped with long, slender, white bristles, the basal teeth or lobes reduced to similar bristles; inflorescences dense, terminal, leafy-bracted; calyx lobes subequal, bristle-tipped; corolla narrowly funnelform, bilabiate to nearly regular, pale blue, lavender, or whitish; stamens exserted or included, the filaments erect or declined; capsule 2- to several-seeded.

KEY TO THE SPECIES

1. Leaves more or less petioled, wedge-shaped, abruptly dilated and 3-lobed at apex, the marginal bristles commonly in pairs; corolla regular or nearly so; stamens not or only slightly exserted, erect or nearly so (2).
1. Leaves sessile, oblong or oblanceolate, seldom dilated at apex, the marginal bristles single; corolla distinctly bilabiate, spotted or streaked with purple; stamens well exserted, somewhat declined (3).
2. Corolla 10 to 18 mm. long, not spotted, the lobes much shorter than the tube 1. *L. setosissima*
2. Corolla 15 to 25 mm. long, purple-spotted along the veins, the lobes nearly as long as the tube .... 2. *L. punctata*
3. Corolla lobes much shorter than the tube, the latter usually not or but slightly surpassing the calyx.. 3. *L. Schottii*
3. Corolla lobes nearly as long as the tube, the latter distinctly surpassing the calyx 4. *L. Matthewsii*

1. Langloisia setosissima (Torr. & Gray) Greene (*Gilia setosissima* Gray). Western Coconino, Mohave, and Yuma counties, and western parts of Maricopa and Pima counties, up to 5,000 feet but usually much lower, rather common in sandy washes, February to June. Idaho to Arizona and southeastern California.

***2. Langloisia punctata** (Coville) Goodding (*Gilia punctata* Munz). Fort Mohave (*Lemmon* in 1884), April. Western Nevada, western Arizona (?), and southeastern California.

Owing to the uncertainty of Lemmon's data of locality, the presence of this species in Arizona needs confirmation.

3. Langloisia Schottii (Torr.) Greene (*Gilia Schottii* Wats.). Mohave, western Maricopa, and Yuma counties, 2,000 feet or lower, sandy washes, March to May. Southern Utah, western Arizona, California, and northwestern Mexico.

***4. Langloisia Matthewsii** (Gray) Greene (*Gilia Matthewsii* Gray). Reported as having been collected at Cottonwood, presumably in Mohave County, Arizona (*Palmer* 404, in part). Arizona (?), Nevada, and California.

11. LOESELIA

Plants perennial, the stems somewhat woody below, with exfoliating bark; herbage glandular-hispidulous; leaves simple, mostly alternate, narrow, spinulose-serrate or -serrulate; flowers solitary or in pairs, closely invested by the scarious, entire or few-toothed bracts; corolla pale violet with darker spots.

1. Loeselia glandulosa (Cav.) G. Don. Santa Rita Mountains, Pima County (*Pringle* in 1884), Bear Valley (Sycamore Canyon), Santa Cruz County (*Hardies & Proctor* 6093, *Darrow & Haskell* 2055), about 4,000 feet, flowering almost throughout the year. Southern Arizona to Central America.

12. POLEMONIUM (241). JACOBS-LADDER

Plants mostly perennial, herbaceous; stems leafy; leaves alternate, pinnate, the uppermost leaflets often confluent; corolla rotate-campanulate, funnelform, or nearly salverform, violet, yellow, or white; stamens equally inserted, the filaments declined and usually hairy at base.

KEY TO THE SPECIES

Contributed by Edgar T. Wherry

1. Corolla tubular-funnelform to nearly salverform, the lobes much shorter than the tube and throat (2).
1. Corolla funnelform-campanulate to rotate-campanulate, the lobes longer than the tube and throat. Leaflets thin, not appearing verticillate; inflorescence somewhat viscid (3).
2. Stems 30 cm. or longer, rather sparsely villous with flaccid hairs, somewhat viscid; leaflets 15 to 25, thin, lanceolate, 10 to 25 mm. long, not appearing verticillate; corolla light yellow, almost salverform, about 40 mm. long. Uppermost leaflets decurrent and confluent; corolla tube narrow, about 3 times as long as the lobes....1. *P. pauciflorum*
2. Stems seldom more than 20 cm. long; herbage copiously viscid-villous; leaflets many more than 25, thickish, narrowly elliptic to nearly orbicular, less than 6 mm. long, appearing verticillate; corolla blue, campanulate-funnelform, less than 25 mm. long 2. *P. viscosum*
3. Flowering stems usually several, slender, not more than 20 mm. long, with not more than 3 leaves; leaflets up to 15 mm. long; inflorescence few-flowered; corolla not more than 10 mm. long. Stems puberulent to sparsely villous with soft hairs; corolla violet blue, with a yellow or white throat....................................3. *P. delicatum*
3. Flowering stems solitary or few, stout, 40 cm. or longer, with numerous leaves; leaflets up to 25 mm. long; inflorescence several- to many-flowered; corolla 12 to 20 mm. long. Uppermost leaflets decurrent and confluent (4).
4. Stem usually pubescent nearly or quite to the base, copiously so above; leaflets elliptic to oblong-ovate, acutish or acute; corolla lobes obtuse or mucronulate 4. *P. foliosissimum*
4. Stem glabrous or nearly so toward base, sparsely pubescent above; leaflets lanceolate, usually narrowly so, acute or acuminate (5).
5. Corolla lobes acuminate, yellow...5. *P. flavum*
5. Corolla lobes acutish, violet..6. *P. filicinum*

1. Polemonium pauciflorum Wats. Chiricahua Mountains, Cochise County (*Blumer* 1626, *Kusche* in 1927, *Kaiser* 49–195), 7,500 to 9,500 feet, along brooks, June to August. Southeastern Arizona and northern Mexico.

The Arizona plant probably belongs to subsp. *typicum* Wherry.

2. Polemonium viscosum Nutt. San Francisco Peaks (Coconino County), 9,000 to 12,000 feet, June to September. Wyoming to Washington and northern Arizona.

Some of the Arizona specimens belong to subsp. *Lemmonii* (Brand) Wherry (*P. Lemmonii* Brand), with more than ⅔ of the length of the filament adnate to the corolla tube, and some to subsp. *genuinum* Wherry, with less than ⅔ of the filament adnate. The type of *P. Lemmonii* was collected on Mount Agassiz (*Lemmon* in 1884).

3. Polemonium delicatum Rydb. (*P. pulcherrimum* Hook. subsp. *delicatum* Brand). San Francisco Peaks (Coconino County), Baldy Peak (Apache County), 10,000 to 11,500 feet, June to August. Idaho to New Mexico and Arizona.

The plant has a mephitic odor and is sometimes called skunk-leaf.

4. Polemonium foliosissimum Gray. Kaibab Plateau (Coconino County), White Mountains (Apache County), moist soil along streams, July. Colorado, Utah, New Mexico, and Arizona.

The Arizona plants seem to belong mainly to subsp. *robustum* (Rydb.) Brand (*P. robustum* Rydb.), with violet corollas 12 to 15 mm. long, included styles, and copious pubescence. A white-flowered form with somewhat exserted styles (subsp. *albiflorum* (Eastw.) Brand ?, *P. albiflorum* Eastw. ?) was collected along Oak Creek, southern Coconino County (*Whiting & Sanders* 756.3998, *Deaver* 2208).

5. Polemonium flavum Greene. Buck Springs (Coconino County), Pinaleno Mountains (Graham County), 7,500 to 9,500 feet, rich, moist soil in coniferous forests, July to September. Southwestern New Mexico and Arizona.

6. Polemonium filicinum Greene. White Mountains (Apache and Greenlee counties), Chiricahua and Huachuca mountains (Cochise County), 7,500 to 9,500 feet, rich, moist soil along brooks, July to September. Southern New Mexico and eastern Arizona.

115. HYDROPHYLLACEAE (242). Water-leaf Family

Plants herbaceous or shrubby; leaves simple or pinnate, alternate (seldom opposite) or mostly basal; flowers perfect, regular or nearly so, 5-merous, mostly in cymes, these often with the branches 1-sided and coiled in the bud (scorpioid or circinate); stamens inserted on the corolla, alternate with its lobes; pistil of 2 united carpels, the ovary superior, 1-celled or more or less completely 2-celled; styles 2 or if 1, then 2-cleft; fruit a longitudinally dehiscent capsule; seeds often pitted or reticulate.

KEY TO THE GENERA

1. Plant an aromatic, somewhat glutinous shrub; leaves evergreen, coriaceous . 9. *Eriodictyon*
1. Plants herbaceous or barely suffrutescent, not glutinous; leaves not evergreen or coriaceous (2).
2. Calyx lobes very unequal, the 3 outer ones suborbicular, cordate, enlarged and veiny in fruit. Inflorescence loosely racemose; stamens included . . . . 7. *Tricardia*
2. Calyx lobes not, or not conspicuously, unequal (3).
3. Plants scapose, the leaves all basal; flowers solitary on elongate, naked peduncles arising near the surface of the ground . . . . 8. *Hesperochiron*
3. Plants not scapose, the stems more or less leafy; flowers not solitary or if so, then not borne as above (4).
4. Ovary 1-celled, the placentas becoming expanded and lining the inner wall of the ovary. Inflorescences not, or very indistinctly, scorpioid (5).
4. Ovary more or less completely 2-celled by intrusion of the placentas (7).
5. Plants perennial; leaves mostly basal, long-petioled, the petioles not dilated at base; inflorescences short, dense, headlike, long-peduncled; stamens much exserted
1. *Hydrophyllum*
5. Plants annual; stems leafy, some or all of the leaves sessile and auriculate-clasping, or the petioles with a dilated more or less clasping base; inflorescences loosely racemose or cymose, or the flowers solitary in the leaf axils; stamens included (6).
6. Herbage and capsules prickly with stout, white, often retrorse or hooked hairs; calyx with a bractlike appendage in each sinus; corolla rotate, 8 to 14 mm. in diameter
2. *Pholistoma*
6. Herbage and capsules not prickly, hirsutulous and often glandular; calyx unappendaged; corolla campanulate, not more than 4 mm. in diameter . . . . 3. *Eucrypta*
7. Stamens inserted at different levels on the corolla tube; flowers axillary, solitary or in small, dense, leafy clusters. Plants annual; leaves simple, entire . . . . 6. *Nama*
7. Stamens inserted at the same level; flowers mostly in 1-sided, racemelike cymes (8).
8. Corolla blue, purple, or white, soon deciduous . . . . 4. *Phacelia*
8. Corolla pale yellow or cream-colored, marcescent-persistent. Flowers on filiform pedicels, becoming pendulous . . . . 5. *Emmenanthe*

1. HYDROPHYLLUM (243). WATER-LEAF

Perennial, pubescent herbs; leaves large, mostly basal or nearly so, long-petioled, pinnate or deeply pinnatifid; flowers in dense, short, terminal clusters; corolla campanulate, pale blue, the lobes appendaged within; filaments hairy; style 2-cleft.

Indians as well as white settlers in the northeastern United States used plants of this genus as potherbs.

1. **Hydrophyllum occidentale** (Wats.) Gray. Oak Creek Canyon (Coconino County), Mazatzal Mountains (Gila County), 5,000 to 6,000 feet, along streams in shade, preferring rich soil, May. Utah to Oregon, central Arizona, and California. Western squaw-lettuce.

2. PHOLISTOMA (244)

Plants annual; stems weak, few-branched, trailing or clambering, retrorsely prickly, usually at least 30 cm. long; leaves runcinate-pinnatifid, these and the calyx hispid; flowers axillary and terminal, solitary or in very few-flowered cymes, the corolla pale blue, with scalelike appendages; calyx accrescent; ovules borne on the axial face of the placenta; mature capsule bristly or prickly.

1. **Pholistoma auritum** (Lindl.) Lilja (*Nemophila aurita* Lindl.). Canyons of the Colorado River (Coconino and Mohave counties), to Maricopa, Pinal, and Pima counties, 3,000 feet or lower, common on rocky slopes, February to April. Arizona and California.

The relatively small-flowered Arizona plant is var. *arizonicum* (Jones) Constance (*Nemophila arizonica* Jones), type from Yucca, Mohave County (*Jones* in 1884).

3. EUCRYPTA (245)

Small, delicate annuals; stems erect or diffuse, rarely more than 25 cm. long; leaves alternate or opposite, pinnatifid to bipinnate with small segments; inflorescences terminal, racemelike, becoming loose and elongate; flowers very small, the corolla without appendages or these minute; ovules borne on both faces of the placenta.

The plants appear in early spring, preferring the partial shade of shrubs and disappearing as the soil dries out.

KEY TO THE SPECIES

1. Leaves simply pinnatifid to entire, the upper ones sessile or nearly so; inflorescence usually stipitate-glandular; seeds all alike, corrugate or tuberculate......1. *E. micrantha*
1. Leaves bipinnatifid, all more or less petiolate; inflorescence not stipitate-glandular; seeds dimorphic, some of them thick and corrugate, others thin (meniscoid) and smooth 2. *E. chrysanthemifolia*

1. **Eucrypta micrantha** (Torr.) Heller (*Ellisia micrantha* Brand). From the Grand Canyon to the Mexican boundary, and from Greenlee County to the Colorado River, 4,000 feet or (usually) lower, common, February to May. Western Texas to southern Utah and southern California.

2. Eucrypta chrysanthemifolia (Benth.) Greene. Central-southern and western Arizona, 3,000 feet or lower, widely distributed and common, February to April. Arizona, California, and Baja California.

The Arizona plants belong to var. *bipinnatifida* (Torr.) Constance (*Ellisia Torreyi* Gray), with fewer and smaller flowers and less-dissected basal leaves than in typical *E. chrysanthemifolia* of coastal California. The type of *E. Torreyi* came from Yampai Valley, Yavapai County (*Newberry* in 1858).

4. PHACELIA (246–252)

Plants herbaceous, annual or perennial, mostly pubescent; leaves mostly alternate, simple and entire to pinnatifid or pinnate; flowers mostly in 1-sided false racemes (modified cymes); corolla narrowly funnelform to broadly campanulate, usually with vertical folds or scales in the tube; style 2-cleft or 2-divided; seeds variously roughened, often boat-shaped.

Contact with species having glandular pubescence, such as *P. crenulata* and *P. pedicellata,* causes dermatitis in susceptible persons. Some of the species are known as caterpillar-weed and wild-heliotrope.

KEY TO THE SPECIES

1. Leaves entire or irregularly dentate or crenate, not pinnate or pinnatifid. Plants annual (except *P. laxiflora*); stems not more than 30 cm. long; flowers in simple, mostly few-flowered racemes; corolla cylindric or cylindric-campanulate; stamens and style not exserted (2).

1. Leaves one or more times pinnate or pinnatifid, or pinnately toothed, or if entire, then the veins very prominent beneath and the stems hispid (12).

2. Stems rather stout, sparingly branched; leaves slightly fleshy; racemes rather dense in fruit; flowers nearly sessile; seeds transversely corrugate. Leaves broadly ovate to suborbicular, subcordate, shallowly (seldom deeply) crenate; corolla about 5 mm. long 1. *P. neglecta*

2. Stems relatively slender, weak, often much branched; leaves not fleshy; racemes usually becoming loose in fruit; flowers mostly distinctly pedicelled; seeds not transversely corrugate (3).

3. Corolla broadly campanulate; style deeply parted or divided; capsule sharply acuminate. Herbage hirsute and puberulent, sparsely glandular in the inflorescence; leaves mostly entire, lanceolate, oblanceolate, or elliptic; corolla 4 to 6 mm. long; seeds alveolate 2. *P. curvipes*

3. Corolla tubular to funnelform-campanulate; style shortly bifid to parted ⅓ of its length; capsule obtuse or truncate and apiculate, or (in *P. rotundifolia* and sometimes in *P. Lemmoni*) acute or acutish (4).

4. Ovules 8 to 16 in each ovary. Leaves entire or undulate-margined (5).

4. Ovules 20 or more in each ovary (6).

5. Stems ascending to nearly erect, glandular-puberulent; leaves broadly ovate to orbicular; racemes loose below..3. *P. demissa*

5. Stems nearly prostrate, glandular-villous and puberulent; leaves oblong or elliptic; racemes dense, subcapitate..4. *P. cephalotes*

6. Stems glandular-villous or glandular-hirsutulous, usually also with nonglandular hairs (7).

6. Stems finely glandular-puberulent, not villous or hirsutulous. Leaves oblong, ovate, or elliptic, to nearly orbicular (9).

7. Leaves oblanceolate or spatulate; corolla persistent after anthesis; seeds shallowly reticulate or nearly smooth. Corolla 3 to 4 mm. long..................5. *P. saxicola*

7. Leaves orbicular or nearly so; corolla not persistent; seeds pitted (8).

8. Plants annual; corolla about 5 mm. long and 2.5 mm. wide, the tube pale yellow
6. *P. rotundifolia*
8. Plants perennial; corolla about 13 mm. long and 8 to 10 mm. wide, white, sometimes fading purple ....7. *P. laxiflora*
9. Pedicels not more than 2 mm. long; corolla 4 to 6 mm. long, the scales 1.5 mm. long or shorter; ovules 70 or more..8. *P. Lemmoni*
9. Pedicels mostly 3 mm. or longer; corolla 7 mm. or longer or, if less than 7 mm., then the ovules not more than 50; corolla scales 2 to 2.5 mm. long (10).
10. Pedicels seldom more than 5 mm. long, shorter than the fruiting calyx...9. *P. pulchella*
10. Pedicels mostly more than 5 mm. long, usually longer than the fruiting calyx (11).
11. Corolla 5 to 9 mm. long; ovules not more than 25....................10. *P. filiformis*
11. Corolla 10 to 15 (exceptionally only 9) mm. long; ovules 35 or more
11. *P. glechomaefolia*
12. Seeds transversely corrugate all around; stamens and style not exserted. Flowers in simple, terminal and axillary, more or less scorpioid racemes, these rather loose in fruit; plants annual, glandular-puberulent and pilose; stems usually less than 30 cm. long; corolla funnelform (13).
12. Seeds corrugate (if at all) only on the margins and keel of the excavated side; stamens and style mostly exserted (15).
13. Corolla 2 to 3 times as long as the calyx, 10 to 15 mm. long, deep blue or violet with a yellow throat. Leaves pinnatifid or pinnatisect with numerous lobes or divisions, these entire or incised; primary leaf lobes or divisions usually less than twice as long as wide; fruiting calyx slightly to ⅓ longer than the capsule; corolla funnelform-campanulate ....12. *P. Fremontii*
13. Corolla little, if at all, longer than the calyx, not more than 5 mm. long, white or purplish (14)
14. Plants few-branched from the base; stems erect or ascending; leaves pinnatifid, the lobes entire or incised; racemes much surpassing the leafy portion of the stem; calyx lobes broadly spatulate..13. *P. affinis*
14. Plants many-branched from the base; stems decumbent or spreading; leaves pinnatifid, pinnatisect, or bipinnatifid; racemes not or little surpassing the leafy portion of the stem; calyx lobes linear or oblanceolate..........................14. *P. Ivesiana*
15. Racemes not circinate or scorpioid, dense, composing a narrow, elongate, spikelike panicle. Plant perennial (sometimes biennial?), not glandular, silvery-sericeous or glabrate; leaves pinnately parted, with numerous linear or narrowly oblong divisions, these often incised or pinnatifid; corolla campanulate, usually violet; seeds not cymbiform ....15. *P. sericea*
15. Racemes circinate or scorpioid, at least until anthesis (16).
16. Stems hispid; leaves simple or pinnate with not more than 5 entire-margined leaflets, the veins very prominent beneath; plants perennial. Corolla whitish or bluish
16. *P. magellanica*
16. Stems not hispid or if so, then the leaves pinnate, pinnatifid, or toothed, with more than 5 leaflets, lobes, or teeth, or the veins not prominent beneath, or the plants not perennial (17).
17. Seeds not excavated or corrugate, deeply pitted; stems weak, usually supported by other vegetation, hirsute or hispid. Leaves once or twice pinnately parted or divided, the divisions variously incised; calyx densely hirsute or hispid, 2 to 4 times as long as the capsule (18).
17. Seeds excavated on one side (cymbiform); stems relatively stout, usually erect or ascending. Plants annual or biennial, usually glandular-viscid, at least in the inflorescence, often ill-scented or if not so, then the herbage silvery-pilose or the corolla lobes erose-denticulate (22).
18. Plants perennial, glandular-pubescent, at least in the inflorescence, and with scattered glistening granules. Leaves not bipinnate (sometimes bipinnatifid), the divisions rather large; racemes dense and relatively short, even in fruit; calyx lobes lanceolate or elliptic; corolla pale blue or whitish; stamens exserted........17. *P. ramosissima*
18. Plants annual, usually with little or no glandular pubescence (19).

19. Calyx conspicuously hispid, the lobes narrowly spatulate and contracted into long, slender claws, after anthesis elongate (8 to 12 mm. long); corolla lavender-purple; leaves usually simply pinnate or pinnatifid, with variously toothed or cleft leaflets or divisions. Stamens not or only slightly exserted (20).
19. Calyx hirsute or villous (exceptionally somewhat hispid), the lobes not contracted into long, slender claws; corolla blue; leaves often bipinnate or bipinnatifid (21).
20. Corolla not more than 7 mm. wide and usually much smaller.........18. *P. cryptantha*
20. Corolla 8 to 16 mm. wide....................................19. *P. vallis-mortae*
21. Scales near the base of the corolla tube with free tips or lobes; stamens commonly not or only moderately exserted.......................................20. *P. distans*
21. Scales wholly adnate to the corolla tube; stamens well exserted.....21. *P. tanacetifolia*
22. Corolla funnelform; inflorescence usually elongate and narrow. Stems stout, erect; leaves oblong-lanceolate, crenate, dentate, or cleft usually not more than halfway to the midrib (23).
22. Corolla campanulate; inflorescence more corymbiform and spreading (25).
23. Seeds not corrugate. Leaves crenate with broad, rounded teeth......22. *P. integrifolia*
23. Seeds transversely corrugate or crenate on the excavated side (24).
24. Leaves shallowly dentate; stems copiously hirsute and tomentulose......23. *P. Palmeri*
24. Leaves deeply serrate-dentate; stems sparsely hirsute and canescent-puberulent 24. *P. serrata*
25. Seeds not corrugate or crenate (26).
25. Seeds transversely corrugate or crenate on the keel or the margins or both, of the excavated side. Corolla normally violet or purple, the lobes entire or crenulate; plants very ill-scented (28).
26. Corolla blue, the lobes entire or nearly so; herbage densely silvery-pilose, soft to the touch, not glandular. Leaves coarsely pinnatifid, with broad, few-cleft or few-toothed divisions ....25. *P. congesta*
26. Corolla white or pale bluish purple, the lobes erose-denticulate; herbage not silvery-pilose (27).
27. Gland-tipped hairs few or none; stems decumbent or prostrate, not leafy to the inflorescence; leaves pinnatifid, the lobes entire, crenate, or irregularly incised; corolla lobes with obtuse or acutish teeth; fruiting racemes rarely more than 4 cm. long; capsule globose ....26. *P. arizonica*
27. Gland-tipped hairs usually numerous on the upper part of the plant; stems leafy to the inflorescence; leaves pinnatisect, the divisions pinnatifid or pinnately incised; corolla lobes with very acute teeth; fruiting racemes 6 to 12 cm. long; capsule short-ovoid 27. *P. neomexicana*
28. Calyx scarious in fruit, about twice as long as the capsule, the lobes conspicuously spatulate and narrowed at base; pedicels filiform, in fruit often longer than the calyx and decurved; corolla lilac or whitish. Stem leaves long-petioled, with broadly ovate to nearly orbicular lobes...................................28. *P. pedicellata*
28. Calyx not becoming scarious, little if at all longer than the capsule, the lobes not conspicuously spatulate or narrowed at base; pedicels shorter than the calyx, straight; corolla normally violet-purple (29).
29. Stamens included or but slightly exserted; stem leaves often long-petioled. Leaves broadly ovate or oblong-ovate, shallowly cleft with broad rounded lobes, but sometimes narrower and pinnatifid...................................29. *P. coerulea*
29. Stamens exserted; stem leaves usually short-petioled or subsessile (30).
30. Leaves oblong, coarsely crenate or cleft not more than halfway to the midrib, seldom deeper ....30. *P. corrugata*
30. Leaves oblong or ovate-oblong, mostly pinnatisect or deeply pinnatifid, with usually acute or acutish, often crenate, leaflets or lobes....................31. *P. crenulata*

1. Phacelia neglecta Jones. Mohave, western Maricopa, western Pima, and Yuma counties, below 1,500 feet, dry, stony soils commonly of volcanic origin, March and April. Southern Nevada, western Arizona, southeastern California, and Baja California.

Corolla whitish or occasionally pale pink and fading blue. Arizona specimens have been referred to *P. pachyphylla* Gray, but this species, apparently limited to California, is a larger plant, the herbage with black-headed stalked glands and the racemes much surpassing the foliage (248, p. 134).

*2. **Phacelia curvipes** Torr. No specimens have been seen from Arizona, but the plant has been collected in the Beaver Dam Mountains, Utah, and this range extends into northern Mohave County. Southern Utah to California.

3. **Phacelia demissa** Gray. Known in Arizona only by the type collection at Fort Defiance, Apache County (*E. Palmer*), and by a collection at Lees Ferry, Coconino County (*Jones* in 1890). Wyoming, Utah, and northeastern Arizona.

4. **Phacelia cephalotes** Gray. Near Holbrook, Navajo County (*Goodding* 8–45). Southern Utah and northeastern Arizona.

5. **Phacelia saxicola** Gray. Grand Canyon (Coconino County), near Kingman, Mohave County (*Lemmon* in 1884, the type collection), 3,500 to 7,000 feet, April to September. Central and southern Utah, Nevada, northwestern Arizona, and southeastern California.

6. **Phacelia rotundifolia** Torr. Grand Canyon (Coconino County) and Mohave, Yavapai, and northern Yuma counties, 2,000 to 5,000 feet, infrequent on rocky talus, March to May. Southern Utah, Nevada, western Arizona, and southeastern California.

7. **Phacelia laxiflora** J. T. Howell. Grand Canyon and Havasu Canyon (Coconino County), near Emory Falls (Mohave County), 650 to 6,000 feet, May to July, type from near Toroweap Point (*Hilend* in 1932). Northern Arizona and southern Nevada.

8. **Phacelia Lemmoni** Gray. Mohave and Yavapai counties, type from Mineral Park, Mohave County (*Lemmon* 3350). Northern Arizona, Nevada, and southern California.

9. **Phacelia pulchella** Gray. Kaibab Plateau (Coconino County) and northern Mohave County, 1,500 to 5,000 feet, April to June. Southern Utah and Nevada, northwestern Arizona, and southeastern California.

Two varieties, var. *typica* J. T. Howell and var. *Gooddingii* (Brand) J. T. Howell, occur in Arizona. The former has the leaves usually entire or repand and a purple corolla limb. The latter has the leaves usually crenate or dentate and a violet corolla limb.

10. **Phacelia filiformis** Brand. Kaibab Plateau and Grand Canyon (Coconino County), 3,000 to 7,500 feet, rare, May to September, type from the Grand Canyon (*MacDougal* 186). Known only from northern Arizona.

11. **Phacelia glechomaefolia** Gray. Peach Springs to the Grand Canyon (*Gray* in 1885, the type collection), Grand Canyon, Coconino County (*Lemmon* in 1884), north of Pierce Ferry, Mohave County (*Nichol* in 1938), 4,000 feet or lower, April to June. Known only from northern Arizona.

Howell (249, pp. 13, 14) mentions other, less typical specimens collected in Havasu Canyon (western Coconino County) and in the Toroweap Valley and at Lake Mead (Mohave County).

12. **Phacelia Fremontii** Torr. Mohave, Yavapai, and northern Maricopa counties, 2,000 to 5,000 feet, plains and mesas, March to June, type from

Yampai Valley (*Newberry* in 1858). Southern Utah, Nevada, northwestern Arizona, and California.

The most beautiful Arizona species, deserving of cultivation as an ornamental. The corolla limb, sky blue to lavender in color, contrasts strongly with the bright-yellow tube and throat.

13. Phacelia affinis Gray. Grand Canyon (Coconino County), and Yavapai, Greenlee, Gila, Maricopa, Santa Cruz, and Pima counties, 2,000 to 4,000 feet, along streams, March to May. New Mexico to Nevada, California, and Baja California.

14. Phacelia Ivesiana Torr. Apache County to northern Mohave County, 4,000 to 7,000 feet, sandy soil, plains and mesas, March to June, type (*Newberry* in 1858) from "Diamond River" (probably Diamond Creek, Mohave or Coconino County). Southern Wyoming and Colorado to Nevada, New Mexico, Arizona, and southeastern California.

Var. *pediculoides* J. T. Howell, distinguished from var. *typica* J. T. Howell by having plumper seeds, these obtuse or truncate at base and with fewer (5 to 7) corrugations, has been collected at Willow Beach (Mohave County), Wickenburg (Maricopa County), and Wellton (Yuma County), 200 to 2,000 feet.

15. Phacelia sericea (Graham) Gray. Near Flagstaff, Coconino County (*Purpus* in 1899), apparently very rare in Arizona. Alberta to British Columbia, south to Colorado, northern Arizona, Nevada, and California.

Purpus's specimens have green and glabrate herbage, pinnately parted leaves, and the style 2 or 3 times as long as the corolla.

16. Phacelia magellanica (Lam.) Cov. Apache County to Coconino County, south to Cochise and Pima counties, 4,000 to 9,500 feet, common in rich, rather moist soil in coniferous forests, descending lower along streams, April to October. Alberta and British Columbia to New Mexico, Arizona, and California; South America.

The writers are indebted to Lincoln Constance for the following note: "The Arizona material of this complex (polyploid) group of plants has customarily been referred to *P. heterophylla* Pursh, or else has been divided between *P. heterophylla* and *P. mutabilis* Greene on the basis of color of foliage and presence or absence of a definite basal rosette. From herbarium material alone it does not seem possible to establish natural units by morphological, geographical, altitudinal, or phenological criteria, or by any readily apparent combination of these. None of the Arizona plants are very similar to topotype *P. heterophylla* of northern Idaho, a stout, virgate biennial or weak perennial. The relationship is undoubtedly closer to *P. mutabilis* of Oregon and California, but the two are not identical, and that entity is itself tetraploid. Both diploids (*K. F. Parker et al.* 6212) and tetraploids (*Gould* 3668) are known to occur in Arizona. Until the group can be examined cytotaxonomically and in the field, therefore, it appears wiser to keep all the material under the all-inclusive binomial, *P. magellanica*."

17. Phacelia ramosissima Dougl. Black Mountains (Mohave County) and Greenlee, Gila, eastern Maricopa, Pinal, and Pima counties, 2,000 to 7,000 feet, among shrubs in canyons, March to May. Washington to Arizona and California.

18. Phacelia cryptantha Greene. Grand Canyon (Coconino County), and Mohave County to Greenlee (?), Gila, eastern Maricopa, and northeastern Pinal counties, 2,500 to 4,500 feet, dry slopes under *Quercus turbinella* and other shrubs, April to June. Nevada, Arizona, and southeastern California.

Var. *derivata* Voss, with somewhat wider sepals and no bristlelike tip to the corolla appendages, occurs at Mount Trumbull, northern Mohave County (*Palmer* 334.5) according to Voss (246, p. 175).

19. Phacelia vallis-mortae Voss. Pierce Ferry to Yucca (Mohave County), 1,000 to 3,000 feet, among desert shrubs, April and May. Southern Nevada, northwestern Arizona, and southeastern California.

20. Phacelia distans Benth. Grand Canyon (Coconino County) and Mohave County to Greenlee, Graham, Gila, Pinal, Pima, and Yuma counties, 1,000 to 4,000 feet, very common under bushes along washes and in the foothills, February to May. Nevada, Arizona, and California.

The delicate foliage and bright-blue flowers are attractive. The plants disappear rapidly as the soil dries out. There is much variation in the dissection of the leaves, some plants having pinnate or pinnatifid leaves with few and large leaflets or divisions, others, bipinnate leaves with many very small leaflets. According to Macbride (*Contrib. Gray Herb.* 49 : 27), however, the Arizona specimens all are referable to var. *australis* Brand, the calyx lobes being linear-lanceolate or oblanceolate and not conspicuously unequal.

21. Phacelia tanacetifolia Benth. Cultivated fields near Tucson (*Thornber* 2535) and Yuma (*L. Swingle* in 1916), doubtless introduced from California.

22. Phacelia integrifolia Torr. Apache County to Coconino and Greenlee counties, 5,000 to 7,000 feet, May to September. Kansas and western Texas to Utah, northern Arizona, and Chihuahua.

23. Phacelia Palmeri Torr. Coconino County and northern Mohave County, 3,000 to 6,000 feet, April to August. Southern Utah, Nevada, and northern Arizona.

Var. *foetida* (Goodding) Brand, with longer leaves and larger seeds than in typical *P. Palmeri,* has been collected at Lees Ferry (*Jones* in 1890) and south of Cameron (*J. T. Howell* 24397).

24. Phacelia serrata Voss. Grand Canyon, San Francisco Peaks, and near Flagstaff (Coconino County), 5,000 to 7,000 feet, July to September, type from the San Francisco Peaks (*Purpus* 8064). Known only from northern Arizona.

25. Phacelia congesta Hook. A collection in the Chiricahua Mountains, Cochise County (*Goodding* 2330), is cited by Voss (247, p. 133), who refers it to var. *rupestris* (Greene) Macbr., characterized by grayish-pubescent herbage and 4-seeded capsules. The variety was collected also at Del Rio, Yavapai County (*Rusby* in 1883). Texas to southeastern Arizona.

26. Phacelia arizonica Gray (*P. Popei* Torr. & Gray var. *arizonica* Voss). Graham, Pinal, Cochise, Santa Cruz, and Pima counties, 1,500 to 5,000 feet, common on plains and mesas, February to May, types from southern Arizona (*Thurber, Greene*). Western Texas to southern Arizona and northern Mexico.

27. Phacelia neomexicana Thurber. Apache County to Coconino and Yavapai counties, 5,500 to 8,500 feet, May to August. Colorado, New Mexico, and Arizona.

Most of the Arizona specimens belong to var. *pseudo-arizonica* (Brand) Voss, type from near Flagstaff (*Purpus* 8030), characterized by low, spreading or decumbent stems, these not or sparingly branched above the base, and by a bluish-purple (exceptionally whitish) corolla. Var. *alba* (Rydb.) Brand, with tall, erect, very leafy, freely branched stems and whitish corollas, occurs in Apache County on open flats at Alpine (*Goodding* 1262) and near Springerville (*Stitt* 1481).

28. Phacelia pedicellata Gray. Grand Canyon (Coconino County), and Mohave, southwestern Yavapai, western Gila, Maricopa, Pinal, and Yuma counties, 1,000 to 3,500 feet, dry, rocky slopes, March to May. Western Arizona, southeastern California, and Baja California.

The plant is glandular-viscid and very ill-scented.

29. Phacelia coerulea Greene. Coconino and Mohave counties to Greenlee, Graham, Santa Cruz, Pima, and Yuma counties, 2,000 to 5,000 feet, rocky slopes, March to June. Western Texas to southern Nevada, southeastern California, and northern Mexico.

Specimens with relatively narrow, pinnatifid leaves are difficult to distinguish from *P. crenulata* Torr. except by the included stamens.

30. Phacelia corrugata A. Nels. Apache County to Coconino County, 5,000 to 7,000 feet, gravelly flats and barren, rocky slopes, April to September. Colorado and Utah to Texas, northern Arizona, and northern Mexico.

Scarcely more than a variety of *P. crenulata* Torr. (*P. crenulata* var. *corrugata* Brand).

31. Phacelia crenulata Torr. Throughout the state except in the extreme northeastern portion, 4,000 feet or lower, very common on plains, mesas, and foothills, February to June. Southern Utah to New Mexico, Arizona, southeastern California, and Baja California.

Arizona's most abundant species, very conspicuous in spring with its rich violet-purple (rarely white) flowers, the plant glandular-viscid and with an unpleasant, somewhat onionlike odor.

The prevailing phase in Arizona is var. *ambigua* (Jones) Macbr., which has the herbage hispid-hirsute with long, very slender hairs in addition to the (sometimes very scanty) glandular pubescence. The type collection of *P. crenulata* Torr., however, has similar pubescence. *P. intermedia* Wooton, a small-seeded form recognized as a species by Voss (247, p. 140) and stated by him to be intermediate between *P. corrugata* and *P. crenulata,* has been collected in most of the central and southern counties. It is difficult to separate this entity satisfactorily from *P. crenulata.* Specimens with the corolla only 3 to 4 mm. long and wide have been collected at Wickenburg, Maricopa County (*Palmer* 626), and near Sacaton, Pinal County (*Kearney* in 1944). The Palmer collection has been referred to *P. minutiflora* Voss (247, p. 144), but it appears to be merely a small-flowered variant of *P. crenulata.* At Sacaton it was found in the midst of a colony of *P. crenulata* with normal-sized flowers.

Phacelia glaberrima (Torr.) J. T. Howell (*Emmenanthe glaberrima* Torr.) was reported by A. Gray (*Syn. Fl.* ed. 2, 2[1]: 171) as having been collected by Newberry along the Flax River, an old name of the Little Colorado. This species is known certainly only from north-central Nevada, and it seems very unlikely that it really occurs in Arizona (250, p. 368).

5. EMMENANTHE

Plants annual, the herbage villous and glandular; leaves sessile or nearly so, often slightly clasping at base, oblong, pinnatifid; flowers pendulous, borne on decurved, filiform pedicels; corolla cream-colored, marcescent-persistent, much longer than the calyx; capsules compressed, oblong; seeds reticulate, compressed, slightly concave on one side.

1. Emmenanthe penduliflora Benth. Mohave, Yavapai, and Gila counties to Pima County, 4,000 feet or lower, on slopes and along streams, usually under bushes, March to May. Southern Utah and Arizona to California.

Whispering-bells, so called from the rustling sound made by the persistent dry corollas.

6. NAMA (253)

Plants (Arizona species) annual; leaves alternate, narrow, entire or very nearly so; calyx deeply cleft or parted; corolla funnelform, usually red-purple; stamens not exserted; styles distinct or partly united.

KEY TO THE SPECIES

1. Sepals united ¼ to ½ of their length, the calyx tube narrow, adnate to the ovary; styles partly united. Stems leafy, hirsute, up to 35 cm. long: section *Zonolacus* 1. *N. stenocarpum*
1. Sepals distinct or nearly so, the calyx free from the ovary; styles distinct or nearly so, at least when dry: section *Eunama* (2).
2. Seeds deeply and regularly pitted, with large alveolas; herbage glandular-puberulent and hirtellous. Stems erect, slender, sparingly branched; leaves linear-elliptic or narrowly spatulate, 1 to 4 mm. wide; corolla about 5 mm. long..........2. *N. dichotomum*
2. Seeds shallowly, or irregularly, or obscurely pitted, often rugulose; herbage (except sometimes in *N. hispidum*) not glandular (3).
3. Leaves mostly clustered at the ends of the branches, and in a basal rosette; herbage pilose or hirsutulous, the hairs soft; stems not more than 15 cm. long, spreading or prostrate (4).
3. Leaves well distributed along the stem; herbage hirsute or hispid, the longer hairs rather stiff; stems often more than 15 cm. long, decumbent to erect (5).
4. Leaves prevailingly rhombic-ovate to obovate, 3 to 6 mm. long; corolla 3 to 5 mm. long 3. *N. pusillum*
4. Leaves prevailingly narrowly spatulate, 10 to 40 mm. long; corolla 8 to 15 mm. long 4. *N. demissum*
5. Stems more or less spreading; stem hairs ascending; corolla 7 to 15 mm. long, the limb usually relatively broad and spreading; seeds mostly less than 0.6 mm. long 5. *N. hispidum*
5. Stems erect, fastigiate; stem hairs (the shorter ones) retrorse; corolla 4 to 7 mm. long, the limb narrow and usually suberect; seeds 0.6 to 0.8 mm. long......6. *N. retrorsum*

1. Nama stenocarpum Gray. Yuma (*Vasey* in 1881). Southern Texas, southern Arizona, southern California, and northern Mexico.

2. Nama dichotomum (Ruiz & Pavon) Choisy. Near Navajo Mountain, Grand Canyon, Sunset Crater, and San Francisco Peaks (Coconino County), Chiricahua Mountains (Cochise County), 5,500 to 7,000 feet, August and September. Colorado, New Mexico, Arizona, and Mexico; South America.

3. Nama pusillum Lemmon. Fort Mohave, Mohave County (*Lemmon* in 1884, the type collection), also at Hemenway Wash on the Nevada side of Lake Mead (*Poyser* 23). Western Arizona, southern Nevada, and southern California.

4. Nama demissum Gray. Mohave, western Gila, Maricopa, Pinal, Pima, and Yuma counties, 3,500 feet or lower, very common on sandy deserts, February to May. Utah, Arizona, southeastern California, and Baja California.

This small plant is conspicuous in spring because of its abundance and the vivid red-purple color of the flowers. Most of the Arizona specimens belong to the diminutive var. *deserti* Brand.

5. Nama hispidum Gray. Almost throughout the state, 5,000 feet or lower, dry plains and mesas, usually in sandy soil, February to June (sometimes autumn). Oklahoma and Colorado to Arizona, southeastern California, and northern Mexico.

KEY TO THE VARIETIES

1. Leaves glandular beneath; herbage hispid............................var. *Mentzelii*
1. Leaves not or but slightly glandular beneath; herbage hirsute or moderately hispid (2).
2. Leaves strongly revolute; pubescence of the herbage soft................var. *revolutum*
2. Leaves flat or moderately revolute; pubescence of the herbage harsh....var. *spathulatum*

Most of the Arizona plants belong to var. *spathulatum* (Torr.) C. L. Hitchc. A collection of var. *revolutum* Jepson at Yuma (*Beard* in 1911) is cited by Hitchcock (253, p. 524). Var. *Mentzelii* Brand is known in Arizona only by a collection from the Gila River bed near Sacaton, Pinal County (*Harrison & Kearney* 8814), where doubtless it grew from seeds brought down the river by flood waters.

6. Nama retrorsum J. T. Howell. Betatakin, Navajo County (*Eastwood & Howell* 6601), Inscription House, Coconino County (*Howell* 24582), Klethla Valley, Coconino County (*Eastwood & Howell* 6525, the type collection), 6,000 to 7,000 feet, sand dunes, June to September. Known only from northeastern Arizona.

7. TRICARDIA

Plants perennial; stems branched from the base; leaves mostly basal or nearly so, spatulate; flowers in short racemes, the corolla with 10 narrow internal appendages, broadly campanulate, white and purple; stamens unequal.

1. Tricardia Watsoni Torr. Beaver Dam, northwestern Mohave County, about 2,000 feet (*Jones* 5024 ai), April. Southern Utah and northwestern Arizona to southeastern California.

8. HESPEROCHIRON

Plants acaulescent, herbaceous, perennial; leaves few in a basal rosette, arising from a thickish, vertical caudex, long-petioled, linear to ovate or spatulate, ciliate; flowers solitary on long peduncles; calyx 5-parted, the lobes oblong-lanceolate, ciliate; corolla rotate-campanulate, 12 to 18 mm. wide, lilac-purple or whitish; style 2-cleft near apex; ovary 1-celled.

1. Hesperochiron pumilus (Dougl.) Porter. North Rim of Grand Canyon, Mormon Lake, south of Flagstaff, and south of Williams (Coconino County), 7,000 to 8,000 feet, moist soil among pines, April and May. Idaho and Washington to northern Arizona and California.

9. ERIODICTYON. Yerba-santa

Shrub up to about 2 m. (6½ feet) high; leaves lanceolate, dentate or denticulate, dark green and resinous above, white-tomentose beneath; flowers numerous in scorpioid, often subcapitate, cymes, these forming terminal panicles; corolla broadly funnelform, deeply lobed, lilac or whitish; styles 2 or if the style 1, then 2-parted; capsules 4-valved.

1. **Eriodictyon angustifolium** Nutt. Southern Coconino and Mohave counties to Greenlee, Graham, Pinal, and Pima counties, 2,000 to 7,000 feet, dry slopes, common in chaparral, April to August. Southern Utah, southern Nevada, and Arizona.

Sometimes known as mountain-balm. According to Mrs. Collom, an infusion of the aromatic leaves is used locally in treating respiratory ailments and is considered by "old-timers" to be very efficacious for sore throat and coughs. *E. californicum* is an official drug plant.

A large-leaved plant from the Pinal Mountains, Gila County (*Jones* in 1890), is the type of var. *amplifolium* Brand.

116. BORAGINACEAE. Borage Family

Contributed by Ivan M. Johnston

Plants herbaceous or shrubby, usually bristly; leaves simple, prevailingly alternate; flowers perfect, regular, solitary or cymose; cymes glomerate-racemose or spicate, frequently unilateral and coiled (scorpioid), usually with bracts between, to one side of, or opposite the flowers; calyx usually deeply lobed, somewhat irregular; corolla 5-lobed, commonly with folds or saccate-intruded appendages in the throat; stamens 5, borne on the corolla tube alternate with the lobes; ovary superior, bicarpellate, usually 4-ovulate, entire or lobed, becoming tough or bony at maturity; fruit commonly breaking up into 4 single-seeded lobes (nutlets); style lobed or entire, seated in the pericarp at the apex of the fruit or borne between the fruit lobes (nutlets) on the receptacle, or on an upward prolongation thereof (gynobase); endosperm absent or scarce.

The classification of this family is based primarily upon the structure of the fruit. In many cases it is very difficult to recognize the genus and almost impossible to obtain a precise identification of the species if the specimens lack mature fruiting structures.

The Boraginaceae are of small importance economically, but the family comprises numerous species that are cultivated as ornamentals, notably in the genera *Heliotropium* (heliotrope), *Anchusa*, *Echium*, and *Myosotis* (forget-me-not).

KEY TO THE GENERA

1. Style 2-cleft; stigmas 2, distinct. Flowers solitary or clustered in the stem forks ..... 1. *Coldenia*
1. Style simple; stigmas united (2).
2. Style springing from the pericarp at apex of the fruit, falling away with the nutlets; stigma annulate, usually surmounted by a sterile conic or cylindric appendange; corolla plaited in the bud ..... 2. *Heliotropium*

2. Style borne between the lobes of the fruit (i.e. the nutlets), and attached to the receptacle or gynobase; stigma capitate, unappendaged; corolla not plaited (3).
3. Mature calyx very irregular, burlike, 3 of the lobes nearly distinct, the others more united and enclosing the fruit, becoming cornute with 7 to 9 coarse barbed appendages; ovules 2 .... 3. *Harpagonella*
3. Mature calyx usually regular or practically so, not armed with hornlike barbed appendages; ovules usually 4 (4).
4. Nutlets stellately spreading, attached at the apical (radicle) end, armed with hooked appendages. Small, slender annuals .... 4. *Pectocarya*
4. Nutlets erect, incurved, or weakly divergent, attached at or below the middle, i. e. toward the cotyledon end (5).
5. Margin of the nutlets with barbed appendages (6).
5. Margin of the nutlets lacking barbed appendages (7).
6. Plants annual; pedicels erect; style surpassing the nutlets .... 5. *Lappula*
6. Plants perennial or biennial; pedicels reflexed; style surpassed by the nutlets 6. *Hackelia*
7. Corolla blue. Nutlets minutely papillate and obscurely rugose; corolla funnelform, the throat greatly surpassing the calyx .... 7. *Mertensia*
7. Corolla white or yellow (8).
8. Nutlets attached above the base along a usually open and generally basally forked ventral groove or slit, or by a triangular opening in the pericarp .... 8. *Cryptantha*
8. Nutlets lacking a distinct ventral groove or opening in the pericarp, this usually replaced by an elevated ventral keel (9).
9. Plants annual; nutlets attached by a caruncular scar borne upon or at the basal end of the ventral keel, the attachment usually lateral or suprabasal; nutlets usually rough (10).
9. Plants perennial; nutlets attached by a broad, rounded, quite basal, noncaruncular attachment; nutlets ovoid, smooth and shiny. Corolla orange or yellow (11).
10. Corolla white; cotyledons entire .... 9. *Plagiobothrys*
10. Corolla orange or yellow; cotyledons 2-cleft .... 10. *Amsinckia*
11. Corolla 1 to 4 cm. long; stamens short, included .... 11. *Lithospermum*
11. Corolla 5 to 6 cm. long; stamens reaching to the corolla sinuses .... 12. *Macromeria*

1. COLDENIA (254)

Low, spreading, fruticulose plants; leaves small and usually with ovate or elliptic, revolute-margined, pinnately veined blades, the veining usually impressed on the upper surface; corolla funnelform or tubular-funnelform, white, pink, or lavender; flowers commonly opening late in the afternoon.

KEY TO THE SPECIES

1. Fruit depressed-globose and unlobed until completely mature, bearing the style on its rounded summit, finally breaking up into quarters to form the nutlets. Leaves ovate to elliptic, white-tomentose, obscurely veined; flowers borne singly in the leaf axils or at the forks of the stem .... 1. *C. canescens*
1. Fruit parted into distinct nutlets even in the bud, bearing the style between the apices of the nutlets (2)
2. Leaves not evidently nerved, lanceolate to linear, usually very conspicuously and pungently setose; base of the petiole expanded, indurate, usually distinctly villous; flowers solitary in the leaf axils; nutlets finely warty, ovate, the inner face somewhat angled 2. *C. hispidissima*
2. Leaves with evident impressed nerves, ovate or obovate to nearly orbicular, lacking conspicuous setae, strigose or merely short-hispid; base of the petiole not expanded, or indurate, or villous; flowers in dense clusters at the forks of the stem; nutlets smooth or merely granulate (3).

3. Plants annual; corolla pink or white; sepals with short, pungent hairs; style surpassed by the calyx; cotyledons horseshoe-shaped..........................3. *C. Nuttallii*
3. Plants perennial; corolla blue or bluish; sepals merely villous; style surpassing the calyx; cotyledons ovate or suborbicular, entire or merely nicked at one end (4).
4. Leaves with about 6 pairs of deeply impressed veins; calyx long-villous within; nutlets elongate, with a somewhat angulate inner face.........................4. *C. plicata*
4. Leaves with only 2 or 3 pairs of shallow veins; calyx glabrous or short-pubescent within; nutlets nearly globose..5. *C. Palmeri*

1. Coldenia canescens DC. Grand Canyon (Coconino County) and Mohave County to Pima and Yuma counties, 3,500 feet or lower, frequent on dry, sunny mesas and slopes, especially in rocky, calcareous soil, March to September. Texas to southeastern California and Mexico.

The distribution in Arizona as stated above is that of typical *C. canescens,* with corollas 5 to 8 mm. long, 4 to 6 mm. wide. In the Castle Dome, Kofa, and Plomosa mountains (Yuma County) there occurs var. *pulchella* Johnst., with corollas much larger (9 to 12 mm. long, 5 to 8 mm. wide) and apparently more deeply colored, type from the Kofa Mountains (*Shreve* 6257). This variety is poorly understood, and field observations are needed before its exact relationship to typical *C. canescens* can be established.

2. Coldenia hispidissima (Torr.) Gray. Northern Apache County and basin of the Colorado River and its tributaries, south to near Holbrook and Snowflake (Navajo County), west to Beaver Dam and Pierce Ferry (Mohave County), 1,000 to 5,000 feet, dunes and dry slopes, April to September. Southern Utah, Nevada, and northern Arizona.

The Arizona plant is var. *latior* Johnst., with broader, more lanceolate leaves than in the typical phase of the species.

3. Coldenia Nuttallii Hook. Eastern Coconino County to northern Mohave County, about 4,500 feet, in dry, sandy places. Wyoming to Washington, northern Arizona, and California; Argentina.

4. Coldenia plicata (Torr.) Coville. Extreme western Arizona and eastward along the Williams, Gila, and Salt rivers and their tributaries to Wikieup, Wickenburg, Tempe, and Florence (Mohave, Yuma, Maricopa, and Pinal counties), mostly below 2,000 feet, sandy flats and dry river bottoms, April to October. Deserts of the *Larrea* belt, southern Nevada and southeastern California to northwestern Mexico.

5. Coldenia Palmeri Gray (*C. brevicalyx* Wats.). Topock and Fort Mohave (Mohave County), Pinacate Plateau, Yuma, and Wellton (Yuma County), 500 feet or lower, dry, sandy places in the *Larrea* belt, March to October. Extreme western Arizona, southeastern California, and adjacent Mexico.

2. HELIOTROPIUM. Heliotrope

Herbs; flowers in scorpioid cymes or borne along the stem, usually between or opposite the leaves; corolla white or purplish; fruit unlobed, at maturity breaking up into 4 nutlets or falling away entire.

KEY TO THE SPECIES

1. Plants entirely glabrous, very succulent..........................3. *H. curassavicum*
1. Plants evidently hairy, not succulent (2).

2. Corolla 8 to 15 mm. wide, with a long-exserted tube; style elongate, many times longer than the stigma..1. *H. convolvulaceum*
2. Corolla 2 to 4 mm. wide, usually with an included tube; style short, about as long as the stigma ....2. *H. phyllostachyum*

1. Heliotropium convolvulaceum (Nutt.) Gray. Apache, Navajo, and eastern Coconino counties, 4,500 to 6,000 feet, dry, sandy places, March to October. Nebraska to Texas, southern Utah, Arizona, and Mexico.

The flowers are very attractive, being large, pure white, and sweet-scented, opening in the late afternoon. The distribution in Arizona as given above is that of the typical plant, with stems and leaves closely strigose. Var. *californicum* (Greene) Johnst., a well-marked Western variant with spreading pungent bristles on the stems and leaves, occurs in similar situations but at lower altitudes and under drier conditions at Beaver Dam (Mohave County) and near Yuma, also in adjacent California and Sonora.

2. Heliotropium phyllostachyum Torr. Central and southeastern Arizona (Graham, Gila, Pinal, Cochise, Santa Cruz, and Pima counties), 4,000 to 5,000 feet, sandy and gravelly soils and grassy slopes, often along streams, August and September, type from near Santa Cruz, Sonora. Southern Arizona, Sonora, and Baja California.

3. Heliotropium curassavicum L. Coconino, western Gila, and eastern Pinal counties to Mohave and Yuma counties, in the valleys of the Colorado, Little Colorado, Virgin, Salt, and Gila rivers, 5,000 feet or lower, moist, saline soil, flowering throughout the year. Widely distributed in the warmer parts of the Western Hemisphere.

These plants are sometimes known as Chinese-pusley and quail-plant. It is reported that the finely powdered dried root was applied to sores and wounds by the Pima Indians.

The distribution in Arizona as given above is that of var. *oculatum* (Heller) Johnst. (*H. oculatum* Heller), with the corolla 3 to 5 (rarely 7) mm. wide, usually becoming distinctly purple or purplish at the throat, and the fruit 1.5 to 2 mm. wide. This variety ranges from southwestern Utah to Baja California. Var. *obovatum* DC. (*H. spathulatum* Rydb.), with the corolla 5 to 10 mm. wide, at most only purplish-tinged at the throat, and the fruit 2.5 mm. wide, has been collected in and near the valley of the Little Colorado River, at Holbrook and Tuba, 4,500 to 6,000 feet, and ranges from Iowa, southwestward to Chihuahua and Arizona, northwestward to Saskatchewan and Washington. It intergrades northward with var. *oculatum* but apparently remains distinct in Arizona. Typical *H. curassavicum* (*H. xerophilum* Cockerell), widely distributed in the American tropics and extending northward to New Mexico, is a more slender, less glaucous plant than var. *oculatum*, with narrower leaves and smaller flowers. It is to be sought in southeastern Arizona.[1]

[1] Ewan (255) recognized *H. spathulatum* Rydb. as a species distinct from *H. curassavicum*, characterized by permanently erect, ascending, or incurved sepals, adherent outer seed coat, and spikes little if at all elongate in fruit; whereas he characterized *H. curassavicum* as having the sepals soon spreading or reflexed, the outer seed coat peeling, and the spikes elongate in fruit. In Ewan's view apparently neither *H. curassavicum* L. nor var. *xerophilum* (Cockerell) Nels. & Macbr. occurs in this state, the Arizona plants being referrable to *H. spathulatum* subsp. *oculatum* (Heller) Ewan.

3. HARPAGONELLA

Slender, spreading annuals; stems fragile, at maturity disarticulating at the nodes; leaves linear; corolla white, inconspicuous; pedicels deflexed at maturity; sepals united, the tips becoming appressed to the stem below the base of the pedicel.

1. Harpagonella Palmeri Gray. Graham County to western Maricopa County, southward through eastern Pima County to northern Sonora, 3,500 feet or lower, gravelly slopes and benches in the *Larrea* belt, frequently under mesquite, March and April.

Only var. *arizonica* Johnst., type from Camp Lowell, Pima County (*Parish* 162), occurs in Arizona. The typical phase, a more slender plant, smaller throughout, is confined to coastal southern California and Baja California.

4. PECTOCARYA (256)

Slender, annual herbs with linear leaves and inconspicuous white flowers; pedicels recurving at maturity; nutlets in divergent pairs.

The species frequently grow together, and some of them are very similar in general appearance. Collections made without care may contain mixtures of 2 or even 3 species. The surprising rarity of hybrids among these associated, closely related species suggests that self-pollination or apogamy may occur. Specialized cleistogamic flowers, producing distinctive fruit, are frequently developed about the base of the plant, abundantly so in *P. heterocarpa*. Detailed observation on the behavior of the species is much needed. These little plants are very abundant in spring in southwestern Arizona, growing mostly in dry, sandy or gravelly soil in the *Larrea* belt. It is said that they are eaten by sheep before the fruits mature.

KEY TO THE SPECIES

1. Plants erect; body of the nutlet distinctly obovate, entire-margined or merely erose or denticulate .... 1. *P. setosa*
1. Plants prostrate or spreading; body of the nutlet linear or oblong (2).
2. Nutlets conspicuously heteromorphous, 2 of them more or less ascending and having distinct, upturned, sparsely toothed or entire margins, the other 2 somewhat recurved and inconspicuously margined; calyx strongly asymmetric; fruit about the base of the plant apparently from cleistogamic flowers, its nutlets reflexed and not margined 2. *P. heterocarpa*
2. Nutlets homomorphous or practically so, all of the nutlets with pectinately lacerate or dentate margins, all spreading or all recurved; calyx nearly regular; fruit about the base of the plant apparently from normal flowers, not much modified (3).
3. Nutlets with a very conspicuous, broad, toothed, cartilaginous margin, the triangular or cuneate teeth evidently united at base, the body of the nutlet straight or only moderately recurved .... 3. *P. platycarpa*
3. Nutlets with a very inconspicuous margin, dissected into distinct, pectinately arranged, subulate teeth, the body of the nutlet becoming very strongly and conspicuously recurved .... 4. *P. recurvata*

1. Pectocarya setosa Gray. Mohave, Yavapai, Gila, Maricopa, and Pinal counties, 2,000 to 5,000 feet, apparently not common, March to June. Idaho and Washington to central Arizona and California.

2. Pectocarya heterocarpa Johnst. Grand Canyon (Coconino County), Mohave County, and southward and southeastward across Arizona, 3,000 feet or lower, March and April. Southwestern Utah, Arizona, southeastern California, and northern Sonora.

3. Pectocarya platycarpa Munz & Johnst. Grand Canyon (Coconino County) and Mohave County to Graham, Cochise, Pima, and Yuma counties, up to 5,000 feet but mostly lower, February to April, type from Camp Lowell, Pima County (*Pringle* in 1884). Southern Utah, Arizona, southeastern California, and northern Sonora.

4. Pectocarya recurvata Johnst. Mohave County southeast to Graham, Cochise, and Pima counties, up to 5,000 feet but mostly lower, February to April, type from near Chandler, Maricopa County (*Harrison & Kearney* 6507). Southwestern Utah, Arizona, southeastern California, and northern Sonora.

This species has not been seen from Yuma County, although several collections of *P. heterocarpa* are at hand from that area. Elsewhere in Arizona species 2, 3, and 4 are frequently found growing together.

5. LAPPULA (257). STICK-SEED

Small annuals with blue or white flowers in bracted racemes; nutlets 4, erect, attached to a slender, elongate gynobase along the length of the well-developed ventral keel; plants of dry, open situations.

KEY TO THE SPECIES

1. Margin of 2 or more of the nutlets obese, completely inflated, bearing a row of very short, terete, barbed appendages seated upon its usually rounded edge...1. *L. texana*
1. Margin of the nutlets consisting of a row of distinct, barbed appendages, or the margin more or less cup-shaped through the obvious partial union of the appendages, the free upper portion of the appendages sometimes inflated, but always appearing as lobes of the margin and not as though merely seated upon it (2).
2. Nutlets with 2 rows of distinct, marginal appendages, the principal row on the rim of the nutlet and the secondary row just outside the rim. Corollas usually larger than in the next species..2. *L. echinata*
2. Nutlets with a single row of distinct or partly united appendages on the rim, with no secondary appendages outside....................................3. *L. Redowskii*

1. Lappula texana (Scheele) Britton (*L. heterosperma* Greene). Apache County to eastern Mohave County, 5,000 to 6,500 feet, April to June, type of *L. heterosperma* from Peach Springs (*Greene* in 1889). Kansas to Idaho, south to Texas, New Mexico, and northern Arizona.

The range in Arizona as given above is that of typical *L. texana,* with heteromorphic nutlets. Var. *coronata* (Greene) Nels. & Macbr. (*L. coronata* Greene), with nutlets all alike in form, is found, mostly at lower elevations, in Graham, Gila, Cochise, and Pima counties, type from mesas near Tucson (*Pringle* in 1884). A very similar form is found in the northern Rocky Mountain states.

2. Lappula echinata Gilib. Schultz Pass, Coconino County (*Whiting* 1173B). Widely distributed in the northern United States and Canada; introduced from Eurasia.

3. Lappula Redowskii (Hornem.) Greene. Widely distributed in Arizona in the *Larrea,* juniper, and pine belts, 1,000 to 8,500 feet, usually in sunny places in disturbed soil, March to September. Western United States; Argentina; Asia.

The species has many forms, but these do not seem to be correlated with geography, and more than one of them may occur in a given locality. The typical plant has the marginal appendages distinct or nearly so on all 4 nutlets. This is less common in Arizona than var. *desertorum* (Greene) Johnst. (*L. leucotricha* Rydb.), in which the appendages are more or less evidently confluent on 1 or more nutlets in each fruit. The type of *L. leucotricha* came from near Tucson (*Toumey* in 1894).

6. HACKELIA (258)

Coarse, perennial or biennial herbs; corolla white or blue, in naked racemes; nutlets attached by a submedial areola to a pyramidal gynobase; plants of meadows, thickets, and pine forests.

KEY TO THE SPECIES

1. Corolla white or at most bluish only about the center, 5 to 10 mm. wide; marginal appendages of the nutlets commonly much united......................1. *H. ursina*

1. Corolla normally blue, 3 to 7 mm. wide; marginal appendages of the nutlets free or united only at base (2).

2. Middle stem leaves tending to be petiolate; branches usually few, elongating, spreading, only very rarely aggregated into a conspicuously cylindric inflorescence; plants slender 2. *H. pinetorum*

2. Middle stem leaves tending to be sessile; branchlets bearing mature cymes numerous, short, together forming a leafy-bracted, elongate terminal panicle; plants coarse, rather strict..3. *H. floribunda*

1. Hackelia ursina (Greene) Johnst. Southern Gila County and Greenlee County to Cochise and Pima counties, 5,000 to 8,000 feet, moist, shaded places in the oak and pine belts, not common, May to September. New Mexico, Arizona, and Chihuahua.

2. Hackelia pinetorum (Greene) Johnst. Bill Williams Mountain (Coconino County), and mountains of Cochise, Santa Cruz, and Pima counties, 6,000 to 9,000 feet, moist, shaded places in the pinyon and pine belts, not common, July and August. New Mexico, Arizona, and Chihuahua.

3. Hackelia floribunda (Lehm.) Johnst. White Mountains (southern Apache County), Flagstaff, San Francisco Peaks, and Grand Canyon (Coconino County), 7,000 feet and higher, stream sides and meadows in the pine and spruce belts. Canada to northern New Mexico and Arizona, and southern Nevada.

7. MERTENSIA (259). BLUEBELLS

Broad-leaved, perennial herbs, glabrous or with inconspicuous, appressed hairs, never bristly; corolla blue (or pink when immature), heterostyled, the throat cup-shaped or broadly cylindric, usually twice as long as the erect or ascending lobes; nutlets minutely papillate and obscurely rugose, ovoid; plants of meadows and woodlands.

KEY TO THE SPECIES

1. Pedicels distinctly strigose; calyx at anthesis cut to the base or nearly so, the margin usually evidently ciliolate; leaves usually glabrous beneath and strigose above ..... 1. *M. franciscana*
1. Pedicels glabrous or at most merely pustulate; calyx at anthesis cut only to beyond the middle or less, the margin not evidently ciliolate; leaves glabrous (2).
2. Plants 10 to 20 cm. high; nutlets with a distinct, coarse, lobulate, reflexed margin; corolla tube glabrous within.......................................2. *M. Macdougalii*
2. Plants 30 to 120 cm. high; nutlets not margined; corolla tube usually hairy within ..... 3. *M. arizonica*

1. Mertensia franciscana Heller. Apache, Navajo, and Coconino counties to the mountains of Cochise County, mostly above 7,000 feet, moist, shaded places in the pine and aspen belts, June to September, type from near Flagstaff (*MacDougal* 232). Western Colorado and southeastern Utah to New Mexico and Arizona.

2. Mertensia Macdougalii Heller. Grand Canyon to Oak Creek Canyon (Coconino County), Pine (Gila County), Fort Whipple (Yavapai County), 6,000 to 9,000 feet, mostly in the pine belt, locally abundant, March to June, type from near Mormon Lake (*MacDougal* 95). Known only from Arizona. The margined nutlets of this species are unique in the genus.

***3. Mertensia arizonica** Greene. This species is based upon specimens labeled as collected in Arizona (*E. Palmer* in 1869), but is known definitely only from central and southwestern Utah. The type is probably mislabeled (259, p. 62), but the species eventually may be found in meadows at the lower edge of the pine belt in extreme northwestern Arizona.

Mertensia Palmeri Nels. & Macbr. also is based upon a Palmer collection labeled as from Arizona. According to Williams (259, p. 43), the type represents *M. paniculata* (Ait.) G. Don of the northern United States and Canada and could not have been collected in Arizona or adjacent states.

8. CRYPTANTHA (260, 261)

Annual, biennial, or perennial herbs, usually bristly; corolla white or yellow; cymes prevailingly scorpioid but sometimes glomerate or loosely racemose, with or without bracts; nutlets smooth, tuberculate, or wrinkled, with rounded, angled, or winged edges, attached through a break in the pericarp along a ventral groove or a more or less triangular lateral areola to a somewhat hemispheric or elongate gynobase.

Several of the species are reported to have value as forage for sheep in Arizona. The popular name "nievitas" is applied to white-flowered species in California. *C. crassisepala* is said to be used by the Hopi Indians in treating boils. Most of the species grow in dry, sandy or gravelly soil.

KEY TO THE SPECIES

1. Plants coarse, biennial or perennial: section *Oreocarya* (2).
1. Plants slender, mostly annual: section *Krynitzkia* (11).
2. Corolla tube elongate, distinctly surpassing the calyx; flowers usually heterostyled (3).
2. Corolla tube short, scarcely if at all surpassing the calyx; flowers not heterostyled (7).
3. Nutlets roughened, tuberculate or muricate, dull. Corolla white; inflorescence a somewhat interrupted cylindric thryse.............................1. *C. fulvocanescens*
3. Nutlets smooth and shiny (4).

4. Corolla white (5).
4. Corolla yellow. Leaves usually more abundantly strigose and paler (6).
5. Stems rather uniformly pubescent throughout; upper leaf surface evenly strigose; inflorescences subcapitate, about 2 cm. long, with sometimes 1 or 2 axillary glomerules below the main inflorescence, inconspicuously bracteate; corolla limb about 8 mm. wide, the tube little surpassing the calyx............................2. *C. capitata*
5. Stems toward base (and the persistent leaf bases) densely sericeous, strigose and hispid above; upper leaf surface glabrous or with a few hairs near the margin; inflorescences narrowly thyrsoid, up to 12 cm. long, conspicuously leafy-bracteate below; corolla limb at least 10 mm. wide, the tube considerably surpassing the calyx.3. *C. semiglabra*
6. Inflorescence an elongate, cylindric thyrse; nutlets lance-ovate, more than 1½ times as long as wide, usually only 1 developing, the margin acute; calyx becoming 8 to 10 mm. long in fruit, with abundant, spreading bristles; pedicels slender, 2 to 4 mm. long; plants 10 to 30 cm. high..4. *C. flava*
6. Inflorescence consisting of a large terminal cluster, capitate in flower, with 1 or more remote, at maturity frequently stalked, much smaller, lateral clusters; nutlets broadly ovate, less than 1½ times as long as wide, all 4 of them usually maturing, the margin narrowly winged; calyx becoming 8 to 14 mm. long in fruit, its hairs not very conspicuous and commonly more or less appressed; pedicels stout, usually 1 to 2 mm. long; plants 20 to 50 cm. high................................5. *C. confertiflora*
7. Nutlets smooth and shiny, not compressed.............................6. *C. Jamesii*
7. Nutlets rugose or tuberculate, strongly compressed (8).
8. Margin of the nutlets with a conspicuous, papery wing; coarse, bristly biennials 40 to 100 cm. high...7. *C. setosissima*
8. Margin of the nutlets not winged, thickened or merely acute; plants less than 40 cm. high (9).
9. Plants biennial; stems from a coarse rosette crowning a short-lived taproot; scorpioid cymes elongating and becoming 3 to 8 cm. long, loosely 6- to 10-flowered 8. *C. virginensis*
9. Plants perennial, with a persistent caudex; cymose branches short, 1 to 2.5 cm. long, more or less corymbose, 2- to 6-flowered (10).
10. Pedicels stout, at first 1 to 2 mm. long, becoming 2 to 4 mm. long in fruit; calyx lobes oblong or oblong-lanceolate, 5 mm. long or less, with inconspicuous, pale bristles; nutlets about 1½ times as long as wide, ovate, asymmetric, the ventral face prominently rugose; stems usually less than 10 cm. high......................9. *C. abata*
10. Pedicels slender, 3 to 5 mm. long, becoming 5 to 10 mm. long in fruit; calyx lobes lance-linear, 5 to 8 mm. long in fruit, with abundant, conspicuous, usually tawny bristles; nutlets about twice as long as wide, weakly asymmetric, the ventral face weakly rugose; stems usually 15 cm. or more high........................10. *C. humilis*
11. Calyx circumscissile at maturity. Low herbs with a compact inflorescence; flowers each with a foliaceous bract...11. *C. circumscissa*
11. Calyx not circumscissile (12).
12. Gynobase protruding beyond the nutlets, bearing a sessile stigma on its apex; root and base of the plant conspicuously charged with a purple dye; small, slender, dichotomous herbs; flowers each with a foliaceous bract..................12. *C. micrantha*
12. Gynobase shorter than the nutlets; style developed; root or herbage only very obscurely if at all charged with a purple dye; plants not conspicuously dichotomous; flowers commonly all or in part bractless (13).
13. Nutlets all smooth and shiny (14).
13. Nutlets all (or some of them) roughened (16).
14. Spikes bracteate; stems reddish; nutlet 1, axial; ovules 2.............16. *C. maritima*
14. Spikes bractless; stems green; nutlets 1 to 4; ovules 4 (15).
15. Calyx broadly conic at base, densely appressed hispid-villous, lacking conspicuous bristles; nutlets 1 to 3; style surpassed by the nutlets................13. *C. gracilis*
15. Calyx rounded at base, hispid or hirsute, inconspicuously strigose along the margin; nutlets 4; style about reaching the tips of the nutlets...............14. *C. Fendleri*

16. Calyx conspicuously recurving, most hirsute on the axial side; nutlet solitary (rarely 2), bent. Ovules 2 .... 15. *C. recurvata*
16. Calyx strict to spreading, not recurving (except sometimes in *C. crassisepala*), most hirsute on the abaxial side; nutlets straight (17).
17. Nutlets distinctly heteromorphous, differing in size, frequently also in the markings and in the firmness of the attachment (18).
17. Nutlets homomorphous or practically so (24).
18. Nutlets 2, one smooth and shiny, the other roughened; ovules 2; stems reddish 16. *C. maritima*
18. Nutlets 4, all more or less roughened; ovules 4; stems not reddish (19).
19. Odd nutlet abaxial; nutlets dark, with pale tuberculations (20).
19. Odd nutlet axial; nutlets and tuberculations usually pale (22).
20. Plants shrubby, much branched, twiggy; inflorescence not scorpioid, loosely racemose; flowers distant and not 2-ranked; pedicels frequently spreading, 1 to 6 mm. long 17. *C. racemosa*
20. Plants herbaceous; inflorescence scorpioid, unilateral, the crowded flowers 2-ranked; pedicels strict or ascending, less than 1 mm. long (21).
21. Nutlets with merely angled margins; style reaching to or only slightly beyond the tip of the odd nutlet; inflorescence naked, except near the base; plants usually prostrate or decumbent .... 18. *C. angustifolia*
21. Nutlets with narrowly winged margins; style coarse, much surpassing the odd nutlet; inflorescence bearing scattered, minute bracts throughout; plants usually erect 19. *C. inaequata*
22. Consimilar nutlets winged. Calyx lobes not noticeably thickened...... 26. *C. pterocarya*
22. Consimilar nutlets not margined (23).
23. Stems with spreading hairs; calyx with a short but distinct pedicel, evidently spreading or even recurving, the base usually not gibbous on one side, the mature lobes (particularly the axial one) usually much thickened and very conspicuous 20. *C. crassisepala*
23. Stems distinctly strigose; calyx sessile, strictly and closely appressed to the rachis, gibbous on the axial side due to a basal prolongation of the odd nutlet, the mature lobes (particularly the axial ones) weakly and inconspicuously thickened 21. *C. dumetorum*
24. Style distinctly surpassing the nutlets (25).
24. Style surpassed by the nutlets or, at most, barely surpassing them (30).
25. Nutlets strongly bent above the base, glossy. Gynobase narrowly pyramidal. 22. *C. pusilla*
25. Nutlets not bent, dull except in some forms of *C. muricata* (26).
26. Groove of the nutlet replaced by a large, more or less excavated, triangular areola occupying much of the ventral surface of the nutlet; gynobase narrowly pyramidal. Nutlets thick, angulate, not margined .... 23. *C. albida*
26. Groove of the nutlet narrow; gynobase subulate (27).
27. Nutlets obscurely rugose, the back high-convex, the face flat, the margin knifelike and inflexed .... 24. *C. costata*
27. Nutlets tuberculate or verrucose, not decidedly plano-convex, the margin spreading, or none (28).
28. Margin of the nutlets merely angled, frequently thickened; nutlets usually glossy and with the back obtusely angled, the body ovate or triangular-ovate.... 28. *C. muricata*
28. Margin of the nutlets more or less winged; nutlets with a rounded back, the body lanceolate to narrowly ovate (29).
29. Mature calyx 2.5 to 3 mm. long, 1.5 to 2 mm. wide, the lobes linear-lanceolate, the margins lacking fulvous hairs; nutlets with an entire margin; style much surpassing the nutlets; leaves acute, very roughly hairy; plants with a distinct, erect, central axis, usually 30 to 60 cm. high .... 25. *C. holoptera*
29. Mature calyx 3 to 5 mm. long, 2.5 to 3.5 mm. wide, the lobes linear-lanceolate to ovate-oblong, the margin usually with short, fulvous hairs; nutlets usually with sinuate to lobulate, winged margins; style shortly surpassing the nutlets; leaves usually rounded

or obtuse at apex, less rough; plants mostly lacking a definite central axis, usually 10 to 30 cm. high .... 26. *C. pterocarya*

30. Margin of the nutlets sharp, knifelike or winged (31).
30. Margin of the nutlets angled or rounded, not sharp (32).
31. Nutlets 4, wing-margined; calyx broad and symmetric .... 26. *C. pterocarya*
31. Nutlets 1 or rarely 2, usually with a knifelike margin; calyx obliquely conic at base 27. *C. utahensis*
32. Nutlets ovate or triangular-ovate, with an acute, usually somewhat thickened margin, the back usually obtusely angled, frequently glossy .... 28. *C. muricata*
32. Nutlets lanceolate, with rounded or obtuse, unthickened margins, the back not angulate, dull (33).
33. Nutlets 1 or rarely 2; style reaching ⅔ or less the height of the nutlets... 29. *C. decipiens*
33. Nutlets 4; style reaching beyond ⅔ the height of the nutlets (34).
34. Plants bristly with mostly spreading hairs .... 30. *C. barbigera*
34. Plants distinctly strigose, but usually with some interspersed, spreading hairs 31. *C. nevadensis*

1. **Cryptantha fulvocanescens** (Wats.) Payson (*C. echinoides* (Jones) Payson). Apache County to northern Mohave County, 4,500 to 6,000 feet, slopes and mesas, May and June. Colorado, Utah, New Mexico, and northeastern Arizona.

The nutlets may be tuberculate with the warts rounded, or, as is most common in Arizona, they may be muricate with the roughenings more or less broadly conic and terminated usually by a single short, stout bristle tip.

2. **Cryptantha capitata** (Eastw.) Johnst. Known only from the Grand Canyon, type from Hermit Trail (*Eastwood* 5969), also North Kaibab Trail (*Eastwood & Howell* 1005) and South Kaibab Trail, 5,500 feet (*Collom* in 1940), April to June.

The species is evidently related to *C. confertiflora*, but the plant is smaller, 10 to 30 cm. high, more sparsely strigose and hence greener. It has the large calyx of its relative but somewhat broader, more herbaceous calyx lobes. The fruit is similar to that of its relative. The inflorescence is also similar but differs in having the small lateral flower clusters very reduced or entirely absent.

3. **Cryptantha semiglabra** Barneby. Near Fredonia, Coconino County, 5,000 feet, on detrital clay hills, May and June (*Ripley & Barneby* 4363, 4829, the type collections, 8513, 8519). Known only from this locality.

4. **Cryptantha flava** (A. Nels.) Payson. Valley of the Little Colorado and eastward, southward to beyond Holbrook (Apache, Navajo, and Coconino counties), 5,000 to 7,000 feet, slopes and mesas, May and June. Wyoming to northwestern New Mexico and northeastern Arizona.

5. **Cryptantha confertiflora** (Greene) Payson. Coconino and Mohave counties, mostly north of the Colorado River, 4,500 to 5,500 feet, in the juniper belt, commonly on limestone, May. Western Utah to northwestern Arizona and California.

6. **Cryptantha Jamesii** (Torr.) Payson (*Oreocarya Lemmoni* Eastw.). Widely distributed in Arizona but most common in the northeastern part, 5,000 to 7,500 feet, in the juniper, oak, and lower pine belts, April to September. Nebraska and Wyoming to Texas, Arizona, and California.

This variable species occurs in a number of floristic areas, but its geographical variants are vague and a successful definition of them has not been achieved. The problem is complicated by the presence of numerous seasonal and ecological forms, which, even in the same region, differ greatly in appearance. Among the variants found in Arizona 2 are worthy of special mention. There is the plant with erect stems 30 to 50 cm. high, which at maturity produces 3 to 7, usually well-developed, scorpioid cymes 4 to 10 cm. long, grouped at apex of the leafy stem, which is commonly 2 to 4 times as long as the basal tuft of leaves. This is var. *multicaulis* (Torr.) Payson (*C. Jamesii* var. *setosa* Johnst.) of which the typical bristly form occurs in the White Mountains and southward to the Santa Catalina Mountains. A strigose variant occurs about Flagstaff.

The most common phase of the species in northern Arizona is var. *cinerea* (Greene) Payson. This has spreading or decumbent stems usually 10 to 20 cm. long, rarely becoming twice the length of the basal tuft of leaves. The stem leaves are fewer and proportionately longer than in var. *multicaulis*, and at maturity most of them bear reduced cymes in their axils. The inflorescence of var. *cinerea* hence becomes proportionately more elongate, but the individual scorpioid cymes are shorter and less perfectly developed than in var. *multicaulis*. This low form has been found in most parts of Apache, Navajo, and Coconino counties. Evidently related to it is a similar but more robust plant, with looser, somewhat tomentose indument, that has been collected near Prescott and Peach Springs (Yavapai and Mohave counties). Material collected by Lemmon in 1884, almost certainly in Mohave County, was described as *O. Lemmoni* Eastw. In southeastern Utah *C. Jamesii* tends to become glabrous and to have nonpersistent basal leaves. From that area the completely glabrous form has been described as *Oreocarya pustulosa* Rydb., and a tall, erect, subglabrescent form as var. *disticha* (Eastw.) Payson. These tendencies are exhibited by some collections from extreme northeastern Arizona.

7. Cryptantha setosissima (Gray) Payson. White Mountains (Apache County) to Bill Williams Mountain (Coconino County) and northward to the Kaibab Plateau, reported also from the Santa Catalina Mountains (Pima County), 6,000 to 8,500 feet, in the pine belt, June to August. South-central Utah and Arizona.

8. Cryptantha virginensis (Jones) Payson. Grand Canyon and Havasu Canyon, Coconino County (*Eastwood* 6019, *Clover* 6392), near Burro Creek, southeastern Mohave County (*Gould* 4248), chiefly on rocky slopes in the pinyon-juniper belt. Southwestern Utah to northern Arizona and California.

9. Cryptantha abata Johnst. (*C. modesta* Payson, non Brand). Monument Valley (Navajo County), south side of Grand Canyon (Coconino County), Beaver Dam and near Burro Creek (Mohave County). Western Utah, adjacent Nevada, and western Arizona.

The plants collected near Burro Creek (*Gould & Darrow* 4248) are large for the species, with stems about 20 cm. long. They grew on a rocky volcanic ridge at about 2,000 feet elevation.

***10. Cryptantha humilis** (Gray) Payson. To be expected in northwestern Arizona, particularly in the valley of the Virgin River. Southwestern Utah to Oregon and California.

11. Cryptantha circumscissa (Hook. & Arn.) Johnst. Mohave and Yavapai counties, 1,500 to 5,000 feet, mostly in the *Larrea* belt, April to June. Wyoming and Colorado to Washington, Arizona, and California; Argentina.

12. Cryptantha micrantha (Torr.) Johnst. Western and southern Arizona, common, chiefly in the *Larrea* belt, occasional northeastward at higher elevations, as at St. Johns (Apache County) and Prescott (Yavapai County), March to June. Western Texas to Oregon and California.

13. Cryptantha gracilis Osterh. Kayenta (Navajo County), Frazier Well and Grand Canyon (Coconino County), Ash Fork (Yavapai County), Littlefield (Mohave County), 1,500 to 7,000 feet, mesas and rocky slopes chiefly in the pinyon-juniper belt, frequently on limestone, May and June. Eastern Colorado to Idaho, northern Arizona, and eastern California.

14. Cryptantha Fendleri (Gray) Greene. Apache County to Yavapai and eastern Mohave counties, 5,000 to 7,000 feet, in the sagebrush-saltbush, juniper, and lower pine belts, June to September. Saskatchewan and eastern Washington south along the Rocky Mountains to northern New Mexico and Arizona.

15. Cryptantha recurvata Coville. Grand Canyon, Coconino County (*Eastwood* 6067), valley of the Virgin River (Mohave County), about 2,000 feet, sandy deserts in the *Larrea* belt, rare, April. Southwestern Utah and northwestern Arizona to southeastern Oregon and eastern California.

16. Cryptantha maritima Greene. Western and southern Arizona, eastward to eastern Maricopa and Pinal counties and central Pima County, commonly not above 2,000 feet, February to April. Southern Nevada and Arizona to California and northwestern Mexico.

The typical plant has evident, coarse, short bristles on the calyx. A rather common variant in which the bristles are hidden by an abundance of spreading, white, silky hairs, var. *pilosa* Johnston, is sporadic within the range of the species.

17. Cryptantha racemosa (Wats.) Greene. Grand Canyon and Havasu Canyon (Coconino County) and Mohave and Yuma counties, mostly below 3,000 feet, dry, rocky slopes in the *Larrea* belt, February to September. Southern Nevada and western Arizona to California and northern Baja California.

Sheltered by large rocks, this species frequently forms a small, much-branched, twiggy bush 30 to 60 cm. high. The persistent stems may become more than 1 cm. thick and distinctly woody. The dead leaves are rather persistent, becoming blanched so that when numerous the plants are conspicuous among dark rocks.

18. Cryptantha angustifolia (Torr.) Greene. Western and southern Arizona, generally below 4,000 feet, *Larrea* desert, abundant, February to June, types from near Yuma (*Thomas, DuBarry*). Western Texas to southern Nevada, southeastern California, and northwestern Mexico.

19. Cryptantha inaequata Johnst. Havasu Canyon (Coconino County), Pierce Ferry and Topock (Mohave County), May to July. Northwestern Arizona, southern Nevada, and eastern California.

20. Cryptantha crassisepala (Torr. & Gray) Greene. Most common in northeastern Arizona (Apache County to eastern Coconino County), known also from Mohave, Yavapai, Greenlee, Graham, Cochise, and eastern Pima counties (near Tucson), commonly 4,000 to 6,500 feet, slopes, mesas, and plains, frequently in dry grasslands, May and June. Southern Colorado and western Texas to southern Utah, Arizona, and northern Mexico.

*21. **Cryptantha dumetorum** Greene. Near Needles, California, possibly in Mohave County, Arizona, *Larrea* desert, not common, usually growing under bushes and scrambling through them. Southern Nevada, Arizona (?), and southeastern California.

22. **Cryptantha pusilla** (Torr. & Gray) Greene. Cochise, Santa Cruz, and Pima counties, 4,000 to 5,500 feet, April to June. Western Texas to southern Arizona and northern Sonora.

23. **Cryptantha albida** (H.B.K.) Johnst. Cochise, Santa Cruz, and eastern Pima counties, mostly 4,000 to 5,000 feet, plains and hillsides, in the *Larrea* and oak belts, and in disturbed places in grasslands, August and September. Western Texas to southeastern Arizona and central Mexico; Argentina.

24. **Cryptantha costata** T. S. Brandeg. Southern Yuma County, *Larrea* desert, rare. Extreme southwestern Arizona and southeastern California, type from Needles, California (*Jones* 3841).

25. **Cryptantha holoptera** (Gray) Macbr. Peach Springs to Lake Mead (Mohave County), south to southern Yuma County, *Larrea* belt, rare, February to April, type from Ehrenberg, Yuma County (*Palmer*). Western Arizona and southeastern California.

26. **Cryptantha pterocarya** (Torr.) Greene. Kayenta (northern Navajo County) and Grand Canyon (Coconino County) to Kingman (Mohave County), south to the Mexican border, mostly below 4,000 feet, March to June. Western Texas to Washington, northern Sonora, and southeastern California.

The typical plant, with fruit composed of 1 wingless and 3 winged nutlets, has been found in Navajo, Mohave, and northwestern Maricopa counties, ranging northward and northwestward from Arizona. The most common and widely distributed variant in Arizona is var. *cycloptera* (Greene) Macbr., type from near Tucson (*Pringle*), with fruit composed of 4 similar, winged nutlets. It ranges in Arizona from Mohave and Yavapai counties southward and southeastward to Graham, Cochise, Pima, and Yuma counties, and has a more southerly distribution outside the state. Var. *stenoloba* Johnston, known only from the Virgin River Valley in Arizona and adjacent Nevada, differs from all other forms of the species in its narrowly linear-lanceolate (rather than ovate or broadly lanceolate) calyx lobes, these 5 to 7 mm. long. Its nutlets are heteromorphous, as in the type of the species.

27. **Cryptantha utahensis** (Gray) Greene. Mohave County, 3,000 feet or lower, gravelly slopes and sandy washes in the *Larrea* belt, March and April. Southwestern Utah and western Arizona to California.

28. **Cryptantha muricata** (Hook. & Arn.) Nels. & Macbr. Known in Arizona from Yucca (Mohave County), Prescott and Skull Valley (Yavapai County), Oracle (Pinal County), and near Tucson (Pima County), 2,000 to 5,000 feet, usually on gravelly slopes in openings among trees and brush, rare, April and May. Southern Utah to southern Arizona and California.

The species is represented in Arizona by var. *denticulata* (Greene) Johnst.

29. Cryptantha decipiens (Jones) Heller. Mohave, Yavapai, and western Gila counties, south to Pima and Yuma counties, 2,000 to 4,000 feet, among bushes in the *Larrea* belt, March to May, type from Yucca, Mohave County (*Jones* in 1884). Southern Nevada, Arizona, and California.

30. Cryptantha barbigera (Gray) Greene (*Krynitzkia mixta* Jones). Southern Apache County, Grand Canyon (Coconino County), and Mohave County, southward and eastward to Greenlee, Cochise, Pima, and Yuma counties, 5,000 feet or lower, most frequent in the deserts, March to June, type of *K. mixta* from the Mescal Mountains (*Jones*). Western New Mexico and southern Utah to southeastern California and northern Sonora.

The common (typical) phase of the species has inconspicuous corollas 1.5 to 3.5 mm. wide. Occurring within the range of the species in central Arizona are plants with more conspicuous corollas 4 to 7 mm. wide. These represent var. *Fergusonae* Macbr., known elsewhere only from the western borders of the Colorado and Mojave deserts in California. Plants resembling *C. barbigera* but with a solitary, smooth, very acuminate nutlet 3 mm. long were collected in the Tule Desert, Yuma County (*Darrow* in 1941).

31. Cryptantha nevadensis Nels. & Kenn. Grand Canyon and Havasu Canyon (Coconino County) and Mohave County to Graham and Pima counties, mostly below 4,000 feet, *Larrea* desert, frequently under bushes, March to May. Southern Utah and Arizona to California.

Typical *C. nevadensis* has very slender, frequently sinuous stems, very slender calyx lobes, and narrowly lanceolate, long-acuminate nutlets. Var. *rigida* Johnst. is a stiffly erect, less slender plant with stiffer, less elongate calyx lobes and less attenuate, lance-ovate nutlets. It occurs throughout most of the range of *C. nevadensis* in Arizona.

9. PLAGIOBOTHRYS (262, 263)

Annual, strigose or bristly herbs with white corollas; nutlets rugose, erect or incurved, attached at or below the middle to a depressed gynobase through a caruncular scar, this decurrent on the lower part of the ventral keel, or situated at the lower end of the keel and sunken below its crest, or elevated to the level of the keel on a more or less well-developed, suprabasal, stipelike, lateral projection from the body of the nutlet; basal leaves opposite or crowded into a rosette.

Some of the California species are known as popcorn-flower. It is said that the plants are grazed by sheep.

KEY TO THE SPECIES

1. Nutlets tessellate with broad, flattened, contiguous, pavementlike tuberculations. Plants erect, hispid, with terminal, bractless, scorpioid cymes .... 1. *P. Jonesii*
1. Nutlets not tessellate, the back wrinkled or with ridges, the tuberculations scattered or none (2).
2. Leaves charged with a purple dye, particularly about the midrib and margins; calyx circumscissile, the lobes short and strongly connivent at maturity. Nutlets incurved 2. *P. arizonicus*
2. Leaves green, lacking a conspicuous purple dye; calyx not circumscissile (3).

3. Basal leaves crowded into a distinct rosette, none opposite; plants slender, erect, loosely branched, not producing flowers near the base; nutlets incurved, contracted at both ends, somewhat cruciform, the transverse ridges very broad and separated by grooves 3. *P. tenellus*
3. Basal leaves distinct or at least not in a well-developed rosette, the lower leaves opposite; plants decumbent or prostrate, frequently floriferous throughout, even in the axils of the lowermost leaves; nutlets not incurved, rounded at base, the ridges well spaced (4).
4. Plants sparsely strigose, very slender; nutlets lanceolate, flattened, attached by a broadly affixed, sessile, ovate scar borne below the level of the ventral keel, the back with low, rounded, irregular ridges .... 4. *P. cognatus*
4. Plants distinctly hispid, rather coarse; nutlets ovoid, not compressed, attached by a scar elevated to the level of the ventral keel and more or less stiped, usually with sharp, narrow ridges (5).
5. Stipe of the nutlet elongate, about equaling the body in length; nutlets commonly united in pairs .... 5. *P. Pringlei*
5. Stipe of the nutlet very short; nutlets distinct .... 6. *P. californicus*

1. Plagiobothrys Jonesii Gray. Grand Canyon (Coconino County) and Mohave, Maricopa, Pinal, Pima, and Yuma counties, mostly below 2,000 feet, dry, sandy and gravelly soils on slopes and in valleys of the *Larrea* belt, March and April. Southwestern Utah to south-central and western Arizona and adjacent California.

2. Plagiobothrys arizonicus (Gray) Greene. Grand Canyon (Coconino County) and Mohave and Yavapai counties, south to Graham, Cochise, Santa Cruz, and Pima counties, 5,000 feet or lower, frequently among bushes or rocks, *Larrea* and oak belts, March to May, types from Tucson (*Greene, Pringle*). Western New Mexico to southern Nevada, California, and northern Sonora.

Sometimes known as blood-weed.

3. Plagiobothrys tenellus (Nutt.) Gray. Known in Arizona from only a few scattered stations in Graham, Gila, Maricopa, Pinal, and Pima counties, mostly in the oak belt, March to May. Idaho and Washington to southern Arizona and southern California.

4. Plagiobothrys cognatus (Greene) Johnst. (*Allocarya cognata* Greene). Flagstaff and vicinity (Coconino County), about 7,000 feet, wet soils in the pine belt, June and July. Utah, Nevada, and northern Arizona.

5. Plagiobothrys Pringlei Greene (*Echidiocarya arizonica* Gray). Verde Mesa (Yavapai County ?), and Maricopa, Pinal, Cochise, and Pima counties, mostly in the *Larrea* belt, February and March, type from Verde Mesa (*Smart*). Central Arizona to northern Sonora.

6. Plagiobothrys californicus Greene. Creek banks near Prescott (Yavapai County), and in the Santa Catalina and Santa Rita mountains, Pima County (*Graham* 3463, 3538), about 4,500 feet, March. Central and southern Arizona and southeastern California.

The Arizona plant is var. *fulvescens* Johnst. (*P. micranthus* A. Nels.), type from Prescott (*A. Nelson* 10232). The species is very closely related to *P. Pringlei*, which it resembles in all details except the separate, unstalked nutlets and the slightly shorter calyx tube. It is possible that collectors have mistaken the plant for the more common and better-known relative and so have failed to collect it. Consequently it may be more common and widely distributed than the few specimens at hand indicate.

10. AMSINCKIA. Fiddle-neck

Bristly, erect herbs with scorpioid cymes of yellow or orange flowers; corolla heterostyled, with an elongate tube, the appendages of the throat reduced or absent; leaves alternate; gynobase pyramidal; plants of dry, open places.

The plants are very abundant on sandy or gravelly soil in western and southern Arizona, and are reported to make good spring forage while young. On the other hand it has been reported recently that horses, cattle, and swine eating the nutlets may develop cirrhosis of the liver.

KEY TO THE SPECIES

1. Nutlets tuberculate, roughened by short ridges and a distinct dorsal keel; calyx 5-lobed, the lobes all distinct; corolla tube 10-nerved below the stamens.......1. *A. intermedia*
1. Nutlets tessellate, the ridges and dorsal keel low and usually rounded; calyx (except in early flowers) 2- to 4-lobed, the broader lobes 2- or 3-dentate at apex; corolla tube 20-nerved below the stamens.......................................2. *A. tessellata*

1. Amsinckia intermedia Fisch. & Meyer (*A. echinata* Gray). Grand Canyon (Coconino County) and Mohave County, southward and eastward to Cochise, Pima, and Yuma counties, chiefly in the *Larrea* belt, March to May. Western New Mexico to California.

Amsinckia echinata (*A. intermedia* var. *echinata* (Gray) Wiggins), based on material from near Fort Mohave (*Cooper* in 1860), is a form having the tuberculations and the dorsal keel of the nutlets elevated, narrow, and fragile. Such plants are frequent in Arizona and adjacent California, but are connected by many transitions to the forms with less prominently roughened nutlets that are typical of *A. intermedia*. Suksdorf described various forms of *A. intermedia* as *A. nana, A. demissa, A. rigida, A. arizonica,* and *A. microphylla,* all based on types collected in Arizona.

2. Amsinckia tessellata Gray (*A. macra* Suksd.). Grand Canyon (Coconino County) and Mohave County to Graham and Pima counties, *Larrea* desert, February to June, type of *A. macra* from Sacaton, Pinal County (*Eastwood* 8025). Eastern Washington to southern and western Arizona and California; Chile and Argentina.

11. LITHOSPERMUM (264). Gromwell, Puccoon

Plants perennial or biennial; corolla yellow or orange, the lobes rounded. The conspicuous corollas may be more or less sterile, and most of the seed may be developed from inconspicuous, cleistogamous flowers produced later in the season.

A purple dye was obtained by the Indians from the roots of these plants.

KEY TO THE SPECIES

1. Corolla trumpet-shaped, 2 to 4 cm. long, the lobes erose or fimbriate; late flowers cleistogamous, obscure but very fertile...............................1. *L. incisum*
1. Corolla funnelform, less than 1.5 cm. long, the lobes entire; cleistogamous flowers usually wanting (2).
2. Basal leaves persistent at anthesis; root biennial or short-lived perennial, usually lacking purple dye; corolla light yellow.......................................2. *L. cobrense*
2. Basal leaves disappearing long before anthesis; root strongly perennial, the crown usually discolored by abundant purple dye; corolla deep yellow or orange....3. *L. multiflorum*

1. Lithospermum incisum Lehm. (*L. linearifolium* Goldie). Apache County to Mohave County, south to Cochise, Santa Cruz, and Pima counties, 4,000 to 7,500 feet, grassy flats and slopes in the saltbush, sagebrush, juniper, and pine belts, March to May and perhaps later. Canada to Illinois, Texas, and Arizona.

The plant has been used medicinally by the Hopi.

2. Lithospermum cobrense Greene. Southern Apache, Navajo, and Coconino counties, south to Cochise, Santa Cruz, and Pima counties, 5,000 to 9,000 feet, mostly in the pine belt, July and August. Western Texas to Arizona and Mexico.

3. Lithospermum multiflorum Torr. (*L. arizonicum* Gandoger). Apache County to Coconino County, south to Cochise and Pima counties, 6,000 to 9,500 feet, gravelly benches and slopes, mostly in the juniper and pine belts, June to September, type of *L. arizonicum* from Flagstaff (*MacDougal* 242). Wyoming to New Mexico and Arizona.

12. MACROMERIA

Coarse, bristly perennials with elongate, hairy, greenish-yellow corollas and more or less acute corolla lobes; plants of thickets and woodlands.

1. Macromeria viridiflora DC. (*M. Thurberi* (Gray) Mackenz.). Mountains of Apache, Navajo, Coconino, Greenlee, Cochise, Santa Cruz, and Pima counties, 6,000 to 9,000 feet, rocky slopes and valleys in pine woods, July to September. New Mexico, Arizona, and northern Mexico.

It is said that the dried leaves and flowers mixed with tobacco are smoked by the Hopi Indians in their "rain-bringing" ceremony.

Myosotis scorpioides L., the forget-me-not of gardens, has been planted at Lindberg (Fulton) Springs, near Flagstaff, and may become established. It has small flowers with a short-salverform (almost rotate) corolla, this sky blue with a yellowish throat. The nutlets are smooth and shining. It is a European plant extensively naturalized in North America in wet places.

117. VERBENACEAE (265). Vervain Family

Herbs or shrubs; stems often 4-angled; leaves simple, opposite or whorled; flowers perfect, in spikes or heads; calyx variously toothed or cleft, or almost entire; corolla sympetalous, usually somewhat bilabiate; stamens 4, inserted on the corolla tube, often in pairs at different heights; style 1, apical or subapical, entire or 2-cleft at apex; fruit a schizocarp of 2 to 4 nutlets (cocci), these separating at maturity, or a drupe containing 1 or 2 stones.

The best-known members of this family are the garden verbena, derived from hybrids among several South American species of *Verbena,* and lemon-verbena (*Aloysia triphylla*), also a native of South America. Species of *Lantana* and *Vitex* are extensively planted as ornamental shrubs in the warmer parts of the United States.[1]

KEY TO THE GENERA

1. Inflorescence determinate and centrifugal, cymose. Flowers slightly irregular; stamens didynamous; fruit coarsely reticulate, pubescent........................6. *Tetraclea*

[1] Dr. Harold N. Moldenke has supplied much valuable information in regard to this family.

1. Inflorescence indeterminate and centripetal, racemose, spicate, or subcapitate (2).
2. Fruit more or less fleshy, drupaceous, not splitting, usually 2-seeded. Plants shrubby ....2. *Lantana*
2. Fruit dry, splitting into 2 or more nutlets at maturity (3).
3. Calyx short, more or less campanulate with 2 to 4 teeth or lobes. Nutlets 2; plants perennial (4).
3. Calyx elongate, cylindric, 5-toothed and 5-ribbed. Plants herbaceous or nearly so (5).
4. Flowers scattered in slender, elongate spikes or spikelike racemes; bractlets narrow, not closely subtending the flowers....3. *Aloysia*
4. Flowers in dense, short-spicate or subcapitate inflorescences; bractlets broad, imbricate, closely subtending the flowers....4. *Phyla*
5. Nutlets 4, these at maturity shorter than the calyx, not beaked....1. *Verbena*
5. Nutlets 2, these at maturity usually equaling or surpassing the calyx, beaked..5. *Bouchea*

1. VERBENA (266). VERVAIN

Plants mostly biennial or perennial; flowers sessile, in elongate spikes, these often forming panicles, or in short, headlike clusters; corolla salverform, the tube often curved; stamens included; style slender, 2-lobed at apex, with 1 lobe stigmatic and the other smooth; nutlets normally 4.

The species, especially in section *Glandularia,* are difficult to distinguish. It is not improbable that natural hybrids are of frequent occurrence in this genus.

KEY TO THE SPECIES

1. Spikes mostly broad and dense; calyx usually more than twice as long as the fruit and constricted or contorted above it; corolla relatively large and showy, bright pink or mauve when fresh; connective of the upper anthers commonly appendaged; sterile lobe of the style distinctly surpassing the stigmatic lobe: section *Glandularia* (2).
1. Spikes mostly slender and elongate after anthesis; calyx seldom more than twice as long as the fruit, not contorted above it; corolla often relatively small and inconspicuous, whitish, blue, or violet; connective not appendaged; sterile lobe of the style usually not surpassing the stigmatic lobe: section *Verbenaca* (7).
2. Corolla tube only slightly longer than the calyx. Herbage conspicuously villous; corolla limb 8 to 9 mm. wide....1. *V. Gooddingii*
2. Corolla tube 1⅓ to 1½ times as long as the calyx (3).
3. Calyx shorter than to equaling the floral bractlets, hirsute, not glandular ....2. *V. bipinnatifida*
3. Calyx longer than the bractlets, commonly somewhat glandular (4).
4. Hairs of the calyx appressed. Leaves tripartite-pinnatifid, the ultimate segments very narrow; spikes 3 to 4 cm. long in fruit....3. *V. tenuisecta*
4. Hairs of the calyx spreading (5).
5. Calyx teeth 2 to 3 mm. long....4. *V. ambrosifolia*
5. Calyx teeth mostly less than 2 mm. long (6).
6. Stems prostrate to decumbent-ascending; calyx sparsely glandular, scarcely viscid ....5. *V. ciliata*
6. Stems erect or ascending; calyx copiously glandular, somewhat viscid....6. *V. Wrightii*
7. Spikes several to numerous, panicled; bracts and bractlets commonly inconspicuous (8).
7. Spikes 1 to 3 at apex of the stems and branches or if more numerous and panicled, then subtended at base by more or less leafy bracts; bractlets often conspicuous (14).
8. Spikes relatively thick and densely flowered, the fruits contiguous or overlapping. Stems erect, tall and stout; leaves oblong-lanceolate to ovate-lanceolate, coarsely serrate, sometimes shallowly incised or subhastately lobed at base; corolla blue or violet (9).
8. Spikes slender, elongate, the fruits usually remote (10).
9. Stems villous-hirsute with spreading hairs; leaves thick, coarsely and prominently rugose-veiny; spikes 7 to 10 mm. thick....7. *V. Macdougalii*

9. Stems short-pilose with appressed or ascending hairs; leaves relatively thin, not prominently rugose-veiny; spikes less than 7 mm. thick........................8. *V. hastata*
10. Leaves merely crenate or serrate. Corolla whitish or bluish, barely surpassing the calyx (11).
10. Leaves once or twice pinnatifid, or 3- to 5-cleft, or deeply incised (12).
11. Leaves distinctly petioled, the lower petioles about 1 cm. long, the blades very scabrous above, more or less ovate (at least the lower ones), short-cuneate at base; stems short-pilose with ascending or subappressed hairs..........................9. *V. scabra*
11. Leaves subsessile or the petioles less than 1 cm. long, the blades not or only slightly scabrous above, oblong-lanceolate, tapering at base; stems hirsute with spreading hairs ....10. *V. carolina*
12. Corolla limb about 1 mm. wide; nutlets not more than 1.5 mm. long 11. *V. Ehrenbergiana*
12. Corolla limb 6 to 7 mm. wide; nutlets 2 to 2.5 mm. long (13).
13. Nutlets little longer than wide; leaves all deeply cleft or incised....12. *V. menthaefolia*
13. Nutlets about twice as long as wide; basal leaves incised-dentate, the middle leaves once or twice pinnatifid, the upper leaves entire or sparingly dentate.........13. *V. Halei*
14. Leaves mostly entire and linear, the lower ones few-toothed. Stems somewhat woody at base; herbage sparsely hispidulous with short, stiff, antrorse hairs; leaves strongly revolute; corolla limb 5 to 7 mm. wide; fruits remote, the nutlets about 3 mm. long 14. *V. perennis*
14. Leaves serrate-dentate to deeply incised or pinnatifid, seldom entire (15).
15. Leaves serrate-dentate or shallowly incised. Stems erect, up to 80 cm. long; leaves thick, coarsely rugose-veiny; spikes thick and very dense.................7. *V. Macdougalii*
15. Leaves deeply incised-dentate, or pinnatifid, or 3-cleft (16).
16. Bractlets conspicuous, somewhat foliaceous, at least at base of the spike (17).
16. Bractlets inconspicuous, not foliaceous (19).
17. Leaves more or less plicate, the veins often whitish toward the margin; corolla limb 4 to 6 mm. wide..15. *V. plicata*
17. Leaves not plicate or noticeably whitish-veined; corolla limb 2 to 3 mm. wide (18).
18. Spikes (including the bractlets) commonly at least 1 cm. wide; herbage coarsely hirsute, not or obscurely glandular; plants often drying blackish...........16. *V. bracteata*
18. Spikes much less than 1 cm. wide; herbage more or less glandular; plants not drying blackish ....17. *V. gracilis*
19. Leaves (at least the lower ones) more or less ovate in outline, not elongate (20).
19. Leaves (at least the lower ones) oblong-lanceolate to narrowly obovate in outline, elongate. Corolla limb mostly 6 mm. or wider (21).
20. Blades more or less plicate, the veins whitish toward the margin; corolla limb 4 to 6 mm. wide ....15. *V. plicata*
20. Blades not noticeably plicate or whitish-veined; corolla limb 2 to 3 mm. wide 17. *V. gracilis*
21. Stems decumbent or ascending; calyx 2.5 to 3 mm. long; corolla limb about 6 mm. wide. Nutlets muricate on the inner (commissural) face..............12. *V. menthaefolia*
21. Stems mostly erect; calyx 3 to 4 mm. long; corolla limb 6 to 15 mm. wide (22).
22. Corolla tube slightly to moderately surpassing the calyx, the limb up to 10 mm. wide; nutlets usually densely white-scaberulous on the inner (commissural) face 18. *V. neomexicana*
22. Corolla tube much surpassing the calyx, the limb up to 15 mm. wide; nutlets very sparsely scaberulous on the inner face...........................19. *V. pinetorum*

1. Verbena Gooddingii Briq. Throughout the state, mostly below 5,000 feet, dry slopes and mesas, flowering almost throughout the year. Southern Texas; Utah, Nevada, Arizona, California, and northwestern Mexico.

This is much the commonest and most widely distributed species in Arizona of the showy-flowered verbenas of section *Glandularia*. Var. *nepetifolia* Tidestrom (*V. arizonica* Briq., *V. verna* A. Nels.), with leaves not cleft or cleft only near the base (several-cleft in typical *V. Gooddingii*), is found in Mohave, Yavapai, Cochise, Santa Cruz, Pima, and Yuma counties. The type of *V. arizonica* Briq. was collected at Yucca, Mohave County (*Jones* 3901), and that of *V. verna* at Diamond Creek, Mohave County (*N. C. Wilson* 95).

2. Verbena bipinnatifida Nutt. Apache, Navajo, and Coconino counties, south to Cochise, Santa Cruz, and Pima counties, 5,000 to 10,000 feet, mostly in open coniferous forests, May to September. Alabama, westward to Texas, South Dakota, Colorado, and Arizona.

Var. *latilobata* Perry, with leaves merely cleft and with relatively broad segments (bipinnatifid and with relatively narrow segments in the typical plant), is found in most of the range of the species in Arizona.

3. Verbena tenuisecta Briq. Roadsides and parkings at Tucson, Pima County (*Gould* 5303), common. Introduced from South America and spreading rapidly in the southern United States.

4. Verbena ambrosifolia Rydb. Apache, Coconino, Graham, and Pima counties. Oklahoma and Texas to Colorado, Arizona, and northern Mexico.

Forma *eglandulosa* Perry, distinguished by absence of glands on the calyx, occurs in Cochise and Pima counties. It differs from *V. bipinnatifida* only in its short bractlets.

5. Verbena ciliata Benth. Apache County to Mohave County, south to Cochise and Pima counties, 2,000 to 7,000 feet, dry plains and mesas and openings in pine forests, April to July. Western Texas to Arizona, south to southern Mexico.

Var. *pubera* (Greene) Perry, with a more prostrate and compact habit of growth and more revolute leaf margins than in typical *V. ciliata,* occurs throughout the range of the species in Arizona.

6. Verbena Wrightii Gray. Apache, Navajo, Coconino, Graham, and Pinal counties, probably elsewhere, but apparently rare in Arizona. Western Texas to Colorado and Arizona.

7. Verbena Macdougalii Heller. Apache, Navajo, and Coconino counties (reported by Moldenke also from Greenlee, Yavapai, and Pima counties), 6,000 to 7,500 feet, mostly in pine forest, June to September, type from Flagstaff (*MacDougal* 249). Southern Wyoming to western Texas, New Mexico, and Arizona.

8. Verbena hastata L. Flagstaff, Coconino County (*MacDougal* 566), near Prescott, Yavapai County (*Coues & Palmer* 279), Camp Verde, Yavapai County (*Toumey* in 1891). Nova Scotia to British Columbia, south to Florida, Nebraska, Arizona, and California.

9. Verbena scabra Vahl. Gila, Pinal, Santa Cruz, and Pima counties, July to October. North Carolina to Florida, west to California and northern Mexico; West Indies.

10. Verbena carolina L. Gila, Cochise, Santa Cruz, and Pima counties, 5,000 to 6,000 feet, along streams, September and October. Florida to Texas; southern Arizona, south to Central America.

11. Verbena Ehrenbergiana Schauer. A specimen in the herbarium of the University of Texas (*Stalmach* 198) is labeled as from Arizona without definite locality. The species is known otherwise only from Mexico.

12. Verbena menthaefolia Benth. Near Yuma (*Jones* in 1906). Western Texas; southern Arizona, southern California, and Mexico.

13. Verbena Halei Small. Near Casa Grande, Pinal County (*Peebles & Harrison* 4224), Menengers Dam, western Pima County (*Goodding & Lusher* 186–45). Alabama to Texas and northeastern Mexico; southern Arizona.

14. Verbena perennis Wooton. Baboquivari Mountains, Pima County (*Jones* 24994). Western Texas, New Mexico, and southern Arizona.

15. Verbena plicata Greene. Yavapai, Cochise, and Pima counties, March to September. Texas to Arizona and northern Mexico.

16. Verbena bracteata Lag. & Rodr. Almost throughout the state, 1,000 to 7,500 feet, waste land and river bottoms, May to September. Widely distributed in North America.

17. Verbena gracilis Desf. (*V. arizonica* Gray). Pinal, Cochise, Santa Cruz, and Pima counties, 4,500 to 6,500 feet, June to October, type of *V. arizonica* Gray from near Fort Huachuca (*Lemmon*). Southern Arizona to southern Mexico.

18. Verbena neomexicana (Gray) Small. Mohave, Yavapai, Gila, Maricopa, Pinal, Cochise, Santa Cruz, and Pima counties, 2,000 to 6,000 feet, foothills and canyons, common, March to October. Texas to southern California and northern Mexico.

Represented in Arizona chiefly by var. *xylopoda* Perry, type from the Santa Catalina Mountains (*Hanson* A1130), which has a slightly glandular inflorescence and a larger corolla than in typical *V. neomexicana.*

19. Verbena pinetorum Moldenke. Santa Rita Mountains, Santa Cruz County (*Berry* in 1904), reported also for Cochise County by Moldenke. Southeastern Arizona and Chihuahua.

2. LANTANA

Shrubs; leaves opposite, petioled, lanceolate to broadly ovate, crenate or serrate; inflorescences mostly axillary, subcapitate, long-peduncled, the flowers subtended by bractlets; corolla regular or nearly so, with a slender tube and a spreading limb; fruit drupelike, the exocarp fleshy or dry, the endocarp hard.

KEY TO THE SPECIES

1. Bractlets narrow, lanceolate or oblong, loosely subtending the flowers; corolla mostly yellow or red, the tube up to 10 mm. long, the limb up to 9 mm. wide; drupes fleshy, black and shiny at maturity; stems often prickly........................1. *L. horrida*
1. Bractlets broadly ovate, closely subtending the flowers; corolla white or pink, the tube up to 5 mm. long, the limb about 3 mm. wide; drupes dry at maturity, the exocarp thin; stems unarmed ....2. *L. macropoda*

1. Lantana horrida H.B.K. Near Sells, western Pima County, about 2,500 feet, bank of a stream, August (*Peebles et al.* 2741, *Harrison* 5822). Mississippi, Texas, southern Arizona, and Mexico.

It is unlikely that the plants at this Arizona locality were introduced by man, but the seeds may have been brought there by migrating birds.

2. Lantana macropoda Torr. La Noria, Santa Cruz County (*Mearns* 1153). Texas, southern Arizona, and northern Mexico.

3. ALOYSIA

Plants shrubby, aromatic; leaves scabrous-strigose above, tomentose or tomentulose beneath; flowers small, in very open, leafy panicles composed of slender, elongate, spikes or spikelike racemes; bractlets narrow, not closely subtending the flowers; calyx tubular-campanulate, deeply 4-lobed, conspicuously villous-hirsute; corolla somewhat irregular, 2-lipped, whitish or bluish.

A. lycioides and *A. Wrightii* are neat and graceful shrubs, responding well to cultivation. They afford browse for livestock and are reported to yield excellent honey.

KEY TO THE SPECIES

1. Leaves entire or sparsely and irregularly denticulate, oblong, elliptic, or obovate, dull green and scarcely rugose above; young stems canescent-puberulent....1. *A. lycioides*
1. Leaves regularly crenate with numerous teeth, ovate or suborbicular, bright green and rugose above, whitish beneath; young stems finely whitish-tomentose....2. *A. Wrightii*

1. Aloysia lycioides Cham. (*Lippia lycioides* Steud., *L. ligustrina* of authors). Near Ruby, Santa Cruz County, about 4,000 feet (*Harrison & King* 6964, *Kaiser* 49–109b). Western Texas to southern Arizona and Mexico; South America.

The Arizona plant is referred to var. *Schulzae* (Standl.) Moldenke, which has broader leaves and coarser pubescence than typical *A. lycioides*.

2. Aloysia Wrightii (Gray) Heller (*Lippia Wrightii* Gray). Grand Canyon (Coconino County) and northern Mohave County to Greenlee, Cochise, Pima, and Yuma counties, 1,500 to 6,000 feet, common on dry, rocky slopes, August to October. Western Texas to southern Nevada, southeastern California, and northern Mexico.

Reported as growing only on northern slopes at low altitudes and only on southern exposures at its higher altitudinal limit.

4. PHYLA

Plants herbaceous, not aromatic; stems procumbent, rooting at the nodes; leaves strigose but scarcely scabrous; inflorescences few, very dense, short-spicate or subcapitate; bractlets broad, imbricate, closely subtending the flowers; calyx campanulate, 2- to 4-toothed, strigose; corolla somewhat 2-lipped, 4-parted.

P. cuneifolia and *P. lanceolata* are efficient soil binders but are nowhere abundant in Arizona. *P. nodiflora* is sometimes used for lawns.

KEY TO THE SPECIES

1. Leaves thickish, somewhat rigid, narrowly oblanceolate to obovate, with 1 to 4 pairs of teeth above the middle; herbage strigillose-canescent, usually densely so (2).
1. Leaves thin, not rigid, with several (often 5 or more) pairs of teeth, these extending usually to the middle or lower; herbage green, sparsely strigose to nearly glabrous (3).
2. Peduncles shorter than to somewhat longer than the subtending leaves (rarely twice as long); fruit oval..1. *P. cuneifolia*
2. Peduncles 1.5 to 4 times as long as the subtending leaves; fruit broadly obovoid 2. *P. incisa*
3. Calyx lobes longer than the tube; leaves lanceolate to rhombic-ovate, mostly widest below the middle, toothed well below the middle........................3. *P. lanceolata*
3. Calyx lobes shorter than the tube; leaves rhombic-obovate or -oblanceolate, widest above the middle, seldom toothed below the middle........................4. *P. nodiflora*

1. **Phyla cuneifolia** (Torr.) Greene (*Lippia cuneifolia* (Torr.) Steud.). Apache County to Coconino, Yavapai, and Cochise counties, and along the Colorado River at Fort Mohave and near Yuma, stream beds and playas, usually in heavy soil, June to August. South Dakota and Wyoming to Texas, Arizona, and northern Mexico.

2. **Phyla incisa** Small (*Lippia incisa* (Small) Tidestrom). White Mountains, Apache County (?) (*Ellis* 4), Gadsden, Yuma County (*Stitt* 1319), and reported from Pima County. Missouri to Colorado, southward and southwestward to northern Mexico and southern California.

In Yuma County it grows as a weed in alfalfa fields.

3. **Phyla lanceolata** (Michx.) Greene (*Lippia lanceolata* Michx.). Tucson, Pima County (*Toumey* in 1892), also at St. Thomas, Nevada, very near the northwestern border of Arizona (*Purpus* 6180). Ontario to Florida, westward to California and northern Mexico.

4. **Phyla nodiflora** (L.) Greene (*Lippia nodiflora* (L.) Michx.). Tucson, Pima County, well established in lawns (*Pultz* 1641), also at Yuma. Tropical and subtropical regions of both hemispheres.

The Arizona plants belong to var. *rosea* (D. Don) Moldenke.

5. BOUCHEA (267)

Plants annual; stem erect, leafy, sparingly branched; leaves long-petioled, the blades oval or ovate, crenate or serrate; flowers in slender elongate spikes; corolla deep violet; nutlets 2.

1. **Bouchea prismatica** (L.) Kuntze. Cochise and Santa Cruz counties, 3,500 to 6,000 feet, infrequent in rich, shaded ground along streams, August to October. Southern Arizona, Mexico, and tropical America.

The species is represented in Arizona chiefly by var. *brevirostra* Grenzebach, with the nutlets short-beaked.

6. TETRACLEA

Perennial herbs, with ashy-green foliage and fine, rough pubescence; leaves ovate or oblong, mostly toothed; flowers in axillary cymes; calyx equally 4- or 5-lobed, the lobes acute or acuminate, longer than the tube, this hemispheric in fruit; corolla at least twice as long as the calyx, the lobes oval, subequal; stamens clearly exserted beyond the corolla, curving up under the upper lobes but not strongly so; nutlets obovate, pitted, hirtellous.

This genus is usually included in the Labiatae, but recent authorities consider it to belong to the Verbenaceae, in which family it was placed at first by Asa Gray.

1. **Tetraclea Coulteri** Gray. Coconino, Yavapai, Greenlee, Maricopa, Cochise, Santa Cruz, and Pima counties, 4,500 feet or lower, April to August. Western Texas to southern Arizona and Mexico.

Specimens from Cochise County may be referable to var. *angustifolia* (Woot. & Standl.) Nels. & Macbr. (*T. angustifolia* Woot. & Standl.), which was described as a more slender, less pubescent plant with narrower calyx lobes, smaller corollas, and more strongly reticulate nutlets; but there seems to be complete intergradation with typical *T. Coulteri.* Specimens from Cameron, Coconino County (*Ripley & Barneby* 4899, *J. T. Howell* 24400), have a 4-lobed calyx, and plants with both 4- and 5-lobed calyxes have been collected elsewhere.

118. LABIATAE. Mint Family

Contributed by Carl Epling

Plants herbaceous, sometimes annual, or perennial and then often spreading by rhizomes, less often woody shrubs or undershrubs; stems usually square; leaves opposite; flowers variously disposed; calyx commonly more or less 2-lipped, the upper 3 teeth more or less joined, the lower pair usually free, or all sometimes equal, the tube sometimes enlarged in fruit; corolla obscurely to (usually) clearly bilabiate, the upper 2 petals usually joined to form an erect, sometimes galeate lip enclosing the stamens, or this sometimes very short and deeply notched, or the 5 lobes sometimes subequal, the lower lip usually spreading, its middle lobe sometimes dipperlike; stamens 4 or 2, usually in 2 unequal pairs, the connective sometimes strongly developed at the expense of the filament, the anthers parallel or divergent, with one theca sometimes completely or partly aborted; style bifid at the apex, arising between the quite distinct lobes of the 4-lobed ovary, or from near the apex of the ovary when the lobes (and the nutlets) are partly united below.

An attractive family of largely aromatic plants, including such notable contributions to the herb garden as mint, sage, lavender, thyme, and rosemary. Nearly all the Arizona Labiatae are good honey plants.

KEY TO THE GENERA

1. Functional stamens 2, with small staminodes sometimes also present (2).
1. Functional stamens 4, staminodes none (6).
2. Stamens appearing jointed, the connective strongly developed, often arcuate and bearing fertile thecae at both ends, or straight and thrust downward into the corolla tube, or the lower end wholly abortive. Apparent calyx teeth 3 . . . . 15. *Salvia*
2. Stamens not as in genus *Salvia*, both thecae fertile and approximate or confluent (3).
3. Corolla nearly regular, small. Flowers in axillary glomerules; herbage glabrous or merely puberulent . . . . 21. *Lycopus*
3. Corolla irregular, distinctly bilabiate (4).
4. Flowers numerous, in dense, subglobose verticils, these forming an interrupted spike or terminal cluster, the glomerules subtended by numerous conspicuous bracts; leaves 3 to 8 cm. long or longer . . . . 16. *Monarda*
4. Flowers 1 to 6 (seldom more), in sessile or pedunculate, axillary clusters, sometimes subtended by small bracteoles, the clusters never dense and subglobose; leaves rarely as long as 2 cm. (5).
5. Stems ashy with small, curled or spreading, not at all feltlike hairs; calyx variously pubescent but not as in genus *Poliomintha*, the lower pair of teeth usually bristly with stiffish hairs along the margin . . . . 17. *Hedeoma*
5. Stems whitened with a dense feltlike tomentum; calyx softly and densely hairy with hairs as long as the calyx teeth . . . . 18. *Poliomintha*
6. Calyx teeth 10, hooked at apex. Flowers in dense, subglobose glomerules; upper lip of the corolla deeply notched . . . . 5. *Marrubium*
6. Calyx teeth 5 or fewer, rarely none (7).
7. Teeth of the calyx none, the lips even and rounded like the lips of a purse. Upper lip of the corolla galeate, completely enclosing the stamens (8).

7. Teeth of the calyx 5 (9).
8. Entire calyx becoming enlarged and bladderlike at maturity, completely enclosing the nutlets ....3. *Salazaria*
8. Upper surface of the calyx bearing a small appendage, this developing at maturity into an erect, conspicuous flap, the calyx at maturity separating into 2 portions longitudinally, the upper, flap-bearing portion deciduous......................4. *Scutellaria*
9. Calyx enlarging into a flaring, veiny funnel 2.5 cm. across. Plants quite glabrous 13. *Moluccella*
9. Calyx enlarging to some extent, but not as in *Moluccella* (10).
10. Lobes of the calyx markedly unequal; flowers in dense, oblong, bracteate spikes (11).
10. Lobes of the calyx somewhat unequal in length, but otherwise all much alike, deltoid, lanceolate, or subulate (12).
11. Upper calyx lip with the middle tooth ovate, twice as broad as the other teeth, all of the teeth spine-tipped; bracts spinose-toothed, hollylike..................8. *Moldavica*
11. Upper calyx lip with the teeth completely joined to form a tricuspidate, squarish lip; bracts ciliate, entire, clasping...9. *Prunella*
12. Hairs of the canescent calyx and foliage branched, intricately tangled; midlobe of the lower lip of the corolla distinctly dipper-shaped. Plants shrubby.........23. *Hyptis*
12. Hairs simple; midlobe of the lower lip of the corolla plane and spreading or at most shallowly cupped (13).
13. Flowers solitary or not more than 3 in the cluster (14).
13. Flowers several or many at each node, forming an interrupted or congested spike, rarely a terminal head. Corolla tube short or if elongate, then narrow and gradually enlarged upward (except in genus *Clinopodium*); stamens attached near the throat of the corolla (16).
14. Flowers solitary and subsessile in the axils of small bracts, disposed in racemes or panicles. Lower pair of stamens attached near the middle of the corolla tube, the latter distended above ....10. *Dracocephalum*
14. Flowers solitary or in pedunculate cymules of 2 or 3, in the axils of the upper leaves. Calyx teeth equaling the calyx tube, or longer; nutlets pitted (15).
15. Flowers solitary in the axils; leaves pinnatifid, or (at least some of them) 3-lobed; upper lip of the corolla much shorter than the lower, which is spreading and obvious 1. *Teucrium*
15. Flowers usually in small cymules, sometimes solitary; leaves entire or somewhat toothed, not at all lobed; limb of the corolla bell-shaped, the lobes subequal....2. *Trichostema*
16. Anther sacs parallel or nearly so (17).
16. Anther sacs divergent or divaricate, forming an angle of about 90 degrees, or placed end to end (20).
17. Leaves 3- to 5-parted, the divisions incised...........................12. *Leonurus*
17. Leaves crenate, serrate, or entire (18).
18. Leaves (at least the lower ones) deltoid-ovate, truncate or cordate at base. Upper lip of the corolla conspicuous and galeate................................6. *Agastache*
18. Leaves elliptic or oblong, narrowed at base, not at all cordate (19)
19. Upper lip of the corolla apparently wanting, very short and scarcely exserted from the calyx, deeply notched; calyx becoming saccate; nutlets pitted. Stamens arched, exserted through the notch of the upper corolla lip.......................1. *Teucrium*
19. Upper lip of the corolla plane, entire, subequal to the other lobes; calyx top-shaped; nutlets smooth ....22. *Mentha*
20. Flowers disposed in 1 or sometimes 2 hemispheric glomerules at the end of each branch (21).
20. Flowers in several glomerules disposed in the axils of the more or less reduced upper leaves and forming interrupted spikes (22).
21. Glomerules often 2; upper calyx teeth joined to the middle; bracts acicular or linear; corolla lobes 4, the upper 2 joined to form an erect lip..............19. *Clinopodium*
21. Glomerules solitary and terminal; calyx teeth subequal; bracts ovate or ovate-lanceolate; corolla lobes 5, subequal.......................................20. *Monardella*
22. Calyx 5-veined; plants annual...11. *Lamium*

22. Calyx with many (usually 14 or more) veins; plants perennial (23).
23. Flower clusters pedunculate; upper lip of the corolla essentially plane, the tube glabrous within; leaves petiolate, truncate or cordate at base; calyx generally 14- or 15-veined, the veins bright green, the intervenous tissue transparent................7. *Nepeta*
23. Flower clusters sessile; upper lip of the corolla cupped, the tube bearing a hairy annulus within; leaves sessile or, if petiolate and truncate at base, then the corolla bright red; calyx with an indefinite number of veins (usually more than 15), the intervenous tissue opaque and green..14. *Stachys*

1. TEUCRIUM. GERMANDER

Perennial herbs up to 1 m. high, with serrate, simple leaves and the flowers in terminal, slender spikes, or else smaller plants, annual or perennial, with at least some of the leaves pinnatifid and with the flowers in the axils of the reduced upper leaves; calyx saccate and toothed, or deeply 5-lobed; corolla pinkish, bluish, or pallid, the upper lip very short, deeply notched, the lower lip conspicuous and spreading, with small, lateral lobes; stamens 4, paired; nutlets roughened.

KEY TO THE SPECIES

1. Leaves merely serrate; flowers in a terminal bracteate spike; calyx saccate, 5-toothed 1. *T. canadense*
1. Leaves laciniate or pinnatifid, less often 3-lobed or some of them entire and linear; flowers in the axils of the reduced upper leaves; calyx deeply lobed, the lobes lanceolate (2).
2. Pedicels up to 40 mm. long; plants perennial......................2. *T. glandulosum*
2. Pedicels 1 to 5 mm. long; plants annual...............................3. *T. cubense*

1. Teucrium canadense L. Navajo, Yavapai, Graham, and Pima counties, 2,500 to 5,000 feet. Canada to Florida, Arizona, and Mexico.

Most of the Arizona specimens belong to var. *angustatum* Gray, which has fine and appressed pubescence, narrow, obscurely serrate leaves, and a nonglandular calyx, type from Camp Grant, Graham County (*Palmer* 177). A collection on Hassayampa Creek, Yavapai County (*Palmer* 166), was one of the types of *T. occidentale* Gray (*T. canadense* var. *occidentale* McClintock & Epling), which has coarser and more spreading pubescence, broader, more deeply serrate leaves, and a glandular calyx.

2. Teucrium glandulosum Kellogg. Chemehuevi (Mohave County), Castle Dome Mountains (Yuma County), 2,000 feet, locally abundant in depressions and arroyos, May to July. Western Arizona, southern California, and Baja California.

An ample collection made recently on Cedros Island, the type locality, permits comparison with the Arizona specimens. The habit and pubescence are very similar. The flowers of the Arizona specimens are somewhat larger and seem paler, but no other differences are apparent.

3. Teucrium cubense Jacq. Maricopa, Pinal, Santa Cruz, Pima, and Yuma counties, 4,000 feet or lower, commonly in wet soil along streams, March to May. Southern Texas to southeastern California; widely distributed in tropical and subtropical America.

The Arizona plant is subsp. *depressum* (Small) McClintock & Epling (*T. depressum* Small).

2. TRICHOSTEMA (268). Blue-curls

Plants herbaceous or suffrutescent, the herbage glandular-puberulent or pilosulous; leaves petioled, the blades entire; flowers axillary, solitary or in loose, few-flowered cymes; calyx lobes approximately equal, about equaling the tube, the latter hemispheric in fruit; stamens nearly straight to strongly arched; nutlets rugose-reticulate, somewhat hirtellous.

KEY TO THE SPECIES

1. Plants annual; leaves lanceolate to narrowly elliptic, the upper ones not greatly reduced; cymes 1- to 3-flowered; corolla pale blue, 3 to 6 mm. long, little surpassing the calyx; stamens straight or nearly so, 2 to 4 mm. long .... 1. *T. brachiatum*
1. Plants perennial, suffrutescent; leaves oval or ovate, the upper ones much reduced and bractlike; cymes 1- to 7-flowered, forming a narrow panicle; corolla lavender or violet and white, 7 to 14 mm. long, greatly surpassing the calyx; stamens strongly arched, 10 to 25 mm. long .... 2. *T. arizonicum*

1. Trichostema brachiatum L. (*Isanthus brachiatus* B.S.P.). Southern Navajo, southern Coconino (?), Gila, and Cochise counties, 5,500 to 6,500 feet, open pine forest, August to October, rare in Arizona. Canada to Florida, Texas, and Arizona.

2. Trichostema arizonicum Gray. Greenlee, Cochise, Santa Cruz, and Pima counties, 3,500 to 6,000 feet, rocky slopes, July to October, type from the Chiricahua Mountains (*Wright* 1541). Southern New Mexico and Arizona, and northern Mexico.

3. SALAZARIA. Bladder-sage

Subspinose shrubs with divaricate branches and inconspicuous leaves; flowers in the axils of small, bractlike leaves; calyx equally 2-lipped, the lips entire, becoming inflated at maturity into a papery bladder enclosing the nutlets; corolla violet and white, tubular, the limb relatively short, the lateral lobes more or less joined with the upper lip to form a galea, this including the stamens and style; stamens 4, paired; nutlets roughened.

1. Salazaria mexicana Torr. Mohave, Yavapai, western Gila, Maricopa, and Yuma counties, usually below 3,000 feet, foothills and washes in the creosote-bush association, reaching the margin of the juniper association, March to May, and October. Western Texas to southern Nevada, Arizona, southern California, and northern Mexico.

This plant is reported to furnish forage for livestock throughout the year in the drier parts of the state. The flowers and bladderlike fruits are attractive.

4. SCUTELLARIA (269). Skull-cap

Small, perennial herbs, either with few stems and slender, spreading rhizomes, or with several stems ascending from a woody caudex; leaves petiolate, entire or crenate-serrate; flowers axillary in the upper part of the plant, or borne in lateral racemes and subtended by leaflike bracts; corolla violet, tubular, the limb relatively short, the lateral lobes more or less joined with the upper lip to form a galea, this including the stamens and style; stamens 4, in pairs, with 1 anther sac abortive in the lower pair; nutlets variously tuberculate.

KEY TO THE SPECIES

1. Plants with spreading slender rhizomes; leaves deltoid-ovate or oblong, crenate-serrate, mostly 3 to 7 cm. long; nutlets buff or straw-colored (2).
1. Plants with tufted stems from a woody caudex; leaves mostly ovate or oval, entire or subentire, mostly 1 to 2.5 cm. long; nutlets dark (3).
2. Galea and tube of the corolla 5 to 7 mm. long; flowers in lateral and terminal, bracteate racemes ....1. *S. lateriflora*
2. Galea and tube of the corolla 13.5 to 21 mm. long; flowers in the axils of the upper leaves ....2. *S. galericulata*
3. Pubescence minute, curled, essentially eglandular; nutlets dark, the protuberances flattened, giving the appearance of a mosaic..............................3. *S. tessellata*
3. Pubescence spreading and obvious, often dense, more or less glandular; nutlets black, evenly and closely tuberculate.....................................4. *S. potosina*

1. Scutellaria lateriflora L. Southern Coconino County and Yavapai County, moist ground, August. Canada to Florida, New Mexico, and central Arizona.

These Arizona stations represent the southwesternmost extension of this species.

2. Scutellaria galericulata L. White Mountains (Apache County), Lakeside (Navajo County), Buck Springs (Coconino County), 6,000 to 9,500 feet, moist ground, June to August. Widely distributed in temperate North America; Eurasia.

This species reaches its southern limit at these Arizona localities.

3. Scutellaria tessellata Epling. Yavapai, Maricopa, Cochise, Santa Cruz, and Pima counties, mostly 4,000 to 6,000 feet, rocky slopes and canyons, August to October, type from the Huachuca Mountains (*Jones* in 1903). Southern New Mexico and Arizona.

An ally of *S. resinosa* Torr. and *S. Wrightii* Gray, separable from them chiefly on the basis of the pubescence and nutlets and, to a less extent, the habit of growth.

4. Scutellaria potosina T. S. Brandeg. Southern Apache, Yavapai, Gila, Maricopa, Pinal, and Cochise counties, 2,500 to 5,500 feet, April to August. Central Arizona to Mexico.

The Arizona plants belong to subsp. *platyphylla* Epling, type from Fish Creek Canyon, Maricopa County (*Harrison* 7778), which has the stems more copiously hirsute and usually taller and the leaves longer than in typical *S. potosina*. Specimens from this state have commonly been referred to *S. Drummondii* Benth., an annual species of Texas with similar pubescence.

5. MARRUBIUM. Horehound

Perennial herbs with densely white-woolly stems and strongly corrugate leaves; flowers crowded in subglobose verticils, these forming interrupted spikes; calyx tubular, with 10 hooked teeth; corolla white, small, the upper lip erect, notched, laterally reflexed; stamens 4, paired; nutlets black, somewhat roughened.

1. Marrubium vulgare L. Throughout the state, a common roadside weed in some places, April to September. Extensively naturalized in the United States; from Europe.

The hooked calyx teeth at maturity cling to wool and clothing. The present use of horehound in medicine is limited almost entirely to a confection for checking coughs and easing sore throat. It formerly was used in domestic medicine for colds and dyspepsia and to expel worms.

6. AGASTACHE (270). GIANT-HYSSOP

Perennial herbs, with short, sparse pubescence; leaves deltoid, crenate, at least at the base of the plant; flowers either in pedunculate cymules, these disposed in a slender, spare panicle, or sessile in verticils, these forming an interrupted or dense spike; calyx tubular, 5-toothed, the teeth usually thin and membranaceous, less often subulate and somewhat rigid, deltoid or attenuate, subequal; corolla pallid or rose-colored, tubular, arcuate, somewhat hooded, the lips subequal; stamens 4, slightly exserted from the corolla tube.

A widespread genus of North America, centered in the southwestern United States, but of Eurasian affinities. The species are apparently localized. The popular name "horsemint" is sometimes applied to these plants as well as to species of *Monarda.*

KEY TO THE SPECIES

1. Calyx tube 2 mm. long or shorter; corolla tube 2.5 to 5 mm. long (2).
1. Calyx tube usually more than 3 mm. long; corolla tube 6 to 30 mm. long (3).
2. Spikes interrupted, the verticils distinct; leaves usually about twice as long as wide; corolla rose-purple, the tube 3 to 5 mm. long . . . . 1. *A. Wrightii*
2. Spikes more or less continuous, at least toward apex; leaves usually about 3 times as long as wide; corolla white, the tube 2.5 to 3 mm. long . . . . 2. *A. micrantha*
3. Leaves linear or linear-lanceolate, the margins entire. Corolla rose-purple, the tube 20 to 30 mm. long . . . . 3. *A. rupestris*
3. Leaves deltoid-ovate or deltoid-lanceolate, the margins crenate or serrate (4).
4. Corolla tube 20 to 30 mm. long. Calyx usually deep rose-purple . . . . 4. *A. Barberi*
4. Corolla tube less than 20 mm. long (5).
5. Corolla usually rose-purple, the tube 6 to 10 mm. long . . . . 5. *A. breviflora*
5. Corolla whitish to rose-purple, the tube 9 to 18 mm. long . . . . 6. *A. pallidiflora*

1. Agastache Wrightii (Greenm.) Woot. & Standl. Greenlee County to Yavapai County, south to Cochise, Santa Cruz, and Pima counties, 4,000 to 6,000 feet, rich soil, canyons and slopes, June to October. Southern New Mexico, Arizona, and northern Mexico.

2. Agastache micrantha (Gray) Woot. & Standl. Springerville, Apache County, 6,000 feet (*Goldman* 2417), Atoscacita (Atascosita?) Spring, Apache County (?) (*Eggleston* 17055), Red Mountain, Coconino County, 6,500 feet (*Phillips* 2842). Western Texas to eastern Arizona and Mexico.

3. Agastache rupestris (Greene) Standl. Mazatzal Mountains and near Payson (Gila County), Galiuro Mountains (Pinal County), Huachuca Mountains (Cochise County), Baboquivari Mountains (Pima County), 4,000 to 7,000 feet, September and October. Southwestern New Mexico and central and southern Arizona.

An apparent hybrid between *A. rupestris* and *A. Wrightii* (270, p. 230) has been collected in Bear Valley or Sycamore Canyon, Santa Cruz County (*Goodding* 4521, *Kearney & Peebles* 14448, *Darrow & Haskell* 2054). The lower leaves are ovate to ovate-lanceolate and more or less crenate, the upper leaves linear-lanceolate and entire. The calyx tube is about 3 mm. long (4 to 6 mm. in *A. rupestris,* 2 to 3 mm. in *A. Wrightii*), and the corolla tube is 10 to 15 mm. long (20 to 30 mm. in *A. rupestris,* 3 to 5 mm. in *A. Wrightii*). Typical *A. rupestris,* however, has not been found in this locality, and the collections cited may represent an undescribed species.

4. Agastache Barberi (Robins.) Epling. Patagonia Mountains (Santa Cruz County), about 5,000 feet (*Peebles & Harrison* 4748, *Kearney & Peebles* 10122), August and September. Southern Arizona and northern Mexico.

Flowers showy for the genus, lavender-purple.

5. Agastache breviflora (Gray) Epling. Mountains of Graham, Cochise, Santa Cruz, and Pima counties, 6,000 to 7,000 feet, July to October, type from the Santa Rita Mountains (*Pringle*). Southern New Mexico and Arizona, and northern Mexico.

6. Agastache pallidiflora (Heller) Rydb. Lukachukai and White mountains (Apache County), and many localities in Coconino County, 7,000 to 10,000 feet, rich, moist soil of coniferous forests, July to October, type from Bill Williams Mountain (*MacDougal* 313). Colorado, New Mexico, and Arizona.

Subsp. *typica* Lint & Epling, with a greenish calyx and a white or lavender-tinged corolla, is found in Apache, Coconino, and Yavapai counties. The species is represented in southern Arizona by subsp. *neomexicana* (Briq.) Lint & Epling, with a rose-purple calyx and corolla, which occurs in the mountains of Graham, Cochise, and Pima counties, 7,500 to 9,500 feet.

7. NEPETA

Plants perennial, herbaceous; leaves soft, ample, usually canescent, truncate or subcordate at base, rather coarsely toothed; flowers in dense, usually pedunculate cymes, these disposed in an interrupted spike; calyx tubular or campanulate, the tube somewhat constricted above, the teeth deltoid-subulate, somewhat spreading, the posterior 3 teeth joined at base, the orifice therefore oblique; corolla white or pinkish, the upper lip erect, notched, laterally reflexed; stamens 4, paired; nutlets smooth, oblong-ovate.

1. Nepeta Cataria L. Lakeside (Navajo County), Kaibab Plateau and near Flagstaff (Coconino County), Prescott (Yavapai County), Chiricahua Mountains (Cochise County), usually at roadsides, July to September. Widely distributed in North America; naturalized from the Mediterranean region.

Catnip. Although having no therapeutic virtue other than that of a mild aromatic, this was formerly an official drug plant. It is reputed to have a quieting effect on the nerves and is used as a mild stimulant, tonic, and emmenagogue. The odor of the plant has a peculiar attraction for cats.

8. MOLDAVICA. Dragon-head

Glabrate, annual or biennial herbs; leaves oblong, sharply and coarsely toothed; flowers in dense, oblong, often leafy spikes, these sometimes interrupted or the lower verticils remote, the subtending bracts spinose along the margins; calyx tubular, strongly veined, the anterior teeth deltoid-lanceolate, spinose at tip, the posterior tooth ovate, twice as broad as the others; corolla blue or purplish pink, scarcely exserted from the calyx, the upper lip erect, notched; stamens 4, paired; nutlets oblong-ovate, smooth.

See also under *Dracocephalum* (below).

1. **Moldavica parviflora** (Nutt.) Britton. Apache, Coconino, Mohave, Yavapai, Greenlee (?), Gila, Maricopa, and Pinal counties, 3,500 to 8,500 feet, pine woods, April to October. Canada to New Mexico and Arizona.

The Havasupai Indians are reported to make flour from the seeds of this plant.

9. PRUNELLA. Self-heal

Small, perennial herbs; leaves oblong, long-petioled, subentire; flowers in dense, terminal, bracteate spikes, the bracts sheathing; calyx 2-lipped, the upper lip truncate, bearing 3 cusps, the lower teeth essentially free; corolla violet, the upper lip galeate, enclosing the stamens; stamens 4, paired; nutlets smooth, ovate.

1. **Prunella vulgaris** L. White Mountains (Apache County), Navajo County, and Kaibab Plateau (Coconino County) to the mountains of Graham and Pima counties, 5,000 to 9,500 feet, moist ground, June to September. Throughout the cooler parts of North America; Eurasia.

The plant was formerly used as a domestic remedy for various disorders.

10. DRACOCEPHALUM. False Dragon-head

Perennial herbs; stems erect, stiffish; leaves glabrous, serrate, sessile; flowers in terminal, showy racemes or panicles; calyx 5-toothed, tubular; corolla rose-colored or pallid, funnelform, with an inflated throat and an arched entire upper lip; stamens 4, in 2 pairs, included in the galea or but slightly exserted, the longer pair attached near the middle of the tube.

A collection at Santa Cruz, Sonora, about 10 miles south of the Arizona boundary (*Wright* 1530), was identified by Gray as *Physostegia virginiana* (L.) Benth. var. *obovata* (Ell.) Gray. It seems to belong to *Dracocephalum Correllii* Lundell (*Physostegia Correllii* Shinners).[1]

11. LAMIUM. Dead-nettle

Plants annual; herbage soft-pubescent; stems low, decumbent at base, with elongate lower internodes; leaves coarsely crenate or incised, the lower ones petioled, cordate, the upper ones sessile; flowers in 1 to 3 dense axillary verticils; calyx teeth 5, nearly equal, awn-pointed; corolla reddish purple, the tube slender, elongate, the throat enlarged, the upper lip galeate, erect or nearly so, hairy on the back, the lower lip spreading, spotted, with the midlobe much larger than the others and contracted at base; stamens 4, ascending under the upper lip of the corolla, the anther cells divaricate, hairy; nutlets truncate at apex.

[1] McClintock (*Leaflets West. Bot.* 5: 171–172. 1949) has proposed restoration of the name *Physostegia* for this American genus. If this proposal is accepted, the generic name *Dracocephalum* should be substituted for *Moldavica*, and *D. parviflorum* Nutt. for *M. parviflora* (Nutt.) Britton.

1. **Lamium amplexicaule** L. Tucson, Pima County, lawns and waste ground, May (*L. Benson* 9424, *Gould* 2935), San Francisco Peaks, Coconino County (*Deaver* 2852). Extensively naturalized in the United States; from Eurasia. Henbit.

12. LEONURUS. MOTHERWORT

Plants herbaceous, tall, puberulent or glabrate; leaves long-petioled, the blades 3-cleft in the upper leaves, 3- to 5-parted with incised divisions in the lower leaves; flowers in axillary clusters much shorter than the leaves; calyx campanulate, 5-veined, the teeth triangular-aristate, somewhat unequal in length; corolla pink, markedly bilabiate, the tube about equaling the calyx, the upper lip erect and slightly concave, conspicuously white-bearded dorsally, the lower lip spreading and 3-lobed; nutlets 3-sided, truncate at apex.

1. **Leonurus Cardiaca** L. Lakeside, Navajo County (*Thornber* 8902), Mormon Lake, Coconino County (*Deaver* in 1928, *Collom* in 1936). Extensively naturalized in North America; from Eurasia.

13. MOLUCCELLA. MOLUCCA-BALM

Plants annual, herbaceous, glabrous; leaves coarsely toothed, rotund, petiolate; flowers several in the axils of the upper leaves; calyx with lobes wholly united into a flaring funnel-shaped structure resembling a small morning-glory flower; corolla white or pinkish, the upper lip concave or galeate, including the stamens; stamens 4, paired; nutlets truncate at apex.

1. **Moluccella laevis** L. Oracle, Pinal County, 5,000 feet (*Oslar* in 1903), also in southern Utah, very near the Arizona border. An occasional escape from gardens in the United States; native of the Mediterranean region.

14. STACHYS (271). BETONY, HEDGE-NETTLE

Perennial herbs; leaves deltoid-ovate or oblong, the upper ones gradually reduced; flowers usually 3 in the axils of leaflike bracts, disposed in interrupted spikes; flowering calyx turbinate, somewhat enlarged at maturity, the teeth more or less deltoid and spinulose at tip; corolla red, pink, or whitish, the tube cylindric, pilose-annulate within below the middle and often constricted at the annulus, even saccate, the upper lip galeate, including the stamens, the lower lip spreading; stamens 4, attached near the middle of the corolla tube, paired; nutlets obovate, smooth or roughened.

The tubers of *S. palustris* are sometimes eaten, and a related species known as Chinese artichoke is extensively cultivated in China and Japan.

KEY TO THE SPECIES

1. Corolla bright red, the tube 18 to 21 mm. long, transversely annulate near the base; leaves deltoid-ovate, petiolate....................................1. *S. coccinea*
1. Corolla white, pallid, or pink, the tube 5.5 to 8.5 mm. long, obliquely annulate below the middle and more or less constricted and saccate on the lower side; leaves oblong, sessile or nearly so (2).
2. Stems clothed with soft, appressed, silvery hairs; leaves prevailingly 8 to 12 mm. wide, appressed-pubescent on both surfaces, the upper surface somewhat silky 2. *S. Rothrockii*
2. Stems clothed with spreading, stiffish hairs; leaves prevailingly 1.5 to 4 cm. wide, thinly clothed with spreading hairs on both surfaces........................3. *S. palustris*

1. Stachys coccinea Jacq. (*S. limitanea* A. Nels.). Southern Apache County to Maricopa County, south to Cochise, Santa Cruz, and Pima counties, 1,500 to 8,000 feet, rich soil, canyons and slopes, March to October, type of *S. limitanea* from near Ruby, Santa Cruz County (*A. & R. Nelson* 1471). Western Texas to southern Arizona and Mexico.

A showy plant, responding readily to cultivation.

2. Stachys Rothrockii Gray. Apache County to eastern Mohave and northern Yavapai counties, 5,000 to 7,000 feet, chiefly in open pine forests, summer, type from Arizona, near Zuni, New Mexico (*Rothrock* 177). Western New Mexico, Arizona, and southwestern Utah.

The plants grow in colonies from deep rootstocks. The wilted corollas are purplish.

3. Stachys palustris L. White Mountains (Apache and Greenlee counties), Flagstaff and Willow Spring (Coconino County), 7,000 to 9,000 feet, moist, shady places, summer. Widely distributed in the cooler parts of North America; Eurasia.

The Rocky Mountain (and Arizona) representative of this widespread complex is subsp. *pilosa* (Nutt.) Epling. The flowers are lavender-pink.

15. SALVIA (272, 273). Sage

Shrubs or herbs of varied habit; flowers in interrupted spikes or terminal heads; calyx 2-lipped, usually laterally compressed, the upper lip commonly entire, less often 3-mucronate or 3-toothed, the lower lip usually 2-toothed; corolla blue, red, or white, tubular, strongly 2-lipped, the upper lip either plane and notched, or galeate and entire; stamens 2, exserted from the corolla tube beyond the limb, or contained within the galea, the connective strongly developed, often more prominent than the filament, bearing a single terminal anther sac (rarely 1 at each end), either straight and projected back into the corolla tube, or geniculate; style usually exserted from the galea or beyond the upper lip; nutlets smooth.

KEY TO THE SPECIES

1. Upper lip of the corolla essentially plane or laterally reflexed and usually notched, not at all galeate; stamens clearly exserted beyond the tube and limb of the corolla (2).
1. Upper lip of the corolla clearly galeate, including the stamens, or these not much exserted (5).
2. Plants annual, with dissected leaves. Lower arm of the connective bearing a fertile anther sac .... 1. *S. Columbariae*
2. Plants shrubby, the leaves not dissected (3).
3. Leaves rugose, deltoid or oblong-elliptic, crenulate, green, thinly hispidulous .... 4. *S. mohavensis*
3. Leaves smooth, obovate, entire, canescent (4).
4. Calyx 5 to 7 mm. long; corolla tube 5 to 10 mm. long, pubescent within above the middle .... 2. *S. carnosa*
4. Calyx 8 to 13 mm. long; corolla tube 15 to 22 mm. long, strongly pilose-annulate below the middle .... 3. *S. pachyphylla*
5. Leaves 10 to 25 cm. long; flowers subtended by conspicuous, persistent, sheathing bracts. A coarse, woolly herb .... 15. *S. aethiopis*
5. Leaves rarely as much as 8 cm. long; bracts deciduous or if persistent, then not sheathing the flowers (6).

6. Perennial herbs; leaves usually pinnate, at least below; stamen connectives bearing fertile thecae at both ends (7).
6. Perennial or annual herbs, or shrubs; leaves simple throughout; lower theca of each stamen wholly abortive (8).
7. Leaves 3-foliolate, or simple by suppression of the lateral pair of smaller leaflets, the terminal leaflet essentially rotund or broadly and obtusely deltoid, coarsely sinuate-dentate .... 5. *S. Davidsonii*
7. Leaves 3- to 5-foliolate, rarely simple by suppression of the lateral leaflets, the terminal leaflet deeply and irregularly toothed or lobed........ 6. *S. Henryi*
8. Stamen connective straight or slightly curved, directed downward into the tube and across it, usually bearing a small, triangular tooth near the middle (9).
8. Stamen connective patently bent at a sharp angle within the corolla tube, the terminal portion assurgent into the throat, often expanded at apex but bearing no theca. Flowers, at least the lower ones, usually 3 or more in the verticil (12).
9. Verticils usually of 3 or more flowers (10).
9. Verticils mostly of 2 opposite flowers (11).
10. Herbage canescent; leaves deltoid-ovate; corolla tube 6.5 to 7 mm. long, much surpassing the calyx .... 12. *S. amissa*
10. Herbage green, sparsely hirtellous; leaves rounded-ovate; corolla tube 3.5 to 4 mm. long, shorter than the calyx........ 13. *S. tiliaefolia*
11. Plants annual; corolla blue........ 11. *S. reflexa*
11. Plants perennial, suffrutescent; corolla lavender-pink or crimson...... 14. *S. Lemmoni*
12. Leaves tomentulose and incanous with minute hairs, at least beneath; verticils crowded (13).
12. Leaves essentially green and glabrous on both surfaces, not at all canescent; verticils spaced at intervals of 1 to 3 cm. or more (14).
13. Calyx densely villous with branched hairs; herbs, woody chiefly at base..... 7. *S. Parryi*
13. Calyx canescent with minute, simple, appressed hairs; shrubs, generally 1 m. high or more........ 8. *S. pinguifolia*
14. Perennial herbs with creeping rootstocks; leaves deltoid-ovate, crenate, commonly 2.5 to 4 cm. wide........ 9. *S. arizonica*
14. Annual herbs; leaves oblong, sharply toothed, rarely more than 1.5 cm. wide
10. *S. subincisa*

1. Salvia Columbariae Benth. Mohave County to Graham, Cochise, and Pima counties (doubtless also Yuma County), 3,500 feet or lower, common in sandy washes, March to May. Southern Nevada, Arizona, and California.

One of the species known as chia. The seeds were utilized by the Indians to make pinole and also mucilaginous poultices. A mucilaginous beverage prepared from the seeds was popular with the Pima Indians. The seeds of other species known as chia are extensively used in Mexico for similar purposes.

2. Salvia carnosa Dougl. Coconino, Mohave, and Yavapai counties, 2,500 to 5,000 feet, sandy soil, plains and washes, spring. Washington to Arizona and California.

Desert sage. A small, compact, much-branched shrub, very ornamental in flower, the sky-blue corollas contrasting with the purple bracts. Although browsed to some extent by livestock its palatability is considered low.

KEY TO THE SUBSPECIES

1. Leaves 2 to 4 mm. wide, oblanceolate or linear........ 1. subsp. *Mearnsii*
1. Leaves 7 mm. or wider (2).
2. Bracts glabrate; leaves 8 to 15 (20) mm. wide........ 2. subsp. *argentea*
2. Bracts thinly hairy outside; leaves 7 to 10 (15) mm. wide........ 3. subsp. *pilosa*

Subsp. *Mearnsii* (Britton) Epling (*Audibertia Mearnsii* Britton) is a little-known plant, found near Fort Verde, Yavapai County (*Mearns* 246, the type collection, *McClintock* in 1947), reported also from Sedona and Jerome Junction (southern Coconino and northern Yavapai counties). In Coconino and Mohave counties occur subsp. *argentea* (Rydb.) Epling (*Audibertiella argentea* Rydb.), type from Mokiak Pass (*Palmer* 395), and subsp. *pilosa* (Gray) Epling (*Audibertia incana* Benth. var. *pilosa* Gray, *Salvia pilosa* Merriam).

3. Salvia pachyphylla Epling. Near Winslow, Navajo County, 5,000 to 5,500 feet (*Jones* in 1929, *Whiting* 756, *Peebles* 14406), eroded slopes, June. Northern Arizona, southern California, and Baja California.

4. Salvia mohavensis Greene. Mohave, western Maricopa, and northern Yuma counties, desert mountain ranges, 1,000 to 4,000 feet, dry, rocky slopes, spring. Western Arizona, southern Nevada, southeastern California, and northwestern Sonora.

A small shrub with a strong, sagelike odor.

5. Salvia Davidsonii Greenm. Havasu Canyon (western Coconino County), and Greenlee, eastern Maricopa, and Cochise counties, among rocks in shade, April to July, type from the Chiricahua Mountains (*Lemmon* 3077). Arizona, probably also in southern New Mexico.

The pubescence of this species, like some other Labiatae, is highly variable in the presence and abundance of elongate slender hairs on stems and leaves. So far as available material indicates, these hairs are infrequent in *S. Davidsonii* and generally represented in *S. Henryi.* In Havasu Canyon specimens have been collected in which the corolla of the first-named species is bluish purple when dried rather than bright red, and is said to have been "pink-mauve" living. The plant is at the same time glabrate, although the red-flowered form collected in the same place is densely pilose. Just what relation exists between these forms remains to be determined.

6. Salvia Henryi Gray. Chiricahua Mountains, Cochise County (*Lemmon* in 1882), Santa Rita Mountains, Pima or Santa Cruz County (*Pringle* in 1884), April to September. Western Texas to southern Arizona and Chihuahua.

7. Salvia Parryi Gray (*S. confinis* Fern.). Cochise, Santa Cruz, and Pima counties, 3,500 to 5,000 feet, April to August, type of *S. confinis* from near Fort Huachuca (*Lemmon* 2861). Southwestern New Mexico, southern Arizona, and northern Sonora.

8. Salvia pinguifolia (Fern.) Woot. & Standl. Greenlee, Graham, Maricopa, Pinal, Cochise, and Pima counties, 2,000 to 7,000 feet, rocky slopes, June to September. Western Texas to southern Arizona and Chihuahua.

A shrub, up to about 1.5 m. (5 feet) high, the flowers blue.

9. Salvia arizonica Gray (*S. arizonica* var. *huachucana* Jones). Mountains of Graham, Cochise, and Pima counties, 7,000 to 9,500 feet, rich, moist soil in forests, July to September, type from Mount Graham (*Rothrock* 407), type of var. *huachucana* from the Huachuca Mountains (*Jones* in 1903). Western Texas and southern Arizona. Flowers indigo blue.

10. Salvia subincisa Benth. Yavapai, Greenlee, Cochise, Santa Cruz, and

Pima counties, up to 5,500 feet, plains and mesas, August and September. Western Texas to Arizona and northern Mexico.

11. Salvia reflexa Hornem. Apache, Navajo, Coconino, and Yavapai counties, 4,000 to 7,000 feet, plains, mesas, rocky slopes, and open pine forest, July to October. North Dakota and Wyoming to Texas, Arizona, and Mexico.

Sometimes called Rocky Mountain sage. Medicinal properties are attributed to this plant, of which infusions are sometimes used in treating malarial and rheumatic fevers and as a tonic and astringent.

12. Salvia amissa Epling (*S. albiflora* Mart. & Gal. var. *Pringlei* Gray). Aravaipa Canyon (Graham County), Fish Creek (eastern Maricopa County), by streams in the Santa Catalina Mountains, Pima County (*Pringle* in 1881, the type collection), 1,500 to 3,000 feet, May to October. Known only from southern Arizona.

The plant collected in Aravaipa Canyon (*Darrow* in 1942) is exceptionally broad-leaved.

13. Salvia tiliaefolia Vahl. Mustang Mountains, Cochise County, 5,500 feet, in deep shade of a limestone cliff (*Darrow* 3628). Marfa, Texas (a street weed, probably introduced), and southeastern Arizona (apparently indigenous) to northern South America.

14. Salvia Lemmoni Gray. Mountains of Cochise and Pima counties, 6,000 to 8,000 feet, rocky slopes and canyons, July to October, type from the Huachuca Mountains (*Lemmon* in 1881). Southern Arizona and northern Mexico.

This species appears to be confluent with a polymorphic complex ranging from Arizona to central and eastern Mexico. Although the Arizona plants have a somewhat different habital aspect, they seem hardly separable from the variety of *S. microphylla* H.B.K. from Chihuahua that was described by Gray as var. *Wislizeni*. The plant has an odor of peppermint, according to Blumer.

15. Salvia aethiopis L. South Rim of the Grand Canyon (Coconino County), Prescott, Yarnell, and Peeples Valley (Yavapai County), fields and roadsides, May to August. Here and there in the United States (Texas, Arizona, Oregon); introduced from the Mediterranean region.

The plant has spread rapidly in Peeples Valley during the past 12 years and has become a pest in overgrazed range land.

16. MONARDA (274). Bee-balm

Perennial or annual herbs; leaves oblong, elliptic, or ovate; flowers crowded in dense, axillary or terminal glomerules, these subtended by an involucre of linear, oblong, or ovate, reflexed or ascending bracts; calyx tubular, the teeth aristate or deltoid; corolla white, yellowish, pink, or rose, the tube longer than the lips, slender, abruptly expanded into a funnel-shaped throat, the upper lip galeate, arched or straight; stamens seated in the corolla throat, included within the galea or somewhat exserted; nutlets smooth, oblong.

Several species of this genus have been used in domestic medicine. *M. menthaefolia* has limited value as a forage plant and is sometimes cultivated by the Hopi Indians, who use it as a potherb, drying the plants for use in winter.

KEY TO THE SPECIES

1. Flowers in terminal heads; petioles seldom more (often less) than 5 mm. long; leaves lanceolate to ovate-lanceolate, 1 to 3 cm. wide; corolla purple......1. *M. menthaefolia*
1. Flowers in axillary and terminal clusters; petioles usually 5 to 10 mm. or longer; leaves oblong or elliptic, commonly 1 to 1.5 cm. wide; corolla whitish (2).
2. Bracts strongly reflexed, usually puberulent and whitish or purple, only the midvein at all prominent ....2. *M. austromontana*
2. Bracts spreading or ascending, glabrous and usually green on the upper surface, 3 of the veins usually prominent......................................3. *M. pectinata*

1. Monarda menthaefolia Graham (*M. fistulosa* L. var. *menthaefolia* Fern.). Apache County to Coconino County, south to Cochise and Pima counties, 5,000 to 8,000 feet, mostly in pine forest, summer. Canada to New Mexico and Arizona.

2. Monarda austromontana Epling. Apache, Navajo, and Gila counties to Cochise, Santa Cruz, and Pima counties, 4,000 to 8,500 feet, mesas and slopes, usually with grasses, late summer. Southwestern New Mexico, Arizona, and northern Mexico.

3. Monarda pectinata Nutt. Apache County to Coconino, Yavapai, and Gila counties, also at Avondale (Maricopa County), where probably from seeds carried down by flood water, commonly 5,000 to 7,000 feet, late summer. Nebraska and Colorado to Texas and Arizona.

17. HEDEOMA (275). MOCK-PENNYROYAL

Small, perennial herbs; leaves simple, essentially sessile; flowers in small cymules in the axils of the upper leaves, these either bractlike or not much reduced; calyx tubular, the teeth relatively short, the upper 3 teeth usually joined below the middle, the lower 2 teeth free, subulate, longer than the upper teeth, bristly with stiffish hairs, the orifice of the calyx more or less hispid-annulate; corolla rose, lavender, or white, tubular; stamens 2, exceeding the corolla tube; nutlets smooth, oblong.

The genus *Hedeoma* is largely confined to the Texas-Arizona region, but occurs also in South America. Probably some of the Arizona species are diaphoretic and stimulant-aromatic like *H. pulegioides*, the American pennyroyal of the eastern United States.

KEY TO THE SPECIES

1. Calyx teeth connivent at maturity, closing the tube, the whole calyx thus tapering from the middle of the tube to the apex. Leaves entire..................4. *H. Drummondii*
1. Calyx teeth usually spreading, the upper ones more or less reflexed, the calyx thus clearly bilabiate (2).
2. Tube of the calyx notably distended at base at maturity, the pouch forming more than half of the length of the tube, the latter about 1/3 as wide as long. Leaves entire or minutely denticulate ....2. *H. nanum*
2. Tube of the calyx only moderately distended, 1/5 to 1/4 as wide as long (3).
3. Leaves dentate, at least in the upper half (4).
3. Leaves entire or essentially so (6).
4. Leaves inconspicuously dentate, with 4 to 6 teeth, the veins inconspicuous; early leaves glabrous ....7. *H. oblongifolium*
4. Leaves conspicuously dentate, with usually more than 6 teeth, the veins conspicuous; early leaves pubescent (5).

5. Calyx tube about 4 times as long as wide; leaves elliptic to rhomboid, usually 2 to 3 times as long as wide. . . . . 5. *H. dentatum*
5. Calyx tube 4.5 to 5 times as long as wide; leaves oval or ovate, usually less than twice as long as wide. . . . . 6. *H. costatum*
6. Leaves linear-lanceolate, pointed, strict, stiffish, the lateral veins straight, rather prominent, parallel . . . . 1. *H. hyssopifolium*
6. Leaves elliptic-lanceolate, oblong, or oval, usually spreading, the lateral veins not markedly straight and parallel (7).
7. Stems prostrate, wiry, 5 to 20 cm. long; corolla tube hirtellous within and somewhat annulate near the middle. . . . . 3. *H. diffusum*
7. Stems erect, 25 to 50 cm. long, stiffish; corolla tube glabrous within or essentially so 7. *H. oblongifolium*

1. Hedeoma hyssopifolium Gray. Southern parts of Apache, Navajo, and Coconino counties to Cochise, Santa Cruz, and Pima counties, 5,000 to 9,500 feet, slopes and canyons, May to October, type from Mount Graham (*Rothrock* 418). New Mexico, Arizona, and northern Mexico.

2. Hedeoma nanum (Torr.) Briq. Coconino and Mohave counties to Cochise, Pima, and Yuma counties, 650 to 5,500 feet, March to October. Western Texas to southeastern California and Mexico.

KEY TO THE SUBSPECIES

1. Calyx tube 3.5 to 4 mm. long; basal leaves present, purple beneath, more than 1 cm. long. Corolla tube 5 to 7 mm. long. . . . . subsp. *macrocalyx*
1. Calyx tube 2.5 to 3 (rarely 4) mm. long; basal leaves usually deciduous (2).
2. Stems forming a close bunch or tuft, 10 to 15 cm. high, sparingly covered with retrorse pubescence above, glabrous below; leaves about 5 mm. long. . . . . . subsp. *californicum*
2. Stems loose, 10 to 35 cm. high, with downward-curling pubescence; leaves usually 5 to 10 mm. long. . . . . subsp. *typicum*

Subsp. *macrocalyx* Stewart occurs in Coconino, Mohave, Yavapai, Gila, Maricopa, Pinal, and Pima counties. Subsp. *californicum* Stewart is known in Arizona only from Toroweap Valley (northern Mohave County). Subsp. *typicum* Stewart occurs throughout the range of the species in Arizona.

3. Hedeoma diffusum Greene (*H. blepharodontum* Greene). Navajo, Coconino, and Yavapai counties, commonly 6,000 to 7,000 feet, May to July, types of *H. diffusum* and *H. blepharodontum* both from Flagstaff (*Rusby* in 1883, *Jones* in 1884). Known only from northern Arizona.

4. Hedeoma Drummondii Benth. Apache County to eastern and northern Mohave County, south to Cochise and Santa Cruz counties, 3,500 to 7,500 feet, April to September. North Dakota and Montana to Texas, Arizona, and northern Mexico.

5. Hedeoma dentatum Torr. Graham, Gila, Cochise, Santa Cruz, and Pima counties, 4,000 to 7,500 feet, August to October. Southern Arizona and northern Mexico.

6. Hedeoma costatum Gray. Bisbee, Cochise County (*Carlson* in 1915). Western Texas to southeastern Arizona and northern Mexico.

Carlson's specimen has relatively small flowers (about 9 mm. long), whereas the flowers are commonly 11 mm. or longer in this species.

7. Hedeoma oblongifolium (Gray) Heller. Coconino and Mohave counties to Greenlee, Cochise, and Pima counties, 2,000 to 8,000 feet, April to September. New Mexico, Arizona, and northern Mexico.

18. POLIOMINTHA

Shrubs, clothed with a minute, feltlike, whitish tomentum; leaves essentially linear or linear-oblong, thickish; inflorescence composed of axillary, sometimes subspicate, 1- to 3-flowered cymules; calyx 15-veined, the tube pilose with silky, spreading hairs, the teeth subequal, more or less connivent, the calyx strongly annulate in the throat; corolla pale blue, the tube strongly but incompletely annulate somewhat above the middle, with coarse, ascending hairs; stamens 2, seated well above the middle of the corolla tube, ascending against the upper lip; nutlets smooth, oblong.

1. Poliomintha incana (Torr.) Gray. Apache, Navajo, and Coconino counties, 4,500 to 6,000 feet, common in sandy deserts, May to September. Western Texas to southern Utah, northern Arizona, southeastern California, and northern Mexico.

An attractive plant, reported as responding well to cultivation. It is said that the Hopi Indians eat the herbage raw or boiled, sometimes drying it for use in winter, and use the flowers for seasoning food.

19. CLINOPODIUM. Wild-basil

Perennial herbs, hirsute with spreading hairs; leaves ovate, subentire; flowers in rather dense cymules, these either terminal and hemispheric or more commonly 2, forming an interrupted spike; calyx tubular, bilabiate, the 3 posterior teeth united to the middle, subulate, somewhat reflexed, the lower teeth free, subulate; corolla rose-colored, the upper lip erect, somewhat concave but scarcely galeate; stamens 4, paired, seated well within the corolla throat; nutlets smooth, ovate.

1. Clinopodium vulgare L. (*Satureja vulgaris* (L.) Fritsch var. *neogaea* Fern.). Coconino County, Chiricahua and Huachuca mountains (Cochise County), Santa Catalina Mountains (Pima County), 6,000 to 7,500 feet, rich, shaded ground, July to September. Canada to North Carolina, New Mexico, and Arizona; Eurasia.

20. MONARDELLA (276)

Perennial herbs, usually forming a low bush with several or numerous ascending branches; leaves elliptic or oblong, entire or denticulate; flowers in terminal, subglobose heads, these subtended by a few bracts similar to the leaves but usually thinner and colored; calyx tubular, the teeth narrowly deltoid, erect, subequal; corolla white and usually purple-punctate, or rose-colored, the tube somewhat exserted from the calyx, the lobes oblong-linear, subequal; stamens 4, exserted; nutlets oblong, smooth.

The genus *Monardella* is inexplicable in terms of a conventional system. The forms of each isolated region have a more or less characteristic aspect difficult or impossible to define, and form a continuous spectrum of variation with the forms of other regions. The genus is predominantly Californian.

KEY TO THE SPECIES

1. Pubescence curled downward; bracts ovate to rotund, usually cupped. . 1. *M. odoratissima*
1. Pubescence spreading; bracts narrowly ovate, spreading. . . . . 2. *M. arizonica*

1. Monardella odoratissima Benth. (*M. parvifolia* Greene). Kaibab Plateau and San Francisco Peaks (Coconino County), Hualpai Mountain (Mohave County), Beaver Creek (Yavapai County), Chiricahua Mountains (Cochise County), 3,500 to 11,000 feet, mostly in coniferous forests, June to September. Montana to Washington, south to New Mexico, Arizona, and California.

2. Monardella arizonica Epling. Black Mountains (Mohave County), near Santa Maria River and near Congress Junction (Yavapai County), Sierra Estrella (Maricopa County), Silver Bell and Quijotoa mountains (Pima County), Kofa Mountains (Yuma County), 2,000 to 4,000 feet, rocky ledges in canyons, spring, type from the Sierra Estrella (*Epling*). Known only from Arizona.

The plant, when fresh, has a strong odor suggesting that of *Pluchea camphorata*. The plant of the Black Mountains may not be conspecific.

21. LYCOPUS. Bugleweed

Plants perennial, glabrous or merely puberulent; stems erect, up to 1 m. high; leaves sessile or nearly so, oblong-lanceolate, sharply serrate; flowers small, in dense, sessile, axillary clusters; calyx campanulate, nearly equaling the corolla, regular or nearly so, with 4 or 5 subulate-aristate teeth; corolla whitish, nearly equally 4-cleft; perfect stamens 2, the others rudimentary; nutlets turbinate, trigonous.

***1. Lycopus lucidus** Turcz. The only collection known to the writer that is labeled as from Arizona (*E. Palmer* in 1869) may have been made elsewhere. Minnesota to British Columbia, south to Kansas, Arizona (?), and California; northern Asia.

Palmer's specimen has continuous, corklike wings on the angles and summit of the nutlet, and the leaves are sessile and merely serrate.

22. MENTHA. Mint

Perennial, aromatic herbs, usually of wet places, with creeping, mat-forming rhizomes; leaves ovate or oblong, serrate; flowers small and pinkish, in dense clusters in the axils of the upper leaves, or in narrow, dense, terminal spikes; calyx equally 5-toothed, the teeth deltoid, acute, usually shorter than the tube; corolla subequally 5-lobed; stamens 4, exserted; nutlets smooth.

Two Old World species, peppermint (*Mentha piperita*) and spearmint (*M. spicata*), are commonly cultivated. Oil of peppermint is used extensively in medicine and confectionery. Spearmint is used fresh for flavoring, and the oil is extracted, chiefly for medicinal use. The Hopi use mint leaves for flavoring mush.

KEY TO THE SPECIES

1. Flowers in axillary clusters. . . . . 1. *M. arvensis*
1. Flowers in terminal spikes (2).

2. Herbage glabrous or nearly so; leaves oblong-lanceolate, acuminate........2. *M. spicata*
2. Herbage villous; leaves broadly ovate to suborbicular, obtuse to acutish
3. *M. rotundifolia*

1. **Mentha arvensis** L. Apache, Navajo, Coconino, Yavapai, Greenlee, and Cochise counties, 5,000 to 9,500 feet, wet places, July to October. A circumpolar, polymorphic species.

The Arizona plants belong to var. *villosa* (Benth.) S. R. Stewart (*M. arvensis* var. *Penardi* Briq., *M. Penardi* Rydb.).

2. **Mentha spicata** L. Prescott and Montezuma Well (Yavapai County), Tonto Basin (Gila County), Santa Catalina Mountains (Pima County), July to September. Extensively naturalized in North America; from Europe.

3. **Mentha rotundifolia** (L.) Hudson. Pomerene, Cochise County (*W. W. Jones* in 1927). A weed of waste land, here and there in the United States; introduced from Europe.

23. HYPTIS

Shrubs, usually canescent, the hairs branched; leaves ovate; flowers several, in axillary or subspicate, globose, woolly cymules, these sometimes paniculate; calyx usually 10-veined, the tube cylindric, enlarging somewhat at maturity, the teeth subequal, subulate; corolla violet, the tube little exserted from the calyx, the upper lip plane or laterally reflexed, the midlobe of the lower lip deeply saccate; stamens 4, declined along the lower lip, paired; nutlets smooth, oblong.

1. **Hyptis Emoryi** Torr. Mohave, Yavapai, Graham, Gila, Maricopa, Pinal, Pima, and Yuma counties, up to 5,000 feet (usually lower), dry, rocky slopes and canyons, flowering almost throughout the year at lower elevations, type from the lower Gila River (*Emory*). Southern Arizona, southern California, and northwestern Mexico.

Desert-lavender. The plant is browsed to a limited extent by livestock. The seeds, like those of certain species of *Salvia*, are used in Mexico as food under the name "chia."

A specimen of *Pycnanthemum californicum* Torr. is labeled as from Garfield, Greenlee County (*Davidson* 611), but so great an eastward extension of range is unlikely, and it is probable that the specimen was collected in California. Of the genera of Labiatae known to occur in Arizona, *Pycnanthemum* is most nearly related to *Monardella*, but in the former the anther sacs are parallel and the corolla is more distinctly bilabiate, with a nearly entire upper lip and a 3-lobed lower lip.

119. SOLANACEAE. Potato Family

Plants herbaceous or shrubby; leaves alternate (sometimes nearly opposite) or fascicled; flowers in umbels, cymes, or panicles, or solitary and lateral, perfect, regular or nearly so, 4- or 5- (rarely 6-) merous; stamens distinct or slightly united by the anthers, the filaments distinct, inserted on the corolla tube alternate with the lobes; style 1; stigma entire or 2-lobed; ovary superior, usually 2-celled; fruit a berry or a capsule.

This large family comprises numerous economically important plants, such as the potato (*Solanum tuberosum*), egg-plant (*S. Melongena*), tomato (*Lycopersicum esculentum*), red-pepper (*Capsicum* spp.), and tobacco (*Nico-*

tiana Tabacum), as well as a number of very poisonous species, among them henbane (*Hyoscyamus niger*) and belladonna (*Atropa Belladonna*), sources of the powerful drugs hyoscyamine and atropine respectively. Many species in the genera *Solanum, Nicotiana, Petunia,* and others are grown as ornamentals.

Tomato plants (*Lycopersicum esculentum*) are found occasionally growing wild (e.g. at Paradise, Cochise County, *Blumer* 2266). But there is no evidence that this species has become established anywhere in the state, and therefore it is not regarded as a member of the flora.

KEY TO THE GENERA

1. Fruit a dry capsule, dehiscent by valves or if bursting irregularly, then the capsule large and spiny (2).
1. Fruit berrylike, indehiscent, commonly fleshy, but (in *Lycium*) sometimes dry and bony (4).
2. Flowers solitary in the forks of the stem; corolla 4 cm. or longer; capsules large, usually spiny, regularly dehiscent or bursting irregularly. Corolla broadly funnelform or trumpet-shaped, white or pale violet .... 8. *Datura*
2. Flowers in terminal inflorescences or, if solitary, then the corolla less than 1 cm. long; capsules small, not spiny, regularly dehiscent (3).
3. Flowers in terminal racemes or panicles; corolla more than 1 cm. long, white, greenish, or yellow .... 9. *Nicotiana*
3. Flowers extra-axillary, solitary; corolla less than 1 cm. long, the limb purple. Stems prostrate, forming mats; leaves oblanceolate or spatulate, not more than 2 cm. long 10. *Petunia*
4. Plants spiny shrubs. Flowers axillary, solitary or in very few-flowered fascicles; calyx scarcely enlarged in fruit; corolla campanulate, funnelform, or salverform; fruit commonly juicy when ripe .... 1. *Lycium*
4. Plants mostly herbaceous or suffrutescent, if shrubby, then not spiny, if spiny, then herbaceous (5).
5. Anthers opening by terminal pores or slits, seldom dehiscent throughout; plants sometimes spiny; inflorescences terminal or extra-axillary .... 7. *Solanum*
5. Anthers completely dehiscent longitudinally; plants not spiny; inflorescences axillary (6).
6. Calyx thick, unaltered in fruit, truncate; berry pungent to the taste .... 6. *Capsicum*
6. Calyx papery, accrescent in fruit, obviously toothed; berry not pungent (7).
7. Flowers in a sessile or pedunculate umbel; calyx saucer-shaped at maturity, not enclosing the fruit .... 5. *Saracha*
7. Flowers solitary; calyx enclosing the fruit at maturity (8).
8. Corolla urceolate .... 2. *Margaranthus*
8. Corolla rotate to funnelform-campanulate (9).
9. Calyx in fruit not angled, the lobes not connivent, leaving the top of the berry exposed 3. *Chamaesaracha*
9. Calyx in fruit angled, the lobes connivent over the berry .... 4. *Physalis*

1. LYCIUM (277, 278). WOLF-BERRY, DESERT-THORN

Plants shrubby, usually spiny; leaves mostly fascicled, entire; flowers chiefly axillary, solitary or in small clusters; calyx campanulate, irregularly toothed or cleft; corolla campanulate, tubular-funnelform, or salverform; stamens 4 or 5; style slender, the stigma capitate or 2-lobed; berry fleshy or dry, globose or ovoid, subtended by the persistent calyx.

These plants, also known as squaw-berry and tomatillo, have contributed to the subsistence of the Indians. Several of the species produce abundant quantities of insipid, slightly bitter, juicy berries that are eaten raw or prepared as a sauce. Probably all the native species furnish winter forage for livestock. They grow commonly along washes and on dry slopes, in desert or semidesert areas. Some of the species tolerate a rather high degree of soil salinity. All the Arizona species, with the possible exception of *L. pallidum*, shed their leaves and become dormant during periods of drought, refoliating quickly when conditions become favorable. They flower regularly in spring, often again after summer rains.

KEY TO THE SPECIES

1. Twigs tomentose when young; fruit transversely constricted, the upper part dry, bony, brown (2).
1. Twigs not tomentose, but sometimes copiously hirtellous; fruit not constricted, a plump, succulent, scarlet berry, or (in *L. californicum*) the endocarp often hardened (3).
2. Fruit constricted below the middle; calyx lobes subulate or linear, equaling to twice as long as the tube; corolla glabrous . . . . 1. *L. macrodon*
2. Fruit constricted above the middle; calyx lobes from ⅓ as long to as long as the tube; corolla densely pubescent to glabrous . . . . 2. *L. Cooperi*
3. Calyx tube densely glandular-pubescent; leaves, except in age, abundantly glandular-pubescent (4).
3. Calyx tube glabrous or nearly so; leaves glabrous or slightly scurfy, rarely minutely puberulent (6).
4. Calyx campanulate to turbinate, the lobes longer than the tube and rounded, to half as long as the tube and acute; berry 7- to 20-seeded; foliage often cinereous. Corolla lobes purple . . . . 5. *L. Parishii*
4. Calyx cylindric, or turbinate in small-flowered forms, the lobes rarely more than ¼ as long as the tube; berry usually many-seeded; foliage not cinereous. Plants sexually dimorphic (5).
5. Filaments densely villous on the lower half of the free portion; corolla lobes pale lavender; flowers mostly pendulous . . . . 6. *L. exsertum*
5. Filaments glabrous or sparsely villous at base of the free portion; corolla lobes purple; flowers not pendulous . . . . 7. *L. Fremontii*
6. Corolla greenish white, the tube 9 to 20 mm. long, greatly expanded above the middle, the mouth 5 to 6 mm. in diameter. Leaves large, pallid; berry glaucous . . 4. *L. pallidum*
6. Corolla ochroleucous, the lobes pale lavender or pinkish, the tube 4 to 15 (rarely 18) mm. long, if greatly expanded then less than 10 mm. long (7).
7. Corolla lobes finely lanate-ciliate. Corolla tube funnelform, 8 to 15 mm. long . 8. *L. Torreyi*
7. Corolla lobes glabrous or sparsely ciliate (8).
8. Corolla lobes half as long as to equaling the tube, normally 4; fruit often with a somewhat hardened endocarp, 2 to 4 mm. long; seeds 2. Leaves very succulent and thickened, often pyriform, less than 10 mm. long; corolla tube 1 to 5 (usually 2 to 3) mm. long 3. *L. californicum*
8. Corolla lobes ⅙ to ⅓ as long as the tube; fruit entirely succulent, 3 to 8 mm. long; seeds several to many (9).
9. Tube of the corolla tubular-funnelform, 4 to 14 mm. long; leaves glabrous or slightly scurfy, rarely pubescent . . . . 9. *L. Andersonii*
9. Tube of the corolla funnelform, often greatly expanded, 4 to 6 (8) mm. long; leaves glabrous or minutely puberulent . . . . 10. *L. Berlandieri*

1. Lycium macrodon Gray. Maricopa, Pinal, western Pima, and Yuma counties, up to 2,500 feet, February to May, the type probably from southern Arizona (*Frémont Expedition,* 1849). Southern Arizona and Sonora.

A large, spiny shrub with shining, mahogany-colored branches, locally common in parts of Pinal County.

2. Lycium Cooperi Gray. Mohave and Yuma counties, up to 3,000 feet, March and April. Southwestern Utah and Arizona to southeastern California.

3. Lycium californicum Nutt. (*L. californicum* var. *arizonicum* Gray). Maricopa, Pinal, Cochise, Pima, and (doubtless) Yuma counties, saline soils, ordinarily at low altitudes but occasionally (in Cochise County), up to 5,000 feet, February and March, type of var. *arizonicum* from Maricopa, Pinal County (*Gray* in 1885). Southern Arizona, California, Sonora, and Baja California, commonly littoral outside of Arizona.

Arizona's smallest species, ordinarily about 0.6 m. (2 feet) high. Berry small, scarlet, of bony hardness except for the thin, succulent, orange-red exocarp.

4. Lycium pallidum Miers. Almost throughout the state, 3,500 to 7,000 feet, April to June. Western Texas to southern Colorado and Utah, Arizona, southern California, and Mexico.

The Indians of northern Arizona ate the fresh berries, and during famines ate a mixture of the dried berries and saline clay.

5. Lycium Parishii Gray (*L. Pringlei* Gray). Southern Mohave, Maricopa, southern Pinal, western Pima, and Yuma counties, up to 2,500 feet, February to April (occasionally autumn-flowering). Southern Arizona, southeastern California, and Sonora.

6. Lycium exsertum Gray. Mohave County to Graham, Cochise, Pima, and Yuma counties, up to 4,000 feet, flowering throughout the year, but mostly in January and February. Southern Arizona and northwestern Mexico.

The fertile, pistillate form with reduced, abortive stamens was described as *L. Fremontii* var. *Bigelovii* Gray, type from Williams River (*Bigelow* in 1854).

7. Lycium Fremontii Gray (*L. gracilipes* Gray). Western and southern Arizona, up to about 2,500 feet, throughout the year, mostly January and February, type of *L. gracilipes* from Williams River (*Palmer* 423). Arizona, southeastern California, and northwestern Mexico.

The abundant, juicy berries produced by this and the preceding species were gathered by the desert Indians for food. Both species are hosts of the destructive root-rot fungus, *Phymatotrichum omnivorum.*

8. Lycium Torreyi Gray (*L. Torreyi* var. *filiforme* Jones). Western and southern parts of the state, mostly below 3,000 feet, March to June, type of var. *filiforme* from Beaver Dam, Mohave County (*Jones* 5015). Western Texas to southern Nevada, southeastern California, and Mexico.

9. Lycium Andersonii Gray. Coconino and Mohave counties, southward to Graham, Cochise, and Yuma counties, up to 5,500 feet, February to April (August and September). Utah and New Mexico to California and northwestern Mexico.

The typical plant has the corolla 7 to 14 mm. long, usually 5-merous, and the leaves 3 to 15 mm. long, usually thickened and succulent. Var. *deserticola* C. L. Hitchc. ex Munz (*L. Andersonii* forma *deserticola* C. L. Hitchc.), with thin, flat leaves up to 35 mm. long and plants usually large and robust, is not uncommon along washes in Maricopa and Yuma counties, below 1,500 feet, and in southeastern California. Var. *Wrightii* Gray, with corolla 4 to 8 mm.

long, usually 4-merous, and leaves 3 to 8 mm. long but occasionally larger, occurs from Greenlee County to Cochise and Yuma counties, and in Sonora. The type of var. *Wrightii* probably came from near Chiricahua, Cochise County (*Wright* 1610, in part).

10. Lycium Berlandieri Dunal. Greenlee, Maricopa, Pinal, Pima, and Yuma counties, up to about 3,000 feet, March to September. Texas to Arizona and Mexico.

Represented in Arizona by var. *parviflorum* (Gray) Terrac. (*L. parviflorum* Gray, *L. Berlandieri* vars. *longistylum* and *brevilobum* C. L. Hitchc.), which is distinguished by stout, leafy branches and flowers mostly 4 to 6 mm. long. The type of var. *longistylum* came from the Santa Catalina Mountains (*Pringle* in 1881).

2. MARGARANTHUS

Small, annual herbs, resembling *Physalis;* stems leafy, branched; leaves petioled, thin, entire or somewhat sinuate, ovate or lanceolate; flowers small, solitary, on slender pedicels; corolla subglobose.

KEY TO THE SPECIES

1. Stems spreading, diffusely branched from the base; corolla apparently white, deeply cleft, the lobes at least as long as the tube .... 1. *M. Lemmoni*
1. Stems erect, not or sparingly branched from the base; corolla greenish yellow or lurid purple, cylindric-urceolate, merely denticulate at the orifice .... 2. *M. solanaceus*

1. Margaranthus Lemmoni Gray. Known only from the type collection in the Huachuca Mountains, Cochise County (*Lemmon* 2847), September.

2. Margaranthus solanaceus Schlecht. Yavapai, Greenlee, and Gila counties to Cochise, Santa Cruz, and Pima counties, 3,500 to 5,500 feet, rich soil in shade, August and September. New Mexico, Arizona, and Mexico.

3. CHAMAESARACHA

Low, perennial herbs; stems leafy, decumbent or prostrate, branched; flowers axillary, solitary on slender pedicels, these recurved or reflexed in fruit; corolla rotate, with pubescent, cushionlike appendages in the throat, alternate with the stamens; berry closely invested but not hidden by the calyx.

KEY TO THE SPECIES

1. Corolla appendages large, contiguous or nearly so, almost filling the throat; herbage not viscid, normally scurfy-puberulent (usually sparsely so) with sessile or short-stalked, white, stellate hairs; leaves sessile or short-petioled, lanceolate or narrowly oblong, mostly pinnatifid; calyx copiously stellate-puberulent, sometimes also with a few longer, simple hairs .... 1. *C. Coronopus*
1. Corolla appendages smaller, more distant, not filling the throat; herbage more or less viscid, sparsely to copiously villous with weak, segmented, unbranched hairs, also copiously glandular-puberulent; leaves distinctly petioled, oblong, ovate, oblanceolate, or obovate, entire or shallowly repand-dentate to pinnatifid; calyx shaggy-villous with mostly simple hairs .... 2. *C. coniodes*

1. Chamaesaracha Coronopus (Dunal) Gray. Apache, Navajo, and Coconino counties, south to Cochise, Pima, and Yuma counties, 2,500 to 7,500 feet, dry plains and mesas, April to September. Kansas to Utah, south to northern Mexico.

The berries are eaten by the Navajo and Hopi Indians. The pubescence is usually as described in the key, but specimens from the Pinacate Plateau, southeastern Yuma County (*Gentry* 3507), and from north of Cameron, Coconino County (*Darrow* 2722), are also sparsely villous with long, weak, segmented, simple or dendritic hairs. The leaves are exceptionally broad in these specimens.

2. Chamaesaracha coniodes (Moric.) Britton (*C. sordida* Gray). Gila, Cochise, and Pima counties, 3,500 to 5,500 feet, dry plains and mesas, often on limestone, March to October. Kansas and Colorado to southeastern Arizona and northern Mexico.

4. PHYSALIS (279). GROUND-CHERRY, HUSK-TOMATO

Annual or perennial herbs (rarely suffrutescent); stems branched, leafy; flowers solitary, on lateral peduncles; calyx 5-toothed or 5-cleft, becoming greatly enlarged, papery, veiny; berry globose, many-seeded; seeds flat.

The berries are more or less edible and are sometimes made into preserves. The Indians eat them, both raw and cooked. Two Old World species, the strawberry ground-cherry (*P. Alkekengi*) and the lantern ground-cherry (*P. Franchetii*), are often grown as ornamentals.

KEY TO THE SPECIES

1. Corolla rotate or rotate-campanulate, purple or whitish. Leaves sinuately denticulate or dentate, or pinnately incised, cuneate at base, long-petioled (2).
1. Corolla campanulate, yellow or yellowish, usually with a brown or purplish eye (4).
2. Plants perennial, sparsely whitish-scurfy, at least on the young parts; leaves oblong to ovate; corolla violet or purple, with a white eye; anthers yellow; seeds thick, coarsely and irregularly rugose on the back ....1. *P. lobata*
2. Plants annual, not scurfy, sparsely puberulent or glabrate; leaves oblong- to ovate-lanceolate; corolla whitish, often with a large yellow eye; anthers purplish; seeds thin, not coarsely rugose (3).
3. Corolla rotate, 15 to 25 mm. wide; leaves deeply dentate to almost laciniate; herbage somewhat glaucous ....2. *P. Wrightii*
3. Corolla rotate-campanulate, not more than 10 mm. wide; leaves sinuate to shallowly dentate; herbage green.......................................3. *P. lanceifolia*
4. Pubescence partly of forked or stellate hairs (these sometimes very few), cinereous, not or very slightly viscid. Plants perennial; leaves broadly ovate or deltoid, acute or acutish at apex, sinuate-dentate, usually shallowly so; corolla open-campanulate 4. *P. Fendleri*
4. Pubescence of simple hairs only (5).
5. Herbage conspicuously pubescent, the stems usually with few to many longer spreading hairs (6).
5. Herbage inconspicuously pubescent or glabrate, the hairs mostly minute, not or scarcely viscid. Plants perennial (9).
6. Pubescence not viscid, not dense, many of the stem hairs long, flat, segmented, tapering; leaves lanceolate or ovate-lanceolate, acute at both ends, entire to coarsely few-toothed. Plants perennial; corolla 15 to 20 mm. wide; fruiting calyx 25 to 35 mm. long 5. *P. lanceolata*
6. Pubescence viscid, copious; leaves ovate, deltoid, or suborbicular (7).
7. Plants annual; main stem erect, stout, seldom branching from the base; leaves thin, sparingly repand-dentate to nearly entire; fruiting calyx conspicuously acuminate. Leaves broadly ovate to suborbicular, often subcordate at base; corolla yellow or greenish with a conspicuous dark eye; stem hairs slender, not noticeably tapering 6. *P. pubescens*

7. Plants perennial; flowering stems from creeping rootstocks, usually diffusely branched from the base; leaves thickish, usually coarsely toothed; fruiting calyx not conspicuously acuminate (8).

8. Leaves commonly at least 5 cm. long, acute or acuminate at apex; herbage pubescent with both short, glandular hairs and long, flat, segmented hairs; corolla at least 15 mm. wide; fruiting calyx 2.5 to 4 cm. long....................... 7. *P. heterophylla*

8. Leaves commonly less than 5 cm. long, obtuse to acute at apex, coarsely toothed to nearly entire; herbage copiously glandular-pilose with very few, if any, long, flat, segmented hairs; corolla commonly less than 15 mm. wide; fruiting calyx not more than 2.5 cm. long... 8. *P. hederaefolia*

9. Herbage sparsely pubescent with short, stiff, appressed or subappressed hairs, or glabrate; leaves lanceolate or oblong-lanceolate (exceptionally oblanceolate); flowering stems sparingly branched, stout, erect from creeping rootstocks. Leaves entire or repand (exceptionally sinuate-dentate), usually acuminate at both ends; corolla 15 to 25 mm. wide, pale yellow with a dark eye; fruiting calyx (20) 25 to 35 mm. long, ovoid, the teeth triangular-lanceolate........................... 9. *P. longifolia*

9. Herbage persistently puberulent; leaves broadly ovate, rounded-deltoid, or suborbicular; flowering stems diffusely branched (10).

10. Leaves thin, coarsely and rather deeply sinuate-dentate; corolla with a distinct eye, greenish yellow, drying purplish............................... 10. *P. versicolor*

10. Leaves thickish, entire to shallowly sinuate-dentate; corolla often without an eye, drying yellow. Plants often suffrutescent; leaves occasionally subcordate 11. *P. crassifolia*

1. Physalis lobata Torr. (? *Chamaesaracha physaloides* Greene). Navajo, Greenlee (?), Maricopa, Pinal, and Pima counties, 1,000 to 5,000 feet, plains, mesas, and roadsides, March to October. Kansas to Texas, southern Nevada, Arizona, and northern Mexico.

Chamaesaracha physaloides, type from the Patagonia Mountains (*Buckminster* in 1881), was referred to *P. lobata* by Gray (*Syn. Fl. N. Amer.* ed 2, 2^1: 437), but Wooton and Standley (*Flora New Mexico,* p. 571) made it a synonym of *P. Wrightii.*

2. Physalis Wrightii Gray. Navajo County to Cochise, Pima, and Yuma counties, 100 to 4,000 feet, fields and along ditches, June to September. Western Texas to southern California and northern Mexico.

A common weed of cultivated land in southern Arizona.

3. Physalis lanceifolia Nees. Sacaton (Pinal County), 1,200 feet (doubtless elsewhere in southern Arizona). Western Texas to southern California, south to South America.

4. Physalis Fendleri Gray. Apache County to Mohave County, south to Cochise, Santa Cruz, and Pima counties, 3,000 to 7,500 feet, dry mesas and slopes, often on limestone with juniper and pinyon, May to August. Colorado and Utah to Arizona, southeastern California, and northern Mexico.

5. Physalis lanceolata Michx. White Mountains (Apache or Greenlee County) and mountains of Cochise, Santa Cruz, and Pima counties, 5,000 to 8,500 feet, August and September. Illinois to South Dakota, Arkansas, and eastern Arizona.

6. Physalis pubescens L. Yavapai, Greenlee, Santa Cruz, and Pima counties, 3,000 to 6,000 feet, mostly along streams in partial shade, August and September. Pennsylvania to Colorado, Florida, and Arizona, southward to Panama.

The Arizona plants, with relatively tall, stout, erect stems not branching from the base, probably are *P. neomexicana* Rydb. but seem scarcely more than varietally distinct from *P. pubescens.*

7. Physalis heterophylla Nees. Navajo and Coconino counties to Pima County, 4,000 to 7,000 feet, June to October. Canada to Florida and Arizona.

8. Physalis hederaefolia Gray. Coconino and Mohave counties to Cochise and Pima counties, 3,000 to 7,000 feet, foothills and plains, April to August. Colorado and Utah to Texas, Arizona, southeastern California, and northern Mexico.

Plant with the aspect of *P. Fendleri,* but distinguishable by the absence of branched hairs and the presence of glandular hairs.

9. Physalis longifolia Nutt. Southern Navajo County and Gila, Cochise, and Pima counties, 2,500 to 5,000 feet, rare in Arizona, April to August. Iowa to Montana, south to Arkansas, Arizona, and northern Mexico.

Specimens from Flagstaff, Coconino County (*Thornber* in 1920), resemble *P. longifolia* except that the fruiting calyx is barely 2 cm. long.

10. Physalis versicolor Rydb. Metcalf (Greenlee County), Sierra Ancha (Gila County), western Pima County, 2,000 to 5,000 feet, mesas and foothills, August and September. Southern Arizona and northern Mexico.

11. Physalis crassifolia Benth. Canyons of the Colorado River in Coconino and Mohave counties, to Pima and Yuma counties, 3,000 feet or lower, dry, rocky slopes, February to October. Southern Utah and Arizona to southeastern California and Baja California.

Arizona's only suffrutescent species and also the most xerophytic one. A form with subcordate leaves, var. *cardiophylla* (Torr.) Gray, is occasional in Arizona.

5. SARACHA (280)

Large, perennial herbs, with a vertically elongate, tuberous root; stems branching, sharply 4-angled; leaves long-petioled, large, very thin, ovate, acuminate at apex, cuneate at base, entire or nearly so; flowers in axillary umbels; corolla rotate, greenish; berry many-seeded, dark purple when mature.

1. Saracha procumbens (Cav.) Ruiz & Pavon (*S. sessilis* Greene). Mountains of Cochise, Santa Cruz, and Pima counties, 3,500 to 5,500 feet, shady canyons in rich soil, August and September, type of *S. sessilis* from the Chiricahua Mountains (*Blumer* in 1907). Southern Arizona to South America.

6. CAPSICUM. Red-pepper

Plants more or less shrubby; stems widely branched; leaves slender-petioled, thin, ovate or lance-ovate, acuminate, entire; peduncles long and slender, often in pairs, spreading or somewhat reflexed; calyx small, shallowly toothed or truncate; corolla rotate, deeply cleft, whitish; fruit short-ovoid or nearly globose, persistent.

1. Capsicum baccatum L. West slope of the Baboquivari Mountains, Pima County, about 4,000 feet, in a canyon (*Peebles et al.* 403, 610, *Gould et al.*

2687), September. Florida to southern Texas, southern Arizona, and south to tropical America.

Bird-pepper, chillipiquin. The very pungent berries are used as a condiment and medicinally as a local stimulant.

7. SOLANUM (281). NIGHTSHADE

Plants herbaceous or suffrutescent, sometimes prickly; leaves petioled, entire to bipinnatifid, often very unequal in size in the pair; flowers mostly lateral (extra-axillary), solitary or in cymes; corolla rotate or rotate-campanulate, 5-toothed to 5-parted; anthers opening by apical pores or short slits; seeds numerous, more or less flattened.

A few of the species are troublesome weeds, and the leaves and unripe fruits of several of them are reputed to be poisonous, e.g. those of the black nightshade (*S. nigrum*), also *S. triflorum*, *S. elaeagnifolium*, and *S. rostratum*. An alkaloid, solanin, is the active principle. However, the fruits of the cultivated garden-huckleberry or wonderberry, a form of *S. nigrum*, have been used for making preserves and desserts.

KEY TO THE SPECIES

1. Fruit closely invested by the accrescent calyx. Herbage and calyx densely armed with long, straight, very sharp, straw-colored spines; plants annual; leaves pinnatifid or bipinnatifid (2).
1. Fruit not invested by the calyx, or loosely so (4).
2. Herbage and corolla not glandular, copiously stellate-pubescent, often also puberulent. Leaf segments broad, very obtuse; corolla yellow, 15 to 25 mm. wide . . 1. *S. rostratum*
2. Herbage and corolla glandular-puberulent, frequently also villous or hirsute with simple or forked, nonglandular hairs (3).
3. Corolla violet; leaf segments broadly ovate or obovate, very obtuse . . . . 2. *S. heterodoxum*
3. Corolla yellow; leaf segments linear or lanceolate, acute or acutish. Spines, when fresh, nearly black at base . . . . 3. *S. Lumholtzianum*
4. Herbage and calyx spiny, or if (exceptionally) unarmed, then densely and minutely whitish-lepidote. Corolla 20 to 30 mm. wide, purple, violet, or nearly white (5).
4. Herbage and calyx not spiny; pubescence never lepidote (6).
5. Leaves pinnatifid or bipinnatifid; spines up to 20 mm. long, often stout; herbage copiously villous, the pubescence more or less glandular; plants annual; calyx loosely investing the fruit . . . . 4. *S. sisymbriifolium*
5. Leaves entire to coarsely sinuate-dentate; spines not more than 5 mm. long, slender, sometimes wanting; herbage densely and minutely whitish-lepidote; plants perennial, with long, deep, creeping rootstocks; calyx not investing the fruit or very loosely investing it at base . . . . 5. *S. elaeagnifolium*
6. Plants with nearly globose tubers and long, slender stolons. Leaves pinnate, with 5 or more leaflets, some of these often very small; herbage pilose, usually sparsely so, with flat, simple, flaccid hairs; corolla 12 to 18 mm. wide (7).
6. Plants not tuberiferous or stoloniferous (8).
7. Leaflets oval, ovate, or obovate; corolla angulately 5-toothed, normally violet; herbage sparsely to copiously pubescent . . . . 6. *S. Fendleri*
7. Leaflets narrowly lanceolate to broadly oblong-lanceolate or occasionally obovate; corolla deeply 5-cleft, normally white; herbage usually very sparsely pubescent, or glabrate . . . . 7. *S. Jamesii*
8. Corolla violet or lilac-purple, rarely white, 20 to 30 mm. wide, angulately 5-lobed; peduncle shorter than the pedicels, the latter with a cupulate thickening at base. Plants perennial, becoming somewhat woody at base, puberulent or soft-pilose, sometimes copiously glandular but usually without glandular hairs; leaves ovate to oblong-lanceolate, entire or undulate, occasionally somewhat auriculate at base . . . 8. *S. Xanti*

8. Corolla white or whitish (sometimes tinged with purple), not more than 20 mm. wide, deeply 5-cleft; pedicels not cupulate at base or very obscurely so (9).
9. Flowers solitary or geminate (exceptionally in threes or fours), on slender, strongly deflexed pedicels, without an evident peduncle; herbage and calyx pubescent with stiff, more or less spreading hairs; seeds radially rugose. Plants annual; leaves ovate to oblong-lanceolate, entire or slightly repand........................9. *S. deflexum*
9. Flowers commonly in umbelliform cymes (sometimes solitary), these borne on peduncles nearly as long as to longer than the pedicels; herbage pubescent with more or less appressed hairs, or with soft and viscid hairs; seeds not radially rugose (10).
10. Leaves deeply pinnatifid, with acute or acutish, narrowly triangular segments. Plants annual, the pubescence appressed, almost scurfy; stems strongly decumbent or prostrate; cymes 1- to 3- (commonly 2-) flowered; corolla less than 10 mm. wide; berry green at maturity..10. *S. triflorum*
10. Leaves entire to sinuate-dentate (11).
11. Plants perennial, often suffrutescent; corolla 10 to 18 mm. wide. Herbage sparsely to densely cinereous-puberulent or short-pilose, the hairs mostly appressed or subappressed; berries black at maturity, many-seeded, persistent.......11. *S. Douglasii*
11. Plants annual or perennial; corolla not more than 8 mm. wide (12).
12. Herbage viscid-villous with spreading hairs; calyx thin, accrescent; berries greenish at maturity. Plants annual....................................12. *S. sarachoides*
12. Herbage sparsely puberulent, strigose, or glabrate; calyx firm, not accrescent; berries black at maturity (13).
13. Leaves firm; calyx lobes all distinct, reflexed at maturity; concretions of the stone cells absent or few, rarely more than 3; plants annual or perennial......13. *S. nodiflorum*
13. Leaves thin; calyx lobes some of them partly fused, unequal in length, not becoming reflexed; concretions always present, usually 4 to 8; plants strictly annual 14. *S. americanum*

1. Solanum rostratum Dunal. Apache County to Coconino County, south to Cochise and Pima counties, 1,000 to 7,000 feet, common on plains and at roadsides in the northern part of the state, perhaps introduced from farther east, June to August. North Dakota and Wyoming to Arizona and Mexico.

Buffalo-bur. This species, believed to be the original host of the Colorado potato-beetle, is considered a pest in range land.

2. Solanum heterodoxum Dunal. Kirkland, Yavapai County (*Peebles et al.* 7422), Guadalupe Mountains, Cochise County (*Darrow et al.* 3576), 4,000 to 4,500 feet, roadsides, October. Western Texas to Arizona and Mexico.

The species is represented in Arizona by var. *novomexicanum* Bartlett (*Androcera novomexicana* Woot. & Standl.), which is described as being more densely pubescent and more spiny and having a larger corolla and stamens than in typical *S. heterodoxum.* It was stated that the spines are brownish yellow in the variety, olivaceous in the typical form.

3. Solanum Lumholtzianum Bartlett. Patagonia, Santa Cruz County (*Harrison & Fulton* 8185), near Arivaca, Pima County (*Kearney & Peebles* 13775), 3,000 to 4,000 feet, sandy soil at roadsides, August to October. Southern Arizona and northern Sonora.

The plants are about 0.6 m. (2 feet high), with long, widespreading branches.

4. Solanum sisymbriifolium Lam. Near the Boyce-Thompson Southwestern Arboretum, Superior, Pinal County (*McLellan* in 1935), perhaps not established. Adventive from tropical America.

The leaves resemble those of watermelon.

5. Solanum elaeagnifolium Cav. Throughout the state, 1,000 to 5,500 feet, fields and roadsides, May to October. Kansas and Colorado to Arizona and California, south to tropical America.

White (or silver) horse-nettle, bull-nettle, trompillo. A troublesome weed in irrigated land, especially in the southern counties, difficult and expensive to eradicate. The crushed berries are added to milk by the Pima Indians in making cheese. A protein-digesting enzyme resembling papain exists in this plant.

6. Solanum Fendleri Gray. Mountains of Greenlee, Gila, Cochise, Santa Cruz, and Pima counties, 6,000 to 9,000 feet, rich soil in open pine forests, July and August. New Mexico and Arizona.

7. Solanum Jamesii Torr. Apache County to Coconino County, south to Cochise and Yavapai counties, 5,500 to 8,500 feet, mostly in coniferous forests, July to September. Colorado and Utah to Texas and Arizona.

Wild-potato, a name applied also to *S. Fendleri.* Both species are related to the cultivated potato and have similar although much smaller tubers. The plants sometimes are found growing wild in the gardens of the Indians, who used them as food. The tubers are cooked by the Hopi with a saline clay and are said to have been used by them also in making yeast.

8. Solanum Xanti Gray. Southern Navajo, southern Coconino, Yavapai, Greenlee, Gila, and eastern Maricopa counties, 3,500 to 5,500 feet, rocky slopes, usually in chaparral, April to November. Arizona, California, and Baja California.

Purple nightshade, a showy plant when in flower.

9. Solanum deflexum Greenm. (*Salpichroa Wrightii* Gray). Cochise, Santa Cruz, and Pima counties, 3,000 to 4,500 feet, not infrequent in sandy soil, August and September, type of *Salpichroa Wrightii* collected "on the Sonoita" (*Wright* 1692). Southern Arizona to Central America.

The berries are milk white at maturity.

10. Solanum triflorum Nutt. Apache County to Coconino County, south to Greenlee, Pinal, and Yavapai counties, 1,000 to 7,000 feet, roadsides and stream beds, May to September. Canada to Kansas, Arizona, and southern California.

The Hopi are reported to plant this species in hills with watermelons, believing that the growth of the latter is thus stimulated.

11. Solanum Douglasii Dunal (*S. arizonicum* Parish). Coconino County, south to Cochise, Santa Cruz, Pima, and Yuma counties, 1,500 to 6,000 feet, common on rocky slopes and in canyons, mostly in chaparral, March to October, type of *S. arizonicum* from Hot Springs, southern Yavapai County (*Toumey* 397). Western New Mexico and Arizona to Oregon and southern California.

12. Solanum sarachoides Sendt. Occasional in Navajo (?), Coconino, Pinal, Cochise, and Pima counties, a weed in cultivated and waste land, 1,500 to 7,000 feet, July to September. Sparingly adventive in the United States; introduced from South America.

Until recently this plant has been mistaken for *S. villosum* Mill.

13. Solanum nodiflorum Jacq. Maricopa and Pinal counties, probably elsewhere in Arizona, roadsides and waste land, summer. Washington to California and Arizona, introduced from tropical America.

Most of the plants referred to *S. nigrum* L. in *Flowering Plants and Ferns of Arizona* doubtless belong to this species.

14. Solanum americanum Mill. (*S. nigrum* of American authors, in large part, non L.). Oak Creek, Coconino County (*Fulton* 9668). Widely distributed in the eastern United States, apparently rare westward.

The collection cited was referred to this species by G. Ledyard Stebbins, Jr. (personal communication), with the notation "appears transitional toward *S. nodiflorum.*"

8. DATURA (282). THORN-APPLE

Coarse, weedy herbs with ill-scented herbage; stems stout, mostly erect, branched; leaves petioled, large, ovate, repand to pinnately lobed; flowers large and showy, short-peduncled, solitary in the forks of the stem, fragrant; calyx cylindric or prismatic, 5-toothed; corolla funnelform, purple to nearly white; fruit a large, globose or ovoid, normally prickly capsule.

All parts of the plants are poisonous, containing various alkaloids, notably atropine (daturine). Children as well as horses, cattle, and sheep have been poisoned by *D. Stramonium,* the common jimson-weed. The roots and other parts of *D. meteloides* are narcotic and are sometimes eaten by the Indians, even the children, to induce visions, a dangerous practice. One of the effects is dilation of the pupil of the eye, the effect being similar to that of belladonna. Contact with these plants is reported to cause dermatitis in susceptible persons.

KEY TO THE SPECIES

1. Fruit erect, regularly dehiscent, 4-valved, ovoid. Corolla not more than 10 cm. long, 5-toothed (2).
1. Fruit nodding, bursting irregularly, globose or nearly so. Leaves repand to sinuate-dentate (3).
2. Spines of the fruit many, subequal, less than 10 mm. long, relatively slender, sometimes much reduced or wanting; leaves repand to coarsely sinuate-toothed; corolla 6 cm. or longer, whitish or purple .... 1. *D. Stramonium*
2. Spines of the fruit relatively few, very unequal, the longer ones more than 10 mm. long, very stout; leaves usually pinnately lobed; corolla not more than 6 cm. long, purple 2. *D. quercifolia*
3. Corolla broadly funnel-shaped, 5-toothed, 15 to 20 cm. long; herbage canescent-puberulent; calyx 8 to 13 (rarely only 6) cm. long; fruit puberulent, not viscid, with slender spines usually less than 1 cm. long; seeds light brown when ripe 3. *D. meteloides*
3. Corolla trumpet-shaped, 10-toothed, usually less than 15 cm. long; herbage green, sparsely puberulent; calyx seldom more than 6 cm. long; fruit viscid-pubescent, with relatively stout spines, the longer ones 1 cm. or longer; seeds black when ripe 4. *D. discolor*

1. Datura Stramonium L. Tonto Creek, Gila County (*Kearney & Harrison* 8364), Paradise, Cochise County (*Blumer* 2267), 5,500 feet, October. Naturalized throughout the United States; from South America or the Eastern Hemisphere.

Jimson-weed. The form with herbage and flowers purplish (*D. Tatula* L.) may also occur in Arizona.

2. Datura quercifolia H.B.K. Big Lue Range (Greenlee County), Douglas and Tombstone (Cochise County), Patagonia and Elgin (Santa Cruz County), 4,000 to 6,000 feet, roadsides, September and October. Texas to southern Arizona and Mexico.

3. Datura meteloides DC. Navajo County to Mohave County, south to Cochise, Santa Cruz, and Pima counties, 1,000 to 6,500 feet, roadsides and along ditches, May to October. Colorado to Texas, Arizona, southern California, and Mexico.

Sacred datura, Indian-apple, tolguacha. With its very large, trumpet-shaped, pale-lavender flowers, this plant is a conspicuous feature of the vegetation. It is used by the Indians for various medicinal purposes, the seeds, it is reported, being sometimes administered to prevent miscarriage.

It has been suggested by Joseph Ewan (*Rhodora* 46: 317–323) that the application of the name *Datura meteloides* to this plant of the southwestern United States is questionable and that the name *D. Wrightii* Regel should perhaps, be substituted.

4. Datura discolor Bernh. Pinal, Maricopa, Cochise, Pima, and Yuma counties, seldom above 2,000 feet, roadsides and waste ground, autumn. Southern Arizona, southeastern California, and Mexico

9. NICOTIANA (283, 284). Tobacco

Plants herbaceous or (1 species) arborescent; leaves sessile or petioled, entire or sinuate-margined; inflorescence terminal, paniculate or racemelike; calyx 5-lobed; corolla tubular to salverform; capsule apically dehiscent, 2- or 4-valved; seeds small, very numerous.

The Arizona species were classified by Goodspeed (283) as follows: *N. glauca* in subgenus *Rustica,* section *Paniculatae; N. trigonophylla* and *N. Palmeri* in subgenus *Petunioides,* section *Trigonophyllae; N. Clevelandi* and *N. attenuata* in subgenus *Petunioides,* section *Acuminatae.*

The leaves of many of the species beside *N. Tabacum* contain nicotine and were smoked by the Indians. *N. trigonophylla* is still used for this purpose, chiefly on ceremonial occasions. Animals usually avoid these plants, but cases of poisoning in cattle, horses, and sheep have been reported. Tree tobacco (*N. glauca*) contains an alkaloid, anabasine, reported to be more efficacious than nicotine in killing certain species of aphid.

KEY TO THE SPECIES

1. Plants shrubby or arborescent; herbage glabrous and very glaucous; corolla yellow, tubular-funnelform, 25 to 50 mm. long, densely pubescent externally, with a very short, erect limb. Flowers diurnal; leaf margins entire or slightly undulate......1. *N. glauca*

1. Plants herbaceous and annual or (in *N. trigonophylla*) sometimes perennial and suffrutescent; herbage pubescent or puberulent, viscid, not glaucous; corolla white or greenish white, with a well-developed, more or less spreading limb (2).

2. Leaves mostly cordate- or auriculate-clasping, sessile or with short broad petioles; corolla tube pubescent externally, often copiously so; flowers diurnal (3).

2. Leaves not cordate or auriculate at base; corolla glabrous or very sparsely pubescent externally, with the hairs mostly confined to the throat and limb; flowers vespertine (4).

3. Calyx lobes not, or but moderately, surpassing the capsule; corolla limb 6 to 8 mm. wide 2. *N. trigonophylla*

3. Calyx lobes much surpassing the capsule; corolla limb 10 to 15 mm. wide..3. *N. Palmeri*

4. Calyx lobes linear-lanceolate, conspicuously unequal, the longer ones in fruit equaling or longer than the calyx tube; stem leaves mostly sessile, prevailingly ovate or lance-ovate; corolla 10 to 18 mm. long.................................4. *N. Clevelandi*
4. Calyx lobes deltoid or deltoid-lanceolate, not conspicuously unequal, all shorter than the calyx tube; stem leaves mostly petioled, prevailingly linear-lanceolate to oblong-lanceolate; corolla 20 to 40 mm. long...............................5. *N. attenuata*

1. Nicotiana glauca Graham. Greenlee, Gila, Maricopa, Pinal, Cochise, Pima, and Yuma counties, reported also from Havasu Canyon (Coconino County) and Lake Mead (Mohave County), common below 3,000 feet, flowering nearly throughout the year. Texas to central California; naturalized from South America.

Tree tobacco. Stems up to 4 m. (13 feet) high. A conspicuous plant in southern Arizona, along streams, ditches, and washes.

2. Nicotiana trigonophylla Dunal. Almost throughout the state, 6,000 feet or (usually) lower, very common along sandy washes, flowering the year around. Western Texas to southern California and Mexico.

The plant is sometimes perennial in southwestern Arizona.

3. Nicotiana Palmeri Gray (*N. trigonophylla* var. *Palmeri* Jones). Yucca (Mohave County), Williams River, Mohave or Yuma County (*Palmer* 433, the type collection), Agua Caliente (western Maricopa County), Kofa Mountains (Yuma County), 2,000 feet or lower, March to May. Western Arizona and southern California.

4. Nicotiana Clevelandi Gray. Fort Mohave (*Cooper* 415) and rather frequent in southern Yuma County, 500 feet or lower, sandy washes, February to April. Western Arizona, southern California, and Baja California.

5. Nicotiana attenuata Torr. Almost throughout the state, 1,000 to 7,000 feet, common along streams and washes, May to October. Utah to Texas, Arizona, and California.

10. PETUNIA

Plants annual, glandular-puberulent; stems prostrate and rooting at the nodes, diffusely branched, forming mats, leafy; leaves narrow, rather fleshy, seldom more than 1 cm. long; flowers solitary, lateral, 4 to 6 mm. long; corolla funnelform, slightly irregular, with a purple limb and a whitish tube.

1. Petunia parviflora Juss. Navajo County to Mohave County, south to Graham, Pima, and Yuma counties, 400 to 5,000 feet, moist soil in beds of streams, and muddy flats, April to September. Southern Florida to California, south to tropical America.

The Arizona plant is a humble relative of the showy cultivated petunias, which are derived from 2 South American species, *P. axillaris* and *P. violacea*, and from hybrids between them.

120. SCROPHULARIACEAE (285, 286). Figwort Family

Plants annual or perennial, a few shrubby, in some genera partially parasitic; leaves opposite, alternate, or mostly basal, simple, entire to pinnately parted; flowers perfect, very irregular to nearly regular; calyx 4- or 5-toothed or -lobed; stamens inserted on the corolla tube, commonly 4, in unequal pairs, a fifth stamen (staminode) often present but nonfunctional, or sometimes only 2 of the stamens perfect, or (in one genus) all 5 of the stamens perfect; style 1, the stigma entire or 2-lobed, the ovary superior, more or less completely 2-celled; fruit a capsule, usually 2-valved and longitudinally dehiscent, sometimes opening by pores or bursting irregularly; seeds usually many, small.

A large and diverse family, comprising many plants that are cultivated as ornamentals. The plants are mostly innocuous, but the Old World foxglove (*Digitalis purpurea*), often grown in the United States as an ornamental, is the source of the drug digitalis, a powerful cardiac stimulant.

KEY TO THE GENERA

1. Anther-bearing stamens 5; corolla nearly regular. Flowers many, in elongate, spikelike, terminal inflorescences; corolla rotate, yellow, the upper lobes external in bud 1. *Verbascum*

1. Anther-bearing stamens 4 or 2 (rarely 5 in genus *Penstemon*); corolla very irregular (bilabiate) to nearly regular (2).

2. Corolla with the lower (anterior) lobes external in the bud (3).

2. Corolla with the upper (posterior) lobes external in the bud (11).

3. Upper corolla lobes flattened or broadly concave, often spreading; anthers all distinct (4).

3. Upper corolla lobes arched, forming a definite galea enclosing the stamens; anthers frequently cohering. Plants often root-parasitic; stamens commonly four (7).

4. Stamens 2; corolla appearing 4-lobed, the upper (posterior) lobes completely united; capsules strongly compressed, little if any longer than wide; plants not parasitic. Corolla blue, purple, or whitish (5).

4. Stamens 4, in pairs; corolla evidently 5-lobed; capsules not compressed, considerably longer than wide; plants root-parasitic (6).

5. Stems leafy, the leaves opposite, at least below the inflorescence; flowers axillary or in loose racemes; corolla nearly regular, rotate or nearly so.............. 17. *Veronica*

5. Stems scapelike, the larger leaves in a basal rosette, the stem leaves reduced and bractlike, alternate; flowers in dense, cylindric, spikelike racemes; corolla irregular (bilabiate), cleft nearly to the base...................................... 18. *Besseya*

6. Pedicels not bracteolate; corolla yellow, the throat ample, glabrous within; filaments and style nearly as long as to longer than the corolla tube; anthers 2-celled; capsule partly exserted .... 19. *Brachystigma*

6. Pedicels bibracteolate; corolla violet or white, the throat narrow, pilose within; filaments and style less than half as long as the corolla tube; anthers 1-celled (by abortion); capsule nearly or quite included in the calyx tube.............. 20. *Buchnera*

7. Anther cells equal, parallel, approximate; seed coat not evidently reticulate (8).

7. Anther cells unequal, separated, the upper (outer) one versatile, the other suspended by its apex and mostly smaller, sometimes sterile; seed coat evidently reticulate. Leaves alternate (9).

8. Leaves opposite; calyx 4-toothed, becoming bladderlike, completely enclosing the fruit; capsule symmetric, both cells dehiscing equally.................... 24. *Rhinanthus*

8. Leaves mostly alternate or basal; calyx cleft on one or both sides, becoming distended, but not bladderlike or completely enclosing the fruit; capsule asymmetric, opening chiefly or wholly on one side................................... 25. *Pedicularis*

9. Calyx 1-lobed (sometimes appearing 2-lobed when the opposite bract is similar); floral bracts and calyx not brightly colored; stamens 2 or 4............. 22. *Cordylanthus*

9. Calyx with 2 or more lobes; floral bracts and calyx (at least the tips) often brightly colored; stamens four (10).

10. Upper corolla lip (galea) much longer than the small, 3-toothed or 3-keeled lower lip; calyx tubular; plants perennial or annual.......................... 21. *Castilleja*

10. Upper corolla lip not, or not greatly, surpassing the inflated, saccate lower lip; calyx tubular-campanulate; plants annual............................23. *Orthocarpus*
11. Stigma punctiform or capitate, entire (see also *Limosella*); seeds smooth, tuberculate, ridged, or winged (12).
11. Stigma (except in *Limosella*) flattened, usually 2-lobed or 2-parted; seeds often reticulate or cross-ribbed, wingless. Leaves opposite or mostly basal; stamens 4, in pairs, or only two (18).
12. Capsule primarily septicidal (dehiscing through the partitions); corolla not saccate or spurred at base; filaments 5, one of them not anther-bearing. Plants perennial (13).
12. Capsule wholly or partly loculicidal (dehiscing between the partitions); corolla gibbous, saccate, or spurred at base, strongly bilabiate; filaments 4 or (in *Collinsia*) a fifth one represented by a minute gland. Corolla throat often closed by a fold or palate (14).
13. Sterile stamen a flat scale approximately as wide as long, this partly adnate to the upper side of the corolla throat; corolla somewhat urceolate, with little distinction of tube and throat...7. *Scrophularia*
13. Sterile stamen an elongate, often bearded filament, not, or not much, shorter than the anther-bearing ones; corolla with tube and throat usually well differentiated 8. *Penstemon*
14. Leaves opposite or whorled; stamens and style enclosed in a keel-shaped fold of the lower corolla lip; capsule regularly dehiscent by valves; seeds few, smooth 6. *Collinsia*
14. Leaves (at least the upper ones) alternate; stamens and style not enclosed in a fold of the lower corolla lip; capsule irregularly dehiscent by pores or transverse chinks; seeds many, angled, winged, tuberculate, or alveolate (15).
15. Anther-bearing stamens 2, the other pair reduced to small abortive filaments; anthers 1-celled, by confluence...3. *Mohavea*
15. Anther-bearing stamens 4, some of the anthers occasionally sterile; anthers more or less completely 2-celled (16).
16. Leaves triangular-hastate, or reniform and deeply lobed. Capsules symmetric, opening nearly or quite their entire width; seeds winged or with corky thickenings 5. *Maurandya*
16. Leaves entire, linear to ovate (17).
17. Corolla tube with a slender spur at base; capsule symmetric, opening nearly or quite its entire width...2. *Linaria*
17. Corolla tube merely gibbous or saccate at base; capsule more or less asymmetric, opening by narrow, subapical perforations or irregular chinks............4. *Antirrhinum*
18. Anthers 1-celled, by confluence; plants subscapose; stigma capitate. Leaves entire, long-petioled; flowers borne on scapelike axillary pedicels; corolla whitish, nearly regular, rotate-campanulate ....15. *Limosella*
18. Anthers distinctly 2-celled; plants caulescent; stigma flattened (19).
19. Cells of the anthers divergent. Pedicels without bractlets (20).
19. Cells of the anthers parallel or nearly so. Sepals almost or quite distinct (21).
20. Sepals united half or more of their length; corolla yellow, pink, or red; anther-bearing stamens 4 (rarely only 2); filaments attached only toward base, entire; capsules loculicidal ....9. *Mimulus*
20. Sepals distinct or nearly so; corolla blue or bluish; anther-bearing stamens 2; filaments attached to the corolla most of their length, the sterile ones each with a knoblike projection; capsules septicidal.................................16. *Lindernia*
21. Corolla campanulate; style bifid near apex. Stems creeping or floating, usually ascending at apex; leaves palmately veined; pedicels without bractlets; sepals very unequal in width; corolla white or bluish; capsule obtuse........................14. *Bacopa*
21. Corolla tubular or funnelform; style entire, the stigma dilated or 2-lobed (22).
22. Leaves pinnatifid; pedicels without bractlets; capsule narrowly lanceolate, acuminate; seeds spirally ridged. Corolla limb purple....................12. *Schistophragma*

22. Leaves entire or merely toothed; pedicels usually bibracteolate; capsule ovate; seeds not spirally ridged (23).
23. Corolla limb violet; anther cells stipitate (each borne on a short arm of the connective). Sepals uniform or nearly so.......................................10. *Stemodia*
23. Corolla limb yellow or whitish; anther cells not stipitate (24).
24. Pedicels bibracteolate at apex; sepals nearly equal in width; corolla limb whitish; anther-bearing stamens 2...11. *Gratiola*
24. Pedicels bibracteolate at base; sepals unequal, the outer ones much broader; corolla limb yellow; anther-bearing stamens 4..........................13. *Mecardonia*

1. VERBASCUM. Mullein

Plants herbaceous, biennial; stems tall, leafy; leaves sessile, clasping or decurrent at base; flowers in elongate spikes or spikelike racemes; corolla yellow, with 5 rounded, slightly unequal lobes; stamens 5, all anther-bearing, the filaments (some or all of them) bearded; style flattened at apex.

Coarse, introduced weeds. The leaves and flowers of *V. Thapsus* have been used medicinally. It is reported that the Hopi Indians dry and smoke them mixed with *Macromeria viridiflora* in treatment of mental aberrations.

KEY TO THE SPECIES

1. Herbage densely woolly-tomentose throughout, not glandular; leaves all with entire or obscurely crenate margins, those of the stem oblanceolate, decurrent; flowers in long, thick, very dense spikes...1. *V. Thapsus*
1. Herbage loosely pilose, glandular in the inflorescence; lower leaves with dentate-serrate margins, those of the stem lanceolate, clasping; flowers in open, elongate, spikelike racemes ....2. *V. virgatum*

1. Verbascum Thapsus L. Here and there in Coconino, Yavapai, Gila, and Cochise counties, 5,000 to 7,000 feet, waste ground and roadsides, summer. Widely distributed in North America; naturalized from Europe.

2. Verbascum virgatum Stokes. Flagstaff (Coconino County), Chiricahua and Mule mountains (Cochise County), 6,000 to 7,000 feet, waste land, late summer and fall. Here and there in North America; adventive from Europe.

2. LINARIA. Toad-flax

Plants herbaceous, glabrous or nearly so, flaxlike in habit and foliage; flowering stems erect, simple or few-branched, leafy; leaves sessile, narrow, entire; flowers in terminal racemes; corolla strongly bilabiate, with a long, slender basal spur (this rarely obsolete), and a prominent palate in the throat; capsule opening near the apex by pores or chinks.

KEY TO THE SPECIES

1. Plants perennial; racemes dense; corolla yellow with an orange-colored palate, 25 to 30 mm. long; seeds winged..1. *L. vulgaris*
1. Plants annual or biennial, with short, sterile basal shoots; racemes slender, becoming elongate; corolla bright blue, not more than 10 mm. long; seeds wingless..2. *L. texana*

1. Linaria vulgaris Mill. Flagstaff and vicinity, Coconino County (*Thornber* in 1930, *Deaver* 909, *Perkins* in 1944), July to September. Widely distributed in waste ground in North America; naturalized from Eurasia.

Common toad-flax, sometimes called butter-and-eggs.

2. Linaria texana Scheele (*L. canadensis* var. *texana* Pennell). Greenlee (?), Graham, Gila, Maricopa, Pinal, Cochise, and Pima counties, 1,500 to 5,000 feet, plains and mesas, February to May. South Carolina to British Columbia, south to southern Mexico.

It is doubtful that *L. texana* is more than a variety of *L. canadensis* (L.) Du Mont. In Arizona specimens the seeds, although tuberculate, are rather sharply angled, and the pedicels are usually sparsely puberulent.

3. MOHAVEA

Plants annual, viscid-villous; leaves alternate, petioled, narrowly to broadly lanceolate, entire; flowers in leafy spikes or racemes; corolla yellow, with a short tube and an ample limb, the lower lip with a relatively small palate, streaked or spotted with red or purple.

KEY TO THE SPECIES

1. Corolla 25 to 35 mm. long, pale yellow, conspicuously marked with numerous, commonly linear, purple spots, the lower lip shallowly cleft (not nearly down to the palate); stems up to 30 cm. long..1. *M. confertiflora*
1. Corolla 15 to 20 mm. long, bright yellow, rather inconspicuously marked with few reddish-brown spots, the lower lip deeply cleft (nearly to the palate); stems not more than 15 cm. long..2. *M. breviflora*

1. Mohavea confertiflora (Benth.) Heller. Mohave and Yuma counties, 2,500 feet or lower, locally abundant in sand and on stony talus slopes, February to April. Nevada, western Arizona, southeastern California, and Baja California.

2. Mohavea breviflora Coville. Northern Mohave County, from near Beaver Dam to 20 miles south of Lake Mead, 2,500 feet or lower, dry sandy or stony slopes, March and April. Nevada, northwestern Arizona, and southeastern California.

4. ANTIRRHINUM (287). SNAPDRAGON

Plants annual or biennial; stems erect or twining; leaves (at least the upper ones) alternate, entire; flowers axillary, solitary or in leafy terminal racemes; corolla strongly bilabiate, with a prominent palate in the throat.

The popular garden snapdragon (*A. majus*) is a native of southern Europe.

KEY TO THE SPECIES

1. Stems climbing by the filiform, tendrillike pedicels, these commonly at least 3 cm. long; corolla bright yellow. Herbage villous or lanate at base of the stem, otherwise glabrous; lower leaves oblong-ovate, the upper ones narrowly lanceolate or linear; corolla conspicuously saccate at base; seeds very irregularly corky-tuberculate and winged
1. *A. filipes*
1. Stems not climbing, commonly erect; corolla not yellow (2).
2. Herbage not viscid-pilose, glandular-puberulent in the inflorescence and sparsely lanate at base of the stem; leaves lanceolate or linear. Corolla white with purple veins; capsule oblique; seeds somewhat winged..............................2. *A. Kingii*
2. Herbage copiously viscid-pilose throughout; leaves ovate (3).
3. Flowers subsessile or on pedicels shorter than the calyx; corolla rose-purple and white (drying violet); capsule nearly globose, not oblique, somewhat didymous, rounded or depressed at apex; seeds 1 mm. or more in greatest diameter, with an elliptic or orbicular, deeply cup-shaped wing much larger than the body......3. *A. cyathiferum*

3. Flowers mostly on pedicels as long or longer than the calyx; corolla violet; capsule oblong-lanceolate or narrowly ovate in outline, very oblique, not at all didymous, attenuate at apex; seeds much less than 1 mm. in greatest diameter, sharply ribbed and muricate. Flowers variable in size..........................4. *A. Nuttallianum*

1. Antirrhinum filipes Gray. Mohave, Maricopa, Pinal, Pima, and Yuma counties, 2,500 feet or lower, sandy plains and slopes, February to May, type from above Fort Mohave, Mohave County (*Newberry* in 1858). Southern Utah and western Arizona to southeastern California.

***2. Antirrhinum Kingii** Wats. This plant is not known definitely to occur in Arizona, but var. *Watsoni* (Vasey & Rose) Munz has been collected in northwestern Sonora.

3. Antirrhinum cyathiferum Benth. Pinal and Yuma counties, 1,500 feet or lower, usually on stony talus slopes, preferring partial shade, January to March. Southwestern Arizona, northwestern Sonora, and Baja California.

4. Antirrhinum Nuttallianum Benth. Yavapai, Pinal, and Pima counties, mostly 4,000 feet or lower, canyons, March to May. Southwestern Arizona, southern California, and Baja California.

A form with most of the flowers nearly sessile was collected by L. N. Gooding in the Baboquivari Mountains. What appears to be a shade form of this species was collected in the Mazatzal Mountains, Gila County, 6,000 feet (*Collom* 1635).

5. MAURANDYA

Perennial herbs of diverse habit; stems twining or procumbent; leaves alternate, petioled, coarsely toothed or hastately lobed; corolla bilabiate, with or without a palate; filaments pubescent or puberulent; capsule irregularly dehiscent near the apex.

KEY TO THE SPECIES

1. Plants densely viscid-villous throughout; stems prostrate, matted, stout, very brittle; leaves cordate or reniform, wider than long, coarsely several-toothed; calyx lobes triangular-ovate. Corolla greenish white, with a narrow, cylindric tube scarcely expanded at apex, the throat open, without a palate...................1. *M. acerifolia*

1. Plants glabrous throughout; stems climbing by the tendrillike petioles and peduncles, slender, not brittle; leaves triangular-hastate, often also cordate; calyx lobes lanceolate (2).

2. Calyx moderately accrescent in fruit, the lower portion not becoming much thickened, or strongly carinate, or reticulate; corolla violet purple or carmine (exceptionally white), the throat partly closed by a large, hairy palate; style slender, moderately dilated below; seeds thick, corky-tuberculate; leaves seldom more than 2 cm. long, mostly as wide as or wider than long..................................2. *M. antirrhiniflora*

2. Calyx greatly accrescent in fruit, the lower portion becoming much thickened, strongly carinate, and reticulate; corolla light blue (?), without a palate; style flattened and broadly triangular below; seeds thin, narrowly winged; leaves up to 6 cm. long, mostly longer than wide...3. *M. Wislizeni*

1. Maurandya acerifolia Pennell. Eastern Maricopa County, side canyons along Salt River, about 2,000 feet, shaded rock ledges and cliffs, the stems often hanging, March to May, type from Fish Creek Canyon (*Peebles et al.* 5246), also reported from near Globe, Gila County (*Ricker* 4084). Known only from southern central Arizona.

Corolla greenish outside, the limb whitish within, the guide lines yellow.

2. Maurandya antirrhiniflora Humb. & Bonpl. (*Maurandella antirrhiniflora* Rothmaler). Grand Canyon (Coconino County), and Mohave County to Cochise, Santa Cruz, Pima, and Yuma counties, 1,500 to 6,000 feet, common on stony slopes, usually among shrubs, April to October. Western Texas to southeastern California and southward.

The showy snapdragonlike flowers make this plant well worth cultivating. It is suited to growing on trellises. There are 2 sharply distinct color forms, lilac or pale violet, and rose-red.

3. Maurandya Wislizeni Engelm. (*Epixiphium Wislizeni* Munz). Greenlee County, at Duncan (*Davidson* 66) and between Duncan and Clifton (*Chapman* in 1947). Western Texas to southeastern Arizona and northern Mexico.

6. COLLINSIA (288)

Small, annual herbs; stems decumbent or erect, widely branched; leaves opposite, oblong to narrowly lanceolate or spatulate; flowers axillary, mostly in whorls, the lower flowers often solitary; corolla small, blue and white, deeply 2-lipped; filaments glabrous; capsule few-seeded.

1. Collinsia parviflora Dougl. Kaibab Plateau and both rims of Grand Canyon (Coconino County), Mt. Trumbull (Mohave County), and in Yavapai and Gila counties, 4,000 to 8,000 feet, moist soil, February to June. Ontario and Michigan to Alaska, south to New Mexico, Arizona, and California.

Newsom (288, p. 291) doubtfully referred to this species a collection in the Chiricahua Mountains, Cochise County (*Lemmon* 3073), in which "the corolla-throat is open, the upper lip not at all reflexed, the tube slender and not gibbous, and the plant quite conspicuously glandular."

7. SCROPHULARIA. Figwort

Coarse, perennial herbs; stems tall, erect or nearly so, leafy; leaves opposite, petioled, ovate to lanceolate, serrate or laciniate; flowers numerous, in ample loose terminal panicles; corolla greenish or dull red, short and broad, short-lobed; capsule 2-valved; seeds many.

KEY TO THE SPECIES

1. Leaves (at least the lower ones) prevailingly triangular-ovate and cordate at base, coarsely and irregularly, often doubly, dentate or laciniate...........1. *S. californica*
1. Leaves prevailingly lanceolate to ovate-lanceolate, mostly truncate, cuneate, or subcordate at base, usually more evenly and shallowly serrate-dentate or crenate...2. *S. parviflora*

1. Scrophularia californica Cham. Pinal Mountains, Gila County, about 4,000 feet (*Harrison* 2090), Santa Rita Mountains, Santa Cruz County, 5,000 feet (*Thornber* in 1903), June. Central Arizona, Oregon, and California.

Although from far outside the main area of this species, the specimens cited closely resemble many from California and Oregon.

2. Scrophularia parviflora Woot. & Standl. (*S. glabrata* Davidson, non Ait., *S. Davidsonii* Pennell). Southern Coconino County and Hualpai Mountain (Mohave County), to Greenlee, Graham, Cochise, and Pima counties,

5,000 to 8,000 feet, common in rich soil in coniferous forests, July to October, type of *S. glabrata* Davidson from Metcalf, Greenlee County (*Davidson* in 1902). Western New Mexico and Arizona.

The stems reach a height of 1.2 m. (4 feet) or more. This species apparently intergrades in Arizona with *S. californica,* which it resembles in its relatively lax inflorescence, but is more like *S. lanceolata* Pursh in the shape of the leaves. The type of *S. glabrata,* as compared with the type of *S. parviflora,* has the herbage more finely puberulent, the leaves thinner and more coarsely toothed, and the flowers smaller (barely 5 mm. long).

S. coccinea Gray, a rare species of southwestern New Mexico, is to be looked for in southeastern Arizona. The corolla is much larger (15 to 20 mm. long) and of a brighter-red color than in *S. californica* and *S. parviflora.*

8. PENSTEMON (289, 290). Beardtongue

Contributed by David D. Keck

Perennial herbs or shrubs; leaves opposite, the upper ones sessile; flowers showy, paniculate; calyx 5-parted; corolla tubular, usually somewhat ventricose, 2-lipped, the upper lip 2-lobed, the lower lip 3-cleft; fertile stamens 4, paired, with arching filaments; sterile filament (staminode) attached to upper side of corolla at junction of tube and throat, extending downward and forward; anthers 2-celled, the cells often confluent; capsule septicidal; seeds numerous, angled.

These plants mostly flower in spring or early summer and are often very showy. One may expect them in light, dry, neutral soils in eroded or mountainous regions throughout the state and at all elevations, although few species are found on the deserts. Some use has been made of them for ornamental plantings, which, on account of the wide range of growth habits and colors available, offer greater opportunities than have been realized thus far. Otherwise the species are of insignificant economic importance, although many of them are browsed, especially the shrubby, evergreen *P. microphyllus.*

KEY TO THE SPECIES

1. Corolla scarlet, carmine, yellow, or (in *P. Parryi*) rose-magenta (2).
1. Corolla whitish, pink, lavender, blue, or blue-purple (never scarlet, carmine, or yellow) or if reddish, then the leaves toothed (11).
2. Corolla sulphur-yellow; fertile filaments strongly pubescent at base. Shrubs up to 2 m. high; leaves less than 2 cm. long, usually elliptic; corolla strongly bilabiate, 10 mm. wide; staminode densely long-bearded........................1. *P. microphyllus*
2. Corolla scarlet to carmine, rarely orange or rose; fertile filaments glabrous (3).
3. Calyces and pedicels obviously glandular-pubescent. Corolla strongly bilabiate (4).
3. Calyces and pedicels glabrous or puberulent, at most very obscurely glandular. Plants strictly herbaceous (6).
4. Leaves filiform, 1 mm. wide, crowded; anthers dehiscent throughout, explanate. Plants woody below; corolla scarlet..2. *P. pinifolius*
4. Leaves much wider, not crowded; anthers not explanate (5).
5. Anther sacs broad, dehiscent from apex to base; leaves linear-attenuate; stems more or less densely puberulent below the inflorescence, entirely herbaceous; corolla orange to dull red ....3. *P. lanceolatus*

5. Anther sacs narrow, dehiscent across their contiguous apices for less than half their length, the lower portion saccate; leaves linear-oblanceolate; stems glabrous below the inflorescence, often slightly woody near the base; corolla bright red
4. *P. Bridgesii*
6. Corolla strongly bilabiate, the prominent lower lip reflexed, the upper lip projecting, scarlet. Stems tall, virgate .... 5. *P. barbatus*
6. Corolla obscurely bilabiate, the lips about equally erect or spreading (7).
7. Anther sacs dehiscent only part way from the free tips, not explanate. Corolla glabrous, scarlet; herbage not at all glaucous .... 6. *P. Eatoni*
7. Anther sacs dehiscent throughout and explanate (8).
8. Corolla glabrous throughout, narrowly tubular, the limb very narrow, scarlet; herbage green or glaucescent. Cauline leaves linear-lanceolate .... 7. *P. subulatus*
8. Corolla glandular externally and internally, the limb broad and flaring; herbage glaucous (9).
9. Glands on the corolla sessile; throat tubular; staminode glabrous (rarely obsoletely bearded). Corolla carmine; cauline leaves lance-oblong: northern Arizona
8. *P. utahensis*
9. Glands on the corolla stalked; throat somewhat ampliate; staminode bearded: southern Arizona (10).
10. Corolla 15 to 20 mm. long, rose-magenta, rather broadly funnelform; cauline leaves narrowly lanceolate to lance-oblong, not blackening on drying .... 9. *P. Parryi*
10. Corolla 20 to 25 mm. long, carmine to scarlet, narrowly funnelform; cauline leaves broadly ovate to oblong-ovate, large, blackening on drying .... 10. *P. superbus*
11. Leaves, at least a few of them, toothed, sometimes (in *P. Jamesii*) very obscurely so. Corolla glandular externally (12).
11. Leaves always entire (18).
12. Leaves obscurely denticulate, none perfoliate; plants not glaucous, less than 60 cm. high; calyx lobes lance-oblong to attenuate, 7 to 12 mm. long. Inflorescence glandular-pubescent; corolla purple or blue-purple (13).
12. Leaves conspicuously toothed, the uppermost leaves connate-perfoliate; plants glaucous or glaucescent (often green in one subspecies of *P. pseudospectabilis*), 60 to 140 cm. high; calyx lobes mostly ovate, 4 to 6 mm. long (15).
13. Corolla villous but not glandular within, the lower lip projecting, exceeding the upper lip; staminode bearded only apically; anther sacs dehiscent throughout but not explanate through the connective .... 11. *P. Whippleanus*
13. Corolla glandular within, the lower lip reflexed, about equaling the upper lip; staminode bearded most of its length; anther sacs peltately explanate (14).
14. Leaves large, ovate, obviously toothed; corolla more than 35 mm. long, the throat abruptly much inflated, not villous within .... 12. *P. Cobaea*
14. Leaves smaller, linear-oblanceolate, some of them obscurely toothed, usually many of them entire; corolla up to 22 mm. long, the inflated throat villous within
13. *P. Jamesii*
15. Corolla abruptly inflated from a tube not longer than the calyx, strongly bilabiate, 10 to 20 mm. wide, white tinged with pink; staminode exserted, uncinate, long-bearded
14. *P. Palmeri*
15. Corolla gradually inflated from a tube twice longer than the calyx, nearly regular, narrower, deep pink to rose-purple; staminode included or barely exserted, straight (16).
16. Throat of the corolla definitely ventricose, 9 to 12 mm. wide; inflorescence usually leafy below, frequently interrupted by long internodes .... 15. *P. Clutei*
16. Throat of the corolla moderately ampliate, 6 to 9 mm. wide; inflorescence not leafy, rarely interrupted (17).
17. Staminode normally glabrous; throat not villous at the lower lip; anther sacs usually not longer than wide .... 16. *P. pseudospectabilis*
17. Staminode bearded; throat sparsely villous at the lower lip; anther sacs longer than wide .... 17. *P. bicolor*

18. Herbage blue-glaucous or obviously glaucescent, glabrous; leaves coriaceous, lanceolate or wider. Staminode bearded (19).
18. Herbage green or, if pubescent, then often grayish (more or less glaucescent in *P. comarrhenus*); leaves mostly thin, relatively narrow (23).
19. Corolla strongly bilabiate, lavender, the throat about 10 mm. wide, pilose at the orifice; inflorescence paniculate, with long internodes, the divergent peduncles 1 to 3 cm. long 18. *P. nudiflorus*
19. Corolla nearly regular, the throat 4 to 7 mm. wide; inflorescence an interrupted thyrse of verticillasters, the peduncles erect and very short, or suppressed (20).
20. Corolla nearly tubular, blue; verticillasters few-flowered and rather open, all very short-bracteate, the inflorescence appearing bare........................19. *P. Fendleri*
20. Corolla more or less ampliate; verticillasters dense, the lower ones prominently bracteate (21).
21. Staminode bearded most of its length with long, fine, pale-yellow hairs, not dilated. Corolla deep blue-purple, nearly glabrous at the orifice; calyx lobes acute, scarious-margined nearly throughout; cauline leaves mostly obtuse or rounded at apex 20. *P. pachyphyllus*
21. Staminode bearded only at and near the apex with relatively short, coarse, deep-yellow hairs, dilated apically (22).
22. Corolla pinkish lavender, glabrous throughout; calyx lobes acuminate, scarious-margined only at base; cauline leaves acuminate........................21. *P. angustifolius*
22. Corolla bright blue, slightly pilose at the orifice; calyx lobes acute to short-acuminate, scarious-margined nearly to the apex; cauline leaves obtuse or rounded, mucronate 22. *P. lentus*
23. Throat of the corolla more or less distinctly 2-ridged within ventrally, the ridges densely hairy about the orifice. Corolla relatively narrow (24)
23. Throat of the corolla rounded ventrally, lightly if at all hairy at the orifice (29).
24. Plants strongly cespitose, not more than 10 cm. high, cinereous-puberulent; leaves up to 2 cm. long, mucronate. Corolla narrowly funnelform, lavender-purple (25).
24. Plants not cespitose, more than 10 cm. high, not cinereous; leaves (at least the basal ones) more than 2 cm. long, not mucronate (26).
25. Stems scarcely creeping; leaves cinereous-whitened with closely appressed hairs, oblanceolate, spatulate-oblong, or obovate; herbage of the inflorescence very obscurely viscid ....23. *P. Thompsoniae*
25. Stems widely creeping; leaves greenish, with spreading hairs, in Arizona linear-oblanceolate to lance-obovate, smaller; herbage of the inflorescence more obviously viscid ....24. *P. caespitosus*
26. Calyx and corolla glandular-puberulent, the latter pale blue, much paler on the often strongly 2-ridged ventral portion, the lower lip exceeding the upper lip 25. *P. oliganthus*
26. Calyx and corolla essentially glabrous, the latter moderately 2-ridged ventrally, the upper and lower lips subequal (27).
27. Leaves and calyx lobes prominently white-margined, thick; inflorescence conspicuously leafy; corolla pinkish, moderately ampliate; staminode glabrous 26. *P. albomarginatus*
27. Leaves not prominently white-margined, thin; calyx lobes more or less hyaline-margined; inflorescence scarcely leafy; corolla blue-purple, tubular; staminode usually densely bearded (28).
28. Stems terminating rootstocks, arising from a basal rosette of leaves; verticillasters dense, the peduncles and pedicels obscure; calyx lobes oblong, caudate-tipped, rather prominently erose, 3.5 to 5.5 mm. long; corolla densely golden-bearded at the palate 27. *P. Rydbergii*
28. Stems densely tufted from a subligneous caudex, leafy throughout but with no definite rosette at base; verticillasters looser, the peduncles and pedicels more obvious; calyx lobes ovate or obovate to rotund, very short-tipped, slightly erose, 2 to 3.5 mm. long; corolla lightly white-bearded at the palate........................28. *P. Watsoni*
29. Plants shrubby or decidedly suffrutescent (sometimes nearly herbaceous in *P. linarioides*); leaves crowded, narrowly linear, mucronate (30).

29. Plants strictly herbaceous; leaves not crowded or mucronate. Corolla large, funnelform, strongly bilabiate (33).
30. Staminode bearded; corolla viscid-puberulent externally, the limb moderately expanded (31).
30. Staminode glabrous; corolla glabrous externally, the limb much expanded (32).
31. Corolla 14 to 24 mm. long, blue-purple; leaves 8 to 20 (or 30) mm. long, not fleshy, grayish or greenish on both surfaces.........................29. *P. linarioides*
31. Corolla 10 mm. long, white tinged with lavender; leaves 5 to 8 (or 10) mm. long, fleshy, white-puberulent on the flat upper surface, deep green and glabrous on the rounded lower surface ....30. *P. discolor*
32. Corolla blue-purple (rarely pinkish) throughout, funnelform, 10 to 12 mm. long, the limb pubescent only at base of the lower lip, scarcely oblique, the throat not curved but ventricose; stamens exserted..............................31. *P. Thurberi*
32. Corolla pink, salverform, 14 to 16 mm. long, the limb white within, densely puberulent on all sides at the orifice, set obliquely on the curved throat; stamens included 32. *P. ambiguus*
33. Inflorescence (at least the corolla) glandular-pubescent; anther sacs obviously spinose-dentate along the suture. Leaves all linear-attenuate; staminode glabrous (34).
33. Inflorescence not at all glandular; anther sacs microscopically denticulate or glabrous along the suture (35).
34. Herbage cinereous throughout; inflorescence densely glandular-pubescent, strict, racemose, the mostly 1-flowered peduncles erect; calyx lobes oblong or oblong-lanceolate, with entire, narrowly or obsoletely scarious margins; flowering mainly in spring 33. *P. dasyphyllus*
34. Herbage glabrate; inflorescence very lightly glandular-puberulent, open, paniculate, the commonly 2- or 3-flowered peduncles divaricate; calyx lobes broadly ovate, with broad erose scarious margins; flowering in late summer..........34. *P. stenophyllus*
35. Anther sacs glabrous or finely scabrid on the sides. Inflorescence strict, secund (36).
35. Anther sacs villous on the sides (37).
36. Sacs of the anthers opening throughout, straight, opposite; staminode glabrous (except in subsp. *arizonicus*); lower lip usually bearded; leaves mostly linear in the Arizona plants (except in subsp. *arizonicus*)............................35. *P. virgatus*
36. Sacs of the anthers opening partially, curved, divaricate; staminode bearded; lower lip glabrous; leaves lanceolate...36. *P. laevis*
37. Corolla pale blue, the tube nearly as long as the throat; lower peduncles divergent, bearing elongate pedicels; anthers usually nearly hidden by hairs..37. *P. comarrhenus*
37. Corolla deep blue, the tube much shorter than the throat; peduncles and pedicels appressed and short, the strict panicle secund; anthers less densely villous.38. *P. strictus*

1. Penstemon microphyllus Gray (*P. Plummerae* Abrams). Gila, Yavapai, and Mohave counties to western Pima and northern Yuma counties, 1,500 to 5,000 feet, desert mountain ranges, often with junipers, March to May, type from Williams River (*Bigelow* in 1853–1854), type of *P. Plummerae* from Mineral Park, Mohave County (*Lemmon* in 1884). Southern and western Arizona, southern California, and northern Baja California.

2. Penstemon pinifolius Greene. Known in Arizona only from the Clifton area (Greenlee County) and the Chiricahua and Swisshelm mountains (Cochise County), rocky summits above 5,000 feet, summer, type from near Clifton (*Greene* in 1880). Southwestern New Mexico, southeastern Arizona, and adjacent Mexico.

3. Penstemon lanceolatus Benth. Mountains of Greenlee, Graham, and Cochise counties, 5,000 to 6,000 feet, occasional, usually in rocky canyons, May and June. Southwestern New Mexico, southeastern Arizona, and northern Mexico.

4. Penstemon Bridgesii Gray. Northern Navajo County and Kaibab Plateau (Coconino County), south to the Sierra Ancha (Gila County), west to Mohave and Yavapai counties, 4,500 to 7,500 feet, occasional in the mountains among pinyons and ponderosa pine, May to September. Southwestern Colorado and western New Mexico to California.

5. Penstemon barbatus (Cav.) Roth. White Mountains (Apache County), northern Navajo County, and Kaibab Plateau (Coconino County), south to the Mexican border, 4,000 to 10,000 feet, common in the mountains in coniferous or in oak woods, June to October. Southern Colorado and Utah to the central highlands of Mexico.

KEY TO THE SUBSPECIES

1. Anthers villous .... subsp. *trichander*
1. Anthers essentially glabrous (2).
2. Lower lip of the corolla bearded .... *P. barbatus* (typical)
2. Lower lip glabrous .... subsp. *Torreyi*

Most of the Arizona plants are typical *P. barbatus*. Subsp. *Torreyi* (Benth.) Keck is much less frequent than in Colorado and New Mexico. In northern Apache County one occasionally finds subsp. *trichander* (Gray) Keck. The species as a whole is variable as to the presence or absence of puberulence on the herbage, all gradations being found at random. Therefore var. *puberulus* Gray, founded on a collection from Guadalupe Canyon, Cochise County (*Thurber* in 1851), is not retained.

Several plants of what is almost certainly a hybrid between *P. barbatus* subsp. *Torreyi* and *P. virgatus* Gray have been found on the north side of the Grand Canyon growing in close proximity to the presumable parental species. A specimen of this hybrid collected by Rose E. Collom has corollas of a deep-violet color.

6. Penstemon Eatoni Gray. Northern Arizona south to Gila and northern Pinal counties, 2,000 to 7,000 feet, common on mesas, in fields, and at roadsides, sandy or clay soils, February to June. Southwestern Colorado to central Arizona and California.

KEY TO THE SUBSPECIES

1. Herbage glabrous throughout .... *P. Eatoni* (typical)
1. Herbage puberulent (2).
2. Stamens included or barely exserted .... subsp. *undosus*
2. Stamens long-exserted .... subsp. *exsertus*

Typical *P. Eatoni* is present only in the northernmost tier of counties and in northern Yavapai County, but is abundant in Utah. In Arizona 2 subspecies occur much more frequently. Subsp. *undosus* (Jones) Keck (*P. coccinatus* Rydb.), type from the Grand Canyon (*MacDougal* 173) is frequent in southern Utah and northern Arizona, occasional in Gila and Pinal counties. Southward, particularly in Gila and Pinal counties, the prevailing entity is subsp. *exsertus* (A. Nels.) Keck (*P. exsertus* A. Nels., *P. amplus* A. Nels.), type from along Salt River (*Nelson* 10624), type of *P. amplus* from Oak Creek Canyon, near Sedona (*A. & R. Nelson* 2075). A hybrid between *P.*

Eatoni subsp. *exsertus* and *P. Palmeri* subsp. *typicus,* collected in Oak Creek Canyon, near Sedona (*A. & R. Nelson* 2076), has been named *P. mirus* A. Nels.

7. Penstemon subulatus Jones. Central Mohave County to Graham, Pima, and Yuma counties, 1,500 to 4,500 feet, stony hillsides, canyons, and mesas, nowhere abundant, February to May, type from Hackberry, Mohave County (*Jones* in 1903). Known only from central and southern Arizona.

This is the Arizona counterpart of the well-known scarlet-bugler (*P. centranthifolius*) of coastal California. It has the most slender corolla tube of all scarlet penstemons except the very distinct and shrubby *P. pinifolius.*

A plant collected near Springerville, Apache County (*L. Benson* 9566), in grassland at an elevation of about 7,000 feet, represents a considerable extension of the geographical and altitudinal ranges of *P. subulatus.* It approaches *P. utahensis* in its wider corolla, but the corolla is glabrous.

8. Penstemon utahensis Eastw. Navajo, Coconino, and Mohave counties, 4,000 to 6,500 feet, uncommon, canyons and mesas, March to May. Southern Utah and northern Arizona to eastern California.

9. Penstemon Parryi Gray (*P. Shantzii* A. Nels., *P. Shantzii* var. *incognitus* A. Nels.). Gila, Pinal, Cochise, Santa Cruz, and Pima counties, 1,500 to 5,000 feet, mountain canyons and well-drained slopes, spring, type from the Gila River (*Parry* in 1852), type of *P. Shantzii* from near Superior, Pinal County (*Nelson* 11263), type of var. *incognitus* from near Sells, Pima County (*A. & R. Nelson* 1373). Southern Arizona and Sonora.

In favorable situations and seasons this plant produces from the base many erect stems up to 4 feet in length, bearing very showy flowers. The species is not uncommon, but the individuals are usually well scattered.

10. Penstemon superbus A. Nels. Greenlee, Graham, and Cochise counties, 3,500 to 5,500 feet, uncommon, rocky canyons and along washes, sandy or gravelly soils, April and May. New Mexico, southeastern Arizona, and Chihuahua.

In habit very similar to *P. Parryi,* with which it apparently intergrades.

11. Penstemon Whippleanus Gray (*P. arizonicus* Heller). San Francisco Peaks (Coconino County), Hualpai Mountain (Mohave County), 6,500 to 12,000 feet, July and August, type of *P. arizonicus* from the crater of the San Francisco Peaks (*MacDougal* in 1898). Southern Montana to New Mexico and northern Arizona.

The flower color of Arizona specimens is of the purple phase rather than lemon yellow or whitish, as it is occasionally elsewhere within the range of the species.

12. Penstemon Cobaea Nutt. (*P. Hansonii* A. Nels.). Flagstaff (Coconino County), apparently escaped from cultivation or introduced (*Hanson* 709, the type of *P. Hansonii*). Southeastern Nebraska to southern Texas.

A showy plant, known in Texas as foxglove, a name properly applied to the Old World *Digitalis purpurea.*

13. Penstemon Jamesii Benth. Apache County to Mohave and Yavapai counties, 4,500 to 7,000 feet, frequent in sandy soils, in the pinyon-juniper and ponderosa pine associations, May and June. Southwestern Colorado, southern Utah, western New Mexico, and northern Arizona.

KEY TO THE SUBSPECIES

1. Corolla 14 to 16 mm. long, 5 to 6.5 mm. wide, scarcely inflated, the lower lip glabrous within .... subsp. *breviculus*
1. Corolla 17 mm. long, or longer, much inflated, the lower lip glandular-pubescent within (2).
2. Corolla 17 to 22 mm. long, 7 to 10 mm. wide .... subsp. *ophianthus*
2. Corolla 25 to 32 mm. long, 10 to 15 mm. wide .... subsp. *typicus*

The staminode in this species is conspicuously bearded and exserted. The corolla is lavender, veined with darker purple. Subsp. *typicus* Keck occurs east of the Continental Divide, in New Mexico and western Texas. The Arizona plants belong to subsp. *ophianthus* (Pennell) Keck (*P. ophianthus* Pennell, *P. pilosigulatus* A. Nels.), type of the latter from Flagstaff (*Hanson* 554), but subsp. *breviculus* Keck is to be looked for in northeastern Arizona since it occurs in northwestern New Mexico and adjacent Colorado.

Penstemon pulchellus Lindl. is to be expected in Cochise County since it is found in southwestern New Mexico. It has thin, finely serrate, lanceolate, nonperfoliate, essentially glabrous leaves, and abruptly much-inflated, showy corollas.

14. Penstemon Palmeri Gray. Coconino, Mohave, and Yavapai counties, 3,500 to 6,500 feet, frequent in washes and at roadsides in the sagebrush and pinyon regions, March to September, type from Skull Valley, Yavapai County (*Coues & Palmer* 228). Utah and Arizona to California.

This is one of the handsomest species of *Penstemon* and is additionally notable for its delicate fragrance. The common plant is subsp. *typicus* Keck with calyces, pedicels, and peduncles glandular-pubescent, but in northernmost Navajo, Coconino, and Mohave counties and in adjacent Utah is found subsp. *eglandulosus* Keck, in which these parts are glabrous.

Penstemon petiolatus T. S. Brandeg. is to be looked for in the northwestern corner of the state. It is a rare plant of the Beaver Dam Mountains, Washington County, Utah, and of southern Nevada, of low, shrubby habit (10 to 20 cm. high), and with dentate, pruinose-puberulent leaves only 10 to 25 mm. long.

15. Penstemon Clutei A. Nels. Coconino County, about 7,000 feet, very local in the region about Sunset Crater, northeast of Flagstaff, in volcanic cinders, June and July, type collected by W. N. Clute in 1923. Known only from Arizona.

16. Penstemon pseudospectabilis Jones. Apache, southern Coconino, and Mohave counties, south and east to Graham, Cochise, Pima, and Yuma counties, 2,000 to 7,000 feet, open land, spring and summer, type from near Chemehuevi, Mohave County (*Jones* in 1903). Southwestern New Mexico to eastern California.

A common and beautiful species, separable into western and eastern subspecies on the presence or absence of glands on the calyces and pedicels. In Mohave, Yuma, and western Pinal and Pima counties and in California is found subsp. *typicus* Keck, with the glands and with usually glaucous herbage. To the eastward grows the much more abundant, greener, and eglandular subsp. *connatifolius* (A. Nels.) Keck, type from the Apache Trail, Gila or Maricopa County (*A. Nelson* 10314). A beautiful form, apparently a first-generation hybrid between subsp. *connatifolius* and *P. Eatoni* subsp. *exsertus,* collected near Superior (*A. Nelson* 11262) was described as *P. Crideri* A. Nels.

The staminode, although normally glabrous in this species, is sparsely bearded toward the apex in a specimen of subsp. *typicus* from the Kofa Mountains, Yuma County (*Harbison* 41.51), and in one from Montezuma Well, Yavapai County (*Deaver* 1991). The latter specimen also has the anther sacs distinctly longer than wide.

17. Penstemon bicolor (T. S. Brandeg.) Clokey & Keck. Portland Mine to Chloride, Mohave County, 2,500 feet (*Kearney & Peebles* 13163), spring, very rare. Southern Nevada and northwestern Arizona.

The Arizona plant is subsp. *roseus* Clokey & Keck. The typical phase of the species has essentially white flowers, and has been found only in Clark County, Nevada. The species as a whole is found associated with cresote-bush on outwash fans and plains.

18. Penstemon nudiflorus Gray. Coconino, Mohave, and Yavapai counties, mountainous regions south of the Grand Canyon, 4,500 to 7,000 feet, dry slopes in ponderosa pine forests, uncommon, summer, type from near Flagstaff (*Lemmon* in 1884). Known only from north-central Arizona.

19. Penstemon Fendleri Torr. & Gray. Guthrie, Greenlee County (*Davidson* 120), and in Cochise County, 4,000 to 5,000 feet, not common, April to June. Oklahoma and Texas to southeastern Arizona and Chihuahua.

20. Penstemon pachyphyllus Gray. Apache, Coconino, and Mohave counties, from the Kaibab Plateau south to Williams, 4,500 to 7,000 feet, dry slopes among ponderosa pine, pinyon, or juniper, May and June. Utah, Nevada, and northern Arizona.

The species is represented in Arizona by subsp. *congestus* (Jones) Keck (*P. congestus* Pennell). Typical *P. pachyphyllus,* from the Uintah Basin, northern Utah, differs from this subspecies in having a very broad, flaring corolla limb. In subsp. *congestus* there is variation in the amount of pubescence on the orifice, in the color of the beard on the staminode, and in the size of the calyx lobes.

21. Penstemon angustifolius Nutt. Apache, Navajo, and Coconino counties, 5,000 to 6,500 feet, mesas and sandy grasslands, frequently on dunes, May and June. North Dakota and eastern Montana to Kansas, New Mexico, and northern Arizona.

The Arizona plants belong to subsp. *venosus* Keck, type from 12 miles northeast of Tuba, Coconino County (*Peebles & Fulton* 11877), found also in southern Utah and northwestern New Mexico. This is the westernmost representative of the variable *P. angustifolius,* which in typical form grows on the high plains east of the Continental Divide. It is replaced in southeastern Colorado, Kansas, and northern New Mexico by subsp. *caudatus* (Heller) Keck, which intergrades completely with typical *P. angustifolius*. The Arizona plant, subsp. *venosus,* is closely related to subsp. *caudatus* but is distinguished by pinkish-lavender instead of blue flowers, herbage not darkening appreciably in drying, and bracts of the inflorescence more prominently venose on both sides. The Hopi Indians are reputed to make a medicine of the roots and call the plant "tci-eq-pi," meaning "snake-plant."

22. Penstemon lentus Pennell. Apache County, in the Lukachukai Mountains (*Goodman & Payson* 2882) and at Fort Defiance (*E. Palmer* 100), 6,000 to 8,500 feet, very rare, dry hills and mesas, usually in sandy soil, June. Southwestern Colorado, southeastern Utah, and northeastern Arizona.

23. Penstemon Thompsoniae (Gray) Rydb. Kaibab Plateau and south (Navajo, Coconino, Mohave, and Yavapai counties), 4,000 to 7,000 feet, light soils with pinyon and juniper, May and June (sometimes also in autumn). Southern Utah, southeastern Nevada, and northern Arizona.

24. Penstemon caespitosus Nutt. Painted Desert and Grand Canyon regions (Navajo and Coconino counties), 4,500 to 7,000 feet, sometimes on limestone soils, June to August. Wyoming, Colorado, Utah, and northern Arizona.

This variable species, which comprises some of the smallest known plants in the genus, is represented in Arizona by subsp. *desertipicti* (A. Nels.) Keck (*P. desertipicti* A. Nels.), type from near Cameron, Coconino County (*Hanson* A177).

25. Penstemon oliganthus Woot. & Standl. White Mountains (Apache County), 8,000 to 9,000 feet, grassy mountain meadows in loamy soil, July to September. Colorado, New Mexico, and eastern Arizona.

A low herb with few stems arising from small rosettes, the flowers often somewhat declined. A collection, labeled as from Flagstaff (*Carter* in 1927), represents a considerable westward extension of the range if the locality is given correctly.

26. Penstemon albomarginatus Jones. Near Yucca (Mohave County), 1,500 to 2,500 feet, sometimes with Joshua trees (*Yucca brevifolia*), rare, April to June. Southern Nevada, western Arizona, and eastern California.

This singular species grows in drifting sand, out of which many stems arise from the long, fleshy root.

27. Penstemon Rydbergii A. Nels. North Rim of the Grand Canyon, Coconino County (*Purchase* 2902, *McHenry* in 1933), 8,500 feet, moist soil, very rare, July, also in the Tunitcha Mountains, San Juan County, New Mexico, so probably in the same range in Apache County, Arizona. Wyoming and eastern Idaho to northern New Mexico and Arizona.

The two Arizona collections combine the characteristics of the small-flowered typical *P. Rydbergii* with the puberulent-calyxed subsp. *aggregatus* (Pennell) Keck. Both are atypical in the naked or nearly naked staminodes. These specimens indicate that at this distant outpost in the range of *P. Rydbergii* a recombination race is found.

28. Penstemon Watsoni Gray. Mokiak Pass, northern Mohave County (*E. Palmer* 377), rare in Arizona. Common in the sagebrush and pinyon belts of Colorado, Utah, and Nevada.

The plant is herbaceous throughout, with several moderately tall stems arising directly from a crownlike base without forming a basal rosette. The sepals are remarkably small, seldom exceeding 3 mm. in length.

29. Penstemon linarioides Gray. Apache County to Mohave County, south to Cochise, Gila, and Yavapai counties, 4,500 to 9,000 feet, often on calcareous soil, June to August. Colorado, Utah, New Mexico, and Arizona.

KEY TO THE SUBSPECIES AND VARIETY

1. Leaves principally oblanceolate....................................subsp. *Maguirei*
1. Leaves essentially linear (2).
2. Leaves glabrous..var. *viridis*
2. Leaves densely puberulent (3).
3. Hairs of the leaf fine, erect or retrorsely spreading......................subsp. *Sileri*
3. Hairs of the leaf flattened, closely appressed (4).
4. Staminode sparsely bearded apically..........................subsp. *coloradoensis*
4. Staminode more densely bearded, with longer hairs, for most of its length (5).
5. Stems ascending from a decumbent rootstock; leaves closely overlapping, heathlike, mostly 1 cm. long....................................subsp. *compactifolius*
5. Stems strictly erect from a compact caudex; leaves more remote, longer...subsp. *typicus*

Arizona is the center of greatest diversity in this variable species, which comprises many subspecies that are both geographically and morphologically separable but intergrade at their points of contact. Subsp. *typicus* Keck is found from Apache and Navajo counties south to Cochise County and east to New Mexico. Subsp. *Maguirei* Keck is local in the Gila River valley (Greenlee County) and in adjacent New Mexico. Subsp. *compactifolius* Keck is locally common in the Flagstaff region. Subsp. *coloradoensis* (A. Nels.) Keck grows at high elevations in the northeastern corner of the state, at Marsh Pass, Navajo County (*Harvey* in 1937), and in the Lukachukai Mountains, Apache County (*Goodman & Payson* 2848), but is much more abundant in adjacent New Mexico and Colorado. Subsp. *Sileri* (Gray) Keck is common from the Kaibab Plateau region (Coconino and Mohave counties) to Yavapai County, and is occasional southeast to the White Mountains and to Cochise County, also common in southern Utah. Var. *viridis* Keck is less frequent than subsp. *Sileri* but occupies the same range as far south as Gila County.

30. Penstemon discolor Keck. Known only from at and near the type locality, Bear Canyon, Santa Catalina Mountains, Pima County (*Shreve* 5319, the type collection, *Parker* 5912), 6,000 to 7,500 feet, June to August.

31. Penstemon Thurberi Torr. (*P. scoparius* A. Nels.). Mohave and Yavapai counties, southeast to Graham, Cochise, and Pima counties, 2,000 to 5,000 feet, open, sandy or stony slopes and scrub-oak thickets, March to August, type of *P. scoparius* from West Wells (*Goodding* 1037). New Mexico, Arizona, California, and Baja California.

This species is ordinarily distinctly set off from the next, but there is some evidence of their mixing in west-central New Mexico.

32. Penstemon ambiguus Torr. Apache, Navajo, and Coconino counties, 4,500 to 6,500 feet, open, sandy mesas and grassland, rather common in the Painted Desert, summer. Kansas and Texas to Nevada and northern Arizona.

The typical plant, with puberulent herbage, occurs farther east. The glabrous Arizona entity, subsp. *laevissimus* Keck, ranges from southwestern Texas to Utah and Nevada. In the Hopi country it has been called cow-tobacco.

33. Penstemon dasyphyllus Gray. Cochise, Santa Cruz, and Pima counties, 3,500 to 5,500 feet, open, gravelly slopes, April to June and rarely again in late summer. Western Texas to southeastern Arizona and Chihuahua.

34. Penstemon stenophyllus Gray (*P. rubescens* Gray). Huachuca and

Patagonia mountains (Cochise and Santa Cruz counties), 4,000 to 5,500 feet, open canyons and slopes, August and September. Southeastern Arizona, Sonora, and Chihuahua.

The type of the species came from Sonora. The type of *P. rubescens* came from near Fort Huachuca, Cochise County (*Lemmon*), and was blue-flowering although Gray mistook the color for red. The flowers of this and the preceding species are a rich violet-blue when fresh.

35. Penstemon virgatus Gray (*P. putus* A. Nels.). Apache County to northeastern Mohave County, south to Cochise, Gila, and Yavapai counties, 5,000 to 11,000 feet, pine woodlands and mountain meadows, summer. New Mexico and Arizona.

The corolla is marked with deep-purple guide lines within the throat and is usually pale violet, but throughout the range of the species occasional plants are found with white flowers. This color variation was the basis of *P. putus*, type from Black River, White Mountains (*Goodding* 1100). *P. virgatus* is a highly variable species as to leaf width, shape of the calyx lobes, presence or absence of puberulence, and corolla size. Two geographic variants are significant. One, subsp. *arizonicus* (Gray) Keck, occurs in the White and Pinaleno mountains, 7,500 to 11,000 feet, the type collected by Rothrock on Mount Graham. It differs from the species in having the staminode bearded, the leaves somewhat broader and oblong-spatulate instead of merely linear-lanceolate, and the calyx lobes broadly scarious-margined and erose. The other subspecies, not yet published, is essentially limited to the Kaibab Plateau, Coconino County, 7,500 to 8,500 feet (*C. L. Hitchcock et al.* 4548), with a station on Navajo Mountain at 9,500 feet, just over the Utah state line. This entity differs from typical *P. virgatus* in having folded, linear-filiform cauline leaves not over 2.5 mm. wide. Intergrades connect each of these units with the species.

***36. Penstemon laevis** Pennell. Not rare just over the state line in Utah, as in Kanab Canyon, and likely to be found in Coconino and Mohave counties, May and June.

This species has blue-purple flowers marked with guide lines as in *P. virgatus*, but its thyrse is somewhat broader and more compact.

Penstemon leiophyllus Pennell, very closely related to *P. laevis* but distinguished by a finely glandular-puberulent inflorescence and a usually glabrous staminode, may also be expected in Mohave and Coconino counties, having been found in Utah within 10 miles of the Arizona state line.

37. Penstemon comarrhenus Gray. Apache, Navajo, and Coconino counties, up to 7,500 feet, with sagebrush and in pine-oak forest, June and July. Western Colorado, Utah, and northeastern Arizona.

What seems to be a hybrid between *P. comarrhenus* and *P. barbatus* subsp. *Torreyi* was collected in Segi Canyon, Navajo County (*Gould & Phillips* 4769).

38. Penstemon strictus Benth. Lukachukai and Chuska mountains, Skeleton Mesa, Segi Canyon (Apache and Navajo counties), 7,000 to about 8,000 feet, uncommon, June and July. Southern Wyoming to northern New Mexico, northeastern Arizona, and Utah.

The Arizona specimens are slightly puberulent instead of glabrous at the very base of the stem, and so may be referable to subsp. *angustus* Pennell. Specimens collected in Canyon de Chelly National Monument, Apache County (*Cronyn* in 1939), and from north of Flagstaff (*Stone* 363) belong to subsp. *strictiformis* (Rydb.) Keck, an entity otherwise limited to southwestern Colorado, characterized by lanceolate, scarious-margined sepals up to 10 mm. long, and broadly lanceolate upper cauline leaves.

9. MIMULUS (291). MONKEY-FLOWER

Plants herbaceous, annual or perennial; stems leafy or scapose; leaves opposite or basal, sessile or petioled, entire or dentate; flowers in leafy terminal racemes, or axillary and solitary; calyx tubular or campanulate, usually 5-angled, sometimes bilabiate; corolla bilabiate or nearly regular, with a pair of longitudinal ridges on the lower side of the throat; stigma 2-lobed, the lobes distinct or united; capsule usually longitudinally dehiscent (2-valved); seeds numerous.

Many of the species have showy, handsome flowers, and some of them are in cultivation. Most of the Arizona species grow in wet soil.

KEY TO THE SPECIES

1. Corolla normally scarlet or carmine, 3 to 5.5 cm. long. Plants perennial, with creeping rootstocks, loosely villous and somewhat viscid with flaccid hairs; leaves sessile with a somewhat clasping base, sharply dentate or serrate; pedicels nearly as long as to much longer than the subtending leaves; calyx tubular-obconic, the teeth nearly equal; corolla bilabiate, the upper lip erect; stamens exserted (2).

1. Corolla not scarlet or carmine, less than 3 cm. long or, if longer, then bright yellow and usually spotted with red (3).

2. Plants stoloniferous; stems not more than 20 cm. long, procumbent or prostrate; leaves ovate or obovate. Corolla tube moderately to greatly surpassing the calyx
1. *M. Eastwoodiae*

2. Plants not stoloniferous; stems usually at least 30 cm. long, erect or ascending; leaves commonly oblong, oblong-ovate, or rhombic-elliptic................2. *M. cardinalis*

3. Pedicels not more than 1/3 as long as the calyx; plants of dry, gravelly slopes, annual. Stems not more than 25 cm. long, erect or ascending, usually branching from the base; leaves entire or nearly so; corolla 15 to 25 mm. long, with a narrow tube and throat and a spreading limb; stamens included (4).

3. Pedicels nearly as long as, to much longer than, the calyx; plants of moist soil (5).

4. Corolla 13 to 16 mm. long, distinctly bilabiate, yellow, sometimes tinged or spotted with reddish purple; herbage glandular-puberulent; leaves lanceolate, oblanceolate, or narrowly elliptic; calyx asymmetric; anthers glabrous.................3. *M. Parryi*

4. Corolla 18 to 23 mm. long, nearly regular, mallow pink with a yellow tube and usually a bright-yellow patch in the throat; herbage villous, usually viscid; leaves lanceolate to broadly ovate or obovate; calyx nearly symmetric; anthers usually hispidulous
4. *M. Bigelovii*

5. Fruiting calyx strongly asymmetric, the upper tooth much longer than the others; corolla distinctly bilabiate (except in *M. pilosus*), yellow, the throat often spotted with red (6).

5. Fruiting calyx not strongly asymmetric, the upper tooth not, or barely longer than the others; corolla only slightly bilabiate, not more than 20 mm. long (10).

6. Corolla throat open (7).

6. Corolla throat partly or wholly closed by a prominent palate. Stems usually erect or ascending; leaves mostly denticulate to sharply and coarsely dentate; fruiting calyx more or less closed by the connivent teeth, the lateral teeth well developed (9).

7. Plants annual; stems usually erect; leaves oblong-lanceolate; calyx deeply cleft, the upper tooth about equaling the tube. Herbage viscid-villous with long, white hairs .... 5. *M. pilosus*
7. Plants perennial; stems strongly decumbent to creeping and rooting at the nodes; leaves oval to suborbicular; calyx shallowly cleft, the teeth all much shorter than the tube (8).
8. Fruiting calyx nearly closed by the connivent teeth, the lateral teeth well developed; corolla lobes erose to laciniate; stems very slender, less than 10 cm. long, closely matted; leaves less than 10 mm. long .... 6. *M. dentilobus*
8. Fruiting calyx open, the lateral teeth very short; corolla lobes entire or nearly so; stems relatively stout, usually more than 10 cm. long, not closely matted; leaves 10 to 50 mm. long .... 7. *M. glabratus*
9. Plants usually perennial, with a rootstock or stolons, or with the flowering stems rooting at the lower nodes; fruiting calyx only partly closed by the infolding of the lower teeth, the upper tooth commonly less than 3 times as long as the others, usually slightly ascending; corolla 1.5 to 4 cm. long .... 8. *M. guttatus*
9. Plants usually annual; fruiting calyx almost completely closed, the upper tooth 3 or more times as long as the others, horizontal or nearly so; corolla commonly less than 2 cm. long .... 9. *M. nasutus*
10. Calyx becoming inflated, not strongly prismatic, campanulate in fruit; leaves mostly distinctly petioled, dentate or denticulate; corolla lobes entire or nearly so. Plants annual, viscid-villous; leaves much shorter than the internodes, broadly ovate; corolla 7 to 15 mm. long, narrowly funnelform, yellow, the throat often dotted or streaked with red .... 10. *M. floribundus*
10. Calyx not becoming inflated (except as distended by the enlarging capsule), strongly prismatic, often with conspicuous, dark-colored ribs; leaves sessile or nearly so, entire or shallowly dentate; corolla lobes usually emarginate or obcordate (11).
11. Plants perennial, with filiform stolons or underground rootstocks and bulbils, scapose to subcaulescent with very short internodes (the stems up to 10 cm. long and bearing none to several pairs of leaves), glabrous or sparsely (seldom copiously) viscid-villous; leaves elliptic, oblanceolate, or obovate; corolla 8 to 20 mm. long. Flowers solitary on erect filiform pedicels much surpassing the leaves; corolla broadly funnelform, yellow, sometimes spotted with reddish brown .... 11. *M. primuloides*
11. Plants annual, caulescent, glandular-puberulent; stems commonly much branched from the base, with well-developed internodes; leaves lanceolate, oblanceolate, or narrowly oblong (exceptionally oval); corolla 5 to 10 mm. long (12).
12. Stems up to 7 cm. long; leaves crowded, nearly as long as to much longer than the internodes; pedicels 3 to 9 mm. long; calyx teeth not ciliate; corolla yellow; stigma lobes unequal .... 12. *M. Suksdorfii*
12. Stems up to 20 cm. long; leaves not crowded, much shorter than the internodes; pedicels 6 to 20 mm. long; calyx teeth usually ciliate; corolla yellow, or with a pink limb; stigma lobes equal .... 13. *M. rubellus*

1. Mimulus Eastwoodiae Rydb. Navajo Indian Reservation (Apache, Navajo, and eastern Coconino counties), 6,000 to 7,000 feet, wet caves and recesses in rock walls. Southeastern Utah and northeastern Arizona.

2. Mimulus cardinalis Dougl. Grand Canyon (Coconino County), and Mohave County, south and southeast to the White Mountains (Apache County) and Cochise and Pima counties, 2,000 to 8,500 feet, along streams, usually in shade, March to October. Utah to Oregon, south to northwestern Mexico.

Crimson monkey-flower. The large orange-red or scarlet flowers are very conspicuous. Var. *verbenaceus* (Greene) Kearney & Peebles (*M. verbenaceus* Greene), with the corolla tube nearly twice as long as the calyx (only moderately exserted in the typical plant), occupies the same range as the latter in Arizona and is about equally common. A yellow-flowered specimen was collected near Lake Mead, Mohave County (*Clover* 4273).

3. Mimulus Parryi Gray. Near Littlefield, northwestern Mohave County, about 2,000 feet (*Maguire* 5003, 5005). Virgin River region, in southwestern Utah, southern Nevada, and northwestern Arizona.

4. Mimulus Bigelovii Gray. Mohave County, from Lake Mead to Williams River, 500 to 2,500 feet, open, sandy places, February to April. Nevada, northwestern Arizona, and southern California.

This small plant, with its disproportionately large, beautifully colored flowers, is well worth cultivating.

5. Mimulus pilosus (Benth.) Wats. (*Mimetanthe pilosa* Greene). Yavapai, Graham, Gila, Pinal, Maricopa, and Pima counties, 1,000 to 4,500 feet, moist, sandy soil along streams, April to August. Nevada and Oregon to Arizona and southern California.

6. Mimulus dentilobus Robins. & Fern. (*M. parvulus* Woot. & Standl.). Santa Rita Mountains, Pima County, 4,500 feet (*Thornber* 505, etc.). Southern New Mexico, southern Arizona, and Sonora.

Thornber's specimens correspond to the descriptions of *M. dentilobus* in the character of the pubescence but seem otherwise more like *M. glabratus.*

7. Mimulus glabratus H.B.K. Northern Apache County and Grand Canyon (Coconino County) to Cochise, Santa Cruz, and Pima counties, 1,500 to 5,500 feet, April to July. Ontario and Manitoba, south to Michigan, Wisconsin, Texas, Nevada, and Mexico.

Most of the Arizona specimens belong to var. *Fremontii* (Benth.) Grant (*M. Geyeri* Torr.), distinguished by having diffusely spreading stems, nearly orbicular leaves, these cuneate to subcordate at base, with the margin entire or merely denticulate, and the corolla not more than 12 mm. long and not spotted within; whereas, in typical *M. glabratus,* the stems are decumbent-ascending, the leaves ovate or oval and often dentate, and the corolla up to 16 mm. long and usually spotted. A collection of typical *M. glabratus* from the Baboquivari Mountains (*Harrison & Kearney* 3783) was cited by Pennell.

8. Mimulus guttatus DC. (*M. prionophyllus* Greene). Apache County to Mohave County, south to Cochise, Santa Cruz, Pima, and northern Yuma counties, 500 to 9,500 feet, abundant in springy places and along brooks, March to September, type of *M. prionophyllus* from Willow Spring, Apache County (*Palmer* 527). Montana to Alaska, south to northern Mexico.

A conspicuous plant with showy, yellow flowers, sometimes used for salad and greens.

Several entities, closely related to *M. guttatus* and represented by specimens collected in Arizona, were regarded as specifically distinct by Pennell (286, pp. 4–6). These are: *M. unimaculatus* Pennell, type from Sierra Ancha, Gila County (*Harrison* 7892); *M. cordatus* Greene, from the Chiricahua Mountains and from near Tucson; *M. puberulus* Greene, from the Lukachukai Mountains, Apache County; and *M. Maguirei* Pennell, type from near Williams, Coconino County (*Maguire et al.* 12214).

9. Mimulus nasutus Greene. Graham, Gila, Pinal, Cochise, and Pima counties, 2,500 to 9,000 feet, wet soil, March to September. Idaho to British Columbia, south to Chihuahua, Arizona, and Baja California.

10. Mimulus floribundus Dougl. White Mountains (Apache County) and Coconino County to Graham and Pima counties, 3,000 to 9,000 feet, wet soil, April to September. Wyoming to British Columbia, south to northern Mexico and California.

11. Mimulus primuloides Benth. Kaibab Plateau (Coconino County), White Mountains (Apache County), 8,000 to 9,500 feet, wet soil about springs, July and August. Idaho to Arizona and southern California.

Plants tending to form mats, with stems rooting at the nodes.

12. Mimulus Suksdorfii Gray. Jumpup Divide, Kaibab Plateau, Coconino County, gravelly flats and slopes, April (*Goodding* 23–49), north of Beaver Dam, Mohave County (*Barkley* 3344), "northern Arizona" (*Lemmon* 3270). British Columbia to California, Colorado, and northern Arizona.

13. Mimulus rubellus Gray. Coconino and Mohave counties to Greenlee, Cochise, Santa Cruz, and Pima counties, 2,000 to 7,500 feet, frequent in sandy soil along streams, March to June (sometimes September). Wyoming to New Mexico, Arizona, and southern California.

10. STEMODIA

Plants herbaceous, glandular-pubescent; stems erect, leafy, simple or sparingly branched; leaves opposite, sessile with a somewhat clasping base, lanceolate, elliptic, or somewhat obovate, serrate; flowers in terminal spikelike racemes or thyrsoid panicles; calyx 5-parted, with narrow lobes; corolla with a narrow tube and a bilabiate, violet-purple limb, the lower lip 3-cleft; stamens 4, all of them anther-bearing; capsule appearing 4-valved (the valves 2-parted).

1. Stemodia durantifolia (L.) Swartz. Gila, Maricopa, Pinal, and Pima counties, 1,000 to 3,500 feet, wet soil along streams, March to October. Southern Texas, southern Arizona, and southern California to tropical America.

The Arizona plants were regarded by Pennell (286, p. 3) as belonging to a different species, *S. arizonica* Pennell, type from the Santa Catalina Mountains (*Pringle* in 1881).

11. GRATIOLA

Low, glandular-pubescent herbs; leaves opposite, sessile, denticulate; flowers axillary, solitary, on long, slender pedicels; calyx 5-parted, with narrow divisions; corolla tubular-funnelform, nearly regular, the limb whitish, shallowly 5-lobed, the tube yellow; sterile stamens rudimentary; capsule 4-valved; seeds many, striate and reticulate.

1. Gratiola neglecta Torr. Coconino County, near Flagstaff (*Lemmon* in 1884), Williams (*Kearney & Peebles* 14005), and Foxboro (*Deaver* 2886), about 7,000 feet, mud flats, July and August. Almost throughout the United States.

12. SCHISTOPHRAGMA

Plants small, annual, glandular-pubescent; leaves opposite, petioled, pinnatifid, the divisions wedge-shaped, usually toothed; flowers axillary, short-peduncled, small; calyx 5-parted, the divisions narrow; corolla obscurely bilabiate, the tube yellowish, the limb violet; stigma 2-lobed; capsule longitudinally dehiscent, appearing 4-valved; seeds numerous.

1. **Schistophragma intermedia** (Gray) Pennell (*Conobea intermedia* Gray). Mountains of Greenlee, Gila, Cochise, Santa Cruz, and Pima counties, 4,000 to 6,000 feet, grassy or wooded slopes, mostly in loose soil, July to September. Southern New Mexico and Arizona, and northern Mexico.

13. MECARDONIA

Small, glabrous, perennial herbs; leaves opposite, short-petioled or nearly sessile, lanceolate, ovate, or obovate, crenate or dentate; flowers axillary, solitary, on long, slender peduncles; corolla funnelform, slightly bilabiate, yellow with dark veins.

1. **Mecardonia vandellioides** (H.B.K.) Pennell (*Pagesia vandellioides* Pennell). Cochise, Santa Cruz, and Pima counties, 3,000 to 4,500 feet, wet, sandy soil along streams, March to September. Southeastern Arizona and Mexico.

14. BACOPA (292)

Aquatic, somewhat succulent herbs; herbage sparsely pilose to nearly glabrous; stems weak, rooting at the nodes; leaves opposite, sessile, entire, broadly obovate or suborbicular, palmately veined; flowers borne on axillary pedicels, often in clusters of 3; calyx 5-parted, the divisions oblong or ovate, unequal in width; corolla whitish or bluish, nearly regular; stamens 4, all antherbearing; style 2-cleft near apex; capsule broadly ovoid, obtuse, many-seeded.

1. **Bacopa rotundifolia** (Michx.) Wettst. Gila River bed between Phoenix and Maricopa, Maricopa County (*Pinney* in 1912). Indiana to Montana, south to Tennessee, Louisiana, Texas, Colorado, and southern Arizona.

15. LIMOSELLA. MUDWORT

Plants small, subscapose, glabrous or nearly glabrous, with slender runners; leaves on long, slender petioles, the blades entire, slightly fleshy; flowers very small, solitary, on long, slender peduncles, the corolla white or purplish; style short, the stigma capitate; capsule 2-celled only at base; seeds many, cross-ribbed.

KEY TO THE SPECIES

1. Leaves elliptic or oval, cuneate at base; corolla lobes acute, glabrous or very nearly so; capsules obovoid; seeds much longer than wide........................1. *L. aquatica*
1. Leaves oblanceolate, attenuate at base; corolla lobes obtuse, puberulent; capsules globose-ovoid; seeds little longer than wide..........................2. *L. pubiflora*

1. **Limosella aquatica** L. Near Flagstaff, Coconino County, 7,500 feet (*Lemmon* in 1884), wet soil, August. Almost throughout North America; Eurasia.

2. **Limosella pubiflora** Pennell. Known only from the type collection in the Chiricahua Mountains, Cochise County (*Peebles & Loomis* 5420).

16. LINDERNIA

Plants small, annual; herbage glabrous; stems leafy, usually diffusely branched; leaves opposite, sessile or the lowest short-petioled, usually denticulate; flowers solitary on slender axillary peduncles, or in short terminal racemes; calyx 5-parted, the lobes narrowly lanceolate; corolla purplish, bilabiate, the upper lip 2-lobed, erect, the lower lip 3-cleft, spreading; anther-bearing stamens 2, the sterile filaments forked; stigma 2-lobed; capsule ellipsoid, about as long as the calyx, many-seeded.

*1. **Lindernia dubia** (L.) Pennell (*Ilysanthes dubia* Barnhart). This plant is not known definitely to occur in Arizona, but a collection from the valley of the Santa Cruz River, Sonora (*Schott*) was referred by Pennell to the typical phase of *L. dubia*. The species is widely distributed in temperate North America.

17. VERONICA (293). SPEEDWELL

Plants herbaceous, annual or perennial, terrestrial or aquatic; leaves opposite or the upper alternate, sessile or short-petioled; flowers small, slightly irregular, axillary, or in usually loose racemes; corolla rotate or broadly campanulate; capsule compressed, often notched at apex.

KEY TO THE SPECIES

1. Plants annual; flowers solitary in the axils. Herbage glandular-puberulent or pilose; floral leaves alternate, the lower ones opposite (2).
1. Plants perennial, with creeping rootstocks; flowers in racemes. Corolla blue or bluish; style much surpassing the notch of the capsule (5).
2. Sepals ovate, shorter than the pedicels, these often longer than the subtending leaves; style usually distinctly surpassing the notch of the capsule; capsules somewhat turgid, with divergent lobes. Stems decumbent or prostrate; leaves broadly ovate, coarsely crenate-dentate; corolla blue (3).
2. Sepals linear or lanceolate, longer than the pedicels, these shorter than the subtending leaves; style not, or barely surpassing the notch; capsules flat, the lobes erect or only slightly divergent. Corolla shorter than to barely surpassing the calyx, not more than 5 mm. wide (4).
3. Corolla much surpassing the calyx; style as long as the capsule............1. *V. persica*
3. Corolla not or barely surpassing the calyx; style shorter than the capsule....2. *V. didyma*
4. Herbage glandular-puberulent to nearly glabrous; leaves linear to oblanceolate, mostly entire; corolla whitish..3. *V. peregrina*
4. Herbage pilose, scarcely glandular; leaves (the lower ones) ovate, dentate; corolla blue-violet ....4. *V. arvensis*
5. Racemes terminal; capsules flat, not orbicular; plants not aquatic; leaves sessile or nearly so, the floral ones alternate, the lower ones mostly opposite (6).
5. Racemes axillary; capsules somewhat turgid, orbicular or nearly so; plants aquatic or semiaquatic; leaves all opposite, except the greatly reduced bracts of the inflorescence (7).
6. Stems decumbent, much branched at base, appressed-puberulent; leaves round-oval to oblong; racemes becoming loose and elongate; corolla little surpassing the calyx, bluish; capsules wider than long, deeply notched..................5. *V. serpyllifolia*
6. Stems erect, simple, these and the inflorescence villous; leaves ovate to oblong; racemes short, rather dense; corolla much surpassing the calyx, sky blue; capsules much longer than wide, shallowly notched................................6. *V. Wormskjoldii*
7. Leaves all short-petioled. Blades elliptic, oblong, or ovate, serrate or serrulate; capsules wider than long..7. *V. americana*
7. Leaves all sessile and cordate-clasping, or the lowest ones short-petioled (8).
8. Sepals acute to acuminate; pedicels ascending; corolla bluish lilac; capsules suborbicular, mostly shorter than the sepals, not or barely notched at apex..8. *V. Anagallis-aquatica*
8. Sepals obtuse to acutish; pedicels divaricate; corolla white or pinkish; capsules round-reniform or obcordate, mostly longer than the sepals, distinctly notched..9. *V. connata*

1. Veronica persica Poir. Oak Creek (Coconino County), Clifton (Greenlee County), Tempe (Maricopa County), Aravaipa Canyon (Pinal County), Tucson (Pima County), a weed in lawns, January to April. Here and there in the United States; introduced from Europe.

2. Veronica polita Fries (*V. didyma* Tenore?). Tucson, Pima County (*Gould* 3478). A weed of lawns, here and there in the United States; introduced from Eurasia.

3. Veronica peregrina L. Throughout the state, up to 9,000 feet but usually much lower, common along streams and washes, March to September. Widely distributed in North America; South America.

The Arizona plants belong to subsp. *xalapensis* (H.B.K.) Pennell, with glandular pubescence.

4. Veronica arvensis L. Fort Huachuca, Cochise County (*Goodding* 140–50), near Tucson, Pima County, in lawns, July (*Pultz* 1650). Extensively naturalized in North America; from Eurasia.

5. Veronica serpyllifolia L. Kaibab Plateau (Coconino County), White Mountains (Apache and Greenlee counties), Pinaleno Mountains (Graham County), Huachuca Mountains (Cochise County), 8,000 to 10,000 feet, summer. Throughout most of North America; Eurasia.

The species is represented in Arizona by var. *borealis* Laestad, a more pubescent, relatively large-flowered variant, regarded by some authorities as specifically distinct (*V. tenella* All., *V. humifusa* Dickson).

6. Veronica Wormskjoldii Roem. & Schult. Baldy Peak (Apache County), San Francisco Peaks (Coconino County), 9,500 to 12,000 feet, July to September. Greenland to Alaska, south in the mountains to New Hampshire, New Mexico, Arizona, and California.

7. Veronica americana (Raf.) Schwein. Apache County to Coconino, Gila, and Pinal counties, 1,500 to 9,500 feet, in and around springs and streams, May to August. Widely distributed in North America. American brooklime.

8. Veronica Anagallis-aquatica L. Apache County to Mohave County, south to Graham, Gila, and Pinal counties, 1,500 to 7,000 feet, habitat similar to that of *V. americana,* March to September. Widely distributed in the Northern Hemisphere; South America.

Water speedwell. Specimens of intermediate character, suggesting hybridization with *V. americana,* have been collected near Camp Verde, Yavapai County (*Goodding* 64–47).

9. Veronica connata Raf. A collection labeled as from Arizona without definite locality (*Palmer* in 1869) and one cited by Pennell (285, p. 370) under subsp. *glaberrima* Pennell, from Beaver Creek, Yavapai County (*Rusby*), are the basis for including this species in the flora of Arizona. The range of the species, as given by Pennell (285, pp. 365, 369), is from Massachusetts and Ontario to Saskatchewan and Washington, south to Pennsylvania, Tennessee, Oklahoma, New Mexico, Arizona, and California.

There is difference of opinion as to the correct name of this species. Fernald (*Gray's Man.,* ed. 8, p. 1284) questioned the applicability of Rafinesque's name and concluded that the American plant is identical with the Eurasian

V. comosa Richter. But Fassett (*Wash. Acad. Sci. Jour.* 37: 353–354) concurred with Pennell in accepting the name *V. connata* Raf.

18. BESSEYA (294)

Plants herbaceous, perennial, subscapose; stem leaves bractlike, much smaller than the large basal leaves; inflorescence terminal, very dense, spikelike, conspicuously bracted; corolla whitish or purplish, the upper lip entire, the lower lip 3-lobed.

KEY TO THE SPECIES

1. Corolla 5 mm. long or longer, conspicuously exserted; capsules emarginate at apex, 5 to 6 mm. long; leaves commonly subcuneate at base, up to 20 cm. long....1. *B. plantaginea*
1. Corolla not more (usually less) than 5 mm. long, moderately exserted; capsules rounded to acutish at apex, not more (usually less) than 5 mm. long; leaves rounded or subcordate at base, not more than 8 cm. long..........................2. *B. arizonica*

1. Besseya plantaginea (James) Rydb. (*B. Gooddingii* Pennell, *Synthyris plantaginea* Benth.). White Mountains (Apache and Greenlee counties), 7,000 to 9,500 feet, moist meadows and pine woods, June to August, type of *B. Gooddingii* from Sitgreaves Camp, White River (*Goodding* 1119). Wyoming to New Mexico and eastern Arizona.

B. Gooddingii is a variant with exceptionally elongate, less pubescent leaves.

2. Besseya arizonica Pennell. White Mountains (Apache County), San Francisco Peaks and vicinity (Coconino County), 7,000 to 9,500 feet, moist meadows and coniferous forests, May to August, type from the San Francisco Peaks (*Leiberg* 5537). Northwestern New Mexico and northern Arizona.

19. BRACHYSTIGMA

Plants perennial, with a large, woody caudex, probably root parasites; herbage retrorsely hispidulous; leaves opposite or nearly so, or in whorls of 3, narrowly linear, entire, often revolute; flowers long-pedicelled, in loose, leafy racemes; corolla large, brownish yellow or orange-yellow, with a short tube scarcely surpassing the calyx, and a large ventricose throat.

1. Brachystigma Wrightii (Gray) Pennell (*Gerardia Wrightii* Gray). Cochise, Santa Cruz, and Pima counties, 5,000 to 7,500 feet, dry slopes and mesas, often among live oaks, August and September, type from between Babocomari, Cochise County, Arizona, and Santa Cruz, Sonora (*Wright*). Southwestern New Mexico, southern Arizona, and northern Mexico.

20. BUCHNERA. Blue-hearts

Perennial, probably root-parasitic; herbage hispid, the hairs pustulate at base; stem erect, virgate; leaves mostly opposite, the basal leaves obovate or oblong, larger than the lanceolate or linear stem leaves, usually sparingly dentate; flowers subsessile, in a slender, elongate, small-bracted spike; corolla dark violet.

1. Buchnera arizonica (Gray) Pennell. Huachuca Mountains (*Lemmon* 2830, the type collection), August. Southeastern Arizona and Mexico.

21. CASTILLEJA. Paint-brush, Painted-cup

Plants annual or perennial, herbaceous or barely suffrutescent, often partially root-parasitic; stems leafy, mostly erect; leaves alternate, sessile, entire to pinnatifid; flowers very irregular, in conspicuously bracted terminal spikes, the bracts usually colored otherwise than green; calyx tubular, cleft above and below; corolla long and narrow, with a long upper lip (galea), the very short lower lip often reduced to teeth or callosities.

Most of the species have bright-red floral bracts that are more conspicuous than the flowers. Where abundant these plants contribute to the forage value of the range for livestock, although it has been discovered that *C. chromosa* takes up large quantities of selenium from certain soils. Plants of this genus are reported to be used medicinally and ceremonially by the Hopi Indians, who ate the flowers of *C. linariaefolia*.

KEY TO THE SPECIES

1. Plants usually annual. Herbage somewhat viscid; stems tall and often unbranched; leaves and bracts entire, the latter scarlet, at least at the tip; corolla yellowish (except sometimes the lower lip), this well developed, much shorter than the galea, thin, not callose, deeply cleft; plants of marshy places (2).
1. Plants perennial (3).
2. Stems usually very slender, sparsely villous or glabrate, at least below; leaves lance-linear; lower lip colored differently from the rest of the corolla, bright red 1. *C. minor*
2. Stems usually rather stout, copiously villous nearly to the base; leaves lanceolate; lower lip not differently colored.......................................2. *C. exilis*
3. Bracts and flowers not highly colored, never bright red; galea from slightly more to considerably less than 3 times as long as the lip, the latter thin (not callose), deeply cleft, with linear or narrowly lanceolate lobes; bracts (and often some of the stem leaves) pinnatifid, with narrow lobes (4).
3. Bracts wholly or partly bright red (rarely yellow or salmon-colored); galea commonly much more than 5 times as long as the thick, more or less callose, green lower lip (5).
4. Corolla much surpassing the calyx, falcate, 35 to 50 mm. long, ochroleucous or purplish, as are sometimes the bracts and calyx; herbage scarcely lanate (except sometimes on the stems below the inflorescence), the hairs mostly short or subappressed. Lower corolla lip conspicuous, cleft nearly to the base....................3. *C. sessiliflora*
4. Corolla not or little surpassing the calyx, not noticeably falcate, not more than 20 mm. long, greenish or yellowish, as are also the bracts and calyx; herbage pilosulous to conspicuously but loosely lanate....................................4. *C. lineata*
5. Bracts entire or shallowly toothed toward the apex. Leaves all entire or nearly so (6).
5. Bracts (some or all of them) deeply incised or pinnatifid. Stems puberulent or short-pilose with mostly retrorse hairs (9).
6. Corolla 3 to 4 cm. long, well exserted. Stems pilose or hirsute with rather stiff, spreading hairs; leaves thin, lax, usually spreading or somewhat reflexed, only the midrib prominent; calyx cleft much deeper before (below) than behind................5. *C. laxa*
6. Corolla 2 to 3 (in *C. integra* sometimes 4) cm. long (7).
7. Stems more or less whitish-tomentose or lanate, the hairs all soft and subappressed; corolla 3 to 4 cm. long. Calyx not cleft much deeper before than behind, the teeth usually rather long and narrow, acute or acutish; leaves glabrous or glabrate above 6. *C. integra*
7. Stems not tomentose or lanate, the longer hairs rather stiff, spreading or retrorse; corolla commonly less than 3 cm. long (8).
8. Leaves thick, rather rigid, very prominently 3-ribbed; stems villous-hirsute; calyx cleft much deeper before than behind; corolla not exserted, about 2 cm. long. . 7. *C. cruenta*

8. Leaves thin, not rigid, the veins not very prominent; stems inconspicuously puberulent or short-pilose; calyx not cleft much deeper before than behind; corolla well exserted, 1.5 to 3 cm. long..8. *C. austromontana*
9. Corolla 3 to 5 cm. long; calyx cleft much deeper before (below) than behind (10).
9. Corolla (except sometimes in *C. confusa*) less than 3 cm. long; calyx not cleft much deeper before than behind (11).
10. Upper calyx lip 4-toothed, the teeth subulate, acute; galea about equaling the corolla tube; leaves (below the inflorescence) mostly entire or, if cleft, then with not more than 2 lateral lobes..9. *C. linariaefolia*
10. Upper calyx lip 2-lobed, the lobes lanceolate or broader, obtuse or obtusish, sometimes dentate; galea usually considerably longer than the corolla tube; leaves (below the inflorescence) mostly deeply cleft or pinnatifid, often with more than 2 lateral lobes 10. *C. patriotica*
11. Stems conspicuously white-lanate, at least when young, sometimes slightly woody below. Leaves (at least the upper ones) usually deeply pinnately cleft, with narrow, divergent lobes; calyx teeth short, broad, obtuse or rounded; corolla 1.5 to 2 cm. long 11. *C. lanata*
11. Stems not white-lanate but sometimes copiously pilose or villous, entirely herbaceous (12).
12. Foliage leaves (some or most of them) pinnatifid, the lobes narrow, divergent; herbage usually grayish, pilose or villous-hirsute..........................12. *C. chromosa*
12. Foliage leaves mostly entire; herbage green, sparsely (seldom copiously) short-pilose to glabrate ....13. *C. confusa*

1. Castilleja minor Gray. Navajo, Coconino, and Mohave counties to Greenlee, Cochise, Santa Cruz, and Pima counties, 3,000 to 7,500 feet, moist soil around springs and along brooks, April to August. New Mexico, Arizona, and northern Mexico.

Although normally annual, the plant is reported to be sometimes perennial. Some of the Arizona specimens approach *C. stenantha* Gray, which has a longer corolla with more exserted galea and lip.

2. Castilleja exilis A. Nels. Navajo Reservation (*Whitehead* in 1916), near Tuba and Moenkopi, Coconino County (*Kearney & Peebles* 12860, *Deaver* 2036), about 5,000 feet, marshy ground, April to September. Montana to Washington, south to northwestern New Mexico, northern Arizona, and Nevada.

3. Castilleja sessiliflora Pursh. Cochise, Santa Cruz, and Pima counties, 4,000 to 5,000 feet, grassy plains, sometimes with *Dasylirion* and *Agave*, May. Illinois to Saskatchewan, south to Missouri, Texas, and southern Arizona.

The corolla varies in color at the same station from ochroleucous to pale purple. The Arizona specimens seem to approach *C. mexicana* (Hemsl.) Gray.

4. Castilleja lineata Greene. White Mountains, Apache County, north of Hannigan (*Kearney & Peebles* 12422), 8,000 to 9,000 feet, along streams and in marshy meadows, July to September. Colorado, New Mexico, and eastern Arizona.

5. Castilleja laxa Gray (*C. retrorsa* Standl., *C. setosa* Pennell). Cochise, Santa Cruz, and Pima counties, 4,000 to 7,500 feet, chaparral and rock ledges, March to November, type of *C. retrorsa* from the Chiricahua Mountains (*Blumer* 2132), type of *C. setosa* from the Santa Catalina Mountains (*Pringle* in 1883). Southern Arizona and northern Mexico.

C. setosa was described by Pennell (286) as differing from *C. laxa* in the more hispid herbage, more stiffly erect stems, and a hirsute calyx more closely investing the corolla.

6. Castilleja integra Gray. Apache County to Mohave County, south to Cochise, Santa Cruz, and Pima counties, 3,000 to 7,500 feet, mostly among oaks and pines, March to September. Colorado, western Texas, New Mexico, Arizona, and northern Mexico.

Var. *gloriosa* (Britton) Cockerell (*C. gloriosa* Britton) is a form with extraordinarily large bracts and flowers (corolla up to 4.5 cm. long), suggesting a polyploid variation. The type of *C. gloriosa* was collected at Fort Verde, Yavapai County (*Mearns* 208).

7. Castilleja cruenta Standl. Known definitely only from the type collection in the Chiricahua Mountains (*Blumer* 2133), but a specimen supposed to have been collected between Fort Huachuca and the San Pedro River (*Mearns* 1539) may belong here. Known only from southeastern Arizona.

No good flowering specimens have been examined, but the vegetative characters are very distinctive.

8. Castilleja austromontana Standl. & Blumer. White Mountains (Apache and Greenlee counties) to Cochise County, west to the Sierra Ancha (Gila County) and the Santa Catalina Mountains (Pima County), 7,000 to 9,500 feet, pine forests, July to September, type from the Rincon Mountains (*Blumer* 3411). Southern New Mexico, southeastern Arizona, and northeastern Sonora.

This species resembles *C. confusa,* but the plant is less rigid and of softer texture. The ranges of these species apparently do not overlap.

9. Castilleja linariaefolia Benth. Apache County to Hualpai Mountain (Mohave County), south to Cochise and Pima counties (3,000?) 5,000 to 10,000 feet, common in juniper, pine, and spruce-fir forests and with sagebrush, April to October. Wyoming to New Mexico, Arizona, southern Nevada, and California.

In the typical plant the stem is glabrous except at base and in the inflorescence. A variant with the stems pubescent throughout (forma *omnipubescens* Pennell, var. *omnipubescens* (Pennell) Clokey) appears to be more abundant in northern Arizona (Navajo, Coconino, and Mohave counties) than the typical plant.

10. Castilleja patriotica Fern. (*C. galeata* A. Nels.). Chiricahua and Huachuca mountains, Cochise County, about 8,000 feet, openings in pine forests, July to September, type of *C. galeata* from the Huachuca Mountains (*Goodding* 1354). Southeastern Arizona and northern Mexico.

Var. *Blumeri* (Standl.) Kearney & Peebles (*C. Blumeri* Standl.) apparently differs only in its smaller flowers, with corolla less than 3 cm. long (3.5 to 4.5 cm. in typical *C. patriotica*). The type of *C. Blumeri* was collected in the Chiricahua Mountains (*Blumer* 143).

11. Castilleja lanata Gray. Greenlee, Graham, and Cochise counties to the Sierra Estrella (Maricopa County) and the Ajo Mountains (Pima County), 2,500 to 7,000 feet, arid, granitic or limestone slopes, flowering throughout the year. Western Texas to southern Arizona and northern Mexico.

12. Castilleja chromosa A. Nels. (*C. eremophila* Woot. & Standl.). Apache County to Mohave County, south to Gila, Pinal, and Maricopa counties, 2,000 to 8,000 feet, common in chaparral and among pines, March to August, type of *C. eremophila* from the Carrizo Mountains, Apache County (*Standley* 7464). Colorado to British Columbia, south to New Mexico, Arizona, and southern California.

13. Castilleja confusa Greene. Apache County to Coconino County, on both sides of the Grand Canyon, 7,000 to 10,000 feet, common in pine forests, June to October. Colorado, northern New Mexico, and northern Arizona.

What appears to be a narrow-leaved form of *C. confusa,* often with yellow to salmon-colored bracts, occurs in wet meadows on the Kaibab Plateau. In *Flowering Plants and Ferns of Arizona* (p. 829) this plant was referred erroneously to *C. flava* Wats. Specimens with the entire leaves of *C. confusa* but resembling *C. chromosa* in their copious pubescence were collected near Mormon Lake (*Sampson* 232).

A species of *Castilleja,* apparently related to *C. lutescens* (Greenm.) Rydb., has been collected in the Phelps Botanical Area, White Mountains (Apache County), where it is rather abundant in wet meadows at 9,500 feet (*Phillips* 3146, 3228, *Phillips & Kearney* 3442). It has the stems villous with very slender, somewhat viscid hairs; the leaves narrowly lanceolate and nearly always entire; the bracts canary yellow with reddish tips, deeply 3-dentate at apex with the middle tooth about twice as wide as the others; the calyx cleft about equally before and behind, the lobes rather deeply bidentate; the galea erect, straight, bright green without, pink within; and the lip thickish, deep green, about 1/5 as long as the galea, 3-ribbed, the ribs extended into slender, acute teeth. This species was published May 29, 1951, as *C. mogollonica* Pennell, *Notul. Nat. Acad. Nat. Sci. Phila.* No. 237.

22. CORDYLANTHUS (295). Club-flower, Bird-beak

Plants annual, probably partially root-parasitic; stems leafy, mostly erect and much branched; leaves alternate, entire to pinnatifid; inflorescence various; bracts entire or parted; calyx with a tubular base and a spathelike lobe, this often opposed by a more or less similar bract, giving the appearance of a 2-lobed calyx; corolla narrow, bilabiate, the lips equal or unequal, the upper lip enclosing the stamens and pistil; capsule compressed.

KEY TO THE SPECIES

1. Flowers in terminal capitate or short-spicate inflorescences, or sometimes solitary at the ends of the branches. Pubescence of the herbage not or obscurely glandular; corolla lips nearly equal; seed coats alveolate (2).

1. Flowers scattered along the branches, 10 to 18 mm. long. Outer bracts 3-cleft (4).

2. Leaves and bracts entire; corolla 15 to 20 mm. long; anthers of the longer stamens 2-celled, of the others 1-celled; herbage loosely villous .... 1. *C. canescens*

2. Leaves (at least some of them) 3- to 5-lobed; outer bracts deeply 3- to 7-lobed, the lobes very narrow; corolla 20 to 30 mm. long; anthers all 2-celled; herbage short-pilose to nearly glabrous (3).

3. Inner bracts of the inflorescence narrowly lanceolate, entire or obscurely 2- or 3-dentate at apex .... 2. *C. Wrightii*

3. Inner bracts broadly lanceolate, 3- to 5-lobed toward apex .... 3. *C. tenuifolius*

4. Corolla bright yellow (drying purplish), the lower lip 1/2 to 2/3 as long as the upper. Herbage hirsute below, villous and often slightly glandular above; anthers commonly 1-celled .... 4. *C. laxiflorus*

4. Corolla mainly pink or lavender, the lower lip more than 2/3 as long as the upper (5).

5. Tips of the outer bracts (and of the leaf lobes) whitish-callose; anthers 1-celled; herbage hirsute or pilose, with few gland-tipped hairs; corolla not yellow-tipped...5. *C. Nevinii*
5. Tips of the outer bracts not callose; anthers mostly 2-celled; herbage copiously glandular-pilose; tip of the corolla yellow..................................6. *C. parviflorus*

*1. **Cordylanthus canescens** Gray. No Arizona specimens have been seen, but the species occurs at St. George, Utah, not far from the northern border. Utah to California.

2. **Cordylanthus Wrightii** Gray. Apache, Navajo, and Coconino counties to Cochise and Pima counties, 5,000 to 7,500 feet, mostly in open pine forests, July to October. Western Texas to Arizona and northern Mexico.

A variant with flowers solitary or 2 (rarely 3) in the cluster, and with glabrate or obscurely glandular-puberulent herbage, var. *pauciflorus* Kearney & Peebles, type from near Tuba, Coconino County (*Kearney & Peebles* 12884), is rather common on sandy plains in Apache, Navajo, and Coconino counties, occurring also in adjacent New Mexico and Utah.

3. **Cordylanthus tenuifolius** Pennell. Coconino County, northern Yavapai County, and northwestern Gila County, 6,500 to 8,000 feet, July to September, type from the Grand Canyon (*Eggleston* 15677a). Known only from northern and central Arizona.

4. **Cordylanthus laxiflorus** Gray. Navajo, Coconino, Yavapai, Gila, and Pima (?) counties, 3,500 to 6,000 feet, dry slopes and mesas, sometimes with *Cupressus glabra,* August to October. Known only from Arizona.

5. **Cordylanthus Nevinii** Gray. Hualpai Mountain, Mohave County (*Goldman* 2995, *Braem* 533, *Kearney & Peebles* 12699), about 6,500 feet, among pines, September. Western Arizona and southern California.

The terminal callosities of the lobes of the leaves and bracts are less conspicuous than in California specimens.

6. **Cordylanthus parviflorus** (Ferris) Wiggins (*Adenostegia parviflora* Ferris). Both sides of the Grand Canyon (Coconino County), Peach Springs and near Wikieup (Mohave County), 2,500 to 7,000 feet, dry, stony slopes and mesas, often with juniper, August and September, type from the Grand Canyon (*Knowlton* 270). Northwestern Arizona, southern Nevada, and southeastern California (?).

The species was described as having the subtending, sepallike bract slightly exceeding the calyx, and the corolla 10 to 11 mm. long. Examination of the type specimen shows, however, that although in most of the flowers the bract is somewhat longer than the calyx, in others the calyx is as long as or even slightly longer than the bract, also that the corollas reach a length of about 15 mm. In most of the Arizona specimens the calyx is longer (often considerably longer) than the bract, and the corollas are 14 to 17 mm. long. In these characters they resemble *C. glandulosus* Pennell & Clokey of southern Nevada, but in that species the flowers are more or less aggregated in small heads, whereas in *C. parviflorus* they are borne singly along and at the ends of the branches.

23. ORTHOCARPUS. Owl-clover

Plants annual; stems leafy, mostly erect; leaves alternate, sessile or nearly so, entire to pinnately parted; inflorescence spicate, usually dense, leafy-bracted, the bracts green or purple; calyx narrowly campanulate; corolla bilabiate, the lips approximately equal, the lower lip entire to trisacculate at apex.

KEY TO THE SPECIES

1. Leaves and bracts pinnately parted, the divisions narrowly linear or filiform; tips of the bracts and calyx lobes purplish pink, as is the corolla; lower lip of the corolla trisacculate at apex, much wider than the galea; stigma large, much wider than the style, depressed-capitate; herbage villous-hirsute, not scabrous..........1. *O. purpurascens*
1. Leaves entire or 3-cleft; tips of the bracts and calyx lobes green; lower lip of the corolla entire or rather obscurely tridentate, not sacculate, at apex; stigma small, scarcely wider than the style; herbage scabrous-puberulent (2).
2. Inflorescence many-flowered, usually dense; leaves mostly entire, linear or lanceolate; corolla yellow; tip of the galea obtuse, not inflexed; lower lip about as long as and not much wider than the galea..2. *O. luteus*
2. Inflorescence few-flowered, loose; leaves mostly 3-cleft, with filiform lobes; corolla purple and white; tip of the galea mucronate, inflexed; lower lip somewhat shorter and much wider than the galea....................................3. *O. purpureo-albus*

1. Orthocarpus purpurascens Benth. Mohave County to Graham, Gila, Maricopa, Pinal, and Pima counties, 1,500 to 4,500 feet, open mesas and slopes, March to May. Western and southern Arizona, California, and Baja California.

Escobita. In favorable seasons extensive areas are bright purple with the flowers of this owl-clover, which is grazed by cattle and sheep. The Arizona plant is var. *Palmeri* Gray, type from near Wickenburg (*Palmer*), distinguished from typical *O. purpurascens* by having the tip of the lower lip purple like the rest of the corolla and often with 1 or more deeper-colored spots instead of yellow or white.

2. Orthocarpus luteus Nutt. Apache, Navajo, Coconino, and Yavapai counties, 7,000 to 9,000 feet, mostly in pine forests, July to September. Manitoba to British Columbia, south to Minnesota, Nebraska, New Mexico, Arizona, and California.

3. Orthocarpus purpureo-albus Gray. Apache, Navajo, Coconino, and Gila counties, 5,500 to 9,000 feet, coniferous forests, July to September. Colorado to Idaho, New Mexico, and Arizona.

24. RHINANTHUS. YELLOW-RATTLE

Plants annual, presumably root-parasitic; stems strictly erect, leafy, 4-angled; leaves opposite, sessile, thickish, rigid, scabrous, lanceolate, sharply serrate; flowers in a rather dense, leafy-bracted, spikelike raceme; corolla yellow, bilabiate, the upper lip arched, the lower lip 3-lobed; anthers hairy; capsule compressed, orbicular, enclosed in the accrescent bladderlike calyx; seeds winged.

The plants have been used as an insecticide.

1. Rhinanthus rigidus Chabert. White Mountains, Apache County (*Eggleston* 17083, *Peebles & Smith* 12493, *Gould & Robinson* 5045), 9,000 to 9,500 feet, August and September. Canada and Alaska to Colorado, eastern Arizona, and Washington.

25. PEDICULARIS. Lousewort, Wood-betony

Plants partially root-parasitic, perennial, herbaceous, caulescent or subacaulescent; leaves alternate or basal, toothed to bipinnatifid; flowers in bracted spikes; corolla strongly bilabiate, narrow, the upper lip compressed on the sides, arched and often beaked at the apex, the lower lip 3-lobed with the midlobe smaller than the lateral ones; capsule oblique, compressed.

Some of the species are known also as duck-bill and fern-leaf.

KEY TO THE SPECIES

1. Galea prolonged into a filiform, recurved beak (curved outward and upward), this as long as or longer than the rest of the corolla. Herbage glabrous; stems erect, moderately leafy; leaves deeply and incisely pinnatifid or bipinnatifid; inflorescence spiciform, cylindric, many-flowered; corolla bright pink to claret red.........1. *P. groenlandica*
1. Galea not beaked or if so, then the beak straight or incurved (curved downward), much shorter than the rest of the corolla (2).
2. Leaves merely crenate or crenate-dentate, often doubly so, linear or linear-lanceolate. Beak of the galea strongly incurved, 4 to 5 mm. long; lower lip of the corolla nearly equaling the galea; stems commonly 30 cm. or longer, very leafy; herbage glabrous or sparsely pubescent; flowers in loose, leafy racemes; corolla 12 to 20 mm. long, white 2. *P. racemosa*
2. Leaves pinnatifid or bipinnatifid (3).
3. Galea falcate, the beak short, stout, conic, straight or slightly incurved; corolla 10 to 15 mm. long, ochroleucous. Stems stout, strict, moderately leafy, up to 50 cm. long; leaves bipinnatifid, with numerous narrow divisions; racemes elongate-spiciform, usually dense, with numerous or many flowers..............................3. *P. Parryi*
3. Galea strongly cucullate at apex, not beaked; corolla 25 mm. long or longer (4).
4. Plants subacaulescent; herbage nearly glabrous below the inflorescence, the latter sparsely villous, relatively few-flowered, and not surpassing the leaves; leaves pinnatifid, with broad, mostly obtuse, crenate-dentate lobes, the teeth conspicuously white-mucronate; corolla purple; anther cells conspicuously aristate at base, the awns projecting like tusks from the hood of the galea..............................4. *P. centranthera*
4. Plants strongly caulescent, the stems stout, very leafy, up to 1.5 m. long; herbage more or less pubescent below the inflorescence, the latter copiously villous, many-flowered, elongate, and greatly surpassing the leaves; leaves pinnate, the primary divisions pinnatifid, lanceolate, acute, the secondary lobes serrate, with setose-tipped teeth; corolla purplish brown (sometimes greenish yellow?); anther cells not aristate 5. *P. Grayi*

1. Pedicularis groenlandica Retz. Baldy Peak, White Mountains, Apache County (*Peebles & Smith* 12503), 10,000 feet, August. Greenland to Alaska, south in the mountains to New Mexico, eastern Arizona, and California.

Elephant-head. The Arizona specimens are of the large-flowered variant, *P. surrecta* Benth., with the galea more than twice as long as the calyx.

2. Pedicularis racemosa Dougl. Baldy Peak, White Mountains, Apache County (*Goodding* 613, *Peebles & Smith* 12498, *Phillips* 3347), 9,500 to 11,000 feet, in deep coniferous forest, July and August. Canada to New Mexico, eastern Arizona, and California.

3. Pedicularis Parryi Gray. Apache, Greenlee, and Coconino counties, especially in the White Mountains and on the San Francisco Peaks, 7,500 to 12,000 feet, moist mountain meadows, June to September. Wyoming to Montana, south to northern New Mexico and northern Arizona.

4. Pedicularis centranthera Gray. Apache, Coconino, and northern Mohave counties, south to Pinal and (probably) Pima counties, 5,000 to 7,500 feet, common in pine forests, April to June. Colorado, Utah, New Mexico, and Arizona.

5. Pedicularis Grayi A. Nels. White Mountains (southern Apache and northern Greenlee counties), Pinaleno Mountains (Graham County), Chiricahua Mountains (Cochise County), 8,000 to 10,000 feet, rich soil in coniferous forests, July and August. Wyoming to New Mexico and eastern Arizona.

This tall, hollow-stemmed species is sometimes cultivated as an ornamental.

121. BIGNONIACEAE. Bignonia Family

Shrubs or small trees; leaves mostly opposite, simple or compound; flowers large and showy, in terminal racemes or panicles; corolla sympetalous, bilabiate; stamens 5, the filaments borne on the corolla, usually only 4 of them anther-bearing; ovary superior, 2-celled; style 1; stigma bilabiate; fruit an elongate, 2-valved capsule; seeds many, large and flat, winged or comose.

Several species of the typical genus *Bignonia* are cultivated as ornamental climbers in the warmer parts of the United States.

KEY TO THE GENERA

1. Leaves simple; seeds comose .... 1. *Chilopsis*
1. Leaves pinnate; seeds with hyaline wings .... 2. *Tecoma*

1. CHILOPSIS. Desert-willow

Small trees or large shrubs, up to 9 m. (30 feet) high; leaves alternate or the lower ones opposite, simple, linear or linear-lanceolate, entire, elongate; corolla white, often tinged, streaked, or spotted with purple; wing of the seed dissected into hairs.

1. Chilopsis linearis (Cav.) Sweet. Coconino and Mohave counties to Greenlee, Cochise, Santa Cruz, Pima, and Yuma counties, up to 4,000 (rarely 6,000) feet but usually lower, mostly along washes in the deserts and foothills, April to August. Western Texas to southern Nevada, Arizona, southern California, and northern Mexico.

The desert-willow is sometimes cultivated as an ornamental for the sake of its attractive catalpalike flowers. It probably would be useful for planting to control soil erosion. It is browsed only where more palatable forage is scarce. The Arizona plants belong mostly to var. *arcuata* Fosberg, with the sterile branchlets nearly or quite glabrous and leaf veins not prominent, but Fosberg (296) mentioned also Arizona specimens approaching var. *glutinosa* (Engelm.) Fosberg, which he distinguished from var. *arcuata* by the glutinous herbage.

2. TECOMA. Trumpet-bush

Shrubs; leaves opposite, pinnate, the leaflets 5 or more, lanceolate, long-acuminate, deeply serrate or laciniate; corolla funnelform-campanulate, bright yellow; seeds flat, with a thin, entire wing.

1. Tecoma stans (L.) H.B.K. Pinal, Cochise, Santa Cruz, and Pima counties, 3,000 to 5,500 feet, dry, stony or gravelly slopes, May to October. Southern New Mexico and Arizona, southward into tropical America.

The plant is much cultivated as an ornamental in the warmer parts of the United States. It is said that the roots are used in Mexico medicinally and for making a sort of beer. Rubber in small quantity is found in the stems and leaves. The plants are browsed by bighorns and doubtless other animals. The Arizona plants, which seldom exceed a height of 2.5 m. (8 feet), belong to var. *angustata* Rehder (*Stenolobium incisum* Rose & Standl., *Tecoma Tronodora* (Loes.) Johnst.), with narrower, more deeply incised, and often more numerous leaflets than in typical *T. stans*.

122. MARTYNIACEAE. Unicorn-plant Family

1. PROBOSCIDEA (297). Unicorn-plant

Coarse, viscid-pubescent, annual herbs; leaves petioled, the lower ones mostly opposite, large, entire to shallowly lobed; flowers few, large and showy, in terminal racemes; calyx somewhat inflated, subtended by 1 or 2 bractlets; corolla sympetalous, somewhat bilabiate; stamens 4, all perfect or 2 of them sterile, the filaments attached to the corolla, the anthers gland-tipped, with divaricate cells; fruit large, somewhat fleshy, imperfectly dehiscent, ending in a long, incurved, hooked, dehiscent beak.

These plants are usually known in Arizona as devils-claws. The young pods are sometimes eaten as a vegetable. The black designs in the baskets made by the Pima and other Arizona Indians are woven with the split mature pods of *P. parviflora*. The plants are regarded as somewhat of a pest on sheep ranges because the hooked beaks of the pods become entangled in the fleece.

KEY TO THE SPECIES

1. Corolla reddish purple to nearly white, often dotted or blotched with red-purple and streaked with yellow, the limb about 2.5 cm. wide, the tube only slightly ventricose; leaves usually longer than wide, nearly entire to shallowly sinuate-lobed 1. *P. parviflora*
1. Corolla yellow or copper-colored, often dotted or splotched with red or brown, the limb 3 to 4 cm. wide; leaves usually wider than long (2).
2. Leaves distinctly lobed; bracts elliptic to broadly oblong; calyx with lobes nearly as long as the tube; corolla tube strongly ventricose........................2. *P. arenaria*
2. Leaves nearly entire to shallowly, sinuately lobed; bracts ovate to suborbicular; calyx with lobes much shorter than the tube; corolla tube scarcely ventricose 3. *P. altheaefolia*

1. **Proboscidea parviflora** (Wooton) Woot. & Standl. (*Martynia parviflora* Wooton). Coconino, Greenlee (?), Graham, and Gila counties to Cochise and Pima counties, 1,000 to 5,000 feet, plains, mesas, and roadsides, April to October. Western Texas to southern Nevada, Arizona, southern California, and northern Mexico.

A collection near Sacaton, Pinal County (*Peebles et al.* 75), differs from the characterization in the key in having the leaves considerably wider than long and the corolla tube strongly ventricose. These are characters of *P. fragrans* (Lindl.) Dcne., but the corolla is too small for that species as described (297, p. 21). The specimen cited has the horns of the fruit exceptionally long, both absolutely and relatively (nearly 30 cm. long and nearly 3 times as long as the body of the fruit).

2. Proboscidea arenaria (Engelm.) Decne. (*Martynia arenaria* Engelm.). Graham, Pinal, Cochise, Santa Cruz, and Pima counties, 3,000 to 4,000 feet, plains and mesas, July to September. Western Texas to southern Arizona and northern Mexico.

Corolla copper-colored outside, yellow within, the throat spotted with purple, the limb streaked with orange, the upper lobes spreading.

3. Proboscidea altheaefolia (Benth.) Decne. (*Martynia altheaefolia* Benth.). Southwest of Kingman, Mohave County (*Ripley & Barneby* 5001), Harqua-Hala Plain, eastern Yuma County (*Ferris* 1318), Yuma, Yuma County (*H. Brown* in 1905), 2,500 feet or lower, July to September. Western Texas to southeastern California and northern Mexico.

Brown's specimen corresponds with the characterization of *P. altheaefolia* (297, pp. 13, 19) except that the calyx lobes are rather more than half as long as the tube. Another collection near Yuma (*Peebles et al.* 4956) seems to be intermediate between this species and *P. arenaria.* It is somewhat doubtful that the latter is specifically distinct from *P. altheaefolia.*

123. OROBANCHACEAE. Broom-rape Family

Plants herbaceous, without chlorophyll, root-parasitic; stems fleshy; leaves alternate, reduced to scales; corolla irregular, bilabiate, the tube narrow, the lower lip 3-lobed; stamens commonly 4, in pairs, inserted on the corolla tube; style 1, elongate; ovary 1-celled; capsule 2-valved; seeds many, very small.

KEY TO THE GENERA

1. Calyx very irregular, spathelike, deeply cleft on the lower side, several-toothed on the upper side; upper lip of the corolla deeply concave....................1. *Conopholis*
1. Calyx nearly regular, the lobes or teeth almost equal; upper lip of the corolla not deeply concave..2. *Orobanche*

1. CONOPHOLIS. Squaw-root

Plants yellowish; stems clustered, covered with imbricate scales; inflorescence spikelike, elongate, not branched, dense, conspicuously bracted, the flowers in several rows; corolla strongly bilabiate, the upper lip arched, emarginate.

1. Conopholis mexicana Gray. Southern Apache County to Cochise and northern Gila counties, 5,000 to 6,000 feet, May and June. New Mexico, Arizona, and Mexico.

The plant grows with and is presumably parasitic on species of *Pinus, Cupressus, Juglans,* and *Quercus.* The plant resembles a cluster of slender pine cones, reaching a length of about 25 cm. in fruit.

2. OROBANCHE (298, 299). Broom-rape

Plants glandular-pilose, purplish brown or yellowish brown; inflorescences loosely fasciculate or densely spikelike; calyx 5-cleft; corolla more or less curved, the upper lip 2-lobed.

These plants are sometimes called cancer-root, in reference to their reputed efficacy in treatment of ulcers by application of the stems to the sore. The Navajo Indians are reported to use a decoction of the plant for this purpose. These plants were eaten by the southwestern Indians. Some of the Old World species are parasitic on clover and other cultivated plants.

KEY TO THE SPECIES

1. Flowers not subtended by bractlets, few, in loose, fasciculate inflorescences; pedicels commonly much longer than the flowers; corolla somewhat falcate: section *Gymnocaulis* ..... 1. *O. fasciculata*
1. Flowers subtended by bractlets, many, in dense, spikelike inflorescences, these sometimes branched below; pedicels none or shorter than the flowers; corolla straight or nearly so, brownish purple and white: section *Myzorrhiza* (2).
2. Corolla lobes rounded..2. *O. multiflora*
2. Corolla lobes narrowed toward the acute or acutish apex..............3. *O. ludoviciana*

1. Orobanche fasciculata Nutt. Apache County to Mohave County, south to Cochise and Pima counties, 4,000 to 8,000 feet, mostly in chaparral and in coniferous forests, May to August. Michigan to British Columbia, south to Texas, Arizona, and Baja California.

Both typical *O. fasciculata,* with herbage and corolla brownish purple, and var. *lutea* (Parry) Achey, with herbage and corolla dull yellow, sometimes tinged with pink, occur in Arizona, the latter being the commoner. Arizona specimens with exceptionally large corollas, therein approaching var. *franciscana* Achey, are not infrequent. A specimen from Mormon Lake, Coconino County (*MacDougal* 78), was referred by Goodman (*Leafl. West Bot.* 5:36) to his var. *subulata,* characterized by exceptionally long calyx lobes. The plants are parasitic on *Artemisia tridentata, Eriogonum Wrightii,* and other plants.

2. Orobanche multiflora Nutt. Apache, Navajo, Mohave, Maricopa, and Pima counties, 1,000 to 6,000 feet, sandy soil, April to September. Wyoming to Washington, south to northern Mexico and southern California.

Most of the Arizona specimens belong to var. *arenosa* (Suksdorf) Munz, with corolla lips not more than 5 mm. long. A collection in the Santa Catalina Mountains (*Pringle* in 1884), with lips 6 to 7 mm. long, represents var. *Pringlei* Munz, otherwise known only from Chihuahua.

3. Orobanche ludoviciana Nutt. Navajo, Coconino, and Mohave counties, south to Santa Cruz, Pima, and Yuma counties, 200 to 7,000 feet, February to September. Western Texas to southern Utah and Nevada, southeastern California, and northwestern Mexico.

Two intergrading varieties occur in Arizona. These are: var. *Cooperi* (Gray) G. Beck, type from Fort Mohave (*Cooper* in 1860), with corolla lips 3 to 6 mm. long and lobes of the lower lip lanceolate, tapering gradually to an acute apex; and var. *latiloba* Munz, with corolla lips 6 to 9 mm. long and lobes of the lower lip oblong or oblong-ovate, narrowed abruptly at apex. The latter variety occurs only in the southern part of the state. The plants are normally parasitic on *Franseria* and other Compositae, but specimens of var. *Cooperi* collected near Flagstaff showed attachment to roots of cacti (*Opuntia, Echinocactus*).

124. LENTIBULARIACEAE. Bladder-wort Family

A small family, interesting because of the adaptations of the plants for capturing and digesting insects.

1. UTRICULARIA. Bladder-wort

Aquatic plants with delicate immersed stems and leaves, only the flowering scapes emersed; leaves pinnately dissected, the segments filiform; leaves and stems bearing small insectivorous bladders, these urn-shaped and with the mouth closed by a lid; flowers several, perfect, in bracted racemes; corolla yellow, bilabiate, conspicuously spurred, the mouth closed by a palate; stamens 2, borne on the corolla; fruit a capsule, bursting irregularly.

1. Utricularia vulgaris L. Head of Black River, White Mountains (Apache County), Kaibab Plateau, San Francisco Peaks, and Dude Lake, Mogollon Escarpment (Coconino County), about 8,000 feet, in shallow water, July. Widely distributed in North America; Eurasia.

125. ACANTHACEAE. Acanthus Family

Plants perennial, herbaceous or shrubby; stems commonly quadrangular; leaves opposite, simple, entire; inflorescences usually bracteate, cymose, racemose, spicate, or the flowers solitary; corolla sympetalous, more or less irregular; anther-bearing stamens 2, or 4 in unequal pairs, the filaments attached to the throat or tube of the corolla; style 1, slender, elongate; stigmas 1 or 2, small; ovary superior, borne on a disk; fruit a 2-celled, elastically dehiscent capsule; seeds few, their stalks (except in *Elytraria*) persisting on the placentas as prominent, hardened, hooklike projections.

Many plants of this family are cultivated as ornamentals, notably the European *Acanthus mollis* and the South African climber, *Thunbergia alata.*[1]

KEY TO THE GENERA

1. Flowers in dense, terminal spikes, closely subtended by imbricate bracts, these cuspidate or aristate at apex. Plants herbaceous; stamens 2 (2).
1. Flowers not in dense spikes, or not closely subtended by imbricate bracts (3).
2. Plants subcaulescent, with the well-developed leaves basal or nearly so and short-lived; floral bracts obscurely veined (only the midrib at all conspicuous), with scarious margins prolonged into 2 winglike teeth on each side of the terminal awn; corolla limb purple .... 1. *Elytraria*
2. Plants caulescent, the stems with several pairs of well-developed leaves; floral bracts prominently 3-ribbed and often with 2 smaller additional veins, not scarious-margined; corolla limb pale yellow, often tinged or spotted with purple .... 7. *Tetramerium*
3. Flowers subtended by a pair of thin, valvelike, closely compressed, nearly orbicular bractlets, these very different from the foliage leaves .... 8. *Dicliptera*
3. Flowers not so subtended (4).
4. Corolla lobes contorted (twisted around each other) in the bud, the corolla slightly bilabiate, violet. Plants herbaceous; stamens 4, the filaments much longer than the 2-celled anthers; stigma linear (5).
4. Corolla lobes overlapping but not contorted in the bud, the corolla strongly bilabiate (6).

[1] E. C. Leonard has supplied valuable information in regard to this family.

5. Flowers axillary, solitary or in few-flowered clusters, sessile or nearly so; corolla less than 3 cm. long; anthers acutish, mucronulate........................2. *Dyschoriste*
5. Flowers in loose terminal panicles, mostly distinctly pedicelled; corolla more than 3 cm. long; anthers obtuse, muticous..3. *Ruellia*
6. Stamens 4, the anthers 1-celled, pubescent, as long as or longer than the filaments; stigma somewhat funnelform, small....................................4. *Berginia*
6. Stamens 2, the anthers 2-celled, much shorter than the filaments; stigma minute, capitate or somewhat flattened (7).
7. Anther cells inserted at the same height or very nearly so, parallel, contiguous, muticous. Plants shrubby or suffruticose (8).
7. Anther cells inserted at different heights (9).
8. Corolla very open, 1 to 1.5 cm. long, the lobes about twice as long as the tube 5. *Carlowrightia*
8. Corolla tubular-funnelform, more than 2 cm. long, the lobes shorter than the tube 6. *Anisacanthus*
9. Corolla white, the tube very slender, 3 cm. long or longer; plants herbaceous or suffrutescent; leaves lanceolate......................................9. *Siphonoglossa*
9. Corolla red; plants shrubby; leaves oval or ovate (10).
10. Lower cell of the anther without a conspicuous basal callus, merely mucronulate 10. *Jacobinia*
10. Lower cell of the anther with a conspicuous, whitish basal callus........11. *Beloperone*

1. ELYTRARIA (300)

Plants herbaceous; scapes numerous, decumbent or spreading, entirely covered with closely imbricate, glumelike bracts; floral bracts like those of the scapes but larger, ciliate with short soft hairs, otherwise glabrous, bluish green; calyx 4-parted; corolla imbricate in the bud, narrowly funnelform.

1. Elytraria imbricata (Vahl) Pers. Cochise, Santa Cruz, and Pima counties, 3,500 to 5,000 feet, mesas and slopes among rocks, April to September. Western Texas to southern Arizona, south to tropical America.

2. DYSCHORISTE (301)

Plants herbaceous, canescent-puberulent; stems decumbent or prostrate; leaves subsessile, oblanceolate or obovate; corolla violet or violet-purple, moderately irregular, convolute in the bud; stamens 4, the 2 pairs somewhat unequal in length; capsule narrowly oblong; seeds 2 to 4.

1. Dyschoriste decumbens (Gray) Kuntze. Cochise, Santa Cruz, and Pima counties, 4,000 to 5,500 feet, plains, mesas, and foothills, in the open or with oaks and junipers, April to October. Western Texas, southern Arizona, and Mexico.

3. RUELLIA

Plants herbaceous, glandular-pilose in the inflorescence; stems mostly erect or ascending, branched; leaves petioled, ovate; calyx deeply 5-cleft or 5-parted; corolla violet, moderately irregular, convolute in the bud; stamens 4, the 2 pairs somewhat unequal in length; capsule narrow, often subclavate; seeds 6 or more.

1. Ruellia nudiflora (Engelm. & Gray) Urban. Cochise, Santa Cruz, and Pima counties, 2,500 to 4,000 feet, canyons and foothills, usually among rocks, May to October. Southern Texas, southern Arizona, and Mexico.

Var. *glabrata* Leonard (*R. glabrata* (Leonard) Tharp & Barkley) is commoner in Arizona than the typical, more pubescent plant. *R. nudiflora* is worth cultivating as an ornamental because of its large, richly colored flowers.

4. BERGINIA

Low shrubs with whitish branches; leaves narrowly lanceolate, entire; flowers subtended by 2 bractlets, in interrupted leafy-bracted spikes; bractlets and calyx lobes narrow, acute, rigid; corolla about 1 cm. long, pink, the limb bilabiate.

***1. Berginia virgata** Harv. No specimens from Arizona have been seen, but the plant has been collected near Altar, Sonora, not far from the southern boundary of the state.

5. CARLOWRIGHTIA

Plants straggling undershrubs with slender branches; leaves entire, linear to ovate-lanceolate; flowers few, in spikes, racemes, or panicles; corolla imbricate in the bud, purple, or white streaked with purple, the tube narrow, the limb 4-cleft; stamens 2; capsule flattened, acuminate at apex, stalked; seeds flat.

KEY TO THE SPECIES

1. Leaves narrowly linear; flowers in narrow, racemelike panicles; corolla purple, the lobes 5 to 7 mm. long; filaments hirsutulous; anthers sagittate; herbage puberulent
1. *C. linearifolia*

1. Leaves lanceolate or ovate-lanceolate; flowers in interrupted spikes, these often forming a very open, few-flowered, leafy panicle; corolla white, the upper (posterior) lobe with a median yellow spot bordered by purple, the lobes 8 to 12 mm. long; filaments glabrous; anthers not sagittate, at most subcordate at base; herbage finely canescent or hirsutulous ....2. *C. arizonica*

1. Carlowrightia linearifolia (Torr.) Gray. Graham County, near Safford and Matthews, 2,500 to 3,000 feet, apparently also at Cascabel, Cochise County (*W. W. Jones* in 1947), mesas and washes, August and September. Western Texas to southeastern Arizona.

2. Carlowrightia arizonica Gray. Graham, Maricopa, Pinal, Cochise, Santa Cruz, Pima, and Yuma counties, 2,500 to 4,000 feet, rather common on dry, stony slopes, April and May, type from Camp Grant, Graham County (*Palmer* in 1867). Southern Arizona and northwestern Mexico.

The plant is browsed by cattle and sheep. There is much variation in appearance in different seasons or at different stages of growth. Typically the flowers are in slender, elongate, interrupted spikes, with the floral leaves reduced to small bracts, but there also occurs a more compact form with flowers scattered in the axils of well-developed leaves.

6. ANISACANTHUS (302)

Shrubs, up to 2.5 m. (8 feet) high, with rather stout branches and whitish, exfoliating bark; leaves short-petioled, lanceolate or ovate-lanceolate; corolla normally brick red, sometimes yellow or orange, the tube long and slender, the limb bilabiate, the lower lip 3-parted; capsule flattened, long-stalked.

1. **Anisacanthus Thurberi** (Torr.) Gray. Southern Apache County to Yavapai County, south to Cochise, Santa Cruz, Pima, and Yuma counties, 2,500 to 5,500 feet, mostly in canyons and along washes, flowering almost throughout the year, but chiefly in spring. Southwestern New Mexico, Arizona, and northern Mexico.

Known as chuparosa and desert-honeysuckle. The plant is browsed by cattle and sheep, especially when other forage is scarce.

7. TETRAMERIUM

Plants herbaceous or suffrutescent; stems several, erect or decumbent, leafy, the old bark exfoliating; foliage leaves and bracts bright green, conspicuously ciliate with long, stiff hairs and usually sparsely hirsute on the veins, often also puberulent; spikes 4-rowed; corolla slightly bilabiate, the tube longer than the limb, the upper lip entire, the lower lip 3-parted.

1. **Tetramerium hispidum** Nees. Graham, Santa Cruz, and Pima counties, 3,000 to 5,000 feet, among rocks and shrubs, usually in partial shade, April to October. Southern Arizona and Mexico.

This plant is reported to be very palatable to livestock.

8. DICLIPTERA

Plants herbaceous; stems erect or decumbent, branched, leafy; flowers each subtended by a pair of valvelike bractlets, solitary in the axils or in few-flowered cymes, the whole inflorescence a loose leafy panicle; corolla rose-purple, deeply bilabiate, the lips entire to shallowly lobed; stamens 2.

KEY TO THE SPECIES

1. Bractlets cordate, subcordate, or slightly cuneate at base, distinct or very nearly so; cymes mostly distinctly pedunculate, the peduncles often surpassing the leaves ....1. *D. resupinata*
1. Bractlets truncate and abruptly cuneate at base, united the length of the wedge-shaped portion; cymes shorter than the leaves, sessile or subsessile....2. *D. pseudoverticillaris*

1. **Dicliptera resupinata** (Vahl) Juss. (*D. Torreyi* Gray). Graham, Pinal, Cochise, and Pima counties, 3,000 to 6,000 feet, rocky slopes and canyons, April to October. Southwestern New Mexico, southern Arizona, and Mexico.

D. Torreyi was based on several collections in southern Arizona.

2. **Dicliptera pseudoverticillaris** Gray. Pima County, south of Tucson (*Thornber* 5351, 5512) and in the Baboquivari Mountains (*Loomis & King* 3255). Southern Arizona and northwestern Mexico.

The Arizona specimens, as compared with the type from Altar Valley, Sonora (*Pringle* in 1884), are less extreme in their characters, approaching *D. resupinata*.

9. SIPHONOGLOSSA

Plants herbaceous or nearly so; stems clustered, usually decumbent, commonly puberulent; leaves short-petioled, lanceolate; flowers clustered in the axils; corolla white, the tube very long and slender, the limb short, the lower lip spreading, deeply 3-lobed.

1. Siphonoglossa longiflora (Torr.) Gray. Pinal and Pima counties, 3,000 to 4,000 feet, rocky slopes and canyons, April to October, type (*Schott*) probably from southern Arizona. Apparently known only from southern Arizona, but doubtless also in Sonora.

Domestic and various wild animals feed upon this plant, especially in times of drought, and it is able to withstand close browsing. The white flowers are vespertine, slightly fragrant, and ephemeral.

10. JACOBINIA

Plants shrubby, up to 1.5 m. (5 feet) high, much branched, the herbage soft-villous; leaves petioled, ovate; flowers in dense, sessile or subsessile, leafy-bracted, axillary clusters; corolla broadly funnelform, bright red, the lower lip deeply 3-lobed; stamens 2, the anther cells unequal.

1. Jacobinia ovata Gray. Near Canyon Lake (Maricopa County), Ajo Mountains (Pima County), Bear Valley or Sycamore Canyon (Santa Cruz County), 1,500 to 3,500 feet, canyons and washes, February to May. Southern Arizona and northern Mexico.

Closely related to *J. candicans* (Nees) Benth. & Hook., to which the Arizona plants were referred in *Flowering Plants and Ferns of Arizona.*

11. BELOPERONE

Shrubs, up to 2 m. high, with spreading, brittle branches and canescent-puberulent herbage; leaves petioled, ovate; flowers in naked racemes; corolla tubular-funnelform, rather dull red, strongly bilabiate, the lower lip shallowly lobed or toothed; stamens 2, exserted, the anther cells unequal.

1. Beloperone californica Benth. Pinal, Maricopa, Pima, and Yuma counties, 1,000 to 4,000 feet, rocky slopes and along washes, flowering most of the year. Southern Arizona, southeastern California, and northwestern Mexico.

Called chuparosa in Sonora. The plant is browsed to some extent by livestock. The flowers are very attractive to hummingbirds, and it is reported that they were eaten by the Papago Indians. The name "honeysuckle" is sometimes used locally for this plant.

126. PLANTAGINACEAE. Plantain Family

1. PLANTAGO (303, 304, 305). Plantain, Indian-wheat

Scapose herbs with the foliage leaves all basal; flowers small, perfect or unisexual, regular, in terminal, long-peduncled, bracted spikes; calyx and corolla 4-divided or 4-lobed, persistent, usually scarious or scarious-margined; stamens 2 or 4, distinct, attached to the corolla tube; style filiform, stigmatic most of its length; ovary superior, 2- to 4-celled; fruit a circumscissile, usually few-seeded capsule.

Some of the native species afford excellent forage for sheep and cattle. The seeds become mucilaginous when wet, and are produced in such quantity where the plants are abundant as to cement the sand grains after rain, forming a thin crust on the surface of the soil. Seeds of the Arizona species that are known as Indian-wheat (*P. insularis, P. Purshii*) are sometimes gathered and used as a substitute for the psyllium seeds of commerce, which are obtained from an Old World species, *Plantago Psyllium*.

KEY TO THE SPECIES

1. Plants more or less dioecious or polygamous, many of the flowers cleistogamous with the corolla remaining closed and its lobes erect or connivent, other flowers with spreading corolla lobes and exserted stamens. Plants annual or biennial; leaves often coarsely dentate or cleft (2).
1. Plants not dioecious, the flowers all perfect, none cleistogamous; corolla lobes spreading or reflexed (5).
2. Leaves and scapes inconspicuously pubescent or glabrate, the hairs short, mostly appressed; leaves filiform or linear, less than 5 mm. wide; spikes loosely flowered, often interrupted below; calyx glabrous; corolla white or pale straw-colored; capsule with 6 or more seeds..1. *P. heterophylla*
2. Leaves and scapes copiously villous or subhirsute; leaves elliptic or oblong-oblanceolate, seldom less than 8 mm. wide; spikes usually densely flowered; calyx villous or subhirsute; corolla buff or orange; capsule 2- to 4-seeded (3).
3. Capsule 3-seeded; plants perennial, flowering in late summer............2. *P. hirtella*
3. Capsule usually 2-seeded; plant annual, flowering in spring (4).
4. Bracts and sepals obtuse; fruiting calyx less than 3 mm. long; mature seeds yellowish brown, less than 2 mm. long, deeply concave on the ventral face......3. *P. virginica*
4. Bracts and sepals acute or apiculate; fruiting calyx 3 to 4 mm. long; mature seeds dark red, 2.5 to 3 mm. long, flat to slightly concave on the ventral face...4. *P. rhodosperma*
5. Leaves lanceolate, oblanceolate, or broader, seldom less than 10 mm. wide; plants glabrous or loosely pubescent, not sericeous or lanate, usually perennial, with a thick caudex (6).
5. Leaves linear or lanceolate, commonly much less than 10 mm. wide; plants copiously silky-villous, sericeous, or lanate (except in *P. Wrightiana*), annual or winter annual. Spikes cylindric at maturity; seeds 2, deeply concave on the ventral face (8).
6. Spikes short-conic, becoming oblong, very dense, 1 to 6 cm. long at maturity. Leaves lanceolate or oblong-lanceolate, entire or denticulate; seeds 2, concave on the ventral face ....5. *P. lanceolata*
6. Spikes cylindric, moderately dense or rather loose, usually 6 cm. long or longer at maturity (7).
7. Leaves broadly ovate, abruptly contracted at base; scapes seldom woolly at base; seeds several or numerous, not more than 1 mm. long, angulate, not concave ventrally 6. *P. major*
7. Leaves lanceolate, oblanceolate, or elliptic, tapering at base; scapes usually woolly at base; seeds 4 or fewer, 2 to 3 mm. long, rounded on the back, somewhat concave ventrally ....7. *P. eriopoda*
8. Bracts subulate or narrowly lanceolate, not or very indistinctly scarious-margined, at least the lower ones commonly longer than the calyx.................8. *P. Purshii*
8. Bracts broadly lanceolate to nearly orbicular, conspicuously scarious-margined, none longer than the calyx (9).
9. Leaves not at all rigid, conspicuously whitish-sericeous or lanate, not noticeably discolored in drying; spikes at maturity 1 to 4 cm. long, seldom more than 4 times as long as wide; lowest bracts much like the sepals, broadly ovate or nearly orbicular, broadly scarious-margined to the apex, the scarious portion forming more than half of the area of the bract; seeds somewhat shiny, reddish brown............9. *P. insularis*
9. Leaves somewhat rigid, turning dark in drying; spikes at maturity commonly 3 to 6 cm. long and more than 4 times as long as wide; lowest bracts unlike the sepals, broadly lanceolate to broadly deltoid, not scarious-margined to the apex, the scarious portion forming less than half of the area of the bract; seeds not shiny (10).
10. Herbage glabrous or sparsely villous; bracts triangular-lanceolate or deltoid; seeds olive brown at maturity......................................10. *P. Wrightiana*
10. Herbage copiously villous or sericeous; bracts ovate-lanceolate; seeds black, or nearly so, at maturity..11. *P. argyraea*

1. **Plantago heterophylla** Nutt. Near Casa Grande (Pinal County), between Pearce and Sunglow (Cochise County), Tucson Mountains and Silver Bell Road (Pima County), 1,500 to 4,500 feet, stream beds, desert, and grassland, February to April. New Jersey to Florida, Texas, southern Arizona, and southern California.

The Arizona specimens are referred doubtfully to this species. In their small number of seeds (6 to 8 per capsule) they approach *P. pusilla* Nutt., which has only 4.

2. **Plantago hirtella** Kunth. Chiricahua and Huachuca mountains (Cochise County), Santa Catalina Mountains (Pima County), 5,500 to 8,000 feet, springy or boggy places along streams, July to September. Southern Arizona to tropical America.

The species is represented in Arizona by var. *mollior* Pilger, more copiously pubescent and with leaf margins more frequently denticulate than in typical *P. hirtella.*

3. **Plantago virginica** L. Coconino, Cochise, and Pima counties (doubtless elsewhere), 2,500 to 7,000 feet, commonly in moist soil, February to April. Connecticut to Michigan and Missouri, south to Florida, Arizona, and southern California.

4. **Plantago rhodosperma** Decne. Yavapai, Gila, Maricopa, Pinal, Cochise, and Pima counties, 1,000 to 6,000 feet, mostly along streams, March to May. Mississippi and Oklahoma to Texas and Arizona.

5. **Plantago lanceolata** L. Apache, Navajo, Coconino, Yavapai, Gila, Pinal, Cochise, and Pima counties, lawns and meadows, April to September. Widely distributed in the United States; naturalized from Europe.

Commonly known as ribwort or buckhorn plantain.

6. **Plantago major** L. Apache, Navajo, and Coconino counties to Cochise, Santa Cruz, and Pima counties, 1,000 to 7,500 feet, moist soil along streams, March to October. Widely distributed in the United States; naturalized from Europe. Common plantain.

7. **Plantago eriopoda** Torr. Kaibab Plateau and North Rim of the Grand Canyon (Coconino County), 8,500 to 9,000 feet (*Kearney & Peebles* 13739, *Collom* in 1942), moist meadows, July and August. Canada to New Mexico, northern Arizona, and California.

In the almost complete absence of basal "wool" and in the size and color of the seeds the Arizona plants approach *P. Tweedyi* Gray, and may represent an undescribed species.

8. **Plantago Purshii** Roem. & Schult. Apache County to Mohave County, south to Cochise, Santa Cruz, and Pima counties, 1,000 to 7,000 feet, dry slopes and mesas, February to July. Canada to Texas, Arizona, and Baja California.

There occur in Arizona both the typical phase, with the basal bracts of the spike not more, usually less, than twice as long as the calyx, and var. *picta* (Morris) Pilger (*P. picta* Morris, *P. xerodea* Morris, *P. ignota* Morris), with the basal bracts 2 to 3 times as long as the calyx. The variety is more common in the southern part of the state. The type of *P. ignota* came from Fort Verde (*Mearns* 199).

9. Plantago insularis Eastw. (*P. fastigiata* Morris). Grand Canyon (Coconino County) and Mohave, Maricopa, Pinal, Pima, and Yuma counties, 3,000 feet or lower, abundant, dry plains and mesas, January to May (occasionally autumn), type of *P. fastigiata* from Tucson (*Toumey* 355a). Southern Utah and Nevada, Arizona, and southern California.

10. Plantago Wrightiana Decne. Near Prescott (Yavapai County), Pinal Mountains (Gila County), about 5,000 feet, openings in pine woods, May to September. Western Texas to central Arizona, Oregon, and California.

11. Plantago argyraea Morris. Navajo, Coconino, and Yavapai counties, 6,000 to 7,500 feet, mostly in pine forest, June to September, type from Castle Creek, Yavapai County (*Toumey* 355c). Western New Mexico and Arizona.

127. RUBIACEAE. Madder Family

Plants annual or perennial, herbaceous or shrubby; leaves simple, entire, opposite or whorled, with stipules, these in one genus (*Galium*) enlarged and leaflike; flowers mostly perfect, regular or very nearly so, usually 4- or 5-merous; calyx tube adnate to the ovary, the limb reduced to teeth or lobes, or obsolete; stamens distinct, borne on the corolla alternate with its lobes; style 1, the stigmas 1 to 4; ovary inferior, 2- to 4-celled; fruit a longitudinally dehiscent or circumscissile, sometimes didymous, few- to many-seeded capsule, or achenelike, or separating at maturity into 2 to 4, usually indehiscent 1-seeded carpels.

A very large, mainly tropical family. Its best-known members are the coffee-tree (*Coffea arabica*), the trees from whose bark quinine is obtained (*Cinchona* spp.), and the formerly important dye plant, madder (*Rubia tinctoria*). The Arizona representatives are of almost no economic importance.

KEY TO THE GENERA

1. Ovules and seeds several in each carpel. Fruits capsular, 2-celled; flowers 4-merous (2).
1. Ovule solitary in each carpel (4).
2. Plants large, suffrutescent or shrubby; leaves lanceolate to ovate; corolla tubular, scarlet or pink .... 3. *Bouvardia*
2. Plants small, herbaceous or barely suffrutescent; leaves linear, narrowly lanceolate, or spatulate; corolla salverform, pink or white (3).
3. Capsule wholly adnate to the calyx tube; seeds angled........ 1. *Oldenlandia*
3. Capsule partly free from the calyx; seeds saucer-shaped........ 2. *Houstonia*
4. Plants shrubby; flowers in dense, globose heads........ 4. *Cephalanthus*
4. Plants herbaceous or (in a few species of *Galium*) suffrutescent; flowers not in globose heads (5).
5. Stipules similar to and nearly as large as the blades, the leaves thus appearing to be in whorls of 4 or more; corolla rotate........ 9. *Galium*
5. Stipules unlike the blades and much smaller; leaves opposite, the floral ones sometimes appearing whorled; corolla salverform or funnelform (6).
6. Flowers on long, slender, divaricate pedicels; fruits hispid with hooked hairs, didymous; stipules small, not forming a sheath around the stem, entire or few-toothed .... 5. *Kelloggia*
6. Flowers sessile or nearly so, in axillary or terminal cymules (rarely solitary in the leaf axils); fruits not hispid with hooked hairs; stipules connate, forming a sheath around the stem, cuspidate or setose (7).

7. Fruit circumscissile, the upper part falling off with the calyx limb; seeds about as wide as long, 4-lobed . . . . 8. *Mitracarpus*
7. Fruit not circumscissile, the carpels separating partly or completely by longitudinal cleavage; seeds elongate, not lobed (8).
8. Calyx limb with the teeth united at base, deciduous at or before the separation of the carpels . . . . 6. *Crusea*
8. Calyx limb of separate or nearly separate teeth, these persistent even after the carpels separate . . . . 7. *Diodia*

1. OLDENLANDIA

Plants annual, small, glabrous; stems slender, erect, often diffusely branched; leaves opposite, narrow; flowers small, all alike, in terminal cymes or solitary in the forks; calyx teeth subulate; corolla whitish or pale blue with a yellow throat, salverform; capsule hemispheric and somewhat quadrangular.

1. Oldenlandia Greenei Gray. Graham, Cochise, Santa Cruz, and Pima counties, 5,000 to 8,000 feet, rich soil in woods, August and September. Southern New Mexico and Arizona (doubtless also in northern Mexico).

2. HOUSTONIA

Plants perennial, often cespitose, herbaceous or barely suffrutescent; stems erect to procumbent; leaves narrow, opposite or appearing fascicled; flowers dimorphic in relative length of the stamens and style; corolla white to deep pink, salverform or funnelform; capsules more or less didymous.

KEY TO THE SPECIES

1. Stems mostly erect or ascending; leaves thin; pedicels erect or ascending in fruit; capsules longer than wide, free from the calyx only about ⅓ of their length 1. *H. nigricans*
1. Stems mostly procumbent or spreading; leaves thickish; pedicels recurved in fruit; capsules wider than long, free from the calyx ⅔ or more of their length (2).
2. Plants rather densely cespitose, subacaulescent, with leaves mostly basal or nearly so and erect (exceptionally distinctly caulescent and leafy-stemmed, with leaves more or less spreading); corolla bright pink, tubular-salverform, the tube 10 to 30 mm. long, 3 to 4 times as long as the lobes, very slender . . . . 2. *H. rubra*
2. Plants loosely cespitose, caulescent, the stems leafy, up to 20 cm. long, the leaves spreading or ascending; corolla white or pinkish, funnelform-salverform, the tube 3 to 4 mm. long, not more than twice as long as the lobes . . . . 3. *H. Wrightii*

1. Houstonia nigricans (Lam.) Fern. (*H. angustifolia* Michx.). Baker Butte, Mogollon Escarpment, Coconino County (*Mearns* 115), near Pine, Gila County (*Peebles & Fulton* 9473, in part), pine forest, May to July. Southern Michigan to Florida, west to Nebraska, central Arizona, and northern Mexico.

2. Houstonia rubra Cav. Navajo, Coconino, Cochise, and Santa Cruz counties (doubtless elsewhere), 4,000 to 6,000 feet, mesas and dry, rocky hills, often in sandy soil, April to August. New Mexico, Arizona, and Mexico.

3. Houstonia Wrightii Gray. Apache County to Coconino County, south to Cochise and Pima counties, 5,000 to 8,500 feet, dry mesas and slopes, among chaparral shrubs, oaks, or pines, and in grassy meadows, common, May to September. Western Texas to Arizona and Mexico.

3. BOUVARDIA

Plants suffrutescent or shrubby, the herbage glabrous or hispidulous; stems branched, the old bark whitish brown; leaves mostly in whorls of 3, lanceolate or ovate-lanceolate, up to 8 cm. long; flowers in mostly terminal cymes, dimorphic in the length of the stamens and the style; calyx lobes subulate, persistent; corolla slender, 2 to 3 cm. long; capsules didymous, subglobose; seeds flat, peltate, winged.

1. **Bouvardia glaberrima** Engelm. (*B. ovata* Gray). Southern Apache County to Cochise, Santa Cruz, and Pima counties, 3,000 to 9,000 feet, canyons and slopes, preferring partial shade, May to October, type of *B. ovata* from between the San Pedro River and Santa Cruz, perhaps in Arizona (*Wright* 1117). Southern New Mexico and Arizona, and northern Mexico.

A handsome shrub, worthy of cultivation, with neat foliage and clusters of bright-red (occasionally pink or white) honeysucklelike flowers. The stems attain a height of about 1 m.

4. CEPHALANTHUS. Button-bush

Shrubs, up to 2.5 m. (8 feet) high; leaves large, opposite or in whorls of 3, broadly lanceolate to oblong-ovate; flowers small, very numerous, in dense, globose, long-peduncled heads, 4-merous; corolla tubular-funnelform, whitish; fruits achenelike, obpyramidal, 2-celled, 1- or 2-seeded.

1. **Cephalanthus occidentalis** L. Apache, Gila, Maricopa, Pinal, and Pima counties, 1,000 to 5,000 feet, wet soil along streams, June to September. Throughout most of temperate North America.

The Arizona plants belong to var. *californicus* Benth., with shorter-petioled, narrower leaves more often in threes than in the typical plant of the eastern United States. This shrub is not palatable to livestock and is reputed poisonous, containing glucosides such as cephalanthine. The bark has been used medicinally. The flowers are attractive to bees.

5. KELLOGGIA

Plants herbaceous, perennial, with slender rootstocks, with the aspect of *Galium;* leaves opposite, sessile, lanceolate; flowers in very open cymose panicles, small; calyx with an obovoid tube and minute teeth; corolla funnelform-salverform, whitish; stamens inserted in the throat of the corolla; stigmas clavate.

1. **Kelloggia galioides** Torr. Northern Apache, Navajo, and Coconino counties, 7,000 to 9,000 feet, rich soil in coniferous forests, June to August. Wyoming to Washington, northern Arizona, and California.

6. CRUSEA

Plants small, annual; stems slender, erect or ascending, simple or sparingly branched; leaves oblong-lanceolate to subulate; flowers small, in terminal or axillary glomerules; calyx lobes 2 to 4, often conspicuously unequal; corolla white or purple, salverform or nearly so; fruit of 2 to 4 obovoid or globose carpels, these separating at maturity from the persistent axis.

KEY TO THE SPECIES

1. Herbage sparsely hirsute; leaves oblong-lanceolate, up to 10 mm. wide, conspicuously several-veined; glomerules of flowers mostly solitary at the ends of the main stem and branches, capitate; calyx lobes attenuate-subulate, more or less unequal in length but not conspicuously different in texture; corolla rose-purple..............1. *C. Wrightii*

1. Herbage glabrous or (commonly) sparsely hispidulous; leaves linear-lanceolate or subulate, not more than 3 mm. wide, with only the midvein apparent; glomerules several, axillary and terminal; calyx lobes very unequal, some of them lanceolate and foliaceous, the others reduced to setaceous, scarious teeth; corolla whitish..2. *C. subulata*

1. **Crusea Wrightii** Gray. Huachuca Mountains, Cochise County (*Lemmon* 2724), Bear Valley (Sycamore Canyon), Santa Cruz County, 3,500 feet (*Kearney & Peebles* 14447), August and September. Southeastern Arizona and northern Mexico.

2. **Crusea subulata** (Pavon) Gray. Chiricahua and Huachuca mountains (Cochise County), Santa Rita Mountains (Pima or Santa Cruz County), 5,000 to 7,500 feet, in woods or in the open, August and September. Southern New Mexico and Arizona; Mexico.

7. DIODIA. BUTTON-WEED

Plants annual; stems erect or diffuse; leaves narrowly lanceolate; stipules fringed with long, stiff bristles; flowers small, in axillary glomerules; corolla funnelform-salverform, pink; fruit obovoid-turbinate, crowned by the persistent calyx lobes.

1. **Diodia teres** Walt. Greenlee, Gila, Cochise, Santa Cruz, and Pima counties, usually about 4,000 to 6,000 feet, dry mesas and slopes, and along streams, often in sandy soil, August and September. Connecticut to Missouri and Arizona, south to Florida and Panama.

Var. *angustata* Gray, with erect, simple or sparingly branched stems, is more common in Arizona than typical *D. teres,* with spreading or procumbent, freely branched stems.

8. MITRACARPUS

Plants annual; stems erect, simple or sparingly branched; leaves opposite, lanceolate; stipules setose; flowers small, in few, very dense, terminal and axillary clusters; calyx with the 2 pairs of lobes very unlike in size and texture, the larger ones equaling or surpassing the whitish corolla; capsule didymous, 2-celled.

1. **Mitracarpus breviflorus** Gray. Cochise, Santa Cruz, and Pima counties, 4,000 to 6,000 feet, dry plains and mesas, August and September. Southern Texas, southern Arizona, and Mexico.

9. GALIUM (306). BEDSTRAW

Plants annual or perennial, herbaceous or suffrutescent; stems angled, often winged, usually weak and reclining or supported on other plants; herbage often retrorsely hispid; leaves appearing whorled, usually narrow; flowers small, perfect or unisexual, in axillary or terminal cymes or glomerules, these often panicled, or the flowers solitary in the axils; fruits didymous (paired), smooth, tuberculate, or covered with straight hairs or hooked bristles, indehiscent; seeds with a deeply concave face.

The plants have been used as remedies for various diseases, but their medicinal value is questionable. The seeds of *G. Aparine* have been used as a substitute for coffee, and the beverage is said to have a distinct coffee flavor.

KEY TO THE SPECIES

(Throughout the key, for the sake of brevity, no distinction is made between the true leaf blades and the scarcely distinguishable foliaceous stipules which together form the whorl. The term "leaves" includes both.)

1. Flowers involucrate (closely subtended by leaflike bracts), sessile or nearly so, solitary in each involucre; fruit slightly fleshy at maturity, granulate or tuberculate, not hairy. Plants glabrous or nearly so; stems deeply grooved, with thick, whitish angles; leaves mostly 4 in the whorl, somewhat rigid, thickish, shiny, mostly narrowly linear, sharply cuspidate, the midrib and margin thick, whitish..........1. *G. microphyllum*

1. Flowers not involucrate, usually distinctly pedicellate; fruit not fleshy at maturity (2).

2. Fruit minutely hispidulous to glabrate. Plants perennial; stems herbaceous above ground (3).

2. Fruit (except sometimes in *G. boreale*) conspicuously hairy, the hairs often as long as the width of the carpel (4).

3. Stems usually rather stout, very rough to the touch; leaves 5 or more in the whorl, the lower ones commonly more than 15 mm. long, narrowly elliptic, lanceolate, or oblanceolate; pedicels less than 5 mm. long; fruit tuberculate or minutely hispidulous 2. *G. asperrimum*

3. Stems slender, not or barely rough to the touch; leaves 4 to 6 in the whorl, commonly less than 15 mm. long; pedicels (some of them) usually more than 5 mm. long; fruit smooth, glabrate ....3. *G. tinctorium*

4. Hairs of the fruit straight, soft, white; stems more or less woody at base. Leaves in whorls of 4 or fewer (5).

4. Hairs of the fruit curved or hooked (except sometimes in *G. boreale*); stems entirely herbaceous above ground (12).

5. Corolla purplish to dark brown-purple; plants not dioecious. Leaves linear, narrowly lanceolate, or somewhat oblanceolate, spreading or ascending, the midrib stout, prominent beneath (6).

5. Corolla white, yellowish, or greenish (rarely purplish); plants (except *G. Collomae*) dioecious (7).

6. Stems and leaves hirtellous, at least near the base of the plant; leaves often somewhat flaccid ....4. *G. Wrightii*

6. Stems and leaves glabrous or puberulent; leaves usually somewhat rigid.5. *G. Rothrockii*

7. Leaves obtuse to acutish at apex, the midrib slender, not very prominent; plants scarcely woody above ground; corolla glabrous. Herbage hirtellous-puberulent to glabrate, scarcely rough to the touch; leaves spreading or reflexed, thin, not rigid, linear or narrowly lanceolate, the lateral veins obsolete......................6. *G. Fendleri*

7. Leaves acute or acuminate, often sharply cuspidate, the midrib very prominent beneath; plants mostly distinctly woody above ground; corolla (except in *G. coloradoense*) pubescent externally. Bark of the older stems pale, usually exfoliating (8).

8. Lateral veins of the leaf obscure or obsolete (9).

8. Lateral veins (1 or more) usually distinct (except in *G. Munzii*) but often very short. Plants suffrutescent (10).

9. Plants more or less shrubby; herbage hispid or hispidulous, rough to the touch; leaves lanceolate to ovate, 10 mm. long or shorter; pedicels erect or ascending.7. *G. stellatum*

9. Plants suffrutescent; herbage glabrous or obscurely puberulent, smooth; leaves linear, 10 to 25 mm. long; pedicels decurved..........................8. *G. coloradoense*

10. Flowers perfect; leaves somewhat flaccid. Herbage hirsute with straight, spreading hairs; leaves up to 6 mm. wide....9. *G. Collomae*
10. Flowers unisexual, the plants dioecious; leaves firm (11).
11. Herbage smooth, glabrous or obscurely puberulent; leaves up to 5 mm. wide, the lateral veins (1 or more) usually evident....10. *G. Watsoni*
11. Herbage copiously rough-pubescent (hispidulous); leaves up to 6 mm. wide, the lateral veins obscure or obsolete....11. *G. Munzii*
12. Leaves 5 or more in the whorl. Stems long, weak, commonly reclining (13).
12. Leaves not more than 4 in the whorl (14).
13. Plants annual, without a rootstock; stems rough to the touch, retrorsely hispid with pricklelike hairs; leaves linear, lanceolate, or oblanceolate; fruit mostly 3 to 4 mm. in transverse diameter at maturity, the hairs stiff, triangular-tuberculate at base, shorter than the transverse diameter of the carpel....12. *G. Aparine*
13. Plants perennial, with slender, elongate, branched rootstocks; stems not or very slightly rough to the touch, hispidulous, pilose, or nearly glabrous; leaves elliptic-lanceolate, broadly oblanceolate, or oblong-ovate, conspicuously cuspidate; fruit less than 3 mm. in transverse diameter at maturity, the hairs not stiff, scarcely enlarged at base, about as long as the transverse diameter of the carpel....13. *G. triflorum*
14. Plants perennial; stems elongate; flowers clustered; peduncles straight (15).
14. Plants annual; stems short, usually less than 25 cm. long; flowers mostly solitary; peduncles more or less curved (16).
15. Flowers in terminal, elongate, often rather dense, many-flowered panicles; herbage puberulent or glabrate; leaves linear to broadly lanceolate, rather stiff and firm, distinctly 3-nerved; corolla bright white; fruit sometimes glabrescent....14. *G. boreale*
15. Flowers in axillary and terminal, few-flowered, very open cymes; herbage puberulent to villous; leaves elliptic to ovate, soft, only the midrib conspicuous; corolla yellowish or brownish; fruit permanently pubescent....15. *G. pilosum*
16. Leaves approximately equal in the whorl; flowers solitary or in twos, nearly sessile, on short sometimes forked branchlets, each flower subtended by a pair of reduced leaves; herbage usually hispidulous....16. *G. proliferum*
16. Leaves very unequal in the whorl; flowers solitary on naked axillary peduncles, these elongate in fruit; herbage glabrous. Cotyledons persistent; stem leaves sometimes reduced to 1 pair....17. *G. bifolium*

1. Galium microphyllum Gray (*G. nitens* Jones, *Relbunium microphyllum* Hemsl.). Greenlee County to Coconino County, south to Cochise, Santa Cruz, and Pima counties, below 5,000 feet, mostly along streams, common, April to October, type of *G. nitens* from Nogales (*Jones* 22352). Western Texas to Arizona and Mexico.

Some authorities recognize *Relbunium* as a genus distinct from *Galium*.

2. Galium asperrimum Gray. Southern Apache and Navajo counties to Cochise, Santa Cruz, and Pima counties, 4,000 to 9,500 feet, common in rich soil in coniferous forests, July to September. New Mexico, Arizona, and Mexico.

3. Galium tinctorium L. Southern Apache County and at Prescott (Yavapai County), 5,000 to 9,500 feet. Labrador to Florida, west to the Pacific Coast.

The Arizona plants presumably belong to var. *subbiflorum* (Wiegand) Fern. A collection in southern Apache County (*E. Palmer* 514) is one of the types of var. *diversifolium* W. F. Wight, which has 5 or 6 leaves in the whorl. Specimens from the North Rim of Grand Canyon (*Collom* in 1950), with stems about 5 cm. long and leaves about 5 mm. long, have been referred to *G. Brandegei* Gray, but it is doubtful that they are more than a reduced form of *G. tinctorium* var. *subbiflorum*.

4. **Galium Wrightii** Gray. Apache County to Coconino County, south to Cochise, Santa Cruz, and Pima counties, 3,500 to 8,500 feet, June to October. Western Texas to Arizona and northern Mexico.

5. **Galium Rothrockii** Gray. Coconino County (both sides of the Grand Canyon) to Graham, Cochise, Santa Cruz, and Pima counties, 3,000 to 8,000 feet, rocky slopes and canyons, common, June to October, type from Camp Crittenden, Santa Cruz County (*Rothrock* 675). Southern New Mexico to southeastern California and Mexico.

Intergrades or hybridizes with the nearly related *G. Wrightii.*

6. **Galium Fendleri** Gray. Greenlee, Graham, Cochise, Santa Cruz, and Pima counties, 8,000 to 9,500 feet, coniferous forests, July and August. New Mexico and southeastern Arizona.

7. **Galium stellatum** Kellogg. Grand Canyon (Coconino County) and Mohave County to Pima and Yuma counties, 3,000 feet or lower, dry, rocky slopes, common, January to May. Southern Utah and Nevada, Arizona, southern California, and Baja California.

The species is represented in Arizona by var. *eremicum* Hilend & Howell.

8. **Galium coloradoense** W. F. Wight. Carrizo Mountains, Apache County (*Standley* 7353). Southwestern Colorado, southeastern Utah, and northeastern Arizona.

Standley's specimen has hispidulous stems and leaf margins.

9. **Galium Collomae** J. T. Howell. Fossil Creek Hill, Mogollon Escarpment, Gila County, 6,500 feet (*Collom* 596, the type collection, and in 1945). Known only by these collections in central Arizona.

10. **Galium Watsoni** (Gray) Heller. Both sides of the Grand Canyon (Coconino County), apparently common, 5,000 to 7,000 feet, Lake Mead (Mohave County), April to September. Idaho and Oregon to northern Arizona.

11. **Galium Munzii** Hilend & Howell. North Kaibab Trail, Grand Canyon, Coconino County (*Eastwood & Howell* 1008, 7087). Northern Arizona, western Nevada, and southern California.

12. **Galium Aparine** L. Navajo County to northern Mohave County, south to Greenlee, Graham, and Pima counties, 2,000 to 8,000 feet, mostly along streams, spring. Widely distributed in the United States; presumably naturalized from Europe.

Goosegrass bedstraw, cleavers. A small-fruited variety, var. *Vaillantii* (DC.) Koch, also has been reported from Arizona.

13. **Galium triflorum** Michx. White Mountains (Apache County), Kaibab Plateau and North Rim of Grand Canyon, Mogollon Escarpment, and Oak Creek Canyon (Coconino County), Beaver Creek (Yavapai County), Pinaleno Mountains (Graham County), Santa Catalina Mountains (Pima County), 4,500 to 8,500 feet, moist, shady places, June to September. Canada and Alaska to Alabama, Arizona, and California.

Sweet-scented bedstraw. The herbage is very fragrant in drying.

Philip J. Leyendecker, Jr. (*Iowa State College Jour. Sci.* 15: 180–181), distinguished forma *glabrum* Leyendecker as having the stem angles glabrous instead of retrorsely hispid, and cited a collection in the White Mountains, Arizona (*Griffiths* 5342).

14. Galium boreale L. Apache County, at Lukachukai Pass (*Eastwood & Howell* 6789) and Mount Baldy (*Phillips* 3264), Navajo County, at Lakeside (*Thornber* 8901), 6,000 to 9,500 feet, July to September. Canada to Pennsylvania, Texas, eastern Arizona, and California; Eurasia.

15. Galium pilosum Ait. Pinaleno Mountains, Graham County, about 7,000 feet (*Peebles et al.* 4504), rich soil in shade, growing with *Circaea,* July. Massachusetts to Indiana, south to Florida and central Texas, and in southeastern Arizona.

The Arizona station is far to the west of the main area occupied by this species.

16. Galium proliferum Gray. Havasu Canyon (Coconino County) and Greenlee, Gila, Pinal, Maricopa, Pima, and Yuma counties, 2,000 to 4,000 feet, canyons and rocky slopes, March to May. Western Texas to southern Utah, Arizona, and northern Mexico.

17. Galium bifolium Wats. Greenland Lake, North Rim of Grand Canyon (*Collom* 1670). Southern Montana to southern British Columbia, south to Colorado, northern Arizona, and California.

128. CAPRIFOLIACEAE. Honeysuckle Family

Plants almost entirely herbaceous, shrubby, or arborescent; leaves opposite, simple or compound; flowers perfect, regular or irregular; calyx with tube wholly adnate to the ovary, the limb represented by 3 to 5 teeth or lobes, or nearly obsolete; corolla rotate, funnelform, or somewhat bilabiate; stamens 4 or 5, distinct, borne on the corolla; style 1, the stigmas 1 to 5; ovary 2- to 5-celled; fruit berrylike, drupelike, or achenelike.

This family includes numerous plants that are cultivated as ornamentals, notably the bush honeysuckles and climbing honeysuckles (*Lonicera* spp.).

KEY TO THE GENERA

1. Leaves pinnately compound; flowers small, very numerous; corolla rotate or saucer-shaped 1. *Sambucus*
1. Leaves simple; flowers relatively large, not very numerous; corolla tubular to funnelform-campanulate (2).
2. Plants low, only slightly woody, with prostrate, creeping stems; flowers in pairs on an elongate, slender, erect peduncle . . . . 3. *Linnaea*
2. Plants shrubby, with erect, climbing, or trailing stems; flowers in short spikes or axillary clusters (3).
3. Corolla regular or very nearly so, funnelform, funnelform-campanulate, or salverform, the tube not gibbous or swollen near the base; ovary 4-celled; berry commonly white, 1- or 2-seeded . . . . 2. *Symphoricarpos*
3. Corolla more or less irregular, with a slightly to pronouncedly bilabiate limb, the tube more or less gibbous or swollen on one side near the base; ovary 2- or 3-celled (the partitions sometimes incomplete); berry not white, often containing more than 2 seeds 4. *Lonicera*

1. SAMBUCUS. Elder, Elderberry

Large shrubs or small trees; stems pithy; leaves large, pinnate, the leaflets lanceolate to ovate; flowers in broad compound cymes or pyramidal cymose panicles; corolla rotate, cream-colored or yellowish; fruits berrylike.

The foliage is browsed by domestic animals and deer. The fruits are edible (except those of the red-fruited *S. racemosa*) and are very attractive to birds. They are sometimes used for making jelly, pies, and wine. It has been reported, however, that the fruits, as well as the flowers, bark, and roots, when eaten raw, may poison animals. All parts of the plants are reputed to have medicinal virtues, but these are probably largely imaginary, although the flowers are said to be diuretic. The Mexican elder (*S. mexicana*), Arizona's most treelike species, is sometimes planted as an ornamental in the southern part of the state.

KEY TO THE SPECIES

1. Inflorescence broadly short-pyramidal, with the axis extended beyond the lowest branches; berries not glaucous. Leaflets lanceolate to oblong-ovate, sharply serrate (2).
1. Inflorescence flat-topped, with elongate, compound rays, the axis not or seldom extended beyond the lowest branches; berries dark blue, commonly glaucous (3).
2. Berries bright red at maturity; branchlets and leaves glabrous or glabrate; leaflets commonly 7; corolla yellowish....................1. *S. racemosa*
2. Berries black at maturity; branchlets and the lower surface of the leaves usually scurfy-puberulent or sparsely villous; leaflets commonly 5; corolla whitish..2. *S. melanocarpa*
3. Leaflets gradually long-acuminate, thin, the larger ones seldom less than 8 cm. long; inflorescences very open, 10 to 30 cm. wide, the primary branches usually divaricate. Leaflets commonly 5 or 7, lanceolate or oblong-lanceolate..........3. *S. neomexicana*
3. Leaflets usually shortly or abruptly acuminate and thickish, commonly less than 8 cm. long; inflorescences somewhat dense, seldom more than 15 cm. wide, the primary branches usually ascending (4).
4. Branchlets and the lower surface of the leaflets finely pubescent, the branchlets densely so. Leaves deciduous; leaflets ovate or ovate-oblong....................4. *S. velutina*
4. Branchlets and leaves glabrous or sparsely pubescent, rarely copiously so (5).
5. Leaves persistent; leaflets 3 to 5 (seldom with a small, additional basal pair), oblong-lanceolate to ovate or somewhat obovate, rather abruptly acuminate, finely and closely serrate; berries moderately glaucous; usually trees, up to 10 m. high....5. *S. mexicana*
5. Leaves deciduous; leaflets 5 to 9 (rarely only 3), lanceolate or oblong-lanceolate, gradually short-acuminate, sharply and often rather deeply serrate; berries very glaucous; commonly large shrubs. Branchlets and leaves glabrous or very nearly so 6. *S. coerulea*

1. Sambucus racemosa L. White Mountains (Apache and Greenlee counties), both sides of the Grand Canyon and San Francisco Peaks (Coconino County), 7,500 to 10,000 feet, moist forests, June and July. Throughout the cooler parts of North America; Eurasia.

The Arizona plants belong to var. *microbotrys* (Rydb.) Kearney & Peebles (*S. microbotrys* Rydb., *S. acuminata* Greene), with the leaves and branchlets glabrous or nearly so. The herbage is strong-scented, and the fruits are reputed to be poisonous. The type of *S. acuminata* came from Mount Agassiz (*Pearson* 330).

2. Sambucus melanocarpa Gray. White Mountains (Apache County), Santa Catalina Mountains (Pima County), reported also from the Kaibab Plateau (Coconino County), 7,500 feet and probably higher, May to July. Alberta and British Columbia to New Mexico, Arizona, and California.

3. Sambucus neomexicana Wooton. Coconino County to Graham, Cochise, and Pima counties, 5,000 to 9,500 feet, mostly along streams, July and August. New Mexico and Arizona.

This shrub attains a height of 2.5 m. (8 feet) and has been reported to reach 6.5 m. (21 feet). Both the typical plant, with branchlets and leaves glabrous or nearly so, and var. *vestita* (Woot. & Standl.) Kearney & Peebles (*S. vestita* Woot. & Standl.), with branchlets and leaves persistently puberulent or tomentulose, are found in Arizona. The variety has been collected in the Pinal, Santa Catalina, and Santa Rita mountains.

4. Sambucus velutina Dur. & Hilg. Hualpai Mountain, Mohave County (*Goldman* 2980, *Kearney & Peebles* 12725), 7,000 to 8,000 feet, pine forest. Western Arizona and California.

5. Sambucus mexicana Presl (*S. coerulea* var. *arizonica* Sarg., *S. coerulea* var. *mexicana* L. Benson). Navajo County to Mohave County, south to Cochise, Santa Cruz, and Pima counties, 1,000 to 4,000 feet, frequent along streams and ditches, March to June. Western Texas to southern California and Mexico.

Tapiro, Mexican elder. The tree attains a height of 10 m. (33 feet) and a trunk diameter of 45 cm. (18 inches). It is the only nonmontane species in the state.

6. Sambucus coerulea Raf. (*S. glauca* Nutt.). Mountains of Apache, Coconino, Yavapai, Cochise, and Santa Cruz counties, 6,500 to 8,500 feet, coniferous forests, June to September. Alberta and British Columbia to Arizona and southern California.

Usually a many-stemmed shrub, up to 6 m. (20 feet) high, sprouting freely from the root.

2. SYMPHORICARPOS (307). SNOWBERRY

Shrubs, with the older bark exfoliating; leaves opposite, simple, without stipules, entire to sinuately lobed; flowers regular or nearly so, the corolla campanulate, funnelform, or salverform, 4- or 5-lobed, pink or white, often pubescent within, with 1 to 5 nectaries at base; ovary 4-celled, 2 of the cells containing 1 large, fertile, pendulous ovule, the other cells containing several small, abortive ovules; berry ovoid, ellipsoid, or subglobose; nutlets normally 2, more or less compressed.

The plants afford valuable browse for livestock and deer. They contain saponin, but in such small quantity that poisoning rarely occurs. The plants are often cultivated for their long-persistent, very ornamental, waxy-looking white fruits, which are eaten by many kinds of birds. The species are difficult to identify in the absence of flowers.

KEY TO THE SPECIES

1. Corolla salverform, 9 to 13 mm. long, with only one small basal nectary, glabrous within; style 4 to 7 mm. long, usually pilose above the middle; free portion of the filaments much shorter than the anthers, or the latter sessile. Leaves oblanceolate, oval, or lanceolate, glaucous, 6 to 15 mm. long, 2 to 5 mm. wide....................1. *S. longiflorus*

1. Corolla long-campanulate or tubular-funnelform, 6 to 13 mm. long, with 5 basal nectaries; style 3 to 5 mm. long, usually glabrous; free portion of the filaments equaling or somewhat longer than the anthers (2).

2. Young twigs completely glabrous (3).

2. Young twigs puberulent or pubescent. Corolla pilose within (4).

3. Corolla long-campanulate, 6 to 7 mm. long, the tube pilose within; leaf buds lanceolate in outline, acuminate; leaves pilosulous, rarely glabrous, normally very glaucous beneath; plants low, spreading..2. *S. Parishii*
3. Corolla tubular-funnelform, 9 to 13 mm. long, the tube sparsely pilose to nearly glabrous within; leaf buds ovate in outline, acute; leaves almost always entirely glabrous; plants erect ....3. *S. oreophilus*
4. Pubescence of the young twigs dense, of straight spreading hairs. Corolla tubular-funnelform, 8 to 10 mm. long; anthers reaching only to the base of the corolla lobes; leaves roundish-oval, dark green, obtuse or obtusish, soft-pubescent on both surfaces, 1 to 3 cm. long, 6 to 18 mm. wide..............................4. *S. rotundifolius*
4. Pubescence of the young twigs not dense or if so, then the hairs short and curved (5).
5. Corolla funnelform-campanulate, 6 to 7 mm. long; young twigs loosely pilosulous, the internodes occasionally glabrous. Low, spreading shrubs; leaves glaucous, pilosulous 2. *S. Parishii*
5. Corolla tubular-funnelform, 8 to 12 mm. long; young twigs tomentulose-puberulent with short, curved hairs (6).
6. Plants erect; leaves puberulent, scarcely paler beneath, the principal veins prominent on the upper surface in dried specimens, the petioles 2 to 4 mm. long; nutlets lanceolate in outline or fusiform, acute or apiculate at base, 5 to 7 mm. long......5. *S. utahensis*
6. Plants trailing; leaves short-pilosulous, paler beneath, the veins obscure on the upper surface, the petioles 1 to 2 mm. long; nutlets elliptic in outline, flattened, acutish at base, 4 to 5 mm. long..6. *S. Palmeri*

1. Symphoricarpos longiflorus Gray. Coconino and Mohave counties, 4,000 to 8,000 feet, foothills, canyons, and pine forests, April to August. Western Colorado and western Texas to Oregon and eastern California.

A *Symphoricarpos* with the corolla salverform, as in *S. longiflorus*, but only about 7 mm. long, light violet, the anthers borne on very short filaments, was collected near the summit of Baboquivari Peak, Pima County, 7,500 feet (*Gould et al.* 2765), flowering in October. This plant may represent an undescribed species.

2. Symphoricarpos Parishii Rydb. Black Mesa (Navajo County), Kaibab Plateau, Grand Canyon, and Flagstaff (Coconino County), 5,000 to 9,000 feet, dry hills, in the open or in chaparral, April to August. Nevada, northern Arizona, and southern California.

Specimens labeled as from Fort Mohave (*Lemmon* in 1884) doubtless were collected at a more elevated station.

3. Symphoricarpos oreophilus Gray. Apache, Navajo, and Coconino counties, south to the mountains of Cochise and Pima counties, 5,500 to 9,000 feet, mostly in pine forests, May to August. Colorado and western Texas to eastern Nevada, Arizona, and northern Sonora.

4. Symphoricarpos rotundifolius Gray. Apache, Navajo, Coconino, Yavapai, western Graham, and Gila counties, 4,000 to 10,000 feet, rocky slopes, May and June. Southern Colorado, New Mexico, and Arizona.

5. Symphoricarpos utahensis Rydb. Carrizo Mountains, Apache County (*Standley* 7384), Navajo Mountain, Coconino County (*Peebles & Smith* 13960), near Pine, Gila County (*MacDougal* 702), 6,000 to 8,000 feet, canyons and slopes, June and July. Wyoming, Colorado, Utah, and Arizona.

A collection on the San Francisco Peaks, 8,500 feet (*Whiting & Sanders* in 1935), probably also belongs to this species.

6. Symphoricarpos Palmeri G. N. Jones. White Mountains (Apache

County), Keam Canyon (Navajo County), Chiricahua Mountains (Cochise County), 7,000 to 8,000 feet, moist slopes and swales, May to July. Southern Colorado and western Texas to eastern Arizona.

3. LINNAEA. Twin-flower

Plants evergreen, nearly herbaceous; stems slender, creeping, forming loose mats; leaves thickish, obovate or nearly orbicular, crenulate; flowers nodding; corolla nearly regular, broadly funnelform, 5-lobed, white tinged with pink; stamens 4, unequal in length; ovary 3-celled; fruit 1-seeded, dry, indehiscent.

1. Linnaea borealis L. White Mountains (Apache County), Kaibab Plateau (Coconino County), deep coniferous forests, June and July. Greenland to Alaska, south to New Jersey, Michigan, New Mexico, northern Arizona, and California; Eurasia.

The name of this beautiful little plant, very rare in Arizona, commemorates the great Linnaeus. The Arizona plants belong to var. *americana* (Forbes) Rehder.

4. LONICERA (308). Honeysuckle

Plants shrubby; stems erect, trailing, or twining; leaves broad, normally entire; flowers in twos or threes on axillary peduncles, or in axillary or terminal whorls, these often forming interrupted, spikelike inflorescences; corolla nearly regular to strongly bilabiate; fruit berrylike.

It is said that the fruits contain saponin and have emetic and cathartic effects. The plants are mostly unpalatable, but some species are browsed. They are, however, regarded with suspicion by stockmen. The fruits are eaten by birds and chipmunks. *L. arizonica,* which resembles the trumpet or coral honeysuckle of the eastern United States (*L. sempervirens* L.), is worthy of cultivation as an ornamental climber.

KEY TO THE SPECIES

1. Flowers in axillary pairs; leaves all petioled, never connate-perfoliate (2).
1. Flowers several or numerous in axillary or terminal, whorllike compound cymes, these sometimes aggregated in spikelike inflorescences; leaves (the uppermost pair or pairs) usually sessile and connate-perfoliate. Stems straggling or clambering (seldom twining); leaves broadly ovate, oval, or nearly orbicular, whitish beneath; fruits red at maturity (4).
2. Stems twining; corolla deeply cleft, the limb strongly bilabiate, the tube slender. Flowers white or tinged with red, fading yellow; fruits normally black at maturity 1. *L. japonica*
2. Stems erect or ascending; corolla shallowly cleft, the limb nearly regular, the tube gibbous or saccate on one side, at base (3).
3. Fruits at maturity black, closely enveloped by the bracts, these glandular-ciliate, broadly ovate, becoming enlarged and rose-red to reddish brown; leaves ovate or oblong, acuminate at apex, paler but green beneath; corolla slightly gibbous, yellow, less than 15 mm. long .... 2. *L. involucrata*
3. Fruits at maturity orange-yellow to bright red (rarely dark blue), not closely enveloped by the bracts, these small, glabrous, subulate, green; leaves oblong, elliptic, ovate, or slightly obovate, obtuse or rounded at apex, whitish beneath; corolla strongly gibbous (almost spurred), yellowish white, usually more than 15 mm. long..... 3. *L. utahensis*
4. Corolla coral red outside (often drying purplish), orange within, the limb short, only slightly bilabiate, not more than ¼ as long as the tube; stamens inserted deep in the corolla tube. Leaves ciliate, up to 7 cm. long; whorl of flowers single, or the whorls

crowded together in a subcapitate spike, this usually distinctly stalked and not closely subtended by the uppermost pair of leaves; corolla tube slightly swollen on one side well above the base; style glabrous or very nearly so . . . . 4. *L. arizonica*

4. Corolla whitish or pale yellow, deeply 2-lipped, the lips nearly as long as the tube, the lower lip spreading or reflexed; stamens inserted near the summit of the corolla tube (5).

5. Bracts more than half as long as the ovary; corolla 10 to 15 mm. long, the tube as long as the limb; stems herbaceous above; leaves commonly glabrous, sometimes pubescent beneath; whorls of flowers several, well separated, forming elongate, interrupted, small-bracted spikes, these terminating the axis and the somewhat elongate branchlets 5. *L. interrupta*

5. Bracts less than half as long as the ovary; corolla usually more than 15 mm. long, the tube usually longer than the limb; stems woody almost throughout; leaves pubescent, at least beneath; whorl of flowers single (if more than one then the whorls so closely crowded as to appear one), borne at the ends of short branchlets and closely subtended by the uppermost pair of leaves . . . . 6. *L. albiflora*

1. Lonicera japonica Thunb. Escaped from cultivation at Douglas and in the Huachuca Mountains, Cochise County (*W. W. Jones* in 1928, *Thornber* 7236). Extensively naturalized in the southeastern United States; native of eastern Asia.

Japanese honeysuckle, common in cultivation and prized for its deliciously fragrant flowers.

2. Lonicera involucrata (Richards.) Banks. Apache, Coconino, Greenlee, Graham, and Cochise counties, especially on the San Francisco Peaks and in the White, Pinaleno, and Chiricahua mountains, 7,500 to 10,500 feet, open coniferous forests, often along streams, June and July. Canada and Alaska to Michigan, Arizona, California, and northern Mexico.

Bear-berry honeysuckle, ink-berry, pigeon-bush, twin-berry. The plants sucker freely and tend to form thickets. The flowers attract hummingbirds.

3. Lonicera utahensis Wats. White Mountains (Apache and Greenlee counties), 9,500 to 11,000 feet, open coniferous forests, gregarious, June and July. Montana to British Columbia, south to New Mexico, Arizona, and northern California.

4. Lonicera arizonica Rehder. Apache County, Coconino County (both sides of the Grand Canyon), and Cochise and Pima counties, 6,000 to 9,000 feet, open coniferous forests, June and July, type from the Rincon Mountains, Pima County (*Pringle* in 1884). Utah, New Mexico, and Arizona.

Arizona honeysuckle. Closely related to *L. ciliosa* Poir., and some of the Arizona specimens (as *Rusby* in 1883) appear to be intermediate.

5. Lonicera interrupta Benth. Gila and Pima counties, especially in the Pinal, Sierra Ancha, Mazatzal, and Santa Catalina mountains, 4,000 to 6,000 feet, in chaparral, usually near streams, May and June. South-central Arizona and California.

Chaparral honeysuckle.

6. Lonicera albiflora Torr. & Gray. Southern Apache County and Gila County to Cochise and Santa Cruz counties, 3,500 to 6,000 feet, along streams, April to June. Arkansas to southern Arizona and northern Mexico.

The species is represented in Arizona by var. *dumosa* (Gray) Rehder (*L. dumosa* Gray).

129. VALERIANACEAE. Valerian Family

Plants herbaceous, annual or perennial; leaves opposite, without stipules; flowers perfect or unisexual, in cymose panicles; calyx tube wholly adnate to the ovary, the limb obsolete or represented by pappuslike bristles; corolla somewhat bilabiate or nearly regular, 4- or 5-lobed; stamens commonly 3, distinct, borne on the corolla; style and stigma 1; ovary 3-celled, with 2 of the cells empty or aborted; fruit achenelike, 1-seeded.

KEY TO THE GENERA

1. Calyx limb obsolete; corolla with a pendent spur at base of the throat......1. *Plectritis*
1. Calyx limb represented by a circle of setiform lobes, these at maturity elongating and becoming a pappuslike ring of plumose bristles surmounting the fruit; corolla not spurred, the tube sometimes gibbous or slightly saccate................2. *Valeriana*

1. PLECTRITIS (309)

Plants annual, glabrous or nearly so; stems erect, sparingly branched; leaves (except the lowest ones) sessile, entire, the upper leaves oblong, the lower leaves spatulate or obovate; flowers small, in dense, terminal or subterminal clusters, these often forming narrow, interrupted panicles; corolla 2-lipped, spurred, pink; fruit commonly short-pilose, keeled on the convex side, winged on the concave (open) side.

KEY TO THE SPECIES

1. Fruit with strongly incurved wings, the opening on the concave side orbicular, deeply cup-shaped ....1. *P. californica*
1. Fruit with spreading, only slightly incurved wings, the opening elliptic or saucer-shaped 2. *P. macroptera*

1. Plectritis californica (Suksd.) Dyal. Prescott (Yavapai County), Mazatzal Mountains (Maricopa or Gila County), Tucson to Redington (Pima County), doubtless elsewhere, in moist soil, April and May. Washington to California and Arizona.

The Arizona plants belong to var. *rubens* (Suksd.) Dyal, with smaller flowers and fruit than in typical *P. californica.*

2. Plectritis macroptera (Suksd.) Rydb. Between Superior and Miami, Gila County, about 5,000 feet, March and April (*Harrison & Kearney* 1522). Washington to California and Arizona.

The collection cited represents var. *patelliformis* (Suksd.) Dyal, with relatively small flowers, a short spur, and small fruits.

2. VALERIANA. Valerian, Tobacco-root

Plants perennial, with rootstocks or tubers; leaves mostly pinnate or pinnately parted; flowers perfect or unisexual; calyx limb of pappuslike bristles, these elongate and spreading in fruit; corolla limb nearly regular, the tube swollen on one side; fruit achenelike, glabrous or puberulent.

The dried plants have a strong, unpleasant, very characteristic odor, which persists for years in herbaria. The roots of *V. edulis* and perhaps of other species were boiled and eaten by the Indians. *V. acutiloba* is reported to have some value as forage. The taxonomy of this genus is confused. The following treatment is necessarily provisional.

KEY TO THE SPECIES

1. Stems from a large vertical caudex, up to 1 m. long, stout, erect; leaves thickish, with several conspicuous veins nearly parallel with the midvein; corolla yellowish or greenish. Basal leaves oblanceolate or spatulate, long-petioled, entire or pinnately parted with few divisions, the stem leaves sessile or nearly so, usually pinnately parted with few, elongate, linear, lanceolate, or spatulate divisions; flowers in an elongate, very open panicle .... 1. *V. edulis*
1. Stems not from a large vertical caudex; leaves thin, with inconspicuous or spreading lateral veins; corolla pink or whitish (2).
2. Rootstock short, tuberlike, usually vertical; leaves very thin and flaccid, all more or less petioled; inflorescence an open, elongate panicle of loosely flowered cymes; flowers unisexual; corolla 1 to 2 mm. long. Leaves pinnate with 3 or more leaflets (the basal ones rarely simple), the leaflets broadly ovate to oblong-lanceolate, coarsely serrate or laciniate; fruits strigose-puberulent .... 2. *V. sorbifolia*
2. Rootstock elongate, usually horizontal; leaves firm, the upper stem leaves sessile or subsessile, the basal ones petioled; inflorescence at anthesis usually short and compact, becoming more open and elongate in fruit; flowers perfect; corolla 3 mm. or longer. Basal leaves mostly simple and entire, the upper stem leaves pinnately cleft or pinnate with very few, narrow, mostly entire divisions or leaflets; fruits glabrous or pubescent (3).
3. Corolla 7 to 13 mm. long, tubular-funnelform, the tube equaling to longer (usually much longer) than the throat and limb; basal leaves broadly elliptic, ovate, or suborbicular, and rounded, subcordate, or short-cuneate at base .... 3. *V. arizonica*
3. Corolla not more than 6 mm. long, funnelform-campanulate, the tube shorter than the throat and limb; basal leaves spatulate, obovate, ovate, or elliptic, and attenuate to rounded or subcordate at base (4).
4. Lateral leaflets of the stem leaves usually narrowly lanceolate and acuminate (sometimes elliptic and obtuse in var. *ovata*); inflorescences usually dense and headlike in flower, compact to open in fruit; corolla 4 to 6 mm. long .... 4. *V. acutiloba*
4. Lateral leaflets of the stem leaves usually broadly lanceolate to elliptic, obtuse to acute (seldom acuminate); inflorescences moderately dense to open in flower, open and often elongate in fruit; corolla 3 to 4 mm. long .... 5. *V. occidentalis*

1. Valeriana edulis Nutt. Apache County to Coconino County and in Yavapai, Graham, and Cochise counties, 7,000 to 9,500 feet, rich, moist soil, usually in open coniferous forests, June to September. Montana and Idaho to New Mexico and Arizona.

The prevailing phase in Arizona, with glabrous fruits, is *V. trachycarpa* Rydb., which does not seem to be more than varietally distinct.

2. Valeriana sorbifolia H.B.K. Mountains of Cochise, Santa Cruz, and Pima counties, 5,500 to 7,000 feet, rich soil in coniferous forests, July to October. Southern Arizona to Central America.

3. Valeriana arizonica Gray. Apache, Navajo, and Coconino counties, to the mountains of Pima County, 4,500 to 8,000 feet, rich, moist soil in coniferous forests, April to July, type from near Prescott (*Palmer*). Southern Utah to southern Arizona.

Specimens from the Santa Catalina Mountains (*Peebles* 4055, *Peebles & Loomis* 7101, *Beckett* 11719) are exceptional in the ciliate-bearded petioles of the stem leaves.

4. Valeriana acutiloba Rydb. Apache County to Coconino County, and in the Santa Rita and (?) Santa Catalina mountains (Santa Cruz and Pima counties), 7,000 to 9,500 feet, May to July. Wyoming to New Mexico and Arizona.

Typical *V. acutiloba* has the basal leaves spatulate, obovate, or elliptic, and attenuate at base. In var. *ovata* (Rydb.) A. Nels. (*V. ovata* Rydb.) the basal leaves are broadly ovate and rounded or subcordate at base. The variety has been collected in the White Mountains, on the San Francisco Peaks, in Oak Creek Canyon, and in the Pinaleno, Huachuca, and Santa Rita mountains (Apache, Coconino, Graham, Cochise, and Santa Cruz counties).

5. Valeriana occidentalis Heller. North Rim of the Grand Canyon (*Collom* 1056), San Francisco Peaks (*Leiberg* 5611), about 8,000 feet. Montana and Idaho to northern Arizona.

The Arizona plants, which have the fruits pubescent, at least when young, are referred doubtfully to this species.

130. CUCURBITACEAE. Gourd Family

Plants herbaceous, annual or perennial; stems with tendrils, trailing or climbing; leaves alternate, simple or compound; flowers mostly unisexual, regular or nearly so, solitary or in racemose or corymbose inflorescences; calyx tube wholly adnate to the ovary; corolla adnate at base to the calyx, usually 5-merous, sympetalous or the petals nearly distinct; stamens 3 or 5, in the latter case usually appearing to be 3, four of the anthers being united in pairs; style 1, the stigmas usually 3; ovary 1- to 4-celled; fruits various.

This family comprises numerous useful plants such as the melons, squashes, pumpkin, cucumber, and gourds. None of the species native to Arizona is of much economic importance, although some of them seem worth cultivating as ornamental climbers.

KEY TO THE GENERA

1. Fruits gourdlike, hard-shelled, at maturity 4 cm. in diameter or larger, not spiny, many-seeded; stems prostrate or trailing, with short, relatively stout, few-coiled tendrils. Leaves large; corolla yellow (2).

1. Fruit not gourdlike or hard-shelled, at maturity not more than 3 cm. in diameter; stems climbing, with long, filiform, many-coiled tendrils. Plants monoecious (except *Ibervillea*); corolla whitish or pale yellow (3).

2. Leaves reniform, wider than long, very shallowly lobed or with merely undulate, dentate margins; staminate flowers in racemes or corymbs; calyx tube cylindric, about as long as the corolla..1. *Apodanthera*

2. Leaves not reniform, longer than wide, or else palmately lobed or parted; staminate flowers solitary; calyx tube campanulate, much shorter than the corolla..4. *Cucurbita*

3. Hypanthium (at least in the pistillate flowers) long and narrowly cylindric; fruits globose, berrylike, smooth, several-seeded (4).

3. Hypanthium short-campanulate or, if elongate-cylindric (in the pistillate flowers of *Brandegea*), then the fruit asymmetric, long-beaked, 1-celled and 1-seeded; fruits not globose and berrylike or if so, then spiny. Corolla of the staminate flowers rotate, whitish (5).

4. Thickened root single; plants dioecious; staminate flowers with a short campanulate hypanthium; seeds rounded at the large end........................2. *Ibervillea*

4. Thickened roots clustered, often forked; plants monoecious; staminate flowers with a long, slender-cylindric hypanthium; seeds truncate at the large end....3. *Tumamoca*

5. Ovary 1-celled; ovule solitary. Corolla of the staminate flowers not more than 6 mm. in diameter (6).

5. Ovary several-celled, or with several placentas; ovules several. Pistillate flowers solitary; fruit spiny (8).

6. Fruit very asymmetric, narrowly obovoid, with a long, slender beak; plants perennial 8. *Brandegea*

6. Fruit symmetric or nearly so, ovoid, not long-beaked; plants annual (7).
7. Pistillate flowers in short-stalked, subcapitate or subumbellate clusters, not hidden by the bracts; fruit spiny; corolla lobes not bicuspidate....................9. *Sicyos*
7. Pistillate flowers solitary or in pairs in the leaf axils, each enveloped and hidden by an infolded, leaflike, 3-lobed and dentate bract; fruit smooth; corolla lobes bicuspidate ....10. *Sicyosperma*
8. Anthers completely fused and appearing as a horizontal ring opening all around; leaves pedately parted or divided...11. *Cyclanthera*
8. Anthers evidently more than 1 although more or less connate, opening longitudinally; leaves not parted or divided, but sometimes deeply cleft (9)
9. Plants perennial; fruit globose; seeds round and turgid...................7. *Marah*
9. Plants annual; fruit ovoid; seeds flat (10).
10. Fruit regularly dehiscent (operculate), the acuminate apical portion coming away as a lid, the spines glandular-hirsute................................5. *Echinopepon*
10. Fruit bursting irregularly at the rounded apex (not operculate), the spines not hirsute 6. *Echinocystis*

1. APODANTHERA. Melon-loco

Plants monoecious, perennial, coarse, with a large, thick root, the herbage harshly appressed-pubescent; flowers few; calyx tube (hypanthium) 1.5 to 2.5 cm. long, the teeth subulate; corolla yellow, the lobes distinct or very nearly so; stamens 3, one with a 1-celled anther and two with the anthers 2-celled, the anthers distinct, sessile; fruit oval, longitudinally ridged.

1. Apodanthera undulata Gray. Southern Yavapai, Pinal, Cochise, Santa Cruz, and Pima counties, 1,500 to 5,500 feet, dry plains and mesas, sometimes on limestone slopes, June to September. Western Texas to southern Arizona and Mexico.

The plant has a disagreeable odor.

2. IBERVILLEA

Plants dioecious, perennial, glabrous or nearly so; stems slender, from a thick, nearly globose root; leaves pedately parted or divided; flowers small, the staminate ones in short racemes or corymbs, the pistillate flowers solitary, axillary; fruit a red, globose berry.

***1. Ibervillea tenuisecta** (Gray) Small. Guadalupe Canyon, Sonora, and reported to occur in this canyon on the Arizona side of the boundary, in the southeastern corner of Cochise County. Western Texas, northern Mexico, and (?) southeastern Arizona.

3. TUMAMOCA

Slender-stemmed, glabrous, monoecious perennials with a cluster of thick but not very large, tuberlike roots; leaves thin, pedately 3-parted, the divisions deeply cleft or parted into narrow segments; staminate flowers in short racemes, the pistillate flowers solitary; calyx tube slender; corolla pale yellow, the lobes narrow; fruit a globose, several-seeded berry, red or yellow at maturity.

1. Tumamoca Macdougalii Rose. Pima County, near Tucson (*MacDougal* in 1908, the type collection, *Wiggins & Rollins* in 1941), and near Gunsight Wells (*Goodding & Reeder* in 1943), dry soil among rocks. Southern Arizona and Sonora.

Ira L. Wiggins, who discovered the Sonoran station for this very rare plant near Carbo, wrote (personal communication): "Root tuberous, each tuber 5 to 12 cm. long; flowers greenish; plants trailing on the ground or climbing 1 to 1.5 m. in a tangled mass in bushes. The tubers smell like decaying cabbage when bruised."

4. CUCURBITA (310)

Coarse, perennial, monoecious herbs with trailing stems from a large, thick root, the herbage harshly appressed-pubescent; flowers few, axillary, solitary; corolla yellow, 3 cm. long or longer, the tube campanulate, much longer than the lobes; stamens with distinct filaments and connate anthers; fruit globose, smooth, gourdlike.

KEY TO THE SPECIES

1. Leaves longer than wide, triangular-ovate, at most shallowly angulate-lobed, acuminate, up to at least 30 cm. long, the upper surface uniformly pubescent....1. *C. foetidissima*
1. Leaves about as wide as long, palmately 5-cleft or 5-parted, the lobes or divisions entire or angulately few-toothed, the upper surface usually conspicuously more pubescent along the veins than elsewhere, the lower surface uniformly pubescent (2).
2. Blades up to 25 cm. long, cleft nearly to the base, with lanceolate, often very narrow lobes, the upper surface dark green, with conspicuous broad bands of white pubescence along the veins, otherwise nearly glabrous; staminate calyx with a cylindric tube 2. *C. digitata*
2. Blades not more than 10 cm. long, cleft not nearly to the base, with deltoid lobes, the upper surface light green, with pubescence less restricted to the veins; staminate calyx with a trumpet-shaped tube..3. *C. palmata*

1. Cucurbita foetidissima H.B.K. Navajo and Coconino counties to Cochise and Pima counties, 1,000 to 7,000 feet, mostly in alluvial soil, often at roadsides, May to August. Missouri and Nebraska to Texas, Arizona, southern California, and Mexico.

Buffalo-gourd, calabazilla. A conspicuous, rank-growing, usually ill-smelling plant, with numerous stems up to 6 m. (20 feet) long and striped, gourdlike fruits about 10 cm. (4 inches) in diameter. The fruits were eaten by the Indians of Arizona cooked, or dried for winter use, and the seeds were eaten in the form of mush. This plant should be useful as a ground cover.

2. Cucurbita digitata Gray. Graham County to southern Yavapai County, south to Cochise, Santa Cruz, Pima, and eastern Yuma counties, 5,000 feet or lower, dry plains and mesas, June to October. Southwestern New Mexico to southeastern California and northern Mexico.

Fruit at maturity pale yellow, striped longitudinally. This species is usually easily distinguished by the characters given in the key, but specimens from the Mazatzal Mountains (*Eastwood* 17498) and from near Yuma (*Wolf* 2287) approach *C. palmata* in the relatively shallowly cleft, broadly lobed leaves, although in the distribution of the pubescence of the upper surface they resemble *C. digitata*. The Mazatzal Mountains are considerably to the east of the known range of *C. palmata*.

3. Cucurbita palmata Wats. Western parts of Mohave and Yuma counties, 3,000 feet or lower, sandy plains and mesas, and rocky slopes, April to September. Western Arizona, southern California, and Baja California.

Sometimes called coyote-melon.

A specimen collected at the head of Lake Mead, Mohave County (*Cutler & Baker* 3588), was identified by L. H. Bailey as *C. californica* Torr., but it may be merely a form of *C. palmata* with relatively short and broad leaf lobes.

5. ECHINOPEPON

Plants monoecious, annual, with slender, climbing stems; leaves shallowly lobed, cordate at base; staminate flowers in long racemes, the pistillate flowers solitary; corolla of the staminate flowers 7 to 8 mm. in diameter, glandular-punctate, the lobes triangular; fruit ovoid, not more than 1.5 cm. in diameter, spiny, opening by an apical lid, commonly 3-celled.

1. **Echinopepon Wrightii** (Gray) Wats. Santa Cruz and Pima (doubtless also Cochise) counties, 3,000 to 4,000 feet, along streams, climbing over shrubs, July to October. Southwestern New Mexico, southern Arizona, and Mexico.

6. ECHINOCYSTIS. Mock-cucumber

Plants monoecious; stems slender, climbing, from an annual root; leaves thin, deeply lobed; flowers of both sexes from the same axils, the staminate ones numerous in long compound racemes, the pistillate flowers few or solitary; staminate corollas not punctate, the lobes ligulate-lanceolate; fruit ovoid, up to 2.5 cm. in diameter, armed with soft spines, irregularly dehiscent.

1. **Echinocystis lobata** (Michx.) Torr. & Gray. Coconino County, at Flagstaff (*Thornber* 8579) and Oak Creek Canyon (*Dearing* in 1943), alluvial soil, September. New Brunswick to Manitoba and southern Idaho, south to Pennsylvania, Texas, northern New Mexico, and Arizona.

The plant is ornamental, and its occurrence in Arizona may be attributable to escape from cultivation.

7. MARAH (311). Big-root, Wild-cucumber

Plants perennial with a very large, tuberlike root; stems climbing; leaves deeply cleft, the lobes triangular or oblong-lanceolate; corolla of the staminate flowers 6 to 10 mm. in diameter, whitish; fruit 2 to 3 cm. in diameter, somewhat fleshy, with stout, smooth spines, bursting irregularly.

The root of a related species, *M. fabacea* (Naud.) Greene, contains 2 glucosides, one cathartic, the other with the property of dilating the pupil of the eye. It is not known whether these constituents are present in the Arizona species.

1. **Marah gilensis** Greene. Greenlee County to Mohave County, south to Pinal and Pima counties, 5,000 feet or lower, common, mostly in thickets along streams, February to April. Southwestern New Mexico and Arizona.

"This wild cucumber sends up succulent shoots or stems very early, sometimes in early March. These shoots grow very rapidly, but in spite of their early appearance appear to be very susceptible to cold. I have known two sets of shoots to be killed down by freezing temperatures in early spring, and the third set to come to perfection later on" (Collom, personal communication).

8. BRANDEGEA

Stems slender, from a thick root, clambering over low shrubs; leaves conspicuously pustulate above, deeply cleft, the lobes narrow, entire or sparingly dentate; staminate flowers in racemes, the pistillate flowers solitary; corolla rotate, deeply 5-lobed, yellowish white; fruit obovoid, long-beaked, sparsely echinate with short, stout spines.

1. **Brandegea Bigelovii** (Wats.) Cogn. Western parts of Maricopa, Pinal, and Pima counties, and in Yuma County, 1,200 feet or lower, not infrequent in sandy soil along washes, January to May (sometimes autumn), type from "the Lower Colorado Valley" (*Bigelow*). Southwestern Arizona, southeastern California, and northwestern Mexico.

The flowers are fragrant.

9. SICYOS. One-seeded bur-cucumber

Plants annual; stems climbing; leaves angulate to deeply cleft; staminate and pistillate inflorescences mostly from the same axils, the staminate ones loose, the pistillate ones dense; corolla of the staminate flowers rotate, whitish; anthers 2 to 5, distinct or united, the filaments united; fruit not fleshy, ovoid, armed with slender, minutely and retrorsely barbed, deciduous spines.

KEY TO THE SPECIES

1. Leaves cleft to the middle or deeper, the lobes oblong, more or less lobulate ... 1. *S. laciniatus*
1. Leaves not cleft to the middle, angulate or very shallowly lobed, the lobes broadly triangular, merely dentate ... 2. *S. ampelophyllus*

1. **Sicyos laciniatus** L. Huachuca Mountains, Cochise County (*Lemmon* in 1882), September. Southeastern Arizona and Mexico.

The Arizona plant belongs to var. *genuinus* Cogn.

2. **Sicyos ampelophyllus** Woot. & Standl. (*S. laciniatus* var. *subinteger* Cogn.). Yavapai, Cochise, and Pima counties (probably elsewhere), 4,000 to 5,500 feet, in shade along streams, August and September. New Mexico and Arizona.

10. SICYOSPERMA

Plants monoecious, annual; stems slender, climbing; leaves thin, angulate or shallowly lobed with broad triangular lobes; inflorescences few-flowered, axillary, the pistillate flowers conspicuously bracteate; staminate corolla rotate, whitish; fruit small, dry, without spines.

1. **Sicyosperma gracile** Gray. Gila, Cochise, Santa Cruz, and Pima counties, 3,500 to 5,500 feet, along streams in partial shade, August and September. Southern Arizona and northern Mexico.

11. CYCLANTHERA

Plants annual, glabrous; stems slender, climbing; leaves mostly pedately compound, the leaflets stalked; flowers of both sexes from the same axils, the staminate flowers in small racemes or panicles, the pistillate flowers solitary; corolla rotate, whitish; fruit dry, narrowly ovoid, somewhat asymmetric, acuminate, 2 to 3 cm. long, armed with long, slender, smooth spines, bursting irregularly.

1. Cyclanthera dissecta (Torr. & Gray) Arn. Santa Rita and Baboquivari mountains (Pima County), 4,000 (to 5,000?) feet, along streams, September and October. Kansas to Louisiana and Texas; southern Arizona and Mexico.

131. CAMPANULACEAE. Bellflower Family

Contributed by Rogers McVaugh

Plants herbaceous or rarely suffrutescent, the juice usually milky; leaves simple, alternate, exstipulate; flowers mostly perfect, normally 5-merous (the carpels 2 to 5, in *Triodanis* the calyx lobes 3 to 5); corolla sympetalous, regular or irregular; stamens distinct or united; style 1; ovary usually inferior or partly so; fruit a capsule; seeds numerous, minute.

KEY TO THE GENERA

1. Corolla regular; anthers and filaments distinct; capsule opening by lateral pores formed by the uplifting of small lids: subfamily Campanuloideae (2).
1. Corolla irregular, usually strongly so; filaments united into a tube; anthers distinct or united; capsule opening apically, by valves or somewhat irregularly: subfamily Lobelioideae (3).
2. Inflorescence loosely branched, at least the terminal flowers of the main axis solitary and long-peduncled; cleistogamous flowers none, all the corollas normally open and expanded .... 1. *Campanula*
2. Inflorescence (at least the upper part) a narrow spike, at least the upper and terminal flowers sessile; cleistogamous flowers usually present (at least the lower flowers), these with vestigial corollas and stamens.................................... 2. *Triodanis*
3. Anthers distinct, all alike; flowers minute, 5 mm. long or less; leaves in a basal rosette; plants of deserts and semideserts, annual.......................... 3. *Nemacladus*
3. Anthers united into a tube, 3 of them longer than the other 2; flowers 10 mm. long or more; leaves mostly cauline; plants of moist situations (4).
4. Corolla tube slit down one side nearly to the base.......................... 4. *Lobelia*
4. Corolla tube entire, not slit down one side............................ 5. *Porterella*

1. CAMPANULA. Bellflower

Plants glabrous or nearly so; upper leaves linear or linear-lanceolate, the basal and lower leaves oblanceolate, spatulate, or ovate, petiolate, sometimes cordate; flowers blue or violet, showy; hypanthium turbinate-obconic; ovary and capsule trilocular.

Several species of this very attractive genus are cultivated as ornamentals.

KEY TO THE SPECIES

1. Calyx lobes entire; mature capsule nodding, the valves at the very base; basal leaves orbicular or nearly so, more or less cordate...................... 1. *C. rotundifolia*
1. Calyx lobes normally with 1 or more callose teeth on each side; mature capsule erect, the valves distinctly above the middle, usually ⅔ to ¾ of the distance from base to apex; basal leaves mostly oblanceolate or obovate, tapering at base........ 2. *C. Parryi*

1. Campanula rotundifolia L. White Mountains (Apache and Greenlee counties), San Francisco Peaks (Coconino County), Pinaleno Mountains (Graham County), 9,000 feet or higher, meadows and rocky slopes, June to September. Boreal regions of North America and Eurasia, south in western North America in the mountains to Nuevo León, eastern Arizona, and northern California.

Harebell, bluebell. An exceedingly widespread and variable species. Numerous segregates have been proposed, but at the present time the most conservative course appears to be to regard them all as phases of one polymorphic species.

2. Campanula Parryi Gray. Lukachukai Mountains (Apache County), Kaibab Plateau to Oak Creek Canyon (Coconino County), 7,000 to 10,000 feet, mountain meadows, July to September. Wyoming and central Utah to northern New Mexico and northern Arizona.

2. TRIODANIS (312)

Plants somewhat hairy; leaves ovate to oblong or suborbicular, sessile or clasping; flowers dimorphic, the earlier ones small and cleistogamous, the later flowers usually petaliferous, showy, purple or violet; hypanthium short-cylindric or elliptic; ovary and capsule regularly trilocular, or sometimes bilocular.

KEY TO THE SPECIES

1. Openings of the capsule linear, 0.2 to 0.4 mm. wide, the elastic cartilage with very narrow, scarious margins; seeds minutely low-tuberculate in longitudinal lines 1. *T. Holzingeri*
1. Openings of the capsule broadly elliptic to rounded, 0.5 to 1.5 mm. wide, the cartilage with broad, rounded, scarious margins; seeds smooth to muriculate (2).
2. Pores at or very near the apex of the capsule; seeds smooth; bracts usually longer than wide, not prominently veined beneath; open corolla usually only 1 and terminal 2. *T. biflora*
2. Pores from 1 mm. below the apex of the capsule to (usually) about midway between base and apex; seeds smooth to muriculate; bracts usually as wide as or wider than long, usually prominently veined beneath; open corollas usually several......3. *T. perfoliata*

1. Triodanis Holzingeri McVaugh (*Specularia Holzingeri* (McVaugh) Fern.). Graham, Cochise, and Pima counties, spring and early summer. Western Missouri to southeastern Wyoming, south to southern Texas; Tennessee; southeastern Arizona.

Formerly confused with *T. perfoliata.*

2. Triodanis biflora (Ruiz & Pavon) Greene (*Specularia biflora* (Ruiz & Pavon) Fisch. & Mey.). Gila County and south, 5,500 feet or lower, April to June. Southern Virginia to Arkansas, southern Arizona, coastal California, and northern Mexico; South America.

3. Triodanis perfoliata (L.). Nieuwl. (*Specularia perfoliata* (L.) A. DC.). Greenlee County to Coconino County and south, 7,500 feet or lower, spring and summer. Southern Ontario to southern British Columbia, south to Florida, Texas, and southwest of the Rockies to northern California, Arizona, and northern Mexico; southern Mexico; West Indies; locally in northern South America.

3. NEMACLADUS (313, 314)

Plants diffusely branched, with delicate, slender stems, 5 to 30 cm. high; cauline leaves reduced to subulate or linear bracts; flowers loosely racemose on all the branches; corolla white or purplish-tinged, more or less bilabiate; ovary and capsule bilocular, or sometimes becoming unilocular.

KEY TO THE SPECIES

1. Capsule 2.3 to 5 mm. long, free from the calyx its entire length; corolla tube 2 to 3 mm. long, usually exceeding the calyx lobes. Corolla (in the Arizona variety) 3 to 3.5 mm. long . . . . 1. *N. longiflorus*
1. Capsule 1.0 to 2.0 (rarely 2.5) mm. long, free from the calyx about half its length; corolla tube not more than 1.5 mm. long, usually exceeded by the calyx lobes (2).
2. Anthers 0.4 to 0.8 mm. long; lower part of the stem silvery gray, shining. Corolla lobes united at the very base only, the lobes much longer than the tube . . . . . . 3. *N. rubescens*
2. Anthers 0.1 to 0.3 (rarely 0.4) mm. long; lower part of the stem purplish or brownish, lacking a silvery-gray sheen (3).
3. Seeds pitted, the pits in about 10 rows of 10 to 12 each; pedicels usually spreading in a graceful double curve; corolla tubular, the tube equaling the lobes or nearly so 2. *N. gracilis*
3. Seeds with distinct, somewhat flattened, longitudinal ridges separated by sharply impressed lines, each ridge divided by 15 to 20 fine transverse lines; pedicels usually somewhat ascending, distinctly stiff; corolla deeply divided, the lobes longer than the tube . . . . 4. *N. glanduliferus*

1. Nemacladus longiflorus Gray. Represented in Arizona by var. *breviflorus* McVaugh, of which the type (*Peebles et al.* 3754) was collected between Tucson and Sells (Pima County), about 2,500 feet, sandy soil. This is the only known Arizona locality.

The variety, as well as the typical phase of the species, occurs most abundantly in the deserts and desert mountains of southern California. The typical phase, with corolla 5.0 to 8.0 mm. long, has not been reported from Arizona.

***2. Nemacladus gracilis** Eastw. This species is included in the flora of Arizona by virtue of a specimen, now in the herbarium of the Academy of Natural Sciences of Philadelphia, labeled "Central Arizona" (*Palmer* 300). The range of the species lies west of the Mojave and Colorado deserts for the most part, and although it is known from Clark County, Nevada, and eastern San Bernardino County, California, there are no authentic modern records of its occurrence in Arizona.

3. Nemacladus rubescens Greene. Western Mohave and Yuma counties, 2,000 feet or lower, dry, gravelly or rocky soil in desert regions, mostly April and May. Nevada, western Arizona, southern California, and Baja California.

A well-marked species, easily distinguished from all others by the silvery-gray stems and the smooth, yellowish-green, nearly entire leaves.

4. Nemacladus glanduliferus Jepson. Throughout Arizona except the northeastern part (absent in Navajo and Apache counties), up to 5,000 feet, sandy deserts and desert mountains, March to May. Southwestern Utah to southern California, south to Sonora and Baja California.

Represented in Arizona by var. *orientalis* McVaugh. This variety is more widely distributed than any other member of the genus. It is characterized by stiffer branches and by usually shorter calyx lobes than in typical *N. glanduliferus*.

4. LOBELIA

Plants erect, more or less strict; inflorescence a single terminal raceme (occasionally with subordinate lateral inflorescences); corolla showy, strongly bilabiate; flowers inverted in anthesis, the pedicel twisted.

Many species of *Lobelia* have beautiful flowers, and several of them are grown as ornamentals.

KEY TO THE SPECIES

1. Corolla normally blue; flower, when straightened, less than 25 mm. long; filament tube 1.5 to 5.0 mm. long (2).
1. Corolla normally red, or red and yellow; flower, when straightened, 25 mm. long or longer; filament tube 18 to 35 mm. long (3).
2. Plants annual or biennial; filament tube 1.5 to 2.3 mm. long, the anthers all densely white-tufted at tip; corolla tube with slits in the sides in addition to the dorsal slit; leaves sharply serrate, broad and often clasping at base....................1. *L. fenestralis*
2. Plants with a perennial rootstock; filament tube 3.5 to 5.0 mm. long, the 2 smaller anthers white-tufted at tip, the 3 larger ones smooth or nearly so; corolla tube entire except for the dorsal slit; cauline leaves shallowly dentate or subentire, narrowed at base, never clasping ....2. *L. anatina*
3. Pedicels elongate, 3 to 10 cm. long in fruit; anther tube yellow or tan, 5 to 9 mm. long 3. *L. laxiflora*
3. Pedicels short, seldom more than 1.5 cm. long in fruit; anther tube bluish gray, 3 to 4 mm. long ....4. *L. Cardinalis*

1. Lobelia fenestralis Cav. Chiricahua and Huachuca mountains (Cochise County), near Patagonia (Santa Cruz County), also a collection by Lemmon labeled as from Oak Creek (Coconino County) but probably collected in southern Arizona, 5,000 to 6,000 feet, meadows and swales, August to November. Western Texas to Arizona, southward to Oaxaca.

2. Lobelia anatina Wimmer. Apache County to Coconino County, south to the mountains of Cochise and Pima counties, 5,500 to 9,000 feet, meadows, marshy places, and stream banks, July to October. Southern New Mexico and Arizona, southward to Durango.

A species quite distinct from *L. gruina* Cav., to which most of the United States material was previously referred.

3. Lobelia laxiflora H.B.K. Bear Valley (Sycamore Canyon), Santa Cruz County (*Gooding* in 1935 and 1936, *Benson* 10950), 4,000 to 5,000 feet, oak woodland, May. Southern Arizona and throughout most of Mexico and Central America.

A polymorphic species. Many so-called species have been segregated from it, but the better course appears to be to regard them all as varieties of the original *L. laxiflora*, the typical phase of which is found in the region of Vera Cruz and southward. The plant of Arizona, at least for the present, is best referred to var. *angustifolia* A.DC.

4. Lobelia Cardinalis L. Throughout most of the state except the low western portion, 3,000 to 7,500 feet, frequent in moist soil, especially along streams, June to October. Widely distributed in the United States, Mexico, and Central America.

Cardinal-flower. Represented in Arizona by subsp. *graminea* (Lam.) McVaugh. Plants of this subspecies had previously been reported as *L. splendens* Willd. or *L. fulgens* Willd.

5. PORTERELLA

Plants annual, diffuse, weak-stemmed; leaves lanceolate to linear, entire or essentially so, the lower leaves often immersed and early deciduous; corolla blue with a yellow eye, showy; flowers inverted in anthesis.

1. Porterella carnosula (Hook. & Arn.) Torr. Coconino County, Mount Agassiz (*Korstian* in 1915) and Fort Valley (*Korstian & Wyman* in 1916, *Fulton* 4373), 7,000 to 9,000 feet, wet meadows, muddy pools, and margins of streams and ponds, June and July. Northern Wyoming to Oregon, south to northern Arizona and northern California.

132. COMPOSITAE. Sunflower Family

Contributed by S. F. Blake

Herbs or shrubs; leaves opposite or alternate, rarely whorled, entire to dissected, never truly compound; flowers borne in a head (this rarely 1-flowered) on a receptacle, surrounded by an involucre of phyllaries (bracts); flowers hermaphrodite, or pistillate, or hermaphrodite with a more or less abortive pistil and functionally staminate, or neutral (with an ovary but lacking a style and stigma); corolla gamopetalous, either regular, tubular, and 5-toothed (rarely 2- to 4-toothed), or bilabiate, or ligulate (flattened, strap-shaped, and usually 2- to 5-toothed), the corolla rarely wanting in the pistillate flowers; stamens (in the hermaphrodite or staminate flowers) almost always 5, united by the anthers or rarely only by the filaments, inserted on the corolla; ovary inferior, 1-celled, with an erect anatropous ovule; style normally 2-branched, the branches stigmatiferous inside, often bearing sterile appendages at apex, the style in functionally staminate flowers often undivided; fruit an achene, with a single, erect, exalbuminous seed, usually bearing a pappus of bristles, of awns, or of scales (paleae or squamellae), this often considered to represent the calyx limb.

The family comprises several well-known vegetables, notably lettuce (*Lactuca sativa*), artichoke (*Cynara Scolymus*), salsify (*Tragopogon porrifolius*), endive (*Cichorium Endivia*), and Jerusalem-artichoke (*Helianthus tuberosus*). Species of goldenrod (*Solidago*), rabbit-brush (*Chrysothamnus*), and Colorado rubber plant (*Hymenoxys*), also of *Aplopappus* and *Guardiola*, contain appreciable quantities of rubber, but commercial exploitation of these plants has not yet been found practicable. Guayule (*Parthenium argentatum*) has been exploited extensively in the wild state in Mexico and is grown to a limited extent in the United States as a source of rubber. The family is rich in oils, resins, and bitter principles, and many species are used in medicine. The well-known insecticide pyrethrum is furnished by the powdered flowers of a few species of *Chrysanthemum*. Many of the family are browsed by grazing animals, but their palatability is generally low because of their resinous or acrid properties. A few species are known to be poisonous to livestock.

Compositae that have wind-blown pollen, especially the ragweeds and their allies, are among the most important causes of hay fever. The ragweeds (*Ambrosia* spp.), so virulent in other parts of the United States, are not sufficiently numerous in Arizona to rank high among the hay-fever plants of the state, their place being taken by the widely distributed and very abundant bur-sages (species of *Franseria*). The wormwoods and sagebrushes (*Artemisia* spp.) probably rank next in importance, especially in northern Arizona. Species of *Dicoria, Hymenoclea,* and *Helianthus* are also suspected of causing this disease.

The flower heads are frequently showy, often beautiful, and Compositae of many genera are in cultivation as ornamentals. Of these, species of *Aster, Calendula, Callistephus* (China-aster), *Chrysanthemum, Coreopsis, Cosmos, Dahlia, Helianthus* (sunflower), *Tagetes,* and *Zinnia* are especially well known.

KEY TO THE PRINCIPAL DIVISIONS

1. Corollas of some or all of the flowers distinctly bilabiate..........................A.
1. Corollas not or obscurely bilabiate (2).
2. Flowers all hermaphrodite (perfect) and with strap-shaped 5-toothed corollas.......B.
2. Flowers, when hermaphrodite, with the corolla tubular, and regular or nearly so. Marginal flowers of the head often pistillate or neutral and often with strap-shaped, 2- or 3-toothed corollas (3).
3. Rays (strap-shaped corollas) none or the ligule vestigial (4).
3. Rays present but sometimes small (6).
4. Pappus none or vestigial...C.
4. Pappus evident on some or all of the achenes (5).
5. Achenes with a pappus of awns or scales (paleae, squamellae) or both, these sometimes united into a low, chaffy crown...D.
5. Achenes with a pappus of capillary bristles, rarely with additional outer scales......E.
6. Pappus of capillary bristles, rarely with a few short outer scales....................F.
6. Pappus none, or of awns or scales (7).
7. Achenes without pappus, or the pappus vestigial.................................G.
7. Achenes (some or all of them) with a pappus of awns, or scales, or both, these separate or united into a crown..H.

KEYS TO THE GENERA

A. Corollas of some or all of the flowers bilabiate.

1. Flowers at the margin of the head pistillate and with strap-shaped 3-toothed corollas, the other flowers perfect and with bilabiate corollas; plants scapose....128. *Chaptalia*
1. Flowers all perfect and with bilabiate corollas; stems leafy (2).
2. Plants herbaceous; corollas pink or whitish; involucre strongly graduated..129. *Perezia*
2. Plants shrubby; corollas yellow; involucre in only 2 distinct series.........130. *Trixis*

B. Flowers all hermaphrodite and with strap-shaped 5-toothed corollas.

1. Achenes without pappus...132. *Atrichoseris*
1. Achenes with pappus (2).
2. Pappus, at least in part, of plumose bristles (3).
2. Pappus of nonplumose bristles or of awns or scales (7).
3. Phyllaries in several strongly graduated series, very obtuse, broadly scarious-margined; receptacle chaffy. Flowers yellow...............................135. *Anisocoma*
3. Phyllaries not in several graduated series but often with a much shorter outer series, not, or narrowly, scarious-margined; receptacle naked (4).
4. Achenes truncate at apex. Plants more or less rushlike; corollas pink 137. *Stephanomeria*
4. Achenes tapering or beaked at apex (5).
5. Plants scapose; leaves hispid. Corollas yellow.......................136. *Leontodon*
5. Plants leafy-stemmed; leaves not hispid (6).

6. Leaves pinnatifid; corollas white or pinkish; involucre with an outer series of short bractlets .... 138. *Rafinesquia*
6. Leaves grasslike, entire; corollas yellow or violet-purple; involucre without short outer bractlets .... 139. *Tragopogon*
7. Pappus, at least in part, of awns or scales (8).
7. Pappus of capillary bristles only (10).
8. Corollas blue. Plants with leafy, branched stems .... 131. *Cichorium*
8. Corollas yellow or orange (9).
9. Plants scapose or subscapose; pappus simple, of bifid paleae tipped with a bristle 133. *Microseris*
9. Plants caulescent, the stems few-leaved; pappus double, the outer series of thin, short scales, the inner series of much longer bristles .... 134. *Krigia*
10. Achenes more or less flattened. Stems leafy; heads in panicles (11).
10. Achenes not flattened (12).
11. Involucre campanulate or hemispheric; achenes not beaked .... 145. *Sonchus*
11. Involucre cylindric or ovoid-cylindric; achenes beaked .... 146. *Lactuca*
12. Achenes not beaked (13).
12. Achenes beaked (16).
13. Pappus quickly deciduous, or with 1 to 8 stiff, persistent bristles. Plants annual 141. *Malacothrix*
13. Pappus persistent (14).
14. Plants rushlike or spinescent; corollas pink .... 147. *Lygodesmia*
14. Plants not rushlike or spinescent; corollas yellow, rarely white or flesh-colored (15).
15. Phyllaries somewhat thickened at base or on the midrib; pappus white ...... 150. *Crepis*
15. Phyllaries not thickened; pappus dingy or brownish, or sometimes white 151. *Hieracium*
16. Pappus deciduous at maturity of the achene (17).
16. Pappus persistent (18).
17. Leaves not crustaceous-margined; plants conspicuously stipitate-glandular above; achenes tapering into the beak, not transversely rugulose .... 142. *Calycoseris*
17. Leaves crustaceous-margined; plants not stipitate-glandular; achenes abruptly beaked, transversely rugulose between the ribs .... 143. *Glyptopleura*
18. Achenes 10- to 15-ribbed or -nerved, not spinulose-muricate; phyllaries imbricate in several graduated series (19).
18. Achenes 4- or 5-ribbed, spinulose-muricate above; phyllaries in 2 unequal series, the outer ones much shorter (20).
19. Corollas rosy or whitish; phyllaries with conspicuous blackish or brownish scarious tips; pappus brownish .... 140. *Pinaropappus*
19. Corollas yellow or orange, sometimes turning purplish in age; phyllaries without dark scarious tips; pappus white or whitish .... 148. *Agoseris*
20. Plants scapose, the scapes naked, 1-headed; pappus whitish, without a woolly ring at base .... 144. *Taraxacum*
20. Plants caulescent, the stems more or less leafy and branched; pappus brownish, with a woolly ring at base .... 149. *Pyrrhopappus*

C. Hermaphrodite flowers with a tubular, regular or nearly regular corolla; rays and pappus none or vestigial.

1. Heads unisexual, monoecious; pistillate heads with 1 to 4 flowers enclosed in a nutlike or burlike involucre, only the style tips exserted (2).
1. Heads not unisexual; involucre not nutlike or burlike (5).
2. Phyllaries of the staminate heads separate. Fruiting involucres burlike, covered with hooked prickles .... 51. *Xanthium*
2. Phyllaries of the staminate heads united (3).
3. Pistillate involucre with several transverse scarious wings; leaves or their lobes linear-filiform .... 48. *Hymenoclea*
3. Pistillate involucre without transverse wings; leaves and their lobes not linear-filiform (4).

4. Fruiting involucre unarmed or with a few teeth or tubercles in a single series below the beak . . . . 49. *Ambrosia*
4. Fruiting involucre with several or many spines in more than 1 series. . . . . . 50. *Franseria*
5. Flowers all hermaphrodite (6).
5. Flowers not all hermaphrodite, the outer ones pistillate and fertile, the inner ones hermaphrodite but often sterile (14).
6. Heads 1- or 2-flowered, aggregated in dense glomerules (7).
6. Heads many-flowered, not in glomerules (8).
7. Phyllaries of the individual heads connate into a toothed tube; leaves ovate, petioled 38. *Lagascea*
7. Phyllaries of the individual heads separate; leaves lanceolate, sessile or subsessile 88. *Flaveria*
8. Receptacle chaffy (9).
8. Receptacle not chaffy (11).
9. Inner phyllaries united to the middle or higher into a cup; achenes thickish, papillate 74. *Thelesperma*
9. Inner phyllaries not united; achenes strongly compressed, very flat, not papillate (10).
10. Achenes strongly ciliate on the margin. . . . . 64. *Encelia*
10. Achenes not evidently ciliate. . . . . 65. *Simsia*
11. Achenes strongly compressed. . . . . 85. *Laphamia*
11. Achenes plump (12).
12. Phyllaries with conspicuous whitish or yellow tips and margins; achenes strongly 4- or 5-angled, usually hirsute on the angles. . . . . 91. *Hymenopappus*
12. Phyllaries not with conspicuous whitish or yellowish tips and margins; achenes otherwise (13).
13. Heads solitary or cymose. . . . . 109. *Matricaria*
13. Heads in spikes or racemes, these often panicled. . . . . 112. *Artemisia*
14. Plants woolly, annual, dwarf (15).
14. Plants not woolly or if so, then perennial (17).
15. Bracts subtending the pistillate flowers completely enclosing them and tipped with a hyaline appendage . . . . 32. *Stylocline*
15. Bracts subtending the pistillate flowers more or less open, not completely enclosing the flowers (16).
16. Pales of the receptacle flattish. . . . . 33. *Evax*
16. Pales of the receptacle, at least the outer ones, boat-shaped. . . . . 34. *Filago*
17. Receptacle not chaffy (18).
17. Receptacle chaffy, at least toward the margin (19).
18. Heads few, solitary at the ends of the stems and branches; phyllaries 2-seriate, approximately equal; outer flowers without corollas; achenes obcompressed, margined 111. *Cotula*
18. Heads usually numerous, in spikes, racemes, or panicles; phyllaries 3- or 4-seriate, the outer ones shorter; outer flowers with corollas; achenes not obcompressed or margined . . . . 112. *Artemisia*
19. Achenes densely long-villous; leaves or their lobes linear-filiform. . . . . 46. *Oxytenia*
19. Achenes not long-villous; leaves or their lobes not linear-filiform (20).
20. Achenes with pectinate or toothed wings. . . . . 47. *Dicoria*
20. Achenes not with pectinate or toothed wings (21).
21. Pistillate flowers 6 to 8; achenes with an acute margin and a terminal apiculation, falling away at maturity with the pales of the opposed disk flowers. . . . 44. *Parthenice*
21. Pistillate flowers 5; achenes otherwise. . . . . 45. *Iva*

D. Hermaphrodite flowers with a tubular, regular or nearly regular corolla; rays none or vestigial; pappus present, of awns or scales or both, these sometimes united into a low, chaffy crown.

1. Receptacle chaffy (2).
1. Receptacle not chaffy, sometimes setose or fimbrillate (8)

2. Pappus of about 15 to 20 plumose awns....77. *Bebbia*
2. Pappus not of plumose awns (3).
3. Pappus of numerous flattened bristles....30. *Baccharis*
3. Pappus of 1 to 4 teeth, squamellae, awns, or paleae (4).
4. Achenes with pectinate or toothed wings....47. *Dicoria*
4. Achenes not with pectinate or toothed wings (5).
5. Awns or teeth of the pappus not retrorsely barbed; plants shrubby (6)
5. Awns or teeth retrorsely barbed; plants herbaceous (7).
6. Achenes somewhat thickened, not notably ciliate on the margin....63. *Flourensia*
6. Achenes very flat, conspicuously ciliate on the margin....64. *Encelia*
7. Inner phyllaries united about to the middle, forming a cup....74. *Thelesperma*
7. Inner phyllaries not united....75. *Bidens*
8. Heads 1-flowered, crowded in dense glomerules. Involucres of the individual heads tubular, 5- or 6-toothed at tip....38. *Lagascea*
8. Heads several- or many-flowered (9).
9. Plants strictly dioecious, the heads on some plants entirely pistillate, on others with hermaphrodite but sterile and functionally staminate flowers....30. *Baccharis*
9. Plants not dioecious (10).
10. Receptacle densely setose all over; thistlelike herbs (11).
10. Receptacle naked, fimbrillate, or sparsely setose; plants not thistlelike (12).
11. Pappus of numerous slender, plumose paleae, united at base and deciduous in a ring 124. *Cynara*
11. Pappus of numerous setae or paleae, not united at base, separately deciduous 126. *Centaurea*
12. Pappus of 2 to 8 caducous awns. Plants usually strongly glutinous....10. *Grindelia*
12. Pappus otherwise (13).
13. Pappus of numerous graduated bristles, the inner ones somewhat flattened and paleaceous. Low shrubs, with crowded, subterete, impressed-punctate leaves 115. *Peucephyllum*
13. Pappus otherwise (14).
14. Leaves and involucre conspicuously punctate with translucent oil glands (15).
14. Leaves and involucre sometimes impressed-punctate, but not with translucent oil glands (16).
15. Pappus of 3 to 6 entire paleae; phyllaries strictly 1-seriate, united essentially to the apex into a toothed cup or tube....103. *Tagetes*
15. Pappus of paleae dissected into bristles, or else of 10 paleae or squamellae; phyllaries more or less distinctly 2-seriate....104. *Dyssodia*
16. Pappus of 12 or more paleae, these nearly or quite as long as the achene (17).
16. Pappus of fewer than 12 paleae or squamellae, or else these much shorter than the achene (20).
17. Herbs; involucre not conspicuously graduated....92. *Hymenothrix*
17. Shrubs; involucre conspicuously graduated (18).
18. Leaves opposite; corollas purple-tinged....5. *Carphochaete*
18. Leaves alternate; corollas yellow (19).
19. Plants very viscid; paleae of the pappus 12 to 16, essentially equal....11. *Vanclevea*
19. Plants not viscid; paleae and bristles of the pappus more numerous, distinctly graduated 14. *Acamptopappus*
20. Heads 5-flowered; pappus of 3 to 5 awns, with or without as many short squamellae, or reduced to a toothed crown....1. *Stevia*
20. Heads with more numerous flowers, or else the pappus otherwise (21).
21. Achenes strongly compressed; pappus of 1 or 2 slender awns and often with a crown of squamellae, or of a crown of lacerate squamellae only (22).
21. Achenes not compressed, or else the pappus otherwise (24).
22. Achenes not evidently ciliate; pappus of 1 or 2 awns....85. *Laphamia*
22. Achenes strongly ciliate on the margin; pappus of a crown of squamellae and often 1 or 2 awns (23).

23. Heads radiate; leaves not hastate-triangular or caudate-acuminate.......84. *Perityle*
23. Heads discoid; leaves hastate-triangular, caudate-acuminate............86. *Pericome*
24. Pappus a low crown (25).
24. Pappus otherwise (26).
25. Plants annual, pineapple-scented; leaves twice or thrice pinnatisect; phyllaries subequal 109. *Matricaria*
25. Plants, shrubby, not pineapple-scented; leaves once pinnatifid; phyllaries somewhat graduated ....113. *Pentzia*
26. Heads mostly 4- to 6-flowered, about 2.5 cm. high....................5. *Carphochaete*
26. Heads with more numerous flowers, or else much smaller (27).
27. Pappus a crown of short, dissected squamellae, or of 5 paleae dissected into bristles (28).
27. Pappus otherwise (29).
28. Glabrous annuals; phyllaries chartaceous, green-tipped; pappus a low, dissected crown 22. *Greenella*
28. Woolly annuals; phyllaries not chartaceous; pappus of 5 paleae dissected into numerous bristles ....98. *Trichoptilium*
29. Phyllaries with a thin, scarious, white, yellow, or purplish margin and tip (30).
29. Phyllaries not with a scarious, colored margin and tip (32).
30. Plants tomentose; pappus of 10 or more paleae or squamellae......91. *Hymenopappus*
30. Plants not tomentose; pappus of 8 paleae or squamellae (31).
31. Heads with fewer than 10 flowers; phyllaries impressed-punctate........90. *Schkuhria*
31. Heads with more than 10 flowers; phyllaries not impressed-punctate........97. *Bahia*
32. Leaves decurrent; corollas fuscous; pappus squamellae about 5, obtuse, much shorter than the achene..100. *Helenium*
32. Leaves not decurrent; corollas not fuscous; pappus paleae or squamellae usually more numerous or longer (33).
33. Corollas yellow ....94. *Eriophyllum*
33. Corollas white, flesh-colored, or purplish (34).
34. Plants scapose, with roundish entire or crenate leaves...........96. *Chamaechaenactis*
34. Plants leafy-stemmed; leaves not roundish, nor entire or subentire (35).
35. Pappus paleae with a strong midrib; leaves lanceolate or linear, entire...93. *Palafoxia*
35. Pappus paleae nerveless or essentially so; leaves, at least in part, toothed to pinnatifid 95. *Chaenactis*

E. Hermaphrodite flowers with a tubular, regular or nearly regular corolla; rays none or vestigial; pappus present, usually of capillary or flattened bristles, rarely paleaceous.

1. Receptacle densely bristly (2).
1. Receptacle naked or chaffy (8).
2. Leaf margins prickly or spiny; pappus bristles united at base and deciduous in a ring (3).
2. Leaf margins not prickly or spiny; pappus bristles not united at base nor deciduous in a ring (6).
3. Pappus bristles merely barbellate (4).
3. Pappus bristles plumose (5).
4. Appendages of the phyllaries entire; leaves not blotched................122. *Carduus*
4. Appendages of the phyllaries spinulose-margined; leaves blotched with white 125. *Silybum*
5. Phyllaries comparatively narrow, at least the outer ones spine-tipped.....123. *Cirsium*
5. Phyllaries broad (about 1 cm. wide), not spine-tipped..................124. *Cynara*
6. Leaves broadly ovate, entire or shallowly dentate, often cordate, all petioled; achenes basally attached. Large coarse herbs with very large basal leaves; involucre burlike, the numerous phyllaries narrow, rigid, ending in a hooked spine.......121. *Arctium*
6. Leaves linear to oblong-lanceolate, sometimes pinnatifid, never cordate, the stem leaves sessile; achenes obliquely attached (7).

7. Achenes compressed or quadrangular, not toothed at apex; pappus not definitely 2-seriate, either setose, or paleaceous, or wanting..................126. *Centaurea*
7. Achenes terete, many-striate, 10-toothed at apex; pappus definitely 2-seriate, the stiff outer bristles longer and naked, the short inner ones fimbriolate. Plant annual, the stems low and branched; heads almost concealed by the outer leafy bracts; corollas yellow ....127. *Cnicus*
8. Phyllaries scarious or hyaline or, in genera *Baccharis* and *Pluchea*, only partly so (9).
8. Phyllaries herbaceous, at least in the center (15).
9. Heads unisexual; plants dioecious (10).
9. Heads with the marginal flowers pistillate, the central flowers perfect (12).
10. Plants large, not tomentose, shrubby, at least at base; heads numerous, in panicles; phyllaries chartaceous, scarious-margined..........................30. *Baccharis*
10. Plants small, tomentose, strictly herbaceous; heads few, in short racemes or corymbs, these sometimes panicled, or the heads rarely solitary; phyllaries strictly scarious (11).
11. Plants strictly dioecious, low, the basal leaves in a rosette, the stem leaves reduced 35. *Antennaria*
11. Plants subdioecious, the pistillate heads usually with a few hermaphrodite flowers in the center; plants usually at least 30 cm. high, without a basal rosette..36. *Anaphalis*
12. Receptacle chaffy except in the center. Plants small, woolly................34. *Filago*
12. Receptacle naked (13).
13. Phyllaries subscarious; corollas purplish; plants not tomentose...........31. *Pluchea*
13. Phyllaries scarious; corollas rarely purplish; plants tomentose (14).
14. Plants subdioecious, the pistillate heads usually with a few central, hermaphrodite flowers ....36. *Anaphalis*
14. Plants not dioecious, the heads all alike and heterogamous............37. *Gnaphalium*
15. Heads unisexual. Plants dioecious...................................30. *Baccharis*
15. Heads not unisexual (16).
16. Plants winter annual, low, depressed, scurfy-pubescent. Leaves broadly ovate or suborbicular ....116. *Psathyrotes*
16. Plants perennial or, if annual, then not low and scurfy-pubescent (17).
17. Phyllaries and leaves bearing conspicuous translucent oil glands (18).
17. Phyllaries and leaves sometimes impressed-glandular but not with translucent oil glands (19).
18. Phyllaries relatively numerous, usually in more than 1 series and with some bractlets at base; pappus bristles basally united in groups....................104. *Dyssodia*
18. Phyllaries few (5 to 8), free, equal, strictly 1-seriate; pappus bristles free, numerous 105. *Porophyllum*
19. Phyllaries proper 4 to 7, in a single series, of equal length (20).
19. Phyllaries more than seven (21).
20. Plants herbaceous; leaves very large, decompound.....................117. *Cacalia*
20. Plants shrubby; leaves narrow, entire..............................118. *Tetradymia*
21. Pappus bristles plumose (22).
21. Pappus bristles not plumose (25).
22. Annual herbs, with deltoid-ovate leaves and cylindric heads, these in a slender, leafless, virgate panicle ....3. *Carminatia*
22. Perennial herbs, or shrubs; leaves and inflorescence otherwise (23).
23. Intricately branched shrubs; corollas yellow...........................77. *Bebbia*
23. Perennial herbs, or shrubby at base, not much branched; corollas whitish (24).
24. Leaves lanceolate or ovate, 3-nerved; heads 9- to 12-flowered.............6. *Brickellia*
24. Leaves narrowly linear or linear-lanceolate, commonly 1-nerved; heads many-flowered 7. *Kuhnia*
25. Plants shrubby; leaves of the branches reduced to scales; involucre strongly graduated, the phyllaries not in vertical ranks (26).
25. Plants otherwise in habit and foliage, or else the phyllaries in distinct vertical ranks (27).
26. Phyllaries lanceolate, acute; achenes silky-pubescent......................27. *Aster*

26. Phyllaries ovate to oblong, obtuse; achenes glabrous............113a. *Lepidospartum*
27. Shrubs; leaves crowded, linear-filiform, subterete, impressed-punctate; phyllaries nearly in 1 series and of equal length, subulate, herbaceous.............115. *Peucephyllum*
27. Herbs or, if shrubby and with subterete, impressed-punctate leaves, then the phyllaries not equal, nor subulate, nor herbaceous (28).
28. Pappus double, the outer series of short scales, the inner series of capillary bristles; white-barked shrubs. Leaves small, the petioles much longer than the blades 2. *Hofmeisteria*
28. Pappus simple, or else the plants herbaceous (29).
29. Heads with the outer flowers pistillate and the central flowers hermaphrodite (30).
29. Heads with all of the flowers hermaphrodite (31).
30. Phyllaries subequal; corollas whitish....................................29. *Conyza*
30. Phyllaries distinctly graduated; corollas purplish.......................31. *Pluchea*
31. Pappus of 2 to 8 caducous, bristlelike awns. Herbs, more or less viscid....10. *Grindelia*
31. Pappus bristles more numerous, not caducous (32).
32. Achenes 5-angled or 5-ribbed; corollas white, pink, blue, or purple. Leaves often opposite[1] ....4. *Eupatorium*
32. Achenes not 5-angled or 5-ribbed, or else the corollas yellow (33).
33. Pappus of 5 paleae dissected above into bristles...................98. *Trichoptilium*
33. Pappus otherwise (34).
34. Outer corollas enlarged and palmate. Low annuals, tomentose below, glandular above 20. *Lessingia*
34. Outer corollas not enlarged (35).
35. Achenes 10-ribbed; involucre usually strongly graduated, the phyllaries striate; leaves often opposite; corollas not yellow................................6. *Brickellia*
35. Achenes not 10-ribbed; phyllaries not striate; leaves alternate; corollas nearly always yellow (36).
36. Phyllaries in a single series of equal length, or with a few much shorter outer bractlets; style tips truncate..119. *Senecio*
36. Phyllaries more or less unequal and imbricate, in more than 1 principal series; style tips not truncate (37).
37. Phyllaries in more or less distinct vertical ranks.................19. *Chrysothamnus*
37. Phyllaries not in vertical ranks (38).
38. Plants woody, or else the leaves spinulose-toothed; phyllaries in 2 or more graduated series, often closely imbricate...................................18. *Aplopappus*
38. Plants herbaceous, the leaves not spinulose-toothed; phyllaries usually subequal, scarcely imbricate ....28. *Erigeron*

F. Hermaphrodite flowers with a tubular, regular or nearly regular corolla; rays evident but sometimes small; pappus of bristles, these mostly capillary, rarely with a few short outer scales.

1. Rays white, pink, purple, or violet (2).
1. Rays yellow or orange (8).
2. Leaves and involucre marked with translucent oil glands...............104. *Dyssodia*
2. Leaves and involucre without translucent oil glands (3).
3. Pappus of 1 or 2 bristlelike awns, these not plumose...................85. *Laphamia*
3. Pappus bristles more numerous, or else single and plumose (4).
4. Ray achenes enveloped by the subtending phyllaries; pappus of the disk achenes of 10 stout, hairy bristles, the hairs on their outer side straight, on the inner side entangled into a woolly mass..81. *Layia*
4. Ray achenes not enveloped by the subtending phyllaries; pappus otherwise (5).
5. Ray flowers with pappus none or vestigial..........................24. *Psilactis*
5. Ray flowers with evident pappus (6).

[1] *Brickellia floribunda,* which might also run down to this point in the key, may be distinguished from any native species of *Eupatorium* by its combination of rather large heads (9 to 12 mm. high) and strongly graduated involucre of obtuse or obtusish phyllaries.

6. Plants dwarf, hispid-hirsute, winter annual; upper leaves closely subtending the heads; pappus of a single plumose bristle and a scarious cup, or of numerous unequal bristles or narrow paleae..25. *Monoptilon*
6. Plants otherwise in habit, or in the pappus (7).
7. Style tips lanceolate or subulate, acute or acuminate; phyllaries usually strongly graduated, often partly herbaceous; rays mostly relatively broad..............27. *Aster*
7. Style tips deltoid, obtuse or rounded; phyllaries usually equal or little graduated; rays mostly very narrow..28. *Erigeron*
8. Pappus of squamellae or paleae dissected into bristles above, but entire at base (9).
8. Pappus otherwise (10).
9. Plants floccose-woolly, annual................................89. *Syntrichopappus*
9. Plants not floccose-woolly......................................104. *Dyssodia*
10. Leaves opposite, at least below (11).
10. Leaves alternate (14).
11. Involucre and leaves with translucent oil glands; leaf margins with a few stiff bristles near the base..106. *Pectis*
11. Involucre and leaves without translucent oil glands; leaves not bristly at base (12).
12. Heads large; pappus of numerous bristles............................114. *Arnica*
12. Heads small; pappus otherwise (13).
13. Rays persistent on the achenes; pappus of 3 short awns or tubercles......53. *Sanvitalia*
13. Rays deciduous; pappus of 1 or 2 bristlelike awns, or wanting..........85. *Laphamia*
14. Pappus of 2 to 8 caducous bristlelike awns; plants glutinous.............10. *Grindelia*
14. Pappus bristles more numerous or persistent, or else the plants not glutinous (15).
15. Pappus wholly of numerous simple and similar capillary bristles (16).
15. Pappus not wholly of numerous simple and similar capillary bristles (19).
16. Phyllaries proper 1-seriate, equal, sometimes with some short outer bractlets; style tips truncate (17).
16. Phyllaries in more than 1 series, usually more or less unequal and graduated; style tips not truncate (18).
17. Plants herbaceous, or a few species shrubby; involucre often calyculate, the proper phyllaries distinct or nearly so; pappus bristles somewhat persistent, seldom barbellate 119. *Senecio*
17. Plants shrubby; involucre not calyculate, the phyllaries united toward base; pappus bristles caducous, conspicuously barbellate or short-plumose..........120. *Euryops*
18. Heads usually small and very numerous, panicled or cymose; phyllaries rarely distinctly herbaceous at apex; plants always herbaceous......................17. *Solidago*
18. Heads usually few and relatively large, if small and panicled, then the plants shrubby; phyllaries often distinctly herbaceous at apex......................18. *Aplopappus*
19. Shrubs; heads small, few-flowered, crowded in small rounded terminal clusters; rays 1 or 2, small..13. *Amphipappus*
19. Herbs; inflorescence otherwise; rays more numerous, conspicuous (20).
20. Achenes dissimilar, those of the rays essentially glabrous and epappose, the disk achenes hairy, their pappus double, the inner series of capillary bristles, the outer series of short bristles or squamellae..................................15. *Heterotheca*
20. Achenes all consimilar and with a consimilar (double) pappus..........16. *Chrysopsis*

G. Hermaphrodite flowers with a tubular, regular or nearly regular corolla; rays evident but sometimes small; pappus none or vestigial.

1. Rays white or exceptionally blue, sometimes with a yellow base (2).
1. Rays yellow, sometimes partly purple or maroon (13).
2. Receptacle naked (3).
2. Receptacle paleaceous (6).
3. Receptacle broad and flattish; phyllaries with a dark-brown submarginal line 110. *Chrysanthemum*
3. Receptacle convex, conic or hemispheric; phyllaries not with a dark-brown submarginal line (4).

4. Achenes oblique, ribbed only on the inner side; leaves dissected into linear-filiform lobes; phyllaries oblong or oval....109. *Matricaria*
4. Achenes prismatic or subterete, ribbed on all sides, or somewhat flattened and 2-nerved; leaves entire to pinnatifid; phyllaries lanceolate (5).
5. Achenes prismatic or subterete, ribbed on all sides....21. *Aphanostephus*
5. Achenes somewhat flattened, 2-nerved....23. *Achaetogeron*
6. Rays sessile, persistent on the achenes, becoming indurated (7).
6. Rays not sessile and persistent on the achenes (8).
7. Involucre strongly graduated; pales of the receptacle not cuspidate....52. *Zinnia*
7. Involucre not distinctly graduated; pales of the receptacle cuspidate....53. *Sanvitalia*
8. Leaves opposite (9).
8. Leaves alternate (11).
9. Ray achenes tightly and completely enclosed by the subtending pales..40. *Melampodium*
9. Ray achenes not tightly and completely enclosed by the subtending pales (10).
10. Rays 1 to 5; leaves roundish-ovate; anthers green....39. *Guardiola*
10. Rays very numerous, small; leaves lanceolate; anthers blackish....55. *Eclipta*
11. Heads very small, numerous, in dense flattish or rounded cymose panicles..108. *Achillea*
11. Heads relatively large, solitary or few (12).
12. Receptacle conic, bearing stiff, narrow acuminate pales above, naked below 107. *Anthemis*
12. Receptacle convex, with broad, blunt, membranaceous pales throughout 108a. *Leucampyx*
13. Receptacle not chaffy (14).
13. Receptacle chaffy, at least toward the margin (23).
14. Phyllaries graduated in several unequal series, closely imbricate (15).
14. Phyllaries in 1 or 2 series, or if pluriseriate, not graduated in length (17).
15. Rays 9 or fewer; heads small, very numerous, in dense, rounded, terminal clusters; involucre ovoid or oblong....8. *Selloa*
15. Rays 12 or more; heads relatively large and few, scattered at the ends of the branchlets; involucre campanulate or hemispheric (16).
16. Leaves entire; phyllaries with narrow herbaceous tips....9. *Xanthocephalum*
16. Leaves pinnately dissected; phyllaries with broad scarious tips....110. *Chrysanthemum*
17. Phyllaries of the outer series about 4, united to the middle or higher into a cup. Heads small, numerous, in terminal cymose panicles....102. *Plummera*
17. Phyllaries separate or essentially so (18).
18. Heads 1- or 2-flowered, in dense, glomerate clusters, these sessile in the forks of the stem, or terminal and leafy-involucrate....88. *Flaveria*
18. Heads several- to many-flowered, solitary on terminal peduncles (19).
19. Plants woolly (20).
19. Plants not woolly (21).
20. Rays persistent, becoming papery....83. *Baileya*
20. Rays not persistent....94. *Eriophyllum*
21. Involucre and leaves with translucent oil glands....106. *Pectis*
21. Involucre and leaves not with translucent oil glands (22).
22. Leaves linear, entire; receptacle conic....87. *Baeria*
22. Leaves broad, dissected; receptacle flat....97. *Bahia*
23. Rays pistillate and fertile; disk flowers hermaphrodite but sterile, with a nearly or quite undivided style and an abortive ovary (24).
23. Rays and disk flowers all fertile (27).
24. Leaves with spinescent lobes and tip, alternate; heads involucrate by spinescent bracts 80. *Hemizonia*
24. Leaves not with spinescent lobes or tip; heads not involucrate by spinescent bracts (25).
25. Ray-achenes not enwrapped by the subtending phyllaries, but falling in connection with them and with the pales of the opposed outer disk flowers. Leaves alternate 41. *Berlandiera*
25. Ray achenes closely enwrapped by the subtending phyllaries (26).
26. Leaves opposite; phyllaries in 2 series; plants not heavy-scented......40. *Melampodium*

26. Leaves alternate, at least the upper ones; phyllaries in 1 series; plants heavy-scented 79. *Madia*
27. Rays sessile and persistent on their achenes, becoming papery (28).
27. Rays not sessile, nor persistent, nor becoming papery (29).
28. Achenes very strongly flattened; phyllaries dry, strongly graduated.........52. *Zinnia*
28. Achenes plump; phyllaries herbaceous at tip, scarcely or not graduated...54. *Heliopsis*
29. Involucre distinctly double, the outer phyllaries herbaceous, the inner ones broader, longer, membranaceous (30).
29. Involucre not double (31).
30. Inner phyllaries free...71. *Coreopsis*
30. Inner phyllaries connate about to the middle or higher................74. *Thelesperma*
31. Plants scapose perennials, with broad, silvery-pubescent, entire leaves and very large, solitary heads...67. *Enceliopsis*
31. Plants leafy-stemmed; leaves and heads otherwise (32).
32. Achenes conspicuously ciliate on the margin, notched at apex, very flat. Plants shrubby 64. *Encelia*
32. Achenes not conspicuously ciliate on the margin (33).
33. Achenes 2-winged ....70. *Verbesina*
33. Achenes not 2-winged, sometimes acutely margined (34).
34. Leaves pinnately parted or dissected.................................57. *Ratibida*
34. Leaves entire to merely toothed or 3-lobed (35).
35. Achenes very flat, notched at apex.....................................65. *Simsia*
35. Achenes more or less thickened, not notched at apex (36).
36. Leaves alternate, ovate, often 3-lobed.................................58. *Zaluzania*
36. Leaves opposite, at least below, linear to lance-ovate, not lobed (37).
37. Ray achenes not enwrapped by the subtending phyllaries................61. *Viguiera*
37. Ray achenes closely enwrapped by the subtending phyllaries...............79. *Madia*

H. Hermaphrodite flowers with a tubular, regular or nearly regular corolla; rays evident but sometimes small; pappus present, of awns, squamellae, or paleae, these sometimes united into a crown.

1. Receptacle paleaceous (2).
1. Receptacle naked or rarely fimbrillate, not paleaceous (31)
2. Rays white, pink, or purple (3).
2. Rays yellow, or partly purple-brown (10).
3. Rays fertile, the disk flowers sterile. Achenes with a narrow callose margin, this adnate at base to the pales of the 2 opposed disk flowers and to the subtending phyllary, at length tearing away from the achene below but remaining attached at apex 43. *Parthenium*
3. Rays and disk flowers all fertile (4).
4. Pappus of 10 bristlelike, hairy awns, the hairs on their outer side straight, on the inner side entangled into a woolly mass.......................................81. *Layia*
4. Pappus otherwise (5).
5. Rays sessile and persistent on their achenes (6).
5. Rays not sessile and persistent on their achenes (7).
6. Involucre strongly graduated; pales of the receptacle not cuspidate........52. *Zinnia*
6. Involucre not distinctly graduated; pales of the receptacle cuspidate.....53. *Sanvitalia*
7. Leaves merely toothed (8).
7. Leaves dissected (9).
8. Rays numerous and very narrow; leaves sessile or subsessile..............55. *Eclipta*
8. Rays 5 or fewer, short and broad; leaves petioled......................78. *Galinsoga*
9. Achenes beaked; plants glabrous...76. *Cosmos*
9. Achenes not beaked; plants more or less woolly....................108a. *Leucampyx*
10. Rays fertile, the disk flowers sterile; achenes of the rays adnate at base to the subtending phyllary and to the pales of the opposed outer disk flowers, the whole falling away together ....42. *Engelmannia*

10. Rays and disk flowers all or mostly fertile, or the rays infertile, or else (genus *Hemizonia*) the achenes of the rays not as in genus *Engelmannia* (11).
11. Rays sessile and persistent on their achenes, becoming papery (12).
11. Rays not persistent or papery (13).
12. Involucre strongly graduated; pales of the receptacle not cuspidate; disk achenes strongly compressed .... 52. *Zinnia*
12. Involucre not distinctly graduated; pales of the receptacle cuspidate; disk achenes 4-angled .... 53. *Sanvitalia*
13. Involucre distinctly double, the outer phyllaries narrow, herbaceous, the inner ones broader, membranaceous (14).
13. Involucre otherwise (17).
14. Inner phyllaries connate to the middle, or higher .... 74. *Thelesperma*
14. Inner phyllaries free essentially to the base (15).
15. Pappus not of retrorsely hispid awns .... 71. *Coreopsis*
15. Pappus of retrorsely hispid awns (16).
16. Achenes dimorphous, the outer ones with a winged or callose margin, the inner ones narrower, wingless, somewhat beaked. Leaves pinnately parted into 3 to 7, entire or few-parted, linear lobes .... 73. *Heterosperma*
16. Achenes not dimorphous, neither winged nor truly beaked .... 75. *Bidens*
17. Achenes dorsoventrally compressed, with a pectinate wing .... 72. *Coreocarpus*
17. Achenes laterally if at all compressed, not with a pectinate wing (18).
18. Plants scapose, perennial, with broad, entire, densely silvery-pubescent leaves
67. *Enceliopsis*
18. Plants not scapose; leaves not silvery-pubescent (19).
19. Pales of the receptacle in a single series inside the rays, united into a toothed cup
80. *Hemizonia*
19. Receptacle paleaceous throughout, the pales not united (20).
20. Leaves, at least the lower ones, parted or pinnately divided (21).
20. Leaves entire or merely toothed (22).
21. Pappus a short chaffy crown; upper leaves entire or 3-lobed .... 56. *Rudbeckia*
21. Pappus of 1 or 2 short awns or teeth, sometimes also with squamellae; leaves, even the upper ones, pinnatifid .... 57. *Ratibida*
22. Rays pistillate (23).
22. Rays neutral (25).
23. Pappus a conspicuous, more or less divided, chaffy crown, often with 1 or 2 awns; leaves alternate .... 59. *Wyethia*
23. Pappus of 1 or 2 awns and sometimes with a very short crown of more or less united squamellae; leaves usually opposite (24).
24. Pappus of 2 awns and several short, more or less united, intermediate squamellae, or the latter obsolete; perennial herbs, with opposite, ovate, subsessile leaves; achenes not with broad white wings .... 69. *Zexmenia*
24. Pappus of 1 or 2 awns, without squamellae; leaves not opposite and subsessile, or else the achenes with broad white wings .... 70. *Verbesina*
25. Achenes more or less thickened (26).
25. Achenes strongly compressed, very flat (28).
26. Pappus of caducous awns or paleae (usually 2, sometimes many); achenes pubescent
62. *Helianthus*
26. Pappus of persistent awns and squamellae, or if wanting, then the achenes glabrous (27).
27. Peduncle noticeably thickened below the head; plants annual, with broad ovate leaves
60. *Tithonia*
27. Peduncle not thickened below the head; plants, if annual, with narrow leaves
61. *Viguiera*
28. Achenes with 2 white wings; leaves opposite, at least the lower ones..... 70. *Verbesina*
28. Achenes not winged but sometimes narrowly white-margined; leaves usually alternate (29).
29. Achenes usually not conspicuously ciliate on the margin; plants herbaceous perennials
68. *Helianthella*

29. Achenes conspicuously long-ciliate on the margin; plants not herbaceous perennials (30).
30. Shrubs; achenes with a narrow white margin, no crown, and 1 or 2 weak awns 64. *Encelia*
30. Annuals; achenes with a conspicuous white margin and crown, and 2 strong awns 66. *Geraea*
31. Involucre and leaves with translucent oil glands (32).
31. Involucre and leaves without translucent oil glands (34).
32. Phyllaries 1-seriate, united almost to the apex into a toothed cup or tube; pappus of 5 or 6 unequal paleae..103. *Tagetes*
32. Phyllaries more or less 2-seriate, or else quite free; pappus otherwise (33).
33. Leaves without stiff spreading bristles at base; phyllaries more or less 2-seriate, often partly united, often with bractlets at base........................104. *Dyssodia*
33. Leaves with a few stiff, spreading bristles at base; phyllaries 1-seriate, free, without bractlets at base..106. *Pectis*
34. Rays white, pink, blue, or purple (35).
34. Rays yellow (43).
35. Achenes compressed, 2-edged or 2-nerved (36).
35. Achenes more or less thickened, 4- or 5-angled (39).
36. Pappus, at least in the disk flowers, of several or many paleae or flattened bristles; phyllaries with scarious margin and tip.........................26. *Townsendia*
36. Pappus of only 1 or 2 bristlelike awns, or a crown of squamellae, or both; phyllaries not scarious-margined (37).
37. Pappus of a minute setulose crown only; leaves entire, alternate......23. *Achaetogeron*
37. Pappus of 1 or 2 weak awns, with or without a crown of squamellae; leaves usually toothed, lobed, or dissected, at least the lower ones opposite (38).
38. Pappus of 1 or 2 awns (these sometimes wanting) and a crown of squamellae; achenes usually callose-margined and ciliate................................84. *Perityle*
38. Pappus of 1 or 2 awns, without squamellae; achenes not callose-margined, not ciliate 85. *Laphamia*
39. Pappus a low, dissected or ciliolate crown only (40).
39. Pappus of distinct awns or paleae, sometimes also with a low crown (41).
40. Leaves toothed to pinnatifid, or the upper ones entire; phyllaries green-centered, scarious-margined ....21. *Aphanostephus*
40. Leaves entire; phyllaries whitish, thick, chartaceous, with a scarious margin and a green tip ....22. *Greenella*
41. Phyllaries with a conspicuous, scarious, whitish or yellowish margin and tip; plants perennial or biennial......................................91. *Hymenopappus*
41. Phyllaries without a conspicuous scarious margin and tip; plants annual (42).
42. Herbage hispid-pilose; upper leaves subtending the heads..............25. *Monoptilon*
42. Herbage woolly; heads not subtended by the upper leaves............94. *Eriophyllum*
43. Phyllaries distinctly graduated, in several series (44).
43. Phyllaries equal or subequal or in 2 unequal series, not distinctly graduated (49).
44. Pappus of 2 to 8 slender caducous awns............................10. *Grindelia*
44. Pappus otherwise (45).
45. Annual herbs; rays 12 to 50..................................9. *Xanthocephalum*
45. Perennial herbs or shrubs; rays 12 or fewer, except in genus *Acamptopappus* (46).
46. Heads comparatively large, solitary. Pappus of numerous, narrowly linear awns and bristles, several-seriate.................................14. *Acamptopappus*
46. Heads small or very small, clustered (47).
47. Pappus of disk achenes a low crown, without distinct paleae or bristles.......8. *Selloa*
47. Pappus of disk achenes of distinct paleae, bristles, or squamellae (48).
48. Disk achenes with a pappus of several straight paleae or squamellae in a single series; involucre not compressed; leaves chiefly linear....................12. *Gutierrezia*
48. Disk achenes with a pappus of numerous, more or less twisted, flattened bristles or narrow paleae; involucre compressed; leaves obovate or elliptic.....13. *Amphipappus*
49. Rays persistent on the achenes, becoming papery (50)

49. Rays not persistent on the achenes (51).
50. Rays 3 to 5, about as wide as long; achenes linear, slightly angled, essentially glabrous 82. *Psilostrophe*
50. Rays 10 or more, much longer than wide; achenes obpyramidal, 5-angled, hirsute 99. *Hymenoxys*
51. Achenes strongly flattened, 2-edged; pappus of 1 or 2 bristlelike awns and sometimes with a crown of squamellae (52).
51. Achenes not flattened or 2-edged; pappus awns or squamellae more numerous (53).
52. Pappus of 1 or 2 awns and with a crown of squamellae; achenes usually callose-margined and ciliate ....84. *Perityle*
52. Pappus of 1 or 2 awns, without squamellae; achenes not callose-margined, not ciliate 85. *Laphamia*
53. Pappus squamellae dissected into numerous bristles, these united at base (54).
53. Pappus squamellae not dissected into bristles (55).
54. Heads short-peduncled, solitary; rays conspicuous..............89. *Syntrichopappus*
54. Heads essentially sessile, clustered; rays inconspicuous..............94. *Eriophyllum*
55. Phyllaries spreading or reflexed, herbaceous (56).
55. Phyllaries erect (57).
56. Receptacle naked; pappus usually much shorter than the achene........100. *Helenium*
56. Receptacle fimbrillate; pappus usually as long as the achene or longer..101. *Gaillardia*
57. Plants low annuals, not woolly; pappus of 2 to 5 lanceolate awns. Leaves opposite, linear, entire ....87. *Baeria*
57. Plants perennial, or woolly, or with more numerous awns or paleae in the pappus (58).
58. Achenes obpyramidal, strongly 4-angled; ray one. Pappus of 8 scarious paleae or squamellae; slender annuals with leaves entire or pinnately divided into filiform lobes 90. *Schkuhria*
58. Achenes not obpyramidal and 4-angled; rays more than 1 (59).
59. Rays 2 to 5, fertile, the disk flowers 6 or 7, sterile. Leaves dissected into filiform lobes; heads very small, very numerous, crowded in dense cymose panicles..102. *Plummera*
59. Both rays and disk flowers fertile, more numerous (60).
60. Pappus of 12 or more lanceolate, awn-tipped paleae; plants nearly glabrous, with dissected leaves; involucre not double............................92. *Hymenothrix*
60. Pappus paleae fewer, or else the plant tomentose, or the leaves not dissected, or the involucre double and of 2 more or less distinct sets of phyllaries (61).
61. Achenes obpyramidal, 5-angled, only 2 or 3 times as long as wide......99. *Hymenoxys*
61. Achenes narrowly obpyramidal, several times as long as wide (62).
62. Phyllaries concave, partly enclosing the ray achenes; plants woolly....94. *Eriophyllum*
62. Phyllaries flat, not at all enclosing the ray achenes; plants not woolly.......97. *Bahia*

1. STEVIA (315)

Herbaceous or shrubby plants; leaves alternate or opposite; heads small, 5-flowered; phyllaries 5 or 6, equal, firm; corollas tubular, 5-toothed, white to purple; achenes slender; pappus of awns or squamellae or both, or reduced to a toothed crown.

KEY TO THE SPECIES

1. Annuals; leaves membranaceous, ovate, more than half as wide as long, serrate. Pappus of 3 awns and 3 very short intermediate squamellae; heads rather loosely cymose; corollas whitish ....1. *S. micrantha*
1. Perennial; leaves usually firm, linear to elliptic, much less than half as wide as long (2).
2. Heads definitely pedicellate, loosely cymose-panicled; corollas purple. Leaves alternate, linear, entire or bluntly toothed...................................2. *S. viscida*
2. Heads subsessile, in dense, fastigiate cymose panicles; corollas white or rose-color (3).
3. Leaves entire, linear-elliptic or lanceolate, opposite, 3-nerved but not strongly veiny; plants suffrutescent or shrubby. Pappus a very short, toothed crown....3. *S. Lemmoni*
3. Leaves serrate, linear to lance-oblong or oblong, often alternate or veiny; plants herbaceous (4).

4. Leaves mostly opposite, not very numerous, strongly veiny beneath, elliptic to elliptic-oblong or lance-oblong, the larger leaves 10 to 20 mm. wide; pappus of squamellae only 4. *S. Plummerae*
4. Leaves mostly alternate, usually very numerous and crowded, narrowly linear-oblanceolate to spatulate-oblanceolate, cuneate, or oval, 3 to 10 (rarely 15) mm. wide; pappus normally of awns and squamellae. . . . . 5. *S. serrata*

1. Stevia micrantha Lag. Mountains of Cochise, Santa Cruz, and Pima counties, 5,000 to 6,500 feet, rich, moist soil in canyons, August to October. Southwestern New Mexico, southern Arizona, and Mexico.

2. Stevia viscida H.B.K. (*S. amabilis* Lemmon). Plains near the Huachuca Mountains, Cochise County (*Lemmon* 2729, the type of *S. amabilis*), September. Southeastern Arizona and Mexico.

3. Stevia Lemmoni Gray. Mountains of Santa Cruz and Pima counties, 2,500 (?) to 5,500 feet, rocky canyons, February to May (September), type from the Santa Catalina Mountains (*Lemmon*). Southern Arizona and northern Mexico.

4. Stevia Plummerae Gray. White Mountains (Apache County), and mountains of Cochise, Santa Cruz, and Pima counties, 6,000 to 8,000 feet, rich soil in canyons, (April) August to October, type from Rucker Valley, Chiricahua Mountains (*Mrs. Lemmon*). New Mexico, Arizona, and northern Mexico.

The flowers are fragrant. The var. *alba* Gray is a form with white flowers, type from the Huachuca Mountains (*Lemmon*).

5. Stevia serrata Cav. Mountains of southern Apache, Graham, Cochise, Santa Cruz, and Pima Counties, 4,500 to 9,000 feet, chiefly in pine forest, (April) July to October. Western Texas to southern Arizona, southward into Mexico.

The typical form has oblanceolate to spatulate leaves. The less common var. *ivaefolia* (Willd.) Robins. has broader, lanceolate to oblong-oblanceolate or oval leaves. A collection in the Pinaleno Mountains (*Darrow et al.* 1164), with many of the achenes exaristate, approaches var. *haplopappa* Robins. of Durango and Aguascalientes, Mexico.

2. HOFMEISTERIA. Arrow-leaf

Much-branched, low shrubs, glandular-puberulous; leaves opposite or alternate, the blades lanceolate or lance-ovate, 2 to 10 mm. long, shorter than the petioles; heads loosely panicled, white, discoid; involucre strongly graduated, of dry, few-ribbed phyllaries, the outer ones with acuminate subherbaceous tips; pappus of about 12 bristles, alternating irregularly with much shorter, narrow squamellae or bristles.

1. Hofmeisteria pluriseta Gray. Mohave, western Pima, and Yuma counties, 3,000 feet or lower, dry, granitic slopes, January to March (sometimes autumn), type from Williams River. Southern Utah and Nevada, western Arizona, southeastern California, and northern Baja California.

3. CARMINATIA

Low, annual herbs; leaves mostly opposite, deltoid-ovate, toothed, thin, slender-petioled; heads discoid, whitish, cylindric, in a terminal, leafless, virgate panicle; involucre graduated, of thin, few-striate phyllaries; achenes slender; pappus of long-plumose bristles.

1. **Carminatia tenuiflora** DC. Big Lue Range (Greenlee County) and mountains of Cochise, Santa Cruz, and Pima counties, 4,000 to 7,000 feet, rich soil in canyons, August to October. Western Texas to southern Arizona, south to Central America.

4. EUPATORIUM. Thoroughwort

Herbs or low shrubs; leaves usually opposite; heads small or medium-sized, usually panicled, discoid, white to lavender, violet, or purple; involucre scarcely to strongly graduated; achenes 5-ribbed; pappus of numerous capillary bristles.

KEY TO THE SPECIES

1. Leaves palmately 3- to 5-cleft, with toothed to pinnatifid divisions; receptacle convex; flowers violet . . . . 1. *E. Greggii*
1. Leaves entire to sharply serrate, not palmately cleft; receptacle flat or essentially so; flowers white to purple (2).
2. Leaves mostly in whorls of 3 to 6. Plants tall, the herbage canescent to nearly glabrous; leaves short-petioled, ovate to oblong-lanceolate, coarsely incurved-serrate; heads numerous, in corymbiform inflorescences, cylindric or narrowly campanulate, usually purplish pink; phyllaries in about 4 series, the outer ones short, very obtuse, herbaceous, the inner ones elongate, somewhat scarious, purplish . . . . 2. *E. maculatum*
2. Leaves opposite (3).
3. Heads 3- to 6-flowered, sessile or subsessile in small clusters at the tips of usually short branches, forming a thyrsoid panicle; leaves narrowly lanceolate or lance-ovate, seldom more than 18 mm. wide, very short-petioled, acuminate . . . . 3. *E. solidaginifolium*
3. Heads 10- to 30-flowered, distinctly pedicellate, not thyrsoid-panicled; leaves usually broader (4).
4. Leaves obtuse, broadly ovate, usually not more than 15 mm. long and wide, entire or crenate-serrate, thickish, on petioles 5 mm. long or less. Plants shrubby, low, puberulous; involucre 3 to 5 mm. high, scarcely half as long as the flowers . . . . . . 4. *E. Wrightii*
4. Leaves acute or acuminate, or if rarely obtuse (*E. Lemmoni*), then the plants strictly herbaceous and with sessile leaves (5).
5. Involucre strongly graduated, the phyllaries in 3 to 5 lengths (6).
5. Involucre of subequal phyllaries, or obscurely graduated (7).
6. Heads small, 5 mm. high or less (excluding the styles), in close clusters at the tips of the stem and branches, the pedicels mostly 4 mm. long or less; achenes 1.5 mm. long; involucre 4 mm. high or less, the inner phyllaries obtuse, the outer ones acute 5. *E. pycnocephalum*
6. Heads larger, 8 to 15 mm. high, loosely cymose, on pedicels mostly 6 to 20 mm. long; achenes about 3 mm. long; involucre 6 to 10 mm. high, the phyllaries all acute or acuminate . . . . 6. *E. Bigelovii*
7. Achenes glabrous, 1 to 1.3 mm. long; leaves narrowly lanceolate, 3 to 5 times as long as wide, long-acuminate; involucre 3 mm. high, the phyllaries acuminate, pubescent 7. *E. pauperculum*
7. Achenes pubescent, (1.5) 2 to 3 mm. long; leaves rhombic- or elliptic-ovate to broadly ovate, not more than about twice as long as wide; involucre 3 to 6 mm. high (8).
8. Corolla lobes glabrous; involucre 3 to 4 mm. high; leaves triangular-ovate to broadly ovate, usually subcordate or truncate at base . . . . 8. *E. herbaceum*
8. Corolla lobes strongly hairy outside; involucre 5 to 7 mm. high; leaves oblong-ovate or rhombic-ovate, cuneate to rounded or rarely subcordate at base (9).
9. Leaves on petioles 1 to 3 cm. long, the blades 3.5 to 7 cm. long, coarsely serrate or crenate-serrate . . . . 9. *E. Rothrockii*
9. Leaves sessile or essentially so, the blades 2 to 3 cm. long, crenate . . . . 10. *E. Lemmoni*

1. **Eupatorium Greggii** Gray. Throughout Cochise County, 3,500 to 6,000 feet, plains and mesas, June to October. Western Texas to southeastern Arizona, southward to Zacatecas.

2. **Eupatorium maculatum** L. Near McNary, southern Apache County (*Deaver* 1732), Lakeside, southern Navajo County (*Pultz* 1705), Sedona, Coconino County (*Beal* in 1947), 4,500 to 7,500 feet, wet soil along streams, August. New Brunswick to British Columbia, south to Florida, New Mexico, and eastern Arizona.

3. **Eupatorium solidaginifolium** Gray. Southeastern Cochise County, and in Pima County from the Rincon to the Ajo Mountains, 3,000 to 5,000 feet, rocky canyons, May to October. Western Texas to southern Arizona and northern Mexico.

4. **Eupatorium Wrightii** Gray. Chiricahua and Huachuca mountains (Cochise County), Mustang Mountains (Santa Cruz County), 5,000 to 6,000 feet, limestone slopes, September and October. Western Texas to southeastern Arizona and northern Mexico.

Plant "very bushy, spreading" (Blumer, ms.).

5. **Eupatorium pycnocephalum** Less. Superstition Mountains (Pinal County), and mountains of Cochise, Santa Cruz, and Pima counties, 3,500 to 5,000 feet, rich soil along streams, May to October. Southern Arizona to South America.

Flowers pale blue or lavender.

6. **Eupatorium Bigelovii** Gray. Apparently known in Arizona only from the type collection on the Gila River (*Bigelow*). Northern Mexico, where it flowers March to October.

7. **Eupatorium pauperculum** Gray (*Piptothrix arizonica* A. Nels.). Greenlee, Graham, Cochise, Santa Cruz, and Pima counties, 3,500 to 5,000 feet, by streams in canyons, March to May, type from the Santa Rita Mountains (*Pringle*), type of *Piptothrix arizonica* from the west side of the Baboquivari Mountains (*A. & R. Nelson* 1567). Southern Arizona and Sonora.

8. **Eupatorium herbaceum** (Gray) Greene (*E. arizonicum* (Gray) Greene). Near Kayenta and Holbrook (Navajo County), Kaibab Plateau and Grand Canyon (Coconino County), Hualpai Mountain (Mohave County), south to the mountains of Greenlee, Cochise, Santa Cruz, and Pima counties, 5,000 to 9,000 feet, mostly in open pine forest, June to October. Utah, New Mexico, Arizona, and northern Mexico.

Flowers white, the plant fragrant in drying.

9. **Eupatorium Rothrockii** Gray (*E. Rothrockii* var. *Shrevei* Robins.). Mountains of southern Apache, Graham, Cochise, Santa Cruz, and Pima counties, 6,500 to 7,500 feet, chiefly in pine forest, July to October, type from Mount Graham (*Rothrock* 740, 741), type of var. *Shrevei* from Ramsey Canyon, Huachuca Mountains (*Shreve* 5017). Southern New Mexico and southeastern Arizona.

10. **Eupatorium Lemmoni** Robins. Chiricahua Mountains, Cochise County (*Lemmon* 316, in part, the type collection). Southeastern Arizona and northern Mexico.

5. CARPHOCHAETE

Low, branching shrubs; leaves opposite, sessile, entire, punctate, linear-elliptic to spatulate-elliptic, 15 to 30 mm. long, with axillary fascicles; heads cylindric, about 25 mm. high, few-flowered, the corollas purplish-tinged; phyllaries rather few, narrow, unequal, subherbaceous, acute to acuminate, densely glandular-punctate; achenes linear, 8- to 10-ribbed; pappus of about 12 narrowly scarious-margined, linear-attenuate awns, these barbellate above.

1. Carphochaete Bigelovii Gray. Mountains of southern Apache, Greenlee, Gila, Cochise, Santa Cruz, and Pima counties, 4,000 to 7,000 feet, rocky slopes and canyons, March to July. Western Texas to southern Arizona and Chihuahua.

The plant is browsed.

6. BRICKELLIA (316)

Herbs or shrubs; leaves opposite or alternate; heads small to medium-sized, discoid, usually whitish, solitary or panicled; involucre usually definitely graduated, the phyllaries generally dryish and striate; achenes 10-ribbed (sometimes only 5-ribbed in *B. Fendleri*); pappus of numerous capillary bristles (plumose only in *B. brachyphylla*).

KEY TO THE SPECIES

1. Heads 3- to 5-flowered (2).
1. Heads (6) 8- to 60-flowered (3).
2. Leaves linear-lanceolate, elongate, 2 to 10 mm. wide, 3 to 13 cm. long....1. *B. longifolia*
2. Leaves lance-ovate, 12 to 28 mm. wide............................2. *B. multiflora*
3. Plants with abundant, very short, budlike branchlets bearing minute, crowded, 4-ranked, scalelike leaves; stem leaves (mostly deciduous) narrowly linear, entire, 3 to 7 cm. long, 2 mm. wide or less....................................3. *B. squamulosa*
3. Plants without budlike branchlets bearing crowded, scalelike leaves; stem leaves persistent, usually more than 2 mm. wide (4).
4. Leaves spatulate or oblanceolate. Intricately branched shrubs; herbage tomentellous; leaves 1 to 3 mm. wide; heads about 26-flowered....................4. *B. frutescens*
4. Leaves linear to broadly ovate (5).
5. Leaves linear to elliptic or lance-ovate, cuneate to rounded at base, subsessile or on petioles 3 mm. long or less (6).
5. Leaves definitely ovate or rhombic-ovate or, if lance-oblong, then with distinctly cordate-clasping base (12).
6. Heads about 40- to 50-flowered, solitary or few and cymose. Leaves elliptic-ovate to lance-linear, usually entire...................................11. *B. oblongifolia*
6. Heads 9- to 26-flowered, numerous, panicled (7).
7. Leaves 5 to 12 mm. long, lanceolate to ovate; achenes finely hispidulous; heads mostly solitary at the tips of slender, minutely leafy peduncles, forming a long, racemiform panicle ....5. *B. scabra*
7. Leaves usually 20 to 60 mm. long; achenes densely pubescent; heads otherwise arranged (8).
8. Pappus bristles plumose; leaves lanceolate or lance-ovate..........7. *B. brachyphylla*
8. Pappus bristles not at all plumose; leaves linear to oblong or lance-oblong, rarely lanceolate (9).
9. Phyllaries few (about 14); heads about 10-flowered..................8. *B. Lemmoni*
9. Phyllaries numerous (20 to 37); heads usually 19- to 26-flowered (10).
10. Plants glandular-pubescent, especially above. Leaves narrowly lanceolate, 12 to 15 mm. wide ....15. *B. amplexicaulis*
10. Plants not glandular-pubescent (11).

11. Heads slender-peduncled, in a loose panicle; leaves linear or narrowly linear-oblong, 3 to 9 mm. wide .... 9. *B. venosa*
11. Heads sessile or subsessile, subspicate on the stem and branches; leaves elliptic, oblong, or lance-oblong, 8 to 20 mm. wide .... 10. *B. Pringlei*
12. Leaves coriaceous, bright green, spiny-toothed; outer phyllaries herbaceous or subcoriaceous, much broader than the inner ones and nearly as long; heads solitary at the tips of the branches, about 50-flowered .... 12. *B. atractyloides*
12. Leaves not coriaceous (except in *B. baccharidea*), not spiny-toothed; outer phyllaries not noticeably broader, and much shorter, than the inner ones; heads panicled, except in *B. simplex* and *B. incana* (13).
13. Herbage (at least when young) densely white-tomentose; heads about 60-flowered. Leaves small, ovate, sessile; heads about 23 mm. high, solitary at tips of stem and branches .... 13. *B. incana*
13. Herbage not densely white-tomentose, if grayish-tomentellous (*B. desertorum*), then the heads only 8- to 12-flowered (14).
14. Leaves lance-oblong to ovate, sessile or very short-petioled, strongly cordate at base (15).
14. Leaves ovate or rhombic-ovate, usually slender-petioled (16).
15. Peduncles hispidulous or short-hirsute, not glandular; leaves ovate or oblong-ovate, distinctly short-petioled .... 14. *B. betonicaefolia*
15. Peduncles densely stipitate-glandular; leaves lance-oblong or ovate-oblong, sessile, strongly cordate-clasping .... 15. *B. amplexicaulis*
16. Blades of the largest leaves not more than 15 mm. long, green (17)
16. Blades of the larger leaves (except sometimes in *B. Coulteri*, *B. baccharidea*, and *B. desertorum*) more than 15 mm. long (18).
17. Stems up to 40 cm. long; heads numerous, mostly solitary at the tips of slender, minutely leafy branches, forming a long racemiform panicle; phyllaries 20 or more, in several series, with conspicuous green tips .... 5. *B. scabra*
17. Stems not more than 25 cm. long; heads solitary, or few in a loosely cymose inflorescence; phyllaries 14 or fewer, in about 2 series, not conspicuously tipped
6. *B. parvula*
18. Leaves sharply 1- to 3-hastate-toothed at base, deltoid- or rhombic-ovate. Heads slender-peduncled, loosely panicled, about 17-flowered .... 16. *B. Coulteri*
18. Leaves not sharply and hastately few-toothed at base (19).
19. Leaves rhombic-ovate, coriaceous, repand-toothed, cuneate at base, 10 to 38 mm. long; phyllaries firm, all rounded or obtuse, stramineous, often resinous. . 17. *B. baccharidea*
19. Leaves ovate, not coriaceous, crenate to dentate, usually cordate or subcordate at base; phyllaries thin (20).
20. Heads solitary or few, about 60-flowered. Leaves triangular-ovate, coarsely toothed; involucre about 13 mm. high .... 24. *B. simplex*
20. Heads numerous, 8- to 38-flowered (21).
21. Outer phyllaries with loose caudate-attenuate herbaceous tips. Leaves thin, triangular-ovate, acuminate, coarsely serrate or crenate-serrate; involucre 10 to 14 mm. high
23. *B. grandiflora*
21. Outer phyllaries not caudate-attenuate (22).
22. Peduncles densely stipitate-glandular; outermost phyllaries (a few of them) herbaceous, usually longer than the next inner series, sometimes surpassing the heads
22. *B. floribunda*
22. Peduncles not stipitate-glandular; outer phyllaries not herbaceous or elongate (23).
23. Heads 28- to 35-flowered. Leaves mostly deltoid- or triangular-ovate, thin; heads slender-peduncled, mostly nodding, in usually few-headed cymose panicles at the tips of the stem and branches .... 21. *B. Fendleri*
23. Heads 8- to 18-flowered (24).
24. Plants herbaceous; leaves 5 to 10 cm. long, acuminate; heads slender-peduncled, the panicles rather loose .... 20. *B. Rusbyi*
24. Plants shrubby; leaves not more than 5 cm. long, obtuse to acute; heads mostly subsessile, clustered in the leaf axils, forming spikelike panicles (25).

25. Phyllaries puberulous on the back; leaves 3 to 13 mm. long and wide, densely grayish-tomentellous .... 18. *B. desertorum*

25. Phyllaries essentially glabrous; leaves 10 to 50 mm. long and wide, green 19. *B. californica*

1. Brickellia longifolia Wats. Cameron, Grand Canyon, and Havasu Canyon (Coconino County), 3,500 to 8,000 feet, September. Southeastern Utah to eastern California and northern Arizona.

2. Brickellia multiflora Kellogg. Coconino County, North Rim of Grand Canyon (*Eastwood & Howell* 7106) and Havasu Canyon (*Clover* 5138), September. Southern Nevada, eastern California, and northwestern Arizona.

3. Brickellia squamulosa Gray. Chiricahua, Mule, and Huachuca mountains (Cochise County), Santa Cruz River valley (Santa Cruz County), 4,000 to 6,000 feet, May to September. Southwestern New Mexico and southeastern Arizona, south to Mexico.

4. Brickellia frutescens Gray. Ajo Mountains, Pima County (*Darrow & Gould* 3841), Kofa Mountains, Yuma County (*Darrow & Gould* 3835), about 3,000 feet, rocky canyon slopes, March. Southern Nevada, southwestern Arizona, southeastern California, and Baja California.

5. Brickellia scabra (Gray) A. Nels. Apache, Navajo, Coconino, and Yuma counties, 3,000 to 5,500 feet, July to October. Wyoming to New Mexico, Arizona, and southern Nevada.

Brickellia Watsonii Robins., of Utah, southeastern Nevada, and southeastern California, will probably be found in Arizona. It is similar to *B. scabra*, but the stem is finely lanulose or crisp-puberulous, whereas in *B. scabra* it is glandular-puberulous.

6. Brickellia parvula Gray. Baboquivari Peak, Pima County (several collections), 6,500 to 7,000 feet. Western Texas and southern Arizona.

This species is related to *B. brachyphylla*, differing in the definitely ovate leaves, few, long-peduncled, loosely cymose or solitary heads, and merely hispidulous pappus bristles.

7. Brickellia brachyphylla Gray. Southern Apache, Greenlee, and Gila counties, 5,500 to 7,000 feet, pine forests, September. Western Texas to southern Colorado and eastern Arizona.

8. Brickellia Lemmoni Gray. Cochise County, in the Chiricahua Mountains (*Lemmon* 306, the type, *Blumer* 1786), and in Rucker Canyon (*Gould & Haskell* 4581, 4583), 6,000 to 6,500 feet, in ponderosa pine forest, August to October. Southeastern Arizona and Chihuahua.

9. Brickellia venosa (Woot. & Standl.) Robins. Greenlee, Graham, Gila, Pinal, Cochise, Santa Cruz, and Pima counties, 4,000 to 5,500 feet, September and October. Southern New Mexico and Arizona, and Chihuahua.

10. Brickellia Pringlei Gray. Huachuca Mountains (Cochise County), Santa Rita and Santa Catalina mountains (Santa Cruz and Pima counties), rich soil in canyons, April and May, type from the Santa Catalina Mountains (*Pringle* in 1881). Southern Arizona, south to Mexico.

11. Brickellia oblongifolia Nutt. Apache County to eastern Mohave County, 4,500 to 8,500 feet, May and June (September). Colorado to New Mexico, west to Nevada and southeastern California.

The species is represented in Arizona by var. *linifolia* (D. C. Eaton) Robins., with few or no glandular hairs on the achenes, whereas, in typical *B. oblongifolia,* all or most of the achene hairs are glandular.

12. Brickellia atractyloides Gray. Grand Canyon (Coconino County) and Mohave County to Greenlee, Pima, and Yuma counties, up to 3,500 feet, March to May (September). Southern Utah and Nevada, southeastern California, and Arizona.

A small, much-branched shrub.

13. Brickellia incana Gray. Mohave County, 1,500 to 3,500 feet, sandy washes, May to October. Southern California, southern Nevada, and western Arizona.

14. Brickellia betonicaefolia Gray. Southern parts of Navajo and Coconino counties, south to Greenlee, Graham, Cochise, Santa Cruz, and Pima counties, 4,500 to 6,500 feet, June to October. New Mexico and central Arizona to Sonora and Chihuahua.

15. Brickellia amplexicaulis Robins. Pinal, Cochise, Santa Cruz, and Pima counties, 3,000 to 5,000 feet, September and October. Southern Arizona and northern Mexico.

The typical plant has ovate-oblong leaves with a cordate-amplexicaul base. Var. *lanceolata* (Gray) Robins., with narrowly lanceolate, scarcely clasping leaves, is known only from Greenlee County, where the type was collected near Clifton (*Greene* in 1880).

16. Brickellia Coulteri Gray. Graham County to Mohave County and in Havasu Canyon (Coconino County), south to Cochise, Pima, and Yuma counties, up to about 4,500 feet, dry, rocky slopes and canyons, common, March to November. Arizona to central Mexico and Baja California.

17. Brickellia baccharidea Gray. Greenlee, Cochise, Santa Cruz, Pima, and Yuma counties, 500 to 5,500 feet, often on limestone, September to November. Western Texas to southwestern Arizona.

18. Brickellia desertorum Coville. Near Pierce Ferry and near Oatman (Mohave County), Kofa Mountains (Yuma County), 1,000 to 4,000 feet, September. Western Arizona, southern Nevada, and southern California.

A small, much-branched shrub.

19. Brickellia californica (Torr. & Gray) Gray. Throughout the state, 3,000 to 7,500 feet, very common, July to November. Colorado to western Texas, west to California, south to Sonora and Baja California.

Called pachaba by the Hopi Indians, who are reported to rub it on the head for headache.

20. Brickellia Rusbyi Gray. Southern Apache and Gila counties to Cochise, Santa Cruz, and Pima counties, 4,500 to 8,500 feet, commonly in pine forest, August to October. New Mexico, southeastern Arizona, Sonora, and Chihuahua.

Plant malodorous.

21. Brickellia Fendleri Gray. Pinaleno Mountains (Graham County), Chiricahua Mountains (Cochise County), 6,000 to 9,500 feet, August and September. New Mexico and southeastern Arizona.

This species is intermediate in achenial characters between *Eupatorium* and *Brickellia,* the achene being 5-angulate, with or without 1 to 5 intermediate secondary ribs.

22. Brickellia floribunda Gray. Southern part of Navajo and Coconino counties and eastern Mohave County, south to Cochise, Santa Cruz, and Pima counties, 3,000 to 5,500 feet, rich soil in canyons, September and October. Southwestern New Mexico, southern Arizona, Sonora, and Chihuahua.

23. Brickellia grandiflora (Hook.) Nutt. (*Coleosanthus umbellatus* Greene). Lukachukai Mountains (Apache County) and Kaibab Plateau (Coconino County) south to the mountains of Cochise, Santa Cruz, and Pima counties, 5,000 to 9,000 feet, rich soil, mostly in coniferous forests, August to October, type of *C. umbellatus* from "the mountain districts of northern Arizona." Missouri and Arkansas to Montana and Washington, south to New Mexico, southern Arizona, California, and northern Baja California.

24. Brickellia simplex Gray. Chiricahua, Huachuca, and Patagonia mountains (Cochise and Santa Cruz counties), 5,000 to 8,000 feet, August and September. Southeastern Arizona, Sonora, and Chihuahua.

7. KUHNIA (317)

Slender perennial herbs; leaves mostly alternate, linear or linear-lanceolate, entire, usually revolute-margined, glandular-punctate; heads rather small, panicled, discoid, whitish; involucre more or less graduated, the phyllaries mostly linear, strongly few-ribbed, narrowly scarious-margined; achenes slender, 10- to 20-nerved; pappus a single series of plumose bristles, usually becoming tawny.

1. Kuhnia rosmarinifolia Vent. Apache, Navajo, and Coconino counties to Cochise, Santa Cruz, and Pima counties, 4,500 to 7,500 feet, mesas and slopes and openings in pine forests, May to October. Texas to Arizona, south to Mexico.

In the typical plant the phyllaries are ciliate, nearly or quite glabrous on the back, not or only slightly graduated, the outer ones abruptly shorter than the inner ones. In var. *chlorolepis* (Woot. & Standl.) Blake (*K. chlorolepis* Woot. & Standl.), known from Greenlee, Pinal, Cochise, and Pima counties, 2,500 to 5,000 feet, the phyllaries are rather densely pilosulous or puberulous on the back and more regularly graduated.

The name *rosmarinifolia* was rejected by Shinners (317) as a "nomen confusum," and the Arizona plants were referred by him to *K. chlorolepis* Woot. & Standl.

Liatris punctata Hook. has been assigned a range extending to Arizona by Rydberg and by Wooton and Standley. There is no specimen from Arizona in the United States National Herbarium or in the herbarium of the New York Botanical Garden, and it seems best to omit the species until a definite specimen is forthcoming. L. O. Gaiser (*Rhodora* 48: 361) stated: "No collections of this [*L. punctata*] or any other species of *Liatris* from the state of Arizona have been found in any herbaria examined." A specimen in the Dudley Herbarium, Stanford University, labeled "Arizona, Brandegee" may have been collected in Colorado.

8. SELLOA

Suffrutescent, branched, glabrous or obscurely puberulent, and glutinous; leaves alternate, oblong-lanceolate to linear, entire, punctate; heads small, inconspicuously radiate, yellow, in dense, rounded, terminal clusters, these panicled; involucre graduated, the phyllaries chartaceous, with short green tips; achenes puberulous, 4- or 5-ribbed; pappus very minute and coroniform, or essentially wanting.

1. Selloa glutinosa Spreng. (*Gymnosperma glutinosum* Less., *G. corymbosum* DC.). Greenlee, Maricopa, Pinal, Cochise, Santa Cruz, Pima, and Yuma counties, 1,000 to 6,000 feet, rocky canyons and slopes, March to December. Texas to southern Arizona, south to Central America.

In Mexico the plant is used in decoctions for treating diarrhea, and the gum is used externally in cases of rheumatism and ulcers.

9. XANTHOCEPHALUM

Annual herbs, nearly or quite glabrous, sometimes glutinous; leaves alternate, linear to lance-oblong; heads radiate, yellow, cymose-panicled or scattered; involucre broad, the phyllaries graduated, chartaceous, green-tipped; pappus of several unequal more or less united awns, or reduced to a setulose crown, sometimes none in the ray achenes.

KEY TO THE SPECIES

1. Leaves chiefly lanceolate, 5 to 20 mm. wide; heads crowded in cymose panicles, the peduncles stipitate-glandular; rays 30 to 50; pappus in the ray achenes none or reduced to an obscure border, in the disk achenes of unequal, basally united squamellae or paleae — 1. *X. gymnospermoides*
1. Leaves chiefly linear, 2 to 4 mm. wide; heads mostly solitary at the tips of cymosely arranged glabrous branches; rays about 10 to 20; pappus a short undulate crown — 2. *X. Wrightii*

1. Xanthocephalum gymnospermoides (Gray) Benth. & Hook. Pinal, Cochise, Santa Cruz, and Pima counties, 1,000 to 5,500 feet, locally abundant in alluvial, often saline soil, August to October, type from the San Pedro (*Wright* 1178). Southeastern Arizona to northern Mexico.

The plant attains a height of 6 feet or more.

2. Xanthocephalum Wrightii Gray. Southern Apache and Navajo counties to the mountains of Cochise, Santa Cruz, and Pima counties, 4,500 to 9,000 feet, openings in pine and oak forest, August to October, type from between Babocomari, Arizona, and Santa Cruz, Sonora (*Wright* 1177). New Mexico and Arizona.

10. GRINDELIA (318). GUM-WEED

Herbs, more or less resinous-viscid; leaves alternate, entire, toothed, or rarely pinnatifid; heads medium-sized, yellow, radiate or discoid; involucre strongly graduated; achenes short, more or less thickened; pappus of 2 to 8 slender caducous paleaceous awns.

Some of the species (including *G. squarrosa*) were formerly official drug plants, being antispasmodic and stomachic, administered in asthma and externally to relieve the irritation caused by poison ivy. The plants are suspected of being toxic to livestock, but are rarely eaten.

KEY TO THE SPECIES

1. Heads discoid .... 1. *G. aphanactis*
1. Heads radiate (2).
2. Phyllaries with strongly spreading or recurved subulate tips .... 2. *G. squarrosa*
2. Phyllaries with appressed or nearly erect triangular tips (3).
3. Leaves all or mostly laciniate-dentate or pinnatifid .... 3. *G. laciniata*
3. Leaves merely serrulate, or sometimes entire .... 4. *G. arizonica*

1. Grindelia aphanactis Rydb. Apache, Navajo, Coconino, Yavapai, and Cochise counties, mostly 5,000 to 7,000 feet, June to October. Southeastern Utah and southern Colorado to western Texas, New Mexico, and eastern Arizona.

2. Grindelia squarrosa (Pursh) Dunal. Navajo, Coconino, Yavapai, and Gila counties, 4,000 to 7,500 feet. Native from Minnesota to Texas and Wyoming, and perhaps farther west, and occurring as an introduction both eastward and westward.

The typical, relatively broad-leaved form of the species has been collected near Jacob Lake, Kaibab Plateau (*Eastwood & Howell* 6427) and near Flagstaff (*Collom* in 1936). Var. *serrulata* (Rydb.) Steyermark, with linear-oblong to oblanceolate leaves, occurs in Navajo, Coconino, Yavapai, and Gila counties, 4,000 to 7,000 feet.

3. Grindelia laciniata Rydb. Southwestern Coconino and northwestern Yavapai counties, at Williams and Seligman, 5,000 to 6,500 feet, June to August. Southeastern Utah and northern Arizona.

4. Grindelia arizonica Gray. Apache, Navajo, and Gila counties, 3,500 to 7,500 feet, openings in pine forests, August to October, type from Black River (*Rothrock* 796). Southwestern New Mexico and eastern Arizona.

Var. *microphylla* Steyermark, with longer pappus awns (5 to 6.5 mm. long), is known only from north of Clifton, Greenlee County (*Davidson* 736, the type).

11. VANCLEVEA

Slender, branching shrubs, white-barked, glabrous, glutinous; leaves alternate, linear-lanceolate, 3- to 5-nerved, entire or slightly toothed, often conduplicate, falcate-recurved; heads medium-sized, discoid, yellow, solitary or cymose; involucre graduated; achenes slender, about 5-ribbed; pappus of about 12 to 16 linear, acuminate, persistent awns.

1. Vanclevea stylosa (Eastw.) Greene. Northern Navajo and eastern Coconino counties, 5,000 to 6,000 feet, September. Southeastern Utah and northeastern Arizona.

12. GUTIERREZIA. Snake-weed

Perennial herbs, sometimes suffrutescent, more or less glutinous; leaves alternate, linear to narrowly oblanceolate, entire; heads small, yellow, radiate, usually numerous and crowded; involucre cylindric to campanulate, the phyllaries chartaceous, scarious-margined, with small green tips; achenes small, oblong or obovoid; pappus of several squamellae or paleae, often shorter in the ray flowers.

Also known as match-weed, resin-weed, broom-weed, and turpentine-weed. Plants of dry, stony plains, mesas, and slopes. The snake-weeds are worthless plants that are not even of much value in retarding soil erosion. They are more or less poisonous to sheep and goats when eaten in quantity, but are unpalatable and are seldom grazed. It is said that *G. microcephala* absorbs selenium in large quantity on certain soils. The carrying capacity of much Southwestern grassland has been greatly reduced by encroachment of the snake-weeds.

KEY TO THE SPECIES

1. Heads tiny, cylindric, about 1 mm. thick; rays 1 or 2; disk flowers 1 to 3 (2).
1. Heads larger, slender-turbinate to subglobose; rays 3 to 12; disk flowers 1 to 12 (3).
2. Heads sessile, in fasciculate glomerules of 2 to 5; ray 1 (very rarely 2); disk flowers 1 or 2 .... 1. *G. lucida*
2. Heads often peduncled, not fasciculate-glomerulate; rays 2; disk flowers 2 or 3 .... 2. *G. linoides*
3. Heads subglobose to broadly turbinate; ray flowers 7 to 14; disk flowers 7 to 24 .... 3. *G. californica*
3. Heads turbinate; flowers of the rays and the disk each 3 to 8, or the disk flowers rarely only 1 or 2 (4).
4. Involucre very slenderly turbinate, 1 to 1.5 mm. thick; rays 4 or 5; disk flowers 1 to 3 .... 4. *G. microcephala*
4. Involucre turbinate, usually 2 mm. thick or more; rays 3 to 8; disk flowers 3 to 8 .... 5. *G. Sarothrae*

1. Gutierrezia lucida Greene. Almost throughout the state, 1,000 to 7,000 feet, June to October. Colorado to Texas, west to Nevada and California, south to Mexico.

2. Gutierrezia linoides Greene. Definitely known only from the Chiricahua Mountains, Cochise County (*Blumer* in 1907, the type). The status of this species is not clear. Benson & Darrow (8, p. 316) made *G. linoides* a synonym of *G. longifolia* Greene, which they reported as occurring in Graham, Cochise, and Pima counties.

3. Gutierrezia californica (DC.) Torr. & Gray (*G. serotina* Greene, *G. polyantha* A. Nels.). Yavapai, Gila, Pinal, Cochise, and Pima counties, 1,000 to 4,000 feet, March to October, type of *G. serotina* from Tucson (*Toumey* in 1892), type of *G. polyantha* from near Tucson (*A. & R. Nelson* 1638). Central coastal California to southern Arizona and Chihuahua.

4. Gutierrezia microcephala (DC.) Gray (*G. Sarothrae* var. *microcephala* L. Benson). Navajo, Coconino, and eastern Mohave counties, south and southeast to Cochise County, 3,500 to 6,500 feet, June to October. Texas to Idaho, south to Arizona and Coahuila.

5. Gutierrezia Sarothrae (Pursh) Britt. & Rusby. Apache County to Mohave County, south to Cochise, Santa Cruz, and Pima counties, 3,000 to 8,000 feet, July to November. Saskatchewan to Kansas, south to northern Mexico and Baja California.

13. AMPHIPAPPUS (319). Chaff-bush

Low, branching shrubs, white-barked; leaves alternate, obovate or elliptic, small, entire; heads small, yellow, few-flowered, crowded in small, rounded, terminal clusters; involucre graduated, the phyllaries broad, blunt, dryish;

rays 1 or 2, small; disk flowers 3 to 6; ray achenes hairy, their pappus of more or less united bristles, awns, or paleae; disk achenes glabrous, their pappus of twisted, hispidulous bristles and narrow paleae.

1. Amphipappus Fremontii Torr. & Gray. Mohave County to northern Maricopa County, 2,000 to 2,500 feet, March to May. Southwestern Utah and northwestern Arizona to eastern California.

The Arizona form is var. *spinosus* (A. Nels.) C. L. Porter, distinguished by having densely scabro-puberulent herbage, whereas in the typical phase of the species the herbage is glabrous or nearly so.

14. ACAMPTOPAPPUS. Golden-head

Low, branching shrubs, white-barked; leaves alternate, spatulate to nearly linear, small, entire; heads medium-sized, discoid, yellow; involucre broad, graduated, the dry, scarious-margined, blunt phyllaries with a greenish subapical spot; achenes turbinate, densely villous; pappus of numerous narrowly linear paleae, some of the outer ones narrower and setiform.

1. Acamptopappus sphaerocephalus (Harv. & Gray) Gray. Northern Mohave County to Graham, Gila, Maricopa, Pima, and Yuma counties, 1,000 to 4,500 feet, dry plains and mesas, April to October. Southern Utah to central Arizona and southern California.

The plant is browsed when better forage is unavailable.

Acamptopappus Shockleyi Gray, of southern Nevada and Inyo County, California, may yet be found in northwestern Arizona. It has radiate heads solitary at the tips of the branches, whereas those of *A. sphaerocephalus* are discoid and borne at the tips of cymosely arranged branchlets.

15. HETEROTHECA. Telegraph-plant

Annual or biennial herbs, hirsute or hispid, glandular above; leaves alternate, the lower ones usually with a foliaceous stipuliform dilation at base of the petiole; heads yellow, medium-sized; involucre graduated, of numerous narrow phyllaries; ray achenes glabrous or slightly pubescent, essentially epappose; disk achenes densely hairy, their pappus double, the outer series of short bristles or setiform squamellae, the inner series of longer capillary bristles.

Plants often large and coarse, browsed by cattle.

KEY TO THE SPECIES

1. Heads relatively small, the disk in fruit (including the pappus) 8 to 10 mm. high; involucre 6 to 8 mm. high; ray achenes glabrous; appendages of the style branches ⅔ as long as the stigmatic region; upper leaves usually with cordate-clasping bases ..... 1. *H. subaxillaris*
1. Heads relatively large, the disk in fruit (including the pappus) 10 to 12 mm. high; involucre 7 to 10 mm. high; ray achenes pubescent, at least on the angles; appendages of the style branches about half as long as the stigmatic region; upper leaves usually narrowed at base ..... 2. *H. grandiflora*

1. Heterotheca subaxillaris (Lam.) Britt. & Rusby. Southern Navajo County to Yavapai County, south to Graham, Cochise, Santa Cruz, and Pima counties, 1,000 to 5,500 feet, abundant and conspicuous along roads and ditches, March to November. Delaware to Kansas, south to Florida, Texas, Arizona, and Mexico.

Sometimes known as camphor-weed because of the odor of the plant. A specimen from near Nogales (*Gould & Robbins* 3619) approaches *H. grandiflora* in having the ray achenes pubescent on the angles.

2. Heterotheca grandiflora Nutt. Near Prescott (Yavapai County), Huachuca Mountains and Mescal (Cochise County), Tucson and in the Santa Rita and Baboquivari mountains (Pima County), 2,500 to 5,500 feet, thickets and chaparral, September and October. California and Arizona.

16. CHRYSOPSIS. Golden-aster

Low, pubescent, perennial herbs; leaves alternate, spatulate to oblong or obovate, entire; heads medium-sized, yellow, radiate, usually few and cymose; involucre graduated; achenes compressed; pappus double in both the ray flowers and the disk flowers, the outer series of short squamellae or squamellate bristles, the inner series of much longer, capillary bristles.

KEY TO THE SPECIES

1. Plants silvery-silky throughout; heads large (disk in fruit 1 to 1.5 cm. high), leafy-bracted, the bracts surpassing the proper involucre and usually also the heads 1. *C. Rutteri*
1. Plants not silvery-silky, or else the heads smaller (disk in fruit 1 cm. high or less) or not leafy-bracted (2).
2. Involucre densely pubescent, its glands obscure or none. Plants canescent or grayish green (3).
2. Involucre distinctly glandular, the longer, eglandular hairs usually few or none (4).
3. Plants canescent; leaves elliptic to obovate, usually not distinctly petioled, subsericeous-canescent .... 2. *C. foliosa*
3. Plants grayish green; leaves usually spatulate or spatulate-obovate and distinctly petioled 3. *C. villosa*
4. Heads leafy-bracted; leaves mostly elliptic or oblong, sessile, green, the middle ones 2 to 4 cm. long, the lowest leaves obovate or oblanceolate. Involucre glandular and hirsute 4. *C. fulcrata*
4. Heads rarely leafy-bracted; leaves chiefly obovate or the lower ones spatulate-oblanceolate, narrowed to a petiolelike base, 2.5 cm. long or less (5).
5. Stem and involucre sparsely glandular........ 5. *C. hispida*
5. Stem and involucre densely glandular, with few long, eglandular hairs...... 6. *C. viscida*

1. Chrysopsis Rutteri (Rothr.) Greene. Western Cochise and Santa Cruz counties, about 5,000 feet, dry plains and mesas, August to October, type from the Sonoita Valley (*Rothrock* 662). Known only from southeastern Arizona.

2. Chrysopsis foliosa Nutt. White Mountains (southern Apache and Navajo counties), near Flagstaff and Tuba (Coconino County), Pinaleno Mountains (Graham County), Pine Creek and Pinal Mountains (Gila County), Galiuro Mountains (Pinal County), 3,000 to 8,500 feet, plains and canyons, August to October. Minnesota to Washington, south to central Arizona.

3. Chrysopsis villosa (Pursh) Nutt. Apache, Navajo, and Coconino counties, south to Pima County, 1,500 to 8,500 feet, dry slopes, mesas, and plains, May to October. Minnesota to Saskatchewan, south to Texas and southern Arizona.

The Hopi Indians are said to use a decoction of the leaves and flowers to relieve pain in the chest.

4. Chrysopsis fulcrata Greene (*C. resinolens* A. Nels.). Coconino, Yavapai, Graham, and Pima counties, 5,000 to 8,000 feet, usually among rocks, June to October. Montana to Texas, New Mexico, and Arizona.

5. Chrysopsis hispida (Hook.) DC. Apache County to northeastern Mohave County, south to Cochise and Pima counties, 2,000 to 7,000 feet, dry rocky slopes, April to October. Saskatchewan to British Columbia, south to southern Arizona and California.

A collection at Springerville, Apache County (*Darrow* in 1937), has numerous small heads, the involucre only about 6 mm. high.

6. Chrysopsis viscida (Gray) Greene. Near Holbrook (Navajo County), Chiricahua and Huachuca mountains (Cochise County), Santa Rita Mountains (Santa Cruz County), 5,000 to 9,500 feet, dry ledges of cliffs, May to October. Colorado to Texas and Arizona.

17. SOLIDAGO. Golden-rod

Perennial herbs; leaves alternate, usually narrow, entire or toothed; heads small, radiate, yellow, in usually racemiform or cymose panicles, often secund on the branches; involucre narrow, the phyllaries more or less graduated, usually thin and dry, sometimes with herbaceous tips; achenes short; pappus of capillary bristles.

Some of the species are reputed to be poisonous to livestock, especially to sheep. The leaves of *S. missouriensis* are said to be eaten as a salad by the Indians of northern Arizona.

KEY TO THE SPECIES

1. Heads in small, rounded cymose clusters at tips of the branches and branchlets; rays small, inconspicuous, more numerous than the disk flowers. Plants glabrous, rather tall, usually with numerous semierect branches, uniformly leafy; leaves linear or narrowly lance-linear, entire, 3-ribbed, not coriaceous....................11. *S. occidentalis*
1. Heads otherwise arranged, or else the plants low and cespitose, with coriaceous leaves; rays fewer than the disk flowers (2).
2. Heads subcylindric, in dense flattish, fastigiate cymes; leaves coriaceous; stems low, cespitose from a branched caudex (3).
2. Heads not cylindric, racemose or panicled, sometimes few and glomerate or cymose, never in dense, fastigiate cymes; leaves not definitely coriaceous; stems not cespitose from a branched caudex (4).
3. Leaves 3-nerved and reticulate, the lower ones oblanceolate or linear-oblanceolate, 2.5 to 7 mm. wide..9. *S. Petradoria*
3. Leaves 1-nerved, narrowly linear, the lower ones about 1 mm. wide.....10. *S. graminea*
4. Stem glabrous or sometimes loosely villous, never densely puberulous (5).
4. Stem densely puberulous or glandular-puberulous, sometimes (in *S. sparsiflora*) sparsely incurved-puberulous (7).
5. Heads smaller (involucre 3 to 5 mm. high), secund on the spreading, recurving, or sometimes erect branches of the usually pyramidal panicle............3. *S. missouriensis*
5. Heads larger (involucre 4 to 6 mm. high), glomerate, racemose, or in a narrow thyrse, not secund on the branches. High montane (6).
6. Phyllaries linear-lanceolate to lanceolate, acuminate to acute, thin; leaves villous-ciliate, especially toward the base...1. *S. ciliosa*
6. Phyllaries oblong, obtuse to acutish, firm; leaves not ciliate..........2. *S. decumbens*

7. Leaves very numerous and nearly uniform, only gradually reduced above, lanceolate or linear-lanceolate, gradually acuminate, strongly triplinerved, usually sharply serrate, sometimes entire (8).
7. Leaves comparatively few and usually distinctly dimorphous (the basal leaves much larger than the middle and upper stem leaves and distinctly petioled) or, if numerous and nearly uniform, then blunt or merely acute, either feather-veined or triplinerved (9).
8. Involucre 3 to 5 mm. high .... 4. *S. altissima*
8. Involucre 2 to 3 mm. high .... 5. *S. canadensis*
9. Phyllaries, at least the outer ones, with definite herbaceous or subherbaceous tips, densely puberulous or stipitate-glandular; leaves nearly uniform, the middle and upper ones elliptic to ovate-elliptic, usually 1 to 2.5 cm. wide, feather-veined or obscurely 3-nerved .... 8. *S. Wrightii*
9. Phyllaries without definite herbaceous tips (sometimes obscurely greenish above), glabrous or rarely slightly puberulous; middle and upper stem leaves usually much smaller than the basal ones, mostly lanceolate to linear or spatulate, seldom as much as 1 cm. wide, usually 3-nerved (10).
10. Phyllaries oblong or the outer ones ovate, very obtuse, firm and substramineous; involucre 4 to 6 mm. high; inflorescence thyrsoid, or with erect branches; herbage cinereous-puberulous .... 6. *S. nana*
10. Phyllaries chiefly linear to linear-oblong or the outer ones usually lanceolate, acute or acutish or the outer ones acuminate, thinner; involucre 3 to 5 mm. high; inflorescence when well developed pyramidal, the heads secund on the more or less recurved branches; herbage greener .... 7. *S. sparsiflora*

1. Solidago ciliosa Greene. Kaibab Plateau and San Francisco Peaks (Coconino County), White Mountains (Apache County), 8,500 to 12,000 feet, July to September, type from the San Francisco Peaks (*Greene* in 1889). Alberta and British Columbia to northern Arizona and California.

2. Solidago decumbens Greene. San Francisco Peaks to Flagstaff and Elden Mountain (Coconino County), White Mountains (northern Greenlee County), 8,000 to 9,500 feet, July and August. British Columbia to Oregon and Arizona.

Plants collected near Flagstaff (*Deaver*) had stems 50 cm. long.

3. Solidago missouriensis Nutt. (*S. Marshallii* Rothr.). Southern Apache and Navajo counties to North Rim of the Grand Canyon (?), Flagstaff, Mormon Lake, and Oak Creek (Coconino County), southward to the mountains of Cochise, Santa Cruz, and Pima counties, 5,000 to 9,000 feet, open pine forest and along streams, June to August, type of *S. Marshallii* from the Chiricahua Mountains (*Rothrock* 530). Michigan and Tennessee to British Columbia, Oregon, and Arizona.

Specimens from Navajo, Coconino, Yavapai, and Cochise counties have been identified by R. C. Friesner as *S. missouriensis* var. *fasciculata* Holz.

4. Solidago altissima L. (*S. canadensis* var. *arizonica* Gray, *S. arizonica* Woot. & Standl.). Coconino, Yavapai, Gila, Cochise, Santa Cruz, and Pima counties, 2,500 to 8,500 feet, August to October, type of *S. canadensis* var. *arizonica* from "Boulder Creek" (*Rothrock* 782). Atlantic coast states to Wyoming and Arizona.

5. Solidago canadensis L. San Francisco Peaks and Grand Canyon (Coconino County), eastern Yavapai and western Gila counties, 3,000 to 8,500 feet, July and August. Newfoundland to Virginia, westward to Montana, Nevada, and central Arizona.

The species is represented in Arizona by var. *gilvocanescens* Rydb. (*S. gilvocanescens* Smyth). The stems reach a height of 2.5 m.

6. Solidago nana Nutt. Coconino County, from the Kaibab Plateau to Oak Creek, 6,500 to 9,000 feet, plains, canyons, and slopes, July to September. Alberta to Nebraska and northern Arizona.

The stems reach a much greater height in Arizona than elsewhere within the range of the species.

7. Solidago sparsiflora Gray (*S. sparsiflora* var. *subcinerea* Gray, *S. trinervata* Greene). Almost throughout the state, 2,000 to 8,500 feet, pine forest, mountain meadows, and chaparral, June to October, type of *S. sparsiflora* from Camp Lowell, Pima County (*Rothrock* 706), type of var. *subcinerea* from Rucker Canyon, Cochise County (*Lemmon*). South Dakota and Wyoming to Texas, New Mexico, and Arizona.

A specimen from the Chiricahua Mountains (*K. F. Parker* 6679) was identified by R. C. Friesner as *S. sparsiflora* × *S. Wrightii.*

8. Solidago Wrightii Gray. Apache, Coconino, Mohave, Yavapai, Pinal, Cochise, and Pima counties, 3,500 to 9,500 feet, mostly in pine forest, August to November. Western Texas to Arizona.

Var. *adenophora* Blake (*S. subviscosa* Greene?) occurs farther south, in the mountains of Graham, Pinal, Cochise, and Pima counties, type from the Santa Catalina Mountains (*Harrison* 3106). It is distinguished by the presence of stipitate glands on the involucre, pedicels, stem, and leaves.

9. Solidago Petradoria Blake (*Petradoria pumila* (Nutt.) Greene). Northeastern Apache County to the Kaibab Plateau and Grand Canyon (Coconino County), near Pipe Springs (Mohave County), 4,500 to 8,000 feet, rocky slopes and canyons, locally common, June to September. Wyoming to Oregon (?), western Texas, northern Arizona, and southeastern California.

It is reported that the Hopi Indians use this plant to alleviate pain in the breast.

10. Solidago graminea (Woot. & Standl.) Blake (*Petradoria graminea* Woot. & Standl.). Kaibab Plateau (Coconino County), about 7,000 feet, in pine and juniper-oak forest, June to September. Southern Utah, northwestern New Mexico, and northern Arizona.

11. Solidago occidentalis (Nutt.) Torr. & Gray. Near Winslow (Navajo County), Tuba (Coconino County), Chiricahua Mountains (Cochise County), 4,500 to about 7,000 feet, August and September. Alberta to British Columbia, south to New Mexico, Arizona, and California.

18. APLOPAPPUS (320)

Herbs or shrubs; leaves alternate, entire to bipinnatifid; heads small to large, usually radiate, yellow, or the rays rarely saffron-colored; involucre usually definitely graduated; achenes cylindric to turbinate; pappus copious, of graduated capillary bristles.

KEY TO THE SPECIES

1. Plants with strictly herbaceous stems but sometimes with a woody caudex. (If the stems exceptionally herbaceous in normally woody species (*A. heterophyllus*, *A. Drummondii*), then the heads discoid and turbinate, and the leaves usually entire.) (2).

1. Plants shrubs or undershrubs, or if almost entirely herbaceous (rarely so in *A. heterophyllus* and *A. Drummondii*), then the heads discoid and turbinate (9).
2. Heads discoid. Leaves closely serrate, the teeth with white, spinescent tips; heads campanulate or hemispheric, many-flowered; plants rarely more than 25 cm. high ....1. *A. Nuttallii*
2. Heads radiate (3).
3. Leaves strongly 3-nerved and veiny, coriaceous, entire; plants with woody branched caudices. Stems low, few-leaved; heads usually solitary (4).
3. Leaves not 3-nerved and veiny; plants not with woody, branched caudices (5).
4. Phyllaries very obtuse to barely acutish, strongly graduated.........7. *A. armerioides*
4. Phyllaries acuminate to acute, usually little graduated..................8. *A. acaulis*
5. Leaves entire or rarely with a few teeth, large, the basal ones lanceolate to obovate, rarely less than 1.5 cm. wide; phyllaries not spinescent-tipped (6).
5. Leaves sharply serrate to bipinnatifid, the teeth or lobes spinescent-tipped, the basal leaves not large; phyllaries spinescent-tipped (7).
6. Heads 1 to 3 per stem; disk 1.5 to 2.5 cm. thick; rays usually saffron-colored or dull orange; stem loosely pilose above.................................5. *A. croceus*
6. Heads several or numerous; disk 1 cm. thick or less; rays yellow; stem hispidulous above ....6. *A. Parryi*
7. Plants annual; involucre strigose or hirsute, obscurely if at all glandular..2. *A. gracilis*
7. Plants perennial; involucre usually conspicuously glandular or tomentose (8).
8. Stems leafy, at least up to the peduncles..........................3. *A. spinulosus*
8. Stems nearly leafless, the leaves reduced and remote, the upper ones scalelike, not more than 5 mm. long...4. *A. junceus*
9. Heads solitary at the tips of the branches, definitely peduncled, radiate, large, the disk 1 cm. high or more; involucre broad, the phyllaries about 3-seriate, not strongly graduated; pappus bright white, about 6 mm. long...............9. *A. linearifolius*
9. Heads cymose or panicled or, if solitary, then not definitely peduncled; involucre usually narrow, often strongly graduated; pappus straw-colored or dull whitish or brownish (10).
10. Heads usually solitary, leafy-bracted, the proper phyllaries about 2-seriate, subequal, 8 to 11 mm. long; style appendages at least twice as long as the stigmatic portion, acuminate. Rays 0 to 6; plants densely glandular..............10. *A. suffruticosus*
10. Heads cymose or panicled or, if solitary, then not leafy-bracted; involucre strongly graduated, or else much shorter; style appendages less than twice as long as the stigmatic portion, or else (in *A. scopulorum*) obtusish (11).
11. Leaves spatulate to broadly obovate; heads discoid. Leaves entire, obtuse or apiculate, 8 to 23 mm. long, 3 to 15 mm. wide; heads small, in small terminal cymes ....11. *A. cuneatus*
11. Leaves linear to linear-lanceolate or cuneate or, if spatulate, then the heads radiate (12).
12. Heads radiate, or exceptionally discoid in *A. laricifolius* (13).
12. Heads discoid (15).
13. Leaves densely glandular-punctate, essentially linear, not more than 2 mm. wide; involucre 2- or 3-seriate, not strongly graduated, 3 to 5 mm. high...12. *A. laricifolius*
13. Leaves obscurely if at all glandular-punctate, often stipitate-glandular or resinous, spatulate to obovate, 2 to 5 mm. wide, cuspidate-pointed; involucre several-seriate, strongly graduated (14).
14. Plants densely stipitate-glandular; leaves spatulate to oblanceolate, 15 to 20 mm. long, 2 to 3 mm. wide..13. *A. Watsoni*
14. Plants glabrous but often glutinous; leaves obovate or spatulate-obovate, 8 to 15 mm. long, 3 to 5 mm. wide......................................14. *A. cervinus*
15. Leaves linear-lanceolate, 3-ribbed, entire or with a minutely spinulose-denticulate margin; style appendages linear, much longer than the stigmatic region, sometimes twice as long; corolla teeth 1.5 to 2 mm. long (16).
15. Leaves linear to cuneate or broader, 1-nerved, entire to toothed or pinnatifid; style appendages deltoid or triangular, shorter than the stigmatic region; corolla teeth not more than 1 mm. long (17).

16. Leaves densely impressed-punctate . . . . 15. *A. salicinus*
16. Leaves not impressed-punctate . . . . 16. *A. scopulorum*
17. Leaves pinnatifid, the lobes linear, several times as long as the breadth of the leaf rachis . . . 17. *A. tenuisectus*
17. Leaves entire or sometimes with a few teeth, occasionally subpinnatifid, the lobes then scarcely longer than the breadth of the leaf rachis (18).
18. Phyllaries with a thickened apex bearing a large rounded gland; leaves entire to laciniate-pinnatifid . . . . 18. *A. acradenius*
18. Phyllaries scarcely thickened at apex, without a distinct gland; leaves usually entire (19).
19. Heads 7- to 15-flowered, the involucre 3.5 to 5 mm. high, the phyllaries usually obscurely if at all green-tipped . . . . 19. *A. heterophyllus*
19. Heads 18- to 30-flowered, the involucre 6 to 8 mm. high, the phyllaries definitely greenish-tipped . . . . 20. *A. Drummondii*

1. Aplopappus Nuttallii Torr. & Gray. Apache, Navajo, and Coconino counties, 5,500 to 8,000 feet, barren, rocky hills and plains, June to October. Saskatchewan to Alberta, south to Nebraska, New Mexico, and northern Arizona.

The Hopi Indians make a tea from the roots, which they administer for coughs.

A collection in House Rock Valley (*Kearney & Peebles* 13632) was stated by Maguire (*Amer. Midland Nat.* 37 : 145) to be intermediate between *Haplopappus Nuttallii* var. *Nuttallii* Maguire and var. *depressus* Maguire, the latter characterized by "low densely cespitose habit . . . and conspicuous long white bristles along the leaf margins."

2. Aplopappus gracilis (Nutt.) Gray. Throughout the state, up to 7,000 feet, dry plains, mesas, and rocky slopes, February to November. Colorado to Texas, Arizona, southeastern California, and Mexico.

3. Aplopappus spinulosus (Pursh) DC. Throughout the state in various forms, March to October. Minnesota to Alberta, south to Texas, Arizona, southeastern California, and northern Mexico.

KEY TO THE VARIETIES

1. Involucre tomentose, not glandular . . . . ssp. *typicus*
1. Involucre glandular (2).
2. Leaves (the lower ones) usually bipinnatifid, the middle ones 0.5 to 2.5 cm. long, serrate to pinnatifid; rays 8 mm. long or less . . . . var. *turbinellus*
2. Leaves merely pinnatifid, the middle ones 2 to 4 cm. long, with remote narrow lobes; rays 8 to 15 mm. long . . . . var. *Gooddingii*

The typical form (ssp. *typicus* H. M. Hall), apparently rare in Arizona, has been collected near Burro Creek (southeastern Mohave County), near Superior (Pinal County), and in western Pima County, 2,000 to 3,000 feet. The most common variety in Arizona, occurring throughout the state up to about 5,000 feet, is var. *turbinellus* (Rydb.) Blake. Var. *Gooddingii* (A. Nels.) Blake, the prevailing variety in western Arizona, ranges from the Grand Canyon and northern Mohave County to Cochise, Pima, and Yuma counties, 3,500 feet or lower.

4. Aplopappus junceus Greene. San Bernardino Ranch, Cochise County (*Mearns* 762). Southern Arizona and California, northwestern Sonora, and northern Baja California.

It is possible that Mearns's specimen (identified by H. M. Hall, not seen by the writer) is incorrectly labeled as to locality. Otherwise it represents a considerable eastward extension of the range of *A. junceus.*

5. **Aplopappus croceus** Gray. The Arizona plants belong to var. *genuflexus* (Greene) Blake (*Pyrrocoma genuflexa* Greene, *P. adsurgens* Greene). San Francisco Peaks to the Mogollon Escarpment (Coconino County), White Mountains (Apache and northern Greenlee counties), 6,000 to 9,500 feet, mountain meadows and openings in coniferous forest, July to October, type of *P. genuflexa* from near Flagstaff (*Toumey* in 1894), type of *P. adsurgens* from Flagstaff (*Rusby* 645). The typical phase of the species ranges from Wyoming to New Mexico and eastern Utah.

6. **Aplopappus Parryi** Gray. Kaibab Plateau, San Francisco Peaks, and Bill Williams Mountain (Coconino County), White Mountains (Apache and northern Greenlee counties), Pinaleno Mountains (Graham County), Chiricahua and Huachuca mountains (Cochise County), Santa Catalina Mountains (Pima County), 8,000 to 11,500 feet, coniferous forests, July to September. Wyoming to New Mexico and Arizona.

7. **Aplopappus armerioides** (Nutt.) Gray. Northern Apache, Navajo, and eastern Coconino counties, about 6,000 feet, mesas, often with pinyon and juniper, May and June. Saskatchewan to Nebraska, New Mexico, and northeastern Arizona.

*8. **Aplopappus acaulis** (Nutt.) Gray. The glabrous var. *glabratus* D. C. Eaton has been reported from northern Arizona. Saskatchewan to Wyoming, Utah, Arizona (?), and California.

9. **Aplopappus linearifolius** DC. The species is represented in Arizona by var. *interior* (Coville) M. E. Jones. Northern Mohave County through Yavapai County to eastern Maricopa County, 2,000 to 5,000 feet, March to June. Southwestern Colorado and Utah to central and western Arizona, southeastern California, and Baja California.

The handsomest species in Arizona, very showy in flower.

*10. **Aplopappus suffruticosus** (Nutt.) Gray. Mountain ridges and slopes, Sierra Nevada of California to Montana, Wyoming, and Nevada (reported from Arizona), July to September.

11. **Aplopappus cuneatus** Gray. Mohave and Yavapai counties to Graham, Cochise, and Pima counties, 3,000 to 7,000 feet, rock ledges, September and October. Arizona to southeastern California and Baja California.

The species is represented in Arizona by var. *spathulatus* (Gray) Blake, with discoid heads.

12. **Aplopappus laricifolius** Gray. Greenlee County to Mohave County, south to Cochise, Pima, and Yuma counties, 3,000 to 6,000 feet, mesas, slopes, and canyons, August to November. Western Texas to Arizona and northern Mexico.

Sometimes called turpentine-brush.

13. **Aplopappus Watsoni** Gray. Grand Canyon, Coconino County (*Eastwood* in 1905 and 1913, *Blake* 9815, 9818), about 7,000 feet, on rocks, September and October. Southern Utah, southern Nevada, and northern Arizona.

14. Aplopappus cervinus Wats. Grand Canyon, Coconino County, 7,000 to 7,500 feet (*Hall* 11190, 11191). Southern Utah and northern Arizona. (Hall's material not seen by the writer).

15. Aplopappus salicinus Blake. Known only from the type locality, Bright Angel Trail, Grand Canyon (*Eastwood* 10, in 1905, the type, in bud in October, and *Eastwood* 3766, in 1913, foot of Bright Angel Trail, in bud in September).

16. Aplopappus scopulorum (Jones) Blake. Both sides of the Grand Canyon, Coconino County (*Eastwood* 3586, *Thornber* 8498, *Eastwood & Howell* 7111), September. Southern Utah and northern Arizona.

In the typical plant the stem and leaves are glabrous except for the ciliolate leaf margin. Var. *hirtellus* Blake, with hirtellous young branches, peduncles, and leaves, has been collected in Waterlily Canyon, 35 miles north of Kayenta, Navajo County (*R. E. Burton* in 1934), and it occurs also in southeastern Utah.

17. Aplopappus tenuisectus (Greene) Blake (*Isocoma tenuisecta* Greene). Greenlee, Graham, Gila, Pinal, Cochise, and Pima counties, 2,000 to 5,500 feet, plains and mesas, August to October, type of *Isocoma tenuisecta* from Tucson (*Smart* in 1867). Southwestern Texas, southern Arizona, and northern Mexico. Burro-weed.

18. Aplopappus acradenius (Greene) Blake. Grand Canyon (Coconino County), and Mohave, Gila, Maricopa, Pinal, Pima, and Yuma counties, up to 4,000 feet, in various habitats, often in saline soil, June to October. Southern Utah to Arizona, southern California, and northern Mexico.

19. Aplopappus heterophyllus (Gray) Blake. Navajo County to eastern Mohave County, south to Cochise, Pima, and Yuma counties, up to 5,000 feet, mesas and plains, often in saline soil, June to October. Colorado to Texas, Arizona, and northern Mexico.

Jimmy-weed, rayless-goldenrod. The plant often occupies overgrazed range land and is a common roadside weed in the irrigated districts. This and doubtless some of the closely related species are generally unpalatable, but when eaten in quantity by cattle cause the disease known as "milk-sickness" or "trembles," which is transmissible through the milk to human beings.

A specimen from the Sierra Ancha (Gila County) approaches *A. acradenius* in having a rather definite gland on the phyllaries.

20. Aplopappus Drummondii (Torr. & Gray) Blake (*Isocoma Rusbyi* Greene). Apache, Navajo, and Coconino counties, 2,000 to 7,000 feet, July to November, type of *Isocoma Rusbyi* from Holbrook (*Rusby* in 1883). Texas to northern Arizona, and northeastern Mexico.

19. CHRYSOTHAMNUS (321). Rabbit-brush

Shrubs; leaves alternate, linear to linear-filiform, sometimes dotted with impressed glands; heads small or medium-sized, discoid, yellow, 4- to 7-flowered, usually panicled; involucre several-seriate, graduated, the phyllaries chartaceous, sometimes herbaceous-tipped, in more or less definite vertical ranks; achenes pubescent or glabrous; pappus of numerous capillary bristles.

The latex of several species contains rubber of fair quality, that of *C. nauseosus* being reported to yield 2.8 per cent on the average and as much as 6.5 per cent in selected individual plants. It has not proved practicable as yet to utilize this source of rubber commercially notwithstanding the great abundance of the plants. The rubber content is reported to be highest in plants growing in saline soil. Several species are browsed to a limited extent, but the genus as a whole is of small forage value. The plants tend to increase on overgrazed land at the expense of more valuable species. Rabbit-brushes are used by the Hopi as one of the "kiva" fuels, and for making wind breaks, arrows, and wicker work. A yellow dye is obtained from the flowers, and a green dye from the inner bark.

KEY TO THE SPECIES

1. Leaves conspicuously punctate with impressed glands, terete or slightly flattened. Plants glabrous throughout (2).
1. Leaves not punctate with impressed glands (3).
2. Phyllaries with the midrib often glandular-thickened for most of its length, but without a roundish terminal gland .... 1. *C. paniculatus*
2. Phyllaries with a large, roundish terminal gland .... 2. *C. teretifolius*
3. Stems covered with a dense, often matted tomentum (4).
3. Stems glabrous to pubescent, but never tomentose (5).
4. Heads in a leafy spikelike panicle, or in racemelike clusters; outer phyllaries acuminate, herbaceous-tipped .... 7. *C. Parryi*
4. Heads cymose, terminal; phyllaries obtuse to acute, not herbaceous-tipped 8. *C. nauseosus*
5. Involucre 9 to 13 mm. high, the phyllaries in very sharply defined vertical ranks, all very acute to cuspidate-acuminate; achenes glabrous to rather sparsely pubescent (6).
5. Involucre 5 to 9 mm. high, the phyllaries in less sharply defined vertical ranks, obtuse to merely acute, or the outer ones acuminate; achenes densely pubescent, rarely merely glandular (7).
6. Plants densely cinereous-puberulent, only 1 to 3 dm. high; leaves narrowly spatulate to oblanceolate or nearly linear .... 3. *C. depressus*
6. Plants glabrous except for the ciliolate leaf margins, taller; leaves linear. 4. *C. pulchellus*
7. Phyllaries obtuse to acute .... 5. *C. viscidiflorus*
7. Phyllaries, at least the outer ones, with slender, acuminate and often cuspidate, greenish tips .... 6. *C. Greenei*

1. Chrysothamnus paniculatus (Gray) H. M. Hall. Near Peach Springs, Kingman, and Black Mountains (Mohave County), Rye Creek (Gila County), 2,500 to 4,000 feet, September to November. Utah, Arizona, southeastern California, and Sonora.

This species is reported to contain 2.5 per cent or more of rubber.

2. Chrysothamnus teretifolius (Dur. & Hilgard) H. M. Hall. Union Pass, Mohave County, 3,500 feet (*Palmer* in 1870), September and October. Southwestern Utah, Nevada, northwestern Arizona, and southern California.

3. Chrysothamnus depressus Nutt. Navajo and Coconino counties to Hualpai Mountain (Mohave County), 5,000 to 7,000 feet, dry, rocky slopes, May to October. Colorado to Nevada, New Mexico, Arizona, and southeastern California.

4. Chrysothamnus pulchellus (Gray) Greene. Navajo County, near Inscription House (*Darrow* 2881) and Hopi Indian Reservation (*Whiting*

2784), Coconino County, northeast of Tuba (*Kearney & Peebles* 12894), 5,500 to 6,500 feet, sandy soil, September. Kansas to Texas, New Mexico, and northeastern Arizona.

The Arizona plants belong to var. *Baileyi* (Woot. & Standl.) Blake, with leaves glabrous on both surfaces, the margins scabrous-ciliolate (entirely glabrous in typical *C. pulchellus*).

5. Chrysothamnus viscidiflorus (Hook.) Nutt. Including its varieties, the species ranges from North Dakota to British Columbia, south to New Mexico, Arizona, and eastern California.

KEY TO THE VARIETIES

1. Achenes merely glandular. Herbage evenly but not densely hispidulous with both conical and glandular-capitate hairs; leaves linear, 10 to 18 mm. long, about 1 mm. wide var. *molestus*
1. Achenes densely pubescent (2).
2. Plants usually more than 40 cm. high; leaves 2 to 5 (rarely only 1) mm. wide. Herbage glabrous or nearly so.................................*C. viscidiflorus* (typical)
2. Plants 10 to 40 cm. high; leaves not more than 2 mm. wide (3).
3. Herbage glabrous or nearly so; leaves 1-nerved, not more than 1 mm. wide; phyllaries without subapical spot..var. *stenophyllus*
3. Herbage more or less hirtellous; leaves mostly 3-nerved, 1 to 2 mm. wide; phyllaries with green, somewhat thickened, subapical spot...........................var. *elegans*

Typical *C. viscidiflorus* is known from northern Apache, Navajo, and Coconino counties, 6,000 to 7,000 feet. Var. *molestus* Blake has been found in the vicinity of the San Francisco Peaks, Coconino County (*Hall* 11199, the type), and on the Hualpai Indian Reservation, northwestern Coconino County, 5,500 feet. Var. *stenophyllus* (Gray) H. M. Hall, the commonest form in Arizona, occurs in Apache, Navajo, and Coconino counties, 5,000 to 6,000 feet, dry mesas and slopes, August to October. Var. *elegans* (Greene) Blake (*C. elegans* Greene) is known in Arizona only from Coconino County, where it has been collected on the Kaibab Plateau, in the Grand Canyon, and eastward to the valley of the Little Colorado, 5,000 to 6,500 feet.

6. Chrysothamnus Greenei (Gray) Greene. Represented in Arizona by var. *filifolius* (Rydb.) Blake (*C. laricinus* Greene), with leaves less than 1 mm. wide and usually only 1 to 2 cm. long. Apache, Navajo, and Coconino counties, 5,000 to 7,000 feet, sandy or stony plains and mesas, July to October, type of *C. laricinus* from the Hopi Reservation–Little Colorado River region (*Hough* 33). Colorado to Nevada, New Mexico, and northeastern Arizona.

7. Chrysothamnus Parryi (Gray) Greene. The species, with its varieties, ranges from Wyoming and Nebraska to New Mexico, Arizona, and California. The Arizona plants belong to var. *nevadensis* (Gray) Kittell. Known from the Kaibab Plateau to Coconino Wash, 9 miles south of the Grand Canyon (Coconino County), and Hopi Indian Reservation (Navajo County), 5,500 to 9,000 feet, in open pine forest, sometimes with sagebrush, August and September.

C. Parryi subsp. *attenuatus* (Jones) Hall & Clements (321, p. 201) also is recorded from Arizona on the basis of *Jones* 6052k from the "Buckskin Mountains," Kaibab Plateau. This specimen, however, is quite indistinguishable from material referred by Hall and Clements to subsp. *nevadensis*.

8. Chrysothamnus nauseosus (Pall.) Britton. In one or another of its forms, this variable species ranges from Saskatchewan to British Columbia, south to western Texas, northern Mexico, and Baja California. The following key to the Arizona varieties is based on Hall and Clements's monograph.

KEY TO THE VARIETIES

1. Involucre glabrous, or the margins of the phyllaries sometimes ciliate (2).
1. Involucre more or less tomentose or tomentulose, at least on the outer phyllaries (5).
2. Achenes glabrous....................var. *abbreviatus*
2. Achenes densely pubescent (3).
3. Leaves linear, 1 to 2 mm. wide, persistent....................var. *graveolens*
3. Leaves linear-filiform, less than 1 mm. wide, soon deciduous in var. *junceus* (4).
4. Plants essentially leafless; phyllaries in very distinct vertical rows; corolla teeth thinly long-pilose, at least when young....................var. *junceus*
4. Plants leafy; phyllaries in less distinct vertical rows; corolla teeth glabrous var. *consimilis*
5. Achenes glabrous (6).
5. Achenes densely pubescent (7).
6. Phyllaries acute or subacuminate; corolla 9 to 10 mm. long..............var. *Bigelovii*
6. Phyllaries obtuse; corolla 10 to 12 mm. long....................var. *glareosus*
7. Branches clothed with a dense, clear-white tomentum. Inner phyllaries obtuse, glabrous var. *latisquameus*
7. Branches with a yellowish or dull-white tomentum (8).
8. Involucre 10 to 12 mm. high, the phyllaries obtuse, in very distinct vertical ranks; corolla 10 to 12 mm. long, the teeth villous....................var. *turbinatus*
8. Involucre 7 to 9 mm. high, the phyllaries acute, the vertical ranks not very sharply defined; corolla 7 to 9 mm. long, the teeth glabrous................var. *gnaphalodes*

Var. *abbreviatus* (Jones) Blake occurs 19 miles west of Cameron and on the north side of the Grand Canyon, Coconino County (*Kearney & Peebles* 12816, *Eastwood & Howell* 7071), about 6,000 feet, dry slopes, September.

Var. *graveolens* (Nutt.) Piper ranges from Apache County to Hualpai Mountain and Pipe Springs (Mohave County), 2,000 to 8,000 feet, mesas and slopes, July to October.

Var. *junceus* (Greene) H. M. Hall (*C. junceus* Greene) is found from Cameron to the Kaibab Plateau and Grand Canyon (Coconino County), near Kingman (Mohave County), and along the Gila River, eastern Greenlee County (*Greene* in 1880, the type collection), 3,500 to 5,500 feet, often on limestone bluffs, September and October. Known only from Arizona.

Var. *consimilis* (Greene) H. M. Hall (*C. consimilis* Greene) occurs in Apache, Navajo, Coconino, eastern Mohave, and Yavapai counties, also in foothills of the Santa Rita and Santa Catalina mountains (Pima County), 4,000 to 7,000 feet, valleys, plains, and mesas, often in saline soil, July to November, type from the base of the San Francisco Peaks (*Greene* in 1895).

Var. *Bigelovii* (Gray) H. M. Hall (*C. moquianus* Greene), extends from northern Apache and northern Navajo counties to the vicinity of Flagstaff (Coconino County), 4,500 to 7,000 feet, dry slopes and mesas, July to September, type of *C. moquianus* from the Hopi Indian Reservation (*Zuck* in 1897).

Var. *glareosus* (Jones) H. M. Hall. A collection on the Little Colorado River (*Thurber*? in 1851) was cited by Hall.

Var. *latisquameus* (Gray) H. M. Hall (*C. arizonicus* Greene) has been collected near Oraibi (northern Navajo County) and is apparently more common in Cochise, Santa Cruz, and eastern Pima counties, 4,000 to 7,000 feet, chiefly along streams, September and October, type of *C. arizonicus* from the Santa Rita Mountains (*Brandegee*).

Var. *turbinatus* (Jones) Blake (*C. turbinatus* (Jones) Rydb.) is known from Adamana and Billings (Apache County) and from the Canaan Ranch, Kaibab Plateau (northern Coconino County), 5,000 to 5,500 feet, sandy saline soil, September and October, type from Canaan Ranch (*Jones* 6066c).

Var. *gnaphalodes* (Greene) H. M. Hall occurs near the Carrizo Mountains (northern Apache County), on the Kaibab Plateau and vicinity of Flagstaff (Coconino County), and in Cochise County, 6,000 to 7,000 feet, gravelly-sandy slopes and mesas, July to September. Some of the Arizona specimens have leaves up to 2.5 mm. wide and approach var. *speciosus* (Nutt.) Hall.

20. LESSINGIA (322)

Low, branching annuals, tomentose below, glandular above; leaves alternate, oblong-obovate to linear, entire or toothed; heads small, solitary, yellow, discoid, but with enlarged and palmately lobed outer corollas; involucre graduated; achenes turbinate, silky; pappus of numerous unequal brown bristles.

1. Lessingia Lemmoni Gray (*L. germanorum* var. *Lemmoni* J. T. Howell). Ash Fork, Yavapai County (*Lemmon* in 1884, the type collection), and near Peach Springs, Kingman, and Santa Claus (Mohave County), 3,500 to 5,000 feet, June and July. West-central Arizona and southern California.

21. APHANOSTEPHUS (323)

Low, somewhat cinereous-puberulous, annuals, branching; leaves linear to spatulate or obovate, the lower ones crenate-toothed or pinnatifid, the upper leaves often entire; heads rather small, solitary at the tips of the stem and branches, the rays white or purple-tinged, the disk yellow; involucre broad, the phyllaries about 3-seriate, somewhat graduated, lanceolate, green-centered with scarious margin and apex; achenes prismatic or subterete, about 10-ribbed; pappus a very short ciliolate-fringed crown.

1. Aphanostephus humilis (Benth.) Gray (*A. arizonicus* Gray). Near Clifton and Guthrie (Greenlee County), near Winkelman, Florence, Sacaton, and Red Rock (Pinal County), Benson (Cochise County), 1,000 to 3,500 feet, river banks and plains, March to September, type of *A. arizonicus* collected on the Gila River (*Rothrock*). Texas to southern Arizona, south to Mexico.

Shinners (323) considered *A. humilis* to be confined to Mexico, and took up Gray's name, *A. arizonicus,* for the Arizona plants.

22. GREENELLA

Low, slender, branching, glabrous, winter annuals; leaves alternate, linear to lanceolate, entire; heads small, scattered or somewhat corymbose, radiate or discoid, the rays when present white; involucre few-seriate, the phyllaries chartaceous, scarious-margined, green-tipped; achenes villous or glabrous; pappus a dissected hyaline crown.

KEY TO THE SPECIES

1. Rays present, white; ovaries and achenes densely canescent-pilosulous. . . .1. *G. arizonica*
1. Rays none; ovaries glabrous or minutely puberulous. . . . . .2. *G. discoidea*

1. Greenella arizonica Gray. Pinal and Pima counties, 1,000 to 4,000 feet, gravelly soil on mesas and plains, March to September, type from near Tucson (*Greene* in 1877). Southern Arizona and northern Mexico.

2. Greenella discoidea Gray. Known only from the type collection in Tanner Canyon, Huachuca Mountains, Cochise County (*Lemmon*), September.

23. ACHAETOGERON

Plants annual (?); stems several, up to 30 cm. high, ascending, leafy except toward the apex, 1- to few-headed, spreading-pilose below, ascending- or appressed-pilose above; leaves spatulate to linear, entire; heads medium-sized, the rays pale blue (sometimes white?), the disk yellow; involucre about 2-seriate, pilose, of thin, lanceolate, subequal phyllaries with a greenish mid-line; achenes compressed, 2-nerved, puberulous; pappus a minute setulose crown.

1. Achaetogeron chihuahuensis Larsen. White Mountains (southern Apache County), up to 9,500 feet, grassy hillsides, locally abundant, July to September. Eastern Arizona and Chihuahua.

The resemblance to a small-headed *Erigeron* is striking.

24. PSILACTIS

Slender, branching annuals, glandular, at least above; leaves alternate, entire to pinnatifid; heads radiate, small, the rays white to purplish, the disk yellow; involucre few-seriate, graduated, the phyllaries herbaceous, at least at tip; ray achenes epappose; disk achenes with a pappus of capillary bristles.

KEY TO THE SPECIES

1. Plants 50 cm. high or more, erect, branched only above; stem below glabrous or hirsute; leaves mostly entire or merely toothed. . . . . .1. *P. asteroides*
1. Plants usually about 30 cm. high or less, diffusely branched from the base; stem glandular-hispidulous throughout, usually also somewhat pilose; leaves mostly laciniate-pinnatifid with spinescent-tipped lobes, sometimes merely toothed. . . . . .2. *P. Coulteri*

1. Psilactis asteroides Gray. Woodruff (Navajo County), Tuba and near Leupp (Coconino County), south to Cochise and Pinal counties, 1,500 to 5,000 feet, river bottoms, marshes, and roadsides, June to October. South-western Texas to Arizona and central Mexico.

2. Psilactis Coulteri Gray. Maricopa, Pinal, Pima, and Yuma counties, 200 to 2,500 feet, river bottoms and roadsides, sometimes in saline soil, February to October. Southern Arizona, southern Nevada, southeastern California, and Sonora.

25. MONOPTILON

Dwarf, hispid-hirsute, winter annuals, diffusely branched; leaves linear-spatulate, small, entire, the upper ones subtending the solitary heads; rays white or rose-tinged, often drying bluish; disk yellow; involucre broad, nearly

1-seriate, the phyllaries equal, subherbaceous, scarious-margined; achenes obovate-oblong, compressed, 2-nerved; pappus of numerous unequal bristles or narrow paleae, or of a scarious cup and a single subplumose bristle.

KEY TO THE SPECIES

1. Pappus of numerous unequal bristles, or of bristles and short narrow paleae
1. *M. bellioides*
1. Pappus of a scarious cup and a single subplumose bristle............2. *M. bellidiforme*

1. Monoptilon bellioides (Gray) H. M. Hall. Mohave, Maricopa, Pinal, Pima, and Yuma counties, 200 to 3,500 feet, sandy or stony mesas and slopes, February to April. Southern Utah, Arizona, southern California, and Sonora.

2. Monoptilon bellidiforme Torr. & Gray. Fort Mohave, Mohave County (*Lemmon* in 1884, mixed with *M. bellioides*), 500 feet, April, also on the Nevada shore of Lake Mead (*Poyser* 37). Southwestern Utah to western Arizona and southeastern California.

26. TOWNSENDIA (324)

Low, perennial, or rarely annual or biennial herbs; leaves spatulate to nearly linear, entire; heads medium-sized or rather large, *Aster*-like, with white, rosy, or violet rays and a yellow disk; involucre graduated, the phyllaries mostly lanceolate, with a narrow, colored, scarious margin; achenes of the disk compressed, usually thick-margined, pubescent with 2-forked or glochidiate hairs; pappus of several or many long awns or paleae, or sometimes of few awns and several squamellae, or of squamellae only, often reduced in the ray achenes.

KEY TO THE SPECIES

1. Plants tall, usually 25 cm. high or more, simple-stemmed; pappus in both the ray flowers and the disk flowers of a few squamellae not longer than the proper tube of the corolla
1. *T. formosa*
1. Plants usually dwarf, or, if nearly 25 cm. high, then much branched; pappus, at least in the disk flowers, much longer than the proper tube of the corolla (2).
2. Plants pulvinate-cespitose perennials, strictly acaulescent; heads comparatively large, strictly sessile among the leaves or very rarely on a bracted peduncle up to 2 cm. long, usually much surpassed by the leaves; involucre (1) 1.5 to 2 cm. high....2. *T. exscapa*
2. Plants with evident leafy stems when fully developed, but the stems sometimes very short; heads smaller, usually pedunculate, normally exceeding the leaves; involucre rarely more than 1 cm. high (3).
3. Plants diffusely much branched from a slender, apparently annual or at most biennial root; pappus of the ray flowers of short scales, very much shorter than that of the disk...3. *T. strigosa*
3. Plants pulvinate-cespitose, from a much-branched, perennial caudex; pappus of the ray flowers of awns, similar to that of the disk, but sometimes only ⅓ as long (4).
4. Pappus of the ray flowers nearly or quite as long as that of the disk; leaves relatively broad, obovate-spatulate; stems cinereous-pubescent.................4. *T. arizonica*
4. Pappus of the ray flowers ⅓ to ⅔ as long as that of the disk; leaves narrowly spatulate; stems very densely white-pubescent.................................5. *T. incana*

1. Townsendia formosa Greene. White Mountains (Apache and northern Greenlee counties), 7,000 to 9,500 feet, grassy slopes and wet meadows, June to September. Southwestern New Mexico and eastern Arizona.

A very beautiful plant.

2. Townsendia exscapa (Richards.) Porter. Coconino, southern Navajo, Yavapai, Gila, Pinal, and Santa Cruz counties, 4,500 to 7,000 feet, open slopes and mesas, April to August. Alberta and Saskatchewan to Texas, Arizona, and Chihuahua.

3. Townsendia strigosa Nutt. Apache, Navajo, Coconino, Yavapai, Graham, and Pima counties, 3,000 to 7,000 feet, dry mesas and slopes, March to September. Wyoming to New Mexico and Arizona.

4. Townsendia arizonica Gray. Navajo, Coconino, eastern Mohave, and northern Yavapai counties, about 5,000 feet, dry plains and mesas, May to August, type material partly from Mount Trumbull, Mohave County (*E. Palmer* 204). Southwestern Colorado, southern Utah, and northern Arizona.

5. Townsendia incana Nutt. Apache County to eastern Mohave and northern Yavapai counties, 5,000 to 7,000 feet, dry, sandy or stony soil, May to September. Wyoming and Utah to New Mexico and central Arizona.

This species and the preceding one probably hybridize, since intermediate specimens occur.

27. ASTER (325, 326)

Perennial herbs, rarely annual or biennial herbs or shrubs; leaves alternate, entire to bipinnatifid; heads medium or large, often showy, the rays white, violet, or purple, very rarely wanting, the disk yellow or rarely white, sometimes turning purple in age; involucre usually definitely graduated, the phyllaries usually with herbaceous tips; achenes hairy or glabrous; pappus of subequal capillary bristles, very rarely with a short outer series of bristles.

Many of the Arizona asters are worthy of cultivation as ornamentals, notably the Mojave aster (*A. abatus*), characterized by often silvery foliage and large heads with lavender or violet rays, a plant well adapted to dry, hot situations. *A. arenosus* and *A. tanacetifolius* are used medicinally by the Hopi Indians. *A. adscendens* and *A. commutatus* are reported to absorb selenium from the soil in quantity sufficient to make them toxic to livestock. The spiny aster or Mexican-devilweed (*A. spinosus*) is sometimes a troublesome weed on canal banks and in pastures where the soil is saline, but its creeping rootstocks make it a useful plant for controlling soil erosion.

KEY TO THE SPECIES

1. Leaves, at least the lower ones, once or twice pinnatifid (2).
1. Leaves entire or merely toothed (4).
2. Heads small, the disk 5 to 7 mm. high; green tips of the phyllaries rhombic or rhombic-lanceolate, usually shorter than the whitish chartaceous base.........16. *A. parvulus*
2. Heads larger, the disk 8 to 12 mm. high; green tips of the phyllaries subulate to lance-triangular, usually equaling or exceeding the whitish chartaceous base (3).
3. Green tips of the phyllaries usually erect, triangular or lance-triangular, sometimes rhombic-lanceolate, not or scarcely narrower than the whitish base...17. *A. tagetinus*
3. Green tips of the phyllaries usually spreading, narrowly subulate or linear-subulate, distinctly narrower than the whitish base....................18. *A. tanacetifolius*
4. Stems low, usually 10 cm. high or less, numerous from a branched woody caudex; leaves very small, less than 1 cm. long, linear or spatulate, or the upper scalelike (5).
4. Stems taller, rarely from a branched caudex; leaves (except in *A. intricatus, A. spinosus,* and *A. riparius*) much larger (6).

5. Leaves all strongly hispid-ciliate and, at least on the upper surface, densely glandular, the lower leaves spatulate, the upper ones linear or linear-spatulate, those of the sterile branches sometimes subulate . . . . . 8. *A. hirtifolius*
5. Leaves rather densely strigose or strigillose, at least the upper ones not or only inconspicuously ciliate, mostly linear or subulate . . . . . 9. *A. arenosus*
6. Upper leaves tiny, scalelike, entire; stems suffrutescent (except in *A. riparius,* this annual). Plants tall (except in *A. riparius*), essentially glabrous; heads medium-sized (7).
6. Upper leaves not tiny and scalelike; stems strictly herbaceous or, if somewhat woody (*A. abatus*), then the heads very large (9).
7. Heads discoid. Stems intricately branched; achenes silky . . . . . 11. *A. intricatus*
7. Heads radiate (8).
8. Rays purple; plants annual, not spiny; achenes pubescent . . . . . 10. *A. riparius*
8. Rays white; plants perennial, often spiny; achenes glabrous . . . . . 12. *A. spinosus*
9. Leaves spiny-toothed; heads very large, about 5 cm. wide or more; involucre 1.5 to 2.5 cm. high . . . . . 13. *A. abatus*
9. Leaves not spiny-toothed; heads smaller; involucre much shorter (10).
10. Phyllaries glabrous on the back, sometimes ciliate on the margin (11).
10. Phyllaries more or less densely glandular or pubescent on the back (18).
11. Plants annual (12).
11. Plants perennial (13).
12. Rays exceeding the not very copious pappus; phyllaries all linear or lance-linear, acuminate, with a scarious margin; heads loosely panicled . . . . . 14. *A. exilis*
12. Rays shorter than the very copious and soft mature pappus; outer phyllaries narrowly oblong or linear-spatulate, obtuse or merely acutish, herbaceous; heads usually numerous and crowded . . . . . 15. *A. frondosus*
13. Rays white, very rarely (in *A. commutatus*) lavender (14).
13. Rays violet or purple, rarely (in a form of *A. foliaceus*) white (15).
14. Plants hairy, not from slender running rootstocks; stem leaves nearly uniform, linear, those of the branches bractlike; heads numerous; phyllaries with a whitish, chartaceous base and an abrupt green tip . . . . . 2. *A. commutatus*
14. Plants glabrous, from slender, running rootstocks; lower leaves spatulate or oblanceolate, the upper ones linear, all entire; heads solitary or few; phyllaries lanceolate or lance-linear, with a green midline (at least above) and a scarious margin 6. *A. Lemmoni*
15. Plants normally tall, 50 cm. high or more (16).
15. Plants normally low (not more than 30 cm. high) and simple or little branched. Outer phyllaries narrowly spatulate to oblong-obovate (17).
16. Outer phyllaries thin, linear or lance-linear, conspicuously green-tipped, more or less spreading . . . . . 4. *A. coerulescens*
16. Outer phyllaries, thick, oblong to ovate, not conspicuously green-tipped, appressed 7. *A. glaucodes*
17. Outer phyllaries narrowly spatulate, about 1 mm. wide; leaves normally narrow (less than 1 cm. wide), not or scarcely clasping . . . . . 3. *A. adscendens*
17. Outer phyllaries broadly oblong-spatulate or oblong-obovate, 2.5 to 5 mm. wide; leaves usually more than 1 cm. wide and definitely clasping . . . . . 5. *A. foliaceus*
18. Rays white; heads usually numerous and small, not more than 1.5 cm. wide, including the rays; phyllaries never glandular, several-seriate, with a whitish, coriaceous base and an abrupt, usually rhombic, green tip, normally cuspidate or mucronate 2. *A. commutatus*
18. Rays lavender or purple; heads usually larger; phyllaries usually glandular (19).
19. Leaves grasslike, entire; phyllaries subherbaceous throughout, or the inner ones narrowly scarious-margined. Leaves elongate-linear or the lower ones narrowly spatulate; involucre densely stipitate-glandular, without other pubescence . . . . . 1. *A. pauciflorus*
19. Leaves not grasslike, usually toothed; phyllaries not subherbaceous throughout (20).
20. Plants perennial; involucre not conspicuously many-ranked. Leaf margins usually hispidulous-ciliate or serrulate (21).

20. Plants annual, biennial, or rarely perennial; involucre conspicuously many-ranked. Phyllaries with a whitish, chartaceous base and an abrupt, often squarrose, herbaceous tip (22).
21. Leaves not glaucous, the lower ones much larger than the upper ones, petioled; involucre pubescent dorsally, not glandular ....3. *A. adscendens*
21. Leaves glaucous or glaucescent, the lower ones not much larger than the upper ones, sessile; involucre, and at least the upper part of the stem, stipitate-glandular. Leaves oblong-lanceolate or -oblanceolate to linear-oblong.................7. *A. glaucodes*
22. Herbaceous tips of the phyllaries comparatively short, usually appressed, rhombic or narrowly triangular, much shorter than the chartaceous whitish base (23).
22. Herbaceous tips of the phyllaries comparatively long, subulate, spreading, often as long as or longer than the chartaceous, whitish base (24).
23. Phyllaries canescent-puberulent, with few or no glandular hairs, their green tips usually narrow-triangular ....19. *A. canescens*
23. Phyllaries conspicuously glandular, their green tips usually rhombic..24. *A. cichoriaceus*
24. Phyllaries densely cinereous-puberulous, scarcely or not at all glandular 20. *A. tephrodes*
24. Phyllaries more or less densely glandular, sometimes also pubescent with eglandular hairs (25).
25. Stem densely glandular-hispid and -hispidulous, essentially without eglandular hairs 21. *A. Bigelovii*
25. Stem with abundant eglandular hairs, sometimes also glandular (26).
26. Plants relatively low (40 cm. high or less); leaves (except the lowest) linear or essentially so, those of the stem commonly 3 (rarely 5) mm. wide or less 22. *A. adenolepis*
26. Plants tall (60 cm. high or more); leaves mostly lanceolate or oblong-lanceolate 23. *A. aquifolius*

1. Aster pauciflorus Nutt. San Bernardino Ranch and valley of the San Pedro River (Cochise County), Santa Cruz River valley and Agua Caliente (Pima County), 2,500 to 4,000 feet, alluvial soil, May to September. Saskatchewan to Texas, New Mexico, southeastern Arizona, and Mexico.

2. Aster commutatus (Torr. & Gray) Gray. Widely distributed in Arizona, 5,000 to 8,000 feet, mostly in the pine belt, August to October. Minnesota to British Columbia, south to Texas, New Mexico, and Arizona.

KEY TO THE VARIETIES

1. Phyllaries subequal, the involucre scarcely graduated. Stems appressed-pubescent *A. commutatus* (typical)
1. Phyllaries conspicuously unequal, the involucre distinctly graduated (2).
2. Stems appressed-pubescent.......................................var. *polycephalus*
2. Stems hispid or hirsutulous with spreading or reflexed hairs..............var. *crassulus*

The typical plant has been collected at Flagstaff and Mormon Lake, Coconino County (*Fulton* 6371, *Collom* 806). Var. *polycephalus* (Rydb.) Blake ranges from the Hopi Indian Reservation and the Kaibab Plateau to the mountains of Cochise, Santa Cruz, and Pima counties. Var. *crassulus* (Rydb.) Blake is the commonest form in Arizona, ranging from northern Apache County to the Kaibab Plateau and south to Cochise and Santa Cruz counties.

3. Aster adscendens Lindl. Lakeside (Navajo County) and widely distributed in Coconino County, 6,500 to 8,500 feet, mountain meadows and openings in coniferous forests, August and September. Saskatchewan to Washington, south to Colorado and northern Arizona.

4. Aster coerulescens DC. Hopi Indian Reservation and White Mountains (Apache and Navajo counties), and throughout Coconino County, south to Cochise and Pima counties, 3,000 to 7,500 feet, marshy places and along streams, August to October. Wisconsin to Alberta, south to Texas, Arizona, and California.

Includes material from Arizona that has been referred to *A. hesperius* Gray, *A. Wootonii* Greene, and *A. foliaceus* var. *Canbyi* Gray. Shinners (*Rhodora* 51: 91–92) has stated recently that the presumed type collection of *A. coerulescens* is *A. praealtus* Poir. (*A. salicifolius* Ait.) and that the correct name for the plant here called *A. coerulescens* is *A. hesperius* Gray.

5. Aster foliaceus Lindl. Kaibab Plateau to the Mogollon Escarpment (Coconino County), White Mountains (Apache and Greenlee counties), Black Mesa (Navajo County), 7,500 to 9,500 feet, coniferous forest and mountain meadows, August and September. Wyoming to British Columbia, south to New Mexico and central Arizona.

The only form known from Arizona is var. *Burkei* Gray (*A. Burkei* (Gray) Howell).

Cronquist (325, p. 454) reported *A. oregonus* (Nutt.) Torr. & Gray, with which he synonymizes *A. Eatoni* (Gray) Howell, from Arizona, citing a collection at Flagstaff (*Purpus* 8113, not seen by the writer).

6. Aster Lemmoni Gray. Chiricahua and Huachuca mountains (Cochise County), Santa Rita Mountains (Santa Cruz County), 6,000 to 7,000 feet, mossy ledges, June to September, type from the Santa Rita Mountains (*Pringle*). Known only from southeastern Arizona.

7. Aster glaucodes Blake (*A. glaucus* (Nutt.) Torr. & Gray). The typical plant, with glabrous stem and involucre, has been collected near Beclabite, Apache County, 6,000 feet (*J. Howell, Jr.* in 1934) and in Havasu Canyon, Coconino County, about 3,000 feet (*Whiting* in 1940, *Deaver* 2135), and ranges from Wyoming, Colorado, and Utah to northern Arizona. Var. *pulcher* Blake, with stems and involucre stipitate-glandular, occurs on the Kaibab Plateau and both rims of the Grand Canyon, in Havasu Canyon, and near Sunset Crater (Coconino County), 6,500 to 8,000 feet, and is known only from southern Utah and northern Arizona.

8. Aster hirtifolius Blake. Apache County to Mohave County, south to Cochise and Pima counties, 3,500 to 7,000 feet, dry slopes and mesas, sometimes in pine woods, March to September. Wyoming to Nevada, south to Texas and Arizona.

This and the next are perhaps only forms of 1 species. (See Shinners, 326).

9. Aster arenosus (Heller) Blake. Apache County to Coconino County, south to Cochise, Santa Cruz, and Pima counties, 3,500 to 7,500 feet, dry mesas and slopes, often on limestone, April to October. Colorado to Texas, Arizona, and Mexico.

10. Aster riparius H.B.K. (*A. sonorae* Gray). West of the Chiricahua Mountains (*Wright* 1163, the type of *A. sonorae*), Willcox, Cochise County, saline soil (*Thornber* in 1930). Southeastern Arizona and Mexico.

11. Aster intricatus (Gray) Blake (*Aster carnosus* Gray). Toroweap Valley, northeastern Mohave County (*Laws* in 1944), and Maricopa, Pinal, Cochise, and Pima counties, 1,000 to 4,000 feet, moist, saline soil, May to October, type of *A. carnosus* from west of the Chiricahua Mountains (*Wright* 1187). Nevada to Arizona and southeastern California.

12. Aster spinosus Benth. Coconino and Mohave counties to Cochise, Pima, and Yuma counties, 150 to 4,000 feet, moist, saline soil, May to October. Texas to Utah, Nevada, Arizona, and California, south to Costa Rica.

13. Aster abatus Blake. Grand Canyon (Coconino County) and Mohave County to western Pima and northern Yuma counties, 2,000 to 3,500 (7,000?) feet, dry rocky slopes and mesas, March to May. Utah, Nevada, western Arizona, and southeastern California.

The very large heads with lavender rays make this a showy plant. The herbage varies from densely arachnoid-tomentose to green and glabrescent.

14. Aster exilis Ell. Eastern Yavapai, Maricopa, Pinal, Cochise, Santa Cruz, Pima, and Yuma counties, 1,000 to 5,500 feet, moist soil along streams and ditches, June to October. South Carolina to Florida, west to California, south to tropical America.

15. Aster frondosus (Nutt.) Torr. & Gray. Northern part of Apache and Navajo counties and eastern Coconino County, 5,000 to 7,000 feet, wet soil or among junipers, September. Wyoming to Oregon, south to northern Arizona and California.

16. Aster parvulus Blake (*A. parviflorus* Gray). Eastern Coconino County (reported as common around Tuba) and Graham, Pinal, Maricopa, Cochise, and Pima counties, 1,000 to 5,000 feet, mesas and plains, February to October. Utah, New Mexico, and Arizona.

Machaeranthera pygmaea (Gray) Woot. & Standl. does not seem specifically distinct from *A. parvulus.*

17. Aster tagetinus (Greene) Blake. Western Coconino and northern Mohave counties to Graham, Cochise, and Pima counties, 1,500 to 4,500 feet, mesas and roadsides, common, March to November, type from Arizona without definite locality (*Wilcox* in 1891). Utah, New Mexico, and Arizona.

18. Aster tanacetifolius H.B.K. (*Machaeranthera Parthenium* Greene). Northern Navajo County and Grand Canyon (Coconino County) to Yavapai, Graham, Cochise, Santa Cruz, and Pima counties, 1,000 to 6,000 (rarely 8,000) feet, roadsides and waste ground, June to October, type of *M. Parthenium* from Davidson Canyon, Empire Mountains, eastern Pima County (*Pringle* in 1884). South Dakota to Alberta, south to Texas, Arizona, and Mexico.

19. Aster canescens Pursh (*Machaeranthera oxylepis* Greene, *M. verna* A. Nels., *M. scoparia* Greene). Northeastern Apache County to Mohave County, south to Pima and Yuma counties, 150 to 8,000 feet, dry soil in open pine forests, chaparral, and roadsides, common, June to November, type of *M. oxylepis* from Apache Pass, Cochise County (*Lemmon* in 1881), type of *M. verna* from Big Bend, Virgin River, Mohave County (*Goodding* 757), type of *M. scoparia* from Turkey Tanks, Coconino County (*Jardine & Hill* in 1911). Colorado to British Columbia, south to Arizona and California.

20. Aster tephrodes (Gray) Blake. Southern Apache and Navajo counties, Havasu Canyon and near Flagstaff (Coconino County), near Kingman (Mohave County), south to Graham, Cochise, Pima, and Yuma counties, 150 to 7,000 feet, mostly in alluvial soil, March to October. New Mexico to Nevada and southern California.

21. Aster Bigelovii Gray. Coconino, Mohave, Yavapai, Greenlee, Graham, Gila, Maricopa, Pinal, and Pima counties, 3,000 to 7,000 feet, canyons and slopes, March to November. Colorado, New Mexico, and Arizona.

This and the similar *A. tephrodes* are very showy and handsome. The plants are browsed freely.

22. Aster adenolepis Blake. Kaibab Plateau and North Rim of the Grand Canyon (Coconino County), 8,000 to 9,000 feet, perhaps also on the South Rim, 7,000 feet, August and September, type from Thompson Canyon (*Jones* 6056b1). Known only from northern Arizona.

A collection on the Kaibab Plateau (*Darrow* 2933) has the middle stem leaves exceptionally wide (up to 5 mm.).

23. Aster aquifolius (Greene) Blake. White Mountains (northern Greenlee County), Cameron to the Grand Canyon (Coconino County), and Hualpai Mountain (Mohave County), south to the mountains of Graham, Gila, and Yavapai counties, 3,000 to 9,500 feet, May to November. New Mexico and Arizona.

24. Aster cichoriaceus (Greene) Blake (*Machaeranthera rigida* Greene, ? *M. Hansonii* A. Nels.). Apache, Navajo, and Coconino counties, 5,000 to 7,500 feet, dry slopes and mesas, June to October, type of *M. rigida* from Keam Canyon, Navajo County (*Zuck*), type of *M. Hansonii* from "Mount Ellen" (Elden Mountain ?), near Flagstaff, 7,500 feet (*Hanson* 814). Colorado, New Mexico, Utah, and Arizona.

28. ERIGERON (327). Fleabane, Wild-daisy

Herbs; leaves alternate, sometimes all basal, usually narrow, entire to pinnatifid; heads small or medium-sized, radiate or rarely discoid, the rays white, pink, or purple, the disk yellow; involucre usually only slightly or not at all graduated, the phyllaries not herbaceous-tipped, but sometimes subherbaceous throughout; achenes usually 2-nerved, sometimes 4- to 10-nerved; pappus usually sparse, of subequal capillary bristles, sometimes with an outer series of short squamellae or bristles; appendages of the style short, triangular, obtuse or rounded.

Many of the species have attractive daisylike heads, with blue, lavender, or white rays. An oil distilled from horseweed, *E. canadensis,* was formerly prescribed for diarrhea and dysentery. This plant is reported to cause irritation of the throat and dermatitis in susceptible persons. When eaten by livestock it may cause colic, but in Arizona the species is not sufficiently abundant to be a serious pest.

KEY TO THE SPECIES

1. Rays very short and inconspicuous, not or only slightly surpassing the pappus, rarely wanting (2).
1. Rays present and conspicuous, much surpassing the pappus (8).

2. Plants perennial, dwarf, rarely more than 15 cm. high, often scapose, the scapes or stems 1- or few-headed (3).
2. Plants annual, normally 30 cm. high or more, leafy-stemmed, several- or many-headed (4).
3. Lower leaves spatulate, ciliate, essentially glabrous on the faces. Stems loosely pilose 1. *E. simplex*
3. Lower leaves narrowly linear-spatulate, hispid like the whole plant......9. *E. concinnus*
4. Plants gray-tomentose, without stiff or glandular hairs. Stem sparsely branched above, the branches nearly naked; lower leaves oblanceolate, toothed......28. *E. eriophyllus*
4. Plants green, with stiff or glandular hairs or both (5).
5. Plants loosely arachnoid-pilose, also glandular, especially on the leaves and heads; leaves sessile and somewhat clasping, lanceolate or oblong-lanceolate, few-toothed; heads medium-sized, about 6 mm. high, numerous, in a narrow leafy racemiform or spikelike panicle ....29. *E. Schiedeanus*
5. Plants hispid, hirsute, or strigose to nearly glabrous, not glandular; leaves not clasping; heads otherwise (6).
6. Heads relatively large (6 to 8 mm. thick); involucre densely hirsutulous; plants relatively low (usually 40 cm. high or less), grayish green, densely hispid and hispidulous 30. *E. linifolius*
6. Heads tiny (about 2 to 4 mm. thick); involucre essentially glabrous, or sparsely hispidulous; plants usually tall (1 m. or more), green, rather sparsely hispid to essentially glabrous. Heads about 4 mm. high, usually very numerous, in an elongate panicle (7).
7. Phyllaries not with colored tips; herbage usually hispid.............31. *E. canadensis*
7. Phyllaries with definite purplish tips; herbage glabrous or nearly so....32. *E. pusillus*
8. Stem leaves all deeply pinnatifid, with several pairs of linear lobes; plants perennial, normally 30 to 50 cm. high, leafy throughout. Rays white or pinkish (9).
8. Stem leaves entire or if rarely pinnatifid, then the plants either low, or not leafy throughout, or annual or biennial (10).
9. Stem hispid or hirsute but not glandular-puberulous..............26. *E. neomexicanus*
9. Stem glandular-puberulous as well as hispid or hirsute...............27. *E. oreophilus*
10. Plants comparatively coarse, large-leaved, and large-headed; stems usually more than 30 cm. high; disk of the head more than 1 cm. wide; middle stem leaves usually more than 5 mm. wide (11).
10. Plants comparatively slender, small-leaved, and small-headed; stems usually less than 30 cm. high; disk of the head usually less than 1 cm. wide; middle stem leaves commonly less than 5 mm. wide (18).
11. Stem very leafy to the inflorescence, the upper leaves not much reduced; heads usually several, the peduncles not elongate (12).
11. Stem more or less naked above, the upper leaves strongly reduced; heads mostly 1 to 3, on comparatively long, mostly naked peduncles (14).
12. Stem glabrous or sometimes sparsely pilose above, not glandular......2. *E. macranthus*
12. Stem densely and finely glandular-puberulent, at least above, often also spreading-hirsute (13).
13. Stem usually rather densely spreading-hirsute as well as glandular-puberulent; stem leaves usually densely ciliate..................................3. *E. platyphyllus*
13. Stem densely and finely glandular-puberulent, at least above the middle, essentially without long hairs; stem leaves not or only very sparsely ciliate........4. *E. superbus*
14. Rays violet or purple...8. *E. formosissimus*
14. Rays white or pink (15).
15. Plants very sparsely hairy or nearly glabrous (16).
15. Plants rather densely hairy (17).
16. Stem leaves not ciliate; stem densely and finely glandular-puberulent above, with very few long hairs or none...4. *E. superbus*
16. Stem leaves ciliate; stem near the apex (the peduncle) spreading-pilose, essentially without glandular hairs...5. *E. Kuschei*
17. Plants green, the pubescence usually appressed, at least on the upper part of the stem; pappus nearly simple, the outer series inconspicuous...................6. *E. Rusbyi*

17. Plants cinereous-pilose with spreading or ascending hairs; pappus distinctly double, the outer series setulose-squamulose .... 7. *E. arizonicus*
18. Leaves twice ternately divided into linear or spatulate lobes; plants essentially scapose, dwarf, 10 cm. high or less .... 11. *E. compositus*
18. Leaves entire or rarely toothed, lobed, or pinnatifid, never ternately divided; stems usually leafy (19).
19. Stem pubescent with widespreading or sometimes deflexed hairs (20).
19. Stem pubescent with appressed or incurved hairs, rarely nearly or quite glabrous (26).
20. Stems diffuse; leaves (some of them, particularly on the sterile branches) often pinnatifid, with 1 to 3 pairs of lobes, small, 2 cm. long or less, the entire ones spatulate to linear. Peduncles short (usually 2.5 cm. long or less) .... 19. *E. Lemmoni*
20. Stems (at least the first flowering stem) erect or suberect; leaves entire or sometimes toothed or pinnatifid (regularly so in *E. lobatus*, very rarely in *E. divergens* and *E. nudiflorus*), usually longer (21).
21. Plants scapose, dwarf, not more than 10 cm. high; leaves ciliate, otherwise essentially glabrous, all basal, spatulate. Plants high-montane .... 1. *E. simplex*
21. Plants leafy-stemmed, normally taller; leaves pubescent on the faces (22).
22. Flowering stems at first scapiform, eventually producing long, spreading, leafy, usually sterile branches from the lower axils .... 21. *E. nudiflorus*
22. Flowering stems usually leafy from the start, not producing long, spreading, leafy, sterile branches from the lower axils (23).
23. Leaves (the lower and usually the middle ones), obovate, oblanceolate, or spatulate in outline (24).
23. Leaves linear to narrowly oblanceolate. Blades entire, or, if a few of the lowest ones rarely few-toothed, then the stems not at all glandular-puberulous (25).
24. Stems glandular-puberulous, at least above, as well as rather sparsely spreading-pilose; leaves mostly obovate in outline, pinnatifid with (3) 5 to 7 lobes; disk 5 to 10 mm. wide .... 24. *E. lobatus*
24. Stems hirtellous or canescent, not glandular; leaves (the basal ones) oblanceolate or spatulate, all entire; disk 9 to 18 mm. wide .... 25. *E. caespitosus*
25. Plants definitely perennial; herbage hispid or coarsely hirsute; rays usually drying white or pink .... 9. *E. concinnus*
25. Plants annual or biennial; herbage puberulous to pilose, the hairs soft; rays usually drying blue or lavender .... 23. *E. divergens*
26. Plants (the first flowering stems) scapiform, eventually producing long, spreading, small-leaved, usually sterile runners from the base or the lower axils .... 20. *E. flagellaris*
26. Plants without long, spreading runners (27).
27. Leaves densely cinereous- or canescent-strigose, sometimes (in *E. utahensis*) only sparsely so (28).
27. Leaves not densely strigose, but sometimes cinereous-pubescent with incurved hairs (30).
28. Stems scapiform, naked except toward the base, 8 cm. high or lower, 1-headed. Plants densely cespitose, often forming large mats; achenes 2-nerved, compressed, pilose-ciliate on the margin, glabrous on the sides .... 16. *E. compactus*
28. Stems leafy at least to above the middle (29).
29. Achenes essentially glabrous, subterete, 8- to 10-nerved; plants 12 to 20 cm. high 14. *E. canus*
29. Achenes pubescent, subquadrangular, 4-nerved; plants usually 30 cm. high or more 15. *E. utahensis*
30. Stems glabrous, striate-angled, tall (about 60 cm. high), simple below the cymose heads. Leaves linear or narrowly linear-oblanceolate, callose-pointed, hispidulous-ciliate, otherwise nearly glabrous .... 18. *E. oxyphyllus*
30. Stems more or less densely pubescent, or else the plants low, not more than 25 cm. high, and the stems 1-headed (31).
31. Plants annual, single-stemmed. Stem rather densely incurved-puberulous 22. *E. Bellidiastrum*
31. Plants perennial, more or less densely cespitose (32).

32. Involucre glabrous; basal leaves linear-spatulate to obovate, some of them usually 3-toothed to pinnatifid. Stem leaves much narrower, linear, entire.....12. *E. Pringlei*
32. Involucre more or less glandular (sometimes very minutely so) and often also pilose or hirsute; leaves all entire (33).
33. Plants entirely glabrous except for the very finely glandular involucre and apex of the peduncles; leaves 1-nerved......................................10. *E. perglaber*
33. Plants more or less strigose-pubescent on the stems and leaves; lowest leaves usually 3-nerved (34).
34. Rays purple; basal leaves essentially glabrous except on the margin; peduncles glandular and spreading-pilose toward the apex...........................13. *E. ursinus*
34. Rays white or purplish-tinged; basal leaves strigose on the faces; peduncles strigose 17. *E. Eatoni*

1. Erigeron simplex Greene. Alpine meadows, San Francisco Peaks, 12,000 feet (*Little* 4656, 4709, 4733, 4777), August and September. Montana to northern New Mexico, northern Arizona, and California.

Some of the material from Arizona is rayless or essentially so, but this feature is not constant even in the same colony.

2. Erigeron macranthus Nutt. (*E. speciosus* (Lindl.) DC. var. *macranthus* (Nutt.) Cronquist). Apache, Navajo, Coconino, Graham, and Cochise counties, 6,000 to 9,500 feet, oak thickets and pine woods, July to October. Alberta and British Columbia, south to New Mexico and Arizona.

3. Erigeron platyphyllus Greene (*E. patens* Greene, *E. foliosissimus* Greene). White Mountains (southern Apache and Navajo counties), San Francisco Peaks and vicinity (Coconino County), Hualpai Mountain (Mohave County), south to the mountains of Cochise and Pima counties, 4,000 to 9,000 feet, rich soil in coniferous forests, July to October, type of *E. patens* from Strawberry Valley, Gila County (*MacDougal* in 1891), type of *E. foliosissimus* from near Fort Huachuca (*Wilcox* 460). New Mexico and Arizona.

4. Erigeron superbus Greene (*E. eldensis* Greene, *E. apiculatus* Greene). North Rim of the Grand Canyon, San Francisco Peaks, and Bill Williams Mountain (Coconino County), Pinaleno Mountains (Graham County), Chiricahua and Huachuca mountains (Cochise County), probably also in the Rincon and Santa Catalina mountains (Pima County), (6,500?) 8,000 to 10,500 feet, coniferous forests, July to September, type of *E. eldensis* from Elden Mountain, Coconino County (*Leiberg* 5837), type of *E. apiculatus* from Mount Graham (*Rothrock* 736). Wyoming to Utah, New Mexico, and Arizona.

5. Erigeron Kuschei Eastw. Known only from Cave Creek, Chiricahua Mountains, Cochise County, 6,000 to 8,000 feet (*Kusche* in 1927, the type collection).

6. Erigeron Rusbyi Gray. Pinaleno Mountains (Graham County), Chiricahua and Huachuca mountains (Cochise County), Santa Catalina Mountains (Pima County), 7,000 to 10,500 feet, coniferous forests, July to October. New Mexico and southeastern Arizona.

Both *Erigeron Rusbyi* and *E. Kuschei* are perhaps only less pubescent forms of *E. arizonicus*. *E. Kuschei*, except for its ciliate leaves, is also strongly suggestive of a dwarf form of *E. superbus*.

7. Erigeron arizonicus Gray (*E. huachucanus* Greene). Huachuca Mountains (Cochise County), where the type of *E. arizonicus* was collected in Tanner Canyon (*Lemmon* 2751), and the type of *E. huachucanus* near Fort Huachuca (*Wilcox* 482), Santa Catalina Mountains, Pima County (*Benson* 9707), 6,500 to 8,000 feet. Known only from southern Arizona.

8. Erigeron formosissimus Greene. Throughout Coconino County, White Mountain region (southern Apache and Navajo and northern Greenlee counties), Pinaleno Mountains (Graham County), Pine (northwestern Gila County), 5,500 to 10,500 feet, coniferous forests, July to September. Western South Dakota, Wyoming, and Utah, to New Mexico and Arizona.

Two varieties are represented in Arizona: var. *typicus* Cronquist (*E. pecosensis* Standl.), with the involucre glandular and more or less densely hirsute, and the uppermost leaves long-hairy or nearly glabrous; and var. *viscidus* (Rydb.) Cronquist (*E. subasper* Greene, *E. Gulielmi* Greene, *E. scaberulus* Greene), with the involucre densely glandular, sometimes also with a few long hairs, and the uppermost leaves glandular-scabrous. The type of *E. subasper* came from the San Francisco Peaks (*Purpus* 8084), the type of *E. Gulielmi* from Bill Williams Mountain (*Palmer* in 1869), and the type of *E. scaberulus* from the White Mountains (*Griffiths* 5353).

9. Erigeron concinnus (Hook. & Arn.) Torr. & Gray (*E. pumilus* Nutt. subsp. *concinnoides* Cronquist). Northern Apache County to northern and eastern Mohave County, south to the Chiricahua Mountains (Cochise County) and the Santa Catalina Mountains (Pima County), 3,000 to 8,000 feet, dry, sandy or stony mesas and slopes, with pine and juniper, April to October. Montana to British Columbia, south to New Mexico, Arizona, and California.

A collection at the Grand Canyon (*Mrs. E. Meire* in 1917) has exceptionally large heads, the disk 12 mm. in diameter. Var. *aphanactis* Gray (*E. aphanactis* (Gray) Greene) with rayless heads, has been collected in Apache County, at the north end of the Carrizo Mountains (*Standley* 7465) and near Rock Point (*Peebles & Smith* 13536). Var. *condensatus* D. C. Eaton, a dwarf, subscapose form, occurs in Navajo County.

10. Erigeron perglaber Blake. Known only from a collection made presumably in Arizona but without definite locality (*E. Palmer* in 1869).

The plant has the appearance of *Erigeron concinnus* but is glabrous.

11. Erigeron compositus Pursh. San Francisco Peaks, Coconino County, 11,500 feet (*Little* 4750). Greenland to Alaska, south to Colorado, northern Arizona, and California.

The Arizona plant is var. *multifidus* (Rydb.) Macbr. & Payson.

12. Erigeron Pringlei Gray. Oak Creek Canyon (Coconino County), Natural Bridge, Pine, Sierra Ancha, and Mazatzal mountains (Gila County), Santa Rita Mountains (Santa Cruz County), apparently also in the Pinaleno Mountains, Graham County (*Pultz* 1158), 5,000 to 9,000 feet, ledges of cliffs and rock crevices in canyons, May to July, type from Mount Wrightson, Santa Cruz County (*Pringle* in 1881). Known only from central and southern Arizona.

13. Erigeron ursinus D.C.Eaton. Kaibab Plateau and North Rim of Grand

Canyon, Coconino County, 8,500 to 9,000 feet (*Cottam* 2675, *Collom* in 1940). Montana and Idaho to Colorado, Utah, northern Arizona, and California.

14. Erigeron canus Gray. Defiance Plateau (Apache County), Keam Canyon (Navajo County), Kaibab Plateau, Grand Canyon, and vicinity of Flagstaff (Coconino County), 6,500 to 7,500 feet, mesas and open pine forest, June and July. South Dakota and Wyoming to Nebraska, New Mexico, and northern Arizona.

15. Erigeron utahensis Gray. Apache County to Mohave County, 3,000 to 7,000 feet, dry, rocky slopes, May to October. Southern Utah and northern Arizona.

Two varieties, both reported from Arizona, were distinguished by Cronquist (327, p. 273). These are var. *tetrapleurus* (by error *tetrapleuris*) (Gray) Cronq. and var. *sparsifolius* (Eastw.) Cronq.

16. Erigeron compactus Blake. Northern Apache County (*Peebles & Smith* 13467, *Turner* in 1935), El Capitan and southeast of Kayenta, northern Navajo County (*Peebles & Fulton* 11921, 11971), 5,500 to 6,000 feet, barren, rocky slopes and mesas, May and June. Utah, Nevada, southeastern California, and northeastern Arizona.

Plants forming compact mounds about 0.3 m. in diameter, attractive in flower, the rays pink. The Arizona plant is var. *consimilis* (Cronq.) Blake (*E. consimilis* Cronq.), with somewhat larger heads and involucre, the phyllaries with more or less spreading instead of appressed hairs, type from 15 miles north of Ganado, Apache County (*Peebles & Smith* 13467).

17. Erigeron Eatoni Gray. Betatakin Canyon (Navajo County), Navajo Mountain, Kaibab Plateau, and Grand Canyon (Coconino County), 7,000 to 8,000 feet, rocky slopes, June. Wyoming to Oregon, northern Arizona, and California.

18. Erigeron oxyphyllus Greene. Black Mountains and near Yucca (Mohave County), Sierra Estrella (Maricopa County), Picacho Mountain (Pinal County), 2,000 to 3,000 feet, dry, rocky slopes, March to May, type from Yucca (*Jones* in 1884). Known only from western and southwestern Arizona.

Stems wandlike, many from a woody crown, the dead stems long-persistent.

19. Erigeron Lemmoni Gray. Fish Creek Canyon, eastern Maricopa County (*Peebles* 7953), Huachuca Mountains, Cochise County (*Lemmon* in 1882), near Tucson, Pima County (*Lemmon* in 1883), 1,500 (to 6,000?) feet, cliffs, May to September, type from Tanner Canyon, Huachuca Mountains (*Lemmon*). Known only from southern Arizona.

20. Erigeron flagellaris Gray (*E. Macdougalii* Heller). Northern Apache, Navajo, and Coconino counties to the mountains of Cochise, Santa Cruz, and Pima counties, 3,000 to 9,500 feet, open coniferous forests and mountain parks, April to September, type of *E. Macdougalii* from the San Francisco Peaks (*MacDougal* 390). South Dakota and Wyoming to Texas, New Mexico, and southern Arizona.

Occasional specimens are suggestive of hybridism between this species and *E. nudiflorus,* combining the habit of one of these species with the pubescence of the other.

21. Erigeron nudiflorus Buckl. (*E. divergens* Torr. & Gray var. *cinereus*

Gray). Apache County to Mohave County, south to Cochise, Santa Cruz, and Pima counties, 4,000 to 7,500 feet, rocky slopes, mesas, and canyons, common, March to July. Colorado and Utah to Texas, New Mexico, Arizona, and northern Mexico.

22. Erigeron Bellidiastrum Nutt. (*E. Eastwoodiae* Woot. & Standl.). Known for Arizona only by a collection in the Carrizo Mountains, Apache County (*Standley* 7433, the type of *E. Eastwoodiae*). South Dakota to Nevada, Texas, and northeastern Arizona.

23. Erigeron divergens Torr. & Gray (? *E. accedens* Greene, *E. gracillimus* Greene, *E. furcatus* Greene). Throughout the state, 1,000 to 9,000 feet, dry, rocky slopes and mesas and open pine woods, very common, February to October, type of *E. accedens* (not examined) from Clifton, Greenlee County (*Davidson* in 1899), type of *E. gracillimus* from Coconino National Forest (*Jardine & Hill* in 1911), type of *E. furcatus* from Kendrick Peak, Coconino County (*Knowlton* 45). South Dakota to British Columbia, south to Texas, New Mexico, Arizona, California, and northern Mexico.

A specimen from Mount Trumbull, Mohave County (*Cottam* 8710), has exceptionally large heads, the disk 1.5 cm. wide, pressed. Cronquist (327, p. 255) cites, under *E. modestus* Gray, a collection from "Montezuma Wells, Maricopa County" (*Purpus* 8278), presumably from Yavapai County. The specimen came from far outside the normal range of *E. modestus,* however, and is probably rather a form of *E. divergens.*

24. Erigeron lobatus A. Nels. Grand Canyon (Coconino County) and Mohave County, south to Graham, Pima, and Yuma counties, 1,500 to 3,000 feet, plains, mesas, and rocky slopes, February to May (October), type from Canyon Lake, Maricopa County (*A. Nelson* 11209). Known only from southern and western Arizona.

25. Erigeron caespitosus Nutt. A collection at Flagstaff, Coconino County (*Jones* in 1884, not seen by the writer), is cited by Cronquist (327, p. 168). Western North Dakota to Alaska, south to New Mexico, northern Arizona, and Washington.

26. Erigeron neomexicanus Gray (*E. delphinifolius* Willd. subsp. *neomexicanus* (Gray) Cronquist var. *euneomexicanus* Cronquist). San Francisco Peaks, Flagstaff, and Walnut Canyon (Coconino County) to the mountains of Greenlee, Graham, Cochise, Santa Cruz, and Pima counties, 4,000 to 9,000 feet, with oaks or pines, August to October. New Mexico and Arizona.

Several specimens from the Chiricahua, Patagonia, and Santa Rita mountains have the pubescence mostly appressed to ascending (not widespreading as in the typical plant) and thus approach *E. delphinifolius* Willd. of Mexico, but in all of them the hairs toward the base of the stem are spreading, and they appear to represent a phase of *E. neomexicanus* rather than *E. delphinifolius,* which has not been recognized in the United States.

27. Erigeron oreophilus Greenm. (*E. delphinifolius* Willd. subsp. *neomexicanus* (Gray) Cronquist var. *oreophilus* (Greenm.) Cronquist). Apache, Navajo, and Coconino counties to Graham, Cochise, Santa Cruz, and Pima counties, 4,500 to 9,500 feet, oak chaparral and open pine forests, May to October. Arizona and northern Mexico.

28. Erigeron eriophyllus Gray (*Conyza eriophylla* Cronquist). On the Sonoita, southwestern Cochise County (*Wright*, the type collection), near Ruby, Santa Cruz County, 4,500 feet (*Kearney & Peebles* 13783, 14914), with live-oaks and grasses, August and September. Apparently known only from these 3 collections in southeastern Arizona.

29. Erigeron Schiedeanus Less. (*Conyza Schiedeana* Cronquist). Kaibab Plateau, San Francisco Peaks, and Flagstaff (Coconino County), White Mountains (Apache County), Pinaleno Mountains (Graham County), Chiricahua Mountains (Cochise County), Rincon and Santa Catalina mountains (Pima County), 7,000 to 9,000 feet, open pine forests, August and September. Southwestern Colorado, New Mexico, Arizona, and Mexico.

30. Erigeron linifolius Willd. Agua Fria River bottom near Avondale, Maricopa County, 1,000 feet (*Peebles et al.* 2462), streets of Tucson (*Thornber* in 1913). Southeastern United States, southern Arizona, and California; a common weed in the warmer parts of the Eastern and Western hemispheres.

Erigeron lonchophyllus Hook., mainly a northern species but known also from New Mexico, Utah, Nevada, and southern California, is represented in the United States National Herbarium by specimens labeled as from northern Arizona (*P. F. Mohr* in 1874). It seems inadvisable to include the species formally in the flora of Arizona until specimens with more definite data become available.

31. Erigeron canadensis L. (*Conyza canadensis* Cronquist). Throughout most of the state except the extreme western portion, 1,000 to 7,000 feet, waste land and cultivated fields, July to October. Widely distributed in North America and South America; naturalized in the Old World.

Horseweed, a coarse unsightly plant. The form occurring in Arizona is much less pubescent than the common Eastern form. The name var. *glabratus* Gray applies to it, but it scarcely appears to merit varietal distinction.

32. Erigeron pusillus Nutt. (*Conyza parva* Cronquist). Superior to Miami, Gila County, 5,000 feet (*Gillespie* 8634), October. Massachusetts to Florida, west to Arizona and south into tropical America.

29. CONYZA

Herbs, similar to some species of *Erigeron* in habit, distinguished only by having the corollas of the outer (pistillate) flowers of the head tubular-filiform, not ligulate.

KEY TO THE SPECIES

1. Leaves merely toothed to coarsely pinnatifid; achenes hispidulous (sometimes also glandular), not puncticulate .... 1. *C. Coulteri*
1. Leaves once or twice pinnately parted into mostly linear lobes; achenes glabrous, puncticulate in lines .... 2. *C. sophiaefolia*

1. Conyza Coulteri Gray. Kaibab Plateau (Coconino County) to Greenlee, Cochise, and Pima counties, 1,500 to 8,000 feet, fields, plains, and river bottoms, April to October. Colorado and Texas to California and Mexico.

Closely similar in appearance to *Erigeron Schiedeanus* Less. and often confused with it. In *C. Coulteri* the corollas of the pistillate flowers are tubular-filiform, without a ligule, and the achenes are only 0.5 to 0.8 mm. long. In *E. Schiedeanus* the pistillate flowers possess a small but definite ligule, and the achenes are 1 to 1.4 mm. long.

2. Conyza sophiaefolia H.B.K. (*C. Coulteri* var. *tenuisecta* Gray). Cochise, Santa Cruz, and Pima counties, 2,500 to 6,000 feet, cultivated fields, dry hills, and plains, August to October, type of var. *tenuisecta* from near Fort Huachuca (*Lemmon* 2753). Southwestern New Mexico, southeastern Arizona, and Mexico.

30. BACCHARIS

Dioecious shrubs, rarely only suffrutescent at base; leaves alternate, entire to toothed; heads usually numerous and panicled, discoid, whitish; involucre graduated, of chartaceous, whitish phyllaries; pistillate heads composed entirely of tubular-filiform pistillate flowers; staminate heads composed entirely of hermaphrodite flowers, with tubular, 5-toothed corollas, infertile; achenes small, 5- to 10-ribbed; pappus in the pistillate flowers of copious capillary bristles, in the staminate flowers of scabrous, often twisted, clavellate bristles.

Most of the species are browsed at times, but their palatability is generally low and some of them (*B. pteronioides, B. sarothroides*) are reputed poisonous to livestock. The widely distributed and very common seep-willow (*B. glutinosa*) is recommended for erosion control along watercourses because of its deep, widespreading root system and its tendency to form thickets. It is said to be readily propagated by cuttings.

KEY TO THE SPECIES

1. Plants low (usually 60 cm. high or less), woody only at or near the base; leaves entire, all small (mostly less than 1 cm. long), narrow (2).
1. Plants either much taller, or definitely woody-stemmed, or the leaves conspicuously toothed (3).
2. Plants glabrous; involucre 6 to 10 mm. high; pistillate pappus deep brown or purplish brown .... 1. *B. Wrightii*
2. Plants densely hispidulous or hirtellous; involucre 6 mm. high or less; pistillate pappus merely brownish-tinged .... 2. *B. brachyphylla*
3. Heads solitary at the tips of very short, leafy branchlets, these racemosely arranged along the branches. Larger leaves mostly obovate or oblanceolate and sharply toothed, those of the flowering branchlets minute and entire .... 3. *B. pteronioides*
3. Heads otherwise arranged (4).
4. Broomlike shrubs, with numerous, erect, strongly sulcate, angled branches; leaves usually essentially absent at flowering time, if present, then usually entire and mostly obovate or spatulate (5).
4. Plants not broomlike (or sometimes somewhat so in *B. Emoryi*); leaves normally present at flowering time, usually toothed (6).
5. Pistillate pappus very short, 3 to 4 mm. long, scarcely or not surpassing the styles; receptacle often rounded or elevated, usually bearing some pales between the flowers, at least in the pistillate heads; staminate heads 2.5 to 4 mm. high.... 4. *B. sergiloides*
5. Pistillate pappus elongate in fruit (10 mm. long or more), much surpassing the styles; receptacle flattish, deeply alveolate but not bearing pales; staminate heads 3.5 to 7 mm. high .... 5. *B. sarothroides*
6. Pistillate pappus in fruit elongate, surpassing the styles by 3 mm. or more (7).
6. Pistillate pappus in fruit not elongate, scarcely or not surpassing the styles (8).
7. Larger leaves cuneate-oblanceolate or oblong-oblanceolate, 3 to 18 (30) mm. wide, distinctly 3-nerved; involucre 4 to 8 mm. high .... 6. *B. Emoryi*
7. Larger leaves mostly linear or narrowly oblanceolate, 3 to 8 mm. wide, usually 1-nerved; involucre 3 to 4 mm. high .... 7. *B. neglecta*
8. Leaves narrowly linear or linear-lanceolate, 1.5 to 8 mm. wide, usually closely, evenly, and sharply spinulose-serrulate .... 8. *B. thesioides*
8. Leaves from lanceolate or elliptic to cuneate-obovate, entire or rather coarsely or unevenly toothed, the teeth not spinulose (9).

9. Leaves cuneate-oblong to oblong-oblanceolate, rather coarsely and unequally toothed, 1.5 to 3.5 cm. long, 4 to 15 mm. wide............................9. *B. Bigelovii*
9. Leaves mostly lanceolate or elliptic-lanceolate, entire or evenly toothed, 5 to 15 cm. long (10).
10. Heads in a cymose panicle terminating the stem; leaves usually toothed.10. *B. glutinosa*
10. Heads in small cymose panicles terminating numerous short, lateral branches as well as the main stem; leaves mostly entire............................11. *B. viminea*

1. **Baccharis Wrightii** Gray. Apache, Navajo, Coconino, and northern Yavapai counties, 5,000 to 6,000 feet, May to July. Kansas and western Texas to Arizona and Durango.

2. **Baccharis brachyphylla** Gray. Graham County to Mohave County, south to Cochise, Santa Cruz, Pima, and Yuma counties, 1,500 to 4,000 feet, May to October, type from "between Conde's Camp and the Chiricahui Mountains" (*Wright* 1199). Arizona, southern California, and Sonora.

3. **Baccharis pteronioides** DC. (*B. ramulosa* (DC.) Gray). Yavapai, Greenlee, Graham, Gila, Maricopa, Pinal, Cochise, Santa Cruz, and Pima counties, 3,500 to 6,000 feet, April to September. Western Texas, southern New Mexico, and Arizona, south to Puebla.

Yerba-de-pasmo. Although the name *ramulosa* has page priority, the name *pteronioides* was chosen by Gray when the two specific names were first combined and must be used under the International Rules.

4. **Baccharis sergiloides** Gray. Grand Canyon (Coconino County) to Mohave, Yavapai, Greenlee, and Maricopa counties, 2,000 to 5,500 feet, flowering nearly throughout the year, the type collected by Emory's expedition along the Gila or Colorado River, probably in Arizona. Utah, Arizona, southeastern California, and Sonora.

5. **Baccharis sarothroides** Gray (*B. arizonica* Eastw.). Mohave, Yavapai, Greenlee, Graham, and Gila counties to Cochise, Santa Cruz, Pima, and Yuma counties, 1,000 to 5,500 feet, hillsides and bottom lands, sometimes in saline soil, September to February, types of *B. arizonica* from Packard, Gila County (*Eastwood* 15832, 15833). Southwestern New Mexico to southern California, northern Mexico, and Baja California.

Broom baccharis, rosin-brush, desert-broom.

6. **Baccharis Emoryi** Gray. Grand Canyon (Coconino County) and Mohave County to Cochise, Pima, and Yuma counties, 500 to 5,000 feet, mostly along streams, September to November, the type collected by Emory's expedition along the Gila River in 1846. Texas to southern California and southern Utah, reported also to occur in northern Mexico.

Plants up to 2 m. high.

7. **Baccharis neglecta** Britton. Cochise County at Dragoon Summit (*Thornber* in 1905) and Huachuca Mountains (*Lemmon* in 1882), Santa Cruz County, near Sonoita, 4,500 feet (*Darrow & Haskell* 2273). Nebraska to Texas, and Arizona to northern Mexico.

8. **Baccharis thesioides** H.B.K. Mogollon Escarpment (Coconino County) and mountains of Greenlee, Graham, Cochise, Santa Cruz, and Pima counties, 4,000 to 8,000 feet, August and September. Western Texas to southern Arizona, south to central Mexico.

Plants up to 2 m. high.

9. Baccharis Bigelovii Gray. Near Douglas and in the Chiricahua, Mule, and Huachuca mountains (Cochise County), 4,500 to 6,000 feet, September. Southwestern Texas, southeastern Arizona, and northern Mexico.

10. Baccharis glutinosa Pers. Grand Canyon and Lees Ferry (Coconino County) and northern Mohave County, to Greenlee, Graham, Cochise, Santa Cruz, Pima, and Yuma counties, up to 5,500 feet but usually lower, very common along watercourses, often forming thickets, March to December. Colorado and Texas to California and Mexico; South America.

Seep-willow, also known as water-willow, water-wally, batamote, and water-motie. Plants up to at least 2.2 m. high.

11. Baccharis viminea DC. Montezuma Castle, Yavapai County, 3,000 feet (*Bowen* in 1940), Willow Springs Mountains (*Griffiths* 3647), near the mouth of the Gila River, Yuma County (*Monnet* 1057), spring and late summer. Southwestern Utah and western Arizona to California. Mule-fat.

31. PLUCHEA. Marsh-fleabane

Herbs or shrubs; leaves alternate; heads small, disciform, in terminal cymes or panicles, the corollas purplish; involucre graduated, the bracts chartaceous to subscarious; receptacle naked; outer flowers pistillate, with a tubular-filiform corolla, the inner hermaphrodite; achenes small, 4- or 5-ribbed; pappus of capillary bristles; anthers caudate at base.

The rank-smelling arrow-weed (*P. sericea*) forms dense thickets in stream beds and in moist saline soil at relatively low elevations. It is browsed by deer and sometimes by cattle and horses. The straight stems are used by the Pima Indians in constructing the roofs and walls of primitive huts, for making baskets, and formerly for arrow shafts. An infusion of the herbage was used as a remedy for sore eyes. The flowers are reported to be an important source of honey in Arizona.

KEY TO THE SPECIES

1. Green, glandular-pubescent annuals; leaves mostly oblong or ovate, toothed; pappus bristles not dilated at tip..1. *P. camphorata*
1. Silky-pubescent shrubs; leaves linear to lanceolate, entire; pappus bristles dilated at tip, especially in the hermaphrodite flowers..............................2. *P. sericea*

1. Pluchea camphorata (L.) DC. Bill Williams Mountain, Coconino County (*Palmer* in 1869), and along the Colorado, Salt, and Gila rivers, in Mohave, Maricopa, Pinal, and Yuma counties, in alluvial, more or less saline soil, September and October. Maine to Texas, Nevada, and California, south to Mexico.

Salt-marsh fleabane. The plant has a strong, camphorlike odor. The specific name *P. camphorata* is here used in the sense of Gray's *Manual*, ed. 7, p. 819.

2. Pluchea sericea (Nutt.) Coville. Almost throughout the state, up to 3,000 feet (seldom higher), very abundant along streams, sometimes in saline soil, flowering chiefly in spring. Texas to Utah, southern California, and northern Mexico.

32. STYLOCLINE

Low, woolly annuals; leaves alternate, linear or spatulate-linear, entire; heads small, clustered, heterogamous, disciform; outer flowers pistillate, with filiform corollas, each subtended by a pale, the 4 or 5 inner flowers hermaphrodite and infertile; pales of the pistillate flowers boat-shaped, tipped with a hyaline appendage, completely enclosing the achenes and deciduous with them; pales subtending the hermaphrodite flowers flattish; pappus none in the fertile flowers, of a few bristles in the hermaphrodite flowers.

KEY TO THE SPECIES

1. Bracts enclosing the pistillate flowers with a narrow, only moderately woolly body, this broadly hyaline-winged throughout its length.................... 1. *S. gnaphalioides*
1. Bracts enclosing the pistillate flowers with a broader, densely long-woolly body produced only at apex into an ovate, hyaline appendage.................... 2. *S. micropoides*

1. Stylocline gnaphalioides Nutt. (*S. arizonica* Coville). Fort Mohave (Mohave County) and at several localities in Maricopa, Pinal, Cochise, and Pima counties, 500 to 4,500 feet, February to May, type of *S. arizonica* from Verde Mesa (*Smart* in 1867). California to Baja California and Arizona.

2. Stylocline micropoides Gray. Mohave, Graham, Maricopa, Pinal, Santa Cruz, and Pima counties, 500 to 4,000 feet, March to May. New Mexico and Utah to southern California.

33. EVAX

Low, diffusely branched, white-woolly annuals; leaves alternate, small, oblanceolate or spatulate, entire; heads small, disciform, clustered in bracted, globose, terminal glomerules; outer flowers pistillate, the inner (2 to 5) flowers hermaphrodite, infertile, all subtended by flattish pales; pappus none.

1. Evax multicaulis DC. Maricopa, Pinal, and Pima counties, 1,500 to 2,500 feet, March to May. Texas and Oklahoma to southern Arizona and northern Mexico.

34. FILAGO

Low, whitish-woolly annuals; leaves alternate, narrow, entire; heads small, glomerate, disciform; outer flowers pistillate, the outermost of these subtended by boat-shaped, open pales, epappose, the others usually without pales and with a pappus of capillary bristles; innermost flowers (2 to 5) hermaphrodite, usually without pales, pappose.

KEY TO THE SPECIES

1. Plants diffuse, with very slender, naked internodes, the leaves practically wanting except for those subtending and conspicuously surpassing the heads; flowers within the inner circle of pales mostly 4 or 5, all hermaphrodite...................... 1. *F. arizonica*
1. Plants erect or diffuse, the stems normally leafy at least below, the leaves subtending the heads only rarely surpassing the latter; flowers within the inner circle of pales about 12 to 20, only about 2 to 4 of them hermaphrodite (2).
2. Plants diffuse or merely ascending; achenes all smooth; outer bracts with a hyaline appendage about as long as the body............................... 2. *F. depressa*
2. Plants normally erect; inner achenes papillose; outer bracts with a hyaline appendage about half as long as the body or less............................. 3. *F. californica*

1. Filago arizonica Gray. Maricopa, Pinal, Pima, and Yuma counties, 1,000 to 2,500 feet, March and April, type from Verde Mesa (*Smart* in 1867). Southern Arizona to southern California and northern Baja California.

2. Filago depressa Gray. Maricopa, Pinal, and Pima counties, 2,000 to 3,500 feet, March and April. Southern Arizona to southern California.

3. Filago californica Nutt. Pierce Ferry and Fort Mohave (Mohave County), and southern Apache, Gila, Maricopa, Pinal, and Pima counties, 500 to 7,000 feet, March to May. Utah to southern Arizona, California, and Baja California.

35. ANTENNARIA. Pussy-toes

Dioecious, dwarf, tomentose, stoloniferous, perennial herbs, rarely suffruticulose; leaves mostly in a basal rosette, small, entire, obovate or spatulate, those of the stem reduced; heads rather small, discoid, strictly staminate or pistillate; involucre strongly graduated, of scarious phyllaries; pistillate corollas filiform; hermaphrodite (functionally staminate) corollas tubular, 5-toothed, whitish; achenes small; pappus of the pistillate flowers of copious capillary bristles, united at base and deciduous in a ring; pappus of the staminate flowers of more or less clavellate and slightly flattened bristles.

In many species of this genus parthenogenetic reproduction is the rule and staminate plants are very rare; they are common, however, in *A. rosulata* and *A. marginata.*

KEY TO THE SPECIES

1. Heads sessile or subsessile among the leaves of the basal rosettes, solitary or 2 or 3 together . . . . 1. *A. rosulata*
1. Heads capitate or closely cymose at the tips of erect stems, these normally 5 cm. high or more (2).
2. Leaves usually soon glabrate and green above; inflorescence finely stipitate-glandular 2. *A. marginata*
2. Leaves persistently tomentose above; inflorescence not glandular (3).
3. Pistillate heads 8 to 10 mm. or more high; basal leaves usually 5 to 9 mm. wide 3. *A. aprica*
3. Pistillate heads less than 8 mm. high; basal leaves usually 4 mm. wide or less (4).
4. Phyllaries blackish green or fuscous toward the base, or nearly throughout 4. *A. umbrinella*
4. Phyllaries greenish or light brown at base, otherwise white . . . . 5. *A. arida*

1. Antennaria rosulata Rydb. Chuska Mountains and near Fort Defiance (Apache County), and from the Kaibab Plateau to the Mogollon Escarpment (Coconino County), 5,500 to 11,000 feet, May to July, co-type from the Mogollon Escarpment (*Mearns* 40). Colorado, Utah, and Arizona.

2. Antennaria marginata Greene (*A. recurva* Greene, *A. marginata* var. *glandulifera* A. Nels.). White Mountains (Apache County) and Coconino County to the mountains of Cochise and Pima counties, 5,000 to 9,500 feet, rocky slopes and ridges, April to July, type of *A. recurva* from Flagstaff (*MacDougal* in 1891), type of *A. marginata* var. *glandulifera* from Schultz Pass, near Flagstaff (*Hanson* 635). Colorado, Utah, and Arizona.

The type of *Antennaria recurva,* which is very immature, is one of the occasional specimens of this species in which the leaves remain tomentose above.

3. **Antennaria aprica** Greene. Apache, Coconino, Yavapai, Graham, Gila, and Pima counties, from the Kaibab Plateau to the southern mountains, 5,000 to 12,000 feet, May to August. Manitoba to British Columbia, south to New Mexico and central Arizona.

4. **Antennaria umbrinella** Rydb. San Francisco Peaks, Coconino County, 11,000 to 12,000 feet (*Little* 4637 (?), 4708, 4718). Montana to British Columbia, south to Colorado, Arizona, and California.

5. **Antennaria arida** E. Nels. Kaibab Plateau, North Rim of Grand Canyon, and San Francisco Peaks (Coconino County), Chiricahua Mountains (Cochise County), 6,500 to 11,000 feet. Montana to New Mexico and Arizona.

36. ANAPHALIS. Pearly Everlasting

Erect, perennial herbs, usually 30 cm. high or more, tomentose, at least on the stem and the lower leaf surface; leaves alternate, linear or linear-oblong, entire, slightly or not at all decurrent; heads in close cymose panicles, discoid or disciform, subdioecious, the pistillate ones usually with a few central hermaphrodite flowers; involucre strongly graduated, of milk-white papery-scarious phyllaries, radiating when dry; achenes short; pappus of capillary bristles, these not united at base.

1. **Anaphalis margaritacea** (L.) Gray. Kaibab Plateau, North Rim of Grand Canyon, San Francisco Peaks, and Oak Creek (Coconino County), 4,500 to 8,500 feet, late summer and autumn. Newfoundland to Alaska, south to Pennsylvania, Kansas, Utah, and northern Arizona; Asia.

The Arizona form of this variable species is nearest var. *subalpina* Gray although not typical.

37. GNAPHALIUM. Cud-weed, Everlasting

Low herbs, more or less woolly; leaves alternate, narrow, entire; heads small, usually numerous, sometimes glomerate, disciform, all the flowers fertile; involucre graduated, of numerous scarious phyllaries; outer flowers numerous, pistillate, the inner ones hermaphrodite; receptacle naked; pappus of capillary bristles.

KEY TO THE SPECIES

1. Pappus bristles united at base and deciduous in a ring; heads spicate. Phyllary tips normally deep purple or brownish .... 10. *G. purpureum*
1. Pappus bristles not united at base, falling separately or in groups, or else the heads not spicate (2).
2. Heads very small, clustered and imbedded in wool, the clusters leafy-bracted; involucre scarcely graduated, the scarious tips of the phyllaries relatively inconspicuous; low annuals, seldom more than 20 cm. high (3).
2. Heads medium-sized (or smaller in *G. luteo-album*), not leafy-bracted; involucre strongly graduated, the phyllaries conspicuously scarious nearly throughout; plants usually 30 cm. high or more (4).
3. Plants thinly but rather closely woolly; leaves linear-spatulate or linear, 1 to 3 mm. wide; inflorescence (when well developed) spiciform .... 8. *G. Grayi*
3. Plants loosely floccose-woolly; leaves spatulate to oblong or obovate, 3 to 8 mm. wide; heads clustered at the tips of the stem and branches, not spicately arranged 9. *G. palustre*
4. Leaves gray-tomentose above as well as beneath, sometimes (in *G. arizonicum*) only thinly so (5).

4. Leaves bright green and strongly glandular-pubescent above, whitish-tomentose beneath, all decurrent (8).
5. Leaves not at all or only obsoletely decurrent. Phyllary tips white or slightly tinged with straw-color, mostly obtuse....................................1. *G. Wrightii*
5. Leaves definitely decurrent, or at least with broad adnate auricles (6).
6. Phyllaries acute or acuminate, brownish- or purple-tinged at tip; heads slender, about 31- to 47-flowered..4. *G. arizonicum*
6. Phyllaries very obtuse, straw-colored, yellowish, or greenish; heads campanulate-subglobose (7).
7. Plants with few or no sterile leafy basal branches; leaves mostly strongly decurrent; heads 4 to 6 mm. high; phyllary tips scarious, yellowish or straw-colored; corollas yellowish; achenes smooth..2. *G. chilense*
7. Plants usually with numerous, sterile, leafy basal branches; leaves clasping but scarcely decurrent; heads 3 to 3.5 mm. high; phyllary tips almost hyaline, whitish, faintly tinged with brownish; corollas purplish; achenes papillose.........3. *G. luteo-album*
8. Phyllary tips pearly white, papery, not shining; stem densely whitish-tomentose 5. *G. leucocephalum*
8. Phyllary tips straw-colored or whitish, thin-scarious, shining; stem glandular-pilose, with or without a thin gray tomentum (9).
9. Heads campanulate-subglobose, 5 to 6 mm. high, about 130- to 150-flowered; phyllary tips straw-colored to pale brownish................................6. *G. Macounii*
9. Heads subcylindric, 4 mm. high, 30- to 35-flowered; phyllary tips whitish..7. *G. Pringlei*

1. Gnaphalium Wrightii Gray. Grand Canyon (Coconino County) and Hackberry (eastern Mohave County) to Greenlee, Cochise, Santa Cruz, and Pima counties, 3,500 to 7,000 feet, dry rocky slopes, May to October. Western Texas to southern California and northern Mexico.

2. Gnaphalium chilense Spreng. Chuska Mountains (Apache County), Rainbow Lodge and Flagstaff (Coconino County), and Pierce Spring (Mohave County), to Cochise, Santa Cruz, Pima, and Yuma counties, 100 to 7,000 feet, along streams, May to October. Montana to Washington, south to Texas, southern Arizona, and California.

3. Gnaphalium luteo-album L. Near Topock, Mohave County, in damp, sandy ground (*Monson* in 1950). Western Arizona and California; adventive from the Old World.

4. Gnaphalium arizonicum Gray. Mogollon Escarpment (Coconino County), near Sonoita (Santa Cruz County), Rincon and Santa Catalina mountains (Pima County), 5,000 to 7,500 feet, pine forests, August to October, type from near Fort Huachuca (*Lemmon*). Arizona and northern Mexico.

5. Gnaphalium leucocephalum Gray. Eastern Maricopa County, Cochise, Santa Cruz, and Pima counties, 2,000 to 5,000 feet, sandy beds of streams, July to October. Southern Arizona, southern California, and Sonora.

Plant with an odor of lemon-verbena, the stems sometimes 50 or more from 1 root.

6. Gnaphalium Macounii Greene. White Mountains (Apache County), Kaibab Plateau (Coconino County), and Hualpai Mountain (Mohave County), to the mountains of Cochise and Pima counties, 5,500 to 10,000 feet, open coniferous forests, July to October. Canada south to West Virginia, Texas, southern Arizona, and northern California.

7. Gnaphalium Pringlei Gray. Mazatzal Mountains (Gila County) and mountains of Cochise, Santa Cruz, and Pima counties, 4,000 to 8,000 feet, canyons, August to October. Southern Arizona and Chihuahua.

8. Gnaphalium Grayi Nels. & Macbr. (*G. strictum* Gray). White Mountains (Apache and Greenlee counties), Kaibab Plateau, San Francisco Peaks, and Flagstaff (Coconino County), 6,000 to 9,500 feet, mountain meadows, August and September. Wyoming to New Mexico and Arizona.

9. Gnaphalium palustre Nutt. Coconino, Yavapai, Gila, Maricopa, Pinal, Pima, and Yuma counties, 1,000 to 5,000 feet, moist soil, April to October. Alberta and British Columbia to New Mexico, southern Arizona, California, and Baja California.

10. Gnaphalium purpureum L. Happy Valley Road (Cochise County) and Rincon, Santa Catalina, and Baboquivari mountains (Pima County), 3,000 to 7,500 feet, March to May. Maine to Kansas and southern Arizona, also British Columbia to California.

38. LAGASCEA (328)

Pubescent shrubs about 1 m. high; leaves opposite, ovate, petioled, acuminate; heads 1-flowered, small, crowded at the tips of the branches in dense glomerules, these subtended by a few herbaceous bracts; involucre of the individual heads tubular, gamophyllous, 5- or 6-toothed; corolla tubular, yellow; achene columnar; pappus a short crown.

1. Lagascea decipiens Hemsl. (*Calhounia Nelsonae* A. Nels.). Oro Blanco Mountains (Santa Cruz County), Baboquivari and Quinlan mountains (Pima County), 3,000 to 4,000 feet, canyons, flowering throughout the year, type of *Calhounia Nelsonae* from the Baboquivari Mountains (*Hanson* 1023). Southern Arizona and Mexico.

39. GUARDIOLA (329)

Branching perennials, up to 1 m. high, glabrous, somewhat glaucous; leaves opposite, roundish-ovate, obtuse, toothed, subcordate, very short-petioled; heads rather small, in terminal cymose clusters, radiate, the corollas white, the anthers green; involucre cylindric, of few, thin, equal phyllaries; rays 1 to 5, fertile; disk flowers hermaphrodite, sterile; receptacle paleaceous; achenes oblong, epappose.

1. Guardiola platyphylla Gray. Huachuca Mountains (Cochise County) to the Santa Catalina and Baboquivari mountains (Pima County), also Slate Creek (western Gila County), 3,000 to 5,000 feet, canyons, February to September, type from between Babocomari, Arizona, and Santa Cruz, Sonora (*Wright* 1236 bis). Southern Arizona and northern Mexico.

A specimen in the United States National Herbarium labeled as collected near Holbrook, Navajo County, by Myrtle Zuck probably came from the vicinity of Tucson.

40. MELAMPODIUM (330)

Plants low, herbaceous or nearly so; leaves opposite, entire to pinnatifid; heads small or medium-sized, radiate; outer phyllaries 4 or 5, herbaceous;

inner phyllaries of the same number as the rays, completely enclosing the ray achenes like a coat and falling with them; rays white or yellow, pistillate, fertile; disk flowers hermaphrodite, sterile; receptacle paleaceous; achenes obovate-oblong; pappus none.

KEY TO THE SPECIES

1. Rays white, often purplish-veined, conspicuous, 6 to 15 mm. long; plants perennial, often slightly suffrutescent; heads 10 to 30 mm. wide, on peduncles usually 2 to 10 cm. long 1. *M. leucanthum*
1. Rays yellow, inconspicuous, usually not more than 2 mm. long; plants annual, strictly herbaceous; heads only 2 to 6 mm. wide, usually sessile or subsessile (2).
2. Fruit (achene with the tightly enwrapping inner phyllary) with a raised hood at apex, this usually prolonged into a recurved, dorsally pilosulous beak up to 2 mm. long 2. *M. longicorne*
2. Fruit not hooded or beaked at apex.................................3. *M. hispidum*

1. Melampodium leucanthum Torr. & Gray. Fort Apache (southern Navajo County) to Mount Trumbull (Mohave County), south to Greenlee, Cochise, and Pima counties, 2,000 to 5,000 feet, common on dry, rocky slopes and mesas, often on limestone, March to November. Kansas to Texas, southern Arizona, and Chihuahua.

A showy and attractive plant.

2. Melampodium longicorne Gray. Mountains of Cochise, Santa Cruz, and Pima counties, 4,000 to 5,500 feet, canyons, often on limestone, August and September. Southeastern Arizona and northern Mexico.

3. Melampodium hispidum H.B.K. Cochise, Santa Cruz, and Pima counties, 4,000 to 5,500 feet, August and September. Southern Arizona and Mexico.

41. BERLANDIERA

Perennial herbs, finely canescent-tomentulose; leaves alternate, lyrate-pinnatifid or merely crenate; heads rather large, radiate, yellow, solitary, long-peduncled; involucre broad, about 3-seriate, the phyllaries broad; rays fertile; disk flowers sterile; receptacle paleaceous; ray achenes strongly flattened, epappose, adnate at base to the subtending phyllary and the spatulate pales of the 2 opposed sterile flowers, the whole falling together.

1. Berlandiera lyrata Benth. Cochise, Santa Cruz, and eastern Pima counties, 4,000 to 5,000 feet, plains and mesas, April to October. Kansas and Arkansas to Texas, southeastern Arizona, and northern Mexico.

Var. *macrophylla* Gray, the type of which was collected in southern Arizona by Lemmon, is occasional in Cochise and Santa Cruz counties. It differs from typical *B. lyrata* in having merely crenate instead of lyrate-pinnatifid leaves. The flower heads of *B. lyrata* are reported to have been used by the Indians as seasoning in foods.

42. ENGELMANNIA

Perennial herbs, rough-pubescent; leaves alternate, deeply pinnatifid; heads medium-sized, yellow, radiate, slender-peduncled; involucre graduated; rays fertile; disk flowers hermaphrodite but sterile; receptacle paleaceous; ray achenes strongly flattened, each adnate at base to the subtending phyllary and the pales of the opposed outer disk flowers, the whole falling away together; pappus in the ray flowers of an unequally lobed or toothed crown, in the disk flowers reduced.

***1. Engelmannia pinnatifida** Nutt. Kansas to Louisiana, west to Colorado and New Mexico, south to northern Mexico (reported from Arizona), dry hills and prairies, May to September.

43. PARTHENIUM

Low, branching, gray-tomentulose shrubs; leaves alternate, pinnatifid, with blunt, roundish lobes; heads small, white, cymose-panicled; rays fertile, very small; disk flowers sterile; achenes small, flattened, surrounded by a narrow callose margin, this adnate at base to the pales of the 2 opposed disk flowers and the subtending phyllary, at length tearing away from the achene below but remaining attached at apex.

1. Parthenium incanum H.B.K. Grand Canyon (Coconino County) to Lake Mead (Mohave County), south to Greenlee (?), Cochise, and Pima counties, 2,500 to 6,000 feet, dry plains and mesas, usually in "caliche" soil, June to October. Western Texas to Arizona and Mexico.

Mariola. A small shrub, reported to be common in the Lower Sonoran zone in the Grand Canyon, and common in southeastern Arizona. The plant is rarely browsed. It contains rubber like that of guayule (*Parthenium argentatum* Gray) but in smaller amount. Guayule, which may be distinguished by its narrowly lanceolate leaves, these entire or laciniate-toothed or lobed with acuminate lobes, and finely silvery-pubescent on both sides, has been cultivated in Arizona as a source of rubber.

44. PARTHENICE

Branching, cinereous-puberulent annuals; leaves alternate, ovate, long-petioled, toothed; heads disciform, numerous, panicled, greenish white; outer flowers pistillate, their corollas tubular, the ligule obsolete; disk flowers hermaphrodite, sterile; achenes dorsoventrally flattened, apiculate, falling away at maturity with the pales of the opposed disk flowers.

1. Parthenice mollis Gray. Patagonia Mountains and near Ruby (Santa Cruz County), Fort Lowell and Santa Rita and Baboquivari mountains (Pima County), 3,500 to 5,000 feet, foothills and canyons, July to December. Southern Arizona and northwestern Mexico, also reported from Colorado and New Mexico.

Stems up to 2 m. high.

45. IVA. Marsh-elder

Herbs; leaves alternate or opposite, entire to dissected; heads small, greenish, panicled or solitary in the axils, disciform; involucre simple, a 4- or 5-lobed cup, or double, with the outer phyllaries 5, broad, herbaceous, the inner ones also 5, membranaceous or scarious; outer flowers 5, pistillate, fertile, their corollas vestigial, the inner flowers hermaphrodite, sterile; achenes obovate, thickened, epappose.

KEY TO THE SPECIES

1. Heads solitary in the upper leaf axils, forming leafy racemes; leaves mostly entire, linear to obovate; plants perennial; involucre simple........................1. *I. axillaris*
1. Heads panicled; leaves serrate or pinnatifid; plants annual; involucre double (2).
2. Leaves 2- or 3-pinnatifid, pubescent but not canescent beneath; stem pubescent throughout; heads loosely panicled, on slender peduncles up to 1 cm. long..2. *I. ambrosiaefolia*
2. Leaves sharply and unequally serrate, canescent beneath, at least when young; stem essentially glabrous below; heads sessile or subsessile, spicate-panicled 3. *I. xanthifolia*

1. Iva axillaris Pursh. Woodruff (Navajo County), Kaibab Plateau, Tuba, Black Falls, and Fredonia (Coconino County), 4,000 to 5,000 feet, cultivated and waste land, sometimes a troublesome weed. Manitoba to British Columbia, south to New Mexico, northern Arizona, and California.

This plant may have been introduced into Arizona, as it seems not to have been collected there before 1913.

2. Iva ambrosiaefolia Gray. Greenlee, Pinal, Cochise, and Pima counties, 1,000 to 5,500 feet, mostly along streams, May to October. Western Texas to southern Arizona and northern Mexico.

3. Iva xanthifolia Nutt. Near Ganado, Apache County (*Griffiths* 5820), Keam Canyon and Kayenta, Navajo County (*Whiting* 854, *Blos* 92), 5,000 to 6,500 feet, along streams and in waste ground, July to October. Saskatchewan to Alberta and Washington, south to Nebraska, New Mexico, and northern Arizona, introduced eastward.

Contact with the plant induces dermatitis in some persons, and the pollen is a cause of hay fever.

46. OXYTENIA

Shrubby; leaves alternate, pinnately parted into 3 or 5 long, filiform lobes, or the upper ones entire; heads numerous, disciform, small, whitish, in dense panicles; outer flowers about 5, pistillate, their corollas vestigial; inner flowers hermaphrodite, sterile; achenes obovoid, long-villous.

1. Oxytenia acerosa Nutt. Holbrook and Kayenta (Navajo County), Willow Spring, Tuba, and Grand Canyon (Coconino County), 3,500 to 6,000 feet, often on saline soil, July to September. Southwestern Colorado and New Mexico to Nevada and southeastern California.

A large, often leafless shrub with rushlike branches. Considered by stockmen to be poisonous to cattle and sheep, but seldom eaten. Cases of human dermatitis have been attributed to contact with this plant.

47. DICORIA

Much-branched, annual herbs, cinereous-strigose or -strigillose; leaves mostly alternate, ovate or suborbicular to lanceolate, toothed or entire; heads small, very numerous, panicled, heterogamous and disciform, or some of them unisexual and staminate; outer phyllaries about 5, small, herbaceous, the inner ones (subtending the 1 or 2 pistillate flowers) subscarious, accrescent, much surpassing the outer phyllaries at maturity; achenes dorsoventrally flattened, oblong, black, with a narrow or broad, toothed or pectinate, crustaceous, whitish wing; pappus none or vestigial.

KEY TO THE SPECIES

1. Mature achenes (solitary) surpassing the subtending phyllary; leaves linear-lanceolate to lance-oblong; wing of the achene pectinately divided into toothed lobes
1. *D. Brandegei*
1. Mature achenes (1 or 2) shorter than the subtending phyllaries; upper leaves elliptic to suborbicular, rarely oblong; wing of the achene merely toothed to pectinately divided
2. *D. canescens*

1. Dicoria Brandegei Gray. Navajo and eastern Coconino counties, doubtless also in Apache County, about 5,000 feet, sandy soil, June to September. Southwestern Colorado, southern Utah, and northeastern Arizona.

It is said that the Indians of northeastern Arizona used the flowers and seeds as food.

2. Dicoria canescens Gray. Maricopa, Pinal, and Yuma counties, doubtless elsewhere, up to 2,500 feet, sandy beds of streams and washes, June to November, type from along the Gila River (*Emory*). Southwestern Utah, Arizona, southeastern California, and Sonora.

Specimens from near Tuba, Coconino County (*Kearney & Peebles* 12854), and from Wupatki National Monument, Coconino County, about 5,000 feet (*Whiting* 5255), with oblong upper leaves and a comparatively narrow wing to the achene, agree with *Dicoria oblongifolia* Rydb. It is not clear, however, that *D. oblongifolia* is specifically distinct from *D. canescens.*

48. HYMENOCLEA. Burro-brush

Low, much-branched shrubs, monoecious or subdioecious; leaves alternate, linear-filiform and entire, or pinnately parted into a few linear-filiform lobes; heads small, those of both sexes usually intermixed in the same leaf axils; pistillate involucre fusiform, beaked, indurate, 1-flowered, with 5 to 12 transverse scarious wings near the middle, completely enclosing the achene; staminate heads with a flattish, 4- to 6-lobed involucre.

Characteristic and abundant shrubs of sandy stream beds and washes, tending to form thickets. Their forage value is small. The conspicuously winged fruiting involucres make these plants rather attractive. The pollen is reported to be an important cause of hay fever.

KEY TO THE SPECIES

1. Wings of the fruiting involucre spirally arranged, 5 to 8 mm. wide.........1. *H. Salsola*
1. Wings of the fruiting involucre in a single whorl (rarely with 1 or 2 additional wings above or below the middle), 1 to 4 mm. wide (2).
2. Wings 5 to 7, flabellate or reniform-orbicular, 2.5 to 4 mm. wide........2. *H. pentalepis*
2. Wings 7 to 12, mostly cuneate or obovate, 1 to 2 mm. wide..............3. *H. monogyra*

1. Hymenoclea Salsola Torr. & Gray. Grand Canyon (Coconino County) and Mohave County to Pima and Yuma counties, up to 4,000 feet but usually lower, sandy washes and rocky slopes, sometimes in saline soil, March and April. Southern Utah, Arizona, southern California, and northwestern Mexico.

A specimen in anthesis from Willow Beach, Mohave County (*Clokey* 5959), is probably *H. fasciculata* A. Nels, but more mature material is to be desired. This species differs from *H. Salsola* in the narrower (3 to 5 mm. wide), mostly erect or appressed wings of the fruit, those of *H. Salsola* being soon widespreading, the lower ones 5 to 8 mm. wide.

2. Hymenoclea pentalepis Rydb. (*H. Salsola* var. *pentalepis* L. Benson, *H. hemidioica* A. Nels.). Topock (Mohave County) and Yavapai, Graham, Maricopa, Pima, and Yuma counties, 2,500 feet or lower, February to April, type of *H. pentalepis* from the Santa Catalina Mountains (*Griffiths* 2630), types of *H. hemidioica* from the Mohawk Mountains, Yuma County (*A. & R. Nelson* 1340, 1341). Western and southern Arizona, southeastern California, and adjacent Mexico.

3. Hymenoclea monogyra Torr. & Gray. Mohave, southern Yavapai, Graham, Gila, Pinal, Maricopa, Cochise, Santa Cruz, and Pima counties, 1,000 to 4,000 feet, usually in sandy soil, September. Western Texas to southern California and northern Mexico.

49. AMBROSIA. Ragweed

Weedy, monoecious herbs; leaves opposite or alternate, lobed or dissected; pistillate heads mostly axillary, 1-flowered, their involucres more or less turbinate, short-beaked, indurate, armed with a few tubercles in a single series around the middle, completely enclosing the achene; staminate heads naked-racemose above the pistillate ones, terminating the stem and branches.

Ragweed pollen is one of the commonest causes of hay fever.

KEY TO THE SPECIES

1. Leaves large, palmately 3- to 5-lobed with serrate lobes (rarely ovate, not lobed, and merely serrate); fruiting involucre 4 to 7 mm. long....................1. *A. aptera*
1. Leaves smaller, pinnatifid or bipinnatifid; fruiting involucre 3 to 3.8 mm. long (2).
2. Perennials, with a running rootstock; leaves thickish, mostly only once pinnatifid 2. *A. psilostachya*
2. Annuals; leaves thin, the lower ones usually twice pinnatifid........3. *A. artemisiifolia*

1. Ambrosia aptera DC. Southern Apache, Navajo, and Coconino counties to Cochise, Santa Cruz, and Pima counties, 2,500 to 8,000 feet, roadsides and bottom lands, July to October. Illinois to Colorado, south to Texas, Arizona, and northern Mexico.

2. Ambrosia psilostachya DC. White Mountain region (Apache and Navajo counties), Flagstaff and Oak Creek (Coconino County), south to Cochise and Pima counties, 2,500 to 7,000 feet, along streams and at roadsides, July to October. Illinois to Saskatchewan and Washington, south to northern Mexico.

3. Ambrosia artemisiifolia L. Near Prescott, Yavapai County (*Kusche* in 1929), Sierra Ancha, Gila County, about 6,000 feet, in a meadow (*Harrison & Kearney* 8317), Sentinel, Maricopa County (*Jones* 25050). A common weed nearly throughout the United States and southern Canada.

50. FRANSERIA. Bur-sage

Plants with the characters of *Ambrosia*, but with the pistillate involucre armed with spines or prickles in more than 1 series, and 1- to 4-flowered.

KEY TO THE SPECIES

1. Plants herbaceous (2).
1. Plants shrubby, at least at base (4).
2. Fruit (mature pistillate involucre) 2 to 4 mm. long, obovoid, armed with about 10 to 20 hooked spines, these 0.8 mm. long or less......................1. *F. confertiflora*

2. Fruit usually 4 to 8 mm. long, the spines longer, rarely hooked (3).
3. Leaves ovate or deltoid in outline, once to thrice pinnatifid, green or merely slightly paler beneath; fruit 4 to 8 mm. long, 1-beaked, armed with 6 to 30 strongly flattened, straight, spreading spines, these 2 to 5 mm. long; annual..................2. *F. acanthicarpa*
3. Leaves mostly oblong in outline, interruptedly bipinnatifid with a strongly toothed or lobed rachis, green above, densely canescent-strigillose beneath; fruit 3.5 to 6 mm. long, 2- or 3-beaked, bearing about 4 to 9 thick-subulate, rarely hooked spines, these 1 to 2 mm. long and flattened only at base; perennial from running rootstocks. 3. *F. discolor*
4. Leaves canescent-pubescent beneath (5).
4. Leaves green beneath or sometimes (in *F. cordifolia*) cinereous or canescent when young (7).
5. Leaves once to thrice pinnately divided into small, mostly ovate or obovate divisions, canescent-strigillose on both faces; fruit 4 to 6 mm. long, bearing about 25 to 40 rigid, flattened, straight spines..4. *F. dumosa*
5. Leaves ovate to oblong, not pinnately divided into small divisions, greenish above, densely canescent-tomentulose beneath; fruit otherwise (6).
6. Leaves subsessile, sinuate-toothed to pinnatifid; fruit fusiform, 8 to 10 mm. long, 1-beaked, glandular and densely long-villous, especially on the 20 or fewer subulate straight spines ....5. *F. eriocentra*
6. Leaves distinctly petioled, serrate or serrulate; fruit turbinate-ovoid or turbinate-subglobose, about 6 mm. long, 2- or 3-beaked, glandular and somewhat pilose-tomentose on the body, at least when young, bearing about 20 or more mostly strongly flattened, often hook-tipped spines.......................................6. *F. deltoidea*
7. Leaves sessile and cordate-clasping, coarsely spinose-toothed, ovate, reticulate. Fruit fusiform or globose-fusiform, 10 to 23 mm. long, 1- or 2-beaked, bearing very numerous hook-tipped spines, these 4 to 6 mm. long...........................9. *F. ilicifolia*
7. Leaves slender-petioled, not spinose-toothed (8).
8. Leaves broadly ovate, about as long as wide, 3-nerved, obtuse or acutish, bluntly toothed, sometimes lobed; fruit turbinate-ovoid or obovoid, 5 to 7 mm. long, bearing about 8 to 20 subulate hook-tipped spines, these 1.5 to 2.5 mm. long, often flattish at base 7. *F. cordifolia*
8. Leaves elongate-triangular or oblong-lanceolate, pinnate-veined, acuminate, coarsely toothed; fruit fusiform (*Xanthium*-like), about 12 mm. long, bearing very numerous slender-subulate hook-tipped spines, these 2 to 3 mm. long, not flattened at base 8. *F. ambrosioides*

1. **Franseria confertiflora** (DC.) Rydb. (*F. incana* Rydb.). Near Flagstaff and in Havasu Canyon (Coconino County) and Kingman (Mohave County), south to Graham, Cochise, Santa Cruz, and Pima counties, 1,000 to 6,500 feet, mesas and slopes, sometimes a weed in cultivated land, April to October, type of *F. incana* from near Fort Huachuca (*Wilcox* in 1892). Oklahoma and Colorado to California, south to northern Mexico.

2. **Franseria acanthicarpa** (Hook.) Coville. Apache County to Mohave, Yavapai, and Greenlee counties, also in Maricopa and Cochise counties, 1,000 to 7,000 feet, dry or moist, sandy soil, June to December. Minnesota to Alberta, south to western Texas, northern Arizona, and California.

Bur-weed. Affords forage for sheep in northern Arizona.

3. **Franseria discolor** Nutt. Winslow (Navajo County) near Flagstaff and Mormon Lake (Coconino County), 5,000 to 7,000 feet, plains and cultivated ground, June to September. Nebraska to Wyoming, south to New Mexico and Arizona.

4. **Franseria dumosa** Gray. Grand Canyon (Coconino County) and Mohave County to Graham, Pinal, Santa Cruz, Pima, and Yuma counties, up to 3,000

feet, very common on dry plains and mesas, April to November. Southern Utah to southeastern California and northwestern Mexico.

White bur-sage. Plants up to 1 m. high, much branched, compact, spinescent. It is reported that horses prefer this plant to all other desert shrubs, as forage. It is sometimes called burro-weed, a name used for various Southwestern plants.

5. Franseria eriocentra Gray. Grand Canyon (Coconino County) and Beaver Dam (Mohave County) to western Gila, Maricopa, and Pinal counties, 1,500 to 5,000 feet, usually in sandy soil, often in washes, locally abundant, April and May. Southern Utah to southeastern California and south-central Arizona.

Woolly bur-sage. Plant up to 1 m. high, with rigid, spreading branches.

6. Franseria deltoidea Torr. Western Gila, Maricopa, Pinal, Pima, and Yuma counties, 1,000 to 3,000 feet, very abundant on plains and mesas, often in nearly pure stands, December to April, type collected by Frémont on the Gila River. South-central Arizona and Sonora.

7. Franseria cordifolia Gray. Eastern Maricopa County and mountains of Santa Cruz, Pima, and Yuma counties, 1,500 to 3,500 feet, locally abundant, canyons, rocky slopes, and washes, February to April, type from mountains near Tucson (*Pringle*). Southern Arizona and northern Mexico.

8. Franseria ambrosioides Cav. Southern Yavapai, Graham, and western Gila counties to Pinal, Pima, and Yuma counties, up to 4,500 feet, common in sandy washes and canyons, March to May. Southern Arizona, southern California, and northern Mexico.

9. Franseria ilicifolia Gray. Desert mountain ranges of southern Yuma County, 1,500 feet or lower, usually in sand, March and April. Southwestern Arizona, southeastern California, and northern Baja California.

Holly-leaf bur-sage. A roundish, densely much-branched evergreen subshrub, the leaves spinescent.

A specimen in young flower collected at Tule Tank, Yuma County (*Gooding & Hardies* in 1938), may represent a hybrid between *F. ambrosioides* and *F. ilicifolia*. The whole plant is densely stipitate-glandular and spreading-hirsute; the leaves (about 10 cm. long, including the winged petiole, and 4 cm. wide) are ovate in outline, deeply pinnatifid with about 2 pairs of sharply toothed lobes, the petiole (2.5 to 3 cm. long) winged to the base, the wing repand-toothed; the teeth are callose-tipped but not cuspidate.

51. XANTHIUM (331, 332). Cocklebur

Monoecious, weedy annuals; leaves alternate, sometimes with triple spines in the axils; hermaphrodite (staminate) heads clustered, borne above the pistillate ones, their involucres with free phyllaries; pistillate involucre burlike, 2-celled, 2-beaked, covered with stiff, hooked prickles.

These are troublesome weeds in pastures and cultivated fields. The spiny burs clot the manes and tails of horses and occasionally cause death of young animals by irritating or clogging the intestinal tract. The seeds and seedlings contain a glucoside, xanthostrumarin, that is poisonous to livestock, especially to swine and poultry. Some of the species yield an extract that is said to have styptic properties.

KEY TO THE SPECIES

1. Leaves lanceolate or lance-ovate, acute or acuminate at both ends, densely whitish-strigillose beneath; axils of the leaves bearing conspicuous, 3-forked spines; fruit about 1 cm. long ....1. *X. spinosum*
1. Leaves deltoid or broadly ovate, usually cordate, green beneath; axils without spines; fruit about 2 cm. long or more ....2. *X. saccharatum*

1. Xanthium spinosum L. Ash Fork to Congress Junction (Yavapai County), also in Santa Cruz and Pima counties, roadsides, August and September. A weed throughout most of the United States and the warmer and temperate parts of the world.

2. Xanthium saccharatum Wallr. Throughout the state, 100 to 6,000 feet, moist, alluvial soil, summer. A common weed throughout the United States and in the Hawaiian Islands, perhaps elsewhere.

52. ZINNIA (333)

Herbs or undershrubs; leaves opposite, entire; heads medium-sized or large, radiate, showy; involucre graduated, of dry phyllaries; receptacle becoming conic or cylindric; rays yellow, white, or red, pistillate, sessile and indurate-persistent on the achenes; disk achenes strongly compressed; pappus none, or of 1 to 4 awns or teeth.

The very popular garden zinnia (*Z. elegans* Jacq.) is a native of Mexico. The Arizona *Z. grandiflora* is an attractive plant, worthy of trial for ornamental borders.

KEY TO THE SPECIES

1. Annuals, usually 30 cm. high or more, single-stemmed; leaves ovate or lanceolate, usually 10 mm. wide or more; rays dark red, greenish on the back............1. *Z. multiflora*
1. Low perennials, normally 20 cm. high or less, much branched from the base, woody below; leaves linear or acerose, not more than 2.5 mm. wide; rays yellow or white (2).
2. Leaves more or less 3-ribbed; rays 8 to 16 mm. long, bright yellow; style branches hispid, with long, acuminate appendages..............................2. *Z. grandiflora*
2. Leaves 1-ribbed; rays usually 12 mm. long or less, light yellow or white; style branches merely hispidulous, with short, bluntish appendages..................3. *Z. pumila*

1. Zinnia multiflora L. Mule and Huachuca mountains (Cochise County), base of the Patagonia Mountains and near Ruby (Santa Cruz County), 4,000 to 5,500 feet, August and September. Florida and West Indies; southern Arizona and Mexico; South America.

2. Zinnia grandiflora Nutt. Navajo County to eastern Mohave County, south to Greenlee, Cochise, and Santa Cruz counties, 4,000 to 6,500 feet, dry slopes and mesas, May to October. Kansas to Nevada, south to Texas, Arizona, and northern Mexico.

3. Zinnia pumila Gray. Central Yavapai County to Cochise and Pima counties, 2,500 to 5,000 feet, dry mesas and slopes, commonly on "caliche" soil, April to October. Texas to southern Arizona and northern Mexico.

Plant woodier and more pubescent than *Z. grandiflora.*

53. SANVITALIA

Slender annuals; leaves opposite, lanceolate or lance-linear, mostly entire; heads small, terminal, radiate, the rays yellow, fading whitish; pales of the receptacle with rigid, cuspidate tips; rays sessile, persistent on the achenes; ray achenes narrowly 4-sulcate, their pappus of 3 short awns or tubercles; disk achenes 4-angled, epappose or nearly so.

The colors of the rays were noted in the field by L. M. Pultz.

1. **Sanvitalia Aberti** Gray. Apache County to eastern Mohave County, south to Cochise, Santa Cruz, and Pima counties, 4,000 to 7,500 feet, dry slopes and mesas and along streams, July to September. Western Texas to Arizona and northern Mexico.

54. HELIOPSIS. Ox-eye

Perennial herbs; leaves opposite, triangular-ovate, usually toothed, petioled; heads solitary, terminal, long-peduncled, radiate, yellow, showy; rays pistillate, persistent on the achenes; achenes short and thick, epappose.

1. **Heliopsis parvifolia** Gray. Mountains of Cochise, Santa Cruz, and Pima counties, 4,000 to 8,000 feet, rich soil in canyons, July to October, type from between Babocomari, Arizona, and Santa Cruz, Sonora (*Wright* 1218). Southwestern Texas to southern Arizona and northern Mexico.

55. ECLIPTA

Low annual herbs; leaves opposite, lanceolate, toothed; heads radiate, white, peduncled in the upper axils; rays numerous, short, narrow; pales of the receptacle bristlelike; achenes short, thick, 3- or 4-angled, truncate, epappose or essentially so.

1. **Eclipta alba** (L.) Hassk. Mohave, Yavapai, Greenlee, Pinal, Pima, and Yuma counties, up to 3,500 feet, along streams and ditches, June to September. Massachusetts to Nebraska, south to Florida, Texas, southern Arizona, California, and South America; widely distributed in the warmer regions of the world.

56. RUDBECKIA. Coneflower

Tall perennials; leaves alternate, the lower ones pinnately divided into few lobes, the upper leaves 3-cleft to entire; heads large, radiate, yellow; involucre herbaceous or subfoliaceous; disk becoming cylindric; achenes 4-angled; pappus a short crown.

1. **Rudbeckia laciniata** L. (*R. umbrosa* Greene). Apache County to Coconino County, south to Cochise and Pima counties, 5,000 to 8,500 feet, rich soil along mountain streams, July to September, type of *R. umbrosa* from Oak Creek, Coconino County (*Pearson* in 1909). Maine to Saskatchewan and Idaho, south to Florida, Colorado, and southern Arizona.

Cut-leaf coneflower. A showy but rather coarse plant, of which a double form known as goldenglow is often cultivated. The plants are reported to be poisonous to cattle, sheep, and swine.

57. RATIBIDA (334, pp. 62–77). Prairie-coneflower

Perennial herbs; leaves alternate, pinnately parted; heads terminal, long-peduncled, showy, the rays yellow or partly brown-purple; disk globose to cylindric; achenes short and broad, compressed, with 1-angled sides; pappus of 1 or 2 teeth, sometimes of squamellae.

KEY TO THE SPECIES

1. Disk soon cylindric, 10 to 40 mm. long, 6 to 10 mm. thick; rays usually 8 to 30 mm. long; peduncles 6 to 25 cm. long . . . . 1. *R. columnaris*
1. Disk subglobose, becoming oblong-ellipsoid, 6 to 13 mm. long, 6 to 9 mm. thick; rays 3 to 8 mm. long; peduncles 1 to 5 cm. long . . . . 2. *R. Tagetes*

1. Ratibida columnaris (Sims) D. Don. Apache, Navajo, Coconino, Greenlee, and Santa Cruz counties, 5,000 to 7,500 feet, plains and openings in pine woods, June to November. Minnesota to British Columbia, south to Tennessee, Colorado, and Arizona.

Var. *pulcherrima* (DC.) D. Don is an unimportant form, with the rays partly or wholly brownish purple (yellow throughout in the typical form). *R. columnaris* is suspected of being poisonous to cattle, but the plant is rarely eaten.

2. Ratibida Tagetes (James) Barnhart. Adamana, Apache County (*Griffiths* 5092), Flagstaff, Coconino County (*Ensor* in 1933, *Deaver* in 1945), 5,000 to 7,000 feet, plains, June to September. Kansas and Colorado to Texas, New Mexico, and northern Arizona.

58. ZALUZANIA (334, pp. 95–114)

Plants suffrutescent, branched, slightly pubescent; leaves alternate, ovate, 3-lobed; heads radiate, yellow, loosely clustered; receptacle conic; rays pistillate; achenes of the disk short, quadrangular, epappose, those of the rays with a few short setae.

1. Zaluzania Grayana Robins. & Greenm. (*Gymnolomia triloba* Gray). Chiricahua and Huachuca mountains, Cochise County (*Lemmon, Pringle, Wilcox*), slopes and canyons, July to September, type of *G. triloba* from the Chiricahua Mountains (*Lemmon*). Southwestern New Mexico, southern Arizona, and Chihuahua.

59. WYETHIA (335). MULES-EARS

Perennial herbs; leaves alternate, linear to oblong, entire or essentially so; heads large, terminal, solitary, yellow, radiate; rays pistillate; achenes rather large, 3- or 4-angled; pappus a chaffy, dentate crown, or divided into a few teeth, or long-awned at the angles.

KEY TO THE SPECIES

1. Leaves uniform, linear or lance-linear, sessile, 0.7 to 2.3 cm. wide, like the stem harshly tuberculate-hispidulous or -hispid; involucre strongly graduated, the phyllaries with an ovate, indurate base, abruptly narrowed into a longer, very narrowly subulate, spreading herbaceous tip . . . . 1. *W. scabra*
1. Leaves not uniform, mainly oblong or elliptic, at least the basal ones petioled, 3 to 9 cm. wide, pilose or hirsute but not tuberculate, much larger than those of the stem; involucre few-seriate, the phyllaries subequal, oblong, oval, or ovate, not abruptly narrowed into a subulate, spreading tip . . . . 2. *W. arizonica*

1. Wyethia scabra Hook. Apache, Navajo, and Coconino counties, 5,000 to 6,000 feet, fairly common on dry slopes and mesas, June to October. Wyoming to east-central Utah, northwestern New Mexico, and northeastern Arizona.

A handsome but rather coarse plant, with numerous stems from a woody base. Said to be used as an emetic by the Hopi and Navajo Indians, but they consider it dangerous. Most of the Arizona plants were referred by Weber (335, pp. 425, 426) to var. *canescens* W. A. Weber, with closely imbricate, uniformly pubescent phyllaries, but a collection from House Rock, Coconino County (*Jones* in 1930), was referred to var. *attenuata* W. A. Weber, with more loosely imbricate, very long-attenuate phyllaries, these with the hairs mostly marginal.

2. Wyethia arizonica Gray. Apache, Navajo, Coconino, and Gila counties, 7,000 to 9,500 feet, slopes and canyons, mostly in pine forest, June to August, type from Bear Springs (*Palmer* in 1869). Colorado, Utah, and northern New Mexico and Arizona.

60. TITHONIA (336)

Annuals; leaves opposite below, alternate above, large, ovate, toothed, petioled; heads rather large, solitary, radiate, orange-yellow, long-peduncled, the peduncle fistulose; rays neutral; achenes obovoid-oblong, thickened; pappus of 1 awn and several squamellae.

1. Tithonia Thurberi Gray. Santa Cruz and Pima counties, from the Patagonia and Santa Rita mountains to the Baboquivari Mountains, 3,000 to 4,500 feet, rich soil near streams, locally abundant, August and September. Southern Arizona and Sonora.

61. VIGUIERA (337)

Herbs or shrubs; leaves opposite (at least below), linear to ovate, usually toothed; heads medium-sized, radiate, yellow; rays neutral; achenes laterally compressed, thickened; pappus of 2 awns and several short squamellae, or sometimes none.

These plants are sometimes known as golden-eye and resin-weed, names applied also to plants of several other genera of Compositae.

KEY TO THE SPECIES

1. Pappus persistent, of 2 awns and several intermediate squamellae; achenes pubescent; leaves ovate, rarely lanceolate (2).
1. Pappus wanting; achenes glabrous; leaves linear to lanceolate or oval (4).
2. Plants shrubby. Leaves mostly opposite, deltoid-ovate, 1.5 to 3.5 cm. long, tuberculate-hispidulous; involucre 5 to 9 mm. high .... 1. *V. deltoidea*
2. Plants strictly herbaceous (3).
3. Leaves usually thin, on slender petioles, these rarely less and usually much more than 10 mm. long; disk of the heads 7 to 10 mm. high; phyllaries with an ovate or oblong-ovate, indurate and ribbed base, and a rather abrupt, linear or sometimes spatulate, herbaceous tip .... 2. *V. dentata*
3. Leaves usually firm, very short-petioled (petioles 1 to 8 mm. long); disk of the heads 11 to 15 mm. high; phyllaries linear-lanceolate to oblong-lanceolate, acuminate, subherbaceous throughout or indurate below .... 3. *V. cordifolia*
4. Plants perennial (5).
4. Plants annual (6).

5. Leaves lance-ovate to linear-lanceolate, 2 to 30 mm. wide..............4. *V. multiflora*
5. Leaves oval to elliptic-oblong, 14 to 23 mm. wide, 2.4 to 5 cm. long..........5. *V. ovalis*
6. Phyllaries merely hispid-ciliate, or sometimes also sparsely hispid on the back
6. *V. ciliata*
6. Phyllaries densely and more or less canescently strigose or strigillose on the back (7).
7. Leaves lanceolate to linear-lanceolate, 4 to 14 mm. wide; heads relatively large, the disk 6 to 14 mm. thick...7. *V. longifolia*
7. Leaves linear or linear-lanceolate, 1.5 to 3 mm. wide; heads relatively small, the disk 6 to 8 mm. thick...8. *V. annua*

1. Viguiera deltoidea Gray. Oatman (Mohave County) and from southern Yavapai and western Gila counties to Pima and Yuma counties, up to 3,500 feet, dry mesas and rocky slopes, January to October. Southern Nevada to southern Arizona, southern California, and northwestern Mexico.

A small, much-branched shrub, of limited value as browse. The species is represented in Arizona by var. *Parishii* (Greene) Vasey & Rose, with harsher pubescence, usually smaller leaves, and often smaller heads than in var. *genuina* Blake.

2. Viguiera dentata (Cav.) Spreng. Southern Apache County, near Prescott (Yavapai County), and Greenlee, Gila, Cochise, Santa Cruz, and Pima counties, 3,000 to 7,000 feet, dry slopes and canyons, fields and ditch banks, occasionally in woods, June to October. Western Texas to Arizona and Mexico.

Var. *lancifolia* Blake, differing in its narrowly lanceolate leaves, less than 2 cm. wide (in the typical plant ovate, up to 8 cm. wide), occurs in the Santa Rita, Santa Catalina, and Baboquivari mountains (Pima County), also in Sonora.

3. Viguiera cordifolia Gray. Fort Apache (southern Navajo County), Pinaleno Mountains (Graham County), and mountains of Cochise, Santa Cruz, and Pima counties, 3,500 to 9,000 feet, dry slopes, canyons, and plains, mostly in pine forest, June to October. Western Texas to eastern and southern Arizona and northern Mexico.

4. Viguiera multiflora (Nutt.) Blake. Apache County to northeastern Mohave County, south to the mountains of Cochise, Santa Cruz, and Pima counties, 4,500 to 9,500 feet, dry slopes and mountain meadows, often in pine forest, common, May to October. Southwestern Montana to New Mexico, southern Arizona, Nevada, and eastern California.

Var. *nevadensis* (A. Nels.) Blake, which differs in its narrowly linear-lancealote, usually strongly revolute-margined leaves, only 2 to 5 mm. wide (in the typical plant the leaves lanceolate or lance-ovate, plane, 6 to 30 mm. wide), occurs throughout most of the range of the species in Arizona.

5. Viguiera ovalis Blake. Cochise County, at Cave Creek, Chiricahua Mountains, about 6,000 feet (*Harrison & Kearney* 6150), and summit of Huachuca Peak, 8,400 feet (*Goodding* 930–49), September. Southwestern New Mexico and southeastern Arizona.

6. Viguiera ciliata (Robins. & Greenm.) Blake. Camp Lowell, Pima County, 2,500 feet (*Thornber* 97), September and October. Southern Utah to eastern New Mexico, southern Arizona, and Sonora.

Thornber's specimen is intermediate between typical *V. ciliata* and var. *hispida* (Gray) Blake.

7. Viguiera longifolia (Robins. & Greenm.) Blake. Coconino County to the mountains of Greenlee, Cochise, Santa Cruz, and Pima counties, 4,500 to 8,000 feet, dry slopes and plains, July to October. Western Texas to Arizona and Mexico.

8. Viguiera annua (Jones) Blake. Apache County to eastern Mohave County, south to Cochise, Santa Cruz, and Pima counties, 2,500 to 7,000 feet, hills, plains, and river bottoms, May to October. Western Texas to Arizona and northern Mexico.

Colors the landscape with brilliant yellow for many miles in northern Yavapai County. It is reported to make good forage for sheep.

62. HELIANTHUS (338). SUNFLOWER

Annual or perennial herbs; leaves opposite or alternate, usually toothed; heads medium-sized to large, usually solitary or few, radiate, the rays yellow, the disk yellow, brown, or purple-brown; involucre more or less herbaceous; achenes oblong, thickened; pappus of 2, rarely many, caducous paleaceous awns.

H. annuus, the state flower of Kansas, is a common and very conspicuous roadside weed in Arizona. In the plains states it is used for ensilage. The seeds of this and other species are eaten by the Southwestern Indians. Cultivated varieties with very large heads (Russian sunflower) are grown for the seeds, which are fed to poultry and other birds. In Russia they are roasted and are a popular delicacy, taking the place of peanuts with us. They yield oil of sunflower, which is used as a hairdressing and sometimes as salad oil. The residue after pressing is of value as a concentrated cattle food. The Hopi Indians extract from the seeds of this and perhaps other species of *Helianthus* purple and black dyes for baskets and textiles and for painting the body in certain of their ceremonies. The white tubers of the Jerusalem-artichoke (*H. tuberosus* L.) are used for human food and as feed for swine. They are rich in sugars, particularly levulose.

KEY TO THE SPECIES

1. Plants strongly glaucous, low, usually less than 0.5 m. high, perennial; phyllaries obtuse to acute, glabrous on the back, white-ciliate, ovate to oblong, closely imbricated, shorter than the disk. . . . . 1. *H. ciliaris*

1. Plants slightly or not at all glaucous, usually taller; phyllaries usually acuminate, more or less pubescent on the back (2).

2. Disk (i. e. the tips of the disk corollas) yellow; plants perennial; leaves elongate-lanceolate, green, not conspicuously hispid. Involucre not conspicuously hispid 2. *H. Nuttallii*

2. Disk (i. e. the tips of the disk corollas) brown or purple-brown; plants annual; leaves usually ovate or lance-ovate (3).

3. Phyllaries narrowly linear or linear-lanceolate, only 1 to 2 mm. wide, light green, usually much surpassing the disk, conspicuously hispid, in about 2 series; pappus of numerous unequal paleae; leaves linear-lanceolate to lance-ovate, hispid with strongly tuberculate-based hairs . . . . 3. *H. anomalus*

3. Phyllaries broader, or else blackish green; pappus normally of 2 awns; leaves otherwise (4).

4. Central pales of the disk densely and conspicuously white-bearded; phyllaries normally lanceolate or linear-lanceolate, gradually acuminate, hispidulous but not or scarcely ciliate; leaves usually lanceolate or lance-ovate, seldom cordate. . . . . . . 4. *H. petiolaris*

4. Central pales of the disk not white-bearded; phyllaries ovate, with an abrupt cirrhuslike acumination, hispidulous and conspicuously ciliate; leaves usually broadly ovate, at least the lower ones usually cordate..............................5. *H. annuus*

1. Helianthus ciliaris DC. Apache County to Mohave County, south to Cochise and Pinal counties, 1,000 to 6,500 feet, often in saline soil, July to October. Texas to northern Arizona and Mexico.

Blueweed. Sometimes a troublesome weed in cultivated land because of the creeping rootstocks.

2. Helianthus Nuttallii Torr. & Gray. Greer, Apache County (*Eggleston* 17091), Tuba, Coconino County (*Kearney & Peebles* 12876), Oak Creek, Coconino or Yavapai County (*Rusby* in 1883), 5,000 to 9,000 feet, marshy places, August and September. Saskatchewan to Alberta, south to New Mexico and northern Arizona.

3. Helianthus anomalus Blake. Monument Valley, Navajo County (*Eastwood & Howell* 6659), Hopi Indian Reservation, Navajo County (*Whiting* 854), Beaver Dam, Mohave County (*Peebles* 13083), 2,000 to 6,000 feet, June to September. Utah and northern Arizona.

4. Helianthus petiolaris Nutt. Apache, Navajo, and Coconino counties, south to Cochise, Santa Cruz, Pima, and Yuma counties, 500 to 7,500 feet, alluvial and cultivated land, common, March to October. Saskatchewan to Missouri and Texas, west to British Columbia and California, occasional farther east as an introduction.

The typical plant is green or greenish and not conspicuously pubescent. Var. *canescens* Gray in its extreme form is densely canescent-strigose or canescent-strigillose on the leaves, stem, and involucre. This variety occurs in Graham, Gila, Pima, and Yuma counties, 3,000 feet or lower, on sandy or gravelly mesas, ranging from Texas to Arizona. Although at times very distinct in appearance, it intergrades with typical *H. petiolaris*. It is also perplexingly close to *H. niveus* (Benth.) T. S. Brandeg. of Baja California.

5. Helianthus annuus L. Nearly throughout the state, 100 to 7,000 feet, roadsides and fields, very common, March to October. Saskatchewan to Texas and westward, cultivated and escaping or becoming established elsewhere throughout the United States.

A sheet of *Helianthus subrhomboideus* Rydb., labeled: "Arizona, Dr. E. Palmer, 1869," is in the United States National Herbarium. The species is not otherwise known from Arizona but is reported from New Mexico. As the data accompanying Palmer's plants of 1869 are not always reliable, it seems best to await additional material before formally including the species in the flora of Arizona.

63. FLOURENSIA (339). TAR-BUSH, VARNISH-BUSH

Resinous, much-branched shrubs; leaves alternate, small, ovate to oval, entire; heads rather small, discoid, yellow, nodding, axillary and terminal; involucre herbaceous; achenes cuneate, laterally compressed but somewhat thickened, villous; pappus of 2 unequal awns.

1. Flourensia cernua DC. Cochise County, 3,500 to 5,000 feet, plains and mesas, often on limestone, locally abundant, July to December. Western Texas to southeastern Arizona and northern Mexico.

A small shrub with hoplike odor and bitter taste, unpalatable to livestock. The leaves and heads are sold in the drug markets of northern Mexico under the name "hojasé" or "hojasén" and are taken in the form of a decoction for indigestion. The plant is sometimes known locally as black-brush, but this name belongs properly to *Coleogyne,* of the rose family.

64. ENCELIA (340, pp. 358–376)

Low, branching shrubs; leaves alternate, ovate or oblong, entire or toothed; heads medium-sized, solitary or panicled, radiate or discoid, yellow, or the disk purple; rays neutral; achenes compressed, very flat, obovate, notched at apex, ciliate, more or less pubescent on the sides; pappus none, or of 1 or 2 weak awns.

KEY TO THE SPECIES

1. Heads cymose or panicled; inflorescence glabrous or essentially so; leaves densely whitish-tomentulose .... 1. *E. farinosa*
1. Heads solitary at the tips of the stem and branches; peduncles pubescent; leaves not tomentulose .... 2. *E. frutescens*

1. Encelia farinosa Gray. Grand Canyon (Coconino County) and Mohave, Greenlee, Gila, Maricopa, Pinal, Pima, and Yuma counties, up to 3,000 feet, very abundant on dry, rocky slopes, November to May. Southwestern Utah (according to Benson and Darrow), southern Nevada, southern and western Arizona, southern California, and northwestern Mexico.

Var. *phenicodonta* (Blake) Johnst., which has dark-purple instead of yellow disk flowers, occurs with the typical plant in the Mohawk and Tule mountains, Yuma County.

Brittle-bush, incienso. The plants reach a height of about 1 m. The stems exude a gum that was chewed by the Indians and also was used as incense in the churches of Baja California. The leaves contain an organic compound that is toxic to other plants, according to Reed Gray and James Bonner (*Amer. Jour. Bot.* 35: 52–57). The plants are reported to be browsed by bighorns.

2. Encelia frutescens Gray. Grand Canyon (Coconino County) and Mohave County, to southern Apache (?), Greenlee, Graham, Pima, and Yuma counties, up to about 4,000 feet, common on rocky slopes and mesas, January to September, type from Agua Caliente, Maricopa County (*Emory* in 1846). Southern Utah to southern California and Arizona.

KEY TO THE VARIETIES

1. Leaves not densely cinereous- or canescent-pubescent (2).
1. Leaves cinereous- or canescent-pubescent. Heads radiate (3).
2. Leaves sparsely tuberculate-hispidulous with conical, tuberculate-based, white hairs, not obviously glandular; peduncles not glandular; involucre hispid, not or only slightly glandular; heads usually discoid .... *E. frutescens* (typical)
2. Leaves tuberculate-hispidulous and also conspicuously glandular; peduncles more or less glandular; involucre densely glandular, sparsely hispidulous; heads radiate .... var. *resinosa*
3. Leaves cinereous with fine appressed hairs, these intermixed with stouter, tuberculate-based, antrorse hairs .... var. *virginensis*
3. Leaves canescent or cinereous with fine, soft, appressed hairs, without longer, tuberculate-based hairs .... var. *actoni*

Var. *resinosa* Jones has been collected near Winslow (type, *Jones* in 1890), in Monument Valley (Navajo County), and along the Little Colorado River near Cameron (Coconino County), 4,500 to 5,500 feet. Var. *virginensis* (A. Nels.) Blake occurs in Coconino, Mohave, Graham, Gila, Pinal, and western Cochise counties. Var. *actoni* (Elmer) Blake has been reported from western Arizona.

Encelia californica Nutt. A collection of this species purporting to be from Fort Mohave (*Cooper* in 1860–1861) has been the only basis for the recording of this species from Arizona. As the plant has not been found by subsequent collectors, and as it is definitely known only from coastal southern California and the west coast of Baja California, it seems almost certain that the locality given for this collection is incorrect. The species may be recognized by its solitary heads, purple disk, and densely soft-pubescent involucre. The plant is reported to cause severe dermatitis in susceptible persons.

65. SIMSIA (340, pp. 376–396)

Annual herbs, pubescent, branched; leaves mostly opposite, ovate, often toothed, petioled; heads numerous, panicled, medium-sized, yellow, radiate; achenes obovate, compressed, very flat, glabrous, epappose.

1. Simsia exaristata Gray. Near Tombstone, Cochise County, 4,500 feet (*Peebles et al.* 3377), valleys, September and October. Western Texas to southeastern Arizona and Mexico.

66. GERAEA (340, pp. 355–357). Desert-sunflower

Annual herbs, hirsute, branching; leaves alternate, oblong, ovate, or obovate, toothed; heads large, showy, yellow, radiate; phyllaries conspicuously white-ciliate; achenes cuneate, strongly compressed, silky-villous, the body black, the narrow, whitish margin continuous with the 2 strong awns, these connected by a low, entire, whitish crown.

1. Geraea canescens Torr. & Gray. Mohave, Maricopa, Pinal, Pima, and Yuma counties, 3,000 feet or lower, common in sandy soil, January to June. Southern Utah to southeastern California, western and southern Arizona, and Sonora.

Var. *paniculata* (Gray) Blake is an unimportant form, with stems paniculately much branched above, bearing very numerous heads only 2 to 2.8 cm. wide, known only from the type collection, at Phoenix (*Pringle* in 1882).

67. ENCELIOPSIS (340, pp. 351–355)

Scapose, perennial herbs, the herbage canescent or silvery-velutinous; leaves obovate to suborbicular, entire, in a basal tuft; heads solitary, large, yellow, radiate, on naked scapes; achenes oblong, strongly compressed, silky-villous, black, with a white cartilaginous border and crown; pappus of 2 short awns, or none.

KEY TO THE SPECIES

1. Pubescence silvery; leaves acute, rhombic-ovate ....1. *E. argophylla*
1. Pubescence dull; leaves obtuse, obovate or suborbicular.................2. *E. nudicaulis*

1. Enceliopsis argophylla (D. C. Eaton) A. Nels. Virgin Narrows and Lake Mead (Mohave County), about 500 feet, dry slopes and sandy washes, April to June. Southern Utah, southern Nevada, and northwestern Arizona.

2. Enceliopsis nudicaulis (Gray) A. Nels. Coconino County, near Navajo Bridge (*Peebles & Parker* 14654, *Mrs. Cantelow* in 1942) and near Fredonia (*Ripley & Barneby* 8511), 3,500 to 4,500 feet, dry slopes, May and June. Idaho to northern Arizona and Nevada.

It is somewhat doubtful that *E. nudicaulis* is distinct as a species from *E. argophylla.*

68. HELIANTHELLA

Perennial herbs; leaves opposite or alternate, lanceolate to ovate, entire; heads radiate, medium-sized or large, with yellow rays and a yellow or purple disk; achenes strongly compressed, obovate, ciliate or eciliate; pappus of fimbriate squamellae and often 2 slender awns.

KEY TO THE SPECIES

1. Disk purple; pales of the receptacle comparatively firm; plants usually much branched above, with several to many small heads........................1. *H. microcephala*
1. Disk yellow; pales of the receptacle soft and mainly scarious; plants usually simple or little branched, the heads solitary or few (2).
2. Heads small, the disk not more than 1.5 cm. wide, the rays 2 cm. long or less; leaves thick and firm, the basal and lower ones not more than 2.3 cm. wide......2. *H. Parryi*
2. Heads large, the disk usually 2 to 3 cm. wide, the rays usually 3 cm. long or more; leaves thin, the basal and lower ones 2.5 to 7.5 cm. wide................3. *H. quinquenervis*

1. Helianthella microcephala Gray. Carrizo Mountains (Apache County), near Fredonia and in the Grand Canyon (Coconino County), near Pipe Springs (Mohave County), 4,500 to 6,000 feet, July and August. Southwestern Colorado, southern Utah, northwestern New Mexico, and northern Arizona.

2. Helianthella Parryi Gray. White Mountains, Apache County (*Griffiths* 5266, *Eggleston* 17051), 8,500 feet, mountainsides, often under aspens, July to September. Wyoming to New Mexico and eastern Arizona.

3. Helianthella quinquenervis (Hook.) Gray (*H. quinquenervis* var. *arizonica* Gray). White Mountains (Apache and Greenlee counties), Kaibab Plateau and Flagstaff (Coconino County), Pinaleno Mountains (Graham County), Chiricahua and Huachuca mountains and Johnston Ranch (Cochise County), Santa Catalina Mountains (Pima County), 5,000 to 10,000 feet, mountain meadows and woods, July to October, types of var. *arizonica* from northern and southern Arizona (*Woodhouse, Lemmon*). South Dakota to Idaho (?), south to New Mexico, Arizona, and Chihuahua.

69. ZEXMENIA (341).

Perennial herbs, the roots tuberous-thickened; leaves opposite, ovate, toothed, subsessile; heads medium-sized, yellow, radiate, solitary, long-peduncled; phyllaries 2- or 3-seriate, essentially equal in length, very broad and blunt; rays pistillate; achenes obovate, compressed, 2-winged; pappus of 2 slender awns and several small intermediate squamellae, or the latter reduced to a mere rim connecting the awns.

1. Zexmenia podocephala Gray. Mountains of Cochise, Santa Cruz, and Pima counties, 4,000 to 5,000 feet, August and September, type from between the San Pedro and the Sonoita (*Wright* 1239 bis). Southwestern New Mexico, southeastern Arizona, and northern Mexico.

70. VERBESINA (342). Crown-beard

Annual or perennial herbs; leaves opposite or alternate, linear to ovate, usually toothed; heads medium-sized or rather large, radiate, yellow; rays pistillate or rarely neutral; achenes strongly compressed, 2-winged; pappus none or of 1 or 2 awns.

KEY TO THE SPECIES

1. Leaves linear or linear-lanceolate, long-acuminate, sessile, 9 to 23 cm. long, 5 to 18 mm. wide, subentire to serrate, strongly veiny beneath; achenes narrowly winged. Pappus awns wanting or rudimentary.................................... 1. *V. longifolia*
1. Leaves lanceolate to ovate or deltoid, obtuse to acuminate, much broader in proportion to length, usually sharply toothed, seldom veiny; achenes mostly broadly winged (2).
2. Leaves sessile, auriculate-clasping, green on both sides; perennial, with usually simple stems and solitary heads on long, naked peduncles.................. 2. *V. Rothrockii*
2. Leaves petioled, densely white-strigose beneath, green or greenish above; annual (sometimes perennial?), often much-branched, and with numerous heads... 3. *V. encelioides*

1. Verbesina longifolia Gray. Mountains of Cochise, Santa Cruz, and Pima counties, 5,000 to 8,000 feet, July to October. Southwestern New Mexico, southeastern Arizona, and northern Mexico.

2. Verbesina Rothrockii Robins. & Greenm. Cochise, Santa Cruz, and Pima counties, 3,500 to 5,500 feet, rocky slopes, May to September, type from Camp Bowie (*Rothrock* 452). Southern New Mexico, southern Arizona, and Coahuila.

3. Verbesina encelioides (Cav.) Benth. & Hook. Apache County to Mohave County, south to Cochise, Pima, and Yuma counties, up to 7,000 feet but usually much lower, a common weed of roadsides and waste ground, April to November. Kansas to Montana, south to Texas, California, and northern Mexico.

Most of the Arizona plants belong to var. *exauriculata* Robins. & Greenm., distinguished from the typical plant by having the petioles not auriculate-dilated at base. This plant is said to have been used by Indians and white pioneers for boils and skin diseases. The Hopis are reported to bathe in water in which this plant has been soaked to relieve the pain of spider bite.

71. COREOPSIS (343). Tickseed

Annual or sometimes perennial herbs; leaves mostly opposite, pinnatisect or dissected; heads showy, the rays yellow or with a brown-purple base; involucre double, the outer phyllaries narrow, herbaceous, the inner ones broad, membranaceous; achenes strongly dorsoventrally compressed, sometimes wing-margined; pappus a small cup, or obsolete.

Several species of this genus are well known as cultivated ornamentals.

KEY TO THE SPECIES

1. Plants scapose or subscapose, usually less than 30 cm. high; heads solitary; rays yellow throughout. Achenes with thick, corky wings........................ 1. *C. Douglasii*
1. Plants leafy-stemmed, usually 30 cm. high or more; heads loosely cymose-panicled; rays yellow with a purple-brown base (2).
2. Achenes wingless .... 2. *C. tinctoria*
2. Achenes with thin wings (3).
3. Achenes about 2 mm. long.. 3. *C. cardaminefolia*
3. Achenes 2.4 to 3 mm. long.. 4. *C. Atkinsoniana*

1. Coreopsis Douglasii (DC.) H. M. Hall. Coconino, Mohave, Graham, Gila, Maricopa, and Pinal counties, 1,500 to 3,500 feet, open places, February to May (sometimes August). Arizona, southern California, and northern Baja California.

2. Coreopsis tinctoria Nutt. "Middle Verde," Yavapai County (*W. W. Jones* 227), probably an escape from cultivation. In low ground, Minnesota and Manitoba to Louisiana, west to British Columbia and New Mexico.

Commonly cultivated in gardens under the name "calliopsis," and frequently escaping both east and west of the range here given. The collection cited is exceptional in having papillate achenes.

3. Coreopsis cardaminefolia (DC.) Torr. & Gray. Near Pinetop (southern Navajo County), Coconino County (several localities), and Yavapai County, 5,000 to 7,500 feet, openings in pine forest, rich, moist soil, June to September. Kansas to Louisiana, north-central Arizona, and northern Mexico.

4. Coreopsis Atkinsoniana Dougl. Coconino County, near the San Francisco Peaks (*Lemmon* 4159), between Williams and the Grand Canyon, about 6,500 feet (*C. L. Hitchcock et al.* 4509). Saskatchewan and British Columbia to South Dakota, northern Arizona, and Oregon.

The Arizona specimens, although technically *C. Atkinsoniana* in size of the achenes, are otherwise like *C. cardaminefolia,* which inhabits the same general area.

72. COREOCARPUS (344, pp. 342–345)

Branching suffrutescent perennials; leaves opposite, pinnately divided into 3 or 5 linear lobes; heads small, cymose, radiate, yellow; involucre of 5 to 8 subequal, 2-seriate, submembranaceous, dark-lined phyllaries, and often with a few much shorter outer bractlets; rays pistillate; achenes oblong, strongly dorsoventrally compressed, with 2 pectinately divided wings; pappus none, or of 1 or 2 retrorsely spinulose awns.

1. Coreocarpus arizonicus (Gray) Blake. Cochise, Santa Cruz, and Pima counties, 3,000 to 5,500 feet, rich soil along streams, January to October, type from Camp Lowell, Pima County (*Lemmon* in 1880). Southern Arizona and northern Mexico.

73. HETEROSPERMA

Low slender annuals; leaves opposite, once or twice pinnately divided into linear lobes; heads small, terminal, radiate, yellow; involucre double, much as in *Coreopsis;* rays fertile; outer achenes oval, incurved, wing-margined, epappose; inner achenes narrower, often infertile, not margined, narrowed into a beak, their pappus of 2 or 3 deciduous awns.

1. Heterosperma pinnatum Cav. Near Flagstaff (Coconino County), near Prescott (Yavapai County), south to Greenlee, Cochise, Santa Cruz, and Pima counties, commoner in southern Arizona, 4,000 to 7,000 feet, rich soil, sometimes on limestone, August and September. Southwestern Texas to Arizona, south to Guatemala.

74. THELESPERMA

Slender perennials; leaves opposite, mostly pinnately parted into a few narrow lobes; heads medium-sized or small, long-peduncled, radiate or discoid, entirely yellow or the disk brownish; involucre double, the outer phyllaries narrow, herbaceous, the inner ones broad, scarious-margined, connate to about the middle or higher; pales of the receptacle broadly scarious-margined; achenes oblong to linear, thickish, more or less papillate; pappus of 2 retrorsely hispid awns, or obsolete.

KEY TO THE SPECIES

1. Lobes of the disk corollas lanceolate or linear, longer than the throat; pappus of 2 triangular, hispid teeth 1 mm. long or longer; plants usually 30 to 60 cm. high, leafy below, naked above, usually branched. Heads normally discoid, rather large, the disk 1 to 1.5 cm. thick; outer phyllaries very short, broadly ovate, rounded
   1. *T. megapotamicum*
1. Lobes of the disk corollas ovate, shorter than the throat; pappus a mere crown, or obsolete; plants normally 40 cm. high or less, very leafy below, with long, naked, erect peduncles (2).
2. Heads relatively large (the disk about 1 to 1.5 cm. thick), normally with conspicuous rays; leaf divisions lanceolate to linear, 1.5 to 6 mm. wide; involucre 7 to 9 mm. high; inner phyllaries with broad and conspicuous scarious margins........2. *T. subnudum*
2. Heads small (the disk 5 to 10 mm. thick), always discoid; leaf divisions filiform, about 0.5 mm. wide; involucre 4 to 5 mm. high; inner phyllaries less conspicuously scarious-margined ....3. *T. longipes*

1. Thelesperma megapotamicum (Spreng.) Kuntze (*T. gracile* (Torr.) Gray). Apache, Navajo, Yavapai, Pinal, Cochise, Santa Cruz, and Pima counties, 4,000 to 7,500 feet, grassy plains and mesas and open woodland, May to October. Nebraska and Wyoming to Utah, south to Texas, Arizona, and Mexico; southern South America.

The Hopi make a tea from the flowers and young leaves, which are dried and then boiled. A reddish-brown dye for baskets and textiles is also obtained by them from this plant.

2. Thelesperma subnudum Gray. Apache County to northern Mohave County, 4,500 to 6,000 feet, dry hills and stream banks, May to September. Colorado and Utah to northern New Mexico and northeastern Arizona.

Sometimes known as Navajo-tea.

3. Thelesperma longipes Gray. Cochise, Santa Cruz, and Pima counties, 4,500 to 6,000 feet, slopes and canyons, often on limestone, June to October. Western Texas to southeastern Arizona and northern Mexico.

Used as a substitute for tea in New Mexico under the name "cota."

Thelesperma simplicifolium Gray has been reported to occur in Arizona but has not been seen by the writer from west of Texas.

75. BIDENS (345). Spanish-needles

Annual or perennial herbs; leaves, at least the lower ones, opposite, entire to dissected; heads medium-sized to large, often showy, usually radiate and yellow, sometimes discoid or with white rays; involucre double, as in *Coreopsis;* achenes linear-fusiform or linear to cuneate, more or less tetragonal or dorsoventrally compressed; pappus of 2 to 4 retrorsely hispid awns, rarely wanting.

These plants are known also as bur-marigold, stick-tight, pitch-forks, beggar-ticks, and water-marigold, most of these names referring to the tendency of the awned fruits to adhere to clothing and the hair of animals.

KEY TO THE SPECIES

1. Achenes obovate or wedge-shaped, either strongly flattened or 3- or 4-angled (2)
1. Achenes narrow, linear or linear-fusiform (3).
2. Leaves sessile, merely serrate, lanceolate or lance-oblong; heads radiate, the rays 1.5 to 3 cm. long, bright yellow; achenes 2- to 4-awned, the margins retrorse-hispidulous 1. *B. laevis*
2. Leaves petioled, pinnate, with 3 to 5 lanceolate, coarsely serrate leaflets; heads discoid, or the rays small and inconspicuous; achenes 2-awned, the marginal hairs mostly antrorse .... 2. *B. frondosa*
3. Achenes not conspicuously elongate, usually little exceeding the involucre; inner phyllaries with conspicuous yellow margins half as wide as the brown center, or wider; rays yellow, conspicuous, 1 cm. long or longer (4).
3. Achenes (at least the inner ones) conspicuously elongate (least so in *B. pilosa*), usually (including the awns) twice as long as the inner phyllaries; inner phyllaries with inconspicuous pale margins; rays wanting or inconspicuous, or if conspicuous, then white (5).
4. Leaves twice or thrice pinnatisect, the lobes narrowly linear, not more than 2 mm. wide 10. *B. ferulaefolia*
4. Leaves lanceolate or oblong, serrate and not lobed, or pinnately parted into 3 or 5 lanceolate or linear divisions at least 6 mm. wide....................... 11. *B. aurea*
5. Leaves all pinnately 3- or 5-parted, with lance-oblong to rhombic-ovate, serrate to incised divisions .... 3. *B. pilosa*
5. Leaves all once or twice pinnately dissected (6).
6. Outer and inner phyllaries more or less densely pilose or hirsute. Heads campanulate; leaves 2 or 3 times dissected into linear lobes..................... 6. *B. tenuisecta*
6. Outer and inner phyllaries glabrous or merely short-ciliate, rarely sparsely pilose (7).
7. Heads (when normally developed) campanulate, more than 13-flowered (8).
7. Heads cylindric or subcylindric, 5- to 13-flowered (9).
8. Outer achenes only moderately shorter than the inner ones and not conspicuously broader; awns usually 3 or 4................................. 4. *B. bipinnata*
8. Outer achenes conspicuously shorter and broader than the inner ones, usually not more than half their length; awns 2, sometimes 3........................ 5. *B. Bigelovii*
9. Leaf divisions filiform or narrowly linear (1 mm. wide or less). Achenes glabrous; heads 5- to 9-flowered... 9. *B. heterosperma*
9. Leaf divisions lanceolate or ovate to linear, usually more than 1 mm. wide (10).
10. Heads subsessile or short-peduncled; achenes glabrous; outer phyllaries 5 to 8 mm. long, or longer, linear-spatulate; leaf divisions mostly linear or linear-oblong, 1.5 to 3.5 mm. wide.. 7. *B. Lemmoni*
10. Heads long-peduncled; achenes antrorse-hispidulous above; outer phyllaries minute, slenderly subulate or filiform-subulate, less than 3 mm. long; leaf divisions broader, mostly lanceolate .... 8. *B. leptocephala*

1. Bidens laevis (L.) B.S.P. (*B. persicaefolia* Greene). Big Lake (southern Apache County), Bradshaw Mountains, Yavapai County (*Toumey* 680, the type of *B. persicaefolia*), near Ruby (Santa Cruz County), Santa Cruz River (Pima ? County), 3,500 to 9,000 feet, in sloughs and shallow lakes. Massachusetts to Florida, Arizona, and California, south to South America.

2. Bidens frondosa L. Near Blue, Greenlee County, 6,000 feet, in a meadow (*Gould & Robinson* 5178). Widely distributed in North America.

3. Bidens pilosa L. Chiricahua, Huachuca, Patagonia, Oro Blanco, and Baboquivari mountains (Cochise, Santa Cruz, and Pima counties), 3,500 to 5,500 feet. Florida to California; tropical and subtropical regions of the world.

Typical *B. pilosa* with discoid heads has been reported from Arizona but has not been seen by the writer. The Arizona plants belong mostly to var. *radiata* Schultz Bip. (*B. leucantha* (L.) Willd.), with 5 or 6 conspicuous white rays 7 to 15 mm. long. Other Arizona specimens are intermediate between var. *radiata* and typical *B. pilosa,* e. g. a collection in the Mule Mountains, Cochise County (*Harrison & Kearney* 6237).

4. **Bidens bipinnata** L. Big Lue Range (Greenlee County), Mule and Huachuca mountains (Cochise County), Tumacacori Mission (Santa Cruz County), 3,000 to 6,000 feet, rich, alluvial soil, August and September. Rhode Island to Kansas, southeastern Arizona, and California, southward to South America; also in the Old World.

Specimens from Arizona closely approach the preceding and the next following species. A collection in the Big Lue Range (*Gould & Haskell* 4037) has achenes with only 2 awns, but there is little difference between the outer and inner achenes.

5. **Bidens Bigelovii** Gray. Southern Apache, Greenlee, Gila, Cochise, and Santa Cruz counties, mostly in the mountains, 3,000 to 6,500 feet, stream banks and moist shady places, July to October. Colorado to western Texas, New Mexico, Arizona, and Sonora.

6. **Bidens tenuisecta** Gray. White Mountains (Apache County), Mormon Lake and near Flagstaff (Coconino County), Chiricahua and Huachuca mountains (Cochise County), Santa Catalina Mountains (Pima County), 6,000 to 8,000 feet, oak-chaparral and open pine forests, July to October. Idaho to western Texas, New Mexico, Arizona, and Chihuahua.

7. **Bidens Lemmoni** Gray. Chiricahua and Huachuca mountains (Cochise County), Patagonia Mountains (Santa Cruz County), Santa Rita Mountains (Pima County), 4,500 to 7,000 feet, September and October, type from Apache Pass (*Lemmon* in 1881). Southern Arizona and Mexico.

8. **Bidens leptocephala** Sherff. Apache, Graham, Gila, Pinal, Cochise, Santa Cruz, and Pima counties, 3,000 to 6,000 feet, mostly along streams, preferring shaded, sandy soil, August to October, type from the Chiricahua Mountains (*Blumer* 1712). Western Texas to Arizona, Chihuahua, and Baja California.

9. **Bidens heterosperma** Gray. San Francisco Peaks (Coconino County) and mountains of Graham, Cochise, Santa Cruz, and Pima counties, 5,500 to 9,000 feet, September. Southern Colorado to New Mexico, Arizona, and ·thern Mexico.

10. **Bidens ferulaefolia** (Jacq.) DC. Near Fort Apache (Apache County), Skull Valley (Yavapai County), Chiricahua and Huachuca mountains (Cochise County), Sonoita Valley (Santa Cruz County), up to 6,500 feet, August and September. Arizona to Guatemala.

The stems are exceptionally tall and stout in specimens from Skull Valley (*Thornber* 8787).

11. **Bidens aurea** (Ait.) Sherff. Cochise, Santa Cruz, and Pima counties, 3,000 to 6,000 feet, moist, sandy soil along streams, spring and autumn. Southern Arizona to Guatemala; adventive or naturalized in France and Italy.

The poorly distinguished var. *Wrightii* (Gray) Sherff, type from between the San Pedro River and Santa Cruz, Sonora (*Wright* 1233 bis), has narrowly lance-linear leaves only 6 to 8 mm. wide, these entire or divided into entire or subentire lance-linear lobes.

76. COSMOS (346)

Slender annuals; leaves opposite, dissected into narrow lobes; heads medium-sized, terminal, the rays white or rosy, the disk yellow; involucre double, as in *Coreopsis* and *Bidens;* achenes fusiform, slender-beaked; pappus of 2 to 4 retrorse-hispid awns.

Two species of *Cosmos, C. bipinnatus* with white to crimson rays and *C. sulphureus* with yellow rays, are common in cultivation.

1. **Cosmos parviflorus** (Jacq.) H.B.K. Southern Apache and Coconino counties to Cochise, Santa Cruz, and Pima counties, apparently rare in northern Arizona, 4,000 to 9,000 feet, open pine forest, hillsides, and canyons, sometimes in cultivated land, July to October. Southeastern Colorado to southwestern Texas, Arizona, and Mexico.

77. BEBBIA

Intricately branched shrubs with slender branches; leaves few, linear, the lower ones opposite; heads yellow, discoid, solitary or few, terminal; involucre strongly graduated; achenes somewhat compressed; pappus of about 20 plumose, bristlelike awns.

1. **Bebbia juncea** (Benth.) Greene. Grand Canyon (Coconino County) and Mohave County to Greenlee, Pima, and Yuma counties, up to 4,000 feet, dry slopes and washes, flowering most of the year. Western Texas to southern Nevada, southern California, and northern Mexico.

The plant is a shrub with rushlike branches. The typical plant, with glabrous stems, is less common in Arizona than var. *aspera* Greene, which has the stems more or less hispidulous with usually tuberculate-based, often deciduous hairs.

A plant collected on Silver Creek, Cochise County (*W. W. Jones* in 1947), has leaves up to 5 cm. long and 7 mm. wide exclusive of the 1 or 2 pairs of divaricate lobes. Similarly lobed leaves characterize *B. juncea* var. *atriplicifolia* (Gray) Johnst. of Baja California, but indications of lobing are found occasionally in var. *aspera.* Jones's specimen is herbaceous or nearly so and may be only a vigorous young shoot of *B. juncea* var. *aspera.*

78. GALINSOGA (347)

Annual herbs; leaves opposite, narrowly lanceolate, or the lower leaves lance-ovate, toothed, petioled; heads small, radiate, the rays white, very small, the disk yellow; achenes small, obovoid, somewhat thickened; pappus of the disk flowers of about 20 fimbriate, subequal squamellae or paleae, in the ray flowers reduced or wanting, in a rare variety wanting in both rays and disk.

1. **Galinsoga semicalva** (Gray) St. John & White. Mountains of southern Apache, Graham, Cochise, and Pima counties, 5,500 to 8,000 feet, rich soil in shade, September and October. Southern New Mexico and Arizona, and northern Mexico.

In the typical plant the disk achenes are finely hispidulous and bear a pappus of about 20 blunt, fringed squamellae or paleae nearly as long as the corolla, and the ray achenes are glabrous or usually somewhat hispidulous above on the inner face and bear a reduced pappus or none. In var. *percalva* Blake, known only from the Santa Rita Mountains, Santa Cruz or Pima County (*Griffiths & Thornber* 162, the type, *Gould* 2608), both ray achenes and disk achenes are glabrous and epappose.

79. MADIA. Tarweed

Plants annual, herbaceous, heavy-scented, the herbage hirsute, glandular above; leaves numerous, alternate, entire, narrowly linear; heads small, in a raceme or a narrow panicle, glomerate at the ends of the stems and branches; phyllaries in 1 series, closely enveloping the ray achenes, strongly carinate; rays tiny, 2 to 5 or sometimes none; ray achenes somewhat curved, 1-nerved on each face; disk achenes 4- or 5-angled; pappus none.

1. Madia glomerata Hook. Kaibab Basin, Coconino County, about 8,000 feet (*Hawbecker* in 1935, *Collom* in 1941), July to September. Saskatchewan to British Columbia, south to northern New Mexico, northern Arizona, and California.

80. HEMIZONIA. Tarweed

Annual herbs; leaves all or mostly alternate, pinnatifid to entire, sometimes spinescent; heads small or medium-sized, radiate, yellow, sometimes leafy-bracted; phyllaries in a single series, each partly enclosing a ray achene in its enfolded base; ray achenes much thickened, epappose, tipped with a very short, oblique beak; disk achenes all or mostly sterile, epappose or with a pappus of several, often connate, squamellae or paleae.

KEY TO THE SPECIES

1. Leaves and their lobes with spinescent tips; heads involucrate by spinose-tipped bracts; receptacle paleaceous throughout; disk achenes epappose.............1. *H. pungens*
1. Leaves and their lobes not spinescent-tipped; heads not involucrate; pales of the receptacle in a single row, connate into a cup; disk achenes with pappus......2. *H. Kelloggii*

1. Hemizonia pungens (Hook. & Arn.) Torr. & Gray. Santa Cruz River bottoms, Tucson, Pima County (*Thornber* in 1903). California; a casual introduction in Arizona.

2. Hemizonia Kelloggii Greene (*H. Wrightii* Gray). Common around habitations on mesas, Tucson, Pima County (*Thornber* 386 in 1903, and in 1905). California; introduced in Arizona.

Two of the 3 sheets examined are true *H. Kelloggii*, with the pappus squamellae partly united into a tube; the other (No. 386) has the squamellae mostly free and represents the form described as *H. Wrightii* Gray.

81. LAYIA

Plants low, annual, pubescent and stipitate-glandular; leaves lanceolate or linear, mostly alternate, the lower ones pinnatifid or incised, the upper leaves entire; heads terminal, showy, the rays white, the disk yellow; receptacle with a series of thin pales between the rays and the outer disk flowers, otherwise naked; ray achenes glabrous, epappose; disk achenes pubescent, their pappus of about 10 stout, villous bristles, the hairs on their outer side straight, on the inner entangled into a woolly mass.

1. Layia glandulosa (Hook.) Hook. & Arn. Navajo County to Mohave County, south to Greenlee (?), Graham, Cochise, Santa Cruz, and Pima counties, up to 5,000 feet, dry slopes and mesas, not abundant, March to June. Idaho and British Columbia to southwestern New Mexico, Arizona, and Baja California.

Plant handsome in flower, with pure white rays.

82. PSILOSTROPHE (348). PAPERFLOWER, PAPER-DAISY

Herbs or shrubs, more or less woolly; leaves alternate, entire or the lower ones pinnatifid; heads small, radiate, yellow; receptacle naked; rays persistent, becoming papery; achenes slender; pappus of 4 to 6 hyaline paleae

KEY TO THE SPECIES

1. Stem and branches densely pannose-tomentose with white wool; leaves linear or narrowly linear-lanceolate, entire; plants definitely shrubby. Heads mostly solitary at the tips of the branches, slender-peduncled .... 1. *P. Cooperi*
1. Stems and branches not pannose-tomentose; leaves (at least the basal ones) usually spatulate to obovate; plants herbaceous (2).
2. Stem and leaves rather densely lanate or tomentose; heads in close cymose clusters at the tips of the stem and branches .... 2. *P. tagetina*
2. Stem and leaves green, thinly pilose or pilosulous; heads usually scattered or in loose cymes .... 3. *P. sparsiflora*

1. Psilostrophe Cooperi (Gray) Greene. Navajo County to Mohave County, south to Cochise, Pima, and Yuma counties, 2,000 to 5,000 feet, mesas and plains, common, flowering throughout the year, type from Fort Mohave (*Cooper* in 1861). Utah and western New Mexico to southern California and northwestern Mexico.

Plant showy and handsome in flower.

2. Psilostrophe tagetina (Nutt.) Greene. Apache, Navajo, Coconino, Greenlee, Graham, and Cochise counties, 4,000 to 7,000 feet, open plains and mesas and pine forests, April to October. Western Texas to eastern Arizona and northern Mexico.

It is reported that sheep are sometimes fatally poisoned by this plant. Var. *grandiflora* (Rydb.) Heiser (*P. grandiflora* Rydb.) occurs in Cochise County, in and near the Chiricahua Mountains, type from near Paradise (*Blumer* 1709); also in southwestern New Mexico. It differs from typical *P. tagetina* in having usually longer peduncles and ray flowers and in being greener and less pubescent. It connects *P. tagetina* with *P. sparsiflora*.

3. Psilostrophe sparsiflora (Gray) A. Nels. (*P. divaricata* Rydb.). Apache County to eastern Mohave and Yavapai counties, and in Cochise County, 4,500 to 7,500 feet, dry mesas, slopes, and open pine forests, abundant in places, May to October, type of *P. divaricata* from the Grand Canyon (*D. T. Allen* in 1897). Utah, New Mexico, Arizona, and Chihuahua.

83. BAILEYA. Desert-marigold

Low floccose-woolly herbs; leaves alternate, pinnatifid to entire; heads solitary or cymose, long-peduncled, radiate, yellow; rays persistent, becoming papery, reflexed in age; achenes striate, epappose.

KEY TO THE SPECIES

1. Rays 5 to 7; heads loosely cymose toward the tips of the branches, relatively small, the disk in flower 6 mm. thick or less..............................1. *B. pauciradiata*
1. Rays about 20 to 50; heads mostly solitary at the tips of the stem and branches, larger, the disk in flower at least 10 mm. thick (2).
2. Stem leafy only at base or below the middle, the peduncles 10 to 20 cm. long 2. *B. multiradiata*
2. Stem leafy to above the middle or nearly to the apex, the peduncles 10 cm. long or less 3. *B. pleniradiata*

1. Baileya pauciradiata Harv. & Gray. Yuma County, 500 feet or lower, sandy deserts, March to May (sometimes October). Southwestern Arizona, southeastern California, and adjacent Sonora and Baja California.

2. Baileya multiradiata Harv. & Gray. Grand Canyon (Coconino County) and Mohave County to Graham, Cochise, Santa Cruz, Pima, and Yuma counties, up to 5,000 feet, very common on sandy plains and mesas, especially at roadsides, March to November. Western Texas to southern Utah and Nevada, southeastern California, and Chihuahua.

The large flower heads are very showy, and this or a similar species is sometimes cultivated in California for the flower trade. It is said that horses crop the heads, but fatal poisoning of sheep and goats eating this plant on overgrazed ranges has been reported.

3. Baileya pleniradiata Harv. & Gray. Mohave County to Cochise, Pima, and Yuma counties, 200 to 6,000 feet, plains and mesas, common in Cochise County, February to November. Western Texas to southern Utah and Nevada, southeastern California, and northern Mexico.

Not always readily distinguishable from *B. multiradiata*.

84. PERITYLE (349)

Herbaceous or suffruticulose, not woolly; leaves toothed or dissected, the lower ones opposite; heads small, radiate, the rays white; achenes strongly compressed, strongly ciliate on the margin, usually with a narrow callose margin; pappus of 1 or 2 awns (except sometimes in *P. Emoryi*), and a crown of squamellae.

KEY TO THE SPECIES

1. Plants annual, more or less stipitate-glandular above (2).
1. Plants perennial, many-stemmed, cinereous-puberulent above, not obviously glandular (3).
2. Achenes without an evident callose margin; awn 1, or wanting............1. *P. Emoryi*
2. Achenes with a conspicuous callose margin; awns 2..................2. *P. microglossa*
3. Leaves triangular in outline, ternately lobed or dissected............5. *P. coronopifolia*
3. Leaves ovate, merely toothed (4).
4. Leaves not impressed-punctate beneath; pappus awns 2, very short....3. *P. spilanthoides*
4. Leaves strongly impressed-punctate beneath; pappus awns 2, nearly as long as the corolla 4. *P. ciliata*

1. Perityle Emoryi Torr. Havasu Canyon (western Coconino County) and Mohave, southern Yavapai, western Gila, Maricopa, Pinal, Pima, and Yuma counties, up to 3,000 feet, common on rocky slopes and cliffs, February to October. Southwestern Arizona, southern California, and northwestern Mexico.

The typical plant has a pappus of squamellae and a slender awn. In var. *nuda* (Torr.) Gray the awn is wanting. The glandular pubescence is sometimes obscure.

2. Perityle microglossa Benth. (*P. microglossa* var. *effusa* Gray). Apparently known to occur in Arizona only in the Santa Catalina Mountains (Pima County), where the type of var. *effusa* was collected (*Pringle* in 1882). Southern Arizona to Nicaragua.

Perityle plumigera Harv. & Gray, based on a plant collected by Thomas Coulter in "California," has been ascribed to Arizona by several authors. Everly (349, p. 390) regarded it as probably from Sonora and synonymized it with *P. californica* Benth.

***3. Perityle spilanthoides** (Schultz Bip.) Rydb. Reported as occurring in Arizona; known otherwise only from northern Mexico.

4. Perityle ciliata (L. H. Dewey) Rydb. Southern Apache, Coconino, Mohave, Yavapai, Gila, and Pima counties, 3,000 to 7,500 feet, rocky slopes, May to November, type from near Pine, Gila County (*MacDougal* 676). Known only from central and southern Arizona.

5. Perityle coronopifolia Gray. Mountains of Greenlee, Graham, Cochise, Santa Cruz, and Pima counties, 4,000 to 7,500 feet, among rocks and on cliffs, often on limestone, June to October. Southern New Mexico and southern Arizona.

85. LAPHAMIA (349)

Plants low, suffruticulose; leaves toothed, lobed, or parted, rarely entire, at least the lower ones opposite; heads small or medium-sized, radiate or discoid, yellow or the rays white; achenes strongly compressed, not ciliate, rarely callose-margined; pappus of 1 or 2 bristlelike awns, or wanting.

KEY TO THE SPECIES

1. Leaves entire or merely toothed (2).
1. Leaves lobed or parted (3).
2. Leaves spatulate or oblanceolate to ovate, entire or repandly 3- to 5-toothed, not more than 8 mm. wide; stems hirtellous.................................1. *L. congesta*
2. Leaves ovate, sharply several-toothed, mostly 10 to 18 mm. wide; stems short-pilose or pilosulous ....2. *L. Palmeri*
3. Heads radiate, the rays white or yellow; stems essentially glabrous (4).
3. Heads discoid; stems pilosulous to villous (5).
4. Larger leaves 3-parted, with short, ovate or oblong segments..............3. *L. gilensis*
4. Leaves dissected into elongate, linear or filiform divisions................4. *L. saxicola*
5. Leaves pedately 3-parted or 3-lobed, with relatively broad, chiefly entire divisions, these up to 2.5 mm. wide..5. *L. gracilis*
5. Leaves lobed or pedately parted, with numerous small ultimate divisions...6. *L. dissecta*

1. Laphamia congesta Jones (*L. Toumeyi* Robins. & Greenm.). Coconino and northern Mohave counties, 2,500 to 8,000 feet, clefts of rocks and dry ground among rocks, June to October, type from below the "Buckskin Mountains," Kaibab Plateau (*Jones* 6063), type of *L. Toumeyi* from the Grand Canyon (*Toumey* in 1892). Known only from northern Arizona.

2. Laphamia Palmeri Gray. Beaver Dam, Mohave County (*E. Palmer* 199, in 1877, the type collection), about 2,000 feet, rocky canyons, July and probably later. Southern Utah and northwestern Arizona.

Var. *tenella* Jones, from crevices of sandstone rocks at Springdale, Utah, distinguished by its loosely villous stems, may occur in Arizona.

3. Laphamia gilensis Jones (*L. arizonica* Eastw., *L. dura* A. Nels.). Southern Gila, northeastern Pinal, and eastern Maricopa counties, about 2,500 feet, clefts in rocks in canyons, April to October, type from Putnam Ranch (*Jones* in 1890), type of *L. arizonica* from Fish Creek (*Eastwood* 8753), type of *L. dura* from Canyon Lake (*A. Nelson* 10323). Known only from south-central Arizona.

4. Laphamia saxicola Eastw. Near Roosevelt Dam (Gila and eastern Maricopa counties), about 2,500 feet (*Eastwood* 17401, the type collection, *Peebles* 9420), Salt River Canyon, northeast of Globe, Gila County, 4,000 feet (*Darrow* in 1940 and 1943), the stems hanging from crevices of cliffs, May and doubtless later. Known only from south-central Arizona.

Darrow's specimens are not typical, the rays being white and the leaves more elongate with fewer segments than in the type.

5. Laphamia gracilis Jones. Nagle Ranch, Kaibab Plateau, 7,000 feet (*Jones* 6050c, type), September. Southern Nevada and northern Arizona.

6. Laphamia dissecta Torr. (*L. Lemmoni* Gray, *L. Lemmoni* var. *pedata* Gray). Mountains of Greenlee, Graham, Cochise, and Pima counties, 3,000 to 7,000 feet, crevices of cliffs, May to October, types of *L. Lemmoni* and *L. Lemmoni* var. *pedata* from near Camp Lowell, Pima County (*Lemmon* in 1880). Southwestern Texas to southeastern Arizona.

86. PERICOME

Perennial herbs, branched, puberulent; leaves mostly opposite, hastate-triangular, caudate-acuminate; heads numerous, cymose-panicled, discoid, yellow; phyllaries lightly connate into a cup; achenes narrow-oblong, strongly compressed, villous-ciliate; pappus a crown of lacerate-ciliate squamellae, sometimes with 1 or 2 awns.

1. Pericome caudata Gray. Mountains of Apache, Navajo, Coconino, Greenlee, Graham, Cochise, Santa Cruz, and Pima counties, 6,000 to 9,000 feet, rich soil in coniferous forests, locally abundant, July to October. Southern Colorado and New Mexico to southern Nevada, California, and Chihuahua.

87. BAERIA

Low, slender annuals, slightly pubescent; leaves opposite, linear, entire; heads small, terminal, radiate, yellow; receptacle conical; achenes linear-clavate, 4-angled; pappus none, or of 2 to 5 lanceolate awns.

1. Baeria chrysostoma Fisch. & Mey. Graham, Gila, Maricopa, Pinal, Cochise, and Pima counties, 1,500 to 4,500 feet, mesas and plains, March to May. Central and southern Arizona to Oregon, California, and Baja California.

Goldfields. In spring extensive areas are sometimes carpeted with the bright-yellow flowers of this plant, which is reported to be cropped by horses. The only form occurring in Arizona is var. *gracilis* (DC.) Hall, characterized by pubescent achenes with a pappus of 2 to 5 lanceolate awns.

88. FLAVERIA (350)

Low, glabrous annuals, dichotomously branched; leaves opposite, lanceolate, toothed, 3-nerved; heads very small, 1- or 2-flowered, densely glomerate, the glomerules sessile in the forks and terminal; phyllaries 1 or 2; ray solitary or none; achenes oblong, 8- to 10-ribbed, glabrous, epappose.

1. **Flaveria trinervia** (Spreng.) C. Mohr. Southern Yavapai, Greenlee, Graham, and Pinal counties (probably elsewhere), 1,000 to 4,000 feet, moist soil at roadsides and on ditch banks, May to November. Florida and Alabama to southern Arizona, southward to South America; probably not indigenous in the United States.

89. SYNTRICHOPAPPUS

Dwarf, floccose-woolly winter annuals; leaves mostly alternate, spatulate to linear, often 3-lobed at tip; heads small, yellow, radiate, solitary at the tips of the branches; paleae of the pappus dissected into numerous bristles, these united only at base.

1. **Syntrichopappus Fremontii** Gray. Peach Springs to Oatman (Mohave County), Ajo Mountains (Pima County), probably also in Yuma County, 3,500 to 5,000 feet, mesas and rocky slopes, March to June. Southern Utah and Nevada, western Arizona, and southern California.

90. SCHKUHRIA (351)

Slender, branching annuals; leaves opposite or alternate, pinnately parted into few filiform lobes, or entire; heads small (the flowers 9 or fewer), discoid or with 1 very small ray; involucre more or less turbinate, the phyllaries only about 4 or 5, strongly impressed-punctate, with scarious, usually purplish tips; achenes shortly obpyramidal, 4-angled, conspicuously ciliate on the angles; pappus of 8 scarious squamellae or paleae, sometimes awned.

1. **Schkuhria Wislizeni** Gray (*Hopkirkia anthemoidea* DC., *Schkuhria anthemoides* [*sic*] Coult., not *S. anthemoidea* Wedd.). Cochise, Santa Cruz, and Pima counties, 4,000 to 6,000 feet, mesas and slopes, July to October. Western Texas to southern Arizona, south to Central America.

KEY TO THE VARIETIES

1. Pappus scales all very obtuse, subequal. . . . . var. *Wrightii*
1. Pappus scales (at least some of them) awned (2).
2. Pappus scales equal or subequal, all or most of them awn-tipped. . . . . var. *frustrata*
2. Pappus scales decidedly unequal, only those on the angles awned. . . *S. Wislizeni* (typical)

Var. *frustrata* Blake occurs in the Chiricahua and Huachuca mountains (Cochise County), near Nogales (Santa Cruz County), and in eastern Pima County. Var. *Wrightii* (Gray) Blake (*S. Wrightii* Gray), is found in Cochise, Santa Cruz, and Pima counties, type from along the Sonoita (*Wright* 1254).

The typical form of *S. Wislizeni* Gray, has been collected in the Chiricahua and Mule mountains (Cochise County) and in eastern Pima County.

91. HYMENOPAPPUS (352, pp. 92–98)

Tomentose herbs; leaves alternate, pinnatifid to dissected, or the lowest leaves in one species entire, often all or most of them basal; heads usually several, cymose, medium-sized, usually discoid, yellow or white, in one species with conspicuous white rays; involucre of 6 to 12 equal, oblong or oval, scarious-margined phyllaries; achenes obpyramidal, 4- or 5-angled, pubescent, often villous; pappus of 10 to 20 mostly obtuse scarious paleae or squamellae or (in *H. mexicanus*) sometimes obsolete.

The root of *H. lugens* is reported to be used by the Hopi as an emetic and in treating toothache.

KEY TO THE SPECIES

1. Heads radiate, the rays white, conspicuous.............................1. *H. radiatus*
1. Heads discoid (2).
2. Leaves entire and oblanceolate or obovate, or once pinnatifid with comparatively broad segments, these 2 to 6 mm. wide.................................2. *H. mexicanus*
2. Leaves once to thrice pinnatifid, with narrowly linear or linear-filiform segments (3).
3. Plants biennial, single-stemmed, tall, 30 to 80 cm. high; corolla teeth more than half as long as the throat. Stem leaves numerous, at least the lower ones not conspicuously smaller than the basal leaves.....................................3. *H. robustus*
3. Plants perennial, usually several-stemmed, lower, rarely more than 30 cm. high; corolla teeth ¼ to ⅓ as long as the throat (4).
4. Plants scapose or subscapose, rarely with 2 or 3 stem leaves...............4. *H. lugens*
4. Plants leafy-stemmed ....5. *H. pauciflorus*

1. Hymenopappus radiatus Rose. Southern Apache, Navajo, and Coconino counties, 5,000 to 7,000 feet, pine forests and open flats, May to July, type from Willow Spring, Apache County (*Palmer* 615, in 1890). New Mexico and Arizona.

2. Hymenopappus mexicanus Gray (*H. obtusifolius* Heller, *H. petaloideus* Rydb.). White Mountains (Apache and Navajo counties), San Francisco Peaks and Flagstaff (Coconino County), south to the mountains of Cochise and Pima counties, 5,000 to 10,000 feet, open coniferous forests, common, June to September, type of *H. obtusifolius* from Fort Valley, near Flagstaff (*MacDougal* 240), type of *H. petaloideus* from the Chiricahua Mountains (*Blumer* 1202). Western Texas to Arizona and Mexico.

3. Hymenopappus robustus Greene. Apache, Navajo, Coconino, Greenlee, Graham, and Gila counties, (2,500?) 3,500 to 6,500 feet, mesas and slopes, April to September. Texas to central Arizona and northern Mexico.

4. Hymenopappus lugens Greene (*H. gloriosus* Heller, *H. scaposus* Rydb., *H. macroglottis* Rydb.). Apache County to eastern Mohave County, south to Santa Cruz and Pima counties, 4,000 to 7,500 feet, dry, rocky slopes and mesas, usually with pines or junipers, common, May to September, type of *H. gloriosus* from Mormon Mountain, Coconino County (*MacDougal* 71), type of *H. scaposus* from near Flagstaff (*MacDougal* 129), type of *H. macroglottis* from Oak Creek (*Rusby* in 1883). Colorado to Nevada, Arizona, and southern California.

5. Hymenopappus pauciflorus Johnst. Northern Apache and Navajo counties and eastern Coconino County, 3,500 to 7,000 feet, dry mesas and slopes, June to September. Southern Utah and northeastern Arizona.

92. HYMENOTHRIX

Slender, annual or perennial herbs, slightly pubescent; leaves alternate, dissected into narrow divisions; heads numerous, cymose-panicled, radiate or discoid, yellow, white, or purple; achenes narrowly obpyramidal, 4- or 5-angled; pappus of 12 to 20 lanceolate hyaline paleae, the costa prolonged into an awn.

KEY TO THE SPECIES

1. Heads radiate, definitely yellow....................................1. *H. Wislizeni*
1. Heads discoid, purple, white, whitish, or pale yellow (2).
2. Petioles incurved-puberulent; lobes of the disk corollas about as long as the throat or a little longer ....2. *H. Loomisii*
2. Petioles (at least the lower ones) usually spreading-hirsute; lobes of the disk corollas several times longer than the very short throat......................3. *H. Wrightii*

1. Hymenothrix Wislizeni Gray (*H. Wislizeni* var. *setiformis* Jones). Greenlee, Graham, Gila, Cochise, Santa Cruz, Pima, and Pinal counties, 2,500 to 5,500 feet, plains, mesas, and washes, preferring sandy soil, June to December, type of var. *setiformis* from Oracle, Pinal County (*Jones* in 1903). Southern New Mexico, southern Arizona, and northern Mexico.

2. Hymenothrix Loomisii Blake (*Hutchinsonia hyalina* Jones). Near Flagstaff (Coconino County), to Kingman (Mohave County) and eastern and southern Yavapai County, 3,500 to 7,500 feet, mesas, plains, and along streams, often abundant at roadsides, June to October, type of *H. Loomisii* from Ash Fork (*Loomis* 3241), type of *Hutchinsonia hyalina* from Peach Springs (*Hutchinson* in 1932). Known only from central Arizona.

3. Hymenothrix Wrightii Gray. Greenlee, southern Navajo, northern Yavapai, and eastern Mohave counties, south to Cochise, Santa Cruz, and Pima counties, 4,000 to 8,000 feet, mostly in the scrub-oak and pine belts, June to November, type from between Babocomari and the Chiricahua Mountains (*Wright* 1266). Southern New Mexico, central and southern Arizona, southern California, and northwestern Mexico.

A densely glandular-viscid form has been collected near Horse Mesa, eastern Maricopa County, 3,000 feet (*K. F. Parker* 6087), and in the Superstition Mountains, Pinal County (*Gillespie* 8601).

93. PALAFOXIA (353)

Annual (or perennial?) herbs; leaves mostly alternate, linear or lanceolate, entire; heads 2 to 2.8 cm. high, discoid, flesh-colored or whitish; achenes slender, linear-tetragonal, about 1 cm. long; pappus of 4 or 5 linear paleae with an excurrent nerve, or reduced in some of the flowers.

1. Palafoxia linearis (Cav.) Lag. Northern Mohave and western Maricopa counties to southern Yuma County, up to 2,000 feet, sandy plains and mesas, February to November. Southern Utah, southern Nevada, western Arizona, southeastern California, and northern Mexico.

This plant, known in California as Spanish-needles, apparently is rapidly spreading eastward along highways in Arizona.

A large, coarse form, var. *gigantea* Jones (var. *arenicola* A. Nels.), 1 to 2 m. high, with larger leaves and heads (the involucre 1.5 to 2 cm. high, the achenes 1.5 to 1.8 cm. long), occurs on sand dunes west of Yuma, in southeastern California.

94. ERIOPHYLLUM (354). WOOLLY-DAISY

Tomentose herbs or shrubs; leaves alternate or opposite, entire to bipinnatifid; heads small, radiate or discoid, yellow, or the rays sometimes white or rosy; achenes slender, 4- or 5-angled; pappus of 4 to 12 squamellae or paleae, rarely wanting.

KEY TO THE SPECIES

1. Suffrutescent, up to 0.5 m. high; leaves pinnatifid or bipinnatifid with very narrow segments. Heads cymosely clustered, pedicellate, radiate, yellow. . 1. *E. confertiflorum*
1. Dwarf annuals; leaves entire or few-toothed or -lobed (2).
2. Heads clustered at the tips of the stem and branches, sessile; pappus of laciniate squamellae (sometimes wanting in the disk flowers); anther tips obtuse (3).
2. Heads solitary at the tips of the stem and branches, pedunculate; pappus of entire or merely erose squamellae; anther tips elongate, linear-subulate (4).
3. Rays present . . . . 2. *E. multicaule*
3. Rays wanting . . . . 3. *E. Pringlei*
4. Rays yellow; squamellae of the pappus equal, short, very blunt, opaque, sometimes wanting . . . . 4. *E. Wallacei*
4. Rays white or rosy; squamellae or paleae of the pappus unequal, the longer ones nearly as long as the corolla, lanceolate, acuminate or awned, not opaque. . . . . . 5. *E. lanosum*

*1. **Eriophyllum confertiflorum** (DC.) Gray. Dry hills, central California to Baja California and northern Mexico (reported from Arizona), April to September.

Eriophyllum confertiflorum is a variable species divided into several by Rydberg, who used the name *E. tenuifolium* (DC.) Rydb. for the plant he recorded from Arizona (*North Amer. Flora* 34: 96).

2. **Eriophyllum multicaule** (DC.) Gray. Near Prescott, Yavapai County (*Palmer* 221), and (according to Gray, *Syn. Fl.* 1^2: 328) at Tucson (Pima County), March to June. Arizona and California.

3. **Eriophyllum Pringlei** Gray. Yucca (Mohave County) and in southwestern Yavapai, Maricopa, and Pima counties, 1,500 to 3,000 feet, gravelly mesas and slopes, March to May. Southern Nevada, western and southern Arizona, and southern California.

4. **Eriophyllum Wallacei** Gray. Mohave County south of the Colorado River, and near Burro Creek (southwestern Yavapai County), 2,000 to 4,000 feet, mesas and plains, March to June. Southern Utah and western Arizona to southern California.

5. **Eriophyllum lanosum** Gray. Northern Mohave County to Greenlee, Graham, Cochise, Santa Cruz, Pima, and Yuma counties, 1,000 to 3,000 feet, dry, gravelly mesas and slopes, common, February to May. Southern Utah, southern Nevada, western Arizona, southeastern California, and Baja California.

95. CHAENACTIS (355)

Low herbs; leaves alternate, entire to bipinnatifid, the blade or its divisions narrow; heads flesh-colored or white (rarely yellow), discoid but sometimes with enlarged outer corollas; achenes linear; pappus of hyaline paleae, these without a midrib.

KEY TO THE SPECIES

1. Phyllaries attenuate into pallid, almost aristiform tips; plants scurfy-puberulent, not tomentose, scarcely or not at all glandular; receptacle with some setaceous pales
1. *C. carphoclinia*
1. Phyllaries obtuse to acuminate, but not attenuate into aristiform tips; plants tomentose or glandular, at least when young; receptacle naked (2).
2. Leaves entire and linear, or once pinnatifid with few unequal lobes; plants soon glabrate (3).
2. Leaves (at least the lower ones) bipinnatifid, with usually numerous lobes; plants more or less persistently tomentose (4).
3. Involucre 8 to 10 mm. high, the phyllaries not loose-tipped; pappus of 4 usually equal paleae, these in a single series; marginal corollas decidedly larger than the others
2. *C. Fremonti*
3. Involucre normally 12 to 17 mm. high, the phyllaries usually loose-tipped; pappus of 4 long paleae and 4 very short outer squamellae; marginal corollas not definitely larger than the others..3. *C. Xantiana*
4. Involucre 10 to 15 mm. high; anthers not exserted; pappus of 4 paleae and 2 to 4 very short outer squamellae; outer phyllaries with loose tips..............4. *C. macrantha*
4. Involucre 6 to 9 (rarely 12) mm. high; anthers exserted; pappus of 4 to 8 paleae, these sometimes unequal but not 2-seriate; outer phyllaries scarcely loose-tipped (5).
5. Pappus of 4 or 5 paleae; plants annual; outer corollas distinctly larger than the others
5. *C. stevioides*
5. Pappus of about 8 paleae; plants perennial or biennial; outer corollas not enlarged
6. *C. Douglasii*

1. Chaenactis carphoclinia Gray. Mohave, Maricopa, Pinal, Pima, and Yuma counties, up to 3,000 feet, plains and mesas, February to May, type from Fort Yuma (*Schott* 617a). Southern Utah to southern Arizona, southeastern California, and northern Baja California.

The type of var. *attenuata* (Gray) Jones, distinguished by having the pappus paleae of all the achenes very short and usually obtuse (those of the central achenes in the typical plant being at least half as long as the corolla and usually acuminate), came from Ehrenberg, Yuma County (*Janvier*). The variety has been collected also at Oatman (Mohave County), at Papago Well (western Pima County), and in the Gila and Pothole mountains (Yuma County).

2. Chaenactis Fremonti Gray. Mohave, Gila, Maricopa, Pinal, and (doubtless) Yuma counties, 1,000 to 3,500 feet, plains and mesas, March to June. Southwestern Utah, southern Nevada, western Arizona, and southeastern California.

3. Chaenactis Xantiana Gray. Havasu Canyon (Coconino County), Virgin Narrows, Hackberry, and Kingman (Mohave County), 2,000 to 3,500 feet, April to June. Eastern Oregon to southern California, western Nevada, and western Arizona.

4. Chaenactis macrantha D. C. Eaton. Near Holbrook (Navajo County), Grand Canyon (Coconino County), Kingman and west and south (Mohave

County), also near Tucson (Pima County), 2,000 to about 3,000 feet, dry plains and slopes, April to June. Southern Utah and Nevada, Arizona, and southeastern California.

5. Chaenactis stevioides Hook. & Arn. Throughout the state, 1,000 to 6,500 feet, very common on dry mesas and plains, February to May. Wyoming to Idaho, south to New Mexico, southern California, and Sonora.

Var. *brachypappa* (Gray) H. M. Hall differs in having the paleae of the pappus in all the flowers blunt and not more than 2 mm. long, whereas in typical *C. stevioides* the central flowers have the paleae at least ⅔ as long as the corolla and usually acute. The variety is not typically developed in Arizona, but an approach to this form is shown by collections in the Sacaton Mountains, Pinal County (*Peebles* 11038), and at Chloride, Mohave County (*Kearney & Peebles* 11204). Var. *Thornberi* Stockwell, which differs in being coarser and having yellow or sometimes cream-colored flowers, those of the other forms being white, is cited from Yavapai, Pinal, Pima, and Santa Cruz counties, type from Wilmot, near Tucson (*Thornber* 385).

6. Chaenactis Douglasii (Hook.) Hook. & Arn. Jacob Lake to Fredonia, northern Coconino County, about 7,000 feet, open pine forest (*Peebles* 13041, 13052), North Rim of Grand Canyon (*Collom*), also just over the line on Navajo Mountain, Utah, 8,500 feet (*Darrow* 2862). Alberta and British Columbia to northern New Mexico, northern Arizona, and California.

Chaenactis Gillespiei Stockwell (355, p. 123), described from a single collection from near Granite Reef Dam, Maricopa County (*Gillespie* 5611), is insufficiently known. It is a thinly lanate, glabrescent annual with bipinnatifid leaves with comparatively few linear segments, involucre 7 to 8.5 mm. high, white flowers, slightly enlarged outer corollas, and a double pappus of 4 long, lanceolate inner paleae and 1 to 4 short outer squamellae.

96. CHAMAECHAENACTIS

Dwarf, scapose, cespitose perennials; leaves small, roundish, crenate or entire, slender-petioled, greenish above, canescent-strigose beneath, 3-nerved; heads medium-sized, discoid, flesh-colored, solitary on short scapes; achenes slender, obpyramidal, about 5-ribbed, villous; pappus of about 8 unequal, blunt, oblong, scarious paleae, these with a thickened midrib.

1. Chamaechaenactis scaposa (Eastw.) Rydb. 15 miles north of Ganado, Apache County, 6,000 feet, May and June (*Peebles & Smith* 13468). Southwestern Wyoming, southwestern Colorado, eastern Utah, and northeastern Arizona.

97. BAHIA

Herbs, rarely suffrutescent; leaves opposite or alternate, entire to dissected; heads radiate or rarely discoid, yellow; phyllaries not punctate; achenes narrow, 4-angled, usually more conspicuously pubescent toward base than above; pappus of several squamellae or paleae, these with a callose-thickened base or midrib, or rarely wanting.

KEY TO THE SPECIES

1. Leaves oblong or elliptic, entire, green at maturity, about 1 cm. wide. .1. *B. oblongifolia*
1. Leaves dissected or lobed, or if entire, then much narrower or else whitened beneath (2).

2. Pappus wanting. Plants normally 30 cm. high or more, glandular, especially above; leaves once to thrice ternately divided into linear to oblong lobes; heads usually numerous and panicled..2. *B. dissecta*
2. Pappus present (3).
3. Heads discoid. Low annuals; leaves entire or 3-cleft, narrowly linear or linear-filiform 3. *B. neomexicana*
3. Heads radiate (4).
4. Stem glandular above (5).
4. Stem strigillose above, not evidently glandular (6).
5. Leaves pedately divided into 3 stipitate, mostly obovate or cuneate divisions, these again lobed or parted, the ultimate divisions broad; squamellae of the pappus in all of the flowers very blunt, with the nerve disappearing below the apex..........4. *B. pedata*
5. Leaves biternately divided into linear segments about 1 mm. wide, or less; squamellae of the pappus in the inner flowers pointed by the excurrent nerve......5. *B. biternata*
6. Heads solitary or few at the tips of the stem and branches, long-peduncled; achenes conspicuously long-hairy at base; leaves opposite below, alternate above, with usually broad segments, sometimes entire; plants usually relatively tall, canescent-puberulent with appressed hairs..8. *B. absinthifolia*
6. Heads usually several (sometimes solitary) at the tips of the stem and branches, on peduncles not more than 2 cm. long; achenes not conspicuously long-hairy at base; leaves nearly all opposite, 3- or 5-parted into linear usually entire divisions, these 2.5 mm. wide or narrower; plants low, much branched, not more than 20 cm. high, green or merely cinereous (7).
7. Achenes hirsutulous, especially toward base; paleae of the pappus lanceolate, acute or acuminate, the stout nerve reaching the apex and usually excurrent..6. *B. Woodhousei*
7. Achenes sessile-glandular only; paleae of the pappus narrowly obovate, blunt, the nerve disappearing below the apex...................................7. *B. oppositifolia*

1. Bahia oblongifolia Gray. Carrizo Mountains, Apache County, dry hills (*Standley* 7363). Southwestern Colorado or southeastern Utah, northwestern New Mexico, and northeastern Arizona.

2. Bahia dissecta (Gray) Britton. Apache County to Mohave County, south to Cochise, Santa Cruz, and Pima counties, 5,000 to 9,000 feet, common in grassland and open pine forests, August to October. Wyoming to northern Mexico and Arizona.

The plant is sometimes called yellow-ragweed. The heads are normally radiate, but a specimen from the Pinaleno Mountains (*Thornber & Shreve* 7788) has the rays obsolete.

3. Bahia neomexicana Gray (*Schkuhria multiflora* Hook. & Arn., not *B. multiflora* Nutt.). Apache, Navajo, Coconino, and Yavapai counties, 5,000 to 7,000 feet, grassland, August and September. Western Texas to Colorado, Arizona, and northern Mexico; South America.

Heiser (351) referred this species to *Schkuhria,* mainly if not entirely because the heads are discoid. This single point is of much less importance than the several features—more numerous flowers, nonpunctate phyllaries, and particularly the slenderly obpyramidal achenes, much more conspicuously pubescent toward base than above—in which it agrees with the generality of species in *Bahia.*

*4. **Bahia pedata** Gray. Hills and rocky slopes, southwestern Texas, southern New Mexico, and (according to Rydberg in *North Amer. Fl.* 34: 36) Arizona, April to September.

5. Bahia biternata Gray. Pinal Mountains and Sierra Ancha (Gila County), Horse Mesa (eastern Maricopa County), near Sacaton (Pinal County), 1,000 to 4,500 feet, among scrub-oaks and in river bottoms, where doubtless carried by flood waters, May to October. Western Texas to central-southern Arizona and Sonora.

6. Bahia Woodhousei Gray. Woodruff, Navajo County (*Ward* in 1901), Red Butte to Rattlesnake Tanks, Coconino County (*Leiberg* 5915), about 5,000 feet, June to September. Northwestern Texas, Colorado, and northern Arizona.

***7. Bahia oppositifolia** (Nutt.) DC. Plains and hillsides, North Dakota to Montana, south to western Texas and New Mexico (reported from Arizona), June to September.

The plant contains a cyanogenic principle, but it is rarely if ever eaten by livestock in sufficient quantity to cause prussic-acid poisoning.

8. Bahia absinthifolia Benth. Cochise County near the Mexican boundary (*Mearns* 756, etc.), mesas and slopes, April to October. Southern Texas and southeastern Arizona to central Mexico.

Much more common in Arizona than the typical plant is var. *dealbata* Gray, which differs in having the leaves merely 3-cleft into lanceolate lobes or entire (in the typical plant the leaves pedately parted into 3 or 5 narrowly linear or lance-linear divisions, these usually again few-lobed). The variety occurs in Greenlee, Graham, Gila, Maricopa, Pinal, Cochise, Santa Cruz, and Pima counties, 2,500 to 5,500 feet, and ranges to western Texas and Chihuahua. It is particularly abundant on shallow, caliche soil around Tucson. Plants more or less intermediate between the variety and typical *B. absinthifolia* have been collected in the Galiuro Mountains (Pinal County), Huachuca Mountains (Cochise County), and Rincon Mountains (Pima County).

98. TRICHOPTILIUM

Low, diffusely branched, floccose-woolly, winter annuals; leaves mostly alternate, oblong to lanceolate, sharply dentate; heads terminal, solitary, slender-peduncled, discoid, yellow; achenes turbinate, 5-nerved, hairy; pappus of 5 paleae, these dissected into numerous bristles.

1. Trichoptilium incisum Gray. Mohave, Maricopa, western Pima, and Yuma counties, up to 2,500 feet, sandy or gravelly mesas and slopes, February to May (sometimes autumn). Southern Nevada, western Arizona, southeastern California, and Baja California.

99. HYMENOXYS (356, 356a)

Annual or perennial herbs; leaves alternate, entire to pinnatifid; heads radiate, yellow; involucre in 2 or more series, the phyllaries often rigid, the outer ones sometimes united at base; pappus of 5 to 12 paleae.

The fragrant-bitterweed (*H. odorata*), and pinguë or pingwing (*H. Richardsoni*) are toxic to livestock, especially to sheep, but are eaten only when other forage is scarce. These plants tend to increase on overgrazed ranges. The latex of some of the species contains rubber, and *H. Richardsoni* var. *floribunda* is known as Colorado rubber-weed. The Hopi are reported to make a stimulating drink from *H. acaulis* var. *arizonica* and to apply the plant locally in alleviating pain, especially in pregnancy. The bark of the roots of *H. Richardsoni* is used as a substitute for chewing gum by Indians in New Mexico. Arizona's larger-headed species are handsome, and *H. acaulis* is cultivated in Europe as an ornamental. In *Flowering Plants and Ferns of Arizona* these plants were treated under the generic name *Actinea.*

KEY TO THE SPECIES

1. Heads solitary (rarely 2 or 3), long-peduncled, on naked scapes or sparsely leafy stems; leaves entire or 3-parted (2).
1. Heads several or many, cymose or cymose-panicled; stems very leafy; leaves almost always deeply pinnatifid (6).
2. Involucre densely and loosely pilose-tomentose with matted hairs. Basal leaves narrowly linear, entire or 3-parted; stem about 10 to 20 cm. high, sparsely leafy; pappus of about 5 lance-attenuate paleae .... 1. *H. Brandegei*
2. Involucre not pilose-tomentose, although sometimes silky-villous (3).
3. Phyllaries lanceolate, the outer ones subherbaceous, acuminate, the inner ones attenuate, indurate, stiff; leaves not punctate. Basal leaves narrowly linear, entire or sometimes 3-lobed; stem 20 to 40 cm. high, sparsely leafy; pappus of 10 lance-attenuate paleae .... 2. *H. Bigelovii*
3. Phyllaries mostly elliptic or oblong, obtuse or merely acute; leaves conspicuously impressed-punctate (4).
4. Plants strictly scapose, the leaves all basal; scapes always simple and 1-headed. Leaves sparsely or rather densely villous: var. *arizonica* .... 3. *H. acaulis*
4. Plants with stems sparsely leafy at least below; stems often few-branched, 1- to 3-headed (5).
5. Leaves spreading, villous to glabrate, bright gray-green: var. *Ivesiana* .... 3. *H. acaulis*
5. Leaves densely to sparingly silky-sericeous with appressed hairs, often silvery 4. *H. argentea*
6. Stems several or numerous from a perennial multicipital caudex, conspicuously long-villous or woolly at base among the petiole bases of the lowest leaves: var. *floribunda* 5. *H. Richardsoni*
6. Stems solitary or few from an annual, biennial, or perennial root or caudex, not long-villous or woolly at base (7).
7. Plants annual, usually branching almost from the base .... 6. *H. odorata*
7. Plants biennial or perennial, usually branching only in the inflorescence (8).
8. Plants subsericeous-canescent; leaves entire or 3-cleft, the blades or their lobes relatively broad, 1.5 to 4 mm. wide .... 7. *H. subintegra*
8. Plants green or, if slightly canescent, then the leaves or their lobes much narrower (9).
9. Leaves either entire or divided into 3 to 5 lobes, the blades or their lobes relatively broad, mostly 2 to 5 mm. wide (10).
9. Leaves all (or nearly all) divided into 3 to 5 narrowly linear or linear-filiform lobes (0.8 to 1.5 mm. wide). Heads usually large (disk 8 to 15 (25) mm. thick) and few (normally 3 to 6, rarely up to 20 per stem), usually on elongate peduncles conspicuously surpassing the leaves (11).
10. Heads usually few (3 to 10 per stem, rarely more) and comparatively large (the disk 10 to 18 mm. thick), on more or less elongate peduncles distinctly surpassing the leaves; leaves divided into 3 or 5 lobes, or the lowest sometimes entire. . 8. *H. Lemmoni*
10. Heads numerous and small (disk 5 to 10 mm. thick), in close, rounded or flattish cymes or cymose panicles, on peduncles scarcely or not surpassing the leaves; leaves entire or the middle ones divided into 3 or 5 lobes .... 9. *H. Rusbyi*
11. Outer phyllaries 5 to 8, very strongly keeled (or even 3-ribbed) especially in fruit, the keels conspicuously decurrent on the peduncles; leaf divisions linear, mostly 0.8 to 1.5 mm. wide; plants apparently perennial, usually without a conspicuous basal rosette of spreading or deflexed leaves or persistent petioles .... 10. *H. quinquesquamata*

11. Outer phyllaries (8) 10 to 14, not strongly keeled, usually with a thickened and sulcate center; leaf divisions usually linear-filiform and less than 1 mm. wide; plants biennial, usually with a conspicuous basal rosette of spreading or deflexed leaves or of persistent petioles..11. *H. Cooperi*

1. **Hymenoxys Brandegei** (Porter) K. F. Parker. Summit of Baldy Peak, Apache County, 11,500 feet (*Bailey* 1449, *Peebles & Smith* 12546, *Phillips* 3341), August and September. Southern Colorado, New Mexico, and eastern Arizona, mostly above timber line.

2. **Hymenoxys Bigelovii** (Gray) K. F. Parker (*Actinea Gaillardia* A. Nels.). White Mountains (Apache and Navajo counties), San Francisco Peaks and southward (Coconino County), Sierra Ancha and Mazatzal Mountains (Gila County), 5,500 to 7,500 feet, mostly in pine forests, April to July, types of *A. Gaillardia* from Flagstaff (*Hanson* A32 and 584). Western New Mexico to central Arizona.

3. **Hymenoxys acaulis** (Pursh) K. F. Parker. Apache County to eastern Mohave County, especially common at the Grand Canyon, 4,000 to 7,000 feet, dry rocky slopes and mesas, mostly with pine or juniper, April to October. Colorado to Nevada, south to New Mexico, north-central Arizona, and southern California.

The species is represented in Arizona mainly by var. *arizonica* (Greene) K. F. Parker, type from Mt. Trumbull, Mohave County (*Palmer* 259, in 1877). In var. *Ivesiana* (Greene) K. F. Parker the basal leaves are narrowly oblanceolate or almost linear, 1 to 3 mm. wide (3 to 6 mm. wide in the typical form).

4. **Hymenoxys argentea** (Gray) K. F. Parker (*Actinea leptoclada* (Gray) Kuntze). 10 miles south of Snowflake, southern Navajo County, in the pinyon-juniper association (*Peebles* 9627), April to October. New Mexico and eastern Arizona.

5. **Hymenoxys Richardsoni** (Hook.) Cockerell. Apache, Navajo, Coconino, and Yavapai counties, 5,000 to 9,000 feet, mostly in pine forest, June to September. Saskatchewan and Alberta to New Mexico and Arizona.

The species is represented in Arizona by the more southern var. *floribunda* (Gray) K. F. Parker (*H. floribunda* Cockerell, *H. floribunda* vars. *arizonica* and *intermedia* Cockerell). The types of vars. *arizonica* and *intermedia* were collected near Flagstaff (*MacDougal* 219 and 359).

6. **Hymenoxys odorata** DC. (*Hymenoxys Davidsonii* (Greene) Cockerell, *H. chrysanthemoides* var. *excurrens* Cockerell). South-central Navajo and eastern Coconino counties, to Greenlee, Cochise, Pima, and Yuma counties, 6,000 feet or lower, moist, alluvial soil, especially abundant along the lower Gila River, January to June, type of *H. Davidsonii* from Clifton (*Davidson* in 1900), type of *H. chrysanthemoides* var. *excurrens* from Yuma (*Vasey* in 1881). Kansas to Texas, west to southeastern California, south into Mexico.

7. **Hymenoxys subintegra** Cockerell. Coconino County, from Navajo Bridge and Cameron to the Kaibab Plateau and to between Williams and Grand Canyon, 5,500 to 8,000 feet, dry soil in the open and in coniferous forests, June to September, type from Nagles Ranch, Kaibab Plateau (*Jones* 6054o). Known only from northern Arizona.

8. Hymenoxys Lemmoni (Greene) Cockerell. Lukachukai Mountains, Apache County, 7,000 feet (*Peebles* 14401), June. Utah, Nevada, northeastern Arizona, and California.

9. Hymenoxys Rusbyi (Gray) Cockerell. Southern Coconino and eastern Yavapai counties to central Gila County, 5,500 to 7,000 feet, dry soil, mostly with juniper, pinyon, and *Gutierrezia,* locally very abundant, July to September. Southwestern New Mexico and central Arizona.

10. Hymenoxys quinquesquamata Rydb. (*H. Cooperi* subsp. *Grayi* Cockerell). Chiricahua and Huachuca mountains (Cochise County), Rincon and Santa Rita mountains (Pima and Santa Cruz counties), 5,000 to 7,000 feet, April to September, type from Carr Peak, Huachuca Mountains (*Goodding* 874), type of *H. Cooperi* subsp. *Grayi* from the Huachuca Mountains (*Lemmon* 2774). Known only from southeastern Arizona.

11. Hymenoxys Cooperi (Gray) Cockerell (*Actinella biennis* Gray, *Hymenoxys Cooperi* var. *argyraea* Cockerell, *H. virgata* A. Nels.). Coconino and northern Mohave counties, 2,000 to 7,000 feet, dry rocky places, May to September, type of *Actinella biennis* from Mokiak Pass (*Palmer* 260 in 1877), type of *Hymenoxys Cooperi* var. *argyraea* from the Grand Canyon (*MacDougal* 189), type of *H. virgata* from Bright Angel Trail, Grand Canyon (*Osterhout* 6991). Southern Utah and Nevada, northern Arizona, and southeastern California.

Var. *canescens* (D. C. Eaton) K. F. Parker has been collected near Jacob Lake, Kaibab Plateau. It differs from the species in its larger heads (disk up to 25 mm. wide), more canescent herbage, and shorter and fewer stems, these branched only in the inflorescence.

100. HELENIUM. Sneeze-weed

Biennial or perennial herbs; leaves alternate, usually narrow, entire or toothed; heads small to large, radiate or discoid, the rays yellow, the disk yellow or purple-brown; involucre about 2-seriate, spreading, at length reflexed; achenes turbinate, 8- to 10-ribbed; pappus of 5 to 8 scarious paleae or squamellae.

KEY TO THE SPECIES

1. Leaves not decurrent; plants subtomentose when young; rays spreading, linear or oval. Leaves thickish, strictly entire, the lowest obovate or oblanceolate, normally 15 cm. long or longer; heads few (usually 3 to 6), large, at least 5 cm. wide across the spreading rays . . . . . 1. *H. Hoopesii*

1. Leaves more or less decurrent; plants not subtomentose; rays soon drooping, cuneate, wanting in one species (2).

2. Rays wanting; squamellae of the pappus very short and blunt; plants annual or biennial 2. *H. Thurberi*

2. Rays present; squamellae of the pappus at least half as long as the corolla, acute or acuminate; plants perennial (3).

3. Leaves essentially uniform, sessile, lanceolate; stem winged essentially throughout 3. *H. autumnale*

3. Leaves not uniform, the basal ones oblanceolate, distinctly larger than the cauline leaves and narrowed into a petioliform base; cauline leaves with an ampliate amplexicaul base, very shortly decurrent; stem not winged throughout . . . . . 4. *H. arizonicum*

1. Helenium Hoopesii Gray. Apache, Coconino, Greenlee, Graham, Cochise, and Pima counties, 7,000 to 11,000 feet, abundant in rich soil in coniferous forests and mountain meadows, June to September. Wyoming to Oregon, south to New Mexico, Arizona, and California.

Orange sneeze-weed, sometimes called owl-claws. The plant contains a toxic glucoside—dugaldin—which causes "spewing sickness" in sheep. It is also poisonous to cattle but is rarely eaten by them.

2. Helenium Thurberi Gray. Pinal, Cochise, Pima, and Yuma counties, 100 to 5,000 feet, marshy places along streams, March to August, types from southern Arizona (*Coulter* 359, *Thurber* 346, *Pringle* 137). Southern Arizona and Mexico.

3. Helenium autumnale L. Skull Valley, Yavapai County (*Thornber* 8792), near Fort Huachuca, Cochise County (*Lemmon* 276). Eastern Canada to British Columbia, south to Florida, Texas, New Mexico, and Arizona.

Fernald (*Rhodora* 45: 489–492) referred the Lemmon collection to var. *canaliculatum* (Lam.) Torr. & Gray, which he distinguished from typical *H. autumnale* as having narrower, more nearly entire leaves and narrower rays, these strongly narrowed and often convolute at base.

4. Helenium arizonicum Blake. Southern Coconino County, near Mormon Lake (*Toumey* 681, the type collection), and near Buck Springs, Alder Lake, and Myrtle Lake, 7,000 to 8,000 feet, August and September. Known only from central Arizona.

Helenium laciniatum Gray, a species of northern Mexico, was reported from southwestern Arizona and southeastern California by Gray (*Syn. Fl.* 1^2: 349), apparently on the basis of specimens collected by Thomas Coulter. The species has not been found in either state by any recent collector, and should be excluded from the flora of the United States until a definite specimen is forthcoming. It is a cinereous-puberulent plant, with pinnatifid or laciniate, somewhat decurrent leaves, radiate heads, and awned pappus paleae more than half as long as the corolla.

101. GAILLARDIA (357). BLANKET-FLOWER

Annual or perennial herbs; leaves alternate, entire to pinnatifid; heads solitary, radiate, showy, the rays yellow or partly purple, the disk yellow or purple; receptacle with subulate or setiform fimbrillae; achenes turbinate, 5-ribbed, villous at least at base; pappus of 5 to 10 scarious paleae, these often awned.

Species of this genus are prized as ornamentals, the cultivated gaillardias being derived mainly from *G. pulchella*.

KEY TO THE SPECIES

1. Corolla teeth long-acuminate, tipped with a long awn or cusp. Plants annual, leafy-stemmed; lower leaves usually lobed, the upper ones mostly oblong-lanceolate, entire; achenes densely silky-pilose on the lower half, but the whole body concealed by the hairs; rays yellow toward the tip, purple at base; disk purple; paleae of the pappus lanceolate, awn-pointed .... 1. *G. pulchella*

1. Corolla teeth merely acute or obtuse, not tipped with an awn or cusp (2).

2. Paleae of the pappus broadly oblong or oval, awnless or with an abrupt awn shorter than the body of the paleae. Plants annual, scapose or subscapose, with deeply pinnatifid or sometimes merely toothed leaves; rays and disk yellow............... 2. *G. arizonica*

2. Paleae of the pappus lanceolate, rather gradually narrowed into an awn (3).

3. Leaves ovate or obovate, entire or subentire, 3-nerved; rays and disk yellow; plants scapose or subscapose, perennial, multicipital.........................3. *G. Parryi*
3. Leaves linear to lanceolate or rarely obovate, when broad always toothed or lobed to pinnatifid; rays yellow, the disk purple; stem usually leafy..........4. *G. pinnatifida*

1. Gaillardia pulchella Foug. (*G. neomexicana* A. Nels.). Yavapai, Greenlee, Graham, Gila, Pinal, Cochise, and Pima counties, mostly 3,500 to 5,500 feet, plains, April to September. Nebraska and Missouri to Louisiana, west to Colorado and southeastern Arizona. Sometimes known as fire-wheel and Indian-blanket.

Biddulph (357) referred all the Arizona material of this species to *G. neomexicana,* which she separated by its biennial or perennial duration, differently colored rays, and more hairy achenes. The achenes are definitely more hairy than those of Texan material (although the hairs are likewise attached below the middle of the achene, not "all over" as she says in her key), but the writer cannot confirm the other alleged distinctions and feels that, as she herself suggested, *G. neomexicana* is better considered "only the western extension of *G. pulchella*" rather than a distinct species.

2. Gaillardia arizonica Gray. Navajo County to Mohave County, south to Cochise, Pima, and Yuma counties, 1,000 to 4,000 feet, plains and mesas, February to July, type from Camp Grant, Graham County (*Palmer* 137). Southern Utah and southern Nevada to southern Arizona.

Var. *Pringlei* (Rydb.) Blake (*G. Pringlei* Rydb., *G. crinita* Rydb.), distinguished by the awned paleae of the pappus (these blunt in the typical form), has been collected below Black Falls, Little Colorado River (eastern Coconino County), and occurs also in Gila, Maricopa, Pinal, and Pima counties. The type of *G. Pringlei* (*Pringle* in 1884) and that of *G. crinita* (*Griffiths* 2386) both came from near Tucson.

3. Gaillardia Parryi Greene. Kaibab Plateau and North Rim of Grand Canyon (Coconino County), Mokiak Pass, Pagumpa Springs, and near Wolf Hole (Mohave County), 4,000 to 8,000 (?) feet, plains and hillsides, May and June. Southern Utah and northern Arizona.

4. Gaillardia pinnatifida Torr. (*G. multiceps* Greene, *G. Mearnsii* Rydb., *G. linearis* Rydb., *G. crassa* Rydb., *G. globosa* A. Nels.). Apache, Navajo, Coconino, and eastern Mohave counties, south to Cochise, Santa Cruz, and Pima counties, 3,500 to 7,000 feet, mesas, plains, and open pine forest, often on limestone, April to October, type of *G. multiceps* from south of Woodruff (*Wooton* in 1892), type of *G. Mearnsii* from Fort Verde (*Mearns* 322), type of *G. crassa* from foothills of the Santa Rita Mountains (*Pringle* in 1884), type of *G. globosa* from near Flagstaff (*MacDougal* 291). Colorado and Utah to Texas, Arizona, and Mexico.

This species is reported to be used by the Hopi Indians as a diuretic. *G. linearis* and *G. multiceps* are forms with narrow, entire leaves. *G. Mearnsii* apparently was based on plants flowering in their first year. Biddulph recognized as distinct *G. multiceps* and *G. Mearnsii,* using for their discrimination characters the writer regards as of no taxonomic significance; she also separated the form with linear, entire leaves as *G. pinnatifida* var. *linearis* (Rydb.) Biddulph.

102. PLUMMERA

Perennial (?) herbs, with aspect of *Hymenoxys Richardsoni* but taller; leaves alternate, divided into filiform lobes; heads very small, cymose-panicled, radiate, yellow, the rays 2 to 5, fertile, the disk flowers 6 or 7, hermaphrodite but sterile; involucre double; ray achenes obovoid, plump, about 15-ribbed, villous; pappus none, or of 2 to 6 oblong squamellae.

KEY TO THE SPECIES

1. Achenes epappose or with a pappus of 2 squamellae, villous with flexuous hairs 1. *P. floribunda*
1. Achenes with a pappus of 4 to 6 oblong or lanceolate squamellae, or the inner ones epappose, villous with straight hairs .... 2. *P. ambigens*

1. Plummera floribunda Gray. Sulphur Springs Valley and Swisshelm, Chiricahua, Dos Cabezas, and Mule mountains (Cochise County), 5,000 to 6,500 feet, July to September, type from Apache Pass (*Lemmon* in 1881). Known only from southeastern Arizona.

2. Plummera ambigens Blake. Known only from the lower slopes of the Pinaleno Mountains (Graham County), where it is abundant at 5,000 to 7,000 feet, in stony, sterile soil, July to October, type from this locality (*Peebles et al.* 4395).

103. TAGETES. Marigold

Herbaceous or suffrutescent; leaves mostly opposite, entire or pinnately divided, dotted with translucent oil glands; heads small or large, radiate or discoid, yellow; phyllaries 1-seriate, united into a toothed cup or tube, dotted with oil glands; achenes slender; pappus of 2 to 6 squamellae and paleae, with or without awns.

Several species, particularly *Tagetes erecta* and *T. patula,* are commonly cultivated as ornamentals under the names "African marigold" and "French marigold."

KEY TO THE SPECIES

1. Tall, suffrutescent perennials; rays large, about 10 mm. long, bright yellow; leaves with 3 to 7 mostly lanceolate divisions, these 4 to 12 mm. wide.............. 1. *T. Lemmoni*
1. Dwarf annuals; rays tiny (or wanting), 2 mm. long or less, pale yellow or whitish; leaves filiform and entire, or of 3 or 5 filiform divisions.................... 2. *T. micrantha*

1. Tagetes Lemmoni Gray. Mountains of Cochise, Santa Cruz, and Pima counties, 4,000 to 8,000 feet, rich, moist soil in canyons, August to October, type from the Huachuca Mountains (*Lemmon* in 1882). Known only from southeastern Arizona.

A handsome but very ill-scented plant.

2. Tagetes micrantha Cav. Southern parts of Apache, Navajo, and Coconino counties, south to Cochise, Santa Cruz, and Pima counties, 6,000 to 8,500 feet, August and September. Western Texas to Arizona and central Mexico.

The oil glands of the leaves and phyllaries are sometimes obscure.

104. DYSSODIA

Annual or perennial herbs, or suffruticose; leaves opposite or alternate, entire to pinnatisect, marked with translucent glands; heads small to rather large, radiate, yellow or orange, or the rays rarely white; involucre usually calyculate (subtended by bractlets), the principal phyllaries equal, usually 2-seriate and united at base, or almost to the apex; pappus of 10 to 15 squamellae or paleae, often tipped with 1 or 3 bristles or dissected into numerous bristles.

Dyssodia papposa and *D. acerosa* are sometimes abundant on overgrazed land and are regarded as range pests. The former is suspected of being poisonous to livestock, but definite information is lacking.

KEY TO THE SPECIES

1. Heads large, the disk 1.2 to 2 cm. high or more. Pappus of 10 to 15 paleae, each dissected into numerous bristles (2).
1. Heads smaller, the disk less than 1 cm. high, or (in *D. acerosa*) up to 1.5 cm. (3).
2. Stem glabrous; leaves pinnately 3- or 5-parted into narrow lobes....1. *D. porophylloides*
2. Stem densely puberulous or hispidulous; larger leaves oval or ovate, merely spinulose-toothed, or with small basal lobes....................................2. *D. Cooperi*
3. Paleae of the pappus each dissected into 5 to 12 capillary bristles, the alternate ones (in *D. acerosa*) dissected into only 3 bristles (4).
3. Paleae of the pappus tipped with only 1 to 3 bristles, or the outer paleae without bristles (6).
4. Leaves entire, linear-filiform, needlelike; plants perennial, the numerous stems or branches woody at base..3. *D. acerosa*
4. Leaves pinnately lobed, not needlelike; plants annual (5).
5. Rays white, conspicuous; involucre essentially naked at base, the phyllaries united ¾ of their length or more, and with numerous small, roundish glands........4. *D. concinna*
5. Rays yellow, small, inconspicuous; involucre subtended by several mostly herbaceous bractlets of about its own length, the phyllaries distinctly in 2 series and united only toward the base, their glands few, linear or elliptic, rather large........5. *D. papposa*
6. Outer paleae (as well as the inner ones) bristle- or awn-tipped...........6. *D. Thurberi*
6. Outer paleae merely acute or obtuse, not awn-tipped (7).
7. Phyllaries of both the inner and the outer series (not the bractlets of the calyculus) united nearly to the apex. Involucre puberulent all over...............7. *D. Hartwegi*
7. Phyllaries of the outer series free-margined to the middle or lower (8).
8. Paleae of the pappus 10, the 5 inner ones 3-toothed and 1-awned, the lateral teeth sometimes aristiform but much shorter than the central awn; outer phyllaries ciliate on the free margins; rays conspicuous, surpassing the pappus.............8. *D. pentachaeta*
8. Paleae of the pappus 20, the 10 inner ones 3-cleft and 3-awned, the lateral awns about half as long as the central awn; outer phyllaries not ciliate; rays inconspicuous, not surpassing the pappus...9. *D. neomexicana*

1. Dyssodia porophylloides Gray. Grand Canyon (Coconino County) and Mohave County to Pima and Yuma counties, 4,000 feet or lower, washes, mesas, and dry, rocky slopes, March to October. Arizona, southern California, Sonora, and Baja California.

The plant has a strong, disagreeable odor.

2. Dyssodia Cooperi Gray (*Clomenocoma laciniata* Rydb.). Hackberry to the Colorado River (Mohave County), 3,500 feet or lower, dry canyons, slopes, and mesas, April to August, type of *Clomenocoma laciniata* from Hackberry (*Jones* in 1884). Western Arizona, southern Nevada, and southeastern California.

3. Dyssodia acerosa DC. North Rim of the Grand Canyon (Coconino County) and Mohave County, south to Cochise and Pima counties, 3,500 to 6,000 (8,000?) feet, dry, rocky slopes and mesas, March to October. Texas to southern Nevada, Arizona, and south to central Mexico.

4. Dyssodia concinna (Gray) Robins. Mesas near Tucson (*Pringle* in 1884, the type collection, *Thornber* 916, etc.), near Quitobaquito and Organ Pipe Cactus National Monument, Pima County (*Gould et al.* 2989, *Supernaugh* in 1950), 1,000 to 2,500 feet, March to May. Southern Arizona and Sonora.

5. Dyssodia papposa (Vent.) Hitchc. Navajo, Coconino, Greenlee, Gila, Cochise, and Santa Cruz counties, 4,500 to 7,000 feet, fields, roadsides, and waste places, August to October. Illinois to Montana, south to Louisiana and Arizona.

6. Dyssodia Thurberi (Gray) A. Nels. Grand Canyon and Havasu Canyon (Coconino County) and northern Mohave County to Pima and Yuma counties, 3,500 feet or lower, dry, rocky slopes and mesas, April to October. Texas to southern Nevada, southeastern California, and northern Mexico.

7. Dyssodia Hartwegi (Gray) Robins. Chiricahua Mountains, Cochise County, about 5,500 feet, on limestone, September and October. Southeastern Arizona to central Mexico.

8. Dyssodia pentachaeta (DC.) Robins. Lees Ferry and Grand Canyon (Coconino County), Tuzigoot National Monument (Yavapai County), and Cochise, Santa Cruz, and Pima counties, 2,500 to 4,500 feet, dry slopes and mesas, March to October. Texas to Arizona and northern Mexico.

9. Dyssodia neomexicana (Gray) Robins. Springerville, Apache County (*William Riley* in 1946). Western New Mexico, eastern Arizona, and Chihuahua.

105. POROPHYLLUM

Glabrous herbs or suffrutescent perennials; leaves opposite or alternate, with conspicuous translucent oil glands in the tissue; heads medium-sized, discoid, whitish or purplish; phyllaries 5 to 8, linear or oblong, 1-seriate, equal, free, with conspicuous, linear oil glands, without accessory bractlets; achenes slender, elongate; pappus of copious, free capillary bristles.

P. gracile, called yerba-del-venado by the Mexicans, is said to be relished by deer and cattle, notwithstanding the strong, unpleasant odor of the plant.

KEY TO THE SPECIES

1. Annuals; leaves oval, thin, mostly 2 to 4 cm. long, with slender petioles about as long as the blades; peduncles thickened toward the apex; heads 2 to 2.5 cm. high
   1. *P. macrocephalum*

1. Perennials, more or less woody toward the base; leaves filiform to narrowly linear, sessile; peduncles not noticeably thickened toward the apex; heads 1.5 to 2 cm. high
   2. *P. gracile*

1. Porophyllum macrocephalum DC. Cochise, Santa Cruz, and Pima counties, from the Peloncillo to the Baboquivari mountains, 3,500 to 5,000 feet, rocky slopes and canyons, August to October. Southern Arizona, south to South America.

2. Porophyllum gracile Benth. (*P. junciforme* Greene, *P. putidum* A. Nels.). Grand Canyon (Coconino County) and Mohave County to western

Cochise, Pima, and Yuma counties, 4,000 feet or lower, dry, rocky slopes and canyons, March to October, type of *P. junciforme* from the Mescal Mountains (*M. E. Jones*), type of *P. putidum* from east of Douglas, Cochise County (*Goodding* 2277). Southern Nevada, Arizona, southeastern California, Sonora, and Baja California.

106. PECTIS (358). Fetid-marigold

Annual or perennial herbs, usually low, slender-stemmed; leaves opposite, entire, dotted with pellucid glands, almost always ciliate with a few stiff bristles toward the base; heads small, radiate, yellow, the rays often purplish beneath; involucre 1-seriate; achenes slender; pappus of numerous bristles or of few awns or paleae, or reduced to a low crown.

With remarkable promptitude after summer rains have fallen, especially in the southern part of the state, the ground is carpeted with the small yellow heads of the strong-smelling chinchweed (*P. papposa*). In New Mexico the flowers of this species are used by the Indians for seasoning meat. It is reported that in Arizona the Hopi Indians use *P. angustifolia,* either raw or dried, for food and seasoning, and extract a dye from the plant.

KEY TO THE SPECIES

1. Pappus of 2 to 6 lanceolate, acuminate, more or less scarious paleae. Low and diffuse annuals; leaves oblanceolate to nearly linear, with scattered oil glands; heads sessile or essentially so, much surpassed by the subtending leaves (2).
1. Pappus of bristles or of stout, not paleaceous awns, or sometimes of short squamellae, rarely reduced to a low crown (3).
2. Phyllaries 5 .... 1. *P. prostrata*
2. Phyllaries 3 .... 2. *P. cylindrica*
3. Pappus of 2 to 6 rigid, subulate, corneous awns, sometimes also with a few short squamellae (4).
3. Pappus various, but not of 2 to 6 rigid subulate awns (7).
4. Plants definitely perennial, from a woody rootstock; stems relatively tall, 30 to 100 cm. high, erect, stiff; leaves normally without basal bristles; pappus usually partly of erect subulate awns and partly of short squamellae .... 3. *P. imberbis*
4. Plants annual (sometimes perennial in *P. Coulteri* ?); stems low and diffuse, or else not rigid; leaves usually with basal bristles; pappus usually wholly of stout subulate awns (5).
5. Pappus of 2 to 6 retrorsely barbed, spreading awns. Plants diffuse; leaves conspicuously bristle-toothed at base .... 4. *P. Coulteri*
5. Pappus of 2 to 4 smooth or antrorse-hispidulous awns (6).
6. Pappus of 2 or 3 smooth, divergent awns; ligules about 1 mm. long .... 5. *P. linifolia*
6. Pappus of 2 to 4 antrorse-hispidulous awns; ligules 4 to 6 mm. long .... 7. *P. filipes*
7. Plants perennial, with a woody rootstock. Stems low, 10 to 20 cm. high, leafy only toward the base, with long, naked peduncles; leaves with 1 to 3 pairs of bristles at base; phyllaries 12 to 15; pappus of the disk flowers of 20 to 40 unequal bristles, that of the ray flowers of 2 bristles and sometimes a few squamellae .... 6. *P. longipes*
7. Plants annual (8).
8. Phyllaries 5. Pappus of 1 to 4 subulate awns and sometimes a few squamellae, or reduced to a crown of squamellae .... 7. *P. filipes*
8. Phyllaries 8 to 10 (9).
9. Heads rather crowded, sessile or peduncled, not obviously surpassing the leaves (10).
9. Heads mostly solitary at the tips of the branches and in the axils and forks of the stem, the peduncles usually considerably surpassing the leaves (11)
10. Leaves not dilated at base; pappus normally of 12 to 18 bristles, sometimes reduced to a crown .... 8. *P. papposa*

10. Leaves dilated at base; pappus a crown of squamellae, with or without 1 or 2 awns
9. *P. angustifolia*

11. Pappus in the ray flowers of 2 to 5 slender awns or bristles, in the disk flowers of numerous bristles at least half as long as the disk corollas............10. *P. Palmeri*

11. Pappus in the ray flowers of 1 or 2 slender awns or bristles, in the disk flowers of numerous short bristles, or reduced to a crown......................11. *P. Rusbyi*

1. Pectis prostrata Cav. Greenlee, Graham, Gila, Cochise, Santa Cruz, and Pima counties, 4,000 to 6,000 feet, along streams and on sandy plains and dry slopes, August to October. Western Texas to southeastern Arizona, south to northern South America; also in Florida, Cuba, and Jamaica.

The phyllaries are glabrous in typical *P. prostrata,* puberulent in var. *urceolata* Fern. (*P. urceolata* Rydb.).

2. Pectis cylindrica (Fern.) Rydb. Yavapai, Greenlee, Pinal, Cochise, Santa Cruz, and Pima counties, 1,500 to 5,000 feet, sandy-gravelly plains and mesas, May to September. New Mexico, Arizona, and Sonora.

The phyllaries in this species vary from glabrous to puberulent.

3. Pectis imberbis Gray. Cochise and Santa Cruz counties, 4,000 to 5,500 feet, August to October, type from "on the Sonoita," probably in southwestern Cochise County (*Wright* 1399). Southern Arizona, Sonora, and Chihuahua.

***4. Pectis Coulteri** Harv. & Gray. Sonora and reported from Arizona.

5. Pectis linifolia L. Cochise, Santa Cruz, and Pima counties, 3,000 to 6,000 feet, shaded canyons and slopes, August and September. Southeastern Arizona to northern South America; West Indies.

6. Pectis longipes Gray. Greenlee, Pinal, Cochise, Santa Cruz, and Pima counties, 3,500 to 5,500 feet, gravelly flats and rocky slopes, April to September, type from between the San Pedro River and Santa Cruz, Sonora (*Wright* 1127). Western Texas to southeastern Arizona and northern Mexico.

7. Pectis filipes Harv. & Gray. Greenlee, Pinal, Cochise, Santa Cruz, and Pima counties, 3,000 to 6,000 feet, sandy plains, mesas, rocky slopes, and along streams, August to October. Western Texas to southern Arizona and northern Mexico.

8. Pectis papposa Harv. & Gray. Greenlee, Coconino, and Mohave counties, south to Graham, Pima, and Yuma counties, 6,000 feet or lower, sandy-gravelly plains and mesas, very common, June to November. New Mexico to California and northern Mexico.

Chinchweed, a host, in Arizona, of the beet leaf hopper.

9. Pectis angustifolia Torr. Southern Apache, Navajo, eastern Coconino, and western Gila counties, 3,500 to 7,000 feet, also reported from the vicinity of Yuma, dry, sandy or gravelly mesas, August and September. Nebraska to Texas, Arizona, and Mexico.

10. Pectis Palmeri Wats. (*P. Mearnsii* Rydb.). Eastern Yavapai, southern Gila, and western Pima counties, 2,500 to 3,500 feet, August to October, type of *P. Mearnsii* from Fort Verde (*Mearns* 184). Southern Arizona, Sonora, and Baja California.

11. Pectis Rusbyi Greene. Yavapai County, about 4,000 feet, August and September, type from Beaver Creek (*Rusby* in 1883). Known only from central Arizona.

107. ANTHEMIS. Camomile

Branching annuals, thinly pilose, rank-scented; leaves alternate, 2- or 3-pinnatisect into narrowly linear, cuspidate-tipped divisions; heads medium-sized, solitary at the tips of the stem and branches, naked-peduncled, the rays 10 to 15, white, the disk yellow; involucre hemispheric, of lance-ovate or oblong, scarious-margined phyllaries; receptacle conic, naked toward the base, bearing stiff, narrow acuminate pales above; achenes subcylindric, 10-ribbed, glandular-roughened; pappus none.

1. Anthemis Cotula L. Coconino, Yavapai, Maricopa, Pinal, Cochise, and Pima counties, 1,000 to 8,000 feet, roadsides and waste places, April to July. Throughout most of the United States and Canada; naturalized from Eurasia.

May-weed, dog-fennel. The powdered flowers are effective against bedbugs, fleas, and flies. A decoction of the leaves may also be employed as an insecticide.

108. ACHILLEA. Yarrow, Milfoil

Perennial herbs, usually less than 0.5 m. high, thinly or densely pilose, sometimes silky-canescent, leafy, with creeping rootstocks; leaves linear to oblong, finely dissected into very numerous, short, linear to ovate, callose-cuspidate divisions not more than 1 mm. wide; heads small, numerous, in a dense, corymbose terminal panicle, both the rays and the disk white or occasionally pink; rays small, roundish, about 2 mm. long; receptacle with scarious chaff; pappus none.

A. Millefolium L., a European species extensively naturalized in the eastern United States, contains achilleine, a drug sometimes used in acute suppression of the menses. It was formerly prescribed as a tonic and in urinary disorders. Mrs. Collom reported that a decoction of the leaves and flowers of *A. lanulosa* is used in family medicine in Arizona.

1. Achillea lanulosa Nutt. Apache, Navajo, and Coconino counties to Cochise and Pima counties, 5,500 to 11,500 feet, mostly in the mountains, common in pine forests, June to September. Manitoba to British Columbia, south to Kansas, New Mexico, Arizona, California, and northern Mexico.

In var. *alpicola* Rydb. (*A. subalpina* Greene, *A. alpicola* Rydb.), seen from the San Francisco Peaks, Coconino County, and Baldy Peak, Apache County, 8,500 to 11,000 feet, the margins of the phyllaries are dark brown, whereas they are straw-colored to pale brown in the typical plant.

Achillea lanulosa, Western yarrow, a certainly indigenous plant, is not clearly distinguishable from the introduced *A. Millefolium*, common in the eastern United States. Specimens from Hualpai Mountain, Mohave County (*U. S. Dept. Interior Range Survey*), from near Prescott (*Shreve* 7767), and from the Chiricahua Mountains (*Blumer* 1340) are suggestive of *A. Millefolium*. A collection from the north end of the Carrizo Mountains, Apache County (*Standley* 7379), has comparatively large heads and large leaves with remote divisions, and has been referred to *A. laxiflora* Pollard & Cockerell, but does not seem to be specifically distinct from other Arizona material.

108a. LEUCAMPYX

Leucampyx Newberryi Gray was reported from Arizona (Wooton & Standley, *Flora New Mexico* p. 723) on the basis of a collection by E. Palmer in 1872, labeled "Arizona" without definite locality. This plant is known certainly only from Colorado and New Mexico, and the Palmer specimen probably was collected in one or the other of these states. Specimens of the very similar *Hymenopappus radiatus* Rose have been mistaken for *L. Newberryi.*

109. MATRICARIA. False-camomile

Annual herbs, usually 30 cm. high or less, sweet-scented, essentially glabrous; leaves alternate, 2- or 3-pinnatisect into linear or linear-filiform divisions; heads medium-sized, radiate or discoid, solitary at the tips of the stem and branches; involucre of subequal, oblong or oval, broadly scarious-margined phyllaries; receptacle conic, naked; achenes small, oblong, oblique, ribbed on the inner side; pappus a low crown, or nearly wanting.

KEY TO THE SPECIES

1. Rays none. Pappus an evident oblique crown; plants pineapple-scented ... 1. *M. matricarioides*
1. Rays present, white (2).
2. Pappus obsolete ... 2. *M. Chamomilla*
2. Pappus a scarious, 1-sided crown ... 3. *M. Courrantiana*

1. Matricaria matricarioides (Less.) Porter. Gila, Maricopa, Pinal, and Pima counties, up to 2,500 feet, roadsides, waste places, and river bottoms, February to April. Native from Alaska to Montana, Arizona, and Baja California, naturalized eastward; adventive in Europe.

Called pineapple-weed, in allusion to the pleasant odor of the plant.

2. Matricaria Chamomilla L. Known in Arizona only by a collection at Phoenix (*Dewey* in 1891). Occasionally adventive in the United States; native of Europe.

3. Matricaria Courrantiana DC. Tucson, Pima County, in a dooryard (*Thornber* in 1913). Texas and Arizona; Mexico and Bolivia; introduced from Spain.

110. CHRYSANTHEMUM

Annual or perennial, glabrous or sparsely pubescent, leafy-stemmed herbs; leaves toothed to tripinnatisect; heads medium-sized, solitary at the tips of the stem and branches, the rays white or yellow, the disk yellow; involucre of scarious-margined phyllaries; receptacle naked, broad, flattish; achenes 5- to 10-ribbed, sometimes narrowly 2- or 3-winged; pappus none.

Several species are in cultivation as ornamentals, and the insecticide pyrethrum is obtained from 2 of them.

KEY TO THE SPECIES

1. Lower leaves obovate or spatulate, toothed or incised, on slender petioles longer than the blades; stem leaves linear to oblanceolate, sessile, incised-serrate to pinnatifid; phyllaries with a dark-brown submarginal area; rays clear white, normally more than 1 cm. long ... 1. *C. Leucanthemum*
1. Leaves all essentially similar, mostly broadly obovate, 2- or 3-pinnatifid, with a winged, pinnatifid or toothed, petiolar base; phyllaries with broad, pale, scarious margin; rays pale yellow, less than 1 cm. long ... 2. *C. coronarium*

1. Chrysanthemum Leucanthemum L. Near Lakeside (southern Navajo County), Mogollon Escarpment and Pinal Mountains (Gila County), fields and roadsides, rare in Arizona, summer and autumn. Extensively naturalized in the United States; introduced from Europe.

The common ox-eye-daisy, represented in Arizona by var. *pinnatifidum* Lecoq & Lam.

2. Chrysanthemum coronarium L. Near Chandler, Maricopa County (*Harrison* 1787), Tucson, Pima County (*Toumey* 710), roadsides and waste ground, spring. Occasional in the United States; introduced from Europe.

Crown-daisy, sometimes cultivated under the name "summer chrysanthemum."

111. COTULA

Low perennials; herbage glabrous or nearly so; stems somewhat succulent, rooting at the lower nodes; leaves entire or pinnatifid, the base forming a sheath around the stem; heads flat, pedunculate, disciform, yellow; phyllaries 2-seriate, nearly equal, membranaceous, greenish, scarious-margined; receptacle naked; flowers of the outer circle pistillate, without corollas; achenes pedicelled, compressed, margined or winged, without pappus.

1. Cotula coronopifolia L. Wikieup, southeastern Mohave County (*Gould & Darrow* 3708), delta of Williams River, northwestern Yuma County (*Monson* in 1949), 400 to 2,000 feet, in wet, saline soil. British Columbia to western Arizona and Baja California; naturalized from South Africa. Brass-buttons.

112. ARTEMISIA (359, 360). Wormwood, Sagebrush

Herbs or shrubs; leaves alternate, entire to once, twice, or thrice pinnatifid; heads small, discoid or disciform, usually very numerous, spicate, racemose, or panicled; pistillate outer flowers, without rays, sometimes present; achenes short, thick, glabrous or merely glandular, or in one species pubescent; pappus none.

The best-known species is *A. tridentata,* the big sagebrush, state flower of Nevada, which covers vast areas in the Great Basin region and adjacent territory. The strong, aromatic odor of the plant is unmistakable and persistent. Many of the species, notably estafiata (*A. frigida*), *A. tridentata, A. filifolia, A. Bigelovii,* and bud-sage (*A. spinescens*), are valuable browse plants, especially in winter and early spring, but some of them are reported to be toxic to domestic animals if eaten in excess. Several species are useful for controlling soil erosion. Some of these plants (*A. filifolia, A. frigida, A. tridentata*) were used medicinally by the Indians and the early white settlers. The drug santonin, a remedy for roundworm, is obtained from *A. mexicana.* A vermifuge is also manufactured from the European *A. Absinthium* L. The Hopi Indians roasted leaves of *A. frigida* with sweet corn to flavor it. The silvery foliage and feathery panicles of some of the sagebrushes, together with their pleasant odor, make them worthy of consideration for cultivation as ornamentals. They are attractive in mixed bouquets.

KEY TO THE SPECIES

1. Plants shrubs or undershrubs (2).
1. Plants herbs (8).
2. Stems less than 20 cm. high; leaves green, sparsely puberulent, not over 5 mm. long, pinnately parted into 3 to 9 linear or oblanceolate lobes.............13. *A. pygmaea*
2. Stems over 20 cm. high; leaves otherwise (3).
3. Leaves twice or thrice pinnatifid into linear or spatulate divisions (4).
3. Leaves entire, 3-dentate, or 3-parted (5).
4. Branches not spinescent; leaves silvery-canescent; achenes glabrous......7. *A. frigida*
4. Branches spinescent; leaves spreading-pilose, not silvery; achenes of the pistillate flowers cobwebby-pubescent ....14. *A. spinescens*
5. Leaves linear-filiform, less than 1 mm. wide, entire or 3-parted..........10. *A. filifolia*
5. Leaves broader, at least the lower ones 3-dentate at apex (6).
6. Pistillate outer flowers (1 or 2) usually present; involucre densely and persistently gray-tomentose throughout, only about 2 mm. high......................6. *A. Bigelovii*
6. Pistillate outer flowers none; involucre usually glabrate, at least on the inner phyllaries, larger (7).
7. Heads very numerous, in dense crowded panicles 1.5 to 7 cm. thick; phyllaries, at least the outer ones, canescent-pubescent; plants usually at least 50 cm. high
11. *A. tridentata*
7. Heads fewer, in less crowded often spikelike or racemelike panicles rarely more than 1.5 cm. thick; phyllaries nearly glabrous, greenish yellow or yellowish brown; plants 10 to 30 cm. high..12. *A. nova*
8. Leaves entire to once pinnatifid (9).
8. Leaves, at least the lower ones, twice to thrice pinnatifid (11).
9. Leaves elongate-linear, entire or some of the lower ones sometimes 3-cleft, usually 5 to 8 cm. long, 1 to 6 mm. wide, glabrous or somewhat pubescent but never tomentose
8. *A. dracunculoides*
9. Leaves usually broader or else usually lobed, toothed, or pinnatifid, always tomentose, at least beneath (10).
10. Leaves from entire to toothed or pinnatisect, the lobes when present 2 mm. wide or wider ....2. *A. ludoviciana*
10. Leaves pinnatisect essentially to the midrib into linear or linear-filiform lobes 1 mm. wide or less..3. *A. Carruthii*
11. Plants essentially glabrous, equably leafy throughout. Leaves twice or thrice pinnatisect into very numerous acute divisions (12).
11. Plants decidedly hairy, at least on the lower leaves or the under leaf surface, the leaves usually much reduced above (13).
12. Panicle dense, spikelike; plants inodorous............................4. *A. biennis*
12. Panicle very loose and broad; plants sweet-scented......................5. *A. annua*
13. Plants densely silky-canescent throughout; receptacle villous............7. *A. frigida*
13. Plants not silky-canescent except sometimes on the lower leaves; receptacle glabrous (14).
14. Leaves green above, canescent-tomentose beneath, bipinnatifid, the divisions not linear-filiform; heads 5 to 7 mm. thick..............................1. *A. franserioides*
14. Leaves about equally pubescent on both sides (sometimes sparsely so), not bicolored, the principal divisions linear-filiform or very narrowly linear; heads about 2 to 3 mm. thick ....9. *A. pacifica*

1. Artemisia franserioides Greene. White Mountains (Apache County), Pinaleno Mountains (Graham County), Chiricahua and Huachuca mountains (Cochise County), Santa Catalina Mountains (Pima County), 8,000 to 10,000 feet, open coniferous forest, August and September. Colorado to New Mexico, southeastern Arizona, and Chihuahua.

2. Artemisia ludoviciana Nutt. Apache and Coconino counties to the mountains of Cochise, Santa Cruz, and Pima counties, 2,500 to 8,500 feet, dry slopes and canyons and open pine forest, often on limestone, August to November. Southern Canada to northern Mexico.

The following treatment is based on the revision by Keck (360).

KEY TO THE SUBSPECIES

1. Principal leaves entire or merely lobed (2).
1. Principal leaves more or less parted or divided (4).
2. Leaves mostly 1 to 2 cm. long, lanceolate, thickish, usually white-tomentose on both sides, the margin often narrowly revolute; panicle open, the heads pendent in slender, more or less leafy racemes.......................................ssp. *albula*
2. Leaves 3 to 10 cm. long, the margin not revolute (3).
3. Panicle usually compact, elongate, not leafy; leaves not thin, usually white-tomentose on both surfaces, rarely bright green above; involucre 3 to 4 mm. high........ssp. *typica*
3. Panicle open, leafy (sometimes contracted); leaves thin, usually bright green above, white-tomentose beneath; involucre 2.5 to 3 mm. high..................ssp. *sulcata*
4. Lower leaves 5 to 10 cm. long, pinnately or ternately divided, the lobes entire, strongly bicolored ....ssp. *mexicana*
4. Lower leaves 2 to 5 cm. long, bipinnately dissected, green and more or less glandular, or moderately tomentose beneath...ssp. *redolens*

Subsp. *typica* Keck (*A. gnaphalodes* Nutt.) is rare in Arizona, being known only from the Grand Canyon (*Eggleston* 15665). Subsp. *sulcata* (Rydb.) Keck (*A. sulcata* Rydb.) occurs throughout central and southern Arizona except in the low western portion. Arizona specimens cited by Fernald (*Rhodora* 47 : 248) under *A. ludoviciana* var. *americana* (Bess.) Fern. were referred by Keck to subsp. *sulcata*. Subsp. *albula* (Wooton) Keck (*A. albula* Wooton) is found throughout the range of the species in Arizona. Specimens of this subspecies were referred previously to *A. gnaphalodes* Nutt. Subsp. *mexicana* (Willd.) Keck (*A. mexicana* Willd.) occurs in Apache, Navajo, and Coconino counties, apparently also in Yavapai, Greenlee, and Cochise counties. Subsp. *redolens* (Gray) Keck is represented from Arizona only by a doubtfully identified collection on Mount Graham (*Kearney & Peebles* 9796) and perhaps by a collection on Turkey Creek Mesa, Cochise County (*Goodding* 136G).

3. Artemisia Carruthii Wood. White Mountains (Apache County), Hopi Indian Reservation (Navajo County), Kaibab Plateau, Grand Canyon, and vicinity of Flagstaff (Coconino County), mostly 6,000 to 9,000 feet, open pine forest, August to November. Missouri to Texas, west to Utah and Arizona.

Var. *Wrightii* (Gray) Blake (*A. Wrightii* Gray), differing from the typical plant only in having the upper surface of the leaves glabrate and green, has a wider range in Arizona, from the Carrizo Mountains (Apache County) to Hualpai Mountain (Mohave County), also in the White, Pinaleno, Pinal, and Chiricahua mountains (Greenlee, Graham, Pinal, and Cochise counties).

Keck (360, p. 440), who did not recognize var. *Wrightii*, cited collections of *A. Carruthii* from Apache, Navajo, Coconino, Mohave, Yavapai, Greenlee, Graham, and Cochise counties.

4. Artemisia biennis Willd. Near Flagstaff, Coconino County (*Pearson* 102, *Thornber* in 1920), Gila River bed, Sacaton, Pinal County (*Harrison* 1969), June to September. Canada to New Jersey, Kentucky, Missouri, Arizona, and California, native in the western part of this range, elsewhere naturalized.

5. Artemisia annua L. Santa Catalina Mountains, Pima County, about 6,000 feet, roadside, August (*Harrison & Kearney* 8134). A weed here and there in the United States; naturalized from Europe. Sweet wormwood.

6. Artemisia Bigelovii Gray (*A. petrophila* Woot. & Standl.). Apache, Navajo, and Coconino counties, 5,000 to 8,000 feet, dry mesas and slopes, sometimes with pinyon, August to October, type of *A. petrophila* from the Carrizo Mountains (*Standley* 7355). Colorado and Utah to Texas, New Mexico, and northern Arizona.

A small, many-stemmed shrub.

7. Artemisia frigida Willd. Apache, Navajo, and Coconino counties, 5,500 to 8,000 feet, dry, stony soil, July to October. Canada and Alaska, south to Texas, New Mexico, and northern Arizona; Siberia.

Estafiata. An undershrub or almost herbaceous, often forming mats, valuable as forage.

8. Artemisia dracunculoides Pursh. Apache County to Mohave County, south to the mountains of Cochise, Santa Cruz, and Pima counties, 3,500 to 9,000 feet, open coniferous forests and chaparral, common, July to November. Manitoba to British Columbia, south to Texas, Arizona, and Baja California.

False-tarragon. Much of the Arizona material, in its small, rather long-peduncled heads (the peduncles up to 5 mm. long) approaches var. *dracunculina* (Wats.) Blake (*A. dracunculina* Wats., *A. glauca* Pall. var. *dracunculina* Fern.), described from Chihuahua. The type of that variety, however, has rather densely pilose stems and leaves, and the only Arizona specimen examined that completely agrees with it is one from Mount Lemmon, Pima County, 6,000 feet (*Harrison & Kearney* 8051).

9. Artemisia pacifica Nutt. Apache, Navajo, Coconino, Yavapai, Greenlee, Gila, and Cochise counties, 5,500 to 8,500 feet, open coniferous forest, July to October. South Dakota to northwestern Canada, south to New Mexico and Arizona.

10. Artemisia filifolia Torr. Apache, Navajo, Coconino, Mohave, Graham, and Cochise counties, 4,000 to 6,000 feet, loose, sandy soil, August to November. Nebraska and Wyoming to Nevada, Texas, Arizona, and northern Mexico.

Sand sagebrush. A small, much-branched shrub.

11. Artemisia tridentata Nutt. Apache, Navajo, and Coconino counties, 5,000 to 8,000 feet, plains, mesas, and rocky slopes, in the open or with pinyon and juniper, common, July to October. South Dakota to British Columbia, south to New Mexico, northern Arizona, and Baja California.

Big sagebrush occurs in nearly pure stands over large areas in central Apache and Navajo counties. It varies greatly in size according to habitat, forming an extensive root system and reaching a height of 2 m. (7 feet) or more where conditions permit. A good growth of this sagebrush indicates a deep, fertile, nonsaline soil.

12. Artemisia nova A. Nels. Navajo Indian Reservation (Apache and Navajo counties), Kaibab Plateau and Grand Canyon (Coconino County), 6,000 to 8,000 feet, dry slopes and mesas, usually in shallow, stony soil, August and September. Montana to New Mexico, northern Arizona, and California.

A small shrub.

13. Artemisia pygmaea Gray. Near Fredonia, northwestern Coconino County, 4,500 feet (*Darrow* 3006, *Ripley & Barneby* 8504). Western Utah, eastern Nevada, and northern Arizona.

Pigmy sagebrush. The plant was described by R. A. Darrow as forming low cushions 1 foot in diameter, occurring on gypseous soil.

14. Artemisia spinescens D. C. Eaton. Northern Apache County, Carrizo Mountains and near Rock Point (*Standley* 7479, *Peebles & Smith* 13545), 5,500 to 6,000 feet, dry slopes and mesas, often in saline soil. Montana to Oregon, south to New Mexico, northeastern Arizona, and California.

Bud-sage, a small rigid, spiny shrub, very resistant to drought and overgrazing.

113. PENTZIA

Low shrubs, the herbage tomentulose-canescent; leaves alternate, pinnatifid, with revolute margins; heads small, solitary, peduncled, discoid; flowers all hermaphrodite and fertile; phyllaries graduated, the inner ones scarious-margined; achenes 5-angled; pappus a low, cup-shaped, scarious crown.

***1. Pentzia incana** (Thunb.) O. Kuntze. Planted by the United States Soil Conservation Service near Safford, Graham County, and reported as tending to seed itself, but probably not yet established as a wild plant in Arizona; native of South Africa.

113a. LEPIDOSPARTUM. Scalebroom

Lepidospartum squamatum Gray was reported by Gray (*Proc. Amer. Acad.* 8: 290) from "Desert of the Colorado, Arizona, 1870. Dr. Palmer." This reference was to a peculiar variety *Linosyris squamata* var. *Palmeri* Gray, now *Lepidospartum squamatum* var. *Palmeri* (Gray) L. C. Wheeler. According to Wheeler (*Rhodora* 40: 320–323), Palmer's plant probably came from near Whitewater, Riverside County, California. Practically all later authors have continued to include Arizona in the range of the species, but not one has cited material from the state. It seems advisable to exclude *Lepidospartum* from the flora of Arizona until actual specimens are forthcoming.

114. ARNICA (361)

Low, pubescent, perennial herbs; leaves opposite; heads solitary or few, rather large, long-peduncled, yellow, radiate; involucre 2-seriate, of thin, subherbaceous, lance-oblong, acute or acuminate, equal phyllaries; pappus of barbellate bristles.

Tincture of arnica is obtained from a European species, *A. montana.*

KEY TO THE SPECIES

1. Lower leaves broadly ovate, cordate; heads almost always solitary.......1. *A. cordifolia*
1. Lower leaves oblong to lance-ovate, tapering at base; heads normally 3 or more
2. *A. foliosa*

1. Arnica cordifolia Hook. Kaibab Plateau, North Rim of the Grand Canyon, and San Francisco Peaks, about 8,500 feet. Alaska to South Dakota, New Mexico, northern Arizona, and California.

2. Arnica foliosa Nutt. (*A. Chamissonis* Less. subsp. *foliosa* Maguire). Kaibab Plateau and North Rim of Grand Canyon, 8,000 to 9,000 feet. Alaska to Colorado, New Mexico, northern Arizona, and California.

115. PEUCEPHYLLUM. Pigmy-cedar

Much-branched shrubs, whitish-barked, resinous-viscid, essentially glabrous, denudate below, very leafy above; leaves alternate, linear-filiform, subterete, obtuse or apiculate, densely glandular-punctate, 8 to 20 mm. long; heads solitary at the tips of the branches, subsessile, yellow, discoid; involucre 2-seriate, of linear-lanceolate, acuminate, subequal phyllaries; achenes silky-pilose; pappus of numerous graduated bristles, the inner ones sometimes narrowly linear, flattened, and paleaceous.

1. Peucephyllum Schottii Gray. Northern Mohave County to southern Yuma County, up to 5,000 feet but commonly lower, dry, rocky slopes, March to June. Southern Nevada, western Arizona, southern California, Sonora, and Baja California.

116. PSATHYROTES

Low, spreading, divaricately branched, annual (or perennial ?), scurfy-tomentose, very leafy herbs; leaves alternate, ovate to deltoid-ovate, toothed or entire, petioled; heads solitary in the forks, small, discoid, nodding or erect, yellow or purplish; involucre 2- or 3-seriate, somewhat graduated, the phyllaries lanceolate to oblong or obovate, at least the outer ones herbaceous, at least above; achenes densely silky-villous; pappus of numerous graduated bristles, becoming yellow-brown in age.

KEY TO THE SPECIES

1. Leaves entire, scurfy-tomentulose and bearing conspicuous, long, many-jointed hairs, especially on the margin and petiole; achenes subcylindric, hispidulous . . . 1. *P. pilifera*
1. Leaves toothed or crenate, the many-jointed hairs, when present, relatively short and inconspicuous; achenes obconic, densely silky-pilose (2).
2. Outer phyllaries obovate, much broader than the inner ones; plants lanate-tomentose as well as scurfy . . . 2. *P. ramosissima*
2. Outer phyllaries lanceolate, lance-ovate, or spatulate, not broader than the inner ones; plants scurfy-tomentose . . . 3. *P. annua*

1. Psathyrotes pilifera Gray. Eastern Coconino County, at Lees Ferry, Navajo Bridge, and Houserock Valley, about 4,000 feet, August and September. Southern Utah and northern Arizona.

2. Psathyrotes ramosissima (Torr.) Gray. Mohave and Yuma counties, 1,500 feet or lower, plains and mesas in gravelly or sandy soil, flowering throughout the year. Southwestern Utah to western Arizona, southeastern California, and northern Baja California.

***3. Psathyrotes annua** (Nutt.) Gray. No specimens from Arizona have been seen, but this species has been collected at St. George, Utah (*E. Palmer* 266), near the Arizona border. Southern Utah to southeastern California and northwestern Mexico.

117. CACALIA. Indian-plantain

Herbaceous perennials, about 1 m. high, woolly-tufted at base, otherwise nearly glabrous; leaves mostly in a basal tuft, broad, long-petioled, 3- or 4-pinnatisect into linear or lance-linear ultimate divisions, the stem leaves few, the upper ones much reduced; heads numerous, small, discoid, white, 5- to 7-flowered, in a dense, terminal corymbiform panicle; phyllaries 5 or 6, less than half as long as the flowers; achenes glabrous; pappus of numerous capillary bristles.

1. Cacalia decomposita Gray. Willow Spring (southern Apache County), Chiricahua and Huachuca mountains (Cochise County), Patagonia Mountains (Santa Cruz County), 5,000 to 8,000 feet, rich, shaded soil, July to September. Southeastern Arizona, southwestern New Mexico, and Sonora.

118. TETRADYMIA. Horse-brush

Shrubs, commonly less than 1 m. high, stiffly much branched, canescent-tomentose throughout; leaves oblanceolate to linear, entire, sessile, callose-tipped, usually less than 1.5 cm. long, often with axillary fascicles; heads medium-sized, discoid, yellow, solitary and axillary or clustered at the tips of the branches; involucre of equal, thick-chartaceous phyllaries, tomentose outside; achenes obovoid, densely silky-pilose; pappus of copious, whitish, capillary bristles.

KEY TO THE SPECIES

1. Primary leaves transformed into stiff, straight, spreading spines; heads solitary, axillary, 5- to 9-flowered; phyllaries 5 or 6 .... 1. *T. axillaris*
1. Primary leaves not transformed into spines; heads clustered at the ends of the branches, 4-flowered; phyllaries 4 .... 2. *T. canescens*

1. Tetradymia axillaris A. Nels. Southwest of Kingman, Mohave County, 2,000 feet (*United States Dept. Interior Range Survey,* in 1938). Utah and western Arizona to California.

2. Tetradymia canescens DC. Navajo County to eastern Mohave County, 6,000 to 7,000 feet, dry, open ground, in sandy or rocky, sometimes saline soil, often abundant, July to October. Montana to New Mexico, northern Arizona, and California.

The Arizona plants belong to var. *inermis* (Nutt.) Gray. This plant, sometimes erroneously called black-sage, is said to be browsed locally by sheep, but its forage value is low and it is suspected of being poisonous. The Hopi are reported to use the leaves and roots as a tonic and for uterine disorders. White-haired animals eating this plant are reported to suffer photosensitization of the skin, resulting in severe dermatitis.

119. SENECIO (362). Groundsel

Herbs or shrubs; leaves alternate, entire to once or twice pinnatifid; heads medium-sized or rather large, yellow, radiate or discoid; involucre essentially 1-seriate, of equal, usually narrow, subherbaceous phyllaries, often with a calyculus of shorter bractlets at base; achenes glabrous or pubescent; pappus of soft, white capillary bristles.

S. longilobus, S. spartioides, and perhaps other species, are poisonous to cattle and horses, less so to sheep, but are seldom eaten when better forage is available. The liver is the organ chiefly affected. *S. cruentus,* of the Canary Islands, is supposed to be the parent of the showy forms cultivated under the name cineraria.

KEY TO THE SPECIES

1. Leaves pinnatilobate with narrowly linear or linear-filiform entire lobes, or entire and narrowly linear or linear-filiform. Plants very leafy throughout, usually suffrutescent (2).
1. Leaves neither pinnatilobate with narrowly linear or linear-filiform entire lobes, nor entire and narrowly linear (5).
2. Leaves entire, narrowly linear, or rarely with a pair of filiform lobes; plants essentially glabrous. Heads numerous, narrowly campanulate or subcylindric, in a close, cymose panicle .... 1. *S. spartioides*
2. Leaves pinnatilobate, or the upper ones often entire, or if all the leaves entire (rarely so in *S. longilobus*), then the plants tomentose (3).
3. Heads subcylindric or narrowly campanulate; phyllaries 8 to 10; plants glabrous or obscurely puberulent .... 2. *S. multicapitatus*
3. Heads broadly campanulate; phyllaries 13 to 21; plants glabrous or tomentose (4).
4. Bracteoles inconspicuous, less than half as long as the involucre; plants permanently tomentose, or sometimes glabrate below .... 3. *S. longilobus*
4. Bracteoles conspicuous, half as long as the involucre or longer; plants glabrous or nearly so .... 4. *S. monoensis*
5. Leaves all or nearly all deeply pinnatifid, with usually toothed or lobed divisions (6).
5. Leaves mostly entire or merely toothed (the stem leaves pinnatifid in *S. hartianus* and sometimes in *S. neomexicanus*) (14).
6. Stem uniformly leafy to the inflorescence; leaves laciniately once or twice pinnatifid, the lobes mostly acute or acuminate; heads small and narrow, the involucre 5 to 7 mm. high, 3 to 5 mm. thick. Phyllaries usually black-tipped (7).
6. Stem with the upper leaves usually much reduced; leaf lobes often blunt; heads larger, the involucre 6 to 10 mm. high, 4 to 10 mm. thick (8).
7. Stems tall, up to 1 meter; leaf lobes elongate, narrow; heads radiate; achenes glabrous or obscurely puberulent .... 5. *S. Macdougalii*
7. Stems low, less than 0.5 meter; leaf lobes short, relatively broad; heads discoid; achenes hirtellous .... 6. *S. vulgaris*
8. Plants coarse and tall (70 to 100 cm. high), somewhat glaucous, glabrous or slightly tomentose in the leaf axils; leaves with comparatively few divisions, the terminal one large, roundish, 2.5 to 5 cm. wide .... 7. *S. quercetorum*
8. Plants smaller and more slender, not evidently glaucous, often tomentose; leaves mostly with numerous divisions, the terminal one rarely more than 2 cm. wide (9).
9. Plants dwarf (10 cm. high or less); heads few; involucre 7 to 10 mm. high 8. *S. franciscanus*
9. Plants taller; heads usually rather numerous; involucre not more than 8 mm. high (10).
10. Achenes hirtellous (11).
10. Achenes glabrous (12).
11. Phyllaries about 13 .... 9. *S. multilobatus*
11. Phyllaries about 21 .... 10. *S. millelobatus*
12. Leaves with the primary divisions again pinnatifid into numerous small divisions 11. *S. lynceus*
12. Leaves not with the primary divisions pinnatifid into numerous small divisions (13).
13. Plants essentially glabrous except for woolly tufts in the leaf axils ...... 12. *S. stygius*
13. Plants more or less tomentose on the leaves and stem .... 13. *S. uintahensis*
14. Heads nodding, discoid. Stems stout, leafy, up to 1 m. high; leaves elongate, lanceolate or oblanceolate, serrate or serrulate with glandular teeth .... 14. *S. Bigelovii*
14. Heads not nodding, usually radiate (15).

15. Leaves suborbicular, about 10 cm. wide, palmately 5- or 7-nerved from the base and conspicuously reticulate, shallowly repand-lobed and repand-dentate. Heads very numerous, small, in a broad panicle. . . . . 15. *S. Seemannii*
15. Leaves much narrower, not palmate-nerved, not evidently reticulate (16).
16. Phyllaries 8, broad, blunt, with a conspicuous, usually bright-yellow margin; plants shrubby, with narrowly lance-linear leaves tapering to the base. . . . . . . 16. *S. salignus*
16. Phyllaries more numerous, narrow, usually acute or acuminate; plants herbaceous or suffrutescent (17).
17. Leaves conspicuously clasping; plants not tomentose (18).
17. Leaves not clasping, or if sometimes so (in *S. arizonicus* and *S. cynthioides*), then the lower leaves tomentose beneath (21).
18. Plants densely glandular-pubescent. . . . . 17. *S. Parryi*
18. Plants glabrous (19).
19. Plants annual; head disciform or discoid; leaves 2 to 6 cm. long, oblong or oblong-ovate . . . . 18. *S. mohavensis*
19. Plants perennial; heads conspicuously radiate (exceptionally discoid in *S. huachucanus*); leaves usually larger (20).
20. Leaves oblong, the middle ones 20 to 25 cm. long, 4 to 6.5 cm. wide (exceptionally only 10 cm. long and 2 cm. wide), closely repand-denticulate; plants herbaceous, simple below the inflorescence. . . . . 19. *S. huachucanus*
20. Leaves lanceolate or linear, 4 to 10 (17) cm. long, rarely more than 1.5 cm. wide, irregularly dentate, chiefly toward the base; plants suffrutescent, branched 20. *S. Lemmoni*
21. Plants glabrous and somewhat glaucous. . . . . 21. *S. Wootonii*
21. Plants tomentose, at least at the base of the stem (22).
22. Heads solitary (very rarely 2 or 3), large, the involucre 10 to 12 mm. high. Basal leaves obovate or oblanceolate, entire or crenate; stem leaves few, small, narrow, entire, bractlike . . . . 22. *S. Actinella*
22. Heads several or many (rarely solitary in *S. werneriaefolius* and *S. neomexicanus*), smaller (23).
23. Basal leaves ovate, 2 to 5.5 cm. wide, often subcordate, closely dentate or dentate-serrate . . . . 23. *S. arizonicus*
23. Basal leaves linear-oblanceolate to obovate, usually much narrower or, if as broad, then not closely dentate-serrate (24).
24. Involucre persistently tomentose, at least toward the base; plants low, rarely more than 20 cm. high; stem naked or with 1 or 2 narrowly subulate bracts. 24. *S. werneriaefolius*
24. Involucre glabrous or soon glabrate (exceptionally persistently tomentose in *S. neomexicanus*); plants usually taller; stems more or less leafy (25).
25. Basal leaves elliptic, oval, or obovate, finely and closely crenate-serrate or serrate; tomentum mostly fugacious. . . . . 25. *S. hartianus*
25. Basal leaves otherwise; tomentum more persistent (26).
26. Basal and stem leaves essentially similar, narrowly oblanceolate (about 5 to 12 cm. long, 4 to 10 mm. wide), entire to remotely toothed, green and glabrescent or glabrate above, gray-tomentulose beneath; stem almost uniformly leafy throughout, or the upper leaves much reduced. . . . . 26. *S. cynthioides*
26. Basal and stem leaves usually dissimilar, at least the basal ones normally broader, often sharply toothed; stem usually with much-reduced leaves above. . . 27. *S. neomexicanus*

1. Senecio spartioides Torr. & Gray. Northern Navajo County, Coconino County, and Hualpai Mountain (Mohave County), 6,500 to 9,000 feet, openings in pine forests, July to October. Nebraska and Wyoming, south to Texas and Arizona.

Broom groundsel. The plant is known to be poisonous to livestock, but is rarely eaten.

2. Senecio multicapitatus Greenm. Apache, Navajo, Coconino, Yavapai, Greenlee, and Gila counties, 5,000 to 7,000 feet, plains, mesas, canyons, and pine forest, May to November. Colorado, Utah, New Mexico, and Arizona.

3. Senecio longilobus Benth. (*S. orthophyllus* Greene, *S. Douglasii* DC. var. *longilobus* L. Benson). Apache County to Mohave County, south to Cochise, Santa Cruz, and Pima counties, 2,500 to 7,500 feet, common on dry plains, mesas, and slopes, and along washes, flowering throughout the year, type of *S. orthophyllus* from Willow Spring, Apache County (*Palmer* 479, in 1890). Colorado and Utah, south to Texas and Mexico.

Thread-leaf groundsel. The United States Department of Agriculture has found *S. longilobus* to be one of the most poisonous of the groundsels, especially to cattle and horses, the leaves of the new growth being most toxic. The primary effect is to produce lesions in the liver. This species is used extensively in the domestic medicine of the Indians.

4. Senecio monoensis Greene (*S. lathyroides* Greene, *S. filicifolius* Greenm., *S. pectinatus* A. Nels., *S. Douglasii* var. *monoensis* Jepson). Almost throughout the state, 1,000 to 6,000 feet, dry, sandy or stony soil on mesas and slopes, and in canyons, often in chaparral, flowering in nearly all months, type of *S. lathyroides* from Pierce Spring (*M. E. Jones* 5077), type of *S. filicifolius* from the Santa Cruz River valley (*Pringle* 316, in 1881), type of *S. pectinatus* from near the Baboquivari Mountains, Pima County (*Hanson* 1020). Utah and Arizona to California and Mexico.

A specimen from Superior (*Whitehead* 269) has a woolly involucre and young branchlets and approaches *S. Douglasii* DC., but other specimens from the same region are definitely *S. monoensis*. The latter is closely related to *S. Douglasii*, differing chiefly in its essentially glabrous character and its more herbaceous stem, and may be no more than a variety.

5. Senecio Macdougalii Heller. Apache County to Mohave County, south to Cochise, Santa Cruz, and Pima counties, 6,500 to 10,500 feet, coniferous forests, July to October, type from Walnut Canyon, Coconino County (*MacDougal* 342). New Mexico and Arizona.

A specimen from the north end of the Carrizo Mountains (*Standley* 7376), identified by Greenman as *S. ambrosioides* Rydb., is the only collection of that species cited by him from Arizona. (See Greenman 362, 2, p. 596). The heads are young, but the number of flowers (6 ray, 30 disk) agrees as well with *S. Macdougalii* as with *S. ambrosioides*.

6. Senecio vulgaris L. Near Tucson, Pima County, in an irrigated pasture (*Matlock* in 1947). Widely distributed as a weed in North America; naturalized from Eurasia.

7. Senecio quercetorum Greene (*S. macropus* Greenm.). Western and southern Coconino County and mountains of Graham, Gila, and Pima counties, 3,500 to 6,000 feet, chiefly in the oak belt, March to May, type from Oak Creek (*Rusby* 672), type of *S. macropus* (*Rusby* 175) from Arizona without definite locality. Known only from Arizona.

8. Senecio franciscanus Greene. San Francisco Peaks (Coconino County), where the type was collected by Greene in 1889, up to 12,000 feet, July and August. Known only from Arizona.

9. Senecio multilobatus Torr. & Gray. Northern Apache, Navajo, and Coconino counties, 6,000 to 7,000 feet, June to August. Wyoming to New Mexico, Arizona, and Nevada.

Senecio multilobatus var. *Standleyi* Greenm. does not seem different from *S. multilobatus,* at least as to the only Arizona specimen seen, from the north end of the Carrizo Mountains (*Standley* 7513). Very small, acaulescent plants apparently of this species were collected in Yavapai County (*Goodding* 115–47).

10. Senecio millelobatus Rydb. Kaibab Plateau, Fredonia, and Oak Creek (Coconino County), near Prescott (Yavapai County), 4,500 to 7,000 feet, also bed of the Agua Fria River (Maricopa County), about 1,000 feet, where doubtless a stray from farther north, March to June. Texas to Arizona and Chihuahua.

11. Senecio lynceus Greene. Northern Navajo County and from the Kaibab Plateau to near Prescott (Coconino and Yavapai counties), 5,500 to 7,500 feet, pine forests, May to October, type not definitely stated but, by implication, from Lynx Creek, Yavapai County (*Rusby* 665, in 1883). Southern Utah and northern Arizona.

12. Senecio stygius Greene (*S. prolixus* Greenm.). Coconino, Mohave, Yavapai, and northern Gila counties, 2,000 to 7,000 feet, near springs, April and May, type of *S. stygius* from the Grand Canyon (*Lemmon* in 1884), type of *S. prolixus* (*Lemmon* 3130) probably a duplicate of this collection. Northwestern Arizona, southern Nevada, and southeastern California.

This species apparently intergrades with *S. lynceus.*

13. Senecio uintahensis A. Nels. Kaibab Plateau, Grand Canyon, and near Flagstaff (Coconino County), about 7,000 feet, June to September. Wyoming to Arizona, west to Oregon and California.

Species 9 to 13 are doubtfully distinct. The plants are very variable in the amount of tomentum, the leaves vary greatly in degree of lobing or dissection, and the characters drawn from the number of phyllaries and pubescence of the achenes seem of little consequence. *Senecio Thornberi* Greenm., type from the San Francisco Peaks (*Rusby* 666), seems intermediate between *S. uintahensis* and *S. neomexicanus* and is not clearly distinct from either.

14. Senecio Bigelovii Gray. San Francisco Peaks to the Mogollon Escarpment (Coconino County), White Mountains (Apache and Greenlee counties), Pinaleno Mountains (Graham County), and mountains of Cochise and Pima counties, 7,000 to 11,000 feet, rich, moist soil in coniferous forests and grassy hillsides, July to September. New Mexico and Arizona.

15. Senecio Seemannii Schultz Bip. Mountains of Cochise, Santa Cruz, and Pima Counties, 4,000 to 6,000 feet, rich, shaded soil, September to November. Southern Arizona and northern Mexico.

Senecio Hartwegi Benth., to which the Arizona plant has generally been referred, is according to Greenman (*Contrib. U. S. Nat. Herb.* 23: 1628) a distinct species of west-central Mexico, with the branches, petioles, and lower leaf surface tomentose.

16. Senecio salignus DC. Cochise, Santa Cruz, and Pima counties, near the Mexican boundary, 2,500 to 5,000 feet, rich soil along streams, February to April. Southern Arizona to Guatemala.

A shrub, up to at least 2 m. (7 feet) high.

17. Senecio Parryi Gray. Garfield (Greenlee County), Pinaleno Mountains

(Graham County), Chiricahua Mountains (Cochise County), Santa Rita and Santa Catalina mountains (Santa Cruz and Pima counties), 5,500 to 7,000 feet, July to September. Western Texas to southern California.

Plant strong-scented.

18. Senecio mohavensis Gray. Pinal, western Pima, and Yuma counties, 2,000 feet or lower, rocky slopes, sandy washes, and bottoms of canyons, February to April. Southwestern Arizona, southeastern California, and Sonora.

19. Senecio huachucanus Gray. Huachuca Mountains, Cochise County (*Lemmon* 2784, the type collection), Santa Rita Mountains (Santa Cruz County), 7,000 to 9,500 feet, August to October. Southeastern Arizona and Chihuahua.

In a collection in the Santa Rita Mountains (*Darrow & Arnold* in 1936) the heads appear to be rayless. Another collection in the same mountains (*K. F. Parker* 5856) has the middle stem leaves only 10 cm. long and 2 cm. wide.

20. Senecio Lemmoni Gray (*S. decorticans* A. Nels.). Yavapai, Gila, Maricopa, Pinal, Pima, and northeastern Yuma counties, 1,500 to 3,500 feet, rocky slopes, usually among shrubs, common, February to May, type of *S. Lemmoni* from the Santa Catalina Mountains (*Lemmon* 389), types of *S. decorticans* from along Salt River, near the Apache Trail (*Nelson* 10309, 11287). Southern Arizona and northern Baja California.

An apparently rayless form was collected in the Ajo Mountains, Pima County (*Nichol* in 1939).

21. Senecio Wootonii Greene (*S. percalvus* A. Nels.). Mountains of Coconino, southern Apache, Greenlee, Graham, Gila, Cochise, and Pima counties, 6,000 to 9,500 feet, coniferous forest, May to September, type of *S. percalvus* from Elden Mountain near Flagstaff, Coconino County (*Hanson* 1572). Colorado, New Mexico, Arizona, and Chihuahua.

22. Senecio Actinella Greene. Apache, Coconino, Greenlee, Gila, and Cochise counties, 5,500 to 9,500 feet, coniferous forest, May to August, type from near Flagstaff (*Rusby* 671). New Mexico, Arizona, and northern Mexico.

S. Actinella var. *mogollonicus* (Greene) Greenm., said to have larger leaves than the typical form, does not seem taxonomically distinct. It is also known from New Mexico.

23. Senecio arizonicus Greene. Oak Creek, Coconino County (*Deaver* 2063), near Prescott (Yavapai County), about 5,500 feet, April to July, type from Lynx Creek (*Rusby* in 1883). Southern New Mexico and central Arizona.

24. Senecio werneriaefolius Gray. Kaibab Plateau and San Francisco Peaks (Coconino County), 9,000 to 10,000 feet (and perhaps higher), June to August. South Dakota, Wyoming, Colorado, Utah, and northern Arizona.

A specimen from the San Francisco Peaks (*Darrow* in 1940) probably is referable to var. *incertus* Greenm., the heads being rayless.

25. Senecio hartianus Heller. Mountains of Apache, Navajo, Coconino, and Gila counties, 6,500 to 8,000 feet, May to August, type from Hart Spring, San Francisco Peaks (*MacDougal* 230). New Mexico and Arizona.

26. Senecio cynthioides Greene. White Mountains (Apache County), 7,500 to 9,500 feet, openings in pine forests, July and August. New Mexico and eastern Arizona.

Although the achenes were described as glabrous by Greenman (362, 5, p. 94), the young achenes in specimens collected by R. A. Darrow are puberulent.

27. Senecio neomexicanus Gray (*S. Toumeyi* Greene, *S. Blumeri* Greene, *S. Encelia* Greene, *S. papagonius* A. Nels.). White Mountains (Apache, Navajo, and Greenlee counties) and vicinity of Flagstaff (Coconino County), to the mountains of Cochise, Santa Cruz, and Pima counties, 3,000 to 9,000 feet, commonly in oak chaparral, sometimes in pine forest, April to August, the type of *S. Toumeyi* from the Chiricahua Mountains (*Toumey* in 1896), that of *S. Blumeri* from the Chiricahua Mountains (*Blumer* in 1907), that of *S. Encelia* from the Pinal Mountains (*Jones* in 1890), that of *S. papagonius* from near "Massacre Camp," between Ruby and the Tucson-Nogales highway (*A. & R. Nelson* 1494). Colorado, New Mexico, and Arizona.

One of the most widely distributed and abundant species of *Senecio* in Arizona. Specimens from the Apache-Verde road, east of Baker Butte (*Coville* 1043), referred to *S. mutabilis* Greene by Greenman (362, 5, p. 49), are not distinguishable from others referred by him to *S. neomexicanus*. *S. neomexicanus* var. *Griffithsii* Greenm. is a form with glabrous achenes (those of the typical form being hirtellous); the type was collected in the Santa Rita Mountains (*Griffiths* 4212).

A remarkable form, glabrescent and with oblanceolate basal leaves that are tridentate at apex, otherwise entire, was collected in the White Mountains (Apache County) between Nutrioso and Alpine (*Ripley & Barneby* 8431).

120. EURYOPS

Shrubs, the herbage glabrous except for woolly tufts in the axils; leaves thickish, very narrow, 1.5 cm. long or less, 3- to several-lobed at apex, the lobes narrowly linear, scarcely broader than the petioles; heads solitary on slender, often clustered peduncles; phyllaries few, broad, scarious-margined; flowers yellow, the rays 5 to 7; achenes 10-ribbed, villous; pappus of soft, strongly barbellate, quickly deciduous bristles.

***1. Euryops multifidus** (L. f.) DC. Planted by the United States Soil Conservation Service in the Tucson area and likely to become naturalized there; native of South Africa.

A plant about 1.2 m. (4 feet) in height and diameter was observed by L. M. Pultz near Continental (Pima County).

121. ARCTIUM. Burdock

Large, coarse, usually biennial herbs; stems branching; leaves very large, thin, broadly ovate, often cordate, entire or shallowly dentate, more or less tomentose beneath; heads in racemelike clusters, medium-sized, rayless; involucre burlike, the phyllaries numerous, very narrow, rigid, ending in stiff, hooked spines; achenes 3-angled, somewhat compressed; pappus of numerous, short, stiff, scabrous, distinct, readily deciduous bristles.

1. Arctium minus Schkuhr. Snowflake, Navajo County (*Thornber?* in 1917). A common weed of waste ground in many parts of the United States; introduced from Europe.

122. CARDUUS

Very similar to *Cirsium,* but the pappus bristles merely barbellate or essentially smooth. Plants biennial; stems 30 to 90 cm. high, winged interruptedly; head solitary, nodding, 3 to 5 cm. wide; corollas purple; phyllaries with coriaceous-herbaceous, spine-tipped, entire-margined appendages.

1. Carduus nutans L. Apache County, between Lukachukai Pass and Fort Defiance (*Eastwood & Howell* 6814), and 8 miles miles south of Greasewood Trading Post (*Goodman & Payson* 3174). Here and there in North America; naturalized from Europe. Musk-thistle.

123. CIRSIUM (363). THISTLE

Biennial or perennial herbs, often woolly, with spiny or prickly leaves and involucre; leaves alternate, toothed or lobed; heads medium or large, discoid, purple, pink, or red, rarely white or greenish yellow; involucre broad, many-seriate, the phyllaries, at least some of them, tipped with spines; receptacle densely bristly; achenes oblong or obovate; pappus of numerous plumose bristles or very narrow paleae, these united at base and deciduous in a ring.

The Navajo and Hopi Indians are reported to use thistles medicinally for various disorders.

KEY TO THE SPECIES

1. Plants dioecious; flowering stems from creeping rootstocks. Heads relatively small, less than 2.5 cm. high; phyllaries with small, weak prickles or nearly muticous, the inner ones with reflexed tips . . . . 1. *C. arvense*

1. Plànts not dioecious; stems not from creeping rootstocks (2).

2. Corollas greenish yellow; middle and inner phyllaries arachnoid-villous with long hairs, and with dilated, submembranaceous, lacerate-fringed tips. Heads usually several on each stem or branch . . . . 2. *C. Parryi*

2. Corollas purple, pink, red, or whitish, or if rarely yellowish (in *C. Drummondii*), then the involucre not as in *C. Parryi* (3).

3. Phyllaries more or less densely and persistently tomentose, the middle ones spreading, the outer ones reflexed . . . . 4. *C. neomexicanum*

3. Phyllaries glabrous or the margin hispidulous or somewhat arachnoid-tomentose, the outer ones not reflexed (4).

4. Prickles of the involucre very small (less than 2 mm. long), mostly appressed (mere cusps rather than prickles). Phyllaries very numerous, with a conspicuous glandular dorsal line; plants glabrate or glabrescent . . . . 5. *C. Wrightii*

4. Prickles of the phyllaries more than 2 mm. long (5).

5. Inner phyllaries with elongate, attenuate, plane, usually bright red or reddish (rarely purple) tips (6).

5. Inner phyllaries with usually more or less dilated and twisted, often erose tips (10).

6. Leaves glabrous or quickly glabrate on both faces (7).

6. Leaves persistently tomentose, at least beneath (8).

7. Spines of the middle phyllaries about 10 mm. long, longer than the body of the phyllaries 6. *C. Rothrockii*

7. Spines of the middle phyllaries not more than 5 mm. long, shorter than the body of the phyllaries . . . . 7. *C. bipinnatum*

8. Spines of the middle phyllaries 1 to 2 cm. long, stout, yellowish . . . . 8. *C. nidulum*

8. Spines of the middle phyllaries shorter, slender (9).

9. Corollas bright red or carmine . . . . 9. *C. arizonicum*

9. Corollas pink-purple . . . . 10. *C. pulchellum*

10. Phyllaries stiffly hispidulous-ciliolate (almost serrulate) on the margin. . 11. *C. Grahami*
10. Phyllaries not hispidulous-ciliolate on the margin, or slightly so in *C. Wheeleri* (11).
11. Heads (1 to 3) very large, 4 to 6 cm. thick; phyllaries broad, conspicuously graduated in many ranks; anthers with elongate, slender-subulate tips. Prickles of the leaves and involucre usually strong and stiff, about 1 cm. long or more, sometimes shorter 12. *C. ochrocentrum*
11. Heads usually several or smaller, or the involucre or the anthers not as in *C. ochrocentrum* (12).
12. Heads normally more than 1, crowded and subsessile at the apex of the stem; corollas not more than tinged with purple . . . . 3. *C. Drummondii*
12. Heads normally solitary at the tips of the stem and branches; corollas rosy purple, rarely ochroleucous (13).
13. Phyllaries with a comparatively broad, ovate or lance-ovate body . . . . 13. *C. Wheeleri*
13. Phyllaries with a narrower, lanceolate or lance-oblong body . . . . 14. *C. undulatum*

1. Cirsium arvense (L.) Scop. Flagstaff, Coconino County (*Thornber* in 1920), Prescott, Yavapai County (*Thornber* in 1936). Extensively naturalized in fields and waste ground in North America; introduced from Europe. Canada thistle.

The Arizona specimens belong to var. *mite* Wimm. & Grab., a less spiny plant and with less deeply laciniate leaves than in typical *C. arvense.*

2. Cirsium Parryi (Gray) Petrak. Between Lukachukai Pass and Fort Defiance, and in the White Mountains (Apache and Greenlee counties), San Francisco Peaks (Coconino County), Pinaleno Mountains (Graham County), 7,500 to 9,500 feet, openings in coniferous forests and grassy meadows, July to September. Colorado, New Mexico, and Arizona.

3. Cirsium Drummondii Torr. & Gray. Between Lukachukai Pass and Fort Defiance (Apache County), White Mountains (Apache and northern Greenlee counties), North Rim of Grand Canyon and San Francisco Peaks and vicinity (Coconino County), 7,000 to 9,500 feet, openings in coniferous forest, July to September. Saskatchewan to British Columbia, south to Arizona and California.

Flowers straw-colored, sometimes tinged with purple.

4. Cirsium neomexicanum Gray (*C. arcuum* A. Nels.). Coconino, Yavapai, and northern Mohave counties, south tc Cochise, Santa Cruz, Pima, and Yuma counties, 1,000 to 6,500 feet, plains, mesas, and foothills, common, March to September, type of *C. arcuum* from Canyon Lake, Maricopa County (*A. & R. Nelson* 1740). Colorado to Nevada, south to New Mexico, Arizona, and southern California.

Flowers lavender. *Cirsium utahense* Petrak, at least as represented in Arizona, does not seem distinct from *C. neomexicanum.*

5. Cirsium Wrightii Gray. San Bernardino, Cochise County (*Wright* 1290, the type collection). Western Texas to southeastern Arizona.

6. Cirsium Rothrockii (Gray) Petrak. Grand Canyon (Coconino County) and in southern Navajo, Gila, and Cochise counties, 4,000 to 6,500 feet, June to September, type from central Arizona (*Rothrock* 289). Utah (?) and Arizona.

Flowers rose-colored to carmine.

7. Cirsium bipinnatum (Eastw.) Rydb. Carrizo Mountains (Apache County), near Tuba (Coconino County), Pinal Mountains (Gila County), perhaps also in the Huachuca Mountains (Cochise County), 5,000 to 6,000 feet, June to August. Colorado, Utah, New Mexico, and Arizona.

8. Cirsium nidulum (Jones) Petrak (? *Carduus Rusbyi* Greene). Coconino County, base of Navajo Mountain, Cameron to Navajo Bridge, and on both rims of the Grand Canyon, 5,500 to 8,000 feet, perhaps also in the Santa Rita Mountains (Santa Cruz or Pima County), June and July, type of *C. Rusbyi* from "southern Arizona" (*Rusby* in 1883). Southern Utah, Nevada, and Arizona.

Specimens intermediate between this species and *C. Rothrockii* have been collected at the Grand Canyon (*MacDougal* 174, *A. E. Hitchcock* 68).

9. Cirsium arizonicum (Gray) Petrak. Flagstaff region (Coconino County), Hualpai Mountain (Mohave County), south to the mountains of Cochise, Santa Cruz, and Pima counties, 3,000 to 7,000 feet, frequent on rocky slopes, often in chaparral, May to October, type from between Babocomari, Arizona, and Santa Cruz, Sonora (*Wright* 1294). Utah (?) and Arizona.

In the Hopi Reservation, on the Kaibab Plateau, and at the Grand Canyon specimens intermediate between this species and *C. nidulum* have been collected. Species 6 to 9 are closely related, and intermediates, presumably of hybrid origin, are not infrequent.

10. Cirsium pulchellum (Greene) Woot. & Standl. Hopi Indian Reservation, Navajo County (*Zuck* in 1897), Tuba, Coconino County (*Harrison & King* 8709, *Shreve* 8953), perhaps also in Yavapai and Gila counties, 5,000 to 6,000 (4,500 to 8,000?) feet, June to August. Colorado, Utah, New Mexico, and Arizona.

Plants, apparently of this species, have been collected at Inscription House (Coconino County), in the Sierra Ancha (Gila County), near Patagonia (Santa Cruz County), and in the Santa Catalina Mountains (Pima County). In the specimens from the Sierra Ancha (*Gould & Hudson* 3777), however, the heads are only about 2.5 cm. wide and the corolla limb is little longer than the tube.

11. Cirsium Grahami Gray. White Mountains (Apache County), Huachuca Mountains (Cochise County), 6,500 to 7,500 feet, openings in pine forest, August to October, type from between the Sonoita and the San Pedro (*Wright* 1296). Southwestern New Mexico, eastern Arizona, and Sonora.

Flowers deep purple. What seems to be this species has been collected also near Flagstaff (Coconino County), in the Pinaleno Mountains (Graham County), and in the Santa Catalina Mountains (Pima County).

12. Cirsium ochrocentrum Gray. Apache, Navajo, Coconino, and northwestern Yavapai counties, south to Cochise, Santa Cruz, and Pima counties, 4,500 to 8,000 feet, open land and in the pinyon-juniper association, often abundant, May to October. Nebraska to Texas and Arizona.

Flowers cream-colored or rose-colored to deep carmine or purple. Arizona specimens that have been identified as *C. megacephalum* (Gray) Cockerell are included here.

13. Cirsium Wheeleri (Gray) Petrak (*C. Blumeri* Petrak). Apache, Navajo, Coconino, Yavapai, Graham, Gila, Cochise, and Pima counties, 5,000 to 9,000 feet, mostly in open pine forest, common, June to October, type of *C. Wheeleri* from near Fort Apache (*Rothrock* 293), type of *C. Blumeri* from Spud Ranch, Rincon Mountains (*Blumer* in 1910). Southern New Mexico and Arizona.

Flowers purple or lavender.

14. Cirsium undulatum (Nutt.) Spreng. Between Lukachukai Pass and Fort Defiance, and in the White Mountains (Apache County), Grand Canyon and Flagstaff region (Coconino County), Chiricahua and Huachuca mountains (Cochise County), Santa Catalina Mountains (Pima County), mostly 7,000 to 9,500 feet, June to October. Michigan to British Columbia, south to Texas and Arizona.

A thistle, apparently related to *Cirsium mohavense* (Greene) Petrak, but with very large lower leaves, up to 60 cm. long and 16 cm. wide, was collected in Havasu Canyon, western Coconino County (*Clover* 4277A).

Seedling plants collected recently in Havasu Canyon (*J. T. Howell* 26629) apparently are *Cirsium vulgare* (Savi) Airy-Shaw (*C. lanceolatum* (L.) Scop.), the bull thistle, an Old World species that is extensively naturalized in the United States. It differs from all of the foregoing species in having the leaves conspicuously decurrent upon the stem, forming interrupted spiny wings.

124. CYNARA. Artichoke, Globe-artichoke

Coarse perennials, whitish-woolly, especially on the lower leaf surface; leaves very large, deeply pinnatisect, with mostly lanceolate, entire or few-toothed, scarcely spiny lobes; heads large (about 6 to 10 cm. thick or more), the phyllaries many-seriate, coriaceous, ovate to oblong, not spiny, glabrous, the inner ones often purplish; receptacle densely setose; corollas purple; pappus of numerous, several-seriate, very narrow, plumose paleae, deciduous in a ring.

1. Cynara Scolymus L. Near Phoenix (Maricopa County), occasional along roadsides and in waste places, May to August. Occasionally or frequently escaping from cultivation in the western United States, most commonly in California, where the artichoke is grown extensively; native of Europe.

125. SILYBUM. Milk-thistle

Coarse, thistlelike herbs; leaves large, lobed or pinnatifid, white-veined and white-blotched; heads large (the disk 2.5 to 5 cm. wide), purple; involucre subglobose, the phyllaries several-seriate, rigid, with spreading, ovate or lanceolate, spinulose-margined and stiffly spine-tipped, coriaceous-herbaceous appendages.

1. Silybum Marianum (L.) Gaertn. Gila, Maricopa, Pinal, Cochise, and Pima counties, occasional at roadsides and in waste ground, May to September. Occasional in much of the United States and adjacent Canada, naturalized in California; native of Europe.

The plant has been used medicinally in Europe. A noxious field weed, difficult to exterminate.

126. CENTAUREA. STAR-THISTLE, KNAP-WEED

Herbs, with entire leaves or the lower ones pinnatifid; heads medium or large, pink, purple, blue, or yellow, rarely white; involucre strongly graduated, the phyllaries appendaged with spines or prickles, or with scarious or hyaline tips; outer flowers sometimes enlarged and falsely radiate; achenes with an oblique attachment; pappus various, setose, paleaceous, or wanting.

The introduced species are objectionable weeds in California, especially in grainfields, because of the prickly involucres. The plants are not yet sufficiently abundant in Arizona to be troublesome.

KEY TO THE SPECIES

1. Phyllaries tipped with a stiff, spreading spine or prickle, 5 mm. long or more, this with 2 or 3 pairs of small prickles at its base (2).
1. Phyllaries not spine-tipped or spine, if present, less than 1 mm. long and without smaller prickles at base (4).
2. Stem leaves pinnatifid, not decurrent; corollas purple; pappus none.....1. *C. Calcitrapa*
2. Stem leaves mostly entire, decurrent on the stem; corollas yellow; pappus present (3).
3. Terminal spine of the phyllary yellowish, comparatively stout, 1.2 to 2 cm. long; plants thinly tomentose ....2. *C. solstitialis*
3. Terminal spine or prickle of the phyllary normally purplish toward the base, rather slender, less than 1 cm. long; plants hispidulous (or occasionally thinly tomentose) 3. *C. melitensis*
4. Phyllaries essentially entire; outer corollas not enlarged, or only inconspicuously so. Corollas rosy purple (5).
4. Phyllaries pectinately fringed, at least toward apex; outer corollas much enlarged (6).
5. Lower leaves deeply pinnatifid, scabrous-pubescent, very much larger than the linear, spinulose-denticulate, upper leaves; involucre (dried) subglobose-ovoid, usually 1 cm. thick or more, the phyllaries firm, greenish white, essentially glabrous, blackish at extreme tip and usually tipped with a mucro less than 1 mm. long......4. *C. salmantica*
5. Lower leaves (mostly withered at flowering time) pinnatifid, more or less tomentose, not greatly larger than the upper, the latter linear, entire; heads small, the involucre less than 1 cm. thick; phyllaries with thin whitish hyaline tips, the outer phyllaries obtuse and essentially glabrous, the inmost long-pointed and densely pilose above..5. *C. Picris*
6. Herbage floccose-lanate, at least when young; heads relatively small, the involucre less than 2 cm. thick; outer phyllaries fringed to the base, or nearly so; corollas normally blue, sometimes white or pink..6. *C. Cyanus*
6. Herbage glabrous or nearly so; heads large, the involucre commonly 2 to 4 cm. thick; outer phyllaries fringed only on the terminal appendage; corollas rosy or purplish, rarely white (7).
7. Appendage of the phyllaries whitish (rarely brownish or purplish), with 4 to 6 pairs of not conspicuously ciliolate lobes, its undivided portion lanceolate......7. *C. americana*
7. Appendage deep brown, with 8 to 12 pairs of light-edged conspicuously ciliolate lobes, its undivided portion broadly triangular or ovate......................8. *C. Rothrockii*

1. Centaurea Calcitrapa L. Near Yuma, Yuma County (*Forbes* in 1906). Here and there in the United States in waste ground; introduced from Europe. Star-thistle.

2. Centaurea solstitialis L. Waste places, Yuma (*Thornber* in 1925, *Stitt* 1248), summer. An occasional weed in much of the United States, naturalized in California; native of Europe.

3. Centaurea melitensis L. Apache, Yavapai, Maricopa, Pinal, Cochise, and Pima counties, waste ground and open, rocky slopes, May to July. Occasional weed in much of the United States, abundant in the West; native of Europe.

A specimen with the spines of *C. melitensis* but with thinly tomentose stems was collected at Montezuma Well (*Schroeder* in 1948).

4. Centaurea salmantica L. Jerome, Yavapai County, a dooryard weed (*Goodding* 5–48, in 1948). Native of the Mediterranean region, known also as an occasional adventive in California.

5. Centaurea Picris Pall. Navajo, Yavapai, Greenlee, Cochise, and Pima counties, waste places, summer. Occasional or naturalized from Michigan and Missouri to Washington, Oregon, and California; native of eastern Europe and Asia.

Russian knap-weed, often a pernicious weed.

6. Centaurea Cyanus L. Tucson, Pima County (*Thornber* 7264), Yuma, Yuma County (*Blackledge* in 1938). An occasional escape from gardens; native of Europe. Cornflower, bluebottle, bachelor's button.

7. Centaurea americana Nutt. White Mountains (*Griffiths* 5398), Fort Apache, Navajo County (*Hough, Thornber* 4605), apparently rare in Arizona, July and August. Missouri and Louisiana to Kansas, eastern Arizona, and northern Mexico.

A very ornamental plant, sometimes cultivated under the name "American basket-flower."

8. Centaurea Rothrockii Greenm. Chiricahua and Huachuca mountains (Cochise County), 6,000 to 8,000 feet, along streams, July to October, type probably from the Chiricahua Mountains (*Rothrock* 527). Southwestern New Mexico and southeastern Arizona to Oaxaca.

Very similar to *C. americana* and equally handsome.

127. CNICUS. Blessed-thistle

Thistlelike annuals with branching stems; herbage soft-pubescent; leaves oblong, sinuate-dentate to pinnatifid, spiny-toothed; heads solitary at the ends of the branches, about 2.5 cm. long, almost hidden by the foliaceous outer bracts; phyllaries imbricate, the outer ones ovate and ending in a simple spine, the inner ones narrower, with tips pinnately spinose; achenes nearly terete, many-striate, 10-toothed at apex; pappus double, the outer of 10 stiff awns longer than the achene, the inner of 10 much shorter awns.

1. Cnicus benedictus L. Yuma (*Coglon* in 1925). Occasional in waste ground in the United States; introduced from Europe.

128. CHAPTALIA (364)

Scapose perennials; leaves pinnatifid or sinuate, glabrate and green above, thinly white-tomentose beneath; scapes several, up to 0.5 m. high, 1-headed; head 2 to 2.5 cm. high; involucre strongly graduated; outer flowers pistillate, with a strap-shaped, 3-toothed purplish corolla, the disk flowers hermaphrodite, whitish, more or less bilabiate; achenes fusiform, beaked; pappus of copious, capillary, whitish or brownish bristles.

KEY TO THE SPECIES

1. Leaves oval-lanceolate, lyrate (the terminal lobe much larger than the few lateral lobes), abruptly contracted into the petiole; peduncle enlarged at apex........1. *C. alsophila*
1. Leaves oblanceolate, crenate-denticulate to repand-crenate, tapering into the petiole; peduncle not noticeably enlarged................................2. *C. leucocephala*

1. Chaptalia alsophila Greene (*C. confinis* Greene, *C. sonchifolia* Greene). White Mountains (southern Apache County), Swisshelm, Chiricahua, and Huachuca mountains (Cochise County), Rincon Mountains (Pima County), 6,500 to 8,000 feet, coniferous forests, rare, May to October, the type of *C. confinis* from the Huachuca Mountains (*Lemmon* 2789), the type of *C. sonchifolia* from the Rincon Mountains (*Nealley* 223). Southwestern New Mexico, southeastern Arizona, and Mexico.

C. alsophila is the Southwestern representative of the widespread tropical and subtropical American *C. nutans* (L.) Polak., differing from that species primarily in the much shorter, stouter, densely hispidulous beak of its achene.

2. Chaptalia leucocephala Greene. White Mountains, southern Apache County, 8,500 to 9,000 feet (*Ripley & Barneby* 8447), Long Valley, Coconino County (*Peebles & Fulton* 9496), pine forest, May to August. Arizona and Mexico.

Burkart (364, p. 546) cited *C. monticola* Greene as a synonym of *C. leucocephala.*

129. PEREZIA (365)

Perennial herbs, with alternate, sessile and often clasping, spinulose-toothed leaves and lavender-pink heads, the stems woolly-tufted at base; involucre strongly graduated; flowers all hermaphrodite, bilabiate, the outer lip 3-toothed, the inner lip 2-parted; achenes subcylindric or fusiform, densely glandular or glandular-hispidulous; pappus of numerous scabrous bristles.

KEY TO THE SPECIES

1. Heads solitary, broad, about 15- to 24-flowered; leaves suborbicular or obovate-suborbicular, 2 to 5 cm. long and about as wide, coarsely and unequally spinulose-dentate; plants usually dwarf, sometimes up to 25 cm. high .... 1. *P. nana*

1. Heads numerous, panicled, narrow, 4- to 11-flowered; leaves mostly ovate, larger, normally much longer than wide, usually more evenly spinulose-denticulate; plants taller, 40 to 150 cm. high (2).

2. Phyllaries obtuse or merely acutish, not (or obscurely) glandular; heads 8- to 11-flowered 2. *P. Wrightii*

2. Phyllaries strongly acuminate, stipitate-glandular; heads 4- to 6-flowered . . 3. *P. Thurberi*

1. Perezia nana Gray. Southern Yavapai, Greenlee, Gila, Maricopa, Pinal, Cochise, and Pima counties, up to 6,000 feet, mesas and slopes, usually under bushes, March to June. Texas to Arizona and northern Mexico.

Desert-holly, an attractive little plant, making a good ground cover where sufficiently abundant. The flowers are delightfully fragrant with an odor of violets. The roots of this species and of *P. Wrightii* yield pipitzahoic acid, which may be used in chemical analysis as an indicator of alkalinity.

2. Perezia Wrightii Gray (*P. arizonica* Gray). Throughout most of the state, not above 6,000 feet, foothills and canyons, common, January to June (sometimes autumn), type of *P. arizonica* collected in Arizona without definite locality (*Palmer*). Western Texas to southern Utah, Arizona, and northern Mexico.

The pink flowers are honey-scented. The root was used as a styptic by the Indians.

3. Perezia Thurberi Gray. Mountains of Cochise, Santa Cruz, and Pima counties, 4,000 to 6,000 feet, slopes and canyons, June to October. Southwestern New Mexico and southern Arizona to central Mexico.

130. TRIXIS (366)

Shrubs; leaves alternate, lanceolate, densely sessile-glandular beneath; heads 9- to 12-flowered, yellow, solitary or clustered; involucre double, the outer series of several linear to elliptic herbaceous bractlets, the inner series of about 8 linear, acuminate phyllaries, these corky-thickened at base; corollas all 2-lipped, the outer lip 3-toothed, the inner lip 2-cleft; achenes densely hispidulous, subrostrate; pappus of numerous straw-colored bristles.

1. Trixis californica Kellogg. Grand Canyon (Coconino County), and from Greenlee and southern Yavapai counties to Cochise, Pima, and Yuma counties, up to 5,000 feet but usually lower, rocky slopes, flowering throughout the year. Western Texas to southern California and northern Mexico.

Browsed to some extent by cattle.

131. CICHORIUM. Chicory

Branched, perennial herbs; lowest leaves runcinate-pinnatifid, those of the stem clasping, toothed, those of the stiff branches minute; heads blue, rarely pink or white, sessile along the branches or at the tips of short, fistulose branchlets; involucre double, the outer series of about 5 short, ovate phyllaries, these corky-thickened at base in age, the inner series of 8 to 10 linear phyllaries; achenes obovoid, somewhat ribbed; pappus a short, toothed crown.

The well-known salad plant endive (*Cichorium Endivia*) belongs to this genus.

1. Cichorium Intybus L. Navajo, Yavapai, and Pima counties, occasional at roadsides and in waste ground, summer. A common weed in the United States and Canada; naturalized from Europe.

The leaves, especially of cultivated varieties, are cooked like spinach or eaten raw in salads; the root is eaten, under the names "barbe" and "witloof," and also furnishes one of the leading adulterants or substitutes for coffee. The large, bright-blue flower heads are very attractive.

132. ATRICHOSERIS

Glabrous, scapose, winter annuals, the scape branched above, several- or many-headed; leaves obovate, spinulose-toothed, 3 to 10 cm. long, often spotted; heads up to 3.5 cm. wide, white; involucre of about 12 to 15 equal, 2-seriate, lance-linear, narrowly scarious-margined phyllaries, with a few much shorter, ovate outer bractlets; achenes oblong, with more or less corky-thickened ribs, epappose.

1. Atrichoseris platyphylla Gray. Mohave and Yuma counties, up to 2,500 feet, sandy or stony slopes and mesas, March and April, type from near Fort Mohave (*Cooper*). Southwestern Utah and western Arizona to southeastern California.

133. MICROSERIS

Subscapose annuals; leaves narrowly linear and entire, or pinnatisect into narrow lobes; heads solitary; corollas yellow, often drying purplish; pappus of 5 linear-lanceolate, scarious, 1-nerved, bifid paleae, as long as the achene or longer, the midrib excurrent as a slender bristle shorter than the body of the palea.

1. **Microseris linearifolia** (DC.) Schultz Bip. Grand Canyon (Coconino County) and Mohave County to Greenlee (?), Cochise, Santa Cruz, and Pima counties, up to 5,000 feet but usually lower, common on plains, mesas, and foothills, March to June. Idaho and Washington to New Mexico, Arizona, and Baja California.

134. KRIGIA (367)

Slender, perennial herbs; leaves chiefly basal, oblanceolate or spatulate, entire or sinuate-dentate; stem leaves few, small, sessile and somewhat clasping; heads few, yellow; involucre of about 9 to 18 thin equal phyllaries; pappus of 10 to 15 small oblong squamellae, and an equal or larger number of much longer capillary bristles.

1. **Krigia biflora** (Walt.) Blake. Willow Spring (southern Apache County), Lakeside (southern Navajo County), Mogollon Escarpment (Coconino County), 6,000 to 7,500 feet, June. New York to Minnesota, south to Georgia, Texas, and Arizona.

135. ANISOCOMA

Low, scapose, winter annuals; leaves pinnatifid; scapes 1-headed; heads pale yellow; involucre strongly graduated, 1.2 to 2.7 cm. high, the phyllaries orbicular (the outer ones) to linear-oblong, broadly rounded to obtuse, thin, appressed, with a usually purplish midline and a pale scarious margin; pappus readily deciduous, of about 5 long and 5 much shorter, bright-white, plumose bristles, inserted within the apical border of the achene.

1. **Anisocoma acaulis** Torr. & Gray. Peach Springs to Kingman, Chloride, and Yucca (Mohave County), 2,000 to 4,000 feet, plains and mesas, March to May. Nevada, northwestern Arizona, and eastern California.

136. LEONTODON. Hawkbit

Low, scapose, perennials; leaves subentire to sinuate-pinnatifid, hispid; scapes 1-headed; head rather small, yellow; achenes fusiform, tapering into a slender beak equaling or shorter than the body, muriculate; pappus of the outer achenes of short, more or less united squamellae, that of the other achenes of about 10 paleaceous-based, long-plumose setae and about as many much shorter, naked setae.

1. **Leontodon nudicaulis** (L.) Banks. Sacaton, Pinal County, in a lawn (*Beckett* 13084), probably not established, summer. An occasional weed in the United States and Canada; introduced from Europe.

137. STEPHANOMERIA

Annual or perennial herbs; leaves linear to oblong, entire to pinnatifid, those on the upper part of the stem usually greatly reduced; heads small, usually panicled, rosy or flesh-colored; involucre of several equal phyllaries and some calyculate bractlets, or more regularly graduated; achenes columnar, 5-angled; bristles of the pappus 1-seriate, plumose at least above, often paleaceous toward the base, sometimes connate into groups.

The names "stick-weed" and "wire-lettuce" have been applied to these plants.

KEY TO THE SPECIES

1. Involucre 9 to 13 mm. high, 10- to 20-flowered (2).
1. Involucre 5 to 9 (rarely 10) mm. high, 3- to 9-flowered (3).
2. Stems branched from the base, more or less spreading, leafy throughout with runcinate-pinnatifid leaves, the branchlets near the heads with greatly reduced, ovate, spinulose-toothed leaves; pappus brownish-tinged, the bristles naked toward the base 1. *S. Parryi*
2. Stems erect, usually branched only above, the leaves at base of the stem runcinate-pinnatifid, the upper leaves much reduced, linear, entire, the heads thus naked-panicled; pappus bright white, the bristles plumose to the base .... 2. *S. Thurberi*
3. Plants annual (4).
3. Plants perennial (5).
4. Pappus of 4 to 6 lanceolate paleae, these plumose on the shorter aristiform tip, not longer than the minutely scabrous achene .... 3. *S. Schottii*
4. Pappus of 5 to 18 bristles, these plumose above the middle, longer than the tuberculate-rugose achene, often paleaceous-dilated at base .... 4. *S. exigua*
5. Pappus brownish-tinged, the bristles naked and merely scabrous toward the base 5. *S. pauciflora*
5. Pappus bright white, the bristles plumose to the base .... 6. *S. tenuifolia*

1. Stephanomeria Parryi Gray. Known in Arizona only by a collection at Kingman, Mohave County, 3,500 feet (*Lemmon* in 1884). Southwestern Utah, northwestern Arizona, and southeastern California, May and June.

2. Stephanomeria Thurberi Gray. Navajo, Coconino, Mohave, Yavapai, Greenlee, Cochise, and Santa Cruz counties, 4,000 to 8,000 feet, open forests of pine, pinyon, and juniper, April to August. New Mexico, Arizona, and northern Mexico.

3. Stephanomeria Schottii Gray. Known only from the original collection made by Schott on the Gila River, Arizona.

4. Stephanomeria exigua Nutt. Throughout the state, 2,000 to 8,000 feet, plains, mesas, and hillsides, often among shrubs, April to September. Wyoming to New Mexico, Arizona, and California.

It is reported that the Navajo Indians use this plant as a diuretic. In the typical phase the pappus consists of 9 to 18 bristles, these often more or less connate at base into about 5 groups. In var. *pentachaeta* (D. C. Eaton) H. M. Hall, which has been collected at Yucca (Mohave County) and near Tucson, the pappus consists of 5 to 7 distinct bristles.

5. Stephanomeria pauciflora (Torr.) A. Nels. Almost throughout the state, 150 to 7,000 feet, dry plains, mesas, and slopes, flowering throughout the year. Kansas to Texas, Arizona, and California.

The Hopi Indians, according to one authority, apply the plant both externally and internally to stimulate milk flow in women.

6. Stephanomeria tenuifolia (Torr.) H. M. Hall. Apache, Navajo, Coconino, and Yavapai counties, apparently also in the Chiricahua Mountains (*Peebles* 5882), 4,500 to 8,000 feet, May to September. Montana to Washington, south to Colorado, Arizona, and California.

Stephanomeria Wrightii Gray was doubtfully recorded by Gray (*Syn. Fl.* 1²: 414) from Arizona on the basis of material collected by Rusby. The identity of these specimens is uncertain, but they do not seem to belong to *S. Wrightii.*

138. RAFINESQUIA

Glabrous or obscurely puberulent branching annuals; leaves mostly pinnatifid; heads rather large, solitary at the tips of the branches and branchlets, white; involucre of about 7 to 15 equal, lanceolate, acuminate, scarious-margined phyllaries and of some much shorter, unequal, outer bractlets; achenes subfusiform, tapering into a beak; pappus of long-plumose setae.

KEY TO THE SPECIES

1. Corollas exceeding the involucre by about 5 mm.; achenes with a slender beak about as long as the body; pappus dull white or brownish white, the bristles plumose essentially to the apex with straight hairs....................................1. *R. californica*
1. Corollas exceeding the involucre by 10 mm. or more; achenes with a more gradually tapering and stouter beak shorter than the body; pappus bright white, the bristles not plumose near the apex, the hairs of the plume softer, subarachnoid..2. *R. neomexicana*

1. Rafinesquia californica Nutt. (*Nemoseris californica* Greene). Mazatzal Mountains (Gila-Maricopa counties), Santa Catalina, Baboquivari, and Ajo mountains (Pima County), probably elsewhere, 3,000 to 4,500 feet, March to May. Southwestern Utah to southern Arizona, California, and northern Baja California.

2. Rafinesquia neomexicana Gray (*Nemoseris neomexicana* Greene). White Mountains (Apache? County), and Mohave, Graham, Gila, Maricopa, Pinal, Pima, and Yuma counties, 200 to 3,000 feet (rarely higher), abundant on plains and mesas, February to July. Western Texas to southern Utah, southern California, and northern Baja California.

One of the conspicuous spring flowers of the more desert parts of the state, sometimes called desert-chicory.

139. TRAGOPOGON. Goats-beard

Biennial or perennial, nearly glabrous herbs; leaves elongate, grasslike, strongly nerved, entire, somewhat clasping; heads large, yellow or purple, solitary on often fistulose peduncles; involucre of 8 to 13 lanceolate, acuminate, equal phyllaries; achenes subfusiform, long-beaked, muricate; pappus of somewhat flattened plumose bristles.

KEY TO THE SPECIES

1. Flowers purple. Phyllaries surpassing the corollas....................1. *T. porrifolius*
1. Flowers yellow (2).
2. Phyllaries 8 or 9, equaling or shorter than the chrome-yellow corollas....2. *T. pratensis*
2. Phyllaries usually 10 to 13, much longer than the lemon-yellow corollas.....3. *T. dubius*

1. Tragopogon porrifolius L. Flagstaff (Coconino County), Prescott (Yavapai County), Santa Cruz Valley (Pima County), waste ground. Adventive here and there in the United States; native of Europe.

The well-known garden vegetable, salsify, oyster-plant, or vegetable-oyster.

2. Tragopogon pratensis L. Coconino County, where common in waste ground near Flagstaff and Williams, also at Lakeside (Navajo County), summer. Widely distributed in the United States and Canada; naturalized from Europe.

3. Tragopogon dubius Scop. Near Betatakin (Navajo County), Grand Canyon, Walnut Canyon, Flagstaff, and Oak Creek Canyon (Coconino County), Montezuma Well (Yavapai County), 3,500 to 7,000 feet, summer. Colorado to Idaho, New Mexico, and Arizona; adventive from Europe.

140. PINAROPAPPUS

Low, perennial herbs, sometimes scapose; leaves entire to runcinate-pinnatifid; heads solitary; receptacle bearing thin, narrow paleae; achenes subfusiform, tapering into a short beak; pappus of copious, brownish, capillary bristles.

1. Pinaropappus roseus Less. Chiricahua and Huachuca mountains (Cochise County), Santa Catalina Mountains (Pima County), 7,500 to 8,000 feet, open, grassy pine forests, July to October. Texas and Arizona, south to Guatemala.

The large heads are white, turning pink in drying. A collection in the Santa Catalina Mountains (*Thornber & Shreve*) is exceptionally broad-leaved.

141. MALACOTHRIX

Annual herbs; leaves toothed to pinnatisect; heads small or medium-sized, yellow or white; involucre of subequal phyllaries and calyculate, or else strongly graduated, the phyllaries narrowly or broadly scarious-margined; achenes columnar, truncate, ribbed; pappus of soft bristles, deciduous more or less in a ring, 1 to 8 of them (rarely none) stiffer and persistent, the achene often also crowned with a ring of minute teeth.

KEY TO THE SPECIES

1. Involucre 12 to 15 mm. high, strongly graduated, the phyllaries 3 to 4 mm. wide, with a linear green-and-purplish midline and a very broad scarious margin, all but the inner phyllaries suborbicular to oval; stem leaves oblong or elliptic to ovate, cordate-clasping, the upper leaves subentire to repand-dentate, rarely laciniate..........1. *M. Coulteri*

1. Involucre 5 to 12 mm. high, calyculate but scarcely graduated, the phyllaries 1.5 mm. wide or less, lanceolate to linear, acute or acuminate, with a narrow subscarious margin; leaves linear to oblong or ovate, mostly pinnatifid (2).

2. Leaf segments linear-filiform, less than 1 mm. wide....................2. *M. glabrata*

2. Leaf segments oblong or triangular to linear-lanceolate, comparatively short, usually toothed (3).

3. Ligules inconspicuous, little exceeding the involucre; heads numerous, panicled, small (the involucre 5 to 8 mm. high); achenes finely 15-ribbed, with 1 (rarely 2) persistent pappus bristles and a crown of minute, white setulose teeth..........3. *M. Clevelandi*

3. Ligules conspicuous, bright yellow, much exceeding the involucre; heads larger, mostly solitary or few at the tips of the branches and branchlets; pappus otherwise (4).

4. Achene cylindric, evenly 15-ribbed, 2 mm. long, the outer coat prolonged into a dark, truncate, unribbed, entire collar, this bearing inside below the apex a denticulate whitish ring and 0 to 2 persistent pappus bristles..........................4. *M. Fendleri*
4. Achene more or less 5-angled (5 of the 15 ribs stronger than the others, sometimes almost winglike, continuous to the apex of the achene or prolonged beyond it into knoblike teeth), the outer coat not prolonged into a collar (5).
5. Persistent bristles of the pappus 2 to 8, with minute teeth between them; 5 of the achene ribs much stronger than the others, almost winglike; phyllaries long-acuminate; achene 3 to 3.5 mm. long..5. *M. Torreyi*
5. Persistent bristles of the pappus none, the achene with a finely denticulate, whitish crown; stronger achene ribs not winglike; phyllaries merely acute or short-acuminate; achene about 2.5 mm. long...................................6. *M. sonchoides*

1. Malacothrix Coulteri Gray. Mohave, Maricopa, Pinal, and Pima counties, 500 to 3,500 feet, March and April. Southwestern Utah to southern Arizona and southern California; Argentina.

Sometimes called snakes-head.

2. Malacothrix glabrata Gray. Coconino and Mohave counties to Greenlee, Graham, Pima, and Yuma counties, up to 7,000 feet but usually much lower, common and abundant on plains and mesas, March to June. Idaho to Arizona and California.

A conspicuous element of the spring flora, sometimes called desert-dandelion.

3. Malacothrix Clevelandi Gray. Grand Canyon (Coconino County), Yucca (Mohave County), Burro Creek to Bagdad (Yavapai County), and Mazatzal Mountains (Gila-Maricopa counties), to the mountains of Pima County, 2,500 to 4,500 feet, mostly along streams, March to May. Southern Arizona and California.

4. Malacothrix Fendleri Gray. Navajo, Yavapai, Greenlee, Graham, Pinal, Cochise, Santa Cruz, and Pima counties, 2,000 to 5,000 feet, sandy plains, mesas, and rocky slopes, March to June. Western Texas to Arizona.

5. Malacothrix Torreyi Gray. Both sides of the Grand Canyon (Coconino County), southwest of Pipe Springs (Mohave County), 5,000 to 7,000 feet, May to July. Utah to Oregon, northern Arizona, and California.

6. Malacothrix sonchoides (Nutt.) Torr. & Gray. Navajo, Coconino, and Mohave counties, especially along the Little Colorado River, apparently also at Patagonia (Santa Cruz County), 1,500 to 6,000 feet, April to June. Nebraska to Idaho, Arizona, and California.

Malacothrix saxatilis var. *tenuifolia* (Nutt.) Torr. & Gray was reported by Gray (*Syn. Fl.* 1²: 423) from Arizona in 1884, but no material from the state has been seen by the author, and it is apparently not recorded by later writers. The locality of Gray's specimen was probably erroneus.

142. CALYCOSERIS

Low, branching, winter annuals, glabrous except for conspicuous stipitate glands on the upper part of the stem and on the involucre; leaves pinnatisect into narrowly linear or filiform divisions; heads white, rosy, or yellow; involucre of equal phyllaries, calyculate; achenes fusiform, 5- or 6-ribbed, tapering into a short beak, this expanded at apex into a shallow, denticulate cup; pappus of numerous hispidulous, white bristles, deciduous in a ring.

KEY TO THE SPECIES

1. Flowers white or rose-colored; achenes tuberculate, not very deeply sulcate between the ribs, dark-colored ....1. *C. Wrightii*
1. Flowers bright yellow; achenes minutely rugulose, not tuberculate, very deeply sulcate between the ribs, light gray..2. *C. Parryi*

1. Calycoseris Wrightii Gray. Mohave County to Greenlee, Graham, Pinal, Cochise, Pima, and Yuma counties, 500 to 4,000 feet, common on plains, mesas, and rocky slopes, March to May. Western Texas to Utah, Arizona, southern California, and northern Mexico.

One of the handsomest of Arizona spring flowers, conspicuous in the more desert areas.

2. Calycoseris Parryi Gray. Grand Canyon (Coconino County), throughout Mohave County, Cottonwood Creek and Bumblebee (Yavapai County), mostly about 3,000 feet, March and April. Southern Utah, Arizona, and southern California.

143. GLYPTOPLEURA

Dwarf, depressed winter annuals; leaves pinnatifid, with a toothed, white, crustaceous margin; heads white or pale yellow, turning pink in drying; involucre of about 7 to 12 equal, lanceolate, scarious-margined phyllaries, with a calyculus of several spatulate bractlets, these crustaceous-margined above and lacerate-toothed or pinnatifid; achenes oblong or columnar, 5-ribbed, cancellate-rugose, at apex produced into a thick, 5-lobed, cuplike border, from which is exserted an abrupt, short beak, this dilated at apex to bear the pappus; pappus copious, soft, white, capillary, deciduous.

1. Glyptopleura setulosa Gray. Fredonia (Coconino County), near Littlefield, Short Creek, Hackberry, and Fort Mohave (Mohave County), 500 (?) to 5,000 feet, in sandy soil, April to June. Southern Utah and northwestern Arizona to southern California.

Glyptopleura marginata D. C. Eaton has been reported from Arizona, but no specimens have been seen by the writer. In *G. marginata* the crustaceous margin of the leaves is broad, with short teeth, the bractlets of the calyculus are pinnatifid above and lacerate-fringed most of the way to the base, and the ligules are short and little exserted. In *G. setulosa* the crustaceous leaf margin is produced into slender teeth longer than the margin itself, the bractlets of the calyculus are lacerate-fringed at the conspicuously dilated apex and naked or merely sparsely setose below it, and the ligules are long-exserted.

144. TARAXACUM (368). DANDELION

Scapose perennial herbs; leaves all basal, runcinate-pinnatifid or sometimes merely toothed or sinuate-lobed; heads solitary, large, yellow, on hollow scapes; involucre double, the outer phyllaries much shorter than the inner ones, often recurved, the inner phyllaries 1-seriate, erect; achenes more or less fusiform, 4- or 5-ribbed, muricate above, prolonged into a slender beak, this bearing the simple capillary pappus.

The common dandelion (*T. officinale*) is often used for greens as a substitute for spinach, and large-leaved cultivated varieties have been developed in Europe. The roots, reputed to have medicinal properties, are sometimes eaten as a salad.

Propagation in this genus is partly or wholly parthenogenetic, which results in the production and preservation in nature of a multitude of closely similar but distinguishable forms. The material available from Arizona is rather scanty, and the treatment here given is tentative. The possible existence of a fourth species is indicated by a dwarf specimen from the San Francisco Peaks (*Little* 4635), but the material available is not sufficient for identification in the present knowledge of the Western species of the genus.

KEY TO THE SPECIES

1. Achenes blackish, the murications comparatively short and blunt; outer phyllaries ovate, appressed or erect, not recurved....................................1. *T. lyratum*
1. Achenes greenish or reddish; outer phyllaries recurved-spreading (2).
2. Achenes bright red or reddish brown, the murications toward the apex very sharp and comparatively long ....2. *T. laevigatum*
2. Achenes greenish, the murications less acute and shorter................3. *T. officinale*

1. Taraxacum lyratum (Ledeb.) DC. San Francisco Peaks, Coconino County (*Knowlton* 142). Greenland and northern Canada, south to Colorado, Utah, and northern Arizona; Asia.

2. Taraxacum laevigatum (Willd.) DC. Lukachukai and White mountains (Apache County), Betatakin (Navajo County), Tuba and Kaibab Plateau to Oak Creek (Coconino County), 5,000 to 11,500 feet, March to August. Nova Scotia to British Columbia, south to Virginia, New Mexico, northern Arizona, and California; naturalized from Europe.

3. Taraxacum officinale Weber. Apache, Navajo, and Coconino counties to Cochise, Santa Cruz, and Pima counties, 2,500 to 9,000 feet, lawns and roadsides, April to September. Northern Canada to Mexico; probably naturalized from Europe.

145. SONCHUS. Sow-thistle

Coarse, weedy annuals, with subentire to pinnatifid, spinulose-toothed leaves and medium-sized, irregularly cymose-panicled, yellow heads; lower leaves usually petioled, the upper ones sessile and strongly clasping; involucre more or less regularly graduated, the phyllaries thin, corky-thickened at base in age; achenes strongly flattened, several-ribbed, not beaked; pappus copious, of soft, white capillary bristles.

KEY TO THE SPECIES

1. Achenes strongly 3- (5-) ribbed on each face, thin-margined, not transversely wrinkled; auricles of the leaf base rounded...1. *S. asper*
1. Achenes striate and also strongly wrinkled transversely, not thin-margined; auricles of the leaf base acute...2. *S. oleraceus*

1. Sonchus asper (L.) Hill. Apache and Coconino counties to Cochise, Santa Cruz, Pima, and Yuma counties, 150 to 8,000 feet, roadsides and waste ground, February to August. An abundant weed nearly throughout North America; naturalized from Europe.

2. Sonchus oleraceus L. Coconino and Mohave counties to Graham, Cochise, Pima, and Yuma counties, roadsides and waste places, March to September. An abundant weed in most parts of North America; naturalized from Europe.

A gum obtained by evaporation of the juice of this plant is said to be a powerful cathartic, and it has been used as a so-called cure for the opium habit.

146. LACTUCA. Lettuce

Annual or perennial herbs, nearly or quite glabrous, leafy-stemmed; leaves variable, from linear and entire to oblong and pinnatifid; heads small or medium-sized, panicled, yellow or blue; involucre rather slender, more or less strongly graduated; achenes strongly flattened, abruptly or gradually beaked; pappus copious, of soft, white capillary bristles.

The garden lettuce is *L. sativa.* A sedative—lactucarium—is obtained from *L. virosa.* Sheep sometimes feed on *L. pulchella,* but the plant is reputed to be slightly poisonous.

KEY TO THE SPECIES

1. Achene lanceolate or lance-oblong, strongly several-ribbed on each face, gradually narrowed into a short, stoutish beak; flowers blue........................1. *L. pulchella*
1. Achene oval or oval-oblong, abruptly narrowed into a slender beak sometimes as long as the body; flowers yellow or purplish (2).
2. Achene small, the body 3 mm. long or less, light gray, about 5-nerved on each side, finely hispidulous above, not rugulose, the beak very slender, longer than the achene; leaves stiffly spinulose on the margin and usually on the midrib beneath......2. *L. Serriola*
2. Achene larger, the body 4 to 6 mm. long, dark brown or blackish, 1-nerved on each side, not hispidulous, finely transverse-rugulose; leaves not stiffly spinulose (3).
3. Achene with a body about 6 mm. long, the beak about half as long; leaves mostly narrowly linear and entire, the lower ones often broader and pinnatifid..3. *L. graminifolia*
3. Achene with a body about 4 mm. long, the beak about as long; leaves all obovate or oblong, at least the lower ones pinnatifid, the upper ones often not lobed
4. *L. ludoviciana*

1. Lactuca pulchella (Pursh) Riddell. Apache, Navajo, Coconino, and Cochise counties, 6,000 to 7,500 feet, summer. Saskatchewan to British Columbia, south to Missouri, New Mexico, Arizona, and California.

2. Lactuca Serriola L. (*L. Scariola* L.). Navajo and Coconino counties, south to Greenlee, Cochise, Pima, and Yuma counties, 1,000 to 7,000 feet, waste land and roadsides, May to September. Abundant in southern Canada and the United States; naturalized from Europe.

Prickly lettuce, a common weed of roadsides and waste ground. Two forms occur commonly—the typical plant, with pinnatifid leaves, and forma *integrifolia* Bogenhard (*L. Scariola* var. *integrata* Gren. & Godr.), with unlobed leaves.

3. Lactuca graminifolia Michx. Apache, Coconino, Graham, Gila, Cochise, Santa Cruz, and Pima counties, 5,000 to 8,000 feet, May to October. South Carolina to Florida, west to Arizona.

4. Lactuca ludoviciana (Nutt.) DC. Southern Coconino, Yavapai, Greenlee, Gila, and Pima counties, 3,000 to 5,500 feet, June to August. Minnesota to Missouri, west to Colorado, Texas, and central Arizona.

147. LYGODESMIA

Annual or perennial herbs, essentially glabrous or merely puberulent, sometimes spinescent; leaves entire or toothed; heads very small to large, the corollas pink or rosy; involucre slender, of few equal phyllaries and a calyculus; achenes subcylindric or linear-prismatic, few-ribbed, not beaked; pappus of numerous capillary bristles, stiffish or soft.

KEY TO THE SPECIES

1. Branches spinescent, rigid, divaricate; stems with tufts of brown wool at base; lower leaves linear, entire, about 3 cm. long or less, the upper ones reduced to scales ... 1. *L. spinosa*
1. Branches not spinescent; stem without tufts of wool at base; leaves otherwise (2).
2. Leaves grasslike, linear, entire, conspicuous, up to 10 cm. long; involucre 18 to 20 mm. high; achenes at least 10 mm. long; plants perennial .... 2. *L. grandiflora*
2. Basal leaves obovate or oblanceolate, repand-toothed or lobed, those of the stem reduced and inconspicuous; involucre about 5 mm. high; achenes about 3 mm. long; plants annual .... 3. *L. exigua*

1. Lygodesmia spinosa Nutt. Coconino County, from the Little Colorado River to Flagstaff and Fredonia, 5,000 to 7,500 feet, August and September. Montana to British Columbia, northern Arizona, and California.

2. Lygodesmia grandiflora (Nutt.) Torr. & Gray. Apache, Navajo, and Coconino counties, 5,000 to 6,500 feet, common on sandy plains, often with grasses, May and June. Wyoming to Idaho, south to New Mexico and central Arizona.

The bright-pink heads are large for the size of the plant. It is reported that the Hopi Indians boil the leaves with meat and with mush and regard the plant as stimulative of milk flow in women.

3. Lygodesmia exigua Gray (*Stephanomeria minima* Jones). Navajo, Coconino, Mohave, Pima, and Yuma counties (probably elsewhere), 1,000 to 5,500 feet, hills and mesas, March to June, the type of *Stephanomeria minima* from Fredonia, Coconino County (*M. E. Jones* in 1929). Colorado to Texas, Arizona, and California.

148. AGOSERIS. Mountain-dandelion

Perennial or rarely annual, scapose herbs; leaves narrow, entire to pinnatifid; heads solitary, medium-sized, on long, naked scapes; involucre usually graduated; receptacle naked; corollas yellow, orange, or purple; achenes subfusiform, ribbed, smooth, beaked; pappus of soft, white, capillary bristles.

The genus is much in need of a thorough revision, and the present key is only tentative.

KEY TO THE SPECIES

1. Plants annual; achene 3 to 4 mm. long, the beak about twice as long; involucre pilose or villous with many-celled, often gland-tipped hairs .... 1. *A. heterophylla*
1. Plants perennial; achene larger; involucre usually glabrous or glabrate, the hairs, when present, not gland-tipped (2).
2. Beak of the achene comparatively stout, nerved throughout, much shorter than the body ... 2. *A. glauca*
2. Beak of the achene slender, not nerved throughout (3).
3. Beak of the achene shorter than the body; corollas orange, becoming purple in age ... 3. *A. aurantiaca*
3. Beak of the achene as long as or longer than the body; corollas light yellow, often becoming purplish in age or on drying .... 4. *A. arizonica*

1. Agoseris heterophylla (Nutt.) Greene. Yavapai, Cochise, and Pima counties, 2,500 to 5,000 feet, plains, mesas, and canyons, March to May. British Columbia to New Mexico, Arizona, California, and northwestern Mexico.

2. Agoseris glauca (Pursh) D. Dietr. Apache County to northeastern Mohave County, south to Graham and Pima counties, 6,500 to 10,000 feet, meadows and open coniferous forests, May to October. British Columbia, south to New Mexico, Arizona, and Nevada.

KEY TO THE VARIETIES

1. Leaves mostly pinnatifid, often retrorsely so. Phyllaries glabrous or nearly so ..... var. *laciniata*
1. Leaves mostly entire or subentire (2).
2. Phyllaries glabrous or nearly so; leaves less than 1 cm. wide, linear or linear-lanceolate ..... var. *parviflora*
2. Phyllaries ciliate and more or less pilose; leaves mostly about 2 cm. wide, lanceolate or oblanceolate ..... var. *dasycephala*

Var. *laciniata* (D. C. Eaton) Smiley, the commonest variety in Arizona, occurs in Apache, Coconino, Mohave, and Yavapai counties. Var. *parviflora* (Nutt.) Rydb. is found chiefly in the southeastern counties, but also in the White Mountains, on the Kaibab Plateau, and near Flagstaff. The presence in Arizona of var. *dasycephala* (Torr. & Gray) Jepson rests upon the uncertain evidence of a collection by E. Palmer in 1869 labeled as from "Arizona" without definite locality, but specimens from the White Mountains (*Goodding* 550, *Phillips* 3201) approach this variety.

3. Agoseris aurantiaca (Hook.) Greene (*A. purpurea* (Gray) Greene). Kaibab Plateau to near Flagstaff (Coconino County), White Mountains (Apache County), near Prescott (Yavapai County), between Pine and Fossil Creek (Gila County), Chiricahua Mountains (Cochise County), Santa Catalina Mountains (Pima County), 5,000 to 9,500 feet, grassy slopes, meadows, and open pine forests, June to August. Alberta and British Columbia to New Mexico and northern Arizona.

4. Agoseris arizonica Greene. Apache, Coconino, Yavapai, Graham (?), Gila, and Pima counties, 5,500 to 11,000 feet, mostly in open pine forests, especially in the Flagstaff region, April to August. Wyoming to New Mexico and Arizona.

Doubtfully distinct from *A. aurantiaca,* or at any rate difficult to separate in herbarium specimens, which mostly lack ripe fruit, and almost never have the color of the flowers noted.

149. PYRRHOPAPPUS. False-dandelion

Nearly glabrous perennials, several-stemmed, the stems sparsely leafy, somewhat branched; leaves entire to pinnatifid, mostly in a basal rosette; heads few, long-peduncled, rather large, yellow; phyllaries narrowly lanceolate, corniculate below the tip, with a calyculus of narrow bractlets less than half as long; achenes subfusiform, about 5-sulcate, tapering into a slender beak longer than the body; pappus copious, of soft brown hairs, surrounded at base by a short villous ring.

1. Pyrrhopappus multicaulis DC. (*P. Rothrockii* Gray). White Mountains (Apache and Navajo counties), Huachuca Mountains (Cochise County), near Canelo (Santa Cruz County), Arivaca (Pima County), 3,500 to 7,000 feet, in moist, sometimes saline soil, apparently rare, June to September, the type of *P. Rothrockii* from Fisch's Ranch, southern Arizona (*Rothrock* 699). Texas to Arizona and Mexico.

150. CREPIS (369, 370). Hawks-beard

Low, perennial herbs, leafy-stemmed or scapose, glabrous to tomentose; leaves mostly in a basal rosette, entire to pinnatifid; heads several or numerous, medium-sized, cymose or panicled, yellow; involucre of narrow, equal phyllaries and some calyculate bractlets; achenes columnar or fusiform, 10- to 20-ribbed; pappus copious, of soft, white, capillary bristles.

KEY TO THE SPECIES

1. Plants glabrous and glaucous, scapose or with a single stem leaf...........1. *C. glauca*
1. Plants more or less pubescent, glandular or tomentose; stem more or less leafy (2).
2. Heads thick-cylindric (5 to 9 mm. thick at anthesis), normally 12- to 25-flowered; plants low, usually less than 30 cm. high; involucre tomentulose and often glandular-hispid; heads usually few..2. *C. occidentalis*
2. Heads slender-cylindric (3 to 5 mm. thick at anthesis), normally 5- to 15-flowered; plants usually taller, more than 30 cm. high; involucre tomentulose or glabrous; heads usually numerous (3).
3. Involucre normally glabrous except for some ciliolation on the outer bractlets; principal phyllaries 5 to 7; heads 5- to 10-flowered...........................3. *C. acuminata*
3. Involucre tomentulose; principal phyllaries 8; heads 10- to 15-flowered..4. *C. intermedia*

1. Crepis glauca (Nutt.) Torr. & Gray (*C. chamaephylla* Woot. & Standl., *C. runcinata* Torr. & Gray ssp. *glauca* (Nutt.) Babcock & Stebbins). Apache and Navajo counties, about 5,000 feet, sometimes in saline soil, June to August, the type of *C. chamaephylla* from the Carrizo Mountains (*Standley* 7419). South Dakota, Saskatchewan, and Idaho, south to Colorado and northeastern Arizona.

A specimen from the White Mountains, Apache County (*Gould & Robinson* 5062a), has been identified by Babcock as *C. runcinata* ssp. *Barberi* (Greenm.) Babc. & Stebb., and another specimen (*Griffiths* 5354) from the same region was formerly identified by Babcock and Stebbins as intermediate between their *C. runcinata* ssp. *glauca* and ssp. *Barberi*. Ssp. *Barberi* differs from ssp. *glauca* in its narrower leaves (0.6 to 1.5 cm. wide) and longer involucres (14 to 16 instead of 8 to 12 mm. high) with longer outer phyllaries. In the Gould & Robinson specimen it further differs in having a slightly tomentulose involucre and leaves that are hispidulous above.

2. Crepis occidentalis Nutt. Keam Canyon (Navajo County), Grand Canyon and Flagstaff to Ash Fork (Coconino County), 6,000 to 7,000 feet, open ground, June. Saskatchewan to British Columbia, south to northern New Mexico, northern Arizona, and California.

Babcock has identified *Eastwood & Howell* 955 from North Rim of Grand Canyon as ssp. *costata* (Gray) Babc. & Stebb., which differs from the typical form principally in having 10- to 14-flowered heads with 8 phyllaries (in the typical form the heads are 18- to 30-flowered, with 10 to 13 phyllaries in the larger heads).

3. Crepis acuminata Nutt. Tunitcha Mountains, Apache County, 9,000 feet (*Goodman & Payson* 2918), "Arizona" without definite locality (*Palmer* in 1869), July and August. Montana to Washington, south to Colorado, northern Arizona, and California.

The species normally has a glabrous involucre. Babcock and Stebbins (369, p. 177) cited a specimen from "Jacobs Pools," Coconino County (*Jaeger* in 1926) as related to their apomictic form *nevadensis,* which has the inner phyllaries lightly tomentulose.

4. Crepis intermedia Gray. Grand Canyon, Coconino County, June and July. Colorado to Alberta, south to northern Arizona and California.

A specimen collected by Goodding in 1916 has exceptionally thick heads. It was identified by Babcock as apm. *arizonica* Babc. & Stebb.

151. HIERACIUM (371). HAWK-WEED

Perennial herbs, sometimes scapose; leaves linear to oblong or obovate, entire or toothed; heads small, yellow, rarely white or rosy; involucre of several equal phyllaries and a few short calyculate bractlets; achenes columnar, or sometimes narrowed upward but not beaked, about 10-ribbed; pappus of brownish or white capillary bristles.

KEY TO THE SPECIES

1. Basal leaves villous-lanate...1. *H. Pringlei*
1. Basal leaves pilose, hirsute, or hispid, not villous-lanate (2).
2. Plants somewhat glaucous, glabrous except for the long-pilose basal leaves and base of the stem, and the sometimes sparsely ciliate lower stem leaves; stem leafy to the inflorescence, the leaves elongate, narrowly linear; corollas rose-colored..2. *H. carneum*
2. Plants usually not glaucous, usually hairy on the stem; stem leaves when present not elongate or narrowly linear; corollas not rose-colored (3).
3. Stem (scape) leafless, or if with 1 or 2 small leaves or bracts, then these not at all clasping; achenes tapering from near the base. Stems rather sparsely setose..3. *H. Fendleri*
3. Stems decidedly leafy; leaves with a broad, more or less clasping base, mostly oblong; achenes columnar, not or scarcely tapering (4).
4. Stems copiously to densely long-hirsute, at least below; leaves remotely denticulate with callose teeth; involucre 8 to 12 mm. high, glabrous, except at base....4. *H. Lemmoni*
4. Stems rather sparsely hirsute below, glabrous above; leaves entire or nearly so; involucre 6 to 8 mm. high, hirsutulous and glandular-puberulent................5. *H. Rusbyi*

1. Hieracium Pringlei Gray. Huachuca Mountains (Cochise County), Santa Rita Mountains (Santa Cruz County), probably about 6,000 feet, June to September, type from the Santa Rita Mountains (*Pringle*). Southwestern New Mexico and southeastern Arizona.

2. Hieracium carneum Greene. Chiricahua and Huachuca mountains (Cochise County), Patagonia Mountains (Santa Cruz County), Rincon and Santa Catalina mountains (Pima County), 6,000 to 7,000 feet, July to September. Southwestern Texas to southeastern Arizona and adjacent Mexico.

3. Hieracium Fendleri Schultz Bip. (*H. Fendleri* var. *discolor* Gray). Springerville (Apache County), Navajo Mountain and Kaibab Plateau (Coconino County), southward to the mountains of Graham, Cochise, and Pima counties, 6,000 to 9,500 feet, mostly in pine forests, May to August. South Dakota to New Mexico, Arizona, and Mexico.

The commonest and most widely distributed species of hawk-weed in Arizona. *H. Fendleri* var. *discolor* was based on specimens collected in the Santa Rita and Huachuca mountains by J. G. Lemmon and C. G. Pringle. The var. *mogollense* Gray, with narrower leaves, shorter hairs of the herbage, and bright-white pappus, has been collected in the White Mountains, Apache County (*Gould & Robinson* 5004).

4. **Hieracium Lemmoni** Gray. Mountains of Cochise and Pima counties, 7,000 to 8,000 feet, July to October, type from Cave Canyon, Huachuca Mountains (*Lemmon*). Southwestern New Mexico and southeastern Arizona.

5. **Hieracium Rusbyi** Greene. Pinaleno Mountains, Graham County, 9,000 feet, on a dry slope (*Pultz* 1146), Chiricahua Mountains, Cochise County, 6,000 to 8,000 feet (*Kusche* in 1927). Western Texas to southeastern Arizona.

Literature Consulted

Note: Mention should be made of two publications that were not available in time to be consulted in writing this book but are sure to be useful to future students of the flora. These are: the second edition of Lyman Benson's "The Cacti of Arizona," issued in 1950 (see reference No. 191a); and "Grasses of Southwestern United States," by Frank W. Gould (*University of Arizona Biol. Sci. Bulletin* No. 7), issued in 1951.

1. Ewan, Joseph, "Bibliography of the Botany of Arizona." *Amer. Midland Nat.* 17: 430–454 (1936).
2. Coville, Frederick V., "The Botanical Explorations of Thomas Coulter in Mexico and California." *Bot. Gaz.* 20: 519–531 (1895).
3. McVaugh, Rogers, and Thomas H. Kearney, "Edward Palmer's Collections in Arizona in 1869, 1876, and 1877. *Amer. Midland Nat.* 29: 768–778 (1943).
4. Shreve, Forrest, "The Vegetation of Arizona." In Kearney, Thomas H., and Robert H. Peebles, *Flowering Plants and Ferns of Arizona.* U. S. Dept. Agric. Misc. Publ. 423 (1942), pp. 10–23.
5. Little, Elbert L., Jr., "Alpine Flora of San Francisco Mountain, Arizona." *Madroño* 6: 65–81 (1941).
6. Clover, Elzada U., and Lois Jotter, "Floristic Studies in the Canyon of the Colorado River and Tributaries." *Amer. Midland Nat.* 32:591–642 (1944).
7. MacDougal, W. B., *Plants of Grand Canyon National Park.* Bul. 10 Grand Canyon Nat. Hist. Assoc. (1947).
8. Benson, Lyman, and Robert A. Darrow, *Manual of Southwestern Desert Trees and Shrubs.* Bul. 15 University of Arizona (1944), illustrated.
9. Phillips, Walter S., "A Check-List of the Ferns of Arizona." *Amer. Fern Jour.* 36: 97–108 (1946); 37: 13–20, 39–51 (1947).
10. Pfeiffer, N. E., "Monograph of the Isoëtaceae." *Ann. Mo. Bot. Gard.* 9: 79–232 (1922).
11. Clausen, Robert T., "A Monograph of the Ophioglossaceae." *Mem. Torrey Bot. Club* 19: 1–177 (1938).
12. Svenson, H. K., "The New World Species of Azolla." *Amer. Fern Jour.* 34: 69–84 (1944).
13. Johnston, Ivan M., "Plants of Coahuila, Eastern Chihuahua, and Adjoining Zacatecas and Durango." *Jour. Arnold Arbor.* 24: 306–339, 375–421 (1943). (*Pteridophyta* by C. A. Weatherby).
14. Tryon, R. M., Jr., "Revision of the Genus Pteridium." *Rhodora* 43: 1–31, 37–67 (1941). (*Contrib. Gray Herb.* 134).
15. Wiggins, Ira L., "Distributional Notes on and a Key to the Species of Cheilanthes in the Sonoran Desert and Certain Adjacent Regions." *Amer. Fern Jour.* 29: 59–69 (1939).
16. Sudworth, George B., *The Pine Trees of the Rocky Mountain Region.* U. S. Dept. Agr. Bul. 460. 47 pp. illus. (1917).
17. Sudworth, George B., *The Spruce and Balsam Fir Trees of the Rocky Mountain Region.* U. S. Dept. Agr. Bul. 327 (1916).
18. Marco, H. F., "Needle Structure as an Aid in Distinguishing Colorado Blue Spruce from Engelmann Spruce." *Bot. Gaz.* 92: 446–449 (1931).

19. Van Dersal, William R., *Native Woody Plants of the United States.* U. S. Dept. Agr. Misc. Pub. 303 (1938).

20. Wolf, Carl B., and Willis E. Wagener, "The New World Cypresses." *El Aliso* 1: 1–444 (1948).

21. Sudworth, George B., *The Cypress and Juniper Trees of the Rocky Mountain Region.* U. S. Dept. Agr. Bul. 207 (1915).

22. Whiting, Albert F., "Juniperus of the Flagstaff Region." *Plateau* 15: 23–32 (1942).

23. Little, Elbert L., Jr., "Older Names for Two Western Species of Juniperus." *Leafl. West. Bot.* 5: 125–132 (1948).

24. Morton, C. V., "Notes on Juniperus." *Rhodora* 43: 344–348 (1941).

25. Fassett, Norman C., "Juniperus virginiana, J. horizontalis and J. scopulorum." *Bul. Torrey Bot. Club* 71: 410–418 (1944).

26. Cutler, Hugh Carson, "Monograph of the North American Species of the Genus Ephedra." *Ann. Mo. Bot. Gard.* 26: 373–428 (1939).

27. Benson, Lyman, "Revision of Status of Southwestern Trees and Shrubs. I. Ephedra." *Amer. Jour. Bot.* 30: 230–233 (1943).

28. Hotchkiss, Neil, and Herbert L. Dozier, "Taxonomy and Distribution of North American Cat-Tails." *Amer. Midland Nat.* 41: 237–254 (1949).

29. Fernald, M. L., "The Linear-Leaved North American Species of Potamogeton, Section Axillares." *Mem. Amer. Acad.* 17: pt. 1. (1932).

30. Ogden, E. C., "The Broad-Leaved Species of Potamogeton of North America North of Mexico." *Rhodora* 45: 57–105, 119–163, 171–214 (1943).

31. Howell, John Thomas, "Remarks on Triglochin concinna." *Leafl. West. Bot.* 5: 13–19 (1947).

32. Fernald, M. L., "The North American Representatives of Alisma Plantago-aquatica." *Rhodora* 48: 86–88 (1946).

33. Smith, Jared G., "North American Species of Sagittaria and Lophotocarpus" (1894). Reprinted in *Ann. Rpt. Mo. Bot. Gard.* 6: 27–64 (1895).

34. Hitchcock, A. S., *Manual of the Grasses of the United States.* U. S. Dept. Agric. Misc. Publ. 200 (1935). 2d ed., revised by Agnes Chase (1950).

35. Church, George L., "A Cytotaxonomic Study of Glyceria and Puccinellia." *Amer. Jour. Bot.* 36: 155–165 (1949).

36. Reeder, John R., "The Status of Distichlis dentata." *Bul. Torrey Bot. Club* 70: 53–57 (1943).

37. Boyle, W. S., "A Cyto-Taxonomic Study of the North American Species of Melica." *Madroño* 8: 1–26 (1945).

38. Chase, Agnes, "Enneapogon Desvauxii and Pappophorum Wrightii, an Agrostological Detective Story." *Madroño* 8: 187–189 (1946).

39. Covas, Guillermo, "Taxonomic Observations on the North American Species of Hordeum." *Madroño* 10: 1–21 (1949).

40. Swallen, Jason R., "The Awnless Annual Species of Muhlenbergia." *Contrib. U. S. National Herbarium* 29: 203–208 (1947).

41. Johnson, B. Lennart, "Cyto-Taxonomic Studies in Oryzopsis." *Bot. Gaz.* 107: 1–32 (1945).

42. Johnson, B. Lennart, "Natural Hybrids between Oryzopsis and Several Species of Stipa." *Amer. Jour. Bot.* 32: 599–608 (1945).

43. Beetle, Alan A., *The Genus Aegopogon Humb. & Bonpl.* Univ. Wyom. Publ. 13: 17–23 (1948).

44. Chase, Agnes, "The North American Species of Paspalum." *Contrib. U. S. National Herbarium* 28: 1–310 (1929).

45. Hitchcock, A. S., and Agnes Chase, "The North American species of Panicum." *Contrib. U. S. National Herbarium* 15: 1–396 (1910).

46. Cutler, Hugh C., and Edgar Anderson, "A Preliminary Survey of the Genus Tripsacum." *Ann. Mo. Bot. Gard.* 28: 249–269 (1941).

47. Friedland, Solomon, "The American Species of Hemicarpha." *Amer. Jour. Bot.* 28: 855–861 (1941).

48. O'Neill, Hugh T., "Cyperus." In Tidestrom, Ivar, and Sister Teresita Kittell, *A Flora of Arizona and New Mexico.* pp. 761–767 (1941).

49. McGivney, Sister M. Vincent de Paul, *A Revision of the Subgenus Eucyperus Found in the United States.* Cath. Univ. Amer. Biol. Ser. No. 26 (1938).

50. Horvat, Sister Mary L., *A Revision of the Subgenus Mariscus Found in the United States.* Cath. Univ. Amer. Biol. Ser. No. 33 (1941).

51. Corcoran, Sister Mary L., *A Revision of the Subgenus Pycreus in North and South America.* Cath. Univ. Amer. Biol. Ser. No. 37 (1941).

52. Beetle, Alan A., "Studies in the Genus Scirpus L." *Amer. Jour. Bot.* 27: 63–64 (1940). *Ibid.* 28: 469–476, 691–700 (1941). *Ibid.* 29: 82–88, 653–656 (1942). *Ibid.* 30: 395–401 (1943).

53. Svenson, H. K., "Monographic Studies in the Genus Eleocharis." *Rhodora* 31: 121–135, 152–163, 167–191, 199–219, 224–242 (1929). *Ibid.* 34: 192–203, 215–227 (1932). *Ibid.* 36: 377–389 (1934). *Ibid.* 39: 210–231, 236–273 (1937). *Ibid.* 41: 1–19, 43–77, 90–110 (1939).

54. Pennell, Francis W., "The Genus Commelina (Plumier) in the United States." *Bul. Torrey Bot. Club* 43: 96–111 (1916).

55. Anderson, E., and R. E. Woodson, "The Species of Tradescantia Indigenous to the United States." *Contrib. Arnold Arbor.* 9: 1–132 (1935).

56. Johnston, Ivan M., "Plants of Coahuila, Eastern Chihuahua, and Adjoining Zacatecas and Durango." *Jour. Arnold Arbor.* 25: 43–83, 133–182, 431–453 (1944).

57. Ownbey, Marion, *The Genus Allium in Arizona.* Research Studies State College Wash. 15: 211–232 (1947).

58. Hoover, Robert F., "A Definition of the Genus Brodiaea." *Bul. Torrey Bot. Club* 66: 161–166 (1939).

59. Hoover, Robert F., "The Genus Dichelostemma." *Amer. Midland Nat.* 24: 463–476 (1940).

60. Hoover, Robert F., "A Systematic Study of Triteleia." *Amer. Midland* Nat. 25: 73–100 (1941). (Genus *Triteleiopsis,* pp. 98–100).

61. Beetle, Dorothy E., "A Monograph of the North American Species of Fritillaria." *Madroño* 7: 133–159 (1944).

62. Ownbey, Marion, "A Monograph of the Genus Calochortus." *Ann. Mo. Bot. Gard.* 27: 371–560 (1940).

63. McKelvey, Susan Delano, *Yuccas of the Southwestern United States.* Part 1, pp. 1–150 (1938). Part 2, pp. 1–192 (1947). (Published by the Arnold Arboretum of Harvard University).

64. Munz, Philip A., "Let's Save the Short-Leaved Joshua Tree." *National Parks Mag* 22: 8–12 (1948).

65. Webber, John Milton, "The Navajo Yucca, a New Species from New Mexico." *Madroño* 8: 105–110 (1945).

66. Trelease, William, "The Desert Group Nolineae." *Proc. Amer. Phil. Soc.* 50: 404–426 (1911).

67. Galway, Desma H., "The North American Species of Smilacina." *Amer. Midland Nat.* 33: 644–666 (1945).

68. Ownbey, Ruth P., "The Liliaceous Genus Polygonatum in North America." *Ann. Mo. Bot. Gard.* 31: 373–413 (1944).

69. Brackett, A., "Revision of the American Species of Hypoxis." *Rhodora* 25: 120–163 (1923).

70. Mulford, A. I., "A Study of the Agaves of the United States." *Ann. Rpt. Mo. Bot. Gard.* 7: 47–100 (1896).

71. Trelease, William, "Agave." In Standley, Paul C., "Trees and Shrubs of Mexico." *Contrib. U. S. National Herbarium* 23: 107–142 (1920).

72. Foster, Robert C., "A Cyto-Taxonomic Survey of the North American Species of Iris." *Contrib. Gray Herb.* 119: 1–80 (1937)

73. Foster, Robert C., "A Revision of the North American Species of Nemastylis Nutt." *Contrib. Gray Herbarium* 155: 26–44 (1945).

74. Correll, Donovan S., "The Genus Habenaria in Western North America." *Leafl. West. Bot.* 3: 233–247 (1943).

75. Sudworth, George B., *Poplars, Principal Tree Willows, and Walnuts of the Rocky Mountain Region.* U. S. Dept. Agr. Tech. Bul. 420 (1934).

76. Goodding, Leslie N., *Willows in Region VIII: Notes on Their Classification, Distribution, and Present Significance, with Suggestions for Their Use in Erosion-Control.* U. S. Dept. Agr. Soil Conserv. Serv. Regional Bul. 65 (1940). (Multigraphed).

77. Ball, Carleton R., "New Varieties and Combinations in Salix." *Jour. Wash. Acad. Sci.* 28: 443–452 (1938)

77a. Ball, Carleton R., "New Combinations in Southwestern Salix." *Jour. Wash. Acad. Sci.* 40: 324–335 (1950).

78. Hermann, F. J., "The Perennial Species of Urtica in the United States East of the Rocky Mountains." *Amer. Midland Nat.* 35: 773–778 (1946).

79. Trelease, William, *The Genus Phoradendron, a Monographic Revision.* Univ. Illinois Bul. 13: 1–224 (1916).

80. Fosberg, F. Raymond, "Notes on Mexican Plants." *Lloydia* 4: 274–280 (1941).

81. Gill, L. S., "Arceuthobium in the United States." *Trans. Conn. Acad. Arts and Sci.* 32: 111–245 (1935).

82. Goodman, George Jones, "A Revision of the North American Species of Chorizanthe." *Ann. Mo. Bot. Gard.* 21: 1–102 (1934).

83. Stokes, Susan G., *The Genus Eriogonum, a Preliminary Study Based on Geographic Distribution.* (1936).

84. Howell, John Thomas, "Eriogonum angulosum and Related Species." *Leafl. West Bot.* 4: 270–279 (1946).

85. Fosberg, F. Raymond, "Eriogonum Abertianum and Its Varieties." *Madroño* 4: 189–194 (1938).

86. Rechinger, K. H., Jr., *The North American Species of Rumex.* Field Mus. Bot. Ser. 17: 1–151 (1937).

87. Small, J. K., "A Monograph of the North American Species of the Genus Polygonum." *Bot. Mem. Columbia Univ.* 1: 1–183 (1895).

88. Aellen, P., "Beitrag zur Systematik der Chenopodium-arten Amerikas." *Repert. Sp. Nov.* 26:31–64, 119–160 (1929).

89. Aellen, P., and T. Just, "Key and Synopsis of the American Species of Chenopodium L." *Amer. Midland Nat.* 30: 47–76 (1943).

90. Hall, Harvey M., and Frederic E. Clements, *The Phylogenetic Method in Taxonomy. Genus Atriplex.* Carnegie Inst. Wash. Pub. 326: 235–346 (1923).

91. Standley, Paul C., "Allioniaceae of the United States, with Notes on Mexican Species." *Contrib. U. S. National Herbarium* 12: 303–389 (1909).

92. Poellnitz, Karl von, "Monographie der Gattung Talinum Adans." *Repert. Sp. Nov.* 35: 1–34 (1934).

93. Maguire, Bassett, "Great Basin Plants III. Caryophyllaceae." *Madroño* 6: 22–26 (1941).

94. Maguire, Bassett, "Studies in the Caryophyllaceae I." *Bul. Torrey Bot. Club* 73: 326 (1946).

95. Maguire, Bassett, "Studies in the Caryophyllaceae II," *Madroño* 8: 258–263 (1946).

96. Maguire, Bassett, "Studies in the Caryophyllaceae III. A Synopsis of the North American Species of Arenaria, Sect. Eremogone Fenzl." *Bul. Torrey Bot. Club.* 74: 38–56 (1947).

97. Rossbach, Ruth P., "Spergularia in North and South America." *Rhodora* 42: 57–83, 105–143, 158–193, 203–213 (1940).

98. Wiggins, Ira L., "The Genus Drymaria in and Adjacent to the Sonoran Desert." *Proc. Calif. Acad. Sci.* ser. 4, 25: 189–213 (1944).

99. Core, Earl L., "The North American Species of Paronychia." *Amer. Midland Nat.* 26: 369–397 (1943).

100. Hitchcock, C. Leo, and Bassett Maguire, *A Revision of the North American Species of Silene.* Univ. Wash. Publ. Biol. 13: 1–73 (1947).

101. Munz, Philip A., "Aquilegia, the Cultivated and Wild Columbines." *Gentes Herbarum* 7: 1–150 (1946).

102. Ewan, Joseph, *A Synopsis of the North American Species of Delphinium.* Univ. Colo. Studies. Ser. D. Physical and Biol. Sci. 2: 55–244 (1945).

103. Erickson, Ralph O., "Taxonomy of Clematis Section Viorna." *Ann. Mo. Bot. Gard.* 30: 1–60 (1943).

104. Benson, Lyman, "A Treatise on the North American Ranunculi." *Amer. Midland Nat.* 40: 1–261 (1948).

105. Boivin, Bernard, "American Thalictra and Their Old World Allies." *Rhodora* 46: 337–377, 391–445, 453–487 (1944).

106. Ownbey, Gerald Bruce, "Monograph of the North American Species of Corydalis." *Ann. Mo. Bot. Gard.* 34: 187–258 (1947).

107. Rollins, Reed C., "The Cruciferous Genus Stanleya." *Lloydia* 2: 109–127 (1939).

108. Payson, Edwin Blake, "A Monographic Study of Thelypodium and Its Immediate Allies." *Ann. Mo. Bot. Gard.* 9: 233–324 (1922).

109. Hitchcock, C. Leo, "The Genus Lepidium in the United States." *Madroño* 3: 265–320 (1936).

110. Payson, Edwin Blake, *The Genus Thlaspi in North America.* Univ. Wyom. Publ. Bot. 1:145–163 (1926).

111. Maguire, Bassett, "Great Basin Plants VII. Cruciferae." *Amer. Midland Nat.* 27: 463–469 (1942).

112. Payson, Edwin Blake, *Species of Sisymbrium Native to North America North of Mexico.* Univ. Wyom. Publ. Bot. 1: 1–27 (1922).

113. Rollins, Reed C., "Generic Revisions in the Cruciferae: Halimolobos." *Contrib. Dudley Herbarium* 3: 241–265 (1943).

114. Rollins, Reed C., "The Cruciferous Genus Dryopetalon." *Contrib. Dudley Herbarium* 3: 199–207 (1941).

115. Rollins, Reed C., "A Revision of Lyrocarpa." *Contrib. Dudley Herbarium* 3: 169–173 (1941).

116. Rollins, Reed C., "The Cruciferous Genus Physaria." *Rhodora* 41: 392–415 (1939).

117. Payson, Edwin Blake, "Monograph of the Genus Lesquerella." *Ann. Mo. Bot. Gard.* 8: 103–236 (1921).

118. Hitchcock, C. Leo, *A Revision of the Drabas of Western North America.* Univ. Wash. Publ. Biol. 11: 1–132 (1941).

119. Detling, LeRoy E., "A Revision of the North American Species of Descurainia." *Amer. Midland Nat.* 22: 481–520 (1939).

120. Rollins, Reed C., "A Monographic Study of Arabis in Western North America." *Rhodora* 43: 289–325, 348–411, 425–481 (1941).

121. Payson, Edwin Blake, *A Synoptical Revision of the Genus Cleomella.* Univ. Wyo. Publ. Bot. 1: 29–46 (1922).

122. Clausen, Robert T., and Charles H. Uhl, "Revision of Sedum Cockerellii and Related Species." *Brittonia* 5: 33–46 (1943).

123. Moran, Reid, "Graptopetalum Bartramii in Chihuahua." *Desert Plant Life* 21: 53–56 (1949).

124. Rosendahl, C. O., F. K. Butters, and O. Lakela, *A Monograph of the Genus Heuchera.* Minn. Studies Plant Sci. 2: 1–180 (1936).

125. Hitchcock, C. Leo, "The Xerophilous Species of Philadelphus in Southwestern North America." *Madroño* 7: 35–56 (1943).

126. Wolf, Carl B., *California Plant Notes.* Occasional Papers Rancho Santa Ana Bot. Gard. 1: 31–43 (1935).

127. Ley, Arline, "A Taxonomic Revision of the Genus Holodiscus (Rosaceae)." *Bul. Torrey Bot. Club* 70: 275–288 (1943).

128. Jones, George Neville, "A Synopsis of the North American Species of Sorbus." *Jour. Arnold Arboretum* 20: 1–43 (1939).

129. McVaugh, Rogers, *Rosaceae of Nevada. Contributions toward a Flora of Nevada.* Bur. Plant Ind. U. S. Dept. Agric. processed No. 22: 1–100 (1942). (pp. 90–96).

130. Clokey, Ira W., "Notes on the Flora of the Charleston Mountains, Clark County, Nevada VI." *Madroño* 8: 56–61 (1945).

131. Jones, George Neville, *American Species of Amelanchier.* Univ. Illinois Biol. Monog. 20: 1–100 (1946).

132. Bailey, L. H., "Species Batorum." *Gentes Herbarum* 5: 1–932 (1941–1945).

133. Keck, David D., "Revision of Horkelia and Ivesia." *Lloydia* 1: 75–142 (1938).

134. Clausen, Jens, David D. Keck, and William M. Hiesey, *Potentilla glandulosa and Its Allies. Potentilla gracilis and Its Allies.* Publ. Carnegie Inst. Wash. 520: 26–175 (1940).

135. Howell, John Thomas, "Studies in Rosaceae, Tribe Potentilleae. 1. A Reconsideration of the Genus Purpusia." *Leafl. West. Bot.* 4: 171–175 (1945).

136. Kearney, Thomas H., "A New Cliff-Rose from Arizona." *Madroño* 7: 16–18 (1943).

137. Erlanson, Eileen Whitehead, "Experimental Data for a Revision of the North American Wild Roses." *Bot. Gaz.* 96: 197–259 (1934).

138. Wiggins, Ira L., "Acacia angustissima (Mill.) Kuntze and Its Near Relatives." *Contrib. Dudley Herbarium* 3: 227–239 (1942).

139. Benson, Lyman, "Revisions of Status of Southwestern Desert Trees and Shrubs. IX. Acacia." *Amer. Jour. Bot.* 30: 237–238 (1943).

139a. Turner, B. L., "Mexican Species of Desmanthus (Leguminosae)." *Field and Lab.* 18: 119–130 (1950).

140. Burkart, A., "Materiales para una monografía del género Prosopis (Leguminosae)." *Darwiniana* 4: 57–128 (1940).

141. Benson, Lyman, "The Mesquites and Screwbeans of the United States." *Amer. Jour. Bot.* 28: 748–754 (1941).

142. Hopkins, Milton, "Cercis in North America." *Rhodora* 44: 193–211 (1942).

143. Johnston, Ivan M., "Parkinsonia and Cercidium." *Contrib. Gray Herbarium* 70: 61–68 (1924).

144. Fisher, E. M., "Revision of the North American Species of Hoffmanseggia." *Contrib. U. S. National Herbarium* 1: 143–150 (1892).

145. Larisey, Mary M., "A Revision of the North American Species of the Genus Thermopsis." *Ann. Mo. Bot. Gard.* 27: 245–258 (1940).

146. Senn, Harold A., "The North American Species of Crotalaria." *Rhodora* 41: 317–367 (1939).

147. Smith, C. P., *A Distributional Catalog of Lupinus in Arizona. Species Lupinorum,* Paper 10 (1939–1940).

148. Martin, James S., "Notes on Trifolium." *Bul. Torrey Bot. Club* 73: 366–369 (1946).

149. Ottley, Alice M., "The American Loti with Special Consideration of a Proposed New Section, Simpeteria." *Brittonia* 5: 81–123 (1944).

150. Palmer, E. J., "Conspectus of the Genus Amorpha." *Jour. Arnold Arboretum* 12: 157–220 (1931).

151. Wiggins, Ira L., "Taxonomic Notes on the Genus Dalea Juss. and Related Genera as Represented in the Sonoran Desert." *Contrib. Dudley Herbarium* 3: 41–64 (1940).

151a. Wood, Carroll E., Jr., "The American Barbistyled Species of Tephrosia." *Rhodora* 51: 193–231, 233–302, 305–364, 369–384 (1949).

152. Jones, Marcus E., *Revision of North American Species of Astragalus* (1923).

153. Barneby, R. C., "Pugillus Astragalorum." I. *Leafl. West. Bot.* 3: 97–144 (1942). II. *Proc. Calif. Acad. Sci.* ser. 4, 25: 147–169 (1944). III. *Leafl. West. Bot.* 4: 49–63 (1944). IV. *Ibid.* 4: 65–147 (1945). V. *Ibid.* 4: 228–238 (1946). VI. *Ibid.* 5: 1–9 (1947). VII. *Amer. Midland Nat.* 37: 421–516 (1947). VIII. *Leafl. West. Bot.* 5: 25–35 (1947). IX. *Ibid.* 5: 82–89 (1948). X. *Amer. Midland Nat.* 41: 496–502 (1949).

154. Beath, O. A., C. S. Gilbert, and H. F. Eppson, "The Use of Indicator Plants in Locating Seleniferous Areas in Western United States." III. "Further Studies." *Amer. Jour. Bot.* 27: 564–573 (1940).

155. Kearney, Thomas H., "Leaf Hairs of Astragalus." *Leafl. West. Bot.* 4: 223–227 (1946).

156. Rollins Reed C., "Studies in the Genus Hedysarum in North America." *Rhodora* 42: 217–239 (1940).

157. Schubert, Bernice G., "Desmodium: Preliminary Studies." *Contrib. Gray Herbarium* 129: 3–31 (1940).

158. Krukoff, B. A., "The American Species of Erythrina." *Brittonia* 3: 205–337 (1939).

159. Krukoff, B. A., "Supplementary Notes on the American Species of Erythrina." I. *Amer. Jour. Bot.* 28: 683–691 (1941). II. *Bul. Torrey Bot. Club* 70: 633–637 (1943).

160. Piper, C. V., "Studies in American Phaseolineae." *Contrib. U. S. National Herbarium* 22: 663–701 (1926).

161. Jones, G. Neville, and Florence Freeman Jones, "A Revision of the Perennial Species of Geranium of the United States and Canada." *Rhodora* 45: 5–26, 32–53 (1943).

162. Muller, Cornelius H., "A Revision of Choisya." *Amer. Midland Nat.* 24: 729–742 (1940).

163. Greene, Edward L., "The Genus Ptelea in the Western and Southwestern United States and Mexico." *Contrib. U. S. National Herbarium* 10: 49–79 (1906).

164. Bullock, A. A., "Notes on the Mexican Species of the Genus Bursera." *Kew Bul. Misc. Inf.* 1936: 346–387 (1936).

165. Wheeler, L. C., "Pedilanthus and Cnidoscolus Proposed for Conservation." *Contrib. Gray Herbarium* 124: 47–52 (1939).

166. Croizat, Leon, "A Study of Manihot in North America." *Jour. Arnold Arboretum* 23: 216–225 (1942).

167. Norton, J. B. S., "North American Species of Euphorbia Section Tithymalus." 60 pp. pl. 11–52. 1899. (preprint from *Mo. Bot. Gard. Ann. Rpt.* 11: 85–144. 1900).

168. Wheeler, Louis C., "Revision of the Euphorbia polycarpa Group of the Southwestern United States and Adjacent Mexico." *Bul. Torrey Bot. Club* 63: 397–416, 429–450 (1936).

169. Wheeler, Louis Cutter, "Euphorbia Subgenus Chamaesyce in Canada and the United States Exclusive of Southern Florida." *Rhodora* 43: 97–154, 168–205, 223–286 (1941).

170. Barkley, Fred Alexander, "A Monographic Study of Rhus and Its Immediate Allies in North and Central America." *Ann. Mo. Bot. Gard.* 24: 265–498 (1937).

171. Ensign, Margaret, "A Revision of the Celastraceous Genus Forsellesia (Glossopetalon)." *Amer. Midland Nat.* 27: 501–511 (1942).

172. St. John, Harold, "Nomenclatorial Changes in Glossopetalon." *Proc. Biol. Soc. Wash.* 55: 109–112 (1942).

173. Keller, Allan C., "Acer glabrum and Its Varieties." *Amer. Midland Nat.* 27: 491–500 (1942).

174. Sherff, Earl Edward, "Some Additions to the Genus Dodonaea L. (Fam. Sapindaceae)." *Amer. Jour. Bot.* 32: 202–214 (1945).

175. Wolf, Carl B., *The North American Species of Rhamnus.* Monog. Rancho Santa Ana Bot. Gard. Bot. Ser. 1: 1–136 (1938).

176. McMinn, Howard E., "A Systematic Study of the Genus Ceanothus." In *Ceanothus* (Santa Barbara Botanic Garden) 131–308 (1942).

177. Bailey, L. H., "Grapes of North America." *Gentes Herbarum* 3: 151–244 (1934).

178. Kearney, Thomas H., *The North American Species of Sphaeralcea Subgenus Eusphaeralcea.* Univ. Calif. Publ. Bot. 19: 1–128 (1935).

179. Wiggins, Ira L., "A Resurrection and Revision of the Genus Iliamna Greene." *Contrib. Dudley Herbarium* 1: 213–229 (1936).

180. Roush, Eva M. Fling, "A Monograph of the Genus Sidalcea." *Ann. Mo. Bot. Gard.* 18: 117–244 (1931).

181. Hochreutiner, B. P. G., "Monographia Generis Anodae." *Ann. Conserv. et Jard. Bot. Genève* 20: 29–68 (1916).

182. Harvey, Margaret, "A Revision of the Genus Fremontia." *Madroño* 7: 100–110 (1943).

183. Fassett, Norman C., "Elatine and Other Aquatics." *Rhodora* 41: 367–376 (1939).

184. Fernald, M. L., "Elatine americana and E. triandra." *Rhodora* 43: 208–211 (1941).

184a. McClintock, Elizabeth, "Studies in California Plants 3. The Tamarisks." *Jour. Calif. Hort. Soc.* 12: 76–83 (1951).

185. Sprague, T. A., "A Revision of Amoreuxia." *Kew Bul. Misc. Inf.* 1922: 97–105 (1922).

186. Morton, C. V., "The Genus Hybanthus in Continental North America." *Contrib. U. S. National Herbarium*, 29: 74–82 (1944).

187. Baird, Viola Brainerd, *Wild Violets of North America*, pp. 1–225 (1942).

188. Killip, Ellsworth P., *The American Species of Passifloraceae.* Field Museum Nat. Hist. Bot. Ser. 19: 1–613 (1938).

189. Darlington, Josephine, "A Monograph of the Genus Mentzelia." *Ann. Mo. Bot. Gard.* 21: 103–226 (1934).

190. Britton, N. L., and J. N. Rose, *The Cactaceae.* Carnegie Inst. Wash. Publ. 248 (1919–1923).

191. Baxter, Edgar Martin, *California Cactus* (1935).

191a. Benson, Lyman. "The Cacti of Arizona." *Univ. Arizona Biol. Sci. Bul.* 5: 1–134 (1940).

192. Marshall, W. Taylor, and Thor Methven Bock, *Cactaceae* (1941).

192a. Clover, Elzada U., and Lois Jotter, "Cacti of the Canyon of the Colorado River and Tributaries." *Bul. Torrey Bot. Club* 68: 409–419 (1941).

193. Benson, Lyman, "A Revision of Some Arizona Cactaceae." *Proc. Calif. Acad. Sci.* ser. 4, 25: 245–267 (1944).

194. Croizat, Leon, "Navajoa, a New Genus in Cactaceae." *Cactus & Succ. Jour.* 15: 88–89 (1943).

195. Munz, Philip A., "Studies in Onagraceae XII. A Revision of the New World Species of Jussiaea." *Darwiniana* 4: 177–284 (1942).

196. Munz, Philip A., "Studies in Onagraceae XIII. The American Species of Ludwigia." *Bul. Torrey Bot. Club* 71: 152–165 (1944).

197. Hilend, Martha, "A Revision of the Genus Zauschneria." *Amer. Jour. Bot.* 16: 58–68 (1929).

198. Munz, Philip A., and C. Leo Hitchcock, "A Study of the Genus Clarkia, with Special Reference to Its Relationship to Godetia." *Bul. Torrey Bot. Club* 56: 181–197 (1929).

199. Hitchcock, C. Leo, "Revision of North American Species of Godetia." *Bot. Gaz.* 89: 321–361 (1930).

200. Munz, Philip A., "Studies in Onagraceae." 1. "A Revision of the Subgenus Chylismia of the Genus Oenothera." *Amer. Jour. Bot.* 15: 223–240 (1928). 2. "Revision of the North American Species of Subgenus Sphaerostigma." *Bot. Gaz.* 85: 233–270 (1928). 3. "A Revision of the Subgenera Taraxia and Eulobus." *Amer. Jour. Bot.* 16: 246–257 (1929). 4. "A Revision of the Subgenera Salpingia and Calylophis." *Ibid.* 16: 702–715 (1929). 5. "The North American Species of the Subgenera Lavauxia and Megapterium." *Ibid.* 17: 358–370 (1930). 6. "The Subgenus Anogra." *Ibid.* 18: 309–327 (1931). 7. "The Subgenus Pachylophis." *Ibid.* 18: 728–738 (1931). 8. "The Subgenus Hartmannia." *Ibid.* 19: 755–765 (1932). 9. "The Subgenus Raimannia." *Ibid.* 22: 645-663 (1935).

200a. Munz, Philip A., "The Oenothera Hookeri Group." *El Aliso* 2: 1–47 (1949).

201. Munz, Philip A., "Studies in Onagraceae. VIII. The Genus Gayophytum." *Amer. Jour. Bot.* 19: 768–778 (1932).

202. Munz, Philip A., "Studies in Onagraceae. XI. A Revision of the Genus Gaura." *Bul. Torrey Bot. Club* 65: 105–122, 211–228 (1938).

203. Mathias, Mildred E., and Lincoln Constance, "Umbelliferae." *North American Flora* 28B: 43–160 (1944); 161–295 (1945).

204. Constance, Lincoln, and Ren Hwa Shan, *The Genus Osmorhiza.* Univ. Calif. Publ. Bot. 23: 111–156 (1948).

205. Mathias, Mildred E., and Lincoln Constance, "A Synopsis of the American Species of Cicuta." *Madroño* 6: 145–151 (1942).

206. Mathias, Mildred E., "Studies in the Umbelliferae III. A Monograph of Cymopterus Including a Critical Study of Related Genera." *Ann. Mo. Bot. Gard.* 17: 213–476 (1930).

207. Mathias, Mildred E., "A Revision of the Genus Lomatium." *Ann. Mo. Bot. Gard.* 25: 225–297 (1938).

208. Mathias, Mildred E., and Lincoln Constance, "A Synopsis of the North American Species of Eryngium." *Amer. Midland Nat.* 25: 361–387 (1941).

209. Thackery, F. A., and M. F. Gilman, "A Rare Parasitic Food Plant of the Southwest." *Smithsonian Inst. Ann. Rpt.* 1930: 409–416 (1931).

210. Copeland, Herbert F., "Observations on the Structure and Classification of the Pyroleae." *Madroño* 9: 65–102 (1947).

211. Camp, W. H., "Aphyllous Forms in Pyrola." *Bul. Torrey Bot. Club* 67: 453–465 (1940).

212. Eastwood, Alice, "A Revision of Arctostaphylos." *Leafl. West. Bot.* 1: 105–127 (1934).

213. Adams, J. E., "A Systematic Study of the Genus Arctostaphylos Adans." *Jour. Elisha Mitchell Sci. Soc.* 56: 1–62 (1940).

214. Camp, W. H., "A Survey of the American Species of Vaccinium Subgenus Euvaccinium." *Brittonia* 4: 205–247 (1942).

215. Williams, Louis O., "Revision of the Western Primulas." *Amer. Midland Nat.* 17: 741–748 (1936).

216. St. John, Harold, "Revision of Certain North American Species of Androsace." *Mem. Victoria Mus.* 126: 45–55 (1922).

217. Robbins, G. Thomas, "North American Species of Androsace." *Amer. Midland Nat.* 32: 137–163 (1944).
218. Clark, Robert B., "A Revision of the Genus Bumelia in the United States." *Ann. Mo. Bot. Gard.* 29: 155–182 (1942).
219. Rehder, Alfred, "The Genus Fraxinus in New Mexico and Arizona." *Proc. Amer. Acad.* 53: 199–212 (1917).
220. Steyermark, Julian A., "A Revision of the Genus Menodora." *Ann. Mo. Bot. Gard.* 19: 87–160 (1932).
221. Card, Hamilton H., "A Revision of the Genus Frasera." *Ann. Mo. Bot. Gard.* 18: 245–282 (1931).
222. St. John, Harold, "Revision of the Genus Swertia (Gentianaceae) of the Americas and the Reduction of Frasera." *Amer. Midland Nat.* 26: 1–29 (1941).
223. Allen, Caroline K., "A Monograph of the American Species of the Genus Halenia." *Ann. Mo. Bot. Gard.* 20: 119–222 (1933).
224. Benson, Lyman, "Revisions of Status of Southwestern Desert Trees and Shrubs II." *Amer. Jour. Bot.* 30: 630–632 (1943).
225. Woodson, Robert E., Jr., "Studies in the Apocynaceae. III. A Monograph of the Genus Amsonia." *Ann. Mo. Bot. Gard.* 15: 379–434 (1928).
226. Woodson, Robert E., Jr., "Studies in the Apocynaceae I." *Ann. Mo. Bot. Gard.* 17: 83–212 (1930).
227. Woodson, Robert E., Jr., "The North American Asclepiadaceae I. Perspective of the Genera." *Ann. Mo. Bot. Gard.* 28: 193–244 (1941).
227a. Holm, Richard W. "The American Species of Sarcostemma R. Br." *Ann. Mo. Bot. Gard.* 37: 477–560 (1950).
228. Yuncker, T. G., *Revision of the North American and West Indian Species of Cuscuta.* Univ. Illinois Biol. Monog. 6: 1–142 (1921).
229. Yuncker, T. G., "The Genus Cuscuta." *Mem. Torrey Bot. Club* 18: 113–331 (1932).
230. Van Ooststroom, S. J., "A Monograph of the Genus Evolvulus." *Meddel. Bot. Mus. Herb. Rijksuniv. Utrecht* 14: 1–267 (1934).
231. House, H. D., "The North American Species of the Genus Ipomoea." *Ann. N. Y. Acad. Sci.* 18: 181–263 (1908).
232. Wherry, Edgar T., "A Provisional Key to the Polemoniaceae." *Bartonia* 20: 14–17 (1940).
233. Mason, Herbert L., "The Taxonomic Status of Microsteris Greene." *Madroño* 6: 122–127 (1941).
234. Wherry, Edgar T., "Microsteris, Phlox, and an Intermediate." *Brittonia* 5: 60–63 (1943).
235. Wherry, Edgar T., "Review of the Genera Collomia and Gymnosteris." *Amer. Midland Nat.* 31: 216–231 (1944).
236. Mason, Herbert L., "The Genus Eriastrum and the Influence of Bentham and Gray upon the Problem of Generic Confusion in Polemoniaceae." *Madroño* 8: 65–91 (1945).
237. Ewan, Joseph, "Linanthastrum, a New West American Genus of Polemoniaceae." *Jour. Wash. Acad. Sci.* 32: 138–141 (1942).
238. Kearney, Thomas H., and Robert H. Peebles, "Gilia multiflora Nutt. and Its Nearest Relatives." *Madroño* 7: 59–63 (1943).
239. Wherry, Edgar T., "The Gilia Aggregata Group." *Bul. Torrey Bot. Club* 73: 194–202 (1946).
240. Mason, Herbert L., and Alva D. Grant, "Some Problems in the Genus Gilia." *Madroño* 9: 201–220 (1948).
241. Wherry, Edgar T., "The Genus Polemonium in America." *Amer. Midland Nat.* 27: 741–760 (1942).
242. Constance, Lincoln, "The Genera of the Tribe Hydrophylleae of the Hydrophyllaceae." *Madroño* 5: 28–33 (1939).
243. Constance, Lincoln, "The Genus Hydrophyllum L." *Amer. Midland Nat.* 27: 710–731 (1942).
244. Constance, Lincoln, "The Genus Pholistoma Lilja." *Bul. Torrey Bot. Club* 66: 341–352 (1939).
245. Constance, Lincoln, "The Genus Eucrypta Nutt." *Lloydia* 1: 143–152 (1938).

246. Voss, John W., "A Revisional Study of the Phacelia hispida Group." *Bul. So. Calif. Acad. Sci.* 33: 169–178 (1935).

247. Voss, John W., "A Revision of the Phacelia crenulata Group for North America." *Bul. Torrey Bot. Club* 64: 81–96, 133–144 (1937).

248. Howell, John Thomas, "Studies in Phacelia." *Leafl. West. Bot.* 3: 95–96, 117–120, 134 (1942).

249. Howell, John Thomas, "Studies in Phacelia—a Revision of Species Related to P. pulchella and P. rotundifolia." *Amer. Midland Nat.* 29: 1–26 (1943).

250. Howell, John Thomas, "A Revision of Phacelia Section Miltitzia." *Proc. Calif. Acad. Sci.* ser. 4, 25: 357–376 (1944).

251. Howell, John Thomas, "Studies in Phacelia. Revision of Species Related to P. Douglasii, P. linearis, and P. Pringlei." *Amer. Midland Nat.* 33: 460–494 (1945).

252. Howell, John Thomas, "A Revision of Phacelia Section Euglypta." *Amer. Midland Nat.* 36: 381–411 (1946).

253. Hitchcock, C. Leo, "A Taxonomic Study of the Genus Nama." *Amer. Jour. Bot.* 20: 415–430, 518–534 (1933).

254. Johnston, Ivan M., "Key to the North American Species of Coldenia." *Proc. Calif. Acad. Sci.* ser. 4, 12: 1139–1141 (1924).

255. Ewan, Joseph, "A Review of the North American Weedy Heliotropes." *Bul. So. Calif. Acad. Sci.* 41: 51–57 (1942).

256. Johnston, Ivan M., "Studies in the Boraginaceae. II. Pectocarya." *Contrib. Gray Herbarium* 70: 34–39 (1924).

257. Johnston, Ivan M., "Studies in the Boraginaceae. II. Lappula." *Contrib. Gray Herbarium* 70: 47–51 (1924).

258. Johnston, Ivan M., "Restoration of the Genus Hackelia." *Contrib. Gray Herbarium* 68: 43–48 (1923).

259. Williams, Louis Otho, "A Monograph of the Genus Mertensia in North America." *Ann. Mo. Bot. Gard.* 24: 17–159 (1937).

260. Payson, Edwin Blake, "A Monograph of the Section Oreocarya of Cryptantha." *Ann. Mo. Bot. Gard.* 14: 211–358 (1927).

261. Johnston, Ivan M., "The North American Species of Cryptantha." *Contrib. Gray Herbarium* 74: 1–114 (1925).

262. Johnston, Ivan M., "Plagiobothrys. A Synopsis and Redefinition of the Genus." *Contrib. Gray Herbarium* 68: 57–80 (1923).

263. Johnston, Ivan M., "The Allocarya Section of Plagiobothrys in the Western United States." *Contrib. Arnold Arboretum* 3: 1–102 (1932).

264. Johnston, Ivan M., "Studies in the Boraginaceae. II. Lithospermum." *Contrib. Gray Herbarium* 70: 18–31 (1924).

265. Moldenke, Harold N., "Verbenaceae." In Lundell, C. L., *Flora of Texas* 3: 13–87 (1942).

266. Perry, Lily M., "A Revision of the North American Species of Verbena." *Ann. Mo. Bot. Gard.* 20: 239–362 (1933).

267. Moldenke, Harold N., "A Monograph of the Genus Bouchea." *Repert. Sp. Nov.* 49: 91–139 (1940).

268. Lewis, Harlan, "A Revision of the Genus Trichostema." *Brittonia* 5: 276–303 (1945).

269. Epling, Carl, *The American Species of Scutellaria.* Univ. Calif. Publ. Bot. 20: 1–146 (1942).

270. Lint, Harold, and Carl Epling, "A Revision of Agastache." *Amer. Midland Nat.* 33: 207–230 (1945).

271. Epling, Carl, "Preliminary Revision of American Stachys." *Repert. Spec. Nov.* 80: 1–75 (1934).

272. Epling, Carl, "The Californian Salvias, a Review of Salvia, Section Audibertia." *Ann. Mo. Bot. Gard.* 25: 95–188 (1938).

273. Epling, Carl, "A Revision of Salvia: Subgenus Calosphace." *Repert. Spec. Nov. Beih.* 110: 1–380 (1938–1939).

274. McClintock, Elizabeth, and Carl Epling, *A Review of the Genus Monarda (Labiatae).* Univ. Calif. Publ. Bot. 20: 147–194 (1942).

275. Epling, Carl, and William S. Stewart, "A Revision of Hedeoma with a Review of Allied Genera." *Repert. Spec. Nov. Beih.* 115: 1–50 (1939).

276. Epling, Carl, "Monograph of the Genus Monardella." *Ann. Mo. Bot. Gard.* 12: 1–106 (1925).

277. Hitchcock, C. Leo, "A Monographic Study of the Genus Lycium of the Western Hemisphere." *Ann. Mo. Bot. Gard.* 19: 179–374 (1932).

278. Wiggins, Ira L., "A Report on Several Species of Lycium from the Southwestern Deserts." *Contrib. Dudley Herbarium* 1: 197–206 (1934).

279. Rydberg, P. A., "The North American Species of Physalis and Related Genera." *Mem. Torrey Bot. Club* 4: 297–372 (1896).

280. Morton, C. V., "Notes on the Genus Saracha." *Proc. Biol. Soc. Wash.* 51: 75–78 (1938).

281. Stebbins, G. Ledyard, Jr., and Elton F. Paddock, "The Solanum nigrum Complex in Pacific North America." *Madroño* 10: 70–81 (1949).

282. Safford, William E., "Synopsis of the Genus Datura." *Jour. Wash. Acad. Sci.* 11: 173–189 (1921).

283. Goodspeed, T. H., *Studies in Nicotiana III. A Taxonomic Organization of the Genus.* Univ. Calif. Publ. Bot. 18: 335–344 (1945).

284. Goodspeed, T. H., "Cytotaxonomy of Nicotiana." *Bot. Rev.* 11: 532–592 (1945).

285. Pennell, Francis W., *The Scrophulariaceae of Temperate Eastern North America.* Monog. Acad. Nat. Sci. Phila. 1: 1–541 (1935).

286. Pennell, Francis W., *New Species of Scrophulariaceae from Arizona.* Notul. Nat. Acad. Nat. Sci. Phila. 43: 1–10 (1940).

287. Munz, Philip A., "The Antirrhinoideae-Antirrhineae of the New World." *Proc. Calif. Acad. Sci.* ser. 4, 15: 323–397 (1926).

288. Newsom, Vesta M., "A Revision of the Genus Collinsia." *Bot. Gaz.* 87: 260–301 (1929).

289. Pennell, Francis W., "Scrophulariaceae of the Central Rocky Mountain States. Penstemon." *Contrib. U. S. National Herbarium* 20: 325–381 (1920).

290. Keck, David D., *Studies in Penstemon I. A Systematic Treatment of the Section Saccanthera.* Univ. Calif. Publ. Bot. 16: 367–426 (1932). II. "The Section Hesperothamnus." *Madroño* 3: 200–219 (1936). IV. "The Section Ericopsis." *Bul. Torrey Bot. Club* 64: 357–381 (1937). V. "The Section Peltanthera." *Amer. Midland Nat.* 18: 790–829 (1937). VI. "The Section Aurator." *Bul. Torrey Bot. Club* 65: 233–255 (1938). VII. "The Subsections Gairdneriani, Deusti, and Arenarii of the Graciles." *Amer. Midland Nat.* 23: 594–616 (1940). VIII. "A Cytotaxonomic Account of the Section Spermunculus." *Ibid.* 33: 128–206 (1945).

291. Grant, Adele Lewis, "A Monograph of the Genus Mimulus." *Ann. Mo. Bot. Gard.* 11: 99–388 (1924).

292. Pennell, Francis W., "Reconsideration of the Bacopa-Herpestis Problem." *Proc. Acad. Nat. Sci. Phila.* 98: 83–98 (1946).

293. Pennell, Francis W., " 'Veronica' in North and South America." *Rhodora* 23: 1–22, 29–41 (1921).

294. Pennell, Francis W., "A Revision of Synthyris and Besseya." *Proc. Acad. Nat. Sci. Phila.* 85: 77–106 (1933).

295. Ferris, Roxana S., "Taxonomy and Distribution of Adenostegia." *Bul. Torrey Bot. Club* 45: 399–423 (1918).

296. Fosberg, F. Raymond, "Varieties of the Desert-Willow, Chilopsis linearis." *Madroño* 3: 362–366 (1936).

297. Van Eseltine, G. P., *A Preliminary Study of the Unicorn Plants.* N. Y. Agr. Expt. Sta. Tech. Bul. 149 (1929).

298. Munz, Philip A., "The North American Species of Orobanche Section Myzorrhiza." *Bul. Torrey Bot. Club* 57: 611–624 (1930).

299. Achey, Daisy M., "A Revision of the Section Gymnocaulis of the Genus Orobanche." *Bul. Torrey Bot. Club* 60: 441–451 (1933).

300. Leonard, E. C., "The American Species of Elytraria." *Jour. Wash. Acad. Sci.* 24: 443–447 (1934).

301. Kobuski, Clarence E., "A Monograph of the American Species of the Genus Dyschoriste." *Ann. Mo. Bot. Gard.* 15: 9–90 (1928).

302. Hagen, Stanley Harlan, "A Revision of the North American Species of Anisacanthus." *Ann. Mo. Bot. Gard.* 28: 385–408 (1941).

303. Poe, Ione, "A Revision of the Plantago patagonica Group of the United States and Canada." *Bul. Torrey Bot. Club* 55: 406–420 (1928).

304. Pilger, Robert, "Plantaginaceae." *Pflanzenreich* IV. 269: 1–466 (1937).

305. Wheeler, Louis C., "Notes on Plantago in the Pacific States." *Amer. Midland Nat.* 20: 331–333 (1938).

306. Hilend, Martha, and John Thomas Howell, "The Genus Galium in Southern California." *Leafl. West. Bot.* 1: 145–168 (1935).

307. Jones, George Neville, "A Monograph of the Genus Symphoricarpos." *Jour. Arnold Arboretum* 21: 201–252 (1940).

308. Rehder, Alfred, "Synopsis of the Genus Lonicera." *Ann. Rpt. Mo. Bot. Gard.* 14: 27–232 (1903).

309. Nielsen, Sarah Dyal, "Systematic Studies in the Valerianaceae." *Amer. Midland Nat.* 42: 480–501 (1949).

310. Bailey, L. H., "Species of Cucurbita." *Gentes Herbarum* 6: 267–322 (1943).

311. Dunn, S. T., "The Genus Marah." *Kew Bul.* 1913: 145–153 (1913).

312. McVaugh, Roger, "The Genus Triodanis Rafinesque, and its Relationships to Specularia and Campanula." *Wrightia* 1: 13–52 (1945).

313. Munz, Philip A., "A Revision of the Genus Nemacladus (Campanulaceae)." *Amer. Jour. Bot.* 11: 233–248 (1924).

314. McVaugh, Rogers, "Some Realignments in the Genus Nemacladus." *Amer. Midland Nat.* 22: 521–550 (1939).

315. Robinson, B. L., "The Stevias of North America." *Contrib. Gray Herbarium* 90: 90–159 (1930).

316. Robinson, B. L., "A Monograph of the Genus Brickellia." *Mem. Gray Herbarium* 1: 1–151 (1917).

317. Shinners, Lloyd H., "Revision of the Genus Kuhnia L." *Wrightia* 1: 122–144 (1946).

318. Steyermark, Julian A., "Studies in Grindelia II. A Monograph of the North American Species of the Genus Grindelia." *Ann. Mo. Bot. Gard.* 21: 433–608 (1934).

319. Porter, C. L., "The Genus Amphipappus Torr. and Gray." *Amer. Jour. Bot.* 30: 481–483 (1943).

320. Hall, Harvey M., *The Genus Haplopappus. A Phylogenetic Study in the Compositae.* Carnegie Inst. Wash. Pub. 389: 1–391 (1928).

321. Hall, Harvey M., and Frederic E. Clements, *The Phylogenetic Method in Taxonomy. Genus Chrysothamnus.* Carnegie Inst. Wash. Publ. 326: 157–234 (1923).

322. Howell, John Thomas, *A Systematic Study of the Genus Lessingia Cham.* Univ. Calif. Publ. Bot. 16: 1–44 (1929).

323. Shinners, Lloyd H., "Revision of the Genus Aphanostephus DC." *Wrightia* 1: 95–121 (1946).

324. Larsen, Esther Louise, "A Revision of the Genus Townsendia." *Ann. Mo. Bot. Gard.* 14: 1–46 (1927).

325. Cronquist, Arthur, "Revision of the Western North American Species of Aster Centering about Aster foliaceus Lindl." *Amer. Midland Nat.* 29: 429–468 (1943).

326. Shinners, Lloyd H., "Revision of the Genus Leucelene Greene." *Wrightia* 1: 82–89 (1946).

327. Cronquist, Arthur, "Revision of the North American Species of Erigeron, North of Mexico." *Brittonia* 6: 121–300 (1947).

328. Robinson, B. L., "Synopsis of the Genus Nocca." (*Contrib. Gray Herbarium* 20) *Proc. Amer. Acad.* 36: 467–471 (1901).

329. Robinson, B. L., "Revision of the Genus Guardiola." *Bul. Torrey Bot. Club* 26: 232–235 (1899).

330. Robinson, B. L., "Synopsis of the Genus Melampodium." (*Contrib. Gray Herbarium* 20) *Proc. Amer. Acad.* 36: 455–466 (1901).

331. Millspaugh, Charles Frederick, and Earl Edward Sherff, *Revision of the North American Species of Xanthium.* Field Mus. Nat. Hist. Bot. Ser. 4: 9–49 (1919).

332. Widder, F. J., "Die Arten der Gattung Xanthium. Beitraege zu einer Monographie." *Repert. Spec. Nov.* 20: 1–221 (1923).

333. Robinson, B. L., and J. M. Greenman, "A Revision of the Genus Zinnia." (*Contrib. Gray Herbarium* 10) *Proc. Amer. Acad.* 32: 14–20 (1896).

334. Sharp, Ward McClintic, "A Critical Study of Certain Epappose Genera of the Heliantheae-Verbesininae of the Natural Family Compositae." *Ann. Mo. Bot. Gard.* 22: 51–152 (1935).

335. Weber, William A., "A Taxonomic and Cytological Study of the Genus Wyethia, Family Compositae." *Amer. Midland Nat.* 35: 400–452 (1946).

336. Blake, S. F., "Revision of the Genus Tithonia." *Contrib. U. S. National Herbarium* 20: 423–436 (1921).

337. Blake, S. F., "A Revision of the Genus Viguiera." *Contrib. Gray Herbarium* 54: 1–205 (1918).

338. Watson, E. E., *Contributions to a Monograph of the Genus Helianthus.* Papers Mich. Acad. Sci. 9: 305–475 (1929).

339. Blake, S. F., "Revision of the Genus Flourensia." *Contrib. U. S. National Herbarium* 20: 393–409 (1921).

340. Blake, S. F., "A Revision of Encelia and Some Allied Genera." (*Contrib. Gray Herbarium* 41) *Proc. Amer. Acad.* 49: 346–396 (1913).

341. Jones, W. W., "A Revision of the Genus Zexmenia." (*Contrib. Gray Herbarium* 30) *Proc. Amer. Acad.* 41: 143–167 (1905).

342. Robinson, B. L., and J. M. Greenman, "Synopsis of the Genus Verbesina, with an Analytical Key to the Species." (*Contrib. Gray Herbarium* 16) *Proc. Amer. Acad.* 34: 534–566 (1899).

343. Sherff, Earl Edward, *Revision of the Genus Coreopsis.* Field Mus. Nat. Hist. Bot. Ser. 11: 279–475 (1936).

344. Blake, S. F., "A Redisposition of the Species Heretofore Referred to Leptosyne." (*Contrib. Gray Herbarium* 41) *Proc. Amer. Acad.* 49: 335–346 (1913).

345. Sherff, Earl Edward, *The Genus Bidens.* Field Museum Nat. Hist. Bot. Ser. 16: 1–709 (1937).

346. Sherff, Earl Edward, *Revision of the Genus Cosmos.* Field Museum Nat. Hist. Bot. Ser. 8: 401–447 (1932).

347. St. John, Harold, and Donald White, "The Genus Galinsoga in North America." *Rhodora* 22: 97–101 (1920).

348. Heiser, Charles B., Jr., "Monograph of Psilostrophe." *Ann. Mo. Bot. Gard.* 31: 279–301 (1944).

349. Everly, Mary Louise, "A Taxonomic Study of the Genus Perityle and Related Genera." *Contrib. Dudley Herbarium* 3: 377–396 (1947).

350. Johnston, J. R., "A Revision of the Genus Flaveria." (*Contrib. Gray Herbarium* 26) *Proc. Amer. Acad.* 39: 279–292 (1903).

351. Heiser, Charles B., Jr., "A Revision of the Genus Schkuhria." *Ann. Mo. Bot. Gard.* 32: 265–278 (1945).

352. Johnston, I. M., "Diagnoses and Notes Relating to the Spermatophytes Chiefly of North America." *Contrib. Gray Herbarium* 68: 80–104 (1923).

353. Baltzer, Elizabeth A., "A Monographic Study of the Genus Palafoxia and Its Immediate Allies." *Ann. Mo. Bot. Gard.* 31: 249–277 (1944).

354. Constance, Lincoln, *A Systematic Study of the Genus Eriophyllum Lag.* Univ. Calif. Publ. Bot. 18: 69–135 (1937).

355. Stockwell, Palmer, "A Revision of the Genus Chaenactis." *Contrib. Dudley Herbarium* 3: 89–167 (1940).

356. Cockerell, T. D. A., "The North American Species of Hymenoxys." *Bul. Torrey Bot. Club* 31: 461–509 (1904).

356a. Parker, K. F., "New combinations in Hymenoxys." *Madroño* 10: 159 (1950).

357. Biddulph, Susan Fry, *A Revision of the Genus Gaillardia.* State Coll. Wash. Research Studies 12: 195–256 (1944).

358. Fernald, M. L., "A Systematic Study of the United States and Mexican Species of Pectis." (*Contrib. Gray Herbarium* 12) *Proc. Amer. Acad.* 33: 57–86 (1897).

359. Hall, Harvey M., and Frederic E. Clements, *The Phylogenetic Method in Taxonomy. Genus Artemisia.* Carnegie Inst. Wash. Publ. 326: 31–156 (1923).

360. Keck, David D., "Revision of the Artemisia vulgaris Complex in North America." *Proc. Calif. Acad. Sci.* ser. 4, 25: 421–468 (1946).

361. Maguire, Bassett, "A Monograph of the Genus Arnica." *Brittonia* 4: 386–527 (1943).

362. Greenman, J. M., "Monograph of the North and Central American Species of the Genus Senecio" Part II. *Ann. Mo. Bot. Gard.* 2: 573–626 (1915); 3: 85–194 (1916); 4: 15–36 (1917); 5: 37–108 (1918). (Not completed.)

363. Petrak, F., "Die nordamerikanischen Arten der Gattung Cirsium." *Bot. Centrbl. Beih.* 35, Abt. 2: 223–567 (1917).

364. Burkart, A., "Estudio del género de Compuestas *Chaptalia* con especial referencia a las especial argentinas." *Darwiniana* 6: 505–594 (1944).

365. Bacigalupi, Rimo, "A Monograph of the Genus Perezia, Section Acourtia, with a Provisional Key to the Section Euperezia." *Contrib. Gray Herbarium* 97: 1–81 (1931).

366. Robinson, B. L., and J. M. Greenman, "Revision of the Mexican and Central American Species of Trixis." (*Contrib. Gray Herbarium* 28) *Proc. Amer. Acad.* 40: 6–14 (1904).

367. Standley, Paul C., "A Revision of the Cichoriaceous Genera Krigia, Cynthia, and Cymbia." *Contrib. U. S. National Herbarium* 13: 351–357 (1911).

368. Sherff, Earl Edward, "North American Species of Taraxacum." *Bot. Gaz.* 70: 329–358 (1920).

369. Babcock, E. B., and G. L. Stebbins, Jr., *The American Species of Crepis. Their Interrelationships and Distribution as Affected by Polyploidy and Apomixis.* Carnegie Inst. Wash. Publ. 504: 1–199 (1938).

370. Babcock, Ernest Brown, *The Genus Crepis.* Univ. Calif. Publ. Bot. 22: 1–1030 (1947).

371. Robinson, B. L., and J. M. Greenman, "Revision of the Mexican and Central American Species of Hieracium." (*Contrib. Gray Herbarium* 28) *Proc. Amer. Acad.* 40: 14–24 (1904).

Glossary

A– (prefix). Without, lacking, as apetalous (without petals).

Abaxial. On the side or face away from the axis, dorsal.

Aberrant. Departing from type, atypical.

Abortion (adj. *abortive*). Nondevelopment or imperfect development of an organ or part.

Abruptly acuminate. Not tapering, having a sharp extremity from a rounded or truncate summit.

Abruptly pinnate. Applied to a compound (pinnate) leaf ending with a pair of leaflets instead of one.

Acaulescent. Without a leafy stem above ground.

Accrescent. Growing larger with age, as often the calyx.

Accumbent. Lying against, as in cotyledons with edges against the radicle.

Acerose. Needle-shaped, like pine leaves.

Achene. A small, dry and hard, indehiscent fruit, usually of one carpel and one-seeded.

Acicular. Needle-shaped, bristlelike.

Acorn. The fruit of oaks.

Acotyledonous. Lacking cotyledons or seed leaves.

Acrid. Sharp- or bitter-tasting, as the sap of many plants.

Acropetal. Produced successively toward the apex, as flowers in an inflorescence.

Acumination (adj. *acuminate*). A tapering point.

Acute. Sharp-pointed but not tapering.

Adherent. Attached to or united with a (usually dissimilar) organ or part.

Adnate. Grown together, as with anthers attached throughout their length to their filaments, or a calyx tube united with the ovary.

Adventive. Introduced recently, scarcely established or naturalized.

Aerial. Parts above ground or out of water.

Aggregate, aggregated. Crowded together, as flowers in a dense inflorescence, drupelets in the compound fruits of mulberry and blackberry.

Alkaloid. An organic base produced in a plant, often with medicinal or poisonous properties, as quinine, strychnine.

Allergic. Susceptible to some foreign substance, especially to pollen causing hay fever and allied diseases.

Alpine. Growing at high elevations, usually above timber line.

Alternate. Leaves, when not opposite or whorled; stamens, when between the petals.

Alternation of generations. In successive, alternate generations sexual, then asexual, especially noticeable in ferns.

Alveola (pl. *alveolae, alveolas*), *alveole* (adj. *alveolate*). Pits in the surface of seeds or receptacles, like cells of a honeycomb.

Amethystine. Amethyst-colored.

Amphibious. Capable of growing both in water and soil.

Amplexicaul. Clasping the stem, as a leaf base or stipule.

Ampliate. Enlarged, expanded.

Anastomosing. With veins uniting so as to form a network.

Anatropous. Of an ovule when in reversed position with the aperture (micropyle) close to the hilum (point of attachment).

Androgynous. With male and female flowers in the same inflorescence; in *Carex* with the staminate flowers above in the spike.

Androphore. A stalk supporting a stamen column or group of stamens.

Angulate. Angled.

Annual. Completing development and dying within a year.

Annulus (adj. *annular, annulate*). A ring-shaped part or organ.

Anterior. In front, away from the axis, dorsal.

Anther. The enlarged, pollen-containing part of a stamen.
Anther cells. The (usually two) locules into which an anther is divided.
Antheriferous. Bearing anthers.
Antheridium (pl. *antheridia*). The male sexual organ of ferns etc., analogous to the anther in flowering plants.
Anther sac. The pollen sac of an anther.
Anthesis. Opening of a flower, time of pollination.
Anthocarp. A structure in which the fruit proper is united with the perianth or receptacle.
Antrorse (adv. *antrorsely*). Directed upward, opposite of retrorse.
Apetalous. Without petals, but the sepals sometimes petallike.
Aphid (pl. *aphides*). A plant louse.
Aphis. A genus of plant lice.
Apical (adv. *apically*). At the apex or summit.
Apiculation (adj. *apiculate*). A short, abrupt point.
Apomixis (adj. *apomictic*). Reproduction without sexual union (*See:* Parthenogenetic).
Appendages (adj. *appendaged, appendiculate*). Outgrowths of various character.
Appressed. Lying flat, as leaves against the stem.
Approximate. Close, as contrasted with remote or separated.
Aquatic. Growing in water, as contrasted with terrestrial.
Arachnoid. Cobwebby, bearing very fine, weak, tangled hairs.
Arboreal, arboreous. Pertaining to trees, treelike, growing upon trees.
Arborescent. Treelike in size and habit of growth.
Archegonium (pl. *archegonia*). The female sexual organ of ferns etc., containing the egg cell.
Arcuate. Moderately curved, like a bow.
Arenicolous. Growing in sand.
Areola (pl. *areolae, areolas*), *areole* (adj. *areolate*). A sharply defined space, such as the spine-bearing area in Cactaceae, spaces between veins etc.
Aril (adj. *arillate*). An expansion of the funicle (attachment of the ovule to the placenta) that more or less envelopes the seed.
Arilliform. Arillike.
Aristate. Awned, having a stiff, bristlelike tip or dorsal appendage.
Aristiform. Awnlike.
Articulated. Jointed to and separating with a clean scar, as often with petioles, leaflets of a compound leaf, etc.
Articulation. A joint or point of union from which an organ may become detached.
Ascending. Directed upward but not erect.
Asexual (adv. *asexually*). Without sex, applied especially to reproduction without union of male and female elements, as by buds, bulblets, etc.
Assurgent. Ascending.
Asymmetric. Irregular, not divisible into equal and like halves, as often in leaves and other parts.
Attenuate. Tapering, gradually narrowed.
Atypical. Not typical.
Auricle (adj. *auriculate*). Earlike basal appendages of leaves, petals, etc.; in grasses etc., these at the junction of blade and sheath.
Autophytic. Independent of associated organisms, having chlorophyll, and not parasitic or saprophytic.
Awn (adj. *awned*). A bristlelike terminal or lateral appendage, especially on the glumes and lemmas of grasses.
Axial. Pertaining to the axis.
Axil. The inner angle where one organ is united with another, especially the angle between leaf and stem.
Axillary. Borne in the axil.
Axis (pl. *axes*, adj. *axial*). The central line around which organs are borne, such as the portion of the stem that supports the inflorescence, and the center, whether or not produced, of a flower.
Banner. The upper, odd or posterior (usually the largest) petal of a papilionaceous (pealike) flower.
Barb (adj. *barbed*). A hair or bristle with retrorse, hooklike projections.
Barbate. Bearded, provided with long hairs.
Barbellate. Provided with short, usually stiff hairs, as the pappus bristles of many Compositae.
Bark. The external tissues, especially in exogenous woody plants, beneath the epidermis and outside the wood and the growth tissue or cambium.
Basal (adv. *basally*). At or very near the true base.
Basifixed. Attached at base, as hairs that are not dolabriform.
Basiscopic. Looking toward the base.
Beak (adj. *beaked*). A more or less narrowed projection, as at apex of petals, fruits, etc.

Bearded. Bearing hairs, usually in tufts.

Berry. A fruit, usually indehiscent, with flesh or pulp surrounding the seed or seeds.

Bi– (prefix). Two–, as biaristate (two-awned), bidentate (two-toothed), bilocular (with two locules or cells), bimucronate (with two mucros), bisulcate (two-furrowed), bivalvate (two-valved).

Bicolored, bicolorous. Of two colors.

Biconvex. Convex on both sides.

Biennial. Requiring two years to complete development, usually flowering and fruiting the second year.

Bifid. Cleft about halfway into two usually equal lobes.

Bifoliolate. Of a compound leaf with two leaflets.

Bifurcate. Two-pronged or -forked.

Bilabiate. Two-lipped, as in many sympetalous, irregular corollas.

Bilateral. Disposed on two sides, as leaves on a branch.

Bipartite. Divided nearly to the base into two divisions.

Bipinnate. Twice-pinnate, as in a compound leaf with the secondary divisions also pinnate.

Bipinnatifid. With the secondary divisions of a pinnatifid leaf also pinnatifid.

Biseriate. In two rows or series, one above the other, as the phyllaries of the involucre in some Compositae.

Biternate. Twice-ternate, with the secondary divisions of a compound (ternate) leaf also in threes.

Biturbinate. Turbinate at both ends, hence broadest near the middle.

Blade. The expanded portion of a leaf as distinguished from the petiole, sheath, or stipules; the corresponding part of a petal as distinguished from the claw.

Brackish. Somewhat saline.

Bract (adj. *bracteal*). A reduced or modified leaf, subtending the divisions of an inflorescence or borne at base of the pedicel of an individual flower.

Bracteate, bracted. Provided with bracts.

Bracteiform. Bractlike, as reduced leaves.

Bracteole (adj. *bracteolate*), *bractlet.* A (usually small) bract subtending the base of an individual flower or borne on the pedicel. The term is sometimes applied to the small outer bracts in the involucre of certain Compositae.

Bridge. The band of tissue connecting adjacent corolla scales in *Cuscuta.*

Bud scales. Dry or sometimes glutinous modified leaves enveloping and protecting a bud.

Bulb. A modified, usually subterranean bud enveloped with scales, as in lilies and onions.

Bulbiferous. Possessing bulbs or bulbils.

Bulbil, bulblet. A small bulb, usually above ground and axillary.

Bulb scale. One of the modified leaves forming the envelope of a bulb.

Bundle. See: Vascular bundle.

Caducous. Falling very early, as leaves or petals.

Calceolate. Shaped like a shoe.

Caliche. A calcareous (limy) hardpan formed under certain soils.

Callose, callous. Thickened and hardened; bearing callosities.

Callosity. A thickened and hardened area.

Callus. A thickening, applied in grasses to the thickened and hardened base either of the lemma below its point of insertion, or of the first glume, or of the rachilla.

Calyculus (adj. *calyculate*). The short outer phyllaries of an involucre, in Compositae, simulating a calyx; an involucel.

Calyx (pl. *calyces, calyxes*). The outer envelope of a flower, usually green and of firmer texture than the corolla.

Campanulate. Bell-shaped.

Canaliculate. Channeled, having a longitudinal groove or grooves.

Cancellate. With crossing lines or ridges, as in a lattice.

Canescent. Gray or hoary, usually referring to pubescence of this character.

Capillary. Hairlike, very fine and slender.

Capitate. Headlike, as is the inflorescence of most Compositae and the stigma of many flowers.

Capsule (adj. *capsular*). A dry dehiscent fruit opening longitudinally along more than one suture, or circumscissily, or by pores; *compare:* Follicle.

Cardiac. Of or pertaining to the heart; a heart or stomach stimulant.

Carina (adj. *carinal, carinate*). A central dorsal ridge or keel.

Carpel (adj. *carpellate*). A simple pistil or unit of a compound pistil, containing the ovule or ovules.

Carpophore. A prolongation of the receptacle between and supporting the carpels, as in Umbelliferae.

Cartilaginous. Firm and tough, like animal cartilage.

Caruncle (adj. *caruncular, carunculate*). A protuberance or appendage of a seed, at or near the hilum.

Caryopsis. The fruit or grain of grasses, indehiscent and with one seed, this usually united with the pericarp.

Catkin. A spike or spikelike inflorescence, in which the flowers are usually of one sex, an ament.

Caudate. Tailed, long-attenuate.

Caudex. The combined stem and root, applied often to the thickened basal parts of stems.

Caulescent. Having a leafy stem above ground, as contrasted with acaulescent and scapose.

Cauline. Pertaining to or borne on the stem, as cauline leaves.

Cell (adj. *cellular*). The mostly microscopic particle of protoplasm constituting the structural unit of organisms, this in plants usually enclosed by walls; also the locule or locules of an anther, ovary, or fruit.

Centrifugal inflorescence. Determinate, developing from the center outward, as in cymes.

Centripetal inflorescence. Indeterminate, developing toward the center, as in corymbs.

Cespitose. Growing in tufts.

Chaff (adj. *chaffy*). Small, dry, membranaceous scales, as on the receptacle of many Compositae.

Chamaephyte. A low-growing plant, with resting buds just above ground.

Chaparral. A vegetation type characterized by a dense growth of shrubs with usually leathery, evergreen leaves.

Chartaceous. Papery.

Chlorophyll. The green substance in plant tissues.

Chromosomes. Minute bodies, usually of a definite number in each species, that develop in the nucleus during cell division and with which the genes or determiners of heredity are associated.

Cilium (pl. *cilia*, adj. *ciliate*). A marginal hair.

Ciliolate. Diminutive of Ciliate, fringed with very short hairs.

Cinereous. Ashy gray.

Circinate. Coiled into a circle or partially so.

Circumscission (adj. *circumscissile*). Transverse dehiscence, as by a circular cut.

Cirrhus. A tendril.

Clambering. Supported on other plants or various objects, but scarcely climbing.

Clasping. Embracing the stem, as leaf bases, bracts, and stipules.

Clathrate. Like a lattice.

Clavate. Club-shaped, enlarged toward apex, as in thickened organs such as pedicels, styles, and capsules.

Clavellate. Diminutive of Clavate, slightly thickened upward.

Claw. The narrowed basal portion of a sepal or petal.

Cleft. Cut about halfway, as in many leaves and petals.

Cleistogamous. Applied to (usually reduced and modified) flowers that are self-fertilized without expanding.

Climbing. Ascending another plant or other object, usually by twining or with the support of holdfasts such as tendrils.

Coalescent. Grown together, referring to similar parts.

Coat. One of the successive layers of a bulb; the covering of a seed.

Cohere (adj. *coherent*). To have (commonly similar) parts united, as coherent petals or stamens.

Coleoptile. The embryo sheath in monocotyledons.

Collar. The expanded apical portion of the style, as in *Pyrola*.

Column. Body formed by the union of stamens or, in orchids, of stamens with the style or stigma; lower portion of the awn in some grasses; *see also:* Stamen column.

Columnar. Shaped like a column.

Coma (adj. *comose*). A tuft of hairs, specifically of hairs at end of seeds, as in *Asclepias*.

Commissure (adj. *commissural*). Inner face of two coherent carpels, as in Umbelliferae.

Complicate. Folded upon itself.

Compound. Opposite of simple, as leaves composed of leaflets, inflorescences with secondary branches, etc.

Compressed. Flattened.

Concavo-convex. Concave on one side, convex on the other.

Concolorous. Uniform in color.

Conduplicate. Folded together lengthwise.

Cone. A strobile, the more or less conic, bracted fruiting structure of Coniferae, also of *Equisetum*.

Confluent. Passing gradually into one another, as confluent leaflets.

Congested. Crowded together, as flowers in a congested inflorescence.

Conic, conical. Cone-shaped, tapering to a point from a broad base.
Connate. United, at least at base.
Connective. Portion of a stamen connecting the lobes of the anther.
Connivent. Coming into contact by convergence but not cohering, as often the petals or the stamens.
Consimilar. Like one another.
Constricted. Drawn in, narrowed and then expanded.
Contiguous. Adjoining, in contact.
Contorted. Twisted, as the petals or corolla lobes in convolute buds.
Contracted. Shortened or narrowed, as a contracted inflorescence.
Converging. Applied especially to veins that approach each other near the leaf margin.
Convolute. Rolled together or around one another as the petals in some buds.
Copious. Abundant, plentiful.
Cordate. Heart-shaped, as in leaves with indented base.
Coriaceous. Leathery, thick and tough.
Corm. The bulblike, fleshy, but apparently solid base of a stem, as in *Crocus.*
Corneous. Horny.
Corniculate. Bearing a hornlike appendage or appendages.
Cornute. Horned or spurred.
Corolla. The inner floral envelope, when different from the calyx in texture and color.
Corona. A structure, usually crownlike, in the throat of a corolla, between the petals and stamens, as in *Narcissus, Silene, Passiflora.*
Coroniform. Crown-shaped, like the pappus of some Compositae.
Corrugate. Strongly and usually regularly wrinkled.
Corrugation. A wrinkle.
Corymb (adj. *corymbose*). A flat-topped or convex open inflorescence, with flowers opening successively toward the center.
Corymbiform. Corymblike.
Costa (pl. *costae,* adj. *costal, costate*). A rib, as a thickened midvein.
Costular. Pertaining to the primary veins in ferns, especially if parallel and prominent.
Cotyledon. The leaf or leaves of the plant embryo.
Creeping. With stems running on or just below the surface of the ground and rooting at the nodes.
Crenate. With rounded teeth, scalloped.
Crenulate. Diminutive of Crenate.
Crest (adj. *crested*). A ridge or other elevation on the surface or apex of an organ.
Crispate, crisped. Curled or crumpled.
Crossbar. A transverse ridge, as on certain seeds.
Cross-pollination. Transfer of pollen from one flower to another.
Crown. Of a tree, the branches and foliage together; of stems, the (usually thickened) bases; *see also:* Caudex and Corona.
Cruciform. Cross-shaped.
Crustaceous. Brittle in texture when dry, as are many lichens.
Cucullate. Hood-shaped.
Culm. The (usually hollow) stem of grasses; also applied to stems of sedges.
Cuneate. Wedge-shaped.
Cupulate. Cup-shaped, as the involucre of an acorn.
Cusp (adj. *cuspidate*). A sharp, rigid, apical point.
Cyanogen (adj. *cyanogenic*). An organic compound, present in the very poisonous hydrocyanic (prussic) acid.
Cyathium (pl. *cyathia*). The specialized ultimate divisions of the inflorescence of *Euphorbia* (see description of that genus).
Cylindric. Elongate and circular in cross section.
Cylindroid. Like a cylinder but elliptic in cross section.
Cymbiform. Boat-shaped, excavated on one side.
Cyme (adj. *cymose*). A (usually broad and flattened) inflorescence with flowers opening successively from the center outward.
Cymule. A small cyme, or a cymelike division of a compound inflorescence.
Cytogenetic. Applied to morphology of the cells, and especially of the chromosomes, in its relationship to inheritance of characters.
Deciduous. Falling, as leaves in autumn, contrasted with evergreen; also when applied to sepals and other organs, contrasted with persistent.
Declined. Turned downward or outward, as the stamens of many flowers.
Decompound. Several times divided, as in repeatedly compound leaves and inflorescences.
Decumbent. Reclining, then ascending.
Decurrent. Prolonged downward, as are some leaves along the stem.
Decurved. Curved downward.
Decussate. In pairs, each pair at a right angle to the pairs immediately above and below.

Deflexed. Bent downward.

Dehiscence (adj. *dehiscent*). The opening, as of anthers and fruits, longitudinally or transversely, and usually in a regular line or lines of cleavage.

Deltoid. Shaped like the Greek letter Δ; equilaterally triangular.

Dendritic. Branched like a tree, as are certain hairs, especially in Cruciferae.

Dentate. Toothed, with teeth usually directed outward, and more pointed than is implied by crenate.

Denticulate. Diminutive of Dentate.

Denudate. Stripped, becoming bare.

Depressed. Flattened, usually with a concave center.

Dermatitis. Inflammation of the skin.

Determinate. Ending with a bud, as the season's growth in some plants; applied also to a centrifugal inflorescence, such as a cyme.

Diadelphous. With stamens in two groups, each more or less united by their filaments, as often in Leguminosae.

Dichotomous. Forked, in pairs, usually referring to branches.

Didymous. In pairs, or two-lobed, as are many fruits.

Didynamous. With stamens in pairs, two of them longer and two shorter, as usually in Labiatae.

Diffuse (adv. *diffusely*). Spreading widely and loosely.

Digitate (adv. *digitately*). Applied to a compound leaf when all the leaflets are clustered at apex of the petiole (as in *Lupinus*) like the fingers of a hand; applied to an inflorescence with branches thus disposed.

Dilated. Expanded or enlarged, usually toward apex.

Dimidiate. Appearing reduced to one half, the other half much smaller or obsolete.

Dimorphic, dimorphous. Having two forms.

Dioecious. Having unisexual flowers, with the two sexes on different individual plants.

Diploid. Having the somatic (body) number of chromosomes, i.e. twice as many as in the germ cells after reduction.

Disarticulate. To become detached at a joint, as leaves at base of the petiole or leaflets at base of the petiolule.

Disciform. Disk-shaped.

Discoid. In Compositae having a rayless head; also the more or less tubular flowers constituting the disk or central part of the head.

Discolored. Of changed color, as in drying; two-colored, usually applied to leaves with the two surfaces differently colored.

Disk. A more or less flat circular body, especially as found in many flowers at the bottom of the calyx and often bearing the stamens and petals; in Compositae the central part of the head as distinguished from the rays.

Dissected. Deeply and usually finely cut or divided, as are many leaves.

Distal (adv. *distally*). Opposite the point of attachment, apical, away from the axis.

Distended. Swollen.

Distichous. In two vertical ranks.

Distinct. Separate, not united.

Diurnal. Referring to flowers opening during the day, as contrasted with nocturnal or vespertine.

Divaricate. Widely spreading.

Divergent. Spreading apart.

Divided. Lobed or segmented to the base.

Division. Segment of a parted or divided leaf or other organ.

Dolabriform. Hatchet-shaped, applied to two-armed hairs attached at or toward the middle and tapering to each end, as contrasted with basifixed.

Dorsal (adv. *dorsally*). Referring to the back or surface away from the axis.

Dorsifixed. Attached on or by the back.

Dorsiventral. Possessing an upper (ventral) and a lower (dorsal) surface.

Dorsoventral axis. The axis through an organ from the dorsal to the ventral side.

Drupe (adj. *drupaceous*). A fruit, fleshy or leathery externally, with a hard stone containing the seed, such as a plum.

Drupelet. A small drupe, sometimes aggregated as in the blackberry.

E– (prefix). Without, lacking, as ebracteate (without bracts), eglandular (without glands), epappose (without pappus).

Echinate. Beset with prickles.

Ecology (adj. *ecological*). Science of the relationships of organisms to their environment.

Ecotype. A variant adapted to only one kind of environment within the range of the species.

Egg cell. The female germ cell or oosphere that, after union with a male germ cell, develops into a new plant.

Ellipsoid. A solid body of elliptic outline.

Elliptic. Nearly oblong, but with curving sides and ends.

Elongate. Drawn out, lengthened.

Emarginate. Notched, usually at apex.

Embryo. The young plant of the next generation still enclosed in the seed or the archegonium.

Embryonic. Pertaining to the embryo; in an early stage.

Embryo sac. The macrospore or cell of the ovule in which the embryo develops.

Emersed. Projecting above the surface of water.

Endemic. Confined to a limited geographic area.

Endocarp. Inner layer of a pericarp.

Endogenous (adv. *endogenously*). Growing mainly by internal multiplication of tissue, as in monocotyledonous plants. (*Compare:* Exogenous.)

Endosperm. The albumen, or portion of a seed within the coats but outside the embryo.

Ensiform. Sword-shaped.

Entire. With a uniform, not indented margin.

Enzyme. An unorganized soluble ferment, such as diastase.

Ephemeral. Of very short duration.

Epidermis (adj. *epidermal*). The external layer or layers of tissue immediately beneath (or including) the cuticle or outermost skin of a stem, leaf, or other organ.

Epigaeous, epigeous. Growing upon or close to the ground; not aquatic.

Epigynous. Borne on the pistil or apparently so, as are sometimes the petals and stamens.

Epiphyte (adj. *epiphytic*). An air plant, growing upon other plants but not parasitic.

Equisetoid. Resembling *Equisetum.*

Equitant. Folded lengthwise over one another, as the ensiform leaves of *Iris.*

Erose. Irregularly indented, as if gnawed.

Exalbuminous. Without endosperm, the embryo occupying the whole cavity of the seed.

Excavated. Concave, as though dug out, as the seeds of some species of *Phacelia, Plantago,* etc.

Excurrent. Applied to nerves prolonged beyond the apex of a leaf or other organ into a mucro, awn, etc.

Excurved. Curved outward.

Exfoliating. Separating in flakes or plates, as often the outer bark.

Exocarp. Outer layer of a pericarp.

Exogenous. Growing by peripheral increase of woody tissue underneath the expanding bark, as in dicotyledonous trees and shrubs.

Expanded. Spread out, fully open.

Explanate. Spread out flat.

Exserted. Projecting beyond, as stamens from a corolla. (*Compare:* Included.)

Exstipulate. Without stipules.

Extra-axillary. Outside of but usually near the axil, as certain flowers.

Extrafloral. Outside the flower, as extrafloral nectaries.

Extrorse. Directed outward, as in the dehiscence of anthers.

Exudate. Anything excreted, as water, salts, resin, wax, etc., often forming a deposit on the surface of a plant.

Faceted. With several plane faces, like those of a crystal, as in some seeds.

Facial. On the face, as distinguished from the edges or sides.

Falcate. Curved like a sickle.

Farinose. Floury, covered with meallike particles, as often in *Chenopodium.*

Fascicle (adj. *fascicled, fasciculate*). A bundle or close cluster of stems, leaves, flowers, etc.

Fastigiate. Clustered and erect, as the branches of some trees.

Faveolate. Alveolate, with a surface like honeycomb, as are the receptacles of many Compositae.

Feather-veined. With secondary veins branching from the midvein.

Fenestrate. Pierced with windowlike holes or transparent areas.

Fertile. Designating flowers that possess functional pistils and are capable of producing seeds; also applied to stamens that have functional pollen.

Fibrillose. Furnished with fine fibers or hairlike appendages.

Fibrous. Possessing or resembling fibers, as fibrous roots.

Filament. The stalk of an anther; a threadlike body, as on the margins of *Yucca* leaves.

Filamentose, filamentous. Furnished with or resembling threads.

Filiferous. Bearing threads, as sometimes the leaf margin.

Filiform. Threadlike.

Fimbriate. Fringed, having elongate, slender processes on the margin.

Fimbrilla (pl. *fimbrillae,* adj. *fimbrillate, fimbriolate*). A diminutive fringe.

Fistulose, fistulous. Hollow, like the scape of an onion.

Flabellate, flabelliform. Fan-shaped.

Flaccid. Flabby, limp, lax.

Flexuose, flexuous. Bent or curved alternately in opposite directions, zigzag.

Floccose. Bearing or covered with tufts of soft, woolly hairs.

Floral. Pertaining to flowers, as floral leaves or bracts.

Floret. A small flower; in grasses, the stamens and pistil with the subtending lemma and palea.

Floriferous. Bearing flowers, flowering abundantly.

Floristic. Relating to vegetation with respect to the component species, their geographic distribution, relative abundance, etc.

Foliage leaves. Ordinary, well-developed leaves as distinguished from bracts, etc.

Foliaceous. Resembling a foliage leaf in texture and size, often applied to bracts and calyx lobes when of this character.

Foliolate. Referring to the number of leaflets in a compound leaf, as bifoliolate, trifoliolate.

Follicle (adj. *follicular*). A dry fruit opening on only one (the ventral) suture.

Fovea (pl. *foveae,* adj. *foveate*). A shallow depression or pit, often glandular.

Frond. The conspicuous leaflike structure or sporophyll in ferns, produced in the nonsexual (sporophytic) generation.

Frutescent, fruticose. Shrubby.

Fruticulose. Slightly shrubby, or small and shrubby.

Fugacious. Soon falling, as sepals and petals that are deciduous very shortly after anthesis.

Fulvous. Tawny, dull yellow.

Functional. Able to perform, not abortive or rudimentary, here used in the sense of fertile, as of stamens containing viable pollen or pistils capable of producing seeds.

Funnelform. Shaped like a funnel, as a corolla gradually expanded upward.

Fuscous. Dusky, brownish gray.

Fusiform. Spindle-shaped, more or less terete and tapering at both ends.

Galea (adj. *galeate*). A helmet-shaped petal or corolla lip, the uppermost or axial one, as in *Aconitum* and *Castilleja;* the term "galeate" also applied to the helmet-shaped carpels of certain Malvaceae.

Gametophyte. The sexual generation of a plant, as the prothallium of ferns, contrasted with the much larger asexual generation or sporophyte.

Gamopetalous. With petals more or less united, sympetalous.

Gamophyllous. With segments more or less united, as in involucres, calyxes, and corollas.

Gamosepalous. With sepals more or less united.

Geminate. In pairs, twin.

Geniculate. Abruptly bent, like the knee joint.

Germ cell. A reproductive cell, usually limited to the female (egg) cell.

Gibbous. Swollen, protuberant, as often the base of the calyx.

Glabrate, glabrescent. Nearly, or tending to become, hairless.

Glabrous. Without pubescence, hairless.

Gland (adj. *glandular*). A secreting structure, either borne on the surface (glandular hairs, etc.) or imbedded in the tissues, as are many oil glands.

Glanduliferous. Bearing glands.

Glaucescent. Somewhat glaucous, or becoming glaucous.

Glaucous. Covered and more or less whitened with bloom.

Globose, globular. Spherical or nearly so.

Glochid (pl. *glochidia,* adj. *glochidiate*). A hair or prickle with retrorse, hooklike projections.

Glomerate. Crowded, as in a headlike inflorescence.

Glomerule (adj. *glomerulate*). A small headlike inflorescence or division of an inflorescence.

Glucoside. One of the complex organic substances in plants that decompose into sugar (glucose) and other products.

Glumaceous. Resembling a glume.

Glume. In grasses the one or two empty, scalelike bracts at base of the spikelets; sometimes applied also to the inflorescence scales of sedges.

Glutinous. Sticky, gluelike.

Graduated. With rows or series of different length, as the phyllaries (bracts) of the involucre in Compositae.

Granules (adj. *granular, granulate, granuliferous*). Small particles, often resinous, on the surface of an organ.

Grenadine. Bright red, the color of pomegranate blossoms.

Guide lines. Lines of different (usually darker) color on the lower side of the throat of insect-pollinated flowers, directed toward the nectary.

Gynaecandrous. Having the staminate flowers below the pistillate ones, as in the spikes of many species of *Carex.*

Gynobase. A prolongation or enlargement of the receptacle or torus on which the pistil or pistils are borne.

Gynoecium, gynecium. The pistil or the pistils collectively.

Gynophore. The stalk or stipe of a pistil when not sessile.

Gypseous. Consisting, wholly or partly, of gypsum.

Habitat. The kind of situation in which a plant grows, as woods, meadows, etc.

Halophyte (adj. *halophytic*). A salt-tolerant plant.

Haploid. Having half of the diploid (somatic) number of chromosomes, as in the germ cells after the reduction division.

Hastate (adv. *hastately*). Similar to Sagittate, but with the basal lobes spreading.

Haustoria. The suckerlike attaching and penetrating organs of parasites, such as *Cuscuta.*

Head. A very dense, usually rounded inflorescence with a very short axis.

Heartwood. The oldest, innermost wood, enclosing the pith.

Hemispheric. Half spherical.

Hemispheroidal. Half spheroidal.

Herb. A plant that is not woody, or not woody above ground.

Herbaceous. Like or pertaining to an herb; of the texture of an ordinary foliage leaf.

Herbage. The leaves and stems, as distinguished from the flowers.

Hermaphrodite. Having functional stamens and pistils in the same flower.

Heterogamous. Having flowers of different sex, as often the ray and disk flowers in the heads of Compositae.

Heteromorphic, heteromorphous. Unlike in form or size, as sometimes the lengths of stamens and pistils on different plants of the same species or the several carpels of a fruit.

Heterosporous. Having spores of different size or shape.

Heterostyled. Having styles of different length or character.

Hexamerous. With the parts in sixes.

Hilum. The scar or "eye" of a seed, marking the point of attachment.

Hirsute. Hairy with rather long and stiff, spreading hairs.

Hirsutulous. Somewhat hirsute.

Hirtellate, hirtellous. Finely or minutely hirsute.

Hispid. Roughly hirsute or bristly.

Hispidulous. Minutely hispid.

Holarctic. Circumpolar, the northern regions of the Eastern and Western hemispheres.

Homochromous. Of uniform color.

Homogeneous. Of the same kind, uniform.

Homomorphic, homomorphous. Of uniform shape and size.

Homosporous. With spores all alike in size and shape.

Hood (adj. *hooded*). A deeply concave organ or part, as certain petals.

Host. A plant giving nourishment to a parasite, either another plant or an animal.

Hyaline. Very thin and translucent.

Hybrid. A plant resulting from cross-fertilization of one species with another.

Hybridization. The production of a hybrid.

Hydathode. A specialized, water-excreting organ, usually in the form of a pore or a hair.

Hydrophilous. Water-loving, referring to the habitat preference.

Hydrophyte. An aquatic plant.

Hypanthium. An enlargement of the receptacle or torus under, and often partly united with, the calyx.

Hypocotyl. The axis of an embryo, between the cotyledons and the radicle.

Hypogynous. Inserted beneath but free from the pistil or pistils.

Imbricate. Overlapping by the edge in regular succession, like tiles.

Immersed. Under water, as the whole or part of an aquatic plant.

Imparipinnate. Pinnate with a single terminal leaflet, odd-pinnate.

Imperfect. With some part or parts undeveloped, as in a unisexual flower.

Implicate. Twisted together or interweaved.

Impressed. Lying below the general surface, as often the leaf veins.

In– (prefix). In or not.

Incanous. Hoary with whitish pubescence.

Incised. Cut sharply and usually irregularly, as a leaf margin.

Included. Not exserted or protruding.

Incumbent. Lying upon, as an embryo with the back of one cotyledon against the radicle, or an anther against the inner face of its filament.

Incurved. Curved inward.

Indefinite (adv. *indefinitely*). Variable in number or too numerous to be counted readily, as often the stamens.

Indehiscent. Not opening, or opening only irregularly or very tardily.

Indeterminate. Applied to an inflorescence with no flower at the farther (distal) end of the axis.

Indigenous. Native, not introduced.

Indument. A covering, as of hairs, etc.

Induplicate. Having the edges folded inward and applied to each other by their outer faces but not overlapping.

Indurate, indurated. Hardened.

Indusium (pl. *indusia*, adj. *indusial, indusiate*). Outgrowth of the epidermis in ferns covering the sori.

Inequilateral. With sides of unequal length, asymmetric.

Inferior. Below, usually applied to an ovary adnate to and appearing as if below the calyx.

Infertile. Not fertile or viable.

Inflated. Swollen, bladderlike.

Inflexed. Bent sharply inward.

Inflorescence. The flowers collectively when not solitary; their arrangement on the axis as in racemes, umbels, etc.

Inframedial. Below the middle.

Infrastamineal. Below the stamens.

Innate. Borne at the apex, as an anther attached to the tip of its filament.

Innovation. A newly formed basal shoot, as in grasses.

Inodorous. Without odor.

Insecticide. A substance poisonous to insects.

Insectivorous. Capturing and digesting insects.

Inter– (prefix). Between.

Intercostal (adv. *intercostally*). Between the ribs or nerves.

Internerve. Space between two nerves, as in the glumes of grasses.

Internode. The space between two successive nodes of a stem.

Interpetiolar. Between the petioles or enclosed by the base of the petiole, as the bud in *Platanus;* also applied to the connate stipules of opposite leaves.

Interrupted. Referring to an inflorescence with intervals, mostly of varying length, between the flowers or cymules.

Intervenous. Relating to the spaces between veins, as intervenous tissue.

Intricate (adv. *intricately*). Tangled together, as the branches of some shrubs.

Introflexion. State of becoming inflexed.

Introrse (adv. *introrsely*). Turned inward, as anthers opening toward the axis and stigmas decurrent on the axial side.

Introversion (adj. *introverted*). Act or condition of being turned inward.

Intrusion. Projection inward, as septa or placentas into the cavity of the ovary.

Inverted. Reversed; opposite to the normal direction.

Investing. Surrounding, covering, or enclosing.

Involucel. A secondary involucre, subtending an ultimate division of an inflorescence, as in Umbelliferae, or a single flower, as in Malvaceae.

Involucre (adj. *involucral, involucrate*). A ring or rings of bracts, united or distinct, surrounding a group of flowers.

Involute. With edges rolled inward.

Iridescent. Showing several colors, as in a rainbow.

Irregular. Applied to flowers when the members of one or more whorls of parts are unlike in size or shape.

Joint. A unit of a segmented stem, especially in Cactaceae (*Opuntia*) ; a node.

Junction membrane. In Polemoniaceae the scarious pleats or folds between the ribs of the calyx.

Keel (adj. *keeled*). A central, dorsal ridge, like the keel of a boat; in Leguminosae the two lowest, more or less united petals, in *Polygala* the single lower petal.

Lacerate. With margins irregularly cut or cleft, as if torn.

Laciniate. Incised, with narrow lobes.

Lamella (adj. *lamellate*). A thin plate.

Lamina. The blade or expanded part of a leaf or leaflike organ.

Laminate. Having or separating into plates or layers.

Lanate, lanuginose. Covered with matted hairs, woolly.

Lanceolate. Narrow and tapering to the apex, broadest below the middle.

Lanulose. Somewhat woolly.

Lateral. At the side, not central or terminal, as lateral veins, leaflets, etc.; also in fruits and seeds, the surfaces between the dorsal and ventral surfaces.

Latex. The milky sap of Euphorbiaceae, Asclepiadaceae, etc., sometimes containing rubber.

Leaves. The (usually thin and expanded) organs, borne laterally on the stem. Strictly, the term "leaf" comprises the petiole and the stipules also, if present, but in practice it is commonly restricted to the blade only.

Leaflets. The divisions of a compound leaf.

Lemma. The scale or bract of a grass spikelet above the glumes, sometimes termed "flowering glume."

Lens-shaped, lenticular. Convex on both sides.

Lepidote. Covered with small scales or scurf.

Liana. A climbing woody plant.

Ligneous. Woody.

Ligulate. Strap-shaped; possessing a ligule.

Ligule. In grasses the (usually thin and scarious) appendage between the sheath and blade of the leaf; in Compositae the (usually expanded) blade of an outermost or ray flower.

Limb. The uppermost, usually expanded, part of a gamophyllous calyx or corolla, as distinguished from the tube and throat.

Linear. Narrow and of nearly uniform width.

Lineate. Marked with (usually darker-colored) lines.

Lingulate. Shaped like a tongue.

Lip. One of the two main divisions of a bilabiate calyx or corolla.

Littoral. Growing on the seashore.

Lobe (adj. *lobed*). A (usually more or less rounded) projection of a deeply indented leaf or other organ.

Lobulate. With small lobes, or shallowly lobed.

Locule (adj. *locular*). A compartment of an anther, ovary, or fruit.

Loculicidal (adv. *loculicidally*). Longitudinally dehiscent on the back, midway between (not through) the septa.

Loment. A leguminous fruit that is more or less constricted between the seeds, the one-seeded segments finally separating.

Longitudinal (adv. *longitudinally*). Lengthwise, as contrasted with transverse.

Lunate. Crescent-shaped.

Lyrate. Pinnatifid, with the rounded terminal lobe much larger than the others.

Macrospore. See: Embryo sac.

Malodorous. Having an unpleasant odor.

Malpighiaceous. Dolabriform, with hairs like those of Malpighiaceae.

Marcescent. Persistent after withering.

Marginal. On or pertaining to the margin of a leaf or other organ.

Mauve. Purplish pink.

Massula (pl. *massulae*). The masses in which the microspores are aggregated in Salviniaceae.

Mealy. See: Farinose.

Medial, median (adv. *medianly*). At or pertaining to the middle.

Megasporangium. The structure bearing or containing the megaspores, in ferns, etc.

Megaspore. The larger (female) spores in ferns, etc.

Megasporophyll. A leaflike structure bearing megaspores in ferns, etc.

Membranaceous, membranous. Thin, pliable.

Meniscoid. Concavo-convex, like a watch crystal.

Mephitic. Having a strong and disagreeable, skunklike odor.

Mericarp. A portion of a fruit with carpels separating at maturity, as the two carpels in Umbelliferae.

Merous. Suffix indicating number of parts, as 5-merous.

Mesic, mesophytic. Requiring a moderate quantity of water, as distinguished from xerophytic and hydrophytic.

Microclimate. An extremely local climate.

Microphyll (adj. *microphyllous*). A small-leaved plant.

Microsporangium. A sporangium producing microspores.

Microspores. The smaller (male) spores of ferns, etc.; pollen grains.

Microsporocarp. The structure producing the microsporangia, in *Azolla.*

Microsporophyll. The leaflike structure in ferns, etc. on which the microsporangia are borne.

Midlobe. The central or terminal lobe.

Midnerve, midrib, midvein. The central nerve, rib, or vein.

Monadelphous. Referring to stamens when all of them are united by their filaments, usually in a tubular structure.

Moniliform. Like a necklace or string of beads.

Monochasial. Applied to a cymose inflorescence with one main axis.

Monochromatic. Of one color.

Monoecious. Having stamens and pistils in separate flowers on the same individual plant.

Monopodial. Applied to a stem having a single, continuous axis.

Montane. Growing in the mountains.

Mottled. Marked with spots colored differently from the general surface.

Mucilaginous. Moist and sticky, like mucilage.

Mucro (adj. *mucronate*). A short, abrupt, often hardened point.

Mucronulate. Diminutive of Mucronate.

Multicipital. With several or numerous stems from a single caudex or taproot.

Multifoliolate. Referring to a compound leaf with numerous leaflets.

Muricate. With surface roughened with sharp, hard tubercles.

Murication. A sharp, hard tubercle.

Muriculate. Diminutive of Muricate.

Muticous. Blunt, not mucronate or awned.

Mycorrhizal. Resembling a mycorrhiza, the threadlike structure formed by the close association of fungus strands with the roots of higher plants.

Naked. Destitute of, as buds without scales, flowers without a perianth, a receptacle without chaff in Compositae, etc.

Nascent. In the formative stage, not completely developed.

Nectariferous. Possessing nectar, the sweet fluid excreted on the surface of flowers and occasionally on other parts of plants.

Nectary. A nectar-secreting gland or the organ containing such glands.

Nerve (adj. *nerved*). An unbranched vein.

Neuter (adj. *neutral*). Without sex, as flowers lacking stamens and pistils, sometimes applied, in Compositae, to flowers having an ovary but without style or stigmas.

Nigrescent. Turning black.

Nocturnal. Flowering at night.

Nodding. Hanging down.

Node (adj. *nodal*). The place on a stem where a leaf is borne, often more or less swollen, as in grasses.

Nodulose. Having small knots or transverse septa.

Non– (prefix). Not, without.

Nut. A one-seeded, indehiscent, usually hard-shelled fruit.

Nutlet. Diminutive of nut, applied to the indehiscent, one-seeded segments of the fruit of Boraginaceae, Labiatae, etc.

Ob– (prefix). Inversely, upside down, as obconic (conic with the pointed end basal), obcordate (cordate with the notch apical), obdeltoid (deltoid with the pointed end basal), oblanceolate (inversely lanceolate), obovate (inversely ovate), obovoid (inversely ovoid), obpyramidal (inversely pyramidal).

Obcompressed. Flattened dorsoventrally instead of laterally.

Obligate. Applied to a parasite incapable of independent existence.

Oblique (adv. *obliquely*). Slanting

Oblong. Longer than wide, with sides parallel or nearly so, but broader than linear.

Obsolete. Almost or quite wanting, very rudimentary.

Obtuse. With the end rounded or blunt.

Ocrea. A stipule or pair of opposite stipules united to form a more or less tubular sheath.

Ochroleucous. Yellowish white.

Odd-pinnate. Pinnate with a single terminal leaflet.

Oil tubes. The oil containers in the fruits of Umbelliferae.

Operculate. Transversely dehiscent, opening by a lid.

Opposed, opposite. In front of, as a stamen in front of a petal; against one another, as two leaves at the same node.

Orbicular. Flat and of circular outline.

Organ. Part of a plant having a definite function, as a leaf, stamen, etc.

Orifice. A mouthlike opening.

Oval. Broadly elliptic.

Ovary. The lower part of the pistil, containing the ovule or ovules.

Ovate. Flat but having the outline of an egg and broadest below the middle.

Ovoid. Egg-shaped.

Ovulate. Referring to the number of ovules, as one-ovulate.

Ovule. One of the structures in the ovary that develop into seeds after fertilization.

Palate. A projection or appendage in the throat of a sympetalous, bilabiate corolla.

Pale. One of the chaffy scales on the receptacle of Compositae.

Palea (pl. *paleae*, adj. *paleaceous*). In grasses the innermost bract of the floret, opposite the lemma and usually two-keeled; in certain Compositae one of the scarious, scalelike segments of the pappus.

Pallid. Somewhat pale, light-colored.

Palmate (adv. *palmately*). Lobed or veined outwardly from the base, like the fingers of a hand.

Palmatifid. Palmately cleft or lobed.

Palmatisect. Palmately divided.

Palustrine. Inhabiting wet ground, marsh-dwelling.

Panicle (adj. *paniculate*). A compound inflorescence, composed of several racemes, corymbs, etc.

Pannose. Feltlike, covered with closely interwoven hairs.

Papain. A protein-digesting enzyme.

Papilionaceous. Butterfly-shaped, as the irregular corolla of the pea and most Leguminosae.

Papilla (pl. *papillae*). A small, soft projection, as on the surface of leaves, etc.

Papillate, papillose. Bearing papillae.

Pappus (adj. *pappose*). The limb at the apex of the achene in Compositae, consisting of hairs, bristles, awns, or scales, presumably a modified calyx.

Parasite (adj. *parasitic*). An organism living in or upon and deriving nourishment from another organism.

Parietal. Borne upon the wall, as placentas that are not axial.

Parted. Cleft nearly to the base.

Parthenogenetic. Developing from the egg cell asexually, without union with a male sex cell.

Parti-colored. Variegated, of different colors.

Pectinate (adv. *pectinately*). Comblike, having narrow, closely set teeth or divisions.

Pedate (adv. *pedately*). Palmate, with the lateral lobes or divisions two-cleft.

Pedicel (adj. *pedicellate, pedicelled*). The stalk of a single flower, or, in Compositae, an ultimate branch of an inflorescence, bearing a head.

Peduncle (adj. *pedunculate, peduncled*). The stalk of a flower cluster, sometimes of a single flower if the latter represents a reduced cluster.

Pellicle. A very thin, skinlike covering.

Pellucid. Transparent or nearly so.

Peltate (adv. *peltately*). Referring to a leaf attached to the petiole by its lower surface, not its base, or to a scale or other structure similarly attached.

Pendent, pendulous. Hanging, drooping.

Pentagonal. Five-angled.

Pentamerous. Of five parts, as a flower with five petals.

Perennial. Living several (usually more than two) years.

Perfect. Applied to a hermaphrodite flower, having functional stamens and pistil; and to a functional stamen or other organ.

Perfoliate. Referring to a leaf with base surrounding the stem, so that the stem apparently passes through it.

Perforate. Pierced with openings or having translucent dots that resemble holes.

Perforation. A windowlike opening.

Perianth. The outer envelope or envelopes of a flower comprising both calyx and corolla, when the latter is present. Term usually employed when there is no marked differentiation of calyx and corolla, as in Cactaceae.

Pericarp. The outer wall of an ovary after fertilization, hence of the fruit.

Perigynium. The more or less flask-shaped bract of *Carex* enclosing the achene.

Perigynous. Applied to petals and stamens when surrounding the pistil or pistils and borne upon a ring or cup in the throat of the hypanthium, as in many Rosaceae.

Peripheral. External, surrounding.

Perispore. The covering of a spore in *Equisetum*, etc.

Persistent. Remaining long-attached, not caducous.

Petal. One of the segments of the inner floral envelope or corolla.

Petaliferous. Bearing petals.

Petaloid. Like a petal in texture or color, as a petaloid sepal or a stamen with an expanded, bladelike filament.

Petiolar. Relating to or borne on the petiole, as a petiolar gland.

Petiole (adj. *petiolate, petioled*). The foot stalk of a leaf.

Petioliform. Resembling a petiole.

Petiolule (adj. *petiolulate*). The foot stalk of a leaflet, in compound leaves.

Phalange. Two or more stamens when partly united by the filaments.

Phyllary (pl. *phyllaries*). One of the bracts of the involucre, in Compositae.

Phytogeography (adj. *phytogeographic*). Plant geography, the science of plant distribution.

Pilose. Hairy with distinct, soft, more or less spreading hairs.

Pilosulous. Diminutive of Pilose.

Pinna (pl. *pinnae*). A primary division of a pinnately decompound leaf.

Pinnate (adv. *pinnately*). Applied to a compound leaf when the leaflets are arranged along the rachis or stalk; and to veins when branching from along the midrib (feather-veined).

Pinnatifid. Pinnately cleft, the clefts not extending to the midrib.

Pinnatilobate. Pinnately lobed.

Pinnatisect. Pinnately divided, the clefts extending to the midrib.

Pinnule. A leaflet or ultimate segment of a compound leaf.

Pistil. The female (central) organ of a flower, comprising the ovary, style when present, and stigma.

Pistillate. Referring to an inflorescence or flower possessing pistils but not functional stamens.

Pith (adj. *pithy*). The central tissue of a (usually exogenous) stem, composed of commonly thin-walled cells.

Placenta (pl. *placentae, placentas*). The structure in an ovary to which the ovules are attached.

Plait. A flattened, usually longitudinal, fold, as in some corollas.

Plano-convex. Flat on one side, convex on the other.

Pleiochasium. A compound cyme with more than two branches at each division.

Plicate. Folded, usually longitudinally, into plaits.

Plumose. Feathered with fine hairs, as often the awns of grasses and the pappus bristles of Compositae.

Plumule. The bud or growing point of an embryo.

Pluricellular. Of several or many cells.

Pluriseriate. In several or many series.

Pod. A generalized term for a dry, usually dehiscent, fruit containing several or many seeds.

Pollen. The microspores of seed plants, contained in the anther, which give rise to the male germ cells.

Pollen grain. An individual pollen cell, the microspore of seed plants.

Pollination. Deposition of pollen upon the stigma.

Polliniferous. Containing pollen.

Pollinium (pl. *pollinia*). Pollen coherent in a mass, as in Orchidaceae and Asclepiadaceae.

Polygamo-dioecious. With some of the plants polygamous, others dioecious in the same species.

Polygamous. With both perfect and unisexual flowers on the same individual plant or on different individuals of the same species.

Polygonal. Having several angles and sides.

Polymorphic. Of several forms, as a highly variable species.

Polyploid. With chromosome number in the somatic (body) nuclei greater than the diploid number, often accompanied by increased size and vigor of the plant.

Pome. A two- to several-celled, inferior, more or less fleshy fruit, as the apple.

Pore. A small but definite opening, as in certain otherwise indehiscent anthers and fruits.

Porrect. Extended outward or horizontally.

Posterior. Behind, toward the axis, ventral.

Prickle. A more or less rigid, often hooked outgrowth of the bark or subepidermal tissue.

Prismatic. With several angles and flat faces, like a prism.

Process. An outgrowth or appendage.

Procumbent. Lying on the ground.

Produced. Drawn out, extended.

Proliferate (adj. *proliferous*). To propagate vegetatively, by buds, offshoots, etc.; to transform flowers into leafy shoots, as often in *Allium* etc.

Prominent. Raised above the surface, as prominent veins.

Prophyllum. A bracteole.

Prostrate. Flat on the ground.

Protein. A complex, nitrogenous, organic substance.

Prothallium (pl. *prothallia*). The small structure representing the gametophytic generation in ferns etc., resulting from germination of a spore and producing the sex organs.

Proximal (adv. *proximally*). Nearest the axis or the base, as contrasted with Distal.

Pruinose. Covered with powdery wax, as the bloom of plums, etc.

Pseudoscape. The more or less elongate subterranean stem of an apparently acaulescent plant, as in certain Umbelliferae.

Puberulence (adj. *puberulent, puberulous*). Fine or minute pubescence.

Pubescence (adj. *pubescent*). Hairiness.

Pulverulent. Covered with a dusty or powdery deposit.

Pulvinate. Cushionlike, having short, very crowded stems.

Punctate. Marked with dots, pits, or translucent glands.

Puncticulate. Diminutive of Punctate.

Punctiform. Reduced to a point, not expanded.

Pungent. Sharply and rigidly pointed; hot to the taste.

Pustule (adj. *pustular, pustulate*). A small blisterlike elevation, as the hair base in many Boraginaceae.

Pustuliferous. Bearing or covered with pustules.

Pyramidal. Shaped like a pyramid, with a broad base and sides sloping to a point.

Pyriform. Pear-shaped.

Quadrangular. Four-angled.
Quadrate. Square.
Quadrilateral. Four-sided.
Quadripinnatifid. Four times pinnatifid.
Raceme (adj. *racemiform, racemose*). An unbranched, indeterminate, more or less elongate inflorescence with pedicelled flowers.
Rachilla. The axis or rachis of the spikelet of grasses.
Rachis. The axis of an inflorescence or of a pinnately compound leaf.
Radial (adv. *radially*). Spreading from the center, developing uniformly around a central axis; in Cactaceae the lateral spines as distinguished from the central spine of an areole.
Radiate (adj. *radiant*). To spread out from a central point, like the rays of a star. In Compositae, having ray flowers, as distinguished from Discoid.
Radical. Pertaining to the root. Basal leaves as distinguished from cauline or stem leaves.
Radicant. Striking root, as often at the nodes of prostrate stems.
Radicle. The primary rootlet of an embryo.
Range. Geographical distribution.
Raphe (adj. *raphal*). In anatropous ovules a cord or ridge of vascular tissue extending from the base of the nucellus or body of the ovule to the placenta.
Ray. One of the arms of a stellate hair; one of the branches of an umbel; one of the outer (more or less strap-shaped) flowers in heads of Compositae.
Receptacle. The axis of a flower; in Compositae the (usually expanded and thickened) axis of the flower head.
Reclining. Bending or curving toward the ground.
Rectangular. With the four sides forming right angles.
Recurved. Curved backward or downward.
Reduced. Not fully developed, smaller than normal.
Reflexed. Bent more or less abruptly downward or backward.
Refoliate. To put forth leaves again, as after rain.
Refracted. Bent sharply backward, from the base.
Regular. Referring to flowers when the members of each whorl of parts, as the petals, are alike in size and shape.
Remote. Distant, scattered, as flowers in an inflorescence.
Reniform. Kidney-shaped, deeply notched at base and broader than long.
Repand (adv. *repandly*). Having a slightly uneven or undulate margin.
Repent. Creeping, with stems prostrate and emitting roots.
Replum. In Cruciferae, the persistent framelike placenta from which the valves separate at maturity of the fruit.
Resin (adj. *resinous*). An organic substance (hydrocarbon) insoluble in water, becoming hard when exposed to air, e.g. turpentine.
Resin canal, resin duct. The tubular components of the wood in which resin is deposited.
Resupinate. Appearing upside down as a result of a twisting of the axis.
Reticulate. Forming a network, as do the veins of many leaves.
Reticulation, reticulum (pl. *reticula*). A network of fibers or veins.
Retrorse (adv. *retrorsely*). Turned backward or downward.
Retuse. Rounded and shallowly notched at apex.
Revolute. Rolled back or under from apex or margin.
Rhizome (adj. *rhizomatous*). A rootstock or prostrate stem, on or just below the ground surface, rooting at the nodes and producing buds.
Rhombic. With equal sides forming oblique angles.
Rhomboid, rhomboidal. Similar to rhombic but with the adjacent sides unequal.
Rib (adj. *ribbed*). A primary vein, especially when thick or prominent.
Rigid. Stiff, not easily bent.
Root. The portion of the plant that absorbs water and nutrients, usually descending from the base of the stem and underground.
Rootstock. See: Rhizome.
Roseate. Tinged with pink.
Rosette. A cluster of basal leaves, more or less symmetrically arranged.
Rostrate. Narrowed to a beaklike point or projection.
Rosulate. In a rosette.
Rotate. Wheel-shaped, usually applied to a circular, nearly flat, short-tubed, sympetalous corolla.
Rotund. Of rounded outline.
Rudiment (adj. *rudimentary*). A much-reduced, nonfunctional organ, a vestige.
Rugose. Wrinkled.

Rugulose. Diminutive of Rugose, slightly or finely wrinkled.

Runcinate. Sharply incised pinnately, with the teeth or lobes retrorse.

Runner. See: Stolon.

Sac (adj. *saccate, saclike*). A pouch or bag.

Sacculate. Diminutive of Saccate, furnished with a small pouch.

Sagittate. Shaped like an arrowhead, with the two basal lobes directed downward and usually acute.

Salient. Projecting outward, as salient teeth or prominent ribs.

Salinity (adj. *saline*). The state of containing salt, as soil salinity.

Salverform. Applied to a sympetalous corolla with limb expanding abruptly and more or less at a right angle to the tube.

Samara. An indehiscent, winged fruit as in maple and ash.

Saponin. A lather-forming, more or less poisonous glucoside.

Saprophyte (adj. *saprophytic*). A plant, often without chlorophyll, living upon dead organic matter.

Scaberulous, scabrellate. Minutely scabrous.

Scabrid. Slightly scabrous.

Scabrous. Rough to the touch.

Scale. A much-reduced leaf or an epidermal outgrowth, usually thin and scarious, applied also to thin appendages of the corolla, as in *Cuscuta,* and to the glumes of the spikelets in Cyperaceae.

Scape (adj. *scapose*). A peduncle or a flower-bearing stem, leafless or with leaves reduced to bracts, arising from the ground.

Scapiform. Like a scape.

Scarious. Thin and dry, not green.

Schizocarp. A fruit that splits septicidally at maturity into separate, usually one-seeded, carpels.

Sclerenchyma (adj. *sclerenchymatous*). Tissue in stems and other organs composed of very thick-walled cells, the bast cells.

Sclerotic. Hardened, stony.

Scorpioid. Like a scorpion's tail, applied to one-sided inflorescences with a coiled axis.

Scurf (adj. *scurfy*). Small branlike scales on the surface of a leaf or other part.

Secondary. Departing from the primary axis or the midrib, as secondary branches and veins.

Secund. One-sided.

Seed. The structure developing from the ovule after fertilization and containing the embryo of the plant of the next generation.

Segment (adj. *segmented*). Part of a compound leaf or other organ, especially if there are several like parts.

Seleniferous. Containing selenium.

Selenium. A poisonous chemical element found in some soils and in plants growing on them.

Self-pollination. Transfer of pollen from the anthers to the stigma of the same flower or another flower on the same individual plant.

Semi– (prefix). Literally half, but used also in the sense of partly or nearly.

Sensitive. Reacting to a touch or other stimulus, as *Mimosa* leaves.

Sepal. One of the segments of the calyx or outer envelope of the flower.

Septicidal. Dehiscing through the partitions so as to divide the fruit more or less completely. (*Compare:* Loculicidal.)

Septum (pl. *septa,* adj. *septate*). A partition, usually longitudinal in fruits, transverse in stems and leaves.

Seriate. Arranged in one or more rows, these either transverse or longitudinal.

Sericeous. Covered with appressed, straight, silky hairs.

Serrate. Saw-edged, with sharp teeth directed forward or upward.

Serrulate. Diminutive of Serrate.

Sessile. Not stalked, without a petiole, pedicel, or stipe.

Seta (pl. *setae,* adj. *setaceous, setiform*). A bristle.

Setose. Bristly, bearing bristles.

Setulose. Diminutive of Setose.

Sheath. A more or less tubular organ, in grasses the lower part of the leaf enclosing the culm, in Polygonaceae the ocrea.

Shoot. A young branch, applied especially to the basal branches, as in *Phlox* etc.

Shrub (adj. *shrubby*). A woody plant, smaller than a tree and several-stemmed, without a trunk.

Sigmoid. Curved twice, in opposite directions, like the letter "S."

Silex (adj. *siliceous*). Silica; a surface deposit like fine sand, as in *Equisetum.*

Silicle. A short silique.

Silique. The fruit of Cruciferae, with the two valves usually separating at maturity from the frame or replum.

Simple. Not compound, as unbranched stems, undivided leaves, pistils with a single carpel, etc.

Sinuate, sinuous (adv. *sinuately*). Having a deeply wavy margin.

Sinus (pl. *sinuses*). An indentation or recess, as between lobes or at base of a leaf.
Solitary. Alone, not clustered.
Somatic. Pertaining to the body, as all cells of a plant except the germ cells.
Sorus (pl. *sori*). Groups of sporangia in ferns.
Spathe (adj. *spathaceous*). A specialized, often large, bract enclosing the flower cluster, especially in Araceae.
Spatulate, spathulate. Like a spatula or spoon, expanded from a narrow basal part and broadest toward apex.
Spermatozoid. The motile male gamete or sex cell in ferns, etc.
Spherical. Perfectly round.
Spheroidal. Nearly spherical.
Spike (Adj. *spicate, spiciform, spikelike*). An unbranched, usually elongate, indeterminate inflorescence, differing from a raceme in having the flowers sessile.
Spikelet. The unit of the inflorescence in grasses, comprising usually two glumes and one or more florets.
Spine (adj. *spinose, spiny*). A sharp-pointed, usually woody structure, being a modified branch, petiole, or stipule. (*Compare:* Thorn.)
Spinescent. Becoming, or somewhat, spiny.
Spiniferous. Bearing thorns or spines.
Spinulose. Diminutive of Spinose.
Spongiose. Spongy, soft.
Sporadic. Occurring here and there, widely scattered.
Sporangiophore. A structure on which a sporangium is borne.
Sporangium (pl. *sporangia*). A (usually saclike) structure in ferns etc. in which the spores are contained.
Spore. A reproductive cell, especially of ferns and other nonflowering plants.
Sporocarp. The capsulelike structure containing the sporangia in *Marsilea* etc.
Sporophyll. The spore-bearing structure in ferns etc., known as a frond when large and leaflike.
Sporophyte. The larger, asexual or diploid generation in ferns etc.
Spur (adj. *spurred*). A slender projection, this hollow and nectariferous in many flowers.
Squamella (pl. *squamellae*, adj. *squamellate*). A small scale, as in the pappus of certain Compositae.
Squamiform. Scalelike.
Squamose. Resembling or bearing scales, scaly.
Squamulose. Diminutive of Squamose.
Squarrose. Rough with outstanding tips or processes.
Stamen. The male or pollen-producing organ in flowering plants.
Stamen column, stamen tube. A stalk or tube bearing the stamens, usually formed by coalescence of the lower part of the filaments, as in many Leguminosae and Malvaceae.
Staminate. Applied to male flowers or inflorescences having stamens only, the pistils, if present, rudimentary and nonfunctional.
Staminode, staminodium (pl. *staminodia*). A reduced or abortive stamen, producing no pollen.
Standard. See: Banner.
Stele. The portion of an axis or stem containing the vascular tissue.
Stellate. Star-shaped, applied mostly to hairs with several radiating arms.
Stem. The axis or portion of a plant above the root on which the leaves and flowers are borne.
Sterile. Not fertile, as branches or shoots not bearing flowers, flowers not producing seeds, and stamens not producing viable pollen.
Stigma (adj. *stigmatic*). The (usually enlarged) apical part of the pistil on which the pollen is deposited and germinates.
Stigmatiferous. Bearing a stigma.
Stipe (adj. *stiped, stipitate*). In ferns the stalk of a frond; the stalk of a pistil or gynecium; the stalk of an anther cell.
Stipe bundle. The vascular bundle or bundles of the stipe in ferns.
Stipel (adj. *stipellate*). The small stipule of a leaflet in compound leaves.
Stipule (adj. *stipular, stipulate*). An appendage of a leaf on each side of its insertion, borne at the base of the petiole when one is present.
Stipuliform. Shaped like a stipule.
Stolon (adj. *stoloniferous*). A basal branch or shoot often rooting at the nodes, a runner.
Stoma (pl. *stomata*, adj. *stomatal*). A minute, specialized structure in the epidermis having an aperture through which exchange of gases with the atmosphere takes place.
Stone. The hardened endocarp or inner portion of a drupe enclosing the seed.
Stramineous. Straw-colored.

Striate. Finely marked with parallel lines, ridges, or furrows.
Strict (adv. *strictly*). Very erect and straight.
Strigose. Bearing stiff, straight, appressed hairs.
Strigillose, strigulose. Diminutive of Strigose.
Strobile. See: Cone.
Style. The (usually slender) portion of a pistil, above the ovary, that supports the stigma.
Stylodium (pl. *stylodia*). The free tips of the styles, in *Geranium;* sometimes applied to stylelike stigmas, as in grasses and Compositae.
Stylopodium. The swelling at base of the styles in Umbelliferae.
Sub– (prefix). Somewhat; below or under.
Submerged, submersed. Under water, immersed.
Subtend (adj. *subtended*). Under or supporting, as a leaf or bract subtends a flower when borne in its axil.
Subterranean. Beneath the surface of the ground.
Subulate. Very narrow and tapering gradually from the base, awl-shaped.
Succulent. Fleshy, thick and juicy.
Sucker. A shoot originating below ground; *see also:* Haustoria.
Suffrutescent. Slightly woody toward base, barely shrubby.
Suffruticose, suffruticulose. Somewhat or slightly shrubby, applied especially to diminutive shrubs.
Sulcate. Furrowed or grooved.
Superior. Above or over; applied to an ovary when free from the perianth.
Supra– (prefix). Above, as suprabasal.
Suture. A seam, the external line of union of the carpels or of their halves, and of dehiscence in fruits.
Symmetric. Regular, capable of division into equal and like parts.
Sympetalous. With petals more or less united, same as gamopetalous.
Sympodial. Referring to a stem when development takes place by a succession of branches rather than by continuation of the original axis.
Taproot. The main or primary root, a continuation of the radicle of the embryo.
Taxonomy (adj. *taxonomic*). The science of classification of plants and animals.
Tendril. A slender organ, a modified leaf or branch, by means of which plants support themselves on other plants or on various objects.
Terete. Cylindric, with a circular cross section, not angled.
Terminal. At the summit or apex.
Ternate (adv. *ternately*). Arranged in threes, as branches, leaf divisions, etc.
Terrestrial. Rooted in soil, as distinguished from aquatic, epiphytic, etc.
Tessellate. Like a tiled floor or a checkerboard.
Testa. The outer (usually hard and shell-like) seed coat.
Tetradynamous. With two of the six stamens shorter than the other four, as in Cruciferae.
Tetragonal. Four-angled.
Tetrahedral. Four-sided.
Thallus (pl. *thalli,* adj. *thalloid*). A vegetative body without evident distinction of stem and leaf.
Theca. See: Anther sac.
Thorn. A shortened and hardened, sharp-pointed branch.
Throat. The more or less expanded upper part of a gamophyllous calyx or corolla between the tube and the limb.
Thyrse (adj. *thyrsoid*). A more or less contracted panicle with the main axis indeterminate and the ultimate flower clusters cymose.
Tissue. The aggregate of mainly similar cells, usually having a definite function, as epidermal tissue, vascular tissue.
Tomentellous, tomentulose. Diminutive of Tomentose.
Tomentum (adj. *tomentose*). A covering of soft, matted hairs.
Tooth. A small projection, shorter than a lobe.
Tortuous. Twisted or bent in different directions.
Torulose. More or less cylindric and constricted at intervals.
Torus. See: Receptacle.
Toxic. Poisonous.
Trailing. Lying on the ground but not rooting at the nodes.
Translator. The body to which the pollinia are attached, in Asclepiadaceae.
Translucent. Nearly transparent.
Transpiration. Emission of water vapor mostly through the stomata.
Transverse (adv. *transversely*). Across, at a right angle to the longitudinal axis.
Tree. A large, woody plant with an evident trunk below the branches.

Tri- (prefix). Three, as triaristate (three-awned).

Trichome. A hair or hairlike extension of an epidermal cell.

Trichotomous. Three-forked, with branches in threes at a node.

Trifid. Three-cleft.

Trifoliolate. With three leaflets.

Trifurcate. Three-forked.

Trigonous. Three-angled.

Trilocular. Having three locules or cells.

Tripinnate. Thrice-pinnate. (*Compare:* Bipinnate.)

Tripinnatifid. Thrice-pinnatifid.

Triplinerved. Having three principal veins from the base or just above it.

Triternate. Thrice-ternate.

Truncate. Even at apex or base as if cut off, not rounded.

Trunk. The main axis of a tree below the branches.

Tube (adj. *tubular*). A cylindric, hollow structure, especially the narrower basal part of a gamophyllous calyx or corolla, often gradually expanded into a throat. *See also:* Stamen column.

Tuber (adj. *tuberous, tuberlike*). A thickened underground branch, bearing buds and serving for storage of reserve food.

Tubercle, tuberculation (adj. *tuberculate*). A small knoblike projection.

Tuberìferous. Bearing tubers.

Tumid. Swollen, inflated.

Tunicate. Having successive, close-fitting coats, as in an onion bulb.

Turbinate. Top-shaped, inverted conic.

Turgid. Not flat; swollen or distended.

Turion. A short, scaly branch of a rhizome.

Twining. Climbing by having the stem winding spirally around the supporting object.

Ultimate. The last or final division, as the ultimate branch or lobe.

Umbel (adj. *umbellate, umbelliform*). An inflorescence, more or less flat-topped, in which all of the pedicels arise at the same point, like the ribs of an umbrella.

Umbellet. A secondary or diminutive umbel.

Umbilicate. Depressed in the center, like a navel.

Umbo (adj. *umbonate*). A central boss or low hump.

Uncinate. Hook-shaped, as are many hairs and prickles.

Undershrub. A low shrub or a plant that is woody only toward base, the stems herbaceous above.

Undulate. With a wavy margin or surface.

Uni- (prefix). One, as uniaristate (one-awned), unicellular (one-celled), unifoliolate (with one leaflet), unilocular (one-celled).

Unilateral. One-sided.

Uniseriate. In one horizontal row or circle, as the phyllaries of the involucre in some Compositae.

Unisexual. Of one sex, as flowers with functional stamens or with functional pistils only.

Urceolate. Urn-shaped or pitcher-shaped, i. e. contracted at or near the mouth.

Utricle (adj. *utricular*). Like an achene, but with a loose (usually thin and membranaceous) outer envelope.

Vallecula (adj. *vallecular*). The grooves between the ridges of a stem or between the ribs of a fruit.

Valve (adj. *valvate*). One of the segments of a longitudinally dehiscent fruit. Valvate also applied to the segments of a flower when not overlapping in the bud.

Vascular. Referring to the tissue in stems, leaves, etc. that contains the elongate tubes or ducts.

Vascular bundle. A group or strand of vascular tissue.

Vegetative. Referring to the stems, leaves, etc. as contrasted with the floral parts of a plant.

Vein. A strand of mostly vascular tissue in a leaf or leaflike organ, especially if branched. (*Compare:* Nerve.)

Velum. A fold of tissue partly enclosing the sporangium in *Isoëtes.*

Velutinous. Velvety, densely covered with fine, soft hairs.

Venation. The system of veins.

Venose, venulose. Veiny.

Ventral (adv. *ventrally*). Relating to the posterior, inner or upper, axial face.

Ventricose. Bellylike, swollen on one side.

Vermifuge. A worm expellant.

Vernation. The arrangement of leaves in the bud.

Verrucose. Warty, covered with small, wart-like swellings or projections.

Versatile. Swinging, referring especially to an anther when attached above its base to the filament.

Verticil (adj. *verticillate*). A whorl or circle of more than two similar parts at a node.

Verticillaster. Two opposite cymes, simulating a verticil of flowers.

Vesicle (adj. *vesicular*). A small, inflated or bladderlike structure.

Vespertine. Opening in the evening.

Vestigial. Applied to the remnant or trace of an organ no longer fully developed, rudimentary.

Villosulous. Diminutive of Villous.

Villous. Pubescent with long and weak but not matted hairs, long-pilose.

Virgate. Wandlike, erect and straight.

Viscid. Sticky.

Viviparous. Sprouting from buds or germinating from seeds still attached to the parent plant.

Whorl. See: Verticil.

Wing. A thin, expanded extension or appendage of a stem, fruit, or seed; one of the lateral petals of the flower of Leguminosae.

Winter annual. A plant germinating in autumn and flowering or fruiting the following spring.

Xeric, xerophilous. Growing in dry places, having a small water requirement.

Xerophyte (adj. *xerophytic*). A xerophilous or desert plant.

Index of Scientific and Common Names

The first page number after each scientific name indicates the principal entry of the respective family, genus, or species. Species are indexed only if the genus is represented in Arizona by more than 10 species. Synonyms are in italic type.

Abies, 56
Abronia, 277, 279
Abutilon, 538
Acacia, 397
Acaciella, 398, 399
Acaciopsis, 398
Acalypha, 506
Acamtopappus, 854
Acanthaceae, 798
Acanthochiton, 267
Acanthogonum, 230
Acanthus family, 798
Acer, 526
Aceraceae, 526
Acerates, 660
Achaetogeron, 867
Achillea, 936
Achyranthes, 269
Achyronychia, 300
Acleisanthes, 271
Aconitum, 310
Actaea, 305
Actinea, 926, 927
Actinella, 928
Adder's-mouth, 202
Adder's tongue, 32
Adder's tongue family, 31
Adelia, 643
Adenostegia, 791
Adiantum, 36
Aegopogon, 122
Aeschynomene, 472
Agaloma, 513
Agastache, 736
Agave, 192
Agoseris, 967
Agrimonia, 391
Agrimony, 391
Agropyron, 92
  arizonicum, 94
  cristatum, 93
  dasystachyum, 93
  desertorum, 93
  pauciflorum, 93
  pseudorepens, 93
  repens, 93
  riparium, 93
  Saundersii, 94
  saxicola, 94
  Scribneri, 94
  Smithii, 93
  spicatum, 94
  subsecundum, 94
  trachycaulum, 93
Agrostideae, 73
Agrostis, 102
Aizoaceae, 280
Alder, 215
Alepidocalyx, 482
Aletes, 613
Alfalfa, 421
Alfilaria, 486
Algerita, 320
Algodoncillo, 553
Alhagi, 471
Alisma, 68
Alismaceae, 68
Alkali Grass, 81
Alkali-sacaton, 112
Allenrolfea, 262
Allionia, 273, 274
Allium, 177
  acuminatum, 179
  Bigelovii, 181
  cernuum, 180
  cristatum, 181
  deserticola, 179
  funiculosum, 179
  Geyeri, 179
  glandulosum, 180
  Gooddingii, 180
  Kunthii, 180
  macropetalum, 179
  neomexicanum, 180
  nevadense, 180
  Palmeri, 181
  Parishii, 181
  Plummerae, 179
  rubrum, 179
  rhizomatum, 180
  sabulicola, 179
Allocarya, 722
All-scale, 258
All-thorn, 558
Alnus, 215
Alopecurus, 103
Aloysia, 729
Alsine, 293
Alternanthera, 269
Alum-root, 363
Amaranth, 265
Amaranthaceae, 264
Amaranth family, 264
Amaranthus, 265
  albus, 266
  blitoides, 266
  caudatus, 267
  cruentus, 267
  fimbriatus, 266
  gracilis, 266
  graecizans, 266
  hybridus, 267
  leucocarpus, 267
  obcordatus, 266
  Palmeri, 266
  Powellii, 267
  Pringlei, 266
  retroflexus, 267
  Torreyi, 266
  venulosus, 266
  viridis, 266
  Wrightii, 267
Amarella, 649

Amaryllidaceae, 191
Amaryllis family, 191
Amber-lily, 177
Ambrosia, 894
Amelanchier, 376
Ammannia, 588
Ammobroma, 629
Ammocodon, 271
Ammoselinum, 610
Amole, 185, 192
Amoreuxia, 557
Amorpha, 431, 429
Amphipappus, 853
Amsinckia, 723
Amsonia, 651
Anacardiaceae, 521
Anacharis, 69
Anagallis, 638
Anaphalis, 887
Androcera, 757
Andropogon, 141
Andropogoneae, 72
Androsace, 636
Androstephium, 182
Anemone, 311
Anemopsis, 207
Anethum, 620
Angelica, 620
Angel-trumpet, 271
Angiospermae, 62
Angle-pod, 665
Anisacanthus, 800
Anisocoma, 959
Anoda, 551
Anogra, 594, 596, 597
Antelope-brush, 391
Antelope-horns, 660
Antelope-sage, 240
Antennaria, 886
Anthemis, 936
Anthericum, 177
Antirrhinum, 765
Anulocaulis, 277
Apache-plume, 387
Aparejo Grass, 106
Apetalae, 203
Aphanostephus, 866
Apiastrum, 609
Apium, 613
Aplopappus, 858
acaulis, 861
acradenius, 862
armerioides, 861
cervinus, 862
croceus, 861
cuneatus, 861
Drummondii, 862
gracilis, 860
heterophyllus, 862
junceus, 860
laricifolius, 861
linearifolius, 861
Nuttallii, 860
Parryi, 861
salicinus, 862
scopulorum, 862
spinulosus, 860
suffruticosus, 861
tenuisectus, 862
Watsoni, 861
Apocynaceae, 651
Apocynum, 654
Apodanthera, 821
Apricot-mallow, 543
Aquilegia, 306
Arabis, 350
Araceae, 165
Aragallus, 470
Aralia, 605
Araliaceae, 605
Arbutus, 633
Arceuthobium, 225
Arctium, 950
Arctomecon, 324
Arctostaphylos, 633
Arenaria, 295
Argemone, 323
Argentina, 381, 384
Argythamnia, 506
Aristida, 118
adscensionis, 120
arizonica, 121
barbata, 120
californica, 119
divaricata, 120
Fendleriana, 120
glabrata, 119
glauca, 120
hamulosa, 120
longiseta, 120
oligantha, 120
Orcuttiana, 119
pansa, 120
Parishii, 121
purpurea, 120
ternipes, 119
Wrightii, 121
Aristolochia, 227
Aristolochiaceae, 227
Arizona-poppy, 492
Arizona-rosewood, 375
Arnica, 942
Arrhenatherum, 100
Arrow-grass, 67
Arrow-grass family, 67
Arrow-head, 68
Arrow-leaf, 843
Arrow-weed, 884
Artemisia, 938
albula, 940
annua, 941
biennis, 941
Bigelovii, 941
Carruthii, 940
dracunculina, 941
dracunculoides, 941
filifolia, 941
franserioides, 939
frigida, 941
glauca, 941
gnaphalodes, 940
ludoviciana, 940
mexicana, 940
nova, 942
pacifica, 941
petrophila, 941
pygmaea, 942
spinescens, 942
sulcata, 940
tridentata, 941
Wrightii, 940
Artichoke, 954
Arum family, 165
Arundo, 89
Asclepiadaceae, 655
Asclepiadeae, 656
Asclepias, 657
albicans, 660
angustifolia, 661
auriculata, 660
brachystephana, 662
capricornu, 660
cryptoceras, 662
Cutleri, 662
demissa, 662
elata, 663
Engelmanniana, 660
erosa, 662
fascicularis, 661
galioides, 661
Hallii, 662
hypoleuca, 660
involucrata, 662
latifolia, 663
Lemmoni, 662
Linaria, 661
linifolia, 661
lonchophylla, 662
macrosperma, 662
macrotis, 661
mexicana, 661
nummularia, 662
nyctaginifolia, 662
quinquedentata, 661

speciosa, 662
subulata, 661
subverticillata, 661
tuberosa, 661
uncialis, 662
viridiflora, 660
Asclepiodora, 660
Ash, 641
Asparagus, 190
Aspen, 207, 209
Aspicarpa, 497
Asplenium, 47
Astephanus, 657
Aster, 869
abatus, 873
adenolepis, 874
adscendens, 871
aquifolius, 874
arenosus, 872
Bigelovii, 874
Burkei, 872
canescens, 873
carnosus, 873
cichoriaceus, 874
coerulescens, 872
commutatus, 871
Eatoni, 872
exilis, 873
foliaceus, 872
frondosus, 873
glaucodes, 872
glaucus, 872
hesperius, 872
hirtifolius, 872
intricatus, 873
Lemmoni, 872
oregonus, 872
parviflorus, 873
parvulus, 873
pauciflorus, 871
riparius, 872
sonorae, 872
spinosus, 873
tagetinus, 873
tanacetifolius, 873
tephrodes, 874
Wootonii, 872
Astragalus, 445
accumbens, 461
agninus, 467
albulus, 461
allochrous, 463
amphioxys, 458
ampullarius, 464
argophyllus, 458
aridus, 463
arizonicus, 468
Arthu-schottii, 467
Astragalus (*cont'd*)
artipes, 462
Beathii, 464
Bigelovii, 469
Blyae, 460
Brandegei, 465
Bryantii, 467
calycosus, 467
canovirens, 457
castaneaeformis, 459
ceramicus, 462
chloridae, 459
cobrensis, 465
confertiflorus, 457
convallarius, 457
crassicarpus, 469
cremnophylax, 461
Crotalariae, 464
desperatus, 461
didymocarpus, 469
diphysus, 466
dispermus, 469
Egglestonii, 465
ensiformis, 468
episcopus, 457
eremiticus, 465
famelicus, 462
gilensis, 461
Hartwegii, 468
Haydenianus, 457
Hosackiae, 461
humillimus, 461
humistratus, 460
hypoxylus, 469
impensus, 456
insularis, 463
intermedius, 459
junciformis, 457
kaibensis, 457
Kentrophyta, 456
lancearius, 457
Layneae, 467
lentiginosus, 465
lonchocarpus, 460
Macdougalii, 466
Matthewsii, 469
micromerius, 461
mitophyllus, 462
moencoppensis, 458
mogollonicus, 469
mokiacensis, 467
musimonum, 462
naturitensis, 470
Newberryi, 459
niveus, 463
nothoxys, 468
Nuttallianus, 468
oophorus, 462
palans, 466
Palmeri, 464
Pattersoni, 464
Peirsonii, 463
pephragmenus, 459
phoenicis, 459
pictus, 462
pinonis, 462
praelongus, 464
Preussii, 464
recurvus, 465
remulcus, 459
Rothrockii, 464
Rusbyi, 465
sabulonum, 464
scaposus, 467
sesquiflorus, 461
Sileranus, 463
sonorae, 461
sophoroides, 458
striatiflorus, 467
subcinereus, 463
tephrodes, 459
tetrapterus, 460
Thompsonae, 469
Thurberi, 464
triquetrus, 463
troglodytus, 458
vaccarum, 468
vespertinus, 458
Wilsonii, 466
wingatanus, 457
Wootoni, 463
yuccanus, 467
zionis, 458
Astrophyllum, 493
Atamisquea, 358
Atelophragma, 465
Athel, 557
Athyrium, 48
Athysanus, 349
Atragene, 312
Atrichoseris, 958
Atriplex, 254
acanthocarpa, 258
argentea, 257
canescens, 259
caput-medusae, 257
collina, 259
confertifolia, 259
corrugata, 258
elegans, 258
fasciculata, 258
Garrettii, 259
Griffithsii, 259
hymenelytra, 259
Jonesii, 258
lentiformis, 259
linearis, 259

Atriplex (*cont'd*)
Nuttallii, 260
obovata, 258
patula, 257
polycarpa, 258
Powellii, 257
rosea, 257
sabulosa, 258
saccaria, 257
semibaccata, 257
Serenana, 258
Thornberi, 258
Wrightii, 258
Audibertia, 742
Audibertiella, 742
Avena, 100
Aveneae, 74
Avens, 386
Avicularia, 246
Ayenia, 555
Azolla, 33

Baccharis, 882
arizonica, 883
Bigelovii, 884
brachyphylla, 883
Emoryi, 883
glutinosa, 884
neglecta, 883
pteronioides, 883
ramulosa, 883
sarothroides, 883
sergiloides, 883
thesioides, 883
viminea, 884
Wrightii, 883
Bachelor's button, 956
Bacopa, 783
Baeria, 917
Bahia, 923
Baileya, 915
Ball-clover, 270
Ball-moss, 166
Bane-berry, 305
Barbarea, 338
Barberry, 320
Barberry family, 320
Barley, 96
Barnyard Grass, 137
Barrel cactus, 573
Basistelma, 663
Basket-flower, American, 956
Bassia, 261
Bastard-toadflax, 227
Batamote, 884
Batidophaca, 460, 461
Batrachium, 314
Bean, 481
Bean subfamily, 409
Bear-berry, 633
Beard Grass, 141
Beardtongue, 768
Bear-grass, 188
Beavertail cactus, 581
Bebbia, 912
Bedstraw, 808
Bee-balm, 743
Beech family, 215
Beehive cactus, 574
Bee-plant, 356
Beggar-ticks, 473, 909
Bellflower, 825
Bellflower family, 825
Bellota, 219
Beloperone, 802
Bent, 102
Bent Grass, 102
Benthamantha, 443
Berberidaceae, 320
Berberis, 320
Berginia, 800
Berlandiera, 890
Bermuda Grass, 124
Bernardia, 506
Berula, 615
Besseya, 786
Betony, 739
Betula, 215
Betulaceae, 214
Bidens, 909
aurea, 911
Bigelovii, 911
bipinnata, 911
ferulaefolia, 911
frondosa, 910
heterosperma, 911
laevis, 910
Lemmoni, 911
leptocephala, 911
leucantha, 911
persicaefolia, 910
pilosa, 910
tenuisecta, 911
Bignoniaceae, 794
Bignonia family, 794
Big-root, 823
Bilderdykia, 246
Bind-weed, 673
Birch, 215
Birch family, 214
Bird-beak, 790
Bird-of-paradise-flower, 409
Bird-pepper, 756
Birdsfoot-trefoil, 427
Birthwort family, 227
Biscuit-root, 621
Bisnaga, 573
Bistorta, 246
Bitter-brush, 391
Bitter-cress, 341
Bitter-root, 288
Bitter-sweet family, 524
Blackberry, 378
Black-brush, 390, 904
Black-sage, 944
Bladder Fern, 44
Bladder-pod, 342
Bladder-sage, 734
Bladder-stem, 238
Bladder-wort, 798
Bladder-wort family, 798
Blanket-flower, 929
Bledo, 266
Blepharoneuron, 114
Blessed-thistle, 956
Blitum, 252
Blood-leaf, 270
Blood-weed, 722
Blooming-sally, 591
Blowout Grass, 87
Bluebell, 826
Bluebells, 713
Blueberry, 634
Bluebottle, 956
Blue-curls, 734
Bluedicks, 182
Blue-eyed-grass, 196
Blue Grass, 82
Blue-hearts, 786
Blue-stem, 141
Blueweed, 903
Bluewood, Mexican, 530
Boerhaavia, 275
caribaea, 276
coccinea, 276
Coulteri, 276
erecta, 276
gracillima, 276
intermedia, 276
megaptera, 277
pterocarpa, 277
purpurascens, 276
spicata, 276
Thornberi, 276
Torreyana, 276
triquetra, 276
universitatis, 276
Wrightii, 276
Bog-orchid, 199
Bommeria, 36
Borage family, 707
Boraginaceae, 707
Botrychium, 31

Bouchea, 730
Bouncing-bet, 303
Bouteloua, 126
aristidoides, 128
barbata, 127
chondrosioides, 128
curtipendula, 129
eludens, 128
eriopoda, 128
filiformis, 129
glandulosa, 128
gracilis, 128
hirsuta, 128
micrantha, 127
Parryi, 127
polystachya, 128
radicosa, 128
Rothrockii, 128
simplex, 127
trifida, 128
Bouvardia, 807
Bowlesia, 609
Box-elder, 527
Box family, 521
Box-leaf, 525
Brachystigma, 786
Bracken, 35
Brandegea, 824
Brass-buttons, 938
Brassica, 337
Brayulinea, 268
Bread-root, 429
Brickellia, 846
amplexicaulis, 849
atractyloides, 849
baccharidea, 849
betonicaefolia, 849
brachyphylla, 848
californica, 849
Coulteri, 849
desertorum, 849
Fendleri, 849
floribunda, 850
frutescens, 848
grandiflora, 850
incana, 849
Lemmoni, 848
longifolia, 848
multiflora, 848
oblongifolia, 848
parvula, 848
Pringlei, 848
Rusbyi, 849
scabra, 848
simplex, 850
squamulosa, 848
venosa, 848
Watsonii, 848
Bristle Grass, 138
Brittle-bush, 904
Brodiaea, 182
Brome, 76
Bromeliaceae, 166
Bromus, 76
anomalus, 77
arizonicus, 77
arvensis, 78
carinatus, 77
catharticus, 77
ciliatus, 77
commutatus, 78
frondosus, 78
inermis, 77
japonicus, 78
latiglumis, 78
madritensis, 78
marginatus, 77
mollis, 78
mucroglumis, 78
Orcuttianus, 78
Porteri, 77
purgans, 78
racemosus, 78
Richardsoni, 77
rigidus, 78
rubens, 78
secalinus, 78
tectorum, 78
texensis, 77
Trinii, 78
unioloides, 77
Brook Grass, 87
Brooklime, American, 785
Broom-rape, 796
Broom-rape family, 796
Broom-sedge, 141
Broom-weed, 853
Buchloë, 129
Buchnera, 786
Buck-brush, 534
Buck-thorn, 531
Buck-thorn family, 529
Buckwheat, 249
Buckwheat-brush, 230, 242
Buckwheat family, 228
Buddleja, 644
Bud-sage, 942
Buffalo-bean, 469
Buffalo-berry, 586
Buffalo-bur, 757
Buffalo-gourd, 822
Buffalo Grass, 129
Bugbane, 305
Bugleweed, 747
Bug-seed, 262
Bulbostylis, 156
Bull-head, 491
Bull-nettle, 758
Bulrush, 151
Bumelia, 640
Bupleurum, 614
Bur-clover, 421
Bur-cucumber,
one-seeded, 824
Burdock, 950
Bur Grass, 121
Bur-marigold, 909
Burnet, 391
Bur-nut, 491
Bur-reed, 64
Bur-reed family, 64
Burro-brush, 390, 893
Burro Grass, 92
Burro-weed, 862, 896
Bur-sage, 894
Bursera, 496
Burseraceae, 496
Burst-wort, 300
Bur-weed, 895
Butter-and-eggs, 764
Buttercup, 314
Butterfly-pea, 479
Butterfly-weed, 661
Button-bush, 807
Button-snakeroot, 623
Button-weed, 808
Button-wood, 371
Buxaceae, 521

Cacalia, 944
Cacao family, 554
Cactaceae, 567
Cactus family, 567
Caesalpinia, 409
Caesalpinioideae, 395, 403
Calabazilla, 822
Calamagrostis, 101
Calamovilfa, 102
Calandrinia, 287
Calhounia, 889
California-buckwheat, 230, 242
California Palm, 164
California-poppy, 323
Calliandra, 396
Calliopsis, 908
Callirhoë, 549
Callisteris, 693
Callitrichaceae, 521
Callitriche, 521
Calochortus, 184
Caltha, 305
Caltrop, 491
Caltrop family, 489

Calycoseris, 963
Calylophis, 594
Calypso, 203
Calyptridium, 289
Camelina, 344
Camel-thorn, 471
Camomile, 936
Camote-de-monte, 441
Camote-de-raton, 408
Campanula, 825
Campanulaceae, 825
Campanuloideae, 825
Camphor-weed, 855
Campion, 301, 303
Canaigre, 245
Canary Grass, 130
Cancer-root, 796
Canchalagua, 645
Cane cactus, 579
Canotia, 526
Caper family, 355
Capparidaceae, 355
Caprifoliaceae, 812
Capsella, 344
Capsicum, 755
Cardamine, 341
Cardaria, 333
Cardinal-flower, 828
Carduus, 951
Careless-weed, 265, 266
Carex, 157
  agrostoides, 161
  albo-nigra, 163
  alma, 161
  aquatilis, 163
  athrostachya, 162
  aurea, 162
  bella, 163
  Bolanderi, 161
  brevior, 162
  canescens, 161
  chalciolepis, 163
  chihuahuensis, 161
  curatorum, 162
  Douglasii, 160
  ebenea, 161
  Eleocharis, 160
  festivella, 161
  filifolia, 162
  geophila, 162
  Hassei, 162
  Haydeniana, 161
  hystricina, 164
  interior, 161
  Kelloggii, 163
  lanuginosa, 163
  leptopoda, 161
  leucodonta, 162
  Meadii, 162
  microptera, 161
  nebraskensis, 163
  occidentalis, 161
  oreocharis, 162
  petasata, 162
  praegracilis, 160
  Rossii, 162
  rostrata, 164
  Rusbyi, 160
  scoparia, 162
  senta, 163
  siccata, 160
  simulata, 160
  specuicola, 163
  spissa, 163
  stipata, 161
  subfusca, 162
  Thurberi, 164
  ultra, 164
  vesicaria, 164
  vulpinoidea, 161
  Wootoni, 162
Carlowrightia, 800
Carminatia, 843
Carnegiea, 569
Carpet-weed, 280
Carpet-weed family, 280
Carphochaete, 846
Carrizo, 89
Carrot, 612
Carrot-leaf, 622
Carum, 614
Caryophyllaceae, 291
Cashew family, 521
Cassia, 404
Castilleja, 787
  austromontana, 789
  Blumeri, 789
  chromosa, 790
  confusa, 790
  cruenta, 789
  eremophila, 790
  exilis, 788
  flava, 790
  galeata, 789
  gloriosa, 789
  integra, 789
  lanata, 789
  laxa, 788
  linariaefolia, 789
  lineata, 788
  lutescens, 790
  mexicana, 788
  minor, 788
  patriotica, 789
  retrorsa, 788
  sessiliflora, 788
  setosa, 788
  stenantha, 788
Castor-bean, 508
Catabrosa, 87
Catchfly, 301
Catchfly-gentian, 650
Catclaw, 398, 400
Caterpillar-weed, 698
Cathestecum, 129
Catnip, 737
Cattail, 63
Cattail family, 63
Cattle-spinach, 258
Caucalis, 612
Caulanthus, 330
Ceanothus, 532
Cedar, 58
Celastraceae, 524
Celery, 613
Celtis, 220
Cenchrus, 140
Cenizo, 259
Centaurea, 955
Centaurium, 645
Centrostegia, 229, 230
Centunculus, 638
Century-plant, 192
Cephalanthus, 807
Cerastium, 293
Cerasus, 394
Ceratophyllaceae, 303
Ceratophyllum, 303
Cercidium, 407
Cercis, 404
Cercocarpus, 388
Cereus, 568–570
Ceterach, 47
Cevallia, 563
Chacate, 403
Chaenactis, 922
Chaff-bush, 853
Chaff-weed, 638
Chain Fern, 49
Chamaebatiaria, 374
Chamaechaenactis, 923
Chamaecrista, 406
Chamaesaracha, 752, 754
Chamaesyce, 513, 518–520
Chamiso, 259
Chamiza, 259
Chaptalia, 956
Charlock, 338
Cheat Grass, 76
Checker-mallow, 548
Cheese-weed, 548
Cheilanthes, 38
  alabamensis, 40
  castanea, 40

Covillei, 39
Eatoni, 40
Feei, 39
Fendleri, 39
lendigera, 39
Lindheimeri, 39
Parryi, 40
Pringlei, 40
pyramidalis, 40
tomentosa, 39
villosa, 39
Wootoni, 39
Wrightii, 40
Chenopodiaceae, 249
Chenopodium, 250
albescens, 254
album, 254
ambrosioides, 252
arizonicum, 253
Berlandieri, 253
Botrys, 252
capitatum, 252
desiccatum, 253
Fremontii, 253
glaucum, 252
graveolens, 252
hians, 253
inamoenum, 253
incanum, 254
incisum, 252
leptophyllum, 253
murale, 253
pratericola, 253
rubrum, 253
salinum, 252
Watsoni, 253
Cherry, 393
Chess, 76
Chia, 741
Chicalote, 323
Chickweed, 293
Chico, 262
Chicory, 958
Chilicote, 480
Chillipiquin, 756
Chilopsis, 794
Chimaphila, 631
Chinch-weed, 934, 935
Chinese-pusley, 710
Chlorideae, 72
Chloris, 125
Choisya, 493
Choke-cherry, 394
Cholla, 579, 585
Choripetalae, 281
Chorispora, 354
Chorizanthe, 229
Christmas cactus, 584
Chrysanthemum, 937
Chrysopsis, 855
Chrysothamnus, 862
Chuchupate, 615
Chufa, 146, 150
Chuparosa, 801, 802
Chylismia, 595, 601, 602
Cichorium, 958
Cicuta, 616
Cimicifuga, 305
Cinquefoil, 381
Circaea, 604
Cirsium, 951
arcuum, 952
arizonicum, 953
arvense, 952
bipinnatum, 953
Blumeri, 953
Drummondii, 952
Grahami, 953
lanceolatum, 954
megacephalum, 953
mohavense, 954
neomexicanum, 952
nidulum, 953
ochrocentrum, 953
Parryi, 952
pulchellum, 953
Rothrockii, 952
undulatum, 954
utahense, 952
vulgare, 954
Wheeleri, 953
Wrightii, 952
Cissus, 536
Cladium, 156
Clammy-weed, 358
Clarkia, 592
Claytonia, 289
Cleavers, 811
Clematis, 311
Clementsia, 359
Cleome, 355
Cleomella, 356
Cliff-brake, 37
Cliff-bush, 367
Cliff-rose, 388
Climbing-milkweed, 663
Clinopodium, 746
Clitoria, 479
Cloak Fern, 40
Clomenocoma, 932
Clover, 442
Club-flower, 790
Cnemidophacos, 457, 458
Cnicus, 956
Cnidoscolus, 509
Coach-whip, 640
Cocculus, 321
Cochlospermaceae, 557
Cochlospermum family, 557
Cocklebur, 896
Cockroach-plant, 651
Cock-spur, 137
Coffee-berry, 532
Coffeeberry-bush, 625
Coffee-bush, 521
Cogswellia, 622
Coldenia, 708
Coleogyne, 390
Coleosanthus, 850
Collinsia, 767
Collomia, 684
Cologania, 479
Colorado-River-hemp, 444
Colorado rubber-weed, 925
Colubrina, 534
Columbine, 306
Comandra, 227
Comarella, 384
Commelina, 167
Commelinaceae, 166
Commicarpus, 277
Compositae, 829
Condalia, 529
Coneflower, 898
Conioselinum, 619
Conium, 614
Conobaea, 783
Conopholis, 796
Conringia, 355
Convolvulaceae, 666
Convolvulus, 673
Convolvulus family, 666
Conyza, 881
Coral-bean, 480
Coral-bells, 365
Corallorhiza, 201
Coral-root, 201
Coral-tree, 480
Corchorus, 536
Cord Grass, 125
Cordylanthus, 790
Coreocarpus, 908
Coreopsis, 907
Coriander, 615
Coriandrum, 615
Corispermum, 262
Corkwing, 617
Cornaceae, 624
Corn-bind, 247
Cornflower, 956
Corn-lily, 176
Cornus, 625
Corona-de-Cristo, 558
Corydalis, 325

Corynopuntia, 583
Coryphantha, 577
Cosmos, 912
Cota, 909
Cotta Grass, 91
Cottea, 91
Cotton, 553
Cotton Grass, 131
Cotton-top, 131
Cottonwood, 207
Cotula, 938
Coursetia, 443
Covena, 182
Covillea, 491
Cowania, 388
Cow-parsnip, 623
Cow-tobacco, 777
Coyote-melon, 822
Crab Grass, 132
Cracca, 443
Crag-lily, 177
Cranesbill, 484
Crassulaceae, 358
Crataegus, 378
Cream-bush, 375
Cream-cups, 322
Creosote-bush, 491
Crepis, 969
Cress, 332, 340
Cressa, 673
Crested-coralroot, 203
Crimson-beak, 403
Crinkle-awn, 144
Crossosoma, 371
Crossosoma family, 371
Crossosomataceae, 371
Crotalaria, 413
Croton, 503
Crowfoot, 314
Crowfoot family, 304
Crowfoot Grass, 124
Crown-beard, 907
Crown-daisy, 938
Crown-of-thorns, 558
Cruciferae, 325
Crucifixion-thorn, 495, 558
Crusea, 807
Crymodes, 314
Cryptantha, 714
  abata, 718
  albida, 720
  angustifolia, 719
  barbigera, 721
  capitata, 717
  circumscissa, 719
  confertiflora, 717
  costata, 720
  crassisepala, 719
  decipiens, 721
  dumetorum, 720
  echinoides, 717
  Fendleri, 719
  flava, 717
  fulvocanescens, 717
  gracilis, 719
  holoptera, 720
  humilis, 718
  inaequata, 719
  Jamesii, 717
  maritima, 719
  micrantha, 719
  modesta, 718
  muricata, 720
  nevadensis, 721
  pterocarya, 720
  pusilla, 720
  racemosa, 719
  recurvata, 719
  semiglabra, 717
  setosissima, 718
  utahensis, 720
  virginensis, 718
Cucurbita, 822
Cucurbitaceae, 820
Cud-weed, 887
Cup Grass, 132
Cuphea, 588
Cupressaceae, 57
Cupressus, 57
Curly-mesquite, 121
Currant, 368
Cuscuta, 666
  applanata, 670
  californica, 668
  campestris, 668
  Cephalanthi, 668
  Coryli, 669
  dentatasquamata, 670
  denticulata, 669
  erosa, 670
  Gronovii, 670
  indecora, 669
  mitraeformis, 670
  nevadensis, 669
  odontolepis, 670
  salina, 669
  tuberculata, 671
  umbellata, 671
Cut Grass, 131
Cyclanthera, 824
Cycloloma, 254
Cylindropuntia, 584, 585
Cymopterus, 617
Cynanchum, 657, 664
Cynara, 954
Cynodon, 124
Cyperaceae, 145
Cyperus, 146
  acuminatus, 149
  albomarginatus, 149
  amabilis, 149
  aristatus, 149
  asper, 151
  cyrtolepis, 149
  difformis, 149
  erythrorhizos, 150
  esculentus, 150
  Fendlerianus, 150
  ferax, 149
  flavus, 151
  hermaphroditus, 151
  inflexus, 149
  laevigatus, 149
  lateriflorus, 149
  manimae, 150
  Mutisii, 151
  niger, 149
  odoratus, 149
  Parishii, 150
  Pringlei, 151
  rotundus, 150
  Rusbyi, 150
  Schweinitzii, 150
  seslerioides, 149
  spectabilis, 150
  subambiguus, 151
  uniflorus, 151
  Wrightii, 151
Cypress, 57
Cypress family, 57
Cypripedium, 198
Cyrtorhyncha, 315
Cystium, 465, 467
Cystopteris, 44

Dactylis, 88
Dactyloctenium, 124
Dalea, 432
  albiflora, 439
  alopecuroides, 437
  amoena, 435
  angulata, 436
  aurea, 438
  brachystachys, 437
  calycosa, 436
  diffusa, 436
  Emoryi, 435
  filiformis, 437
  formosa, 438
  Fremontii, 435
  Grayi, 438
  Greggii, 438
  Hutchinsoniae, 437
  Jamesii, 437

lachnostachys, 436
laevigata, 438
Lagopus, 437
lanata, 437
Lemmoni, 437
leporina, 437
Lumholtzii, 438
mollis, 436
nana, 438
neomexicana, 436
nummularia, 431
Ordiae, 439
Parryi, 436
pogonathera, 438
polyadenia, 435
polygonoides, 437
Pringlei, 438
pulchra, 439
Schottii, 435
scoparia, 436
spinosa, 435
tentaculoides, 439
terminalis, 437
Thompsonae, 436
urceolata, 437
Whitingi, 436
Wislizeni, 438
Wrightii, 438
Dallis Grass, 133
Dandelion, 964
Danthonia, 101
Darnel, 97
Dasiphora, 381, 383
Dasyliron, 189
Datil, 185
Datura, 759
Daucus, 612
Dayflower, 167
Dead-nettle, 738
Death-camas, 176
Deer-brier, 534
Deer-browse, 389
Deer-brush, 533
Deer-clover, 425
Deerhorn cholla, 585
Deer-nut, 521
Deers-ears, 650
Deer-vetch, 425
Delphinium, 307
Deschampsia, 99
Descurainia, 349
Desert-almond, 394
Desert-broom, 883
Desert-chicory, 961
Desert-dandelion, 963
Desert-holly, 259, 957
Desert-honeysuckle, 801
Desert-lavender, 748
Desert-lily, 177
Desert-mallow, 543
Desert-marigold, 915
Desert-plume, 328
Desert-poppy, 324
Desert-sunflower, 905
Desert-thorn, 749
Desert-trumpet, 238
Desert-willow, 794
Desmanthus, 401
Desmodium, 473
angustifolium, 474
annuum, 475
arizonicum, 476
batocaulon, 476
Bigelovii, 475
cinerascens, 476
exiguum, 475
Grahami, 475
gramineum, 474
intortum, 475
Metcalfei, 476
neomexicanum, 475
procumbens, 475
psilocarpum, 475
psilophyllum, 475
retinens, 475
Rosei, 475
scopulorum, 475
sonorae, 475
Wigginsii, 475
Wislizeni, 475
Wrightii, 475
Devils-claws, 398, 795
Dew-berry, 380
Dichelostemma, 182
Dichondra, 671
Dicliptera, 801
Dicoria, 892
Dicotyledoneae, 203
Dicrophyllum, 517
Digitaria, 132
Diholcos, 457
Dill, 620
Diodia, 808
Diphysa, 443
Diplotaxis, 337
Disporum, 191
Distichlis, 88
Ditaxis, 505
Dithyrea, 341
Dock, 243
Dodder, 666
Dodecatheon, 638
Dodonaea, 528
Dogbane, 654
Dogbane family, 651
Dog-fennel, 936
Dogwood, 625
Dogwood family, 624
Dondia, 263
Douglas-fir, 55
Dove-weed, 504
Draba, 345
albertina, 347
asprella, 346
aurea, 346
brachycarpa, 347
crassifolia, 347
cuneifolia, 347
Helleriana, 346
patens, 347
petrophila, 346
platycarpa, 348
rectifructa, 347
reptans, 347
sonorae, 348
spectabilis, 346
Standleyi, 346
viridis, 346
Dracocephalum, 738
Dragon-head, 737
Drop-seed, 112
Drymaria, 299
Drymocallis, 381, 383
Dryopetalon, 339
Dryopteris, 45, 46
Duck-bill, 793
Duckweed, 165
Duckweed family, 165
Dudleya, 360
Dyschoriste, 799
Dyssodia, 932

Echeveria, 360
Echidiocarya, 722
Echinocactus, 572, 573–575
Echinocereus, 570
Echinochloa, 137
Echinocystis, 823
Echinodorus, 68
Echinomastus, 574
Echinopepon, 823
Echinopsilon, 261
Eclipta, 898
Elaeagnaceae, 586
Elaeagnus, 586
Elatinaceae, 556
Elatine, 556
Elder, 812
Elderberry, 812
Eleocharis, 153
acicularis, 155
bella, 155
cancellata, 155
caribaea, 154

Eleocharis (*cont'd*)
Engelmanni, 154
macrostachya, 154
montana, 154
montevidensis, 155
nodulosa, 154
Parishii, 155
parvula, 154
pauciflora, 154
radicans, 155
rostellata, 155
Elephant-head, 793
Elephant-tree, 496
Eleusine, 123
Elks-lip, 305
Elkweed, 649
Ellisia, 697, 698
Elm family, 220
Elodea, 70
Elymus, 94
Elyonurus, 144
Elytraria, 799
Emmenanthe, 705
Emplectocladus, 394
Encelia, 904
Enceliopsis, 905
Enchanters-nightshade, 604
Engelmannia, 890
Enneapogon, 91
Ephedra, 60
Ephedraceae, 60
Epicampes, 111, 112
Epilobium, 590
Epipactis, 200
Epixiphium, 767
Equisetaceae, 30
Equisetum, 30
Eragrostis, 85
arida, 87
Barrelieri, 86
cilianensis, 86
curvula, 87
diffusa, 86
echinochloidea, 87
intermedia, 87
Lehmanniana, 87
lutescens, 86
mexicana, 87
neomexicana, 87
obtusiflora, 87
Orcuttiana, 86
pectinacea, 86
pilosa, 86
spectabilis, 87
Eremalche, 548
Eremocrinum, 177
Erianthus, 140
Eriastrum, 685
Ericaceae, 629
Erigeron, 874
accedens, 880
aphanactis, 878
apiculatus, 877
arizonicus, 878
Bellidiastrum, 880
caespitosus, 880
canadensis, 881
canus, 879
compactus, 879
compositus, 878
concinnus, 878
consimilis, 879
delphinifolius, 880
divergens, 880
Eastwoodiae, 880
Eatoni, 879
eldensis, 877
eriophyllus, 881
flagellaris, 879
foliosissimus, 877
formosissimus, 878
furcatus, 880
gracillimus, 880
Gulielmi, 878
huachucanus, 878
Kuschei, 877
Lemmoni, 879
linifolius, 881
lobatus, 880
lonchophyllus, 881
Macdougalii, 879
macranthus, 877
modestus, 880
neomexicanus, 880
nudiflorus, 879
oreophilus, 880
oxyphyllus, 879
patens, 877
pecosensis, 878
perglaber, 878
platyphyllus, 877
Pringlei, 878
pumilus, 878
pusillus, 881
Rusbyi, 877
scaberulus, 878
Schiedeanus, 881
simplex, 877
speciosus, 877
subasper, 878
superbus, 877
ursinus, 878
utahensis, 879
Eriochloa, 132
Eriodictyon, 707
Eriogonum, 230
Abertianum, 237
alatum, 239
angulosum, 236, 237
arcuatum, 240
arizonicum, 237, 239
aureum, 242
Bakeri, 240
capillare, 239
cernuum, 238
Clutei, 238
cognatum, 240
corymbosum, 242
cyclosepalum, 237
Darrovii, 236
deflexum, 239
densum, 236
divaricatum, 236
effusum, 242, 243
exaltatum, 239
fasciculatum, 242
Ferrissii, 240
flavum, 240
flexum, 237
glandulosum, 237
gracillimum, 237
Heermanni, 242
hieracifolium, 239
Hookeri, 238
Howellii, 242
inflatum, 237
insigne, 239
Jamesii, 240
Jonesii, 242
juncinellum, 236
lachnogynum, 241
lanosum, 242
leptocladon, 242
leptophyllum, 243
Macdougalii, 243
maculatum, 236
Mearnsii, 243
microthecum, 242, 243
nidularium, 236
Ordii, 238
ovalifolium, 240
pallidum, 242
Palmeri, 241
panduratum, 239
Parryi, 238
pharnaceoides, 237
pinetorum, 237
Plumatella, 241
polifolium, 242
polycladon, 236
Pringlei, 241
puberulum, 236
pulchrum, 243
pulvinatum, 241
pusillum, 238

racemosum, 241
reniforme, 238
Ripleyi, 240
rotundifolium, 238
salsuginosum, 237
Shockleyi, 241
Simpsoni, 243
stellatum, 240
subreniforme, 239
sulcatum, 242
Thomasii, 237
Thompsonae, 241
Thurberi, 239
trachygonum, 241
trichopes, 238
triste, 239
turbinatum, 239
umbellatum, 240
villiflorum, 241
vimineum, 236
vineum, 240
Wetherillii, 239
Wrightii, 241
zionis, 242
Eriophyllum, 921
Erodium, 486
Eruca, 338
Eryngium, 623
Eryngo, 623
Erysimum, 353
Erythrina, 480
Erythrobalanus, 216
Erythrocoma, 387
Eschscholtzia, 323
Escoba-colorada, 436
Escobita, 792
Estafiata, 941
Esula, 512
Eucnide, 567
Eucrypta, 697
Eulobus, 594, 599
Eulophus, 615
Eupatorium, 844
Euphorbia, 511
Abramsiana, 520
albomarginata, 518
alta, 516
arizonica, 519
bilobata, 517
brachycera, 516
capitellata, 518
Chamaesula, 516
Chamaesyce, 520
dentata, 515
dictyosperma, 516
eremica, 517
eriantha, 515
exstipulata, 517
Fendleri, 519
flagelliformis, 518
florida, 518
glyptosperma, 520
gracillima, 519
heterophylla, 516
hirta, 518
hyssopifolia, 518
incisa, 516
indivisa, 520
Jonesii, 518
Lathyris, 516
lurida, 517
maculata, 518
marginata, 517
melanadenia, 519
micromera, 520
ocellata, 518
odontadenia, 516
Palmeri, 517
Parryi, 518
pediculifera, 519
Peplus, 517
platysperma, 517
Plummerae, 517
polycarpa, 519
prostrata, 520
pseudoserpyllifolia, 520
pycnanthema, 518
radians, 515
revoluta, 519
robusta, 517
Rusbyi, 518
schizoloba, 516
serpens, 519
serpyllifolia, 520
serrula, 520
setiloba, 520
spathulata, 516
stictospora, 520
supina, 520
trachysperma, 518
vermiculata, 518
vermiformis, 519
versicolor, 519
yaquiana, 516
Euphorbiaceae, 501
Eurotia, 260
Euryops, 950
Eustoma, 650
Evax, 885
Evening-primrose, 593
Evening-primrose family, 589
Evening-snow, 687
Everlasting, 887
Evolvulus, 672
Eysenhardtia, 432
Fagaceae, 215
Fagonia, 490
Fagopyrum, 249
Fairy-duster, 396
Fallugia, 387
Fall Witch Grass, 132
False Buffalo Grass, 129
False-camomile, 937
False-dandelion, 968
False-dragonhead, 738
False-flax, 344
False-hellebore, 176
False-indigo, 431
False-mesquite, 396
False-Solomon-seal, 190
False-tarragon, 941
Feather-geranium, 252
Feather Grass, 116
Fendlera, 367
Fendlerella, 368
Fennel, 612
Fern-bush, 374
Fern family, 34
Fern-leaf, 793
Ferns and fern allies, 27
Ferocactus, 573
Fescue, 79
Festuca, 79
arizonica, 80
Eastwoodae, 80
elatior, 80
Grayi, 80
idahoensis, 80
megalura, 80
myuros, 80
octoflora, 80
ovina, 80
pacifica, 80
reflexa, 80
rubra, 80
sororia, 80
Thurberi, 80
Festuceae, 74
Festucoideae, 71
Fetid-marigold, 934
Fiddle-neck, 723
Field-pansy, 560
Figwort, 767
Figwort family, 761
Filago, 885
Filaree, 486
Fimbristylis, 155
Finger Grass, 125
Fir, 56
Fire-weed, 591
Fire-wheel, 930
Fishhook cactus, 576
Flag, 195

Flannel-bush, 554
Flat-sedge, 146
Flaveria, 918
Flax, 488
Flax family, 488
Fleabane, 874
Fleur-de-lis, 195
Flourensia, 903
Flowering plants, 49
Fluff Grass, 90
Foam-bush, 375
Foeniculum, 612
Forestiera, 643
Forget-me-not, 724
Forsellesia, 526
Fountain Grass, 140
Fouquieria, 639
Fouquieriaceae, 639
Four-o'clock, 271
Four-o'clock family, 270
Foxglove, 773
Fox-tail, 103
Foxtail-millet, 138
Fragaria, 380
Fragrant-bitterweed, 925
Franseria, 894
Frasera, 650
Fraxinus, 641
Fremontia, 554
Fremontodendron, 554
Fringe-pod, 348
Fritillaria, 183
Fritillary, 183
Froelichia, 268
Frogs-bit family, 69
Fumarioideae, 322
Funastrum, 663

Gaillardia, 929
Galactia, 480
Galinsoga, 912
Galium, 808
  Aparine, 811
  asperrimum, 810
  bifolium, 812
  boreale, 812
  Brandegei, 810
  Collomae, 811
  coloradoense, 811
  Fendleri, 811
  microphyllum, 810
  Munzii, 811
  nitens, 810
  pilosum, 812
  proliferum, 812
  Rothrockii, 811
  stellatum, 811
  tinctorium, 810
  triflorum, 811
  Watsoni, 811
  Wrightii, 811
Galleta, 121
Galpinsia, 599, 600
Gama Grass, 145
Garrya, 624
Gastridium, 105
Gaura, 603
Gayoides, 540
Gayophytum, 602
Gentian, 646
Gentiana, 646
  affinis, 648
  Amarella, 649
  barbellata, 648
  Bigelovii, 648
  elegans, 648
  Forwoodii, 648
  Fremontii, 648
  grandis, 648
  heterosepala, 648
  interrupta, 648
  microcalyx, 649
  monantha, 648
  Parryi, 648
  scopulorum, 649
  strictiflora, 649
  superba, 648
  tenella, 648
  thermalis, 648
  Wislizeni, 649
  Wrightii, 648
Gentianaceae, 645
Gentian family, 645
Geoprumnon, 469
Geraea, 905
Geraniaceae, 484
Geranium, 484
Geranium family, 484
Gerardia, 786
Germander, 733
Geum, 386
Giant cactus, 569
Giant-hyssop, 736
Giant-reed, 89
Gilia, 687
  acerosa, 690
  achilleaefolia, 691
  aggregata, 692
  arenaria, 691
  arizonica, 693
  Ashtonae, 687
  aurea, 687
  Bigelovii, 687
  Breweri, 684
  Dactylophyllum, 687
  demissa, 687
  depressa, 690
  dichotoma, 687
  eremica, 685
  filifolia, 685
  filiformis, 690
  flavocincta, 691
  floribunda, 686
  gilioides, 690
  gracilis, 683
  Gunnisoni, 690
  hutchinsifolia, 691
  inconspicua, 691
  Jonesii, 687
  latifolia, 690
  laxiflora, 692
  leptomeria, 691
  longiflora, 692
  Macombii, 693
  Matthewsii, 694
  minima, 685
  multicaulis, 691
  multiflora, 692
  Nuttallii, 686
  ochroleuca, 691
  polyantha, 692
  polycladon, 690
  pumila, 690
  punctata, 694
  pungens, 686
  rigidula, 690
  Schottii, 694
  scopulorum, 691
  setosissima, 694
  sinuata, 691
  subnuda, 692
  superba, 692
  tenuiflora, 691
  tenuituba, 693
  Thurberi, 693
Ginseng family, 605
Glandularia, 725
Globe-amaranth, 269
Globe-artichoke, 954
Globe-mallow, 540
Glossopetalon, 525
Glyceria, 81
Glycyrrhiza, 471
Glyptopleura, 964
Gnaphalium, 887
Goat-nut, 521
Goats-beard, 961
Godetia, 592
Golden-aster, 855
Golden-eye, 900
Golden-head, 854
Golden-pea, 413
Golden-rod, 856
Goldentop, 88

Goldeye-grass, 192
Gold Fern, 36
Gold-fields, 918
Gold-poppy, 323
Golondrina, 511
Gomphocarpus, 660
Gomphrena, 269
Gonolobeae, 656
Gonolobus, 665
Goodyera, 201
Goose-berry, 368
Goose-foot, 251
Goose-foot family, 249
Goose Grass, 123
Gossypium, 553
Gourd family, 820
Grama, 126
Gramineae, 70
Grandfathers-beard, 387
Granjeno, 220
Grape, 535
Grape family, 534
Grape-fern, 31
Graptopetalum, 360
Grass family, 70
Grass-nuts, 182
Grass-of-Parnassus, 366
Gratiola, 782
Grayia, 260
Gray-thorn, 530
Grease-bush, 525
Grease-wood, 262, 491
Greenella, 866
Green-gentian, 649
Green-molly, 261
Grindelia, 851
Gromwell, 723
Grossularia, 368
Ground-cedar, 59
Ground-cherry, 753
Ground-plum, 469
Groundsel, 944
Grusonia, 584
Guardiola, 889
Guayule, 891
Gum-weed, 851
Gutierrezia, 852
Guttiferae, 556
Gymnocarpium, 45
Gymnolomia, 899
Gymnosperma, 851
Gymnospermae, 50

Habenaria, 199
Hackberry, 220
Hackelia, 713
Hackelochloa, 144
Hair Grass, 99
Halenia, 650
Halimolobos, 337
Halliophytum, 503
Haloragaceae, 604
Hamosa, 467–469
Haplopappus (See:
Aplopappus)
Haplophyton, 651
Harebell, 826
Hares-ear-mustard, 355
Harpagonella, 711
Hartmannia, 593, 598, 599
Hawkbit, 959
Hawks-beard, 969
Hawk-weed, 970
Hawthorn, 378
Heather family, 629
Hedeoma, 744
Hedgehog cactus, 570
Hedge-nettle, 739
Hedge-parsley, 611
Hedysarum, 471
Helenium, 928
Helianthella, 906
Helianthus, 902
Heliopsis, 898
Heliotrope, 709
Heliotropium, 709
Helleborine, 200
Hemicarpha, 146
Hemizonia, 913
Heracleum, 623
Hermannia, 555
Herniaria, 300
Heron-bill, 486
Hesperidanthus, 336
Hesperocallis, 177
Hesperochiron, 706
Hesperonia, 272
Heteranthera, 168
Heteropogon, 143
Heterosperma, 908
Heterotheca, 854
Heuchera, 363
Hexalectris, 203
Hibiscus, 552
Hieracium, 970
Hierba-de-la-cucaracha, 651
Hierochloë, 130
Hilaria, 121
Himalaya-berry, 380
Hippuris, 605
Hoffmanseggia, 408
Hofmeisteria, 843
Hog-potato, 408
Holacantha, 495
Holcus, 101
Holly Fern, 44
Holly-grape, 320
Hollyhock, 541
Holodiscus, 375
Holy Grass, 130
Homalobus, 457
Honeysuckle, 816
Honeysuckle family, 812
Hop, 221
Hopbush, 528
Hop-hornbeam, 214
Hopkirkia, 918
Hop-sage, 260
Hop-tree, 494
Hordeae, 72
Hordeum, 96
Horehound, 735
Horned-pondweed, 67
Hornwort, 303
Hornwort family, 303
Horse-bean, 407
Horse-brush, 944
Horsemint, 736
Horse-nettle, 758
Horsetail, 30
Horsetail family, 30
Horseweed, 881
Horsfordia, 540
Hosackia, 427, 428
Houstonia, 806
Huegelia, 685
Huisache, 399
Hummingbird-trumpet, 590
Humulus, 221
Husk-tomato, 753
Hutchinsia, 344
Hutchinsonia, 920
Hybanthus, 559
Hydastylus, 197
Hydrocharitaceae, 69
Hydrocotyle, 609
Hydrophyllaceae, 696
Hydrophyllum, 697
Hymenoclea, 893
Hymenopappus, 919
Hymenothrix, 920
Hymenoxys, 925
acaulis, 927
argentea, 927
Bigelovii, 927
Brandegei, 927
chrysanthemoides, 927
Cooperi, 928
Davidsonii, 927
floribunda, 927
Lemmoni, 928
odorata, 927
quinquesquamata, 928
Richardsoni, 927

Hymenoxys (*cont'd*)
  Rusbyi, 928
  subintegra, 927
  virgata, 928
Hypericum, 556
Hypopitys, 630
Hypoxis, 192
Hyptis, 748

Ibervillea, 821
Iliamna, 547
Ilysanthes, 784
Imperata, 140
Incienso, 904
Indian-apple, 760
Indian-bean, 480
Indian-blanket, 930
Indian-chickweed, 280
Indian Grass, 143
Indian-hemp, 654
Indian-lettuce, 290
Indian-mallow, 538
Indian-millet, 115
Indian-pea, 469
Indianpipe-weed, 238
Indian-plantain, 944
Indian-root, 227, 621
Indian-wheat, 802
Indigo, 428
Indigo-bush, 431, 432
Indigofera, 428
Ink-berry, 817
Iodine-bush, 262
Ionoxalis, 487
Ipomoea, 675
  barbatisepala, 677
  cardiophylla, 678
  coccinea, 676
  costellata, 677
  cuneifolia, 677
  desertorum, 678
  egregia, 677
  futilis, 677
  heterophylla, 677
  hirsutula, 678
  Lemmoni, 677
  leptotoma, 677
  Lindheimeri, 677
  longifolia, 676
  muricata, 677
  patens, 677
  Plummerae, 677
  purpurea, 678
  tenuiloba, 677
  Thurberi, 676
  triloba, 678
Ipomopsis, 688
Iresine, 270

Iridaceae, 195
Iris, 195
Iris family, 195
Ironwood, 442
Isanthus, 734
Isocoma, 862
Isoëtaceae, 29
Isoëtes, 29
Iva, 891
Ivesia, 382, 384, 385

Jackass-clover, 357
Jacobinia, 802
Jacobs-ladder, 694
Jacquemontia, 673
Jamesia, 367
Janusia, 497
Jatropha, 508
Jerusalem-oak, 252
Jerusalem-thorn, 406
Jimmy-weed, 862
Jimson-weed, 759
Johnson Grass, 142
Jointfir, 60
Jointfir family, 60
Jojoba, 521
Jonesiella, 464
Joshua-tree, 187
Juglandaceae, 213
Juglans, 213
Jumping cholla, 584
Juncaceae, 168
Juncaginaceae, 67
Junco, 558
Junco family, 558
Juncus, 168
  acuminatus, 172
  acutus, 171
  arizonicus, 172
  badius, 173
  balticus, 171
  brunnescens, 173
  bufonius, 171
  confusus, 171
  Cooperi, 171
  Drummondii, 170
  Dudleyi, 171
  effusus, 170
  ensifolius, 173
  interior, 171
  longistylis, 172
  macer, 171
  macrophyllus, 172
  marginatus, 172
  Mertensianus, 173
  mexicanus, 171
  neomexicanus, 172
  nodosus, 172
  parous, 173
  saximontanus, 173
  setosus, 172
  sphaerocarpus, 171
  tenuis, 171
  Torreyi, 172
  Tracyi, 173
  truncatus, 173
  xiphioides, 173
June Grass, 98
Jungle-rice, 137
Juniper, 58
Juniperus, 58
Jussiaea, 590

Kallstroemia, 491
Kelloggia, 807
Kentrophyta, 456
Kidney-wood, 432
Kinnikinnick, 633
Knap-weed, 955
Knot Grass, 133
Knotweed, 246
Kochia, 261
Koeberlinia, 558
Koeberliniaceae, 558
Koeleria, 98
Krameria, 403
Krigia, 959
Krynitzkia, 714, 721
Kuhnia, 850

Labiatae, 731
Lace-pod, 348
Lachnostoma, 665
Lactuca, 966
Lady Fern, 48
Lady's-slipper, 198
Lady's-thumb, 246
Lady's-tresses, 200
Lagascea, 889
Lamarckia, 88
Lambs-quarters, 254
Lamium, 738
Lamprophragma, 330
Langloisia, 693
Lantana, 728
Laphamia, 916
Lappula, 712
Larkspur, 307
Larrea, 491
Lathyrus, 477
Lavauxia, 593, 599
Layia, 913
Leather-flower, 312
Lechuguilla, 192
Leersia, 131
Leguminosae, 395
Lemaireocereus, 569

Lemna, 165
Lemnaceae, 165
Lemon-weed, 430
Lennoaceae, 629
Lennoa family, 629
Lens-scale, 259
Lentibulariaceae, 798
Leontodon, 959
Leonurus, 739
Lepidium, 332
Lepidobalanus, 216
Lepidospartum, 942
Leptochloa, 123
Leptodactylon, 685
Leptoloma, 132
Leptotaenia, 622
Lesquerella, 342
Lessingia, 866
Lettuce, 966
Leucampyx, 937
Leucocrinum, 177
Lewisia, 288
Liatris, 850
Licorice, 471
Ligusticum, 615
Lilaeopsis, 617
Liliaceae, 174
Lilium, 183
Lily, 183
Lily family, 174
Limnia, 290
Limnorchis, 199
Limonium, 639
Limosella, 783
Linaceae, 488
Linanthastrum, 686
Linanthus, 686
Linaria, 764
Linden family, 536
Lindernia, 783
Linnaea, 816
Linosyris, 942
Linum, 488
Lip Fern, 38
Lippia, 729, 730
Listera, 200
Lithophragma, 365
Lithosperum, 723
Lizard-tail family, 207
Loasaceae, 562
Loasa family, 562
Lobelia, 828
Lobelioideae, 825
Loco, 468
Loco-weed, 445, 470
Locust, 441
Loeflingia, 300
Loeselia, 694
Loganiaceae, 644
Logania family, 644
Lolium, 97
Lomatium, 621
Lonchophaca, 457, 460
Lonicera, 816
Loosestrife, 588, 637
Loosestrife family, 587
Lophocereus, 570
Lophotocarpus, 68
Loranthaceae, 223
Lotus, 425
  alamosanus, 427
  americanus, 427
  corniculatus, 427
  Greenei, 428
  hamatus, 427
  humilis, 427
  humistratus, 427
  longebracteatus, 428
  Mearnsii, 428
  neomexicanus, 428
  oblongifolius, 427
  oroboides, 428
  puberulus, 428
  Purshianus, 427
  rigidus, 427
  salsuginosus, 427
  tomentellus, 427
  Torreyi, 427
  utahensis, 428
  Wrightii, 427
Lousewort, 793
Love Grass, 85
Love-lies-bleeding, 267
Lucerne, 421
Ludwigia, 590
Lupine, 414
Lupinus, 414
  aduncus, 419
  alpestris, 420
  argenteus, 418
  aridus, 420
  arizonicus, 416
  barbiger, 418
  bicolor, 417
  Blumeri, 420
  brevicaulis, 417
  capitatus, 417
  chihuahuensis, 418
  concinnus, 417
  Cutleri, 420
  Greenei, 419
  Hillii, 419
  huahucanus, 418
  ingratus, 419
  Kingii, 417
  latifolius, 418
  Lemmonii, 419
  lepidus, 420
  Marcusianus, 419
  micensis, 417
  odoratus, 417
  Orcuttii, 417
  Osterhoutianus, 418
  Palmeri, 419
  Parishii, 418
  platanophilus, 418
  polyphyllus, 418
  pusillus, 417
  rubens, 417
  sericeus, 418
  Shockleyi, 418
  Sitgreavesii, 420
  sparsiflorus, 416
  succulentus, 416
Luzula, 173
Lychnis, 303
Lycium, 749
Lycopersicum, 749
Lycopus, 747
Lycurus, 104
Lygodesmia, 966
Lyrocarpa, 341
Lysiloma, 397
Lysimachia, 637
Lythraceae, 587
Lythrum, 588

Machaeranthera, 873, 874
Macromeria, 724
Macrosiphonia, 653
Madder family, 805
Madia, 913
Madrone, 633
Madroño, 633
Mahonia, 320
Maidenhair Fern, 36
Malacothrix, 962
Mala-mujer, 509
Malaxis, 202
Malcolmia, 354
Mal de ojos, 541
Male Fern, 47
Mallow, 548
Mallow family, 536
Malpighiaceae, 497
Malpighia family, 497
Malva, 548
Malvaceae, 536
Malvastrum, 547
Mammillaria, 576
  aggregata, 577
  arizonica, 577
  chlorantha, 577
  deserti, 577

Mammillaria (*cont'd*)
  Engelmannii, 577
  fasciculata, 578
  Heyderi, 578
  Macdougalii, 578
  Mainae, 577
  microcarpa, 578
  Oliviae, 578
  recurvata, 577
  robustispina, 577
  tetrancistra, 577
  vivipara, 577
  Wilcoxii, 578
Manihot, 509
Manna Grass, 81
Manzanita, 633
Maple, 526
Maple family, 526
Marah, 823
Mares-tail, 605
Margaranthus, 752
Marigold, 931
Mariola, 891
Mariposa, 184
Marrubium, 735
Marsh-elder, 891
Marsh-fleabane, 884
Marsh-marigold, 305
Marsh-purslane, 590
Marsilea, 33
Marsileaceae, 33
Martynia, 795, 796
Martyniaceae, 795
Match-weed, 853
Matelea, 665, 666
Matricaria, 937
Matthiola, 354
Maurandella, 767
Maurandya, 766
May-weed, 936
Meadow-rue, 319
Mecardonia, 783
Medicago, 420
Medick, 420
Megapterium, 593
Meibomia, 475, 476
Melampodium, 889
Melica, 89
Melic Grass, 89
Melilotus, 421
Mellichampia, 664
Meloncilla, 550
Melon-loco, 821
Menispermaceae, 321
Menodora, 643
Mentha, 747
Mentzelia, 564
Meriolix, 600
Mertensia, 713
Mescal, 192
Mesquite, 401
Mesquitilla, 396
Metastelma, 663
Mexican-devilweed, 869
Mexican-poppy, 492
Mexican-star, 183
Mexican-tea, 252
Mexican-thistle, 624
Micranthes, 363
Microchloa, 124
Microrhamnus, 530
Microseris, 959
Microsisymbrium, 330
Microsteris, 683
Microstylis, 202
Mignonette family, 358
Milfoil, 936
Milk-thistle, 954
Milk-vetch, 445
Milkweed, 657
Milkweed family, 655
Milk-wort, 498
Milk-wort family, 497
Milla, 183
Mimetanthe, 781
Mimosa, 399
Mimosa subfamily, 395
Mimosoideae, 395
Mimulus, 779
  Bigelovii, 781
  cardinalis, 780
  cordatus, 781
  dentilobus, 781
  Eastwoodiae, 780
  floribundus, 782
  Geyeri, 781
  glabratus, 781
  guttatus, 781
  Maguirei, 781
  nasutus, 781
  Parryi, 781
  parvulus, 781
  pilosus, 781
  primuloides, 782
  prionophyllus, 781
  puberulus, 781
  rubellus, 782
  Suksdorfii, 782
  unimaculatus, 781
  verbenaceus, 780
Miners-lettuce, 290
Mint, 747
Mint family, 731
Mirabilis, 271
Mistletoe, 224
Mistletoe family, 223
Mitracarpus, 808
Mock-cucumber, 823
Mock-locust, 431
Mock-orange, 366
Mock-pennyroyal, 744
Mohavea, 765
Mojave-stinkweed, 357
Moldavica, 737
Mollugo, 280
Mollugophytum, 299
Molucca-balm, 739
Molucella, 739
Monarda, 743
Monardella, 746
Moneses, 632
Monkey-flower, 779
Monks-hood, 310
Monnina, 501
Monocotyledoneae, 62
Monolepis, 254
Monoptilon, 867
Monotropa, 630
Montia, 289
Moonseed family, 321
Moraceae, 221
Morning-glory, 675
Mortonia, 525
Morus, 221
Motherwort, 739
Mountain-ash, 375
Mountain-balm, 707
Mountain-dandelion, 967
Mountain-heath, 632
Mountain-laurel, 523
Mountain-lilac, 532
Mountain-lover, 525
Mountain-mahogany, 388
Mountain-sorrel, 246
Mountain-spray, 375
Mouse-ear-chickweed, 293
Mouse-tail, 313
Mud-plantain, 168
Mudwort, 783
Muhlenbergia, 105
  abata, 109
  andina, 110
  appressa, 109
  arenacea, 109
  arenicola, 111
  arizonica, 111
  asperifolia, 109
  brevis, 109
  ciliata, 108
  curtifolia, 110
  depauperata, 109
  dubia, 112
  dubioides, 111
  dumosa, 109
  eludens, 108

Emersleyi, 112
filiculmis, 111
filiformis, 109
fragilis, 108
glauca, 110
gracilis, 111
huachucana, 110
longiligula, 111
mexicana, 110
microsperma, 109
minutissima, 108
montana, 111
monticola, 111
mundula, 110
pauciflora, 111
pectinata, 108
polycaulis, 111
Porteri, 111
pulcherrima, 109
pungens, 109
racemosa, 110
repens, 109
Richardsonis, 110
rigens, 110
rigida, 112
Schreberi, 112
sinuosa, 108
sylvatica, 110
texana, 108
Thurberi, 110
Torreyi, 111
utilis, 109
Vaseyana, 112
virescens, 111
Wolfii, 108
Wrightii, 110
xerophila, 112
Muhly, 105
Mulberry, 221
Mulberry family, 221
Mule-fat, 884
Mules-ears, 899
Mullein, 764
Munroa, 129
Musk-thistle, 951
Mustard, 338
Mustard family, 325
Mutton Grass, 83
Myosotis, 724
Myosurus, 313
Myriophyllum, 604

Naiad, 67
Naiad family, 67
Nailwort, 300
Najadaceae, 67
Najas, 67
Nama, 705
Nasturtium, 340
Natal Grass, 138
Navajoa, 575
Navajo-tea, 909
Navarettia, 684
Needle-and-thread, 117
Needle Grass, 116
Nemacaulis, 229
Nemacladus, 827
Nemastylis, 196
Nemophila, 697
Nemoseris, 961
Neomammillaria, 577, 578
Nepeta, 737
Nettle, 222
Nettle family, 222
Nicotiana, 760
Nievitas, 714
Niggerhead cactus, 573
Night-blooming cereus, 568
Nightshade, 756
Ninebark, 374
Nissolia, 472
Nit Grass, 105
Nitrophila, 250
Nogal, 214
Nolina, 188
None-such, 421
Nose-burn, 507
Notholaena, 37, 40
Nothoscordum, 181
Nut-grass, 146, 150
Nyctaginaceae, 270

Oak, 216
Oak Fern, 45
Oat, 100
Oat Grass, 101
Ocean-spray, 375
Ocotillo, 639
Ocotillo family, 639
Odostemon, 320
Oenothera, 593
albicaulis, 596
Boothii, 600
brachycarpa, 599
brevipes, 601
caespitosa, 598
cardiophylla, 601
cavernae, 598
chamaenerioides, 600
clavaeformis, 601
Clutei, 596
contorta, 600
coronopifolia, 596
decorticans, 600
deltoides, 596
deserti, 600
flava, 599
Greggii, 600
Hartwegii, 599
Hookeri, 595
irrigua, 595
Kunthiana, 599
laciniata, 596
lavandulaefolia, 599
leptocarpa, 599
longissima, 596
micrantha, 600
multijuga, 601
neomexicana, 597
pallida, 597
pallidula, 601
Parryi, 602
primiveris, 598
procera, 596
refracta, 600
rosea, 599
runcinata, 597
scapoidea, 601
serrulata, 600
speciosa, 598
strigosa, 596
taraxacoides, 599
trichocalyx, 598
Wrightii, 599
Oldenlandia, 806
Oldman-whiskers, 387
Old-witch Grass, 136
Oleaceae, 640
Oleaster, 586
Oleaster family, 586
Oligomeris, 358
Olive family, 640
Olneya, 442
Onagra, 593
Onagraceae, 589
Onion, 177
Onion Grass, 89
Ophioglossaceae, 31
Ophioglossum, 32
Opuntia, 578
abyssi, 584
acanthocarpa, 585
angustata, 583
arbuscula, 584
aurea, 581
basilaris, 581
Bigelovii, 584
canada, 583
chlorotica, 582
clavata, 583
delicata, 582
discata, 583
echinocarpa, 584
Engelmannii, 583

Opuntia (*cont'd*)
erinacea, 582
flavescens, 583
fragilis, 581
fulgida, 585
gilvescens, 583
Gosseliniana, 582
hualpaensis, 585
hystricina, 582
Kunzei, 584
laevis, 583
leptocaulis, 584
Loomisii, 582
macrocentra, 583
mojavensis, 583
Parishii, 583
phaeacantha, 583
plumbea, 582
polyacantha, 581
procumbens, 583
pulchella, 583
Rafinesquii, 582
ramosissima, 584
rhodantha, 582
santa-rita, 582
spinosior, 585
Stanlyi, 583
tenuispina, 583
tetracantha, 584
Thornberi, 585
tortispina, 583
ursina, 582
versicolor, 585
vivipara, 584
Whipplei, 585
Wrightiana, 584
Orache, 254
Orchard Grass, 88
Orchidaceae, 197
Orchis family, 197
Oregon-boxwood, 525
Oreobroma, 288
Oreocarya, 714, 717, 718
Oreoxis, 613, 617
Organpipe cactus, 569
Orobanchaceae, 796
Orobanche, 796
Orpine family, 358
Orthocarpus, 791
Oryzeae, 72
Oryzopsis, 115
Osha, 615
Osmorhiza, 610
Ostrya, 214
Our-Lord's candle, 188
Owl-claws, 929
Owl-clover, 791
Oxalidaceae, 487
Oxalis, 487
Ox-eye, 898
Oxeye-daisy, 938
Oxybaphus, 273
Oxypolis, 621
Oxyria, 246
Oxytenia, 892
Oxytheca, 230
Oxytropis, 470
Oyster-plant, 962

Pachaba, 849
Pachylophis, 593, 598
Pachystima, 525
Padus, 394
Pagesia, 783
Paint-brush, 787
Painted-cup, 787
Palafoxia, 920
Palmae, 164
Palm family, 164
Palmilla, 188
Palo-blanco, 220
Palo-de-hierro, 442
Palo-verde, 407, 526
Pancake-pear, 582
Paniceae, 71
Panicoideae, 71
Panicum, 134
antidotale, 137
arizonicum, 135
barbipulvinatum, 136
bulbosum, 136
capillare, 136
dichotomiflorum, 135
fasciculatum, 135
Hallii, 137
hirticaule, 136
huachucae, 136
lepidulum, 137
miliaceum, 136
obtusum, 137
pampinosum, 136
plenum, 137
Scribnerianum, 136
sonorum, 135
stramineum, 136
tennesseense, 136
texanum, 135
Urvilleanum, 137
virgatum, 137
Papaveraceae, 322
Papaveroideae, 322
Paper-daisy, 914
Paper-flower, 914
Papilionoideae, 395, 409
Pappophorum, 91
Pappus Grass, 91
Parietaria, 223
Parkinsonia, 406
Parnassia, 366
Paronychia, 300
Parosela, 436, 438
Parrot-feather, 605
Parryella, 430
Parsley family, 606
Parsnip, 623
Parthenice, 891
Parthenium, 891
Parthenocissus, 535
Partridge-pea, 406
Paspalum, 133
Passiflora, 562
Passifloraceae, 562
Passionflower, 562
Passionflower family, 562
Pastinaca, 623
Patata (Patota), 254
Paxistima, 525
Pea-bush, 432
Peach, 393
Pea family, 395
Pearly everlasting, 887
Pearl-wort, 295
Peavine, 477
Pectis, 934
angustifolia, 935
Coulteri, 935
cylindrica, 935
filipes, 935
imberbis, 935
linifolia, 935
longipes, 935
Mearnsii, 935
Palmeri, 935
papposa, 935
prostrata, 935
Rusbyi, 935
urceolata, 935
Pectocarya, 711
Pedicularis, 793
Pediomelum, 429, 430
Peganum, 490
Pellaea, 37
Pellitory, 223
Pencil cholla, 584
Pencil-flower, 472
Peniocereus, 568
Pennellia, 330
Pennisetum, 139
Penny-cress, 335
Penstemon, 768
albomarginatus, 776
ambiguus, 777
amplus, 772
angustifolius, 775
arizonicus, 773

barbatus, 771
bicolor, 775
Bridgesii, 771
caespitosus, 776
Clutei, 774
Cobaea, 773
coccinatus, 772
comarrhenus, 778
congestus, 775
Crideri, 775
dasyphyllus, 776
desertipicti, 776
discolor, 777
Eatoni, 772
exsertus, 772
Fendleri, 775
Hansonii, 773
Jamesii, 773
laevis, 778
lanceolatus, 771
leiophyllus, 778
lentus, 776
linarioides, 776
microphyllus, 771
mirus, 773
nudiflorus, 775
oliganthus, 776
ophianthus, 774
pachyphyllus, 775
Palmeri, 774
Parryi, 773
petiolatus, 774
pilosigulatus, 774
pinifolius, 771
Plummerae, 771
pseudospectabilis, 774
pulchellus, 774
putus, 778
rubescens, 777
Rydbergii, 776
scoparius, 777
Shantzii, 773
stenophyllus, 777
strictus, 778
subulatus, 773
superbus, 773
Thompsoniae, 776
Thurberi, 777
utahensis, 773
virgatus, 778
Watsoni, 776
Whippleanus, 773
Pentzia, 942
Pepper-grass, 332
Pepperwort, 33
Pepperwort family, 33
Peraphyllum, 378
Perezia, 957
Pericome, 917
Perideridia, 614
Peritoma, 356
Perityle, 915
Persicaria, 246, 248
Petalonyx, 563
Petalostemum, 439
Peteria, 441
Petradoria, 858
Petrophytum, 374
Petunia, 761
Peucedanum, 619
Peucephyllum, 943
Phaca, 462–464
Phacelia, 698
affinis, 702
arizonica, 703
cephalotes, 701
coerulea, 704
congesta, 703
corrugata, 704
crenulata, 704
cryptantha, 703
curvipes, 701
demissa, 701
distans, 703
filiformis, 701
Fremontii, 701
glaberrima, 704
glechomaefolia, 701
heterophylla, 702
integrifolia, 703
intermedia, 704
Ivesiana, 702
laxiflora, 701
Lemmoni, 701
magellanica, 702
minutiflora, 704
mutabilis, 702
neglecta, 701
neomexicana, 703
pachyphylla, 701
Palmeri, 703
pedicellata, 704
Popei, 703
pulchella, 701
ramosissima, 702
rotundifolia, 701
saxicola, 701
sericea, 702
serrata, 703
tanacetifolia, 703
vallis-mortae, 703
Phalarideae, 72
Phalaris, 130
Phanerophlebia, 45
Phaseolus, 481
Phellopterus, 618
Phellosperma, 577
Pherotrichis, 665
Philadelphus, 366
Phleum, 105
Phlox, 680
acerba, 681
amabilis, 682
austromontana, 681
caespitosa, 681
Cluteana, 682
Covillei, 681
diffusa, 681
gilioides, 681
gladiformis, 681
gracilis, 683
Grayi, 683
griseola, 681
Hoodii, 681
longifolia, 683
nana, 682
Nelsonii, 682
Stansburyi, 682
tenuifolia, 681
triovulata, 682
visenda, 683
Woodhousei, 682
Phlox family, 678
Pholistoma, 697
Phoradendron, 224
Phragmites, 89
Phyla, 729
Phyllodoce, 632
Physalis, 753
crassifolia, 755
Fendleri, 754
hederaefolia, 755
heterophylla, 755
lanceifolia, 754
lanceolata, 754
lobata, 754
longifolia, 755
neomexicana, 755
pubescens, 754
versicolor, 755
Wrightii, 754
Physaria, 342
Physocarpus, 374
Physostegia, 738
Phytolacca, 280
Phytolaccaceae, 279
Picea, 54
Pickerel-weed family, 168
Pickle-weed, 262
Pigeon-berry, 279, 532
Pigeon-bush, 817
Pigmy-cedar, 943
Pigmy-weed, 361
Pig-weed, 265, 281

Pilostyles, 227
Pimpernel, 638
Pinaceae, 50
Pinaropappus, 962
Pincushion cactus, 576
Pine, 50
Pineapple family, 166
Pineapple-weed, 937
Pine-drops, 630
Pine-dropseed, 114
Pine family, 50
Pinesap, 630
Pingüe, 925
Pingwing, 925
Pink family, 291
Pinus, 50
Pinyon, 51, 52
Pinyon rice Grass, 115
Pipsissewa, 631
Piptochaetium, 115
Piptothrix, 845
Pisophaca, 462
Pistia, 165
Pitahaya, 570
Pitch-forks, 909
Pityrogramma, 36
Plagiobothrys, 721
Plane-tree, 371
Plane-tree family, 371
Plantaginaceae, 802
Plantago, 802
  argyraea, 805
  eriopoda, 804
  fastigiata, 805
  heterophylla, 804
  hirtella, 804
  ignota, 804
  insularis, 805
  lanceolata, 804
  major, 804
  picta, 804
  Purshii, 804
  pusilla, 804
  rhodosperma, 804
  Tweedyi, 804
  virginica, 804
  Wrightiana, 805
  xerodea, 804
Plantain, 802
Plantain family, 802
Platanaceae, 371
Platanus, 371
Platystemon, 322
Plectritis, 818
Pleurisy-root, 661
Pluchea, 884
Plum, 393
Plumbaginaceae, 639
Plumbago, 639
Plumbago family, 639
Plume Grass, 140
Plummera, 931
Poa, 82
  annua, 84
  arida, 84
  Bigelovii, 83
  bulbosa, 85
  Canbyi, 85
  compressa, 84
  Fendleriana, 84
  glaucifolia, 84
  interior, 84
  longiligula, 85
  nevadensis, 85
  palustris, 84
  pratensis, 84
  reflexa, 84
  rupicola, 85
  secunda, 85
Poinciana, 409
Poinsettia, 512, 515, 516
Poison-hemlock, 614
Poison-ivy, 522
Poison-oak, 522
Poison-vetch, 445
Pokeberry, 280
Pokeberry family, 279
Poke-weed, 280
Polanisia, 358
Polemoniaceae, 678
Polemonium, 694
Poliomintha, 746
Polygala, 498
  acanthoclada, 500
  alba, 500
  arizonae, 500
  Barbeyana, 499
  glochidiata, 500
  hemipterocarpa, 501
  Lindheimeri, 500
  longa, 500
  macradenia, 499
  obscura, 500
  orthotricha, 500
  piliophora, 499
  racemosa, 499
  reducta, 499
  Rusbyi, 500
  scoparioides, 501
  subspinosa, 500
  Tweedyi, 500
Polygalaceae, 497
Polygonaceae, 228
Polygonatum, 191
Polygonum, 246
  amphibium, 248
  argyrocoleon, 247
  aviculare, 247
  bistortoides, 248
  coccineum, 248
  Convolvulus, 247
  Douglasii, 247
  fusiforme, 248
  incanum, 248
  Kelloggii, 247
  lapathifolium, 248
  natans, 248
  pensylvanicum, 248
  Persicaria, 248
  punctatum, 248
  ramosissimum, 248
  sawatchense, 248
  viviparum, 248
  Watsoni, 247
Polypodiaceae, 34
Polypodium, 42
Polypody, 42
Polypogon, 104
Polystichum, 44
Pomatophytum, 39
Pondweed, 65
Pondweed family, 64
Pontederiaceae, 168
Poorman's-weatherglass, 638
Popcorn-flower, 721
Poplar, 207
Poppy family, 322
Poppy-mallow, 549
Populus, 207
Porcupine Grass, 116
Porophyllum, 933
Porterella, 829
Portulaca, 290
Portulaca family, 285
Portulacaceae, 285
Potamogeton, 65
Potamogetonaceae, 64
Potato family, 748
Potentilla, 381
  albiflora, 386
  Anserina, 384
  arguta, 383
  arizonica, 384
  atrorubens, 384
  biennis, 384
  concinna, 386
  concinnaeformis, 386
  convallaria, 383
  crinita, 385
  diversifolia, 385
  fruticosa, 383
  glandulosa, 383
  glaucophylla, 385

Hippiana, 385
Lemmoni, 385
Macdougalii, 383
millegrana, 384
modesta, 386
monspeliensis, 384
multifoliolata, 384
norvegica, 384
Osterhoutii, 383
pensylvanica, 385
plattensis, 384
propinqua, 385
pulcherrima, 385
ramulosa, 386
rivalis, 384
sabulosa, 384
sanguinea, 384
Sibbaldi, 383
subviscosa, 386
Thurberi, 384
viscidula, 386
Wheeleri, 386
Powder-horn, 293
Prairie-clover, 439
Prairie-coneflower, 898
Prairie-mallow, 548
Prickle-poppy, 323
Prickly-pear, 579
Primrose, 635
Primrose family, 635
Primula, 635
Primulaceae, 635
Proboscidea, 795
Prosopis, 401
Protobalanus, 216
Prunella, 738
Prunus, 393
Psathyrotes, 943
Pseudocymopterus, 619
Pseudotsuga, 55
Psilactis, 867
Psilostrophe, 914
Psoralea, 429
Psoralidium, 429, 430
Psorothamnus, 436
Ptelea, 494
Pteridium, 35
Pteridophyta, 27
Pterophacos, 460
Pterospora, 630
Pterostegia, 228
Pteryxia, 618
Puccinellia, 81
Puccoon, 723
Puncture-vine, 491
Purpusia, 381, 383
Purshia, 391
Purslane, 291
Pusley, 291
Pussy-toes, 886
Pycnanthemum, 748
Pyrola, 631
Pyrrhopappus, 968
Pyrrocoma, 861

Quack Grass, 92
Quail-brush, 259
Quail-plant, 710
Quamoclidion, 272
Quamoclit, 676
Quelite, 266
Quelite-salado, 263
Quercus, 216
arizonica, 218
diversicolor, 217
Emoryi, 219
Fendleri, 219
Gambelii, 219
grisea, 218
hypoleuca, 219
hypoleucoides, 219
oblongifolia, 217
Palmeri, 219
pungens, 218
reticulata, 217
submollis, 219
subobtusifolia, 219
subturbinella, 218
Toumeyi, 218
turbinella, 218
undulata, 219
Wilcoxii, 219
Quillwort, 29
Quillwort family, 29
Quinine-bush, 388, 625
Quinine-plant, 521

Rabbit-brush, 862
Rabbit-foot Grass, 104
Radish, 338
Rafflesiaceae, 227
Rafflesia family, 227
Rafinesquia, 961
Ragweed, 894
Raimannia, 594
Rainbow cactus, 571
Ramischia, 631
Ranunculaceae, 304
Ranunculus, 314
aquatilis, 316
arizonicus, 319
Bongardi, 317
cardiophyllus, 318
circinatus, 316
Collomae, 316
Cymbalaria, 316
eremogenes, 316
Eschscholtzii, 317
eximius, 317
Flammula, 317
glaberrimus, 316
hydrocharoides, 317
inamoenus, 318
juniperinus, 316
Macounii, 317
macranthus, 317
oreogenes, 316
pedatifidus, 318
pensylvanicus, 317
sceleratus, 316
stolonifer, 317
trichophyllus, 316
uncinatus, 317
Raphanus, 338
Raspberry, 378
Ratany, 403
Rathbunia, 570
Ratibida, 898
Rattle-box, 413
Rattlesnake-plantain, 201
Rattlesnake-weed, 511
Rattle-weed, 445
Ravenna Grass, 141
Rayless-goldenrod, 862
Red-and-yellow-pea, 425
Redbud, 404
Redfieldia, 87
Red-maids, 288
Red-pepper, 755
Red-root, 266
Red-sage, 261
Red-scale, 257
Reed, 89
Reed Grass, 101
Reina de la noche, 568
Rein-orchis, 199
Relbunium, 810
Rescue Grass, 76
Resedaceae, 358
Resin-weed, 853, 900
Reverchonia, 502
Rhamnaceae, 529
Rhamnus, 531
Rhinanthus, 792
Rhodes Grass, 125
Rhus, 522
Rhynchelytrum, 138
Rhynchosia, 481
Ribes, 368
aureum, 369
cereum, 370
divaricatum, 370
inebrians, 370
inerme, 370

Ribes (*cont'd*)
leptanthum, 370
montigenum, 369
pinetorum, 370
quercetorum, 370
velutinum, 370
viscosissimum, 370
Wolfii, 370
Ribesia, 368
Ribwort, 804
Rice Grass, 115
Ricinus, 508
Ring Grass, 105
Rivina, 279
Robinia, 441
Rock-cress, 350
Rocket-salad, 338
Rock-jasmine, 636
Rock-mat, 374
Rock-nettle, 567
Rock-purslane, 287
Rock-spiraea, 375
Roquette, 338
Rorippa, 339
Rosa, 392
Rosaceae, 372
Rosary-bean, 481
Rose, 392
Rose family, 372
Rose-mallow, 552
Rosin-brush, 883
Rotala, 587
Rothrockia, 666
Rouge-plant, 279
Rubiaceae, 805
Rubus, 378
Rudbeckia, 898
Rue family, 493
Ruellia, 799
Rumex, 243
Acetosella, 244
altissimus, 244
arizonicus, 245
californicus, 244
conglomeratus, 245
crispus, 245
ellipticus, 244
fueginus, 245
hymenosepalus, 245
mexicanus, 244
obtusifolius, 245
occidentalis, 245
orthoneurus, 245
triangulivalvis, 244
violascens, 245
Rush, 168
Rush family, 168
Russian-olive, 586
Russian-thistle, 264
Rutaceae, 493
Rutosma, 494
Rye, 94
Rye Grass, 97

Sacahuista, 189
Sacaton, 112
Sage, 740, 258
Sagebrush, 938, 258
Sageretia, 531
Sagina, 295
Sagittaria, 68
Saguaro, 569
Salazaria, 734
Salicaceae, 207
Salix, 209
amygdaloides, 212
Bebbiana, 211
Bonplandiana, 213
caudata, 212
exigua, 211
Geyeriana, 212
Gooddingii, 212
irrorata, 212
laevigata, 212
lasiandra, 213
lasiolepis, 212
ligulifolia, 213
lutea, 213
nigra, 212
pseudocordata, 213
pseudomonticola, 213
Scouleriana, 211
taxifolia, 211
Toumeyi, 213
Wrightii, 213
Salpichroa, 758
Salpingia, 594
Salsify, 962
Salsola, 264
Salt-bush, 254
Salt Grass, 88
Salvia, 740
aethiopis, 743
albiflora, 743
amissa, 743
arizonica, 742
carnosa, 741
Columbariae, 741
confinis, 742
Davidsonii, 742
Henryi, 742
Lemmoni, 743
microphylla, 743
mohavensis, 742
pachyphylla, 742
Parryi, 742
pilosa, 742
pinguifolia, 742
reflexa, 742
subincisa, 742
tiliaefolia, 743
Salviniaceae, 32
Sambucus, 812
Samolus, 637
Sandalwood family, 226
Sand-berry, 633
Sand-bur, 140
Sand-corn, 176
Sand-cress, 289
Sand-food, 629
Sand-lily, 177
Sandpaper-plant, 563
Sand-reed, 102
Sand-root, 629
Sand-spurry, 298
Sand-verbena, 277
Sandwort, 295
Sangre-de-Cristo, 509
Sangre-de-drago, 509
Sanguisorba, 391
Santalaceae, 226
Santa-rita cactus, 582
Sanvitalia, 897
Sapindaceae, 528
Sapindus, 528
Sapium, 511
Saponaria, 303
Sapotaceae, 640
Sapote family, 640
Saracha, 755
Sarcobatus, 262
Sarcostemma, 664
Satin-tail, 140
Satureja, 746
Saururaceae, 207
Saw-grass, 156
Saxifraga, 362
Saxifragaceae, 361
Saxifrage, 362
Saxifrage family, 361
Scalebroom, 942
Scandix, 611
Schedonnardus, 125
Schismus, 98
Schistophragma, 783
Schkuhria, 918, 924
Schmaltzia, 524
Scirpus, 151
Sclerocactus, 574
Scleropogon, 92
Scoke, 280
Scouring-rush, 30
Screwbean, 402
Scrophularia, 767
Scrophulariaceae, 761

Scurf-pea, 429
Scutellaria, 734
Sea-lavender, 639
Sea-purslane, 281
Secale, 94
Sedge, 157
Sedge family, 145
Sedum, 359
Seed-box, 590
Seep-weed, 263
Seep-willow, 884
Sego-lily, 184
Selaginella, 28
Selaginellaceae, 28
Selaginella family, 28
Self-heal, 738
Selinocarpus, 271
Selloa, 851
Seneca Grass, 130
Senecio, 944
  Actinella, 949
  ambrosioides, 947
  arizonicus, 949
  Bigelovii, 948
  Blumeri, 950
  cynthioides, 949
  decorticans, 949
  Douglasii, 947
  Encelia, 950
  filicifolius, 947
  franciscanus, 947
  hartianus, 949
  Hartwegi, 948
  huachucanus, 949
  lathyroides, 947
  Lemmoni, 949
  longilobus, 947
  lynceus, 948
  Macdougalii, 947
  macropus, 947
  millelobatus, 948
  mohavensis, 949
  monoensis, 947
  multicapitatus, 946
  multilobatus, 947
  mutabilis, 950
  neomexicanus, 950
  orthophyllus, 947
  papagonius, 950
  Parryi, 948
  percalvus, 949
  prolixus, 948
  quercetorum, 947
  salignus, 948
  Seemannii, 948
  spartioides, 946
  stygius, 948
  Thornberi, 948
  Toumeyi, 950
  uintahensis, 948
  vulgaris, 947
  werneriaefolius, 949
  Wootonii, 949
Senegalia, 397
Senna, 404
Senna subfamily, 403
Sensitive joint-vetch, 472
Sensitive-pea, 406
Service-berry, 376
Sesban, 444
Sesbania, 444
Sesuvium, 281
Setaria, 138
Shad-scale, 259
Shamrock, 424
Sheep-fat, 259
Sheep sorrel, 244
Sheep-sour, 487
Shepherdia, 586
Shepherds-purse, 344
Shield Fern, 46
Shinleaf, 631
Shooting-star, 638
Sibbaldia, 381, 383
Sicyos, 824
Sicyosperma, 824
Sida, 549
Sidalcea, 548
Sidanoda, 552
Sidebells-pyrola, 631
Sieversia, 387
Silene, 301
Silk-tassel, 624
Silver-dollar cactus, 582
Silverweed, 384
Silybum, 954
Simaroubaceae, 495
Simarouba family, 495
Simmondsia, 521
Simsia, 905
Sinapis, 338
Sinita, 570
Siphonella, 686
Siphonoglossa, 801
Sisymbrium, 335
Sisyrinchium, 196
Sitanion, 95
Sium, 616
Skeleton-weed, 239
Skull-cap, 734
Skunk-bush, 523
Skunk-cabbage, 177
Skunk-leaf, 695
Skunk-weed, 356
Skyrocket, 692
Sleepy Grass, 116
Slimwood, 640
Small-mistletoe, 225
Smartweed, 246
Smilacina, 190
Smoke-thorn, 435
Smoke-tree, 435
Smother-weed, 261
Snail-seed, 321
Snake-cotton, 268
Snakeroot, 305
Snakes-head, 963
Snake-weed, 852
Snapdragon, 765
Sneeze-weed, 928
Snowberry, 814
Snow-on-the-mountain, 511, 517
Soapberry, 528
Soapberry family, 528
Soap-weed, 185
Soapwort, 303
Solanaceae, 748
Solanum, 756
  americanum, 759
  arizonicum, 758
  deflexum, 758
  Douglasii, 758
  elaeagnifolium, 758
  Fendleri, 758
  heterodoxum, 757
  Jamesii, 758
  Lumholtzianum, 757
  nigrum, 759
  nodiflorum, 759
  rostratum, 757
  sarachoides, 758
  sisymbriifolium, 757
  triflorum, 758
  villosum, 758
  Xanti, 758
Solidago, 856
  altissima, 857
  arizonica, 857
  canadensis, 857
  ciliosa, 857
  decumbens, 857
  gilvocanescens, 858
  graminea, 858
  Marshallii, 857
  missouriensis, 857
  nana, 858
  occidentalis, 858
  Petradoria, 858
  sparsiflora, 858
  subviscosa, 858
  trinervata, 858
  Wrightii, 858
Solomon-seal, 191

Sonchus, 965
Sophora, 412
Sorbus, 375
Sore-eye-poppy, 541
Sorghastrum, 143
Sorghum, 142
Sorrel, 243
Sotol, 189
Sour-clover, 421
Sour-grass, 387
Sow-thistle, 965
Spanish-bayonet, 185
Spanish-clover, 427
Spanish-needles, 909, 921
Spanish-tea, 252
Sparganiaceae, 64
Sparganium, 64
Spartina, 125
Spear Grass, 116
Spearmint, 747
Spectacle-pod, 341
Specularia, 826
Speedwell, 784
Spergularia, 298
Spermatophyta, 49
Spermolepis, 610
Sphaeralcea, 540
  ambigua, 543
  angustifolia, 545
  arida, 543
  arizonica, 546
  coccinea, 546
  Coulteri, 542
  cuspidata, 545
  digitata, 546
  Emoryi, 542
  Fendleri, 544
  grossulariaefolia, 546
  incana, 544
  laxa, 543
  leiocarpa, 544
  leptophylla, 546
  lobata, 545
  Orcuttii, 542
  parvifolia, 546
  pedata, 546
  pumila, 545
  Rusbyi, 543
  subhastata, 545
  variabilis, 543
  Wrightii, 544
Sphaerostigma, 594, 600
Sphenopholis, 98
Sphinctospermum, 444
Spice-bush, 431
Spider-flower, 355
Spider Grass, 119
Spiderling, 275
Spiderwort, 167
Spiderwort family, 166
Spikenard, American, 606
Spike-rush, 153
Spineless cactus, 583
Spiraea, 374
Spiranthes, 200
Spirodela, 165
Spleenwort, 47
Sporobolus, 112
  airoides, 114
  argutus, 113
  asper, 113
  contractus, 114
  cryptandrus, 114
  flexuosus, 114
  giganteus, 114
  interruptus, 114
  microspermus, 108
  Nealleyi, 114
  neglectus, 113
  patens, 113
  pulvinatus, 113
  pyramidatus, 113
  ramulosus, 108
  texanus, 114
  vaginiflorus, 113
  Wrightii, 114
Sprangle-top, 123
Spring-beauty, 289
Spruce, 54
Spurge, 511
Spurge family, 501
Spur-gentian, 650
Squaw-apple, 378
Squaw-berry, 749
Squaw-bush, 523, 530
Squaw-cabbage, 330
Squaw-lettuce, western, 697
Squaw-root, 796
Squirrel-tail, 95, 96
Stachys, 739
Staghorn cholla, 585
Stanleya, 328
Stanleyella, 329
Starflower, 190
Star-glory, 676
Star-leaf, 493
Star-lily, 177
Star-thistle, 955
Starwort, 292
Steironema, 638
Stellaria, 292
Stemodia, 782
Stenolobium, 795
Stephanomeria, 960, 967
Sterculiaceae, 554
Stevia, 842
Stick-leaf, 564
Stick-seed, 712
Stick-tight, 909
Stick-weed, 960
Stillingia, 510
Stink-grass, 85
Stinking-willow, 431
Stipa, 116
  arida, 118
  columbiana, 118
  comata, 117
  coronata, 117
  eminens, 117
  Lettermani, 118
  lobata, 117
  neomexicana, 117
  Pringlei, 117
  robusta, 118
  Scribneri, 117
  speciosa, 117
  viridula, 118
St. Johns-wort, 556
Stonecrop, 359
Strawberry, 380
Strawberry-blite, 252
Strawberry cactus, 570
Streptanthella, 331
Streptanthus, 331
Streptopus, 191
Strombocarpa, 402
Strophostyles, 483
Sylocline, 885
Stylosanthes, 472
Suaeda, 263
Sudan Grass, 142
Sugar-berry, 220
Sugar-bush, 523
Sulphur-flower, 240
Sumac, 522
Summer-cypress, 261
Summer-poppy, 492
Sun-drops, 593
Sunflower, 902
Sunflower family, 829
Swainsona, 444
Sweet-cicely, 610
Sweet-clover, 421
Sweet Grass, 130
Sweet-root, 610
Sweet-vetch, 471
Swertia, 649
Switch Grass, 134
Switch-sorrel, 528
Sycamore, 371
Sympetalae, 626
Symphoricarpos, 814
Synthyris, 786
Syntrichopappus, 918
Syringa, 366

Tagetes, 931
Talinum, 286
Tall-oat Grass, 101
Tamaricaceae, 557
Tamarix, 557
Tamarix family, 557
Tangle-head, 143
Tansy-mustard, 349
Tapiro, 814
Taraxacum, 964
Tar-bush, 903
Tarweed, 913
Tecoma, 794
Teddybear cactus, 584
Telegraph-plant, 854
Tepary bean, 483
Tephrosia, 440
Tesota, 442
Tetraclea, 730
Tetracoccus, 503
Tetradymia, 944
Tetramerium, 801
Teucrium, 733
Texas-timothy, 105
Thalictrum, 319
Thamnosma, 494
Thelesperma, 908
Thelypodium, 329
Thelypteris, 45
Thermopsis, 413
Thimble-berry, 378
Thistle, 951
Thlaspi, 335
Thorn-apple, 759
Thornbera, 439
Thoroughwort, 844
Three-awn, 118
Thurberia, 553
Thysanocarpus, 348
Tick-clover, 473
Tickseed, 907
Tidestromia, 268
Tiliaceae, 536
Tillaea, 361
Tillandsia, 166
Timothy, 105
Tithonia, 900
Tithymalus, 516, 517
Tium, 465
Toad-flax, 764
Tobacco, 760
Tobacco-root, 818
Tobosa, 122
Tolguacha, 760
Tomatillo, 749
Tomato, 749
Torch-wood family, 496
Torilis, 611
Tornillo, 402
Torreyochloa, 82
Toumeya, 575, 576
Townsendia, 868
Toxicodendron, 522
Trachypogon, 144
Tradescantia, 167
Tragia, 507
Tragopogon, 961
Tragus, 121
Trailing-four-o'clock, 274
Trautvetteria, 314
Trianthema, 281
Tribulus, 491
Tricardia, 706
Trichachne, 131
Trichloris, 126
Tricholaena, 138
Trichoptilium, 925
Trichostema, 734
Tridens, 90
Trifolium, 422
  albopurpureum, 424
  amabile, 424
  andinum, 423
  arizonicum, 424
  dubium, 424
  Fendleri, 424
  fistulosum, 424
  gracilentum, 424
  gymnocarpum, 424
  hybridum, 425
  involucratum, 424
  lacerum, 424
  longicaule, 424
  longipes, 425
  microcephalum, 424
  neurophyllum, 425
  Ortegae, 424
  pinetorum, 424
  pratense, 424
  repens, 425
  Rusbyi, 425
  subcaulescens, 424
  variegatum, 424
  Wildenovii, 424
  Wormskjoldii, 424
Triglochin, 67
Triodanis, 826
Triodia, 90
Tripsacum, 145
Tripterocalyx, 279
Trisetum, 99
Triteleia, 182
Triteliopsis, 181
Trixis, 958
Trompillo, 758
Trumpet-bush, 794
Tumamoca, 821
Tumble Grass, 125
Tumble-mustard, 336
Tuna, 579
Turkey-foot, 141
Turpentine-broom, 494
Turpentine-brush, 861
Turpentine-weed, 853
Turritis, 352
Tway-blade, 200
Twin-berry, 817
Twin-flower, 816
Twin-pod, 342
Twisted-stalk, 191
Twist-flower, 331
Typha, 63
Typhaceae, 63

Ulmaceae, 220
Umbelliferae, 606
Unicorn-plant, 795
Unicorn-plant family, 795
Uniola, 88
Urtica, 222
Urticaceae, 222
Utahia, 575
Utricularia, 798

Vaccaria, 303
Vaccinium, 634
Vachellia, 398
Valerian, 818
Valeriana, 818
Valerianaceae, 818
Valerian family, 818
Vanclevea, 852
Vanilla Grass, 130
Varnish-bush, 903
Vauquelinia, 374
Vegetable-oyster, 962
Velvet Grass, 101
Venus' comb, 611
Veratrum, 176
Verbascum, 764
Verbena, 725
  ambrosifolia, 727
  arizonica, 726, 728
  bipinnatifida, 727
  bracteata, 728
  carolina, 727
  ciliata, 727
  Ehrenbergiana, 727
  Gooddingii, 726
  gracilis, 728
  Halei, 728
  hastata, 727
  Macdougalii, 727
  menthaefolia, 728

Verbena (*cont'd*)
neomexicana, 728
perennis, 728
plicata, 728
pinetorum, 728
scabra, 727
tenuisecta, 727
verna, 726
Wrightii, 727
Verbenaca, 725
Verbenaceae, 724
Verbesina, 907
Veronica, 784
Vervain, 725
Vervain family, 724
Vetch, 476
Vicia, 476
Viguiera, 900
Vincetoxicum, 665
Vine-mesquite, 134
Viola, 559
Violaceae, 559
Violet, 559
Violet family, 559
Viorna, 312
Virginia-creeper, 535
Vitaceae, 534
Vitis, 535

Wafer-parsnip, 617
Wait-a-bit, 400
Wait-a-minute, 400
Walnut, 213
Walnut family, 213
Waltheria, 555
Washingtonia, 164
Water-cress, 340
Water-fern family, 32
Water-hemlock, 616
Water-leaf, 697
Water-leaf family, 696
Water-lettuce, 165
Water-marigold, 909
Water-milfoil, 604
Water-milfoil family, 604
Water-motie, 884
Water-parsnip, 615, 616
Water-pennywort, 609
Water-pimpernel, 637
Water-plantain, 68
Water-plaintain family, 68
Water-stargrass, 168
Water-starwort, 521
Water-starwort family, 521
Water-wally, 884
Water-weed, 69
Water-willow, 884
Water-wort, 556
Water-wort family, 556
Wedelia, 274
Wedge Grass, 98
Wedge-scale, 98
Western-wallflower, 354
Wheat Grass, 92
Whipplea, 368
Whispering-bells, 705
White-sage, 261
White-thistle, 259
White-thorn, 399
Whortleberry, 634
Wilcoxia, 569
Wild-basil, 746
Wild-buckwheat, 230
Wild-candytuft, 335
Wild-caraway, 615
Wild-carrot, 622
Wild-cucumber, 823
Wild-daisy, 874
Wild-hazel, 521
Wild-heliotrope, 698
Wild-hollyhock, 547
Wild-lilac, 532
Wild-olive, 643
Wild-potato, 758
Wild-rhubarb, 245
Wild-rye, 94
Willow, 209
Willow family, 207
Willow-weed, 590
Wind-flower, 311
Winter-cress, 338
Winter-fat, 260
Wintergreen, 631
Wire-lettuce, 960
Wislizenia, 357
Wolf-berry, 749
Wolf-tail, 104
Wood-betony, 793
Woodland-star, 365
Wood-rush, 173
Woodsia, 43
Wood-sorrel, 487
Wood-sorrel family, 487
Woodwardia, 49
Woolly-daisy, 921
Wormwood, 938
Wyethia, 899

Xanthium, 896
Xanthocephalum, 851
Xanthoxalis, 487
Xylophacos, 458–460

Yampa, 615
Yard Grass, 123
Yarrow, 936
Yellow-ragweed, 924
Yellow-rattle, 792
Yellow waterweed, 590
Yerba-de-fleche, 511
Yerba-de-la-rabia, 271
Yerba-del-venado, 933
Yerba-de-pasmo, 883
Yerba-mansa, 207
Yerba-santa, 707
Yucca, 185
angustissima, 188
arizonica, 187
baccata, 187
Baileyi, 188
brevifolia, 187
confinis, 187
elata, 188
Harrimaniae, 188
kanabensis, 188
mohavensis, 187
navajoa, 188
Newberryi, 188
schidigera, 187
Schottii, 187
Standleyi, 188
Thornberi, 187
utahensis, 188
verdiensis, 188
Whipplei, 188

Zaluzania, 899
Zannichellia, 67
Zauschneria, 590
Zephyranthes, 192
Zephyr-lily, 192
Zexmenia, 906
Zigadenus, 176
Zinnia, 897
Zorillo, 493
Zornia, 473
Zoysieae, 71
Zuckia, 260
Zygadenus (see Zigadenus)
Zygophyllaceae, 489
Zygophyllidium, 517

Supplement (1960)

by JOHN THOMAS HOWELL

ELIZABETH McCLINTOCK and collaborators

PTERIDOPHYTA

FERNS AND FERN ALLIES

Contributed by C. V. Morton

1. SELAGINELLACEAE. Selaginella Family

1. SELAGINELLA

There is a recent revision: "Selaginella rupestris and its allies," by Rolla M. Tryon (461). The following revised key to the Arizona species (pp. 28, 29) is adapted from this work.

1. Stems erect or ascending, with rhizophores (aerial roots) produced only near the base, or rarely with a few stems prostrate throughout; basal branch buds present. Leaves with the base abruptly adnate and distinct from the stem in color on all branches (2).
1. Stems prostrate to decumbent or irregularly ascending, with rhizophores produced throughout (at least on the primary stem); basal branch buds rarely and irregularly present (3).
2. Spores normal (megaspores present; microspores well developed, 38 to 64μ diam.); leaves gradually attenuate into an elongate seta 0.7 mm. long or more . 6. *S. rupincola*
2. Spores abortive (megaspores not found; microspores diminutive, averaging 22μ diam.); leaves abruptly narrowed into a shorter seta (about 0.4 mm. long) . . 7. *S. neomexicana*
3. Leaves of the branches with the base abruptly adnate and distinct from the stem in color on all sides, those of the main stem slightly decurrent; setae absent or reduced in size and only about 0.2 mm. long . . . . 4. *S. mutica*
3. Under leaves (at least) strongly decurrent, the upper leaves decurrent or not (4).
4. Leafy stems radially symmetrical to definitely dorsiventral, the zone of green leaves about equal on all sides of the stem; branch tips straight or slightly curled in the dormant condition; upper leaves with the base usually decurrent (5).
4. Leafy stems strongly dorsiventral, the zone of green leaves well developed on the upper side of the stem, very short to absent on the under side; branch tips involute in the dormant condition; upper leaves with the base abruptly adnate, distinct from the stem in color (rarely a distinct stem ridge present) (6).
5. Upper and under leaves unequal in length on the same portion of the stem, the under definitely longer; stems forming compact mats with discrete branches; base of the setae of the sporophylls often strongly broadened and flattened . . . . 3. *S. densa*
5. Upper and under leaves essentially or quite equal in length; stems forming open, spreading mats with intricate branching; base of setae of sporophylls not or only slightly broadened and flattened . . . . 5. *S. Underwoodii*
6. Setae delicate, filiform, tortuous, mostly soon deciduous (or sometimes present only in the bud) . . . . 1. *S. eremophila*
6. Setae stout, straight, mostly or entirely persistent . . . . 2. *S. arizonica*

3. Selaginella densa Rydb. (p. 28). The Arizona specimens, all from the top of Mount Baldy, are reported by Tryon (461, p. 58) to be somewhat intermediate between var. *densa* and var. *scopulorum* (Maxon) Tryon.

4. Selaginella mutica D. C. Eaton (p. 28). Most Arizona specimens represent var. *mutica* (with muticous leaves essentially lacking a seta), but Tryon reports var. *limitanea* Weatherby (with setate leaves, the seta short but evident) from a single collection (Paradise, Cochise County, *Ferriss* in March, 1904).

5. Selaginella Underwoodii Hieron. (p. 29). Tryon considers that var. *dolichotricha* Weatherby is not worthy of nomenclatural recognition, since, although it is distinct in some areas, it merges insensibly in others with var. *Underwoodii.*

Another species that occurs in southwestern Utah near the Arizona border and which may be found in Arizona eventually is *S. utahensis* Flowers (*Amer. Fern Jour.* 39:83, —1949). It is like *S. Watsonii* in being a cushion-forming species, but differs in having the leaves muticous (without setae) or essentially so; both these species may be distinguished from *S. Underwoodii* (which forms loose spreading mats) by their cushion-forming habit.

7. Selaginella neomexicana Maxon (p. 29). Considered by Tryon to be a naturally occurring hybrid *S. mutica* × *S. rupincola,* on the basis of its sterility and intermediate morphology.

3. EQUISETACEAE. Horsetail Family

1. EQUISETUM. Horsetail

2. Equisetum laevigatum A. Braun (p. 30). Delete New York, New Jersey, and Guatemala from the general range, which can then be stated as follows: Ontario and Michigan to British Columbia, south to Ohio, Oklahoma, Texas, and California; Mexico.

Add the following paragraph:

A recent study by Emily L. Hartman (404) recognized two subspecies—subsp. *laevigatum* being characterized by having the supporting tissue (collenchyma) behind the stem-ridges elongate internally and intruded through the parenchyma so as to almost or quite reach the subtending vascular bundle, and subsp. *Funstonii* (A. A. Eaton) Hartman (*E. Funstonii* A. A. Eaton, *E. kansanum* Schaffn.) in which this collenchyma is not so extended. A still more recent study by Richard L. Hauke indicates that this anatomical feature does not delimit natural groups.

3. Equisetum hyemale L. (p. 30). Replace the last paragraph (top of p. 31) by the following:

The Arizona plant is var. *affine* (Engelm.) A. A. Eaton (*E. prealtum* Raf.).

Certain aberrant specimens formerly referred to *E. hyemale* (as var. *intermedium* A. A. Eaton) or to *E. laevigatum* have been shown by Dr. Hauke to be the naturally occurring hybrid *E. hyemale* var. *affine* × *E. laevigatum.* Specimens of the hybrid mostly resemble *E. laevigatum,* but have some of the sheath-teeth persistent, some sheaths with black basal bands like *E. hyemale,* and some cones more prominently apiculate than is usual in *E. laevigatum.* The diagnostic feature is the aborted or malformed and sterile spores. For this hybrid the name *Equisetum* × *Ferrissii* Clute is available. The hybrid has been found in the Chiricahua, Santa Rita, and Huachuca mountains and near Flagstaff; it is of general occurrence where the parent species occur together.

7. POLYPODIACEAE. Fern Family

5. PELLAEA. Cliff-brake

1. Pellaea limitanea (Maxon) Morton (p. 37). In a recent publication, Rolla M. Tryon (462) has retained this (*op. cit.*, p. 86) and *P. Jonesii* (Maxon) Morton (*op. cit.*, p. 83) in the genus *Notholaena.* They belong to a group that is aberrant in either *Pellaea* or *Notholaena;* a proper generic disposition awaits further study.

2. Pellaea Jonesii (Maxon) Morton (p. 37). See note above.

3. Pellaea atropurpurea (L.) Link. The var. *simplex* (Butters) Morton (p. 38) is referred by A. F. Tryon (460, p. 142) to *P. glabella* var. *simplex* Butters; she considers it possibly of hybrid origin, *i.e., P. atropurpurea* × *P. glabella* var. *occidentalis.*

6. Pellaea Wrightiana Hook. (p. 38). Better called now *P. ternifolia* (Cav.) Link var. *Wrightiana* (Hook.) A. F. Tryon. In its best-developed form, *P. Wrightiana* seems quite different in its elongate pinnae and several pairs of pinnules, but many specimens are somewhat intermediate. Mrs. Tryon has identified a few specimens from Santa Cruz and Cochise counties as var. *ternifolia* (460, p. 153) but it is likely that they are only growth forms of var. *Wrightiana,* which sometimes simulates var. *ternifolia* closely.

7. NOTHOLAENA

In the recent treatment of American *Notholaena* by Rolla M. Tryon (462), there are no essential differences from the treatment in *Arizona Flora* (pp. 40, 42) except for the inclusion of two species, *N. limitanea* Maxon and *N. Jonesii* Maxon, in *Notholaena* rather than in *Pellaea,* as noted above.

Modify division 7 of the key as follows:

7. Blades broadly pentagonal, acute; stipes and rachises reddish brown; pinnules glandular on the upper surface, their margins crenulate and rather slightly revolute
7. *N. californica*

7. Blades narrowly triangular, long-acuminate; stipes and rachises blackish; pinnules glabrate above, their margins entire, strongly revolute.............. 8. *N. neglecta*

12. CYRTOMIUM

1. Cyrtomium auriculatum (Underw.) Morton. This generic and specific name replaces *Phanerophlebia auriculata* Underw. (p. 45; 438, p. 54).

13. THELYPTERIS

1. Thelypteris puberula (Baker) Morton. Recently Morton (439, p. 138) has regarded *T. augescens* (Link) Munz & Johnst. var. *puberula* (Fée) Munz & Johnst. (p. 45) as specifically distinct from *T. augescens,* in which case the correct name of the Arizona plant is *T. puberula* (Baker) Morton.

SPERMATOPHYTA

FLOWERING PLANTS

Gymnospermae

8. PINACEAE. Pine Family

1. PINUS. Pine

6. Pinus strobiformis Engelm. in Wisliz. *P. flexilis* James var. *reflexa* Engelm. in Rothr. *P. Ayacahuite* Ehrenb. var. *brachyptera* Shaw. Southwestern white pine, Mexican white pine. To replace *P. reflexa* (Engelm. in Rothr.) Engelm. (p. 53). In defining the name *P. strobiformis* we are following Little (424, p. 266) rather than Loock (425, pp. 107–116) who places *P. strobiformis* as a synonym of *P. Ayacahuite* var. *brachyptera* and treats *P. reflexa* as a separate species.

7. Pinus leiophylla Schiede & Deppe in Schlect. & Cham. var. **chihuahuana** (Engelm. in Wisliz.) Shaw. Chihuahua Pine. To replace *P. chihuahuana* Engelm. in Wisliz. (p. 53; Little, 424, p. 267).

9. Pinus Engelmannii Carr. Apache pine, Arizona longleaf pine. To replace *P. latifolia* Sarg. (p. 54; Little, 424, p. 265).

3. PSEUDOTSUGA. Douglas-fir

1. Pseudotsuga Menziesii (Mirbel) Franco var. **glauca** (Beissner in Jäger & Beissner) Franco. Rocky Mountain Douglas-fir; blue Douglas-fir. To replace *Pseudotsuga taxifolia* (Poiret) Britton var. *glauca* (Beissner) Sudworth (pp. 55, 56; Little, 424, p. 307).

Angiospermae

MONOCOTYLEDONEAE

13. POTAMOGETONACEAE. Pondweed Family

1. POTAMOGETON. Pondweed

Add as species 1a, to precede 1b. *P. pectinatus* L. (p. 66):

1a. Potamogeton crispus L. Reported from an irrigation tank near Camp Verde, Yavapai County, by J. A. McCleary (433, p. 155). Introduced from Europe and widely naturalized in North America. Differs from all Arizona pondweeds with "leaves all much alike and immersed" (p. 65) in its broadly linear to oblong and sharply serrulate leaf blades.

15. JUNCAGINACEAE.[1] Arrow-grass Family

1. TRIGLOCHIN. Arrow-grass

1. Triglochin debilis (Jones) Löve & Löve. To replace *T. concinna* Burtt Davy var. *debilis* (Jones) J. T. Howell (p. 67). The chromosomes in *T. concinna* number 2n = 48, whereas in *T. debilis,* 2n = 96 (Löve & Löve, 426, p. 160).

16. ALISMACEAE. Water-plantain Family

Add new genus and species (pp. 68, 69):

3. ECHINODORUS. Burhead

1. Echinodorus Berteroi (Sprengel) Fassett. Reported by Charles T. Mason (428, p. 64) from the Imperial National Wildlife Refuge, 1.5 miles north of Imperial Dam, Yuma County. Central and southern California east to the Atlantic and south to Mexico and the West Indies. From all other species of *Alismaceae* in Arizona, Bertero's burhead can be distinguished as a plant in which all flowers are perfect and in which the achenes (in many series) form a dense burlike head.

17. HYDROCHARITACEAE. Frogs-bit Family

1. Stem leafy, the leaves short (0.5 to 3.5 cm. long), in numerous whorls of 3 or 4 ..... 1. *Elodea*
1. Stem short and condensed (or prolonged in stolons), bearing a tuft of elongate ribbon-like leaves 1.5 to 20 dm. long ..... 2. *Vallisneria*

1. ELODEA. Water-weed

Elodea should replace *Anacharis* (pp. 69, 70):

1. Elodea densa (Planch.) Caspary. *Anacharis densa* (Planch.) Marie-Victorin.

2. Elodea canadensis Michx. *Anacharis canadensis* (Michx.) Planch.

2. VALLISNERIA. Tapegrass

1. Vallisneria americana Michx. A genus and species new to Arizona and the western states, reported by J. A. McCleary (432, p. 153) as "growing very extensively for a distance of over a mile in a canal of the Salt River Valley irrigation system," in Maricopa County. Widespread in eastern North America.

18. GRAMINEAE. Grass Family

Contributed by Jason R. Swallen

1. BROMUS. Brome

Substitute the following for divisions 7 to 10 in the key to the species (p. 76), and species 6 to 10 (pp. 77, 78).

6. Rhizomes wanting (7).

[1] Scheuchzeriaceae, family name proposed for conservation by A. A. Bullock (*Taxon* 7: 30, 163, —1958) against *Juncagineae* (*Juncaginaceae*) and *Triglochineae* (*Triglochinaceae*).

7. Lemmas densely ciliate on the margins, the back glabrous or nearly so
6. *B. Richardsonii*
7. Lemmas evenly pubescent or pilose over the back, or rarely glabrous (8).
8. First glume 1-nerved, or sometimes faintly 3-nerved (9).
8. First glume 3-nerved (10).
9. Panicles with long or short, flexuous or drooping branches; lower sheaths lanate pubescent, sometimes nearly glabrous ....7. *B. lanatipes*
9. Panicles narrow, the branches short, stiffly ascending to spreading...9. *B. Orcuttianus*
10. Glumes pubescent or pilose; panicles small, drooping, the branches usually densely pilose ....8. *B. Porteri*
10. Glumes glabrous; panicles ample, with long arcuate spreading branches, these scabrous on the angles ....10. *B. frondosus*

6. Bromus Richardsonii Link. Apache, Coconino, Yavapai, Graham, Cochise, and Pima counties, up to 11,000 feet, moist woods and rocky slopes, July to October. British Columbia southward to western Texas, New Mexico, and Arizona. Formerly included under *B. ciliatus* L. (p. 77), a distinct species which apparently does not occur in Arizona. *Bromus mucroglumis* Wagnon (464, p. 472) seems to be composed of rather diverse elements, which are probably best referred to *B. Richardsonii.*

7. Bromus lanatipes (Shear) Rydb. Apache, Coconino, Greenlee, Cochise, and Pima counties, mostly 6,500 to 9,500 feet, rocky slopes, mountain meadows, and moist pinelands, July to September. Colorado, northwestern Oklahoma, Texas, Arizona, and New Mexico. A form with sheaths glabrous or nearly so is f. *glaber* Wagnon.

Specimens referred to *B. texensis* (Shear) Hitchc. (p. 77) in the first edition have been reidentified as a nearly glabrous form of *B. lanatipes. Bromus texensis* is an annual species, whereas these specimens are definitely perennial.

8. Bromus Porteri (Coult.) Nash. Apache, Coconino, Greenlee, and Pima counties, open woods and mountain meadows, 2,000 to 10,000 feet, July to October. Manitoba to British Columbia, south to New Mexico, Arizona, and Nevada. Formerly included under *B. anomalus* (p. 77), a distinct Mexican species which extends into southwestern Texas.

9. Bromus Orcuttianus Vasey (p. 78). The Arizona specimen (*Lemmon* in 1883) is fragmentary, but apparently is this species. Since it is far out of range, however, the record should be considered doubtful.

10. Bromus frondosus (Shear) Woot. & Standl. (p. 78). Greenlee, Cochise, Santa Cruz, and Pima counties, 3,500 to 7,500 feet, rocky or shady slopes, and openings in pine-oak forests. Southern Utah, New Mexico, Arizona, and northern Mexico.

22a. AEGILOPS. Goatgrass

1. Aegilops cylindrica Host. Grand Canyon Village, Coconino County (*Whiting* in 1941); Prescott, Yavapai County (*Passey* in 1955); and Pinetop, Navajo County (*Michaels* in 1955). This European weed has become troublesome in fields and pastures. *Aegilops* is distinguished from *Agropyron* (p. 92) and *Triticum,* to which it is closely related, by the turgid, cylindrical spikelets.

26. HORDEUM. Barley

8. Hordeum vulgare L., cultivated barley. Common roadside weed, Casa Grande, Pinal County (*K. F. Parker* 8241). Readily distinguished from *H. arizonicum* Covas (p. 97), by the conspicuous sheath auricles, the continuous rachis, and the sessile lateral spikelets.

45. MUHLENBERGIA. Muhly

8a. Muhlenbergia tenuifolia (H.B.K.) Trin. Dragoon Mountains, Cochise County (*L. N. Goodding* 44–54). Southern Arizona and Mexico. This species is allied with *M. microsperma* (p. 109) but has longer, acute or awn-pointed, unequal glumes, and shorter florets, 2 to 2.5 mm. long.

61a. BECKMANNIA. Sloughgrass

1. Beckmannia syzigachne (Steud.) Fernald. Greenland Lake, north rim of Grand Canyon, 8,500 feet (*Merkle* 411), Coconino County; near Alpine, Apache County, *Isaacson*. *Beckmannia* is similar to *Spartina* (p. 125) in that the articulation is below the glumes, the spikelets falling entire, but differs in having the glumes equal, inflated, usually longer than the 1 or 2 florets.

72. DIGITARIA. Crab Grass

2. Digitaria Ischaemum (Schreb.) Schreb. ex Muhl. Tucson, Pima County, a common weed in lawns (*K. F. Parker* 8427). Introduced from Eurasia. This species is readily distinguished from *D. sanguinalis* (p. 132) by the glabrous sheaths, more compact habit, and smaller spikelets 2 mm. long. The first glume, when present, is a small nerveless scale.

79. SETARIA. Bristle Grass

8. Setaria macrostachya H.B.K. (p. 139). Pima and Santa Cruz counties, 2,500 to 4,000 feet, rocky soil, August to October. Texas, Arizona, and Mexico. *Harrison & Kearney* 8018, the only record of *S. Scheelei* (Steud.) Hitchc. in Arizona (p. 139), has been reidentified as *S. macrostachya.*

8a. Setaria leucopila (Scribn. & Merr.) K. Schum. Apache, Coconino, Greenlee, Yavapai, Cochise, Santa Cruz, and Pima counties, 2,000 to 6,000 feet, rocky soil, May to October. Texas to Colorado, Arizona, and Mexico. Included under *S. macrostachya* in the first edition, but has the sterile palea one-half to three-fourths as long as the flat or concave fertile palea rather than equalling the convex fertile palea.

80. PENNISETUM

2. Pennisetum ciliare (L.) Link. Near Oracle, Pinal County (*K. F. Parker* 7752). Introduced from India. Differs from *P. setaceum* (Forsk.) Chiov. (p. 140) in the much shorter, denser panicles 2 to 10 cm. long, and shorter bristles, 5 to 10 mm. long, subtending the spikelets.

84. ANDROPOGON. Broom-sedge

9a. Andropogon Wrightii Hack. Canelo Hills, Santa Cruz County, 5,000 feet (*Reynolds* in 1954, *R. White* 153). New Mexico, Arizona, and northern Mexico. This species can be distinguished from the other species of the Sect.

Amphilophis in Arizona by the stiffly ascending or appressed racemes and nearly glabrous spikelets. The first glume is sometimes pitted.

9b. Andropogon Ischaemum L. Santa Catalina Mountains, Pima County, 7,500 feet (*K. F. Parker* 8093). Introduced from the Old World. Differs from *A. barbinodis* Lag. and *A. saccharoides* Swartz, in the more slender, usually ascending culms and nodding racemes and in the pedicellate spikelets as large as the sessile ones.

19. CYPERACEAE. Sedge Family

2. CYPERUS. Flat-sedge

16a. Cyperus strigosus L. Reported by Charles T. Mason (429) from Sabino Canyon, Santa Catalina Mountains, Pima County. As the flat-sedges are keyed on pages 146 to 149, *C. strigosus* comes nearest to *C. Parishii* Britton from which it may be distinguished by its perennial tufted non-stoloniferous habit, the swollen and cormlike bases of the culms, its yellowish scales, and the narrowly oblong achenes. It is found throughout much of the United States and southern Canada.

8. CAREX

Contributed by F. J. Hermann

21a. Carex Bonplandii Kunth. Tonto National Forest, Gila County, turbinella oak chaparral, on coarse granitic soil, 3,600 feet, Rock Creek, Three Bar Game Management Area (*C. P. Pase* 946). A species of the Andes of Colombia, Peru, and Ecuador, its previously known range extending southward to Bolivia and northward in the high mountains of Costa Rica and Vera Cruz. The central Arizona record extends its range northwestward by 1,300 miles.

This sedge is distinguished from the densely cespitose *C. subfusca* (p. 162) by having long, slender rootstocks and narrower (scarcely more than 1 mm. rather than 1.5 mm. wide) perigynia in which the margins, only weakly serrulate and smooth below the middle, are almost unwinged so that they approach closely the condition in Section *Stellulatae.*

39a. Carex serratodens W. Boott. Tonto National Forest, Gila County, same locality as that of *C. Bonplandii* (*C. P. Pase* 947). Previously known only from California and southern Oregon, this record extends the known range of the species 400 miles eastward.

Distinguished from *C. Meadii* (p. 162) and *C. specuicola* (p. 163) by having broader lateral spikes (5 to 8 mm. wide rather than 4 mm. or less) and the culms lateral (that is arising from without the tufts of leaves of the previous year) rather than central, and not clothed at the base with the dried leaves of the previous year.

22. LEMNACEAE. Duckweed Family

2. LEMNA. Duckweed

5a. Lemna aequinoctialis Welwitsch. This tropical or subtropical duckweed has been reported from El Hambre charco, Papago Indian Reservation, Pima County, based on *Goodding* 412–45, the only known collection in the United States (cf. Mason, 429). It is characterized (N. E. Brown in Thiselton-Dyer *Fl. Trop. Afr.* 8:201, 203) by its elliptic, somewhat fleshy, very small thalli (less than 2 mm. long) that are "obtusely keeled longitudinally on the upper face." It is not clear how the Arizona plant might differ from *L. minima* Phil. which is rather widespread in the New World.

26. JUNCACEAE. Rush Family

Contributed by F. J. Hermann

1. JUNCUS. Rush

17. Juncus acuminatus Michx. (p. 172). Forma *sphaerocephalus* F. J. Hermann (Leafl. West. Bot. 8:12, 13), characterized by globose, many-flowered heads, has been collected in the Chiricahua Mountains, Cochise Co. (*Lemmon* 309). It occurs occasionally across the United States.

27. LILIACEAE. Lily Family

6. ALLIUM. Onion

Contributed by Marion Ownbey

1a. Allium Geyeri Wats. var. **tenerum** Jones. This is the name to be used for the asexual bulbiliferous plants that have been called *A. rubrum* Osterh. (p. 179).

7. Allium rhizomatum Woot. & Standl. The reduction of *A. rhizomatum* to synonymy under the Mexican species *A. glandulosum* Link & Otto (p. 180) is unwarranted. In addition to its pale flowers, *A. rhizomatum* is distinguished by its 3-lobed, apically 3-grooved ovary and lack of septal glands. *Allium glandulosum* is red-flowered and has a nearly cylindrical, apically rounded ovary which secretes 3 prominent drops of nectar from septal glands. The last feature does not show in herbarium specimens.

16. YUCCA. Soap-weed, Spanish-bayonet, Datil

By T. H. Kearney (MS.)

A recent revision by J. M. Webber (467) departs in many particulars from the treatment in *Arizona Flora* (pp. 185–188). *Yucca Thornberi* McKelvey (No. 6) and *Y. confinis* McKelvey (under No. 5) were thought to have originated as hybrids between *Y. baccata* Torr. (No. 4) and *Y. arizonica* McKelvey (No. 5). *Yucca Newberryi* McKelvey (No. 7) was reduced to synonymy under *Y. Whipplei* Torr. *Yucca Harrimaniae* Trel. (No. 8) was given as a synonym, in part, of *Y. neomexicana* Woot. & Standl., which apparently is not found in Arizona. The plant referred to *Y. Harrimaniae* in *Arizona Flora* is probably *Y. Baileyi* Woot. & Standl., of which *Y. Standleyi* McKelvey was cited by Webber as a synonym. *Yucca navajoa* J. M. Webber (No. 13) was reduced to a variety of *Y. Baileyi*. *Yucca utahensis* McKelvey (No. 10), *Y. verdiensis* McKelvey (No. 11), and perhaps *Y. kanabensis* McKelvey (No. 14) were considered as products of hybridization between *Y. elata* Engelm. (No. 9) and some other species. Little (424, p. 434), however, cited *Y. verdiensis* as a synonym of *Y. elata* Engelm.

Webber's key is here abstracted as to taxa that occur in Arizona.

1. Fruit indehiscent (2).
1. Fruit dehiscent (6).
2. Leaves small and fine, stiletto-like, the margin minutely denticulate; perianth segments

thick and fleshy; fruit spongy and dry. . . . . *Y. brevifolia*
2. Leaves large, coarse, swordlike, the margin entire or with free fibers; perianth segments thin; fruit fleshy (3).
3. Pistil 18 to 40 mm. long; plants commonly fruticose or arborescent, with few stems and with leaves in a rather open clump (4).
3. Pistil 45 to 90 mm. long; plants acaulescent or caulescent, with numerous stems or with leaves in rather dense clumps (5).
4. Plants caulescent, with rather tall, erect or assurgent stems; leaf blades narrow (18 mm.) the margins with few, fine fibers. . . . . *Y. arizonica*
4. Plants acaulescent or with short, procumbent stems; leaf blades broad (38 mm.), the margins with coarse, often curly fibers. . . . . *Y. baccata*
5. Leaf margins without free fibers, or in age, with a few fine straight fibers; ovary rather slender . . . . *Y. Schottii*
5. Leaf margins with free fibers, these rather coarse and curly; ovary rather stout *Y. schidigera*[2]
6. Margin of the leaf corneous, completely or partly denticulate, yellow, brown, or greenish yellow; stigma capitate . . . . *Y. Whipplei*
6. Margin of the leaf thin, paperlike, soon finely filiferous, white (rarely green) (7).
7. Plants arborescent, with thick, long stems; clusters of leaves commonly large. . *Y. elata*
7. Plants acaulescent or rarely with thin, short stems up to 1 m. tall; clusters of leaves commonly small (8).
8. Capsule usually deeply constricted. . . . . *Y. angustissima*
8. Capsule rarely, and then only slightly, constricted. . . . . *Y. Baileyi*

DICOTYLEDONEAE

Series 1. Apetalae

32. SALICACEAE. Willow Family

1. POPULUS. Cottonwood, Poplar, Aspen

4a. Populus nigra L. var. **italica** Muenchh. Lombardy poplar. Extensively planted in Arizona, its suberect branches distinguishing it from all native poplars. It was collected near Oraibi, Navajo County, apparently growing wild, but can scarcely be counted as an established member of the flora since it is sterile and spreads only by root-shoots.

2. SALIX. Willow

16. Salix padophylla Rydb. This name, published in 1901, to replace *S. pseudomonticola* Ball (p. 213), published in 1921, as pointed out by Little (424, p. 392).

34. BETULACEAE. Birch Family

2. BETULA. Birch

1. Betula occidentalis Hook. This name for the water birch is to replace *B. fontinalis* Sarg. (p. 215), a synonym.

35. FAGACEAE. Beech Family

1. QUERCUS. Oak

Contributed by Cornelius H. Muller

1. Quercus rugosa Née. *Q. reticulata* Humb. & Bonpl., *Q. diversicolor* Trel. (p. 217). This polymorphic species extends to Guatemala; none of the

[2] The name *schidigera,* although it was accepted by McKelvey (63), by Benson and Darrow (378), and by Webber (467), was rejected by Little (424) as a *nomen subnudum* in favor of *mohavensis.*

many proposed segregates is truly distinguishable. The type of Née's name is representative of the most common and widespread form of the species. In the Chiricahua Mountains in Cochise County this species has hybridized to a limited extent with *Q. arizonica.*

3. Quercus arizonica Sarg. (p. 218). This species hybridizes with both *Q. grisea* Liebm. (No. 4) and *Q. rugosa* Née in the Chiricahua Mountains, Cochise County.

6. Quercus turbinella Greene (p. 218). *Q. subturbinella* Trel. Although most of the Californian population is distinguishable as a subspecies, the Arizona population and that of Baja California (the type locality) are identical. This species has hybridized locally with *Q. ajoensis* in the Castle Dome and Kofa mountains in Yuma County.

6a. Quercus ajoensis C. H. Mull. Pima, Pinal, and Gila counties, between 2,500 and 4,500 feet elevation, the type from the Ajo Mountains, Pima County.

This species is known only in Arizona but probably occurs also in adjacent Sonora. It is most similar to *Q. turbinella* from which it is distinguished by its very elongate marginal leaf spines, the waxy-glaucous leaf surface, much more sparse pubescence, and the presence of microscopic papillae on the epidermis of the lower leaf surface. *Quercus ajoensis* and *Q. turbinella* have hybridized locally, especially in Yuma County where neither parent now exists but intermediates between them occur in the Castle Dome and Kofa mountains.

10. Quercus Dunnii Kell. *Q. Palmeri* Engelm. (p. 219). Cochise, Coconino, Gila, Graham, Mohave, and Yavapai counties, 3,500 to 6,000 feet elevation, the type from Baja California and scattered in southern California. Although *Q. Palmeri* was written as a binomial and is the earlier name, Engelmann referred to it as a subspecies of *Q. chrysolepis,* and *Q. Dunnii* is therefore the first specific name available for this species. There is ample room for doubt that Engelmann intended the name *Palmeri* as a trinomial, for in the same publication he used the category *varietas* for this purpose in trinomial combination.

Sargent (452) treated all the Arizona material of this group as *Q. chrysolepis* whereas Rydberg (451) treated it all as *Q. Wilcoxii* (a form of *Q. chrysolepis* suggestive of *Q. Dunnii*). Current investigations by J. M. Tucker (unpublished) have resulted in the first recognition of the fact that this group of oaks is represented in Arizona by two species, *Q. chrysolepis* and *Q. Dunnii.* Tucker has found that where these two occur in contact, the upper elevations harbor *Q. chrysolepis* but only *Q. Dunnii* occurs at the lower elevations, and the intermediate elevations frequently carry both species and an elaborate series of hybrids between them. The broader, thicker leaves with few elongate-spinose teeth and the stiff, divaricate branches of this shrub distinguish it from the arboreal species following.

10a. Quercus chrysolepis Liebm. *Q. Wilcoxii* Rydb. Cochise, Coconino, Gila, Mohave, Pima, and Yavapai counties, between 5,500 and 7,000 feet elevation, the type from coastal California reaching Sonora, Baja California, Oregon, and Nevada. Tucker's examination of the type of *Q. Wilcoxii* reveals it to belong here rather than with *Q. Dunnii.*

43. POLYGONACEAE. Buckwheat Family

5. ERIOGONUM. Wild-buckwheat

13. Eriogonum flexum Jones. This name to replace that of *E. glandulosum* (Nutt.) Nutt. (p. 237), a species that is not known to occur in Arizona (cf. Howell, 407, p. 37). Outside of northern Arizona, *E. flexum* is known also in Colorado and Utah. It is remarkable for its geniculate peduncle and its 3-lobed involucre.

17a. Eriogonum Parishii Wats. On the roadside, Crown King, Yavapai County, *Beaty* in 1951, the only collection known from Arizona. The Parish buckwheat is otherwise found in southern California and northern Baja California. It differs from *E. pusillum* and *E. reniforme* (species with puberulent perianths) in the hirsute (not lanate) pubescence of the basal leaves.

6. RUMEX. Dock, Sorrel

Rumex occidentalis (No. 9, p. 245) is to be replaced by the following two species:

9. Rumex nematopodus Rech. f. Known in Arizona only from the Huachuca Mountains, Cochise County (*Jones* in 1903, the type). Also in New Mexico and Chihuahua.

9a. Rumex densiflorus Osterh. Known in Arizona only from the White Mountains, Apache County. East and north in the Rocky Mountains to Wyoming.

The following key prepared by Dr. Kearney (418, p. 68) distinguishes these two species:

1. Flowering stems from a vertical root; longest pedicels 3 to 4 times as long as the perianth; valves in fruit 4 to 5 mm. long, 3 to 4 mm. wide........9. *R. nematopodus*
1. Flowering stems from a black, horizontal rootstock, this covered with rootlets; pedicels (as described) 1 to 1½ times as long as the perianth; valves in fruit 5 to 6 mm. long and wide ....9a. *R. densiflorus*

8. POLYGONUM. Knotweed, Smartweed

4a. Polygonum aviculare L. var. **littorale** (Link) W. D. J. Koch. *P. buxiforme* Small. Specimens with relatively broad, obtusish leaves are referable to this variety, which is considered indigenous in North America, whereas typical *P. aviculare* is naturalized from Europe.

44. CHENOPODIACEAE. Goose-foot Family

2. CHENOPODIUM. Goose-foot

2. Chenopodium graveolens Willd. var. **neomexicanum** (Aellen) Aellen. To replace *C. incisum* Poir. var. *neomexicanum* Aellen (p. 252) acc. Wahl (464a, p. 7).

The following key to species No. 7 to No. 16 has been contributed by Herbert A. Wahl:

7. Pericarp rugose or alveolate (8).
7. Pericarp smooth or nearly so (13).

8. Surface of the pericarp with irregular markings, seeds sharp-edged, the pericarp tightly adherent. Leaves thickish, deltoid to rhombic-ovate, coarsely and often sharply dentate or laciniate ....7. *C. murale*
8. Surface of the pericarp regularly and finely alveolate; pericarp adherent (9).
9. Herbage densely farinose, whitish; seeds subglobose; pericarp often whitened; stems much branched from the base, stout, the branches decumbent or spreading-ascending; leaves scarcely longer than wide, thickish; plants very ill-scented . . 8. *C. Watsoni*
9. Herbage moderately farinose to glabrescent, green; seeds flattened; stems usually not much branched from the base, the branches mostly erect or ascending at a narrow angle; leaves as long as to considerably longer than wide, thin or moderately thickened; plants not or only moderately ill-scented (10).
10. Stems slender; glomerules of flowers small and rather widely spaced, the inflorescence open; perianth parts not closely investing the fruit, spreading at full maturity, their midribs low-keeled; leaves entire, or subhastate with one pair of short teeth (11).
10. Stems stout; glomerules relatively large and crowded, the inflorescence usually dense; perianth parts closely investing the fruit, more or less connivent even at full maturity, their midribs strongly keeled or winged; leaves usually toothed above the basal lobes (12).
11. Seeds 1.2 to 1.5 mm. broad; leaves thin, the lobes usually acute; plants freely branching; inflorescences moderately leafy........................9a. *C. neomexicanum*
11. Seeds 1.0 to 1.1 mm. broad; leaves thin or firm, broadly triangular, usually with bipartite basal lobes; plants upright, with short basal branches and terminal leafless inflorescence ....9b. *C. Palmeri*
12. Seeds 1.0 to 1.3 mm. broad; leaves firm, with basal lobes, the terminal lobe elongate, toothed, the sides tending to be parallel half their length or nearly to the tip
10a. *C. Berlandieri* var. *sinuatum*
12. Seeds 1.2 to 1.5 mm. broad; leaves thin to coriaceous, variable in shape but the larger ones tending to be deltoid, with or without basal lobes and tapering gradually to the tip ....10b. *C. Berlandieri* var. *Zschackei*
13. Leaves linear, entire, 1-nerved; seeds 0.8 to 1.0 mm. broad; pericarp adherent
11. *C. leptophyllum*
13. Primary leaves lanceolate or broader, with 3 or more nerves (14).
14. Larger leaves chiefly 2 to 3 times as long as broad, irregularly toothed throughout, with more than a single pair of basal teeth or bipartite lobes; seeds 1.1 to 1.5 mm. broad, the pericarp chiefly adherent but often not strongly so..........16. *C. album*
14. Leaves entire or only with basal teeth or lobes, or (in *C. Fremontii*) occasionally with low teeth above the basal lobes if broader than ovate (15).
15. Pericarp strongly separable (16).
15. Pericarp adherent (20).
16. Seeds maturing mixed in the glomerules, so that mature seeds may be present before all flowers in the same glomerule have reached anthesis; perianth parts exposing half or more of the utricle when closed over it (17).
16. Seeds maturing relatively uniformly or in close succession in the glomerules; perianth parts largely or entirely hiding utricle when closed over it (18).
17. Leaf blades thin, deltoid, 1.0 to 4.0 (5.0) cm. broad, hardly longer than broad, usually with basal lobes; seeds chiefly 1.1 to 1.5 mm. broad..............14. *C. Fremontii*
17. Leaf blades thin or firm, ovate or narrower, 1.5 cm. broad or less, only occasionally with basal lobes; seeds chiefly 1.0 to 1.3 mm. broad..............14a. *C. atrovirens*
18. Leaf blades thick, chiefly less than twice as long as broad, prominently and usually sharply lobed at the base; utricle entirely covered by perianth lobes; plants profusely branched, usually low and bushy and pyramidal in shape....14b. *C. incanum*
18. Leaf blades thick or moderately thin, chiefly 3 or 4 more times as long as broad, lobed or unlobed at the base; utricles entirely covered or partially exposed by the perianth lobes; plants upright and strict or variously branched, usually not pyramidal (19).
19. Leaves lanceolate to oblong-elliptic, chiefly unlobed; perianth lobes relatively greenish, obtuse, usually curving inward over the fruit and with a prominent keel; plants upright-spreading to low and diffuse..............13a. *C. desiccatum* var. *desiccatum*

19. Leaves lanceolate to ovate-lanceolate, lobed or unlobed; perianth lobes relatively yellowish, acute and ascending, less prominently keeled; plants erect, often strict
13b. *C. desiccatum* var. *leptophylloides*

20. Leaves thin, ovate-lanceolate or broader; seeds 1.2 to 1.5 mm. broad
14c. *C. incognitum*

20. Leaves firm, ovate-lanceolate or narrower; seeds 1.0 mm. broad...........12. *C. hians*

Delete *C. arizonicum* Standley (No. 9, p. 253) and substitute the two following:

9a. Chenopodium neomexicanum Standley. Cochise County in Huachuca and Chiricahua mountains.

9b. Chenopodium Palmeri Standley. *C. arizonicum* Standley (type from the Santa Rita Mountains). Maricopa, Pima, and Apache counties.

10. Chenopodium Berlandieri Moq. (p. 253). Represented in Arizona by var. *sinuatum* (Murr) Wahl and var. *Zschackei* (Murr) Murr. According to Dr. Wahl (in litt.), *C. Berlandieri* is the most common and widespread midwestern and western species.

13. Chenopodium desiccatum A. Nels. *C. pratericola* Rydb. (p. 253). Represented in Arizona by var. *desiccatum* (*C. pratericola* subsp. *desiccatum* Aellen) and var. *leptophylloides* (Murr) Wahl (*C. pratericola* var. *leptophylloides* Aellen).

Chenopodium Fremontii as treated on pages 253 and 254 comprises the following four species:

14. Chenopodium Fremontii Wats. Apache County to Mohave County, south to Pima County.

14a. Chenopodium atrovirens Rydb. Apache County.

14b. Chenopodium incanum (Wats.) Heller. *C. Fremontii* Wats. var. *incanum* Wats. Mohave and Pima counties.

14c. Chenopodium incognitum Wahl. Apache, Navajo, and Coconino counties.

Delete No. 15, *Chenopodium albescens* Small (p. 254).

45. AMARANTHACEAE. Amaranth Family

1. AMARANTHUS. Amaranth

4. Amaranthus Torreyi (Gray) Benth. (p. 266). This name has been regarded as an ambiguous name by Dr. Jonathan Sauer (453, p. 38) who would replace it in part by *A. Watsoni* Standl. (a dioecious species), a collection of which he cites from Yuma (*op. cit.*, p. 39). However, as Dr. Kearney (418, p. 69) points out, *A. Torreyi* in *Arizona Flora* is treated as a monoecious species and the identity of the Arizona plant so-called is still to be determined.

2. ACANTHOCHITON

1. Acanthochiton Wrightii Torr. Add the following synonym and reference: *Amaranthus Acanthochiton* Sauer (453, p. 44).

Series 2. Choripetalae

49. PORTULACACEAE. Purslane Family

7. PORTULACA

1. Portulaca umbraticola H.B.K. According to Legrand (421, p. 2), *P. coronata* Small (p. 291) and *P. lanceolata* Engelm. are synonyms of *P. umbraticola* H.B.K. which would range from the southern United States to South America.

50. CARYOPHYLLACEAE. Pink Family

1. STELLARIA. Starwort

4. Stellaria gonomischa Boivin. This name is applicable to North American plants (Boivin, *Svensk Bot. Tidskr.* 50:113, 114), whereas *S. umbellata* Turcz. (p. 293) is for Asiatic plants.

2. CERASTIUM

4. Cerastium nutans Raf. Add the following note by Dr. Kearney (415, p. 173): "*Cerastium adsurgens* Greene . . . appears to be a narrow-leaved form of *C. nutans* Raf., to judge from a specimen from New Mexico identified by Greene himself as *C. adsurgens*." The second sentence in the note in small print (p. 295) is to be deleted.

4. ARENARIA. Sandwort

5. Arenaria lanuginosa (Michx.) Rohrb. subsp. **saxosa** (Gray) Maguire. This name replaces both *A. confusa* Rydb. (p. 296) and *A. saxosa* Gray (p. 297) according to Dr. Bassett Maguire (427, p. 498).

52. RANUNCULACEAE. Buttercup Family

9. MYOSURUS. Mouse-tail

2. Myosurus nitidus Eastw. To replace *M. Egglestonii* Woot. & Standl. (p. 314; Campbell, 380, p. 401).

3. Myosurus aristatus Benth. (p. 314). Arizona plants of this species belong to subsp. *montanus* (Campbell) Stone (*M. minimus* L. subsp. *montanus* Campbell). Their relationship to *M. minimus* and *M. aristatus* has been discussed by H. L. Mason and D. E. Stone (430, p. 503).

11. RANUNCULUS. Buttercup, Crowfoot

Contributed by Lyman Benson

15. Ranunculus pedatifidus J. E. Smith var. **affinis** (R. Br.) L. Benson. This varietal name applies to the North American plants of this species. In Utah and northern Arizona the plant occurs "in modified form" (p. 318; Benson, 377, p. 355).

55. PAPAVERACEAE. Poppy Family

3. ARGEMONE. Prickle-poppy

The following key and disposition of taxa are adapted from the monograph of the genus by Gerald B. Ownbey (441).

1. Leaf surfaces above and below and stems usually closely prickly; lower cauline leaves lobed to about one-half of distance to midrib (2).
1. Leaf surfaces prickly almost exclusively on primary and secondary veins, sometimes essentially smooth; stems more sparsely prickly; lower cauline leaves mostly lobed to four-fifths or more of distance to midrib (3).
2. Largest capsular spines compound, *i.e.*, with few to many smaller spines or prickles arising from the basal portion; prickles on buds often compound

1a. *A. munita* subsp. *rotundata*
2. Largest capsular spines simple; prickles on bud simple. . .1b. *A. munita* subsp. *argentea*
3. Uppermost leaves usually definitely clasping (4).
3. Uppermost leaves not definitely clasping (5).
4. Latex bright yellow when fresh; stamens 120 or less; stigma yellowish green or purplish 2a. *A. pleiacantha* subsp. *ambigua*
4. Latex orange when fresh; stamens 150 or more; stigma purple 5. *A. corymbosa* subsp. *arenicola*
5. Buds usually subspheric, sepal horns smooth. . . . . .3. *A. gracilenta*
5. Buds usually oblong, elliptic, or obovate, sepal horns nearly always prickly (6).
6. Sepal horns terete, 12 to 15 mm. long. . . . . .4. *A. arizonica*
6. Sepal horns usually flattened on adaxial side, 5 to 12 mm. long 2b. *A. pleiacantha* subsp. *pleiacantha*

1a. Argemone munita Dur. & Hilg. subsp. **rotundata** (Rydb.) G. B. Ownbey. Open desert slopes and foothills to middle elevations, Coconino and Mohave counties. Northwestern Arizona to California, Nevada, and Utah.

1b. Argemone munita Dur. & Hilg. subsp. **argentea** G. B. Ownbey. Sandy soil of desert slopes and washes, Mohave County. Deserts of southern Nevada and southeastern California.

2a. Argemone pleiacantha Greene subsp. **ambigua** G. B. Ownbey. Well-drained gravelly soils, 3,500 to 6,000 feet, central Arizona (Coconino, Yavapai, Gila, and Pinal counties). The type (*Ownbey & Ownbey* 1827) was collected 10.3 miles northeast of Prescott, Yavapai County.

2b. Argemone pleiacantha Greene subsp. **pleiacantha.** Washes, slopes, and plains, often in disturbed soil, Yavapai and Pima counties east to Apache and Cochise counties. Southwestern New Mexico, northern Mexico.

3. Argemone gracilenta Greene. Sandy and gravelly soil of desert slopes and washes, central and southwestern Arizona (Yavapai to Pima and Yuma counties), Sonora and Baja California, Mexico.

4. Argemone arizonica G. B. Ownbey. Rocky slopes on north and south walls of the Grand Canyon, Grand Canyon National Park, Coconino County. The type (*Eastwood & Howell* 1003) was collected on Kaibab Trail to Roaring Springs.

5. Argemone corymbosa Greene subsp. **arenicola** G. B. Ownbey. In loose or stable red sands, northeastern Coconino County; north to Utah.

6. FUMARIA. Fumitory

1. Fumaria parviflora Lam. Occurring as a weed in Tucson, Pima County; a native of the Old World. This genus differs from *Corydalis* (p. 325), the only other member of the fumitory subfamily of the Poppy Family in Arizona, in having globose, 1-seeded, indehiscent fruits and whitish, purplish-tipped corollas.

56. CRUCIFERAE. Mustard Family

6a. CARDARIA

1. Cardaria Draba (L.) Desv. *Lepidium Draba* L. (p. 333). Hoary cress, a noxious weed of agricultural lands, should be known under the generic name *Cardaria*. This genus may be distinguished from *Lepidium* by its silicles which are turgid, subglobose, and nearly or quite indehiscent.

8. SISYMBRIUM

2a. Sisymbrium Kearneyi Rollins. Known only from Grand Canyon National Monument, Mohave County; the type, *McClintock 52–481a.* Of the several species in *Arizona Flora,* this most closely resembles *S. elegans* (Jones) Payson (p. 336) from which it may be distinguished by its lower pinnatifid leaves and by its erect siliques. In *S. elegans* (not yet known from Arizona), the lower leaves are entire and the siliques widely spreading.

11. BRASSICA

6a. Brassica Tournefortii Gouan. A weed of roadsides and cultivated land in the Yuma area, according to Charles T. Mason (429). The plant is a native of the Old World, and among the Arizona species of *Brassica* it is most similar to *B. juncea* (L.) Coss. (p. 338) from which it differs in its hirsute pubescence of leaves and lower stems, its narrower petals (less than 2 mm. wide), and its smaller seeds (about 1 mm. in diameter).

11a. ERUCA

1. Eruca sativa Mill. *Brassica Eruca* L. (p. 338). The so-called garden-rocket should be separated from *Brassica,* from which it may be readily distinguished by the two rows of seeds in each cell of the silique.

17. DITHYREA. Spectacle-pod

2. Dithyrea Wislizeni Engelm. According to R. C. Rollins (447), var. *Griffithsii* (Woot. & Standl.) Payson (p. 342) is not distinguishable as a separate entity from *D. Wislizeni.*

29a. MATTHIOLA

1. Matthiola bicornis (Sibth. & Smith) DC. This native of the eastern Mediterranean region, given only incidental mention on page 354, is common on the mesas near Tucson (where it was reported by Thornber as early as 1914), and it has also been collected near Holbrook, Navajo County. In the key to the genera of *Cruciferae, Matthiola* would come next to *Chorispora* from which it differs in the much shorter, more deeply cleft beak of the capsule, the body of which is scarcely or not at all torulose.

60. SAXIFRAGACEAE. Saxifrage Family

5. PHILADELPHUS. Mock-orange

1. Philadelphus microphyllus Gray (pp. 366, 367). The *P. microphyllus* complex has been given a novel and very different treatment by Dr. Shiu-ying Hu (409) and the following analysis of her work was prepared by Dr. Kearney (MS.):

"Dr. Hu recognized as species *P. microphyllus* Gray, *P. occidentalis* A. Nels., *P. argenteus* Rydb., *P. argyrocalyx* Woot., and *P. stramineus* Rydb., but no Arizona specimens of the two latter taxa were cited. Two new varieties of *P. microphyllus* were described, var. *linearis* S. Y. Hu and var. *ovatus* S. Y. Hu, both occurring in the Lukachukai Mountains, northern Apache County (*Goldman* 2928, type of var. *linearis,* and *Goldman* 2913, referred to var. *ovatus*). Collections from the Chiricahua Mountains, Cochise County (*Burrall* 2079, *Kusche* in 1937) were referred to *P. Palmeri*

Rydb., from the Santa Catalina Mountains, Pima County (*Rehder* 470) to *P. maculatus* (C. L. Hitchc.) S. Y. Hu, several from the Pinaleno and Chiricahua mountains to *P. madrensis* Hemsl., and several from the Santa Catalina Mountains to *P. crinitus* (C. L. Hitchc.) S. Y. Hu. A collection from Grand Canyon National Park (*Gilstrap* 13) was cited under the narrow-leaved *P. occidentalis* var. *minutus* (Rydb.) S. Y. Hu."

63. ROSACEAE. Rose Family

21. PRUNUS. Plum, Cherry

4. Prunus serotina Ehrh. subsp. **virens** (Woot. & Standl.) McVaugh. According to McVaugh (434) this name replaces *P. virens* Woot. & Standl. (pp. 394, 395). In Arizona, the subspecies is represented by two varieties: var. *virens* and var. *rufula* (Woot. & Standl.) McVaugh [*P. virens* var. *rufula* (Woot. & Standl.) Sarg., which is described on p. 395].

64. LEGUMINOSAE. Pea Family

17. LUPINUS. Lupine

8a. Lupinus pusillus Pursh subsp. **intermontanus** (Heller) Dunn. According to D. B. Dunn (392), all Arizona collections referred to *L. pusillus* (var. *pusillus*) or var. *intermontanus* (Heller) C. P. Sm. in *Arizona Flora* (p. 417) are referable to subsp. *intermontanus*.

8b. Lupinus pusillus Pursh subsp. **rubens** (Rydb.) Dunn. According to Dr. Dunn (*loc. cit.*) this trinomial should replace *L. rubens* Rydb. (p. 417).

13a. Lupinus volutans Greene. Between Frazier Well and Hualpai Hilltop, Coconino County. This species, mentioned in the paragraph in small type (p. 420), is found at scattered stations from southeastern Oregon through Nevada to northwestern Arizona. It is part of the *L. lepidus* group. In Arizona *L. volutans* resembles *L. huachucanus* Jones (p. 418) from which it differs in the longer hairs of the stems and petioles (up to 3 mm. long), racemes scarcely surpassing the foliage, and larger corollas (12 to 13 mm. long).

16a. Lupinus caudatus Kell. Specimens from Kaibab Plateau, in which the leaflets are pubescent on the upper side and the calyx is more or less spurred, have been removed from *L. argenteus* Pursh (p. 418) and referred to *L. caudatus* by Dr. Kearney (418, pp. 62, 63). Dunn (*Contrib. Fl. Nevada* 39:46,—1956) cites Arizona in the distribution of *L. caudatus*.

21. Lupinus neomexicanus Greene. According to D. B. Dunn (in litt.), *L. neomexicanus* Greene and *L. Blumeri* Greene (p. 420) are synonymous and, since *L. neomexicanus* is the older name, it should be used.

38. SPHAEROPHYSA

1. Sphaerophysa Salsula (Pall.) DC. This name to replace *Swainsona Salsula* (p. 445).

39. ASTRAGALUS

Contributed by R. C. Barneby

4. Delete this entry (*A. convallarius,* p. 457) and replace by the two following.

4. Astragalus xiphoides (Barneby) Barneby. *A. convallarius* var. *xiphoides* Barneby. Desert mesas, on sandstone, about 5,200 feet, May, known only from Navajo County, the type from near Holbrook (*Ripley & Barneby* 5246).

This and the next are sparsely leafy switch-plants with small loosely racemose flowers, the banner abruptly recurved as in *A. convallarius* Greene, although even the lowest stipules are free. The upper leaves of *A. xiphoides* are reduced to the threadlike rachis and the ripe pods are flat and bladelike, 2.5 to 4 cm. long, 4 to 5.5 mm. wide.

4a. Astragalus titanophilus Barneby. *A. convallarius* var. *foliolatus* Barneby. Sandy soil on the limestone plateau about the common boundary point of Coconino, Mohave, and Yavapai counties, locally common in juniper woodland and desert grassland at 4,450 to 6,000 feet, known only from northwestern Arizona, the type from near Frazier Well, Coconino County (*Ripley & Barneby* 5229).

Differs from the preceding in having all leaves pinnately foliolate and in the narrower pod (3 to 4.5 mm. wide) only a little laterally flattened when ripe.

5a. Astragalus flexuosus Dougl. Northern and east-central Arizona to New Mexico, north, mostly east of the Continental Divide, to Alberta and southern Manitoba.

A widely dispersed and complex species, distinguished from other Arizona *Astragali* by its subterranean caudex, connate lower stipules, diffuse stems, and small pinkish flowers giving rise to declined, sessile or subsessile, unilocular pods of thin texture, the cavity traversed by pulpy filaments. The typical variety extends south and west from its main range in the Cordillera and Great Plains into southeast Utah and New Mexico but is not yet recorded from Arizona. It has the banner recurved through about 45° and the keel 5.5 to 8 mm. long; its pods are narrowly oblong or linear in outline, straight or curved either way, 12 to 24 mm. long and 2.5 to 4.5 mm. in diameter. The var. *Greenei* (Gray) Barneby (*A. Greenei* Gray, *A. gracilentus* var. *Greenei* Jones, *Pisophaca Greenei* Rydb.), differing ideally from var. *flexuosus* in its tumid or definitely inflated pod of ovate or broadly elliptic outline, mostly 5 to 9 mm. in diameter, is common locally at 4,400 to 5,700 feet in juniper-grassland and on gullied slopes under cliffs in the upper Salt River Valley, Gila County, and extends into southwest and central New Mexico. A third variety, var. *Diehlii* (Jones) Barneby (*A. Diehlii* Jones, *Pisophaca Diehlii* Rydb.), characterized by a very small flower with abruptly recurved banner and keel only about 5 mm. long, represents *A. flexuosus* at low elevations on the Colorado Plateau in southwestern Colorado, southeastern Utah, and northern Navajo County, Arizona (Second Mesa, Hopi Indian Reservation, *Michaels* 10 in 1955), where it is locally plentiful on sandy valley floors and open plains, associated with red and white sandstones. Its pod is similar to that of var. *flexuosus* but averages shorter (11 to 15 mm.); it is usually a little curved downward and keeled ventrally by the blunt suture.

7. Delete this entry (p. 457) and replace by the following.

7. Astragalus bisulcatus (Hook.) Gray. Northern Arizona to New Mexico and Alberta.

A coarse, ill-scented *Astragalus*, its presence indicating selenium in the soil. It is highly toxic, but not sufficiently abundant in Arizona to cause much trouble. Two forms of the species, which is readily recognized by its pendulous, stipitate, unilocular pod with ventral suture prominent but depressed and lying in a double groove, are known to occur in Arizona: var. *bisulcatus* (*Diholcos bisulcatus* Rydb., *A. Haydenianus* var. *major* Jones), with usually purplish flowers of moderate length (the banner 10 to 15 mm. long) and relatively long, cross-reticulate but not rugulose pods (the body 10 to 15 mm. long); it is widely dispersed over the Great Plains and extends westward over the Colorado Plateau into southwestern Utah and Coconino County, Arizona; and var. *Haydenianus* (Gray) Jones (*A. Haydenianus* Gray, *Diholcos Haydenianus* Rydb.), a plant of the west slope of the Rocky Mountains from southern Wyoming to northern New Mexico, entering Arizona in northern Apache County (Piute Rock, *Starr* 90), distinguished by its smaller, whitish, often very numerous flowers (the banner 8 to 11 mm. long) and shorter, transversely wrinkled pods less than 10 mm. long.

8. Delete this entry (p. 457) and replace by the following:

8. Astragalus flavus Nutt. Wyoming to northwestern New Mexico and northern Arizona.

A polymorphic species, of which two intergradient varieties occur in Arizona: var. *flavus* (*A. flaviflorus* Sheld., *Cnemidophacos flavus* Rydb.), with calyx tube about 2.5 to 4 mm. wide and pale yellow petals, found mostly above 5,500 feet from Coconino County northward and eastward; and var. *candicans* Gray (*A. confertiflorus* Gray, *Cnemidŏphacos confertiflorus* Rydb.), with calyx tube 2 to 2.5 mm. wide and petals whitish or lilac-tinged when fresh, dispersed between 2,000 and 5,900 feet from Mohave County, Arizona, north into central Utah and southern Nevada.

14. Astragalus amphioxys Gray (p. 458). In addition to the varieties mentioned, the var. *modestus* Barneby (374a), distinguished by its small flowers, the banner only 13 to 16 mm. (not 18 to 28 mm.) long, occurs in the piñon belt in Mohave County, Arizona, north and west of the Grand Canyon, and in southern Nevada, the type from Toroweap Valley (*McClintock* 52–193). The report of *A. musimonum* Barneby (p. 462) from Mokiak Pass, Mohave County, was based on specimens of var. *modestus* misidentified by the writer.

17a. Astragalus coccineus Brandg. *Xylophacos coccineus* Heller. Rocky slopes in the piñon-juniper zone, mostly on granite, March to May, northern Mohave and western Colorado deserts, California, and extreme northern Baja California, and isolated on the Kofa Mountains, Yuma County, Arizona (*H. B. Crandell* in 1952).

Similar in growth habit and in the large woolly pods to *A. Newberryi* Gray (p. 459), but distinguished by its narrow, little-recurved scarlet petals, the banner and keel of about equal length (3 to 3.5 cm.), the wings slightly shorter.

20. Astragalus humistratus Gray (p. 460). The var. *sonorae* as defined in *Arizona Flora* consists of two forms: genuine var. *sonorae* (Gray) Jones, with relatively long (1 to 2 cm.) and narrow pod dorsally sulcate from base up to the beak, and 18 to 26 ovules; and var. *humivagans* (Rydb.) Barneby (*Batidophaca humivagans* Rydb.), with comparatively short pod of plumper outline, sulcate at base only or not at all, and 10 to 16 ovules. The former is found at 4,500 to 6,600 feet in the Gila watershed of southeastern Arizona, extending into western New Mexico and northern Sonora, the latter at 5,000 to 8,000 feet from the White Mountains, Apache County, west across north and north-central Arizona to southeastern Nevada and southern Utah. The var. *crispulus* Barneby is now known to be common locally at 7,250 to 8,150 feet in pine forests of the White Mountains, and has been traced into adjoining New Mexico.

25. Astragalus sesquiflorus Wats. (p. 461). The Arizona plant has been distinguished as var. *brevipes* Barneby, the type from Betatakin, Navajo County (*J. T. Howell* 24631); it extends just into southeastern Utah. The var. *sesquiflorus,* with larger flowers and acute leaflets, is still known only from the base (at 5,000 to 5,500 feet) of Zion Escarpment in Kane County, Utah, but is to be sought in adjoining Arizona.

27a. Astragalus monumentalis Barneby. *Astragalus* sp., *Arizona Flora,* p. 470. Near Kayenta, Navajo County (*Peebles & Fulton* 11928). Northeastern Arizona and southeastern Utah, in the piñon zone, 5,000 to 6,100 feet, commonly in crevices of sandstone rimrock, April to June.

A dwarf tufted plant, with large veiny overlapping stipules in contact opposite the petiole but not united, about 9 to 15 small leaflets to the leaf, and 3 to 9 ascending, racemose, purplish pink flowers on subscapose peduncles. The pod is ascending, humistrate, linear-oblong but lunately incurved, compressed-triquetrous, brightly mottled, bilocular, 1.8 to 3 cm. long. A related species, *A. naturitensis* Pays., known at present only from southwestern Colorado, is to be looked for in similar habitats in northern Arizona. It resembles *A. monumentalis* closely, but the flowers are bicolored (purple and white) and the pod is shorter and unilocular.

***28. Astragalus musimonum** Barneby (p. 462). Delete this entry (cf. note on 14. *A. amphioxys,* above).

29. Delete this entry and replace by the following:

29. Astragalus Hallii Gray. Rocky Mountains of Colorado, south and west to central and southeastern Arizona and western New Mexico.

The Arizona plant is var. *fallax* (Wats.) Barneby (*A. fallax* Wats., not Fisch., *A. gracilentus* var. *fallax* Jones, A. *famelicus* Sheld.), of which the distribution is given correctly, under the last synonym, in *Arizona Flora* (p. 462). A collection from Richville, Apache County (*Richey* in 1950) was identified by me and subsequently listed by Kearney as typical *A. Hallii,* but is better interpreted as one of several intergradient forms which link the two varieties. In Arizona the pod of var. *fallax* varies from subsessile to stipitate (the stipe up to 4.5 mm. long), and the valves may be either strigulose or villosulous; the petals are sometimes bright purple as in var. *Hallii,* but more often whitish tipped or suffused with lurid lilac.

32. Astragalus ceramicus Sheld. (p. 462). Despite great superficial variation in size and outline of the pod and in development of lateral leaflets in some or all leaves, it is probably correct to refer all Arizona material of the species to var. *ceramicus* (including var. *magnus* Jones, var. *angustus* Jones, and the var. *filifolius* of *Arizona Flora*). The entity so formed is distinguished from genuine var. *filifolius* (Gray) Hermann, which barely extends west of the Continental Divide, by its vesture of largely dolabriform hairs, small flowers (calyx mostly 3 to 4, not 4 to 6 mm. long), and 12 to 16, not 20 to 30 ovules.

36. Delete this entry (p. 463) and replace by the following:

36. Astragalus fucatus Barneby (374a). Sandy plains, desert washes, sometimes on dunes, associated with the red sandstones of the Colorado Plateau, 4,500 to 6,200 feet, May to July, locally common in the drainage of the lower San Juan and Little Colorado rivers in extreme eastern Coconino and northern Navajo and Apache counties, Arizona, and in adjoining Utah, the type from near Hotevila, Navajo County (*Peebles* 13392).

Named only recently but long known as *A. subcinereus,* of which the type, however, belongs to the species next in order, the *A. Sileranus* of *Arizona Flora* (p. 463). Similar in most technical characters to genuine *A. subcinereus* but readily recognized by the closely appressed vesture of the calyx and fruit, the acutely triangular, beaklike keel tip, and the firmer although more strongly inflated pod (12 to 22 mm., not 6 to 13 mm. in diameter) enclosing 21 to 32 (not 10 to 20) ovules or seeds.

37. Delete this entry (p. 463) and replace by the following:

37. Astragalus subcinereus Gray. *Phaca subcinerea* Rydb., *A. Sileranus* Jones, *Phaca Silerana* Rydb. Dry hillsides and parklike openings in pine and piñon forests, or in oak thickets, mostly between 6,000 and 9,000 feet, May to September, strongly calciphile with us, locally abundant, southwestern Utah and southeastern Nevada, south in Arizona to the region of Mokiak Pass and the Kaibab Plateau, northern Mohave and Coconino counties, the type from Mokiak Pass (*Palmer* 117 in 1877). The change of name explained in the preceding entry.

45. Astragalus Preussii Gray (p. 464). Two distinct species are confused under this name in *Arizona Flora.* The range of genuine *A. Preussii* (with stipitate fruit) and its var. *laxiflorus* (with subsessile fruit) are given correctly, but the so-called var. *latus,* differing in its nodding, whitish (not ascending, purple) flowers and semibilocular pods of fleshy, ultimately subligneous texture represents a variety of the next following.

49. Astragalus praelongus Sheld. (p. 464). The distribution and synonymy as given in *Arizona Flora* are substantially correct for var. *praelongus.* Its sessile or nearly sessile pod varies somewhat in orientation, size, and outline, and the calyx teeth vary greatly in length, but these features are not correlated, and there seems to be no basis for varietal segregation. In northern Apache and Navajo counties and adjoining Utah, at 3,700 to 5,400 feet, mostly on red or white sandstone, *A. praelongus* is represented by var. *lonchopus* Barneby (374a), the type from near Rattlesnake Mines, Carrizo Mountains, Apache County (*U. S. Geological Survey* no. 511), easily recognized by its erect, narrowly fusiform pod contracted at base into a stout stipe 4.5 to 8 mm. long. The variety is mentioned in *Arizona Flora* as *A. Preussii* var. *latus* (cf. the preceding entry). In all forms of *A. praelongus* the fleshy pods become woody and remain attached to the withered stems after dehiscence, often persisting over winter until new growth from the base is well advanced.

54a. Astragalus scopulorum Porter. *Tium scopulorum* Rydb. Upper Salt River valley, southern Navajo and adjoining Gila counties, and on Black Mesa, Apache County, in piñon forest or oak-brush, 4,800 to 7,250 feet, May to August. Rocky Mountains of Colorado and north-central New Mexico, westward interruptedly to central Utah and eastern Arizona.

A rather handsome *Astragalus,* easily recognized by subterranean caudex, connate lower stipules, comparatively large nodding ochroleucous flowers, and pendulous stipitate trigonously compressed glabrous pod divided into two chambers by a complete septum.

55a. Astragalus straturensis Jones. *Atelophragma straturense* Rydb., *Tium atratiforme* Rydb. Nixon Springs, Mount Trumbull, northern Mohave County, 6,500 feet (*Gould* 1716). Southwestern Utah and adjoining Nevada and Arizona.

Similar in habit to *A. Rusbyi* Greene (with which Arizona material was at first confused) and *A. recurvus* Greene, but differing in its free lower stipules and bright amethyst-purple (violet when dry) flowers. In *A. straturensis* the wing petals are about as long as the banner (not greatly exserted as in *A. Rusbyi*), and the pod is gently curved inward (not downward, as in *A. recurvus*). In contrast to these two species confined to volcanic soils in central Arizona, *A. straturensis* is strongly calciphile.

63a. Astragalus Emoryanus (Rydb.) Barneby. *Hamosa Emoryana* Rydb. Grand Canyon, Coconino County (Bright Angel Trail, 3,500 feet, *Darrow* in 1938), apparently greatly isolated in Arizona and possibly introduced from farther east. Central New Mexico to southern Texas, northern Chihuahua, and Nuevo Leon.

A slender annual, easily mistaken for some form of the polymorphic *A. Nuttallianus,* but distinguished from Arizona forms of the latter by its rounded keel tip combined with truncate-retuse leaflets in all the leaves. The pod of *A. Nuttallianus* persists on the raceme until it turns black, and dehisces apically, whereas that of *A. Emoryanus* is straw-colored when ripe and is readily deciduous, discharging its seeds by dehiscence at both ends after falling.

45. AESCHYNOMENE

1. Aeschynomene villosa Poir. This name to replace *A. americana* (p. 472) for the Arizona plants according to V. E. Rudd (450, p. 33).

49. VICIA. Vetch

5. Vicia exigua Nutt. (p. 477). Two names in earlier literature that have been associated with the slender vetch in Arizona are *V. Thurberi* Wats. and *V. producta* Rydb. Both are regarded as synonyms of *V. exigua* by F. J. Hermann (404 b, p. 57).

50. LATHYRUS. PEAVINE

Contributed by C. Leo Hitchcock

1a. Lathyrus latifolius L. (to precede 1b. *Lathyrus eucosmus* Butters & St. John, p. 478). This European species often escapes and becomes established for varying lengths of time. It is known from many areas in North America, our record from Prescott, Yavapai County (*Kuntze* in 1896). It is recognizable because of the single pair of leaflets and woody winged stems.

2a. Lathyrus zionis C. L. Hitchc. This name is to replace *L. brachycalyx* Rydb. in Arizona except for report from Dos Cabezas Mountains, Cochise County (p. 478).

2b. Lathyrus pauciflorus Fern. var. **utahensis** (Jones) Peck. Known from the Carrizo Mountains, Apache Co. (*Standley* 7383). This is a large-flowered plant in many ways similar to *L. brachycalyx* Rydb. and *L. eucosmus* Butters & St. John. In both of those species the keel petal is subequal to the wing petals; in *L. pauciflorus,* on the other hand, the keel is 2 to 4 mm. shorter than the wings.

3. Lathyrus leucanthus Rydb. var. **laetivirens** (Greene ex. Rydb.) C. L. Hitchc. (405, p. 39). This name is to replace *L. laetivirens* Greene (p. 478).

70. SIMAROUBACEAE. SIMAROUBA FAMILY

2. AILANTHUS. TREE OF HEAVEN

1. Ailanthus altissima (Mill.) Swingle. Pecks Lake, Yavapai County. Tree with malodorous foliage and flowers, large pinnate leaves, panicles of small yellowish green flowers, and winged fruits. It is extensively planted as a shade tree in the United States, sometimes escaping into roadsides and wasteland. Native of China.

74. EUPHORBIACEAE. SPURGE FAMILY

Contributed by Louis C. Wheeler

2. TETRACOCCUS

1. Tetracoccus fasciculatus (Watson) Croizat var. **Hallii** (T. S. Brandegee) Dressler (391, p. 57). *T. Hallii* T. S. Brandegee (p. 503).

4. ARGYTHAMNIA

1. Argythamnia Clariana Jepson. Western Arizona, southeastern California. Replaces *Argythamnia adenophora* Gray [*Ditaxis adenophora* (Gray) Pax & Hoffmann, p. 505], considered by Dr. John Ingram to be a separate species.

2. Argythamnia mercurialina (Nutt.) Muell. Arg. *Ditaxis mercurialina* (Nutt.) Coult. (p. 505).

3. Argythamnia cyanophylla (Woot. & Standl.) Ingram. *Ditaxis cyanophylla* Woot. & Standl. (p. 505).

4. Argythamnia Brandegei Millsp. var. **intonsa** (I. M. Johnston) Ingram. *Ditaxis Brandegei* (Millsp.) Rose & Standl. var. *intonsa* I. M. Johnston (p. 506).

5. Argythamnia lanceolata (Benth.) Muell. Arg. *Ditaxis lanceolata* (Benth.) Pax & Hoffmann (p. 506).

6. Argythamnia neomexicana Muell. Arg. *Ditaxis neomexicana* (Muell. Arg.) Heller (p. 506).

7. Argythamnia serrata (Torrey) Muell. Arg. *Ditaxis serrata* (Torrey) Heller (p. 506).

14. EUPHORBIA. Spurge

5. Euphorbia spathulata Lam. (p. 516). Delete "perhaps also on the North Rim of the Grand Canyon (*Eastwood & Howell* 962)." That specimen belongs to *E. lurida* Engelm.

46. Euphorbia prostrata Aiton replaces *E. Chamaesyce* L. (p. 520) which is a European plant.

75. CALLITRICHACEAE. Water-starwort Family

1. CALLITRICHE. Water-starwort

1. Fruits distinctly longer than wide and winged at apex .... *C. verna*
1. Fruits not or but slightly longer than wide, not winged or very obscurely so *C. heterophylla*

1. Callitriche verna L. emend. Kutz. *C. palustris* (p. 521). Chuska Mountains, northern Apache County, Flagstaff, Coconino County, Tucson and Santa Catalina mountains, Pima County. Throughout most of North America, possibly extending into Mexico.

2. Callitriche heterophylla Pursh emend. Darby. Kaibab Plateau, Coconino County, 8,000 feet, northern Gila County, Fort Lowell and Sabino Canyon, Pima County.

The Arizona plants belong to var. *heterophylla* which ranges from Nova Scotia to Washington, south to Guatemala; West Indies; southern South America. Fassett (394, p. 177) limited var. *Bolanderi* (Hegelm.) Fassett (*C. Bolanderi* Hegelm., p. 521) to British Columbia and the Pacific coast states.

80. SAPINDACEAE. Soapberry Family

2. DODONAEA. Hopbush

1. Dodonaea viscosa Jacq. (p. 528). Benson and Darrow (378, p. 257) referred the Arizona plants to *D. viscosa* var. *angustifolia* (L. f.) Benth.

3. CARDIOSPERMUM

1. Cardiospermum Halicacabum L. Balloon Vine. Rocky slope, Coyote Mountains, Pima County (Mason, 428, p. 64). Widely distributed in tropical and subtropical regions. Sometimes cultivated as an ornamental. Herbaceous vines, with bipinnate leaves, small flowers, capsules showy, 3-angled, inflated, seeds black, about size of a pea.

80a. MELIACEAE

1. MELIA. Chinaberry, Umbrella Tree

1. Melia Azedarach L. Pecks Lake, Yavapai County. Tree with large bipinnate leaves, showy panicles of lavender flowers, fleshy single-seeded fruit. It is often planted as a shade tree in the warmer parts of the United States; occasionally escaping. Native of southeastern Asia.

84. MALVACEAE. Mallow Family

1. ABUTILON. Indian-mallow

7. Abutilon Palmeri Gray (p. 539). Delete *A. Parishii* Watson as a synonym.

7a. Abutilon Parishii Watson. Santa Catalina Mountains, Pima County; Mazatzal Mountains, Gila County. Known only from these Arizona localities. Distinguished from *A. Palmeri* in having corollas only about 10 mm. long (15 to 25 mm. in *A. Palmeri*), a more paniculate inflorescence, strongly discolorous and usually more acuminate leaves, and the longer stem hairs mostly reflexed.

2. BOGENHARDIA

1. Bogenhardia crispa (L.) Kearney. *Abutilon crispum* (L.) Sweet, *Gayoides crispum* (L.) Small (p. 540).

6. MALVASTRUM

Transfer species 2 and 3 (p. 548) to:

6a. EREMALCHE (417)

1. Eremalche rotundifolia (Gray) Greene. *Malvastrum rotundifolium* Gray.

2. Eremalche exilis (Gray) Greene. *Malvastrum exile* Gray.

Eremalche, a genus of annual plants having purple to whitish petals and carpels without appendages and with thin lateral walls, is considered by Kearney to be distinct from *Malvastrum*. In Arizona, *Malvastrum* is a shrubby plant, in which the flowers have orange-yellow petals and the carpels have coriaceous lateral walls and two dorsal appendages.

10. SIDA

8. Sida rhombifolia L. Sycamore Canyon near Ruby, Santa Cruz County, 3,500 feet, rocky slopes (*Darrow & Haskell 2223*). In this species the leaves are more or less rhombic and cuneate at the base, finely dentate or serrate, and the carpels are long-aristate to nearly blunt. These characters distinguish it from No. 7, *S. tragiaefolia* Gray (p. 551) in which the leaves are ovate or oblong-ovate, truncate or subcordate at base, coarsely serrate, and the carpels bear 2 short awns.

85. STERCULIACEAE. Cacao Family

1. FREMONTODENDRON

1. Fremontodendron californicum (Torrey) Coville. *Fremontia californica* Torrey (p. 554).

88. TAMARICACEAE. Tamarix Family

1. TAMARIX

1. Tamarix pentandra Pall. (p. 557). Benson and Darrow (378, p. 267) refer this plant to *T. gallica* L. var. *pycnostachys* Ledeb. *Tamarix* is a difficult genus which requires further study in the Old Word where it is native, and until that is done the identity of this widespread naturalized species will remain in doubt.

91. VIOLACEAE. Violet Family

2. VIOLA. Violet

7. Viola aurea Kellogg subsp. **arizonensis** Baker & Clausen (374, pp. 11, 12). North Peak, Mazatzal Mountains, 6,000 feet (*Collom* 48 in 1948) and Sierra Ancha (*Crooks et al.*, in 1939), Gila County. This species, mainly Californian, is known in Arizona only by these two collections. It is closely related to *V. purpurea* Kellogg (p. 561) which does not occur in Arizona.

93. LOASACEAE

3. MENTZELIA. Stick-leaf

7a. Mentzelia Veatchiana Kellogg. *M. albicaulis* Dougl. var. *Veatchiana* (Kellogg) Urban & Gilg. (p. 566). Thompson and Lewis (459, p. 103) recognize this as specifically different from *M. albicaulis*. The two differ in chromosome number, in *M. albicaulis* n = 27 and in *M. Veatchiana* n = 18. They may be distinguished by petal size: the petals of *M. Veatchiana* are 7 to 12 mm. long whereas those of *M. albicaulis* are 3 to 6 mm. long.

9. Mentzelia pumila (Nutt.) Torrey & Gray. Thompson and Lewis evidently consider *M. pumila* var. *multiflora* (Nutt.) Urban & Gilg. (p. 566), to be a species, *M. multiflora* (Nutt.) Gray, since they record a chromosome count for it but do not mention its relationship to *M. pumila.*

5. SYMPETELEIA

1. Sympeteleia rupestris (Baillon) Gray. *Loasella rupestris* Baillon. Cabeza Prieta Tanks on Cabeza Prieta Game Range, Yuma County, *Gale Monson.* Extreme southern California, Baja California, and Sonora. A hirsute annual much like *Eucnide* in habit but differing from it in the smaller flowers having petals united to form a distinct tube.

94. CACTACEAE. Cactus Family

Benson (376) included within the genus *Echinocactus* the species in the following genera: *Ferocactus, Echinomastus, Sclerocactus, Utahia, Toumeya,* and *Navajoa* (pp. 573–576). This treatment is similar to that of Peebles in *Flowering Plants and Ferns of Arizona.* However, Peebles chose to recognize these segregate genera in his later treatment of the family for *Arizona Flora.*

97. ONAGRACEAE. Evening-primrose Family

Contributed by Philip A. Munz

3. ZAUSCHNERIA. Hummingbird-trumpet

1. Zauschneria californica Presl subsp. **latifolia** (Hook.) Keck. *Z. latifolia* (Hook.) Greene var. *arizonica* (Davidson) Hilend (p. 590).

4. EPILOBIUM. Willow-weed

8. Epilobium adenocaulon Hausskn. (p. 592). Both var. *adenocaulon* and var. *Parishii* (Trelease) Munz (*E. californicum* Hausskn., No. 9, p. 592) are widely distributed in Arizona. Var. *adenocaulon* is glandular pubescent in the inflorescence, whereas var. *Parishii* has nonglandular, somewhat appressed hairs.

9a. Epilobium ciliatum Raf. *E. adenocaulon* var. *perplexans* Trel. This differs from *E. adenocaulon* in its thinner, less firm, paler green, more definitely petioled leaves; papillae on seeds more conical, less rounded; coma less persistent.

5. CLARKIA (422)

1. Clarkia rhomoidea Dougl. (p. 592).

2. Clarkia purpurea (Curt.) Nels & Macbr. subsp. **quadrivulnera** (Dougl.) Lewis & Lewis. *Godetia quadrivulnera* (Dougl.) Spach. (p. 592).

3. Clarkia epilobioides (Nutt.) Nels. & Macbr. *Godetia epilobioides* Nutt. (p. 593).

7. OENOTHERA. EVENING-PRIMROSE, SUN-DROPS

Following species No. 32 (p. 601) add:

32a. Oenothera pterosperma Wats. *Chylismia pterosperma* Small. Grand Canyon National Monument, northeastern Mohave County, 4,500 to 5,000 feet. Southern Utah to eastern Oregon, northern Arizona to southeastern California. A low delicate annual, the seeds with an incurved wing so as to appear boat-shaped, cellular-pubescent, flowers pinkish, the petals barely 2 mm. long.

34. Oenothera pallidula (Munz) Munz (p. 601). Plants quite intermediate between *O. clavaeformis* Torr. & Frém. and *O. brevipes* Gray are found in Mohave and Yuma counties where the ranges of the two species overlap. They probably represent hybrids.

98. HALORAGACEAE. WATER-MILFOIL FAMILY

1. MYRIOPHYLLUM. WATER-MILFOIL

2. Myriophyllum spicatum L. subsp. **exalbescens** (Fern.) Hult. *M. exalbescens* Fern., *M. spicatum* var. *exalbescens* (Fern.) Jeps. (p. 605). Patten (445) concluded that *M. exalbescens* represents a geographical variant of the circumboreal *M. spicatum* complex.

100. UMBELLIFERAE. PARSLEY FAMILY

Contributed by Mildred E. Mathias and Lincoln Constance

6. OSMORHIZA. SWEET-ROOT, SWEET-CICELY

3. Osmorhiza depauperata Phil. *O. obtusa* (Coulter & Rose) Fern. (p. 611).

31. LOMATIUM. BISCUIT-ROOT, INDIAN-ROOT

5. Lomatium mohavense Coulter & Rose (p. 622). Grand View (3 miles west), Grand Canyon, Coconino County, *Newlon* 858. This species is now definitely known to be a member of the Arizona flora.

14. BUPLEURUM

1. Bupleurum subovatum Link. Replaces *B. rotundifolium* L. (p. 614). Tucson, Pima County, and near Cave Creek, Maricopa County. Naturalized in the eastern United States. Native of Mediterranean region. The Maricopa County collection was reported by McCleary (433, p. 155). Two collections from Tucson have been called *B. rotundifolium:* that of Toumey in 1892 and a more recent one reported by Mason (427, p. 64). All Arizona material has been identified as *B. subovatum* by Mathias.

Series 3. Sympetalae

102. LENNOACEAE. Lennoa Family

2. PHOLISMA

1. Pholisma arenarium Nutt. Sand dunes near Parker, Mohave County, *Cowan*. Colorado and Mohave deserts, and coastal southern California to Baja California.

This curious root parasite differs from *Ammobroma sonorae* Torrey in having the flowers in a dense ovoid or cylindric, sometimes branched spike, instead of saucer-shaped head, and sepals which are linear and glabrous whereas in *Ammobroma* they are filiform and plumose.

104. PRIMULACEAE. Primrose Family

7. DODECATHEON. Shooting-star

1. Dodecatheon Ellisiae Standley (p. 639). Thompson (458) reduced this species to a subspecies of *D. dentatum* Hook. According to him, *D. dentatum* has two subspecies separated by technical characters of the stamens. They are, however, widely separated geographically, subsp. *dentatum* being found in Washington, northern Oregon and northern Idaho, and subsp. *Ellisiae* much to the south in Arizona, New Mexico, and Utah.

2. Dodecatheon pulchellum (Raf.) Merrill replaces *D. radicatum* Greene (p. 639) which becomes a synonym. The Arizona plants belong to the widely distributed subsp. *pulchellum*.

3. Dodecatheon alpinum (A. Gray) Greene (p. 639). The Arizona plants belong to subsp. *majus* H. J. Thompson.

108. OLEACEAE. Olive Family

FRAXINUS. Ash

1. Fraxinus anomala Torr. Add as synomyms: *F. anomala* var. *Lowellii* (Sargent) Little and *F. Lowellii* Sargent, deleting species No. 2, p. 642.

Both Little (423, pp. 370, 371) and Miller (436, p. 50) discuss the variation in leaf shape and number of leaflets of *F. Lowellii*. This variation falls within the variability of *F. anomala* as expressed throughout its geographic distribution.

3a. Fraxinus dipetala Hooker & Arnott. Two Petal Ash. Coconino County. California eastward to southern Nevada and Utah, and northwestern Arizona. Distinguished from all other North American species by having only two petals.

4. Fraxinus Gooddingii Little. Goodding Ash. Santa Cruz County, 3,600 to 5,000 feet, northeastern Sonora, Mexico. Little (423, pp. 373–375) designated as holotype a specimen collected in Bear Valley (Sycamore Canyon), Santa Cruz County, by L. N. Goodding in 1936. In *Arizona Flora* this specimen was cited under *F. Greggii* (p. 642), a species of southwestern Texas and northeastern Mexico.

5. Fraxinus pennsylvanica Marshall subsp. **velutina** (Torrey) G. N. Miller. To replace *F. velutina* Torrey, and its varieties, var. *glabra* Rehder, var. *Toumeyi* (Britton) Rehder, and var. *coriacea* (Watson) Rehder. Velvet Ash.

6. Fraxinus papillosa Lingelsheim. Cochise and Santa Cruz counties, mountainous areas 5,200 to 7,700 feet. Mexico, southern Arizona, and New Mexico. Distinguished from other Arizona species by its leaflets which are sessile and papillose beneath.

2. FORESTIERA. Adelia

1. Forestiera Shrevei Standley. Replaces *F. phillyreoides* (Benth.) Torr. (p. 643), a species of very different plants of the high deserts of central Mexico according to M. C. Johnston (413, p. 145).

110. GENTIANACEAE. Gentian Family

2. GENTIANA. Gentian

The following names and references should be added to those species of *Gentiana* that have recently been studied monographically under the generic name of *Gentianella* (cf. Gillett, 396).

2. Gentiana tenella Rottb. (p. 648). *Gentianella tenella* (Rottb.) Börner subsp. *tenella,* Gillett (396, pp. 263–265).

7. Gentiana barbellata Engelm. (p. 648). *Gentianella barbellata* (Engelm.) J. M. Gillett (396, pp. 230, 231).

8. Gentiana thermalis Kuntze (p. 648). *Gentianella detonsa* (Rottb.) G. Don subsp. *elegans* (A. Nels.) J. M. Gillett (396, p. 217).

9. Gentiana grandis (Gray) Holm (p. 648). *Gentianella detonsa* (Rottb.) G. Don subsp. *superba* (Greene) J. M. Gillett (396, p. 219).

10. Gentiana heterosepala Engelm. (p. 648). *Gentianella Amarella* (L.) Börner subsp. *heterosepala* (Engelm.) J. M. Gillett (396, p. 251).

11. Gentiana Wrightii Gray (p. 648). *Gentianella Amarella* (L.) Börner subsp. *Wrightii* (Gray) J. M. Gillett (396, p. 259).

12. Gentiana Wislizeni Engelm. (p. 649). *Gentianella Wislizeni* (Engelm.) J. M. Gillett (396, p. 235).

13. Gentiana microcalyx Lemmon (p. 649). *Gentianella microcalyx* (Lemmon) J. M. Gillett (396, p. 246).

14. Gentiana strictiflora (Rydb.) A. Nels. (p. 649). *Gentianella Amarella* (L.) Börner subsp. *acuta* (Michx.) J. M. Gillett (396, p. 253).

5. EUSTOMA

The synopsis of *Eustoma* by L. H. Shinners (455) was based on specimens preserved in two Texas herbaria, and since these contained no material from Arizona, he does not include this state within the range of *E. exaltatum* (L.) Salisbury ex G. Don.

6. MENYANTHES

1. Menyanthes trifoliata L. Woolsey Lake, Apache County, about 8,000 feet, and at other lakes in the same general area (432, p. 153). This semi-aquatic plant is readily distinguished from all other Arizona *Gentianaceae* (pp. 645–650) by its trifoliolate, long-petiolate, basal or alternate leaves, and racemose flowers (415, p. 168).

112. ASCLEPIADACEAE. Milkweed Family

2. ASCLEPIAS (470). Milkweed

2. Asclepias Engelmanniana Woodson (p. 660). Prairies and swales, open sandy hillsides, draws, washes and bottoms. Gila and Mohave counties.

2a. Asclepias Rusbyi (Vail) Woodson. *A. Engelmanniana* Woodson var. *Rusbyi* (Vail) Kearney (p. 660). Rocky slopes in open oak forest, Coconino, Gila, and Mohave counties.

4. Asclepias asperula (Decne.) Woodson. *A. capricornu* Woodson (p. 660). The Arizona plants belong to subsp. *asperula.*

7. Asclepias tuberosa L. Arizona material of this species is referred by Woodson (470) to subsp. *terminalis* Woodson. More recently (personal communication) he has called it subsp. *interior.*

10. Asclepias fascicularis Decne. (p. 661). Apparently this species does not occur in Arizona. Woodson believes that Arizona specimens referred to it belong to *A. subverticillata* (No. 9, p. 661).

10. ROTHROCKIA

1. Rothrockia cordifolia Gray (p. 666). *Matelea cordifolia* Woodson. Canyon Diablo, Ajo Mountains, 5,000 feet, Pima County, *Supernaugh* in 1952. This plant is now to be included in the flora of Arizona.

114. POLEMONIACEAE. Phlox Family

7. LINANTHASTRUM

1. Linanthastrum Nuttallii (Gray) Ewan (p. 686). *Linanthus Nuttallii* (Gray) Greene acc. Mason (430a, p. 431) and acc. V. Grant (402, p. 145).

8. LINANTHUS

Add the following as an accepted species:

5. Linanthus Jonesii (Gray) Greene. *Gilia Bigelovii* Gray var. *Jonesii* Brand. The Arizona plant is discussed under *L. Bigelovii* (Gray) Greene (p. 687).

9. GILIA

4. Gilia gilioides (Benth.) Greene (p. 690). The following synonym and reference are to be added: *Allophyllum gilioides* (Benth.) A. & V. Grant (399, p. 105).

5. Gilia polycladon Torr. (p. 690). *Ipomopsis polycladon* (Torr.) V. Grant (401, p. 361).

6. Gilia pumila Nutt. (p. 690). *Ipomopsis pumila* (Nutt.) V. Grant (*loc. cit.*).

7. Gilia Gunnisoni Torr. & Gray (p. 690). *Ipomopsis Gunnisoni* (Torr. & Gray) V. Grant (*loc. cit.*).

9. Delete No. 9 (p. 691), *G. multicaulis* Benth., and replace under No. 8 with *G. achilleaefolia* Benth. subsp. *multicaulis* (Benth.) V. & A. Grant. Specimens from Greenlee and Gila counties have been examined by Dr. Verne Grant who writes that the Arizona material is far out of range and might possibly be introduced in Arizona.

10a. Gilia hutchinsifolia Rydb. *G. leptomeria* Gray subsp. *rubella* (Brand) Mason & A. Grant (p. 691). This species, reported from sandy places in northern Arizona, differs from *G. leptomeria* in its more deeply cut, mostly bipinnatifid basal leaves and in its corollas 8 to 14 mm. long (in *G. leptomeria* only 4.5 to 6.5 mm. long).

11a. Gilia stellata Heller. Reported from dry sandy slopes in Arizona (Mason and A. Grant, 431, p. 471), the star gilia is closely related to *G. scopulorum* Jones (p. 691) from which it may be distinguished by its short corolla tube which does not exceed the calyx (but which is long-exserted in *G. scopulorum*).

12. Gilia transmontana (Mason & A. Grant) A. & V. Grant. To replace *G. ochroleuca* Jones subsp. *transmontana* Mason & A. Grant (p. 691).

13. Gilia flavocincta A. Nels. *G. ophthalmoides* Brand subsp. *flavocincta* (A. Nels.) A. & V. Grant. This and the two following replace *G. tenuiflora* as reported for Arizona (p. 691), *G. tenuiflora* being known only from coastal California (400, p. 244). *Gilia ophthalmoides* subsp. *australis* A. & V. Grant (400, p. 263) is a small-flowered variant of *G. flavocincta* according to Dr. Verne Grant (in litt.).

14a. Gilia ophthalmoides Brand. According to Dr. Verne Grant (in litt.), *G. ophthalmoides* in Arizona is the plant treated under the name *G. ophthalmoides* subsp. *Clokeyi* in the study of the "cobwebby gilias" (400, pp. 260 to 262, map 11), and an unnamed plant related to *G. ophthalmoides* is the Arizona plant that has been called *G. inconspicua* (Smith) Sweet (A. and V. Grant, 400, p. 249). *Gilia ophthalmoides* is found in sandy places in the piñon belt in desert mountains and is not uncommon in northern Arizona.

14b. Gilia Clokeyi Mason. In sandy places in the desert below the piñon belt in Mohave and Navajo counties. Eastern California, southern Nevada, southern Utah, and northwestern New Mexico.

14c. Gilia mexicana A. & V. Grant. Pima, Cochise, and Pinal counties, Arizona, east to New Mexico, and south to Chihuahua, Mexico.

14d. Gilia minor A. & V. Grant. Sandy places in the desert or on desert borders, in Arizona known from Wickenburg and Aguila in Maricopa County (V. Grant in litt.).

The six preceding species (Nos. 12, 13, 14a–d), together with *G. sinuata* Dougl. (No. 14), belong to the "cobwebby gilias," *Gilia* sect. *Arachnion* of A. & V. Grant. The following key, adapted from their work (400) and prepared with assistance from both Alva and Verne Grant, will distinguish the Arizona plants of this complex which are annuals with the lower stems or leaves arachnoid-woolly.

1. Stems at base glabrous or rarely glandular-pubescent.................. 14. *G. sinuata*
1. Stems at base sparsely to densely cobwebby-pubescent (2).
2. Capsule broadly ovoid to globular (3).
2. Capsule oblongish (6).
3. Calyx glandular.. 12. *G. transmontana*
3. Calyx glabrous or sparsely pubescent (4).
4. Calyx lobes acute; corolla 3.5 to 5 mm. long, the tube included in the calyx
14b. *G. Clokeyi*

4. Calyx lobes acuminate; corolla exceeding 5 mm., the tube exserted from the calyx (or sometimes included in small-flowered forms of *G. flavocincta*) (5).
5. Corolla 7.5 to 30 mm. long, lower throat pale yellow, limb pinkish violet; lower part of plant lightly to moderately pubescent............................13. *G. flavocincta*
5. Corolla 5.5 to 12 mm. long; the throat bright yellow, limb pale pink to white; lower part of plant moderately to densely pubescent....................14a. *G. ophthalmoides*
6. Inflorescence loosely branching; calyx glabrous or sparsely pubescent; corolla throat white with yellow spots...14c. *G. mexicana*
6. Inflorescence strictly branching; calyx glandular-dotted; corolla throat purple with yellow spots...14d. *G. minor*

16. Gilia longiflora (Torr.) G. Don (p. 692). *Ipomopsis longiflora* (Torr.) V. Grant (401, p. 361).

17. Gilia laxiflora (Coult.) Osterh. (p. 692). *Ipomopsis laxiflora* (Coult.) V. Grant (*loc. cit.*).

18. Gilia multiflora Nutt. (p. 692). *Ipomopsis multiflora* (Nutt.) V. Grant (401, p. 357).

19. Gilia polyantha Rydb. (p. 692). *Ipomopsis polyantha* (Rydb.) V. Grant (*loc. cit.*).

20. Gilia aggregata (Pursh) Sprengel (p. 692). *Ipomopsis aggregata* (Pursh) V. Grant (401, p. 361).

21. Gilia Macombii Torr. (p. 693). *Ipomopsis Macombii* (Torr.) V. Grant (*loc. cit.*).

22. Gilia Thurberi Torr. (p. 693). *Ipomopsis Thurberi* (Torr.) V. Grant (*loc. cit.*).

An additional species in this series is:

20a. Gilia tenuituba Rydb. *Ipomopsis tenuituba* (Rydb.) V. Grant (*loc. cit.*). The Arizona plant (p. 693) is forma *macrosiphon* Wherry [*G. aggregata* var. *macrosiphon* (Wherry) Kearney & Peebles].

115. HYDROPHYLLACEAE. Water-leaf Family

4. PHACELIA

14a. Phacelia pediculoides (J. T. Howell) Constance in Cave & Constance. This name is to replace *P. Ivesiana* Torr. var. *pediculoides* J. T. Howell (p. 702). According to Cave and Constance (382, table 2, p. 243), n = 11 in *P. Ivesiana* and n = 23 in *P. pediculoides.*

6. NAMA

7. Nama Rothrockii Gray. Reported from near Mount Trumbull, Mohave County; known otherwise only from California. Distinguished from all other Arizona species of *Nama* (pp. 705, 706) by its perennial habit, its thickish, coarsely serrate-dentate leaves, and its densely subcapitate, many-flowered, terminal inflorescences.

116. BORAGINACEAE. Borage Family

1a. CORDIA

1. Cordia parvifolia A. DC. *C. Greggii* Torrey. Gravel ridge, among *Larrea* bushes, about 17 miles south of Tucson, Pima County, *Altfillisch* in 1951. A single plant, an old, many-stemmed shrub about 6 feet tall, was found at this station. However, there did not appear to be any evidence that it was planted.

Both *Cordia* and *Coldenia* have twice cleft styles, but in *Cordia* each style is again divided so that there are four style branches and stigmas whereas in *Coldenia* there are two stigmas. *Cordia parviflora* is a larger shrub than any of the species of *Coldenia,* its white flowers are larger than in *Coldenia,* and its fruit is drupaceous.

11. LITHOSPERMUM. Gromwell, Puccoon

Two species of *Lithospermum* have been added to the flora of Arizona by I. M. Johnston (410):

4. Lithospermum confine I. M. Johnston. Chiricahua Mine, Cochise County (*Blumer* 1796), 8 miles north of Metcalf, Greenlee County (*Maguire, et al.,* 11805). Western Texas and northeastern Mexico.

5. Lithospermum viride Greene. Southwestern United States (Arizona to Texas) and northeastern Mexico. Although Johnston includes Arizona within the range of this species, he did not cite specimens or localities from there. However, it has been collected in Guadelupe Canyon, southwestern New Mexico, and so its occurrence in adjacent Arizona is very probable. The following key will serve to distinguish the five Arizona species (Kearney, 419, p. 5):

1. Flowers (the showy ones) heterostylic or dimorphic, the 2 types of flowers differing in the height at which the stamens are affixed in the corolla tube. Cleistogamous flowers usually wanting (2).
1. Flowers (the showy ones) not heterostylic, monomorphic (3).
2. Stems not arising from a leaf rosette, the lower cauline leaves very much smaller than the middle and upper ones; corolla tube glabrous within............. *L. multiflorum*
2. Stems arising from a well-developed leaf rosette, the basal leaves much larger than the middle and upper cauline ones; corolla tube hairy within, at least in the long-styled flowers.. *L. cobrense*
3. Pollen ellipsoidal or subglobose; cleistogamous flowers absent; stem leaves 10 to 35 mm. wide, evidently veined. Corolla yellow or yellowish, with entire, spreading or recurved lobes.. *L. viride*
3. Pollen perfectly globose; cleistogamous flowers always present; stem leaves not more, usually less than 10 mm. wide, obscurely veined (4).
4. Corolla lobes erose or fimbriate; fruiting calyx usually nutant or cernuous; cleistogamous flowers abundant... *L. incisum*
4. Corolla lobes entire; fruiting calyx erect; flowers commonly nearly all cleistogamous *L. confine*

12. MACROMERIA

1. Macromeria viridiflora DC. (p. 724). According to I. M. Johnston (411), both varieties of this species [var. *viridiflora* and var. *Thurberi* (Gray) I. M. Johnston] occur in Arizona.

118. LABIATAE. Mint Family

2. TRICHOSTEMA. Blue-curls

1a. Trichostema micranthum Gray. Near Mount Trumbull, about 6,000 feet, Mohave County, *Jaeger* in 1941. Southern California and northern Baja California. This species is an annual as is *T. brachiatum* L. (p. 734). In *T. micranthum* the corolla tube is curved upward and the leaves are 2 to 6 mm. wide, but in *T. brachiatum* the corolla tube is straight and the leaves are 4 to 16 mm. wide.

14. STACHYS

4. Stachys agraria Cham. & Schlecht. (pp. 740). In the region of McNary and Lakeside (the precise locality not recorded), southern Apache and Navajo counties. Texas, Mexico, and Guatemala. Its small flowers (corolla tube not more than 4 mm. long) distinguish it from the other species in Arizona in which the corolla tube is more than 5.5 mm. long.

15. SALVIA. Sage

Replace the subspecies of *Salvia carnosa* Dougl. (pp. 741, 742) with the following:

2a. Salvia Dorrii (Kellogg) Abrams subsp. **Dorrii.** *S. carnosa* subsp. *pilosa* (Gray) Epling.

2b. Salvia Dorrii (Kellogg) Abrams subsp. **Mearnsii** (Britt.) McClintock. *S. carnosa* subsp. *Mearnsii* (Britt.) Epling.

2c. Salvia Dorrii (Kellogg) Abrams subsp. **argentea** (Rydb.) Munz. *S. carnosa* subsp. *argentea* (Rydb.) Epling.

20. MONARDELLA

1. Monardella odoratissima Benth. (p. 747). Delete *M. parvifolia* Greene as synonym in line 2 and add that the species is represented in Arizona by subsp. *parvifolia* (Greene) Epling (*M. parvifolia* Greene).

21. LYCOPUS. Bugleweed

2. Lycopus americanus Muhl. Wet soil, Lakeside, 7,000 feet, southern Navajo County. Almost throughout the United States and Canada. This species has leaves which are petiolate, deeply incised or pinnatifid, and nutlets which are entire at their apices. These characters distinguish it from *L. lucidus* (p. 747), reported from Arizona on the basis of a Palmer collection but not otherwise known from the state.

119. SOLANACEAE. Potato Family

11. SALPICHROA

1. Salpichroa rhomboidea (Gill. & Hook.) Miers. Cocks Eggs. Common weed in yards, Tucson, Pima County. Cultivated and tending to become naturalized in California and southern Arizona. Native of Argentina. It may be distinguished from other Arizona members of the family by its climbing habit. It is a perennial from a fleshy root with green flexuous branches, white flowers with small urn-shaped corollas, and a many-seeded berry.

120. SCROPHULARIACEAE. Figwort Family

2. LINARIA. Toad-flax

2. Linaria dalmatica Miller. Casual garden escape, campus of Arizona State College, Flagstaff. Native of the Mediterranean region. A handsome yellow-flowered plant which differs from *L. vulgaris* Miller (p. 764) in having broader, ovate or ovate-lanceolate, cordate-clasping leaves and larger flowers, the corolla 3.5 to 4 cm. long.

4. ANTIRRHINUM. SNAPDRAGON

2. Antirrhinum Kingii Watson. Toroweap Valley and inner gorge of the Colorado River, both in the Grand Canyon National Monument, northeastern Mohave County (*McClintock* 52–424, 52–556). This species tentatively included in *Arizona Flora* (p. 766), is now definitely known to occur in the state.

9. MIMULUS. MONKEY-FLOWER

8. Mimulus guttatus DC. (p. 781). According to Campbell (379), var. *guttatus* and var. *gracilis* (Gray) Campbell occur in Arizona. Var. *gracilis* is characterized by recurved pedicels, commonly geniculate stems, and frequently scorpioid young inflorescences.

8a. Mimulus Tilingii Regel. Campbell (379) included Arizona within the range of this species on the basis of a single collection, *Leiberg* 5721, made on "Smith Creek," presumably on the San Francisco Peaks, Coconino County. Campbell used several quantitative rather than qualitative characters for differentiating this species from *M. guttatus:* the fewer flowers on each stem, the upper and lower leaves similar in shape and size, and the small fibrous roots in *M. Tilingii,* in contrast with numerous flowers, reduced upper stem leaves, and well-developed fibrous root system in *M. guttatus.*

14. BACOPA

1. Bacopa Eisenii (Kellogg) Pennell. Gila River bed between Phoenix and Maricopa, Maricopa County. Central California and western Nevada. *B. rotundifolia* (Michx.) Wettst. (p. 783) is a species of the central United States.

126. PLANTAGINACEAE. PLANTAIN FAMILY

1. PLANTAGO. PLANTAIN, INDIAN-WHEAT

7. Plantago Tweedyi Gray. Kearney (419) suggested that the Arizona plants be referred to this species rather than to *P. eriopoda* Torr. (p. 804).

127. RUBIACEAE. MADDER FAMILY

9. GALIUM. BEDSTRAW

2. Galium asperrimum Gray. Dempster (389) considered this to be related to *G. mexicanum* H.B.K. and designated the Arizona plants as *G. mexicanum* var. *asperulum* (Gray) Dempster.

5. Galium Wrightii Gray var. **Rothrockii** (Gray) Ehrendorfer (393). *G. Rothrockii* Gray (p. 811).

11. Galium Munzii Hilend & Howell (p. 811). Glabrous specimens of this otherwise rough hispidulous species have been designated as forma *glabrum* Ehrendorfer (393). This taxon occurs in the Grand Canyon, Coconino County, in eastern California, and in southern Nevada and Utah. Dempster considered the relationship of this taxon to be with *G. Mathewsii* A. Gray and designated it as var. *magnifolium* Dempster (389); however, later she raised it to specific rank as *G. magnifolium* (Dempster) Dempster (390).

128. CAPRIFOLIACEAE. Honeysuckle Family

1. SAMBUCUS. Elder, Elderberry

1. Sambucus microbotrys Rydb. *S. racemosa* L. var. *microbotrys* (Rydb.) Kearney & Peebles.

3. Sambucus neomexicana Wooton (p. 813). This was reduced in 1953 by Little (424, p. 396) to synonymy under *S. glauca* Nutt., a much more widespread species.

6. Sambucus glauca Nutt. This should replace the name *S. coerulea* Raf., which according to Little (424, p. 397), was very briefly and inadequately described without specimens.

129. VALERIANACEAE. Valerian Family

1. PLECTRITIS

According to D. H. Morey (437), the names of Arizona *Plectritis* should be:

1. Plectritis ciliosa (Greene) Jepson subsp. **ciliosa.** *P. californica* (Suksd.) Dyal (p. 818).

1a. Plectritis ciliosa (Greene) Jepson subsp. **insignis** (Suksd.) Morey. *P. macroptera* (Suksd.) Dyal var. *patelliformis* (Suksd.) Dyal (p. 818).

2. VALERIANA. Valerian, Tobacco-root

1. Valeriana edulis Nutt. (p. 819). Represented in Arizona by subsp. *edulis* (Meyer, 435, pp. 425, 426).

2. Valeriana sorbifolia H.B.K. (p. 819). Represented in Arizona by var. *sorbifolia* (Meyer, 435, pp. 476, 477).

3. Valeriana arizonica Gray (p. 819). Add as synonyms: *V. acutiloba* Rydb. var. *ovata* (Rydb.) A. Nels. and *V. ovata* Rydb. which in *Arizona Flora* are cited under *V. acutiloba* (p. 820).

4. Valeriana capitata Pall. ex Link subsp. **acutiloba** (Rydb.) F. G. Meyer (435, p. 407) replaces *V. acutiloba* Rydb. (p. 819).

131. CAMPANULACEAE. Bellflower Family

3. NEMACLADUS

2. Nemacladus gracilis Eastw. Yucca, western Mohave County, 2,300 feet, *Gould & Darrow* 4311, part, and 4312. Tentatively included in *Arizona Flora* (p. 827), this species is now known to occur in the state.

132. COMPOSITAE. Sunflower Family

15. HETEROTHECA. Telegraph-plant

1. Heterotheca subaxillaris (Lam.) Britt. & Rusby (p. 854). Burdette L. Wagenknecht (463a) does not include Arizona within the distribution of this species which he stated has a range along the Atlantic and Gulf coasts from Delaware to northeastern Mexico. He described a new species, *H. psammophila* Wagenknecht, the type of which was collected near Sedona, Yavapai County. This perhaps would replace *H. subaxillaris* of the *Arizona Flora.* (The first part only of Wagenknecht's paper appeared while this Supplement was in press.)

17. SOLIDAGO. Golden-rod

1. **Solidago multiradiata** Ait. This name, used by Cronquist (385, p. 309) and by Keck (420, p. 296), should replace *S. ciliosa* Greene (p. 857).

9. **Solidago Petradoria** Blake (p. 858). Keck (420, p. 296) treats this under the generic name *Petradoria.* If this course were followed in Arizona, *S. graminea* (Woot. & Standl.) Blake (p. 858) would also be referred to that genus as *Petradoria graminea* Woot. & Standl.

26. TOWNSENDIA

3a. **Townsendia annua** Beaman (375, p. 132). Widespread and rather common in Arizona from Apache County to Coconino County south to Graham and Pinal counties; also in Utah, southwestern Colorado, New Mexico, and Texas. Differs from *T. strigosa* Nutt. (p. 869) in its annual habit, shorter achenes (2 to 2.8 mm. long), shorter disk pappus (1.8 to 3 mm. long), and other characters.

Townsendia arizonica Gray (p. 869) is referred by Beaman to synonymy under *T. incana* Nutt. (*op. cit.,* p. 123).

27. ASTER

Cronquist and Keck (386) have recently "reconstituted" the genus *Machaeranthera* which, if accepted in *Arizona Flora,* would affect ten of the twenty-four names in the genus *Aster* (pp. 873, 874) as follows:

13. **Aster abatus** Blake. *Machaeranthera tortifolia* (Torr. & Gray) Cronquist & Keck.

16. **Aster parvulus** Blake. *Machaeranthera parviflora* Gray.

17. **Aster tagetinus** (Greene) Blake. *Machaeranthera tagetina* Greene.

18. **Aster tanacetifolius** H.B.K. *Machaeranthera tanacetifolia* (H.B.K.) Nees.

19. **Aster canescens** Pursh. *Machaeranthera canescens* (Pursh) Gray.

20. **Aster tephrodes** (Gray) Blake. *Machaeranthera tephrodes* (Gray) Greene.

21. **Aster Bigelovii** Gray. *Machaeranthera Bigelovii* (Gray) Greene.

22. **Aster adenolepis** Blake. *Machaeranthera mucronata* Greene.

23. **Aster aquifolius** (Greene ex Woot. & Standl.) Blake. *Machaeranthera aquifolia* Greene ex Woot. & Standl. This was not included by Cronquist and Keck among "the more evident species" that they treated.

24. **Aster linearis** (Greene) Cory. *Machaeranthera linearis* Greene; *M. cichoriacea* Greene; *Aster cichoriaceus* (Greene) Blake (p. 874).

31. PLUCHEA. Marsh-fleabane

1. **Pluchea purpurascens** (Sw.) DC. var. **purpurascens.** Godfrey (397, pp. 255–257) places in this taxon the Arizona specimens included in *P. camphorata* (L.) DC. (p. 884). *Pluchea camphorata* is a species of the southeastern and middle western United States, but *P. purpurascens* var. *purpurascens* occurs also in the southwestern United States and southward into Mexico, Central America, northern South America, and the West Indies.

35. ANTENNARIA. Pussy-toes

3. Antennaria parvifolia Nutt. This name should replace *A. aprica* Greene (p. 887) according to Cronquist (385, p. 38) and C. W. Sharsmith (454, p. 482).

42. ENGELMANNIA

1. Engelmannia pinnatifida Nutt. (p. 891). Window Rock, northern Apache County, 6,500 feet, *Deaver* 4836. Apparently the first definite record of the occurrence of this plant in Arizona.

43. PARTHENIUM

2. Parthenium confertum Gray var. **lyratum** (Gray) Rollins. *P. lyratum* (Gray) Gray. Reported by Mason (429a, p. 93) from a collection made along Sala Ranch Road, northeast of Tombstone, Cochise County (*Goodding* 433–58). New Mexico, Texas, and northeastern Mexico. Distinguished from *P. incanum* (p. 891) by its herbaceous, rather than frutescent habit.

85. LAPHAMIA

Lloyd H. Shinners (455a) has recently called attention to the tenuous differences between this genus and *Perityle* and proposes that the two be united. Although he takes the "logical step of making the necessary transfers to accommodate all accepted species under *Perityle*" his list of transfers does not account for all taxa included under *Laphamia* in *Arizona Flora*, pp. 916, 917.

91. HYMENOPAPPUS

The following key and notes are taken from the monographic study of the genus *Hymenopappus* L'Hér. by Billie L. Turner (463). The key replaces the third and fourth dichotomies of the key on page 919, and the notes replace species 3 to 5 on pages 919 and 920.

3. Plants biennial, the roots normally bearing a single crown
3. *H. flavescens* var. *cano-tomentosus*
3. Plants perennial, the roots bearing a branched crown; stem leaves few, less than 12 (4).
4. Corollas 2 to 3 mm. long; flowers in head less than 30; phyllaries 3 to 6 mm. long
4a. *H. filifolius* var. *pauciflorus*
4. Corollas 3 to 7 mm. long; flowers in head 20 to 70; phyllaries 6 to 14 mm. long (5).
5. Anthers 3 to 4 mm. long; corolla 4 to 7 mm. long (6).
5. Anthers 2 to 3 mm. long; corolla 2.5 to 4.5 mm. long (7).
6. Stem leaves usually more than 2; phyllaries 8 to 14 mm. long
4b. *H. filifolius* var. *megacephalus*
6. Stem leaves 3 or less; phyllaries 6 to 10 mm. long..........4c. *H. filifolius* var. *lugens*
7. Hairs on achenes 1 mm. long or less; pappus 1.5 to 3 mm. long; stem leaves 2 or less
4d. *H. filifolius* var. *nanus*
7. Hairs on achenes 1 to 3 mm. long; pappus 1.5 to 2 mm. long; stem leaves usually 2 to 4
4e. *H. filifolius* var. *cinereus*

3. Hymenopappus flavescens Gray var. **cano-tomentosus** Gray. *H. robustus* Greene (p. 919). Sandy or gravelly soil in eastern Arizona (Apache, Navajo, Gila, Graham, and Pima counties); east to Texas and south to Chihuahua, Mexico.

4a. Hymenopappus filifolius Hook. var. **pauciflorus** (Johnst.) B. L. Turner. *H. pauciflorus* Johnst. (p. 920). Sandy flats and slopes, northeastern

Arizona (Coconino, Navajo, and Apache counties); north to Utah and southwestern Colorado.

4b. Hymenopappus filifolius Hook. var. **megacephalus** B. L. Turner. Sandy or gravelly slopes, northeastern Arizona (Coconino, Navajo, and Apache counties); west to Nevada and southeastern California, and north and east to Utah and Colorado.

4c. Hymenopappus filifolius Hook. var. **lugens** (Greene) Jeps. *H. lugens* Greene (p. 919). Wooded mountain slopes in sandy or gravelly soil, widespread in Arizona from Apache County to Mohave County and south to Santa Cruz County; west to southern California and northern Baja California, north and east to Utah and New Mexico.

4d. Hymenopappus filifolius Hook. var. **nanus** (Rydb.) B. L. Turner. *H. nanus* Rydb. Slopes of hills and mountains in rocky soil, northwestern Arizona (Mohave County); west to eastern California, north in Nevada and Utah.

4e. Hymenopappus filifolius Hook. var. **cinereus** (Rydb.) Johnst. *H. cinereus* Rydb. Sandy places about sandstone rocks, northeastern Arizona (Coconino, Navajo, and Apache counties); north to Utah and Colorado, east to Texas.

94. ERIOPHYLLUM. Woolly-daisy

Since Carlquist (381) has expressed the opinion that the genus *Antheropeas* should be segregated from *Eriophyllum,* the following synonymy should be noted for *Arizona Flora* (p. 921):

4. Eriophyllum Wallacei Gray. *Antheropeas Wallacei* (Gray) Rydb.

5. Eriophyllum lanosum Gray. *Antheropeas lanosum* (Gray) Rydb.

99. HYMENOXYS

3a. Hymenoxys Ivesiana (Greene) K. F. Parker. *H. acaulis* (Pursh) K. F. Parker var. *Ivesiana* (Greene) K. F. Parker. North central Arizona. Colorado, New Mexico, and Utah. In *Arizona Flora* this was treated as a variety of *H. acaulis* (p. 927). More recently Parker (444a, pp. 92, 93) has treated it as a distinct species, distinguishing it from *H. acaulis* var. *arizonica* by its stems, usually few branched, bearing 1 to 6 reduced leaves, whereas *H. acaulis* var. *arizonica* is strictly scapose with the leaves all basal. She also points out that *H. Ivesiana* differs in chromosome number from *H. acaulis,* the former being diploid, 2n = 30, whereas the latter is tetraploid, 2n = 60.

8. Hymenoxys helenioides (Rydb.) Cockerell replaces *H. Lemmoni* (Greene) Cockerell of *Arizona Flora* (p. 928). According to Parker (444a, p. 93) *H. Lemmoni* does not occur in Arizona. Known in Arizona only from the Lukachukai Mountains, Apache County. Colorado, Utah.

111. COTULA

2. Cotula australis (Sieber) Hook. f. Sandy-silt soil, Papago Park near Tempe, Maricopa County, *E. R. Blakely* 1981. Native of Australia. Differs from *C. coronopifolia* L. (p. 938) in having finely pinnate leaves and smaller heads, 2 to 5 mm. broad. In *C. coronopifolia* the leaves are entire or pinnatifid and the heads are 8 to 10 mm. broad.

112. ARTEMISIA

12. Artemisia arbuscula Nutt. subsp. **nova** (A. Nels.) G. H. Ward (465, pp. 183–185). *Artemisia nova* A. Nels. (p. 942).

116. PSATHYROTES

3. Psathyrotes annua (Nutt.) Gray. Toroweap Valley, northeastern Mohave County, 4,000 to 5,000 feet, *McClintock* 52–589a, *Cottam* 13340. These collections establish this plant as a member of the flora of Arizona (p. 943).

123. CIRSIUM. THISTLE

3. Cirsium Drummondii Torr. & Gray (p. 952). This name is treated both by Cronquist (385, p. 138) and by Howell (408, p. 532) as a synonym of *C. foliosum* (Hook.) DC. The complex of "leafy thistles" is in need of critical systematic study.

4a. Cirsium utahense Petrak (p. 952). The Utah thistle has recently been distinguished by Howell (408, pp. 516, 524) from the New Mexico thistle by the following characters:

1. Heads usually broader than long, hemispheric; phyllaries tipped by spines mostly 5 to 10 mm. long, the outer phyllaries often reflexed.................. *C. neomexicanum*
1. Heads subglobose; spines on phyllaries mostly 3 to 7 mm. long, the outer phyllaries spreading or ascending... *C. utahense*

Although the type locality of the Utah thistle is in southern Utah (Silver Reef, Washington County), it is not certain that the plant occurs in Arizona. It should be watched for, particularly in the northern part of the state.

7a. Cirsium Rydbergii Petrak. Black Mesa, northern Apache County, 6,000 feet, *Deaver* 3675. Southwestern Utah. Resembles *C. bipinnatum* (Eastw.) Rydb. (p. 953), from which it differs in its glabrous or glabrescent leaves and stems, rounded and clasping leaf bases, and outer phyllaries appressed or ascending (widely spreading or deflexed in *C. bipinnatum*).

15. Cirsium vulgare (Savi) Tenore (p. 954). The occurrence of the Old World bull thistle as a weed in pastures in Havasu Canyon, Coconino County, has now been confirmed.

126. CENTAUREA. STAR-THISTLE, KNAP-WEED

5. Centaurea repens L. This name for the Russian knap-weed should replace *C. Picris* Pall. (p. 956).

133. MICROSERIS

2. Microseris heterocarpa (Nutt.) K. Chambers. San Carlos Indian Reservation, Gila County, *A. & R. Nelson* 1848. California and Baja California. This species differs from *M. linearifolia* (p. 959) in the stouter and usually longer awn of the paleae and in the beakless achenes.

139. TRAGOPOGON. GOATS-BEARD

Ownbey (442) reported natural hybridization between the three species, *Tragopogon dubius* Scop., *T. porrifolius* L., and *T. pratensis* L., where they occur together in southeastern Washington and adjacent Idaho. Similar hybrids may be looked for in Arizona in localities where two of these species are found together.

140. PINAROPAPPUS

1. **Pinaropappus roseus** Less. var. **foliosus** (Heller) Shinners. Arizona plants are placed by Shinners (Field and Laboratory 19:48) in this taxon which he characterized as having the stems leafy to above the middle and leaves up to 12 cm. long. Var. *roseus*, according to Shinners, occurs only east of the Continental Divide.

141. MALACOTHRIX

2. **Malacothrix glabrata** Gray (p. 963). Williams (469, p. 500) reduced this to varietal status as *M. californica* DC. var. *glabrata* Eaton.

5. **Malacothrix Torreyi** Gray (p. 963). Williams (469, p. 503) reduced this to varietal status as *M. sonchoides* (Nutt.) Torr. & Gray var. *Torreyi* (Gray) Williams.

6. **Malacothrix sonchoides** (Nutt.) Torr. & Gray. According to Williams (469, p. 503), var. *sonchoides* occurs in Arizona. She distinguished between var. *Torreyi* and var. *sonchoides* on the basis of persistent setae on the achenes and glandular hairs on the stems and leaves in var. *Torreyi,* whereas in var. *sonchoides* the pappus setae are deciduous and the leaves and stems are usually glabrous.

Literature Consulted for Supplement

(The number preceding each entry continues the enumeration of literature references given on pages 973 to 986. There the references were arranged according to the systematic position of the group that was treated; here the references are arranged alphabetically by author.)

372. Abrams, LeRoy, *Illustrated Flora of the Pacific States.* Volume III. Stanford University Press (1951).

373. Avery, Amos G., Sophie Satina, Jacob Rietsema, *Blakeslee: The Genus Datura.* Ronald Press Co., New York (1959).

374. Baker, Milo S., "Studies in Western Violets. VII." *Madroño* 12:8–18 (1953).

374a. Barneby, R. C., "New and Critical Arizona Astragali." Leafl. West. Bot. vol. 9:89–91 (1960).

375. Beaman, John H., "The Systematics and Evolution of Townsendia (Compositae)." *Contrib. Gray Herbarium* 183:1–151 (1957).

376. Benson, Lyman, *The Cacti of Arizona.* University of Arizona Press, Tucson (1950).

377. Benson, Lyman, "Supplement to a Treatise on the North American Ranunculi." *Amer. Midland Nat.* 52:328–369 (1954).

378. Benson, Lyman, and Robert A. Darrow, *The Trees and Shrubs of the Southwestern Deserts,* ed. 2. University of Arizona Press, Tucson; University of New Mexico Press, Albuquerque (1954).

379. Campbell, Gloria R., "Mimulus guttatus and Related Species." *El Aliso* 2:319–337 (1950).

380. Campbell, Gloria R., "The Genus Myosurus in North America." *El Aliso* 2:389–402 (1952).

381. Carlquist, Sherwin, "On the Generic Limits of Eriophyllum (Compositae) and Related Genera." *Madroño* 13:226–239 (1956).

382. Cave, M. S., and L. Constance, "Chromosome numbers in the Hydrophyllaceae: V." Univ. Calif. Publ. Bot. 30:233–258 (1959).

383. Chambers, Kenton L., "Biosystematic Study of the Annual Species of Microseris." *Contrib. Dudley Herbarium* 4:207–312 (1955).

384. Croizat, L., "New or Critical Euphorbiaceae from the Americas." *Jour. Arnold Arbor.* 26:181–196 (1945).

385. Cronquist, Arthur, "Compositae," Part 5 of *Vascular Plants of the Pacific Northwest* by C. L. Hitchcock, A. Cronquist, M. Ownbey, and J. W. Thompson. University of Washington Press, Seattle (1955).

386. Cronquist, Arthur, and David D. Keck, "A Reconstitution of the Genus Machaeranthera." *Brittonia* 9:231–239 (1957).

387. Dayton, William A., "Some More Notes on United States Ashes." *Jour. Wash. Acad. Sci.* 44:385–390 (1954).

388. Dempster, Lauramay T., "Dimorphism in the Fruits of Plectritis, and Its Taxonomic Implications." *Brittonia* 10:14–28 (1958).

389. Dempster, Lauramay T., "New Names and Combinations in the Genus Galium."

Brittonia 10:181–192 (1958).

390. Dempster, Lauramay T., "A Re-evaluation of Galium multiflorum and Related Taxa." *Brittonia* 11:105–122 (1959).

391. Dressler, Robert L., "The Genus Tetracoccus (Euphorbiaceae)." *Rhodora* 56:45–61 (1954).

392. Dunn, David B., "Lupinus pusillus and Its Relationship." *Amer. Midland Nat.* 62:500–510 (1959).

393. Ehrendorfer, Friedrich, "Survey of the Galium multiflorum Complex in Western North America." *Contrib. Dudley Herbarium* 5:1–36 (1956).

394. Fassett, Norman C., "Callitriche in the New World." *Rhodora* 53:137–155, 161–182, 185–194, 209–222 (1951).

395. Ferris, Roxana Stinchfield, Vol. IV of L. R. Abrams and R. S. Ferris, *Illustrated Flora of the Pacific States.* Stanford University Press (1960).

396. Gillett, John M., "A Revision of the North American Species of Gentianella Moench," *Ann. Mo. Bot. Gard.* 44:195–269 (1957).

397. Godfrey, R. K., "Pluchea, Section Stylimnus, in North America." *Jour. Elisha Mitchell Sci. Soc.* 68:238–272 (1952).

398. Goodspeed, Thomas Harper, *The Genus Nicotiana.* Chronica Botanica Company, Waltham, Mass. (1954).

399. Grant, Alva, and Verne Grant, "The Genus Allophyllum (Polemoniaceae)." *El Aliso* 3:93–110 (1955).

400. Grant, Alva, and Verne Grant, "Genetic and Taxonomic Studies in Gilia. VIII. The Cobwebby Gilias." *El Aliso* 3:203–287 (1956).

401. Grant, Verne, "A Synopsis of Ipomopsis." *El Aliso* 3:351–362 (1956).

402. Grant, Verne, *Natural History of the Phlox Family.* Martinus Nijhoff, The Hague (1959).

403. Grant, Verne, and Alva Grant, "Genetic and Taxonomic Studies in Gilia. X. Conspectus of the Subgenus Gilia." *El Aliso* 3:297–300 (1956).

404. Hartman, Emily L., "The Taxonomy and Ecology of the Equisetum laevigatum Complex." *Trans. Kansas Acad. Sci.* 61:125–148 (1958).

404a. Hermann, F. J., "An Andean and a Californian Sedge in Arizona." *Leafl. West. Bot.* 9:86, 87 (1960).

404b. Hermann, F. J., *Vetches of the United States—Native, Naturalized, and Cultivated.* U. S. Dept. Agriculture Handbook No. 168 (1960).

405. Hitchcock, C. Leo, "A Revision of the North American Species of Lathyrus." Univ. Wash. Publ. Biol. 15:1–104 (1952).

406. Hitchcock, C. Leo, Arthur Cronquist, Marion Ownbey, and J. W. Thompson, *Vascular Plants of the Pacific Northwest,* Part 4 (1959), Part 5 (1955). University of Washington Press, Seattle.

407. Howell, J. T., "Eriogonum Notes V: E. glandulosum, with a New Variety." *Leafl. West. Bot.* 8:37–39 (1956).

408. Howell, John Thomas, "Cynareae" in L. R. Abrams and R. S. Ferris, *Illustrated Flora of the Pacific States,* Vol. IV, pp. 506–548 (1960).

409. Hu, Shiu-ying, "A Monograph of the Genus Philadelphus [Section Microphyllus]." *Jour. Arnold Arbor.* 37:15–35 (1956).

409a. Ingram, John, "New Species and New Combinations in the Genus Argythamnia." *Bul. Torrey Bot. Club* 80:420–423 (1953).

410. Johnston, Ivan M., "Studies in the Boraginaceae XXIII. A Survey of the Genus Lithospermum." *Jour. Arnold Arbor.* 33:299–363 (1952).

411. Johnston, Ivan M., "Studies in the Boraginaceae XXVI. Further Revaluations of the Genera of the Lithospermeae." *Jour. Arnold Arbor.* 35:1–81 (1954).

412. Johnston, Marshall C., "The Texas Species of Dyssodia (Compositae)." *Field and Laboratory* 24:60–69 (1956).

413. Johnston, Marshall C., "Synopsis of the United States Species of Forestiera (Oleaceae)." *Southwestern Naturalist* 2:140–151 (1957).

414. Kearney, Thomas H., Robert H. Peebles, and collaborators, *Arizona Flora.* University of California Press, Berkeley and Los Angeles (1951).

415. Kearney, Thomas H., "Further Additions to the Known Flora of Arizona." *Leafl.*

West. Bot. 7:165–175 (1954).

416. Kearney, Thomas H., "Malvastrum, A. Gray—a Redefinition of the Genus." *Leafl. West. Bot.* 7:238–241 (1955).

417. Kearney, Thomas H., "Notes on Malvaceae VIII. Eremalche." *Madroño* 13:241–243 (1956).

418. Kearney, Thomas H., "Arizona Plant Records III." *Leafl. West. Bot.* 8:61–80 (1956).

419. Kearney, Thomas H., Elizabeth McClintock, and Kittie F. Parker, "Recent Additions to the Known Flora of Arizona." *Leafl. West. Bot.* 7:1–11 (1953).

420. Keck, David D., "Solidago" and "Petradoria" [and others] in P. A. Munz and D. D. Keck, *A California Flora*, pp. 1184–1187 (1959); and in L. R. Abrams and R. S. Ferris, *Illustrated Flora of the Pacific States*, Vol. IV, pp. 291–296 (1960).

421. Legrand, Diego, "Revisando tipos de Portulaca." *Comunic. Bot. Mus. Hist. Nat. Montevideo* 2(24):1–10 (1952).

422. Lewis, H., and M. E. Lewis, "The Genus Clarkia." Univ. Calif. Publ. Bot. 20:241–392 (1955).

423. Little, Elbert L., Jr., "Notes on Fraxinus (Ash) in the United States." *Jour. Wash. Acad. Sci.* 42:369–380 (1952).

424. Little, Elbert L., Jr., *Check List of Native and Naturalized Trees of the United States (including Alaska)*. U.S. Dept. Agriculture Handbook No. 41 (1953).

425. Loock, E. E. M., *The Pines of Mexico and British Honduras*. Union So. Afr. Dept. Forestry Bul. No. 35, Pretoria (1950).

426. Löve, Åskell, and Doris Löve, "Biosystematics of Triglochin maritimum Agg." *Naturaliste Canadien* 85:156–165 (1958).

427. Maguire, Bassett, "Studies in the Caryophyllaceae V. Arenaria in America North of Mexico." *Amer. Midland Nat.* 46:493–511 (1951).

428. Mason, Charles T., Jr., "Notes on the Flora of Arizona." *Madroño* 15:64 (1959).

429. Mason, Charles T., Jr., "Notes on the Flora of Arizona. II." *Leafl. West. Bot.* vol. 9:87, 88 (1960).

429a. Mason, Charles T., Jr., "A Parthenium New to Arizona." *Leafl. West. Bot.* 9:93 (1960).

430. Mason, Herbert L., *A Flora of the Marshes of California*. University of California Press (1957).

430a. Mason, Herbert Louis, "Linanthus" in Abrams' *Illustrated Flora of the Pacific States* 3:413–431 (1951).

431. Mason, Herbert Louis, and Alva Day Grant, "Gilia" in Abrams' *Illustrated Flora of the Pacific States* 3:456–474 (1951).

432. McCleary, James A., "Notes on Arizona Plants." *Southwestern Naturalist* 2:152–154 (1957).

433. McCleary, James A., "New Arizona Plant Records." *Southwestern Naturalist* 4:154, 155 (1959).

434. McVaugh, Rogers, "A Revision of the North American Black Cherries (Prunus serotina Ehrh., and Relatives)." *Brittonia* 7:279–315 (1951).

435. Meyer, Frederick G., "Valeriana in North America and the West Indies (Valerianaceae)." *Ann. Mo. Bot. Gard.* 38:377–503 (1951).

436. Miller, Gertrude N., "The Genus Fraxinus, the Ashes, in North America, North of Mexico." Cornell University, Agric. Exp. Station, Ithaca, Memoir No. 335 (1955).

437. Morey, Dennison H., "Changes in Nomenclature in the Genus Plectritis." *Contrib. Dudley Herbarium* 5:119–121 (1959).

438. Morton, C. V., "The Proper Generic Name of the Holly Fern." *Amer. Fern Jour.* 47:52–55 (1957).

439. Morton, C. V., "The Californian Species of Thelypteris." *Amer. Fern Jour.* 48:136–142 (1958).

440. Munz, Philip A., in collaboration with David D. Keck, *A California Flora*. University of California Press, Berkeley and Los Angeles (1959).

441. Ownbey, Gerald B., "Monograph of the Genus Argemone for North America and the West Indies." *Mem. Torrey Bot. Club* 21:1–159 (1958).

442. Ownbey, Marion, "Natural Hybridization and Amphiploidy in the Genus Tragopogon." *Amer. Jour. Bot.* 37:487–499 (1950).

443. Ownbey, Marion, and Hannah C. Aase, "Cytotaxonomic Studies in Allium I. The Allium canadense Alliance." *Research Studies of the State College of Washington, Monographic Supplement* 1:1–106 (1956).

444. Parker, Kittie F., *Arizona Ranch, Farm, and Garden Weeds.* Agricultural Extension Service, Circular 265, Univ. Arizona, Tucson (1958).

444a. Parker, Kittie F., "Two Species of Hymenoxys (Compositae) New for Arizona." *Leafl. West. Bot.* 9:92, 93 (1960).

445. Patten, Bernard C., Jr., "The Status of Some American Species of Myriophyllum." *Rhodora* 56:213–225 (1954).

446. Rogers, David J., "A Revision of Stillingia in the New World." *Ann. Mo. Bot. Gard.* 38:207–259 (1951).

446a. Rollins, Reed C., "The Guayule Rubber Plant and Its Relatives." *Contrib. Gray Herbarium* 172:1–73 (1950).

447. Rollins, Reed C., "The Genetic Evaluation of a Taxonomic Character in Dithyrea (Cruciferae)." *Rhodora* 60: 145–152 (1958).

448. Rothmaler, Werner, "Notes on Western Antirrhineae." *Leafl. West. Bot.* 7:113–117 (1954).

449. Rothmaler, W., "Taxonomische Monographie der Gattung Antirrhinum." *Fedde's Repertorium Beiheft* 136, pp. 1–124 (1956).

450. Rudd, Velva E., "The American Species of Aeschynomene." *Contrib. U. S. National Herbarium* 32:1–172 (1955).

451. Rydberg, P. A., "The Oaks of the Continental Divide North of Mexico." *Bul. N. Y. Bot. Gard.* 2:187–233 (1901).

452. Sargent, C. S., *The Silva of North America,* Vol. 8(1895).

453. Sauer, Jonathan, "Revision of the Dioecious Amaranths." *Madroño* 13:5–46 (1955).

454. Sharsmith, Carl William, "Antennaria" in L. R. Abrams and R. S. Ferris, *Illustrated Flora of the Pacific States,* Vol. IV, pp. 474–485 (1960).

455. Shinners, Lloyd H., "Synopsis of the Genus Eustoma (Gentianaceae)." *Southwestern Naturalist* 2:38–43 (1957).

455a. Shinners, Lloyd H., "Species of Laphamia Transferred to Perityle (Compositae-Helenieae)." *Southwestern Naturalist* 4: 204, 205 (1959).

456. Stocking, Kenneth M., "Some Considerations of the Genera Echinocystis and Echinopepon in the United States and Northern Mexico." *Madroño* 13:84–100 (1955).

457. Stocking, Kenneth M., "Some Taxonomic and Ecological Considerations on the Genus Marah (Cucurbitaceae)." *Madroño* 13:113–140 (1955).

458. Thompson, Henry J., "The Biosystematics of Dodecatheon." *Contrib. Dudley Herbarium* 4:73–154 (1953).

459. Thompson, Henry J., and Harlan Lewis, "Chromosome Numbers in Mentzelia (Loasaceae)." *Madroño* 13:102–107 (1955).

460. Tryon, Alice F., "A Revision of the Fern Genus Pellaea Section Pellaea." *Ann. Mo. Bot. Gard.* 44:125–193 (1957).

461. Tryon, Rolla M., "Selaginella rupestris and Its Allies." *Ann. Mo. Bot. Gard.* 42:1–99 (1955).

462. Tryon, Rolla M., "A Revision of the American Species of Notholaena." *Contrib. Gray Herbarium* 179:1–106 (1956).

463. Turner, Billie L., "A Cytotaxonomic Study of the Genus Hymenopappus (Compositae)." *Rhodora* 58:163–186, 208–242, 250–269, 295–308 (1956).

463a. Wagenknecht, Burdette L., "Revision of Heterotheca, Section Heterotheca (Compositae)." *Rhodora* 62:61–76 (1960).

464. Wagnon, H. Keith, "A Revision of the Genus Bromus, Section Bromopsis, of North America." *Brittonia* 7:415–480 (1952).

464a. Wahl, Herbert A., "A Preliminary Study of the Genus Chenopodium in North America." *Bartonia* No. 27:1–46 (1954).

465. Ward, George H., "Artemisia, Section Seriphidium, in North America." *Contrib. Dudley Herbarium* 4:155–205 (1953).

466. Waterfall, U. T., "A revision of Eucnide." *Rhodora* 61:231–243 (1959).
467. Webber, John Milton, *Yuccas of the Southwest.* U. S. Dept. Agri. Monograph No. 17 (1953).
468. Wherry, Edgar T., *The Genus Phlox.* Morris Arboretum Monographs III (1955).
469. Williams, Elizabeth W., "The Genus Malacothrix (Compositae)." *Amer. Midland Nat.* 58:494–512 (1957).
470. Woodson, Robert E., Jr., "The North American Species of Asclepias, L." *Ann. Mo. Bot. Gard.* 41:1–211 (1954).

Index to Supplement

Abutilon, 1060
Acanthochiton, 1048
Aegilops, 1040
Aeschynomene, 1057
Agropyron, 1040
Ailanthus, 1058
Alismaceae, 1039
Allium, 1043
Allophyllum, 1065
Amaranthaceae, 1048
Amaranthus, 1048
Ammobroma, 1063
Anacharis, 1039
Andropogon, 1041, 1042
Antennaria, 1073
Antheropeas, 1074
Antirrhinum, 1070
Arenaria, 1049
Argemone, 1049, 1050
Argythamnia, 1058, 1059
Artemisia, 1075
Asclepiadaceae, 1065
Asclepias, 1065
Aster, 1072
Astragalus, 1052
  amphioxys, 1054
  bisulcatus, 1054
  ceramicus, 1055
  coccineus, 1054
  confertiflorus, 1054
  convallarius, 1052, 1053
  Diehlii, 1053
  Emoryanus, 1057
  famelicus, 1055
  flaviflorus, 1054
  flavus, 1054
  flexuosus, 1053
  fucatus, 1056
  gracilentus, 1053, 1055
  Greenei, 1053
  Hallii, 1055
  Haydenianus, 1054
  humistratus, 1054
  monumentalis, 1055
  musimonum, 1054, 1055
  naturitensis, 1055
  Newberryi, 1054
  Nuttallianus, 1057
  praelongus, 1056
  Preussii, 1056
  recurvus, 1057
  Rusbyi, 1057
  scopulorum, 1057
  sesquiflorus, 1055
  Sileranus, 1056
  straturensis, 1057
  subcinereus, 1056
  titanophilus, 1053
  xiphoides, 1053
Atelophragma, 1057

Bacopa, 1070
Batidophaca, 1055
Beckmannia, 1041
Betula, 1044
Betulaceae, 1044
Bogenhardia, 1060
Boraginaceae, 1067
Brassica, 1051
Bromus, 1039
Bupleurum, 1062

Cactaceae, 1061
Callitrichaceae, 1059
Callitriche, 1059
Campanulaceae, 1071
Caprifoliaceae, 1071
Cardaria, 1050
Cardiospermum, 1059
Carex, 1042
Caryophyllaceae, 1049
Centaurea, 1075
Cerastium, 1049
Chenopodiaceae, 1046
Chenopodium, 1046
  albescens, 1048
  arizonicum, 1048
  atrovirens, 1048
  Berlandieri, 1048
  desiccatum, 1048
  Fremontii, 1048
  graveolens, 1046
  incanum, 1048
  incisum, 1046
  incognitum, 1048
  neomexicanum, 1048
  Palmeri, 1048
  pratericola, 1048
Cirsium, 1075
Clarkia, 1062
Cnemidophacos, 1054
Coldenia, 1067
Compositae, 1071
Cordia, 1067
Corydalis, 1050
Cotula, 1074
Cruciferae, 1050
Cyperaceae, 1042
Cyperus, 1042
Cyrtomium, 1037

Digitaria, 1041
Diholcos, 1054
Ditaxis, 1058, 1059
Dithyrea, 1051
Dodecatheon, 1063
Dodonaea, 1059

Echinocactus, 1061
Echinodorus, 1039
Echinomastus, 1061
Elodea, 1039
Engelmannia, 1073
Epilobium, 1061
Equisetaceae, 1036
Equisetum, 1036
Eremalche, 1060
Eriogonum, 1046
Eriophyllum, 1074
Eruca, 1051
Eucnide, 1061
Euphorbia, 1059
Euphorbiaceae, 1058
Eustoma, 1064

Fagaceae, 1044
Ferocactus, 1061
Forestiera, 1064
Fraxinus, 1063, 1064
Fremontia, 1060
Fremontodendron, 1060
Fumaria, 1050

Galium, 1070
Gayoides, 1060
Gentiana, 1064
Gentianaceae, 1064
Gentianella, 1064
Gilia, 1065
  achilleaefolia, 1065
  aggregata, 1067
  Clokeyi, 1066
  flavocincta, 1066
  gilioides, 1065
  Gunnisoni, 1065
  hutchinsifolia, 1066
  inconspicua, 1066
  laxiflora, 1067
  leptomeria, 1066
  longiflora, 1067
  Macombii, 1067
  mexicana, 1066
  minor, 1066
  multicaulis, 1065
  multiflora, 1067
  ochroleuca, 1066
  ophthalmoides, 1066
  polyantha, 1067
  polycladon, 1065
  pumila, 1065
  scopulorum, 1066
  stellata, 1066
  tenuiflora, 1066
  tenuituba, 1067
  Thurberi, 1067
  transmontana, 1066
Godetia, 1062
Gramineae, 1039

Haloragaceae, 1062
Hamosa, 1057
Heterotheca, 1071
Hordeum, 1041
Hydrocharitaceae, 1039
Hydrophyllaceae, 1067
Hymenopappus, 1073
Hymenoxys, 1074

Ipomopsis, 1065, 1067

Juncaceae, 1043
Juncaginaceae, 1039
Juncus, 1043

Labiatae, 1068
Laphamia, 1073
Lathyrus, 1058
Leguminosae, 1052
Lemna, 1042
Lemnaceae, 1042
Lennoaceae, 1063
Lepidium, 1050
Liliaceae, 1043
Linanthastrum, 1065
Linanthus, 1065
Linaria, 1069
Lithospermum, 1068
Loasaceae, 1061
Loasella, 1061
Lomatium, 1062
Lupinus, 1052
Lycopus, 1069

Machaeranthera, 1072
Macromeria, 1068
Malacothrix, 1076
Malvaceae, 1060
Malvastrum, 1060
Matelea, 1065
Matthiola, 1051
Melia, 1059
Meliaceae, 1059
Mentzelia, 1061
Menyanthes, 1064
Microseris, 1075
Mimulus, 1070
Monardella, 1069
Muhlenbergia, 1041
Myosurus, 1049
Myriophyllum, 1062

Nama, 1067
Navajoa, 1061
Nemacladus, 1071
Notholaena, 1037

Oenothera, 1062
Oleaceae, 1063
Onagraceae, 1061
Osmorhiza, 1062

Papaveraceae, 1049
Parthenium, 1073
Pellaea, 1037
Pennisetum, 1041
Phaca, 1056
Phacelia, 1067
Philadelphus, 1051
Pholisma, 1063
Pinaceae, 1038
Pinaropappus, 1076
Pinus, 1038
Pisophaca, 1053

Plantaginaceae, 1070
Plantago, 1070
Plectritis, 1071
Pluchea, 1072
Polemoniaceae, 1065
Polygonaceae, 1046
Polygonum, 1046
Polypodiaceae, 1037
Populus, 1044
Portulaca, 1048, 1049
Portulacaceae, 1048
Potamogeton, 1038
Potamogetonaceae, 1038
Primulaceae, 1063
Prunus, 1052
Psathyrotes, 1075
Pseudotsuga, 1038

Quercus, 1044, 1045

Ranunculaceae, 1049
Ranunculus, 1049
Rosaceae, 1052
Rothrockia, 1065
Rubiaceae, 1070
Rumex, 1046

Salicaceae, 1044
Salix, 1044
Salpichroa, 1069
Salvia, 1069
Sambucus, 1071
Sapindaceae, 1059
Saxifragaceae, 1051
Scheuchzeriaceae, 1039
Sclerocactus, 1061
Scrophulariaceae, 1069
Selaginella, 1035
Selaginellaceae, 1035
Setaria, 1041
Sida, 1060
Simaroubaceae, 1058
Sisymbrium, 1051
Solanaceae, 1069
Solidago, 1072
Spartina, 1041
Sphaerophysa, 1052
Stachys, 1069
Stellaria, 1049
Sterculiaceae, 1060
Sympeteleia, 1061

Tamaricaceae, 1060
Tamarix, 1060
Tetracoccus, 1058
Thelypteris, 1037
Tium, 1057
Toumeya, 1061
Townsendia, 1072